CONCEPTS OF GENETICS

Third Edition

Robert J. Brooker

McGraw Hill Education

CONCEPTS OF GENETICS

Published by McGraw-Hill Education, 2 Penn Plaza, New York, NY 10121. Copyright © 2019 by McGraw-Hill Education. All rights reserved. Printed in the United States of America. No part of this publication may be reproduced or distributed in any form or by any means, or stored in a database or retrieval system, without the prior written consent of McGraw-Hill Education, including, but not limited to, in any network or other electronic storage or transmission, or broadcast for distance learning.

Some ancillaries, including electronic and print components, may not be available to customers outside the United States.

This book is printed on acid-free paper.

2 3 4 5 6 7 8 9 BRP 21 20 19

ISBN 978-1-260-28858-2
MHID 1-260-28858-7

Cover Image: ©BLUEXHAND/iStock/Getty Images Plus, ©Radius Images/Alamy

All credits appearing on page or at the end of the book are considered to be an extension of the copyright page.

The Internet addresses listed in the text were accurate at the time of publication. The inclusion of a website does not indicate an endorsement by the authors or McGraw-Hill Education, and McGraw-Hill Education does not guarantee the accuracy of the information presented at these sites.

mheducation.com/highered

ABOUT THE AUTHOR

Robert J. Brooker is a professor in the Department of Genetics, Cell Biology, and Development and the Department of Biology Teaching and Learning at the University of Minnesota–Minneapolis. He received his B.A. in biology from Wittenberg University in 1978 and his Ph.D. in genetics from Yale University in 1983. At Harvard, he conducted postdoctoral studies on the lactose permease, which is the product of the *lacY* gene of the *lac* operon. He continued his work on transporters at the University of Minnesota. Dr. Brooker's laboratory has also investigated the structure, function, and regulation of manganese and iron transporters found in bacteria and *C. elegans*. At the University of Minnesota, he teaches undergraduate courses in biology, genetics, and cell biology.

Courtesy of Robert Brooker

DEDICATION

To my wife, Deborah, and our children,
Daniel, Nathan, and Sarah

BRIEF CONTENTS

PART I
1 Overview of Genetics 1

PART II
2 Reproduction and Chromosome Transmission 19
3 Mendelian Inheritance 41
4 Sex Determination and Sex Chromosomes 72
5 Extensions of Mendelian Inheritance 89
6 Extranuclear Inheritance, Imprinting, and Maternal Effect 113
7. Genetic Linkage and Mapping in Eukaryotes 131
8 Variation in Chromosome Structure and Number 156
9 Genetics of Bacteria 185
10 Genetics of Viruses 207

PART III
11 Molecular Structure of DNA and RNA 221
12 Molecular Structure of Chromosomes and Transposition 240
13 DNA Replication and Recombination 266

PART IV
14 Gene Transcription and RNA Modification 296
15 Translation of mRNA 325
16 Gene Regulation in Bacteria 355
17 Gene Regulation in Eukaryotes 380
18 Non-Coding RNAs 412
19 Gene Mutation and DNA Repair 431

PART V
20 Molecular Technologies 459
21 Genomics 491

PART VI
22 Medical Genetics and Cancer 518
23 Population Genetics 554
24 Quantitative Genetics 584

TABLE OF CONTENTS

Preface viii

A Visual Guide to Concepts of Genetics xv

PART I
INTRODUCTION 1

1 OVERVIEW OF GENETICS 1

1.1 The Molecular Expression of Genes 3
1.2 The Relationship Between Genes and Traits 6
1.3 Fields of Genetics 12
1.4 The Science of Genetics 14

PART II
PATTERNS OF INHERITANCE 19

2 REPRODUCTION AND CHROMOSOME TRANSMISSION 19

2.1 General Features of Chromosomes 19
2.2 Cell Division 23
2.3 Mitosis and Cytokinesis 26
2.4 Meiosis 30
2.5 Sexual Reproduction 34

3 MENDELIAN INHERITANCE 41

3.1 Mendel's Study of Pea Plants 42
3.2 Law of Segregation 45
3.3 Law of Independent Assortment 50
3.4 Chromosome Theory of Inheritance 55
3.5 Studying Inheritance Patterns in Humans 58
3.6 Probability and Statistics 59

4 SEX DETERMINATION AND SEX CHROMOSOMES 72

4.1 Mechanisms of Sex Determination Among Various Species 72
4.2 Dosage Compensation and X-Chromosome Inactivation in Mammals 75
4.3 Properties of the X and Y Chromosomes in Mammals 79
4.4 Transmission Patterns for X-Linked Genes 80

5 EXTENSIONS OF MENDELIAN INHERITANCE 89

5.1 Overview of Simple Inheritance Patterns 89
5.2 Dominant and Recessive Alleles 91
5.3 Environmental Effects On Gene Expression 93
5.4 Incomplete Dominance, Overdominance, and Codominance 95
5.5 Sex-Influenced and Sex-Limited Inheritance 99
5.6 Lethal Alleles 101
5.7 Understanding Complex Phenotypes Caused By Mutations in Single Genes 102
5.8 Gene Interaction 104

6 EXTRANUCLEAR INHERITANCE, IMPRINTING, AND MATERNAL EFFECT 113

6.1 Extranuclear Inheritance: Chloroplasts 114
6.2 Extranuclear Inheritance: Mitochondria 117
6.3 Theory of Endosymbiosis 119
6.4 Epigenetics: Imprinting 120
6.5 Maternal Effect 124

7 GENETIC LINKAGE AND MAPPING IN EUKARYOTES 131

7.1 Overview of Linkage 131
7.2 Relationship Between Linkage and Crossing Over 133
7.3 Genetic Mapping in Plants and Animals 139
7.4 Mitotic Recombination 146

8 VARIATION IN CHROMOSOME STRUCTURE AND NUMBER 156

8.1 Microscopic Examination of Eukaryotic Chromosomes 156
8.2 Changes in Chromosome Structure: An Overview 159
8.3 Deletions and Duplications 160
8.4 Inversions and Translocations 164
8.5 Changes in Chromosome Number: An Overview 168
8.6 Variation in the Number of Chromosomes Within A Set: Aneuploidy 170
8.7 Variation in the Number of Sets of Chromosomes 172
8.8 Mechanisms That Produce Variation in Chromosome Number 175

9 GENETICS OF BACTERIA 185

9.1 Overview of Genetic Transfer in Bacteria 185
9.2 Bacterial Conjugation 187
9.3 Conjugation and Mapping via Hfr Strains 191
9.4 Bacterial Transduction 196
9.5 Bacterial Transformation 199
9.6 Medical Relevance of Horizontal Gene Transfer 200

10 GENETICS OF VIRUSES 207

10.1 Virus Structure and Genetic Composition 207
10.2 Viral Reproductive Cycles 212

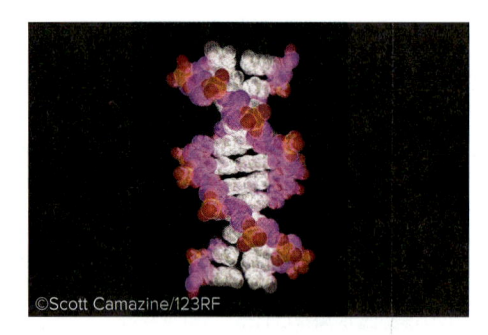

PART III
MOLECULAR STRUCTURE AND REPLICATION OF THE GENETIC MATERIAL 221

11 MOLECULAR STRUCTURE OF DNA AND RNA 221

11.1 Identification of DNA as the Genetic Material 221
11.2 Overview of DNA and RNA Structure 224
11.3 Nucleotide Structure 225
11.4 Structure of a DNA Strand 227
11.5 Discovery of the Double Helix 228
11.6 Structure of the DNA Double Helix 230
11.7 RNA Structure 234

12 MOLECULAR STRUCTURE OF CHROMOSOMES AND TRANSPOSITION 240

12.1 Organization of Functional Sites Along Bacterial Chromosomes 240
12.2 Structure of Bacterial Chromosomes 241
12.3 Organization of Functional Sites Along Eukaryotic Chromosomes 245
12.4 Sizes of Eukaryotic Genomes and Repetitive Sequences 246
12.5 Transposition 247
12.6 Structure of Eukaryotic Chromosomes in Nondividing Cells 253
12.7 Structure of Eukaryotic Chromosomes During Cell Division 259

13 DNA REPLICATION AND RECOMBINATION 266

13.1 Structural Overview of DNA Replication 266
13.2 Bacterial DNA Replication: The Formation of Two Replication Forks at the Origin of Replication 270
13.3 Bacterial DNA Replication: Synthesis of New DNA Strands 273
13.4 Bacterial DNA Replication: Chemistry and Accuracy 278
13.5 Eukaryotic DNA Replication 280
13.6 Homologous Recombination 285

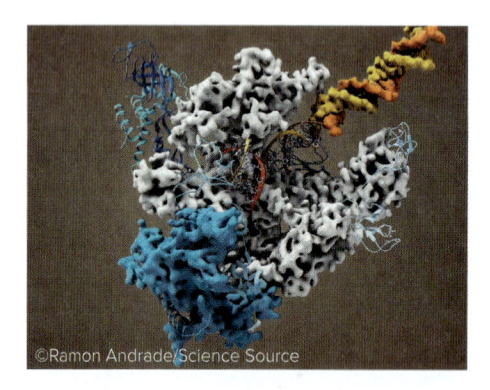

PART IV
MOLECULAR PROPERTIES OF GENES 296

14 GENE TRANSCRIPTION AND RNA MODIFICATION 296

14.1 Overview of Transcription 297
14.2 Transcription in Bacteria 299
14.3 Transcription in Eukaryotes 304
14.4 RNA Modification 309
14.5 A Comparison of Transcription and RNA Modification in Bacteria and Eukaryotes 319

15 TRANSLATION OF mRNA 325

15.1 The Genetic Basis for Protein Synthesis 325
15.2 The Relationship Between the Genetic Code and Protein Synthesis 328
15.3 Experimental Determination of the Genetic Code 334
15.4 Structure and Function of tRNA 338
15.5 Ribosome Structure and Assembly 341
15.6 Stages of Translation 343

16 GENE REGULATION IN BACTERIA 355

16.1 Overview of Transcriptional Regulation 355
16.2 Regulation of the *lac* Operon 358
16.3 Regulation of the *trp* Operon 368
16.4 Translational and Posttranslational Regulation 372
16.5 Riboswitches 373

17 GENE REGULATION IN EUKARYOTES 380

17.1 Regulatory Transcription Factors 381
17.2 Chromatin Remodeling, Histone Variants, and Histone Modification 386
17.3 DNA Methylation 391
17.4 Overview of Epigenetics 394
17.5 Epigenetics and Development 397
17.6 Epigenetics and Environmental Agents 402
17.7 Regulation of Translation and RNA Stability 405

18 NON-CODING RNAs 412

18.1 Overview of Non-Coding RNAs 413
18.2 Non-Coding RNAs: Effects on Chromatin Structure and Transcription 416
18.3 Non-Coding RNAs: Effects on Translation and mRNA Degradation 417
18.4 Non-Coding RNAs and Protein Targeting 421

- 18.5 Non-Coding RNAs and Genome Defense 422
- 18.6 Role of Non-Coding RNAs in Human Diseases 426

19 GENE MUTATION AND DNA REPAIR 431

- 19.1 Effects of Mutations on Gene Structure and Function 432
- 19.2 Random Nature of Mutations 439
- 19.3 Spontaneous Mutations 440
- 19.4 Induced Mutations 445
- 19.5 DNA Repair 449

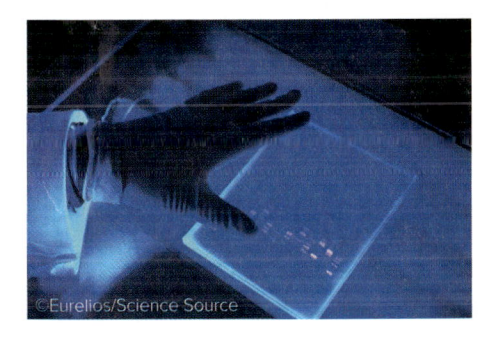

PART V
GENETIC TECHNOLOGIES 459

20 MOLECULAR TECHNOLOGIES 459

- 20.1 Gene Cloning Using Vectors 460
- 20.2 Polymerase Chain Reaction 466
- 20.3 Reproductive Cloning and Stem Cells 472
- 20.4 DNA Sequencing 477
- 20.5 Gene Editing 479
- 20.6 Blotting Methods To Detect Gene Products 482
- 20.7 Analyzing DNA- and RNA-Binding Proteins 484

21 GENOMICS 491

- 21.1 Overview of Chromosome Mapping 492
- 21.2 In situ Hybridization 492
- 21.3 Molecular Markers 495
- 21.4 Genome-Sequencing Projects 498
- 21.5 Metagenomics 505
- 21.6 Functional Genomics 507

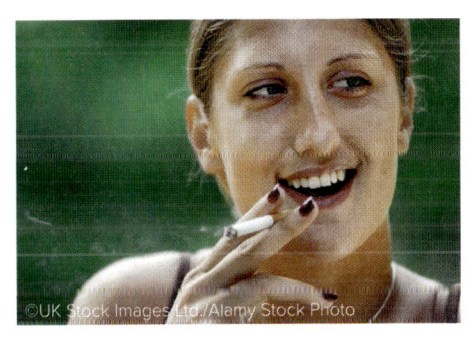

PART VI
GENETIC ANALYSIS OF INDIVIDUALS AND POPULATIONS 518

22 MEDICAL GENETICS AND CANCER 518

- 22.1 Inheritance Patterns of Genetic Diseases 519
- 22.2 Detection of Disease-Causing Alleles via Haplotypes 525
- 22.3 Genetic Testing and Screening 529
- 22.4 Overview of Cancer 531
- 22.5 Oncogenes 532
- 22.6 Tumor-Suppressor Genes 536
- 22.7 Role of Epigenetics in Cancer 542
- 22.8 Personalized Medicine 544

23 POPULATION GENETICS 554

- 23.1 Genes in Populations and the Hardy-Weinberg Equation 554
- 23.2 Overview of Microevolution 560
- 23.3 Natural Selection 561
- 23.4 Genetic Drift 567
- 23.5 Migration 570
- 23.6 Nonrandom Mating 571
- 23.7 Sources of New Genetic Variation 572

24 QUANTITATIVE GENETICS 584

- 24.1 Overview of Quantitative Traits 584
- 24.2 Statistical Methods for Evaluating Quantitative Traits 586
- 24.3 Polygenic Inheritance 590
- 24.4 Identification of Genes That Control Quantitative Traits 592
- 24.5 Heritability 595
- 24.6 Selective Breeding 600

Appendix A A-1

Appendix B A-7

Glossary G-1

Index I-1

PREFACE

Based on discussions with instructors from many institutions, I have learned that most instructors want a broad textbook that clearly explains concepts in a way that is interesting, accurate, concise, and up-to-date. *Concepts of Genetics* has been written to achieve these goals. It is intended for students who want to gain a conceptual grasp of the various fields of genetics. The content reflects current trends in genetics, and the pedagogy is based on educational research. In particular, a large amount of formative assessment is woven into the content. As an author, researcher, and teacher, I want a textbook that gets students actively involved in learning genetics. To achieve this goal, I have worked with a talented team of editors, illustrators, and media specialists who have helped me to make the third edition of *Concepts of Genetics* a fun learning tool.

FLIPPING THE CLASSROOM

A trend in science education is the phenomenon that is sometimes called "flipping the classroom." This phrase refers to the idea that some of the activities that used to be done in class are now done out of class, and vice versa. For example, instead of spending the entire class time lecturing about the textbook and other materials, some of the class time is spent engaging students in various activities, such as problem solving, working through case studies, and designing experiments. This approach is called active learning. For many instructors, the classroom has become more learner centered rather than teacher centered. A learner-centered classroom provides a rich environment in which students can interact with each other and with their instructors. Instructors and fellow students often provide formative assessment—immediate feedback that helps students understand if their learning is on the right track.

What are some advantages of active learning? Educational studies reveal that active learning usually promotes greater learning gains. In addition, active learning often focuses on skill development rather than the memorization of facts that are easily forgotten. Students become trained to "think like scientists" and to develop a skill set that enables them to apply scientific reasoning.

A common concern among instructors who are beginning to try out active learning is that they think they will have to teach their students less material. However, this may not be the case. Although students may be provided with online lectures, "flipping the classroom" typically gives students more responsibility for understanding the textbook material on their own. Along these lines, *Concepts of Genetics,* Third Edition, is intended to provide students with a resource that can be effectively used out of the classroom. Several key pedagogical features include the following:

- **Learning Outcomes** Each section of every chapter begins with a set of learning outcomes. These outcomes help students understand what they should be able to do if they have mastered the material in that section.
- **Formative Assessment** When students are expected to learn textbook material on their own, it is imperative that they are given formative assessment on a regular basis so they can gauge whether or not they are mastering the material. Formative assessment is a major feature of this textbook and is bolstered by McGraw-Hill Connect®—a state-of-the art digital assignment and assessment platform. In *Concepts of Genetics,* Third Edition, formative assessment is provided in multiple ways.

  1. Each section of every chapter ends with multiple-choice questions. Formative assessment at the end of each section allows students to evaluate their mastery of the material before moving on to the next section.
  2. Most figures have concept check questions so students can determine if they understand the key points in the figure.
  3. Extensive end-of-chapter questions continue to provide students with feedback regarding their mastery of the material.
  4. A feature called **Genetic TIPS** provides a consistent approach to help students solve problems in genetics. This approach has three components: First, the student is made aware of the *T*opic at hand. Second, the question is evaluated with regard to the *I*nformation that is available to the student. Finally, the student is guided through a *P*roblem-*S*olving *S*trategy to tackle the question.
  5. Additional questions, including questions that pertain to every feature investigation, are available to the student in Connect: http://successinhighered.com/genetics-molecular-biology.
  6. The textbook material is supported by digital learning tools found in Connect. Questions and activities are assignable in Connect. Assignments due before class time or following an in-class activity help students prepare or review.
  7. McGraw-Hill SmartBook is an adaptive learning tool available in Connect that has been shown to strengthen recall and increase retention so that students can move beyond memorizing and truly learn the material.

- **Chapter Organization** In genetics, it is sometimes easy to "lose the forest for the trees." Genetics can be a dense subject. To circumvent this difficulty, the content in *Concepts of Genetics* has been organized to foster a better appreciation for the big picture of genetic principles. The chapters are divided into several sections, and each section ends with a summary that touches on the main points. As mentioned, multiple-choice questions at the end of each section are also intended to help students grasp the broader concepts in genetics. Finally, the end of each chapter contains a summary, which allows students to connect the concepts that were learned in each section.

- **Interactive Exercises** Working with education specialists, the author has crafted interactive exercises in which the students can make their own choices in problem-solving activities and predict what the outcomes will be. Many of these exercises are focused on inheritance patterns and human genetic diseases. (For example, see Chapters 5 and 22.) In addition, there are many interactive exercises for the molecular chapters. These types of exercises engage students in the learning process. The interactive exercises are found online, and the corresponding material in the chapter is indicated with an Interactive Exercise icon.

- **Animations** Our media specialists have created over 50 animations for a variety of genetic processes. These animations were made specifically for this textbook and use the art from the textbook. The animations literally make many of the figures in the textbook "come to life." The animations are found online and the corresponding material in the chapter is indicated with an Online Animation icon.

An effective textbook needs to accomplish three goals: First, it needs to provide comprehensive, accurate, and up-to-date content in its field. Second, it needs to expose students to the techniques and skills they will need to become successful in that field. And finally, it should inspire students so they want to study the material. The hard work that has gone into the third edition of *Concepts of Genetics* has been aimed at achieving all three of these goals. Furthermore, the pedagogy of *Concepts of Genetics* has been designed to foster student learning. Instead of being a collection of "facts and figures," *Concepts of Genetics*, Third Edition, by Robert Brooker, is intended to be an engaging and motivating textbook in which formative assessment allows students to move ahead and learn the material in a productive way. We welcome your feedback so we can make future editions even better!

HOW WE EVALUATED YOUR NEEDS

ORGANIZATION

In surveying many genetics instructors, it became apparent that most people fall into two camps: **Mendel first** versus **Molecular first**. I have taught genetics both ways. As a teaching tool, this textbook has been written with these different teaching strategies in mind. The organization and content lend themselves to various teaching formats.

Chapters 2 through 10 are largely inheritance chapters, whereas Chapters 23 and 24 examine population and quantitative genetics. The bulk of the molecular genetics is found in Chapters 11 through 22, although I have tried to weave a fair amount of molecular genetics into Chapters 2 through 10 as well. The information in Chapters 11 through 22 does not assume that a student has already covered Chapters 2 through 10. Actually, each chapter is written with the perspective that instructors may want to vary the order of their chapters to fit their students' needs.

For those who like to discuss inheritance patterns first, a common strategy would be to cover Chapters 1 through 10 first, and then possibly 23 and 24. (However, many instructors like to cover quantitative and population genetics at the end. Either way works fine.) The more molecular and technical aspects of genetics would then be covered in Chapters 11 through 22. Alternatively, if you like the "Molecular first" approach, you would probably cover Chapter 1, then skip to Chapters 11 through 22, then return to Chapters 2 through 10, and then cover Chapters 23 and 24 at the end of the course. This textbook was written in such a way that either strategy works well.

ACCURACY

Both the publisher and I acknowledge that inaccuracies can be a source of frustration for both the instructor and students. Therefore, throughout the writing and production of this textbook we have worked very hard to catch and correct errors during each phase of development and production.

Each chapter has been reviewed by faculty members who teach the course or conduct research in genetics or both. In addition, a developmental editor has gone through the material to check for accuracy in art and consistency between the text and art. When the problem sets were first developed, we had a team of students work through all of the problems and one developmental editor also checked them. The author personally checked every question and answer when the chapters were completed for this edition.

ILLUSTRATIONS

In surveying students whom I teach, I often hear it said that most of their learning comes from studying the figures. Likewise, instructors frequently use the illustrations from a textbook as a central teaching tool. For these reasons, a great amount of effort has gone into the illustrations. The illustrations are created with four goals in mind:

1. **Completeness** For most figures, it should be possible to understand an experiment or genetic concept by looking at the illustration alone. Students have complained that it is difficult to understand the content of an illustration if they have to keep switching back and forth between the figure and text. In cases where an illustration shows the steps in a scientific process, the steps are described in brief statements that allow the students to understand the whole process (e.g., see Figure 17.10). Likewise, such illustrations should make it easier for instructors to explain these processes in the classroom.

2. **Clarity** The figures have been extensively reviewed by students and instructors. This has helped us to avoid drawing things that may be confusing or unclear. Aside from being unmistakably drawn, all new elements within each figure are clearly labeled.
3. **Consistency** Before we began to draw the figures, we generated a style sheet that contained recurring elements that are found in many places in the textbook. Examples include the DNA double helix, DNA polymerase, and fruit flies. We agreed on the best way(s) to draw these elements and also what colors they should be. Therefore, as students and instructors progress through this textbook, they become accustomed to the way things should look.
4. **Realism** An important emphasis of this textbook is to make each figure as realistic as possible. When drawing macroscopic elements (e.g., fruit flies, pea plants), the illustrations are based on real images, not on cartoonlike simplifications. Our most challenging goal, and one that we feel has been achieved most successfully, is the realism of our molecular drawings. Whenever possible, we have tried to depict molecular elements according to their actual structures, if such structures are known. For example, the ways we have drawn RNA polymerase, DNA polymerase, DNA helicase, and ribosomes are based on their crystal structures. When a student sees a figure in this textbook that illustrates an event in transcription, RNA polymerase is depicted in a way that is as realistic as possible (e.g., Figure 14.8 below).

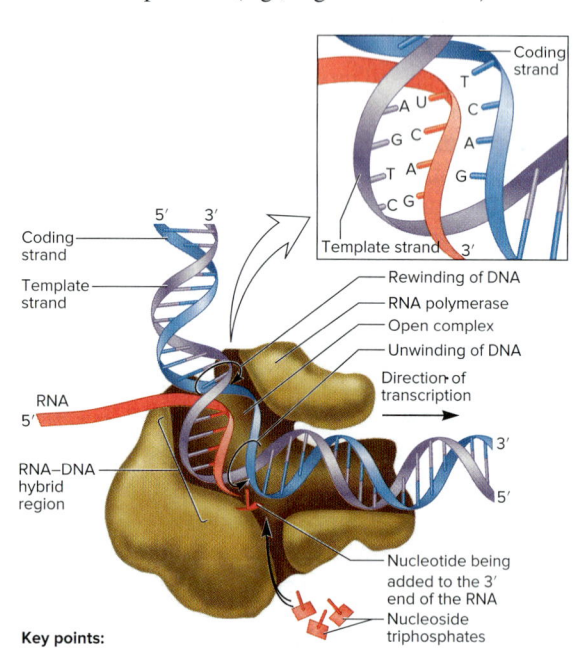

Key points:
- RNA polymerase slides along the DNA, creating an open complex as it moves.
- The DNA strand known as the template strand is used to make a complementary copy of RNA, resulting in an RNA–DNA hybrid.
- RNA polymerase moves along the template strand in a 3′ to 5′ direction, and RNA is synthesized in a 5′ to 3′ direction using nucleoside triphosphates as precursors. Pyrophosphate is released (not shown).
- The complementarity rule is the same as the AT/GC rule except that U is substituted for T in the RNA.

FEATURE EXPERIMENTS

Many chapters have one or two experiments that are presented according to the scientific method. These experiments are integrated within the chapters and flow with the rest of the text. As you are reading the experiments, you will simultaneously explore the scientific method and the genetic principles that have been discovered using this approach. For students, I hope this textbook helps you to see the fundamental connection between scientific analysis and principles. For both students and instructors, I expect that this strategy makes genetics much more fun to explore.

WRITING STYLE

Motivation in learning often stems from enjoyment. If you enjoy what you're reading, you are more likely to spend longer amounts of time with it and focus your attention more crisply. The writing style of this book is meant to be interesting, down to earth, and easy to follow. Each section of every chapter begins with an overview of the contents of that section, usually with a table or figure that summarizes the broad points. The section then examines how those broad points were discovered experimentally, as well as explaining many of the finer scientific details. Important terms appear in the text in a boldface font. These terms are also found at the end of the chapter and in the glossary.

There are various ways to make a genetics book interesting and inspiring. The subject matter itself is pretty amazing, so it's not difficult to build on that. In addition to describing the concepts and experiments in ways that motivate students, it is important to draw on examples that bring the concepts to life. In a genetics book, many of these examples come from the medical realm. This textbook contains lots of examples of human diseases that convey some of the underlying principles of genetics. Students often say they remember certain genetic concepts because they remember how defects in certain genes can cause disease. For example, defects in DNA repair genes cause a higher predisposition to develop cancer. In addition, I have tried to be evenhanded in providing examples from the microbial and plant world. Finally, students are often interested in applications of genetics that affect their everyday lives. Because we frequently hear about genetics in the news, it's inspiring for students to learn the underlying basis for such technologies. Chapters 20 and 21 are devoted to genetic technologies, and applications of these and other technologies are found throughout this textbook. By the end of their genetics course, students should come away with a greater appreciation for the influence of genetics in their lives.

SIGNIFICANT CONTENT CHANGES TO THE THIRD EDITION

- A new feature called ***Genetic TIPS*** was added to the second edition. Many of these problem-solving activities have been refined based on student and instructor feedback.

Examples of Specific Content Changes to Individual Chapters

In addition to the usual updates to material based on new research information, other additions and changes to the third edition have been made, as described next.

- Chapter 5. Extensions of Mendelian Inheritance: A new section has been added that explores how development plays a role in producing certain traits. In particular, having dark fur on the back and white fur on the underside, a trait that is observed in certain breeds of dogs and other animals, is discussed in the context of the migration of melanocyte precursor cells during embryonic development (see Figure 5.13).
- Chapter 7. Genetic Linkage and Mapping in Eukaryotes: Based on student feedback, the discussion of linkage has been improved by increased emphasis on the description of how crossing over is related to the production of recombinant offspring and by the revision of certain figures (e.g., Figure 7.8).
- Chapter 10. Genetics of Viruses: New information about Zika virus has been added.
- Chapter 12. Molecular Structure of Chromosomes and Transposition: The discussion of the mechanism of bacterial chromosome compaction has been updated with a new figure (see Figure 12.3). The topic of transposable elements has been moved to this chapter because of their impact on chromosome structure and genome size.
- Chapter 13. DNA Replication and Recombination: Based on instructor feedback, the topic of homologous recombination has been added to this chapter so that it appears before many of the molecular topics in subsequent chapters.
- Chapter 14. Gene Transcription and RNA Modification: The topic of alternative splicing has been moved to this chapter so that it immediately follows the discussion of splicing.
- Chapter 17. Gene Regulation in Eukaryotes: The discussion of epigenetics has been expanded from one to three sections, which begins with an overview and then focuses on how epigenetic changes are programmed during development and how environmental factors cause epigenetic changes.
- *NEW!* Chapter 18. Non-Coding RNAs: In the past decade or so, technical advances have allowed researchers to identify and study the functions of RNA molecules that do not encode proteins. This new chapter has been added to this edition to focus on this critical topic in molecular genetics.
- Chapter 20. Molecular Technologies: Discussion of CRISPR-Cas technology for gene editing has been added to this chapter (see Figure 20.18), and the topics of reproductive cloning and stem cells have been moved to this chapter from a later chapter.
- Chapter 21. Genomics: The coverage of the topics of genomics and functional genomics has been streamlined and combined in this chapter. The relatively new method of RNA sequencing (RNA-Seq) has been added (see Figure 21.14).
- Chapter 22. Medical Genetics and Cancer: This chapter now covers the use of genome-wide association studies (GWAS) to identify genes associated with human diseases (see Figure 22.8). The section on the genetic basis of cancer has been subdivided into four sections: 22.4 Overview of Cancer, 22.5 Oncogenes, 22.6 Tumor-Suppressor Genes, and 22.7 Role of Epigenetics and Cancer.
- Chapter 24. Quantitative Genetics: The discussion of identifying genes involved in quantitative traits has been set apart in its own section.

SUGGESTIONS WELCOME!

It seems very appropriate to use the word *evolution* to describe the continued development of this textbook. I welcome any and all comments. The refinement of any science textbook requires input from instructors and their students. These include comments regarding writing, illustrations, supplements, factual content, and topics that may need greater or less emphasis. You are invited to contact me at:

Dr. Rob Brooker
Dept. of Genetics, Cell Biology, and Development
University of Minnesota
6-160 Jackson Hall
321 Church St
Minneapolis, MN 55455
brook005@umn.edu

connect®

McGraw Hill Education

Students—study more efficiently, retain more and achieve better outcomes. Instructors—focus on what you love—teaching.

SUCCESSFUL SEMESTERS INCLUDE CONNECT

For Instructors

You're in the driver's seat.

Want to build your own course? No problem. Prefer to use our turnkey, prebuilt course? Easy. Want to make changes throughout the semester? Sure. And you'll save time with Connect's auto-grading too.

65% Less Time Grading

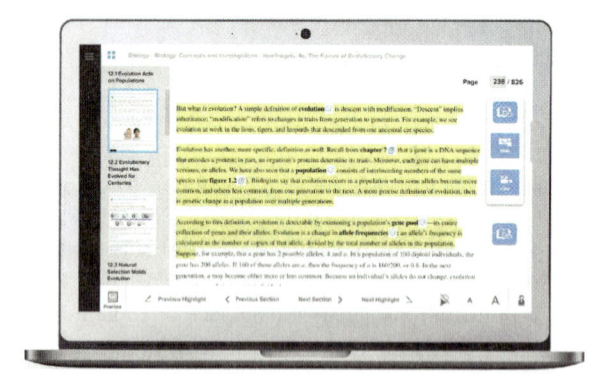

They'll thank you for it.

Adaptive study resources like SmartBook® help your students be better prepared in less time. You can transform your class time from dull definitions to dynamic debates. Hear from your peers about the benefits of Connect at **www.mheducation.com/highered/connect**

Make it simple, make it affordable.

Connect makes it easy with seamless integration using any of the major Learning Management Systems—Blackboard®, Canvas, and D2L, among others—to let you organize your course in one convenient location. Give your students access to digital materials at a discount with our inclusive access program. Ask your McGraw-Hill representative for more information.

©Hill Street Studios/Tobin Rogers/Blend Images LLC

Solutions for your challenges.

A product isn't a solution. Real solutions are affordable, reliable, and come with training and ongoing support when you need it and how you want it. Our Customer Experience Group can also help you troubleshoot tech problems—although Connect's 99% uptime means you might not need to call them. See for yourself at **status.mheducation.com**

FOR STUDENTS

Effective, efficient studying.

Connect helps you be more productive with your study time and get better grades using tools like SmartBook, which highlights key concepts and creates a personalized study plan. Connect sets you up for success, so you walk into class with confidence and walk out with better grades.

©Shutterstock/wavebreakmedia

> "I really liked this app—it made it easy to study when you don't have your textbook in front of you."
>
> - Jordan Cunningham,
> Eastern Washington University

Study anytime, anywhere.

Download the free ReadAnywhere app and access your online eBook when it's convenient, even if you're offline. And since the app automatically syncs with your eBook in Connect, all of your notes are available every time you open it. Find out more at **www.mheducation.com/readanywhere**

No surprises.

The Connect Calendar and Reports tools keep you on track with the work you need to get done and your assignment scores. Life gets busy, Connect tools help you keep learning through it all.

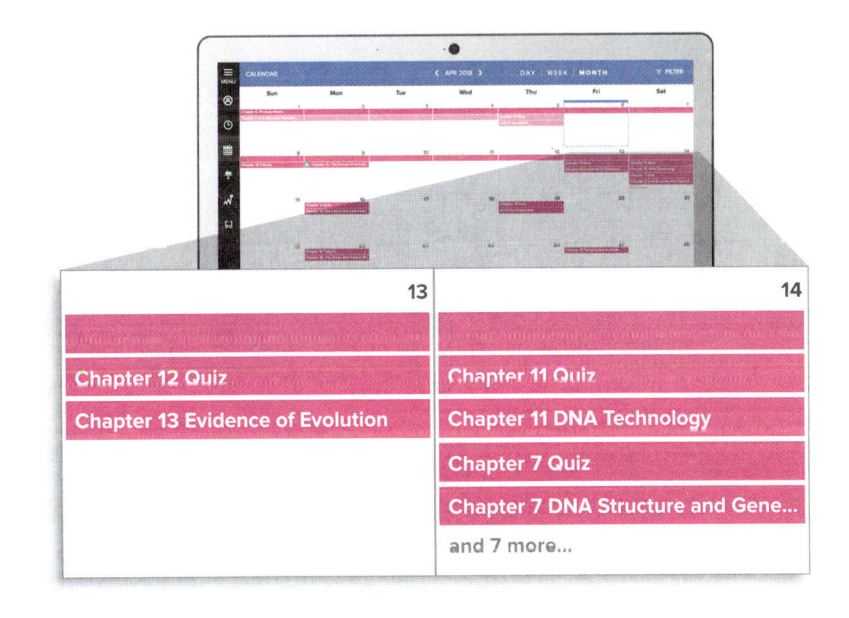

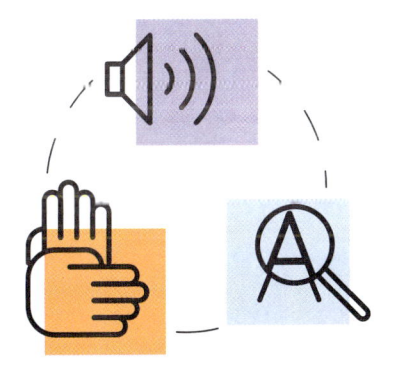

Learning for everyone.

McGraw-Hill works directly with Accessibility Services Departments and faculty to meet the learning needs of all students. Please contact your Accessibility Services office and ask them to email accessibility@mheducation.com, or visit **www.mheducation.com/about/accessibility.html** for more information.

ACKNOWLEDGMENTS

The production of a textbook is truly a collaborative effort, and I am greatly indebted to a variety of people. This textbook has gone through multiple rounds of rigorous revision that involved the input of faculty, students, editors, and educational and media specialists. Their collective contributions are reflected in the final outcome.

Let me begin by acknowledging the many people at McGraw-Hill Education whose efforts are amazing. My highest praise goes to Andrew Urban (Portfolio Manager), who managed many aspects of this project. I also would like to thank Elizabeth Sievers (Senior Product Developer) for her patience in overseeing this project and her contributions to the digital components. Liz is the glue that holds textbook development and textbook production together. Other people at McGraw-Hill have played key roles in producing an actual book and the supplements that go along with it. In particular, Jessica Portz (Content Project Manager) has done a superb job of managing the components that need to be assembled to produce a book. I would also like to thank Lori Hancock (Content Licensing Specialist), who acted as an interface between me and the photo company. In addition, my gratitude goes to David Hash (Designer), who provided much input into the internal design of the book as well as creating an awesome cover. Finally, I would like to thank Kelly Brown (Marketing Manager), whose major efforts begin once the book is published!

With regard to the content of the book, Deborah Brooker (Freelance Developmental Editor) has worked closely with me in developing the art for this textbook for all of the editions. She has scrutinized each figure for clarity and logic. I would also like to thank Jane Hoover (Freelance Copy Editor) for her superb copy editing. Her crisp understanding of the material allows her to edit it in a meaningful way, which has significantly improved the text's clarity.

I would also like to extend my thanks to everyone at MPS Limited, including the many artists who have played important roles in developing the art for this textbook. Also, the folks at MPS Limited worked with great care in the paging of the book, making sure that the figures and relevant text are as close to each other as possible. Likewise, the people at Aptara have done a great job of locating many of the photographs that have been used in this textbook.

Finally, I want to thank the many scientists who reviewed the chapters of this textbook and previous editions. Their broad insights and constructive suggestions were an important factor that shaped its final content and organization. I am truly grateful for their time and effort.

A Visual Guide to
CONCEPTS OF GENETICS

LEARNING THROUGH EXPERIMENTATION

Many chapters contain an experiment that is presented according to the scientific method. These experiments are integrated within the chapters and flow with the rest of the chapter material. As you read the experiments, which can be hypothesis-testing or discovery-based science, you will simultaneously explore the scientific method and the genetic principles learned from this approach.

BACKGROUND OBSERVATIONS

Each experiment begins with a description of the information that led researchers to study a hypothesis-driven or discovery-based problem. Detailed information about the researchers and the experimental challenges they faced help students to understand actual research.

THE HYPOTHESIS OR THE GOAL

The student is given a possible explanation for the observed phenomenon that will be tested or the question researchers were hoping to answer. This section reinforces the scientific method and allows students to experience the process for themselves.

Mendel Followed the Outcome of a Single Character for Two Generations

Prior to conducting his studies, Mendel did not have a hypothesis to explain the formation of hybrids. However, his educational background led him to realize that a quantitative analysis of crosses might uncover mathematical relationships that would otherwise be mysterious. His experiments were designed to determine the relationships that govern hereditary traits. This rationale is called an **empirical approach.** Laws that are deduced from an empirical approach are known as **empirical laws.**

Mendel's experimental procedure is shown in **Figure 3.5**. He began with true-breeding plants that differed with regard to a single character. These plants are termed the **parental generation**

▶ THE GOAL (DISCOVERY-BASED SCIENCE)

Mendel speculated that the inheritance pattern for a single character may follow quantitative natural laws. The goal of this experiment was to uncover such laws.

TESTING THE HYPOTHESIS OR ACHIEVING THE GOAL

This section illustrates the experimental process, including the actual steps followed by scientists to test their hypothesis or study a question. Science comes alive for students with this detailed look at experimentation.

▶ **THE GOAL (DISCOVERY-BASED SCIENCE)**
Mendel speculated that the inheritance pattern for a single character may follow quantitative natural laws. The goal of this experiment was to uncover such laws.

▶ **ACHIEVING THE GOAL — FIGURE 3.5** Mendel's analysis of single-factor crosses.

Starting material: Mendel began his experiments with true-breeding pea plants that varied with regard to only one of seven different characters (see Figure 3.4).

Experimental level

P plants

Tall × Dwarf

Note: The P cross produces seeds that are part of the F₁...

F₁ seeds

Conceptual level

$TT \times tt$

All Tt

1. For each of seven characters, Mendel cross-fertilized two different true-breeding strains. Keep in mind that each cross involved two plants that differed in regard to only one of the seven characters studied. The illustration at the right shows one cross between a tall and dwarf plant. This is called a P (parental) cross.

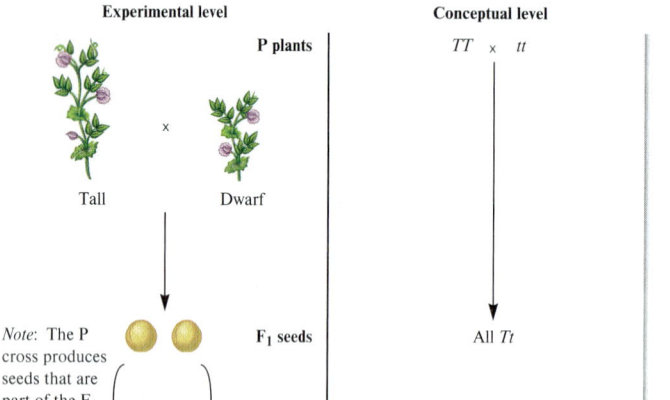

2. Collect the F₁ generation seeds. The following spring, plant the seeds and allow the plants to grow. These are the plants of the F₁ generation.

THE DATA

Actual data from the original research paper help students understand how real-life research results are reported. Each experiment's results are discussed in the context of the larger genetic principle to help students understand the implications and importance of the research.

▶ **THE DATA**

P cross	F₁ generation	F₂ generation	Ratio of traits in F₂ generation
Tall × dwarf height	All tall	787 tall, 277 dwarf	2.84:1
Purple × white flowers	All purple	705 purple, 224 white	3.15:1
Axial × terminal flowers	All axial	651 axial, 207 terminal	3.14:1
Yellow × green seeds	All yellow	6022 yellow, 2001 green	3.01:1
Round × wrinkled seeds	All round	5474 round, 1850 wrinkled	2.96:1
Green × yellow pods	All green	428 green,	

▶ **INTERPRETING THE DATA**

The data shown in the table above are the results of producing an F₁ generation via cross-fertilization and an F₂ generation via self-fertilization of the F₁ monohybrids. A quantitative analysis of these data allowed Mendel to propose three important ideas:

1. Mendel's data argued strongly against a blending mechanism of inheritance. In all seven cases, the F₁ generation displayed traits that were distinctly like one of the two parents rather than traits intermediate in character. His first proposal was that one variant of a character is **dominant** to another variant. For example, the variant of green pods is dominant to that of yellow pods. The term **recessive** is used to describe a variant that is masked by the presence of a dominant trait but reappears in subsequent generations. Yellow pods and dwarf height are examples of recessive variants. They can also be referred to as recessive traits.

2. When a true-breeding plant with a dominant trait was crossed to a true-breeding plant with a recessive trait, the dominant trait was always observed in the F₁ generation. In the F₂ generation, some offspring displayed the dominant trait, but a smaller proportion showed the recessive trait. How did Mendel explain this observation? Because the recessive trait appeared in the F₂ generation, he made a second proposal—the genetic determinants of traits are passed along as "unit factors" from generation to generation. His data were consistent with a **particulate theory of inheritance,** in which the genes that govern traits are inherited as discrete units that remain unchanged as they are passed from parent to offspring. Mendel called them unit factors, but we now call them genes (from the Greek, *genesis*, meaning "birth," or *genos*, meaning "origin").

INTERPRETING THE DATA

This discussion, which examines whether the experimental data supported or disproved the hypothesis or provided new information to propose a hypothesis, gives students an appreciation for scientific interpretation.

xvi

Learning–Assessment–Problem Solving

These study tools and problems are crafted to aid students in reviewing key information in the text, assess their understanding, and develop problem-solving skills.

18.1 OVERVIEW OF NON-CODING RNAs

Learning Outcomes:

1. Describe the ability of ncRNAs to bind to other molecules and macromolecules.
2. Outline the general functions of ncRNAs.
3. Define *ribozyme*.
4. List several examples of ncRNAs, and describe their functions.

LEARNING OUTCOMES

Each section begins with one or more Learning Outcomes. These allow a student to appreciate the skills and knowledge they will gain if they master the material.

REVIEWING THE KEY CONCEPTS

These bulleted lists at the end of each section help students identify important concepts. Students should understand these concepts before moving on to the next section.

COMPREHENSION QUESTIONS

Multiple choice questions found at the end of each section allow students an opportunity to test their knowledge of key information and concepts. This helps students better identify what they know and don't know before tackling more concepts. Answers are provided at the end of the chapter.

4.2 REVIEWING THE KEY CONCEPTS

- Dosage compensation often occurs in species in which males and females differ in their sex chromosomes (see Table 4.1).
- In mammals, the process of X-chromosome inactivation (XCI) in females compensates for the single X chromosome found in males. The inactivated X chromosome is called a Barr body. The process can lead to a variegated phenotype, such as a calico cat (see Figure 4.5).
- After it occurs during embryonic development, X-chromosome inactivation is maintained when somatic cells divide (see Figure 4.6).
- X-chromosome inactivation is controlled by the X-inactivation center (Xic) that contains the *Xist* gene. The three phases of XCI are initiation, spreading, and maintenance phases (see Figure 4.7).

4.2 COMPREHENSION QUESTIONS

1. In fruit flies, dosage compensation is achieved by
   a. X-chromosome inactivation
   b. doubling the expression of genes on the single X chromosome in the male.
   c. decreasing the expression of genes on each X chromosome in the female to one-half.
   d. all of the above.
2. According to the Lyon hypothesis,
   a. one of the X chromosomes is converted to a Barr body in somatic cells of female mammals.
   b. one of the X chromosomes is converted to a Barr body in all cells of female mammals.
   c. both of the X chromosomes are converted to Barr bodies in somatic cells of female mammals.
   d. both of the X chromosomes are converted to Barr bodies in all cells of female mammals.
3. Which of the following is *not* a phase of XCI?
   a. Initiation
   b. Spreading
   c. Maintenance
   d. Erasure

GENES → TRAITS

Because genetics is such a broad discipline ranging from the molecular level to populations, many students have trouble connecting the concepts they learn in molecular genetics with the traits that occur at the level of an organism. To make this connection more meaningful, certain figures have a "Genes→Traits" feature that reminds students that molecular and cellular phenomena ultimately lead to traits observed in organisms.

CONCEPT CHECK QUESTIONS

Students can test their knowledge and understanding with Concept Check questions that are associated with the figure legends. These questions often go beyond simple recall of information and ask students to apply or interpret information presented in the illustrations.

FIGURE 4.6 The mechanism of X-chromosome inactivation.
Genes ⟶ Traits The top of this figure represents a mass of several cells that compose the early embryo. Initially, both X chromosomes are active. At an early stage of embryonic development, random inactivation of one X chromosome occurs in each cell. This inactivation pattern is maintained as the embryo matures into an adult.

Concept Check: At which stage of development does X-chromosome inactivation initially occur?

Genetic TIPS

The Question: If a diploid cell contains four chromosomes (i.e., two per set), how many possible random arrangements of homologs could occur during metaphase of meiosis I?

Topic: What topic in genetics does this question address?

The topic is meiosis. More specifically, the question is about metaphase of meiosis I.

Information: What information do you know based on the question and your understanding of the topic?

From the question, you know a cell that started with two pairs of homologous chromosomes has entered meiosis and is now in metaphase of meiosis I. From your understanding of the topic, you may remember that tetrads align along the metaphase plate (see Figure 2.11). The orientations of the homologs within the tetrads are random.

Problem-Solving Strategies: Make a drawing. Make a calculation.

One strategy to solve this problem is to make a drawing in which the homologs are in different colors, such as red and blue. Note: The spindle poles are labeled A and B in the drawing below. The alignment occurs relative to the spindle poles.

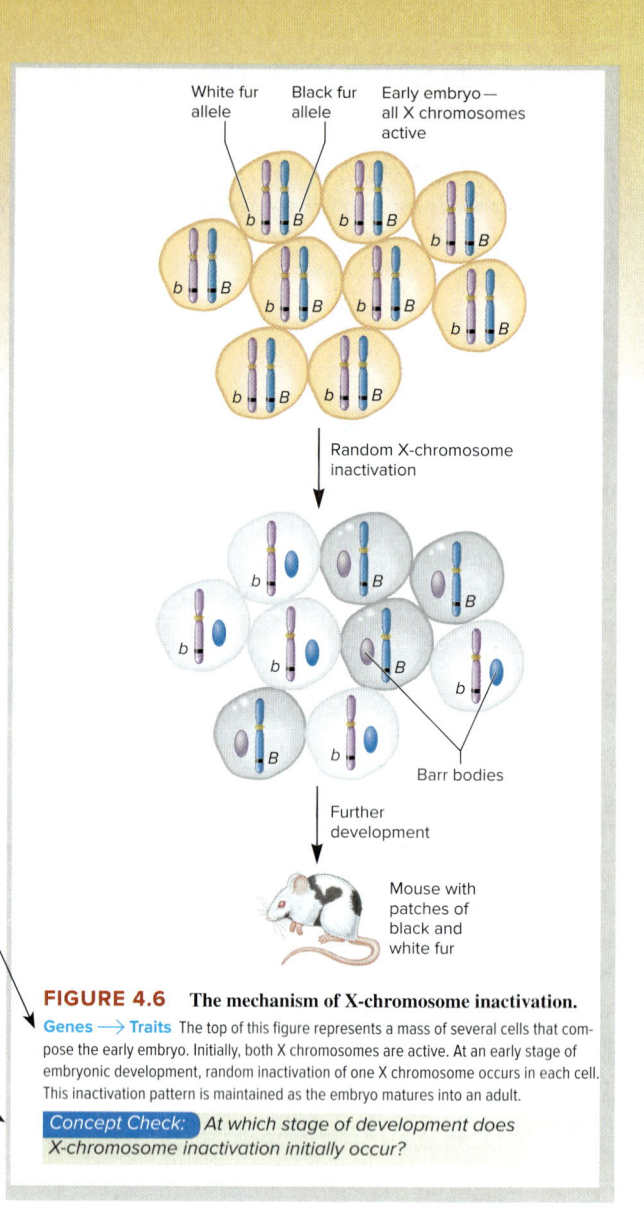

GENETIC TIPS

Problem solving is a skill that genetics students need to master. Genetic TIPS provides a consistent approach to help students solve problems in genetics. This approach has three components: First, the student is made aware of the Topic at hand. Second, the question is evaluated with regard to the Information that is available to the student. Finally, the student is guided through a Problem-Solving Strategy to tackle the question. More Genetic TIPS are presented at the end of the chapter, allowing for additional practice in strengthening problem-solving skills.

End-of-Chapter Support Materials

These study tools and problems are crafted to aid students in reviewing key information in the text and developing a wide range of problem-solving skills. They also develop a student's cognitive, writing, analytical, computational, and collaborative abilities.

KEY TERMS

Providing the key terms from the chapter enhances student development of vital vocabulary necessary for the understanding and application of chapter content. Important terms are boldfaced throughout the chapter and page referenced at the end of each chapter for reflective study.

CHAPTER SUMMARY

These bulleted summaries, which are organized by section, emphasize the main concepts of the chapter to provide students with a thorough review of the main topics covered.

KEY TERMS

Page 114. nuclear genes, extranuclear inheritance (cytoplasmic inheritance), nucleoid, chloroplast DNA (cpDNA)
Page 115. reciprocal cross, maternal inheritance, heteroplasmy
Page 116. heterogamous
Page 117. mitochondrial DNA (mtDNA)
Page 118. paternal leakage
Page 119. endosymbiosis, endosymbiosis theory
Page 120. epigenetic inheritance, genomic imprinting (imprinting), monoallelic expression
Page 122. DNA methylation, imprinting control region (ICR)
Page 124. maternal effect

CHAPTER SUMMARY

- Non-Mendelian inheritance refers to inheritance patterns that do not obey Mendel's laws of inheritance.

6.1 Extranuclear Inheritance: Chloroplasts

- Extranuclear inheritance is the inheritance of genes that are found in chloroplasts or mitochondria.
- Chloroplasts have circular chromosomes in a nucleoid. These circular chromosomes carry many genes but far fewer compared with the number on chromosomes in the cell nucleus (see Figure 6.1, Table 6.1).
- Maternal inheritance occurs when organelles, such as chloroplasts, are transmitted via the egg (see Figure 6.2).
- Heteroplasmy of chloroplasts can result in a variegated pheno-

6.3 Theory of Endosymbiosis

- Chloroplasts and mitochondria were derived from ancient endosymbiotic relationships (see Figure 6.6).

6.4 Epigenetics: Imprinting

- Epigenetic inheritance is an inheritance pattern in which a gene or chromosome is modified so as to alter gene expression, but the modification is not permanent over the course of many generations. An example is imprinting, in which an offspring expresses a gene that is inherited from one parent but not both (see Figures 6.7, 6.8).
- DNA methylation at an imprinting control region is the marking

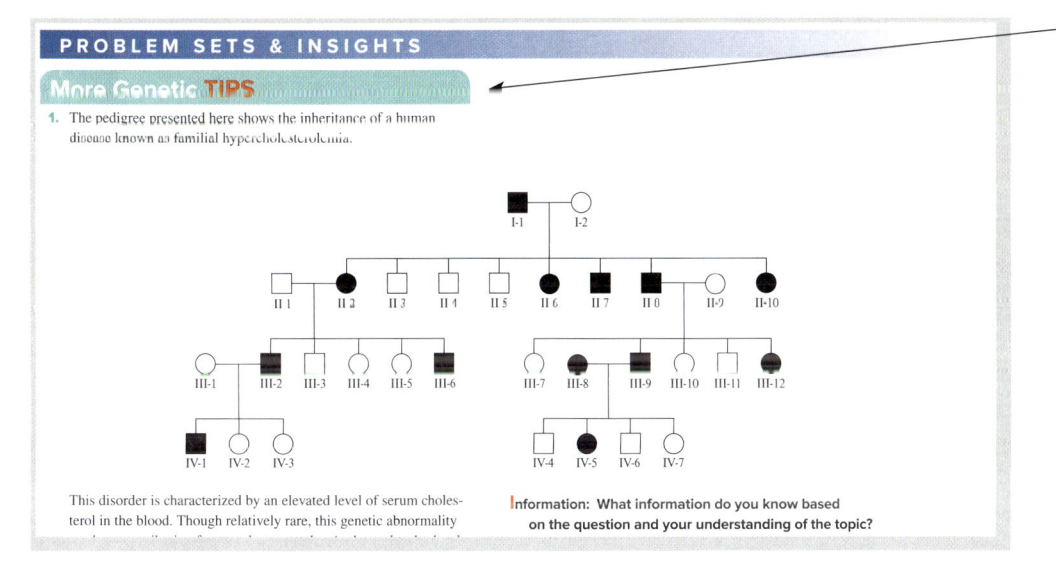

MORE GENETIC TIPS

*Like the Genetic TIPS within the chapter, these problems provide more practice in developing problem-solving skills before the students work through more problems unaided. The Genetic TIPS help the student identify the primary question (the **T**opic), evaluate the question based on the student's knowledge of the topic (**I**nformation), and then the student is guided through the solution revealing a **P**roblem-**S**olving **S**trategy. These provide a reference for when students encounter similar problems later.*

PROBLEM SETS & INSIGHTS

More Genetic TIPS

1. The pedigree presented here shows the inheritance of a human disease known as familial hypercholesterolemia.

This disorder is characterized by an elevated level of serum cholesterol in the blood. Though relatively rare, this genetic abnormality

Information: What information do you know based on the question and your understanding of the topic?

xix

CONCEPTUAL QUESTIONS

These questions test the understanding of basic genetic principles. The student is given many questions with a wide range of difficulty. Some require critical-thinking skills, and some require the student to write coherent answers in an essay form.

Application and Experimental Questions

E1. Describe three advantages of using pea plants as an experimental organism.

E2. Explain the technical differences between a cross-fertilization experiment and a self-fertilization experiment.

E3. How long did it take Mendel to complete the experiment in Figure 3.5?

E4. For all seven characters described in the data of Figure 3.5, Mendel allowed the F_2 plants to self-fertilize. He found that when F_2 plants with recessive traits were crossed to each other, they always bred true. However, when F_2 plants with dominant traits were crossed, some bred true but others did not. A summary of Mendel's results is shown in the following table.

The Ratio of True-Breeding and Non-True-Breeding Parents of the F_2 Generation

F_2 Parents	True-Breeding	Non-True-Breeding	Ratio
Round	193	372	1:1.93
Yellow	166	353	1:2.13
Gray	36	64	1:1.78
Smooth	29	71	1:2.45
Green	40	60	1:1.5
Axial	33	67	1:2.08
Tall	28	72	1:2.57
TOTAL:	525	1059	1:2.02

When considering the data in this table, keep in mind that they describe the characteristics of the F_2 generation parents that had

Conceptual Questions

C1. The process of binary fission begins with a single mother cell ends with two daughter cells. Would you expect the mother an daughter cells to be genetically identical? Explain why or why not.

C2. What is a homolog? With regard to genes and alleles, how are homologs similar to and different from each other?

C3. What is a sister chromatid? Are sister chromatids genetically s lar or identical? Explain.

C4. With regard to sister chromatids, which phase of mitosis is the organization phase, and which is the separation phase?

C5. A species is diploid with three chromosomes per set. Make a d ing that shows what the chromosomes would look like in the G and G_2 phases of the cell cycle.

C6. How does the attachment of kinetochore microtubules to the k ochore differ in metaphase of meiosis I from metaphase of mit Discuss what you think would happen if a sister chromatid wa attached to a kinetochore microtubule.

C7. For the following events, specify whether they occur during m sis, meiosis I, or meiosis II:

a. Separation of conjoined chromatids within a pair of sister

APPLICATION AND EXPERIMENTAL QUESTIONS

These questions test the ability to analyze data, design experiments, or appreciate the relevance of experimental techniques.

QUESTIONS FOR STUDENT DISCUSSION/ COLLABORATION

These questions encourage students to consider broad concepts and practical problems. Some questions require a substantial amount of computational activities, which can be worked on as a group.

Questions for Student Discussion/Collaboration

1. Consider this cross in pea plants: Tt Rr yy Aa × Tt rr Yy Aa, where T = tall, t = dwarf, R = round, r = wrinkled, Y = yellow, y = green, A = axial, a = terminal. What is the expected phenotypic outcome of this cross? Have one group of students solve this problem by making one big Punnett square, and have another group solve it by making four single-gene Punnett squares and using the multiplication method. Time each other to see who gets done first.

or dwarf with terminal flowers and the fourth offspring will be tall with axial flowers? Discuss what operation(s) (e.g., product rule or binomial expansion equation) you used and in what order they were used.

3. Consider this four-factor cross: Tt Rr yy Aa × Tt RR Yy aa, where T = tall, t = dwarf, R = round, r = wrinkled, Y = yellow, y = green, A = axial, a = terminal. What is the probability that the first three plants

xx

PART I INTRODUCTION

1

CHAPTER OUTLINE

1.1 The Molecular Expression of Genes
1.2 The Relationship Between Genes and Traits
1.3 Fields of Genetics
1.4 The Science of Genetics

CC (for "carbon copy" or "copy cat"), the first cloned pet. In 2002, the cat shown here was produced by cloning, a procedure described in Chapter 20.
©Texas A&M University/Getty Images

OVERVIEW OF GENETICS

Hardly a week goes by without a major news story involving a genetic breakthrough. The increasing pace of genetic discoveries has become staggering. The Human Genome Project is a case in point. This project began in the United States in 1990, when the National Institutes of Health and the Department of Energy joined forces with international partners to decipher the massive amount of information contained in our **genome**—the **deoxyribonucleic acid (DNA)** found within all of our chromosomes (**Figure 1.1**). Remarkably, in only a decade, the researchers working on this project determined the DNA sequence of 90% of the human genome. The completed sequence, published in 2003, has an accuracy greater than 99.99%; fewer than one mistake was made in every 10,000 base pairs (bp)!

In 2008, a more massive undertaking, called the 1000 Genomes Project, was launched, with the goal of establishing a detailed understanding of human genetic variation. In this international project, researchers set out to determine the DNA sequence of at least 1000 anonymous participants from around the globe. In 2015, the sequencing of over 2500 genomes was described in the journal *Nature*.

Studying the human genome allows us to explore fundamental details about ourselves at the molecular level. The results of human genome projects have shed considerable light on basic questions, such as how many genes we have, how genes direct the activities of living cells, how species evolve, how single cells develop into complex tissues, and how defective genes cause disease. Furthermore, such understanding may lend itself to improvements in modern medicine by providing better diagnoses of diseases and allowing the development of new treatments for them.

A controversial example of a genetic technology is mammalian cloning. In 1997, Ian Wilmut and his colleagues produced clones of sheep, using mammary cells from an adult animal (**Figure 1.2**). More recently, such cloning has been achieved in several mammalian species, including cows, mice, goats, pigs, and cats. In 2002, the first pet was cloned, a cat named CC (for "carbon copy" or "copy cat"; see the photo at the beginning of the chapter). The cloning of mammals provides the potential for many practical applications. Cloning of livestock would enable farmers to use cells from their best individuals to create genetically homogeneous herds. This could be advantageous in terms of agricultural yield, although such a genetically homogeneous herd may be more susceptible to certain diseases. However, people have become greatly concerned with the possibility of human cloning. As discussed in Chapter 20, this prospect has raised serious ethical questions. Within the past few years, legislative bills have been introduced that involve bans on human cloning.

1

DNA, the molecule of life

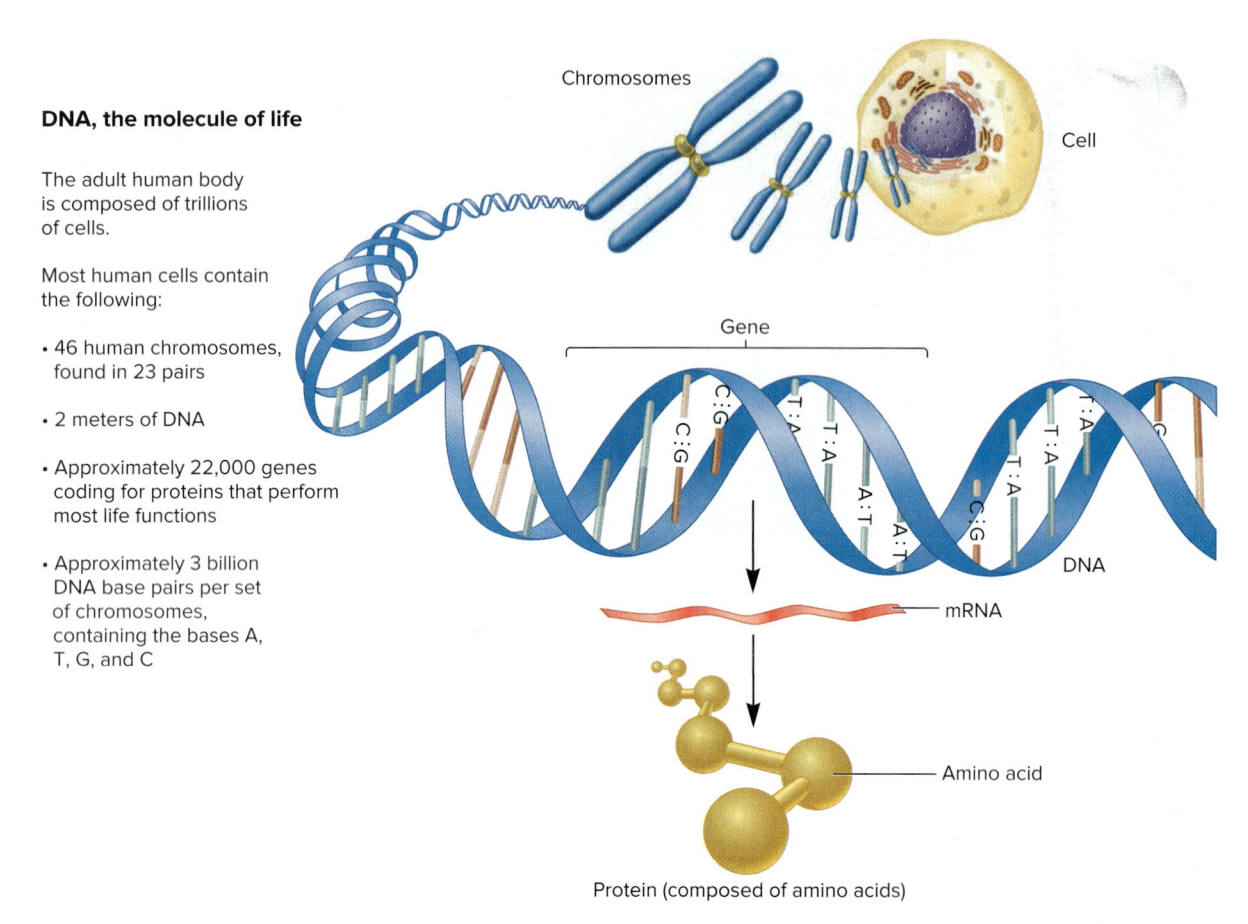

The adult human body is composed of trillions of cells.

Most human cells contain the following:

- 46 human chromosomes, found in 23 pairs
- 2 meters of DNA
- Approximately 22,000 genes coding for proteins that perform most life functions
- Approximately 3 billion DNA base pairs per set of chromosomes, containing the bases A, T, G, and C

FIGURE 1.1 **The Human Genome Project.** The human genome is a complete set of human chromosomes. People have two sets of chromosomes, one set from each parent. Collectively, each set of chromosomes is composed of a DNA sequence that is approximately 3 billion nucleotide base pairs long. Estimates suggest that each set contains about 22,000 protein-encoding genes. This figure emphasizes the DNA found in the cell nucleus. Humans also have a small amount of DNA in their mitochondria, which has also been sequenced.

Concept Check: How might a better understanding of our genes be used in the field of medicine?

FIGURE 1.2 **The cloning of a mammal.** The lamb on the left is Dolly, the first mammal to be cloned. She was cloned from a cell of a Finn Dorset (a white-faced sheep). The sheep on the right is Dolly's surrogate mother, a Blackface ewe. A description of how Dolly was produced is presented in Chapter 20.
©R. Scott Horner KRT/Newscom

Concept Check: What ethical issues may be associated with human cloning?

Finally, genetic technologies provide the means of modifying the traits of animals and plants in ways that would have been unimaginable just a few decades ago. **Figure 1.3a** shows a bizarre example in which scientists introduced a gene from jellyfish into mice. Certain species of jellyfish emit a "green glow" produced by a gene that encodes a bioluminescent protein called green fluorescent protein (GFP). When exposed to blue or ultraviolet (UV) light, the protein emits a striking green-colored light. Scientists were able to clone the *GFP* gene from a sample of jellyfish cells and then introduce this gene into laboratory mice. The green fluorescent protein is made throughout the cells of their bodies. As a result, their skin, eyes, and organs give off an eerie green glow when exposed to UV light.

The expression of green fluorescent protein allows researchers to identify particular proteins in cells or specific body parts. For example, Andrea Crisanti and colleagues have altered mosquitoes to express GFP only in the gonads of males (**Figure 1.3b**). This enables the researchers to distinguish males from females and sort mosquitoes by sex. Why is this useful? The ability to rapidly

(a) GFP expressed in mice

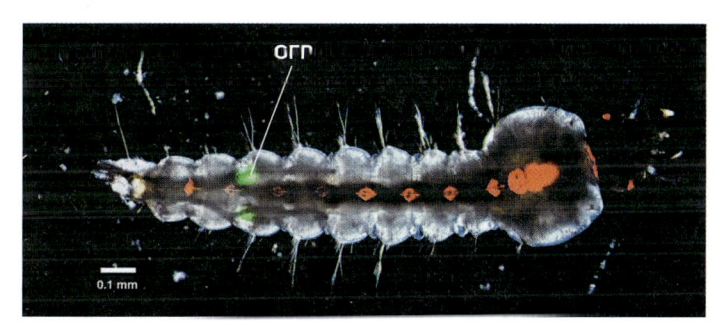

(b) GFP expressed in the gonads of a male mosquito

FIGURE 1.3 **The introduction of a jellyfish gene into laboratory mice and mosquitoes.** (a) A gene that naturally occurs in certain jellyfish encodes a protein called green fluorescent protein (GFP). The *GFP* gene was cloned and introduced into mice. When these mice are exposed to ultraviolet light, GFP emits a bright green color. These mice glow green, just like the jellyfish! (b) *GFP* was introduced next to a gene sequence that causes the expression of GFP only in the gonads of male mosquitoes. The resulting green glow allows researchers to identify and sort males from females.
(a) ©Eye of Science/Science Source; (b) Courtesy of Flaminia Catteruccia, Jason Benton and Andrea Crisanti

Concept Check: Why is it useful to sort male mosquitoes from female mosquitoes?

sort mosquitoes by sex makes it possible to produce populations of sterile males and then release the sterile males without the risk of releasing additional females. The release of sterile males may be an effective means of controlling mosquito populations because females breed only once. Mating with a sterile male prevents a female from producing offspring. In 2008, Osamu Shimomura, Martin Chalfie, and Roger Tsien received the Nobel Prize in chemistry for the discovery and the development of GFP, which has become a widely used tool in biology.

Overall, as we move forward in the twenty-first century, the excitement level in the field of genetics is high, perhaps higher than it has ever been. Nevertheless, new genetic knowledge and technologies will also create many ethical and societal challenges. In this chapter, we begin with an overview of genetics and then explore the various fields of genetics and their experimental approaches.

1.1 THE MOLECULAR EXPRESSION OF GENES

Learning Outcomes:
1. Describe the biochemical composition of cells.
2. Outline how DNA stores the information to make proteins.
3. Explain how proteins are largely responsible for cell structure and function.

Genetics is the branch of biology that deals with heredity and variation. It stands as the unifying discipline in biology by allowing us to understand how life can exist at all levels of complexity, ranging from the molecular to the population level. Genetic variation is the root of the natural diversity that we observe among members of the same species and among different species.

Genetics is centered on the study of genes. A gene is classically defined as a unit of heredity, but such a vague definition does not do justice to the exciting characteristics of genes as intricate molecular units that manifest themselves as critical contributors to cell structure and function.

- At the molecular level, a **gene** is a segment of DNA that has the information to produce a functional product. The functional product of most genes is a polypeptide—a linear sequence of amino acids that folds into units that constitute proteins.
- Genes are commonly described according to the way they affect **traits**, which are the characteristics of an organism. In humans, for example, we observe traits such as eye color, hair texture, and height. An ongoing theme of this textbook is the relationship between genes and traits. As an organism grows and develops, its collection of genes provides a blueprint that determines its characteristics.

In this section, we will examine the general features of life with an emphasis on the molecular level. Genetics is the common thread that explains the existence of life and its continuity from generation to generation. For most students, this chapter should serve as a cohesive review of topics they learned in other introductory courses such as general biology. Even so, it is usually helpful to see the "big picture" of genetics before delving into the finer details that are covered in Chapters 2 through 24.

Living Cells Are Composed of Biochemicals

To fully understand the relationship between genes and traits, we need to begin with an examination of the composition of living organisms. Every cell is constructed from intricately organized chemical substances. Small organic molecules such as glucose and amino acids are produced by the linkage of atoms via chemical bonds. The chemical properties of organic molecules are essential for cell vitality in two key ways.

- First, the breaking of chemical bonds during the degradation of small molecules provides energy to drive cellular processes.
- A second important function of these small organic molecules is their role as the building blocks for the synthesis of larger molecules. Four important categories of larger cellular molecules are **nucleic acids** (i.e., DNA and RNA), **proteins, carbohydrates,** and **lipids.** Three of these—nucleic acids, proteins, and carbohydrates—form **macromolecules** that are composed of many repeating units of smaller building blocks. Proteins, RNA, and carbohydrates can be made from hundreds or even thousands of repeating building blocks. DNA is the largest macromolecule found in living cells. A single DNA molecule can be composed of a linear sequence of hundreds of millions of building blocks called nucleotides!

The formation of cellular structures relies on the interactions of molecules and macromolecules. **Figure 1.4** illustrates this concept.

- Nucleotides are small organic molecules.
- Nucleotides are linked to each other and form the building blocks of DNA, which is a macromolecule.
- DNA is a component of chromosomes, which also contain proteins that contribute to chromosome structure.
- Within a eukaryotic cell, the chromosomes are contained in a compartment called the cell nucleus. The nucleus is bounded by a double membrane composed of lipids and proteins that shields the chromosomes from the rest of the cell. The nucleus is an example of an **organelle**—a membrane-bound compartment with a specialized function. The cell nucleus protects the chromosomes from mechanical damage and provides a single compartment for genetic activities such as gene transcription.
- Finally, cellular molecules, macromolecules, and organelles are organized to make a complete living cell.

Each Cell Contains Many Different Proteins That Determine Cell Structure and Function

To a great extent, the characteristics of a cell depend on the types of proteins that it makes. The entire collection of proteins that a cell makes at a given time is called its **proteome.** As we will learn throughout this textbook, proteins are the "workhorses" of all living cells. The range of functions among different types of proteins is truly remarkable. Some examples include the following:

- Proteins help determine the shape and structure of a given cell. For example, the protein known as tubulin can assemble into large structures known as microtubules, which provide the cell with internal structure and organization.
- Proteins are inserted into cell membranes and aid in the transport of ions and small molecules across the membrane.

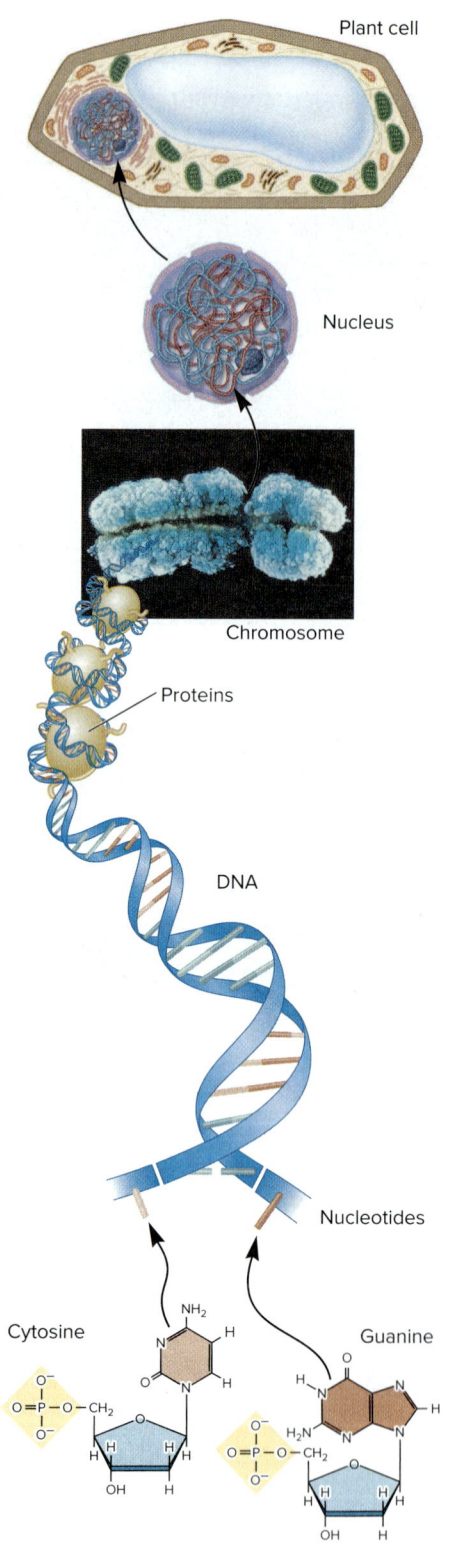

FIGURE 1.4 **Molecular organization of a living cell.** Cellular structures are constructed from smaller building blocks. In this example, DNA is formed from the linkage of nucleotides, producing a very long macromolecule. The DNA associates with proteins to form a chromosome. The chromosomes are located within a membrane-bound organelle called the nucleus, which, along with many different types of organelles, is found within a complete cell.
©Biophoto Associates/Science Source

Concept Check: *Is DNA a small molecule, a macromolecule, or an organelle?*

- Proteins may also function as biological motors. An interesting case is the protein known as myosin, which is involved in the contractile properties of muscle cells.
- Within multicellular organisms, certain proteins function in cell-to-cell recognition and signaling. For example, hormones such as insulin are secreted by endocrine cells and bind to the insulin receptor protein found within the plasma membrane of target cells.
- **Enzymes,** which accelerate chemical reactions, are a particularly important category of proteins. Some enzymes play a role in the breakdown of molecules or macromolecules into smaller units. These enzymes are important in the utilization of energy.

Molecular biologists have come to realize that the functions of proteins underlie the cellular characteristics of every organism. At the molecular level, proteins can be viewed as the active participants in the enterprise of life.

DNA Stores the Information for Protein Synthesis

As mentioned, the genetic material of living organisms is composed of a substance called deoxyribonucleic acid, abbreviated DNA. The DNA stores the information needed for the synthesis of all proteins. In other words, the main function of the genetic blueprint is to code for the production of proteins in the correct cell, at the proper time, and in suitable amounts. This task is extremely complicated because living cells make thousands of different proteins. Genetic analyses have shown that a typical bacterium can make a few thousand different proteins, and estimates of the numbers of proteins produced by complex eukaryotes range in the tens of thousands.

DNA's ability to store information is based on its structure.

- DNA is composed of a linear sequence of **nucleotides,** each of which contains one of four nitrogen-containing bases: adenine (A), thymine (T), guanine (G), or cytosine (C).
- The linear order of these bases along a DNA molecule contains information similar to the way that groups of letters of the alphabet represent words. For example, the "meaning" of the sequence of bases ATGGGCCTTAGC differs from that of TTTAAGCTTGCC.
- DNA sequences within most genes contain the information to direct the order of amino acids within **polypeptides** according to the **genetic code.** In the code, a three-base sequence, called a **codon,** specifies one particular **amino acid** among the 20 possible choices.
- The sequence of amino acids in a polypeptide causes it to fold into a particular structure; one or more polypeptides form a functional protein.

In this way, the DNA can store the information to specify the proteins made by an organism.

DNA Sequence	Amino Acid Sequence
ATG GGC CTT AGC	Methionine Glycine Leucine Serine
TTT AAG CTT GCC	Phenylalanine Lysine Leucine Alanine

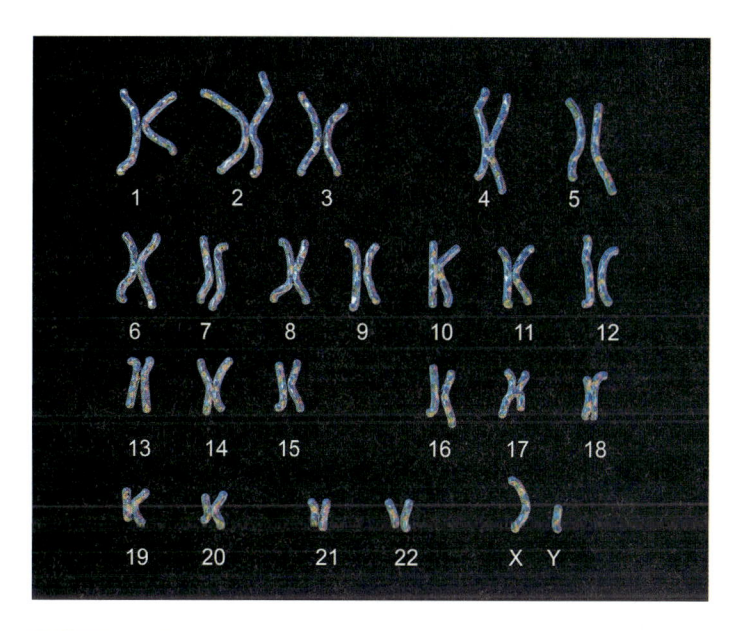

FIGURE 1.5 A micrograph of the 46 chromosomes found in a cell from a human male.
©Kateryna Kon/Shutterstock

Concept Check: Which types of macromolecules are found in chromosomes?

In living cells, DNA is found within large structures known as **chromosomes. Figure 1.5** is a micrograph of the 46 chromosomes in a cell from a human male, which are found in pairs. The DNA of an average human chromosome is an extraordinarily long, linear, double-stranded structure that contains well over a hundred million nucleotides. Along the immense length of a chromosome, the genetic information is parceled into functional units known as genes. An average-sized human chromosome is expected to carry about 1000 different genes.

The Information in DNA Is Accessed During the Process of Gene Expression

To synthesize its proteins, a cell must be able to access the information that is stored within its DNA. The process of using a gene sequence to affect the characteristics of cells and organisms is referred to as **gene expression.** At the molecular level, the information is accessed in a stepwise process (**Figure 1.6**).

1. In the first step, known as **transcription,** the DNA sequence within a gene is copied into a nucleotide sequence of **ribonucleic acid (RNA).** Most genes encode RNAs that contain the information for the synthesis of a particular polypeptide. This type of RNA is called **messenger RNA (mRNA).**
2. During the process of **translation,** the sequence of nucleotides in an mRNA provides the information (using the genetic code) to produce the amino acid sequence of a polypeptide.
3. A polypeptide folds into a three-dimensional structure. As mentioned, a protein is a functional unit. Some proteins are composed of a single polypeptide, and other proteins consist of two or more polypeptides.
4. The functioning of proteins largely determines cell structure and function.

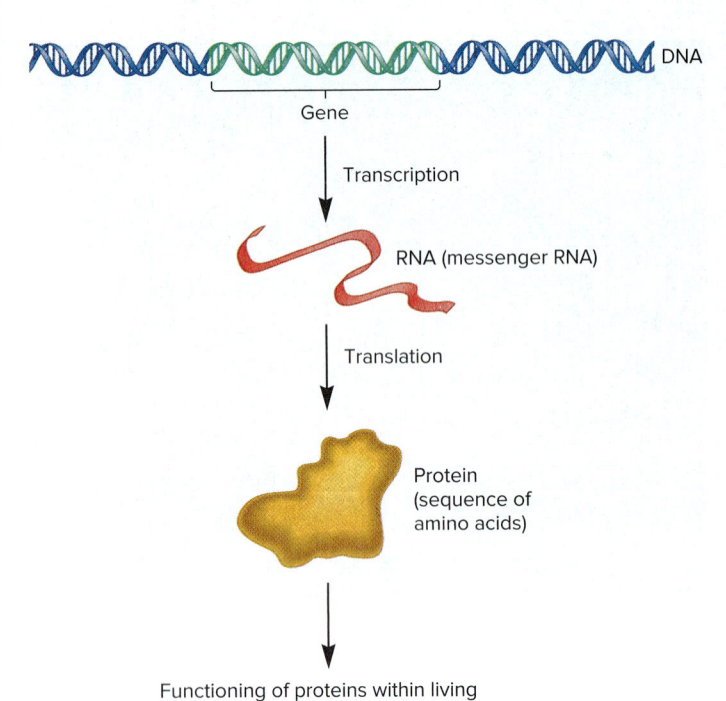

FIGURE 1.6 Gene expression at the molecular level. The expression of a gene is a multistep process. During transcription, one of the DNA strands is used as a template to make an RNA strand. During translation, the RNA strand is used to specify the sequence of amino acids within a polypeptide. One or more polypeptides produce a functional protein, thereby influencing an organism's traits.

Concept Check: Where is the information to make a polypeptide stored?

1.1 REVIEWING THE KEY CONCEPTS

- Living cells are composed of nucleic acids (DNA and RNA), proteins, carbohydrates, and lipids. The proteome largely determines the structure and function of cells (see Figure 1.4).
- DNA, which is found within chromosomes, stores the information to make proteins (see Figure 1.5).
- Most genes encode polypeptides that are units within functional proteins. Gene expression at the molecular level involves transcription to produce mRNA and translation to produce a polypeptide (see Figure 1.6).

1.1 COMPREHENSION QUESTIONS

1. Which of the following is *not* a constituent of a cell's proteome?
   a. An enzyme
   b. A motor protein
   c. A receptor in the plasma membrane
   d. An mRNA

2. A gene is a segment of DNA that has the information to produce a functional product. The functional product of most genes is
   a. DNA.
   b. mRNA.
   c. a polypeptide.
   d. none of the above.

3. The function of the genetic code is to
   a. promote transcription.
   b. specify the amino acids within a polypeptide.
   c. alter the sequence of DNA.
   d. do none of the above.

4. The process of transcription directly results in the synthesis of
   a. DNA.
   b. RNA.
   c. a polypeptide.
   d. all of the above.

1.2 THE RELATIONSHIP BETWEEN GENES AND TRAITS

Learning Outcomes:
1. Outline how the expression of genes leads to an organism's traits.
2. Define *genetic variation*.
3. Discuss the relationship between genes, traits, and the environment.
4. Describe how genes are transmitted in sexually reproducing species.
5. Describe the process of evolution.

A trait is any characteristic that an organism displays. In genetics, we can place traits into different categories.

- **Morphological traits** affect the appearance, form, and structure of an organism. The color of a flower and the height of a pea plant are morphological traits. Geneticists frequently study these types of traits because they are easy to evaluate. For example, an experimenter can simply look at a plant and tell if it has red or white flowers.
- **Physiological traits** affect the ability of an organism to function. For example, the rate at which a bacterium metabolizes a sugar such as lactose is a physiological trait. Like morphological traits, physiological traits are controlled, in part, by the expression of genes.
- **Behavioral traits** affect the ways an organism responds to its environment. An example is the mating calls of bird species. In animals, the nervous system plays a key role in governing such traits.

In this section, we will examine the relationship between the expression of genes and an organism's traits.

The Molecular Expression of Genes Within Cells Leads to an Organism's Traits

A complicated, yet very exciting, aspect of genetics is that our observations and theories span four levels of biological organization: molecules, cells, organisms, and populations. This broad scope can make it difficult to appreciate the relationship between genes and traits. To understand this connection, we need to relate the following four phenomena:

1. As we learned in Section 1.1, genes are expressed at the **molecular level.** In other words, gene transcription and translation lead to the production of a particular protein, which is a molecular process.
2. Proteins often function at the **cellular level.** The function of a protein within a cell affects the structure and workings of that cell.
3. An organism's traits are determined by the characteristics of its cells. We do not have microscopic vision, yet when we view morphological traits, we are really observing the properties of an individual's cells. For example, a red flower has its color because its cells make a red pigment. The trait of red flower color is an observation at the **organism level,** yet the trait is rooted in the molecular characteristics of the organism's cells.
4. A **species** is a group of organisms that maintains a distinctive set of attributes in nature. The occurrence of a trait within a species is an observation at the **population level.** Along with learning how a trait occurs, we also want to understand why a trait becomes prevalent in a particular species. In many cases, researchers discover that a trait predominates within a population because it promotes the reproductive success of the members of the population.

As a schematic example to illustrate the four levels of genetics, **Figure 1.7** shows the trait of pigmentation in a species of butterflies. One member of this species is light-colored and the other is very dark. Let's consider how we can explain this trait at the molecular, cellular, organism, and population levels.

1. At the molecular level, we need to understand the nature of the gene or genes that govern this trait. As shown in Figure 1.7a, a gene, which we will call the pigmentation gene, is responsible for the amount of pigment produced. The pigmentation gene can exist in two different forms called **alleles.** In this example, one allele confers a dark pigmentation and one causes a light pigmentation. Each of these alleles encodes a protein that functions as a pigment-synthesizing enzyme. However, the DNA sequences of the two alleles differ slightly from each other. This difference in the DNA sequence leads to a variation in the structure and function of the respective pigmentation enzymes.
2. At the cellular level (Figure 1.7b), the functional differences between the pigmentation enzymes affect the amount of pigment produced. The allele causing dark pigmentation, which is shown on the left, encodes an enzyme that functions very well. Therefore, when this gene is expressed in the cells of the wings, a large amount of pigment is made. By comparison, the allele causing light pigmentation encodes an enzyme that functions poorly. Therefore, when this allele is the only pigmentation gene expressed, little pigment is made.

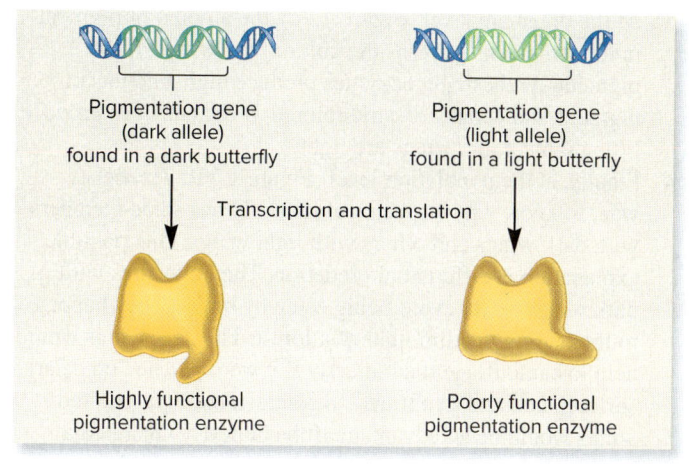

(a) Molecular level

(b) Cellular level

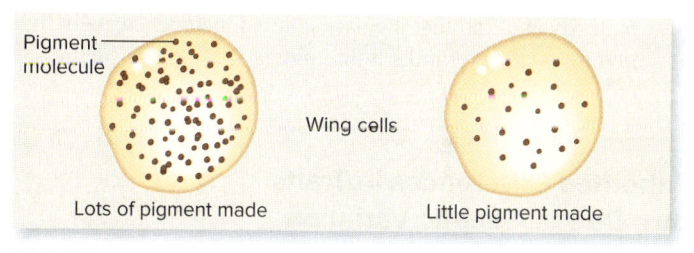

(c) Organism level

(d) Population level

FIGURE 1.7 The relationship between genes and traits at the (a) molecular, (b) cellular, (c) organism, and (d) population levels.

Concept Check: Which butterfly has a more active pigment-synthesizing enzyme, the light- or dark-colored one?

3. At the organism level (Figure 1.7c), the amount of pigment in the wing cells governs the color of the wings. If the pigment-synthesizing enzymes produce high amounts of pigment, the wings are dark-colored; if the enzymes produce little pigment, the wings are light.
4. Finally, at the population level (Figure 1.7d), geneticists want to know why a species of butterfly has some members with dark wings and others with light wings. One possible explanation is differential predation. The butterflies with dark wings might avoid being eaten by birds if they happen to live within the dim light of a forest. The dark wings would help to camouflage the butterfly if it were perched on a dark surface such as a tree trunk. In contrast, the light-colored wings would be an advantage if the butterfly inhabited a brightly lit meadow. Under these conditions, a bird might be less likely to notice a light-colored butterfly that was perched on a sunlit surface. A geneticist might study this species of butterfly and find that the dark-colored members usually live in forested areas and the light-colored members reside in unforested regions.

Inherited Differences in Traits Are Due to Genetic Variation

In Figure 1.7, we considered how gene expression can lead to variation in a trait of an organism, specifically, dark- versus light-colored wings in butterflies. Variation in traits among members of the same species is very common. For example, some people have black hair, and others have brown hair; some petunias have white flowers, but others have purple flowers. These are examples of **genetic variation.** This term describes the differences in inherited traits among individuals within a population.

In large populations that occupy a wide geographic range, genetic variation can be quite striking. Morphological differences have often led geneticists to misidentify two members of the same species as belonging to separate species. As an example, **Figure 1.8** shows two dyeing poison frogs that are members of the same species, *Dendrobates tinctorius*. They display dramatic differences in their markings. Such contrasting forms within a single species are termed **morphs.** You can easily imagine how someone might mistakenly conclude that these frogs are not members of the same species.

Changes in the nucleotide sequence of DNA underlie the genetic variation that we see among individuals. Throughout this textbook, we will routinely examine how variation in the genetic material results in changes in the outcome of traits. At the molecular level, genetic variation can be attributed to different types of modifications.

- Small or large differences can occur within gene sequences. When such changes initially occur, they are called **gene mutations,** which are heritable changes in the genetic material. Gene mutations result in genetic variation in which a gene is found in two or more alleles, as previously described in Figure 1.7. In many cases, gene mutations alter the expression or function of a protein that a gene specifies.

FIGURE 1.8 Two dyeing poison frogs (*Dendrobates tinctorius*) showing different morphs within a single species.
(a) ©Natalia Kuzmina/Shutterstock; (b) ©Valt Ahyppo/Shutterstock

Concept Check: Why do these two frogs look so different?

- Major alterations can also occur in the structure of a chromosome. A large segment of a chromosome can be lost, rearranged, or reattached to another chromosome.
- Variation may also occur in the total number of chromosomes. In some cases, an organism may inherit one too many or one too few chromosomes. In other cases, it may inherit an extra set of chromosomes.

Variations within the sequences of genes are a common source of genetic variation among members of the same species. In humans, familiar examples of sequence variation involve genes for eye color, hair texture, and skin pigmentation. Chromosome variation—a change in chromosome structure or number (or both)—is also found, but this type of change is often detrimental. Many human genetic disorders are the result of chromosomal alterations. An example is Down syndrome, which is due to the presence of an extra chromosome (**Figure 1.9a**). By comparison, chromosome variation in plants is common and often results in plants with superior characteristics, such as increased resistance to disease. Plant breeders have frequently exploited this observation. Cultivated varieties of wheat, for example, have many more chromosomes than the wild species (**Figure 1.9b**).

Traits Are Governed by Genes and by the Environment

In our discussion thus far, we have considered the role that genes play in the outcome of traits. Another critical factor is the **environment**—the surroundings in which an organism exists. A variety of factors in an organism's environment profoundly affect its morphological and physiological features. For example, a person's diet greatly influences many traits, such as height, weight, and even intelligence. Likewise, the amount of sunlight a plant receives affects its growth rate and the color of its flowers. The term **norm of reaction** refers to the effects of environmental variation on an individual's traits.

FIGURE 1.9 Examples of chromosome variation. (a) A person with Down syndrome. She has 47 chromosomes rather than the common number of 46, because she has an extra copy of chromosome 21. (b) A wheat plant. Bread wheat is derived from the contributions of three related species with two sets of chromosomes each, producing an organism with six sets of chromosomes.
(a) ©Stockbyte/Alamy Stock Photo; (b) ©Pixtal/age fotostock

Concept Check: Are these examples of gene mutations, variation in chromosome structure, or variation in chromosome number?

FIGURE 1.10 Environmental influence on the outcome of PKU. This girl with PKU has developed normally because she followed a diet that is very low in phenylalanine.
©Noah Goodrich/Newscom

Concept Check: What would have been the consequences if this girl had followed a standard diet, which contains a higher amount of phenylalanine?

External influences may dictate the way that genetic variation is manifested in an individual. An interesting example is the human genetic disease **phenylketonuria (PKU).** Humans have a gene that encodes an enzyme known as phenylalanine hydroxylase. Most people have two functional copies of this gene. People with one or two functional copies of the gene can eat foods containing the amino acid phenylalanine and metabolize it properly.

A rare variation in the sequence of the phenylalanine hydroxylase gene results in a nonfunctional version of this protein. Individuals with two copies of this rare, inactive allele cannot metabolize phenylalanine properly. Such individuals represent about 1 in 8000 births in the United States. When given a standard diet containing phenylalanine, individuals with this disorder are unable to break down this amino acid. Phenylalanine accumulates and is converted into phenylketones, which are detected in the urine. PKU individuals manifest a variety of detrimental traits, including mental impairment, underdeveloped teeth, and foul-smelling urine. In contrast, when PKU individuals are identified at birth and raised on a restricted diet that is low in phenylalanine, they develop normally (**Figure 1.10**). Fortunately, through routine newborn screening, most affected babies in the United States are now diagnosed and treated early. PKU provides a dramatic example of how the environment and an individual's genes can interact to influence the traits of the organism.

During Reproduction, Genes Are Passed from Parent to Offspring

Now that we have considered how genes and the environment govern the outcome of traits, we can turn to the issue of inheritance. How are traits passed from parents to offspring? The foundation for our understanding of inheritance came from the studies of pea plants by Gregor Mendel in the nineteenth century. His work revealed that genetic determinants, which we now call genes, are passed from parent to offspring as discrete units. We can predict the outcome of many genetic crosses based on Mendel's laws of inheritance.

The inheritance patterns identified by Mendel can be explained by the existence of chromosomes and their behavior during cell division.

- Like Mendel's pea plants, sexually reproducing species are commonly **diploid.** This means that their cells contain two copies of each chromosome, one from each parent. The two copies are called **homologs** of each other.
- Because genes are located within chromosomes, diploid organisms have two copies of most genes. Humans, for example, have 46 chromosomes, which are found in homologous pairs (**Figure 1.11a**). With the exception of the sex chromosomes (X and Y), each homologous pair contains the same kinds of genes. For example, both copies of human chromosome 12 carry the gene that encodes phenylalanine hydroxylase, which was discussed previously. Therefore, an individual has two copies of this gene that may or may not be identical alleles.
- Most cells of the human body that are not directly involved in sexual reproduction contain 46 chromosomes. These cells are called **somatic cells.** In contrast, the **gametes**—sperm and egg cells—contain half that number (23) and are termed **haploid** (**Figure 1.11b**).
- The union of gametes during fertilization restores the diploid number of chromosomes. The primary advantage of sexual reproduction is that it enhances genetic variation. For example, a tall person with blue eyes and a short person with

10   CHAPTER 1 :: OVERVIEW OF GENETICS

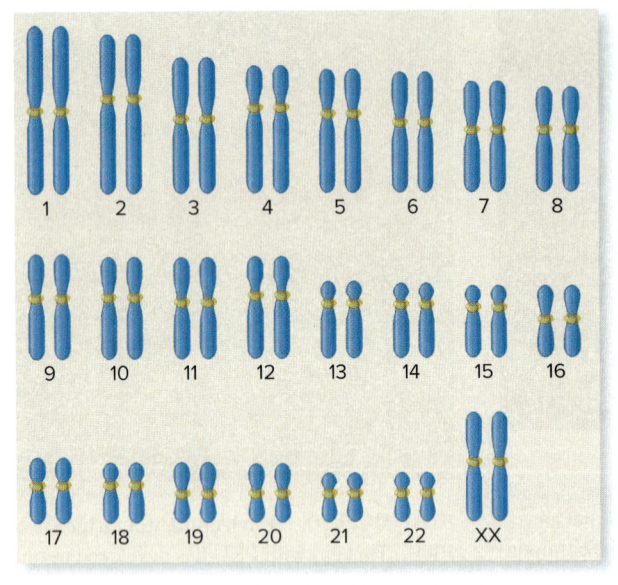

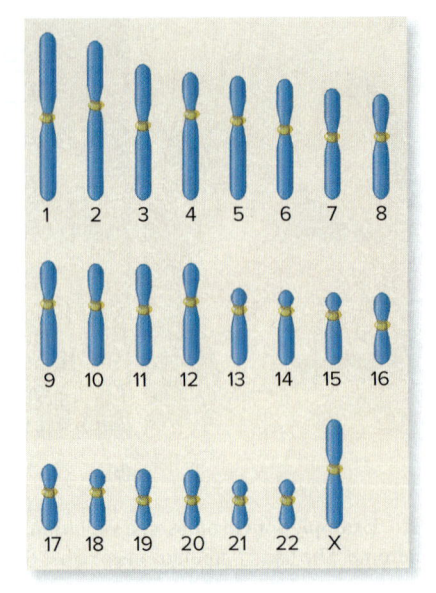

(a) Chromosomal composition found in most female human cells (46 chromosomes)

(b) Chromosomal composition found in a human gamete (23 chromosomes)

FIGURE 1.11   The complement of human chromosomes in somatic cells and gametes. (a) A schematic drawing of the 46 chromosomes of a human. With the exception of the sex chromosomes, these are always found in homologous pairs in somatic cells, such as skin or nerve cells. (b) The chromosomal composition of a gamete, which contains only 23 chromosomes, one from each pair. This gamete contains an X chromosome. Half of the gametes from human males contain a Y chromosome instead of an X chromosome.

Concept Check: The leaf cells of a corn plant contain 20 chromosomes each. How many chromosomes are found in a gamete made by a corn plant?

brown eyes may have short offspring with blue eyes or tall offspring with brown eyes. Therefore, sexual reproduction can result in new combinations of two or more traits that differ from those of either parent.

The Genetic Composition of a Species Evolves over the Course of Many Generations

As we have just seen, sexual reproduction has the potential to enhance genetic variation. This can be an advantage for a population of individuals as they struggle to survive and compete within their natural environment. The term **biological evolution,** or simply, **evolution,** refers to the phenomenon that the genetic makeup of a population can change from one generation to the next.

As proposed by Charles Darwin, the members of a species are in competition with one another for essential resources. Random genetic changes (i.e., mutations) occasionally occur within an individual's genes, and sometimes these changes lead to a modification of traits that promote reproductive success. For example, over the course of many generations, random gene mutations have lengthened the snout of the anteater, enabling it to feed on ants located in the ground. When a mutation creates a new allele that is beneficial, the allele may become prevalent in future generations because the individuals carrying the allele are more likely to survive and reproduce and pass the beneficial allele to their offspring. This process is known as **natural selection.** In this way, a species becomes better adapted to survive and reproduce in its native environment.

Over a long period of time, the accumulation of many genetic changes may lead to rather striking modifications in a species' characteristics. As an example, **Figure 1.12** depicts the evolution of the modern-day horse. A variety of morphological changes occurred, including an increase in size, fewer toes, and modified jaw structure. The changes can be attributed to natural selection producing adaptations to changing global climates. Over North America, where much of horse evolution occurred, large areas of dense forests were replaced with grasslands. The increase in size and changes in foot structure enabled horses to escape predators more easily and travel greater distances in search of food. The changes seen in horses' teeth are consistent with a shift from eating tender leaves to eating grasses and other types of vegetation that are more abrasive and require more chewing.

1.2 REVIEWING THE KEY CONCEPTS

- Genetics spans the molecular, cellular, organism, and population levels (see Figure 1.7).
- Genetic variation underlies variation in traits. In addition, the environment plays a key role (see Figures 1.8–1.10).
- During reproduction, genetic material is passed from parents to offspring. In many species, somatic cells are diploid and have two sets of chromosomes, whereas gametes are haploid and have a single set (see Figure 1.11).
- Evolution refers to a change in the genetic composition of a population from one generation to the next (see Figure 1.12).

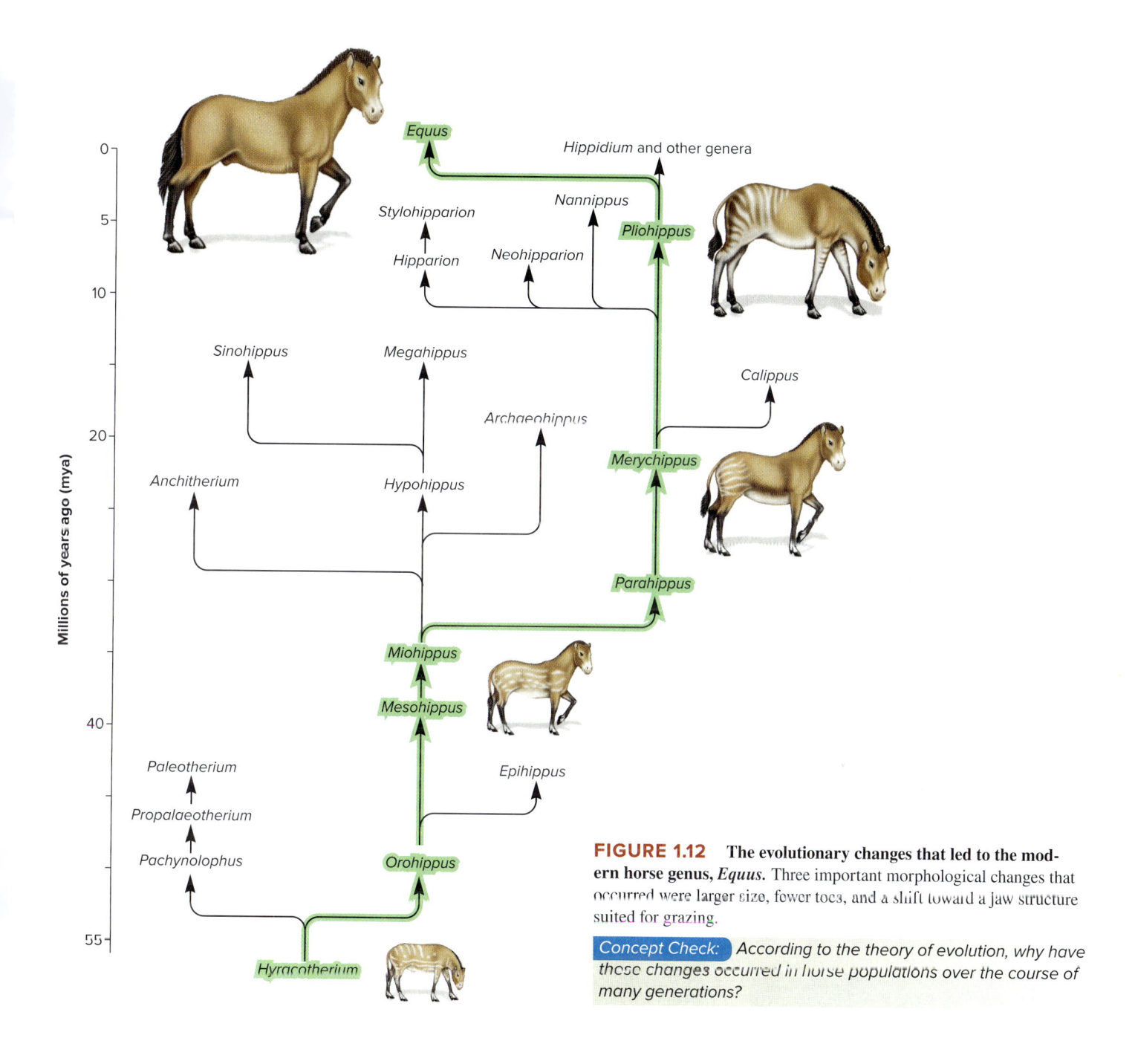

FIGURE 1.12 The evolutionary changes that led to the modern horse genus, *Equus*. Three important morphological changes that occurred were larger size, fewer toes, and a shift toward a jaw structure suited for grazing.

Concept Check: According to the theory of evolution, why have these changes occurred in horse populations over the course of many generations?

1.2 COMPREHENSION QUESTIONS

1. Gene expression can be viewed at which of the following levels?
   a. Molecular and cellular levels
   b. Organism level
   c. Population level
   d. All of the above

2. Variation in the traits of organisms may be attributable to
   a. gene mutations.
   b. alterations in chromosome structure.
   c. variation in chromosome number.
   d. all of the above.

3. A human skin cell has 46 chromosomes. A human sperm cell has
   a. 23.
   b. 46.
   c. 92.
   d. None of the above is the number of chromosomes in a sperm cell.

4. Evolutionary change caused by natural selection results in species with
   a. greater complexity.
   b. less complexity.
   c. greater reproductive success in their native environment.
   d. the ability to survive longer.

1.3 FIELDS OF GENETICS

Learning Outcome:
1. Compare and contrast the three major fields of genetics: transmission, molecular, and population genetics.

Genetics is a broad discipline encompassing molecular, cellular, organism, and population biology. Many scientists who are interested in genetics have been trained in supporting disciplines such as biochemistry, biophysics, cell biology, mathematics, microbiology, population biology, ecology, agriculture, and medicine. Experimentally, geneticists often focus their efforts on **model organisms**—organisms studied by many different researchers so they can compare their results and determine scientific principles that apply more broadly to other species. **Figure 1.13** shows some examples of model organisms, including *Escherichia coli* (a bacterium), *Saccharomyces cerevisiae* (a yeast), *Drosophila melanogaster* (fruit fly), *Caenorhabditis elegans* (a nematode worm), *Mus musculus* (mouse), and *Arabidopsis thaliana* (a flowering plant). Model organisms offer experimental advantages over other species. For example, *E. coli* is a very simple organism that can be easily grown in the laboratory. By limiting their work to a few such model organisms, researchers can more easily unravel the genetic mechanisms that govern the traits of a given species. Furthermore, the genes found in model organisms often function in a similar way to those found in humans.

The study of genetics has been traditionally divided into three areas—transmission, molecular, and population genetics—although there is some overlap of these three fields. In this section, we will examine the general questions that scientists in these areas are attempting to answer.

Transmission Genetics Explores the Inheritance Patterns of Traits as They Are Passed from Parents to Offspring

A scientist working in the field of transmission genetics examines the relationship between the transmission of genes from parent to offspring and the outcome of the offspring's traits. For example, how can two brown-eyed parents produce a blue-eyed child? Or why do tall parents tend to produce tall children, but not always? Our modern understanding of transmission genetics began with the studies of Gregor Mendel. His work provided the conceptual framework for transmission genetics. In particular, he originated the idea that genetic determinants, which we now call genes, are passed as discrete units from parents to offspring via sperm and egg cells. Since Mendel's pioneering studies of the 1860s, our knowledge of genetic transmission has greatly increased. Many patterns of genetic transmission are more complex than the simple Mendelian patterns that are described in Chapter 3. The additional complexities of transmission genetics are examined in Chapters 4 through 10.

Experimentally, the fundamental technique used by a transmission geneticist is the genetic cross. A **genetic cross** involves

(a) *Escherichia coli*

(b) *Saccharomyces cerevisiae*

(c) *Drosophila melanogaster*

(d) *Caenorhabditis elegans*

(e) *Mus musculus*

(f) *Arabidopsis thaliana*

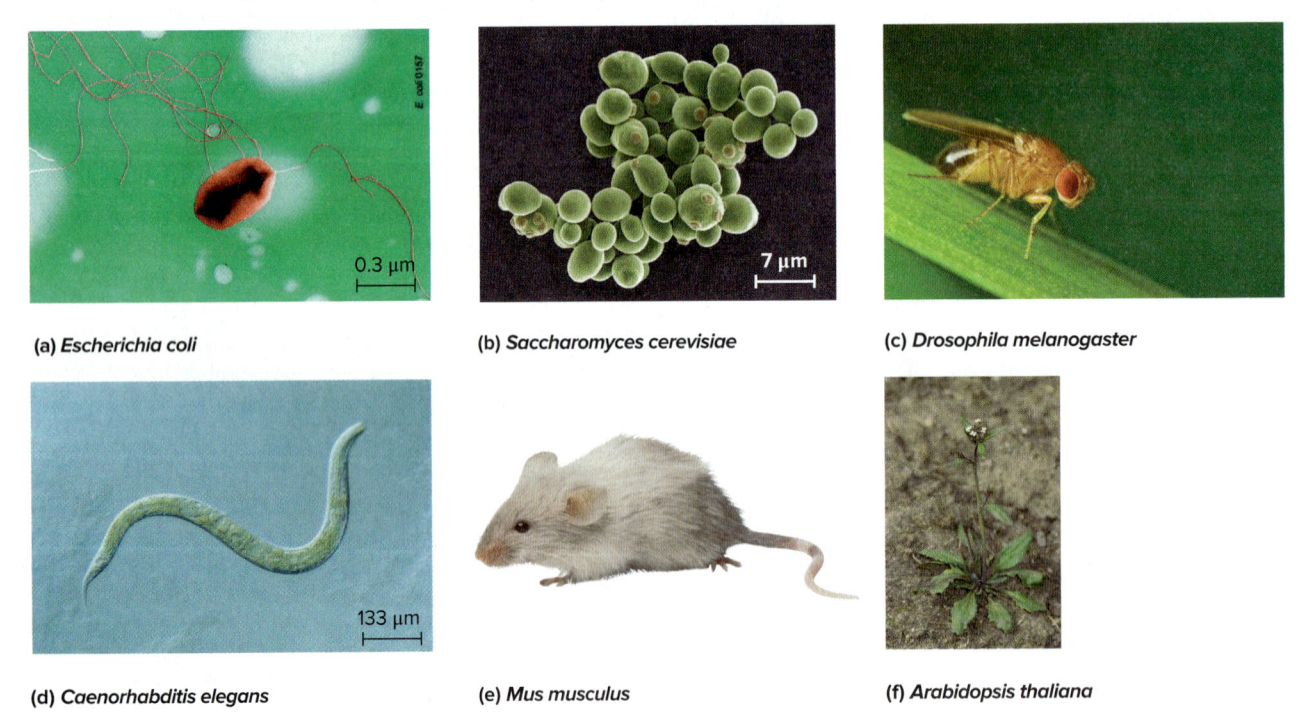

FIGURE 1.13 Examples of model organisms studied by geneticists. (a) *Escherichia coli* (a bacterium), (b) *Saccharomyces cerevisiae* (a yeast), (c) *Drosophila melanogaster* (fruit fly), (d) *Caenorhabditis elegans* (a nematode worm), (e) *Mus musculus* (mouse), and (f) *Arabidopsis thaliana* (a flowering plant).
(a) Source: CDC/Peggy S. Hayes & Elizabeth H. White, M.S; (b) ©Science Photo Library/Alamy Stock Photo; (c) ©janeff/Getty Images; (d) ©Sinclair Stammers/Science Source; (e) ©G.K. & Vikki Hart/Getty Images; (f) ©WILDLIFE GmbH/Alamy Stock Photo

Concept Check: Can you think of another example of a model organism?

breeding two selected individuals and then analyzing their offspring in an attempt to understand how traits are passed from parents to offspring. In the case of experimental organisms, the researcher chooses two parents with particular traits and then categorizes the offspring according to the traits they possess. In many cases, this analysis is quantitative in nature. For example, an experimenter may cross two tall pea plants and obtain 100 offspring that fall into two categories: 75 tall and 25 dwarf. As we will see in Chapter 3, the ratio of tall to dwarf offspring (3:1) provides important information concerning the inheritance pattern of the height trait.

Throughout Chapters 2 to 10, we will learn how researchers try to answer many fundamental questions concerning the passage of genetic material from cell to cell and the passage of traits from parents to offspring. Here are some of these questions:

How are chromosomes transmitted during cell division and gamete formation? **Chapter 2**

What are the common patterns of inheritance for genes? **Chapters 3–5**

Are there unusual patterns of inheritance that cannot be explained by the simple transmission of genes located on chromosomes in the cell nucleus? **Chapter 6**

When two or more genes are located on the same chromosome, how is the pattern of inheritance affected? **Chapter 7**

How do variations in chromosome structure or chromosome number occur, and how are they transmitted from parents to offspring? **Chapter 8**

How are genes transmitted by bacterial species? **Chapter 9**

How do viruses proliferate? **Chapter 10**

Molecular Genetics Focuses on a Biochemical Understanding of the Hereditary Material

The goal of molecular genetics, as the name of the field implies, is to understand how the genetic material works at the molecular level. In other words, molecular geneticists want to understand the molecular features of DNA and how these features underlie the expression of genes. The experiments of molecular geneticists are usually conducted within the confines of a laboratory. Their efforts frequently progress to a detailed analysis of DNA, RNA, and proteins, using a variety of techniques that are described throughout Parts III, IV, and V of this textbook.

Molecular geneticists often study mutant genes that have abnormal function. This is called a **genetic approach** to the study of a research question. In many cases, researchers analyze the effect of a gene mutation that eliminates the function of a gene. This is called a **loss-of-function mutation,** and the resulting gene is called a **loss-of-function allele.** Studying the effect of such a mutation often reveals the role of the functional, nonmutant gene. For example, let's suppose that a particular plant species produces purple flowers. If a loss-of-function mutation within a given gene causes a plant of that species to produce white flowers, one would suspect that the role of the functional gene involves the production of purple pigmentation.

Studies within molecular genetics interface with other disciplines such as biochemistry, biophysics, and cell biology. In addition, advances within molecular genetics have shed considerable light on the areas of transmission and population genetics. Our quest to understand molecular genetics has spawned a variety of modern molecular technologies and computer-based approaches. Furthermore, discoveries within molecular genetics have had widespread applications in agriculture, medicine, and biotechnology.

The following are some general questions within the field of molecular genetics:

What are the molecular structures of DNA and RNA? **Chapter 11**

What is the composition and conformation of chromosomes? **Chapter 12**

How is the genetic material copied? **Chapter 13**

How are genes expressed at the molecular level? **Chapters 14, 15**

How is gene expression regulated so that it occurs under the appropriate conditions, in the appropriate cell type, and at the correct stage of development? **Chapters 16, 17**

What are the roles of RNA molecules that do not encode polypeptides? **Chapter 18**

What is the molecular nature of mutations? How are mutations repaired? **Chapter 19**

How have genetic technologies advanced our understanding of genetics? **Chapter 20**

What is the genetic composition and function of whole genomes? **Chapter 21**

Population Genetics Is Concerned with Genetic Variation and Its Role in Evolution

The foundations of population genetics arose during the first few decades of the twentieth century. Although many scientists of this era did not accept the findings of Mendel or Darwin, the theories of population genetics provided a compelling way to connect the two viewpoints. Mendel's work and that of many succeeding geneticists gave insight into the nature of genes and how they are transmitted from parents to offspring. The theory of evolution by natural selection proposed by Darwin provided a biological explanation for the variation in characteristics observed among the members of a species. To relate these two phenomena, population geneticists have developed mathematical theories to explain the prevalence of certain alleles within populations of individuals. The work of population geneticists helps us understand how processes such as natural selection have resulted in the prevalence of individuals that carry particular alleles.

Population geneticists are particularly interested in genetic variation and how that variation is related to an organism's environment. In this field, the frequencies of alleles within a population are of central importance. The following are some general questions in population genetics:

What is the underlying relationship between genes and genetic diseases? **Chapter 22**

Why are two or more different alleles of a gene maintained in a population? **Chapter 23**

What factors alter the prevalence of alleles within a population? **Chapter 23**

What are the contributions of genetics and environment in the outcome of a trait? **Chapter 24**

How do genetics and the environment influence quantitative traits, such as size and weight? **Chapter 24**

1.3 REVIEWING THE KEY CONCEPTS

- Model organisms are studied by many different researchers so they can compare their results and determine scientific principles that apply more broadly to other species (see Figure 1.13).
- Genetics is traditionally divided into transmission genetics, molecular genetics, and population genetics, though overlap occurs among these fields.

1.3 COMPREHENSION QUESTIONS

1. Which of the following is *not* a model organism?
   a. *Mus musculus* (laboratory mouse)
   b. *Escherichia coli* (a bacterium)
   c. *Saccharomyces cerevisiae* (a yeast)
   d. *Sciurus carolinensis* (gray squirrel)
2. A person studying the rate of transcription of a particular gene is working in the field of
   a. molecular genetics.
   b. transmission genetics.
   c. population genetics.
   d. None of the above is correct.

1.4 THE SCIENCE OF GENETICS

Learning Outcomes:
1. Describe what makes genetics an experimental science.
2. Outline different strategies for solving problems in genetics.

Science is a way of knowing about our natural world. The science of genetics allows us to understand how the expression of our genes produces the traits that we possess. In this section, we will consider how scientists attempt to answer questions via experimentation. We will also consider general approaches for solving problems.

Genetics Is an Experimental Science

Regardless of what field of genetics they work in, researchers typically follow two general types of scientific approaches: hypothesis testing and discovery-based science. In **hypothesis testing,** also called the **scientific method,** scientists follow a series of steps to reach verifiable conclusions about the world. Although scientists arrive at their theories in different ways, the scientific method provides a way to validate (or invalidate) a particular hypothesis. Alternatively, research may also involve the collection of data without a preconceived hypothesis. For example, researchers might analyze the genes found in cancer cells to identify those genes that have become mutant. In this case, the scientists may not have a hypothesis about which particular genes may be involved. The collection and analysis of data without the need for a preconceived hypothesis is called **discovery-based science** or, simply, discovery science.

In traditional science textbooks, the emphasis often lies on the product of science. That is, many textbooks are aimed primarily at teaching the student about the observations scientists have made and the hypotheses they have proposed to explain these observations. Along the way, the student is provided with many bits and pieces of experimental techniques and data. Although this textbook provides you with many observations and hypotheses, it attempts to go one step further. Many of the following chapters contain one or two figures presenting experiments that have been "dissected" into five individual components to help you to understand the entire scientific process. The five steps are as follows:

1. Background information is provided so that you can appreciate observations that were known prior to conducting the experiment.
2. Most experiments involve hypothesis testing. In those cases, the figure presenting the experiment states the hypothesis the scientists were trying to test. In other words, what scientific question was the researcher trying to answer?
3. Next, the figure follows the experimental steps the scientists took to test the hypothesis. The steps necessary to carry out the experiment are listed in the order in which they were conducted. The figure presents parallel illustrations labeled "Experimental level" and "Conceptual level." The experimental level helps you to understand the techniques that were used. The conceptual level helps you to understand what is actually happening at each step in the procedure.
4. The raw data from the experiment are then presented.
5. Last, an interpretation of the data is offered within the text.

The rationale behind this approach is that it enables you to see the experimental process from beginning to end. Hopefully, you will find this a more interesting and rewarding way to learn about genetics. As you read through the chapters, the experiments will help you to see the relationship between science and scientific theories.

As a student of genetics, you will be given the opportunity to involve your mind in the experimental process. As you are reading an experiment, you may find yourself thinking about alternative approaches and hypotheses. Different people can view the same data and arrive at very different conclusions. As you progress through the experiments in this book, you will enjoy genetics far more if you try to develop your own skills at formulating hypotheses, designing experiments, and interpreting data. Also, some of the questions in the problem sets are aimed at refining these skills.

Genetic TIPS Will Help You to Improve Your Problem-Solving Skills

As you progress through this textbook, your learning will involve two general goals:

- *You will gather foundational knowledge.* In other words, you will be able to describe core concepts in genetics. For example, you will be able to explain how DNA replication occurs and describe the proteins that are involved in this process.

- *You will develop problem-solving skills that allow you to apply that foundational knowledge in different ways.* For example, you will learn how to use statistics to determine if a genetic hypothesis is consistent with experimental data.

The combination of foundational knowledge and problem-solving skills will enable you not only to understand genetics, but also to apply your knowledge in different situations. To help you develop these skills, Chapters 2 through 24 contain solved problems named **Genetic TIPS,** which stands for **T**opic, **I**nformation, and **P**roblem-solving **S**trategy. These solved problems follow a consistent pattern.

Genetic TIPS

The Question: All of the Genetic **TIPS** begin with a question. As an example, let's consider the following question:

The coding strand of DNA in a segment of a gene is as follows: ATG GGC CTT AGC. This strand carries the information to make a region of a polypeptide with the amino acid sequence methionine-glycine-leucine-serine. What would be the consequences if a mutation changed the second cytosine (C) in this sequence to an adenine (A)?

Topic: What topic in genetics does this question address?

The topic is gene expression. More specifically, the question is about the relationship between a gene sequence and the genetic code.

Information: What information do you know based on the question and your understanding of the topic?

In the question, you are given the base sequence of a short segment of a gene and told that one of the bases has been changed. From your understanding of the topic, you may remember that a polypeptide sequence is determined by reading the mRNA (transcribed from a gene) in groups of three bases called codons.

Problem-Solving Strategy: Compare and contrast.

One strategy to solve this problem is to compare the mRNA sequence (transcribed from this gene) before and after the mutation:

Original: AUG GGC CUU AGC
Mutant: AUG GGC **A**UU AGC
↑

Answer: The mutation has changed the sequence of bases in the mRNA so that the third codon has changed from CUU to AUU (see arrow). Because codons specify amino acids, this may change the third amino acid to something else. Note: If you look ahead to Chapter 15 (see Table 15.1), you will see that CUU specifies leucine, whereas AUU specifies isoleucine. Therefore, you would predict that the mutation would change the third amino acid from leucine to isoleucine.

Throughout Chapters 2 through 24, each chapter will contain several Genetic TIPS. Some of these will be within the chapter itself and some will precede the problem sets that are at the end of each chapter. Though there are many different problem-solving strategies, Genetic TIPS will focus on ten strategies that will help you to solve problems. You will see these ten strategies over and over again as you progress through the textbook:

1. *Define key terms.* In some cases, a question may be difficult to understand because you don't know the meaning of one or more key terms in the question. If so, you will need to begin your problem solving by defining such terms, either by looking them up in the glossary or by using the index to find the location in the text where the key terms are explained.
2. *Make a drawing.* Genetic problems are often difficult to solve in your head. Making a drawing may make a big difference in your ability to see the solution.
3. *Predict the outcome.* Geneticists may want to predict the outcome of an experiment. For example, in Chapters 3 through 6, you will learn about different ways to predict the outcome of genetic crosses. Becoming familiar with these methods will help you to predict the outcomes of particular experiments.
4. *Compare and contrast.* Making a direct comparison between two things, such as two RNA sequences, may help you to understand how they are similar and how they are different.
5. *Relate structure and function.* A recurring theme in biology and genetics is that structure determines function. This relationship holds true at many levels of biology, including the molecular, microscopic, and macroscopic levels. For some questions, you will need to understand how certain structural features are related to their biological functions.
6. *Describe the steps.* At first, some questions may be difficult to understand because they may involve mechanisms that occur in a series of several steps. Sometimes, if you sort out the steps, you may identify the key step that you need to understand to solve the problem.
7. *Propose a hypothesis.* A hypothesis is an attempt to explain an observation or data. Hypotheses may be made in many forms, including statements, models, equations, and diagrams.
8. *Design an experiment.* Experimental design lies at the heart of science. In many cases, an experiment begins with some type of starting material(s), such as strains of organisms or purified molecules, and then the starting materials are subjected to a series of steps. The experiments featured throughout the textbook will also help you refine the skill of designing experiments.
9. *Analyze data.* Because genetics is an experimental science, many problems involve the analysis of data, which are the product of experiments. A variety of different statistical methods are used to analyze data and make conclusions about what the data mean.
10. *Make a calculation.* Genetics is a quantitative science. Researchers have devised mathematical relationships to understand and predict genetic phenomena. Becoming familiar with these mathematical relationships will help you to better understand genetic concepts and to make predictions.

For most problems throughout this textbook, one or more of these strategies may help you to arrive at the correct solution. Genetic TIPS will provide you with practice at applying these ten problem-solving strategies.

1.4 REVIEWING THE KEY CONCEPTS

- Researchers in genetics carry out hypothesis testing or discovery-based science.
- Genetic TIPS are aimed at improving your ability to solve problems.

1.4 COMPREHENSION QUESTION

1. The scientific method involves which of the following?
   a. The collection of observations and the formulation of a hypothesis
   b. Experimentation
   c. Data analysis and interpretation
   d. All of the above

KEY TERMS

Page 1. genome, deoxyribonucleic acid (DNA)
Page 3. genetics, gene, traits
Page 4. nucleic acids, proteins, carbohydrates, lipids, macromolecules, organelle, proteome
Page 5. enzymes, nucleotides, polypeptides, genetic code, codon, amino acid, chromosomes, gene expression, transcription, ribonucleic acid (RNA), messenger RNA (mRNA), translation
Page 6. morphological traits, physiological traits, behavioral traits
Page 7. molecular level, cellular level, organism level, species, population level, alleles
Page 8. genetic variation, morphs, gene mutations, environment, norm of reaction
Page 9. phenylketonuria (PKU), diploid, homologs, somatic cells, gametes, haploid
Page 10. biological evolution, evolution, natural selection
Page 12. model organisms, genetic cross
Page 13. genetic approach, loss-of-function mutation, loss-of-function allele
Page 14. hypothesis testing, scientific method, discovery-based science

CHAPTER SUMMARY

- The complete genetic composition of a cell is called a genome. The genome encodes all of the proteins a cell can make. Many key discoveries in genetics are related to the study of genes and genomes (see Figures 1.1–1.3).

1.1 The Molecular Expression of Genes

- Living cells are composed of nucleic acids (DNA and RNA), proteins, carbohydrates, and lipids. The proteome largely determines the structure and function of cells (see Figure 1.4).
- DNA, which is found within chromosomes, stores the information to make proteins (see Figure 1.5).
- Most genes encode polypeptides that are units within functional proteins. Gene expression at the molecular level involves transcription to produce mRNA and translation to produce a polypeptide (see Figure 1.6).

1.2 The Relationship Between Genes and Traits

- Genetics spans the molecular, cellular, organism, and population levels (see Figure 1.7).
- Genetic variation underlies variation in traits. In addition, the environment plays a key role (see Figures 1.8–1.10).
- During reproduction, genetic material is passed from parents to offspring. In many species, somatic cells are diploid and have two sets of chromosomes, whereas gametes are haploid and have a single set (see Figure 1.11).
- Evolution refers to a change in the genetic composition of a population from one generation to the next (see Figure 1.12).

1.3 Fields of Genetics

- Model organisms are studied by many different researchers so they can compare their results and determine scientific principles that apply more broadly to other species (see Figure 1.13).
- Genetics is traditionally divided into transmission genetics, molecular genetics, and population genetics, though overlap occurs among these fields.

1.4 The Science of Genetics

- Researchers in genetics carry out hypothesis testing or discovery-based science.
- Genetic TIPS are aimed at improving your ability to solve problems.

PROBLEM SETS & INSIGHTS

More Genetic TIPS

1. Most genes encode proteins. Explain how proteins produce an organism's traits. Provide examples.

   Topic: What topic in genetics does this question address?

   The topic is the relationship between genes and traits. More specifically, the question is about how proteins, which are encoded by genes, produce an organism's traits.

   Information: What information do you know based on the question and your understanding of the topic?

   In the question, you are reminded that most genes encode proteins and that proteins play a role in producing an organism's traits. From your understanding of the topic, you may remember that proteins carry out a variety of functions that are critical to cell structure and function.

Problem-Solving Strategy: Relate structure and function.

One strategy for you to solve this problem is to consider the relationship between protein structure and function. Think about examples in which the structure and function of proteins govern the structure and function of living cells. Also, consider how the structure and functions of cells determine an organism's traits.

Answer: The structure and function of proteins govern the structure and function of living cells. For example, specific proteins help determine the shape and structure of a given cell. The protein known as tubulin can assemble into large structures known as microtubules, which provide the cell with internal structure and organization. The proteins that a cell makes are largely responsible for the cell's structure and function. For example, the proteins made by a nerve cell cause the cell to be very elongated and to be able to transmit signals from one cell to another. The structure of a nerve cell provides animals with many traits, such as the ability to sense the temperature of their environment and the ability to send signals to their muscles to promote movement.

2. A human gene called the *CFTR* gene (for cystic fibrosis transmembrane regulator) encodes a protein that functions in the transport of chloride ions across the cell membrane. Most people have two copies of a functional *CFTR* gene and do not have cystic fibrosis. However, a mutant version of the *CFTR* gene is found in some people. If a person has two mutant copies of the gene, he or she develops the disease known as cystic fibrosis. Are the following examples a description of genetics at the molecular, cellular, organism, or population level?

A. People with cystic fibrosis have lung problems due to a buildup of thick mucus in their lungs.

B. The mutant *CFTR* gene encodes a defective chloride transporter.

C. A defect in the chloride transporter causes a salt imbalance in lung cells.

D. Scientists have wondered why the mutant *CFTR* gene is relatively common. In fact, it is the most common mutant gene that causes a severe disease in people of northern European descent. Usually, mutant genes that cause severe diseases are relatively rare. One possible explanation why cystic fibrosis is so common is that people who have one copy of the functional *CFTR* gene and one copy of the mutant gene may be more resistant to diarrheal diseases such as cholera. Therefore, even though individuals with two mutant copies are very sick, people with one mutant copy and one functional copy might have a survival advantage over people with two functional copies of the gene.

Topic: What topic in genetics does this question address?

The topic is how genetics can be viewed at different levels, ranging from the molecular to the population level.

Information: What information do you know based on the question and your understanding of the topic?

The question describes the disease called cystic fibrosis. Parts A through D give descriptions of various aspects of the disease. From your understanding of the topic, you may remember that genetics can be viewed at the molecular, cellular, organism, and population level. This concept is described in Figure 1.7.

Problem-Solving Strategies: Make a drawing. Compare and contrast.

One strategy to solve this problem is to make a drawing of the descriptions of parts A through D and decide if you are drawing something at the molecular, cellular, organism, or population levels. For example, if you drew the description in part B, you would be drawing a protein, which is a molecule. If you drew the description of part C, you would be drawing a cell and indicating that a salt imbalance occurs. Another strategy to solve this problem would be to compare and contrast parts A, B, C, and D with each other. For example, if you compared part A and part D, you might realize that part A is describing something in one person, whereas part D is describing the occurrence of the mutant gene in multiple people.

Answer:

A. Organism. This is a description of a trait at the level of an entire individual.

B. Molecular. This is a description of a gene and the protein it encodes.

C. Cellular. This is a description of how protein function affects the cell.

D. Population. This is a possible explanation why two alleles of the gene occur within a population.

Conceptual Questions

C1. At the molecular level, what is a gene? Where are genes located?

C2. Briefly explain how gene expression occurs at the molecular level.

C3. A human gene called the β-globin gene encodes a polypeptide that functions as a subunit of the protein known as hemoglobin. Hemoglobin carries oxygen within red blood cells. In human populations, the β-globin gene can be found as the more common allele called the Hb^A allele, but it can also be found as the Hb^S allele. Individuals who have two copies of the Hb^S allele have the disease called sickle cell disease. Are the following examples a description of genetics at the molecular, cellular, organism, or population level?

A. The Hb^S allele encodes a polypeptide that functions slightly differently from the polypeptide encoded by the Hb^A allele.

B. If an individual has two copies of the Hb^S allele, that person's red blood cells take on a sickle shape.

C. Individuals who have two copies of the Hb^A allele do not have sickle cell disease, but they are not resistant to malaria. People who have one Hb^A allele and one Hb^S allele do not have sickle cell disease, and they are resistant to malaria. People who have two copies of the Hb^S allele have sickle cell disease, and this disease may significantly shorten their lives.

D. Individuals with sickle cell disease have anemia because their red blood cells are easily destroyed by the body.

C4. What is meant by the term *genetic variation*? Give two examples of genetic variation not discussed in this chapter. What causes genetic variation at the molecular level?

C5. What is the cause of Down syndrome?

C6. Your textbook describes how the trait of phenylketonuria (PKU) is greatly influenced by the environment. Pick a trait of your favorite plant, and explain how genetics and the environment may play important roles.

C7. What is meant by the term *diploid*? Which cells of the human body are diploid, and which cells are not?

C8. What is a DNA sequence?

C9. What is the genetic code?

C10. Explain the relationship between each of these pairs of genetic terms:

A. Gene and trait

B. Gene and chromosome

C. Allele and gene

D. DNA sequence and amino acid sequence

C11. With regard to biological evolution, which of the following statements is incorrect? Explain why.

A. During its lifetime, an animal evolves to become better adapted to its environment.

B. The process of biological evolution has produced species that are better adapted to their environments.

C. When an animal is better adapted to its environment, the process of natural selection makes it more likely for that animal to reproduce.

C12. What are the primary interests of researchers working in the following fields of genetics?

A. Transmission genetics

B. Molecular genetics

C. Population genetics

Application and Experimental Questions

E1. Pick any example of a genetic technology and describe how it has directly affected your life.

E2. What is a genetic cross?

E3. The technique known as DNA sequencing (described in Chapter 20) enables researchers to determine the DNA sequence of genes. Would this technique be used primarily by transmission geneticists, molecular geneticists, or population geneticists?

E4. Figure 1.5 shows a micrograph of chromosomes from a normal human cell. If you created this type of display using cells from a person with Down syndrome, what would you expect to see?

E5. Many organisms are studied by geneticists. Of the following species, do you think it is more likely for them to be studied by a transmission geneticist, a molecular geneticist, or a population geneticist? Explain your answer. Note: More than one answer may be possible for a given species.

A. Dogs

B. *E. coli*

C. Fruit flies

D. Leopards

E. Corn

E6. Pick any trait you like in any species of wild plant or animal. The trait must somehow vary among different members of the species. For example, some butterflies have dark wings and others have light wings (see Figure 1.7). Note: When picking a trait to answer this question, do not pick the trait of wing color in butterflies.

A. Discuss all of the background information that you already have (from personal observations) regarding this trait.

B. Propose a hypothesis that explains the genetic variation within the species. For example, in the case of the butterflies, your hypothesis might be that the dark butterflies survive better in dark forests, and the light butterflies survive better in sunlit fields.

C. Describe the experimental steps you would follow to test your hypothesis.

D. Describe the possible data you might collect.

E. Interpret your data.

Answers to Comprehension Questions

1.1: d, c, b, b

1.2: d, d, a, c

1.3: d, a

1.4: d

Note: All answers appear in Connect; the answers to even-numbered questions and all Concept Check questions are in Appendix B.

PART II PATTERNS OF INHERITANCE

2

CHAPTER OUTLINE

- 2.1 General Features of Chromosomes
- 2.2 Cell Division
- 2.3 Mitosis and Cytokinesis
- 2.4 Meiosis
- 2.5 Sexual Reproduction

Chromosome sorting during cell division. When eukaryotic cells divide, they replicate and sort their chromosomes (shown in light blue), so that each cell receives the correct number.
©Conly L. Rieder

REPRODUCTION AND CHROMOSOME TRANSMISSION

Reproduction is the biological process by which new cells or new organisms are produced. In this chapter, we will first survey reproduction at the cellular level, paying close attention to the inheritance of chromosomes. An examination of chromosomes at the microscopic level provides us with insights regarding the inheritance patterns of traits, which we will consider in Chapter 3. To appreciate this relationship, we will examine how cells distribute their chromosomes during the process of cell division. We will see that in bacteria and most unicellular eukaryotes, simple cell division provides a way to reproduce asexually. Then we will explore a form of cell division called meiosis that produces cells with half the number of chromosomes. This form of cell division is needed for sexual reproduction, which is the formation of a new individual following the union of two gametes. This chapter will end with a discussion of how sexual reproduction occurs in animals and plants.

2.1 GENERAL FEATURES OF CHROMOSOMES

Learning Outcomes:
1. Define the term *chromosome*.
2. Outline key differences between prokaryotic and eukaryotic cells.
3. Describe the procedure for making a karyotype.
4. Compare and contrast the similarities and differences between homologous chromosomes.

Chromosomes are structures within living cells that contain the genetic material. Genes are physically located within chromosomes. Biochemically, each chromosome contains a very long segment of DNA, which is the genetic material, and proteins, which are bound to the DNA and provide it with an organized structure. In eukaryotic cells, this complex between DNA and proteins is called **chromatin.** In this chapter, we will focus on the cellular

19

mechanics of chromosome transmission to better understand the patterns of gene transmission that we will consider in Chapters 3 through 7. In particular, we will examine how chromosomes are copied and sorted into newly made cells. In later chapters, particularly Chapters 11 and 12, we will examine the molecular features of chromosomes in greater detail.

Before we begin a description of chromosome transmission, we need to consider the distinctive cellular differences between prokaryotic and eukaryotic species. Bacteria and archaea are referred to as **prokaryotes,** from the Greek meaning "prenucleus," because their chromosomes are not contained within a membrane-bound nucleus of the cell. Prokaryotes usually have a single type of circular chromosome in a region of the cytoplasm called the **nucleoid** (**Figure 2.1a**). The cytoplasm is enclosed by a plasma membrane that regulates the uptake of nutrients and the excretion of waste products. Outside the plasma membrane is a rigid cell wall that protects the cell from breakage. Certain species of bacteria also have an outer membrane located beyond the cell wall.

Eukaryotes, from the Greek meaning "true nucleus," include some simple species, such as single-celled protists and some fungi (such as yeast), and more complex multicellular species, such as plants, animals, and other fungi. The cells of eukaryotic species have internal membranes that enclose highly specialized compartments (**Figure 2.1b**). These compartments form membrane-bound

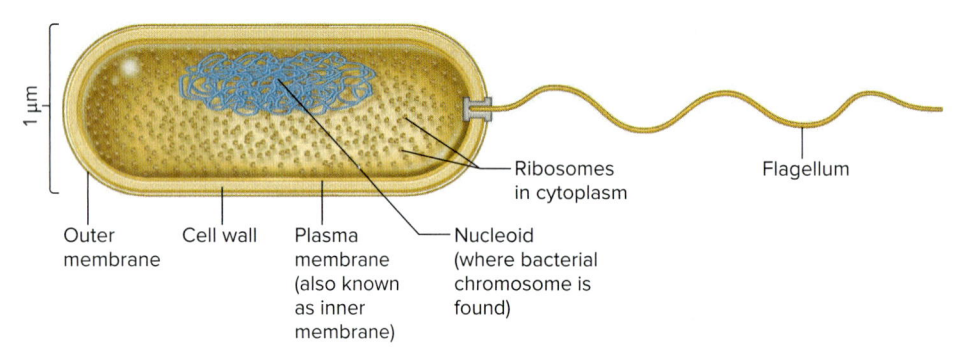

(a) Prokaryotic cell

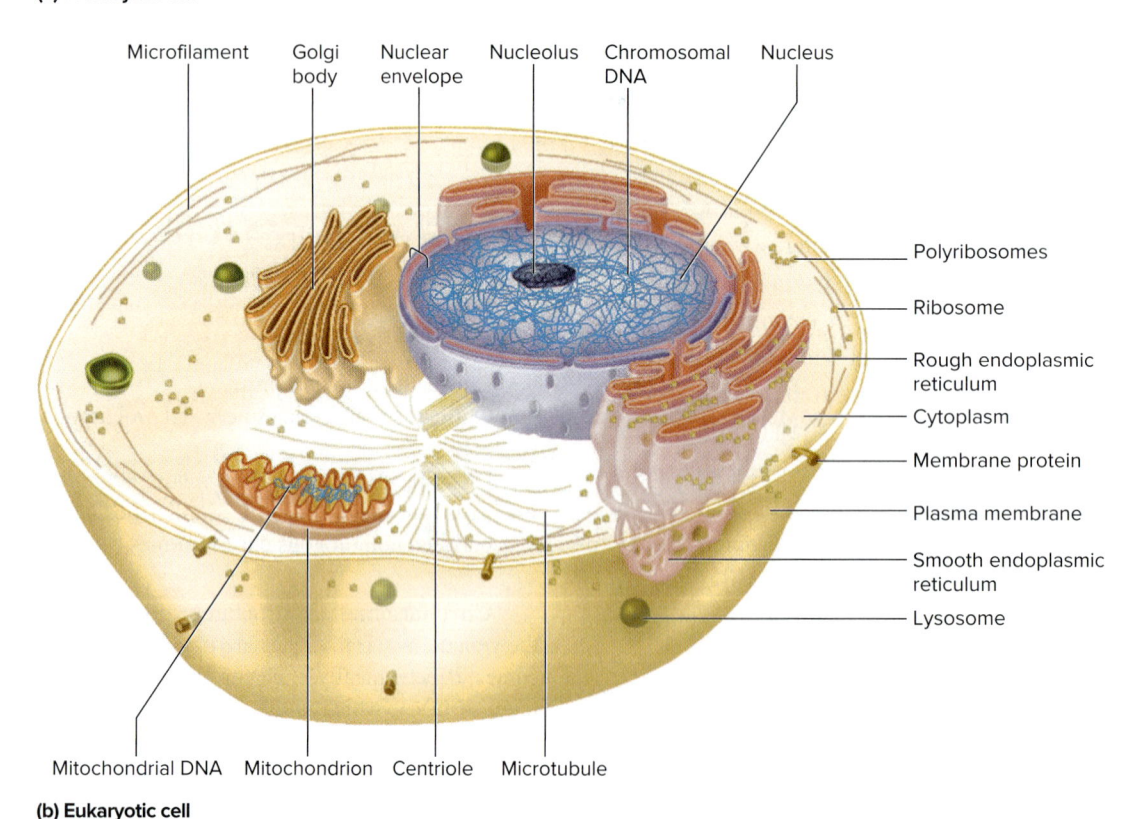

(b) Eukaryotic cell

FIGURE 2.1 **The basic organization of cells.** (a) A bacterial cell. The example shown here is typical of a bacterium such as *Escherichia coli*, which has an outer membrane. (b) A eukaryotic cell. The example shown here is a typical animal cell.

Concept Check: Eukaryotic cells exhibit compartmentalization. What does this mean?

organelles with specific functions. For example, a particularly conspicuous organelle is the **nucleus,** which is bounded by two membranes that constitute the nuclear envelope. Most of the genetic material is found within chromosomes that are located in the nucleus. In addition to the nucleus, certain organelles in eukaryotic cells contain a small amount of their own DNA. These include the mitochondrion, which plays a role in ATP synthesis, and, in plant cells, the chloroplast, which functions in photosynthesis. The DNA found in these organelles is referred to as extranuclear, or extrachromosomal, DNA to distinguish it from the DNA that is found in the cell nucleus. We will examine the role of mitochondrial and chloroplast DNA in Chapter 6.

In this section, we will focus on the composition of chromosomes found in the nucleus of eukaryotic cells. As you will learn, eukaryotic species contain genetic material that comes in sets of linear chromosomes.

Eukaryotic Chromosomes Are Examined Cytologically to Yield a Karyotype

Insights into inheritance patterns have been gained by observing chromosomes under the microscope. **Cytogenetics** is the field of genetics that involves the microscopic examination of chromosomes. The most basic observation that a **cytogeneticist** can make is to examine the chromosomal composition of a particular cell. For eukaryotic species, this is usually accomplished by observing the chromosomes as they are found in actively dividing cells. When a cell is preparing to divide, the chromosomes become more tightly coiled, which shortens them, thereby increasing their diameter. The consequence of this shortening is that distinctive shapes and numbers of chromosomes become visible with a light microscope. Each species has a particular chromosome composition. For example, most human cells contain 23 pairs of chromosomes, for a total of 46. On rare occasions, some individuals may inherit an abnormal number of chromosomes or a chromosome with an abnormal structure. Such abnormalities can often be detected by a microscopic examination of the chromosomes within actively dividing cells. In addition, a cytogeneticist may examine chromosomes as a way to distinguish two closely related species.

Figure 2.2a shows the general procedure for preparing human chromosomes to be viewed by microscopy. In this example, the cells were obtained from a sample of human blood; more specifically, the chromosomes within leukocytes (a type of white blood cell) were examined. Blood cells are a type of **somatic cell,** which is any cell of the body that is not a **gamete** or a precursor to a gamete. As discussed later, gametes are involved with sexual reproduction.

1. After the blood cells have been removed from the body, they are treated with one chemical that stimulates them to begin cell division and another chemical that halts cell division during mitosis, which is described later in this chapter.
2. As shown in Figure 2.2a, these actively dividing cells are subjected to centrifugation to concentrate them. The concentrated preparation is then mixed with a hypotonic solution that makes the cells swell. This swelling causes the chromosomes to spread out within the cell, thereby making it easier to see each individual chromosome.
3. After a second centrifugation step, the cells, which are at the bottom of the tube, are treated with a fixative that chemically freezes them so the chromosomes can no longer move around. The cells are then treated with a chemical dye that binds to the chromosomes and stains them. As discussed in greater detail in Chapter 8, this gives chromosomes a distinctive banding pattern that greatly enhances geneticists' ability to visualize and uniquely identify them (look ahead to Figure 8.1c, d). The cells are then placed on a slide and viewed with a light microscope.

In a cytogenetics laboratory, the microscopes are equipped with a camera that can photograph the chromosomes. In recent years, advances in technology have allowed cytogeneticists to view microscopic images on a computer screen (**Figure 2.2b**). On the screen, the chromosomes can be organized in a standard way, usually from largest to smallest. As seen in **Figure 2.2c**, the human chromosomes are lined up, and a number is given to designate each type of chromosome. An exception is the sex chromosomes, which are designated with the letters X and Y. An organized representation of the chromosomes within a cell is called a **karyotype.** A karyotype reveals how many chromosomes are found within an actively dividing somatic cell.

Eukaryotic Chromosomes Are Inherited in Sets

Most eukaryotic species are **diploid** or have a diploid phase to their life cycles, which means that each type of chromosome is a member of a pair. A diploid cell has two sets of chromosomes. In humans, for example, most somatic cells have 46 chromosomes—two sets of 23 each. Other diploid species, however, have different numbers of chromosomes in their somatic cells. For example, the dog has 39 chromosomes per set (78 total), the fruit fly has 4 chromosomes per set (8 total), and the tomato has 12 per set (24 total).

When a species is diploid, the members of a pair of chromosomes are called **homologs;** each type of chromosome is found in a homologous pair. As shown in Figure 2.2c, a human somatic cell has two copies of chromosome 1, two copies of chromosome 2, and so forth. Within each pair, the chromosome on the left is a homolog to the one on the right, and vice versa. In each pair, one chromosome was inherited from the mother and its homolog was inherited from the father. The two chromosomes in a homologous pair are nearly identical in size, have the same banding pattern, and contain a similar composition of genetic material. If a particular gene is found on one copy of a chromosome, it is also found on the other homolog. However, the two homologs may carry different versions of a given gene, which are called **alleles.** In Chapter 3, we will see that some alleles are dominant, meaning that they mask the expression of alleles that are recessive. As an example, let's consider a gene in humans, called *OCA2,* which is one of a few different genes that affect eye color. The *OCA2* gene is located on chromosome 15 and comes in alleles that result in brown or blue eyes. In a person with brown eyes, one copy of chromosome 15 may carry a dominant brown allele, whereas its

22   CHAPTER 2 :: REPRODUCTION AND CHROMOSOME TRANSMISSION

A sample of blood is collected and treated with chemicals that stimulate the cells to divide. Colchicine is added because it disrupts spindle formation and stops cells in mitosis where the chromosomes are highly compacted. The cells are then subjected to centrifugation.

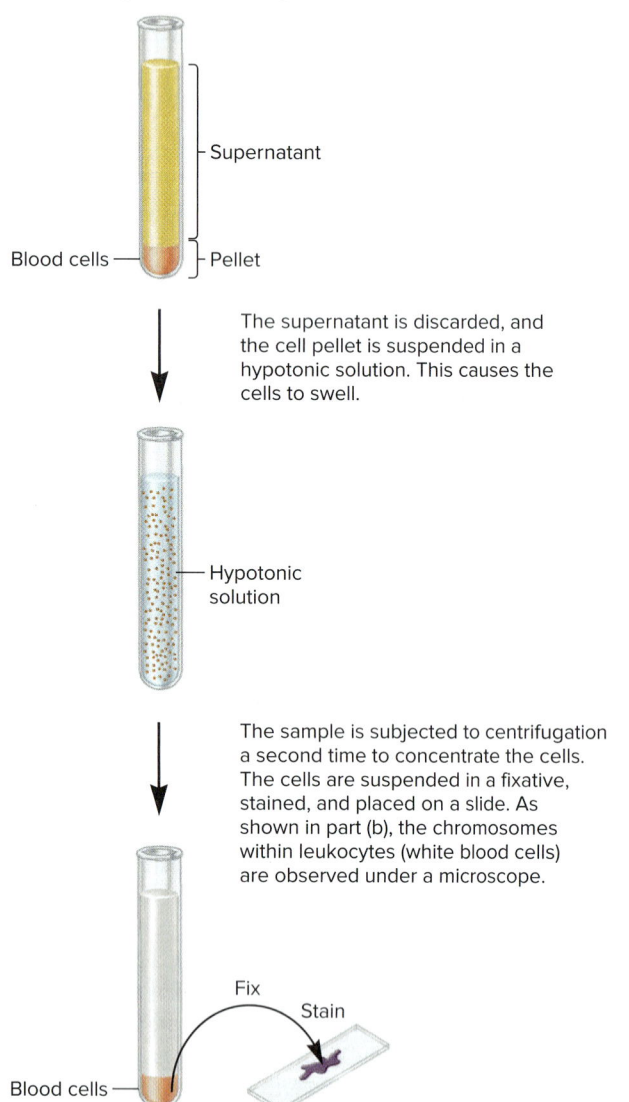

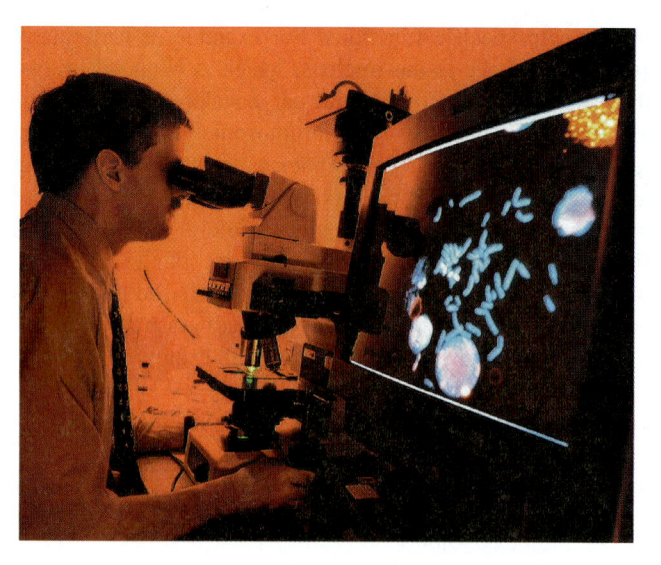

Supernatant
Blood cells — Pellet

The supernatant is discarded, and the cell pellet is suspended in a hypotonic solution. This causes the cells to swell.

Hypotonic solution

The sample is subjected to centrifugation a second time to concentrate the cells. The cells are suspended in a fixative, stained, and placed on a slide. As shown in part (b), the chromosomes within leukocytes (white blood cells) are observed under a microscope.

Fix Stain
Blood cells

(a) Preparing cells for a karyotype

(b) The slide is viewed by a light microscope; the sample is seen on a computer screen. The chromosomes can be arranged electronically on the screen.

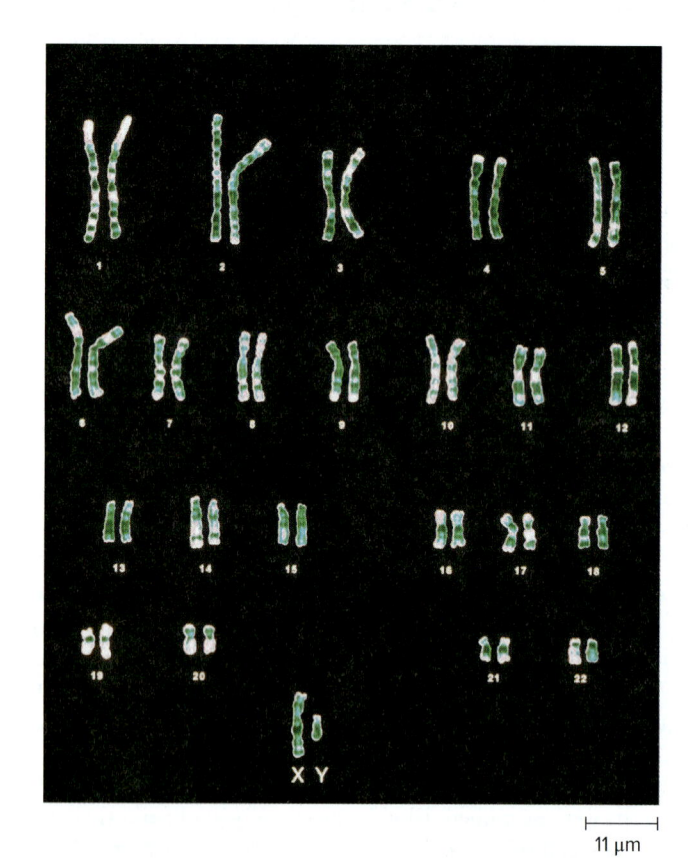

11 μm

(c) For a diploid human cell, two complete sets of chromosomes from a single cell constitute a karyotype of that cell.

FIGURE 2.2 The procedure for making a human karyotype.
(b) ©David Parker/Science Source; (c) ©Leonard Lessin/Science Source

Concept Check: How do you think the end results would be affected if the cells were not treated with a hypotonic solution?

homolog may carry a recessive blue allele. One copy was inherited from the mother and the other from the father.

At the molecular level, how similar are homologous chromosomes? The answer is that the sequence of bases of one homolog usually differs by less than 1% compared with the sequence of the other homolog. For example, the DNA sequence of chromosome 1 that you inherited from your mother is more than 99% identical to the sequence of chromosome 1 that you inherited from your father. Nevertheless, it should be emphasized that the sequences are not identical. The slight differences in DNA sequences provide the allelic differences in genes. Again, if we use the eye color gene as an example, a slight difference in DNA sequence distinguishes the brown and blue alleles. However, the striking similarities between homologous chromosomes do not apply to the pair of sex chromosomes—X and Y. These chromosomes differ in size and genetic composition. Certain genes that are found on the X chromosome are not found on the Y chromosome, and vice versa. The X and Y chromosomes are not considered homologous chromosomes, though they do have short regions of homology.

Figure 2.3 shows two homologous chromosomes with three different genes. An individual carrying these two chromosomes is **homozygous** for the dominant allele of gene *A*, which means that both homologs carry the same allele. The individual is **heterozygous**, *Bb*, for the second gene, meaning that the homologs carry different alleles. For the third gene, the individual is homozygous for a recessive allele, *c*. The physical location of a gene is called its **locus** (plural: **loci**). As seen in Figure 2.3, for example, the locus of gene *C* is toward one end of this chromosome, whereas the locus of gene *B* is more in the middle.

2.1 REVIEWING THE KEY CONCEPTS

- Chromosomes are structures that contain the genetic material, which is DNA.
- Prokaryotic cells are simple and lack cell compartmentalization, whereas eukaryotic cells contain a cell nucleus and other compartments (see Figure 2.1).
- Chromosomes can be examined under the microscope. An organized representation of the chromosomes from a single cell is called a karyotype (see Figure 2.2).
- In eukaryotic species, the chromosomes are found in sets. Eukaryotic cells are often diploid, which means that each type of chromosome occurs in a homologous pair (see Figure 2.3).

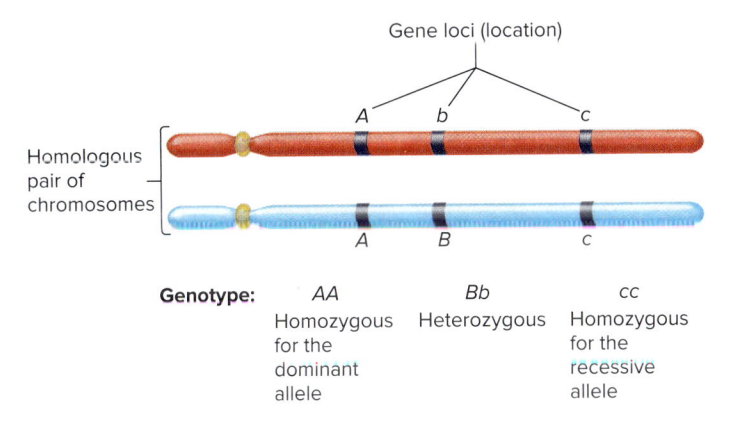

FIGURE 2.3 A comparison of homologous chromosomes. Each pair of homologous chromosomes carries the same types of genes, but, as shown here, the alleles may or may not be different.

Concept Check: How are homologs similar to each other and how are they different?

2.1 COMPREHENSION QUESTIONS

1. Which of the following is *not* found in a prokaryotic cell?
   a. Plasma membrane
   b. Ribosome
   c. Cell nucleus
   d. Cytoplasm

2. When preparing a karyotype, which of the following steps is conducted?
   a. Treat the cells with a chemical that causes them to begin cell division.
   b. Treat the cells with a hypotonic solution that causes them to swell.
   c. Expose the cells to chemical dyes that bind to the chromosomes and stain them.
   d. All of the above steps are carried out.

3. How many sets of chromosomes are found in a human somatic cell, and how many chromosomes are within one set?
   a. 2 sets, with 23 in each set
   b. 23 sets, with 2 in each set
   c. 1 set, with 23 in each set
   d. 23 sets, with 1 in each set

2.2 CELL DIVISION

Learning Outcomes:
1. Describe the process of binary fission in bacteria.
2. Outline the phases of the eukaryotic cell cycle.

Now that we have an appreciation for the chromosomal composition of living cells, we can consider how chromosomes are copied and transmitted when cells divide. One purpose of cell division is **asexual reproduction**. In this process, a preexisting cell divides to produce two new cells. By convention, the original cell is usually called the mother cell, and the two new cells are the daughter cells. When species are unicellular, the mother cell is judged to be one individual, and the two daughter cells are two new separate organisms. Asexual reproduction is how bacterial cells proliferate. In addition, certain unicellular eukaryotes, such as the amoeba and baker's yeast (*Saccharomyces cerevisiae*), can reproduce asexually.

Another purpose of cell division is multicellularity. Species such as plants, animals, most fungi, and some protists are derived from a single cell that has undergone repeated cell divisions. Humans, for example, begin as a single fertilized egg; repeated cell divisions produce an adult with many trillions of cells. The precise transmission of chromosomes during every cell division is critical so that all cells of the body receive the correct amount of genetic material.

In this section, we will consider how the process of cell division requires the duplication, organization, and distribution of the chromosomes. In bacteria, which have a single circular chromosome, the division process is relatively simple. Prior to cell division, bacteria duplicate their circular chromosome; they then distribute a copy into each of the two daughter cells. This process, known as binary fission, is described first. Eukaryotes have multiple numbers of chromosomes that occur as sets. Compared to the process of binary fission in bacteria, this added complexity requires a more complicated sorting process to ensure that each newly made cell receives the correct number and types of chromosomes. In this section, we will examine how eukaryotic cell division follows a series of stages called the cell cycle.

Bacteria Reproduce Asexually by Binary Fission

As discussed earlier (see Figure 2.1a), bacterial species are typically unicellular, although individual bacteria may associate with each other to form pairs, chains, or clumps. Unlike eukaryotes, which have their chromosomes in a separate nucleus, the circular chromosomes of bacteria are in direct contact with the cytoplasm. In Chapter 12, we will consider the molecular structure of bacterial chromosomes in greater detail.

The capacity of bacteria to divide is really quite astounding. Some species, such as *Escherichia coli,* a common bacterium of the intestine, can divide every 20 to 30 minutes. As shown in **Figure 2.4**, bacteria reproduce by a process called **binary fission.**

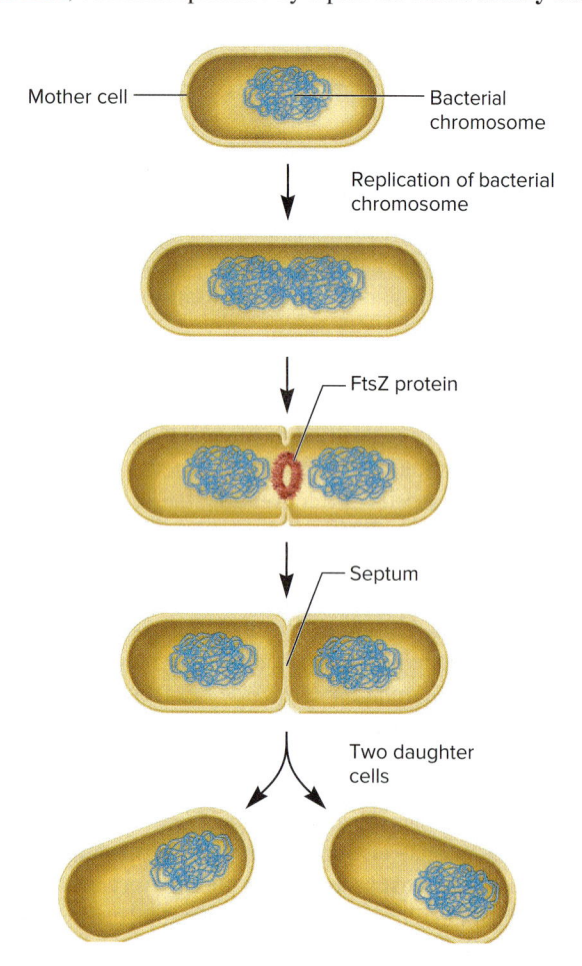

FIGURE 2.4 **Binary fission: the process by which bacterial cells divide.** Prior to division, the chromosome replicates to produce two identical copies. These two copies segregate from each other, with one copy going to each daughter cell.

Concept Check: What is the function of the FtsZ protein during binary fission?

1. Prior to cell division, bacterial cells copy, or replicate, their chromosomal DNA. This produces two identical copies of the genetic material, as shown at the top of Figure 2.4.
2. A protein called FtsZ assembles into a ring at the future site of the septum. FtsZ is thought to be the first protein to move to this division site.
3. FtsZ recruits other proteins that produce a septum, which is a new cell wall between the daughter cells. FtsZ is evolutionarily related to a eukaryotic protein called tubulin. As discussed later in this chapter, tubulin is the main component of microtubules, which play a key role in chromosome sorting in eukaryotes. Both FtsZ and tubulin form structures that provide cells with organization and play key roles in cell division.
4. As a result of binary fission, a bacterial cell called the mother cell has divided into two daughter cells. Each daughter cell receives a copy of the chromosomal genetic material. Except when rare mutations occur, the daughter cells are usually genetically identical because they contain exact copies of the genetic material from the mother cell.

Binary fission is an asexual form of reproduction because it does not involve genetic contributions from two different gametes. On occasion, bacteria can exchange small pieces of genetic material with each other. We will consider some interesting mechanisms of such genetic exchange in Chapter 9.

Eukaryotic Cells Advance Through a Cell Cycle to Produce Genetically Identical Daughter Cells

The common outcome of eukaryotic cell division is to produce two daughter cells that have the same number and types of chromosomes as the original mother cell. This requires a replication and division process that is more complicated than simple binary fission. Eukaryotic cells that are destined to divide advance through a series of phases known as the **cell cycle** (**Figure 2.5**). These phases are named G for gap, S for synthesis (of the genetic material), and M for mitosis. There are two G phases: G_1 and G_2. The term *gap* originally described the gaps between S phase and mitosis in which it was not microscopically apparent that significant changes were occurring in the cell. However, we now know that both gap phases are critical periods in the cell cycle that involve many molecular changes. In actively dividing cells, the G_1, S, and G_2 phases are collectively known as **interphase.** In addition, cells may remain permanently, or for long periods of time, in a phase of the cell cycle called G_0. A cell in the G_0 phase is either temporarily not advancing through the cell cycle or, in the case of terminally differentiated cells, such as most nerve cells in an adult mammal, will never divide again. In other words, the G_0 phase is a nondividing stage.

Let's consider the key steps in these four phases.

1. During the G_1 **phase,** a cell may prepare to divide. Depending on the cell type and the conditions it encounters, a cell in the G_1 phase may accumulate molecular changes and reach a **restriction point** and thereby be committed on

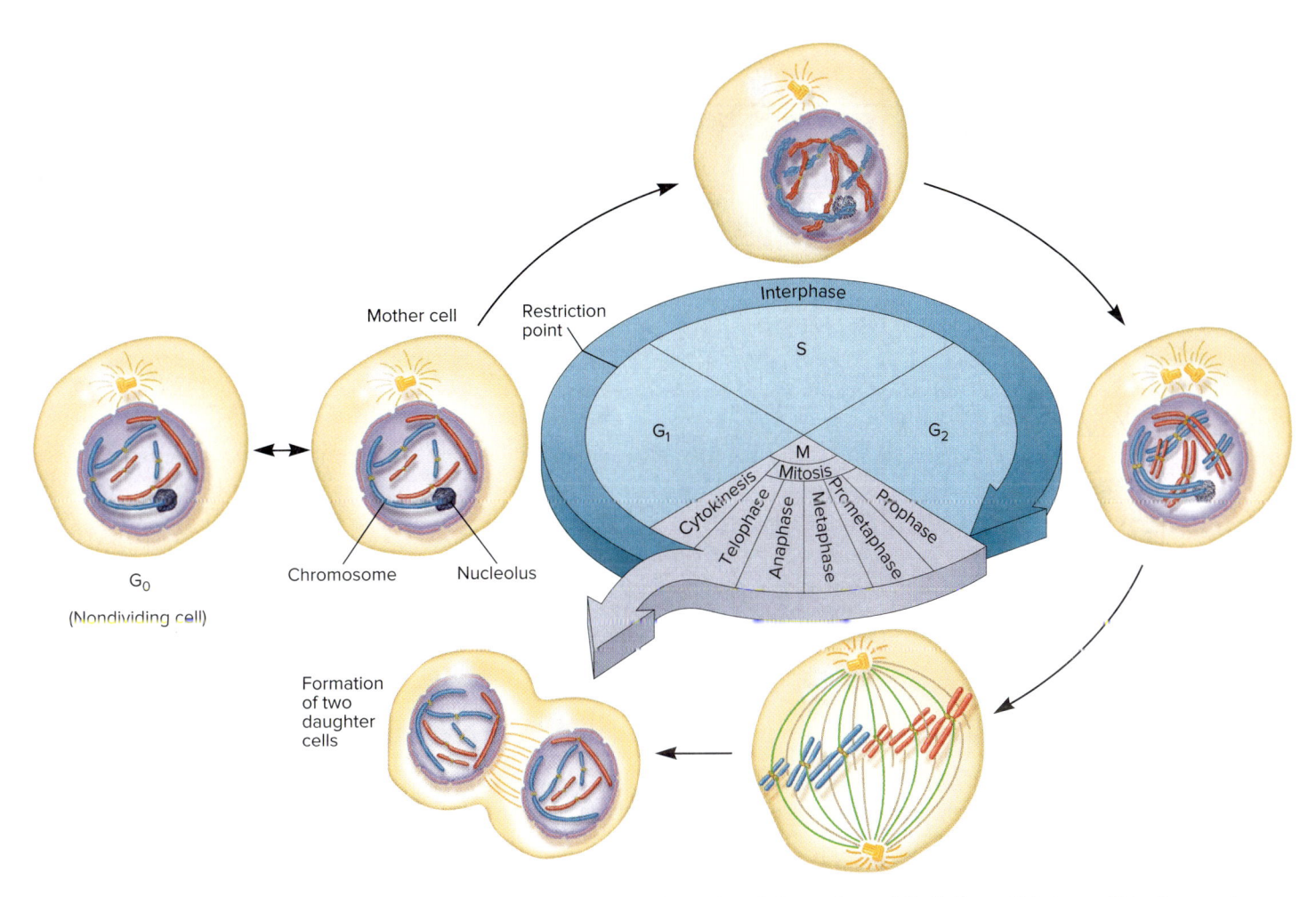

FIGURE 2.5 The eukaryotic cell cycle. Dividing cells advance through a series of phases, denoted G_1, S, G_2, and M phases. This diagram shows the advancement of a cell through mitosis to produce two daughter cells. The original mother cell had three pairs of chromosomes, for a total of six individual chromosomes. By the G_2 phase, these have replicated to yield 12 chromatids found in six pairs of sister chromatids. After mitosis and cytokinesis are completed, each of the two daughter cells contains six individual chromosomes, just like the mother cell. Note: The chromosomes in G_0, G_1, S, and G_2 phases are not condensed (look ahead to Figure 2.8a). In this drawing, they are shown partially condensed so they can be easily counted.

Concept Check: What is the difference between the G_0 and G_1 phases?

a pathway that leads to cell division. After a cell has reached a restriction point, it will advance through the remainder of G_1, and then proceed through the S, G_2, and M phases to complete the cell cycle.

2. Once past the restriction point, the cell then advances to the **S phase,** during which the chromosomes are replicated. After replication, the two copies are called **chromatids.** They are joined to each other at a region of DNA called the **centromere** to form a unit known as a pair of **sister chromatids,** or a **dyad** (**Figure 2.6**). A single chromatid within a dyad is called a **monad.** An unreplicated chromosome can also be called a monad. When S phase is completed, a cell has twice as many chromatids as it had chromosomes in the G_1 phase. The **kinetochore** is a group of proteins that are bound to the centromere. These proteins help to hold the sister chromatids together and also play a role in chromosome sorting, as discussed later.

3. During the **G_2 phase,** the cell accumulates the materials that are necessary for nuclear and cell division.

4. Finally, the cell advances into the **M phase** of the cell cycle, when **mitosis** occurs. The primary purpose of mitosis is to distribute the replicated chromosomes, dividing one cell nucleus into two nuclei, so that each daughter cell receives the same complement of chromosomes. For example, a human cell in the G_2 phase has 92 chromatids, which are found in 46 pairs. During mitosis, these pairs of chromatids are separated and sorted to produce two nuclei that contain 46 chromosomes each.

5. Two daughter cells are formed by a process called cytokinesis.

2.2 REVIEWING THE KEY CONCEPTS

- Bacteria divide by binary fission (see Figure 2.4).
- To divide, eukaryotic cells advance through a cell cycle (see Figure 2.5).
- Prior to cell division, eukaryotic chromosomes are replicated to form sister chromatids (see Figure 2.6).

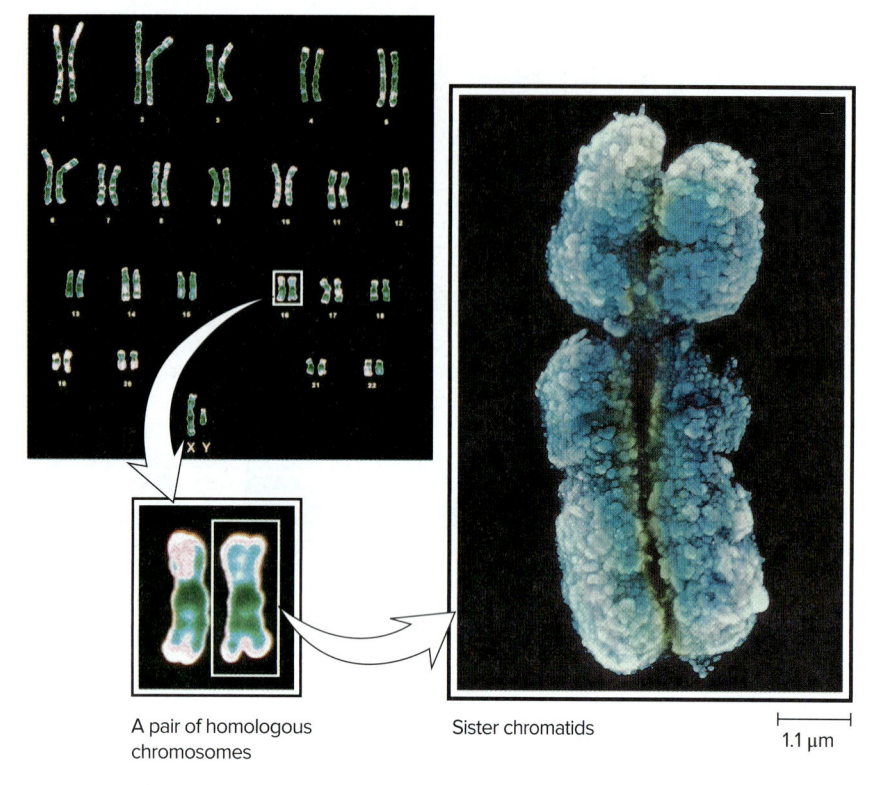

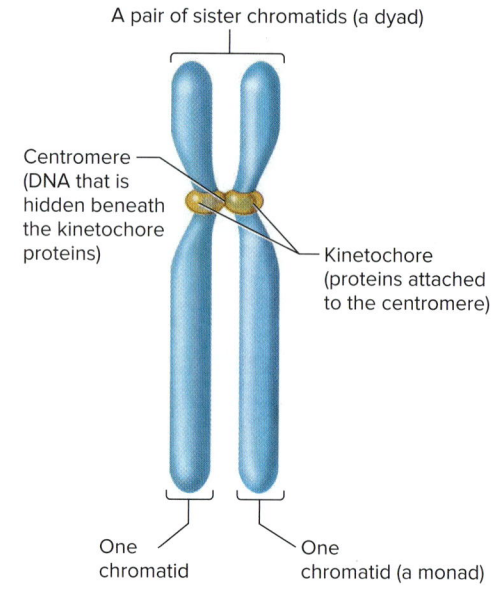

(a) **Homologous chromosomes and sister chromatids**

(b) **Schematic drawing of sister chromatids**

FIGURE 2.6 Chromosomes following DNA replication. (a) The photomicrograph at the upper left shows a human karyotype. The photomicrograph on the right shows a chromosome in the form of a pair of sister chromatids. This chromosome is in the metaphase stage of mitosis, which is described later in the chapter. Note: Each of the 46 chromosomes that are viewed in a human karyotype (upper left) is actually a pair of sister chromatids. Look closely at the white rectangular boxes in the two insets. (b) A schematic drawing of sister chromatids. This structure has two chromatids that lie side by side. As seen here, each chromatid is a distinct unit. The two chromatids are held together by kinetochore proteins that bind to each other and to the centromere of each chromatid.

(a.1, 2) ©Leonard Lessin/Science Source; (a.3) ©Biophoto Associates/Science Source

Concept Check: What is the difference between homologs and the chromatids within a pair of sister chromatids?

2.2 COMPREHENSION QUESTIONS

1. Binary fission
   a. is a form of asexual reproduction.
   b. is a way for bacteria to reproduce.
   c. begins with a single mother cell and produces two genetically identical daughter cells.
   d. is all of the above.

2. Which of the following is the correct order of phases of the eukaryotic cell cycle?
   a. G_1, G_2, S, M
   b. G_1, S, G_2, M
   c. G_1, G_2, M, S
   d. G_1, S, M, G_2

3. What is a critical event that occurs during S phase of the cell cycle?
   a. Cells decide whether or not to divide.
   b. DNA replication produces pairs of sister chromatids.
   c. The chromosomes condense.
   d. The single nucleus is divided into two nuclei.

2.3 MITOSIS AND CYTOKINESIS

Learning Outcomes:
1. Describe the structure and function of the mitotic spindle.
2. List and describe the phases of mitosis.
3. Outline the key differences between cytokinesis in animal and plant cells.

As we have seen, eukaryotic cell division involves a cell cycle in which the chromosomes are replicated and later sorted so that each daughter cell receives the same amount of genetic material. This process ensures genetic consistency from one cell generation to the next. In this section, we will examine the stages of mitosis and cytokinesis in greater detail.

The Mitotic Spindle Apparatus Organizes and Sorts Eukaryotic Chromosomes

Before we discuss the events of mitosis, let's first consider the structure of the **mitotic spindle apparatus** (also known simply

2.3 MITOSIS AND CYTOKINESIS

each other play a role in the separation of the two poles. They help to "push" the poles away from each other.
- The kinetochore microtubules have attachments to kinetochores, which are protein complexes bound to the centromeres of individual chromosomes.

The mitotic spindle allows cells to organize and separate chromosomes so that each daughter cell receives the same complement of chromosomes. This sorting process, known as mitosis, is described next.

The Transmission of Chromosomes During the Division of Eukaryotic Cells Involves a Process Known as Mitosis

In **Figure 2.8**, the process of mitosis is shown for a diploid animal cell. In the simplified diagrams below the micrographs in this figure, the original mother cell contains six chromosomes; it is diploid (2n) and has three chromosomes per set (n = 3). One set is shown in blue, and the homologous set is red. As discussed next, mitosis is subdivided into phases known as prophase, prometaphase, metaphase, anaphase, and telophase.

Prophase Prior to mitosis, the cells are in interphase, during which the chromosomes are **decondensed**—less tightly compacted—and found in the nucleus (Figure 2.8a). At the start of mitosis, in **prophase**, the chromosomes have already replicated to produce 12 chromatids that are joined as six pairs of sister chromatids (Figure 2.8b). As prophase proceeds, the nuclear membrane begins to dissociate into small vesicles. At the same time, the chromatids **condense** into more compact structures that are readily visible by light microscopy. The mitotic spindle also begins to form, and the nucleolus, which is the site of ribosome assembly, disappears.

Prometaphase As mitosis advances from prophase to prometaphase, the centrosomes move to opposite ends of the cell and demarcate two spindle poles, one within each of the future daughter cells. Once the nuclear membrane has dissociated into vesicles, the spindle fibers can interact with the sister chromatids. This interaction occurs in a phase of mitosis called **prometaphase** (Figure 2.8c). How do sister chromatids become attached to the spindle? Initially, microtubules are rapidly formed and can be seen growing out from the two poles. As a microtubule grows, if its end happens to make contact with a kinetochore, the end is said to be captured and remains firmly attached to the kinetochore. This random process is how sister chromatids become attached to kinetochore microtubules. Alternatively, if the end of a microtubule does not collide with a kinetochore, the microtubule eventually depolymerizes and retracts to the centrosome. As the end of prometaphase nears, the kinetochore on a pair of sister chromatids is attached to kinetochore microtubules from opposite poles. As these events are occurring, the sister chromatids are seen to undergo jerky movements as they are tugged, back and forth, between the two poles. By the end of prometaphase, the mitotic spindle is completely formed.

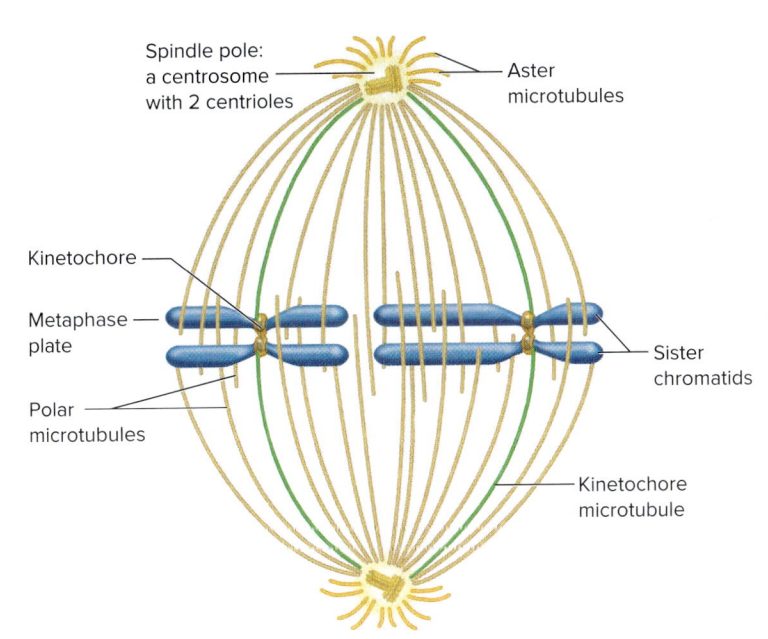

FIGURE 2.7 **The structure of the mitotic spindle in a typical animal cell.** A single centrosome duplicates during S phase, and the two centrosomes separate at the beginning of M phase. The mitotic spindle is formed from microtubules that are rooted in the centrosomes. Each centrosome is located at a spindle pole. The aster microtubules emanate away from the region between the poles. They help to position the spindle within the cell and are used as reference points for cell division. However, aster microtubules are not found in many species, such as plants. The polar microtubules project into the region between the two poles; they play a role in pole separation. The kinetochore microtubules are attached to the kinetochore of sister chromatids.

> **Concept Check:** *Where are the two ends of a kinetochore microtubule?*

as the **mitotic spindle**), which is involved in the organization and sorting of chromosomes (**Figure 2.7**). The spindle apparatus is formed from **microtubule-organizing centers (MTOCs),** which are structures found in eukaryotic cells from which microtubules grow. Microtubules are produced from the rapid polymerization of tubulin proteins. In animal cells, the mitotic spindle is formed from two MTOCs called **centrosomes.** Each centrosome is located at a **spindle pole.** A pair of **centrioles** at right angles to each other is found within each centrosome of animal cells. Centrosomes and centrioles are found in animal cells but not in all eukaryotic species. For example, plant cells do not have centrosomes. Instead, the nuclear envelope functions as an MTOC for spindle formation in plant cells.

The mitotic spindle of a typical animal cell has three types of microtubules (see Figure 2.7).

- The aster microtubules emanate outward from the centrosome toward the plasma membrane. They are important for the positioning of the spindle apparatus within the cell and later in the process of cytokinesis.
- The polar microtubules project toward the region where the chromosomes are found during mitosis—the region between the two spindle poles. Polar microtubules that overlap with

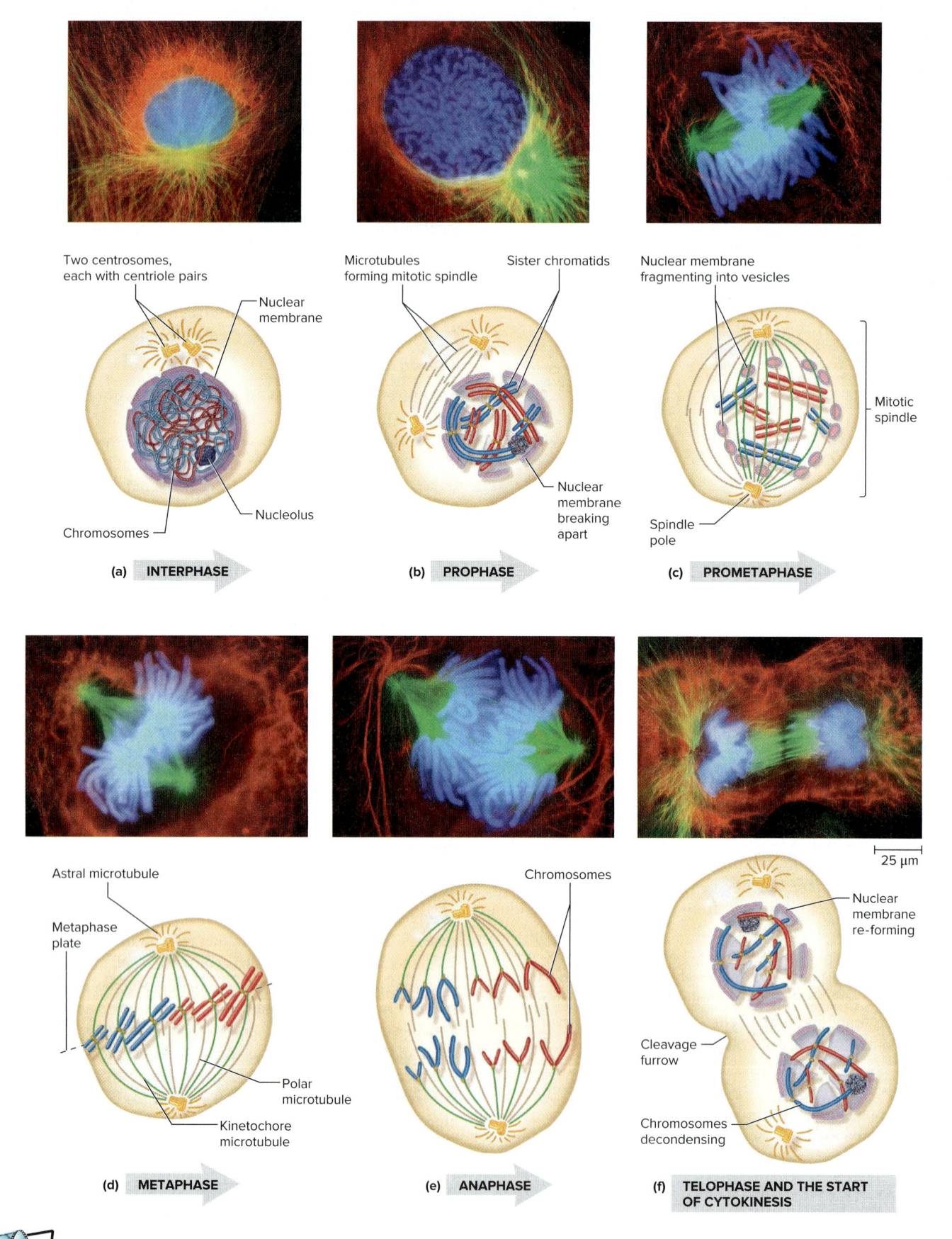

FIGURE 2.8 The process of mitosis in an animal cell. The upper rows of micrographs illustrate cells of a fish embryo advancing through mitosis. The chromosomes are stained in blue and the spindle is green. Below the micrographs are schematic drawings that emphasize the sorting and separation of the chromosomes. In the drawings, the original diploid cell is shown with six chromosomes (three in each set). At the start of mitosis, these have already replicated into 12 chromatids. After cytokinesis is completed, the final result is two daughter cells, each containing six chromosomes.

©Conly L. Rieder

Concept Check: During which phase are sister chromatids separated and sent to opposite poles?

2.3 MITOSIS AND CYTOKINESIS

Metaphase Eventually, the pairs of sister chromatids align themselves along a plane called the **metaphase plate.** As shown in Figure 2.8d, when this alignment is complete, the cell is in **metaphase** of mitosis. At this point, each pair of chromatids (each dyad) is attached to both poles by kinetochore microtubules. The pairs of sister chromatids have become organized into a single row along the metaphase plate. When this organizational process is finished, the chromatids can be equally distributed into two daughter cells.

Anaphase At **anaphase,** the connection that is responsible for holding the pairs of chromatids together is broken (Figure 2.8e). Each chromatid or monad, now an individual chromosome, is linked to only one of the two poles. As anaphase proceeds, the chromosomes move toward the pole to which they are attached. This movement is due to a shortening of the kinetochore microtubules. In addition, the two poles themselves move farther apart due to the elongation of the polar microtubules, which slide in opposite directions as a result of the actions of motor proteins.

Telophase During **telophase,** the chromosomes reach their respective poles and decondense. The nuclear membrane now reforms to produce two separate nuclei. In Figure 2.8f, this process has produced two nuclei that contain six chromosomes each. The nucleoli have also reappeared.

Cytokinesis In most cases, mitosis is quickly followed by **cytokinesis,** in which the two nuclei are segregated into separate daughter cells. Likewise, cytokinesis also segregates cell organelles such as mitochondria and chloroplasts into daughter cells. In animal cells, cytokinesis begins shortly after anaphase. A contractile ring, composed of myosin motor proteins and actin filaments, assembles adjacent to the plasma membrane. Myosin hydrolyzes ATP, which shortens the ring and thereby constricts the plasma membrane to form a **cleavage furrow** that ingresses, or moves inward (**Figure 2.9a**). Ingression continues until a midbody structure is formed that physically pinches one cell into two.

In plants, the two daughter cells are separated by the formation of a **cell plate** (**Figure 2.9b**). At the end of anaphase, Golgi-derived vesicles carrying cell wall materials are transported to the equator of a dividing cell. These vesicles are directed to their locations via microtubules and actin filaments that serve as tracks for vesicle movement. The fusion of these vesicles gives rise to the cell plate, which is a membrane-bound compartment. The cell plate begins in the middle of the cell and expands until it attaches to the mother cell's wall. Once this attachment has taken place, the cell plate undergoes a process of maturation and eventually separates the mother cell into two daughter cells.

Outcome of Mitotic Cell Division Mitosis and cytokinesis ultimately produce two daughter cells having the same number of chromosomes as the mother cell. Barring rare mutations, the two daughter cells are genetically identical to each other and to the mother cell from which they were derived. The critical consequence of this sorting process is to ensure genetic consistency from one somatic cell to the next. The development of multicellularity relies on the repeated process of mitosis and cytokinesis. In diploid organisms that are multicellular, most of the somatic cells are diploid and genetically identical to each other.

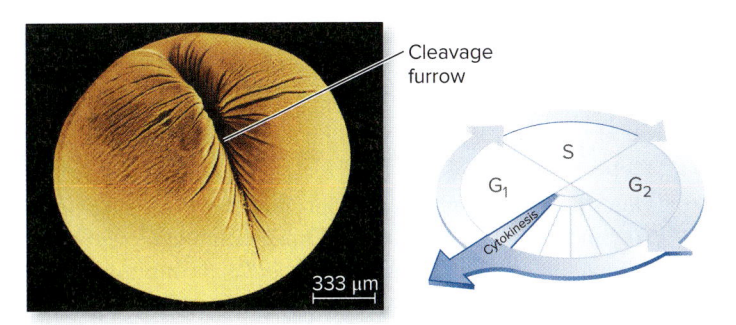

(a) Cleavage of an animal cell

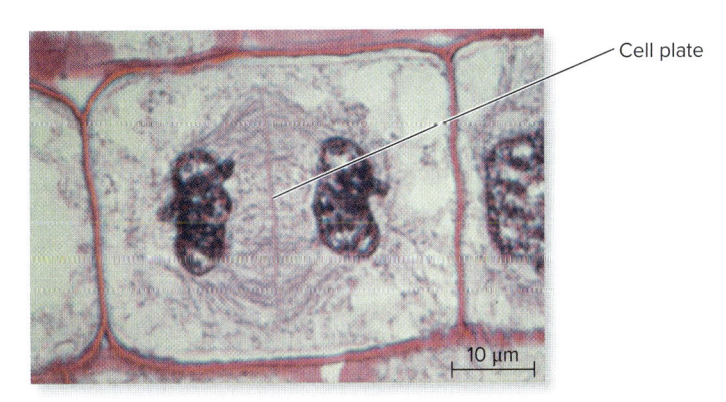

(b) Formation of a cell plate in a plant cell

FIGURE 2.9 Cytokinesis in animal and plant cells. (a) In an animal cell, cytokinesis involves the formation of a cleavage furrow. (b) In a plant cell, cytokinesis occurs via the formation of a cell plate between the two daughter cells.

(a) ©Don W. Fawcett/Science Source; (b) ©Ed Reschke

Concept Check: What causes the cleavage furrow to ingress?

Genetic TIPS

The Question: What are the functional roles of the mitotic spindle in an animal cell? Explain how these functions are related to the three types of microtubules: aster, polar, and kinetochore microtubules.

Topic: What topic in genetics does this question address?

The topic is mitosis. More specifically, the question is about the role of the mitotic spindle.

Information: What information do you know based on the question and your understanding of the topic?

From the question, you know there are three types of microtubules. From your understanding of the topic, you may remember the structure of the mitotic spindle, which is shown in Figure 2.7. Also, Figure 2.8 describes how the spindle plays different roles during mitosis.

Problem-Solving Strategy: Define key terms. Describe the steps.

One strategy to begin solving this problem is to make sure you understand the key terms. In particular, you may want to look up the meaning of *mitotic spindle* and *microtubules*, if you don't already know what those terms mean. After you understand the key terms, a problem-solving strategy is to describe the steps of mitosis, and think about the roles of the types of microtubules in the various steps. These steps are shown in Figure 2.8. You may also want to refer back to Figure 2.7 to appreciate the structure of the mitotic spindle.

Answer:
- The mitotic spindle is involved in sorting the chromosomes and promoting the division of one cell into two daughter cells.
- The polar microtubules overlap with each other and push the poles apart during anaphase.
- The aster microtubules help to orient the spindle in the cell and play a role in cytokinesis.
- The kinetochore microtubules attach to chromosomes and aid in their sorting. They are needed to align the chromosomes at the metaphase plate and to pull the chromosomes to the poles during anaphase.

2.3 REVIEWING THE KEY CONCEPTS

- Chromosome sorting in eukaryotes is achieved via a mitotic spindle apparatus (see Figure 2.7).
- A common way for eukaryotic cells to divide is by mitosis and cytokinesis. Mitosis is divided into prophase, prometaphase, metaphase, anaphase, and telophase. During cytokinesis, animal cells divide by a cleavage furrow and plant cells form a cell plate (see Figures 2.8, 2.9).

2.3 COMPREHENSION QUESTIONS

1. During which phase of mitosis does the nuclear envelop re-form?
    a. Prometaphase   c. Anaphase
    b. Metaphase   d. Telophase
2. Which phase of mitosis is depicted in the drawing below?

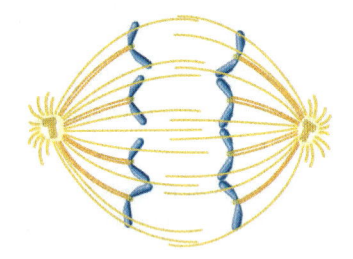

    a. Prophase   d. Anaphase
    b. Prometaphase   e. Telophase
    c. Metaphase

2.4 MEIOSIS

Learning Outcomes:
1. List and describe the phases of meiosis.
2. Compare and contrast the key differences between mitosis and meiosis.

In the previous section, we considered how eukaryotic cells can divide by mitosis and cytokinesis so that a mother cell produces two genetically identical daughter cells. Diploid eukaryotic cells may also divide by an alternative process called **meiosis** (from the Greek meaning "less"). During meiosis, **haploid** cells, which contain a single set of chromosomes, are produced from a cell that was originally diploid. For this to occur, the chromosomes must be correctly sorted and distributed in a way that reduces the chromosome number to half its original value. In the case of humans, for example, each gamete must receive half of the total number of chromosomes. But not just any 23 chromosomes will do—a gamete must receive one chromosome from each of the 23 pairs. In this section, we examine how the phases of meiosis lead to the formation of cells with a haploid complement of chromosomes.

Meiosis Produces Cells That Are Haploid

The process of meiosis bears striking similarities to mitosis. Like mitosis, meiosis begins after a cell has advanced through the G_1, S, and G_2 phases of the cell cycle. However, meiosis involves two successive divisions rather than one (as in mitosis). Prior to meiosis, the chromosomes are replicated in S phase to produce pairs of sister chromatids. This single replication event is then followed by two sequential cell divisions: called meiosis I and II. As in mitosis, each of these divisions is subdivided into prophase, prometaphase, metaphase, anaphase, and telophase.

Prophase of Meiosis I Figure 2.10 emphasizes some of the important events that occur during prophase of meiosis I, which is further subdivided into stages known as leptotene, zygotene, pachytene, diplotene, and diakinesis.

1. During the **leptotene** stage, the replicated chromosomes begin to condense and become visible with a light microscope.
2. Unlike prophase of mitosis, the **zygotene** stage of prophase of meiosis I involves a recognition process known as **synapsis,** in which the homologous chromosomes recognize each other and begin to align themselves along their entire lengths. As this occurs, a **synaptonemal complex** forms between the homologs. This complex is thought to promote the binding of homologs to each other. However, the synaptonemal complex may not be required for the pairing of homologous chromosomes, because some species, such as *Aspergillus nidulans* and *Schizosaccharomyces pombe*, completely lack such a complex, yet their chromosomes synapse correctly.
3. At **pachytene,** the homologs have become completely aligned. The associated chromatids are known as **bivalents.** Each bivalent contains two pairs of sister chromatids, or a total of four chromatids. A bivalent is also called a **tetrad** (from the prefix *tetra-*, meaning "four") because it is composed of four

STAGES OF PROPHASE OF MEIOSIS I

LEPTOTENE — Nuclear membrane. Replicated chromosomes condense.

ZYGOTENE — Bivalent forming. Synaptonemal complex forming. Synapsis begins.

PACHYTENE — A bivalent has formed and crossing over has occurred.

DIPLOTENE — Chiasma. Synaptonemal complex dissociates.

DIAKINESIS — Nuclear membrane breaking apart. End of prophase I

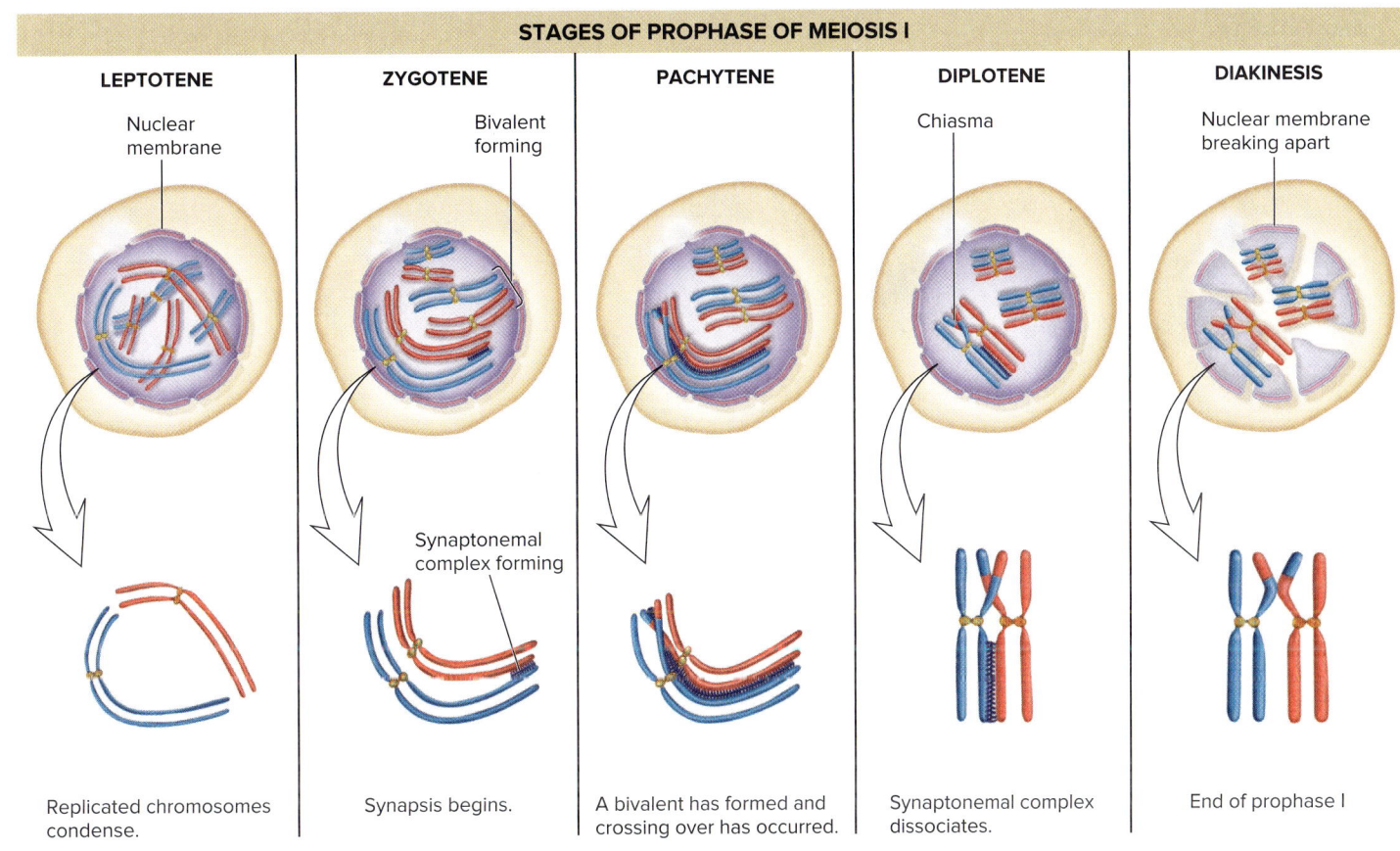

FIGURE 2.10 The events that occur during prophase of meiosis I. Note: for simplicity, the nucleolus is not shown in this figure.

Concept Check: What is the end result of crossing over?

---

chromatids—that is, four monads. Prior to the pachytene stage, when synapsis is complete, an event known as crossing over usually occurs. **Crossing over** involves a physical exchange of chromosome pieces that results in an exchange of genetic information. Depending on the size of the chromosome and the species, an average eukaryotic chromosome incurs from a couple to a couple of dozen crossovers. During spermatogenesis in humans, for example, an average chromosome undergoes slightly more than 2 crossovers, whereas chromosomes in certain plant species may undergo 20 or more crossovers. In Figure 2.10, crossing over has occurred at a single site between two of the larger chromatids. The connection that results from crossing over is called a **chiasma** (plural: **chiasmata**), because it physically resembles the Greek letter chi, χ. We will consider the genetic consequences of crossing over in Chapter 7 and the molecular process of crossing over in Chapter 13.

4. By the end of the **diplotene** stage, the synaptonemal complex has largely disappeared. The chromatids within a bivalent pull apart slightly, and microscopically it becomes easier to see that a bivalent is actually composed of four chromatids.
5. In the last stage of prophase of meiosis I, **diakinesis,** the synaptonemal complex completely disappears.

Prometaphase of Meiosis I Figure 2.10 emphasizes the pairing and crossing over that occurs during prophase of meiosis I. In **Figure 2.11**, we turn our attention to the general events in meiosis.

Prophase of meiosis I is followed by prometaphase, in which the spindle apparatus is completely formed and the chromatids are attached via kinetochore microtubules.

Metaphase of Meiosis I At metaphase of meiosis I, the bivalents are organized along the metaphase plate. However, their pattern of alignment is strikingly different from that observed during mitosis (refer back to Figure 2.8d). Before we consider the rest of meiosis I, a particularly critical feature for you to appreciate is how the bivalents are aligned along the metaphase plate. In particular, the pairs of sister chromatids (the dyads) are aligned in a double row rather than a single row, as occurs in mitosis. Furthermore, the arrangement of sister chromatids within this double row is random with regard to the blue and red homologs. In Figure 2.11, one of the blue homologs is above the metaphase plate and the other two are below, whereas one of the red homologs is below the metaphase plate and other two are above.

In an organism that produces many gametes, meiosis in other cells can produce a different arrangement of homologs—three blues above and none below, or none above and three below, and so on. Because most eukaryotic species have several chromosomes per set, the sister chromatids can be randomly aligned along the metaphase plate in many possible ways. For example, consider humans, who have 23 chromosomes per set. The possible number of different random alignments equals 2^n, where n equals the

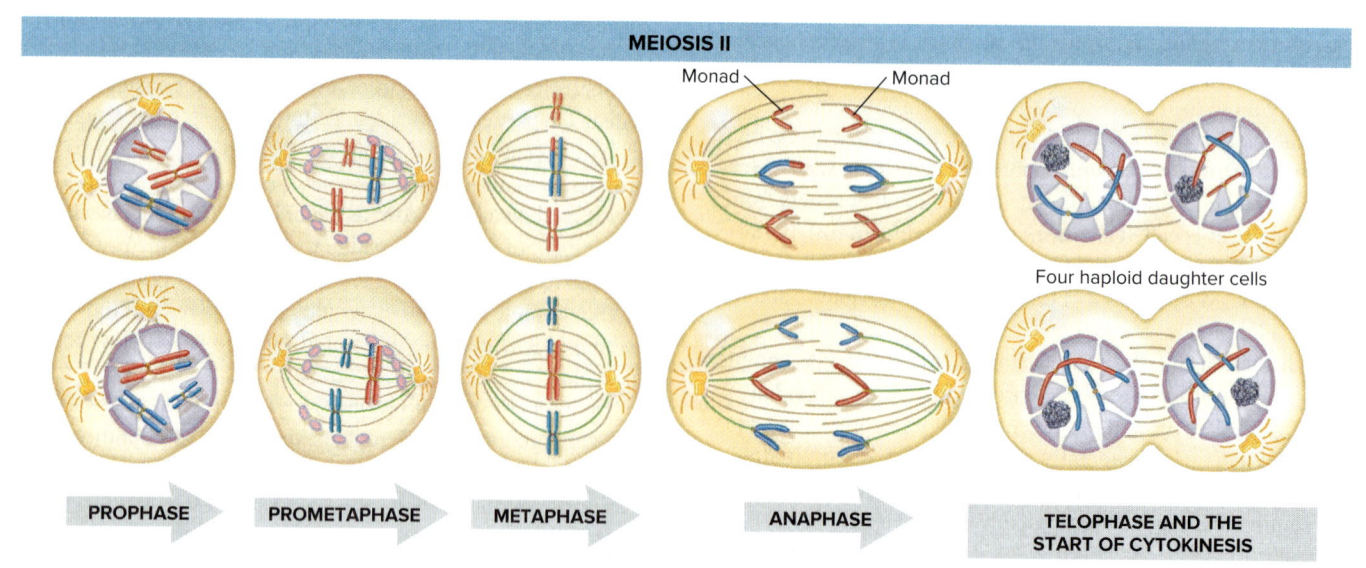

FIGURE 2.11 The stages of meiosis in an animal cell. See text for details.

Concept Check: How do the four cells at the end of meiosis differ from the original mother cell?

2.4 MEIOSIS

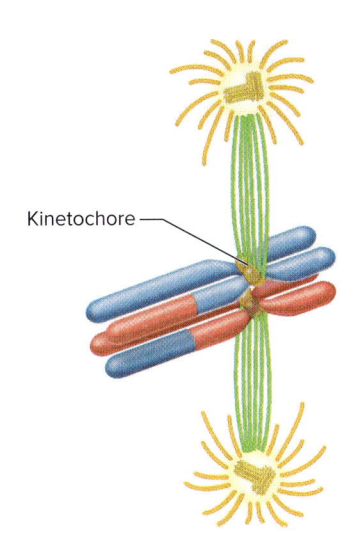

FIGURE 2.12 Attachment of the kinetochore microtubules to replicated chromosomes during meiosis. The kinetochore microtubules from a given pole are attached to one pair of chromatids in a bivalent, but not both. Therefore, each pair of sister chromatids is attached to only one pole.

Concept Check: How is this attachment of chromosomes to kinetochore microtubules different from their attachment during metaphase of mitosis?

number of chromosomes per set. Thus, in humans, there are 2^{23}, or over 8 million, possibilities! Because the homologs are genetically similar but not identical, we see from this calculation that the random alignment of homologous chromosomes provides a mechanism to promote a vast amount of genetic diversity.

In addition to the random arrangement of homologs within a double row, a second distinctive feature of metaphase of meiosis I is the attachment of kinetochore microtubules to the sister chromatids (**Figure 2.12**). One pair of sister chromatids is linked to one of the poles, and the homologous pair is linked to the opposite pole. This arrangement is quite different from the kinetochore attachment sites during mitosis in which a pair of sister chromatids is linked to both poles (see Figure 2.8d).

Anaphase of Meiosis I During anaphase of meiosis I, the two pairs of sister chromatids within a bivalent separate from each other (see Figure 2.11). However, the connection that holds sister chromatids together does not break. Instead, each joined pair of chromatids migrates to one pole, and the homologous pair of chromatids moves to the opposite pole. Another way of saying this is that the two dyads within a tetrad separate from each other and migrate to opposite poles.

Telophase of Meiosis I and Cytokinesis Finally, at telophase of meiosis I, the sister chromatids have reached their respective poles, and decondensation occurs in many, but not all, species. The nuclear membrane may re-form to produce two separate nuclei. In the example shown in Figure 2.11, the end result of meiosis I is two cells, each with three pairs of sister chromatids. It is a reduction division. The original diploid cell had its chromosomes in homologous pairs, but the two cells produced at the end of meiosis I are considered to be haploid; they do not have pairs of homologous chromosomes.

Meiosis II The sorting events that occur during meiosis II are similar to those that occur during mitosis, but the starting point is different. For a diploid organism with six chromosomes, mitosis begins with 12 chromatids that are joined as six pairs of sister chromatids (refer back to Figure 2.8). In other words, mitosis begins with six dyads in this case. By comparison, the two cells that begin meiosis II each have six chromatids that are joined as three pairs of sister chromatids; meiosis II begins with three dyads. Otherwise, the steps that occur during prophase, prometaphase, metaphase, anaphase, and telophase of meiosis II are analogous to a mitotic division.

Meiosis Versus Mitosis If we compare the outcome of meiosis (see Figure 2.11) to that of mitosis (see Figure 2.8), the results are quite different. In these examples, mitosis produced two diploid daughter cells with six chromosomes each, whereas meiosis produced four haploid daughter cells with three chromosomes each. In other words, meiosis has halved the number of chromosomes per cell. **Table 2.1** compares the key differences among mitosis, meiosis I, and meiosis II.

With regard to alleles, the results of mitosis and meiosis are also different. The daughter cells produced by mitosis are genetically identical. However, the haploid cells produced by meiosis are not genetically identical to each other because they contain only one homologous chromosome from each pair. In Chapter 3, we will consider how gametes may differ in the alleles that they carry on their homologous chromosomes.

TABLE 2.1

A Comparison of Mitosis, Meiosis I, and Meiosis II

Phase	Event	Mitosis	Meiosis I	Meiosis II
Prophase	Synapsis	No	Yes	No
Prophase	Crossing over	Rarely	Commonly	Rarely
Prometaphase	Attachment to the poles	A pair of sister chromatids to both poles	A pair of sister chromatids to one pole	A pair of sister chromatids to both poles
Metaphase	Alignment along the metaphase plate	Sister chromatids	Bivalents	Sister chromatids
Anaphase	Separation of:	Sister chromatids	Bivalents	Sister chromatids
End result		Two diploid cells		Four haploid cells

Genetic TIPS

The Question: If a diploid cell contains four chromosomes (i.e., two per set), how many possible random arrangements of homologs could occur during metaphase of meiosis I?

Topic: What topic in genetics does this question address?

The topic is meiosis. More specifically, the question is about metaphase of meiosis I.

Information: What information do you know based on the question and your understanding of the topic?

From the question, you know a cell that started with two pairs of homologous chromosomes has entered meiosis and is now in metaphase of meiosis I. From your understanding of the topic, you may remember that tetrads align along the metaphase plate (see Figure 2.11). The orientations of the homologs within the tetrads are random.

Problem-Solving Strategies: Make a drawing. Make a calculation.

One strategy to solve this problem is to make a drawing in which the homologs are in different colors, such as red and blue. Note: The spindle poles are labeled A and B in the drawing below. The alignment occurs relative to the spindle poles.

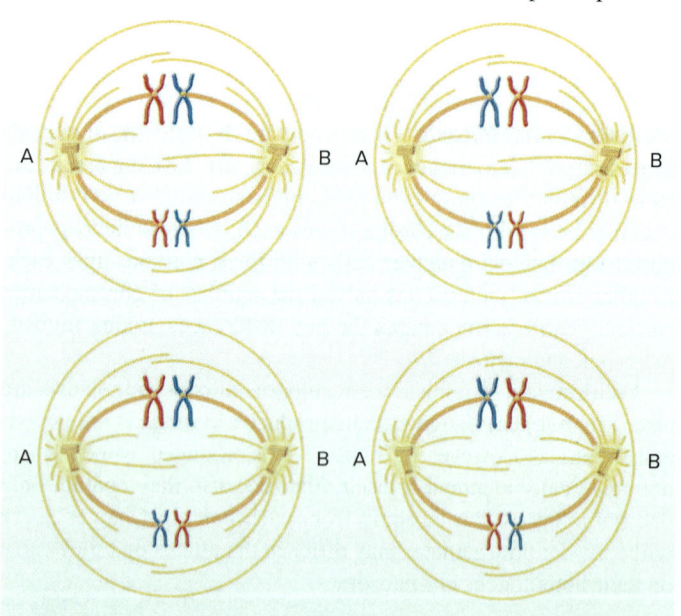

Another strategy is to make a calculation in which the number of different random alignments equals 2^n, where n equals the number of chromosomes per set.

Answer: As seen in the drawing, the number of random alignments is 4. As mentioned in the text, the number of different random alignments equals 2^n. So the possible number of arrangements in this case is 2^2, which is 4.

2.4 REVIEWING THE KEY CONCEPTS

- Another way for eukaryotic cells to divide is via meiosis, which produces four haploid cells. During prophase of meiosis I, homologs synapse, and crossing over may occur (see Figures 2.10–2.12).

2.4 COMPREHENSION QUESTIONS

1. When does crossing over usually occur, and what is the end result?
   a. It occurs during prophase of meiosis I, and the end result is the exchange of pieces between homologous chromosomes.
   b. It occurs during prometaphase of meiosis I, and the end result is the exchange of pieces between homologous chromosomes.
   c. It occurs during prophase of meiosis I, and the end result is the separation of sister chromatids.
   d. It occurs during prometaphase of meiosis I, and the end result is the separation of sister chromatids.

2. Which phase of meiosis is depicted in the following drawing?

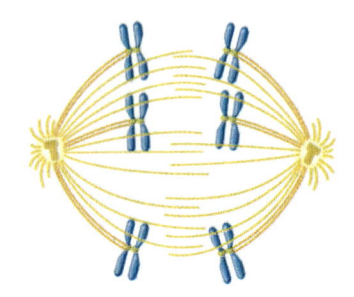

   a. Metaphase of meiosis I
   b. Metaphase of meiosis II
   c. Anaphase of meiosis I
   d. Anaphase of meiosis II

2.5 SEXUAL REPRODUCTION

Learning Outcomes:

1. Define *sexual reproduction*.
2. Describe how animals make sperm and egg cells.
3. Explain how plants alternate beween haploid and diploid generations.

In the previous section, we considered how a diploid cell divides by meiosis to produce cells with half the genetic material of the original mother cell. This process is critical for sexual reproduction, which is a common way for eukaryotic organisms to produce offspring. During **sexual reproduction,** gametes are made that contain half the amount of an organism's genetic material. These gametes fuse with each other in the process of fertilization to begin the life of a new organism.

Gametes are highly specialized cells that are produced by a process called **gametogenesis.** As discussed previously, gametes are typically haploid, which means they contain half the number of chromosomes of diploid cells. Haploid cells are represented by $1n$ and diploid cells by $2n$, where n refers to a set of chromosomes. A haploid gamete contains a single set of chromosomes, whereas a diploid cell has two sets. For example, a diploid human cell contains two sets of chromosomes, for a total of 46, but a human gamete (sperm or egg cell) contains only a single set of 23 chromosomes.

Some simple eukaryotic species are **isogamous,** which means that the gametes are morphologically similar. Examples of isogamous organisms include many species of fungi and algae. Most eukaryotic species, however, are **heterogamous**—they produce two morphologically different types of gametes. Male gametes, or

sperm cells, are relatively small and usually travel far distances to reach the female gamete—the **egg cell,** or **ovum.** The mobility of the sperm is an important characteristic, making it likely that it will come in close proximity to the egg cell. The sperm of most animal species contain a single flagellum that enables them to swim. The sperm of ferns and nonvascular plants such as bryophytes may have multiple flagella. In flowering plants, however, the sperm are contained within pollen grains. Pollen is a small mobile structure that can be carried by the wind or on the feet or hairs of insects. In flowering plants, sperm are delivered to egg cells via pollen tubes. Compared with sperm cells, an egg cell is usually very large and nonmotile. In animal species, the egg stores a large amount of nutrients to nourish the growing embryo. In this section, we will examine how sperm and egg cells are made in animal and plant species.

In Animals, Spermatogenesis Produces Four Haploid Sperm Cells and Oogenesis Produces a Single Haploid Egg Cell

In male animals, **spermatogenesis,** the production of sperm, occurs within glands known as the testes. The testes contain spermatogonial cells that divide by mitosis to produce two cells. One of these remains a spermatogonial cell, and the other cell becomes a primary spermatocyte. As shown in **Figure 2.13a,** the spermatocyte advances through meiosis I and meiosis II to produce four haploid cells, which are known as spermatids. These cells then mature into sperm cells. The structure of a sperm cell includes a long flagellum and a head. The head of the sperm contains little more than a haploid nucleus and an organelle, known as an acrosome, at its tip. The acrosome contains digestive enzymes that are released when a sperm meets an egg cell. These enzymes enable the sperm to penetrate the outer protective layers of the egg and gain entry into the egg cell's cytosol. In animal species without a mating season, sperm production is a continuous process in mature males. A mature human male, for example, produces several hundred million sperm each day.

In female animals, **oogenesis,** the production of egg cells, occurs within specialized diploid cells of the ovary known as oogonia. Quite early in the development of the ovary, the oogonia initiate meiosis to produce primary oocytes. For example, in humans, approximately 1 million primary oocytes per ovary are produced before birth. These primary oocytes are arrested—enter a dormant phase—at prophase of meiosis I, and remain at this stage until the female becomes sexually mature. At maturity, primary oocytes are

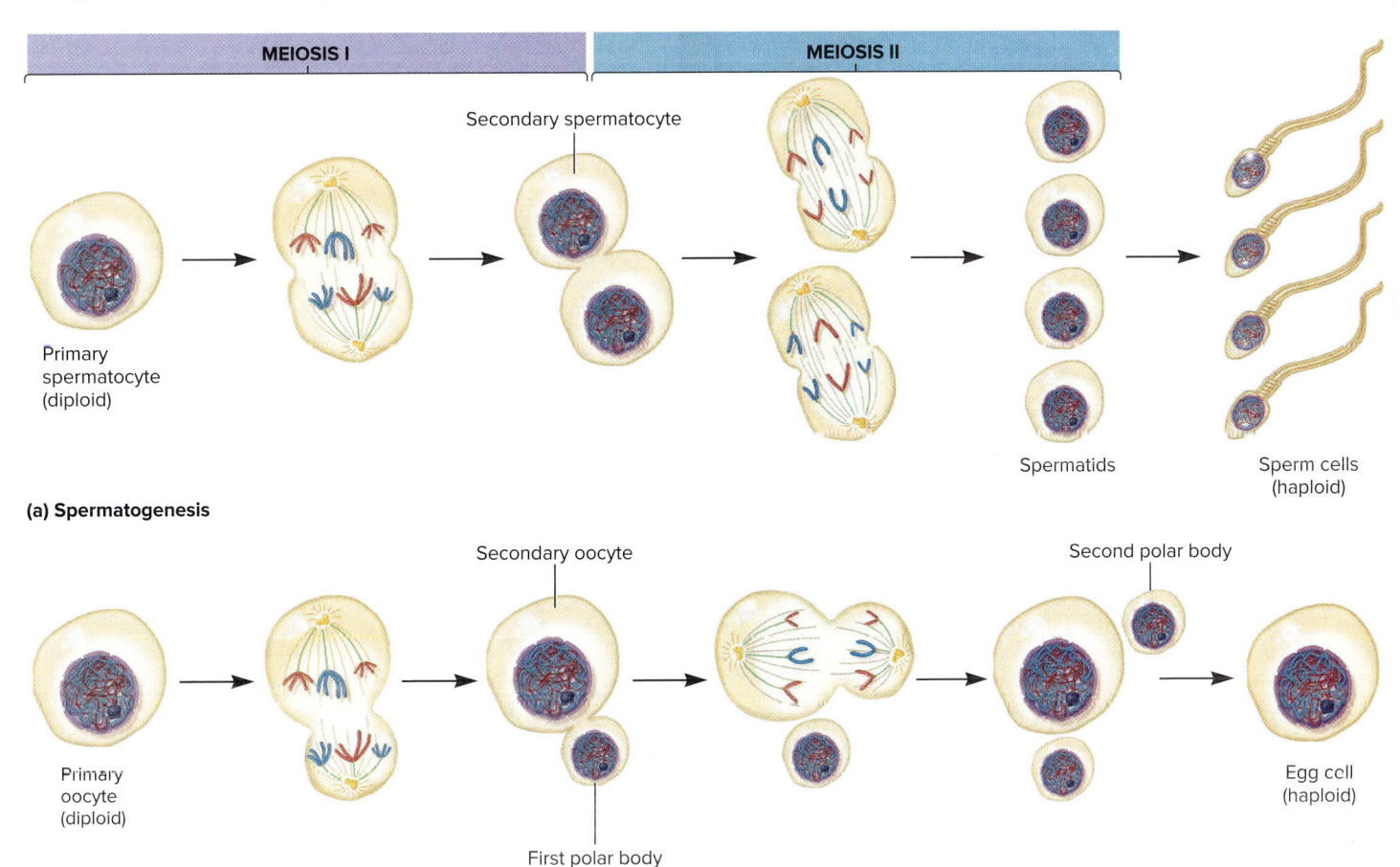

FIGURE 2.13 Gametogenesis in animals. (a) Spermatogenesis. A diploid spermatocyte undergoes meiosis to produce four haploid (1n) spermatids. These differentiate during spermatogenesis to become mature sperm. (b) Oogenesis. A diploid oocyte undergoes meiosis to produce one haploid egg cell and two or three polar bodies. In some species, the first polar body divides; in other species, it does not. Because of asymmetrical cytokinesis, the amount of cytoplasm the egg receives is maximized. The polar bodies degenerate.

Concept Check: What are polar bodies?

periodically activated to advance through the remaining stages of oocyte development.

During oocyte maturation, meiosis produces only one cell that is destined to become an egg, as opposed to the four gametes produced from each primary spermatocyte during spermatogenesis. How does oogenesis occur? As shown in **Figure 2.13b**, the first meiotic division is asymmetrical and produces a secondary oocyte and a much smaller cell, known as a polar body. Most of the cytoplasm is retained by the secondary oocyte and very little by the polar body, allowing the oocyte to become a larger cell with more stored nutrients. The secondary oocyte then begins meiosis II. In mammals, the secondary oocyte is released from the ovary—an event called ovulation—and travels down the oviduct toward the uterus. During this journey, if a sperm cell penetrates the secondary oocyte, it is stimulated to complete meiosis II; in this case, the secondary oocyte produces a haploid egg and a second polar body. The haploid egg and sperm nuclei then unite to create the diploid nucleus of a new individual.

Plant Species Alternate Between Haploid (Gametophyte) and Diploid (Sporophyte) Generations

Most species of animals are diploid, and their haploid gametes are considered to be a specialized type of cell. By comparison, the life cycles of plant species alternate between haploid and diploid generations. The haploid generation is called the **gametophyte,** whereas the diploid generation is called the **sporophyte.** Meiosis produces haploid cells called spores, which divide by mitosis to produce the gametophyte. In simpler plants, such as mosses, a haploid spore can produce a large multicellular gametophyte by repeated mitoses and cellular divisions. In flowering plants, however, spores develop into gametophytes that contain only a few cells. In this case, the organism that we think of as a plant is the sporophyte, whereas the gametophyte is very inconspicuous. The gametophytes of most plant species are small structures produced within the much larger sporophyte. Certain cells within the haploid gametophytes then become specialized as haploid gametes.

Figure 2.14 provides an overview of gametophyte development and gametogenesis in flowering plants, using as an example a flower from an angiosperm (a plant that produces seeds within an ovary). Meiosis occurs within two different structures of the sporophyte: the anthers and the ovaries, which produce male and female gametophytes, respectively.

In the anther, diploid cells called microsporocytes undergo meiosis to produce four haploid microspores. These separate into individual microspores. In many flowering plants, each microspore undergoes mitosis to produce a two-celled structure containing one tube cell and one generative cell, both of which are haploid. This structure differentiates into a **pollen grain,** which is the male gametophyte with a thick cell wall. Later, the generative cell undergoes a mitotic cell division to produce two haploid sperm cells. In most plant species, this division occurs only if the pollen grain germinates—if it lands on a stigma and forms a pollen tube (look ahead to Figure 3.2c).

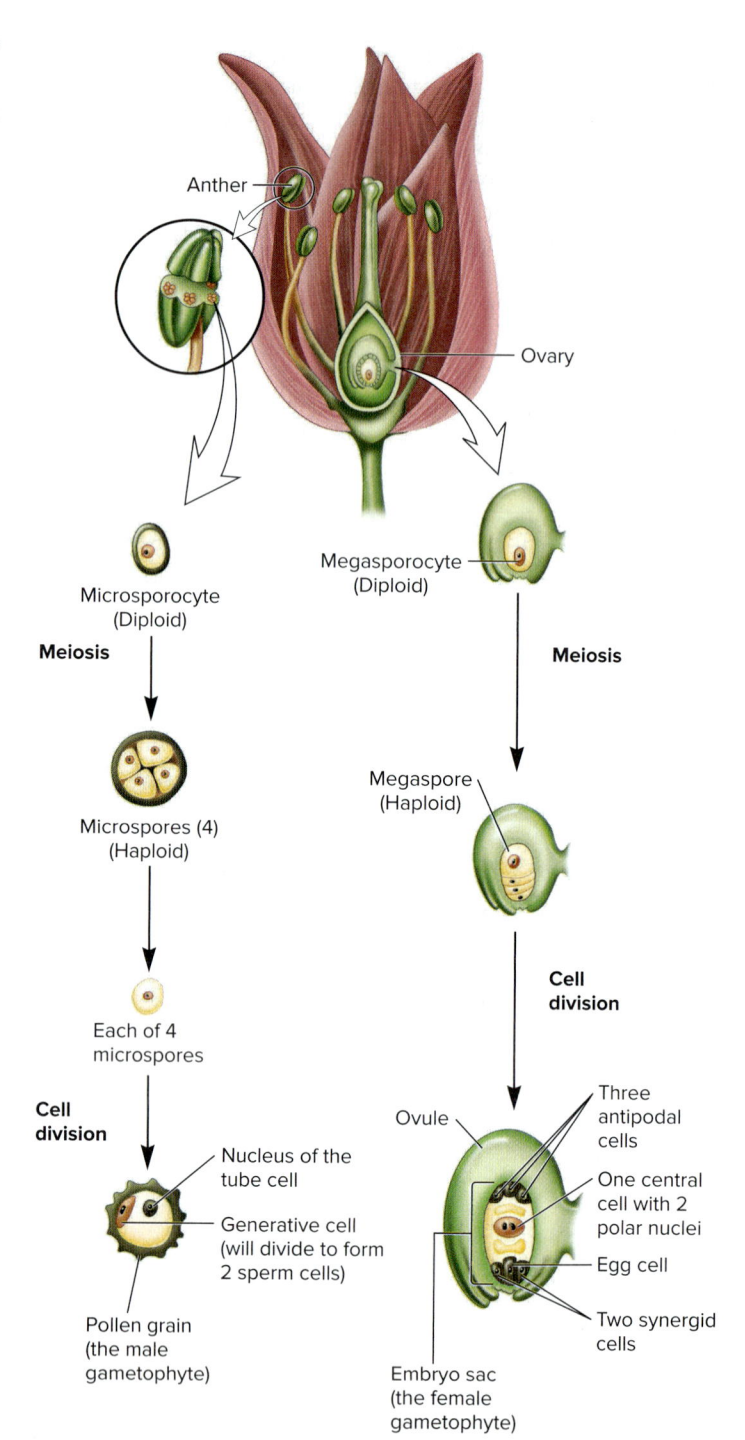

FIGURE 2.14 The formation of male and female gametes by the gametophytes of angiosperms (flowering plants).

Concept Check: Are all of the cell nuclei in the embryo sac haploid or is just the egg haploid?

By comparison, female gametophytes are produced within ovules found in the plant ovaries. A type of cell known as a megasporocyte undergoes meiosis to produce four haploid megaspores. Three of the four megaspores degenerate. The remaining haploid megaspore then undergoes three successive mitotic divisions accompanied by asymmetrical cytokinesis to produce seven individual cells—one egg cell, two synergids, three antipodals, and one central cell. This seven-celled structure, also known as the **embryo sac,** is the mature female gametophyte. Each embryo sac is contained within an ovule.

For fertilization to occur, specialized cells within the male and female gametophytes must meet. The steps of plant fertilization are described in Chapter 3.

When comparing animals and plants, it's interesting to consider how gametes are made. Animals produce gametes by meiosis. In contrast, meiosis in plants produce spores that develop into gametophytes. The gametophyte of plants is a haploid multicellular organism that is produced by mitotic cellular divisions of a haploid spore. Within the multicellular gametophyte, certain cells become specialized as gametes.

2.5 REVIEWING THE KEY CONCEPTS

- Animals produce gametes via spermatogenesis and oogenesis (see Figure 2.13).

- Plants exhibit alternation of generations between a diploid sporophyte and a haploid gametophyte. The gametophyte produces gametes (see Figure 2.14).

2.5 COMPREHENSION QUESTIONS

1. In animals, a key difference between spermatogenesis and oogenesis is
   a. only oogenesis involves meiosis.
   b. only spermatogenesis involves meiosis.
   c. spermatogenesis produces four sperm, whereas oogenesis produces only one egg cell.
   d. None of the above describes a difference between the two processes.
2. Which of the following statements regarding plants is *false*?
   a. Meiosis within anthers produces spores that develop into pollen.
   b. Meiosis within ovules produces spores that develop into an embryo sac.
   c. The male gametophyte is a pollen grain, and the female gametophyte is an embyro sac.
   d. Meiosis directly produces sperm and egg cells in plants.

KEY TERMS

Page 19. reproduction, chromosomes, chromatin
Page 20. prokaryotes, nucleoid, eukaryotes
Page 21. organelles, nucleus, cytogenetics, cytogeneticist, somatic cell, gamete, karyotype, diploid, homologs, alleles
Page 23. homozygous, heterozygous, locus (loci), asexual reproduction
Page 24. binary fission, cell cycle, interphase, G_1 phase, restriction point
Page 25. S phase, chromatids, centromere, sister chromatids, dyad, monad, kinetochore, G_2 phase, M phase, mitosis
Page 26. mitotic spindle apparatus
Page 27. mitotic spindle, microtubule-organizing centers (MTOCs), centrosomes, spindle pole, centrioles, decondensed, prophase, condense, prometaphase

Page 29. metaphase plate, metaphase, anaphase, telophase, cytokinesis, cleavage furrow, cell plate
Page 30. meiosis, haploid, leptotene, zygotene, synapsis, synaptonemal complex, pachytene, bivalents, tetrad
Page 31. crossing over, chiasma (chiasmata), diplotene, diakinesis
Page 34. sexual reproduction, gametogenesis, isogamous, heterogamous
Page 35. sperm cells, egg cell, ovum, spermatogenesis, oogenesis
Page 36. gametophyte, sporophyte, pollen grain
Page 37. embryo sac

CHAPTER SUMMARY

2.1 General Features of Chromosomes

- Chromosomes are structures that contain the genetic material, which is DNA.
- Prokaryotic cells are simple and lack cell compartmentalization, whereas eukaryotic cells contain a cell nucleus and other compartments (see Figure 2.1).
- Chromosomes can be examined under the microscope. An organized representation of the chromosomes from a single cell is called a karyotype (see Figure 2.2).
- In eukaryotic species, the chromosomes are found in sets. Eukaryotic cells are often diploid, which means that each type of chromosome occurs in a homologous pair (see Figure 2.3).

2.2 Cell Division

- Bacteria divide by binary fission (see Figure 2.4).
- To divide, eukaryotic cells advance through a cell cycle (see Figure 2.5).
- Prior to cell division, eukaryotic chromosomes are replicated to form sister chromatids (see Figure 2.6).

2.3 Mitosis and Cytokinesis

- Chromosome sorting in eukaryotes is achieved via a mitotic spindle apparatus (see Figure 2.7).
- A common way for eukaryotic cells to divide is by mitosis and cytokinesis. Mitosis is divided into prophase, prometaphase,

2.4 Meiosis

- Another way for eukaryotic cells to divide is via meiosis, which produces four haploid cells. During prophase of meiosis I, homologs synapse and crossing over may occur (see Figures 2.10–2.13).

2.5 Sexual Reproduction

- Animals produce gametes via spermatogenesis and oogenesis (see Figure 2.13).
- Plants exhibit alternation of generations between a diploid sporophyte and a haploid gametophyte. The gametophyte produces gametes (see Figure 2.14).

PROBLEM SETS & INSIGHTS

More Genetic TIPS

1. Describe how the kinetochore microtubules attach the sister chromatids to the spindle poles during metaphase of mitosis, meiosis I, and meiosis II.

Topic: *What topic in genetics does this question address?*

The topic is cell division. More specifically, the question is about how chromosomes are attached to the spindle during metaphase of mitosis or meiosis.

Information: *What information do you know based on the question and your understanding of the topic?*

From the question, you know that you are supposed to describe how the chromosomes are attached in three different circumstances: metaphase of mitosis, metaphase of meiosis I, and metaphase of meiosis II. From your understanding of the topic, you may remember that the attachments of sister chromatids is distinctly different in meiosis I compared to mitosis and meiosis II. Compare Figures 2.8 and 2.11.

Problem-Solving Strategies: *Make a drawing. Compare and contrast.*

One strategy to solve this problem is to make a drawing. If you make drawings, like those shown in Figures 2.8 and 2.11, you may appreciate that tetrads and bivalents are attached differently to the kinetochore microtubules. Another strategy is to compare and contrast what happens during meiosis I compared to mitosis and meiosis II. During meiosis I, tetrads align along the metaphase plate, and during anaphase, bivalents move toward the poles. During mitosis and meioisis II, bivalents align along the metaphase plate, and during anaphase, sister chromatids separate and each chromatid moves to a pole.

Answer: During mitosis and meiosis II, each pair of sister chromatids is attached by kinetochore microtubules to both poles. By comparison, during metaphase of meiosis I, the sister chromatids line up as a tetrad. Within a bivalent, kinetochore microtubules attach a pair of sister chromatids to just one pole.

2. What are the key differences in anaphase when comparing mitosis, meiosis I, and meiosis II?

Topic: *What topic in genetics does this question address?*

The topic is cell division. More specifically, the question is about the events that occur during anaphase.

Information: *What information do you know based on the question and your understanding of the topic?*

From the question, you know you are supposed to distinguish the key differences among anaphase of mitosis, anaphase of meiosis I, and anaphase of meiosis II. From your understanding of the topic, you may remember that the separation of chromosomes is distinctly different in meiosis I compared to mitosis and meiosis II. Compare Figures 2.8 and 2.11.

Problem-Solving Strategies: *Make a drawing. Compare and contrast.*

Just as for the previous problem, one strategy to solve this problem is to make a drawing. If you make drawings, like those shown in Figures 2.8 and 2.11, you may appreciate that bivalents move to opposite poles during anaphase of meiosis I, whereas individual chromatids move to opposite poles during anaphase of mitosis and meiosis II. Another strategy is to compare and contrast what happens during meiosis I compared to mitosis and meiosis II. During meiosis I, anaphase does not involve the splitting of centromeres, whereas centromeres split during mitosis and meioisis II, thereby separating sister chromatids.

Answer: During mitosis and meiosis II, the centromeres split and individual chromatids move to their respective poles. In meiosis I the centromeres do not split. Instead, the bivalents separate and pairs of sister chromatids move to opposite poles.

3. A diploid cell has eight chromosomes, four per set. In the following diagram, in what phase of mitosis, meiosis I, or meiosis II, is this cell?

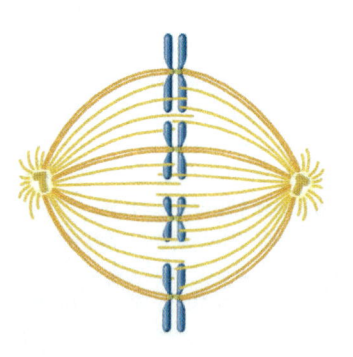

Topic: *What topic in genetics does this question address?*

The topic is cell division. More specifically, the question is asking you to look at a drawing and discern which phase of cell division a particular cell is in.

Information: *What information do you know based on the question and your understanding of the topic?*

In the question, you are given a diagram of a cell at a particular phase of the cell cycle. This cell is derived from a mother cell with four pairs of chromosomes. From your understanding of the topic, you may remember the various phases of mitosis, meiosis I, and meiosis II, which are described in Figures 2.8 and 2.11. If so, you may initially realize that the cell is in metaphase.

Problem-Solving Strategy: *Describe the steps.*

To solve this problem, you may need to describe the steps, starting with a mother cell that has four pairs of chromosomes. Keep in mind that a mother cell with four pairs of chromosomes has eight chromosomes during G_1, which then replicate to form eight pairs of sister chromatids during S phase. Therefore, at the beginning of M phase, this mother cell will have eight pairs of sister chromatids. During mitosis, the mother cell will align eight pairs of sister chromatids in a single row at metaphase. During meiosis I, the mother cell will align four tetrads along the metaphase plate. During meiosis II, two cells will align four pairs of sister chromatids along the metaphase plate.

Answer: The cell is in metaphase of meiosis II. You can tell because the chromosomes are lined up in a single row along the metaphase plate, and the cell has only four pairs of sister chromatids. If the diagram was showing mitosis, the cell would have eight pairs of sister chromatids in a single row. If it was showing meiosis I, tetrads would be aligned along the metaphase plate.

Conceptual Questions

C1. The process of binary fission begins with a single mother cell and ends with two daughter cells. Would you expect the mother and daughter cells to be genetically identical? Explain why or why not.

C2. What is a homolog? With regard to genes and alleles, how are homologs similar to and different from each other?

C3. What is a sister chromatid? Are sister chromatids genetically similar or identical? Explain.

C4. With regard to sister chromatids, which phase of mitosis is the organization phase, and which is the separation phase?

C5. A species is diploid with three chromosomes per set. Make a drawing that shows what the chromosomes would look like in the G_1 and G_2 phases of the cell cycle.

C6. How does the attachment of kinetochore microtubules to the kinetochore differ in metaphase of meiosis I from metaphase of mitosis? Discuss what you think would happen if a sister chromatid was not attached to a kinetochore microtubule.

C7. For the following events, specify whether they occur during mitosis, meiosis I, or meiosis II:

  a. Separation of conjoined chromatids within a pair of sister chromatids

  b. Pairing of homologous chromosomes

  c. Alignment of chromatids along the metaphase plate

  d. Attachment of sister chromatids to both poles

C8. Describe the key events during meiosis that result in a 50% reduction in the amount of genetic material per cell.

C9. A cell is diploid and contains three chromosomes per set. Draw the arrangement of chromosomes during metaphase of mitosis and metaphase of meiosis I and II. In your drawing, make one set of chromosomes dark and the other lighter.

C10. The alignment of homologs along the metaphase plate during metaphase of meiosis I is random. In your own words, explain what this means.

C11. A eukaryotic cell is diploid, containing 10 chromosomes (5 in each set). For mitosis and meiosis, how many daughter cells will be produced, and how many chromosomes will each one contain?

C12. If a diploid cell contains six chromosomes (i.e., three per set), how many possible random arrangements of homologs could occur during metaphase of meiosis I?

C13. A cell has four pairs of chromosomes. Assuming that crossing over does not occur, what is the probability that a gamete will contain all of the paternal chromosomes? If n equals the number of chromosomes in a set, which of the following expressions can be used to calculate the probability that a gamete will receive all of the paternal chromosomes: $(1/2)^n$, $(1/2)^{n-1}$, or $n^{1/2}$?

C14. With regard to question C13, how would the phenomenon of crossing over affect the results? In other words, would the probability of a gamete inheriting only paternal chromosomes be higher or lower? Explain your answer.

C15. Eukaryotic cells must sort their chromosomes during mitosis so that each daughter cell receives the correct number of chromosomes. Why don't bacteria need to sort their chromosomes?

C16. Why is it necessary for the chromosomes to condense during mitosis and meiosis? What do you think might happen if the chromosomes were not condensed?

C17. Nine-banded armadillos almost always give birth to four offspring that are genetically identical quadruplets. Explain how you think this happens.

C18. A diploid species contains four chromosomes per set for a total of eight chromosomes in its somatic cells. Draw the cell as it would look in prophase of meiosis II and prophase of mitosis. Discuss how prophase of meiosis II and prophase of mitosis differ from each other, and explain how the difference originates.

C19. Explain why the products of meiosis may not be genetically identical, whereas the products of mitosis are.

C20. The period between meiosis I and meiosis II is called interphase II. Does DNA replication take place during interphase II?

C21. List several ways in which telophase appears to be the reverse of prophase and prometaphase.

C22. Corn has 10 chromosomes per set, and the sporophyte of the species is diploid. If you performed a karyotype, what is the total number of chromosomes you would expect to see in the following types of corn cells?

A. A leaf cell

B. The sperm nucleus of a pollen grain

C. An endosperm cell after fertilization

D. A root cell

C23. The arctic fox has 50 chromosomes (25 per set), and the common red fox has 38 chromosomes (19 per set). These species can interbreed to produce viable but infertile offspring. How many chromosomes will the offspring have? What problems do you think may occur during meiosis that would explain the offspring's infertility?

C24. Describe the cellular differences between male and female gametes.

C25. At puberty, the testes contain a finite number of cells and produce an enormous number of sperm cells during the life span of a man. Explain why testes do not run out of spermatogonial cells.

C26. Describe the timing of meiosis I and II during human oogenesis.

C27. A phenotypically normal woman with an abnormally long chromosome 13 (and a normal homolog of chromosome 13) has offspring with a phenotypically normal man with an abnormally short chromosome 11 (and a normal homolog of chromosome 11). What is the probability of producing an offspring that will have both a long chromosome 13 and a short chromosome 11? If such a child is produced, what is the probability that this child would eventually pass both abnormal chromosomes to one of her or his offspring?

Application and Experimental Questions

E1. When studying living cells in a laboratory, researchers sometimes use drugs as a way to make cells remain at a particular stage of the cell cycle. For example, aphidicolin is an antibiotic that inhibits DNA synthesis in eukaryotic cells and causes them to remain in the G_1 phase because they cannot replicate their DNA. In what phase of the cell cycle—G_1, S, G_2, prophase, metaphase, anaphase, or telophase—would you expect somatic cells to stay if they were treated with each of the following types of drug?

A. A drug that inhibits microtubule formation

B. A drug that allows microtubules to form but prevents them from shortening

C. A drug that inhibits cytokinesis

D. A drug that prevents chromosomal condensation

E2. With regard to thickness and length, what do you think the chromosomes would look like if you microscopically examined them during interphase? How would that compare with their appearance during metaphase?

E3. A rare form of dwarfism that also includes hearing loss was found to run in a particular family. It is inherited in a dominant manner. It was discovered that an affected individual had one normal copy of chromosome 15 and one abnormal copy of chromosome 15 that was unusually long. How would you determine if the unusually long chromosome 15 was causing this disorder?

Questions for Student Discussion/Collaboration

1. A diploid eukaryotic cell has 10 chromosomes (5 per set). As a group, take turns having one student draw the cell as it would look during a phase of mitosis, meiosis I, or meiosis II; then have the other students guess which phase it is.

2. Discuss the advantages and disadvantages of asexual versus sexual reproduction.

Answers to Comprehension Questions

2.1: c, d, a

2.2: d, b, b

2.3: d, d

2.4: a, c

2.5: c, d

Note: All answers appear in Connect; the answers to even-numbered questions and all Concept Check questions are in Appendix B.

CHAPTER OUTLINE

- 3.1 Mendel's Study of Pea Plants
- 3.2 Law of Segregation
- 3.3 Law of Independent Assortment
- 3.4 Chromosome Theory of Inheritance
- 3.5 Studying Inheritance Patterns in Humans
- 3.6 Probability and Statistics

The garden pea, studied by Mendel.
©Zigzag Mountain Art/Alamy Stock Photo

MENDELIAN INHERITANCE

An appreciation for the concept of heredity can be traced far back in human history. Hippocrates, a Greek physician, was the first person to provide an explanation for hereditary traits (around 400 B.C.E.). He suggested that "seeds" are produced by all parts of the body and then collected and transmitted to the offspring at the time of conception. Furthermore, he hypothesized that these seeds cause certain traits of the offspring to resemble those of the parents. This idea, known as **pangenesis,** was the first attempt to explain the transmission of hereditary traits from generation to generation.

The first systematic studies of genetic crosses were carried out by German botanist Joseph Kölreuter from 1761 to 1766. In crosses between different strains of tobacco plants, he found that the offspring were usually intermediate in appearance between the two parents. This led Kölreuter to conclude that both parents make equal genetic contributions to their offspring. Furthermore, his observations were consistent with the **blending hypothesis of inheritance.** According to this view, the factors that dictate hereditary traits could blend together from generation to generation. The blended traits are then passed to the next generation. The popular view before the 1860s, which combined the notions of pangenesis and blending inheritance, was that hereditary traits were rather malleable and could change and blend over the course of one or two generations. However, the pioneering work of Gregor Mendel would prove instrumental in refuting this viewpoint.

In this chapter, we will first examine the outcome of Mendel's crosses of pea plants. We begin our inquiry into genetics here because the inheritance patterns observed in peas are fundamentally related to inheritance patterns found in other eukaryotic species, such as corn, fruit flies, mice, and humans. We will discover how Mendel's insights into the patterns of inheritance in pea plants revealed some simple rules that govern the process of inheritance. In Chapters 4 through 8, we will explore more complex patterns of inheritance and also consider the role that chromosomes play as the carriers of the genetic material.

In the last section of this chapter, we will become familiar with general concepts in probability and statistics. How are statistical methods useful? Probability calculations allow us to predict the outcomes of simple genetic crosses, as well as the outcomes of more complicated crosses described in later chapters. In addition, we will learn how to use statistics to test the validity of genetic hypotheses that attempt to explain the inheritance patterns of traits.

3.1 MENDEL'S STUDY OF PEA PLANTS

Learning Outcomes:
1. Describe the characteristics of pea plants that make them a suitable organism to study genetically.
2. Outline the steps that Mendel followed to make crosses between different strains of pea plants.
3. List the seven characteristics of pea plants that Mendel chose to study.

Gregor Johann Mendel, born in 1822, is now remembered as a pioneer of genetics (**Figure 3.1**). He grew up on a farm in Hynčice (formerly Heinzendorf) in northern Moravia, which was then a part of Austria and is now a part of the Czech Republic. Instead of becoming a farmer, however, Mendel entered the Augustinian monastery of St. Thomas, completed his studies for the priesthood, and was ordained in 1847. Soon after becoming a priest, Mendel worked for a short time as a substitute teacher. To continue that role, he needed to obtain a teaching license from the government. Surprisingly, he failed the licensing exam due to poor answers in the areas of physics and natural history.

Therefore, Mendel then enrolled at the University of Vienna to expand his knowledge in these two areas. Mendel's training in physics and mathematics taught him to perceive the world as an orderly place, governed by natural laws. In his studies, Mendel learned that these natural laws could be stated as simple mathematical relationships.

In 1856, Mendel began his historic studies on pea plants. For 8 years, he grew and crossed thousands of pea plants on a small 23- by 115-foot plot. He kept meticulously accurate records that included quantitative data concerning the outcomes of his crosses. He published his work, entitled *Experiments on Plant Hybrids*, in 1866. This paper was largely ignored by scientists at that time, possibly because of its title, which gave little indication of its contents. Another reason his work went unrecognized could be tied to a lack of understanding of chromosomes and their transmission, a topic we discussed in Chapter 2. Nevertheless, Mendel's groundbreaking work allowed him to propose the natural laws that now provide a framework for our understanding of genetics.

Prior to his death in 1884, Mendel reflected, "My scientific work has brought me a great deal of satisfaction and I am convinced that it will be appreciated before long by the whole world." Sixteen years later, in 1900, the work of Mendel was independently rediscovered by three biologists with an interest in plant genetics: Hugo de Vries of Holland, Carl Correns of Germany, and Erich von Tschermak of Austria. Within a few years, the influence of Mendel's studies was felt around the world. In this section, we will examine Mendel's experiments and consider their great significance in the field of genetics.

Mendel Chose Pea Plants as His Experimental Organism

Mendel's study of genetics grew out of his interest in ornamental flowers. Prior to his work with pea plants, many plant breeders had conducted experiments aimed at obtaining flowers with new varieties of colors. When two distinct individuals with different characteristics are bred, or **crossed,** to each other—a process called a **hybridization** experiment—their offspring are referred to as **hybrids.** For example, a hybridization experiment could involve a cross between a purple-flowered plant and a white-flowered plant. Mendel was particularly intrigued, in such experiments, by the consistency with which offspring of subsequent generations showed characteristics of one or the other parent. His intellectual foundation in physics and the natural sciences led him to consider that this regularity might be rooted in natural laws that could be expressed mathematically. To uncover these laws, he realized that he needed to carry out quantitative experiments in which the numbers of offspring carrying certain traits were carefully recorded and analyzed.

Mendel chose the garden pea, *Pisum sativum*, to investigate the natural laws that govern plant hybrids. The reproductive features of this plant are shown in **Figure 3.2a, b**.

- The term **gamete** is used to describe haploid reproductive cells that can unite to form a zygote.

FIGURE 3.1 Gregor Johann Mendel (1822–1884).
Source: National Library of Medicine

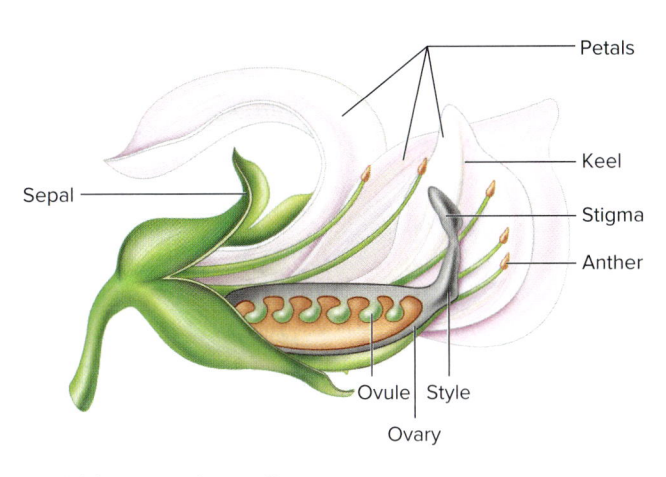

(a) Structure of a pea flower

(b) A flowering pea plant

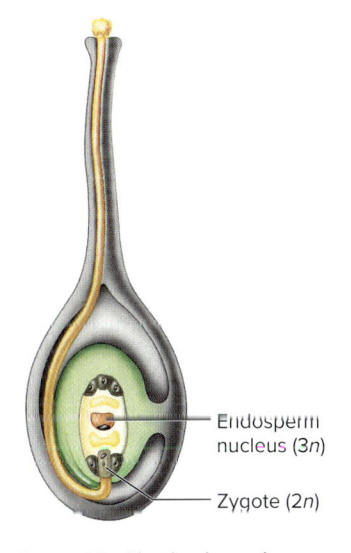

(c) Pollination and fertilization in angiosperms

FIGURE 3.2 Flower structure and pollination in pea plants. (a) The pea flower produces both pollen and egg cells. The pollen grains are produced within the anthers, and the egg cells are produced within the ovules that are contained within the ovary. A modified petal called a keel encloses the anthers and ovaries. (b) Photograph of a flowering pea plant. (c) A pollen grain must first land on the stigma. After this occurs, the pollen grain sends out a long tube through which two sperm cells travel toward an ovule to reach an egg cell. The fusion between a sperm and an egg cell results in fertilization and creates a zygote. A second sperm fuses with a central cell containing two polar nuclei to create the endosperm. The endosperm provides nutritive material for the developing embryo.
©np-e07/Getty Images

Concept Check: Prior to fertilization, where is the male gamete located?

- In plants, male gametes (**sperm**) are produced within **pollen grains** formed in the **anthers.**
- The female gametes (**eggs**) are contained within **ovules** that form in the **ovaries.**
- For fertilization to occur, a pollen grain lands on the **stigma,** which stimulates the growth of a pollen tube. This enables sperm cells to enter the stigma and migrate toward an ovule. Fertilization occurs when a sperm enters the micropyle, an opening in the ovule wall, and fuses with an egg cell.

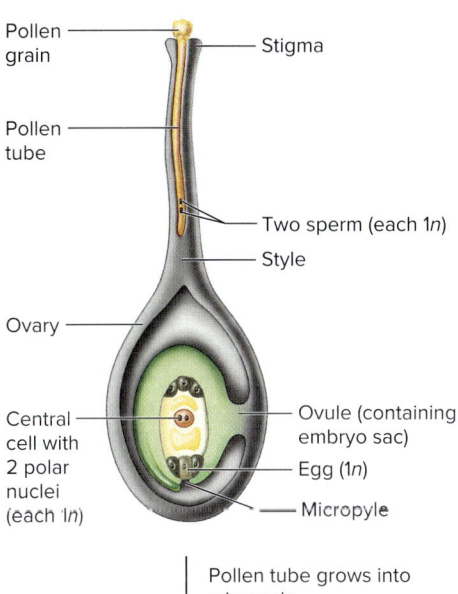

Mendel recognized that a key advantage of garden peas is the ease of making crosses. In flowering plants, sexual reproduction occurs by a pollination event (**Figure 3.2c**).

- In some experiments, Mendel allowed plants to reproduce by **self-fertilization,** which means that the pollen and eggs are derived from the same plant. In peas, a modified petal known as the keel covers the reproductive structures of a flower. Because of this covering, pea plants naturally reproduce by self-fertilization. Usually, pollination occurs even before the flower opens.

- In other experiments, Mendel wanted to make crosses between different plants. How did he accomplish this goal? Fortunately, pea plants contain relatively large flowers that are easy to manipulate, making it possible to make crosses between two particular plants and study their outcomes. This process, known as **cross-fertilization,** requires that the pollen from one plant be placed on the stigma of another plant (**Figure 3.3**). Mendel was able to pry open immature flowers and remove the anthers before they produced pollen. Therefore, these flowers could not self-fertilize. He then obtained pollen from another plant by gently touching its mature anthers with a paintbrush. Mendel applied this pollen to the stigma of the flower that already had its anthers removed.

Mendel Studied Seven Characters That Bred True

Another advantage of garden peas is that they were available in several distinct varieties, which differed with regard to their morphological characteristics.

- General characteristics of an organism are called **characters.** For example, seed color is a character of peas and eye color is a character of humans.
- The term **trait,** or **variant,** is typically used to describe the specific properties of a character. For example, yellow color is a trait (or variant) of peas and blue eye color is a trait (or variant) found in some people.

Over the course of 2 years, Mendel tested his varieties of peas to determine if their characters bred true. *Breeding true* means that a trait does not vary in appearance from generation to generation. For example, if the seeds from a pea plant were yellow, the next generation would also produce yellow seeds. Likewise, if these offspring were allowed to self-fertilize, all of their offspring would also produce yellow seeds, and so on.

- A variety that continues to produce the same characteristic after several generations of self-fertilization is called a **true-breeding line,** or **strain.**

Mendel concentrated his efforts on the analysis of characteristics that were clearly distinguishable among different true-breeding lines. **Figure 3.4** illustrates the seven characters that Mendel eventually chose to follow in his breeding experiments. All seven were found in two variants. For example, one characteristic he followed was height, which was found in two variants: tall and dwarf plants.

3.1 REVIEWING THE KEY CONCEPTS

- Mendel chose pea plants as his experimental organism (see Figures 3.1, 3.2).
- Using pea plants made it easy for Mendel to carry out self-fertilization or cross-fertilization experiments (see Figure 3.3).

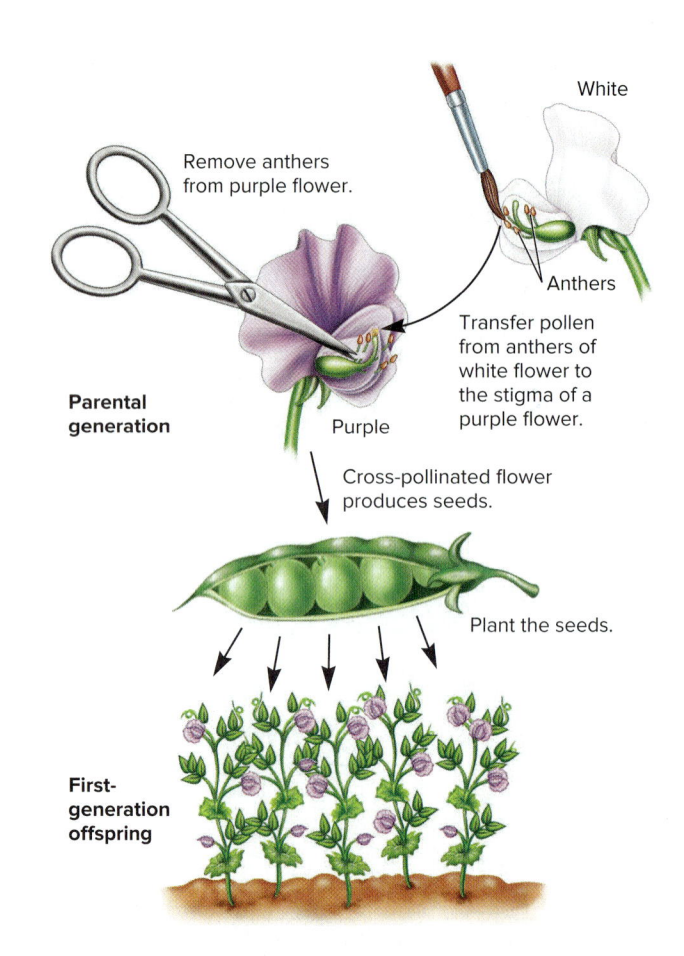

FIGURE 3.3 **How Mendel cross-fertilized two different pea plants.** This illustration depicts a cross between a plant with purple flowers and another plant with white flowers. The offspring from this cross are the result of pollination of the purple flower using pollen from a white flower.

Concept Check: In this experiment, which plant, the white- or purple-flowered plant, is providing the egg cells, and which is providing the sperm cells?

- Another advantage of using pea plants is they were available in several varieties in which a character existed in two distinct variants (see Figure 3.4).

3.1 COMPREHENSION QUESTIONS

1. Experimental advantages of using pea plants include which of the following?
   a. They came in several different varieties.
   b. They were capable of self-fertilization.
   c. They were easy to cross.
   d. All of the above were advantages.

2. The term *cross* refers to an experiment in which
   a. the gametes come from different individuals.
   b. the gametes come from a single flower of the same individual.
   c. the gametes come from different flowers of the same individual.
   d. Both a and c are true.

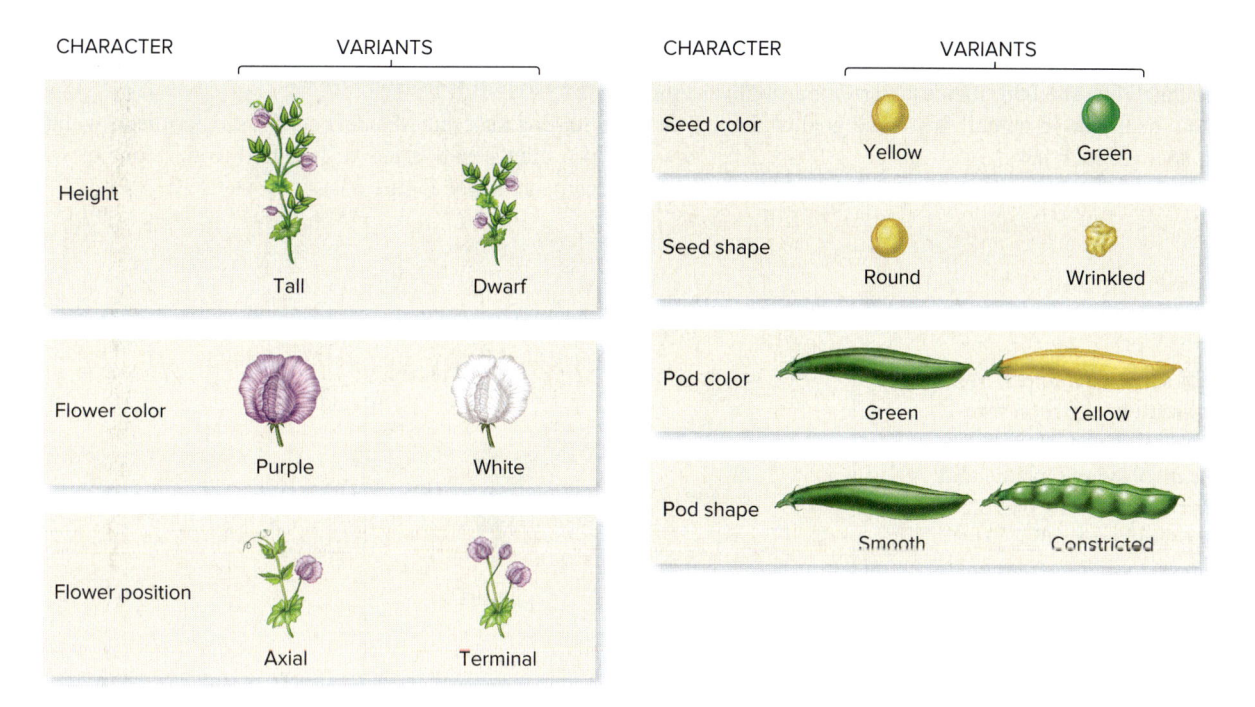

FIGURE 3.4 An illustration of the seven characters that Mendel studied. Each character was found as two variants that were decisively different from each other.

Concept Check: What do we mean when we say a strain is true-breeding?

3. To avoid self-fertilization in his pea plants, Mendel had to
   a. spray the plants with a chemical that damaged the pollen.
   b. remove the anthers from immature flowers.
   c. grow the plants in a greenhouse that did not contain pollinators (e.g., bees).
   d. do all of the above.

3.2 LAW OF SEGREGATION

Learning Outcomes:
1. Analyze Mendel's experiments involving single-factor crosses.
2. State Mendel's *law of segregation*, and explain how it is related to gamete formation and fertilization.
3. Predict the outcome of a single-factor cross or self-fertilization experiment using a Punnett square.

As discussed in the previous section, Mendel was able to carry out self-fertilization or cross-fertilization experiments with his pea plants. In this section, we will examine how he studied the inheritance of characters by crossing variants to each other. A cross in which an experimenter is observing only one character is called a **single-factor cross.** When the two parents are different variants for a given character, this type of cross produces single-character hybrids, also known as **monohybrids.** As you will learn, this type of experimental approach led Mendel to propose the law of segregation.

Mendel Followed the Outcome of a Single Character for Two Generations

Prior to conducting his studies, Mendel did not have a hypothesis to explain the formation of hybrids. However, his educational background led him to realize that a quantitative analysis of crosses might uncover mathematical relationships that would otherwise be mysterious. His experiments were designed to determine the relationships that govern hereditary traits. This rationale is called an **empirical approach.** Laws that are deduced from an empirical approach are known as **empirical laws.**

Mendel's experimental procedure is shown in **Figure 3.5**. He began with true-breeding plants that differed with regard to a single character. These plants are termed the **parental generation,** or **P generation.** The true-breeding parents were crossed to each other—a process called a P cross—producing offspring called the F_1 **generation,** or first filial generation (from the Latin *filius* meaning "son"). As seen in the data, all plants of the F_1 generation showed the phenotype of one parent but not the other. This prompted Mendel to follow the transmission of this character for one additional generation. To do so, the plants of the F_1 generation were allowed to self-fertilize to produce a second generation, called the F_2 **generation,** or second filial generation.

46   CHAPTER 3 :: MENDELIAN INHERITANCE

▶ **THE GOAL (DISCOVERY-BASED SCIENCE)**

Mendel speculated that the inheritance pattern for a single character may follow quantitative natural laws. The goal of this experiment was to uncover such laws.

▶ **ACHIEVING THE GOAL — FIGURE 3.5** Mendel's analysis of single-factor crosses.

Starting material: Mendel began his experiments with true-breeding pea plants that varied with regard to only one of seven different characters (see Figure 3.4).

Experimental level   **Conceptual level**

1. For each of seven characters, Mendel cross-fertilized two different true-breeding strains. Keep in mind that each cross involved two plants that differed in regard to only one of the seven characters studied. The illustration at the right shows one cross between a tall and dwarf plant. This is called a P (parental) cross.

   P plants: Tall × Dwarf

   $TT \times tt$

2. Collect the F_1 generation seeds. The following spring, plant the seeds and allow the plants to grow. These are the plants of the F_1 generation.

   Note: The P cross produces seeds that are part of the F_1 generation.

   F_1 seeds → F_1 plants: Tall, Tall

   All Tt

   Tt

3. Allow the F_1 generation plants to self-fertilize. This produces seeds that are part of the F_2 generation.

   Self-fertilization → F_2 seeds

   Self-fertilization

   $TT + 2\,Tt + tt$

4. Collect the F_2 generation seeds and plant them the following spring to obtain the F_2 generation plants.

   F_2 plants: Tall, Tall, Dwarf, Tall

5. Analyze the traits found in each generation.

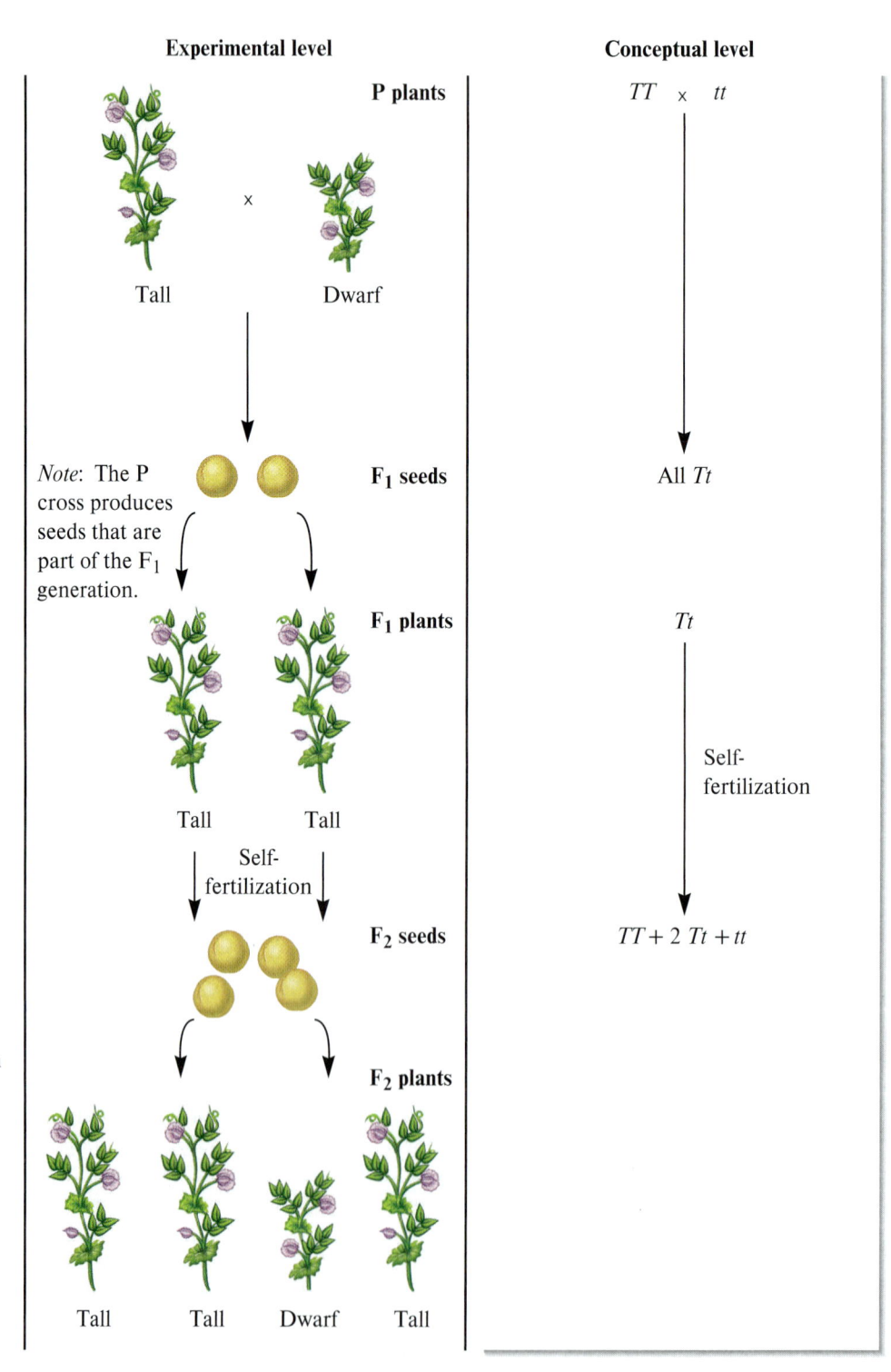

▶ THE DATA

P cross	F_1 generation	F_2 generation	Ratio of traits in F_2 generation
Tall × dwarf height	All tall	787 tall, 277 dwarf	2.84:1
Purple × white flowers	All purple	705 purple, 224 white	3.15:1
Axial × terminal flowers	All axial	651 axial, 207 terminal	3.14:1
Yellow × green seeds	All yellow	6022 yellow, 2001 green	3.01:1
Round × wrinkled seeds	All round	5474 round, 1850 wrinkled	2.96:1
Green × yellow pods	All green	428 green, 152 yellow	2.82:1
Smooth × constricted pods	All smooth	882 smooth, 299 constricted	2.95:1
Total	All dominant	14,949 dominant, 5010 recessive	2.98:1

Source: Mendel, Gregor (1866) *Versüche uber Pflanzenhybriden. Verhandlungen des naturforschenden Vereines in Brünn, Bd IV für das Jahr 1865,* Abhandlungen, 3–47.

▶ INTERPRETING THE DATA

The data shown in the table above are the results of producing an F_1 generation via cross-fertilization and an F_2 generation via self-fertilization of the F_1 monohybrids. A quantitative analysis of these data allowed Mendel to propose three important ideas:

1. Mendel's data argued strongly against a blending mechanism of inheritance. In all seven cases, the F_1 generation displayed traits that were distinctly like one of the two parents rather than traits intermediate in character. His first proposal was that one variant of a character is **dominant** to another variant. For example, the variant of green pods is dominant to that of yellow pods. The term **recessive** is used to describe a variant that is masked by the presence of a dominant trait but reappears in subsequent generations. Yellow pods and dwarf height are examples of recessive variants. They can also be referred to as recessive traits.

2. When a true-breeding plant with a dominant trait was crossed to a true-breeding plant with a recessive trait, the dominant trait was always observed in the F_1 generation. In the F_2 generation, some offspring displayed the dominant trait, but a smaller proportion showed the recessive trait. How did Mendel explain this observation? Because the recessive trait appeared in the F_2 generation, he made a second proposal—the genetic determinants of traits are passed along as "unit factors" from generation to generation. His data were consistent with a **particulate theory of inheritance,** in which the genes that govern traits are inherited as discrete units that remain unchanged as they are passed from parent to offspring. Mendel called them unit factors, but we now call them genes (from the Greek, *genesis*, meaning "birth," or *genos*, meaning "origin").

3. When Mendel compared the numbers of dominant and recessive traits in the F_2 generation, he noticed a recurring pattern. Within experimental variation, he always observed approximately a 3:1 ratio between the dominant trait and the recessive trait. Mendel was the first scientist to apply this type of quantitative analysis in a biological experiment. As described next, this quantitative approach allowed him to make a third proposal—genes **segregate** from each other during the process that gives rise to gametes.

Mendel's 3:1 Phenotypic Ratio Is Consistent with the Law of Segregation

Mendel's research was aimed at understanding the laws that govern the inheritance of traits. At that time, scientists did not understand the molecular composition of the genetic material or its mode of transmission during gamete formation and fertilization. We now know that the genetic material is composed of deoxyribonucleic acid (DNA), a component of chromosomes. Each chromosome contains hundreds or thousands of shorter segments that function as genes—a term that was originally coined by the Danish botanist Wilhelm Johannsen in 1909. A **gene** is defined as a unit of heredity that may influence the outcome of an organism's traits. Each of the seven pea plant characters that Mendel studied is influenced by a different gene.

As discussed in Chapter 2, most eukaryotic species, such as pea plants and humans, have their genetic material organized into pairs of chromosomes. For this reason, eukaryotes have two copies of most genes. These copies may be the same or they may differ. The term **allele** (from the Latin *alius* meaning "other") refers to a different form of the same gene. With this modern knowledge, the results shown in Figure 3.5 are consistent with the idea that each parent transmits only one copy of each gene (i.e., one allele) to each offspring. **Mendel's law of segregation** states:

The two copies of a gene segregate (or separate) from each other during transmission from parent to offspring.

Therefore, only one copy of each gene is found in a gamete. At fertilization, two gametes combine randomly, potentially producing different allelic combinations.

Let's use Mendel's cross of tall and dwarf pea plants to illustrate how genes are passed from parents to offspring (**Figure 3.6**). The letters *T* and *t* are used to represent the alleles of the gene that determines plant height. By convention, the uppercase letter represents the dominant allele (*T* for tall height, in this case), and the recessive allele is represented by the same letter in lowercase (*t*, for dwarf height). For the P cross, both parents are true-breeding plants. Therefore, we know that each has identical copies of the height gene.

- When an individual possesses two identical copies of a gene, the individual is said to be **homozygous** with respect to that gene. (The prefix *homo-* means "like," and the suffix *-zygo* means "pair.") In the P cross, the tall plant is homozygous for the tall allele *T*, and the dwarf plant is homozygous for the dwarf allele *t*.

- In contrast, the F_1 generation is **heterozygous,** with the genotype *Tt*, because every individual carries one copy of the tall allele and one copy of the dwarf allele. A heterozygous individual carries different alleles of a gene. (The prefix *hetero-* means "different.")

48   CHAPTER 3 :: MENDELIAN INHERITANCE

- The term **phenotype** refers to an observable characteristic of an organism. In the P generation, the plants exhibit a phenotype that is either tall or dwarf. In the F_1 generation, their phenotypes are tall because they have a copy of the dominant tall allele.

The law of segregation predicts that the phenotypes of the F_2 generation will be tall and dwarf in a ratio of 3:1 (see Figure 3.6). Because the parents of the F_2 generation are heterozygous, each gamete can transmit either a *T* allele or a *t* allele to a particular offspring, but not both. Therefore, *TT*, *Tt*, and *tt* are the possible genotypes of the F_2 generation. By randomly combining these alleles, the genotypes are produced in a 1:2:1 ratio. Because *TT* and *Tt* both produce tall phenotypes, a 3:1 phenotypic ratio is observed in the F_2 generation.

A Punnett Square Can Be Used to Predict the Outcome of a Cross or Self-Fertilization Experiment

An easy way to predict the outcome of simple genetic crosses and self-fertilization experiments is to use a **Punnett square,** a method originally proposed by Reginald Punnett. To construct a Punnett square, you must know the genotypes of the parents. With this information, the Punnett square enables you to predict the types of offspring the parents are expected to produce and in what proportions.

1. *Write down the genotypes of both parents.* (*In a self-fertilization experiment, a single parent provides the sperm and egg cells.*) Let's consider an example in which a heterozygous tall plant is crossed to another heterozygous tall plant. The plant providing the sperm (via pollen) is considered the male parent, and the plant providing the eggs is the female parent.

   Male parent: *Tt*

   Female parent: *Tt*

2. *Write down the possible gametes that each parent can make.* Remember that the law of segregation tells us that a gamete can carry only one copy of each gene.

   Male gametes: *T* or *t*

   Female gametes: *T* or *t*

3. *Create an empty Punnett square.* In the examples shown in this textbook, the number of columns equals the number of male gametes, and the number of rows equals the number of female gametes. Our example has two rows and two columns. Place the male gametes across the top of the Punnett square and the female gametes along the side.

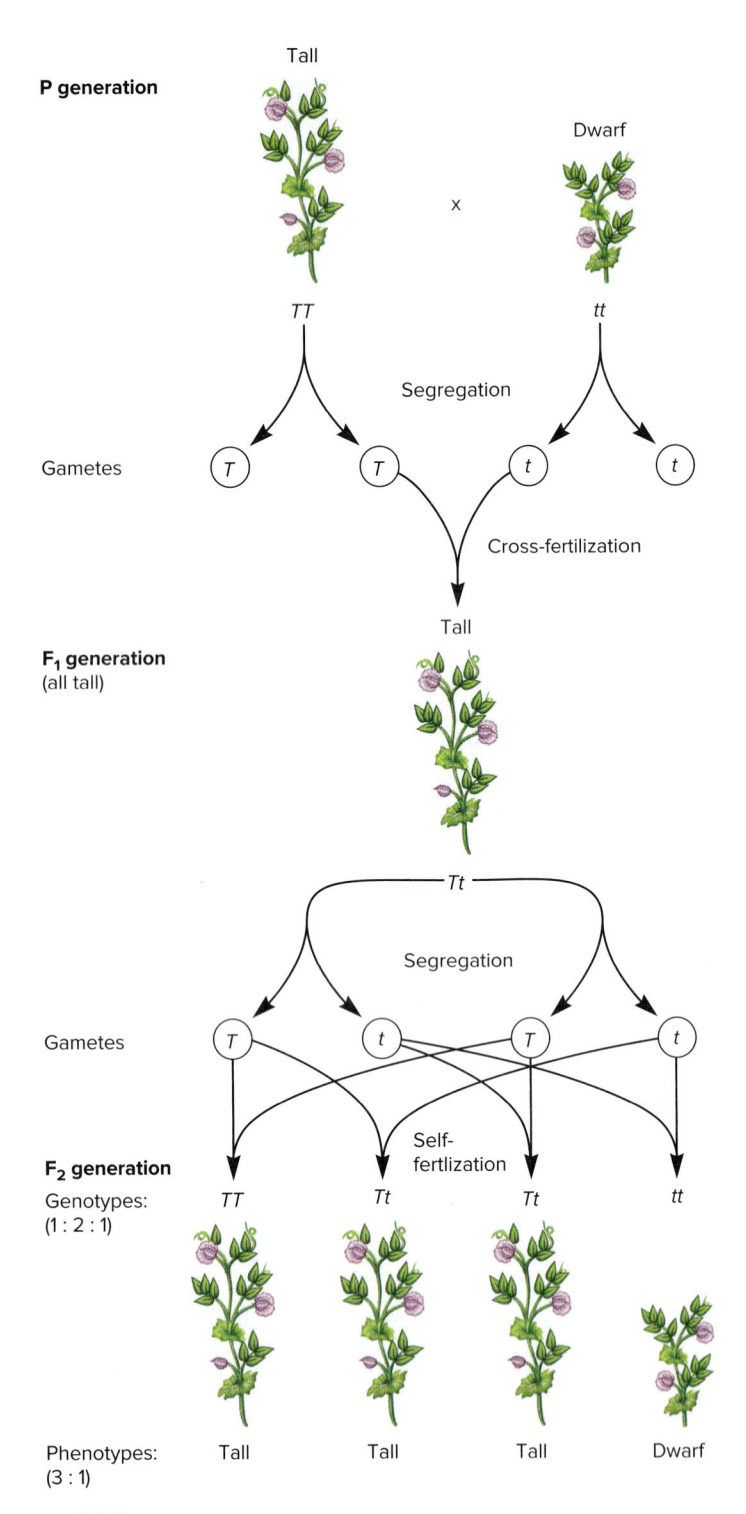

FIGURE 3.6 Mendel's law of segregation. This illustration shows a cross between a true-breeding tall plant and a true-breeding dwarf plant and the subsequent segregation of the tall (*T*) and dwarf (*t*) alleles in the F_1 and F_2 generations.

Concept Check: With regard to the *T* and *t* alleles, explain what the word segregation means.

- The term **genotype** refers to the genetic composition of an individual. *TT* and *tt* are the genotypes of the P generation in this experiment, and *Tt* is the genotype of the F_1 generation.

4. *Fill in the possible genotypes of the offspring by combining the alleles of the gametes in the empty boxes.*

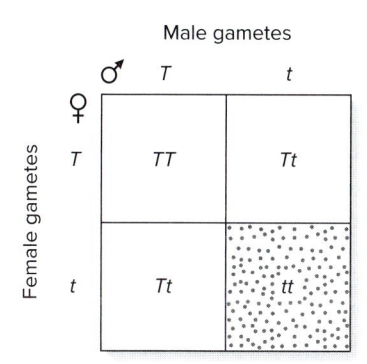

	♂ T	t
♀ T	TT	Tt
t	Tt	tt

(Male gametes across top; Female gametes down side)

5. *Determine the relative proportions of genotypes and phenotypes of the offspring.* The genotypes are obtained directly from the Punnett square. They are contained within the boxes that have been filled in. In this example, the genotypes are *TT*, *Tt*, and *tt* in a 1:2:1 ratio. To determine the phenotypes, you must know the dominant/recessive relationship between the alleles. For plant height, we know that *T* (tall) is dominant to *t* (dwarf). The genotypes *TT* and *Tt* are tall, whereas the genotype *tt* is dwarf. Therefore, our Punnett square shows us that the ratio of phenotypes is 3:1, or 3 tall plants : 1 dwarf plant.

Genetic TIPS

The Question: One pea plant that is heterozygous with regard to flower color (purple is dominant to white) is crossed to a plant with white flowers. What are the predicted outcomes of genotypes and phenotypes for the offspring?

Topic: What topic in genetics does this question address? The topic is Mendelian inheritance. More specifically, the question is about a single-factor cross.

Information: What information do you know based on the question and your understanding of the topic? From the question, you know that one plant is heterozygous for flower color. If *P* is the purple allele and *p* the white allele, the genotype of this plant is *Pp*. The other plant exhibits the recessive phenotype, so its genotype must be *pp*. From your understanding of the topic, you may remember that alleles segregate during gamete formation and parents each pass one allele to their offspring; the two alleles combine at fertilization.

Problem-Solving Strategy: Predict the outcome. One strategy to solve this type of problem is to use a Punnett square to predict the outcome of the cross. The Punnett square is shown next.

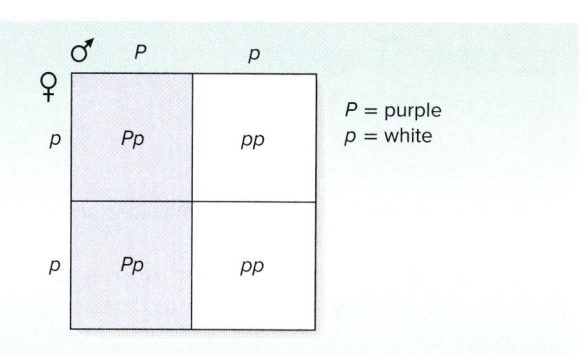

	♂ P	p
♀ p	Pp	pp
p	Pp	pp

P = purple
p = white

Answer: The ratio of offspring genotypes is 1 *Pp* : 1 *pp*. The ratio of the phenotypes is 1 purple : 1 white.

3.2 REVIEWING THE KEY CONCEPTS

- Mendel conducted single-factor crosses in which he followed the variants for a single character (see Figure 3.5).
- The results of his single-factor crosses showed that the dominant trait was observed in all individuals in the F_1 generation and displayed in a 3:1 ratio in the F_2 generation.
- Based on the results of his single-factor crosses, Mendel proposed three key ideas regarding inheritance: (1) Traits may be dominant or recessive. (2) Genes are passed unaltered from generation to generation. (3) The two copies of a given gene segregate (or separate) from each other during transmission from parent to offspring. This third idea is known as the law of segregation (see Figure 3.6).
- A Punnett square can be used to predict the outcome of a single-factor cross or self-fertilization experiment.

3.2 COMPREHENSION QUESTIONS

1. A pea plant is *Tt*. Which of the following statements is correct?
   a. Its genotype is *Tt*, and its phenotype is dwarf.
   b. Its phenotype is *Tt*, and its genotype is dwarf.
   c. Its genotype is *Tt*, and its phenotype is tall.
   d. Its phenotype is *Tt*, and its genotype is tall.

2. A *Tt* plant is crossed to a *tt* plant. What is the expected ratio of phenotypes for offspring from this cross?
   a. 3 tall : 1 dwarf
   b. 1 tall : 1 dwarf
   c. 1 tall : 3 dwarf
   d. 2 tall : 1 dwarf

3. You may need to refer back to Chapter 2 to answer this question. If a plant is *Tt*, at which stage do the *T* and *t* alleles segregate from each other?
   a. Anaphase of mitosis
   b. Anaphase of meiosis I
   c. Anaphase of meiosis II
   d. None of the above is the stage at which segregation occurs.

3.3 LAW OF INDEPENDENT ASSORTMENT

Learning Outcomes:
1. Analyze Mendel's experiments involving two-factor crosses.
2. State Mendel's *law of independent assortment*.
3. Predict the outcome of two-factor crosses using a Punnett square.

Though his experiments as described in Figure 3.5 revealed important ideas regarding hereditary laws, Mendel realized that additional insights might be uncovered if he conducted more complicated experiments. In this section, we will examine how he conducted crosses in which he simultaneously investigated the pattern of inheritance for two different characters. In other words, he carried out **two-factor crosses** in which he followed the inheritance of two different characters within the same groups of individuals. These experiments led to the formulation of a second law called the law of independent assortment.

Mendel Also Analyzed Crosses Involving Two Different Characters

Let's consider one of Mendel's experiments in which one of the characters was seed shape, found in round or wrinkled variants, and the second character was seed color, which existed as yellow and green variants. In this two-factor cross, Mendel followed the inheritance pattern for both characters simultaneously.

What results are possible from a two-factor cross? Keep in mind that the results of Figure 3.5 have already shown us that a gamete carries only one allele for each gene.

- One possibility is that the genetic determinants for two different characters are always linked to each other and inherited as a single unit (**Figure 3.7a**). If this were the case, the F_1 offspring could produce only two types of gametes: *RY* and *ry*.
- A second possibility is that they are not linked and can assort themselves independently into haploid gametes (**Figure 3.7b**). According to the independent-assortment hypothesis, an F_1 offspring could produce four types of gametes: *RY*, *Ry*, *rY*, and *ry*.

The experimental protocol for one of Mendel's two-factor crosses is shown in **Figure 3.8**. He began with two different strains of true-breeding pea plants that were different with regard to two characters: seed shape and seed color. In this example, one plant was produced from seeds that were round and yellow; the other plant from seeds that were wrinkled and green. When these plants were crossed, the seeds, which contain the plant embryo, are considered part of the F_1 generation. As expected, all F_1 seeds displayed a phenotype of round and yellow. This phenotype was observed because round and yellow are dominant traits. It is the F_2 generation that supports the independent-assortment model and refutes the linkage model.

▶ **THE HYPOTHESES**

The inheritance pattern for two different characters follows one or more quantitative natural laws. Two possible hypotheses are described in Figure 3.7.

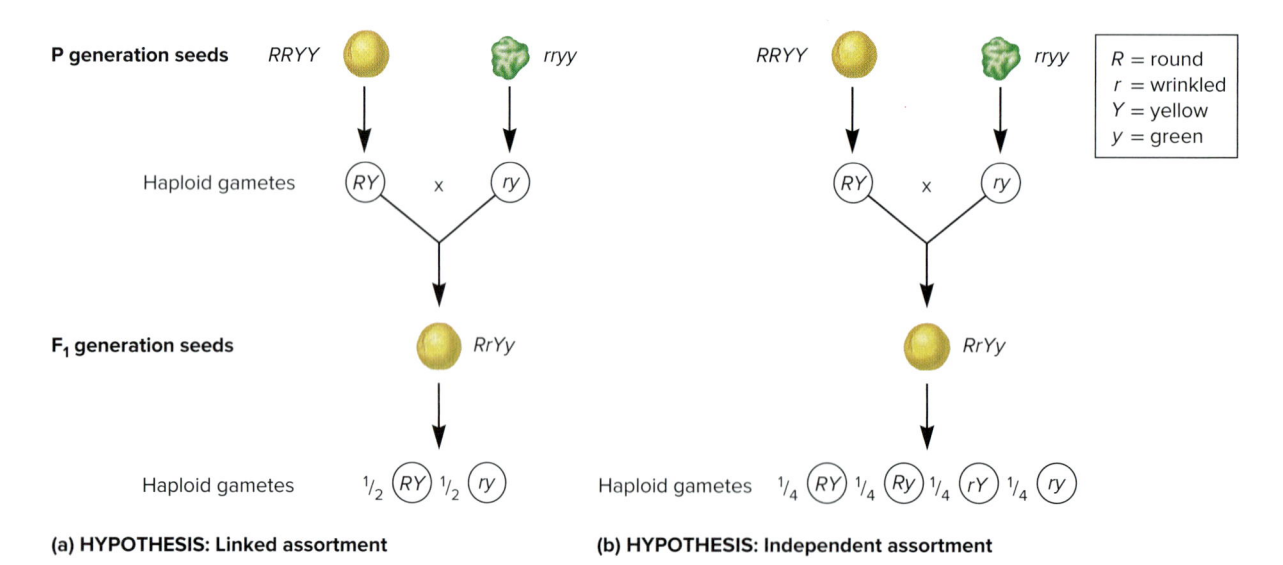

FIGURE 3.7 Two hypotheses to explain how two different genes assort during gamete formation. (a) According to the hypothesis of linked assortment, the two genes always stay associated with each other. (b) In contrast, the independent-assortment hypothesis proposes that the two different genes randomly segregate into haploid cells.

Concept Check: According to the hypothesis of linked assortment shown here, what is linked? Are two different genes linked, or are two different alleles of the same gene linked, or both?

3.3 LAW OF INDEPENDENT ASSORTMENT

▶ **TESTING THE HYPOTHESES** — **FIGURE 3.8** Mendel's analysis of two-factor crosses.

Starting material: In this experiment, Mendel began with true-breeding pea strains that were different with regard to two characters. One strain produced round, yellow seeds (*RRYY*); the other strain produced wrinkled, green seeds (*rryy*).

1. Cross the two true-breeding strains to each other. This produces F_1 generation seeds.

2. Collect many seeds and record their phenotype.

3. F_1 seeds are planted and grown, and the F_1 plants are allowed to self-fertilize. This produces seeds that are part of the F_2 generation.

4. Analyze the characteristics found in the F_2 generation seeds.

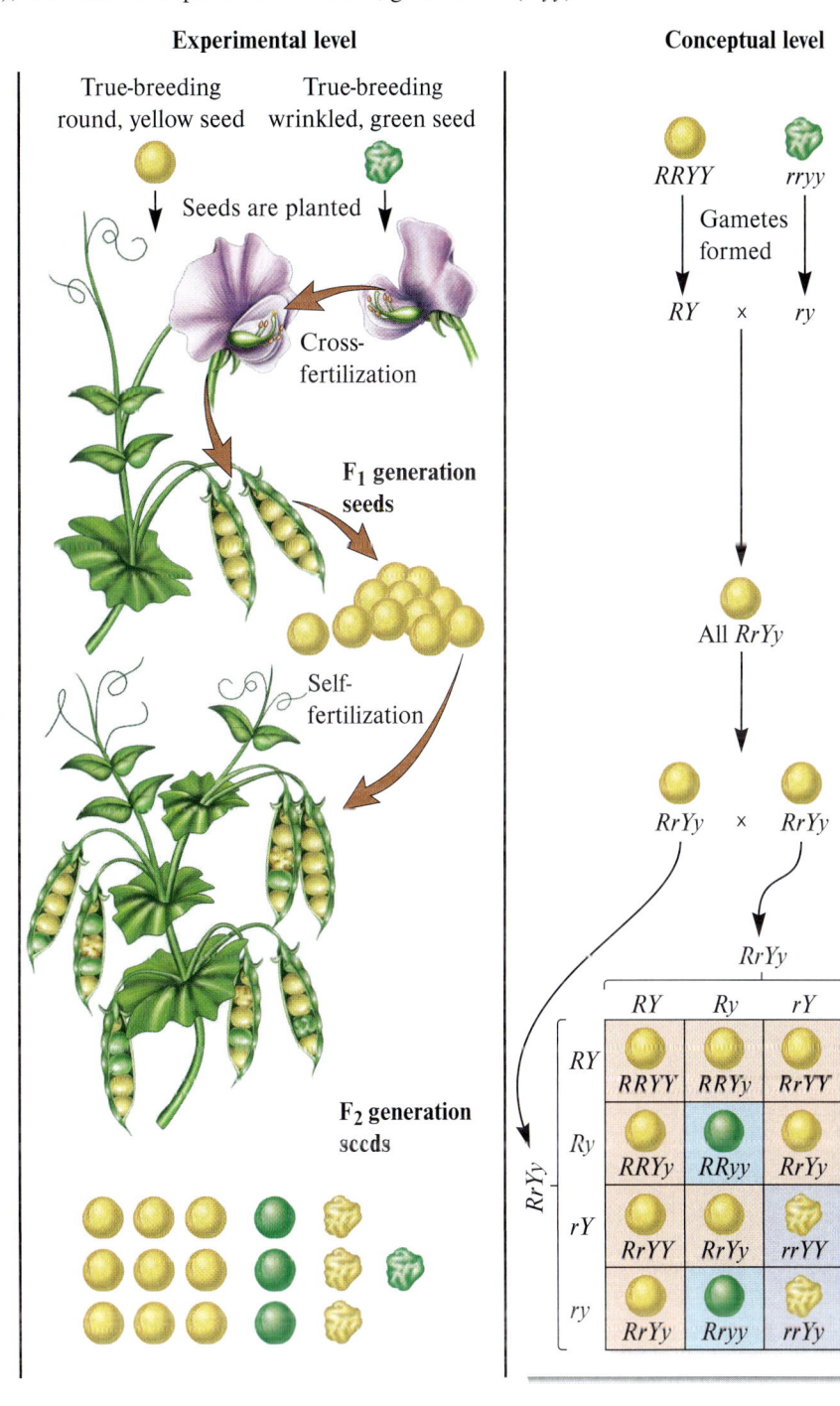

▶ **THE DATA**

P cross	F_1 generation	F_2 generation
Round, yellow × wrinkled, green seeds	All round, yellow	315 round, yellow seeds
		108 round, green seeds
		101 wrinkled, yellow seeds
		32 wrinkled, green seeds

▶ **INTERPRETING THE DATA**

As seen in the data, the F_2 generation had four categories of seeds. Those that were round and green or wrinkled and yellow are called **nonparentals,** because these combinations of traits were not found in the true-breeding plants of the parental generation. The occurrence of nonparental variants contradicts the linked-assortment hypothesis. According to that model, the *R* and *Y* alleles should be linked together and so should the *r* and *y* alleles.

If this were the case, the F_1 plants could only produce gametes that were *RY* or *ry*. These would combine to produce *RRYY* (round, yellow), *RrYy* (round, yellow), or *rryy* (wrinkled, green) seeds in a 1:2:1 ratio. Nonparental seeds could not be produced. However, Mendel did not obtain this result. Instead, he observed a phenotypic ratio of 9:3:3:1 in the F_2 generation.

Mendel's Two-Factor Crosses Led to the Law of Independent Assortment

Mendel's results from many two-factor crosses rejected the hypothesis of linked assortment and, instead, supported the hypothesis that different characters assort themselves independently. Using the modern notion of genes, **Mendel's law of independent assortment** states:

> *Two different genes randomly assort their alleles during the formation of haploid cells.*

In other words, the allele for one gene is found within a resulting gamete independently of whether the allele for a different gene is found in the same gamete. Using the example given in Figure 3.8, the round and wrinkled alleles are assorted into haploid gametes independently of the yellow and green alleles. Therefore, a heterozygous *RrYy* parent can produce four different gametes—*RY*, *Ry*, *rY*, and *ry*—in equal proportions.

In an F_1 self-fertilization experiment, any two gametes can combine randomly during fertilization. This allows for 4^2, or 16, possible offspring, although some offspring are genetically identical to each other. As shown in **Figure 3.9**, these 16 possible combinations result in seeds with the following phenotypes: 9 round, yellow; 3 round, green; 3 wrinkled, yellow; and 1 wrinkled, green. This 9:3:3:1 ratio is the expected outcome when heterozygotes for two traits are allowed to self-fertilize. Mendel was clever enough to realize that the data for his two-factor crosses were close to a 9:3:3:1 ratio. In Figure 3.8, for example, his F_1 generation produced F_2 seeds with the following characteristics: 315 round, yellow seeds; 108 round, green seeds; 101 wrinkled, yellow seeds; and 32 wrinkled, green seeds. The phenotypic ratio of the F_2 generation is 9.8:3.2:3.4:1.0. Within experimental error, Mendel's data approximated the predicted 9:3:3:1 ratio for the F_2 generation.

The law of independent assortment held true for two-factor crosses involving the characters that Mendel studied in pea plants. However, in other cases, the inheritance pattern of two different genes is consistent with the linkage model described in Figure 3.7a. In Chapter 7, we will examine the inheritance of genes that are linked to each other because they are physically within the same chromosome. As we will see, linked genes do not assort independently.

A Punnett Square Can Be Used to Solve Independent-Assortment Problems

As depicted in Figure 3.8, we can make a Punnett square to predict the outcome of experiments involving two or more genes that assort independently. Let's see how such a Punnett square is made by considering a cross between two plants that are heterozygous for height and seed color (**Figure 3.10**). This cross is $TtYy \times TtYy$. When we construct a Punnett square for this cross, we must keep

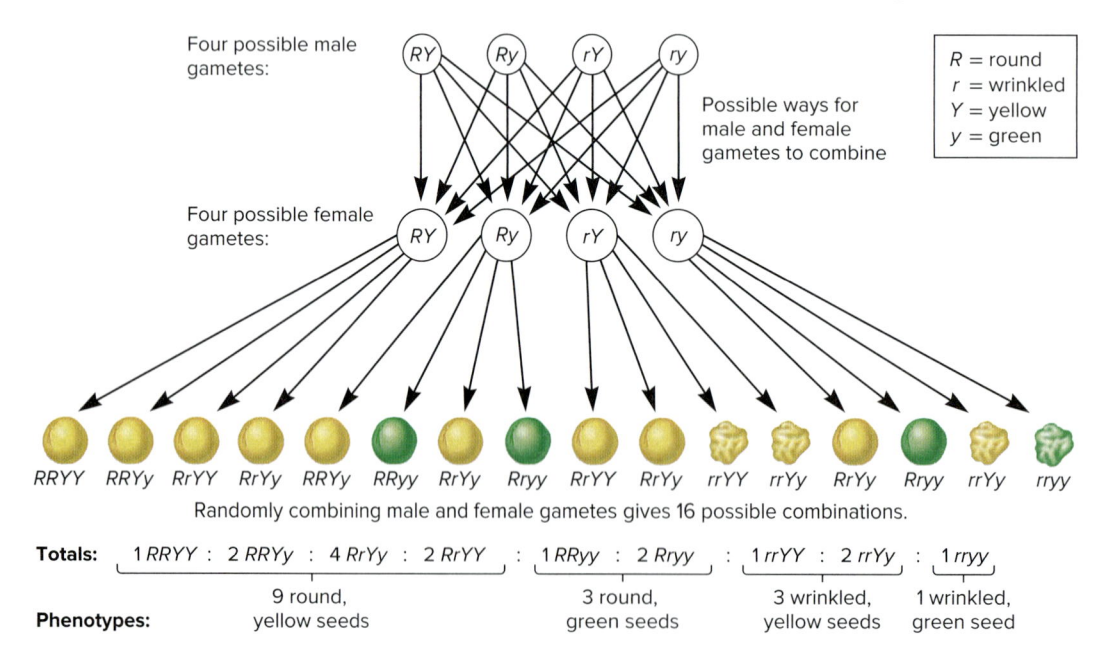

FIGURE 3.9 Mendel's law of independent assortment.

Genes → Traits This self-fertilization experiment involves a parent that is heterozygous for seed shape and seed color (*RrYy*). Four types of male gametes are possible: *RY*, *Ry*, *rY*, and *ry*. Likewise, four types of female gametes are possible: *RY*, *Ry*, *rY*, and *ry*. These four types of gametes are the result of the independent assortment of the seed shape and seed color alleles relative to each other. During fertilization, any one of the four types of male gametes can combine with any one of the four types of female gametes. This results in 16 offspring, each one containing two copies of the seed shape gene and two copies of the seed color gene.

Concept Check: *Why does independent assortment promote genetic variation?*

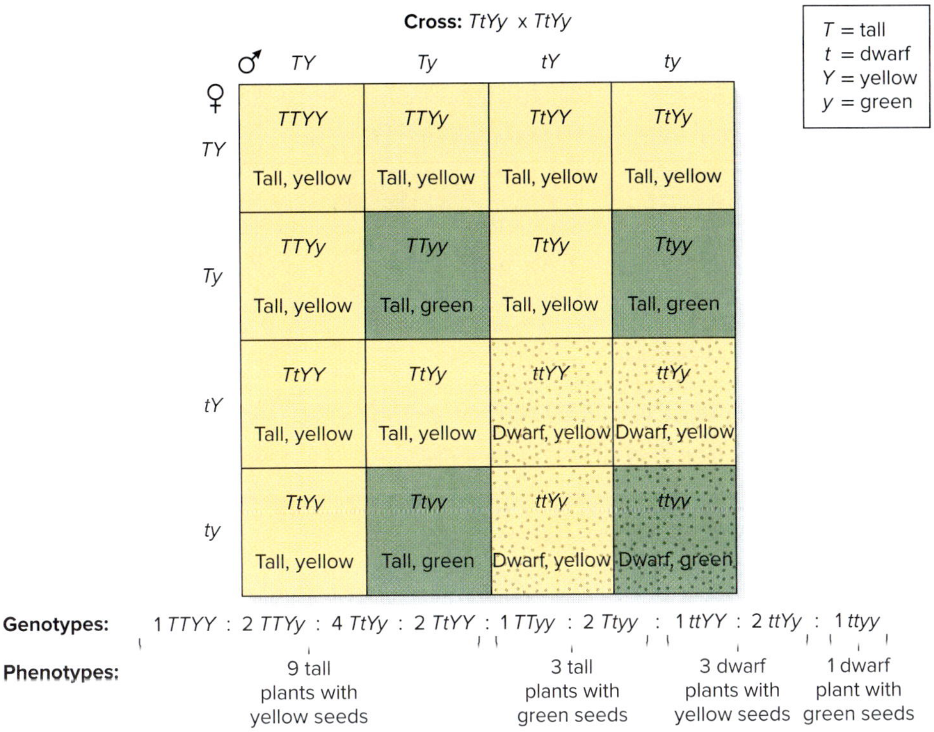

Cross: TtYy × TtYy

	TY	Ty	tY	ty
TY	TTYY Tall, yellow	TTYy Tall, yellow	TtYY Tall, yellow	TtYy Tall, yellow
Ty	TTYy Tall, yellow	TTyy Tall, green	TtYy Tall, yellow	Ttyy Tall, green
tY	TtYY Tall, yellow	TtYy Tall, yellow	ttYY Dwarf, yellow	ttYy Dwarf, yellow
ty	TtYy Tall, yellow	Ttyy Tall, green	ttYy Dwarf, yellow	ttyy Dwarf, green

T = tall
t = dwarf
Y = yellow
y = green

Genotypes: 1 TTYY : 2 TTYy : 4 TtYy : 2 TtYY : 1 TTyy : 2 Ttyy : 1 ttYY : 2 ttYy : 1 ttyy

Phenotypes: 9 tall plants with yellow seeds : 3 tall plants with green seeds : 3 dwarf plants with yellow seeds : 1 dwarf plant with green seeds

FIGURE 3.10 A Punnett square for a two-factor cross. The Punnett square shown here involves a cross between two pea plants that are heterozygous for height and seed color. The cross is TtYy × TtYy.

Concept Check: If a parent plant is Ttyy, how many different types of gametes can it make?

---

in mind that each gamete has a single allele for each of two genes. In this example, the four possible gametes from each parent are

TY, Ty, tY, and ty

In this two-factor cross, we need to make a Punnett square containing 16 boxes. The phenotypes of the resulting offspring are predicted to occur in a ratio of 9:3:3:1.

The Multiplication and Forked-Line Methods Can Also Be Used to Solve Independent-Assortment Problems

In crosses involving three or more genes, the construction of a single large Punnett square becomes very unwieldy. For example, in a three-factor cross between two pea plants that are Tt Rr Yy, each parent can make 2^3, or 8, possible gametes. Therefore, the Punnett square must contain 8 × 8 = 64 boxes. As a more reasonable alternative, we can consider each gene separately and then algebraically combine them by multiplying together the expected outcomes for each gene. Two methods for doing this kind of analysis are termed the multiplication method and the forked-line method. To illustrate these methods, let's consider the following question:

Two pea plants are heterozygous for three genes (Tt Rr Yy), where T = tall, t = dwarf, R = round seeds, r = wrinkled seeds, Y = yellow seeds, and y = green seeds. If these plants are crossed to each other, what are the predicted phenotypes of the offspring, and what fraction of the offspring will occur in each category?

You could solve this problem by constructing a large Punnett square and filling in the boxes. However, in this case, eight different male gametes and eight different female gametes are possible: *TRY, TRy, TrY, tRY, trY, Try, tRy,* and *try*. It would be rather time-consuming to construct and fill in this Punnett square, which would contain 64 boxes. As an alternative, we can consider each gene separately and then algebraically combine them by multiplying together the expected phenotypic outcomes for each gene. In the cross *Tt Rr Yy × Tt Rr Yy*, a Punnett square can be made for each gene (**Figure 3.11a**).

According to the **multiplication method,** we can simply use the product rule (discussed later in Section 3.6) and multiply these three combinations together (**Figure 3.11b**). Even though the multiplication steps are also somewhat tedious, this approach is much easier than making a Punnett square with 64 boxes, filling them in, deducing each phenotype, and then adding them up! A second approach analogous to the multiplication method is the **forked-line method.** In this case, the genetic proportions are determined by multiplying together the probabilities of each phenotype (**Figure 3.11c**).

3.3 REVIEWING THE KEY CONCEPTS

- By conducting two-factor crosses, Mendel formulated the law of independent assortment, which states that two different genes randomly assort their alleles during the formation of haploid cells (see Figures 3.8, 3.9).
- A Punnett square can be used to predict the outcomes of two-factor crosses (see Figure 3.10).

FIGURE 3.11

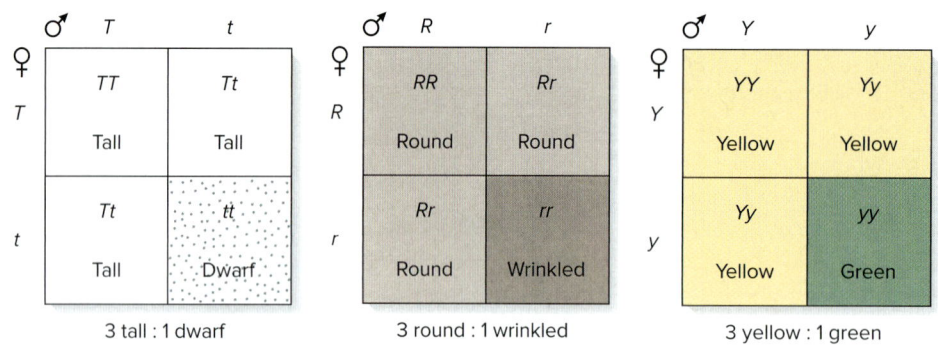

(a) Punnett squares for each gene

♀\♂	T	t
T	TT Tall	Tt Tall
t	Tt Tall	tt Dwarf

3 tall : 1 dwarf

♀\♂	R	r
R	RR Round	Rr Round
r	Rr Round	rr Wrinkled

3 round : 1 wrinkled

♀\♂	Y	y
Y	YY Yellow	Yy Yellow
y	Yy Yellow	yy Green

3 yellow : 1 green

P = (3 tall + 1 dwarf)(3 round + 1 wrinkled)(3 yellow + 1 green)

First, multiply (3 tall + 1 dwarf) times (3 round + 1 wrinkled)

(3 tall + 1 dwarf)(3 round + 1 wrinkled) = 9 tall, round + 3 tall, wrinkled + 3 dwarf, round + 1 dwarf, wrinkled

Next, multiply this product by (3 yellow + 1 green).

P = (9 tall, round + 3 tall, wrinkled + 3 dwarf, round + 1 dwarf, wrinkled) (3 yellow + 1 green) = 27 tall, round, yellow + 9 tall, round, green + 9 tall, wrinkled, yellow + 3 tall, wrinkled, green + 9 dwarf, round, yellow + 3 dwarf, round, green + 3 dwarf, wrinkled, yellow + 1 dwarf, wrinkled, green

(b) Multiplication method

Tall or dwarf	Round or wrinkled	Yellow or green	Observed product	Phenotype
3/4 tall	3/4 round	3/4 yellow	(3/4)(3/4)(3/4) = 27/64	tall, round, yellow
		1/4 green	(3/4)(3/4)(1/4) = 9/64	tall, round, green
	1/4 wrinkled	3/4 yellow	(3/4)(1/4)(3/4) = 9/64	tall, wrinkled, yellow
		1/4 green	(3/4)(1/4)(1/4) = 3/64	tall, wrinkled, green
1/4 dwarf	3/4 round	3/4 yellow	(1/4)(3/4)(3/4) = 9/64	dwarf, round, yellow
		1/4 green	(1/4)(3/4)(1/4) = 3/64	dwarf, round, green
	1/4 wrinkled	3/4 yellow	(1/4)(1/4)(3/4) = 3/64	dwarf, wrinkled, yellow
		1/4 green	(1/4)(1/4)(1/4) = 1/64	dwarf, wrinkled, green

(c) Forked-line method

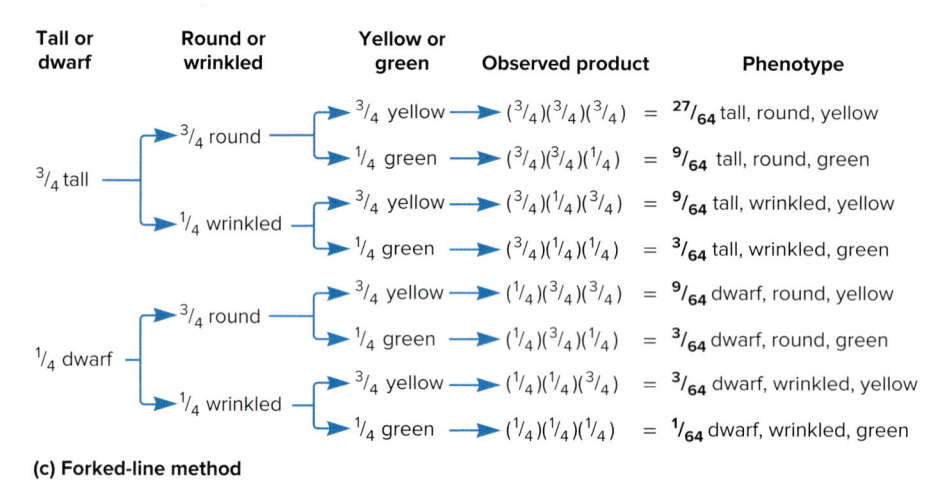

FIGURE 3.11 Two alternative ways to predict the outcome of a three-factor cross. The cross is $Tt\ Rr\ Yy \times Tt\ Rr\ Yy$. (a) For both methods, each gene is treated separately. These three Punnett squares predict the outcome for each gene. (b) Multiplication method. (c) Forked-line method.

- The multiplication and forked-line methods are used to predict the outcomes of crosses involving three or more genes (see Figure 3.11).

3.3 COMPREHENSION QUESTIONS

1. A pea plant has the genotype $rrYy$. How many different types of gametes can it make and in what proportions?
   a. 1 rr : 1 Yy
   b. 1 rY : 1 ry
   c. 3 rY : 1 ry
   d. 1 RY : 1 rY : 1 Ry : 1 ry

2. A cross is made between a pea plant that is $RrYy$ and one that is $rrYy$. What is the predicted outcome of the seed phenotypes?
   a. 9 round, yellow : 3 round, green : 3 wrinkled, yellow : 1 wrinkled green
   b. 3 round, yellow : 3 round, green : 1 wrinkled, yellow : 1 wrinkled green
   c. 3 round, yellow : 1 round, green : 3 wrinkled, yellow : 1 wrinkled green
   d. 1 round, yellow : 1 round, green : 1 wrinkled, yellow : 1 wrinkled green

3. In a population of wild squirrels, most of them have gray fur but an occasional squirrel is completely white. If we let P and p represent dominant and recessive alleles, respectively, of a gene that encodes an enzyme necessary for pigment formation, which of the following statements do you think is most likely to be correct?
   a. The white squirrels are pp, and the p allele is a loss-of-function allele.
   b. The gray squirrels are pp, and the p allele is a loss-of-function allele.
   c. The white squirrels are PP, and the P allele is a loss-of-function allele.
   d. The gray squirrels are PP, and the P allele is a loss-of-function allele.

3.4 CHROMOSOME THEORY OF INHERITANCE

Learning Outcomes:
1. List the key tenets of the chromosome theory of inheritance.
2. Explain the relationship between meiosis and Mendel's laws of inheritance.

Thus far, we have considered how Mendel's crosses in pea plants led to two important laws of inheritance. In Chapter 2, we examined how chromosomes are transmitted during cell division and gamete formation. In this section, we will explore how chromosomal transmission is related to the patterns of inheritance observed by Mendel. This relationship, known as the chromosome theory of inheritance, was a major breakthrough in our understanding of genetics because it established the framework for understanding how chromosomes carry and transmit the genetic determinants that govern the outcome of traits.

The Chromosome Theory of Inheritance Relates the Behavior of Chromosomes to the Mendelian Inheritance of Traits

The chromosome theory of inheritance dramatically unfolded as a result of three lines of scientific inquiry (**Table 3.1**). One avenue concerned Mendel's breeding studies, in which he analyzed the transmission of traits from parent to offspring. A second line of inquiry focused on the biochemical basis for heredity. A Swiss botanist, Carl Nägeli, and a German biologist, August Weismann, championed the idea that a substance found in living cells is responsible for the transmission of traits from parents to offspring. Nägeli also suggested that both parents contribute equal amounts of this substance to their offspring. Several scientists, including Oscar Hertwig, Eduard Strasburger, and Walther Flemming, conducted studies suggesting that the chromosomes are the carriers of the genetic material. We now know the DNA within the chromosomes is the genetic material. Finally, the third line of evidence involved the microscopic examination of the processes of fertilization, mitosis, and meiosis. Researchers became increasingly aware that the characteristics of organisms are rooted in the continuity of cells during the life of an organism and from one generation to the next. When the work of Mendel was rediscovered, several scientists noted striking parallels between the segregation and assortment of traits noted by Mendel and the behavior of chromosomes during meiosis. Among them were Theodor Boveri, a German biologist, and Walter Sutton at Columbia University. They independently proposed the chromosome theory of inheritance, which was a milestone in our understanding of genetics.

According to the **chromosome theory of inheritance,** the inheritance patterns of traits can be explained by the transmission patterns of chromosomes during meiosis and fertilization. This theory is based on a few fundamental principles:

1. Chromosomes contain the genetic material that is transmitted from parent to offspring and from cell to cell.

2. Chromosomes are replicated and passed along, generation after generation, from parent to offspring. They are also passed from cell to cell during the development of a

TABLE 3.1
Chronology for the Development and Proof of the Chromosome Theory of Inheritance

Year	Event
1866	Gregor Mendel: Analyzed the transmission of traits from parents to offspring and showed that it follows a pattern of segregation and independent assortment.
1876–77	Oscar Hertwig and Hermann Fol: Observed that the nucleus of the sperm enters the egg during animal cell fertilization.
1877	Eduard Strasburger: Observed that the sperm nucleus of plants (and no detectable cytoplasm) enters the egg during plant fertilization.
1878	Walther Flemming: Described mitosis in careful detail.
1883	Carl Nägeli and August Weismann: Proposed the existence of a genetic material, which Nägeli called idioplasm and Weismann called germ plasm.
1883	Wilhelm Roux: Proposed that the most important event of mitosis is the equal partitioning of nuclear qualities to the daughter cells.
1883	Edouard van Beneden: Showed that gametes contain half the number of chromosomes and that fertilization restores the diploid number.
1884–1885	Hertwig, Strasburger, and August Weismann: Proposed that chromosomes are carriers of the genetic material.
1889	Theodor Boveri: Showed that enucleated sea urchin eggs that are fertilized by sperm from a different species develop into larvae that have characteristics that coincide with the sperm's species.
1900	Hugo de Vries, Carl Correns, and Erich von Tschermak: Rediscovered Mendel's work, while analyzing inheritance patterns in other plants.
1901	Thomas Montgomery: Determined that maternal and paternal chromosomes pair with each other during meiosis.
1901	C. E. McClung: Discovered that sex determination in insects is related to differences in chromosome composition.
1902	Theodor Boveri: Showed that when sea urchin eggs were fertilized by two sperm, the abnormal development of the embryo was related to an abnormal number of chromosomes.
1903	Walter Sutton: Showed that even though the chromosomes seem to disappear during interphase, they do not actually disintegrate. Instead, he argued that chromosomes must retain their continuity and individuality from one cell division to the next.
1902–1903	Theodor Boveri and Walter Sutton: Independently proposed tenets of the chromosome theory of inheritance. Some historians primarily credit this theory to Sutton
1910	Thomas Hunt Morgan: Showed that a genetic trait (i.e., white-eyed phenotype in *Drosophila*) is linked to a particular chromosome.
1913	E. Eleanor Carothers: Demonstrated that homologous pairs of chromosomes show independent assortment.
1916	Calvin Bridges: Studied chromosomal abnormalities in *Drosophila* as a way to confirm the chromosome theory of inheritance.

Source: Voeller, Bruce (1968) *The Chromosome Theory of Inheritance: Classic Papers in Development and Heredity.* New York, NY: Appleton-Century-Crofts, 229.

multicellular organism. Each type of chromosome retains its individuality during cell division and gamete formation.

3. The nuclei of most eukaryotic cells contain chromosomes that are found in homologous pairs—they are diploid. One member of each pair is inherited from the mother, the other from the father. At meiosis, one of the two members of each pair segregates into one daughter nucleus, and the homolog segregates into the other daughter nucleus. Gametes contain one set of chromosomes—they are haploid.
4. During the formation of haploid cells, different types of (nonhomologous) chromosomes segregate independently of each other.
5. Each parent contributes one set of chromosomes to its offspring. The maternal and paternal sets of homologous chromosomes are functionally equivalent; each set carries a full complement of genes.

The Law of Segregation Is Explained by the Separation of Homologs during Meiosis

The chromosome theory of inheritance allows us to see the relationship between Mendel's laws and chromosome transmission. As shown in **Figure 3.12**, Mendel's law of segregation can be explained by the homologous pairing and segregation of chromosomes during meiosis. This figure depicts the behavior of a pair of homologous chromosomes that carry a gene for seed color in pea plants. One of the chromosomes carries a dominant allele that confers yellow seed color, whereas the homologous chromosome carries a recessive allele that confers green color. A heterozygous individual passes along only one of these alleles to each offspring. In other words, a gamete may contain the yellow allele or the green allele, but not both. Because homologous chromosomes segregate from each other, a gamete contains only one copy of each type of chromosome.

The Law of Independent Assortment Is Explained by the Random Alignment of Homologs during Meiosis

How is the law of independent assortment explained by the behavior of chromosomes? **Figure 3.13** considers the segregation of two types of chromosomes, each carrying a different gene.

- One pair of chromosomes carries the gene for seed color: The yellow (Y) allele is on one chromosome, and the green (y) allele is on the homolog.
- The other pair of (smaller) chromosomes carries the gene for seed shape: One copy has the round (R) allele, and the homolog carries the wrinkled (r) allele.
- At metaphase of meiosis I, the different types of chromosomes are randomly aligned along the metaphase plate. On the left, the R allele has sorted with the y allele, whereas the r allele has sorted with the Y allele. On the right, the opposite situation has occurred. Therefore, the random alignment of chromatid pairs during meiosis I can lead to an independent assortment of genes that are found on nonhomologous chromosomes.

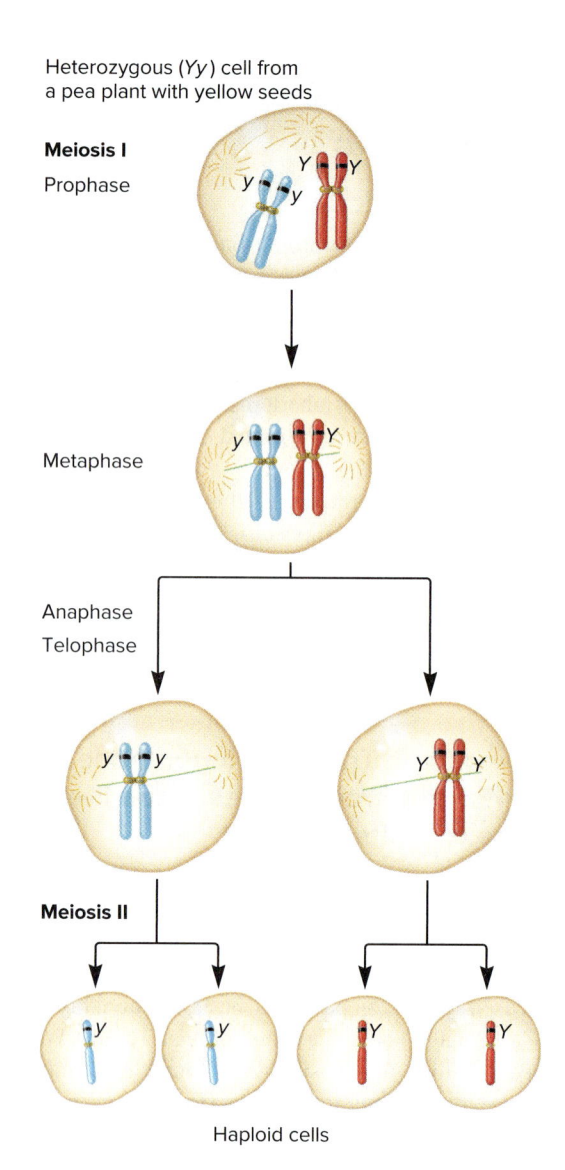

FIGURE 3.12 **Mendel's law of segregation can be explained by the segregation of homologs during meiosis.** The two copies of a gene are located on homologous chromosomes. In this example using pea seed color, the two alleles are Y (yellow) and y (green). During meiosis, the homologous chromosomes segregate from each other, leading to segregation of the two alleles into separate gametes.

Genes ⟶ Traits The gene for seed color exists in two alleles, Y (yellow) and y (green). During meiosis, the homologous chromosomes that carry these alleles segregate from each other. The resulting cells receive the Y or y allele, but not both. When two gametes unite during fertilization, the alleles they carry determine the traits of the resulting offspring. In this case, they determine seed color, producing yellow or green seeds.

Concept Check: At which stage do homologous chromosomes separate from each other?

An important consequence of the law of independent assortment is that a single individual can produce a vast array of genetically different gametes. As discussed in Chapter 2, diploid species have pairs of homologous chromosomes, which may differ with respect to the alleles they carry. When an offspring receives a combination of alleles that differs from those in its parents, this phenomenon is termed **genetic recombination.** One mechanism that accounts for

3.4 CHROMOSOME THEORY OF INHERITANCE   57

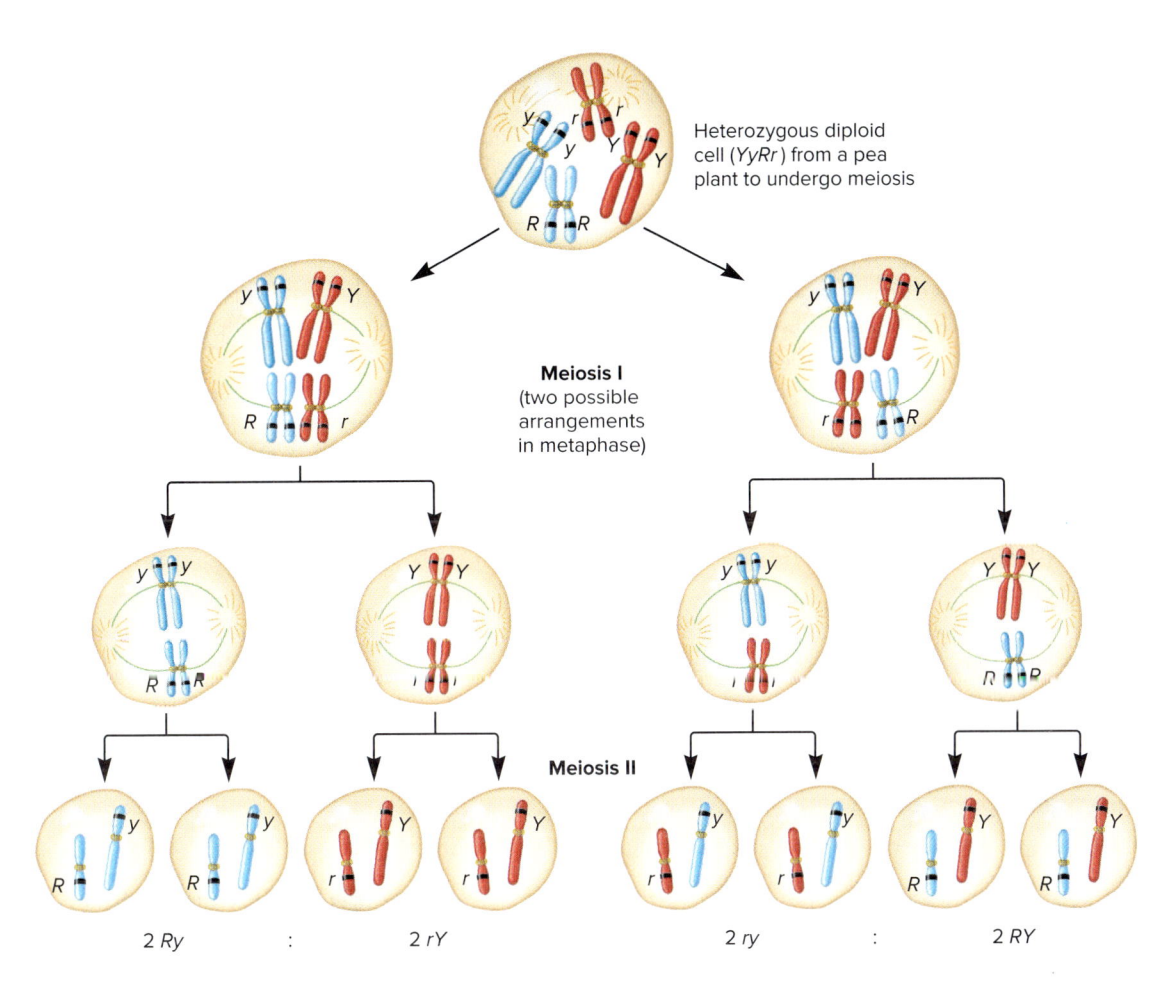

FIGURE 3.13 Mendel's law of independent assortment can be explained by the random alignment of bivalents during metaphase of meiosis I. This figure shows the assortment of two genes located on two different chromosomes, using pea seed color and shape as an example (*YyRr*). During metaphase of meiosis I, different possible arrangements of the homologs within bivalents can lead to different combinations of alleles in the resulting gametes. For example, on the left, the dominant *R* allele has sorted with the recessive *y* allele; on the right, the dominant *R* allele has sorted with the dominant *Y* allele.

Genes → Traits Most species have several different chromosomes that carry many different genes. In this example, the gene for seed color exists in two alleles, *Y* (yellow) and *y* (green), and the gene for seed shape is found as *R* (round) and *r* (wrinkled) alleles. The two genes are found on different (nonhomologous) chromosomes. During meiosis, the homologous chromosomes that carry these alleles segregate from each other. In addition, the chromosomes carrying the *Y* or *y* alleles independently assort from the chromosomes carrying the *R* or *r* alleles. As shown here, this provides a reassortment of alleles, potentially creating combinations of alleles that are different from the parental combinations. When two gametes unite during fertilization, the alleles they carry affect the traits of the resulting offspring.

Concept Check: Let's suppose a pea plant is heterozygous for three genes, *Tt Yy Rr*, and each gene is on a different chromosome. How many different ways could the three pairs of homologous chromosomes line up during metaphase of meiosis I?

genetic recombination is independent assortment. A second mechanism, discussed in Chapters 2 and 7, is crossing over, which can reassort alleles that happen to be linked along the same chromosome.

The phenomenon of independent assortment is rooted in the random pattern in which the homologs assort themselves during the process of meiosis (see Figure 3.13). A species containing a large number of homologous chromosomes has the potential for creating an enormous amount of genetic diversity. For example, human cells contain 23 pairs of chromosomes. These pairs can randomly assort into gametes during meiosis. The number of different gametes an individual can make equals 2^n, where *n* is the number of pairs of chromosomes. Therefore, a human can make 2^{23}, or over 8 million, possible gametes, due to independent assortment. The capacity to make so many genetically different gametes enables a species to produce individuals with many different combinations of traits. This variety of phenotypes allows environmental factors to select for those combinations of traits that favor reproductive success.

3.4 REVIEWING THE KEY CONCEPTS

- The chromosome theory of inheritance describes how the transmission of chromosomes can explain Mendel's laws (see Table 3.1).
- Mendel's law of segregation is explained by the separation of homologs during meiosis (see Figure 3.12).
- Mendel's law of independent assortment is explained by the random alignment of different chromosomes during metaphase of meiosis I (see Figure 3.13).

3.4 COMPREHENSION QUESTIONS

1. Which of the following is *not* one of the tenets of the chromosome theory of inheritance?
   a. Chromosomes contain the genetic material that is transmitted from parent to offspring and from cell to cell.
   b. Chromosomes are replicated and passed along, generation after generation, from parent to offspring.
   c. Chromosome replication occurs during S phase of the cell cycle.
   d. Each parent contributes one set of chromosomes to its offspring.

2. A pea plant has the genotype *TtRr*. The independent assortment of these two genes occurs at _____ because chromosomes with the _____ alleles line up independently of those with the _____ alleles.
   a. metaphase of meiosis I, *T* and *t*, *R* and *r*
   b. metaphase of meiosis I, *T* and *R*, *t* and *r*
   c. metaphase of meiosis II, *T* and *t*, *R* and *r*
   d. metaphase of meiosis II, *T* and *R*, *t* and *r*

3.5 STUDYING INHERITANCE PATTERNS IN HUMANS

Learning Outcomes:
1. Describe the features of a pedigree.
2. Analyze a pedigree to determine if a trait or disease is dominant or recessive.

Before we end our discussion of simple Mendelian traits, let's address the question of how we can analyze inheritance patterns among humans. In his experiments, Mendel selectively made crosses and then analyzed a large number of offspring. When studying human traits, however, researchers cannot control parental crosses. Instead, they must rely on the information contained within family trees, also called pedigrees, which are charts representing family relationships. This type of approach, known as **pedigree analysis,** is aimed at determining the type of inheritance pattern that a gene follows. Although this method may be less definitive than performing experiments like Mendel's, a pedigree analysis can often provide important clues concerning the pattern of inheritance of traits within human families. An expanded discussion of human pedigrees is provided in Chapter 22, which concerns the inheritance patterns of many human diseases.

In order to discuss the applications of pedigree analyses, we need to understand the organization and symbols of a pedigree (**Figure 3.14**).

- The oldest generation is at the top of the pedigree, and the most recent generation is at the bottom. Vertical lines connect each succeeding generation.
- A male is depicted as a square and a female as a circle.
- A man and woman who produce one or more offspring are directly connected by a horizontal line.
- A vertical line connects parents with their offspring.
- If parents produce two or more offspring, the group of siblings (brothers and/or sisters) is denoted by two or more squares or circles projecting downward from the same horizontal line.

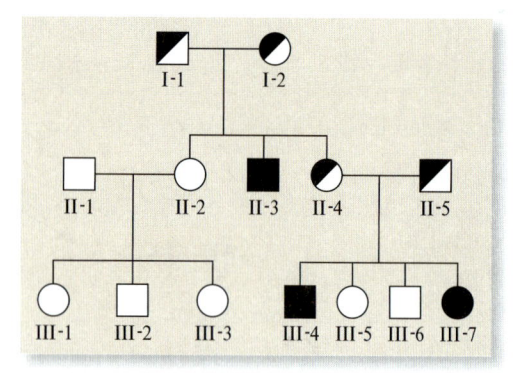

(a) Human pedigree showing inheritance pattern for cystic fibrosis

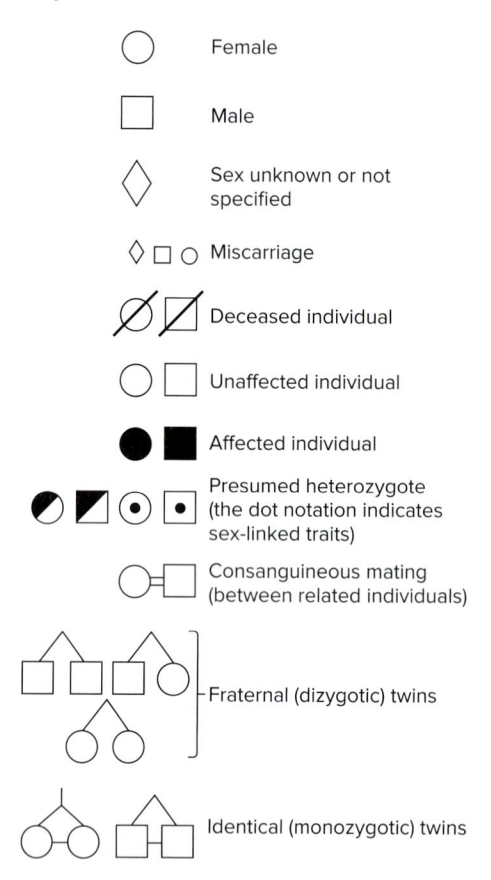

○ Female
□ Male
◇ Sex unknown or not specified
◇ □ ○ Miscarriage
⊘ ⊠ Deceased individual
○ □ Unaffected individual
● ■ Affected individual
◐ ◩ ⊙ ⊡ Presumed heterozygote (the dot notation indicates sex-linked traits)
○═□ Consanguineous mating (between related individuals)
Fraternal (dizygotic) twins
Identical (monozygotic) twins

(b) Symbols used in a human pedigree

FIGURE 3.14 Pedigree analysis. (a) A family pedigree in which some of the members are affected with cystic fibrosis. Individuals I-1, I-2, II-4, and II-5 are depicted as presumed heterozygotes because they produce affected offspring. (b) The symbols used in a pedigree analysis. *Note:* In most pedigrees shown in this textbook, such as those found in the problem sets, the heterozygotes are not shown as half-filled symbols. Most pedigrees throughout the book show individuals' phenotypes— open symbols are unaffected individuals and filled (closed) symbols are affected individuals.

Concept Check: What are the two different meanings of horizontal lines in a pedigree?

When a pedigree involves the transmission of a human trait or disease, affected individuals are depicted by filled symbols (in this case, filled with black) that distinguish them from unaffected individuals. Each generation is given a roman numeral designation, and individuals within the same generation are numbered from left to right. A few examples of the genetic relationships in Figure 3.14a are described here:

- Individuals I-1 and I-2 are the grandparents of III-1, III-2, III-3, III-4, III-5, III-6, and III-7.
- Individuals III-1, III-2, and III-3 are brother and sisters.
- Individuals II-3, III-4, and III-7 are affected by a genetic disease.

The symbols in Figure 3.14 depict certain individuals, such as I-1, I-2, II-4, and II-5, as presumed heterozygotes because they are unaffected with the recessive genetic disease (cystic fibrosis) but produce homozygous offspring that are affected. However, in many pedigrees, such as those found in the problem sets at the end of the chapter, the symbols reflect only phenotypes. In most pedigrees, affected individuals are shown with filled symbols, and unaffected individuals, including those that might be heterozygous for a recessive disease, are depicted with open symbols.

Pedigree analysis is commonly used to determine the inheritance pattern of human genetic diseases. Human geneticists are routinely interested in knowing whether a genetic disease is inherited as a recessive or dominant trait. One way to discern the dominant/recessive relationship between two alleles is by a pedigree analysis. Genes that play a role in disease may exist as common (wild-type) alleles or as mutant alleles that cause disease symptoms. If the disease follows a simple Mendelian pattern of inheritance and is caused by a recessive allele, an individual must inherit two copies of the mutant allele to exhibit the disease. Therefore, a recessive pattern of inheritance makes two important predictions. First, two heterozygous (unaffected) individuals will, on average, have 1/4 of their offspring affected. Second, all offspring of two affected individuals will be affected. Alternatively, a dominant trait predicts that affected individuals will have inherited the gene from at least one affected parent except in rare cases when a new mutation has occurred during gamete formation.

The pedigree in Figure 3.14a illustrates inheritance of a human genetic disease known as cystic fibrosis (CF). Among people of northern European descent, approximately 3% of the population are heterozygous carriers of the recessive allele. In homozygotes, the disease symptoms include abnormalities of the pancreas, intestine, sweat glands, and lungs. These abnormalities are caused by an imbalance of ions across the plasma membrane. In the lungs, this leads to a buildup of thick, sticky mucus. Respiratory problems may lead to early death, although modern treatments have greatly increased the life span of CF patients. In the late 1980s, the gene for CF was identified. The *CFTR* gene encodes a protein called the cystic fibrosis transmembrane conductance regulator (CFTR). This protein regulates the ion balance across the cell membrane in tissues of the pancreas, intestine, sweat glands, and lungs. The mutant allele causing CF alters the encoded CFTR protein. The altered CFTR protein is not correctly inserted into the plasma membrane, resulting in a decreased function that causes the ionic imbalance. As seen in the pedigree, the pattern of affected and unaffected individuals is consistent with

a recessive mode of inheritance. Two unaffected individuals can produce an affected offspring. Although not shown in this pedigree, a recessive mode of inheritance is also characterized by the observation that two affected individuals produce 100% affected offspring. However, for human genetic diseases that limit survival or fertility (or both), there may never be cases where two affected individuals produce offspring.

3.5 REVIEWING THE KEY CONCEPTS

- Human inheritance patterns are determined by analyzing family trees known as pedigrees (see Figure 3.14).

3.5 COMPREHENSION QUESTIONS

1. Which of the following would *not* be observed in a pedigree if a genetic disorder was inherited in a recessive manner?
    a. Two unaffected parents have an affected offspring.
    b. Two affected parents have an unaffected offspring.
    c. One affected and one unaffected parent have an unaffected offspring.
    d. All of the above are possible for a recessive disorder.
2. For the pedigree shown here, which pattern(s) of inheritance is/are possible? Affected individuals are shown with a filled symbol.

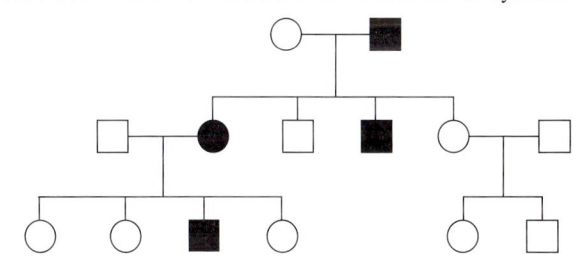

    a. Recessive
    b. Dominant
    c. Both recessive and dominant
    d. Neither recessive nor dominant

3.6 PROBABILITY AND STATISTICS

Learning Outcomes:
1. Define *probability*.
2. Predict the outcomes of crosses using the product rule and binomial expansion equation.
3. Evaluate the validity of a hypothesis using a chi square test.

A powerful application of Mendel's work is that the laws of inheritance can be used to predict the outcomes of genetic crosses. In agriculture, for example, plant and animal breeders are concerned with the types of offspring their crosses will produce. This information is used to produce commercially important crops and livestock. In addition, people are often interested in predicting the characteristics of the children they may have. This may be particularly important to individuals who carry alleles that cause inherited diseases. Of course, we cannot see into the future and definitively predict what will happen. Nevertheless, genetic counselors can help couples to predict the likelihood of having an affected child.

This probability is one factor that may influence a couple's decision to have children.

In this section, we will see how probability calculations are used in genetic problems to predict the outcomes of crosses. To compute probability, we will consider two mathematical operations known as the product rule and the binomial expansion equation. These methods allow us to determine the probability that a cross between two individuals will produce a particular outcome. To apply these operations, we must have knowledge regarding the genotypes of the parents and the pattern of inheritance of a given trait.

Probability calculations can also be used in hypothesis testing. In many situations, a researcher wants to discern the genotypes and patterns of inheritance for traits that are not yet understood. A traditional approach to this problem is to conduct crosses and then analyze their outcomes. The proportions of offspring may provide important clues that allow the experimenter to propose a hypothesis, based on the quantitative laws of inheritance, that explains the transmission of the trait from parents to offspring. Statistical methods, such as the chi square test, can then be used to evaluate how well the observed data from crosses fit the expected data. We will end this chapter with an example that applies the chi square test to a genetic cross.

Probability Is the Likelihood That an Outcome Will Occur

The chance that an outcome will occur in the future is called the outcome's **probability.** For example, if you flip a coin, the probability is 0.50, or 50%, that the head side will be showing when the coin lands. Probability depends on the number of possible outcomes. In this case, two possible outcomes (heads or tails) are equally likely. This allows us to predict a 50% chance that a coin flip will produce heads. The general formula for probability (P) is

$$\text{Probability} = \frac{\text{Number of times an outcome will occur}}{\text{Total number of possible outcomes}}$$

Thus, the probability that a coin flip will result in heads is

$$P_{\text{heads}} = 1 \text{ heads} / (1 \text{ heads} + 1 \text{ tails}) = 1/2 = 50\%$$

In genetic problems, we are often interested in the probability that a particular type of offspring will be produced. Recall that when two heterozygous tall pea plants (Tt) are crossed, the phenotypic ratio of the offspring is 3 tall to 1 dwarf. This information can be used to calculate the probability for either type of offspring.

$$\text{Probability} = \frac{\text{Expected number of individuals with a given phenotype}}{\text{Total number of individuals}}$$

$$P_{\text{tall}} = 3 \text{ tall} / (3 \text{ tall} + 1 \text{ dwarf}) = 3/4 = 75\%$$

$$P_{\text{dwarf}} = 1 \text{ dwarf} / (3 \text{ tall} + 1 \text{ dwarf}) = 1/4 = 25\%$$

The probability is 75% for offspring that are tall and 25% for offspring that are dwarf. When we add together the probabilities of all possible outcomes (tall and dwarf), we should get a sum of 100% (here, 75% + 25% = 100%).

A probability calculation allows us to predict the likelihood that an outcome will occur in the future. The accuracy of this prediction, however, depends to a great extent on the size of the sample. For example, if we toss a coin six times, our probability prediction suggests that 50% of the time we should get heads (i.e., three heads and three tails). In this small sample size, however, we would not be too surprised if we came up with four heads and two tails. Each time we toss a coin, there is a random chance that it will be heads or tails. The deviation between the observed and expected outcomes is called the **random sampling error.** In a small sample of coin tosses, the error between the predicted percentage of heads and the actual percentage observed may be quite large. By comparison, if we flipped a coin 1000 times, the percentage of heads would be fairly close to the predicted 50% value. In a larger sample, we expect the random sampling error to be a much smaller percentage.

The Product Rule Can Be Used to Predict the Probability of Independent Outcomes

We can use probability to make predictions regarding the likelihood of two or more independent outcomes from a genetic cross. When we say that outcomes are independent, we mean that the occurrence of one outcome does not affect the probability of another outcome. As an example, let's consider a rare, recessive human trait known as congenital analgesia. Persons with this trait can distinguish between sharp and dull, and hot and cold, but do not perceive extremes of sensation as being painful. The first case of congenital analgesia, described in 1932, was a man who made his living entertaining the public as a "human pincushion."

For a phenotypically unaffected couple, in which both parents are heterozygous, Pp (where P is the common allele and p is the recessive allele causing congenital analgesia), we can ask: What is the probability that the couple's first three offspring will have congenital analgesia? To answer this question, we can use the **product rule,** which states:

The probability that two or more independent outcomes will occur is equal to the product of their individual probabilities.

A strategy for solving this type of problem is shown here.

The Cross: $Pp \times Pp$

The Question: What is the probability that the couple's first three offspring will have congenital analgesia?

Step 1. *Calculate the individual probability of this phenotype.* As described previously, this is accomplished using a Punnett square.

The probability of an affected offspring is 1/4.

Step 2. *Multiply the individual probabilities.* In this case, we are asking about the first three offspring, and so we multiply 1/4 three times.

$$1/4 \times 1/4 \times 1/4 = 1/64 = 0.016, \text{ or } 1.6\%$$

In this case, the probability that the first three offspring will have this trait is 0.016. We predict that 1.6% of the time the first three offspring will all have congenital analgesia when both parents are heterozygotes. In this example, the phenotypes of the first, second, and third offspring are independent outcomes. The phenotype of the first offspring does not have an effect on the phenotype of the second or third offspring, and vice versa.

In the problem described here, we have used the product rule to determine the probability that the first three offspring will all have the same phenotype (congenital analgesia). We can also apply the rule to predict the probability of a sequence of outcomes that involves combinations of different offspring. For example, consider this question: What is the probability that the first offspring will be unaffected, the second offspring will have congenital analgesia, and the third offspring will be unaffected? Again, to solve this problem, begin by calculating the individual probability of each phenotype.

Unaffected = 3/4

Congenital analgesia = 1/4

The probability that these three phenotypes will occur in this specified order is

$$3/4 \times 1/4 \times 3/4 = 9/64 = 0.14, \text{ or } 14\%$$

In other words, this sequence of outcomes is expected to occur only 14% of the time.

The product rule can also be used to predict the outcome of a cross involving two or more genes. Let's suppose an individual with the genotype *Aa Bb CC* was crossed to an individual with the genotype *Aa bb Cc*. We could ask: What is the probability that an offspring will have the genotype *AA bb Cc*? If the three genes independently assort, the probability of inheriting alleles for each gene is independent of the probability for other two genes. Therefore, we can separately calculate the probability of the desired outcome for each gene.

Cross: *Aa Bb CC* × *Aa bb Cc*

Probability that an offspring will be *AA* = 1/4, or 0.25

Probability that an offspring will be *bb* = 1/2, or 0.5

Probability that an offspring will be *Cc* = 1/2, or 0.5

We can use the product rule to determine the probability that an offspring will be *AA bb Cc*.

$$P = (0.25)(0.5)(0.5) = 0.0625, \text{ or } 6.25\%$$

The Binomial Expansion Equation Can Be Used to Predict the Probability of an Unordered Combination of Outcomes

Another predictive problem in genetics is to determine the probability that a certain proportion of offspring will be produced with particular characteristics; in such a case, they can be produced in an unspecified order. For example, we can consider a group of children produced by two heterozygous brown-eyed (*Bb*) individuals. We can ask: What is the probability that two out of five children will have blue eyes?

In this case, we are not concerned with the order in which the offspring are born. Instead, we are only concerned with the final numbers of blue-eyed and brown-eyed offspring. One possible outcome would be the following: firstborn child with blue eyes, second child with blue eyes, and then the next three with brown eyes. Another possible outcome could be firstborn child with brown eyes, second with blue eyes, third with brown eyes, fourth with blue eyes, and fifth with brown eyes. Both of these scenarios would result in two offspring with blue eyes and three with brown eyes. In fact, several other ways to have such a family could occur.

To solve this type of question, the **binomial expansion equation** can be used. This equation represents all of the possibilities for a given set of two unordered events.

$$P = \frac{n!}{x!(n-x)!} p^x q^{n-x}$$

where

P = the probability that the unordered combination of outcomes will occur

n = total number of outcomes

x = number of outcomes in one category (e.g., blue eyes)

p = individual probability of x

q = individual probability of the other category (e.g., brown eyes)

Note: In this case, $p + q = 1$.

The symbol ! denotes a factorial. The factorial $n!$ is the product of all integers from n down to 1. For example, $4! = 4 \times 3 \times 2 \times 1 = 24$. An exception is 0!, which equals 1.

The use of the binomial expansion equation is described next.

The Cross: *Bb* × *Bb*

The Question: What is the probability that two out of five offspring will have blue eyes?

Step 1. *Calculate the individual probabilities of the blue-eye and brown-eye phenotypes.* If we construct a Punnett square, we find the probability of blue eyes is 1/4 and the probability of brown eyes is 3/4:

$$p = 1/4$$
$$q = 3/4$$

Step 2. *Specify the number of outcomes in category x (in this case, blue eyes) and the total number of outcomes.* In this example, the number of outcomes in category x is two blue-eyed children among a total number of five.

$$x = 2$$
$$n = 5$$

Step 3. *Substitute the values for p, q, x, and n in the binomial expansion equation.*

$$P = \frac{n!}{x!(n-x)!} p^x q^{n-x}$$

$$P = \frac{5!}{2!(5-2)!} (1/4)^2 (3/4)^{5-2}$$

$$P = \frac{5 \times 4 \times 3 \times 2 \times 1}{(2 \times 1)(3 \times 2 \times 1)} (1/16)(27/64)$$

$$P = 0.26, \text{ or } 26\%$$

Thus, the probability is 0.26 that two out of five offspring will have blue eyes. In other words, 26% of the time we expect a $Bb \times Bb$ cross yielding five offspring to have two blue-eyed children and three brown-eyed children.

If a cross involves more than two categories of offspring, we can use an expanded version of this approach that uses the **multinomial expansion equation.** A general expression for this equation is

$$P = \frac{n!}{a!b!c!\cdots} p^a q^b r^c \cdots$$

where P = the probability that a combination of unordered outcomes will occur.

n = total number of outcomes

$$a + b + c + \cdots = n$$

$$p + q + r + \cdots = 1$$

(p is the likelihood of a, q is the likelihood of b, r is the likelihood of c, and so on).

The multinomial expansion equation can be useful in many genetic problems where more than two combinations of offspring are possible. For example, this formula can be used to solve problems involving a two-factor cross in which there are four categories of offspring.

The Chi Square Test Is Used to Test the Validity of a Genetic Hypothesis

Let's now consider a different issue in genetic problems, namely **hypothesis testing.** Our goal here is to determine if the data from genetic crosses are consistent with a particular pattern of inheritance. For example, a geneticist may study the inheritance of body color and wing shape in fruit flies over the course of two generations. The following question may be asked about the F_2 generation: Do the observed numbers of offspring agree with the predicted numbers based on Mendel's laws of segregation and independent assortment? As we will see in Chapters 4 through 8, not all traits follow a simple Mendelian pattern of inheritance. Some genes do not segregate and independently assort themselves in the same way that Mendel's seven characters did in pea plants.

To distinguish between inheritance patterns that obey Mendel's laws versus those that do not, a conventional strategy is to make crosses and then quantitatively analyze the offspring. Based on the observed outcome, an experimenter may make a tentative hypothesis. For example, the data may seem to obey Mendel's laws. Hypothesis testing provides an objective, statistical method to evaluate whether the observed data really agree with the hypothesis. In other words, we use statistical methods to determine whether the data that have been gathered from crosses are consistent with predictions based on quantitative laws of inheritance.

The rationale behind a statistical approach is to evaluate the **goodness of fit** between the observed data and the data that are predicted from a hypothesis. This is called a **null hypothesis** when it assumes there is no real difference between the observed and expected values. Any actual differences that occur are presumed to be due to random sampling error. If the observed and predicted data are very similar, we can conclude that the hypothesis is consistent with the observed outcome. In this case, it is reasonable to accept the null hypothesis. However, it should be emphasized that this does not prove a hypothesis is correct. Statistical methods can never prove that a hypothesis is correct. They can provide insight about whether or not the observed data seem reasonably consistent with the hypothesis. Alternative hypotheses, perhaps even ones that the experimenter has failed to realize, may also be consistent with the data. In some cases, statistical methods may reveal a poor fit between hypothesis and data. In other words, a high deviation is found between the observed and expected values. If this occurs, the null hypothesis is rejected. Hopefully, the experimenter can subsequently propose an alternative hypothesis that has a better fit with the data.

One commonly used statistical method for determining goodness of fit is the **chi square test** ("chi square" is often written χ^2). We can use the chi square test to analyze population data in which the members of the population fall into different categories. We typically have this kind of data when we evaluate the outcomes of genetic crosses, because these usually produce a population of offspring that differ with regard to phenotypes. The general formula for the chi square test is

$$\chi^2 = \Sigma \frac{(O - E)^2}{E}$$

where

O = observed data in each category

E = expected data in each category based on the experimenter's hypothesis

Σ means to sum the data values for each category. For example, if the population data fell into two categories, the chi square calculation would be

$$\chi^2 = \frac{(O_1 - E_1)^2}{E_1} + \frac{(O_2 - E_2)^2}{E_2}$$

We can use the chi square test to determine if a genetic hypothesis is consistent with the observed outcome of a genetic cross. The strategy described next provides a step-by-step outline for applying the chi square test. In this problem, the experimenter wants to determine if a two-factor cross obeys Mendel's laws. The experimental organism is *Drosophila melanogaster* (the common fruit fly), and the two characters involve wing shape and body color. Straight wing shape and curved wing shape are designated by c^+ and c, respectively; gray body color and ebony body color are designated by e^+ and e, respectively. Note: In certain species, such as *D. melanogaster*, the convention is to designate the common (wild-type) allele with a plus sign. Recessive mutant alleles are designated with lowercase letters and dominant mutant alleles with capital letters.

The Cross: A true-breeding fly with straight wings and a gray body ($c^+c^+e^+e^+$) is crossed to a true-breeding fly with curved wings and an ebony body (*ccee*). The flies of the F_1 generation are then allowed to mate with each other to produce an F_2 generation.

The Outcome:

F_1 generation:	All offspring have straight wings and gray bodies
F_2 generation:	193 straight wings, gray bodies
	69 straight wings, ebony bodies
	64 curved wings, gray bodies
	26 curved wings, ebony bodies
Total:	352

Step 1. *Propose a hypothesis that allows us to calculate the expected values based on Mendel's laws.* The F_1 generation suggests that the trait of straight wings is dominant to curved wings and gray body coloration is dominant to ebony. Looking at the F_2 generation, it appears that the data show a 9:3:3:1 ratio. If so, this is consistent with an independent assortment of the two characters.

Based on these observations, the hypothesis is

Straight (c^+) is dominant to curved (c), and gray (e^+) is dominant to ebony (e). The two characters segregate and assort independently from generation to generation.

Step 2. *Based on the hypothesis, calculate the expected values of the four phenotypes.* We first need to calculate the individual probabilities of the four phenotypes. According to our hypothesis, the ratio for the F_2 generation should be 9:3:3:1. Therefore, the expected probabilities are

9/16 = straight wings, gray bodies
3/16 = straight wings, ebony bodies
3/16 = curved wings, gray bodies
1/16 = curved wings, ebony bodies

The observed F_2 generation contained a total of 352 individuals. Our next step is to calculate the expected numbers of each type of offspring when the total equals 352. This can be accomplished by multiplying each individual probability by 352.

9/16 × 352 = 198 (expected number with straight wings, gray bodies)
3/16 × 352 = 66 (expected number with straight wings, ebony bodies)
3/16 × 352 = 66 (expected number with curved wings, gray bodies)
1/16 × 352 = 22 (expected number with curved wings, ebony bodies)

Step 3. *Apply the chi square formula, using the data for the expected values that have been calculated in step 2.* In this case, the data include four categories, and thus the sum has four terms.

$$\chi^2 = \frac{(O_1 - E_1)^2}{E_1} + \frac{(O_2 - E_2)^2}{E_2} + \frac{(O_3 - E_3)^2}{E_3} + \frac{(O_4 - E_4)^2}{E_4}$$

$$\chi^2 = \frac{(193 - 198)^2}{198} + \frac{(69 - 66)^2}{66} + \frac{(64 - 66)^2}{66} + \frac{(26 - 22)^2}{22}$$

$$\chi^2 = 0.13 + 0.14 + 0.06 + 0.73 = 1.06$$

Step 4. *Interpret the calculated chi square value. This is done using a chi square table.*

In order to interpret the chi square value we obtained, we must understand how to use **Table 3.2.** The probabilities, called ***P* values,** listed in the chi square table allow us to determine the likelihood that the amount of variation indicated by a given chi square value is due to random chance alone, based on a particular hypothesis. For example, let's consider a value (0.00393) listed in row 1. (The meaning of the rows will be explained shortly.) Chi square values that are equal to or greater than 0.00393 are expected to occur 95% of the time when a hypothesis is correct. In other words, 95 out of 100 times we expect that random chance alone will produce a deviation between the experimental data and the hypothesized model that is equal to or greater than 0.00393. A low chi square value indicates a high probability that the observed deviation could be due to random chance alone. By comparison, chi square values that are equal to or greater than 3.841 are expected to occur less than 5% of the time due to random sampling error. If a high chi square value is obtained, an experimenter becomes suspicious that the high deviation has occurred because the null hypothesis is incorrect and that the deviation is not attributed to chance alone.

A common convention is to reject the null hypothesis if the chi square value results in a probability that is less than 0.05 (less than 5%) or if the probability is less than 0.01 (less than 1%). These are called the 5% and 1% significance levels, respectively. Which level is better to choose? The choice is somewhat subjective. If you choose a 5% level rather than a 1% level, a disadvantage is that you are more likely to reject a hypothesis that happens to be correct. Even so, choosing a 5% level rather than a 1% level has the advantage that you are less likely to accept an incorrect hypothesis.

For our problem involving flies with straight or curved wings and gray or ebony bodies, we have calculated a chi square value of 1.06. Before we can determine the probability that this deviation occurred as a matter of random chance, we must first determine the degrees of freedom (*df*) in this experiment. The **degrees of freedom** is a measure of the number of categories that are independent of each other. When phenotype categories are derived from a Punnett square, it is typically $n - 1$, where n equals the total number of categories. In our fruit fly problem, $n = 4$ (the categories are the phenotypes: straight wings and gray body; straight wings and ebony body; curved wings and gray body; and curved wings and ebony body); thus, the degrees of freedom equals 3.* We now have sufficient information to interpret our chi square value of 1.06.

With $df = 3$, the chi square value of 1.06 we obtained is slightly greater than 1.005, which gives a *P* value of 0.80, or 80%. What does this *P* value mean? If the hypothesis is correct, chi

---

*If the hypothesis had assumed that the law of segregation is obeyed, the degrees of freedom would be 1 (see Chapter 7).

TABLE 3.2
Chi Square Values and Probability

Degrees of Freedom	P = 0.99	0.95	0.80	0.50	0.20	Null Hypothesis Rejected 0.05	0.01
1.	0.000157	0.00393	0.0642	0.455	1.642	3.841	6.635
2.	0.020	0.103	0.446	1.386	3.219	5.991	9.210
3.	0.115	0.352	1.005	2.366	4.642	7.815	11.345
4.	0.297	0.711	1.649	3.357	5.989	9.488	13.277
5.	0.554	1.145	2.343	4.351	7.289	11.070	15.086
6.	0.872	1.635	3.070	5.348	8.558	12.592	16.812
7.	1.239	2.167	3.822	6.346	9.803	14.067	18.475
8.	1.646	2.733	4.594	7.344	11.030	15.507	20.090
9.	2.088	3.325	5.380	8.343	12.242	16.919	21.666
10.	2.558	3.940	6.179	9.342	13.442	18.307	23.209
15.	5.229	7.261	10.307	14.339	19.311	24.996	30.578
20.	8.260	10.851	14.578	19.337	25.038	31.410	37.566
25.	11.524	14.611	18.940	24.337	30.675	37.652	44.314
30.	14.953	18.493	23.364	29.336	36.250	43.773	50.892

Source: Fisher, Ronald and Yates, Frank (1943) *Statistical Tables for Biological, Agricultural and Medical Research.* London, U.K.: Oliver and Boyd, 98.

square values equal to or greater than 1.005 are expected to occur 80% of the time based on random chance alone. To reject the null hypothesis at the 5% significance level, the chi square would have to be greater than 7.815. Because it is actually far less than this value, we accept that the hypothesis is correct.

Again, keep in mind that the chi square test does not prove a hypothesis is correct. It is a statistical method for evaluating whether the data and hypothesis have a good fit.

3.6 REVIEWING THE KEY CONCEPTS

- Probability is the number of times an outcome occurs divided by the total number of outcomes.
- According to the product rule, the probability that two or more independent outcomes will occur is equal to the product of their individual probabilities. This rule can be used to predict the outcome of crosses involving two or more genes.
- The binomial expansion equation is used to predict the probability of a given set of two unordered outcomes.
- The chi square test is used to test the validity of a hypothesis (see Table 3.2).

3.6 COMPREHENSION QUESTIONS

1. A cross is made between *AA Bb Cc Dd* and *Aa Bb cc dd* individuals. Rather than making a very large Punnett square, which statistical operation could you use to solve this problem, and what would be the probability that the cross produces an offspring that is *AA bb Cc dd*?
   a. Product rule, 1/32
   b. Product rule, 1/4
   c. Binomial expansion, 1/32
   d. Binomial expansion, 1/4

2. In dogs, brown fur color (*B*) is dominant to white (*b*). A cross is made between two heterozygotes for fur color. If the litter contains six pups, what is the probability that half of them will be white?
   a. 0.066, or 6.6%
   b. 0.13, or 13%
   c. 0.25, or 25%
   d. 0.26, or 26%

3. Which of the following operations could be used for hypothesis testing?
   a. Product rule
   b. Binomial expansion
   c. Product rule and the binomial expansion
   d. Chi square test

KEY TERMS

Page 41. pangenesis, blending hypothesis of inheritance
Page 42. crossed, hybridization, hybrids, gamete
Page 43. sperm, pollen grains, anthers, eggs, ovules, ovaries, stigma, self-fertilization
Page 44. cross-fertilization, characters, trait, variant, true-breeding line, strain
Page 45. single-factor cross, monohybrids, empirical approach, empirical laws, parental generation (P generation), F_1 generation, F_2 generation
Page 47. dominant, recessive, particulate theory of inheritance, segregate, gene, allele, Mendel's law of segregation, homozygous, heterozygous
Page 48. genotype, phenotype, Punnett square
Page 50. two-factor crosses
Page 51. nonparentals
Page 52. Mendel's law of independent assortment
Page 53. multiplication method, forked-line method
Page 55. chromosome theory of inheritance
Page 56. genetic recombination
Page 58. pedigree analysis
Page 60. probability, random sampling error, product rule
Page 61. binomial expansion equation
Page 62. multinomial expansion equation, hypothesis testing, goodness of fit, null hypothesis, chi square test
Page 63. *P* values, degrees of freedom

CHAPTER SUMMARY

- Early ideas regarding inheritance of traits included pangenesis and the blending hypothesis of inheritance. These ideas were later refuted by the work of Mendel.

3.1 Mendel's Study of Pea Plants

- Mendel chose pea plants as his experimental organism (see Figures 3.1, 3.2).
- Using pea plants, it was easy for Mendel to carry out self-fertilization or cross-fertilization experiments (see Figure 3.3).
- Another advantage of using pea plants is they were available in several varieties in which a character existed in two distinct variants (see Figure 3.4).

3.2 Law of Segregation

- Mendel conducted single-factor crosses in which he followed the variants for a single character (see Figure 3.5).
- The results of his single-factor crosses showed that the dominant trait was observed in all individuals in the F_1 generation and displayed in a 3:1 ratio in the F_2 generation.
- Based on the results of his single-factor crosses, Mendel proposed three key ideas regarding inheritance: (1) Traits may be dominant or recessive. (2) Genes are passed unaltered from generation to generation. (3) The two copies of a given gene segregate (or separate) from each other during transmission from parent to offspring. This third idea is known as the law of segregation (see Figure 3.6).
- A Punnett square can be used to predict the outcome of a cross or self-fertilization experiment.

3.3 Law of Independent Assortment

- By conducting two-factor crosses, Mendel formulated the law of independent assortment, which states that two different genes randomly assort their alleles during the formation of haploid cells (see Figures 3.7–3.9).
- A Punnett square can be used to predict the outcomes of two-factor crosses (see Figure 3.10).
- The multiplication and forked-line methods are used to predict the outcomes of crosses involving three or more genes (see Figure 3.11).

3.4 Chromosome Theory of Inheritance

- The chromosome theory of inheritance describes how the transmission of chromosomes can explain Mendel's laws (see Table 3.1).
- Mendel's law of segregation is explained by the separation of homologs during meiosis (see Figure 3.12).
- Mendel's law of independent assortment is explained by the random alignment of different chromosomes during metaphase of meiosis I (see Figure 3.13).

3.5 Studying Inheritance Patterns in Humans

- Human inheritance patterns are determined by analyzing family trees known as pedigrees (see Figure 3.14).

3.6 Probability and Statistics

- Probability is the number of times an outcome occurs divided by the total number of outcomes.
- According to the product rule, the probability that two or more independent outcomes will occur is equal to the product of their individual probabilities. This rule can be used to predict the outcome of crosses involving two or more genes.
- The binomial expansion equation is used to predict the probability of a given set of two unordered outcomes.
- The chi square test is used to test the validity of a hypothesis (see Table 3.2).

PROBLEM SETS & INSIGHTS

More Genetic TIPS

1. As described in this chapter, a human disease known as cystic fibrosis is inherited as a recessive trait. Two unaffected individuals have a first child with the disease. What is the probability that their next two children will *not* have the disease?

Topic: What topic in genetics does this question address?

The topic is Mendelian inheritance. More specifically, the question is about a single-factor cross involving cystic fibrosis.

Information: What information do you know based on the question and your understanding of the topic?

From the question, you know that both parents are unaffected, but they produced an affected offspring who must be homozgyous for the recessive allele. Therefore, both parents must be heterozygotes. If C is the common (non-disease-causing) allele and c is the disease-causing allele, the genotype of each parent must be Cc. From your understanding of the topic, you may remember how to use a Punnett square to predict the outcome of a cross. You may also realize that the phenotypes of offspring are independent outcomes and that the product rule can be used to solve this type of problem.

Problem-Solving Strategies: Predict the outcome. Make a calculation.

As shown next, you can make a Punnett square to predict the ratio of affected to unaffected offspring.

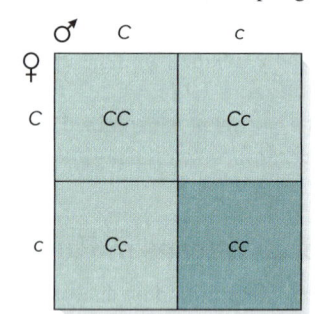

	C	c
C	CC	Cc
c	Cc	cc

C = common allele
c = cystic fibrosis allele

To calculate the probability that the parents will have two unaffected offspring in a row, you need to consider two things. First, you need to know the probability of having an unaffected offspring. This probability can be deduced from the Punnett square. You then use the product rule to calculate the likelihood of having two unaffected offspring in a row.

Answer: The genotypes of the offspring are in the ratio 1 CC : 2 Cc : 1 cc. The ratio of the phenotypes is 3 unaffected : 1 affected with cystic fibrosis.

The probability of a single unaffected offspring is

$$P_{\text{unaffected}} = 3/(3+1) = 3/4$$

To obtain the probability of having two unaffected offspring in a row (i.e., in a specified order), you apply the product rule.

$$3/4 \times 3/4 = 9/16 = 0.56, \text{ or } 56\%$$

The chance that the couple's next two children will be unaffected is 56%.

2. In dogs, black fur color is dominant to white. Two heterozygous black dogs are mated. What is the probability of the following combinations of offspring?

A. A litter of six pups, four with black fur and two with white fur

B. A first litter of six pups, four with black fur and two with white fur, and then a second litter of seven pups, five with black fur and two with white fur

Topic: What topic in genetics does this question address?

The topic is Mendelian inheritance. More specifically, the question is about a single-factor cross involving fur color in dogs.

Information: What information do you know based on the question and your understanding of the topic?

From the question, you know that black fur is dominant to white and that the parents are heterozygotes. If B is the black allele and b is the white allele, the genotype of each parent must be Bb. From your understanding of the topic, you may remember how to use a Punnett square to predict the outcome of a cross. You may also realize that each litter is an unordered combination of two different outcomes and, therefore, you can use the binomial expansion equation to calculate the probability for each litter.

Problem-Solving Strategies: Predict the outcome. Make a calculation.

To begin this problem, you need to know the probability of producing black offspring compared to white offspring. This can be deduced from a Punnett square, which is shown next.

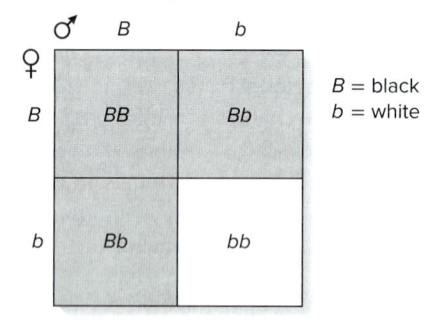

	B	b
B	BB	Bb
b	Bb	bb

B = black
b = white

For part A of the question, you can derive probabilities for black and white fur from the Punnett square, and then use those values in the binomial expansion equation. For part B, you need two types of calculations. To determine the probability of each litter occurring, you can use the binomial expansion equation. Because each litter is an independent outcome, you can multiply the probability of the first litter times the probability of the second litter to get the probability of both litters occurring in this order.

Answer: From the Punnett square, you can deduce that the probability of black fur is 3/4, or 0.75, and the probabilty of white fur is 1/4, or 0.25.

A. Because this is an unordered combination of outcomes, you use the binomial expansion equation, where $n = 6$, $x = 4$, $p = 0.75$ (probability of black), and $q = 0.25$ (probability of white).

The answer is that such a litter will occur 0.297, or 29.7%, of the time.

B. The two litters occur in a row, so they are independent outcomes. Therefore, you use the product rule, and multiply the probability of the first litter times the probability of the second litter. You need to use the binomial expansion equation for each litter.

(binomial expansion of the first litter)(binomial expansion of the second litter)

For the first litter, $n = 6$, $x = 4$, $p = 0.75$, $q = 0.25$. For the second litter, $n = 7$, $x = 5$, $p = 0.75$, $q = 0.25$.

The answer is that two such litters will occur in this order 0.092, or 9.2%, of the time.

3. A cross was made between a plant that has blue flowers and purple seeds and a plant with white flowers and green seeds. The F_1 generation was then allowed to self-fertilize. The following data were obtained:

F_1 generation:	All offspring have blue flowers with purple seeds.
F_2 generation:	208 blue flowers, purple seeds
	13 blue flowers, green seeds
	19 white flowers, purple seeds
	60 white flowers, green seeds
Total:	300

Start with the hypothesis that blue flowers and purple seeds are dominant traits and that the two genes assort independently. Calculate a chi square value. What does this value mean with regard to your hypothesis? If you decide to reject your hypothesis, which aspect of the hypothesis do you think is incorrect (i.e., that blue flowers and purple seeds are dominant traits or that the two genes assort independently)?

Topic: What topic in genetics does this question address?

The topic is hypothesis testing. More specifically, the question is about evaluating the dominant/recessive relationships of two genes and determining if they are obeying the law of independent assortment.

Information: What information do you know based on the question and your understanding of the topic?

From the question, you know the outcome of a two-factor cross. You are given a starting hypothesis. From your understanding of the topic, you may remember that this type of experiment should produce a 9:3:3:1 ratio of the four types of offspring, according to the law of independent assortment. Alternatively, you could set up a Punnett square to predict the outcome for the F_2 generation. You may also remember that a chi square test can be used to evaluate the validity of a hypothesis.

Problem-Solving Strategy: Analyze data.

One strategy to solve this problem is to analyze the data using the chi square test, which compares observed and expected data. The expected data are predicted from the hypothesis.

Answer: The hypothesis is that blue flowers and purple seeds are dominant traits and they are governed by two genes that assort independently. According to this hypothesis, the F_2 generation should display a ratio of 9 blue flowers, purple seeds : 3 blue flowers, green seeds : 3 white flowers, purple seeds : 1 white flower, green seeds. Because a total of 300 offspring were produced, the expected numbers are

$9/16 \times 300 = 169$ blue flowers, purple seeds

$3/16 \times 300 = 56$ blue flowers, green seeds

$3/16 \times 300 = 56$ white flowers, purple seeds

$1/16 \times 300 = 19$ white flowers, green seeds

$$\chi^2 = \frac{(208-169)^2}{169} + \frac{(13-56)^2}{56} + \frac{(19-56)^2}{56} + \frac{(60-19)^2}{19}$$

$$\chi^2 = 154.9$$

Looking up this value in the chi square table under 3 degrees of freedom, you see that it is much higher than would be expected 1% of the time by chance alone. Therefore, you reject the hypothesis. The idea that the two genes are assorting independently seems to be incorrect. The F_1 generation supports the idea that blue flowers and purple seeds are dominant traits. Note: We will discuss why independent assortment may not occur in Chapter 7.

Conceptual Questions

C1. Why did Mendel's work refute the idea of blending inheritance?

C2. What is the difference between cross-fertilization and self-fertilization?

C3. Describe the difference between genotype and phenotype. Give three examples. Is it possible for two individuals to have the same phenotype but different genotypes?

C4. With regard to genotypes, what is a true-breeding organism?

C5. How can you determine whether an organism is heterozygous or homozygous for a dominant trait?

C6. In your own words, describe what Mendel's law of segregation means. Do not use the word *segregation* in your answer.

C7. Based on genes in pea plants that we have considered in this chapter, which statement(s) is/are incorrect?

A. The gene causing tall plants is an allele of the gene causing dwarf plants.

B. The gene causing tall plants is an allele of the gene causing purple flowers.

C. The alleles causing tall plants and purple flowers are dominant.

C8. For a cross between a heterozygous tall pea plant and a dwarf plant, predict the ratios of genotypes and phenotypes in the offspring.

C9. Do you know the genotype of an individual with a recessive trait or a dominant trait? Explain your answer.

C10. A cross is made between a pea plant that has constricted pods (a recessive trait; smooth is dominant) and is heterozygous for seed color (yellow is dominant to green) and a plant that is heterozygous for both pod texture and seed color. Construct a Punnett square that depicts this cross. What are the predicted outcomes of genotypes and phenotypes of the offspring?

C11. A pea plant that is heterozygous with regard to seed color (yellow is dominant to green) is allowed to self-fertilize. What are the predicted outcomes of genotypes and phenotypes of the offspring?

C12. Describe the significance of nonparentals with regard to the law of independent assortment. In other words, explain how the appearance of nonparentals refutes the hypothesis of linked assortment.

C13. For the following two pedigrees, describe what you think is the more likely inheritance pattern (dominant or recessive). Explain your reasoning. Filled (black) symbols indicate affected individuals.

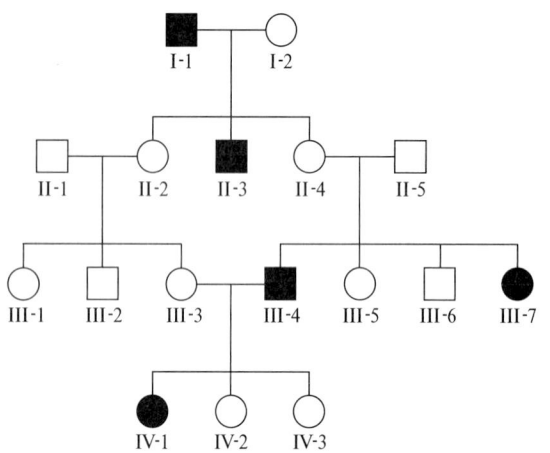

(a)

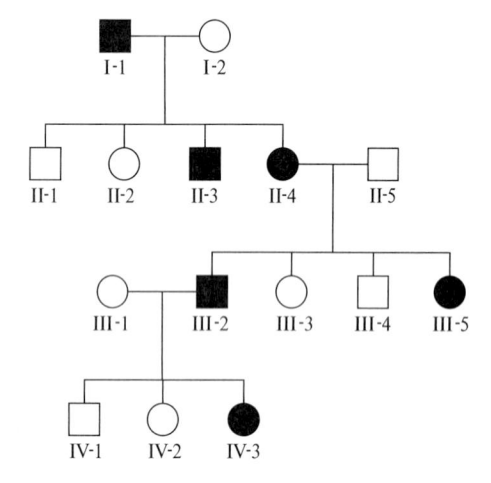

(b)

C14. Ectrodactyly, also known as "lobster claw syndrome," is a recessive disorder in humans. If a phenotypically unaffected couple produces an affected offspring, what are the following probabilities?

A. Both parents are heterozygotes.

B. The next offspring is a heterozygote.

C. The next three offspring will be phenotypically unaffected.

D. Any two out of the next three offspring will be phenotypically unaffected.

C15. Identical twins are produced from the same sperm and egg (which splits after the first mitotic division), whereas fraternal twins are produced from separate sperm and separate egg cells. If two parents with brown eyes (a dominant trait) produce one twin boy with blue eyes, what are the following probabilities?

A. If the other twin is identical, he will have blue eyes.

B. If the other twin is fraternal, he or she will have blue eyes.

C. If the other twin is fraternal, he or she will transmit the blue eye allele to his or her offspring.

D. The parents are both heterozygotes.

C16. In cocker spaniels, solid coat color is dominant over spotted coat color. If two heterozygous dogs were crossed to each other, what would be the probability of the following combinations of offspring?

A. A litter of five pups, four with solid fur and one with spotted fur

B. A first litter of six pups, four with solid fur and two with spotted fur, and then a second litter of five pups, all with solid fur

C. A first litter of five pups, the firstborn with solid fur, and then among the next four, three with solid fur and one with spotted fur, and then a second litter of seven pups in which the firstborn is spotted, the second born is spotted, and the remaining five are composed of four solid and one spotted animal

D. A litter of six pups, the firstborn with solid fur, the second born spotted, and among the remaining four pups, two with spotted fur and two with solid fur

C17. Crosses were made between a white male dog and two different black females. The first female gave birth to eight black pups, and the second female gave birth to four white and three black pups. What are the likely genotypes of the male parent and the two female parents? Explain whether you are uncertain about any of the genotypes.

C18. In humans, the allele for brown eye color (B) is dominant to that for blue eye color (b). If two heterozygous parents produce children, what are the following probabilities?

A. The first two children have blue eyes.

B. Among a total of four children, two have blue eyes and the other two have brown eyes.

C. The first child has blue eyes, and the next two have brown eyes.

C19. Albinism, a condition characterized by a partial or total lack of skin pigment, is a recessive human trait. If a phenotypically unaffected couple produced an albino child, what is the probability that their next child will be albino?

C20. A true-breeding tall plant was crossed to a dwarf plant. Tallness is a dominant trait. The F_1 individuals were allowed to self-fertilize. What are the following probabilities for the F_2 generation?

A. The first plant is dwarf.

B. The first plant is dwarf or tall.

C. The first three plants are tall.

D. For any seven plants, three are tall and four are dwarf.

E. The first plant is tall, and then among the next four, two are tall and the other two are dwarf.

C21. For pea plants with the following genotypes, list the possible gametes that each plant can make.

A. *TT Yy Rr*

B. *Tt YY rr*

C. *Tt Yy Rr*

D. *tt Yy rr*

C22. An individual has the genotype *Aa Bb Cc* and makes an abnormal gamete with the genotype *AaBc*. Does this gamete violate the law of independent assortment or the law of segregation (or both)? Explain your answer.

C23. In people with maple syrup urine disease, the body is unable to metabolize the amino acids leucine, isoleucine, and valine. One of the symptoms is that the urine smells like maple syrup. An unaffected couple produced six children in the following order: unaffected daughter, affected daughter, unaffected son, unaffected son, affected son, and unaffected son. The youngest unaffected son and an unaffected woman have three children in the following order: affected daughter, unaffected daughter, and unaffected son. Draw a pedigree that describes this family. What type of inheritance (dominant or recessive) would you propose to explain maple syrup urine disease?

C24. Marfan syndrome is a rare inherited human disorder characterized by unusually long limbs and digits plus defects in the heart (especially the aorta) and the eyes, among other symptoms. Following is a pedigree for this disorder. Affected individuals are shown with filled (black) symbols. What type of inheritance pattern do you think is most likely?

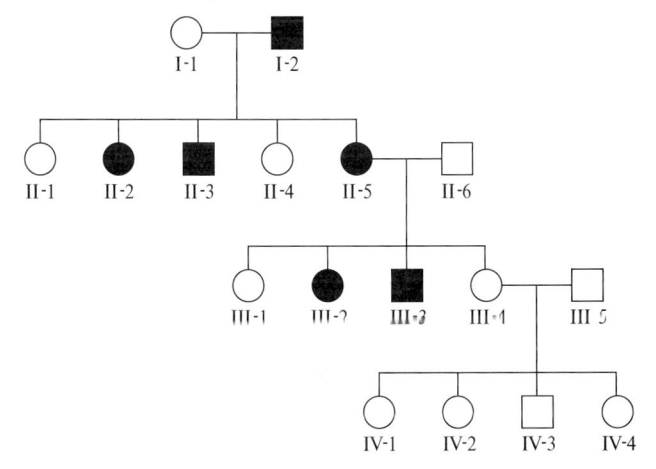

C25. A true-breeding pea plant with round and green seeds was crossed to a true-breeding plant with wrinkled and yellow seeds. Round and yellow seeds are the dominant traits. The F_1 plants were allowed to self-fertilize. What are the following probabilities for the F_2 generation?

A. A wrinkled, yellow seed

B. Three out of three seeds that are round and yellow

C. Five seeds in the following combination: two round, yellow, one round green, and two wrinkled green

D. A seed that is not round and yellow

C26. A true-breeding tall pea plant was crossed to a true-breeding dwarf plant. What is the probability that an F_1 individual will be true-breeding? What is the probability that an F_1 individual will be a true-breeding tall plant?

C27. What are the expected phenotypic ratios from the following cross: *Tt Rr yy Aa × Tt rr YY Aa*, where *T* = tall, *t* = dwarf, *R* = round, *r* = wrinkled, *Y* = yellow, *y* = green, *A* = axial, *a* = terminal; *T*, *R*, *Y*, and *A* are dominant alleles. *Hint*: See Figure 3.11 for help.

C28. On rare occasions, an organism may have three copies of a gene (instead of the usual number of two copies). The alleles for the gene usually segregate so that a gamete contains one or two copies of the gene. Let's suppose that a rare pea plant has three copies of the height gene. Its genotype is *TTt*. The plant is also heterozygous for the seed color gene, *Yy*. How many types of gametes can this plant make, and in what proportions? (Assume that it is equally likely that a gamete will contain one or two copies of the height gene.)

C29. Honeybees are unusual in that male bees (drones) have only one copy of each gene, but female bees have two copies of their genes. This difference arises because drones develop from eggs that have not been fertilized by sperm cells. In bees, the trait of long wings is dominant over short wings, and the trait of black eyes is dominant over white eyes. If a drone with short wings and black eyes was mated to a queen bee that is heterozygous for both genes, what are the predicted genotypes and phenotypes of male and female offspring? What are the phenotypic ratios if we assume an equal number of male and female offspring?

C30. A pea plant that is dwarf with green, wrinkled seeds is crossed to a true-breeding plant that is tall with yellow, round seeds. The F_1 generation is then allowed to self-fertilize. What types of gametes, and in what proportions, will the F_1 generation make? What will be the ratios of genotypes and phenotypes of the F_2 generation?

C31. A true-breeding plant with round and green seeds is crossed to a true-breeding plant with wrinkled and yellow seeds. The F_1 plants are allowed to self-fertilize. What is the probability of obtaining the following plants in the F_2 generation: two that have round, yellow seeds; one with round, green seeds; and two with wrinkled, green seeds?

C32. Woolly hair is a rare dominant trait found in people of Scandinavian descent in which the hair resembles the wool of a sheep. A male with woolly hair, who has a mother with straight hair, moves to an island that is inhabited by people who are not of Scandinavian descent. Assuming that no other Scandinavians immigrate to the island, what is the probability that a great-grandchild of this male will have woolly hair? (*Hint:* You may want to draw a pedigree to help you figure this out.) If this woolly-haired male has eight great-grandchildren, what is the probability that one out of eight will have woolly hair?

C33. Huntington disease is a rare dominant trait that causes neurodegeneration later in life. A man in his thirties, who already has three children, discovers that his mother has Huntington disease though his father is unaffected. What are the following probabilities?

A. That the man in his thirties will develop Huntington disease

B. That his first child will develop Huntington disease

C. That one out of three of his children will develop Huntington disease

C34. A woman with achondroplasia (a dominant form of dwarfism) and a phenotypically unaffected man have seven children, all of whom have achondroplasia. What is the probability of producing such a family if this woman is a heterozygote? What is the probability that the woman is a heterozygote if her eighth child does not have this disorder?

Application and Experimental Questions

E1. Describe three advantages of using pea plants as an experimental organism.

E2. Explain the technical differences between a cross-fertilization experiment and a self-fertilization experiment.

E3. How long did it take Mendel to complete the experiment in Figure 3.5?

E4. For all seven characters described in the data of Figure 3.5, Mendel allowed the F_2 plants to self-fertilize. He found that when F_2 plants with recessive traits were crossed to each other, they always bred true. However, when F_2 plants with dominant traits were crossed, some bred true but others did not. A summary of Mendel's results is shown in the following table.

The Ratio of True-Breeding and Non-True-Breeding Parents of the F_2 Generation

F_2 Parents	True-Breeding	Non-True-Breeding	Ratio
Round	193	372	1:1.93
Yellow	166	353	1:2.13
Gray	36	64	1:1.78
Smooth	29	71	1:2.45
Green	40	60	1:1.5
Axial	33	67	1:2.08
Tall	28	72	1:2.57
TOTAL:	525	1059	1:2.02

When considering the data in this table, keep in mind that they describe the characteristics of the F_2 generation parents that had displayed a dominant phenotype. These data were deduced by analyzing the outcome of the F_3 generation. Based on Mendel's laws, explain why the ratios were approximately 1:2.

E5. From the point of view of crosses and data collection, what are the experimental differences between single-factor and two-factor crosses?

E6. As in many animals, albino coat color is a recessive trait in guinea pigs. Researchers removed the ovaries from an albino female guinea pig and then transplanted ovaries from a true-breeding black guinea pig. They then mated this albino female (with the transplanted ovaries) to an albino male. The albino female produced three offspring. What were their coat colors? Explain the results.

E7. The fungus *Melampsora lini* causes a disease known as flax rust. Different strains of *M. lini* cause varying degrees of the disease. Conversely, different strains of flax are resistant or sensitive to the various varieties of the fungus. The Bombay variety of flax is resistant to *M. lini* strain 22 but sensitive to *M. lini* strain 24. A strain of flax called 770B is just the opposite; it is resistant to strain 24 but sensitive to strain 22. When 770B was crossed to Bombay, all F_1 individuals were resistant to both strain 22 and strain 24. When F_1 individuals were self-fertilized, the following data were obtained:

   43 resistant to strain 22 but sensitive to strain 24

   9 sensitive to strain 22 and strain 24

   32 sensitive to strain 22 but resistant to strain 24

   110 resistant to strain 22 and strain 24

Explain the inheritance pattern for flax resistance and sensitivity to *M. lini* strains.

E8. For Mendel's data shown in Figure 3.8, conduct a chi square analysis to determine if the data agree with Mendel's law of independent assortment.

E9. Would it be possible to deduce the law of independent assortment by analyzing the results of single-factor crosses? Explain your answer.

E10. In fruit flies, curved wings are recessive to straight wings, and ebony body is recessive to gray body. A cross was made of true-breeding flies with curved wings and gray bodies and flies with straight wings and ebony bodies. The F_1 offspring were then mated to flies with curved wings and ebony bodies to produce an F_2 generation.

A. Diagram the genotypes of this cross, starting with the parental generation and ending with the F_2 generation.

B. What is the predicted phenotypic ratio of the F_2 generation?

C. Let's suppose the following data were obtained for the F_2 generation:

   114 curved wings, ebony body

   105 curved wings, gray body

   111 straight wings, gray body

   114 straight wings, ebony body

Conduct a chi square analysis to determine if the experimental data are consistent with the expected outcome based on Mendel's laws.

E11. A recessive allele in mice results in an unusually long neck. Sometimes, during early embryonic development, the long neck causes the embryo to die. An experimenter began with a population of true-breeding mice with normal necks and true-breeding mice with long necks. Crosses were made between these two populations to produce an F_1 generation of mice with normal necks. The F_1 mice were then mated to each other to obtain an F_2 generation. For the mice that were born alive, the following data were obtained:

   522 mice with normal necks

   62 mice with long necks

What percentage of homozygous mice (that would have had long necks if they had survived) died during embryonic development?

E12. The data in Figure 3.5 show the results of the F_2 generation for seven of Mendel's experiments. Conduct a chi square analysis to determine if these data are consistent with the law of segregation.

E13. Let's suppose you conducted an experiment involving genetic crosses and calculated a chi square value of 1.005. There were four categories of offspring (i.e., the degrees of freedom equaled 3). Explain what the 1.005 value means. Your answer should include the phrase "80% of the time."

E14. A tall pea plant with axial flowers was crossed to a dwarf plant with terminal flowers. Tallness and axial flowers are dominant traits. The following offspring were obtained: 27 tall, axial flowers; 23 tall, terminal flowers; 28 dwarf, axial flowers; and 25 dwarf, terminal flowers. What are the genotypes of the parents?

E15. A cross was made between two strains of plants that are agriculturally important. One strain was disease-resistant but herbicide-sensitive; the other strain was disease-sensitive but herbicide-resistant. A plant breeder crossed the two plants and then allowed the F_1 generation to self-fertilize. The following data were obtained:

F_1 generation: All offspring are disease-sensitive and herbicide-resistant.

F_2 generation: 157 disease-sensitive, herbicide-resistant

                    57 disease-sensitive, herbicide-sensitive

                    54 disease-resistant, herbicide-resistant

                    20 disease-resistant, herbicide-sensitive

Total: 288

Formulate a hypothesis that you think is consistent with the observed data. Test the goodness of fit between the data and your hypothesis using a chi square test. Explain what the chi square results mean.

E16. Discuss why crosses (i.e., the experiments of Mendel) and the microscopic observations of chromosomes during mitosis and meiosis were both needed to deduce the chromosome theory of inheritance.

Questions for Student Discussion/Collaboration

1. Consider this cross in pea plants: *Tt Rr yy Aa* × *Tt rr Yy Aa*, where T = tall, t = dwarf, R = round, r = wrinkled, Y = yellow, y = green, A = axial, a = terminal. What is the expected phenotypic outcome of this cross? Have one group of students solve this problem by making one big Punnett square, and have another group solve it by making four single-gene Punnett squares and using the multiplication method. Time each other to see who gets done first.

2. A cross was made between two pea plants, *TtAa* and *Ttaa*, where T = tall, t = dwarf, A = axial, and a = terminal. What is the probability that the first three offspring will be tall with axial flowers or dwarf with terminal flowers and the fourth offspring will be tall with axial flowers? Discuss what operation(s) (e.g., product rule or binomial expansion equation) you used and in what order they were used.

3. Consider this four-factor cross: *Tt Rr yy Aa* × *Tt RR Yy aa*, where T = tall, t = dwarf, R = round, r = wrinkled, Y = yellow, y = green, A = axial, a = terminal. What is the probability that the first three plants will have round seeds? What is the easiest way to solve this problem?

Answers to Comprehension Questions

3.1: d, a, b

3.2: c, b, b

3.3: b, c, a

3.4: c, a

3.5: b, c

3.6: a, b (use the binomial expansion), d

Note: All answers appear in Connect; the answers to even-numbered questions and all Concept Check questions are in Appendix B

4

CHAPTER OUTLINE

- 4.1 Mechanisms of Sex Determination Among Various Species
- 4.2 Dosage Compensation and X-Chromosome Inactivation in Mammals
- 4.3 Properties of the X and Y Chromosomes in Mammals
- 4.4 Transmission Patterns for X-Linked Genes

Opposite sexes. Most species of animals, such as these cardinals, are found in two sexes (here the male is on the right and the female on the left).
©Steve Byland/Shutterstock

SEX DETERMINATION AND SEX CHROMOSOMES

In Chapter 2, we examined the process of sexual reproduction, in which two gametes fuse with each other to begin the life of a new individual. Within a population, sexual reproduction enhances genetic diversity because the genetic material of offspring comes from two sources. For most species of animals and some species of plants, sexual reproduction is carried out by individuals of opposite sexes—females and males. The underlying mechanism by which an individual develops into a female or a male is called **sex determination.** As we will see, a variety of mechanisms promote this process.

For some species, females and males differ in their genomes. For example, you are probably already familiar with the idea that people differ with regard to X and Y chromosomes. Females are XX and males are XY, which means that females have two copies of the X chromosome, whereas males have one X and one Y chromosome. Because these chromosomes carry different genes, chromosomal differences between the sexes also result in unique phenotypes and inheritance patterns that differ from those that we discussed in Chapter 3. In this chapter, we will explore how genes that are located on the X chromosome exhibit a characteristic pattern of inheritance.

4.1 MECHANISMS OF SEX DETERMINATION AMONG VARIOUS SPECIES

Learning Outcome:
1. Outline different mechanisms of sex determination.

After gametes fuse with each other during fertilization, what factors determine whether the resulting zygote and then embryo develops into a female or a male? Researchers have studied the process of sex determination in a wide range of species and discovered that several different mechanisms exist. In this section, we will explore some common mechanisms of sex determination in animals and plants.

Sex Differences May Depend on the Presence of Sex Chromosomes

According to the chromosome theory of inheritance, which we discussed in Chapter 3, chromosomes carry the genes that determine an organism's traits. Not surprisingly, sex determination in some species is due to the presence of particular chromosomes.

In 1901, Clarence McClung, who studied grasshoppers, was the first to suggest that male and female sexes are due to the inheritance of particular chromosomes. Since McClung's initial observations, we now know that a pair of chromosomes, called the **sex chromosomes,** determines sex in many different species. Some examples are described in **Figure 4.1**.

X-Y System In the X-Y system of sex determination, which operates in mammals, the male carries one X chromosome and one Y chromosome, whereas the female has two X chromosomes (Figure 4.1a). In this case, the male is called the **heterogametic sex.** Two types of sperm are produced: one that carries only the X chromosome, and another that carries the Y. The female is the **homogametic sex,** because all eggs carry a single X chromosome. The 46 chromosomes found in humans consist of 1 pair of sex chromosomes and 22 pairs of **autosomes**—chromosomes that are not sex chromosomes. In the human male, each of the four sperm produced during gametogenesis has 23 chromosomes. Two sperm carry an X chromosome, and the other two have a Y chromosome. The sex of the offspring is determined by whether the sperm that fertilizes the egg carries an X or a Y chromosome.

What causes an offspring to develop into a male or female? One possibility is that two X chromosomes are required for female development. A second possibility is that the Y chromosome promotes male development. In the case of mammals, the second possibility is correct. This is known from the analysis of rare individuals who carry chromosomal abnormalities. For example, mistakes that occasionally occur during meiosis may produce an individual who carries two X chromosomes and one Y chromosome. Such an individual develops into a male. In addition, people are sometimes born with a single X chromosome and no other sex chromosome. Such an individual becomes a female. The chromosomal basis for sex determination in mammals is rooted in a particular gene on the Y chromosome. The presence of a gene on the Y chromosome called the *SRY* gene causes maleness.

X-0 System Another mechanism of sex determination that involves sex chromosomes is the X-0 system that operates in many insects (Figure 4.1b). In some insect species, the male has only one sex chromosome (the X) and is designated X0, whereas the female has a pair (two X's). In other insect species, such as *Drosophila melanogaster*, the male is XY. For both types of insect species (i.e., X0 or XY males, and XX females), the ratio between X chromosomes and the number of autosomal sets determines sex. If a fly has one X chromosome and is diploid (2*n*) for the autosomes, the ratio is 1/2, or 0.5. This fly becomes a male even if it does not receive a Y chromosome. In contrast to the X-Y system of mammals, the Y chromosome in the X-0 system does not determine maleness. If a fly receives two X chromosomes and is diploid, the ratio is 2/2, or 1.0, and the fly becomes a female.

Z-W system For the Z-W system, which determines sex in birds and some fish, the male is ZZ and the female is ZW (Figure 4.1c). The letters Z and W are used to distinguish these types of sex chromosomes from those found in the X-Y pattern of sex determination of other species. In the Z-W system, the male is the homogametic sex, and the female is heterogametic.

Sex Differences May Depend on the Number of Sets of Chromosomes

Another interesting mechanism of sex determination, known as the **haplodiploid system,** is found in bees, wasps, and ants (**Figure 4.2**). For example, in honeybees, the male, which is called a drone, is produced from an unfertilized haploid egg. A male honeybee has a single set of 16 chromosomes. By comparison, female honeybees, both worker bees and queen bees, are produced from fertilized eggs and therefore are diploid. They contain two sets of chromosomes, for a total of 32. In this case, only females are produced by sexual reproduction.

Sex Differences May Depend on the Environment

Although sex in many species of animals is determined by chromosomes, other mechanisms are also known. In certain reptiles and fish,

(a) X–Y system in mammals
44 + XY ♂
44 + XX ♀

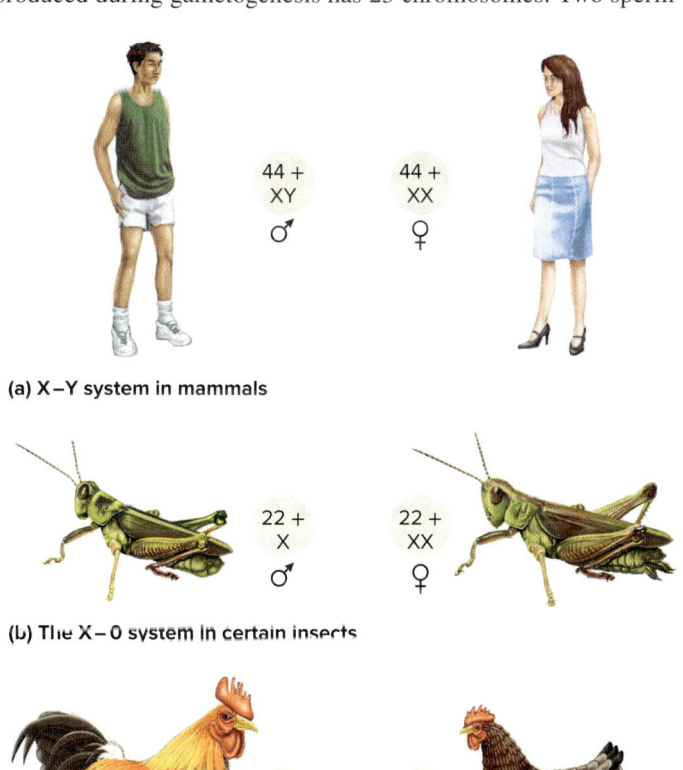

(b) The X–0 system in certain insects
22 + X ♂
22 + XX ♀

(c) The Z–W system in birds
76 + ZZ ♂
76 + ZW ♀

FIGURE 4.1 Sex determination via the presence of sex chromosomes. See text for a description.

Genes → Traits Certain genes that are found on the sex chromosomes play a key role in the development of sex (male vs. female). For example, in mammals, a gene on the Y chromosome initiates male development. In the X-0 system, the ratio of X chromosomes to the sets of autosomes plays a key role in governing the pathway of development to produce a male or female.

Concept Check: What is the difference between the X-Y and X-0 systems of sex determination?

74   CHAPTER 4 :: SEX DETERMINATION AND SEX CHROMOSOMES

Male honeybee (drone)
Haploid – 16 chromosomes

Female honeybee
Diploid – 32 chromosomes

FIGURE 4.2 The haplodiploid mechanism of sex determination. In this system, males are haploid, whereas females are diploid.
(a) Source: USGS Bee Inventory and Monitoring Lab. Photo by Sue Boo; (b) Source: Rob Flynn/USDA

Concept Check: *Is the male bee produced by sexual reproduction? Explain.*

(a) Sex determination via temperature: American alligator (*A. mississippiensis*)

(b) Sex determination via behavior: Clownfish (*Amphiprion ocellaris*)

FIGURE 4.3 Sex determination caused by environmental factors. (a) In the alligator, temperature determines whether an individual develops into a female or male. (b) In clownfish, males can change into females due to behavioral changes that occur when a dominant female dies.
(a) Source: NASA; (b) ©Krzysztof Odziomek/Getty Images

Concept Check: *How might global warming affect alligator populations?*

sex is controlled by environmental factors such as temperature. For example, in the American alligator (*Alligator mississippiensis*), temperature controls sex development (**Figure 4.3a**). When fertilized eggs of this alligator species are incubated at 33°C, nearly 100% of them produce male individuals. In contrast, eggs incubated at a temperature a few degrees below 33°C produce nearly all females, whereas those incubated a few degrees above 33°C produce about 95% females.

Another way that sex can be environmentally determined is via behavior. Clownfish of the genus *Amphiprion* are coral reef fish that live among anemones on the ocean floor (**Figure 4.3b**). One anemone typically harbors a harem of clownfish consisting of a large female, a medium-sized reproductive male, and small nonreproductive juveniles. Clownfish are **protandrous hermaphrodites**—they can switch from male to female! When the female of a harem dies, the reproductive male changes sex to become a female and the largest of the juveniles matures into a reproductive male. Unlike male and female humans, the opposite sexes of clownfish are not determined by chromosome differences. Male and female clownfish have the same chromosomal composition.

How can a clownfish switch from female to male? A juvenile clownfish has both male and female immature sexual organs. Hormone levels, particularly those of an androgen called testosterone and an estrogen called estradiol, control the expression of particular genes. In nature, the first sexual change that usually happens is that a juvenile clownfish becomes a male. This occurs when the testosterone level becomes high, which promotes the expression of genes that encode proteins that cause the male organs to mature. Later, when the female of the harem dies, the estradiol level in the reproductive male becomes high and testosterone is decreased. This alters gene expression in a way that leads to the synthesis of some new proteins and prevents the synthesis of others. When this occurs, the female organs grow and the male reproductive system degenerates. The male fish becomes female.

(a) American holly (*I. opaca*)

(b) Female and male flowers on separate individuals in white campion (*S. latifolia*)

FIGURE 4.4 Examples of dioecious plants in which individuals produce only male gametophytes or only female gametophytes. **(a)** American holly (*Ilex opaca*). The female sporophyte, which produces red berries, is shown here. **(b)** White campion (*Silene latifolia*), which is often studied by researchers.

(a) ©valentino cazzanti/Shutterstock; (b, c) ©Arco Images GmbH/Alamy Stock Photo

Concept Check: Which are the opposite sexes in dioecious plants—the sporophytes or the gametophytes?

What factor determines the hormone levels in clownfish? A female seems to control the other clownfish in the harem through aggressive dominance, thereby preventing the formation of other females. This aggressive behavior suppresses an area of the brain in the other clownfish that is responsible for the production of certain hormones that are needed to promote female development. If a clownfish is left by itself in an aquarium, it will automatically develop into a female because this suppression does not occur.

Dioecious Plant Species Have Opposite Sexes

In most flowering plants, a single diploid individual (a sporophyte) produces both female and male gametophytes, which are haploid and contain egg or sperm cells, respectively (see Chapter 2, Figure 2.14). However, some plant species are **dioecious**, which means that some individuals produce only male gametophytes, whereas others produce only female gametophytes. These include hollies (**Figure 4.4a**), willows, and ginkgo trees. The genetics of sex determination in dioecious plant species is beginning to be understood. To study this process, many researchers have focused their attention on the white campion, *Silene latifolia*, which is a relatively small dioecious plant with a short generation time (**Figure 4.4b**). In this species, sex chromosomes, designated X and Y, are responsible for sex determination. The male plant has X and Y chromosomes, whereas the female plant is XX. Sex chromosomes are also found in other plant species such as papaya and spinach. However, in other dioecious species, cytological examination of the chromosomes does not reveal distinct types of sex chromosomes. Even so, in these plant species, the male plants usually appear to be the heterogametic sex.

4.1 REVIEWING THE KEY CONCEPTS

- In the X-Y, X-0, and Z-W systems, sex is determined by the presence and number of particular sex chromosomes (see Figure 4.1).
- In the haplodiploid system, sex is determined by the number of sets of chromosomes (see Figure 4.2).
- In some species, such as alligators and clownfish, sex is determined by environmental factors (see Figure 4.3).
- Dioecious plants exist as individuals that produce only pollen and those that produce only eggs. Sex chromosomes sometimes determine sex in these species (see Figure 4.4).

4.1 COMPREHENSION QUESTIONS

1. Among different species, sex may be determined by
   a. differences in sex chromosomes.
   b. differences in the number of sets of chromosomes.
   c. environmental factors.
   d. all of the above.

2. In mammals, sex is determined by
   a. the *SRY* gene on the Y chromosome.
   b. having two copies of the X chromosome.
   c. having one copy of the X chromosome.
   d. both a and c.

3. An abnormal fruit fly has two sets of autosomes and is XXY. Such a fly is
   a. a male.
   b. a female.
   c. a hermaphrodite.
   d. none of the above.

4.2 DOSAGE COMPENSATION AND X-CHROMOSOME INACTIVATION IN MAMMALS

Learning Outcomes:
1. Compare and contrast the mechanisms of dosage compensation in different animal species.
2. Describe the process of X-chromosome inactivation in mammals.
3. Explain how X-chromosome inactivation may affect the phenotype of female mammals.

Dosage compensation refers to the phenomenon in which the level of expression of many genes on the sex chromosomes (e.g., the X chromosome) is similar in both sexes, even though males and females have a different complement of sex chromosomes. This term was coined in 1932 by Hermann Muller to explain the effects of eye color mutations in *Drosophila*. Muller observed that female flies homozygous for certain alleles on the X chromosome had a phenotype similar to that of hemizygous males, which have only one copy of the gene. He noted that an allele on the X chromosome conferring an apricot eye color produces a very similar phenotype in a female carrying two copies of the gene and in a male with just one. In contrast, a female that has one copy of the apricot allele and a deletion of the apricot allele on the other X chromosome has eyes of paler color. Therefore, one copy of the allele in the female is not equivalent to one copy of the allele in the male. Instead, two copies of the allele in the female produce a phenotype that is similar to that produced by one copy in the male. In other words, the difference in gene dosage—two copies in females versus one copy in males—is being compensated at the level of gene expression. In this section, we will explore how this occurs in different species of animals.

Dosage Compensation Is Necessary in Some Species to Ensure Genetic Equality Between the Sexes

Dosage compensation has been studied extensively in mammals, *Drosophila*, and *Caenorhabditis elegans* (a nematode). Depending on the species, dosage compensation occurs via different mechanisms (Table 4.1).

- Female mammals equalize the expression of genes on the X chromosome by turning off one of their two X chromosomes. This process is known as **X-chromosome inactivation (XCI).**
- In *Drosophila*, the male accomplishes dosage compensation by doubling the expression of most genes on the X chromosome.
- In *C. elegans*, the XX animal is a hermaphrodite that produces both sperm and egg cells, and an animal carrying a single X chromosome is a male that produces only sperm. The XX hermaphrodite diminishes the expression of genes on each X chromosome to approximately 50%.

- In birds, the Z chromosome is a large chromosome, usually the fourth or fifth largest, and contains many genes. The W chromosome is generally a much smaller chromosome containing a high proportion of repeat-sequence DNA that does not encode genes. Male birds are ZZ and females are ZW. Though dosage compensation is not well understood in birds, most evidence suggests that some Z-linked genes may be dosage-compensated, but many of them are not.

Dosage Compensation Occurs in Female Mammals by the Inactivation of One X Chromosome

In 1961, Mary Lyon proposed that dosage compensation in mammals occurs by the inactivation of a single X chromosome in females. This proposal brought together two lines of study. The first evidence came from cytological studies. In 1949, Murray Barr and Ewart Bertram identified a highly condensed structure in the interphase nuclei of somatic cells in female cats that was not found in male cats. This structure became known as the **Barr body** (Figure 4.5a). In 1960, Susumu Ohno correctly proposed that the Barr body is a highly condensed X chromosome.

In addition to this cytological evidence, Lyon was also familiar with examples in which the coat color of a mammal had a variegated pattern. Figure 4.5b is a photo of a calico cat, which is a female that is heterozygous for a gene on the X chromosome that can occur as an orange or a black allele. (The cat's white underside is due to a dominant allele in a different gene.) The orange and black patches are randomly distributed in different female individuals. The calico pattern does not occur in male cats, but similar kinds of mosaic patterns have been identified in the female mouse. Lyon suggested that both the Barr body and the calico pattern are the result of XCI in the cells of female mammals.

The proposed mechanism of XCI, known as the **Lyon hypothesis,** is schematically illustrated in Figure 4.6. This example involves a white and black variegated coat found in certain strains of mice. As shown here, a female mouse has inherited an X chromosome from its mother that carries an allele conferring

TABLE 4.1
Mechanisms of Dosage Compensation Among Different Species

	Sex Chromosomes in:		
Species	Females	Males	Mechanism of Compensation
Placental mammals	XX	XY	One of the X chromosomes in the somatic cells of females is inactivated. In certain species, the paternal X chromosome is inactivated, and in other species, such as humans, either the maternal or paternal X chromosome is randomly inactivated throughout the somatic cells of females.
Marsupial mammals	XX	XY	The paternally derived X chromosome is inactivated in the somatic cells of females.
Drosophila melanogaster	XX	XY	The level of expression of genes on the X chromosome in males is doubled.
Caenorhabditis elegans	XX*	X0	The level of expression of genes on each X chromosome in hermaphrodites is decreased to 50% of the level occurring on the X chromosome in males.

*In *C. elegans*, an XX individual is a hermaphrodite, not a female.

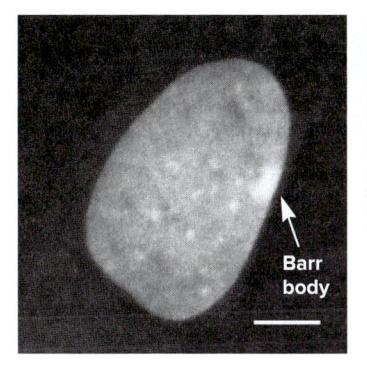

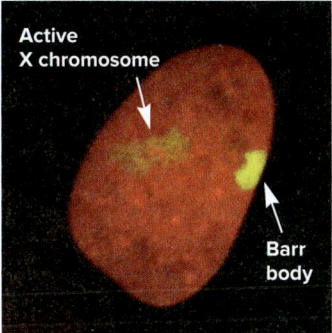

(a) Nucleus with a Barr body

(b) A calico cat

FIGURE 4.5 X-chromosome inactivation in female mammals. (a) The left micrograph shows the Barr body on the periphery of a human nucleus after staining with a DNA-specific dye. Because it is compact, the Barr body is more brightly stained. The white scale bar is 5 μm. The right micrograph shows the same nucleus using a yellow fluorescent probe that recognizes the X chromosome. The Barr body is more compact than the active X chromosome, which is to the left of the Barr body. (b) The fur pattern of a calico cat.
Genes → Traits The pattern of black and orange fur on this cat is due to random X-chromosome inactivation during embryonic development. The orange patches of fur are due to the inactivation of the X chromosome that carries a black allele; the black patches are due to the inactivation of the X chromosome that carries the orange allele. In general, only heterozygous female cats can be calico. A rare exception is a male cat (XXY) that has an abnormal composition of sex chromosomes.
(a) Courtesy of I. Solovei, University of Munich (LMU); (b) ©Tim Davis/Science Source

Concept Check: Why is the Barr body more brightly stained by DNA-specific dye in a cell nucleus?

white coat color (X^b). The X chromosome from its father carries a black coat color allele (X^B).

1. Initially, both X chromosomes are active.
2. At an early stage of embryonic development, one of the two X chromosomes is randomly inactivated in each somatic cell and becomes a Barr body. During inactivation, the chromosomal DNA of the inactivated X chromosome becomes highly compacted into a Barr body, so most of the genes on that X chromosome cannot be expressed.

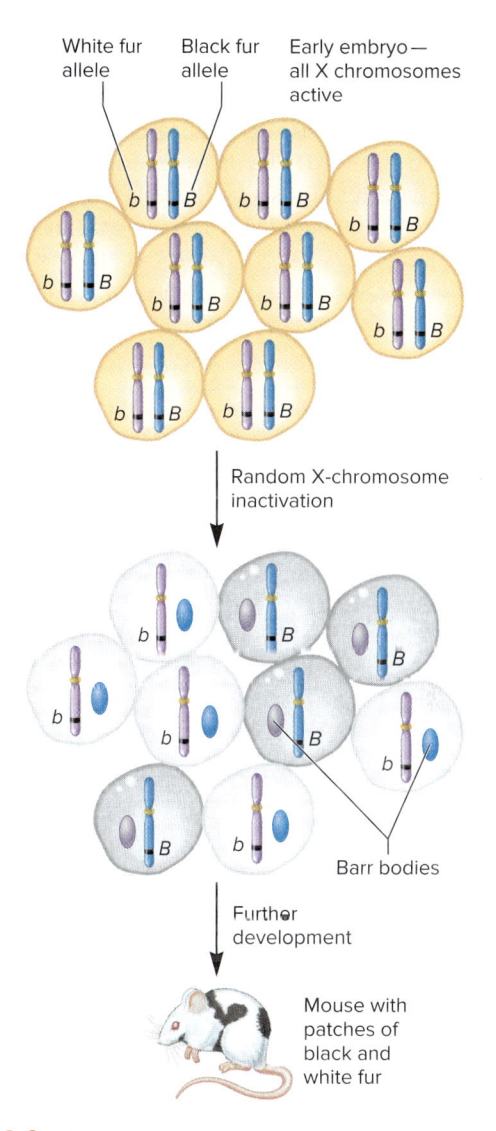

FIGURE 4.6 The mechanism of X-chromosome inactivation.
Genes → Traits The top of this figure represents a mass of several cells that compose the early embryo. Initially, both X chromosomes are active. At an early stage of embryonic development, random inactivation of one X chromosome occurs in each cell. This inactivation pattern is maintained as the embryo matures into an adult.

Concept Check: At which stage of development does X-chromosome inactivation initially occur?

3. As the embryo continues to grow and mature, each embryonic cell divides and may eventually give rise to billions of cells in the adult animal. The epithelial (skin) cells that are derived from an embryonic cell that had the X^B chromosome inactivated will produce a patch of white fur because the pattern of XCI is maintained in subsequent cell divisions. Alternatively, another embryonic cell may have the other X chromosome inactivated (i.e., X^b). The epithelial cells derived from this embryonic cell produce a patch of black fur.
4. Because the primary event of XCI is a random process that occurs at an early stage of development, the result is an animal with some patches of white fur and other patches of black fur. This is the basis of the variegated phenotype.

Mammals Maintain One Active X Chromosome in Their Somatic Cells

Since the Lyon hypothesis was confirmed, the genetic control of X-chromosome inactivation has been investigated further by many laboratories. Research has shown that mammalian somatic cells possess the ability to count the X chromosomes they contain and allow only one of them to remain active. How was this determined? A key observation came from comparisons of the chromosome composition of people who were born with normal or abnormal numbers of sex chromosomes.

Phenotype	Chromosome Composition	Number of X Chromosomes	Number of Barr Bodies
Normal female	XX	2	1
Normal male	XY	1	0
Turner syndrome (female)	X0	1	0
Triple X syndrome (female)	XXX	3	2
Klinefelter syndrome (male)	XXY	2	1

In normal females, two X chromosomes are counted and one is inactivated, whereas in males, one X chromosome is counted and it is not inactivated. If the number of X chromosomes exceeds two, as in triple X syndrome, additional X chromosomes are converted to Barr bodies.

X-Chromosome Inactivation in Mammals Depends on the X-Inactivation Center and the *Xist* Gene

A short region on the X chromosome called the **X-inactivation center** (**Xic**—pronounced "zic") plays a critical role in X-chromosome inactivation. Eeva Therman and Klaus Patau identified the Xic from its key role in XCI. The counting of human X chromosomes is accomplished by counting the number of Xics. A Xic must be found on an X chromosome for inactivation to occur. Therman and Patau discovered that if one of the two X chromosomes in a female is missing its Xic due to a chromosome mutation, a cell counts only one Xic and X-chromosome inactivation does not occur. Having two active X chromosomes is a lethal condition for a human female embryo.

Let's consider how the molecular expression of certain genes controls X-chromosome inactivation. The expression of a specific gene within the Xic is required for the compaction of the X chromosome into a Barr body. This gene, discovered in 1991, is named *Xist* (for X-inactive specific transcript). The *Xist* gene on the inactivated X chromosome is active, which is unusual because most other genes on the inactivated X chromosome are silenced. The *Xist* gene product is an RNA molecule that does not encode a protein. Instead, the role of the *Xist* RNA is to coat the X chromosome and inactivate it. After coating, other proteins associate with the *Xist* RNA and promote chromosomal compaction into a Barr body.

X-Chromosome Inactivation Occurs in Three Phases: Initiation, Spreading, and Maintenance

The process of XCI can be divided into three phases: initiation, spreading, and maintenance (**Figure 4.7**).

- During initiation, which occurs during embryonic development, one of the X chromosomes remains active, and the other is chosen to be inactivated.
- During the spreading phase, the chosen X chromosome is inactivated. This spreading requires the expression of the *Xist* gene. The *Xist* RNA coats the inactivated X chromosome and recruits proteins that promote compaction. The spreading phase is so named because inactivation begins near the Xic and spreads in both directions along the X chromosome.
- Once the initiation and spreading phases occur for a given X chromosome, the inactivated X chromosome is maintained as a Barr body during future cell divisions. When a cell divides, the Barr body is replicated, and both copies remain compacted. This maintenance phase continues from the embryonic stage through adulthood.

Some genes on the inactivated X chromosome are expressed in the somatic cells of adult female mammals. These genes are said to escape the effects of XCI. In humans, up to 1/4 of the genes on the X chromosome may escape inactivation to some degree. Many of these genes occur in clusters. Among these are the pseudoautosomal genes found on both the X and Y chromosomes in the regions of homology, which are described in Section 4.3. Dosage compensation is not necessary for pseudoautosomal genes because they are located on both the X and Y chromosomes.

Genetic TIPS

The Question: A cat is born with two X chromosomes and one Y chromosome. One of the X chromosomes carries the black fur allele and the other carries the orange fur allele. Would you expect this cat to be a male or female? Would it be calico?

Topic: What topic(s) in genetics does this question address?

The topics are sex determination and X-chromosome inactivation.

Information: What information do you know based on the question and your understanding of the topic?

From the question, you know the composition of sex chromosomes in a cat and the fur color alleles carried on the X chromosomes. From your understanding of the topics, you may remember that the Y chromosome determines maleness in mammals, and that X-chromosome inactivation occurs and only one X chromosome remains active in somatic cells.

Problem-Solving Strategy: Predict the outcome.

With regard to sex determination, you would predict that the cat is a male because the Y chromosome causes maleness. You would also predict that random X-chromosome inactivation would occur in this cat's somatic cells, because the cells contain two X chromosomes. The cat would be heterozygous for the orange and black fur color alleles, resulting in some patches of orange fur and some patches of black.

Answer: It would be a male cat with a calico coat.

Initiation: Occurs during embryonic development. The number of X-inactivation centers (Xics) is counted and one of the X chromosomes remains active and the other is targeted for inactivation.

Spreading: Occurs during embryonic development. It begins at the Xic and progresses toward both ends until the entire chromosome is inactivated. The *Xist* gene encodes an RNA that coats the X chromosome and recruits proteins that promote its compaction into a Barr body.

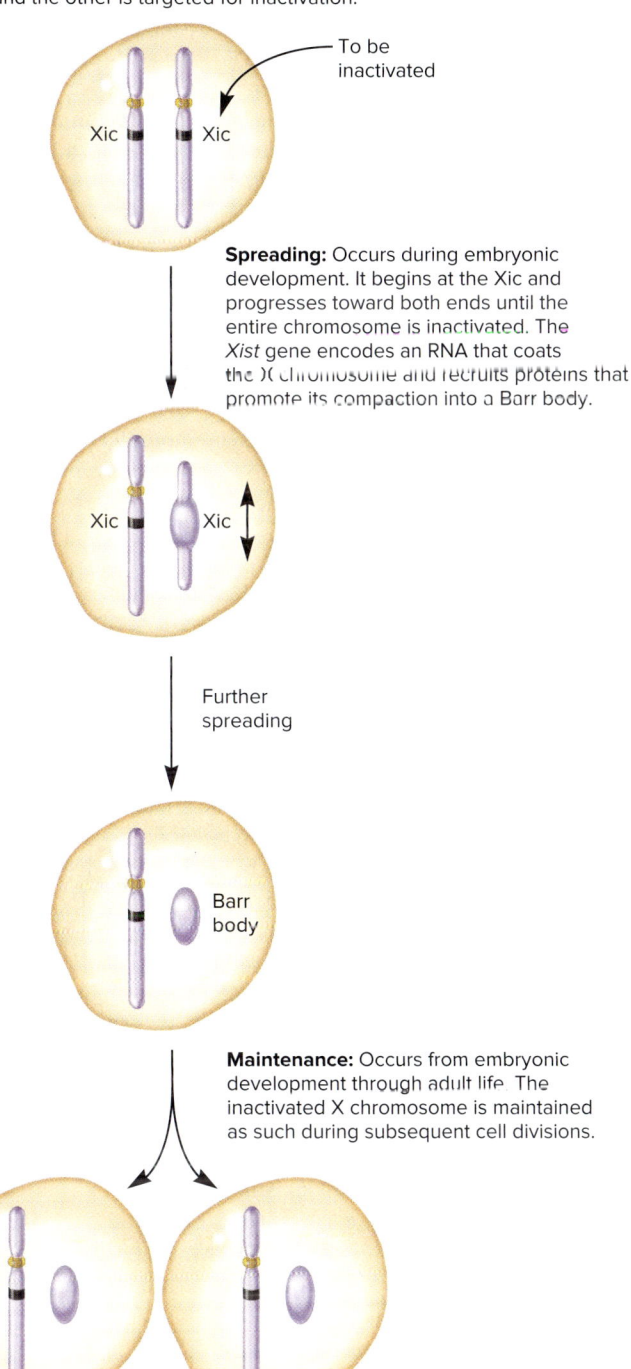

Further spreading

Barr body

Maintenance: Occurs from embryonic development through adult life. The inactivated X chromosome is maintained as such during subsequent cell divisions.

FIGURE 4.7 The phases of X-chromosome inactivation.

Concept Check: Which of these phases occurs in an adult female?

4.2 REVIEWING THE KEY CONCEPTS

- Dosage compensation often occurs in species in which males and females differ in their sex chromosomes (see Table 4.1).
- In mammals, the process of X-chromosome inactivation (XCI) in females compensates for the single X chromosome found in males. The inactivated X chromosome is called a Barr body. The process can lead to a variegated phenotype, such as a calico cat (see Figure 4.5).
- After it occurs during embryonic development, X-chromosome inactivation is maintained when somatic cells divide (see Figure 4.6).
- X-chromosome inactivation is controlled by the X-inactivation center (Xic) that contains the *Xist* gene. The three phases of XCI are initiation, spreading, and maintenance phases (see Figure 4.7).

4.2 COMPREHENSION QUESTIONS

1. In fruit flies, dosage compensation is achieved by
    a. X-chromosome inactivation.
    b. doubling the expression of genes on the single X chromosome in the male.
    c. decreasing the expression of genes on each X chromosome in the female to one-half.
    d. all of the above.
2. According to the Lyon hypothesis,
    a. one of the X chromosomes is converted to a Barr body in somatic cells of female mammals.
    b. one of the X chromosomes is converted to a Barr body in all cells of female mammals.
    c. both of the X chromosomes are converted to Barr bodies in somatic cells of female mammals.
    d. both of the X chromosomes are converted to Barr bodies in all cells of female mammals.
3. Which of the following is *not* a phase of XCI?
    a. Initiation   c. Maintenance
    b. Spreading   d. Erasure

4.3 PROPERTIES OF THE X AND Y CHROMOSOMES IN MAMMALS

Learning Outcomes:
1. Compare and contrast the features of the X and Y chromosomes in mammals.
2. Explain how pseudoautosomal inheritance occurs.

As discussed in the first section of this chapter, sex in mammals is determined by the presence of the Y chromosome, which carries the *SRY* gene. The X and Y chromosomes also differ in other ways. The X chromosome is typically much larger than the Y and carries more genes. For example, in humans, researchers estimate that the X chromosome carries about 800 protein-encoding genes, whereas the Y chromosome has about 50.

- Genes that are found on only one sex chromosome, but not both, are called **sex-linked genes.**
- **X-linked genes** are found only on the X chromosome.

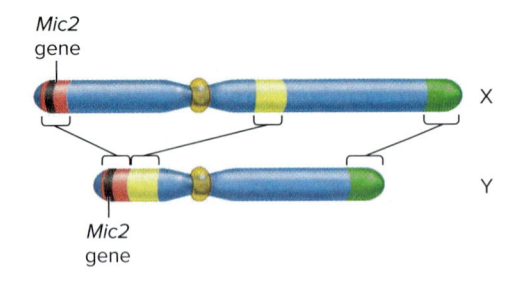

FIGURE 4.8 **A comparison of the homologous and nonhomologous regions of the X and Y chromosomes in humans.** The brackets show three regions of homology between the X and Y chromosome. A few pseudoautosomal genes, such as *Mic2*, are found on both the X and Y chromosomes in these small regions of homology.

- **Y-linked genes,** or **holandric genes,** are found only on the Y chromosome.
- **Pseudoautosomal genes** are found on both the X and Y chromosomes.

As shown in **Figure 4.8**, the human sex chromosomes have three homologous regions. These regions, which are evolutionarily related, promote the necessary pairing of the X and Y chromosomes that occurs during meiosis I of spermatogenesis. Relatively few genes are located in these homologous regions. One example is a human gene called *Mic2*, which encodes a cell-surface protein. The *Mic2* gene is found on both the X and Y chromosomes. It follows a pattern of inheritance called **pseudoautosomal inheritance.** The term *pseudoautosomal* refers to the idea that the inheritance pattern of the *Mic2* gene is the same as the inheritance pattern of a gene located on an autosome even though the *Mic2* gene is actually located on the sex chromosomes. As in autosomal inheritance, males have two copies of pseudoautosomally inherited genes, and they can transmit the genes to both daughters and sons.

By comparison, genes that are found only on the X or Y chromosome exhibit transmission patterns that are quite different from genes located on an autosome. A Y-linked inheritance pattern is very distinctive—the gene is transmitted only from fathers to sons. By comparison, transmission patterns involving X-linked genes are more complex because females inherit two X chromosomes, whereas males receive only one X chromosome, from their mother. We will consider the complexities of X-linked inheritance patterns next.

4.3 REVIEWING THE KEY CONCEPTS

- Chromosomes that differ between males and females are termed sex chromosomes and carry sex-linked genes.
- X-linked genes are found only on the X chromosome, whereas Y-linked genes are found only on the Y chromosome. Pseudoautosomal genes are found on both the X and Y chromosomes in regions of homology (see Figure 4.8).

4.3 COMPREHENSION QUESTIONS

1. A Y-linked gene is passed from
    a. father to son.
    b. father to daughter.
    c. father to daughter or son.
    d. mother to son.

4.4 TRANSMISSION PATTERNS FOR X-LINKED GENES

Learning Outcomes:

1. Analyze the results of Morgan's experiment, which showed that a gene affecting eye color in fruit flies is located on the X chromosome.
2. Predict the outcome of crosses involving X-linked genes.

In the first section of this chapter, we discussed how sex determination in certain species is controlled by sex chromosomes. In fruit flies and mammals, a female is XX, whereas a male is XY. The inheritance pattern of X-linked genes, known as **X-linked inheritance,** shows certain distinctive features. For example, males transmit X-linked genes only to their daughters, and sons receive their X-linked genes from their mothers. The term **hemizygous** is used to describe the single copy of an X-linked gene in the male. A male mammal or fruit fly is said to be hemizygous for X-linked genes. Because males of certain species, such as humans, have a single copy of the X chromosome, another distinctive feature of X-linked inheritance is that males are more likely to be affected by rare, recessive X-linked disorders. We will consider the medical implications of X-linked inheritance in Chapter 22. In this section, we will examine X-linked inheritance in fruit flies and mammals.

Morgan's Experiments Showed a Connection Between a Genetic Trait and the Inheritance of a Sex Chromosome in *Drosophila*

In the early 1900s, Thomas Hunt Morgan carried out the first study that confirmed the location of a gene on a particular chromosome. In this experiment, he showed that a gene affecting eye color in fruit flies is located on the X chromosome. Morgan was trained as an embryologist, and much of his early research involved descriptive and experimental work in that field. He was particularly interested in ways that organisms change. He wrote, "The most distinctive problem of zoological work is the change in form that animals undergo, both in the course of their development from the egg (embryology) and in their development in time (evolution)." Throughout his life, he usually had dozens of different experiments going on simultaneously, many of them unrelated to each other. He jokingly said there are three kinds of experiments—those that are foolish, those that are damn foolish, and those that are worse than that!

In one of his most famous studies, Morgan engaged one of his graduate students to rear fruit flies (*Drosophila melanogaster*) in the dark, hoping to produce flies whose eyes would atrophy from disuse and disappear in future generations. Even after many consecutive generations, however, the flies appeared to have no noticeable changes despite repeated attempts at inducing mutations by treatments with agents such as X-rays and radium. After 2 years, Morgan finally obtained an interesting result when a true-breeding line of *Drosophila* produced a male fruit fly with white eyes rather than the common (wild-type) red eyes. Because this had been a true-breeding line of flies, this white-eyed male must have arisen from a new mutation that converted a red-eye allele (denoted w^+) into a white-eye allele (denoted w). Morgan is said to have carried this fly home with him in a jar, put it by his bedside at night while he slept, and then taken it back to the laboratory during the day.

Much like Mendel, Morgan studied the inheritance of this white-eye trait by making crosses and quantitatively analyzing their outcome. In the experiment described in **Figure 4.9**, he began with his white-eyed male and crossed it to a true-breeding red-eyed female. All of the F_1 offspring had red eyes, indicating that red is dominant to white. The F_1 offspring were then mated to each other to obtain an F_2 generation.

▶ **THE GOAL (DISCOVERY-BASED SCIENCE)**

This is an example of discovery-based science rather than hypothesis testing. In this case, a quantitative analysis of genetic crosses may reveal the inheritance pattern for the white-eye allele.

▶ **ACHIEVING THE GOAL** — **FIGURE 4.9** Inheritance pattern of an X-linked trait in fruit flies.

Starting material: A true-breeding line of red-eyed fruit flies plus one white-eyed male fly that was discovered in Morgan's collection of flies.

Concept Check. What is the key result that suggests an X-linked inheritance pattern?

Experimental level

1. Cross the white-eyed male to a true-breeding red-eyed female.

   P generation

2. Record the results of the F_1 generation. This involves noting the eye color and sex of many offspring.

   F_1 generation

3. Cross F_1 offspring with each other to obtain F_2 offspring. Also record the eye color and sex of the F_2 offspring.

   F_2 generation

4. In a separate experiment, perform a testcross between a white-eyed male from the F_2 generation and a red-eyed female from the F_1 generation. Record the results.

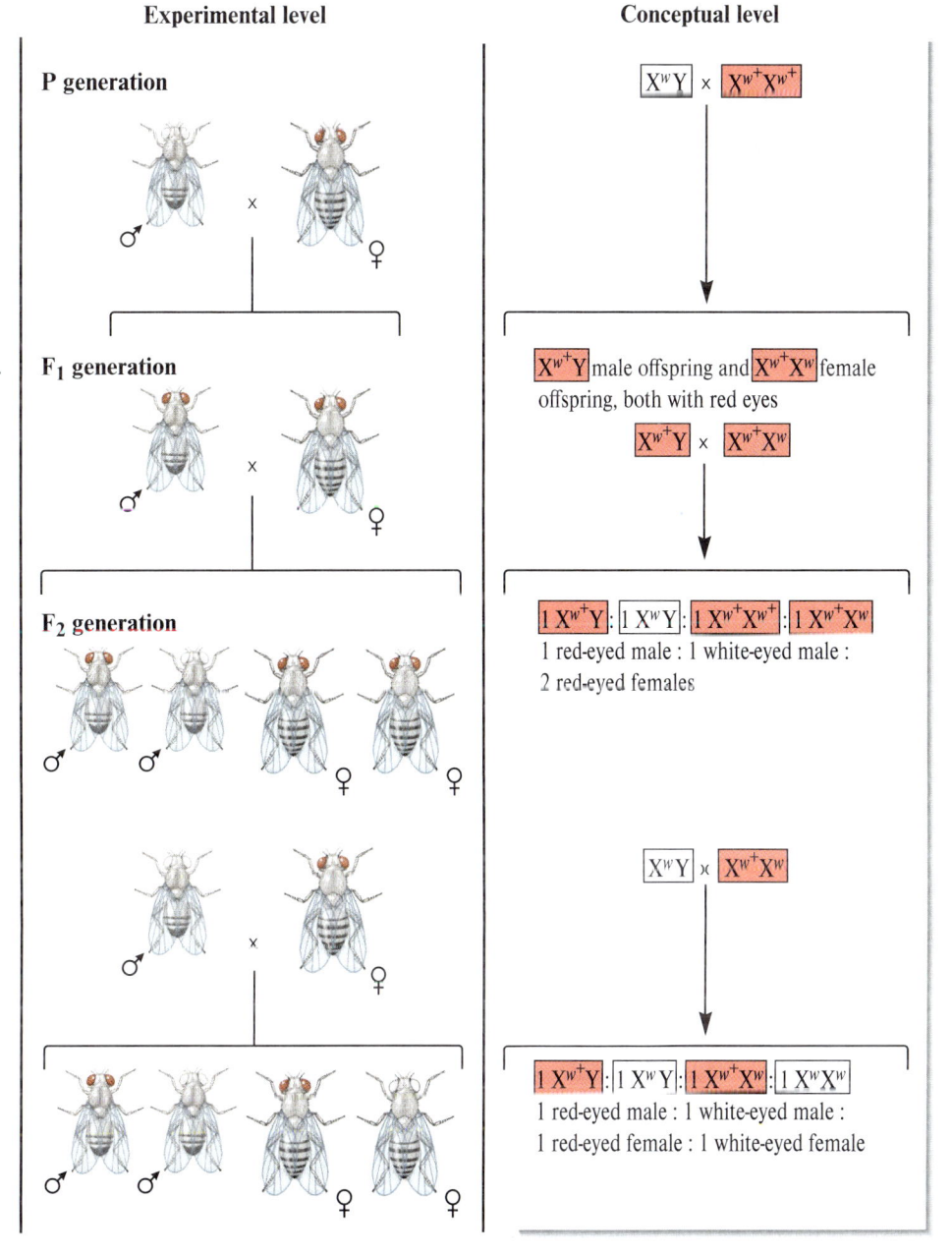

Conceptual level

$X^w Y$ × $X^{w^+} X^{w^+}$

$X^{w^+} Y$ male offspring and $X^{w^+} X^w$ female offspring, both with red eyes

$X^{w^+} Y$ × $X^{w^+} X^w$

$1 X^{w^+} Y : 1 X^w Y : 1 X^{w^+} X^{w^+} : 1 X^{w^+} X^w$
1 red-eyed male : 1 white-eyed male : 2 red-eyed females

$X^w Y$ × $X^{w^+} X^w$

$1 X^{w^+} Y : 1 X^w Y : 1 X^{w^+} X^w : 1 X^w X^w$
1 red-eyed male : 1 white-eyed male : 1 red-eyed female : 1 white-eyed female

▶ THE DATA

Cross	Results	
Original white-eyed male to a red-eyed female	F_1 generation:	All red-eyed flies
F_1 male to F_1 females	F_2 generation:	2459 red-eyed females 1011 red-eyed males 0 white females 782 white-eyed males
White-eyed males to F_1 females	Testcross:	129 red-eyed females 132 red-eyed males 88 white-eyed females 86 white-eyed males

Source: Morgan, T. H. (1940) Sex Limited Inheritance in *Drosophila. Science*, Vol. 32, 120–122.

▶ INTERPRETING THE DATA

As seen in the data, the F_2 generation consisted of 2459 red-eyed females, 1011 red-eyed males, and 782 white-eyed males. Most notably, no white-eyed female offspring were observed in the F_2 generation. These results suggested that the pattern of transmission from parent to offspring depends on the sex of the offspring and on the alleles that they carry. As shown in the Punnett square here, the data are consistent with the idea that the eye color alleles are located on the X chromosome:

F_1 male is $X^{w+}Y$
F_1 female is $X^{w+}X^w$

	♂ X^{w+}	Y
♀ X^{w+}	$X^{w+}X^{w+}$ Red, female	$X^{w+}Y$ Red, male
X^w	$X^{w+}X^w$ Red, female	X^wY White, male

The Punnett square predicts that the F_2 generation will not have any white-eyed females. This prediction was confirmed experimentally. These results indicated that the eye color alleles are located on the X chromosome. As mentioned earlier, genes that are physically located within the X chromosome are called X-linked genes, or **X-linked alleles.** However, it should also be pointed out that the experimental ratio of red eyes to white eyes in the F_2 generation is (2459 + 1011):782, which equals 4.4:1. This ratio deviates significantly from the predicted ratio of 3:1. How can this discrepancy be explained? Later work revealed that the lower-than-expected number of white-eyed flies is due to their decreased survival rate.

Morgan also conducted a **testcross** (see step 4, Figure 4.9) in which an individual with a dominant phenotype and unknown genotype is crossed to an individual with a recessive phenotype. In this case, he mated F_1 red-eyed females to white-eyed males. This cross produced red-eyed males and females in approximately equal numbers, and white-eyed males and females in approximately equal numbers. The testcross data are also consistent with an X-linked pattern of inheritance. As shown in the following Punnett square, a 1:1:1:1 ratio is predicted for this testcross:

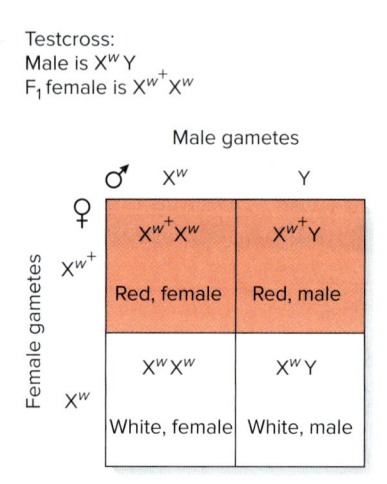

Testcross:
Male is X^wY
F_1 female is $X^{w+}X^w$

	♂ X^w	Y
♀ X^{w+}	$X^{w+}X^w$ Red, female	$X^{w+}Y$ Red, male
X^w	X^wX^w White, female	X^wY White, male

The observed data were 129:132:88:86, which is a ratio of 1.5:1.5:1:1. Again, the lower-than-expected numbers of white-eyed males and females can be explained by a lower survival rate for white-eyed flies. In his own interpretation, Morgan concluded that red eye color and X (a sex factor that is present in two copies in the female) are combined and have never existed apart. In other words, this gene for eye color is on the X chromosome. In 1933, Morgan received the Nobel Prize in physiology or medicine.

The Inheritance Pattern of X-Linked Genes Can Be Revealed by Reciprocal Crosses

X-linked patterns are also observed in mammalian species. As an example, let's consider a human disease known as Duchenne muscular dystrophy (DMD), which was first described by the French neurologist Guillaume Duchenne in the 1860s. Affected individuals show signs of muscle weakness as early as age 3. The disease gradually weakens the skeletal muscles and eventually affects the heart and breathing muscles. Survival is rare beyond the early 30s. The gene for DMD, found on the X chromosome, encodes a protein called dystrophin that is required inside muscle cells for structural support. Dystrophin is thought to strengthen muscle cells by anchoring elements of the internal cytoskeleton to the plasma membrane. Without it, the plasma membrane becomes permeable and may rupture.

DMD follows an inheritance pattern that is called **X-linked recessive**—the allele causing the disease is recessive and located on the X chromosome. In the pedigree shown in **Figure 4.10**, several males are affected by this disorder, as indicated by filled squares. The mothers of these males are presumed heterozygotes for this X-linked recessive allele. This recessive disorder is very rare among females because daughters would have to inherit a copy of the mutant allele from both their mother and an affected father.

4.4 TRANSMISSION PATTERNS FOR X-LINKED GENES

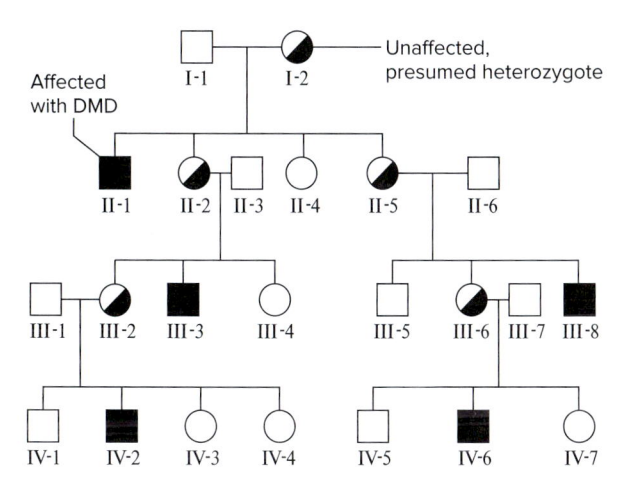

FIGURE 4.10 **A human pedigree for Duchenne muscular dystrophy, an X-linked recessive trait.** Affected individuals are shown with filled symbols. Females who are unaffected with the disease but have affected sons are presumed to be heterozygous carriers, shown with half-filled symbols.

Concept Check: *What features of this pedigree indicate that the allele for Duchenne muscular dystrophy is X linked?*

and has survived to reproductive age. When setting up a Punnett square involving X-linked traits, we must consider the alleles on the X chromosome as well as the observation that males may transmit a Y chromosome instead of the X chromosome. The X chromosomes from the female and male are designated with their corresponding alleles.

- The male makes two types of gametes: one that carries the X chromosome and one that carries the Y. The Punnett square must also include the Y chromosome even though this chromosome does not carry any X-linked genes.
- As seen on the left side of Figure 4.11b, the heterozygous female produces X^D and X^d gametes.
- When the Punnett square is filled in, it predicts the X-linked genotypes and sexes of the offspring. As seen on the left side, none of the offspring from this cross are affected with the disorder, although all female offspring are carriers.
- The right side of Figure 4.11b shows a **reciprocal cross**—a second cross in which the sexes and phenotypes are reversed. In this case, an affected female animal is crossed to an unaffected male. This cross produces female offspring that are carriers, and all male offspring will be affected with muscular dystrophy.

By comparing the two Punnett squares, we see that the outcome of the reciprocal cross yields different results. This is expected with X-linked genes, because the male transmits the gene only to female offspring, whereas the female transmits an X chromosome to both male and female offspring. Because the male parent does not transmit the X chromosome to his sons, he does not contribute to their X-linked phenotypes. This explains why X-linked traits do not behave equally in reciprocal crosses. Experimentally, the observation that reciprocal crosses do not yield the same results is an important clue that a trait may be X-linked.

X-linked muscular dystrophy has also been found in certain breeds of dogs such as golden retrievers (**Figure 4.11a**). Like humans, the mutation occurs in the dystrophin gene, and the symptoms include severe weakness and muscle atrophy that begin at about 6 to 8 weeks of age. Many dogs that inherit this disorder die within the first year of life, though some can live 3 to 5 years and reproduce.

Figure 4.11b (left side) considers a cross between an unaffected female dog with two copies of the wild-type gene and a male dog with muscular dystrophy that carries the mutant allele

(a) Male golden retriever with X-linked muscular dystrophy

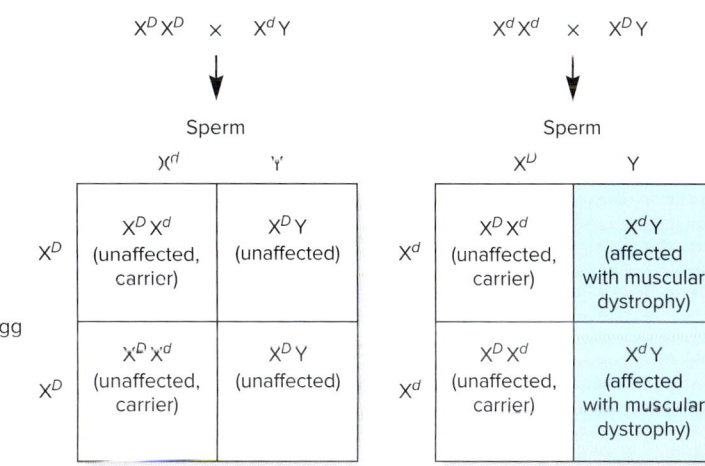

(b) Examples of X-linked muscular dystrophy inheritance patterns

FIGURE 4.11 **X-linked muscular dystrophy in dogs.** (a) The male golden retriever shown here has the disease. (b) The Punnett square on the left shows a cross between an unaffected female and an affected male. The one on the right shows a reciprocal cross between an affected female and an unaffected male. *D* represents the common (non-disease-causing) allele for the dystrophin gene, and *d* is the mutant allele that causes a defect in dystrophin function.

Courtesy of Dr. Joseph Kornegay

Concept Check: *Explain why the reciprocal cross yields a different result from that obtained in the first cross.*

Genetic TIPS

The Question: Calvin Bridges, who worked in Morgan's lab, made crosses involving the inheritance of X-linked traits in fruit flies. One of his experiments concerned two different X-linked genes affecting eye color and wing length. For the eye color gene, the red-eye allele (w^+) is dominant to the white-eye allele (w). A second X-linked trait is wing length; the allele called *miniature* is recessive to the normal allele. In this case, m represents the *miniature* allele and m^+ the normal allele, which is designated long wings. A male fly carrying a *miniature* allele on its single X chromosome has small (miniature) wings. A female must be homozygous, mm, in order to have miniature wings.

Bridges made a cross between $X^{w,m^+} X^{w,m^+}$ female flies (white eyes and long wings) and $X^{w^+,m}$ Y male flies (red eyes and miniature wings). He then examined the eyes, wings, and sexes of thousands of offspring. As expected, most of the offspring were females with red eyes and long wings or males with white eyes and long wings. On rare occasions (approximately 1 out of 1700 flies), however, he also obtained female offspring with white eyes or males with red eyes. He also noted the wing length in these flies and then cytologically examined their chromosome composition using a microscope. The following results were obtained:

Offspring	Eye Color	Wing Length	Sex Chromosomes
Expected females	Red	Long	XX
Expected males	White	Long	XY
Unexpected females (rare)	White	Long	XXY
Unexpected males (rare)	Red	Miniature	X0

Source: Bridges, Calvin (1916) Non-Disjunction as Proof of the Chromosome Theory of Heredity, *Genetics*, vol. 1, no. 2, 107–163.

Explain how the unexpected female and male offspring were produced.

Topic: What topic in genetics does this question address?

The topic is X-linked inheritance—more specifically, the topic is about X-linked inheritance and its relationship to abnormalities in chromosome number.

Information: What information do you know based on the question and your understanding of the topic?

From the question, you know the outcome of a cross involving $X^{w,m^+} X^{w,m^+}$ female flies and $X^{w^+,m}$ Y male flies. From your understanding of the topic, you may remember that females normally transmit an X chromosome to both daughters and sons, whereas males normally transmit an X to their daughters and a Y to their sons. In fruit flies, sex is determined by the ratio of the number of X chromosomes to the number of sets of autosomes. It is not determined by the presence of the Y chromosome (see Figure 4.1).

Problem-Solving Strategies: Predict the outcome. Compare and contrast.

One strategy to solve this problem is to compare and contrast the outcomes of this cross if the female transmits the wrong number of chromosomes or if the male transmits the wrong number.

The cross is $X^{w,m^+} X^{w,m^+} \times X^{w^+,m}$ Y

Female Transmits	Male Transmits	Offspring Genotype	Offspring Phenotype
$X^{w,m^+} X^{w,m^+}$	Y	$X^{w,m^+} X^{w,m^+}$ Y	White eyes, long wings, female
$X^{w,m^+} X^{w,m^+}$	$X^{w^+,m}$	$X^{w,m^+} X^{w,m^+} X^{w^+,m}$	Red eyes, long wings, female
No sex chromosomes	Y	Y	Inviable
No sex chromosomes	$X^{w^+,m}$	$X^{w^+,m}$	Red eyes, miniature wings, male
X^{w,m^+}	$X^{w^+,m}$ Y	$X^{w,m^+} X^{w^+,m}$ Y	Red eyes, long wings, female
X^{w,m^+}	No sex chromosomes	X^{w,m^+}	White eyes, long wings, male

Answer: The white-eyed female flies were due to the union between an abnormal XX female gamete and a normal Y male gamete. The unexpected male offspring had only one X chromosome and no Y. These male offspring were due to the union between an abnormal egg without any X chromosome and a normal sperm containing one X chromosome. The wing length of the unexpected males was a particularly significant result. The red-eyed males had miniature wings. As noted by Bridges, this means that they inherited their X chromosome from their father rather than their mother.

4.4 REVIEWING THE KEY CONCEPTS

- Morgan's work showed that a gene affecting eye color in fruit flies is inherited on the X chromosome (see Figure 4.9).
- X-linked inheritance is also observed in mammals. An example is Duchenne muscular dystrophy, which is an X-linked recessive disorder. For X-linked alleles, reciprocal crosses yield different results (see Figures 4.10, 4.11).

4.4 COMPREHENSION QUESTIONS

1. In Morgan's experiments, a key observation of the F_2 generation that suggested an X-linked pattern of inheritance was
   a. a 3:1 ratio of red-eyed to white-eyed flies.
   b. the only white-eyed flies were males.
   c. the only white-eyed flies were females.
   d. the only red-eyed flies were males.

2. Hemophilia is a blood-clotting disorder in humans that follows an X-linked recessive pattern of inheritance. A man with hemophilia and a woman without hemophilia have a daughter with hemophilia. If you let H represent the common (non-disease-causing) allele and h the hemophilia allele, what are the genotypes of the parents?
   a. Mother is $X^H X^h$ and father is $X^h Y$.
   b. Mother is $X^h X^h$ and father is $X^h Y$.
   c. Mother is $X^h X^h$ and father is $X^H Y$.
   d. Mother is $X^H X^h$ and father is $X^H Y$.

3. A cross is made between a white-eyed female fruit fly and a red-eyed male. What would be the reciprocal cross?
   a. Female is $X^w X^w$ and male is $X^w Y$.
   b. Female is $X^{w+} X^{w+}$ and father is $X^{w+} Y$.
   c. Mother is $X^{w+} X^{w+}$ and father is $X^w Y$.
   d. Mother is $X^w X^w$ and father is $X^{w+} Y$.

KEY TERMS

Page 72. sex determination
Page 73. sex chromosomes, heterogametic sex, homogametic sex, autosomes, haplodiploid system
Page 74. protandrous hermaphrodites
Page 75. dioecious, dosage compensation
Page 76. X-chromosome inactivation (XCI), Barr body, Lyon hypothesis
Page 78. X-inactivation center (Xic)
Page 79. sex-linked genes, X-linked genes
Page 80. Y-linked genes (holandric genes), pseudoautosomal genes, pseudoautosomal inheritance, X-linked inheritance, hemizygous
Page 82. X-linked alleles, testcross, X-linked recessive
Page 83. reciprocal cross

CHAPTER SUMMARY

4.1 Mechanisms of Sex Determination Among Various Species

- In the X-Y, X-0, and Z-W systems, sex is determined by the presence and number of particular sex chromosomes (see Figure 4.1).
- In the haplodiploid system, sex is determined by the number of sets of chromosomes (see Figure 4.2).
- In some species, such as alligators and clownfish, sex is determined by environmental factors (see Figure 4.3).
- Dioecious plants exist as individuals that produce only pollen and those that produce only eggs. Sex chromosomes sometimes determine sex in these species (see Figure 4.4).

4.2 Dosage Compensation and X-Chromosome Inactivation in Mammals

- Dosage compensation often occurs in species in which males and females differ in their sex chromosomes (see Table 4.1).
- In mammals, the process of X-chromosome inactivation (XCI) in females compensates for the single X chromosome found in males. The inactivated X chromosome is called a Barr body. The process can lead to a variegated phenotype, such as a calico cat (see Figure 4.5).

- After it occurs during embryonic development, X-chromosome inactivation is maintained when somatic cells divide (see Figure 4.6).
- X-chromosome inactivation is controlled by the X-inactivation center (Xic) that contains the *Xist* gene. The three phases of XCI are initiation, spreading, and maintenance phases (see Figure 4.7).

4.3 Properties of the X and Y Chromosomes in Mammals

- Chromosomes that differ between males and females are termed sex chromosomes and carry sex-linked genes.
- X-linked genes are found only on the X chromosome, whereas Y-linked genes are found only on the Y chromosome. Pseudoautosomal genes are found on both the X and Y chromosomes in regions of homology (see Figure 4.8).

4.4 Transmission Patterns for X-Linked Genes

- Morgan's work showed that a gene affecting eye color in fruit flies is inherited on the X chromosome (see Figure 4.9).
- X-linked inheritance is also observed in mammals. An example is Duchenne muscular dystrophy, which is an X-linked recessive disorder. For X-linked alleles, reciprocal crosses yield different results (see Figures 4.10, 4.11).

PROBLEM SETS & INSIGHTS

More Genetic TIPS

1. A human male named Phillip has an X chromosome that is missing its Xic. Is this caused by a new mutation (one that occurred during gametogenesis), or could it be the result of a mutation that occurred in an earlier generation and is found in the somatic cells of one of his parents? Explain your answer. How would this mutation affect his ability to produce viable offspring?

Topic: What topic in genetics does this question address?

The topic is X-chromosome inactivation. More specifically, the question is about the effects of a mutation that removes the Xic from one of the X chromosomes.

Information: What information do you know based on the question and your understanding of the topic?

From the question, you know that Phillip has an X chromosome that is missing its Xic. From your understanding of the topic, you may remember that an X chromosome that is missing its Xic will not be inactivated. This is a lethal condition in females.

Problem-Solving Strategies: Predict the outcome. Compare and contrast.

One strategy to solve this problem is to compare the outcomes if the missing Xic occurred in a previous generation (denoted with an asterisk) or if it occurred as a new mutation during oogenesis or spermatogenesis in one of Phillip's parents. Let's compare the following crosses, in which X* indicates an X chromosome that is missing its Xic:

X*X × XY: Mother would have been inviable, so this cross is not possible.

XX × X*Y: Father could not pass X* to his son.

XX × XY (new mutation during oogenesis): A son with a missing Xic could be produced.

XX × XY (new mutation during spermatogenesis): Father could not pass X* to his son.

Answer: The missing Xic must be due to a new mutation that occurred during oogenesis in Phillip's mother. Phillip will pass his X chromosome to his daughters and his Y chromosome to his sons. He cannot produce living daughters, because a missing Xic is lethal in females. Phillip can produce living sons.

2. An unaffected woman (i.e., without disease symptoms), who is heterozygous for the recessive X-linked allele causing Duchenne muscular dystrophy has children with a man with the common (non-disease-causing) allele. What is the probability that this couple will have an unaffected son?

Topic: What topic in genetics does this question address?

The topic is X-linked inheritance. More specifically, the question is about the inheritance of Duchenne muscular dystrophy.

Information: What information do you know based on the question and your understanding of the topic?

From the question, you know a woman is heterozygous for the disease-causing allele, which is X-linked. The man carries the common, or non-disease-causing, allele. From your understanding of the topic, you may remember that females inherit two copies of X-linked genes, one from each parent, whereas males inherit the X chromosome only from their mother.

Problem-Solving Strategy: Predict the outcome.

One strategy to solve this problem is to set up a Punnett square in which D represents the common allele and d is the recessive allele that causes Duchenne muscular dystrophy. The mother is heterozygous, and the father has the common allele.

	X^D	Y
X^D	$X^D X^D$	$X^D Y$
X^d	$X^D X^d$	$X^d Y$

Phenotype ratio is

2 unaffected daughters :
1 unaffected son :
1 affected son

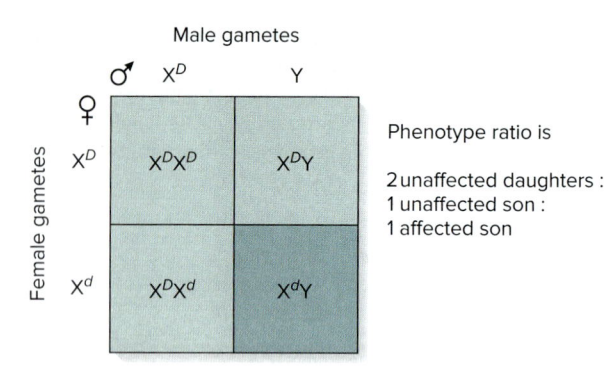

Answer: Four genotypes are possible for the couple's children, one of whom is an unaffected son. Therefore, the probability of an unaffected son is 1/4.

3. Red-green color blindness is inherited as a recessive X-linked trait in humans. A woman named Selena has parents who are not color-blind and a color-blind brother. What is the probability that Selena's first child will be a color-blind son? Assume that her first child could be a male or a female.

Topic: What topic in genetics does this question address?

The topic is X-linked inheritance.

Information: What information do you know based on the question and your understanding of the topic?

From the question, you know that Selena has a father who carries the common allele, which does not cause color blindness. Selena's mother must have been heterozygous because she had a color-blind son (Selena's brother). From your understanding of the topic, you may remember that females inherit two copies of X-linked genes, one from each parent, whereas males inherit the X chromosome only from their mother.

Problem-Solving Strategies: Predict the outcome. Make a calculation.

The first thing you need to do is to determine the likelihood that Selena carries the color-blind allele. One strategy to solve this problem is to set up a Punnett square in which C represents the common allele and c is the recessive allele that causes red-green color blindness. Selena's mother is heterozygous, and her father has the common allele.

	X^C	Y
X^C	$X^C X^C$	$X^C Y$
X^c	$X^C X^c$	$X^c Y$

Phenotype ratio is

2 unaffected daughters :
1 unaffected son :
1 color-blind son

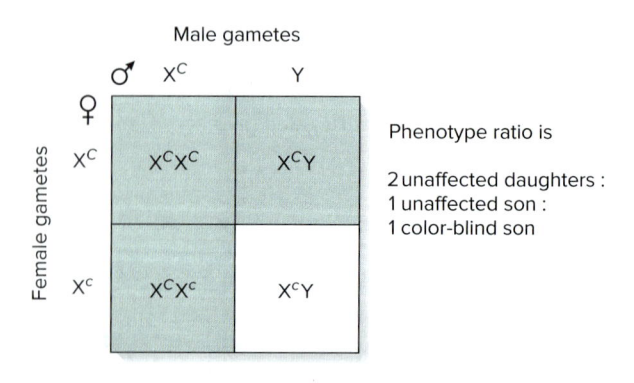

After you have determined the likelihood that Selena carries the color-blind allele, you can use the product rule to determine the likelihood that she will pass that allele to her son because these are independent outcomes.

P = (probability that Selena carries the color-blind allele)(probability that she will pass it to a son)

Answer: From the Punnett square, there is a 50% chance that Selena is a carrier. If she has children, 25% will be affected sons if she is a carrier. However, there is only a 50% chance that she is a carrier. We multiply 50% times 25%, which equals $0.5 \times 0.25 = 0.125$, or a 12.5% chance.

Conceptual Questions

C1. In the X-Y and Z-W systems of sex determination, which sex is the heterogametic sex, the male or female?

C2. Let's suppose that a gene affecting pigmentation is found on the X chromosome (in mammals or insects) or the Z chromosome (in birds) but not on the Y or W chromosome. It is found on an autosome in bees. This gene exists in two alleles: D (dark) is dominant to d (light). What would be the phenotypic results of crosses between true-breeding dark females and true-breeding light males, and the reciprocal crosses involving true-breeding light females and true-breeding dark males, in the following species? Refer back to Figures 4.1 and 4.2 for the mechanism of sex determination in these species.

　A. Birds

　B. Fruit flies

　C. Bees

　D. Humans

C3. Assuming that such a fly would be viable, what would be the sex of a fruit fly with each of the following chromosomal compositions?

　A. One X chromosome and two sets of autosomes

　B. Two X chromosomes, one Y chromosome, and two sets of autosomes

　C. Two X chromosomes and four sets of autosomes

　D. Four X chromosomes, two Y chromosomes, and four sets of autosomes

C4. What would be the sex of a human with each of the following numbers of sex chromosomes?

　A. XXX

　B. X (also denoted X0)

　C. XYY

　D. XXY

C5. With regard to the numbers of sex chromosomes, explain why dosage compensation is necessary.

C6. What is a Barr body? How is its structure different from other chromosomes in the cell? How does the structure of a Barr body affect the level of X-linked gene expression?

C7. Among different species, describe three distinct mechanisms for accomplishing dosage compensation.

C8. Describe when X-chromosome inactivation occurs and how this leads to phenotypic results at the organism level. In your answer, you should explain why X-chromosome inactivation causes results such as variegated coat patterns in mammals. Why do two different calico cats have their patches of orange and black fur in different places? Explain whether or not a variegated coat pattern due to X-chromosome inactivation could occur in marsupials.

C9. Describe the molecular process of X-chromosome inactivation. This description should include the three phases of inactivation and the role of the Xic. Explain what happens to the X chromosomes during embryogenesis, in adult somatic cells, and during oogenesis.

C10. On rare occasions, a human male is born who is somewhat feminized compared with most males. Microscopic examination of the cells of one such individual revealed that he has a single Barr body in each cell. What is the chromosomal composition of this individual?

C11. How many Barr bodies would you expect to find in a human with each of the following abnormal compositions of sex chromosomes?

　A. XXY

　B. XYY

　C. XXX

　D. X0 (a person with just a single X chromosome)

C12. Certain forms of human color blindness are inherited as an X-linked recessive trait. Hemizygous males are color-blind, but heterozygous females are not. However, heterozygous females sometimes have partial color blindness.

　A. Discuss why heterozygous females may have partial color blindness.

　B. Doctors identified an unusual case in which a heterozygous female was color-blind in her right eye but had normal color vision in her left eye. Explain how this might have occurred.

C13. A black female cat ($X^B X^B$) and an orange male cat ($X^O Y$) were mated to each other and produced a male cat that was calico. Which sex chromosomes did this male offspring inherit from its mother and father? Remember that the presence of the Y chromosome determines maleness in mammals.

C14. What is the spreading phase of X-chromosome inactivation? Why do you think it is called a spreading phase? Discuss the role of the *Xist* gene in the spreading phase of X-chromosome inactivation.

C15. A human disease known as vitamin D-resistant rickets is inherited as an X-linked dominant trait. If a male with the disease produces children with a female who does not have the disease, what is the expected ratio of affected and unaffected offspring?

C16. Hemophilia is an X-linked recessive disorder in humans. If a heterozygous woman has children with an unaffected man, what are the probabilities that she will have the following combinations of children?

A. An affected son

B. Four unaffected offspring in a row

C. An unaffected daughter or son

D. Two out of five offspring that are affected

C17. Incontinentia pigmenti is a rare, X-linked dominant disorder in humans that is characterized by swirls of pigment in the skin. If an affected female, who had an unaffected father, has children with an unaffected male, what are the predicted ratios of affected and unaffected sons and daughters?

Application and Experimental Questions

E1. Discuss the experimental observations that Lyon brought together in proposing her hypothesis concerning X-chromosome inactivation. In your own words, explain how these observations were consistent with her hypothesis.

E2. The *Mic2* gene in humans is present on both the X and Y chromosomes. Let's suppose the *Mic2* gene exists in a dominant *Mic2* allele, which results in a functional cell-surface protein, and a recessive *mic2* allele, which results in a defective cell-surface protein. Using molecular techniques, it is possible to identify homozygous and heterozygous individuals. By following the transmission of the *Mic2* and *mic2* alleles in a large human pedigree, would you be able to distinguish between pseudoautosomal inheritance and autosomal inheritance? Explain your answer.

E3. In Morgan's experiments, which result do you think is the most convincing piece of evidence pointing to X-linkage of the eye color gene? Explain your answer.

E4. In his original studies, summarized in Figure 4.9, Morgan first suggested that the original white-eyed male had two copies of the white-eye allele. In this problem, let's assume that he meant the fly was $X^w Y^w$ instead of $X^w Y$. Are his data in Figure 4.9 consistent with this hypothesis? What crosses would need to be made to rule out the possibility that the Y chromosome carries a copy of the eye color gene?

E5. How would you set up crosses to determine if a gene is Y-linked versus X-linked?

E6. Occasionally during meiosis, a mistake can happen whereby a gamete receives zero or two sex chromosomes rather than one. Bridges made a cross between white-eyed female flies and red-eyed male flies. As you would expect, most of the offspring were red-eyed females and white-eyed males. On rare occasions, however, he found a white-eyed female or a red-eyed male. These rare flies were not due to new gene mutations but instead were due to mistakes during meiosis in the parent flies. Consider the mechanism of sex determination in fruit flies and propose how this could happen. In your answer, describe the sex chromosome composition of the rare flies.

E7. White-eyed flies have a lower survival rate than red-eyed flies. Based on Morgan's data in Figure 4.9, what percentage of white-eyed flies survived compared with red-eyed flies, assuming 100% survival of red-eyed flies?

E8. A cross was made between female flies with white eyes and miniature wings (both X-linked recessive traits) to male flies with red eyes and long wings. On rare occasions, female offspring were produced with white eyes. If we assume these females are due to errors in meiosis, what would be their most likely chromosomal composition? What would be their wing length?

E9. Experimentally, how do you think researchers were able to determine that the Y chromosome causes maleness in mammals, whereas the ratio of X chromosomes to the sets of autosomes causes sex determination in fruit flies?

Questions for Student Discussion/Collaboration

1. Sex determination among different species is caused by a variety of mechanisms. With regard to evolution, discuss why you think this has happened. During the evolution of life on Earth over the past 4 billion years, do you think the phenomenon of opposite sexes is a relatively recent event?

2. In the experiment described in Figure 4.9, Morgan obtained a white-eyed male fly in a population containing many red-eyed flies that he thought were true-breeding. The figure says that he crossed this fly with a red-eyed female, and all the offspring had red eyes. But actually this is not quite true. Morgan observed 1237 red-eyed flies and 3 white-eyed males. Provide two or more explanations why he obtained 3 white-eyed males in the F_1 generation.

Answers to Comprehension Questions

4.1: d, a, b

4.2: b, a, d

4.3: a

4.4: b, a, c

Note: All answers appear at the website in Connect; the answers to even-numbered questions and all Concept Check questions are in Appendix B.

5

CHAPTER OUTLINE

- 5.1 Overview of Simple Inheritance Patterns
- 5.2 Dominant and Recessive Alleles
- 5.3 Environmental Effects on Gene Expression
- 5.4 Incomplete Dominance, Overdominance, and Codominance
- 5.5 Sex-Influenced and Sex-Limited Inheritance
- 5.6 Lethal Alleles
- 5.7 Understanding Complex Phenotypes Caused by Mutations in Single Genes
- 5.8 Gene Interaction

Inheritance patterns and alleles. In the petunia, multiple alleles can result in flowers with several different colors, such as the ones shown here.
©dimitriosp/123RF

EXTENSIONS OF MENDELIAN INHERITANCE

The term **Mendelian inheritance** refers to inheritance patterns that obey two laws: the law of segregation and the law of independent assortment. Until now, we have mainly considered traits that are affected by a single gene that is found in two different alleles. In these cases, one allele is dominant to the other. This type of inheritance is sometimes called **simple Mendelian inheritance** because the observed ratios in the offspring clearly obey Mendel's laws. For example, when two different true-breeding pea plants are crossed (e.g., tall and dwarf), and the F_1 generation is allowed to self-fertilize, the F_2 generation shows a 3:1 phenotypic ratio of tall to dwarf offspring. In Chapter 4, we also considered patterns of inheritance when genes are found on sex chromosomes.

In this chapter, we will extend our understanding of Mendelian inheritance in several ways. First, we will take a closer look at simple dominant/recessive patterns of inheritance in order to understand why some alleles are recessive whereas others are dominant. We will then consider situations in which alleles do not show a simple dominant/recessive relationship. Geneticists have discovered an amazing diversity of mechanisms by which alleles affect the outcome of traits. Many alleles don't produce the ratios of offspring that are expected with simple Mendelian inheritance. This does not mean that Mendel was wrong. Rather, the inheritance patterns of many traits are more complex and interesting than he had realized. In this chapter, we will examine how the outcome of a trait may be influenced by a variety of factors, such as the level of protein expression, the sex of the individual, the presence of multiple alleles of a given gene, and environmental effects. At the end of this chapter, we will explore how two different genes can contribute to the outcome of a single trait. Later, in Chapters 6 and 7, we will examine eukaryotic inheritance patterns that actually violate the law of segregation or the law of independent assortment.

5.1 OVERVIEW OF SIMPLE INHERITANCE PATTERNS

Learning Outcomes:

1. Compare and contrast the different types of Mendelian inheritance patterns involving single genes.
2. Describe the molecular mechanisms that account for the Mendelian inheritance patterns involving single genes.

Before we delve more deeply into inheritance patterns, let's first compare a variety of inheritance patterns (Table 5.1). We have already discussed two of these—simple dominant/recessive inheritance and X-linked inheritance—in Chapters 3 and 4, respectively. These various patterns occur because the outcome of a trait may be governed by two or more alleles in several different ways. Geneticists want to understand Mendelian inheritance for two reasons.

- One goal is to predict the outcome of crosses. Many of the inheritance patterns described in Table 5.1 do not produce a 3:1 phenotypic ratio when two heterozygotes produce offspring. In this chapter, we will consider how allelic interactions produce ratios that may differ from a simple Mendelian pattern.
- A second goal is to understand how the molecular expression of genes can account for an individual's phenotype. In other words, what is the underlying relationship between molecular genetics—the expression of genes to produce functional proteins—and the traits of individuals that inherit the genes? The remaining sections of this chapter will explore several patterns of inheritance from a molecular perspective.

5.1 REVIEWING THE KEY CONCEPTS

- Mendelian inheritance patterns obey Mendel's laws.
- Several inheritance patterns involving single genes differ from those observed by Mendel (see Table 5.1).

5.1 COMPREHENSION QUESTION

1. Which of the following statements is *false*?
   a. Mendelian inheritance patterns obey the law of segregation.
   b. Mendelian inheritance patterns obey the law of independent assortment.
   c. All inheritance patterns show a simple dominant/recessive relationship.
   d. None of the above is false.

TABLE 5.1

Types of Mendelian Inheritance Patterns Involving Single Genes

Type	Description
Simple Mendelian	**Inheritance:** As described in Chapter 3, this term is commonly applied to the inheritance of alleles that obey Mendel's laws and follow a strict dominant/recessive relationship. In this chapter, we will see that some genes can be found in three or more alleles, making the dominant/recessive relationship more complex. **Molecular:** The dominant allele encodes a functional protein, and 50% of the protein is sufficient to produce the dominant trait.
X-linked	**Inheritance:** As described in Chapter 4, this pattern involves the inheritance of genes that are located on the X chromosome. In mammals and fruit flies, males have a single copy of X-linked genes, whereas females have two copies. **Molecular:** If a pair of X-linked alleles shows a simple dominant/recessive relationship, the dominant allele encodes a functional protein, and 50% of the protein is sufficient to produce the dominant trait (in a heterozygous female).
Incomplete penetrance	**Inheritance:** This pattern occurs when a dominant phenotype is not expressed even though an individual carries a dominant allele. An example is an individual who carries the polydactyly allele but has a normal number of fingers and toes. **Molecular:** Even though a dominant gene may be present, the protein encoded by the gene may not exert its effects. This can be due to environmental influences or to other genes that may encode proteins that counteract the effects of the dominant allele.
Incomplete dominance	**Inheritance:** This pattern occurs when the heterozygote has a phenotype that is intermediate between either corresponding homozygote. For example, a cross between homozygous red-flowered and homozygous white-flowered parents produces heterozygous offspring with pink flowers. **Molecular:** 50% of a functional protein is not sufficient to produce the same trait as in a homozygote with 100% of that protein.
Overdominance	**Inheritance:** This pattern occurs when the heterozygote has a trait that is more beneficial than either homozygote. **Molecular:** Three common ways that heterozygotes may gain benefits: (1) Their cells may have increased resistance to infection by microorganisms; (2) they may produce more forms of protein dimers with enhanced function; or (3) they may produce proteins that function under a wider range of conditions.
Codominance	**Inheritance:** This pattern occurs when the heterozygote expresses both alleles simultaneously without forming an intermediate phenotype. For example, in blood typing, an individual carrying the *A* and *B* alleles has an AB blood type. **Molecular:** The codominant alleles encode proteins that function slightly differently from each other, and the function of each protein in the heterozygote affects the phenotype uniquely.
Sex-influenced inheritance	**Inheritance:** This pattern refers to the effect of sex on the phenotype of the individual. Some alleles are recessive in one sex and dominant in the opposite sex. **Molecular:** Sex hormones may regulate the molecular expression of genes. This regulation can influence the phenotypic effects of alleles.
Sex-limited inheritance	**Inheritance:** In this pattern, a trait occurs in only one of the two sexes. An example is breast development in mammals. **Molecular:** Sex hormones may regulate the molecular expression of genes. This regulation can influence the phenotypic effects of alleles. In this pattern of inheritance, sex hormones that are primarily produced in only one sex are essential for an individual to display a particular phenotype.
Lethal alleles	**Inheritance:** A lethal allele is one that has the potential of causing the death of an organism. **Molecular:** Lethal alleles are most commonly loss-of-function alleles that encode proteins that are necessary for survival. In some cases, such an allele may be due to a mutation in a nonessential gene that changes a protein so that it functions with abnormal and detrimental consequences.

5.2 DOMINANT AND RECESSIVE ALLELES

Learning Outcomes:
1. Define *wild-type allele* and *genetic polymorphism*.
2. Explain the common underlying mechanisms for recessive and dominant alleles.
3. Describe how traits can exhibit incomplete penetrance and vary in their expressivity.

In Chapters 3 and 4, we examined patterns of inheritance that showed a simple dominant/recessive relationship. This means that a heterozygote exhibits the dominant trait. In this section, we will take a closer look at why an allele may be dominant or recessive and discuss how dominant alleles may not always exert their effects.

Recessive Alleles Often Cause a Reduction in the Amount or Function of the Encoded Proteins

For any given gene, geneticists refer to prevalent alleles in a natural population as **wild-type alleles.** In large populations, more than one wild-type allele may occur—a phenomenon known as **genetic polymorphism.** For example, **Figure 5.1** illustrates a striking example of polymorphism in the elderflower orchid, *Dactylorhiza sambucina*. Throughout the range of this species in Europe, both yellow- and red-flowered individuals are prevalent. Both colors are considered wild type. At the molecular level, a wild-type allele typically encodes a protein that is made in the proper amount and functions normally. As discussed in Chapter 23, wild-type alleles tend to promote the reproductive success of organisms in their native environments.

FIGURE 5.1 An example of genetic polymorphism. Both yellow and red flowers are common in natural populations of the elderflower orchid, *Dactylorhiza sambucina*, and both are considered wild type.
©blickwinkel/Alamy Stock Photo

Concept Check: Why are both of these colors considered to be wild type?

In addition, random mutations occur in populations and alter preexisting alleles. Geneticists sometimes refer to these kinds of alleles as **mutant alleles** to distinguish them from the more common wild-type alleles. Because random mutations are more likely to disrupt gene function, mutant alleles are often defective in their ability to express a functional protein. Such mutant alleles tend to be rare in natural populations. They are typically, but not always, inherited in a recessive fashion.

Among Mendel's seven characters, discussed in Chapter 3, the wild-type alleles are those that produce tall plants, purple flowers, axial flowers, yellow seeds, round seeds, green pods, and smooth pods (refer back to Figure 3.4). The mutant alleles result in dwarf plants, white flowers, terminal flowers, green seeds, wrinkled seeds, yellow pods, and constricted pods. You may have already noticed that the seven wild-type alleles are dominant over the seven mutant alleles. Likewise, red eyes and long wings are examples of traits produced by wild-type alleles in *Drosophila*, and white eyes and miniature wings are due to recessive mutant alleles.

The idea that recessive alleles usually cause a substantial decrease in the expression of a functional protein is supported by the analysis of many human genetic diseases. Keep in mind that a genetic disease is usually caused by a mutant allele. **Table 5.2** lists several examples of human genetic diseases in which the recessive allele fails to produce a specific cellular protein in its active form. In many cases, molecular techniques have enabled researchers to study these genes and determine the differences between the wild-type and mutant alleles. They have found that the recessive allele usually contains a mutation that causes a defect in the synthesis of a fully functional protein.

To understand why many defective mutant alleles are inherited recessively, we need to take a quantitative look at protein function. With the exception of sex-linked genes, diploid individuals have two copies of every gene. In a simple dominant/recessive relationship, the recessive allele does not affect the phenotype of the heterozygote. In other words, a single copy of the dominant allele is sufficient to mask the effects of the recessive allele. If the recessive allele cannot produce a functional protein, how do we explain the wild-type phenotype of the heterozygote?

- One common explanation is that 50% of the functional protein is adequate to provide the wild-type phenotype (**Figure 5.2**). In the example shown in the figure, the *PP* homozygote and *Pp* heterozygote each make sufficient amounts of the functional protein to yield purple flowers. This means that the homozygous individual makes twice as much of the wild-type protein as it really needs to produce purple flowers. Therefore, if the amount is reduced to 50%, as in the heterozygote, there is still enough of this protein to accomplish whatever cellular function it performs.
- A second possible explanation for other dominant alleles is that the heterozygote actually produces more than 50% of the functional protein. Due to gene regulation, the expression of the normal gene may be increased, or up-regulated, in the heterozygote to compensate for the lack of function of the defective allele. The topic of gene regulation is discussed in Chapters 16 and 17.

TABLE 5.2
Examples of Recessive Human Diseases

Disease	Protein That Is Produced by the Normal Gene*	Description
Phenylketonuria	Phenylalanine hydroxylase	Inability to metabolize phenylalanine. The disease can be prevented by following a phenylalanine-free diet. If the diet is not followed early in life, the result can be severe mental impairment and other symptoms.
Albinism	Tyrosinase	Lack of pigmentation in the skin, eyes, and hair.
Tay-Sachs disease	Hexosaminidase A	Defect in lipid metabolism. Leads to paralysis, blindness, and early death.
Sandhoff disease	Hexosaminidase B	Defect in lipid metabolism. Muscle weakness in infancy, early blindness, and progressive mental and motor deterioration.
Cystic fibrosis	Chloride transporter	Inability to regulate ion balance across epithelial cells. Leads to production of thick lung mucus and chronic lung infections.
Lesch-Nyhan syndrome	Hypoxanthine-guanine phosphoribosyl transferase	Inability to metabolize purines, which are bases found in DNA and RNA. Leads to self-mutilation behavior, poor motor skills, and usually mental impairment and kidney failure.

*Individuals who exhibit the disease are either homozygous for a recessive allele or hemizygous (for X-linked genes in human males). The disease symptoms result from a defect in the amount or function of the normal protein.

Dominant (functional) allele: *P* (purple)
Recessive (defective) allele: *p* (white)

Genotype	PP	Pp	pp
Amount of functional protein P	100%	50%	0%
Phenotype	Purple	Purple	White
Simple dominant/recessive relationship			

FIGURE 5.2 A comparison of protein levels among homozygous and heterozygous genotypes: *PP*, *Pp*, and *pp*.

Genes → Traits In a simple dominant/recessive relationship, 50% of the protein encoded by one copy of the dominant allele in the heterozygote is sufficient to produce the wild-type phenotype, in this case, purple flowers. A complete lack of the functional protein results in white flowers.

Concept Check: Does a PP individual produce more of the protein encoded by the P gene than is necessary for the purple color?

Dominant Mutant Alleles Usually Exert Their Effects in One of Three Ways

Though dominant mutant alleles are much less common than recessive alleles, they do occur in natural populations. How can a mutant allele be dominant over a wild-type allele? One of three mechanisms accounts for most dominant mutant alleles: a gain-of-function mutation, a dominant-negative mutation, or haploinsufficiency.

- **Gain-of-function mutations** change the gene or the protein encoded by a gene so that it gains a new or abnormal function. For example, a mutant gene may be overexpressed, thereby producing too much of the encoded protein.

- **Dominant-negative mutations** change a protein such that the mutant protein acts antagonistically to the normal protein. In a heterozygote, the mutant protein counteracts the effects of the normal protein, thereby altering the phenotype.

- In **haploinsufficiency,** the dominant mutant allele is a loss-of-function allele. Haploinsufficiency is used to describe patterns of inheritance in which a heterozygote (with one functional allele and one inactive allele) exhibits an abnormal or disease phenotype. An example in humans is a condition called polydactyly, in which a heterozygous individual has extra fingers or toes, which is discussed next.

Traits May Skip a Generation Due to Incomplete Penetrance and Vary in Their Expressivity

As we have seen, dominant alleles are expected to influence the outcome of a trait when they are present in heterozygotes. Occasionally, however, this may not occur.

- The phenomenon called **incomplete penetrance** is a situation in which an allele that is expected to cause a particular phenotype does not.

Figure 5.3a presents a human pedigree for a dominant trait known as polydactyly. This trait causes the affected individual to have additional fingers or toes (or both) (**Figure 5.3b**). Polydactyly is due to an autosomal dominant allele—the allele occurs in a gene located on an autosome (not a sex chromosome), and a single copy of this allele is sufficient to cause this condition. Sometimes, however, individuals carry the dominant allele but do not exhibit the trait. In Figure 5.3a, individual III-2 has inherited the polydactyly allele from his mother and passed the allele to a daughter and son. However, individual III-2 does not actually exhibit the trait himself, even though he is a heterozygote. In the case of polydactyly, the dominant allele does not always "penetrate" into the phenotype of the individual. Alternatively, for recessive traits, incomplete penetrance occurs if a homozygote carrying the recessive allele does not exhibit the recessive trait.

has a high expressivity of this trait, whereas a person with a single extra digit has a low expressivity.

How do we explain incomplete penetrance and variable expressivity? Although the answer may not always be understood, the range of phenotypes is often due to two factors.

- The environment may affect the outcome of the phenotype.
- One or more modifier genes may also affect the phenotype. For example, a modifier gene may affect the expression of the gene associated with polydactly and thereby influence the number of fingers or toes.

5.2 REVIEWING THE KEY CONCEPTS

- Wild-type alleles are those that are prevalent in a natural population. When a gene exists in two or more wild-type alleles, this is genetic polymorphism (see Figure 5.1).
- Recessive alleles are often due to mutations that result in a reduction or loss of function of the encoded protein (see Figure 5.2 and Table 5.2).
- Dominant mutant alleles are most commonly caused by gain-of-function mutations, dominant-negative mutations, or haploinsufficiency.
- Incomplete penetrance occurs when an allele that is expected to be expressed is not expressed (see Figure 5.3).
- Traits may vary in their expressivity—the degree to which they are expressed.

5.2 COMPREHENSION QUESTIONS

1. Which of the following phenotypes is *not* due to a wild type allele?
   a. Yellow-flowered elderflower orchid
   b. Red-flowered elderflower orchid
   c. A gray elephant
   d. An albino (white) elephant

2. Dominant alleles may result from a mutation that causes
   a. the overexpression of a gene or its protein product.
   b. a protein to inhibit the function of a normal protein.
   c. a protein to be inactive, and 50% of the normal protein is insufficient for a normal phenotype.
   d. all of the above.

3. Polydactyly is a condition in which a person has extra fingers and/or toes. It is caused by a dominant allele. If a person carries this allele but does not have any extra fingers or toes, this is an example of
   a. haploinsufficiency.
   b. a dominant-negative mutation.
   c. incomplete penetrance.
   d. a gain-of-function mutation.

5.3 ENVIRONMENTAL EFFECTS ON GENE EXPRESSION

Learning Outcomes:
1. Discuss the role of the environment with regard to an individual's traits.
2. Define *norm of reaction*.

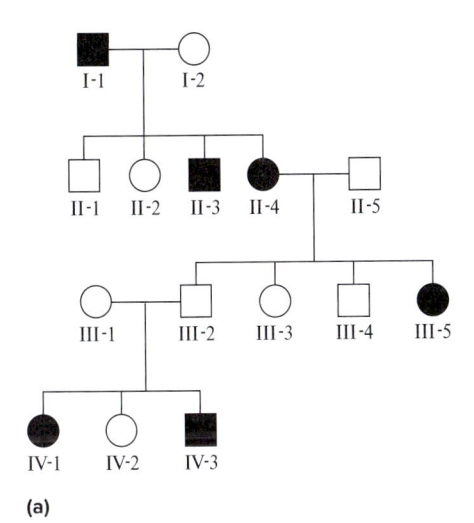

FIGURE 5.3 Polydactyly, a dominant trait that shows incomplete penetrance. (a) A family pedigree. Affected individuals are shown in black. Notice that offspring IV-1 and IV-3 have inherited the trait from a parent, III-2, who is heterozygous but does not exhibit polydactyly. (b) Antonio Alfonseca, a baseball player with polydactyly.
(b) ©Bob Shanley/Newscom

Concept Check: Which individual(s) in this pedigree exhibit(s) the effect of incomplete penetrance?

The measure of penetrance is described at the population level. For example, if 60% of the heterozygotes carrying a dominant allele exhibit the trait, we say that this trait is 60% penetrant. At the individual level, the trait is either present or not.

Another term used to describe the outcome of a trait is the degree to which the trait is expressed, or its **expressivity.** In the case of polydactyly, the number of extra digits can vary. For example, one individual may have an extra toe on only one foot, whereas a second individual may have extra digits on both the hands and feet. Using genetic terminology, a person with several extra digits

Throughout this book, our study of genetics tends to focus on the roles of genes in the outcome of traits. In addition to genetic variation, environmental conditions have a great effect on the phenotype of the individual. For example, the arctic fox (*Alopex lagopus*) goes through two color phases. During the cold winter, the arctic fox is primarily white, but in the warmer summer, it is mostly brown (**Figure 5.4a**). The fox has temperature-sensitive alleles affecting fur color, which are found among many species of mammals.

A dramatic example of the relationship between environment and phenotype can be seen in the human genetic disease known as phenylketonuria (PKU). This autosomal recessive disease is caused by a defect in a gene that encodes the enzyme phenylalanine hydroxylase. Homozygous individuals with this defective allele are unable to metabolize the amino acid phenylalanine properly. When given a standard diet containing phenylalanine, which is found in most protein-rich foods, PKU individuals manifest a variety of detrimental traits, including mental impairment, underdeveloped teeth, and foul-smelling urine. In contrast, when PKU individuals are diagnosed early and follow a restricted diet free of phenylalanine, they develop properly (**Figure 5.4b**).

Since the 1960s, testing methods have been developed that can determine if an individual is lacking the phenylalanine hydroxylase enzyme. These tests permit the identification of infants who have PKU. Their diets can then be modified before the harmful effects of phenylalanine ingestion have occurred. As a result of government legislation, more than 90% of infants born in the United States are now tested for PKU. This test prevents a great deal of human suffering and is cost-effective. In the United States, the annual cost of PKU testing is estimated to be a few million dollars, whereas the cost of treating severely affected individuals with the disease would be hundreds of millions of dollars.

The arctic fox and individuals with PKU provide examples of the dramatic effects of different environmental conditions. When considering the environment, geneticists often examine a range of conditions, rather than simply observing phenotypes under two different conditions. The term **norm of reaction** refers to the effects of environmental variation on a phenotype. Specifically, it is the phenotypic range seen in individuals with a particular genotype. To evaluate the norm of reaction, researchers begin with true-breeding strains that have the same genotypes and subject them to different environmental conditions. As an example, let's consider facet number in the eyes of fruit flies, *Drosophila melanogaster*. This species has compound eyes composed of many individual facets. **Figure 5.4c** shows the norm of reaction for facet

(a) Arctic fox in winter and summer

(b) Healthy person with PKU

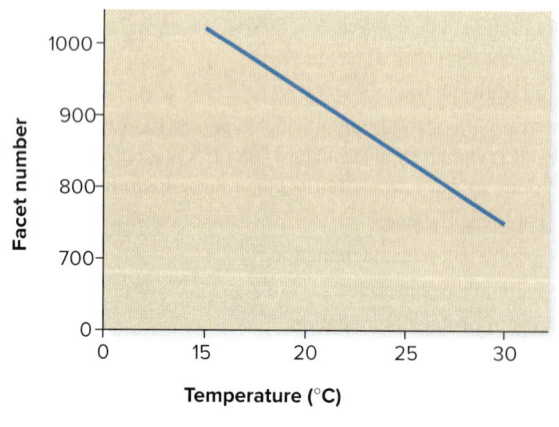

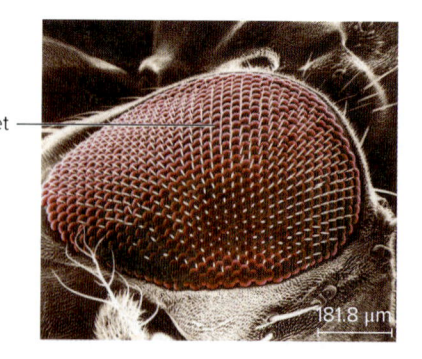

(c) Norm of reaction

FIGURE 5.4 Variation in the expression of traits due to environmental effects. (a) The arctic fox in winter and summer. (b) A person with PKU who has followed a restricted diet and developed properly. (c) Norm of reaction. In this experiment, fertilized eggs from a population of genetically identical fruit flies (*Drosophila melanogaster*) were allowed to develop into adults at different environmental temperatures. The graph shows the relationship between temperature (an environmental factor) and facet number in the eyes of the resulting adult flies. The micrograph shows the many facets of an eye of *D. melanogaster*.
(a: left) ©Dmitry Deshevykh/Getty Images; (a: right) ©Ondrej Prosicky/Shutterstock; (b) ©Sally Haugen/Virginia Schuett, www.pkunews.org; (c: right) ©Science History Images/Alamy Stock Photo

Concept Check: *What are the two main factors that determine an organism's traits?*

number in genetically identical fruit flies that developed from fertilized eggs at different temperatures. As shown in the graph, the facet number varies with changes in temperature. At a higher temperature (30°C), the facet number is approximately 750, whereas at a lower temperature (15°C), it is over 1000.

5.3 REVIEWING THE KEY CONCEPTS

- The outcome of traits is also influenced by the environment (see Figure 5.4).

5.3 COMPREHENSION QUESTION

1. The outcome of an individual's traits is controlled by
   a. genes.
   b. the environment.
   c. genes and the environment.
   d. neither genes nor the environment.

5.4 INCOMPLETE DOMINANCE, OVERDOMINANCE, AND CODOMINANCE

Learning Outcomes:
1. Predict the outcome of crosses involving incomplete dominance, overdominance, and codominance.
2. Explain the underlying molecular mechanisms of incomplete dominance, overdominance, and codominance.

Thus far, we have considered inheritance patterns that follow a simple dominant/recessive inheritance pattern. In these cases, the heterozygote exhibits a phenotype that is the same as a homozygote that carries two copies of the dominant allele but different from the homozygote carrying two copies of the recessive allele. In this section, we will examine three different inheritance patterns in which the heterozygote shows a phenotype that is different from both types of homozygotes.

Incomplete Dominance Occurs When Two Alleles Produce an Intermediate Phenotype

Although many alleles display a simple dominant/recessive relationship, some do not.

- **Incomplete dominance** is a condition in which the phenotype of a heterozygote is intermediate between the corresponding homozygous individuals.

In 1905, German botanist Carl Correns first observed this phenomenon in the colors of the flowers of the four-o'clock plant (*Mirabilis jalapa*). **Figure 5.5** describes Correns's experiment, in which a homozygous red-flowered four-o'clock plant was crossed to a homozygous white-flowered plant. The wild-type allele for red flower color is designated C^R and the allele for white flower color is C^W. As shown in the figure, the offspring had pink flowers. If these F_1 offspring were allowed to self-fertilize, the F_2 generation consisted of 1/4 red-flowered plants, 1/2 pink-flowered plants, and 1/4 white-flowered plants. The pink plants in the F_2 generation were heterozygotes with an intermediate phenotype. As shown in the Punnett square in Figure 5.5, the F_2 generation displayed a 1:2:1 phenotypic ratio, which is different from the 3:1 ratio observed for simple Mendelian inheritance.

In Figure 5.5, incomplete dominance occurs because a heterozygote has an intermediate phenotype. At the molecular level, the allele that causes a white phenotype is expected to result in a lack of a functional protein required for pigmentation. Depending on the effects of gene regulation, the heterozygotes may produce only 50% of the normal protein, but this amount is not sufficient to produce the same phenotype as the $C^R C^R$ homozygote, which may make twice as much of this protein. In this example, a reasonable explanation is that 50% of the functional protein cannot accomplish the same level of pigment synthesis that 100% of the protein can.

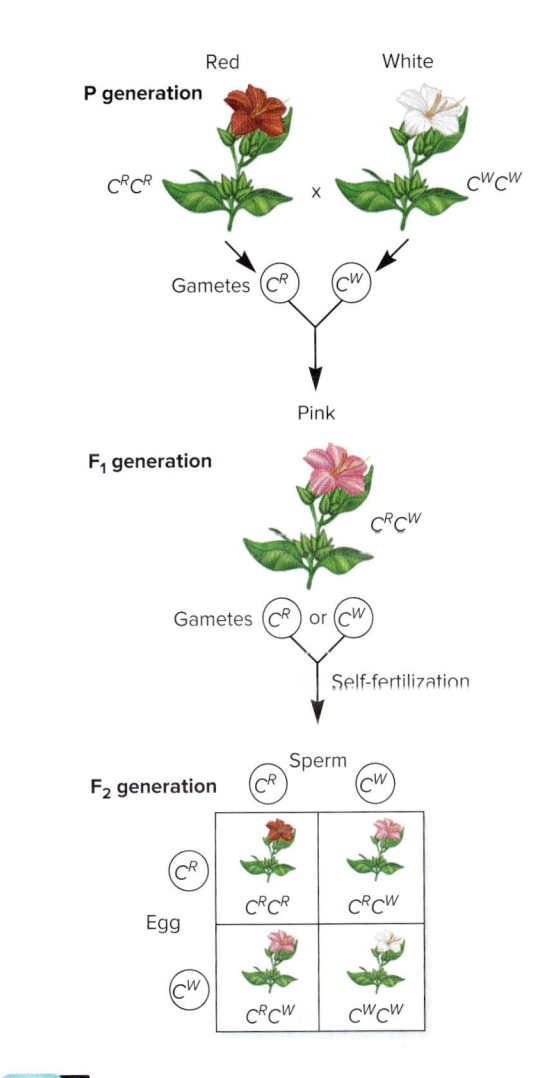

FIGURE 5.5 Incomplete dominance in the four-o'clock plant, *Mirabilis jalapa*.

Genes → Traits When two different homozygotes ($C^R C^R$ and $C^W C^W$) are crossed, the resulting heterozygote, $C^R C^W$, has an intermediate phenotype of pink flowers. In the heterozygote, 50% of the functional protein encoded by the C^R allele is not sufficient to produce a red phenotype.

Concept Check: At the molecular level, what is the explanation for why the four-o'clock flowers are pink instead of red?

Our Conclusions About Dominance May Depend on the Level of Examination

Our opinion of whether a trait is dominant or incompletely dominant may depend on how closely we examine the trait in the individual. The more closely we look, the more likely we are to discover that the heterozygote is not quite the same as the wild-type homozygote. For example, Mendel studied the characteristic of pea seed shape and visually concluded that the *RR* and *Rr* genotypes produced round seeds and the *rr* genotype produced wrinkled seeds. The peculiar morphology of the wrinkled seed is caused by a large decrease in the amount of starch deposition in the seed due to a defective *r* allele. More recently, other scientists have dissected round and wrinkled seeds and examined their contents under a microscope. They have found that round seeds from heterozygotes actually contain an intermediate number of starch grains compared with seeds from the corresponding homozygotes (**Figure 5.6**). Within the seed, an intermediate amount of the functional protein is not enough to produce as many starch grains as in the homozygote carrying two copies of the *R* allele. Even so, at the level of our unaided eyes, heterozygotes produce seeds that appear to be round. With regard to phenotypes, the *R* allele is dominant to the *r* allele at the level of visual examination, but the *R* and *r* alleles show incomplete dominance at the level of starch biosynthesis.

Overdominance Occurs When Heterozygotes Have Greater Reproductive Success

As we have seen, the environment plays a key role in the outcome of traits. For certain genes, heterozygotes may display characteristics that are more beneficial for their survival in a particular environment. Such heterozygotes may be more likely to survive and reproduce. For example, a heterozygote may be larger, more disease-resistant, or better able to withstand harsh environmental conditions.

- The phenomenon in which a heterozygote has greater reproductive success compared with either of the corresponding homozygotes is called **overdominance,** or **heterozygote advantage.**

A well-documented example of overdominance involves a human allele that causes sickle cell disease in homozygous individuals. This disease is an autosomal recessive disorder in which the affected individual produces an altered form of the protein hemoglobin, which carries oxygen within red blood cells. Most people carry the Hb^A allele and make hemoglobin A. Individuals affected with sickle cell disease are homozygous for the Hb^S allele and produce only hemoglobin S. This causes their red blood cells to deform into a sickle shape under conditions of low oxygen concentration (**Figure 5.7a, b**). The sickling phenomenon reduces the life span of these cells to only a few weeks, compared with a normal span of 4 months, and therefore, anemia results. In addition, abnormal sickled cells can become clogged in the capillaries throughout the body, leading to localized areas of oxygen depletion. Such an event causes pain and sometimes tissue and organ damage. For these reasons, the homozygous $Hb^S Hb^S$ individual usually has a shortened life span relative to an individual producing hemoglobin A.

In spite of the harmful consequences to homozygotes, the sickle cell allele has been found at a fairly high frequency among human populations that are exposed to malaria. The protozoan genus that causes malaria, *Plasmodium,* spends part of its life cycle within the *Anopheles* mosquito and another part within the red blood cells of humans who have been bitten by an infected mosquito. However, red blood cells of heterozygotes, $Hb^A Hb^S$, are likely to rupture when infected by this parasite, thereby preventing the parasite from propagating. People who are heterozygous have better resistance to malaria than do $Hb^A Hb^A$ homozygotes, and they do not suffer the ill effects of sickle cell disease. Therefore, even though the homozygous $Hb^S Hb^S$ condition is detrimental, the greater survival of the heterozygote has selected for the presence of the Hb^S allele within populations where malaria is prevalent. When viewing survival in such a region, overdominance explains the prevalence of the sickle cell allele. In Chapter 23, we will consider the role that natural selection plays in maintaining alleles that are beneficial to the heterozygote but harmful to the homozygote.

Figure 5.7c illustrates the predicted outcome when two heterozygotes have children. In this example, 1/4 of the offspring are $Hb^A Hb^A$ (unaffected, not malaria-resistant), 1/2 are $Hb^A Hb^S$ (unaffected, malaria-resistant), and 1/4 are $Hb^S Hb^S$ (have sickle cell disease). This 1:2:1 ratio deviates from a simple Mendelian 3:1 phenotypic ratio.

Overdominance is usually due to two alleles that produce proteins with slightly different amino acid sequences. How can we explain the observation that two protein variants in the heterozygote produce a more favorable phenotype? There are three common explanations.

Dominant (functional) allele: *R* (round)
Recessive (defective) allele: *r* (wrinkled)

Genotype	RR	Rr	rr
Amount of functional (starch-producing) protein	100%	50%	0%
Phenotype			
With unaided eye (simple dominant/recessive relationship)	Round	Round	Wrinkled
With microscope (incomplete dominance)			

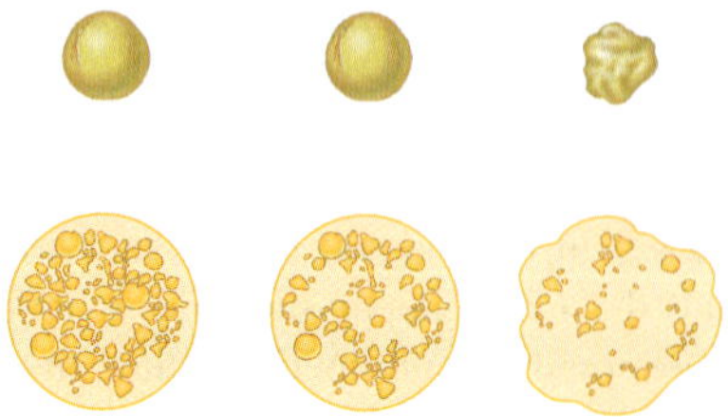

FIGURE 5.6 A comparison of phenotypes at the macroscopic and microscopic levels.

Genes → Traits This illustration shows the effects of a heterozygote having only 50% of the functional protein needed for starch production. The seed from the heterozygote appears to be as round as that of the homozygote carrying the *R* allele, but when examined microscopically, it has produced only half the amount of starch.

Concept Check: *At which level is incomplete dominance more likely to be observed—at the molecular/cellular level or at the organism level?*

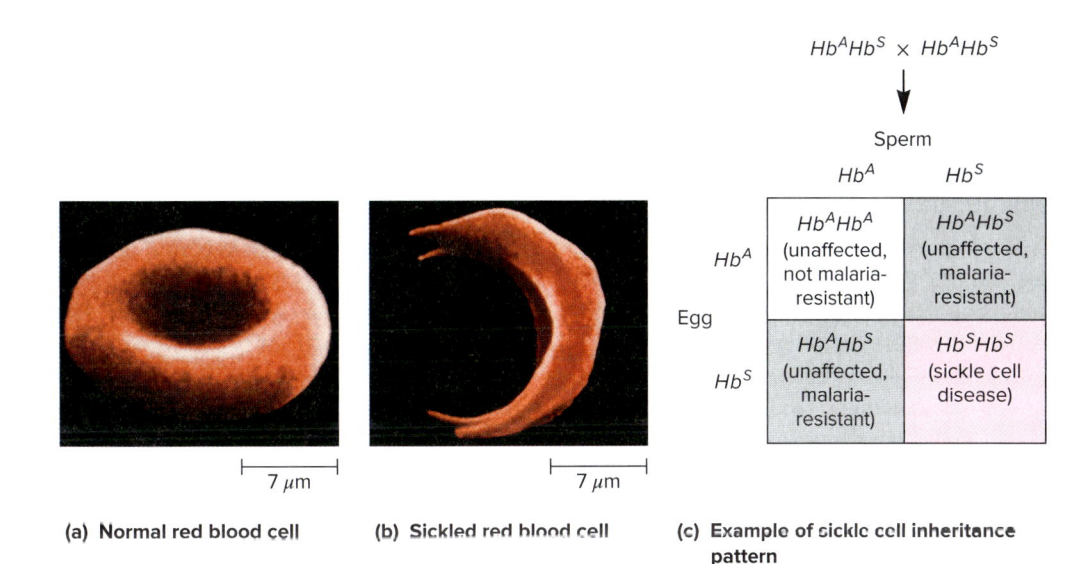

5.4 INCOMPLETE DOMINANCE, OVERDOMINANCE, AND CODOMINANCE

$Hb^A Hb^S \times Hb^A Hb^S$

(a) Normal red blood cell

(b) Sickled red blood cell

(c) Example of sickle cell inheritance pattern

	Sperm Hb^A	Hb^S
Egg Hb^A	$Hb^A Hb^A$ (unaffected, not malaria-resistant)	$Hb^A Hb^S$ (unaffected, malaria-resistant)
Hb^S	$Hb^A Hb^S$ (unaffected, malaria-resistant)	$Hb^S Hb^S$ (sickle cell disease)

FIGURE 5.7 Inheritance of sickle cell disease. A comparison of (a) normal red blood cells and (b) those from a person with sickle cell disease. (c) The outcome of a cross between two heterozygous individuals.
(a, b) ©Mary Martin/SPL/Science Source

Concept Check: Why does the heterozygote have an advantage?

Disease Resistance The heterozygote may exhibit increased disease resistance. In the case of sickle cell disease, the phenotype is related to the infectivity of *Plasmodium* (**Figure 5.8a**). In the heterozygote, the infectious agent is less likely to propagate within red blood cells. Interestingly, researchers have speculated that other alleles in humans may confer disease resistance in the heterozygous condition but are detrimental in the homozygous state. These include PKU, in which the heterozygous fetus may be resistant to miscarriage caused by a fungal toxin, and Tay-Sachs disease, in which the heterozygote may be resistant to tuberculosis.

Subunit Composition of Proteins In some cases, a protein functions as a complex of multiple subunits; each subunit is composed of one polypeptide. A protein composed of two subunits is called a dimer. When both subunits are encoded by the same gene, the protein is a **homodimer**. The prefix *homo-* means that the subunits come from the same type of gene, although the gene may exist in different alleles. **Figure 5.8b** considers a situation in which a gene exists in two alleles that encode polypeptides designated A1 and A2. Homozygous individuals can produce only A1A1 or A2A2 homodimers, whereas a heterozygote can also produce an A1A2 homodimer. Thus, heterozygotes produce three forms of the homodimer, homozygotes only one. For some proteins, A1A2 homodimers may have better functional activity because they are more stable or able to function under a wider range of conditions. The greater activity of the homodimer protein may be the underlying reason why a heterozygote has characteristics superior to either homozygote.

Differences in Protein Function In some cases, the proteins encoded by different alleles may exhibit functional differences. For example, let's suppose that a gene encodes a metabolic enzyme that can be found in two forms (corresponding to the two alleles), one that functions better at a lower temperature and another that functions optimally at a higher temperature (**Figure 5.8c**). The

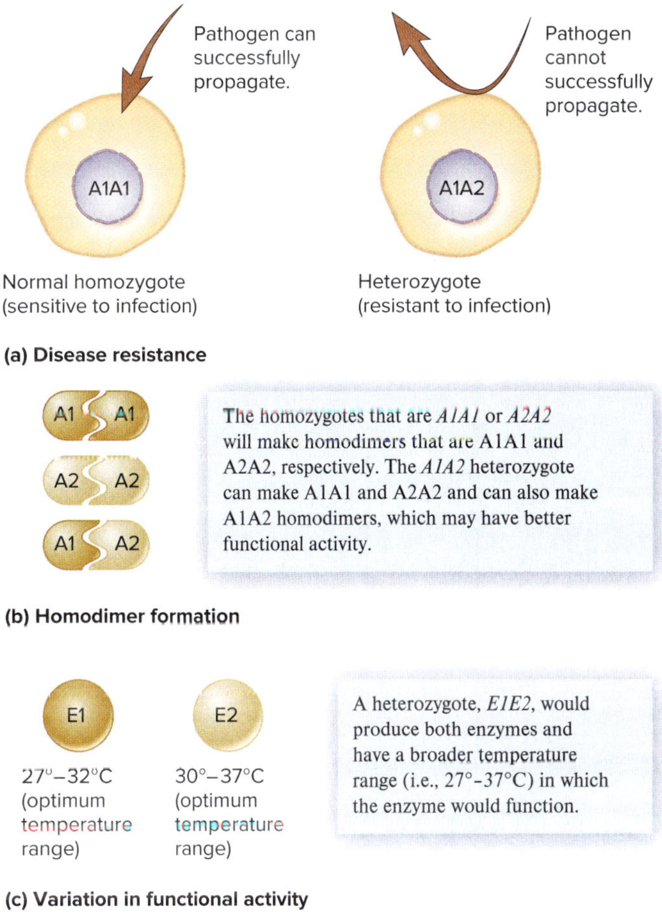

Normal homozygote (sensitive to infection) — Pathogen can successfully propagate. A1A1

Heterozygote (resistant to infection) — Pathogen cannot successfully propagate. A1A2

(a) Disease resistance

The homozygotes that are *A1A1* or *A2A2* will make homodimers that are A1A1 and A2A2, respectively. The *A1A2* heterozygote can make A1A1 and A2A2 and can also make A1A2 homodimers, which may have better functional activity.

(b) Homodimer formation

E1 — 27°–32°C (optimum temperature range)

E2 — 30°–37°C (optimum temperature range)

A heterozygote, *E1E2*, would produce both enzymes and have a broader temperature range (i.e., 27°–37°C) in which the enzyme would function.

(c) Variation in functional activity

FIGURE 5.8 Three possible explanations for overdominance at the molecular level.

Concept Check: Which of these three scenarios explains overdominance with regard to the sickle cell allele?

heterozygote, which makes a mixture of both enzymes, may be at an advantage under a wider temperature range than either of the corresponding homozygotes.

Alleles of the ABO Blood Group Can Be Dominant, Recessive, or Codominant

Thus far, we have considered examples in which a gene exists in two different alleles. As researchers have probed genes at the molecular level within natural populations of organisms, they have discovered that most genes exist in **multiple alleles.** Within a population, genes are typically found in three or more alleles. The ABO group of antigens, which determine blood type in humans, are produced in the body under the control of multiple alleles, which exhibit a relationship called codominance. To understand this concept, we first need to examine the molecular characteristics of human blood types. The plasma membranes of red blood cells have groups of interconnected sugars—oligosaccharides—that act as surface antigens (**Figure 5.9a**). Antigens are molecular structures that are recognized by antibodies produced by the immune system.

The synthesis of these surface antigens is controlled by three alleles, designated i, I^A, and I^B.

- The i allele is recessive to both I^A and I^B. A person who is homozygous ii has type O blood and produces a relatively short oligosaccharide, which is called H antigen.
- A homozygous $I^A I^A$ or heterozygous $I^A i$ individual has type A blood. The red blood cells of this individual contain the surface antigen known as A.
- A homozygous $I^B I^B$ or heterozygous $I^B i$ individual produces surface antigen B.
- A person who is $I^A I^B$ has the blood type AB and expresses both surface antigens A and B. The phenomenon in which

(a) ABO blood type

(b) Example of the ABO inheritance pattern

(c) Formation of A and B antigen by glycosyl transferase

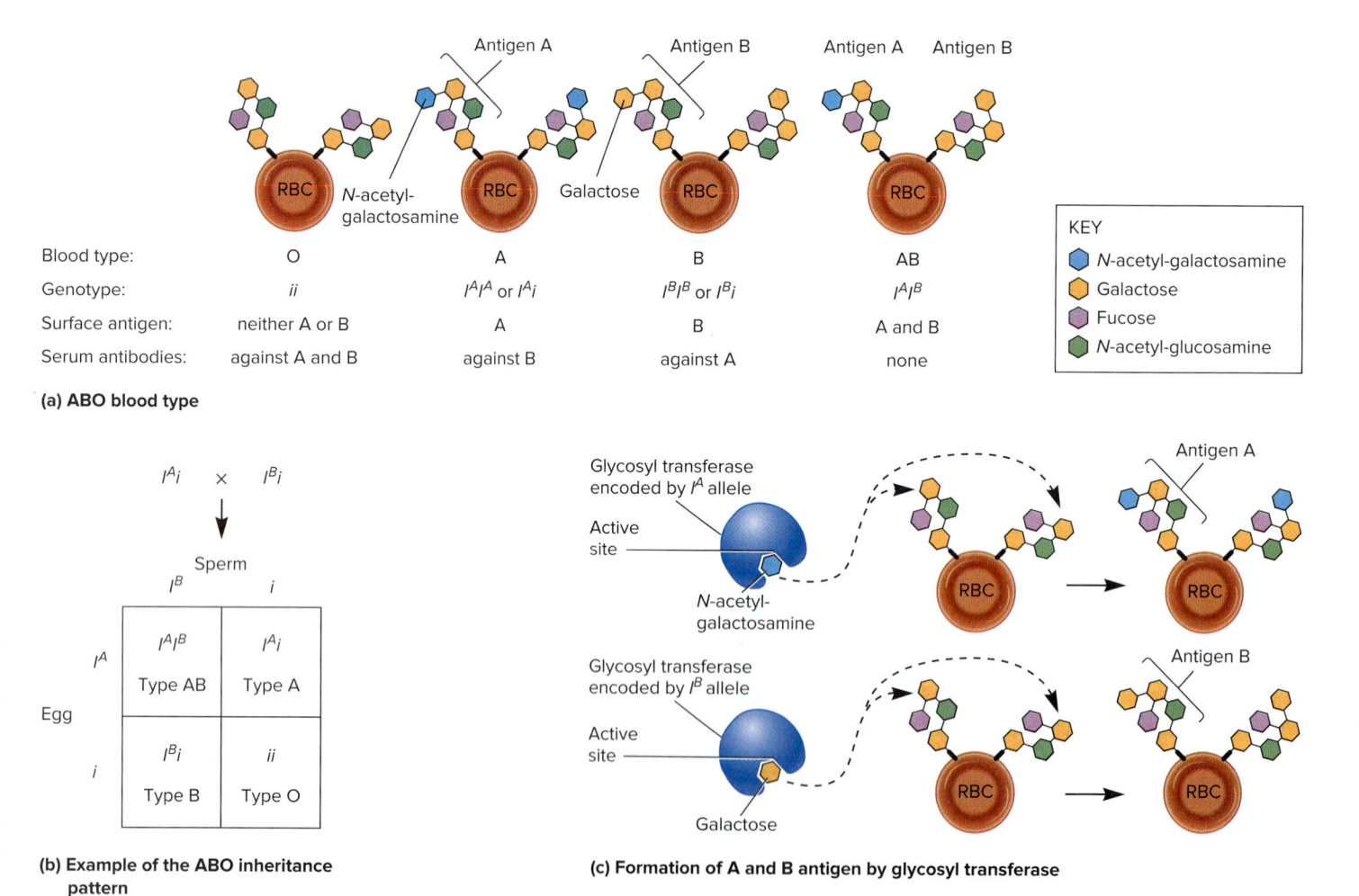

FIGURE 5.9 **ABO blood type.** (a) A schematic representation of blood type at the cellular level. Note: This is not drawn to scale. A red blood cell is much larger than any oligosaccharide on its surface. (b) The predicted offspring from parents who are $I^A i$ and $I^B i$. (c) The glycosyl transferase encoded by the I^A and I^B alleles recognizes different sugars due to changes in its active site. The i allele results in a nonfunctional enzyme.

Concept Check: Which allele is an example of a loss-of-function allele?

two alleles are both expressed in the heterozygous individual is called **codominance.** In this case, the I^A and I^B alleles are codominant to each other.

As an example of the inheritance of blood type, let's consider the possible offspring between two parents who are $I^A i$ and $I^B i$ (**Figure 5.9b**). The $I^A i$ parent makes I^A and i gametes, and the $I^B i$ parent makes I^B and i gametes. These combine to produce $I^A I^B$, $I^A i$, $I^B i$, and ii offspring in a 1:1:1:1 ratio. The resulting blood types are AB, A, B, and O, respectively.

Biochemists have analyzed the oligosaccharides on the surfaces of cells of differing blood types. In type O, the tree is smaller than type A or type B because a sugar has not been attached to a specific site on the tree. This idea is schematically shown in Figure 5.9a. How do we explain this difference at the molecular level? The gene that determines ABO blood type encodes an enzyme called glycosyl transferase that attaches a sugar to the oligosaccharide.

- The i allele carries a mutation that renders this enzyme inactive, which prevents the attachment of an additional sugar.
- The glycosyl transferase encoded by the I^A allele recognizes uridine diphosphate N-acetylgalactosamine (UDP-GalNAc) and attaches GalNAc to the oligosaccharide (**Figure 5.9c**). GalNAc is symbolized as a blue hexagon. This produces the structure of surface antigen A.
- The glycosyl transferase encoded by the I^B allele recognizes UDP-galactose and attaches galactose to the oligosaccharide. Galactose is symbolized as an orange hexagon. This produces the molecular structure of surface antigen B.
- A person with type AB blood makes both types of enzymes and thereby produces trees with both types of sugar attached.

A small difference in the structure of the oligosaccharide, namely, a GalNAc in antigen A versus galactose in antigen B, explains why the two antigens are different from each other at the molecular level. These differences enable them to be recognized by different antibodies. A person who has blood type A makes antibodies to blood type B (refer back to Figure 5.9a). The antibodies against blood type B require a galactose in the oligosaccharide for their proper recognition. Their antibodies do not recognize and destroy their own blood cells, but they do recognize and destroy the blood cells from a type B person.

5.4 REVIEWING THE KEY CONCEPTS

- Incomplete dominance is an inheritance pattern in which the heterozygote has an intermediate phenotype (see Figure 5.5).
- Whether we consider an allele to be dominant or incompletely dominant may depend on how closely we examine the phenotype (see Figure 5.6).
- Overdominance is an inheritance pattern in which the heterozygote has greater reproductive success (see Figures 5.7, 5.8).
- Many genes exist in multiple alleles in a population.
- Some alleles, such as those that produce A and B blood antigens, are codominant (see Figure 5.9).

5.4 COMPREHENSION QUESTIONS

1. A pink-flowered four-o'clock is crossed to a red-flowered plant. What is the expected outcome for the offspring?
   a. All pink
   b. All red
   c. 1 red : 2 pink : 1 white
   d. 1 red : 1 pink

2. A person with type AB blood has a child with a person with type O blood. What are the possible blood types of the child?
   a. A and B
   b. A, B, and O
   c. A, B, AB, and O
   d. O only

5.5 SEX-INFLUENCED AND SEX-LIMITED INHERITANCE

Learning Outcomes:
1. Compare and contrast sex-influenced inheritance and sex-limited inheritance.
2. Predict the outcome of crosses for sex-influenced inheritance.

In Chapter 4, we examined how the transmission pattern of sex-linked genes depends on whether the gene is on the X or Y chromosome and on the sex of the offspring. Sex can influence traits in other ways as well.

- The term **sex-influenced inheritance** refers to the phenomenon in which an allele is dominant in one sex but recessive in the opposite sex. Therefore, sex influence is a phenomenon of heterozygotes.

Sex-influenced inheritance should not be confused with sex-linked inheritance. The genes that govern sex-influenced traits are autosomal, not on the X or Y chromosome. Researchers once thought that human pattern baldness, which is characterized by hair loss on the front and top of the head but not on the sides, is an example of sex-influenced inheritance. However, recent research indicates that mutations in the androgen receptor gene, which is located on the X chromosome, often play a key role in pattern baldness. Therefore, pattern baldness typically follows an X-linked pattern of inheritance.

An example of sex-influenced inheritance is found in cattle. Certain breeds exhibit scurs, which are small horn-like growths on the frontal bone in the same locations where horns (in other breeds of cattle) would grow (**Figure 5.10**). This trait appears to be controlled by a single gene that exists in Sc and sc alleles. The trait is dominant in males and recessive in females:

Genotype	Phenotype	
	Males	Females
ScSc	Scurs	Scurs
Scsc	Scurs	No scurs (hornless)
scsc	No scurs	No scurs

Genetic TIPS

The Question: Scurs is an example of a sex-influenced trait in cattle that is dominant in males and recessive in females. A male and a female, neither of which has scurs, produce a male offspring with scurs. What are the genotypes of the parents?

Topic: What topic in genetics does this question address?

The topic is Mendelian inheritance. More specifically, the question is about sex-influenced inheritance.

Information: What information do you know based on the question and your understanding of the topic?

From the question, you know that a male and a female without scurs produce a male offspring with scurs. From your understanding of the topic, you may remember that the unique feature of sex-influenced inheritance is that a trait is dominant in one sex and recessive in the other.

Problem-Solving Strategy: Predict the outcome.

One strategy to solve this type of problem is to use a Punnett square to predict the possible outcomes of this cross. Because the father does not have scurs, you know he must be homozygous, *scsc*. Otherwise, he would have scurs. A female without scurs can be either *Scsc* or *scsc*. Therefore, two Punnett squares are possible, which are shown next.

♂	*sc*	*sc*
♀		
Sc	*Scsc*	*Scsc*
sc	*scsc*	*scsc*

♂	*sc*	*sc*
♀		
sc	*scsc*	*scsc*
sc	*scsc*	*scsc*

Answer: Because she has produced a male offspring with scurs (see the gray-shaded boxes), you know that she must be *Scsc* in order to pass the *Sc* allele to her son.

Some Traits Are Limited to One Sex

Another way in which sex affects an organism's phenotype is through **sex-limited inheritance,** in which a trait occurs in only one of the two sexes. The genes that influence sex-limited traits may be autosomal or sex-linked. In humans, examples of sex-limited traits are the presence of ovaries in females and the presence of testes in males. Due to these two sex-limited traits, mature females can produce only eggs, whereas mature males produce only sperm.

FIGURE 5.10 Scurs in cattle, an example of a sex-influenced trait.
Courtesy of Sheila M. Schmutz, Ph.D.

Concept Check: What is the phenotype of a female cow that is heterozyous for the scurs alleles?

Sex-limited traits are responsible for **sexual dimorphism,** in which members of the opposite sex have different morphological features. This phenomenon is common among many animal species and is often striking among various species of birds in which the male has more ornate plumage than the female. As shown in **Figure 5.11**, roosters have a larger comb and wattles and longer neck and tail feathers than do hens. These sex-limited features are never observed in normal hens.

(a) Hen   (b) Rooster

FIGURE 5.11 Differences in morphological features between female and male chickens, an example of sex-limited inheritance.
(a) ©Pixtal/age fotostock; (b) ©Image Source/Getty Images

Concept Check: What is the molecular explanation for sex-limited inheritance?

5.5 REVIEWING THE KEY CONCEPTS

- For sex-influenced traits such as scurs in cattle, heterozygous males and females have different phenotypes (see Figure 5.10).
- Sex-limited traits are expressed in only one sex, thereby resulting in sexual dimorphism (see Figure 5.11).

5.5 COMPREHENSION QUESTION

1. A cow with scurs and a bull with no scurs have an offspring. This offspring could be
   a. a female with scurs or a male with scurs.
   b. a female with no scurs or a male with scurs.
   c. a female with scurs or a male with no scurs.
   d. a female with no scurs or a male with no scurs.

5.6 LETHAL ALLELES

Learning Outcomes:
1. Describe the different types of lethal alleles.
2. Predict how lethal alleles may affect the outcome of a cross.

Let's now turn our attention to alleles that have the most detrimental effect on phenotype—those that result in death.

- An allele that has the potential to cause the death of an organism is called a **lethal allele.** Such alleles are usually inherited in a recessive manner.
- When the absence of a specific protein results in a lethal phenotype, the gene that encodes the protein is considered an **essential gene,** one that must be present for survival. Though the proportion varies by species, researchers estimate that approximately 1/3 of all genes are essential genes.
- By comparison, **nonessential genes** are not absolutely required for survival, although they are likely to be beneficial to the organism. A loss-of-function mutation in a nonessential gene does not usually cause death. On rare occasions, however, a nonessential gene may acquire a mutation that causes the gene product to be abnormally expressed in a way that may interfere with normal cell function and lead to a lethal phenotype.

Not all lethal mutations occur in essential genes, although the great majority do.

Many lethal alleles prevent cell division, thereby causing an organism to die at a very early stage. Others, however, may only exert their effects later in life. For example, a human genetic disease known as Huntington disease is caused by a dominant allele. The disease is characterized by a progressive degeneration of the nervous system, dementia, and early death. The age when these symptoms appear, or the **age of onset,** is usually between 30 and 50.

Other lethal alleles may kill an organism only when certain environmental conditions prevail. Such **conditional lethal alleles** have been extensively studied in experimental organisms. For example, some conditional lethal alleles cause an organism to die only in a particular temperature range. These alleles, called **temperature-sensitive (ts) lethal alleles,** have been observed in many organisms, including *Drosophila*. A ts lethal allele may be fatal for a developing larva at a high temperature (30°C), but the larva survives if grown at a lower temperature (22°C). Temperature-sensitive lethal alleles are typically caused by mutations that alter the structure of the encoded protein so it does not function correctly at the nonpermissive temperature or becomes unfolded and is rapidly degraded. Conditional lethal alleles may also be identified when an individual is exposed to a particular agent in the environment. For example, people with a defect in the gene that encodes the enzyme glucose-6-phosphate dehydrogenase (G-6-PD) have a negative reaction to the ingestion of fava beans. This reaction can lead to an acute hemolytic syndrome with a 10% mortality rate if not treated properly.

Finally, it is surprising that certain lethal alleles act only in some individuals. These are called **semilethal alleles.** Of course, any particular individual cannot be semidead. However, within a population, a semilethal allele will cause some individuals to die but not all of them. The reasons for semilethality are not always understood, but environmental conditions and the actions of other genes within the organism may help to prevent the detrimental effects of certain semilethal alleles. An example of a semilethal allele is the X-linked white-eye allele in fruit flies, which was described in Chapter 4 (see Figure 4.9). Depending on the growth conditions, approximately 1/4 to 1/3 of the flies that would be expected to exhibit the white-eyed trait die prematurely.

In some cases, a lethal allele may produce ratios that seemingly deviate from Mendelian ratios. An example is an allele in a breed of cats known as Manx, which originated on the Isle of Man (**Figure 5.12a**). The Manx cat carries a dominant mutation that affects the spine. This mutation shortens the tail, resulting in a range of tail lengths from normal to tailless. When two Manx cats are crossed to each other, the ratio of offspring is 1 normal to 2 Manx. How do we explain the 1:2 ratio? The answer is that about 1/4 of the offspring are homozygous for the dominant mutant allele, and they die during early embryonic development (**Figure 5.12b**). In this case, the Manx phenotype is dominant, whereas the lethal phenotype occurs only in the homozygous condition.

5.6 REVIEWING THE KEY CONCEPTS

- Lethal alleles most commonly occur in essential genes.
- Lethal alleles may result in inheritance patterns that yield unexpected ratios (see Figure 5.12).

5.6 COMPREHENSION QUESTION

1. The Manx phenotype in cats is caused by a dominant allele that is lethal in the homozygous state. A Manx cat is crossed to a normal (non-Manx) cat. What is the expected outcome for the surviving offspring?
   a. All Manx
   b. All normal
   c. 1 normal : 1 Manx
   d. 1 normal : 2 Manx

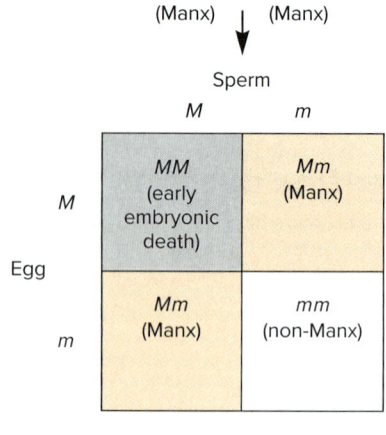

(a) A Manx cat

(b) Example of a Manx inheritance pattern

FIGURE 5.12 The Manx cat, which carries a lethal allele. (a) Photo of a Manx cat, which typically has a shortened tail. (b) Outcome of a cross between two Manx cats. Animals that are homozygous for the dominant Manx allele (*M*) die during early embryonic development.

(a) ©Juniors Bildarchiv GmbH/Alamy Stock Photo

Concept Check: Why do you think the *Mm* heterozygote survives with a developmental abnormality, whereas the *MM* homozygote dies?

5.7 UNDERSTANDING COMPLEX PHENOTYPES CAUSED BY MUTATIONS IN SINGLE GENES

Learning Outcomes:

1. Explain the phenomenon of pleiotropy.
2. Describe how embryonic development determines certain coat patterns in animals.

Thus far, we have considered a variety of examples in which mutations in a single gene affect the outcome of a single trait. For some of these examples, such as ABO blood typing, flower color in the four-o'clock plant, and sickle cell disease, the relationship between a mutation in a gene and its effect on a single trait is relatively easy to understand. However, for other genes, the phenotypic effects may be more complex, and researchers may have to dig deeper to understand how a mutation in a single gene can produce several effects on phenotype. In this section, we will consider how the expression of a single gene can have multiple effects throughout the body and see how an understanding of embryonic development can explain certain complex phenotypes.

Pleiotropy Occurs When the Expression of a Single Gene Has Two or More Phenotypic Effects

Although we tend to discuss genes within the context of how they influence a single trait, most genes actually have multiple effects throughout a cell or throughout a multicellular organism. The multiple effects of a single gene on the phenotype of an organism is called **pleiotropy**. Pleiotropy occurs for several reasons, including the following:

- The expression of a single gene can affect cell function in more than one way. For example, a defect in a microtubule protein may affect cell division and cell movement.
- A gene may be expressed in different cell types in a multicellular organism. For example, a gene may be expressed in a muscle cell and also expressed in a nerve cell.
- A gene may be expressed at different stages of development. For example, a gene may be expressed during embryonic development and also expressed in the adult.

In all or nearly all cases, the expression of a gene is pleiotropic with regard to the characteristics of an organism. The expression of any given gene influences the expression of many other genes in the genome, and vice versa. Pleiotropy is revealed when researchers study the effects of gene mutations. As an example of a pleiotropic mutation, let's consider cystic fibrosis, which is a recessive human disorder. In the late 1980s, the gene for cystic fibrosis was identified. It encodes a protein called the cystic fibrosis transmembrane conductance regulator (CFTR), which regulates ionic balance by allowing the transport of chloride ions (Cl^-) across epithelial cell membranes.

The mutation that causes cystic fibrosis diminishes the function of this Cl^- transporter, which affects several parts of the body in different ways. Because the movement of Cl^- affects water transport across membranes, the most severe symptom of cystic fibrosis is thick mucus in the lungs that occurs because of a water imbalance. This thickened mucus results in difficulty breathing and frequent lung infections. Thick mucus can also block the tubes that carry digestive enzymes from the pancreas to the small intestine. Without these enzymes, certain nutrients are not properly absorbed by the body, resulting in poor weight gain. In sweat glands, a functional Cl⁻ transporter is needed to recycle salt out of the glands and back into the skin before it can be lost to the outside world. Persons with cystic fibrosis have excessively salty sweat due to their inability to recycle salt back into their skin cells; a common test for cystic fibrosis is the measurement of salt on the skin. Taken together, these symptoms indicate that a defect in CFTR has multiple effects throughout the body.

Certain Coat-Color Patterns in Dogs Are Determined by Events During Embryonic Development

Many breeds of dogs and other mammals have a coat-color pattern, called white spotting, in which portions of an animal's fur lack pigmentation (see **Figure 5.13**). The spotting gene exists in multiple alleles that affect the amount of pigmentation of an animal's fur. The S^+ allele results in full pigmentation (no white spotting), whereas other alleles vary with regard to the amount of pigmentation produced; possible patterns include Irish spotting (s^I), and extreme-white spotting (s^w). In many breeds, the areas that are white include the legs, belly, neck and the tip of the tail (compare parts a and b of Figure 5.13a and b). For decades, researchers were baffled by this coat pattern in which some areas are pigmented and others are white.

Fur pigmentation is dependent on melanocytes, which are cells located in the bottom layer of the skin's epidermis and in the hair follicles. Melanocytes produce the pigment, melanin, which is found in the skin, eyes, and hair. If melanin is not produced within hair follicles, a dog's fur remains white. An understanding of melanocyte development provides insight regarding the intriguing phenotype of white spotting.

During embryogenesis, melanocyte precursor cells, called melanoblasts, originate in the neural crest, which is a temporary group of cells that are associated with each other only during embryonic development. The neural crest is located dorsal to the neural tube, which gives rise to the brain and spinal cord. Melanoblasts originate only in the part of the neural crest that is located in the trunk region of the embryo (between the neck and the tail). From there, the melanoblasts migrate to other parts of the body (Figure 5.13c). As they migrate, the melanoblasts proliferate, which allows some of them to travel longer distances and reach more ventral regions of the embryo. Once the melanoblasts reach their final destination, they continue to proliferate and differentiate into pigment-producing melanocytes in places such as the epidermis and hair follicles.

Researchers speculate that the alleles conferring the white-spotting phenotype cause a decrease in the number of melanocytes due to failure of melanoblast migration, proliferation, and/or survival during embryonic development. In 2007, a genome-wide association study determined that the spotting gene encodes a protein called microphthalmia-associated transcription factor (MITF). (Genome-wide association studies are described in Chapter 22.) In mice and humans, MITF expression is needed for proper migration, proliferation, and survival of melanoblasts. Reduced expression of MITF is expected to decrease the number of melanocytes in adult animals. Because the melanoblasts begin their journey from the neural crest in the trunk, this reduced expression of MITF causes the regions of the body in the adult that are farthest away from the spinal cord to contain fewer melanocytes and therefore more likely to be white.

5.7 REVIEWING THE KEY CONCEPTS

- Single genes usually exhibit pleiotropy, which means that they exert multiple phenotypic effects.
- Some phenotypes are best understood within the context of development (see Figure 5.13).

5.7 COMPREHENSION QUESTION

1. Which of the following is a possible explanation for pleiotropy?
    a. The expression of a single gene can affect cell function in more than one way.
    b. A gene may be expressed in different cell types in a multicellular organism.
    c. A gene may be expressed at different stages of development.
    d. All of the above are possible explanations.

(a) Irish spotting

(b) Extreme-white spotting

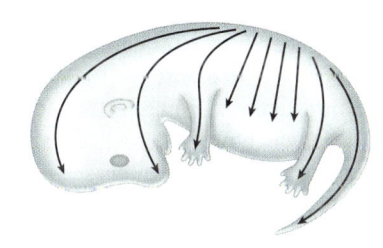

(c) Migration of melanoblasts during embryonic development

FIGURE 5.13 **White-spotting phenotype in dogs.** (a) This animal shows a typical Irish spotting pattern seen for an $s^I s^I$ homozygote or an $s^I s^w$ heterozygote. (b) The extreme-white spotting pattern of an $s^w s^w$ homozygote. (c) During embryonic development, melanoblasts migrate away from the trunk region of the neural crest. *Source (parts a and b):* Figure 1 in Annu. Rev. Anim. Biosci. 2013. 1:125–156.

(a, b) ©Deanna Vout

5.8 GENE INTERACTION

Learning Outcomes:
1. Define *gene interaction*.
2. Compare and contrast epistasis, complementation, modifying genes, and gene redundancy.
3. Predict the outcome of crosses that exhibit epistasis, complementation, and gene redundancy.

Thus far, we have considered the effects of a single gene on the outcome of a trait. This approach helps us to understand the various ways that alleles can influence traits. Researchers often examine the effects of a single gene on the outcome of a single trait as a way to simplify the genetic analysis. For example, Mendel studied one gene that affected the height of pea plants—tall versus dwarf alleles. Actually, many other genes in pea plants also affect height, but Mendel did not happen to study variants in those other height genes. How then did Mendel study the effects of a single gene? The answer lies in the genotypes of his strains. Although many genes affect the height of pea plants, Mendel chose true-breeding strains that differed with regard to only one of those genes. As a hypothetical example, let's suppose that pea plants have 10 genes affecting height, which we will call *K, L, M, N, O, P, Q, R, S,* and *T*. The genotypes of two hypothetical strains of pea plants may be

Tall strain: KK LL MM NN OO PP QQ RR SS TT
Dwarf strain: KK LL MM NN OO PP QQ RR SS tt

In this example, the alleles affecting height may differ at only a single gene. One strain is *TT* and the other is *tt*, and this accounts for the difference in their height. If we make crosses between these tall and dwarf strains, the genotypes of the F_2 offspring will differ with regard to only one gene; the other nine genes will be identical in all of them. This approach allows a researcher to study the effects of a single gene even though many genes may affect a single trait.

Researchers now appreciate that essentially all traits are affected by the contributions of many genes. Morphological features such as height, weight, growth rate, and pigmentation are all affected by the expression of many different genes in combination with environmental factors. In this section, we will further our understanding of genetics by considering how the allelic variants of two different genes affect a single trait. This phenomenon is known as **gene interaction**. Table 5.3 considers examples of inheritance patterns in which two different genes interact to influence the outcome of particular traits. In this section, we will examine some of these examples in greater detail.

A Gene Interaction Can Exhibit Epistasis and Complementation

William Bateson and Reginald Punnett discovered an unexpected gene interaction when studying crosses involving the sweet

TABLE 5.3
Types of Mendelian Inheritance Patterns Involving Two Genes

Type	Description
Epistasis	An inheritance pattern in which the alleles of one gene mask the phenotypic effects of the alleles of a different gene.
Complementation	A phenomenon in which two different parents that express the same or similar recessive phenotypes produce offspring with a wild-type phenotype.
Modifying genes	A phenomenon in which an allele of one gene modifies the phenotypic outcome of the alleles of a different gene.
Gene redundancy	A pattern in which the loss of function in a single gene has no phenotypic effect, but the loss of function of two genes has an effect. Functionality of only one of the two genes is necessary for a normal phenotype; the genes are functionally redundant.

pea, *Lathyrus odoratus*. The wild sweet pea has purple flowers. However, these researchers obtained several true-breeding mutant varieties with white flowers. Not surprisingly, when they crossed a true-breeding purple-flowered plant to a true-breeding white-flowered plant, the F_1 generation had all purple-flowered plants and the F_2 generation (produced by self-fertilization of the F_1 generation) consisted of purple- and white-flowered plants in a 3:1 ratio.

A surprising result came in an experiment in which Bateson and Punnett crossed two different varieties of white-flowered plants (**Figure 5.14**). All of the F_1 generation plants had purple flowers! The researchers then allowed the F_1 offspring to self-fertilize. The F_2 generation resulted in purple- and white-flowered plants in a ratio of 9 purple to 7 white. From this result, Bateson and Punnett deduced that two different genes were involved, with the following relationship:

- *C* (one purple-color-producing) allele is dominant to *c* (white).
- *P* (another purple-color-producing) allele is dominant to *p* (white).
- *cc* masks the *P* allele and *pp* masks the *C* allele, producing white color.

When the alleles of one gene mask the phenotypic effects of the alleles of another gene at a different locus, the phenomenon is called **epistasis**. Geneticists describe epistasis relative to a particular phenotype. If possible, geneticists use the wild-type phenotype as their reference phenotype when describing an epistatic interaction. In the case of sweet peas, purple flowers are wild-type. Homozygosity for the white allele of one gene masks the expression of the purple-producing allele of another gene. In other words, the *cc* genotype is epistatic to a purple phenotype, and the *pp* genotype is also epistatic to a purple phenotype. At the level of genotypes, *cc* is epistatic to *PP* or *Pp*, and *pp* is epistatic to *CC* or *Cc*. This is an example intermediate. Two copies of the recessive allele (*cc*) result in a lack of production of this enzyme in the homozygote. Gene *P* encodes a functional enzyme P, which converts the colorless intermediate into the purple pigment. Like the *c* allele, the recessive *p* allele encodes a defective enzyme P. If a plant is homozygous for either recessive allele (*cc* or *pp*), it cannot make any functional enzyme C or enzyme P, respectively. When one of these enzymes is missing, purple pigment cannot be made, and the flowers remain white.

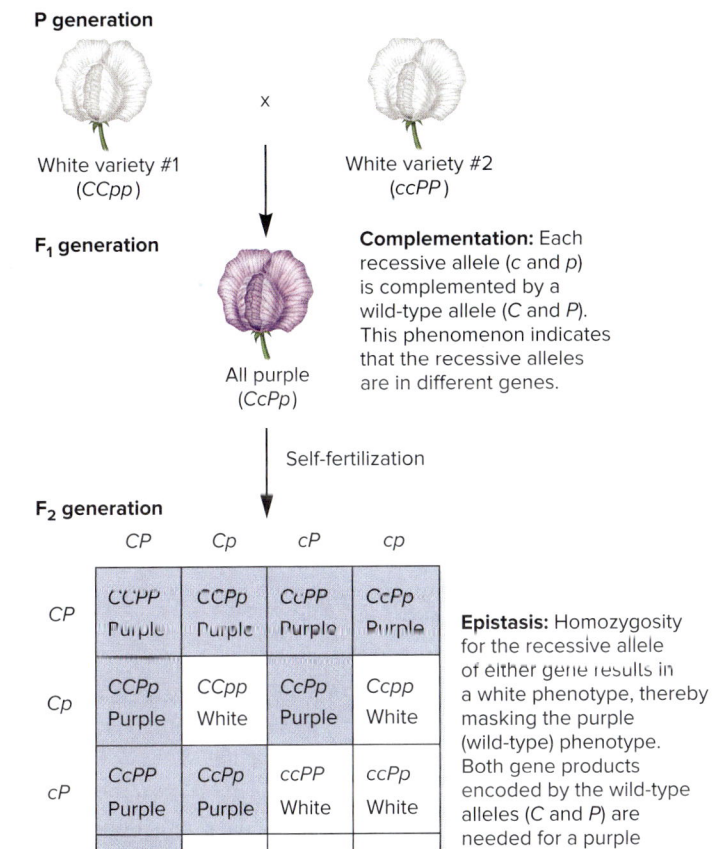

The parental cross shown in Figure 5.14 illustrates another genetic phenomenon called **complementation.** This term refers to the production of offspring with a wild-type phenotype from parents that both display the same or similar recessive phenotype. In this example, purple-flowered F$_1$ offspring were obtained from two white-flowered parents. Complementation typically occurs because the recessive phenotype in the parents is due to homozygosity at two different genes. In our sweet pea example, one parent is *CCpp* and the other is *ccPP*. In the F$_1$ offspring, the *C* and *P* alleles, which are wild-type and dominant, complement the *c* and *p* alleles, which are recessive. The offspring must have one wild-type allele of both genes to display the wild-type phenotype. Why is complementation an important experimental observation? When geneticists observe complementation in a genetic cross, this outcome suggests that the recessive phenotype in the two parent strains is caused by mutant alleles in two different genes.

Due to Gene Redundancy, Loss-of-Function Alleles May Have No Effect on Phenotype

In some cases, two or more different genes may carry out the same or similar functions. When one of these genes is nonfunctional due to a mutation, the other gene or genes can still carry out its function and therefore no phenotypic effect of the mutation is observed. Geneticists may attribute this observation to **gene redundancy**—the phenomenon in which one gene can compensate for the loss of function of another gene.

Gene redundancy may be due to different underlying causes.

- One common cause is gene duplication. Certain genes have been duplicated during evolution, so a species may have two or more copies of similar genes. These copies, which are not identical due to the accumulation of random changes during evolution, are called **paralogs.** When one gene is missing, a paralog may be able to carry out the missing function.
- Alternatively, gene redundancy may involve proteins that are involved in a common cellular function. When one of the proteins is missing due to a faulty gene, the function of another protein may be increased to compensate for the missing protein and thereby overcome the defect.

Let's explore the consequences of gene redundancy in a genetic cross. George Shull conducted one of the first studies that illustrated the phenomenon of gene redundancy. His work involved *Capsella bursa-pastoris,* a weed known as shepherd's

FIGURE 5.14 A cross between two different white varieties of the sweet pea.

Concept Check: What do the terms epistasis and complementation mean?

of **recessive epistasis.** As seen in Figure 5.14, this epistatic interaction produces only two phenotypes—purple or white flowers—in a 9:7 ratio.

Epistasis often occurs because two (or more) different proteins participate in a common function. For example, two or more proteins may be part of an enzymatic pathway leading to the formation of a single product. To illustrate this idea, let's consider the formation of a purple pigment in the sweet pea flower:

Colorless precursor →(Enzyme C)→ Colorless intermediate →(Enzyme P)→ **Purple pigment**

In this example, a colorless precursor molecule must be acted on by two different enzymes to produce the purple pigment. Gene *C* encodes a functional protein called enzyme C, which converts the colorless precursor into a colorless

106 CHAPTER 5 :: EXTENSIONS OF MENDELIAN INHERITANCE

P generation

TTVV (Triangular) × ttvv (Ovate)

F₁ generation

TtVv — All triangular

F₁ (TtVv) × F₁ (TtVv)

F₂ generation

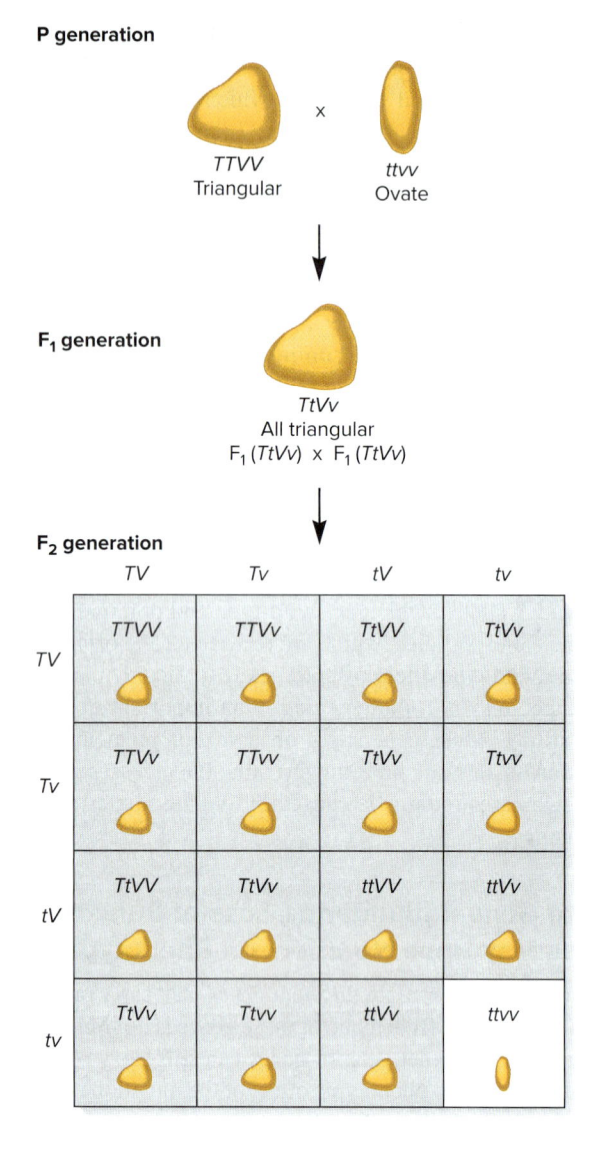

	TV	Tv	tV	tv
TV	TTVV	TTVv	TtVV	TtVv
Tv	TTVv	TTvv	TtVv	Ttvv
tV	TtVV	TtVv	ttVV	ttVv
tv	TtVv	Ttvv	ttVv	ttvv

FIGURE 5.15 **Inheritance of seed capsule shape in *Capsella bursa-pastoris* (shepherd's purse), an example of gene redundancy.** In this case, triangular seed capsule shape requires a dominant allele in one of two genes. The *T* and *V* alleles are redundant.

Concept Check: At the molecular level (with regard to loss-of-function alleles), explain why the ttvv homozygote has an ovate seed capsule.

purse, a member of the mustard family. The trait he followed was the shape of the seed capsule, which is commonly triangular (**Figure 5.15**). Strains producing smaller ovate capsules are homozygous for loss-of-function alleles in two different genes (*ttvv*). When Shull crossed a true-breeding plant with triangular capsules to a plant having ovate capsules, the F₁ generation all had triangular capsules. When the F₁ plants were self-fertilized, a surprising result came in the F₂ generation. Shull observed a 15:1 ratio of plants having triangular capsules to ovate capsules. The result can be explained by gene redundancy. Having at least one functional copy of either gene (*T* or *V*) is sufficient to produce the triangular phenotype. *T* and *V* are functional alleles of redundant genes. Only one of them is necessary for a triangular shape. When the functions of both genes are lost, as in the *ttvv* homozygote, the capsule becomes smaller and ovate.

5.8 REVIEWING THE KEY CONCEPTS

- A gene interaction is a phenomenon in which two or more genes affect a single phenotype (see Table 5.3).
- Epistasis occurs when the allele of one gene masks the phenotypic expression of the alleles of a different gene. Complementation refers to the production of offspring with a wild-type phenotype from parents that both display the same or similar recessive phenotype (see Figure 5.14).
- Two different genes may have redundant functions (see Figure 5.15).

5.8 COMPREHENSION QUESTIONS

1. Two different strains of sweet peas are true-breeding and have white flowers. When plants of these two strains are crossed, the F₁ offspring all have purple flowers. This outcome is due to
   a. epistasis.
   b. complementation.
   c. incomplete dominance.
   d. incomplete penetrance.
2. If the F₁ offspring from question 1 are allowed to self-fertilize, what is the expected outcome for the F₂ offspring?
   a. all white
   b. all purple
   c. 3 purple: 1 white
   d. 9 purple: 7 white

KEY TERMS

Page 89. Mendelian inheritance, simple Mendelian inheritance
Page 91. wild-type alleles, genetic polymorphism, mutant alleles
Page 92. gain-of-function mutations, dominant-negative mutations, haploinsufficiency, incomplete penetrance
Page 93. expressivity
Page 94. norm of reaction
Page 95. incomplete dominance
Page 96. overdominance (heterozygote advantage)
Page 97. homodimer
Page 98. multiple alleles

Page 99. codominance, sex-influenced inheritance
Page 100. sex-limited inheritance, sexual dimorphism
Page 101. lethal allele, essential gene, nonessential genes, age of onset, conditional lethal alleles, temperature-sensitive (ts) lethal alleles, semilethal alleles
Page 102. pleiotropy
Page 104. gene interaction, epistasis
Page 105. recessive epistasis, complementation, gene redundancy, paralogs

CHAPTER SUMMARY

- Mendelian inheritance patterns obey Mendel's laws.

5.1 Overview of Simple Inheritance Patterns

- Several inheritance patterns involving single genes differ from those observed by Mendel (see Table 5.1).

5.2 Dominant and Recessive Alleles

- Wild-type alleles are those that are prevalent in a natural population. When a gene exists in two or more wild-type alleles, this is genetic polymorphism (see Figure 5.1).
- Recessive alleles are often due to mutations that result in a reduction or loss of function of the encoded protein (see Figure 5.2 and Table 5.2).
- Dominant mutant alleles are most commonly caused by gain-of-function mutations, dominant-negative mutations, or haploinsufficiency.
- Incomplete penetrance occurs when an allele that is expected to be expressed is not expressed (see Figure 5.3).
- Traits may vary in their expressivity.

5.3 Environmental Effects on Gene Expression

- The outcome of traits is also influenced by the environment (see Figure 5.4).

5.4 Incomplete Dominance, Overdominance, and Codominance

- Incomplete dominance is an inheritance pattern in which the heterozygote has an intermediate phenotype (see Figure 5.5).
- Whether we consider an allele to be dominant or incompletely dominant may depend on how closely we examine the phenotype (see Figure 5.6).
- Overdominance is an inheritance pattern in which the heterozygote has greater reproductive success (see Figures 5.7, 5.8).
- Many genes exist in multiple alleles in a population.
- Some alleles, such as those that produce A and B blood antigens, are codominant (see Figure 5.9).

5.5 Sex-Influenced and Sex-Limited Inheritance

- For sex-influenced traits, such as scurs in cattle, heterozygous males and females have different phenotypes (see Figure 5.10).
- Sex-limited traits are expressed in only one sex, thereby resulting in sexual dimorphism (see Figure 5.11).

5.6 Lethal Alleles

- Lethal alleles most commonly occur in essential genes.
- Lethal alleles may result in inheritance patterns that yield unexpected ratios (see Figure 5.12).

5.7 Understanding Complex Phenotypes Caused by Mutations in Single Genes

- Single genes usually exhibit pleiotropy, which means that they exert multiple phenotypic effects.
- Some phenotypes are best understood within the context of development (see Figure 5.13).

5.8 Gene Interaction

- A gene interaction is a situation in which two or more genes affect a single phenotype (see Table 5.3).
- Epistasis occurs when the allele of one gene masks the phenotypic expression of the alleles of a different gene. Complementation refers to the production of offspring with a wild-type phenotype from parents that both display the same or similar recessive phenotype (see Figure 5.14).
- Two different genes may have redundant functions (see Figure 5.15).

PROBLEM SETS & INSIGHTS

More Genetic TIPS

1. In Ayrshire cattle, the spotting of the coat can be either red and white or mahogany and white. The mahogany-and-white phenotype is caused by the allele S^M (where the letter S indicates spotting, and the superscript M stands for mahogany). The red-and-white phenotype is controlled by the allele S^R (where the superscript R stands for red). The table below shows the relationships between genotype and phenotype for males and females:

Genotype	Phenotype	
	Females	Males
$S^M S^M$	Mahogany and white	Mahogany and white
$S^M S^R$	Red and white	Mahogany and white
$S^R S^R$	Red and white	Red and white

Explain the pattern of inheritance.

Topic: **What topic in genetics does this question address?**

The topic concerns different patterns of Mendelian inheritance. More specifically, the question asks you to determine which of several patterns (described in Table 5.1) is demonstrated by the spotting of the coat in Ayrshire cattle.

Information: **What information do you know based on the question and your understanding of the topic?**

From the question, you know the relationship between genotype, phenotype, and sex with regard to spotting colors in Ayrshire cattle. From your understanding of the topic, you may remember that certain inheritance patterns may result in differences between males and females.

Problem-Solving Strategy: Compare and contrast.

One strategy to solve this problem is to compare and contrast these results with the inheritance patterns described in Table 5.1. This allows you to rule out certain patterns. The information displayed in the table is not consistent with simple Mendelian inheritance because male and female heterozygotes differ in phenotypes. Because you are not given a pedigree, you don't have evidence for incomplete penetrance. The pattern is not incomplete dominance or overdominance because the heterozygote does not have an intermediate phenotype or greater reproductive success, respectively. The pattern does not exhibit codominance because the heterozygote is not expressing two phenotypes uniquely. It can't be X-linked because the male has two copies of the gene. It is not sex-limited inheritance because neither phenotype is unique to a particular sex. Finally, the alleles are not lethal.

Answer: The inheritance pattern for this trait is sex-influenced inheritance. The S^M allele is dominant in males but recessive in females, whereas the S^R allele is dominant in females but recessive in males.

2. The following pedigree involves a single gene causing an inherited disease. Assuming that incomplete penetrance is *not* occurring, indicate which of the following modes of inheritance is/are possible and explain why. (Affected individuals are shown with filled symbols.)

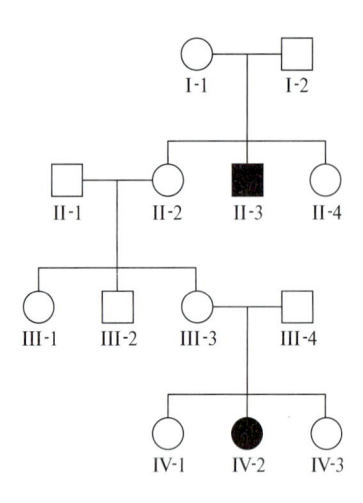

A. Recessive
B. Dominant
C. X-linked, recessive
D. Sex-influenced, dominant in females
E. Sex-limited, recessive in females

Topic: What topic in genetics does this question address?

The topic concerns different patterns of Mendelian inheritance. More specifically, the aim of the question is to determine which of five patterns is/are possible, based on the data in a human pedigree.

Information: What information do you know based on the question and your understanding of the topic?

In the question, you are given a pedigree involving a human genetic disorder. From your understanding of the topic, you may remember how the five patterns of inheritance listed in the question differ from each other.

Problem-Solving Strategy: Analyze data. Predict the outcome.

To solve this problem, you need to analyze the pedigree and determine if the offspring produced by each set of parents are consistent with any of the five modes of inheritance. In other words, you need to predict what types of offspring each set of parents could produce for each of the five inheritance patterns. This analysis allows you to rule out certain patterns.

Answer:

A. It could be recessive.

B. It cannot be dominant (and completely penetrant) because both affected offspring have two unaffected parents.

C. It cannot be X-linked recessive because individual IV-2 is an affected female. An affected female would have to inherit the disease-causing allele from both parents. Because males are hemizyous for X-linked traits, the father of IV-2 would also have to be affected. However, III-4 is unaffected.

D. It cannot be sex-influenced and dominant in females because individual II-3 is an affected male and therefore would have to be homozygous for the disease-causing allele. If so, his mother (individual I-1) would have to carry at least one copy of the disease-causing allele, and she would be affected. However, she is not affected with the disease.

E. It cannot be sex-limited because individual II-3 is an affected male and individual IV-2 is an affected female.

3. Two pink-flowered four-o'clock plants were crossed to each other. What is the probability that a group of six offspring from this cross will be composed of one pink-, two white-, and three red-flowered plants?

Topic: What topic in genetics does this question address?

The topic is Mendelian inheritance. More specifically, the question is about incomplete dominance in four-o'clock plants.

Information: What information do you know based on the question and your understanding of the topic?

From the question, you know that two pink-flowered plants are crossed to each other. From your understanding of the topic, you may remember that pink-flowered plants are heterozygous and show an intermediate phenotype. Also, from Chapter 3, you may recall that the multinomial expansion equation is used to solve problems involving three or more categories of offspring.

Problem-Solving Strategies: Predict the outcome. Make a calculation.

To begin to solve this problem, you need to know the probabilities of producing pink-, white-, and red-flowered offspring. These can be deduced from a Punnett square, which is shown below. The cross is $C^R C^W \times C^R C^W$.

♀ \ ♂	C^R	C^W
C^R	$C^R C^R$ Red	$C^R C^W$ Pink
C^W	$C^R C^W$ Pink	$C^W C^W$ White

Next, you can use the probabilities from the Punnett square in the multinomial expansion equation.

Answer: From the Punnett square, the phenotypic ratio is 1 red : 2 pink : 1 white. In other words, 1/4 are expected to be red, 1/2 pink, and 1/4 white.

$$P = \frac{n!}{a!b!c!} p^a q^b r^c$$

where

n = total number of offspring = 6
a = number of reds = 3
p = probability of red = (1/4)
b = number of pinks = 1
q = probability of pink = (1/2)
c = number of whites = 2
r = probability of white = (1/4)

If you substitute these values into the equation, you obtain

$$P = \frac{6!}{3!1!2!} (1/4)^3 (1/2)^1 (1/4)^2$$

$P = 0.029$, or 2.9%

This result means that 2.9% of the time you expect a group of six offspring from this cross to consist of three with red flowers, one with pink flowers, and two with white flowers.

4. As described in Figure 5.9, a gene in humans that occurs as the i, I^A, and I^B alleles encodes a glycosyl transferase that is involved in attaching galactose or N-acetyl-galactosamine to an oligosaccharide on the surface of red blood cells. In addition, another gene, called the H gene, encodes a different glycosyl transferase that is needed to attach the sugar fucose onto the oligosaccharide and thereby make H antigen, which is the antigen found in type O people.

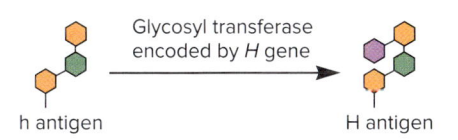

```
                Glycosyl transferase
                encoded by H gene
    [h antigen]  ———————————————→  [H antigen]
```

This gene exists as the common allele, H, and as a very rare, recessive loss-of-function allele, h. An individual who is hh is unable to attach fucose, and the resulting oligosaccharide, which is smaller, is called h antigen. A woman has type O blood that makes H antigen. She gives birth to a daughter with type B blood. The daughter's genotype is $I^B i$. Her I^B allele is *not* due to a new mutation. Surprisingly, the biological father of this daughter does *not* have type B or type AB blood. How is this possible? In your answer, describe the genotypes of the mother and father.

Topic: **What topic in genetics does this question address?**

The topic is concerned with Mendelian patterns of inheritance. More specifically, the question is asking you to form a hypothesis to explain an unexpected genetic outcome.

Information: **What information do you know based on the question and your understanding of the topic?**

From the question, you have learned about a gene that encodes a glycosyl transferase that attaches fucose onto the oligosaccharide on the surface of red blood cells. You also are given the blood types of a mother, father, and daughter, and you are given the genotype of the daughter. From your understanding of the topic, you may remember that another gene is involved with blood type (see Figure 5.9). You know that i is recessive to I^B. You also know that the father must have passed the I^B allele to his daughter, even though he doesn't have type B or type AB blood, which initially seems mysterious.

Problem-Solving **S**trategies: **Relate structure and function. Propose a hypothesis. Predict the outcome.**

It appears that the daughter inherited the I^B allele from her father. Your task is to understand why the father is not expressing the I^B allele even though it appears that he carries it. Because the expression of the dominant I^B allele is being masked, this is an example of epistasis. One strategy to begin to solve this problem is to think about the relationship between the structure of the oligosaccharide and the functions of the two types of glycosyl transferases. The glycosyl transferase encoded by the I^B allele recognizes H antigen and attaches a galactose to the oligosaccharide. The structure of H antigen is important so the glycosyl transferase encoded by the I^B allele can recognize the oligosaccharide and attach an additional galactose. If the oligosaccharide is smaller because it is missing fucose, it will not be recognized by the glycosyl transferase encoded by the I^B allele. To predict the outcome of this cross, you could assume that the mother is $HHii$ and the father is hh and carries at least one copy of the I^B allele. If that were the case, the daughter would be heterozygous, Hh, and make H antigen.

Answer: One hypothesis to explain these results is that the father is hh and that the glycosyl transferase that attaches galactose is unable to recognize the smaller oligosaccharide on the surface of his red blood cells. Another way of saying this is that the hh genotype is epistatic to I^B. In this case, the mother is $HHii$ and the father could be either $hhI^B i$, $hhI^B I^B$, or $hhI^A I^B$ and passed the I^B allele to his daughter. The daughter's genotype is $HhI^B i$.

Conceptual Questions

C1. Describe the differences among dominance, incomplete dominance, codominance, and overdominance.

C2. Discuss the differences among sex-influenced, sex-limited, and sex-linked inheritance. Give examples.

C3. What is meant by a gene interaction? How can a gene interaction be explained at the molecular level?

C4. Let's suppose a recessive allele encodes a completely defective protein. If the functional allele is dominant, what does that tell you about the amount of the functional protein that is sufficient to cause the phenotype? What if the allele shows incomplete dominance?

C5. A nectarine is a peach without the fuzz. The difference is controlled by a single gene that is found in two alleles, D and d. At the molecular level, does it make more sense to you that the nectarine is homozygous for the recessive allele or that the peach is homozygous for the recessive allele? Explain your reasoning.

C6. An allele in *Drosophila* produces a "star-eye" trait in the heterozygous individual. (It is not X-linked.) However, the star-eye allele is lethal in homozygotes. What would be the ratio and phenotypes of surviving flies if star-eyed flies were crossed to each other?

C7. The serum from one individual (individual 1) is known to agglutinate the red blood cells from a second individual (individual 2). List the pairwise combinations of possible genotypes that individuals 1 and 2 could have. If individual 1 is the parent of individual 2, what are his or her possible genotypes?

C8. Which blood phenotype(s) (A, B, AB, and/or O) provide(s) an unambiguous genotype? Is it possible for a couple to produce a family of children with all four blood types? If so, what must be the genotypes of the parents?

C9. A woman with type B blood has a child with type O blood. What are the possible genotypes and blood types of the father?

C10. A type A woman is the daughter of a type O father and a type A mother. If she has children with a type AB man, what are the probabilities of the following offspring?

  A. A type AB child

  B. A type O child

  C. The first three children with type AB blood

  D. A family containing two children with type B blood and one child with type AB

C11. In Shorthorn cattle, coat color is controlled by a single gene that can exist as a red allele (H^R) or a white allele (H^W). Heterozygotes ($H^R H^W$) have a color called roan that looks less red than the color of $H^R H^R$ homozygotes. However, when examined carefully, the roan phenotype is actually due to a mixture of completely red hairs and completely white hairs. Should this effect be called incomplete dominance, codominance, or something else? Explain your reasoning.

C12. In chickens, the Leghorn variety has white feathers due to an autosomal dominant allele. Silkies have white feathers due to a recessive allele in a second (different) gene. If a true-breeding white Leghorn is crossed to a true-breeding white Silkie, what is the expected phenotype of the F_1 generation? If members of the F_1 generation are mated to each other, what is the expected phenotypic outcome of the F_2 generation? Assume the chickens in the parental generation are homozygous for the white allele of one gene and homozygous for the brown allele of the other gene. In subsequent generations, nonwhite birds will be brown.

C13. Propose the most likely mode of inheritance (autosomal dominant, autosomal recessive, or X-linked recessive) for the following pedigree:

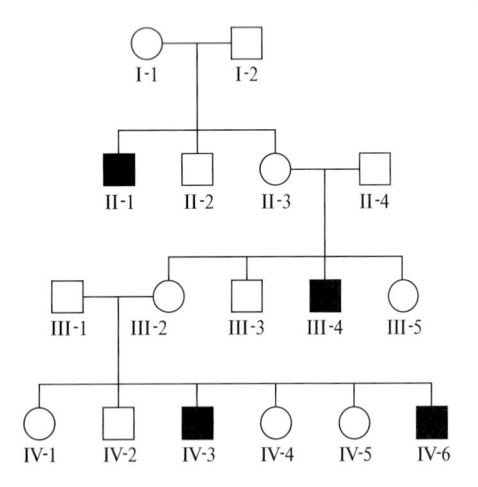

Affected individuals are shown with filled (black) symbols.

C14. With regard to scurs in cattle (a sex-influenced trait), a cow with no scurs whose mother had scurs had offspring with a bull with scurs whose father had no scurs. What is the probability of each of the following for their offspring?

  A. Their first offspring will not have scurs.

  B. Their first offspring will be a male with no scurs.

  C. Their first three offspring will be females with no scurs.

C15. In the pedigree shown here for a trait determined by a single gene (affected individuals are shown as filled symbols), state whether it would be possible for the trait to be inherited in each of the following ways:

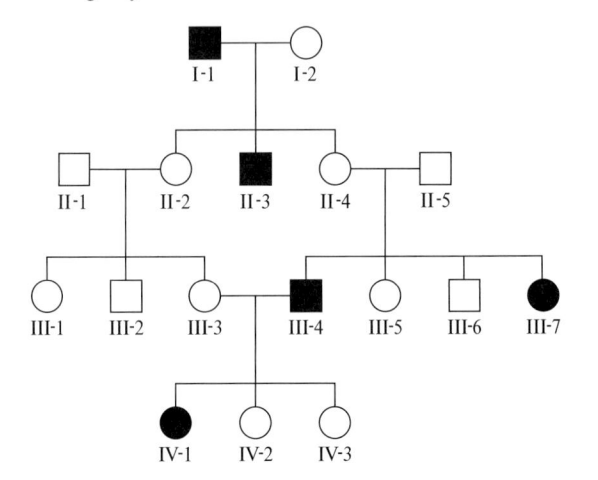

  A. Recessive

  B. X-linked recessive

  C. Dominant, complete penetrance

  D. Sex-influenced, dominant in males

  E. Sex-limited

  F. Dominant, incomplete penetrance

C16. Let's suppose you have pedigree data from thousands of different families involving a particular genetic disease that is fairly rare. How would you decide whether the disease is inherited as a recessive trait as opposed to being dominant with incomplete penetrance?

C17. Compare phenotypes at the molecular, cellular, and organism levels for individuals who are homozygous for the hemoglobin A allele, $Hb^A Hb^A$, and the sickle cell allele, $Hb^S Hb^S$.

C18. A very rare dominant allele that causes the little finger to be crooked has a penetrance value of 80%. In other words, 80% of heterozygotes carrying the allele have a crooked little finger. If a homozygous unaffected person has children with a heterozygote carrying this mutant allele, what is the probability that an offspring will have little fingers that are crooked?

C19. A sex-influenced trait in humans is one that affects the length of the index finger. A short allele is dominant in males and recessive in females. Heterozygous males have an index finger that is significantly shorter than the ring finger. In contrast, the long allele is dominant in females and recessive in males. The gene affecting index finger length is located on an autosome. A woman with short index fingers has children with a man who has long index fingers. They produce five children in the following order: female, male, male, female, male. The oldest female offspring marries a man with long fingers and then has one son. The youngest male among the five children marries a woman with short index fingers, and then they have two sons. Draw the pedigree for this family. Indicate the phenotypes of every individual (filled symbols for individuals with short index fingers and open symbols for individuals with long index fingers).

C20. In horses, three coat-color patterns are termed cremello (beige), chestnut (brown), and palomino (golden with light mane and tail). If two palomino horses are mated, they produce about 1/4 cremello, 1/4 chestnut, and 1/2 palomino offspring. In contrast, cremello horses and chestnut horses breed true. (In other words, two cremello horses will produce only cremello offspring and two chestnut horses will produce only chestnut offspring.) Explain this pattern of inheritance.

Application and Experimental Questions

E1. A seed dealer wants to sell four-o'clock seeds that will produce only plants with red, white, or pink flowers. Explain how this could be done.

E2. Mexican hairless dogs have little hair and few teeth. When a Mexican hairless is mated to another breed of dog, about half of the puppies are hairless. When two Mexican hairless dogs are mated to each other, about 1/3 of the surviving puppies have hair, and about 2/3 of the surviving puppies are hairless. However, about two out of eight puppies from this type of cross are born grossly deformed and do not survive. Explain this pattern of inheritance.

E3. In chickens, some varieties have feathered shanks (legs), but others do not. In a cross between a Black Langhans (feathered shanks) and Buff Rocks (unfeathered shanks), the shanks of the F_1 generation are all feathered. When chickens from the F_1 generation are crossed, the F_2 generation contains chickens with feathered shanks to unfeathered shanks in a ratio of 15:1. Suggest an explanation for this result.

E4. In sheep, the formation of horns is a sex-influenced trait; the allele that results in horns is dominant in males and recessive in females. Females must be homozygous for the horned allele to have horns. A horned ram was crossed to a polled (unhorned) ewe, and the first offspring they produced was a horned ewe. What are the genotypes of the parents?

E5. A particular breed of dog can have long hair or short hair. When true-breeding long-haired animals were crossed to true-breeding short-haired animals, the offspring all had long hair. The F_2 generation consisted of a 3:1 ratio of long- to short-haired animals. A second gene affects the texture of the hair. The two variants are wiry hair and straight hair. F_1 offspring from a cross of these two varieties all had wiry hair, and F_2 offspring showed a 3:1 ratio of wiry- to straight-haired puppies. Recently, a breeder of the short-, wiry-haired dogs discovered a female puppy that was albino. Similarly, another breeder of the long-, straight-haired dogs found a male puppy that was albino. The albino trait is due to a recessive allele. The two breeders got together and mated the two dogs. Surprisingly, all the puppies in the litter had black hair. How would you explain this result?

E6. In the clover butterfly, males are always yellow, but females can be yellow or white. In females, white is a dominant allele. Two yellow butterflies were crossed to yield an F_1 generation consisting of 50% yellow males, 25% yellow females, and 25% white females. Describe how this trait is inherited, and identify the genotypes of the parents.

E7. Duroc Jersey pigs are typically red, but a sandy variation also occurs. When two different strains of true-breeding sandy pigs were crossed to each other, they produced F_1 offspring that were red. When these F_1 offspring were crossed to each other, they produced red, sandy, and white pigs in a 9:6:1 ratio. Explain this pattern of inheritance.

E8. In certain species of summer squash, fruit color is determined by two interacting genes. A dominant allele, W, determines white color, and a recessive allele (w) allows the fruit to be colored. In a homozygous ww individual, a second gene determines fruit color: G (green) is dominant to g (yellow). A white squash and a yellow squash were crossed, and the F_1 generation yielded approximately 50% white fruit and 50% green fruit. What are the genotypes of the parents?

E9. Certain species of summer squash exist in long, spherical, or disk shapes. When a true-breeding long-shaped strain was crossed to a true-breeding disk-shaped strain, all of the F_1 offspring were disk-shaped. When the F_1 offspring were allowed to self-fertilize, the F_2 generation consisted of a ratio of 9 disk-shaped to 6 round-shaped to 1 long-shaped. Assuming that the shape of summer squash is governed by two different genes, with each gene existing in two alleles, propose a mechanism to account for this 9:6:1 ratio.

E10. In a species of plant, two genes control flower color. The red allele (R) is dominant to the white allele (r); the color producing allele (C) is dominant to the non-color-producing allele (c). You suspect that either an rr homozygote or a cc homozygote will produce white flowers. In other words, rr is epistatic to C, and cc is epistatic to R. To test your hypothesis, you allowed heterozygous plants (RrCc) to self-fertilize and counted the phenotypes of the offspring. You obtained the following data: 201 plants with red flowers and 144 with white flowers. Conduct a chi square test to see if your observed data are consistent with your hypothesis.

E11. In *Drosophila*, red eyes is the wild-type phenotype. Several different genes (with each gene existing in two or more alleles) are known to affect eye color. One allele causes purple eyes, and a different allele causes sepia eyes. Both of these alleles are recessive to the red-eye allele. When flies with purple eyes were crossed to flies with sepia eyes, all of the F_1 offspring had red eyes. When the F_1 offspring were allowed to mate with each other, the following results were obtained:

146 purple eyes

151 sepia eyes

50 purplish sepia eyes

444 red eyes

Explain this pattern of inheritance. Conduct a chi square test to see if the experimental data fit your hypothesis.

Questions for Student Discussion/Collaboration

1. Let's suppose a gene exists as a functional wild-type allele and a nonfunctional mutant allele. At the organism level, the wild-type allele is dominant. In a heterozygote, discuss whether dominance occurs at the cellular or molecular level. Discuss examples in which the existence of dominance depends on the level of examination.

2. In oats, the color of the chaff is determined by a two-gene interaction. When a true-breeding black plant was crossed to a true-breeding white plant, the F_1 generation was composed of all black plants. When the F_1 offspring were crossed to each other, the phenotypic ratio of the F_2 generation was 12 black to 3 gray to 1 white. First, construct a Punnett square that accounts for this pattern of inheritance. Which genotypes produce the gray phenotype? Second, at the level of protein function, how would you explain this type of inheritance?

Answers to Comprehension Questions

5.1: c

5.2: d, d, c

5.3: c

5.4: d, a

5.5: b

5.6: c

5.7: d

5.8: b, d

Note: All answers appear in Connect; the answers to even-numbered questions and all of the Concept Check questions are in Appendix B.

6

CHAPTER OUTLINE

- 6.1 Extranuclear Inheritance: Chloroplasts
- 6.2 Extranuclear Inheritance: Mitochondria
- 6.3 Theory of Endosymbiosis
- 6.4 Epigenetics: Imprinting
- 6.5 Maternal Effect

Shell coiling in the water snail, Lymnaea peregra. In this species of snails, some shells coil to the left, and others coil to the right. This variation in phenotype is due to an inheritance pattern called the maternal effect.
Courtesy of John Mendenhall, University of Texas at Austin

EXTRANUCLEAR INHERITANCE, IMPRINTING, AND MATERNAL EFFECT

Mendelian inheritance patterns involve genes that directly influence the outcome of an offspring's traits and obey Mendel's laws. To predict phenotype, we must consider several factors. These include the dominant/recessive relationship of alleles, gene interactions that may affect the expression of a single trait, and the roles that sex and the environment play in influencing the individual's phenotype. Once these factors are understood, we can predict the phenotypes of offspring from their genotypes. Genes that follow a Mendelian inheritance pattern conform to four rules:

1. The genes obey Mendel's law of segregation.
2. Except in the case of rare mutations, the genes are passed unaltered from generation to generation.
3. The expression of the genes in the offspring directly influences their traits.
4. For crosses involving two or more genes, the genes obey Mendel's law of independent assortment.

Most genes in eukaryotic species follow a Mendelian pattern of inheritance. However, many genes do not. In this chapter and the following chapter, we will examine several additional and even bizarre types of inheritance patterns that deviate from a Mendelian pattern because one of these four rules is broken. We begin this chapter by analyzing inheritance patterns that arise because some genetic material is not located in the cell nucleus. Certain organelles, such as mitochondria and chloroplasts, contain their own genetic material. We will consider examples in which traits are determined by genes within these organelles. These genes do not conform to rule 1; they do not obey the law of segregation.

We then turn to epigenetic inheritance, which breaks rule number 2, because the genes are altered in the offspring. As we will see, in certain types of epigenetic inheritance, genes are methylated and this process alters their expression. Finally, in the last section of this chapter, we will consider a pattern called maternal effect in which rule 3 is broken; the expression of the genes in the mother's cells determines an offspring's traits. In Chapter 7, we will examine inheritance patterns that do not obey rule 4, the law of independent assortment.

6.1 EXTRANUCLEAR INHERITANCE: CHLOROPLASTS

Learning Outcomes:
1. Define *extranuclear inheritance*.
2. Describe the general features of chloroplast genomes.
3. Predict the outcome of crosses involving genetic variation in chloroplast genomes.

TABLE 6.1
Genetic Composition of Chloroplasts

Organism(s)	Organelle	Nucleoids per Organelle	Total Number of Chromosomes per Organelle
Chlamydomonas	Chloroplast	~15	~80
Euglena	Chloroplast	20–34	100–300
Flowering plants	Chloroplast	12–25	~60

Source: Gillham, N. W. (1994) *Organelle Genes and Genomes*. New York, NY: Oxford University Press, 424.

In previous chapters, we have considered several types of Mendelian inheritance patterns in eukaryotic species. All of these inheritance patterns involve **nuclear genes**—genes located on chromosomes that are in the cell nucleus. One cause of non-Mendelian inheritance patterns involves genes that are not located in the cell nucleus. In eukaryotic species, the most biologically important example of extranuclear inheritance is due to genetic material in cellular organelles. In addition to the cell nucleus, the chloroplasts and mitochondria contain their own DNA. Because these organelles are found within the cytoplasm of the cells, the inheritance of organellar genetic material is called **extranuclear inheritance** (the prefix *extra-* means "outside of"), or **cytoplasmic inheritance**. In this section, we will examine the genetic composition of chloroplasts and explore the patterns of transmission of this organelle from parent to offspring.

Chloroplasts Contain Circular Chromosomes with Many Genes

The genetic material of chloroplasts is located inside the organelle in a region known as the **nucleoid** (**Figure 6.1**). The genome is a single circular chromosome, although a nucleoid may contain multiple copies of this chromosome. In addition, a chloroplast often has more than one nucleoid. **Table 6.1** describes the genetic composition of chloroplasts for a few selected organisms.

Among algae and plants, substantial variation is found in the sizes of chloroplast chromosomes. A typical chloroplast genome is approximately 100,000 to 200,000 bp in length. As an example, let's consider the **chloroplast DNA (cpDNA)** of the tobacco plant:

- It is a circular DNA molecule that contains 156,000 bp.
- The cpDNA carries between 110 and 120 different genes. By comparison, the genome inside the nucleus of a tobacco cell contains tens of thousands of genes.

- Some chloroplast genes encode ribosomal RNAs and transfer RNAs, which are required for protein synthesis inside the chloroplast. Other chloroplast genes encode proteins involved with photosynthesis. These proteins are made inside the chloroplast. However, many chloroplast proteins are encoded by genes found in the plant cell nucleus. These nuclear-encoded proteins are made in the cytosol and contain chloroplast-targeting signals that direct them into the chloroplasts.

Extranuclear Inheritance Produces Non-Mendelian Results in Reciprocal Crosses

In diploid eukaryotic species, most genes within the nucleus obey a Mendelian pattern of inheritance because the homologous pairs of chromosomes segregate during meiosis. Except for genes that determine sex-linked traits, offspring inherit one copy of each gene from both the maternal and paternal parents. The sorting of chromosomes during meiosis explains the inheritance patterns of nuclear genes, which are genes on chromosomes in the cell nucleus. By comparison, the inheritance of extranuclear genetic material does not display a Mendelian pattern. Chloroplasts and mitochondria are not sorted during meiosis and therefore do not segregate into gametes in the same way as nuclear chromosomes.

In 1909, Carl Correns discovered a trait that showed a non-Mendelian pattern of inheritance involving pigmentation in *Mirabilis jalapa* (the four-o'clock plant). Leaves can be green, white, or variegated with both green and white sectors. Correns conducted crosses in which the parent providing the eggs had a

FIGURE 6.1 **Nucleoids within a chloroplast.** The chloroplast chromosomes are found within the nucleoid of the organelle.

From: Gibbs, SP., Mak, R., Ng, R., & Slankis, T. (1974) The Chloroplast Nucleoid In Ochromonas Danica. II. Evidence For An Increase In Plastid DNA During Greening. *J Cell Sci.* 16(3):579-91. Fig. 1. By permission of the Company of Biologists Limited

Concept Check: How is a nucleoid different from a cell nucleus?

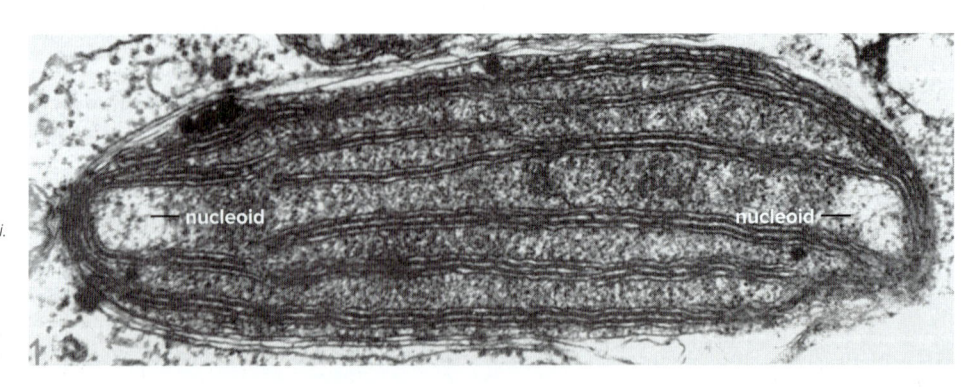

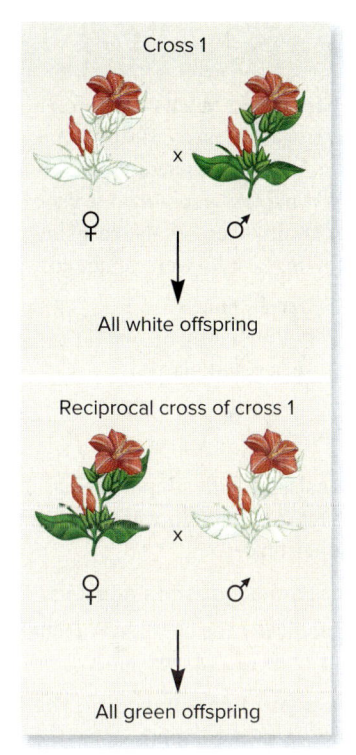

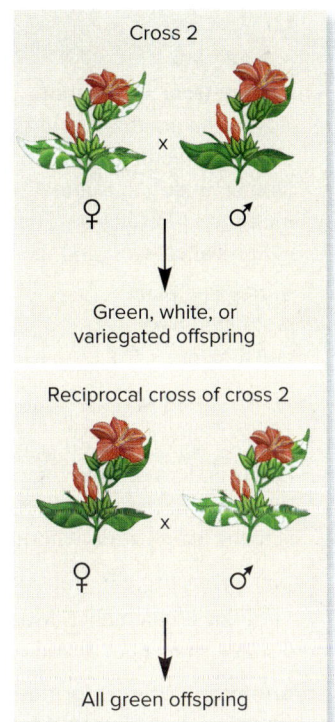

FIGURE 6.2 Maternal inheritance in the four-o'clock plant, *Mirabilis jalapa*. The reciprocal crosses of four-o'clock plants by Carl Correns consisted of a pair of crosses between white-leaved and green-leaved plants and a second pair of crosses between variegated-leaved and green-leaved plants.

Genes → Traits In this example, the white phenotype is due to chloroplasts that carry a mutant allele that diminishes green pigmentation. The variegated phenotype is due to a mixture of chloroplasts, some of which carry the normal (green) allele and some of which carry the white allele. In the crosses shown here, the parent providing the eggs determines the phenotypes of the offspring. This outcome is due to maternal inheritance. The egg contains the chloroplasts that are inherited by the offspring. (Note: The defective chloroplasts that give rise to white sectors are not completely defective in chlorophyll synthesis. Therefore, entirely white plants can survive, though they are smaller than green or variegated plants.)

Concept Check: What is a reciprocal cross?

different phenotype than that of the parent providing the pollen (**Figure 6.2**). In each case, he then carried out a **reciprocal cross** in which the phenotypes of the maternal and paternal parents were reversed.

- If the female parent had white pigmentation, all offspring had white leaves.
- If the female was green, all offspring were green.
- When the female was variegated, the offspring could be green, white, or variegated.

The pattern of inheritance observed by Correns is a type of extranuclear inheritance called **maternal inheritance.** Chloroplasts are a type of plastid that makes chlorophyll, a green photosynthetic pigment. Maternal inheritance occurs because the chloroplasts are inherited only through the cytoplasm of the egg. The sperm cells within pollen grains of *M. jalapa* do not transmit chloroplasts to the offspring.

The phenotypes of leaves can be explained by the types of chloroplasts within the leaf cells. The green phenotype, which is the wild-type condition, is due to the presence of normal chloroplasts that make green pigment. By comparison, the white phenotype is due to a mutation in a gene within the chloroplast DNA that diminishes the synthesis of green pigment. A cell may contain both types of chloroplasts, a condition known as **heteroplasmy.** A leaf cell containing both types of chloroplasts is green because the normal chloroplasts produce green pigment.

How does a variegated phenotype occur? **Figure 6.3** considers the leaf of a plant that began from a fertilized egg that contained both types of chloroplasts (i.e., a heteroplasmic cell). As a plant grows, the two types of chloroplasts are irregularly

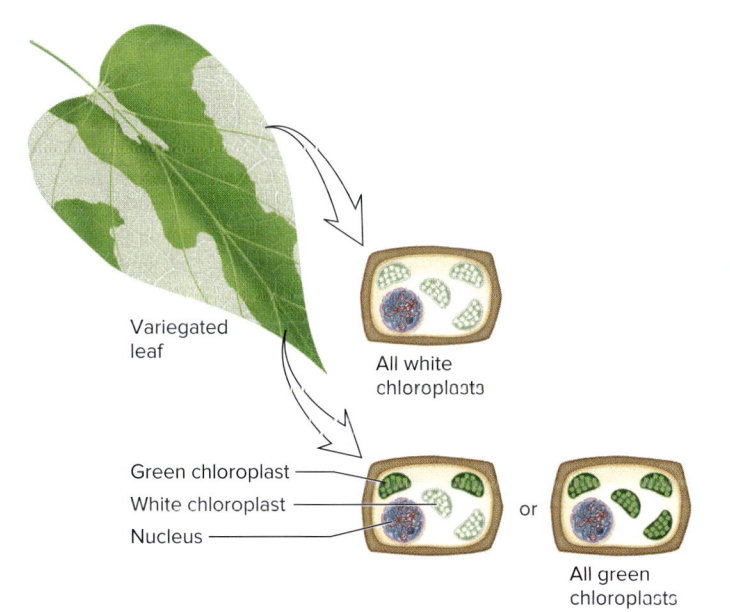

FIGURE 6.3 A cellular explanation of the variegated phenotype in *Mirabilis jalapa*. This plant inherited two types of chloroplasts—those that produce green pigment and those that are defective. As the plant grows, the two types of chloroplasts are irregularly distributed to daughter cells. On occasion, a leaf cell may receive only the chloroplasts that are defective at making green pigment. Such a cell continues to divide and produces a sector of the leaf that is entirely white. Cells that contain both types of chloroplasts or cells that contain only green chloroplasts produce green tissue, which may be adjacent to a sector of white tissue. This is the basis for the variegated phenotype of the leaves.

Concept Check: During growth, can a patch of tissue with a white phenotype give rise to a patch with a green phenotype? Explain.

distributed to daughter cells. On occasion, a cell may receive only the chloroplasts that are defective in making green pigment. Such a cell continues to divide and produces a sector of the plant that is entirely white. In this way, the variegated phenotype is produced. Similarly, if we consider the results of Figure 6.2, a female parent that is variegated may transmit green, white, or a mixture of these types of chloroplasts to the egg cell, thereby producing green, white, or variegated offspring, respectively.

The Pattern of Inheritance of Chloroplasts Varies Among Different Species

The inheritance of traits via genetic material within chloroplasts is now a well-established phenomenon that geneticists have investigated in many different species. In **heterogamous** species, two morphologically different types of gametes are made—eggs and sperm. Egg cells tend to be large and provide most of the cytoplasm to the zygotes, whereas the sperm are small and often provide little more than a nucleus. Therefore, chloroplasts are most often inherited from the maternal parent. However, this is not always the case. Table 6.2 describes the inheritance patterns of chloroplasts in selected species.

For extranuclear genes, you cannot use a Punnett square to predict an offspring's phenotype. Instead, you need two pieces of information:

- You need to know if the offspring inherits the gene from the mother, the father, or both.
- You need to know if heteroplasmy is present in the parent(s) who transmits the extranuclear gene.

With this information, you can predict an offspring's phenotype. For example, let's consider an extranuclear gene that affects leaf pigmentation. If the gene is inherited from the mother, and if the mother has white leaves, you can predict that the mother will transmit the white allele to her offspring and all of them will have white leaves. Alternatively, the mother may exhibit heteroplasmy and have variegated leaves. In this case, the offspring could have green, white, or variegated leaves.

TABLE 6.2
Transmission of Chloroplasts Among Different Organisms

Organism	Organelle	Transmission
Chlamydomonas	Chloroplasts	Inherited from the parent with the mt^+ mating type
Plants		
Angiosperms	Chloroplasts	Often maternal inheritance, although biparental inheritance* is found among some species
Gymnosperms	Chloroplasts	Usually paternal inheritance**

*A pattern in which the organelle is transmitted from both parents.
**A pattern in which the organelle is transmitted only via sperm cells within pollen grains.

Genetic TIPS

The Question: One strain of periwinkle plants has green leaves and another strain has variegated leaves. Both strains are true-breeding. You do not know if the phenotypic difference is due to a mutation of a nuclear gene or to a gene found in chloroplasts. The two strains were analyzed using reciprocal crosses, and the following results were obtained:

Variegated plant pollinated by a green plant
↓
125 variegated leaves

Green plant pollinated by a variegated plant
↓
123 green leaves

Is this pattern of inheritance consistent with simple Mendelian inheritance, with green leaves dominant to variegated, or is it consistent with maternal inheritance?

Topic: What topic in genetics does this question address?

The topic is inheritance. More specifically, the question is about distinguishing nuclear and extranuclear inheritance patterns.

Information: What information do you know based on the question and your understanding of the topic?

From the question, you know there are two strains of periwinkles, green and variegated. From your understanding of the topic, you may remember that some genes are in the nucleus, whereas others are found in organelles, such as chloroplasts. Because nuclear genes segregate differently from chloroplast genes, one way to distinguish these inheritance patterns is to make reciprocal crosses.

Problem-Solving Strategies: Analyze data. Predict the outcome. Compare and contrast.

Reciprocal crosses may yield different results depending on the mode of inheritance. For example, if the gene is a nuclear gene and the green allele is dominant, you would predict that all of the F_1 offspring will be green-leaved. On the other hand, if the gene is in the chloroplasts and follows maternal inheritance, the phenotype of the offspring will depend on which plant contributed the egg.

Answer: The results of the reciprocal crosses are consistent with maternal inheritance because the phenotype of the offspring correlates with inheriting the gene from the plant contributing the egg cells.

6.1 REVIEWING THE KEY CONCEPTS

- Extranuclear inheritance is the inheritance of genes that are found in mitochondria or chloroplasts.
- Chloroplasts have circular chromosomes in a nucleoid. These circular chromosomes carry many genes but far fewer compared with the number on chromosomes in the cell nucleus (see Figure 6.1, Table 6.1).
- Maternal inheritance occurs when organelles, such as chloroplasts, are transmitted via the egg (see Figure 6.2).

- Heteroplasmy of chloroplasts can result in a variegated phenotype (see Figure 6.3).
- The transmission patterns of chloroplasts vary among different organisms (see Table 6.2).

6.1 COMPREHENSION QUESTIONS

1. Extranuclear inheritance occurs due to
   a. chromosomes that may become detached from the spindle apparatus during meiosis.
   b. genetic material that is found in chloroplasts and mitochondria.
   c. mutations that disrupt the integrity of the nuclear membrane.
   d. none of the above.
2. The genetic material of a chloroplast is found in the _____ and consists of _____.
   a. nucleus, several genes
   b. nucleus, one or more copies of a circular chromosome
   c. nucleoid, a few genes
   d. nucleoid, one or more copies of a circular chromosome
3. A cross is made between a green four-o'clock plant and a variegated one. If the variegated plant provides the pollen, the expected outcome for the phenotypes of the offspring will be
   a. all plants with green leaves.
   b. 3 plants with green leaves to 1 plant with variegated leaves.
   c. 3 plants with green leaves to 1 plant with white leaves.
   d. some plants with green leaves, some with variegated leaves, and some with white leaves.

6.2 EXTRANUCLEAR INHERITANCE: MITOCHONDRIA

Learning Outcomes:

1. Describe the general features of mitochondrial genomes.
2. Predict the outcome of crosses involving genetic variation in mitochondrial genomes.
3. Explain how mutations in mitochondrial genes cause human diseases.

As we have seen, chloroplasts contain their own genetic material, and their transmission pattern is different from a traditional Mendelian pattern. A similar situation exists for mitochondria. In this section, we will examine the genetic composition of mitochondria and explore the pattern of transmission of this organelle from parent to offspring.

Mitochondria Also Contain Circular Chromosomes with Many Genes

Like chloroplasts, the genetic material of mitochondria is located inside the organelle in a region known as the nucleoid (**Figure 6.4**). The genome is a single circular chromosome, although a nucleoid usually contains multiple copies of this chromosome. In addition, a mitochondrion often has more than one nucleoid. In mice, for example, each mitochondrion has one to three nucleoids, with each nucleoid containing five to six copies of the circular mitochondrial genome. However, this number varies depending on the

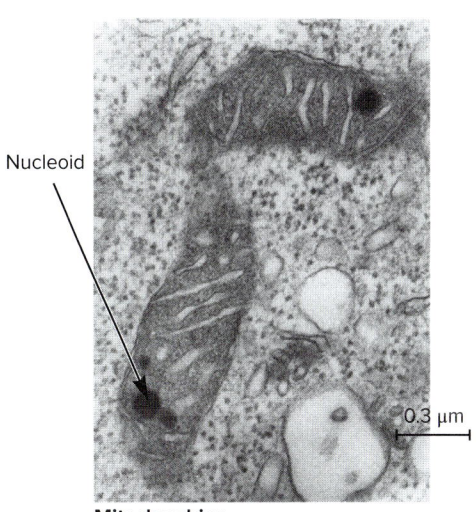

FIGURE 6.4 **Nucleoid within a mitochondrion.** The mitochondrial chromosomes are found within the nucleoid region of the organelle.

From: Prachar J., "Mouse and human mitochondrial nucleoid—detailed structure in relation to function," *Gen Physiol Biophys.* 2010 Jun, 29(2):160-174. Fig 3A

Concept Check: *Within one nucleoid, is the mitochondrial chromosome found in a single copy or multiple copies?*

TABLE 6.3
Genetic Composition of Mitochondria

Organism	Organelle	Nucleoids per Organelle	Total Number of Chromosomes per Organelle
Tetrahymena	Mitochondrion	1	6–8
Mouse	Mitochondrion	1–3	5–6

Source: Gillham, N. W. (1994) *Organelle Genes and Genomes*. New York, NY: Oxford University Press, 424.

type of cell and the stage of development. **Table 6.3** describes the genetic composition of mitochondria for two selected organisms.

Besides variation in copy number, the sizes of mitochondrial genomes also vary greatly among different species. A 400-fold variation is found in the sizes of mitochondrial chromosomes. In general, the mitochondrial genomes of animal species tend to be fairly small; those of fungi and protists are intermediate in size; and those of plant cells tend to be fairly large.

Figure 6.5 shows a map of human **mitochondrial DNA (mtDNA).**

- Each copy of the mitochondrial chromosome consists of a circular DNA molecule that is only 17,000 bp in length. This size is less than 1% of a typical bacterial chromosome.
- The human mtDNA carries relatively few genes. Thirteen genes encode polypeptides that function within the mitochondrion. In addition, the mtDNA carries genes that encode ribosomal RNA and transfer RNA, which are needed for the synthesis of the 13 polypeptides that are encoded by the mtDNA.
- The primary role of mitochondria is to provide cells with the bulk of their adenosine triphosphate (ATP), which is used as an energy source to drive cellular reactions. The 13 polypeptides synthesized in mitochondria are subunits of proteins that

FIGURE 6.5 A genetic map of human mitochondrial DNA (mtDNA). This diagram illustrates the locations of many genes along the circular mitochondrial chromosome. The genes shown in red encode transfer RNAs. For example, tRNA^{Arg} encodes a tRNA that carries arginine. The genes that encode ribosomal RNA are shown in light brown. The remaining genes encode proteins that function within the mitochondrion. The mitochondrial genomes from numerous species have been determined.

Concept Check: Why do mitochondria need genes that encode rRNAs and tRNAs?

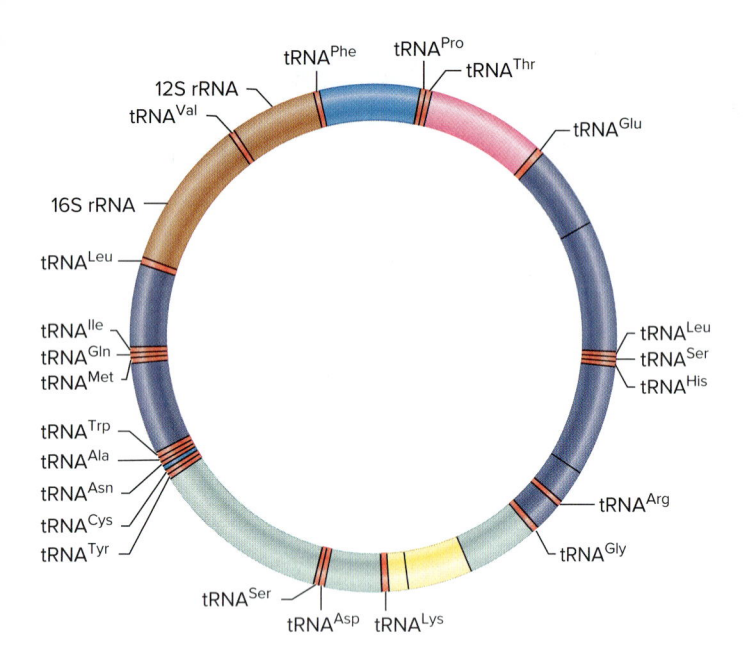

function in a process known as oxidative phosphorylation, in which mitochondria use oxygen and synthesize ATP. However, mitochondria require many additional proteins to carry out oxidative phosphorylation and other mitochondrial functions. Most mitochondrial proteins are encoded by genes within the cell nucleus and contain mitochondrial-targeting signals.

The Transmission of Mitochondria Usually Follows a Maternal Inheritance Pattern

In heterogamous species that make female and male gametes, mitochondria are usually inherited via egg cells. Therefore, the female parent passes mitochondrial genes to her offspring, but the male parent does not. However, this is not always the case. **Table 6.4** describes the inheritance patterns of mitochondria in selected organisms.

In species in which maternal inheritance is generally observed, the paternal parent may occasionally provide mitochondria via the sperm. This phenomenon, called **paternal leakage,** occurs in many species that primarily exhibit maternal inheritance of their organelles. In the mouse, for example, approximately 1 to 4 paternal mitochondria are inherited for every 100,000 maternal mitochondria per generation of offspring. Most offspring do not inherit any paternal mitochondria, but a rare individual may inherit a mitochondrion from the sperm.

Many Human Diseases Are Caused by Mitochondrial Mutations

Researchers have identified many human diseases that are caused by mutations in mtDNA. These diseases can occur in two ways.

- Mitochondrial mutations that cause disease may be transmitted from mother to offspring. Such diseases follow a maternal inheritance pattern, because mtDNA is transmitted from mother to offspring via the cytoplasm of the egg.
- Mitochondrial mutations may occur in somatic cells and accumulate as a person ages. Researchers have discovered that mitochondria are particularly susceptible to DNA damage. When more oxygen is consumed than is actually used to make ATP, mitochondria tend to produce free radicals that damage mtDNA. Unlike nuclear DNA, mitochondrial DNA has very limited repair abilities and almost no protective ability against free-radical damage.

Table 6.5 describes several mitochondrial diseases that have been discovered in humans and are caused by mutations in mitochondrial genes. Over 200 diseases associated with defective mitochondria have been discovered. These are usually chronic degenerative disorders that affect cells requiring a high level of ATP, such as nerve and muscle cells. For example, Leber hereditary optic neuropathy (LHON) affects the optic nerve and may lead to the progressive loss of vision in one or both eyes. LHON can be caused by a defective mutation in one of several different mitochondrial genes. Researchers are still investigating how a defect in these mitochondrial genes produces the symptoms of this disease.

TABLE 6.4
Transmission of Mitochondria Among Different Organisms

Organism(s)	Organelle	Transmission
Mammals	Mitochondria	Maternal inheritance
S. cerevisiae	Mitochondria	Biparental inheritance
Molds	Mitochondria	Usually maternal inheritance; paternal inheritance has been found in the genus *Allomyces*
Chlamydomonas	Mitochondria	Inherited from the parent with the *mt*$^-$ mating type
Plants		
Angiosperms	Mitochondria	Often maternal inheritance, although biparental inheritance is found among some species
Gymnosperms	Mitochondria	Usually paternal inheritance

6.3 THEORY OF ENDOSYMBIOSIS

TABLE 6.5
Examples of Human Mitochondrial Diseases

Disease	Mitochondrial Gene Mutated
Leber hereditary optic neuropathy	A mutation in one of several mitochondrial genes that encode electron transport chain proteins: ND1, ND2, CO1, ND4, ND5, ND6, and cytb
Neurogenic muscle weakness	A mutation in the *ATPase6* gene that encodes a subunit of the mitochondrial ATP-synthetase, which is required for ATP synthesis
Mitochondrial myopathy	A mutation in a gene that encodes a tRNA for leucine
Maternal myopathy and cardiomyopathy	A mutation in a gene that encodes a tRNA for leucine

An important factor in mitochondrial disease is heteroplasmy, a condition in which a cell contains a mixed population of mitochondria. Within a single cell, some mitochondria may carry a disease-causing mutation, whereas others may not. As cells divide, mutant and normal mitochondria randomly segregate into the resulting daughter cells. Some daughter cells may receive a high ratio of mutant to normal mitochondria, whereas others may receive a low ratio. The ratio of mutant to normal mitochondria must exceed a certain threshold value before disease symptoms are observed.

6.2 REVIEWING THE KEY CONCEPTS

- Mitochondria have circular chromosomes in a nucleoid (see Figures 6.4, 6.5, Table 6.3).
- Mitochondria are usually transmitted by a maternal inheritance pattern, but transmission patterns vary among different organisms (see Table 6.4).
- Many human diseases are caused by mutations in mitochondrial DNA (see Table 6.5).

6.2 COMPREHENSION QUESTIONS

1. Human mitochondrial DNA
   a. is found in the nucleoid of the mitochondrion.
   b. is found in a circular chromosome.
   c. encodes rRNAs, tRNAs, and several proteins.
   d. All of the above describe human mitochondrial DNA.
2. In most species, the transmission of mitochondria follows
   a. a maternal inheritance pattern.
   b. a paternal inheritance pattern.
   c. a biparental inheritance pattern.
   d. none of the above.
3. Some human diseases are caused by mutations in mitochondrial genes. Which of the following statements is *false*?
   a. Mitochondrial diseases may follow a maternal inheritance pattern.
   b. Mutations associated with mitochondrial diseases often affect cells with a high demand for ATP.
   c. The symptoms associated with mitochondrial diseases tend to improve with age.
   d. Heteroplasmy plays a key role in the severity of mitochondrial disease symptoms.

6.3 THEORY OF ENDOSYMBIOSIS

Learning Outcome:
1. Describe the endosymbiosis theory.

The idea that the nucleus, chloroplasts, and mitochondria contain their own separate genetic material may at first seem puzzling. Wouldn't it be simpler to have all of the genetic material in one place in the cell? The underlying reason for the distinct genomes of chloroplasts and mitochondria can be traced back to their evolutionary origin, which is thought to have involved a symbiotic association.

A symbiotic relationship occurs when two different species live together in a close association. The symbiont is the smaller of the two species; the host is the larger. The term **endosymbiosis** describes a symbiotic relationship in which the symbiont actually lives inside (*endo-* means "inside") the host. In 1883, Andreas Schimper proposed that chloroplasts were descended from an endosymbiotic relationship between cyanobacteria and eukaryotic cells. This idea, now known as the **endosymbiosis theory,** indicates that the ancient origin of chloroplasts was initiated when a cyanobacterium took up residence within a primordial eukaryotic cell (**Figure 6.6**).

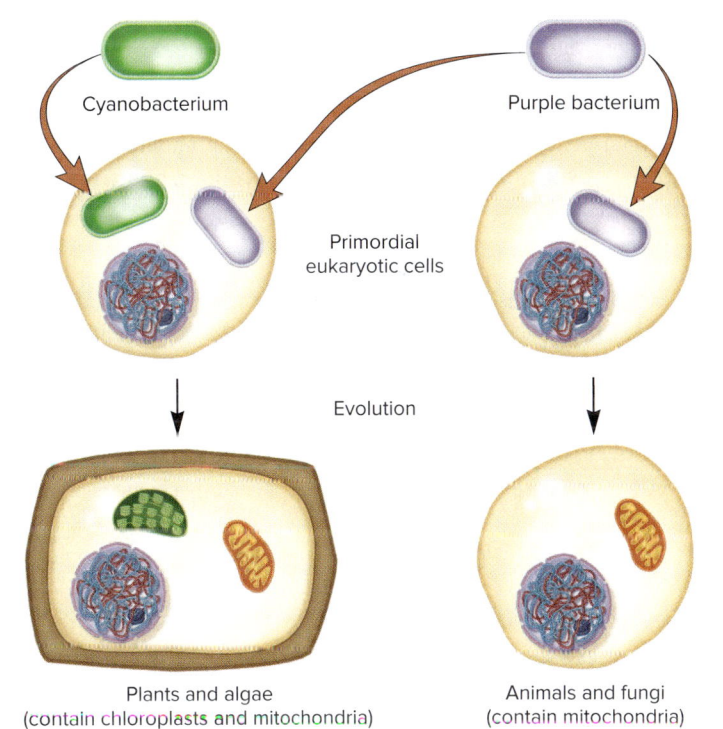

FIGURE 6.6 **The endosymbiotic origin of chloroplasts and mitochondria.** According to the endosymbiosis theory, chloroplasts descended from an endosymbiotic relationship between cyanobacteria and eukaryotic cells. This relationship arose when a cyanobacterium took up residence within a primordial eukaryotic cell. Over the course of evolution, the intracellular bacterial cell gradually changed its characteristics, eventually becoming a chloroplast. Similarly, mitochondria are derived from an endosymbiotic relationship between purple bacteria and eukaryotic cells.

Concept Check: How have chloroplasts and mitochondria changed since the initial endosymbiosis events, which occurred hundreds of millions of years ago?

Over the course of evolution, the characteristics of the intracellular bacterial cell gradually changed to those of a chloroplast. In 1922, Ivan Wallin also proposed an endosymbiotic origin for mitochondria.

The endosymbiosis theory proposes that the relationship provided eukaryotic cells with useful cellular characteristics. Chloroplasts were derived from cyanobacteria, a bacterial species that is capable of photosynthesis. The ability to carry out photosynthesis enabled algal and plant cells to use the energy from sunlight. By comparison, mitochondria are thought to have been derived from a different type of bacteria known as gram-negative nonsulfur purple bacteria. In this case, the endosymbiotic relationship enabled eukaryotic cells to synthesize greater amounts of ATP. It is less clear how the relationship would have been beneficial to cyanobacteria or purple bacteria, though the cytosol of a eukaryotic cell may have provided a stable environment with an adequate supply of nutrients.

During the evolution of eukaryotic species, most genes that were originally found in the genome of the primordial cyanobacteria and purple bacteria have been lost or transferred from the organelles to the nucleus. Such changes have occurred many times throughout evolution, so modern chloroplasts and mitochondria have lost most of the genes that are still found in present-day cyanobacteria and purple bacteria. For example, about 1500 genes have been transferred from the mitochondrial genome to the nuclear genome of mammals. This unidirectional gene transfer from organelles to the nucleus helps to explain why the organellar genomes now contain relatively few genes.

6.3 REVIEWING THE KEY CONCEPTS

- Chloroplasts and mitochondria were derived from ancient endosymbiotic relationships (see Figure 6.6).

6.3 COMPREHENSION QUESTION

1. Chloroplasts and mitochondria evolved from endosymbiotic relationships involving
    a. purple bacteria and cyanobacteria, respectively.
    b. cyanobacteria and purple bacteria, respectively.
    c. cyanobacteria.
    d. purple bacteria.

6.4 EPIGENETICS: IMPRINTING

Learning Outcomes:
1. Define *epigenetic inheritance* and *genomic imprinting*.
2. Predict the outcome of crosses involving imprinted genes.
3. Explain the molecular mechanism of imprinting.

We now turn to a non-Mendelian pattern of inheritance that involves nuclear genes. **Epigenetic inheritance** is a pattern in which a modification occurs to a nuclear gene or chromosome that alters gene expression, but is not permanent over the course of many generations. Epigenetic inheritance patterns are the result of DNA and chromosomal modifications that occur during oogenesis, spermatogenesis, or early stages of embryogenesis. Once they are initiated during these early stages, epigenetic changes alter the expression of particular genes in a way that may be fixed during an individual's lifetime. Therefore, epigenetic changes can permanently affect the phenotype of the individual. However, epigenetic modifications are not permanent over the course of many generations, and they do not change the actual DNA sequence. For example, a gene may undergo an epigenetic change that inactivates it for the lifetime of an individual. However, when this individual makes gametes, the gene may become activated and remain operative during the lifetime of an offspring who inherits the active gene.

We already considered one example of epigenetic inheritance, X-chromosome inactivation, in Chapter 4 (see Figures 4.5–4.7). During this process, one of the two X chromosomes in each somatic cell of female mammals is chosen for random inactivation during embryonic development. This may lead to a variegated phenotype in females. In this section, we will examine another form of epigenetic inheritance called **genomic imprinting,** or simply **imprinting,** in which a modification occurs to a nuclear gene that alters gene expression, but is not permanent over the course of many generations.

The Expression of an Imprinted Gene Depends on the Sex of the Parent from Which the Gene Was Inherited

The term *imprinting* implies a type of marking process that has a memory. For example, newly hatched birds identify marks on their parents that allow them to distinguish their parents from other individuals. The term *genomic imprinting* refers to an analogous situation in which a segment of DNA is marked, and that mark is retained and recognized throughout the life of the organism inheriting the marked DNA. The phenotypes caused by imprinted genes follow a non-Mendelian pattern of inheritance because the marking process causes the offspring to distinguish between maternally and paternally inherited alleles. Depending on how the genes are marked, each offspring expresses only one of the two alleles. This phenomenon is termed **monoallelic expression.**

To understand genomic imprinting, let's consider a specific example. In the mouse, a gene designated *Igf2* encodes a protein growth hormone called insulin-like growth factor 2.

- Imprinting results in the expression of the paternal *Igf2* allele, but not the maternal allele.
- The paternal allele is transcribed into RNA, but the maternal allele is transcriptionally silent.

With regard to phenotype, a functional *Igf2* gene is necessary for normal size. A loss-of-function allele of this gene, designated *Igf2*$^-$, is defective with regard to synthesis of a functional Igf2 protein. This may cause a mouse to be a dwarf, but the occurrence of dwarfism depends on whether the mutant allele is inherited from the male or female parent, as shown in **Figure 6.7**.

- On the left side, an offspring has inherited the *Igf2* allele from its father and the *Igf2⁻* allele from its mother. Due to imprinting, only the *Igf2* allele is expressed in the offspring. Therefore, this mouse grows to a normal size.
- In the reciprocal cross on the right side, an individual has inherited the *Igf2⁻* allele from its father and the *Igf2* allele from its mother. In this case, the *Igf2* allele is not expressed. In this mouse, the *Igf2⁻* allele is transcribed into mRNA, but the mutation renders the Igf2 protein defective. Therefore, the offspring on the right has a dwarf phenotype.
- Both offspring shown in Figure 6.7 have the same genotype; they are heterozygous for the *Igf2* alleles (i.e., *Igf2 Igf2⁻*). They are phenotypically different, however, because only the paternally inherited allele is expressed.

For imprinted genes, you cannot use the Punnett square approach to predict an offspring's phenotype. Instead, you need two pieces of information.

- You need to know if the offspring expresses the allele that is inherited from the mother or the father.
- You need to know which allele was inherited from the mother and which allele was inherited from the father.

With this information, you can predict an offspring's phenotype. As an example, let's consider the *Igf2* gene. We know that the allele inherited from the father is expressed. If the offspring inherits the *Igf2* allele from the father, it will be a normal size. If the offspring inherits the *Igf2⁻* allele from the father, it will be dwarf.

Igf2⁻ Igf2⁻ (mother's genotype) × *Igf2 Igf2* (father's genotype)

Igf2 Igf2 (mother's genotype) × *Igf2⁻ Igf2⁻* (father's genotype)

Igf2 ● *Igf2⁻* ▲
Normal offspring
(Only the *Igf2* allele is expressed in somatic cells of this heterozygous offspring.)

Igf2 ▲ *Igf2⁻* ●
Dwarf offspring
(Only the *Igf2⁻* allele is expressed in somatic cells of this heterozygous offspring.)

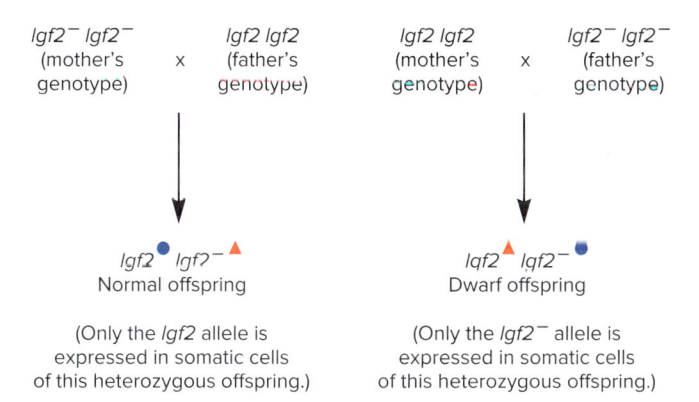

▲ Denotes an allele that is silent in the offspring
● Denotes an allele that is expressed in the offspring

FIGURE 6.7 **An example of genomic imprinting in the mouse.** In the cross on the left, a homozygous male with the functional *Igf2* allele is crossed to a homozygous female carrying a defective allele, designated *Igf2⁻*. An offspring is heterozygous and normal size because the paternal allele is active. In the reciprocal cross on the right, a homozygous male carrying the defective allele is crossed to a homozygous female carrying two functional alleles. In this case, the offspring is heterozygous and dwarf. This is because the paternal allele is defective due to a mutation and the maternal allele is not expressed. The photograph shows normal-size (left) and dwarf littermates (right) derived from a cross between a wild-type female and a heterozygous male carrying a loss-of-function *Igf2* allele (courtesy of A. Efstratiadis). The loss-of-function allele was created using gene knockout methods.
Courtesy of Dr. Argiris Efstratiadis

Concept Check: What would be the outcome of a cross between a heterozygous female and a male that carries two functional copies of the Igf2 gene?

The Imprint Is Established During Gametogenesis

At the cellular level, imprinting is a process that can be divided into three stages: (1) the establishment of the imprint during gametogenesis, (2) the maintenance of the imprint during embryogenesis and in adult somatic cells, and (3) the erasure and reestablishment of the imprint in the germ cells. These stages are described in **Figure 6.8**, which shows the imprinting of the *Igf2* gene.

1. The two mice shown here have inherited the *Igf2* allele from their father and the *Igf2⁻* allele from their mother. Imprinting was established during gametogenesis in the parents of these mice.
2. Due to imprinting, both mice express the *Igf2* allele in their somatic cells, and the pattern of imprinting is maintained in the somatic cells throughout development.
3. In the germ-line cells that give rise to gametes (i.e., sperm or eggs), the imprint is erased; it is reestablished according to the sex of the animal. The female mouse on the left transmits only transcriptionally inactive alleles to offspring. The male mouse on the right transmits transcriptionally active alleles. However, because this male is a heterozygote, it transmits either a functionally active *Igf2* allele or a functionally defective mutant allele (*Igf2⁻*). If this heterozygous male transmits the *Igf2* allele to an offspring, the offspring will be normal size. In contrast, an *Igf2⁻* allele, which is inherited from a male mouse, can be expressed into mRNA (i.e., it is transcriptionally active), but it does not produce a functional Igf2 protein due to the deleterious mutation that created the *Igf2⁻* allele. Therefore, a dwarf phenotype results.

As seen in Figure 6.8, genomic imprinting is permanent in the somatic cells of an animal, but the marking of alleles can be altered from generation to generation. For example, the female mouse on the left has an active copy of the *Igf2* allele, but any allele this female transmits to its offspring is transcriptionally inactive.

Establishment of the imprint
In this example, imprinting occurs during gametogenesis in the *Igf2* gene, which exists in the *Igf2* allele from the male and the *Igf2⁻* allele from the female. This imprinting occurs so that only the paternal allele is expressed.

Maintenance of the imprint
After fertilization, the imprint pattern is maintained throughout development. In this example, the maternal *Igf2⁻* allele will not be expressed in the somatic cells. Note that the offspring on the left is a female and the one on the right is a male; both are normal in size.

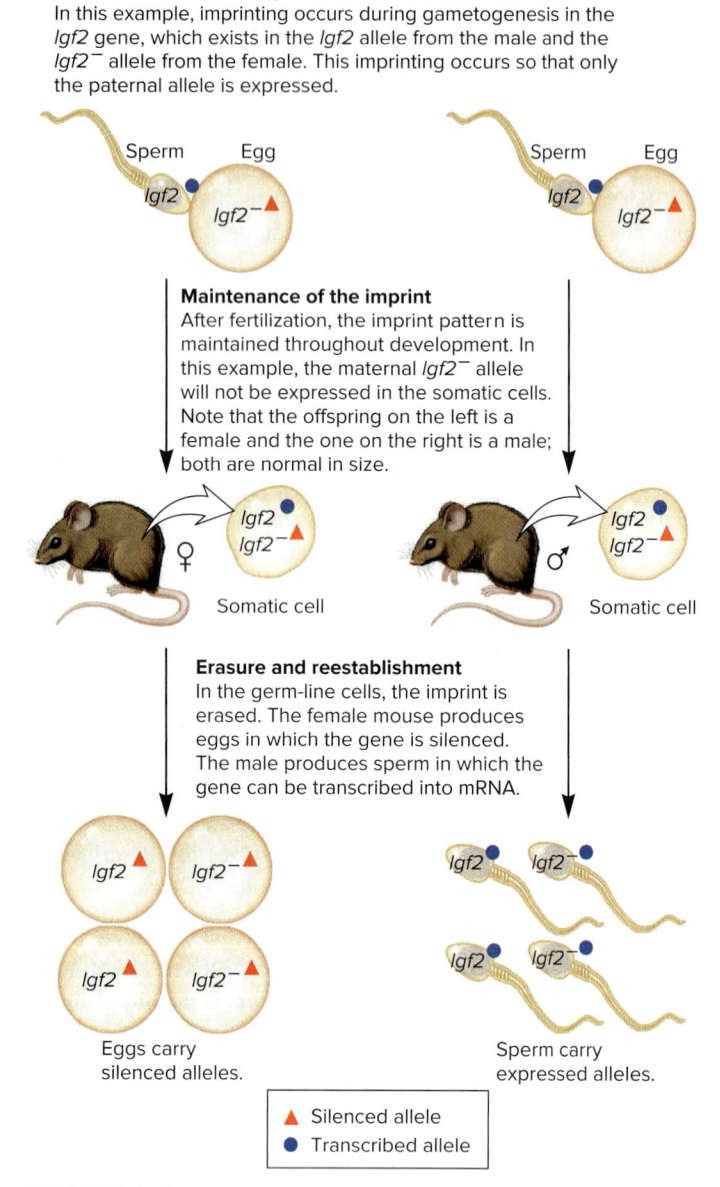

Erasure and reestablishment
In the germ-line cells, the imprint is erased. The female mouse produces eggs in which the gene is silenced. The male produces sperm in which the gene can be transcribed into mRNA.

▲ Silenced allele
● Transcribed allele

FIGURE 6.8 **Genomic imprinting during gametogenesis.** This example involves a mouse gene, *Igf2*, which is found in two alleles designated *Igf2* and *Igf2⁻*. The left side shows a female mouse that was produced from a sperm carrying the *Igf2* allele and an egg carrying the *Igf2⁻* allele. In the somatic cells of this female animal, the *Igf2* allele is active. However, when this female produces eggs, both alleles are transcriptionally inactive when they are transmitted to offspring. The right side of this figure shows a male mouse that was also produced from a sperm carrying the *Igf2* allele and an egg carrying the *Igf2⁻* allele. In the somatic cells of this male animal, the *Igf2* allele is active. However, the sperm from this male carry either a functionally active *Igf2* allele or a functionally defective *Igf2⁻* allele.

Concept Check: *Explain why the erasure stage of imprinting is necessary in eggs.*

Genomic imprinting occurs in several species, including numerous insects, mammals, and flowering plants. Depending on the species, imprinting may involve a single gene, a part of a chromosome, an entire chromosome, or even all of the chromosomes from one parent. Genomic imprinting can also be involved in the process of X-chromosome inactivation, described in Chapter 4. In certain species, imprinting plays a role in the choice of which X chromosome will be inactivated. For example, in marsupials, the paternal X chromosome is marked so that it is always inactivated in the somatic cells of females. In marsupials, X-chromosome inactivation is not random; the maternal X chromosome is always active.

The Imprinting of Genes and Chromosomes Is a Molecular Marking Process That Involves DNA Methylation

As we have seen, genomic imprinting must involve a marking process. A particular gene or chromosome must be marked differently during spermatogenesis compared to oogenesis. After fertilization takes place, this differential marking affects the expression of particular genes. What is the molecular explanation for genomic imprinting? As discussed in Chapter 17 (see Figure 17.11), **DNA methylation**—the attachment of a methyl group onto a cytosine base in DNA—is a common way that eukaryotic genes may be regulated. Research indicates that genomic imprinting involves an **imprinting control region (ICR)** that is located near the imprinted gene. The ICR contains binding sites for one or more proteins that regulate the transcription of the imprinted gene.

Let's consider the methylation process from one generation to the next. In the example shown in **Figure 6.9**, the paternally inherited allele for a particular gene is methylated, but the maternally inherited allele is not.

- A female (left side) and male (right side) have inherited a methylated ICR from their father and an unmethylated ICR from their mother. This pattern of imprinting is maintained in the somatic cells of both individuals.
- When the female makes gametes, the imprinting is erased (demethylated) during early oogenesis, so the female passes an unmethylated ICR to her offspring.
- In the male, the imprinting is also erased during early spermatogenesis, but then *de novo* (new) methylation occurs in both ICRs. Therefore, the male transmits a methylated gene to his offspring.

Many Human Genes Are Imprinted, and Some Are Associated with Human Diseases

Genomic imprinting is a fairly new and exciting area of research. About 100 genes have been shown to be imprinted in humans. Examples are presented in **Table 6.6**. A few human genetic diseases, such as Angelman syndrome (AS), involve imprinted genes. Individuals with Angelman syndrome are thin and hyperactive, have unusual seizures, and exhibit repetitive symmetrical muscle movements and mental deficiencies. The syndrome is caused by a

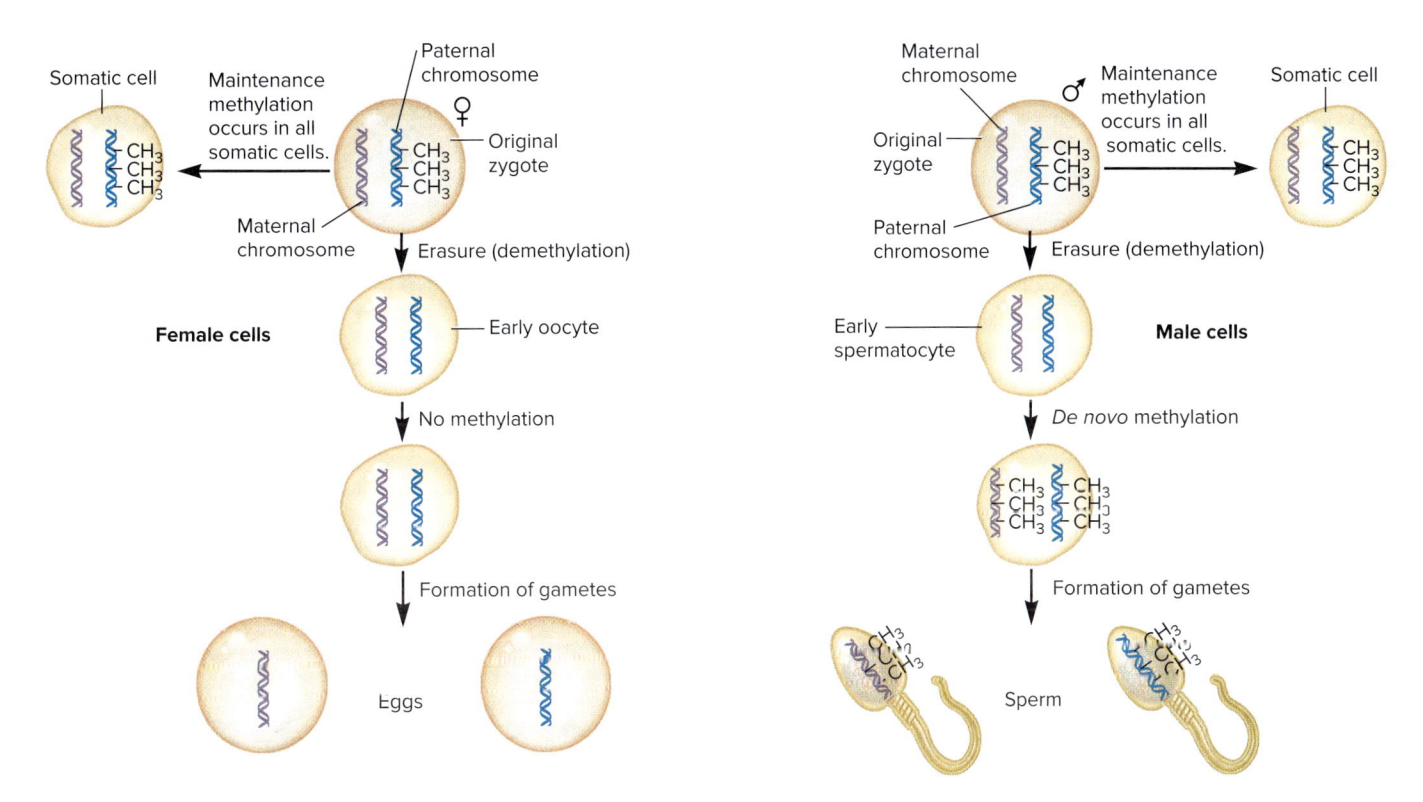

FIGURE 6.9. **The pattern of ICR methylation from one generation to the next.** In this example, a male and a female offspring have inherited a methylated imprinting control region (ICR) and unmethylated ICR from their father and mother, respectively. Maintenance methylation retains the imprinting in somatic cells during embryogenesis and in adulthood. Erasure (demethylation) occurs in cells that are destined to become gametes. In this example, *de novo* methylation occurs only in cells that are destined to become sperm. Haploid male gametes transmit a methylated ICR, whereas haploid female gametes transmit an unmethylated ICR.

Concept Check: *What is the difference between maintenance methylation and de novo methylation? In what cell types (somatic cells or germ-line cells) does each process occur?*

TABLE 6.6

Examples of Human Genes That Are Imprinted*

Gene	Allele Expressed	Function
WT1	Maternal	Wilms tumor-suppressor gene; suppresses cell growth
INS	Paternal	Insulin; hormone involved in cell growth and metabolism
Igf2	Paternal	Insulin-like growth factor 2; similar to insulin
Igf2R	Maternal	Receptor for insulin-like growth factor 2
UBE3A	Maternal	Targets other proteins for degradation
SNRPN	Paternal	Splicing factor
Gabrb	Maternal	Neurotransmitter receptor

*Researchers estimate that approximately 1–2% of human genes are subjected to genomic imprinting, but only about 100 have actually been demonstrated to be imprinted.

rare loss-of-function mutation of the *UBE3A* gene (see Table 6.6). This gene is silenced during sperm formation due to imprinting. A functional copy of the *UBE3A* gene is needed to prevent Angelman syndrome. Therefore, if a defective *UBE3A* gene is inherited from the mother, the offspring will have Angelman syndrome because the father will have transmitted a silenced copy of the gene.

6.4 REVIEWING THE KEY CONCEPTS

- Epigenetic inheritance is an inheritance pattern in which a gene or chromosome is modified so as to alter gene expression, but the modification is not permanent over the course of many generations. An example is imprinting, in which an offspring expresses a gene that is inherited from one parent but not both (see Figures 6.7, 6.8).
- DNA methylation at an imprinting control region is the marking process that causes imprinting (see Figure 6.9).
- Imprinting has been identified in many mammalian genes and may play a role in some human diseases (Table 6.6).

6.4 COMPREHENSION QUESTIONS

1. In mice, the copy of the *Igf2* gene that is inherited from the mother is never expressed in her offspring. This happens because the *Igf2* gene from the mother
   a. always undergoes a mutation that inactivates its function.
   b. is deleted during oogenesis.
   c. is deleted during embryonic development.
   d. is not transcribed in the somatic cells of the offspring.
2. A female mouse that is *Igf2 Igf2⁻* is crossed to a male that is also *Igf2 Igf2⁻*. The expected outcome for the phenotypes of the offspring is
   a. all normal size.
   b. all dwarf.
   c. 1 normal size : 1 dwarf.
   d. 3 normal size : 1 dwarf.
3. The marking process for genomic imprinting initially occurs during
   a. gametogenesis.
   b. fertilization.
   c. embryonic development.
   d. adulthood.

6.5 MATERNAL EFFECT

Learning Outcomes:
1. Define *maternal effect*.
2. Predict the outcome of crosses for genes that exhibit a maternal effect pattern of inheritance.
3. Explain the mechanism of maternal effect inheritance at the molecular and cellular level.

Maternal effect refers to an inheritance pattern for certain nuclear genes in which the genotype of the mother directly determines the phenotype of her offspring. (Note that maternal effect should not be confused with maternal inheritance.) Surprisingly, for maternal effect genes, the genotypes of the father and of the offspring themselves do not affect the phenotype of the offspring. Therefore, you cannot use a Punnett square to predict the phenotype of the offspring. We will see that maternal effect inheritance is explained by the accumulation of gene products that the mother provides to her developing eggs.

The Genotype of the Mother Determines the Phenotype of the Offspring for Maternal Effect Genes

The first example of a maternal effect gene was studied in the 1920s by Arthur Boycott and involved morphological features of the water snail, *Lymnaea peregra*. In this species, the shell and internal organs can be arranged in either a right-handed (dextral) or left-handed (sinistral) direction (see the chapter-opening photo). The dextral orientation is more common and is dominant to the sinistral orientation. **Figure 6.10** describes the results of a genetic analysis carried out by Boycott. In this experiment, he began with two different true-breeding strains of snails with either a dextral or sinistral morphology.

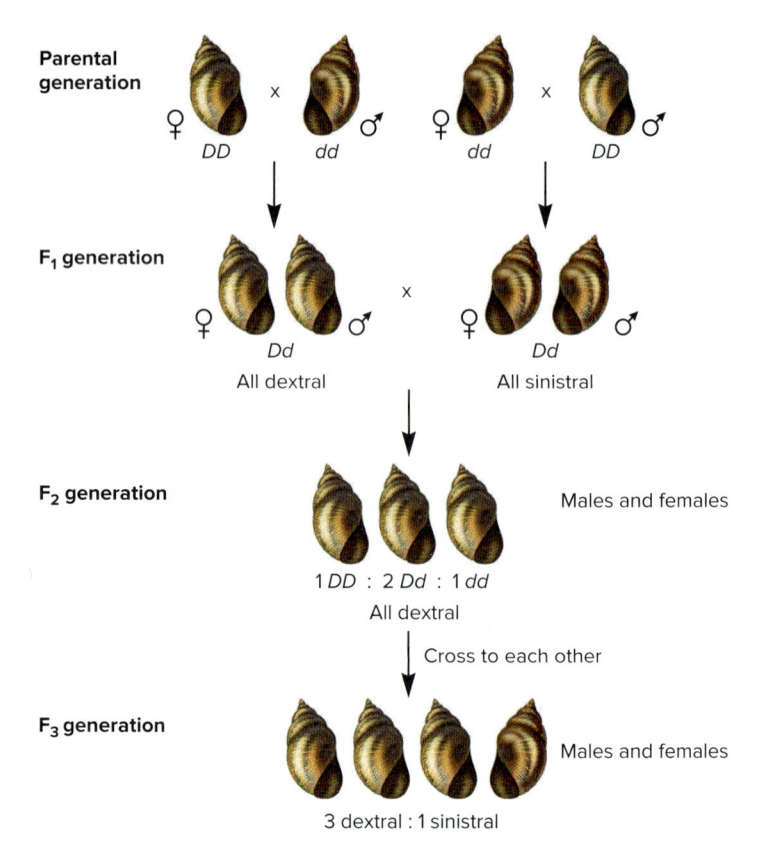

FIGURE 6.10 Experiment showing the inheritance pattern of snail coiling. In this experiment, *D* (dextral) is dominant to *d* (sinistral). The genotype of the mother determines the phenotype of the offspring. This phenomenon is known as the maternal effect. In this case, a *DD* or *Dd* mother produces dextral offspring, and a *dd* mother produces sinistral offspring. The genotypes of the father and offspring do not affect the offspring's phenotype.

Concept Check: Explain why all of the offspring in the F_2 generation are dextral even though some of them are *dd*.

F₁ Generation Many combinations of crosses produced results that could not be explained by a Mendelian pattern of inheritance.

- When a dextral female (*DD*) was crossed to a sinistral male (*dd*), all F_1 offspring were dextral.
- In the reciprocal cross, where a sinistral female (*dd*) was crossed to a dextral male (*DD*), all F_1 offspring were sinistral.

Taken together, these results contradict a Mendelian pattern of inheritance.

How can we explain the unusual results obtained in Figure 6.10? Alfred Sturtevant proposed the idea that snail coiling is due to a maternal effect gene that exists as a dextral (*D*) or sinistral (*d*) allele. His conclusions were drawn from the inheritance patterns of the F_2 and F_3 generations.

F₂ Generation The genotype of the F_1 generation is expected to be heterozygous (*Dd*). When these F_1 individuals were crossed to each other, a genotypic ratio of 1 DD : 2 Dd : 1 dd is predicted for the F_2 generation. Because the D allele is dominant to the d allele, a 3:1 phenotypic ratio of dextral to sinistral snails should be produced according to a Mendelian pattern of inheritance. Instead of this predicted phenotypic ratio, however, the F_2 generation was composed of all dextral snails. This incongruity with Mendelian inheritance is due to the maternal effect. The phenotype of the offspring depended solely on the genotype of the mother. The F_1 mothers were Dd. The D allele in the mothers is dominant to the d allele and caused the offspring to be dextral, even if the offspring's genotype was dd!

F_3 Generation When the members of the F_2 generation were crossed, the F_3 generation exhibited a 3:1 ratio of dextral to sinistral snails. This ratio corresponds to the genotypes of the F_2 females, which were the mothers of the F_3 generation. The ratio of F_2 females was 1 DD : 2 Dd : 1 dd. The DD and Dd females produced dextral offspring, whereas the dd females produced sinistral offspring. This explains the 3:1 ratio of dextral and sinistral offspring in the F_3 generation.

Female Gametes Receive Gene Products from the Mother That Affect Early Developmental Stages of the Embryo

At the molecular and cellular level, the non-Mendelian inheritance pattern of maternal effect genes can be explained by the process of oogenesis in female animals (**Figure 6.11a**). As an animal oocyte (egg) matures, many surrounding maternal cells called nurse cells provide the egg with nutrients and other materials. In Figure 6.11a, a female is heterozygous for the snail-coiling maternal effect gene, with the alleles designated D and d. Depending on the outcome of meiosis, the haploid egg may receive the D allele or the d allele, but not both. The surrounding nurse cells, however, produce both D and d gene products (mRNA and/or proteins). These gene products are then transported into the egg. As shown here, the egg has received both the D gene product and the d gene product. These gene products persist for a significant time after the egg has been fertilized and embryonic development has begun. In this way, the gene products of the nurse cells, which reflect the genotype of the mother, influence the early developmental stages of the embryo.

Now that we understand the relationship between oogenesis and maternal effect genes, let's reconsider the topic of snail coiling.

- As shown in **Figure 6.11b**, a female snail that is DD transmits only the D gene product to the egg. During the early stages of embryonic development, this gene product causes the egg cleavage to occur in a way that promotes a right-handed body plan.
- A heterozygous female transmits both D and d gene products. Because the D allele is dominant, the maternal effect also causes a right-handed body plan.
- A dd mother contributes only the d gene product that promotes a left-handed body plan, even if the egg is fertilized by a sperm carrying a D allele. The sperm's genotype is irrelevant, because the expression of the sperm's gene will occur too late.

The origin of dextral and sinistral coiling can be traced to the orientation of the mitotic spindle at the two- to four-cell stage of embryonic development. The dextral and sinistral snails develop as mirror images of each other (**Figure 6.11c**).

Since these initial studies, researchers have found that maternal effect genes encode proteins that are important in the early steps of embryogenesis. The accumulation of maternal gene products in the egg allows embryogenesis to proceed quickly after fertilization. Maternal effect genes often play a role in cell division, cleavage pattern, and body axis orientation. Therefore, defective alleles in maternal effect genes tend to have a dramatic effect on the phenotype of the offspring, altering major features of morphology, often with dire consequences.

Genetic TIPS

The Question: A female snail has offspring that all coil to the right. What are the possible genotypes of this female snail?

Topic: What topic in genetics does this question address?

The topic is non-Mendelian inheritance. More specifically, the question is about maternal effect inheritance.

Information: What information do you know based on the question and your understanding of the topic?

From the question, you know a female snail has offspring that all coil to the right. From your understanding of the topic, you may remember that this trait shows a maternal effect pattern of inheritance and that the mother's genotype determines the offspring's phenotype.

Problem-Solving Strategy: Predict the outcome.

A strategy to solve this problem is to predict the outcome for each possible genotype of the mother. For maternal effect genes, you cannot use a Punnett square to predict the phenotype of the offspring. Instead, to predict their phenotype, you have to know the mother's genotype. In this question, you already know the offspring's phenotype. With this information, you can deduce the possible genotype(s) of the mother. Because the offspring coil to the right, you may realize that the mother must have at least one D allele. If the mother was DD, all of her offspring would coil to the right. If the mother was Dd, all of her offspring would coil to the right, because D is dominant. If the mother was dd, all of her offspring would coil to the left.

Answer: The mother's genotype could be either DD or Dd.

6.5 REVIEWING THE KEY CONCEPTS

- Maternal effect is an inheritance pattern in which the genotype of the mother determines the phenotype of the offspring (see Figure 6.10).
- Maternal effect inheritance occurs because the gene products of maternal effect genes are transferred from nurse cells to the oocyte. These gene products affect early stages of development (see Figure 6.11).

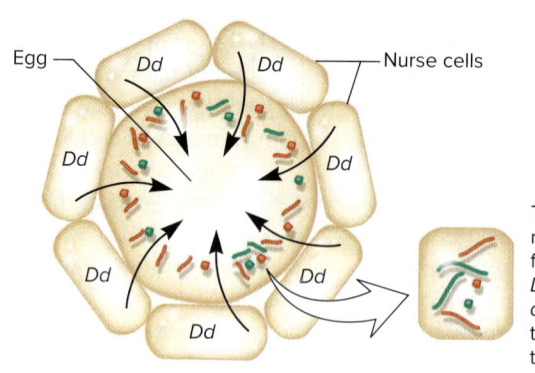

(a) Transfer of gene products from nurse cells to egg

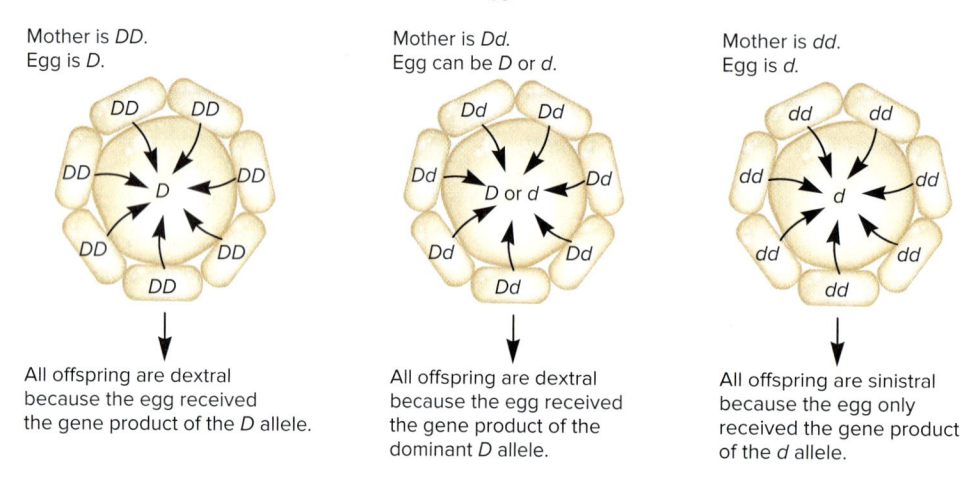

The nurse cells express mRNA and/or protein from genes of the *D* allele (green) and the *d* allele (red) and transfer those products to the egg.

Mother is *DD*. Egg is *D*.

All offspring are dextral because the egg received the gene product of the *D* allele.

Mother is *Dd*. Egg can be *D* or *d*.

All offspring are dextral because the egg received the gene product of the dominant *D* allele.

Mother is *dd*. Egg is *d*.

All offspring are sinistral because the egg only received the gene product of the *d* allele.

(b) Maternal effect in snail coiling

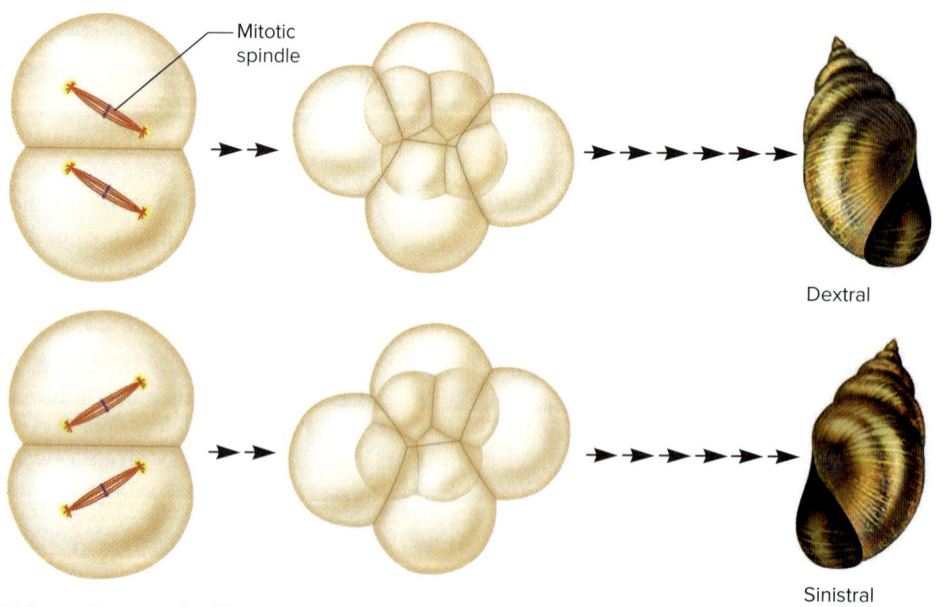

(c) An explanation of coiling direction at the cellular level

FIGURE 6.11 **The mechanism of maternal effect in snail coiling.** (a) Transfer of gene products from nurse cells to an egg. The nurse cells are heterozygous (*Dd*). Both the *D* and *d* alleles are activated in the nurse cells to produce *D* and *d* gene products (mRNA or proteins, or both). These products are transported into the cytoplasm of the egg, where they accumulate to significant amounts. (b) Explanation of the maternal effect in snail coiling. (c) The direction of snail coiling is determined by differences in the cleavage planes during early embryonic development.

Genes → Traits If the nurse cells are *DD* or *Dd*, they will transfer the *D* gene product to the egg, thereby causing the offspring to be dextral. If the nurse cells are *dd*, only the *d* gene product is transferred to the egg, so the resulting offspring will be sinistral.

Concept Check: If a mother snail is heterozygous, *Dd*, which gene products will the egg cell receive?

6.5 COMPREHENSION QUESTIONS

1. A female snail that coils to the left has offspring that coil to the right. What are the genotypes of this mother and of the maternal grandmother of the offspring, respectively?
   a. *dd*, *DD*
   b. *Dd*, *Dd*
   c. *dd*, *Dd*
   d. *Dd*, *dd*

2. What is the explanation for maternal effect inheritance at the molecular and cellular level?
   a. The father's gene is silenced at fertilization.
   b. During oogenesis, nurse cells transfer gene products to the oocyte.
   c. The gene products from nurse cells are needed during the very early stages of development.
   d. Both b and c are correct.

KEY TERMS

Page 114. nuclear genes, extranuclear inheritance (cytoplasmic inheritance), nucleoid, chloroplast DNA (cpDNA)
Page 115. reciprocal cross, maternal inheritance, heteroplasmy
Page 116. heterogamous
Page 117. mitochondrial DNA (mtDNA)
Page 118. paternal leakage
Page 119. endosymbiosis, endosymbiosis theory
Page 120. epigenetic inheritance, genomic imprinting (imprinting), monoallelic expression
Page 122. DNA methylation, imprinting control region (ICR)
Page 124. maternal effect

CHAPTER SUMMARY

- Non-Mendelian inheritance refers to inheritance patterns that do not obey Mendel's laws of inheritance.

6.1 Extranuclear Inheritance: Chloroplasts

- Extranuclear inheritance is the inheritance of genes that are found in chloroplasts or mitochondria.
- Chloroplasts have circular chromosomes in a nucleoid. These circular chromosomes carry many genes but far fewer compared with the number on chromosomes in the cell nucleus (see Figure 6.1, Table 6.1).
- Maternal inheritance occurs when organelles, such as chloroplasts, are transmitted via the egg (see Figure 6.2).
- Heteroplasmy of chloroplasts can result in a variegated phenotype (see Figure 6.3).
- The transmission patterns of chloroplasts vary among different organisms (see Table 6.2).

6.2 Extranuclear Inheritance: Mitochondria

- Mitochondria have circular chromosomes in a nucleoid (see Figures 6.4, 6.5, Table 6.3).
- Mitochondria are usually transmitted by a maternal inheritance pattern, but transmission patterns vary among different organisms (see Table 6.4).
- Many human diseases are caused by mutations in mitochondrial DNA (see Table 6.5).

6.3 Theory of Endosymbiosis

- Chloroplasts and mitochondria were derived from ancient endosymbiotic relationships (see Figure 6.6).

6.4 Epigenetics: Imprinting

- Epigenetic inheritance is an inheritance pattern in which a gene or chromosome is modified so as to alter gene expression, but the modification is not permanent over the course of many generations. An example is imprinting, in which an offspring expresses a gene that is inherited from one parent but not both (see Figures 6.7, 6.8).
- DNA methylation at an imprinting control region is the marking process that causes imprinting (see Figure 6.9).
- Imprinting has been identified in many mammalian genes and may play a role in some human diseases (Table 6.6).

6.5 Maternal Effect

- Maternal effect is an inheritance pattern in which the genotype of the mother determines the phenotype of the offspring (see Figure 6.10).
- Maternal effect inheritance occurs because the gene products of maternal effect genes are transferred from nurse cells to the oocyte. These gene products affect early stages of development (see Figure 6.11).

PROBLEM SETS & INSIGHTS

More Genetic TIPS

1. Let's suppose you are a horticulturist and have focused your work on developing strains of petunias that have interesting characteristics. Although most of your petunia strains have green leaves, you have recently identified an interesting plant with variegated leaves. How would you determine if this trait is the result of nuclear or extranuclear inheritance?

   Topic: What topic in genetics does this question address?
   The topic is inheritance. More specifically, the question is about distinguishing nuclear and extranuclear inheritance patterns.

Information: What information do you know based on the question and your understanding of the topic?

From the question, you know that you have identified a plant with variegated leaves. From your understanding of the topic, you may remember that some genes are in the cell nucleus, whereas others are found in organelles, such as chloroplasts and mitochondria. Variation in nuclear or organellar genes could be responsible for this trait. Because nuclear genes segregate differently from organellar genes, one way to distinguish their inheritance patterns is to make reciprocal crosses.

Problem-Solving Strategy: Design an experiment.

To begin the design of this experiment, you need to consider your starting strains of plants. In this case, you have strains with green and variegated leaves. You could begin this experiment with a true-breeding strain with green leaves and a true-breeding strain with variegated leaves. The experimental approach would be to make crosses and reciprocal crosses and follow the outcomes for two or more generations.

Answer:

1. Start with two true-breeding strains, one with green leaves and one with variegated leaves.

2. Cross the strains to each other. For example, take the pollen from the green-leaved plants to pollinate the variegated plant. Make the reciprocal cross in which you take the pollen from the variegated plant to pollinate the green-leaved plant.

3. Observe the phenotypes of the F_1 generation.

4. Take the F_1 generation plants and do the same types of crosses and reciprocal crosses as described in step 2.

Overall, these experiments may yield different results depending on the mode of inheritance. For example, if the gene is a nuclear gene and the green allele is dominant, you expect all of the F_1 offspring to be green-leaved, and the F_2 generation to have a 3:1 ratio of green-leaved to variegated-leaved plants. On the other hand, if the gene is in the chloroplasts and follows maternal inheritance, the phenotype of offspring will depend on which plant contributed the egg.

2. What are two general mechanisms by which mitochondrial diseases may occur? Which mechanism would you expect to occur later in life?

Topic: What topic in genetics does this question address?

The topic is extranuclear inheritance. More specifically, the question is about how mutations in mitochondrial genes can cause disease.

Information: What information do you know based on the question and your understanding of the topic?

In the question, you are asked to identify two different ways that mitochondrial diseases may occur. From your understanding of the topic, you may remember that such diseases can be inherited or they can be due to new mutations.

Problem-Solving Strategy: Compare and contrast.

One strategy to solve this problem is to directly compare your knowledge of inherited mitochondrial diseases with those that occur as a result of new mutations.

Answer: In some cases, mitochondrial mutations that cause human disease are transmitted from mother to offspring. This mechanism is a maternal inheritance pattern. A second way that mitochondrial diseases occur is via somatic mutations. Mitochondrial mutations may occur in somatic cells and accumulate as a person ages. The second mechanism is cumulative and is more likely to occur later in life.

3. A maternal effect gene in *Drosophila*, called *torso*, occurs as a functional allele (*torso*$^+$) and a nonfunctional, recessive allele (*torso*$^-$) that prevents the correct development of anterior- and posterior-most structures. A wild-type male (*torso*$^+$*torso*$^+$) is crossed to a female of unknown genotype. This cross produces offspring (larva) that are all missing their anterior- and posterior-most structures and therefore die during early development. What are the genotype and the phenotype of the female fly in this cross? What are the genotypes and phenotypes of the female fly's parents?

Topic: What topic in genetics does this question address?

The topic is non-Mendelian inheritance. More specifically, the question is about maternal effect inheritance.

Information: What information do you know based on the question and your understanding of the topic?

From the question, you know that a gene called *torso* can exist as a recessive allele that prevents the proper development of anterior- and posterior-most structures. You are also given the results of a cross in which all of the offspring are missing their anterior- and posterior-most structures, and therefore die at the larval stage. From your understanding of the topic, you may remember that when a trait shows a maternal effect pattern of inheritance, the mother's genotype determines the offspring's phenotype.

Problem-Solving Strategy: Predict the outcome.

A strategy to solve this problem is to relate the outcome of the cross to the mother's genotype. For maternal effect genes, you cannot use a Punnett square to predict the phenotype of the offspring. Instead, you have to know the mother's genotype. From the question, you already know the offspring's phenotype. With this information, you can deduce the possible genotype(s) of their mother. Because the offspring are missing their anterior- and posterior-most structures, you may realize that the mother must be homozygous for the recessive *torso*$^-$ allele because her genotype determines her offspring's phenotype.

Answer: Because the cross produced only abnormal offspring that were missing their anterior- and posterior-most structures, the mother of the abnormal offspring must have been homozygous, *torso*$^-$*torso*$^-$. However, her phenotype with regard to body structure must be normal because she was able to reproduce. As shown below, the mother of the abnormal offspring had a mother that was heterozygous for the *torso* alleles and a father that was either heterozygous or homozygous for the *torso*$^-$ allele.

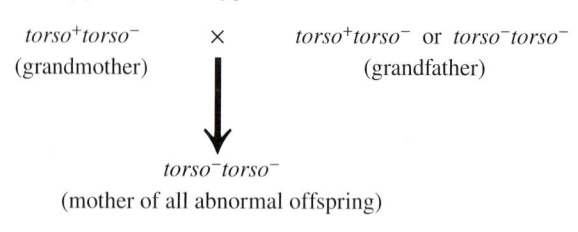

$$torso^+torso^- \quad \times \quad torso^+torso^- \text{ or } torso^-torso^-$$
$$\text{(grandmother)} \quad \quad \quad \text{(grandfather)}$$
$$\downarrow$$
$$torso^-torso^-$$
$$\text{(mother of all abnormal offspring)}$$

The mother of the abnormal offspring is phenotypically normal because her mother was heterozygous and provided the gene product of the *torso*⁺ allele from the nurse cells. However, this homozygous female will produce only abnormal offspring because she cannot provide them with the functional *torso*⁺ gene product.

Conceptual Questions

C1. What is extranuclear inheritance? Describe three examples.

C2. Among different species, does extranuclear inheritance always follow a maternal inheritance pattern? Why or why not?

C3. Extranuclear inheritance often correlates with maternal inheritance. Even so, paternal leakage may occur. What is paternal leakage? If a cross produced 200 offspring and the level of mitochondrial paternal leakage was 3%, how many offspring would be expected to contain paternal mitochondria?

C4. Discuss the structure and organization of the mitochondrial and chloroplast genomes. How large are they, how many genes do they contain, and how many copies of the genome are found in each organelle?

C5. Explain the likely evolutionary origin of chloroplast and mitochondrial genomes. How have the sizes of the chloroplast and mitochondrial genomes changed since their origin? How did this occur?

C6. Is each of the following traits or diseases determined by nuclear genes?

   A. Snail coiling pattern

   B. Dwarfism in mice due to a mutation in *Igf2*

   C. Variegated leaf color in the four-o'clock plant

   D. Leber hereditary optic neuropathy

C7. Acute murine leukemia virus (AMLV) causes leukemia in mice. This virus is easily passed from mother to offspring through the mother's milk. (Note: Even though newborn offspring acquire the virus, they may not develop leukemia until much later in life. Testing can determine if an animal carries the virus.) Describe how the development of leukemia due to AMLV resembles a maternal inheritance pattern. How could you determine that this form of leukemia is not caused by extranuclear inheritance?

C8. Describe how a biparental pattern of extranuclear inheritance could resemble a Mendelian pattern of inheritance for a particular gene. How would they differ?

C9. According to the endosymbiosis theory, chloroplasts and mitochondria are derived from bacteria that took up residence within eukaryotic cells. At one time, prior to being taken up by eukaryotic cells, these bacteria were free-living organisms. However, we cannot take a chloroplast or mitochondrion out of a living eukaryotic cell and get it to survive and replicate on its own. Why not?

C10. Define the term *genomic imprinting*, and give two examples.

C11. When does the erasure and reestablishment phase of genomic imprinting occur? Explain why it is necessary to erase an imprint and then reestablish it according to the sex of the parent.

C12. In what types of cells would you expect *de novo* methylation to occur? In what cell types would it not occur?

C13. Describe the inheritance pattern of maternal effect genes. Explain how the maternal effect occurs at the molecular and cellular level. What are the expected functional roles of the proteins that are encoded by maternal effect genes?

C14. A maternal effect gene exists in a dominant *N* (functional) allele and a recessive *n* (nonfunctional) allele. What are the ratios of genotypes and phenotypes for the offspring of the following crosses?

   A. *nn* female × *NN* male

   B. *NN* female × *nn* male

   C. *Nn* female × *Nn* male

C15. A *Drosophila* embryo dies during early embryogenesis due to a recessive allele of a maternal effect gene called *bicoid*. The wild-type allele is designated *bicoid*⁺. What are the genotypes and phenotypes of the embryo's mother and maternal grandparents?

C16. For Mendelian inheritance, the nuclear genotype (i.e., the alleles found on chromosomes in the cell nucleus) directly influences an offspring's traits. In contrast, for non-Mendelian inheritance patterns, the offspring's phenotype cannot be reliably predicted solely from its genotype. For each of the following traits, what do you need to know to predict the phenotypic outcome?

   A. Dwarfism due to a mutant *Igf2* allele

   B. Snail coiling direction

   C. Leber hereditary optic neuropathy

C17. Suppose a maternal effect gene exists as a functional dominant allele and a nonfunctional recessive allele that causes a disorder. A mother with the disorder produces offspring that are all without the disorder. Explain the genotype of the mother.

C18. Suppose that a maternal effect gene affects the anterior morphology in houseflies. The gene exists in a dominant allele, *H*, and a recessive allele, *h*, which causes a small head. A female fly with a normal size head is mated to a true-breeding male with a small head. All of the offspring have small heads. What are the genotypes of the mother and offspring? Explain your answer.

C19. Explain why maternal effect genes exert their effects during the early stages of development.

Application and Experimental Questions

E1. A variegated trait in plants is analyzed using reciprocal crosses. The following results are obtained.

Variegated female × Normal male   Normal female × Variegated male
          ↓   ↓
1024 variegated + 52 normal   1113 normal + 61 variegated

Explain this pattern of inheritance.

E2. Two male mice, male A and male B, are both normal size. Male A was from a litter that contained half phenotypically normal size mice and half dwarf mice. The mother of male A was known to be homozygous for the functional *Igf2* allele. Male B was from a litter of eight mice that were all phenotypically normal size. The parents of male B were a normal size male and a dwarf female. Male A and male B were put into a cage with two female mice, female A and female B. Female A is dwarf, and female B is normal size. The parents of these two females were unknown, although it was known that they were from the same litter. The mice were allowed to mate with each other, and the following data were obtained:

Female A gave birth to three dwarf babies and four normal size babies.

Female B gave birth to four normal size babies and two dwarf babies.

Which male(s) mated with female A and female B? Explain.

E3. Figure 6.10 illustrates an example of a maternal effect gene. Explain how Sturtevant deduced a maternal effect gene based on the F_2 and F_3 generations.

E4. Chapter 20 describes two blotting methods (i.e., Northern blotting and Western blotting) that are used to detect specific gene products. Northern blotting detects RNA, and Western blotting detects proteins. Suppose that a female fruit fly is heterozygous for a maternal effect gene, gene B. The female is *Bb*. The dominant allele, *B*, encodes a functional mRNA that is 550 nucleotides long. A recessive allele, *b*, encodes a shorter mRNA that is 375 nucleotides long. (Allele *b* is due to a deletion within this gene.) How could you use one or both of the blotting methods to show that nurse cells transfer gene products from gene *B* to developing eggs? You may assume that you can dissect the ovaries of fruit flies and isolate eggs separately from nurse cells. In your answer, describe your expected results.

E5. Suppose that a trait observed in some mice is an unusually abnormally long tail. You initially have a true-breeding strain with normal-length tails and a true-breeding strain with long tails. You then make the following types of crosses:

Cross 1: When true-breeding females with normal tails are crossed to true-breeding males with long tails, all F_1 offspring have long tails.

Cross 2: When true-breeding females with long tails are crossed to true-breeding males with normal tails, all F_1 offspring have normal tails.

Cross 3: When F_1 females from cross 1 are crossed to true-breeding males with normal tails, all offspring have normal tails.

Cross 4: When F_1 males from cross 1 are crossed to true-breeding females with long tails, half of the offspring have normal tails and half have long tails.

Explain the pattern of inheritance of this trait.

E6. You have a female snail that coils to the right, but you do not know its genotype. You may assume that right coiling (*D*) is dominant to left coiling (*d*). You also have male snails of known genotype. How would you determine the genotype of this female snail? In your answer, describe your expected results depending on whether the female is *DD*, *Dd*, or *dd*.

Questions for Student Discussion/Collaboration

1. During the course of evolution, most organellar genes have been transferred from the chloroplast or mitochondrial genome to the nuclear genome. Discuss possible reasons why this may have occurred. In other words, what are possible selective advantages to having genes in the nucleus?

2. Recessive maternal effect genes are identified in flies (for example) when a phenotypically normal mother cannot produce any normal offspring. Because all of the offspring die before reproducing, this female fly cannot be used to produce a strain of heterozygous flies that could be used in future studies. How would you identify heterozygous individuals that are carrying a recessive maternal effect allele? How would you maintain this strain of flies in a laboratory over many generations?

Answers to Comprehension Questions

6.1: b, d, a

6.2: d, a, c

6.3: b

6.4: d, c, a

6.5: d, d

Note: All answers appear in Connect; the answers to even-numbered questions and all Concept Check questions are in Appendix B.

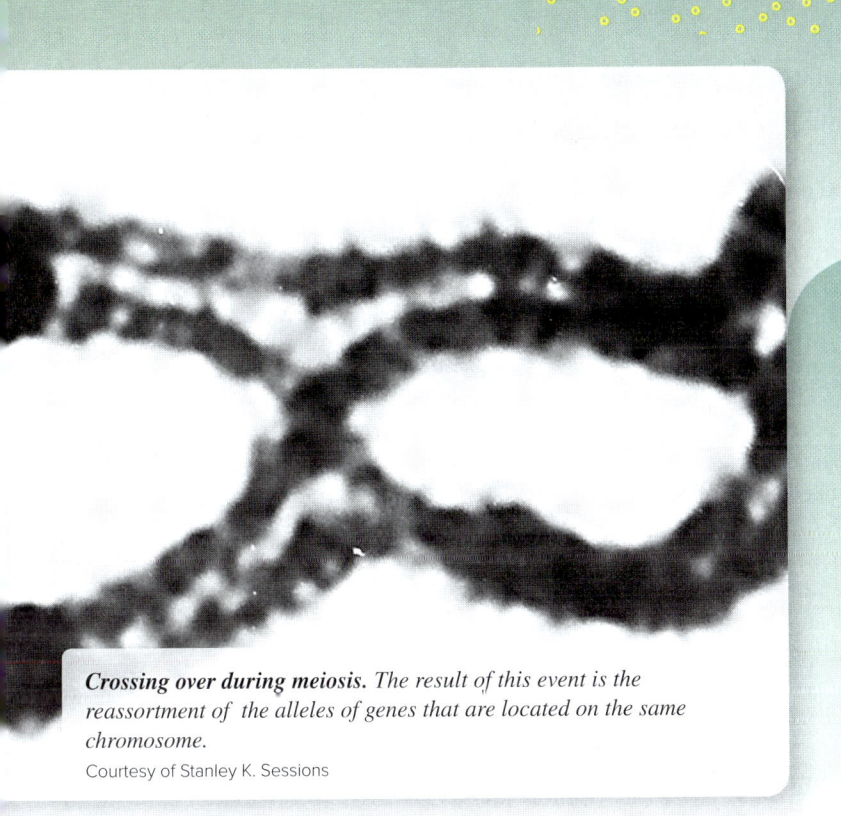

Crossing over during meiosis. The result of this event is the reassortment of the alleles of genes that are located on the same chromosome.
Courtesy of Stanley K. Sessions

GENETIC LINKAGE AND MAPPING IN EUKARYOTES

CHAPTER OUTLINE

7.1 Overview of Linkage
7.2 Relationship Between Linkage and Crossing Over
7.3 Genetic Mapping in Plants and Animals
7.4 Mitotic Recombination

In Chapter 3, we focused on Mendel's laws of inheritance. According to these principles, we expect that two different genes will segregate and independently assort themselves during the process that creates haploid cells. After Mendel's work was rediscovered at the turn of the twentieth century, chromosomes were identified as the cellular structures that carry genes. The chromosome theory of inheritance explained how the transmission patterns of chromosomes are responsible for the passage of genes from parents to offspring.

When geneticists first realized that chromosomes contain the genetic material, they began to suspect that discrepancies might sometimes occur between the law of independent assortment of genes and the behavior of chromosomes during meiosis. In particular, geneticists assumed that each species of organism must contain thousands of different genes, yet cytological studies revealed that most species have at most a few dozen chromosomes. Therefore, it seemed likely, and turned out to be true, that each chromosome carries many hundreds or even thousands of different genes. The transmission of genes located close to each other on the same chromosome violates the law of independent assortment.

In this chapter, we will consider patterns of inheritance that occur when different genes are situated on the same chromosome. In addition, we will briefly explore how the data from genetic crosses are used to construct a **genetic map**—a diagram that describes the order of genes along a chromosome. Newer strategies for gene mapping are described in Chapter 22. However, an understanding of traditional mapping studies, as described in this chapter, will strengthen our appreciation for these newer molecular approaches. More importantly, traditional mapping studies further illustrate how the location of two or more genes on the same chromosome can affect the patterns of gene transmission from parents to offspring.

7.1 OVERVIEW OF LINKAGE

Learning Outcomes:
1. Define *genetic linkage*.
2. Explain how linkage affects the outcome of crosses.

In eukaryotic species, each linear chromosome is composed of a very long segment of DNA and many different proteins. A chromosome carries many individual functional units—called genes—that influence an organism's traits. A typical chromosome is expected to contain many hundreds or perhaps a few thousand different genes.

- The term **synteny** means that two or more genes are located on the same chromosome. Genes that are syntenic are physically linked to each other, because each eukaryotic chromosome contains a single, continuous, linear molecule of DNA.
- **Genetic linkage,** or simply **linkage,** is the phenomenon in which genes that are close together on the same chromosome tend to be transmitted as a unit. For this reason, linkage has an influence on inheritance patterns.

Chromosomes are sometimes called **linkage groups,** because a chromosome contains a group of genes that are physically linked together. In each species, the number of linkage groups equals the number of chromosome types. For example, human somatic cells have 46 chromosomes, which are composed of 22 types of autosomes that come in pairs plus one pair of sex chromosomes, X and Y. Therefore, humans have 22 autosomal linkage groups and an X chromosome linkage group, and human males also have a Y chromosome linkage group. In addition, the human mitochondrial genome is another linkage group.

Geneticists are often interested in the transmission of two or more characters in a genetic cross. When a geneticist follows the variants of two different characters in a cross, this is called a **two-factor cross;** when three characters are followed, it is a **three-factor cross;** and so on. The outcome of a two-factor or three-factor cross depends on whether or not the genes are linked to each other on the same chromosome. Next, we will examine how linkage affects the transmission patterns of two characters. Later sections will examine crosses involving three different genes.

Bateson and Punnett Discovered Two Characters That Did Not Assort Independently

An early study indicating that some characters may not assort independently was carried out by William Bateson and Reginald Punnett in 1905. According to Mendel's law of independent assortment, a two-factor cross between two individuals that are heterozygous for two genes should yield a 9:3:3:1 phenotypic ratio among the offspring. However, a surprising result occurred when Bateson and Punnett conducted a cross of sweet peas involving two different characters: flower color and pollen shape.

As seen in **Figure 7.1**, they began by crossing a true-breeding strain with purple flowers (*PP*) and long pollen (*LL*) to a true-breeding strain with red flowers (*pp*) and round pollen (*ll*). This yielded an F_1 generation of plants that all had purple flowers and long pollen (*PpLl*). An unexpected result came from the F_2 generation. Even though the F_2 generation had four different phenotypic categories, the observed numbers of offspring with the various phenotypes did not conform to the expected 9:3:3:1 ratio. Bateson and Punnett found that the F_2 generation had a much greater proportion of the two phenotypes found in the P generation—purple flowers with long pollen and red flowers with round pollen. Therefore, they suggested that the transmission of these two characters from the P generation to the F_2 generation was somehow coupled; that is, the alleles were not assorted in an independent manner. However, Bateson and Punnett did not realize that this coupling was due to the linkage of the flower color gene and the pollen shape gene on the same chromosome.

P generation

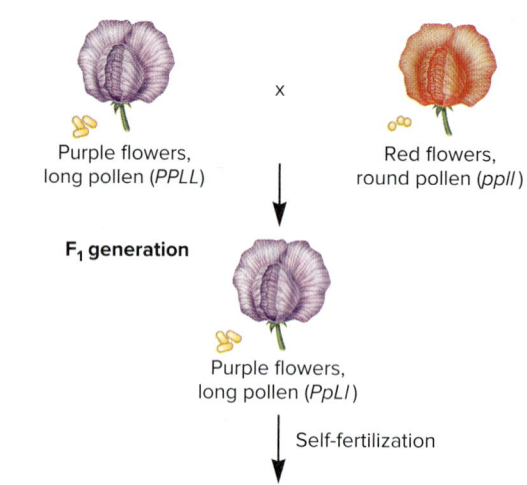

Purple flowers, long pollen (*PPLL*) × Red flowers, round pollen (*ppll*)

F_1 generation

Purple flowers, long pollen (*PpLl*)

Self-fertilization

F_2 generation	Observed number	Ratio	Expected number	Ratio
Purple flowers, long pollen	296	15.6	240	9
Purple flowers, round pollen	19	1.0	80	3
Red flowers, long pollen	27	1.4	80	3
Red flowers, round pollen	85	4.5	27	1

FIGURE 7.1 An experiment of Bateson and Punnett with sweet peas, showing that independent assortment does not always occur. Note: The expected numbers are rounded to the nearest whole number.

Concept Check: Which types of offspring in the F_2 generation in this experiment are found in excess, according to Mendel's law of independent assortment?

7.1 REVIEWING THE KEY CONCEPTS

- Linkage has two related meanings. First, it refers to genes that are located on the same chromosome. Second, linkage also means that the alleles of two or more genes tend to be transmitted as a unit because they are relatively close on the same chromosome.
- Bateson and Punnett discovered the first example of genetic linkage in sweet peas (see Figure 7.1).

7.1 COMPREHENSION QUESTIONS

1. Genetic linkage occurs because
   a. genes that are on the same chromosome may affect the same trait.
   b. genes that are close together on the same chromosome tend to be transmitted together to offspring.
   c. genes that are on different chromosomes are independently assorted.
   d. None of the above explains why genetic linkage occurs.

2. In the experiment by Bateson and Punnett, which of the following observations suggested genetic linkage in the sweet pea?
   a. A 9:3:3:1 ratio was observed in the F_2 offspring.
   b. A 9:3:3:1 ratio was not observed in the F_2 offspring.
   c. An unusually high number of F_2 offspring had the phenotypes of the P generation.
   d. Both b and c suggested genetic linkage.

7.2 RELATIONSHIP BETWEEN LINKAGE AND CROSSING OVER

Learning Outcomes:
1. Describe how crossing over can change the arrangements of alleles along a chromosome.
2. Explain how the distance between linked genes affects the proportions of recombinant and nonrecombinant offspring.
3. Apply a chi square test to distinguish between linkage and independent assortment.
4. Analyze the data of Creighton and McClintock and explain how it indicated that recombinant offspring carry chromosomes that are the result of crossing over.

Even though the alleles for different genes may be linked on the same chromosome, the linkage can be altered during meiosis. In diploid eukaryotic species, homologous chromosomes can exchange pieces with each other, a phenomenon called **crossing over.** This event most commonly occurs during prophase of meiosis I. As discussed in Chapter 2, the replicated chromosomes, known as sister chromatids, associate with the homologous sister chromatids to form a structure known as a **bivalent.** A bivalent is composed of two pairs of sister chromatids. In prophase of meiosis I, it is common for a sister chromatid of one pair to cross over with a sister chromatid from the homologous pair (refer back to Figure 2.10). In this section, we will consider how crossing over affects the pattern of inheritance for genes linked on the same chromosome. In Chapter 13, we will consider the molecular events that cause crossing over to occur.

Crossing Over May Produce Recombinant Genotypes

Figure 7.2 considers meiosis when two genes are linked on the same chromosome. One of the original chromosomes carries the *B* and *A* alleles, whereas the homolog carries the *b* and *a* alleles. In Figure 7.2a, no crossing over has occurred. Therefore, the resulting haploid cells contain the same combination of alleles as the original chromosomes. Two haploid cells carry the dominant *B* and *A* alleles, and the other two carry the recessive *b* and *a* alleles. These are called **nonrecombinant** cells. The arrangement of linked alleles has not been altered from those found in the original cell.

In contrast, Figure 7.2b illustrates what can happen when crossing over occurs. The two lowermost haploid cells contain combinations of alleles, namely, *B* and *a* in one and *b* and *A* in the other, which differ from those in the original chromosomes. In these two cells, the grouping of linked alleles has changed. The haploid cells carrying the *B* and *a* alleles, or the *b* and *A* alleles, are called **recombinant** cells. If such haploid cells are gametes that participate in fertilization, the resulting offspring are called recombinant offspring.

When offspring inherit a combination of two or more alleles or traits that are different from those in either of their parents, this event is known as **genetic recombination**. It commonly occurs in two ways:

1. When two or more genes are linked on the same chromosome, crossing over during meiosis can result in genetic recombination.
2. When two or more genes are on different chromosomes, the independent assortment of those chromosomes during meiosis can result in genetic recombination.

In this chapter, our definition of recombinant offspring is the following:

Recombinant offspring are produced by the exchange of DNA between two homologous chromosomes during meiosis in one or both parents, leading to a novel combination of genetic material.

In other words, recombinant offspring carry chromosomes that are the product of crossing over.

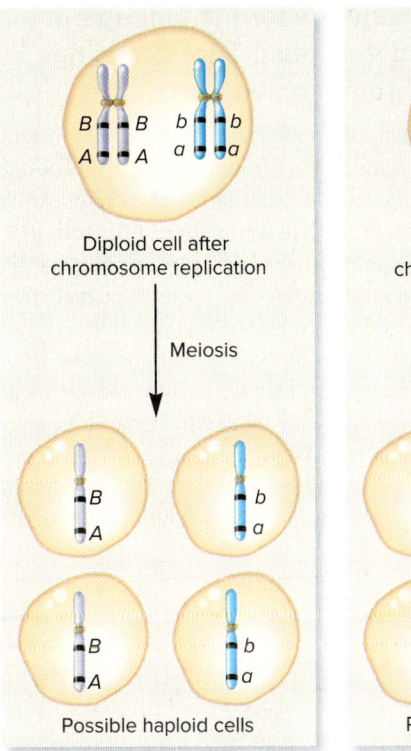

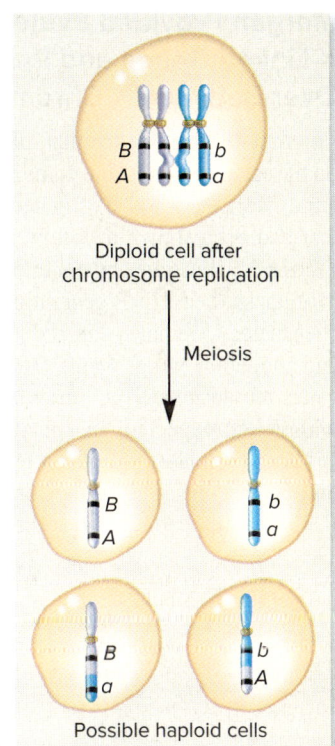

(a) Without crossing over, linked alleles segregate together.

(b) Crossing over can reassort linked alleles.

FIGURE 7.2 Consequences of crossing over during meiosis. (a) In the absence of crossing over, the *A* and *B* alleles and the *a* and *b* alleles are maintained in the same arrangement found in the original chromosomes. (b) Crossing over has occurred in the region between the two genes, producing two recombinant haploid cells with a new combination of alleles.

Concept Check: If a crossover began in the short region between gene *A* and the tip of the chromosome, would this event affect the arrangement of the *A* and *B* alleles?

Morgan Provided Evidence for the Linkage of X-Linked Genes and Proposed That Crossing Over Between X Chromosomes Can Occur

The first direct evidence that different genes are physically located on the same chromosome came in 1911 from the studies of Thomas Hunt Morgan, who investigated the inheritance of different traits that had been shown to follow an X-linked pattern of inheritance. **Figure 7.3** illustrates an experiment involving three characters that Morgan studied. His P-generation crosses were wild-type male fruit flies mated to females that had yellow bodies (yy), white eyes (ww), and miniature wings (mm). The wild-type alleles for these three genes are designated y^+ (gray body), w^+ (red eyes), and m^+ (long wings). As expected, the phenotypes of the F_1 generation were wild-type females and males with yellow bodies, white eyes, and miniature wings. The linkage of these genes was revealed when the F_1 flies were mated to each other and the F_2 generation examined.

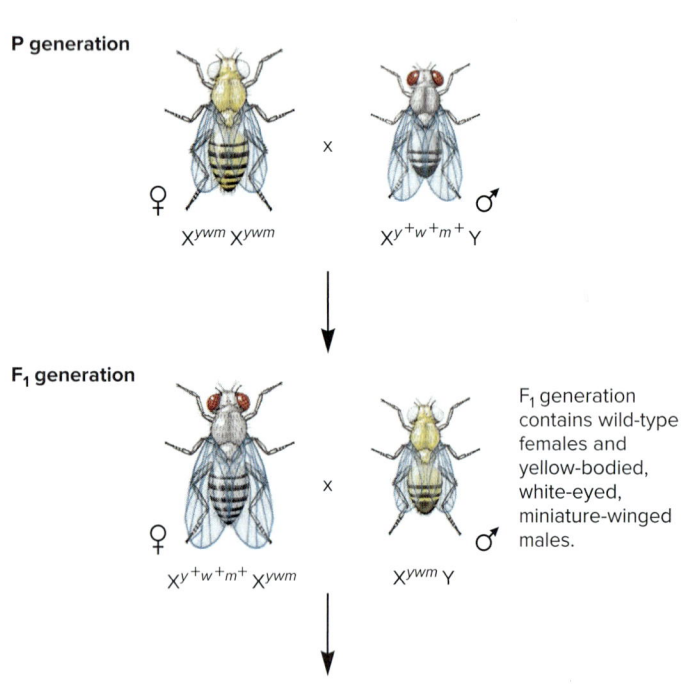

Instead of equal proportions of the eight possible phenotypes, Morgan observed much higher proportions of the combinations of traits found in the P generation.

- He observed 758 flies with gray bodies, red eyes, and long wings and 700 flies with yellow bodies, white eyes, and miniature wings.
- The combination of gray body, red eyes, and long wings was found in the males of the P generation, and the combination of yellow body, white eyes, and miniature wings was the same as the females of the P generation.
- Morgan's explanation for the higher proportions of these combinations was that all three genes are located on the X chromosome and, therefore, tend to be transmitted together as a unit.

However, to fully account for the data shown in Figure 7.3, Morgan needed to explain why a significant proportion of the F_2 generation had a recombinant arrangements of alleles. Along with the two nonrecombinant phenotypes, five other phenotypic combinations appeared that were not found in the P generation. How did Morgan explain these data? He considered the studies conducted in 1909 by the cytologist Frans Alfons Janssens, who observed chiasmata under the microscope and proposed that crossing over involves a physical exchange between homologous chromosomes. Morgan shrewdly realized that crossing over between homologous X chromosomes was consistent with his data.

- Morgan hypothesized that the genes for body color, eye color, and wing length are all located on the same chromosome, namely, the X chromosome. Therefore, the alleles for all three genes are most likely to be inherited together.
- Due to crossing over, Morgan also proposed that, in the female, the homologous X chromosomes can exchange pieces of chromosomes and create new (recombinant) arrangements of alleles and traits in the F_2 generation.

To appreciate Morgan's proposals, let's simplify his data and consider only two of the three genes, those that affect body color and eye color. If we use the data from Figure 7.3, the following totals are obtained:

Gray body, red eyes	1159	
Yellow body, white eyes	1017	
Gray body, white eyes	17	Recombinant offspring
Yellow body, red eyes	12	
Total	2205	

P generation

$X^{ywm} X^{ywm}$ ♀ × $X^{y^+w^+m^+} Y$ ♂

F_1 generation

$X^{y^+w^+m^+} X^{ywm}$ ♀ × $X^{ywm} Y$ ♂

F_1 generation contains wild-type females and yellow-bodied, white-eyed, miniature-winged males.

F_2 generation	Females	Males	Total
Gray body, red eyes, long wings	439	319	758
Gray body, red eyes, miniature wings	208	193	401
Gray body, white eyes, long wings	1	0	1
Gray body, white eyes, miniature wings	5	11	16
Yellow body, red eyes, long wings	7	5	12
Yellow body, red eyes, miniature wings	0	0	0
Yellow body, white eyes, long wings	178	139	317
Yellow body, white eyes, miniature wings	365	335	700

FIGURE 7.3 Morgan's three-factor cross involving three X-linked traits in *Drosophila*.

Genes → Traits Three genes that govern body color, eye color, and wing length are all found on the X chromosome of fruit flies. Therefore, the offspring tend to inherit a nonrecombinant pattern of alleles ($y^+w^+m^+$ or ywm). Figure 7.5 explains how single and double crossovers can create recombinant combinations of alleles.

Concept Check: Of the eight possible phenotypic combinations in the F_2 generation, which ones are the product of a single crossover?

Figure 7.4 shows how Morgan's proposals could account for these data. The nonrecombinant offspring with gray bodies and red eyes or yellow bodies and white eyes were produced when no crossing over had occurred between the two genes (Figure 7.4a). This was the more common situation. By comparison, crossing over could alter the arrangement of alleles along each chromosome and account for the recombinant offspring (Figure 7.4b). Why were there relatively few recombinant

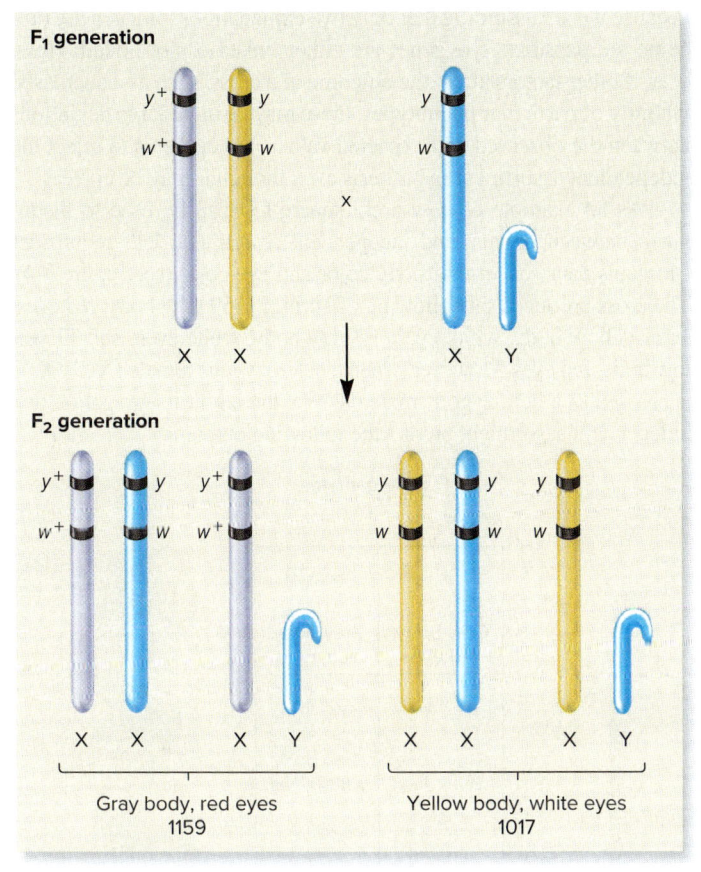

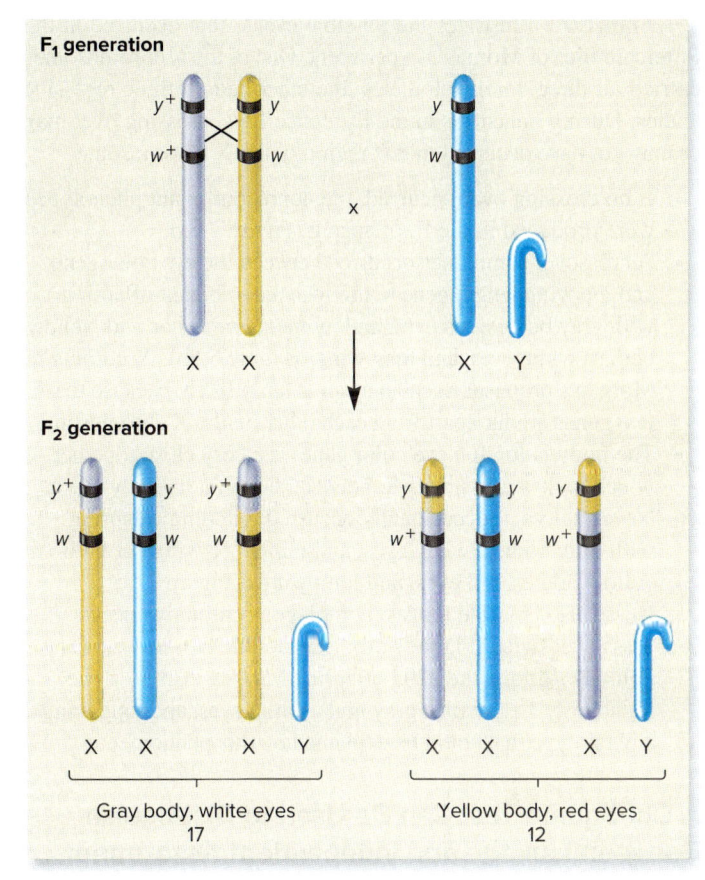

7.2 RELATIONSHIP BETWEEN LINKAGE AND CROSSING OVER

(a) No crossing over, nonrecombinant offspring

F₁ generation → F₂ generation

Gray body, red eyes: 1159
Yellow body, white eyes: 1017

(b) Crossing over, recombinant offspring

F₁ generation → F₂ generation

Gray body, white eyes: 17
Yellow body, red eyes: 12

FIGURE 7.4 **Morgan's explanation for nonrecombinant and recombinant offspring.** As described in Chapter 2, crossing over actually occurs when chromosomes are bivalents, that is, homologous pairs of sister chromatids aligned with each other. For simplicity, this figure shows only two X chromosomes (one from each homolog) rather than four chromatids, which would be present during the bivalent stage of meiosis. Also note that this figure shows only a portion of the X chromosome. A map of the entire X chromosome is shown in Figure 7.7.

Concept Check: *Why are the nonrecombinant offspring more common than the recombinant offspring?*

---

offspring? These two genes are very close together on the same chromosome, which makes it unlikely that a crossover would be initiated between them. As described next, the distance between two genes is an important factor that determines the relative proportions of recombinant offspring.

The Likelihood of Crossing Over Between Two Genes Depends on the Distance Between Them

In the experiment of Figure 7.3, Morgan also noticed a quantitative difference between the numbers of recombinant offspring involving body color and eye color versus those involving eye color and wing length. This quantitative difference is revealed by reorganizing the data of Figure 7.3 by pairs of genes.

Gray body, red eyes	1159
Yellow body, white eyes	1017
Gray body, white eyes	17 ⎫ Recombinant
Yellow body, red eyes	12 ⎭ offspring
Total	2205

Red eyes, long wings	770
White eyes, miniature wings	716
Red eyes, miniature wings	401 ⎫ Recombinant
White eyes, long wings	318 ⎭ offspring
Total	2205

Morgan found a substantial difference between the numbers of recombinant offspring when pairs of genes were considered separately. Recombinant offspring involving only eye color and wing length were fairly common (401 + 318). In sharp contrast, recombinant offspring for body color and eye color were quite rare (17 + 12).

How did Morgan explain these data? He proposed the following:

- The likelihood of crossing over depends on the distance between two genes.
- If two genes are far apart from each other, crossing over is more likely to occur between them than between two genes that are close together.

Figure 7.5 illustrates the possible events that occurred in the F₁ female flies of Morgan's experiment. One of the X chromosomes carried all three dominant alleles; the other had all three recessive alleles. During oogenesis in the F₁ female flies, crossing over may or may not have occurred in this region of the X chromosome.

- If no crossing over occurred, the nonrecombinant phenotypes were produced in the F₂ offspring (Figure 7.5a).
- A crossover sometimes occurred between the eye color gene and the wing length gene to produce recombinant offspring with gray bodies, red eyes, and miniature wings or with yellow bodies, white eyes, and long wings (Figure 7.5b). According to Morgan's proposal, such an event is fairly likely because these two genes are far apart from each other on the X chromosome.
- The body color and eye color genes are very close together, which makes crossing over between them an unlikely event. Nevertheless, it occasionally occurred, yielding offspring with gray bodies, white eyes, and miniature wings or with yellow bodies, red eyes, and long wings (Figure 7.5c).
- It was also possible for two homologous chromosomes to cross over twice (Figure 7.5d). This double crossover is very unlikely. Among the 2205 offspring Morgan examined, he found only 1 fly with a gray body, white eyes, and long wings, a phenotype that could be explained by this phenomenon.

A Chi Square Test Can Be Used to Distinguish Between Linkage and Independent Assortment

Now that we have an appreciation for linkage and the production of recombinant offspring, let's consider how an experimenter can objectively decide whether two genes are linked or assort independently. In Chapter 3, we used a chi square test to evaluate the goodness of fit between a genetic hypothesis and observed experimental data. This method can be employed to determine if the outcome of a two-factor cross is consistent with linkage or independent assortment.

To conduct a chi square test, we must first propose a hypothesis. For a two-factor cross, the standard hypothesis is that the two genes are not linked. This hypothesis is chosen even if the observed data suggest linkage, because an independent assortment hypothesis allows us to calculate the expected number of offspring based on the genotypes of the parents and the law of independent assortment. In contrast, for two linked genes that have not been previously mapped, we cannot calculate the expected number of offspring from a genetic cross because we do not know how likely it is for a crossover to occur between the two genes. Without expected numbers of nonrecombinant and recombinant offspring, we cannot conduct a chi square test. Therefore, we begin with the hypothesis that the genes are not linked. Recall from Chapter 3 that the hypothesis we are testing is called a **null hypothesis**, because it assumes there is no real difference between the observed and expected values. The goal is to determine whether or not the data fit the hypothesis. If the chi square value is low and we cannot reject the null hypothesis, we infer that the genes assort independently. On the other hand, if the chi square value is so high that our hypothesis is rejected, we accept the alternative hypothesis, namely, that the genes are linked.

Of course, a statistical analysis cannot prove that a hypothesis is true. If the chi square value is high, we accept the linkage hypothesis because we are assuming that only two explanations for a genetic outcome are possible: The genes are either linked or not linked. However, if other factors affect the outcome of a cross, such as a decreased viability of particular phenotypes, these may result in large deviations between the observed and expected values and cause us to reject the independent assortment hypothesis even though it may be correct.

As an example of how a chi square test can be used to distinguish between linkage and independent assortment, let's reconsider Morgan's data concerning body color and eye color (see Figure 7.4). The cross produced the following offspring: 1159 gray body, red eyes; 1017 yellow body, white eyes; 17 gray body, white eyes; and 12 yellow body, red eyes. However, when a heterozygous female ($X^{y+w+}X^{yw}$) is crossed to a hemizygous male ($X^{yw}Y$), the laws of segregation and independent assortment predict the following outcome:

F₁ male gametes

	X^{yw}	Y
X^{y+w+}	$X^{y+w+}X^{yw}$ Gray body, red eyes	$X^{y+w+}Y$ Gray body, red eyes
X^{y+w}	$X^{y+w}X^{yw}$ Gray body, white eyes	$X^{y+w}Y$ Gray body, white eyes
X^{yw+}	$X^{yw+}X^{yw}$ Yellow body, red eyes	$X^{yw+}Y$ Yellow body, red eyes
X^{yw}	$X^{yw}X^{yw}$ Yellow body, white eyes	$X^{yw}Y$ Yellow body, white eyes

F₁ female gametes

Mendel's laws predict a 1:1:1:1 ratio among the four phenotypes. The observed data obviously seem to conflict with this expected outcome. Nevertheless, we stick to the strategy just discussed. We begin with the hypothesis that the two genes are not linked and then conduct a chi square test to see if the data fit this hypothesis. If the data do not fit, we reject the idea that the genes assort independently and conclude that the genes are linked.

A step-by-step procedure for applying the chi square test to distinguish between linkage and independent assortment is as follows:

Step 1. *Propose a hypothesis.* Even though the observed data appear inconsistent with this hypothesis, we propose that the two genes for eye color and body color obey Mendel's law of independent assortment. This hypothesis allows us to calculate expected values. Because the data seem to conflict with this hypothesis, we actually anticipate that the chi square test will allow us to reject the independent assortment hypothesis in favor of a linkage hypothesis. We also assume that the alleles follow the law of segregation and the four phenotypes are equally viable.

Step 2. *Based on the hypothesis, calculate the expected values of each of the four phenotypes.* Each phenotype has an equal probability of occurring (see the Punnett square

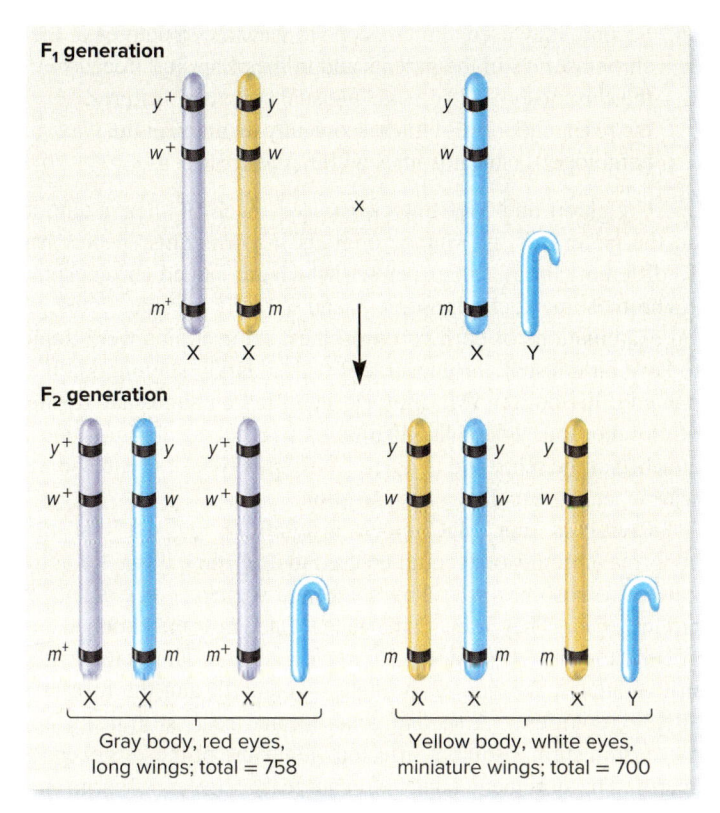

7.2 RELATIONSHIP BETWEEN LINKAGE AND CROSSING OVER 137

(a) **No crossing over in this region, very common**

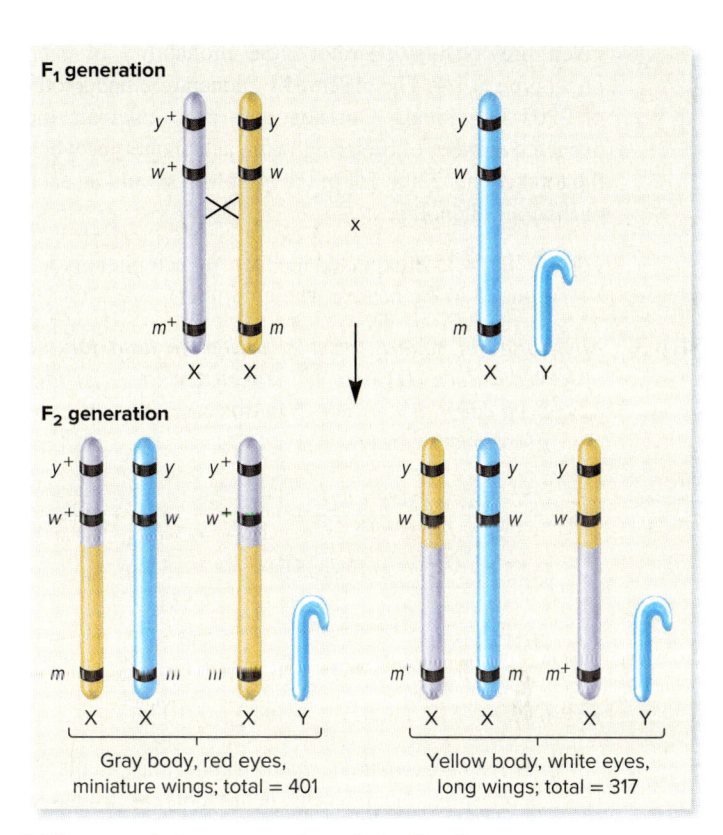

F₁ generation; F₂ generation — Gray body, red eyes, long wings; total = 758. Yellow body, white eyes, miniature wings; total = 700.

(b) **Crossover between eye color and wing length genes, fairly common**

F₁ generation; F₂ generation — Gray body, red eyes, miniature wings; total = 401. Yellow body, white eyes, long wings; total = 317.

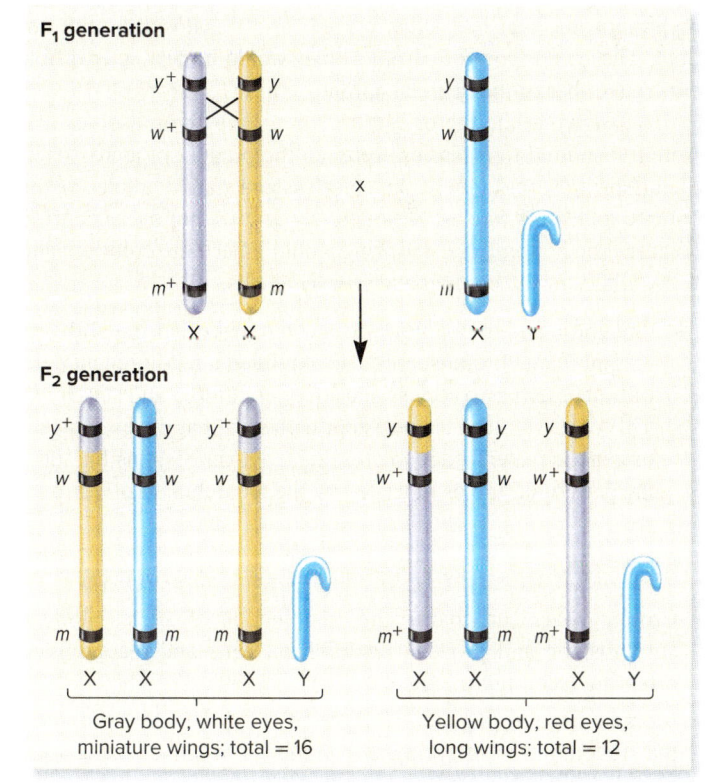

(c) **Crossover between body color and eye color genes, uncommon**

F₁ generation; F₂ generation — Gray body, white eyes, miniature wings; total = 16. Yellow body, red eyes, long wings; total = 12.

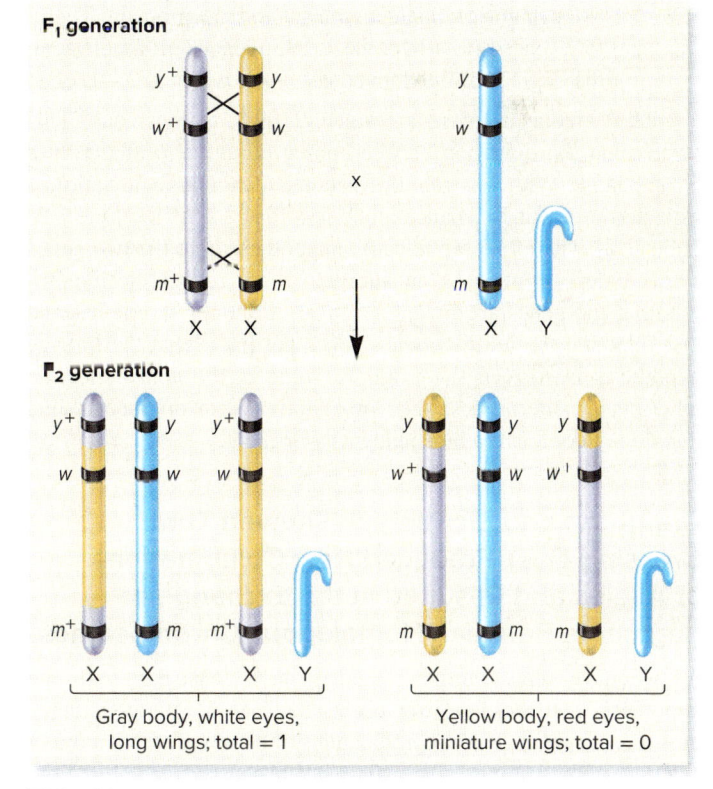

(d) **Double crossover, very uncommon**

F₁ generation; F₂ generation — Gray body, white eyes, long wings; total = 1. Yellow body, red eyes, miniature wings; total = 0.

FIGURE 7.5 **Morgan's explanation for different proportions of recombinant offspring.** Crossing over is more likely to occur between two genes that are relatively far apart than between two genes that are very close together. A double crossover is particularly uncommon.

Concept Check: Why are the types of F₂ offspring shown in part (b) more numerous than those shown in part (c)?

given previously). Therefore, the probability of each phenotype is 1/4. The observed F_2 generation had a total of 2205 individuals. Our next step is to calculate the expected number of offspring with each phenotype when the total equals 2205; 1/4 of the offspring should be each of the four phenotypes:

1/4 × 2205 = 551 (expected number of each phenotype, rounded to the nearest whole number)

Step 3. *Apply the chi square formula, using the data for the observed values (O) and the expected values (E) that have been calculated in step 2. In this case, the data consist of four phenotypes.*

$$\chi^2 = \frac{(O_1 - E_1)^2}{E_1} + \frac{(O_2 - E_2)^2}{E_2} + \frac{(O_3 - E_3)^2}{E_3} + \frac{(O_4 - E_4)^2}{E_4}$$

$$\chi^2 = \frac{(1159 - 551)^2}{551} + \frac{(17 - 551)^2}{551} + \frac{(12 - 551)^2}{551} + \frac{(1017 - 551)^2}{551}$$

$$\chi^2 = 670.9 + 517.5 + 527.3 + 394.1 = 2109.8$$

Step 4. *Interpret the calculated chi square value.* The calculated value is interpreted using a chi square table, as discussed in Chapter 3. The four phenotypes are based on the law of segregation and the law of independent assortment. By itself, the law of independent assortment predicts only two categories: nonrecombinant and recombinant. Therefore, based on a hypothesis of independent assortment, the degrees of freedom equals $n - 1$, which is $2 - 1$, or 1.

The calculated chi square value, 2109.8, is enormous! This means that the deviation between observed and expected values is very large. With 1 degree of freedom, such a large deviation is expected to occur by chance alone less than 1% of the time (see Table 3.2). Therefore, we reject the hypothesis that the two genes assort independently. As an alternative, we accept the hypothesis that the genes are linked. Even so, it should be emphasized that rejecting the null hypothesis does not prove that the linkage hypothesis is correct. For example, some of the non-Mendelian inheritance patterns described in Chapter 6 can produce results that do not conform to independent assortment.

Creighton and McClintock Showed That Crossing Over Produced New Combinations of Alleles and Resulted in the Exchange of Segments Between Homologous Chromosomes

As we have seen, Morgan's studies were consistent with the hypothesis that crossing over occurs between homologous chromosomes to produce new combinations of alleles. To obtain direct evidence that crossing over can result in genetic recombination, Harriet Creighton and Barbara McClintock used an interesting strategy involving parallel observations.

- They first made crosses involving two linked genes to produce nonrecombinant and recombinant offspring.

- Second, they used a microscope to view the structures of the chromosomes in the parents and in the offspring. Because the chromosomes had some unusual structural features, the researchers could microscopically distinguish the two homologous chromosomes within a pair.

Creighton and McClintock focused much of their attention on the pattern of inheritance of traits in corn. This species has 10 different chromosomes per set, which are named chromosome 1, chromosome 2, chromosome 3, and so on. In previous cytological examinations of corn chromosomes, some strains were found to have an unusual chromosome 9 with a darkly staining knob at one end. In addition, McClintock identified an abnormal version of chromosome 9 that also had an extra piece of chromosome 8 attached at the other end (**Figure 7.6a**). This type of chromosomal rearrangement is called a translocation.

Creighton and McClintock insightfully realized that this abnormal chromosome could be used to determine if two homologous chromosomes physically exchange segments as a result of crossing over. They knew that a gene located near the knobbed end of chromosome 9 provided color to corn kernels. This gene existed in two alleles, the dominant allele *C* (colored) and the recessive allele *c* (colorless). A second gene, located near the translocated piece from chromosome 8, affected the texture of the kernel endosperm. The dominant allele *Wx* caused starchy endosperm, and the recessive *wx* allele caused waxy endosperm. Creighton and McClintock reasoned that a crossover involving a normal chromosome 9 and a knobbed/translocated chromosome 9 would produce a chromosome that had either a knob or a translocation, but not both. These two types of chromosomes would be distinctly different from either of the original chromosomes (**Figure 7.6b**).

As shown at the top of **Figure 7.6c**, if the chromosomes that were transmitted to offspring were not the product of a crossover, the offspring would inherit the colorless and starchy alleles or the colored and waxy alleles. Alternatively, if crossing over occurred (Figure 7.6c bottom), the offspring would inherit the colorless and waxy alleles or the colored and starchy alleles.

- Creighton and McClintock discovered that offspring with recombinant chromosomes also exhibited a recombinant pair of traits, such as colorless and waxy or colored and starchy.

- This observation was consistent with idea that the exchange of genetic material due to crossing over was responsible for the pattern of alleles inherited by the recombinant offspring.

7.2 REVIEWING THE KEY CONCEPTS

- Crossing over can change the combination of alleles along a chromosome and produce recombinant cells and recombinant offspring (see Figure 7.2).

- Morgan discovered genetic linkage in *Drosophila* and proposed that recombinant offspring are produced when crossing over occurs during meiosis (see Figures 7.3, 7.4).

- When genes are linked, the relative proportions of recombinant offspring depend on the distance between the genes (see Figure 7.5).

- A chi square test can be used to judge whether or not two genes assort independently.

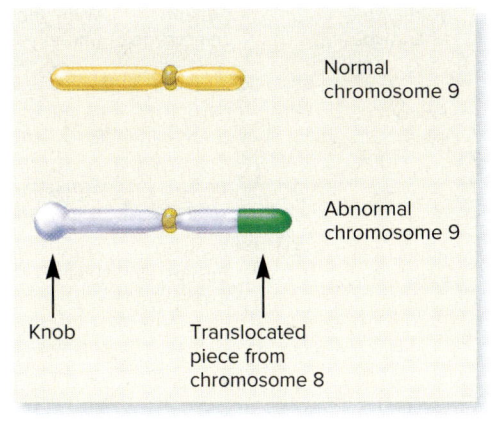

7.3 GENETIC MAPPING IN PLANTS AND ANIMALS

FIGURE 7.6 Crossing over between a normal and abnormal chromosome 9 in corn. (a) A normal chromosome 9 in corn is compared with an abnormal chromosome 9 that contains a knob at one end and a translocation at the opposite end. (b) The morphological features of the chromosomes differ if crossing over occurs. In this case, a crossover produces a chromosome that contains only a knob at one end and another chromosome that contains only a translocation at the other end. (c) The combination of alleles inherited by the offspring also differ if crossing over occurs.

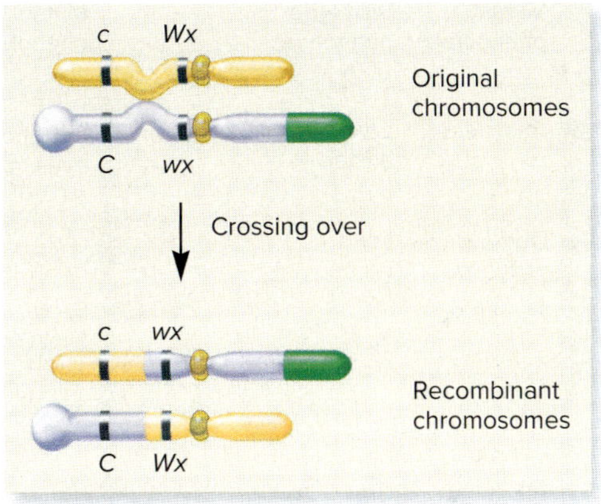

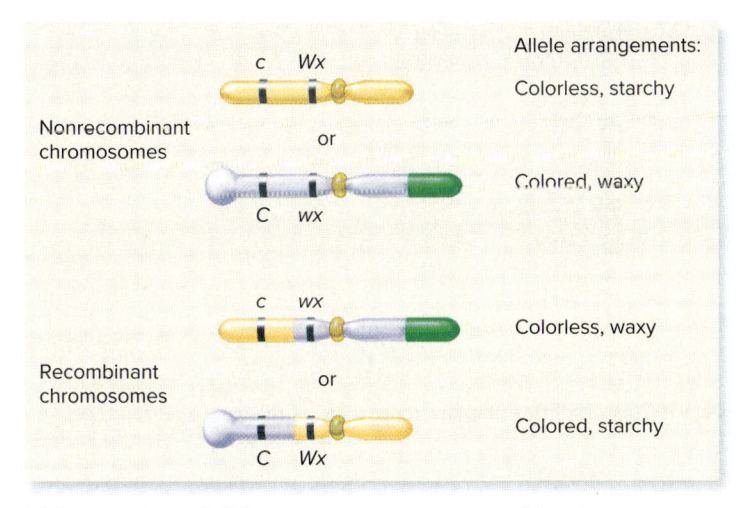

Concept Check: In this experiment, what are the two types of characteristics that crossing over can change? Hint: One type is seen only with a microscope, whereas the other type can be seen with the unaided eye.

- Creighton and McClintock were able to correlate the formation of recombinant offspring with the presence of chromosomes that had exchanged pieces due to crossing over (see Figure 7.6).

7.2 COMPREHENSION QUESTIONS

1. With regard to linked genes on the same chromosome, which of the following statements is *false*?
   a. Crossing over is needed to produce nonrecombinant offspring.
   b. Crossing over is needed to produce recombinant offspring.
   c. Crossing over is more likely to separate alleles if they are far apart on the same chromosome.
   d. Crossing over that separates linked alleles occurs during prophase of meiosis I.

2. Morgan observed a higher number of recombinant offspring involving eye color and wing length (401 + 318) than those involving body color and eye color (17 + 12). These results occurred because
   a. the genes affecting eye color and wing length are farther apart on the X chromosome than are the genes affecting body color and eye color.
   b. the genes affecting eye color and wing length are closer together on the X chromosome than are the genes affecting body color and eye color.
   c. the gene affecting wing length is not on the X chromosome.
   d. the gene affecting body color is not on the X chromosome.

3. For a chi square test involving genes that may be linked, which of the following statements is correct?
   a. An independent assortment hypothesis is not proposed because the data usually suggest linkage.
   b. An independent assortment hypothesis is proposed because it allows the expected numbers of offspring to be calculated.
   c. A large chi square value suggests that the observed and expected data are in good agreement.
   d. The hypothesis is rejected when the chi square value is very low.

7.3 GENETIC MAPPING IN PLANTS AND ANIMALS

Learning Outcomes:
1. Describe why genetic mapping is useful.
2. Calculate the map distance between linked genes using data from a testcross.

The purpose of **genetic mapping**, also known as gene mapping, or chromosome mapping, is to determine the linear order and distance of separation among genes that are linked to each other on

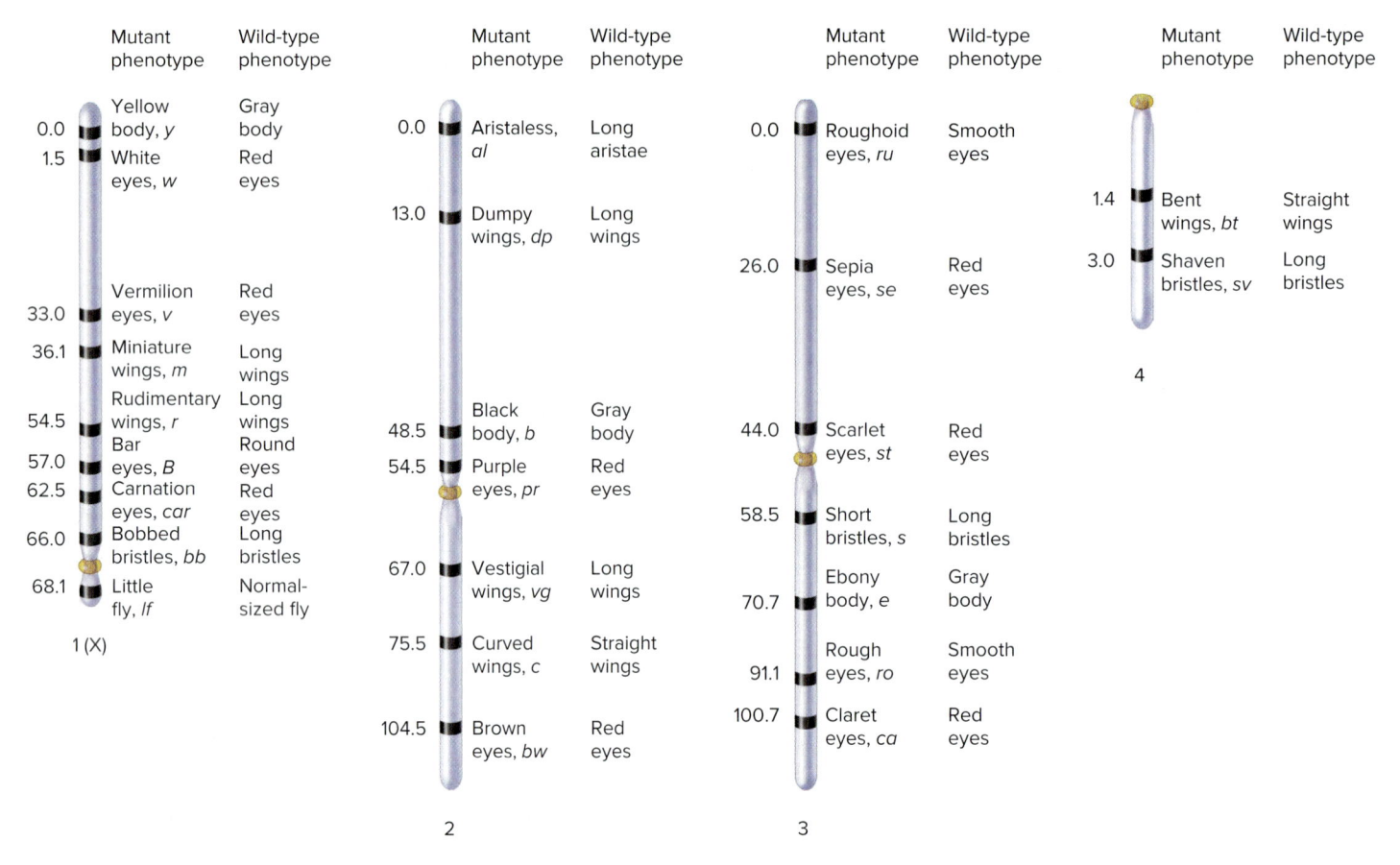

FIGURE 7.7 A simplified genetic linkage map of *Drosophila melanogaster.* This simplified map illustrates a few of the many thousands of genes that have been identified in this organism.

Concept Check: List five reasons why genetic maps are useful.

the same chromosome. **Figure 7.7** presents a simplified genetic map of *Drosophila melanogaster,* depicting the locations of many different genes along the individual chromosomes. As shown here, each gene has its own unique **locus**—the site where the gene is found within a particular chromosome. For example, the gene designated *brown eyes* (*bw*), which affects eye color, is located near one end of chromosome 2. The gene designated *black body* (*b*), which affects body color, is found near the middle of the same chromosome.

Why is genetic mapping useful?

- It allows geneticists to understand the overall complexity and organization of the genome of a particular species.
- The known locus of a gene within a genetic map can help molecular geneticists to clone that gene and thereby obtain greater information about its molecular features.
- Genetic maps are useful from an evolutionary point of view. A comparison of the genetic maps for different species can improve our understanding of the evolutionary relationships among those species.
- Many human genes that play a role in diseases have been genetically mapped. This information can be used to diagnose and perhaps someday treat inherited human diseases.
- Genetic maps are gaining increasing importance in agriculture. A genetic map can provide plant and animal breeders with helpful information for improving agriculturally important strains through selective breeding programs.

In this section, we will examine traditional genetic mapping techniques that involve an analysis of crosses of individuals that are heterozygous for two or more genes. The frequency of recombinant offspring due to crossing over provides a way to deduce the linear order of genes along a chromosome. As shown in Figure 7.7, a depiction of the linear arrangement of genes is known as a **genetic linkage map.** This mapping technique has been useful for analyzing organisms that are easily crossed and produce a large number of offspring in a short period of time. Genetic linkage maps have been constructed for several plant species and certain species of animals, including *Drosophila.* For many organisms, however, traditional mapping approaches are difficult due to long generation times or the inability to carry out experimental crosses (as with humans). Fortunately, many alternative methods of gene mapping have been developed to replace the need to carry out crosses. As described in Chapter 21, molecular approaches are increasingly used to map genes.

The Frequency of Recombination Between Two Genes Is Used to Compute the Map Distance Between Them on a Chromosome

Genetic mapping allows us to estimate the relative distances between linked genes based on the likelihood that a crossover will occur between them. If two genes are very close together on the same chromosome, a crossover is unlikely to begin in the region between them. However, if two genes are very far apart, a crossover is more likely to be initiated between them and thereby recombine the alleles of the two genes. Experimentally, the basis for genetic mapping is that the percentage of recombinant offspring is correlated with the distance between two genes. If two genes are far apart, many recombinant offspring will be produced. However, if two genes are close together, very few recombinant offspring will be observed.

To interpret a genetic mapping experiment, the experimenter must know if the characteristics of an offspring are due to crossing over during meiosis in a parent. This is accomplished by conducting a **testcross**. Most testcrosses are between an individual that is heterozygous for two or more genes and an individual that is recessive and homozygous for the same genes. The goal of the testcross is to determine if recombination has occurred during meiosis in the heterozygous parent. Therefore, genetic mapping is based on the level of recombination that occurs in just one parent—the heterozygote. In a testcross, new combinations of alleles cannot occur in the gametes of the other parent, the one that is homozygous for the genes being studied.

Figure 7.8 illustrates the strategy for conducting a testcross to distinguish between nonrecombinant and recombinant offspring. The experiment begins with a true-breeding P generation, and the two genes of interest are linked genes affecting bristle length and body color in fruit flies. The dominant (wild-type) alleles are s^+ (long bristles) and e^+ (gray body), and the recessive alleles are s (short bristles) and e (ebony body). Because the P generation is true-breeding, the experimenter knows the arrangement of linked alleles. In one parent, s^+ is linked to e^+, and in the other parent, s is linked to e. Therefore, in the F_1 offspring, we know that the s^+ and e^+ alleles are located on one chromosome and the corresponding s and e alleles are located on the homologous chromosome. In the testcross, the F_1 heterozygote is crossed to an individual that is homozygous for the recessive alleles of the two genes (ssee).

Now let's take a look at the four possible types of F_2 offspring. Their phenotypes are long bristles, gray body; short bristles, ebony body; long bristles, ebony body; and short bristles, gray body. All four types of F_2 offspring have inherited a chromosome carrying the s and e alleles from their homozygous parent, which is the blue chromosome shown on the right in each pair of chromosomes. Now, consider the other chromosome in each pair. The offspring with long bristles and gray bodies have inherited a chromosome carrying the s^+ and e^+ alleles from the heterozygous parent. This chromosome is not the product of a crossover. Therefore, this type of offspring is nonrecombinant. The offspring with short bristles and ebony bodies have inherited a chromosome carrying the s and e alleles from the heterozygous parent. Again, this chromosome is not the product of a crossover.

The two types of recombinant F_2 offspring, however, can be produced only if crossing over occurred in the region between the two linked genes. The offspring with long bristles and ebony bodies and those with short bristles and gray bodies have inherited a chromosome that is the product of a crossover during oogenesis in the F_1 female. A key point for you to observe is that the recombinant offspring of the F_2 generation must carry a chromosome that is the product of a crossover. As noted in Figure 7.8, there are fewer recombinant offspring than nonrecombinant offspring.

The amount of recombination can be used as an estimate of the physical distance between two genes on the same chromosome. The **map distance** is defined as the number of recombinant offspring divided by the total number of offspring, multiplied by 100. We can calculate the map distance between the bristle length gene and the body color gene using this formula:

$$\text{Map distance} = \frac{\text{Number of recombinant offspring}}{\text{Total number of offspring}} \times 100$$

where

recombinant offspring are those offspring that were produced by a crossover in the heterozygous parent.

If we apply this equation to the data of Figure 7.8

$$\text{Map distance} = \frac{76 + 75}{537 + 542 + 76 + 75} \times 100$$

$$= 12.3 \text{ map units}$$

The units of map distance are called **map units (mu),** or sometimes **centiMorgans (cM)** in honor of Thomas Hunt Morgan. One map unit is equivalent to a 1% frequency of recombination. In this example, we conclude that the bristle length and body color genes are 12.3 mu apart from each other on the same chromosome.

As the percentage of recombinant offspring approaches 50%, this value becomes a progressively more inaccurate measure of actual map distance. To correct for this inaccuracy, geneticists have developed **mapping functions,** which describe the quantitative relationship between recombination frequencies and physical distances along chromosomes (**Figure 7.9**). What is the basis for this inaccuracy? When the distance between two genes is large, the likelihood of multiple crossovers in the region between them causes the observed number of recombinant offspring to underestimate the physical distance.

Multiple crossovers set a quantitative limit on the relationship between map distance and the percentage of recombinant offspring. Even though two different genes can be on the same chromosome and more than 50 mu apart, a testcross is expected to yield a maximum of only 50% recombinant offspring. What accounts for this 50% limit? The answer lies in the pattern of multiple crossovers. A single crossover in the region between two genes produces only 50% recombinant chromosomes (see Figure 7.2b). Therefore, to exceed a 50% recombinant level, it seems necessary

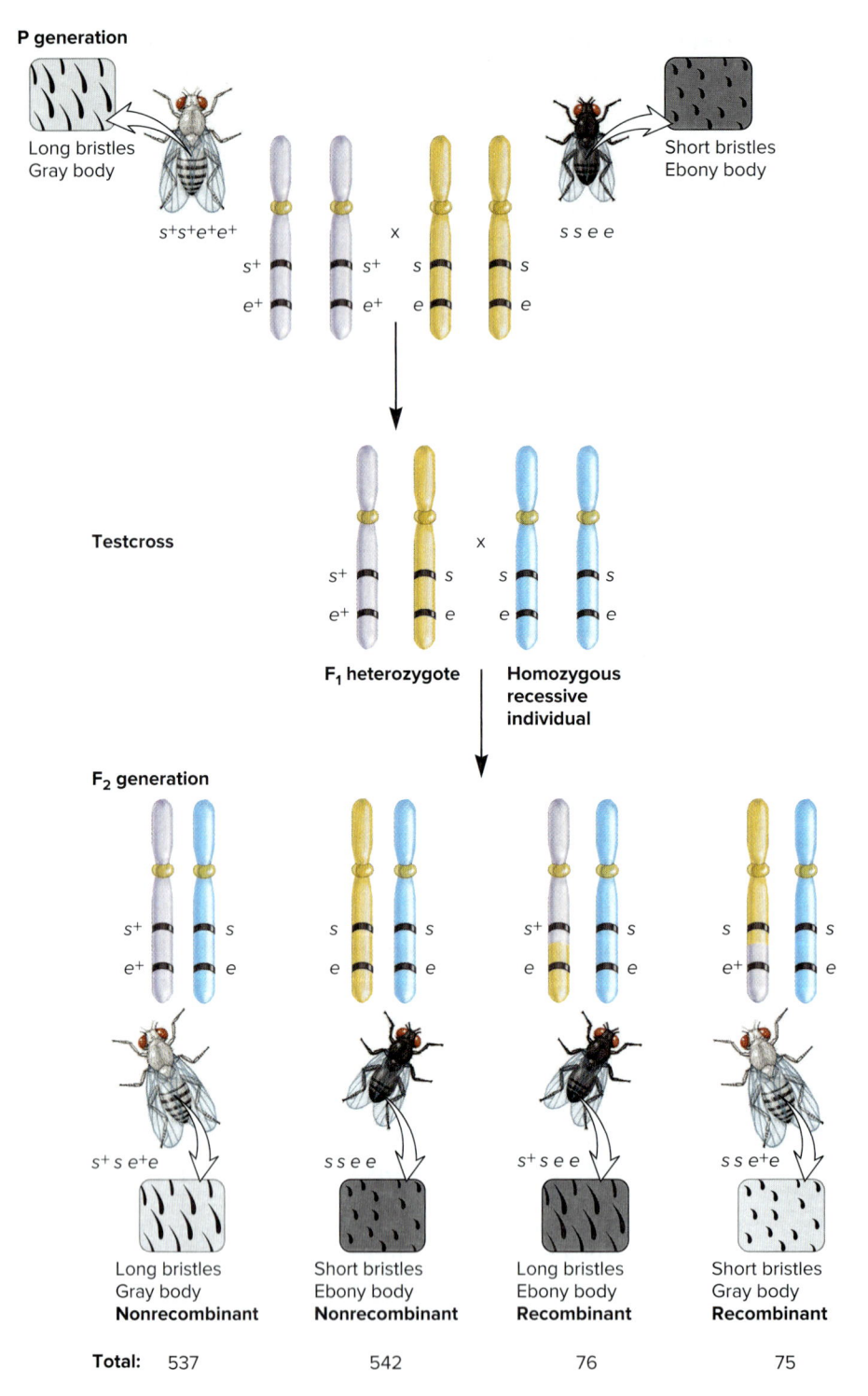

142   CHAPTER 7 :: GENETIC LINKAGE AND MAPPING IN EUKARYOTES

FIGURE 7.8 **Use of a testcross to distinguish between nonrecombinant and recombinant offspring.** The P generation consists of flies from two different true-breeding strains, and the cross produces an F_1 heterozygote. In the testcross, an F_1 female that is heterozygous for both genes (s^+se^+e) is crossed to a male that is homozygous recessive for short bristles (*ss*) and ebony body (*ee*). The F_2 recombinant offspring carry a chromosome that is the product of a crossover. Note: Crossing over does not occur during sperm formation in *Drosophila*, which is unusual among eukaryotes. Therefore, the heterozygote in a testcross involving *Drosophila* must be the female.

Concept Check: When and in which fly or flies did crossing over occur in order to produce the recombinant offspring?

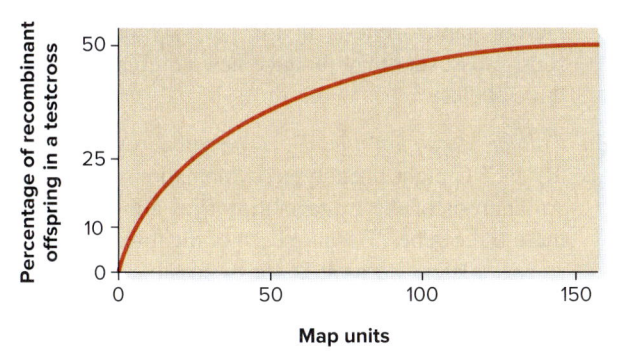

FIGURE 7.9 **Relationship between the percentage of recombinant offspring observed in a testcross and the actual map distance between genes.** The y-axis depicts the percentage of recombinant offspring that would be observed in a two-factor testcross. The actual map distance, shown on the x-axis, is calculated by analyzing the percentages of recombinant offspring from a series of many two-factor crosses involving closely linked genes. Even though two genes may be more than 50 mu apart, the percentage of recombinant offspring will not exceed 50%.

Concept Check: What phenomenon explains why the maximum percentage of recombinant offspring does not exceed 50%?

to have multiple crossovers within a tetrad. However, let's consider double crossovers. As shown in the figure with question 4 of More Genetic TIPS at the end of the chapter, a double crossover between two genes could involve four, three, or two chromatids, which would yield 100%, 50%, or 0% recombinants, respectively. Because all of these double crossovers are equally likely, we take the average of them to determine the maximum recombination frequency. This average equals 50%. Therefore, when two different genes are more than 50 mu apart, they follow the law of independent assortment in a testcross, and only 50% recombinants are observed.

Genetic TIPS

The Question: In the mapping example in Figure 7.8, all of the dominant alleles were on one chromosome and the recessive alleles were on the homolog. Let's consider a two-factor cross in which the dominant allele for one gene is on one chromosome, but the dominant allele for a second gene is on the homolog. A cross is made between *AAbb* and *aaBB* parents. The F_1 offspring are *AaBb*. The F_1 offspring are then testcrossed to *aabb* individuals. Which F_2 offspring are recombinant?

Topic: What topic in genetics does this question address? The topic is linkage and genetic mapping. More specifically, the question is about identifying recombinant offspring in a two-factor testcross.

Information: What information do you know based on the question and your understanding of the topic? From the question, you know that a cross involves a P generation that is *AAbb* and *aaBB*, and then the F_1 heterozygotes are testcrossed to *aabb* individuals. From your understanding of the topic, you may remember that F_2 recombinant offspring are produced by crossing over in the F_1 heterozygotes.

Problem-Solving Strategy: Make a drawing. Predict the outcome. In solving linkage problems, it can be very helpful to draw the chromosomes in the F_1 heterozygote and thereby deduce the possible haploid cells that an F_1 heterozygote can produce. Such a drawing is shown below.

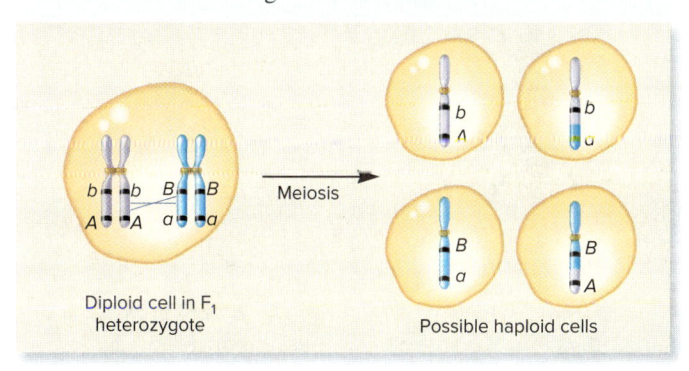

A is initially linked to *b*, and *a* is linked to *B*. If crossing over occurs in the region between these two genes, it will produce chromosomes in which *A* is linked to *B* and *a* is linked to *b* (see the drawing above). All F_2 offspring will inherit *ab* from the *aabb* parent. The F_2 nonrecombinant offspring, which are not the result of a crossover, will inherit *Ab* or *aB* from the F_1 heterozygous parent. They will be *Aabb* and *aaBb*. The F_2 recombinant offspring, which are produced by a crossover, will inherit *AB* or *ab* from the F_1 parent. They will be *AaBb* and *aabb*.

Answer: The F_2 recombinant offspring are those with the genotypes *AaBb* and *aabb*. Note: You might be surprised that the recombinant offspring in this example have genotypes that are the same as their parents that were involved in the testcross. It is important to remember that in this chapter we define recombinant offspring as offspring that have inherited a chromosome that is the product of a crossover.

Three-Factor Crosses Can Be Used to Determine the Order and Distance Between Linked Genes

Thus far, we have considered the construction of genetic maps using two-factor testcrosses to compute map distance. The data from three-factor crosses can yield additional information about map distance and gene order. In a three-factor cross, the experimenter crosses two individuals that differ in three characters.

144   CHAPTER 7 :: GENETIC LINKAGE AND MAPPING IN EUKARYOTES

The following experiment outlines a common strategy for using three-factor crosses to map genes. In this experiment, the P generation consists of fruit flies that differ in body color, eye color, and wing shape. We begin with true-breeding lines so we know which alleles are initially linked to each other on the same chromosome.

Step 1. *Cross two true-breeding strains that differ with regard to three genes.* In this example, we cross a fly that has a black body (*bb*), purple eyes (*prpr*), and vestigial wings (*vgvg*) to a homozygous wild-type fly with a gray body (*b⁺b⁺*), red eyes (*pr⁺pr⁺*), and long wings (*vg⁺vg⁺*):

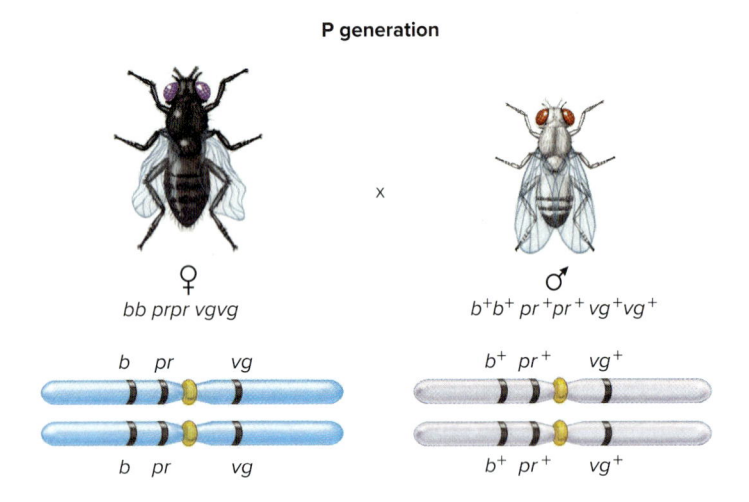

P generation

♀ *bb prpr vgvg* × ♂ *b⁺b⁺ pr⁺pr⁺ vg⁺vg⁺*

The goal in this step is to obtain F₁ individuals that are heterozygous for all three genes. In the F₁ heterozygotes, all dominant alleles are located on one chromosome, and all recessive alleles are on the other homologous chromosome.

Step 2. *Perform a testcross by mating F₁ female heterozygotes to male flies that are homozygous recessive for all three alleles (bb prpr vgvg).*

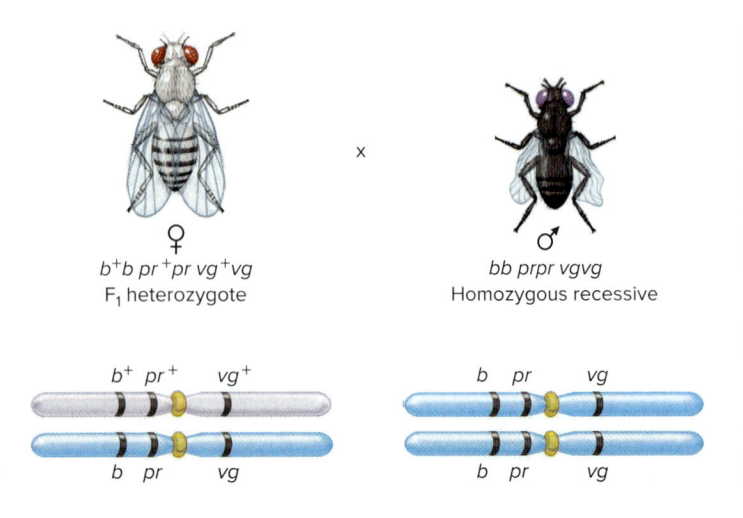

♀ *b⁺b pr⁺pr vg⁺vg* F₁ heterozygote × ♂ *bb prpr vgvg* Homozygous recessive

During gametogenesis in the heterozygous female F₁ flies, crossovers may produce new combinations of the three alleles.

Step 3. *Collect data for the F₂ generation.* As shown in **Table 7.1**, eight phenotypic combinations are possible. An analysis of the F₂ generation flies allows us to map these three genes. Because each of the three genes exists as two alleles, we have 2³, or 8, possible combinations of offspring. If these alleles assorted independently, all eight combinations would occur in equal proportions. However, we see from the data that the proportions of the eight phenotypes are far from equal.

The genotypes of the P generation correspond to two phenotypes: gray body, red eyes, long wings; and black body, purple eyes, vestigial wings. In crosses involving linked genes, the nonrecombinant phenotypes occur most frequently in the offspring. The remaining six phenotypes in the F₂ generation, which are due to crossing over, are recombinants.

A double crossover is always expected to cause the least frequent types of offspring. Two of the phenotypes—gray body, purple eyes, long wings; and black body, red eyes, vestigial wings—arose from a double crossover between two pairs of genes. Also, the combination of traits involved in the double crossover

TABLE 7.1
Data from a Three-Factor Cross (see Steps 2 and 3)

Phenotype	Number of Observed Offspring (males and females)	Chromosome Inherited from F₁ Female
Gray body, red eyes, long wings	411	*b⁺ pr⁺ vg⁺*
Gray body, red eyes, vestigial wings	61	*b⁺ pr⁺ vg*
Gray body, purple eyes, long wings	2	*b⁺ pr vg⁺*
Gray body, purple eyes, vestigial wings	30	*b⁺ pr vg*
Black body, red eyes, long wings	28	*b pr⁺ vg⁺*
Black body, red eyes, vestigial wings	1	*b pr⁺ vg*
Black body, purple eyes, long wings	60	*b pr vg⁺*
Black body, purple eyes, vestigial wings	412	*b pr vg*
Total	1005	

tells us which gene is in the middle along the chromosome. When a chromatid undergoes a double crossover, the gene in the middle becomes separated from the other two genes at either end.

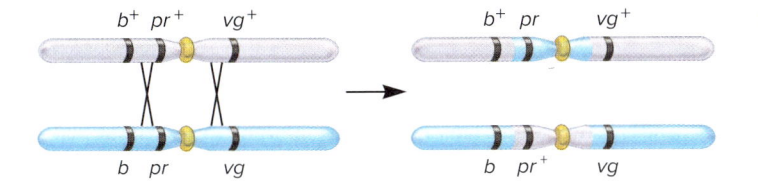

In the types of offspring resulting from a double crossover, the recessive purple eye allele is separated from the other two recessive alleles. When such an offspring is mated to a homozygous recessive fly in the testcross, this yields flies with either gray bodies, purple eyes, and long wings or black bodies, red eyes, and vestigial wings. This observation indicates that the gene for eye color lies between the genes for body color and wing shape.

Step 4. *Calculate the map distance between pairs of genes.* To do this, we need to understand which gene combinations are nonrecombinant and which are recombinant. The recombinant offspring are due to crossing over in the heterozygous female parent. If you look back at step 2, you can see the arrangement of alleles in the heterozygous female parent in the absence of crossing over. Let's consider this arrangement with regard to gene pairs:

b^+ is linked to pr^+, and b is linked to pr
pr^+ is linked to vg^+, and pr is linked to vg
b^+ is linked to vg^+, and b is linked to vg

With regard to body color and eye color, the recombinant offspring have gray bodies and purple eyes (2 + 30) or black bodies and red eyes (28 + 1). As shown along the right side of Table 7.1, these offspring were produced by crossovers in the female parents. The total number of these recombinant offspring is 61. The map distance between the body color and eye color genes is

$$\text{Map distance} = \frac{61}{944 + 61} \times 100 = 6.1 \text{ mu}$$

With regard to eye color and wing shape, the recombinant offspring have red eyes and vestigial wings (61 + 1) or purple eyes and long wings (2 + 60). The total number is 124. The map distance between the eye color and wing shape genes is

$$\text{Map distance} = \frac{124}{881 + 124} \times 100 = 12.3 \text{ mu}$$

With regard to body color and wing shape, the recombinant offspring have gray bodies and vestigial wings (61 + 30) or black bodies and long wings (28 + 60). The total number is 179. The map distance between the body color and wing shape genes is

$$\text{Map distance} = \frac{179}{826 + 179} \times 100 = 17.8 \text{ mu}$$

Step 5. *Construct the map.* Based on the map unit calculation, the body color (b) and wing shape (vg) genes are farthest apart. The eye color gene (pr) must lie in the middle. As mentioned earlier, this order of genes is also confirmed by the pattern of traits found in the double crossovers. To construct the map, we use the distances between the genes that are closest together.

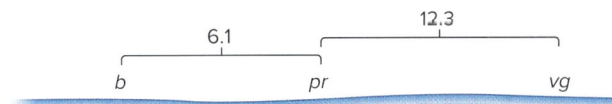

In our example, we have placed the body color gene first and the wing shape gene last. The data also are consistent with a map in which the wing shape gene comes first and the body color gene comes last. In detailed genetic maps, the locations of genes are mapped relative to the centromere.

You may have noticed that our calculations underestimate the distance between the body color and wing shape genes. We obtained a value of 17.8 mu even though the distance seems to be 18.4 mu when we add together the distance between body color and eye color genes (6.1 mu) and the distance between eye color and wing shape genes (12.3 mu). What accounts for this discrepancy? The answer is double crossovers. If you look at the data in Table 7.1, the offspring with gray bodies, purple eyes, long wings and those with black bodies, red eyes, vestigial wings are due to a double crossover. From a phenotypic perspective, these offspring are not recombinant with regard to the body color and wing shape alleles. Even so, we know that they arose from a double crossover between these two genes. Therefore, we should consider these crossovers when calculating the distance between the body color and wing shape genes. In this case, three offspring (2 + 1) were due to double crossovers. Because they are double crossovers, we multiply 2 times the number of double crossovers (2 + 1) and add this number to our previous value of recombinant offspring:

$$\text{Map distance} = \frac{179 + 2(2 + 1)}{826 + 179} \times 100 = 18.4 \text{ mu}$$

7.3 REVIEWING THE KEY CONCEPTS

- A genetic linkage map is a diagram that portrays the order and relative spacing of genes along one or more chromosomes (see Figure 7.7).
- A testcross can be performed to map the distance between two or more linked genes (see Figure 7.8).

146 CHAPTER 7 :: GENETIC LINKAGE AND MAPPING IN EUKARYOTES

- Due to the effects of multiple crossovers, the map distance between two genes obtained from a testcross cannot exceed 50% (see Figure 7.9).
- The data from a three-factor cross can be used to map the three genes (see Table 7.1).

7.3 COMPREHENSION QUESTIONS

Answer the multiple-choice questions based on the following experiment:

P generation: True-breeding flies with red eyes and long wings were crossed to flies with white eyes and miniature wings. All F_1 offspring had red eyes and long wings.

The F_1 female flies were then crossed to males with white eyes and miniature wings. The following results were obtained for the F_2 generation:

    129 red eyes, long wings
    133 white eyes, miniature wings
     71 red eyes, miniature wings
     67 white eyes, long wings

1. What is/are the phenotypes of the recombinant offspring of the F_2 generation?
   a. red eyes, long wings
   b. white eyes, miniature wings
   c. red eyes, long wings and white eyes, miniature wings
   d. red eyes, miniature wings and white eyes, long wings

2. The recombinant offspring of the F_2 generation were produced by crossing over that occurred
   a. during spermatogenesis in the P generation males.
   b. during oogenesis in the P generation females.
   c. during spermatogenesis in the F_1 males.
   d. during oogenesis in the F_1 females.

3. What is the map distance between the two genes for eye color and wing length?
   a. 32.3 mu
   b. 34.5 mu
   c. 16.2 mu
   d. 17.3 mu

7.4 MITOTIC RECOMBINATION

Learning Outcome:

1. Describe the process of mitotic recombination, and explain how it can produce a twin spot.

Thus far, we have considered how the arrangement of linked alleles along a chromosome can be rearranged by crossing over. This event can produce cells and offspring with a recombinant set of traits. In these cases, crossing over occurs during meiosis, when the homologous chromosomes replicate and form bivalents.

In multicellular organisms, the union of egg and sperm is followed by many cellular divisions, which occur in conjunction with mitotic divisions of the cell nuclei. As discussed in Chapter 2, mitosis normally does not involve the homologous pairing of chromosomes to form bivalents. Therefore, crossing over during mitosis is expected to occur much less frequently than during meiosis. Nevertheless, it does happen on rare occasions. Mitotic crossing over may produce a pair of recombinant chromosomes that have a new combination of alleles, an event known as **mitotic recombination.** If this recombination occurs during an early stage of embryonic development, the daughter cells containing one recombinant and one nonrecombinant chromosome will continue to divide many times to produce a patch of tissue in the adult. This may result in a portion of tissue with characteristics that are different from those of the rest of the organism.

In 1936, Curt Stern identified unusual patches on the bodies of flies from certain *Drosophila* strains. He was working with strains carrying X-linked alleles affecting body color and bristle morphology (**Figure 7.10**). A recessive allele confers yellow body color (*y*), and another recessive allele causes shorter body bristles that look singed (*sn*). The corresponding wild-type alleles result in gray body color (y^+) and long bristles (sn^+). Females with the genotype $y^+y\ sn^+sn$ are expected to have gray bodies and long bristles. This was generally the case. However, when Stern carefully observed the bodies of these female flies under a low-power microscope, he occasionally noticed places in which two adjacent regions were different from the rest of the body—a twin spot. He concluded that twin spotting was too frequent to be explained by the random positioning of two independent single spots that happened to occur close together. How then did Stern explain the phenomenon of twin spotting? He proposed that twin spots are due to a single mitotic recombination within one cell during embryonic development.

As shown in Figure 7.10, the X chromosomes of the fertilized egg were $y^+\ sn$ and $y\ sn^+$. During development, a rare crossover occurred during mitosis to produce two adjacent daughter cells that were $y^+y^+\ snsn$ and $yy\ sn^+sn^+$. As embryonic development proceeded, the cell on the left continued to divide to produce many cells, eventually producing a patch on the body that had gray color with singed bristles. The daughter cell next to it produced a patch of body that was yellow with long bristles. These two adjacent patches—a twin spot—were surrounded by cells that were $y^+y\ sn^+sn$ and had gray color and long bristles. Twin spots provide evidence that mitotic recombination occasionally occurs.

7.4 REVIEWING THE KEY CONCEPTS

- Mitotic recombination occurs on rare occasions and may produce twin spots (see Figure 7.10).

7.4 COMPREHENSION QUESTION

1. The process of mitotic recombination involves the
   a. exchange of chromosomal regions between homologs during gamete formation.
   b. exchange of chromosomal regions between homologs during the division of somatic cells.
   c. reassortment of alleles that occurs at fertilization.
   d. reassortment of alleles that occurs during gamete formation.

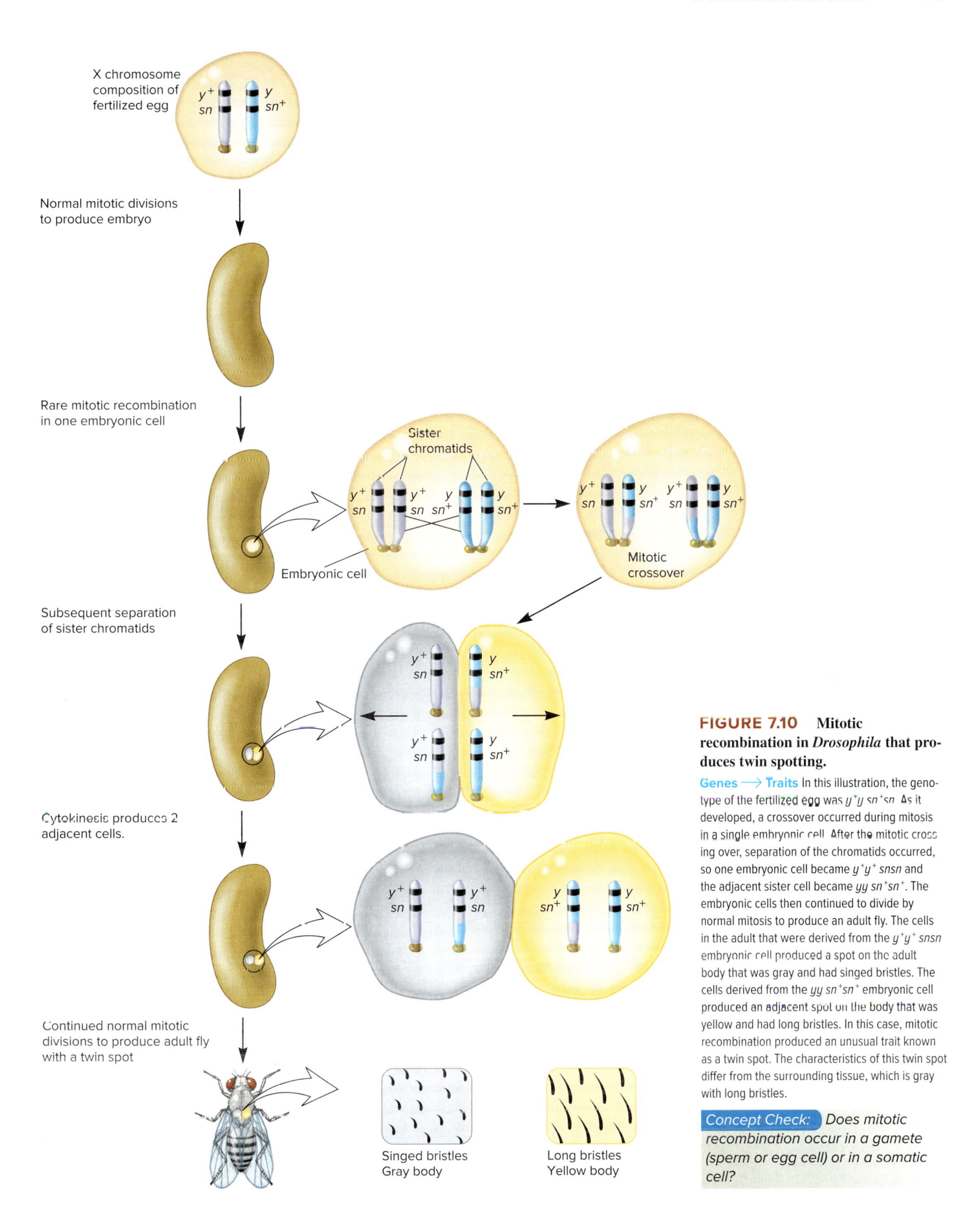

7.4 MITOTIC RECOMBINATION   147

X chromosome composition of fertilized egg

Normal mitotic divisions to produce embryo

Rare mitotic recombination in one embryonic cell

Sister chromatids

Embryonic cell

Mitotic crossover

Subsequent separation of sister chromatids

Cytokinesis produces 2 adjacent cells.

Continued normal mitotic divisions to produce adult fly with a twin spot

Singed bristles
Gray body

Long bristles
Yellow body

FIGURE 7.10 Mitotic recombination in *Drosophila* that produces twin spotting.

Genes → Traits In this illustration, the genotype of the fertilized egg was $y^+y\ sn^+sn$. As it developed, a crossover occurred during mitosis in a single embryonic cell. After the mitotic crossing over, separation of the chromatids occurred, so one embryonic cell became $y^+y^+\ snsn$ and the adjacent sister cell became $yy\ sn^+sn^+$. The embryonic cells then continued to divide by normal mitosis to produce an adult fly. The cells in the adult that were derived from the $y^+y^+\ snsn$ embryonic cell produced a spot on the adult body that was gray and had singed bristles. The cells derived from the $yy\ sn^+sn^+$ embryonic cell produced an adjacent spot on the body that was yellow and had long bristles. In this case, mitotic recombination produced an unusual trait known as a twin spot. The characteristics of this twin spot differ from the surrounding tissue, which is gray with long bristles.

Concept Check: Does mitotic recombination occur in a gamete (sperm or egg cell) or in a somatic cell?

KEY TERMS

Page 131. genetic map
Page 132. synteny, genetic linkage (linkage), linkage groups, two-factor cross, three-factor cross
Page 133. crossing over, bivalent, nonrecombinant, recombinant, genetic recombination
Page 136. null hypothesis
Page 139. genetic mapping
Page 140. locus, genetic linkage map
Page 141. testcross, map distance, map units (mu), centiMorgans (cM), mapping functions
Page 146. mitotic recombination

CHAPTER SUMMARY

7.1 Overview of Linkage

- Linkage has two related meanings. First, it refers to genes that are located on the same chromosome. Second, linkage also means that the alleles of two or more genes tend to be transmitted as a unit because they are relatively close on the same chromosome.
- Bateson and Punnett discovered the first example of genetic linkage in sweet peas (see Figure 7.1).

7.2 Relationship Between Linkage and Crossing Over

- Crossing over can change the combination of alleles along a chromosome and produce recombinant cells and recombinant offspring (see Figure 7.2).
- Morgan discovered genetic linkage in *Drosophila* and proposed that recombinant offspring are produced when crossing over occurs during meiosis (see Figures 7.3, 7.4).
- When genes are linked, the relative proportions of recombinant offspring depend on the distance between the genes (see Figure 7.5).
- A chi square test can be used to judge whether or not two genes assort independently.
- Creighton and McClintock were able to correlate the formation of recombinant offspring with the presence of chromosomes that had exchanged pieces due to crossing over (see Figure 7.6).

7.3 Genetic Mapping in Plants and Animals

- A genetic linkage map is a diagram that portrays the order and relative spacing of genes along one or more chromosomes (see Figure 7.7).
- A testcross can be performed to map the distance between two or more linked genes (see Figure 7.8).
- Due to the effects of multiple crossovers, the map distance between two genes obtained from a testcross cannot exceed 50% (see Figure 7.9).
- The data from a three-factor cross can be used to map the three genes (see Table 7.1).

7.4 Mitotic Recombination

- Mitotic recombination occurs on rare occasions and may produce twin spots (see Figure 7.9).

PROBLEM SETS & INSIGHTS

More Genetic TIPS

1. In the garden pea, orange pods (*orp*) are recessive to green pods (*Orp*), and sensitivity to pea mosaic virus (*mo*) is recessive to resistance to the virus (*Mo*). A plant with orange pods and sensitivity to the virus was crossed to a true-breeding plant with green pods and resistance to the virus. The F_1 plants were then testcrossed to plants with orange pods and sensitivity to the virus. The following results were obtained:

    160 orange pods, virus-sensitive
    165 green pods, virus-resistant
    36 orange pods, virus-resistant
    39 green pods, virus-sensitive
    400 total

    A. Conduct a chi square test to see if these two genes are linked.

    B. If they are linked, calculate the map distance between the two genes.

Topic: What topic in genetics does this question address?

The topic is linkage. More specifically, the question asks you to determine if the outcome of a testcross is consistent with linkage or independent assortment.

Information: What information do you know based on the question and your understanding of the topic?

From the question, you know the outcome of a two-factor cross and a subsequent testcross. From your understanding of the topic, you may remember that linked genes do not independently assort. A chi square test can be used to evaluate whether two genes are likely to be linked. You may also recall the formula for calculating map distance if two genes are linked.

Problem-Solving Strategy: Propose a hypothesis. Predict the outcome. Analyze data. Make a calculation.

To answer part A of this question, one strategy is to follow the four steps for conducting a chi square test, which are described in Section 7.2.

Step 1. *Propose a hypothesis.* In this case, your hypothesis is that the genes are not linked. This hypothesis allows you to predict the outcome of the testcross based on independent assortment.

Step 2. *Based on the hypothesis, calculate the expected value of each of the four phenotypes.* The testcross is *Orporp Momo × orporp momo*.

For this testcross, a 1:1:1:1 ratio of the four phenotypes is predicted if the genes were not linked. In other words, the phenotypes of the offspring should be in these proportions: 1/4 orange pods, virus-sensitive; 1/4, green pods, virus-resistant; 1/4, orange pods, virus-resistant; and 1/4 green pods, virus-sensitive. Because a total of 400 offspring were produced, your hypothesis predicts 100 offspring with each phenotype.

Step 3. *Apply the chi square formula, using the data for the observed values (O) and the expected values (E) that have been calculated in step 2.* In this case, the expected values are 100, as calculated in step 2, and the observed values consist of the numbers of offspring with the four phenotypes, given in the question.

$$\chi^2 = \frac{(O_1 - E_1)^2}{E_1} + \frac{(O_2 - E_2)^2}{E_2} + \frac{(O_3 - E_3)^2}{E_3} + \frac{(O_4 - E_4)^2}{E_4}$$

$$\chi^2 = \frac{(160 - 100)^2}{100} + \frac{(165 - 100)^2}{100} + \frac{(36 - 100)^2}{100} + \frac{(39 - 100)^2}{100}$$

$$\chi^2 = 36 + 42.3 + 41 + 37.2 = 156.5$$

Step 4. *Interpret the calculated chi square value.* The calculated chi square value is quite large. This indicates that the deviation between observed and expected values is very high. For 1 degree of freedom in Table 3.2, such a large deviation is expected to occur by chance alone less than 1% of the time. Therefore, you reject the hypothesis that the genes assort independently. As an alternative, you may infer that the two genes are linked.

To answer part B, use the equation given in Section 7.3 to calculate the map distance between the two genes.

$$\text{Map distance} = \frac{\text{Number of recombinant offspring}}{\text{Total number of offspring}} \times 100$$

$$= \frac{36 + 39}{36 + 39 + 160 + 165} \times 100$$

$$= 18.8 \text{ mu}$$

Answer:

A. You reject the hypothesis that the genes are not linked.

B. The genes are approximately 18.8 mu apart.

2. Two recessive traits in mice—droopy ears and flaky tail—are caused by mutations in two genes that are located 6 mu apart on the same chromosome. A true-breeding mouse with normal ears (*De*) and a flaky tail (*ft*) was crossed to a true-breeding mouse with droopy ears (*de*) and a normal tail (*Ft*). The F_1 offspring had normal ears and a normal tail. The F_1 offspring were then testcrossed to mice with droopy ears and flaky tails. What are the phenotypes of the F_2 recombinants? If this testcross produced 100 offspring, what is the expected outcome of phenotypes?

Topic: **What topic in genetics does this question address?**

The topic is linkage. More specifically, the question asks you to predict the outcome of a testcross when you know the distance between two linked genes.

Information: **What information do you know based on the question and your understanding of the topic?**

From the question, you know that two genes are 6 mu apart on the same chromosome. You also know the genotypes of the P generation and the phenotypes of the F_1 offspring. From your understanding of the topic, you may remember that the map distance is a measure of the percentage of recombinant offspring.

Problem-Solving **S**trategies: **Make a drawing. Predict the outcome. Make a calculation.**

In solving linkage problems, you may find it helpful to make a drawing of the chromosomes to help you distinguish the recombinant and nonrecombinant offspring. The recombinant offspring are produced by a crossover in the F_1 parent. The testcross is shown below, with the F_1 parent on the left side.

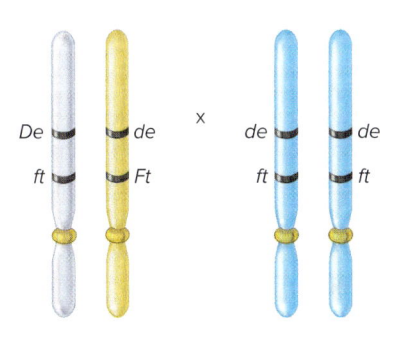

The nonrecombinant offspring are as follows:

Dede ftft   normal ears, flaky tail

dede Ftft   droopy ears, normal tail

The recombinant offspring are as follows:

dede ftft   droopy ears, flaky tail

Dede Ftft   normal ears, normal tail

Note: In this testcross, both dominant alleles were not on the same chromosome in the F_1 parent. This situation is different from that in many previous problems. Crossing over produces chromosomes in which one chromosome carries both dominant alleles and the other carries both recessive alleles. Because the two genes are located 6 mu apart on the same chromosome, 6% of the offspring will be recombinants.

Answer:

The expected outcome of phenotypes for 100 offspring is

3 droopy ears, flaky tail

3 normal ears, normal tail

47 normal ears, flaky tail

47 droopy ears, normal tail

3. The following X-linked traits are found in fruit flies: vermilion eyes are recessive to red eyes, miniature wings are recessive to long wings, and sable body is recessive to gray body. A cross was made between wild-type males with red eyes, long wings, and gray bodies and females with vermilion eyes, miniature wings, and sable bodies. The heterozygous female offspring from this cross, which had red eyes, long wings, and gray bodies, were then crossed to males with vermilion eyes, miniature wings, and sable bodies. The following results were obtained for the F_2 generation (including both males and females):

1320 vermilion eyes, miniature wings, sable body

1346 red eyes, long wings, gray body

102 vermilion eyes, miniature wings, gray body

90 red eyes, long wings, sable body

42 vermilion eyes, long wings, gray body

48 red eyes, miniature wings, sable body

2 vermilion eyes, long wings, sable body

1 red eyes, miniature wings, gray body

Calculate the distances separating the three genes.

Topic: What topic in genetics does this question address?

The topic is linkage. More specifically, you are asked to calculate the map distances separating three linked genes.

Information: What information do you know based on the question and your understanding of the topic?

From the question, you know that the three genes are on the X chromosome. You also know the results of a three-factor testcross. The F_1 offspring have all of the dominant alleles on one chromosome and all of the recessive alleles on another chromosome. (Make a drawing of the chromosomes in these crosses if this isn't obvious.) From your understanding of the topic, you may remember that the map distance is a measure of the percentage of recombinant offspring. In a three-factor cross, double crossovers result in the rarest category of offspring, in which the gene in the middle becomes separated from the genes at the ends.

Problem-Solving Strategy: Analyze data. Make a drawing. Make a calculation.

The first step is to make a drawing that depicts the order of the three genes. You can do this by evaluating the pattern of inheritance for the double crossovers. The double crossovers occur with the lowest frequency. Thus, the double crossovers produced the phenotypes vermilion eyes, long wings, sable body; and red eyes, miniature wings, gray body. Compared with the nonrecombinant patterns of alleles (vermilion eyes, miniature wings, sable body; and red eyes, long wings, gray body), the gene for wing length has been reassorted. Two flies have long wings associated with vermilion eyes and sable body, and one fly has miniature wings associated with red eyes and gray body. Taken together, these results indicate that the wing length gene is found in between the eye color and body color genes.

v   m   s

You can now calculate the distance between the genes for eye color and wing length and between those for wing length and body color. To do this, you can consider the data on numbers of offspring according to gene pairs, first for eye color and wing length:

vermilion eyes, miniature wings = 1320 + 102 = 1422

red eyes, long wings = 1346 + 90 = 1436

vermilion eyes, long wings = 42 + 2 = 44

red eyes, miniature wings = 48 + 1 = 49

The recombinants are vermilion eyes, long wings and red eyes, miniature wings. The map distance between these two genes is

(44 + 49)/(1422 + 1436 + 44 + 49) × 100 = 3.2 mu

Likewise, for the other gene pair of wing length and body color, the numbers of offspring are:

miniature wings, sable body = 1320 + 48 = 1368

long wings, gray body = 1346 + 42 = 1388

miniature wings, gray body = 102 + 1 = 103

long wings, sable body = 90 + 2 = 92

The recombinants are miniature wings, gray body and long wings, sable body. The map distance between these two genes is

(103 + 92)/(1368 + 1388 + 103 + 92) × 100 = 6.6 mu

Answer:

With these data, you can produce the following genetic map:

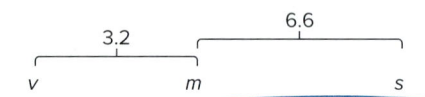

```
        3.2        6.6
      |-----|  |---------|
        v        m        s
```

4. As described in Figure 7.9, a limit exists for the relationship between map distance and the percentage of recombinant offspring. Even though two genes on the same chromosome may be much more than 50 mu apart, we do not expect to obtain greater than 50% recombinant offspring in a testcross. You may be wondering why this is so. The answer lies in the pattern of multiple crossovers. At prophase of meiosis I, a single crossover in the region between two genes produces only 50% recombinant chromosomes (see Figure 7.2b). Therefore, to exceed a 50% recombinant level, it would seem necessary to have multiple crossovers within a bivalent.

Let's suppose that two genes are far apart on the same chromosome. A testcross is made between a heterozygous individual, *AaBb*, and a homozygous individual, *aabb*. In the heterozygous individual, the dominant alleles (*A* and *B*) are linked on the same chromosome, and the recessive alleles (*a* and *b*) are linked on the same chromosome. What are all of the possible double crossovers (between two, three, or four chromatids)? What is the average number of recombinant offspring, assuming an equal probability of occurrence for all of the double-crossovers?

Topic: What topic in genetics does this question address?

The topic is linkage. More specifically, the question asks you to determine how double crossovers affect the maximum map distance in a testcross.

Information: What information do you know based on the question and your understanding of the topic?

In the question, you are reminded that the maximum percentage of recombinant offspring in a testcross is 50%. From your understanding of the topic, you may remember that the map distance is a measure of the percentage of recombinant offspring, which are produced by crossing over.

Problem-Solving Strategies: Make a drawing. Predict the outcome. Make a calculation.

One strategy to solve this problem is to make a drawing that shows how the sister chromatids within a bivalent could cross over. A double crossover affecting the two genes could involve two chromatids, three chromatids, or four chromatids. From the drawing, you should be able to predict the outcomes of the different types of double crossovers, and then calculate the average percentage.

Answer:

The drawing that follows considers two crossovers that occur in the region between the two genes. A double crossover between the two genes could involve two chromatids, three chromatids, or four chromatids. Because a bivalent is composed of two pairs of homologs, a double crossover between homologs could occur in several possible ways. In each part of the drawing, the crossover on the right occurred first. Because all of these double crossovers are equally probable, we take the average of them to determine the maximum number of recombinants. This average equals 50%.

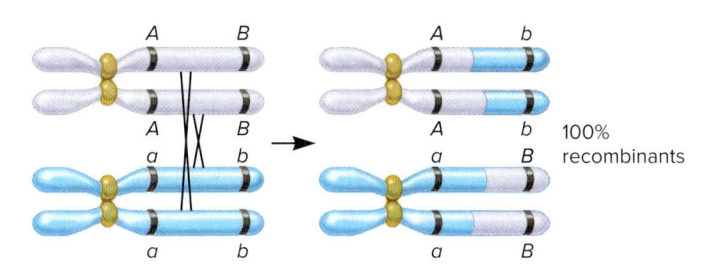

Double crossover (involving 4 chromatids) — 100% recombinants

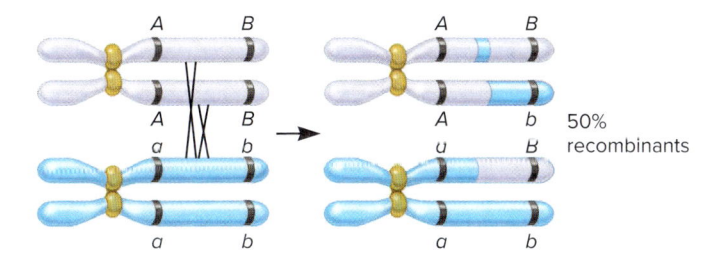

Double crossover (involving 3 chromatids) — 50% recombinants

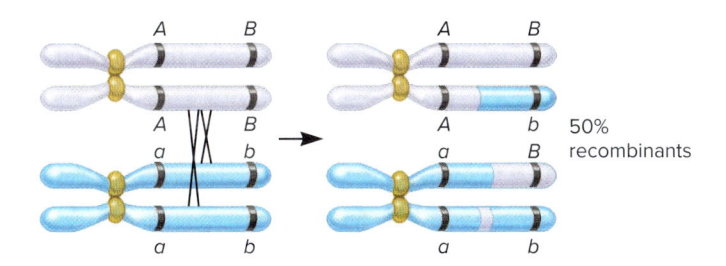

Double crossover (involving 3 chromatids) — 50% recombinants

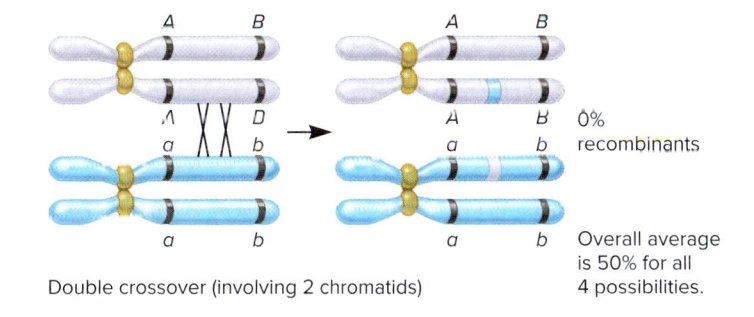

Double crossover (involving 2 chromatids) — 0% recombinants

Overall average is 50% for all 4 possibilities.

Conceptual Questions

C1. What is the difference in meaning between the terms *genetic recombination* and *crossing over*?

C2. When a chi square test is applied to solve a linkage problem, explain why an independent assortment hypothesis is proposed.

C3. A crossover has occurred in the bivalent shown in the drawing on the right. If a second crossover occurs in the same region between these two genes, which two chromatids would be involved to produce the following outcomes?

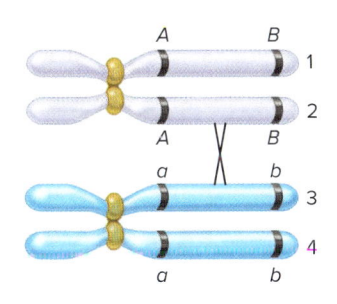

A. 100% recombinants

B. 0% recombinants

C. 50% recombinants

C4. A crossover has occurred in the bivalent shown below:

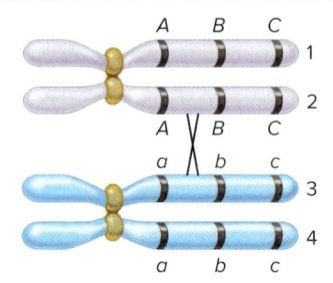

What is the outcome of this single crossover? If a second crossover occurs somewhere between *A* and *C*, explain which two chromatids would be involved and where the crossover would occur (i.e., between which two genes) to produce each of the following arrangements of alleles along the chromosomes:

A. *ABC, AbC, aBc,* and *abc*

B. *Abc, Abc, aBC,* and *aBC*

C. *ABc, Abc, aBC,* and *abC*

D. *ABC, ABC, abc,* and *abc*

C5. A diploid organism has a total of 14 chromosomes and about 20,000 genes per haploid genome. Approximately how many genes are in each linkage group?

C6. If you try to throw a basketball into a basket, the likelihood of succeeding depends on the size of the basket. It is more likely that you will get the ball into the basket if the basket is bigger. In your own words, explain how this analogy applies to the idea that the likelihood of crossing over is greater when two genes are far apart compared to when they are close together.

C7. By conducting testcrosses, researchers have found that the sweet pea has seven linkage groups. How many chromosomes would you expect to find in leaf cells of sweet pea plants?

C8. In humans, a rare dominant disorder known as nail-patella syndrome causes abnormalities in the fingernails, toenails, and kneecaps. Researchers have examined family pedigrees with regard to this disorder and have also examined the blood types of individuals within each pedigree. (A description of blood genotypes is found in Chapter 5.) In the pedigree below, individuals affected with nail-patella syndrome are shown as filled symbols. The genotype of each individual with regard to ABO blood type is also shown. Does this pedigree suggest any linkage between the gene that causes nail-patella syndrome and the gene that determines blood type?

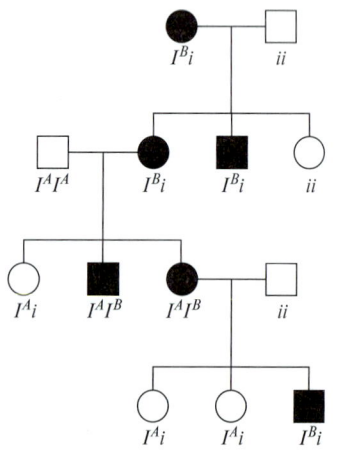

C9. When true-breeding mice with brown fur and short tails (*BBtt*) were crossed to true-breeding mice with white fur and long tails (*bbTT*), all F_1 offspring had brown fur and long tails. The F_1 offspring were crossed to mice with white fur and short tails. What are the possible phenotypes of the F_2 offspring? Which F_2 offspring are recombinant, and which are nonrecombinant? What are the ratios of the phenotypes of the F_2 offspring if independent assortment is taking place? How are the ratios affected by linkage?

C10. Though we often think of genes in terms of the phenotypes they produce (e.g., curly leaves, flaky tail, brown eyes), the molecular function of most genes is to encode proteins. Many cellular proteins function as enzymes. The table that follows gives the map distances between six different genes that encode six different enzymes: *Ada*, adenosine deaminase; *Hao-1*, hydroxyacid oxidase-1; *Hdc*, histidine decarboxylase; *Odc-2*, ornithine decarboxylase-2; *Sdh-1*, sorbitol dehydrogenase-1; and *Ass-1*, arginosuccinate synthetase-1.

Map distances between two genes:

	Ada	Hao-1	Hdc	Odc-2	Sdh-1	Ass-1
Ada		14		8	28	
Hao-1	14		9		14	
Hdc		9		15	5	
Odc-2	8		15			63
Sdh-1	28	14	5			43
Ass-1				63	43	

Construct a genetic map that shows the locations of all six genes.

C11. If the likelihood of a single crossover in a particular chromosomal region is 10%, what is the theoretical likelihood of a double or triple crossover in that same region?

C12. In most two-factor crosses involving linked genes, we cannot tell if a double crossover between the two genes has occurred because the offspring inherit the nonrecombinant pattern of alleles. How does the inability to detect double crossovers affect the calculation of map distance? Is map distance underestimated or overestimated because of our inability to detect double crossovers? Explain your answer.

C13. Researchers have discovered that some regions of chromosomes are much more likely than others to cross over. We might call such a region a "hot spot" for crossing over. Let's suppose that two genes, gene *A* and gene *B*, are 5,000,000 bp apart on the same chromosome. Genes *A* and *B* are in a hot spot for crossing over. Two other genes, let's call them gene *C* and gene *D*, are also 5,000,000 bp apart but are not in a hot spot for crossing over. If we conducted two-factor crosses to compute the map distance between genes *A* and *B* and other two-factor crosses to compute the map distance between genes *C* and *D*, would the calculated map distance between *A* and *B* be the same as that between *C* and *D*? Explain.

C14. What is mitotic recombination? A heterozygous individual (*Bb*) with brown eyes has one eye with a small patch of blue. Provide two or more explanations for how the blue patch may have occurred.

C15. Mitotic recombination can occasionally produce a twin spot. Let's suppose an animal is heterozygous for two genes that govern fur color and length: One gene affects pigmentation, with dark pigmentation (*A*) dominant to albino (*a*); the other gene affects hair length, with long hair (*L*) dominant to short hair (*l*). The two genes are linked on the same chromosome. Let's assume that the animal is *AaLl*; *A* is linked to *l*, and *a* is linked to *L*. Draw the chromosomes labeled with these alleles, and explain how mitotic recombination could produce a twin spot with one spot having albino pigmentation and long fur and the other having dark pigmentation and short fur.

Application and Experimental Questions (Including Many Mapping Questions)

E1. Figure 7.1 shows the first experimental results that indicated linkage between two different genes. Conduct a chi square test to see if the genes are really linked and the data could not be explained by independent assortment.

E2. In the experiment of Figure 7.6, the researchers followed the inheritance pattern of chromosomes that were abnormal at both ends to correlate genetic recombination with the physical exchange of chromosome pieces. Is it necessary to analyze a chromosome that is abnormal at both ends, or could the researchers have used a strain with two abnormal versions of chromosome 9, one with a knob at one end and its homolog with a translocation at the other end?

E3. How would you determine that genes in mammals are in the Y chromosome linkage group? Is it possible to conduct crosses (let's say in mice) to map the distances between genes on the Y chromosome? Explain.

E4. Explain the rationale behind a testcross. Is it necessary for one of the parents to be homozygous recessive for the genes of interest? In the heterozygous parent in a testcross, must all of the dominant alleles be linked on the same chromosome and all of the recessive alleles be linked on the homolog?

E5. In your own words, explain why a testcross cannot produce more than 50% recombinant offspring. When a testcross does produce 50% recombinant offspring, what does this result mean?

E6. Explain why the percentage of recombinant offspring in a testcross is a more accurate measure of map distance when two genes are close together. When two genes are far apart, is the percentage of recombinant offspring an underestimate or overestimate of the actual map distance?

E7. If two genes are more than 50 mu apart, how would you ever be able to show experimentally that they are located on the same chromosome?

E8. From the three-factor crosses of Figure 7.3, Morgan realized that crossing over was more frequent between the eye color and wing length genes than between the body color and eye color genes. Explain how he determined this.

E9. Two genes are located on the same chromosome and are known to be 12 mu apart. An *AABB* individual was crossed to an *aabb* individual to produce *AaBb* offspring. The *AaBb* offspring were then testcrossed to *aabb* individuals.

A. If the testcross produced 1000 offspring, what are the predicted numbers of offspring with each of these four genotypes: *AaBb*, *Aabb*, *aaBb*, and *aabb*?

B. What would be the predicted numbers of offspring with the four genotypes if the P generation had been *AAbb* and *aaBB* instead of *AABB* and *aabb*?

E10. Two genes, designated *A* and *B*, are located 10 mu from each other. A third gene, designated *C*, is located 15 mu from *B* and 5 mu from *A*. The P generation consisting of *AA bb CC* and *aa BB cc* individuals were crossed to each other. The F_1 heterozygotes were then testcrossed to *aa bb cc* individuals. Assuming that no double crossovers occur, what percentage of offspring would you expect with each of the following genotypes?

A. *Aa Bb Cc*

B. *aa Bb Cc*

C. *Aa bb cc*

E11. Two genes in tomatoes are 61 mu apart; normal fruit (*F*) is dominant to fasciated fruit (*f*), and normal number of leaves (*Lf*) is dominant to leafy (*lf*). A true-breeding plant with a normal number of leaves and fruit was crossed to a leafy plant with fasciated fruit. The F_1 offspring were then crossed to leafy plants with fasciated fruit. If this cross produced 600 offspring, what are the expected numbers of plants in each of the four possible categories: normal leaf number, normal fruit; normal leaf number, fasciated fruit; leafy, normal fruit; and leafy, fasciated fruit?

E12. In tomato plants, three genes are linked on the same chromosome. Tall is dominant to dwarf, skin that is smooth is dominant to skin that is peachy, and fruit with a normal tomato shape is dominant to fruit with an oblate (flattened) shape. A plant that is true-breeding for the dominant traits was crossed to a dwarf plant with peachy skin and oblate fruit. The F_1 plants were then testcrossed to dwarf plants with peachy skin and oblate fruit. The following results were obtained:

151 tall, smooth, normal

33 tall, smooth, oblate

11 tall, peach, oblate

2 tall, peach, normal

155 dwarf, peach, oblate

29 dwarf, peach, normal

12 dwarf, smooth, normal

0 dwarf, smooth, oblate

Construct a genetic map that shows the order of these three genes and the distances between them.

E13. A trait in garden peas involves the curling of leaves. A two-factor cross was made involving a plant with yellow pods and curling leaves to a wild-type plant with green pods and normal leaves. All of the F_1 offspring had green pods and normal leaves. The F_1

plants were then crossed to plants with yellow pods and curling leaves. The following results were obtained:

117 green pods, normal leaves

115 yellow pods, curling leaves

78 green pods, curling leaves

80 yellow pods, normal leaves

A. Conduct a chi square test to determine if these two genes are linked.

B. If they are linked, calculate the map distance between the two genes. How accurate do you think this calculated distance is?

E14. In mice, the gene that encodes the enzyme inosine triphosphatase is 12 mu from the gene that encodes the enzyme ornithine decarboxylase. Suppose you have identified a strain of mice homozygous for a defective inosine triphosphatase allele that does not produce any of this enzyme and also homozygous for a defective ornithine decarboxylase allele. In other words, this strain of mice cannot make either enzyme. You cross this homozygous recessive strain to a normal strain of mice to produce heterozygotes. The heterozygotes are then crossed to the strain that cannot produce either enzyme. What is the probability of obtaining a mouse that cannot make either enzyme?

E15. In the garden pea, several different genes affect pod characteristics. A gene affecting pod color (green is dominant to yellow) is approximately 7 mu away from a gene affecting pod width (wide is dominant to narrow). Both genes are located on chromosome 5. A third gene, located on chromosome 4, affects pod length (long is dominant to short). A true-breeding wild-type plant (green, wide, long pods) was crossed to a plant with yellow, narrow, short pods. The F_1 offspring were then testcrossed to plants with yellow, narrow, short pods. If the testcross produced 800 offspring, what are the expected numbers of the eight possible phenotypes?

E16. A sex-influenced trait is dominant in males of a certain species of animal and causes bushy tails. The same trait is recessive in females. Fur color is not sex-influenced. Yellow fur is dominant to white fur. A true-breeding female with a bushy tail and yellow fur was crossed to a white male without a bushy tail (i.e., a normal tail). The F_1 females were then crossed to white males without bushy tails. The following results were obtained:

Males	Females
28 normal tails, yellow	102 normal tails, yellow
72 normal tails, white	96 normal tails, white
68 bushy tails, yellow	0 bushy tails, yellow
29 bushy tails, white	0 bushy tails, white

A. Conduct a chi square test to determine if these two genes are linked.

B. If the genes are linked, calculate the map distance between them. Explain which data you used in your calculation.

E17. Three recessive traits in garden pea plants are as follows: yellow pods are recessive to green pods; bluish green seedlings are recessive to green seedlings; creeper (a plant that cannot stand up) is recessive to normal. A true-breeding normal plant with green pods and green seedlings was crossed to a creeper with yellow pods and bluish green seedlings. The F_1 plants were then crossed to creepers with yellow pods and bluish green seedlings. The following results were obtained:

2059 green pods, green seedlings, normal

151 green pods, green seedlings, creeper

281 green pods, bluish green seedlings, normal

15 green pods, bluish green seedlings, creeper

2041 yellow pods, bluish green seedlings, creeper

157 yellow pods, bluish green seedlings, normal

282 yellow pods, green seedlings, creeper

11 yellow pods, green seedlings, normal

Construct a genetic map that indicates the map distances between these three genes.

E18. Three autosomal genes are linked along the same chromosome. The distance between gene A and gene B is 7 mu, the distance between B and C is 11 mu, and the distance between A and C is 4 mu. An individual that is $AA\ bb\ CC$ was crossed to an individual that is $aa\ BB\ cc$ to produce heterozygous F_1 offspring. The F_1 offspring were then crossed to homozygous $aa\ bb\ cc$ individuals to produce F_2 offspring.

A. Draw the arrangement of the alleles on the chromosomes in the parents and in the F_1 offspring.

B. Where would a crossover have to occur to produce an F_2 offspring that was heterozygous for all three genes?

C. If we assume that no double crossovers occur, what percentage of F_2 offspring is likely to be homozygous for all three genes?

E19. Let's suppose that two different X-linked genes exist in mice, designated with the letters N and L. Gene N exists in a dominant, normal allele and in a recessive allele, n, that is lethal. Similarly, gene L exists in a dominant, normal allele and in a recessive allele, l, that is lethal. Heterozygous females are normal, but males that carry either recessive allele are born dead. Explain whether or not it would be possible to map the distance between these two genes by making crosses and analyzing the number of living and dead offspring. You may assume that you have strains of mice in which females are heterozygous for one or both genes.

Questions for Student Discussion/Collaboration

1. In mice, a dominant allele that causes a short tail is located on chromosome 2. On chromosome 3, a recessive allele causing droopy ears is 6 mu away from another recessive allele that causes a flaky tail. A recessive allele that causes a jerker (uncoordinated) phenotype is located on chromosome 4. A jerker mouse with droopy ears and a short, flaky tail was crossed to a normal mouse. All F_1 generation mice were phenotypically normal, except they had short tails. These F_1 mice were then testcrossed to jerker mice with droopy ears and long, flaky tails. If this testcross produced 400 offspring, what are the expected numbers of the 16 possible phenotypic categories?

2. In Chapter 4, we discussed the observation that the X and Y chromosomes have some genes in common. These genes are inherited in a pseudoautosomal pattern. With this phenomenon in mind, discuss whether or not the X and Y chromosomes are really distinct linkage groups.

3. Mendel studied seven traits in pea plants, and the garden pea happens to have seven different chromosomes. It has been pointed out that Mendel was very lucky not to have conducted crosses involving two traits governed by genes that are closely linked on the same chromosome because the results would have confounded his law of independent assortment. It has even been suggested that Mendel may not have published data involving traits that were linked! An article by Stig Blixt ("Why Didn't Gregor Mendel Find Linkage?" *Nature* 256:206, 1975) considers this issue. Look up this article and discuss why Mendel did not find linkage.

Answers to Comprehension Questions

7.1: b, d

7.2: a, a, b

7.3: d, d, b

7.4: b

Note: All answers appear at the website in Connect; the answers to even-numbered questions and all Concept Check questions are in Appendix B.

8

CHAPTER OUTLINE

- 8.1 Microscopic Examination of Eukaryotic Chromosomes
- 8.2 Changes in Chromosome Structure: An Overview
- 8.3 Deletions and Duplications
- 8.4 Inversions and Translocations
- 8.5 Changes in Chromosome Number: An Overview
- 8.6 Variation in the Number of Chromosomes Within a Set: Aneuploidy
- 8.7 Variation in the Number of Sets of Chromosomes
- 8.8 Mechanisms That Produce Variation in Chromosome Number

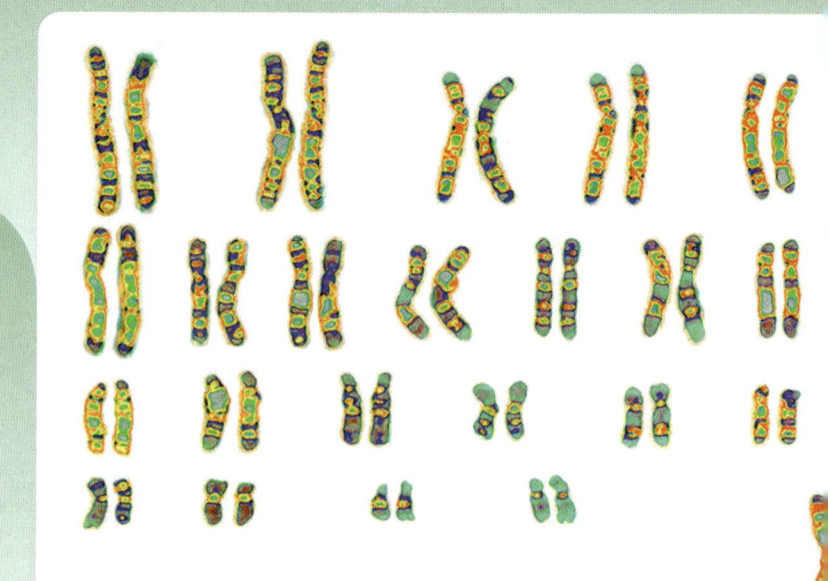

The chromosome composition of humans. Somatic cells in humans contain 46 chromosomes, which come in 23 pairs.

©BSIP/Phototake

VARIATION IN CHROMOSOME STRUCTURE AND NUMBER

The term **genetic variation** refers to genetic differences among members of the same species or among different species. Throughout Chapters 2 through 7, we focused primarily on variation in specific genes, which is called **allelic variation.** In this chapter, our emphasis will shift to larger types of genetic changes that affect the structure or number of eukaryotic chromosomes. These larger alterations may affect the expression of many genes, thereby influencing phenotypes. Variations in chromosome structure and number are of great importance in the field of genetics because they are critical in the evolution of new species and have widespread medical relevance. In addition, agricultural geneticists have discovered that such variations can lead to the development of new strains of crops, which may be quite profitable.

We begin this chapter by exploring how the structure of a eukaryotic chromosome can be modified, either by altering the total amount of genetic material or by rearranging the order of genes along a chromosome. Such changes may often be detected microscopically. The rest of the chapter is concerned with changes in the total number of chromosomes. We will explore how variation in chromosome number occurs and consider examples in which it has significant phenotypic consequences. We will also examine how changes in chromosome number can be induced through experimental treatments and how these approaches have applications in agriculture.

8.1 MICROSCOPIC EXAMINATION OF EUKARYOTIC CHROMOSOMES

Learning Outcome:

1. Describe the characteristics that are used to classify and identify chromosomes.

To identify changes in chromosome structure and number, researchers need to start with a reference point: the chromosomal composition of most members of a given species. For example, most people have two sets of chromosomes with 23 specific chromosomes in each set, resulting in a total of 46 chromosomes. On relatively rare occasions, however, a person may have a chromosomal composition different from that of most other people. Such a person may have a chromosome that has an unusual structure or may have too few or too many chromosomes. **Cytogeneticists**—scientists who study chromosomes microscopically—examine the chromosomes from many members of a given species to determine the common chromosomal composition and to identify rare individuals that show variation in chromosome structure and/or number. In addition, cytogeneticists may be interested in analyzing the chromosomal compositions of two or more species to see how they are similar and how they are different.

Because chromosomes are more compact and microscopically visible during cell division, cytogeneticists usually analyze chromosomes in actively dividing cells. **Figure 8.1a** shows micrographs of chromosomes from individuals of three species: a human, a fruit fly, and a corn plant. As seen here, a cell from a human has 46 chromosomes (23 pairs), that from a fruit fly has 8 chromosomes (4 pairs), and one from corn has 20 chromosomes (10 pairs). Except for the sex chromosomes, which differ between males and females, most members of the same species have very similar chromosomes. By comparison, the chromosomal compositions of distantly related species, such as humans and fruit flies, may be very different. A total of 46 chromosomes is the usual number for humans, whereas 8 chromosomes is the norm for fruit flies.

Cytogeneticists have various ways to classify and identify chromosomes. The three most commonly used features are location of the centromere, size, and banding patterns that are revealed when the chromosomes are treated with stains. As shown in **Figure 8.1b**, chromosomes are classified with regard to centromeric location as follows:

- **metacentric,** the centromere is near the middle;
- **submetacentric,** the centromere is slightly off center;
- **acrocentric,** the centromere is significantly off center but not at the end; or
- **telocentric,** the centromere is at one end.

Because the centromere is never exactly in the center of a chromosome, each chromosome has a short arm and a long arm. For human chromosomes, the short arm is designated with the letter p (for the French, petite), and the long arm is designated with the letter q. In the case of telocentric chromosomes, the short arm may be nearly nonexistent.

A **karyotype** is a micrograph in which all of the chromosomes within a single cell have been arranged in a standard fashion. **Figure 8.1c** shows a human karyotype. The procedure for making a karyotype is described in Chapter 2 (see Figure 2.2a). As seen in Figure 8.1c, the chromosomes are aligned with the short arms on top and the long arms on the bottom. By convention, the chromosomes are numbered according to their size, with the largest chromosomes having the smallest numbers. For example, human chromosomes 1, 2, and 3 are relatively large, whereas 21 and 22 are the two smallest. An exception to the numbering system involves the sex chromosomes, which are designated with letters (for humans, X and Y).

Because different chromosomes often have similar sizes and centromeric locations (e.g., compare human chromosomes 8, 9, and 10), geneticists must use additional methods to accurately identify each type of chromosome within a karyotype. For detailed identification, chromosomes are treated with stains to produce characteristic banding patterns. Several different staining procedures are used by cytogeneticists to identify specific chromosomes. An example is the procedure that produces **G bands,** as shown in Figure 8.1c. In this procedure, chromosomes are treated with mild heat or with proteolytic enzymes that partially digest chromosomal proteins. When exposed to the dye called Giemsa, named after its inventor Gustav Giemsa, some chromosomal regions bind the dye heavily and produce dark bands. In other regions, the stain hardly binds at all and light bands result. Though the mechanism of staining is not completely understood, the dark bands are thought to represent regions that are more tightly compacted. As shown in Figure 8.1c and d, the alternating pattern of G bands is a unique feature for each chromosome.

In the case of human chromosomes, approximately 300 G bands can usually be distinguished during metaphase. A larger number of G bands (in the range of 800) can be observed in prometaphase because the chromosomes are less compacted than metaphase chromosomes. **Figure 8.1d** shows the conventional numbering system that is used to designate G bands along a set of human chromosomes. The left chromatid in each pair of sister chromatids shows the expected banding pattern during metaphase, and the right chromatid shows the banding pattern as it appears during prometaphase.

Why is the banding pattern of eukaryotic chromosomes useful?

- When stained, individual chromosomes can be distinguished from each other, even if they have similar sizes and centromeric locations. For example, compare the differences in banding patterns between human chromosomes 8 and 9 (Figure 8.1d). These differences permit us to distinguish these two chromosomes even though their sizes and centromeric locations are very similar.
- Banding patterns are used to detect changes in chromosome structure. Chromosomal rearrangements or changes in the total amount of genetic material are more easily detected in banded chromosomes.
- Chromosome banding is used to assess evolutionary relationships between species. Research studies have shown that the similarity of chromosome banding patterns is a good measure of genetic relatedness.

8.1 REVIEWING THE KEY CONCEPTS

- Among members of the same species and different species, natural variation exists with regard to chromosome structure and number. Three features of chromosomes that aid in their identification are centromeric location, size, and banding pattern (see Figure 8.1).

158   CHAPTER 8 :: VARIATION IN CHROMOSOME STRUCTURE AND NUMBER

(a) Micrographs of metaphase chromosomes — Human, Fruit fly, Corn

(b) A comparison of centromeric locations — Metacentric, Submetacentric, Acrocentric, Telocentric

(c) Giemsa staining of human chromosomes

(d) Conventional numbering system of G bands in human chromosomes

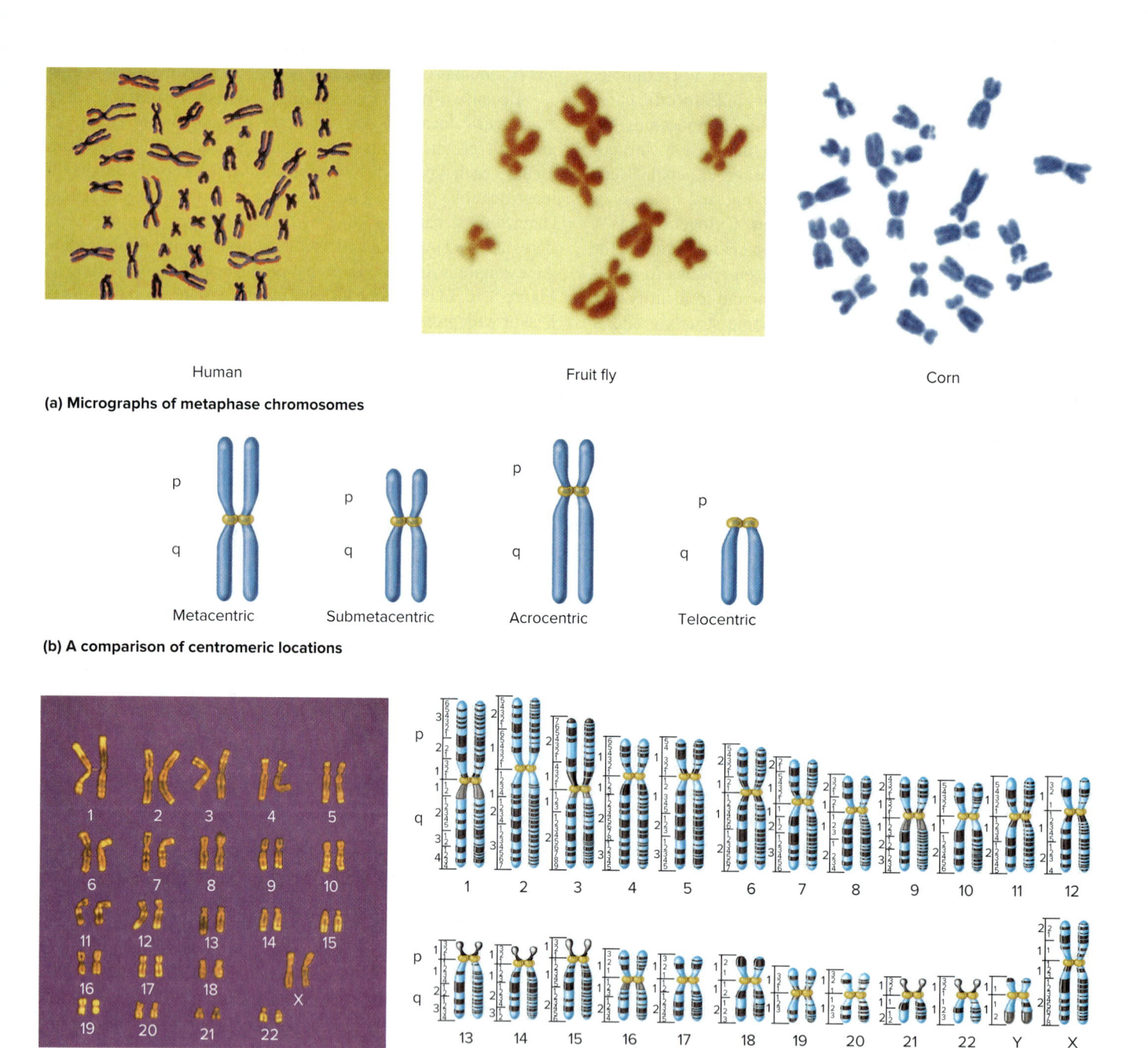

FIGURE 8.1 **Features of chromosomes under the microscope.** (a) Micrographs of chromosomes from a human, a fruit fly, and corn. (b) A comparison of centromeric locations. Centromeres can be metacentric, submetacentric, acrocentric (near one end), or telocentric (at the end). (c) Human chromosomes that have been treated with Giemsa stain. The micrograph has been colorized so that the banding patterns can be more easily discerned. (d) The conventional numbering of bands in Giemsa-stained human chromosomes. Each chromosome is divided into broad regions, which then are subdivided into smaller regions. The numbers increase as regions get farther away from the centromere. For example, if you take a look at the left chromatid of chromosome 1, the uppermost dark band is at a location designated p35. The banding patterns of chromatids change as the chromatids condense. The left chromatid of each pair of sister chromatids shows the banding pattern of a chromatid in metaphase, and the right chromatid shows the banding pattern as it would appear in prometaphase. Note: In prometaphase, the chromatids are less compacted than in metaphase.

a: (1) ©Scott Camazine/Science Source; (2) ©Michael Abbey/Science Source; (3) ©Carlos R Carvalho/Universidade Federal de Viçosa; c: ©C.N.R.I./Phototake

Concept Check: Why is it useful to stain chromosomes?

8.1 COMPREHENSION QUESTIONS

1. A chromosome that is metacentric has its centromere
   a. at the very tip.
   b. near one end, but not at the very tip.
   c. near the middle.
   d. at two distinct locations.
2. Staining eukaryotic chromosomes is useful because it makes it possible to
   a. distinguish chromosomes that are similar in size and centromeric locations.
   b. identify changes in chromosome structure.
   c. explore evolutionary relationships among different species.
   d. do all of the above.

8.2 CHANGES IN CHROMOSOME STRUCTURE: AN OVERVIEW

Learning Outcome:
1. Compare and contrast the four types of changes in chromosome structure.

Now that we understand that chromosomes typically come in a variety of shapes and sizes, let's consider how the structures of normal chromosomes can be modified by mutation. In some cases, the total amount of genetic material within a single chromosome can be increased or decreased significantly. Alternatively, the genetic material in one or more chromosomes may be rearranged without affecting the total amount of material. As shown in **Figure 8.2**, these mutations are categorized as deletions, duplications, inversions, and translocations.

- When a **deletion** occurs, a segment of chromosomal material is missing. In other words, the affected chromosome is deficient in a significant amount of genetic material. The term **deficiency** is also used to describe a missing region of a chromosome.
- In a **duplication,** a section of a chromosome is repeated more than once within the chromosome.
- An **inversion** involves a change in the direction of the genetic material along a single chromosome. For example, in Figure 8.2c, a segment of one chromosome has been inverted, so the order of four G bands is opposite to what it was originally.
- A **translocation** occurs when one segment of a chromosome becomes attached to a different chromosome or to a different part of the same chromosome. A **simple translocation** occurs when a single piece of chromosome is attached to another chromosome. In a **reciprocal translocation,** two different chromosomes exchange pieces, thereby altering both of them.

Deletions and duplications are changes in the total amount of genetic material within a single chromosome. Inversions and translocations are chromosomal rearrangements.

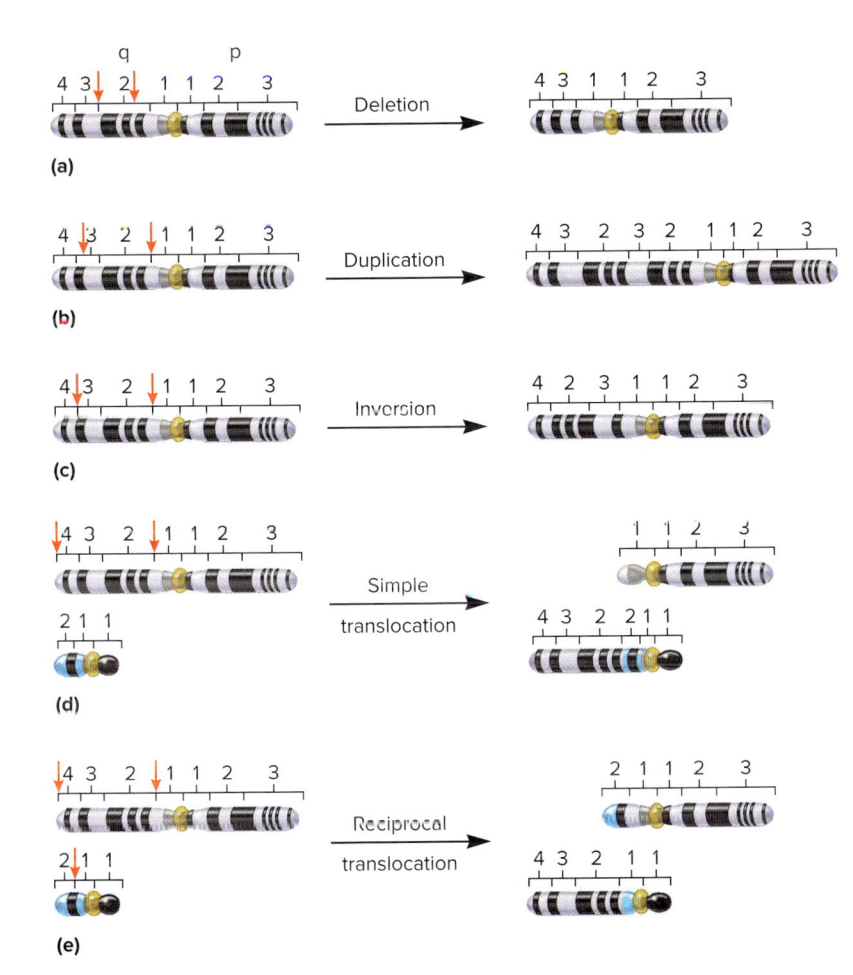

FIGURE 8.2 Types of changes in chromosome structure. The large chromosome shown throughout is human chromosome 1. The smaller chromosome seen in (d) and (e) is human chromosome 21. The red arrows indicate the ends of the affected portion. (**a**) A deletion removes a large portion of the q2 region. (**b**) A duplication doubles the q2-q3 region. (**c**) An inversion inverts the q2-q3 region. (**d**) The q2-q4 region of chromosome 1 is translocated to chromosome 21. A region of a chromosome cannot be attached directly to the tip of another chromosome because telomeres at the tips of chromosomes prevent such an event. In this example, a small piece at the end of chromosome 21 must be removed for the q2-q4 region of chromosome 1 to be attached to chromosome 21. (**e**) The q2-q4 region of chromosome 1 is exchanged with most of the q1-q2 region of chromosome 21.

Concept Check: Which of these changes in chromosome structure alter the total amount of genetic material?

8.2 REVIEWING THE KEY CONCEPTS

- Variations in chromosome structure include deletions, duplications, inversions, and translocations (see Figure 8.2).

8.2 COMPREHENSION QUESTION

1. A change in chromosome structure that does *not* involve a change in the total amount of genetic material is
   a. a deletion.
   b. a duplication.
   c. an inversion.
   d. none of the above.

8.3 DELETIONS AND DUPLICATIONS

Learning Outcomes:
1. Explain how deletions and duplications occur.
2. Describe how deletions and duplications may affect the phenotype of an organism.
3. Define *copy number variation*.

As we have seen, deletions and duplications alter the total amount of genetic material within a chromosome. In this section, we will examine how these changes occur, how they are detected experimentally, and how they affect the phenotypes of the individuals who inherit them.

Loss of Genetic Material Due to a Deletion Tends to Be Detrimental to an Organism

A chromosomal deletion occurs when a chromosome breaks in one or more places and a fragment of the chromosome is lost. In **Figure 8.3a**, a normal chromosome has broken into two separate pieces. The piece without the centromere is missing in future daughter cells because it usually does not find its way into the nucleus following mitosis and is degraded in the cytosol. The other piece is a chromosome with a **terminal deletion.** In **Figure 8.3b**, a chromosome has broken in two places to produce three chromosomal fragments. The central fragment is lost, and the two outer pieces reattach to each other. This process creates a chromosome with an **interstitial deletion.** Deletions can also be created when recombination takes place at incorrect locations between two homologous chromosomes. The products of this type of aberrant recombination event are one chromosome with a deletion and another chromosome with a duplication. This process is examined later in this section.

The phenotypic consequences of a chromosomal deletion depend on the size of the deletion and whether it includes genes or portions of genes that are vital to the development of the organism. When deletions have a phenotypic effect, it is usually detrimental. Larger deletions tend to be more harmful because more genes are missing. Many examples are known in which deletions affect phenotype. For example, a human genetic disease known as cri-du-chat syndrome is caused by a deletion in a segment of the short arm of human chromosome 5 (**Figure 8.4a**). Individuals who carry a single copy of this abnormal chromosome along with a normal chromosome 5 display an array of abnormalities, including mental deficiencies, unique facial anomalies, and, in infancy, an unusual catlike cry, which is the meaning of the French name for the syndrome (**Figure 8.4b**).

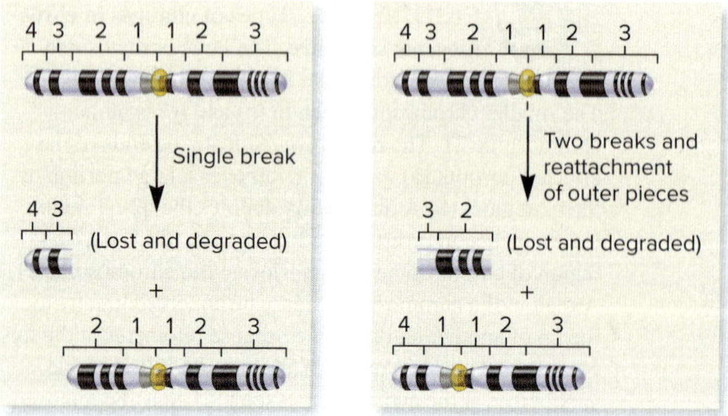

(a) Terminal deletion

(b) Interstitial deletion

FIGURE 8.3 Production of terminal and interstitial deletions. This illustration shows the production of deletions in human chromosome 1.

Concept Check: What is the reason why a chromosomal fragment is eventually lost and degraded?

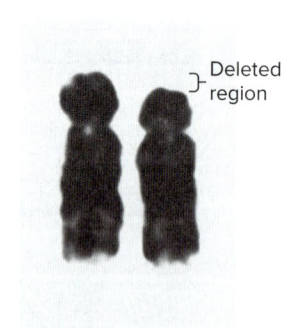

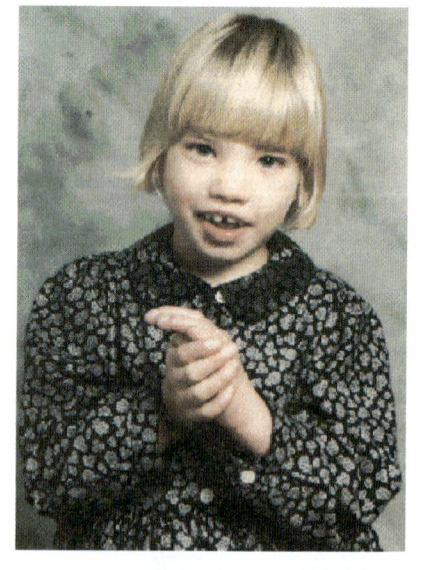

(a) Chromosome 5

(b) A child with cri-du-chat syndrome

FIGURE 8.4 Cri-du-chat syndrome. (a) Chromosome 5 from the karyotype of an individual with this disorder. A section of the short arm of chromosome 5 is missing. (b) An affected individual.

Genes → Traits Compared with an individual who has two copies of each gene on chromosome 5, an individual with cri-du-chat syndrome has only one copy of the genes that are located within the missing segment. This genetic imbalance (one versus two copies of many genes on chromosome 5) causes the phenotypic characteristics of this disorder, which include a catlike cry in infancy, short stature, characteristic facial anomalies (e.g., a triangular face, almond-shaped eyes, broad nasal bridge, and low-set ears), and microencephaly (a smaller-than-normal brain).

(a) ©Biophoto Associates/Science Source; (b) ©Jeff Noneley

Duplications Tend to Be Less Harmful Than Deletions

Duplications result in extra genetic material. They may be caused by abnormal crossover events. During meiosis, crossing over usually occurs after homologous chromosomes have properly aligned with each other. On rare occasions, however, a crossover may occur at misaligned sites on homologs (**Figure 8.5**). What causes the misalignment? In some cases, a chromosome may carry two or more homologous segments of DNA that have identical or similar sequences. These are called **repetitive sequences** because they occur multiple times. An example of a repetitive sequence is a transposable element (see Chapter 12). In Figure 8.5, the repetitive sequence on the right (in the upper chromatid) has lined up with the repetitive sequence on the left (in the lower chromatid). A crossover then occurs. This is called **nonallelic homologous recombination** because it has occurred at homologous sites (such as repetitive sequences), but the sites are not alleles of the same gene. The result is that one chromatid has an internal duplication and another chromatid has a deletion. In Figure 8.5, the chromosome with the extra genetic material carries a **gene duplication**, because the number of copies of gene C has been increased from one to two. In most cases, gene duplications happen as rare, sporadic events during the evolution of species. Shortly, we will consider how multiple copies of genes can evolve into a family of genes with specialized functions.

As with deletions, the phenotypic consequences of duplications tend to be correlated with size. Duplications are more likely to have phenotypic effects if they involve a large piece of a chromosome. In general, small duplications are less likely to have harmful effects than are deletions of comparable size. This observation suggests that having only one copy of a gene is more harmful than having three copies. In humans, relatively few well-defined syndromes are caused by small chromosomal duplications. An example is Charcot-Marie-Tooth disease (type 1A), a peripheral neuropathy characterized by numbness in the hands and feet that is caused by a small duplication on the short arm of chromosome 17.

Duplications Provide Additional Material for Gene Evolution, Sometimes Leading to the Formation of Gene Families

In contrast to the gene duplication that causes Charcot-Marie-Tooth disease, the majority of small chromosomal duplications have no phenotypic effect. Nevertheless, they are vitally important because they provide raw material for the addition of more genes into a species' chromosomes. Over the course of many generations, this can lead to the formation of a **gene family** consisting of two or more genes that are similar to each other. As shown in **Figure 8.6**, the members of a gene family are derived from the same ancestral gene. Over time, two copies of an ancestral gene can accumulate different

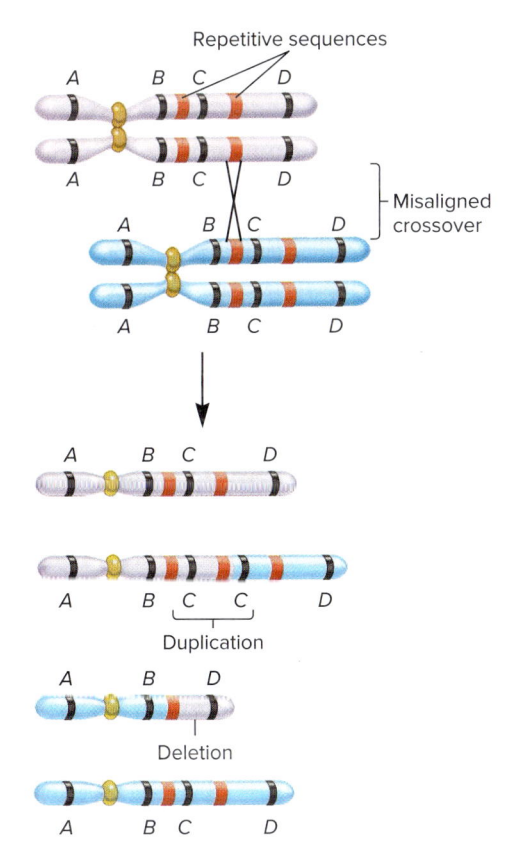

FIGURE 8.5 **Nonallelic homologous recombination, leading to a duplication and a deletion.** Repetitive sequences (shown in red) have promoted the misalignment of homologous chromosomes. A crossover has occurred at sites between genes C and D in one chromatid and between genes B and C in another chromatid. After crossing over is completed, one chromatid contains a duplication, and the other contains a deletion.

Concept Check: In this example, what is the underlying cause of nonallelic homologous recombination?

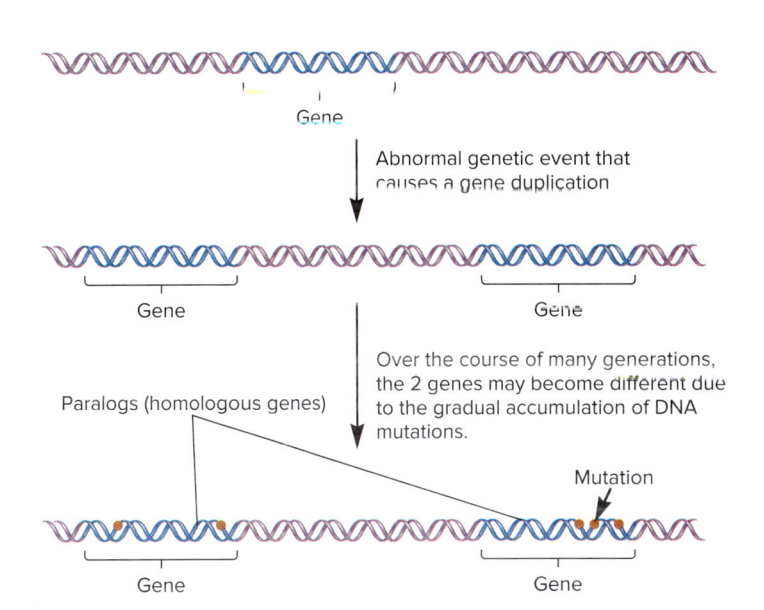

FIGURE 8.6 **Gene duplication and the evolution of paralogs.** An abnormal crossover event like the one described in Figure 8.5 leads to a gene duplication. Over time, each gene accumulates different mutations.

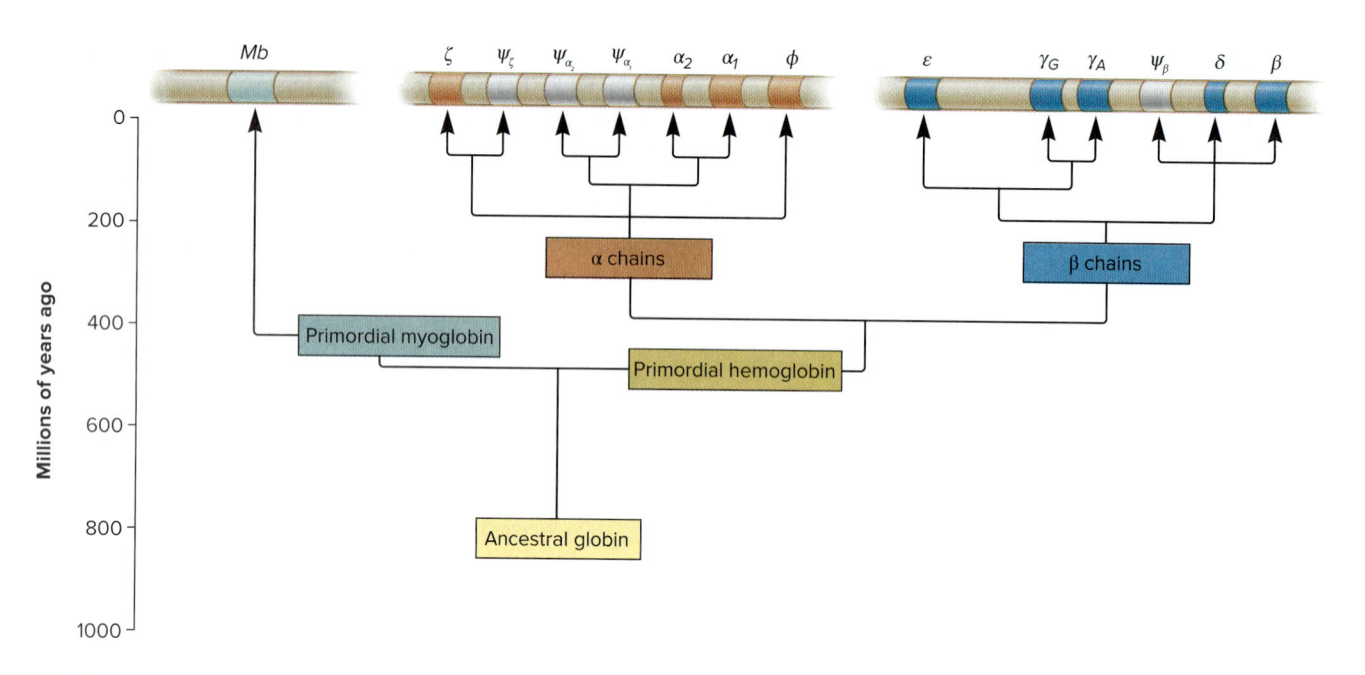

FIGURE 8.7 **The evolution of the globin gene family in humans.** The globin gene family evolved from a single ancestral globin gene. The first gene duplication produced two genes that accumulated mutations and became the genes encoding myoglobin (on chromosome 22) and the group of hemoglobins. The primordial hemoglobin gene then duplicated to produce several α-chain and β-chain genes, which are found on chromosomes 16 and 11, respectively. The four genes shown in gray are nonfunctional pseudogenes.

mutations. Therefore, after many generations, the two genes are similar but not identical. During evolution, this type of event can occur several times, creating a family of many similar genes.

When two or more genes are derived from a single ancestral gene, the genes are said to be **homologous.** Homologous genes within a single species are called **paralogs** and constitute a gene family. A well-studied example of a gene family is shown in **Figure 8.7**, which illustrates the evolution of the globin gene family found in humans. The globin genes encode polypeptides that are subunits of proteins that function in oxygen binding. For example, hemoglobin is a protein found in red blood cells; its function is to carry oxygen throughout the body. The globin gene family is composed of 14 paralogs that were originally derived from a single ancestral globin gene. According to an evolutionary analysis, the ancestral globin gene first duplicated about 500 million years ago and became separate genes encoding myoglobin and the hemoglobin group of genes. The primordial hemoglobin gene duplicated into an α-chain gene and a β-chain gene, which subsequently duplicated to produce several genes located on chromosomes 16 and 11, respectively. Currently, 14 globin genes are found on three different human chromosomes.

Why is it advantageous to have a family of globin genes? Although all globin polypeptides are subunits of proteins that play a role in oxygen binding, the accumulation of different mutations in the various family members has produced globins that are more specialized in their function. For example, myoglobin is better at binding and storing oxygen in muscle cells, and the hemoglobins are better at binding and transporting oxygen via the red blood cells. Also, different globin genes are expressed during different stages of human development. The ε-globin and ζ-globin genes are expressed very early in embryonic life, whereas the α-globin and γ-globin genes are expressed during the second and third trimesters of gestation. Following birth, the α-globin genes remain turned on, but the γ-globin genes are turned off and the β-globin gene is turned on. These differences in the expression of the globin genes reflect the differences in the oxygen transport needs of humans during the embryonic, fetal, and postpartum stages of life.

Copy Number Variation Is Relatively Common Among Members of the Same Species

The term **copy number variation (CNV)** refers to a type of structural variation in which a segment of DNA that is 1000 bp or more in length exhibits copy number differences among members of the same species. Genes that were assumed to always occur in two copies in a diploid cell have now been found to be present in one, three, or more than three copies. One possibility is that some members of a species may carry a chromosome that is missing a particular gene or part of a gene. Alternatively, CNV may be due to a duplication. For example, some members of a diploid species may have one copy of gene A on both homologs of a chromosome, and thereby have two copies of the gene (**Figure 8.8a**). By comparison, other members of the same species might have one copy of gene A on a particular chromosome and two copies on its homolog for a total of three copies (**Figure 8.8b**). The homolog with two copies of gene A is said to have undergone a **segmental duplication.**

In the past 10 years, researchers have discovered that copy number variation is relatively common in animal and plant species. Though the analysis of structural variation is a relatively new area of investigation, researchers estimate that 1% to 10% of a genome may show CNV within a typical species of animal or plant.

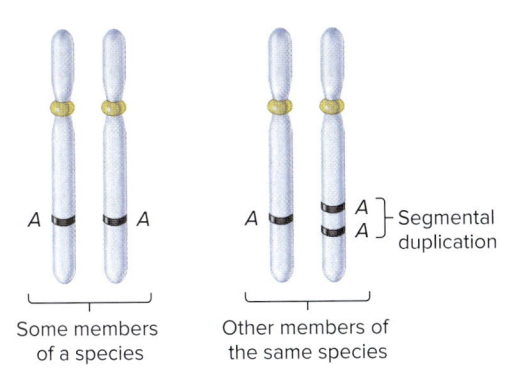

FIGURE 8.8 An example of copy number variation. In part (a), some members of a species have two copies of gene A, whereas in part (b) other members of the species have three copies.

Most CNV is inherited and happened in the past, but CNV may also be caused by new mutations. A variety of mechanisms may bring about copy number variation. One common cause is nonallelic homologous recombination, which was described in Figure 8.5. This type of event can produce a chromosome with a duplication or deletion, thereby altering the copy number of genes. Researchers also speculate that the proliferation of transposable elements, which are described in Chapter 12, may increase the copy number of DNA segments. A third mechanism that underlies CNV may involve errors in DNA replication, which is described in Chapter 13.

What are the phenotypic consequences of CNV? In many cases, CNV has no obvious phenotypic consequences. However, recent medical research is revealing that some CNV in humans is associated with specific diseases. For example, particular types of CNV are associated with schizophrenia, autism, and certain forms of learning disabilities. In addition, CNV may affect susceptibility to infectious diseases. An example is the human *CCL3* gene, which encodes a chemokine protein that is involved in immunity. In human populations, the copy number of this gene varies from one to six. In people infected with HIV (human immunodeficiency virus), copy number variation of *CCL3* may affect the progression of AIDS (acquired immune deficiency syndrome). Individuals with a higher copy number of *CCL3* produce more chemokine protein and often show a slower advancement of AIDS. Finally, another reason why researchers are interested in copy number variation is its possible relationship to cancer, which is discussed in Chapter 22.

Genetic TIPS

The Question: A son and his mother both have an inherited disorder that affects the nervous system. How would you determine if the disorder is caused by a change in chromosome structure, such as a deletion or duplication?

Topic: What topic in genetics does this question address? The topic is how to determine if a genetic disorder is caused by a change in chromosome structure.

Information: What information do you know based on the question and your understanding of the topic? From the question, you know that a son and his mother have an inherited disorder. From your understanding of the topic, you may remember that some disorders are caused by variation in chromosome structure, but some are caused by single gene mutations. You may also recall that karyotyping can detect deletions and duplications.

Problem-Solving Strategy: Design an experiment. Compare and contrast. To solve this problem, you need to design an experiment in which you analyzed the chromosomes in the affected individuals. You also need to examine the chromosomes in unaffected individuals as a control.

Answer:
- Obtain a sample of cells, such as leukocytes, from the son and his mother. As a control, obtain cells from the father and any unaffected siblings.
- Subject the samples to karyotyping, as described in Chapter 2 (Figure 2.2).
- Compare the chromosomes of affected and unaffected individuals.

Expected results: If the disorder is caused by a change in chromosome structure, you would expect to find a change, such as a deletion or duplication, in both the son and his mother, but you would not see the change in the father or any unaffected siblings.

8.3 REVIEWING THE KEY CONCEPTS

- Chromosome breaks can create terminal or interstitial deletions. Some deletions are associated with human genetic disorders, such as cri-du-chat syndrome (see Figures 8.3, 8.4).
- Nonallelic homologous recombination can produce gene duplications and deletions. Over time, gene duplications can lead to the formation of gene families, such as the globin gene family (see Figures 8.5–8.7).
- Copy number variation (CNV) among members of a species is fairly common. In humans, CNV is associated with certain diseases (see Figure 8.8).

8.3 COMPREHENSION QUESTIONS

1. Which of the following statements is correct?
   a. If a deletion and duplication are the same size, the deletion is more likely to be harmful.
   b. If a deletion and duplication are the same size, the duplication is more likely to be harmful.
   c. If a deletion and duplication are the same size, the likelihood of causing harm is about the same.
   d. A deletion is always harmful, whereas a duplication is always beneficial.

2. With regard to gene duplications, which of the following statement(s) is/are correct?
   a. Gene duplications may be caused by nonallelic homologous recombination.
   b. Large gene duplications are more likely to be harmful than smaller ones.
   c. Gene duplications are responsible for creating gene families that encode proteins with similar and specialized functions.
   d. All of the above statements are correct.

8.4 INVERSIONS AND TRANSLOCATIONS

Learning Outcomes:
1. Define *pericentric inversion* and *paracentric inversion*.
2. Diagram the production of abnormal chromosomes due to crossing over in inversion heterozygotes.
3. Explain two mechanisms that result in reciprocal translocations.
4. Describe how reciprocal translocations align during meiosis and how they segregate.

As discussed earlier in this chapter, inversions and translocations are types of chromosomal rearrangements. In this section, we will explore how they occur and how they may affect an individual's phenotype and fertility.

Inversions Often Occur Without Phenotypic Consequences

A chromosome with an inversion contains a segment that has been flipped so that it runs in the opposite direction (**Figure 8.9**). Geneticists classify inversions according to the location of the centromere.

- If the centromere lies within the inverted region of the chromosome, the inverted region is known as a **pericentric inversion** (Figure 8.9b).
- Alternatively, if the centromere is found outside the inverted region, the inverted region is called a **paracentric inversion** (Figure 8.9c).

A B C D E F G H I
(a) Normal chromosome

A B C G F E D H I
Inverted region
(b) Pericentric inversion

A E D C B F G H I
Inverted region
(c) Paracentric inversion

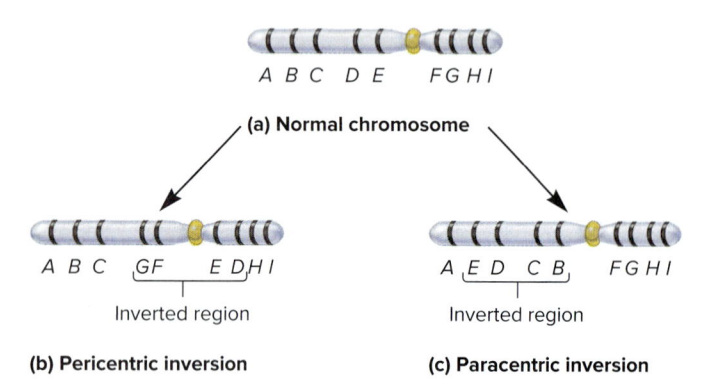

FIGURE 8.9 **Types of inversions.** (a) A normal chromosome with the genes ordered from *A* through *I*. A pericentric inversion (b) includes the centromere, whereas a paracentric inversion (c) does not.

When a chromosome contains an inversion, the total amount of genetic material remains the same as in a normal chromosome. Therefore, the great majority of inversions do not have any phenotypic consequences. In rare cases, however, an inversion can alter the phenotype of an individual. Whether or not this occurs is related to the boundaries of the inverted segment. When an inversion occurs, the chromosome is broken in two places, and the center piece flips around to produce the inversion.

- If either breakpoint occurs within a vital gene, the function of the gene is expected to be disrupted, possibly producing a phenotypic effect. For example, some people with hemophilia (type A) have inherited an X-linked inversion in which a breakpoint has inactivated the gene for factor VIII, a blood-clotting protein.
- In other cases, an inversion (or translocation) may reposition a gene on a chromosome in a way that alters its normal level of expression. This is a type of **position effect**—a change in phenotype that occurs when the position of a gene changes from one chromosomal site to a different location. This topic is also discussed in Chapter 19 (see Figure 19.2).

Because inversions seem like an unusual genetic phenomenon, it is perhaps surprising that they are found in human populations in significant numbers. About 2% of the human population carries inversions that are detectable with a light microscope. In most cases, such individuals show no phenotypic effect and live their lives without knowing they carry an inversion. In a few cases, however, an individual with a chromosomal inversion may produce offspring with phenotypic abnormalities. This event may prompt a physician to request a microscopic examination of the individual's chromosomes. In this way, phenotypically unaffected individuals may discover they have a chromosome with an inversion.

Inversion Heterozygotes May Produce Abnormal Chromosomes Due to Crossing Over

An individual carrying one normal copy of a chromosome and one copy with an inversion is known as an **inversion heterozygote.** Such an individual, though possibly not affected phenotypically, may have a high probability of producing gametes that are abnormal in their total genetic content.

The underlying cause of gamete abnormality is the phenomenon of crossing over within the inverted region. During meiosis I, pairs of homologous sister chromatids synapse with each other. **Figure 8.10** illustrates how this occurs in an inversion heterozygote. In order for the normal chromosome and inversion chromosome to synapse properly, an **inversion loop** must form that permits the homologous genes on both chromosomes to align next to each other despite the inverted region. If a crossover occurs within the inversion loop, highly abnormal chromosomes are produced. A crossover is more likely to occur in this region if the inversion is large. Therefore, individuals carrying large inversions are more likely to produce abnormal gametes.

The consequences of this type of crossover depend on whether the inversion is pericentric or paracentric. Figure 8.10a shows a crossover in the inversion loop when one of the homologs

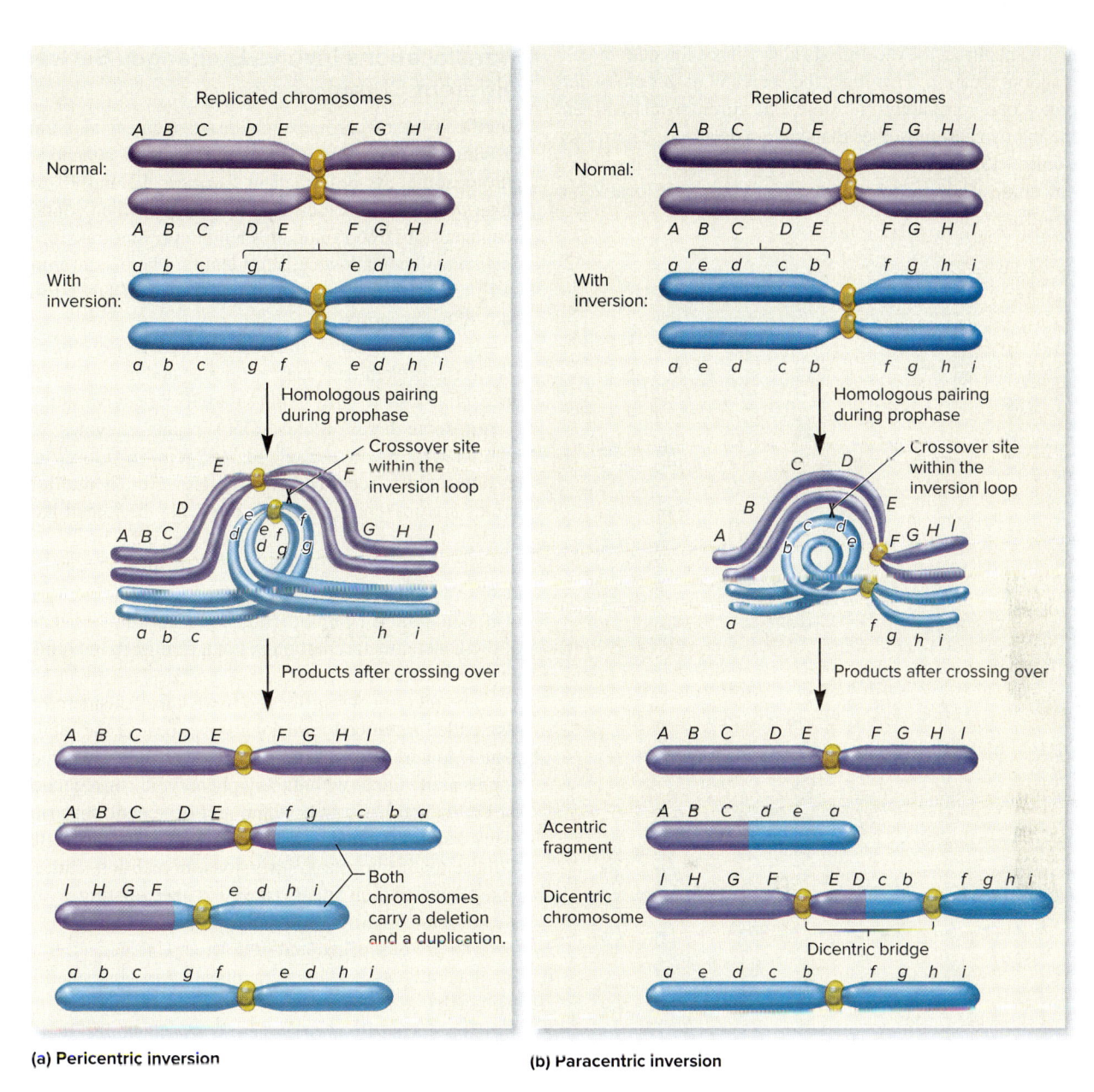

FIGURE 8.10 The consequences of crossing over in the inversion loop. (a) Crossover within a pericentric inversion. (b) Crossover within a paracentric inversion.

Concept Check: Explain why the homologous chromosomes of an inversion heterozygote can synapse only if an inversion loop forms.

has a pericentric inversion, in which the centromere lies within the inverted region of the chromosome. This event consists of a single crossover that involves only two of the four sister chromatids. Following the completion of meiosis, this single crossover yields two chromosomes that have a segment that is deleted and a different segment that is duplicated. In this example, one of the chromosomes is missing genes H and I and has an extra copy of genes A, B, and C. The other chromosome has the opposite situation; it is missing genes A, B, and C and has an extra copy of genes H and I. These abnormal chromosomes may result in gametes that are inviable. Alternatively, if these abnormal chromosomes are passed to offspring, they are likely to produce phenotypic abnormalities, depending on the amount and nature of the duplicated and deleted genetic material. A large deletion is likely to be lethal.

Figure 8.10b shows the outcome of a crossover involving a paracentric inversion, in which the centromere lies outside the inverted region. This single crossover event produces a very strange outcome. One chromosome, called a **dicentric** chromosome, contains two centromeres. The region connecting the two centromeres in such a chromosome is a **dicentric bridge.** The crossover also produces a piece of a chromosome without any centromere—an **acentric fragment**—which is lost and degraded in subsequent cell divisions. The dicentric chromosome is a temporary structure. If the two centromeres try to move toward opposite poles during anaphase, the dicentric bridge will be forced to break at some random location. Therefore, the net result of this crossover is to produce one normal chromosome, one chromosome with an inversion, and two chromosomes that contain deletions. These two chromosomes with deletions result from the breakage of the dicentric chromosome. They are missing the genes that were located on the acentric fragment.

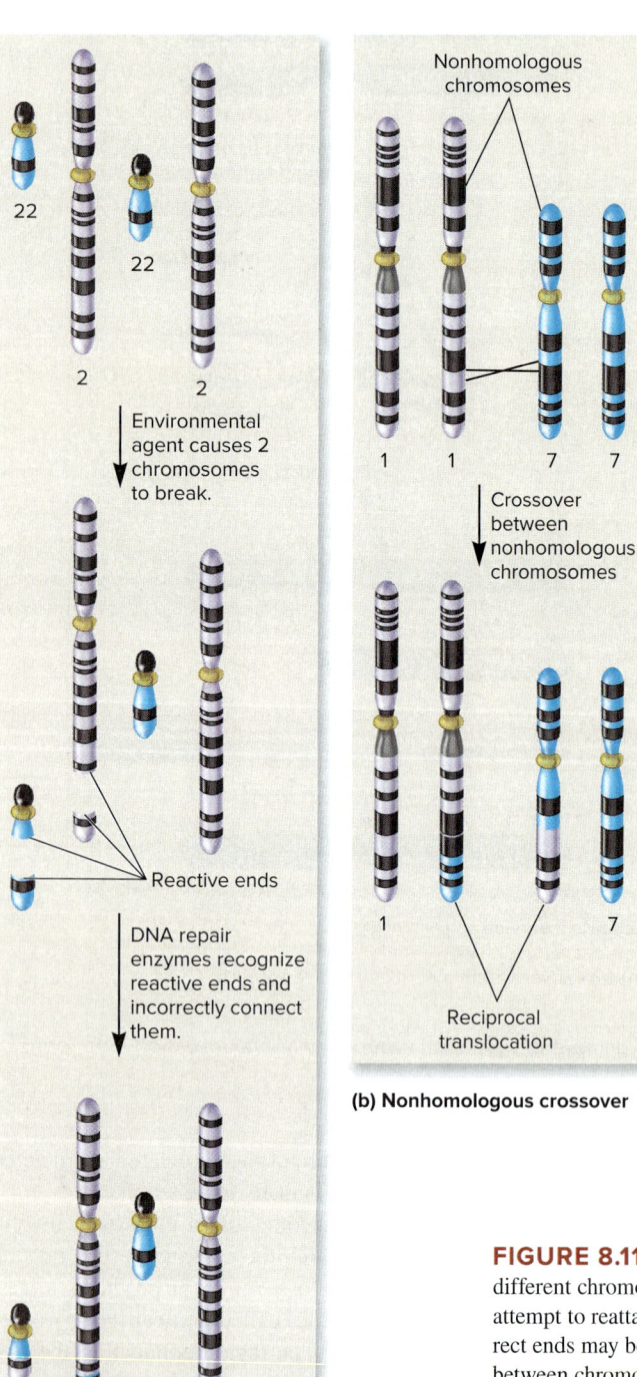

(a) Chromosomal breakage and DNA repair

(b) Nonhomologous crossover

Translocations Involve Exchanges Between Different Chromosomes

Another type of chromosomal rearrangement is a translocation, in which a piece from one chromosome is attached to another chromosome. As described in Chapters 12 and 13, the ends of eukaryotic chromosomes have **telomeres,** which contain repeated sequences of DNA. Telomeres tend to prevent translocations from occurring. They allow cells to identify where a chromosome ends and prevent the attachment of chromosomal DNA to the natural ends of a chromosome.

If cells are exposed to agents that cause chromosomes to break, the broken ends lack telomeres and are said to be reactive. A reactive end readily binds to another reactive end. If a single chromosome break occurs, DNA repair enzymes will usually recognize the two reactive ends and join them back together; the chromosome is repaired properly. However, if multiple chromosomes are broken, the reactive ends may be joined incorrectly to produce a reciprocal translocation (**Figure 8.11a**).

As shown in **Figure 8.11b**, a reciprocal translocation can also be produced when two nonhomologous chromosomes cross over. This type of rare aberrant event results in a rearrangement of the genetic material, though not a change in the total amount of genetic material.

The reciprocal translocations we have considered thus far are also called **balanced translocations** because the total amount of genetic material is not altered. Like inversions, balanced translocations usually occur without any phenotypic consequences because the individual has a normal amount of genetic material. In a few cases, balanced translocations can result in position effects similar to those that can occur in inversions. In addition, carriers of a balanced translocation are at risk of having offspring with an **unbalanced translocation,** in which significant portions of genetic material are duplicated and/or deleted. Unbalanced translocations are generally associated with phenotypic abnormalities or may even be lethal.

The most common type of translocation that occurs in humans is a **Robertsonian translocation,** named after William Robertson, who first described this type of rearrangement in grasshoppers. A Robertsonian translocation arises from breaks near the centromeres of two nonhomologous acrocentric chromosomes. In the example shown in **Figure 8.12**, the short arm of one acrocentric chromosome (chromosome 14) is exchanged with the long arm of a different acrocentric chromosome (chromosome 21). One product is a chromosome that has the long arms from

FIGURE 8.11 Two mechanisms that cause a reciprocal translocation. (a) When two different chromosomes break, the reactive ends are recognized by DNA repair enzymes, which attempt to reattach them. If two different chromosomes are broken at the same time, the incorrect ends may become attached to each other. (b) A nonhomologous crossover has occurred between chromosome 1 and chromosome 7. This crossover yields two chromosomes that carry translocations.

Concept Check: Which of these two mechanisms might be promoted by the presence of the same repetitive sequence, such as a transposable element, in many places in a species' genome?

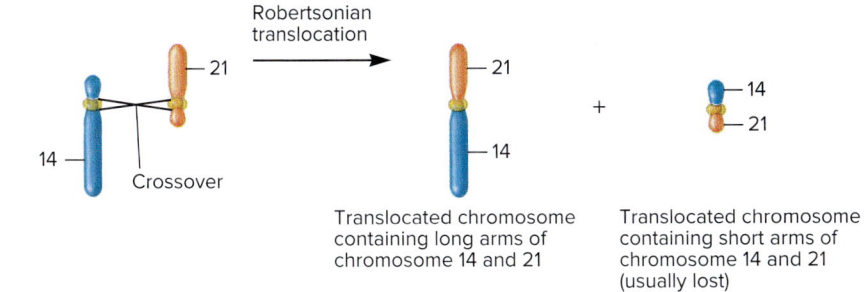

FIGURE 8.12 Robertsonian translocation. In this example, the long arm of chromosome 21 is exchanged with the short arm of chromosome 14.

both chromosomes. The other product is a very small chromosome containing the two short arms of these chromosomes; this small chromosome is usually lost during segregation. In humans, Robertsonian translocations occur at a frequency of approximately 1 in 900 live births. They involve only the acrocentric chromosomes 13, 14, 15, 21, and 22.

Individuals with Reciprocal Translocations May Produce Abnormal Gametes Due to the Pairing and Segregation of Chromosomes

Individuals who carry balanced translocations have a greater risk of producing gametes with unbalanced combinations of chromosomes. Whether or not this occurs depends on the segregation pattern during meiosis I (**Figure 8.13**). In this example, the parent carries a reciprocal translocation and is likely to be unaffected phenotypically. During meiosis, the homologous chromosomes attempt to synapse with each other. Because of the translocations, the pairing of homologous regions leads to the formation of an unusual structure that contains four pairs of sister chromatids (i.e., eight chromatids), termed a **translocation cross**.

To understand the segregation of translocated chromosomes, pay close attention to the centromeres, which are numbered in Figure 8.13. For these translocated chromosomes, the expected segregation pattern is governed by the centromeres. Each haploid gamete should receive one centromere located on chromosome 1 and one centromere located on chromosome 2.

- One possibility is alternate segregation. As shown in Figure 8.13a, this occurs when the chromosomes diagonal to each other within the translocation cross sort into the same cell. One daughter cell receives two normal chromosomes, and the other cell gets two translocated chromosomes. Following meiosis II, four haploid cells are produced: two have normal chromosomes, and two have reciprocal (balanced) translocations.
- Another possible segregation pattern is called adjacent-1 segregation (Figure 8.13b). This occurs when adjacent chromosomes (one with each type of centromere) segregate into the same cell. Following anaphase of meiosis I, each daughter cell receives one normal chromosome and one translocated chromosome. After meiosis II is completed, four haploid cells are produced, all of which are genetically unbalanced, because part of one chromosome has been deleted and part of another has been duplicated. If these haploid cells give rise to gametes that unite with a normal gamete, the zygote is expected to be abnormal genetically and possibly phenotypically.
- On very rare occasions, adjacent-2 segregation can occur (Figure 8.13c). In this case, the centromeres do not segregate as they should. One daughter cell has received both copies of the centromere on chromosome 1; the other, both copies of the centromere on chromosome 2. This rare segregation pattern also yields four abnormal haploid cells that contain an unbalanced combination of chromosomes.

Alternate and adjacent-1 segregation patterns are the likely outcomes when an individual carries a reciprocal translocation. Depending on the sizes of the translocated segments, both types may be equally likely to occur. In many cases, the haploid cells from adjacent-1 segregation are not viable, thereby lowering the fertility of the parent. This condition is called **semisterility**.

8.4 REVIEWING THE KEY CONCEPTS

- Inversions can be pericentric or paracentric. In an inversion heterozygote, crossing over within the inversion loop can create deletions and duplications in the resulting chromosomes (see Figures 8.9, 8.10).
- Two mechanisms can produce translocations: chromosome breakage and subsequent repair, and nonhomologous crossing over (see Figure 8.11).
- A Robertsonian translocation results in a chromosome in which the long arms of two different acrocentric chromosomes become joined (see Figure 8.12).
- Individuals that carry a balanced translocation may have a high probability of producing unbalanced gametes, depending on the segregation pattern during meiosis I (see Figure 8.13).

8.4 COMPREHENSION QUESTIONS

1. A paracentric inversion
   a. includes the centromere within the inverted region.
   b. does not include the centromere within the inverted region.
   c. has two adjacent inverted regions.
   d. is an inverted region at the very end of a chromosome.

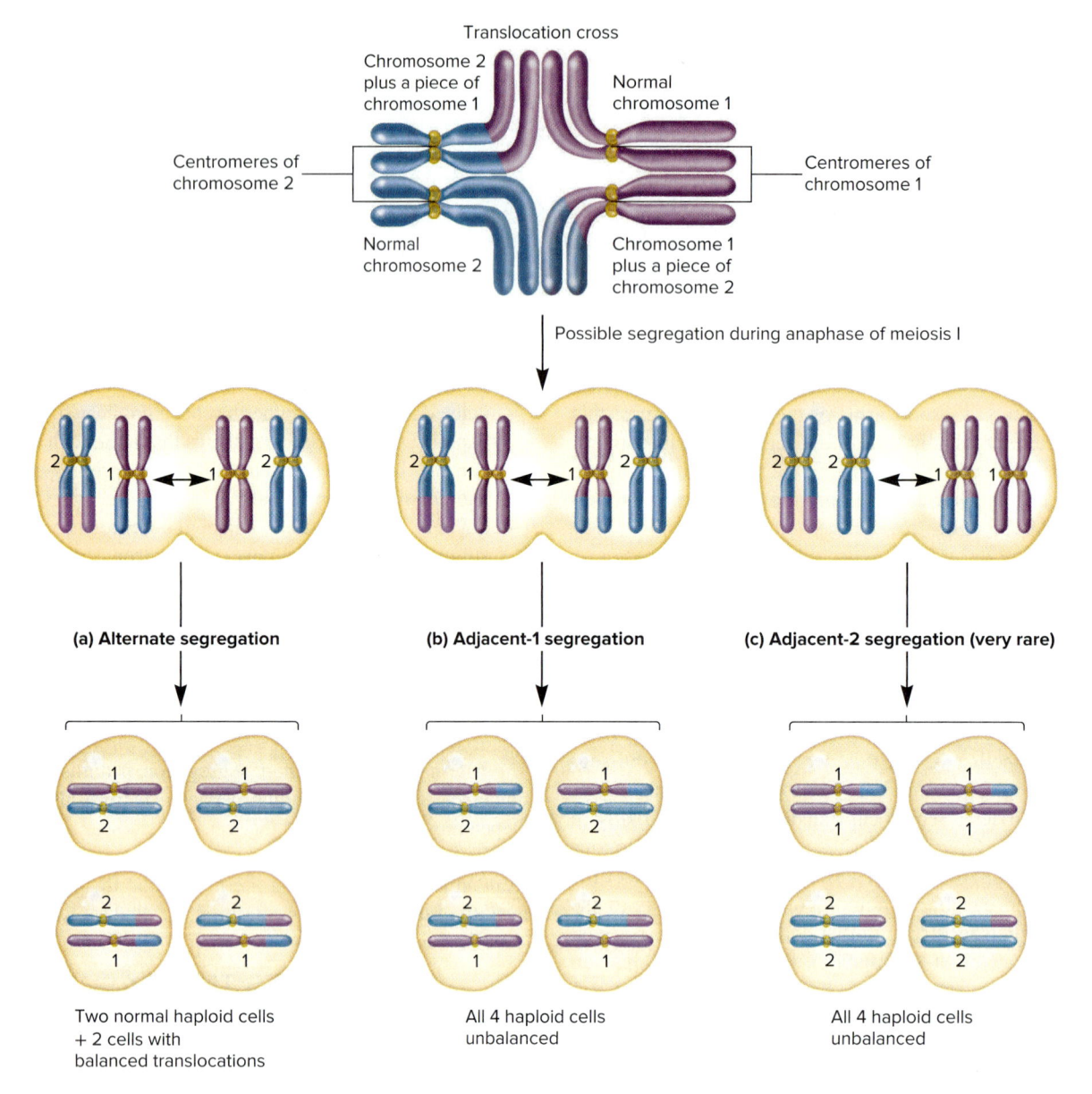

FIGURE 8.13 **Meiotic segregation of chromosomes with a reciprocal translocation.** Follow the numbered centromeres through each process. **(a)** Alternate segregation gives rise to balanced haploid cells, whereas **(b)** adjacent-1 and **(c)** adjacent-2 segregation produce haploid cells with an unbalanced amount of genetic material.

Concept Check: Explain why these chromosomes form a translocation cross during prophase of meiosis I.

2. Due to crossing over within an inversion loop, a heterozygote with a pericentric inversion may produce gametes that carry
   a. a deletion.
   b. a duplication.
   c. a translocation.
   d. both a deletion and a duplication.

3. A mechanism that can produce a translocation is
   a. the joining of reactive ends when two different chromosomes break.
   b. crossing over between nonhomologous chromosomes.
   c. crossing over between homologous chromosomes.
   d. either a or b.

8.5 CHANGES IN CHROMOSOME NUMBER: AN OVERVIEW

Learning Outcomes:
1. Define *euploid* and *aneuploid*.
2. Compare and contrast polyploidy and aneuploidy.

As we saw in previous sections of this chapter, chromosome structure can be altered in a variety of ways. Likewise, the total number of chromosomes can vary. Eukaryotic species typically contain several chromosomes that are inherited as one or more sets. Variations in chromosome number can be categorized in two ways: variation in the number of sets of chromosomes and variation in the number of particular chromosomes within a set.

Organisms that are **euploid** have a chromosome number that is an exact multiple of a chromosome set. In *Drosophila melanogaster*, for example, a normal individual has 8 chromosomes. The species is diploid, having two sets of 4 chromosomes each (**Figure 8.14a**). A normal fruit fly is euploid because 8 chromosomes divided by 4 chromosomes per set equals two exact sets. On rare occasions, an abnormal fruit fly can be produced with 12 chromosomes, having three sets of 4 chromosomes each. This alteration produces a **triploid** fruit fly with 12 chromosomes. Such a fly is also euploid because it has exactly three sets of chromosomes. Organisms with three or more sets of chromosomes are also called **polyploid** (**Figure 8.14b**). Geneticists use the letter n to represent a set of chromosomes. A diploid organism is referred to as $2n$, a triploid organism as $3n$, a **tetraploid** organism as $4n$, and so on.

A second way in which chromosome number can vary occurs when an organism is **aneuploid**. Such variation involves an alteration in the number of particular chromosomes, so the total number of chromosomes is not an exact multiple of a set. For example, an abnormal fruit fly could have 9 chromosomes instead of 8 because it has three copies of chromosome 2 instead of the usual number of two copies (**Figure 8.14c**). Such an animal is said to have trisomy 2 or to be **trisomic**. Instead of being perfectly diploid ($2n$), a trisomic animal is $2n + 1$. By comparison, a fruit fly could be lacking a single chromosome, such as chromosome 1, and contain a total of seven chromosomes ($2n - 1$). This animal is **monosomic** and is described as having monosomy 1.

8.5 REVIEWING THE KEY CONCEPTS

- Variation in the number of chromosomes may involve changes in the number of sets (euploidy) or changes in the number of particular chromosomes within a set (aneuploidy) (see Figure 8.14).

8.5 COMPREHENSION QUESTION

1. Humans have 23 chromosomes per set. A person with 45 chromosomes can be described as being
   a. euploid.
   b. aneuploid.
   c. monoploid.
   d. trisomic.

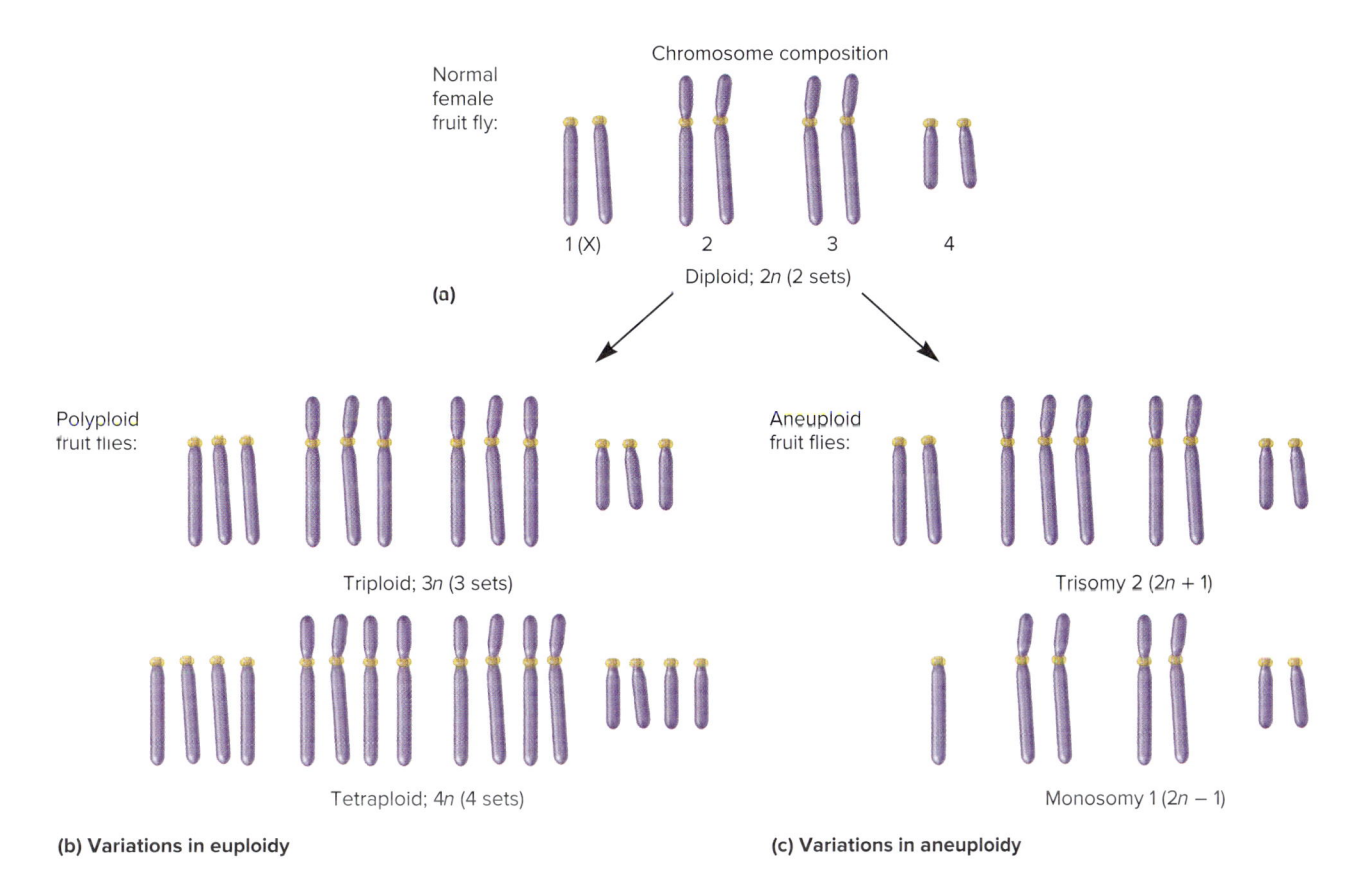

FIGURE 8.14 Types of variation in chromosome number. (a) The normal diploid number of chromosomes in *Drosophila*. (b) Examples of polyploidy. (c) Examples of aneuploidy.

Concept Check: What terms can be used to describe a fruit fly that has a total of 7 chromosomes because it is missing one copy of chromosome 3?

8.6 VARIATION IN THE NUMBER OF CHROMOSOMES WITHIN A SET: ANEUPLOIDY

Learning Outcomes:
1. Explain why aneuploidy usually has a detrimental effect on phenotype.
2. Describe examples of aneuploidy in humans.

In this section, we will consider several examples of aneuploidy. As you will learn, this is generally regarded as an abnormal condition that usually has a negative effect on phenotype.

Aneuploidy Causes an Imbalance in Gene Expression That Is Often Detrimental to the Phenotype of the Individual

The phenotype of every eukaryotic species is influenced by thousands of different genes. In humans, for example, a single set of chromosomes has approximately 22,000 different protein-encoding genes. To produce a phenotypically normal individual, intricate coordination has to occur in the expression of thousands of genes. In the case of humans and other diploid species, evolution has resulted in a developmental process that works correctly when somatic cells have two copies of each chromosome. In other words, when a human is diploid, the balance of gene expression among many different genes usually produces a person with a normal phenotype.

Aneuploidy commonly causes an abnormal phenotype. To understand why, let's consider the relationship between gene expression and chromosome number in a species that has three pairs of chromosomes (**Figure 8.15**). The level of gene expression is influenced by the number of genes per cell. Compared with a diploid cell, if a cell has a chromosome that is present in three copies instead of two, more of the gene product is typically made. For example, a gene present in three copies instead of two may produce 150% of the gene product, though that number may vary due to effects of gene regulation. Alternatively, if only one copy of a gene is present due to a missing chromosome, less of the gene product is usually made, perhaps only 50% as much. Therefore, in trisomic and monosomic individuals, an imbalance occurs between the level of gene expression for the chromosomes found in pairs and the level of gene expression for the chromosomes that are not.

At first glance, the difference in gene expression between euploid and aneuploid individuals may not seem terribly dramatic. Keep in mind, however, that a eukaryotic chromosome carries hundreds or even thousands of different genes. Therefore, when an organism is trisomic or monosomic, many gene products occur in excessive or deficient amounts. This imbalance among many genes appears to underlie the phenotypic abnormalities that aneuploidy frequently causes. In most cases, these effects are detrimental and produce an individual that is less likely to survive than a euploid individual.

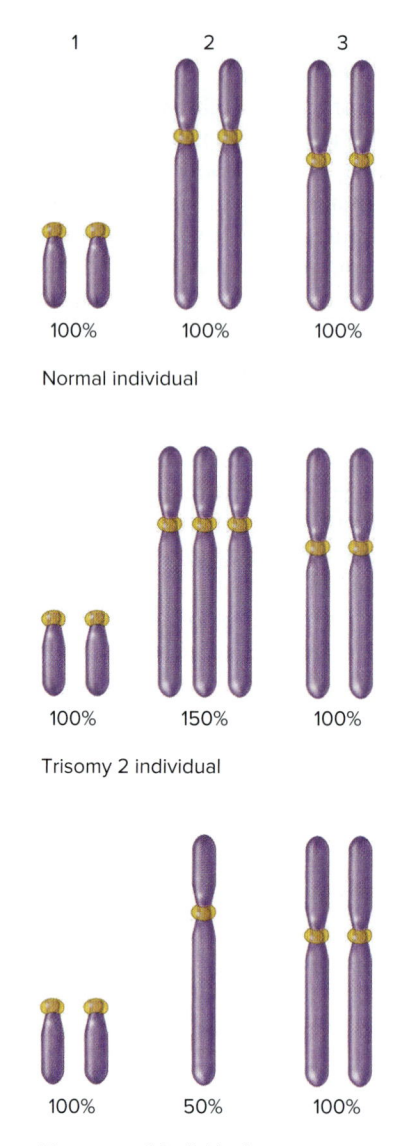

FIGURE 8.15 **Imbalance of gene products in trisomic and monosomic individuals.** Aneuploidy of chromosome 2 (i.e., trisomy and monosomy) leads to an imbalance in the amount of gene products from chromosome 2 compared with the amounts from chromosomes 1 and 3.

Concept Check: Describe the imbalance in gene products that occurs in an individual with monosomy 2.

Aneuploidy in Humans Causes Detrimental Phenotypes

A key reason why geneticists are so interested in aneuploidy is its relationship to certain inherited disorders in humans. Even though most people are born with 46 chromosomes, alterations in chromosome number occur fairly frequently during gamete formation. About 5% to 10% of all fertilized human eggs result in an embryo with an abnormality in chromosome number! In most cases, such an embryo does not develop properly and results in a spontaneous abortion very early in pregnancy. Approximately 50% of all spontaneous abortions are due to alterations in chromosome number.

In some cases, an abnormality in chromosome number produces an offspring that survives to birth or longer. Several human disorders involve abnormalities in chromosome number. The most common are trisomies of chromosome 13, 18, or 21 and abnormalities in the number of the sex chromosomes (**Table 8.1**). Most of the known trisomies involve chromosomes that are relatively small and carry fewer genes compared to larger chromosomes. Trisomies of the other human autosomes and monosomies of all autosomes are presumed to produce a lethal phenotype, and many have been found in spontaneously aborted embryos and fetuses. For example, all possible human trisomies have been found in spontaneously aborted embryos except trisomy 1. It is believed that trisomy 1 is lethal at such an early stage that it prevents the successful implantation of the embryo.

Variation in the number of X chromosomes, unlike that of other large chromosomes, is often nonlethal. The survival of trisomy X individuals may be explained by X-chromosome inactivation, which is described in Chapter 4. In an individual with more than one X chromosome, all additional X chromosomes are converted to Barr bodies in the somatic cells of adult tissues. In an individual with trisomy X, for example, two out of three X chromosomes are converted to inactive Barr bodies. Unlike the level of expression for autosomal genes, the normal level of expression for X-linked genes is from a single X chromosome. In other words, the correct level of mammalian gene expression results from two copies of each autosomal gene and one copy of each X-linked gene. This explains how the expression of X-linked genes in males (XY) can be maintained at the same levels as in females (XX). It may also explain why trisomy X is not a lethal condition.

The phenotypic effects noted in Table 8.1 involving sex chromosomal abnormalities may be due to the expression of X-linked genes prior to embryonic X-chromosome inactivation or to the expression of genes on the inactivated X chromosome. As described in Chapter 4, pseudoautosomal genes and some other genes on the inactivated X chromosome are expressed in humans. Having one or three copies of the sex chromosomes may result in an underexpression or overexpression of these X-linked genes, respectively.

Human abnormalities in chromosome number are influenced by the age of the parents. Older parents are more likely to produce children with abnormalities in chromosome number. Down syndrome provides an example. This syndrome was first described by English physician John Langdon Down in 1866. It occurs at a frequency of about 1 in 800 live births. The symptoms are described in Table 8.1. The common form of this disorder is caused by the inheritance of three copies of chromosome 21. Therefore, Down syndrome is also known as trisomy 21. The incidence of Down syndrome rises with the age of either parent. In males, however, the rise occurs relatively late in life, usually past the age when most men have children. By comparison, the likelihood of having a child with Down syndrome rises dramatically during a woman's reproductive age (**Figure 8.16**). The association between maternal age and Down syndrome was discovered by L.S. Penrose in 1933, even before the chromosomal basis for the disorder was identified by French scientist Jérôme Lejeune in 1959. Down syndrome is most commonly caused by **nondisjunction,** which means that the chromosomes do not segregate properly. (Nondisjunction is discussed later in this chapter.) In this case, nondisjunction of chromosome 21 usually occurs during meiosis I in the oocyte.

Different hypotheses have been proposed to explain the relationship between maternal age and Down syndrome. One popular idea suggests that it may be due to the age of the oocytes. Human primary oocytes are produced within the ovaries of the female fetus prior to birth, are arrested at prophase of meiosis I, and remain in this stage until the time of ovulation. Therefore, as a woman ages, her primary oocytes have been in prophase I for a progressively longer period of time. This added length of time may contribute to an increased frequency of nondisjunction, though direct evidence to support this

TABLE 8.1
Aneuploid Conditions in Humans

Condition	Frequency	Syndrome	Characteristics
Autosomal			
Trisomy 13	1/15,000	Patau	Mental and physical deficiencies, wide variety of defects in organs, large triangular nose, early death
Trisomy 18	1/6000	Edward	Mental and physical deficiencies, facial abnormalities, extreme muscle tone, early death
Trisomy 21	1/800	Down	Mental deficiencies, abnormal pattern of palm creases, slanted eyes, flattened face, short stature
Sex Chromosomal			
XXY	1/1000	Klinefelter	Sexual immaturity (no sperm), breast swelling (males)
XYY	1/1000	Jacobs	Tall and thin (males)
XXX	1/1500	Triple X	Tall and thin, menstrual irregularity (females)
X0	1/5000	Turner	Short stature, webbed neck, sexually undeveloped (females)

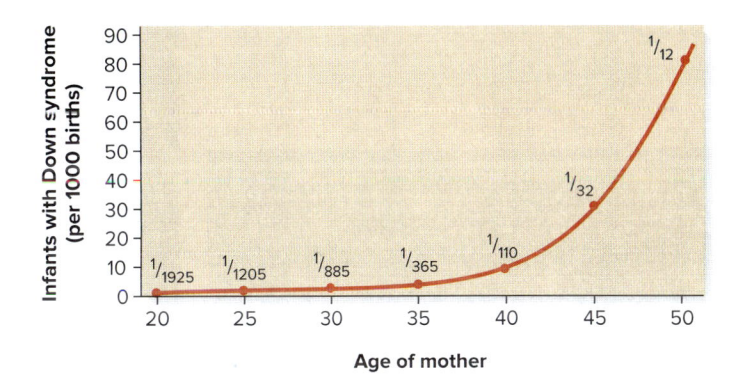

FIGURE 8.16 The incidence of Down syndrome births according to the age of the mother. The *y*-axis shows the number of infants born with Down syndrome per 1000 live births, and the *x*-axis plots the age of the mother at the time of birth. The data points indicate the fraction of live offspring born with Down syndrome.

claim is lacking. About 5% of the time, Down syndrome is due to an extra paternal chromosome. Prenatal tests can determine if a fetus has Down syndrome and some other genetic abnormalities. The topic of genetic testing is discussed in Chapter 22.

8.6 REVIEWING THE KEY CONCEPTS

- Aneuploidy is often detrimental due to an imbalance in gene expression. Down syndrome, an example of aneuploidy in humans, increases in frequency with maternal age (see Figures 8.15, 8.16, Table 8.1).

8.6 COMPREHENSION QUESTIONS

1. In a trisomic individual, such as a person with trisomy 21 (Down syndrome), a genetic imbalance occurs because
   a. genes on chromosome 21 are overexpressed.
   b. genes on chromosome 21 are underexpressed.
   c. genes on the other chromosomes are overexpressed.
   d. genes on the other chromosomes are underexpressed.
2. Humans with aneuploidy who survive usually have incorrect numbers of chromosome 13, 18, or 21 or the sex chromosomes. A possible explanation why these abnormalities permit survival is because
   a. the chromosomes have clusters of genes that aid in embryonic growth.
   b. the chromosomes are small and carry relatively few genes.
   c. an X-chromosome is inactivated.
   d. of both b and c.

8.7 VARIATION IN THE NUMBER OF SETS OF CHROMOSOMES

Learning Outcomes:
1. Describe examples in animals that involve variation in euploidy.
2. Define *endopolyploidy*.
3. Outline the process of polytene chromosome formation.
4. Discuss the effects of polyploidy among plant species and its impact in agriculture.

We now turn our attention to changes in the number of sets of chromosomes, referred to as variations in euploidy. In some cases, such changes are detrimental. However, in many species, particularly plants, additional sets of chromosomes are very common and are often beneficial with regard to an individual's phenotype.

Variations in Euploidy Occur Naturally in a Few Animal Species

Most species of animals are diploid. In some cases, changes in euploidy are not well tolerated. For example, polyploidy in mammals is often a lethal condition. However, many examples of naturally occurring variations in euploidy occur. In **haplodiploid** species, which includes many species of bees, wasps, and ants, one of the sexes is haploid, usually the male, and the other is diploid (see Chapter 4, Figure 4.2). For example, male bees, which are called drones, contain a single set of chromosomes. They are produced from unfertilized eggs. By comparison, female bees are produced from fertilized eggs and are diploid.

Many examples of vertebrate polyploid animals have been discovered. Interestingly, on several occasions, animals that are morphologically very similar can be found as a diploid species as well as a separate polyploid species. This situation occurs among certain amphibians and reptiles. **Figure 8.17** shows photographs of a diploid frog and a tetraploid (4n) frog. As you can see, they look nearly indistinguishable from each other. Their difference can be revealed only by an examination of the chromosome number in the somatic cells of the animals and by their mating calls—*Hyla chrysoscelis* has a faster trill rate than *Hyla versicolor*.

Variations in Euploidy Occur in Certain Tissues Within Animals

Thus far, we have considered variations in chromosome number that occur at fertilization, so all somatic cells of an individual contain this variation. In many animals, certain tissues of the body

(a) *Hyla chrysoscelis*   (b) *Hyla versicolor*

FIGURE 8.17 Differences in euploidy in two closely related frog species. The frog in (a) is diploid, whereas the frog in (b) is tetraploid.

Genes → Traits Though similar in appearance, these two species differ in their number of chromosome sets. At the level of gene expression, this observation suggests that the number of copies of each gene (two versus four) does not critically affect the phenotype of these two species.

©A. B. Sheldon

display normal variations in the number of sets of chromosomes. Diploid animals sometimes produce tissues that are polyploid. For example, the cells of the human liver can vary to a great degree in their number of sets of chromosomes. Liver cells contain nuclei that can be triploid, tetraploid, and even octaploid ($8n$). The occurrence of polyploid cells in organisms that are otherwise diploid is known as **endopolyploidy.** What is the biological significance of endopolyploidy? One possibility is that the increase in chromosome number in certain cells may enhance their ability to produce specific gene products that are needed in great abundance.

An unusual example of natural variation in the euploidy of somatic cells occurs in *Drosophila* and some other insects. Within certain tissues, such as the salivary glands, the chromosomes undergo repeated rounds of chromosome replication without cellular division. For example, in the salivary gland cells of *Drosophila*, the pairs of chromosomes double approximately nine times ($2^9 = 512$). **Figure 8.18a** illustrates how repeated rounds of chromosomal replication produce a bundle of chromosomes that lie together in a parallel fashion. This bundle, termed a **polytene chromosome,** was first observed by E. G. Balbiani in 1881. Later, in the 1930s, Theophilus Painter and colleagues recognized that the size and morphology of polytene chromosomes provided geneticists with unique opportunities to study chromosome structure and gene organization.

Figure 8.18b shows a micrograph of a polytene chromosome. The structure of polytene chromosomes differs from the multiple sets observed in other forms of endopolyploidy because the replicated chromosomes remain attached to each other. Prior to the formation of polytene chromosomes, *Drosophila* cells contain 8 chromosomes (two sets of 4 chromosomes each; see Figure 8.14a). In the salivary gland cells, the homologous chromosomes synapse with each other and replicate to form a polytene structure. During this process, the four types of chromosomes aggregate to form a single structure with several polytene arms. The central point where the chromosomes aggregate is known as the **chromocenter.** Each of the four types of chromosome is attached to the chromocenter near its centromere. The X and Y chromosomes and chromosome 4 are telocentric, and chromosomes 2 and 3 are metacentric. Therefore, chromosomes 2 and 3 have two arms that radiate from the chromocenter, whereas the X and Y and chromosome 4 have a single arm projecting from the chromocenter (**Figure 8.18c**).

Variations in Euploidy Are Common in Plants

We now turn our attention to variations of euploidy that occur in plants. Compared with animals, plants more commonly exhibit polyploidy. Among ferns and flowering plants, at least 30% to

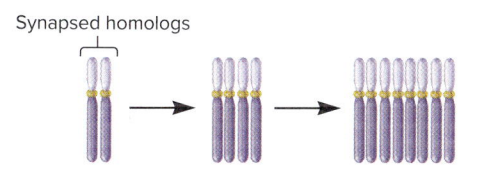

(a) Repeated chromosome replication produces polytene chromosome.

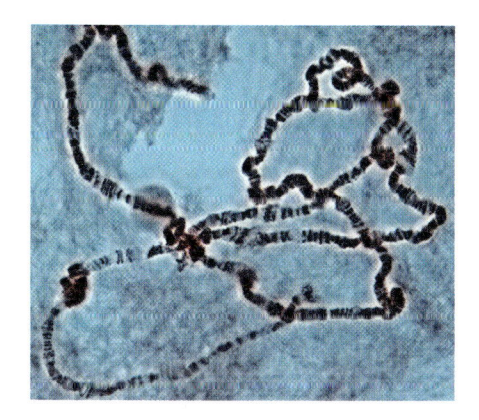

(b) A polytene chromosome

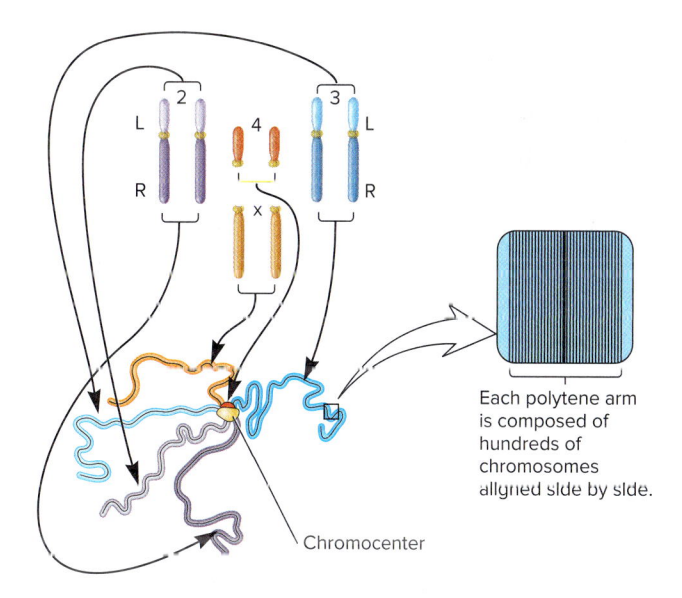

(c) Relationship between a polytene chromosome and regular *Drosophila* chromosomes

Each polytene arm is composed of hundreds of chromosomes aligned side by side.

Chromocenter

FIGURE 8.18 Polytene chromosomes in *Drosophila*. (a) A schematic illustration of the formation of polytene chromosomes. Several rounds of repeated replication without cellular division result in a bundle of sister chromatids that lie side by side. Both homologs also lie parallel to each other. This replication does not occur in highly condensed, heterochromatic DNA near the centromere. (b) A photograph of a polytene chromosome. (c) This drawing shows the relationship between the four pairs of chromosomes and the formation of a polytene chromosome in a salivary gland cell of *Drosophila*. The heterochromatic regions of the chromosomes aggregate at the chromocenter, and the arms of the chromosomes project outward. In chromosomes with two arms, the short arm is labeled L and the long arm is labeled R.

(b) ©David M. Phillips/Science Source

Concept Check: Approximately how many copies of chromosome 2 are found in a polytene chromosome of Drosophila?

35% of species are polyploid. Polyploidy is also important in agriculture. Many of the fruits and grains we eat are produced from polyploid plants. For example, the species of wheat that we use to make bread, *Triticum aestivum*, is a hexaploid (6*n*) that arose from the union of diploid genomes from three closely related species (**Figure 8.19**). Different species of strawberries are diploid, tetraploid, hexaploid, and even octaploid!

In many instances, polyploid strains of plants display outstanding agricultural characteristics. They are often larger in size and more robust. These traits are clearly advantageous in the production of food. In addition, polyploid plants tend to exhibit a greater adaptability, which allows them to withstand harsher environmental conditions. Also, polyploid ornamental plants often produce larger flowers than their diploid counterparts.

Polyploid plants having an odd number of chromosome sets, such as triploids (3*n*) or pentaploids (5*n*), usually cannot reproduce. Why are they sterile? The sterility arises because these plants produce highly aneuploid gametes, which are nonviable. During prophase of meiosis I, homologous pairs of sister chromatids form bivalents. However, organisms with an odd number of chromosomes, such as three, display an unequal separation of homologous chromosomes during anaphase of meiosis I (**Figure 8.20**). An odd number cannot be divided equally between two daughter cells. For each type of chromosome, a daughter cell randomly gets one or two copies. For example, one daughter cell might receive one copy of chromosome 1, two copies of chromosome 2, two copies of chromosome 3, one copy of chromosome 4, and so forth. For a triploid species containing many different chromosomes in a set, meiosis is very unlikely to produce a daughter cell that is euploid. If we assume that a daughter cell receives either one copy or two copies of each kind of chromosome, the probability that meiosis will produce a cell that is perfectly haploid or diploid is $(1/2)^{n-1}$, where *n* is the number of chromosomes in a set. As an example, in a triploid organism containing 20 chromosomes per set, the probability of producing a haploid or diploid cell is 0.000001907, or 1 in 524,288. Thus, meiosis is almost certain to produce cells that contain one copy of some chromosomes and two copies of the others. This high probability of aneuploidy underlies the reason for triploid sterility.

Though sterility is generally a detrimental trait, it can be desirable agriculturally because it may result in a seedless fruit. For example, domestic bananas and seedless watermelons are triploid varieties. The domestic banana was originally derived from a seed-producing diploid species and has been asexually propagated by humans via cuttings. The small black spots in the center of a domestic banana are degenerate seeds. In the case of flowers, the seedless phenotype can also be beneficial. Seed producers such as Burpee have developed triploid varieties of flowering plants

Cultivated wheat, a hexaploid species

FIGURE 8.19 **Example of a polyploid plant.** Cultivated wheat, *Triticum aestivum*, is a hexaploid. It was derived from three different diploid species of grasses that originally were found in the Middle East and were cultivated by ancient farmers in that region.

Genes → Traits An increase in chromosome number from diploid to tetraploid or hexaploid affects the phenotype of the individual. In the case of many plant species, a polyploid individual is larger and more robust than its diploid counterpart. This suggests that, for plants, having additional copies of each gene is somewhat better than having two copies of each gene. The situation is rather different in animals. Tetraploidy in animals may have little effect (as in Figure 8.17b), but it is also common for polyploidy in animals to be detrimental.

©Jim Steinberg/Science Source

Concept Check: *What are some common advantages of polyploidy in plants?*

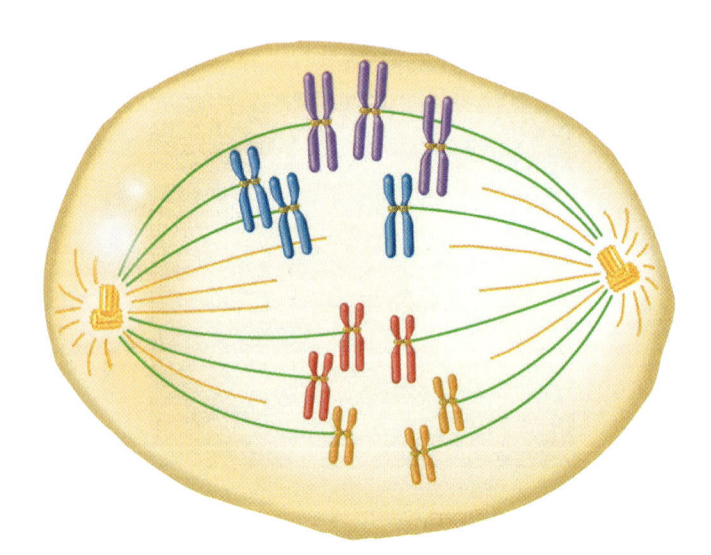

FIGURE 8.20 **Schematic representation of anaphase of meiosis I in a triploid organism containing three sets of four chromosomes.** In this example, the homologous chromosomes (three each) do not evenly separate during anaphase. Each cell receives one copy of some chromosomes and two copies of other chromosomes. This produces aneuploid gametes.

Concept Check: *Explain why a triploid individual is usually infertile.*

such as marigolds. Because the triploid marigolds are sterile and unable to set seed, more of their energy goes into flower production. According to Burpee, "They bloom and bloom, unweakened by seed bearing."

8.7 REVIEWING THE KEY CONCEPTS

- Among animals, variation in euploidy is relatively rare, though it does occur. Some tissues within a diploid animal may exhibit polyploidy. An example is the polytene chromosome found in salivary gland cells of *Drosophila* (see Figures 8.17, 8.18).
- Polyploidy in plants is relatively common and has many advantages for agriculture. Triploid plants are usually seedless because they cannot segregate their chromosomes evenly during meiosis (see Figures 8.19, 8.20).

8.7 COMPREHENSION QUESTIONS

1. The term *endopolyploidy* refers to the phenomenon of having
   a. too many chromosomes.
   b. extra chromosomes inside the cell nucleus.
   c. extra sets of chromosomes in certain cells in the body.
   d. extra sets of chromosomes in gametes.
2. In agriculture, an advantage of triploidy in plants is that the plants are
   a. more fertile.
   b. often seedless.
   c. always disease resistant.
   d. all of the above.

8.8 MECHANISMS THAT PRODUCE VARIATION IN CHROMOSOME NUMBER

Learning Outcomes:

1. Describe the mechanisms of meiotic nondisjunction and mitotic nondisjunction and their possible phenotypic consequences.
2. Compare and contrast autopolyploidy, alloploidy, and allopolyploidy.
3. Describe how colchicine is used to produce polyploid species.

As we have seen, variations in chromosome number are fairly widespread and usually have a significant effect on the phenotypes of plants and animals. For these reasons, researchers have wanted to understand the cellular mechanisms that cause variations in chromosome number. In some cases, a change in chromosome number is the result of nondisjunction. The term *nondisjunction* refers to an event in which the chromosomes do not segregate properly. As we will see, it may be caused by an improper separation of homologous pairs in a bivalent in meiosis or a failure of the centromeres to disconnect during mitosis.

Meiotic nondisjunction can produce cells that have too many or too few chromosomes. If such a cell gives rise to a gamete that fuses with a normal gamete during fertilization, the resulting offspring will have an abnormal chromosome number in all of its cells. An abnormal nondisjunction event also may occur after fertilization in one of the somatic cells of the body. This second mechanism is known as **mitotic nondisjunction.** When this occurs during embryonic stages of development, it may lead to a patch of tissue in the organism that has an altered chromosome number. A third common way in which the chromosome number of an organism can vary is by interspecies crosses. An **alloploid** organism contains sets of chromosomes from two or more different species. This term refers to the occurrence of chromosome sets (ploidy) from the genomes of different (allo) species. In this section, we will examine these three mechanisms in greater detail.

Meiotic Nondisjunction Can Produce Aneuploidy or Polyploidy

The normal process of meiosis is described in Chapter 2 (see Figures 2.10 and 2.11). Nondisjunction during meiosis can occur during anaphase of meiosis I or meiosis II. If it happens during meiosis I, an entire bivalent migrates to one pole (**Figure 8.21a**). Following the completion of meiosis, the four resulting haploid cells are abnormal. If nondisjunction occurs during anaphase of meiosis II (**Figure 8.21b**), the net result is two abnormal and two normal haploid cells. If a gamete that is missing a chromosome is viable and participates in fertilization, the resulting offspring is monosomic for the missing chromosome. Alternatively, if a gamete carrying an extra chromosome unites with a normal gamete, the offspring will be trisomic.

In rare cases, all of the chromosomes may undergo nondisjunction and migrate to one of the daughter cells. The net result of **complete nondisjunction** is a diploid cell and a cell without any chromosomes. The cell without chromosomes is nonviable, but the diploid cell might participate in fertilization with a normal haploid gamete to produce a triploid individual. Therefore, complete nondisjunction can produce individuals that are polyploid.

Mitotic Nondisjunction or Chromosome Loss Can Produce a Patch of Tissue with an Altered Chromosome Number

Abnormalities in chromosome number occasionally occur after fertilization takes place. In this case, the abnormal event happens during mitosis rather than meiosis. One possibility is that the sister chromatids separate improperly, so one daughter cell has three copies of a chromosome, whereas the other daughter cell has only one (**Figure 8.22a**). Alternatively, the sister chromatids can separate during anaphase of mitosis, but one of the chromosomes is improperly attached to the spindle apparatus and so does not migrate to a pole (**Figure 8.22b**). A chromosome will be degraded if it is left outside the nucleus when the nuclear membrane reforms. In this case, one of the daughter cells has two copies of that chromosome, whereas the other has only one.

When genetic abnormalities occur after fertilization, the organism contains a subset of cells that are genetically different from those of the rest of the organism. This condition is referred to as **mosaicism.** The size and location of the mosaic region depend on the timing and location of the original abnormal event. If a

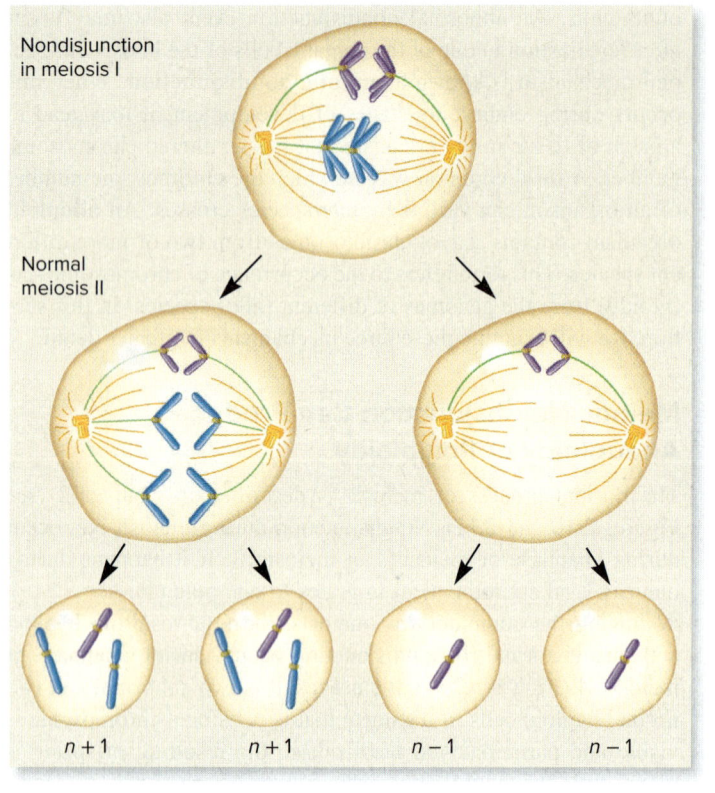

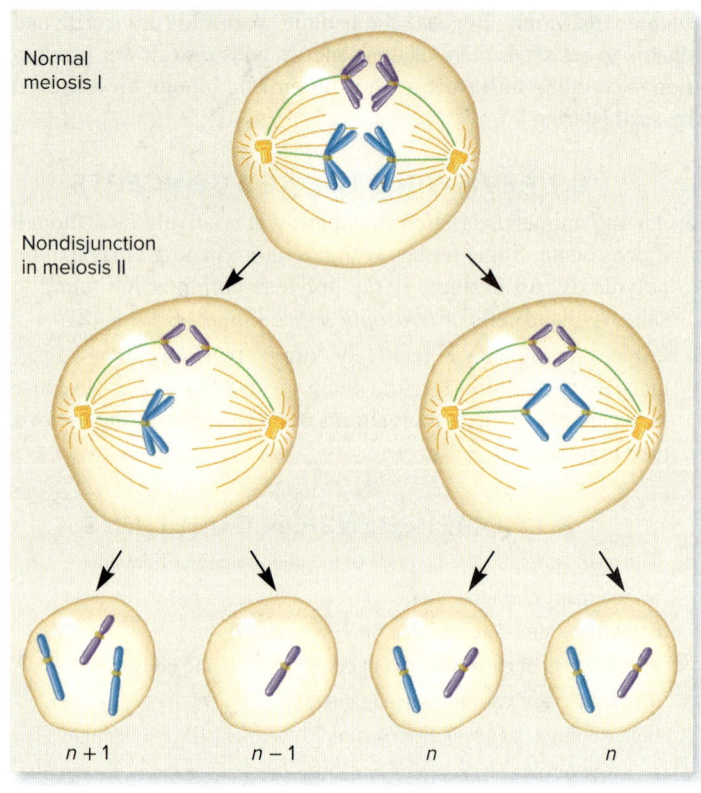

176   CHAPTER 8 :: VARIATION IN CHROMOSOME STRUCTURE AND NUMBER

(a) Nondisjunction in meiosis I

(b) Nondisjunction in meiosis II

FIGURE 8.21 **Nondisjunction during meiosis I and II.** The chromosomes shown in purple are behaving properly during meiosis I and II, so each haploid cell receives one copy of such a chromosome. The chromosomes shown in blue are not disjoining correctly. In (a), nondisjunction occurs in meiosis I, so the resulting four cells receive either two copies of the blue chromosome or zero copies. In (b), nondisjunction occurs during meiosis II, so one cell has two blue chromosomes and another cell has zero. The remaining two cells are normal.

Concept Check: *Explain what the word nondisjunction means.*

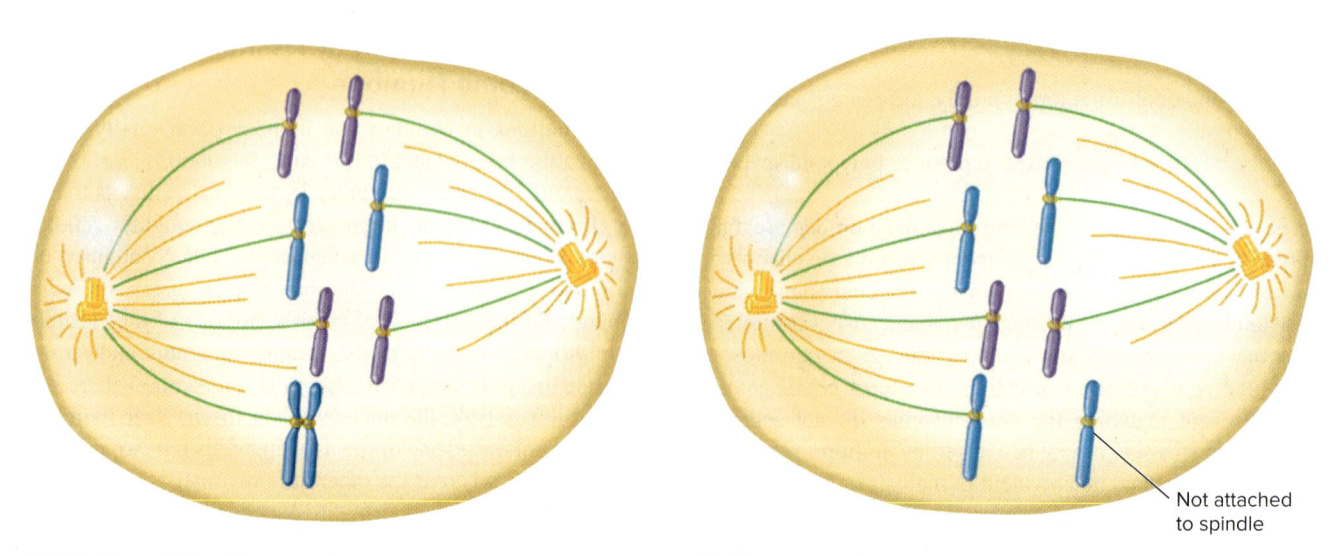

(a) Mitotic nondisjunction

(b) Chromosome loss

FIGURE 8.22 **Nondisjunction and chromosome loss during mitosis in somatic cells.** (a) Mitotic nondisjunction produces trisomic and monosomic daughter cells. (b) Chromosome loss produces normal and monosomic daughter cells.

genetic alteration happens very early in the embryonic development of an organism, the abnormal cell will be the precursor for a large section of the organism.

Changes in Euploidy Occur by Autopolyploidy, Alloploidy, and Allopolyploidy

Different mechanisms account for changes in the number of chromosome sets among natural populations of plants and animals (**Figure 8.23**). As previously mentioned, complete nondisjunction, due to a general defect in the spindle apparatus, can produce an individual with one or more extra sets of chromosomes. This individual is known as an **autopolyploid** (Figure 8.23a). The prefix *auto-* (meaning "self") and the term *polyploid* (meaning "many sets of chromosomes") refer to an increase in the number of chromosome sets within a single species.

Another common mechanism for change in chromosome number, called **alloploidy,** is a result of interspecies crosses.

- An alloploid that has one set of chromosomes from two different species is called an **allodiploid.** An interspecies cross is most likely to occur between species that are close evolutionary relatives (Figure 8.23b). For example, closely related species of grasses may interbreed to produce allodiploids.
- As shown in Figure 8.23c, an **allopolyploid** contains two (or more) sets of chromosomes from two (or more) species. In the example shown, the **allotetraploid** contains two complete sets of chromosomes from two different species, for a total of four sets.

In nature, allotetraploids usually arise from allodiploids. This can occur when a somatic cell in an allodiploid undergoes complete nondisjunction to create an allotetraploid cell. In plants, such a cell can continue to grow and produce a section of the plant that

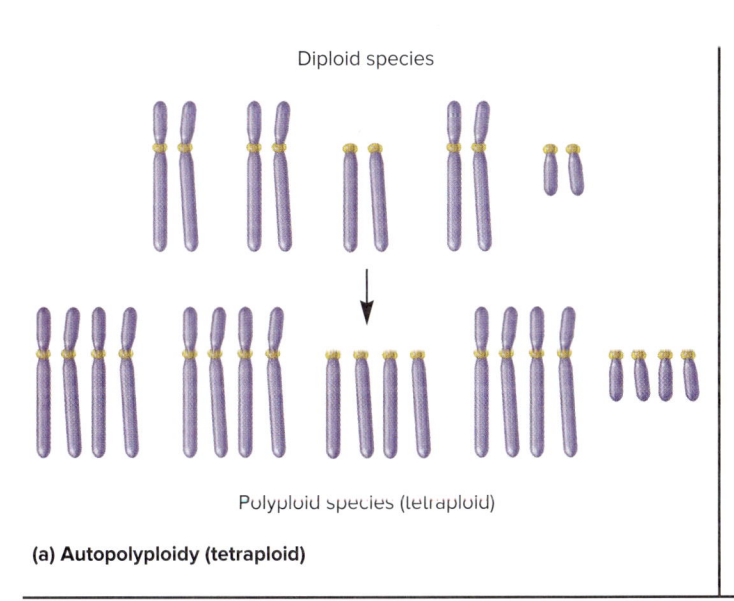

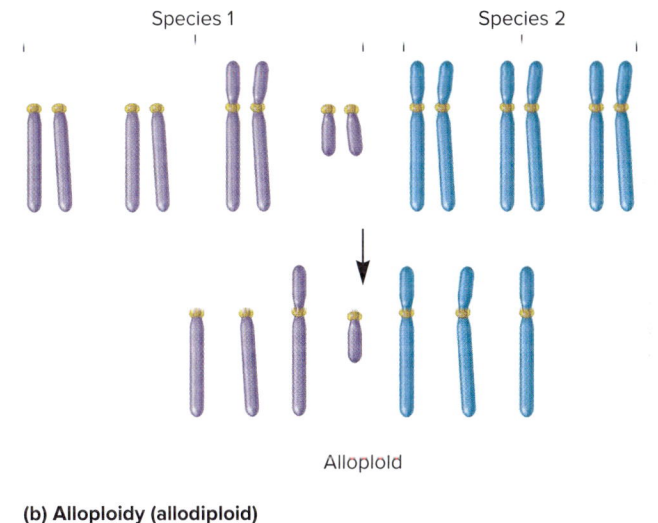

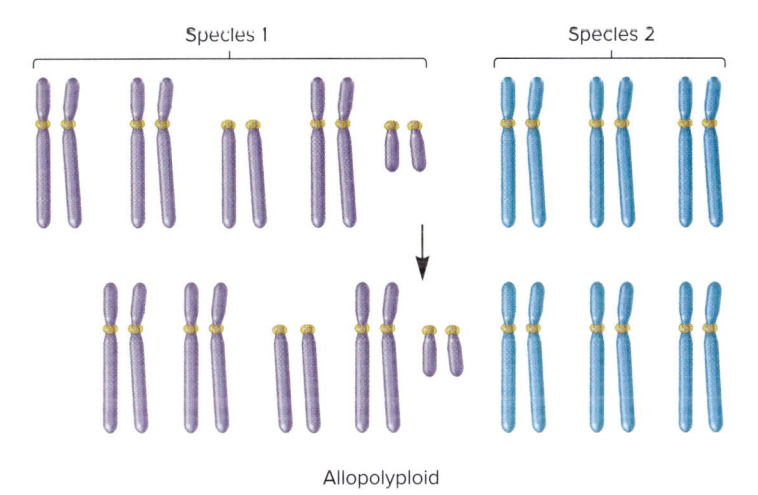

(a) Autopolyploidy (tetraploid)

(b) Alloploidy (allodiploid)

(c) Allopolyploidy (allotetraploid)

FIGURE 8.23 A comparison of autopolyploidy, alloploidy, and allopolyploidy.

Concept Check: What is the key difference between autopolyploidy and allopolyploidy?

is allotetraploid. If this part of the plant produces seeds by self-pollination, the seeds give rise to allotetraploid offspring. Cultivated wheat (refer back to Figure 8.19) is a plant for which two species must have interbred to create an allotetraploid, and then a third species interbred with the allotetraploid to create an allohexaploid.

Experimental Treatments Can Promote Polyploidy

Because polyploid and allopolyploid plants often exhibit desirable traits, the development of polyploids is of considerable interest among plant breeders. Experimental studies on the ability of environmental agents to promote polyploidy began in the early 1900s. Since that time, various treatments have been shown to promote nondisjunction, thereby leading to polyploidy. These include abrupt temperature changes during the initial stages of seedling growth and the treatment of plants with chemical agents that interfere with the formation of the spindle apparatus.

The drug colchicine is commonly used for promoting polyploidy. Once inside the cell, colchicine binds to tubulin (a protein found in the spindle apparatus) and inhibits microtubule assembly, thereby interfering with normal chromosome segregation during mitosis or meiosis. In 1937, Alfred Blakeslee and Amos Avery applied colchicine to plant tissue and found that high doses of the drug caused complete mitotic nondisjunction and produced polyploidy in plant cells. Colchicine can be applied to seeds, young embryos, or rapidly growing regions of a plant (**Figure 8.24**). This application may produce aneuploidy, which is usually an undesirable outcome, but it often produces polyploid cells, which may grow faster than the surrounding diploid tissue. In a diploid plant, colchicine may cause complete mitotic nondisjunction, yielding tetraploid ($4n$) cells. As the tetraploid cells continue to divide, they generate a portion of the plant that is often morphologically distinguishable from the remainder. For example, a tetraploid stem may have a larger diameter and produce larger leaves and flowers.

Because individual plants can be propagated asexually from pieces of plant tissue (i.e., cuttings), the polyploid portion of the plant can be removed, treated with the proper growth hormones, and grown as a separate plant. Alternatively, the tetraploid region of a plant may have flowers that produce seeds by self-pollination. For many plant species, a tetraploid flower produces diploid pollen and eggs, which can combine to produce tetraploid offspring. In this way, the use of colchicine provides a straightforward method of producing polyploid strains of plants.

FIGURE 8.24 Use of colchicine to promote polyploidy in plants. Colchicine interferes with the mitotic spindle apparatus and promotes nondisjunction. If complete nondisjunction occurs in a diploid cell, a daughter cell will be formed that is tetraploid. Such a tetraploid cell may continue to divide and produce a segment of the plant with more robust characteristics. This segment may be cut from the rest of the plant and rooted. In this way, a tetraploid plant can be propagated.

8.8 REVIEWING THE KEY CONCEPTS

- Meiotic nondisjunction, mitotic nondisjunction, and chromosome loss can result in changes in chromosome number (see Figures 8.21–8.22).
- Autopolyploidy is an increased number of sets of chromosomes within a single species. Interspecies matings result in alloploids. Alloploids with multiple sets of chromosomes from each species are allopolyploids (see Figure 8.23).
- Various agents, including the drug colchicine, are used to promote nondisjunction, thereby producing polyploid plants (see Figure 8.24).

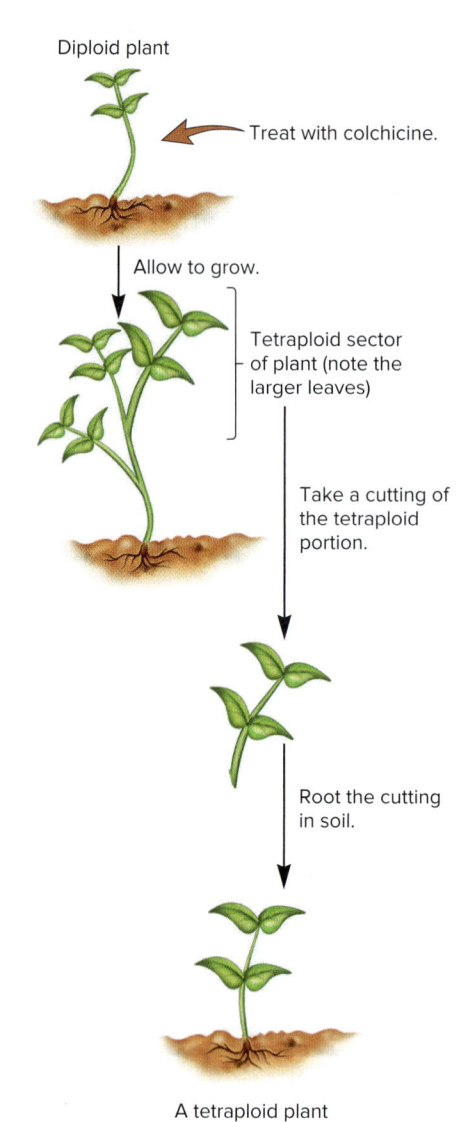

8.8 COMPREHENSION QUESTIONS

1. In a diploid species, complete nondisjunction during meiosis I may produce a viable cell that is
    a. trisomic.
    b. haploid.
    c. diploid.
    d. triploid.
2. The somatic cells of an allotetraploid contain
    a. one set of chromosomes from four different species.
    b. two sets of chromosomes from two different species.
    c. four sets of chromosomes from one species.
    d. one set of chromosomes from two different species.

KEY TERMS

Page 156. genetic variation, allelic variation
Page 157. cytogeneticists, metacentric, submetacentric, acrocentric, telocentric, karyotype, G bands
Page 159. deletion, deficiency, duplication, inversion, translocation, simple translocation, reciprocal translocation
Page 160. terminal deletion, interstitial deletion
Page 161. repetitive sequences, nonallelic homologous recombination, gene duplication, gene family
Page 162. homologous, paralogs, copy number variation (CNV), segmental duplication
Page 164. pericentric inversion, paracentric inversion, position effect, inversion heterozygote, inversion loop
Page 165. dicentric, dicentric bridge, acentric fragment
Page 166. telomeres, balanced translocation, unbalanced translocation, Robertsonian translocation
Page 167. translocation cross, semisterility
Page 169. euploid, triploid, polyploid, tetraploid, aneuploid, trisomic, monosomic
Page 171. nondisjunction
Page 172. haplodiploid
Page 173. endopolyploidy, polytene chromosome, chromocenter
Page 175. meiotic nondisjunction, mitotic nondisjunction, alloploid, complete nondisjunction, mosaicism,
Page 177. autopolyploid, alloploidy, allodiploid, allopolyploid, allotetraploid

CHAPTER SUMMARY

8.1 Microscopic Examination of Eukaryotic Chromosomes

- Among members of the same species and different species, natural variation exists with regard to chromosome structure and number. Three features of chromosomes that aid in their identification are centromeric location, size, and banding pattern (see Figure 8.1).

8.2 Changes in Chromosome Structure: An Overview

- Variations in chromosome structure include deletions, duplications, inversions, and translocations (see Figure 8.2).

8.3 Deletions and Duplications

- Chromosome breaks can create terminal or interstitial deletions. Some deletions are associated with human genetic disorders, such as cri-du-chat syndrome (see Figures 8.3, 8.4).
- Nonallelic homologous recombination can produce gene duplications and deletions. Over time, gene duplications can lead to the formation of gene families, such as the globin gene family (see Figures 8.5–8.7).
- Copy number variation (CNV) among members of a species is fairly common. In humans, CNV is associated with certain diseases (see Figure 8.8).

8.4 Inversions and Translocations

- Inversions can be pericentric or paracentric. In an inversion heterozygote, crossing over within the inversion loop can create deletions and duplications in the resulting chromosomes (see Figures 8.9, 8.10).
- Two mechanisms can produce translocations: chromosome breakage and subsequent repair, and nonhomologous crossing over (see Figure 8.11).
- A Robertsonian translocation results in a chromosome in which the long arms of two different acrocentric chromosomes become joined (see Figure 8.12).
- Individuals that carry a balanced translocation may have a high probability of producing unbalanced gametes, depending on the segregation pattern during meiosis I (see Figure 8.13).

8.5 Changes in Chromosome Number: An Overview

- Variation in the number of chromosomes may involve changes in the number of sets (euploidy) or changes in the number of particular chromosomes within a set (aneuploidy) (see Figure 8.14).

8.6 Variation in the Number of Chromosomes Within a Set: Aneuploidy

- Aneuploidy is often detrimental due to an imbalance in gene expression. Down syndrome, an example of aneuploidy in humans, increases in frequency with maternal age (see Figures 8.15, 8.16, Table 8.1).

8.7 Variation in the Number of Sets of Chromosomes

- Among animals, variation in euploidy is relatively rare, though it does occur. Some tissues within a diploid animal may exhibit polyploidy. An example is the polytene chromosome found in salivary gland cells of *Drosophila* (see Figures 8.17, 8.18).
- Polyploidy in plants is relatively common and has many advantages for agriculture. Triploid plants are usually seedless because they cannot segregate their chromosomes evenly during meiosis (see Figures 8.19, 8.20).

8.8 Mechanisms That Produce Variation in Chromosome Number

- Meiotic nondisjunction, mitotic nondisjunction, and chromosome loss can result in changes in chromosome number (see Figures 8.21–8.22).

CHAPTER 8 :: VARIATION IN CHROMOSOME STRUCTURE AND NUMBER

- Autopolyploidy is an increased number of sets of chromosomes within a single species. Interspecies matings result in alloploids. Alloploids with multiple sets of chromosomes from each species are allopolyploids (see Figures 8.23).

- Various agents, including the drug colchicine, are used to promote nondisjunction, thereby producing polyploid plants (see Figure 8.24).

PROBLEM SETS & INSIGHTS

More Genetic TIPS

1. What is a gene family? Discuss the common and unique features of the members of the globin gene family found in humans.

Topic: What topic in genetics does this question address?

The topic is gene families. More specifically, the question is about the globin gene family in humans.

Information: What information do you know based on the question and based on your understanding of the topic?

In the question, you are reminded that humans carry a globin gene family in their genome. From your understanding of the topic, you may remember that gene families are produced by gene duplication events. The duplications are followed by the accumulation of mutations in each family member, which may alter the gene's function and make it more specialized.

Problem-Solving Strategy: Define key terms. Compare and contrast.

To begin to solve this problem, you may want to define *gene family* and briefly describe how a gene family is produced. You then can compare and contrast the general features of the members of the globin gene family.

Answer: A gene family is produced when a single gene is copied one or more times by a duplication event. This duplication may occur because of a misaligned crossover, which produces a chromosome with a deletion and another chromosome with a gene duplication (see Figure 8.5). This type of duplication may occur several times to produce many copies of a particular gene. Over time, each member of a gene family accumulates mutations, which may subtly alter its function.

All members of the globin gene family bind oxygen. Myoglobin tends to bind oxygen more tightly; therefore, it is good at storing oxygen. Hemoglobin binds oxygen more loosely, so it can transport oxygen throughout the body (via red blood cells) and release it to the tissues that need it. The polypeptides that form hemoglobins are expressed in red blood cells, whereas the myoglobin gene is expressed in many different cell types. The expression pattern of the globin genes changes during different stages of development. The ε-globin and ζ-globin genes are expressed in the early embryo. They are turned off near the end of the first trimester, and then the γ-globin genes exhibit their maximal expression during the second and third trimesters of gestation. Following birth, the γ-globin genes are silenced, and the β-globin gene is expressed for the rest of a person's life. These differences in the expression of the globin genes reflect the differences in the oxygen transport needs of humans during different stages of life. Overall, the evolution of gene families has resulted in gene products that are better suited to particular tissues or stages of development.

2. An inversion heterozygote has the following inversion chromosome:

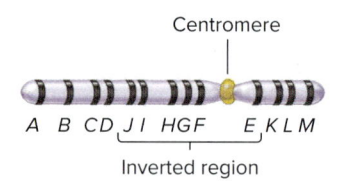

[Diagram of a chromosome with labels: A B CD JI HGF E KLM, Centromere, Inverted region]

What is the result if crossing over occurs between genes *F* and *G* on the inverted chromosome and a normal chromosome in this individual?

Topic: What topic in genetics does this question address?

The topic is changes in chromosome structure. More specifically, the question is about the consequences of crossing over when one of the chromosomes carries an inversion.

Information: What information do you know based on the question and your understanding of the topic?

From the question, you know the features of a chromosome carrying an inversion and the site of a crossover between the inverted chromosome and a normal chromosome. From your understanding of the topic, you may remember that this is a pericentric inversion. For crossing over to occur, a loop must form so that the inverted and normal chromosomes can pair up.

Problem-Solving Strategy: Make a drawing. Predict the outcome.

One strategy to solve this problem is to make a drawing that aligns the inverted and normal chromosomes and includes the crossover site.

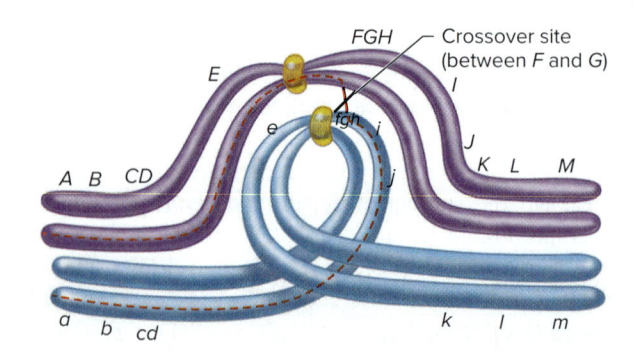

[Diagram of paired chromosomes forming an inversion loop with labels: A B CD, E, FGH, I, J, K, L, M; a b cd, e, fgh, i, j, k, l, m; Crossover site (between F and G)]

To determine the products of the crossover, start at one end of the normal (purple) chromosome (at gene A) with a pencil and trace along the chromosome, as shown by the red dashed line. At the crossover site, your pencil shifts to the inverted (blue) chromosome. Keep your pencil moving in the same direction as you go through the crossover site. (Just before the crossover site, your pencil will be moving away from the centromere. Keep it moving away from the centromere when you shift to the blue chromosome.) When your pencil gets to the end of the blue chromosome, it will have traced one of the chromosomes that is produced, which has a duplication and a deletion. Next, start at the other end of the normal chromosome and do the same thing. This will yield the other chromosome, which has a different duplication and deletion.

Answer: The resulting products are four chromosomes. One chromosome is normal, one has an inversion, and two have a duplication and a deletion. The two chromosomes with duplicated and deleted parts are shown here:

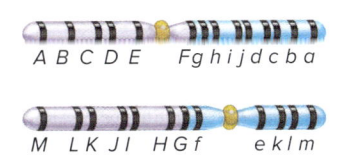

A B C D E   F g h i j d c b a

M L K J I   H G f   e k l m

3. In humans, the number of chromosomes per set is 23. Even though the following conditions are lethal, what would be the total number of chromosomes for an individual with each condition?

   A. Trisomy 22
   B. Monosomy 11
   C. Triploidy

Topic: What topic in genetics does this question address?

The topic is variation in chromosome number. More specifically, the question is about predicting the number of chromosomes in individuals with different types of chromosomal abnormalities.

Information: What information do you know based on the question and your understanding of the topic?

From the question, you know that humans have 23 chromosomes per set, and you are given the chromosomal composition for three conditions that are due to variation in chromosome number. From your understanding of the topic, you may remember that humans are diploid and you may recall the consequences of trisomy, monosomy, and triploidy.

Problem-Solving Strategy: Define key terms. Make a calculation.

One strategy to begin to solve this problem is to define *trisomy, monosomy,* and *triploidy*. If n represents a set of chromosomes, trisomy is $2n + 1$, monosomy is $2n - 1$, and triploidy is $3n$. To calculate the numbers of chromosomes in three individuals with these conditions, you substitute 23 for n. For example, $2n + 1$ is $2(23) + 1$, which equals 47.

Answer:

   A. 47 (the diploid number, 46, plus 1)
   B. 45 (the diploid number, 46, minus 1)
   C. 69 (3 times 23)

4. A diploid species with 44 chromosomes (i.e., 22 per set) is crossed to another diploid species with 38 chromosomes (i.e., 19 per set). How many chromosomes are produced in an allodiploid or allotetraploid from this cross? Would you expect the offspring to be sterile or fertile?

Topic: What topic in genetics does this question address?

The topic is variation in chromosome number. More specifically, the question is about calculating the number of chromosomes in different types of alloploids, and then predicting if they would be sterile or fertile.

Information: What information do you know based on the question and your understanding of the topic?

From the question, you know that both species are diploid; one has 22 chromosomes per set and the other has 19. From your understanding of the topic, you may remember that allodiploids have one set of chromosomes from each species and that allotetraploids have two sets from each species. Also, you may recall that individuals with even numbers of homologous chromosomes are usually fertile.

Problem-Solving Strategy: Define key terms. Make a calculation. Predict the outcome.

One strategy to begin to solve this problem is to define *allodiploid* and *allotetraploid*. As mentioned, an allodiploid has one set of chromosomes from each species and an allotetraploid has two sets from each species. If n_1 represents a set of chromosomes from one species and n_2 represents a set of chromosomes from another species, an allodiploid will be $n_1 + n_2$, whereas an allotetraploid will be $2n_1 + 2n_2$. The allotetraploid will have its chromosomes in homologous pairs.

Answer: The allodiploid will have $22 + 19 = 41$ chromosomes. This individual is likely to be sterile, because not all of the chromosomes would have homologous partners to pair with during meiosis. This represents aneuploidy, which usually causes sterility. An allotetraploid will have $44 + 38 = 82$ chromosomes. Because each chromosome would have a homologous partner, the allotetraploid is likely to be fertile.

Conceptual Questions

C1. Which changes in chromosome structure cause a change in the total amount of genetic material, and which do not?

C2. Explain why small deletions and duplications are less likely than large ones to have a detrimental effect on an individual's phenotype. If a small deletion within a single chromosome happens to have a phenotypic effect, what would you conclude about the genes in the region affected by the deletion?

C3. How does a chromosomal duplication occur?

C4. What is a gene family? How are gene families produced over time? With regard to gene function, what is the biological significance of a gene family?

C5. Following a gene duplication, two genes will accumulate different mutations, causing them to have slightly different sequences. In Figure 8.7, which pair of genes would you expect to have more similar sequences: α_1 and α_2 or ψ_{α_1} and α_2? Explain your answer.

C6. Two chromosomes have the following orders for their genes:

Normal: *A B C* centromere *D E F G H I*

Abnormal: *A B G F E D* centromere *C H I*

Does the abnormal chromosome have a pericentric or a paracentric inversion? Draw a sketch showing how these two chromosomes would pair during prophase of meiosis I.

C7. An inversion heterozygote has the following inversion chromosome:

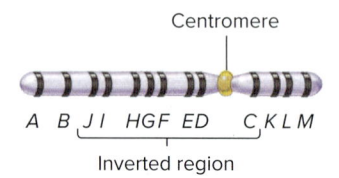

A B J I HGF ED C K L M (Inverted region: J I HGF ED)

What would be the products if a crossover occurred between genes *H* and *I* on the inverted chromosome and a normal chromosome?

C8. An inversion heterozygote has the following inversion chromosome:

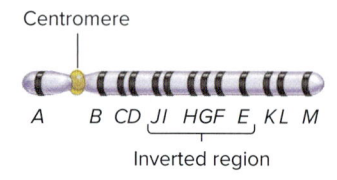

A B CD J I HGF E KL M (Inverted region: CD J I HGF E)

What would be the products if a crossover occurred between genes *H* and *I* on the inverted chromosome and a normal chromosome?

C9. Explain why inversions and reciprocal translocations do not usually cause a phenotypic effect. In a few cases, however, they do. Explain how.

C10. An individual has the following reciprocal translocation:

(Chromosomes with genes A, B, H, C, D, E and H, B, C, D, E and H, I, J, K, L, M and A, I, J, K, L, M)

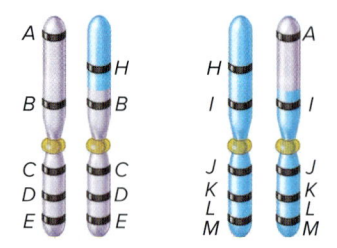

What would be the outcome of alternate segregation and of adjacent-1 segregation?

C11. A phenotypically normal individual has the following combinations of normal and abnormal chromosomes:

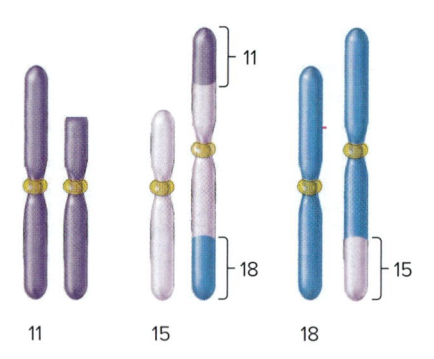

(Chromosomes labeled 11, 15, 18 with abnormal combinations labeled 11, 18, 15)

The normal chromosome is shown on the left in each pair. Suggest a series of events (breaks, translocations, crossovers, etc.) that may have produced these combinations of chromosomes.

C12. Two phenotypically normal parents produce a phenotypically abnormal child in which chromosome 5 is missing part of its long arm but has a piece of chromosome 7 attached to it. The child also has one normal copy of chromosome 5 and two normal copies of chromosome 7. With regard to chromosomes 5 and 7, what do you think are the chromosomal compositions of the parents?

C13. With regard to the segregation of centromeres, why is adjacent-2 segregation less frequent than alternate or adjacent-1 segregation?

C14. Which of the following types of chromosomal changes would you expect to have phenotypic consequences? Explain your choices.

A. Pericentric inversion

B. Reciprocal translocation

C. Deletion

D. Unbalanced translocation

C15. Explain why a translocation cross occurs during metaphase of meiosis I when a cell contains a reciprocal translocation.

C16. A phenotypically abnormal individual has a phenotypically normal father with an inversion on one copy of chromosome 7 and a phenotypically normal mother without any changes in chromosome structure. The orders of genes along the two copies of chromosome 7 in the father are as follows:

R T D M centromere *P U X Z C*   (normal chromosome 7)

R T D U P centromere *M X Z C*   (inversion chromosome 7)

The phenotypically abnormal offspring has a chromosome 7 with the following order of genes:

R T D M centromere *P U D T R*

Using a sketch, explain how this chromosome was formed. In your answer, explain where the crossover occurred (i.e., between which two genes).

C17. A diploid fruit fly has 8 chromosomes. How many chromosomes would a fly with each of the following chromosomal compositions have?

A. Tetraploid

B. Trisomy 2

C. Monosomy 3

D. 3n

E. 4n + 1

C18. A person is born with one X chromosome, zero Y chromosomes, trisomy 21, and two copies of the other chromosomes. How many chromosomes does this person have altogether? Explain whether this person is euploid or aneuploid.

C19. Aneuploidy is typically detrimental, whereas polyploidy is sometimes beneficial, particularly in plants. Discuss why you think this is the case.

C20. Explain how aneuploidy, deletions, and duplications cause genetic imbalances. Why do you think that deletions and monosomies are more detrimental than duplications and trisomies?

C21. Why do you think humans with trisomy 13, 18, or 21 can survive but other trisomies are lethal? Even though an X chromosome is large, aneuploidy of this chromosome is also tolerated. Explain why.

C22. A zookeeper has collected a male and female lizard that look like they belong to the same species. They mate with each other and produce phenotypically normal offspring. However, the offspring are sterile. Suggest one or more explanations for their sterility.

C23. What is endopolyploidy? What is its biological significance?

C24. What is mosaicism? How is it produced?

C25. Explain how polytene chromosomes of *Drosophila* are produced and how they form a six-armed structure.

C26. Describe some of the advantages of polyploid plants. What are the consequences of having an odd number of chromosome sets?

C27. A diploid fruit fly has 8 chromosomes. Which of the following terms should *not* be used to describe a fruit fly with four sets of chromosomes?

   A. Polyploid   D. Tetraploid
   B. Aneuploid   E. 4n
   C. Euploid

C28. Which of the following terms should *not* be used to describe a human with three copies of chromosome 12?

   A. Polyploid   D. Euploid
   B. Triploid   E. 2n + 1
   C. Aneuploid   F. Trisomy 12

C29. The kidney bean plant, *Phaseolus vulgaris*, is a diploid species containing a total of 22 chromosomes in somatic cells. How many possible types of trisomic individuals could be produced in this species?

C30. A triploid plant has 18 chromosomes (i.e., 6 chromosomes per set). If we assume a gamete has an equal probability of receiving one or two copies of each of the 6 types of chromosome, what are the odds of this plant producing a haploid or a diploid gamete? What are the odds of producing an aneuploid gamete? If the plant is allowed to self-fertilize, what are the odds of producing a euploid offspring?

C31. Describe three naturally occurring ways that the chromosome number can change.

C32. Meiotic nondisjunction is much more likely than mitotic nondisjunction. Based on this observation, would you conclude that meiotic nondisjunction is usually due to nondisjunction during meiosis I or meiosis II? Explain your reasoning.

C33. A woman who is heterozygous, *Bb*, has brown eyes; *B* (brown) is the dominant allele, and *b* (blue) is recessive. One of her eyes, however, has a patch of blue color. Give three different explanations for how this might have occurred.

C34. Meiotic nondisjunction usually occurs during meiosis I. What is not separating properly: bivalents or sister chromatids? What is not separating properly during mitotic nondisjunction?

C35. Table 8.1 shows that Turner syndrome occurs when an individual inherits one X chromosome but lacks a second sex chromosome. Can Turner syndrome be due to nondisjunction during oogenesis, spermatogenesis, or both? If a phenotypically normal couple has a color-blind child (due to a recessive X-linked allele) with Turner syndrome, did nondisjunction occur during oogenesis or spermatogenesis in this child's parents? Explain your answer.

C36. Male honeybees, which are haploid, produce sperm by meiosis. Explain what unusual event (compared with other animals) must occur during spermatogenesis in honeybees to produce sperm. Does this unusual event occur during meiosis I or meiosis II?

Application and Experimental Questions

E1. With regard to the analysis of chromosome structure, explain the experimental advantage that polytene chromosomes offer. Discuss why changes in chromosome structure are more easily detected in polytene chromosomes than in ordinary chromosomes.

E2. Describe how colchicine can be used to alter chromosome number.

E3. Female fruit flies homozygous for the X-linked white-eye allele are crossed to males with red eyes. On very rare occasions, an offspring of such a cross is a male with red eyes. Assuming these rare offspring are not due to a new mutation in one of the mother's X chromosomes that converted the white-eye allele into a red-eye allele, explain how a red-eyed male arises.

E4. A cytogeneticist has collected tissue samples from members of a certain butterfly species. Some of the butterflies were located in Canada, and others were found in Mexico. Through karyotyping, the cytogeneticist discovered that chromosome 5 of the Canadian butterflies had a large inversion compared with chromosome 5 of the Mexican butterflies. The Canadian butterflies were inversion homozygotes, whereas the Mexican butterflies had two normal copies of chromosome 5.

   A. Explain whether a mating between Canadian and Mexican butterflies would produce phenotypically normal offspring.

   B. Explain whether the offspring of a cross between Canadian and Mexican butterflies would be fertile.

E5. While conducting field studies on a chain of islands, you karyotype two phenotypically identical groups of turtles, which are found on different islands. The turtles on one island have 24 chromosomes, but those on another island have 48 chromosomes. How would you explain this observation? How do you think the turtles with 48 chromosomes came into being? If you crossed the two types of turtles, would you expect the offspring to be phenotypically normal? Would you expect them to be fertile? Explain.

E6. It is an exciting time to be a plant breeder because so many options are available for the development of new types of agriculturally useful plants. Let's suppose you wish to develop a seedless tomato that can grow in a very hot climate and is resistant to a viral pathogen that commonly infects tomato plants. At your disposal, you have a seed-bearing tomato strain that is heat-resistant and produces great-tasting tomatoes. You also have a wild strain of tomato plants (which have lousy-tasting tomatoes) that is resistant to the viral pathogen. Suggest a series of steps you might follow to produce a great-tasting, seedless tomato that is resistant to heat and the viral pathogen.

E7. What are G-bands? Discuss how G-bands are useful in the analysis of chromosome structure.

E8. A female fruit fly has one normal X chromosome and one X chromosome with a deletion. The deletion occurred in the middle of the X chromosome and removed about 10% of the entire length of the X chromosome. Suppose you stained and observed the chromosomes in salivary gland cells of this female fruit fly. Draw the polytene arm of the X chromosome. Explain your drawing.

E9. Describe two different experimental strategies that could be used to create an allotetraploid from two different diploid species of plants.

Questions for Student Discussion/Collaboration

1. A chromosome that was involved in a reciprocal translocation also has an inversion. In addition, the cell contains two normal chromosomes.

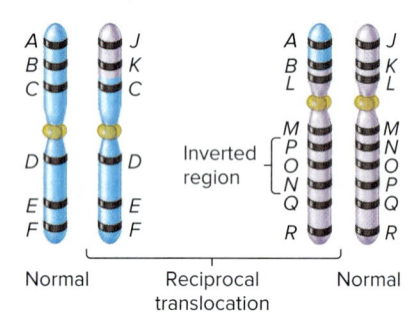

   Make a drawing that shows how these chromosomes will pair during metaphase of meiosis I.

2. Besides the ones mentioned in this textbook, look for other examples of variations in euploidy. Perhaps you might look in more advanced textbooks concerning population genetics, ecology, or other related fields. Discuss the phenotypic consequences of these changes.

3. Cell biology textbooks often discuss cellular proteins encoded by genes that are members of a gene family. Examples of such proteins include myosins and glucose transporters. Look through a cell biology textbook and identify some proteins encoded by members of gene families. Discuss the importance of gene families at the cellular level.

4. Discuss how variation in chromosome number has been useful in agriculture.

Answers to Comprehension Questions

8.1: c, d
8.2: c
8.3: a, d
8.4: b, d, d
8.5: b
8.6: a, d
8.7: c, b
8.8: c, b

Note: All answers appear at the website in Connect; the answers to even-numbered questions and all Concept Check questions are in Appendix B.

CHAPTER OUTLINE

- 9.1 Overview of Genetic Transfer in Bacteria
- 9.2 Bacterial Conjugation
- 9.3 Conjugation and Mapping via Hfr Strains
- 9.4 Bacterial Transduction
- 9.5 Bacterial Transformation
- 9.6 Medical Relevance of Horizontal Gene Transfer

Conjugating bacteria. The bacteria shown here are transferring genetic material by a process called conjugation.
©Eye of Science/Science Source

GENETICS OF BACTERIA

Thus far, in Chapters 1 through 8, we have primarily focused on eukaryotes, which include protists, fungi, plants, and animals. In Chapters 9 and 10, we will turn our attention to bacteria and viruses, respectively. One reason why researchers are so interested in bacteria and viruses is their impact on health. Infectious diseases caused by these agents are a leading cause of human death, accounting for one-quarter to one-third of deaths worldwide. The spread of infectious diseases results from human behavior, and in recent times it has been accelerated by increased trade and travel and the inappropriate use of antibiotic drugs. An alarming increase in more deadly strains of bacteria and viruses has occurred over the past few decades. Since 1980, the number of deaths in the United States due to infectious diseases has approximately doubled.

In this chapter, we will focus on the genetic analysis of bacteria. Like their eukaryotic counterparts, bacteria often possess allelic differences that affect their cellular traits. Common allelic variations among bacteria involve traits such as sensitivity to antibiotics and differences in nutrient requirements for growth. In these cases, the allelic differences are between different strains of bacteria, because any given bacterium is usually haploid for a particular gene. In fact, the haploid nature of bacteria is a feature that makes it easier to identify mutations that produce phenotypes such as altered nutritional requirements. Loss-of-function mutations, which are often recessive in diploid eukaryotes, are not masked by dominant alleles in haploid species. Throughout this chapter, we will consider interesting experiments that examine bacterial strains with allelic differences.

Compared with eukaryotes, a striking difference in prokaryotic species is their mode of reproduction. Because bacteria reproduce asexually, researchers do not use crosses in genetic analyses of bacterial species. Instead, they rely on a similar mechanism, called genetic transfer, in which a segment of bacterial DNA is transferred from one bacterium to another. In this chapter, we will explore different routes of genetic transfer and see how researchers have used this process to map the locations of genes along the chromosome of a few bacterial species. We will also consider the medical relevance of bacterial gene transfer.

9.1 OVERVIEW OF GENETIC TRANSFER IN BACTERIA

Learning Outcome:

1. Compare and contrast the three mechanisms of genetic transfer in bacteria.

CHAPTER 9 :: GENETICS OF BACTERIA

In Chapter 2, we considered how bacteria reproduce asexually by a process called binary fission (refer back to Figure 2.4). This process produces two daughter cells that are genetically identical to the mother cell. In addition, a segment of DNA from one bacterial cell can occasionally be transferred to another cell. This process is called **genetic transfer.** Why is genetic transfer an advantage? Like sexual reproduction in eukaryotes, genetic transfer in bacteria enhances the genetic diversity of bacterial species. For example, a bacterial cell carrying a gene that provides antibiotic resistance may transfer this gene to another bacterial cell, enabling that cell to survive exposure to the antibiotic.

Bacteria can transfer genetic material naturally in three ways (**Table 9.1**).

- **Conjugation** involves a direct physical interaction between two bacterial cells. One bacterium acts as a donor and transfers genetic material to a recipient cell.
- During **transduction,** a virus infects a bacterium and then transfers bacterial genetic material from that bacterium to another.
- **Transformation** is a process in which genetic material is released into the environment when a bacterial cell dies. This material then binds to a living bacterial cell, which can take it up.

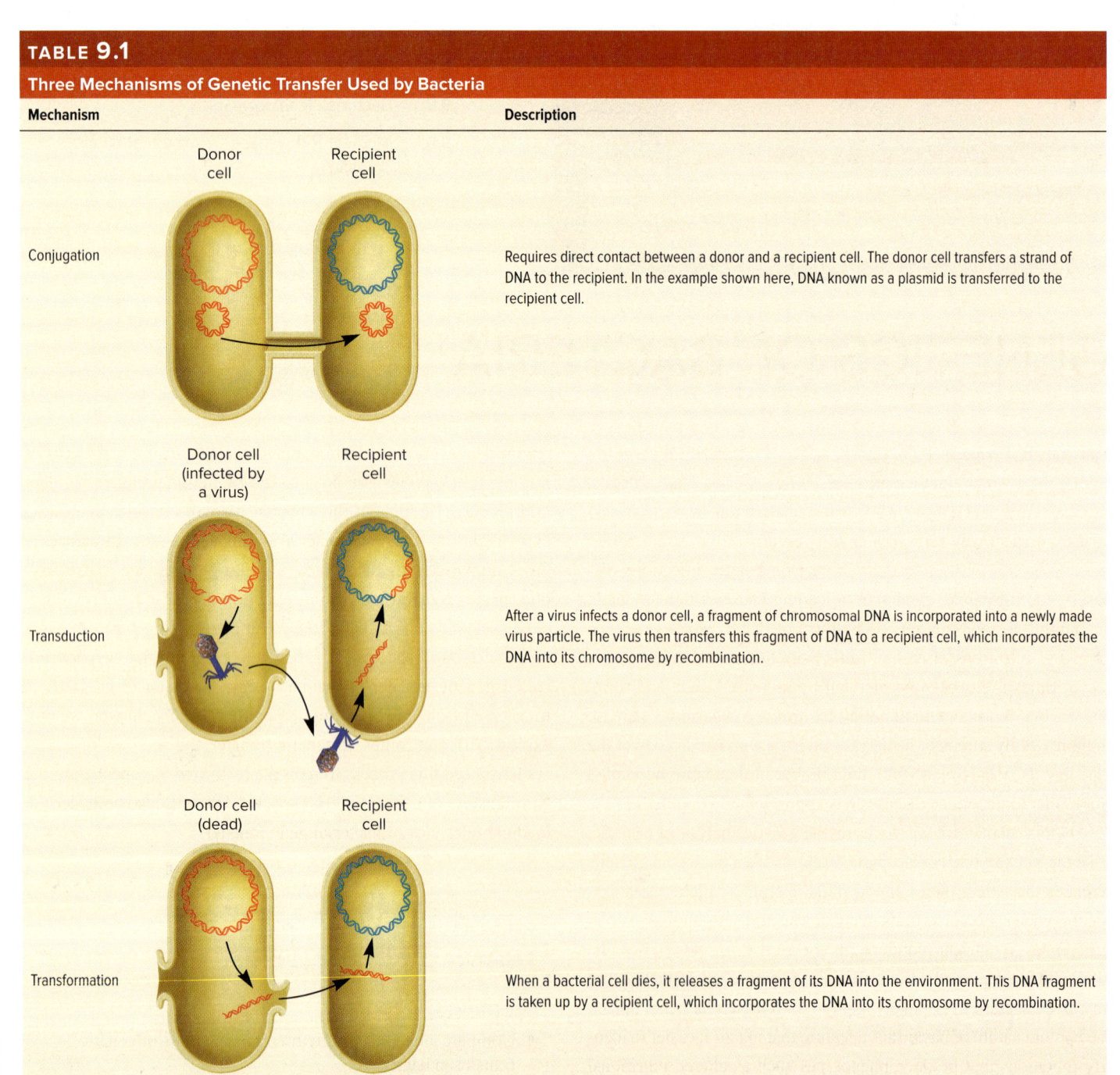

TABLE 9.1
Three Mechanisms of Genetic Transfer Used by Bacteria

Mechanism	Description
Conjugation	Requires direct contact between a donor and a recipient cell. The donor cell transfers a strand of DNA to the recipient. In the example shown here, DNA known as a plasmid is transferred to the recipient cell.
Transduction	After a virus infects a donor cell, a fragment of chromosomal DNA is incorporated into a newly made virus particle. The virus then transfers this fragment of DNA to a recipient cell, which incorporates the DNA into its chromosome by recombination.
Transformation	When a bacterial cell dies, it releases a fragment of its DNA into the environment. This DNA fragment is taken up by a recipient cell, which incorporates the DNA into its chromosome by recombination.

These three mechanisms of genetic transfer have been extensively investigated in research laboratories, and their molecular mechanisms continue to be studied with great interest. In later sections of this chapter, we will examine these three mechanisms in greater detail.

We will also explore how genetic transfer between bacterial cells has provided unique ways to accurately map bacterial genes. The mapping methods described in this chapter have been largely replaced by molecular approaches described in Chapter 21. Even so, the mapping of bacterial genes serves to illuminate the mechanisms by which genes are transferred between bacterial cells and also helps us to appreciate the strategies of newer mapping approaches.

9.1 REVIEWING THE KEY CONCEPTS

- Three general mechanisms for genetic transfer used by various species of bacteria are conjugation, transduction, and transformation (see Table 9.1).

9.1 COMPREHENSION QUESTION

1. A form of genetic transfer that involves the uptake of a fragment of DNA from the environment is called
   a. conjugation.
   b. transduction.
   c. transformation.
   b. all of the above.

9.2 BACTERIAL CONJUGATION

Learning Outcomes:

1. Analyze the work of Lederberg and Tatum and that of Davis, and explain how the data indicated that some strains of bacteria can transfer genetic material via direct physical contact.
2. Outline the steps of conjugation via F factors.
3. Compare and contrast different types of plasmids.

As described briefly in Table 9.1, conjugation involves the direct transfer of genetic material from one bacterial cell to another. In this section, we will examine the steps in this process at the molecular and cellular levels.

Bacteria Can Transfer Genetic Material During Conjugation

The natural ability of one bacterial cell to transfer genetic material to another bacterial cell was first recognized by Joshua Lederberg and Edward Tatum in 1946. They were studying strains of *Escherichia coli* that had different nutritional requirements for growth. A **minimal medium** is a growth medium that contains the essential nutrients for a wild-type (nonmutant) bacterial species to grow. Researchers often study bacterial strains that harbor mutations and cannot grow on minimal media.

- A strain that cannot synthesize a particular nutrient and needs that nutrient to be supplemented in its growth medium

is called an **auxotroph.** For example, a strain that cannot make the amino acid methionine does not grow on a minimal medium. Such a strain needs to have methionine added to its growth medium and is called a methionine auxotroph.
- By comparison, a strain that can make this amino acid is termed a methionine prototroph. A **prototroph** does not need a particular nutrient in its growth medium.

The experiment in **Figure 9.1** considers one *E. coli* strain, designated $met^-\ bio^-\ thr^+\ leu^+\ thi^+$, which required the addition to its growth medium of one amino acid, methionine (met), and one vitamin, biotin (bio). This strain did not require the amino acids threonine (thr) and leucine (leu) or the vitamin thiamine (thi)

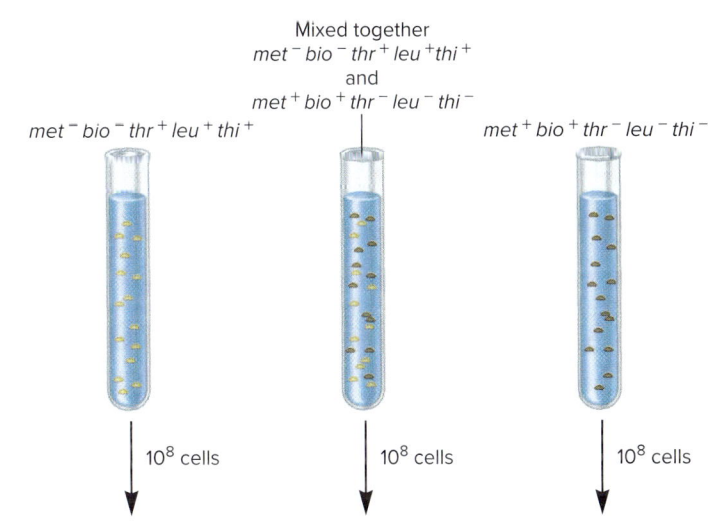

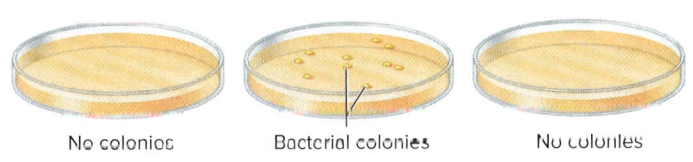

FIGURE 9.1 Experiment of Lederberg and Tatum demonstrating genetic transfer during conjugation in *E. coli*. When plated on a growth medium lacking amino acids, biotin, and thiamine, the $met^-\ bio^-\ thr^+\ leu^+\ thi^+$ and $met^+\ bio^+\ thr^-\ leu^-\ thi^-$ strains were unable to grow. However, if the two strains were mixed together and then plated, some colonies were observed. These colonies were due to the transfer of genetic material between these two strains by conjugation. Note: In bacteria, it is common to give genes a three-letter name (shown in italics) that is related to the function of the gene. A plus superscript ($^+$) indicates a functional gene, and a minus superscript ($^-$) indicates a mutation that has caused the gene or gene product to be inactive. In some cases, several genes have related functions. These may have the same three-letter name followed by different capital letters. For example, different genes involved with leucine biosynthesis may be called *leuA*, *leuB*, *leuC*, and so on. In the experiment described here, the genes involved in leucine biosynthesis were not distinguished, so the gene is simply referred to as leu^+ (for a functional gene) and leu^- (for a nonfunctional gene).

Concept Check: Describe how genetic transfer can explain the growth of colonies on the middle plate.

188   CHAPTER 9 :: GENETICS OF BACTERIA

to be added to the growth medium. Another *E. coli* strain, designated met^+ bio^+ thr^- leu^- thi^-, had just the opposite requirements. It was an auxotroph for threonine, leucine, and thiamine, but a prototroph for methionine and biotin. These differences in nutritional requirements correspond to variations in the genetic material of the two strains. The first strain had two defective genes that would encode enzymes necessary for methionine and biotin synthesis. The second strain had three defective genes that would be needed to make threonine, leucine, and thiamine.

Figure 9.1 compares the results when the two strains were mixed together and when they were not mixed. In each case, about 100 million (10^8) cells were applied to plates on a growth medium lacking amino acids, biotin, and thiamine.

- Without mixing, when 10^8 met^- bio^- thr^+ leu^+ thi^+ cells were plated, no colonies were observed to grow. This result is expected because the medium did not contain methionine or biotin.
- Similarly, when 10^8 met^+ bio^+ thr^- leu^- thi^- cells were plated, no colonies were observed because threonine, leucine, and thiamine were missing from this growth medium.
- When the two strains were mixed together and then 10^8 cells plated, approximately 10 bacterial colonies formed.

A bacterial colony is derived from a single bacterial cell by many successive cell divisions. Because bacterial colonies were observed to grow after mixing, the genotype of the cells within these colonies must have been met^+ bio^+ thr^+ leu^+ thi^+. How could this genotype occur? Because no colonies were observed on either plate in which the two strains were not mixed, Lederberg and Tatum concluded that it was not due to mutations that converted met^- bio^- to met^+ bio^+ or to mutations that converted thr^- leu^- thi^- to thr^+ leu^+ thi^+. Instead, they hypothesized that some genetic material was transferred between the two strains.

- One possibility is that the genetic material providing the ability to synthesize methionine and biotin (met^+ bio^+) was transferred to the met^- bio^- thr^+ leu^+ thi^+ strain.
- Alternatively, the ability to synthesize threonine, leucine, and thiamine (thr^+ leu^+ thi^+) may have been transferred to the met^+ bio^+ thr^- leu^- thi^- cells.

The results of this experiment did not distinguish between these two possibilities.

Conjugation Requires Direct Physical Contact

In 1950, Bernard Davis conducted experiments showing that two strains of bacteria must make physical contact with each other to transfer genetic material. The apparatus he used, known as a U-tube, is shown in **Figure 9.2**. At the bottom of the U-tube is a filter with pores small enough to allow the passage of genetic material (i.e., DNA molecules) but too small to permit the passage of bacterial cells.

1. On one side of the filter, Davis added a bacterial strain with a certain combination of nutritional requirements (the met^- bio^- thr^+ leu^+ thi^+ strain).

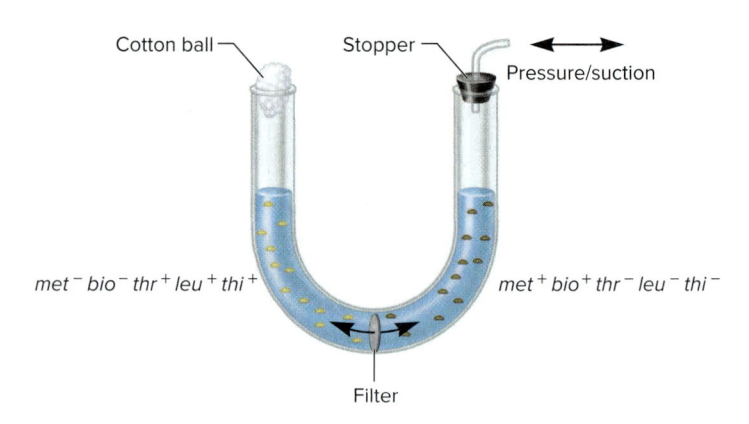

FIGURE 9.2   A U-tube apparatus like that used by Davis. The fluid in the tube is forced through the filter by alternating suction and pressure. However, the pores in the filter are too small for the passage of bacteria.

Concept Check: With regard to studying the mechanism of conjugation, what is the purpose of using a U-tube?

2. On the other side, he added a different bacterial strain (the met^+ bio^+ thr^- leu^- thi^- strain).
3. The application of alternating pressure and suction promoted the movement of liquid through the filter. Because the bacteria were too large to pass through the pores, the movement of liquid did not allow the two types of bacterial strains to mix with each other. However, any genetic material that was released from a bacterium could pass through the filter.
4. After incubation in a U-tube, bacteria from either side of the tube were placed on a medium that selected for the growth of cells that were met^+ bio^+ thr^+ leu^+ thi^+. This minimal medium lacked methionine, biotin, threonine, leucine, and thiamine, but contained all other nutrients essential for growth.
5. In this case, no bacterial colonies grew on the plates. The experiment showed that, without physical contact, the two bacterial strains did not transfer genetic material to one another.

The term *conjugation* is now used to describe the natural process of genetic transfer between bacterial cells that requires direct cell-to-cell contact. It is also called bacterial mating. Many, but not all, species of bacteria can conjugate.

An F⁺ Strain Transfers an F Factor to an F⁻ Strain During Conjugation

We now know that certain donor strains of *E. coli* contain a small circular segment of genetic material known as an **F factor** (for fertility factor) in addition to their circular chromosome. Strains of *E. coli* that contain an F factor are designated F⁺, whereas strains without an F factor are termed F⁻. In recent years, the molecular details of the conjugation process have been extensively studied. Though the mechanisms vary somewhat from one bacterial species to another, some general themes have emerged. F factors carry several genes

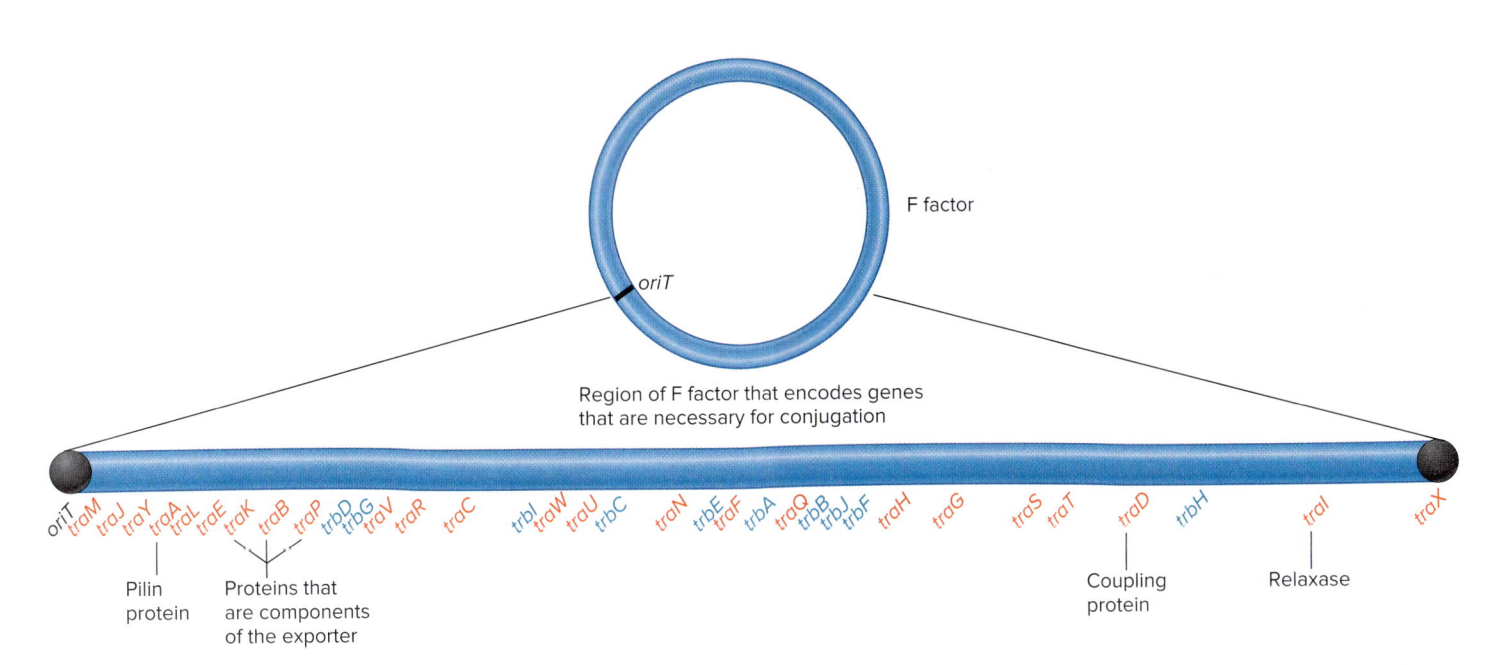

FIGURE 9.3 Genes on the F factor that play a role during conjugation. A region of the F factor contains a few dozen genes that play a role in the conjugation process. Because they play a role in the transfer of DNA from donor to recipient cell, the genes are designated with the three-letter name of *tra* or *trb*, followed by a capital letter. (Note: The *tr-* prefix stands for "transfer.") The *tra* genes are shown in red, and the *trb* genes are shown in blue. The functions of a few examples are indicated. The origin of transfer is designated *oriT*.

Concept Check: Would this circular DNA molecule be found in an F^+ or F^- cell?

that are required for conjugation to occur. For example, **Figure 9.3** shows the arrangement of genes on the F factor found in certain strains of *E. coli*. The proteins encoded by these genes are needed to transfer a strand of DNA from the donor cell to a recipient cell.

Contact between donor and recipient cells is a key step that initiates the conjugation process. **Sex pili** (singular: **pilus**) are made by F^+ strains (**Figure 9.4a**). The gene encoding the pilin protein (*traA*) is located on the F factor. The pili act as attachment sites that promote the binding of bacteria to each other. In this way, an F^+ strain makes physical contact with an F^- strain. In certain species, such as *E. coli*, long pili project from F^+ cells and attempt to make contact with nearby F^- cells.

Figure 9.4b shows the molecular events that occur during conjugation in *E. coli*.

1. Once contact is made, the pili shorten and thereby draw the donor and recipient cells closer together. A **conjugation bridge** is later formed between the two cells, which provides a passageway for DNA transfer.
2. Genes within the F factor encode a protein complex called the **relaxosome**. This complex first recognizes a DNA sequence in the F factor known as the **origin of transfer**. Upon recognition, the relaxosome cuts one DNA strand at that site in the F factor.
3. The relaxosome also catalyzes the separation of the DNA strands, and only the cut DNA strand, called **T DNA**, is transferred to the recipient cell. As the DNA strands separate, most of the proteins within the relaxosome are released, but a protein called relaxase remains bound to one end of the cut DNA strand.
4. The next phase of conjugation involves the export of the nucleoprotein complex from the donor cell to the recipient cell. To begin this process, the DNA/relaxase complex is recognized by a coupling factor that promotes the entry of the nucleoprotein into the exporter, a complex of proteins that spans both inner and outer membranes of the donor cell. In bacterial species, this complex is formed from 10 to 15 different proteins that are encoded by genes within the F factor.
5. Once the DNA/relaxase complex is pumped out of the donor cell, it travels through the conjugation bridge and then into the recipient cell. The other strand of the F factor DNA remains in the donor cell, where DNA replication restores this DNA to its original double-stranded condition. After the recipient cell receives a single strand of the F-factor DNA, relaxase catalyzes the joining of the ends of this linear molecule to form a circular molecule. This single-stranded DNA is replicated in the recipient cell to become double-stranded.

The result of conjugation is that the recipient cell has acquired an F factor, converting it from an F^- to an F^+ cell. The genetic composition of the donor cell has not changed.

Bacteria May Contain Different Types of Plasmids

An F factor is one type of DNA that exists independently of the chromosomal DNA. The more general term for this type of structure is a **plasmid**. Most known plasmids are circular, although some are linear. Plasmids occur naturally in many strains of bacteria and in a few types of eukaryotic cells such as yeast.

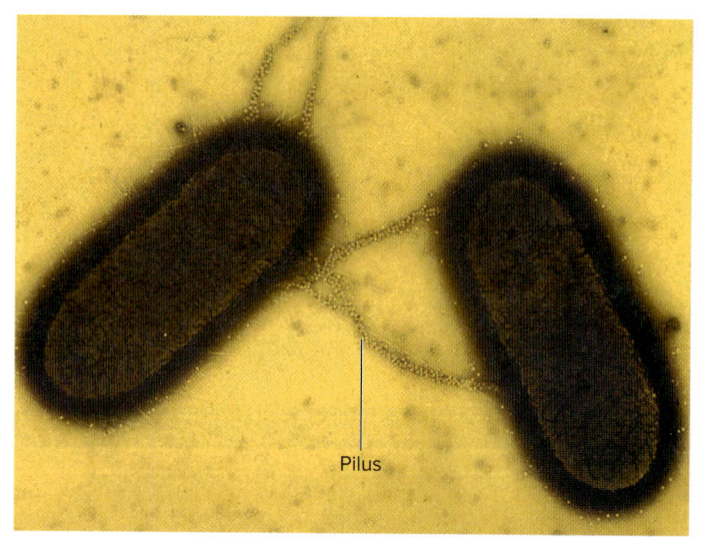

(a) Conjugating E. coli

FIGURE 9.4 **The transfer of an F factor during bacterial conjugation.** (a) Two *E. coli* cells in the act of conjugation. The cell on the left is F$^+$, and the one on the right is F$^-$. The two cells make contact with each other via sex pili that are made by the F$^+$ cell. (b) The mechanism of transfer. The end result is that both cells have an F factor.
©Dr. L. Caro/SPL/Science Source

Concept Check: What are the functions of relaxase, coupling factor, and the exporter in the process of conjugation?

The smallest plasmids consist of just a few thousand base pairs (bp) and carry only a gene or two; the largest are in the range of 100,000 to 500,000 bp and carry several dozen or even hundreds of genes. Some plasmids, such as F factors, can integrate into a chromosome. These plasmids are also called **episomes.**

A plasmid has its own origin of replication that allows it to be replicated independently of the bacterial chromosome. The DNA sequence of the origin of replication influences how many copies of the plasmid are found within a cell. Some origins are said to be very strong because they result in many copies of the plasmid, perhaps as many as 100 per cell. Other origins of replication have sequences that are described as much weaker, because the number of copies created is relatively low, such as one or two per cell.

Why do bacteria have plasmids? Plasmids are not usually necessary for bacterial survival. However, in many cases, certain genes within a plasmid provide some type of growth advantage to the cell. By studying plasmids in many different species, researchers have discovered that most plasmids fall into a few different categories:

- Fertility plasmids, also known as F factors, allow bacteria to conjugate with each other.
- Resistance plasmids, also known as R factors, contain genes that confer resistance against antibiotics and other types of toxins.
- Degradative plasmids carry genes that enable the bacterium to utilize an unusual substance. For example, a degradative plasmid may carry genes that allow a bacterium to metabolize an organic solvent such as toluene.

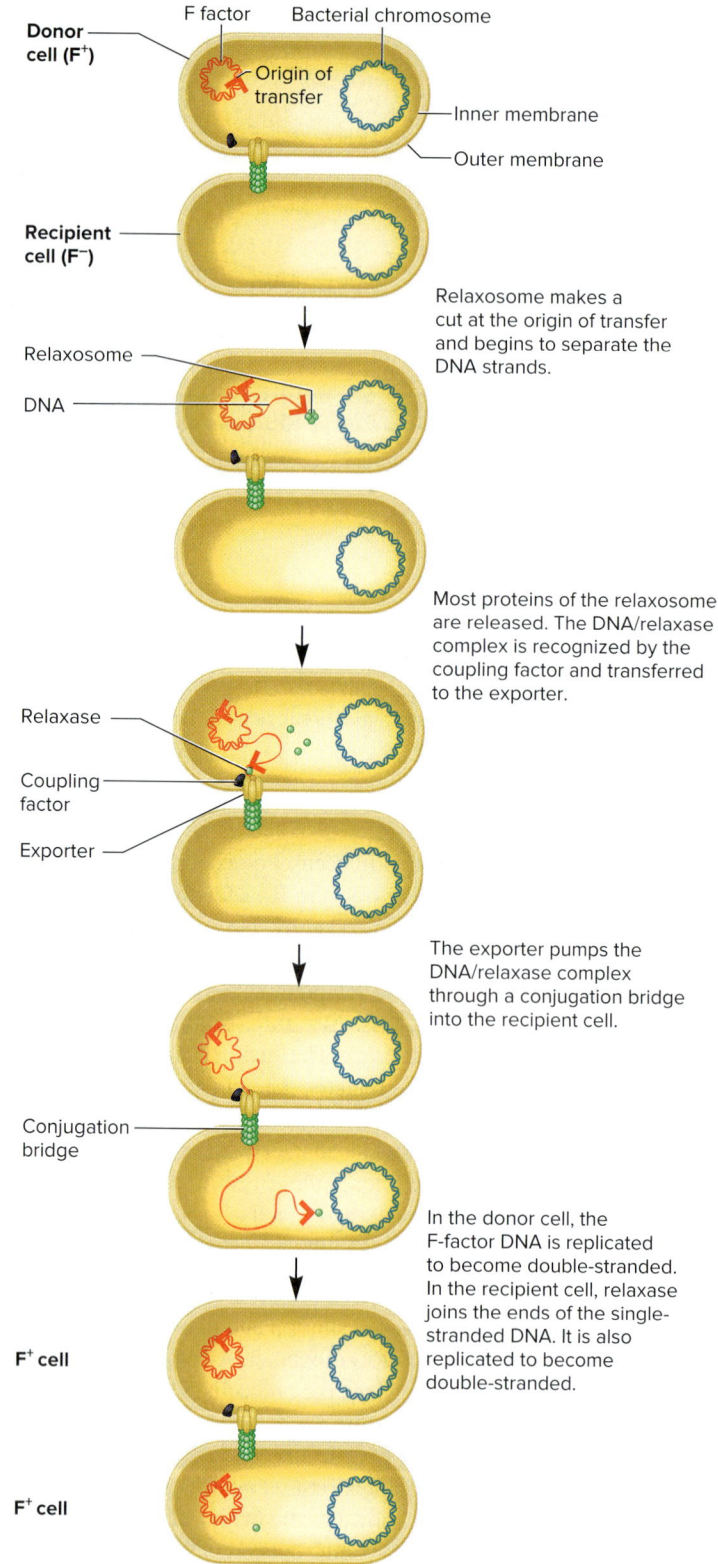

(b) Transfer of an F factor via conjugation

- Col-plasmids contain genes that encode colicins, which are proteins that kill other bacteria.
- Virulence plasmids carry genes that turn a bacterium into a pathogenic strain.

9.2 REVIEWING THE KEY CONCEPTS

- Lederberg and Tatum discovered conjugation in *E. coli* by analyzing auxotrophic strains (see Figure 9.1).
- Using a U-tube apparatus, Davis showed that conjugation requires cell-to-cell contact (see Figure 9.2).
- Certain strains of bacteria have F factors, which they can transfer via conjugation in a series of steps (see Figures 9.3, 9.4).
- Bacteria may contain different types of plasmids.

9.2 COMPREHENSION QUESTIONS

1. A bacterial cell with an F factor conjugates with an F⁻ cell. Following conjugation, the two cells will be
   a. F⁺.
   b. F⁻.
   c. one F⁺ and one F⁻.
   d. none of the above.
2. Which of the following is a type of plasmid?
   a. F factor (fertility factor)
   b. R factor (resistance plasmid)
   c. Virulence plasmid
   d. All of the above are types of plasmids.

9.3 CONJUGATION AND MAPPING VIA Hfr STRAINS

Learning Outcomes:
1. Explain how an Hfr strain is produced.
2. Describe how bacterial cells of an Hfr strain can transfer genes to recipient cells.
3. Construct a genetic map using data from conjugation experiments.

Thus far, we have considered how conjugation may involve the transfer of an F factor from a donor to a recipient cell. In addition to this form of conjugation, certain donor strains of *E. coli*, called **Hfr strains** (for **h**igh **f**requency of **r**ecombination), are capable of a different type of conjugation. In this section, we will examine how Hfr strains are formed and how they transfer genes to recipient cells. We will also explore the use of Hfr strains to map genes along the *E. coli* chromosome.

Hfr Strains Have an F Factor Integrated into the Bacterial Chromosome

Luca Cavalli-Sforza discovered a strain of *E. coli* that was very efficient at transferring many chromosomal genes to recipient F⁻ strains. Cavalli-Sforza designated this bacterial strain an Hfr strain. How is an Hfr strain formed? As shown in **Figure 9.5a**, an F factor may align with a region found in the bacterial chromosome. Due to recombination, which is described in Chapter 13, the F factor may integrate into the bacterial chromosome. In the example shown in Figure 9.5a, the F factor has integrated next to a *lac*⁺ gene. F factors can integrate into several different sites that are scattered around the *E. coli* chromosome.

Occasionally, the integrated F factor in an Hfr strain is excised from the bacterial chromosome. This process involves the looping out of the F-factor DNA from the chromosome, which is followed by recombination that releases the F factor from the chromosome (**Figure 9.5b**). In the example shown in Figure 9.5b, the excision is imprecise. This produces an F factor that carries a portion of the bacterial chromosome and leaves behind some of the F-factor DNA in the bacterial chromosome. F factors that carry a portion of the bacterial chromosome are called **F′ factors** (read "F prime factors"). For example, in the experiment of Lederberg and Tatum described earlier in this chapter (see Figure 9.1), an F′ factor carrying the *bio*⁺ and *met*⁺ genes or an F′ factor carrying the *thr*⁺, *leu*⁺, and *thi*⁺ genes may have been transferred to the recipient strain. Therefore, conjugation may introduce new genes into the recipient strain, thereby altering its genotype.

Hfr Strains Can Transfer a Portion of the Bacterial Chromosome to Recipient Cells

William Hayes, who independently isolated another Hfr strain, demonstrated that conjugation between an Hfr cell and an F⁻ cell involves the transfer of a portion of the bacterial chromosome from the Hfr cell to the F⁻ cell (**Figure 9.6**).

1. One of the DNA strands is cut, or nicked, at the origin of transfer, which determines the starting point and direction of this transfer process. This cut site is the starting point at which the Hfr chromosome enters the F⁻ recipient cell.
2. From this starting point, a strand of the DNA of the Hfr chromosome begins to enter the F⁻ cell in a linear manner. The transfer process occurs in conjunction with chromosomal replication, so the Hfr cell retains its original chromosomal composition. About 1.5 to 2 hours is required for the entire Hfr chromosome to pass into the F⁻ cell. Because most conjugations do not last that long, usually only a portion of the Hfr chromosome is transmitted to the F⁻ cell.
3. Once inside the F⁻ cell, the chromosomal material from the Hfr cell can swap, or recombine, with the homologous region of the recipient cell's chromosome. (Chapter 13 describes the process of homologous recombination.)

How does this process affect the recipient cell? As illustrated in Figure 9.6, this recombination may provide the recipient cell with a new combination of alleles. In this example, the recipient strain was originally *lac*⁻ (unable to metabolize lactose) and *pro*⁻ (unable to synthesize proline). As shown in Figure 9.6, an important feature of Hfr conjugation is that the bacterial chromosome is transferred linearly to the recipient strain. In this example, *lac*⁺ is always transferred first, and *pro*⁺ is transferred later.

- If conjugation occurs for a short time, the recipient cell will receive a short segment of chromosomal DNA from the donor. In this case, the recipient cell becomes *lac*⁺ but remains *pro*⁻.
- If the conjugation is prolonged, the recipient cell will receive a longer segment of chromosomal DNA from the donor. After a longer conjugation, the recipient becomes *lac*⁺ and *pro*⁺.

192 CHAPTER 9 :: GENETICS OF BACTERIA

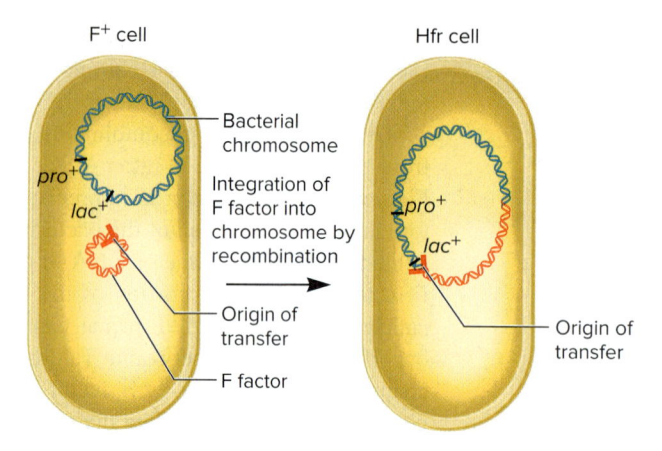

(a) When an F factor integrates into the chromosome, it creates an Hfr cell.

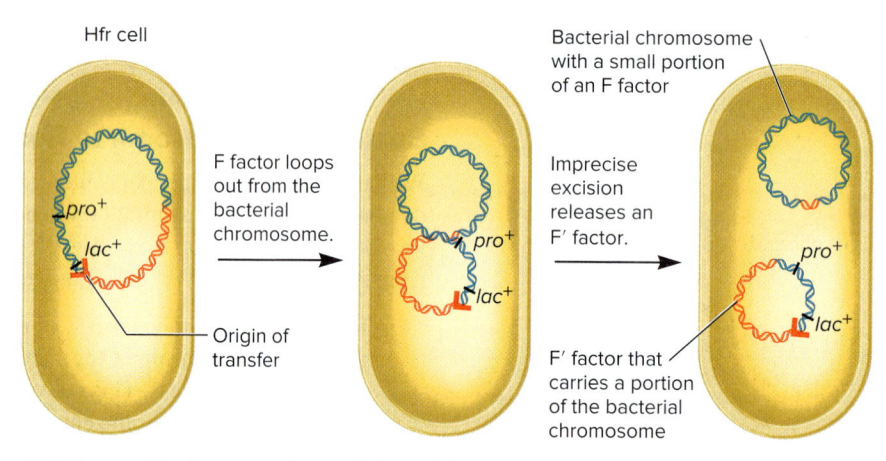

(b) When an F factor excises imprecisely, an F′ factor is created.

FIGURE 9.5 **Integration of an F factor to form an Hfr cell and its subsequent excision to form an F′ factor.** (a) An Hfr cell is created when an F factor integrates into the bacterial chromosome. (b) When an F factor is imprecisely excised, an F′ factor is created, and it carries a portion of the bacterial chromosome.

Concept Check: How is an F′ factor different from an F factor?

In any particular Hfr strain, the origin of transfer has a specific orientation that promotes either a counterclockwise or a clockwise transfer of genes. Among different Hfr strains, the origin of transfer may be located in different regions of the chromosome. Therefore, the order of genetic transfer depends on the location and orientation of the origin of transfer. For example, a different Hfr strain could have its origin of transfer next to pro^+ and transfer pro^+ first and then lac^+.

Conjugation Experiments Can Map Genes Along the *E. coli* Chromosome

The first genetic mapping experiments in bacteria were carried out by Elie Wollman and François Jacob in the 1950s. At the time of their studies, these researchers were aware of previous microbiological studies concerning **bacteriophages**—viruses that bind to bacterial cells and subsequently infect them. Those studies showed that bacteriophages can be sheared from the surface of *E. coli* cells if the cells are spun in a blender. In this treatment, the bacteriophages are detached from the surface of the bacterial cells, but the bacteria themselves remain healthy and viable. Wollman and Jacob reasoned that a blender treatment could also be used to separate bacterial cells that were in the act of conjugation without killing them. This technique is known as an **interrupted mating.**

The rationale behind Wollman and Jacob's mapping strategy is that the time it takes for genes to enter a donor cell is directly related to their order along the bacterial chromosome. They hypothesized that the chromosome of the donor strain in an Hfr conjugation is transferred in a linear manner to the recipient strain. If so, the order of genes along the chromosome can be deduced by determining the time it takes various genes to enter the recipient strain. Assuming the Hfr chromosome is transferred linearly, Wollman and Jacob realized that interruptions of conjugation at different times would lead to various lengths of the Hfr chromosome being transferred to the F⁻ recipient cell. If two bacterial cells conjugated for a short period of time, only a small

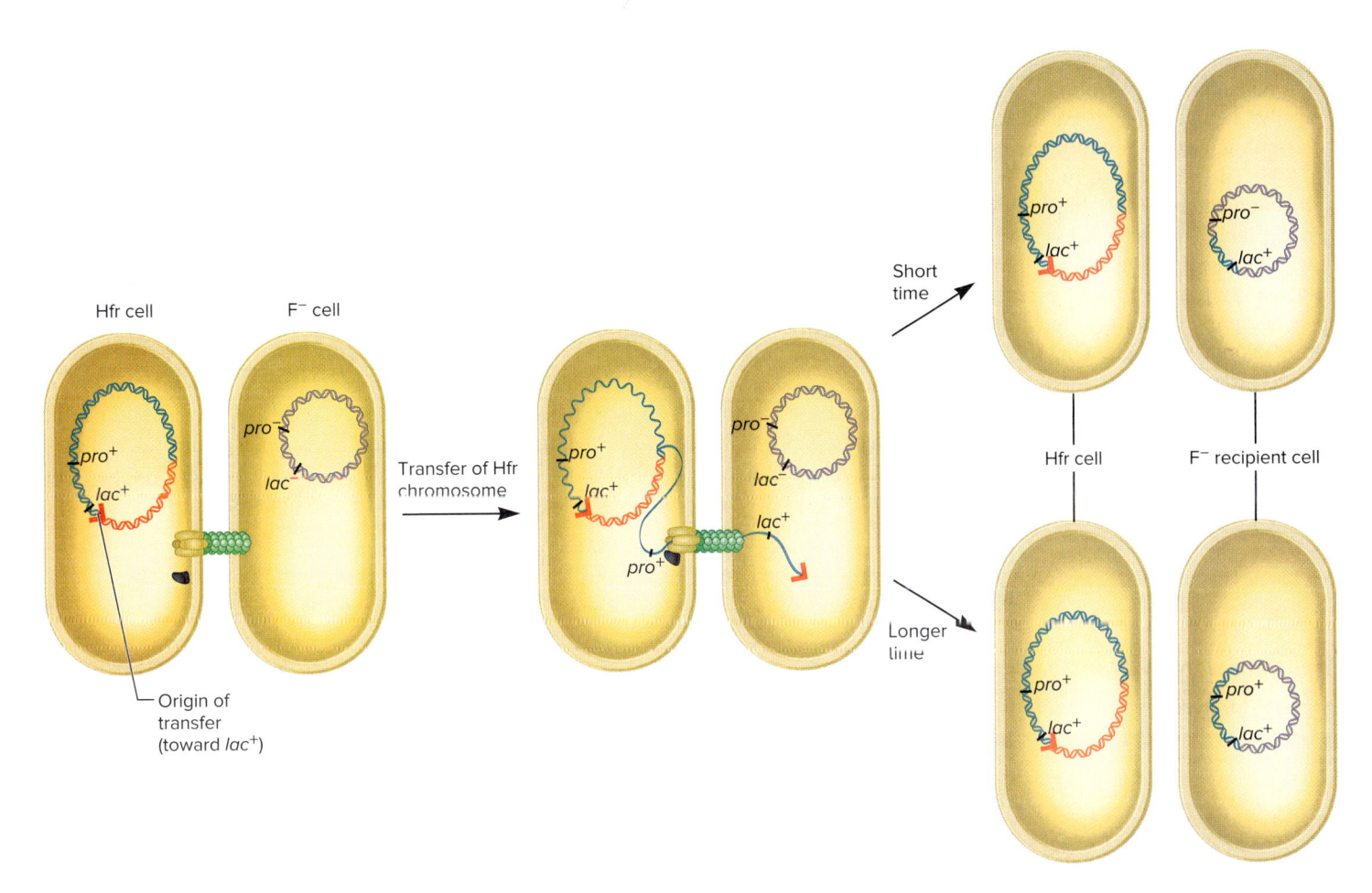

FIGURE 9.6 **Transfer of bacterial genes from an Hfr cell to an F⁻ cell.** The transfer of the bacterial chromosome begins at the origin of transfer and then proceeds around the circular chromosome. After a segment of chromosome has been transferred to the F⁻ recipient cell, it recombines with the recipient cell's chromosome.

Genes → Trait The F⁻ recipient cell was originally *lac*⁻ (unable to metabolize lactose) and *pro*⁻ (unable to synthesize proline). If conjugation occurs for a short period of time, the recipient cell acquires *lac*⁺, allowing it to metabolize lactose. If conjugation occurs for a longer period of time, the recipient cell also acquires *pro*⁺, enabling it to synthesize proline.

Concept Check: With regard to the timing of conjugation, explain why the recipient cell at the top right in this figure is *pro*⁻ whereas the recipient cell at the bottom right is *pro*⁺.

segment of the Hfr chromosome would be transferred to the recipient bacterium. However, if the bacterial cells were allowed to conjugate for a longer period before being interrupted, a longer segment of the Hfr chromosome could be transferred (see Figure 9.6). By determining which genes were transferred during short conjugations and which required longer times, Wollman and Jacob were able to deduce the order of particular genes along the *E. coli* chromosome.

As shown in the experiment of **Figure 9.7**, Wollman and Jacob began with two *E. coli* strains. The donor (Hfr) strain had the following genetic composition:

thr⁺: able to synthesize threonine, an essential amino acid for growth

leu⁺: able to synthesize leucine, an essential amino acid for growth

*azi*ˢ: sensitive (*s* stands for "sensitive") to killing by azide (a toxic chemical)

*ton*ˢ: sensitive to killing by bacteriophage T1

lac⁺: able to metabolize lactose and use it for growth

gal⁺: able to metabolize galactose and use it for growth

*str*ˢ: sensitive to killing by streptomycin (an antibiotic)

The recipient (F⁻) strain had the opposite genotype: *thr*⁻ *leu*⁻ *azi*ʳ *ton*ʳ *lac*⁻ *gal*⁻ *str*ʳ (*r* stands for "resistant"). Before the experiment, Wollman and Jacob already knew that the *thr*⁺ gene was transferred first, followed by the *leu*⁺ gene, and both were transferred relatively soon (5–10 minutes) after conjugation began. Their main goal in this experiment was to determine the times at which the other genes (*azi*ˢ *ton*ˢ *lac*⁺ *gal*⁺) were transferred to the recipient strain. The transfer of the *str*ˢ gene was not examined because streptomycin was used to kill the donor strain following conjugation.

Before discussing the interpretation of this experiment, let's consider how Wollman and Jacob monitored genetic transfer. To determine if particular genes had been transferred after conjugation, they took the conjugated bacterial cells and first plated them on a growth medium that lacked threonine (thr) and leucine (leu) but contained streptomycin (str). On these plates, the original donor and recipient strains could not grow because the donor strain was streptomycin-sensitive and the recipient strain required threonine and leucine. However, recipient cells into which the donor strain had transferred chromosomal DNA carrying the thr^+ and leu^+ genes would be able to grow.

To determine the order of genetic transfer of the azi^s, ton^s, lac^+, and gal^+ genes, Wollman and Jacob picked colonies from the first plates and restreaked them on media that contained azide or bacteriophage T1 or on media that contained lactose or galactose as the sole source of energy for growth. The plates were incubated overnight to observe the formation of visible bacterial growth. Whether or not the bacteria could grow depended on their genotypes. For example, a cell that is azi^s cannot grow on a medium containing azide, and a cell that is lac^- cannot grow on a medium containing lactose as the carbon source for growth. By comparison, a cell that is azi^r and lac^+ can grow on both types of media.

▶ THE GOAL (DISCOVERY-BASED SCIENCE)

The chromosome of the donor strain in an Hfr conjugation is transferred in a linear manner to the recipient strain. The order of genes along the chromosome can be deduced by determining the time various genes take to enter the recipient strain.

▶ ACHIEVING THE GOAL — FIGURE 9.7 The use of conjugation to map the order of genes along the *E. coli* chromosome.

Starting materials: The two *E. coli* strains already described, one Hfr strain ($thr^+ \ leu^+ \ azi^s \ ton^s \ lac^+ \ gal^+ \ str^s$) and one F⁻ strain ($thr^- \ leu^- \ azi^r \ ton^r \ lac^- \ gal^- \ str^r$).

Experimental level

1. Mix together a large number of Hfr donor and F⁻ recipient cells.

   Flask with bacteria

2. After different periods of time, take a sample of cells and interrupt conjugation in a blender.

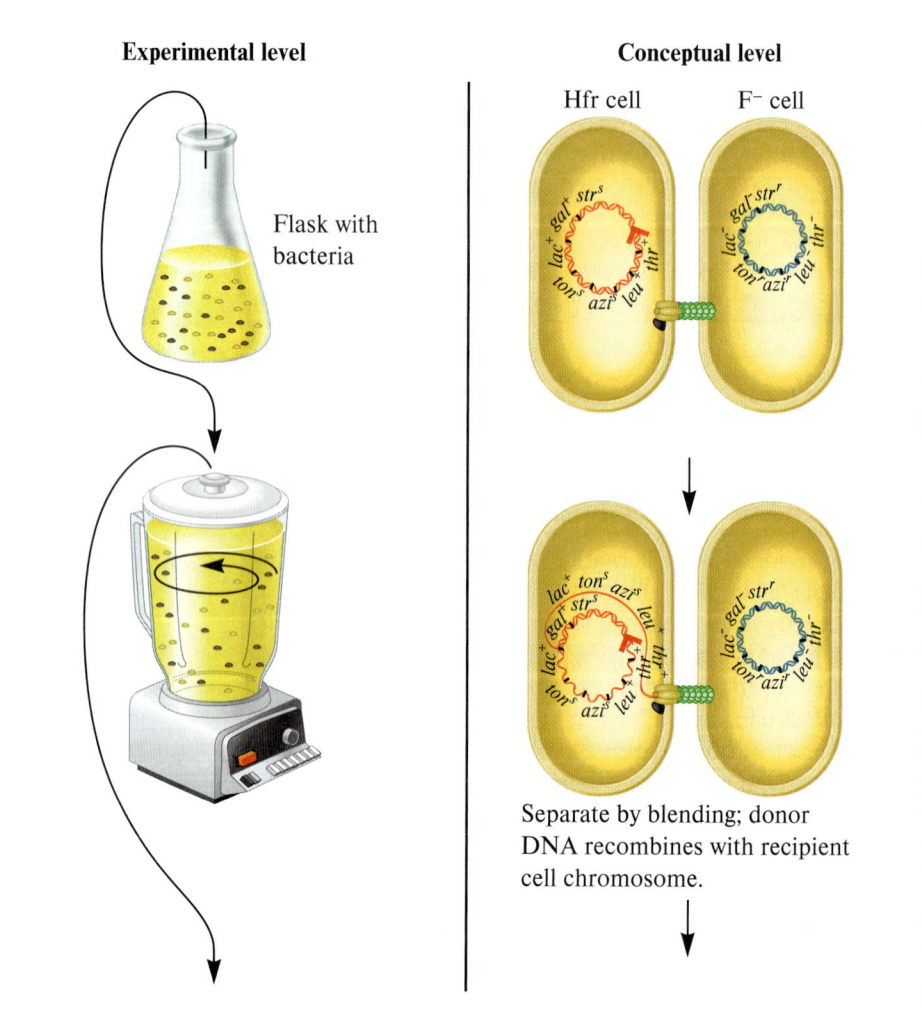

Conceptual level

Hfr cell   F⁻ cell

Separate by blending; donor DNA recombines with recipient cell chromosome.

3. Plate the cells on solid growth medium lacking threonine and leucine but containing streptomycin. Note: The general methods for growing bacteria in a laboratory are described in Appendix A.

4. Pick each surviving colony, which would have to be thr^+ leu^+ str^r, and test to see if it is sensitive to killing by azide, sensitive to infection by T1 bacteriophage, and able to metabolize lactose or galactose.

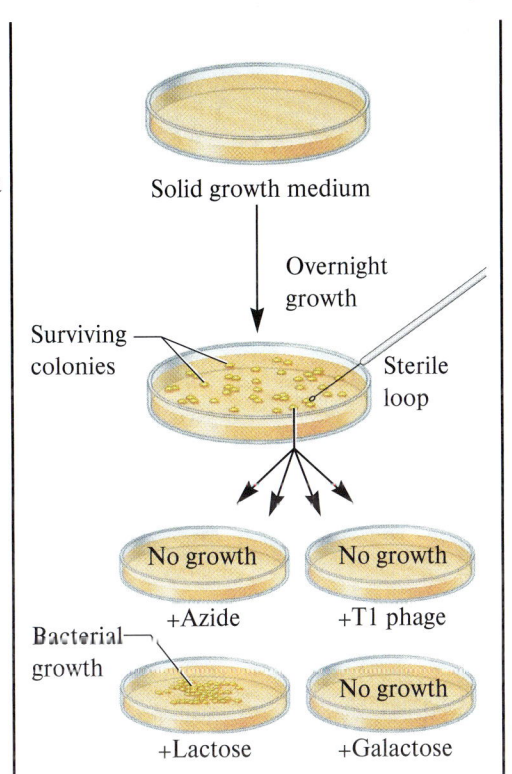

In this conceptual example, the cells have been incubated about 20 minutes.

Cannot survive on plates with streptomycin

Can survive on plates with streptomycin

Additional tests

The conclusion is that the colony that was picked contained cells with a genotype of thr^+ leu^+ azi^s ton^s lac^+ gal^- str^r.

▶ THE DATA

Minutes That Bacterial Cells Were Allowed to Conjugate Before Blender Treatment	Percentage of Surviving Bacterial Colonies with the Following Genotypes:				
	thr^+ leu^+	azi^s	ton^s	lac^+	gal^+
5*	—*	—	—	—	—
10	100	12	3	0	0
15	100	70	31	0	0
20	100	88	71	12	0
25	100	92	80	28	0.6
30	100	90	75	36	5
40	100	90	75	38	20
50	100	91	78	42	27
60	100	91	78	42	27

*There were no surviving colonies within the first 5 minutes of conjugation.

Source: Jacob, François and Wollman, Elie (1961) *Sexuality and the Genetics of Bacteria.* New York, NY: Academic Press.

▶ INTERPRETING THE DATA

Now let's discuss the data shown in the table. After the first plating, all colonies were composed of cells in which the thr^+ and leu^+ alleles had been transferred to the F$^-$ recipient strain, which was already streptomycin resistant. As seen in the data, 5 minutes was not sufficient time to transfer the thr^+ and leu^+ alleles because no surviving colonies were observed. After 10 minutes or longer, however, surviving bacterial colonies with the thr^+ leu^+ genotype were obtained. To determine the order of the remaining genes (azi^s, ton^s, lac^+, and gal^+), each surviving colony was tested to see if it was sensitive to killing by azide, sensitive to killing by T1 bacteriophage, able to use lactose for growth, or able to use galactose for growth. The likelihood of surviving colonies depended on whether the azi^s, ton^s, lac^+, and gal^+ genes were close to the origin of transfer or farther away. For example, when cells were allowed to conjugate for 25 minutes, 80% carried the ton^s gene, whereas only 0.6% carried the gal^+ gene. These results indicate that the ton^s gene is closer to the origin of transfer than the gal^+ gene is. When Wollman and Jacob reviewed all of the data, a consistent pattern emerged. The gene that conferred sensitivity to azide (azi^s) was transferred first, followed by ton^s, lac^+, and finally, gal^+. From these data, as well as results from other experiments, Wollman and Jacob constructed a genetic map that depicted the order of these genes along the *E. coli* chromosome.

| thr | leu | azi | ton | lac | gal |

This work provided bacterial geneticists with the first method for mapping the order of genes along the bacterial chromosome.

A Genetic Map of the *E. coli* Chromosome Has Been Obtained from Many Conjugation Studies

Conjugation experiments have been used to map more than 1000 genes along the circular *E. coli* chromosome. A map of that chromosome is shown in **Figure 9.8**. This simplified map shows the

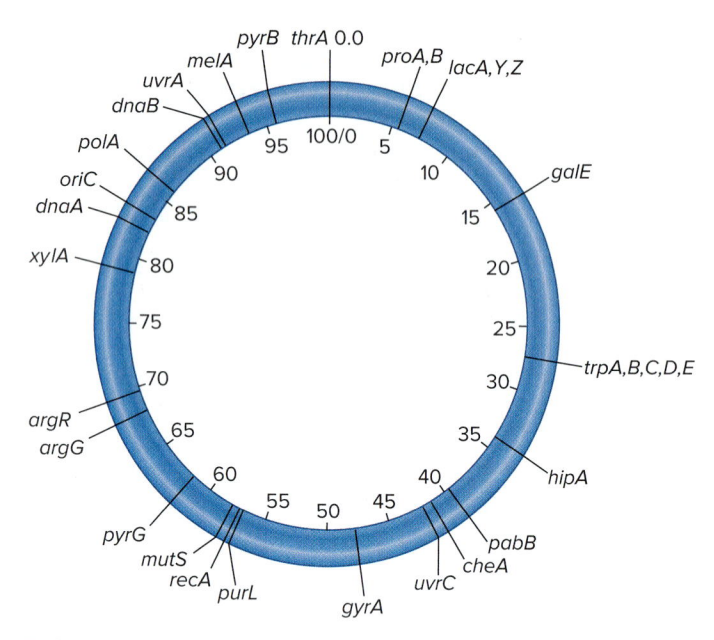

FIGURE 9.8 A simplified genetic map of the *E. coli* chromosome indicating the positions of several genes. *E. coli* has a circular chromosome with about 4377 different genes. This map shows the locations of some of them. The map is scaled in units of minutes and proceeds in a clockwise direction. The starting point on the map is the gene *thrA*.

Concept Check: Why is the scale of this map in minutes?

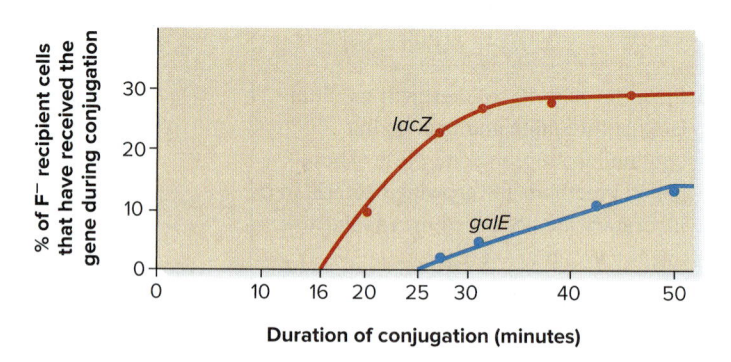

FIGURE 9.9 Time course of an interrupted *E. coli* conjugation experiment. By extrapolating the data back to the origin, the approximate time of entry of the *lacZ* gene is found to be 16 minutes; that of the *galE* gene is 25 minutes. Therefore, the distance between these two genes is 9 minutes.

Concept Check: Which of these two genes is closer to the origin of transfer?

locations of only a few dozen genes. Because the chromosome is circular, we must arbitrarily assign a starting point on the map, in this case, the gene *thrA*. Researchers scale genetic maps from bacterial conjugation studies in units of **minutes.** This unit refers to the relative time it takes for genes to first enter an F^- recipient strain during a conjugation experiment. The *E. coli* genetic map shown in Figure 9.8 is 100 minutes long, which is approximately the time that it takes to transfer the complete chromosome during an Hfr conjugation.

The distance between two genes is determined by comparing their times of entry during a conjugation experiment. As shown in **Figure 9.9**, the time of entry is found by conducting conjugation experiments that proceed for different time intervals before interruption. We compute the time of entry by extrapolating the data back to the *x*-axis. In this experiment, the time of entry of the *lacZ* gene was approximately 16 minutes, and that of the *galE* gene was 25 minutes. Therefore, these two genes are approximately 9 minutes apart from each other along the *E. coli* chromosome.

9.3 REVIEWING THE KEY CONCEPTS

- Hfr strains are formed when an F factor integrates into the bacterial chromosome. An imprecise excision produces an F′ factor that carries a portion of the bacterial chromosome (see Figure 9.5).
- Hfr strains can transfer a portion of the bacterial chromosome to a recipient cell during conjugation (see Figure 9.6).
- Wollman and Jacob showed that conjugation can be used to map the locations of genes along the bacterial chromosome, thereby creating a genetic map (see Figures 9.7–9.9).

9.3 COMPREHENSION QUESTIONS

1. With regard to conjugation, a key difference between F^+ and Hfr cells is that an Hfr cell
   a. is unable to conjugate.
   b. transfers a plasmid to the recipient cell.
   c. transfers a portion of the bacterial chromosome to the recipient cell.
   d. becomes an F^- cell after conjugation.

2. In mapping experiments, _____ strains are conjugated with F^- strains. The distance between two genes is determined by comparing their _____ during a conjugation experiment.
   a. F^+, times of entry
   b. Hfr, times of entry
   c. F^+, expression levels
   d. Hfr, expression levels

9.4 BACTERIAL TRANSDUCTION

Learning Outcomes:
1. Outline the steps of bacterial transduction.
2. Explain how transduction can be used to map genes.

We now turn to a second method of genetic transfer, one that involves bacteriophages (also known as phages), which are viruses that infect bacterial cells. Following infection of a bacterium, new viral particles are made, which are then released from the cell in an event called lysis. In this section, we will examine how mistakes in the bacteriophage reproductive cycle can lead to the transfer of genetic material from one bacterial cell to another.

Bacteriophages Transfer Genetic Material from One Bacterial Cell to Another via Transduction

As noted earlier in this chapter, the ability of phages to transfer genetic material between bacteria is called transduction.

Examples of phages that can transfer bacterial chromosomal DNA from one bacterium to another are the P22 and P1 phages, which infect the bacterial species *Salmonella typhimurium* and *E. coli*, respectively. The P22 and P1 phages can follow either the lytic or lysogenic cycle. (Look ahead to Chapter 10, Figure 10.4, for a description of these two cycles.) In the lytic cycle, when the phage infects the bacterial cell, the bacterial chromosome becomes fragmented into small pieces of DNA. The phage DNA directs the synthesis of more phage DNA and proteins, which then assemble to make new phages, and the bacterial DNA is degraded.

How does a bacteriophage transfer bacterial chromosomal genes from one cell to another? The process is shown in **Figure 9.10**.

1. A phage infects a bacterial cell by injecting its genetic material into the cell.
2. Following infection, the bacterial chromosomal DNA is digested into fragments.
3. New phage proteins and DNA are made and begin to assemble into new phages. Occasionally, a mistake can happen in which a random piece of bacterial DNA assembles with phage proteins. This creates a transducing phage that contains bacterial chromosomal DNA. In this example, the bacterial DNA carries the *his*⁺ gene.
4. The bacterial cell is lysed and releases the newly made phages into the environment.
5. Following release, the transducing phage can bind to a living bacterial cell and inject its genetic material into the bacterium.
6. The DNA fragment, which was derived from the chromosomal DNA of the first bacterium, can then recombine with the recipient cell's chromosome.

In this case, the recipient bacterium has been changed from a cell that was *his*⁻ (unable to synthesize histidine) to a cell that is *his*⁺ (able to synthesize histidine).

Cotransduction Can Be Used to Map Genes That Are Within 2 Minutes of Each Other

Can transduction be used to map the distance between bacterial genes? The answer is yes, but only if the genes are relatively close together. During transduction, P1 phages cannot package pieces that are greater than 2–2.5% of the entire length of the *E. coli* chromosome, and P22 phages cannot package pieces that are greater than 1% of the length of the *S. typhimurium* chromosome. If two genes are close together along the chromosome, a bacteriophage may package a single piece of the chromosome that carries both genes and transfer that piece to another bacterium. This phenomenon is called **cotransduction.** The likelihood that two genes will be cotransduced depends on how close together they lie. If two genes are far apart along a bacterial chromosome, they will never be cotransduced because the bacteriophage cannot physically package a DNA fragment that consists of more than 1–2.5% of the bacterial chromosome. In genetic mapping studies, cotransduction is used to determine the order and distance between genes that lie fairly close to each other.

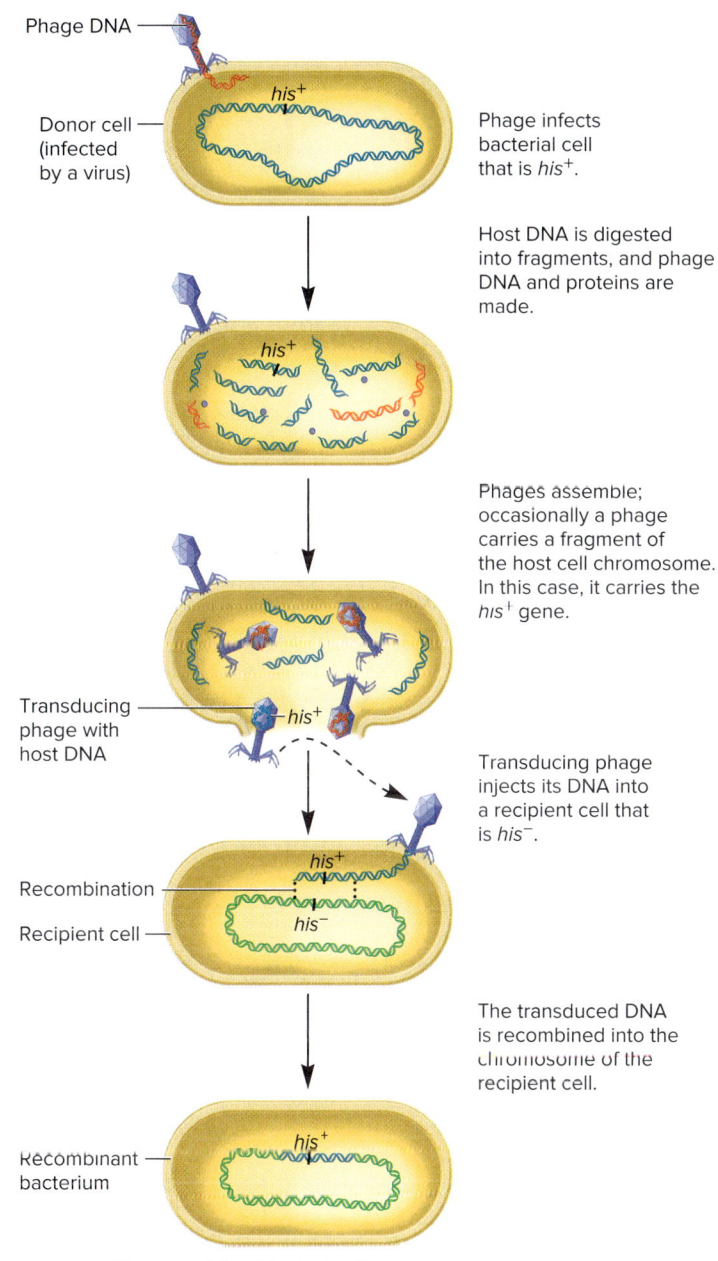

FIGURE 9.10 Transduction in bacteria.

Genes → Traits The transducing phage introduced DNA into a recipient cell that was originally *his*⁻ (unable to synthesize histidine). During transduction, this cell received a segment of bacterial chromosomal DNA that carried *his*⁺. Following recombination, the recipient cell's genotype was changed to *his*⁺, and thus it became able to synthesize histidine.

Concept Check: *Transduction is sometimes described as a mistake in the bacteriophage reproductive cycle. Explain how it can be viewed as a mistake.*

To map genes using cotransduction, a researcher selects for the transduction of one gene and then monitors whether or not a second gene is cotransduced along with it. As an example, let's consider a donor strain of *E. coli* that is *arg*⁺ *met*⁺ *str*ˢ (able to synthesize arginine and methionine but sensitive to killing by streptomycin) and a recipient strain that is *arg*⁻ *met*⁻ *str*ʳ (**Figure 9.11**).

1. The donor strain is infected with P1 bacteriophage.
2. Some of the *E. coli* cells are lysed by P1 and new phages are released. As noted in Figure 9.10, some of the phages in the P1 lysate may carry random pieces of bacterial chromosomal DNA.
3. This P1 lysate is mixed with the recipient cells.
4. After allowing sufficient time for transduction, the recipient cells are plated on a growth medium that contains arginine and streptomycin but not methionine. Therefore, these plates select for the growth of cells in which the *met*⁺ gene has been transferred to the recipient strain, but they do not select for the growth of cells in which the *arg*⁺ gene has been transferred, because the growth media contain arginine.
5. To determine if the *arg*⁺ gene has been cotransduced with the *met*⁺ gene, a sample of cells from each bacterial colony can be picked up with a wire loop and restreaked on media that lack both amino acids. If the cells from a colony can grow, they must have also obtained the *arg*⁺ gene during transduction. In other words, cotransduction of both the *arg*⁺ and *met*⁺ genes has occurred. Alternatively, if the cells from the restreaked colony do not grow, they must have received only the *met*⁺ gene during transduction. Data from this type of experiment are shown at the bottom of Figure 9.11. These data indicate a cotransduction frequency of 21/50 = 0.42, or 42%.

In 1966, Tai Te Wu derived a mathematical expression that relates cotransduction frequency with map distance obtained from conjugation experiments:

$$\text{Cotransduction frequency} = (1 - d/L)^3$$

where

d = distance between two genes in minutes
L = the size of the chromosomal pieces (in minutes) that the phage carries during transduction. (For P1 transduction, this size is approximately 2% of the bacterial chromosome, which equals about 2 minutes.)

This equation assumes that the bacteriophages randomly package pieces of the bacterial chromosome that are similar in size. Depending on the type of phage used in a transduction experiment, this assumption may not always be valid. Nevertheless, this equation has

FIGURE 9.11 **The steps in a cotransduction experiment.** The donor strain, which is *arg*⁺ *met*⁺ *str*ˢ (able to synthesize arginine and methionine but sensitive to streptomycin), is infected with phage P1. Some of the cells are lysed by P1, and this P1 lysate is mixed with cells of the recipient strain, which are *arg*⁻ *met*⁻ *str*ʳ. P1 phages in this lysate may carry fragments of the donor cell's chromosome, and the P1 phage may inject that DNA into the recipient cells. To identify recipient cells that have received the *met*⁺ gene from the donor strain, the recipient cells are plated on a growth medium that contains arginine and streptomycin but not methionine. To determine if the *arg*⁺ gene has also been cotransduced, cells from each bacterial colony (on the plates containing arginine and streptomycin) are lifted with a sterile wire loop and streaked on a medium that lacks both amino acids. If the cells can grow, cotransduction of the *arg*⁺ and *met*⁺ genes has occurred. Alternatively, if the cells of the restreaked colony do not grow, they must have received only the *met*⁺ gene during transduction.

Concept Check: If the *arg*⁺ and *met*⁺ genes were very far apart on the bacterial chromosome, how would the results have been different?

Results

Selected gene	Nonselected gene	Number of colonies that grew on medium + arginine	Number of colonies that grew on medium − arginine	Cotransduction frequency
met⁺	*arg*⁺	50	21	0.42

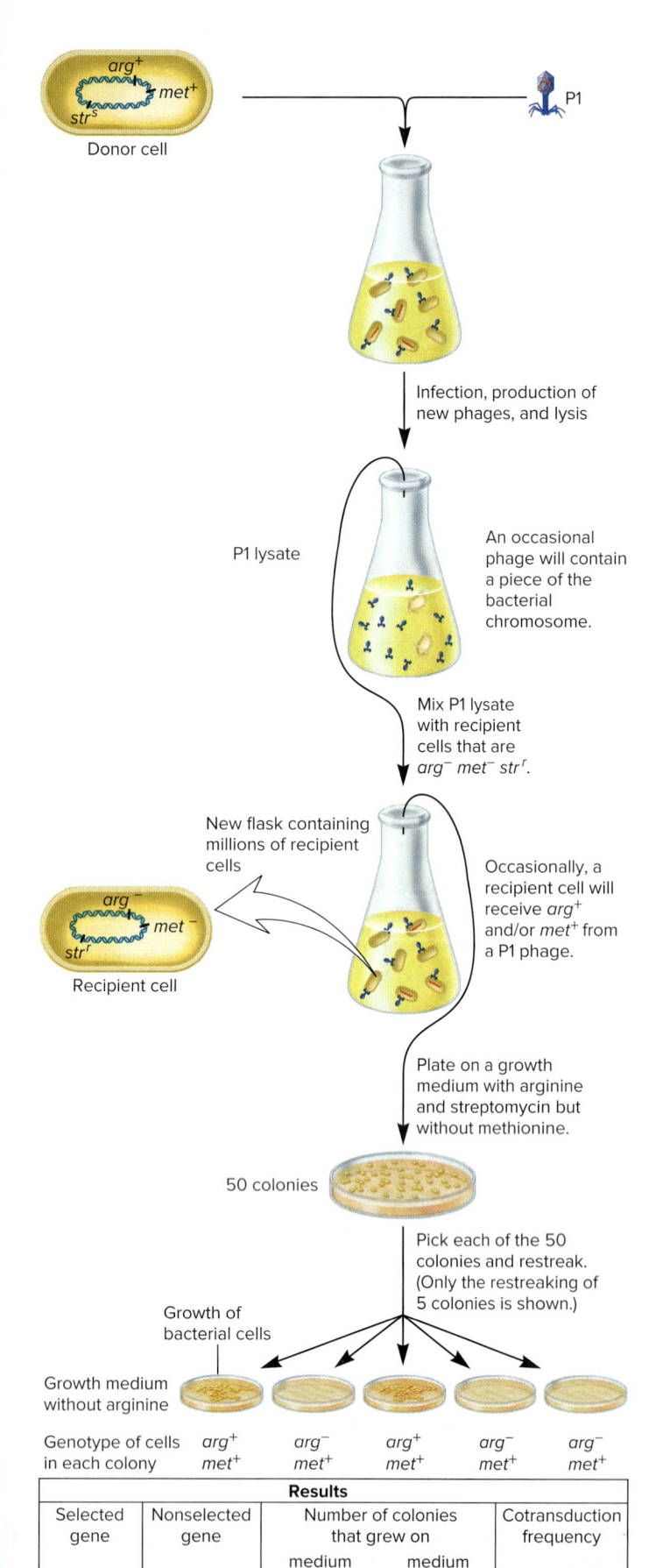

been fairly reliable in computing map distance from P1 transduction experiments with *E. coli*. We can use this equation to estimate the distance between the two genes described in Figure 9.11.

$$0.42 = (1 - d/2)^3$$
$$(1 - d/2) = \sqrt[3]{0.42}$$
$$1 - d/2 = 0.75$$
$$d/2 = 0.25$$
$$d = 0.5 \text{ minute}$$

This calculation tells us that the distance between the met^+ and arg^+ genes is approximately 0.5 minute.

Historically, genetic mapping studies of bacteria often involved data from both conjugation and transduction experiments. Conjugation has been used to determine the relative order and distance of genes, particularly those that are far apart along the chromosome. In comparison, transduction experiments can provide fairly accurate mapping data for genes that are close together.

9.4 REVIEWING THE KEY CONCEPTS

- During transduction, a portion of a bacterial chromosome is transferred to a recipient cell via a bacteriophage (see Figure 9.10).
- A cotransduction experiment can be used to map genes that are close together on a bacterial chromosome (see Figure 9.11).

9.4 COMPREHENSION QUESTIONS

1. During transduction via a P1 phage,
   a. any small fragment of the bacterial chromosome may be transferred to another bacterium by the phage.
   b. two bacterial cells must make direct contact for genetic transfer to occur.
   c. the same phage moves from one bacterial cell to the next.
   d. the bacterial chromosome is transferred to a recipient cell in a linear manner.
2. Cotransduction may be used to map bacterial genes that are
   a. far apart on the bacterial chromosome.
   b. close together on the bacterial chromosome.
   c. both b and c.
   d. neither b or c.

9.5 BACTERIAL TRANSFORMATION

Learning Outcome:

1. Outline the steps of bacterial transformation.

A third mechanism for the transfer of genetic material from one bacterium to another is transformation. This process was first discovered by Frederick Griffith in 1928 while working with strains of *Streptococcus pneumoniae* (formerly known as *Diplococcus pneumoniae*, or pneumococcus). During transformation, a living bacterial cell takes up DNA that is released from a dead bacterium. This DNA may then recombine into the living bacterium's chromosome, producing a bacterium with genetic material that it has received from the dead bacterium. (This experiment is discussed in detail in Chapter 11; see Figure 11.1.)

- Transformation can occur as a natural process that has evolved in certain bacteria, in which case it is called **natural transformation.**
- Transformation may occur due to experimental treatments in which the bacterial cells are forced to take up DNA. This experimental approach is termed **artificial transformation.** For example, a technique known as electroporation, in which an electric current causes the uptake of DNA, is used by researchers to promote the transport of DNA into a bacterial cell.

Since the initial studies of Griffith, we have learned a great deal about the events that occur in natural transformation. This form of genetic transfer has been reported in a wide variety of bacterial species. Bacterial cells that are able to take up DNA are known as **competent cells.** Those that can take up DNA naturally carry genes that encode proteins called **competence factors.** These proteins facilitate the binding of DNA fragments to the cell surface, the uptake of the DNA into the cytoplasm, and its subsequent incorporation into the bacterial chromosome. Temperature, ionic conditions, and nutrient availability can affect whether or not a bacterium is competent to take up genetic material from its environment. These conditions influence the expression of the genes that encode competence factors.

In recent years, geneticists have unraveled some of the steps that occur when competent bacterial cells are transformed by genetic material in their environment. **Figure 9.12** describes the steps of transformation.

1. A large fragment of DNA binds to the surface of the bacterial cell. Competent cells express DNA receptors that promote such binding.
2. Before entering the cell, this large piece of chromosomal DNA must be cut into smaller fragments. This cutting is accomplished by an extracellular bacterial enzyme known as an endonuclease, which makes occasional random cuts in the long piece of chromosomal DNA. At this stage, the DNA fragments are composed of double-stranded DNA.
3. To enter the cell, one of the DNA strands is degraded and the other strand enters the bacterial cytoplasm via an uptake system, which is structurally similar to the one described earlier in this chapter for conjugation (shown in Figure 9.4b), but is involved with DNA uptake rather than export.
4. To be stably inherited, the DNA strand must be incorporated into the bacterial chromosome. If the DNA strand has a sequence that is similar to a region of DNA in the bacterial chromosome, the DNA may be incorporated into the chromosome by a process known as **homologous recombination,** discussed in detail in Chapter 13. For this to occur, the single-stranded DNA aligns itself with a homologous location on the bacterial chromosome. In the example shown in Figure 9.12, the foreign DNA carries a functional lys^+ gene that aligns itself with a nonfunctional (mutant) lys^- gene already present within the bacterial chromosome.
5. The foreign DNA then recombines with one of the strands in the bacterial chromosome of the competent cell. In other words,

FIGURE 9.12 **The steps of bacterial transformation.** In this example, a fragment of DNA carrying a *lys*⁺ gene enters the competent cell and recombines with the chromosome, transforming the bacterium from *lys*⁻ to *lys*⁺.

Genes → Traits Bacterial transformation can also lead to new traits for the recipient cell. The recipient cell was *lys*⁻ (unable to synthesize the amino acid lysine). Following transformation, it became *lys*⁺. The recipient bacterial cell was transformed into a cell that can synthesize lysine and grow on a medium that lacks this amino acid. Before transformation, the recipient *lys*⁻ cell would not have been able to grow on a medium lacking lysine.

> **Concept Check:** *If the recipient cell did not already have a lys⁻ gene, could the lys⁺ DNA become incorporated into the bacterial chromosome? Explain.*

Alternatively, a DNA fragment that has entered a cell may not be homologous to any genes that are already found in the bacterial chromosome. In this case, the DNA strand may be incorporated at a random site in the chromosome. This process is known as **nonhomologous recombination.**

9.5 REVIEWING THE KEY CONCEPTS

- During transformation, a segment of DNA is taken up by a bacterial cell and then is incorporated into the bacterial chromosome (see Figure 9.12).

9.5 COMPREHENSION QUESTION

1. During bacterial transformation, what happens to a fragment of DNA before it enters the cell?
   a. It is cut into smaller pieces.
   b. One of the DNA strands is degraded.
   c. Both a and b occur.
   d. Neither a nor b occurs.

9.6 MEDICAL RELEVANCE OF HORIZONTAL GENE TRANSFER

Learning Outcomes:
1. Define *horizontal gene transfer*.
2. Explain the impact of bacterial horizontal gene transfer in medicine.

Horizontal gene transfer is a process in which an organism incorporates genetic material from another organism without being the offspring of that organism. Conjugation, transduction, and transformation are examples of horizontal gene transfer. This process can occur between members of the same species or members of different species. A key reason why horizontal gene transfer is important is its medical relevance. One area of great concern is the phenomenon of antibiotic resistance. Antibiotics are widely prescribed to treat bacterial infections in humans. They are also used in agriculture to control bacterial diseases. Unfortunately, the widespread use of antibiotics has increased the prevalence of antibiotic-resistant strains of bacteria—strains that have a selective advantage over those that are susceptible to antibiotics. Resistant strains carry genes that counteract the action of antibiotics in various ways. A resistance gene may encode a protein that breaks

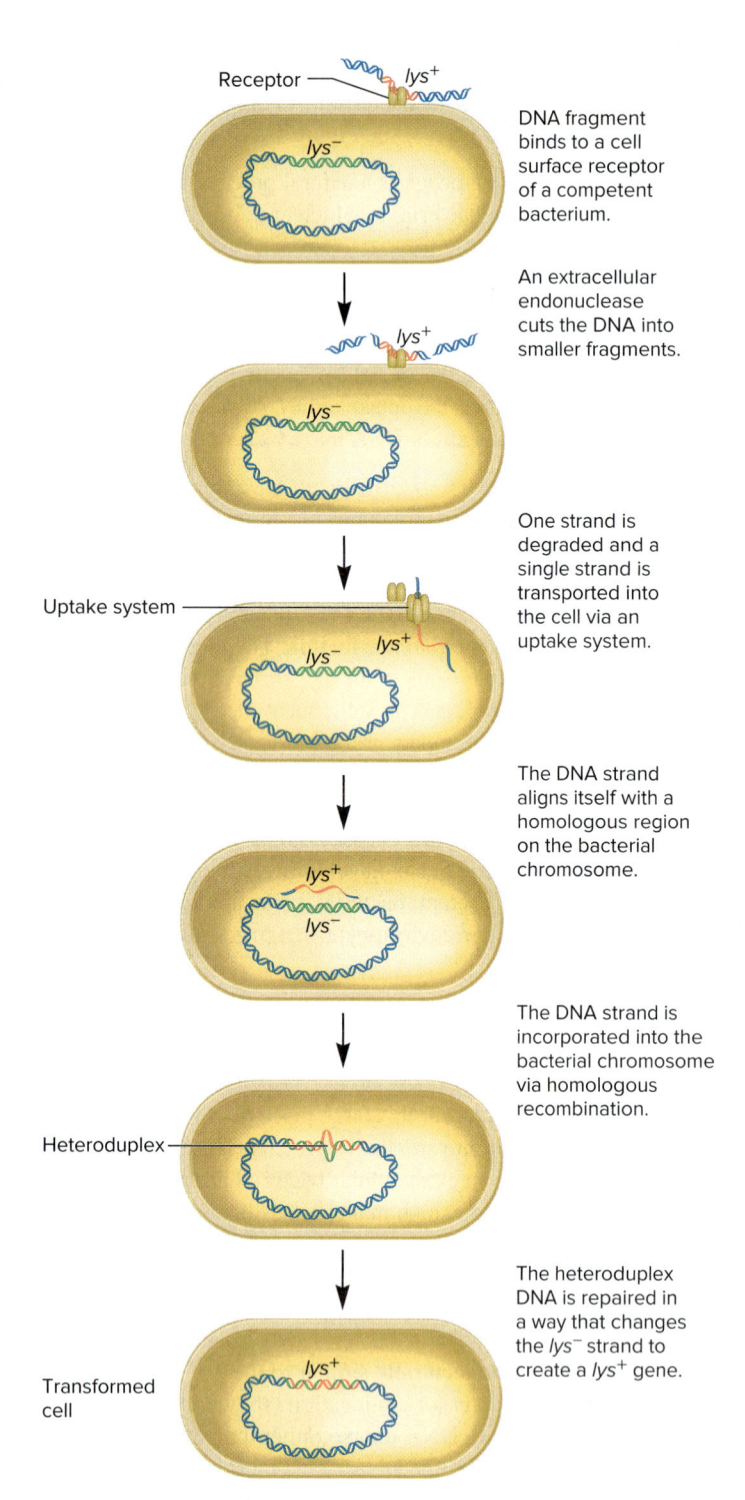

the foreign DNA replaces one of the chromosomal strands of DNA, which is subsequently degraded. During homologous recombination, alignment of the *lys*⁻ and the *lys*⁺ alleles results in a region of double-stranded DNA called a **heteroduplex,** which contains one or more base sequence mismatches.

6. The heteroduplex exists only temporarily. DNA repair enzymes in the recipient cell recognize the heteroduplex and repair it. In this example, the heteroduplex has been repaired by eliminating the mutation that caused the *lys*⁻ genotype, thereby creating a *lys*⁺ gene. Therefore, the recipient cell has been transformed from a *lys*⁻ strain to a *lys*⁺ strain.

down the antibiotic, pumps it out of the cell, or prevents it from inhibiting cellular processes.

The term **acquired antibiotic resistance** refers to the common phenomenon of a previously susceptible strain becoming resistant to a specific antibiotic. This change may result from genetic alterations in the bacterial genome, but it is often due to the horizontal transfer of resistance genes from a resistant strain. As reported in the news media, antibiotic resistance has increased dramatically worldwide over the past few decades, with resistant strains emerging in almost all pathogenic strains of bacteria. As an example, some *Staphylococcus aureus* strains have developed resistance to methicillin and all penicillins. Evidence suggests that these methicillin-resistant strains of *Staphlococcus aureus* (MRSA, pronounced "mersa") acquired the methicillin-resistance gene by horizontal gene transfer, possibly from a strain of *Enterococcus faecalis*. MRSA strains cause skin infections that are more difficult to treat than staph infections caused by nonresistant strains of *S. aureus*.

9.6 REVIEWING THE KEY CONCEPTS

- Horizontal gene transfer is a process in which an organism incorporates genetic material from another organism without being the offspring of that organism. It can be accomplished by conjugation, transduction, or transformation.
- Many strains of bacteria have acquired antibiotic resistance by horizontal gene transfer.

9.6 COMPREHENSION QUESTION

1. Which of the following is an example of horizontal gene transfer?
   a. The transfer of a gene from one strain of *E. coli* to a different strain via conjugation
   b. The transfer of a gene from one strain of *E. coli* to a different strain via transduction
   c. The transfer of an antibiotic resistance gene from *E. coli* to *Salmonella typhimurium* via transformation
   d. All of the above are examples of horizontal gene transfer.

KEY TERMS

Page 186. genetic transfer, conjugation, transduction, transformation
Page 187. minimal medium, auxotroph, prototroph
Page 188. F factor
Page 189. sex pili, conjugation bridge, relaxosome, origin of transfer, T DNA, plasmid
Page 190. episomes
Page 191. Hfr strains, F′ factors

Page 192. bateriophages, interrupted mating
Page 196. minutes
Page 197. cotransduction
Page 199. natural transformation, artificial transformation, competent cells, competence factors, homologous recombination
Page 200. heteroduplex, nonhomologous recombination, horizontal gene transfer
Page 201. acquired antibiotic resistance

CHAPTER SUMMARY

9.1 Overview of Genetic Transfer in Bacteria

- Three general mechanisms for genetic transfer in various species of bacteria are conjugation, transduction, and transformation (see Table 9.1).

9.2 Bacterial Conjugation

- Lederberg and Tatum discovered conjugation in *E. coli* by analyzing auxotrophic strains (see Figure 9.1).
- Using a U-tube apparatus, Davis showed that conjugation requires cell-to-cell contact (see Figure 9.2).
- Certain strains of bacteria have F factors, which they can transfer via conjugation in a series of steps (see Figures 9.3, 9.4).
- Bacteria may contain different types of plasmids.

9.3 Conjugation and Mapping via Hfr Strains

- Hfr strains are formed when an F factor integrates into the bacterial chromosome. An imprecise excision produces an F′ factor that carries a portion of the bacterial chromosome (see Figure 9.5).
- Hfr strains can transfer a portion of the bacterial chromosome to a recipient cell during conjugation (see Figure 9.6).

- Wollman and Jacob showed that conjugation can be used to map the locations of genes along the bacterial chromosome, thereby creating a genetic map (see Figures 9.7, 9.9).

9.4 Bacterial Transduction

- During transduction, a portion of a bacterial chromosome is transferred to a recipient cell via a bacteriophage (see Figure 9.10).
- A cotransduction experiment can be used to map genes that are close together on a bacterial chromosome (see Figure 9.11).

9.5 Bacterial Transformation

- During transformation, a segment of DNA is taken up by a bacterial cell and then is incorporated into the bacterial chromosome (see Figure 9.12).

9.6 Medical Relevance of Horizontal Gene Transfer

- Horizontal gene transfer is a process in which an organism incorporates genetic material from another organism without being the offspring of that organism. It can be accomplished by conjugation, transduction, or transformation.
- Many strains of bacteria have acquired antibiotic resistance by horizontal gene transfer.

PROBLEM SETS & INSIGHTS

More Genetic TIPS

1. In *E. coli*, the gene $bioD^+$ encodes an enzyme involved in biotin synthesis, and $galK^+$ encodes an enzyme involved in galactose utilization. An *E. coli* strain that contained wild-type versions of both genes was infected with P1, and then a P1 lysate was obtained. This lysate was used to transduce (infect) a strain that was $bioD^-$ and $galK^-$. The cells were plated on media containing galactose as the sole carbon source for growth, in order to select for transduction of the $galK^+$ gene. These media also were supplemented with biotin. The resulting colonies were then restreaked on media that lacked biotin to see if the $bioD^+$ gene had been cotransduced. The following results were obtained:

Selected Gene	Non-selected Gene	Number of Colonies That Grew on: Galactose + Biotin	Galactose − Biotin	Cotransduction Frequency
$galK^+$	$bioD^+$	80	10	0.125

How far apart are these two genes?

Topic: What topic in genetics does this question address?

The topic is bacterial transduction. More specifically, the question is about computing the distance between two genes using data from a cotransduction experiment.

Information: What information do you know based on the question and your understanding of the topic?

From the question, you know that a cotransduction experiment was conducted in which the cells were initially placed on media with galactose and biotin. Of the 80 colonies that grew, only 10 of them grew when restreaked on media that lacked biotin. From your understanding of the topic, you may remember that map distance can be computed using this equation:

Cotransduction frequency = $(1 - d/2)^3$

Problem-Solving Strategy: Make a calculation.

To solve this problem, you first need to know the contransduction frequency, which is 10 (the number of cotransductants) divided by 80 (the total number of colonies): $10/80 = 0.125$

Cotransduction frequency = $(1 - d/2)^3$

$$0.125 = (1 - d/2)^3$$
$$1 - d/2 = \sqrt[3]{0.125}$$
$$1 - d/2 = 0.5$$
$$d/2 = 1 - 0.5$$
$$d = 1.0 \text{ minute}$$

Answer: The two genes are approximately 1 minute apart on the *E. coli* chromosome.

2. By conducting conjugation experiments between Hfr and recipient strains, Wollman and Jacob mapped the order of many bacterial genes. Throughout the course of their studies, they identified several different Hfr strains in which the F-factor DNA had been integrated at different places along the bacterial chromosome. A sample of their experimental results is shown in the following table:

Order of Transfer of Several Different Bacterial Genes

Hfr Strain	First								Last
H	thr	leu	azi	ton	pro	lac	gal	str	met
1	leu	thr	met	str	gal	lac	pro	ton	azi
2	pro	ton	azi	leu	thr	met	str	gal	lac
3	lac	pro	ton	azi	leu	thr	met	str	gal
4	met	str	gal	lac	pro	ton	azi	leu	thr
5	met	thr	leu	azi	ton	pro	lac	gal	str
6	met	thr	leu	azi	ton	pro	lac	gal	str
7	ton	azi	leu	thr	met	str	gal	lac	pro

A. Explain how these results are consistent with the idea that the bacterial chromosome is circular.

B. Draw a map of the bacterial chromosome that shows the order of genes and the locations of the origins of transfer among these different Hfr strains.

Topic: What topic in genetics does this question address?

The topic is genetic mapping. More specifically, the question is about using data from conjugation experiments to construct a genetic map.

Information: What information do you know based on the question and your understanding of the topic?

From the question, you know the order of gene transfer from several conjugation experiments. From your understanding of the topic, you may remember that genes are transferred linearly, from donor to recipient cell, starting at the origin of transfer.

Problem-Solving Strategy: Analyze data. Compare and contrast. Make a drawing.

One strategy to solve this problem is to compare the orders of transfer in these conjugation experiments, which involve eight strains and nine different genes.

Answer:

A. In the data for the different *Hfr* strains, the order of the nine genes is always the same or the reverse of that order. For example, *HfrH* and *Hfr4* transfer the same genes, but their orders are reversed relative to each other. In addition, the *Hfr* strains showed an overlapping pattern of transfer with regard to the origin. For example, *Hfr1* and *Hfr2* had the same order of genes, but *Hfr1* began with *leu* and ended with *azi*, whereas *Hfr2* began with *pro* and ended with *lac*. From these findings, Wollman and Jacob concluded that the segment of DNA that was the origin of transfer had been inserted at different points within a circular *E. coli* chromosome in different *Hfr* strains. A circular chromosome explains how the results can be overlapping. The researchers also concluded that the origin can be inserted in either orientation, so the direction of gene transfer can be clockwise or counterclockwise around the circular bacterial chromosome.

B. A genetic map consistent with these results is shown here.

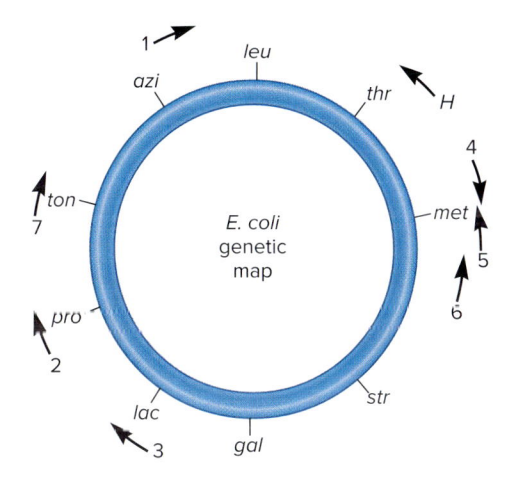

3. An *Hfr* strain that is *leuA*⁺ and *thiL*⁺ was conjugated to a strain that is *leuA*⁻ and *thiL*⁻. The conjugation was interrupted at different time points, and the percentage of recombinants for each gene was determined by streaking on media that lacked either leucine or thiamine. The results are shown in the following graph.

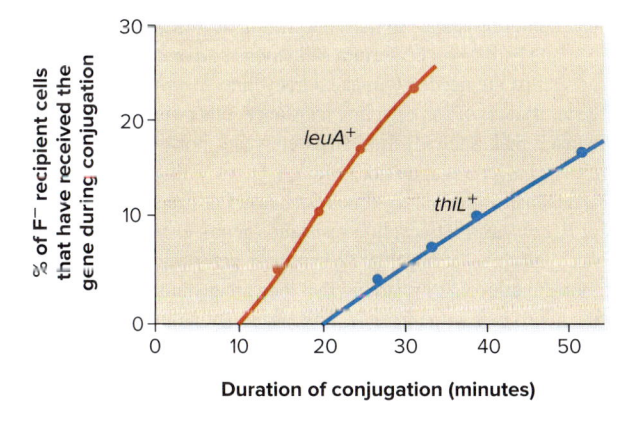

What is the map distance (in minutes) between these two genes?

Topic: What topic in genetics does this question address?

The topic is bacterial genetic mapping via a conjugation experiment. More specifically, the question asks you to use the experimental results to find the map distance between two genes.

Information: What information do you know based on the question and your understanding of the topic?

In the question, you are given the data from a conjugation experiment involving two genes. From your understanding of the topic, you may remember that genes are transferred linearly during conjugation, so the order of transfer is determined by the location of the gene relative to the origin of transfer.

Problem-Solving Strategy: Make a calculation.

One strategy to solve this problem is by extrapolating the data points to the *x*-axis to determine the time of entry. For *leuA*⁺, they extrapolate back to 10 minutes. For *thiL*⁺, they extrapolate back to 20 minutes.

Answer: The distance between the two genes is approximately 10 minutes.

4. Genetic transfer via transformation can also be used to map genes along the bacterial chromosome. In this approach, fragments of chromosomal DNA are isolated from one bacterial strain and used to transform another strain. In this type of experiment, the recipient cell is exposed to a fairly low concentration of donor DNA, making it unlikely that the recipient bacterium will take up more than one fragment of DNA. If two genes are transferred to the recipient strain, this is called cotransformation. Because bacteria take up only a single fragment of DNA, cotransformation is likely only when two genes are fairly close together and found on the same DNA fragment. The experimenter examines the transformed bacteria to see if they have incorporated two or more different genes. One way to calculate the map distance is to use the equation that we used for cotransduction data, but substitute cotransformation frequency for cotransduction frequency:

Cotransformation frequency = $(1 - d/L)^3$

The researcher needs to experimentally determine the value of *L* by running the DNA on a gel and estimating the average size of the DNA fragments.

In a cotransformation experiment, a researcher has isolated DNA from an *araB*⁺ and *leuD*⁺ donor strain of *E. coli*. This DNA was transformed into a recipient strain that was *araB*⁻ and *leuD*⁻. Following transformation, the cells were plated on a medium containing arabinose and leucine. On this medium, only bacteria that are *araB*⁺ can grow. The bacteria can be either *leuD*⁺ or *leuD*⁻ because leucine is provided in the medium. Colonies that grew on this medium were then restreaked on a medium that contained arabinose but lacked leucine. Only *araB*⁺ and *leuD*⁺ cells could grow on these secondary plates. Following this protocol, the researcher obtained the following results:

Number of colonies growing on arabinose plus leucine medium: 57

Number of colonies that grew when restreaked on an arabinose medium without leucine: 42

What is the map distance between these two genes? Assume that the average size of the DNA fragments was about 2% of the bacterial chromosome. In other words, *L* equals 2 minutes, which is the same as 2%.

Topic: What topic in genetics does this question address?

The topic is genetic mapping via a transformation experiment. More specifically, the question asks you to use the experimental results to find the map distance between two genes.

Information: What information do you know based on the question and your understanding of the topic?

From the question, you have learned that you can use the same genetic mapping equation for transformation that was used for transduction. You are also given the results of a cotransformation experiment and told that the average DNA fragment was 2 minutes long. From your understanding of the topic, you may remember that cotransformation can only occur if two genes are on the same DNA fragment and, therefore, are fairly close together.

Problem-Solving Strategy: Make a calculation.

One strategy to solve this problem is to follow the same strategy as for a cotransduction experiment, except that the researcher must determine the average size of DNA fragments that are taken up by the bacterial cells. This would correspond to the value of L in a cotransduction experiment. In this problem, $L = 2$. The cotransformation frequency is the number of cotransformants (42) divided by the total number of transformants (57).

$$\text{Cotransformation frequency} = (1 - d/L)^3$$

$$42/57 = (1 - d/2)^3$$

$$d = 0.2 \text{ minute}$$

Answer: The distance between *araB* and *leuD* is approximately 0.2 minute.

Conceptual Questions

C1. The terms *conjugation, transduction,* and *transformation* are used to describe three different natural forms of genetic transfer between bacterial cells. Briefly discuss the similarities and differences among these processes.

C2. Conjugation is sometimes called bacterial mating. Is it a form of sexual reproduction? Explain.

C3. If you mix together an equal number of F^+ and F^- cells, how would you expect the proportions to change over time? In other words, do you expect an increase in the relative proportion of F^+ or of F^- cells? Explain your answer.

C4. What is the difference between an F^+ and an Hfr strain? Which type of strain do you expect to transfer many bacterial genes to recipient cells?

C5. What is the role of the origin of transfer during F^+-mediated and Hfr-mediated conjugation? What is the significance of the direction of transfer in Hfr-mediated conjugation?

C6. What is the role of sex pili during conjugation?

C7. Think about the structure and transmission of F factors, and discuss how you think F factors may have originated.

C8. What is cotransduction? What determines the likelihood that two genes will be cotransduced?

C9. When bacteriophage P1 causes *E. coli* to lyse, the resulting material is called a P1 lysate. What type of genetic material would be found in most of the P1 phages in the lysate? What kind of genetic material is occasionally found within a P1 phage?

C10. As described in Figure 9.10, host DNA is digested into small pieces, which are occasionally assembled with phage proteins, creating a phage with bacterial chromosomal DNA. If the breakage of the chromosomal DNA is not random (i.e., it is more likely to break at certain spots as opposed to other spots), how might such nonrandom breakage affect cotransduction frequency?

C11. Describe the steps that occur during bacterial transformation. What is a competent cell? What factors may determine whether a cell will be competent?

C12. Which form of bacterial genetic transfer does not require recombination with the bacterial chromosome?

C13. Researchers who study the molecular mechanism of transformation have identified many proteins in bacteria that function in the uptake of DNA from the environment and its recombination into the host cell's chromosome. This means that bacteria have evolved molecular mechanisms for the purpose of transformation by extracellular DNA. What advantage(s) does a bacterium gain from importing DNA from the environment and/or incorporating it into its chromosome?

C14. Antibiotics such as tetracycline, streptomycin, and bacitracin are small organic molecules that are synthesized by particular species of bacteria. Microbiologists have hypothesized that the reason why certain bacteria make antibiotics is to kill other species that occupy the same environment. Bacteria that produce an antibiotic may be able to kill competing species. Eliminating competitors provides more resources for the antibiotic-producing bacteria. In addition, bacteria that have the genes necessary for antibiotic biosynthesis contain genes that confer resistance to the same antibiotic. For example, tetracycline is made by the soil bacterium *Streptomyces aureofaciens*. Besides the genes that are needed to make tetracycline, *S. aureofaciens* also has genes that confer tetracycline resistance; otherwise, it would kill itself when it makes tetracycline. In recent years, however, many other species of bacteria that do not synthesize tetracycline have acquired the genes that confer tetracycline resistance. For example, certain strains of *E. coli* carry tetracycline-resistance genes, even though *E. coli* does not synthesize tetracycline. When these genes were analyzed at the molecular level, it was found that they are evolutionarily related to the genes in *S. aureofaciens*. This observation indicates that the genes from *S. aureofaciens* have been transferred to *E. coli*.

A. What form of genetic transfer (i.e., conjugation, transduction, or transformation) is the most likely mechanism of interspecies genetic transfer?

B. Because *S. aureofaciens* is a nonpathogenic soil bacterium and *E. coli* is a bacterium found in the intestinal tract, do you think the genetic transfer was direct, or do you think it may have occurred in multiple steps (i.e., from *S. aureofaciens* to other bacterial species and then to *E. coli*)?

C. How could the widespread use of antibiotics to treat diseases have contributed to the proliferation of many bacterial species that are resistant to antibiotics?

Application and Experimental Questions

E1. In the experiment of Figure 9.1, a *met⁻ bio⁻ thr⁺ leu⁺ thi⁺* cell could become *met⁺ bio⁺ thr⁺ leu⁺ thi⁺* by a (rare) double mutation that converts the *met⁻ bio⁻* genetic material into *met⁺ bio⁺*. Likewise, a *met⁺ bio⁺ thr⁻ leu⁻ thi⁻* cell could become *met⁺ bio⁺ thr⁺ leu⁺ thi⁺* by three mutations that convert the *thr⁻ leu⁻ thi⁻* genetic material into *thr⁺ leu⁺ thi⁺*. From the results of

Figure 9.1, how do you know that the occurrence of 10 *met⁺ bio⁺ thr⁺ leu⁺ thi⁺* colonies is not due to these rare double or triple mutations?

E2. In the experiment of Figure 9.1, Lederberg and Tatum could not discern whether *met⁺ bio⁺* genetic material was transferred to the *met⁻ bio⁻ thr⁺ leu⁺ thi⁺* strain or if *thr⁺ leu⁺ thi⁺* genetic material was transferred to the *met⁺ bio⁺ thr⁻ leu⁻ thi⁻* strain. Let's suppose that one strain is streptomycin-resistant (say, *met⁺ bio⁺ thr⁻ leu⁻ thi⁻*) and the other strain is sensitive to streptomycin. Describe an experiment that could determine whether the *met⁺ bio⁺* genetic material was transferred to the *met⁻ bio⁻ thr⁺ leu⁺ thi⁺* strain or the *thr⁺ leu⁺ thi⁺* genetic material was transferred to the *met⁺ bio⁺ thr⁻ leu⁻ thi⁻* strain.

E3. Explain how a U-tube apparatus can distinguish between genetic transfer involving conjugation and genetic transfer involving transduction. Do you think a U-tube could be used to distinguish between transduction and transformation?

E4. What is an interrupted mating experiment? What type of experimental information can be obtained from this type of study? Why is it necessary to interrupt conjugation?

E5. In a conjugation experiment, what is meant by the time of entry? How is the time of entry determined experimentally?

E6. In your laboratory, you have an F⁻ strain of *E. coli* that is resistant to streptomycin and is unable to metabolize lactose, but it can metabolize glucose. Therefore, this strain can grow on media that contain glucose and streptomycin, but it cannot grow on media containing only lactose. A researcher has sent you two *E. coli* strains in two separate tubes. One strain, let's call it strain A, has an F factor that carries the genes that are required for lactose metabolism. On its chromosome, it also has the genes that are required for glucose metabolism. However, it is sensitive to streptomycin. This strain can grow on media containing lactose or glucose, but it cannot grow if streptomycin is added to the media. The second strain, let's call it strain B, is an F⁻ strain. On its chromosome, it has the genes that are required for lactose and glucose metabolism. Strain B is also sensitive to streptomycin. Unfortunately, when strains A and B were sent to you, the labels had fallen off the tubes. Describe how you could determine which tubes contain strain A and strain B.

E7. As mentioned in question 2 of More Genetic TIPS, origins of transfer can be located in many different places on a bacterial chromosome, and their direction of transfer can be clockwise or counterclockwise. Let's suppose a researcher conjugated six different Hfr strains that were *thr⁺ leu⁺ tonˢ strʳ azi⁵ lac⁺ gal⁺ pro⁺ met⁺* to an F⁻ strain that was *thr⁻ leu⁻ tonʳ strˢ aziʳ lac⁻ gal⁻ pro⁻ met⁻*, and obtained the following results:

Strain	Order of Gene Transfer
1	*tonˢ aziˢ leu⁺ thr⁺ met⁺ strʳ gal⁺ lac⁺ pro⁺*
2	*leu⁺ aziˢ tonˢ pro⁺ lac⁺ gal⁺ strʳ met⁺ thr⁺*
3	*lac⁺ gal⁺ strʳ met⁺ thr⁺ leu⁺ aziˢ tonˢ pro⁺*
4	*leu⁺ thr⁺ met⁺ strʳ gal⁺ lac⁺ pro⁺ tonˢ aziˢ*
5	*tonˢ pro⁺ lac⁺ gal⁺ strʳ met⁺ thr⁺ leu⁺ aziˢ*
6	*met⁺ strʳ gal⁺ lac⁺ pro⁺ tonˢ aziˢ leu⁺ thr⁺*

Draw a map of the circular *E. coli* chromosome that shows the locations and orientations of the origins of transfer in these six Hfr strains.

E8. An Hfr strain that is *hisE⁺* and *pheA⁺* was conjugated to a strain that is *hisE⁻* and *pheA⁻*. The conjugation was interrupted at different times, and the percentage of recombinants for each gene was determined by streaking on media that lacked either histidine or phenylalanine. The following results were obtained:

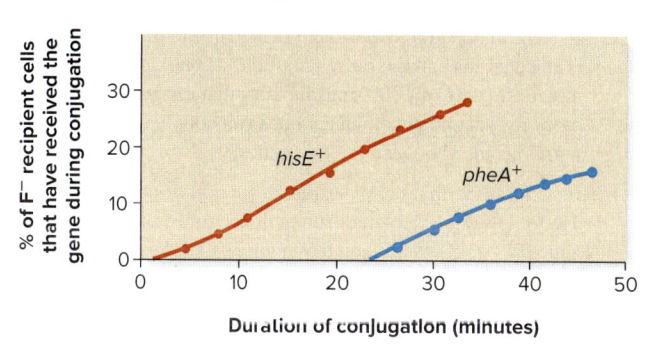

A. Determine the map distance (in minutes) between these two genes.

B. In a previous experiment, it was found that *hisE* is 4 minutes away from *pabB* and that *PheA* is 17 minutes from *pabD*. Draw a genetic map showing the locations of all three genes.

E9. Acridine orange is a chemical that inhibits the replication of F-factor DNA but does not affect the replication of chromosomal DNA, even if the chromosomal DNA is that of an Hfr strain. Let's suppose that you have an *E. coli* strain that is unable to metabolize lactose and has an F factor that carries a streptomycin-resistant gene. You also have an F⁻ strain of *E. coli* that is sensitive to streptomycin and has the genes that allow the bacterium to metabolize lactose. This second strain can grow on lactose-containing media. How would you generate an Hfr strain that is resistant to streptomycin and can metabolize lactose? (Hint: F factors occasionally integrate into the chromosome to become Hfr strains, and occasionally Hfr strains excise their DNA from the chromosome to become F⁺ strains that carry an F' factor.)

E10. In a P1 transduction experiment, the P1 lysate contains phages that carry pieces of the host chromosomal DNA, but the lysate also contains broken pieces of chromosomal DNA (see Figure 9.10). If a P1 lysate is used to transfer chromosomal DNA to another bacterium, how could you show experimentally that the recombinant bacterium has been transduced (i.e., has taken up a P1 phage with a piece of chromosomal DNA inside) versus transformed (i.e., taken up a piece of chromosomal DNA that is not within a P1 phage coat)?

E11. Can you devise an experimental strategy to get a P1 phage to transduce the entire genome of phage λ from one strain of bacterium to another strain? (Note: The general features of bacteriophage reproductive cycles are described in Chapter 10.) Phage λ has a genome size of 48,502 bp (about 1% of the size of the *E. coli* chromosome) and can follow the lytic or lysogenic reproductive cycle. Growth of *E. coli* on minimal growth medium favors the lysogenic reproductive cycle, whereas growth on rich media and/or under UV light promotes the lytic cycle.

E12. Let's suppose a new strain of P1 has been identified that packages larger pieces of the *E. coli* chromosome, pieces that are 5 minutes long. If two genes are 0.7 minute apart along the *E. coli* chromosome, what would be the cotransduction frequency using a normal strain of P1 and using the new strain of P1 that packages larger pieces? What would be the experimental advantage of using this new P1 strain?

E13. If two bacterial genes are 0.6 minute apart on the bacterial chromosome, what cotransduction frequency would you expect to observe in a cotransduction experiment using P1?

E14. In an experiment involving P1 transduction, the cotransduction frequency was 0.53. How far apart are the two genes?

E15. In a cotransduction experiment using P1, the transfer of one gene is selected for and the presence of the second gene is then determined. If 0 out of 1000 transductants that carry the first gene also carry the second gene, what would you conclude about the minimum distance between the two genes?

E16. In a cotransformation experiment (see question 4 of More Genetic TIPS), DNA was isolated from a donor strain that was $proA^+$ and $strC^+$ and sensitive to tetracycline. (The $proA$ and $strC$ genes confer the ability to synthesize proline and confer streptomycin resistance, respectively.) A recipient strain is $proA^-$ and $strC^-$ and is resistant to tetracycline. After transformation, the bacteria were first streaked on a medium containing proline, streptomycin, and tetracycline. Colonies were then restreaked on a medium containing streptomycin and tetracycline. (Note: Both types of media had carbon and nitrogen sources for growth.) The following results were obtained:

70 colonies grew on the medium containing proline, streptomycin, and tetracycline, but only 2 of these 70 colonies grew when restreaked on the medium containing streptomycin and tetracycline but lacking proline.

A. If we assume the average size of the DNA fragments is 2 minutes, how far apart are these two genes?

B. What would you expect the cotransformation frequency to be if the average size of the DNA fragments was 4 minutes and the two genes were 1.4 minutes apart?

Questions for Student Discussion/Collaboration

1. Discuss the advantages of the genetic analysis of bacteria. Make a list of the types of allelic differences among bacteria that are suitable for genetic analyses.

2. Of the three types of genetic transfer, discuss which one(s) is/are more likely to occur between members of different species. Discuss some of the potential consequences of interspecies genetic transfer.

Answers to Comprehension Questions

9.1: c

9.2: a, d

9.3: c, b

9.4: a, b

9.5: c

9.6: d

Note: All answers appear in Connect; the answers to even-numbered questions and all Concept Check questions are in Appendix B.

10

CHAPTER OUTLINE

10.1 Virus Structure and Genetic Composition
10.2 Viral Reproductive Cycles

A plant infected with tobacco mosaic virus.
©Nigel Cattlin/Science Source

GENETICS OF VIRUSES

Viruses are nonliving particles with nucleic acid genomes. Why are viruses considered nonliving? They do not exhibit all of the properties associated with living organisms. For example, viruses are not composed of cells, and by themselves, they do not carry out metabolism, use energy, maintain homeostasis, or even reproduce. A virus or its genetic material must be taken up by a living cell to replicate.

The first virus to be discovered was tobacco mosaic virus (TMV). This virus infects many species of plants and causes mosaic-like patterns on the leaves in which normal-colored patches are interspersed with light green or yellowish patches (see the chapter-opening photo). TMV damages leaves, flowers, and fruit, but almost never kills the plant. In 1883, the German scientist Adolf Mayer determined that this disease could be spread by spraying the sap from one plant onto another. By subjecting this sap to filtration, the Russian scientist Dmitri Ivanovski demonstrated that the disease-causing agent was not a bacterium. Sap that had been passed through filters with pores small enough to prevent the passage of bacterial cells was still able to spread the disease. At first, some researchers suggested the agent was a chemical toxin. However, the Dutch botanist Martinus Beijerinck ruled out this possibility by showing that sap could continue to transmit the disease after many plant generations. A toxin would have been diluted after many generations, whereas Beijerinck's results indicated the disease agent was multiplying in the plant. Around the same time, animal viruses were discovered in connection with a livestock infection called foot-and-mouth disease. In 1900, the first human virus, the virus that causes yellow fever, was identified. Since that time, researchers have identified over 120 different viruses that infect humans!

For many decades, microbiologists, geneticists, and molecular biologists have taken great interest in the structure, genetic composition, and replication of viruses. All organisms are susceptible to infection by viruses. Once a cell is infected, the genetic material of a virus orchestrates a series of events that ultimately leads to the production of new virus particles. In this chapter, we will first examine the structure and genetic composition of viruses and then explore viral reproductive cycles.

10.1 VIRUS STRUCTURE AND GENETIC COMPOSITION

Learning Outcomes:
1. Compare and contrast viruses with regard to host range, structure, and genome.
2. Analyze the results of an experiment indicating that the genome of tobacco mosaic virus is RNA.

207

A **virus** is a small infectious particle that consists of one or more nucleic acid molecules enclosed in a protein coat. Researchers have identified and studied thousands of different viruses. In this section, we will examine their general properties.

Viruses Differ in their Host Range, Structure, and Genome Composition

Although all viruses share some similarities, such as small size and the reliance on a living cell for replication, they vary greatly in their characteristics, including their host range, structure, and genome composition. Some of the major differences are described next, and characteristics of selected viruses are shown in **Table 10.1**.

Differences in Host Range A cell that is infected by a virus is called a **host cell,** and a species that can be infected by a specific virus is called a host species for that virus. Viruses differ greatly in their **host range**—the number of species and cell types they can infect. Table 10.1 lists a few examples of viruses with widely different ranges of host species. Tobacco mosaic virus, which we discussed at the beginning of this chapter, has a broad host range. TMV is known to infect hundreds of different plant species. By comparison, other viruses have a narrow host range, with some infecting only a single species. Furthermore, a virus may infect only certain cell types in a host species. For example, influenza virus specifically infects lung cells.

Structural Differences Although the existence of viruses was postulated in the 1890s, viruses were not observed until the 1930s, when the electron microscope was invented. Viruses cannot be resolved by even the best light microscope. Most of them are smaller than the wavelength of visible light. Viruses range in diameter from about 20 nm to 400 nm (1 nanometer = 10^{-9} meter). For comparison, a typical bacterium is 1000 nm in diameter, and the diameter of most eukaryotic cells is 10 to 1000 times that of a bacterium. Adenoviruses, which cause infections of the respiratory and gastrointestinal tracts, have an average diameter of 75 nm. Over 50 million adenoviruses could fit into an average-sized human cell.

What are the common structural features of all viruses? As shown in **Figure 10.1**, all viruses have a protein coat called a **capsid** that encloses a genome consisting of one or more molecules of DNA or RNA. Capsids are composed of one or more types of protein subunits called **capsomers.** Capsids have a variety of shapes, including helical and polyhedral. Figure 10.1a shows the structure of TMV, which has a helical capsid made of identical capsomers. Figure 10.1b shows an adenovirus, which has a polyhedral capsid. Protein fibers with a terminal knob are located at the corners of the polyhedral capsid. Many viruses that infect animal cells, such as the influenza virus shown in Figure 10.1c, have a **viral envelope** enclosing the capsid. The envelope consists of a lipid bilayer that is derived from the plasma membrane of the host cell and is embedded with virally encoded spike glycoproteins, also called spikes or peplomers.

In addition to encasing and protecting the genetic material, the capsid and envelope enable viruses to infect their hosts. In many viruses, the capsids or envelopes have specialized proteins, including protein fibers with a knob (Figure 10.1b) or spike glycoproteins

TABLE 10.1
Hosts and Characteristics of Selected Viruses

Virus or Group of Viruses	Host	Effect on Host	Nucleic Acid*	Genome Size (kb)†	Number of Genes†
Phage fd	*E. coli*	Slows growth	ssDNA	6.4	10
Phage λ	*E. coli*	Can exist harmlessly in the host cell or cause lysis	dsDNA	48.5	36
Phage T4	*E. coli*	Causes lysis	dsDNA	169	288
Phage Qß	*E. coli*	Slows growth	ssRNA	4.2	4
Tobacco mosaic virus (TMV)	Many plants	Causes mottling and necrosis of leaves and other plant parts	ssRNA	6.4	6
Baculoviruses	Insects	Most baculoviruses are species-specific; they usually kill the insect.	dsDNA	133.9	154
Parvovirus	Mammals	Causes respiratory, flulike symptoms	ssDNA	5.0	5
Influenza virus	Mammals	Causes classic flu symptoms—fever, cough, sore throat, and headache	ssRNA	13.5	11
Epstein-Barr virus	Humans	Causes mononucleosis, with fever, sore throat, and fatigue	dsDNA	172	80
Adenovirus	Humans	Causes respiratory symptoms and diarrhea	dsDNA	34	35
Herpes simplex type II	Humans	Causes blistering sores around the genital region	dsDNA	158.4	77
HIV	Humans	Causes AIDS, an immunodeficiency syndrome eventually leading to death	ssRNA	9.7	9

*The abbreviations ss and ds refer to single-stranded and double-stranded, respectively.
†Several of the viruses listed in this table are found in different strains that show variation with regard to genome size and number of genes. The numbers reported in this table are typical values. The abbreviation kb refers to kilobase, which equals 1000 bases.

10.1 VIRUS STRUCTURE AND GENETIC COMPOSITION   209

(a) Tobacco mosaic virus, a nonenveloped virus with a helical capsid

Helical capsid
Protein subunit (capsomer)
Nucleic acid (RNA)

(b) Adenovirus, a nonenveloped virus with a polyhedral capsid and protein fibers with a knob

Polyhedral capsid
Capsomer
Nucleic acid (DNA)
Protein fiber with a knob

(c) Influenza virus, an enveloped virus with spikes

Polyhedral capsid
Viral envelope
Nucleic acid (RNA)
Spike glycoproteins

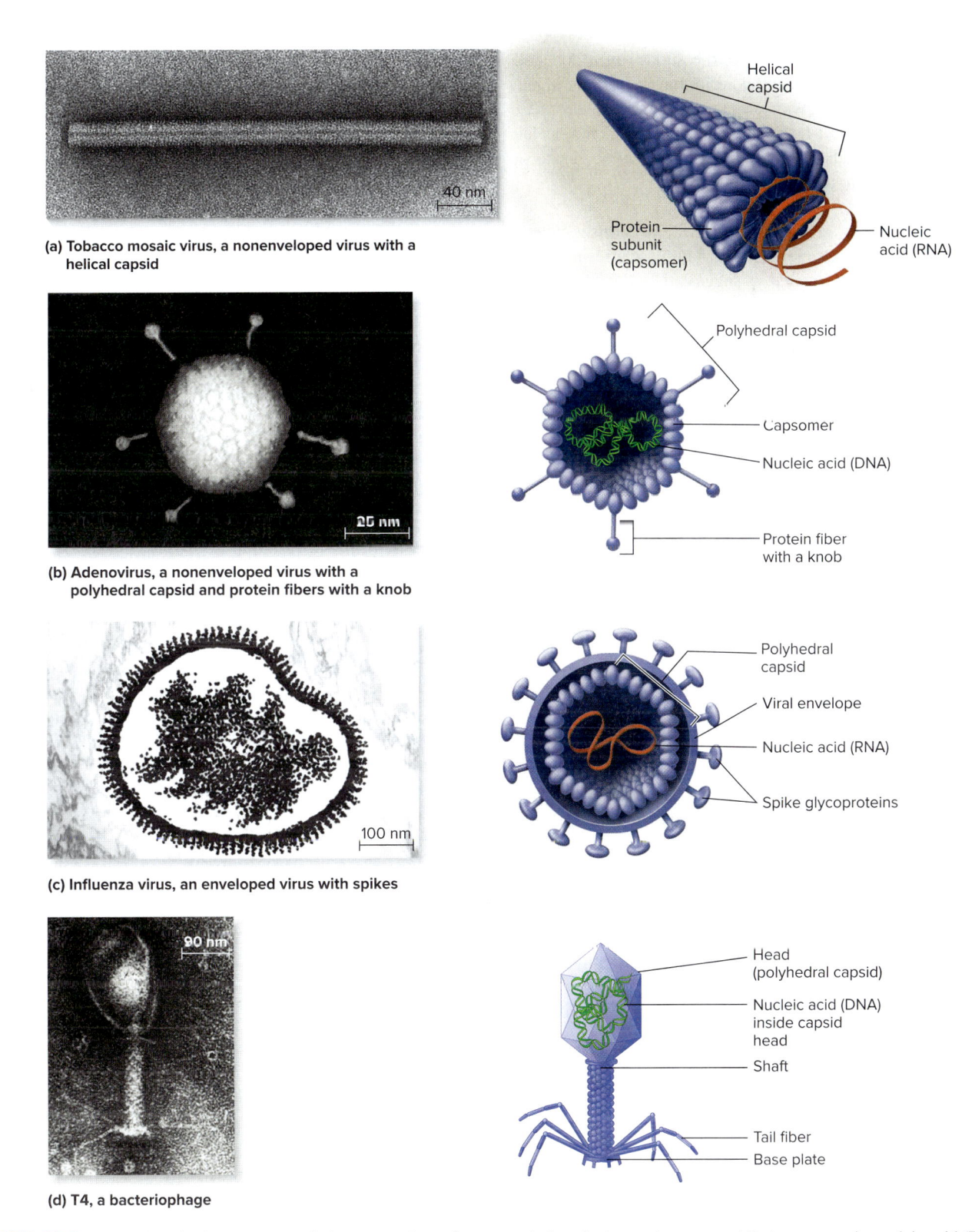

(d) T4, a bacteriophage

Head (polyhedral capsid)
Nucleic acid (DNA) inside capsid head
Shaft
Tail fiber
Base plate

FIGURE 10.1 **Variations in the structure of viruses, as shown by transmission electron microscopy.** All viruses contain nucleic acid (DNA or RNA) surrounded by a protein capsid. They may or may not have an outer envelope surrounding the capsid. **(a)** Tobacco mosaic virus (TMV) has a capsid made of 2130 identical protein subunits, helically arranged around a strand of RNA. **(b)** Adenoviruses have polyhedral capsids containing protein fibers with a knob. **(c)** Many animal viruses, including the influenza virus, have an envelope composed of a lipid bilayer and spike glycoproteins. The lipid bilayer is obtained from the host cell when the virus buds from the plasma membrane. **(d)** Some bacteriophages, such as T4, have capsids with accessory structures, such as tail fibers and base plates, that facilitate invasion of a bacterial cell.

(a) ©Omikron/Science Source; (b) ©Dr. Linda M. Stannard, University of Cape Town/SPL/Science Source; (c) ©Chris Bjornberg/Science Source; (d) ©Omikron/Science Source

Concept Check: What features vary among different types of viruses?

(Figure 10.1c), that help them bind to the surface of a host cell. Viruses that infect bacteria, called **bacteriophages,** or **phages,** may have more complex capsids, with accessory structures that are used for anchoring the virus to a host cell and injecting the viral nucleic acid (Figure 10.1d). As discussed later, the tail fibers of a bacteriophage attach the virus to the bacterial cell wall.

Genome Differences   The genetic material in a virus is called a **viral genome.** The composition of viral genomes varies markedly among different types of viruses, as suggested by the examples in Table 10.1. The nucleic acid in some viruses is DNA, whereas in others it is RNA. These are referred to as DNA viruses and RNA viruses, respectively. It is striking that some viruses use RNA for their genome, whereas all living organisms use DNA. In some viruses, the nucleic acid is single-stranded, whereas in others, it is double-stranded. The genome can be linear or circular, depending on the type of virus. Some kinds of viruses have more than one copy of the genome.

Viral genomes also vary considerably in size, ranging from a few thousand to more than a hundred thousand nucleotides in length (see Table 10.1). For example, the genome of phage Qβ is only a few thousand nucleotides in length and contains only a few genes. Other viruses, particularly those with a complex structure, such as phage T4, contain many more genes. These extra genes encode many different proteins that are involved in the formation of the elaborate structure of a phage, as shown in Figure 10.1d.

The Genome of Tobacco Mosaic Virus Is Composed of RNA

We now know that bacteria, archaea, protists, fungi, plants, and animals all use DNA as their genetic material. In 1956, Alfred Gierer and Gerhard Schramm isolated RNA from tobacco mosaic virus (TMV), which infects plant cells. When this purified RNA was applied to plant tissue, the plants developed the same types of lesions that occurred when they were exposed to intact tobacco mosaic viruses. Gierer and Schramm correctly concluded that the viral genome of TMV is composed of RNA.

To further confirm that TMV uses RNA as its genetic material, Heinz Fraenkel-Conrat and Beatrice Singer conducted additional research that involved different strains of TMV. They focused their efforts on the wild-type strain and a mutant strain called the Holmes ribgrass (HR) strain. The two strains differed in two ways.

- They caused significantly different symptoms when they infected plants. In particular, the wild-type strain produced a mottled area with yellow and green irregularly shaped lesions on infected leaves (see the chapter-opening photo), whereas the HR strain often produced streaks along the veins and ringlike markings on other parts of the leaves.
- The capsid protein in the HR strain had two amino acids, histidine and methionine, which were not found in the wild-type capsid protein.

Previous experiments had shown that purified capsid proteins and purified RNA molecules from TMVs can be mixed together and will self-assemble into intact viruses. Such a procedure is referred to as a reconstitution experiment because intact viruses are made from their individual parts. In the experiment described in **Figure 10.2**, Fraenkel-Conrat and Singer mixed wild-type RNA with HR proteins or HR RNA with wild-type proteins and then placed the reconstituted viruses onto tobacco leaves. Following infection, they then observed the symptoms caused by the viruses and also analyzed the amino acid composition of the proteins of viruses produced after the infection.

▶ **THE HYPOTHESIS**

RNA is the genetic material of TMV.

▶ **TESTING THE HYPOTHESIS** — **FIGURE 10.2**   Evidence that RNA is the genetic material of TMV.

Starting material: Purified preparations of RNA and proteins from wild-type TMV and from the Holmes ribgrass (HR) strain of TMV.

1. Mix together wild-type RNA and HR proteins or HR RNA and wild-type proteins. Allow time for the RNA and proteins to assemble into intact viruses. These are called reconstituted viruses.

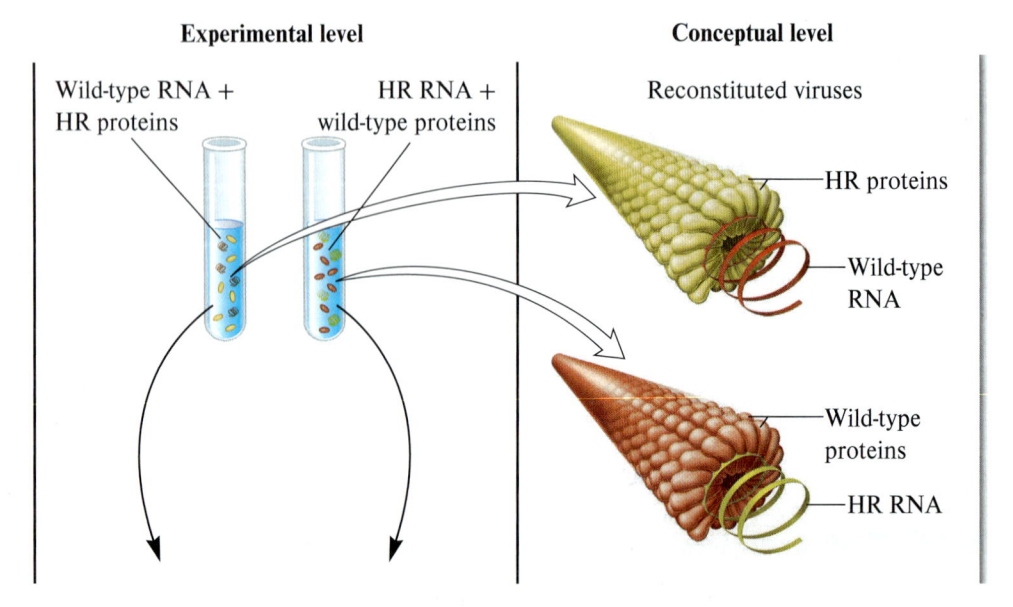

2. Inoculate a small amount of reconstituted viruses onto healthy tobacco leaves. Allow time for infection to occur.

3. Observe the types of lesions that form on the leaves.

4. Take plant tissue containing viral lesions and isolate newly made viral proteins. This is done by extracting the protein with mild alkali.

5. Determine the amino acid composition of the newly made viral proteins. This involves hydrolyzing the proteins into individual amino acids and then separating the amino acids by chromatography.

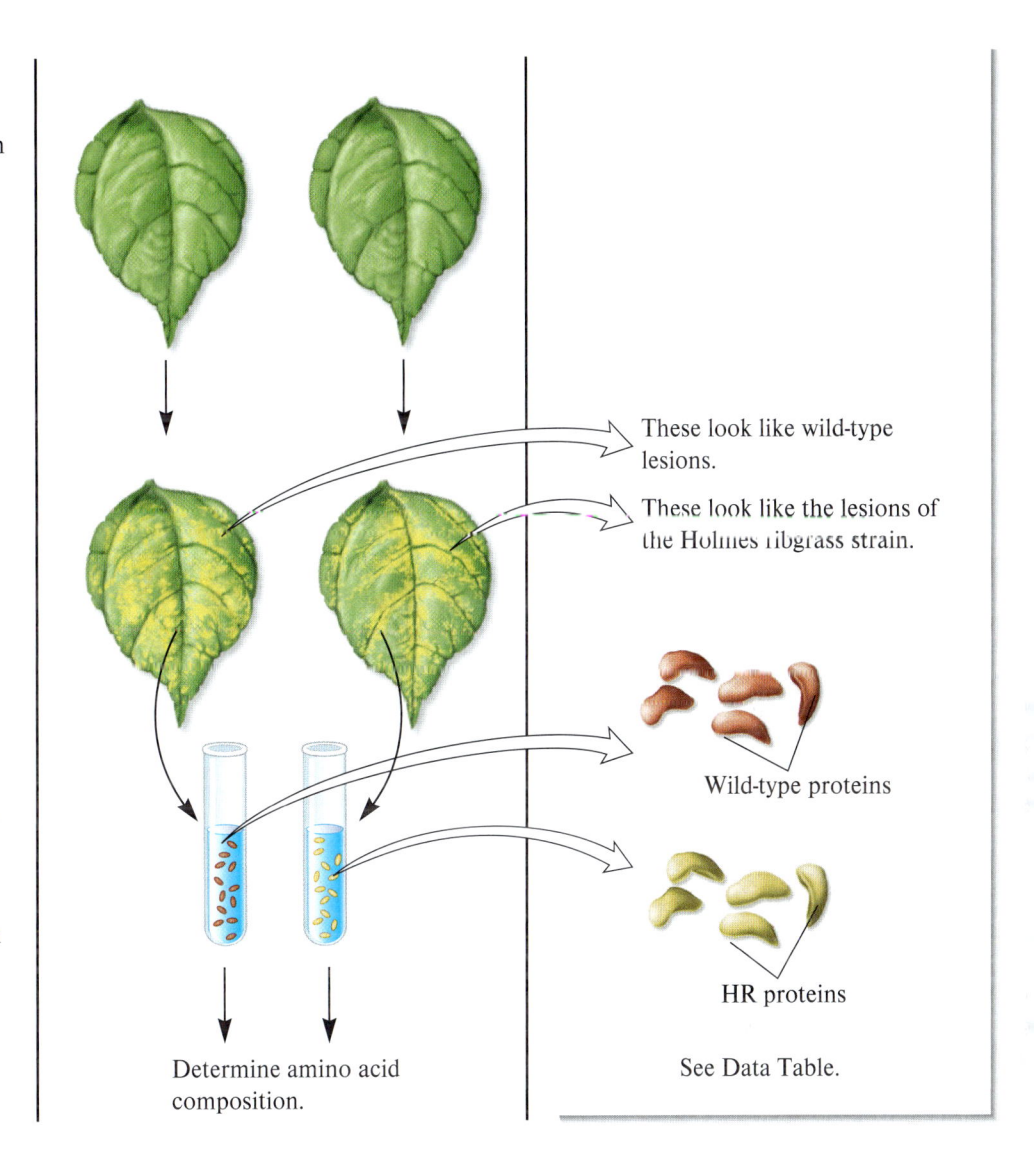

These look like wild-type lesions.

These look like the lesions of the Holmes ribgrass strain.

Wild-type proteins

HR proteins

Determine amino acid composition.

See Data Table.

▶ THE DATA

Source: Data adapted from Fraenkel-Conrat, H. & Singer, B. (1957) Virus Reconstitution II. Combination of Protein and Nucleic Acid from Different Strains, *Biochimica et Biophysica Acta*, vol. 24, no.3, 540–548.

Composition of Reconstituted Virus Placed on Tobacco Leaves	Symptoms on Tobacco Leaves	Amino Acids Found in Newly Made Viral Proteins Following Infection: Methionine	Histidine
Wild-type RNA and HR protein	Like wild-type TMV	No	No
HR RNA and wild-type protein	Like HR TMV	Yes	Yes

▶ INTERPRETING THE DATA

As seen in the data, the outcome of infection depended on the RNA that was found in the reconstituted virus but not the protein. If wild-type RNA was used, the leaves developed symptoms that were typical of wild-type TMV, and the capsid proteins of newly made viruses lacked methionine or histidine. In contrast, if the reconstituted viruses had HR RNA, the symptoms were those of the HR TMV strain, and the newly made capsid proteins contained both methionine and histidine. Taken together, these results are consistent with the hypothesis that the RNA component of TMV is its genetic material.

10.1 REVIEWING THE KEY CONCEPTS

- Tobacco mosaic virus (TMV) was the first virus to be discovered. It infects many species of plants.
- Viruses vary with regard to their host range, structure, and genome composition (see Table 10.1, Figure 10.1).
- The production of reconstituted viruses confirmed that the genome of TMV is composed of RNA (see Figure 10.2).

10.1 COMPREHENSION QUESTIONS

1. What is a common feature found in all viruses?
   a. An envelope
   b. DNA
   c. Nucleic acid surrounded by a protein capsid
   d. All of the above are common features.

2. Viral genomes can be
   a. DNA or RNA.
   b. single-stranded or double-stranded.
   c. linear or circular.
   d. all of the above.

10.2 VIRAL REPRODUCTIVE CYCLES

Learning Outcomes:
1. Compare and contrast the reproductive cycles of phage λ and HIV.
2. Define *latency*, and explain how it occurs for phage λ and HIV.
3. Describe the properties of emerging viruses.

When a virus infects a host cell, the expression of the viral genome leads to a series of steps, called a **viral reproductive cycle,** that results in the production of new viruses. The details of the steps differ among various types of viruses, and even the same virus may have the capacity to follow alternative cycles. Even so, by studying the reproductive cycles of hundreds of different viruses, researchers have determined that the viral reproductive cycle consists of five or six basic steps. In this section, we will examine these basic steps for two different viruses, and consider how new viruses can quickly spread through a population.

Viruses Follow a Reproductive Cycle That Leads to the Synthesis of New Viruses

To illustrate the general features of viral reproductive cycles, **Figure 10.3** considers the basic steps for two types of viruses. Figure 10.3a shows the cycle of **phage λ** (lambda), a bacteriophage with double-stranded DNA as its genome, and Figure 10.3b depicts the cycle of **human immunodeficiency virus (HIV),** an enveloped virus that contains single-stranded RNA and infects humans. The descriptions that follow compare the reproductive cycles of these two very different viruses.

Step 1: *Attachment* In the first step of a viral reproductive cycle, the virus must attach to the surface of a host cell. This attachment is usually specific for one or just a few types of cells because proteins in the virus recognize and bind to specific molecules on the cell surface. In the case of phage λ, the tail fibers bind to proteins in the outer cell membrane of the bacterium *E. coli*. In the case of HIV, spike glycoproteins in the viral envelope bind to receptors in the plasma membrane of human blood cells called helper T cells.

Step 2: *Entry* After attachment, the viral genome enters the host cell. Attachment of phage λ stimulates a conformational change in the phage coat proteins, so that the shaft contracts, and the phage injects its DNA into the bacterial cytoplasm. In contrast, the envelope of HIV fuses with the plasma membrane of the host cell, so both the capsid and its contents are released into the cytosol. Some of the HIV capsid proteins are then removed by host cell enzymes, a process called uncoating. This releases two copies of the viral RNA as well as molecules of enzymes called reverse transcriptase and integrase, which will be needed for step 3.

Once a viral genome has entered the cell, specific viral genes may be expressed immediately due to the action of host cell enzymes and ribosomes. Expression of these key genes leads quickly to either step 3 or step 4 of the reproductive cycle, depending on the type of virus. The genome of some viruses, including both phage λ and HIV, can integrate into a chromosome of the host cell. For such viruses, the cycle may proceed from step 2 to step 3 as described next, delaying the production of new viruses. Alternatively, the cycle may proceed directly from step 2 to step 4 and quickly lead to the production of new viruses.

Step 3: *Integration* Viruses capable of integration carry a gene that encodes an enzyme called **integrase.** For phage λ, the integrase gene is expressed soon after entry and the integrase protein is made. For HIV, integrase proteins are already contained within the virus itself. The function of integrase is to cut the host's chromosomal DNA and insert the viral genome into the chromosome. In the case of phage λ, the double-stranded DNA that entered the cell can be directly integrated into the double-stranded DNA of the host chromosome. Once integrated, the phage DNA in a bacterium is called a **prophage.** When the phage DNA exists as a prophage, the viral reproductive cycle is called the **lysogenic cycle.** As discussed later, new phages are not made during the lysogenic cycle, and the host cell is not destroyed. On occasion, a prophage can be excised from the bacterial chromosome, and the cycle proceeds to step 4.

How can an RNA virus integrate its genome into the host cell's DNA? For this to occur, the viral genome must be copied into DNA. HIV accomplishes this by means of a viral enzyme called **reverse transcriptase,** which is carried within the capsid and released into the host cell along with the viral RNA. Reverse transcriptase uses the viral RNA strand to make a complementary strand of DNA, and it then uses that DNA strand as a template to make double-stranded viral DNA. This process is called reverse transcription because it is the reverse of the usual transcription process, in which a DNA strand is used to make a complementary strand of RNA. The viral double-stranded DNA enters the host cell nucleus and is inserted into a host chromosome via integrase. Once integrated, the viral DNA in a eukaryotic cell is called a **provirus.** Viruses that follow this mechanism are called retroviruses.

Step 4: *Synthesis of Viral Components* The production of new viruses by a host cell involves the replication of the viral genome and the synthesis of viral proteins that make up the capsid. When phage λ has been integrated into the host chromosome, the prophage must be excised as described in step 3 before synthesis of new viral components can occur. A viral enzyme called excisionase is required for this process. Following excision, host cell enzymes make many copies of the phage DNA and transcribe the genes within these copies into mRNA. Host cell ribosomes translate this viral mRNA into viral proteins. The expression of phage genes also leads to the degradation of the host chromosomal DNA.

In the case of HIV, the DNA provirus is not excised from the host chromosome. Instead, it is transcribed in the nucleus to produce many copies of viral RNA. These viral RNA molecules enter the cytosol, where they are used to make viral proteins and serve as the genome for new viral particles.

Step 5: *Viral Assembly* After all of the necessary components have been synthesized, they must be assembled into new viruses. Some viruses with a simple structure self-assemble, meaning that viral components spontaneously bind to each other to form a complete virus particle. An example of a self-assembling virus is TMV, which we examined earlier (see Figure 10.1a). The capsid proteins assemble around the RNA genome, which lies inside the hollow capsid (see question 1 in More Genetic TIPS at the end of the chapter). This assembly process can occur in vitro if purified capsid proteins and RNA are mixed together.

Other viruses, including the two featured in Figure 10.3, do not self-assemble. The correct assembly of phage λ requires the help of noncapsid proteins not found in the completed phage particle. Some of these noncapsid proteins function as enzymes that modify capsid proteins, and others serve as scaffolding for the assembly of the capsid. The assembly of an HIV virus occurs in two stages. First, capsid proteins assemble around two molecules of viral RNA along with molecules of reverse transcriptase and integrase. As this is occurring, the newly formed capsid acquires its outer envelope in a budding process. This second phase of assembly happens during step 6, as the virus is released from the cell.

Step 6: *Release* The last step of a viral reproductive cycle is the release of new viruses from the host cell. The release of bacteriophages is a dramatic event. Because bacteria are surrounded by a rigid cell wall, the phages must burst, or lyse, their host cell in order to escape. After the phages have been assembled, a phage-encoded enzyme called lysozyme digests the bacterial cell wall, causing the cell to burst. Lysis releases many new phages into the environment, where they can infect other bacteria and begin the cycle again. Collectively, steps 1, 2, 4, 5, and 6 are called the **lytic cycle** because they lead to cell lysis.

The release of enveloped viruses from an animal cell is far less dramatic. This type of virus escapes by a mechanism called budding that does not lyse the cell. In the case of HIV, a newly assembled virus particle associates with a portion of the plasma membrane containing HIV spike glycoproteins. The membrane enfolds the viral capsid and eventually buds from the surface of the cell. This is how the virus acquires its envelope, which is a piece of host cell membrane studded with viral glycoproteins.

Latency in Bacteriophages As we saw in step 3, viruses can integrate their genomes into a host chromosome. In some cases, the prophage or provirus may remain inactive, or **latent**, for a long time. Most of the viral genes are silent during latency, and the viral reproductive cycle does not progress to step 4. Latency in bacteriophages is also called **lysogeny.** When this occurs, both the prophage and its host cell are said to be lysogenic. When a lysogenic bacterium prepares to divide, it copies the prophage DNA along with its own DNA, so each daughter cell inherits a copy of the prophage. A prophage can be replicated repeatedly in this way without killing the host cell or producing new phage particles. As mentioned earlier, this process is called the lysogenic cycle.

Many bacteriophages can alternate between lysogenic and lytic cycles (**Figure 10.4**). A bacteriophage that can spend some of its time in the lysogenic cycle is called a temperate phage. Phage λ is an example of a **temperate phage.** Upon infection, it can either enter the lysogenic cycle or proceed directly to the lytic cycle. Other phages, called **virulent phages,** can only follow a lytic cycle. The genome of a virulent phage is not capable of integration into a host chromosome. Phage T4, which we examined earlier in this chapter (see Figure 10.1d), is a virulent phage that infects *E. coli*. Unlike phage λ, which may coexist harmlessly with *E. coli*, T4 always lyses the infected cell.

For phages such as phage λ that can follow either cycle, environmental conditions influence whether or not viral DNA is integrated into a host chromosome and how long the virus remains in the lysogenic cycle. If nutrients are readily available, phage λ usually proceeds directly to the lytic cycle after its DNA enters the cell. Alternatively, if nutrients are in short supply, the lysogenic cycle is often favored because sufficient material may not be available to make new viruses. If more nutrients become available later, this may cause the prophage to become activated. At this point, the viral reproductive cycle switches to the lytic cycle, and new viruses are made and released.

Latency in Human Viruses Latency among human viruses can occur in two different ways. For HIV, latency occurs because the virus has integrated into the host genome and may remain dormant for long periods of time. Alternatively a viral genome can exist as an **episome**—a genetic element that can replicate independently of the chromosomal DNA but also can occasionally integrate into chromosomal DNA. Examples of viral genomes that can exist as episomes include different types of herpesviruses that cause cold sores (herpes simplex type I), genital herpes (herpes simplex type II), and chickenpox (varicella-zoster virus). A person infected with a given type of herpesvirus may have periodic outbreaks of disease symptoms when the virus switches from the latent, episomal form to the active form that produces new virus particles.

As an example, let's consider the herpesvirus called varicella-zoster virus. The initial infection by this virus causes chickenpox, after which the virus may remain latent for many years as an episome. The disease called shingles occurs when varicella-zoster virus switches from the latent state and starts making new virus particles. It usually occurs in adults over the age of 60 or in people with weakened immune systems. Shingles begins as a painful rash that eventually erupts into blisters. The blisters follow the path of the nerve cells that carry the latent varicella-zoster virus, and often form a ring around the patient's body. The term *shingles* is derived from a Latin word meaning "girdle," referring to the observation that the blisters girdle a part of the body.

214 CHAPTER 10 :: GENETICS OF VIRUSES

1 Attachment:
The phage binds specifically to proteins in the outer bacterial cell membrane.

2 Entry:
The phage injects its DNA into the bacterial cytoplasm.

3 Integration:
Phage DNA may integrate into the bacterial chromosome via integrase. The host cell carrying a prophage may then undergo repeated divisions, which is called the lysogenic cycle. To end the lysogenic cycle and switch to the lytic cycle, the phage DNA is excised. Alternatively, the reproductive cycle may completely skip the lysogenic cycle and proceed directly to step 4.

(a) Reproductive cycle of phage λ

1 Attachment:
Spike glycoproteins bind to receptors on the host-cell plasma membrane.

2 Entry:
The viral envelope fuses with the host-cell plasma membrane, releasing the capsid and its contents into the cytosol. Some capsid proteins are removed by cellular enzymes, a process called uncoating. This releases the RNA, reverse transcriptase, and integrase into the cytosol.

3 Integration:
Viral RNA is reverse transcribed into double-stranded DNA and then integrated into the host cell chromosome, via integrase. The integrated provirus may remain latent for a long period of time.

(b) Reproductive cycle of HIV

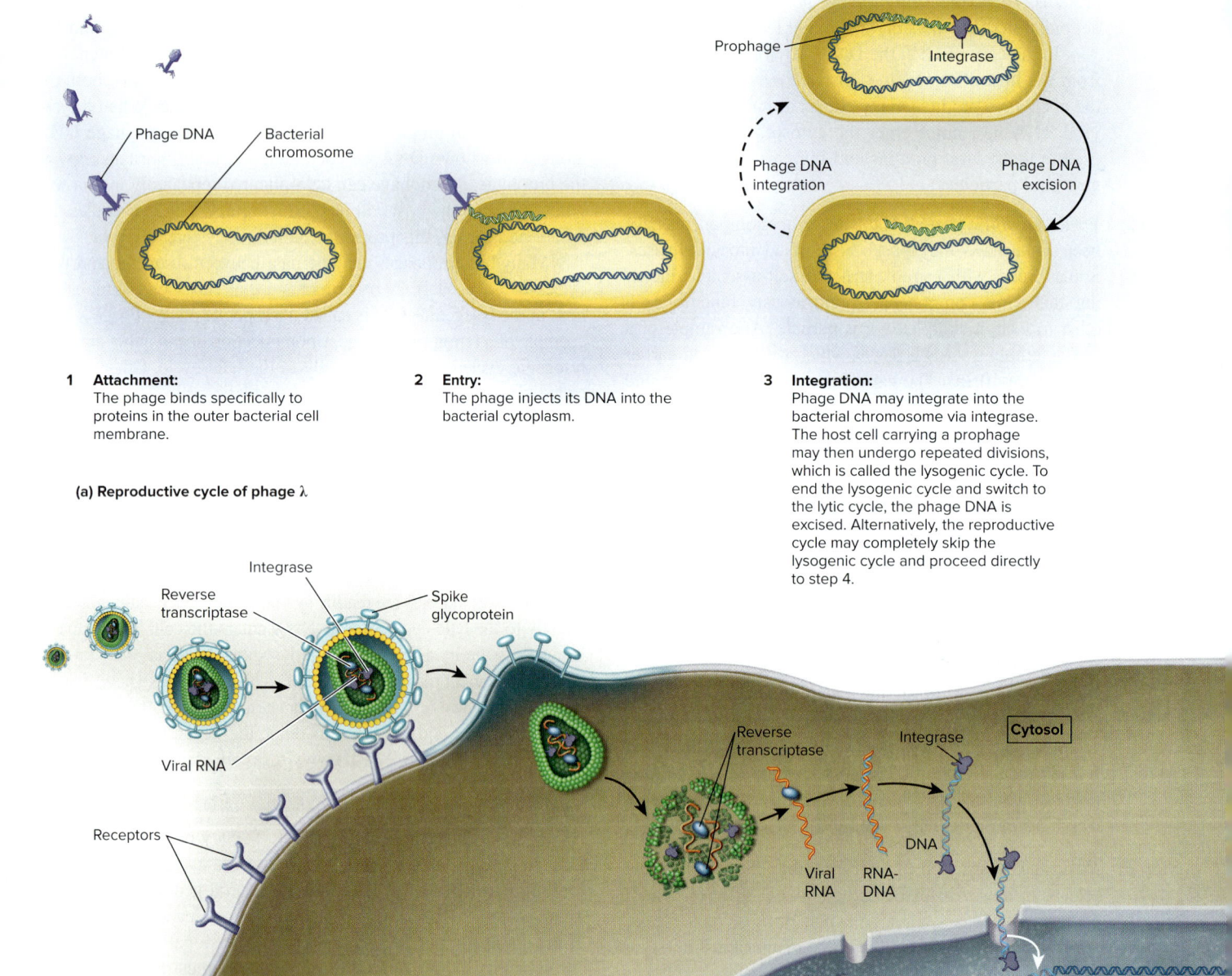

FIGURE 10.3 Comparison of the steps of two viral reproductive cycles. (a) The reproductive cycle of phage λ, a bacteriophage with a double-stranded DNA genome. (b) The reproductive cycle of HIV, an enveloped virus with a single-stranded RNA genome that infects humans.

Concept Check: During which step of the reproductive cycle can a virus remain latent?

10.2 VIRAL REPRODUCTIVE CYCLES   215

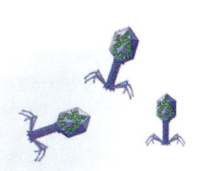

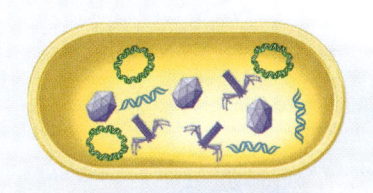

4 Synthesis of viral components:
In the lytic cycle, phage DNA directs the synthesis of viral components. During this process, the phage DNA circularizes, and the host chromosomal DNA is digested into fragments.

5 Viral assembly:
Phage components are assembled with the help of noncapsid proteins to make many new phages.

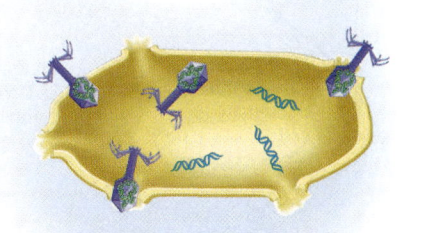

6 Release:
The viral enzyme called lysozyme causes cell lysis, and new phages are released from the broken cell.

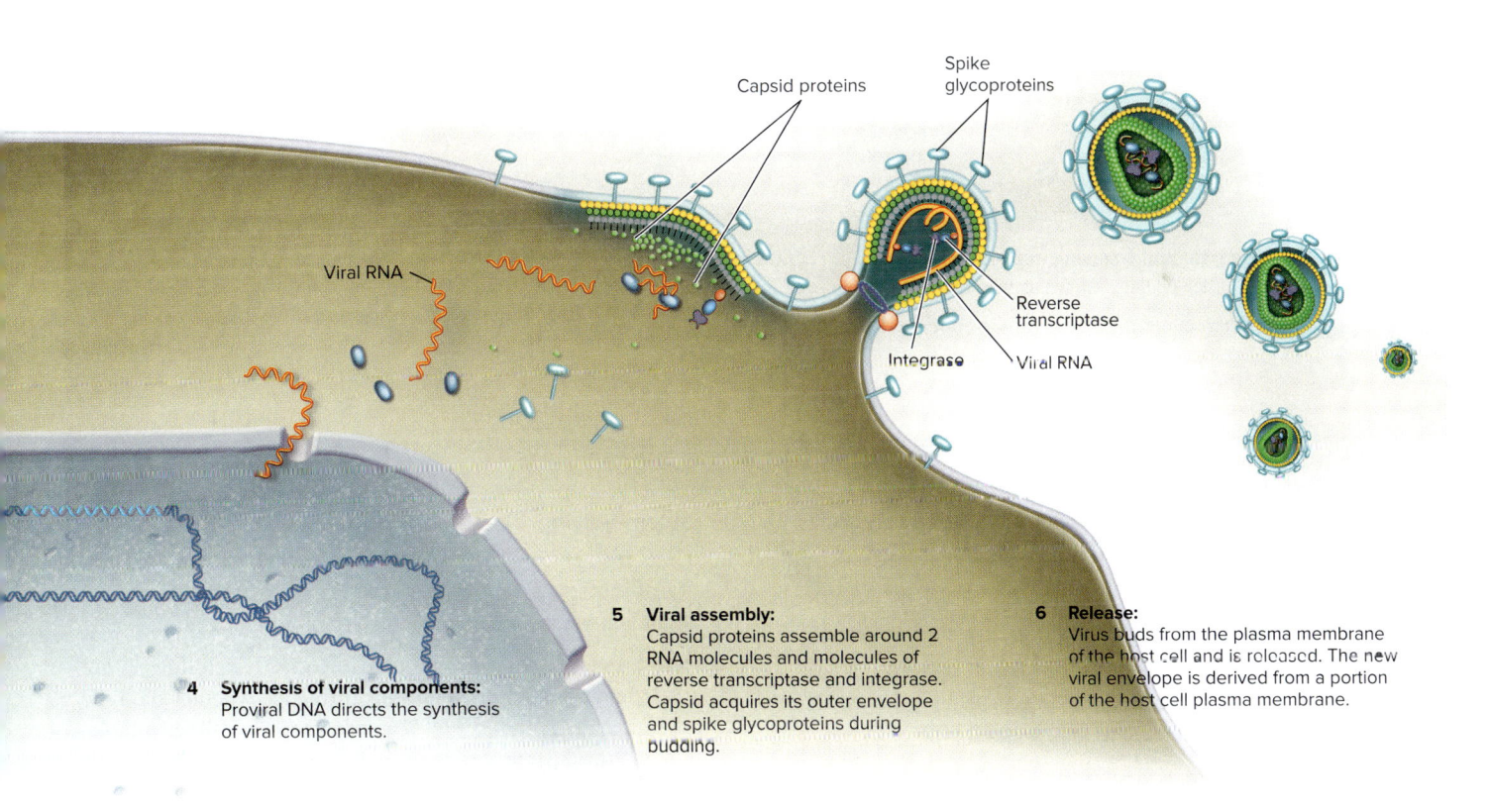

Capsid proteins

Spike glycoproteins

Viral RNA

Reverse transcriptase

Integrase

Viral RNA

4 Synthesis of viral components:
Proviral DNA directs the synthesis of viral components.

5 Viral assembly:
Capsid proteins assemble around 2 RNA molecules and molecules of reverse transcriptase and integrase. Capsid acquires its outer envelope and spike glycoproteins during budding.

6 Release:
Virus buds from the plasma membrane of the host cell and is released. The new viral envelope is derived from a portion of the host cell plasma membrane.

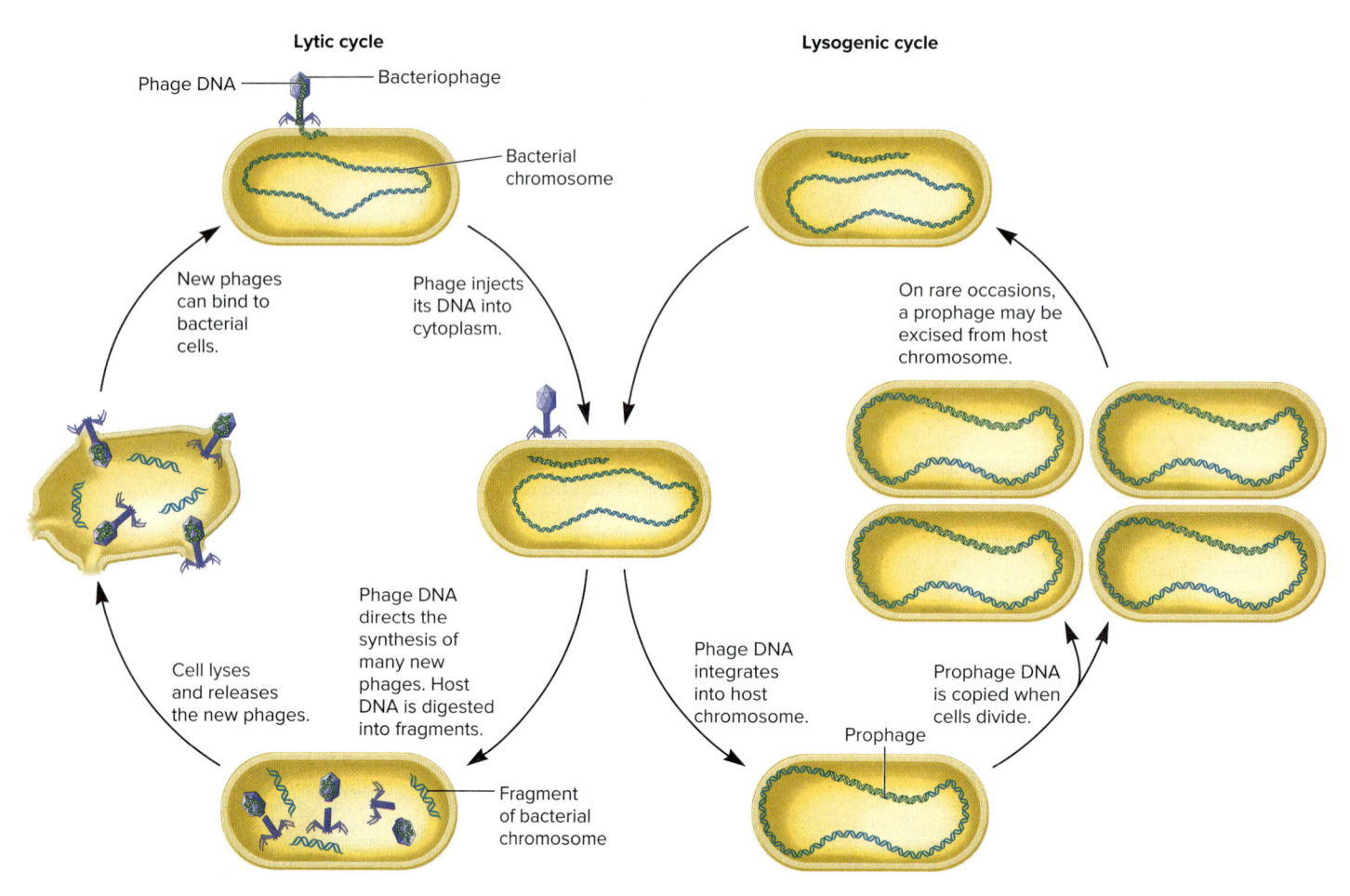

FIGURE 10.4 **The lytic and lysogenic reproductive cycles of certain bacteriophages.** Some bacteriophages, such as temperate phages, can follow both cycles. Other phages, known as virulent phages, can only follow a lytic cycle.

Concept Check: Which cycle produces new phage particles?

Genetic TIPS

The Question: From the perspective of a bacteriophage, what is the advantage of being able to follow either a lytic or a lysogenic cycle?

Topic: What topic in genetics does this question address? The topic is the lytic and lysogenic cycles of bacteriophages. More specifically, the question asks you to identify the advantage to a phage of being able to follow either type of cycle.

Information: What information do you know based on the question and your understanding of the topic? From the question, you know that some viruses can follow either a lytic or a lysogenic cycle. From your understanding of the topic, you may remember that the lytic cycle involves the synthesis of new viruses, whereas the lysogenic cycle involves the integration of the viral DNA into the host chromosome. The choice between the two cycles may be driven by the availability of nutrients to the host cell.

Problem-Solving Strategy: Compare and contrast. One strategy to solve this problem is to compare the ability of the host cell to make new viruses when nutrients are abundant versus when nutrients are limiting.

Answer: For phages such as phage λ that can follow either cycle, environmental conditions influence whether the lytic or lysogenic cycle occurs. If nutrients are readily available, phage λ usually proceeds directly to the lytic cycle after its DNA enters the cell. This allows the phage to rapidly proliferate. Alternatively, if nutrients are in short supply, the lysogenic cycle is usually favored because sufficient nutrients to make new viruses are not available. If more nutrients become available later, the prophage may become activated, and the viral reproductive cycle will switch to the lytic cycle. Therefore, under more favorable conditions, new viruses are made and released.

Emerging Viruses, Such as HIV and Zika Virus, Have Arisen Recently and May Rapidly Spread Through a Population

A primary reason researchers have been interested in viral reproductive cycles is the ability of many viruses to cause diseases in humans and other hosts. Some examples of viruses that infect humans were presented earlier in Table 10.1. **Emerging viruses** are viruses that have arisen recently and are more likely to cause infection compared with previous strains. Such viruses may lead to a significant loss of human life and often cause public alarm.

- New strains of influenza virus arise fairly regularly due to new mutations. An example is the strain H1N1, also called swine flu (**Figure 10.5**). It was originally called swine flu because laboratory testing revealed that it carries two genes that are normally found in flu viruses that infect pigs in Europe and Asia. The Centers for Disease Control and Prevention (CDC) recommends vaccination as the first and most important step in preventing infection by H1N1 and other strains of influenza. However, in the United States, despite vaccination campaigns that aim to minimize deaths from seasonal influenza, over 30,000 people die annually from this disease.

- Another example of an emerging virus is Zika virus, a type of flavivirus. It is an enveloped virus with a genome composed of single-stranded RNA. Its name comes from the Zika Forest of Uganda, where the virus was first isolated in 1947. Zika virus is primarily spread by mosquitoes of the genus *Aedes*. The most common symptoms of a Zika infection are fever, rash, joint pain, and conjunctivitis (inflammation of membranes around the eyes). In most people, the illness is usually mild, but in rare cases, a Zika infection in an adult can cause a more serious illness called Guillain-Barré syndrome. In addition, Zika virus infection during pregnancy can cause the child to be born with a serious birth defect called microcephaly, as well as other severe brain defects. The Zika virus has spread globally from Africa into Asia, South America, and North America. Though estimates for infection rates vary, some epidemiologists predict that millions of people will become infected with the virus in the coming years.

- During the past few decades, the most devastating example of an emerging virus has been human immunodeficiency virus (HIV), the causative agent of acquired immune deficiency syndrome (AIDS). Research studies suggest that HIV is a mutant form of a virus found in chimpanzees in West Africa, which is called simian immunodeficiency virus (SIV). It was most likely transmitted to humans and mutated into HIV when humans hunted chimpanzees for meat and came into contact with their infected blood. Scientists speculate that humans were first infected in the early 1900s.

HIV is primarily spread by sexual contact between infected and uninfected individuals, but it can also be spread by the transfusion of HIV-infected blood, by the sharing of needles among drug users, and from infected mother to unborn child. The total number of deaths due to AIDS between 1981 and the end of 2015 was over 40 million; more than 650,000 of these deaths occurred in the United States. During 2015, around 2 million adults and children became infected with HIV worldwide; approximately 1 in every 100 adults between ages 15 and 49 is infected. In the United States, about 50,000 new HIV infections occur each year; 70% of those infections are in men and 30% in women.

The symptoms of AIDS result from viral destruction of helper T cells, a type of white blood cell that plays an essential role in the immune system of humans. **Figure 10.6** shows HIV virus particles invading a helper T cell, which normally interacts with other cells of the immune system to facilitate the production of antibodies and other molecules that target and kill foreign invaders of the body. When large numbers of helper T cells are destroyed by HIV, the function of the immune system is seriously compromised, and the individual becomes susceptible to opportunistic infections—infectious diseases that would not normally occur in a healthy person. For example, *Pneumocystis jiroveci*, a fungus that causes pneumonia, is easily destroyed by a normal immune system. However, in people with AIDS, infection by this fungus can be fatal. An insidious feature of HIV replication, which is outlined in Figure 10.3b, is that reverse transcriptase, the enzyme that

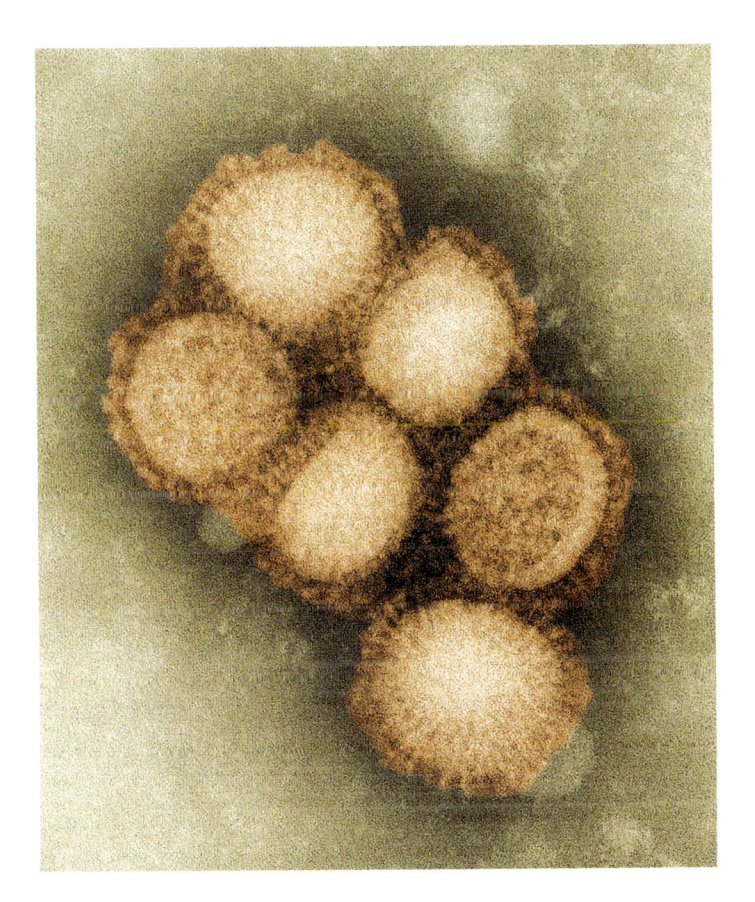

FIGURE 10.5 **A micrograph of influenza virus, strain H1N1.** This virus, which infects people, is a relatively new strain of influenza virus that causes classic flu symptoms.
Source: CDC/C. S. Goldsmith and A. Balish

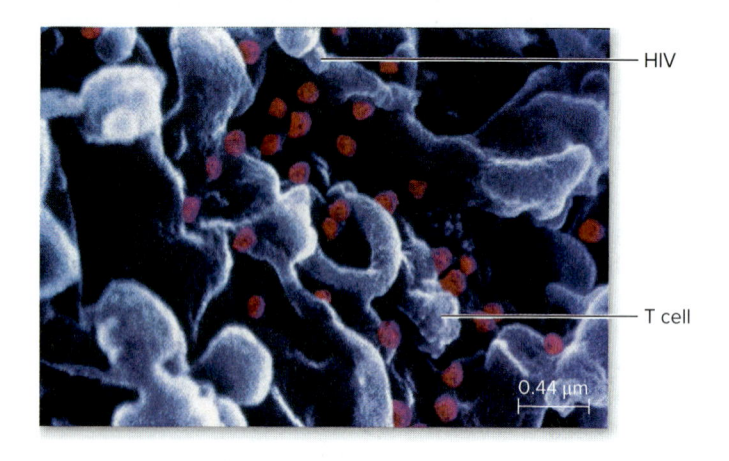

FIGURE 10.6 Micrograph of HIV invading a human helper T cell. This is a colorized scanning electron micrograph. The surface of the T cell is purple, and HIV particles are red.
©Richard J. Green/Science Source

Concept Check: What is an emerging virus?

copies the RNA genome into DNA, lacks a proofreading function. In Chapter 13, we will discuss how DNA polymerase can identify and remove mismatched nucleotides in newly synthesized DNA. Because reverse transcriptase lacks this function, it makes more errors and thereby tends to create mutant strains of HIV. This undermines the ability of the body to combat HIV because mutant strains may not be destroyed by the body's defenses. In addition, mutant strains of HIV may be resistant to antiviral drugs.

10.2 REVIEWING THE KEY CONCEPTS

- The viral reproductive cycle consists of five or six basic steps, including attachment, entry, integration, synthesis, assembly, and release (see Figure 10.3).
- Some bacteriophages can alternate between two reproductive cycles: the lytic and lysogenic cycles (see Figure 10.4).
- Emerging viruses are those that have arisen recently and are more likely than previous strains to cause infection (see Figure 10.5).
- The disease AIDS is caused by human immunodeficiency virus (HIV). This virus is a retrovirus whose reproductive cycle involves the integration of the viral genome into a chromosome in the host cell (see Figure 10.6).

10.2 COMPREHENSION QUESTIONS

1. Which of the following is a common order of the steps in a viral reproductive cycle? (Note: Integration is an optional step.)
    a. Entry, integration, attachment, synthesis of viral components, viral assembly, viral release
    b. Entry, integration, synthesis of viral components, viral assembly, attachment, viral release
    c. Attachment, entry, integration, synthesis of viral components, viral assembly, viral release
    d. Attachment, entry, integration, viral assembly, synthesis of viral components, viral release
2. An emerging virus is
    a. a virus that has arisen recently and is more likely to cause infection than previous strains.
    b. HIV that causes AIDS.
    c. a strain of influenza virus called H1N1 that causes the flu.
    d. all of the above.

KEY TERMS

Page 208. virus, host cell, host range, capsid, capsomers, viral envelope
Page 210. bacteriophages (phages), viral genome
Page 212. viral reproductive cycle, phage λ, human immunodeficiency virus (HIV), integrase, prophage, lysogenic cycle, reverse transcriptase, provirus
Page 213. lytic cycle, latent, lysogeny, temperate phage, virulent phage, episome
Page 217. emerging virus

CHAPTER SUMMARY

10.1 Virus Structure and Genetic Composition

- Tobacco mosaic virus (TMV) was the first virus to be discovered. It infects many species of plants.
- Viruses vary with regard to their host range, structure, and genome composition (see Table 10.1, Figure 10.1).
- The production of reconstituted viruses confirmed that the genome of TMV is composed of RNA (see Figure 10.2).

10.2 Viral Reproductive Cycles

- The viral reproductive cycle consists of five or six basic steps, including attachment, entry, integration, synthesis, assembly, and release (see Figure 10.3).

- Some bacteriophages can alternate between two reproductive cycles: the lytic and lysogenic cycles (see Figure 10.4).
- Emerging viruses are those that have arisen recently and are more likely than previous strains to cause infection (see Figure 10.5).
- The disease AIDS is caused by human immunodeficiency virus (HIV). This virus is a retrovirus whose reproductive cycle involves the integration of the viral genome into a chromosome in the host cell (see Figure 10.6).

PROBLEM SETS & INSIGHTS

More Genetic TIPS

1. Discuss how the components of viruses assemble to make new virus particles. What is the difference between a virus that can self-assemble and one that cannot? Give examples.

Topic: What topic in genetics does this question address?

The topic is the synthesis of new viruses. More specifically, the question is about the assembly of components into viral particles.

Information: What information do you know based on the question and your understanding of the topic?

From the question, you know that viruses are composed of various components that must assemble to make a virus particle. From your understanding of the topic, you may remember that viral assembly occurs in a series of steps.

Problem-Solving Strategy: Describe the steps.

One strategy to solve this problem is to break down the assembly of viruses into a series of steps.

Answer: Some viruses are composed of only nucleic acid (DNA or RNA) surrounded by a capsid of proteins. First, the viral nucleic acid and viral proteins are made by the host cells. Next, the proteins bind to each other to form a capsid that surrounds the viral nucleic acid. For viruses that also have an envelope, such as HIV, the third step is the budding of the virus particle from the cell, during which it acquires its envelope.

Many viruses with a simple structure are able to self-assemble, meaning that viral components spontaneously bind to each other to form a complete virus particle. An example of a self-assembling virus is the tobacco mosaic virus (TMV). The capsid proteins assemble around the RNA genome, which lies inside the hollow capsid.

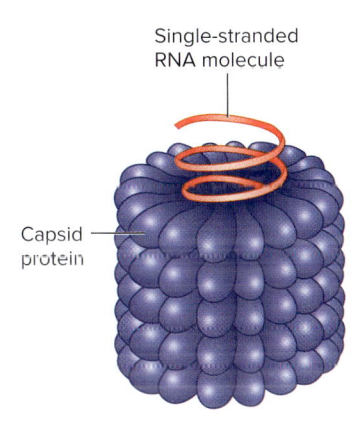

Other viruses do not self-assemble. For example, the assembly of phage λ requires the help of noncapsid proteins not found in the completed viral particle. Some of these noncapsid proteins function as enzymes that modify capsid proteins, whereas others provide scaffolding for the assembly of the capsid.

2. Viruses may be latent for a long period of time. For example, HIV may be latent for many years, during which new viruses are not made. What are three different mechanisms of viral latency?

Topic: What topic in genetics does this question address?

The topic is latency in viruses. More specifically, the question asks you to explain how latency occurs.

Information: What information do you know based on the question and your understanding of the topic?

From the question, you know that viruses may become latent, during which time they do not produce new viruses. From your understanding of the topic, you may remember that latency can occur in different ways depending on the type of virus.

Problem-Solving Strategy: Compare and contrast.

One strategy to solve this problem is to consider how different types of viruses can become latent and contrast different mechanisms.

Answer: Some bacteriophages enter the lysogenic cycle in which the phage DNA becomes incorporated into the bacterial chromosome as a prophage. In this case, the phage DNA is directly incorporated into the host-cell chromosome. Eukaryotic viruses can become latent in other ways. Retroviruses, which have an RNA genome, are reverse transcribed and the viral DNA is incorporated into a host-cell chromosome as a provirus. Alternatively, herpesviruses can exist as episomes that remain latent for long periods of time.

3. What is an emerging virus? Give two examples. Propose an experiment to explain how an emerging virus could arise.

Topic: What topic in genetics does this question address?

The topic is emerging viruses. More specifically, the question asks you to describe what an emerging virus is and propose an experiment to explain how it might come into existence.

Information: What information do you know based on the question and your understanding of the topic?

From your understanding of the topic, you may remember that emerging viruses have arisen recently and have the potential to cause widespread infection and disease. You may also remember that emerging viruses arise via mutations in existing viruses.

Problem-Solving Strategy: Define key terms. Design an experiment.

To begin to solve this problem, you need to define a key term: emerging virus. To design an experiment that explains how such viruses arise, you may want to consider that new forms of viruses generally come from the genetic alteration of preexisting viruses. Another aspect of viruses to consider is that the base sequences of many viruses have already been determined.

Answer: An emerging virus is a virus that has arisen recently and is more likely than previous strains to cause infection. Examples include HIV and the influenza strain called H1N1.

The rationale behind designing an experiment is based on the premise that emerging viruses arise from genetic alterations in preexisting viruses.

1. Determine the base sequence of the emerging virus that you are interested in. (Note: The topic of DNA sequencing is described in Chapter 20.)
2. Compare the base sequence with the known sequences of other viruses; the expectation is that your emerging virus will be closely related to some other virus that is already known.
3. Analyze the differences in base sequences between your emerging virus and to its closest relative.

Results: You may identify changes in particular viral genes that may have altered the infectivity of the virus. For example, a mutation could occur that alters a viral protein in a way that allows the virus to bind more easily to host cells and enter them.

Conceptual Questions

C1. Explain why viruses are considered nonliving.

C2. What structural features are common to all viruses? Which features are found only in certain types of viruses?

C3. What are the similarities and differences among viral genomes?

C4. What is a viral envelope? Describe how it is made.

C5. What do the terms *host cell* and *host range* mean?

C6. Describe why the attachment step in a viral reproductive cycle is usually specific for one or just a few cell types.

C7. Compare and contrast the entry step of the viral reproductive cycles of phage λ and HIV.

C8. What is the role of reverse transcriptase in the reproductive cycle of HIV and other retroviruses?

C9. Describe how lytic bacteriophages are released from their host cells.

C10. What is the difference between a temperate phage and a virulent phage?

C11. What are a prophage, a provirus, and an episome? What is their common role in a viral reproductive cycle?

Application and Experimental Questions

E1. Discuss how researchers determined that TMV is a virus that causes damage to plants.

E2. What technique must be used to visualize a virus?

E3. What is a reconstituted virus?

E4. Following the infection of healthy tobacco leaves by reconstituted viruses, what two characteristics did Fraenkel-Conrat and Singer analyze? Explain how their results were consistent with the idea that the RNA of TMV is responsible for the traits of the virus.

E5. Certain drugs to combat human viral diseases affect spike glycoproteins in the viral envelope. Discuss how you think such drugs may prevent viral infection.

E6. Some drugs that inhibit HIV proliferation are inhibitors of HIV protease, an enzyme that cleaves certain HIV proteins into smaller, functional units. Explain how these drugs would help to stop the spread of HIV.

Questions for Student Discussion/Collaboration

1. Outline the similarities and differences between the reproductive cycles of phage λ and HIV.

2. Discuss ways that you might distinguish whether an infectious disease in mice is caused by a virus or by a bacterium.

Answers to Comprehension Questions

10.1: c, d

10.2: c, d

Note: All answers appear in Connect; the answers to even-numbered questions and all Concept Check questions are in Appendix B.

PART III MOLECULAR STRUCTURE AND
REPLICATION OF THE GENETIC MATERIAL

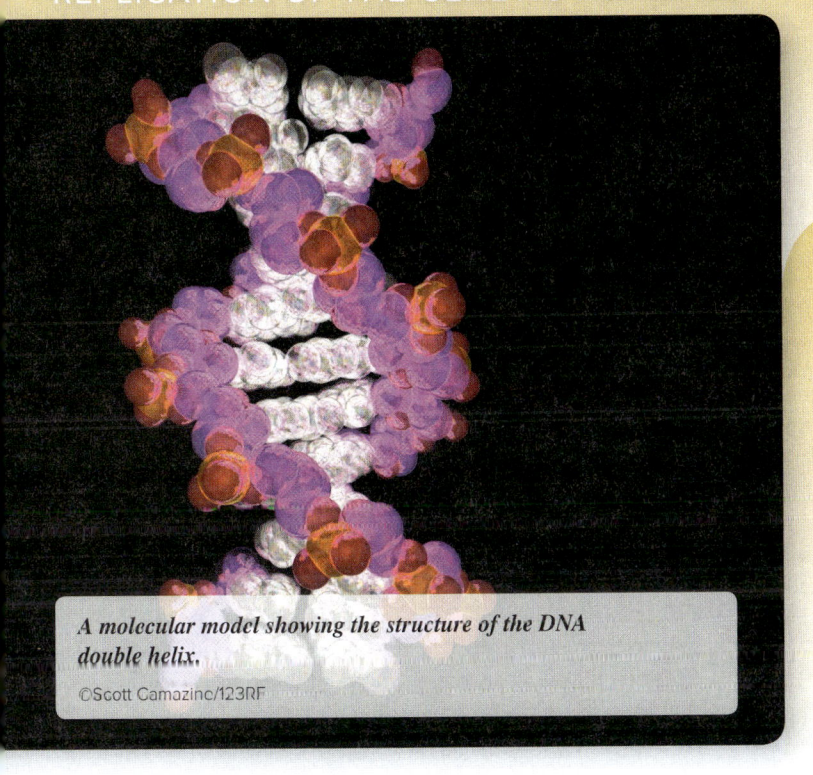

11

A molecular model showing the structure of the DNA double helix.
©Scott Camazine/123RF

CHAPTER OUTLINE

- 11.1 Identification of DNA as the Genetic Material
- 11.2 Overview of DNA and RNA Structure
- 11.3 Nucleotide Structure
- 11.4 Structure of a DNA Strand
- 11.5 Discovery of the Double Helix
- 11.6 Structure of the DNA Double Helix
- 11.7 RNA Structure

MOLECULAR STRUCTURE OF DNA AND RNA

In Chapters 2 through 10, we focused on the relationship between the inheritance of genes and chromosomes and the outcome of an organism's traits. In this chapter, we will shift our attention to **molecular genetics**—the study of DNA structure and function at the molecular level. An exciting goal of molecular genetics is to use our knowledge of DNA structure to understand how DNA functions as the genetic material. Using molecular techniques, researchers have determined the organization of many genes. This information, in turn, has helped us understand how the expression of genes governs the outcome of an individual's inherited traits.

The past several decades have seen dramatic advances in techniques and approaches used to investigate and even to alter the genetic material. These advances have greatly expanded our understanding of molecular genetics and also have provided key insights into the mechanisms underlying transmission and population genetics. Molecular genetic technology is also widely used in supporting disciplines such as biochemistry, cell biology, and microbiology.

To a large extent, our understanding of genetics comes from our knowledge of the molecular structure of **DNA (deoxyribonucleic acid)** and **RNA (ribonucleic acid)**. In this chapter, we will begin by considering classic experiments that showed that DNA is the genetic material. We will then survey the molecular features of DNA and RNA that underlie their function.

11.1 IDENTIFICATION OF DNA AS THE GENETIC MATERIAL

Learning Outcomes:
1. Describe the four criteria that the genetic material must meet.
2. Analyze the results of (1) Griffiths, (2) Avery, MacLeod, and McCarty, and (3) Hershey and Chase and explain how they indicate that DNA is the genetic material.

To fulfill its role, the genetic material must meet four criteria:

1. **Information:** The genetic material must contain the information necessary to construct an entire organism. In other words, it must provide the blueprint to determine the inherited traits of an organism.
2. **Transmission:** During reproduction, the genetic material must be passed from parents to offspring.

221

3. **Replication:** Because the genetic material is passed from parents to offspring, and from mother cell to daughter cells during cell division, it must be copied.
4. **Variation:** Within any species, a significant amount of phenotypic variability occurs. For example, Mendel studied several characteristics in pea plants that varied among different plants. These included height (tall versus dwarf) and seed color (yellow versus green). Therefore, the genetic material must also vary in ways that can account for the known phenotypic differences within each species.

Recall from Chapter 3 that, in the 1880s, August Weismann and Carl Nägeli championed the idea that a chemical substance within living cells is responsible for the transmission of traits from parents to offspring and that subsequent experimentation demonstrated that chromosomes are the carriers of the genetic material. Nevertheless, the story was not complete because chromosomes contain both DNA and proteins. Also, RNA is found in the vicinity of chromosomes. Therefore, further research was needed to precisely identify the genetic material. In this section, we will examine the first experimental attempts to achieve this goal.

Griffith's Experiments Indicated That Genetic Material Can Transform Pneumococcus

Some early work in microbiology was important in developing an experimental strategy to identify the genetic material. Frederick Griffith studied a type of bacterium known then as pneumococci and now classified as *Streptococcus pneumoniae*. Certain strains of *S. pneumoniae* secrete a polysaccharide capsule, whereas other strains do not. When streaked onto petri plates containing a solid growth medium, capsule-secreting strains have a smooth colony morphology, whereas those strains unable to secrete a capsule produce a colony with a rough appearance.

The different forms of *S. pneumoniae* also affect their virulence, or ability to cause disease. When smooth strains of *S. pneumoniae* infect a mouse, the capsule allows the bacteria to escape attack by the mouse's immune system. As a result, the bacteria can grow and eventually kill the mouse. In contrast, the nonencapsulated (rough) bacteria are destroyed by the animal's immune system.

In 1928, Griffith conducted experiments that involved the injection of live and/or heat-killed bacteria into mice. He then observed whether or not the bacteria caused a lethal infection. Griffith was working with two strains of *S. pneumoniae*, type S (for "smooth") and type R (for "rough").

- When injected into a live mouse, the type S bacteria proliferated within the mouse's bloodstream and ultimately killed the mouse (**Figure 11.1a**). Following the death of the mouse, Griffith found many type S bacteria within the mouse's blood.
- In contrast, when type R bacteria were injected into a mouse, the mouse lived (**Figure 11.1b**).
- To verify that the proliferation of the type S bacteria was causing the death of the mouse, Griffith killed the type S

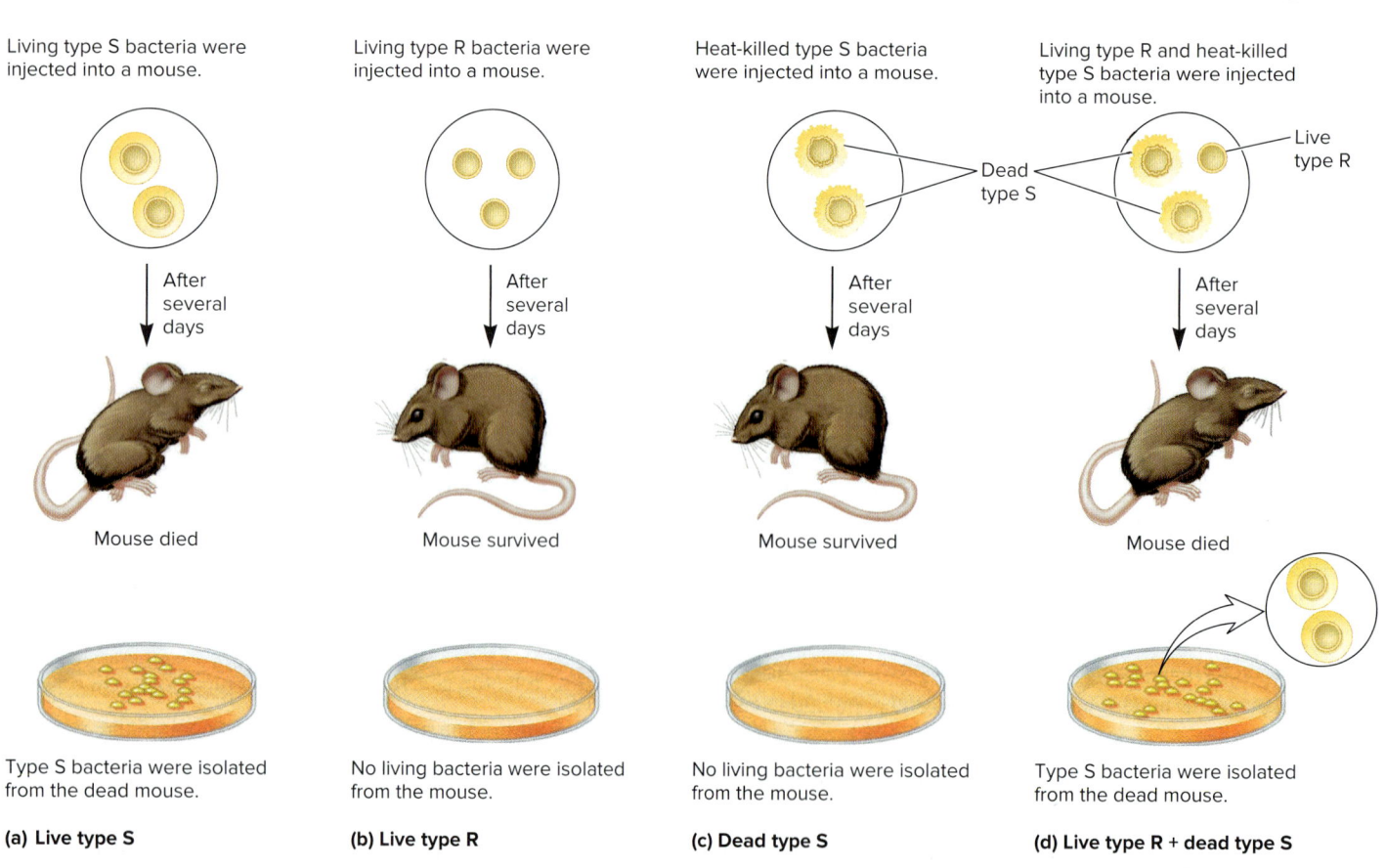

FIGURE 11.1 Griffith's experiments on genetic transformation in pneumococcus.

Concept Check: Explain why the mouse in part (d) died.

bacteria with heat treatment before injecting them into a mouse. In this case, the mouse also survived (**Figure 11.1c**).

- The critical and unexpected result was obtained in the experiment shown in **Figure 11.1d**. In this experiment, live type R bacteria were mixed with heat-killed type S bacteria. Several days after receiving an injection of this mixture, the mouse died. Furthermore, extracts from tissues of the dead mouse were found to contain many living type S bacteria, indicating that type S bacteria were reproducing within the mouse!

What can account for the results shown in Figure 11.1d? Because living type R bacteria alone could not proliferate and kill the mouse (Figure 11.1b), the interpretation of the result in Figure 11.1d is that something from the dead type S bacteria was transforming the type R bacteria into type S bacteria. Griffith called this process **transformation,** and the unidentified substance causing this to occur was termed the transforming principle. The steps of bacterial transformation are described in Chapter 9 (see Figure 9.12).

At this point, let's look at what Griffith's observations mean in terms of the four criteria for the genetic material. The transformed bacteria acquired the *information* to make a capsule. Among different strains, *variation* exists in the ability to create a capsule and to cause mortality in mice. The genetic material that is necessary to create a capsule must be *replicated* so that it can be *transmitted* from mother to daughter cells during cell division. Taken together, these observations are consistent with the idea that the formation of a capsule is governed by the bacteria's genetic material, meeting the four criteria described previously. Griffith's experiments showed that some genetic material from the dead bacteria was transferred to the living bacteria and provided them with a new trait. However, Griffith did not know what the transforming substance was.

Avery, MacLeod, and McCarty Showed That DNA Is the Substance That Transforms Bacteria

Important scientific discoveries often take place when researchers recognize that someone else's experimental observations can be used to address a particular scientific question. Oswald Avery, Colin MacLeod, and Maclyn McCarty realized that Griffith's observations could be used as part of an experimental strategy to identify the genetic material. They asked: What substance is being transferred from the dead type S bacteria to the live type R? To answer this question, they incorporated additional biochemical techniques into their experimental methods.

At the time of these experiments in the 1940s, researchers already knew that DNA, RNA, proteins, and carbohydrates are major constituents of living cells. To separate these components and to determine if any of them is the genetic material, Avery, MacLeod, and McCarty used established biochemical purification procedures and prepared extracts from type S bacterial strains that contained each type of these molecules. After many repeated attempts with different types of extracts, they discovered that only one of the extracts, namely, the one that contained purified DNA, was able to convert type R bacteria into type S. As shown in **Figure 11.2**, when this extract was mixed with type R bacteria, some of the bacteria were converted to type S. However,

if no DNA extract was added, no type S bacterial colonies were observed on the petri plates.

A biochemist might point out that a DNA extract may not be 100% pure. In fact, any purified extract might contain small traces of some other substances. Therefore, one can argue that a small amount of contaminating material in the DNA extract might actually be the genetic material. The most likely contaminating substance in this case would be RNA or protein. To further verify that the DNA in the extract was responsible for the transformation, Avery, MacLeod, and McCarty treated samples of the DNA extract with enzymes that digest DNA (called **DNase**), RNA (**RNase**), or protein (**protease**) (see Figure 11.2).

- When the DNA extracts were treated with RNase or protease, they still converted type R bacteria into type S. These results indicated that any remaining RNA or protein in the extract was not acting as the genetic material.
- However, when the extract was treated with DNase, it lost its ability to convert type R into type S bacteria.

Taken together, these results indicated that any remaining RNA or protein in the extract was not acting as the genetic material. Instead, they indicated that DNA is the genetic material. A more elegant way of saying this is that the transforming principle is DNA.

Hershey and Chase Provided Evidence That DNA Is the Genetic Material of T2 Phage

A second set of experimental results indicating that DNA is the genetic material came from the studies of Alfred Hershey and Martha Chase in 1952. Their research centered on the study of a virus known as T2. This virus infects *Escherichia coli* bacterial cells and is therefore known as a **bacteriophage,** or simply a **phage.** The T2 phage consists of genetic material that is packaged inside a phage coat. During infection, the phage coat remains attached on the outside of the bacterium. Only the genetic material enters the cytoplasm of the cell. From a molecular point of view, this virus is rather simple, because it is composed of only two types of macromolecules: DNA and proteins.

Hershey and Chase asked: What is the biochemical composition of the genetic material that enters the bacterial cell during infection? They used radioisotopes to distinguish proteins from DNA. Sulfur atoms are found in proteins but not in DNA, whereas phosphorus atoms are found in DNA but not in phage proteins. Therefore, ^{35}S (a radioisotope of sulfur) and ^{32}P (a radioisotope of phosphorus) were used to specifically label phage proteins and DNA, respectively. After phages were given sufficient time to infect bacterial cells, the researchers separated the phage coats from the bacterial cells. They determined that most of the ^{32}P had entered the bacterial cells whereas most of the ^{35}S remained outside the cells. These results were consistent with the idea that the genetic material of bacteriophages is DNA not proteins.

We now know that DNA is the genetic material of all bacteria, archaea, protists, fungi, plants, and animals. Many viruses, such as T2, also have DNA as their genetic material. However, as discussed in Chapter 10, the genetic material of some viruses is RNA, rather than DNA.

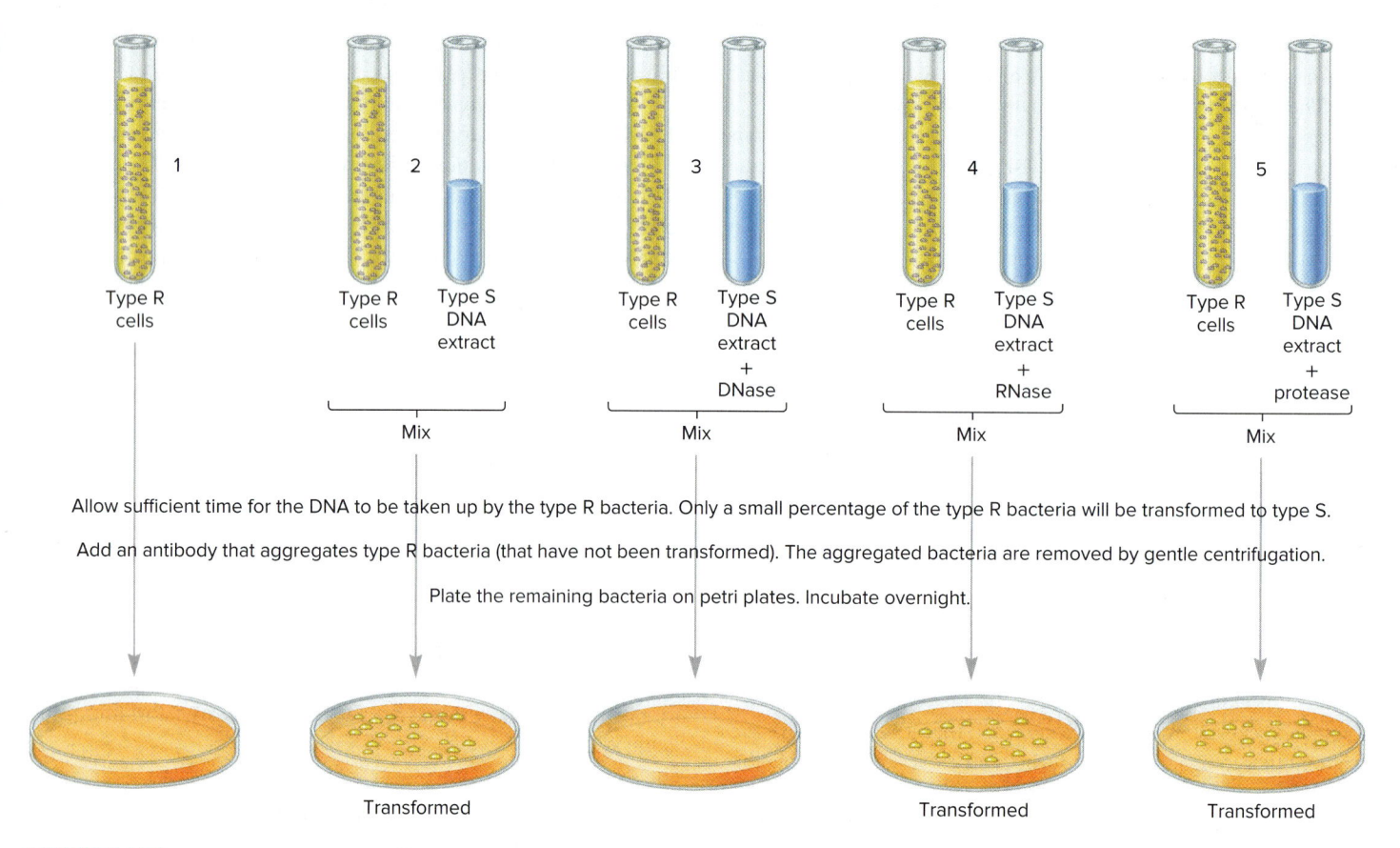

FIGURE 11.2 **Experimental protocol used by Avery, MacLeod, and McCarty to identify the transforming principle.** Samples of *S. pneumoniae* cells were either not exposed to type S DNA extract (tube 1) or exposed to type S DNA extract (tubes 2–5). Tubes 3, 4, and 5 also contained DNase, RNase, or protease, respectively. After incubation, the cells were exposed to antibodies, which are molecules that can specifically recognize the molecular structure of macromolecules. In this experiment, the antibodies recognized the cell surface of type R bacteria and caused them to clump together. The clumped bacteria were removed by a gentle centrifugation step. Only the bacteria that were not recognized by the antibody (namely, the type S bacteria) remained in the supernatant. The cells in the supernatant were plated on solid growth media. After overnight incubation, visible colonies will be observed if type S bacteria were in the supernatant.

Concept Check: What was the purpose of adding RNase or protease to a DNA extract?

11.1 REVIEWING THE KEY CONCEPTS

- To fulfill its role, genetic material must meet four criteria: information, transmission, replication, and variation.
- Griffith showed that the genetic material from type S bacteria could transform type R bacteria into type S (see Figure 11.1).
- Avery, MacLeod, and McCarty discovered that the transforming substance is DNA (see Figure 11.2).
- Hershey and Chase determined that the genetic material of T2 phage is DNA.

11.1 COMPREHENSION QUESTION

1. In the experiment of Avery, McLeod, and McCarty, the addition of RNase or protease to a DNA extract
   a. prevented the conversion of type S bacteria into type R bacteria.
   b. allowed the conversion of type S bacteria into type R bacteria.
   c. prevented the conversion of type R bacteria into type S bacteria.
   d. allowed the conversion of type R bacteria into type S bacteria.

11.2 OVERVIEW OF DNA AND RNA STRUCTURE

Learning Outcomes:
1. Define *nucleic acid*.
2. Describe the four levels of complexity of DNA.

DNA and its molecular cousin, RNA, are known as **nucleic acids.** This term is derived from the discovery of DNA by Friedrich Miescher in 1869. He identified a previously unknown phosphorus-containing substance that was isolated from the nuclei of white blood cells found in waste surgical bandages. He named this substance nuclein. As the structure of DNA and RNA became better understood, they were found to be acidic molecules, which means they release hydrogen ions (H^+) in solution and have a net negative charge at neutral pH. Thus, the name *nucleic acid* was coined.

Both DNA and RNA are macromolecules composed of smaller building blocks. To fully appreciate their structures, we need to consider four levels of complexity (**Figure 11.3**):

1. **Nucleotides** form the repeating structural unit of nucleic acids.
2. Nucleotides are linked together in a linear manner to form a **strand** of DNA or RNA.
3. Two strands of DNA (or sometimes RNA) interact with each other to form a **double helix.**
4. The three-dimensional structure of DNA results from the folding and bending of the double helix. Within living cells, DNA is associated with a wide variety of proteins that influence its structure. Chapter 12 examines the roles of these proteins in creating the three-dimensional structure of DNA found within **chromosomes.**

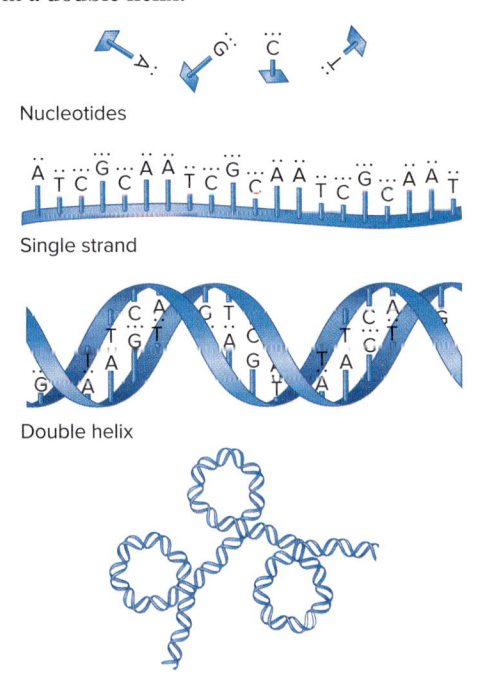

11.2 REVIEWING THE KEY CONCEPTS

- DNA and RNA are types of nucleic acids.
- In DNA, nucleotides are linked together to form strands, which then form a double helix that is found within chromosomes (see Figure 11.3).

11.2 COMPREHENSION QUESTION

1. Going from simple to complex, which of the following is the proper order for the structure of DNA?
    a. Nucleotide, double helix, DNA strand, chromosome
    b. Nucleotide, chromosome, double helix, DNA strand
    c. Nucleotide, DNA strand, double helix, chromosome
    d. Chromosome, nucleotide, DNA strand, double helix

11.3 NUCLEOTIDE STRUCTURE

Learning Outcomes:
1. Describe the structure of a nucleotide.
2. Compare and contrast the structures of nucleotides found in DNA and RNA.

The nucleotide is the repeating structural unit of both DNA and RNA. A nucleotide has three components: at least one phosphate group, a pentose sugar, and a nitrogenous base. As shown in **Figure 11.4**, nucleotides can vary with regard to the sugar and the nitrogenous base.

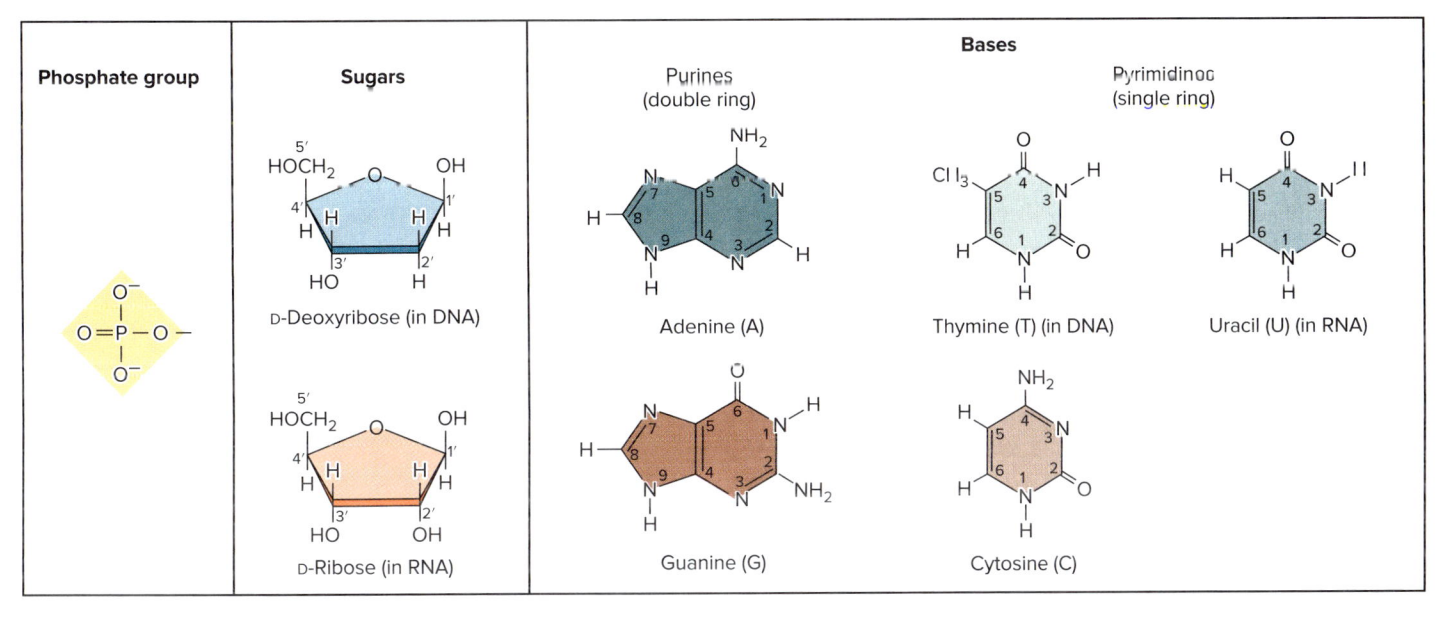

FIGURE 11.3 Levels of nucleic acid structure.

FIGURE 11.4 The components of nucleotides. The three components of a nucleotide are one or more phosphate groups, a sugar, and a base. The bases are purines (adenine and guanine) and pyrimidines (thymine, cytosine, and uracil).

Concept Check: Which of these components of nucleotides are not found in DNA?

226 CHAPTER 11 :: MOLECULAR STRUCTURE OF DNA AND RNA

- The two types of sugars are **deoxyribose** and **ribose,** which are found in DNA and RNA, respectively.
- The five different bases are subdivided into two categories: the **purines** and the **pyrimidines.**
- The purine bases, **adenine (A)** and **guanine (G),** contain a double-ring structure; the pyrimidine bases, **thymine (T), cytosine (C),** and **uracil (U),** contain a single-ring structure.
- The base thymine is found in DNA, whereas uracil is found in RNA instead of thymine. Adenine, guanine, and cytosine occur in both DNA and RNA.

As shown in Figure 11.4, the bases and sugars have a standard numbering system. The nitrogen and carbon atoms found in the ring structure of the bases are given numbers 1 through 9 for the purines and 1 through 6 for the pyrimidines. In comparison, the numbers given to the five carbons found in the sugars have primes, such as 1′, to distinguish them from the numbers used for the bases.

Figure 11.5 shows the repeating units of nucleotides found in DNA and RNA. The locations of the attachment sites of the base and phosphate to the sugar molecule are important to the nucleotide's function. In the sugar ring, carbon atoms are numbered in a clockwise direction, beginning with a carbon atom adjacent to the ring oxygen atom. The fifth carbon is outside the ring structure. In a single nucleotide, the base is always attached to the 1′ carbon atom, and one or more phosphate groups are attached at the 5′ position. As discussed later, the —OH group attached to the 3′ carbon is important in allowing nucleotides to form covalent linkages with each other.

The terminology used to describe nucleic acid units is based on three structural features: the type of base, the type of sugar, and the number of phosphate groups.

- When a base is attached to only a sugar, we call this pair a **nucleoside.** If adenine is attached to ribose, this nucleoside is called adenosine (**Figure 11.6**). Nucleosides containing guanine, thymine, cytosine, or uracil are called guanosine, thymidine, cytidine, and uridine, respectively. When only the bases are attached to deoxyribose, they are called deoxyadenosine, deoxyguanosine, deoxythymidine, and deoxycytidine.
- The covalent attachment of one or more phosphate molecules to a nucleoside creates a nucleotide. If a nucleotide contains adenine, ribose, and one phosphate, it is adenosine monophosphate, abbreviated AMP. If a nucleotide contains adenine, ribose, and three phosphate groups, it is called adenosine triphosphate, or ATP. A nucleotide composed of adenine, deoxyribose, and three phosphate groups is called deoxyadenosine triphosphate (dATP).

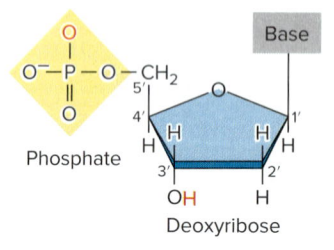

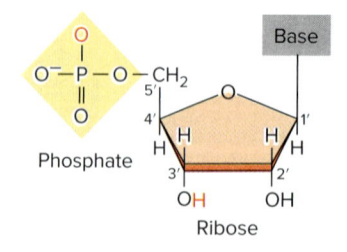

FIGURE 11.5 The structure of nucleotides found in (a) DNA and (b) RNA. DNA contains deoxyribose as its sugar and the bases A, T, G, and C. RNA contains ribose as its sugar and the bases A, U, G, and C. In a DNA or RNA strand, the oxygen on the 3′ carbon is linked to the phosphorus atom of phosphate in the adjacent nucleotide. The two atoms (O and H) shown in red are found within individual nucleotides but not when nucleotides are joined together to make strands of DNA and RNA.

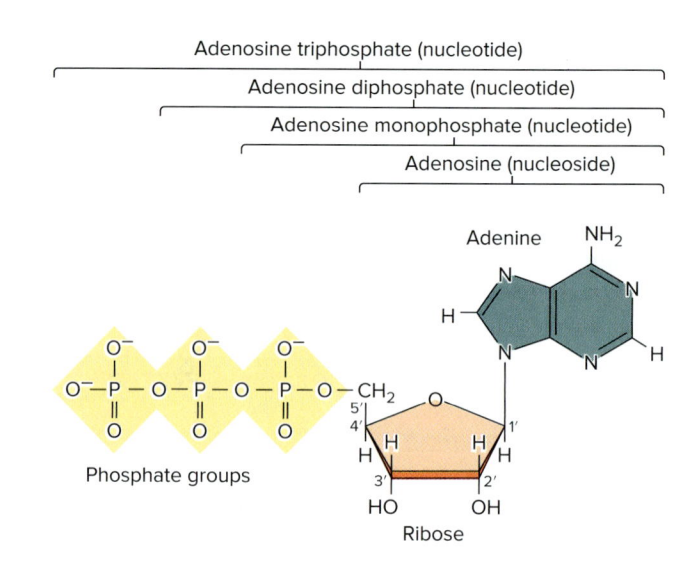

FIGURE 11.6 A comparison of the structures of an adenine-containing nucleoside and nucleotides. A nucleoside is composed of a sugar and base, whereas a nucleotide has a sugar, a base, and one or more phosphates.

Genetic TIPS

The Question: A molecule contains adenine, deoxyribose, and one phosphate. Is it a nucleoside or a nucleotide? Would it be found in DNA or RNA?

Topic: What topic in genetics does this question address?
The topic is the structure of nucleosides and nucleotides and which types are found in DNA and RNA.

Information: What information do you know based on the question and your understanding of the topic?
In the question, you are given the components of a molecule. From your understanding of the topic, you may remember that a nucleoside contains a sugar and a base, whereas a nucleotide contains a sugar, a base, and one or more phosphate groups.

Problem-Solving Strategy: Define key terms.
As mentioned, nucleosides contain a sugar and a base, whereas nucleotides contain a sugar, a base, and one or more phosphate groups.

Answer: The molecule is a nucleotide. It contains a sugar, a base, and a phosphate. Because it contains deoxyribose, it could be found in DNA, but not in RNA.

11.3 REVIEWING THE KEY CONCEPTS

- A nucleotide is composed of one or more phosphates, a sugar, and a base. The purine bases are adenine and guanine, whereas the pyrimidine bases are thymine (DNA only), cytosine, and uracil (RNA only). The sugar deoxyribose is found in DNA, whereas ribose is in RNA (see Figures 11.4, 11.5).
- A nucleoside contains a sugar and a base, whereas a nucleotide contains a sugar, a base, and one or more phosphate groups (see Figure 11.6).

11.3 COMPREHENSION QUESTIONS

1. Which of the following could be the components of a single nucleotide found in DNA?
   a. Deoxyribose, adenine, and thymine
   b. Ribose, phosphate, and cytosine
   c. Deoxyribose, phosphate, and thymine
   d. Ribose, phosphate, and uracil
2. A key difference between the nucleotides in DNA and those in RNA is that
   a. DNA has phosphate but RNA does not.
   b. DNA has deoxyribose, but RNA has ribose.
   c. DNA has thymine, but RNA has uracil.
   d. Both b and c are correct.

11.4 STRUCTURE OF A DNA STRAND

Learning Outcome:
1. Describe the structural features of a DNA strand.

A strand of DNA (or RNA) has many nucleotides that are covalently attached to each other in a linear fashion. **Figure 11.7** depicts a short strand of DNA with four nucleotides. A few structural features are worth noting.

- The linkage of nucleotides to each other involves an ester bond between a phosphate group on one nucleotide and the sugar molecule on the adjacent nucleotide.
- Another way of viewing this linkage is to notice that a phosphate group connects two sugar molecules. This linkage in DNA (or RNA) strands is called a **phosphodiester linkage.**
- The phosphates and sugar molecules form the **backbone** of the strand. The bases project from the backbone. The backbone is negatively charged due to a negative charge on each phosphate.
- A phosphodiester linkage involves a phosphate attachment to the 3′ carbon in one nucleotide and to the 5′ carbon in the next nucleotide. In a strand, all sugar molecules are oriented in the same direction.
- A DNA strand has a **directionality** based on the orientation of the sugar molecules within that strand. In Figure 11.7, the direction of the strand is 5′ to 3′ proceeding from top to bottom.

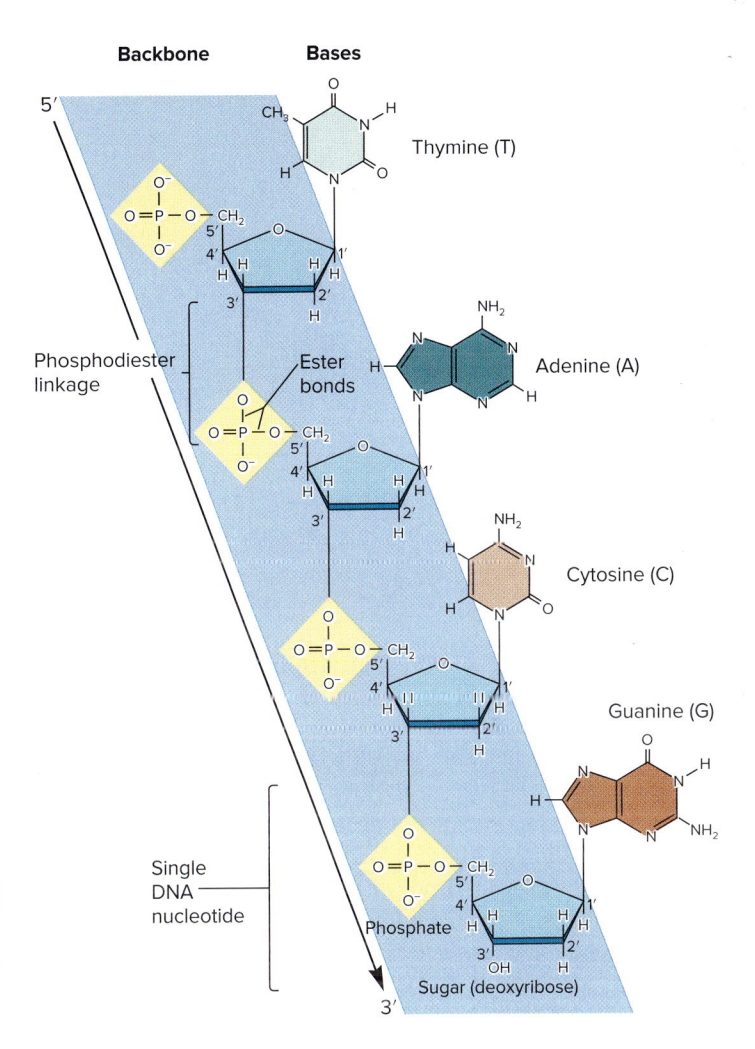

FIGURE 11.7 **A short strand of DNA containing four nucleotides.** Nucleotides are covalently linked together to form a strand of DNA.

Concept Check: Which components of nucleotides form the backbone of a DNA strand?

A critical aspect regarding DNA and RNA structure is that a strand contains a specific sequence of bases. In Figure 11.7, the sequence of bases is thymine–adenine–cytosine–guanine, abbreviated TACG. Furthermore, to show the directionality, the abbreviation for this sequence is written 5′–TACG–3′. Because the nucleotides within a strand are covalently attached to each other, the sequence of bases cannot shuffle around and become rearranged. Therefore, the sequence of bases in a DNA strand remains the same over time, except in rare cases when mutations occur. As we will see throughout this text, the sequence of bases within DNA and RNA is the defining feature that allows them to carry information.

11.4 REVIEWING THE KEY CONCEPTS

- In a DNA strand, nucleotides are covalently attached to one another via phosphodiester linkages (see Figure 11.7).

11.4 COMPREHENSION QUESTION

1. In a DNA strand, a phosphate connects a 3′ carbon atom in one nucleotide to
   a. a 5′ carbon in an adjacent nucleotide.
   b. a 3′ carbon in an adjacent nucleotide.
   c. a base in an adjacent nucleotide.
   d. none of the above.

11.5 DISCOVERY OF THE DOUBLE HELIX

Learning Outcome:
1. Outline the key experiments that led to the discovery of the DNA double helix.

A major discovery in molecular genetics was made in 1953 by James Watson and Francis Crick. At that time, DNA was already known to be composed of nucleotides. However, it was not understood how the nucleotides are bonded together to form the structure of DNA. Watson and Crick committed themselves to determining the structure of DNA because they felt this knowledge was needed to understand the functioning of genes. Other researchers, such as Rosalind Franklin and Maurice Wilkins, shared this view. Before we examine the characteristics of the double helix, let's consider the events that provided the scientific framework for Watson and Crick's breakthrough.

A Few Key Events Led to the Discovery of the Double-Helix Structure

In the early 1950s, Linus Pauling proposed that regions of proteins can fold into a particular type of secondary structure. To elucidate this structure, Pauling built large models by linking together simple ball-and-stick units (**Figure 11.8**). By carefully scaling the objects in his models, he could visualize whether atoms fit together properly in a complex three-dimensional structure. Is this approach still used today? The answer is yes, except that today researchers construct their three-dimensional models on computers. As we will see, Watson and Crick also used ball-and-stick modeling to solve the structure of the DNA double helix. Interestingly, they were well aware that Pauling might figure out the structure of DNA before they did. This awareness provided a stimulating rivalry among the researchers.

A second development that led to the elucidation of the double helix was improvement in the use of X-ray diffraction. When a purified substance, such as DNA, is subjected to X-rays, it produces a well-defined diffraction pattern if the molecular structure has a repeating pattern. An interpretation of the diffraction pattern (using mathematical theory) can ultimately provide information concerning the structure of the molecule. Rosalind Franklin (**Figure 11.9a**), working in the same laboratory as Maurice Wilkins, used X-ray diffraction to study wet DNA fibers. Franklin made marked advances in X-ray diffraction techniques while working with DNA. A diffraction pattern of Franklin's DNA fibers is shown in **Figure 11.9b**. It suggested key structural features of DNA.

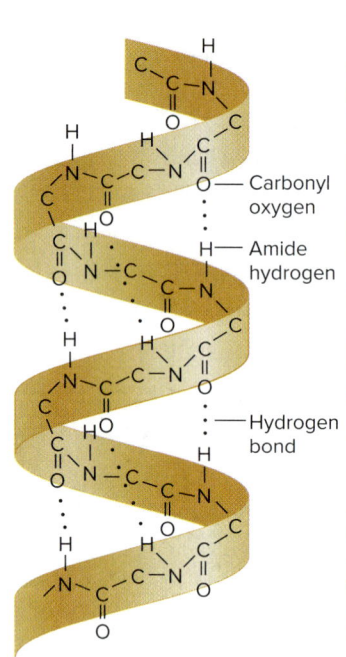

(a) An α helix in a protein   (b) Linus Pauling

FIGURE 11.8 The use of models by Linus Pauling.
(b) ©Tom Hollyman/Science Source

Concept Check: Why is modeling useful?

- The pattern was consistent with a helical structure.
- The diameter of the helical structure was too wide to be only a single-stranded helix.
- The diffraction pattern indicated that the helix contains about 10 base pairs (bp) per complete turn.

These observations were instrumental in solving the structure of DNA.

Chargaff Found That DNA Has a Biochemical Composition in Which the Amounts of A and T Are Equal, as Are the Amounts of G and C

Another piece of information that led to the discovery of the double-helix structure came from the studies of Erwin Chargaff. In the 1940s and 1950s, he pioneered many of the biochemical techniques for the isolation, purification, and measurement of nucleic acids from living cells. This was not an easy undertaking, because the biochemical composition of living cells is complex. At the time of Chargaff's work, researchers already knew that the building blocks of DNA are nucleotides containing the bases adenine, thymine, guanine, or cytosine. Chargaff analyzed the base composition of DNA, which was isolated from many different species. He expected that the results might provide important clues concerning the structure of DNA.

11.5 DISCOVERY OF THE DOUBLE HELIX

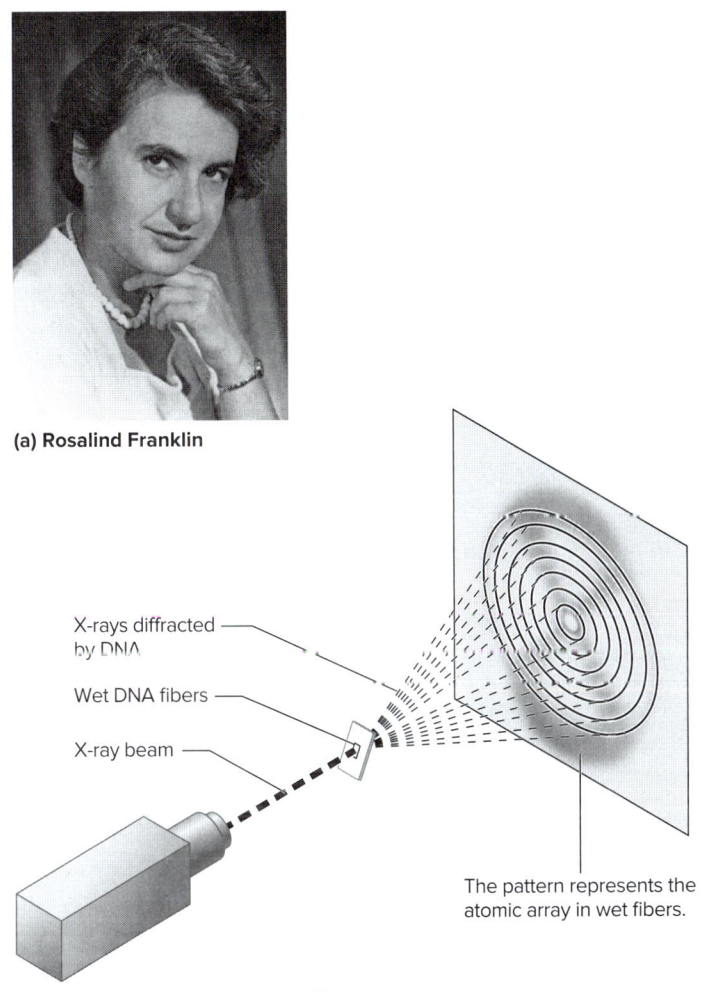

(a) Rosalind Franklin

(b) X-ray diffraction of wet DNA fibers

FIGURE 11.9 X-ray diffraction of DNA.
(a) ©World History Archive/Alamy Stock Photo

TABLE 11.1
Base Content in the DNA from a Variety of Organisms

Organism	Adenine	Thymine	Guanine	Cytosine
Escherichia coli	26.0	23.9	24.9	25.2
Streptococcus pneumoniae	29.8	31.6	20.5	18.0
Yeast	31.7	32.6	18.3	17.4
Turtle red blood cells	28.7	27.9	22.0	21.3
Salmon sperm	29.7	29.1	20.8	20.4
Chicken red blood cells	28.0	28.4	22.0	21.6
Human liver cells	30.3	30.3	19.5	19.9

Percentage of Bases (based on molarity)

*When the base compositions from different tissues within the same species were measured, similar results were obtained. These data were compiled from several sources. See E. Chargaff and J. Davidson, Eds. (1995), *The Nucleic Acids*, Academic Press, New York.

The data shown in **Table 11.1** are only a small sampling of Chargaff's results. The compelling observation was that the amount of adenine was similar to that of thymine, and the amount of guanine was similar to cytosine. The idea that the amount of A in DNA equals the amount of T, and the amount of G equals that of C, is known as **Chargaff's rule.**

These results were not sufficient to propose a model for the structure of DNA. However, they provided the important clue that DNA is structured so that each molecule of adenine interacts with thymine, and each molecule of guanine interacts with cytosine. A DNA structure in which A binds to T and G binds to C would explain the equal amounts of A and T and of G and C observed in Chargaff's experiments. As we will see, this observation became crucial evidence that Watson and Crick used to elucidate the structure of the double helix.

Watson and Crick Deduced the Double-Helix Structure of DNA

Watson and Crick assumed that DNA is composed of nucleotides that are linked together in a linear fashion. They also assumed that the chemical linkage between two nucleotides is always the same.

With these ideas in mind, they tried to build ball-and-stick models that incorporated the known experimental observations. During this time, Franklin had produced even clearer X-ray diffraction patterns, which provided greater detail concerning the relative locations of the bases and backbone of DNA. This major breakthrough suggested a two-strand interaction that was helical.

In their model building, Watson and Crick's emphasis shifted to models containing the two backbones on the outside of the model, with the bases projecting toward each other. At first, a structure was considered in which the bases form hydrogen bonds with the identical base in the opposite strand (A to A, T to T, G to G, and C to C). However, the model building revealed that the bases could not fit together this way. The final hurdle was overcome when it was realized that the hydrogen bonding of adenine to thymine was structurally similar to that of guanine to cytosine. With interactions between A and T and between G and C, the ball-and-stick model showed that the two strands fit together properly. This ball-and-stick model, shown in **Figure 11.10**, was consistent with all of the known data regarding DNA structure.

For their work, Watson, Crick, and Wilkins were awarded the 1962 Nobel Prize in physiology or medicine. The contribution of Franklin to the discovery of the double helix was also critical and has been acknowledged in several books and articles. Franklin was independently trying to solve the structure of DNA. However, Wilkins, who worked in the same laboratory, shared Franklin's X-ray diffraction data with Watson and Crick, presumably without her knowledge. This important information helped them solve the structure of DNA, which was published in the journal *Nature* in April 1953. Though she was not given credit in the original publication of the double-helix structure, Franklin's key contribution became known in later years. Unfortunately, however, Rosalind Franklin died in 1958, and the Nobel Prize is not awarded posthumously.

230   CHAPTER 11 :: MOLECULAR STRUCTURE OF DNA AND RNA

(a) Watson and Crick

FIGURE 11.10   Watson and Crick and their model of the DNA double helix. (a) James Watson is on the left and Francis Crick on the right. (b) The molecular model they originally proposed for the double helix. Each strand contains a sugar-phosphate backbone. Between the strands, A hydrogen bonds to T, and G hydrogen bonds with C.
(a) ©A. Barrington Brown/Science Source; (b) ©Hulton Archives/Getty Images

11.5 REVIEWING THE KEY CONCEPTS

- Pauling used ball-and-stick models to deduce the secondary structure of a protein (see Figure 11.8).
- Franklin performed X-ray diffraction studies that helped to determine the structure of DNA (see Figure 11.9).
- Chargaff determined that, in DNA, the amounts of A and T are equal, and so are the amounts of G and C (see Table 11.1).
- Watson and Crick deduced the structure of DNA by building models that were consistent with the available data (see Figure 11.10).

11.5 COMPREHENSION QUESTIONS

1. Evidence that led to the discovery of the DNA double helix includes
   a. the determination of structures using ball-and-stick models.
   b. the X-ray diffraction data of Franklin.
   c. the base composition data of Chargaff.
   d. all of the above.
2. Chargaff's analysis of the base composition of DNA is consistent with base pairing between
   a. A and G, and T and C.
   b. A and A, G and G, T and T, C and C.
   c. A and T, and G and C.
   d. A and C, and T and G.

(b) Original model of the DNA double helix

11.6 STRUCTURE OF THE DNA DOUBLE HELIX

Learning Outcomes:
1. Outline the key structural features of the DNA double helix.
2. Compare and contrast B DNA and Z DNA.

As we have seen, the discovery of the DNA double helix required different kinds of evidence. In this section, we will examine the double-helix structure in greater detail.

The Molecular Structure of the DNA Double Helix Has Several Key Features

The general structural features of the double helix are shown in **Figures 11.11** and **11.12**.

- In a DNA double helix, two DNA strands are twisted together around a common axis to form a structure that resembles a spiral staircase.

11.6 STRUCTURE OF THE DNA DOUBLE HELIX

Key Features

- Two strands of DNA form a right-handed double helix.
- The bases in opposite strands hydrogen bond according to the AT/GC rule.
- The 2 strands are antiparallel with regard to their 5' to 3' directionality.
- There are 10.0 nucleotides in each strand per complete 360° turn of the helix.

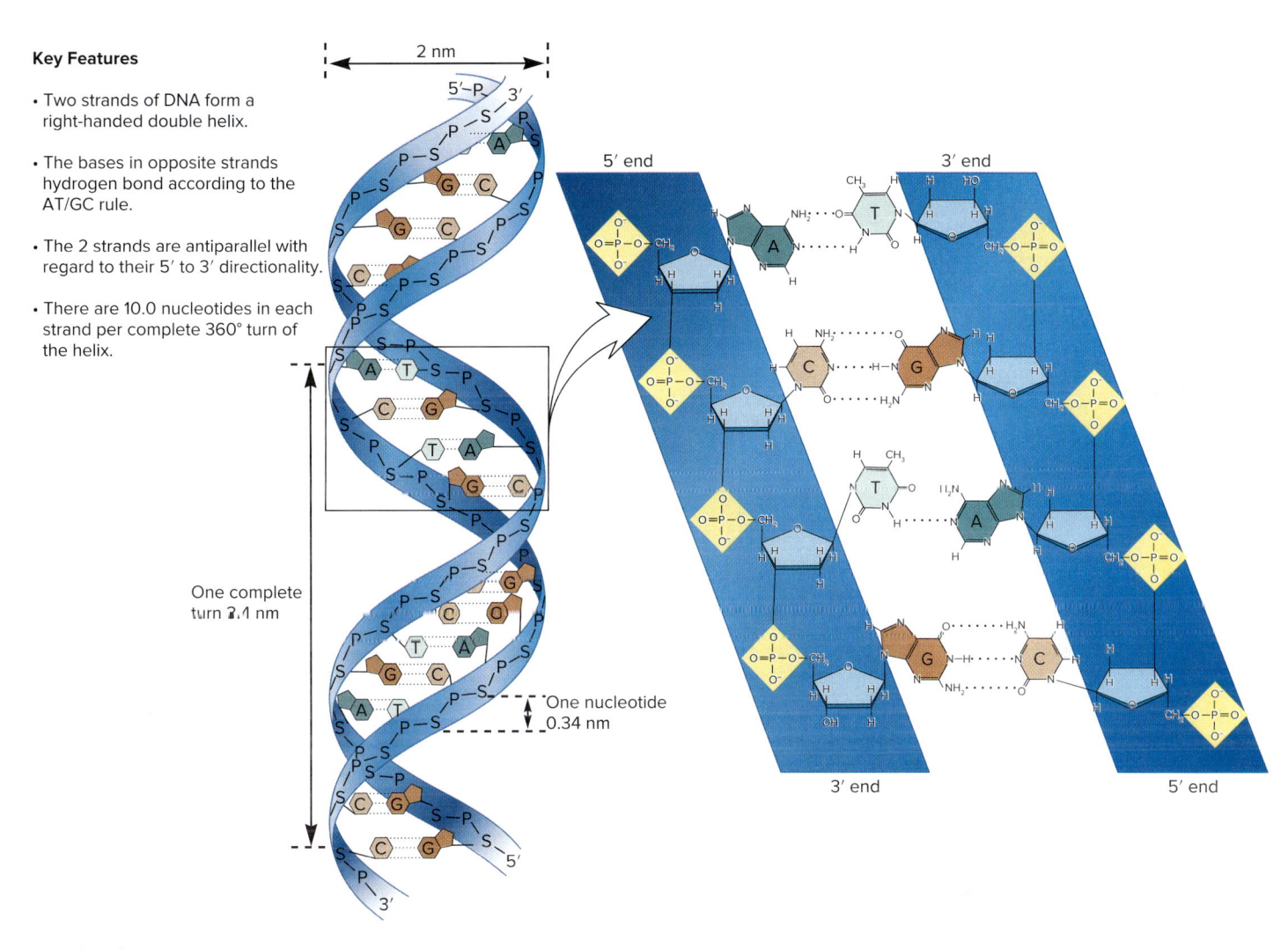

FIGURE 11.11 Key features of the structure of the double helix. Note: In the inset, the planes of the bases and sugars are shown parallel to each other in order to depict the hydrogen bonding between the bases. In an actual DNA molecule, the bases are rotated about 90° so their planes face each other, as shown in Figure 11.12a.

Concept Check: What holds the DNA strands together?

- This double-stranded structure is stabilized by **base pairs (bp),** bases in opposite strands of DNA that are hydrogen-bonded to each other (Figure 11.11). If you count bases along one strand, once you have passed 10 bp, you will have gone 360° around the backbone. The linear distance of a complete turn is 3.4 nm; each base pair traverses 0.34 nm.
- A distinguishing feature of the hydrogen bonding between base pairs is its specificity. An adenine in one strand hydrogen bonds with a thymine in the opposite strand, or a guanine hydrogen bonds with a cytosine. This **AT/GC rule** indicates that purines (A and G) always bond with pyrimidines (T and C). This pairing of bases keeps the width of the double helix relatively constant.
- Three hydrogen bonds occur between G and C but only two between A and T. For this reason, DNA sequences that have a high proportion of G and C tend to form more stable double-stranded structures.

- Due to the AT/GC rule, we can predict the sequence in one DNA strand if the sequence in the opposite strand is known. For example, if one DNA strand is 5'–ATGGCGGATTT–3', the opposite strand has to be 3'–TACCGCCTAAA–5'. In genetic terms, we say that these two sequences are **complementary** to each other or that the two sequences exhibit complementarity.
- As mentioned, two DNA strands have an opposite orientation with regard to their 5' and 3' ends. This is called an **antiparallel** arrangement (Figure 11.11).
- Within DNA, the bases are oriented so that the flattened regions are facing each other—an arrangement referred to as **base stacking** (Figure 11.12a). In other words, if you think of the bases as flat plates, these plates are stacked on top of each other in the double-stranded DNA structure. This is a favorable interaction because the bases are relatively nonpolar and can interact with each other. Base stacking is a structural

232   CHAPTER 11 :: MOLECULAR STRUCTURE OF DNA AND RNA

(a) Ball-and-stick model of DNA

(b) Space-filling model of DNA

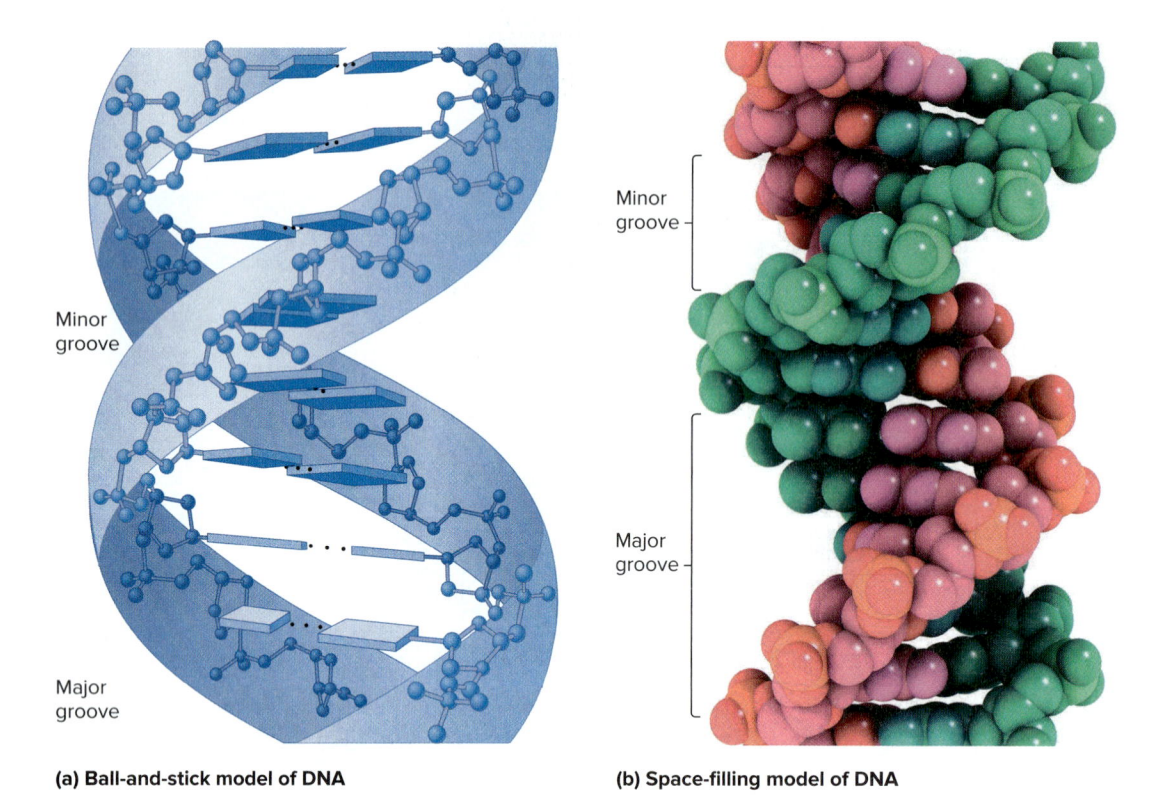

FIGURE 11.12 **Two models of the double helix.** (a) Ball-and-stick model of the double helix. The deoxyribose-phosphate backbone is shown in detail, whereas the bases are depicted as flattened rectangles. (b) Space-filling model of the double helix.
(b) ©Laguna Design/Science Source

Concept Check: Describe the major and minor grooves.

feature that stabilizes the double helix by excluding water molecules, which are polar.
- By convention, the DNA double helix shown in Figure 11.12a spirals in a direction that is called right-handed. To understand this terminology, imagine that a double helix is laid on your desk; one end of the helix is close to you, and the other end is at the opposite side of the desk. As it spirals away from you, a right-handed helix turns in a clockwise direction. By comparison, a left-handed helix would spiral in a counterclockwise direction. Both strands in Figure 11.12a spiral in a right-handed direction.
- Figure 11.12b is a space-filling model of DNA in which the atoms are represented by spheres. Biochemists use the term *grooves* to describe the indentations where atoms of the bases are in contact with water in the surrounding cellular fluid. As you travel around the DNA helix, the structure of DNA has two grooves: a wider **major groove** and a narrower **minor groove.**

As discussed in later chapters, proteins can bind to DNA and affect its conformation and function. For example, some proteins can hydrogen bond to the bases within the major groove. This hydrogen bonding can be very precise, allowing a protein to interact with a particular sequence of bases. In this way, a protein can recognize a specific gene and then affect its ability to be transcribed. We will consider such proteins in Chapters 14, 16, and 17. Alternatively, other proteins bind to the DNA backbone. For example, histone proteins, which are discussed in Chapter 12, form ionic interactions with the negatively charged phosphates in the DNA backbone. The histones are important for the proper compaction of DNA in eukaryotic cells and also play a role in gene transcription.

DNA Can Form Alternative Types of Double Helices

The DNA double helix can form different types of structures. **Figure 11.13** compares the structures of **B DNA** and **Z DNA.** The highly detailed structures shown here were deduced by performing X-ray crystallography on short segments of DNA. B DNA is the predominant form of DNA found in living cells. However, under certain conditions, the two strands of DNA can twist into Z DNA, which differs significantly from B DNA. B DNA is a right-handed helix, whereas Z DNA has a left-handed conformation. In addition, the helical backbone in Z DNA appears to zigzag slightly as it winds itself around the double-helix structure. The numbers of base pairs per 360° turn are 10.0 in B DNA and 12.0 in Z DNA. In B DNA, the bases tend to be centrally located, and the hydrogen bonds between base pairs are oriented relatively perpendicular to the central axis. In contrast, the bases in Z DNA are substantially tilted relative to the central axis.

What is the biological significance of Z DNA? Accumulating evidence suggests a possible biological role for Z DNA in the process of transcription. Recent research has identified cellular proteins that specifically recognize Z DNA. In 2005, Alexander Rich and colleagues reported that the Z-DNA-binding region of

11.6 STRUCTURE OF THE DNA DOUBLE HELIX   233

one such protein played a role in regulating the transcription of particular genes. In addition, other research has suggested that short regions of Z DNA may play a role in chromosome structure by affecting the level of compaction.

Genetic TIPS

The Question: A double-stranded molecule of B DNA contains 340 nucleotides. How many complete turns occur in this double helix?

Topic: What topic in genetics does this question address? The topic is DNA structure. More specifically, the question is about determining the number of turns in a DNA molecule.

Information: What information do you know based on the question and your understanding of the topic? From the question, you know that a DNA molecule contains 340 nucleotides. From your understanding of the topic, you may remember that B DNA is double-stranded and contains 10 bp per turn.

Problem-Solving Strategy: Make a calculation. To solve this problem, the first thing you need to do is to determine the number of base pairs. A molecule that has 340 nucleotides will contain 340/2, or 170 bp. To determine the number of turns, we divide 170 bp by 10 bp/turn to calculate the number of turns.

Answer: This DNA molecule contains 17 complete turns.

11.6 REVIEWING THE KEY CONCEPTS

- DNA is usually a right-handed double helix in which A hydrogen bonds to T and G hydrogen bonds to C. The two strands are antiparallel and contain 10 bp per turn. Base stacking stabilizes the double helix by excluding water (see Figure 11.11).
- The double helix of DNA has a major groove and a minor groove (see Figure 11.12).
- B DNA is the major form of DNA found in living cells. Z DNA is an alternative conformation for DNA (see Figure 11.13).

11.6 COMPREHENSION QUESTIONS

1. Which of the following is *not* a feature of the DNA double helix?
   a. It obeys the AT/GC rule.
   b. The DNA strands are antiparallel.
   c. The structure is stabilized by base stacking.
   d. All of the above are features of the DNA double helix.
2. A groove in a DNA double helix refers to
   a. the indentations where the bases are in contact with the surrounding water.
   b. the interactions between bases in the DNA.
   c. the spiral structure of the DNA.
   d. all of the above.

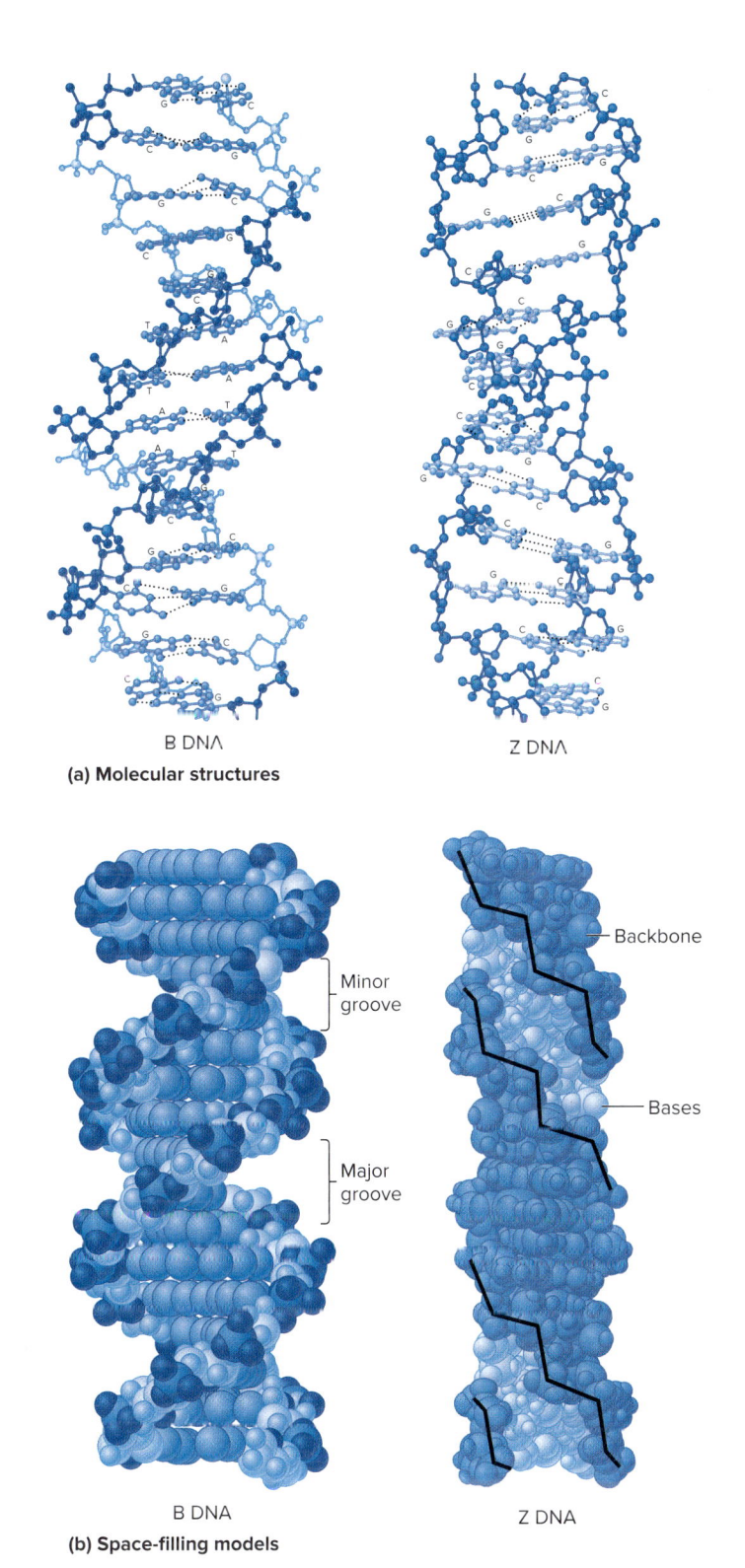

FIGURE 11.13 Comparison of the structures of B DNA and Z DNA. (a) The highly detailed structures shown here were deduced by performing X-ray crystallography on short segments of DNA. (b) Space-filling models of the B-DNA and Z-DNA structures. In the case of Z DNA, the black lines connect the phosphate groups in the DNA backbone. As seen here, they travel along the backbone in a zigzag pattern.

Illustration, Irving Geis. Image from Irving Geiss Collection/Howard Hughes Medical Institute. Rights owned by HHMI. Not to be reproduced without permission.

Concept Check: What are the structural differences between B DNA and Z DNA?

3. A key difference between B DNA and Z DNA is that
   a. B DNA is right-handed, whereas Z DNA is left-handed.
   b. B DNA obeys the AT/GC rule, whereas Z DNA does not.
   c. Z DNA allows ribose in its structure, whereas B DNA uses deoxyribose.
   d. Z DNA allows uracil in its structure, whereas B DNA uses thymine.

11.7 RNA STRUCTURE

Learning Outcome:
1. Outline the key structural features of RNA.

Let's now turn our attention to RNA structure, which has many similarities with DNA structure. The structure of an RNA strand is much like that of a DNA strand (**Figure 11.14**). Strands of RNA are typically several hundred or several thousand nucleotides in length, which is much shorter than chromosomal DNA. During transcription, DNA is used as a template to make a copy of single-stranded RNA. In most cases, only one of the two DNA strands is used as a template for RNA synthesis. Therefore, only one complementary strand of RNA is usually made. Nevertheless, relatively short sequences within one RNA molecule or between two separate RNA molecules can form double-stranded regions.

- RNA double helices are antiparallel and right-handed, with 11 to 12 bp per turn.

In places where an RNA molecule is double-stranded, complementary regions form base pairs between A and U and between G and C. Depending on the relative locations of these complementary regions, different types of structural patterns are possible (**Figure 11.15**). These include bulge loops, internal loops, multibranched junctions, and stem-loops (also called hairpins).

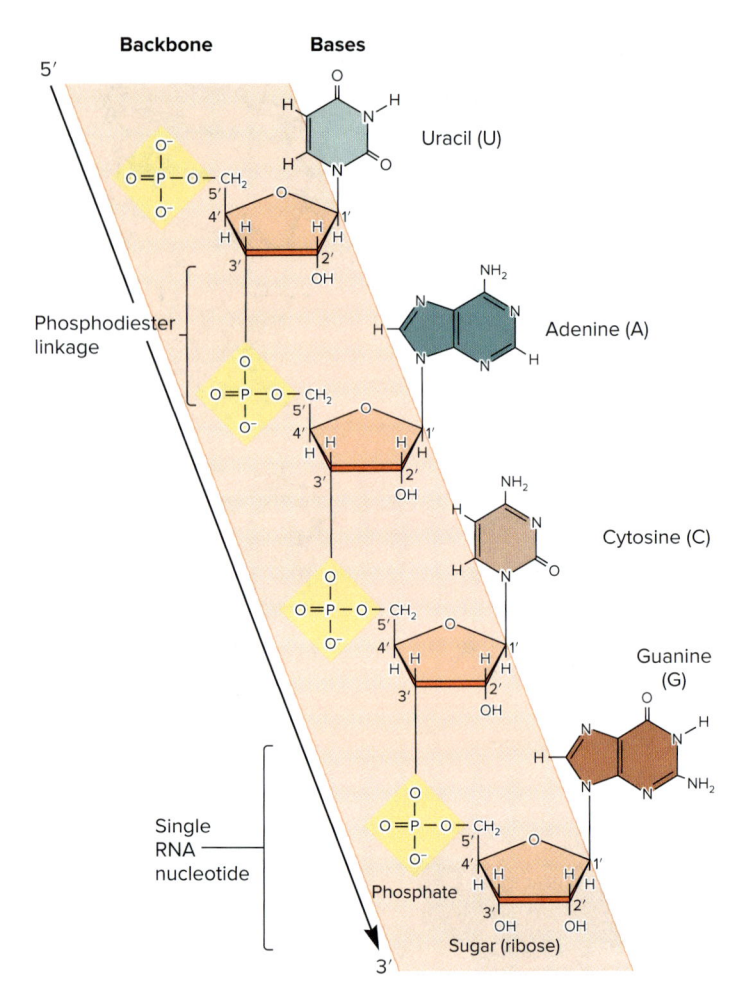

FIGURE 11.14 A strand of RNA. This structure is very similar to a DNA strand (see Figure 11.7), except that the sugar is ribose instead of deoxyribose, and uracil is substituted for thymine.

Concept Check: What types of bonds hold nucleotides together in an RNA strand?

(a) Bulge loop (b) Internal loop (c) Multibranched junction (d) Stem-loop

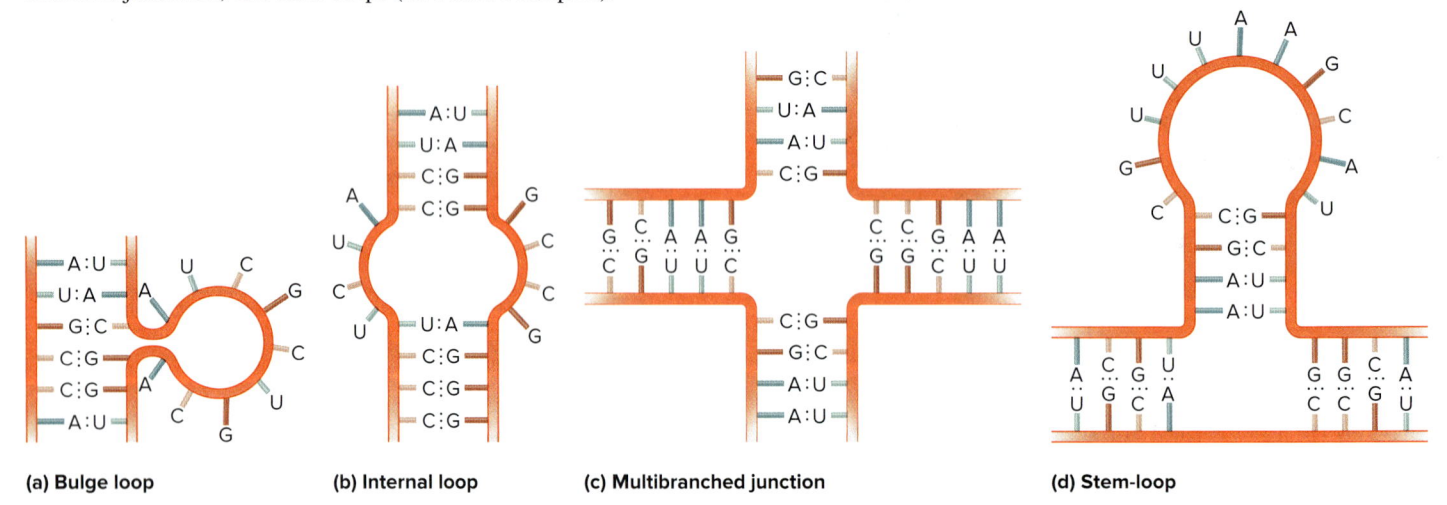

FIGURE 11.15 Possible structures of RNA molecules. The double-stranded regions are formed when hydrogen bonds (represented by two or three dots) connect complementary bases. Loops are noncomplementary regions that are not hydrogen bonded with complementary bases. Double-stranded RNA structures can form within a single RNA molecule or between two separate RNA molecules. Note: Though not shown here, double-stranded regions are in a helical conformation.

Concept Check: What are the base-pairing rules for RNA?

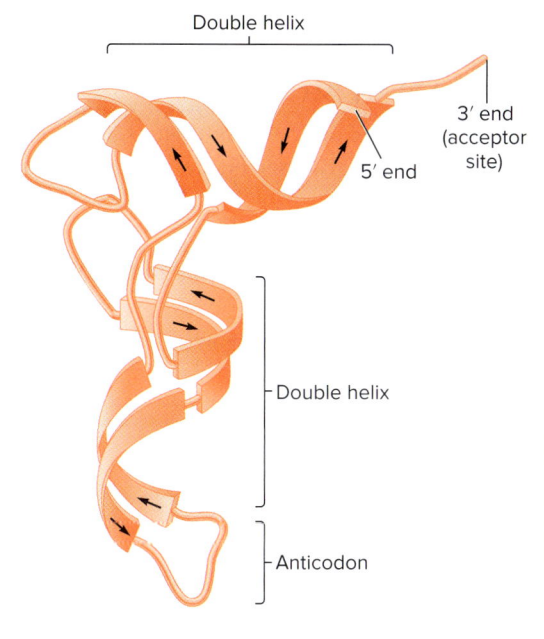

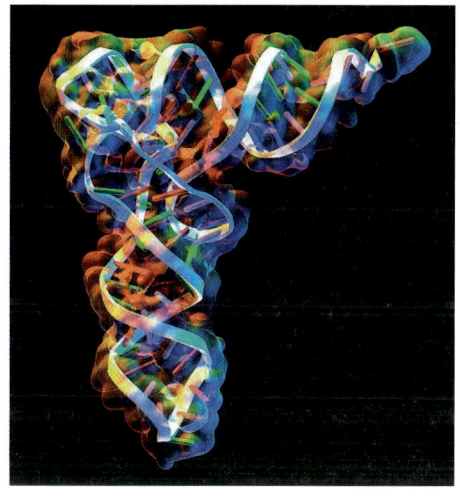

FIGURE 11.16 The structure of tRNA^{Phe}, the transfer RNA molecule that carries phenylalanine. (a) The double-stranded regions of the molecule are shown as antiparallel ribbons. (b) A space-filling model of tRNA^{Phe}.
(b) ©Alfred Pasieka/Science Source

(a) Ribbon model

(b) Space-filling model

Many factors contribute to RNA structure.

- After folding, some parts of RNA molecules may become double-stranded because the base sequences are complementary. The complementary regions are held together by connecting hydrogen bonds. Other regions of RNA molecules may remain single-stranded.
- The structure of RNA is stabilized by hydrogen bonding between base pairs, stacking between bases, and hydrogen bonding between bases and backbone regions.
- In addition, interactions with ions, small molecules, and large proteins may influence RNA structure.

Figure 11.16 depicts the structure of a transfer RNA molecule known as tRNA^{Phe}, which is a tRNA molecule that carries the amino acid phenylalanine (Phe). It was the first naturally occurring RNA to have its structure elucidated. This RNA molecule has several double-stranded and single-stranded regions.

The folding of RNA into a three-dimensional structure is important for its function. For example, as discussed in Chapter 15, a tRNA molecule has two key functional sites—an anticodon and a 3′ acceptor site—that play important roles in translation. In a folded tRNA molecule, these sites are exposed on the surface of the molecule so that they can perform their roles (see Figure 11.16a). Many other examples are known in which RNA folding is key to the molecule's structure and function. These include the folding of ribosomal RNAs (rRNAs), which are important components of ribosomes, and ribozymes, which are RNA molecules with catalytic function.

11.7 REVIEWING THE KEY CONCEPTS

- RNA is composed of a strand of nucleotides (see Figure 11.14).
- RNA can form double-stranded helical regions and folds into a three-dimensional structure (see Figures 11.15, 11.16).

11.7 COMPREHENSION QUESTION

1. A double-stranded region of RNA
   a. forms a helical structure.
   b. obeys the AU/GC rule.
   c. may promote the formation of a structure such as a bulge loop or a stem-loop.
   d. does all of the above.

KEY TERMS

Page 221. molecular genetics, DNA (deoxyribonucleic acid), RNA (ribonucleic acid)
Page 223. transformation, DNase, RNase, protease, bacteriophage (phage)
Page 224. nucleic acids
Page 225. nucleotides, strand, double helix, chromosomes
Page 226. deoxyribose, ribose, purines, pyrimidines, adenine (A), guanine (G), thymine (T), cytosine (C), uracil (U), nucleoside
Page 227. phosphodiester linkage, backbone, directionality
Page 229. Chargaff's rule
Page 231. base pairs (bp), AT/GC rule, complementary, antiparallel, base stacking
Page 232. major groove, minor groove, B DNA, Z DNA

CHAPTER SUMMARY

- Molecular genetics is the study of DNA structure and function at the molecular level.

11.1 Identification of DNA as the Genetic Material

- To fulfill its role, genetic material must meet four criteria: information, transmission, replication, and variation.
- Griffith showed that the genetic material from type S bacteria could transform type R bacteria into type S (see Figure 11.1).
- Avery, MacLeod, and McCarty discovered that the transforming substance is DNA (see Figure 11.2).
- Hershey and Chase determined that the genetic material of T2 phage is DNA.

11.2 Overview of DNA and RNA Structure

- DNA and RNA are types of nucleic acids.
- In DNA, nucleotides are linked together to form strands, which then form a double helix that is found within chromosomes (see Figure 11.3).

11.3 Nucleotide Structure

- A nucleotide is composed of one or more phosphates, a sugar, and a base. The purine bases are adenine and guanine, whereas the pyrimidine bases are thymine (DNA only), cytosine, and uracil (RNA only). The sugar deoxyribose is found in DNA, whereas ribose is in RNA (see Figures 11.4, 11.5).
- A nucleoside contains a sugar and a base, whereas a nucleotide contains a sugar, a base, and one or more phosphate groups (see Figure 11.6).

11.4 Structure of a DNA Strand

- In a DNA strand, nucleotides are covalently attached to one another via phosphodiester linkages (see Figure 11.7).

11.5 Discovery of the Double Helix

- Pauling used ball-and-stick models to deduce the secondary structure of a protein (see Figure 11.8).
- Franklin performed X-ray diffraction studies that helped to determine the structure of DNA (see Figure 11.9).
- Chargaff determined that, in DNA, the amounts of A and T are equal, and so are the amounts of G and C (see Table 11.1).
- Watson and Crick deduced the structure of DNA by building models that were consistent with the available data (see Figure 11.10).

11.6 Structure of the DNA Double Helix

- DNA is usually a right-handed double helix in which A hydrogen bonds to T and G hydrogen bonds to C. The two strands are antiparallel and contain 10 bp per turn. Base stacking stabilizes the double helix by excluding water (see Figure 11.11).
- The double helix of DNA has a major groove and a minor groove (see Figure 11.12).
- B DNA is the major form of DNA found in living cells. Z DNA is an alternative conformation for DNA (see Figure 11.13).

11.7 RNA Structure

- RNA is composed of a strand of nucleotides (see Figure 11.14).
- RNA can form double-stranded helical regions and folds into a three-dimensional structure (see Figures 11.15, 11.16).

PROBLEM SETS & INSIGHTS

More Genetic TIPS

1. A hypothetical sequence of an RNA molecule is
   5′–AUUUGCCCUAGCAAACGUAGCAAACG–3′.

   Using two out of the three underlined parts of this sequence, draw two possible stem-loop structures that might form in this RNA.

   Topic: What topic in genetics does this question address?

   The topic is about RNA structure. More specifically, the question is about the ability of RNA to form stem-loop structures.

   Information: What information do you know based on the question and your understanding of the topic?

   From the question, you know the base sequence of an RNA molecule and are given choices regarding parts of that sequence that could interact to form stem-loops. From your understanding of the topic, you may remember that RNA molecules can form double-stranded regions that are antiparallel and complementary.

   Problem-Solving Strategy: Make a drawing.

   One strategy to solve this problem is to make a drawing that shows how parts of the given base sequence can bind to each other because they are antiparallel and complementary.

   Answer:

   ```
              CU                              G C A A A
            C   A                           A       C
            C : G                           U       G
            C : G                or         C       U
            G : C                           C   A
            U : A                           C : G
            U : A                           G : C
            U : A                           U : A
          A     C                           U : A
       5′-     GUAGCAAACG – 3′            A     C
                                        5′-     G – 3′
   ```

2. Within living cells, many different proteins play important functional roles by binding to DNA. As described throughout this text, the dynamic interactions between nucleic acids and proteins lie at the heart of molecular genetics. Some proteins bind to DNA but not in a sequence-specific manner. For example, histones are proteins important in the formation of chromosome structure. The positively charged histone proteins bind to the negatively charged phosphate groups in DNA. In addition, several other proteins interact with DNA but do not require a specific nucleotide sequence to carry out their function. For example, DNA polymerase, which catalyzes the synthesis of new DNA strands, does not bind to DNA in a sequence-dependent manner. By comparison, many other proteins do interact with nucleic acids in a sequence-dependent way. This means that a specific sequence of bases can provide a structure that is recognized by a particular protein. Throughout this text, the functions of many of these proteins will be described. Some examples include transcription factors that affect the rate of transcription and proteins that bind to origins of replication. With regard to the three-dimensional structure of DNA, where would you expect DNA-binding proteins to bind if they recognize a specific base sequence? What about DNA-binding proteins that do not recognize a base sequence?

Topic: **What topic in genetics does this question address?**

The topic is DNA-binding proteins. More specifically, the question is about deciding where on a DNA molecule a protein will bind if it recognizes a specific base sequence.

Information: **What information do you know based on the question and your understanding of the topic?**

From the question, you know that proteins can bind to specific places on a DNA molecule, such as the backbone or a particular sequence of bases. From your understanding of the topic, you may remember that the bases are accessible along the major and minor grooves.

Problem-Solving **S**trategy: **Relate structure and function.**

One strategy to solve this problem is to relate the function of a protein to the structure of DNA. In the first case, a protein functions by binding to a base sequence in the DNA. In the second case, it binds to the DNA but does not recognize a specific sequence of bases.

Answer: DNA-binding proteins that recognize a base sequence must bind into the major or minor groove of the DNA, which is where the bases are accessible to a DNA-binding protein. Most DNA-binding proteins, which recognize a base sequence, fit into the major groove. By comparison, other DNA-binding proteins, such as histones, which do not recognize a base sequence, typically bind to the DNA backbone.

3. The formation of a double-stranded structure must obey the rule that adenine hydrogen bonds to thymine (or uracil) and cytosine hydrogen bonds to guanine. Based on your understanding of genetics (from this course or a general biology course), discuss reasons why complementarity is an important feature of DNA and RNA structure and function.

Topic: **What topic in genetics does this question address?**

The topic is the importance of complementarity with regard to DNA and RNA structure and function.

Information: **What information do you know based on the question and your understanding of the topic?**

In the question, you are reminded that complementary bases in both DNA and RNA can pair with each other to form double-stranded regions. From your understanding of the topic, you may recall some examples in which the formation of base pairs is important for DNA and RNA function. Some of these examples are found in earlier chapters, such as Chapter 1, and you may have learned other examples in a previous course such as general biology.

Problem-Solving **S**trategy: **Relate structure and function.**

One strategy to solve this problem is to relate the structure of double-stranded regions to different aspects of DNA and RNA function.

Answer: Here are a few examples:
- Complementarity underlies DNA function because it provides the basis for the synthesis of new strands of DNA and RNA. During replication, the synthesis of the new DNA strands occurs in such a way that adenine hydrogen bonds to thymine, and cytosine hydrogen bonds to guanine.
- The ability to transcribe DNA into RNA is based on complementarity. During transcription, one strand of DNA is used as a template to make a complementary strand of RNA.
- The folding of RNA into a particular structure is driven by the hydrogen bonding of complementary regions. This event is necessary to produce functionally active tRNA molecules.
- Another way that complementarity can be functionally important is in promoting the interaction of two separate RNA molecules. During translation, codons in mRNA bind to the anticodons in tRNA (see Chapter 15). This binding is due to complementarity.

Conceptual Questions

C1. What is the meaning of the term *genetic material*?

C2. After the DNA from type S bacteria is exposed to type R bacteria, list all of the steps that you think must occur for the bacteria to start making a capsule.

C3. Look up the meaning of the word *transformation* in a dictionary and explain whether it is an appropriate word to describe the transfer of genetic material from one organism to another.

C4. What are the three components of a nucleotide? With regard to the 5′ and 3′ positions on a sugar molecule, how are nucleotides linked together to form a strand of DNA?

C5. Draw the structures of guanine, guanosine, and deoxyguanosine triphosphate.

C6. Draw the structure of a phosphodiester linkage.

C7. Describe how bases interact with each other in the double helix. This description should include the concepts of complementarity, hydrogen bonding, and base stacking.

C8. If one DNA strand is 5′–GGCATTACACTAGGCCT–3′, what is the sequence of the complementary strand?

C9. What is meant by the term *DNA sequence*?

C10. Make a side-by-side drawing of two DNA helices, one with 10 bp per 360° turn and the other with 15 bp per 360° turn.

C11. Discuss the differences in the structural features of B DNA and Z DNA.

C12. What part(s) of a nucleotide (namely, phosphate, sugar, and/or base) is/are found in the major and minor grooves of double-stranded DNA, and what part(s) is/are found in the DNA backbone? If a DNA-binding protein does not recognize a specific nucleotide sequence, do you expect that it binds to the major groove, the minor groove, or the DNA backbone? Explain.

C13. List the structural differences between DNA and RNA.

C14. Draw the structure of deoxyribose and number the carbon atoms. Describe the numbering of the carbon atoms in deoxyribose with regard to the directionality of a DNA strand. In a DNA double helix, what does the term *antiparallel* mean?

C15. Write a sequence of an RNA molecule that could form a stem-loop with 24 nucleotides in the stem and 16 nucleotides in the loop.

C16. Compare the structural features of a double-stranded RNA molecule with those of a DNA double helix.

C17. Which of the following DNA double helices would be more difficult to separate into single-stranded molecules by treatment with heat, which breaks hydrogen bonds?

   A. GGCGTACCAGCGCAT
      CCGCATGGTCGCGTA

   B. ATACGATTTACGAGA
      TATGCTAAATGCTCT

   Explain your choice.

C18. What structural feature allows DNA to store information?

C19. Discuss the structural significance of complementarity in DNA and in RNA.

C20. The amount of G plus C in an organism's DNA is 64% of the total base content of that DNA. What are the percentages of A, T, G, and C?

C21. A DNA-binding protein recognizes the following double-stranded sequence:

   5′–GCCCGGGC–3′
   3′–CGGGCCCG–5′

   This type of double-stranded structure could also occur within the stem region of an RNA stem-loop. Discuss the structural differences between RNA and DNA that might prevent the DNA-binding protein from recognizing a double-stranded RNA molecule.

C22. Within a protein, certain amino acids are positively charged (e.g., lysine and arginine), some are negatively charged (e.g., glutamate and aspartate), some are polar but uncharged, and some are nonpolar. If you knew that a DNA-binding protein was recognizing the DNA backbone rather than a base sequence, which amino acids in the protein would be good candidates for interacting with the DNA?

C23. A double-stranded DNA molecule contains 560 nucleotides. How many complete turns occur in this double helix?

C24. As the minor and major grooves wind around a DNA double helix, do they ever intersect each other, or do they always run parallel to each other?

C25. What chemical group (phosphate group, hydroxyl group, or a nitrogenous base) is found at the 3′ end of a DNA strand? What group is found at the 5′ end?

C26. The base composition of an RNA virus was analyzed and found to be 14.1% A, 14.0% U, 36.2% G, and 35.7% C. Would you conclude that the viral genetic material is single-stranded RNA or double-stranded RNA?

C27. The genetic material found within some viruses is single-stranded DNA. Would this genetic material contain equal amounts of A and T and equal amounts of G and C?

C28. A medium-sized human chromosome contains about 100 million base pairs. If the DNA was stretched out in a linear manner, how long would it be?

C29. A double-stranded DNA molecule is 1 cm long, and the percentage of adenine is 15%. How many cytosines does this DNA molecule contain?

C30. Could single-stranded DNA form a stem-loop structure? Why or why not?

C31. As described in Chapter 17, the methylation of cytosine bases can have an important effect on gene expression. For example, such methylation may inhibit the transcription of genes. A methylated cytosine base has the following structure:

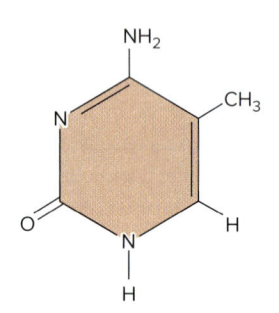

Would you expect the methylation of cytosine to affect the hydrogen bonding between cytosine and guanine in a DNA double helix? Why or why not? (Hint: See Figure 11.11 for help.) Look back at question 2 in More Genetic TIPS and speculate about how methylation could affect gene expression.

C32. An RNA molecule has the following sequence:

   Region 1   Region 2   Region 3
   5′–CAUCCAUCCAUUCCCCAUCCGAUAAGGGGAAUGGAUCCGAAUGGAUAAC–3′

   Parts of region 1 can form a stem-loop with region 2 and with region 3. Can region 1 form a stem-loop with region 2 and region 3 at the same time? Why or why not? Which stem-loop would you predict to be more stable: a region 1/region 2 interaction or a region 1/region 3 interaction? Explain your choice.

Application and Experimental Questions

E1. Genetic material acts as a blueprint for an organism's traits. Explain how the experiments of Griffith indicated that genetic material was transferred to the type R bacteria.

E2. With regard to the experiment described in Figure 11.2, answer the following:

 A. List several possible reasons why only a small percentage of the type R bacteria were converted to type S.

 B. Explain why an antibody was used to remove the bacteria that were not transformed. What would the results look like, in all five cases, if the antibody/centrifugation step had not been included in the experimental procedure?

 C. The DNA extract was treated with DNase, RNase, or protease. Why was this done? (In other words, what were the researchers trying to determine?)

E3. An interesting trait that some bacteria exhibit is resistance to being killed by antibiotics. For example, certain strains of bacteria are resistant to tetracycline, whereas other strains are sensitive to tetracycline. Describe an experiment to determine if tetracycline resistance is an inherited trait encoded by the DNA of the resistant strain.

E4. The type of model building used by Pauling and by Watson and Crick involved the use of ball-and-stick units. Now we can do model building on a computer screen. Discuss potential advantages of using computers in molecular model building.

E5. With regard to Chargaff's data described in Table 11.1, would his experiments have been convincing if they had been done on DNA from only one species? Discuss.

Questions for Student Discussion/Collaboration

1. Try to propose structures for a genetic material that are substantially different from the double helix. Remember that the genetic material must have a way to store information and a way to be faithfully replicated.

2. How might you provide evidence that DNA is the genetic material in mice?

Answers to Comprehension Questions

11.1: d

11.2: c

11.3: c, d

11.4: a

11.5: d, c

11.6: d, a, a

11.7: d

Note: All answers appear in Connect; the answers to even-numbered questions and all Concept Check questions are in Appendix B.

12

CHAPTER OUTLINE

- 12.1 Organization of Functional Sites Along Bacterial Chromosomes
- 12.2 Structure of Bacterial Chromosomes
- 12.3 Organization of Functional Sites Along Eukaryotic Chromosomes
- 12.4 Sizes of Eukaryotic Genomes and Repetitive Sequences
- 12.5 Transposition
- 12.6 Structure of Eukaryotic Chromosomes in Nondividing Cells
- 12.7 Structure of Eukaryotic Chromosomes During Cell Division

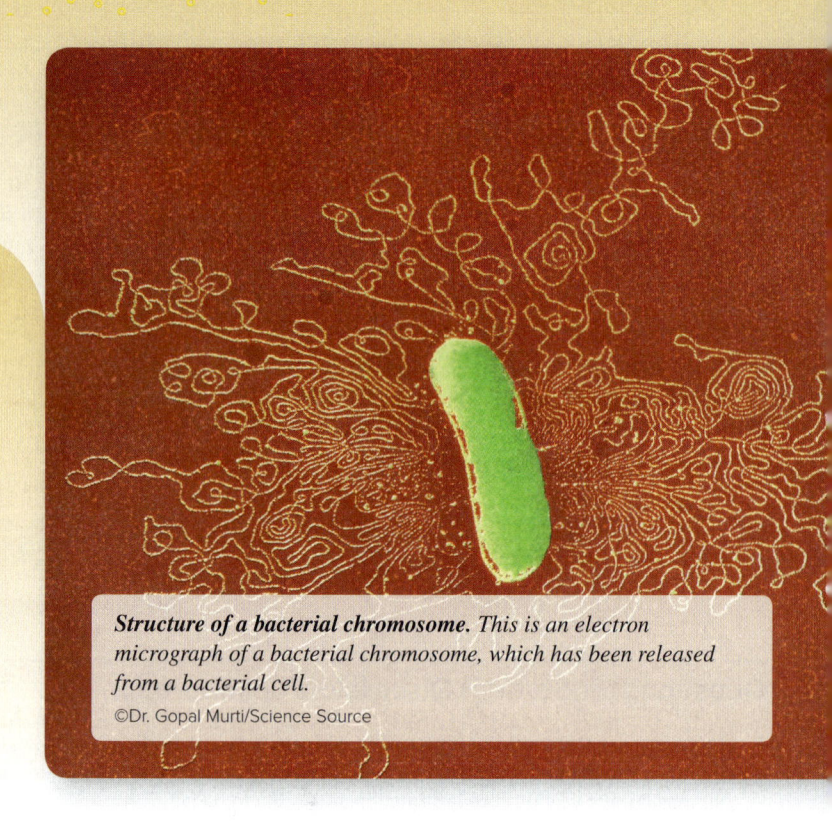

Structure of a bacterial chromosome. This is an electron micrograph of a bacterial chromosome, which has been released from a bacterial cell.
©Dr. Gopal Murti/Science Source

MOLECULAR STRUCTURE OF CHROMOSOMES AND TRANSPOSITION

Chromosomes are the structures within living cells that contain the genetic material. The term **genome** refers to all of the types of genetic material found in a cell or a species. For bacteria, the genome is typically a single circular chromosome. For eukaryotes, the nuclear genome refers to one complete set of chromosomes that resides in the cell nucleus. In other words, the haploid complement of chromosomes is considered a nuclear genome. Eukaryotes have a mitochondrial genome, and plants also have a chloroplast genome. Unless otherwise noted, the term *eukaryotic genome* refers to the nuclear genome.

The primary function of the genetic material is to store the information needed to produce the characteristics of an organism. As we saw in Chapter 11, the sequence of bases in a DNA molecule can store information. To fulfill their role at the molecular level, chromosomal regions facilitate four important processes:

- the synthesis of RNA and cellular proteins
- the replication of chromosomes
- the proper segregation of chromosomes
- the compaction of chromosomes so they can fit within living cells

In this chapter, we will examine three features of bacterial and eukaryotic chromosomes. First, we will consider the general organization of functional sites along a chromosome. Second, we will consider the molecular structures of chromosomes. As you will learn, different mechanisms are responsible for making chromosomes more compact so they can fit within living cells. Third, we will consider transposition, the process by which small segments of DNA called transposable elements can move themselves to multiple locations within the chromosomal DNA and accumulate in large numbers.

12.1 ORGANIZATION OF FUNCTIONAL SITES ALONG BACTERIAL CHROMOSOMES

Learning Outcome:

1. Describe the organization of functional sites along bacterial chromosomes.

12.2 STRUCTURE OF BACTERIAL CHROMOSOMES

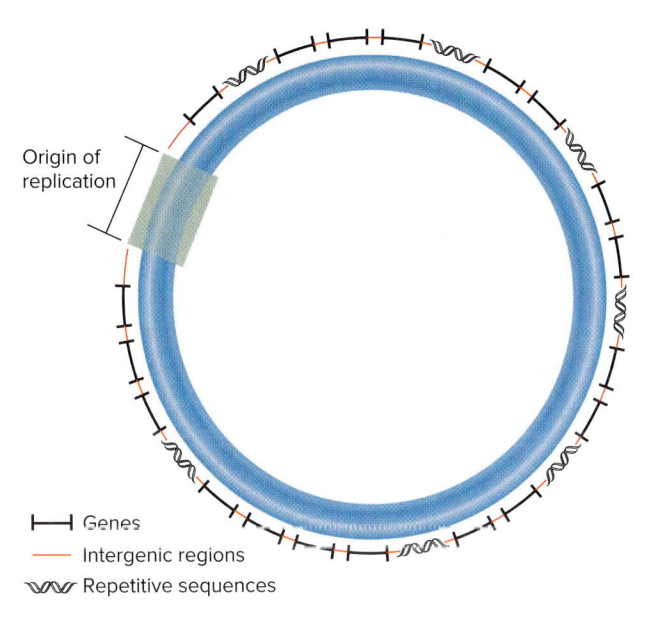

Key features:

- Most, but not all, bacterial species contain circular chromosomal DNA.
- Most bacterial species contain a single type of chromosome, but it may be present in multiple copies.
- A typical chromosome is a few million base pairs in length.
- Several thousand different genes are interspersed throughout the chromosome. The short regions between adjacent genes are called intergenic regions.
- One origin of replication is required to initiate DNA replication.
- Repetitive sequences may be interspersed throughout the chromosome.

FIGURE 12.1 Organization of sequences in bacterial chromosomal DNA.

Concept Check: What types of sequences constitute most of a bacterial genome?

Figure 12.1 shows the general features of a bacterial chromosome and the organization of functional sites along it.

- Bacterial chromosomal DNA is usually a circular molecule, though some bacteria have linear chromosomes.
- A typical chromosome is a few million base pairs (bp) in length. For example, the chromosome of *Escherichia coli* has approximately 4.6 million bp, and that of the *Haemophilus influenzae* has roughly 1.8 million bp.
- Most bacteria contain a single type of chromosome, though it may be present in multiple copies.
- A bacterial chromosome commonly has a few thousand different genes. These genes are interspersed throughout the entire chromosome. **Protein-encoding genes** (also called **structural genes**)—nucleotide sequences that encode proteins—account for the majority of bacterial DNA. The nontranscribed regions of DNA located between adjacent genes are termed **intergenic regions**.
- Bacterial chromosomes have one **origin of replication,** a sequence that is a few hundred base pairs in length. This nucleotide sequence functions as an initiation site for the assembly of several proteins required for DNA replication.
- A variety of **repetitive sequences** have been identified in many bacterial species. These sequences are found in multiple copies and are usually interspersed within the intergenic regions throughout the bacterial chromosome. Repetitive sequences may play a role in a variety of genetic processes, including DNA folding, gene regulation, and genetic recombination. As discussed later in this chapter, some repetitive sequences are transposable elements that can move throughout the genome.

12.1 REVIEWING THE KEY CONCEPTS

- Bacterial chromosomes are typically circular and have an origin of replication and a few thousand genes (see Figure 12.1).

12.1 COMPREHENSION QUESTION

1. A bacterial chromosome typically contains
   a. a few thousand genes.
   b. one origin of replication.
   c. some repetitive sequences.
   d. all of the above.

12.2 STRUCTURE OF BACTERIAL CHROMOSOMES

Learning Outcome:

1. Outline the two processes that make a bacterial chromosome more compact.

Inside a bacterial cell, a chromosome is highly compacted and found within a region of the cell known as a **nucleoid.** Although a bacterial cell usually contains a single type of chromosome, more than one copy of that chromosome may be found within the cell. Depending on the growth conditions, bacteria may have one to four identical chromosomes per cell. In addition, the number of copies varies depending on the bacterial species. As shown in **Figure 12.2,** each chromosome is found within its own distinct nucleoid within the cell. Unlike the eukaryotic nucleus, the bacterial nucleoid is not a separate cellular compartment surrounded by a membrane. Rather, the DNA in a nucleoid is in direct contact with the cytoplasm of the cell. In this section, we will explore the structure of bacterial chromosomes and the processes by which they are compacted to fit within the nucleoid region.

The Formation of Chromosomal Loops Helps Make the Bacterial Chromosome More Compact

To fit within the bacterial cell, the chromosomal DNA must be compacted about 1000-fold. The mechanism of bacterial chromosome compaction is not entirely understood, and it may vary

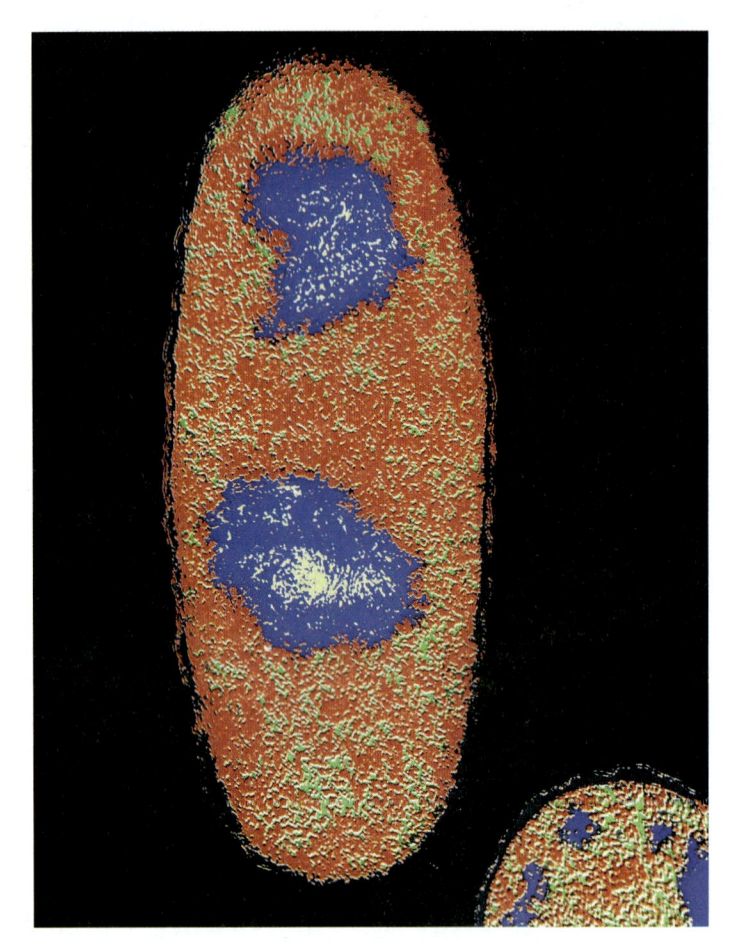

242   CHAPTER 12 :: MOLECULAR STRUCTURE OF CHROMOSOMES AND TRANSPOSITION

FIGURE 12.2 **The localization of nucleoids within *Bacillus subtilis*.** The nucleoids are fluorescently labeled in blue and seen as bright, oval-shaped regions within the bacterial cytoplasm. Note that two or more nucleoids may be found within a cell.
©M. Wurtz/Biozentrum, University of Basel/Science Source

Concept Check: How many nucleoids are in this bacterial cell?

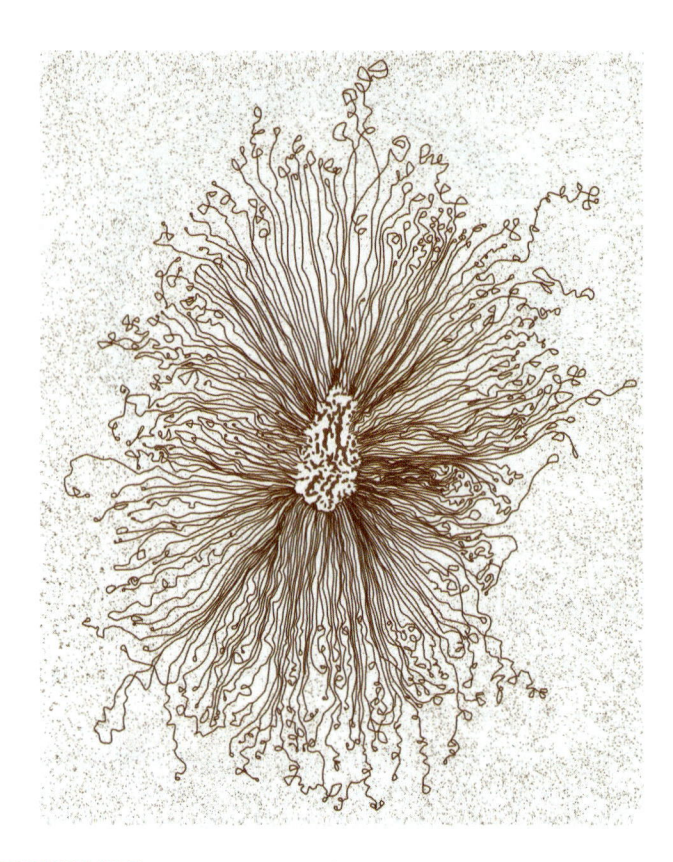

FIGURE 12.3 **Core and loops of a bacterial chromosome.** This is a schematic drawing of an *E. coli* chromosome that has been extracted from a cell and viewed by electron microscopy. The core is in the center with many loops (microdomains) emanating from it. Not all bacterial species have their chromosomes organized into microdomains and macrodomains.
Source: Adapted from Wang, Xindan, Llopis, Paula Montero, and Rudner, David Z. (2013) Organization and Segregation of Bacterial Chromosomes, *Nature Reviews Genetics*, vol. 14, no. 3, 191-203.

among different bacterial species. **Figure 12.3** shows a schematic drawing of a chromosome that has been removed from an *E. coli* cell. As the drawing shows, the chromosome has a central core with many loops emanating from that core.

- The loops that emanate from the core, which are called **microdomains,** are typically 10,000 base pairs (10 kbp) in length. An *E. coli* chromosome is expected to have about 400 to 500 microdomains. The lengths and boundaries of these microdomains are thought to be dynamic, changing in response to environmental conditions.
- In *E. coli*, many adjacent microdomains are further organized into macrodomains that are about 800 to 1000 kbp in length; each macrodomain contains about 80 to 100 microdomains. The macrodomains are not evident in Figure 12.3.

To form microdomains and macrodomains, bacteria use a set of DNA-binding proteins called **nucleoid-associated proteins (NAPs)** that facilitate chromosome compaction and organization. These proteins either bend the DNA or act as bridges that cause different regions of DNA to bind to each other. NAPs also facilitate chromosome segregation and play a role in gene regulation. Examples of NAPs include <u>h</u>istone-like <u>n</u>ucleoid <u>s</u>tructuring (H-NS) proteins and <u>s</u>tructural <u>m</u>aintenance of <u>c</u>hromosomes (SMC) proteins. SMCs are also found in eukaryotes.

DNA Supercoiling Further Compacts the Bacterial Chromosome

Because DNA is a long, thin molecule, twisting forces can dramatically change its conformation. This effect is similar to what happens when you twist a rubber band. If twisted in one direction, a rubber band eventually coils itself into a compact structure as it absorbs the energy applied by the twisting motion. Because the two strands within DNA already coil around each other, the formation of additional coils due to twisting forces is referred to as **DNA supercoiling.**

How do twisting forces affect DNA structure? **Figure 12.4** illustrates four possibilities. In Figure 12.4a, a double-stranded DNA molecule with five complete turns is anchored between two plates. In this hypothetical example, the ends of the DNA molecule

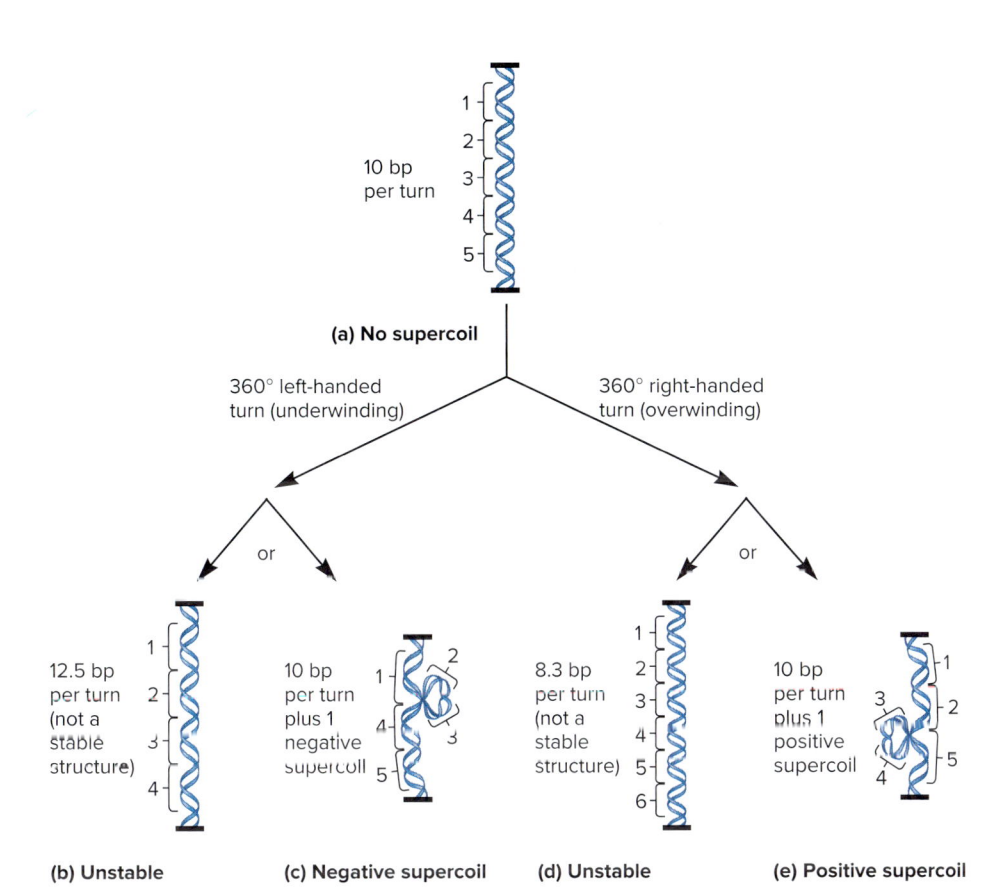

FIGURE 12.4 Schematic representation of DNA supercoiling. In this example, the DNA in (a) is anchored between two plates and given a twist as noted by the arrows. A left-handed twist (underwinding) can produce either (b) fewer turns or (c) a negative supercoil. A right-handed twist (overwinding) can produce (d) more turns or (e) a positive supercoil. The structures shown in (b) and (d) are unstable.

cannot rotate freely. Both underwinding and overwinding of the DNA double helix can induce supercoiling of the helix. Because B DNA is a right-handed helix, underwinding is a left-handed twisting motion, and overwinding is a right-handed twist. Along the left side of Figure 12.4, one of the plates has been given a turn in the direction that tends to unwind the helix. As the helix absorbs this force, two things can happen. Underwinding can cause:

- fewer turns (Figure 12.4b)
- the formation of a negative supercoil (Figure 12.4c)

On the right side of Figure 12.4, one of the plates has been given a right-handed turn, which overwinds the double helix. Overwinding can cause:

- more turns (Figure 12.4d)
- the formation of a positive supercoil (Figure 12.4e)

The DNA conformations shown in Figure 12.4a, c, and e differ only with regard to supercoiling. These three DNA conformations are referred to as **topoisomers** of each other. The DNA conformations shown in Figure 12.4b and d are not structurally favorable and do not occur in living cells.

Chromosome Function Is Influenced by DNA Supercoiling

The chromosomal DNA in bacteria is negatively supercoiled. In the chromosome of *E. coli*, about one negative supercoil occurs per 40 turns of the double helix. Negative supercoiling has several important consequences. As already mentioned, the supercoiling of chromosomal DNA makes it much more compact. Therefore, supercoiling helps to greatly decrease the size of the bacterial chromosome. In addition, negative supercoiling also affects DNA function. To understand how it does so, remember that negative supercoiling is due to an underwinding force on the DNA. Therefore, negative supercoiling creates tension on the DNA strands that may be released by their separation (**Figure 12.5**). Although most of the chromosomal DNA is negatively supercoiled and compact, the force of negative supercoiling may promote DNA strand separation in small regions. This enhances genetic activities such as replication and transcription, which require separation of the DNA strands.

How does bacterial DNA become supercoiled? In 1976, Martin Gellert and colleagues discovered the enzyme **DNA gyrase**, also known as topoisomerase II. This enzyme introduces negative supercoils (or relaxes positive supercoils) using energy from ATP. In addition, DNA gyrase in bacteria and topoisomerase II in eukaryotes can untangle DNA molecules. For example, circular DNA molecules are sometimes intertwined following DNA replication. Such interlocked molecules can be separated by topoisomerase II.

A second type of enzyme, **topoisomerase I**, can relax negative supercoils. This enzyme can bind to a negatively supercoiled region and introduce a break in one of the DNA strands. After one DNA strand has been broken, the DNA molecule can rotate to relieve the tension that is caused by negative supercoiling. The broken strand is then resealed. The competing actions of DNA gyrase and topoisomerase I govern the overall supercoiling of the bacterial DNA.

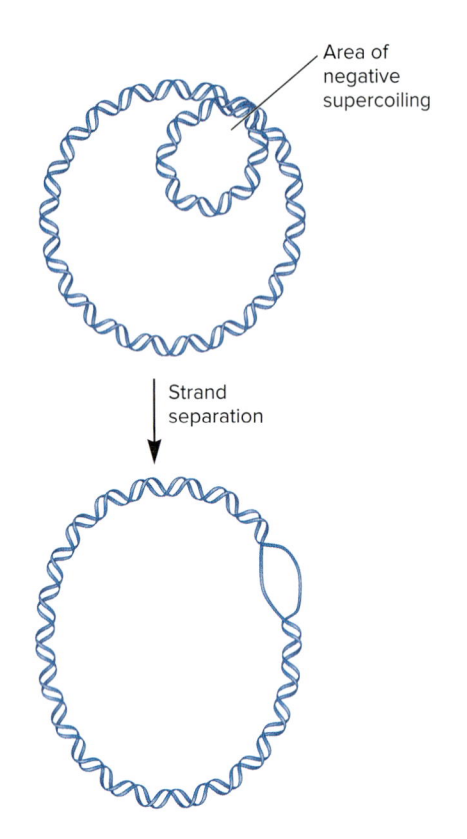

FIGURE 12.5 Negative supercoiling promotes strand separation.

Concept Check: Why is strand separation sometimes beneficial?

The ability of DNA gyrase to introduce negative supercoils into DNA is critical for bacteria to survive. For this reason, much research has been aimed at identifying drugs that specifically block this enzyme's function as a way to cure or alleviate diseases caused by bacteria. Two main classes—quinolones and coumarins—inhibit bacterial topoisomerases, thereby blocking bacterial cell growth. These drugs do not inhibit eukaryotic topoisomerases, which are structurally different from their bacterial counterparts. This finding has been the basis for the production of many drugs with important antibacterial applications. An example is ciprofloxacin (known also by the brand name Cipro), which is used to treat a wide spectrum of bacterial diseases, including anthrax.

Genetic TIPS

The Question: As noted in Chapter 11, 1 bp of DNA is approximately 0.34 nm in length. A bacterial chromosome is about 4 million bp in length and is organized into about 400 microdomains that are about 10,000 bp in length. The size of the bacterial cytoplasm, such as that of *E. coli*, is roughly 0.5 μm wide and 1.0 μm long.

A. If it was stretched out linearly, how long (in micrometers) would one microdomain be?

B. If a bacterial chromosomal microdomain was circular, what would be its diameter? (Note: Circumference = πD, where D is the diameter of the circle.)

C. Is the diameter of the circular microdomain calculated in part B small enough to fit inside a bacterium?

Topic: What topic in genetics does this question address?

The topic is the dimensions of a bacterial chromosome.

Information: What information do you know based on the question and your understanding of the topic?

In the question, you are reminded of the length of 1 bp of DNA, and that a bacterial chromosome is about 4 milllion bp in length. One microdomain is about 10,000 bp. You are also told that the bacterial cytoplasm is about 0.5 μm wide and 1.0 μm long.

Problem-Solving Strategy: Make a calculation.

For part A, you simply multiply 10,000 times 0.34 nm, which is the length of one bp. For part B, you use the equation that is given. The circumference is the linear length of the DNA. For part C, you compare the answer to part B to the dimensions of the bacterial cytoplasm.

Answer:

A. One loop is 10,000 bp. One base pair is 0.34 nm = 0.00034 μm. If you multiply these two values, you obtain (10,000) (0.00034) = 3.4 μm

B. Circumference = πD

 3.4 μm = πD

 D = 1.1 μm

C. No, it is a little too big to fit inside a bacterium such as *E. coli*. NAPs and supercoiling are needed to make the microdomains much more compact so that a single chromosome can occupy a nucleoid within a bacterial cell.

12.2 REVIEWING THE KEY CONCEPTS

- A bacterial chromosome is found in a nucleoid of a bacterial cell (see Figure 12.2).
- Bacterial chromosomes are made more compact by the formation of microdomains and macrodomains and by DNA supercoiling (see Figures 12.3, 12.4).
- Negative DNA supercoiling can promote DNA strand separation (see Figure 12.5).
- DNA gyrase (topoisomerase II) is a bacterial enzyme that introduces negative supercoils.
- Topoisomerase I relaxes negative supercoils.

12.2 COMPREHENSION QUESTIONS

1. Mechanisms that make the bacterial chromosome more compact include
    a. the formation of loop domains.
    b. DNA supercoiling.

c. crossing over.
   d. both a and b.
2. Negative supercoiling may enhance activities like transcription and DNA replication because it
   a. allows the binding of proteins to the major groove.
   b. promotes DNA strand separation.
   c. makes the DNA more compact.
   d. does all of the above.
3. DNA gyrase
   a. promotes negative supercoiling.
   b. relaxes positive supercoils.
   c. relaxes negative supercoiling.
   d. does both a and b.

12.3 ORGANIZATION OF FUNCTIONAL SITES ALONG EUKARYOTIC CHROMOSOMES

Learning Outcome:
1. Describe the organization of functional sites along a eukaryotic chromosome.

Figure 12.6 shows the general features of a eukaryotic chromosome and the organization of functional sites along it.

- Each eukaryotic chromosome contains a long, linear DNA molecule.
- The amount of DNA in a eukaryotic chromosome is tens of millions to hundreds of millions bp in length.
- Eukaryotic species have one or more sets of chromosomes in the cell nucleus; each set is composed of several different linear chromosomes (refer back to Figure 8.1). Humans, for example, have two sets of 23 chromosomes each, for a total of 46.
- A eukaryotic chromosome carries hundreds to a few thousand different genes. For example, the human X chromosome has about 1100 genes. A eukaryotic gene is several thousand to tens of thousands base pairs in length. In less complex eukaryotes such as yeast, genes are relatively short and primarily contain nucleotide sequences that encode the amino acid sequences within proteins. In more complex eukaryotes such as mammals and flowering plants, protein-encoding genes tend to be much longer due to the presence of noncoding intervening sequences called **introns**. Introns range in size from less than 100 bp to more than 10,000 bp. The presence of large introns can greatly increase the lengths of eukaryotic genes.
- Eukaryotic chromosomes contain many origins of replication, interspersed approximately every 100,000 bp. The function of origins of replication is discussed in Chapter 13.
- **Centromeres** are regions that play a role in the proper segregation of chromosomes during mitosis and meiosis. In most eukaryotic species, each chromosome contains a single centromere, which usually appears as a constricted region

Key features:

- Eukaryotic chromosomes are usually linear.
- Eukaryotic chromosomes occur in sets. Many species are diploid, which means that somatic cells contain 2 sets of chromosomes.
- A typical chromosome is tens of millions to hundreds of millions of base pairs in length.
- Genes are interspersed throughout the chromosome. A typical chromosome contains between a few hundred and several thousand different genes.
- Each chromosome contains many origins of replication that are interspersed about every 100,000 bp.
- Each chromosome contains a centromere that forms a recognition site for the kinetochore proteins.
- Telomeres contain specialized sequences located at both ends of the linear chromosome.
- Repetitive sequences are commonly found near centromeric and telomeric regions, but they may also be interspersed throughout the chromosome.

⊢⊣ Genes
〰️ Repetitive sequences

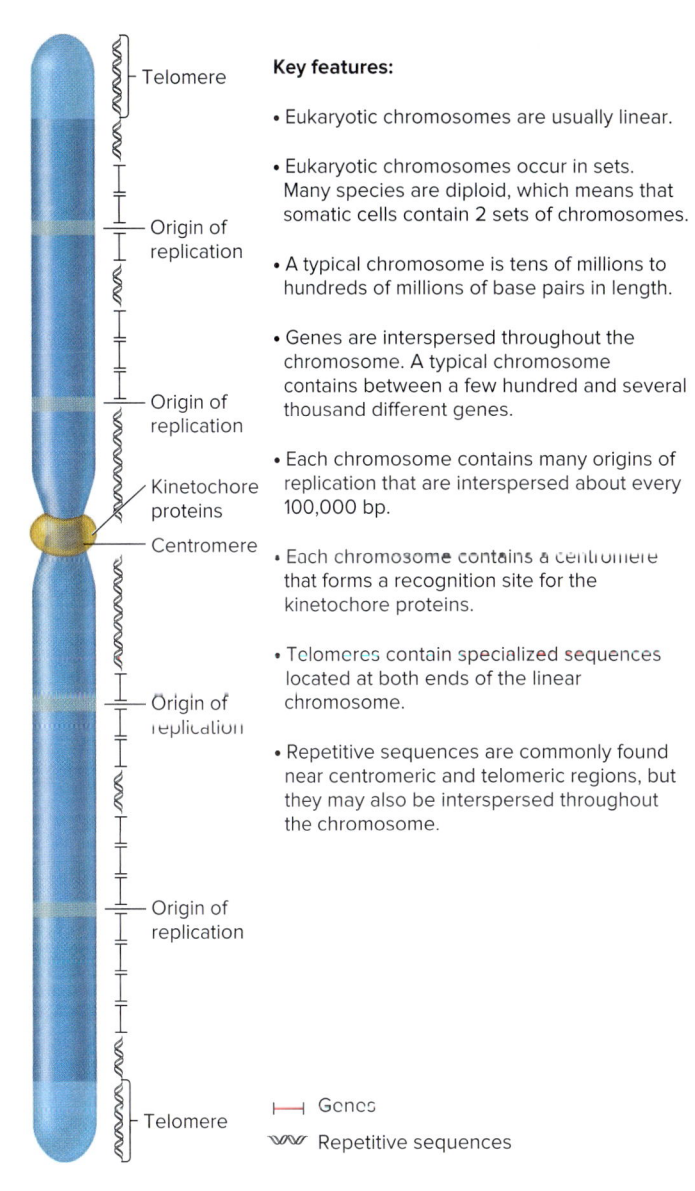

FIGURE 12.6 Organization of eukaryotic chromosomes. This organization applies to chromosomes in the cell nucleus. Mitochondrial and chloroplast chromosomes are described in Chapter 6.

Concept Check: What are some differences between the types of sequences found in eukaryotic chromosomes versus those in bacterial chromosomes?

of a mitotic chromosome. In certain yeast species, such as *Saccharomyces cerevisiae*, the centromere has a defined DNA sequence that is about 125 bp in length. This type of centromere is called a point centromere. By comparison, the centromeres found in more complex eukaryotes are much larger and contain many copies of tandemly repeated DNA sequences. These are called regional centromeres, which can range in length from several thousand base pairs to over a million. The repeated DNA sequences within regional centromeres by themselves are not necessary or sufficient to form a functional centromere. Instead, other biochemical properties are needed to make a functional centromere.

For example, a distinctive feature of all eukaryotic centromeres is that histone H3, which is discussed later in this chapter, is replaced with a histone variant called CENP-A. (Histone variants are discussed in Chapter 17.)

- The **kinetochore** is composed of a group of proteins that link the centromere to the spindle apparatus during mitosis and meiosis, ensuring the proper segregation of the chromosomes to each daughter cell.
- At the ends of linear chromosomes are found specialized regions known as **telomeres.** Telomeres serve several important functions in the replication and stability of chromosomes. As discussed in Chapter 8, telomeres prevent chromosomal rearrangements such as translocations. In addition, they prevent chromosome shortening in two ways. First, the telomeres protect chromosomes from digestion via enzymes called exonucleases that recognize the ends of DNA. Second, an unusual form of DNA replication occurs at the telomere to ensure that eukaryotic chromosomes do not become shortened with each round of DNA replication (see Chapter 13).
- As discussed later in this chapter, eukaryotic chromosomes carry a variety of **repetitive sequences**—sequences that are repeated multiple times throughout the genome.

12.3 REVIEWING THE KEY CONCEPTS

- Eukaryotic chromosomes are usually linear and contain multiple origins of replication, a centromere, telomeres, repetitive sequences, and many genes (see Figure 12.6).

12.3 COMPREHENSION QUESTION

1. The chromosomes of eukaryotes typically contain
   a. a few hundred to several thousand different genes.
   b. multiple origins of replication.
   c. a centromere.
   d. telomeres at their ends.
   e. all of the above.

12.4 SIZES OF EUKARYOTIC GENOMES AND REPETITIVE SEQUENCES

Learning Outcomes:

1. Describe the variation in size of eukaryotic genomes.
2. Define *repetitive sequence*, and explain how repetitive sequences affect genome sizes.

The total amount of DNA in cells of eukaryotic species is usually much greater than the amount in bacterial cells. In addition, eukaryotic genomes contain many more genes than their bacterial counterparts. In this section, we will examine the sizes of eukaryotic genomes and consider how repetitive sequences may greatly contribute to their overall size.

The Sizes of Eukaryotic Genomes Vary Substantially

Different eukaryotic species vary dramatically in the size of their genomes (**Figure 12.7a**; note that the graph uses a log scale). In general, more complex species have larger genome sizes. For example, the genome size of a mammal is much larger than that of a yeast. One reason is because mammalian genomes carry more genes than a yeast genome. However, the relationship between species' complexity and their number of genes is by no means linear (look ahead to Chapter 21, Table 21.3).

In many cases, variation in genome size is not related to the complexity of the species. As shown in **Figure 12.7b** and **c**, two closely related species of salamander, *Plethodon richmondi* and *Plethodon larselli*, differ considerably in genome size. The genome of *P. larselli* is more than twice as large as the genome of *P. richmondi*. However, the genome of *P. larselli* probably doesn't contain more genes. How do we explain the difference in genome size? The additional DNA in *P. larselli* is due to the accumulation

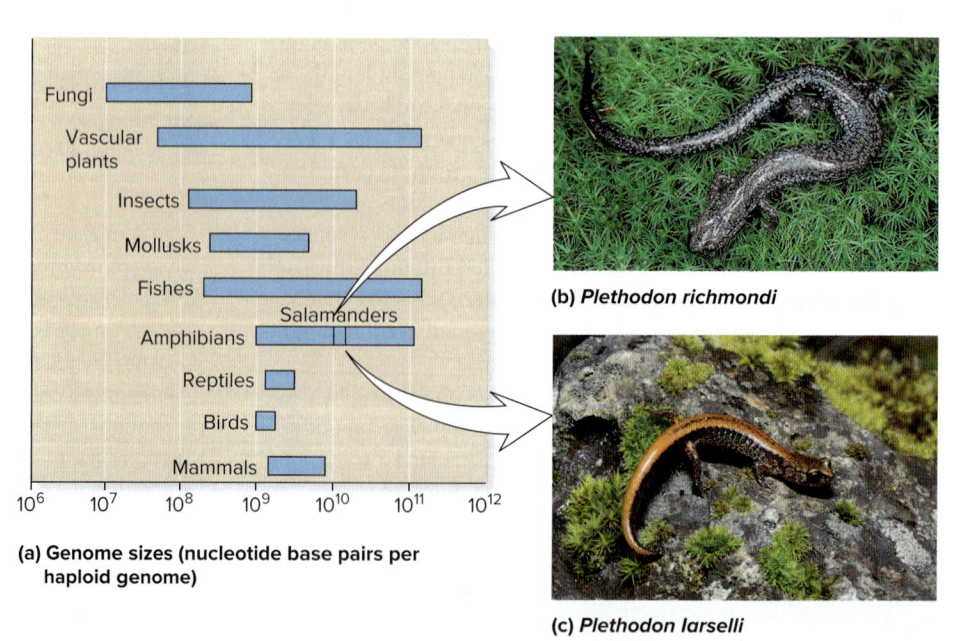

FIGURE 12.7 **Haploid genome sizes among groups of eukaryotic species.** (a) Ranges of genome sizes among different groups of eukaryotes. (b) A species of salamander, *Plethodon richmondi*, and (c) a close relative, *Plethodon larselli*. The genome of *P. larselli* is over twice as large as that of *P. richmondi*.
Source: Data in part (a) from Gregory, T. Ryan, Nichol, J. A., Tamm, H., et al. (2007) Eukaryotic Genome Size Databases, *Nucleic Acids Research,* vol. 35, D332-D338.

Genes → Traits The two species of salamander shown here have very similar traits, even though the genome of *P. larselli* is over twice as large as that of *P. richmondi*. However, the genome of *P. larselli* is not likely to contain more genes. Rather, the additional DNA is due to the accumulation of short repetitive DNA sequences that do not contain functional genes genes and are present in many copies.
(b) ©Ann & Rob Simpson; (c) ©Gary Nafis

Concept Check: What are two reasons for the wide variation in genome sizes among eukaryotic species?

(a) Genome sizes (nucleotide base pairs per haploid genome)

(b) *Plethodon richmondi*

(c) *Plethodon larselli*

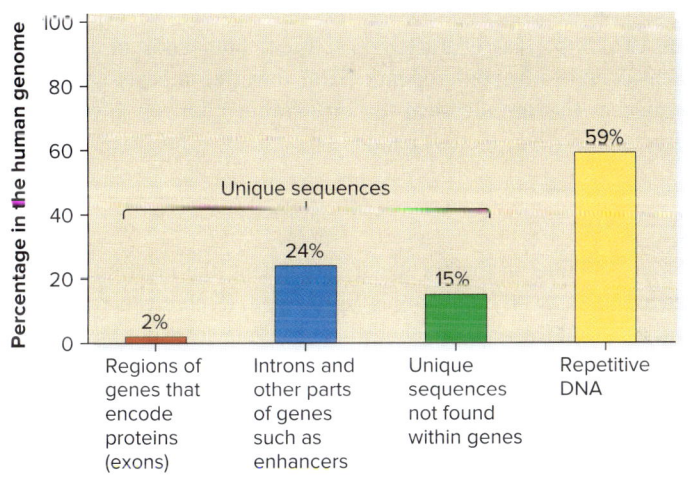

of repetitive DNA sequences present in many copies. In some species, these repetitive sequences have reached enormous levels. Such highly repetitive sequences do not encode proteins, and their function remains a matter of controversy and great interest. The structure and significance of repetitive DNA is discussed next.

The Genomes of Eukaryotes Contain Sequences That Are Unique, Moderately Repetitive, or Highly Repetitive

The term **sequence complexity** refers to the number of times a particular base sequence appears throughout the genome of a species.

- Unique (or nonrepetitive) sequences are those found once or a few times within a haploid genome. Protein-encoding genes are typically unique sequences of DNA. The vast majority of proteins in eukaryotic cells are encoded by genes present in one or a few copies. In the case of humans, unique sequences make up roughly 41% of the entire genome (**Figure 12.8**). These sequences include exons, which encode polypeptides, and unique noncoding DNA such as sequences of introns and other chromosomal regions.
- **Moderately repetitive sequences** are found a few hundred to several thousand times in a genome. In a few cases, moderately repetitive sequences are multiple copies of the same gene. For example, the genes that encode ribosomal RNA (rRNA) are found in many copies. Cells need a large amount of rRNA for making ribosomes, and production of this amount is facilitated by having multiple copies of the genes that encode rRNA. By comparison, many moderately repetitive sequences do not play a functional role and are derived from **transposable elements (TE)**—short segments of DNA that have the ability to move within a genome. This category of repetitive sequences is discussed in greater detail later in this chapter.
- **Highly repetitive sequences** are found tens of thousands or even millions of times throughout a genome. Some of these are transposable elements. A widely studied example is the *Alu* family of sequences found in humans and other primates. The *Alu* sequence is approximately 300 bp long and is present in about 1,000,000 copies in the human genome. Aside from transposable elements, some highly repetitive sequences are clustered together in a **tandem array,** also known as a tandem repeat. In a tandem array, a very short nucleotide sequence is repeated many times in a row. For example, in *Drosophila*, one particular tandem array carries two related sequences, AATAT and AATATAT:

AATATAATATAATATAATATAATATATAATAT
TTATATTATATTATATTATATTATATATTATA

Such tandem arrays are commonly found in centromeric regions of chromosomes and can be quite long, sometimes more than 1,000,000 bp in length!

What is the functional significance of highly repetitive sequences? Whether highly repetitive sequences have any significant function is controversial. Some experiments in *Drosophila* indicate that highly repetitive sequences may be important in the proper segregation of chromosomes during meiosis. It is not yet clear if highly repetitive DNA plays the same role in other species. The sequences within highly repetitive DNA vary greatly from species to species. Likewise, the amount of highly repetitive DNA can vary a great deal even among closely related species (as noted in the discussion of Figure 12.7b and c).

12.4 REVIEWING THE KEY CONCEPTS

- The genome sizes of eukaryotes vary greatly. Some of this variation is due to the accumulation of repetitive sequences (see Figure 12.7).
- The human genome contains about 41% unique sequences and 59% repetitive sequences (see Figure 12.8).

12.4 COMPREHENSION QUESTION

1. Which of the following is an example of moderately repetitive sequences?
   a. Genes that encode rRNAs
   b. Most protein-encoding genes
   c. Both a and b
   d. None of the above

12.5 TRANSPOSITION

Learning Outcomes:
1. Describe the organization of sequences within different types of transposable elements.
2. Explain how transposons and retrotransposons move to new locations in a genome.
3. Discuss the effects of transposable elements on gene function.

Genome sequencing projects have revealed that a sizable portion of many species' genomes are composed of repetitive sequences. In many cases, the repetitive sequences are due to transposition, which involves the integration of small segments of DNA into a

FIGURE 12.8 Relative amounts of unique and repetitive DNA sequences in the human genome.

chromosome. Transposition can occur at many different locations within the genome. As noted in the preceding section, the DNA segments that transpose themselves are known as transposable elements (TEs). TEs have sometimes been referred to as "jumping genes" because they are inherently mobile.

Transposable elements were first identified by Barbara McClintock in the early 1950s during her classic studies with corn plants. Since that time, geneticists have discovered many different types of TEs in organisms as diverse as bacteria, fungi, plants, and animals. The advent of molecular technology has allowed scientists to better understand the characteristics of TEs that enable them to be mobile. In this section, we will examine the characteristics of TEs and explore the mechanisms that explain how they move. We will also discuss the biological significance of TEs.

Transposable Elements Move by Different Transposition Pathways

Since the pioneering studies of McClintock, many different TEs have been found in bacteria, fungi, plants, and animals. Different types of transposition pathways have been identified.

Simple Transposition In **simple transposition,** the TE is removed from its original site and transferred to a new target site (**Figure 12.9a**). This mechanism is called a cut-and-paste mechanism because the element is cut out of its original site and pasted into a new one. Transposable elements that move via simple transposition are widely found in bacterial and eukaryotic species. Such TEs are also called **transposons.**

Retrotransposition Another type of transposable element moves via an RNA intermediate. This mechanism of transposition, termed **retrotransposition,** occurs in eukaryotic species, where it is very common (**Figure 12.9b**). Transposable elements that move via retrotransposition are known as **retrotransposons,** or **retroelements.** In retrotransposition, the element is transcribed into RNA. An enzyme called reverse transcriptase uses the RNA as a template to synthesize a DNA molecule that is integrated into a new region of the genome. Retrotransposons increase in number during retrotransposition.

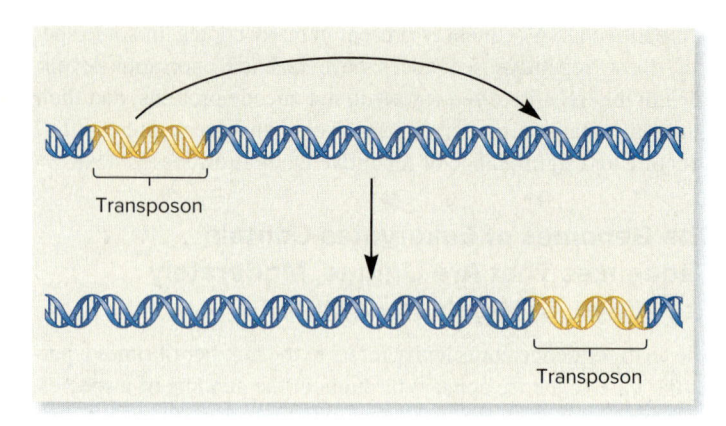

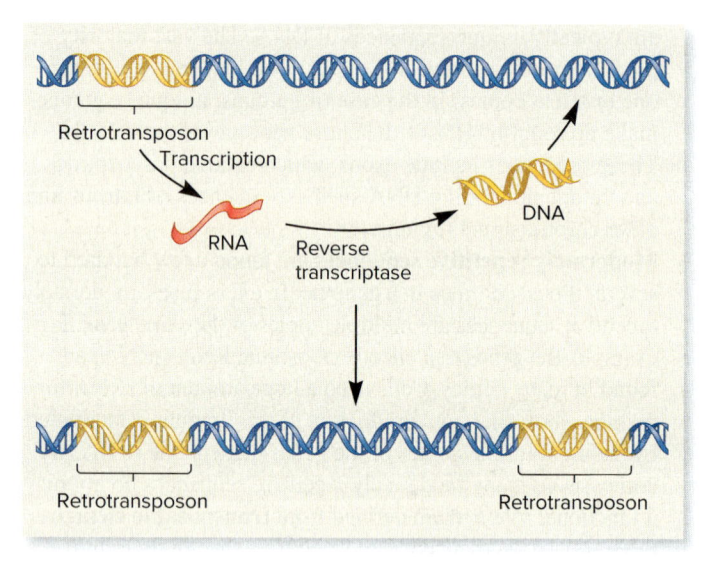

Each Type of Transposable Element Has a Characteristic Pattern of DNA Sequences

Research on TEs from many species has established that DNA sequences within them are organized in several different ways. **Figure 12.10** describes a few of those ways, although many variations are possible. All TEs are flanked by **direct repeats (DRs),** also called **target-site duplications,** which are identical nucleotide sequences that are oriented in the same direction and repeated. Direct repeats are adjacent to both ends of any TE.

Insertion Elements The simplest TE is known as an **insertion element (IS element).** As shown in Figure 12.10a, an IS element

(a) Simple transposition

(b) Retrotransposition

FIGURE 12.9 Different mechanisms of transposition.

Concept Check: Which of these mechanisms causes the TE to increase in number?

has two important characteristics. First, both ends of the element contain **inverted repeats (IRs).** Inverted repeats are DNA sequences that are identical (or very similar) but run in opposite directions, such as the following:

5′–CTGACTCTT–3′   and   5′–AAGAGTCAG–3′
3′–GACTGAGAA–5′   3′–TTCTCAGTC–5′

Depending on the particular IS element, the inverted repeats range from 9 to 40 bp in length. In addition, IS elements may contain a central region that encodes the enzyme **transposase,** which catalyzes the transposition event.

Simple Transposons By comparison, a **simple transposon** carries one or more genes that are not required for transposition. For example, the simple transposon shown in Figure 12.10a carries an antibiotic-resistance gene.

12.5 TRANSPOSITION 249

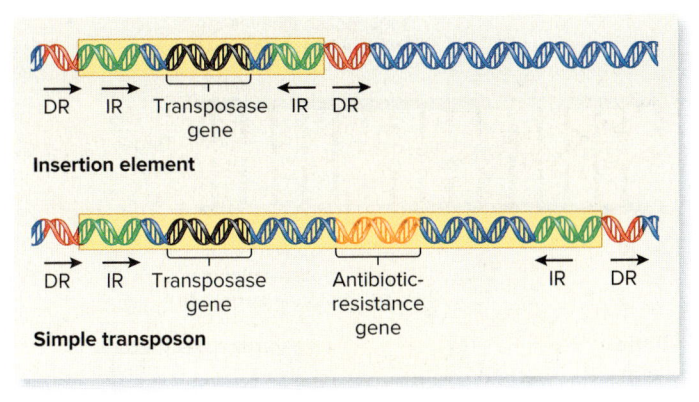

(a) Elements that move by simple transposition

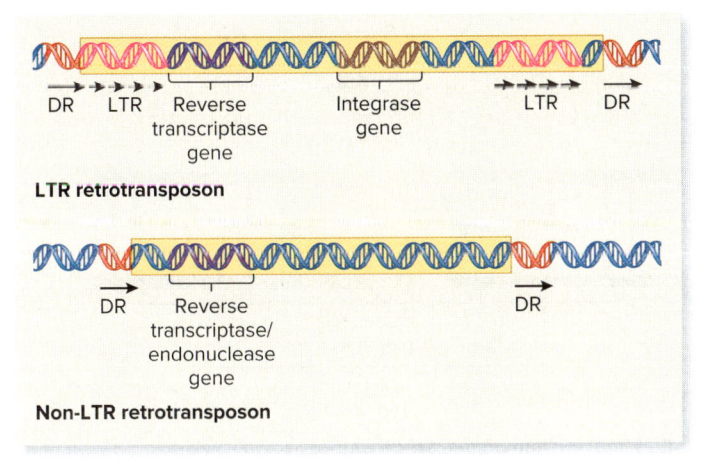

(b) Elements that move by retrotransposition (via an RNA intermediate)

FIGURE 12.10 Common organizations of DNA sequences in transposable elements. Direct repeats (DRs) are identical sequences found on both sides of all TEs. Inverted repeats (IRs) are at the ends of transposons. Long terminal repeats (LTRs) are regions containing a large number of tandem repeats.

The organization of retrotransposons varies greatly. They are categorized based on their evolutionary relationship to retroviruses. As described in Chapter 10, retroviruses are RNA viruses that make a DNA copy that integrates into the host's genome.

LTR Retrotransposons LTR retrotransposons are evolutionarily related to known retroviruses. These TEs have retained the ability to move around the genome, though, in most cases, they do not produce mature viral particles. LTR retrotransposons are so named because they contain **long terminal repeats (LTRs)** at both ends (Figure 12.10b). The LTRs are typically a few hundred base pairs in length. Like their viral counterparts, LTR retrotransposons encode proteins such as reverse transcriptase and integrase that are needed for the retrotransposition process.

Non-LTR Retrotransposons By comparison, the sequences of **non-LTR retrotransposons** appear less like those of retroviruses. These retrotransposons may contain a gene that encodes a protein that has both reverse transcriptase and endonuclease function (see Figure 12.10b). As discussed later, these functions are needed for retrotransposition. Some non-LTR retrotransposons are evolutionarily derived from normal eukaryotic genes. For example, the *Alu* family of repetitive sequences found in humans is derived from a single ancestral gene known as the *7SL RNA* gene (a component of the complex called signal recognition particle, which directs newly made proteins to the endoplasmic reticulum). The *Alu* retrotransposon has been copied by retrotransposition many times, and the current number of copies in the human genome is approximately 1 million.

Transposable elements are considered to be complete or **autonomous elements** when they contain all of the information necessary for transposition or retrotransposition to take place. However, TEs are often incomplete or nonautonomous. A **nonautonomous element** typically lacks a gene that encodes transposase or reverse transcriptase, which is necessary for transposition to occur. For example, the *Ds* element (described later in Table 12.1) is a nonautonomous element because it lacks a transposase-encoding gene. An element that is similar to *Ds* but contains a functional transposase-encoding gene is called the *Ac* element, which stands for Activator element. An *Ac* element provides a transposase-encoding gene that enables *Ds* to transpose. Therefore, nonautonomous TEs such as *Ds* can transpose only when *Ac* is present at another region in the genome.

Transposase Catalyzes the Excision and Insertion of Transposons

Now that we have considered the typical organization of TEs, let's examine the steps of the transposition process. The enzyme transposase catalyzes the removal of a TE from its original site in the chromosome and its subsequent insertion at another location. A general scheme for simple transposition is shown in **Figure 12.11a**.

1. Transposase monomers first bind to the inverted repeat sequences at the ends of the TE.
2. The monomers then dimerize, which brings the inverted repeats close together.
3. The DNA is cleaved between the inverted and direct repeats, excising the TE from its original site within the chromosome.
4. Transposase carries the TE to a new site and cleaves the target DNA sequence at staggered recognition sites. The TE is then inserted and ligated to the target DNA.

As shown in **Figure 12.11b**, the ligation of the transposable element into its new site initially leaves short gaps in the target DNA. Notice that the DNA sequences in these gaps are complementary to each other (in this case, ATGCT and TACGA). Therefore, when they are filled in by DNA gap repair synthesis, the repair produces direct repeats that flank both ends of the TE. These direct repeats are common features found adjacent to all TEs (see Figure 12.10).

Although the transposition process depicted in Figure 12.11 does not directly alter the number of TEs, simple transposition

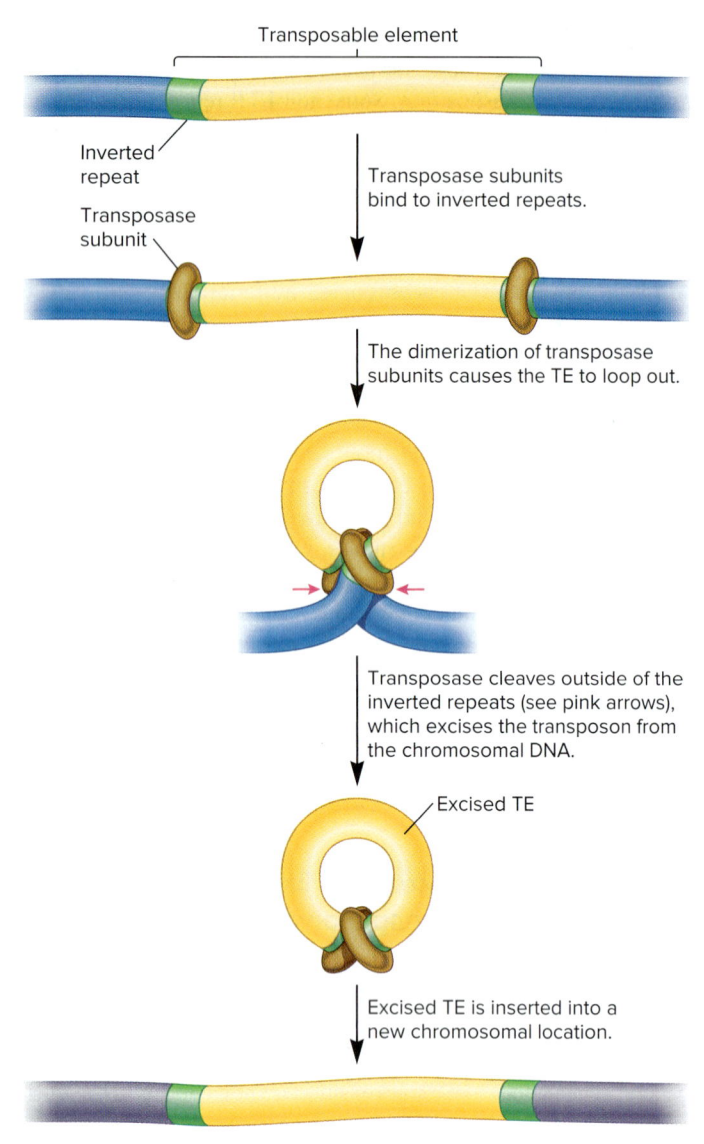

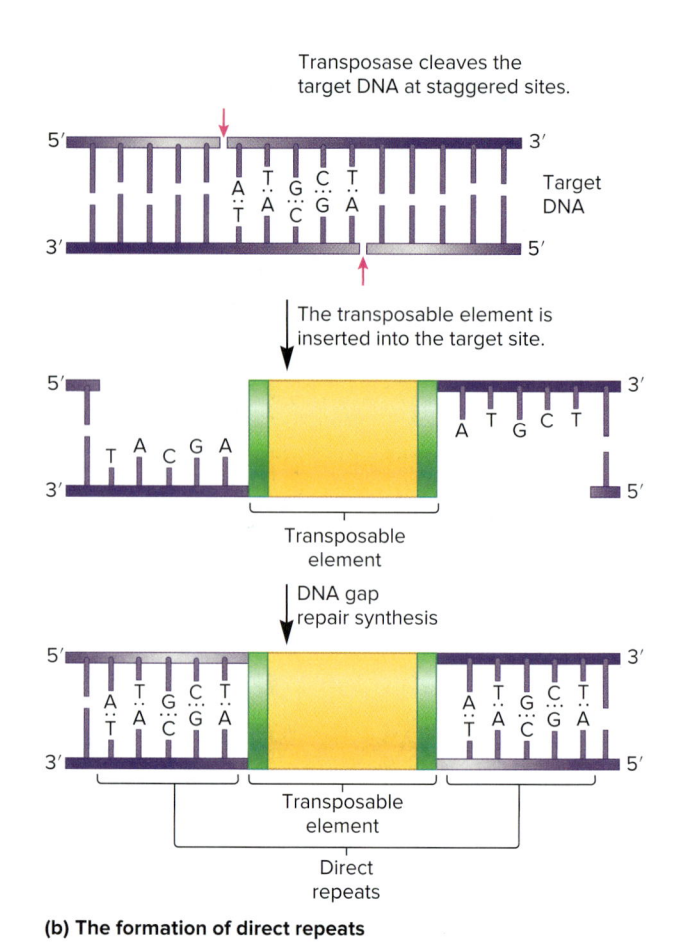

FIGURE 12.11 Simple transposition. (a) Transposase removes the TE from its original site and inserts it into a new site. (b) A closer look at the insertion process.

is known to increase their numbers in genomes, in some cases to fairly high levels. How can this happen? The answer is that transposition often occurs around the time of DNA replication, which is described in Chapter 13. After a site containing a TE has been replicated, one of the TEs may transpose from its original location into a second region that has not yet replicated (**Figure 12.12**). After this second region has been replicated, two TEs will be found in one of the chromosomes and one TE in the other chromosome. In this way, simple transposition can lead to an increase in TEs. We will discuss the biological significance of transposon proliferation later in this section.

Retrotransposons Use Reverse Transcriptase for Retrotransposition

Thus far, we have considered how transposons can move throughout a genome. By comparison, retrotransposons use an RNA intermediate in their transposition mechanism. Let's consider how LTR retrotransposons proliferate. As shown in **Figure 12.13**, the transposition mechanism of LTR retrotransposons requires two key enzymes: reverse transcriptase and integrase. In this example, a yeast cell already contains a retrotransposon known as *Ty* within its genome.

1. The retrotransposon is transcribed into RNA.
2. In a series of steps, **reverse transcriptase** uses this RNA as a template to synthesize a double-stranded DNA molecule.
3. The LTRs at the ends of the double-stranded DNA are then recognized by **integrase,** which makes staggered cuts at a target site in the host chromosome and catalyzes the insertion of the TE into this site.

The integration of a retrotransposon can occur at multiple sites within the genome. (Figure 12.13 shows the integration occurring at two locations.) Furthermore, because a single retrotransposon can be copied into many RNA transcripts, retrotransposons may accumulate rapidly within a genome.

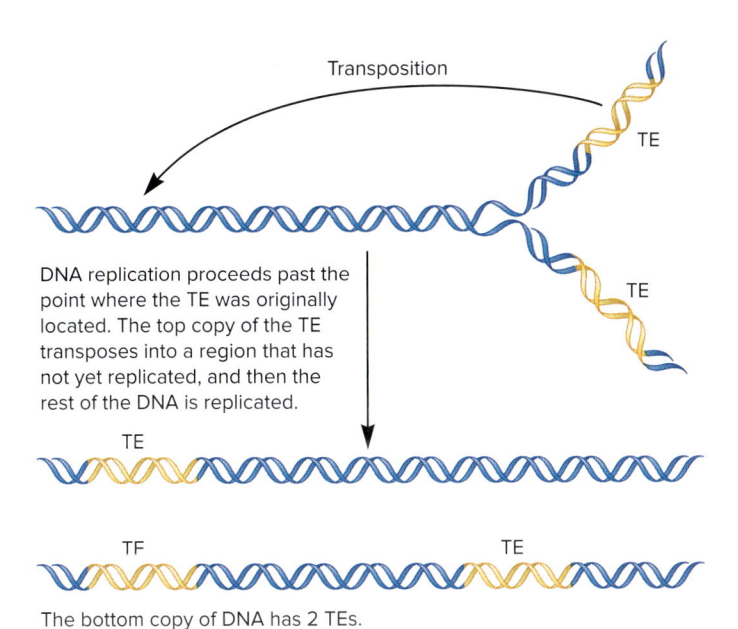

FIGURE 12.12 Increase in the number of copies of a transposable element (TE) via simple transposition. In this example, a TE that has already been replicated transposes to a new site that has not yet replicated. Following the completion of DNA replication, the TE has increased in number.

Transposable Elements May Have Important Influences on Mutation and Evolution

Over the past few decades, researchers have found that TEs probably occur in the genomes of all species. **Table 12.1** describes a few TEs that have been studied in great detail. As discussed earlier in this chapter, the genomes of eukaryotic species typically contain moderately and highly repetitive sequences. In some cases, these repetitive sequences are due to the proliferation of TEs.

The relative abundance of TEs varies widely among different species. As shown in **Table 12.2**, TEs can be quite prevalent in amphibians, mammals, and flowering plants, but tend to be less abundant in simpler organisms such as bacteria and yeast. The biological significance of TEs in the evolution of prokaryotic and eukaryotic species remains a matter of debate. According to

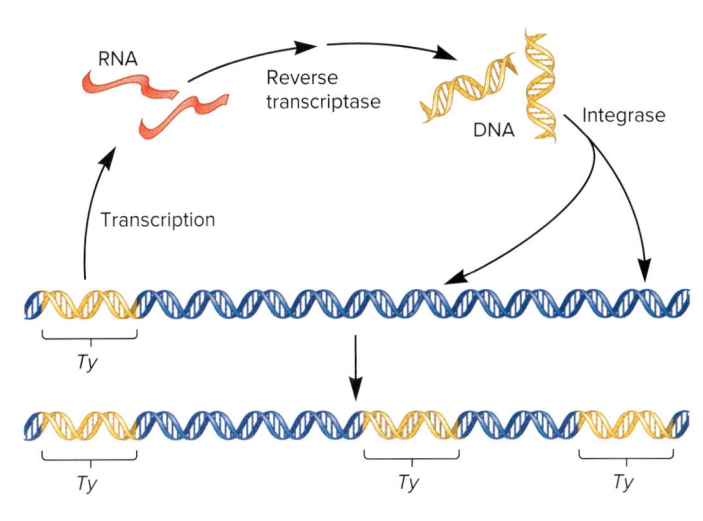

FIGURE 12.13 Retrotransposition of an LTR retrotransposon via reverse transcriptase and integrase.

Concept Check: What is the function of reverse transcriptase?

TABLE 12.1
Examples of Transposable Elements

Element	Type	Approximate Length (bp)	Description
Bacteria			
IS1	Transposon	768	An insertion element that is commonly found in five to eight copies in *E. coli*
Tn10	Transposon	9300	One of many different bacterial transposons that carries antibiotic resistance
Yeast			
Ty elements	Retrotransposon	6300	Found in *S. cerevisiae* in about 35 copies per genome
Drosophila			
P elements	Transposon	500–3000	A transposon that may be found in 30–50 copies in P strains of *Drosophila*. It is absent from M strains.
Humans			
Alu sequence	Retrotransposon	300	A retrotransposon found in about 1 million copies in the human genome
L1	Retrotransposon	6500	A retrotransposon found in about 500,000 copies in the human genome
Plants			
Ac/Ds	Transposon	4500	*Ac* is an autonomous transposon found in corn and other plant species. It carries a transposase-encoding gene. *Ds* is a nonautonomous version that lacks a functional transposase-encoding gene.

TABLE 12.2
Abundance of Transposable Elements in the Genomes of Selected Species

Species	Percentage of the Total Genome Composed of TEs*
Frog (*Xenopus laevis*)	77
Corn (*Zea mays*)	60
Human (*Homo sapiens*)	45
Mouse (*Mus musculus*)	40
Fruit fly (*Drosophila melanogaster*)	20
Nematode (*Caenorhabditis elegans*)	12
Yeast (*Saccharomyces cerevisiae*)	4
Bacterium (*Escherichia coli*)	0.3

*In some cases, the abundance of TEs may vary somewhat among different strains of the same species. The values reported here are typical values.

TABLE 12.3
Possible Consequences of Transposition

Consequence	Cause
Chromosome Structure	
Chromosome breakage	Excision of a TE
Chromosomal rearrangements	Homologous recombination between TEs located at different positions in the genome
Gene Expression	
Mutation	Incorrect excision of TEs
Gene inactivation	Insertion of a TE into a gene
Alteration in gene regulation	Transposition of a gene next to regulatory sequences or the transposition of regulatory sequences next to a gene
Alteration in the exon content of a gene	Insertion of exons into the coding sequence of a gene via TEs, a phenomenon called exon shuffling
Gene duplications	Insertion of a gene into a transposon that transposes to another site in the genome

the **selfish DNA hypothesis,** TEs exist because they have characteristics that allow them to multiply within the chromosomal DNA of living cells. In other words, they resemble parasites in the sense that they inhabit a cell without offering any selective advantage to the organism. They can proliferate as long as they do not harm the organism to the extent that they significantly disrupt survival.

Alternatively, other geneticists have argued that most transposition events are deleterious. Therefore, TEs would be eliminated from the genome by natural selection if they did not also offer a compensating advantage. Several potential advantages have been suggested. For example, TEs may cause greater genetic variability by promoting recombination. In addition, bacterial TEs often carry an antibiotic-resistance gene that provides the organism with a survival advantage. Researchers have also suggested that transposition may cause the insertion of exons from one gene into another gene, thereby producing a new gene with novel function(s). This phenomenon is called **exon shuffling.**

This controversy remains unresolved, but it is clear that TEs can rapidly enter the genome of an organism and proliferate quickly. In *Drosophila melanogaster*, for example, a TE known as the P element was probably introduced into this species in the 1950s. Laboratory stocks of *D. melanogaster* collected prior to this time do not contain P elements. Remarkably, in the last 50 years, the P element has expanded throughout *D. melanogaster* populations worldwide. The only strains without the P element are laboratory strains collected prior to the 1950s. This observation underscores the surprising ability of TEs to infiltrate a population of organisms quickly.

Transposable elements have a variety of effects on chromosome structure and gene expression (**Table 12.3**). Many of these outcomes are likely to be harmful. Usually, transposition is a relatively rare event that occurs only in a few individuals under certain conditions. Agents such as radiation, chemical mutagens, and hormones stimulate the movement of TEs. As described in Chapter 18, prokaryotes and eukaryotes have mechanisms involving non-coding RNAs that prevent the movement of TEs.

When it is not carefully regulated, transposition is likely to be detrimental. For example, in *D. melanogaster*, if females that lack P elements (M strain females) are crossed with males that contain numerous P elements (P strain males), the egg cells allow the P elements inherited via the sperm to transpose at a high rate. The resulting hybrid offspring exhibit a variety of abnormalities, which include high rates of sterility, mutation, and chromosome breakage. This deleterious outcome, which is called **hybrid dysgenesis,** occurs because the P elements were able to insert into a variety of locations in the genome.

12.5 REVIEWING THE KEY CONCEPTS

- Transposable elements can move via simple transposition (cut-and-paste) and retrotransposition (see Figure 12.9).
- Each type of transposable element has its own pattern of DNA sequences, which includes direct repeats (see Figure 12.10).
- Transposase catalyzes the excision and insertion of transposons, which move via a DNA intermediate (see Figure 12.11).
- Simple transposition can increase the copy number of a transposon if it occurs just after a transposon has been replicated (see Figure 12.12).

- Retrotransposition of LTR retrotransposons occurs via reverse transcriptase and integrase (see Figure 12.13).
- Many different transposons are found among living organisms. Their abundance varies among different species (see Tables 12.1, 12.2).
- Transposition can have a variety of effects on chromosome structure and gene expression (see Table 12.3).

12.5 COMPREHENSION QUESTIONS

1. Which of the following types of transposable elements move via RNA intermediate?
   a. Insertion elements
   b. Simple transposons
   c. Retrotransposons
   d. All of the above
2. The function of transposase is
   a. to recognize inverted repeats.
   b. to remove a TE from its original site.
   c. to insert a TE into a new site.
   d. to do all of the above.
3. According to the selfish DNA hypothesis, TEs exist because
   a. they offer the host a selective advantage.
   b. they have characteristics that allow them to multiply within the chromosomal DNA of living cells.
   c. they promote the expression of certain beneficial genes.
   d. of all of the above.

12.6 STRUCTURE OF EUKARYOTIC CHROMOSOMES IN NONDIVIDING CELLS

Learning Outcomes:
1. Define *chromatin*.
2. Describe the structures of nucleosomes, the 30-nm fiber, and radial loop domains.
3. Analyze the results of Noll, and explain how they support the beads-on-a-string model.
4. Describe what a chromosome territory is.

A distinguishing feature of eukaryotic cells is that their chromosomes are located within a cellular compartment known as the **nucleus**. The DNA within a typical eukaryotic chromosome is a single, linear, double-stranded molecule that may be hundreds of millions of base pairs in length. If the DNA from a single set of human chromosomes was stretched from end to end, the length would be over 1 meter! By comparison, most eukaryotic cells are only 10 to 100 μm in diameter, and the cell nucleus is only about 2 to 4 μm in diameter. Therefore, the DNA in a eukaryotic cell must be folded and compacted to a staggering extent to fit inside the nucleus. In eukaryotic chromosomes, as in bacterial chromosomes, this is accomplished by the binding of the DNA to many different cellular proteins. The DNA-protein complex found within eukaryotic chromosomes is termed **chromatin**. In recent years, it has become increasingly clear that the proteins bound to chromosomal DNA are subject to change over the life of the cell. These changes in protein composition, in turn, affect the degree of compaction of the chromatin. In this section, we will consider how chromosomes are compacted during interphase—the period of the cell cycle that includes the G_1, S, and G_2 phases. In the following section, we will examine the additional compaction that is necessary to produce the highly condensed chromosomes present during M phase.

Linear DNA Wraps Around Histone Proteins to Form Nucleosomes

The repeating structural unit within eukaryotic chromatin is the **nucleosome**—a double-stranded segment of DNA wrapped around an octamer of **histone proteins** (**Figure 12.14a**).

- Each octamer contains eight histone subunits, two copies each of four different histone proteins: H2A, H2B, H3, and H4. These are called the core histone proteins.
- Each of the histone proteins consists of a globular domain and a flexible, charged amino terminus called an amino terminal tail.
- The DNA lies on the surface and makes 1.65 negative superhelical turns around the histone octamer. Positively charged lysine and arginine amino acids in the histone proteins play a major role in binding to the negatively charged phosphate groups along the DNA backbone.
- The amount of DNA required to wrap around the histone octamer is 146 or 147 bp.
- At its widest point, a single nucleosome is about 11 nm in diameter.
- In 1997, Timothy Richmond and colleagues determined the structure of a nucleosome by X-ray crystallography (**Figure 12.14b**).

The chromatin of eukaryotic cells contains a repeating pattern in which the nucleosomes are connected by linker regions of DNA that vary in length from 20 to 100 bp, depending on the species and cell type. It has been suggested that the overall structure of connected nucleosomes resembles beads on a string. This structure shortens the length of the DNA molecule about sevenfold.

Another histone protein, H1, is found in most eukaryotic cells and is called the linker histone. It binds to the DNA in the linker region between nucleosomes and may help to compact adjacent nucleosomes (**Figure 12.14c**). The linker histone is less tightly bound to the DNA than are the core histones. In addition, nonhistone proteins bound to the linker region play a role in the organization and compaction of chromosomes, and their presence may affect the expression of nearby genes.

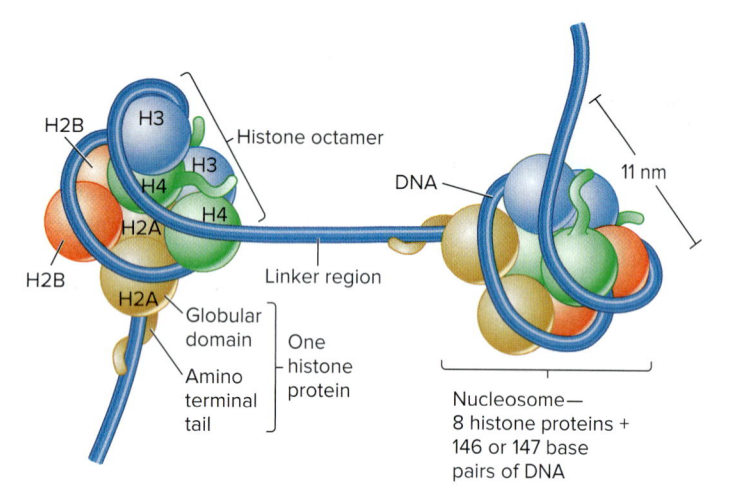

(a) Nucleosomes showing core histone proteins

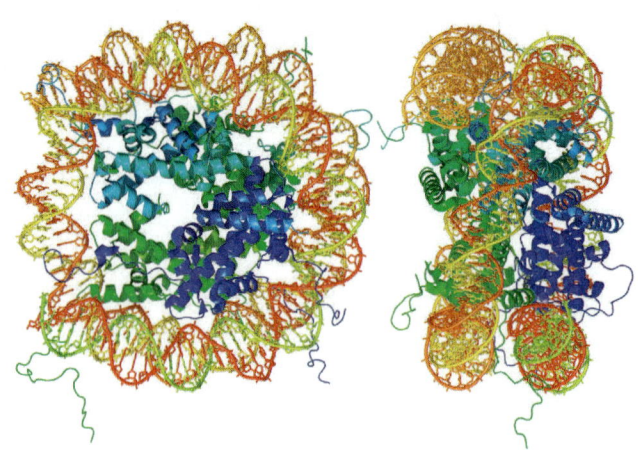

(b) Molecular model for nucleosome structure

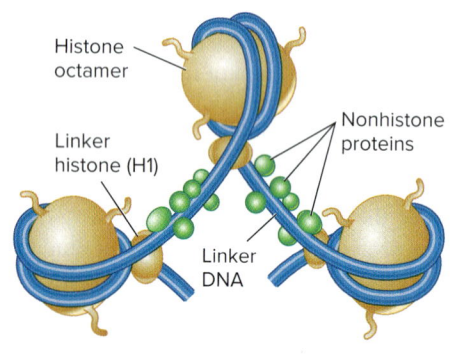

(c) Nucleosomes showing linker histones and nonhistone proteins

FIGURE 12.14 Nucleosome structure. **(a)** A nucleosome consists of 146 or 147 bp of DNA wrapped around an octamer of core histone proteins. **(b)** A model for the structure of a nucleosome as determined by X-ray crystallography. This drawing shows two views of a nucleosome that are at right angles to each other. **(c)** The linker region of DNA connects adjacent nucleosomes. The linker histone H1 and nonhistone proteins also bind to this linker region.
(b) ©Laguna Design/SPL/Science Source

Concept Check: What is the diameter of a nucleosome?

The Repeating Nucleosome Structure is Revealed by Digestion of the Linker Region

The model of nucleosome structure was originally proposed by Roger Kornberg in 1974. He based his proposal on several observations. Biochemical experiments had shown that chromatin contains a ratio of one molecule of each of the four core histone proteins (namely, H2A, H2B, H3, and H4) per 100 bp of DNA. Approximately one H1 protein was found per 200 bp of DNA. In addition, purified core histone proteins were observed to bind to each other via specific pairwise interactions. Subsequent X-ray diffraction studies showed that chromatin is composed of a repeating pattern of smaller units. Finally, electron microscopy of chromatin fibers revealed a diameter of approximately 11 nm. Taken together, these observations led Kornberg to propose a model in which the DNA double helix is wrapped around an octamer of core histone proteins. Including the linker region, this structure involves about 200 bp of DNA.

Markus Noll decided to test Kornberg's model by digesting chromatin with DNase I, an enzyme that cuts the DNA backbone. He then accurately measured the molecular mass of the DNA fragments by gel electrophoresis. Noll assumed that the linker region of DNA is more accessible to DNase I and, therefore, DNase I is more likely to make cuts in the linker region than in the 146-bp region that is tightly bound to the core histones. If this was correct, incubation with DNase I was expected to make cuts in the linker region, thereby producing DNA pieces approximately 200 bp in length. The size of the DNA fragments might vary somewhat because the linker region is not of constant length and because the cut within the linker region may occur at different sites.

Figure 12.15 describes Noll's experimental protocol. He began with nuclei from rat liver cells and incubated them with low, medium, or high concentrations of DNase I. The DNA was extracted into an aqueous phase and then loaded onto an agarose gel that separated the fragments according to their molecular mass. The DNA fragments within the gel were stained with a UV-sensitive dye, ethidium bromide, which made it possible to view the DNA fragments under UV illumination.

▶ **THE HYPOTHESIS**

The aim of this experiment was to test the beads-on-a-string model for chromatin structure. According to this model, DNase I should preferentially cut the DNA in the linker region, thereby producing DNA pieces that are about 200 bp in length.

12.6 STRUCTURE OF EUKARYOTIC CHROMOSOMES IN NONDIVIDING CELLS

▶ **TESTING THE HYPOTHESIS — FIGURE 12.15** DNase I cuts chromatin into repeating units containing 200 bp of DNA.
Starting material: Nuclei from rat liver cells.

1. Incubate the nuclei with low, medium, and high concentrations of DNase I. The conceptual level illustrates a low DNase I concentration.

2. Isolate the DNA. This involves dissolving the nuclear membrane with detergent and treating the sample with the organic solvent phenol.

3. Load the DNA into a well of an agarose gel and run the gel to separate the DNA pieces according to size. On this gel, also load DNA fragments of known molecular mass (marker lane).

4. Visualize the DNA fragments by staining the DNA with ethidium bromide, a dye that binds to DNA and is fluorescent when excited by UV light.

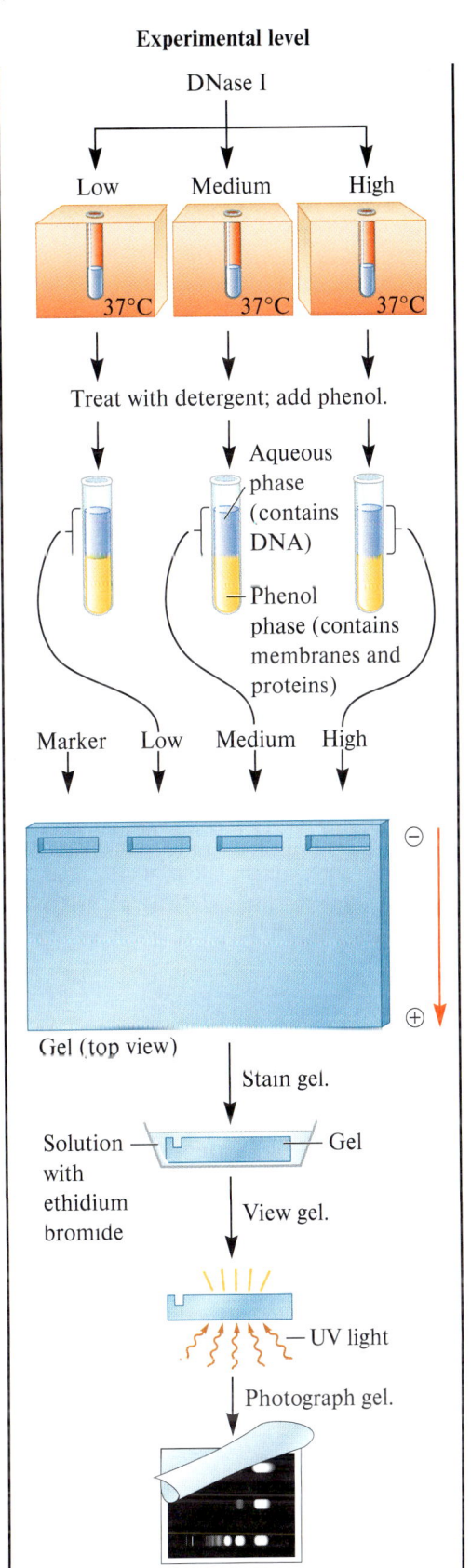

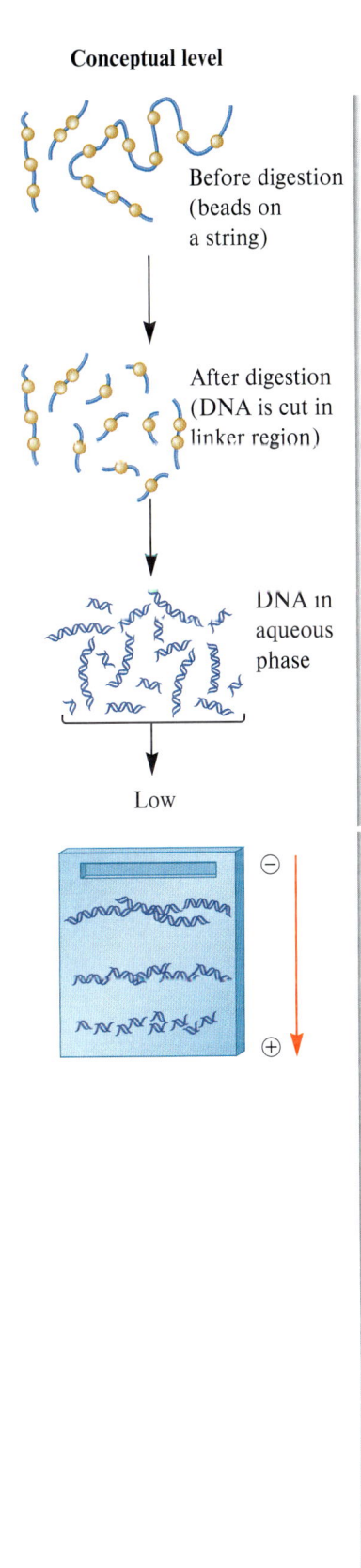

THE DATA

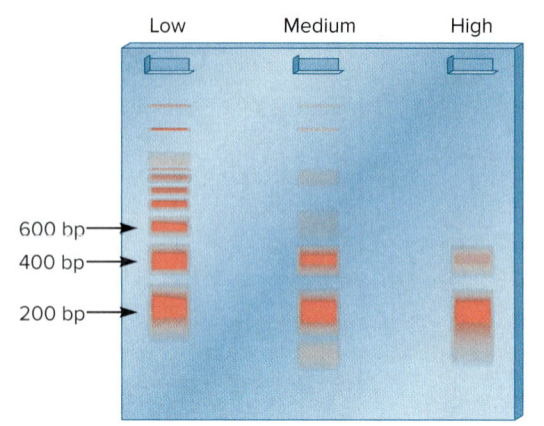

DNase concentration: 30 units mL^{-1} 150 units mL^{-1} 600 units mL^{-1}

Note: The marker lane is omitted from this drawing.

Source: Adapted from Noll, M. (1974) Subunit Structure of Chromatin, *Nature*, vol. 251, 249-251.

INTERPRETING THE DATA

As shown in the data, at a high DNase I concentration, the entire sample of chromosomal DNA was digested into fragments of approximately 200 bp in length. This result is predicted by the beads-on-a-string model. Furthermore, at a low or medium DNase I concentration, longer pieces were observed, and these were in multiples of 200 bp (400, 600, etc.). How do we explain these longer pieces? They occurred because occasional linker regions remained uncut at a low or medium DNase I concentration. For example, if one linker region was not cut, a DNA piece would contain two nucleosomes and be 400 bp in length. If two consecutive linker regions were not cut, a piece with three nucleosomes containing about 600 bp of DNA would result. Taken together, these results strongly supported the nucleosome model for chromatin structure.

Nucleosomes Become Closely Associated to Form a 30-nm Fiber

In eukaryotic chromatin, nucleosomes associate with each other to form a more compact structure that is 30 nm in diameter, known as the **30-nm fiber** (**Figure 12.16a**). The 30-nm fiber shortens the total length of DNA another sevenfold. The structure of the 30-nm fiber has proven difficult to determine, because the conformation of the DNA may be substantially altered when it is extracted from living cells. Most models for the structure of the 30-nm fiber fall into two main classes.

- The solenoid model suggests a helical structure in which contact between nucleosomes produces a symmetrically compact structure within the 30-nm fiber (**Figure 12.16b**). This type of model is still favored by some researchers in the field.
- An alternative model, the zigzag model, advocated by Rachel Horowitz, Christopher Woodcock, and others, is based on techniques such as cryoelectron microscopy (electron microscopy at low temperature). According to the zigzag model, linker regions within the 30-nm structure are variably bent and twisted, and little face-to-face contact occurs between nucleosomes (**Figure 12.16c**). At this level of compaction, the overall picture of chromatin that emerges is an irregular, fluctuating, three-dimensional zigzag structure with stable nucleosome units connected by deformable linker regions.

In 2005, Timothy Richmond and colleagues were the first to determine the crystal structure of a segment of DNA containing four nucleosomes. The structure revealed that the linker region of DNA zigzags back and forth between each nucleosome, a feature consistent with the zigzag model.

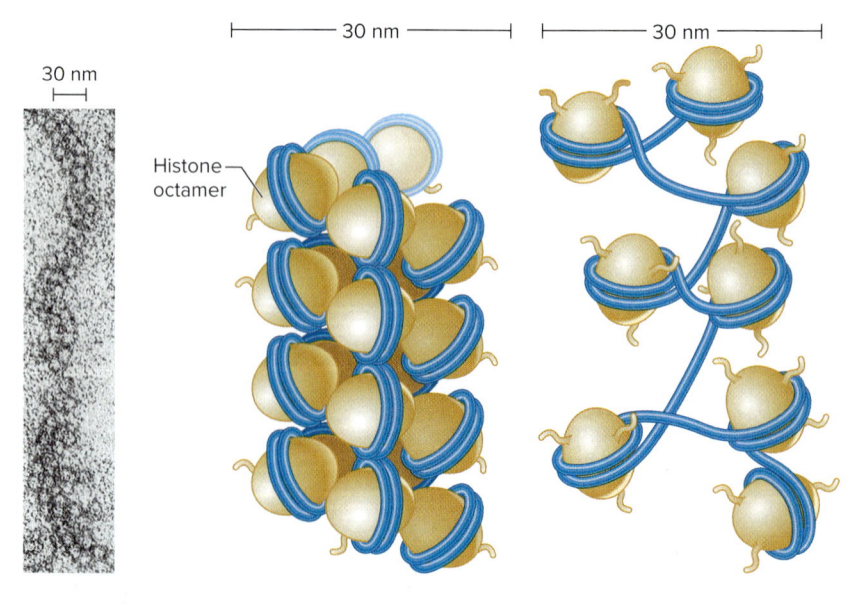

FIGURE 12.16 The 30-nm fiber. (a) A photomicrograph of the 30-nm fiber. (b) In the solenoid model, the nucleosomes are packed in a spiral configuration. (c) In the zigzag model, the linker DNA forms a more irregular structure, and less contact occurs between adjacent nucleosomes. The zigzag model is consistent with more recent data regarding chromatin conformation.
Courtesy of Jerome Rattner/University of Calgary

(a) Micrograph of a 30-nm fiber   (b) Solenoid model   (c) Zigzag model

Concept Check: Describe the distinguishing features of the solenoid and zigzag models.

12.6 STRUCTURE OF EUKARYOTIC CHROMOSOMES IN NONDIVIDING CELLS

Chromosomes Are Further Compacted by Anchoring of the 30-nm Fiber into Radial Loop Domains Along the Nuclear Matrix

Thus far, we have examined two mechanisms that compact eukaryotic DNA: the wrapping of DNA within nucleosomes and the arrangement of nucleosomes to form a 30-nm fiber. Taken together, these two processes shorten the DNA nearly 50-fold. A third level of compaction involves interactions between the 30-nm fibers and a filamentous network of proteins in the nucleus called the **nuclear matrix**, which consists of two parts (**Figure 12.17a–c**).

- The **nuclear lamina** is a collection of fibers that line the inner nuclear membrane. These fibers are composed of intermediate filament proteins.
- The **internal nuclear matrix** is connected to the nuclear lamina and fills the interior of the nucleus. The internal nuclear matrix, whose structure and functional role remain controversial, is hypothesized to be an intricate fine network of irregular protein fibers with many other proteins bound to them.

The proteins of the nuclear matrix are involved in compacting the DNA into **radial loop domains**, similar to the microdomains of the bacterial chromosome. During interphase, chromatin is organized into loops, often 25,000 to 200,000 bp in size, which are anchored to the nuclear matrix. The chromosomal DNA of eukaryotic species contains sequences called **matrix-attachment regions (MARs)**, which are also called scaffold-attachment regions (SARs). They are interspersed at regular intervals throughout the genome. The MARs bind to specific proteins in the nuclear matrix, thus forming chromosomal loops (**Figure 12.17d**).

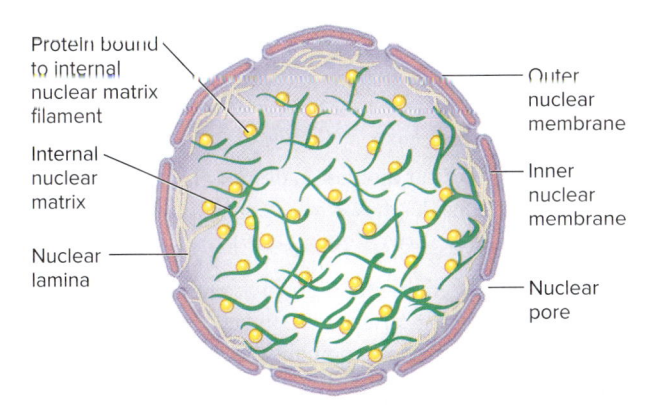

(a) Proteins that form the nuclear matrix

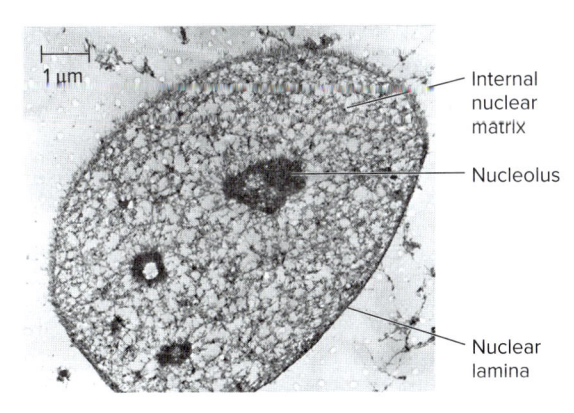

(b) Micrograph of nucleus with chromatin removed

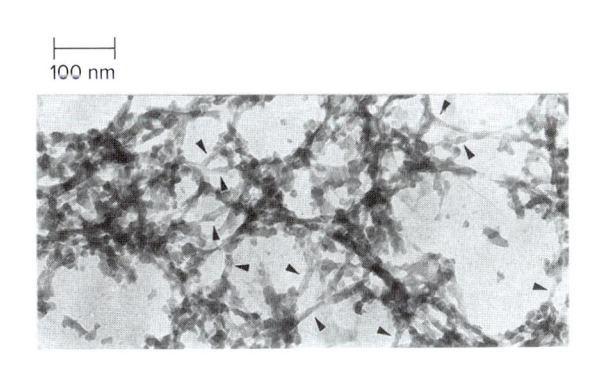

(c) Micrograph showing a close-up of nuclear matrix

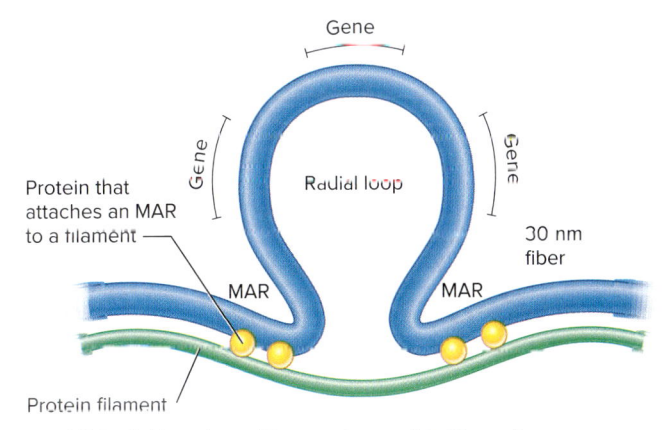

(d) Radial loop bound to a nuclear matrix filament

FIGURE 12.17 Structure of the nuclear matrix. (a) This schematic drawing shows the arrangement of the matrix within a cell nucleus. The nuclear lamina, which lines the inner nuclear membrane, is a collection of fibrous proteins that line the inner nuclear membrane. The internal nuclear matrix is composed of protein filaments (depicted in green) that may be interconnected. These fibers also have many other proteins bound to them (depicted in orange). (b and c) Micrographs of the matrix after the DNA has been removed. (d) The matrix-attachment regions (MARs), which contain a high percentage of A and T bases, bind to the nuclear matrix and create radial loops. Formation of the loops results in a further compaction of eukaryotic chromosomal DNA.

(b, c) Nickerson et al., "The nuclear matrix revealed by eluting chromatin from a cross-linked nucleus." PNAS, 94(9):4446-4450. Fig. 3 Copyright (1997) National Academy of Sciences, U.S.A. Jerome Rattner/University of Calgary

Concept Check: What is the function of the nuclear matrix?

258   CHAPTER 12 :: MOLECULAR STRUCTURE OF CHROMOSOMES AND TRANSPOSITION

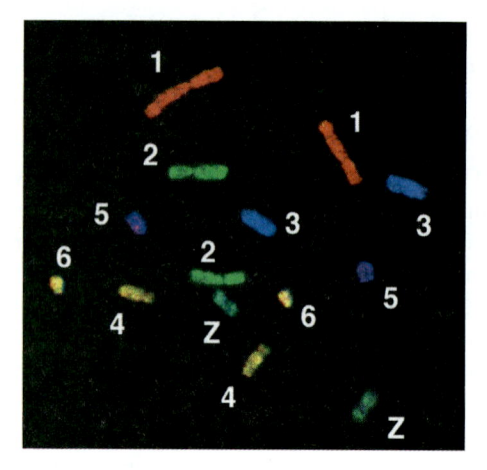

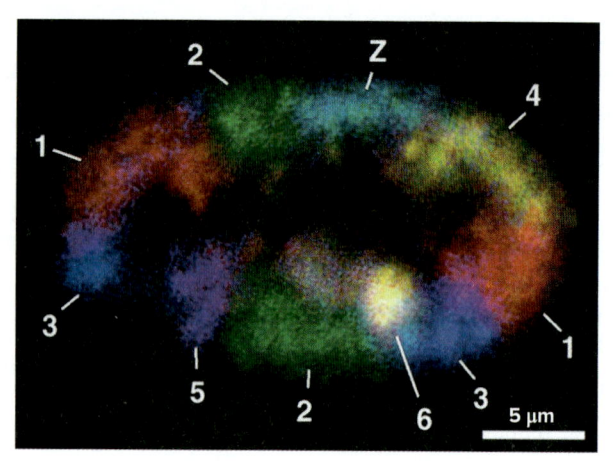

(a) Metaphase chromosomes   (b) Chromosomes in the cell nucleus during interphase

FIGURE 12.18 **Chromosome territories in the cell nucleus.** (a) Several metaphase chromosomes from chicken cells were labeled with chromosome-specific fluorescent molecules. Each of seven types of chicken chromosomes (i.e., 1, 2, 3, 4, 5, 6, and Z) is labeled a different color. (b) The same molecules were used to label interphase chromosomes in the cell nucleus. Each of these chromosomes occupies its own distinct, nonoverlapping territory within the cell nucleus. (Note: Chicken cells are diploid, with two copies of each chromosome. The Z chromosome is a sex chromosome.)
Courtesy of Felix Habermann and Irina Solovei, University of Munich (LMU, Biocenter)

Concept Check: What is a chromosome territory?

During Interphase, Each Chromosome Occupies Its Own Distinct Territory in the Cell Nucleus

Why is the attachment of radial loop domains to the nuclear matrix important? In addition to being involved in compaction, the nuclear matrix serves to organize the chromosomes within the nucleus. Each chromosome in the cell nucleus is located in a distinct **chromosome territory.** As shown in studies by Thomas Cremer, Christoph Cremer, and others, these territories can be viewed when interphase cells are exposed to multiple fluorescent molecules that recognize specific sequences on particular chromosomes. **Figure 12.18** illustrates an experiment in which chicken cells were exposed to a mixture of fluorescent molecules that recognize specific sites along several of the larger chromosomes found in this species (*Gallus gallus*). Figure 12.18a shows the chromosomes in metaphase. The fluorescent molecules label each type of metaphase chromosome with a different color. Figure 12.18b shows the use of the same labeling molecules during interphase, when the chromosomes are less condensed and found in the cell nucleus. As seen here, each chromosome occupies its own distinct territory.

Before ending our discussion of interphase chromosome compaction, let's consider how the compaction level of interphase chromosomes may vary (**Figure 12.19**).

- Less compacted regions, known as **euchromatin,** are usually capable of gene transcription. In euchromatin, the 30-nm fiber forms radial loop domains, as shown in Figure 12.17d.
- In **heterochromatin,** these radial loop domains become compacted even further. These more compacted regions of the chromosome are usually transcriptionally inactive.
- A typical eukaryotic chromosome contains regions of heterochromatin and regions of euchromatin during interphase. Heterochromatin is most abundant in the centromeric regions of the chromosome and, to a lesser extent, in the telomeric regions.

12.6 REVIEWING THE KEY CONCEPTS

- The term *chromatin* refers to the DNA-protein complex found within eukaryotic chromosomes.
- Eukaryotic DNA wraps around an octamer of core histone proteins to form a nucleosome (see Figure 12.14).

Telomere   Centromere   Telomere

Euchromatin (30-nm fiber in radial loops)   Heterochromatin (greater compaction of the radial loops)

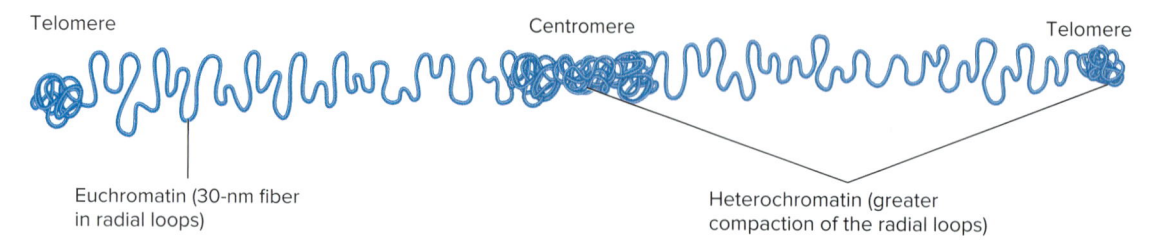

FIGURE 12.19 **Chromatin structure during interphase.** Heterochromatic regions are more highly condensed and tend to be localized in centromeric and telomeric regions.

Concept Check: Would you expect to find active genes in regions of heterochromatin or euchromatin?

- Noll tested Kornberg's nucleosome model by digesting eukaryotic chromatin with varying concentrations of DNase I (see Figure 12.15).
- Nucleosomes are arranged to form a 30-nm fiber. Solenoid and zigzag models of its structure have been proposed (see Figure 12.16).
- Chromatin is further compacted by the attachment of 30-nm fibers to protein filaments to form radial loop domains (see Figure 12.17).
- Within the cell nucleus, each eukaryotic chromosome occupies its own unique chromosome territory (see Figure 12.18).
- In nondividing cells, each chromosome has highly compacted regions called heterochromatin and less compacted regions called euchromatin (see Figure 12.19).

12.6 COMPREHENSION QUESTIONS

1. What are the components of a single nucleosome?
   a. About 146 bp of DNA and four core histone proteins
   b. About 146 bp of DNA and eight core histone proteins
   c. About 200 bp of DNA and four core histone proteins
   d. About 200 bp of DNA and eight core histone proteins

2. In Noll's experiment to test the beads-on-a-string model, exposure of nuclei from rat liver cells to a low concentration of DNase I resulted in
   a. a single band of DNA with a size of approximately 200 bp.
   b. several bands of DNA in multiples of 200 bp.
   c. a single band of DNA with a size of 100 bp.
   d. several bands of DNA in multiples of 100 bp.

3. With regard to the 30-nm fiber, a key difference between the solenoid and zigzag models is
   a. the solenoid model suggests a helical structure.
   b. the zigzag model suggests a more irregular pattern of nucleosomes.
   c. the zigzag model does not include nucleosomes.
   d. Both a and b describe key differences between the models.

4. A chromosome territory is a region
   a. along a chromosome where many genes are clustered.
   b. along a chromosome where the nucleosomes are close together.
   c. in a cell nucleus where a single chromosome is located.
   d. in a cell nucleus where multiple chromosomes are located.

12.7 STRUCTURE OF EUKARYOTIC CHROMOSOMES DURING CELL DIVISION

Learning Outcome:
1. Describe the levels of compaction that lead to a metaphase chromosome.

As described in Chapter 2, when eukaryotic cells prepare to divide, the chromosomes become very condensed or compacted. This aids in their proper sorting during M phase. The highly compacted chromosomes that are viewed during metaphase of M phase are called metaphase chromosomes. **Figure 12.20** illustrates the levels of compaction that lead to a metaphase chromosome.

- The first level of compaction involves the formation of nucleosomes.
- The nucleosomes interact in a three-dimensional zigzag or solenoid structure to form a 30-nm fiber.
- The 30-nm fibers form radial loop domains by anchoring to the protein filaments of the nuclear matrix. The average distance that loops radiate from this protein scaffold is approximately 300 nm.
- This structure is further compacted via additional folding of the radial loop domains and the protein scaffold. This compaction greatly shortens the overall length of a chromosome and produces a diameter of approximately 700 nm. Two parallel chromatids have a larger diameter of approximately 1400 nm but a much shorter length compared with interphase chromosomes.

Highly condensed metaphase chromosomes undergo little gene transcription because it is difficult for transcription proteins to gain access to the compacted DNA. Therefore, most transcriptional activity ceases during M phase, although a few specific genes may be transcribed. M phase is usually a short period of the cell cycle.

In highly condensed chromosomes, such as those found in metaphase, the radial loops are highly compacted and remain anchored to the protein scaffold. **Figure 12.21a** shows a human metaphase chromosome. In this condition, the radial loops of DNA are in a very compact configuration. If this chromosome is treated with a high concentration of salt to remove both the core and linker histones, the highly compact configuration is lost, but the bottoms of the elongated loops remain attached to the scaffold composed of nonhistone proteins. In **Figure 12.21b**, an arrow points to an elongated DNA loop emanating from the darkly staining scaffold. Remarkably, the scaffold retains the shape of the original metaphase chromosome even though the DNA loops have become greatly elongated. These results illustrate that the structure of metaphase chromosomes is determined by the nuclear matrix proteins, which form the scaffold, and by the histones, which are needed to compact the radial loops.

12.7 REVIEWING THE KEY CONCEPTS
- Chromatin compaction occurs due to the formation of nucleosomes, followed by the formation of the 30nm-fiber and then radial loop domains. A metaphase chromosome is highly compacted due to the further compaction of radial loops (see Figure 12.20).
- Both histone and nonhistone proteins are important for the compaction of metaphase chromosomes (see Figure 12.21).

12.7 COMPREHENSION QUESTION
1. The compaction leading to a metaphase chromosome involves which of the following?
   a. The formation of nucleosomes
   b. The formation of the 30-nm fiber
   c. Anchoring and further compaction of the radial loops
   d. All of the above

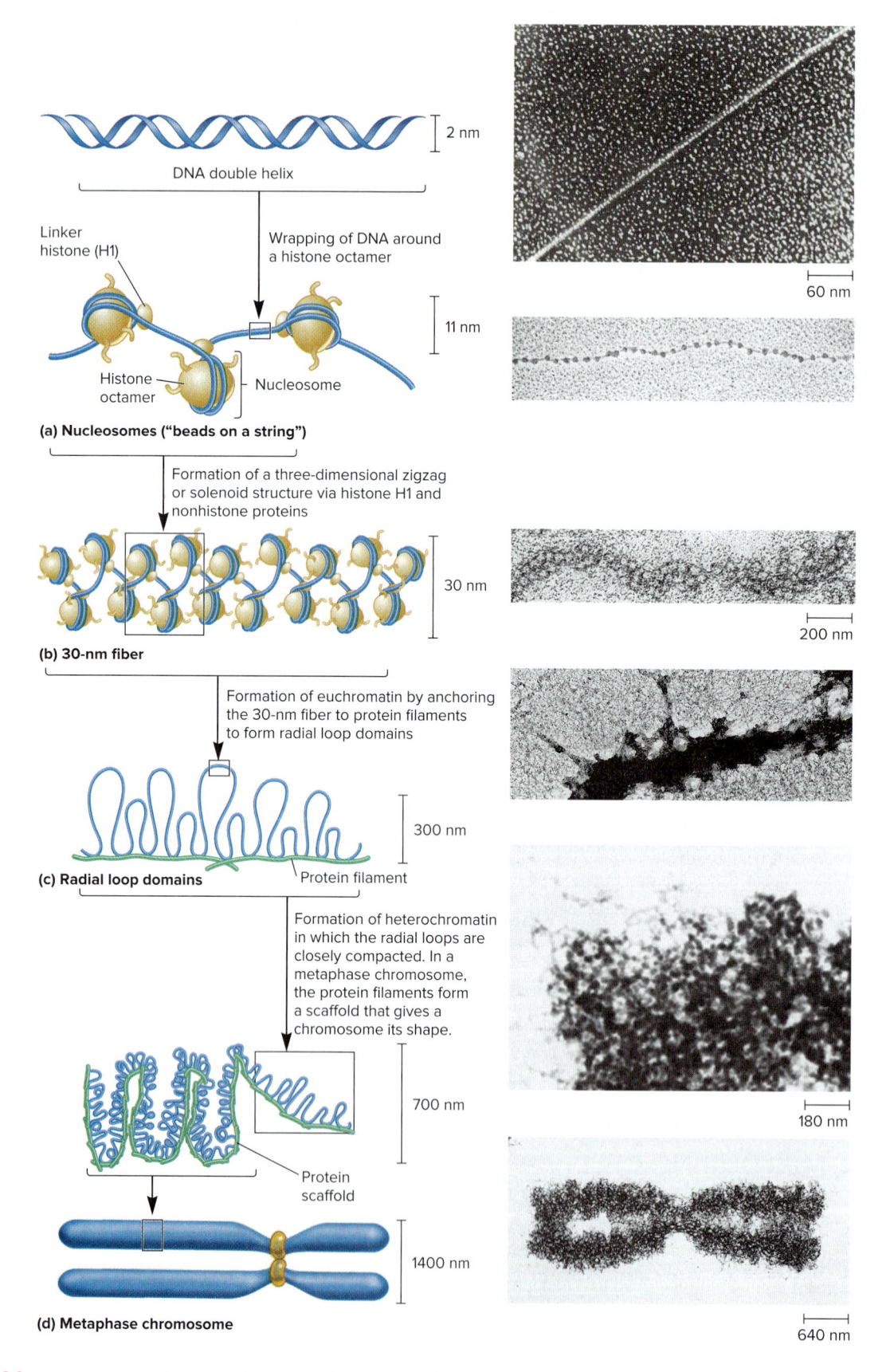

260 CHAPTER 12 :: MOLECULAR STRUCTURE OF CHROMOSOMES AND TRANSPOSITION

FIGURE 12.20 The steps in eukaryotic chromosomal compaction leading to the metaphase chromosome.

(a) ©Dr. Barbara A. Hamkalo; (b) Courtesy of Jerome Rattner/University of Calgary; (c) ©Dr. James Paulson, Ph.D.; (d) Courtesy of Peter Engelhardt/Department of Virology, Haartman Institute

Concept Check: Describe the structural change that converts a region that is 300 nm in diameter to one that is 700 nm in diameter.

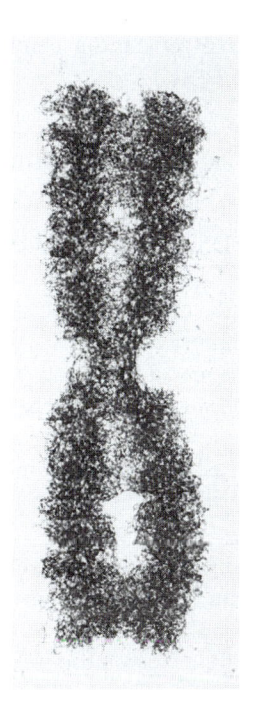

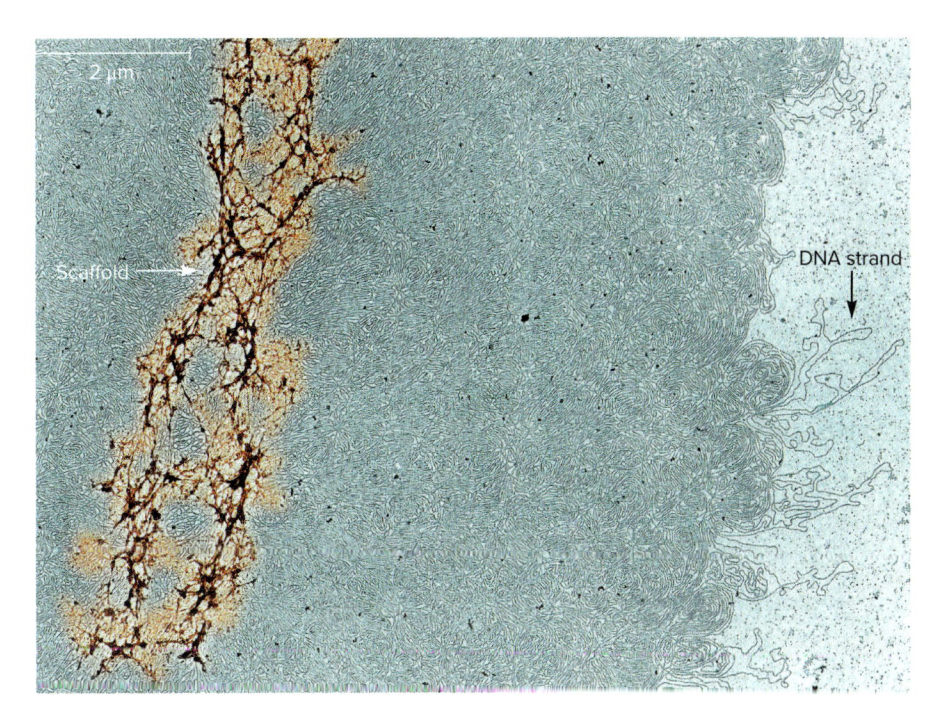

(a) Metaphase chromosome   (b) Metaphase chromosome treated with high salt to remove histone proteins

FIGURE 12.21 The importance of histone proteins and scaffolding proteins in the compaction of eukaryotic chromosomes. (a) A metaphase chromosome. (b) A metaphase chromosome following treatment with a highly concentrated salt solution to remove the histone proteins. The arrow on the left points to the scaffold (composed of nonhistone proteins), which anchors the bases of the radial loops. The arrow on the right points to an elongated strand of DNA.
(a) Courtesy of Peter Engelhardt/Department of Virology, Haartman Institute; (b) ©Don W. Fawcett/Science Source

KEY TERMS

Page 240. chromosomes, genome
Page 241. protein-encoding genes, structural genes, intergenic regions, origin of replication, repetitive sequences, nucleoid
Page 242. microdomains, nucleoid-associated proteins (NAPs), DNA supercoiling
Page 243. topoisomers, DNA gyrase, topoisomerase I
Page 245. introns, centromeres
Page 246. kinetochore, telomeres, repetitive sequences
Page 247. sequence complexity, moderately repetitive sequences, transposable elements (TE), highly repetitive sequences, tandem array
Page 248. simple transposition, transposons, retrotransposition, retrotransposons, retroelements, direct repeats (DRs), target-site duplications, insertion element (IS element), inverted repeats (IRs), transposase, simple transposon
Page 249. LTR retrotransposons, long terminal repeats (LTRs), non-LTR retrotransposons, autonomous elements, nonautonomous element
Page 250. reverse transcriptase, integrase
Page 252. selfish DNA hypothesis, exon shuffling, hybrid dysgenesis
Page 253. nucleus, chromatin, nucleosome, histone proteins
Page 256. 30-nm fiber
Page 257. nuclear matrix, nuclear lamina, internal nuclear matrix, radial loop domains, matrix-attachment regions (MARs)
Page 258. chromosome territory, euchromatin, heterochromatin

CHAPTER SUMMARY

- Chromosomes contain the genetic material, which is DNA. The complete genetic complement of a cell or species is called a genome.

12.1 Organization of Functional Sites Along Bacterial Chromosomes

- Bacterial chromosomes are typically circular and have an origin of replication and a few thousand genes (see Figure 12.1).

12.2 Structure of Bacterial Chromosomes

- A bacterial chromosome is found in a nucleoid of a bacterial cell (see Figure 12.2).
- Bacterial chromosomes are made more compact by the formation of microdomains and macrodomains and by DNA supercoiling (see Figures 12.3, 12.4).
- Negative DNA supercoiling can promote DNA strand separation (see Figure 12.5).

- DNA gyrase (topoisomerase II) is a bacterial enzyme that introduces negative supercoils.
- Topoisomerase I relaxes negative supercoils.

12.3 Organization of Functional Sites Along Eukaryotic Chromosomes

- Eukaryotic chromosomes are usually linear and contain multiple origins of replication, a centromere, telomeres, repetitive sequences, and many genes (see Figure 12.6).

12.4 Sizes of Eukaryotic Genomes and Repetitive Sequences

- The genome sizes of eukaryotes vary greatly. Some of this variation is due to the accumulation of repetitive sequences (see Figure 12.7).
- The human genome contains about 41% unique DNA sequences and 59% repetitive DNA sequences (see Figure 12.8).

12.5 Transposition

- Transposable elements can move via simple transposition (cut-and-paste) and retrotransposition (see Figure 12.9).
- Each type of transposable element has its own pattern of DNA sequences, which includes direct repeats (see Figure 12.10).
- Transposase catalyzes the excision and insertion of transposons, which move via a DNA intermediate (see Figure 12.11).
- Simple transposition can increase the copy number of a transposon if it occurs just after a transposon has been replicated (see Figure 12.12).
- Retrotransposition of LTR retrotransposons occurs via reverse transcriptase and integrase (see Figure 12.13).
- Many different transposons are found among living organisms. Their abundance varies among different species (see Tables 12.1, 12.2).
- Transposition can have a variety of effects on chromosome structure and gene expression (see Table 12.3).

12.6 Structure of Eukaryotic Chromosomes in Nondividing Cells

- The term *chromatin* refers to the DNA-protein complex found within eukaryotic chromosomes.
- Eukaryotic DNA wraps around an octamer of core histone proteins to form a nucleosome (see Figure 12.14).
- Noll tested Kornberg's nucleosome model by digesting eukaryotic chromatin with varying concentrations of DNase I (see Figure 12.15).
- Nucleosomes are further compacted to form a 30-nm fiber. Solenoid and zigzag models of its structure have been proposed (see Figure 12.16).
- Chromatin is further compacted by the attachment of 30-nm fibers to protein filaments to form radial loop domains (see Figure 12.17).
- Within the cell nucleus, each eukaryotic chromosome occupies its own unique chromosome territory (see Figure 12.18).
- In nondividing cells, each chromosome has highly compacted regions called heterochromatin and less compacted regions called euchromatin (see Figure 12.19).

12.7 Structure of Eukaryotic Chromosomes During Cell Division

- Chromatin compaction occurs due to the formation of nucleosomes, followed by the formation of a 30-nm fiber and radial loop domains. A metaphase chromosome is highly compacted due to the further compaction of radial loops (see Figure 12.20).
- Both histone and nonhistone proteins are important for the compaction of metaphase chromosomes (see Figure 12.21).

PROBLEM SETS & INSIGHTS

More Genetic TIPS

1. Suppose that a bacterial DNA molecule is given a left-handed twist. How does this affect the structure and function of the DNA?

Topic: What topic in genetics does this question address?

The topic is DNA supercoiling. More specifically, the question is about the effects of giving a DNA molecule a left-handed twist.

Information: What information do you know based on the question and your understanding of the topic?

From the question, you know that a DNA molecule has been given a left-handed twist. From your understanding of the topic, you may remember that DNA forms a right-handed double helix.

Problem-Solving Strategy: Relate structure and function.

One strategy to solve this problem is to begin with DNA structure. Because DNA is right-handed, a left-handed twist could have either of two potential effects. It could add a negative supercoil, or it could promote DNA strand separation. With regard to function, negative supercoiling makes the DNA more compact, and strand separation makes the DNA strands more accessible.

Answer: Negative supercoiling makes the bacterial chromosome more compact, so it fits better within the cell. Alternatively, a left-handed twist promotes strand separation and thereby enhances DNA functions such as replication and transcription.

2. Describe the differences beween unique DNA and highly repetitive sequences in DNA.

Topic: What topic in genetics does this question address?

The topic is the complexity of DNA sequences. More specifically, the question is about the differences between unique and highly repetitive sequences.

Information: What information do you know based on the question and your understanding of the topic?

From the question, you know that some sequences in DNA are unique and some are highly repetitive. From your understanding of the topic, you may remember that unique DNA occurs once per haploid genome, whereas highly repetitive DNA occurs multiple times.

Problem-Solving Strategy: Compare and contrast.

To answer this question, you can compare and contrast the features of unique and highly repetitive sequences.

Answer: Unique DNA occurs once per haploid genome. Many genes in a genome are unique. By comparison, a highly repetitive sequence as its name suggests, is repeated many times, from tens of thousands to millions of times throughout a genome. It can be interspersed in the genome or found clustered in a tandem array, in which a short nucleotide sequence is repeated many times in a row.

3. To hold bacterial DNA in a more compact configuration, specific proteins must bind to the DNA and stabilize its conformation. Several different proteins are involved in this process. Some of these proteins, such as H-NS, have been referred to as "histone-like" because of their functional similarity to the histone proteins found in eukaryotes. Based on your knowledge of eukaryotic histone proteins, what biochemical properties would you expect bacterial histone-like proteins to have?

Topic: What topic in genetics does this question address?

The topic is DNA compaction. More specifically, the question is asking you to predict how bacterial proteins could make a bacterial chromosome more compact.

Information: What information do you know based on the question and your understanding of the topic?

From the question, you know that bacterial chromosomes have histone-like proteins. From your understanding of eukaryotic chromosomes, you may recall that the negatively charged DNA backbone wraps around positively charged core histone proteins.

Problem-Solving Strategy: Relate structure and function.

One strategy to solve this problem is to consider the structures of nucleosomes, in which DNA is wrapped around core histone proteins.

Answer: The histone-like proteins have the properties expected for proteins involved in DNA folding. They are all small proteins found in relative abundance within the bacterial cell. In some cases, the histone-like proteins are biochemically similar to eukaryotic histones. For example, they tend to be basic (positively charged) and bind to the negatively charged DNA backbone in a non-sequence-dependent fashion.

4. If you assume the average length of a DNA linker region is 50 bp, approximately how many nucleosomes could be found in the haploid human genome, which contains 3 billion bp?

Topic: What topic in genetics does this question address?

The topic is chromosome structure. More specifically, the question is about the number of nucleosomes in a haploid human genome.

Information: What information do you know based on the question and your understanding of the topic?

From the question, you know that the haploid human genome has about 3 billion bp, and you are asked to assume that the average length of a DNA linker is 50 bp. From your understanding of the topic, you may remember that 146 bp of DNA is found within one nucleosome.

Problem-Solving Strategy: Make a calculation.

The repeating unit is a nucleosome with 146 bp of DNA plus a linker region with 50 bp, for a total of 196 bp. To determine the maximum number of nucleosomes, you divide 3 billion by 196.

Answer: $3,000,000,000/196 = 15,306,122$, or about 15.3 million. Note: This is the maximum number. The actual number is somewhat less because a small percentage of eukaryotic DNA occurs in nucleosome-free regions, a topic discussed in Chapter 17.

Conceptual Questions

C1. What is a bacterial nucleoid? With regard to cellular membranes, what is the difference between a bacterial nucleoid and a eukaryotic nucleus?

C2. In Part II of this text, we considered inheritance patterns for diploid eukaryotic species. Bacterial cells frequently contain two or more nucleoids. With regard to genes and alleles, how is a bacterium that contains two nucleoids similar to a diploid eukaryotic cell, and how is it different?

C3. Describe the mechanisms by which bacterial DNA becomes compacted.

C4. Why is DNA supercoiling called supercoiling rather than just coiling? Why is positive supercoiling called overwinding and negative supercoiling called underwinding? How would you define the terms *positive* and *negative supercoiling* for Z DNA (described in Chapter 11)?

C5. Coumarins and quinolones are two classes of drugs that inhibit bacterial growth by directly inhibiting DNA gyrase. Discuss two reasons why inhibiting DNA gyrase might inhibit bacterial growth.

C6. Take two pieces of string that are each approximately 10 inches long, and create a double helix by wrapping them around each other to make 10 complete turns. Tape one end of the strings to a table, and then twist the strings three times (360° each time) in a right-handed direction. Note: As you are looking down at the strings from above, a right-handed twist is in the clockwise direction.

    A. Did the three twists create more or fewer turns in your double helix? How many turns does your double helix have after you twisted it?

    B. Is your double helix right-handed or left-handed? Explain your answer.

    C. Did the three twists create any supercoils?

    D. If you had coated your double helix with rubber cement and allowed the cement to dry before making the three additional right-handed twists, would the rubber cement make it more or less likely that the twists would create supercoiling? Would a pair of cemented strings be more or less like a real DNA double helix than an uncemented pair of strings? Explain your answer.

C7. Try to explain the function of DNA gyrase with a drawing.

C8. How are two topoisomers different from each other? How are they the same?

C9. On rare occasions, a chromosome can suffer a small deletion that removes the centromere. When this occurs, the chromosome usually is not found within subsequent daughter cells. Explain why a chromosome without a centromere is not transmitted very efficiently from mother to daughter cells. (Note: If a chromosome is located outside the nucleus after telophase, it is degraded.)

C10. What is the function of a centromere? At what stage of the cell cycle would you expect the centromere to be most important?

C11. Why does transposition produce direct repeats in the chromosomal DNA?

C12. Which types of TEs have the greatest potential for proliferation: insertion elements, simple transposons, or retrotransposons? Explain your choice.

C13. Do you consider TEs to be mutagens? Explain.

C14. This chapter describes different types of TEs: insertion elements, simple transposons, LTR retrotransposons, and non-LTR retrotransposons. Which of these four types of TEs have the following features?

   A. Require reverse transcriptase to transpose
   B. Require transposase to transpose
   C. Are flanked by direct repeats
   D. Have inverted repeats

C15. What features distinguish a transposon from a retrotransposon? How are their sequences different, and how are their mechanisms of transposition different?

C16. What is the difference between an autonomous and a nonautonomous TE? Is it possible for nonautonomous TEs to move? If it is possible, explain how.

C17. The occurrence of multiple transposons within the genome of organisms has been suggested as a possible cause of chromosomal rearrangements such as deletions, duplications, translocations, and inversions. How could the occurrence of transposons promote these kinds of structural rearrangements?

C18. Describe the structures of a nucleosome and a 30-nm fiber.

C19. Beginning with the G_1 phase of the cell cycle, describe the level of compaction of the eukaryotic chromosome. How does the level of compaction change as the cell progresses through the cell cycle? Why is it necessary to further compact the chromatin during mitosis?

C20. Draw a picture depicting the binding between the nuclear matrix and a MAR.

C21. Compare heterochromatin and euchromatin. What are the differences between them?

C22. Compare the structure and cell localization of chromosomes during interphase and M phase.

C23. What types of genetic activities occur during interphase? Explain why these activities cannot occur during M phase.

C24. Let's assume the linker region of DNA averages 54 bp in length. How many molecules of H2A would you expect to find in a DNA sample that is 46,000 bp in length?

C25. What are the roles of the core histone proteins versus the role of histone H1 in the compaction of eukaryotic DNA?

C26. A typical eukaryotic chromosome found in humans has about 100 million bp of DNA. As noted in Chapter 11, 1 DNA base pair has a linear length of 0.34 nm.

   A. What is the linear length of the DNA in a typical human chromosome in micrometers?
   B. What is the linear length of a 30-nm fiber of a typical human chromosome?
   C. Based on your calculation in part B, would a typical human chromosome fit inside the nucleus (with a diameter of 5 μm) if the 30-nm fiber were stretched out in a linear manner? If not, explain how a typical human chromosome fits inside the nucleus during interphase.

C27. Discuss the differences between the compaction levels of metaphase chromosomes and those of interphase chromosomes. When would you expect gene transcription and DNA replication to take place, during M phase or interphase? Explain why.

Application and Experimental Questions

E1. Two circular DNA molecules, molecule A and molecule B, are topoisomers of each other. When viewed under the electron microscope, molecule A appears more compact than molecule B. The level of gene transcription is much lower for molecule A. Which of the following three possibilities could account for these observations?

   First possibility: Molecule A has three positive supercoils, and molecule B has three negative supercoils.

   Second possibility: Molecule A has four positive supercoils, and molecule B has one negative supercoil.

   Third possibility: Molecule A has zero supercoils, and molecule B has three negative supercoils.

E2. Let's suppose that you have isolated DNA from a cell and have viewed it under a microscope. It looks supercoiled. What experiment would you perform to determine if it is positively or negatively supercoiled? In your answer, describe your expected results. You may assume that you have purified topoisomerases at your disposal.

E3. We seem to know more about the structure of eukaryotic chromosomal DNA than bacterial DNA. Discuss why you think this is so, and list several experimental procedures that have yielded important information concerning the compaction of eukaryotic chromatin.

E4. When chromatin is treated with a salt solution of moderate concentration, the linker histone H1 is removed. A higher salt concentration removes the rest of the histone proteins (see Figure 12.21b). If the experiment presented in Figure 12.15 were carried out after the DNA was treated with a moderately or highly concentrated salt solution, what would be the expected results?

E5. Let's suppose you have isolated chromatin from some bizarre eukaryote that has a DNA linker region that is usually 300 to 350 bp in length. The nucleosome structure is the same as in other eukaryotes. If you digested this eukaryotic organism's chromatin with a high concentration of DNase I, what would be your expected results?

E6. If you were given a sample of chromosomal DNA and asked to determine if it is bacterial or eukaryotic, what experiment would you perform, and what would be your expected results?

E7. Consider how histone proteins bind to DNA and then explain why a salt solution with a high concentration can remove them from DNA (as shown in Figure 12.21b).

E8. In Chapter 21, the technique of fluorescence in situ hybridization (FISH) is described. This is another method for examining sequence complexity within a genome. With this technique, a particular DNA sequence, such as a particular gene sequence, can be detected within an intact chromosome by using a fluorescently labeled DNA molecule that is complementary to the sequence. For example, let's consider the β-globin gene, which is found on human chromosome 11. A labeled DNA molecule complementary to the β-globin gene binds to that gene and shows up as a brightly colored spot on human chromosome 11. In this way, researchers can detect where the β-globin gene is located within a set of chromosomes. Because the β-globin gene is unique and because human cells are diploid (i.e., have two copies of each chromosome), a FISH experiment shows two bright spots per cell; the labeled DNA molecule binds to each copy of chromosome 11. What would you expect to see if you used the following types of labeled DNA molecules?

A. A labeled DNA molecule complementary to the *Alu* sequence

B. A labeled DNA molecule complementary to a tandem array near the centromere of the X chromosome

Questions for Student Discussion/Collaboration

1. Bacterial and eukaryotic chromosomes are very compact. Discuss the advantages and disadvantages of a compact chromosomal structure.

2. The prevalence of highly repetitive sequences seems rather strange to many geneticists. Do they seem strange to you? Why or why not? Discuss whether or not you think they have an important function.

3. Discuss and make a list of the similarities and differences between bacterial and eukaryotic chromosomes.

Answers to Comprehension Questions

12.1: d

12.2: d, b, d

12.3: e

12.4: a

12.5: c, d, b

12.6: b, b, d, c

12.7: d

Note: All answers appear in Connect; the answers to even-numbered questions and all Concept Check questions are in Appendix B.

13

CHAPTER OUTLINE

- **13.1** Structural Overview of DNA Replication
- **13.2** Bacterial DNA Replication: The Formation of Two Replication Forks at the Origin of Replication
- **13.3** Bacterial DNA Replication: Synthesis of New DNA Strands
- **13.4** Bacterial DNA Replication: Chemistry and Accuracy
- **13.5** Eukaryotic DNA Replication
- **13.6** Homologous Recombination

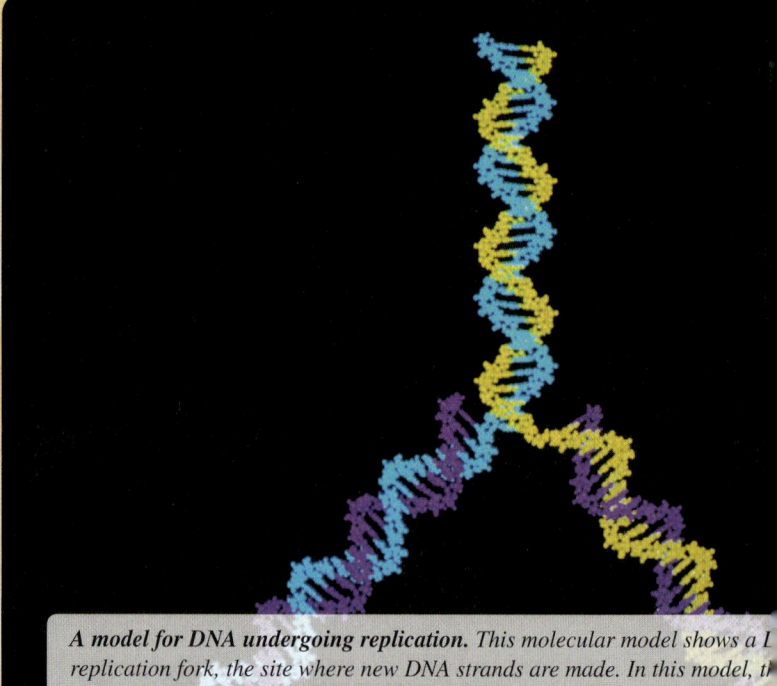

A model for DNA undergoing replication. This molecular model shows a D replication fork, the site where new DNA strands are made. In this model, t original DNA is yellow and blue. The newly made strands are purple.
©Clive Freeman/Science Source

DNA REPLICATION AND RECOMBINATION

As discussed throughout Chapters 2 to 10, genetic material is transmitted from parents to offspring and from cell to cell. For transmission to occur, the genetic material must be copied. During this process, known as **DNA replication,** the original DNA strands are used as templates for the synthesis of new DNA strands. We will begin this chapter with a consideration of the structural features of the double helix that underlie the replication process. Next, we will examine how chromosomes are replicated within living cells, addressing the following questions: Where does DNA replication begin, how does it proceed, and where does it end? We will first consider how DNA replication occurs within bacterial cells, and then turn our attention to the unique features of the replication of eukaryotic DNA. At the molecular level, it is rather remarkable that the replication of chromosomal DNA occurs very quickly, very accurately, and at the appropriate time in the life of the cell. For this to happen, many proteins play vital roles.

In this chapter, we will examine the mechanism of DNA replication and consider the functions of several proteins involved in the process. We will also explore the mechanism of homologous recombination, in which segments of DNA are swapped between chromosomes. Homologous recombination occurs when chromosomes crossover during meiosis (refer back to Chapter 2).

13.1 STRUCTURAL OVERVIEW OF DNA REPLICATION

Learning Outcomes:

1. Describe the structural features of DNA that enable it to be replicated.
2. Analyze the experiment of Meselson and Stahl, and explain how the results were consistent with the semiconservative model of DNA replication.

Because they bear directly on the replication process, let's begin by recalling two important structural features of DNA from Chapter 11. First, DNA is a double helix composed of two DNA strands. Second, the bases in the two DNA strands bind to each other in a very specific way. Adenine (A) in one DNA strand hydrogen bonds with thymine (T) in the opposite strand, and guanine (G) hydrogen bonds with cytosine (C). This structural feature known as the AT/GC rule is the basis for the complementarity of the base sequences

13.1 STRUCTURAL OVERVIEW OF DNA REPLICATION

in double-stranded DNA. In this section, we will consider how the structure of the DNA double helix provides the basis for DNA replication.

Existing DNA Strands Act as Templates for the Synthesis of New Strands

As shown in **Figure 13.1a**, DNA replication relies on the complementarity of DNA strands, based on the AT/GC rule.

- During the replication process, the two complementary strands of DNA come apart and serve as **template strands,** or **parental strands,** for the synthesis of two new strands of DNA.

- After the double helix has separated, individual nucleotides have access to the template strands. Hydrogen bonding between individual nucleotides and the template strands must obey the AT/GC rule.
- To complete the replication process, a covalent bond is formed between the phosphate of one nucleotide and the sugar of the previous nucleotide. The two newly made strands are referred to as the **daughter strands.**
- The base sequences are identical in both double-stranded molecules after replication (**Figure 13.1b**). That is, DNA is replicated in such a way that both copies retain the same information—the same base sequence—as in the original molecule.

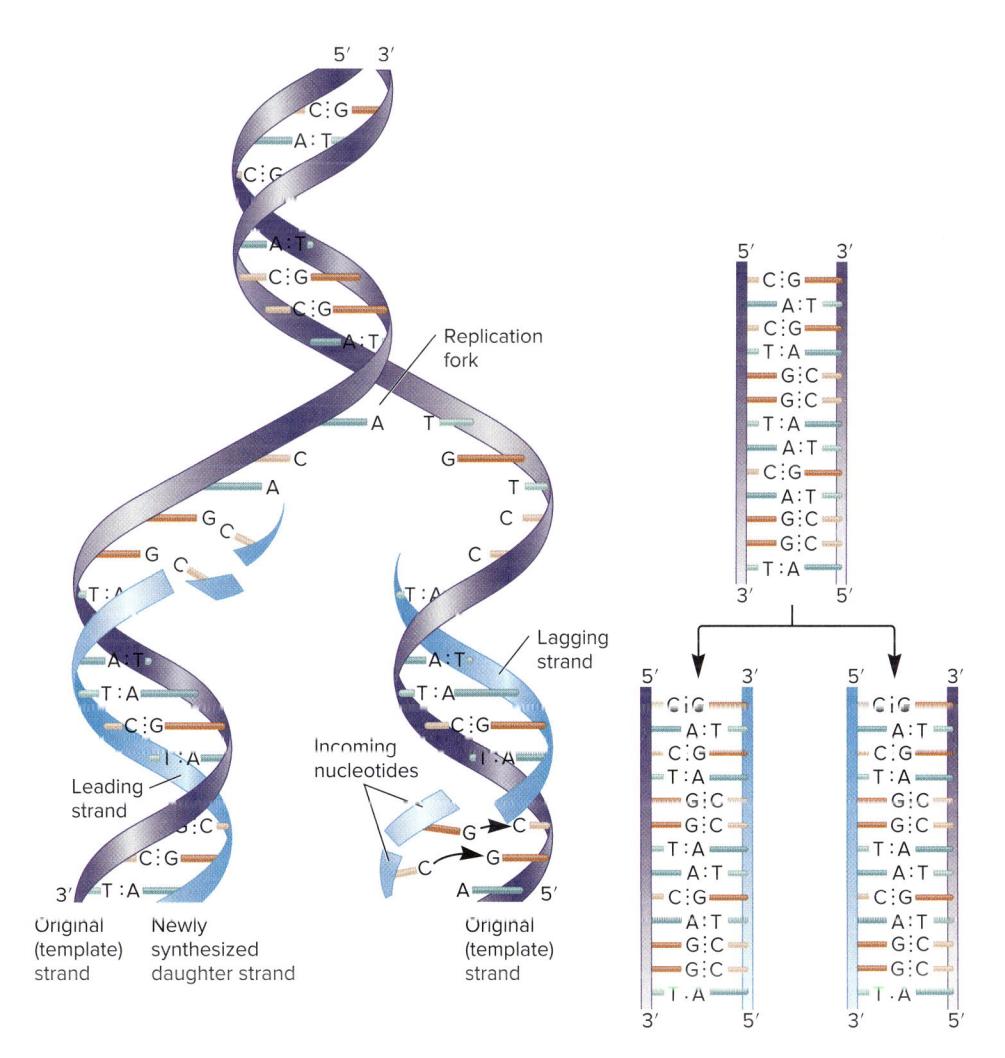

(a) The mechanism of DNA replication

(b) The products of replication

FIGURE 13.1 The structural basis for DNA replication. (a) The mechanism of DNA replication as originally proposed by Watson and Crick. As we will see, the synthesis of one newly made strand (the leading strand) occurs in the direction toward the replication fork, whereas the synthesis of the other newly made strand (the lagging strand) occurs in small segments away from the replication fork. (b) DNA replication produces two copies of DNA with the same base sequence as the original DNA molecule.

Genes → Traits DNA replication allows a cell to copy its genetic material. For example, the genetic material is copied during the formation of new sperm and egg cells. When sperm and egg cells unite during fertilization, this genetic material carries genes that provide the inherited traits of the resulting offspring.

Concept Check: What features of DNA structure enable it to be replicated?

Three Different Models Were Proposed to Describe the End Result of DNA Replication

Scientists in the late 1950s considered three different mechanisms to explain the end result of DNA replication. These mechanisms are shown in **Figure 13.2.**

- In the **conservative model,** both parental strands of DNA remain together following DNA replication. In this model, the original arrangement of parental strands is completely conserved, and the two newly made daughter strands also remain together following replication.
- In the **semiconservative model,** the double-stranded DNA is half conserved following the replication process. In other words, the newly made double-stranded DNA contains one parental strand and one daughter strand.
- In the **dispersive model,** segments of parental DNA and newly made DNA are interspersed in both strands following the replication process.

The semiconservative model, shown in Figure 13.2b, is the correct mechanism. How did scientists determine this? In 1958, Matthew Meselson and Franklin Stahl devised a method to experimentally distinguish newly made daughter strands from the original parental strands. Their technique involved labeling DNA with a heavy isotope of nitrogen (^{15}N). Nitrogen, which is found within the bases of DNA, occurs in both a heavy and a light (^{14}N) form.

Prior to their experiment, Meselson and Stahl grew *E. coli* cells in the presence of ^{15}N for many generations. This produced a population of cells in which all of the DNA was heavy-labeled. At the start of their experiment, shown in **Figure 13.3** (generation 0), they switched the bacteria to a medium that contained only ^{14}N and then collected samples of cells after various time points. Under the growth conditions they employed, 30 minutes is the time required for one doubling, or one generation time. Because the bacteria were doubling in a medium that contained only ^{14}N, all of the newly made DNA strands were labeled with light nitrogen, but the original strands remained in the heavy form.

Meselson and Stahl then analyzed the density of the DNA by centrifugation, using a cesium chloride (CsCl) gradient. (The procedure of gradient centrifugation is described in Appendix A.) If both DNA strands contained ^{14}N, the DNA would have a light density and sediment near the top of the tube. If one strand contained ^{14}N and the other strand contained ^{15}N, the DNA would be half-heavy and have an intermediate density. Finally, if both strands contained ^{15}N, the DNA would be heavy and move closer to the bottom of the centrifuge tube.

▶ **THE HYPOTHESIS**

Based on Watson's and Crick's ideas, the hypothesis was that DNA replication is semiconservative. Figure 13.2 also shows the two alternative models.

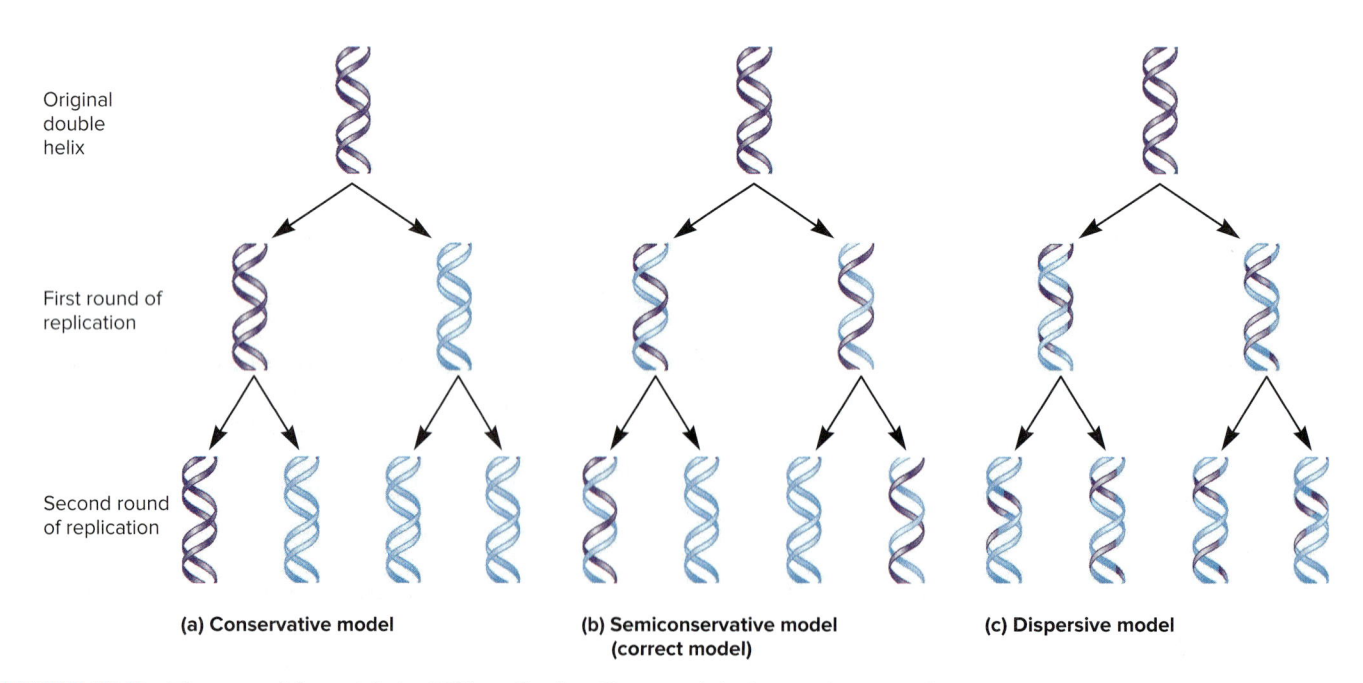

FIGURE 13.2 **Three possible models for DNA replication.** The two original parental strands of DNA are shown in purple, and the newly made strands after one and two generations are shown in light blue.

Concept Check: *Which of these models is the correct one?*

13.1 STRUCTURAL OVERVIEW OF DNA REPLICATION

▶ **TESTING THE HYPOTHESIS** — **FIGURE 13.3** Evidence that DNA replication is semiconservative.
Starting material: A strain of *E. coli* that has been grown for many generations in the presence of ^{15}N. All of the bases in the DNA are labeled with ^{15}N.

Experimental level

Conceptual level

1. Add an excess of ^{14}N-containing compounds to the growth medium so all of the newly made DNA will contain ^{14}N.

2. Incubate the cells for various lengths of time. Note: The ^{15}N-labeled DNA is shown in purple, and the ^{14}N-labeled DNA is shown in blue.

3. Lyse the cells by adding lysozyme and detergent, which disrupt the bacterial cell wall and cell membrane, respectively.

4. Load a sample of the lysate onto a CsCl gradient. (Note: The average density of DNA is around 1.7 g/cm³, which is sufficiently different from other cellular macromolecules.)

5. Centrifuge the gradients until the DNA molecules reach their equilibrium densities.

6. DNA within the gradient can be observed under a UV light.

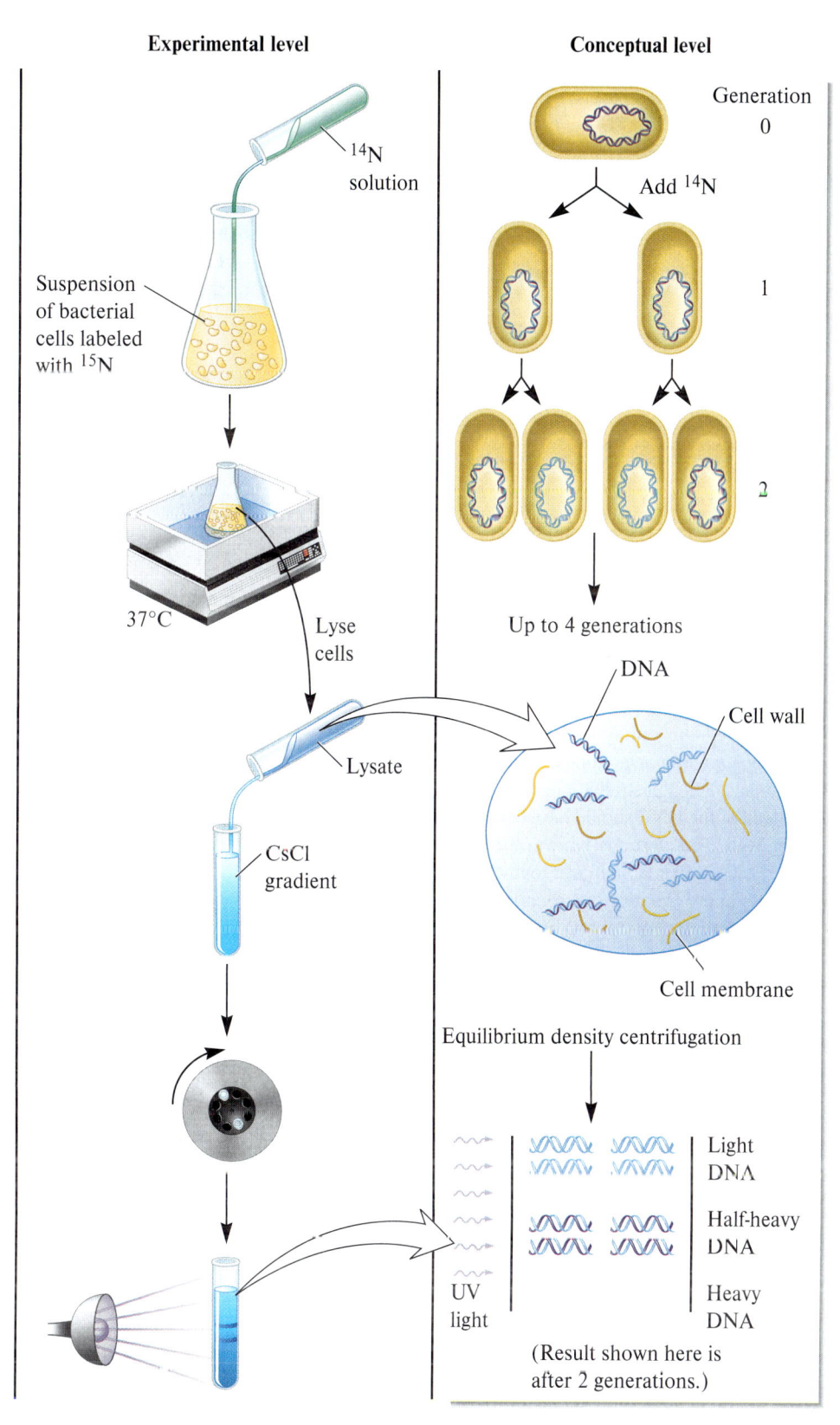

(Result shown here is after 2 generations.)

▶ THE DATA

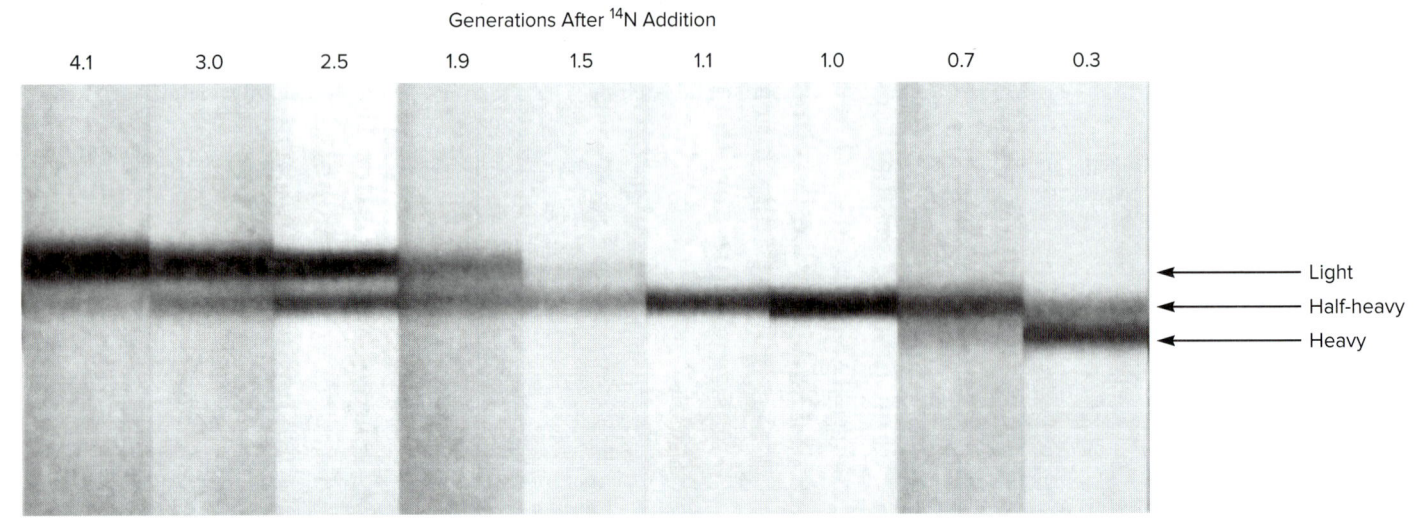

Generations After ^{14}N Addition

4.1   3.0   2.5   1.9   1.5   1.1   1.0   0.7   0.3

← Light
← Half-heavy
← Heavy

Source: Data from Meselson, M., & Stahl, F. W. (1958) The Replication of DNA in *Escherichia coli*, *Proceedings of the National Academy of Sciences of the United States of America*, vol. 44, 671–682.
©Matthew Meselson

▶ INTERPRETING THE DATA

As seen in the data of Figure 13.3, after one round of DNA replication (i.e., after one generation), all of the DNA sedimented at a density that was half-heavy. Both the semiconservative and dispersive models are consistent with this result. In contrast, the conservative model predicts two separate DNA types: a light type and a heavy type. Because all of the DNA sedimented as a single band, this model was disproved. According to the semiconservative model, the replicated DNA would contain one original strand (a heavy strand) and a newly made daughter strand (a light strand). Likewise, in a dispersive model, all of the DNA should have been half-heavy after one generation as well. To determine which of these two remaining models is correct, Meselson and Stahl had to investigate future generations.

After approximately two rounds of DNA replication (i.e., after 1.9 generations), a mixture of light DNA and half-heavy DNA was observed. This result was consistent with the semiconservative model of DNA replication, because some DNA molecules contained all light DNA, and other molecules were half-heavy (see Figure 13.2b). The dispersive model predicts that after two generations, the heavy nitrogen would be evenly dispersed among four strands, each strand containing 1/4 heavy nitrogen and 3/4 light nitrogen (see Figure 13.2c). However, this result was not obtained. Instead, the results of the Meselson and Stahl experiment provided compelling evidence in favor of only the semiconservative model for DNA replication.

13.1 REVIEWING THE KEY CONCEPTS

- DNA replication occurs when the strands of DNA unwind and each strand is used as a template to make a new strand according to the AT/GC rule. The resulting DNA molecules have the same base sequence as the original DNA (see Figure 13.1).
- By labeling DNA with heavy and light isotopes of nitrogen and using centrifugation, Meselson and Stahl showed that DNA replication is semiconservative (see Figures 13.2, 13.3).

13.1 COMPREHENSION QUESTIONS

1. The complementarity of DNA strands is based on
   a. the chemical properties of a phosphodiester linkage.
   b. the binding of proteins to the DNA.
   c. the AT/GC rule.
   d. none of the above.
2. To make a new DNA strand, which of the following is/are necessary?
   a. A template strand
   b. Nucleotides
   c. Heavy nitrogen
   d. Both a and b
3. The model that correctly describes the process of DNA replication is the
   a. conservative model.
   b. semiconservative model.
   c. dispersive model.
   d. Any of the above models describes the process correctly.

13.2 BACTERIAL DNA REPLICATION: THE FORMATION OF TWO REPLICATION FORKS AT THE ORIGIN OF REPLICATION

Learning Outcomes:
1. Describe the key features of a bacterial origin of replication.
2. Explain how DnaA protein initiates DNA replication.

Thus far, we have considered how a complementary, double-stranded structure underlies the ability of DNA to be copied. In addition, the experiments of Meselson and Stahl showed that DNA replication results in two double helices, each one containing an original parental strand and a newly made daughter strand. We now turn our attention to how DNA replication actually occurs within living cells. Much research has focused on understanding the process in the bacterium *Escherichia coli*. In this section, we will examine how bacterial DNA replication begins.

Bacterial Chromosomes Contain a Single Origin of Replication

Figure 13.4 presents an overview of the process of bacterial chromosomal replication. The site on the bacterial chromosome where DNA synthesis begins is known as the **origin of replication.** Bacterial chromosomes have a single origin of replication. The synthesis of new daughter strands is initiated within the origin and proceeds in both directions, or **bidirectionally,** around the bacterial chromosome. A **replication fork** is the site where the parental DNA strands have separated and new daughter strands are being made. Two replication forks move in opposite directions outward from the origin and eventually meet each other on the opposite side of the bacterial chromosome when replication is completed.

Replication Is Initiated at the Origin of Replication

Considerable research has focused on the origin of replication in *E. coli*. This origin is named *oriC* for <u>ori</u>gin of <u>C</u>hromosomal replication (**Figure 13.5**).

1. DNA replication begins with the binding of **DnaA proteins** to sequences within the origin of replication known as **DnaA boxes.** The DnaA boxes serve as recognition sites for the binding of the DnaA proteins; they bind to the five DnaA boxes in *oriC* to initiate DNA replication. DnaA proteins also bind to each other to form a complex (**Figure 13.6**).
2. Other DNA-binding proteins not shown in Figure 13.6 cause the DNA to bend around the complex of DnaA proteins, which results in the separation of the strands at the **AT-rich region.** Note: Because only two hydrogen bonds form between the bases A and T, whereas three hydrogen bonds occur between G and C, the DNA strands are more easily separated at an AT-rich region.
3. Following separation at the AT-rich region, the DnaA proteins, with the help of the DnaC protein, recruit the protein **DNA helicase** (also called DnaB protein) to this site. When DNA helicase encounters a double-stranded region, it breaks the hydrogen bonds between the two strands, thereby generating two single strands.
4. Two DNA helicases begin strand separation within the *oriC* region and continue to separate the DNA strands beyond the origin. These proteins use the energy from ATP hydrolysis to catalyze the separation of the double-stranded DNA. In *E. coli*, DNA helicases bind to single-stranded DNA and travel along the DNA in a 5′ to 3′ direction to keep each replication fork moving. As shown at the bottom of Figure 13.6, the action of DNA helicases promotes the movement of two replication forks outward from *oriC* in opposite directions, a process termed **bidirectional replication.**

The **GATC methylation sites** within *oriC* help to regulate the replication process. Prior to DNA replication, these sites are methylated in both strands. This full methylation of the GATC sites facilitates the initiation of DNA replication at the origin. Following DNA replication, the newly made strands are not methylated, because adenine rather than methyladenine is incorporated into the daughter strands. The initiation of DNA replication at the origin does not readily occur until after it has become fully methylated. Because it takes several minutes to methylate the GATC sequences within this region, DNA replication does not occur again too quickly.

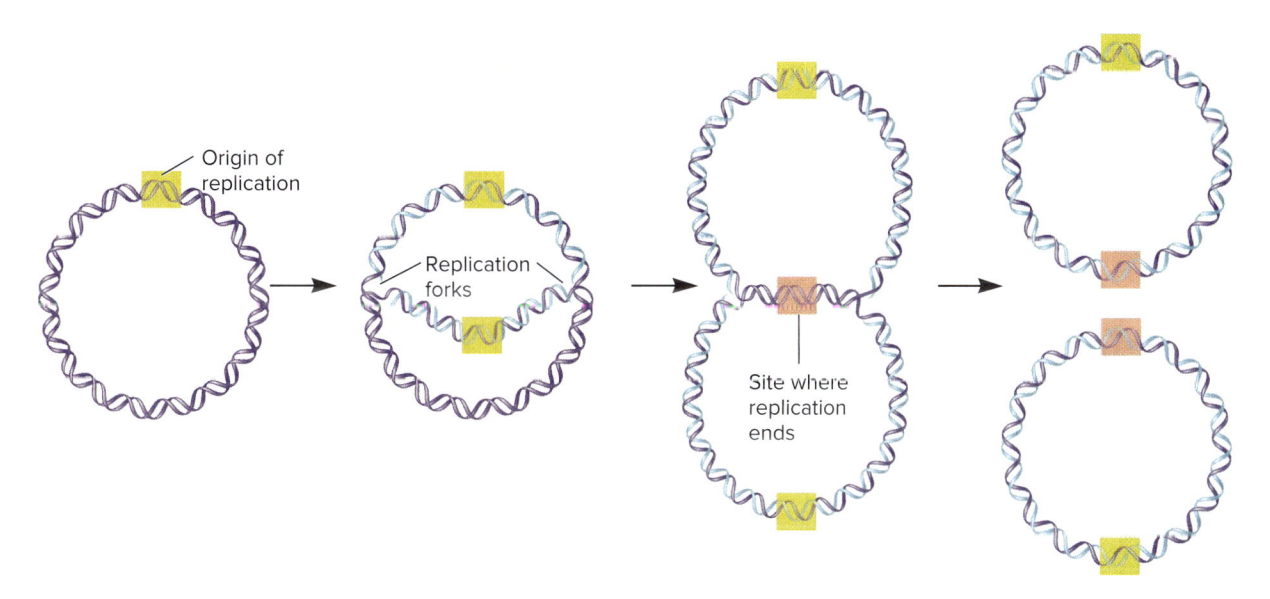

FIGURE 13.4 **The process of bacterial chromosome replication.** An overview of the process of bacterial chromosome replication. Note: The original DNA is purple, and newly made DNA is blue.

272   CHAPTER 13 :: DNA REPLICATION AND RECOMBINATION

FIGURE 13.5 The sequence of *oriC* in *E. coli*. The AT-rich region is composed of three tandem repeats that are 13 bp long and highlighted in blue. The five DnaA boxes are highlighted in orange. The GATC methylation sites are underlined.

Concept Check: What are the functions of the AT-rich region and DnaA boxes?

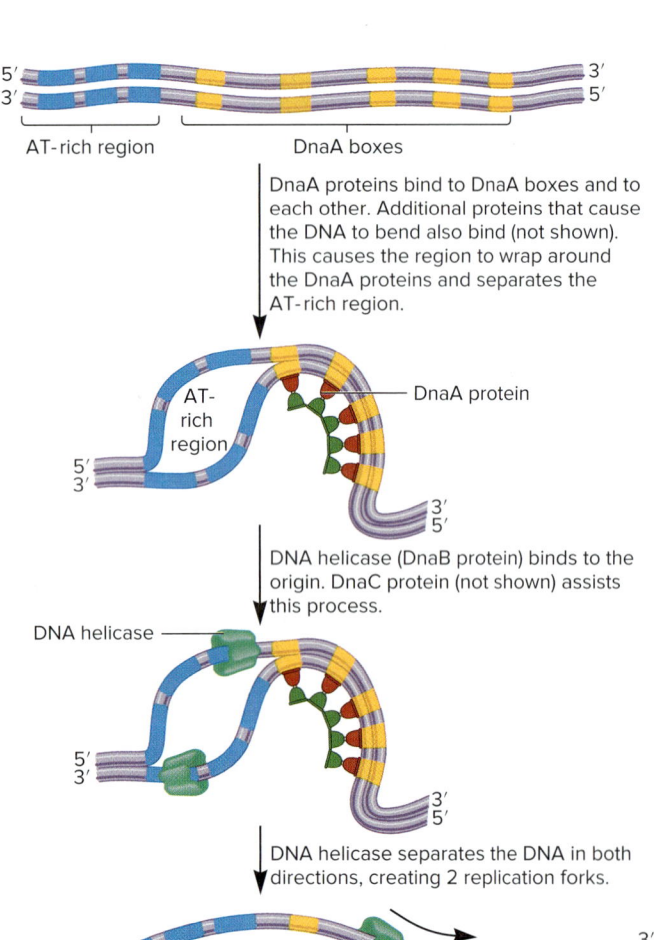

13.2 REVIEWING THE KEY CONCEPTS

- Bacterial DNA replication begins at a single origin of replication and proceeds bidirectionally around the circular chromosome (see Figure 13.4).
- In *E. coli*, DNA replication is initiated when DnaA proteins bind to five DnaA boxes at the origin of replication and cause separation of the strands at the AT-rich region. DNA helicases continue to separate the DNA strands, thereby promoting the movement of two replication forks (see Figures 13.5, 13.6).

13.2 COMPREHENSION QUESTIONS

1. A site in a chromosome where DNA replication begins is
   a. a promoter.
   b. an origin of replication.
   c. an operator.
   d. a replication fork.

FIGURE 13.6 The events that occur at *oriC* to initiate the DNA replication process. To initiate DNA replication, DnaA proteins bind to sequences in the five DnaA boxes, which causes the DNA strands to separate at the AT-rich region. The DnaA protein has a DNA-binding domain, shown in red, which binds to the DnaA boxes. It also has a domain called an oligomerization domain, shown in green, which promotes the binding of DnaA proteins to each other. DnaA and DnaC proteins (not shown) recruit DNA helicase (DnaB) to this region. Each DNA helicase is composed of six subunits, which form a ring around one DNA strand and travel in the 5′ to 3′ direction. As shown here, the movement of two DNA helicases separates the DNA strands beyond the *oriC* region.

Concept Check: How many replication forks are formed at the origin of replication?

2. The origin of replication in *E. coli* contains
   a. an AT-rich region.
   b. DnaA boxes.
   c. GATC methylation sites.
   d. all of the above.

13.3 BACTERIAL DNA REPLICATION: SYNTHESIS OF NEW DNA STRANDS

Learning Outcomes:
1. Describe how helicase, topoisomerase, and single-strand binding protein are important for the unwinding of the DNA double helix.
2. Explain how primase, DNA polymerase, and DNA ligase function in synthesizing new strands of DNA at the replication fork.
3. Compare and contrast the synthesis of the leading and lagging strands.

As we have seen, bacterial DNA replication begins at the origin of replication. The synthesis of new DNA strands is a stepwise process in which many proteins participate. In this section, we will examine how DNA strands are made at a replication fork.

DNA Replication Occurs at the Replication Fork

Figure 13.7 provides an overview of the molecular events that occur as one of the two replication forks moves around the bacterial chromosome. **Table 13.1** summarizes the functions of the major proteins involved in *E. coli* DNA replication.

Unwinding of the Helix To act as the templates for DNA replication, the strands of a double helix must separate.

- The function of DNA helicase is to break the hydrogen bonds between base pairs and thereby unwind the strands. This action generates positive supercoiling ahead of each replication fork.
- As shown in Figure 13.7, an enzyme known as a **topoisomerase II,** also called **DNA gyrase,** travels in front of DNA helicase and relaxes positive supercoiling.
- After the two parental DNA strands have been separated and the supercoiling relaxed, the strands must be kept in that state until the complementary daughter strands have been made. What prevents the DNA strands from coming back together? DNA replication requires **single-strand binding proteins,** which bind to the strands of parental DNA and prevent them from re-forming a double helix. In this way, the bases within the parental strands are kept in an exposed condition that enables them to hydrogen bond with individual nucleotides.

Functions of key proteins involved with bacterial DNA replication

- DNA helicase breaks the hydrogen bonds between the DNA strands.
- Topoisomerase II alleviates positive supercoiling.
- Single-strand binding proteins keep the parental strands apart.
- Primase synthesizes an RNA primer.
- DNA polymerase III synthesizes a daughter strand of DNA.
- DNA polymerase I excises the RNA primers and fills in with DNA (not shown).
- DNA ligase covalently links the Okazaki fragments together.

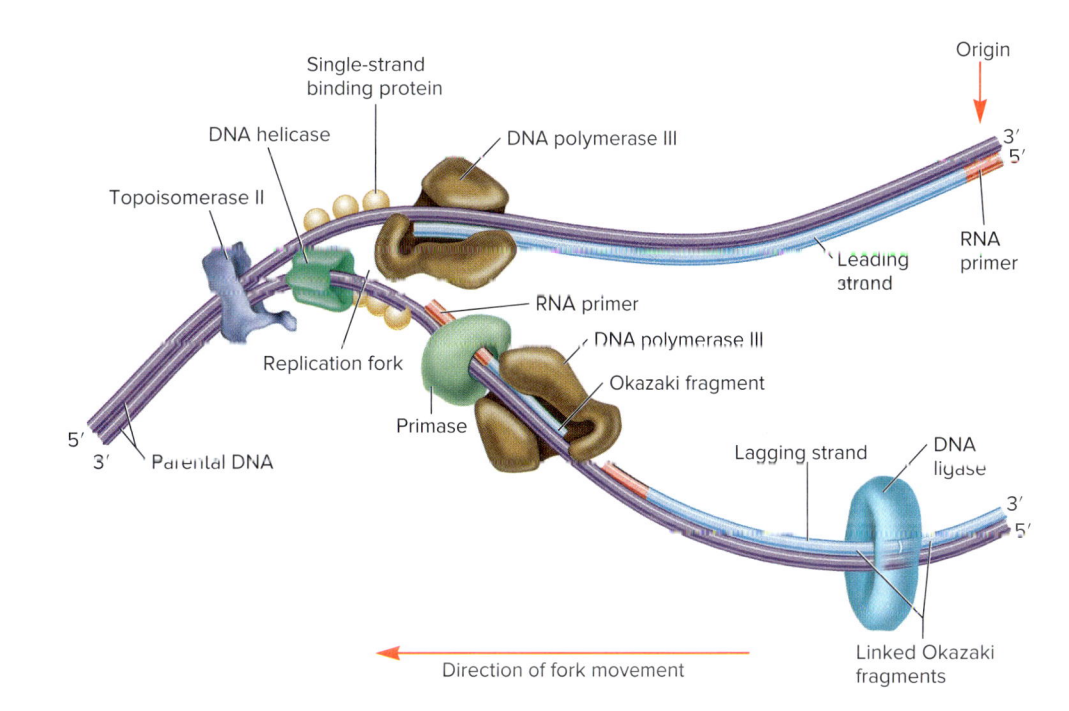

FIGURE 13.7 **The proteins involved with bacterial DNA replication at the replication fork.** Note: The drawing of DNA polymerase III depicts the catalytic subunit that synthesizes DNA. DNA polymerase I, which removes the RNA primers and fills in the vacant region with DNA, is not shown. Also, as shown later in Figure 13.12, DNA polymerase III functions as a dimer. It is shown here as a monomer for simplicity.

Concept Check: Why is primase needed for DNA replication?

274 CHAPTER 13 :: DNA REPLICATION AND RECOMBINATION

TABLE 13.1

Proteins Involved in *E. coli* DNA Replication

Common Name	Function
DnaA protein	Binds to DnaA boxes within the origin to initiate DNA replication
DnaC protein	Aids DnaA in the recruitment of DNA helicase to the origin
DNA helicase (DnaB)	Separates double-stranded DNA
Topoisomerase	Removes positive supercoiling ahead of the replication fork
Single-strand binding protein	Binds to single-stranded DNA and prevents it from re-forming a double-stranded structure
Primase	Synthesizes short RNA primers
DNA polymerase III	Synthesizes DNA in the leading and lagging strands
DNA polymerase I	Removes RNA primers, fills in gaps with DNA
DNA ligase	Covalently attaches adjacent Okazaki fragments

Synthesis of RNA Primers via Primase The next event in DNA replication is the synthesis of short strands of RNA (rather than DNA) called **RNA primers.** These strands of RNA are synthesized by the linkage of ribonucleotides via an enzyme known as **primase.** This enzyme synthesizes short strands of RNA, typically 10 to 12 nucleotides in length. These short RNA strands start, or prime, the process of DNA replication.

- In the **leading strand,** which is made in the same direction as the replication fork is moving in Figure 13.7, a single primer is made at the origin of replication.
- In the **lagging strand,** which is made in short fragments in the opposite direction from the movement of the replication fork, multiple primers are made.

As discussed later, the RNA primers are eventually removed.

Synthesis of DNA via DNA Polymerase An enzyme known as **DNA polymerase** is responsible for synthesizing the DNA of the leading and lagging strands. This enzyme catalyzes the formation of covalent bonds between adjacent nucleotides, thereby making the new daughter strands. In *E. coli*, five distinct proteins function as DNA polymerases and are designated polymerase I, II, III, IV, and V. DNA polymerases I and III are involved in normal DNA replication, whereas DNA polymerases II, IV, and V play a role in DNA repair and the replication of damaged DNA.

DNA polymerase III is responsible for most of the DNA replication (see Figure 13.7). It is a large enzyme consisting of 10 different subunits that play various roles in the DNA replication process. One subunit, called the α subunit, catalyzes the bond formation between adjacent nucleotides, and the remaining nine subunits fulfill other functions. The complex of all 10 subunits together is called DNA polymerase III holoenzyme. By comparison, DNA polymerase I is composed of a single subunit. Its role during DNA replication is to remove the RNA primers and fill in the vacant regions with DNA.

Though the various DNA polymerases in *E. coli* and other bacterial species vary in their subunit composition, several common structural features have emerged. The catalytic subunit of all DNA polymerases has a structure that resembles a human hand. As shown in **Figure 13.8**, the template DNA is threaded through the palm of the hand, with the thumb and fingers wrapped around the DNA. Incoming deoxyribonucleoside triphosphates (dNTPs) enter the catalytic

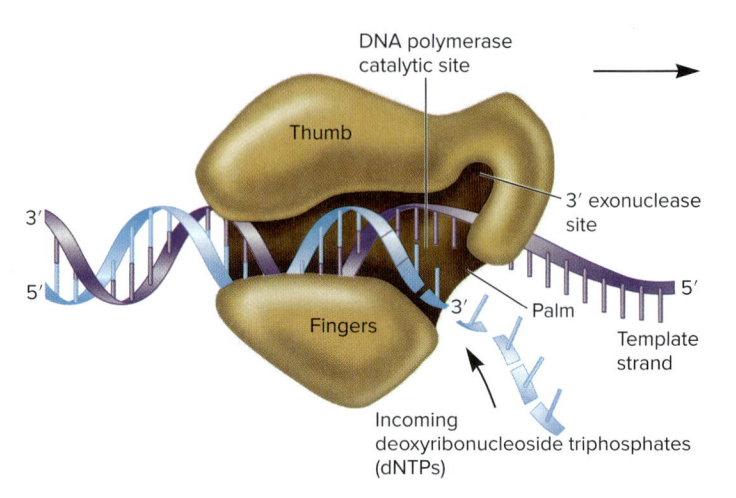

(a) Schematic side view of DNA polymerase III

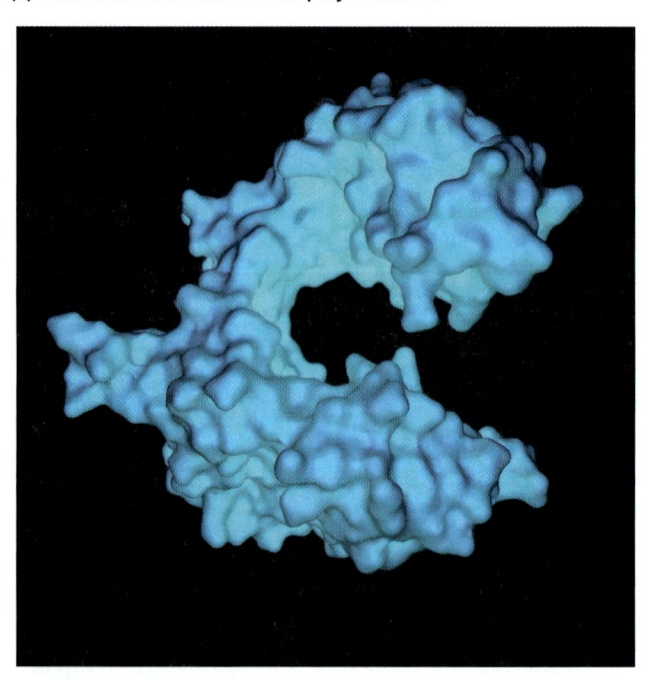

(b) Molecular model for DNA polymerase

FIGURE 13.8 The structure and function of DNA polymerase III. (a) DNA polymerase slides along the template strand as it synthesizes a new strand. As shown later in Figure 13.13, deoxyribonucleoside triphosphates (dNTPs) initially bind to the template strand and deoxyribonucleoside monophosphates are connected in the 5′ to 3′ direction. The catalytic subunit of DNA polymerase III resembles a hand that is wrapped around the template strand. Thus, its movement, along the template strand is similar to a hand that is sliding along a rope. (b) A model of the molecular structure of DNA polymerase.

(b) ©Luk Cox/Alamy Stock Photo

Concept Check: Is the template strand read in the 5′ to 3′ or the 3′ to 5′ direction?

site, binding to the template strand according to the AT/GC rule. As described later, deoxyribonucleoside monophosphates are covalently attached to the 3′ end of the growing strand, and pyrophosphate is released. DNA polymerase also contains a 3′ exonuclease site that removes mismatched bases, which is also described later.

The Synthesis of the Leading and Lagging Strands Is Distinctly Different

As researchers began to unravel the function of DNA polymerase, two features seemed unusual (**Figure 13.9**).

1. DNA polymerase cannot begin DNA synthesis by linking together the first two individual nucleotides. Rather, this type of enzyme can elongate only a preexisting strand starting with an RNA primer or an existing DNA strand (Figure 13.9a).
2. DNA polymerase can attach nucleotides only in the 5′ to 3′ direction, not in the 3′ to 5′ direction (Figure 13.9b).

Due to these unusual features of DNA replication, the synthesis of the leading and lagging strands is distinctly different (**Figure 13.10**).

- The synthesis of RNA primers by primase allows DNA polymerase III to begin the synthesis of complementary daughter strands of DNA at the origin of replication. DNA polymerase III catalyzes the attachment of nucleotides to the 3′ end of each primer, in a 5′ to 3′ direction.

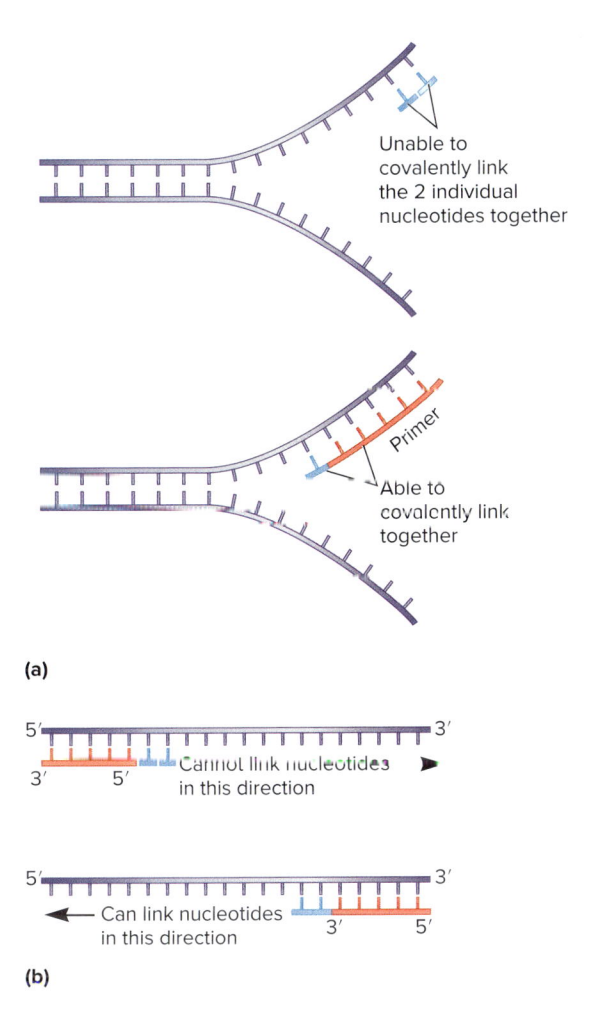

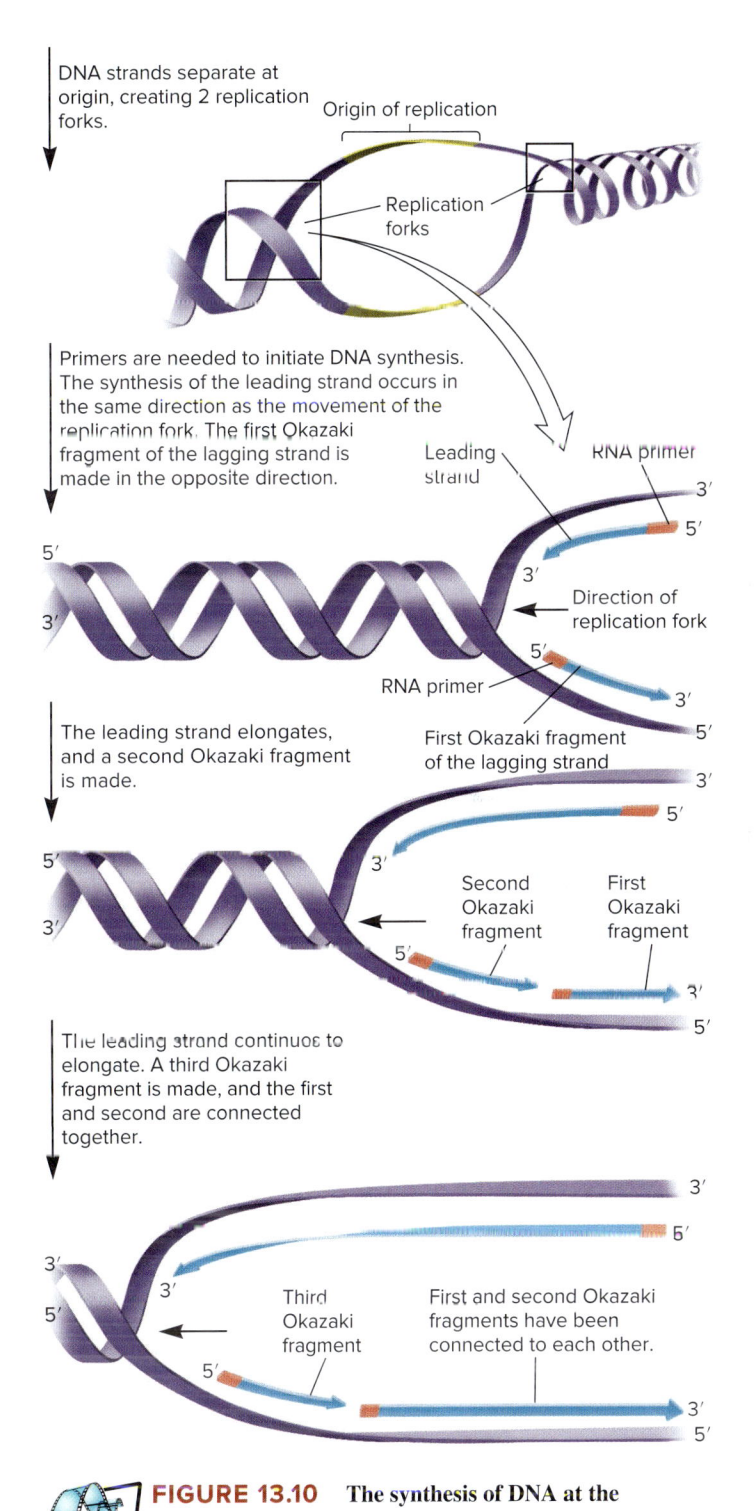

FIGURE 13.9 Unusual features of the functioning of DNA polymerase. (a) DNA polymerase can elongate a strand only from an RNA primer or an existing DNA strand. (b) DNA polymerase can attach nucleotides only in a 5′ to 3′ direction. Note: The template strand is read in the opposite, 3′ to 5′, direction.

FIGURE 13.10 The synthesis of DNA at the replication fork.

Concept Check: Describe the differences between the syntheses of the leading and lagging strands.

- In the leading strand, one RNA primer is made at the origin, and then DNA polymerase III can attach nucleotides in a 5′ to 3′ direction as it slides toward the opening of the replication fork. The synthesis of the leading strand is therefore continuous.
- In the lagging strand, the synthesis of DNA also proceeds in a 5′ to 3′ direction, but moving away from the replication fork. In the lagging strand, RNA primers must repeatedly initiate the synthesis of short segments of DNA. Therefore, the synthesis has to be discontinuous. These fragments in bacteria are typically 1000 to 2000 nucleotides long. (In eukaryotes, the fragments are shorter—100 to 200 nucleotides.) Each fragment contains a short RNA primer at the 5′ end, which is made by primase. The remainder of the fragment is a strand of DNA made by DNA polymerase III. The DNA fragments made in this manner are known as **Okazaki fragments,** after Reiji and Tsuneko Okazaki, who initially discovered them in the late 1960s.
- In *E. coli*, the RNA primers are removed by the action of DNA polymerase I. This enzyme has a 5′ to 3′ exonuclease activity, which means that DNA polymerase I digests away the RNA primers in a 5′ to 3′ direction, leaving a vacant area.
- DNA polymerase I then synthesizes DNA to fill in this region. It uses the 3′ end of an adjacent Okazaki fragment as a primer. For example, in Figure 13.10, DNA polymerase I will remove the RNA primer from the first Okazaki fragment and then synthesize DNA in the vacant region by attaching nucleotides to the 3′ end of the second Okazaki fragment.
- After the gap has been completely filled in with DNA, a covalent bond is still missing between the last nucleotide added by DNA polymerase I and the adjacent DNA strand that had been previously made by DNA polymerase III. An enzyme known as **DNA ligase** catalyzes a covalent bond between adjacent fragments to complete the replication process in the lagging strand.

Now that we have seen how the leading and lagging strands are made, **Figure 13.11** shows how new strands are constructed from a single origin of replication. To the left of the origin, the top strand is made continuously, whereas to the right of the origin, it is made in Okazaki fragments. By comparison, the synthesis of the bottom strand is just the opposite. To the left of the origin, this is made in Okazaki fragments and to the right of the origin, the synthesis is continuous.

Certain Enzymes of DNA Replication Bind to Each Other to Form a Complex

Figure 13.12 provides a more three-dimensional view of the DNA replication process. DNA helicase and primase are physically bound to each other to form a complex known as a **primosome.** This complex leads the way at the replication fork. The primosome tracks along the DNA, separating the parental strands and synthesizing RNA primers at regular intervals along the lagging strand. Being associated together within a complex allows DNA helicase and primase to better coordinate their actions.

The primosome is physically associated with two DNA polymerase holoenzymes to form a **replisome.** As shown in Figure 13.12, two DNA polymerase III proteins act in concert to replicate the leading and lagging strands. The term **dimeric DNA polymerase** is used to describe two DNA polymerase holoenzymes that move as a unit during DNA replication. To facilitate this movement, the lagging strand is looped out with respect to the DNA polymerase that synthesizes the lagging strand. This loop allows the lagging-strand polymerase to make DNA in a 5′ to 3′ direction yet move as a unit with the leading-strand polymerase. Interestingly, when the lagging-strand polymerase reaches the end of an Okazaki fragment, it must be released from the template DNA and "hop" to the RNA primer that is closest to the fork. Similarly, after primase synthesizes an RNA primer in the 5′ to 3′ direction, it must hop over the primer and synthesize the next primer closer to the replication fork.

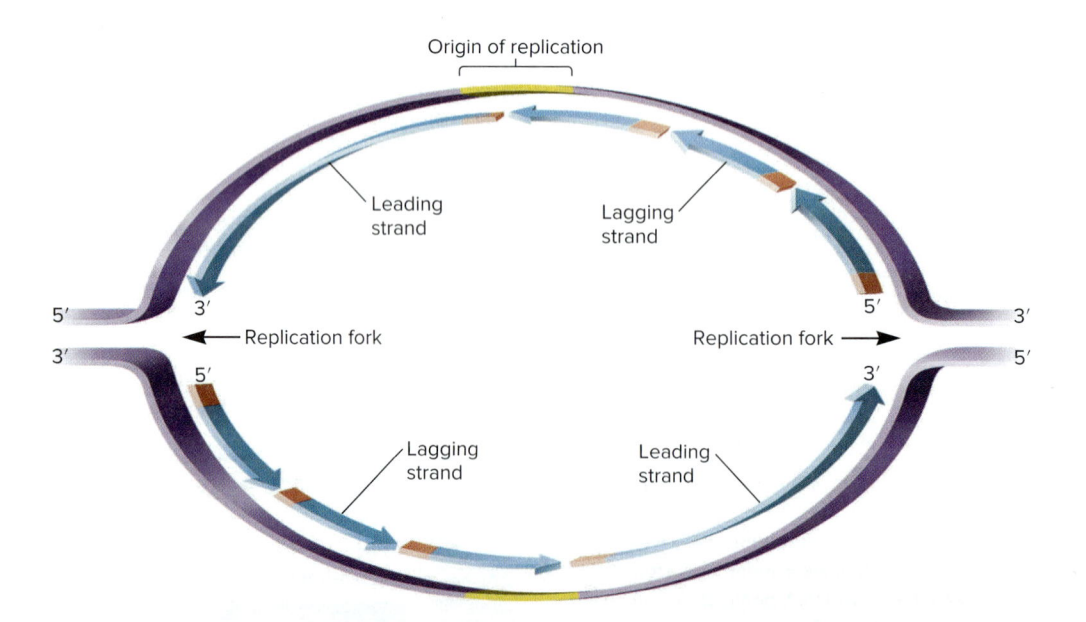

FIGURE 13.11 The synthesis of the leading and lagging strands outward from a single origin of replication.

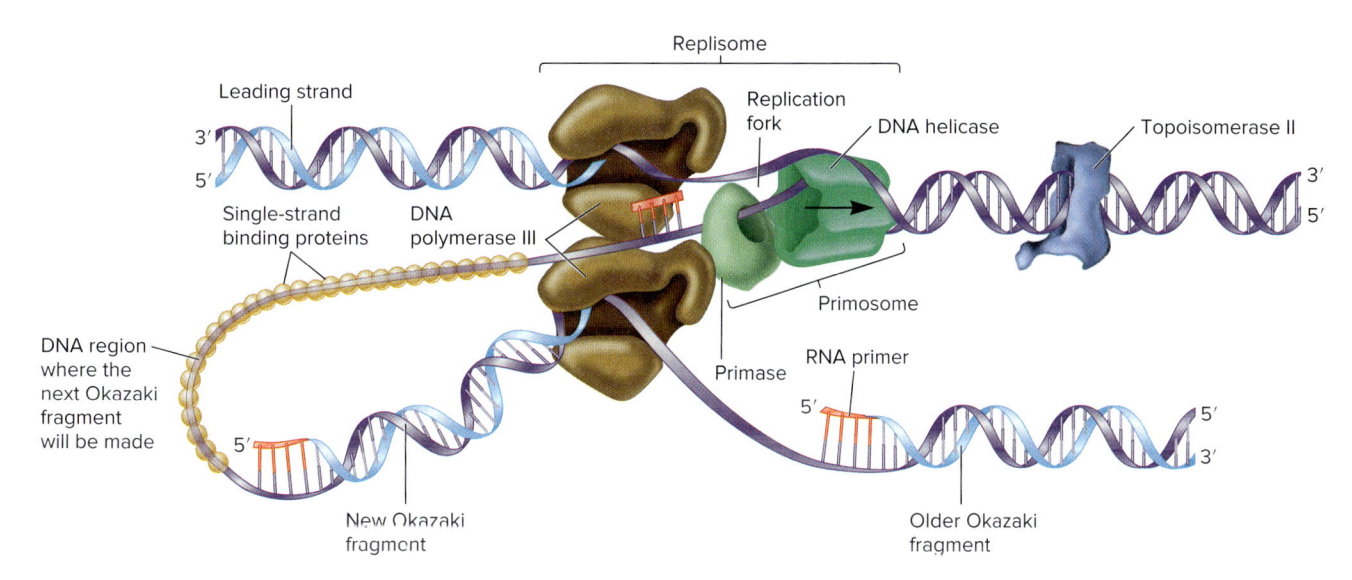

FIGURE 13.12 A three-dimensional view of DNA replication. DNA helicase and primase associate together to form a primosome. The primosome associates with two DNA polymerase holoenzymes to form a replisome.

Genetic TIPS

The Question: What are the similarities and differences in the synthesis of DNA in the leading and lagging strands in *E. coli*?

Topic: What topic in genetics does this question address?

The topic is DNA replication. More specifically, the question is about comparing the synthesis of the leading and lagging strands.

Information: What information do you know based on the question and your understanding of the topic?

From the question, you know that DNA replication occurs in the lagging and leading strands. From your understanding of the topic, you may remember that the leading strand is made continuously, in the direction of the replication fork, whereas the lagging strand is made as Okazaki fragments in the direction away from the fork.

Problem-Solving Strategy: Compare and contrast. Describe the steps.

One strategy to solve this problem is to compare your knowledge of DNA replication of the leading and lagging strands. It may be helpful to break down the process into its individual steps.

Answer: The leading strand is primed once at the origin, and then DNA polymerase III synthesizes DNA continuously in the direction of the replication fork. The DNA is made in the 5′ to 3′ direction. In the lagging strand, many short pieces of DNA (Okazaki fragments) are made. This requires many RNA primers. The primers are removed by DNA polymerase I, which then fills in the gaps with DNA. Like the leading strand, the lagging strand is made in the 5′ to 3′ direction. DNA ligase then covalently connects the Okazaki fragments together.

13.3 REVIEWING THE KEY CONCEPTS

- At each replication fork, DNA helicase unwinds the DNA and topisomerase alleviates positive supercoiling. Single-strand binding proteins coat the DNA to prevent the strands from coming back together. Primase synthesizes RNA primers, and DNA polymerase synthesizes complementary strands of DNA. DNA ligase joins Okazaki fragments by catalyzing the formation of covalent bonds (see Figure 13.7, Table 13.1).
- DNA polymerase III is an enzyme in *E. coli* with several subunits. The catalytic subunit wraps around the DNA like a hand (see Figure 13.8).
- DNA polymerase needs an RNA primer or an existing DNA strand to synthesize DNA and makes a new DNA strand in a 5′ to 3′ direction (see Figure 13.9).
- During DNA synthesis, the leading strand is made continuously in the direction of the replication fork, whereas the lagging strand is made as Okazaki fragments in the direction away from the fork (see Figures 13.10, 13.11).
- A primosome is a complex between DNA helicase and primase. A replisome is a complex between a primosome and dimeric DNA polymerase III (see Figure 13.12).

13.3 COMPREHENSION QUESTIONS

1. The enzyme known as _____ uses _____ and separates the DNA strands at the replication fork.
   a. helicase, ATP
   b. helicase, GTP
   c. gyrase, ATP
   d. gyrase, GTP

2. In the lagging strand, DNA is made in the direction _____ the replication fork and is made as _____.
   a. toward, one continuous strand
   b. away from, one continuous strand
   c. toward, Okazaki fragments
   d. away from, Okazaki fragments

13.4 BACTERIAL DNA REPLICATION: CHEMISTRY AND ACCURACY

Learning Outcomes:
1. Describe how nucleotides are connected to a growing DNA strand.
2. Define *processivity*.
3. Explain the proofreading function of DNA polymerase.

In Sections 13.2 and 13.3, we examined the origin of replication and considered how DNA strands are synthesized. In this section, we will take a closer look at the process in which nucleotides are attached to a growing DNA strand and discuss the amazing accuracy of DNA replication.

DNA Polymerase III Is a Processive Enzyme That Uses Deoxyribonucleoside Triphosphates

Figure 13.13 describes the chemistry of DNA replication.

- The nucleotide about to be attached to the growing strand is a dNTP containing three phosphate groups attached at the 5′–carbon atom of deoxyribose.
- The dNTP binds to the template strand according to the AT/GC rule.
- The 3′–OH group on the nucleotide at the end of the growing strand reacts with the phosphate group adjacent to the sugar on the incoming nucleotide. Pyrophosphate (PP$_i$) is released. The breakage of a covalent bond between two phosphates in a dNTP is a highly exergonic reaction that provides the energy to form a covalent (ester) bond between the sugar at the 3′ end of the DNA strand and the phosphate of the incoming nucleotide.
- The formation of this covalent bond causes the newly made strand to grow in the 5′ to 3′ direction.

As noted in Chapter 11 (refer back to Figure 11.7), a phosphodiester linkage is the linkage between a phosphate and two sugar molecules. As its name implies, a phosphodiester linkage involves two ester bonds. In comparison, as a DNA strand grows, a single covalent (ester) bond is formed between adjacent nucleotides (see Figure 13.13). The other ester bond in the phosphodiester linkage—the bond between the 5′-oxygen and phosphorus—is already present in the incoming nucleotide.

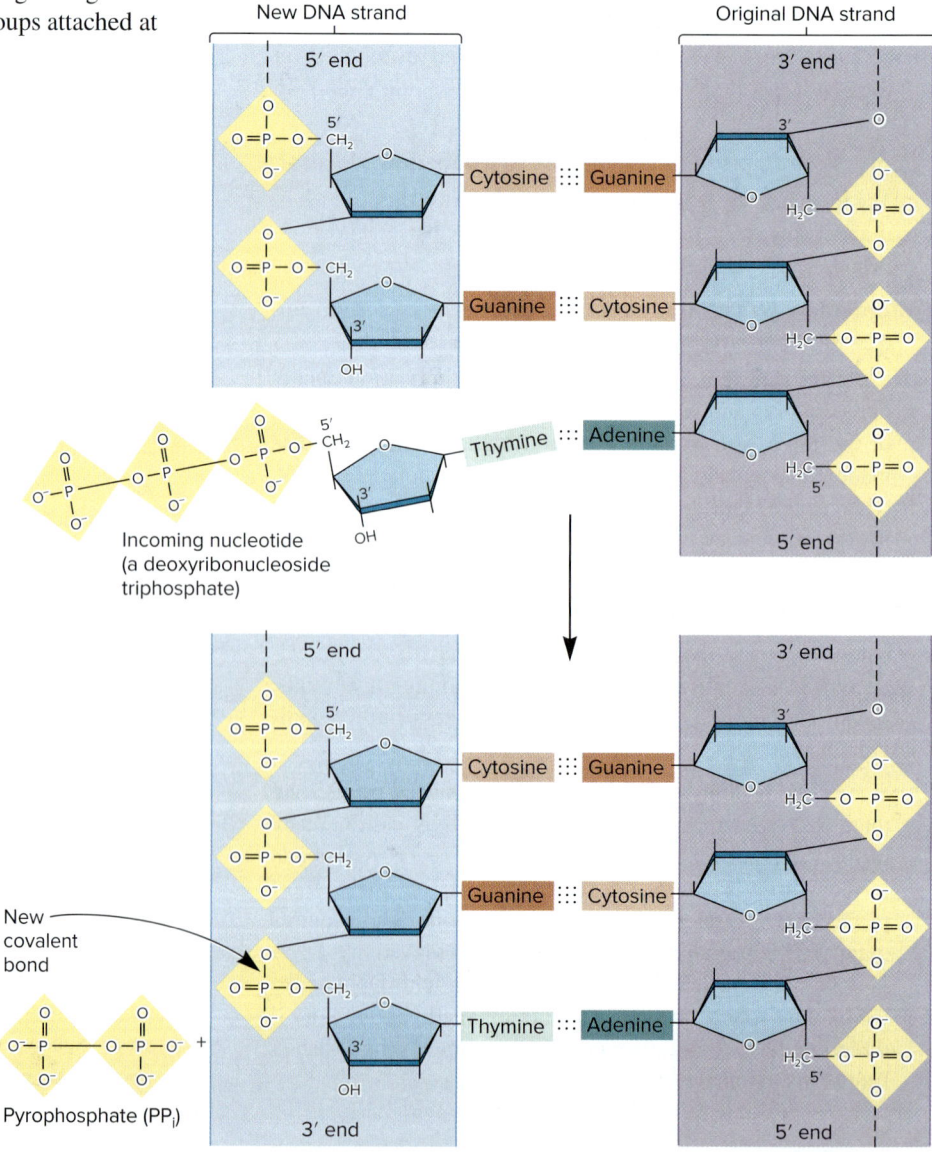

FIGURE 13.13 **The attachment of a nucleotide to a DNA strand.** An incoming deoxyribonucleoside triphosphate (dNTP) is cleaved to form a deoxyribonucleoside monophosphate and pyrophosphate (PP$_i$). The energy released from this exergonic reaction allows the deoxyribonucleoside monophosphate to form a covalent (ester) bond at the 3′ end of the growing strand. This reaction is catalyzed by DNA polymerase.

Concept Check: Does the oxygen in the newly made ester bond come from the phosphate or from the sugar?

DNA polymerase catalyzes the covalent attachment of nucleotides with great speed. In *E. coli*, DNA polymerase III attaches approximately 750 nucleotides per second! DNA polymerase III can catalyze the synthesis of the daughter strands so quickly because it is a **processive enzyme.** This means that it does not dissociate from the growing strand after it has catalyzed the covalent joining of two nucleotides. Rather, as depicted in Figure 13.8a, it remains clamped to the DNA template strand and slides along the template as it catalyzes the synthesis of the daughter strand. One subunit of DNA polymerase III, called the β subunit, or clamp protein, promotes the association of the holoenzyme with the DNA as it glides along the template strand.

The effects of processivity are really quite remarkable. In the absence of the β subunit, DNA polymerase can synthesize DNA at a rate of approximately 20 nucleotides per second. On average, it falls off the DNA template after about 10 nucleotides have been linked together. By comparison, when the β subunit is present, as it is in the holoenzyme, the synthesis rate is approximately 750 nucleotides per second. In the leading strand, DNA polymerase III has been estimated to synthesize a segment of DNA that is over 500,000 nucleotides in length before it falls off.

DNA Replication Is Very Accurate

With replication occurring so rapidly, you might imagine that mistakes could happen in which the wrong nucleotide is incorporated into the growing daughter strand. Although mistakes do happen during DNA replication, they are extraordinarily rare. In the case of DNA synthesis via DNA polymerase III, only one mistake is made per 100 million nucleotides. Therefore, DNA synthesis occurs with a high degree of accuracy, or **fidelity.**

Why is the fidelity so high? Three factors play a role.

- The hydrogen bonding between G and C or A and T is much more stable than that between mismatched pairs of bases. However, this stability accounts for only part of the fidelity, because mismatching due to stability considerations would account for 1 mistake per 1000 nucleotides.
- The active site of DNA polymerase preferentially catalyzes the attachment of nucleotides when the correct bases are located in opposite strands. Helix distortions caused by mispairing usually prevent an incorrect nucleotide from properly occupying the active site of DNA polymerase. By comparison, the active site undergoes a change in shape when the correct nucleotide binds, a process called induced fit, which is necessary for catalysis. The ability of correct nucleotides to cause induced fit decreases the error rate to a range from 1 in 100,000 to 1 in 1 million.
- DNA polymerase decreases the error rate by the enzymatic removal of mismatched nucleotides. As shown in **Figure 13.14**, DNA polymerase can detect a mismatched nucleotide and remove it from the daughter strand. This occurs by exonuclease cleavage of the bonds between adjacent nucleotides at the 3′ end of the newly made strand. The ability of DNA polymerase to remove mismatched bases by this mechanism is called **proofreading.** Proofreading

FIGURE 13.14 The proofreading function of DNA polymerase. When a base pair mismatch is found, the end of the newly made strand is shifted into the 3′ exonuclease site. The DNA is digested in the 3′ to 5′ direction to release the incorrect nucleotide, and replication is resumed in the 5′ to 3′ direction.

occurs by the removal of nucleotides in the 3′ to 5′ direction at the 3′ exonuclease site. After the mismatched nucleotide is removed, DNA polymerase resumes DNA synthesis in the 5′ to 3′ direction.

13.4 REVIEWING THE KEY CONCEPTS

- DNA polymerase III is a processive enzyme that uses deoxyribonucleoside triphosphates (dNTPs) to make new DNA strands (see Figure 13.13).
- The high fidelity of DNA replication is a result of (1) the stability of hydrogen bonding between the correct pairs of bases, (2) the phenomenon of induced fit, and (3) the proofreading ability of DNA polymerase (see Figure 13.14).

13.4 COMPREHENSION QUESTIONS

1. DNA polymerase III is a processive enzyme, which means that
    a. it does not dissociate from the growing strand after it has attached a nucleotide to the 3′ end.
    b. it makes a new strand very quickly.
    c. it proceeds toward the opening of the replication fork.
    d. it copies DNA with relatively few errors.
2. The proofreading function of DNA polymerase involves the recognition of a _____ and the removal of a short segment of DNA in the _____ direction.
    a. missing base, 5′ to 3′
    b. base pair mismatch, 5′ to 3′
    c. missing base, 3′ to 5′
    d. base pair mismatch, 3′ to 5′

13.5 EUKARYOTIC DNA REPLICATION

Learning Outcomes:
1. Compare and contrast the origins of replication in bacteria and eukaryotes.
2. Outline the functions of different DNA polymerases in eukaryotes.
3. Describe how RNA primers are removed in eukaryotes.
4. Explain how DNA replication occurs at the ends of eukaryotic chromosomes.

Extensive similarities exist between the general features of DNA replication in prokaryotes and eukaryotes. For example, DNA helicases, topoisomerases, single-strand binding proteins, primases, DNA polymerases, and DNA ligases—the types of bacterial proteins described in Table 13.1—have also been identified in eukaryotes. Nevertheless, at the molecular level, eukaryotic DNA replication appears to be substantially more complex. These additional intricacies of eukaryotic DNA replication are related to several features of eukaryotic cells. In particular, eukaryotic cells have larger, linear chromosomes with multiple origins of replication; the chromatin is tightly packed within nucleosomes; and cell-cycle regulation is much more complicated. This section emphasizes some of the unique features of eukaryotic DNA replication.

Initiation Occurs at Multiple Origins of Replication on Linear Eukaryotic Chromosomes

Eukaryotes have long, linear chromosomes with multiple origins of replication, which enable the DNA to be replicated in a reasonable

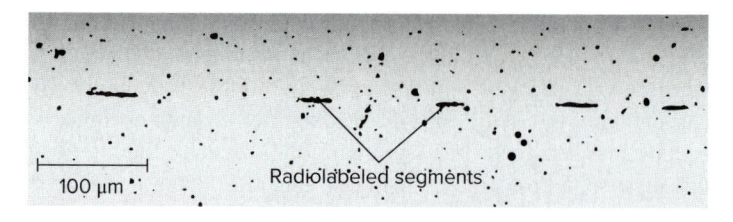

FIGURE 13.15 **Evidence for multiple origins of replication in eukaryotic chromosomes.** In this experiment, cells were given a pulse/chase of ^{3}H-thymidine and unlabeled thymidine. The chromosomes were isolated and subjected to autoradiography. In this schematic illustration, radiolabeled segments were interspersed among nonlabeled segments, indicating that eukaryotic chromosomes contain multiple origins of replication.

length of time. In 1968, Joel Huberman and Arthur Riggs provided evidence for multiple origins of replication (**Figure 13.15**).

1. They added a pulse of a radiolabeled deoxyribonucleoside (^{3}H-thymidine) to a culture of actively dividing cells. The radiolabeled deoxyribothymidine was taken up by the cells and incorporated into their newly made DNA strands for a brief period.
2. The cells were then exposed to a high concentration of unlabeled deoxyribothymidine, which prevented the further incorporation of radiolabeled deoxriboythymidine.
3. The chromosomes were then isolated from the cells and subjected to autoradiography.
4. As seen in Figure 13.15, radiolabeled segments were interspersed among nonlabeled segments. This result is consistent with the hypothesis that eukaryotic chromosomes contain multiple origins of replication.

As shown schematically in **Figure 13.16**, eukaryotic DNA replication proceeds bidirectionally from many origins of replication during S phase of the cell cycle. The multiple replication forks eventually make contact with each other to complete the replication process.

The molecular features of some eukaryotic origins of replication may have some similarities to the origins found in bacteria. At the molecular level, eukaryotic origins of replication have been extensively studied in the yeast *Saccharomyces cerevisiae*. They are named **ARS elements** (for <u>a</u>utonomously <u>r</u>eplicating <u>s</u>equence). ARS elements, which are about 50 bp in length, are necessary for initiation of chromosome replication. ARS elements have unique features in their DNA sequences.

- They contain a higher percentage of A and T bases than the rest of the chromosomal DNA.
- They contain a copy of the ARS consensus sequence, ATTTAT(A or G)TTTA, along with additional elements that enhance origin function.

In animals, the critical features that define origins of replication are not completely understood. In many species, origins are not determined by particular DNA sequences. Rather, origins of replication in animals may be due to specific features of chromatin structure, including protein modifications.

13.5 EUKARYOTIC DNA REPLICATION 281

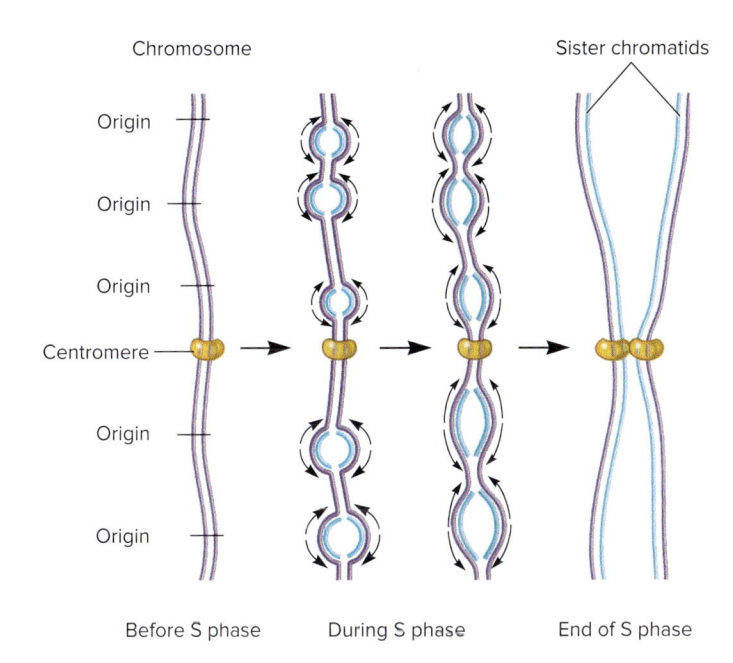

FIGURE 13.16 The replication of eukaryotic chromosomes. At the beginning of the S phase of the cell cycle, eukaryotic chromosome replication begins from multiple origins of replication. As S phase continues, the replication forks move bidirectionally to replicate the DNA. By the end of S phase, all of the replication forks have merged. The net result is two sister chromatids attached to each other at the centromere.

Concept Check: Why do eukaryotes need multiple origins of replication?

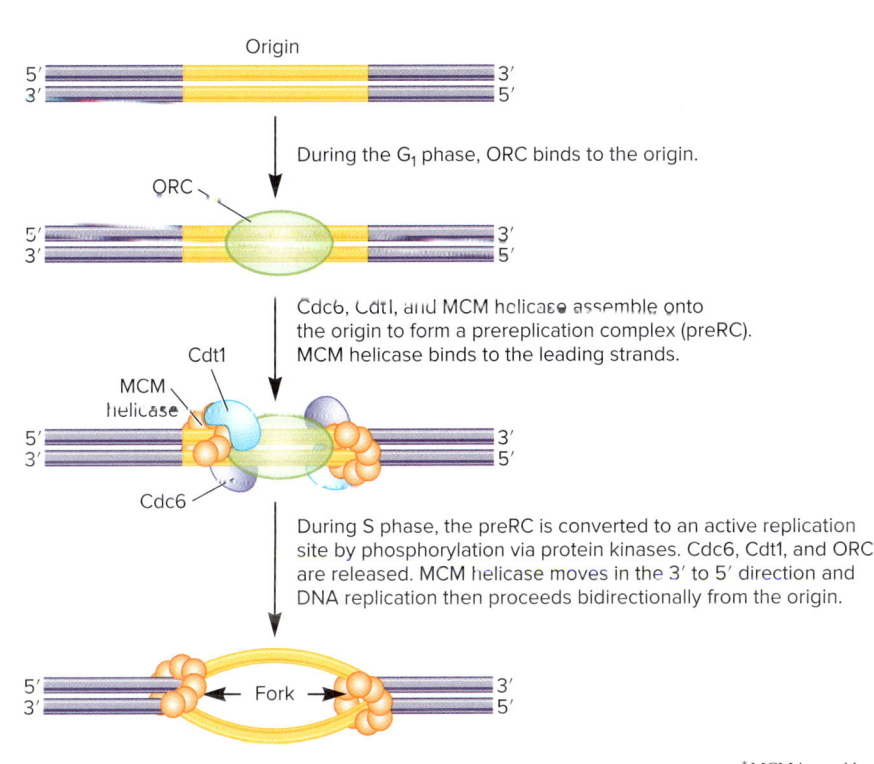

The initiation of DNA replication in eukaryotes is a multistep process. Initiation requires the assembly of a **prereplication complex (preRC)** during the G_1 phase of the cell cycle (**Figure 13.17**). Part of the preRC is a group of proteins called the **origin recognition complex (ORC)**, which acts as the first initiator of preRC assembly. ORC promotes the binding of Cdc6, Cdt1, and a group of six proteins called **MCM helicase.*** The binding of MCM helicase to the leading strand completes a process called **DNA replication licensing.** Those origins with MCM helicases are able to begin the process of DNA synthesis.

As S phase approaches, the preRC is converted to an active replication site by phosphorylation via protein kinases. This phosphorylation promotes the release of Cdc6, Cdt1, and ORC and the assembly of additional replication factors and DNA polymerases. (These additional replication factors and DNA polymerases are not shown in Figure 13.17.) The MCM helicase moves in the 3′ to 5′ direction, and DNA replication then proceeds bidirectionally from the origin.

Eukaryotes Have Many Different DNA Polymerases

Eukaryotes have many types of DNA polymerases. For example, mammalian cells have well over a dozen different DNA polymerases (**Table 13.2**).

- DNA polymerase α (alpha) is the only eukaryotic polymerase that associates with primase. The functional role of the DNA polymerase α/primase complex is to synthesize a short RNA–DNA primer of approximately 10 RNA nucleotides followed by 20 to 30 DNA nucleotides.

FIGURE 13.17 The formation of a prereplication complex in eukaryotes. This is a simplified model; more proteins are involved in this process than are shown here.

*MCM is an abbreviation for minichromosome maintenance. The genes encoding MCM proteins were originally identified in mutant yeast strains that are defective in the maintenance of minichromosomes in the cell. MCM proteins have since been shown to play a role in DNA replication.

TABLE 13.2
Eukaryotic DNA Polymerases

Polymerase Types*	Function
α	Initiates DNA replication in conjunction with primase
ε	Replication of the leading strand during S phase
δ	Replication of the lagging strand during S phase
γ	Replication of mitochondrial DNA
η, κ, ι, ξ (translesion-replicating polymerases)	Replication of damaged DNA
α, β, δ, ε, σ, λ, μ, φ, θ, η	DNA repair or other functions†

*The designations are those of the mammalian enzymes.
†Many DNA polymerases have dual functions. For example, DNA polymerases α, δ, and ε are involved in the replication of normal DNA and also play a role in DNA repair.

- This short RNA–DNA strand is then used by DNA polymerase ε (epsilon) or δ (delta) for the processive elongation of the leading and lagging strands, respectively. For this to happen, the DNA polymerase α/primase complex dissociates from the replication fork and is exchanged for DNA polymerase ε or δ.
- DNA polymerase ε is primarily involved with leading-strand synthesis.
- DNA polymerase δ is responsible for lagging-strand synthesis.
- DNA polymerase γ (gamma) is responsible for the replication of mitochondrial DNA.
- Several DNA polymerases are **translesion-replicating polymerases.** When DNA polymerase α, δ, and ε encounter abnormalities in DNA structure, such as abnormal bases or covalent crosslinks, they may be unable to replicate over the aberration. When this occurs, a translesion-replicating polymerase is attracted to the damaged DNA, and its special properties enable it to synthesize a complementary strand over the abnormal region. The various translesion-replicating polymerases are able to replicate over different kinds of DNA damage. For example, DNA polymerase κ can replicate over damage caused by benzo[a]pyrene, a hydrocarbon present in cigarette smoke, whereas DNA polymerase η can replicate over thymine dimers, which are caused by UV light.
- Several DNA polymerases are involved with DNA repair, a topic that will be examined in Chapter 19. For example, DNA polymerase β is not involved in the replication of normal DNA, but it plays an important role in removing incorrect bases from damaged DNA.

Flap Endonuclease Removes RNA Primers During Eukaryotic DNA Replication

Another key difference between bacterial and eukaryotic DNA replication is the way that RNA primers are removed. As discussed earlier in this chapter, bacterial RNA primers are removed by DNA polymerase I. By comparison, a DNA polymerase enzyme does not play this role in eukaryotes. Instead, an enzyme called **flap endonuclease** is primarily responsible for RNA primer removal.

Flap endonuclease gets its name because it removes small RNA flaps that are generated by the action of DNA polymerase δ (**Figure 13.18**).

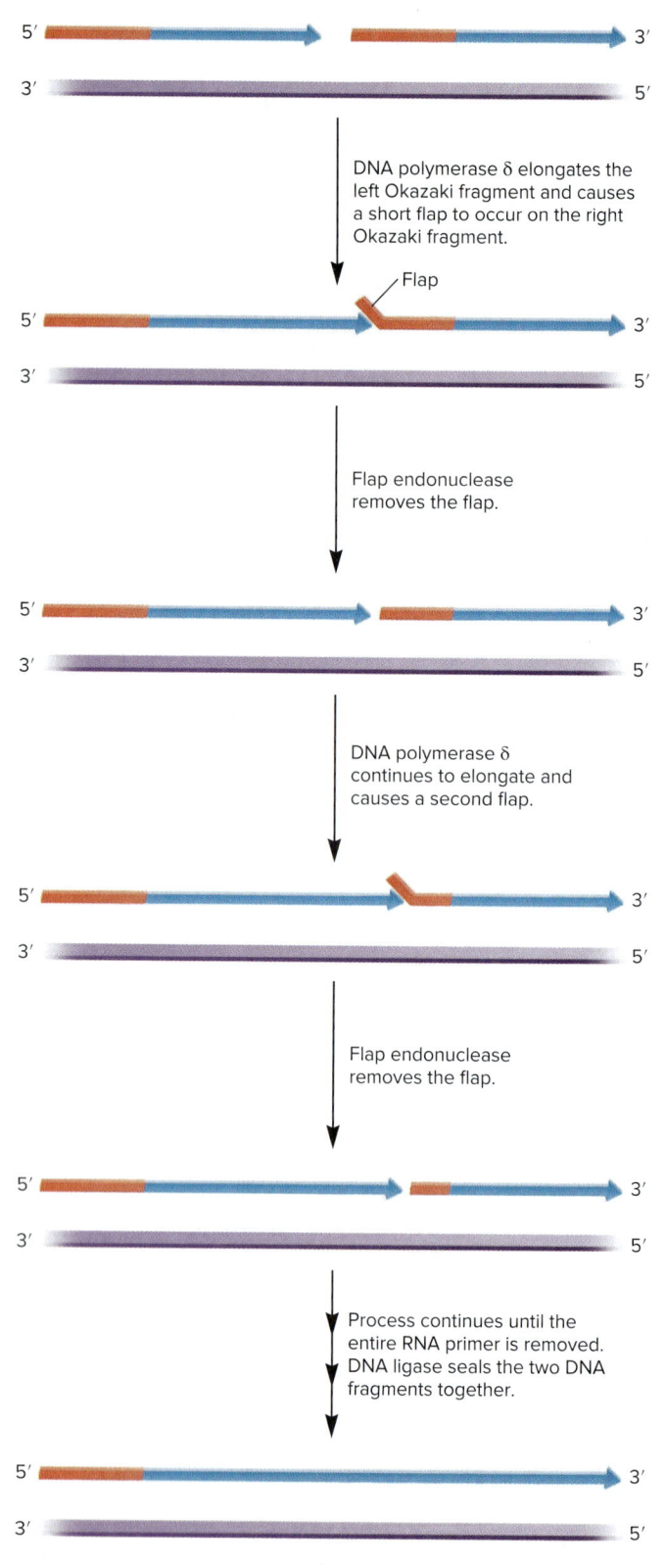

FIGURE 13.18 Removal of an RNA primer by flap endonuclease.

1. DNA polymerase δ elongates the left Okazaki fragment until it runs into the RNA primer of the adjacent Okazaki fragment on the right. This causes a portion of the RNA primer to form a short flap, which is removed by flap endonuclease.
2. As DNA polymerase δ continues to elongate the DNA, short flaps continue to be generated, which are sequentially removed by flap endonuclease.
3. Eventually, all of the RNA primer is removed, and DNA ligase can seal the DNA fragments together.

Though flap endonuclease is thought to be the primary agent for RNA primer removal in eukaryotes, it is unable to remove a flap that is too long. In such cases, the long flap is cleaved by the enzyme called Dna2 nuclease/helicase. This enzyme can cut a long flap, thereby generating a short flap that is then removed via flap endonuclease.

TABLE 13.3
Telomeric Repeat Sequences Within Selected Organisms

Group	Example	Telomeric Repeat Sequence
Mammals	Humans	TTAGGG
Slime molds	*Physarum, Didymium*	TTAGGG
	Dictyostelium	AG$_{(1-8)}$
Filamentous fungi	*Neurospora*	TTAGGG
Budding yeast	*Saccharomyces cerevisiae*	TG$_{(1-3)}$
Ciliates	*Tetrahymena*	TTGGGG
	Paramecium	TTGGG(T/G)
	Euplotes	TTTTGGGG
Flowering plants	*Arabidopsis*	TTTAGGG

The Ends of Eukaryotic Chromosomes Are Replicated by Telomerase

Linear eukaryotic chromosomes have **telomeres** at both ends. The term *telomere* refers to the telomeric sequences within the DNA and the special proteins that are bound to these sequences. Telomeric sequences consist of a moderately repetitive tandem array and a 3′ overhang region that is 12 to 16 nucleotides in length (**Figure 13.19**).

The tandem array that occurs within telomeres has been studied in a wide variety of eukaryotic organisms. A common feature is that the telomeric sequence contains several guanine nucleotides and often many thymine nucleotides (**Table 13.3**). Depending on the species and the cell type, this sequence can be tandemly repeated up to several hundred times in the telomere region.

One reason telomeric repeat sequences are needed is that DNA polymerase is unable to replicate the 3′ ends of DNA strands. Why is DNA polymerase unable to replicate this region? The answer lies in the two unusual functional features of this enzyme. As discussed previously, DNA polymerase synthesizes DNA only in a 5′ to 3′ direction, and it cannot link together the first two individual nucleotides; it can elongate only preexisting strands (refer back to Figure 13.9). These two features of DNA polymerase function pose a problem at the 3′ ends of linear chromosomes. As shown in **Figure 13.20**, the 3′ end of a DNA strand cannot be replicated by DNA polymerase

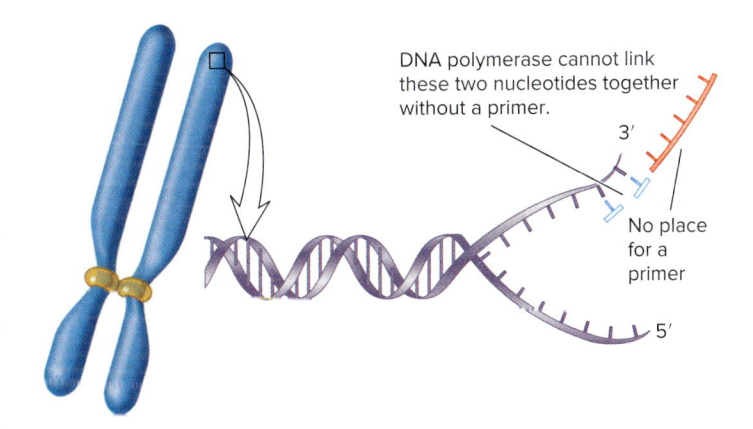

FIGURE 13.20 **The replication problem at the ends of linear chromosomes.** DNA polymerase cannot synthesize a DNA strand that is complementary to the 3′ end because a primer cannot be made upstream from this site.

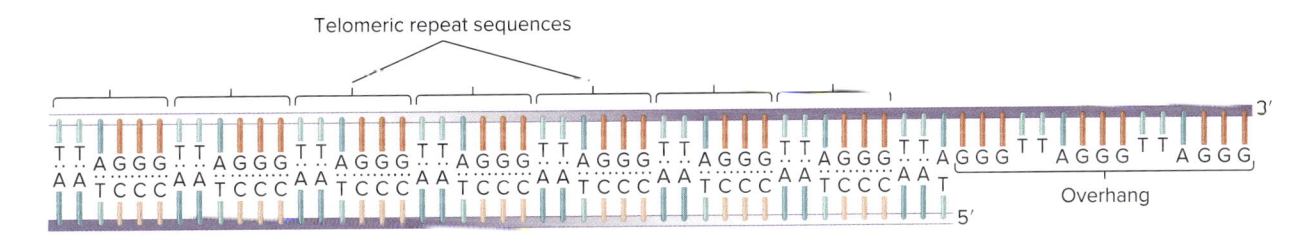

FIGURE 13.19 **General structure of telomeric sequences.** The telomere DNA consists of a tandem array with an overhang consisting of 12- to 16-nucleotides.

because a primer cannot be made upstream from this point. Therefore, if this problem was not solved, the chromosome would become progressively shorter with each round of DNA replication.

To prevent the loss of genetic information due to chromosome shortening, additional DNA sequences are attached to the ends of telomeres. In 1984, Carol Greider and Elizabeth Blackburn discovered an enzyme called **telomerase** that prevents chromosome shortening. These researchers received the 2009 Nobel Prize in physiology or medicine for their discovery. Telomerase recognizes the sequences at the ends of eukaryotic chromosomes and synthesizes additional repeats of telomeric sequences. The telomerase enzyme contains both protein subunits and RNA. The RNA part of telomerase contains a sequence complementary to the DNA sequence found in the telomeric repeat. **Figure 13.21** shows the interesting mechanism by which telomerase works.

1. Telomerase contains both protein subunits and RNA. The RNA part of telomerase, known as **telomerase RNA component (TERC)**, contains a sequence complementary to that found in the telomeric repeat sequence. This allows telomerase to bind to the 3′ overhang region of the telomere.
2. Following binding, the RNA sequence beyond the binding site functions as a template allowing the synthesis of a six-nucleotide sequence at the end of the DNA strand. This synthesis is called polymerization, because it is analogous to the function of DNA polymerase. It is catalyzed by two identical protein subunits called **telomerase reverse transcriptase (TERT)**. TERT's name indicates that it uses an RNA template to synthesize DNA.
3. Following polymerization, telomerase then moves to the new end of this DNA strand, a process called translocation.

This binding–polymerization–translocation cycle occurs many times in a row, thereby greatly lengthening the 3′ end of the DNA strand in the telomeric region. The complementary strand is then synthesized by primase, DNA polymerase, and DNA ligase.

13.5 REVIEWING THE KEY CONCEPTS

- Eukaryotic chromosomes contain multiple origins of replication. Part of the prereplication complex is formed from a group of six proteins called the origin recognition complex (ORC). The binding of MCM helicase completes a process called DNA replication licensing (see Figures 13.15–13.17).
- Eukaryotes have many different DNA polymerases with specialized roles. Different types of DNA polymerases switch with each other during the process of DNA replication (see Table 13.2).
- Flap endonuclease is an enzyme that removes RNA primers from Okazaki fragments (see Figure 13.18).
- The ends of eukaryotic chromosomes contain telomeres, which are composed of short repeat sequences and proteins (see Figure 13.19, Table 13.3).

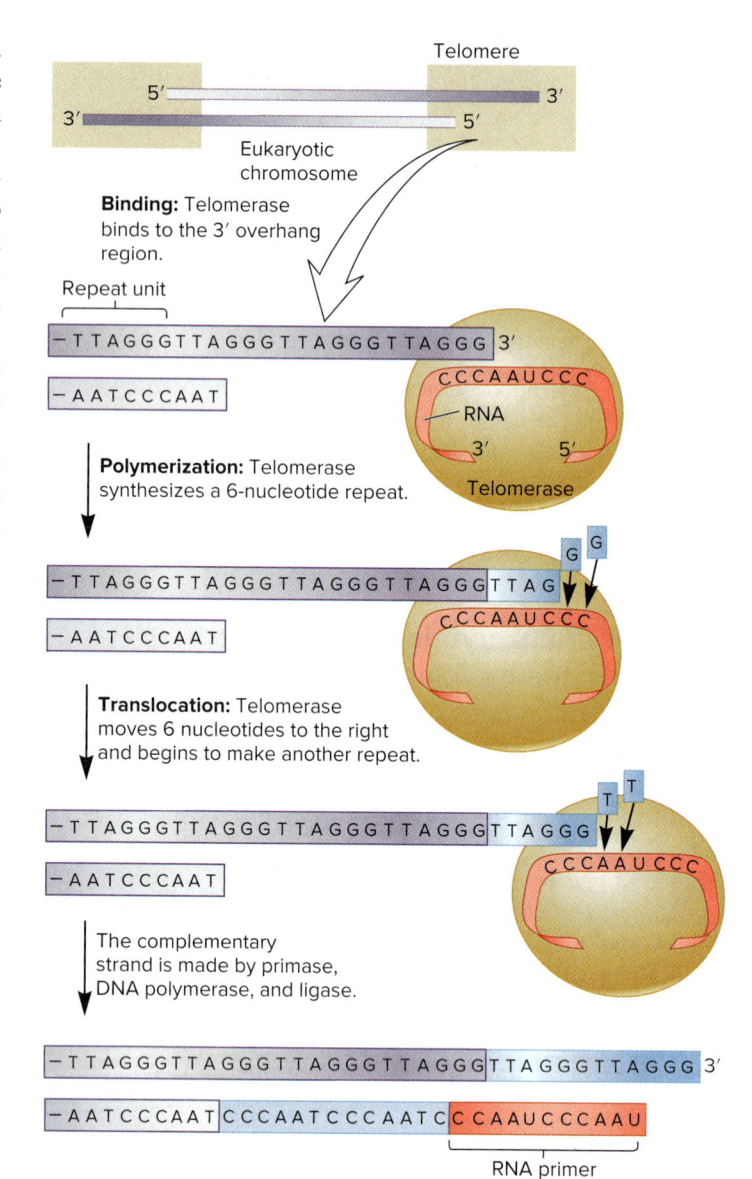

FIGURE 13.21 The enzymatic action of telomerase. A short, three-nucleotide segment of RNA within telomerase allows it to bind to the 3′ overhang. The adjacent part of the RNA is used as a template to make a short, six-nucleotide repeat of DNA. After the repeat is made, telomerase moves six nucleotides to the right and then synthesizes another repeat. This process is repeated many times to lengthen the top strand shown in this figure. The bottom strand is made by DNA polymerase, using an RNA primer at the end of the chromosome that is complementary to the telomeric repeat sequence in the top strand. DNA polymerase fills in the region, which is sealed by ligase.

Concept Check: How many times would telomerase have to bind to a different site in the telomere to make a segment of DNA that is 36 nucleotides in length?

- DNA polymerase is unable to replicate the very end of a eukaryotic chromosome (see Figure 13.20).
- Telomerase uses a short RNA molecule as a template to add repeat sequences onto telomeres (see Figure 13.21).

13.5 COMPREHENSION QUESTIONS

1. In eukaryotes, DNA replication is initiated at an origin of replication by
   a. DnaA.
   b. the origin recognition complex.
   c. DNA polymerase δ.
   d. MCM helicase.

2. Which of the following statements regarding DNA polymerases in eukaryotes is *not* correct?
   a. DNA polymerase α synthesizes a short RNA–DNA primer.
   b. DNA polymerases δ and ε synthesize most of the lagging and leading strands, respectively.
   c. Translesion-replicating DNA polymerases can replicate over damaged DNA.
   d. All of the above statements are correct.

3. In eukaryotes, RNA primers are primarily removed by
   a. DNA polymerase I.
   b. DNA polymerase α.
   c. flap endonuclease.
   d. helicase.

4. To synthesize DNA, what does telomerase use as a template?
   a. It uses the DNA in the 3′ overhang region.
   b. It uses RNA that is already a component of telomerase.
   c. No template is used.
   d. Both a and b are correct.

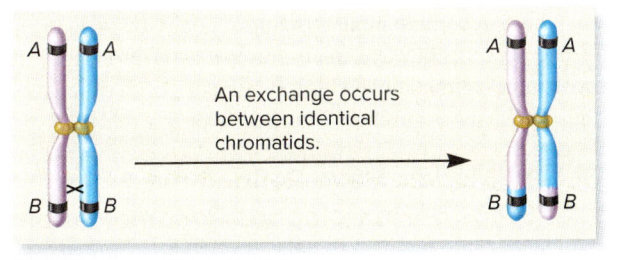

13.6 HOMOLOGOUS RECOMBINATION

Learning Outcomes:
1. Describe the Holiday model and the double-strand break model for homologous recombination.
2. Explain how gene conversion can occur via mismatch repair and DNA gap repair synthesis.

Homologous recombination is the process whereby DNA segments that are similar or identical to each other break and rejoin to form a new combination. This formation of new combinations occurs when chromosomes cross over during meiosis, as described in Chapter 2. Not only does homologous recombination enhance genetic diversity, it also helps to repair DNA and ensures the proper segregation of chromosomes.

Homologous recombination involves an exchange of DNA segments that are similar or identical in their DNA sequences. Eukaryotic chromosomes that have similar or identical sequences frequently participate in crossing over during meiosis I and occasionally during mitosis. Crossing over involves the alignment of a pair of homologous chromosomes, followed by the breakage of two chromatids at analogous locations, and the subsequent exchange of the corresponding segments (refer to Figure 2.10).

Figure 13.22 shows two types of crossing over that may occur between replicated chromosomes in a diploid species.

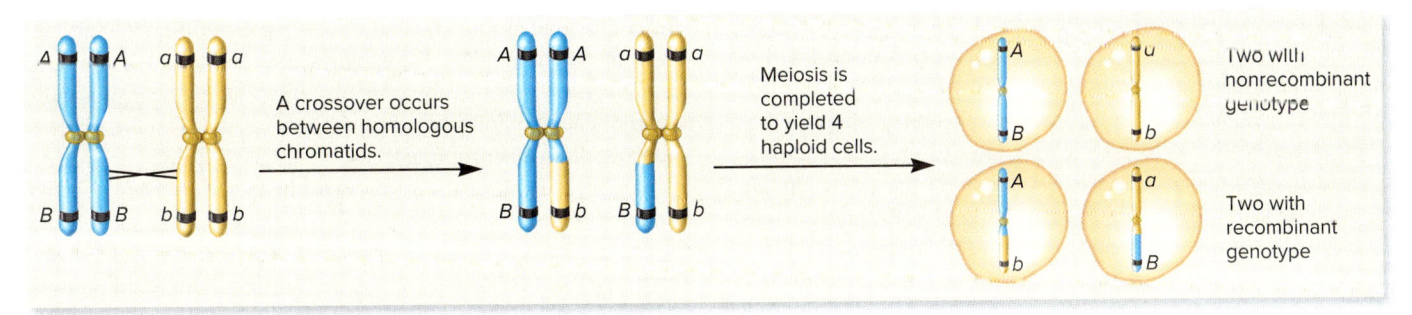

(a) Sister chromatid exchange

(b) Recombination between homologous chromosomes during meiosis

FIGURE 13.22 Crossing over between eukaryotic chromosomes. (a) Sister chromatid exchange occurs when genetically identical chromatids cross over. (b) Homologous recombination can also occur when homologous chromatids cross over. This form of homologous recombination may lead to a new combination of alleles, which is called a recombinant genotype.

Genes → Traits Homologous recombination is particularly important when we consider the relationships between multiple genes and multiple traits. For example, if the X chromosome in a female fruit fly carries alleles for red eyes and gray body, and its homolog carries alleles for white eyes and yellow body, homologous recombination could produce recombinant chromosomes that carry alleles for red eyes and yellow body or that carry alleles for white eyes and gray body. Therefore, new combinations of two or more alleles can arise when homologous recombination takes place.

Concept Check: What is the advantage of genetic recombination, which is depicted in part (b)?

- When crossing over takes place between sister chromatids, the process is called **sister chromatid exchange (SCE).** Because sister chromatids are genetically identical, SCE does not produce a new combination of alleles (Figure 13.22a).
- Crossing over commonly occurs between homologous chromosomes during meiosis. As shown in Figure 13.22b, this form of homologous recombination may produce new combinations of alleles in the resulting chromosomes. In this second case, homologous recombination has resulted in **genetic recombination,** which refers to the shuffling of genetic material to create a new combination of alleles that differs from the original combination. Homologous recombination is an important mechanism for fostering genetic variation.

Bacteria are usually haploid. They do not have pairs of homologous chromosomes. Even so, bacteria can undergo homologous recombination. How can the exchange of DNA segments occur in a haploid organism? First, bacteria may have more than one copy of a chromosome per cell, though the copies are usually identical. These copies can exchange genetic material via homologous recombination. Second, during DNA replication, the replicated regions may also undergo homologous recombination. In bacteria, homologous recombination is particularly important in the repair of DNA segments that have been damaged. In this section, we will focus on the molecular mechanisms that underlie homologous recombination.

The Holliday Model Describes a Molecular Mechanism for the Recombination Process

Geneticists have learned a great deal from the analysis of fungi, because certain species produce asci (singular, ascus), which are sacs that contain the products of a single meiosis. When two haploid fungi that differ at a single gene are crossed to each other, the ascus is expected to contain an equal proportion of each genotype. For example, if a pigmented strain of *Neurospora* producing orange spores is crossed to an albino strain producing white spores, the resulting group of eight cells, called an octad, should contain four orange spores and four white spores. In 1934, H. Zickler noticed that unequal proportions of the spores sometimes occurred within asci. He occasionally observed octads with six orange spores and two white spores, or six white spores and two orange spores.

Zickler used the term **gene conversion** to describe the phenomenon in which one allele is converted to the allele on the homologous chromosome. Subsequent studies by several researchers confirmed this phenomenon in yeast and *Neurospora*. Gene conversion occurred at too high a rate to be explained by new mutations. In addition, research showed that gene conversion often occurs in a chromosomal region where a crossover has taken place.

Based on studies involving gene conversion, Robin Holliday proposed a model in 1964 to explain the molecular steps that occur during homologous recombination. We will first consider the steps in the Holliday model and then consider a more recent model. Later, we will examine how the Holliday model can explain the phenomenon of gene conversion.

The **Holliday model** is shown in **Figure 13.23a**.

1. At the beginning of the process, two homologous chromatids are aligned with each other. According to the model, a break or nick occurs at identical sites in one strand of each of the two homologous chromatids.
2. The strands then invade the opposite helices and base-pair with the complementary strands. This event is followed by covalent linkages to create a **Holliday junction.**
3. The cross in the Holliday junction can migrate in a lateral direction. As it does so, a DNA strand in one helix is swapped for a DNA strand in the other helix. This process is called **branch migration** because the junction connecting the two double helices migrates laterally. Because the DNA sequences in the homologous chromosomes are similar but may not be identical, the swapping of the DNA strands during branch migration may produce a **heteroduplex,** a region in the double-stranded DNA that contains base mismatches.
4. The last event in the recombination process is called **resolution** because it involves the breakage and rejoining of two DNA strands to produce two separate chromosomes (look at the bottom of Figure 13.23a). If breakage occurs in the same two DNA strands that were originally nicked at the beginning of this process, the subsequent joining of strands produces nonrecombinant chromosomes, each with a heteroduplex region. Alternatively, if breakage occurs in the strands that were not originally nicked, the rejoining process results in recombinant chromosomes, also with heteroduplex regions.

The Holliday model can account for the general properties of recombinant chromosomes formed during eukaryotic meiosis. Particularly convincing evidence came from electron microscopy studies in which recombination structures could be visualized. **Figure 13.23b** shows an electron micrograph of two DNA fragments in the process of recombination. This structure has been called a chi form because its shape is similar to the Greek letter χ (chi).

More Recent Models Have Refined the Molecular Steps of Homologous Recombination

As more detailed studies of homologous recombination have become available, certain steps in the Holliday model have been reconsidered. In particular, more recent models have modified the initiation phase of recombination. Researchers now think that homologous recombination is not likely to involve nicks at identical sites in one strand of each homologous chromatid. Instead, it is more likely for a DNA helix to incur a break in both strands of one chromatid or a single nick. Both of these types of changes have been shown to initiate homologous recombination. Therefore, newer models have tried to incorporate these experimental observations. Jack Szostak, Terry Orr-Weaver, Rodney Rothstein, and Franklin Stahl proposed that a double-strand break initiates the recombination process. This is called the **double-strand break model.** Though recombination may occur via more than one mechanism, research indicates that double-strand breaks commonly promote homologous recombination during meiosis and during DNA repair.

Figure 13.24 shows the general steps in the double-strand break model.

1. The top chromosome has experienced a double-strand break.
2. A small region near the break is degraded.
3. Strand degradation generates a single-stranded DNA segment that can invade the intact double helix. The strand

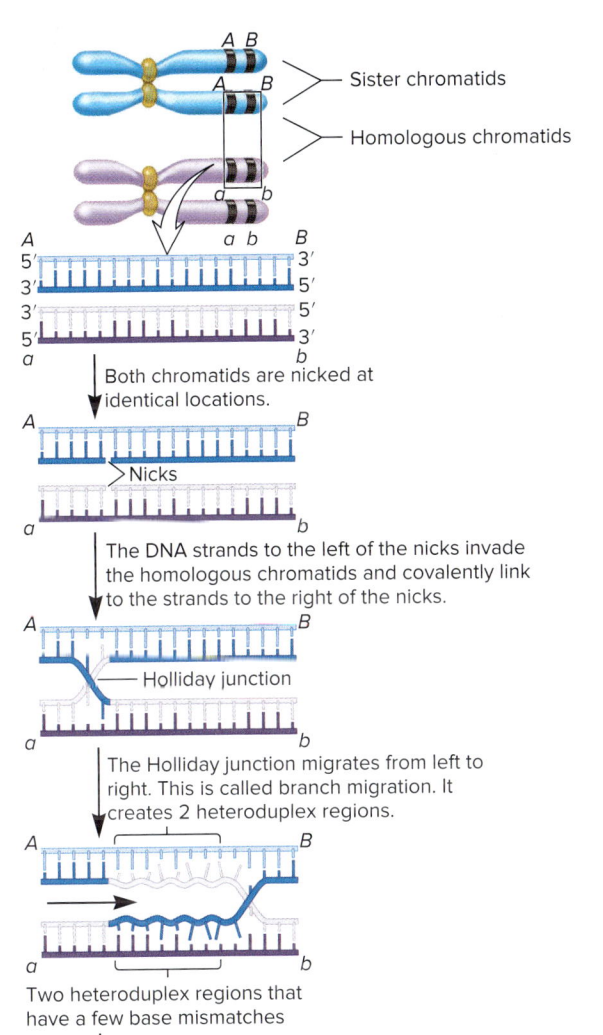

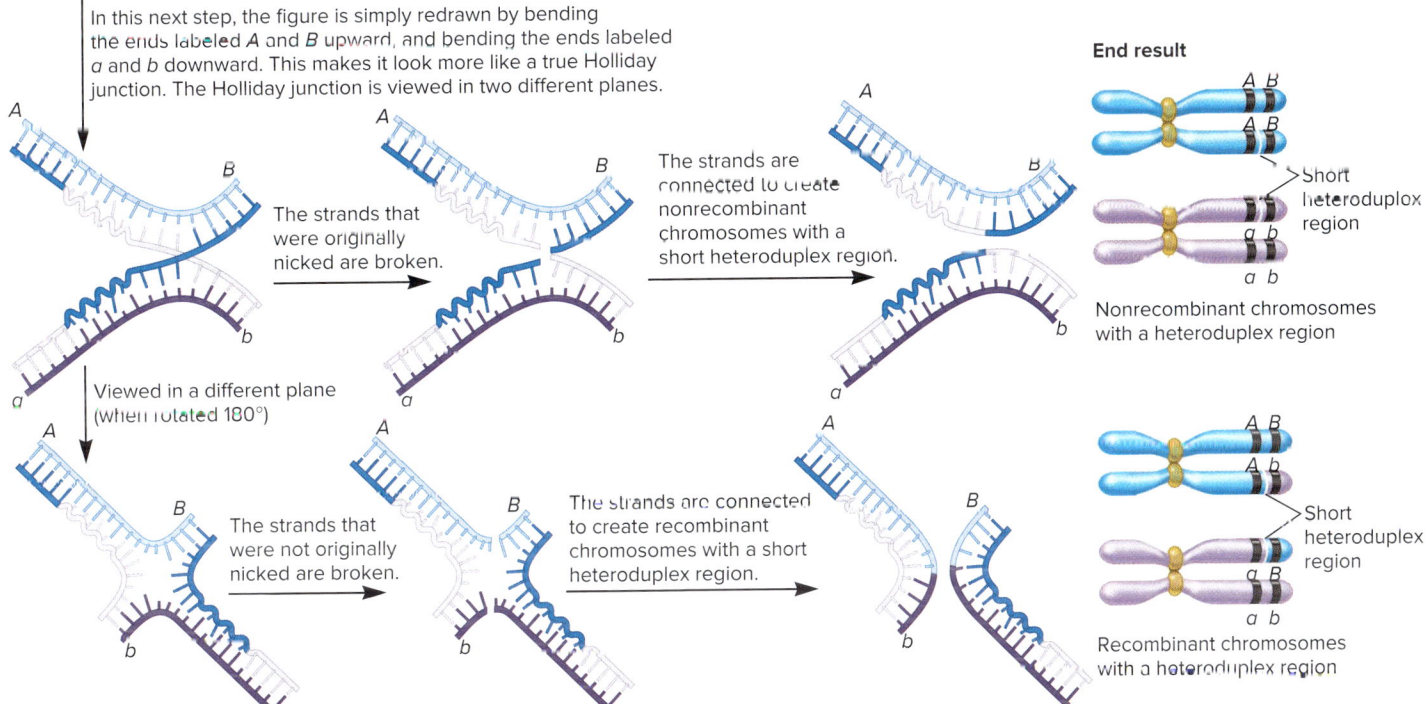

13.6 HOMOLOGOUS RECOMBINATION 287

FIGURE 13.23 The Holliday model for homologous recombination. The Holliday model is adapted from R. Holliday (1964) A mechanism for gene conversion in fungi. *Genet Res 5*, 282–304.

Concept Check: Explain why a heteroduplex region may be produced after branch migration occurs.

(a) The Holliday model for homologous recombination

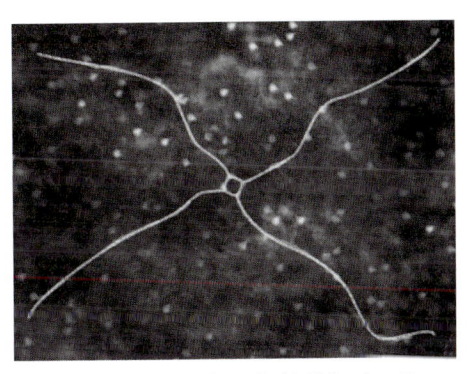

(b) A schematic drawing of a Holliday junction

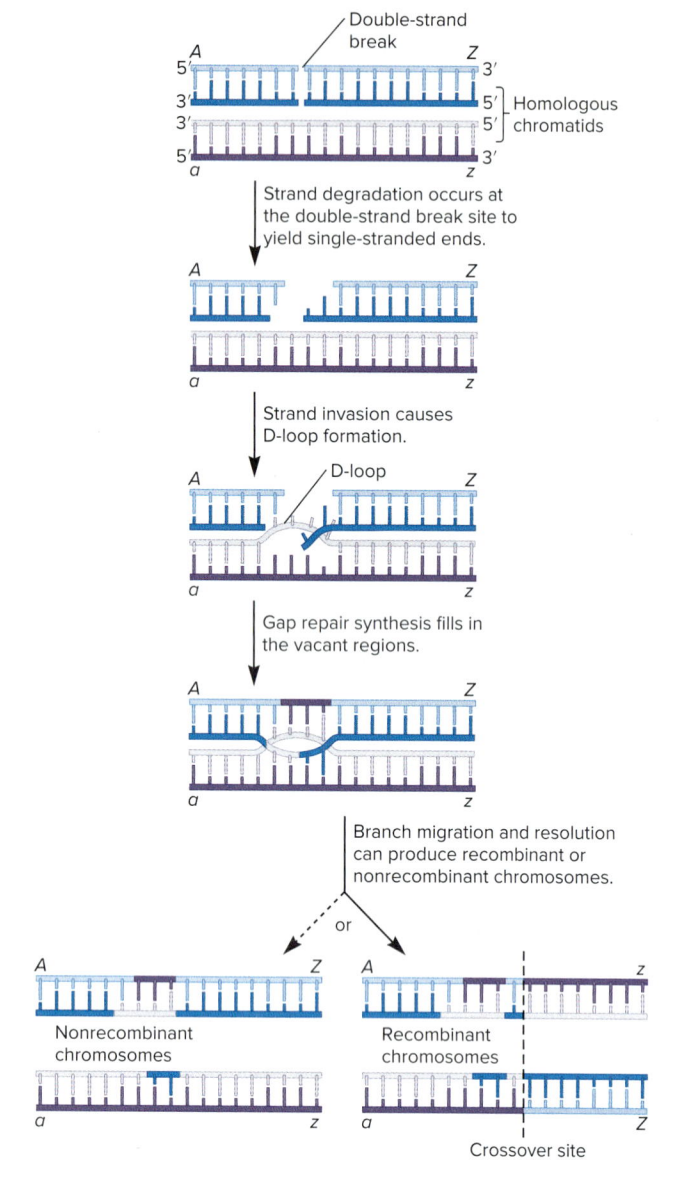

FIGURE 13.24 A simplified version of the double-strand break model. For simplicity, this illustration does not include the formation of heteroduplexes. The dashed arrow indicates that the pathway to the left may be less favored.

Concept Check: Describe the structure and location of a D-loop.

TABLE 13.4
E. coli Proteins That Play a Role in Homologous Recombination

Protein	Description
RecBCD	A complex of three proteins that tracks along the DNA and recognizes double-strand breaks. The complex partially degrades the double-stranded regions to generate single-stranded regions that can participate in strand invasion. RecBCD is also involved in loading RecA onto single-stranded DNA.
Single-strand binding protein	Coats broken ends of chromosomes and prevents excessive strand degradation.
RecA	Binds to single-stranded DNA and promotes strand invasion, which enables homologous strands to find each other. It also promotes the displacement of the complementary strand to generate a D-loop.
RuvABC	This protein complex binds to Holliday junctions. RuvAB promotes branch migration. RuvC is an endonuclease that cuts the crossed or uncrossed strands to resolve Holliday junctions into separate chromosomes.
RecG	RecG protein can also promote branch migration of Holliday junctions.

Various Proteins Are Necessary to Facilitate Homologous Recombination

The homologous recombination process requires the participation of many proteins that catalyze different steps in the recombination pathway. Homologous recombination is found in all species, and the types of proteins that participate in the steps outlined in Figure 13.24 are very similar. The cells of any given species may have more than one molecular mechanism to carry out homologous recombination. This process is best understood in *Escherichia coli*. **Table 13.4** summarizes some of the *E. coli* proteins that play critical roles in one recombination pathway observed in this species. Though discussing them is beyond the scope of this text, *E. coli* has other pathways to carry out homologous recombination.

Gene Conversion May Result from DNA Mismatch Repair or DNA Gap Repair Synthesis

As mentioned earlier, homologous recombination can lead to gene conversion, in which one allele is converted to the allele on the homologous chromosome. The original Holliday model was based on this phenomenon.

How can homologous recombination account for gene conversion? Researchers have identified two possible ways this can occur.

DNA Mismatch Repair As shown in **Figure 13.25**, DNA mismatch repair of a heteroduplex may result in gene conversion. In this example, the two original chromosomes had different alleles due to a single base-pair difference in their DNA sequences, as shown at the top of the figure. During recombination, branch migration has occurred across this region, thereby creating two heteroduplexes with base mismatches. As described in Chapter 19,

displaced by the invading segment forms a structure called a displacement loop (D-loop).

4. After the D-loop is formed, two regions have a gap in the DNA. How is the problem fixed? DNA synthesis occurs in the relatively short gaps where a DNA strand is missing. This DNA synthesis is called **DNA gap repair synthesis.**

5. After DNA gap repair synthesis is completed, two Holliday junctions are produced. Depending on the way these are resolved, the end result is nonrecombinant or recombinant chromosomes, each containing a short heteroduplex. In eukaryotes such as yeast, evidence suggests that certain proteins bound to Holliday junctions may regulate the resolution step in a way that favors the formation of recombinant chromosomes rather than nonrecombinant chromosomes.

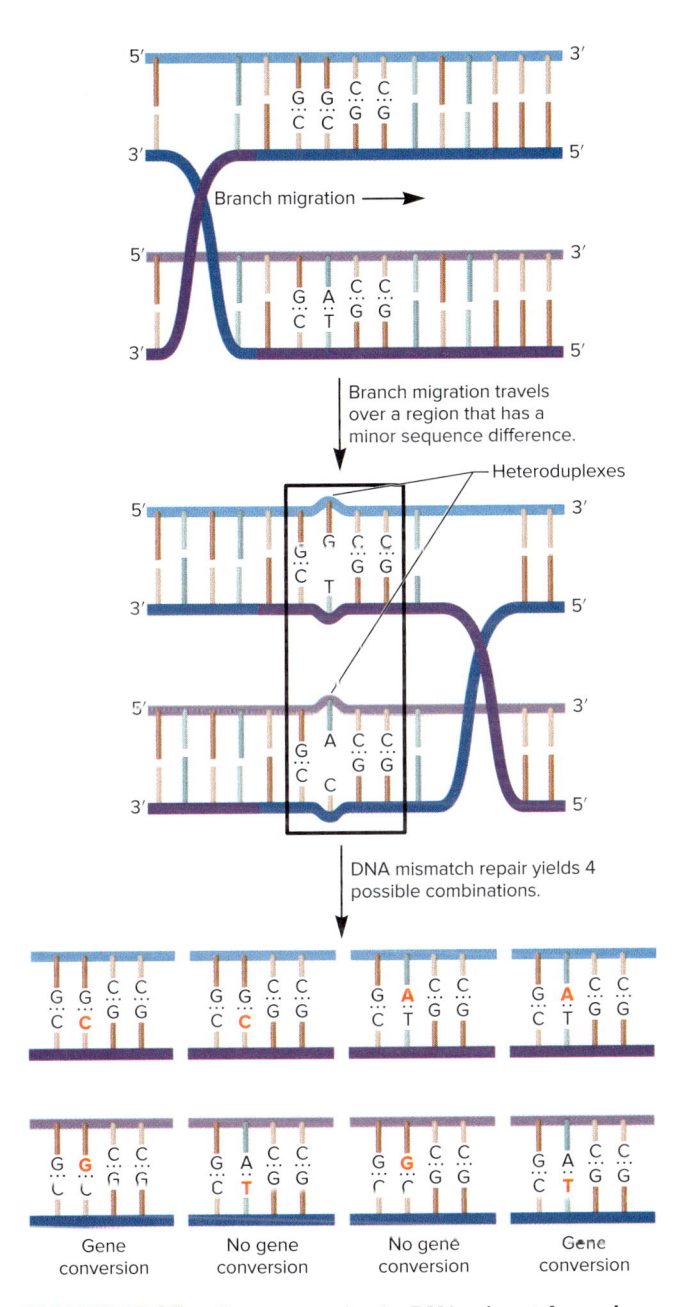

FIGURE 13.25 **Gene conversion by DNA mismatch repair.**
A branch migrates past a homologous region that contains a slightly different DNA sequence. This produces two heteroduplexes: DNA double helices with mismatches. The mismatches can be repaired in four possible ways by the mismatch repair system described in Chapter 19. Two of these ways result in gene conversion. The repaired base is shown in red.

DNA mismatches are recognized by DNA repair systems and repaired to yield a double helix that obeys the AT/GC rule. These two mismatches can be repaired in four possible ways. As shown here, two possibilities produce no gene conversion, whereas the other two lead to gene conversion.

DNA Gap Repair Synthesis A second mechanism that can result in gene conversion is DNA gap repair synthesis. **Figure 13.26** illustrates how gap repair synthesis can lead to gene conversion according to the double-strand break model. The top

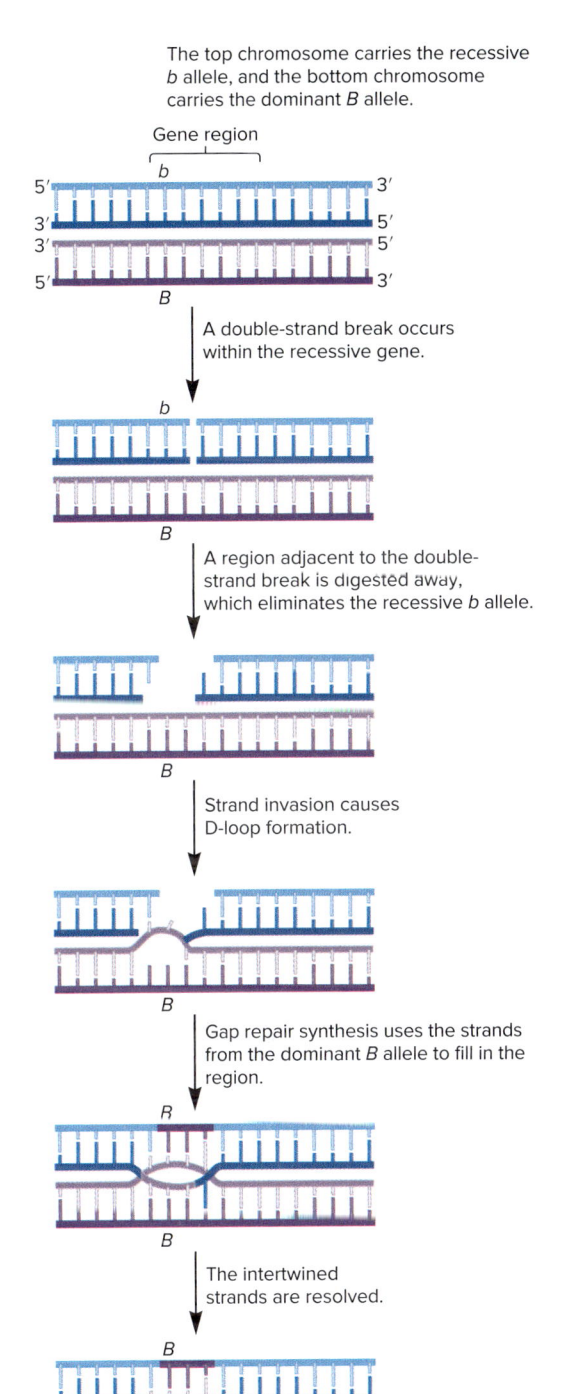

FIGURE 13.26 **Gene conversion by DNA gap repair synthesis according to the double-strand break model.** A gene is found in two alleles, designated B and b. A double-strand break occurs in the DNA encoding the b allele. Both of these DNA strands are digested away, thereby eliminating the b allele. A complementary DNA strand encoding the B allele migrates to this region and provides the template to synthesize a new double-stranded region. Following resolution, both DNA double helices carry the B allele.

Concept Check: Explain what happened to the b allele that allowed gene conversion to occur.

chromosome, which carries the recessive *b* allele, has incurred a double-strand break in this gene. A gap is created by the digestion of the DNA in the double helix. This digestion eliminates the *b* allele. The two template strands used in gap repair synthesis are from one homologous chromatid. This helix carries the dominant *B* allele. Therefore, after gap repair synthesis takes place, the top chromosome carries the *B* allele, as does the bottom chromosome. Gene conversion has changed the recessive *b* allele to a dominant *B* allele.

13.6 REVIEWING THE KEY CONCEPTS

- Homologous recombination involves an exchange of DNA segments that are similar or identical in their DNA sequences. It can occur between sister chromatids or between homologous chromosomes (see Figure 13.22).
- The Holliday model describes the molecular steps that occur during homologous recombination between homologous chromosomes (see Figure 13.23).
- The initiation of homologous recombination usually occurs with a double-strand break (see Figure 13.24).
- Several different proteins are involved in homologous recombination (see Table 13.4).
- Two different mechanisms, DNA mismatch repair and DNA gap repair synthesis, can result in gene conversion during homologous recombination (see Figures 13.25, 13.26).

13.6 COMPREHENSION QUESTIONS

1. Homologous recombination refers to the exchange of DNA segments that are
   a. similar or identical in their DNA sequences.
   b. in close proximity to one another.
   c. broken due to ionizing radiation.
   d. misaligned along a chromosome.

2. During the molecular process of homologous recombination between homologous chromosomes,
   a. a Holliday junction forms.
   b. branch migration occurs.
   c. a heteroduplex region forms.
   d. all of the above occur.

3. A key difference between the original Holliday model and the double-strand break model is the way that
   a. the DNA strands are initially broken.
   b. branch migration occurs.
   c. a heteroduplex is formed.
   d. resolution occurs.

4. Which of the following mechanisms can cause gene conversion?
   a. DNA mismatch repair
   b. DNA gap repair
   c. Resolution of a Holliday junction
   d. Both a and b can result in gene conversion.

KEY TERMS

Page 266. DNA replication
Page 267. template strands (parental strands), daughter strands
Page 268. conservative model, semiconservative model, dispersive model
Page 271. origin of replication, bidirectionally, replication fork, DnaA proteins, DnaA boxes, AT-rich region, DNA helicase, bidirectional replication, GATC methylation sites
Page 273. topoisomerase II (DNA gyrase), single-strand binding proteins
Page 274. RNA primers, primase, leading strand, lagging strand, DNA polymerase
Page 276. Okazaki fragments, DNA ligase, primosome, replisome, dimeric DNA polymerase
Page 279. processive enzyme, fidelity, proofreading
Page 280. ARS elements
Page 281. prereplication complex (preRC), origin recognition complex (ORC), MCM helicase, DNA replication licensing
Page 282. translesion-replicating polymerases, flap endonuclease
Page 283. telomeres
Page 284. telomerase, telomerase RNA component (TERC), telomerase reverse transcriptase (TERT)
Page 285. homologous recombination
Page 286. sister chromatid exchange (SCE), genetic recombination, gene conversion, Holliday model, Holliday junction, branch migration, heteroduplex, resolution, double-strand break model
Page 288. DNA gap repair synthesis

CHAPTER SUMMARY

- DNA replication is the process in which existing DNA strands are used to make new DNA strands.

13.1 Structural Overview of DNA Replication

- DNA replication occurs when the strands of DNA unwind and each strand is used as a template to make a new strand according to the AT/GC rule. The resulting DNA molecules have the same base sequence as the original DNA (see Figure 13.1).
- By labeling DNA with heavy and light isotopes of nitrogen and using centrifugation, Meselson and Stahl showed that DNA replication is semiconservative (see Figures 13.2, 13.3).

13.2 Bacterial DNA Replication: The Formation of Two Replication Forks at the Origin of Replication

- Bacterial DNA replication begins at a single origin of replication and proceeds bidirectionally around the circular chromosome (see Figure 13.4).
- In *E. coli*, DNA replication is initiated when DnaA proteins bind to five DnaA boxes at the origin of replication and cause separation of the strands at the AT-rich region. DNA helicases continue to separate the DNA strands, thereby promoting the movement of two replication forks (see Figures 13.5, 13.6).

13.3 Bacterial DNA Replication: Synthesis of New DNA Strands

- At each replication fork, DNA helicase unwinds the DNA and topoisomerase II alleviates positive supercoiling. Single-strand binding proteins coat the DNA to prevent the strands from coming back together. Primase synthesizes RNA primers, and DNA polymerase synthesizes complementary strands of DNA. DNA ligase joins Okazaki fragments by catalyzing the formation of covalent bonds (see Figure 13.7, Table 13.1).
- DNA polymerase III is an enzyme in *E. coli* with several subunits. The catalytic subunit wraps around the DNA like a hand (see Figure 13.8).
- DNA polymerase needs an RNA primer or an existing DNA strand to synthesize DNA and make a new DNA strand in a 5′ to 3′ direction (see Figure 13.9).
- During DNA synthesis, the leading strand is made continuously in the direction of the replication fork, whereas the lagging strand is made as Okazaki fragments in the direction away from the fork (see Figures 13.10, 13.11).
- A primosome is a complex between DNA helicase and primase. A replisome is a complex between a primosome and dimeric DNA polymerase III (see Figure 13.12).

13.4 Bacterial DNA Replication: Chemistry and Accuracy

- DNA polymerase III is a processive enzyme that uses deoxyribonucleoside triphosphates (dNTPs) to make new DNA strands (see Figure 13.13).
- The high fidelity of DNA replication is a result of (1) the stability of hydrogen bonding between the correct pairs of bases, (2) the phenomenon of induced fit, and (3) the proofreading ability of DNA polymerase (see Figure 13.14).

13.5 Eukaryotic DNA Replication

- Eukaryotic chromosomes contain multiple origins of replication. Part of the prereplication complex is formed from a group of six proteins called the origin recognition complex (ORC). The binding of MCM helicase completes a process called DNA replication licensing (see Figures 13.15–13.17).
- Eukaryotes have many different DNA polymerases with specialized roles. Different types of DNA polymerases switch with each other during the process of DNA replication (see Table 13.2).
- Flap endonuclease is an enzyme that removes RNA primers from Okazaki fragments (see Figure 13.18).
- The ends of eukaryotic chromosomes contain telomeres, which are composed of short repeat sequences and proteins (see Figure 13.19, Table 13.3).
- DNA polymerase is unable to replicate the very end of a eukaryotic chromosome (see Figure 13.20).
- Telomerase uses a short RNA molecule as a template to add repeat sequences onto telomeres (see Figure 13.21).

13.6 Homologous Recombination

- Homologous recombination is an exchange of DNA segments that are similar or identical in their DNA sequences. It can occur between sister chromatids or between homologous chromosomes (see Figure 13.22).
- The Holliday model describes the molecular steps that occur during homologous recombination between homologous chromosomes (see Figure 13.23).
- The initiation of homologous recombination usually occurs with a double-strand break (see Figure 13.24).
- Several different proteins are involved in homologous recombination (see Table 13.4).
- Two different mechanisms, DNA mismatch repair and DNA gap repair synthesis, can result in gene conversion during homologous recombination (see Figures 13.25, 13.26).

PROBLEM SETS & INSIGHTS

More Genetic TIPS

1. Describe three factors that account for the high fidelity of DNA replication. Discuss the quantitative contributions of each of the three.

Topic: What topic in genetics does this question address?

The topic is the high fidelity of DNA replication. More specifically, the question asks you to describe and quantify the factors that contribute to that high fidelity.

Information: What information do you know based on the question and your understanding of the topic?

From the question, you know that three factors account for the high fidelity of DNA replication. From your understanding of the topic, you may recall that these are proper base pairing, induced fit, and proofreading.

CHAPTER 13 :: DNA REPLICATION AND RECOMBINATION

Problem-Solving Strategies: Relate structure and function.

One strategy to solve this problem is to consider how the structures of DNA and DNA polymerase are related to the fidelity of DNA replication.

Answer: First: A-T and G-C pairs are more likely to form than are other types of base pairs. This base pairing limits mistakes to around 1 mistake per 1000 nucleotides added.

Second: Induced fit by DNA polymerase prevents covalent bond formation unless the proper nucleotides are in place. This increases fidelity another 100- to 1000-fold, to a range from 1 error in 100,000 to 1 error in 1 million.

Third: Exonuclease proofreading increases fidelity another 100- to 1000-fold, to about 1 error per 100 million nucleotides added.

2. Summarize the process of chromosomal DNA replication in *E. coli*.

Topic: What topic in genetics does this question address?

The topic is DNA replication in bacteria. More specifically, the questions asks you to summarize the replication process in *E. coli*.

Information: What information do you know based on the question and your understanding of the topic?

From the question, you know that DNA replication occurs in *E. coli*. From your understanding of the topic, you may recall that DNA replication begins at an origin of replication and then two replication forks proceed bidirectionally around the circular chromosome.

Problem-Solving Strategies: Describe the steps.

DNA replication is a complicated process. One strategy to solve this problem is to break it down into its individual steps.

Answer:

Step 1. DnaA proteins bind to the origin of replication, resulting in the separation of the strands at the AT-rich region.

Step 2. DNA helicase breaks the hydrogen bonds between the DNA strands, topoisomerase II alleviates positive supercoiling, and single-strand binding proteins hold the parental strands apart. Two replication forks move bidirectionally outward from the origin.

Step 3. At each of the two forks, primase synthesizes one RNA primer in the leading strand and many RNA primers in the lagging strand. DNA polymerase III then synthesizes the daughter strands of DNA. In the lagging strand, many short segments of DNA (Okazaki fragments) are made. DNA polymerase I removes the RNA primers and fills in with DNA, and DNA ligase covalently links the Okazaki fragments together.

Step 4. The processes described in steps 2 and 3 continue until the two replication forks reach each other on the other side of the circular bacterial chromosome.

3. In the following schematic drawing of a Holliday junction, one chromatid is shown in red, and the homologous chromatid is shown in blue. The red chromatid carries a dominant allele labeled *A* and a recessive allele labeled *b*, whereas the blue chromatid carries a recessive allele labeled *a* and a dominant allele labeled *B*.

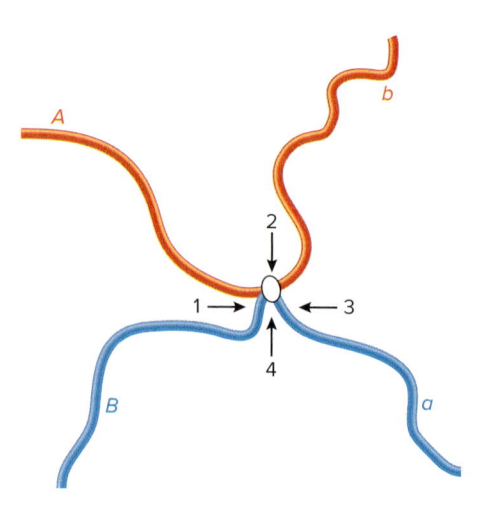

Where would the DNA strands have to be cut to produce recombinant chromosomes? Would they be cut at sites 1 and 3, or at sites 2 and 4? What would be the genotypes of the two recombinant chromosomes?

Topic: What topic in genetics does this question address?

The topic is homologous recombination. More specifically, the question is about the resolution steps of the process.

Information: What information do you know based on the question and your understanding of the topic?

In the question, you are given a drawing of a Holliday junction. From your understanding of the topic, you may remember that resolution can occur in two ways; either the strands that were originally nicked are cut again, or the strands that were not nicked are cut. The second type of event results in recombinant chromosomes with heteroduplex regions.

Problem-Solving Strategy: Describe the steps. Predict the outcome.

One strategy to solve this problem is to consider the two ways that the DNA strands can be cut. Compare the drawing given in this question with Figure 13.23.

Answer: The cutting would have to occur at the sites labeled 2 and 4. This would leave the *A* allele and the *B* allele on the same chromosome. The *a* allele in the other homolog would be on the same chromosome as the *b* allele. In other words, one recombinant chromosome would have *AB*, and the homolog would have *ab*.

Conceptual Questions

C1. What key structural features of the DNA molecule underlie its ability to be accurately replicated?

C2. With regard to DNA replication, define the term *bidirectional replication*.

C3. Which of the following statements is *not* true? Explain why.

   A. A DNA strand can serve as a template strand on many occasions.

   B. Following semiconservative DNA replication, one strand is a newly made daughter strand and the other strand is a parental strand.

   C. A DNA double helix may contain two strands of DNA that were made at the same time.

   D. A DNA double helix obeys the AT/GC rule.

   E. A DNA double helix could contain one strand that is 10 generations older than its complementary strand.

C4. The compound known as nitrous acid is a reactive chemical that replaces amino groups ($-NH_2$) with keto groups ($=O$). When nitrous acid reacts with the bases in DNA, it can change cytosine to uracil and change adenine to hypoxanthine. A DNA double helix has the following sequence:

   TTGGATGCTGG
   AACCTACGACC

   A. What would be the sequence of this double helix immediately after reaction with nitrous acid? Let the letter H represent hypoxanthine and U represent uracil.

   B. Let's suppose this DNA was treated with nitrous acid. The nitrous acid was then removed, and the DNA was replicated for two generations. What would be the sequences of the DNA products after the DNA had replicated twice? Your answer should contain the sequences of four double helices. Note: During DNA replication, uracil hydrogen bonds with adenine, and hypoxanthine hydrogen bonds with cytosine.

C5. One way that bacterial cells regulate DNA replication is through GATC methylation sites within the origin of replication. Would this mechanism work if the DNA was conservatively (rather than semiconservatively) replicated?

C6. The chromosome of *E. coli* contains about 4.6 million bp. How long will it take to replicate its DNA? Assuming that DNA polymerase III is the primary enzyme involved and that it can actively proofread during DNA synthesis, how many base pair mistakes will be made in one round of DNA replication in a bacterial population containing 1000 bacteria?

C7. Here are two strands of DNA:

   ———————————DNA polymerase—>

   The one on the bottom is a template strand, and the one on the top is being synthesized by DNA polymerase in the direction shown by the arrow. Label the 5′ and 3′ ends of the top and bottom strands.

C8. A DNA strand has the following sequence:

   5′–GATCCCGATCCGCATACATTTACCAGATCACCACC–3′

   In which direction would DNA polymerase slide along this strand (from left to right or from right to left)? If this strand was used as a template by DNA polymerase, what would be the sequence of the newly made strand? Indicate the 5′ and 3′ ends of the newly made strand.

C9. List and briefly describe the three types of functionally important sequences within bacterial origins of replication.

C10. As shown in Figure 13.5, five DnaA boxes are found within the origin of replication in *E. coli*. Take a look at these five sequences carefully.

   A. Are the sequences of the five DnaA boxes very similar to each other? (*Hint:* Remember that DNA is double-stranded; think about these sequences in the forward and reverse directions.)

   B. What is the most common sequence for a DnaA box? In other words, what is the most common base in the first position, second position, and so on until the ninth position? The most common sequence is called the consensus sequence.

   C. The *E. coli* chromosome is about 4.6 million bp long. Based on random chance, is it likely that the consensus sequence for a DnaA box occurs elsewhere in the *E. coli* chromosome? If so, why aren't there multiple origins of replication in *E. coli*?

C11. In the following drawing, the top strand is the template DNA, and the bottom strand shows the lagging strand prior to the action of DNA polymerase I. The lagging strand contains three Okazaki fragments. The RNA primers, which are shown in red, have not yet been removed.

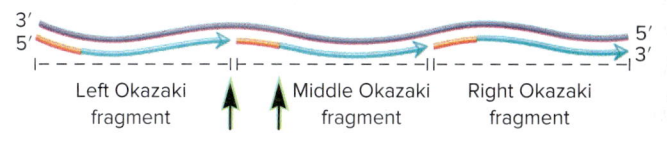

   A. Which Okazaki fragment was made first, the one on the left or the one on the right?

   B. Which RNA primer will be the first one to be removed by DNA polymerase I, the primer on the left or the primer on the right? For this primer to be removed by DNA polymerase I and for the gap to be filled in, is it necessary for the Okazaki fragment in the middle to have already been synthesized? Explain.

   C. Let's consider how DNA ligase connects the left Okazaki fragment with the middle Okazaki fragment. After DNA polymerase I removes the middle RNA primer and fills in the gap with DNA, where does DNA ligase function? See the arrows on either side of the middle RNA primer. Is ligase needed at the left arrow, at the right arrow, or both?

   D. When connecting two Okazaki fragments, DNA ligase needs to use ATP as a source of energy to catalyze this reaction. Explain why DNA ligase needs another source of energy to connect two nucleotides, but DNA polymerase needs nothing more than the incoming nucleotide and the existing DNA strand. Note: You may want to refer to Figure 13.13 to answer this question.

C12. Describe the three important functions of the DnaA protein.

C13. Draw a picture that illustrates how DNA helicase works.

C14. What is an Okazaki fragment? In which strand of replicating DNA are Okazaki fragments found? Based on the properties of DNA polymerase, why is it necessary to make these fragments?

C15. Explain the proofreading function of DNA polymerase.

C16. What is a processive enzyme? Explain why processivity is an important feature of DNA polymerase.

C17. What enzymatic features of DNA polymerase prevent it from replicating one of the DNA strands at the ends of linear chromosomes? Compared with DNA polymerase, how is telomerase different in its ability to synthesize a DNA strand? What does telomerase use as its template for the synthesis of a DNA strand? How does the use of this template result in a telomere sequence that is tandemly repetitive?

C18. As shown in Figure 13.21, telomerase attaches additional DNA, six nucleotides at a time, to the ends of eukaryotic chromosomes. However, it works in only one DNA strand. Describe how the opposite strand is replicated.

C19. If a eukaryotic chromosome has 25 origins of replication, how many replication forks does it have at the beginning of DNA replication?

C20. A diagram of a linear chromosome is shown here:

5′–A————————————————B–3′
3′–C————————————————D–5′

The end of each strand is labeled with A, B, C, or D. Which ends could *not* be replicated by DNA polymerase? Why not?

C21. As discussed in Chapter 10, some viruses contain RNA as their genetic material. Certain RNA viruses can exist as proviruses in which the viral genetic material has been inserted into the chromosomal DNA of the host cell. For this to happen, the viral RNA must be copied into a strand of DNA. An enzyme called reverse transcriptase, encoded by the viral genome, copies the viral RNA into a complementary strand of DNA. The strand of DNA is then used as a template to make a double-stranded DNA molecule. This double-stranded DNA molecule is then inserted into the chromosomal DNA, where it may exist as a provirus for a long period of time.

A. How is the function of reverse transcriptase similar to the function of telomerase?

B. Unlike DNA polymerase, reverse transcriptase does not have a proofreading function. How might this affect the proliferation of the virus?

C22. Telomeres contain a 3′ overhang region, as shown in Figure 13.19. Does telomerase require a 3′ overhang to replicate the telomere region? Explain.

C23. Describe the similarities and differences between homologous recombination involving sister chromatid exchange (SCE) and that involving homologs. Would you expect the same types of proteins to be involved in both processes? Explain.

C24. The molecular mechanism of SCE is similar to homologous recombination between homologs except that the two segments of DNA are sister chromatids instead of homologous chromatids. If branch migration occurs during SCE, will a heteroduplex be formed? Explain why or why not. Can gene conversion occur during sister chromatid exchange?

C25. What two molecular mechanisms can result in gene conversion? Do both occur in the double-strand break model?

C26. What are recombinant chromosomes? How do they differ from the original chromosomes from which they are derived?

C27. In the Holliday model for homologous recombination (see Figure 13.23), the resolution steps can produce recombinant or nonrecombinant chromosomes. Explain how this can occur.

C28. What is gene conversion?

C29. Make a list of the differences between the Holliday model and the double-strand break model.

C30. In recombinant chromosomes, where is gene conversion likely to take place: near the breakpoint or far away from the breakpoint? Explain.

C31. According to the double-strand break model, does gene conversion necessarily involve DNA mismatch repair? Explain.

Application and Experimental Questions

E1. Answer the following questions pertaining to the experiment of Figure 13.3.

A. What would be the expected results if the Meselson and Stahl experiment was carried out for four or five generations?

B. What would be the expected results of the Meselson and Stahl experiment after three generations if the mechanism of DNA replication was dispersive?

C. Considering the data from the experiment, explain why three different bands (i.e., light, half-heavy, and heavy) can be observed in the CsCl gradient.

E2. An absent-minded researcher follows the steps of Figure 13.3 and, when viewing the gradient under UV light, does not see any bands at all. Which of the following mistakes could account for this observation? Explain how.

A. The researcher forgot to add ^{14}N-containing compounds.

B. The researcher forgot to add lysozyme.

C. The researcher forgot to turn on the UV lamp.

E3. Figure 13.4b shows an autoradiograph of a replicating bacterial chromosome. If you analyzed many replicating chromosomes, what types of information could you learn about the mechanism of DNA replication?

E4. The technique of dideoxy sequencing of DNA is described in Chapter 20. The technique relies on the use of dideoxyribonucleotides (shown in Figures 20.5 and 20.16). A dideoxyribonucleotide has a hydrogen atom attached to the 3′-carbon atom instead of an –OH group. When a dideoxyribonucleotide is incorporated into a newly made strand, the strand cannot grow any longer. Explain why.

E5. Another technique described in Chapter 20 is polymerase chain reaction (PCR) (see Figure 20.5), which is based on our understanding of DNA replication. In this method, a small amount of double-stranded template DNA is mixed with a high concentration of primers. Nucleotides and DNA polymerase are also added. The template DNA strands are separated by heat treatment, and when the temperature is lowered, the primers can bind to the single-stranded DNA, and then DNA polymerase replicates the DNA. This increases the amount of DNA made from the primers. This cycle of steps (i.e., heat treatment, lower temperature, allow DNA replication to occur) is repeated over and over. Because the cycle is repeated many times, this method is called a chain reaction. It is called a polymerase chain reaction because DNA polymerase is the enzyme needed to increase the amount of DNA with each cycle. In a PCR experiment, the template DNA is placed in a tube, and the primers, nucleotides, and DNA polymerase are added to the tube. The tube is then placed in a machine called a thermocycler, which raises and lowers the temperature. During one cycle, the temperature is raised (e.g., to 95°C) for a brief period to separate the DNA strands and then lowered (e.g., to 60°C) to allow the primers to bind. The sample is incubated at a slightly higher temperature for a few minutes to allow DNA replication to proceed. In a typical PCR experiment, the tube may be left in the thermocycler for 25 to 30 cycles.

A. Why is DNA helicase not needed in a PCR experiment?

B. How is the sequence of each primer important in a PCR experiment? Do the two primers recognize the same strand or opposite strands?

C. The DNA polymerase used in PCR experiments is isolated from thermophilic bacteria. Why is this kind of polymerase used?

D. If a tube initially contained 10 copies of double-stranded DNA, how many copies of double-stranded DNA (in the region flanked by the two primers) would be obtained after 27 cycles?

E6. A gene exists in two alleles, B and b. The gene is 1123 bp in length, and the B and b alleles exhibit single base pair differences at six different sites. If you determined experimentally that gene conversion changed the b allele into the B allele, do you think that it is more likely to have occurred via DNA mismatch repair or DNA gap repair synthesis? Explain your choice.

Questions for Student Discussion/Collaboration

1. The complementarity of its two strands is the underlying reason that DNA can be accurately copied. Propose alternative chemical structures that could be accurately copied.

2. DNA replication is fast, virtually error-free, and coordinated with cell division. Discuss which of these three features you think is the most important.

Answers to Comprehension Questions

13.1: c, d, b

13.2: b, d

13.3: a, d

13.4: a, d

13.5: b, d, c, b

13.6: a, d, a, d

Note: All answers appear in Connect; the answers to even-numbered questions and all Concept Check questions are in Appendix B.

14

PART IV MOLECULAR PROPERTIES OF GENES

CHAPTER OUTLINE

- 14.1 Overview of Transcription
- 14.2 Transcription in Bacteria
- 14.3 Transcription in Eukaryotes
- 14.4 RNA Modification
- 14.5 A Comparison of Transcription and RNA Modification in Bacteria and Eukaryotes

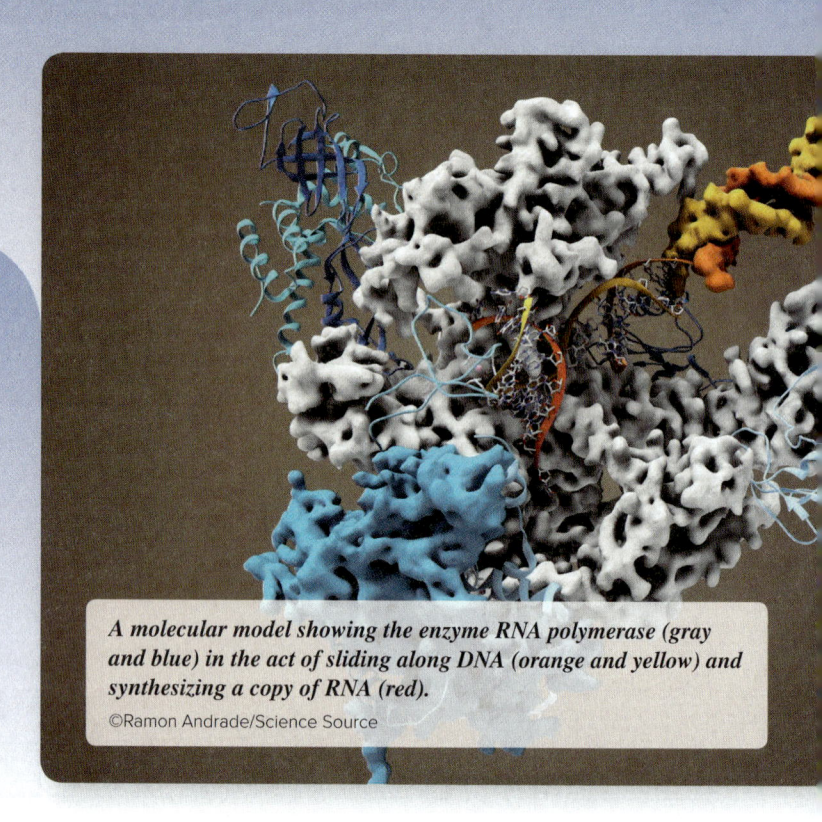

A molecular model showing the enzyme RNA polymerase (gray and blue) in the act of sliding along DNA (orange and yellow) and synthesizing a copy of RNA (red).
©Ramon Andrade/Science Source

GENE TRANSCRIPTION AND RNA MODIFICATION

The primary function of the genetic material, which is DNA, is to store the information necessary to create a living organism. The information is contained within units called genes. At the molecular level, a **gene** is defined as a segment of DNA that is used to make a functional product—either an RNA molecule or a polypeptide. How is the information within a gene accessed? The first step in this process is called **transcription,** which literally means the act or process of making a copy. In genetics, this term refers to the process of synthesizing RNA from a DNA sequence (**Figure 14.1**). The structure of DNA is not altered as a result of transcription. Rather, the DNA base sequence has only been accessed to make a copy in the form of RNA. Therefore, the same DNA can continue to store information. DNA replication, which was discussed in Chapter 13, provides a mechanism for copying that information so it can be transmitted from cell to cell and from parents to offspring.

Protein-encoding genes (also called **structural genes**) encode the amino acid sequence of a polypeptide. When a protein-encoding gene is transcribed, the first product is an RNA molecule known as **messenger RNA (mRNA).** During polypeptide synthesis—a process called **translation**—the sequence of nucleotides within the mRNA determines the sequence of amino acids in a polypeptide. One or more polypeptides then assemble into a functional protein. The structures and functions of proteins largely determine an organism's traits. The model depicted in Figure 14.1, which is called the **central dogma of genetics** (also called the central dogma of molecular biology), was first enunciated by Francis Crick in 1958. The central dogma of genetics specifies that the usual flow of genetic information is from DNA to mRNA to polypeptide. It forms a cornerstone of our understanding of genetics at the molecular level.

In this chapter, we begin to study the molecular steps in **gene expression,** the overall process by which the information within a gene is used to produce a functional product such as a polypeptide. Our emphasis will be on transcription and the modifications that may occur to an RNA transcript after it has been made. In Chapter 15, we will examine the process of translation, and Chapters 16 and 17 will focus on how gene expression is regulated at the molecular level.

296

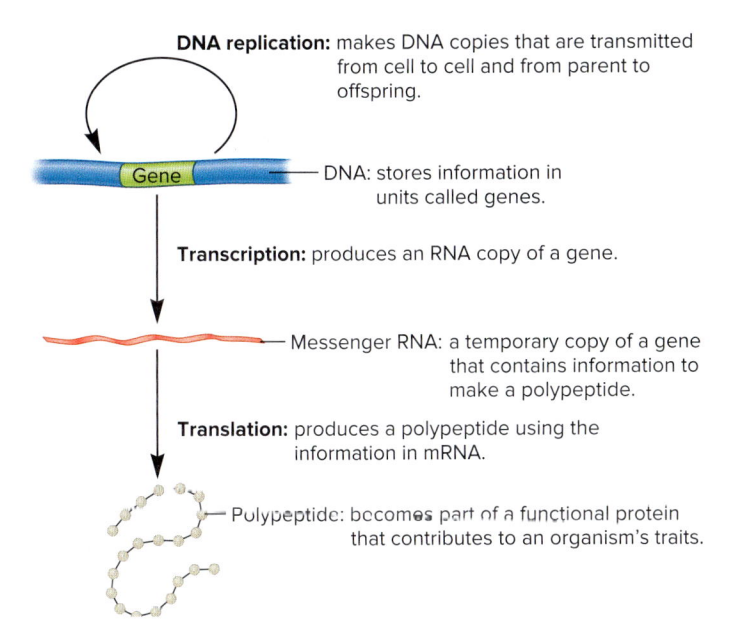

14.1 OVERVIEW OF TRANSCRIPTION

Learning Outcomes:
1. Describe the organization of a protein-encoding gene and its mRNA transcript.
2. Outline the three stages of transcription.

One key concept regarding the process of transcription is that short base sequences define the beginning and ending of a gene and also play a role in regulating the level of RNA synthesis. In this section, we begin by examining the sequences that determine where transcription starts and ends; we also briefly consider DNA sequences, called regulatory sites, that influence whether a gene is turned on or off. A second important concept is the role of proteins in transcription. DNA sequences, in and of themselves, just exist. For genes to be actively transcribed, proteins must recognize particular DNA sequences and act on them in a way that affects the transcription process. Later in this section, we will consider how proteins participate in the general steps of transcription.

Gene Expression Requires Base Sequences That Perform Different Functional Roles

Figure 14.2 shows a common organization of base sequences needed to create a protein-encoding gene that functions in a

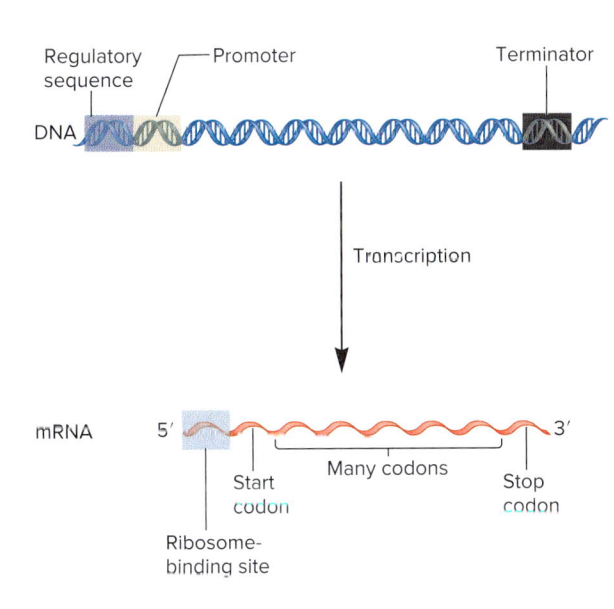

FIGURE 14.1 **The central dogma of genetics.** The usual flow of genetic information is from DNA to mRNA to polypeptide. Note: The direction of informational flow shown in this figure is the most common direction, but exceptions occur. For example, RNA viruses and certain transposable elements use an enzyme called reverse transcriptase to make a copy of DNA from RNA.

DNA replication: makes DNA copies that are transmitted from cell to cell and from parent to offspring.

DNA: stores information in units called genes.

Transcription: produces an RNA copy of a gene.

Messenger RNA: a temporary copy of a gene that contains information to make a polypeptide.

Translation: produces a polypeptide using the information in mRNA.

Polypeptide: becomes part of a functional protein that contributes to an organism's traits.

DNA:
- **Regulatory sequence:** site for the binding of regulatory proteins; the role of regulatory proteins is to influence the rate of transcription. Regulatory sequences can be found in a variety of locations.
- **Promoter:** site for RNA polymerase binding; signals the beginning of transcription.
- **Terminator:** signals the end of transcription.

mRNA:
- **Ribosome-binding site:** site for ribosome binding; translation begins near this site in the mRNA. In eukaryotes, the ribosome scans the mRNA for a start codon.
- **Start codon:** specifies the first amino acid in a polypeptide sequence, usually a formylmethionine (in bacteria) or a methionine (in eukaryotes).
- **Codons:** 3-nucleotide sequences within the mRNA that specify particular amino acids. The sequence of codons within mRNA determines the sequence of amino acids within a polypeptide.
- **Stop codon:** specifies the end of polypeptide synthesis.
- Bacterial mRNA may be polycistronic, which means it encodes two or more polypeptides.

FIGURE 14.2 **Organization of sequences of a bacterial gene and its mRNA transcript.** This figure depicts the general organization of sequences that are needed to create a functional gene that encodes an mRNA.

Concept Check: If a mutation changed the start codon into a stop codon, would the mutation affect the length of the mRNA? Explain.

bacterium such as *E. coli*. Each type of base sequence performs its role during a particular stage of gene expression.

- The **promoter** provides a site to begin transcription.
- The **terminator** specifies the end of transcription.

These two sequences cause RNA synthesis to occur within a defined location. As shown in Figure 14.2, the DNA is transcribed into RNA from the promoter to the terminator.

Proteins called **transcription factors** recognize base sequences in the DNA and control the rate of transcription. Some transcription factors bind directly to the promoter and facilitate transcription. Other transcription factors recognize **regulatory sequences,** or **regulatory elements**—short stretches of DNA involved in the regulation of transcription. Certain transcription factors bind to regulatory sequences and increase the rate of transcription, whereas others inhibit transcription.

Base sequences within an mRNA are used during the translation process.

- In bacteria, a short sequence within the mRNA, the **ribosome-binding site,** provides a location for the ribosome to bind and begin translation.

- Each mRNA contains a series of **codons,** read as groups of three nucleotides, which contain the information for a polypeptide's sequence.
- The first codon, which is very close to the ribosome-binding site, is the **start codon.**
- A **stop codon** signals the end of translation.

The Three Stages of Transcription Are Initiation, Elongation, and Termination

Transcription occurs in three stages: **initiation; elongation,** or synthesis of the RNA transcript; and **termination** (**Figure 14.3**). These steps involve protein-DNA interactions in which proteins such as **RNA polymerase,** the enzyme that synthesizes RNA, interacts with DNA sequences. What causes transcription to begin? The initiation stage in the transcription process is a recognition step. The sequence of bases within the promoter is recognized by one or more transcription factors. The specific binding of transcription factors to the promoter identifies the starting site for transcription.

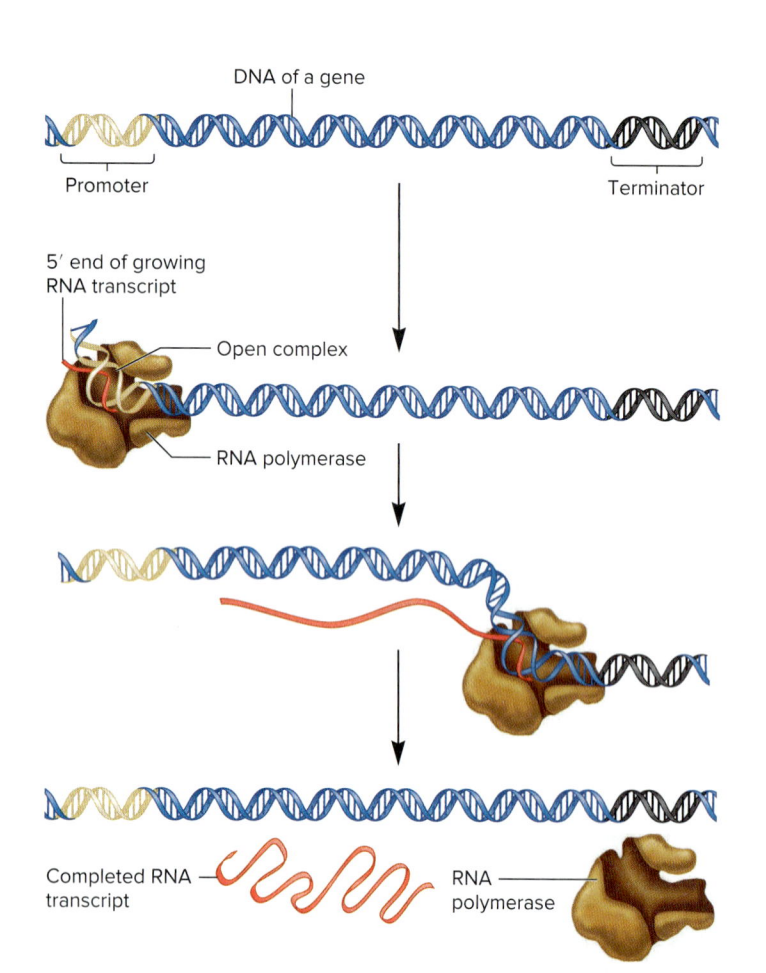

Initiation: The promoter functions as a recognition site for transcription factors (not shown). The transcription factor(s) enables RNA polymerase to bind to the promoter. Following binding, the DNA is denatured into a bubble known as the open complex.

Elongation/synthesis of the RNA transcript: RNA polymerase slides along the DNA in an open complex to synthesize RNA.

Termination: A terminator is reached that causes RNA polymerase and the RNA transcript to dissociate from the DNA.

FIGURE 14.3 Stages of transcription.

Genes → Traits The ability of genes to produce an organism's traits relies on the molecular process of gene expression. Transcription is the first step in gene expression. During transcription, the gene's sequence within the DNA is used as a template to make a complementary copy of RNA. In Chapter 15, we examine how the sequence in mRNA is translated into a polypeptide. After polypeptides are made within a living cell, they fold and become units within functional proteins that govern an organism's traits.

The transcription factor(s) and RNA polymerase first bind to the promoter when the DNA is in the form of a double helix. For transcription to occur, the DNA strands must be separated. This allows one of the two strands to be used as a template for the synthesis of a complementary strand of RNA. This synthesis occurs as RNA polymerase slides along the DNA, forming a small bubble-like structure known as the **open complex.** Eventually, RNA polymerase reaches a terminator, which causes both RNA polymerase and the newly made RNA transcript to dissociate from the DNA.

14.1 REVIEWING THE KEY CONCEPTS

- A gene is an organization of DNA sequences. A promoter signals the start of transcription, and a terminator signals the end. Regulatory sequences control the rate of transcription. For genes that encode polypeptides, the gene sequence also specifies a start codon, a stop codon, and many codons in between. Bacterial genes also specify a ribosome-binding sequence (see Figure 14.2).
- Transcription occurs in three phases called initiation, elongation, and termination (see Figure 14.3)

14.1 COMPREHENSION QUESTIONS

1. Which of the following base sequences is/are used during transcription?
   a. Promoter and terminator
   b. Start and stop codons
   c. Ribosome-binding sequence
   d. Both a and b
2. The three stages of transcription are
   a. initiation, ribosome-binding, and termination.
   b. elongation, ribosome-binding, and termination.
   c. initiation, elongation, and termination.
   d. initiation, regulation, termination.

14.2 TRANSCRIPTION IN BACTERIA

Learning Outcomes:

1. Describe the characteristics of a bacterial promoter.
2. Explain how RNA polymerase transcribes a bacterial gene.
3. Compare and contrast two mechanisms for transcriptional termination in bacteria.

In 1960, Matthew Meselson and François Jacob found that proteins are synthesized on ribosomes. One year later, Jacob and his colleague Jacques Monod proposed that a certain type of RNA acts as a genetic messenger (from the DNA to the ribosome) to provide the information for protein synthesis. They hypothesized that this RNA, which they called messenger RNA (mRNA), is transcribed from the nucleotide sequence within DNA and then directs the synthesis of particular polypeptides. This proposal was a remarkable insight, considering that it was made before the actual isolation and characterization of mRNA molecules in vitro. In 1961, the hypothesis was confirmed by Sydney Brenner in collaboration with Jacob and Meselson. They found that when a virus infects a bacterial cell, a virus-specific RNA is made and rapidly associates with preexisting ribosomes in the cell.

Since these pioneering studies, a great deal has been learned about the molecular features of bacterial gene transcription. Much of our knowledge comes from studies in *E. coli*. In this section, we will examine the three steps in the gene transcription process as they occur in bacteria.

A Promoter Is a Short Sequence of DNA That Is Necessary to Initiate Transcription

The type of DNA sequence known as the promoter gets its name from the idea that it "promotes" gene expression. More precisely, this sequence of bases directs the exact location for the initiation of RNA transcription. Most of the promoter is located just ahead of, or upstream from, the site where transcription of a gene actually begins. By convention, the bases in a promoter sequence are numbered in relation to the **transcriptional start site** (**Figure 14.4**). This site is the first base used as a template for RNA transcription and is denoted +1. The bases preceding this site are numbered in a negative direction. No base is numbered zero. Therefore, most of the promoter is labeled with negative numbers that describe the number of bases preceding the beginning of transcription.

Although the promoter may encompass a DNA sequence several dozen nucleotides in length, short **sequence elements** are particularly critical for promoter recognition. By comparing the sequence of DNA bases within many promoters, researchers have learned that certain sequences of bases are necessary for a promoter to be functional. In many promoters found in *E. coli* and similar species, two sequence elements are important. These are located

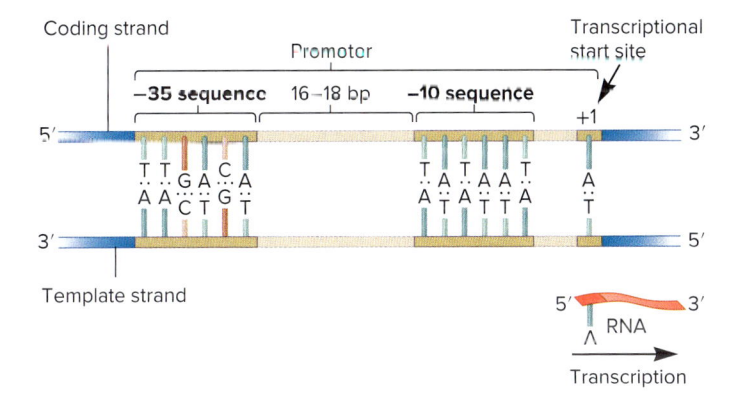

FIGURE 14.4 **The conventional numbering system of promoters.** The first nucleotide that acts as a template for transcription is designated +1. The numbering of nucleotides to the left of this spot is in a negative direction, whereas the numbering to the right is in a positive direction. For example, the nucleotide that is immediately to the left of the +1 nucleotide is numbered –1, and the nucleotide to the right of the +1 nucleotide is numbered +2. There is no zero nucleotide in this numbering system. In many bacterial promoters, sequence elements at the –35 and –10 locations play a key role in promoting transcription.

at approximately the −35 and −10 sites in the promoter (see Figure 14.4). The sequence in the top DNA strand at the −35 site is 5′–TTGACA–3′, and the one at the −10 site is 5′–TATAAT–3′. The TATAAT sequence is sometimes called the **Pribnow box** after David Pribnow, who initially discovered it in 1975.

Sequences within DNA, such as those found in promoters or regulatory elements, vary among different genes. For example, **Figure 14.5** illustrates the sequences found in several different *E. coli* promoters. The most commonly occurring bases within a sequence element form the **consensus sequence.** This sequence is efficiently recognized by proteins that initiate transcription. For many bacterial genes, a strong correlation is found between the maximal rate of RNA transcription and the degree to which the −35 and −10 sequences agree with the consensus sequences. Mutations in the −35 or −10 sequences that lessen their similarity to the consensus sequences typically slow down the rate of transcription.

Bacterial Transcription Is Initiated When RNA Polymerase Holoenzyme Binds at a Promoter

Thus far, we have considered the DNA sequences that constitute a functional promoter. Let's now turn our attention to the proteins that recognize those sequences and carry out the transcription process. The enzyme that catalyzes the synthesis of RNA is RNA polymerase.

- In *E. coli*, the **core enzyme** is composed of five subunits: $\alpha_2\beta\beta'\omega$.
- The association of a sixth subunit, **sigma (σ) factor,** with the core enzyme creates what is referred to as RNA polymerase **holoenzyme.**
- The two α subunits are important in the proper assembly of the holoenzyme and in the process of binding to DNA.
- The β and β′ subunits are also needed for binding to the DNA, and they carry out the catalytic synthesis of RNA.
- The ω (omega) subunit is important for the proper assembly of the core enzyme.
- The holoenzyme is required to initiate transcription; the primary role of σ factor is to recognize the promoter. Proteins such as σ factor that influence the function of RNA polymerase are types of transcription factors.

After RNA polymerase holoenzyme is assembled into its six subunits, it binds loosely to the DNA and then slides along the DNA, much as a train rolls down the tracks. How is a promoter identified? When the holoenzyme encounters a promoter, σ factor recognizes both the −35 and −10 sequences. The σ factor protein contains a structure called a **helix-turn-helix motif** that can bind tightly to these sequences. Alpha (α) helices within the protein fit into the major groove of the DNA double helix and form hydrogen bonds with the bases. This phenomenon of molecular recognition is shown in **Figure 14.6**. Hydrogen bonding occurs between nucleotides in the −35 and −10 sequences of the promoter and amino acid side chains in the helix-turn-helix motif of σ factor.

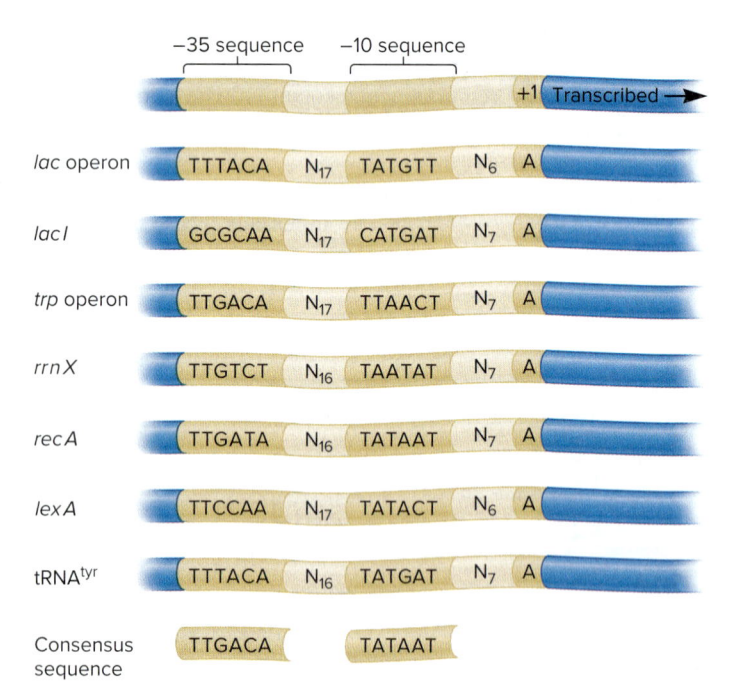

FIGURE 14.5 Examples of −35 and −10 sequences within a variety of bacterial promoters. This figure shows the −35 and −10 sequences for one DNA strand found in seven bacterial promoters. The consensus sequence is at the bottom. The spacer regions contain the designated number of nucleotides between the −35 and −10 sequences or between the −10 sequence and the transcriptional start site. For example, N_{17} means there are 17 nucleotides between the end of the −35 sequence and the beginning of the −10 sequence.

Concept Check: *What does the term consensus sequence mean?*

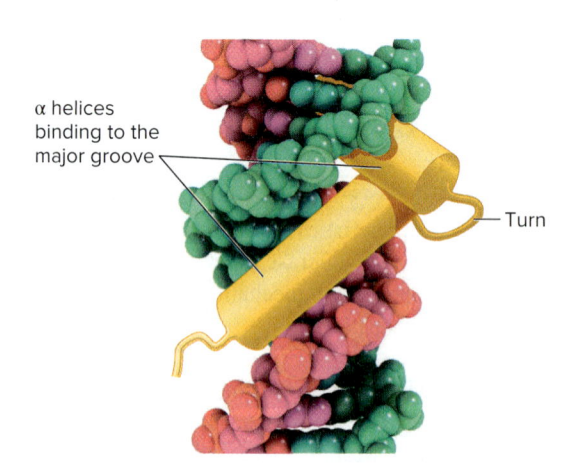

FIGURE 14.6 The binding of σ factor protein to the DNA of a promoter. Part of the protein contains two α helices connected by a turn, termed a helix-turn-helix motif. Two α helices of the protein can fit within the major groove of the DNA. Amino acid side chains within the α helices form hydrogen bonds with the bases in the DNA.
©Laguna Design/Science Source

Concept Check: *Why is it necessary for portions of the sigma (σ) factor protein to fit into the major groove?*

14.2 TRANSCRIPTION IN BACTERIA

As shown in **Figure 14.7**, the process of transcription is initiated when σ factor within the holoenzyme binds to the promoter to form the **closed complex.** In this closed complex, the strands of DNA in the promoter are not separated. For transcription to begin, the double-stranded DNA must be unwound to form an open complex, as described in the preceding section of this chapter. This unwinding begins at the TATAAT sequence at the −10 site, which contains only A-T base pairs, as shown in Figure 14.4. A-T base pairs form only two hydrogen bonds, whereas G-C pairs form three. Therefore, DNA in an AT-rich region is more easily separated because fewer hydrogen bonds must be broken. A short strand of RNA is made within the open complex, and then σ factor is released from the core enzyme. The release of σ factor marks the transition to the elongation phase of transcription. The core enzyme may now slide along the DNA to synthesize a strand of RNA.

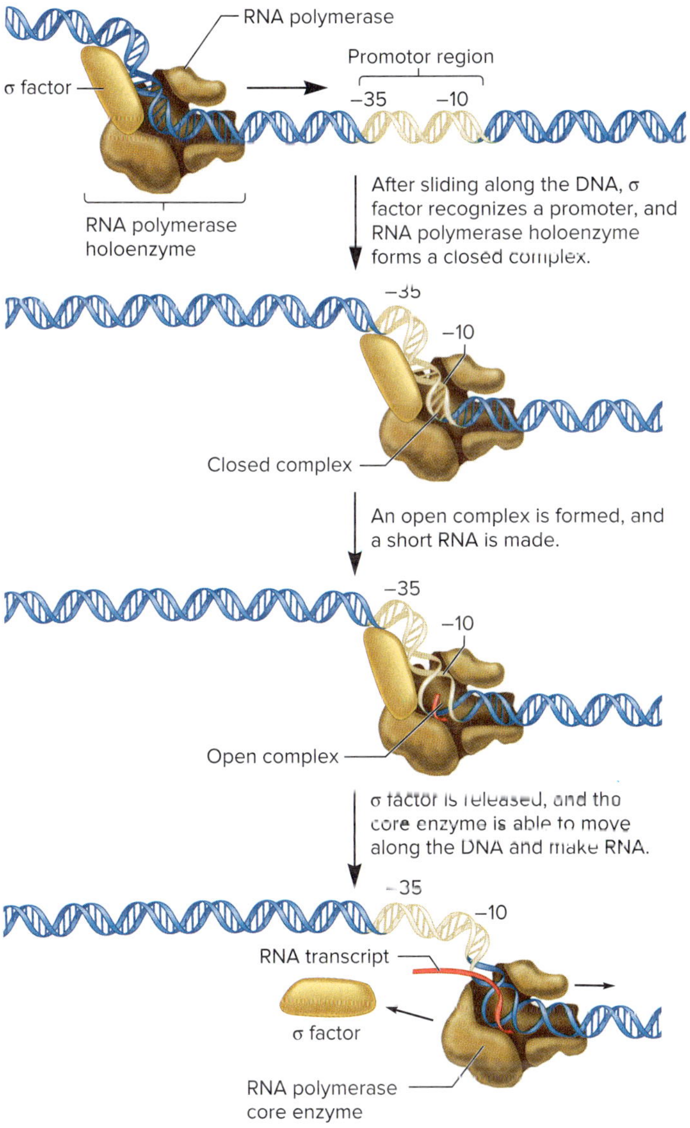

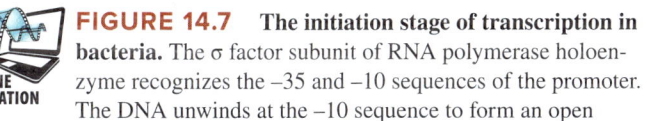

The RNA Transcript Is Synthesized During the Elongation Stage

After the initiation stage of transcription is completed, the RNA transcript is made during the elongation stage.

- During the synthesis of the RNA transcript, RNA polymerase moves along the DNA, causing it to unwind (**Figure 14.8**).
- The DNA strand used as a template for RNA synthesis is termed the **template strand** (also called the antisense strand).

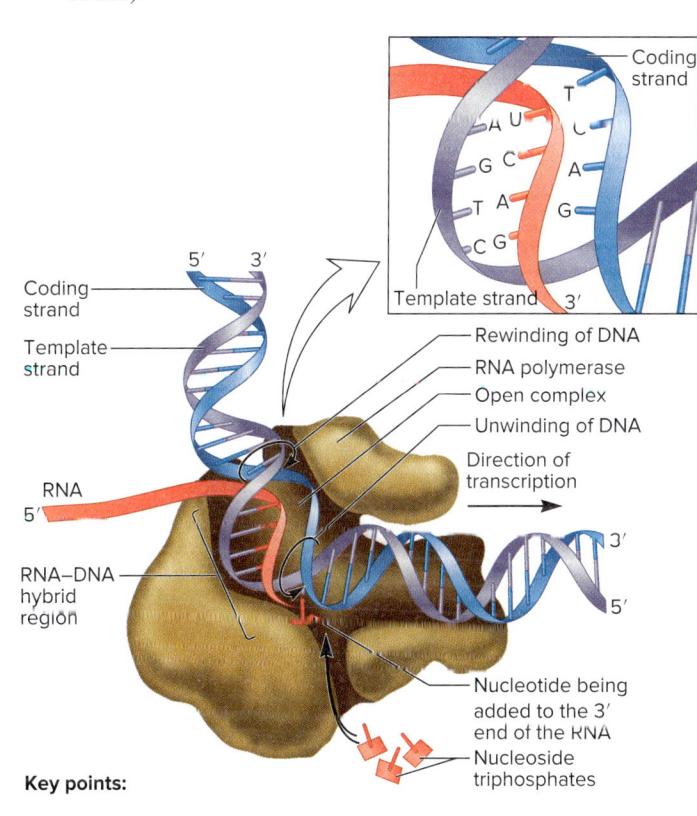

FIGURE 14.7 The initiation stage of transcription in bacteria. The σ factor subunit of RNA polymerase holoenzyme recognizes the −35 and −10 sequences of the promoter. The DNA unwinds at the −10 sequence to form an open complex, and a short RNA is made. Then, σ factor dissociates from the holoenzyme, and RNA polymerase core enzyme can proceed along the DNA, synthesizing RNA and forming an open complex as it goes.

Concept Check: What feature of the −10 sequence makes it easy to unwind?

Key points:

- RNA polymerase slides along the DNA, creating an open complex as it moves.
- The DNA strand known as the template strand is used to make a complementary copy of RNA, resulting in an RNA–DNA hybrid.
- RNA polymerase moves along the template strand in a 3′ to 5′ direction, and RNA is synthesized in a 5′ to 3′ direction using nucleoside triphosphates as precursors. Pyrophosphate is released (not shown).
- The complementarity rule is the same as the AT/GC rule except that U is substituted for T in the RNA.

FIGURE 14.8 Synthesis of the RNA transcript.

- The opposite DNA strand is the **coding strand** (also called the sense strand); it has the same sequence as the RNA transcript being synthesized except that T in the DNA corresponds to U in the RNA.
- As it moves along the DNA, the open complex formed by the action of RNA polymerase is approximately 17 bp long.
- On average, the rate of RNA synthesis is about 43 nucleotides per second.
- Behind the open complex, the DNA rewinds back into a double helix.

As summarized in Figure 14.8, the chemistry of transcription by RNA polymerase is similar to the synthesis of DNA via DNA polymerase, which is discussed in Chapter 13. RNA polymerase always connects nucleotides in the 5′ to 3′ direction. During this process, RNA polymerase catalyzes the formation of a bond between the 5′ phosphate group on one nucleotide and the 3′ —OH group on the previous nucleotide. The complementarity rule is similar to the AT/GC rule, except that uracil substitutes for thymine in the RNA. In other words, RNA synthesis obeys an $A_{DNA}U_{RNA}/T_{DNA}A_{RNA}/G_{DNA}C_{RNA}/C_{DNA}G_{RNA}$ rule.

In the case of the transcription of multiple genes within a chromosome, the direction of transcription and the DNA strand used as a template vary among different genes. **Figure 14.9** shows three genes adjacent to each other within a chromosome. Genes A and B are transcribed from left to right, and the bottom DNA strand is used as the template. By comparison, gene C is transcribed from right to left, and the top DNA strand is used as the template. Note that in all three cases, the template strand is read in the 3′ to 5′ direction, and the synthesis of the RNA transcript occurs in a 5′ to 3′ direction.

Transcription Is Terminated by Either an RNA-Binding Protein or an Intrinsic Terminator

The end of RNA synthesis is referred to as termination. Prior to termination, the hydrogen bonding between the DNA and RNA within the open complex is of central importance in preventing dissociation of RNA polymerase from the template strand. Termination occurs when this short RNA–DNA hybrid region is forced to separate, thereby releasing the newly made RNA transcript. In *E. coli*, two different mechanisms for termination have been identified.

ρ-Dependent Termination In **ρ-dependent termination**, the termination process requires two sequences. First, a sequence upstream from the terminator, called the *rut* site for rho utilization site, acts as a recognition site for the binding of **rho (ρ) protein** (**Figure 14.10**). How does ρ protein facilitate termination? It functions as a helicase, an enzyme that can separate RNA–DNA hybrid regions. After the *rut* site is synthesized in the RNA, ρ protein binds to the RNA and moves in the direction of RNA polymerase.

The second component of ρ-dependent termination is the site where termination actually takes place. At this terminator site, the DNA encodes an RNA sequence containing several GC base pairs that form a stem-loop. This stem-loop forms almost immediately after the RNA sequence is synthesized and quickly binds to RNA polymerase. This binding results in a conformational change that causes RNA polymerase to pause in its synthesis of RNA. The pause allows ρ protein to catch up to the stem-loop, pass through it, and break the hydrogen bonds between the DNA and RNA within the open complex. When this occurs, the completed RNA strand is separated from the DNA along with RNA polymerase.

ρ-Independent Termination The process of **ρ-independent termination** does not require ρ protein. In this case, the terminator is composed of two adjacent nucleotide sequences that function within the RNA (**Figure 14.11**). One is a uracil-rich sequence located at the 3′ end of the RNA. The second sequence is adjacent to the uracil-rich sequence and promotes the formation of a stem-loop. As shown in Figure 14.11, the formation of the stem-loop causes RNA polymerase to pause in its synthesis of RNA. This pausing is stabilized by other proteins that bind to RNA polymerase. For example, a protein called NusA, which is bound to RNA polymerase, promotes pausing at stem-loops. At the precise time that RNA polymerase pauses, the uracil-rich sequence in the RNA transcript is bound to the DNA template strand. As previously mentioned, the hydrogen bonding of RNA to DNA keeps RNA polymerase clamped onto the DNA. However, the binding of this uracil-rich sequence to the DNA template strand is relatively

FIGURE 14.9 The transcription of three different genes found in the same chromosome. RNA polymerase synthesizes each RNA transcript in a 5′ to 3′ direction, sliding along a DNA template strand in a 3′ to 5′ direction. However, which strand is used as the template strand varies from gene to gene. For example, genes A and B are transcribed from the bottom strand, but gene C is transcribed from the top strand.

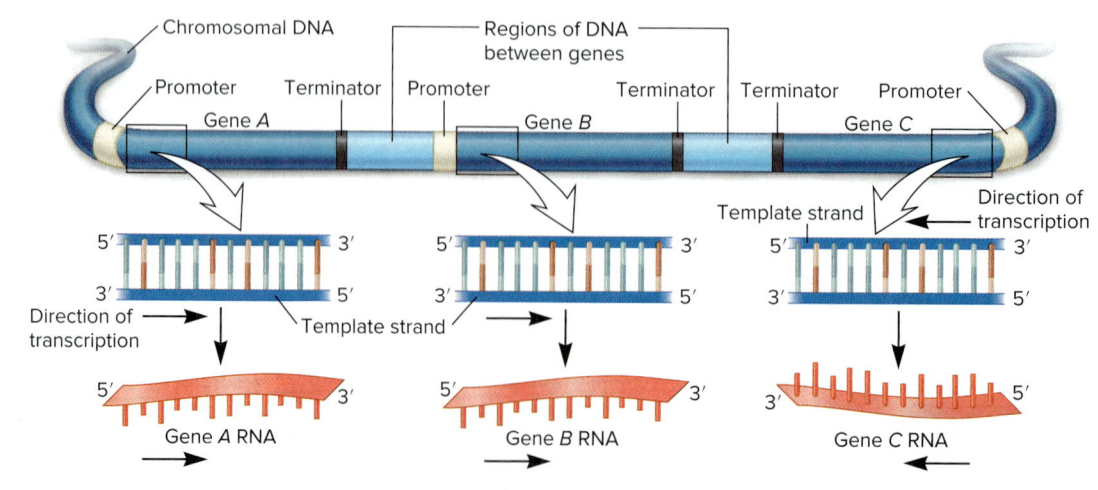

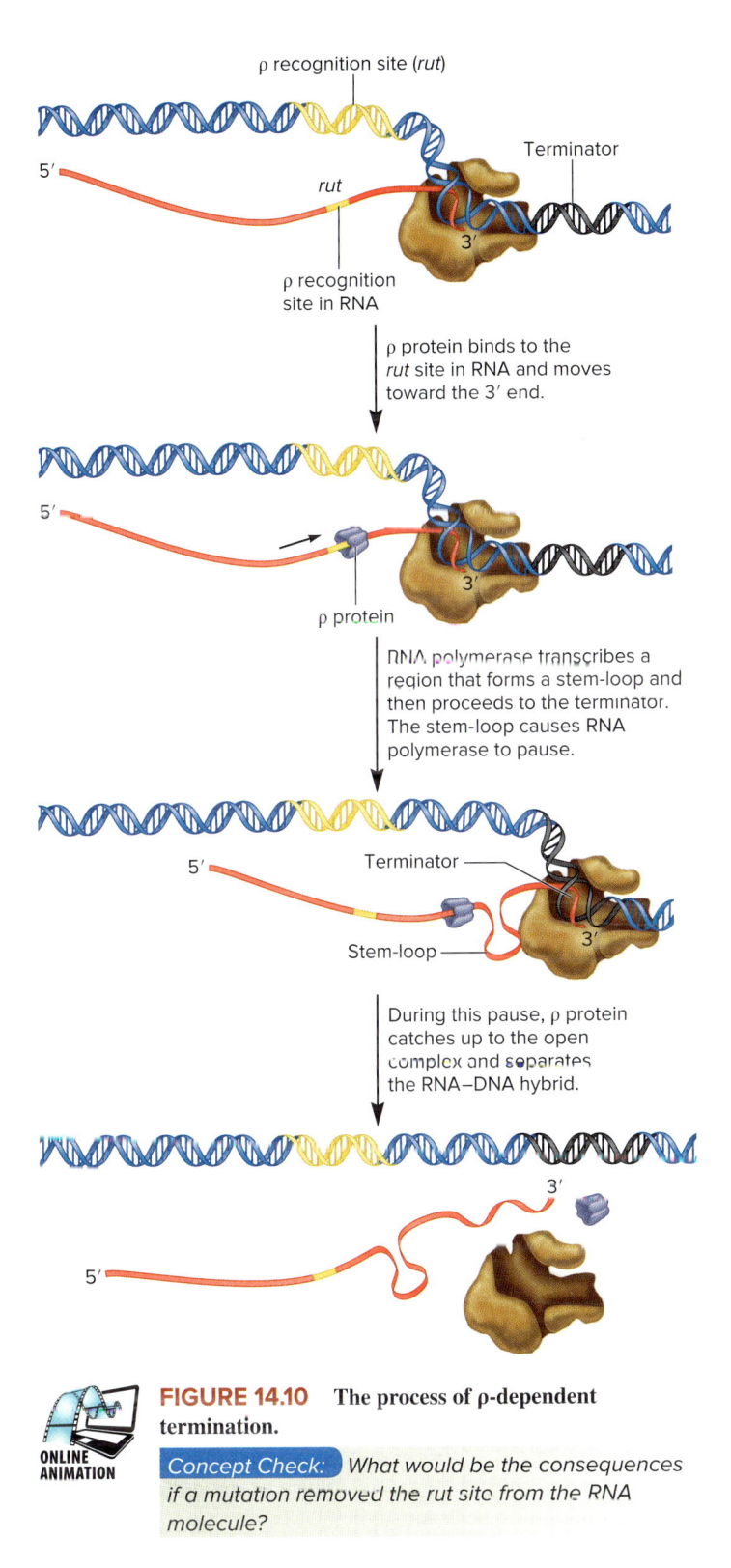

14.2 TRANSCRIPTION IN BACTERIA 303

FIGURE 14.10 The process of ρ-dependent termination.

Concept Check: What would be the consequences if a mutation removed the *rut* site from the RNA molecule?

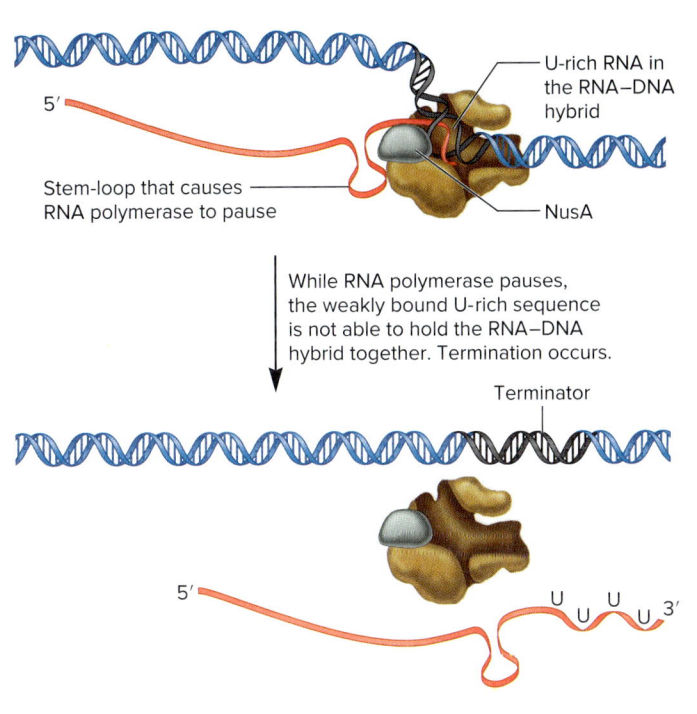

FIGURE 14.11 The process of ρ-independent (or intrinsic) termination. When RNA polymerase reaches the end of the gene, it transcribes a uracil-rich sequence. As this uracil-rich sequence is transcribed, a stem-loop forms just upstream from the open complex. The formation of this stem-loop causes RNA polymerase to pause in its synthesis of the transcript. This pausing is stabilized by the protein NusA, which binds near the region where RNA exits the open complex. While it is pausing, the RNA in the RNA–DNA hybrid is a uracil-rich sequence. Because hydrogen bonds between U and A are relatively weak interactions, the transcript and RNA polymerase dissociate from the DNA.

Concept Check: Why is NusA important for this termination process?

weak, causing the RNA transcript to spontaneously dissociate from the DNA and cease further transcription. Because ρ-independent termination does not require a protein (the ρ protein) to physically remove the RNA transcript from the DNA, it is also referred to as **intrinsic termination**. In *E. coli*, about half of the genes show intrinsic termination, and the other half are terminated by ρ protein.

Genetic TIPS

The Question: The technique of Northern blotting, which is described in Chapter 20, is used to determine how much RNA is transcribed from any particular gene. Figure 14.5 shows the sequences of bacterial promoters from several genes. Let's focus on the *lac* operon promoter. The transcription of the *lac* operon is induced when *E. coli* cells are exposed to lactose. You can use the technique of gene mutagenesis, also described in Chapter 20, to change the *lac* operon promoter sequence in any way you like. How would you determine if the similarity of a gene's promoter sequence to the consensus sequence is an important factor affecting the level of gene transcription?

Topic: What topic in genetics does this question address? The topic is transcription. More specifically, the question is about the importance of the similarity of the promoter sequence to the consensus sequence.

Information: What information do you know based on the question and your understanding of the topic?

From the question, you know that Northern blotting can be used to measure the amounts of RNA transcribed from a gene, and gene mutagenesis can be used to alter a gene's sequence. From your understanding of the topic, you may remember that the consensus sequence is efficiently recognized by proteins that initiate transcription.

Problem-Solving Strategy: Design an experiment.

One strategy to solve this problem is to design an experiment that compares *lac* operon promoters that differ in their similarity to the consensus sequence.

Answer: Starting material is a (wild-type) strain of *E. coli* that carries a normal *lac* operon.

1. Use gene mutagenesis to create different *E. coli* strains in which the *lac* operon promoter sequence is altered to become either more similar to the consensus sequence or less similar to the consensus sequence.

2. Grow the wild-type and mutant strains and induce transcription by adding lactose.

3. Determine the resulting amounts of *lac* operon RNA using Northern blotting.

Expected results: Mutations that made the promoter sequence more like the consensus sequence will result in higher amounts of *lac* operon RNA compared to the wild-type strain. The bands on the gel would appear darker. Mutations that made the promoter sequence less like the consensus sequence will result in lower amounts of that RNA.

14.2 REVIEWING THE KEY CONCEPTS

- Many bacterial promoters have sequence elements at the −35 and −10 sites. The transcriptional start site is at +1 (see Figures 14.4, 14.5).
- During the initiation phase of transcription in *E. coli*, sigma (σ) factor, which is bound to RNA polymerase, binds into the major groove of DNA and recognizes sequence elements at the promoter. This process forms a closed complex. Following the formation of an open complex, σ factor is released (see Figures 14.6, 14.7).
- During the elongation phase of transcription, RNA polymerase slides along the DNA and maintains an open complex as it goes. RNA is made in the 5′ to 3′ direction, and base pairing follows a complementarity rule similar to the AT/GC rule (see Figure 14.8).
- In a given chromosome, the strand that is used as the template strand varies from gene to gene (see Figure 14.9).
- Transcriptional termination in *E. coli* occurs by a ρ-dependent or ρ-independent mechanism (see Figures 14.10, 14.11).

14.2 COMPREHENSION QUESTIONS

1. Within a promoter, a transcriptional start site is
   a. located at the −35 sequence and is recognized by σ factor.
   b. located at the −35 sequence and is where the first base is used as a template for RNA transcription.
   c. located at the +1 site and is recognized by σ factor.
   d. located at the +1 site and is where the first base is used as a template for RNA transcription.

2. For the following five sequences, what is the consensus sequence?

   5′-GGGAGCG-3′
   5′-GAGAGCG-3′
   5′-GAGTGCG-3′
   5′-GAGAACG-3′
   5′-GAGAGCA-3′

   a. 5′-GGGAGCG-3′
   b. 5′-GAGAGCG-3′
   c. 5′-GAGTGCG-3′
   d. 5′-GAGAACG-3′

3. Sigma (σ) factor is needed during which stage(s) of transcription?
   a. Initiation
   b. Elongation
   c. Termination
   d. All of the above

4. A uracil-rich sequence occurs at the end of the RNA in
   a. ρ-dependent termination.
   b. ρ-independent termination.
   c. both a and b.
   d. none of the above.

14.3 TRANSCRIPTION IN EUKARYOTES

Learning Outcomes:
1. List the functions of the three types of RNA polymerases in eukaryotes.
2. Describe the characteristics of a eukaryotic promoter for a protein-encoding gene.
3. Explain how general transcription factors and RNA polymerase assemble at the promoter and form an open complex.
4. Compare and contrast two possible mechanisms for transcriptional termination in eukaryotes.

Many of the basic features of gene transcription are very similar in bacterial and eukaryotic species. Much of our understanding of transcription has come from studies in *Saccharomyces cerevisiae* (baker's yeast) and other eukaryotic species, including mammals. In general, gene transcription in eukaryotes is more complex than it is in bacteria. Eukaryotic cells are larger than bacterial cells and contain a variety of compartments known as organelles. This added level of cellular complexity requires that eukaryotes make many more proteins and consequently have many more protein-encoding genes. In addition, most eukaryotic species are multicellular, being composed of many different cell types. Multicellularity adds a requirement that genes must be transcribed in the correct type of cell and during the proper stage of development. Therefore, in any given eukaryotic species, the transcription of the thousands of different genes that an organism possesses requires appropriate timing and coordination. In this section, we will examine the basic features of gene transcription in eukaryotes. We will focus on the proteins that are needed to make an RNA transcript. In addition, an important factor that affects eukaryotic gene transcription is chromatin structure. Eukaryotic gene transcription requires changes in the positions and structures of nucleosomes. However, because these changes are important for regulating transcription, they are described in Chapter 17 rather than in this chapter.

Eukaryotes Have Multiple RNA Polymerases That Are Structurally Similar to the Bacterial Enzyme

The genetic material within the nucleus of a eukaryotic cell is transcribed by three different RNA polymerase enzymes—designated RNA polymerase I, II, and III. What are the roles of these enzymes? Each of the three RNA polymerases transcribes different categories of genes.

- RNA polymerase I transcribes all of the genes for ribosomal RNAs (rRNAs), except for 5S rRNA.
- RNA polymerase II transcribes all protein-encoding genes. Therefore, it is responsible for the synthesis of all mRNAs. It also transcribes the genes for most snRNAs, which are needed for RNA splicing (discussed later in this chapter). In addition, it transcribes several types of genes that produce other non-coding RNAs (described in Chapter 18), including most long non-coding RNAs, microRNAs, and snoRNAs.
- RNA polymerase III transcribes all the genes for tRNAs and the gene for 5S rRNA. To a much lesser extent than RNA polymerase II, it also transcribes a few genes that produce other non-coding RNAs, such as snRNAs, long non-coding RNAs, microRNAs, and snoRNAs.

All three RNA polymerases are structurally very similar to each other and are composed of many subunits. They contain two large catalytic subunits similar to the β and β′ subunits of bacterial RNA polymerase. The structures of RNA polymerase from a few different species have been determined by X-ray crystallography. A remarkable similarity exists between the bacterial enzyme and its eukaryotic counterparts. **Figure 14.12a** compares the structures of a bacterial RNA polymerase with RNA polymerase II from yeast. As you can see, the two enzymes have similar structures. Interestingly, this structure provides a way to envision how the transcription process works. As seen in **Figure 14.12b**, DNA enters the enzyme through the jaw and lies on a surface within RNA polymerase termed the bridge. The part of the enzyme called the clamp is thought to control the movement of the DNA through RNA polymerase. A wall in the enzyme forces the RNA–DNA hybrid to make a right-angle turn. This bend facilitates the ability of nucleotides to bind to the template strand. Mg^{2+} is located at the catalytic site, which is precisely at the 3′ end of the growing RNA strand. Nucleoside triphosphates (NTPs) enter the catalytic site via

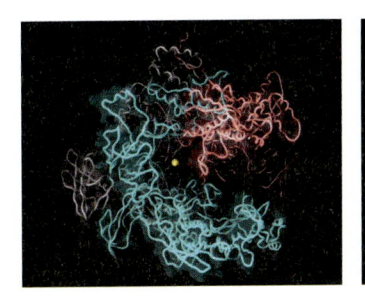

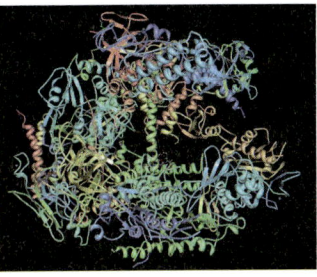

(a) Structure of a bacterial RNA polymerase

Structure of a eukaryotic RNA polymerase II (yeast)

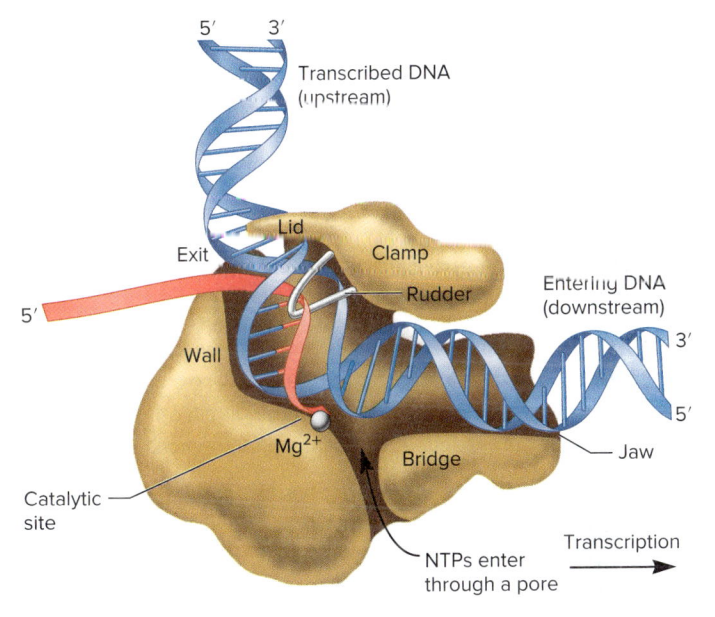

(b) Schematic structure of RNA polymerase

FIGURE 14.12 Structure and molecular function of RNA polymerase. (a) A comparison of the crystal structures of a bacterial RNA polymerase (left) and a eukaryotic RNA polymerase II (right). The bacterial enzyme is from *Thermus aquaticus*. The eukaryotic enzyme is from *Saccharomyces cerevisiae*. (b) A mechanism for transcription based on the enzyme's structure. In this diagram, the direction of transcription is from left to right. The double-stranded DNA enters the polymerase along a bridge surface that is between the jaw and clamp. At a region termed the wall, the RNA–DNA hybrid is forced to make a right-angle turn, which enables nucleotides to bind to the template strand of the DNA. Mg^{2+} is located at the catalytic site. Nucleoside triphosphates (NTPs) enter the catalytic site via a pore and bind to the template strand. At the catalytic site, the nucleotides are covalently attached to the 3′ end of the RNA. As RNA polymerase slides along the template strand, a small region of the protein termed the rudder separates the RNA–DNA hybrid. The single-stranded RNA then exits under a small lid.
(a: left) ©Seth Darst; (a: right) ©David Goodsell/RCSB PDB

a pore. The correct nucleotide binds to the template DNA and is covalently attached to the 3′ end. As RNA polymerase slides along the template strand, a rudder, which is about 9 bp away from the 3′ end of the RNA, forces the RNA–DNA hybrid apart. The single-stranded RNA then exits under a small lid.

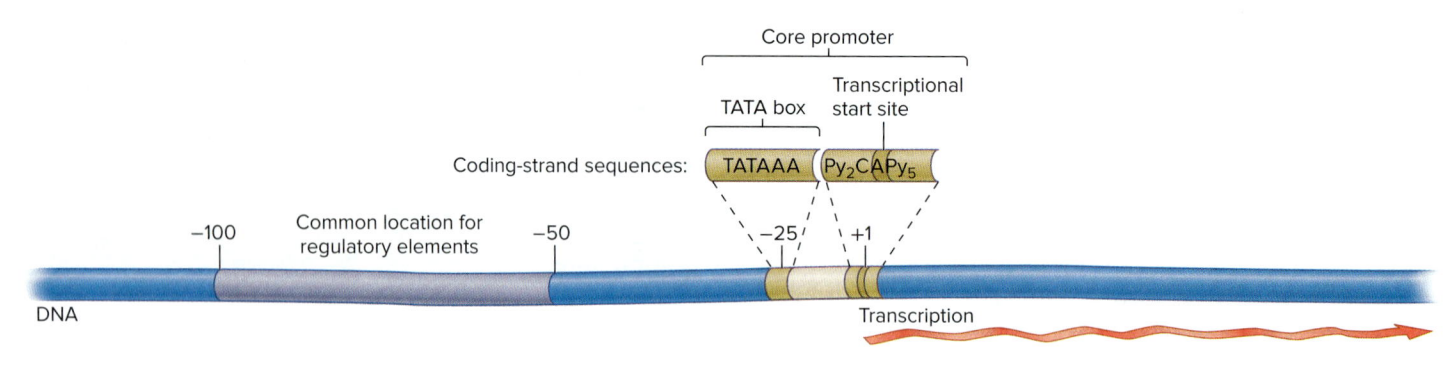

FIGURE 14.13 A common pattern found within the promoter of protein-encoding genes recognized by RNA polymerase II. The transcriptional start site usually occurs at adenine; two pyrimidines (Py; cytosine or thymine) and a cytosine precede this adenine, and five pyrimidines (Py) follow it. A TATA box is approximately 25 bp upstream from the start site. However, the sequences that constitute eukaryotic promoters are quite diverse, and not all protein-encoding genes have a TATA box. Regulatory elements, which are discussed in Chapter 17, vary in their locations but are often found in the −50 to −100 region. The patterns of core promoters for RNA polymerase I and III are quite different from this one. A single upstream regulatory element is involved in the binding of RNA polymerase I to its promoter, whereas two regulatory elements, called A and B boxes, facilitate the binding of RNA polymerase III.

Concept Check: What is the functional role of the TATA box?

Eukaryotic Protein-Encoding Genes Have a Core Promoter and Regulatory Elements

In eukaryotic protein-encoding genes, at least three features are found in most promoters: regulatory elements, a TATA box, and a transcriptional start site. **Figure 14.13** shows a common pattern of sequences found within the promoters of eukaryotic protein-encoding genes.

- The **core promoter** is a relatively short DNA sequence that is necessary for transcription to take place. It consists of a TATAAA sequence called the **TATA box** and the transcriptional start site, where transcription begins.
- The TATA box, which is usually about 25 bp upstream from the transcriptional start site, is important in determining the precise starting point for transcription. If it is missing from the core promoter, the transcription start site becomes undefined, and transcription may start at a variety of different locations.
- The core promoter, by itself, produces a low level of transcription. This is termed **basal transcription.**

Regulatory elements are short DNA sequences that affect the ability of RNA polymerase to recognize the core promoter and begin the process of transcription. These elements are recognized by transcription factors—proteins that influence the rate of transcription. There are two categories of regulatory elements.

- Activating sequences, known as **enhancers,*** are needed to stimulate transcription. In the absence of enhancer sequences, most eukaryotic genes have very low levels of basal transcription.
- Under certain conditions, it may be necessary to prevent transcription of a given gene. This is accomplished via **silencers**—DNA sequences that inhibit transcription.

*Some geneticists use the term *enhancer* to describe a genetic regulatory element that binds activators and/or repressors. In this text, we will used the term *silencer* to describe elements that bind repressors to avoid confusion with regard to the function of these regulatory elements.

- As seen in Figure 14.13, a common location for regulatory elements is the −50 to −100 region. However, the locations of regulatory elements vary considerably among different eukaryotic genes. These elements can be far away from the core promoter yet strongly influence the ability of RNA polymerase to initiate transcription.

DNA sequences such as the TATA box, enhancers, and silencers exert their effects only on a particular gene. They are called *cis*-**acting elements.** The term *cis* comes from chemistry nomenclature and means "next to." *Cis*-acting elements, though possibly far away from the core promoter, are always found within the same chromosome as the genes they regulate. By comparison, the transcription factors that bind to such elements are called ***trans*-acting factors** (the term *trans* means "across from"). The transcription factors that control the expression of a gene are themselves encoded by genes; regulatory genes that encode transcription factors may be far away from the genes they control. When a gene encoding a *trans*-acting factor is expressed, the transcription factor protein that is made can diffuse throughout the cell and bind to its appropriate *cis*-acting element. Let's now turn our attention to the function of such proteins.

Transcription of Eukaryotic Protein-Encoding Genes Is Initiated When RNA Polymerase II and General Transcription Factors Bind to a Promoter Sequence

Thus far, we have considered the DNA sequences that play a role in the promoter of eukaryotic protein-encoding genes. By studying transcription in a variety of eukaryotic species, researchers have discovered that three categories of proteins are needed for basal transcription at the core promoter: RNA polymerase II, general transcription factors, and a complex called mediator (**Table 14.1**).

14.3 TRANSCRIPTION IN EUKARYOTES

TABLE 14.1
Proteins Needed for Transcription via the Core Promoter of Eukaryotic Protein-Encoding Genes

RNA polymerase II: The enzyme that catalyzes the linkage of ribonucleotides in the 5′ to 3′ direction, using DNA as a template. Eukaryotic RNA polymerase II proteins are usually composed of 12 subunits. The two largest subunits are structurally similar to the β and β′ subunits found in *E. coli* RNA polymerase.

General transcription factors:

- **TFIID:** Composed of TATA-binding protein (TBP) and other TBP-associated factors (TAFs). Recognizes the TATA box of eukaryotic protein-encoding gene promoters.
- **TFIIB:** Binds to TFIID and then enables RNA polymerase II to bind to the core promoter. Also promotes TFIIF binding.
- **TFIIF:** Binds to RNA polymerase II and plays a role in its ability to bind to TFIIB and the core promoter. Also plays a role in the ability of TFIIE and TFIIH to bind to RNA polymerase II.
- **TFIIE:** Plays a role in the formation or the maintenance (or both) of the open complex. It may exert its effects by facilitating the binding of TFIIH to RNA polymerase II and regulating the activity of TFIIH.
- **TFIIH:** A multisubunit protein that has multiple roles. First, certain subunits act as helicases and promote the formation of the open complex. Other subunits phosphorylate the carboxyl terminal domain (CTD) of RNA polymerase II, which releases its interaction with TFIIB, thereby allowing RNA polymerase II to proceed to the elongation phase.

Mediator: A multisubunit complex that mediates the effects of regulatory transcription factors on the function of RNA polymerase II. Though mediator typically has certain core subunits, many of its subunits vary, depending on the cell type and environmental conditions. The ability of mediator to affect the function of RNA polymerase II is thought to occur via the CTD of RNA polymerase II. Mediator can influence the ability of TFIIH to phosphorylate CTD, and subunits within mediator itself have the ability to phosphorylate CTD. Because CTD phosphorylation is needed to release RNA polymerase II from TFIIB, mediator plays a key role in the ability of RNA polymerase II to switch from the initiation to the elongation stage of transcription.

Five different proteins called **general transcription factors (GTFs)** are always needed for RNA polymerase II to initiate transcription of protein-encoding genes. **Figure 14.14** describes the assembly of GTFs and RNA polymerase II at the TATA box. As shown in the figure, a series of interactions leads to the formation of the open complex.

1. Transcription factor IID (TFIID) first binds to the TATA box, thereby playing a critical role in the recognition of the core promoter. TFIID is composed of several subunits, including TATA-binding protein (TBP), which directly binds to the TATA box, and several other proteins called TBP-associated factors (TAFs).
2. After TFIID binds to the TATA box, it associates with TFIIB.
3. TFIIB promotes the binding of RNA polymerase II and TFIIF to the core promoter.
4. TFIIE and TFIIH bind to the complex. Their binding completes the assembly of proteins to form a closed complex, also known as a **preinitiation complex.**

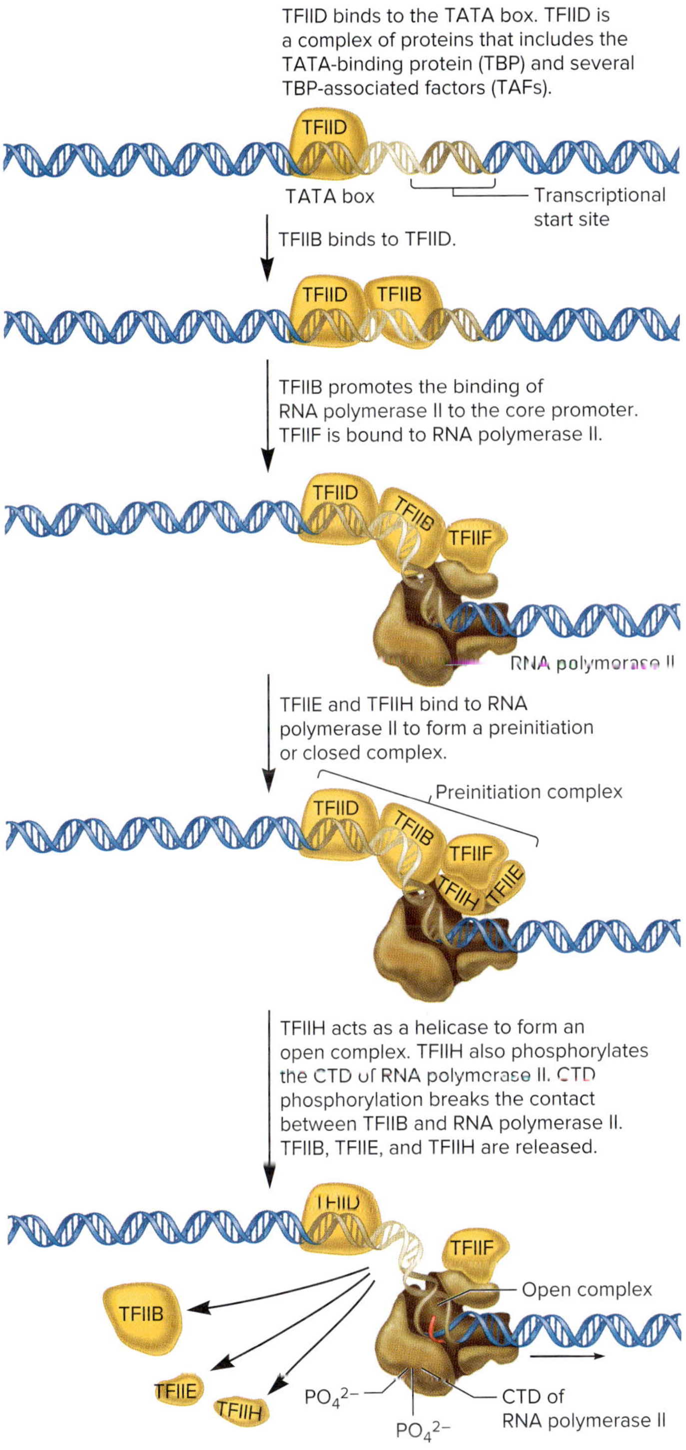

FIGURE 14.14 Steps leading to the formation of the open complex. The strands of the DNA are separated at the TATA box to form an open complex. In this diagram, the open complex has moved to the transcriptional start site, which is usually about 25 bp away from the TATA box.

Concept Check: *Why is carboxyl terminal domain (CTD) phosphorylation functionally important?*

5. TFIIH plays a major role in the formation of the open complex. TFIIH has several subunits that perform different functions. Certain subunits act as helicases, which break the hydrogen bonds between the two strands of the DNA and thereby promote the formation of the open complex. Another subunit hydrolyzes ATP and phosphorylates a domain in RNA polymerase II known as the carboxyl terminal domain (CTD).

6. Phosphorylation of the CTD releases the contact between RNA polymerase II and TFIIB. Next, TFIIB, TFIIE, and TFIIH dissociate, and RNA polymerase II is free to proceed to the elongation stage of transcription.

In addition to GTFs and RNA polymerase II, another component required for transcription is a large protein complex termed mediator. This complex was discovered by Roger Kornberg and colleagues in 1990. In 2006, Kornberg was awarded the Nobel Prize in chemistry for his studies regarding the molecular basis of eukaryotic transcription. **Mediator** derives its name from the observation that it mediates interactions between RNA polymerase II and regulatory transcription factors that bind to enhancers or silencers. It serves as an interface between RNA polymerase II and many, diverse regulatory signals. The subunit composition of mediator is quite complex and variable. The core subunits form an elliptically shaped complex that partially wraps around RNA polymerase II. Mediator itself may phosphorylate the CTD of RNA polymerase II, and it may regulate the ability of TFIIH to phosphorylate the CTD. Therefore, it plays a pivotal role in the switch between transcriptional initiation and elongation.

Transcriptional Termination of RNA Polymerase II Occurs After the 3′ End of the Transcript Is Cleaved Near the Polyadenylation Signal Sequence

As discussed later in this chapter, eukaryotic mRNAs are modified by cleavage near their 3′ end and the subsequent attachment of a string of adenine nucleotides (look ahead at Figure 14.24). This process, which is called polyadenylation, requires a polyadenylation signal sequence that directs the cleavage of the pre-mRNA. Transcription via RNA polymerase II typically terminates about 500 to 2000 nucleotides downstream from the polyadenylation signal sequence.

Figure 14.15 shows a simplified scheme for the transcriptional termination of RNA polymerase II. After RNA polymerase II has transcribed the polyadenylation signal sequence, the RNA is cleaved just downstream from this sequence. This cleavage occurs before transcriptional termination. Two models have been proposed for transcriptional termination.

- According to the allosteric model, RNA polymerase II becomes destabilized after it has transcribed the polyadenylation signal sequence, and it eventually dissociates from the DNA. This destabilization may be caused by the loss of proteins that function as elongation factors or by the binding of proteins that function as termination factors.

- A second model, called the torpedo model, suggests that RNA polymerase II is physically removed from the DNA. According to this model, the region of RNA that is downstream from the polyadenylation signal sequence is cleaved by an exonuclease that degrades the transcript in the 5′ to 3′ direction. When the exonuclease catches up to RNA polymerase II, this causes RNA polymerase II to dissociate from the DNA.

Which of these two models is correct? Additional research is needed, but the results of studies over the past few years have provided evidence that the two models are not mutually exclusive. Therefore, both mechanisms may play a role in transcriptional termination.

14.3 REVIEWING THE KEY CONCEPTS

- Eukaryotes use RNA polymerases I, II, and III to transcribe different categories of genes. Prokaryotic and eukaryotic RNA polymerases have similar structures (see Figure 14.12).
- Eukaryotic promoters have a core promoter and regulatory elements such as enhancers and silencers (see Figure 14.13).
- Transcription of protein-encoding genes in eukaryotes requires RNA polymerase II, five general transcription factors, and mediator. The five general transcription factors and RNA polymerase assemble together to form an open complex (see Table 14.1, Figure 14.14).
- Transcriptional termination of RNA polymerase II may proceed according to the allosteric or torpedo model (see Figure 14.15).

14.3 COMPREHENSION QUESTIONS

1. Which RNA polymerase in eukaryotes is responsible for the transcription of genes that encode proteins?
   a. RNA polymerase I
   b. RNA polymerase II
   c. RNA polymerase III
   d. All of the above transcribe protein-encoding genes.

2. An enhancer is a _____ that _____ the rate of transcription.
   a. *trans*-acting factor, increases
   b. *trans*-acting factor, decreases
   c. *cis*-acting element, increases
   d. *cis*-acting element, decreases

3. Transcription of protein-encoding genes in eukaryotes requires
   a. five general transcription factors.
   b. RNA polymerase II.
   c. a DNA sequence containing a TATA box and transcriptional start site.
   d. all of the above.

4. With regard to transcriptional termination in eukaryotes, which model(s) suggest(s) that RNA polymerase is physically removed from the DNA by an exonuclease?
   a. Allosteric model
   b. Torpedo model
   c. Both models
   d. Neither model

14.4 RNA MODIFICATION 309

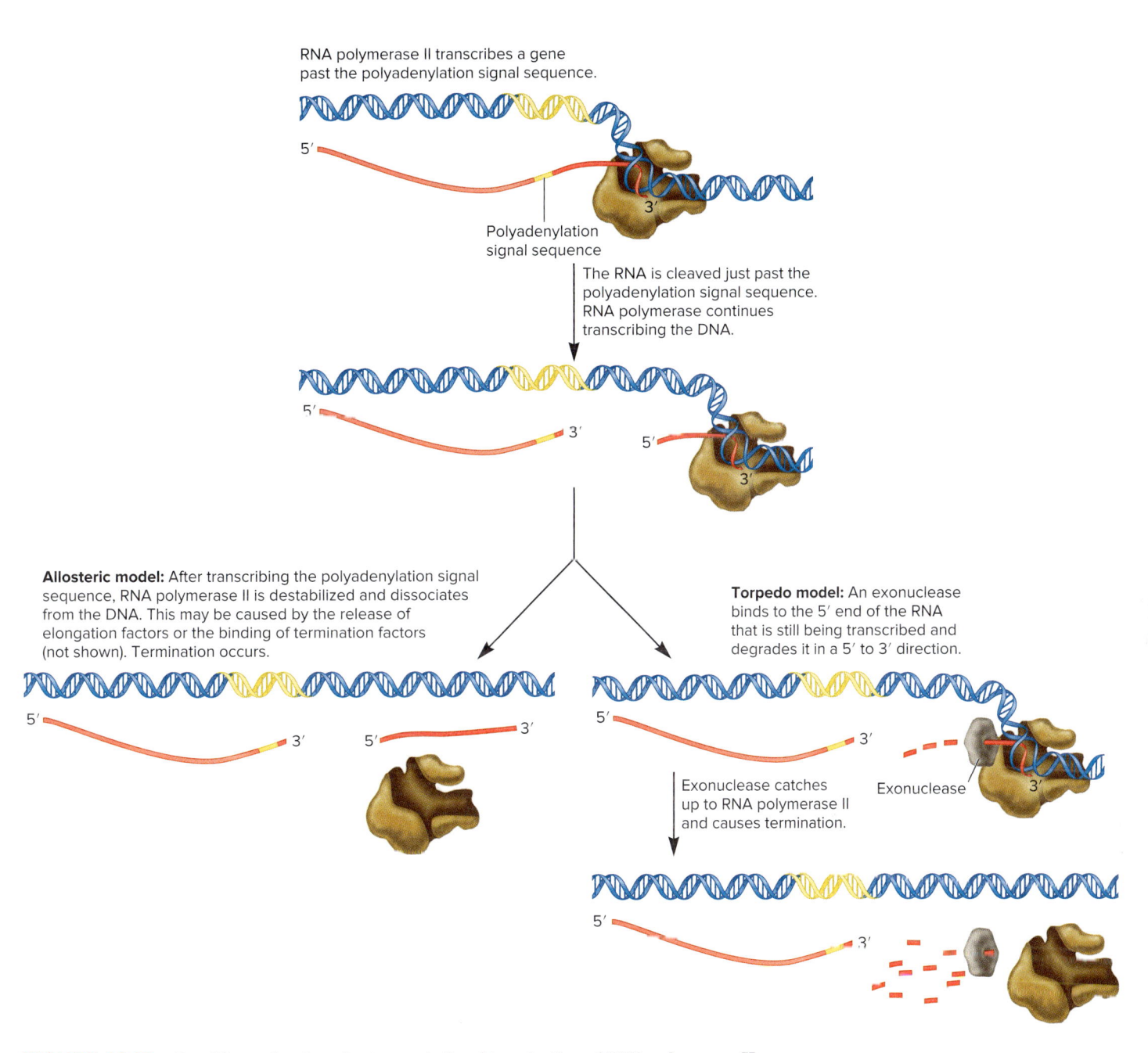

FIGURE 14.15 Possible mechanisms for transcriptional termination of RNA polymerase II.

14.4 RNA MODIFICATION

Learning Outcomes:
1. List the different types of RNA modifications.
2. Describe the processing of ribosomal RNAs and tRNAs.
3. Compare and contrast different mechanisms of RNA splicing.
4. Explain how alternative splicing occurs and what advantage it provides.
5. Explain how eukaryotic mRNAs are modified to have a cap and a tail.
6. Describe the process of RNA editing.

During the 1960s and 1970s, studies in bacteria established the physical structure of the gene. The analysis of bacterial genes showed that the sequence of DNA within the coding strand corresponds to the sequence of nucleotides in the mRNA, except that T is replaced with U. During translation, the sequence of codons in the mRNA is read, providing the instructions for the correct amino acid sequence in a polypeptide. The one-to-one correspondence between the sequence of codons in the DNA-coding strand and the amino acid sequence of the polypeptide has been termed the **colinearity** of gene expression.

The situation changed dramatically in the late 1970s, when the tools became available to study eukaryotic genes at the molecular level. The scientific community was astonished by the discovery that eukaryotic protein-encoding genes are not always colinear with their functional mRNAs. Instead, the coding sequences within many eukaryotic genes are separated by DNA sequences that are not translated into proteins. The coding sequences are found within **exons,** which are regions that are contained within mature mRNA. By comparison, the sequences that are found between the exons are called **intervening sequences,** or **introns.** During transcription, an RNA is made corresponding to the entire gene sequence. Subsequently, as it matures, the sequences in the RNA that correspond to the introns are removed and the exons are connected, or spliced, together. This process is called **RNA splicing.** Since the 1970s, research has revealed that splicing is common in eukaryotic species. Splicing occurs occasionally in bacteria as well.

Aside from splicing, research has also shown that RNA transcripts can be modified in several other ways. **Table 14.2** describes the general types of RNA modifications. In this section, we will examine the molecular mechanisms that account for several types of RNA modifications and consider why they are functionally important.

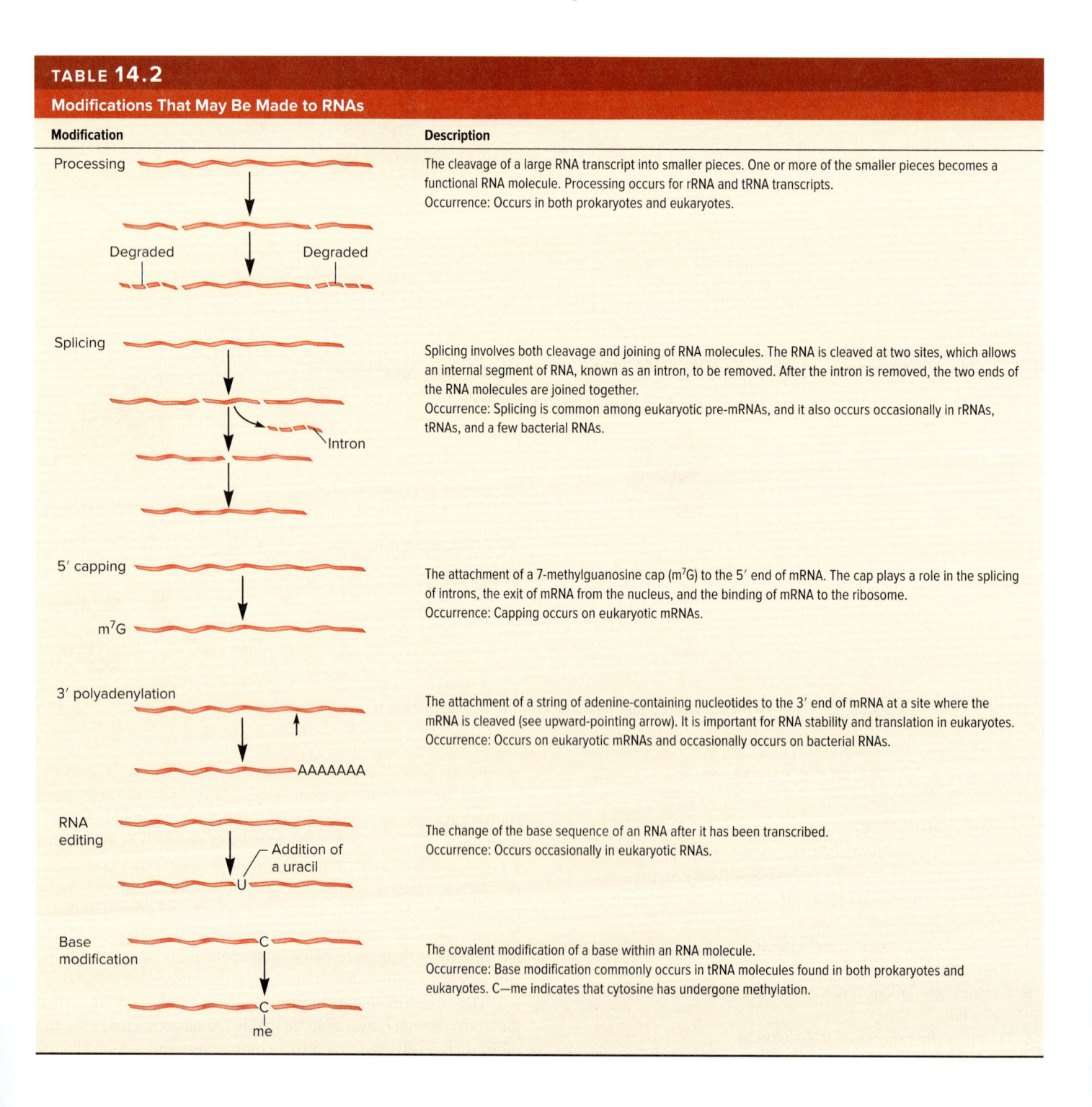

TABLE 14.2
Modifications That May Be Made to RNAs

Modification	Description
Processing	The cleavage of a large RNA transcript into smaller pieces. One or more of the smaller pieces becomes a functional RNA molecule. Processing occurs for rRNA and tRNA transcripts. Occurrence: Occurs in both prokaryotes and eukaryotes.
Splicing	Splicing involves both cleavage and joining of RNA molecules. The RNA is cleaved at two sites, which allows an internal segment of RNA, known as an intron, to be removed. After the intron is removed, the two ends of the RNA molecules are joined together. Occurrence: Splicing is common among eukaryotic pre-mRNAs, and it also occurs occasionally in rRNAs, tRNAs, and a few bacterial RNAs.
5′ capping	The attachment of a 7-methylguanosine cap (m^{7}G) to the 5′ end of mRNA. The cap plays a role in the splicing of introns, the exit of mRNA from the nucleus, and the binding of mRNA to the ribosome. Occurrence: Capping occurs on eukaryotic mRNAs.
3′ polyadenylation	The attachment of a string of adenine-containing nucleotides to the 3′ end of mRNA at a site where the mRNA is cleaved (see upward-pointing arrow). It is important for RNA stability and translation in eukaryotes. Occurrence: Occurs on eukaryotic mRNAs and occasionally occurs on bacterial RNAs.
RNA editing	The change of the base sequence of an RNA after it has been transcribed. Occurrence: Occurs occasionally in eukaryotic RNAs.
Base modification	The covalent modification of a base within an RNA molecule. Occurrence: Base modification commonly occurs in tRNA molecules found in both prokaryotes and eukaryotes. C—me indicates that cytosine has undergone methylation.

Some Large RNA Transcripts Are Cleaved into Smaller Functional Transcripts

For many non-protein-encoding genes, the RNA transcript initially made during gene transcription is processed or cleaved into smaller pieces. As an example, **Figure 14.16** shows the processing of mammalian ribosomal RNA (rRNA). The rRNA gene is transcribed by RNA polymerase I, resulting in a long primary transcript known as 45S rRNA. The designation 45S refers to the sedimentation characteristics of this transcript in Svedberg units(S). Following synthesis of the 45S rRNA, cleavage occurs at several points to produce three fragments: termed 18S, 5.8S, and 28S rRNA. These are functional rRNA molecules that play a key role in forming the structure of the ribosome. In eukaryotes, the cleavage of 45S rRNA into smaller rRNAs and the assembly of ribosomal subunits occur in a structure within the cell nucleus known as the **nucleolus**.

The production of tRNA molecules requires processing via exonucleases and endonucleases.

- An **exonuclease** cleaves a covalent bond between two nucleotides at one end of a strand. Starting at one end, an exonuclease digests a strand, one nucleotide at a time. Some exonucleases begin this digestion only from the 3' end, traveling in the 3' to 5' direction, whereas others begin only at the 5' end and digest in the 5' to 3' direction.
- By comparison, an **endonuclease** cleaves the bond between two adjacent nucleotides within a strand.

Like ribosomal RNAs, tRNAs are synthesized as large precursor tRNAs that must be cleaved to produce mature, functional tRNAs that carry amino acids. This processing has been studied extensively in *E. coli*. **Figure 14.17** shows the processing of a precursor tRNA, which involves the action of two endonucleases and one exonuclease.

1. The precursor tRNA is recognized by RNaseP, which is an endonuclease that cuts the precursor tRNA. The action of RNaseP produces the correct 5' end of the mature tRNA.

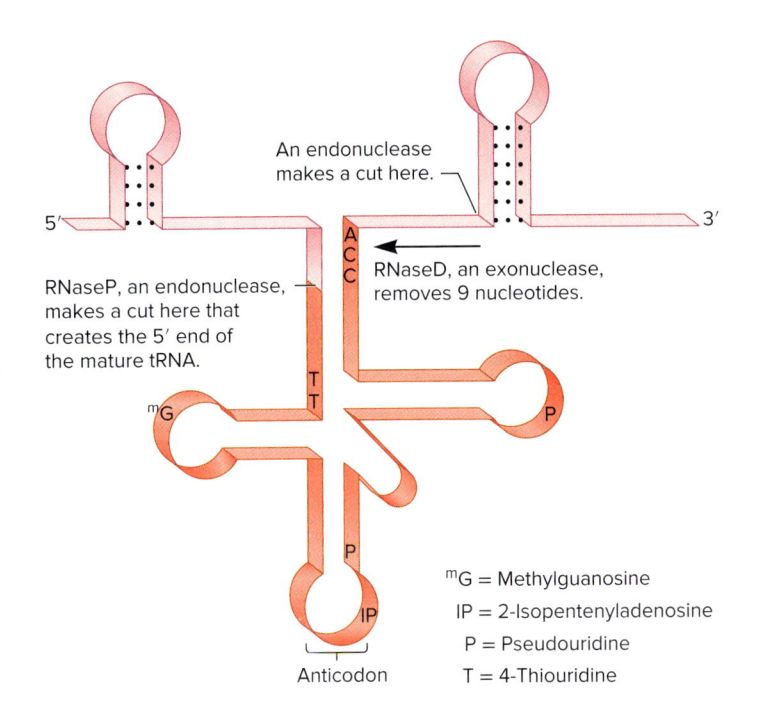

FIGURE 14.17 The processing of a precursor tRNA molecule in *E. coli*. RNaseP is an endonuclease that makes a cut that creates the 5' end of the mature tRNA. To produce the 3' end of a mature tRNA, an endonuclease makes a cut, and then the exonuclease RNaseD removes nine nucleotides at the 3' end. In addition to these cleavage steps, several bases are modified to other bases, as schematically indicated. A similar type of precursor tRNA processing occurs in eukaryotes.

Concept Check: What is the difference between an endonuclease and an exonuclease?

2. A different endonuclease cleaves the precursor tRNA to remove a 170-nucleotide segment from the 3' end.
3. Next, an exonuclease called RNaseD binds to the 3' end and digests the RNA in the 3' to 5' direction. When it reaches an ACC sequence, the exonuclease stops digesting the precursor tRNA molecule. Therefore, all tRNAs in *E. coli* have an ACC sequence at their 3' ends.
4. Finally, certain bases in tRNA molecules may be covalently modified to alter their structure. The functional importance of modified bases in tRNAs is discussed in Chapter 15.

As researchers studied tRNA processing, they discovered certain catalytic features that were very unusual and exciting, changing the way biologists view the actions of catalysts. RNaseP has been found to contain both RNA and protein subunits. In 1983, Sidney Altman and colleagues made the surprising discovery that the RNA portion of RNaseP, not the protein subunit, contains the catalytic ability to cleave the precursor tRNA. RNaseP is an example of a **ribozyme,** an RNA molecule with catalytic activity. Prior to the study of RNaseP and the identification of self-splicing RNAs (discussed later in this section), biochemists had staunchly believed that only enzymes, which are protein molecules, could function as biological catalysts.

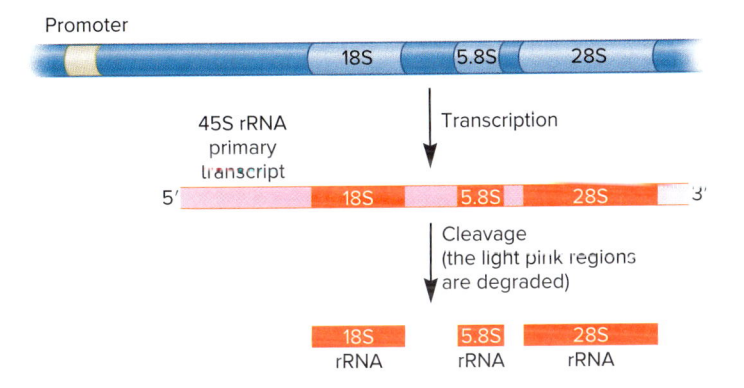

FIGURE 14.16 The processing of ribosomal RNA in **eukaryotes.** The large ribosomal RNA gene is transcribed into a long 45S rRNA primary transcript. This transcript is cleaved to produce 18S, 5.8S, and 28S rRNA molecules, which become associated with protein subunits in the ribosome. This processing occurs within the nucleolus.

Different Splicing Mechanisms Remove Introns

Although the discovery of ribozymes was very surprising, the observation that tRNA and rRNA transcripts are processed to a smaller form did not seem unusual to geneticists and biochemists, because the cleavage of RNA was similar to the cleavage that can occur for other macromolecules such as DNA and proteins. In sharp contrast, when RNA splicing was detected in the 1970s, it was a novel concept. Splicing involves cleavage at two sites. An intron is removed, and—in a unique step—the remaining fragments are hooked back together again.

Since the discovery of introns, the investigations of many research groups have shown that most protein-encoding genes in complex eukaryotes contain one or more introns. Less commonly, introns can occur within genes that encode tRNAs and rRNAs. At the molecular level, different RNA-splicing mechanisms have been identified. In the three examples shown in **Figure 14.18**, splicing leads to the removal of the intron RNA and the covalent connection of the exon RNA.

The splicing of **group I** and **group II introns** occurs via **self-splicing**—splicing that does not require the aid of other catalysts. Instead, the RNA functions as its own ribozyme. Introns of groups I and II differ in the ways that they are removed and the exons are connected. Group I introns that occur within the rRNA of *Tetrahymena* (a protozoan) have been studied extensively by Thomas Cech

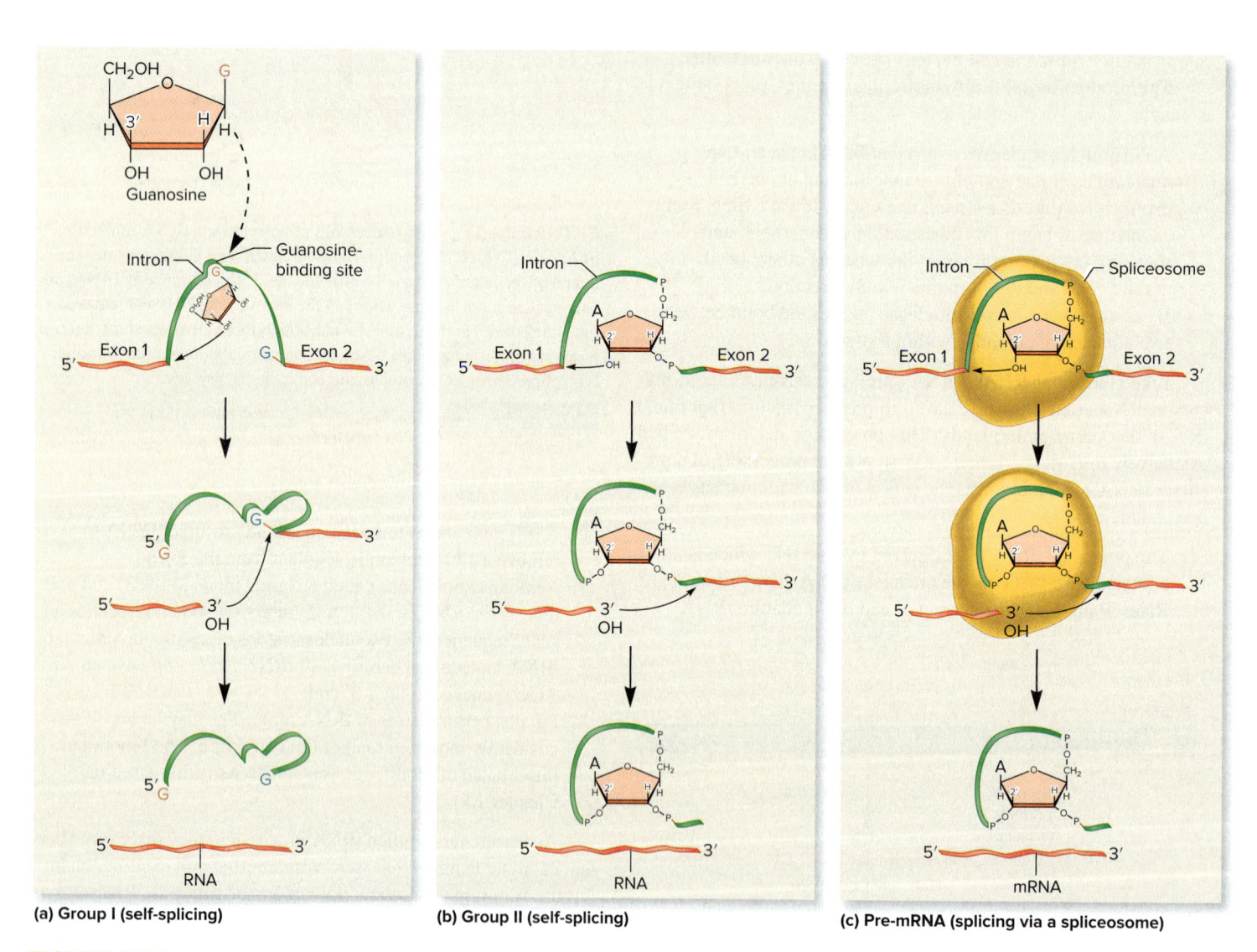

FIGURE 14.18 **Mechanisms of RNA splicing.** Group I and II introns are self-splicing. (a) The splicing of group I introns involves the binding of a free guanosine to a site within the intron, leading to the cleavage of RNA at the 3′ end of exon 1. The bond between a different guanine nucleotide (in the intron strand) and the 5′ end of exon 2 is cleaved. The 3′ end of exon 1 then forms a covalent bond with the 5′ end of exon 2. (b) In group II introns, self-splicing occurs by a similar mechanism, except that the 2′—OH group on an adenine nucleotide (already within the intron) begins the catalytic process. (c) Pre-mRNA splicing requires the aid of a complex known as a spliceosome.

Concept Check: Which of these three mechanisms is very common in eukaryotes?

and colleagues. In this organism, the splicing process involves the binding of a single guanosine to a guanosine-binding site within the intron (Figure 14.18a).

1. The guanosine breaks the bond between the first exon and the intron. As this occurs, the guanosine becomes attached to the 5' end of the intron.
2. The 3' —OH group of exon 1 then breaks the bond next to a different nucleotide (in this example, a guanine, G) that lies at the boundary between the end of the intron and exon 2; exon 1 forms a phosphoester bond with the 5' end of exon 2.
3. The intron RNA is subsequently degraded.

In this example, the RNA molecule functions as its own ribozyme, because it splices itself without the aid of a catalytic protein.

With group II introns, a similar splicing mechanism occurs, except the 2' —OH group on ribose in an adenine (A) nucleotide already within the intron strand begins the catalytic process (Figure 14.18b). Experimentally, self-splicing of group I and group II introns can occur in vitro without the addition of any proteins. However, in a living cell, proteins known as **maturases** often enhance the rate of such self-splicing.

In eukaryotes, the transcription of protein-encoding genes produces a long transcript known as **pre-mRNA,** which is located within the nucleus. This pre-mRNA is usually altered by splicing and other modifications before it exits the nucleus. Unlike group I and II introns, which may undergo self-splicing, pre-mRNA splicing requires the aid of a complex known as a **spliceosome** (Figure 14.18c). As discussed shortly, the spliceosome is needed to recognize the boundaries of the intron and to properly remove it.

Table 14.3 summarizes the occurrence of introns among the genes of different groups of organisms. The splicing of group I and II introns is relatively uncommon. By comparison, pre-mRNA splicing is a widespread phenomenon among complex eukaryotes. In mammals and flowering plants, most protein-encoding genes have at least one intron that can be located anywhere within the gene. For example, an average human gene has about eight introns.

Pre-mRNA Splicing Occurs by the Action of a Spliceosome

The spliceosome is a large complex that splices pre-mRNA in eukaryotes. It is composed of several subunits known as **small nuclear riboproteins** (**snRNPs,** pronounced "snurps"). Each snRNP contains small nuclear RNA (snRNA) and a set of proteins. During splicing, the subunits of a spliceosome carry out several functions. First, spliceosome subunits bind to an intron sequence and precisely recognize the intron-exon boundaries. In addition, the spliceosome must hold the pre-mRNA in the correct configuration to ensure the splicing together of the exons. And finally, the spliceosome catalyzes the chemical reactions that cause the intron to be removed and the exons to be covalently linked.

Intron RNA is defined by particular sequences within the intron and at the intron-exon boundaries. The consensus sequences for the splicing of mammalian pre-mRNA are shown in **Figure 14.19**. The bases most commonly found at these sites—those that are highly conserved evolutionarily—are shown in bold. The 5' and 3' splice sites occur at the ends of the intron, whereas the branch site is somewhere in the middle. These sites are recognized by subunits of the spliceosome.

TABLE 14.3
Occurrence of Introns

Type of Intron	Mechanism of Removal	Occurrence
Group I	Self-splicing	Found in rRNA genes within the nucleus of *Tetrahymena* and other simple eukaryotes. Found in a few protein-encoding, tRNA, and rRNA genes within mitochondrial DNA (in fungi and plants) and in chloroplast DNA. Found very rarely in tRNA genes within bacteria.
Group II	Self-splicing	Found in a few protein-encoding, tRNA, and rRNA genes within mitochondrial DNA (in fungi and plants) and in chloroplast DNA. Also found rarely in bacterial genes.
Pre-mRNA	Spliceosome	Very commonly found in protein-encoding genes within the nucleus of eukaryotic cells.

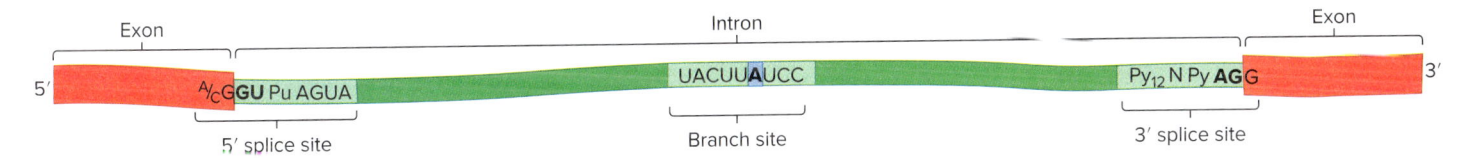

FIGURE 14.19 Consensus sequences for pre-mRNA splicing in complex eukaryotes. Consensus sequences exist at the intron-exon boundaries and at a branch site found within the intron itself. The adenine nucleotide shown in blue in this figure corresponds to the adenine nucleotide at the branch site in Figure 14.20. The nucleotides shown in bold are highly conserved. Designations: A/C = A or C, Pu = purine, Py = pyrimidine, N = any of the four bases.

The molecular mechanism of pre-mRNA splicing is depicted in **Figure 14.20**.

1. The snRNP designated U1 binds to the 5′ splice site, and U2 binds to the branch site.

2. This is followed by the binding of a trimer of three snRNPs: a U4/U6 dimer plus U5. Due to this binding, the intron loops outward, and the two exons are brought closer together.

3. The 5′ splice site is then cut, and the 5′ end of the intron becomes covalently attached to the 2′—OH group of a specific adenine nucleotide in the branch site. U1 and U4 are released.

4. In the final step, the 3′ splice site is cut, and then the exons are covalently attached to each other. The three snRNPs—U2, U5, and U6—remain attached to the intron, which is in a lariat configuration. Eventually, the intron is degraded, and the snRNPs are used again to splice other pre-mRNAs.

Evidence is accumulating that certain snRNA molecules within the spliceosome play a catalytic role in the removal of introns and the connection of exons. In other words, snRNAs may function as ribozymes that cleave the RNA at the exon-intron boundaries and connect the remaining exons. Researchers have speculated that RNA molecules within U2 and U6 may have this catalytic function.

Alternative Splicing Regulates Which Exons Occur in an RNA Transcript, Allowing Different Polypeptides to Be Made from the Same Protein-Encoding Gene

When it was first discovered, the phenomenon of splicing seemed like a rather wasteful process. During transcription, energy is used to synthesize intron sequences. Likewise, energy is also used to remove introns via a spliceosome. This observation intrigued many geneticists, because natural selection tends to eliminate wasteful processes. Therefore, many geneticists expected to find that pre-mRNA splicing has one or more important biological roles. In recent years, one very important biological role, termed **alternative splicing,** has become apparent. Alternative splicing refers to the phenomenon that a pre-mRNA can be spliced in more than one way.

What is the advantage of alternative splicing? To understand the biological effects of alternative splicing, remember that the sequence of amino acids within a polypeptide determines the structure and function of a protein. Alternative splicing produces two or more polypeptides from the same gene that have differences in their amino acid sequences, leading to possible changes in their functions. In most cases, the alternative versions of the protein have similar functions, because most of their amino acid sequences are identical to each other. Nevertheless, alternative splicing produces differences in amino acid sequences that provide each polypeptide with its own unique characteristics. Because alternative splicing allows two or more different polypeptide sequences to be derived from a single gene, some geneticists have speculated that an important advantage of this process is that it allows an organism to carry fewer genes in its genome.

The degree of splicing and alternative splicing varies greatly among different species. Baker's yeast (*S. cerevisiae*), for example, has about 6300 genes, and approximately 300 (i.e., ~5%) encode

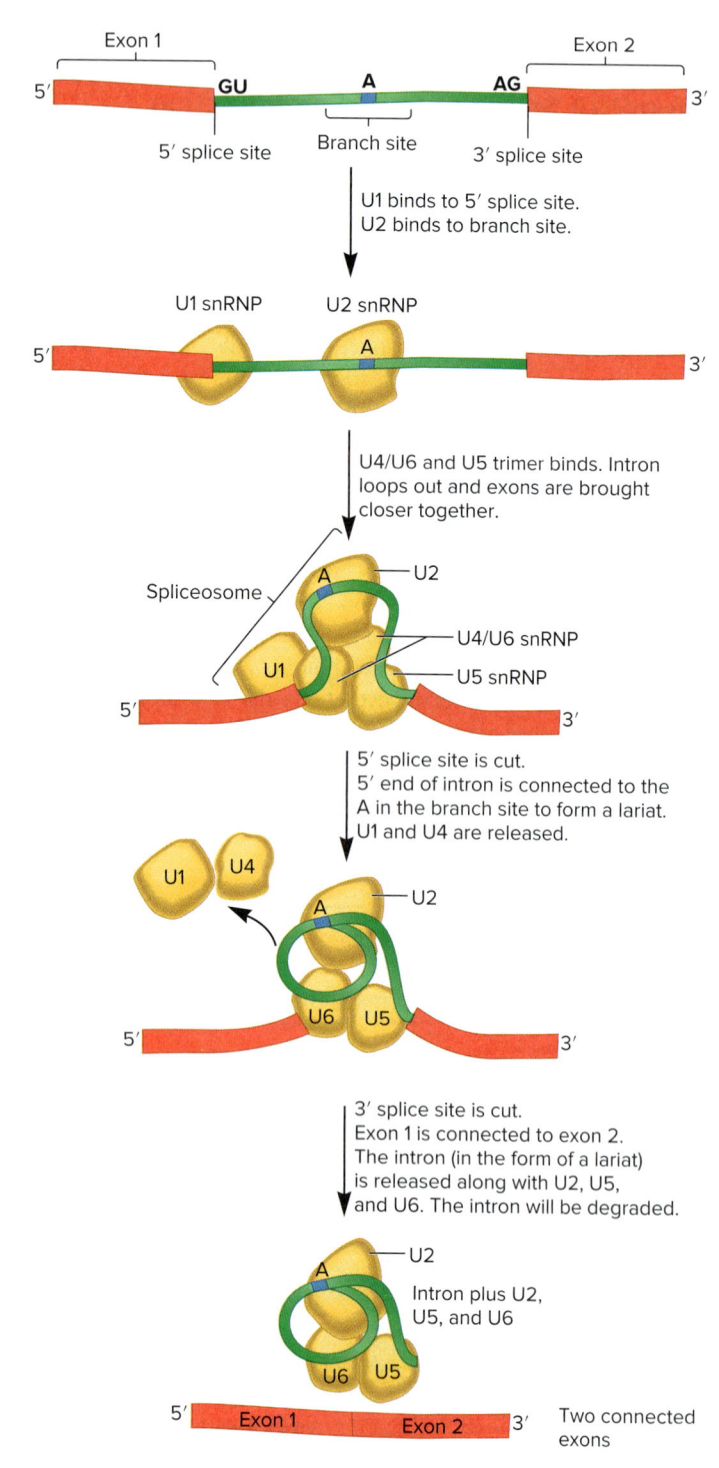

FIGURE 14.20 Splicing of pre-mRNA via a spliceosome.

Concept Check: Describe the roles of snRNPs in the splicing process.

pre-mRNAs that are spliced. Of these, only a few have been shown to be alternatively spliced. Therefore, in this unicellular eukaryote, alternative splicing is not a major mechanism for generating protein diversity. In comparison, complex multicellular organisms rely much more heavily on alternative splicing. Humans have approximately 22,000 different protein-encoding genes, and most of these contain one or more introns. Recent estimates suggest that about 70% of all human pre-mRNAs are alternatively spliced. Furthermore, certain pre-mRNAs are alternatively spliced to an extraordinary extent. Some pre-mRNAs can be alternatively spliced to produce dozens of different mRNAs. This level of alternative splicing provides a much greater potential for human cells to produce protein diversity.

Figure 14.21 considers an example of alternative splicing for a mammalian gene that encodes a protein known as α-tropomyosin. This protein functions in the regulation of cell contraction. It is located along the thin filaments found in smooth muscle cells, such as those in the uterus and small intestine, and in striated muscle cells that are found in cardiac and skeletal muscle. The protein α-tropomyosin is also synthesized in many types of nonmuscle cells but in lower amounts. Within a multicellular organism, different types of cells must regulate their contractibility in subtly different ways. One way this variation in function may be accomplished is by the production of different forms of α-tropomyosin.

The intron-exon structure of rat α-tropomyosin pre-mRNA and two alternative ways that the pre-mRNA can be spliced are shown in Figure 14.21.

- The pre-mRNA contains 14 exons, 6 of which are **constitutive exons** (shown in red), which are always found in the mature mRNA from all cell types. Presumably, constitutive exons encode polypeptide segments of the α-tropomyosin protein that are necessary for its general structure and function.
- By comparison, **alternative exons** (shown in green) are not always found in the mRNA after splicing has occurred. The polypeptide sequences encoded by alternative exons may subtly change the function of α-tropomyosin to meet the needs of the cell type in which it is found. For example, Figure 14.21 shows the predominant splicing products found in smooth muscle cells and striated muscle cells. Exon 2 encodes a segment of the α-tropomyosin protein that alters its function to make it suitable for smooth muscle cells. By comparison, the α-tropomyosin mRNA found in striated muscle cells does not include exon 2. Instead, this mRNA contains exon 3, which is more suitable for that cell type.

Alternative splicing is not a random event. Rather, the specific pattern of splicing is regulated in any given cell. The molecular mechanism for the regulation of alternative splicing involves proteins known as **splicing factors.** Such splicing factors play a key role in the choice of particular splice sites. **SR proteins** are an example of a type of splicing factor. SR proteins contain a domain at their carboxyl-terminal end that is rich in serines (S) and arginines (R) and is involved in protein-protein recognition. They also contain an RNA-binding domain at their amino-terminal end.

As shown earlier, in Figure 14.20, components of the spliceosome recognize the 5′ and 3′ splice sites and then remove the intervening intron. The key effect of splicing factors is to modulate

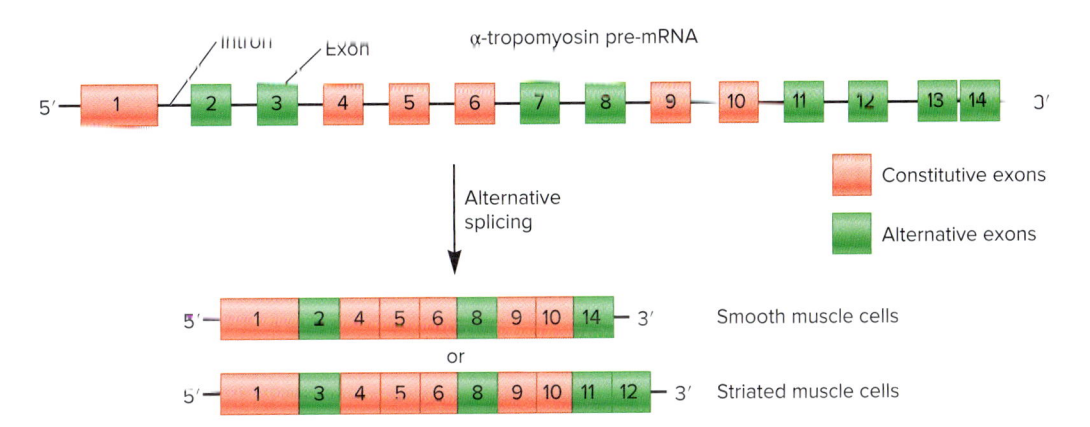

FIGURE 14.21 Alternative ways that the rat α-tropomyosin pre-mRNA can be spliced. The top part of this figure depicts the structure of the rat α-tropomyosin pre-mRNA. Exons are shown as colored boxes, and introns as connecting black lines. The lower part of the figure describes the final mRNA products in smooth and striated muscle cells. Note: Exon 8 is found in the final mRNA of smooth and striated muscle cells, but not in the mature mRNA for α-tropomyosin in certain other cell types.

Genes → Traits α-tropomyosin functions in the regulation of cell contraction in muscle and nonmuscle cells. Alternative splicing of the pre-mRNA provides a way to vary contractibility in different types of cells by modifying the function of α-tropomyosin. As shown here, the alternatively spliced versions of the pre-mRNA produce α-tropomyosin proteins that differ from each other in their structure (i.e., amino acid sequence). These alternatively spliced versions vary in function to meet the needs of the cell type in which they are found. For example, the sequence of exons 1–2–4–5–6–8–9–10–14 produces an α-tropomyosin protein that functions suitably in smooth muscle cells. Overall, alternative splicing affects the traits of an organism by allowing a single gene to encode several versions of a protein, each optimally suited to the cell type in which it is made.

the ability of the spliceosome to choose 5′ and 3′ splice sites. This can occur in two ways:

- Some splicing factors act as repressors that inhibit the ability of the spliceosome to recognize a splice site. In **Figure 14.22a**, a splicing repressor binds to a 3′ splice site and prevents the spliceosome from recognizing the site. Instead, the spliceosome binds to the next available 3′ splice site. The splicing repressor causes exon 2 to be spliced out of the pre-mRNA and not included in the mature mRNA, an event called **exon skipping**.
- Alternatively, other splicing factors enhance the ability of the spliceosome to recognize particular splice sites. In **Figure 14.22b**, splicing enhancers bind to the 3′ and 5′ splice sites that flank exon 3, which results in the inclusion of exon 3 in the mature mRNA.

Alternative splicing in different tissues is thought to occur because each cell type has its own characteristic concentration of many kinds of splicing factors. Furthermore, splicing factors may be regulated by the binding of small effector molecules, protein-protein interactions, and covalent modifications. Overall, the differences in the composition of splicing factors and the regulation of their activities form the basis for alternative splicing outcomes.

The Ends of Eukaryotic Pre-mRNAs Have a 5′ Cap and a 3′ Tail

In addition to splicing, pre-mRNAs in eukaryotes are also modified at their 5′ and 3′ ends. At their 5′ end, most mature mRNAs have a 7-methylguanosine covalently attached—an event known as **capping**. Capping occurs while the pre-mRNA is being made by

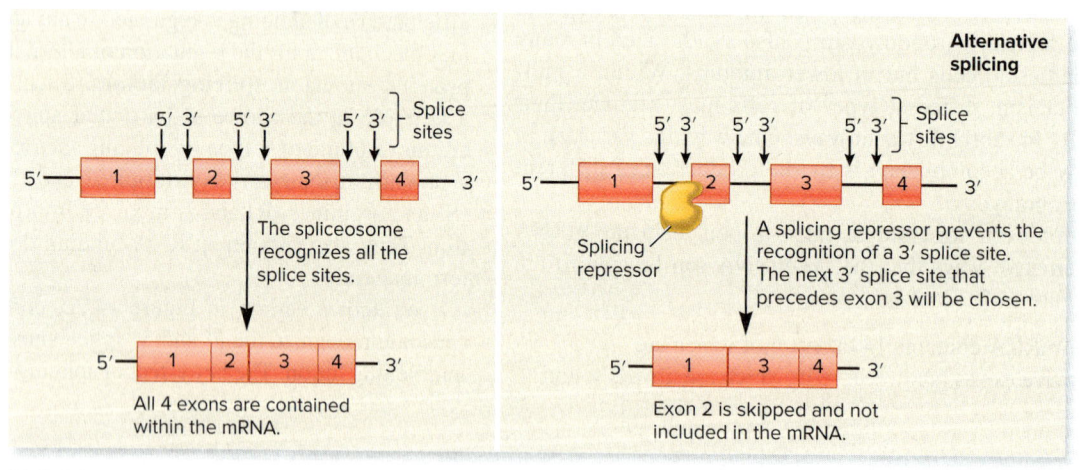

FIGURE 14.22 The roles of splicing factors during alternative splicing. (a) Splicing factors can act as repressors to prevent the recognition of splice sites. In this example, the presence of the splicing repressor causes exon 2 to be skipped and thus not included in the mRNA. (b) Other splicing factors can enhance the recognition of splice sites. In this example, the splicing enhancers promote the recognition of sites that flank exon 3, thereby causing its inclusion in the mRNA.

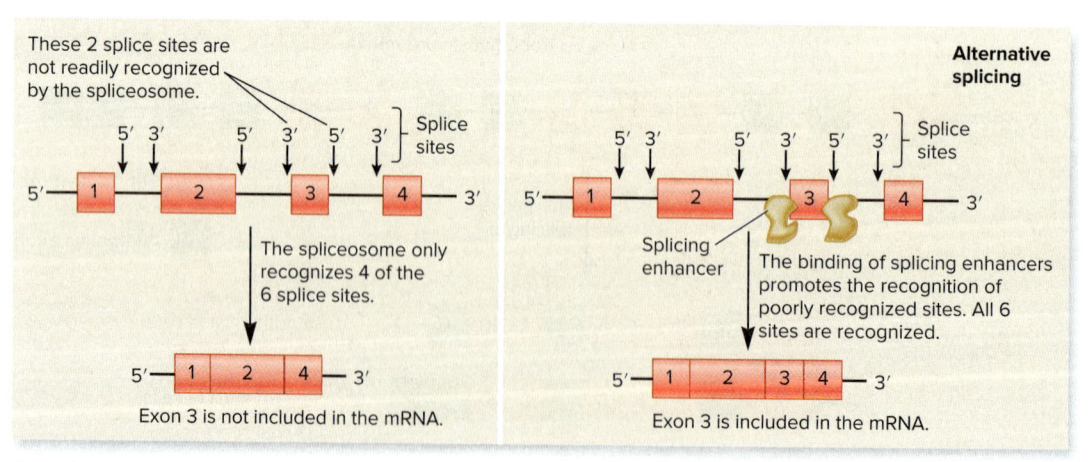

Concept Check: *A pre-mRNA with seven exons and six introns is recognized by just one splicing repressor that binds to the 3′ end of the third intron. The third intron is located between exon 3 and exon 4. After splicing in the presence of the splicing repressor is complete, would you expect the mRNA to contain exon 3 and/or exon 4?*

14.4 RNA MODIFICATION 317

RNA polymerase II, usually when the transcript is only 20 to 25 nucleotides in length. As shown in **Figure 14.23**, capping is a three-step process.

1. The nucleotide at the 5′ end of the transcript has three phosphate groups. An enzyme called RNA 5′-triphosphatase removes one of the phosphates.
2. A second enzyme, guanylyltransferase, uses guanosine triphosphate (GTP) to attach a guanosine monophosphate (GMP) to the 5′ end.
3. Finally, a methyltransferase attaches a methyl group to the base guanine.

What are the functions of the 7-methylguanosine cap? The cap structure is recognized by cap-binding proteins, which perform various roles.

- Cap-binding proteins are required for the proper exit of most mRNAs from the nucleus.
- The cap structure is recognized by initiation factors that are needed during the early stages of translation.
- The cap structure may be important in the efficient splicing of introns, particularly the intron that is closest to the 5′ end.

Let's now turn our attention to the 3′ end of the RNA molecule. At that end, most mature mRNAs have a string of adenine nucleotides, referred to as a **polyA tail**, which is important for mRNA stability, the exit of mRNA from the nucleus, and in the synthesis of polypeptides. The polyA tail is not encoded in the gene sequence. Instead, it is added enzymatically after the pre-mRNA has been completely transcribed—a process termed **polyadenylation**.

To acquire a polyA tail, the pre-mRNA contains a polyadenylation signal sequence near its 3′ end. In mammals, the consensus sequence is AAUAAA. This sequence is downstream (toward the 3′ end) from the stop codon in the pre-mRNA. The steps required to synthesize a polyA tail are shown in **Figure 14.24**.

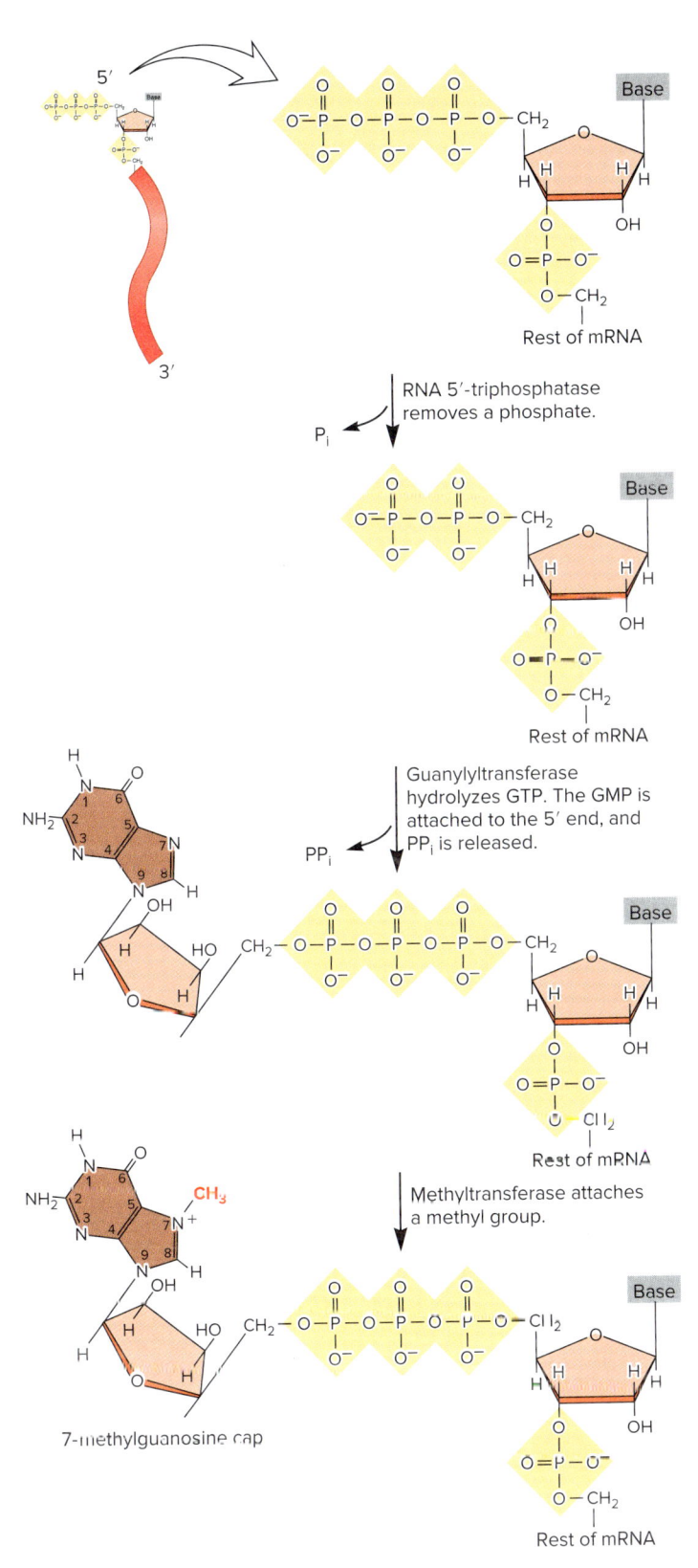

FIGURE 14.23 **Attachment of a 7-methylguanosine cap to the 5′ end of mRNA.** When the transcript is about 20 to 25 nucleotides in length, RNA 5′-triphosphatase removes one of the three phosphates, and then a second enzyme, guanylyltransferase, attaches GMP to the 5′ end. Finally, a methyltransferase attaches a methyl group to the base guanine.

Concept Check: What are three functional roles of the 7-methylguanosine cap?

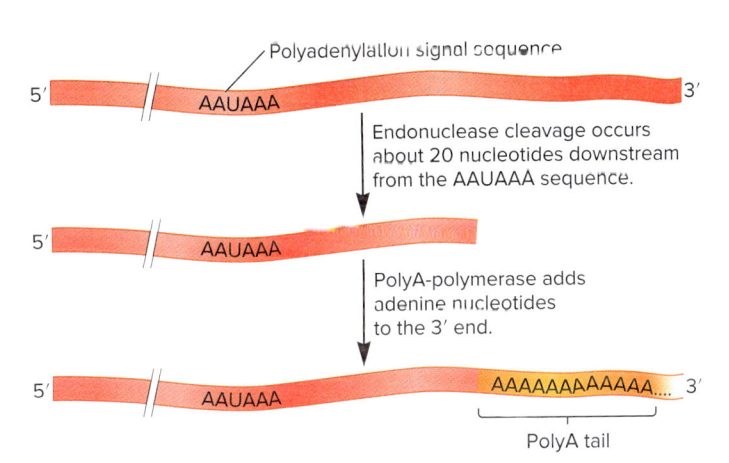

FIGURE 14.24 **Attachment of a polyA tail.** First, an endonuclease cuts the RNA at a location that is about 20 nucleotides downstream from the AAUAAA polyadenylation signal sequence, making the RNA shorter at its 3′ end. Adenine-containing nucleotides are then attached, one at a time, to the 3′ end by the enzyme polyA-polymerase.

1. An endonuclease recognizes the signal sequence and cleaves the pre-mRNA at a location that is about 20 nucleotides beyond the 3′ end of the AAUAAA sequence. The fragment beyond the 3′ cut is degraded.
2. Next, an enzyme known as polyA-polymerase attaches many adenine-containing nucleotides.

The length of the polyA tail varies among different mRNAs, from a few dozen to several hundred adenine nucleotides. A long polyA tail increases the stability of mRNA in eukaryotes. In contrast, some RNAs in bacteria also acquire a polyA tail, but the polyA tail of bacterial RNAs usually promotes degradation.

The Nucleotide Sequence of RNA Can Be Modified by RNA Editing

The term **RNA editing** refers to the process of making a change in the nucleotide sequence of an RNA molecule that involves additions or deletions of particular nucleotides or conversion of one type of base to another, such as a cytosine to a uracil. In the case of mRNAs, editing can have various effects, such as generating start codons, generating stop codons, and changing the coding sequence for a polypeptide.

The phenomenon of RNA editing was first discovered in trypanosomes, the protists that cause sleeping sickness. Since that time, however, RNA editing has been shown to occur in various organisms and in a variety of ways, although its functional significance is only slowly emerging (**Table 14.4**). In the specific case of trypanosomes, the editing process involves the addition or deletion of one or more uracil nucleotides in an RNA.

A more widespread mechanism of RNA editing involves changes of one type of base to another. In this form of editing, a base in the RNA is deaminated—an amino group is removed from the base. When cytosine is deaminated, uracil is formed, and when adenine is deaminated, hypoxanthine (H) is formed (**Figure 14.25**). Hypoxanthine is recognized as guanine during translation.

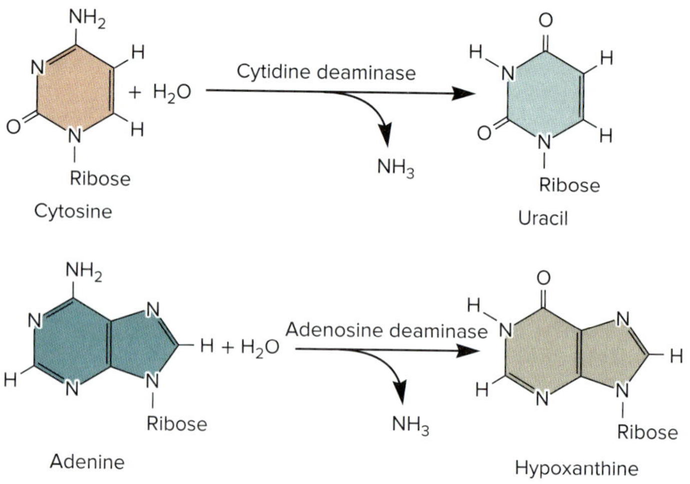

FIGURE 14.25 RNA editing by deamination. A cytidine deaminase can remove an amino group from cytosine, thereby creating uracil. An adenosine deaminase can remove an amino group from adenine to make hypoxanthine.

Concept Check: What is a functional consequence of RNA editing?

An example of RNA editing that occurs in mammals involves an mRNA that encodes a protein called apolipoprotein B. In the liver, translation of an unedited mRNA produces apolipoprotein B-100, a protein that is essential for the transport of cholesterol in the blood. In intestinal cells, the mRNA may be edited so that a single C is changed to a U. What is the significance of this base substitution? This change converts a glutamine codon (CAA) to a stop codon (UAA), thereby resulting in a shorter apolipoprotein. In this case, RNA editing produces an apolipoprotein B with an altered structure. Therefore, RNA editing can produce two proteins from the same gene, much like the phenomenon of alternative splicing.

How widespread is RNA editing that involves C to U and A to H substitutions? In invertebrates such as *Drosophila*, researchers estimate that 50 to 100 pre-mRNAs are edited in a way that changes the RNA coding sequence. In mammals, the pre-mRNAs from fewer than 25 genes are currently known to be edited.

14.4 REVIEWING THE KEY CONCEPTS

- RNA transcripts can be modified in a variety of ways, which include processing, splicing, capping at the 5′ end, adding a polyA tail at the 3′ end, RNA editing, and base modification (see Table 14.2).
- Certain RNA molecules such as rRNAs and tRNAs are processed via cleavage steps to yield smaller, functional molecules (see Figures 14.16, 14.17).
- Group I and group II introns are removed by self-splicing. Pre-mRNA introns are removed via a spliceosome (see Table 14.3, Figure 14.18).

TABLE 14.4
Examples of RNA Editing

Organism	Type of Editing	Found In
Trypanosomes (protozoa)	Primarily additions but occasionally deletions of uracil nucleotides	Many mitochondrial mRNAs
Land plants	C-to-U conversion	Many mitochondrial and chloroplast mRNAs, tRNAs, and rRNAs
Mammals	C-to-U conversion	Apolipoprotein B mRNA
	A-to-H* conversion	Glutamate receptor mRNA, many tRNAs
Drosophila	A-to-H conversion	mRNA for calcium and sodium channels

*H stands for hypoxanthine. Note: When hypoxanthine is attached to ribose, it is called inosine.

- The spliceosome is a multisubunit structure that recognizes intron sequences and removes them from pre-mRNA (see Figures 14.19, 14.20).
- During alternative splicing, proteins called splicing factors regulate which exons are included in a mature mRNA (see Figures 14.21, 14.22).
- In eukaryotes, mRNA is given a 7-methylguanosine cap at the 5′ end and a polyA tail at the 3′ end (see Figures 14.23, 14.24).
- RNA editing changes the base sequence of an RNA after it has been synthesized (see Table 14.4, Figure 14.25).

14.4 COMPREHENSION QUESTIONS

1. Which of the following are examples of RNA modifications?
   a. Splicing
   b. Capping with 7-methylguanosine
   c. Adding a polyA tail
   d. All of the above are examples of RNA modifications.
2. A ribozyme is
   a. a complex between RNA and a protein.
   b. an RNA that encodes a protein that functions as an enzyme.
   c. an RNA molecule with catalytic function.
   d. a protein that degrades RNA molecules.
3. Which of the following statements about the spliceosome is *false*?
   a. A spliceosome splices pre-mRNA molecules.
   b. A spliceosome removes exons from RNA molecules.
   c. A spliceosome is composed of snRNPs.
   d. A spliceosome recognizes the exon-intron boundaries and the branch site.
4. Which of the following is a function of the 7-methylguanosine cap?
   a. Exit of mRNA from the nucleus
   b. Efficient splicing of pre-mRNA
   c. Initiation of translation
   d. All of the above are functions of the cap.

14.5 A COMPARISON OF TRANSCRIPTION AND RNA MODIFICATION IN BACTERIA AND EUKARYOTES

Learning Outcome:
1. Compare and contrast the processes of transcription and RNA modification in bacteria and eukaryotes.

Throughout this chapter, we have considered the processes of transcription and RNA modification. Many similarities have been noted between bacteria and eukaryotes. However, these processes are more complex in eukaryotes than in their bacterial counterparts. **Table 14.5** summarizes many of the key differences.

14.5 COMPREHENSION QUESTION

1. Which of the following is *not* a key difference between bacteria and eukaryotes?
   a. Initiation of transcription requires more proteins in eukaryotes.
   b. A 7-methylguanosine cap is added only to eukaryotic mRNAs.
   c. Splicing is common in complex eukaryotes but not in bacteria.
   d. All of the above are key differences.

TABLE 14.5
Key Differences Between Transcription and RNA Modification in Bacteria and Eukaryotes*

Component	Bacteria	Eukaryotes
Promoter	Consists of −35 and −10 sequences	For protein-encoding genes, the core promoter often consists of a TATA box and a transcriptional start site.
RNA polymerase	A single RNA polymerase	Three types of RNA polymerases; RNA polymerase II transcribes protein-encoding genes.
Initiation	Sigma factor is needed for promoter recognition.	Five general transcription factors assemble at the core promoter.
Elongation	Requires the release of sigma factor	Mediator controls the switch to the elongation phase.
Termination	Is either ρ-dependent or ρ-independent	According to the allosteric or torpedo model
Splicing	Very rare; self-splicing	Commonly occurs in protein-encoding pre-mRNAs in complex eukaryotes via a spliceosome; self-splicing occurs rarely.
Capping	Does not occur	Addition of 7-methylguanosine cap
Tailing	Added to 3′ end; promotes degradation	Added to the 3′ end; promotes stability
RNA editing	Does not occur	Occurs occasionally

*Note: This table does not include the process of gene regulation, which is described in Chapters 16 and 17.

KEY TERMS

Page 296. gene, transcription, Protein-encoding genes (structural genes), messenger RNA (mRNA), translation, central dogma of genetics, gene expression,

Page 298. promoter, terminator, transcription factors, regulatory sequences (regulatory elements), ribosome-binding site, codons, start codon, stop codon, initiation, elongation, termination, RNA polymerase

Page 299. open complex, transcriptional start site, sequence elements

Page 300. Pribnow box, consensus sequence, core enzyme, sigma (σ) factor, holoenzyme, helix-turn-helix motif

Page 301. closed complex, template strand

Page 302. coding strand, ρ-dependent termination, rho (ρ) protein, ρ-independent termination (intrinsic termination)

Page 306. core promoter, TATA box, basal transcription, enhancers, silencers, *cis*-acting elements, *trans*-acting factors

Page 307. general transcription factors (GTFs), preinitiation complex

Page 308. mediator

Page 309. colinearity

Page 310. exons, intervening sequences (introns), RNA splicing

Page 311. nucleolus, exonuclease, endonuclease, ribozyme

Page 312. group I introns, group II introns, self-splicing

Page 313. maturases, pre-mRNA, spliceosome, small nuclear riboproteins (snRNPs)

Page 314. alternative splicing

Page 315. constitutive exons, alternative exons, splicing factors, SR proteins

Page 316. exon skipping, capping

Page 317. polyA tail, polyadenylation

Page 318. RNA editing

CHAPTER SUMMARY

- According to the central dogma of genetics, DNA is transcribed into mRNA, and mRNA is translated into a polypeptide. DNA replication allows the DNA to be passed from cell to cell and from parent to offspring (see Figure 14.1).

14.1 Overview of Transcription

- A gene is an organization of DNA sequences. A promoter signals the start of transcription, and a terminator signals the end. Regulatory sequences control the rate of transcription. For genes that encode polypeptides, the gene sequence also specifies a start codon, a stop codon, and many codons in between. Bacterial genes also specify a ribosomal binding sequence (see Figure 14.2).
- Transcription occurs in three phases called initiation, elongation, and termination (see Figure 14.3).

14.2 Transcription in Bacteria

- Many bacterial promoters have sequence elements at the −35 and −10 sites. The transcriptional start site is at +1 (see Figures 14.4, 14.5).
- During the initiation phase of transcription in *E. coli*, sigma (σ) factor, which is bound to RNA polymerase, binds into the major groove of DNA and recognizes sequence elements at the promoter. This process forms a closed complex. Following the formation of an open complex, σ factor is released (see Figures 14.6, 14.7).
- During the elongation phase of transcription, RNA polymerase slides along the DNA and maintains an open complex as it goes. RNA is made in the 5′ to 3′ direction, and base pairing follows a complementarity rule similar to the AT/GC rule (see Figure 14.8).
- In a given chromosome, the strand that is used as the template strand varies from gene to gene (see Figure 14.9).

- Transcriptional termination in *E. coli* occurs by a ρ-dependent or ρ-independent mechanism (see Figures 14.10, 14.11).

14.3 Transcription in Eukaryotes

- Eukaryotes use RNA polymerases I, II, and III to transcribe different categories of genes. Prokaryotic and eukaryotic RNA polymerases have similar structures (see Figure 14.12).
- Eukaryotic promoters have a core promoter and regulatory elements such as enhancers and silencers (see Figure 14.13).
- Transcription of protein-encoding genes in eukaryotes requires RNA polymerase II, five general transcription factors, and mediator. The five general transcription factors and RNA polymerase assemble together to form an open complex (see Table 14.1, Figure 14.14).
- Transcriptional termination of RNA polymerase II may proceed according to the allosteric or torpedo model (see Figure 14.15).

14.4 RNA Modification

- RNA transcripts can be modified in a variety of ways, which include processing, splicing, capping at the 5′ end, adding a polyA tail at the 3′ end, RNA editing, and base modification (see Table 14.2).
- Certain RNA molecules such as ribosomal RNAs and precursor tRNAs are processed via cleavage steps to yield smaller, functional molecules (see Figures 14.16, 14.17).
- Group I and group II introns are removed by self-splicing. Pre-mRNA introns are removed via a spliceosome (see Table 14.3, Figure 14.18).
- The spliceosome is a multisubunit structure that recognizes intron sequences and removes them from pre-mRNA (see Figures 14.19, 14.20).
- During alternative splicing, proteins called splicing factors regulate which exons are included in a mature mRNA (see Figures 14.21, 14.22).

PROBLEM SETS & INSIGHTS   *321*

- In eukaryotes, mRNA is given a 7-methylguanosine cap at the 5′ end and a polyA tail at the 3′ end (see Figures 14.23, 14.24).
- RNA editing changes the base sequence of an RNA after it has been synthesized (see Table 14.4, Figure 14.25).

14.5 A Comparison of Transcription and RNA Modification in Bacteria and Eukaryotes

- Several key differences have been found between transcription and RNA modification in bacteria and eukaryotes (see Table 14.5).

PROBLEM SETS & INSIGHTS

More Genetic TIPS

1. Describe the important events that occur during gene transcription in bacteria. What proteins play critical roles in the three stages?

Topic: What topic in genetics does this question address?

The topic is gene transcription in bacteria. More specifically, the question is about the critical roles played by proteins in that process.

Information: What information do you know based on the question and your understanding of the topic?

In the question, you are reminded that gene transcription in bacteria has three stages. From your understanding of the topic, you may remember that the stages are called initiation, elongation, and termination.

Problem-Solving Strategy: Describe the steps.

One strategy to solve this problem is to break down transcription into its three stages and describe each one separately.

Answer: The three stages are initiation, elongation, and termination.

Initiation: RNA polymerase holoenzyme slides along the DNA until σ factor recognizes a promoter. Next, σ factor binds tightly to this sequence, forming a closed complex. The DNA strands are then separated to form a bubble-like structure known as the open complex.

Elongation: After σ factor is released, RNA polymerase core enzyme slides along the DNA, synthesizing RNA as it goes. The α subunits of RNA polymerase keep the enzyme bound to the DNA, while the β subunits are responsible for binding and for the catalytic synthesis of RNA. The ω (omega) subunit is also important for the proper assembly of the core enzyme. During elongation, RNA is made according to the AU/GC rule, with nucleotides being added in the 5′ to 3′ direction.

Termination: RNA polymerase eventually reaches a sequence at the end of the gene that signals the end of transcription. In ρ-independent termination, the properties of the termination sequences in the DNA are sufficient to cause termination. In ρ-dependent termination, ρ protein recognizes a sequence within the RNA, binds there, and travels toward RNA polymerase. When the formation of a stem-loop causes RNA polymerase to pause, ρ protein catches up and separates the RNA–DNA hybrid, releasing RNA polymerase.

2. The consensus sequence for the –35 sequence of a bacterial promoter is 5′–TTGACA–3′. The –35 sequence of a particular bacterial gene is 5′–TTAACA–3′. A mutation changes the fifth base in this sequence from a C to a G. Would you expect this mutation to increase or decrease the rate of transcription of the gene?

Topic: What topic in genetics does this question address?

The topic is gene transcription in bacteria. More specifically, the question is about the effects of a mutation in the promoter.

Information: What information do you know based on the question and your understanding of the topic?

In the question, you are given the consensus sequence for the –35 sequence of a bacterial promoter and information regarding a mutation in a bacterial gene's promoter. From your understanding of the topic, you may recall that the consensus sequence is efficiently recognized by proteins that initiate transcription.

Problem-Solving Strategy: Compare and contrast. Predict the outcome.

One way to solve this problem is to compare the –35 sequences of the nonmutant and mutant promoters and contrast them with the consensus sequence. If the mutation makes the –35 sequence more like the consensus sequence, it will increase the rate of transcription, whereas if it makes the –35 sequence less like the consensus sequence, it will slow down the rate of transcription.

Consensus sequence: 5′–TTGACA–3′

Nonmutant promoter: 5′–TTAACA–3′

Mutant promoter: 5′–TTAAGA–3′

The bases that are different from the consensus sequence are shown in red.

Answer: The mutation is predicted to decrease the rate of transcription. The mutant promoter has two bases that differ from those in the consensus sequence, whereas the nonmutant promoter has only one.

3. When RNA polymerase transcribes DNA, only one of the two DNA strands is used as a template. Referring to Figure 14.4, explain how RNA polymerase determines which DNA strand is the template strand.

Topic: What topic in genetics does this question address?

The topic is transcription. More specifically, the question is about how RNA polymerase identifies the template strand.

Information: What information do you know based on the question and your understanding of the topic?

From the question, you know that only one DNA strand is used as a template. From your understanding of the topic, you may remember that specific base sequences form a promoter that determines the starting point for transcription.

Problem-Solving Strategy: Make a drawing. Relate structure and function.

One strategy to begin solving this problem is to make a drawing of a bacterial promoter that is based on Figure 14.4. You want to relate the structure of the promoter sequence to the function of RNA polymerase in choosing the correct strand as the template strand.

Answer: The binding of σ factor and RNA polymerase depends on the sequence of the promoter. RNA polymerase binds to the promoter in such a way that the –35 sequence TTGACA and the –10 sequence TATAAT are within the coding strand, whereas the –35 sequence AACTGT and the –10 sequence ATATTA are within the template strand.

Conceptual Questions

C1. Explain the central dogma of genetics.

C2. In bacteria, what event marks the end of the initiation stage of transcription?

C3. What is the meaning of the term *consensus sequence*? Give an example. Describe the locations of consensus sequences within bacterial promoters. What are their functions?

C4. What is the consensus sequence of the following six DNA sequences?

GGCATTGACT

GCCATTGTCA

CGCATAGTCA

GGAAATGGGA

GGCTTTGTCA

GGCATAGTCA

C5. Mutations in bacterial promoters may increase or decrease the rate of gene transcription. Promoter mutations that increase transcription are termed *up-promoter* mutations, and those that decrease transcription are termed *down-promoter* mutations. The sequence of the –10 site of the promoter for the *lac* operon is TATGTT (see Figure 14.5). Would you expect each of the following mutations to be an up-promoter or down-promoter mutation?

A. TATGTT to TATATT

B. TATGTT to TTTGTT

C. TATGTT to TATGAT

C6. In the examples shown in Figure 14.5, which positions of the –35 sequence (i.e., first, second, third, fourth, fifth, or sixth) are more tolerant of changes? Do you think these positions play a more or less important role in the binding of σ factor? Explain why.

C7. In Chapter 11, we considered the dimensions of the double helix (see Figure 11.11). In an α helix of a protein, there are 3.6 amino acids per complete turn. Each amino acid advances the α helix by 0.15 nm; a complete turn of an α helix is 0.54 nm in length. As shown in Figure 14.6, two α helices of a transcription factor occupy the major groove of the DNA. According to Figure 14.6, estimate the number of amino acids that bind to this region. How many complete turns of the α helices occupy the major groove of DNA?

C8. A mutation within a gene changes the start codon to a stop codon. How will this mutation affect the transcription of this gene?

C9. What is the subunit composition of bacterial RNA polymerase holoenzyme? What are the functional roles of the subunits?

C10. At the molecular level, describe how σ factor recognizes bacterial promoters. Be specific about the structure of σ factor and the type of chemical bonding.

C11. Let's suppose a mutation changes the consensus sequence at the –35 site in a way that inhibits binding of σ factor. Explain how a mutation could inhibit σ factor from binding to the DNA. Look at Figure 14.5 and describe two specific base substitutions you think would inhibit the binding of σ factor. Explain why your base substitutions would have this effect.

C12. What is the complementarity rule that governs the synthesis of an RNA molecule during transcription? An RNA transcript has the following sequence:

5′–GGCAUGCAUUACGGCAUCACACUAGGGAUC–3′

What is the sequence of the template and coding strands of the DNA that encodes this RNA? On which side (5′ or 3′) of the template strand is the promoter located?

C13. Describe the movement of the open complex along the DNA.

C14. Describe what happens to the chemical bonding interactions when transcriptional termination occurs. Be specific about the type of chemical bonding.

C15. Discuss the differences between ρ-dependent and ρ-independent termination.

C16. In Chapter 13, we discussed the function of DNA helicase, which is involved in DNA replication. Discuss how the functions of helicase and ρ protein are similar and how they are different.

C17. Discuss the similarities and differences between RNA polymerase and DNA polymerase (described in Chapter 13).

C18. Mutations that occur at the end of a gene may alter the sequence of the gene and prevent transcriptional termination.

A. What types of mutations would prevent ρ-independent termination?

B. What types of mutations would prevent ρ-dependent termination?

C. If a mutation prevented transcriptional termination at the end of a gene, where would gene transcription end? Or would it end?

C19. If the following RNA polymerases were missing from a eukaryotic cell, what types of genes would *not* be transcribed?

A. RNA polymerase I

B. RNA polymerase II

C. RNA polymerase III

C20. What sequence elements are found within the core promoter of protein-encoding genes in eukaryotes? Describe their locations and specific functions.

C21. For each of the following transcription factors, explain how eukaryotic transcriptional initiation would be affected if it were missing.

  A. TFIIB
  B. TFIID
  C. TFIIH

C22. Describe the allosteric and torpedo models for transcriptional termination of RNA polymerase II. Which model is more similar to ρ-dependent termination in bacteria, and which model is more similar to ρ-independent termination?

C23. Which eukaryotic transcription factor(s) shown in Figure 14.14 plays a role that is equivalent to that of σ factor in bacterial cells?

C24. The initiation phase of eukaryotic transcription via RNA polymerase II is considered an assembly and disassembly process. Which types of biochemical interactions—hydrogen bonding, ionic bonding, covalent bonding, and/or hydrophobic interactions—would you expect to drive the assembly and disassembly process? How would temperature and salt concentration affect assembly and disassembly?

C25. A eukaryotic protein-encoding gene contains two introns and three exons: exon 1–intron 1–exon 2–intron 2–exon 3. The 5′ splice site at the boundary between exon 2 and intron 2 has been eliminated by a small deletion in the gene. Describe how the pre-mRNA encoded by this mutant gene will be spliced. Indicate which introns and exons will be found in the mRNA after splicing occurs.

C26. Describe the processing events that occur during the production of tRNA in *E. coli*.

C27. Describe the structure and function of a spliceosome. Speculate why the spliceosome subunits contain snRNA. In other words, what do you think is/are the functional role(s) of snRNA during splicing?

C28. What is the unique feature of ribozyme function? Give two examples described in this chapter.

C29. What does it mean to say that gene expression is colinear?

C30. What is meant by the term *self-splicing*? What types of introns are self-splicing?

C31. In eukaryotes, what types of modifications occur to pre-mRNAs?

C32. What is alternative splicing? What is its biological significance?

C33. The processing of ribosomal RNA in eukaryotes is shown in Figure 14.16. Why is this called cleavage or processing but not splicing?

C34. In the splicing of group I introns shown in Figure 14.18, does the 5′ end of the intron have a phosphate group? Explain.

C35. According to the mechanism shown in Figure 14.20, several snRNPs play different roles in the splicing of pre-mRNA. Identify the snRNP that recognizes each of the following sites:

  A. 5′ splice site
  B. 3′ splice site
  C. Branch site

C36. After the intron (which is in a lariat configuration) is released during pre-mRNA splicing, a brief moment occurs before the two exons are connected to each other. Which snRNP(s) hold(s) the exons in place so they can be covalently connected to each other?

C37. A lariat contains a closed loop and a linear end. An intron has the following sequence: 5′–GUPuAGUA–60 nucleotides–UACUUAUCC–100 nucleotides–Py$_{12}$NPyAG–3′. Which sequence would be found within the closed loop of the lariat, the 60-nucleotide sequence or the 100-nucleotide sequence?

Application and Experimental Questions

E1. A research group has sequenced the cDNA and genomic DNA for a particular gene. The cDNA is derived from mRNA, so it does not contain introns. Here are the DNA sequences.

cDNA:

5′–ATTGCATCCAGCGTATACTATCTCGGGCCCAATTAAT-
GCCAGCGGCCAGACTATCACCCAACTCGGTTACCTACTAG-
TATATCCCATATACTAGCATATATTTTACCCATAATTTGTGTGT-
GGGTATACAGTATAATCATATA–3′

Genomic DNA (contains one intron):

5′–ATTGCATCCAGCGTATACTATCTCGGGCCCAAT-
TAATGCCAGCGGCCAGACTATCACCCAACTCG-
GCCCACCCCCCAGGTTTACACAGTCATACCATACA-
TACAAAAATCGCAGTTACTTATCCCAAAAAAACCTAG-
ATACCCCACATACTATTAACTCTTTCTTTCTAGGTTACCTAC-
TAGTATATCCCATATACTAGCATATATTTTACCCATAATTTGT-
GTGTGGGTATACAGTATAATCATATA–3′

Indicate where the intron is located. Does the intron contain the normal consensus sequences for splicing, based on those shown in Figure 14.19? Underline the splice site sequences, and indicate whether or not they match the consensus sequence.

E2. Chapter 20 describes a technique known as Northern blotting that is used to detect RNA transcribed from a particular gene. In this method, a specific RNA is detected using a short segment of cloned DNA that is labeled. The labeled DNA is complementary to the RNA that the researcher wishes to detect. After the RNA is run on a gel and blotted onto nylon paper, the labeled DNA is added and the RNA of interest is visualized as a dark (labeled) band. As shown here, Northern blotting can be used to determine the amount of a particular RNA transcribed in a given cell type. If one type of cell produces twice as much of a particular mRNA as another type of cell does, the band will appear twice as intense. Also, the method can distinguish whether alternative RNA splicing has occurred to produce an RNA that has a different molecular mass.

Northern blot

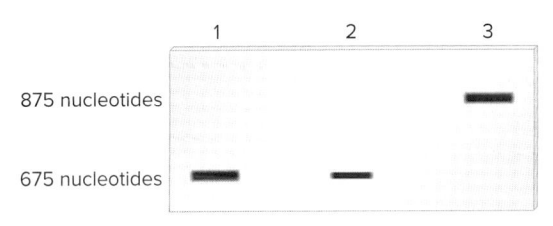

Lane 1 is a sample of RNA isolated from nerve cells.

Lane 2 is a sample of RNA isolated from kidney cells. Nerve cells produce twice as much of this RNA as do kidney cells.

Lane 3 is a sample of RNA isolated from spleen cells. Spleen cells produce an alternatively spliced version of this RNA that is about 200 nucleotides longer than the RNA produced in nerve and kidney cells.

Let's suppose a researcher is interested in the effects of mutations on the expression of a particular protein-encoding gene in eukaryotes. The gene has one intron that is 450 nucleotides long. After this intron is removed from the pre-mRNA, the mRNA transcript is 1100 nucleotides in length. Diploid somatic cells have two copies of this gene. Make a drawing that shows the expected results of a Northern blot using mRNA from the cytosol of somatic cells, which were obtained from the following individuals:

Lane 1: A normal individual

Lane 2: A homozygote for a deletion that removed the –50 to –100 region of the gene that encodes this mRNA

Lane 3: A heterozygote in which one gene is normal and the other gene had a deletion that removed the –50 to –100 region

Lane 4: A homozygote for a mutation that introduced an early stop codon into the middle of the coding sequence of the gene

Lane 5: A homozygote for a two-nucleotide deletion that removed the AG sequence at the 3′ splice site

E3. An electrophoretic mobility shift assay (EMSA) can be used to study the binding of proteins to a segment of DNA. This method is described in Chapter 20. When a protein binds to a segment of DNA, it slows the movement of the DNA through a gel, so the DNA appears at a higher point in the gel, as shown in the following example:

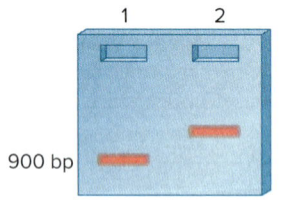

Lane 1: 900-bp fragment alone

Lane 2: 900-bp fragment plus a protein that binds to the 900-bp fragment

In this example, the segment of DNA is 900 bp in length, and the binding of a protein causes the DNA to appear at a higher point in the gel. Assuming that this 900-bp fragment of DNA contains a eukaryotic promoter for a protein-encoding gene, draw a gel that shows the relative locations of the 900-bp fragment under the following conditions:

Lane 1: 900 bp plus TFIID

Lane 2: 900 bp plus TFIIB

Lane 3: 900 bp plus TFIID and TFIIB

Lane 4: 900 bp plus TFIIB and RNA polymerase II

Lane 5: 900 bp plus TFIID, TFIIB, and RNA polymerase II/TFIIF

E4. As described in Chapter 20 and in experimental question E3, an electrophoretic mobility shift assay can be used to determine whether a protein binds to DNA. This method can also determine whether a protein binds to RNA. For each of the following combinations, would you expect the migration of the RNA to be shifted to a higher point in the gel due to the binding of the protein?

A. mRNA from a gene that is terminated in a ρ-independent manner plus ρ protein

B. mRNA from a gene that is terminated in a ρ-dependent manner plus ρ protein

C. pre-mRNA from a protein-encoding gene that contains two introns plus the snRNP called U1

D. Mature mRNA from a protein-encoding gene that contains two introns plus the snRNP called U1

Questions for Student Discussion/Collaboration

1. Based on your knowledge of introns and pre-mRNA splicing, discuss whether or not you think alternative splicing fully explains the existence of introns. Can you think of other possible reasons to explain their existence?

2. Discuss the types of RNA transcripts and the functional roles they play. Why do some RNAs form complexes with protein subunits?

Answers to Comprehension Questions

14.1: a, c

14.2: d, b, a, b

14.3: b, c, d, b

14.4: d, c, b, d

14.5: d

Note: All answers appear in Connect; the answers to even-numbered questions and all Concept Check questions are in Appendix B.

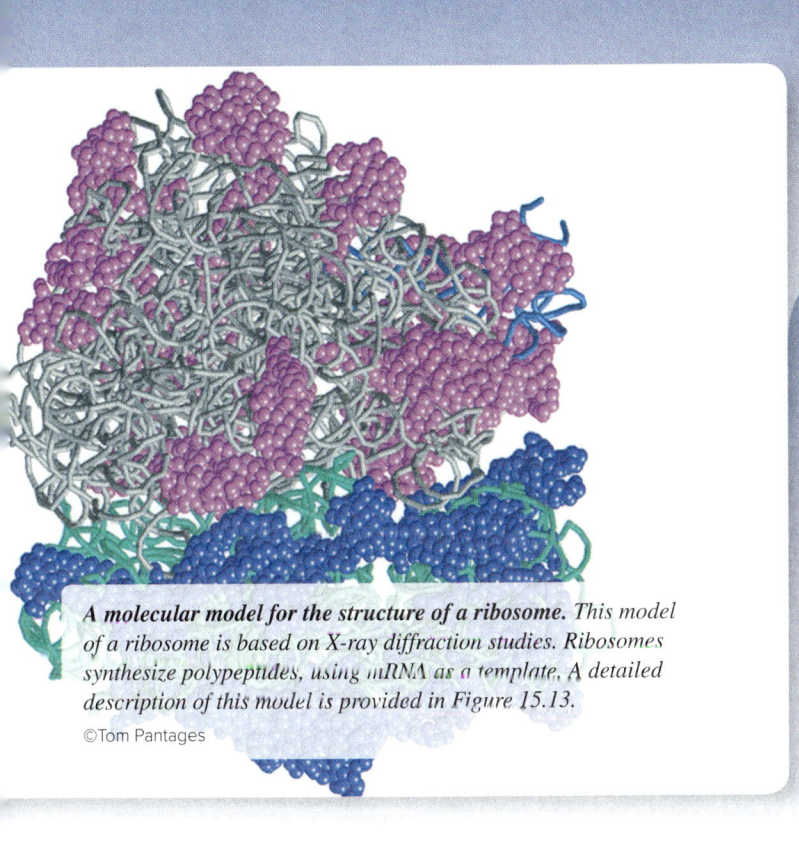

15

CHAPTER OUTLINE

- 15.1 The Genetic Basis for Protein Synthesis
- 15.2 The Relationship Between the Genetic Code and Protein Synthesis
- 15.3 Experimental Determination of the Genetic Code
- 15.4 Structure and Function of tRNA
- 15.5 Ribosome Structure and Assembly
- 15.6 Stages of Translation

A molecular model for the structure of a ribosome. This model of a ribosome is based on X-ray diffraction studies. Ribosomes synthesize polypeptides, using mRNA as a template. A detailed description of this model is provided in Figure 15.13.
©Tom Pantages

TRANSLATION OF mRNA

Translation is the process in which the sequence of codons within mRNA provides the information to synthesize the sequence of amino acids that constitute a polypeptide. One or more polypeptides then fold and assemble to create a functional protein. In this chapter, we will explore the current state of knowledge regarding translation, focusing on the specific molecular interactions responsible for this process. During the past few decades, the concerted efforts of geneticists, cell biologists, and biochemists have profoundly advanced our understanding of translation. Even so, many questions remain unanswered, and this process continues to be an exciting area of investigation.

We will begin by considering the classic experiments that revealed that the purpose of some genes is to encode proteins that function as enzymes. Next, we examine how the genetic code is used to decipher the information within mRNA to produce a polypeptide with a specific amino acid sequence. The rest of the chapter is devoted to understanding translation at the molecular level as it occurs in living cells. To accomplish this, we will need to examine the cellular components—including many different proteins, RNAs, and small molecules—that are involved in the translation process. We will consider the structure and function of tRNA molecules, which act as the translators of the genetic information within mRNA, and then examine the composition of ribosomes. Finally, we will explore the three stages of translation and compare differences in the translation process between bacterial cells and eukaryotic cells.

15.1 THE GENETIC BASIS FOR PROTEIN SYNTHESIS

Learning Outcomes:

1. Explain how the work of Garrod indicated that some genes encode enzymes.
2. Analyze the experiments of Beadle and Tatum, and explain how their results were consistent with the idea that certain genes encode a single enzyme.

Proteins are critically important as active participants in cell structure and function. The primary role of DNA is to store the information needed for the synthesis of all the proteins that an organism makes. As we discussed in Chapter 14, genes that encode an amino acid sequence are known as **protein-encoding genes,** also called **structural genes.** The RNA transcribed from protein-encoding genes is called **messenger RNA (mRNA).** The main function of the genetic material is to encode the production of cellular proteins in the correct cell, at the proper time, and in suitable amounts. This is an extremely complicated task because living cells make thousands of different proteins. Genetic analyses have shown that a typical bacterium can make a few thousand different proteins, and estimates for eukaryotes range from several thousand in simple eukaryotic organisms, such as yeast, to tens of thousands in plants and animals. In this section, we will consider early experiments that showed that the role of some genes is to encode enzymes.

Garrod Proposed That Some Genes Code for the Production of a Single Enzyme

The idea that a relationship exists between genes and the production of proteins was first suggested at the beginning of the twentieth century by Archibald Garrod, a British physician. Prior to Garrod's studies, biochemists had studied many metabolic pathways within living cells. These pathways consist of a series of metabolic conversions of one molecule to another, each step catalyzed by a specific enzyme. Each enzyme is a distinctly different protein that catalyzes a particular chemical reaction. **Figure 15.1** illustrates part of the metabolic pathway for the degradation of phenylalanine, an amino acid commonly found in human diets. The enzyme phenylalanine hydroxylase catalyzes the conversion of phenylalanine to tyrosine, and a different enzyme, tyrosine aminotransferase, converts tyrosine into p-hydroxyphenylpyruvic acid, and so on. In all of the steps shown in Figure 15.1, a specific enzyme catalyzes a single type of chemical reaction.

Garrod studied patients who had defects in their ability to metabolize certain compounds. He was particularly interested in the inherited disease known as **alkaptonuria.** In this disorder, the patient's body accumulates abnormal levels of homogentisic acid (also called alkapton), which is excreted in the urine, causing it to appear black on exposure to air. In addition, the disease is characterized by bluish black discoloration of cartilage and skin (ochronosis). Garrod proposed that the accumulation of homogentisic acid in these patients is due to a missing enzyme, namely, homogentisic acid oxidase (see Figure 15.1).

How did Garrod realize that certain genes encode enzymes? He already knew that alkaptonuria is an inherited trait that follows an autosomal recessive pattern of inheritance. Therefore, an individual with alkaptonuria must have inherited the mutant (defective) gene that causes this disorder from both parents. From these observations, Garrod proposed that a relationship exists between the inheritance of the trait and the inheritance of a defective enzyme. Namely, if an individual inherited the mutant gene (which causes a loss of enzyme function), she or he would not produce any normal enzyme and would be unable to metabolize homogentisic acid. Garrod described alkaptonuria as an **inborn error of metabolism.** This hypothesis was the first suggestion that a connection exists between the function of genes and the production of enzymes. At the time of Garrod's studies, this idea was particularly insightful, because the structure and function of the genetic material were completely unknown.

Beadle and Tatum's Experiments with *Neurospora* Led Them to Propose the One-Gene/One-Enzyme Hypothesis

In the early 1940s, George Beadle and Edward Tatum were also interested in the relationship among genes, enzymes, and traits. They developed an experimental system for investigating

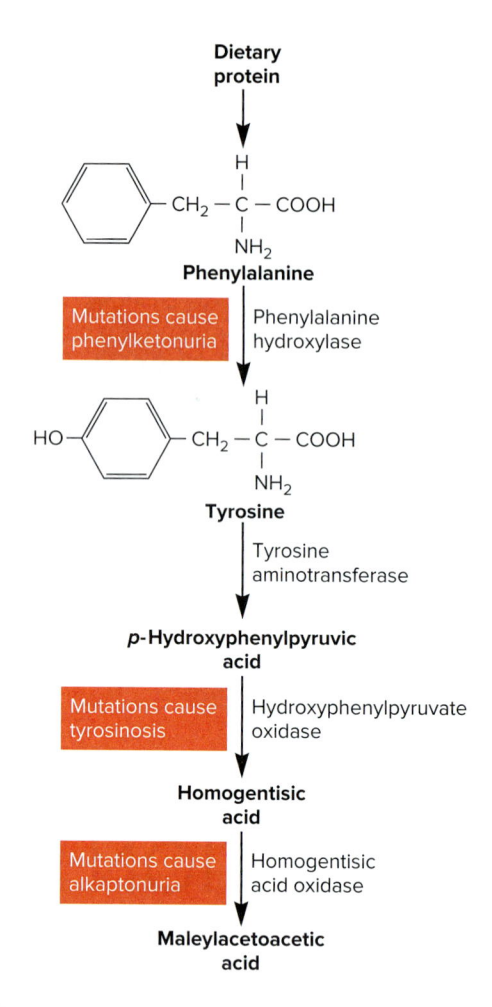

FIGURE 15.1 The metabolic pathway of phenylalanine breakdown. This diagram shows part of the pathway of phenylalanine metabolism, which consists of enzymes that successively convert one molecule to another. Certain human genetic diseases (noted in red boxes) are caused when enzymes in this pathway are missing or defective.

Genes ⟶ Traits When a person inherits two defective copies of the gene that encodes homogentisic acid oxidase, he or she cannot convert homogentisic acid into maleylacetoacetic acid. Such a person accumulates large amounts of homogentisic acid in the body and has other symptoms of the disease known as alkaptonuria. Similarly, if a person has two mutant (defective) alleles of the gene encoding phenylalanine hydroxylase, he or she is unable to synthesize the enzyme phenylalanine hydroxylase and has the disease called phenylketonuria (PKU).

Concept Check: Which disease occurs when homogentisic acid oxidase is defective?

the connection between genes and the production of particular enzymes. Consistent with Garrod's hypothesis, the underlying assumption behind their approach was that a relationship exists between genes and the production of enzymes. However, the quantitative nature of this relationship was unclear. In particular, they asked, "Does one gene control the production of one enzyme, or does one gene control the synthesis of many enzymes involved in a complex biochemical pathway?"

At the time of their studies, many geneticists were trying to understand the nature of the gene by studying morphological traits. However, Beadle and Tatum realized that morphological traits are likely to be based on systems of biochemical reactions so complex as to make analysis exceedingly difficult. Therefore, they turned their genetic studies to the analysis of simple nutritional requirements in *Neurospora crassa*, a common bread mold. *Neurospora* can be easily grown in the laboratory and has few nutritional requirements: a carbon source (sugar), inorganic salts, and the vitamin biotin. Normal *Neurospora* cells produce many different enzymes that can synthesize the organic molecules, such as amino acids and vitamins, that are essential for growth.

Beadle and Tatum wanted to understand how enzymes are controlled by genes. They isolated several different mutant strains that required methionine for growth. They hypothesized that each mutant strain might be blocked at only a single step in the consecutive series of reactions that lead to methionine synthesis. To test this hypothesis, the strains were examined for their ability to grow in the presence of O-acetylhomoserine, cystathionine, homocysteine, or methionine. O-Acetylhomoserine, cystathionine, and homocysteine are intermediates in the synthesis of methionine from homoserine.

A simplified depiction of the results is shown in **Figure 15.2a**. The wild-type strain could grow on minimal growth media that contained the minimum set of nutrients that is required for growth. The minimal media did not contain O-acetylhomoserine, cystathionine, homocysteine, or methionine. Based on their growth properties on media supplemented with O-acetylhomoserine, cystathionine, homocysteine, or methionine, the mutant strains that had been originally identified as requiring methionine for growth could be placed into four groups designated strains 1, 2, 3, and 4 in this figure.

- A strain 1 mutant was missing enzyme 1, needed for the conversion of homoserine into O-acetylhomoserine. The cells could grow only if O-acetylhomoserine, cystathionine, homocysteine, or methionine was added to the growth medium.
- A strain 2 mutant was missing the second enzyme in this pathway, which is needed for the conversion of O-acetylhomoserine into cystathionine.
- A strain 3 mutant was unable to convert cystathionine into homocysteine.
- A strain 4 mutant could not make methionine from homocysteine.

Based on these results, the researchers could order the enzymes into a biochemical pathway as depicted in **Figure 15.2b**. Taken together, the analysis of these mutants allowed Beadle and Tatum to conclude that a single gene controlled the synthesis of a single enzyme. This was referred to as the **one-gene/one-enzyme hypothesis.**

In later decades, this hypothesis had to be modified in four ways.

1. Enzymes are only one category of proteins. All proteins are encoded by genes, and many of them do not function as enzymes.

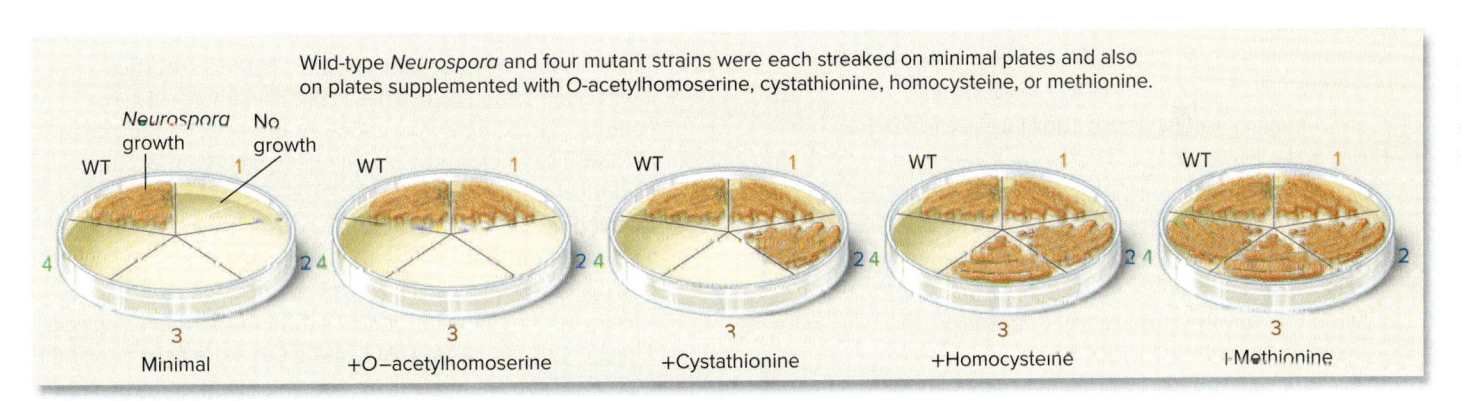

(a) Growth of strains on minimal and supplemented growth media

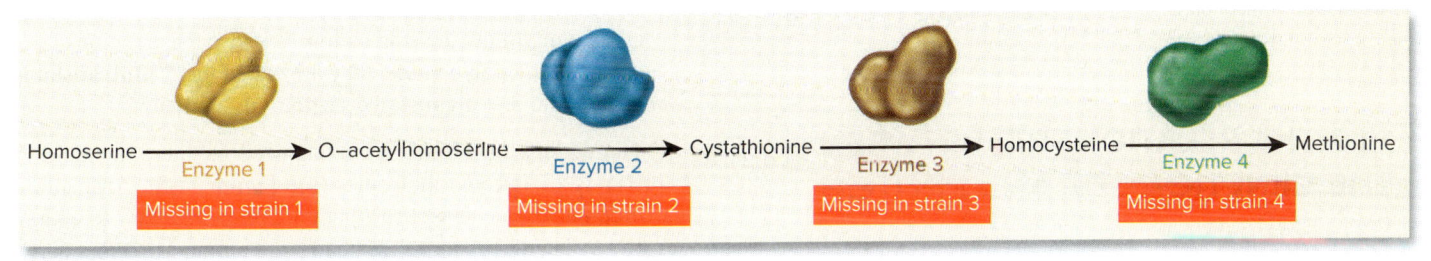

(b) Simplified pathway for methionine biosynthesis

FIGURE 15.2 An example of an experiment that supported Beadle and Tatum's one-gene/one-enzyme hypothesis. (a) Growth of wild-type (WT) and mutant strains on minimal growth media or in the presence of O-acetylhomoserine, cystathionine, homocysteine, or methionine. (b) A simplified pathway for methionine biosynthesis. Note: Homoserine is made by *Neurospora* via enzymes and precursor molecules not discussed in this experiment.

Concept Check: What enzymatic function is missing in the strain 2 mutants?

2. Some proteins are composed of two or more different polypeptides. The term **polypeptide** refers to a structure; it is a linear sequence of amino acids. By comparison, the term **protein** denotes function. Some proteins are composed of one polypeptide. In such cases, a single gene does encode a single protein. In other cases, however, a functional protein is composed of two or more different polypeptides.
3. One gene can encode two or more polypeptides due to alternative splicing or RNA editing (see Chapter 14).
4. A fourth reason why the one-gene/one-enzyme hypothesis needed revision is that we now know that many genes do not encode polypeptides. As we will discuss in Chapter 18, several types of genes specify non-coding RNA molecules.

15.1 REVIEWING THE KEY CONCEPTS

- Garrod studied the disease called alkaptonuria and suggested that some genes encode enzymes (see Figure 15.1).
- Beadle and Tatum studied *Neurospora* mutants that were altered in their nutritional requirements and hypothesized that one gene encodes one enzyme. This one-gene/one-enzyme hypothesis was later modified because: (1) some proteins are not enzymes; (2) some proteins are composed of two or more different polypeptides; (3) one gene can encode two or more polypeptides due to alternative splicing or RNA editing; and (4) some genes encode RNAs that are not translated into polypeptides (see Figure 15.2).

15.1 COMPREHENSION QUESTIONS

1. An inborn error of metabolism is caused by
   a. a mutation in a gene that causes an enzyme to be inactive.
   b. a mutation in a gene that occurs in somatic cells.
   c. the consumption of foods that disrupt metabolic processes.
   d. any of the above.
2. The reason why Beadle and Tatum observed four different categories of mutants that could not grow on media without methionine is
   a. the enzyme involved in methionine biosynthesis is composed of four different subunits.
   b. the enzyme involved in methionine biosynthesis is present in four copies in the *Neurospora* genome.
   c. four different enzymes are involved in a pathway for methionine biosynthesis.
   d. a lack of methionine biosynthesis can inhibit *Neurospora* growth in four different ways.

15.2 THE RELATIONSHIP BETWEEN THE GENETIC CODE AND PROTEIN SYNTHESIS

Learning Outcomes:
1. Outline how the information within DNA is used to make mRNA and a polypeptide.
2. Explain the function of the genetic code.
3. List a few exceptions to the genetic code.
4. Describe the four levels of protein structure.
5. Compare and contrast the functions of several types of proteins.

Thus far, we have considered experiments that led to the conclusion that some genes encode enzymes. Further research indicated that most genes encode polypeptides that form functional units within proteins. The sequence of a protein-encoding gene provides a template for the synthesis of mRNA, which in turn contains the information to synthesize a polypeptide. In this section, we will examine the general features of the genetic code—the sequence of bases in a codon that specifies an amino acid or the end of translation. In addition, we will look at the biochemistry of polypeptide synthesis and explore the structure and function of proteins. Ultimately, proteins are largely responsible for determining the characteristics of living cells and an organism's traits.

During Translation, the Codons in mRNA Provide the Information to Make a Polypeptide with a Specific Amino Acid Sequence

Why have researchers given the name *translation* to the process of polypeptide synthesis? At the molecular level, translation involves an interpretation of one language—the language of mRNA, a nucleotide sequence—into the language of proteins—an amino acid sequence. The ability of mRNA to be translated into a specific sequence of amino acids relies on the **genetic code.** The sequence of bases within an mRNA molecule provides coded information that is read in groups of three nucleotides known as codons (**Figure 15.3**).

- The sequence of three bases in most codons specifies a particular amino acid. These codons are termed **sense codons.** For example, the codon AGC specifies the amino acid serine, whereas the codon GGG encodes the amino acid glycine.
- The codon AUG, which specifies methionine, is used as a **start codon;** it is usually the first codon that begins a polypeptide sequence. The AUG codon is also used to specify additional methionines within the coding sequence.
- Three codons—UAA, UAG, and UGA—are used to end the process of translation and are known as **stop codons.** They are also called **termination codons,** or **nonsense codons.**
- An mRNA molecule also has regions that precede the start codon and follow the stop codon. Because these regions do not encode a polypeptide, they are called the **5′-untranslated region** and **3′-untranslated region,** respectively.

The codons in mRNA are recognized by the anticodons in **transfer RNA (tRNA)** molecules (see Figure 15.3). **Anticodons** are three-nucleotide sequences that are complementary to codons in mRNA. The tRNA molecules carry the amino acids that correspond to the codons in the mRNA. In this way, the order of codons in mRNA dictates the order of amino acids within a polypeptide.

The genetic code is composed of 64 different codons, as shown in **Table 15.1**. Because polypeptides are composed of 20 different kinds of amino acids, a minimum of 20 codons is needed to specify all the amino acids. With four types of bases in mRNA

15.2 THE RELATIONSHIP BETWEEN THE GENETIC CODE AND PROTEIN SYNTHESIS

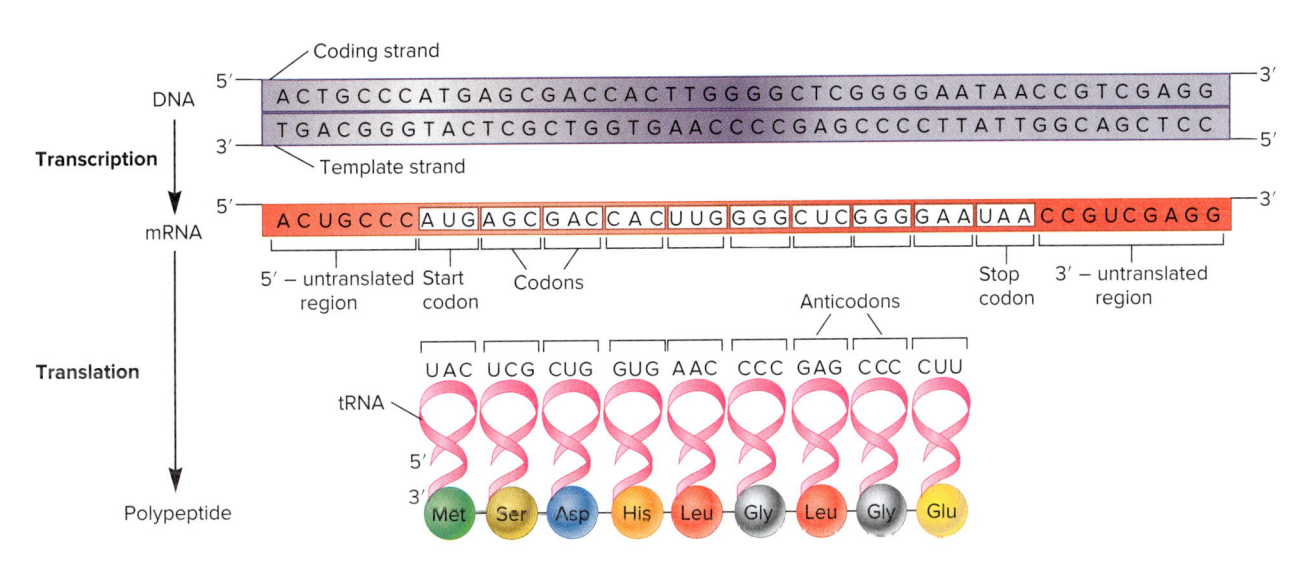

FIGURE 15.3 The relationships among the DNA-coding sequence, mRNA codons, tRNA anticodons, and amino acids in a polypeptide. The sequence of nucleotides within DNA is transcribed to make a complementary sequence of nucleotides within mRNA. This sequence of nucleotides in mRNA is translated into a sequence of amino acids of a polypeptide. tRNA molecules act as intermediates in this translation process.

Concept Check: Describe the role of DNA in the synthesis of a polypeptide.

TABLE 15.1

The Genetic Code

First base	Second base: U	Second base: C	Second base: A	Second base: G	Third base
U	UUU, UUC Phenylalanine (Phe); UUA, UUG Leucine (Leu)	UCU, UCC, UCA, UCG Serine (Ser)	UAU, UAC Tyrosine (Tyr); UAA Stop codon, UAG Stop codon	UGU, UGC Cysteine (Cys); UGA Stop codon; UGG Tryptophan (Trp)	U C A G
C	CUU, CUC, CUA, CUG Leucine (Leu)	CCU, CCC, CCA, CCG Proline (Pro)	CAU, CAC Histidine (His); CAA, CAG Glutamine (Gln)	CGU, CGC, CGA, CGG Arginine (Arg)	U C A G
A	AUU, AUC, AUA Isoleucine (Ile); AUG Methionine (Met); start codon	ACU, ACC, ACA, ACG Threonine (Thr)	AAU, AAC Asparagine (Asn); AAA, AAG Lysine (Lys)	AGU, AGC Serine (Ser); AGA, AGG Arginine (Arg)	U C A G
G	GUU, GUC, GUA, GUG Valine (Val)	GCU, GCC, GCA, GCG Alanine (Ala)	GAU, GAC Aspartic acid (Asp); GAA, GAG Glutamic acid (Glu)	GGU, GGC, GGA, GGG Glycine (Gly)	U C A G

(A, U, G, and C), a genetic code containing two bases in a codon would not be sufficient because it would specify only 4^2, or 16, possible types of amino acids. By comparison, a three-base codon system could form 4^3, or 64, different codons. Because the number of possible codons exceeds 20—which is the number of different types of amino acids—the genetic code is said to contain **degeneracy**. This means that more than one codon can specify the same amino acid. For example, the codons GGU, GGC, GGA, and GGG all specify the amino acid glycine. Such codons are termed **synonymous codons**. In most instances of synonymous codons, the third base in the codon is the base that varies. The third base is sometimes referred to as the **wobble base**. This term is derived from the idea that the complementary base in the tRNA can "wobble" a bit during the recognition of the third base of the codon in mRNA. The significance of the wobble base will be discussed later in this chapter.

The start codon (AUG) defines the **reading frame** of an mRNA—a sequence of codons determined by reading bases in groups of three, beginning with the start codon as a frame of reference. This concept is best understood with a few examples.

The following mRNA sequence encodes a short polypeptide with seven amino acids:

5′–AUGCCCGGAGGCACCGUCCAAU–3′
Met – Pro – Gly – Gly – Thr – Val – Gln

If we remove one base (C) adjacent to the start codon, this changes the reading frame to produce a different polypeptide sequence:

5′–AUGCCGGAGGCACCGUCCAAU–3′
Met – Pro – Glu – Ala – Pro – Ser – Asn

Alternatively, if we remove three bases (CCC) next to the start codon, the resulting polypeptide has the same reading frame as the first polypeptide, though one amino acid (Pro, proline) has been deleted:

5′–AUGGGAGGCACCGUCCAAU–3′
Met – Gly – Gly – Thr – Val – Gln

Genetic TIPS

The Question: An mRNA has the following base sequence:

5′–CAGGCGGCGAUGGACAAUAAAGCGGGCCUGUAAGC–3′

Identify the start codon, and determine the complete amino acid sequence that will be translated from this mRNA.

Topic: What topic in genetics does this question address?

The topic is translation. More specifically, the question is about predicting a polypeptide sequence based on an mRNA sequence.

Information: What information do you know based on the question and your understanding of the topic?

From the question, you know the base sequence of an mRNA. From your understanding of the topic, you may remember that the genetic code determines the amino acid sequence of a polypeptide. Furthermore, certain codons function as start and stop codons.

Problem-Solving Strategy: Predict the outcome.

One strategy to solve this problem is to first identify the start codon (AUG) and then determine the adjacent codons. You need to use Table 15.1 to solve this problem.

Answer: The start codon is AUG. The amino acid sequence is shown below the mRNA sequence:

5′–CAGGCGGCGAUGGACAAUAAAGCGGGCCUGUAAGC–3′
Met Asp Asn Lys Ala Gly Leu Stop

Exceptions to the Genetic Code Include the Incorporation of Selenocysteine and Pyrrolysine into Polypeptides

From the analysis of many different species, including bacteria, archaea, protists, fungi, plants, and animals, researchers have found that the genetic code is nearly universal. This means that most features of the genetic code are used by all living organisms. However, a few exceptions to the genetic code have been noted (Table 15.2). In vertebrates, the mitochondrial genetic code differs from the nuclear genetic code in certain ways, for example, AUA codes for methionine and UGA codes for tryptophan.

Selenocysteine (Sec) and **pyrrolysine** (Pyl) are sometimes called the twenty-first and twenty-second amino acids in polypeptides. Their structures are shown later in this section, in Figure 15.5f. Selenocysteine is found in several enzymes involved in oxidation-reduction reactions in bacteria, archaea, and eukaryotes. Pyrrolysine is found in a few enzymes of methane-producing archaea. Selenocysteine and pyrrolysine are encoded by the codons UGA and UAG, respectively, which usually function as stop codons. Like the standard 20 amino acids, selenocysteine and pyrrolysine are bound to tRNAs that specifically carry them to the ribosomes for their incorporation into polypeptides. The anticodon of the tRNA that carries selenocysteine is complementary to a UGA codon, and the tRNA that carries pyrrolysine has an anticodon that is complementary to UAG.

How does a UGA or UAG codon occasionally specify the incorporation of selenocysteine or pyrrolysine, respectively? In the case of a codon that specifies selenocysteine, the UGA codon is followed by a sequence called the selenocysteine insertion sequence (SECIS), which forms a stem-loop. In bacteria, a SECIS may be located immediately following the UGA codon, whereas a SECIS may be further downstream in the 3′-untranslated region of the mRNA in archaea and eukaryotes. The SECIS is recognized by proteins that favor the binding of a UGA codon to a tRNA carrying selenocysteine instead of the binding of release factors that are needed for polypeptide termination. Similarly, pyrrolysine incorporation may involve sequences downstream from a UAG codon that form a stem-loop.

TABLE 15.2

Examples of Exceptions to the Genetic Code*

Codon	Universal Meaning	Exception
AUA	Isoleucine	Methionine in yeast and vertebrate mitochondria
UGA	Stop	Tryptophan in vertebrate mitochondria
CUU, CUA, CUC, CUG	Leucine	Threonine in yeast mitochondria
UAA, UAG	Stop	Glutamine in ciliated protozoa
UGA	Stop	Selenocysteine in certain genes found in bacteria, archaea, and eukaryotes
UAC	Stop	Pyrrolysine in certain genes found in methane-producing archaea

*Several other exceptions, sporadically found among various species, are also known.

A Polypeptide Has Directionality from Its Amino-Terminal to Its Carboxyl-Terminal End

Let's now turn our attention to polypeptide biochemistry. Polypeptide synthesis has a directionality that parallels the order of codons in the mRNA. As a polypeptide is made, a **peptide bond** is formed between the carboxyl group in the last amino acid of the polypeptide and the amino group in the amino acid being added. As shown in **Figure 15.4a**, this occurs via a condensation reaction that releases a water molecule. The newest amino acid added to a growing polypeptide always has a free carboxyl group. **Figure 15.4b** compares the sequence of a very short polypeptide with the mRNA that encodes it.

- The first amino acid is said to be at the **amino-terminal end**, or **N-terminus**, of the polypeptide. An amino group (NH_3^+) is found at this site. The term *N-terminus* refers to the presence of a nitrogen atom (N) at this end. The first amino acid is specified by a codon that is near the 5′ end of the mRNA.
- The last amino acid in a completed polypeptide is located at the **carboxyl-terminal end**, or **C-terminus**. A carboxyl group (COO^-) is always found at this site in a polypeptide. This last amino acid is specified by a codon that is closer to the 3′ end of the mRNA.

The Amino Acid Sequences of Polypeptides Determine the Structure and Function of Proteins

Now that we have examined how mRNAs encode polypeptides, let's consider the structure and function of the gene product, namely, polypeptides. **Figure 15.5** shows the 22 different amino acids that may be found within polypeptides. Each amino

(a) Attachment of an amino acid to a polypeptide

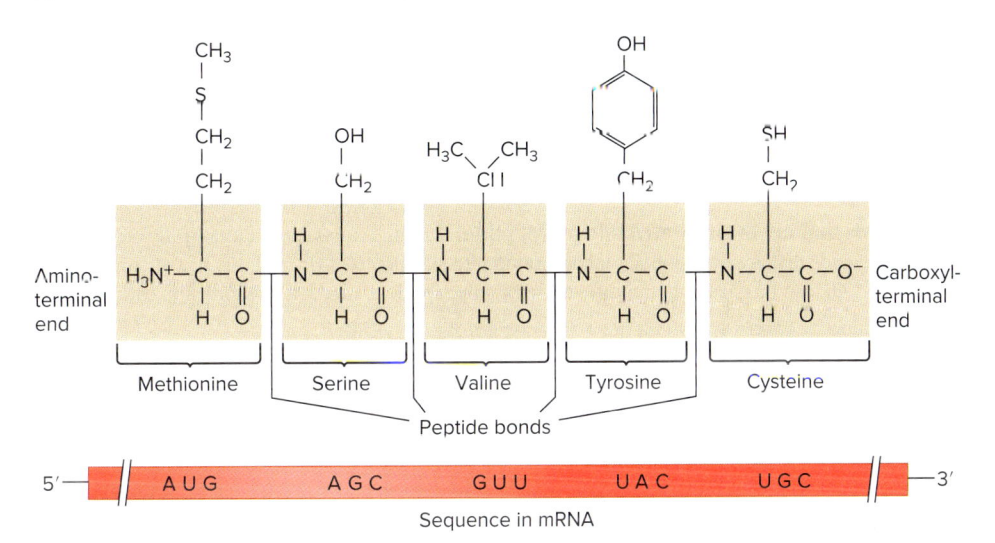

(b) Directionality in a polypeptide and mRNA

FIGURE 15.4 The directionality of polypeptide synthesis. (a) An amino acid is connected to a growing polypeptide via a condensation reaction that releases a water molecule. The letter R is a general designation for an amino acid side chain. (b) The first amino acid in a polypeptide (usually methionine) is located at the amino-terminal end, and the last amino acid is at the carboxyl-terminal end. Therefore, the directionality of amino acids in a polypeptide is from the amino-terminal end to the carboxyl-terminal end, which corresponds to the 5′ to 3′ orientation of codons in mRNA.

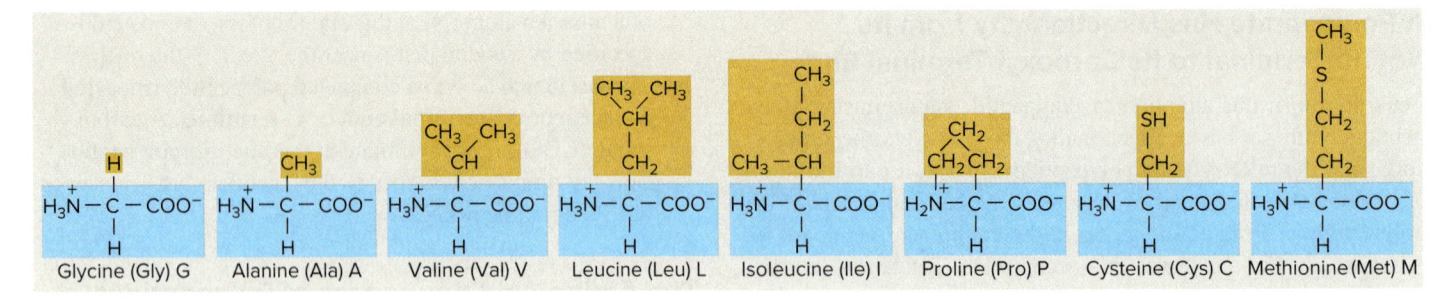

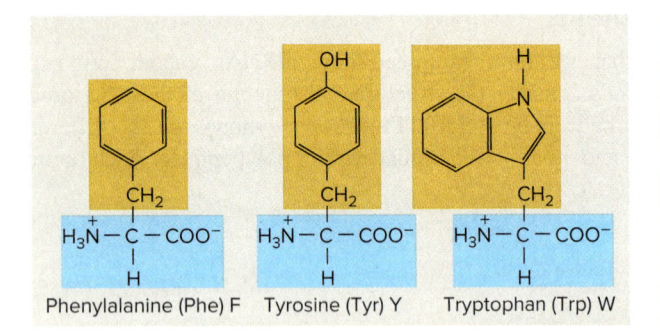

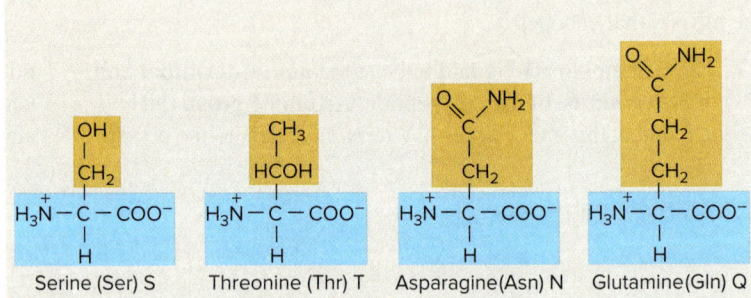

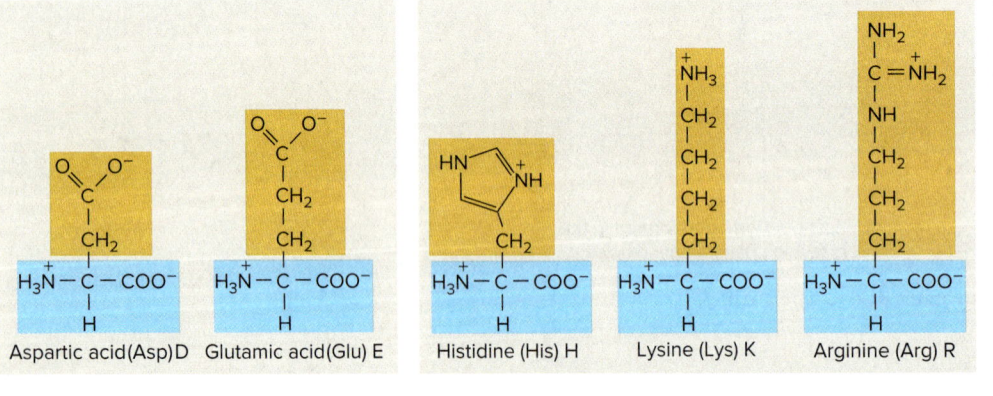

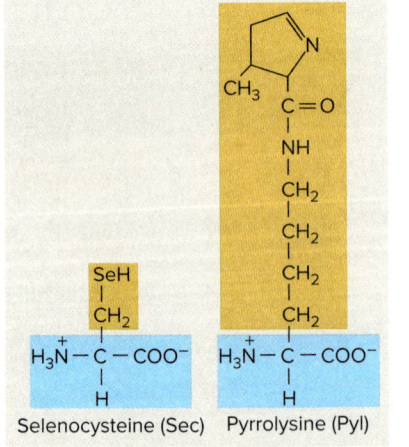

FIGURE 15.5 The amino acids that are incorporated into polypeptides during translation. Parts (a) through (e) show the 20 standard amino acids, and part (f) shows two amino acids that are occasionally incorporated into polypeptides through the use of stop codons (see Table 15.2). The structures of amino acid side chains can also be covalently modified after a polypeptide is made, a phenomenon called posttranslational modification.

Concept Check: Which two amino acids do you think are the least soluble in water?

acid contains a unique **side chain,** or **R group,** that has its own particular chemical properties. For example, aliphatic and aromatic amino acids are relatively nonpolar, which means they are less likely to associate with water. These hydrophobic (meaning "water-fearing") amino acids are often buried within the interior of a folded protein. In contrast, the polar amino acids are hydrophilic ("water-loving") and are more likely to be on the surface of a protein, where they can favorably interact with the surrounding water of cells or extracellular fluids. The chemical properties of the amino acids and their sequences in a polypeptide are critical factors that determine the unique structure of that polypeptide.

Primary Structure Following gene transcription and mRNA translation, the end result is a polypeptide with a defined amino acid sequence. This sequence is the **primary structure** of a polypeptide. The primary structure of a typical polypeptide may be a few hundred or even a couple of thousand amino acids in length. To become a functional unit, most polypeptides quickly

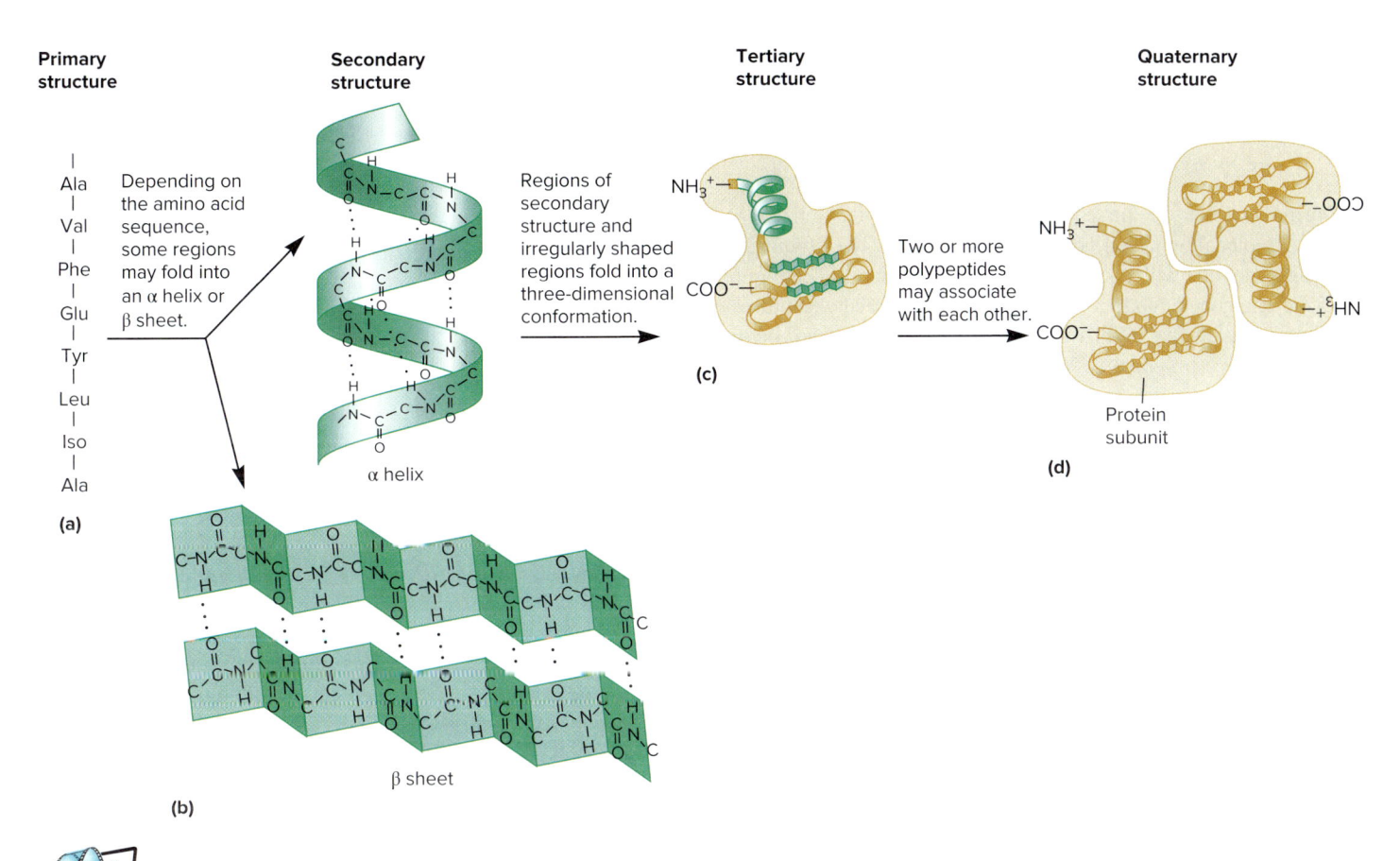

FIGURE 15.6 Levels of structures formed in proteins. (a) The primary structure of a polypeptide is its amino acid sequence. (b) Certain regions of a primary structure fold into a secondary structure; the two types of secondary structures are called α helices and β sheets. (c) Both of these secondary structures can be found within the tertiary structure of a polypeptide. (d) Some polypeptides associate with each other to form a protein with a quaternary structure.

Concept Check: What type of bonding is responsible for the formation of the two types of secondary structures?

adopt a compact three-dimensional structure. The folding process begins while the polypeptide is still being translated. The progression from the primary structure of a polypeptide to the three-dimensional structure of a protein is dictated by the amino acid sequence of the polypeptide. In particular, the chemical properties of the amino acid side chains play a central role in determining the folding pattern of a protein. In addition, the folding of some polypeptides is aided by **chaperones**—proteins that bind to polypeptides and facilitate their proper folding.

Secondary Structure The folding of polypeptides is governed by their primary structure and occurs in multiple stages (**Figure 15.6**). The first stage involves the formation of a regular, repeating shape known as a **secondary structure.** The two main types of secondary structures are the **α helix** and the **β sheet** (Figure 15.6b). A single polypeptide may have some regions that fold into an α helix and other regions that fold into a β sheet. Because of the geometry of secondary structures, certain amino acids are good candidates to form an α helix, whereas others are more likely to be found in a β-sheet conformation. Secondary structures within polypeptides are primarily stabilized by the formation of hydrogen bonds between atoms that are located in the polypeptide backbone. In addition, some regions do not form a repeating secondary structure. Such regions have shapes that look very irregular in their structure because they do not follow a repeating folding pattern.

Tertiary Structure The short regions of secondary structure within a polypeptide are folded relative to each other to make the **tertiary structure** of the polypeptide. As shown in Figure 15.6c, α-helical regions and β-sheet regions are connected by irregularly shaped segments to determine this tertiary structure. The folding of a polypeptide into its secondary and then tertiary conformation can usually occur spontaneously because the process is thermodynamically favorable. The structure is determined by various interactions, including the tendency of hydrophobic amino acids to avoid water, ionic interactions among charged amino acids, hydrogen bonding among amino acids in the folded polypeptide, and weak bonding known as van der Waals interactions.

Quaternary Structure A protein is a functional unit that can be composed of one or more polypeptides. Some proteins are composed of a single polypeptide. Many proteins, however, are composed of two or more polypeptides that associate with each other to form a **quaternary structure** (Figure 15.6d). The individual polypeptides are called **subunits** of the resulting functional protein, and each of them has its own tertiary structure.

15.2 REVIEWING THE KEY CONCEPTS

- During translation, the codons in mRNA are recognized by tRNA molecules, which act as intermediates in the production of a polypeptide with a specific amino acid sequence (see Figure 15.3). The genetic code refers to the relationship between three-base codons in the mRNA and the amino acids that are incorporated into a polypeptide. One codon (AUG) is a start codon, which determines the reading frame of the mRNA. Three codons (UAA, UAG, and UGA) function as stop codons (see Table 15.1).
- The genetic code is largely universal, but some exceptions are known to occur (see Table 15.2).
- A polypeptide is made by the formation of peptide bonds between adjacent amino acids. Each polypeptide has a directionality from its amino-terminal end to its carboxyl-terminal end, which parallels the arrangement of codons in mRNA in the 5′ to 3′ direction (see Figure 15.4).
- Amino acids differ in their side-chain structure (see Figure 15.5).
- Protein structure can be viewed at different levels: primary structure (sequence of amino acids), secondary structure (repeating folding patterns such as the α helix and the β sheet), tertiary structure (additional folding), and quaternary structure (the binding of multiple subunits to each other) (see Figure 15.6).

15.2 COMPREHENSION QUESTIONS

1. What is the genetic code?
   a. The relationship between a three-base codon sequence and an amino acid or the end of translation
   b. The entire base sequence of an mRNA molecule
   c. The entire sequence from the promoter to the terminator of a gene
   d. The binding of tRNA to mRNA
2. The reading frame begins with a _____ and is read _____.
   a. promoter, one base at a time
   b. promoter, in groups of three bases
   c. start codon, one base at a time
   d. start codon, in groups of three bases
3. The fourth codon in an mRNA is GGG, which specifies glycine. If we assume that no amino acids are removed from the polypeptide, which of the following statements is *correct*?
   a. The third amino acid from the N-terminus is glycine.
   b. The fourth amino acid from the N-terminus is glycine.
   c. The third amino acid from the C-terminus is glycine.
   d. The fourth amino acid from the C-terminus is glycine.
4. A type of secondary structure found in proteins is
   a. an α helix.
   b. a β sheet.
   c. Both a and b are secondary structures in proteins.
   d. Neither a nor b is a secondary structure of proteins.

15.3 EXPERIMENTAL DETERMINATION OF THE GENETIC CODE

Learning Outcome:

1. Compare and contrast the experiments of (1) Nirenberg and Matthaei, (2) Khorana, and (3) Nirenberg and Leder that were instrumental in deciphering the genetic code.

In the previous section, we examined how the genetic code determines the amino acid sequence of a polypeptide. In this section, we will consider the experimental approaches that deduced the genetic code.

Synthetic RNA Helped to Determine the Genetic Code

Having learned that the genetic code is read in triplets, how did scientists determine the functions of the 64 codons of the genetic code? During the early 1960s, three research groups headed by Marshall Nirenberg, Severo Ochoa, and H. Gobind Khorana set out to decipher the genetic code. Though they used different methods, all of these groups used synthetic mRNA in their experimental approaches to "crack the code." We first consider the work of Nirenberg and his colleagues. Prior to their studies, several laboratories had already determined that extracts from bacterial cells, containing a mixture of components including ribosomes, tRNAs, and other factors required for translation, are able to synthesize polypeptides if mRNA and amino acids are added. This mixture is termed a **cell-free translation system** (also called an in vitro translation system). If radiolabeled amino acids are added to a cell-free translation system, the synthesized polypeptides are radiolabeled and easy to detect.

To decipher the genetic code, Nirenberg and colleagues needed to gather information regarding the relationship between mRNA composition and polypeptide composition. To accomplish this goal, they made mRNA molecules of a known base composition, added them to a cell-free translation system, and then analyzed the amino acid composition of the resultant polypeptides. For example, if an mRNA molecule consisted of a string of adenine-containing nucleotides (e.g., 5′–AAAAAAAAAAAAAAAA–3′), researchers could add this polyA mRNA to a cell-free translation system in order to answer this question: "Which amino acid is specified by a codon that contains only adenine nucleotides?" (As Table 15.1 shows, it is lysine.)

Before discussing the details of this type of experiment, let's consider how the synthetic mRNA molecules were made. To synthesize mRNA, an enzyme known as polynucleotide phosphorylase was used. In the presence of excess ribonucleoside diphosphates, also called nucleoside diphosphates (NDPs), this enzyme catalyzes the covalent linkage of nucleotides to make a polymer of RNA. Because it does not use a template, the order of the nucleotides is random. For example, if only uracil-containing diphosphates (UDPs) are added, then a polyU mRNA (5′–UUUUUUUUUUUUUUU–3′) is made. If nucleotides containing two different bases, such as uracil and guanine, are added, then the phosphorylase makes a random polymer containing both nucleotides (5′–GGGUGUGUGGUGGGUG–3′).

An experimenter can control the amounts of the nucleotides that are added. For example, if 70% G and 30% U are mixed together with polynucleotide phosphorylase, the predicted amounts of the codons within the random polymer are as follows:

Codon Possibilities	Percentage in the Random Polymer
GGG	$0.7 \times 0.7 \times 0.7 = 0.34 = 34\%$
GGU	$0.7 \times 0.7 \times 0.3 = 0.15 = 15\%$
GUU	$0.7 \times 0.3 \times 0.3 = 0.06 = 6\%$
UUU	$0.3 \times 0.3 \times 0.3 = 0.03 = 3\%$
UUG	$0.3 \times 0.3 \times 0.7 = 0.06 = 6\%$
UGG	$0.3 \times 0.7 \times 0.7 = 0.15 = 15\%$
UGU	$0.3 \times 0.7 \times 0.3 = 0.06 = 6\%$
GUG	$0.7 \times 0.3 \times 0.7 = 0.15 = 15\%$
	100%

The first experiment that demonstrated the ability to synthesize polypeptides from synthetic mRNA was performed by Marshall Nirenberg and J. Heinrich Matthaei in 1961. As shown in **Figure 15.7**, a cell-free translation system was added to 20 different tubes. An mRNA template made using polynucleotide phosphorylase was then added to each tube. In this example, the mRNA was made from 70% G and 30% U. Next, the 20 amino acids were added to each tube, but each tube differed with regard to which of the amino acids was radiolabeled. For example, radiolabeled glycine is found in only one of the 20 tubes. The tubes were incubated for a sufficient length of time to allow translation to occur. The newly made polypeptides were then precipitated onto a filter by treatment with trichloroacetic acid. This step precipitates polypeptides but not individual amino acids. A washing step caused amino acids that had not been incorporated into polypeptides to pass through the filter. Finally, the amount of radioactivity trapped on the filter was determined by liquid scintillation counting.

▶ **THE GOAL**

The researchers assumed that the sequence of bases in mRNA determines the incorporation of specific amino acids into a polypeptide. The purpose of this experiment was to provide information to help decipher the relationship between base composition and particular amino acids.

▶ **ACHIEVING THE GOAL** — **FIGURE 15.7** Elucidation of the genetic code.

Starting material: A cell-free translation system that can synthesize polypeptides if mRNA and amino acids are added.

1. Place the cell-free translation system into 20 tubes.

2. To each tube, add random mRNA polymers of G and U made via polynucleotide phosphorylase using 70% G and 30% U.

3. Add a different radiolabeled amino acid to each tube, and add the other 19 nonradiolabeled amino acids. In this example, glycine was radiolabeled.

4. Incubate for 60 minutes to allow translation to occur.

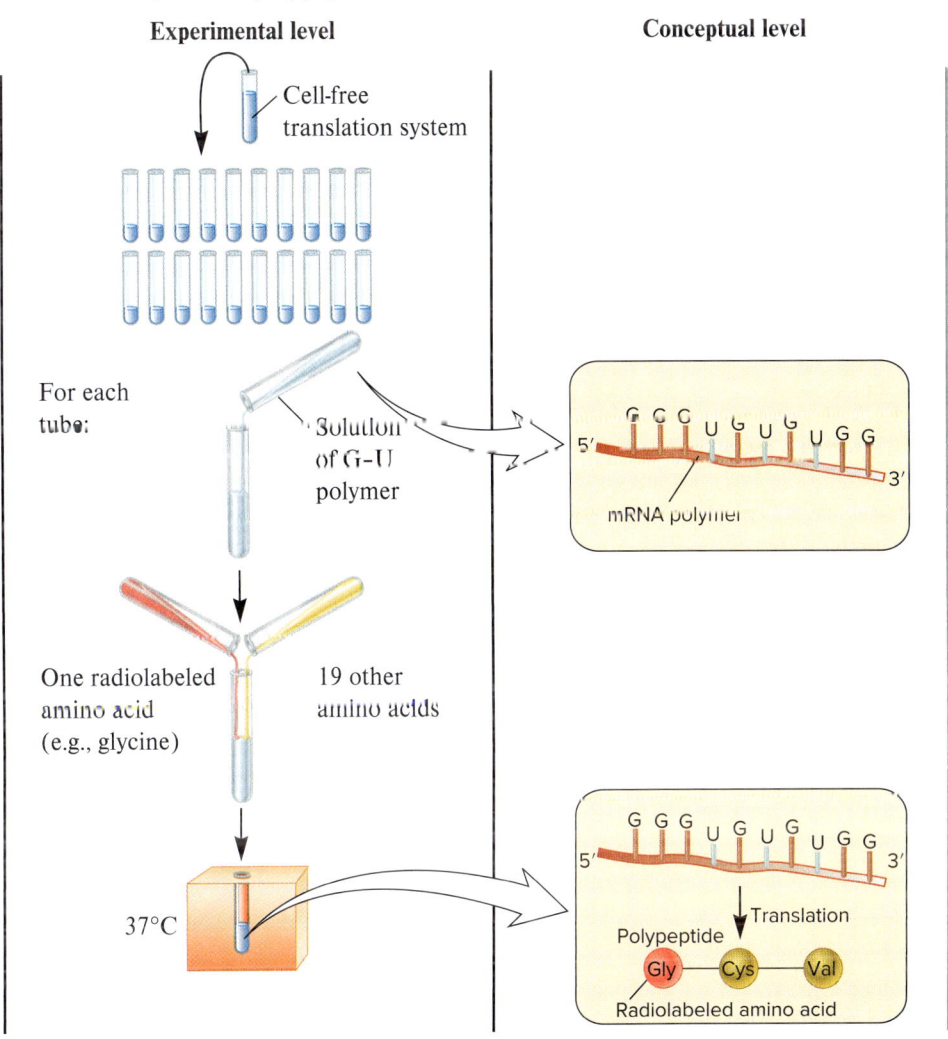

5. Add 15% trichloroacetic acid (TCA), which precipitates polypeptides but not amino acids.

6. Capture the precipitated polypeptides on a filter. Note: Amino acids that were not incorporated into polypeptides pass through the filter.

7. Count the radioactivity on the filter in a scintillation counter (see Appendix A for a description).

8. Calculate the amount of radiolabeled amino acids in the precipitated polypeptides.

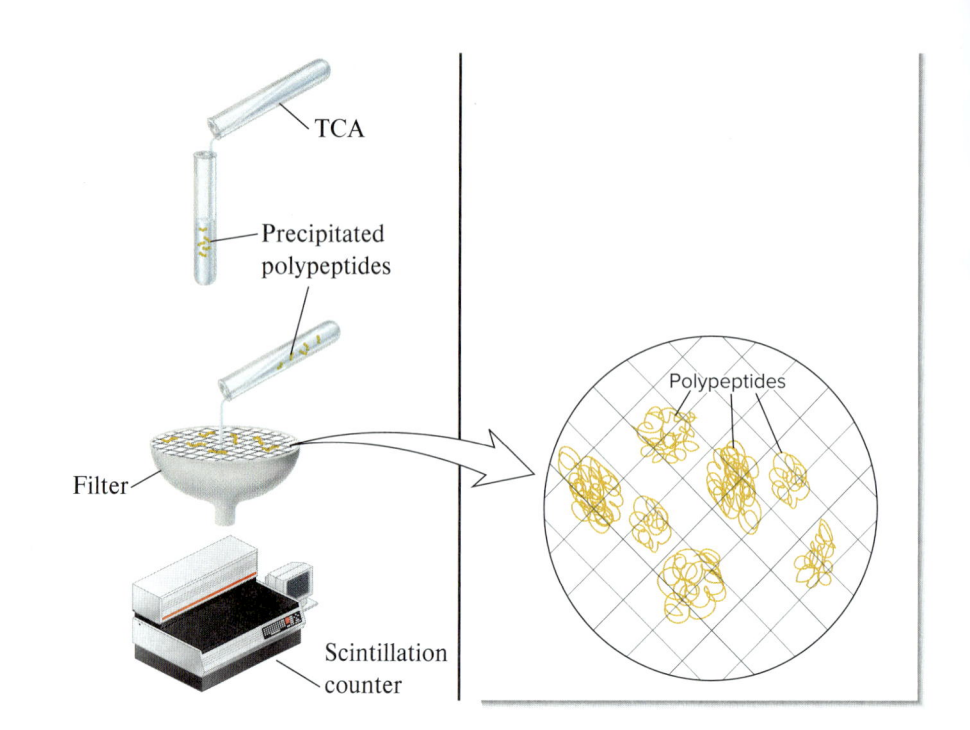

▶ **THE DATA**

Radiolabeled Amino Acid Added	Relative Amount of Radiolabeled Amino Acid Incorporated into Translated Polypeptides (% of total)
Alanine	0
Arginine	0
Asparagine	0
Aspartic acid	0
Cysteine	6
Glutamic acid	0
Glutamine	0
Glycine	49
Histidine	0
Isoleucine	0
Leucine	6
Lysine	0
Methionine	0
Phenylalanine	3
Proline	0
Serine	0
Threonine	0
Tryptophan	15
Tyrosine	0
Valine	21

Source: Adapted from Nirenberg, Marshall W., & Matthaei, J. H. (1961) The Dependence of Cell-Free Protein Synthesis in E. Coli Upon Naturally Occurring or Synthetic Polyribonucleotide *Proceedings of the National Academy of Sciences of the United States of America*, vol. 47, 1588-1602.

▶ **INTERPRETING THE DATA**

The genetic code was not deciphered in a single experiment such as the one described here. Furthermore, this kind of experiment yields information regarding only the nucleotide content of codons, not the specific order of bases within a single codon. According to the calculation previously described, codons should occur in the following percentages: 34% GGG, 15% GGU, 6% GUU, 3% UUU, 6% UUG, 15% UGG, 6% UGU, and 15% GUG. Because 6% of the radiolabeling was due to the incorporation of cysteine, these results are consistent with the idea that a cysteine codon contains two U's and one G. However, this single experiment does not tell us that a cysteine codon is UGU. Based on these data alone, a cysteine codon could be UUG, GUU, or UGU. By comparing many different RNA polymers, the laboratories of Nirenberg and Ochoa established patterns between the specific base sequences of codons and the amino acids they encode. In their first experiments, Nirenberg and Matthaei showed that a random polymer containing only uracil produced a polypeptide containing only phenylalanine. From this result, they inferred that UUU specifies phenylalanine. This idea is consistent with the results shown in the data table. In the random 70% G and 30% U polymer, 3% of the codons are UUU. Likewise, 3% of the amino acids within the polypeptides were found to be phenylalanine. Also, in some cases, the percentage of a given amino acid is due to multiple codons. We now know that the value of 49% for glycine is due to two codons: GGG (34%) and GGU (15%).

The Use of RNA Copolymers and the Triplet-Binding Assay Also Helped to Crack the Genetic Code

In the 1960s, H. Gobind Khorana and colleagues developed a novel method to synthesize RNA. They first created short RNA molecules, two to four nucleotides in length, that had a defined sequence. For example, RNA molecules with the sequence

5′–UG–3′ were synthesized chemically. These short RNAs were then linked together enzymatically, in a 5′ to 3′ manner, to create long copolymers with the sequence

5′–UGUGUGUGUGUGUGUGUGUGUGUGUG–3′

This type of molecule is called a copolymer because it is made from the linkage of several smaller molecules. Depending on the reading frame, such a copolymer contains two different codons: UGU and GUG. In a cell-free translation system like the one described in Figure 15.7, this copolymer produced polypeptides containing cysteine and valine. If these results were compared with the experimental approach of Nirenberg, which showed that a cysteine codon is composed of two U's and one G (see Figure 15.7), the results indicate that UGU specifies cysteine and GUG specifies valine. **Table 15.3** summarizes some of the copolymers that were made using this approach and the amino acids that were incorporated into polypeptides.

Finally, another method that helped to decipher the genetic code also involved the chemical synthesis of short RNA molecules. In 1964, Marshall Nirenberg and Philip Leder discovered that RNA molecules containing three nucleotides—a triplet—can cause a ribosome to bind a tRNA. In other words, the RNA triplet acted like a codon. Ribosomes were able to bind RNA triplets, and then a tRNA with the appropriate anticodon could subsequently bind to the ribosome. To establish the relationship between triplet sequences and specific amino acids, samples containing ribosomes and a particular triplet were exposed to tRNAs with different radiolabeled amino acids.

1. In one experiment, the researchers began with a sample of ribosomes that were mixed with 5′–CCC–3′ triplets (**Figure 15.8**).

2. Portions of this sample were then added to 20 different tubes that had tRNAs with different radiolabeled amino acids. For example, one tube contained radiolabeled histidine, a second tube had radiolabeled proline, a third tube contained radiolabeled glycine, and so on. Only one radiolabeled amino acid was added to each tube.

TABLE 15.3

Examples of Copolymers That Were Analyzed by Khorana and Colleagues

Synthetic RNA*	Codon Possibilities	Amino Acids Incorporated into Polypeptides
UC	UCU, CUC	Serine, leucine
AG	AGA, GAG	Arginine, glutamic acid
UG	UGU, GUG	Cysteine, valine
AC	ACA, CAC	Threonine, histidine
UUC	UUC, UCU, CUU	Phenylalanine, serine, leucine
AAG	AAG, AGA, GAA	Lysine, arginine, glutamic acid
UUG	UUG, UGU, GUU	Leucine, cysteine, valine
CAA	CAA, AAC, ACA	Glutamine, asparagine, threonine
UAUC	UAU, AUC, UCU, CUA	Tyrosine, isoleucine, serine, leucine
UUAC	UUA, UAC, ACU, CUU	Leucine, tyrosine, threonine

*The synthetic RNAs were linked together to make copolymers.

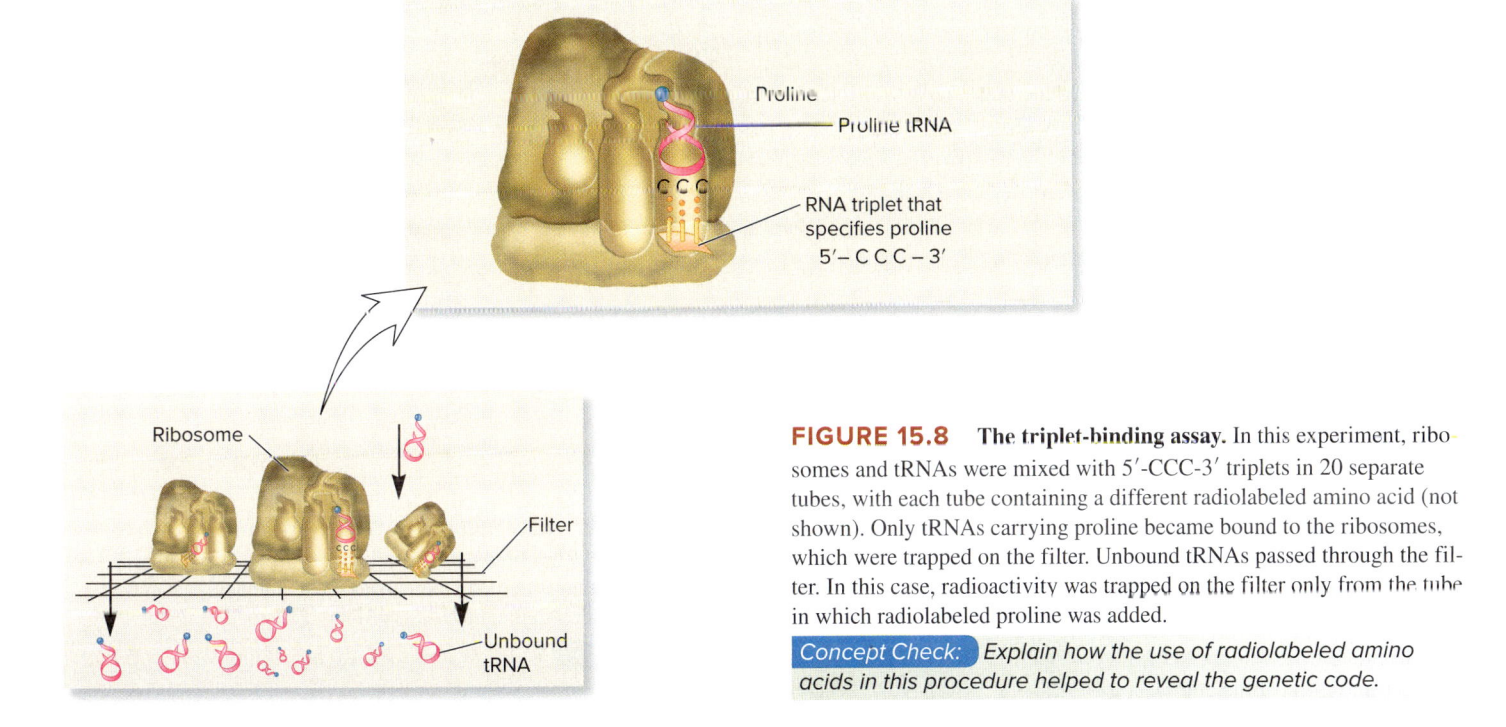

FIGURE 15.8 **The triplet-binding assay.** In this experiment, ribosomes and tRNAs were mixed with 5′-CCC-3′ triplets in 20 separate tubes, with each tube containing a different radiolabeled amino acid (not shown). Only tRNAs carrying proline became bound to the ribosomes, which were trapped on the filter. Unbound tRNAs passed through the filter. In this case, radioactivity was trapped on the filter only from the tube in which radiolabeled proline was added.

Concept Check: Explain how the use of radiolabeled amino acids in this procedure helped to reveal the genetic code.

3. After allowing sufficient time for tRNAs to bind to the ribosomes, the samples were filtered; only the large ribosomes and anything bound to them were trapped on the filter. Unbound tRNAs passed through the filter.
4. Next, the researchers determined the amount of radioactivity trapped on each filter. If the filter contained a large amount of radioactivity, the results indicated that the added triplet encoded the amino acid that was radiolabeled.

Using the triplet-binding assay, Nirenberg and Leder were able to establish relationships between particular triplet sequences and the binding of tRNAs carrying specific (radiolabeled) amino acids. In the case of the 5′–CCC–3′ triplet, they determined that tRNAs carrying radiolabeled proline were bound to the ribosomes. Unfortunately, in some cases, a triplet could not promote sufficient tRNA binding to yield unambiguous results. Nevertheless, the triplet-binding assay was an important tool in the identification of the majority of codons.

15.3 REVIEWING THE KEY CONCEPTS

- Nirenberg and colleagues used synthetic RNA and a cell-free translation system to decipher the genetic code (see Figure 15.7).
- Other methods of deciphering the genetic code included the synthesis of copolymers by Khorana and the triplet-binding assays conducted by Nirenberg and Leder (see Table 15.3, Figure 15.8).

15.3 COMPREHENSION QUESTIONS

1. Let's suppose a researcher mixed together nucleotides with the following percentage of bases: 30% G, 30% C, and 40% A. If RNA was made via polynucleotide phosphorylase, what percentage of the codons would be 5′-GGC-3′?
   a. 30%
   b. 9%
   c. 2.7%
   d. 0%
2. In the triplet-binding assay of Nirenberg and Leder, an RNA triplet composed of three bases was able to cause the
   a. translation of a polypeptide.
   b. binding of a tRNA carrying the appropriate amino acid.
   c. termination of translation.
   d. release of the amino acid from the tRNA.

15.4 STRUCTURE AND FUNCTION OF tRNA

Learning Outcomes:
1. Describe the specificity between the amino acid carried by a tRNA and a codon in mRNA.
2. Describe the key structural features of a tRNA molecule.
3. Explain how an amino acid is attached to a tRNA via aminoacyl-tRNA synthetase.
4. Outline the wobble rules.

Thus far, we have considered the general features of translation and surveyed the structure and functional significance of proteins. The rest of this chapter is devoted to a molecular description of translation as it occurs in living cells. Biochemical studies of protein synthesis and tRNA molecules began in the 1950s. As work progressed toward an understanding of translation, research revealed that different kinds of RNA molecules are involved in the incorporation of amino acids into growing polypeptides. Francis Crick proposed the **adaptor hypothesis.** According to this idea, the position of an amino acid within a polypeptide is determined by the binding between the mRNA and an adaptor molecule carrying a specific amino acid. Later, work by Paul Zamecnik and Mahlon Hoagland suggested that the adaptor molecule is transfer RNA (tRNA). During translation, a tRNA has two functions: (1) It recognizes a three-base codon sequence in mRNA, and (2) it carries an amino acid specific for that codon. In this section, we will examine the structure and function of tRNA molecules.

The Function of a tRNA Depends on the Specificity Between the Amino Acid It Carries and Its Anticodon

The adaptor hypothesis proposes that tRNA molecules recognize the codons within mRNA and carry the correct amino acids to the site of polypeptide synthesis. During mRNA-tRNA recognition, the anticodon in a tRNA molecule binds to a codon in mRNA due to their complementary sequences (**Figure 15.9**). The codon/anticodon binding occurs in an antiparallel manner. Importantly, the anticodon in the tRNA is complementary to the codon for the amino acid that it carries. For example, if the anticodon in the tRNA is 3′–AAG–5′, it is complementary to a 5′–UUC–3′ codon. According to the genetic code described earlier in this chapter, the UUC codon specifies phenylalanine. Therefore, the tRNA with a 3′–AAG–5′ anticodon must carry a phenylalanine. As another

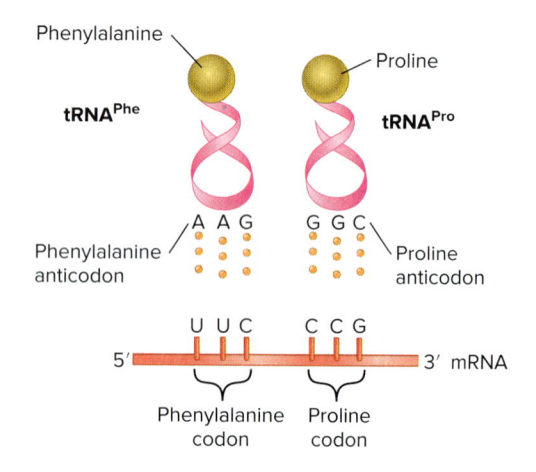

FIGURE 15.9 Recognition between tRNAs and mRNA. The anticodon in the tRNA binds to a complementary sequence in the mRNA. At its other end, the tRNA carries the amino acid that corresponds to the codon in the mRNA according to the genetic code.

Concept Check: What are the two key functional sites of a tRNA molecule?

example, if the tRNA has a 3′–GGC–5′ anticodon, it is complementary to a 5′–CCG–3′ codon that specifies proline. This tRNA must carry proline.

Recall that the genetic code has 64 codons. Of these, 61 are sense codons that specify the 20 amino acids. Therefore, to synthesize proteins, a cell must produce many different tRNA molecules having specific anticodon sequences. To do so, the chromosomal DNA contains many distinct genes that encode tRNA molecules with different sequences. According to the adaptor hypothesis, the anticodon in a tRNA specifies the type of amino acid that it carries. Due to this specificity, tRNA molecules are named according to the amino acid they carry. For example, a tRNA that carries phenylalanine is called tRNA^{Phe}, whereas a tRNA that carries proline is tRNA^{Pro}.

Common Structural Features Are Shared by All tRNAs

To understand how tRNAs are able to carry the correct amino acids during translation, researchers have examined the structural characteristics of these molecules in great detail. Though a cell makes many different tRNAs, all tRNAs share common structural features (**Figure 15.10**).

- As originally proposed by Robert Holley in 1965, the secondary structure of a tRNA exhibits a cloverleaf pattern with three stem-loops and a few variable sites (locations with additional nucleotides not found in all tRNA molecules).
- A tRNA also has an acceptor stem with a 3′ single-stranded region. The acceptor stem is where an amino acid becomes attached (see the inset in the figure).
- A conventional numbering system for the nucleotides within a tRNA molecule begins at the 5′ end and proceeds toward the 3′ end. The anticodon is located in the second loop region.

The three-dimensional, or tertiary, structure of tRNA molecules involves additional folding of the secondary structure. In the tertiary structure of tRNA, the stem-loops are folded into a much more compact molecule. The ability of RNA molecules to form stem-loops and the tertiary folding of tRNA molecules are described in Chapter 11 (see Figures 11.15 and 11.16). In addition to A, U, G, and C bases, tRNA molecules commonly contain modified bases. For example, Figure 15.10 illustrates a tRNA that contains several modified bases. Among many different species, researchers have found that more than 80 different base modifications can occur in tRNA molecules. We will explore the significance of modified bases in codon recognition later in this section.

Aminoacyl-tRNA Synthetases Charge tRNAs by Attaching the Appropriate Amino Acid

To function correctly, each type of tRNA must have the appropriate amino acid attached to its 3′ end. How does an amino acid get attached to a tRNA with the correct anticodon? Enzymes in the cell

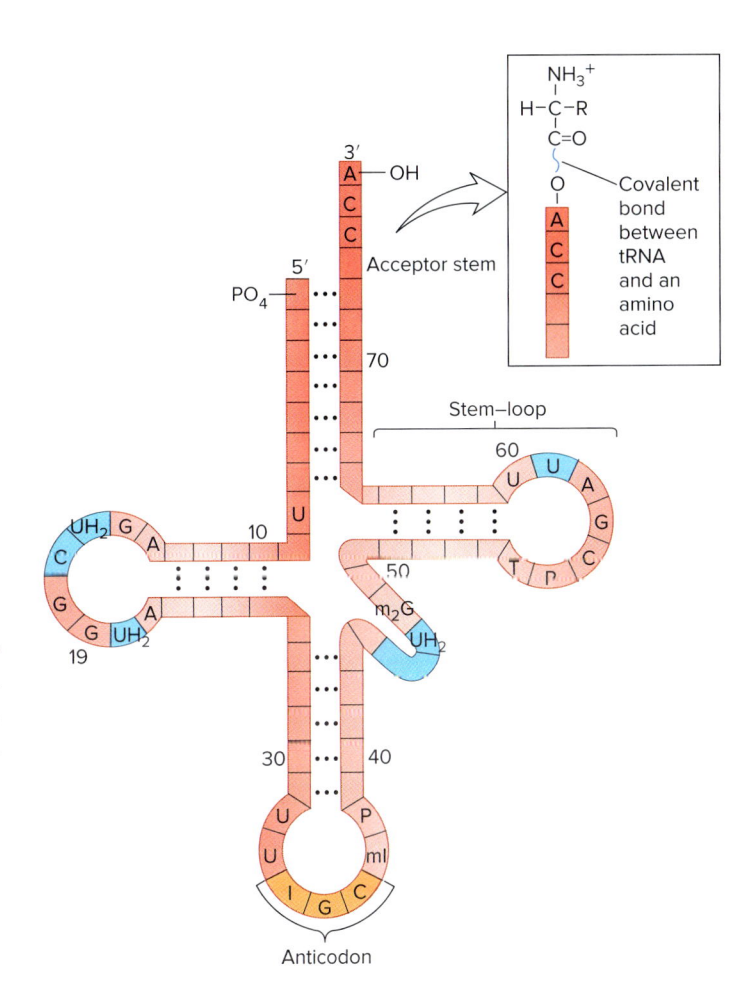

FIGURE 15.10 Secondary structure of tRNA. The conventional numbering of nucleotides within a tRNA begins at the 5′ end and proceeds toward the 3′ end. In all tRNAs, the nucleotides at the 3′ end contain the sequence CCA. Certain locations can have additional nucleotides not found in all tRNA molecules. These variable sites are shown in blue. The figure also shows the locations of a few modified bases specifically found in a yeast tRNA that carries alanine. The modified bases are as follows: I – inosine, mI – methylinosine, T – ribothymidine, UH$_2$ = dihydrouridine, m$_2$G – dimethylguanosine, and P – pseudouridine. The inset shows an amino acid covalently attached to the 3′ end of a tRNA. Note: The 5′ to 3′ orientation of the anticodon in this drawing (left to right) is opposite to that of tRNAs shown in other drawings, such as Figure 15.9.

known as **aminoacyl-tRNA synthetases** catalyze the attachment of amino acids to tRNA molecules. Cells produce 20 different aminoacyl-tRNA synthetase enzymes, one for each of the 20 distinct amino acids. Each aminoacyl-tRNA synthetase is named for the specific amino acid it attaches to tRNA. For example, alanyl-tRNA synthetase recognizes a tRNA with an alanine anticodon—tRNA^{Ala}—and attaches an alanine to it.

Aminoacyl-tRNA synthetases catalyze a chemical reaction involving three different molecules: an amino acid, a tRNA molecule, and ATP (**Figure 15.11**).

1. In the first step of the reaction, a synthetase recognizes a specific amino acid and also ATP.
2. The ATP is hydrolyzed, resulting in the attachment of AMP to the amino acid and the release of pyrophosphate.

3. The correct tRNA then binds to the synthetase and the amino acid becomes covalently attached to the 3′ end of the tRNA molecule at the acceptor stem; AMP is released.
4. Finally, the tRNA with its attached amino acid is released from the enzyme.

At this stage, the tRNA is called a **charged tRNA,** or an **aminoacyl-tRNA.** In a charged tRNA molecule, the amino acid is attached to the 3′ end of the tRNA by a covalent bond (see the inset in Figure 15.10).

The ability of aminoacyl-tRNA synthetases to recognize tRNAs has sometimes been called the "second genetic code." This recognition process is necessary to maintain the fidelity of genetic information. The frequency of error for aminoacyl-tRNA synthetases is less than 10^{-5}. In other words, the wrong amino acid is attached to a tRNA less than once in 100,000 times!

As mentioned previously, tRNA molecules frequently contain bases within their structure that have been chemically modified. These modified bases can have important effects on tRNA function. For example, modified bases within tRNA molecules affect the rate of translation and the recognition of tRNAs by aminoacyl-tRNA synthetases. Positions 34 and 37 contain the largest variety of modified nucleotides; position 34 is the first base (in the 5′ to 3′ direction) in the anticodon that matches the third base in the codon of mRNA (refer to Figure 15.10). As discussed next, a modified base at position 34 can have important effects on codon-anticodon recognition.

The Wobble Rules Describe MIsmatches That Are Allowed at the Third Position in Codon-Anticodon Pairing

Having considered the structure and function of tRNA molecules, let's reexamine some subtle features of the genetic code. As discussed earlier, the genetic code contains degeneracy, which means that more than one codon can specify the same amino acid. Degeneracy usually occurs at the third position in a codon (**Figure 15.12a**). For example, valine is specified by GUU, GUC, GUA, and GUG. In all four cases, the first two bases are G and U. The third base, however, can be U, C, A, or G. To explain this pattern of degeneracy, Francis Crick proposed in 1966 that it is due to "wobble" at the third position in the codon-anticodon recognition process. According to the **wobble rules,** the first two positions pair strictly according to the AU/GC rule. However, the third position can tolerate certain types of mismatches (see **Figure 15.12b**). This proposal suggested that the base at the third position in the codon does not have to hydrogen bond as precisely with the corresponding base in the anticodon.

Because of the wobble rules, some flexibility is observed in the recognition between a codon and an anticodon during the process of translation. When two or more tRNAs that differ at the wobble position are able to recognize the same codon, they are termed **isoacceptor tRNAs.** As an example, tRNAs with an anticodon of 3′–CCA–5′ or 3′–CCG–5′ can recognize a codon with the sequence of 5′–GGU–3′. In addition, the wobble rules enable a single type of tRNA to recognize more than one codon. For example, a tRNA with an anticodon sequence of 3′–AAG–5′ can recognize a 5′–UUC–3′ and a 5′–UUU–3′ codon. The 5′–UUC–3′ codon is a perfect match with this tRNA. The 5′–UUU–3′ codon is mismatched according to the

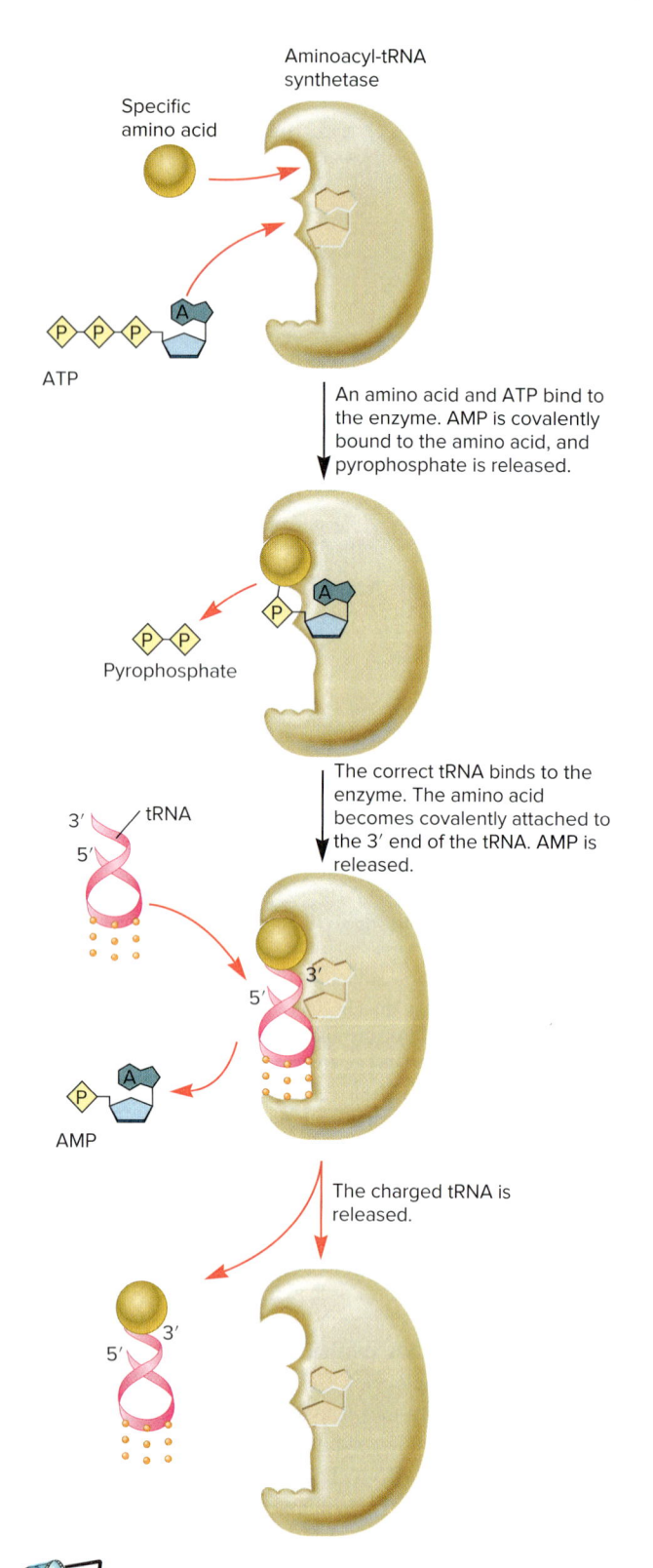

FIGURE 15.11 Catalytic function of aminoacyl-tRNA synthetase. Aminoacyl-tRNA synthetase has binding sites for a specific amino acid, ATP, and a particular tRNA. In the first step, the enzyme catalyzes the covalent attachment of AMP to an amino acid, yielding an activated amino acid. In the second step, the activated amino acid is attached to the appropriate tRNA.

Concept Check: *What is the difference between a charged tRNA and an uncharged tRNA?*

FIGURE 15.12 Wobble position and base-pairing rules.

(a) The wobble position occurs between the first base (meaning the first base in the 5′ to 3′ direction) in the anticodon and the third base in the mRNA codon. (b) The revised wobble rules are slightly different from those originally proposed by Crick. The standard bases found in RNA are G, C, A, and U. In addition, the structures of bases in tRNAs may be modified. Some modified bases that may occur in the wobble position in tRNA are I = inosine; xm^5s^2U = 5-methyl-2-thiouridine; xm^5Um = 5-methyl-2′-O-methyluridine; Um = 2′-O-methyluridine; xm^5U = 5-methyluridine; xo^5U = 5-hydroxyuridine; k^2C = lysidine (a cytosine derivative).

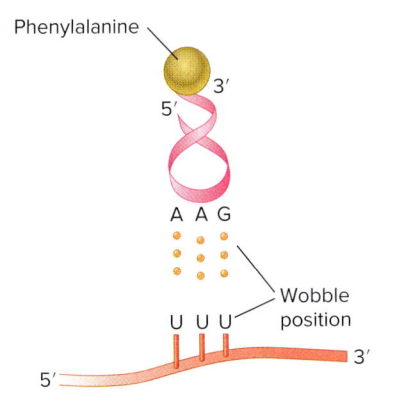

Third base of mRNA codon:	Base in anticodon can be:
A	U, I, xm^5s^2U, xm^5Um, Um, xm^5U, xo^5U, k^2C
U	A, G, U, I, xo^5U
G	C, A, U, xo^5U
C	G, A, I

(a) Location of wobble position
(b) Revised wobble rules

Concept Check: How do the wobble rules affect the total number of different tRNAs that are needed to carry out translation?

---

standard RNA-RNA hybridization rules (namely, G in the anticodon is mismatched to U in the codon), but the two can fit according to the wobble rules described in Figure 15.12b. Likewise, the modification of the wobble base to an inosine in the tRNA allows it to recognize three different codons. At the cellular level, the ability of a single tRNA to recognize more than one codon makes it unnecessary for a cell to make 61 different tRNA molecules with anticodons that are complementary to the 61 possible sense codons. *E. coli* cells, for example, make a population of tRNA molecules that have just 40 different anticodon sequences.

15.4 REVIEWING THE KEY CONCEPTS

- The anticodon in a tRNA is complementary to a codon in mRNA. The tRNA carries a specific amino acid that corresponds to the codon in the mRNA according to the genetic code (see Figure 15.9).
- The secondary structure of tRNA resembles a cloverleaf. The anticodon is in the second loop, and the amino acid is attached to the 3′ end (see Figure 15.10).
- An aminoacyl-tRNA synthetase is one of a group of enzymes that attaches the correct amino acid to a tRNA. The resulting tRNA is called a charged tRNA, or an aminoacyl-tRNA (see Figure 15.11).
- Mismatches are allowed between the pairing of tRNAs and mRNA according to the wobble rules (see Figure 15.12).

15.4 COMPREHENSION QUESTIONS

1. If a tRNA has an anticodon with the sequence 5′-CAG-3′, which amino acid does it carry?
   a. Aspartic acid
   b. Leucine
   c. Valine
   d. Glutamine
2. The anticodon of a tRNA is located in the
   a. 3′ single-stranded region of the acceptor stem.
   b. loop of the first stem-loop.
   c. loop of the second stem-loop.
   d. loop of the third stem-loop.
3. An enzyme known as _____ attaches an amino acid to the _____ of a tRNA, thereby producing _____.
   a. aminoacyl-tRNA synthetase, anticodon, a charged tRNA
   b. aminoacyl-tRNA synthetase, 3′ single-stranded region of the acceptor stem, a charged tRNA
   c. polynucleotide phosphorylase, anticodon, a charged tRNA
   d. polynucleotide phosphorylase, anticodon, an aminoacyl tRNA

15.5 RIBOSOME STRUCTURE AND ASSEMBLY

Learning Outcome:
1. Outline the structural features of ribosomes.

In Section 15.4, we examined how the structure and function of tRNA molecules are important in translation. According to the adaptor hypothesis, tRNAs bind to mRNA due to complementarity between the anticodons and codons. Concurrently, the tRNA molecules have the correct amino acid attached to their 3′ ends.

To synthesize a polypeptide, additional events must occur. In particular, the bond between the 3′ end of the tRNA and the amino acid must be broken, and a peptide bond must be formed between the adjacent amino acids. To facilitate these events, translation occurs on the surface of a macromolecular complex known as the **ribosome**. The ribosome can be thought of as the macromolecular arena where translation takes place. In this section, we begin by outlining the biochemical compositions of ribosomes in bacterial and eukaryotic cells. We will then examine the key functional sites on ribosomes for the translation process.

Bacterial and Eukaryotic Ribosomes Are Assembled from rRNA and Proteins

Bacterial cells have one type of ribosome that is found within the cytoplasm. Eukaryotic cells contain biochemically distinct ribosomes in different cellular locations. The most abundant type of ribosome functions in the cytosol, which is the region of the eukaryotic cell that is inside the plasma membrane but outside the membrane-bound organelles. Besides the cytosolic ribosomes, all eukaryotic cells have ribosomes within the mitochondria. In addition, plant cells and algae have ribosomes in their chloroplasts. The compositions of mitochondrial and chloroplast ribosomes are quite different from that of the cytosolic ribosomes. Unless otherwise noted, the term *eukaryotic ribosome* refers to ribosomes in the cytosol, not to those found within organelles. Likewise, the description of eukaryotic translation refers to translation via cytosolic ribosomes.

Each ribosome is composed of structures called the large and small subunits. This term is perhaps misleading because each ribosomal subunit itself is formed from the assembly of many different proteins and RNA molecules called **ribosomal RNA,** or **rRNA.** In bacterial ribosomes, the 30S subunit is formed from the assembly of 21 different ribosomal proteins and a 16S rRNA molecule; the 50S subunit contains 34 different proteins and 5S and 23S rRNA molecules (**Table 15.4**). The designations 30S and 50S refer to the rate at which these subunits sediment when subjected to a centrifugal force. This rate is described as a sedimentation coefficient in Svedberg units (S), in honor of Theodor Svedberg, who invented the ultracentrifuge. Together, the 30S and 50S subunits form a 70S ribosome. (Note: Svedberg units do not add up linearly.) In bacteria, the ribosomal proteins and rRNA molecules are synthesized in the cytoplasm, and the ribosomal subunits are assembled there.

The synthesis of eukaryotic rRNA occurs within the nucleus, and the ribosomal proteins are made in the cytosol, where translation takes place. The 40S subunit is composed of 33 proteins and an 18S rRNA; the 60S subunit is made of 49 proteins and 5S, 5.8S, and 28S rRNAs (see Table 15.4). The assembly of the rRNAs and ribosomal proteins to make the 40S and 60S subunits occurs within the **nucleolus,** a region of the nucleus specialized for this purpose. The 40S and 60S subunits are then exported into the cytosol, where they associate to form an 80S ribosome during translation.

Components of Ribosomal Subunits Form Functional Sites for Translation

To understand the structure and function of the ribosome at the molecular level, researchers must determine the locations and functional roles of the individual ribosomal proteins and rRNAs. In recent years, many advances have been made toward a molecular understanding of ribosomes. Microscopic and biophysical methods have been used to study ribosome structure. An electron micrograph of bacterial ribosomes is shown in **Figure 15.13a.** In this example, many ribosomes are translating a single mRNA. The term **polyribosome,** or **polysome,** is used to describe an mRNA transcript that has many bound ribosomes in the act of translation.

More recently, a few research groups have succeeded in crystallizing ribosomal subunits and even intact ribosomes. This is

TABLE 15.4
A Comparison of Ribosome Composition in Bacteria and Eukaryotes

	Small Subunit	Large Subunit	Assembled Ribosome
Bacterial			
Sedimentation coefficient	30S	50S	70S
Number of proteins	21	34	55
rRNA molecule(s)	16S rRNA	5S rRNA, 23S rRNA	16S rRNA, 5S rRNA, 23S rRNA
Eukaryotic			
Sedimentation coefficient	40S	60S	80S
Number of proteins	33	49	82
rRNA molecule(s)	18S rRNA	5S rRNA, 5.8S rRNA, 28S rRNA	18S rRNA, 5S rRNA, 5.8S rRNA, 28S rRNA

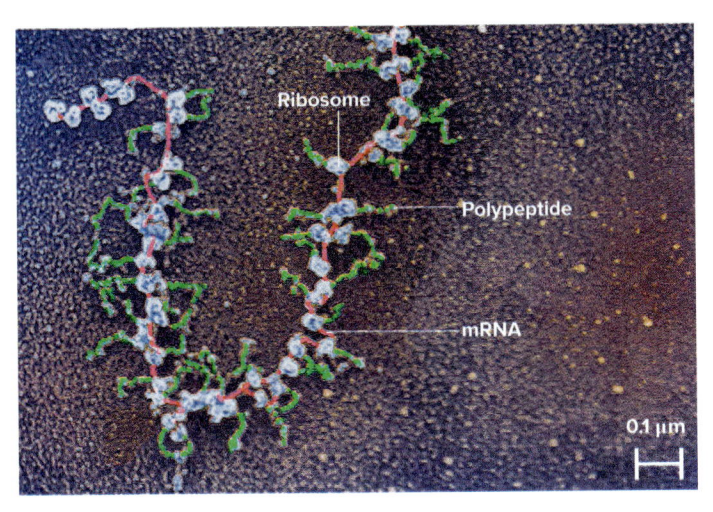

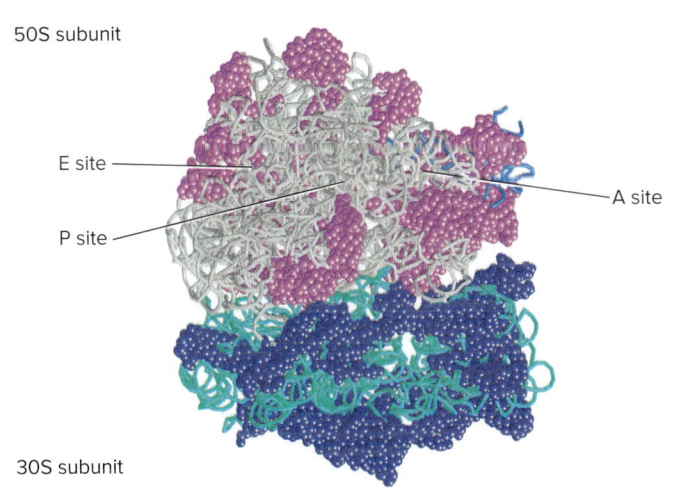

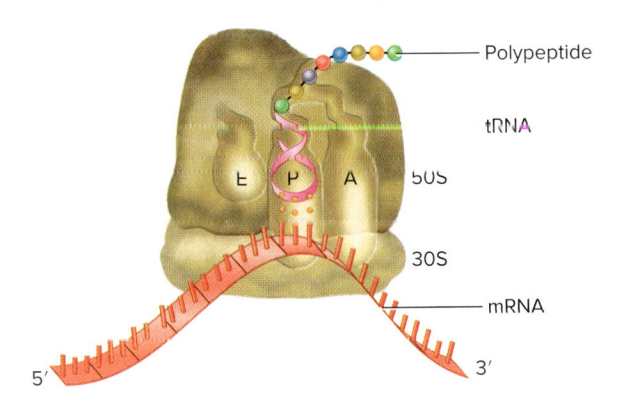

an amazing technical feat, because it is difficult to find the right conditions under which large macromolecules form highly ordered crystals. **Figure 15.13b** shows the crystal structure of bacterial ribosomal subunits. The overall shape of each subunit is largely determined by the structure of the rRNAs, which constitute most of the mass of the ribosome. The interface between the 30S and 50S subunits is primarily composed of rRNA. Ribosomal proteins cluster on the outer surface of the ribosome and on the periphery of the interface.

During bacterial translation, the mRNA lies on the surface of the 30S subunit within a space between the 30S and 50S subunits. As the polypeptide is being synthesized, it exits through a channel within the 50S subunit (**Figure 15.13c**). Ribosomes contain discrete sites where tRNAs bind and the polypeptide is synthesized. In 1964, James Watson was the first to propose a two-site model for tRNA binding to the ribosome. These sites are known as the **peptidyl site (P site)** and **aminoacyl site (A site)**. In 1981, Knud Nierhaus, Hans Sternbach, and Hans-Jorg Rheinberger proposed a three-site model. This model incorporated the observation that uncharged tRNA molecules can bind to a site on the ribosome that is distinct from the P and A sites. This third site is now known as the **exit site (E site)**. The locations of the E, P, and A sites are shown in Figure 15.13c. In the next section, we will examine the roles of these sites during the stages of translation.

15.5 REVIEWING THE KEY CONCEPTS

- Ribosomes are the site of polypeptide synthesis. The small and large subunits of ribosomes are composed of rRNAs and multiple proteins (see Table 15.4).
- A ribosome contains A (aminoacyl), P (peptidyl), and E (exit) sites, which are occupied by tRNA molecules (see Figure 15.13).

15.5 COMPREHENSION QUESTIONS

1. Each ribosomal subunit is composed of
   a. multiple proteins.
   b. rRNA.
   c. tRNA.
   d. both a and b.
2. The site(s) on a ribosome where tRNA molecules may be located include
   a. the A site.
   b. the P site.
   c. the E site.
   d. all of the above.

FIGURE 15.13 Ribosomal structure. (a) A colorized transmission electron micrograph of ribosomes attached to a bacterial mRNA and synthesizing polypeptides. Ribosomes are blue, mRNA is red, and polypeptides are green. (b) Crystal structure of the 50S and 30S subunits in bacterial ribosomes. The rRNA is shown as gray strands (50S subunit) and turquoise strands (30S subunit), and proteins are violet (50S subunit) and navy blue (30S subunit). (c) A model depicting the sites where tRNA and mRNA bind to an intact ribosome. The mRNA lies on the surface of the 30S subunit. The E, P, and A sites are formed at the interface between the large and small subunits. The growing polypeptide exits through a hole in the 50S subunit.
(a) ©Elena Kiseleva/Science Source; (b) ©Tom Pantages

15.6 STAGES OF TRANSLATION

Learning Outcomes:
1. Outline the three stages of translation.
2. Describe the steps that occur during the initiation, elongation, and termination stages of translation.
3. Compare and contrast bacterial and eukaryotic translation.

Like transcription, the process of translation can be viewed as occurring in three stages: initiation, elongation, and termination. **Figure 15.14** presents an overview of these stages. During **initiation,** the ribosomal subunits, mRNA, and the first tRNA assemble to form a complex. After the initiation complex is formed, the ribosome slides along the mRNA in the 5′ to 3′ direction, moving over the codons. This is the **elongation** stage of translation. As the ribosome moves, tRNA molecules sequentially bind to the mRNA in the ribosome, bringing with them the appropriate amino acids. Therefore, amino acids are linked in the order dictated by the codon sequence in the mRNA. Finally, a stop codon is reached, signaling the **termination** of translation. At this point, disassembly occurs, and the newly made polypeptide is released. In this section, we will examine the components required for the translation process and consider their functional roles during the three stages of translation.

The Initiation Stage Involves the Binding of mRNA and the Initiator tRNA to the Ribosomal Subunits

During initiation, an mRNA and the first tRNA bind to the ribosomal subunits. A specific tRNA functions as the **initiator tRNA,** which recognizes the start codon in the mRNA. In bacteria, the initiator tRNA, which is also designated tRNA^{fMet}, carries a methionine that has been covalently modified to *N*-formylmethionine. In this modification, a formyl group (—CHO) is attached to the nitrogen atom in methionine after the methionine has been attached to the tRNA.

Figure 15.15 describes the initiation stage of translation in bacteria, during which the mRNA, tRNA^{fMet}, and ribosomal subunits associate with each other to form an initiation complex. The formation of this complex requires the participation of three initiation factors: IF1, IF2, and IF3.

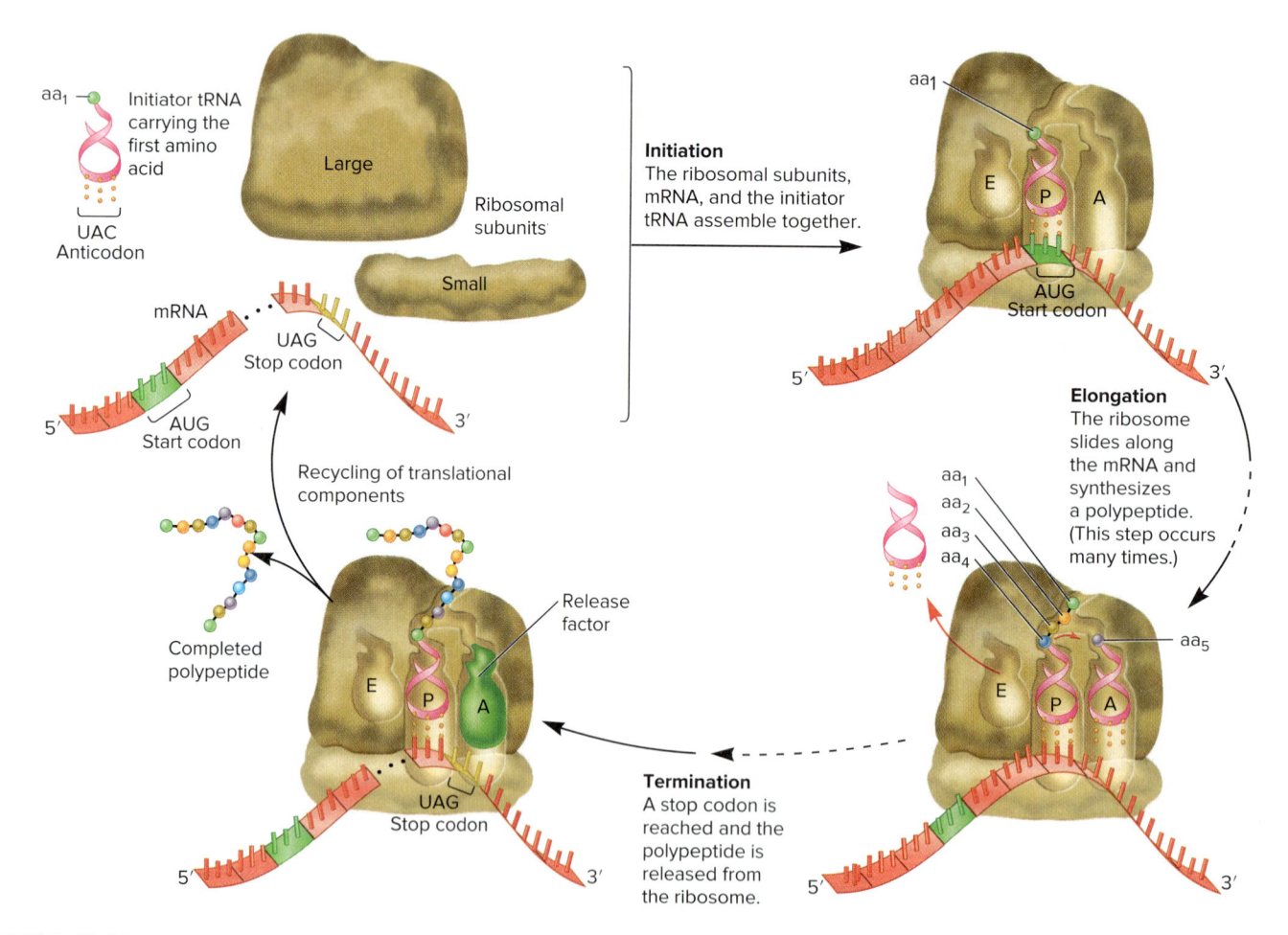

FIGURE 15.14 Overview of the stages of translation. The initiation stage involves the assembly of the ribosomal subunits, mRNA, and the initiator tRNA carrying the first amino acid. During elongation, the ribosome slides along the mRNA and synthesizes a polypeptide. Translation ends when a stop codon is reached and the polypeptide is released from the ribosome. (Note: In this and succeeding figures in this chapter, the ribosomes are drawn schematically to emphasize different aspects of the translation process. The structures of ribosomes are illustrated in Figure 15.13.)

Genes → Traits The ability of genes to produce an organism's traits relies on the molecular process of gene expression. During translation, the codon sequence within mRNA (which is derived from a gene sequence during transcription) is translated into a polypeptide sequence. After polypeptides are made within a living cell, they function as components of proteins to govern an organism's traits. For example, once the β-globin polypeptide is made, it functions within the hemoglobin protein and provides red blood cells with the ability to carry oxygen, a vital trait for survival. Translation allows functional proteins to be made within living cells.

Concept Check: Describe the roles that mRNA plays in the three stages of translation.

15.6 STAGES OF TRANSLATION 345

Figure (left column, diagram labels):

IF1 and IF3 bind to the 30S subunit.
— 30S subunit

The mRNA binds to the 30S subunit. The Shine-Dalgarno sequence is complementary to a portion of the 16S rRNA.

Portion of 16S rRNA
Shine-Dalgarno sequence (actually 9 nucleotides long)
Start codon
5′ ... 3′

IF2, bound to GTP, promotes the binding of the initiator tRNA to the start codon in the P site.

tRNA^{fMet}
GTP
Initiator tRNA

IF1 and IF3 are released.
IF2 hydrolyzes its GTP and is released.
The 50S subunit associates.

tRNA^{fMet}
E P A
70S initiation complex
5′ ... 3′

FIGURE 15.15 The initiation stage of translation in bacteria.

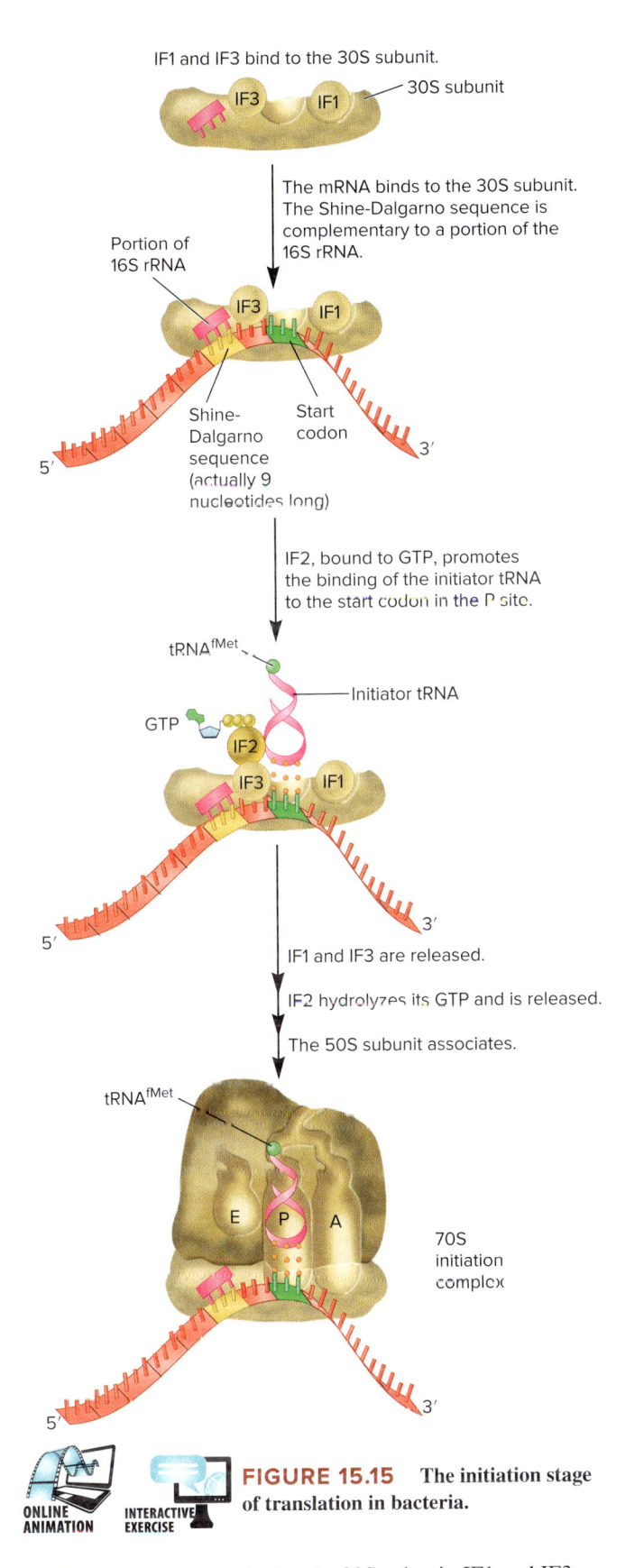

1. First, IF1 and IF3 bind to the 30S subunit. IF1 and IF3 prevent the association of the 50S subunit.
2. Next, the mRNA binds to the 30S subunit. This binding is facilitated by a nine-nucleotide sequence within the bacterial mRNA called the **Shine-Dalgarno sequence.** The location of this sequence is shown in Figure 15.15 and in more detail in **Figure 15.16**. How does the Shine-Dalgarno sequence facilitate the binding of mRNA to the ribosome? The Shine-Dalgarno sequence is complementary to a short sequence within the 16S rRNA, which promotes the hydrogen bonding of the mRNA to the 30S subunit.
3. A tRNA^{fMet} binds to the mRNA that is already attached to the 30S subunit. This step requires the function of IF2, which uses GTP. The tRNA^{fMet} binds to the start codon, which is typically a few nucleotides downstream from the Shine-Dalgarno sequence. The start codon is usually AUG, but in some cases it can be GUG or UUG. Even when the start codon is GUG (which normally encodes valine) or UUG (which normally encodes leucine), the first amino acid in the polypeptide is still a formylmethionine because only a tRNA^{fMet} can initiate translation. During or after translation of the entire polypeptide, the formyl group or the entire formylmethionine may be removed. Therefore, some polypeptides may not have formylmethionine or methionine as their first amino acid.
4. After the mRNA and tRNA^{fMet} have become bound to the 30S subunit, IF1 and IF3 are released, and then IF2 hydrolyzes its GTP and is also released. This allows the 50S ribosomal subunit to associate with the 30S subunit. Much later, after translation is completed, IF1 binding is necessary to dissociate the 50S and 30S ribosomal subunits so that the 30S subunit can reinitiate with another mRNA molecule.

In eukaryotes, the assembly of the initiation complex has similarities to what occurs in bacteria. However, as described in **Table 15.5**, additional factors are required for the initiation process. Note that the initiation factors are designated eIF (for eukaryotic Initiation Factor) to distinguish them from bacterial initiation factors. The initiator tRNA in eukaryotes carries methionine rather than formylmethionine, as in bacteria. A eukaryotic initiation factor, eIF2, binds directly to tRNA^{Met} to recruit it to the 40S subunit. Eukaryotic mRNAs do not have a Shine-Dalgarno sequence. How then are eukaryotic mRNAs recognized by the ribosome? The mRNA is recognized by eIF4, which is a multiprotein complex that recognizes the 7-methylguanosine cap and facilitates the binding of the mRNA to the 40S subunit.

The identification of the correct AUG start codon in eukaryotes occurs in a different way than it does in bacteria. After the initial binding of mRNA to the ribosome, the next step is locating an AUG start codon that is somewhere downstream from the 5′ cap. In 1986, Marilyn Kozak proposed that the ribosome begins at the 5′ end and then scans along the mRNA in the 3′ direction in search of an AUG start codon. In many, but not all cases, the ribosome uses the first AUG codon that it encounters as a start codon. When a start codon is identified, the 60S subunit assembles onto the 40S subunit with the aid of eIF5, forming the initiation complex.

By analyzing the sequences of many eukaryotic mRNAs, researchers have found that not all AUG codons near the 5′ end of mRNA can function as start codons. In some cases, the scanning

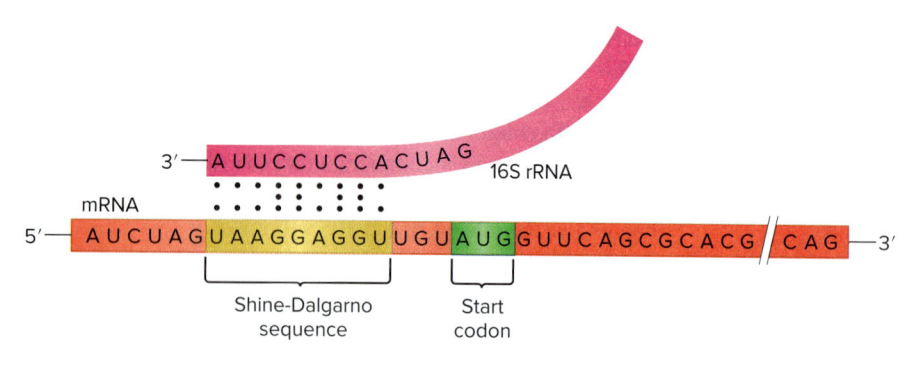

FIGURE 15.16 The locations of the Shine-Dalgarno sequence and the start codon in bacterial mRNA. The Shine-Dalgarno sequence is complementary to a sequence in the 16S rRNA. It hydrogen bonds with the 16S rRNA to promote initiation. The start codon is typically a few nucleotides downstream from the Shine-Dalgarno sequence.

Concept Check: Why does bacterial mRNA bind specifically to the small ribosomal subunit?

TABLE 15.5
A Simplified Comparison of Translational Protein Factors in Bacteria and Eukaryotes

Bacterial Factors	Eukaryotic Factors*	Function
Initiation Factors		
	eIF4	Involved with the recognition of the 7-methylguanosine cap and the binding of the mRNA to the small ribosomal subunit
IF1, IF3	eIF1, eIF3, eIF6	Prevent the association between the small and large ribosomal subunits and favor their dissociation
IF2	eIF2	Promote the binding of the initiator tRNA to the small ribosomal subunit
	eIF5	Helps to dissociate the other elongation factors, which allows the large ribosomal subunit to bind
Elongation Factors		
EF-Tu	eEF1α	Involved in the binding of tRNAs to the A site
EF-Ts	eEF1βγ	Nucleotide exchange factors required for the functioning of EF-Tu and eEF1α, respectively
EF-G	eEF2	Required for translocation
Release Factors		
RF1, RF2	eRF1	Recognize a stop codon and trigger the cleavage of the polypeptide from the tRNA
RF3	eRF3	GTPases that are also involved in termination

*Eukaryotic translation factors are typically composed of multiple proteins.

ribosome passes over the first AUG codon and chooses an AUG farther along within the mRNA. The sequence of bases around the AUG codon plays an important role in determining whether or not it is selected as the start codon by a scanning ribosome. The consensus sequence for optimal start codon recognition in complex eukaryotes, such as vertebrates and vascular plants, is shown here.

					Start Codon		
G	C	C	(A/G)	C	C	A	U	G	G
−6	−5	−4	−3	−2	−1	+1	+2	+3	+4

Aside from an AUG codon itself, a guanine at the +4 position and a purine, preferably an adenine, at the −3 position are the most important sites for start codon selection. These rules for optimal translation initiation are called **Kozak's rules.**

Polypeptide Synthesis Occurs During the Elongation Stage

During the elongation stage of translation, amino acids are added, one at a time, to a growing polypeptide (**Figure 15.17**).

1. At the top of Figure 15.17, a short polypeptide is already attached to the tRNA at the P site of the ribosome.
2. A charged tRNA carrying a single amino acid binds to the A site. This binding occurs because the anticodon in the tRNA is complementary to the codon in the mRNA. The hydrolysis of GTP by the elongation factor EF-Tu provides energy for the binding of a tRNA to the A site. The 16S rRNA can detect when an incorrect tRNA is bound at the A site and prevents elongation until the mispaired tRNA is released from the A site. An incorrect amino acid is incorporated into a growing polypeptide at a low rate of approximately 1 mistake per 10,000 amino acids, or 10^{-4}.
3. The next step of elongation is a reaction called **peptidyl transfer**—the polypeptide is removed from the tRNA in the P site and transferred to the amino acid at the A site. This transfer is accompanied by the formation of a peptide bond between the amino acid at the A site and the polypeptide, lengthening the polypeptide by one amino acid. The peptidyl transfer reaction is catalyzed by a component of the 50S subunit known as **peptidyl transferase,** which is composed of several proteins and rRNA. The 23S rRNA catalyzes the bond formation between adjacent amino acids. In other words, the ribosome is a ribozyme!
4. After the peptidyl transfer reaction is complete, the ribosome moves, or translocates, to the next codon in the mRNA. This moves the tRNAs at the P and A sites to the E and P sites, respectively.

15.6 STAGES OF TRANSLATION   347

5. Finally, the uncharged tRNA exits the E site. You should notice that the next codon in the mRNA is now exposed in the unoccupied A site. At this point, a charged tRNA can enter the empty A site, and the same series of steps can add the next amino acid to the growing polypeptide.

As you may have realized, the A, P, and E sites are named for the role of the tRNA that is usually found there. The A site binds an aminoacyl-tRNA (also called a charged tRNA), the P site usually contains the peptidyl-tRNA (a tRNA with an attached peptide), and the E site is where the uncharged tRNA exits.

Even though elongation is rather complex, it occurs at a remarkable rate. Under normal cellular conditions, a polypeptide can elongate at a rate of 15 to 20 amino acids per second in bacteria or 2 to 6 amino acids per second in eukaryotes!

Termination Occurs When a Stop Codon Is Reached in the mRNA

The final stage of translation, known as termination, occurs when a stop codon is reached in the mRNA. In most species, the three stop codons are UAA, UAG, and UGA. The stop codons are recognized by proteins known as **release factors** (see Table 15.5). Interestingly, the three-dimensional structures of release factor proteins are "molecular mimics" that resemble the structure of tRNAs. Release factors can specifically bind to a stop codon sequence. In bacteria, RF1 recognizes UAA and UAG, and RF2 recognizes UGA and UAA. A third release factor, RF3, is also required. In eukaryotes, a single release factor, eRF1, recognizes all three stop codons, and eRF3 is also required for termination.

Figure 15.18 illustrates the termination stage of translation in bacteria. At the top of this figure, the completed polypeptide is attached to a tRNA in the P site. A stop codon is located at the A site.

1. In the first step, RF1 or RF2 binds to the stop codon at the A site and RF3 (not shown) binds at a different location on the ribosome.
2. After RF1 (or RF2) and RF3 have bound, the bond between the polypeptide and the tRNA is hydrolyzed. The polypeptide and tRNA are then released from the ribosome.
3. The final step in translational termination is the disassembly of ribosomal subunits, mRNA, and the release factors.

Bacterial Translation Can Begin Before Transcription Is Completed

Microscopic, biochemical, and genetic studies have shown that the translation of a bacterial mRNA begins before the mRNA transcript is completed. In other words, as soon as an mRNA strand is long enough, a ribosome attaches to the 5′ end and begins translation, even before RNA polymerase has reached the transcriptional termination site within the gene. This phenomenon in bacterial cells is referred to as coupling between transcription and translation. Note that coupling of these processes does not usually occur in eukaryotes, because transcription takes place in the nucleus of eukaryotic cells, whereas translation occurs in the cytosol.

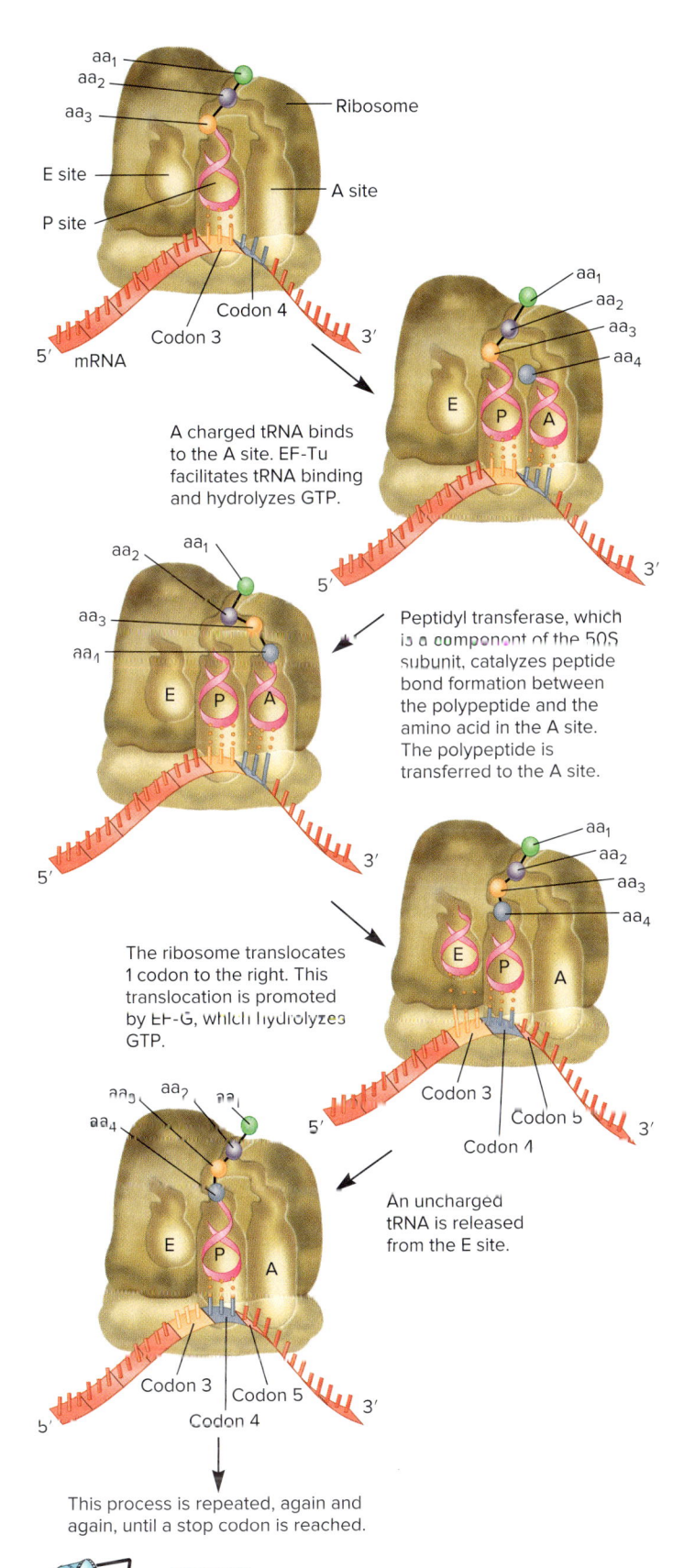

FIGURE 15.17 The elongation stage of translation in bacteria.

Concept Check: What is the role of peptidyl transferase during the elongation stage?

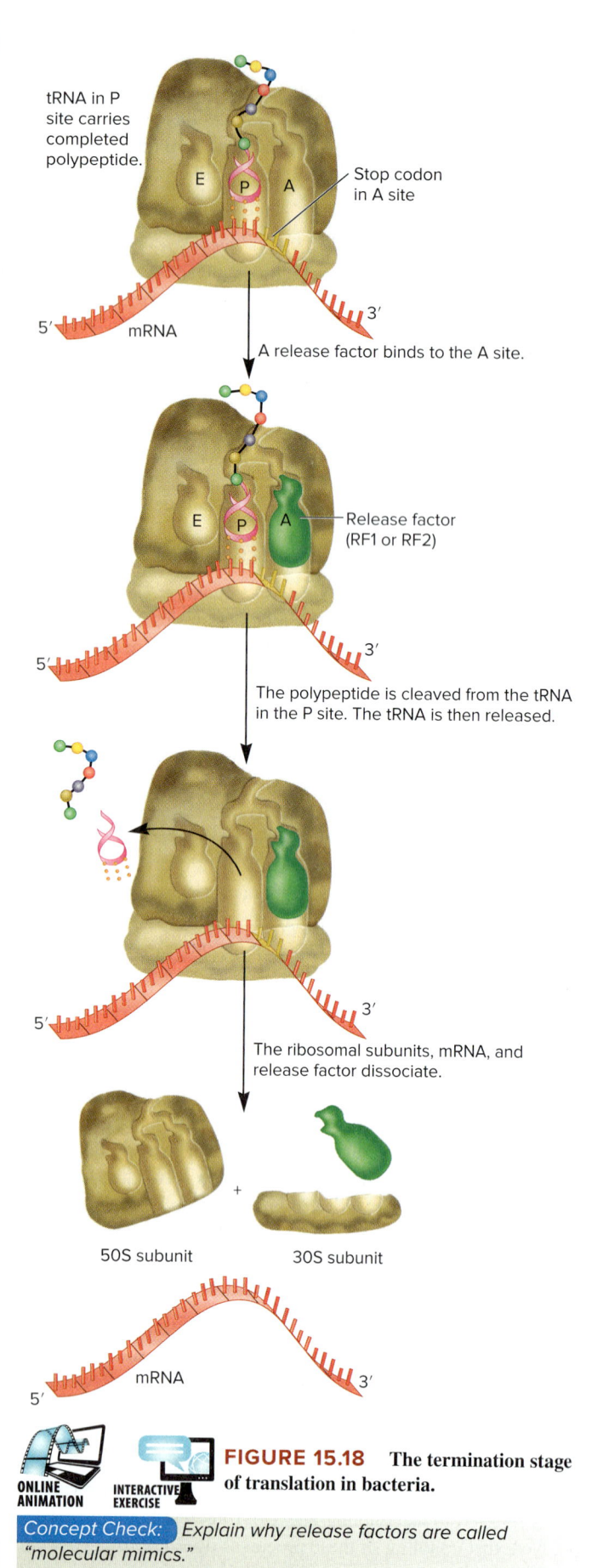

Bacterial and Eukaryotic Translation Show Similarities and Differences

Throughout this chapter, we have compared translation in bacteria and eukaryotic organisms. The general steps of translation are similar in all forms of life, but we have also seen some striking differences between bacteria and eukaryotes. **Table 15.6** compares translation between these groups.

Antibiotics That Inhibit Bacterial Translation Are Used to Treat Bacterial Diseases

Many different diseases that affect people and domesticated animals are caused by pathogenic bacteria. An **antibiotic** is any substance produced by a microorganism that inhibits the growth of other microorganisms, such as pathogenic bacteria. Most antibiotics are small organic molecules, with masses less than 2000 daltons. In some cases, antibiotics exert their effect because they inhibit or interfere with bacterial translation. Because the components of translation are somewhat different between bacteria and eukaryotes, some antibiotics inhibit bacterial translation without affecting eukaryotic translation. Therefore, they can be used to treat bacterial infections in humans, pets, and livestock. **Table 15.7** describes a few examples.

15.6 REVIEWING THE KEY CONCEPTS

- The three stages of translation are initiation, elongation, and termination (see Figure 15.14).
- During the initiation stage of translation, the mRNA, initiator tRNA, and ribosomal subunits assemble. Initiation factors are involved in the process. In bacteria, the Shine-Dalgarno sequence promotes the binding of the mRNA to the small ribosomal subunit (see Figures 15.15, 15.16, Table 15.5).
- Start codon selection in complex eukaryotes follows Kozak's rules.
- During elongation, tRNAs bring amino acids to the A site and a series of peptidyl transfers creates a polypeptide. At each step, the polypeptide is transferred from the P site to the A site. The tRNAs are released from the E site. Elongation factors are involved in this process (see Figure 15.17).
- During termination, a release factor binds to a stop codon in the A site. This promotes the cleavage of the polypeptide from the tRNA and the subsequent disassembly of the tRNA, mRNA, and ribosomal subunits (see Figure 15.18).
- Bacterial translation can begin before transcription is completed.
- Bacterial and eukaryotic translation show many similarities and differences (see Table 15.6).
- Many antibiotics work by inhibiting translation (see Table 15.7).

15.6 COMPREHENSION QUESTIONS

1. During the initiation stage of translation in bacteria, which of the following events occur(s)?
   a. IF1 and IF3 bind to the 30S subunit.
   b. The mRNA binds to the 30S subunit, and tRNA^{fMet} binds to the start codon in the mRNA.
   c. IF2 hydrolyzes its GTP and is released; the 50S subunit binds to the 30S subunit.
   d. All of the above events occur.

FIGURE 15.18 The termination stage of translation in bacteria.

Concept Check: *Explain why release factors are called "molecular mimics."*

TABLE 15.6
Comparison of Bacterial and Eukaryotic Translation

	Bacterial	Eukaryotic

Ribosome composition:	70S ribosomes:
	30S subunit—
	  21 proteins + 1 rRNA
	50S subunit—
	  34 proteins + 2 rRNAs
		80S ribosomes:
		40S subunit—
		  33 proteins + 1 rRNA
		60S subunit—
		  49 proteins + 3 rRNAs

Initiator tRNA:	tRNA^{fMet}	tRNA^{Met}
Formation of the initiation complex:	Requires IF1, IF2, and IF3	Requires more initiation factors than in bacterial translation
Initial binding of mRNA to the ribosome:	Requires a Shine-Dalgarno sequence	Requires a 7-methylguanosine cap
Selection of a start codon:	AUG, GUG, or UUG located just downstream from the Shine-Dalgarno sequence	According to Kozak's rules
Elongation rate:	Typically 15 to 20 amino acids per second	Typically 2 to 6 amino acids per second
Termination:	Requires RF1, RF2, and RF3	Requires eRF1 and eRF3
Coupled to transcription:	Yes	No

TABLE 15.7
Mechanisms of Inhibition of Bacterial Translation via Selected Antibiotics

Antibiotic	Description
Chloraphenical	Blocks elongation by acting as competitive inhibitor of peptidyl transferase.
Erythromycin	Binds to the 23S rRNA and blocks elongation by interfering with the translocation step.
Puromycin	Binds to the A site and causes premature release of the polypeptide. This early termination of translation results in polypeptides that are shorter than normal.
Tetracycline	Blocks elongation by inhibiting the binding of aminoacyl tRNAs to the ribosome.
Streptomycin	Interferes with normal pairing between aminoacyl tRNAs and codons. This causes misreading, and thereby produces abnormal proteins.

2. Kozak's rules determine
   a. the choice of the start codon in complex eukaryotes.
   b. the choice of the start codon in bacteria.
   c. the site in the mRNA where translation ends.
   d. how fast the mRNA is translated.
3. During the peptidyl transfer reaction, the polypeptide, which is attached to a tRNA in the _____ becomes bound via _____ to an amino acid attached to a tRNA in the _____.
   a. A site, several hydrogen bonds, P site
   b. A site, a peptide bond, P site
   c. P site, a peptide bond, A site
   d. P site, several hydrogen bonds, A site
4. A release factor is referred to as a "molecular mimic" because its structure is similar to
   a. a ribosome.
   b. an mRNA.
   c. a tRNA.
   d. an elongation factor.

KEY TERMS

Page 325. translation, protein-encoding genes (structural genes), messenger RNA (mRNA)
Page 326. alkaptonuria, inborn error of metabolism
Page 327. one-gene/one-enzyme hypothesis
Page 328. polypeptide, protein, genetic code, sense codons, start codon, stop codons (termination or nonsense codons), 5′-untranslated region, 3′-untranslated region, transfer RNA (tRNA), anticodons
Page 329. degeneracy, synonymous codons, wobble base, reading frame
Page 330. selenocysteine, pyrrolysine
Page 331. peptide bond, amino-terminal end (N-terminus), carboxyl-terminal end (C-terminus)
Page 332. side chain (R group), primary structure
Page 333. chaperones, secondary structure, α helix, β sheet, tertiary structure
Page 334. quaternary structure, subunits, cell-free translation system
Page 338. adaptor hypothesis
Page 339. aminoacyl-tRNA synthetases

Page 340. charged tRNA (aminoacyl-tRNA), wobble rules, isoacceptor tRNAs
Page 341. ribosome
Page 342. ribosomal RNA (rRNA), nucleolus, polyribosome (polysome)
Page 343. peptidyl site (P site), aminoacyl site (A site), exit site (E site)
Page 344. initiation, elongation, termination, initiator tRNA
Page 345. Shine-Dalgarno sequence
Page 346. Kozak's rules, peptidyl transfer, peptidyl transferase
Page 347. release factors
Page 348. antibiotic

CHAPTER SUMMARY

- Cellular proteins are made via the translation of mRNAs.

15.1 The Genetic Basis for Protein Synthesis

- Garrod studied the disease called alkaptonuria and suggested that some genes encode enzymes (see Figure 15.1).
- Beadle and Tatum studied *Neurospora* mutants that were altered in their nutritional requirements and hypothesized that one gene encodes one enzyme. This one-gene/one-enzyme hypothesis was later modified because: (1) some proteins are not enzymes; (2) some proteins are composed of two or more different polypeptides; (3) one gene can encode two or more polypeptides due to alternative splicing or RNA editing; and (4) some genes encode RNAs that are not translated into polypeptides (see Figure 15.2).

15.2 The Relationship Between the Genetic Code and Protein Synthesis

- During translation, the codons in mRNA are recognized by tRNA molecules, which act as intermediates in the production of a polypeptide with a specific amino acid sequence (see Figure 15.3).
- The genetic code refers to the relationship between three-base codons in the mRNA and the amino acids that are incorporated into a polypeptide. One codon (AUG) is a start codon, which determines the reading frame of the mRNA. Three codons (UAA, UAG, and UGA) function as stop codons (see Table 15.1).
- The genetic code is largely universal, but some exceptions are known to occur (see Table 15.2).
- A polypeptide is made by the formation of peptide bonds between adjacent amino acids. Each polypeptide has a directionality from its amino-terminal end to its carboxyl-terminal end, which parallels the arrangement of codons in mRNA in the 5′ to 3′ direction (see Figure 15.4).
- Amino acids differ in their side-chain structure (see Figure 15.5).
- Protein structure can be viewed at different levels: primary structure (sequence of amino acids), secondary structure (repeating folding patterns such as the α helix and the β sheet), tertiary structure (additional folding), and quaternary structure (the binding of multiple subunits to each other) (see Figure 15.6).

15.3 Experimental Determination of the Genetic Code

- Nirenberg and colleagues used synthetic RNA and a cell-free translation system to decipher the genetic code (see Figure 15.7).
- Other methods of deciphering the genetic code included the synthesis of copolymers by Khorana and the triplet-binding assays conducted by Nirenberg and Leder (see Table 15.3, Figure 15.8).

15.4 Structure and Function of tRNA

- The anticodon in a tRNA is complementary to a codon in mRNA. The tRNA carries a specific amino acid that corresponds to the codon in the mRNA according to the genetic code (see Figure 15.9).
- The secondary structure of tRNA resembles a cloverleaf. The anticodon is in the second loop, and the amino acid is attached to the 3′ end (see Figure 15.10).
- An aminoacyl-tRNA synthetase is one of a group of enzymes that attaches the correct amino acid to a tRNA. The resulting tRNA is called a charged tRNA, or an aminoacyl-tRNA (see Figure 15.11).
- Mismatches are allowed between the pairing of tRNAs and mRNA according to the wobble rules (see Figure 15.12).

15.5 Ribosome Structure and Assembly

- Ribosomes are the site of polypeptide synthesis. The small and large subunits of ribosomes are composed of rRNAs and multiple proteins (see Table 15.4).
- A ribosome contains A (aminoacyl), P (peptidyl), and E (exit) sites, which are occupied by tRNA molecules (see Figure 15.13).

15.6 Stages of Translation

- The three stages of translation are initiation, elongation, and termination (see Figure 15.14).
- During the initiation stage of translation, the mRNA, initiator tRNA, and ribosomal subunits assemble. Initiation factors are involved in the process. In bacteria, the Shine-Dalgarno sequence promotes the binding of the mRNA to the small ribosomal subunit (see Figures 15.15, 15.16, Table 15.5).
- Start codon selection in complex eukaryotes follows Kozak's rules.
- During elongation, tRNAs bring amino acids to the A site and a series of peptidyl transfers creates a polypeptide. At each step, the polypeptide is transferred from the P site to the A site. The tRNAs are released from the E site. Elongation factors are involved in this process (see Figure 15.17).
- During termination, a release factor binds to a stop codon in the A site. This promotes the cleavage of the polypeptide from the tRNA and the subsequent disassembly of the tRNA, mRNA, and ribosomal subunits (see Figure 15.18).
- Bacterial translation can begin before transcription is completed.
- Bacterial and eukaryotic translation show many similarities and differences (see Table 15.6).
- Many antibiotics work by inhibiting translation (see Table 15.7).

PROBLEM SETS & INSIGHTS

More Genetic TIPS

1. The first amino acid in a purified bacterial protein is methionine. The start codon in the mRNA is GUG, which codes for valine. Why isn't the first amino acid formylmethionine or valine?

Topic: What topic in genetics does this question address?

The topic is translation. More specifically, the question is about what determines the first amino acid in a polypeptide.

Information: What information do you know based on the question and your understanding of the topic?

From the question, you know that the start codon in an mRNA is GUG, but the first amino acid in the resulting polypeptide is methionine. From your understanding of the topic, you may remember that the initiator tRNA carries the first amino acid in a polypeptide.

Problem-Solving Strategy: Describe the steps.

One strategy to solve this problem is to describe the steps of translation. This may help you to see how the first amino acid gets incorporated into a polypeptide.

Answer: During polypeptide synthesis, the first amino acid is carried by the initiator tRNA. This initiator tRNA always carries formylmethionine even when the start codon is GUG (valine) or UUG (leucine). The formyl group can be removed later to yield methionine as the first amino acid.

2. A tRNA has the anticodon sequence 3′–CAG–5′. What amino acid does it carry?

Topic: What topic in genetics does this question address?

The topic is translation. More specifically, the question asks you to determine the amino acid that a tRNA carries.

Information: What information do you know based on the question and your understanding of the topic?

From the question, you know that a tRNA has an anticodon that is 3′–CAG–5′. From your understanding of the topic, you may remember that the anticodon and codon are complementary and antiparallel.

Problem-Solving Strategy: Make a drawing.

One strategy to solve this problem is to make a drawing showing how the anticodon in a tRNA binds to a codon in an mRNA.

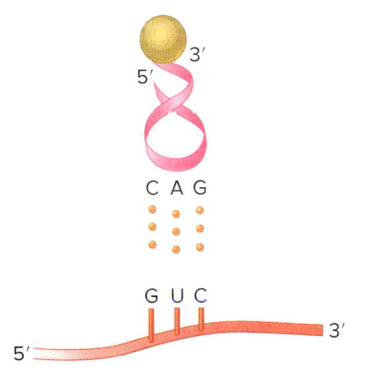

Answer: An anticodon that is 3′–CAG–5′ is complementary to a codon with the sequence 5′–GUC–3′. According to the genetic code, this codon specifies the amino acid valine. Therefore, this tRNA must carry valine at its acceptor stem.

3. An antibiotic is a drug that kills or inhibits the growth of microorganisms. The use of antibiotics has been of great importance in the battle against many infectious diseases caused by microorganisms. The mode of action for many antibiotics is to inhibit the translation process within bacterial cells. Certain antibiotics selectively bind to bacterial (70S) ribosomes but do not inhibit eukaryotic (80S) ribosomes. Why would an antibiotic bind to a bacterial ribosome but not to a eukaryotic ribosome? Why does this binding inhibit growth?

Topic: What topic in genetics does this question address?

The topic is translation. More specifically, the question is about how antibiotics inhibit translation in bacteria but not in eukaryotes.

Information: What information do you know based on the question and your understanding of the topic?

From the question, you know that some antibiotics bind to bacterial ribosomes but not eukaryotic ribosomes. From your understanding of the topic, you may remember that ribosomes must function correctly for translation to occur.

Problem-Solving Strategy: Relate structure and function.

One strategy to solve this problem is to consider how the structure of bacterial ribosomes may differ from that of eukaryotic ribosomes and how such structural differences may affect the ability of antibiotics to bind to ribosomal components. The binding of an antibiotic may inhibit a key step in translation (see Table 15.7).

Answer: Bacterial ribosomes have proteins and rRNAs whose composition differs from those of eukaryotic ribosomes, and certain antibiotics can recognize these different components and bind specifically to bacterial ribosomes, thereby interfering with the process of translation. In other words, the surface of a bacterial ribosome must be somewhat different from the surface of a eukaryotic ribosome so that the antibiotic molecules are able to bind only to the surfaces of bacterial ribosomes. If a bacterial cell is exposed to certain antibiotics, it cannot synthesize new polypeptides because the antibiotic inhibits ribosome function. Because polypeptides form functional proteins needed for processes such as cell division, the bacterium is unable to grow and proliferate.

Conceptual Questions

C1. An mRNA has the following sequence:

5′–GGCGAUGGGCAAUAAACCGGGCCAGUAAGC–3′

Identify the start codon, and determine the complete amino acid sequence that will be translated from this mRNA.

C2. What does it mean when we say that the genetic code contains degeneracy? Discuss the universality of the genetic code.

C3. According to the adaptor hypothesis, is each the following statements true or false?

A. The sequence of an anticodon in a tRNA directly recognizes a codon sequence in mRNA, with some allowance for wobble.

B. The amino acid attached to the tRNA directly recognizes a codon sequence in mRNA.

C. The amino acid attached to the tRNA affects the binding of the tRNA to a codon sequence in mRNA.

C4. Researchers have isolated strains of bacteria that carry mutations within genes that encode tRNAs. These mutations can change the sequence of the anticodon. For example, a normal tRNA^{Trp} gene encodes a tRNA with the anticodon 3′–ACC–5′. A mutation can change this sequence to 3′–CCC–5′. When this mutation occurs, the tRNA still carries a tryptophan at its 3′ acceptor stem, even though the anticodon sequence has been altered.

A. How would this mutation affect the synthesis of polypeptides within the bacterium?

B. What does this mutation tell you about the recognition between tryptophanyl-tRNA synthetase and tRNA^{Trp}? Does the enzyme primarily recognize the anticodon or not?

C5. The covalent attachment of an amino acid to a tRNA is an endergonic reaction. In other words, an input of energy is required for the reaction to proceed. Where does the energy come from to attach amino acids to tRNA molecules?

C6. The wobble rules for tRNA-mRNA pairing are shown in Figure 15.12b. If we assume that the tRNAs do not contain modified bases, what is the minimum number of tRNAs needed to recognize the codons for each of the following amino acids?

A. Leucine

B. Methionine

C. Serine

C7. How many different sequences of mRNA could encode a peptide with the sequence proline-glycine-methionine-serine?

C8. If a tRNA molecule carries a glutamic acid, what are the two possible anticodon sequences that it could contain? Be specific about the 5′ and 3′ ends.

C9. A tRNA has an anticodon sequence 3′–GGU–5′. What amino acid does it carry?

C10. If a tRNA has an anticodon sequence 3′–CCI–5′, what codon(s) can it recognize?

C11. Describe the anticodon of a single tRNA that could recognize the codons 5′–AAC–3′ and 5′–AAU–3′. What type(s) of base modification to this tRNA would allow it to also recognize 5′–AAA–3′?

C12. Describe the structural features that all tRNA molecules have in common.

C13. In the tertiary structure of tRNA, where is the anticodon region relative to the attachment site for the amino acid? Are these located adjacent to each other?

C14. What is the role of aminoacyl-tRNA synthetase? The ability of aminoacyl-tRNA synthetases to recognize tRNAs has sometimes been called the "second genetic code." Why has the function of this type of enzyme been described this way?

C15. What is an activated amino acid?

C16. Discuss the significance of modified bases within tRNA molecules.

C17. How and when does formylmethionine become attached to the initiator tRNA in bacteria?

C18. Is it necessary for a cell to make 61 different tRNA molecules, corresponding to the 61 codons for amino acids? Explain your answer.

C19. List the components required for translation. Describe the relative sizes of these different components. In other words, which components are small molecules, macromolecules, or assemblies of macromolecules?

C20. Describe the components of eukaryotic ribosomal subunits and the location where the assembly of the subunits occurs within living cells.

C21. The term *subunit* can be used in a variety of ways. What is the difference between a protein subunit and a ribosomal subunit?

C22. Do the following events during bacterial translation occur primarily within the 30S subunit, within the 50S subunit, or at the interface between these two ribosomal subunits?

A. mRNA-tRNA recognition

B. Peptidyl transfer reaction

C. Exit of the polypeptide from the ribosome

D. Binding of initiation factors IF1, IF2, and IF3

C23. What are the three stages of translation? Discuss the main events that occur during these three stages.

C24. Describe the sequence in bacterial mRNA that promotes recognition by the 30S subunit.

C25. For each of the following initiation factors, how would eukaryotic initiation of translation be affected if it were missing?

A. eIF2

B. eIF4

C. eIF5

C26. How does a eukaryotic ribosome select its start codon? Describe the sequences in eukaryotic mRNAs that provide an optimal context for a start codon.

C27. Rank each of the following sequences in order (from best to worst) as sequences that could be used to initiate translation according to Kozak's rules.

GACGCCAUGG
GCCUCCAUGC
GCCAUCAAGG
GCCACCAUGG

C28. Explain the functional roles of the A, P, and E sites during translation.

C29. An mRNA has the following sequence:

5′–AUG UAC UAU GGG GCG UAA–3′.

Describe the amino acid sequence of the polypeptide that would be encoded by this mRNA. Be specific about the amino-terminal and carboxyl-terminal ends.

C30. Which steps during the translation of bacterial mRNA involve an interaction between complementary strands of RNA?

C31. What is the function of the nucleolus?

C32. In which of the ribosomal sites, the A site, P site, and/or E site, could the following be found?

A. A tRNA without an amino acid attached

B. A tRNA with a polypeptide attached

C. A tRNA with a single amino acid attached

C33. What is a polysome?

C34. Referring to Figure 15.17, explain why the ribosome translocates along the mRNA in the 5′ to 3′ direction rather than the 3′ to 5′ direction.

C35. Lactose permease, a protein of *E. coli*, is composed of a single polypeptide that is 417 amino acids in length. By convention, the amino acids within a polypeptide are numbered from the amino-terminal end to the carboxyl-terminal end. Are the following questions about lactose permease true or false?

A. Because the sixty-fourth amino acid is glycine and the sixty-eighth amino acid is aspartic acid, the codon for glycine, 64, is closer to the 3′ end of the mRNA than the codon for aspartic acid, 68.

B. The mRNA that encodes lactose permease must be greater than 1241 nucleotides in length.

C36. An mRNA encodes a polypeptide that is 312 amino acids in length. The fifty-third codon in this polypeptide is a tryptophan codon. A mutation in the gene that encodes this polypeptide changes this tryptophan codon into a stop codon. How many amino acids occur in the resulting polypeptide: 52, 53, 259, or 260?

C37. Explain what is meant by the *coupling* of transcription and translation. Does coupling occur in bacterial and/or eukaryotic cells? Explain.

Application and Experimental Questions

E1. In the experiment of Figure 15.7, what would be the predicted amounts of amino acids incorporated into polypeptides if the RNA was a random polymer containing 50% C and 50% G?

E2. Polypeptides can be translated in vitro. Would a bacterial mRNA be translated in vitro by eukaryotic ribosomes? Would a eukaryotic mRNA be translated in vitro by bacterial ribosomes? Why or why not?

E3. Discuss how the elucidation of the structure of the ribosome can help us to understand its function.

E4. Chapter 20 describes a blotting method known as Western blotting, which is used to detect the production of a polypeptide that is translated from a particular mRNA. In this method, a protein is detected using an antibody that specifically recognizes and binds to the protein's amino acid sequence. The antibody acts as a probe to detect the presence of the protein. In a Western blotting experiment, gel electrophoresis is used to separate a mixture of proteins according to their molecular masses. After the antibody has bound to the protein of interest within a blot of a gel, the protein is visualized as a dark band. For example, an antibody that recognizes the β-globin polypeptide could be used to specifically detect the β-globin polypeptide in a blot. As shown here, the method of Western blotting can be used to determine the amount and relative size of a particular protein that is produced in a given cell type.

Western blot

1 2 3

Lane 1 is a sample of proteins isolated from normal red blood cells.

Lane 2 is a sample of proteins isolated from kidney cells. Kidney cells do not produce β globin polypeptide.

Lane 3 is a sample of proteins isolated from red blood cells from a patient with β-thalassemia. This patient is homozygous for a mutation that results in the shortening of the β-globin polypeptide.

Here is the question: A protein called troponin contains 334 amino acids. Because each amino acid weighs 120 daltons (Da) (on average), the molecular mass of this protein is about 40,000 Da, or 40 kDa. Troponin functions in muscle cells, and it is not expressed in nerve cells. Draw the expected results of a Western blot for the following samples:

Lane 1: Proteins isolated from muscle cells

Lane 2: Proteins isolated from nerve cells

Lane 3: Proteins isolated from the muscle cells of an individual who is homozygous for a mutation that introduces a stop codon at codon 177 in the gene that encodes troponin

E5. The protein known as tyrosinase is needed to make certain types of pigments. Tyrosinase is composed of a single polypeptide with 511 amino acids. The molecular mass of this protein is approximately 61,300 Da, or 61.3 kDa. People who carry two defective copies of the tyrosinase gene have the condition known as albinism. They are unable to make pigment in the skin, eyes, and hair. Western blotting is used to detect proteins that are translated from a particular mRNA. This method is described in Chapter 20 and also in experimental question E4. Skin samples were collected from a pigmented individual (lane 1) and from three unrelated albino individuals (lanes 2, 3, and 4) and subjected to a Western blot analysis using an antibody that recognizes tyrosinase.

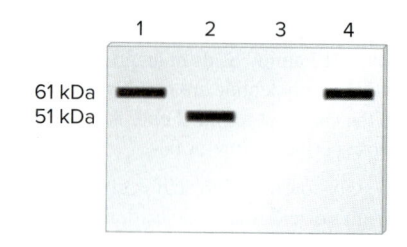

Explain the possible cause of albinism in the three albino individuals.

E6. Although 61 codons specify the 20 amino acids, most species display a codon bias. This means that certain codons are used much more frequently than other codons. For example, UUA, UUG, CUU, CUC, CUA, and CUG all specify leucine. In yeast, however, the UUG codon is used to specify leucine approximately 80% of the time.

A. The experiment of Figure 15.7 shows the use of a cell-free translation system. In this experiment, the RNA that was used for translation was chemically synthesized. Instead of using chemically synthesized RNA, researchers can isolate mRNA from living cells and then add the mRNA to the cell-free translation system. If a researcher isolated mRNA from kangaroo cells and then added it to a cell-free translation system that came from yeast cells, how might the phenomenon of codon bias affect the production of proteins?

B. Discuss potential advantages and disadvantages of codon bias for translation in a given species.

Questions for Student Discussion/Collaboration

1. Discuss why you think the ribosomes need to contain so many proteins and rRNA molecules. Does it seem like a waste of cellular energy to make such a large structure so that translation can occur?
2. Discuss and make a list of the similarities and differences in the events that occur during the initiation, elongation, and termination stages of transcription (see Chapter 14) and translation (Chapter 15).
3. Which events during translation involve molecular recognition between base sequences within different RNAs? Which events involve recognition between different protein molecules?

Answers to Comprehension Questions

15.1: a, c
15.2: a, d, b, c
15.3: c, b
15.4: b, c, b
15.5: d, d
15.6: d, a, c, c

Note: All answers appear in Connect; the answers to even-numbered questions and all Concept Check questions are in Appendix B.

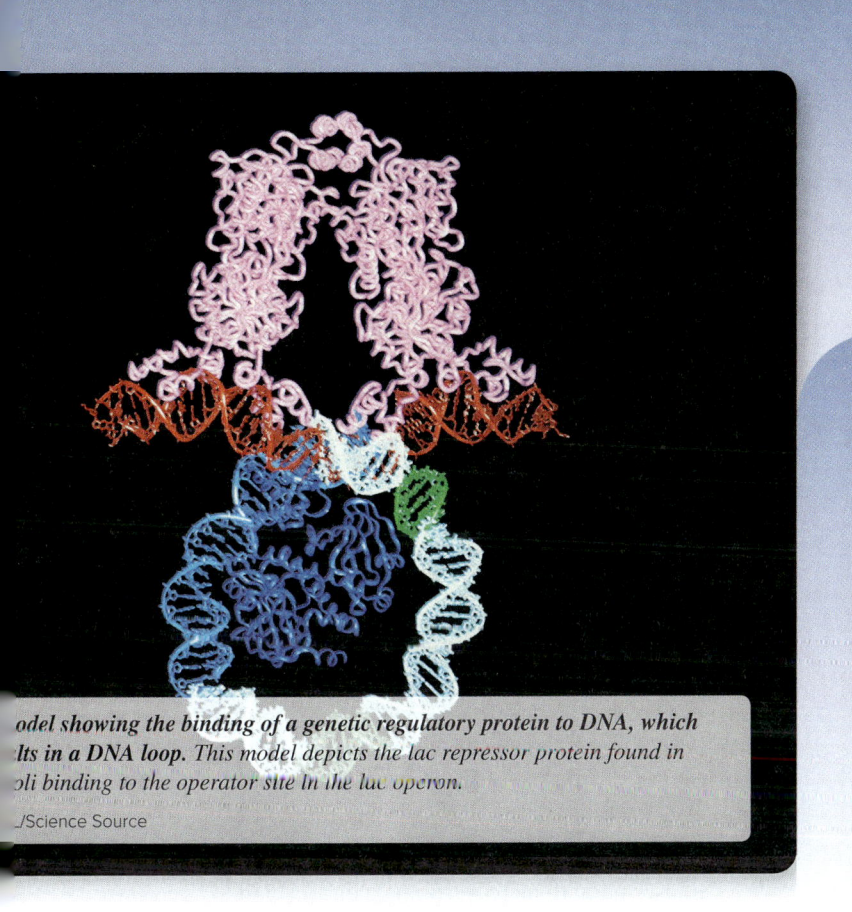

16

CHAPTER OUTLINE

- 16.1 Overview of Transcriptional Regulation
- 16.2 Regulation of the *lac* Operon
- 16.3 Regulation of the *trp* Operon
- 16.4 Translational and Posttranslational Regulation
- 16.5 Riboswitches

odel showing the binding of a genetic regulatory protein to DNA, which lts in a DNA loop. This model depicts the lac repressor protein found in oli binding to the operator site in the lac operon.
/Science Source

GENE REGULATION IN BACTERIA

The expression of protein-encoding genes ultimately leads to the production of functional proteins. As we discussed in Chapters 14 and 15, gene expression is a multistep process that proceeds from transcription to translation, and it may involve posttranslational effects on protein function. **Gene regulation** is the phenomenon in which the level of gene expression can vary under different conditions. In comparison, unregulated genes have essentially constant levels of expression in all conditions over time. Unregulated genes are also called **constitutive genes.** With regard to protein-encoding genes, those that are constitutive encode proteins that are continuously needed for the survival of a bacterium. In contrast, the majority of genes are regulated so that the proteins they encode can be produced at the proper times and in the proper amounts.

A key benefit of gene regulation is that the encoded proteins are produced only when they are required. Therefore, the cell avoids wasting valuable energy making proteins it does not need. From the viewpoint of natural selection, this enables a bacterium to compete as efficiently as possible for limited resources. Gene regulation is particularly important because bacteria exist in an environment that is frequently changing with regard to temperature, nutrients, and many other factors.

The following are a few common processes regulated at the genetic level:

1. *Metabolism:* Some proteins function in the metabolism of small molecules. For example, certain enzymes are needed for a bacterium to metabolize particular sugars. These enzymes are required only when the bacterium is exposed to such sugars in its environment.
2. *Response to environmental stress:* Certain proteins help a bacterium to survive an environmental stress such as osmotic shock or heat shock. These proteins are required only when the bacterium is confronted with the stress.
3. *Cell division:* Some proteins are needed for cell division. These are necessary only when the bacterial cell is getting ready to divide.

In this chapter, we will examine the molecular mechanisms that account for these types of gene regulation. As shown in **Figure 16.1**, gene regulation in bacteria can occur at any step in the pathway of gene expression.

16.1 OVERVIEW OF TRANSCRIPTIONAL REGULATION

Learning Outcomes:
1. Describe the functions of activators and repressors.
2. Explain how small effector molecules affect the functions of activators and repressors.

355

REGULATION OF GENE EXPRESSION

Transcription
- Gene → Genetic regulatory proteins bind to the DNA and control the rate of transcription.
- In attenuation, transcription terminates soon after it has begun due to the formation of a transcriptional terminator.

Translation
- mRNA → Translational repressor proteins can bind to the mRNA and prevent translation from starting.
- Riboswitches can produce an mRNA conformation that prevents translation from starting.
- Antisense RNA can bind to the mRNA and prevent translation from starting.

Posttranslation
- Protein → In feedback inhibition, the product of a metabolic pathway inhibits the first enzyme in the pathway.
- Covalent modifications to the structure of a protein can alter its function.
- Functional protein

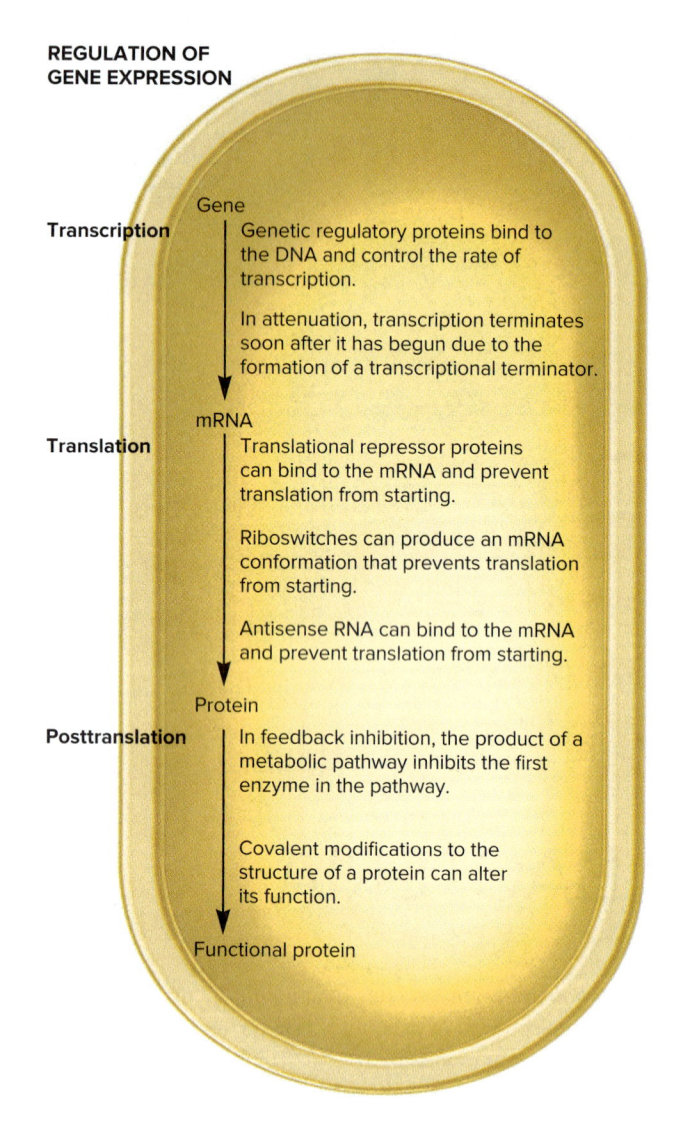

FIGURE 16.1 Common points where regulation of gene expression occurs in bacteria.

Concept Check: What is an advantage of gene regulation?

In bacteria, the most common way to regulate gene expression is by influencing the rate at which transcription is initiated. Although we frequently refer to genes as being "turned on or off," it is more accurate to say that the level of gene expression is increased or decreased. At the level of transcription, this means that the rate of RNA synthesis can be increased or decreased.

In most cases, transcriptional regulation involves the actions of regulatory proteins that can bind the DNA and affect the rate of transcription of one or more nearby genes. Two types of regulatory proteins are common.

- A **repressor** is a regulatory protein that binds to the DNA and inhibits transcription.
- An **activator** is a regulatory protein that increases the rate of transcription.
- Transcriptional regulation by a repressor is termed **negative control.**
- Regulation by an activator is considered to be **positive control.**

In conjunction with regulatory proteins, small effector molecules often play a critical role in transcriptional regulation. However, small effector molecules do not bind directly to the DNA to alter transcription. Rather, an effector molecule exerts its effects by binding to an activator or repressor. The binding of the effector molecule causes a conformational change in the regulatory protein, usually influencing whether or not the protein can bind to the DNA. Genetic regulatory proteins that respond to small effector molecules have two functional domains.

- One domain is a site where the protein binds to the DNA.
- The other domain is the binding site for the effector molecule.

Regulatory proteins are given names describing how they affect transcription when they are bound to the DNA (repressor or activator). In contrast, small effector molecules are given names that describe how they affect transcription when they are present in the cell at a sufficient concentration to exert their effect (**Figure 16.2**).

- An **inducer** is a small effector molecule that causes transcription to increase. Some inducers bind to a repressor protein and prevent it from binding to the DNA, whereas others bind to an activator protein and cause it to bind to the DNA. In either case, the transcription rate is increased.
- Genes that are regulated by inducers are called **inducible genes.**
- A **corepressor** is a small molecule that binds to a repressor protein, thereby causing the protein to bind to the DNA.
- An **inhibitor** binds to an activator protein and prevents it from binding to the DNA.
- Genes that are regulated by corepressors or inhibitors are termed **repressible genes.**

16.1 REVIEWING THE KEY CONCEPTS

- Repressors and activators are regulatory proteins that exert negative control or positive control, respectively. Small effector molecules that control the functions of repressors and activators include inducers, inhibitors, and corepressors (see Figure 16.2).

16.1 COMPREHENSION QUESTIONS

1. A repressor is a _____ that _____ transcription.
   a. small effector molecule, inhibits
   b. small effector molecule, enhances
   c. regulatory protein, inhibits
   d. regulatory protein, enhances

2. Which of the following combinations would cause transcription to increase?
   a. A repressor plus an inducer
   b. A repressor plus a corepressor
   c. An activator plus an inhibitor
   d. None of the above would increase transcription.

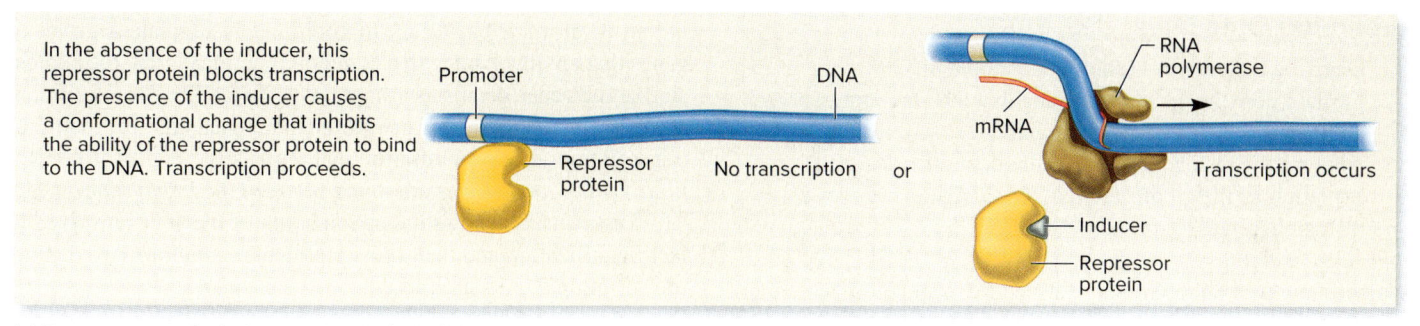

16.1 OVERVIEW OF TRANSCRIPTIONAL REGULATION   357

(a) **Repressor protein, inducer molecule, inducible gene**

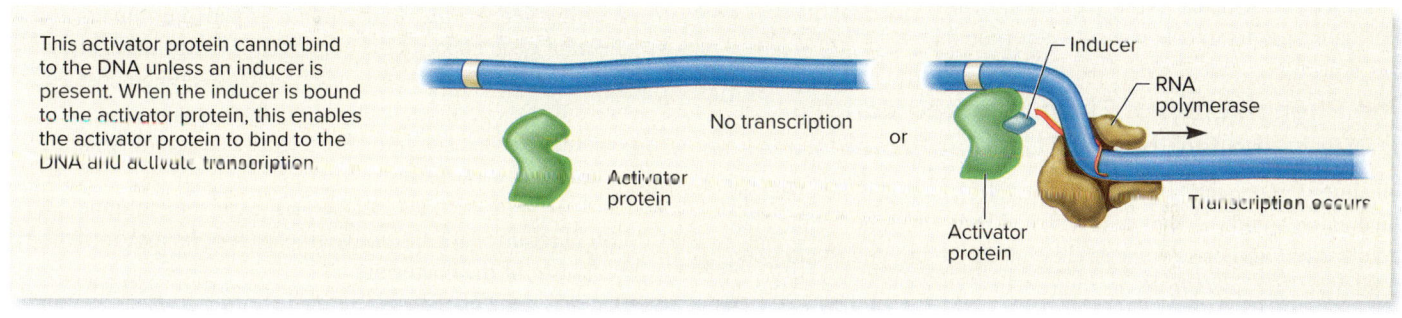

In the absence of the inducer, this repressor protein blocks transcription. The presence of the inducer causes a conformational change that inhibits the ability of the repressor protein to bind to the DNA. Transcription proceeds.

(b) **Activator protein, inducer molecule, inducible gene**

This activator protein cannot bind to the DNA unless an inducer is present. When the inducer is bound to the activator protein, this enables the activator protein to bind to the DNA and activate transcription.

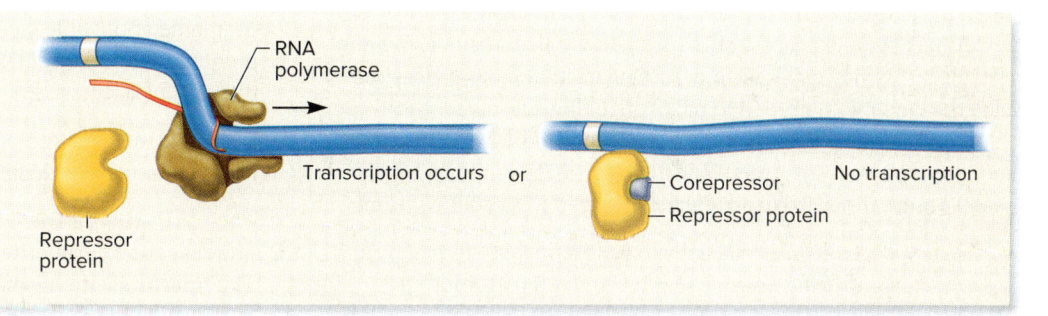

(c) **Repressor protein, corepressor molecule, repressible gene**

In the absence of a corepressor, this repressor protein will not bind to the DNA. Therefore, transcription can occur. When the corepressor is bound to the repressor protein, a conformational change occurs that allows the repressor to bind to the DNA and inhibit transcription.

(d) **Activator protein, inhibitor molecule, repressible gene**

This activator protein will bind to the DNA without the aid of an effector molecule. The presence of an inhibitor causes a conformational change that inhibits the ability of the activator protein to bind to the DNA. This inhibits transcription.

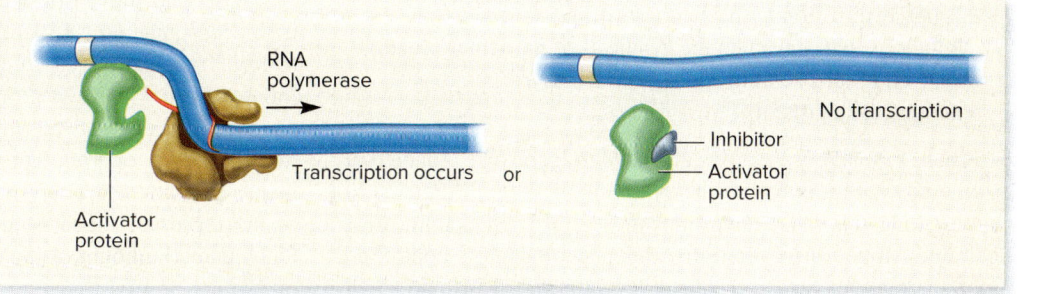

FIGURE 16.2 **Binding sites on a genetic regulatory protein.** In these examples, a regulatory protein has two binding sites: one for a small effector molecule and one for DNA. The binding of the small effector molecule changes the conformation of the regulatory protein, which alters the structure of the DNA-binding site, thereby influencing whether the protein can bind to the DNA.

Concept Check: *Which of these are genetic regulatory proteins and which are small effector molecules?*

16.2 REGULATION OF THE *lac* OPERON

Learning Outcomes:
1. Describe the organization of the *lac* operon.
2. Explain how the *lac* operon is regulated by lac repressor and by catabolite activator protein.
3. Analyze the results of Jacob, Monod, and Pardee, and explain how they indicated that the *lacI* gene encodes a diffusible repressor protein.

We will now turn our attention to a specific example of gene regulation that is found in *E. coli*. This example involves genes that play a role in the utilization of lactose, which is a sugar found in milk. As you will learn, the regulation of these genes involves a repressor protein and also an activator protein.

The Phenomenon of Enzyme Adaptation Is Due to the Synthesis of Cellular Proteins

Our initial understanding of gene regulation can be traced back to the creative minds of François Jacob and Jacques Monod at the Pasteur Institute in Paris, France. Their research into genes and gene regulation stemmed from an interest in the phenomenon known as **enzyme adaptation**, which had been identified at the turn of the twentieth century. Enzyme adaptation refers to the observation that a particular enzyme appears within a living cell only after the cell has been exposed to the substrate for that enzyme. When a bacterium is not exposed to a particular substance, it does not make the enzyme(s) needed to metabolize that substance.

To investigate this phenomenon, Jacob and Monod focused their attention on lactose metabolism in *E. coli*. Several key experimental observations led to an understanding of this genetic system:

- The exposure of bacterial cells to lactose increased the levels of lactose-utilizing enzymes by 1000- to 10,000-fold.
- The activity of these enzymes, which are proteins, was due to the increased synthesis of the enzymes in the cell.
- The removal of lactose from the environment caused an abrupt termination in the synthesis of the enzymes.
- Mutations that prevented the synthesis of particular proteins involved in lactose utilization showed that a separate gene encoded each protein.

These critical observations indicated to Jacob and Monod that enzyme adaptation is due to the synthesis of specific proteins in response to lactose in the environment. Next, we will examine how Jacob and Monod discovered that this phenomenon is controlled by the interactions between genetic regulatory proteins and small effector molecules. In other words, we will see that enzyme adaptation is due to the transcriptional regulation of genes.

The *lac* Operon Encodes Proteins Involved in Lactose Metabolism

In bacteria, it is common for a few genes to be arranged together in an **operon**—a group of two or more genes that are transcribed from a single promoter. An operon encodes a **polycistronic mRNA**, an mRNA that contains the sequences of two or more genes. Why do operons occur in bacteria? One biological advantage of an operon is that it allows a bacterium to coordinately regulate a group of genes that are involved with a common functional goal; the expression of the genes occurs as a single unit. To facilitate transcription, an operon is flanked by a **promoter** that signals the beginning of transcription and a **terminator** that signals the end of transcription. Two or more genes are found between these two sequences.

Figure 16.3a shows the organization of the genes involved in lactose utilization and their transcriptional regulation. Two distinct transcriptional units are present.

- One of these units known as the *lac* operon, contains a CAP site; promoter (*lacP*); operator site (*lacO*); three protein-encoding genes, *lacZ*, *lacY*, and *lacA*; and a terminator.
- The *lacZ* gene encodes the enzyme β-galactosidase, an enzyme that cleaves lactose into galactose and glucose. As a side reaction, β-galactosidase also converts a small percentage of lactose into allolactose, a structurally similar sugar (**Figure 16.3b**).
- The *lacY* gene encodes lactose permease, a membrane protein required for the active transport of lactose into the cytoplasm of the bacterium.
- The *lacA* gene encodes galactoside transacetylase, an enzyme that covalently modifies lactose and lactose analogs by attaching hydrophobic acetyl groups. The acetylation of nonmetabolizable lactose analogs prevents their toxic buildup within the bacterial cytoplasm by allowing them to diffuse out of the cell.

The CAP site and the operator site are short DNA segments that function in gene regulation.

- The **CAP site** is a DNA sequence recognized by an activator protein called **catabolite activator protein (CAP)**.
- The **operator site** (also known simply as the **operator**) is a sequence of bases that provides a binding site for a repressor protein.

A second transcriptional unit involved in genetic regulation is the *lacI* gene (see Figure 16.3a), which is not part of the *lac* operon. The *lacI* gene, which is constitutively expressed at fairly low levels, has its own promoter, the *i* promoter. The *lacI* gene encodes **lac repressor,** a protein that is important for the regulation of the *lac* operon. This repressor functions as a homotetramer, a protein composed of four identical subunits. The amount of lac repressor made is approximately 10 molecules per cell. Only a small amount of lac repressor is needed to repress the *lac* operon.

The *lac* Operon Is Regulated by a Repressor Protein

The *lac* operon can be transcriptionally regulated in more than one way. The first mechanism we will examine is one that is inducible and under negative control. As shown in **Figure 16.4**, this form of regulation involves lac repressor protein, which binds to the

16.2 REGULATION OF THE *lac* OPERON 359

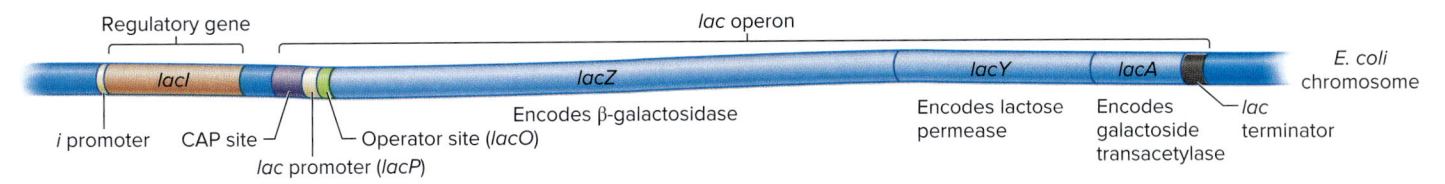

(a) Organization of DNA sequences in the *lac* region of the *E. coli* chromosome

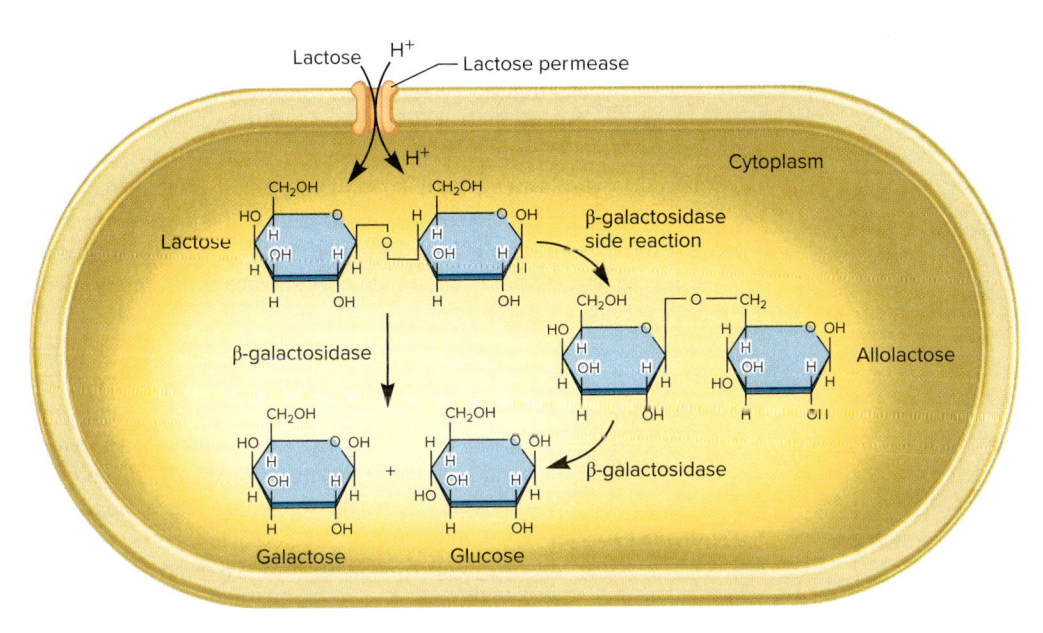

(b) Functions of lactose permease and β-galactosidase

FIGURE 16.3 Organization of the *lac* operon and other genes involved with lactose metabolism in *E. coli*. (a) The CAP site is the binding site for the catabolite activator protein (CAP). The operator site is a binding site for lac repressor. The promoter (*lacP*) is responsible for the transcription of the *lacZ*, *lacY*, and *lacA* genes as a single unit, which ends at the *lac* terminator. The *i* promoter is responsible for the transcription of the *lacI* gene. (b) Lactose permease allows the uptake of lactose into the bacterial cytoplasm. It cotransports lactose with H⁺. Because bacteria maintain an H⁺ gradient across their cytoplasmic membrane, this cotransport permits the active accumulation of lactose against a gradient. β-galactosidase is a cytoplasmic enzyme that cleaves lactose and related compounds into galactose and glucose. As a minor side reaction, β-galactosidase also converts lactose into allolactose. Allolactose can also be broken down into galactose and glucose.

Concept Check: Which genes are under the control of the lac promoter?

sequence of nucleotides found within the *lac* operator site. Once bound, lac repressor prevents RNA polymerase from transcribing the *lacZ*, *lacY*, and *lacA* genes (Figure 16.4a). The binding of the repressor to the operator site is a reversible process. In the absence of allolactose, lac repressor is bound to the operator site most of the time.

The ability of lac repressor to bind to the operator site depends on whether or not allolactose is bound to it. Each of the repressor protein's four subunits has a single binding site for allolactose, the inducer. How does a small molecule like allolactose exert its effects? When allolactose binds to the repressor, a conformational change occurs in lac repressor that prevents it from binding to the operator site. Under these conditions, RNA polymerase is free to transcribe the operon (Figure 16.4b). In genetic terms, we say that the operon has been **induced.** The action of a small effector molecule, such as allolactose, is called **allosteric regulation.** Allosteric proteins have at least two binding sites.

The effector molecule binds to the protein's **allosteric site,** which is a site other than the protein's active site. In the case of lac repressor, the active site is the part of the protein that binds to the DNA.

Rare mutations in the *lacI* gene that alter the regulation of the *lac* operon reveal that lac repressor is composed of a protein domain that binds to the DNA and another domain that contains the allolactose-binding site. As shown later in Figure 16.7, researchers have identified *lacI⁻* mutations that result in the constitutive expression of the *lac* operon, which means that it is expressed in the presence and absence of lactose. Such mutations may result in an inability to synthesize any repressor protein or they may produce a repressor protein that is unable to bind to the DNA at the *lac* operator site. If lac repressor is unable to bind to the DNA, the *lac* operon cannot be repressed. By comparison, *lacI*ˢ mutations have the opposite effect—the *lac* operon cannot be induced even in the presence of lactose. These mutations, which are called

360 CHAPTER 16 :: GENE REGULATION IN BACTERIA

In the absence of the inducer allolactose, the repressor protein is tightly bound to the operator site, thereby inhibiting the ability of RNA polymerase to transcribe the operon.

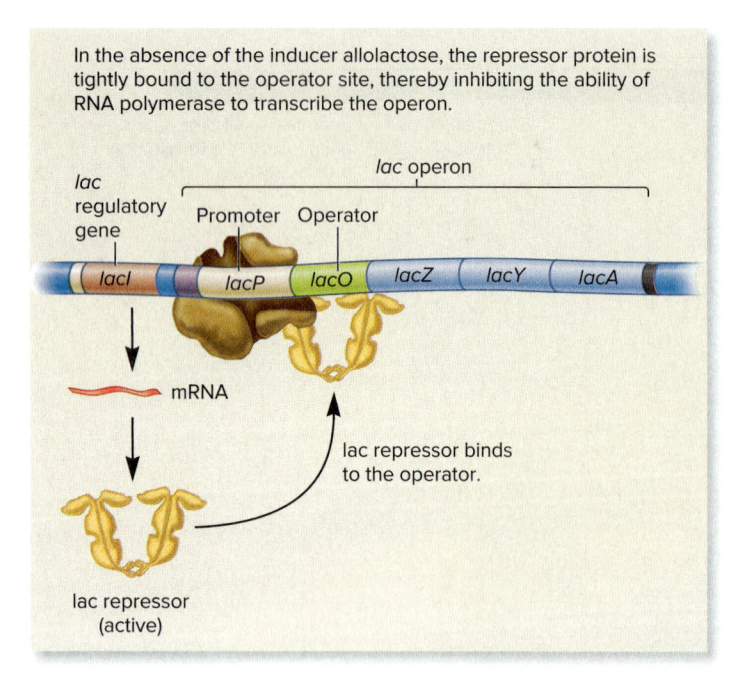

[Diagram showing lac regulatory gene (lacI), Promoter (lacP), Operator (lacO), lacZ, lacY, lacA, with mRNA and lac repressor (active) binding to the operator.]

(a) No lactose in the environment

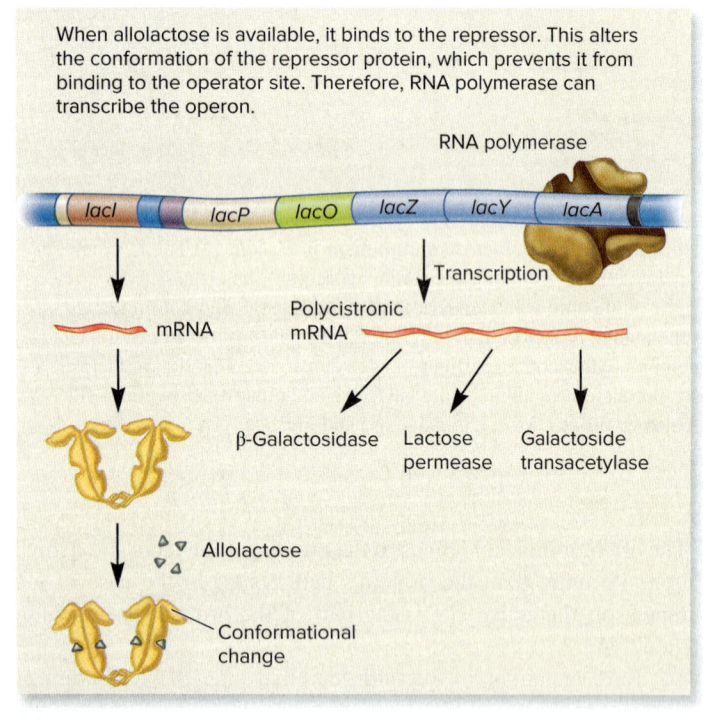

When allolactose is available, it binds to the repressor. This alters the conformation of the repressor protein, which prevents it from binding to the operator site. Therefore, RNA polymerase can transcribe the operon.

[Diagram showing RNA polymerase, lacI, lacP, lacO, lacZ, lacY, lacA, Transcription, Polycistronic mRNA, β-Galactosidase, Lactose permease, Galactoside transacetylase, Allolactose, Conformational change.]

(b) Lactose present

FIGURE 16.4 **Mechanism of induction of the *lac* operon.** Note: The CAP site is not labeled in these diagrams, but it is shown in purple.

The Regulation of the *lac* Operon Allows a Bacterium to Respond to Environmental Change

To better appreciate *lac* operon regulation at the cellular level, let's consider the process as it occurs over time. **Figure 16.5** illustrates the effects of external lactose on the regulation of the *lac* operon. In the absence of lactose, no inducer is available to bind to lac repressor. Therefore, lac repressor binds to the operator site and inhibits transcription. In reality, the repressor does not completely inhibit transcription, so very small amounts of β-galactosidase, lactose permease, and transacetylase are made. However, the levels are far too low to enable the bacterium to readily use lactose. When the bacterium is exposed to lactose, a small amount can be transported into the cytoplasm via lactose permease, and β-galactosidase converts some of that lactose to allolactose. As this occurs, the cytoplasmic level of allolactose gradually rises; eventually, allolactose binds to lac repressor. The binding of allolactose promotes a conformational change that prevents the repressor from binding to the *lac* operator site, thereby allowing transcription of the *lacZ*, *lacY*, and *lacA* genes to occur. Translation of the encoded polypeptides produces the proteins needed for lactose uptake and metabolism.

To understand how the induction process is shut off in a lactose-depleted environment, let's consider the interaction between allolactose and lac repressor. The repressor protein has a measurable affinity for allolactose. The binding of allolactose to lac repressor is reversible. The likelihood that allolactose will bind to the repressor depends on the allolactose concentration. During induction of the operon, the concentration of allolactose rises and approaches the affinity for the repressor protein. This makes it likely that allolactose will bind to lac repressor, thereby causing the repressor to be released from the operator site.

Later on, however, the bacterial cell metabolizes the sugars, thereby lowering the concentration of allolactose below its affinity for the repressor. At this point, allolactose is unlikely to be bound to lac repressor. When allolactose is released, lac repressor returns to the conformation that binds to the operator site. In this way, the binding of the repressor shuts down the *lac* operon when lactose is depleted from the environment. The mRNA and proteins encoded by the *lac* operon are eventually degraded (see Figure 16.5).

The *lacI* Gene Encodes a Diffusible Repressor Protein

Now that we have examined the function of the *lac* operon, let's consider one of the experimental approaches that was used to elucidate its regulation. In the 1950s, Jacob, Monod, and their colleague Arthur Pardee identified a few rare mutant strains of bacteria that had abnormal lactose adaptation. As mentioned earlier, one type of mutant, designated *lacI*⁻, resulted in the constitutive expression of the *lac* operon even in the absence of lactose. As shown in **Figure 16.6a**, the correct explanation is that a loss-of-function mutation in the *lacI* gene prevented lac repressor from binding to the *lac* operator site and inhibiting transcription. At the time of this work, however, the function of lac repressor was not yet known. Instead, the researchers incorrectly hypothesized that the *lacI*⁻ mutation resulted in the synthesis of an internal inducer,

super-repressor mutations, typically occur in the domain that binds allolactose. The mutation usually results in a lac repressor protein that cannot bind allolactose. If lac repressor is unable to bind allolactose, it will remain bound to the *lac* operator site and therefore induction cannot occur.

16.2 REGULATION OF THE *lac* OPERON   361

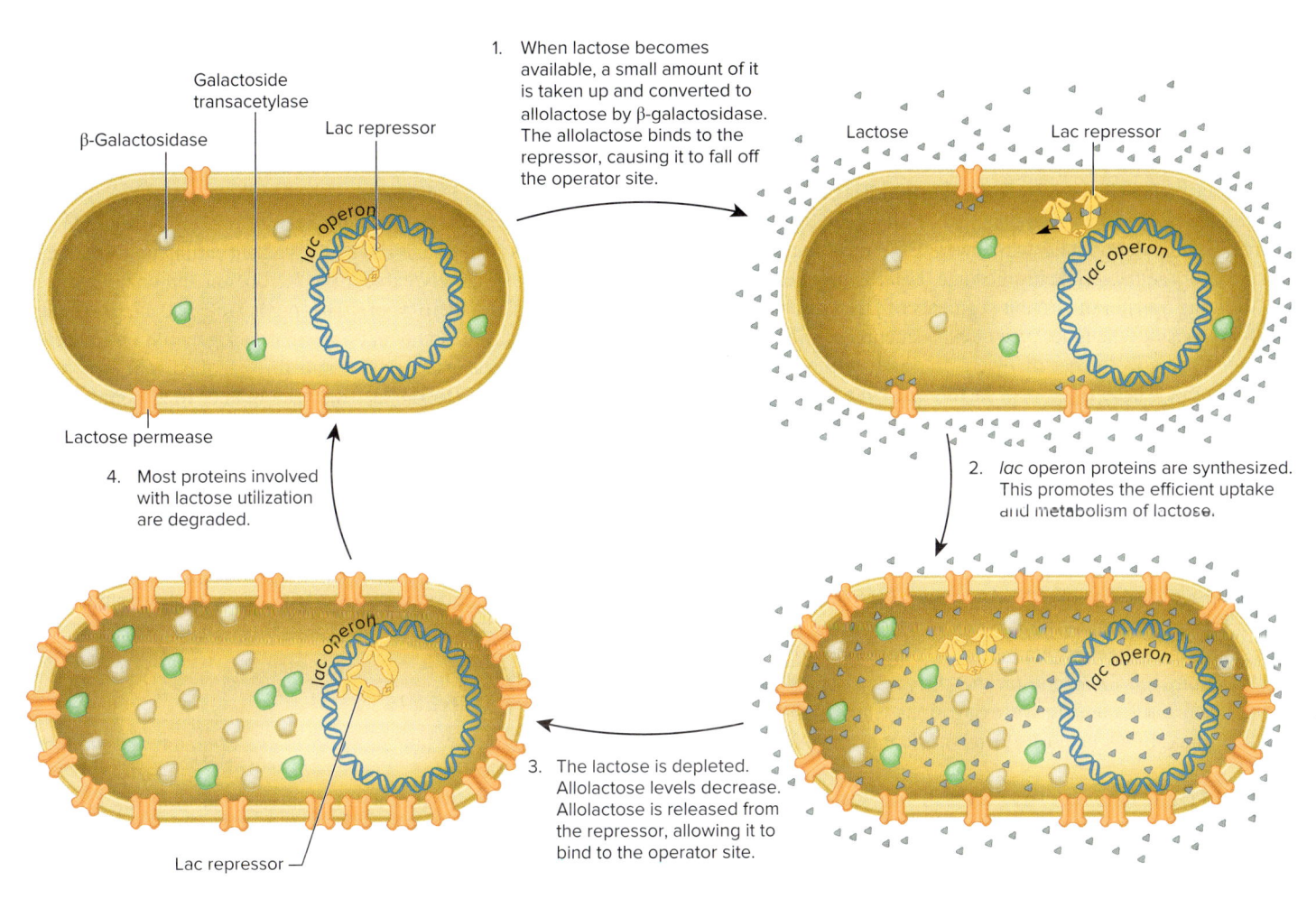

FIGURE 16.5   **The cycle of *lac* operon induction and repression.**

1. When lactose becomes available, a small amount of it is taken up and converted to allolactose by β-galactosidase. The allolactose binds to the repressor, causing it to fall off the operator site.

2. *lac* operon proteins are synthesized. This promotes the efficient uptake and metabolism of lactose.

3. The lactose is depleted. Allolactose levels decrease. Allolactose is released from the repressor, allowing it to bind to the operator site.

4. Most proteins involved with lactose utilization are degraded.

Genes → Traits   The genes of the *lac* operon provide the bacterium with the trait of being able to metabolize lactose in the environment. When lactose is present, the genes of the *lac* operon are induced, and the bacterial cell can efficiently metabolize this sugar. When lactose is absent, these genes are repressed, so the bacterium does not waste its energy expressing them. Note: The proteins involved with lactose utilization are fairly stable, but they will eventually be degraded.

Concept Check: Under what conditions is lac repressor bound to the lac operon?

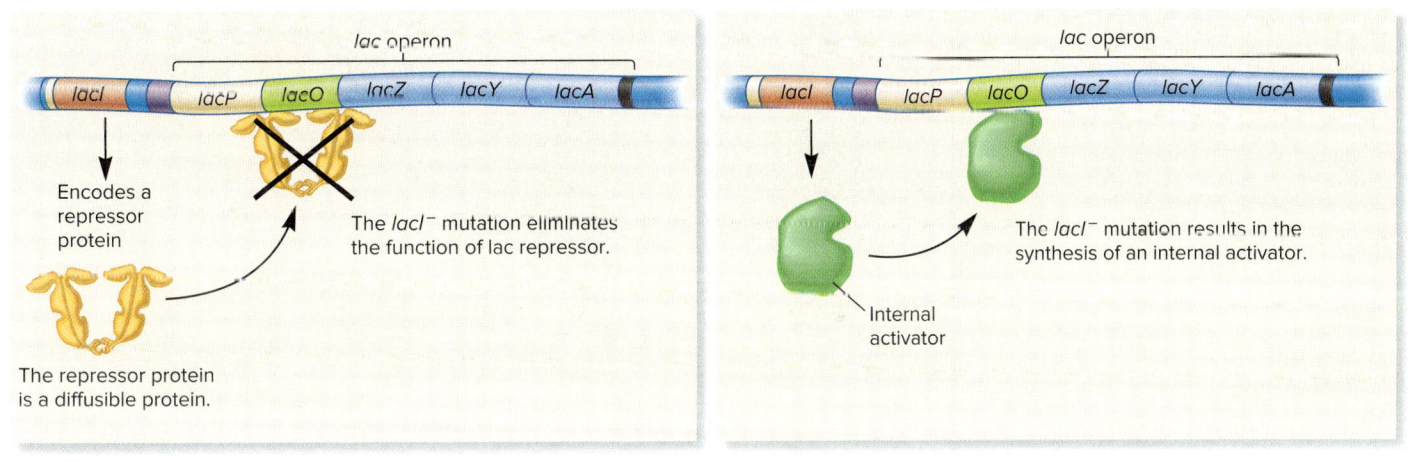

(a) Correct explanation

(b) Internal activator hypothesis

FIGURE 16.6   **Alternative hypotheses to explain how a *lacI⁻* mutation could cause the constitutive expression of the *lac* operon.**
(a) The correct explanation in which the *lacI⁻* mutation eliminates the function of lac repressor, which prevents it from repressing the *lac* operon. (b) The internal inducer hypothesis of Jacob, Monod, and Pardee. In this case, the *lacI⁻* mutation results in the synthesis of an internal inducer that turns on the *lac* operon.

making it unnecessary for cells to be exposed to lactose for induction (**Figure 16.6b**).

To further explore the nature of this mutation, Jacob, Monod, and Pardee applied a genetic approach. In order to understand their approach, let's briefly consider the process of bacterial conjugation (described in Chapter 9). The earliest studies of Jacob, Monod, and Pardee in 1959 involved conjugations between recipient cells, termed F⁻, and donor cells, which were Hfr strains that transferred a portion of the bacterial chromosome. Later experiments in 1961 involved the transfer of circular segments of DNA known as F factors. We consider the latter type of experiment here. Sometimes an F factor also carries genes that were originally within the bacterial chromosome. These types of F factors are called F′ factors (read as "F prime factors"). In their studies, Jacob, Monod, and Pardee identified F′ factors that carried the *lacI* gene and the *lac* operon. These F′ factors can be transferred from one cell to another by bacterial conjugation. A strain of bacteria containing F′ factor genes is called a **merozygote,** or partial diploid.

The production of merozygotes was instrumental in allowing Jacob, Monod, and Pardee to determine the function of the *lacI* gene. This experimental approach was based on two key points. First, the two *lacI* genes in a merozygote may be different alleles. For example, the *lacI* gene on the chromosome may be a *lacI*⁻ allele that causes constitutive expression, whereas the *lacI* gene on the F′ factor may be normal (*lac*⁺). Second, the genes on the F′ factor and the genes on the bacterial chromosome are not physically adjacent to each other. As we now know, the expression of the *lacI* gene on an F′ factor should produce repressor proteins that can diffuse within the cell and eventually bind to the operator site of the *lac* operon located on the chromosome and also to the operator site on an F′ factor.

Figure 16.7 shows one experiment of Jacob, Monod, and Pardee in which they analyzed a *lacI*⁻ mutant strain that was already known to constitutively express the *lac* operon and compared it to the corresponding merozygote. The merozygote had a *lacI*⁻ mutant gene on the chromosome and a normal *lacI* gene on an F′ factor. These two strains were grown and then divided into two tubes each. In half of the tubes, lactose was omitted. In the other tubes, the strains were incubated with lactose to determine if lactose was needed to induce the expression of the operon. The cells were subjected to sonication, which caused them to release β-galactosidase. Next, β-*o*-nitrophenylgalactoside (β-ONPG) was added. This molecule is colorless, but β-galactosidase cleaves it into a product that has a yellow color. Therefore, the amount of yellow color produced in a given amount of time is a measure of the amount of β-galactosidase that is being expressed from the *lac* operon.

▶ **THE HYPOTHESIS**

The *lacI*⁻ mutation results in the synthesis of an internal inducer.

▶ **TESTING THE HYPOTHESIS** — **FIGURE 16.7**   Evidence that the *lacI* gene encodes a diffusible repressor protein.

Starting material: The genotype of the mutant strain was *lacI*⁻ *lacZ*⁺ *lacY*⁺ *lacA*⁺. The merozygote strain had an F′ factor that was *lacI*⁺ *lacZ*⁺ *lacY*⁺ *lacA*⁺, which had been introduced into the mutant strain via conjugation.

1. Grow mutant strain and merozygote strain separately.

2. Divide each strain into two tubes.

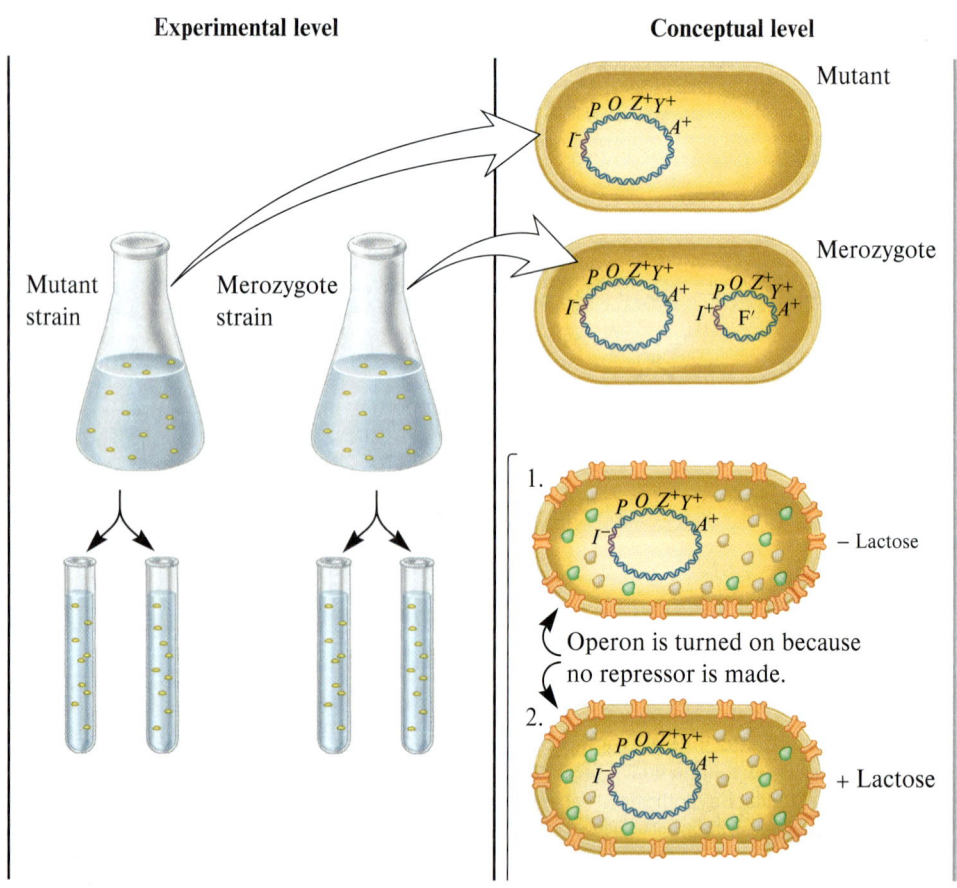

16.2 REGULATION OF THE *lac* OPERON 363

3. To one of the two tubes, add lactose.

4. Incubate the cells long enough to allow *lac* operon induction.

5. Lyse the cells with a sonicator. This allows β-galactosidase to escape from the cells.

6. Add β-*o*-nitrophenylgalactoside (β-ONPG). This is a colorless compound. If β-galactosidase is present, it will cleave the compound to produce galactose and *o*-nitrophenol (O-NP). *o*-Nitrophenol has a yellow color. The deeper the yellow color, the more β-galactosidase was produced.

7. Incubate the sonicated cells to allow β-galactosidase time to cleave β-*o*-nitrophenylgalactoside.

8. Measure the yellow color produced with a spectrophotometer. (See Appendix A for a description of spectrophotometry.)

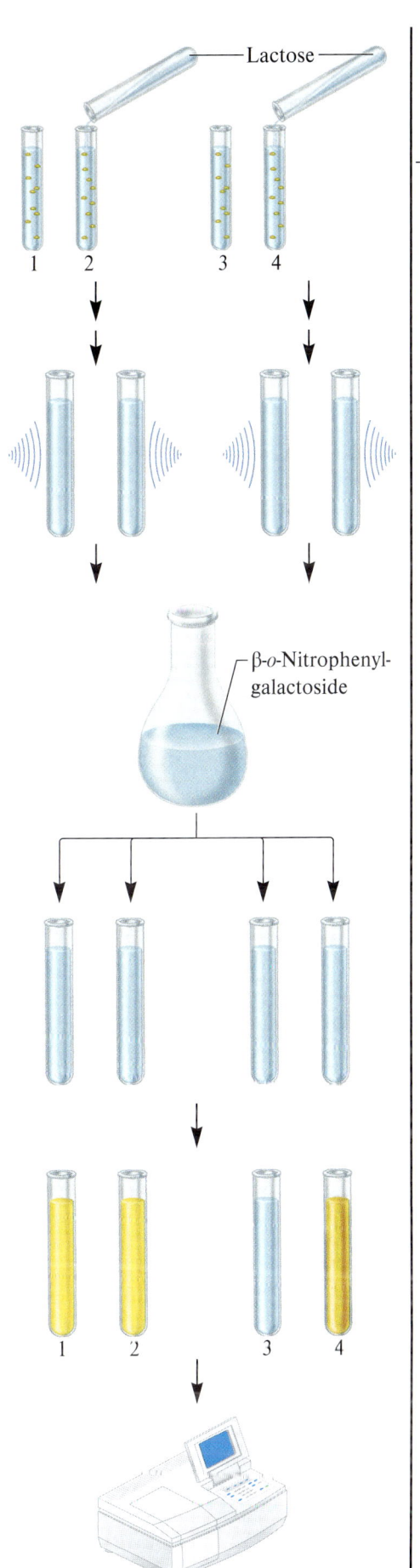

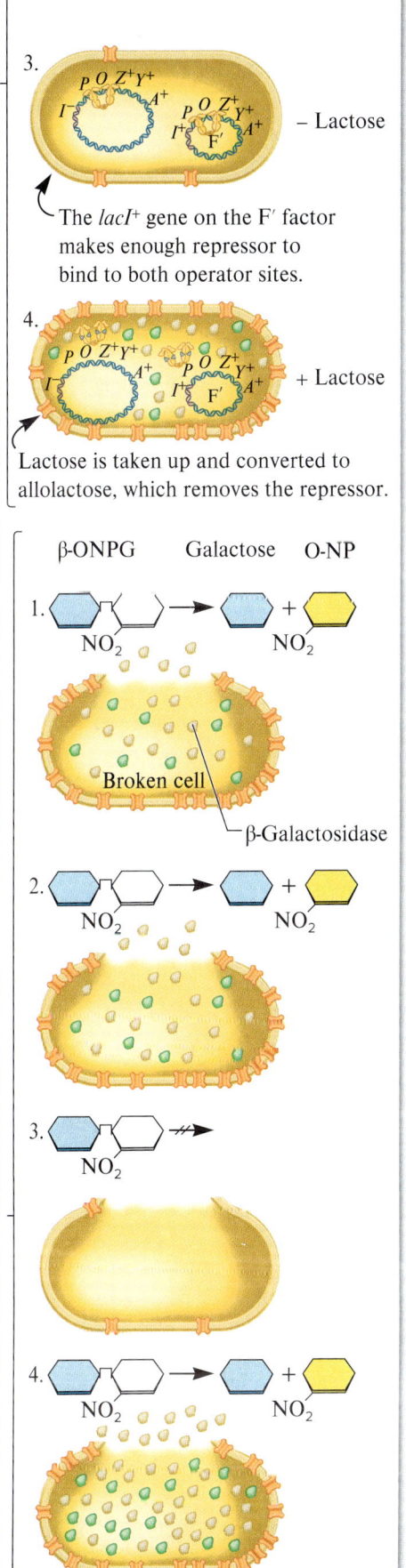

THE DATA

Strain	Addition of Lactose	Amount of β-Galactosidase (% of parent strain)
Mutant	No	100%
Mutant	Yes	100%
Merozygote	No	<1%
Merozygote	Yes	220%

Source: Data from Jacob, F. & Monod, J. (1961) Genetic Regulatory Mechanisms in the Synthesis of Proteins, *Journal of Molecular Biology*, vol. 3, 318-356.

INTERPRETING THE DATA

As seen in the data, the amount of β-galactosidase produced by the original mutant strain was the same in the presence or absence of lactose. This result is expected because the expression of β-galactosidase in the *lacI*⁻ mutant strain was already known to be constitutive. In other words, the presence of lactose was not needed to induce the operon due to a defective *lacI* gene. With the merozygote strain, however, a different result was obtained. In the absence of lactose, the *lac* operons were repressed—even the operon on the bacterial chromosome. How do we explain these results? Because the normal *lacI* gene on the F′ factor was not physically located next to the chromosomal *lac* operon, this result is consistent with the idea that the *lacI* gene codes for a repressor protein that can diffuse throughout the cell and bind to any *lac* operon. The hypothesis that the *lacI*⁻ mutation resulted in the synthesis of an internal inducer was rejected. If that hypothesis had been correct, the merozygote strain would have still made an internal inducer, and the *lac* operons in the merozygote would have been expressed in the absence of lactose. This result was not obtained.

The interactions between regulatory proteins and DNA sequences observed in this experiment led to the definition of two genetic terms. A **trans-effect** is a form of genetic regulation that can occur even though two DNA segments are not physically adjacent. The action of lac repressor on the *lac* operon is a *trans*-effect. A regulatory protein, such as lac repressor, is called a **trans-acting factor.** In contrast, a **cis-acting element** is a DNA segment that must be adjacent to the gene(s) that it regulates, and it is said to have a **cis-effect** on gene expression. The *lac* operator site is an example of a *cis*-acting element. A *trans*-effect is mediated by genes that encode regulatory proteins, whereas a *cis*-effect is mediated by DNA sequences that are bound by regulatory proteins.

Jacob and Monod also isolated constitutive mutants that affected the operator site, *lacO*. **Table 16.1** summarizes the effects of mutations based on their location in the *lacI* regulatory gene or in *lacO* and their analysis in merozygotes. As seen here, a loss-of-function mutation in a gene encoding a repressor protein has the same effect as a mutation in an operator site that cannot bind a repressor protein. In both cases, the genes of the *lac* operon are constitutively expressed. In a merozygote, however, the results are quite different. When a normal *lacI* gene and a normal *lac* operon are introduced into a cell harboring a defective *lacI* gene, the normal *lacI* gene can regulate both operons. In contrast, when a *lac* operon with a normal operator site is introduced into a cell with a defective operator site, the operon with the defective operator site continues to be expressed without lactose present. Overall, a mutation in a *trans*-acting factor can be complemented by the introduction of a second gene with normal function. However, a mutation in a *cis*-acting element is not affected by the introduction of another *cis*-acting element with normal function into the cell.

TABLE 16.1
A Comparison of Loss-of-Function Mutations in the *lacI* Gene or in the Operator Site

Chromosome	F′ factor	Expression of the *lac* Operon (%) With Lactose	Without Lactose
Wild type	None	100	<1
lacI⁻	None	100	100
lacO⁻	None	100	100
lacI⁻	*lacI*⁺ and a normal *lac* operon	200	<1
lacO⁻	*lacI*⁺ and a normal *lac* operon	200	100

The *lac* Operon Is Also Regulated by an Activator Protein

The *lac* operon can be transcriptionally regulated in a second way, known as **catabolite repression.** (As we will see, this is a somewhat imprecise term.) This form of transcriptional regulation is influenced by the presence of glucose, which is a catabolite—a substance that is broken down inside the cell. The presence of glucose ultimately leads to repression of the *lac* operon. When exposed to both glucose and lactose, *E. coli* cells first use glucose, and catabolite repression prevents the use of lactose. Why is this an advantage? The explanation is efficiency. The bacterium does not have to express all of the genes necessary for both glucose and lactose metabolism. If the glucose is used up, catabolite repression is alleviated, and the bacterium then expresses the *lac* operon. The sequential use of two sugars by a bacterium, which is called **diauxic growth,** is a common phenomenon among many bacterial species. Typically, glucose, a more commonly encountered sugar, is metabolized preferentially, and then a second sugar is metabolized only after glucose has been depleted from the environment.

Glucose, however, is not itself the small effector molecule that binds directly to a genetic regulatory protein. Instead, this form of regulation involves a small effector molecule, **cyclic-AMP (cAMP),** which is produced from ATP via an enzyme known as adenylyl cyclase. When a bacterium is exposed to glucose, the transport of glucose into the cell stimulates a signaling pathway that causes the intracellular concentration of cAMP to decrease because the pathway inhibits adenylyl cyclase, the enzyme needed for cAMP synthesis. The effect of cAMP on the *lac* operon is mediated by the activator protein called catabolite activator protein (CAP). CAP is composed of two subunits, each of which binds one molecule of cAMP.

Figure 16.8 considers how the interplay between lac repressor and CAP determines whether the *lac* operon is expressed in the presence or absence of lactose and/or glucose.

- When only lactose is present, cAMP levels are high (Figure 16.8a). The cAMP binds to CAP, and then CAP binds to the CAP site. A domain in CAP interacts with RNA polymerase, which facilitates the binding of RNA polymerase to the promoter. In the presence of lactose, lac repressor is not bound to the operator site, so transcription can proceed at a high rate.
- In the absence of both lactose and glucose, cAMP levels are also high (Figure 16.8b). Under these conditions, however, the binding of lac repressor inhibits transcription even though CAP is bound to the DNA. Therefore, the transcription rate is very low.
- Figure 16.8c considers the situation in which both sugars are present. The presence of lactose causes lac repressor to be inactive, which prevents it from binding to the operator site. Even so, the presence of glucose decreases cAMP levels so that cAMP is released from CAP, which prevents CAP from binding to the CAP site. Because CAP is not bound to the CAP site, the transcription of the *lac* operon is low in the presence of both sugars.
- Figure 16.8d illustrates what happens when only glucose is present. The transcription of the *lac* operon is very low because lac repressor is bound to the operator site and CAP is not bound to the CAP site due to low cAMP levels.

The effect of glucose, called catabolite repression, may seem like a puzzling way to describe this process because this regulation involves the action of an inducer (cAMP) and an activator protein (CAP), not a repressor. The term was coined before the action of the cAMP-CAP complex was understood. At that time, the primary observation was that glucose (a catabolite) inhibited (repressed) lactose metabolism.

Further Studies Have Revealed That the *lac* Operon Has Three Operator Sites for lac Repressor

Our traditional view of the regulation of the *lac* operon has been modified as we gain a greater understanding of the molecular process. In particular, detailed genetic and crystallographic studies have shown that the binding of lac repressor is more complex than originally realized. The site in the *lac* operon commonly called the operator site was first identified by mutations that prevented lac repressor from binding. These mutations, called *lacO*− or *lacO*^C mutations, resulted in the constitutive expression of the *lac* operon even in strains that make a normal lac repressor protein.

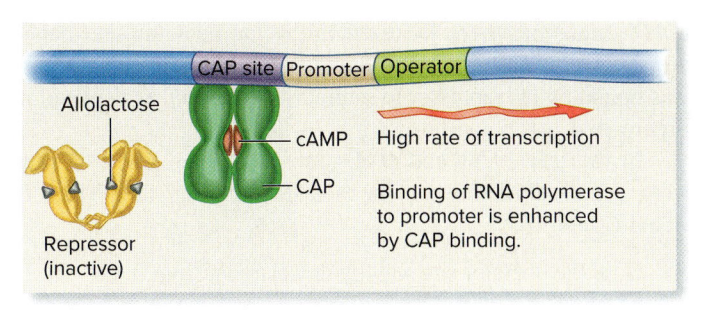

(a) Lactose, no glucose (high cAMP)

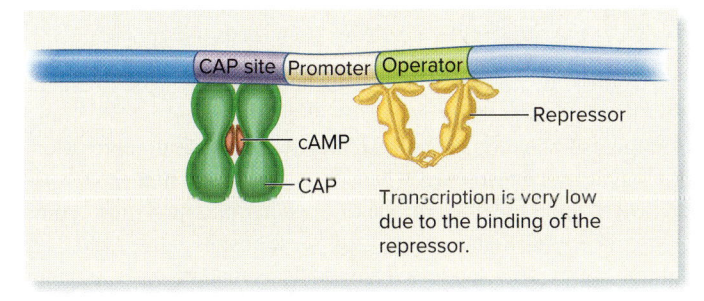

(b) No lactose or glucose (high cAMP)

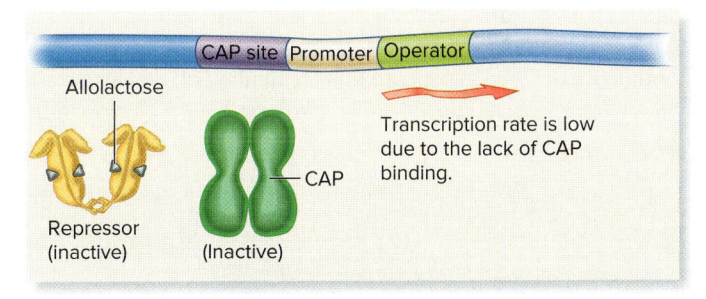

(c) Lactose and glucose (low cAMP)

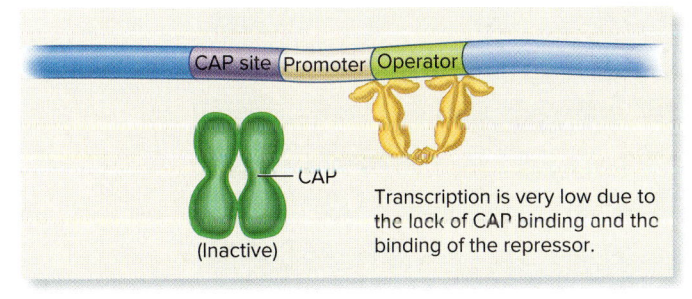

(d) Glucose, no lactose (low cAMP)

FIGURE 16.8 The roles of lac repressor and catabolite activator protein (CAP) in the regulation of the *lac* operon. This figure illustrates how the *lac* operon is regulated depending on its exposure to lactose or glucose.

Genes → Traits The mechanism of catabolite repression provides the bacterium with the trait of being able to choose between two sugars. When exposed to both glucose and lactose, the bacterium metabolizes glucose first. After the glucose is used up, the bacterium expresses the genes necessary for lactose metabolism. This trait allows the bacterium to more efficiently use sugars from its environment.

Concept Check: Why is it beneficial to the bacterium to regulate the lac operon with both a repressor protein and an activator protein?

$LacO^C$ mutations were localized in the *lac* operator site that is now known as O_1. This led to the view that a single operator site was bound by lac repressor to inhibit transcription, as depicted in Figure 16.4.

In the late 1970s and 1980s, two additional operator sites were identified. As shown at the top of **Figure 16.9**, these sites are called O_2 and O_3.

- O_1 is the operator site slightly downstream from the promoter.
- O_2 is located farther downstream in the *lacZ* coding sequence.
- O_3 is located slightly upstream from the promoter.

Studies by Benno Müller-Hill and his colleagues revealed how the three operators work. As shown in Figure 16.9 (fourth example from the top), if both O_2 and O_3 are missing, repression is dramatically reduced even when O_1 is present. When O_1 is missing, even in the presence of one of the other operator sites, repression is nearly abolished. How were these results interpreted? The data of Figure 16.9 supported a hypothesis that lac repressor must bind to O_1 and either O_2 or O_3 to cause full repression. According to this view, lac repressor can readily bind to O_1 and O_2, or to O_1 and O_3, but not to O_2 and O_3. Look at Figure 16.9 and you will notice that the operator sites are a fair distance from each other. For this reason, it was proposed that the binding of lac repressor to two operator sites requires the DNA to form a loop. A loop in the DNA brings the operator sites closer together, thereby facilitating the binding of the repressor protein (**Figure 16.10a**).

In 1996, the proposal that lac repressor binds to two operator sites was confirmed by studies in which lac repressor was crystallized by Mitchell Lewis and his colleagues. The repressor is a tetramer of four identical subunits. The crystal structure revealed that each dimer within the tetramer recognizes one operator site. **Figure 16.10b** is a molecular model illustrating the binding of lac repressor to the O_1 and O_3 sites. The amino acid side chains in the protein interact directly with bases in the major groove of the DNA double helix. This is how genetic regulatory proteins recognize specific DNA sequences. Because each dimer within the tetramer recognizes a single operator site, the association of two dimers to form a tetramer requires that the two operator sites be close to each other. For this to occur, a loop must form in the DNA. The formation of this loop dramatically inhibits the ability of RNA polymerase to slide past the O_1 site and transcribe the operon.

Figure 16.10b also shows the binding of the cAMP-CAP complex to the CAP site (see the blue protein within the loop). A particularly striking observation is that the binding of the cAMP-CAP complex to the DNA causes a 90° bend in the DNA structure. When the repressor is active—not bound to allolactose—the cAMP-CAP complex facilitates the binding of lac repressor to the O_1 and O_3 sites. When the repressor is inactive, the bending of the DNA also appears to be important in the ability of RNA polymerase to initiate transcription slightly downstream from the bend.

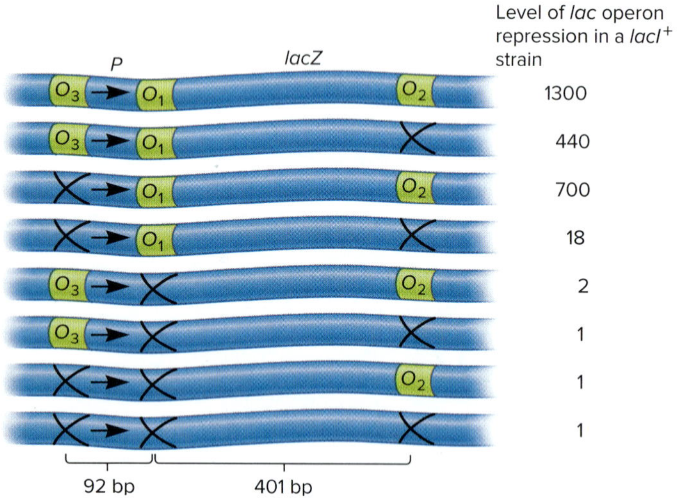

Operator configuration	Level of *lac* operon repression in a *lacI*⁺ strain
O_3, O_1, O_2 present	1300
O_3, O_1 present; O_2 missing	440
O_1, O_2 present; O_3 missing	700
O_1 present; O_2, O_3 missing	18
O_3, O_2 present; O_1 missing	2
O_3 present; O_1, O_2 missing	1
O_2 present; O_1, O_3 missing	1
All missing	1

92 bp / 401 bp

FIGURE 16.9 **The identification of three *lac* operator sites.** The top of this figure shows the locations of three *lac* operator sites, designated O_1, O_2, and O_3. The *lac* operator site shown in previous figures is the same as O_1. The arrows indicate the starting site for transcription. Missing operator sites are represented with an X. When all three operator sites are present, the repression of the *lac* operon is 1300-fold; this means there is 1/1300 the level of expression as compared to when lactose is present. This figure also shows the amount of repression when one or more operator sites are removed. A repression value of 1.0 indicates that no repression is occurring.

Concept Check: Which data provide the strongest evidence that O_1 is not the only operator site?

Genetic TIPS

The Question: Let's suppose you have isolated a mutant strain of *E. coli* in which the *lac* operon is constitutively expressed. In other words, the operon is turned on in the presence or absence of lactose. One possibility is that the mutation may block the transcription of the *lacI* gene, thereby preventing the synthesis of lac repressor. A second possibility is that the mutation could alter the sequence of the *lac* operator in a way that prevents lac repressor from binding to the operator. How would you distinguish between these two possibilities?

Topic: What topic in genetics does this question address?

The topic is gene regulation. More specifically, the question is about how a mutation could potentially alter the expression of the *lac* operon.

Information: What information do you know based on the question and your understanding of the topic?

From the question, you know that a mutation may either inhibit the expression of *lacI* or alter *lacO* in a way that prevents lac repressor from binding. From your understanding of the topic, you may remember that *lacI* exhibits a *trans*-effect because it is a diffusible protein, whereas *lacO* exhibits a *cis*-effect.

Problem-Solving Strategy: Design an experiment.

One strategy to solve this problem is to design an experiment that can distinguish between a mutation that has a *cis*-effect versus one that has a *trans*-effect. The use of merozygotes is one way to accomplish that goal.

Answer: Starting materials: The constitutive strain and a merozygote strain that carries a normal *lac* operon and a normal *lacI* gene on an F′ factor (see Figure 16.7).

1. Place strains into separate tubes with or without lactose.
2. Allow induction to occur.
3. Burst the cells.
4. Add β-ONPG, and measure the intensity of yellow color produced.

Expected results: If the mutation is in *lacI*, the repressor encoded on the F′ factor will inhibit the expression of the *lac* operon on the chromosome and the one on the F′ factor. There will be very little yellow color in the absence of lactose in the merozygote strain. (This was the result obtained in Figure 16.7.) Alternatively, if the mutation is in *lacO*, the *lac* operon on the chromosome will still be turned on even in the absence of lactose (also see Table 16.1). The merozygote will produce a high level of yellow color in the absence of lactose.

16.2 REVIEWING THE KEY CONCEPTS

- Enzyme adaptation is the phenomenon in which an enzyme appears within a living cell only after the cell has been exposed to the substrate for that enzyme.
- The *lac* operon encodes a polycistronic mRNA for proteins that are involved with the uptake and metabolism of lactose (see Figure 16.3).
- Lac repressor binds to the *lac* operator and inhibits transcription. Allolactose binds to the repressor and causes a conformational change that prevents the repressor from binding to the operator site. This event induces transcription (see Figure 16.4).
- The regulation of the *lac* operon enables the bacterium *E. coli* to respond to changes in the level of lactose in its environment (see Figure 16.5).
- Jacob, Monod, and Pardee constructed merozygotes to show that the *lacI* gene encodes a diffusible repressor protein (see Figures 16.6, 16.7).
- Mutations in the *lacI* gene and the *lac* operator site may have different effects in a merozygote (see Table 16.1).
- CAP is an activator protein for the *lac* operon. It binds to the CAP site when cAMP levels are high. Glucose inhibits cAMP levels, thereby inhibiting the *lac* operon (see Figure 16.8).

(a) DNA loops caused by the binding of lac repressor

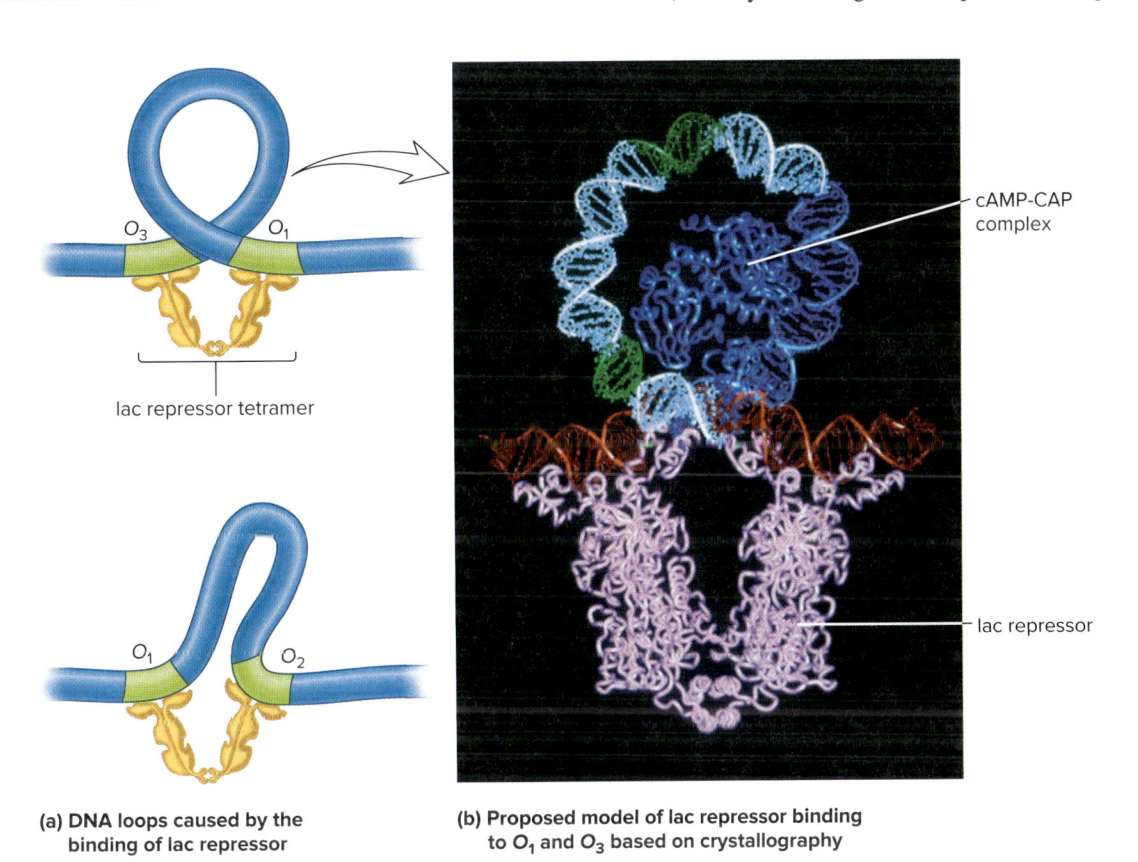

(b) Proposed model of lac repressor binding to O_1 and O_3 based on crystallography

FIGURE 16.10 **The binding of lac repressor to two operator sites.** (a) The binding of lac repressor protein to the O_1 and O_3 (top) or to the O_1 and O_2 (bottom) operator sites. Because the two sites are far apart, a loop must form in the DNA. (b) A molecular model for the binding of lac repressor to O_1 and O_3. Each repressor dimer binds to one operator site, so the repressor tetramer brings the two operator sites together. This causes the formation of a DNA loop in the intervening region. Note that the DNA loop contains the –35 and –10 sequences (shown in green) (refer back to Figure 14.4). These sequences are recognized by σ factor of RNA polymerase. This loop also contains the binding site for the cAMP-CAP complex, which is the blue protein within the loop.

©SPL/Science Source

- Lac repressor, which is a tetramer, binds to two operator sites, either to O_1 and O_2 or to O_1 and O_3, causing a loop to form in the DNA (see Figures 16.9, 16.10).

16.2 COMPREHENSION QUESTIONS

1. What is an operon?
   a. A site in the DNA where a regulatory protein binds
   b. A group of genes under the control of a single promoter
   c. An mRNA that encodes several genes
   d. All of the above are true of an operon.
2. The binding of _____ to lac repressor causes lac repressor to _____ to the operator site and thereby _____ transcription.
   a. glucose, bind, inhibits
   b. allolactose, bind, inhibits
   c. glucose, not bind, increases
   d. allolactose, not bind, increases
3. On its chromosome, an *E. coli* cell has a genotype of *lacI⁻ lacZ⁺ lacY⁺ lacA⁺*. It has an F′ factor with a genotype of *lacI⁺ lacZ⁺ lacY⁺ lacA⁺*. What is the expected level of expression of the *lac* operon genes (*lacZ⁺ lacY⁺ lacA⁺*) in the absence of lactose?
   a. Both *lac* operons will be expressed.
   b. Neither *lac* operon will be expressed.
   c. Only the chromosomal *lac* operon will be expressed.
   d. Only the *lac* operon on the F′ factor will be expressed.
4. When an *E. coli* cell is exposed to glucose, how does this affect the regulation of the *lac* operon via CAP?
   a. cAMP binds to CAP and transcription is increased.
   b. cAMP binds to CAP and transcription is decreased.
   c. cAMP does not bind to CAP and transcription is increased.
   d. cAMP does not bind to CAP and transcription is decreased.

16.3 REGULATION OF THE *trp* OPERON

Learning Outcomes:

1. Describe the organization of the *trp* operon.
2. Explain how the *trp* operon is regulated by trp repressor and by attenuation.

We now turn our attention to a second operon in *E. coli* called the *trp* operon (pronounced "trip"), which encodes enzymes involved with the synthesis of the amino acid tryptophan. Like the *lac* operon, the *trp* operon is regulated by a repressor protein. In addition, the *trp* operon is regulated by another mechanism called **attenuation,** in which transcription begins but is stopped prematurely, or "attenuated," before most of the *trp* operon is transcribed.

The *trp* Operon Is Regulated by a Repressor Protein

The *trp* operon contains six genes—*trpL, trpE, trpD, trpC, trpB,* and *trpA* (**Figure 16.11a**).

- The *trpE, trpD, trpC, trpB,* and *trpA* genes encode enzymes involved in tryptophan biosynthesis.
- *trpL* plays a regulatory role.
- The *trpR* gene, which is not part of the *trp* operon, encodes the **trp repressor** protein.
- When tryptophan levels within the cell are very low, trp repressor cannot bind to the operator site. Under these conditions, RNA polymerase transcribes the *trp* operon (Figure 16.11a). In this way, the cell expresses the genes required for the synthesis of tryptophan.
- When the tryptophan levels within the cell become high, tryptophan acts as a corepressor that binds to the trp repressor protein. This causes a conformational change in trp repressor that allows it to bind to the *trp* operator site (**Figure 16.11b**). This binding inhibits the ability of RNA polymerase to transcribe the operon. Therefore, when a high level of tryptophan is present within the cell—when the cell does not need to make more tryptophan—the *trp* operon is turned off.

The *trp* Operon Is Also Regulated by Attenuation

In addition to regulation via trp repressor, the *trp* operon is also regulated by a mechanism called attenuation (**Figure 16.11c**). Attenuation can occur in bacteria because the processes of transcription and translation are coupled. During attenuation, transcription actually begins, but it is terminated before the entire mRNA is made. When attenuation occurs, the mRNA from the *trp* operon is made as a short piece that terminates at the **attenuator sequence,** which is just downstream from the *trpL* gene (see Figure 16.11c). Because this short mRNA has been terminated before RNA polymerase has transcribed the *trpE, trpD, trpC, trpB,* and *trpA* genes, it does not encode the proteins required for tryptophan biosynthesis. In this way, attenuation inhibits the further production of tryptophan in the cell.

Let's now consider how attenuation happens. The segment of the *trp* operon immediately downstream from the operator site plays a critical role during attenuation. The first gene in the *trp* operon is the *trpL* gene, which encodes a peptide containing 14 amino acids called the leader peptide. As shown in **Figure 16.12**, two features are key in the attenuation mechanism.

- First, two tryptophan (Trp) codons are found within the mRNA that encodes the *trp* leader peptide. What is the role of these codons? As we will see later, these two codons provide a way to sense whether or not the bacterium has sufficient tryptophan to synthesize its proteins.
- Second, the mRNA can form stem-loops. The types of stem-loops that are formed determine whether or not attenuation occurs.

Different combinations of stem-loops are possible due to interactions among four regions within the RNA transcript (see the color key in Figure 16.12).

- Region 2 is complementary to region 1 and also to region 3.
- Region 3 is complementary to region 2 as well as to region 4.
- A particular segment of RNA can participate in the formation of only one stem-loop. For example, if region

16.3 REGULATION OF THE *trp* OPERON

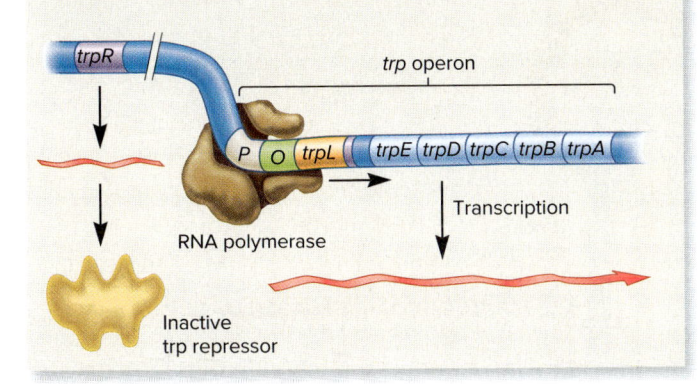

(a) Low tryptophan levels, transcription of the entire *trp* operon occurs

When tryptophan levels are low, tryptophan does not bind to the trp repressor protein, which prevents the repressor protein from binding to the operator site. Under these conditions, RNA polymerase can transcribe the operon, which leads to the expression of the *trpE*, *trpD*, *trpC*, *trpB*, and *trpA* genes. These genes encode enzymes involved in tryptophan biosynthesis.

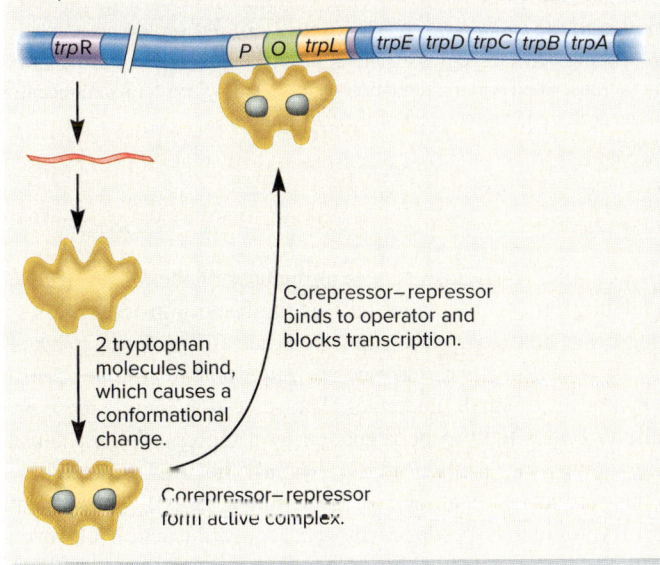

(b) High tryptophan levels, repression occurs

When tryptophan levels are high, tryptophan acts as a corepressor that binds to the trp repressor protein. The tryptophan trp repressor complex then binds to the operator site to inhibit transcription.

- 2 tryptophan molecules bind, which causes a conformational change.
- Corepressor–repressor form active complex.
- Corepressor–repressor binds to operator and blocks transcription.

(c) High tryptophan levels, attenuation occurs

Another mechanism of regulation is attenuation. When attenuation occurs, the RNA is transcribed only to the attenuator sequence, and then transcription is terminated.

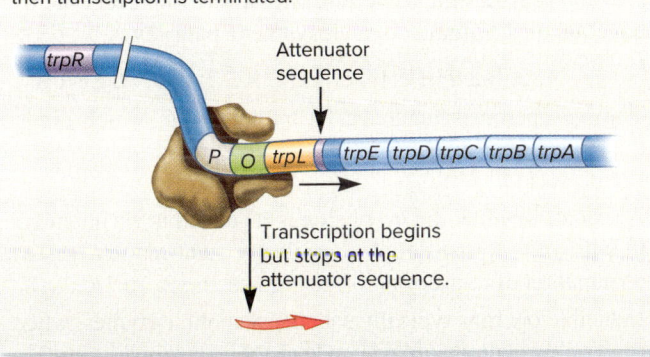

Transcription begins but stops at the attenuator sequence.

---

2 forms a stem-loop with region 1, it cannot (at the same time) form a stem-loop with region 3.

- Though three stem-loops are possible, the 3–4 stem-loop is functionally unique. The 3–4 stem-loop in combination with the U-rich attenuator sequence results in intrinsic termination, also called ρ-independent termination, as described in Chapter 14. Therefore, the formation of the 3–4 stem-loop causes RNA polymerase to pause, and the U-rich sequence dissociates from the DNA. This terminates transcription at the U-rich attenuator.

Conditions that favor the formation of the 3–4 stem-loop ultimately rely on the translation of the *trpL* gene. As shown in **Figure 16.13**, three scenarios are possible.

- In Figure 16.13a, translation is not coupled with transcription. Region 1 rapidly hydrogen bonds to region 2, and region 3 is left to hydrogen bond to region 4. Therefore, the terminator stem-loop forms, and transcription is terminated just past the *trpL* gene at the U-rich attenuator.
- In Figure 16.13b, coupled transcription and translation occur under conditions in which the tryptophan concentration is low. When tryptophan levels are low, the cell cannot make a sufficient amount of charged tRNA^{Trp}. The ribosome pauses at the Trp codons in the *trpL* mRNA because it is waiting for charged tRNA^{Trp}. This pause occurs in such a way that the ribosome shields region 1 of the mRNA. This prevents region 1 from hydrogen bonding to region 2. As an alternative, region 2 hydrogen bonds to region 3. Therefore, the 3–4 stem-loop cannot form. Under these conditions, attenuation does not occur, and RNA polymerase transcribes the rest of the operon. This ultimately enables the bacterium to make more tryptophan.
- In Figure 16.13c, coupled transcription and translation occur under conditions in which a sufficient amount of tryptophan is present in the cell. In this case, translation of the *trpL* mRNA progresses to its stop codon, where the ribosome pauses. The pausing at the stop codon prevents region 2 from hydrogen bonding with any region, thereby enabling region 3 to hydrogen bond with region 4. As in Figure 16.13a, this terminates transcription. Of course, keep in mind that the *trpL* mRNA contains two tryptophan codons. For the ribosome to smoothly progress to the *trpL* stop codon, enough charged tRNA^{Trp} must be available to translate this mRNA. It follows that the bacterium must have a sufficient amount of tryptophan. Under these conditions, the rest of the transcription of the operon is terminated.

Attenuation is a mechanism to regulate transcription that occurs with several other operons involved with amino acid biosynthesis. In all cases, the mRNAs that encode the leader peptides are rich in codons for the particular amino acid that is synthesized

FIGURE 16.11 Organization of the *trp* operon and regulation via trp repressor and attenuation.

Concept Check: How does tryptophan affect the function of trp repressor?

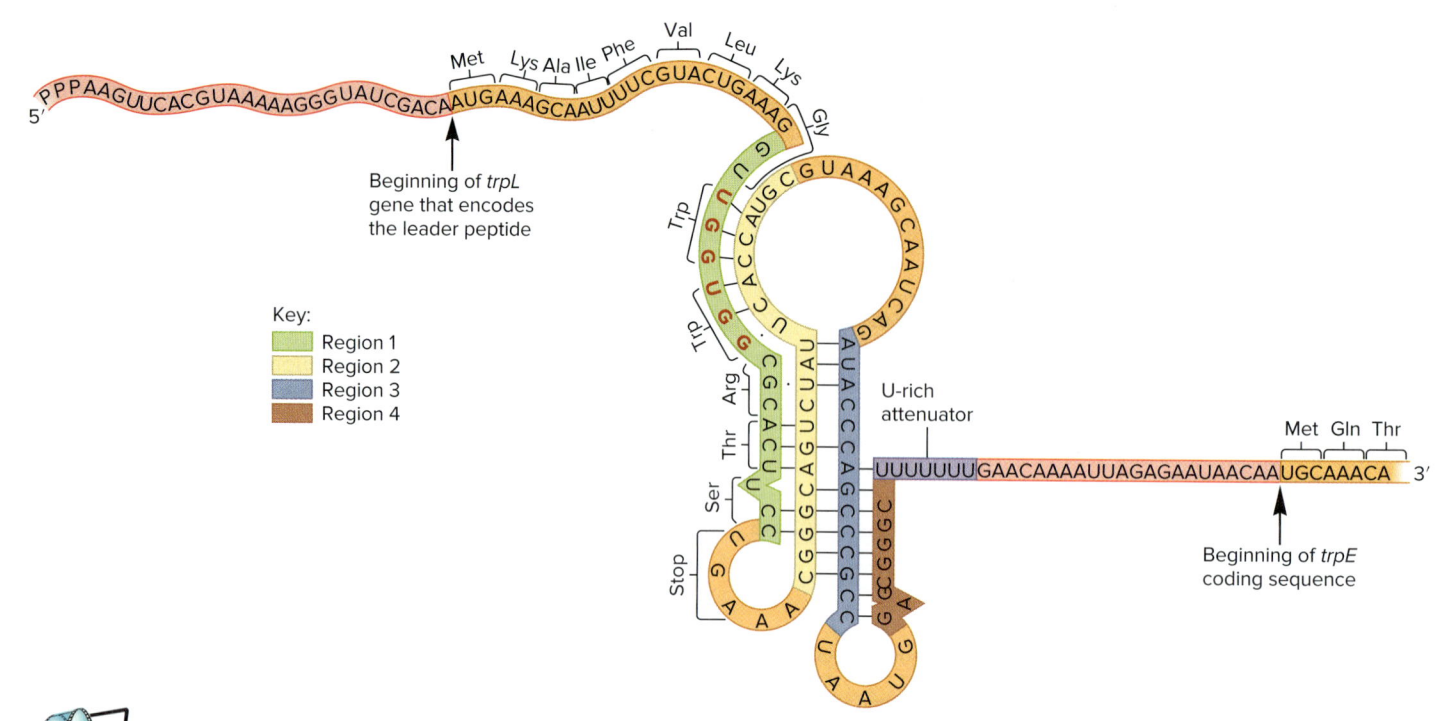

FIGURE 16.12 Sequence of the mRNA produced at the beginning of the *trp* operon. As shown here, this mRNA has several regions that are complementary to each other. The possible hydrogen bonding between regions 1 and 2, 2 and 3, and 3 and 4 is also shown. The last U in the purple attenuator sequence is the last nucleotide that is transcribed when attenuation occurs. At low tryptophan concentrations, however, transcription occurs beyond the end of *trpL* and proceeds through the *trpE* gene and the rest of the *trp* operon.

Concept Check: What type of bonding interactions cause stem-loops to form?

by the enzymes encoded by the particular operon. For example, the mRNA that encodes the leader peptide of the histidine operon has seven histidine codons in its sequence, and the mRNA for the leader peptide of the leucine operon has four leucine codons. Like the *trp* operon, these other operons have alternative stem-loops, one of which is a transcriptional terminator.

Inducible Operons Usually Encode Catabolic Enzymes, and Repressible Operons Encode Anabolic Enzymes

Thus far, we have seen that bacterial genes can be transcriptionally regulated in a positive or negative way—and sometimes both. The *lac* operon is an inducible operon regulated by sugar molecules that activate transcription. By comparison, the *trp* operon is a repressible operon regulated by tryptophan, a corepressor that binds to the repressor and turns the operon off. In addition, an abundance of charged tRNA^{Trp} in the cytoplasm can turn the *trp* operon off via attenuation.

By studying the genetic regulation of many operons, geneticists have discovered a general trend concerning inducible versus repressible regulation. When the genes in an operon encode proteins that function in the catabolism or breakdown of a substance, they are usually regulated in an inducible manner, which means that a small effector molecule induces their expression. The substance to be broken down or a related compound often acts as the inducer.

For example, allolactose acts as an inducer of the *lac* operon. An inducible form of regulation allows the bacterium to express the appropriate genes only when they are needed to break down lactose.

In contrast, other enzymes are important for the anabolism, or synthesis, of small molecules. The genes that encode these anabolic enzymes tend to be regulated by a repressible mechanism. The corepressor or inhibitor is commonly the small molecule that is the product of the enzymes' biosynthetic activities. For example, tryptophan is produced by the sequential action of several enzymes that are encoded by the *trp* operon. Tryptophan itself acts as a corepressor that binds to trp repressor when the intracellular levels of tryptophan become relatively high. This mechanism turns off the genes required for tryptophan biosynthesis when enough of this amino acid has been made. Therefore, genetic regulation via repression provides the bacterium with a way to prevent the overproduction of the product of a biosynthetic pathway.

16.3 REVIEWING THE KEY CONCEPTS

- The *trp* operon is regulated by trp repressor, which binds to the *trp* operator when tryptophan, a corepressor, is present (see Figure 16.11).
- A second way that the *trp* operon is regulated is via attenuation, in which the formation of a terminator stem-loop causes early termination of transcription (see Figures 16.12, 16.13).
- Inducible operons typically encode catabolic enzymes, whereas repressible operons often encode anabolic enzymes.

16.3 REGULATION OF THE *trp* OPERON

When translation is not coupled with transcription, region 1 hydrogen bonds to region 2 and region 3 hydrogen bonds to region 4. Because a 3–4 terminator stem-loop forms, transcription will be terminated at the U-rich attenuator.

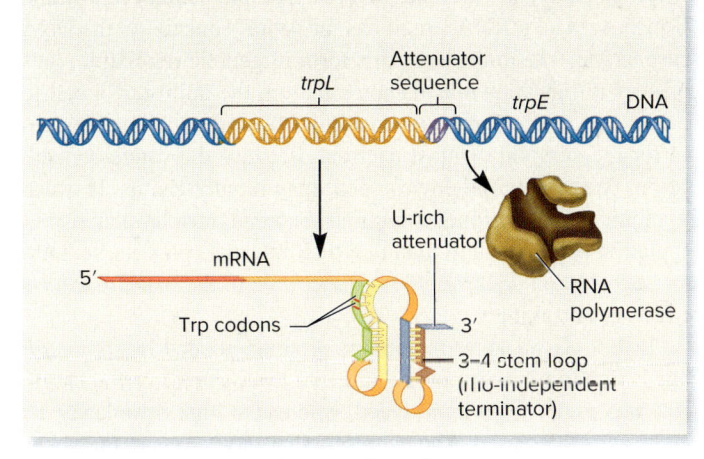

(a) No translation, 1–2 and 3–4 stem-loops form

Coupled transcription and translation occur under conditions in which the tryptophan concentration is very low. The ribosome pauses at the Trp codons in the *trpL* gene because insufficient amounts of charged tRNA^{Trp} are present. This pause blocks region 1 of the mRNA, so region 2 can hydrogen bond only with region 3. When this happens, the 3–4 stem-loop structure cannot form. Attenuation does not occur, and RNA polymerase transcribes the rest of the operon.

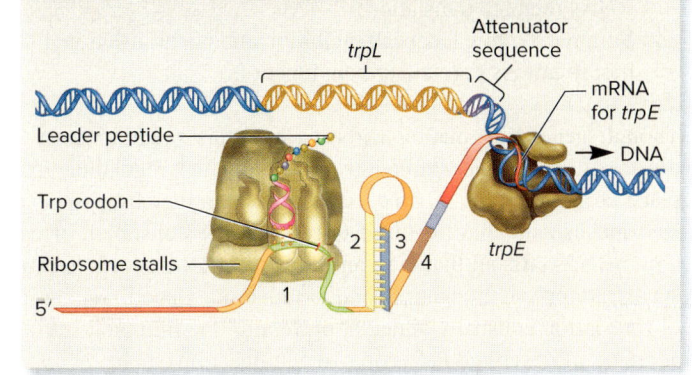

(b) Low tryptophan levels, 2–3 stem-loop forms

Coupled transcription and translation occur under conditions in which a sufficient amount of tryptophan is present in the cell. Translation of the *trpL* gene progresses to its stop codon, where the ribosome pauses. This blocks region 2 from hydrogen bonding with any region and thereby enables region 3 to hydrogen bond with region 4. This terminates transcription at the U-rich attenuator.

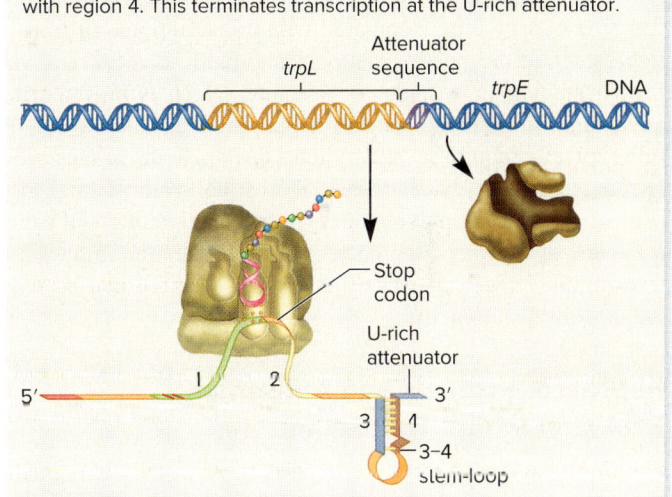

(c) High tryptophan levels, 3–4 stem-loop forms

FIGURE 16.13 Possible stem-loops formed in *trpL* mRNA under different conditions of translation. Attenuation occurs in parts (**a**) and (**c**) due to the formation of a 3–4 stem-loop.

Concept Check: Explain how the presence of tryptophan favors the formation of the 3–4 stem-loop.

16.3 COMPREHENSION QUESTIONS

1. When tryptophan binds to trp repressor, this causes trp repressor to _____ to the *trp* operator and _____ transcription.
    a. bind, inhibit
    b. not bind, inhibit
    c. bind, activate
    d. not bind, activate

2. During attenuation, when tryptophan levels are high, the _____ stem-loop forms and transcription _____ the *trpL* gene.
    a. 2–3, ends just past
    b. 3–4, ends just past
    c. 1–2, continues beyond
    d. 3–4, continues beyond

3. Operons involved with the biosynthesis of molecules, such as amino acids, are most likely to be regulated in which of the following ways?
    a. The product of the biosynthetic pathway represses transcription.
    b. The product of the biosynthetic pathway activates transcription.
    c. A precursor of the biosynthetic pathway represses transcription.
    d. A precursor of the biosynthetic pathway activates transcription.

16.4 TRANSLATIONAL AND POSTTRANSLATIONAL REGULATION

Learning Outcomes:
1. Explain how translational regulatory proteins and antisense RNAs regulate translation.
2. Summarize how feedback inhibition and posttranslational modifications regulate protein function.

Though genetic regulation in bacteria occurs predominantly at transcription, many examples are known in which regulation takes place at a later stage in gene expression. In some cases, specialized mechanisms have evolved to regulate the translation of certain mRNAs. Recall that the translation of mRNA occurs in three stages: initiation, elongation, and termination. Genetic regulation of translation is usually aimed at preventing the initiation step. In addition, as described in Section 16.5, translation can be regulated by riboswitches.

To fully understand how proteins influence an organism's traits, researchers have investigated how the functions of proteins are regulated. The term **posttranslational regulation** refers to the functional control of proteins that are already present in the cell rather than regulation of transcription or translation. Posttranslational regulation can either activate or inhibit the function of a protein. Compared with transcriptional or translational regulation, posttranslational regulation can be relatively fast, occurring in a matter of seconds, which is an important advantage. In contrast, transcriptional and translational regulation typically require several minutes or even hours to take effect because these two mechanisms involve the synthesis and turnover of mRNA and polypeptides. In this section, we will examine some of the ways that bacteria can regulate the initiation of translation, as well as ways that protein function can be regulated posttranslationally.

Repressor Proteins and Antisense RNA Can Inhibit Translation

For some bacterial genes, the translation of mRNA is regulated by the binding of proteins or other RNA molecules that influence the ability of ribosomes to translate the mRNA into a polypeptide. A **translational regulatory protein** recognizes sequences within the mRNA, much as transcription factors recognize DNA sequences. In most cases, translational regulatory proteins act to inhibit translation. These are known as **translational repressors.** When a translational repressor protein binds to the mRNA, it can inhibit translational initiation in one of two ways.

- One possibility is that it can bind in the vicinity of the Shine-Dalgarno sequence and/or the start codon, thereby blocking the ribosome's ability to initiate translation in this region.
- Alternatively, the repressor protein may bind to the mRNA at a location that is not near the Shine-Dalgarno sequence or the start codon, but the binding stabilizes an mRNA secondary structure that prevents initiation.

Translational repression is also a form of genetic regulation found in eukaryotic species, and we will consider specific examples in Chapter 17.

A second way to regulate translation is via the synthesis of **antisense RNA,** an RNA strand that is complementary to a strand of mRNA. (The mRNA strand has the same sequence as the DNA sense strand.) To understand this form of genetic regulation, let's consider a trait known as osmoregulation, the ability of a cell to control the amount of water inside it. Osmoregulation is essential for the survival of most bacteria. Because the solute concentrations in the external environment may rapidly change between hypotonic and hypertonic conditions, bacteria must have an osmoregulation mechanism to maintain their internal cell volume. Otherwise, bacterial cells would be susceptible to the harmful effects of lysis or shrinking.

In *E. coli,* an outer membrane protein encoded by the *ompF* gene is important in osmoregulation. At low osmolarity, the OmpF protein is preferentially produced, whereas at high osmolarity, its synthesis is decreased. The expression of another gene, known as *micF,* is responsible for inhibiting the expression of the *ompF* gene at high osmolarity. The name *mic* stands for mRNA-interfering complementary RNA. As shown in **Figure 16.14**, the *micF* RNA is complementary to the *ompF* mRNA; it is an antisense strand of RNA. When the *micF* gene is transcribed, its RNA product binds to the *ompF* mRNA via hydrogen bonding between their complementary regions. The binding of the *micF* RNA to the *ompF* mRNA prevents the *ompF* mRNA from being translated. The RNA transcribed from the *micF* gene is called antisense RNA because it is complementary to the *ompF* mRNA, which is a sense strand of mRNA that encodes a polypeptide. The *micF* RNA does not encode a polypeptide.

Posttranslational Regulation Can Occur Via Feedback Inhibition and Covalent Modifications

Let's now turn our attention to ways that protein function is regulated posttranslationally. A common mechanism for regulating the

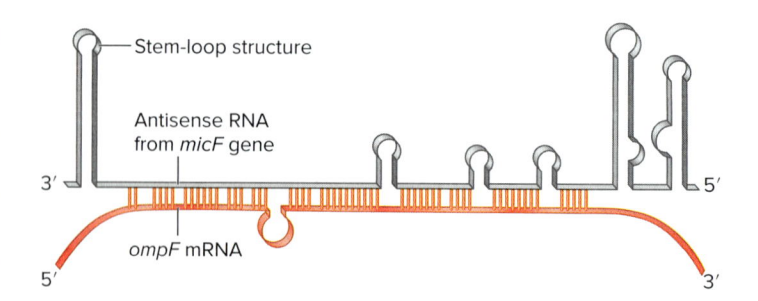

FIGURE 16.14 **The double-stranded RNA structure formed between the *micF* antisense RNA and the *ompF* mRNA.** Because they have regions that are complementary to each other, the *micF* antisense RNA hydrogen bonds with the *ompF* mRNA to form a double-stranded structure that prevents the *ompF* mRNA from being translated.

Concept Check: How does micF antisense RNA affect the translation of ompF mRNA?

activity of metabolic enzymes is **feedback inhibition.** The synthesis of many cellular molecules such as amino acids, vitamins, and nucleotides occurs via the action of a series of enzymes that convert precursor molecules to particular products. During feedback inhibition, the final product in a metabolic pathway inhibits an enzyme that acts early in the pathway.

Figure 16.15 depicts feedback inhibition in a metabolic pathway. Enzyme 1 is an example of an **allosteric enzyme,** an enzyme that contains two different binding sites (lac repressor is an allosteric protein, but not an enzyme). The catalytic site is responsible for the binding of the substrate and its conversion to intermediate 1. The second site is a regulatory, or allosteric, site. This site binds the final product of the metabolic pathway. When bound to the regulatory site, the final product causes a conformational change that inhibits the catalytic ability of enzyme 1.

To appreciate feedback inhibition at the cellular level, we can consider the relationship between the product concentration and the regulatory site on enzyme 1. As the final product is made within the cell, its concentration gradually increases. Once the final product concentration has reached a level that is similar to the product's affinity for enzyme 1, the product is likely to bind to the regulatory site on enzyme 1 and inhibit its function. In this way, the net result is that the final product of a metabolic pathway inhibits the further synthesis of more product. Under these conditions, the concentration of the final product has reached a level sufficient for the cell's needs.

A second strategy to control the function of proteins is the covalent modification of their structure, a process called **posttranslational covalent modification.** Certain types of modifications are involved primarily in the assembly and construction of a functional protein. These alterations include proteolytic processing; disulfide bond formation; and the attachment of prosthetic groups, sugars, or lipids. These are typically irreversible changes required to produce a functional protein. In contrast, other types of modifications, such as phosphorylation ($-PO_4$), acetylation ($-COCH_3$), and methylation ($-CH_3$), are often reversible modifications that transiently affect the function of a protein.

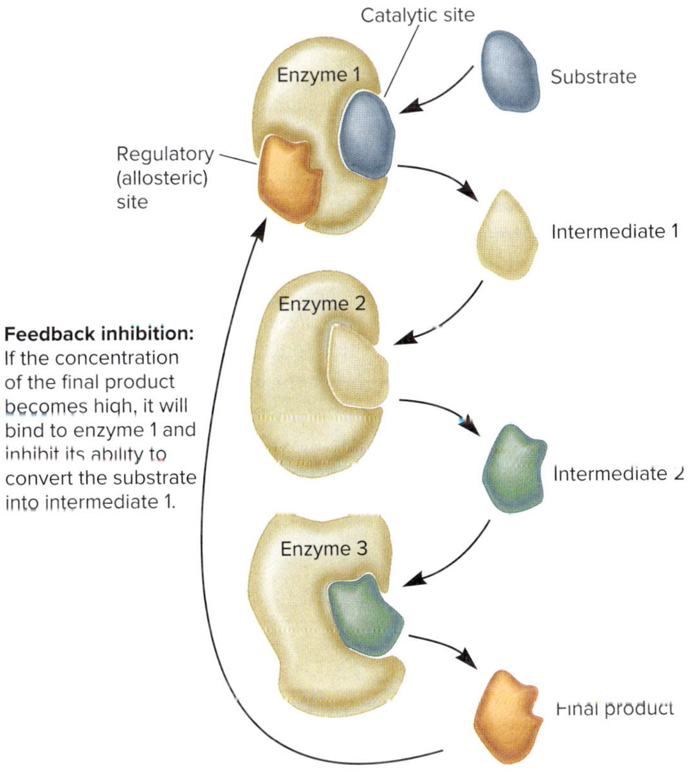

16.4 REVIEWING THE KEY CONCEPTS

- Translation can be regulated by translational regulatory proteins and by antisense RNA (see Figure 16.14).
- Posttranslational control of protein function may involve feedback inhibition and posttranslational covalent modifications (see Figure 16.15).

16.4 COMPREHENSION QUESTIONS

1. Translation can be regulated by
   a. translational repressors.
   b. antisense RNA.
   c. attenuation.
   d. both a and b.
2. An example of a posttranslational covalent modification that may affect protein function is
   a. phosphorylation.
   b. acetylation.
   c. methylation.
   d. Any of the above can affect protein function.

FIGURE 16.15 **Feedback inhibition in a metabolic pathway.** The substrate is converted to a product by the sequential action of three different enzymes. Enzyme 1 has a catalytic site that recognizes the substrate; it also has a regulatory site that recognizes the final product. When the final product binds to the regulatory site, it inhibits the ability of enzyme 1 to convert the substrate into intermediate 1.

Concept Check: Why is feedback inhibition an advantage to a bacterium?

16.5 RIBOSWITCHES

Learning Outcome:

1. Explain how riboswitches can regulate transcription and translation.

In 2001 and 2002, researchers in a few different laboratories discovered a mechanism of gene regulation called a **riboswitch.** In this form of regulation, an RNA molecule can exist in two different secondary conformations. One conformation is active, and the other inhibits gene expression. The conversion from one conformation to the other is due to the binding of a small molecule. As described in **Table 16.2**, a riboswitch can regulate transcription, translation, RNA stability, or splicing.

Riboswitches are widespread in bacteria. Researchers estimate that 3% to 5% of all bacterial genes may be regulated by riboswitches. The bacterial genes subject to this form of regulation

TABLE 16.2
Types of Riboswitches

Type of Regulation	Description
Transcription	The 5′ region of an mRNA may exist in one conformation that forms a terminator stem-loop which causes attenuation of transcription. The other conformation does not form such a terminator and is completely transcribed.
Translation	The 5′ region of an mRNA may exist in one conformation in which the Shine-Dalgarno sequence cannot be recognized by the ribosome, whereas the other conformation has an accessible Shine-Dalgarno sequence that allows the mRNA to be translated.
RNA stability	One mRNA conformation may be stable, whereas the other conformation acts as a ribozyme that causes self-degradation.
Splicing	In eukaryotes, one pre-mRNA conformation may be spliced in one way, whereas another conformation is spliced in a different way.

For example, in *Bacillus subtilis*, the majority of genes involved in TPP synthesis are found within the *thi* operon, which contains seven genes.

The *thi* operon in *B. subtilis* is regulated by a riboswitch that controls transcription (**Figure 16.16**). As the polycistronic mRNA for the *thi* operon is being made, the 5′ end quickly folds into a secondary structure.

- When TPP levels are low, the secondary structure has a stem-loop called an antiterminator, which prevents the formation of the terminator stem-loop. Under these conditions, transcription of the entire *thi* operon occurs. In this way, the bacterium is able to make more TPP, which is in short supply.
- When TPP levels are high, TPP binds to the RNA and causes a change in its secondary structure. As shown on the right side of Figure 16.16, a terminator stem-loop forms instead of the antiterminator stem-loop. The terminator stem-loop abruptly stops transcription, thereby inhibiting the production of the enzymes that are needed to make more TPP. Similar to the effect of the *trp* operon, discussed earlier in this chapter, this is an example of attenuation.

are associated with the biosynthesis of purines, amino acids, vitamins, and other essential molecules. Riboswitches are also found in archaea, algae, fungi, and plants. In this section, we will examine two examples of riboswitches in bacteria. The first example shows how a riboswitch can regulate transcription, and the second example involves translational regulation.

A Riboswitch Can Regulate Transcription

Thiamin, also called vitamin B$_1$, is an important organic molecule for bacteria, archaea, and eukaryotes. The active form of this vitamin is thiamin pyrophosphate (TPP). TPP is an essential coenzyme for the functioning of a variety of enzymes, such as certain enzymes in the citric acid cycle. In bacteria, TPP is made via biosynthetic enzymes that are encoded in the bacterial genome.

A Riboswitch Can Regulate Translation

In gram-positive bacteria, such as *B. subtilis*, the regulation of TPP biosynthetic enzymes occurs via a riboswitch that controls transcription. However, in gram-negative bacteria, such as *E. coli*, a structurally similar riboswitch regulates translation. **Figure 16.17** shows how a riboswitch can regulate translation. In *E. coli*, the *thiMD* operon encodes two enzymes involved with TPP biosynthesis. Recall from Chapter 15 that the Shine-Dalgarno sequence in an mRNA causes the mRNA to bind to a ribosome (refer back to Figure 15.16).

- When TPP levels are low, the 5′ end of the mRNA folds into a structure that contains a stem-loop called the Shine-Dalgarno antisequestor. If this stem-loop forms, the Shine-Dalgarno sequence is accessible, which allows the mRNA

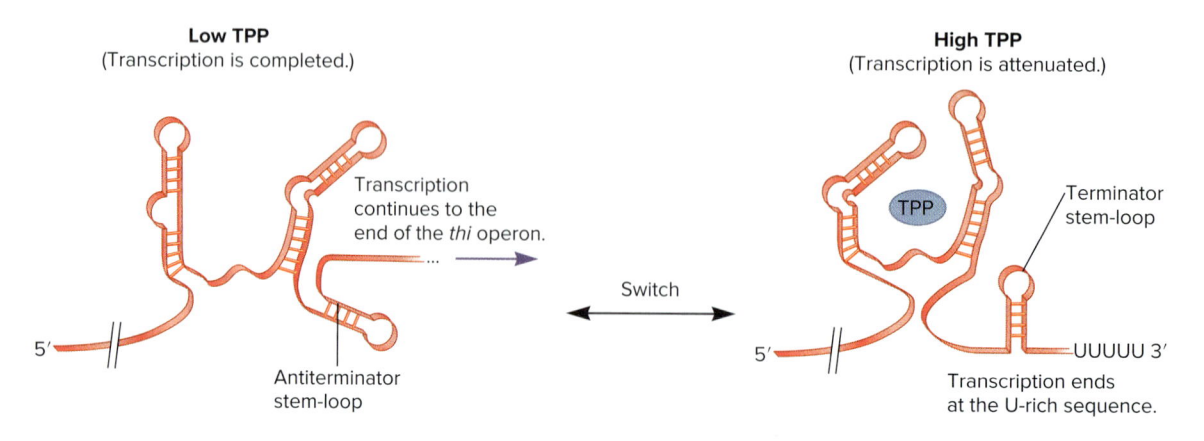

FIGURE 16.16 The TPP riboswitch in *Bacillus subtilis* that regulates transcription.

Concept Check: Which RNA conformation favors transcription—the form with the antiterminator stem-loop or the form with the terminator stem-loop?

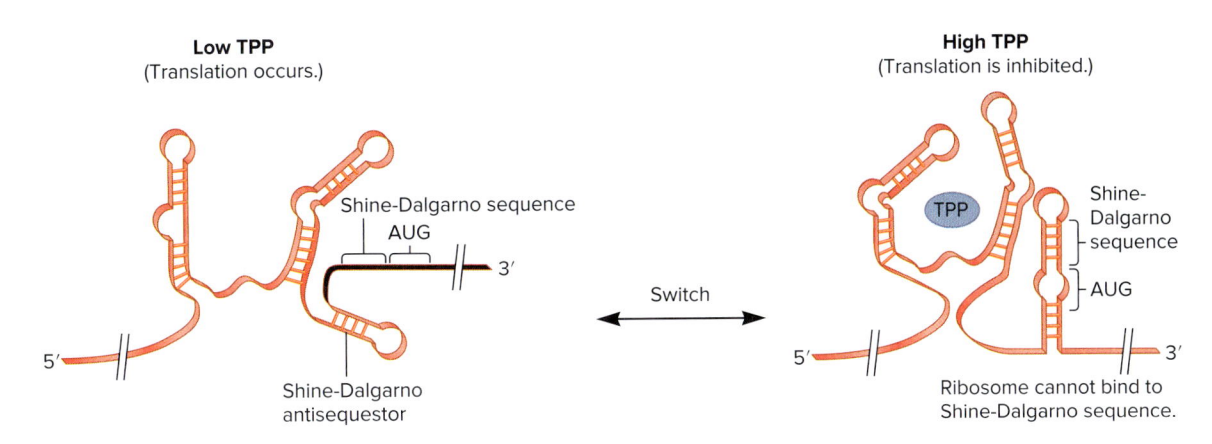

FIGURE 16.17 **The TPP riboswitch in *E. coli* that regulates translation.** Note: In both cases, the *thiMD* operon is completely transcribed.

Concept Check: Which RNA conformation favors translation—the form with the Shine-Dalgarno antisequestor or the form in which the Shine-Dalgarno sequence is within a stem-loop?

to bind to the ribosome. Therefore, the mRNA is translated when TPP is in short supply.
- When TPP levels are high, TPP binds to the RNA and causes a change in its secondary structure. As shown on the right side of Figure 16.17, a stem-loop forms that contains the Shine-Dalgarno sequence and the start codon. The formation of this stem-loop sequesters the Shine-Dalgarno sequence, thereby preventing ribosomal binding. This blocks the translation of the enzymes that are needed to make more TPP.

While comparing Figures 16.16 and 16.17, it is worth noting that the 5' end of the *thiMD* mRNA of *E. coli* can exist in two different secondary structures that greatly resemble those that occur at the 5' end of the *thi* operon of *B. subtilis*. However, the effects on regulation are quite different. The TPP riboswitch in *E. coli* controls translation, whereas the TPP riboswitch in *B. subtilis* controls transcription.

16.5 REVIEWING THE KEY CONCEPTS

- A riboswitch is a mechanism of gene regulation in which an RNA can exist in two different secondary conformations. The conversion from one conformation to the other is due to the binding of a small molecule.
- A riboswitch can regulate transcription, translation, RNA stability, or splicing (see Table 16.2, Figures 16.16, 16.17).

16.5 COMPREHENSION QUESTION

1. For a riboswitch that controls transcription, the binding of a small molecule such as TPP controls whether the mRNA
   a. has an antiterminator or terminator stem-loop.
   b. has a Shine-Dalgarno antisequestor or the Shine-Dalgarno sequence within a stem-loop.
   c. is degraded from its 5' end.
   d. has both a and b.

KEY TERMS

Page 355. gene regulation, constitutive genes
Page 356. repressor, activator, negative control, positive control, inducer, inducible genes, corepressor, inhibitor, repressible genes
Page 358. enzyme adaptation, operon, polycistronic mRNA, promoter, terminator, CAP site, catabolite activator protein (CAP), operator site (operator), lac repressor
Page 359. induced, allosteric regulation, allosteric site
Page 362. merozygote
Page 364. *trans*-effect, *trans*-acting factor, *cis*-acting element, *cis*-effect, catabolite repression, diauxic growth, cyclic-AMP (cAMP)
Page 368. attenuation, trp repressor, attenuator sequence
Page 372. posttranslational regulation, translational regulatory protein, translational repressors, antisense RNA
Page 373. feedback inhibition, allosteric enzyme, posttranslational covalent modification, riboswitch

CHAPTER SUMMARY

- Gene regulation is the phenomenon in which the level of gene expression can vary under different conditions. By comparison, constitutive genes are expressed at relatively constant levels. Gene regulation can occur during transcription, during translation, or posttranslationally (see Figure 16.1).

16.1 Overview of Transcriptional Regulation

- Repressors and activators are regulatory proteins that exert negative control and positive control, respectively. Small effector molecules that control the function of repressors and activators include inducers, inhibitors, and corepressors (see Figure 16.2).

16.2 Regulation of the *lac* Operon

- Enzyme adaptation is the phenomenon in which an enzyme appears within a living cell only after the cell has been exposed to the substrate for that enzyme.
- The *lac* operon encodes a polycistronic mRNA for proteins that are involved with the uptake and metabolism of lactose (see Figure 16.3).
- Lac repressor binds to the *lac* operator and inhibits transcription. Allolactose binds to the repressor and causes a conformational change that prevents the repressor from binding to the operator site. This event induces transcription (see Figure 16.4).
- The regulation of the *lac* operon enables the bacterium *E. coli* to respond to changes in the level of lactose in its environment (see Figure 16.5).
- Jacob, Monod, and Pardee constructed merozygotes to show that the *lacI* gene encodes a diffusible repressor protein (see Figures 16.6, 16.7).
- Mutations in the *lacI* gene and the *lac* operator site may have different effects in a merozygote (see Table 16.1).
- CAP is an activator protein for the *lac* operon. It binds to the CAP site when cAMP levels are high. Glucose inhibits cAMP levels, thereby inhibiting the *lac* operon (see Figure 16.8).
- Lac repressor, which is a tetramer, binds to two operator sites, either to O_1 and O_2 or to O_1 and O_3, causing a loop to form in the DNA (see Figures 16.9, 16.10).

16.3 Regulation of the *trp* Operon

- The *trp* operon is regulated by trp repressor, which binds to the *trp* operator when tryptophan, a corepressor, is present (see Figure 16.11).
- A second way that the *trp* operon is regulated is via attenuation, in which the formation of a terminator stem-loop causes early termination of transcription (see Figures 16.12, 16.13).
- Inducible operons typically encode catabolic enzymes, whereas repressible operons often encode anabolic enzymes.

16.4 Translational and Posttranslational Regulation

- Translation can be regulated by translational regulatory proteins and by antisense RNA (see Figure 16.14).
- Posttranslational control of protein function may involve feedback inhibition and posttranslational covalent modifications (see Figure 16.15).

16.5 Riboswitches

- A riboswitch is a mechanism of gene regulation in which an RNA can exist in two different secondary conformations. The conversion from one conformation to the other is due to the binding of a small molecule.
- A riboswitch can regulate transcription, translation, RNA stability, or splicing (see Table 16.2, Figures 16.16, 16.17).

PROBLEM SETS & INSIGHTS

More Genetic TIPS

1. Researchers have identified mutations in the promoter region of the *lacI* gene that make it more difficult for the *lac* operon to be induced. These are called *lacI*^{Q} mutants, because their effect is that a greater quantity of lac repressor is made. Explain why an increased transcription of the *lacI* gene makes it more difficult to induce the *lac* operon.

Topic: What topic in genetics does this question address?

The topic is gene regulation. More specifically, the question is about how the amount of a repressor protein will affect transcription.

Information: What information do you know based on the question and your understanding of the topic?

From the question, you know that certain mutations result in an overproduction of lac repressor. From your understanding of the topic, you may remember that the binding of lac repressor to the *lac* operator inhibits transcription.

Problem-Solving Strategy: Relate structure and function. Predict the outcome.

One strategy to solve this problem is to consider the structure and function of lac repressor. It forms a tetramer and then binds to the operator in the absence of allolactose. When allolactose binds to the repressor, the result is a conformational change that causes lac repressor to not bind to the operator.

Answer: An increase in the amount of lac repressor makes it easier for tetramers to form that will repress the *lac* operon. When the cell is exposed to lactose, allolactose levels slowly rise. Some of the allolactose binds to lac repressor and causes it to be released from the operator sites. If many more lac repressor proteins accumulate within the cell, more allolactose is needed to ensure that no unoccupied repressor proteins can repress the operon.

2. Explain how the pausing of the ribosome in the presence or absence of tryptophan affects the formation of a terminator (3–4) stem-loop, and describe how this affects transcription.

Topic: What topic in genetics does this question address?

The topic is gene regulation. More specifically, the question is about a form of gene regulation called attenuation.

Information: What information do you know based on the question and your understanding of the topic?

In the question, you are reminded that tryptophan affects the formation of a terminator stem-loop. From your understanding of the topic, you may remember that the location where a ribosome pauses can influence which types of stem-loops can form. The three possible types are 1–2, 2–3, and 3–4. The formation of a 2–3 stem-loop prevents the formation of a 3–4 stem-loop.

Problem-Solving Strategy: Make a drawing. Compare and contrast.

One strategy to solve this problem is to make two drawings similar to those shown in Figure 16.13, parts (b) and (c). In part (b), the ribosome is shielding region 1, and in part (c), it is shielding region 2. Compare the two drawings and think about which stem-loops are able to form.

Answer: The key issue is the location where the ribosome pauses. In the absence of tryptophan, it pauses over the Trp codons in the *trpL* mRNA. Pausing at this site shields region 1 in the mRNA. Because region 1 is unavailable to hydrogen bond with region 2, region 2 hydrogen bonds with region 3. Therefore, region 3 cannot form a terminator stem-loop with region 4. Alternatively, if tryptophan levels in the cell are sufficient, the ribosome pauses over the stop codon in the *trpL* mRNA. In this case, the ribosome shields region 2. Therefore, regions 3 and 4 hydrogen bond with each other to form a terminator stem-loop, which abruptly halts the continued transcription of the *trp* operon.

3. The 3' region of the TPP riboswitch in *Bacillus subtilis* is very similar to the TPP riboswitch in *E. coli*. However, the riboswitch in *B. subtilis* regulates transcription, whereas the one in *E. coli* regulates translation. What is the role of the 5' region in both riboswitches? How can one riboswitch regulate transcription while the other regulates translation?

Topic: What topic in genetics does this question address?

The topic is riboswitches. More specifically, the question is about TPP riboswitches found in two different bacteria.

Information: What information do you know based on the question and your understanding of the topic?

From the question, you know that *B. subtilis* and *E. coli* have TPP riboswitches with similar 5' regions, yet one riboswitch regulates transcription while the other regulates translation. From your understanding of the topic, you may remember that some riboswitches function by forming stem-loops that may affect transcription or translation.

Problem-Solving Strategy: Relate structure and function. Compare and contrast.

One strategy to solve this problem is to first consider the structure and function of the 5' region and then compare its effects in the two types of riboswitches.

Answer: With regard to structure, the 5' region of both riboswitches contains a binding site for TPP. The binding of TPP alters the structure of the 5' region in a way that transmits a subsequent alteration in RNA structure beyond this 5' region, toward the 3' end. In the case of the *B. subtilis* TPP riboswitch, the subsequent alteration in RNA structure causes a terminator-stem loop to form, which ends transcription. In this way, the *B. subtilis* riboswitch controls transcription. By comparison, the binding of TPP to the *E. coli* TPP riboswitch causes the Shine-Dalgarno sequence and AUG start codon to be sequestered in a stem-loop, which inhibits translation. Therefore, the *E. coli* riboswitch controls translation.

Conceptual Questions

C1. What is the difference between a constitutive gene and a regulated gene?

C2. In general, why is it important to regulate genes? Discuss examples of situations in which it would be advantageous for a bacterial cell to regulate genes.

C3. If a gene is repressible and under positive control, what kind of effector molecule and regulatory protein are involved in its regulation? Explain how the binding of the effector molecule affects the regulatory protein.

C4. Transcriptional regulation often involves a regulatory protein that binds to a segment of DNA and a small effector molecule that binds to the regulatory protein. Do each of the following terms apply to a regulatory protein, a segment of DNA, or a small effector molecule?

    A. Repressor   E. Activator

    B. Inducer   F. Attenuator

    C. Operator site   G. Inhibitor

    D. Corepressor

C5. An operon is repressible—a small effector molecule turns off its transcription. Which combination(s) of small effector molecule and regulatory protein could be involved in this process?

    A. An inducer plus a repressor

    B. A corepressor plus a repressor

    C. An inhibitor plus an activator

    D. An inducer plus an activator

C6. Some mutations have a *cis*-effect on gene expression, whereas others have a *trans*-effect. Explain the molecular differences between these two types of mutations. Which type of mutation can be complemented in a merozygote experiment?

C7. What is enzyme adaptation? From a genetic point of view, how does it occur?

C8. In the *lac* operon, how would gene expression be affected if each one of the following segments was missing?

    A. *lac* operon promoter

    B. Operator site

    C. *lacA* gene

C9. If an abnormal repressor protein could still bind allolactose, but the binding of allolactose did not alter the conformation of the repressor protein, how would this affect the expression of the *lac* operon?

C10. What is diauxic growth? Explain the roles of cAMP and the catabolite activator protein (CAP) in this process.

C11. Mutations may have an effect on the expression of the *lac* operon and the *trp* operon. Would the following mutations have a *cis*- or *trans*-effect on the expression of the protein-encoding genes in the operon?

378 CHAPTER 16 :: GENE REGULATION IN BACTERIA

    A. A mutation in the operator site that prevents lac repressor from binding to it

    B. A mutation in the *lacI* gene that prevents lac repressor from binding to DNA

    C. A mutation in the *trpL* gene that prevents attenuation

C12. Would a mutation that inactivated lac repressor and prevented it from binding to the *lac* operator site result in the constitutive expression of the *lac* operon under all conditions? Explain. What is the disadvantage to a bacterium of having a constitutive *lac* operon?

C13. What is meant by the term *attenuation*? Is it an example of gene regulation during transcription or translation? Explain your answer.

C14. As shown in Figure 16.12, four regions within the *trpL* mRNA can form stem-loops. Let's suppose that mutations have been previously identified that prevent the ability of a particular region to form a stem-loop with a complementary region. For example, a region 1 mutant cannot form a 1–2 stem-loop, but it can still form a 2–3 or 3–4 stem-loop. Likewise, a region 4 mutant can form a 1–2 or 2–3 stem-loop but not a 3–4 stem-loop. Under each of the following sets of conditions, would attenuation occur?

    A. Region 1 is mutant, tryptophan is high, and translation is not occurring.

    B. Region 2 is mutant, tryptophan is low, and translation is occurring.

    C. Region 3 is mutant, tryptophan is high, and translation is not occurring.

    D. Region 4 is mutant, tryptophan is low, and translation is not occurring.

C15. As described in Chapter 15, enzymes known as aminoacyl-tRNA synthetases are responsible for attaching amino acids to tRNAs. Let's suppose that in a mutant bacterium tryptophanyl-tRNA synthetase has a reduced ability to attach tryptophan to tRNA; its activity is only 10% of that found in a normal bacterium. How would that affect attenuation of the *trp* operon? Would the operon be more or less likely to be attenuated? Explain your answer.

C16. The combination of a 3–4 stem-loop and a U-rich attenuator sequence in the *trp* operon (see Figure 16.12) is an example of a ρ-independent terminator. The mechanism of ρ-independent termination is described in Chapter 14. Would you expect attenuation to occur if the tryptophan levels were high and mutations changed the attenuator sequence from UUUUUUUU to UGGUUGUC? Explain why or why not.

C17. Mutations in genes that encode tRNAs can create tRNAs that recognize stop codons. Because stop codons are sometimes called nonsense codons, these types of mutations that affect tRNAs are called nonsense suppressors. For example, a normal tRNA^{Gly} has an anticodon sequence CCU that recognizes a glycine codon in mRNA (GGA) and puts in a glycine during translation. However, a mutation in the gene that encodes tRNA^{Gly} could change the anticodon to ACU. This mutant tRNA^{Gly} would still carry glycine, but it would recognize the stop codon UGA. Would this mutation affect attenuation of the *trp* operon? Explain why or why not. Note: To answer this question, you need to look carefully at Figure 16.12 and see if you can identify any stop codons that may exist beyond the UGA stop codon that is located after region 1.

C18. Translational control is usually aimed at preventing the initiation of translation. With regard to cellular efficiency, why do you think this is the case?

C19. What is antisense RNA? How does it affect the translation of a complementary mRNA?

C20. A species of bacteria can synthesize the amino acid histidine, so they do not require histidine in their growth medium. A key enzyme, which we will call histidine synthetase, is necessary for histidine biosynthesis. When these bacteria are given histidine in their growth medium, they stop synthesizing histidine intracellularly. Based on this observation alone, propose three different regulatory mechanisms to explain why histidine biosynthesis ceases when histidine is in the growth medium. To explore this phenomenon further, you measure the amount of intracellular histidine synthetase when cells are grown in the presence and absence of histidine. In both conditions, the amount of this protein is identical. Which mechanism of regulation is consistent with this observation?

C21. Using three examples, describe how allosteric sites are important in the function of genetic regulatory proteins.

C22. How are the actions of lac repressor and trp repressor similar and how are they different with regard to their binding to operator sites, their effects on transcription, and the influences of small effector molecules?

C23. Transcriptional repressor proteins (e.g., lac repressor), antisense RNA, and feedback inhibition are three different mechanisms that turn off the expression of genes and gene products. Which of these three mechanisms will be most effective in each of the following situations?

    A. Shutting down the synthesis of a polypeptide

    B. Shutting down the synthesis of mRNA

    C. Shutting off the function of a protein

For your answers to parts A–C that list more than one mechanism, which mechanism will be the fastest or the most efficient?

Application and Experimental Questions

E1. Answer the following questions regarding the experiment of Figure 16.7.

    A. Why was β-ONPG used? Why was no yellow color observed in one of the four tubes? Can you propose alternative methods for measuring the level of expression of the *lac* operon?

    B. The yellow color as measured by spectrophotometry was twice as intense for the mated strain as for the parent strain. Why was this result obtained?

E2. Chapter 20 describes a blotting method known as Northern blotting, which can be used to detect RNA transcribed from a particular gene or a particular operon. In this method, a specific RNA is detected by using a short segment of labeled DNA, which is complementary to the RNA that the researcher wishes to detect. After the labeled DNA binds to the RNA within a blot of a gel, the RNA is visualized as a dark band. For example, a segment of labeled DNA complementary to the mRNA of the *lac* operon could be used to specifically detect the *lac* operon mRNA on a gel blot. As shown here, the method of Northern blotting can be used to determine the amount of a particular RNA transcribed under different types of growth conditions. In this Northern blot, bacteria containing a normal *lac* operon were grown under different types of conditions, and then the mRNA was isolated from the cells and subjected to Northern blotting, using a segment of labeled DNA that is complementary to the mRNA of the *lac* operon.

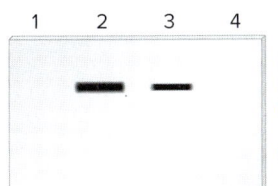

1 2 3 4

Lane 1. Growth in medium containing glucose
Lane 2. Growth in medium containing lactose
Lane 3. Growth in medium containing glucose and lactose
Lane 4. Growth in medium that doesn't contain glucose or lactose

Based on your understanding of the regulation of the *lac* operon, explain these results. Which is more effective at shutting down the *lac* operon, the binding of *lac* repressor or the removal of CAP? Explain your answer based on the results shown in the Northern blot.

E3. As described in experimental question E2 and also in Chapter 20, the technique of Northern blotting can be used to detect the level of transcription of a specific RNA. Draw the results you would expect from a Northern blot if bacteria were grown in a medium containing lactose (and no glucose) but had the following mutations:

Lane 1. Normal strain

Lane 2. Strain with a mutation that inactivates lac repressor

Lane 3. Strain with a mutation that prevents allolactose from binding to lac repressor

Lane 4. Strain with a mutation that inactivates CAP

How would your results differ if these bacterial strains were grown in media that did not contain lactose or glucose?

E4. An absentminded researcher follows the protocol described in Figure 16.7 and (at the end of the experiment) does not observe any yellow color in any of the tubes. Yikes! Which of the following mistakes could account for this observation?

A. Forgot to sonicate the cells

B. Forgot to add lactose to two of the tubes

C. Forgot to add β-ONPG to the four tubes

E5. Explain how the data shown in Figure 16.9 indicate that two operator sites are necessary for repression of the *lac* operon. What would the results have been if all three operator sites were required for the binding of lac repressor?

E6. A mutant strain has a defective *lac* operator site that results in the constitutive expression of the *lac* operon. Outline an experiment you would carry out to demonstrate that the operator site must be physically adjacent to the genes that it influences. Based on your knowledge of the *lac* operon, describe the results you would expect.

E7. Suppose that you have isolated a mutant strain of *E. coli* in which the *lac* operon is constitutively expressed. To understand the nature of this defect, you create a merozygote in which the mutant strain contains an F' factor with a normal *lac* operon and a normal *lacI* gene. You then compare the mutant strain and the merozygote with regard to their β-galactosidase activities in the presence and absence of lactose. You obtain the following results:

	Addition of Lactose	Amount of β-Galactosidase (% of mutant strain in the presence of lactose)
Mutant	No	100
Mutant	Yes	100
Merozygote	No	100
Merozygote	Yes	200

Explain the nature of the defect in the mutant strain.

Questions for Student Discussion/Collaboration

1. Discuss the advantages and disadvantages of genetic regulation at the different points identified in Figure 16.1.

2. Looking at Figure 16.10, discuss possible "molecular ways" that the cAMP-CAP complex and lac repressor may influence RNA polymerase function. In other words, try to explain how the bending and looping in DNA may affect the ability of RNA polymerase to initiate transcription.

Answers to Comprehension Questions

16.1: c, a

16.2: b, d, b, d

16.3: a, b, a

16.4: d, d

16.5: a

Note: All answers appear in Connect; the answers to even-numbered questions and all Concept Check questions are in Appendix B.

17

CHAPTER OUTLINE

- 17.1 Regulatory Transcription Factors
- 17.2 Chromatin Remodeling, Histone Variants, and Histone Modification
- 17.3 DNA Methylation
- 17.4 Overview of Epigenetics
- 17.5 Epigenetics and Development
- 17.6 Epigenetics and Environmental Agents
- 17.7 Regulation of Translation and RNA Stability

X-chromosome inactivation in female mammals. This micrograph shows the nucleus from a human female cell in which a yellow fluorescent probe has been used to highlight the X chromosomes. The Barr body is more compact than the active X chromosome, which is to the left of the Barr body. The compaction of the Barr body is due to epigenetic modifications.

Courtesy of I. Solovei, University of Munich (LMU)

GENE REGULATION IN EUKARYOTES

As discussed in Chapter 16, **gene regulation** refers to the phenomenon whereby the level of gene expression can be controlled so that genes can be expressed at high or low levels. The ability to regulate genes provides many benefits to eukaryotic organisms, a category that includes protists, fungi, plants, and animals. Like their prokaryotic counterparts, eukaryotic cells need to adapt to changes in their environment. For example, eukaryotic cells respond to changes in nutrient availability by enzyme adaptation, much as prokaryotic cells do. Eukaryotic cells also respond to environmental stresses such as ultraviolet (UV) radiation by inducing genes that provide protection against harmful environmental agents. An example is the ability of humans to develop a tan. The tanning response helps to protect a person's skin cells against the damaging effects of UV rays.

Among plants and animals, multicellularity and a more complex cell structure demand a much greater level of gene regulation. The life cycle of complex eukaryotic organisms involves a progression through developmental stages that lead to a mature organism. Some genes are expressed only during early stages of development, such as the embryonic stage, whereas others are expressed in the adult. In addition, complex eukaryotic species are composed of many different tissues that contain a variety of cell types. Gene regulation is necessary to ensure the differences in structure and function among distinct cell types. It is amazing that the various cells within a multicellular organism usually contain the same genetic material, yet phenotypically may be quite different. For example, the appearance of a human nerve cell seems about as similar to a muscle cell as an amoeba is to a paramecium. In spite of these phenotypic differences, a human nerve cell and muscle cell actually contain the same complement of human chromosomes. Nerve and muscle cells look strikingly different because of gene regulation rather than differences in DNA content. Many genes are expressed in the nerve cell and not in the muscle cell, and vice versa.

The molecular mechanisms that underlie gene regulation in eukaryotes bear many similarities to the ways that bacteria regulate their genes. As in bacteria, regulation in eukaryotes can occur at any step in the pathway of gene expression (**Figure 17.1**). Most of this chapter is focused on gene regulation at the level of transcription, an important form of control. We will also consider how eukaryotes regulate translation and RNA stability.

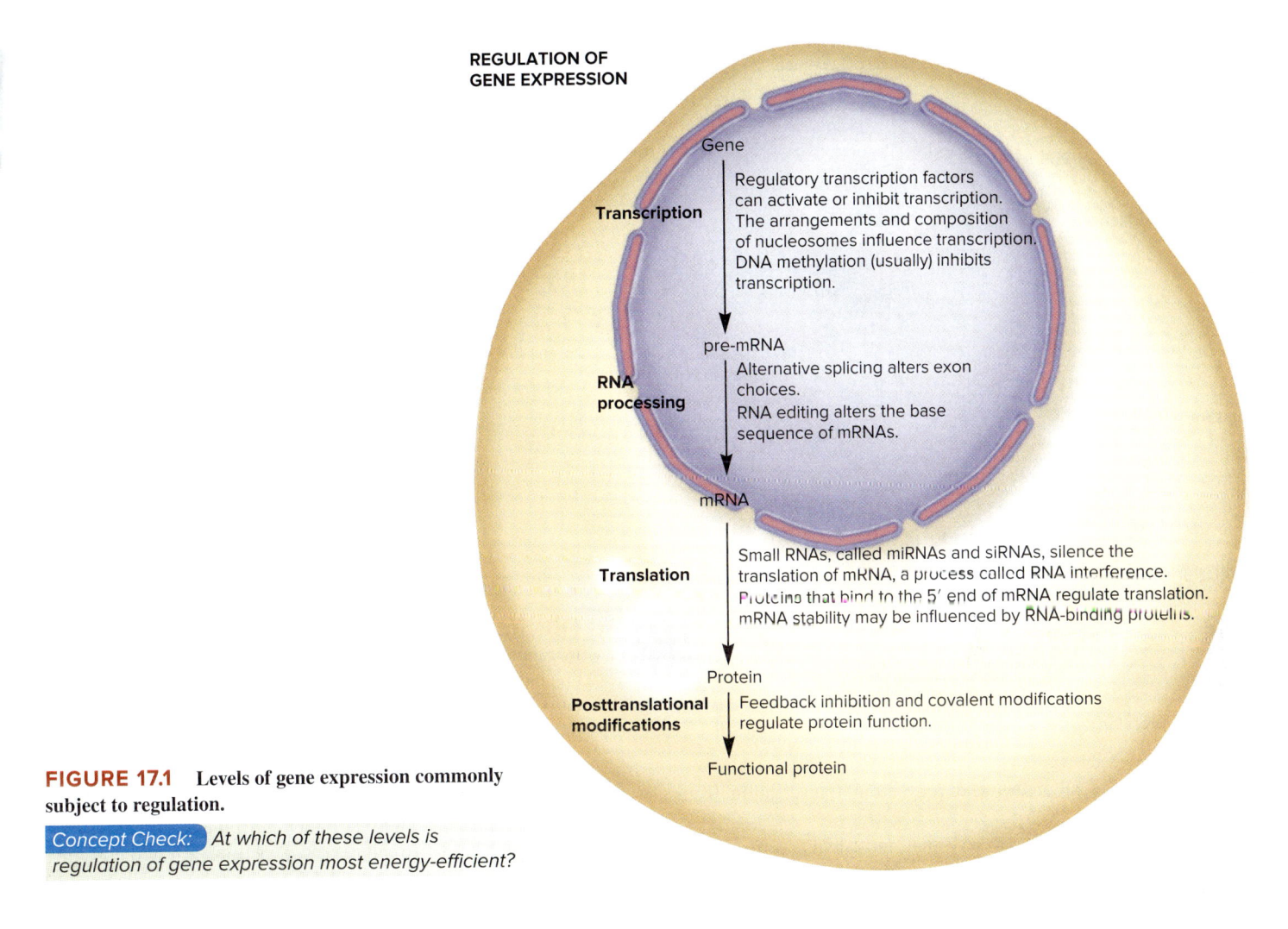

FIGURE 17.1 Levels of gene expression commonly subject to regulation.

Concept Check: At which of these levels is regulation of gene expression most energy-efficient?

17.1 REGULATORY TRANSCRIPTION FACTORS

Learning Outcomes:

1. Distinguish between general and regulatory transcription factors.
2. List the factors that contribute to combinatorial control.
3. Explain how a regulatory transcription factor exerts its effects via TFIID or mediator.
4. Describe three ways that the function of a regulatory transcription factor can be modulated.
5. Outline the steps whereby glucocorticoid receptors regulate genes.

The term **transcription factor** is broadly used to describe a category of proteins that influence the ability of RNA polymerase to transcribe a given gene. We will focus our attention on transcription factors that affect the ability of RNA polymerase to begin the transcription process. Such transcription factors can regulate the binding of RNA polymerase to the core promoter and/or control the switch from the initiation to the elongation stage of transcription. Two types of transcription factors play a key role in these processes. In Chapter 14, we considered **general transcription factors (GTFs)**, which are required for the binding of RNA polymerase to the core promoter and its progression to the elongation stage. General transcription factors are necessary for a basal level of transcription. In addition, eukaryotic cells possess a diverse array of **regulatory transcription factors** that serve to regulate the rate of transcription of target genes.

As discussed in this section, some regulatory transcription factors exert their effects by influencing the ability of RNA polymerase to begin transcription of a particular gene. They recognize *cis* acting elements that are often located in the vicinity of the core promoter but may be relatively far away. These DNA sequences are analogous to the operator sites that control bacterial promoters. In eukaryotes, these DNA sequences are generally known as **regulatory elements,** or **control elements.** When a regulatory transcription factor binds to a regulatory element, it affects the transcription of an associated gene. For example, the binding of regulatory transcription factors may enhance the rate of transcription (**Figure 17.2a**). Such a transcription factor is termed an **activator,** and the sequence it binds to is called an **enhancer.** Alternatively, regulatory transcription factors may act as **repressors** by binding

FIGURE 17.2 Overview of transcriptional regulation by regulatory transcription factors. A regulatory transcription factor can act as either **(a)** an activator to increase the rate of transcription or **(b)** a repressor to decrease the rate of transcription.

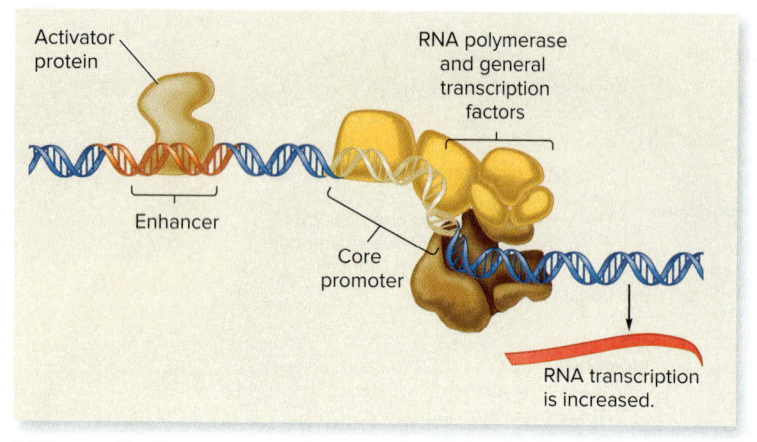

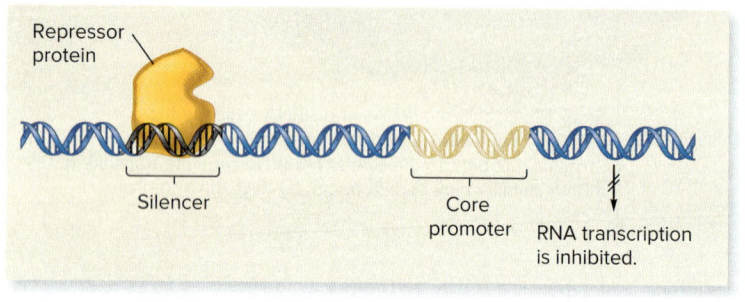

to elements called **silencers** and preventing transcription from occurring (**Figure 17.2b**).*

Researchers have discovered that most eukaryotic genes, particularly those found in multicellular species, are regulated by many factors. This phenomenon is called **combinatorial control** because the combination of many factors determines the expression of any given gene. At the level of transcription, the following are factors that commonly contribute to combinatorial control:

1. One or more activator proteins may stimulate the ability of RNA polymerase to initiate transcription.
2. One or more repressor proteins may inhibit the ability of RNA polymerase to initiate transcription.
3. The function of activators and repressors may be modulated in a variety of ways, including the binding of small effector molecules, protein-protein interactions, and covalent modifications.
4. Regulatory proteins may alter the composition or arrangements of nucleosomes in the vicinity of a promoter, thereby affecting transcription.
5. DNA methylation may inhibit transcription, either by preventing the binding of an activator protein or by recruiting proteins that cause the chromatin to become more compact.

All five of these factors can contribute to the regulation of a single gene, or possibly only three or four will play a role. In most cases, transcriptional regulation is aimed at controlling the initiation of transcription at the promoter. In this section, we will survey the first three factors that may contribute to combinatorial control of transcription. In the next two sections, we will consider the fourth and fifth factors.

Regulatory Transcription Factors Recognize Regulatory Elements That Function as Enhancers or Silencers

As mentioned previously, when the binding of a regulatory transcription factor to a regulatory element increases transcription, the regulatory element is known as an enhancer. Such elements can stimulate transcription 10- to 1000-fold, a phenomenon known as **up regulation.** Alternatively, regulatory elements that serve to inhibit transcription are called silencers, and their action is called **down regulation.**

Many regulatory elements are **orientation-independent,** or **bidirectional.** Such a regulatory element can function in the forward or reverse orientation. For example, let's consider an enhancer with a forward orientation as shown below:

$$5'-GATA-3'$$
$$3'-CTAT-5'$$

This enhancer is also bound by a regulatory transcription factor and enhances transcription even when it is rotated 180° and oriented in the reverse orientation:

$$5'-TATC-3'$$
$$3'-ATAG-5'$$

---

*Some geneticists use the term *enhancer* to describe a genetic element that binds either activator or repressor proteins. In this text, we will use *enhancer* to describe elements that bind activators and *silencer* to describe elements that bind repressors, to avoid confusion.

Striking variation is also found in the location of regulatory elements relative to a gene's promoter. Regulatory elements are often located in a region within 200 base pairs (bp) upstream from the promoter site. However, they can be quite distant from the promoter, even 100,000 bp away, yet exert strong effects on the ability of RNA polymerase to initiate transcription at the core promoter! Regulatory elements were first discovered by Susumu Tonegawa and coworkers in the 1980s. While studying genes that play a role in immunity, these researchers identified a region that is far away from the core promoter, but is needed for high levels of transcription to take place. In some cases, regulatory elements are located downstream from the promoter site and may even be found within introns, the noncoding parts of genes. As you may imagine, the variation in the orientation and location of regulatory elements profoundly complicates the efforts of geneticists to identify those elements that affect the expression of any given gene.

Regulatory Transcription Factors May Exert Their Effects Through TFIID or Mediator

Different mechanisms have been discovered that explain how a regulatory transcription factor can bind to a regulatory element, thereby affecting gene transcription. For most genes, more than one mechanism is involved. The net effect of a regulatory transcription factor is to influence the ability of RNA polymerase to transcribe a given gene. However, most regulatory transcription factors do not bind directly to RNA polymerase. How then do these regulatory transcription factors exert their effects? For protein-encoding genes in eukaryotes, regulatory transcription factors commonly influence the function of RNA polymerase II by interacting with other proteins that directly control RNA polymerase II. Two protein complexes that communicate the effects of regulatory transcription factors are transcription factor IID (TFIID) and mediator.

TFIID As discussed in Chapter 14, **TFIID** is a general transcription factor that binds to the TATA box and is needed to recruit RNA polymerase II to the core promoter. Some regulatory transcription factors bind to a regulatory element and then influence the function of TFIID. Activator proteins can enhance the ability of TFIID to initiate transcription. One possibility is that activator proteins might help recruit TFIID to the TATA box, or they might enhance the function of TFIID in a way that facilitates its ability to recruit RNA polymerase II. In some cases, activator proteins exert their effects by interacting with **coactivators**—proteins that increase the rate of transcription but do not directly bind to the DNA itself. This type of activation is shown in **Figure 17.3a**. In contrast, repressors inhibit the function of TFIID. They could exert their effects by preventing the binding of TFIID to the TATA box (**Figure 17.3b**) or by inhibiting the ability of TFIID to recruit RNA polymerase II to the core promoter.

Mediator A second way that regulatory transcription factors control RNA polymerase II is via mediator—a protein complex discovered by Roger Kornberg and colleagues in 1990. The name **mediator** refers to the observation that this complex mediates the interaction between RNA polymerase II and regulatory transcription factors. As discussed in Chapter 14, mediator controls the ability of RNA polymerase II to progress to the elongation stage of transcription. Transcriptional activators stimulate the ability of mediator to facilitate the switch between the initiation and elongation stages, whereas repressors have the opposite effect. In the example shown in **Figure 17.4**, an activator binds to a distant enhancer element. The activator protein and mediator are brought together by the formation of a loop within the intervening DNA.

A third way that regulatory transcription factors can influence transcription is by recruiting proteins that affect nucleosome positions and compositions to the promoter region. For example, certain transcriptional activators recruit proteins that facilitate the conversion of chromatin to a conformation that is more accessible to RNA polymerase. We will return to this topic later in this chapter.

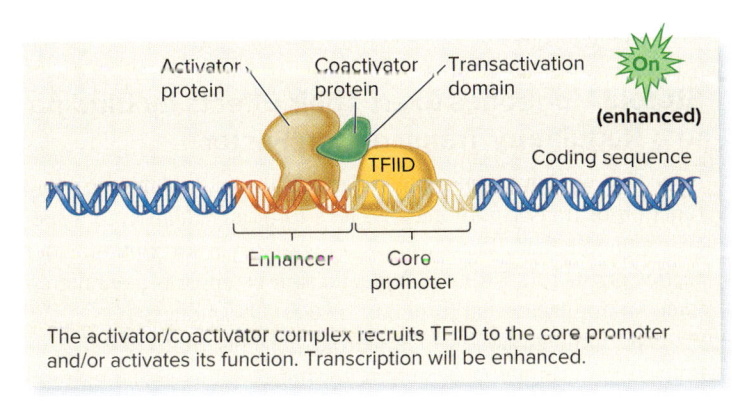

(a) Transcriptional activation via TFIID

The activator/coactivator complex recruits TFIID to the core promoter and/or activates its function. Transcription will be enhanced.

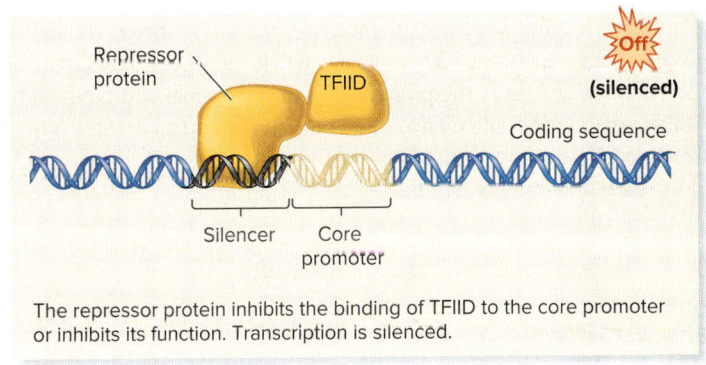

(b) Transcriptional repression via TFIID

The repressor protein inhibits the binding of TFIID to the core promoter or inhibits its function. Transcription is silenced.

FIGURE 17.3 **Effects of regulatory transcription factors on TFIID.** (a) Some activators stimulate the function of TFIID, thereby enhancing transcription. In this example, the activator interacts with a coactivator that directly binds to TFIID and stimulates its function. (b) Regulatory transcription factors may also function as repressors. In the example shown here, the repressor binds to a silencer and inhibits the ability of TFIID to bind to the TATA box at the core promoter.

Concept Check: If a repressor prevents TFIID from binding to the TATA box, why does this inhibit transcription?

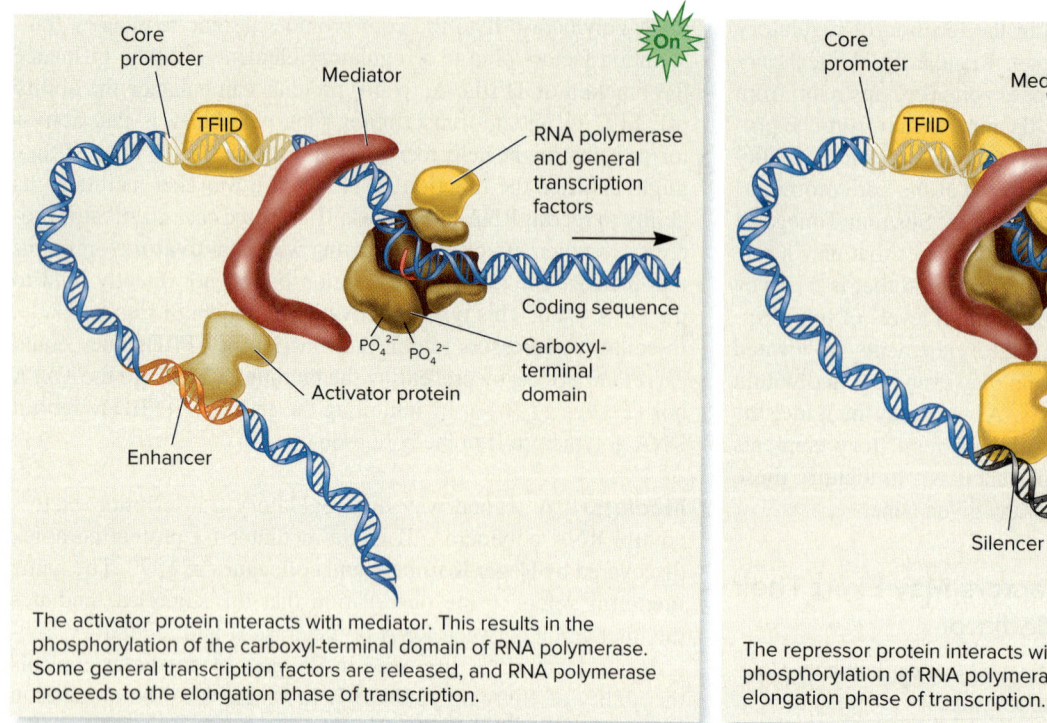

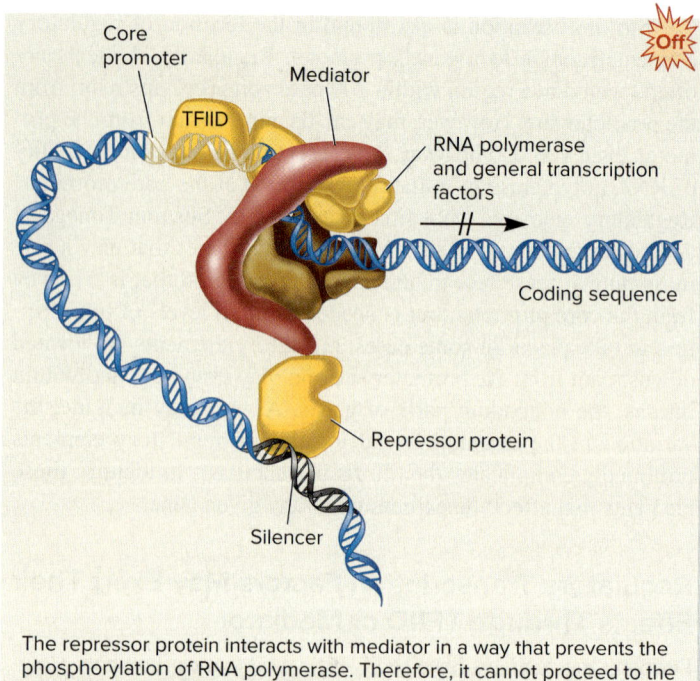

(a) Transcriptional activation via mediator

The activator protein interacts with mediator. This results in the phosphorylation of the carboxyl-terminal domain of RNA polymerase. Some general transcription factors are released, and RNA polymerase proceeds to the elongation phase of transcription.

(b) Transcriptional repression via mediator

The repressor protein interacts with mediator in a way that prevents the phosphorylation of RNA polymerase. Therefore, it cannot proceed to the elongation phase of transcription.

FIGURE 17.4 Effects of regulatory transcription factors on mediator. Some activators interact with mediator in a way that causes RNA polymerase to proceed to the elongation phase of transcription. (a) In this example, the enhancer is relatively far away from the core promoter. Therefore, a loop must form in the DNA so that the activator can interact with mediator. (b) Alternatively, repressors can interact with mediator, thereby preventing RNA polymerase from proceeding to the elongation phase of transcription.

Concept Check: When an activator protein interacts with mediator, how does this affect the function of RNA polymerase?

The Function of Regulatory Transcription Factor Proteins Can Be Modulated in Three Ways

Thus far, we have considered the structures of regulatory transcription factors and the molecular mechanisms that account for their abilities to control transcription. The functions of the regulatory transcription factors themselves must also be modulated. Why is this necessary? The answer is that the genes they control must be turned on at the proper time, in the correct cell type, and under the appropriate environmental conditions. Eukaryotes have evolved different ways to modulate the functions of these proteins.

The functions of regulatory transcription factor proteins are controlled in three common ways:

- The binding of small effector molecules
- Protein-protein interactions
- Covalent modifications.

Figure 17.5 depicts these three mechanisms of modulating the functions of regulatory transcription factors. Usually, one or more of these modulating effects are important in determining whether a transcription factor can bind to the DNA or influence transcription by RNA polymerase. For example, a small effector molecule may bind to a regulatory transcription factor and promote its binding to DNA (Figure 17.5a). We will see that steroid hormones function in this manner. Another important mechanism of modulation is via protein-protein interactions (Figure 17.5b). The formation of dimers is a fairly common means of controlling transcription. Finally, the function of a regulatory transcription factor can be affected by covalent modifications such as the attachment of a phosphate group (Figure 17.5c).

Steroid Hormones Exert Their Effects by Binding to a Regulatory Transcription Factor

Now that we have a general understanding of the structure and function of transcription factors, let's turn our attention to a specific example that illustrates how a regulatory transcription factor carries out its role within living cells. Our example is a regulatory transcription factor that responds to steroid hormones.

The ultimate action of a steroid hormone is to affect gene transcription. In animals, steroid hormones act as signaling molecules that are synthesized by endocrine glands and secreted into the bloodstream. The hormones are then taken up by cells that respond to these substances in different ways. For example, glucocorticoid hormones influence nutrient metabolism in most body cells. Other steroid hormones, such as estrogen and testosterone, promote the development of secondary sexual characteristics and play a role in reproduction.

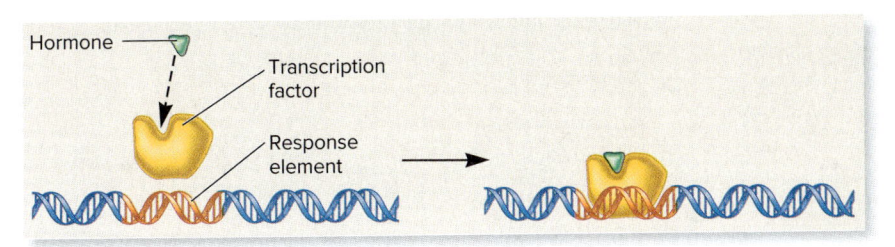

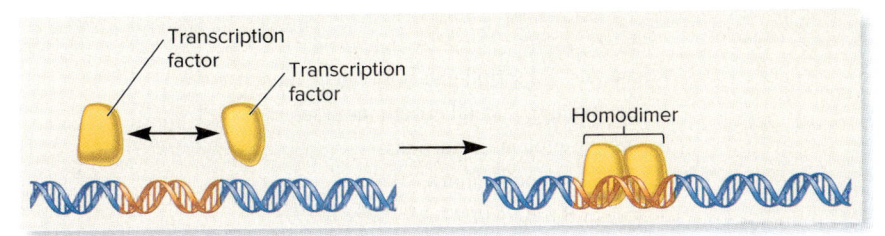

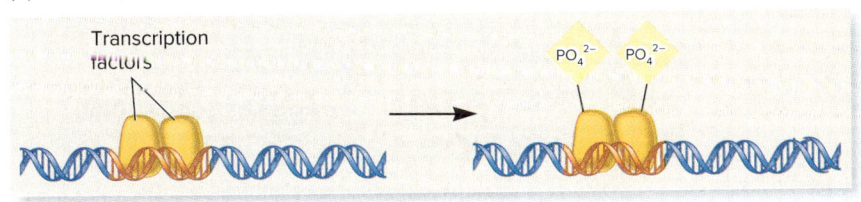

FIGURE 17.5 Common ways to modulate the function of regulatory transcription factors. (a) The binding of an effector molecule such as a hormone may influence the ability of a transcription factor to bind to the DNA. (b) Protein–protein interactions among transcription factor proteins may influence their functions. (c) Covalent modifications such as phosphorylation may alter transcription factor function.

Figure 17.6 shows the stepwise action of glucocorticoid hormones, which are produced in mammals.

1. In this example, the hormone enters the cytosol of a cell by diffusing through the plasma membrane.
2. Once inside, the hormone specifically binds to a **glucocorticoid receptor.** Note: Prior to hormone binding, the glucocorticoid receptor is complexed with proteins known as heat shock proteins (HSP), one example being HSP90.
3. After the hormone binds to the glucocorticoid receptor, HSP90 is released, thereby exposing a nuclear localization signal (NLS) that directs the receptor protein into the nucleus.
4. Two glucocorticoid receptors form a homodimer and then travel through a nuclear pore into the nucleus.
5. In the nucleus, the glucocorticoid receptor homodimer binds to a glucocorticoid response element (GRE) with two copies of the following consensus sequence:

$$5'-AGRACA-3'$$
$$3'-TCYTGT-5'$$

where R is a purine and Y is pyrimidine. A GRE is found next to many genes and functions as an enhancer.
6. The binding of the glucocorticoid receptor homodimer to a GRE activates the transcription of the nearby gene, eventually leading to the synthesis of the encoded protein.

Mammalian cells usually have a large number of glucocorticoid receptors within the cytoplasm. Because GREs are located near dozens of different genes, the uptake of many hormone molecules can activate many glucocorticoid receptors, thereby stimulating the transcription of many different genes. For this reason, a cell can respond to the presence of the hormone in a very complex way. Glucocorticoid hormones stimulate many genes that encode proteins involved in several different cellular processes, including the synthesis of glucose, the breakdown of proteins, and the mobilization of fats. Although the genes are not physically adjacent to each other, the regulation of multiple genes via glucocorticoid hormones is much like the ability of bacterial operons to simultaneously control the expression of several genes.

17.1 REVIEWING THE KEY CONCEPTS

- Regulatory transcription factors can be activators that bind to enhancers and increase transcription, or they can be repressors that bind to silencers and inhibit transcription (see Figure 17.2).
- Combinatorial control means that the expression of a gene is regulated by a variety of factors.
- Regulatory elements are usually orientation-independent.
- Regulatory transcription factors may exert their effects by interacting with TFIID (a general transcription factor) or mediator (see Figures 17.3, 17.4).
- The functions of regulatory transcription factors can be modulated by small effector molecules, protein-protein interactions, and covalent modifications (see Figure 17.5).
- Steroid hormones, such as glucocorticoids, bind to receptors that function as transcriptional activators (see Figure 17.6).

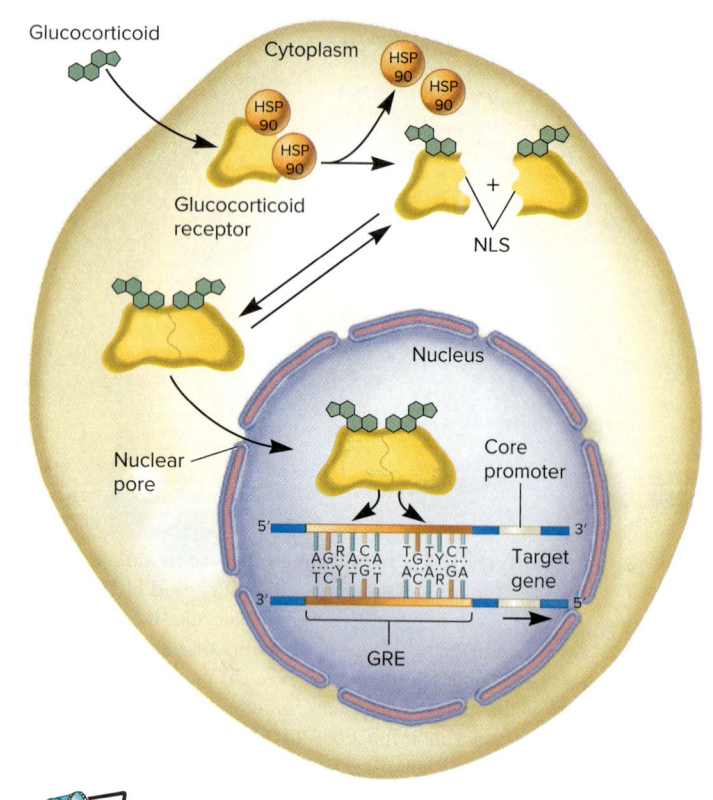

FIGURE 17.6 The action of glucocorticoid hormones. Once inside the cell, the glucocorticoid hormone binds to the glucocorticoid receptor, which then releases a protein known as HSP90. This exposes a nuclear localization signal (NLS). Two glucocorticoid receptors then form a homodimer and travel into the nucleus, where the dimer binds to a glucocorticoid response element (GRE) that is next to a particular gene. The binding of the dimer to the GRE activates the transcription of the adjacent target gene.

Genes → Traits Glucocorticoid hormones are produced by the endocrine glands in response to fasting and activity. They enable the body to regulate its metabolism properly. When glucocorticoids are produced, they are taken into cells and bind to glucocorticoid receptors. This eventually leads to the activation of genes that encode proteins involved in the synthesis of glucose, the breakdown of proteins, and the mobilization of fats.

Concept Check: Explain why the glucocorticoid receptor binds specifically next to the core promoter of certain genes, but not next to the core promoter of most genes.

17.1 COMPREHENSION QUESTIONS

1. Combinatorial control refers to the phenomenon that
   a. transcription factors always combine with each other when regulating genes.
   b. the combination of several factors determines the expression of any given gene.
   c. small effector molecules and regulatory transcription factors are found in many different combinations.
   d. genes and regulatory transcription factors must combine with each other during gene regulation.

2. A bidirectional enhancer has the following sequence:

   5′–GTCA–3′
   3′–CAGT–5′

   Which of the following sequences would also be a functional enhancer?
   a. 5′–ACTG–3′
      3′–TGAC–5′
   b. 5′–TGAC–3′
      3′–ACTG–5′
   c. 3′–GTCA–5′
      5′–CAGT–3′
   d. 3′–TGAC–5′
      5′–ACTG–3′

3. Regulatory transcription factors can be modulated by
   a. the binding of small effector molecules.
   b. protein-protein interactions.
   c. covalent modifications.
   d. any of the above.

17.2 CHROMATIN REMODELING, HISTONE VARIANTS, AND HISTONE MODIFICATION

Learning Outcomes:
1. Describe how chromatin-remodeling complexes alter nucleosomes.
2. Define *histone variant,* and explain why histone variants are functionally important.
3. Explain how histone modifications affect transcription.
4. Summarize the steps that occur for transcriptional activation of a eukaryotic gene.

In Chapter 12, we considered how eukaryotic DNA is complexed with proteins, such as histones, to form chromatin. The term **ATP-dependent chromatin remodeling,** or simply **chromatin remodeling,** refers to dynamic changes in the structure of chromatin that occur during the life of a cell. These changes range from local alterations in the positioning of one or a few nucleosomes to larger changes that affect chromatin structure over a longer distance. Chromatin remodeling is carried out by ATP-dependent chromatin-remodeling complexes, which are a set of diverse multiprotein machines that reposition and restructure nucleosomes.

In eukaryotes, changes in nucleosome position and histone composition are key features of gene regulation. The regulation of transcription depends not only on the activity of regulatory transcription factors that influence RNA polymerase via TFIID or mediator to turn genes on or off, but must also involve changes in chromatin structure that affect the ability of transcription factors to gain access to and bind their target sequences in the promoter region. If the chromatin is in a **closed conformation,** transcription may be difficult or impossible. By comparison, chromatin that is in an **open conformation** is more easily accessible to transcription factors and RNA polymerase, allowing transcription to occur. Although the closed and open conformations may be affected by the relative compaction of a chromosomal region, researchers have recently determined that histone composition and the precise positioning of nucleosomes at or near promoters often play a key role in eukaryotic gene regulation. In this section, we examine the molecular mechanisms that explain how changes in chromatin structure control the regulation of eukaryotic genes.

Chromatin-Remodeling Complexes Alter the Positions and Compositions of Nucleosomes

In recent years, geneticists have been trying to identify the steps that promote the interconversion between the closed and open conformations of chromatin. Nucleosomes have been shown to have different positions in cells that normally express a particular gene compared with cells in which the gene is inactive. For example, in red blood cells that express the β-globin gene, an alteration in nucleosome positioning occurs in the promoter region from nucleotide −500 to nucleotide +200. This alteration is thought to be an important step in the process of expressing the β-globin gene. Based on the analysis of many genes, researchers have discovered that a key role of some transcriptional activators is to orchestrate changes in chromatin structure from the closed to the open conformation by altering nucleosomes.

One way to change chromatin structure is through ATP-dependent chromatin remodeling. In this process, the energy derived from ATP hydrolysis is used to drive changes in the locations and/or compositions of nucleosomes, thereby making the DNA more or less amenable to transcription. Therefore, chromatin remodeling is important for both the activation and repression of transcription.

The remodeling process is carried out by an ATP-dependent chromatin-remodeling complex, a protein complex that recognizes nucleosomes and uses ATP to alter their configuration. All chromatin-remodeling complexes have a catalytic ATPase subunit called **DNA translocase**; this ATPase subunit, similar to what is found in motor proteins, moves along the DNA. Eukaryotes have multiple families of chromatin remodelers. Some common families of ATP-dependent chromatin-remodeling complexes include the SWI/SNF family, the ISWI family, the INO80 family, and the Mi-2 family. The names of these remodelers sometimes refer to the effects of mutations in genes that encode them. For example, the abbreviations SWI and SNF refer to the effects that occur in yeast when these remodeling complexes are defective. SWI mutants are defective in mating-type switching, and SNF mutations create a sucrose nonfermenting phenotype.

How do chromatin remodelers change chromatin structure? Three effects are possible.

- One result of ATP-dependent chromatin remodeling is a change in the positions of nucleosomes (**Figure 17.7a**). This may involve shifts in nucleosomes to new locations or changes in the relative spacing of nucleosomes over a long stretch of DNA.
- A second effect is that remodelers may evict histones from the DNA, thereby creating gaps where nucleosomes are not found (**Figure 17.7b**).
- A third possibility is that remodelers may change the composition of nucleosomes by removing standard histones and replacing them with histone variants (**Figure 17.7c**). The functions of histone variants are described next.

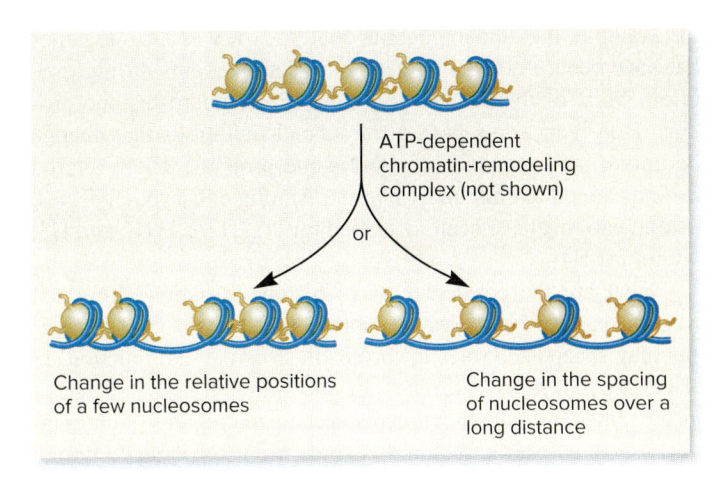

(a) Change in nucleosome position

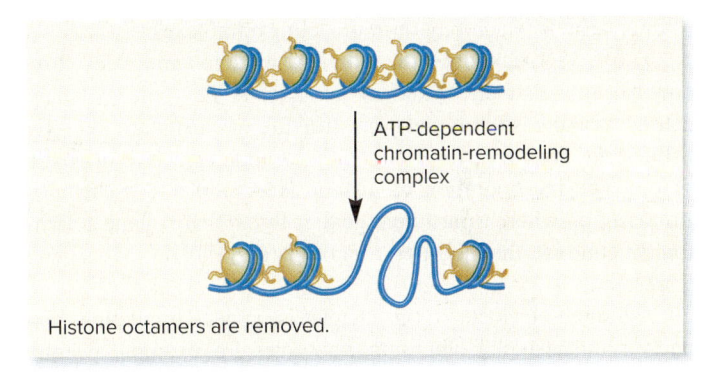

(b) Histone eviction

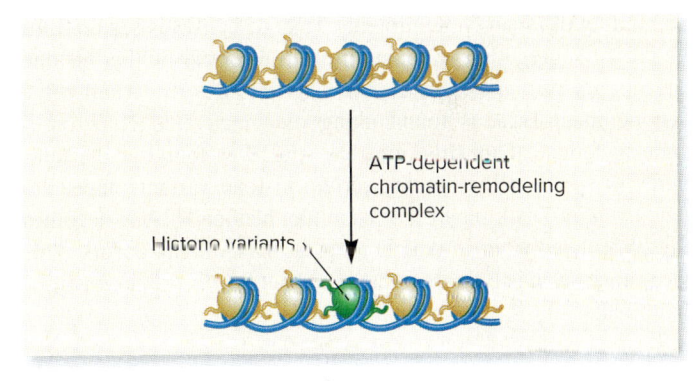

(c) Replacement with histone variants

FIGURE 17.7 ATP-dependent chromatin remodeling. ATP-dependent chromatin-remodeling complexes may **(a)** change the locations of nucleosomes, **(b)** remove histones from the DNA, or **(c)** replace core histones with histone variants.

Concept Check: How might nucleosome eviction affect transcription?

Histone Variants Play Specialized Roles in Chromatin Structure and Function

As discussed in Chapter 12, the genes that encode histones H1, H2A, H2B, H3, and H4 are moderately repetitive. The total number of histone-encoding genes varies from species to species.

For example, the human genome contains over 70 histone genes that have been produced by gene duplication events during evolution. Most of the histone genes encode standard histone proteins. However, a few have accumulated mutations that change the amino acid sequence of the histone proteins. These altered histones are called **histone variants.** Among eukaryotic species, histone variants have been identified for H1, H2A, H2B, and H3, but not for H4.

What are the consequences of histone variation? Research over the past two decades has shown that certain histone variants play specialized roles in chromatin structure and function. In all eukaryotes, histone variants are incorporated into a subset of nucleosomes to create functionally specialized regions of chromatin. In most cases, the standard histones are incorporated into the nucleosomes while new DNA is synthesized during S phase of the cell cycle. Later, some of the standard histones are replaced by histone variants via chromatin-remodeling complexes.

Table 17.1 describes the standard histones and a few histone variants that are found in humans. A key role of many histone variants is to regulate the structure of chromatin, thereby influencing gene transcription. Such variants can have opposite effects. The incorporation of histone H2A.Bbd into a chromosomal region where a particular gene is found favors gene activation. In contrast, the incorporation of histone H1^0 represses gene expression.

Although our focus in this chapter is on gene regulation, histone variants also play other important roles. For example, histone cenH3 (also called CENP-A), which is a variant of histone H3, is found at the centromere of each chromosome and functions in the binding of kinetochore proteins. Histone cenH3 is required for the proper segregation of eukaryotic chromosomes. Other histone variants are primarily found at specialized sites in certain cells. Histone macroH2A is found along the inactivated X chromosome in female mammals, whereas spH2B is found at the telomeres in sperm cells. Finally, certain histone variants appear to play a role in DNA repair. For example, histone H2A.X becomes phosphorylated where a double-stranded DNA break occurs. This phosphorylation is thought to be important for the proper repair of that break.

The Histone Code Also Controls Gene Transcription

As described in Chapter 12, each of the core histone proteins consists of a globular domain and a flexible, charged amino terminus called an amino-terminal tail (refer back to Figure 12.14a). The DNA wraps around the globular domains, and the amino-terminal tails protrude from the chromatin. In recent years, researchers have discovered that particular amino acids in the amino-terminal tails of standard histones and histone variants are subject to several types of covalent modifications, including acetylation, methylation, and phosphorylation. Over 50 different histone-modifying enzymes have been identified in mammals that selectively modify amino-terminal tails of histones. **Figure 17.8a** shows examples of sites in the tails of H2A, H2B, H3, and H4 that can be modified.

TABLE 17.1

Standard Human Histones and Examples of Histone Variants

Histone	Type	Number of Genes in Humans	Function
H1	Standard	11	Standard linker histone*
H1^0	Variant	1	Linker histone associated with chromatin compaction and gene repression
H2A	Standard	15	Standard core histone
MacroH2A	Variant	1	Core histone that is abundant on the inactivated X chromosome in female mammals; plays a role in chromatin compaction
H2A.Z	Variant	1	Core histone that is usually found in nucleosomes that flank the transcriptional start site of promoters; plays a role in gene transcription
H2A.Bbd	Variant	1	Core histone that promotes the open conformation of chromatin; plays a role in gene activation
H2A.X	Variant	1	Plays a role in DNA repair
H2B	Standard	17	Standard core histone
spH2B	Variant	1	Core histone found in the telomeres of sperm cells
H3	Standard	10	Standard core histone
cenH3	Variant	1	Core histone found at centromeres; involved with the binding of kinetochore proteins
H3.3	Variant	2	Core histone that promotes the open conformation of chromatin; plays a role in gene activation
H4	Standard	14	Standard core histone

*H1 in mammals is found in five subtypes.

How do histone modifications affect the level of transcription? First, they may directly influence interactions between histones and the DNA. For example, positively charged lysines within the core histone proteins can be acetylated by a type of enzyme called **histone acetyltransferase.** The attachment of the acetyl group (—COCH$_3$) eliminates the positive charge on the lysine side chain, thereby disrupting the electrostatic attraction between the histone protein and the negatively charged DNA backbone (**Figure 17.8b**).

In addition, histone modifications occur in patterns that are recognized by proteins. According to the **histone code hypothesis,** proposed by Brian Strahl, C. David Allis, and Bryan Turner in 2000, the pattern of histone modification acts much like a language or code in specifying alterations in chromatin structure. For example, one pattern might involve phosphorylation of the serine at the first position in H2A and acetylation of the fifth and

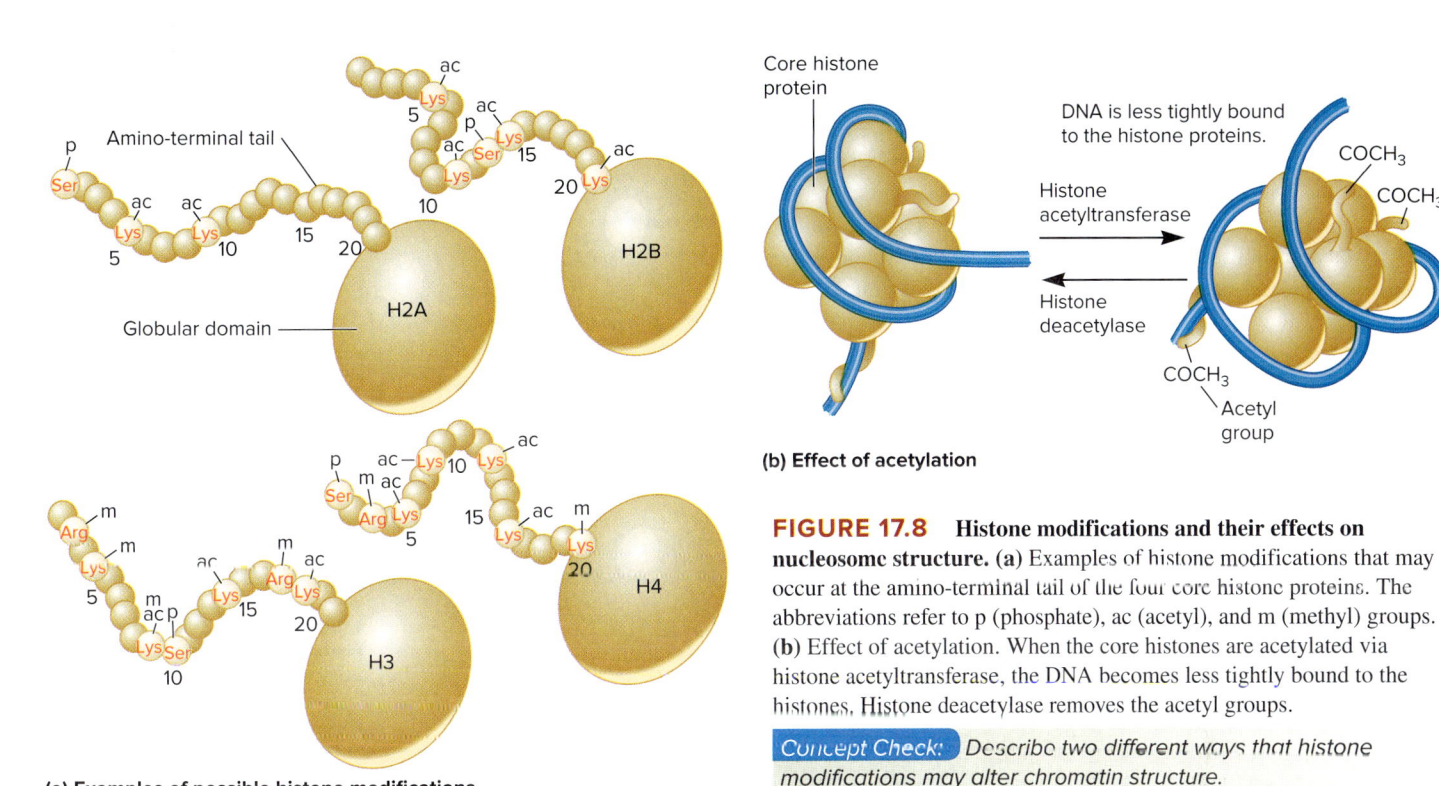

17.2 CHROMATIN REMODELING, HISTONE VARIANTS, AND HISTONE MODIFICATION

(a) Examples of possible histone modifications

(b) Effect of acetylation

FIGURE 17.8 Histone modifications and their effects on nucleosome structure. (a) Examples of histone modifications that may occur at the amino-terminal tail of the four core histone proteins. The abbreviations refer to p (phosphate), ac (acetyl), and m (methyl) groups. (b) Effect of acetylation. When the core histones are acetylated via histone acetyltransferase, the DNA becomes less tightly bound to the histones. Histone deacetylase removes the acetyl groups.

Concept Check: Describe two different ways that histone modifications may alter chromatin structure.

eighth amino acids in H4, which are lysines. A different pattern could involve acetylation of the fifth amino acid, a lysine, in H2B and methylation of the third amino acid in H4, which is an arginine.

The pattern of covalent modifications to the amino-terminal tails provides binding sites for proteins that subsequently affect the degree of transcription. One pattern of histone modification may attract proteins that inhibit transcription, which would silence the transcription of genes in the region. A different combination of histone modifications may attract proteins, such as chromatin-remodeling complexes, which would serve to alter the positions of nucleosomes in a way that promotes gene transcription. For example, the acetylation of histones attracts certain chromatin remodelers that can shift or evict nucleosomes, thereby aiding in the transcription of genes. Overall, the histone code is thought to play an important role in determining whether the information within the genomes of eukaryotic species is accessed. Researchers are trying to decipher the effects of the covalent modifications that make up the histone code.

Eukaryotic Genes Are Flanked by Nucleosome-Free Regions and Well-Positioned Nucleosomes

Studies over the last 15 years or so have revealed that many eukaryotic genes show a common pattern of nucleosome organization (**Figure 17.9**). For active genes or those genes that can be activated, the core promoter is found at a **nucleosome-free region (NFR)**, which is a region of DNA where nucleosomes are not found. An NFR is typically 150 bp in length. Although the NFR may be required for transcription, it is not, by itself, sufficient for gene activation. At any given time in the life of a cell, many genes that contain an NFR are not being actively transcribed.

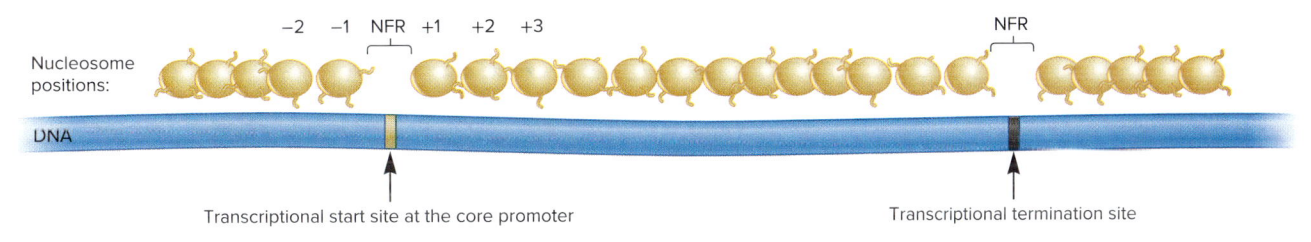

A nucleosome-free region (NFR) is found at the beginning and end of many genes. Nucleosomes tend to be precisely positioned near the beginning and end of a gene but are less regularly distributed elsewhere.

FIGURE 17.9 Nucleosome arrangements and composition in the vicinity of a protein-encoding gene.

Concept Check: Why is an NFR needed at the core promoter for transcription to occur?

The NFR at the transcription start site (TSS) is flanked by two well-positioned nucleosomes that are termed the −1 and +1 nucleosomes. In yeast, the TSS is usually at the boundary between the NFR and the +1 nucleosome. However, in animals, the TSS is about 60 bp farther upstream and within the NFR. The +1 nucleosome typically contains histone variants H2A.Z and H3.3. Depending on the species and the gene, these variants may also be found in the −1 nucleosome and in some of the nucleosomes that immediately follow the +1 nucleosome in the transcribed region. For example, the +2 nucleosome is likely to contain H2A.Z, but not as likely as the +1 nucleosome. Similarly, the +3 nucleosome is likely to contain H2A.Z, but not as likely as the +2 nucleosome.

The nucleosomes downstream from the +1 nucleosome tend to be evenly spaced near the beginning of a eukaryotic gene, but their spacing becomes less regular farther downstream. The end of many eukaryotic genes has a well-positioned nucleosome that is followed by an NFR. This arrangement at the end of genes may be important for transcriptional termination.

Transcriptional Activation Involves Changes in Nucleosome Positions and Composition and Histone Modifications

A key role of certain transcriptional activators is to recruit chromatin-remodeling complexes and histone-modifying enzymes to the promoter region. Though the order of recruitment may differ among specific transcriptional activators, this appears to be critical for transcriptional initiation and elongation. **Figure 17.10** presents a general scheme for how transcriptional activators may facilitate transcription in a eukaryotic gene, such as a gene found in yeast.

1. A transcriptional activator binds to an enhancer in the NFR.
2. The activator then recruits chromatin-remodeling complexes and histone-modifying enzymes to this region. The chromatin remodelers may shift nucleosomes or temporarily evict nucleosomes from the promoter region. Nucleosomes containing the histone variant H2A.Z, which are typically found at the +1 nucleosome, are thought to be more easily removed from the DNA than those containing the standard histone H2A. Histone-modifying enzymes, such as histone acetyltransferase, covalently modify histone proteins and may affect nucleosome contact with the DNA.
3. The actions of chromatin-remodeling complexes and histone-modifying enzymes facilitate the binding of general transcription factors and RNA polymerase II to the core promoter, thereby allowing the formation of a preinitiation complex.

FIGURE 17.10 A simplified model for the transcriptional activation of a eukaryotic gene. The activation involves changes in nucleosome arrangements and histone modifications.

Concept Check: Explain why histone eviction or displacement is needed for the elongation phase of transcription to occur.

Many genes are flanked by nucleosome-free regions (NFR) and well-positioned nucleosomes.

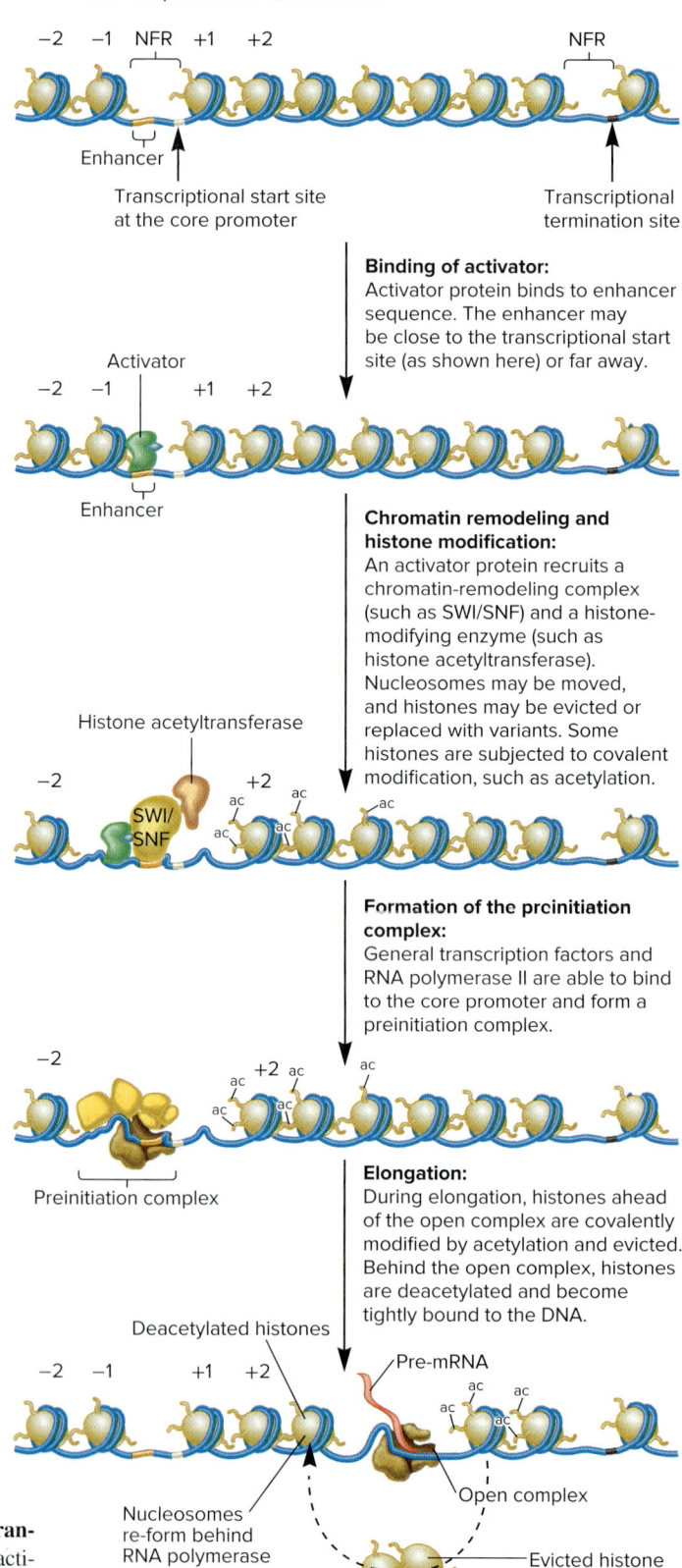

Binding of activator: Activator protein binds to enhancer sequence. The enhancer may be close to the transcriptional start site (as shown here) or far away.

Chromatin remodeling and histone modification: An activator protein recruits a chromatin-remodeling complex (such as SWI/SNF) and a histone-modifying enzyme (such as histone acetyltransferase). Nucleosomes may be moved, and histones may be evicted or replaced with variants. Some histones are subjected to covalent modification, such as acetylation.

Formation of the preinitiation complex: General transcription factors and RNA polymerase II are able to bind to the core promoter and form a preinitiation complex.

Elongation: During elongation, histones ahead of the open complex are covalently modified by acetylation and evicted. Behind the open complex, histones are deacetylated and become tightly bound to the DNA.

4. For elongation to occur, histones are evicted, partially displaced, or destabilized so that RNA polymerase II can move along the DNA. Evicted histones are transferred to histone chaperones, which are proteins that bind histones and aid in the assembly of histones. Assembled histones are then placed back on the DNA behind the moving RNA polymerase II.

Genetic TIPS

The Question: Discuss the roles of histones in eukaryotic transcription. How might histones inhibit transcription, and how are histones modified or moved to allow transcription to occur?

Topic: What topic in genetics does this question address? The topic is the roles of histones in eukaryotic transcription.

Information: What information do you know based on the question and your understanding of the topic?

From the question, you know that histones may inhibit transcription. From your understanding of the topic, you may remember that histones may be moved, evicted, or replaced with histone variants by chromatin-remodeling complexes. You also may recall that histones can undergo covalent modifications that alter their functional properties.

Problem-Solving Strategy: Describe the steps.

Eukaryotic transcription is a complex process. One strategy to solve this problem is to look at each step in the process, which is shown in Figure 17.10, and consider how histones could affect each step.

Answer:

1. Histones are removed at the beginning of a gene to allow transcription to occur. Activators are able to bind to this nucleosome-free region.
2. Chromatin-remodeling complexes remove more histones to allow the preinitiation complex to form.
3. Histone variants are placed near the transcriptional start site; such variants are thought to be more easily removed during transcription.
4. Histones are covalently modified. Some modifications, such as acetylation, make it easier for the histones to be removed so that transcription can take place. Other modifications, such as deacetylation, allow the histones to rebind to the DNA after RNA polymerase has passed.
5. A nucleosome-free region is also at the end of a eukaryotic gene, which may facilitate transcriptional termination.

17.2 REVIEWING THE KEY CONCEPTS

- Chromatin remodeling occurs via ATP-dependent chromatin-remodeling complexes that alter the position and composition of nucleosomes (see Figure 17.7).
- Histone variants, which have amino acid sequences that differ slightly from the standard histones, play specialized roles in chromatin structure and function (see Table 17.1).
- Amino-terminal histone tails are subject to covalent modifications that act as a histone code for the binding of proteins that affect chromatin structure and gene expression (see Figure 17.8).
- Many eukaryotic genes are flanked by a nucleosome-free regions (NFRs) and well-positioned nucleosomes (see Figure 17.9).
- Transcriptional activation involves changes in nucleosome positions and composition and histone modifications (see Figure 17.10).

17.2 COMPREHENSION QUESTIONS

1. A chromatin-remodeling complex may
   a. change the locations of nucleosomes.
   b. evict nucleosomes from DNA.
   c. replace standard histones with histone variants.
   d. do all of the above.

2. According to the histone code hypothesis, the pattern of histone modifications acts like a language that
   a. influences chromatin structure.
   b. promotes transcriptional termination.
   c. inhibits the elongation of RNA polymerase.
   d. does all of the above.

3. Which of the following characteristics is typical of a eukaryotic gene that can be transcribed?
   a. The core promoter is wrapped around a nucleosome.
   b. The core promoter is found in a nucleosome-free region.
   c. The terminator is wrapped around a nucleosome.
   d. None of the above characteristics is typical of a eukaryotic gene.

4. Transcriptional activation of eukaryotic genes involves which of the following events?
   a. Changes in nucleosome locations
   b. Changes in histone composition within nucleosomes
   c. Histone modifications
   d. All of the above

17.3 DNA METHYLATION

Learning Outcomes:

1. Define *DNA methylation*, and explain how it affects transcription.
2. Explain how DNA methylation is heritable.

We now turn our attention to a regulatory mechanism that usually silences gene expression. As discussed in previous chapters, DNA structure can be modified by the covalent attachment of methyl groups, a mechanism called **DNA methylation**. This process is common in some, but not all, eukaryotic species. For example, yeast and *Drosophila* have little or no detectable methylation of their DNA, whereas DNA methylation in vertebrates and plants is relatively abundant. In mammals, approximately 2–7% of the DNA is methylated. In this section, we will examine how DNA methylation occurs and how it can control gene expression.

DNA Methylation Occurs on the Cytosine Base and Usually Inhibits Gene Transcription

As shown in **Figure 17.11**, eukaryotic DNA methylation occurs via an enzyme called **DNA methyltransferase,** which attaches a methyl group to the carbon at the number 5 position of the cytosine base, forming 5-methylcytosine. The sequence that is methylated is shown here.

$$\begin{array}{c} CH_3 \\ | \\ 5'-CG-3' \\ 3'-GC-5' \\ | \\ CH_3 \end{array}$$

Note that this sequence contains cytosines in both strands. Methylation of the cytosine in both strands is termed full methylation, whereas methylation of the cytosine in only one strand is called hemimethylation. The methyl group on cytosine protrudes into the major groove of the DNA and may affect the binding of proteins into that groove.

DNA methylation usually inhibits the initiation of transcription, particularly when it occurs in the vicinity of the promoter. In vertebrates and plants, **CpG islands** occur near many promoters of genes. (Note: CpG refers to a dinucleotide of C and G in DNA that is connected by a phosphodiester linkage.) These CpG islands are commonly 1000 to 2000 bp in length and contain a high number of CpG sites. In the case of **housekeeping genes**—genes that encode proteins required in most cells of a multicellular organism—the cytosine bases in the CpG islands are unmethylated. Therefore, housekeeping genes tend to be expressed in most cell types. By comparison, other genes are highly regulated and may be expressed only in a particular cell type. These are termed **tissue-specific genes.** In some cases, it has been found that the expression of tissue-specific genes may be silenced by the methylation of CpG islands. In this way, DNA methylation is thought to play an important role in the silencing of tissue-specific genes to prevent them from being expressed in the wrong tissue. Methylation can affect transcription in two general ways, as described next.

Alteration in the Binding of Regulatory Transcription Factors   Methylation of CpG islands may prevent or enhance the binding of regulatory transcription factors to the promoter region. For example, methylated CG sequences may prevent the binding of an activator protein to an enhancer element, presumably because methyl groups protrude into the major groove of the DNA (**Figure 17.12a**). The inability of an activator protein to bind to the DNA inhibits the initiation of transcription.

Binding of Methyl-CpG-Binding Proteins   A second way that methylation inhibits transcription is via proteins known as **methyl-CpG-binding proteins,** which bind to methylated sequences (**Figure 17.12b**). These proteins contain a methyl-binding domain that specifically recognizes a methylated CG sequence. Once bound to the DNA, the methyl-CpG-binding proteins recruit to

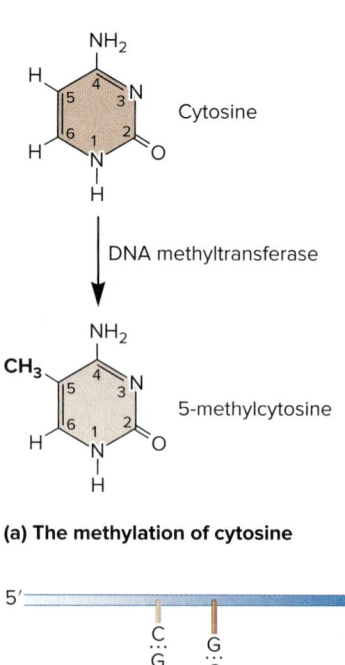

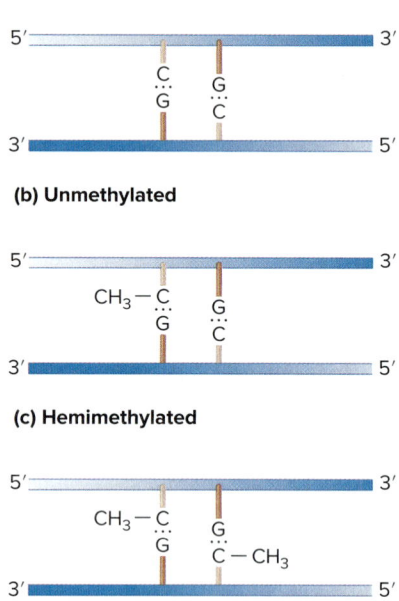

FIGURE 17.11   DNA methylation on cytosine bases. **(a)** Methylation occurs via an enzyme known as DNA methyltransferase, which attaches a methyl group to the number 5 carbon on cytosine. The CG sequence can be **(b)** unmethylated, **(c)** hemimethylated, or **(d)** fully methylated.

the region other proteins that inhibit transcription. For example, methyl-CpG-binding proteins may recruit histone deacetylase to a methylated CpG island near a promoter. Histone deacetylation removes acetyl groups from the histone proteins, which makes it more difficult for nucleosomes to be removed from the DNA. In this way, deacetylation tends to inhibit transcription.

DNA Methylation Is Heritable

Methylated DNA sequences are inherited during cell division. Experimentally, if fully methylated DNA is introduced into a plant or vertebrate cell, the DNA will remain fully methylated even in subsequently produced daughter cells. However, if the same sequence of nonmethylated DNA is introduced into a cell, it will

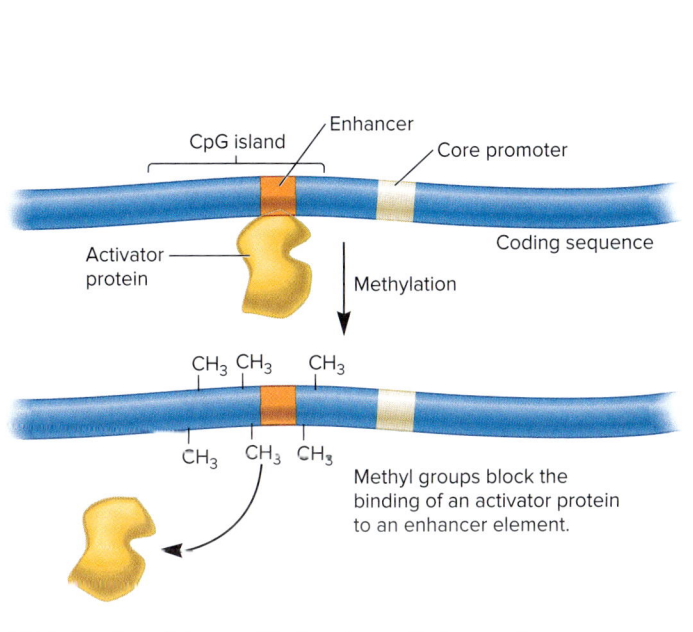

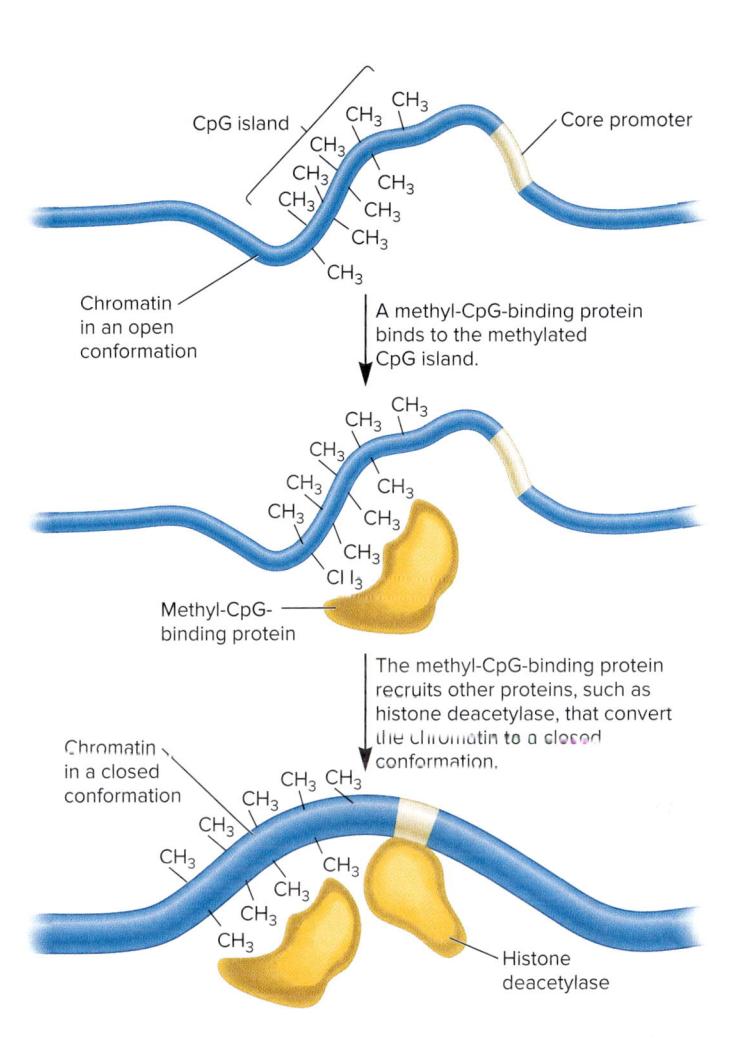

17.3 DNA METHYLATION

FIGURE 17.12 Transcriptional silencing via methylation. (a) The methylation of a CpG island may inhibit the binding of transcriptional activators to the promoter region. (b) The binding of a methyl-CpG-binding protein to a CpG island may lead to the recruitment of proteins, such as histone deacetylase, that convert chromatin to a closed conformation and thus suppress transcription.

Concept Check: Explain why the events shown in part (a) inhibit transcription.

(a) Methylation inhibits the binding of an activator protein.

(b) Methyl-CpG-binding protein recruits other proteins that change the chromatin to a closed conformation.

---

remain nonmethylated in the daughter cells. These observations indicate that the pattern of methylation is retained following DNA replication and, therefore, is inherited in future daughter cells.

How can methylation be inherited from cell to cell? **Figure 17.13** illustrates a molecular model that explains this process, which was originally proposed by Arthur Riggs, Robin Holliday, and J. E. Pugh in 1975.

1. The DNA in a particular cell may become methylated by **de novo methylation**—the methylation of DNA that was previously unmethylated.
2. When a fully methylated segment of DNA replicates in preparation for cell division, the newly made daughter strands contain unmethylated cytosines. Because only one strand is methylated, such DNA is said to be hemimethylated.
3. This hemimethylated DNA is recognized by DNA methyltransferase, which makes it fully methylated. This process is called **maintenance methylation,** because it preserves the methylated condition in future cells. However, maintenance methylation does not act on unmethylated DNA.

Overall, maintenance methylation appears to be an efficient process that routinely occurs within vertebrate and plant cells. By comparison, de novo methylation and demethylation are infrequent and highly regulated events. According to this view, the initial methylation or demethylation of a given gene can be regulated so that it occurs in a specific cell type or stage of development. Once methylation has occurred, it can then be transmitted from mother to daughter cells via maintenance methylation.

17.3 REVIEWING THE KEY CONCEPTS

- DNA methylation in eukaryotes is the attachment of a methyl group to a cytosine base via DNA methyltransferase (see Figure 17.11).
- The methylation of CpG islands near promoters usually silences transcription. Methylation may affect the binding of regulatory transcription factors or it may inhibit transcription via methyl-CpG-binding proteins (see Figure 17.12).
- Maintenance methylation preserves a methylation pattern in daughter cells following cell division (see Figure 17.13).

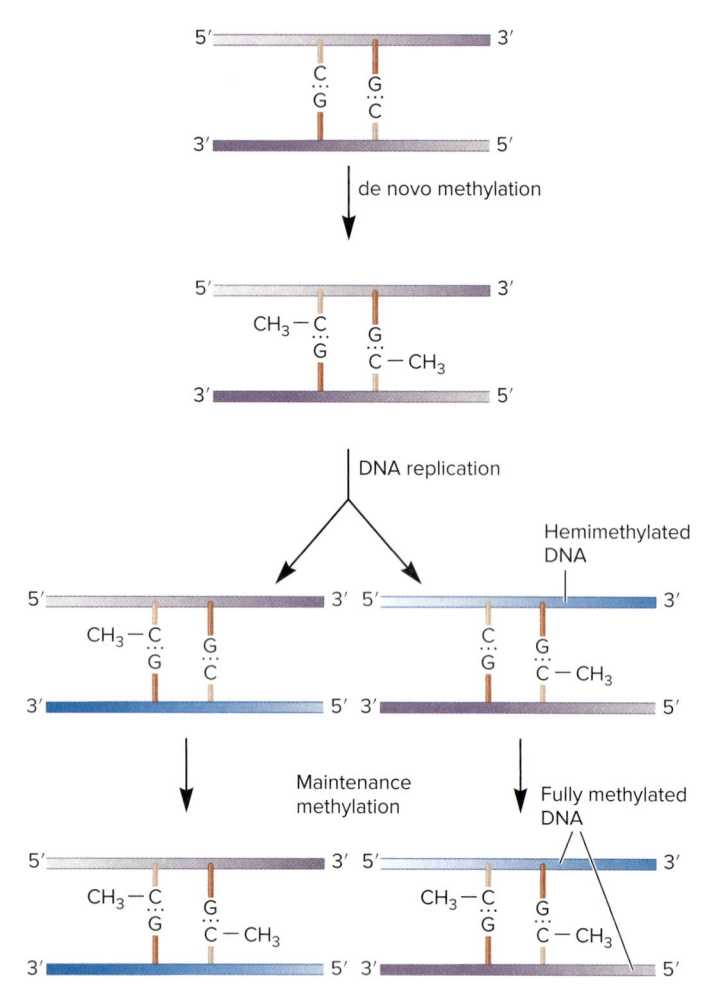

FIGURE 17.13 A molecular model for the inheritance of DNA methylation. The DNA initially undergoes de novo methylation, which is a rare, highly regulated event. Once this occurs, DNA replication produces hemimethylated DNA molecules, which are then fully methylated by DNA methyltransferase. This process, called maintenance methylation, is a routine event that is expected to occur for all hemimethylated DNA.

Concept Check: What is the difference between de novo methylation and maintenance methylation?

17.3 COMPREHENSION QUESTIONS

1. How can methylation affect transcription?
   a. It may prevent the binding of regulatory transcription factors.
   b. It may enhance the binding of regulatory transcription factors.
   c. It may promote the binding of methyl-CpG-binding proteins, which inhibit transcription.
   d. All of the above are possible ways for methylation to affect transcription.
2. The process in which completely unmethylated DNA becomes methylated is called
   a. maintenance methylation.
   b. de novo methylation.
   c. primary methylation.
   d. demethylation.

17.4 OVERVIEW OF EPIGENETICS

Learning Outcomes:
1. Define *epigenetics* and *epigenetics inheritance*.
2. Outline the types of molecular changes that underlie epigenetic gene regulation.
3. Compare and contrast epigenetic changes that are programmed during development versus those that are caused by environmental agents.

The term *epigenetics* was coined by Conrad Waddington in 1941. The prefix *epi-*, which means "over," suggests that some types of changes in gene expression are at a level that goes beyond changes in DNA sequences. How do geneticists distinguish epigenetic effects from other types of gene regulation, such as those described earlier in this chapter? An epigenetic effect begins with an initial event that causes a change in gene expression. For example, DNA methylation may inhibit transcription. However, for this to be an epigenetic effect, the change must be passed from cell to cell and must not involve a change in the sequence of DNA.

Thus, a key feature of an epigenetic effect is the long-term maintenance of a change in gene expression. As an example, let's consider muscle cells in humans. Some genes in the human genome should not be expressed in muscle cells. During embryonic development, these genes are inhibited by epigenetic changes such as DNA methylation in embryonic cells that will give rise to muscle cells. As the embryo grows and eventually becomes an adult, the epigenetic changes are passed from cell to cell so that adult muscle cells do not express these inhibited genes.

Some epigenetic changes, such as those involving the silencing of genes in muscle cells, are relatively permanent during the life of a single individual. Alternatively, other epigenetic changes may be reversible during the life of an individual, or they may be reversible from one generation to the next. For example, a gene that is silenced in one individual may be active in the offspring of that individual.

Although researchers are still debating the proper definition, one way to define epigenetics is the following.

- **Epigenetics** is the study of mechanisms that lead to changes in gene expression that can be passed from cell to cell and are reversible, but do not involve a change in the sequence of DNA. This type of change may also be called an **epimutation**—a heritable change in gene expression that does not alter the sequence of DNA.

In multicellular species that reproduce via gametes (i.e., sperm and egg cells), an epigenetic change that is passed from parent to offspring is called **epigenetic inheritance,** or **transgenerational epigenetic inheritance.** For example, as discussed in Chapter 6, genomic imprinting is an epigenetic change that is passed from parent to offspring. However, not all epigenetic changes fall into this category. For example, a person may be exposed to an environmental agent in cigarette smoke that causes an epigenetic change in a lung cell that is subsequently transmitted from cell to cell and promotes lung cancer. Such a change would not be transmitted to offspring.

Different Types of Molecular Changes Underlie Epigenetic Gene Regulation

The molecular mechanisms that promote epigenetic gene regulation are the subject of a large amount of recent research. The most common types of molecular changes that underlie epigenetic control are DNA methylation, chromatin remodeling, **covalent histone modification,** the localization of histone variants, and **feedback loops** (**Table 17.2**). These types of changes can also be involved in transient (nonepigenetic) gene regulation. The details of the first four mechanisms were examined earlier in this chapter. In Sections 17.5 and 17.6, we will explore specific examples in which epigenetic gene regulation occurs by those mechanisms. In some cases, epigenetic changes stimulate the transcription of a given gene and in other cases, they repress gene transcription.

Epigenetic Changes May Be Targeted to Specific Genes by Transcription Factors or Non-Coding RNAs

How are specific genes or chromosomes targeted for the types of epigenetic changes described in Table 17.2? The answer to this question is not well understood, but researchers are beginning to uncover a few types of mechanisms. In some cases, transcription factors may bind to a specific gene and initiate a series of events that leads to an epigenetic modification. For example, particular transcription factors in stem cells initiate epigenetic modifications that cause the cells to follow a specific pathway of development. For this to occur, the transcription factors recognize specific sites in the genome and recruit proteins to those sites, such as histone-modifying enzymes and DNA methyltransferase. This recruitment leads to epigenetic changes, such as changes in chromatin structure and DNA methylation. A simplified illustration of this process is presented in **Figure 17.14a**.

In other cases, **non-coding RNAs (ncRNAs)**—RNAs that do not encode polypeptides—are involved in establishing an epigenetic modification. Chapter 18 is devoted to the topic of ncRNAs and explores several examples in which ncRNAs facilitate epigenetic changes. Later in this chapter, we will consider how X-chromosome inactivation is mediated by an ncRNA. In some cases, ncRNAs act as bridges between specific sites in the DNA and proteins that alter chromatin or DNA structure, such as histone-modifying enzymes and DNA methyltransferase (**Figure 17.14b**).

Epigenetic Gene Regulation May Occur as a Programmed Developmental Change or Be Caused by Environmental Agents

Many epigenetic modifications that regulate gene expression are programmed changes that occur at specific stages of development (**Table 17.3**). For example, in previous chapters, we examined genomic imprinting (Chapter 6) and X-chromosome inactivation (Chapter 4). Genomic imprinting of the *Igf2* gene occurs during gametogenesis—the maternal allele is silenced, whereas the paternal allele remains active. By comparison, X-chromosome inactivation occurs during embryogenesis in female mammals. In early embryonic cells, one of the X chromosomes of a female is inactivated and forms a Barr body, whereas the other remains active. This pattern is maintained as the cells divide and eventually form an adult organism. Similarly, the differentiation of specific cell types, such as muscle cells and neurons, involves epigenetic modifications. During embryonic development, certain genes undergo epigenetic changes that affect their expression throughout the rest of development. For example, in an embryonic cell that is destined to give rise to muscle cells, a large number of genes that should not be expressed in muscle cells undergo epigenetic modifications that prevent their expression; such changes persist through adulthood.

As indicated in Table 17.3, a particularly surprising discovery in the field of epigenetics is that a wide range of environmental agents have epigenetic effects, although researchers are often uncertain whether such environmental agents are responsible for altering phenotype. Even so, many recent studies have suggested that environmentally induced changes in an organism's characteristics are rooted in epigenetic changes that alter gene expression. For example, several studies have indicated that temperature changes have epigenetic effects. In certain species of flowering plants, a process known as vernalization occurs, in which flowering in the spring requires exposure to colder temperatures during the previous winter. Researchers studying this process in *Arabidopsis* have

TABLE 17.2
Molecular Mechanisms That Underlie Epigenetic Gene Regulation

Type of Modification	Description
DNA methylation	Methyl groups may be attached to cytosine bases in DNA. When methylation occurs near promoters, transcription is usually inhibited.
Chromatin remodeling	Nucleosomes may be moved to new positions or evicted. When such changes occur in the vicinity of promoters, the level of transcription may be altered. Also, larger-scale changes in chromatin structure may occur, such as those that happen during X-chromosome inactivation in female mammals.
Covalent histone modification	Specific amino acid side chains found in the amino-terminal tails of histones can be covalently modified. For example, they can be acetylated or phosphorylated. Such modifications may enhance or inhibit transcription.
Localization of histone variants	Histone variants may become localized to specific positions, such as near the promoters of genes, and affect transcription.
Feedback loop	The activation of a gene that encodes a transcription factor may result in a feedback loop in which that transcription factor continues to stimulate its own expression.

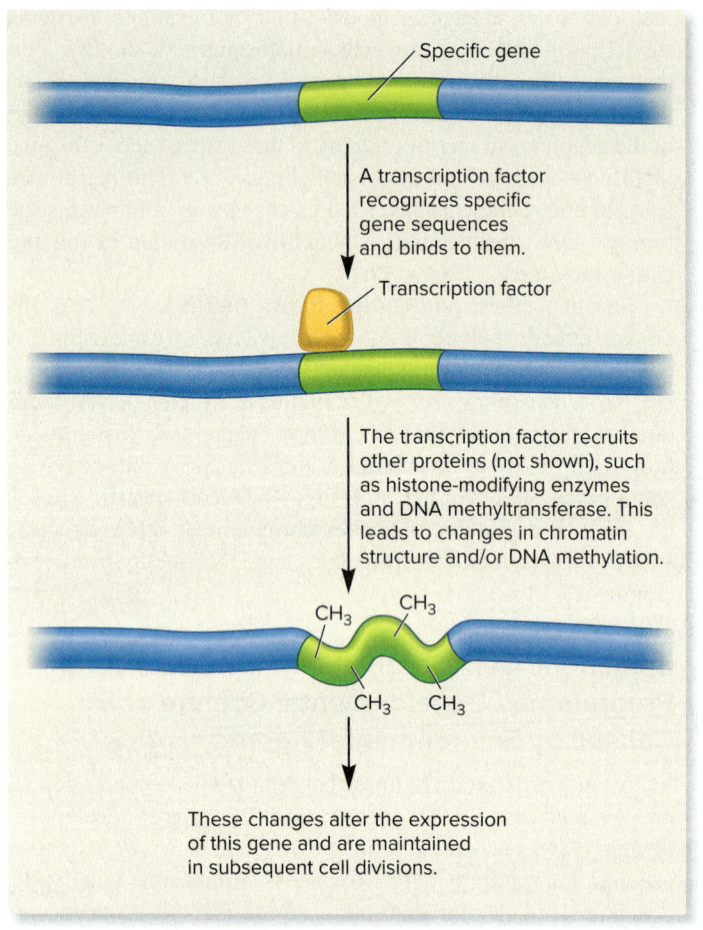

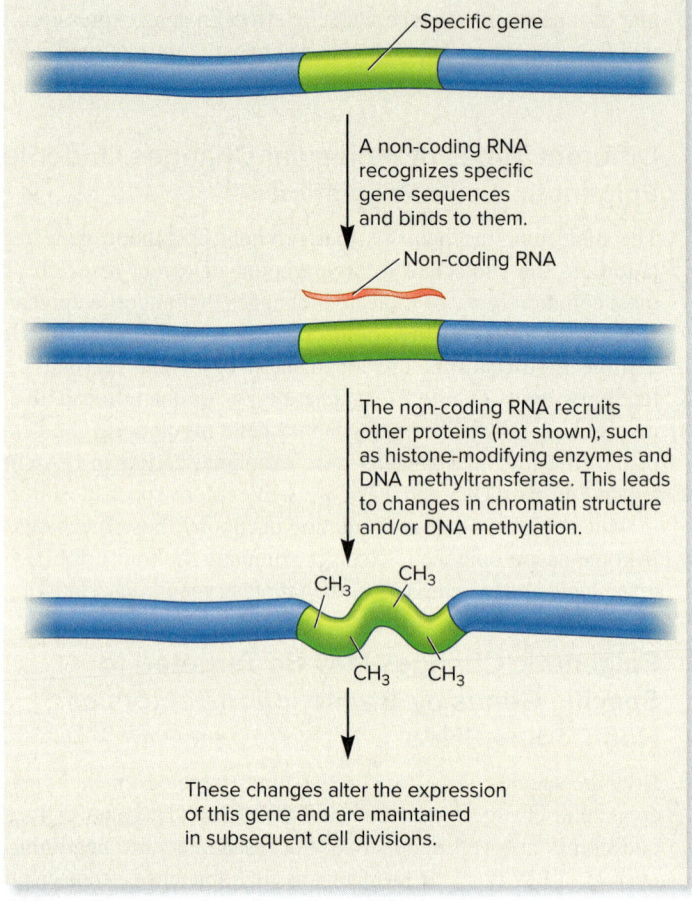

(a) Targeting a gene for epigenetic modification by a transcription factor

(b) Targeting a gene for epigenetic modification by a non-coding RNA

FIGURE 17.14 **Establishing epigenetic modifications.** Two common ways are **(a)** via transcription factors and **(b)** via non-coding RNAs.

TABLE 17.3
Factors That Promote Epigenetic Changes

Factor	Examples
Programmed Changes During Development	
Genomic imprinting	Certain genes, such as the *Igf2* gene discussed in Chapter 6, undergo different patterns of DNA methylation during oogenesis and spermatogenesis. Such patterns affect whether the maternal or paternal allele is expressed in offspring.
X-chromosome inactivation	As described in Chapter 4 and later in this chapter, X-chromosome inactivation occurs during embryogenesis in female mammals.
Cell differentiation	The differentiation of cells into particular cell types involves epigenetic changes such as DNA methylation and covalent histone modification.
Environmental Agents	
Temperature	In some species of flowering plants, cold winter temperatures cause specific types of covalent histone modifications that affect the expression of specific genes the following spring. This process may be necessary for seed germination or flowering in the spring.
Diet	The different diets of queen and worker bees alter DNA methylation patterns, which affects the expression of many genes. Such effects may underlie the different body types of queen and worker bees.
Toxins	Cigarette smoke contains a variety of toxins that affect DNA methylation and covalent histone modifications in lung cells. These epigenetic changes may play a role in the development of lung cancer. In addition, metals, such as cadmium and nickel, and certain chemicals found in pesticides and herbicides, cause epigenetic changes that can affect gene expression.

discovered that vernalization involves covalent histone modifications of specific genes, which persist from winter to spring.

A second environmental factor that can have an epigenetic effect is diet. A striking example is found in honeybees (*Apis mellifora*). Female bees have two alternative body types—queen bees and worker bees. These distinct body types are caused by dietary differences. Only larvae that are persistently fed royal jelly develop into queens. Researchers have determined that patterns of DNA methylation are quite different in queen bees and worker bees and that the pattern of methylation affects the expression of many genes.

A third environmental factor that is of great interest to many geneticists is environmental toxins that can cause epigenetic changes. In humans, exposure to tobacco smoke has been shown to alter DNA methylation and covalent histone modifications of specific genes in lung cells. As discussed in Chapter 22 (see Section 22.7), such changes may play a role in the development of cancer.

17.4 REVIEWING THE KEY CONCEPTS

- Epigenetics can be defined as the study of mechanisms that lead to changes in gene expression that are passed from cell to cell and are reversible but do not involve a change in the sequence of DNA. The transmission of epigenetic changes from one generation to the next is referred to as epigenetic inheritance.
- The most common types of molecular changes that underlie epigenetic control are DNA methylation, chromatin remodeling, covalent histone modification, localization of histone variants, and feedback loops (see Table 17.2).
- Epigenetic changes can be established by transcription factors or non-coding RNAs (see Figure 17.14).
- Some epigenetic changes are programmed during development and others are caused by environmental agents (see Table 17.3).

17.4 COMPREHENSION QUESTIONS

1. Which of the following are examples of molecular changes that can have an epigenetic effect on gene expression?
   a. Chromatin remodeling
   b. Covalent histone modification
   c. Localization of histone variants
   d. DNA methylation
   e. Feedback loops
   f. All of the above can have an epigenetic effect on gene expression.
2. An epigenetic modification to a specific gene may initially be established by
   a. a transcription factor.
   b. a non-coding RNA.
   c. both a and b.
   d. none of the above.
3. Epigenetic changes may
   a. be programmed during development.
   b. be caused by environmental changes.
   c. involve changes in the DNA sequence of a gene.
   d. be both a and b.

17.5 EPIGENETICS AND DEVELOPMENT

Learning Outcomes:
1. Describe the mechanism of genomic imprinting of the *Igf2* gene in mammals.
2. Outline the process of X-chromosome inactivation.
3. Explain how epigenetic modifications are involved in developmental changes that lead to the formation of specific cell types.

Beginning with gametes—sperm and egg cells—**development** in multicellular species involves a series of genetically programmed stages in which a fertilized egg becomes an embryo and eventually develops into an adult. Over the past few decades, researchers have determined that epigenetic changes play key roles in the process of development in animals and plants. At the molecular level, cells in the adult are able to "remember" events that happened much earlier in development. For example, in Chapter 6, we examined genomic imprinting. For the *Igf2* gene in mammals, an offspring expresses the copy that was inherited from the father, but not the copy that was inherited from the mother. From an epigenetic perspective, the cells in the offspring are able to "remember" an event that occurred during gamete formation in their parents. Likewise, during embryonic development, cells become destined to embark on pathways that lead to particular cell types. For example, an embryonic cell may give rise to a lineage of daughter cells that become a group of muscle cells. The muscle cells in the adult "remember" an event that occurred during embryonic development. In this section, we will explore the epigenetic mechanisms that explain how cells can remember events that occurred during specific stages of development.

Genomic Imprinting Occurs During Gamete Formation

We previously considered the imprinting of the *Igf2* gene in mice as an example of epigenetic inheritance (refer back to Figure 6.7). This gene encodes a protein that is required for proper growth.

The molecular mechanism of *Igf2* imprinting is due to different patterns of methylation during oogenesis and spermatogenesis (**Figure 17.15**). The *Igf2* gene is located next to another gene called *H19*. The function of the *H19* gene is not well understood, but it appears to play a role in some forms of cancer. Methylation may occur at a site called the **imprinting control region (ICR)** that is located between the *H19* and *Igf2* genes. A second site called a **differentially methylated region (DMR)** may also be methylated. During oogenesis (see upper right in Figure 17.15), methylation does not occur at either site. A protein called the CTC-binding factor (CTCF) binds to a DNA sequence containing CTC (cytosine-thymine-cytosine) that is found in both the ICR and DMR. The CTCFs bound to these sites may bind to each other to form a loop in the DNA.

Oogenesis How can this loop affect the expression of *Igf2*? To understand how, we need to consider the effects of an enhancer that is located next to the *H19* gene. Even though it is fairly far away, this enhancer can stimulate transcription of the *Igf2* gene. However, when

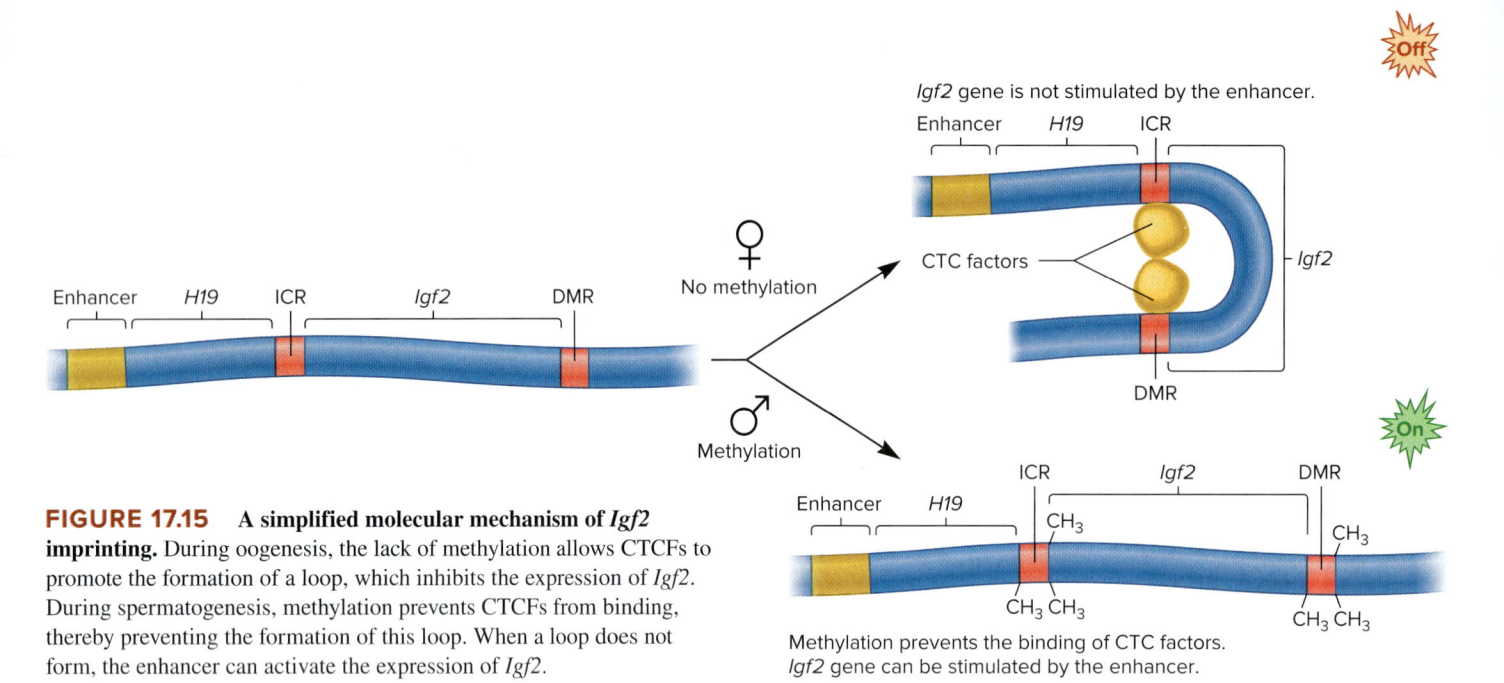

FIGURE 17.15 A simplified molecular mechanism of *Igf2* imprinting. During oogenesis, the lack of methylation allows CTCFs to promote the formation of a loop, which inhibits the expression of *Igf2*. During spermatogenesis, methylation prevents CTCFs from binding, thereby preventing the formation of this loop. When a loop does not form, the enhancer can activate the expression of *Igf2*.

a loop forms due to the interactions between two CTCFs, the change in chromatin structure prevents the enhancer from stimulating *Igf2*. Under these conditions, the maternal allele of *Igf2* is turned off.

Spermatogenesis Alternatively, if the ICR and DMR are methylated, which occurs during sperm formation, CTCFs are unable to bind to these sites (see bottom right of Figure 17.15). This prevents loop formation, which allows the enhancer to stimulate the *Igf2* gene. Therefore, the paternally inherited *Igf2* allele is transcriptionally activated. The methylation that occurs during sperm formation is de novo methylation, which is the methylation of a completely unmethylated site. Following this de novo methylation, this methylation pattern is maintained in the somatic cells of offspring due to maintenance methylation, which is the methylation of hemimethylated sites (refer back to Figure 17.13).

As we have just seen, DNA methylation causes the *Igf2* gene to be transcriptionally active. However, it is worth noting that DNA methylation more commonly has the opposite effect. As discussed in Section 17.3, DNA methylation at CpG islands in the vicinity of promoters for other genes usually inhibits their transcription.

X-Chromosome Inactivation in Mammals Occurs During Embryogenesis

As described in Chapter 4, **X-chromosome inactivation (XCI)** occurs in female mammals. One of the two X chromosomes in somatic cells is inactivated and becomes a condensed Barr body. Because females are XX and males are XY, the process of XCI achieves dosage compensation—both females and males express a single copy of most X-linked genes. In addition, XCI is responsible for traits such as the calico coat pattern observed in certain female cats (refer back to Figure 4.5).

XCI is an epigenetic phenomenon. During early embryonic development in female mammals, one of the X chromosomes in each somatic cell is randomly chosen for inactivation. After this occurs, the same X chromosome is maintained in an inactivated state during subsequent cell divisions (refer back to Figure 4.6). A region of the X chromosome called the X-inactivation center (Xic) plays a key role in this process (**Figure 17.16a**). Within the Xic are two genes called *Xist*, for X inactive-specific transcript, and *Tsix*. *Xist* is expressed from the inactivated X chromosome, whereas *Tsix* is expressed from the active X chromosome. The two genes are transcribed in opposite directions. (Note: *Tsix* is *Xist* spelled backward; this naming reflects the opposite direction of transcription of the two genes.)

The mechanism of XCI is not completely understood, and researchers are still investigating key aspects of the process. **Figure 17.16b** shows a simplified model of how XCI may occur at the molecular level, but some of these steps have not been firmly established. Prior to XCI, transcription factors called

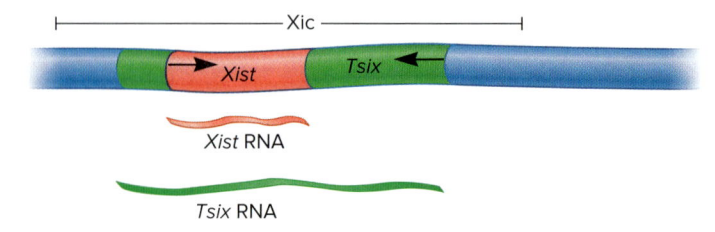

(a) The X-inactivation center (Xic)

FIGURE 17.16 The process of X-chromosome inactivation. Note: This is a simplified description of XCI. Several more proteins and ncRNAs that are not included in this figure are involved in the process.

Concept Check: In X-chromosome inactivation, when is the choice made as to which X chromosome is inactivated? Does this choice occur in embryonic cells, in adult somatic cells, or both?

17.5 EPIGENETICS AND DEVELOPMENT 399

Before X inactivation:
Pluripotency factors bind and stimulate transcription from *Tsix* and inhibit transcription from *Xist*. Both X chromosomes are active (Xa).

X chromosome pairing:
CTCF binds to the Xic and the X chromosomes pair with each other at the *Tsix* gene.

Choosing the active and inactive X chromosome:
Pluripotency factors and CTCFs shift to one of the X chromosomes. This chromosome expresses *Tsix* and remains the active X chromosome (Xa). The other chromosome can now express *Xist* and becomes the inactivated X chromosome (Xi).

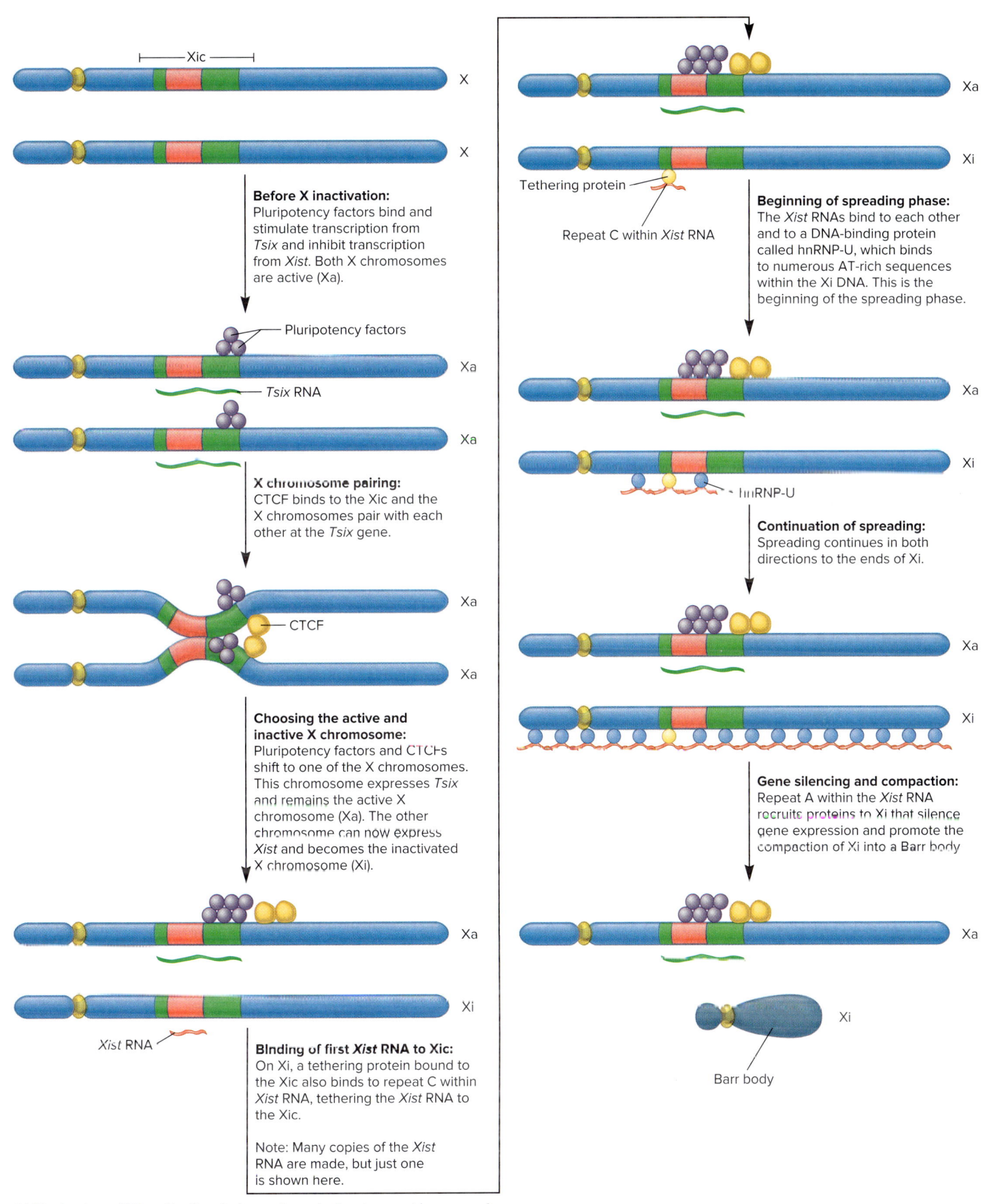

Binding of first *Xist* RNA to Xic:
On Xi, a tethering protein bound to the Xic also binds to repeat C within *Xist* RNA, tethering the *Xist* RNA to the Xic.

Note: Many copies of the *Xist* RNA are made, but just one is shown here.

Beginning of spreading phase:
The *Xist* RNAs bind to each other and to a DNA-binding protein called hnRNP-U, which binds to numerous AT-rich sequences within the Xi DNA. This is the beginning of the spreading phase.

Continuation of spreading:
Spreading continues in both directions to the ends of Xi.

Gene silencing and compaction:
Repeat A within the *Xist* RNA recruits proteins to Xi that silence gene expression and promote the compaction of Xi into a Barr body.

(b) Mechanism of X inactivation during embryonic development in mammals

pluripotency factors stimulate the expression of *Tsix*. The expression of the *Tsix* gene from both X chromosomes inhibits the expression of the *Xist* gene in multiple ways. For example, the *Tsix* RNA may associate with DNA methylation machinery to silence the *Xist* promoter. At this very early embryonic stage, both X chromosomes are active (Xa).

An early event in XCI is the pairing of the two X chromosomes, which occurs briefly (for less than an hour). This pairing, which happens only in embryonic cells, occurs at the onset of XCI, but it is not clear if it is required for choosing one X chromosome to be inactivated. Pairing begins at the *Tsix* gene via a complex that also includes the pluripotency factors and CTCFs (see Figure 17.16b). As you may recall from Figure 17.15, CTCFs bind to unmethylated sequences; in this case, they bind to an unmethylated sequence in the Xic.

The choosing of the active and inactive X chromosome occurs when the pluripotency factors and CTCFs, which were previously bound to both X chromosomes, shift entirely to one of the X chromosomes. Because the pluripotency factors stimulate the expression of the *Tsix* gene, the X chromosome to which they shift is chosen as the active X chromosome (Xa). By comparison, on the other X chromosome, the *Tsix* gene is not expressed, which permits the expression of the *Xist* gene. This chromosome is the one that becomes the inactivated X chromosome (Xi).

The next phase of X-chromosome inactivation, called the spreading phase, involves the coating of Xi with *Xist* RNA, which is a 17-kb non-coding RNA (in humans) that is transcribed exclusively from Xi. This RNA consists of six repetitive sequences (called repeats A–F). Repeat A is located at the 5′ end and is necessary for later events that lead to the formation of a Barr body. Repeat C is necessary for the binding of *Xist* RNA to Xi. The process of spreading is not well understood, but appears to begin at the Xic. In the model shown in Figure 17.16, a tethering protein binds to a DNA site in the Xic and also binds to repeat C in one *Xist* RNA molecule, tethering it to the Xic. After this occurs, subsequent *Xist* RNAs bind to each other and to a protein called hnRNP-U, which binds to numerous sites along Xi.

After the spreading phase, the first repeat sequence (repeat A) within the *Xist* RNA recruits proteins to Xi, thereby leading to epigenetic changes.

- *Xist* RNA recruits protein complexes to Xi that cause covalent modifications of specific sites in histone tails. For example, *Xist* RNA forms a complex with PRC2, which is described later in this chapter (look ahead to Figure 17.17). The Xist-PRC2 complexes cover the whole X chromosome. PRC2 modifies histones, and these histone changes are recognized by other proteins that promote changes in chromatin structure.
- A histone variant, called macroH2A, is incorporated into nucleosomes at many sites along Xi.
- Another key event that occurs after *Xist* RNA coating is the recruitment of DNA methyltransferases to Xi, which leads to DNA methylation.

Collectively, covalent histone modifications, the incorporation of macroH2A into nucleosomes, and the methylation of many CpG islands are thought to play key roles in the silencing of genes on Xi and its compaction into a Barr body. These epigenetic changes in Xi are then maintained in subsequent cell divisions.

The Development of Body Parts and the Formation of Specific Cell Types in Multicellular Organisms Involve Epigenetic Gene Regulation

The process of development involves a series of genetically programmed stages in which a fertilized egg becomes a multicellular adult with well-defined body parts composed of specific types of cells. Over the past three decades, researchers have become increasingly aware that epigenetics plays a key role in the establishment and maintenance of body parts and particular cell types. The process of development involves differential gene regulation in which certain genes are expressed in one cell type but not in another. For example, certain genes that are expressed in muscle cells are not expressed in neurons, and vice versa.

How are genes activated in one cell type and repressed in another? A key mechanism is epigenetic gene regulation. During embryonic development, many genes undergo epigenetic changes that enable them to be transcribed or cause them to be permanently repressed. Such epigenetic changes are then transmitted during subsequent cell divisions. For example, an embryonic cell that will eventually give rise to muscle tissue is programmed to undergo epigenetic modifications that will enable the transcription of muscle-specific genes and repress the transcription of genes that should not be expressed in muscle cells.

Researchers have discovered that two competing groups of protein complexes—the **trithorax group (TrxG)** and the **polycomb group (PcG)**—are key regulators of epigenetic changes that are programmed during development. TrxG complexes are involved with gene activation, whereas PcG complexes cause gene repression. Both types of complexes are found in multicellular species, such as animals and plants, where they are required for proper development. TrxG and PcG complexes were discovered in genetic studies of *Drosophila*. The names *trithorax* and *polycomb* refer to altered body parts associated with mutations in genes that encode TrxG and PcG proteins, respectively.

TrxG and PcG complexes regulate many different genes, particularly those that encode transcription factors that control developmental changes and cell differentiation. For example, PcG complexes regulate *Hox* genes, which are involved in specifying the structures that form along the anteroposterior axis in animals.

At the molecular level, a key function of specific proteins within TrxG and PcG complexes is the covalent modification of histones. For example, a component of TrxG recognizes histone H3 and attaches three methyl groups to a lysine at position 4—a process called **trimethylation.** (Note: This is methylation of a histone protein, not methylation of DNA.) The methylated histone is said to have been marked; the mark is abbreviated H3K4me3 and is called an activating mark. By comparison, a component of certain PcG complexes recognizes histone H3 and trimethylates a lysine at position 27. This mark (H3K27me3) is a repressive mark. Multiple marks are made by TrxG and PcG proteins. However, the precise role of this epigenetic marking process in gene activation or gene repression is not completely understood and is being actively investigated.

Figure 17.17 is a simplified molecular model of how a gene may be targeted for epigenetic silencing by PcG complexes. Keep

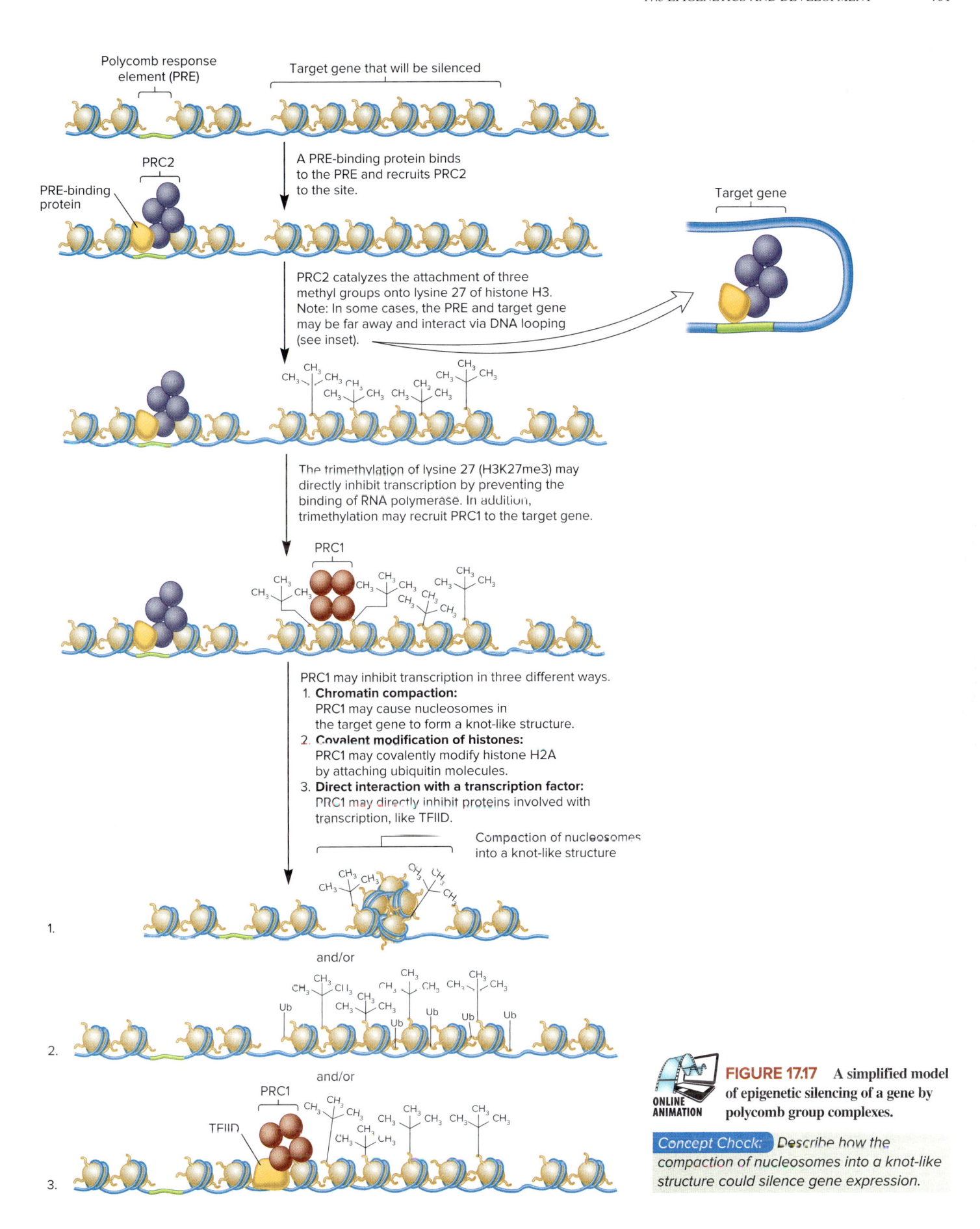

17.5 EPIGENETICS AND DEVELOPMENT   401

Polycomb response element (PRE)

Target gene that will be silenced

A PRE-binding protein binds to the PRE and recruits PRC2 to the site.

PRE-binding protein

PRC2

PRC2 catalyzes the attachment of three methyl groups onto lysine 27 of histone H3. Note: In some cases, the PRE and target gene may be far away and interact via DNA looping (see inset).

Target gene

The trimethylation of lysine 27 (H3K27me3) may directly inhibit transcription by preventing the binding of RNA polymerase. In addition, trimethylation may recruit PRC1 to the target gene.

PRC1

PRC1 may inhibit transcription in three different ways.
1. **Chromatin compaction:** PRC1 may cause nucleosomes in the target gene to form a knot-like structure.
2. **Covalent modification of histones:** PRC1 may covalently modify histone H2A by attaching ubiquitin molecules.
3. **Direct interaction with a transcription factor:** PRC1 may directly inhibit proteins involved with transcription, like TFIID.

Compaction of nucleosomes into a knot-like structure

1.

and/or

2.

and/or

PRC1
TFIID

3.

FIGURE 17.17 A simplified model of epigenetic silencing of a gene by polycomb group complexes.

Concept Check: Describe how the compaction of nucleosomes into a knot-like structure could silence gene expression.

in mind that some of these steps are not entirely understood and that the details of silencing may not be the same for different genes. This model considers the actions of two different types of PcG complexes, which are named polycomb repressive complex 1 and 2 (PRC1 and PRC2). The first step involves the binding of PRC2 to a chromosomal site that is near a gene controlled by PcG complexes. In *Drosophila*, PRC2 binds to a DNA element called a **polycomb response element (PRE)**. The response element is initially recognized by a PRE-binding protein, which then recruits PRC2 to the site. In some cases, the PRE and target gene are far apart and are brought close together by the formation of a DNA loop (see the inset in Figure 17.17). In mammals, the binding of PRC2 to specific genes may occur at PREs or CpG islands, or noncoding RNAs may recruit PRC2 to specific genes.

After PRC2 binds to a chromosomal site, a protein within that PcG complex catalyzes the covalent modification of histone H3 (see Figure 17.17). As mentioned, certain PcG complexes recognize histone H3 and trimethylate a lysine at position 27. The effect of trimethylation is not completely understood, but different effects are possible. First, it may inhibit the binding of RNA polymerase to a promoter, thereby inhibiting transcription. Second, experimental evidence suggests that trimethylation promotes the binding of PRC1, which can also inhibit transcription. However, H3K27me3 modifications may not always be required for PRC1 binding.

How do PRC1 complexes silence gene expression? Three mechanisms have been proposed, which are not mutually exclusive. The first mechanism involves chromatin compaction. PRC1 can catalyze the aggregation of nucleosomes into a more compact knot-like structure, which would silence gene expression (see Figure 17.17). A second mechanism involves another covalent modification—the attachment of a ubiquitin molecule to histone H2A. Though this covalent modification is associated with the silencing of many genes, the molecular mechanism by which it represses transcription is not understood. Finally, a third possibility is a direct interaction with a transcription factor. In the model shown in Figure 17.17, PRC1 interacts with TFIID (refer back to Figure 17.3), thereby inhibiting transcription.

After chromosomal regions have undergone epigenetic changes due to the actions of PcG complexes, these changes are maintained during subsequent cell divisions. In this way, epigenetic changes that occur during embryonic development can be transmitted to a population of cells that gives rise to a particular type of tissue, such as muscle tissue. The molecular mechanism (or mechanisms) by which such epigenetic changes are maintained during subsequent cell divisions is not well understood. One possibility is that PcG complexes may remain bound to their chromosomal sites during the process of DNA replication, thereby facilitating covalent histone modification and changes in chromatin structure after replication has occurred.

17.5 REVIEWING THE KEY CONCEPTS

- The imprinting of the *Igf2* gene in mammals takes place during gametogenesis and involves DNA methylation that occurs during spermatogenesis but not during oogenesis (see Figure 17.15).
- X-chromosome inactivation in female mammals involves epigenetic changes that are initiated during embryogenesis and are maintained throughout the rest of development (see Figure 17.16).

- The trithorax and polycomb groups are protein complexes that promote epigenetic changes important in development and in the formation of specific cell types (see Figure 17.17).

17.5 COMPREHENSION QUESTIONS

1. For the *Igf2* gene, where do de novo methylation and maintenance methylation occur?
    a. De novo methylation occurs in sperm, and maintenance methylation occurs in egg cells.
    b. De novo methylation occurs in egg cells, and maintenance methylation occurs in sperm cells.
    c. De novo methylation occurs in sperm, and maintenance methylation occurs in somatic cells of offspring.
    d. De novo methylation occurs in egg cells, and maintenance methylation occurs in somatic cells of offspring.

2. For XCI to occur, where are the *Xist* and *Tsix* genes expressed?
    a. *Xist* is expressed only on Xa, and *Tsix* is expressed only on Xi.
    b. *Xist* is expressed only on Xi, and *Tsix* is expressed only on Xa.
    c. *Xist* is expressed only on Xa, and *Tsix* is expressed only on Xa.
    d. *Xist* is expressed only on Xi, and *Tsix* is expressed only on Xi.

3. Which of the following mechanisms may be involved when PRC1 complexes silence gene expression?
    a. The compaction of nucleosomes
    b. The attachment of ubiquitin to histone proteins
    c. The direct inhibition of transcription factors, such as TFIID
    d. Any of the above may be involved in silencing of gene expression by PRC1 complexes.

17.6 EPIGENETICS AND ENVIRONMENTAL AGENTS

Learning Outcomes:
1. Explain how coat color in mice is epigenetically modified by dietary factors.
2. Describe the evidence that the development of honeybee queens is due to exposure to royal jelly.

One of the most active fields in genetics is the study of how certain environmental agents cause epigenetic changes that affect gene expression. Two areas that have received a great deal of attention are the effects of diet on epigenetic modifications and the potential effects of toxic agents, such as carcinogens (cancer-causing agents).

Exposure to Environmental Agents at Early Stages of Development May Cause Epigenetic Changes That Affect Phenotype

A striking example of how the environment can promote epigenetic changes is illustrated by studies of the *Agouti* gene (also designated *A*) found in mice. This gene encodes the Agouti signaling peptide that controls the deposition of yellow pigment in developing hairs. In wild-type mice (*AA*), the expression of this gene promotes the synthesis of pheomelanin, a yellow pigment. During the growth of a hair, melanocytes (pigment-producing cells) within a hair follicle initially make eumelanin, which is black. The transient expression of the *Agouti* gene causes the cells to express pheomelanin. The melanoncytes then revert back to making black pigment. The result is a band of yellow pigment sandwiched between layers of black pigment, which gives a brown color. The yellow pigment is not synthesized near the tip of the hair, so the hairs of wild-type mice are brown with black tips.

Researchers have identified many mutations that affect the expression of the *Agouti* gene. For example, mice that are homozygous for a loss-of-function mutation (*aa*) have black fur because pheomelanin is not made. Alternatively, a gain-of-function mutation that causes the *Agouti* gene to be overexpressed results in a mouse with yellow fur. One such mutation is designated A^{vy} (*A* for *Agouti*, *v* for *viable*, and *y* for *yellow*; the letter *v* was used because some mutations of the *Agouti* gene are lethal). By characterizing the A^{vy} allele at the molecular level, researchers determined that it is created by the insertion of a transposable element (TE) upstream from the normal promoter of the *Agouti* gene (**Figure 17.18a**). (TEs are described in Chapter 12.) The TE carries an active promoter that causes the overexpression of the *Agouti* gene.

An intriguing observation about mice carrying the A^{vy} allele is that they exhibit a wide phenotypic variation, ranging from yellow to mottled to pseudo-agouti (**Figure 17.18b**). Why should mice with the same genotype show such a wide range of phenotypic variation? Although the answer is not entirely understood, researchers have speculated that TEs are particularly sensitive to epigenetic modifications. In the case of the A^{vy} allele, epigenetic modifications may affect the function of the promoter within the TE that is responsible for overexpressing the *Agouti* gene. For example, DNA methylation could inhibit this promoter. Furthermore, a variety of environmental factors may cause such epigenetic changes to occur. The sensitivity of TEs to epigenetic modifications together with variation in environmental factors may explain the phenotypic variation seen in these mice.

One environmental factor that may affect epigenetic modification is diet. With regard to the A^{vy} allele, the exposure of pregnant female mice to different types of diets can have a significant effect on the phenotypes of the resulting offspring. For example, in 2003, Robert Waterland and Randy Jirtle conducted a study in which they investigated the effects of certain dietary supplements. Their goal was to determine if nutrients that are known to affect DNA methylation would alter the expression of the *Agouti* gene and thereby affect coat color. When DNA is methylated, DNA methyltransferase removes a methyl group from *S*-adenosyl methionine and transfers it to a cytosine base in DNA. A variety of nutrients can increase the synthesis of *S*-adenosyl methionine in cells. These include folic acid, vitamin B_{12}, betaine, and choline chloride. Waterland and Jirtle divided female mice into a control group that was fed a normal diet and an experimental group that was fed a diet supplemented with folic acid, vitamin B_{12}, betaine, and choline chloride. Both groups were fed their respective diets before and during pregnancy and up to the stage of weaning. Offspring carrying the A^{vy} allele were then analyzed with regard to their coat color and levels of DNA methylation.

As expected, a range of coat colors was observed among the offspring (**Figure 17.18c**). However, the offspring of females that had been fed a supplemented diet tended to have darker coats. For example, over 25% of the offspring with heavily mottled coats had mothers that were fed a supplemented diet (red bars), whereas less than 10% had mothers that were given a normal diet (blue bars).

The coat colors of the offspring largely correlated with the degree of methylation that occurred at CpG islands in the TE—offspring with darker coats had greater levels of DNA methylation (**Figure 17.18d**). How do we explain these results? In the mice that are more yellow, the TE has undergone very little methylation. Therefore, the promoter remains active, thereby leading to the transcription of the *Agouti* gene and the overproduction of yellow pigment. By contrast, the TE in the darker mice has undergone extensive methylation. Such methylation is expected to inhibit the overexpression of the *Agouti* gene, thereby preventing the overproduction of yellow pigment and resulting in a darker coat.

The Different Body Types of Queen and Worker Honeybees Depend on Factors in Royal Jelly That Promote Epigenetic Changes

Evidence that diet may affect DNA methylation also comes from the study of honeybees (*Apis mellifera*). Female honeybees are of two types: queen bees and worker bees (**Figure 17.19**). Queens are larger, live for years, and produce up to 2000 eggs each day. By comparison, the smaller worker bees are sterile, typically live only for weeks, and engage in specialized types of work, which include the cleaning and constructing of comb cells, nurturing larvae, guarding the hive entrance, and foraging for pollen and nectar. The striking differences between queens and worker bees are largely caused by differences in their diets. Certain worker bees, called nurse bees, produce a secretion called royal jelly from glands in their mouths. All female larvae are initially fed royal jelly, but those that are bathed in royal jelly throughout their entire larval development and fed it into adulthood become queens. In contrast, female larvae that are weaned at an early stage of development and switched to a diet of pollen and nectar become worker bees.

In 2008, a study conducted by Ryszard Maleszka and colleagues indicated that DNA methylation may play a role in controlling developmental pathways that result in differing morphologies for queens and worker bees. Bee larvae were fed a diet that should produce worker bees. These larvae were injected with a substance that inhibits DNA methyltransferase. The result was that most of them became queen bees with fully developed ovaries! While other factors may contribute to the development of queens, these results are consistent with the hypothesis that royal jelly may contain a substance that inhibits DNA methylation. Such inhibition is thought to allow the expression of genes that contribute to the development of traits that are observed in queen bees.

404   CHAPTER 17 :: GENE REGULATION IN EUKARYOTES

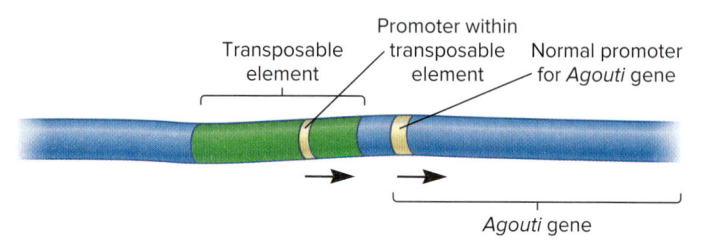

(a) The insertion of a transposable element to create the A^{vy} allele

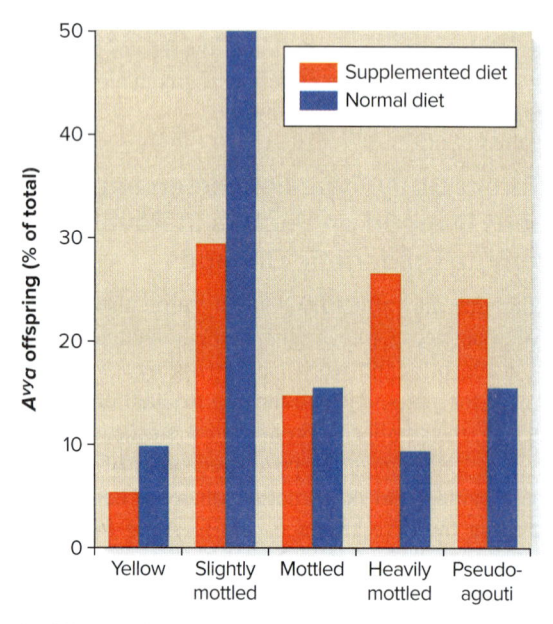

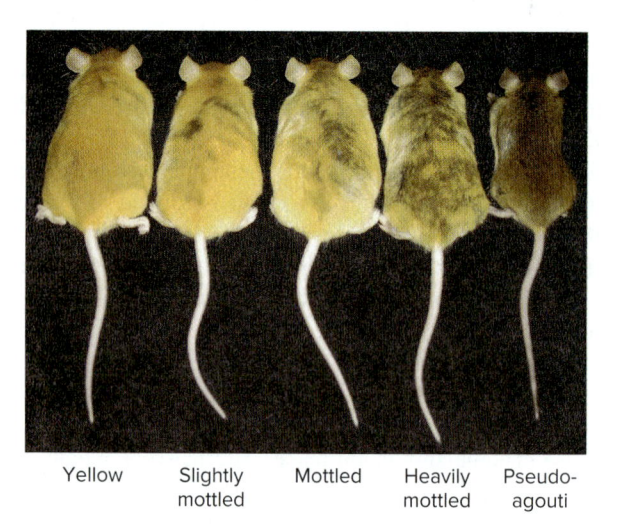

(b) Range in coat-color phenotypes in $A^{vy}a$ mice due to epigenetic changes

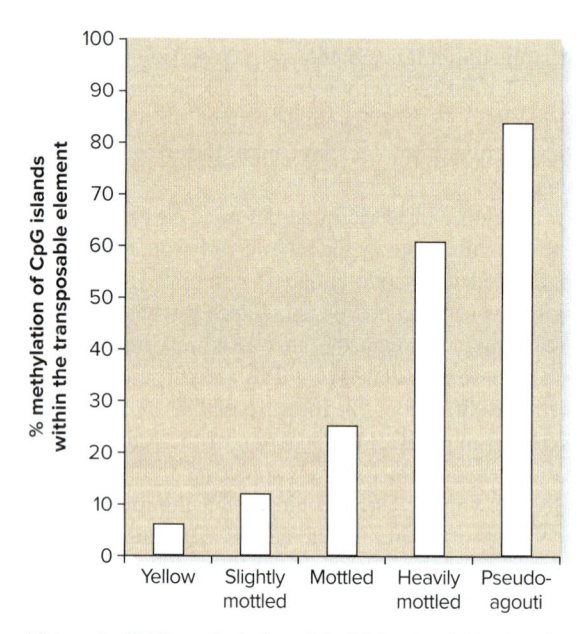

Yellow Slightly mottled Mottled Heavily mottled Pseudo-agouti

(c) Effect of diet on coat color

(d) Level of DNA methylation of CpG islands within the TE among mice with different coat colors

FIGURE 17.18 **Dietary effects on coat color in mice.** (a) A mutation in the *Agouti* gene, designated A^{vy}, is caused by the insertion of a TE upstream from the normal *Agouti* promoter. The TE promoter is very active and causes the overexpression of the *Agouti* gene. (b) Mice carrying the A^{vy} allele exhibit a range of phenotypes. The mice shown here are heterozygotes, $A^{vy}a$; they carry the mutant A^{vy} allele and a loss-of-function allele (*a*). (c) Effects of diet supplementation on coat color. Blue bars represent offspring from females given a normal diet, and red bars represent offspring from females given a supplemented diet. (d) DNA methylation patterns among mice with different coat colors. The samples to determine DNA methylation were obtained from cells in the tail. (a), (c), (d) Source: Data from Waterland, R. A., and Jirtle, R. L. (2003) Transposable Elements: Targets for Early Nutritional Effects on Epigenetic Gene Regulation, *Molecular and Cellular Biology*, vol. 23, 5293–5300. (b) From: D.C. Dolinoy et al. (2006) Maternal genistein alters coat color and protects A^{vy} mouse offspring from obesity by modifying the fetal epigenome, *Environ Health Perspect.* vol. 114, 567–572. Reproduced with permission

Genes → Traits The mice shown in part (b) are genetically identical. Their differences in coat color are due to epigenetic modifications that occur during early stages of development.

17.6 REVIEWING THE KEY CONCEPTS

- During early stages of development, dietary factors can cause epigenetic changes that affect phenotype (see Figures 17.18, 17.19).

17.6 COMPREHENSION QUESTION

1. When mice carrying the A^{vy} allele exhibit a darker coat, this phenotype is thought to be caused by dietary factors that result in

   a. a greater level of DNA methylation and a decrease in the expression of the *Agouti* gene.

FIGURE 17.19 **Dietary effects on honeybee development.** Female honeybees that are fed royal jelly throughout the entire larval stage and into adulthood develop into queen bees. The larger queen bee is shown with a blue disk labeled 68. By comparison, those larvae that do not receive this diet become smaller worker bees. These differences in development are caused by epigenetic modifications.
©Andia/Alamy Stock Photo

Concept Check: Are queen and worker bees genetically different from each other?

b. a lower level of DNA methylation and a decrease in the expression of the *Agouti* gene.
c. a greater level of DNA methylation and the overexpression of the *Agouti* gene.
d. a lower level of DNA methylation and the overexpression of the *Agouti* gene.

17.7 REGULATION OF TRANSLATION AND RNA STABILITY

Learning Outcome:
1. Explain how the translation of mRNA and RNA stability may be controlled by RNA-binding proteins.

Thus far, we have considered a variety of mechanisms that regulate the level of gene transcription. These mechanisms control the amount of RNA transcribed from a given gene. In this section, we will see how the process of mRNA translation may be regulated by RNA-binding proteins that prevent ribosomes from initiating the translation process or affect RNA stability.

Iron Assimilation in Mammals Is Regulated by an RNA-Binding Protein That Affects Translation and mRNA Stability

Specific mRNAs are sometimes regulated by RNA-binding proteins that bind to elements within the non-coding region of the

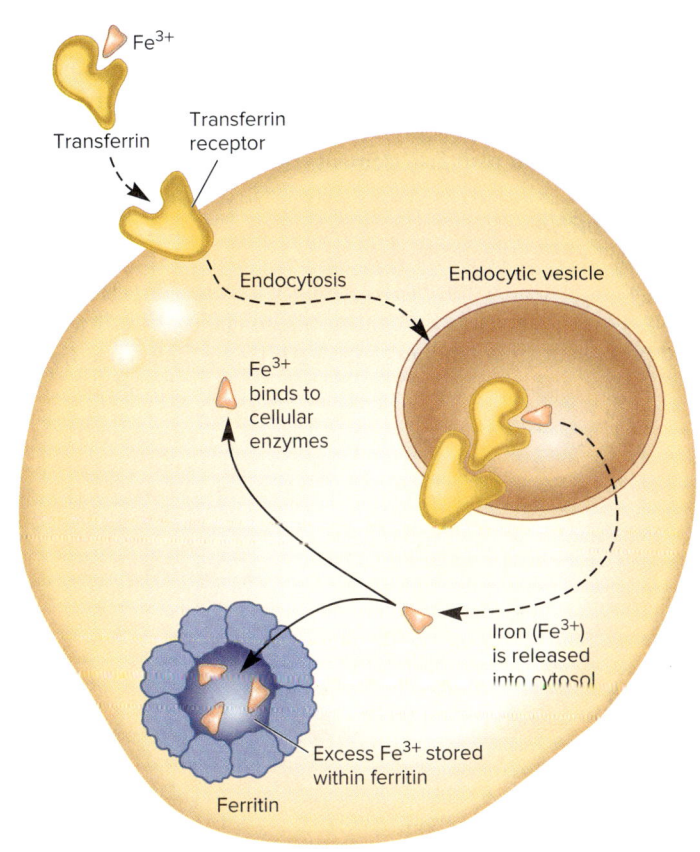

FIGURE 17.20 The uptake of iron (Fe^{3+}) into mammalian cells.

mRNA and directly affect translational initiation or RNA stability. The regulation of iron assimilation provides a well-studied example in which both of these phenomena occur. Before discussing this form of translational control, let's consider the biology of iron metabolism.

Iron is an essential element for the survival of living organisms because it is required for the function of many different proteins. The pathway by which mammalian cells take up iron is depicted in **Figure 17.20**.

1. Iron (Fe^{3+}) ingested by an animal is absorbed into the bloodstream and becomes bound to transferrin, a protein that carries iron in the blood. The transferrin-Fe^{3+} complex is recognized by a transferrin receptor on the surface of cells.
2. The complex binds to the receptor and then is transported into the cytosol by endocytosis, forming an endocytic vesicle.
3. Once inside, the iron is then released from transferrin. At this stage, Fe^{3+} may bind to cellular proteins that require iron for their activity. Alternatively, if too much iron is present, the excess iron is stored within a hollow, spherical protein known as ferritin. The storage of excess iron within ferritin helps to prevent the toxic buildup of too much iron within the cell.

Because iron is a vital yet potentially toxic substance, mammals have evolved an interesting way to regulate its assimilation.

The two mRNAs that encode ferritin and the transferrin receptor are both influenced by an RNA-binding protein known as the **iron regulatory protein (IRP)**. How does IRP exert its effects? This protein binds to a regulatory element within these two mRNAs known as the **iron response element (IRE)**.

Ferritin mRNA The ferritin mRNA has an IRE in its 5′-untranslated region (5′-UTR). When IRP binds to this IRE, it inhibits the translation of the ferritin mRNA (**Figure 17.21a**, left). However, when iron is abundant in the cytosol, the iron binds directly to IRP and prevents it from binding to the IRE. Under these conditions, the ferritin mRNA is translated to make more ferritin protein (Figure 17.21a, right), which prevents the toxic buildup of iron within the cytosol.

Transferrin Receptor mRNA The transferrin receptor mRNA also contains an IRE, but it is located in the 3′-UTR. When IRP binds to this IRE, it does not inhibit translation. Instead, the binding of IRP increases the stability of the mRNA by blocking the action of RNA-degrading enzymes. This leads to increased amounts of transferrin receptor mRNA within the cell when the cytosolic levels of iron are very low (**Figure 17.21b**, left).

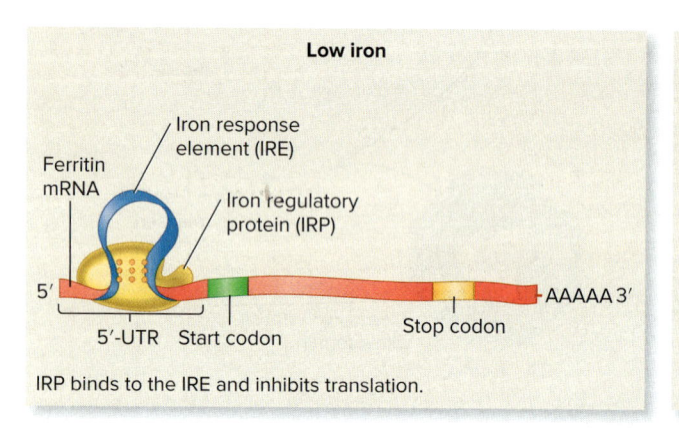

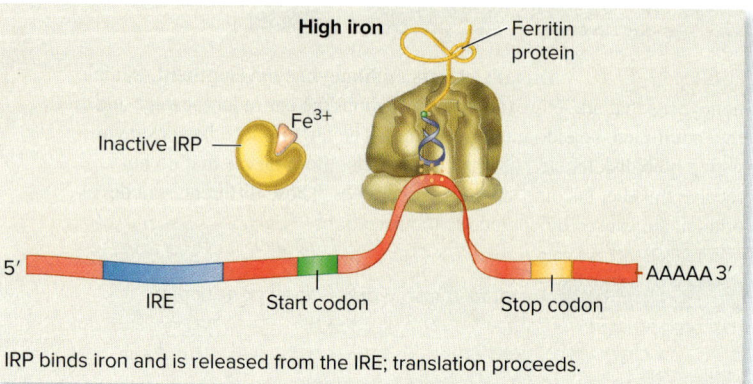

(a) Regulation of translation: ferritin mRNA

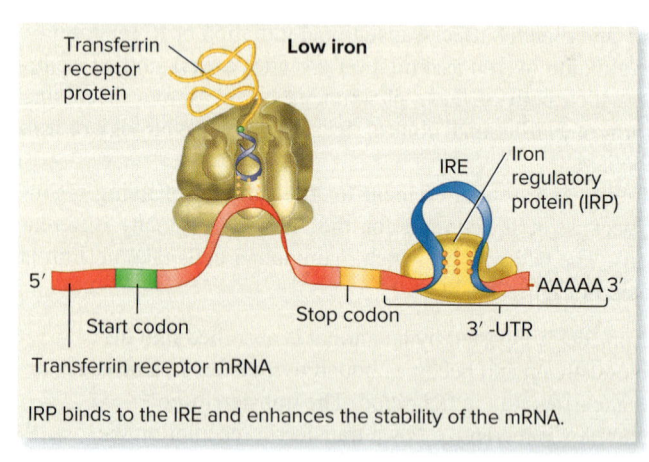

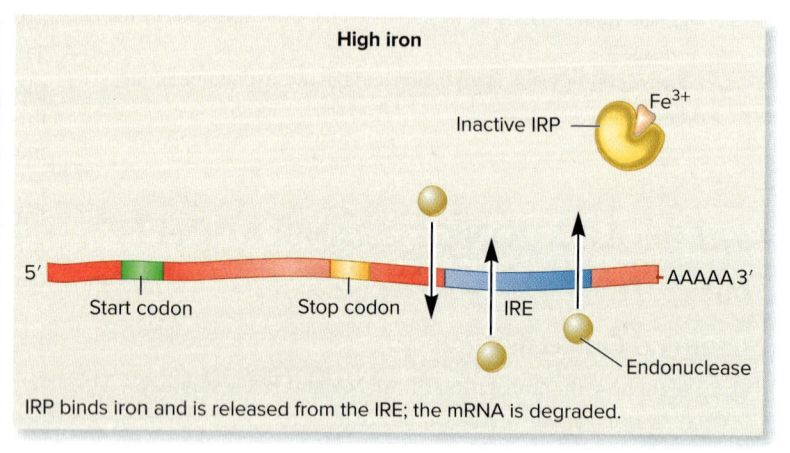

(b) Regulation of RNA stability: transferrin receptor mRNA

FIGURE 17.21 **The regulation of iron assimilation genes by IRP and IRE.** (a) When the Fe^{3+} concentration is low, the binding of the iron regulatory protein (IRP) to the iron response element (IRE) in the 5′-UTR of ferritin mRNA inhibits translation (left). When the Fe^{3+} concentration is high and Fe^{3+} binds to IRP, the IRP is removed from the ferritin mRNA, so translation can proceed (right). (b) The binding of IRP to IRE in the 3′-UTR of the transferrin receptor mRNA enhances the stability of the mRNA and leads to a higher concentration of this mRNA. Therefore, more transferrin receptor proteins are made when the Fe^{3+} concentration is low (left). When the Fe^{3+} concentration is high and iron binds to IRP, the IRP dissociates from the IRE, and the transferrin receptor mRNA is rapidly degraded (right).

Genes → Traits This form of translational control allows cells to use iron appropriately. When the cellular concentration of iron is low, the stability of the transferrin receptor mRNA is increased, thereby enhancing the ability of cells to synthesize more receptors and thereby take up more iron. Also, the translation of ferritin mRNA is inhibited, because ferritin is not needed to store excess iron. By comparison, when the cellular concentration of iron is high, the translation of ferritin mRNA is enhanced. This leads to the synthesis of ferritin and prevents the toxic buildup of iron. When iron is high, the transferrin receptor mRNA is degraded, which decreases further uptake of iron.

Concept Check: *If a mutation prevented IRP from binding to the IRE in the ferritin mRNA, how would the mutation affect the regulation of ferritin synthesis? Do you think there would be too much or too little ferritin?*

Under these conditions, more transferrin receptor is made. This promotes the uptake of iron when it is in short supply. In contrast, when iron is abundant within the cytosol, IRP is removed from the transferrin receptor mRNA, and the mRNA becomes rapidly degraded (Figure 17.21b, right). The degradation of the mRNA leads to a decrease in the amount of transferrin receptor, thereby helping to prevent the uptake of too much iron into the cell.

17.7 REVIEWING THE KEY CONCEPTS

- The regulation of iron uptake and storage is necessary to prevent the toxic buildup of iron. When iron levels are low, the binding of iron regulatory protein (IRP) to an iron regulatory element (IRE) found in the 5'-untranslated region of ferritin mRNA inhibits translation. By comparison, the binding of IRP to an IRE in the 3'-untranslated region of the transferrin receptor mRNA promotes the mRNA's stability when iron levels are low (see Figures 17.20, 17.21).

17.7 COMPREHENSION QUESTION

1. The binding of iron regulatory protein (IRP) to the iron response element (IRE) in the 5'-untranslated region of the ferritin mRNA results in
   a. inhibition of translation of the ferritin mRNA.
   b. stimulation of translation of the ferritin mRNA.
   c. degradation of the ferritin mRNA.
   d. both a and c.

KEY TERMS

Page 380. gene regulation
Page 381. transcription factor, general transcription factors (GTFs), regulatory transcription factors, regulatory elements, control elements, activator, enhancer, repressors
Page 382. silencers, combinatorial control, up regulation, down regulation, orientation-independent, bidirectional
Page 383. TFIID, coactivators, mediator
Page 385. glucocorticoid receptor
Page 386. ATP-dependent chromatin remodeling, chromatin remodeling, closed conformation, open conformation
Page 387. DNA translocase
Page 388. histone variants, histone acetyltransferase, histone code hypothesis
Page 389. nucleosome-free region (NFR)
Page 391. DNA methylation
Page 392. DNA methyltransferase, CpG islands, housekeeping genes, tissue-specific genes, methyl-CpG-binding proteins
Page 393. de novo methylation, maintenance methylation
Page 394. epigenetics, epimutation, epigenetic inheritance, trans-generational epigenetic inheritance
Page 395. covalent histone modification, feedback loops, non-coding RNAs (ncRNAs)
Page 397. development, imprinting control region (ICR), differentially methylated region (DMR)
Page 398. X-chromosome inactivation (XCI)
Page 400. pluripotency factors, trithorax group (TrxG), polycomb group (PcG), trimethylation
Page 402. polycomb response element (PRE)
Page 406. iron regulatory protein (IRP), iron response element (IRE)

CHAPTER SUMMARY

- Gene regulation refers to the phenomenon whereby the level of gene expression can be controlled so that genes can be expressed at high or low levels. Gene regulation can occur at any step in the pathway to a functional protein (see Figure 17.1).

17.1 Regulatory Transcription Factors

- Regulatory transcription factors can be activators that bind to enhancers and increase transcription, or they can be repressors that bind to silencers and inhibit transcription (see Figure 17.2).
- Combinatorial control means that the expression of a gene is regulated by a variety of factors.
- Regulatory elements are usually orientation-independent.
- Regulatory transcription factors may exert their effects by interacting with TFIID (a general transcription factor) or mediator (see Figures 17.3, 17.4).
- The functions of regulatory transcription factors can be modulated by small effector molecules, protein-protein interactions, and covalent modifications (see Figure 17.5).
- Steroid hormones, such as glucocorticoids, bind to receptors that function as transcriptional activators (see Figure 17.6).

17.2 Chromatin Remodeling, Histone Variants, and Histone Modification

- Chromatin remodeling occurs via ATP-dependent chromatin remodeling complexes that alter the position and composition of nucleosomes (see Figure 17.7).
- Histone variants, which have amino acid sequences that differ slightly from the standard histones, play specialized roles in chromatin structure and function (see Table 17.1).
- Amino-terminal histone tails are subject to covalent modifications that act as a histone code for the binding of proteins that affect chromatin structure and gene expression (see Figure 17.8).
- Many eukaryotic genes are flanked by nucleosome-free regions (NFRs) and well-positioned nucleosomes (see Figure 17.9).
- Gene transcription involves changes in nucleosome positions and composition and histone modifications (see Figure 17.10).

17.3 DNA Methylation

- DNA methylation in eukaryotes is the attachment of a methyl group to a cytosine base via DNA methyltransferase (see Figure 17.11).
- The methylation of CpG islands near promoters usually silences transcription. Methylation may affect the binding of regulatory transcription factors, or it may inhibit transcription via methyl-CpG-binding proteins (see Figure 17.12).
- Maintenance methylation preserves a methylation pattern in daughter cells following cell division (see Figure 17.13).

17.4 Overview of Epigenetics

- Epigenetics can be defined as the study of mechanisms that lead to changes in gene expression that are passed from cell to cell and are reversible but do not involve a change in the sequence of DNA. The transmission of epigenetic changes from one generation to the next is referred to as epigenetic inheritance.
- The most common types of molecular changes that underlie epigenetic control are DNA methylation, chromatin remodeling, covalent histone modification, localization of histone variants, and feedback loops (see Table 17.2).
- Epigenetic changes can be established by transcription factors or non-coding RNAs (see Figure 17.14).
- Some epigenetic changes are programmed during development and others are caused by environmental agents (see Table 17.3).

17.5 Epigenetics and Development

- The imprinting of the *Igf2* gene in mammals takes place during gametogenesis and involves DNA methylation that occurs during spermatogenesis but not during oogenesis (see Figure 17.15).
- X-chromosome inactivation in female mammals involves epigenetic changes that are initiated during embryogenesis and are maintained throughout the rest of development (see Figure 17.16).
- The trithorax and polycomb groups are protein complexes that promote epigenetic changes important in development and in the formation of specific cell types (see Figure 17.17).

17.6 Epigenetics and Environmental Agents

- During early stages of development, dietary factors can cause epigenetic changes that affect phenotype (see Figures 17.18, 17.19).

17.7 Regulation of Translation and RNA Stability

- The regulation of iron uptake and storage is needed to prevent the toxic buildup of iron. When iron levels are low, the binding of iron regulatory protein (IRP) to an iron regulatory element (IRE) found in the 5′-untranslated region of ferritin mRNA inhibits translation. By comparison, the binding of IRP to an IRE in the 3′-untranslated region of the transferrin receptor mRNA promotes the mRNA's stability when iron levels are low (see Figures 17.20, 17.21).

PROBLEM SETS & INSIGHTS

More Genetic TIPS

1. The glucorticoid response element (GRE) has two copies of the sequence 5′-AGRACA-3′, where R is a purine. Given a human genome size of about 3 billion bp, how many times would you expect this sequence to occur, due to random chance? Would the glucocorticoid receptor bind to all of those sites? Why or why not?

Topic: What topic in genetics does this question address?

The topic is the frequency of a sequence in the human genome and whether the glucocorticoid receptor would bind to all of them.

Information: What information do you know based on the question and your understanding of the topic?

From the question, you know that a GRE has two copies of a certain sequence and are reminded of the size of the human genome. From your understanding of the topic, you may remember that the glucocorticoid receptor forms a dimer and binds to two adjacent copies of this sequence.

Problem-Solving Strategy: Make a calculation.

One strategy to solve this problem is to first consider the likelihood of the given six-base sequence occuring in the human genome, based on random chance. You can use the product rule, which is discussed in Chapter 3. The likelihood of any particular base in a given position is 1/4 (because there are 4 types of bases) and the likelihood of a purine is 1/2. If you assume random base arrangements, the likelihood of the six-base sequence is (1/4)(1/4)(1/2)(1/4)(1/4)(1/4), which equals 1/2048. In other words, this sequence would occur every 2048 bp by random chance alone.

Answer: If you divide 3 billion by 2048, you find that the sequence should occur about 1,464,844 times in the human genome! However, for the glucocorticoid receptor to bind, two copies of this sequence must be close together. So, it would not bind to most of them.

2. A drug called garcinol, isolated from *Garcinia indica* (a fruit-bearing tree commonly known as kokum), is a potent inhibitor of histone acetyltransferase. Would you expect this drug to enhance or inhibit transcriptional initiation and/or elongation?

Topic: What topic in genetics does this question address?

The topic is the role of histone modifications in eukaryotic transcription. More specifically, the question is about how the attachment of acetyl groups affects transcriptional initiation and elongation.

Information: What information do you know based on the question and your understanding of the topic?

From the question, you know that garcinol inhibits histone acetyltransferase. From your understanding of the topic, you may remember that acetylated histones do not bind as tightly to DNA and are more easily removed than are nonacetylated histones.

Problem-Solving Strategy: Relate structure and function. Predict the outcome.

One strategy to solve this problem is to first consider how the drug will affect the ability of histones to bind to the DNA. You can then relate that effect to what is involved in the processes of transcriptional initiation and elongation.

Answer: Garcinol will inibit the acetylation of histones. Histones that are not acetylated bind more tightly to DNA. This tighter binding will inhibit transcriptional initiation and elongation, which both require the removal of histones.

3. A person carries a mutation in which the IRP has a lower affinity for iron compared to a person without the mutation. Would the person with the mutation be more or less likely to take up enough iron into his or her cells? Would the person be better or worse at dealing with an iron overdose?

Topic: What topic in genetics does this question address?

The topic is the regulation of mRNAs that have an IRE. More specifically, the question is about ferritin and transferrin receptor RNAs, and how a mutation could affect their expression.

Information: What information do you know based on the question and your understanding of the topic?

From the question, you know that a mutation lowers the affinity of IRP for an IRE. From your understanding of the topic, you may remember that the binding of IRP to the ferritin RNA inhibits translation, whereas the binding of IRP to the transferrin receptor mRNA enhances its stability. The binding of iron to IRP releases it from the IRE.

Problem-Solving Strategy: Compare and contrast. Predict the outcome.

One strategy to solve this problem is to first compare how IRP controls these two types of mRNA. If IRP has a lower affinity for iron, it will be more likely to bind to the IRE; a higher concentration of iron will be needed to remove it from the IRE.

Answer: With regard to the transferrin receptor mRNA, the IRP would be more likely to be bound to the IRE in the person with the mutation compared to someone without the mutation. Because IRP stabilizes the transferrin receptor mRNA, the person with the mutation would make more transferrin receptor, and you would predict that his or her cells would take up more iron than those of someone without the mutation. With regard to the ferritin mRNA, this mRNA would be more likely to be inhibited in the person carrying this mutation. This means that this person's cells would produce less ferritin, and therefore would have a diminished capacity to deal with an iron overdose.

Conceptual Questions

C1. Discuss the common points of control in eukaryotic gene regulation.

C2. Discuss the structure and function of regulatory elements. Where are they located relative to the core promoter?

C3. What is meant by the term *transcription factor modulation*? List three general ways this can occur.

C4. What are the functions of transcriptional activator proteins and repressor proteins? Explain how they work at the molecular level.

C5. Is each of the following statements true or false?

A. An enhancer is a type of regulatory element.

B. A core promoter is a type of regulatory element.

C. Regulatory transcription factors bind to regulatory elements.

D. An enhancer may cause the down regulation of transcription.

C6. Describe the steps that need to occur for the glucocorticoid receptor to bind to a GRE.

C7. Suppose that a mutation in the glucocorticoid receptor does not prevent the binding of the glucocorticoid hormone to the protein but prevents the ability of the receptor to activate transcription. Make a list of all the possible defects that may explain why transcription cannot be activated.

C8. The glucocorticoid receptor is an example of a transcriptional activator that binds to response elements and activates transcription. How could the function of the glucocorticoid receptor be shut off? (Note: The answer to this question is not directly described in this chapter. You have to rely on your understanding of the functioning of other proteins that are modulated by the binding of effector molecules, such as lac repressor.)

C9. An enhancer, located upstream from a gene, has the following sequence:

   5′–GTAG–3′
   3′–CATC–5′

This enhancer is orientation independent. Which of the following sequences also works as an enhancer?

A. 5′–CTAC–3′
   3′–GATG–5′

B. 5′–GATG–3′
   3′–CTAC–5′

C. 5′–CATC–3′
   3′–GTAG–5′

C10. Briefly describe three ways that ATP-dependent chromatin-remodeling complexes may change chromatin structure.

C11. What is a histone variant?

C12. Explain how the acetylation of core histones may loosen chromatin packing.

C13. What is meant by the term *histone code*? With regard to gene regulation, what is the proposed role of the histone code?

C14. What is a nucleosome-free region? Where are such regions typically found in a genome? How are nucleosome-free regions thought to be functionally important?

C15. Histones are thought to be displaced as RNA polymerase is transcribing a gene. What would be the potentially harmful consequences if histones were not put back onto a gene after RNA polymerase had passed?

C16. How does histone acetyltransferase affect nucleosome structure and transcription?

C17. What is DNA methylation? When we say that DNA methylation is heritable, what do we mean? How is it passed from a mother to a daughter cell?

C18. Let's suppose that a vertebrate organism carries a mutation that causes some cells that normally differentiate into nerve cells to differentiate into muscle cells. A molecular analysis reveals that this mutation is in a gene that encodes a DNA methyltransferase. Explain how an alteration in a DNA methyltransferase could produce this phenotype.

C19. What is a CpG island? Where would you expect one to be located? How does the methylation of CpG islands affect gene expression?

C20. Define *epigenetics*. Are all epigenetic changes passed from parent to offspring? Explain.

C21. List and briefly describe five types of molecular mechanisms that may underlie epigenetic gene regulation.

C22. Explain how DNA methylation and the formation of a DNA loop control the expression of the *Igf2* gene in mammals. How is this gene imprinted so that only the paternal copy is expressed in offspring?

C23. Let's suppose a mutation removes the ICR next to the *Igf2* gene. If this mutation is inherited from the mother, will the *Igf2* gene (from the mother) be silenced or expressed? Explain.

C24. Outline the molecular steps in the process of X-chromosome inactivation (XCI). Which step plays a key role in choosing which of the X chromosomes will remain active and which will be inactivated.

C25. Following X-chromosome inactivation, most of the genes on the inactivated X chromosome are silenced. Explain how. Name one gene that is not silenced.

C26. In general, explain how epigenetic modifications are an important mechanism for developmental changes that lead to specialized body parts and cell types. How do the protein complexes called the trithorax and polycomb groups participate in this process?

C27. What are the contrasting roles of trithorax and polycomb groups during development in animals and plants?

C28. Describe the molecular steps by which polycomb groups cause epigenetic gene silencing.

C29. With regard to development, what dire consequences would result if polycomb groups did not function properly?

C30. Using coat color in mice and the development of female honeybees as examples, explain how dietary factors cause epigenetic modifications and thereby lead to phenotypic effects.

C31. Describe how the binding of iron regulatory protein to an IRE affects the mRNAs for ferritin and the transferrin receptor. How does iron (Fe^{3+}) influence this process?

C32. Suppose that a person is homozygous for a mutation in the *IRP* gene that changed the structure of the IRP in such a way that it could not bind iron, but it could still bind to IREs. How would this mutation affect the regulation of ferritin and transferrin receptor mRNAs? Do you think such a person would need more iron in his or her diet than normal individuals? Do you think that excess iron in the diet would be more toxic than it would be for normal individuals? Explain your answer.

C33. In response to potentially toxic substances (e.g., high levels of iron), eukaryotic cells often use translational or posttranslational regulatory mechanisms to prevent cell death, rather than using transcriptional regulatory mechanisms. Explain why.

Application and Experimental Questions

E1. Researchers can isolate a sample of cells, such as skin fibroblasts, and grow them in the laboratory. This procedure is called a cell culture (see Appendix A). A cell culture can be exposed to a sample of DNA. If the cells are treated with agents that make their membranes permeable to DNA, the cells may take up the DNA and incorporate it into their chromosomes. This process is called transformation, or transfection. Experimenters transformed human skin fibroblasts with methylated DNA and then allowed the fibroblasts to divide for several cellular generations. The DNA in the daughter cells was then isolated, and the segment that corresponded to the transformed DNA was examined. This DNA segment in the daughter cells was also found to be methylated. However, if the original skin fibroblasts were transformed with unmethylated DNA, the DNA found in the daughter cells was also unmethylated. Do fibroblasts undergo de novo methylation, maintenance methylation, or both? Explain your answer.

E2. Restriction enzymes, discussed in Chapter 20, are enzymes that recognize a particular DNA sequence and cleave the DNA (along the DNA backbone) at that site. The restriction enzyme known as *Not*I recognizes the sequence

5′–GCGGCCGC–3′
3′–CGCCGGCG–5′

However, if the cytosines in this sequence have been methylated, *Not*I will not cleave the DNA at this site. For this reason, *Not*I is commonly used to investigate the methylation state of CpG islands.

A researcher has studied a gene, which we will call gene *T*, that is found in corn. This gene encodes a transporter involved in the uptake of phosphate from the soil. A CpG island is located near the core promoter of gene *T*. The CpG island has a single *Not*I site. The arrangement of gene *T* is shown here.

	CpG island	Core promoter	Coding sequence for gene *T*
Sal I	1500 bp	*Not* I 3800 bp	*Eco*RI

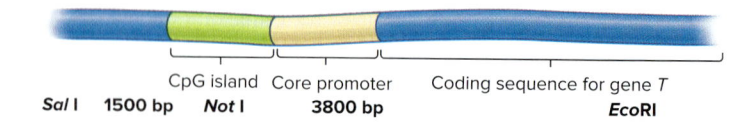

A *Sal*I restriction site is located upstream from the CpG island, and an *Eco*RI restriction site is located near the end of the coding sequence for gene *T*. The distance between the *Sal*I and *Not*I sites is 1500 bp, and the distance between the *Not*I and *Eco*RI sites is 3800 bp. No other sites for *Sal*I, *Not*I, or *Eco*RI are found in this region.

Here is the question. Let's suppose a researcher has isolated DNA samples from four different tissues in a corn plant: the leaf, the tassel, a section of stem, and a section of root. The DNA was then digested with all three restriction enzymes, separated by gel electrophoresis, and then probed with a labeled DNA fragment complementary to the gene *T* coding sequence. The results are shown here.

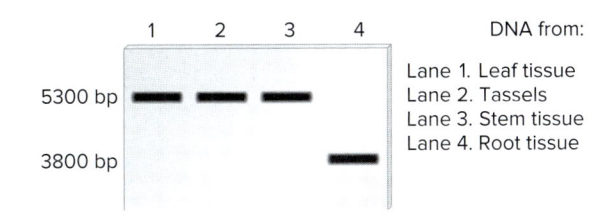

	1	2	3	4	DNA from:
5300 bp	■	■	■		Lane 1. Leaf tissue
					Lane 2. Tassels
					Lane 3. Stem tissue
3800 bp				■	Lane 4. Root tissue

In which type of tissue is the CpG island methylated? Does this make sense based on the function of the protein encoded by gene *T*?

E3. An electrophoretic mobility shift assay (EMSA) can be used to determine if a protein binds to a segment of DNA. When a segment of DNA is bound by a protein, its mobility will be slowed down, and the DNA band will be shifted to a place that is higher in the gel. In the EMSA whose results are shown here, a cloned gene fragment that is 750 bp in length contains a regulatory element that is recognized by a transcription factor called protein X. Previous experiments have shown that the presence of hormone X results in transcriptional activation by protein X.

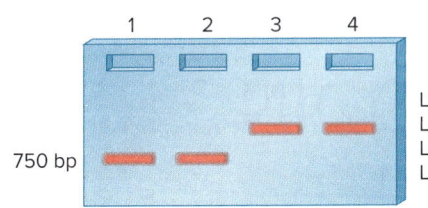

	1	2	3	4	
			■	■	Lane 1. No additions
					Lane 2. Plus hormone X
750 bp	■	■			Lane 3. Plus protein X
					Lane 4. Plus protein X and hormone X

Explain the action of hormone X.

E4. In the experiments described in Figure 17.18, explain the relationship between coat color and DNA methylation. How is coat color related to the diet of the mother?

E5. 5-Azocytidine is an inhibitor of DNA methyltransferase. If it were fed to female mice during pregnancy, explain how you think it would affect the coat color of offspring carrying the A^{vy} allele.

E6. A research study indicated that an agent in cigarette smoke caused the silencing of a tumor suppressor gene called *p53*. However, upon sequencing, no mutation was found in the DNA sequence for this gene. Give two possible explanations for these results.

E7. Let's suppose you were interested in developing drugs to prevent epigenetic changes that may contribute to cancer. What cellular proteins would be the target of your drugs? What possible side effects might your drugs cause?

E8. Chapter 20 describes a blotting method known as Northern blotting, in which a short segment of labeled DNA is used to detect RNA that is transcribed from a particular gene. The labeled DNA is complementary to the RNA that the researcher wishes to detect. After the labeled DNA binds to the RNA within a blot of a gel, the RNA is visualized as a dark band. The method of Northern blotting can be used to determine the amount of a particular RNA transcribed in a given cell type. If one type of cell produces twice as much of a particular mRNA compared to another cell, the band appears twice as intense.

Suppose that a researcher had a segment of labeled DNA that was complementary to the ferritin mRNA. This labeled DNA can be used to specifically detect the amount of ferritin mRNA on a gel. The researcher began with two flasks of human skin cells. One flask contained a very low concentration of iron, and the other flask had a high concentration. The mRNA was isolated from these cells and then subjected to Northern blotting, using labeled DNA that is complementary to the ferritin mRNA. The sample loaded in lane 1 was from the cells grown in a low concentration of iron, and the sample in lane 2 was from the cells grown in a high concentration of iron. Three Northern blots are shown below, but only one of them is correct. Based on your understanding of ferritin mRNA regulation, which blot (a, b, or c) would be the expected result? Explain. Which blot (a, b, or c) would be the expected result if the segment of labeled DNA was complementary to the transferrin receptor mRNA?

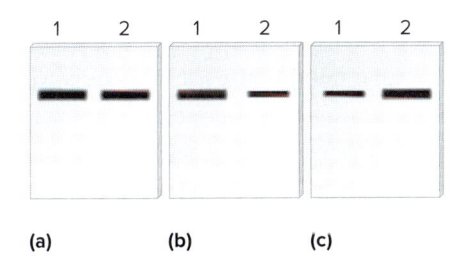

1 2	1 2	1 2
▬ ▬	▬ ▬	▬ ▬
(a)	(b)	(c)

Questions for Student Discussion/Collaboration

1. Explain how DNA methylation can regulate gene expression in a tissue-specific way. When and where would de novo methylation occur, and when would demethylation occur? What would occur in the cells that give rise to eggs and sperm?

2. Go to the PubMed website and search the words *epigenetic* and *cancer*. Scan through the journal articles you retrieve and make a list of environmental agents that may cause epigenetic changes that contribute to cancer.

Answers to Comprehension Questions

17.1: b, b, d

17.2: d, a, b, d

17.3: d, b

17.4: f, c, d

17.5: c, d, b

17.6: a

17.7: a

Note: All answers appear in Connect; the answers to even-numbered questions and all Concept Check questions are in Appendix B.

18

CHAPTER OUTLINE

- 18.1 Overview of Non-Coding RNAs
- 18.2 Non-Coding RNAs: Effects on Chromatin Structure and Transcription
- 18.3 Non-Coding RNAs: Effects on Translation and mRNA Degradation
- 18.4 Non-Coding RNAs and Protein Targeting
- 18.5 Non-Coding RNAs and Genome Defense
- 18.6 Role of Non-Coding RNAs in Human Diseases

Non-coding RNAs and heart repair. The muscles of the mammalian heart have poor regenerating abilities. Researchers have identified several non-coding RNAs that stimulate cardiac muscle regeneration. This heart, from a mouse that was treated with a type of non-coding RNA called a microRNA, showed a significant increase in proliferating cells. This discovery may lead to new therapies to help heart attack victims regenerate new cardiac muscle.

©Mauro Giacca, Ana Eulalio, Miguel Mano

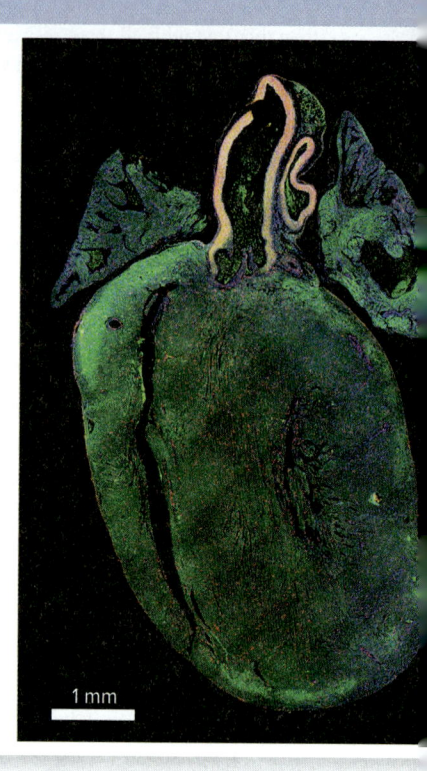

NON-CODING RNAs

In Chapters 14 and 15, we focused our attention on gene expression at the molecular level. The emphasis was on protein-encoding genes, which are transcribed into mRNA. During translation, the information within mRNAs is used to make polypeptides, which then assemble into functional proteins. The human genome has about 22,000 protein-encoding genes. In contrast, other genes are transcribed into **non-coding RNAs (ncRNAs),** which are RNA molecules that do not encode polypeptides. In humans, the number of genes that specify non-coding RNAs is still difficult to measure and remains a matter of controversy. Estimates range from several thousand to tens of thousands.

In the past, educators have tended to emphasize DNA and proteins in the teaching of genetics at the molecular level. Although DNA, RNA, and proteins are all key molecular players in living cells, a historical bias has existed against RNA. With a few exceptions, the educational exploration of RNA has been limited to its role in making proteins (as discussed in Chapter 15).

The purpose of the chapter is to lessen the bias against RNA. New molecular tools have enabled researchers to discover that ncRNAs perform a spectacular array of cellular functions in bacteria, archaea, and eukaryotes, including important roles in a variety of processes, such as DNA replication, chromatin modification, transcription, translation, and genome defense. In most cell types, ncRNAs are more abundant than mRNAs. For example, in a typical human cell, only about 20% of transcription involves the production of mRNAs, whereas 80% of it is associated with making ncRNAs! This observation underscores the importance of RNA in the enterprise of life, and indicates why it deserves greater recognition. Furthermore, abnormalities in ncRNAs are associated with a wide range of human diseases, including cancer, neurological disorders, and cardiovascular diseases.

In this chapter, we will begin with an overview of the general properties of ncRNAs, and then examine specific examples of how they perform their functions. We will end the chapter by considering the role of ncRNAs in different human diseases.

18.1 OVERVIEW OF NON-CODING RNAs

Learning Outcomes:
1. Describe the ability of ncRNAs to bind to other molecules and macromolecules.
2. Outline the general functions of ncRNAs.
3. Define *ribozyme*.
4. List several examples of ncRNAs, and describe their functions.

The study of ncRNAs is rapidly expanding. The functions of many ncRNAs are not yet known, and researchers speculate that many others have yet to be discovered. Also, due to the relatively young age of this field, not all researchers agree on the names of certain ncRNAs or their primary functions. Even so, some broad themes are beginning to emerge. In this section, we will survey the general features of ncRNAs. In later sections of the chapter, we will discuss specific examples in greater detail.

ncRNAs Can Bind to Different Types of Molecules

The ability of ncRNAs to carry out an amazing array of functions is largely related to their ability to bind to different types of molecules. **Figure 18.1a** shows four common types of molecules that are recognized by ncRNAs. Some ncRNAs bind to DNA or to another RNA through complementary base pairing. This binding allows the ncRNAs to affect processes such as transcription and translation. In addition, ncRNAs can bind to proteins or small molecules.

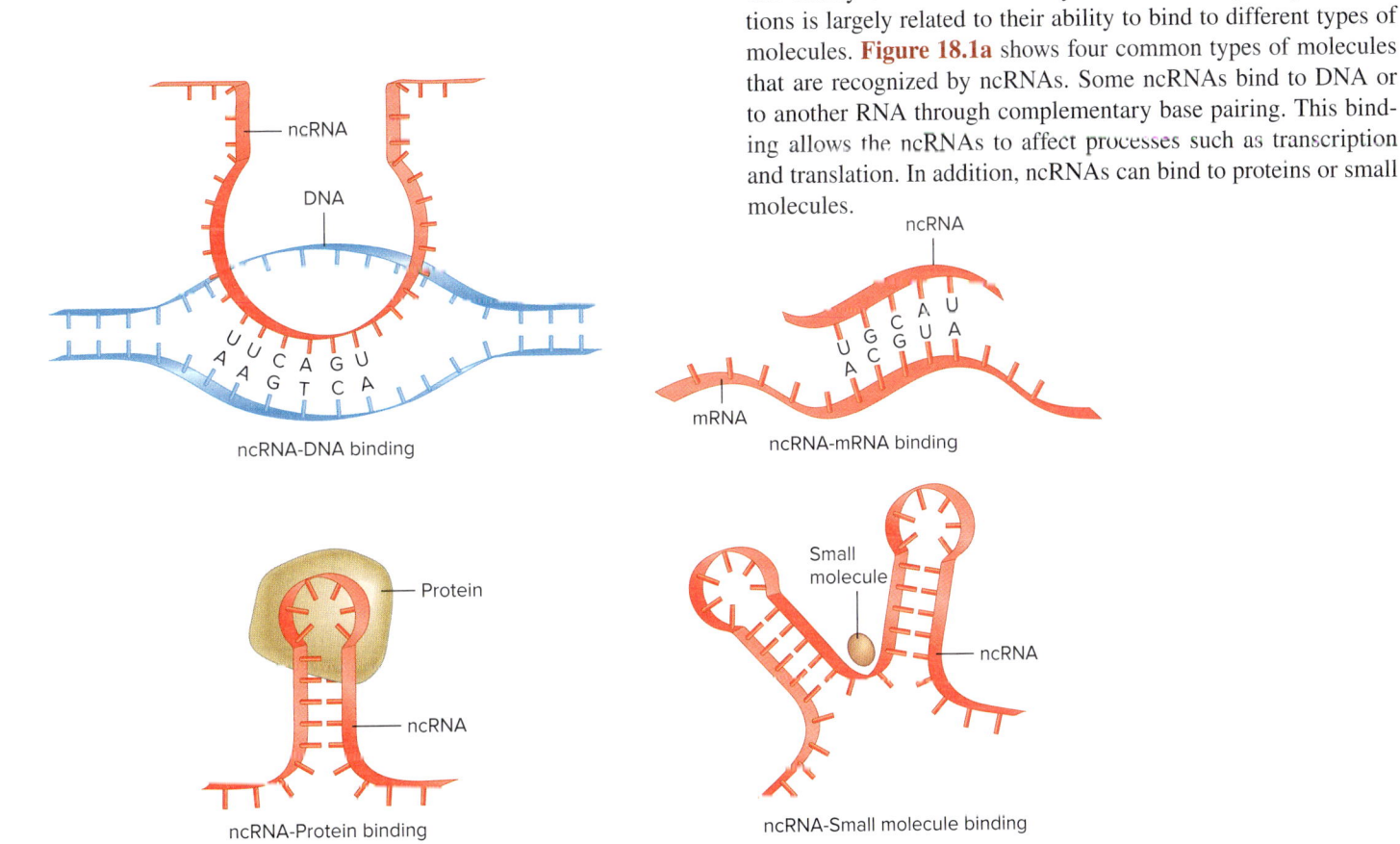

FIGURE 18.1 Ability of ncRNAs to bind to other molecules. (a) ncRNA molecules can bind to DNA, mRNA, proteins, or small molecules. (b) Some ncRNAs have multiple binding sites that can interact with several different molecules, such as proteins.

Concept Check: What types of molecules can bind to a non-coding RNA?

(a) Common binding interactions between ncRNA and other molecules

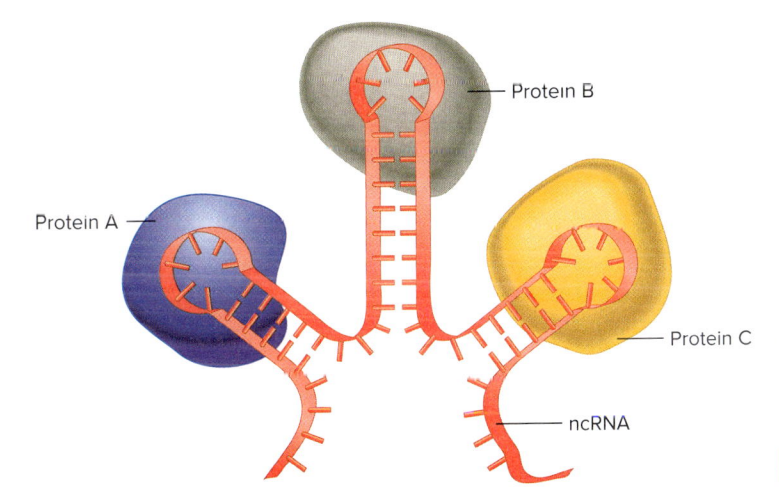

(b) Multiple binding sites in a single ncRNA

As described in Chapter 11, RNA molecules can form stem-loop structures, which may bind to pockets on the surface of proteins. Furthermore, multiple stem-loops may form a binding site for a small molecule. In some cases, a single ncRNA may contain multiple binding sites. The presence of multiple binding sites in an ncRNA facilitates the formation of a large structure composed of multiple molecules, such as the one consisting of an ncRNA and three different proteins shown in **Figure 18.1b**.

ncRNAs Can Perform a Diverse Set of Functions

In recent decades, researchers have uncovered many examples in which ncRNAs play a critical role in different biological processes. Let's first consider six general functions that ncRNAs can perform.

Scaffold Some ncRNAs contain binding sites for multiple components, such as a group of different proteins. In this way, an ncRNA can act as a scaffold for the formation of a complex, as shown in Figure 18.1b.

Guide Another function related to having multiple binding sites is that some ncRNAs can guide one molecule to a specific location in a cell. For example, an ncRNA may bind to a protein and guide it to a specific site in the cell's DNA that is part of a particular gene. This specificity occurs because the ncRNA has a binding site for the protein and another binding site for the DNA.

Alteration of Protein Function or Stability When an ncRNA binds to a protein, it can alter that protein's structure, which in turn, can have a variety of effects. The binding of the ncRNA may affect:

- the ability of the protein to act as a catalyst
- the ability of the protein to bind to another molecule, such as another protein, DNA, or RNA
- the stability of the protein

Ribozyme Another interesting feature of some ncRNAs is that they function as **ribozymes,** which are RNA molecules with catalytic function. For example, in Chapter 14, we considered RNaseP (refer back to Figure 14.17). RNaseP recognizes tRNA molecules. The RNA component of RNaseP functions as an endonuclease that cleaves a tRNA, reducing its size.

Blocker An ncRNA may physically prevent or block a cellular process from happening. For example, as described in Chapter 16, an antisense RNA is a type of ncRNA that is complementary to an mRNA. When an antisense RNA binds to an mRNA in the vicinity of the start codon, it blocks the ability of the mRNA to be translated (refer back to Figure 16.14).

Decoy Some ncRNAs recognize other ncRNAs and sequester them, thereby preventing them from working. For example, a decoy ncRNA may bind to a different ncRNA called a microRNA (miRNA), which is described later in this chapter. The function of an miRNA is to inhibit the translation of a particular mRNA. However, as shown below, if a decoy RNA binds to an miRNA, it is unable to carry out its function. When multiple decoy ncRNAs are found in a cell, they collectively act like a sponge by binding to the miRNAs and preventing them from functioning.

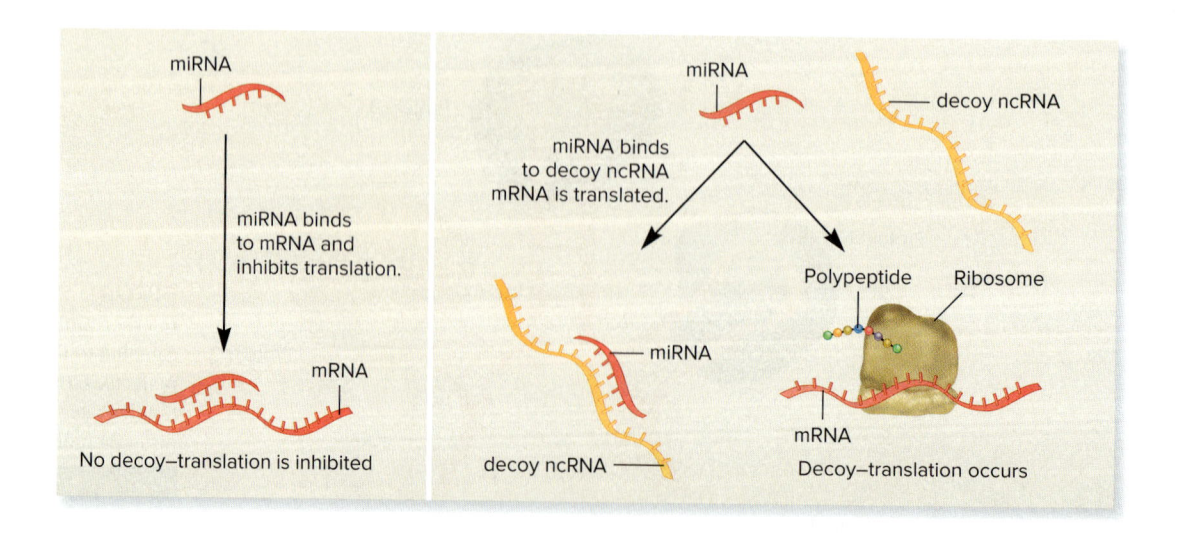

TABLE 18.1
Examples of Non-Coding RNA Molecules

Type of ncRNA	Plays a role in:	lncRNA or Small Regulatory RNA?	Described in:	Description
Telomerase RNA component (TERC)	DNA replication	lncRNA	Chapter 13	TERC facilitates the binding of telomerase to the telomere and acts as a template for DNA replication.
X inactive-specific transcript (*Xist* RNA)	Chromatin structure, transcription	lncRNA	Chapter 17	*Xist* RNA coats one of the X chromosomes in female mammals and plays a role in its compaction and resulting inactivation.
Hox transcript antisense intergenic RNA (HOTAIR)	Chromatin structure, transcription	lncRNA	Section 18.2	HOTAIR alters chromatin structure and thereby represses transcription by guiding histone-modifying enzymes to target genes.
RNaseP RNA	Processing of tRNA molecules	lncRNA	Chapter 14	RNaseP is involved in the processing of tRNA molecules. The RNaseP RNA is the catalytic component.
Small nuclear RNA (snRNA)	Splicing	Small regulatory RNA	Chapter 14	snRNAs associate with proteins to form subunits of the spliceosome, which splices pre-mRNAs in eukaryotes.
Transfer RNA (tRNA)	Translation	Small regulatory RNA	Chapter 15	tRNA molecules recognize mRNA codons during translation and carry the appropriate amino acid.
Ribosomal RNA (rRNA)	Translation	Variable*	Chapter 15	rRNAs are components of ribosomes, which are the site of polypeptide synthesis.
Antisense RNA	Translation	Variable	Chapter 16	An antisense RNA is complementary to an mRNA. The binding of the antisense RNA to the mRNA blocks translation.
MicroRNA (miRNA), small-interfering RNA (siRNA)	Translation and RNA degradation	Small regulatory RNA	Section 18.3	miRNAs and siRNAs regulate the expression and degradation of mRNAs.
RNA component of signal recognition particle (SRP RNA)	Protein targeting and secretion	lncRNA	Section 18.4	An SRP is composed of one ncRNA and one or more proteins. In prokaryotes, SRP directs certain polypeptides synthesized in the cell to the plasma membrane. In eukaryotes, it directs them to the endoplasmic reticulum.
CRISPR RNA (crRNA)	Genome defense	Small regulatory RNA	Section 18.5	crRNA, found in prokaryotes, guides an endonuclease to foreign DNA, such as the DNA of a bacteriophage. Some CRISPR systems can target RNA.
PIWI-interacting RNA (piRNA)	Genome defense	Small regulatory RNA	Section 18.5	pi-RNA associates with PIWI proteins and prevents the movement of transposable elements.

*These types of ncRNA may be shorter or longer than 200 nucleotides.

The Functions of Some ncRNAs Are Understood

Table 18.1 describes several examples of ncRNAs that have been well characterized. Some of these are described in other chapters. In the remaining sections of this chapter, we will focus on the functions of ncRNAs that have not been discussed elsewhere in the text.

As shown in Table 18.1, ncRNAs are broadly categorized according to their length. **Long non-coding RNAs (lncRNAs)** are RNA molecules that are longer than 200 nucleotides. This arbitrary limit distinguishes lncRNAs from **small regulatory RNAs** (also called short ncRNAs), which are shorter than 200 nucleotides. An example is a microRNA, which is usually 20–25 nucleotides in length.

18.1 REVIEWING THE KEY CONCEPTS

- Non-coding RNAs (ncRNAs) bind to different types of molecules, including DNA, other RNAs, proteins, and small molecules (see Figure 18.1).
- An ncRNA can perform any of several functions: provide a scaffold, act as a guide, alter protein function or stability, function as a ribozyme, function as a blocker, and/or act as a decoy.
- With regard to cell structure and function, ncRNAs play a role in DNA replication, chromatin structure, transcription, translation, splicing, RNA degradation, RNA modification, protein secretion, and genome defense (see Table 18.1).

18.1 COMPREHENSION QUESTIONS

1. Which of the following can bind to ncRNAs?
   a. DNA
   b. RNA
   c. Proteins
   d. Small molecules
   e. All of the above

2. When an ncRNA functions as a decoy, it
   a. contains binding sites for many different proteins, thereby promoting the formation of a large complex.

b. recognizes other ncRNAs and sequesters them, thereby preventing them from working.
c. can alter its conformation, allowing it to switch between active and inactivate conformations.
d. may physically prevent or block a cellular process from happening.

18.2 NON-CODING RNAs: EFFECTS ON CHROMATIN STRUCTURE AND TRANSCRIPTION

Learning Outcome:
1. Explain how the ncRNA known as HOTAIR plays a role in gene repression.

<u>Ho</u>x <u>t</u>ranscript <u>a</u>ntisense <u>i</u>ntergenic <u>R</u>NA, known as HOTAIR, is a recently discovered ncRNA that alters chromatin structure and thereby represses gene transcription. The gene that encodes HOTAIR is located on human chromosome 12 within a cluster of genes called the *HoxC* genes. (*Hox* genes play a role in embryonic development in animals.) HOTAIR is a 2.2-kb-long ncRNA that is transcribed from the opposite (antisense) strand with respect to the *HoxC* genes. Though the name HOTAIR refers to the antisense direction of its transcription, this ncRNA does not function like the antisense RNAs described in Chapter 16, which bind directly to mRNAs (refer back to Figure 16.14). Instead, researchers have discovered that HOTAIR acts as a scaffold that guides two histone-modifying enzymes to their target genes.

Figure 18.2 shows a simplified version of the mechanism by which HOTAIR represses gene transcription. HOTAIR acts as a scaffold for the binding of two histone-modifying enzymes known as <u>p</u>olycomb <u>r</u>epressive <u>c</u>omplex <u>2</u> (PRC2) and <u>l</u>ysine-<u>s</u>pecific <u>d</u>emethylase <u>1</u> (LSD1). PRC2 binds to the 5′ end of HOTAIR, and LSD1 binds to the 3′ end. HOTAIR then guides PRC2 and LSD1 to a target gene by binding to a region near the gene that contain many purines, which is called a GA-rich region. For example, HOTAIR binds to a GA-rich region that is next to a specific *HoxD* gene on human chromosome 2. A portion of HOTAIR is complementary to this GA-rich region.

The next event involves histone modifications. As described in Chapter 17, PRC2 functions as a histone methylase that trimethylates lysine 27 on histone H3. LSD1 forms a complex with other proteins and demethylates mono- and dimethylated lysines. In particular, it removes methyl groups from lysine 4 on histone H3. Although the mechanism of gene repression is not well understood, these histone modifications (H3K27 trimethylation and H3K4 demethylation) may inhibit transcription in two ways:

- The modifications may directly inhibit the ability of RNA polymerase to transcribe the target gene. For example, these histone modifications may prevent RNA polymerase from forming a preinitiation complex.

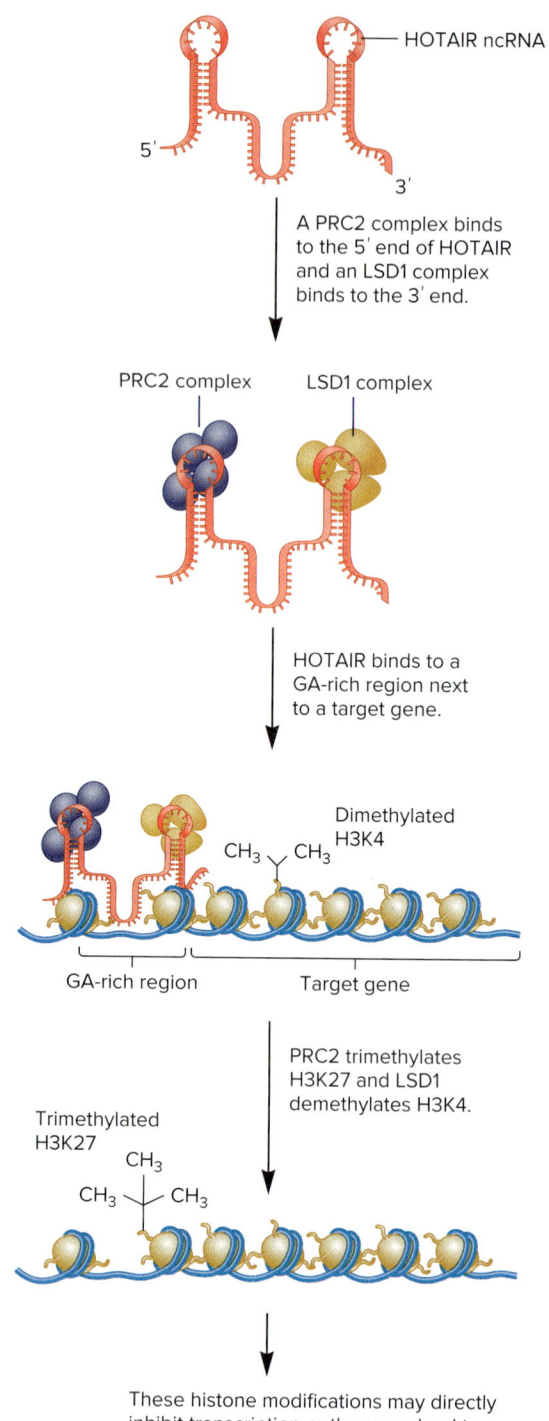

FIGURE 18.2 Simplified mechanism of repression of transcription by HOTAIR. This is just one proposed role of HOTAIR. This ncRNA is known to interact with other proteins as well. Note: For simplicity, this drawing shows just one trimethylation event and one demethylation event. The PRC2 and LSD1 complexes would trimethylate and demethylate many lysine residues, respectively. Also, the structure of HOTAIR is schematic. The actual structure is more complicated than the one shown here.

- Rather than directly affecting transcription, the histone modifications carried out by PRC2 and LSD1 may attract other chromatin-modifying enzymes to the target gene. For example, they may attract a histone deacetylase to the target gene, which would foster a closed chromatin conformation.

Of great interest to researchers investigating HOTAIR is its role in human disease. As discussed later in this chapter, certain types of cancer, such as breast cancer, may develop when HOTAIR is not functioning properly.

18.2 REVIEWING THE KEY CONCEPTS

- HOTAIR is a long ncRNA found in humans that regulates transcription by guiding PRC2 and LSD1 to particular genes. These complexes produce histone modifications that silence the genes (see Figure 18.2).

18.2 COMPREHENSION QUESTION

1. Which of the following functions does HOTAIR perform?
   a. Decoy
   b. Scaffold
   c. Guide
   d. Both b and c

18.3 NON-CODING RNAs: EFFECTS ON TRANSLATION AND mRNA DEGRADATION

Learning Outcomes:
1. Analyze the experimental evidence that double-stranded RNA is more potent than antisense RNA at inhibiting mRNA.
2. Outline the steps of RNA interference.

In the previous section, we considered how an ncRNA can affect the process of transcription. In recent years, researchers have discovered that ncRNAs often exert their effects on RNA molecules that are already made. In this section, we will explore how ncRNAs affect the ability of mRNAs to be translated or degraded.

Fire and Mello Showed That Double-Stranded RNA Is More Potent Than Antisense RNA at Silencing mRNA

Specific mRNAs can be targeted for translational inhibition or degradation by a mechanism involving double-stranded RNA. This mechanism was discovered during research involving plants and the nematode worm *Caenorhabditis elegans*. The study described here involved an examination of gene expression in *C. elegans*.

Prior to this work, researchers had often introduced antisense RNA into cells as a way to inhibit mRNA translation. Because antisense RNA is complementary to mRNA, the antisense RNA binds to the mRNA, thereby preventing translation. Oddly, in other experiments, researchers introduced sense RNA (RNA with the same sequence as mRNA) into cells instead, and, unexpectedly, this also inhibited mRNA expression. Another curious observation was that the effects of antisense RNA often persisted for a very long time, much longer than would have been predicted by the relatively short half-lives of most RNA molecules in the cell. These two unusual observations caused Andrew Fire, Craig Mello, and colleagues to investigate how the introduction of RNA into cells inhibits mRNA.

C. elegans was used as the experimental organism because it is relatively easy to inject with RNA and the expression of many of its genes had already been established. In this study, published in 1998, Fire and Mello investigated the effects of both antisense and sense RNA on the expression of specific mRNAs in *C. elegans*. In **Figure 18.3**, we focus on one of their experiments involving an mRNA encoded by a gene called *mex-3*. This mRNA had already been shown to be made in high amounts in early embryos of *C. elegans*.

Prior to this work, the *mex-3* gene had been identified and inserted into a plasmid. The process of inserting genes into plasmids is described in Chapter 20 (look ahead to Figure 20.2). Let's first look at the plasmid shown at the top of the conceptual level column in step 1 of Figure 18.3. When RNA polymerase, nucleotides, and this plasmid are mixed together in a test tube, *mex-3* mRNA is made; this mRNA is called the sense strand. In living cells, the sense strand is used to make the mex-3 protein. Fire and Mello also switched the location of the promoter to the other end of the gene and transcribed the opposite strand, which is called the antisense strand. The sense and antisense strands are complementary to each other.

Next, Fire and Mello injected these RNAs into the gonads of *C. elegans*. Into some worms, they injected antisense RNA alone. Alternatively, they mixed sense and antisense RNA, which formed double-stranded RNA, and injected this double-stranded RNA into the gonads of other worms. They also used uninjected worms as controls. After injection, the RNA was taken up by eggs, which later developed into embryos. To determine the amount of *mex-3* mRNA present, Fire and Mello incubated the embryos with a segment of labeled DNA that was complementary to this mRNA. The labeled DNA bound to *mex-3* mRNA could be observed under the microscope. After this incubation step, any labeled DNA that was not bound to this mRNA was washed away.

▶ THE GOAL (DISCOVERY-BASED SCIENCE)

The goal was to further understand how the experimental injection of RNA was responsible for the silencing of particular mRNAs.

ACHIEVING THE GOAL — FIGURE 18.3 Injection of antisense and double-stranded RNAs into *C. elegans* to compare their effects on mRNA silencing.

Starting material: The researchers used *C. elegans* as their model organism. They also had the cloned *mex-3* gene, which had been previously shown to be highly expressed in the early embryos of this worm.

Experimental level | **Conceptual level**

1. Make sense and antisense *mex-3* RNA in vitro using cloned genes for *mex-3* with promoters on either side of the gene. RNA polymerase and nucleotides are added to synthesize the RNAs.

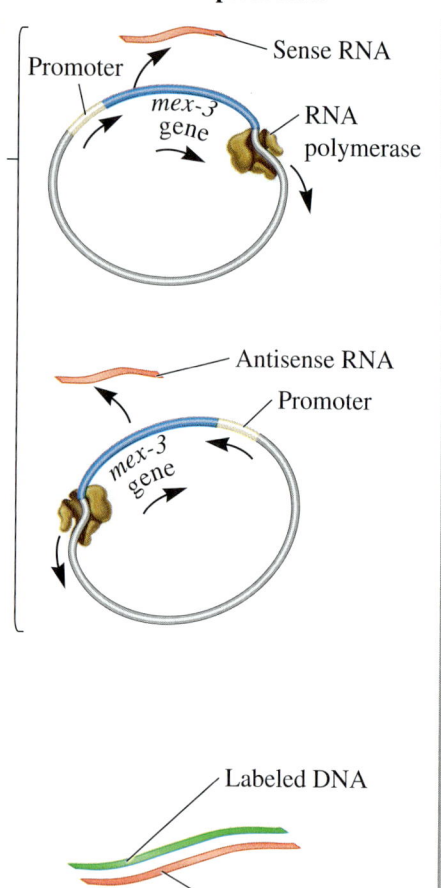

2. Inject either *mex-3* antisense RNA or a mixture of *mex-3* sense and antisense RNA into the gonads of *C. elegans*. This RNA is taken up by the eggs and early embryos. As a control, do not inject any RNA.

3. Incubate and then subject early embryos to *in situ* hybridization. In this method, a labeled segment of DNA is added that is complementary to *mex-3* mRNA. If cells express *mex-3*, the mRNA in the cells will bind to the DNA and become labeled. After incubation with a labeled DNA, the cells are washed to remove unbound DNA.

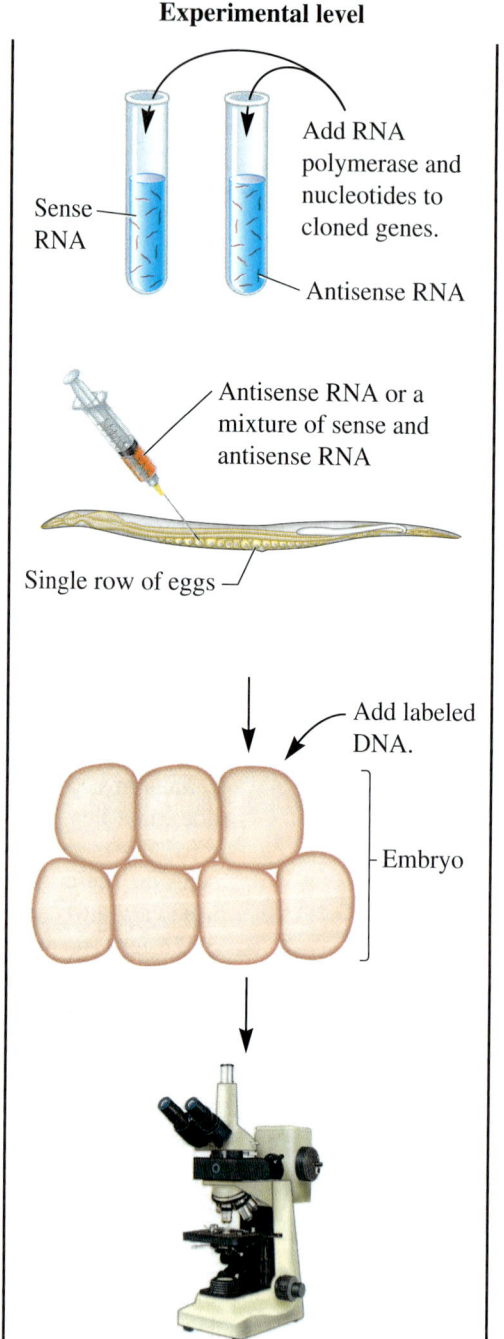

4. Observe embryos under the microscope.

THE DATA

Control  | Injected with *mex-3* antisense RNA 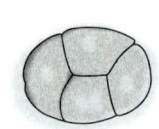 | Injected with double-stranded RNA (both *mex-3* sense and antisense RNAs)

Source: Adapted from Fire, A., Xu, S, Montgomery, M. K., Kostas, S. A., Driver, S. E., & Mello, C. C. (1998) Potent and Specific Genetic Interference by Double-Stranded RNA in *Caenorhabditis elegans*, *Nature*, 391, 806–811.

INTERPRETING THE DATA

As seen in the data, the control embryos were very darkly labeled, as shown by the green color. These results indicated that the control embryos contained a high amount of *mex-3* mRNA, which was known from previous research. In the embryos that had received antisense RNA, *mex-3* mRNA levels were decreased, but detectable, as shown by the light-green color. Remarkably, in embryos that received double-stranded

RNA, no *mex-3* mRNA was detected! These results indicated that double-stranded RNA is more potent at silencing mRNA than is antisense RNA. In this case, the double-stranded RNA caused the *mex-3* mRNA to be degraded. Fire and Mello used the term **RNA interference (RNAi)** to describe the phenomenon in which double-stranded RNA causes the silencing of mRNA. This surprising observation led these researchers to investigate the underlying molecular mechanism that accounts for this phenomenon, as described next.

RNA Interference Is Mediated by MicroRNAs or Small-Interfering RNAs via an RNA-Induced Silencing Complex

RNA interference is found in most eukaryotic species. The RNAs that promote interference are of two types: microRNAs and small-interfering RNAs. **MicroRNAs (miRNAs)** are ncRNAs that are transcribed from endogenous eukaryotic genes—genes that are normally found in the genome. They play key roles in regulating gene expression, particularly during embryonic development in animals and plants. Most commonly, a single type of miRNA inhibits the translation of several different mRNAs. An miRNA and an mRNA bind to each other because they have sequences that are partially complementary. In humans, nearly 2000 genes encode miRNAs, though that number may be an underestimate. Research suggests that 60% of human protein-encoding genes are regulated by miRNAs.

By comparison, **small-interfering RNAs (siRNAs)** are ncRNAs that usually originate from sources that are exogenous, which means they are not normally made by cells. The sources of siRNAs can be viruses that infect a cell, or researchers can make siRNAs to study gene function experimentally, as Fire and Mello did in their experiment (see Figure 18.3). In most cases, an siRNA is a perfect match to a single type of mRNA. The functioning of siRNAs is thought to play a key role in preventing certain types of viral infections. In addition, siRNAs have become important experimental tools in molecular biology.

How do miRNAs and siRNAs cause the silencing of specific mRNAs? **Figure 18.4** shows how an miRNA or siRNA achieves RNA interference. An miRNA is first synthesized as a pri-miRNA (for primary-miRNA) in the nucleus. Due to complementary base pairing, the pri-miRNA forms a hairpin structure (a stem-loop) with long single-stranded 5′ and 3′ ends. The pri-miRNA is recognized in the nucleus by two proteins, Drosha and DGCR8, and is cleaved at both ends. The result is a 70-nucleotide RNA molecule that is called a pre-miRNA (for precursor-miRNA; not to be confused with pri-miRNA). The pre-miRNA is then exported from the nucleus with the aid of a protein called exportin 5.

As shown in Figure 18.4, siRNAs do not go through the processing events that occur in the nucleus. Instead, precursor-siRNAs (pre-siRNAs) are usually derived from viral RNAs or may be made by researchers and taken up by cells. For example, in the work of Fire and Mello shown in Figure 18.3, the double-stranded *mex-3* RNA is an example of a pre-siRNA. In Figure 18.4, the pre-siRNA is formed from two complementary RNA molecules that base-pair with each other.

In the cytosol, both pre-miRNAs and pre-siRNAs are cut by an endonuclease called dicer (see Figure 18.4), releasing a double-stranded RNA molecule that is 20–25 bp long. This double-stranded RNA associates with proteins to form a complex called the **RNA-induced silencing complex (RISC).** One of the RNA strands is degraded. The remaining single-stranded miRNA or siRNA is complementary to specific mRNAs that will be silenced. The miRNA or siRNA acts as a guide that causes RISC to recognize and bind to such mRNA molecules.

After RISC binds to an mRNA, any of the following three outcomes may result:

- RISC may inhibit translation without degrading the mRNA. This outcome is more common for miRNAs, which often are only partially complementary to their target mRNAs.
- The RISC-mRNA complex may remain in a cellular structure called a **processing body (P-body),** where it can be stored and later reused. In this case, the inhibition of translation by an miRNA is only temporary.
- RISC may direct the degradation of the mRNA. One of the proteins in RISC, which is called Argonaute, can cleave the mRNA. This outcome usually occurs with siRNAs, which are typically a perfect match to their target mRNAs.

In the first and third cases, the expression of the mRNA is permanently silenced. The effect is termed RNA interference because the miRNA or siRNA interferes with the proper expression of an mRNA. In 2006, Fire and Mello received the Nobel Prize in physiology or medicine for their discovery of this mechanism.

RNA interference is believed to have at least two benefits:

- RNAi an important means of gene regulation. When genes encoding pri-miRNAs are turned on, the production of miRNAs silences the expression of specific mRNAs.
- RNAi provides a defense against viruses. This mechanism is widely used by plants to prevent viral infections.

In addition to mRNA degradation, siRNAs can inhibit transcription by causing chromatin modifications. This effect is also seen with another type of small regulatory RNA, called PIWI-interacting RNA (piRNA). The ability of piRNAs to inhibit transcription is described later in this chapter (look ahead to Figure 18.7).

18.3 REVIEWING THE KEY CONCEPTS

- Fire and Mello showed that double-stranded RNA is more potent at silencing mRNA than is antisense RNA (see Figure 18.3).
- RNA interference is a mechanism of RNA silencing in which miRNA or siRNA becomes part of an RNA-induced silencing complex (RISC) that inhibits the translation of a specific mRNA or causes its degradation (see Figure 18.4).

18.3 COMPREHENSION QUESTION

1. The process of RNA interference may lead to
   a. the degradation of an mRNA.
   b. the inhibition of translation of an mRNA.
   c. the synthesis of an mRNA.
   d. both a and b.

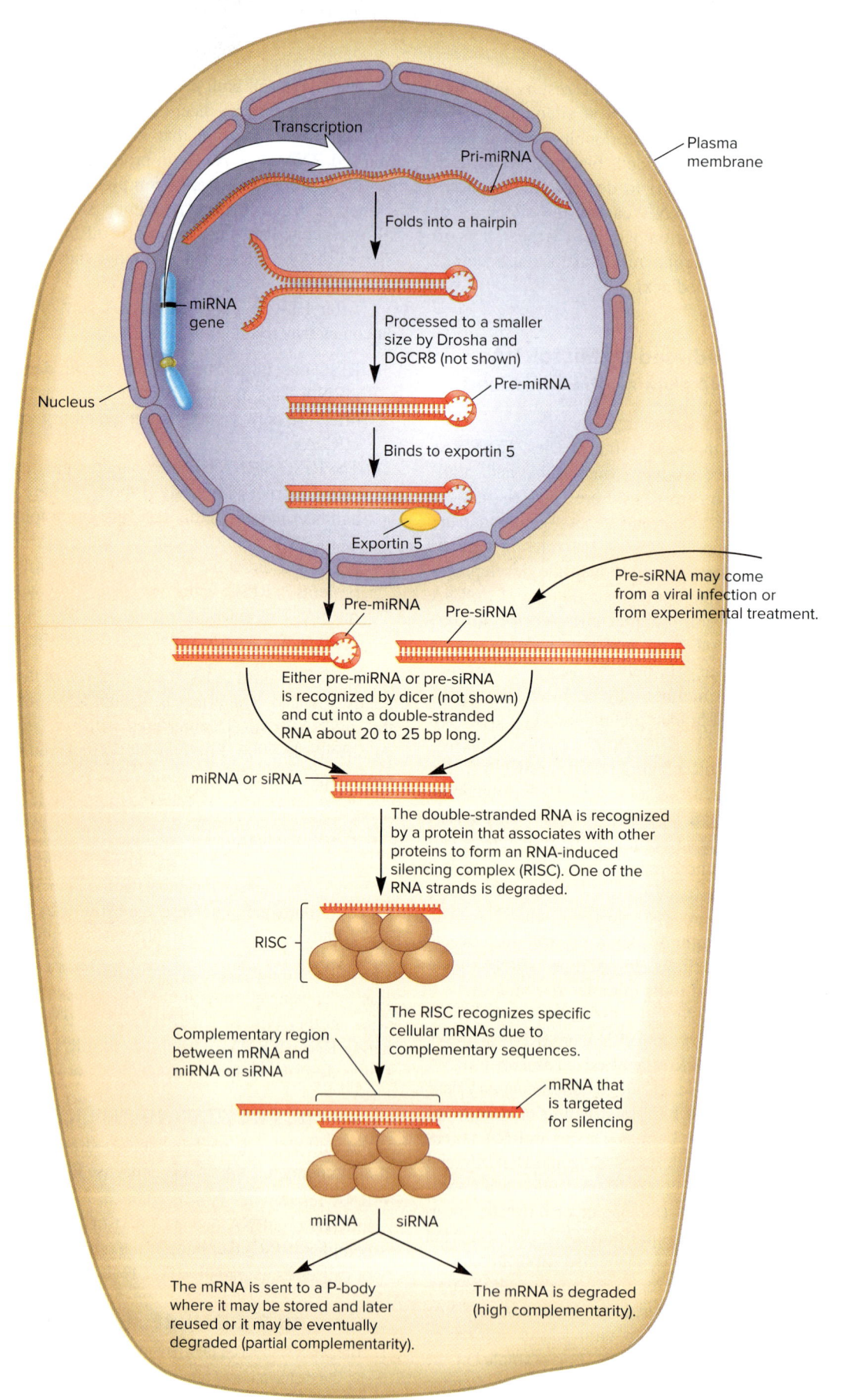

FIGURE 18.4 Mechanism of RNA interference.

Concept Check: Explain why RISC binds to a specific mRNA. What type of bonding occurs?

18.4 NON-CODING RNAs AND PROTEIN TARGETING

Learning Outcome:
1. Describe the function of SRP, and explain the roles of SRP RNA with regard to SRP function.

To carry out their functions, proteins need to be targeted to a particular location. For example, some proteins function extracellularly and need to be secreted from the cell. For such proteins to be secreted, they must first be targeted to the plasma membrane in bacteria and archaea or to the endoplasmic reticulum (ER) membrane in eukaryotes. This targeting process is facilitated by an RNA-protein complex called **signal recognition particle (SRP).** In bacteria, SRP is composed of one ncRNA and one protein. In eukaryotes, SRP is composed of one ncRNA and six different proteins.

Figure 18.5 shows how SRP works in eukaryotes. To be directed to the ER membrane, a polypeptide must contain a sorting signal called an **ER signal sequence,** which is a sequence of about 6–12 amino acids that are predominantly hydrophobic and usually located near the N-terminus. As the ribosome is making the polypeptide in the cytosol, the ER signal sequence emerges from the ribosome and is recognized by a protein in SRP. The binding of SRP pauses translation. SRP then binds to an SRP receptor in the ER membrane, which docks the ribosome over a channel. For this binding to occur, proteins within SRP and the SRP receptor must also be bound by GTP. Next, these GTP-binding proteins hydrolyze their GTP, which causes the release of SRP from the SRP receptor and the polypeptide. Once SRP is released, translation resumes and the growing polypeptide is threaded through a channel to cross the ER membrane. In the case of a secreted protein, the newly made polypeptide travels through the Golgi apparatus and then to the plasma membrane, where it is released outside of the cell.

Researchers have identified at least two key roles for SRP RNA:

1. SRP RNA provides a scaffold for the binding of SRP proteins.
2. After SRP binds to the SRP receptor in the ER membrane, the SRP RNA stimulates proteins within both SRP and the SRP receptor to hydrolyze GTP. In other words, SRP RNA alters the structures of these proteins to enhance their GTPase activities. This stimulation is essential for the release of SRP.

18.4 REVIEWING THE KEY CONCEPTS

- Signal recognition particle (SRP), which is composed of one or more proteins and an ncRNA, plays a role in the targeting of proteins to the plasma membrane of prokaryotic cells or to the ER membrane of eukaryotic cells (see Figure 18.5).

18.4 COMPREHENSION QUESTION

1. Which of the following is a function of SRP?
    a. Pausing translation of polypeptides with an ER signal sequence
    b. Binding to an SRP receptor in the ER membrane
    c. Docking the ribosome over a channel
    d. All of the above are functions of SRP.

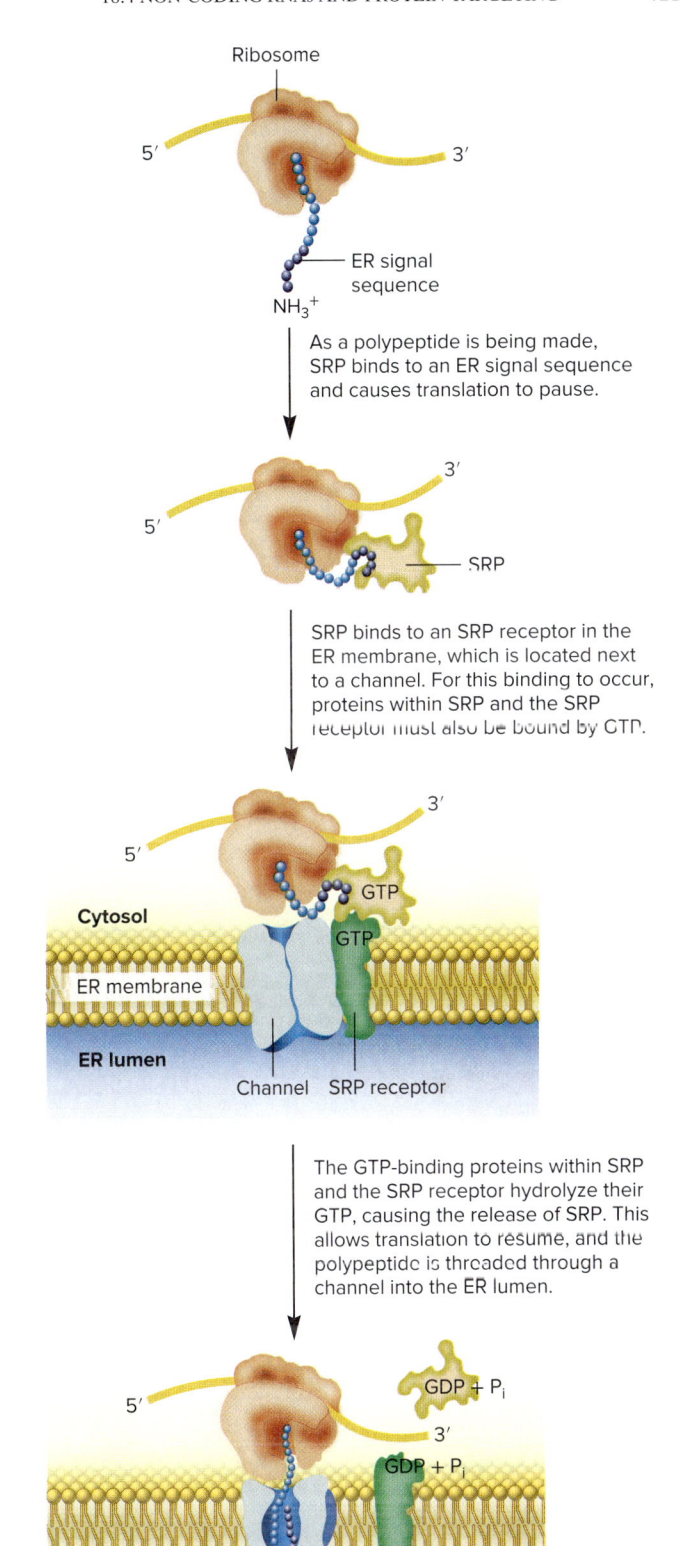

FIGURE 18.5 Targeting of polypeptides to the endoplasmic reticulum membrane via SRP. In eukaryotes, several categories of proteins are first targeted to the ER membrane via SRP. These include secreted proteins as well as proteins that are destined to stay in the ER, Golgi apparatus, lysosomes, or vacuoles.

Concept Check: Why is GTP necessary for this process?

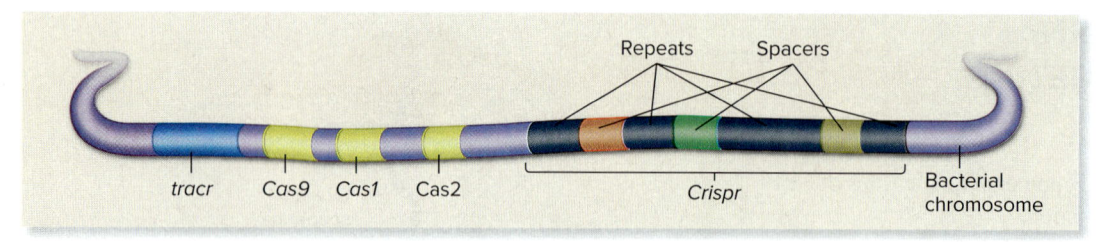

FIGURE 18.6 The CRISPR-Cas system of genome defense in prokaryotes. The system shown here is a type II system, which is found in certain bacterial species but not in archaeal species. (a) Organization of the CRISPR-Cas system in a bacterial chromosome. This drawing shows a typical organization, but it can vary among different species. (b–d) A simplified mechanism for the functioning of the CRISPR-Cas system. The defense occurs in three phases called adaptation, expression, and interference. Parts (c) and (d) are on the next page.

Concept Check: Which component of the CRISPR-Cas system directly recognizes the bacteriophage DNA?

(a) Simplified organization of the CRISPR-Cas system in the bacterial chromosome

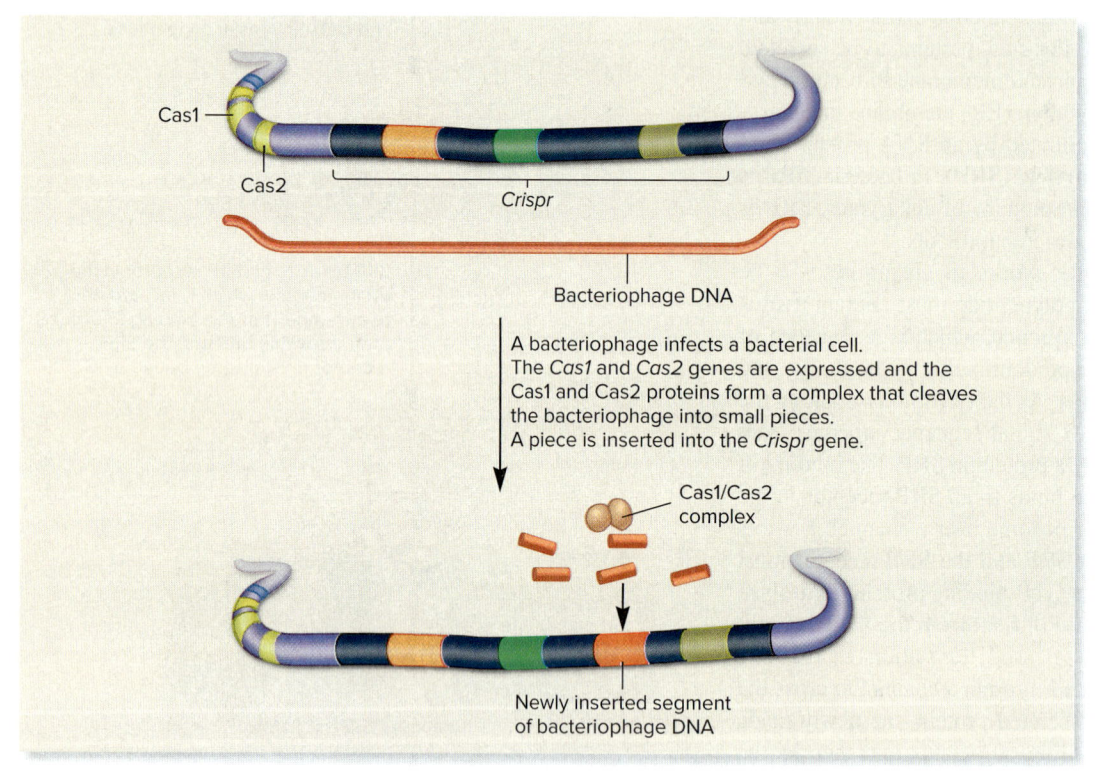

(b) Adaptation

18.5 NON-CODING RNAs AND GENOME DEFENSE

Learning Outcomes:
1. Explain how the CRISPR-Cas system defends bacteria against bacteriophages.
2. Outline the two ways that piRNAs and PIWI proteins inhibit transposable elements.

Much like the immune system found in vertebrates, some prokaryotes have a system, called the **CRISPR-Cas system** that defends against foreign invaders. In many prokaryotes, CRISPR-Cas systems provide an effective defense against bacteriophages (discussed in Chapter 10), plasmids (discussed in Chapter 9), and transposons (discussed in Chapter 12). ncRNAs play a key role in these systems. About half of all bacterial species and most archaeal species have a CRISPR-Cas system. Three general types are known. In this section, we will focus on the type II system and its role in providing bacteria with defense against bacteriophages.

We will also consider another defense system found in animals that involves a type of ncRNA called PIWI-interacting RNA. As you will learn, this ncRNA interacts with PIWI proteins and inhibits the movement of transposable elements.

The CRISPR-Cas System Provides Bacteria with Defense Against Bacteriophages

In 1993, Francisco Mojica and colleagues were the first to recognize that different species of prokaryotes have a site in their chromosome, now called the CRISPR locus, which contains a series of repeated sequences. In 2005, by analyzing the DNA sequences of the CRISPR locus in a variety of prokaryotes, Mojica, Giles Vergnaud, and Alexander Bolotin independently proposed that it provides protection against bacteriophage infection. This hypothesis was based on the observation that the CRISPR locus contains segments that are derived from bacteriophage DNA. Their hypothesis was confirmed in 2007 by Philippe Horvath and colleagues, who showed experimentally that the CRISPR-Cas system provides defense against bacterophage infection.

Figure 18.6a shows a common organization of the CRISPR-Cas system (also called the CRISPR locus) in a bacterial

The genes encoding pre-crRNA and tracrRNA are transcribed.

tracrRNAs bind to pre-crRNA due to complementary base pairing at the repeats. The pre-crRNA is cleaved into several crRNAs.

The *Cas9* gene is also expressed. Each tracrRNA-crRNA complex binds to a Cas9 protein. (Only one is shown.)

(c) Expression

The tracrRNA-crRNA-Cas9 complex binds to bacteriophage DNA due to complementary base pairing at the spacer.

Cas9 cleaves the bacteriophage DNA into pieces, thereby inactivating it.

(d) Interference

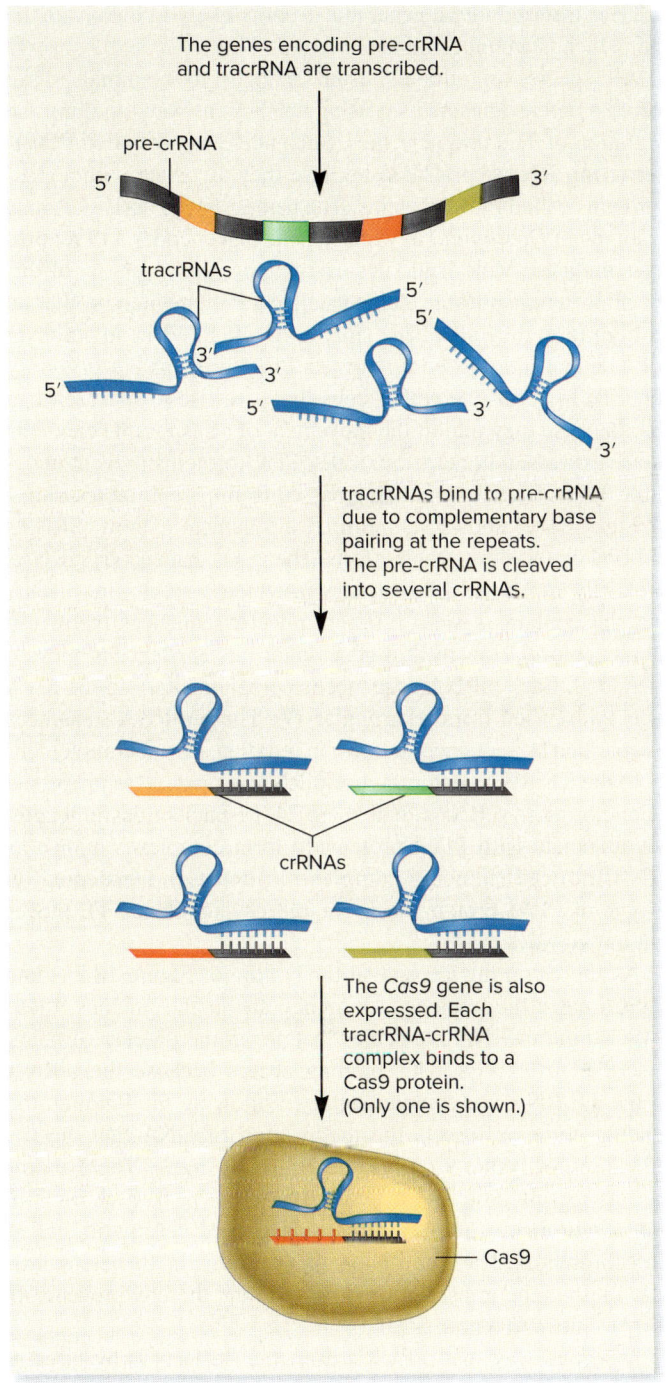

FIGURE 18.6 *continued*.

chromosome. In this example, the system has five genes: *tracr, Cas9, Cas1, Cas2,* and *Crispr*. A key feature of the *Crispr* gene is a group of clustered, regularly interspaced, short, palindromic repeats; hence the name CRISPR. The repeats are interspersed with short, unique sequences, which are called spacers. The CRISPR-Cas type II system also employs a gene that encodes an ncRNA called tracrRNA and a few protein-encoding *Crispr*-associated genes, (*Cas* genes), which are usually adjacent to the *Crispr* gene. These genes are needed to mediate the defense against bacteriophages.

The CRISPR-Cas system is considered an adaptive defense system because a bacterial cell must first be exposed to an agent, such as a bacteriophage, to elicit a response. The defense mechanism has three phases: adaptation, expression, and interference.

Adaptation The process of adaptation (also called spacer acquisition) occurs after a bacterial cell has been exposed to a bacteriophage. The proteins encoded by the *Cas1* and *Cas2* genes form a complex that recognizes the bacteriophage DNA as being foreign and cleaves it into small pieces. As shown in **Figure 18.6b**, a piece of bacteriophage DNA, usually between 20 and 50 bp in length, is inserted into the *Crispr* gene. The mechanism of insertion is not entirely understood. The newly inserted piece of bacteriophage DNA is called a spacer, because it acts as a space between adjacent repeats. The different spacers found in the *Crispr* gene of modern bacterial species are derived from past bacteriophage infections. Each spacer provides a bacterium with defense against a particular bacteriophage. Once a bacterial cell has become adapted to a particular bacteriophage, it will pass this trait on to its daughter cells.

Cleaving the bacteriophage DNA into pieces during the adaptation phase can protect a bacterial cell, because this action inactivates the phage. However, a more effective way of destroying phages is provided by the expression and interference phases of this defense system.

Expression If a bacterial cell has already been adapted to a bacteriophage, a subsequent bacteriophage infection will result in the expression phase, in which the system gets ready for action by expressing the *Crispr, tracr,* and *Cas9* genes (**Figure 18.6c**). The *Crispr* gene is transcribed from a single promoter and produces a long ncRNA called pre-crRNA. The gene encoding the tracrRNA is also transcribed, which produces many molecules of tracrRNA, another kind of ncRNA. A region of the tracrRNA is complementary to the repeat sequences of the pre-crRNA. Several molecules of tracrRNA base-pair with the pre-crRNA. The pre-crRNA is then cleaved into many small molecules, now called crRNAs. Each crRNA is attached to a tracrRNA. A region of the tracrRNA is recognized by the Cas9 protein. The tracrRNA acts as a guide that causes the tracrRNA-crRNA complex to bind to a Cas9 protein.

Interference After the tracrRNA-crRNA-Cas9 complex is formed, the bacterial cell is ready to destroy the bacteriophage DNA. This phase is called interference, because it resembles the process of RNA interference described earlier in this chapter (refer back to Figure 18.4). Each spacer within a crRNA is complementary to one of the strands of the bacteriophage DNA. Therefore, the crRNA acts as a guide that causes the tracrRNA-crRNA-Cas9 complex to bind to that strand (**Figure 18.6d**). After binding, the Cas9 protein functions as an endonuclease that makes double-strand breaks in the bacteriophage DNA. This cleavage inactivates the phage, and thereby prevents its proliferation.

After discovering the CRISPR-Cas system in prokaryotes, researchers have been able to modify certain components of this system and use them to mutate genes in living cells. We will consider this technology in Chapter 20.

PIWI-interacting RNAs Silence Transposable Elements in Animals

As described in Chapter 12, **transposable elements (TEs)** are segments of DNA that can become integrated into chromosomes. Various types of TEs are found in all species. Transposition is the process in which a TE is inserted into a new site in the genome. If a TE is inserted into a gene, this event is likely to inactivate the gene, which is an undesirable outcome.

All species of animals, from sponges to mammals, have evolved a defense mechanism using ncRNAs as way to inhibit the integration of TEs into new sites. Such ncRNAs are called **PIWI-interacting RNAs (piRNAs)** because they associate with a class of proteins called PIWI proteins. The name *PIWI* is derived from a phenotype in *Drosophila* in which a gene that encodes a PIWI protein is mutated. Male fruit flies carrying a mutant version of this gene allow a transposable element called a P-element to proliferate at a high rate; they also have smaller testes. PIWI is an acronym for P-element induced wimpy testes. PIWI proteins are primarily expressed in germ-line cells—cells that give rise to sperm or egg cells. Because they inhibit the movement of transposable elements and are expressed in germ-line cells, PIWI proteins prevent undesirable TE insertions from being passed from parent to offspring.

Figure 18.7 shows a simplified mechanism of how piRNAs and PIWI proteins exert their effects. The genes that encode piRNAs are usually organized into clusters in which a single pre-piRNA transcript contains several different piRNA sequences (shown in different colors). How piRNA genes came into existence is not well understood. Each piRNA sequence is complementary to the RNA that is transcribed from a particular TE. These piRNA sequences are separated by sequences (shown in red) that are not complementary to the TE RNA. After the pre-piRNA is made, it is processed into one or more piRNAs, which are 24–31 nucleotides in length. They associate with PIWI proteins to form complexes known as **piRNA-induced silencing complexes (piRISCs)**. Depending on the type of PIWI protein, such complexes can silence TE movement in two ways.

As shown on the left side at the bottom of Figure 18.7, some piRISCs enter the nucleus. After they are in the nucleus, these piRISCs bind to RNA molecules that are in the process of being transcribed from TEs. This binding occurs because the piRNA and the TE RNA are complementary. Following the binding, the silencing complex directs the methylation of DNA and the trimethylation of lysine 9 on histone H3 (H3K9me3). These modifications recruit proteins that convert loosely packed chromatin, called euchromatin, into more densely packed chromatin, called heterochromatin. The conversion of euchromatin to heterochromatin inhibits the transcription of the TE and thereby prevents its ability to move to a new site.

A second way that piRISCs exert their effects is by directly inhibiting RNAs. The mechanism of this inhibition is very similar to the mechanism of mRNA silencing via siRNAs described earlier in this chapter (look back at Figure 18.4). However, the processing of pre-piRNAs does not involve dicer. As shown on the right side at the bottom of Figure 18.7, the piRISC binds to the TE RNA in the cytosol and an Argonaute protein within piRISC cuts it into pieces, thereby inactivating it.

In recent years, researchers have discovered that piRNAs are an abundant class of ncRNAs. They are thought to be more diverse than any other known class of cellular RNAs and to constitute the largest class of ncRNAs. Although their main function is to prevent the movement and integration of TEs, piRNAs are also known to regulate a few protein-coding genes.

18.5 NON-CODING RNAs AND GENOME DEFENSE   425

A pre-piRNA is transcribed in the nucleus and exported to the cytosol.

Pre-piRNA

piRNA sequences (Note: Each piRNA sequence is complementary to an mRNA from a transposable element.)

The pre-piRNA is cleaved at multiple locations, which produces several piRNAs that are usually 24 to 31 nucleotides long.

piRNAs

Each piRNA associates with PIWI proteins to form a piRNA-induced silencing complex (piRISC). Note: Only one piRISC is shown here.

piRISC

piRISC travels into the nucleus and binds to RNA that is being transcribed from a TE due to complementary base paring.

piRISC binds to TE RNA in the cytosol due to complementary base pairing.

or

TE RNA

Segment of a transposable element

piRISC directs the methylation of DNA and the trimethylation of lysine 9 on histone H3.

TE RNA

piRISC directs the cleavage of the TE RNA, thereby inactivating it.

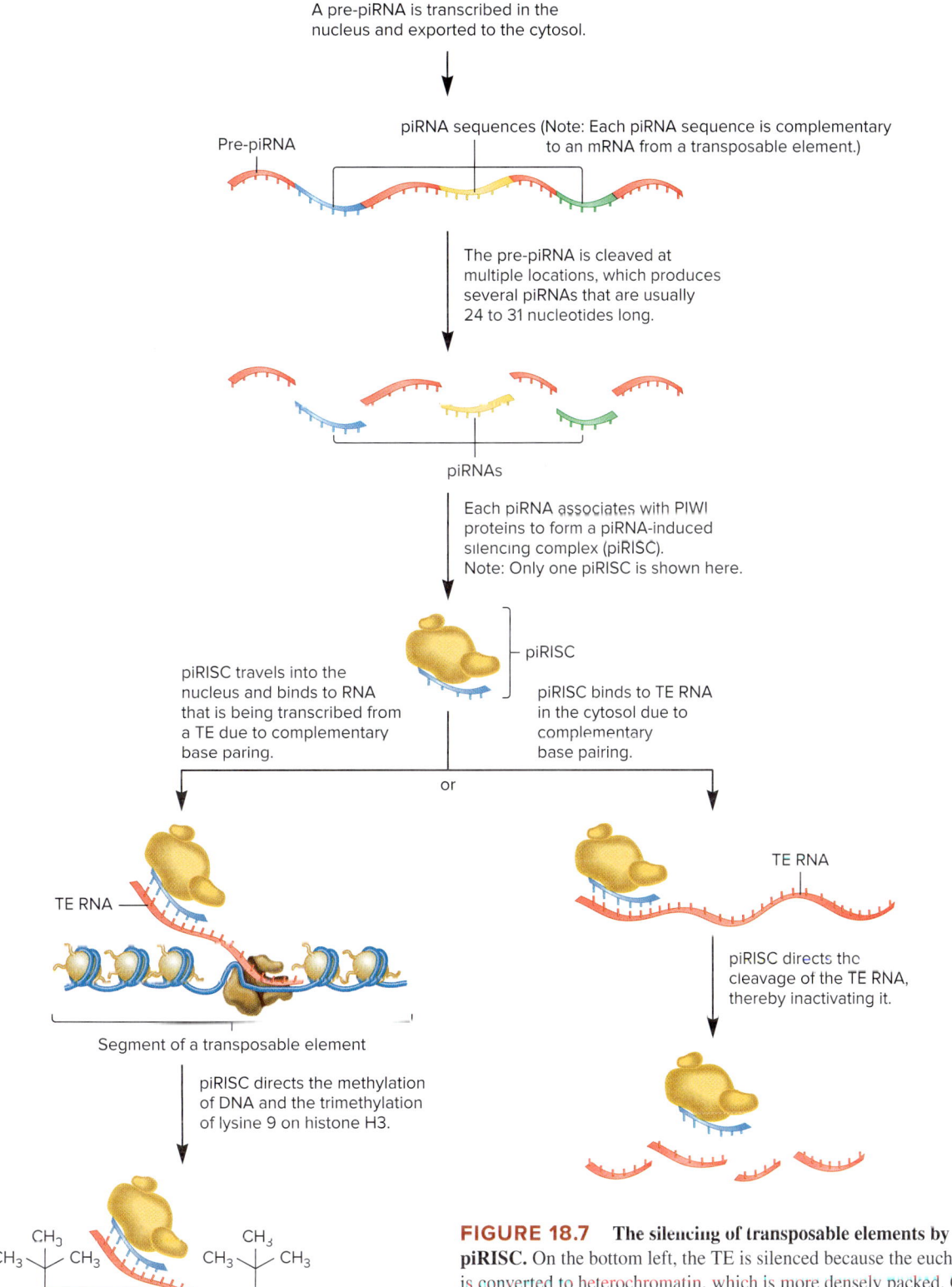

These modifications recruit proteins that convert euchromatin to heterochromatin, which stops the transcription of the TE.

FIGURE 18.7 **The silencing of transposable elements by piRISC.** On the bottom left, the TE is silenced because the euchromatin is converted to heterochromatin, which is more densely packed. On the bottom right, the TE is silenced because the TE RNA is cleaved into several pieces. Note: Some researchers refer to the silencing complex that inhibits transcription as RITS (RNA-induced transcriptional silencing complex) to distinguish it from the complex that degrades RNA, which is called RISC.

Concept Check: What are the two ways in which piRNAs and PIWI proteins prevent the movement of transposable elements?

18.5 REVIEWING THE KEY CONCEPTS

- The CRISPR-Cas system found in bacteria and archaea provides defense against bacteriophages, plasmids, and transposons. The defense mechanism has three phases: adaptation, expression, and interference (see Figure 18.6).
- PIWI-interacting RNAs (piRNAs) silence transposable elements in animals (see Figure 18.7).

18.5 COMPREHENSION QUESTIONS

1. Which of the following components are needed for the adaptation phase of the CRISPR-Cas system?
   a. crRNA and Cas1
   b. crRNA and Cas2
   c. crRNA and Cas9
   d. Cas1 and Cas2
2. In the CRISPR-Cas system, what does tracrRNA bind to?
   a. crRNA and Cas1
   b. crRNA and Cas2
   c. crRNA and Cas9
   d. Cas1 and Cas2
3. Which of the following is a function of the piRISC?
   a. Inhibits transcription of TEs
   b. Causes the degradation of TE RNA
   c. Causes chromosome breakage
   d. Both a and b are functions of the piRISC.

18.6 ROLE OF NON-CODING RNAs IN HUMAN DISEASES

Learning Outcomes:

1. List examples in which ncRNAs are associated with human diseases.
2. Explain how ncRNAs may be targets in treating human diseases.

During the past two decades, researchers have discovered that abnormalities in the expression of ncRNAs are associated with a wide range of human diseases. In 2001, the gene encoding an RNA molecule that is a component of RNaseMRP was the first ncRNA gene shown to be associated with a genetic disease. RNaseMRP is an RNA-protein complex involved in processing some rRNAs, mRNAs, and mitochondrial RNAs. The RNA component is a ribozyme. Mutations in this gene cause a disorder called cartilage-hair hypoplasia (CHH). CHH patients have a short stature, underdeveloped hair, and short limbs. In addition, they have a predisposition to develop lymphomas and other cancers, and suffer from defective T-cell immunity. Since the identification of the first CHH-associated mutations in 2001, many ncRNAs have been shown to play a key role in human diseases. Researchers speculate that they have seen only the "tip of the iceberg" with regard to identifying the roles of ncRNAs in human pathology. In this section, we will consider examples in which ncRNAs are associated with human diseases and explore how they may be targets in treating diseases.

ncRNAs Play a Role in Many Forms of Cancer and Other Human Diseases

As we have seen throughout this chapter, ncRNAs play important roles in regulating chromatin modification, gene transcription, mRNA translation, and protein function. When certain ncRNAs are expressed abnormally, that is, at too high or too low a level, disease conditions are known to occur. Such abnormal expression levels can be caused by mutations in specific genes or by epigenetic changes that alter the expression of genes that encode ncRNAs. Several examples of human diseases associated with the abnormal expression of ncRNAs are listed in **Table 18.2**. We will focus on the roles of ncRNAs in the development of cancer, neurological disorders, and cardiovascular diseases.

ncRNAs and Cancer The roles of ncRNAs in cancer have been most thoroughly studied with respect to miRNAs. In nearly all forms of human cancer, levels of expression of particular miRNAs differ between normal and cancer cells. In some cases, the genes that encode certain miRNAs act as oncogenes; their overexpression promotes cancer. In other cases, the genes behave like tumor-suppressor genes, because a lower level of expression of particular miRNAs allows tumor growth.

A well-studied example of the role of miRNAs in cancer involves a group of several different miRNAs called the miR-200 family. These miRNAs are often involved in cancers that are derived from epithelial cells, such as skin or intestinal cells. Low levels of expression of miR-200 members have been associated with many types of cancer, including bladder cancer, melanoma, stomach cancer, and colorectal cancer.

TABLE 18.2
Examples of Non-Coding RNAS Associated with Human Diseases

ncRNAs or Processing Proteins Associated with Diseases	Disease(s)*
Group of miRNAs called the miR-200 family	Several types of cancer, including bladder cancer, melanoma, stomach cancer, and colorectal cancer
HOTAIR	Several types of cancer, including breast cancer, lung cancer, and colorectal cancer
piRNAs/PIWI proteins	Certain types of testicular cancer
Drosha	Familial amyotrophic lateral sclerosis
Many miRNAs	Alzheimer disease
Many miRNAs	Multiple sclerosis
An miRNA called miR-1	Heart arrhythmias
Many different miRNAs	Damage to heart tissue due to heart failure
Several different miRNAs, including miR-10a, miR-145, and miR-143	Formation of arterial plaques
snoRNAs	Lung cancer

*The diseases listed here show an association with abnormal levels of the particular ncRNAs or miRNA-processing proteins. In many cases, it is not yet clear if the disease symptoms are caused, in part, by the abnormal levels of ncRNAs or proteins or if the abnormal levels are a consequence of the disease symptoms.

The miR-200 family plays an essential role in tumor suppression by inhibiting an event called the epithelial-mesenchymal transition (EMT), which is the initiating step of metastasis. (The process of metastasis is described in Chapter 22.) The EMT occurs as part of normal embryonic development, and it shares many similarities with cancer progression. During the EMT, cells lose their adhesion to neighboring cells. This loss of adhesion is associated with a decrease in expression of E-cadherin, which is a membrane protein that adheres adjacent cells to each other. When E-cadherin levels are low, cells can more easily move to new sites in the body. In an adult, when the miR-200 family of miRNAs is expressed at normal levels, the miRNAs inhibit the EMT. This inhibition maintains a normal level of E-caderin and thereby prevents metastasis.

Though they have been less well studied, lncRNAs are also associated with particular types of human cancers. HOTAIR, which was discussed in Section 18.2, is an lncRNA that is highly expressed in a variety of cancers, including breast cancer, lung cancer, and colorectal cancer. In this regard, HOTAIR behaves like an oncogene. High levels of HOTAIR expression in primary breast tumors are a significant predictor of metastasis and death. HOTAIR is known to interact with a variety of cellular components, but the mechanism by which it promotes cancer is not well understood.

ncRNAs and Neurological Disorders Many miRNAs are essential for the proper development and functioning of the nervous system. Approximately 70% of all miRNAs are expressed in the brain, and many of them are specific to neurons. miRNAs are involved in neuron growth and the overall development of the nervous system. Abnormal levels of expression of miRNAs have been associated with nearly all neurological disorders in which they have been investigated!

Table 18.2 describes some examples in which the expression of miRNAs has been altered and associated with neurological disorders. The changes involve two categories of genes: genes that encode miRNAs and genes that encode proteins that are involved in processing miRNAs

- Specific miRNAs have been linked to particular neurological diseases. For example, in Alzheimer disease, abnormally expressed miRNAs are thought to be involved in down-regulating the expression of the enzyme β-secretase, which leads to the overproduction of certain β-amyloid peptides. miRNAs are also known to control the inflammatory process that leads to the development of multiple sclerosis.
- Mutations or epigenetic changes may alter the expression of genes that encode miRNA-processing proteins, such as Drosha, dicer, and others (refer back to Figure 18.4). For example, mutations in components of Drosha cause up to 50% of all cases of familial amyotrophic lateral sclerosis (ALS) (sometimes called Lou Gehrig disease). These mutations result in a generalized decrease in the processing of many different miRNAs.

ncRNAs and Cardiovascular Diseases Abnormalities in miRNA levels have been linked to several cardiovascular diseases.

- A particular miRNA called miR-1 is associated with the development of heart arrhythmias—irregularities in the rate or rhythm of the heartbeat. This miRNA regulates the expression of genes that encode ion channel proteins, which are important for proper signaling between cardiac muscle cells.
- Cardiac tissue from heart failure patients has a distinctly different expression pattern of many different miRNAs compared to healthy cardiac tissue.
- Particular miRNAs appear to play a role in cardiovascular disease. The formation of arterial plaques is associated with abnormal expression levels of several miRNAs, including miR-10a, miR-145, and miR-143.

Alterations in ncRNAs May Be Used to Combat Certain Diseases

As we have seen, abnormal levels of ncRNAs and miRNA-processing proteins have been linked to a wide variety of human diseases, and the list is growing at a rapid rate. Historically, most drugs that are used to treat human diseases bind to cellular proteins. This bias developed because we have a much better understanding of how proteins affect cell structure and function than we do about the effects of ncRNAs. However, as we learn more about the functions of ncRNAs and their roles in human diseases, researchers are beginning to conduct studies to determine if ncRNAs and miRNA-processing components may be effective targets for novel treatment therapies. So far, much of this work has centered on miRNAs and cancer. Although the development of this methodology is still in its infancy, researchers are optimistic that the approach may be applicable to other types of diseases and other categories of ncRNAs.

Therapies That Inhibit miRNA Function One novel strategy for treating certain forms of cancer is the use of **anti-miRNA oligonucleotides (AMOs)** to inhibit miRNA function. An AMO is complementary to a specific miRNA and can base-pair with it. This action can block the ability of the miRNA to function and/or cause it to be degraded. Some AMOs contain chemical modifications that cause them to bind more tightly to their complementary miRNAs. Examples include the following:

- **Locked nucleic acids (LNAs)** contain a ribose sugar that has an extra bridge connecting the $2'$ oxygen and $4'$ carbon. The bridge locks the ribose in a conformation that causes it to bind more tightly to the complementary miRNAs.
- **Antagomirs** have one or more base modifications that may promote a stronger binding to the complementary miRNAs.

Studies in mice have shown that AMOs may prevent certain types of cancer. For example, in a mouse strain that develops mammary tumors, the intravenous injection of antagomirs that bind to a particular miRNA called miR-10b inhibited the onset of metastasis and suppressed the dissemination of cancer cells to the lungs. However, the silencing of a single miRNA might not be sufficient to inhibit the growth of most types of cancer cells. Recent research suggests that several miRNAs may need to be simultaneously inhibited to slow or prevent the progression of cancer.

Therapies That Restore miRNA Function Some miRNAs behave like tumor suppressors. When their function is decreased, cancer may occur. For this reason, researchers are developing strategies to restore the function of miRNAs that have been downregulated in cancer cells. This approach has been called miRNA replacement therapy. Three examples of this therapy are as follows:

- A viral vector was used to restore a particular miRNA called miR-26a in mice with a type of liver cancer called hepatocellular carcinoma. This therapy inhibited cancer progression.
- Another approach for increasing the levels of miRNAs is to target the miRNA-processing machinery. The drug enoxacin enhances the miRNA-processing machinery in certain cancer cells and thereby results in an increase of many miRNAs. Recent research has shown that this drug may inhibit the growth of cancer. In mice, the drug did not affect miRNA levels in normal cells and was not associated with harmful side effects.
- In some cancers, miRNA expression is decreased due to epigenetic changes such as CpG methylation and histone deacetylation. Therefore, a third approach for restoring miRNAs is the use of DNA demethylating agents and histone deacetylase inhibitors. These compounds reverse the epigenetic silencing of miRNAs that behave as tumor suppressors. Such agents have been shown to slow the progression of certain forms of cancer and have received clinical approval for use in treating those cancers.

18.6 REVIEWING THE KEY CONCEPTS

- Abnormalities in the expression of ncRNAs have been associated with many diseases, including cancer, neurological disorders, and cardiovascular diseases (Table 18.2).
- In the future, therapies to treat certain diseases may involve the inhibition or restoration of ncRNA function.

18.6 COMPREHENSION QUESTIONS

1. Abnormalities in the expression of ncRNAs are associated with
   a. many forms of cancer.
   b. neurological disorders.
   c. cardiovascular diseases.
   d. all of the above.
2. Let's suppose that the overexpression of a particular miRNA was associated with pancreatic cancer. Which of the following agents might be effective in treating this type of cancer?
   a. Enoxacin
   b. A DNA demethylating agent
   c. An anti-miRNA oligonucleotide
   d. Both a and b

KEY TERMS

Page 412. non-coding RNAs (ncRNAs)
Page 414. ribozyme
Page 415. long non-coding RNAs (lncRNAs), small regulatory RNAs
Page 419. RNA interference (RNAi), microRNAs (miRNAs), small-interfering RNAs (siRNAs), RNA-induced silencing complex (RISC), processing body (P-body)
Page 421. signal recognition particle (SRP), ER signal sequence
Page 422. CRISPR-Cas system
Page 424. transposable elements (TEs), PIWI-interacting RNAs (piRNAs), piRNA-induced silencing complexes (piRISCs)
Page 427. anti-miRNA oligonucleotides (AMOs), locked nucleic acids (LNAs), antagomirs

CHAPTER SUMMARY

- Non-coding RNAs (ncRNAs) are RNA molecules that do not encode polypeptides.

18.1 Overview of Non-Coding RNAs

- ncRNAs bind to different types of molecules, including DNA, other RNAs, proteins, and small molecules (see Figure 18.1).
- An ncRNA can perform any of several functions: provide a scaffold, act as a guide, alter protein function or stability, function as a ribozyme, function as a blocker, and/or act as a decoy.
- With regard to cell structure and function, ncRNAs play a role in DNA replication, chromatin structure, transcription, translation, splicing, RNA degradation, RNA modification, protein secretion, and genome defense (see Table 18.1).

18.2 Non-Coding RNAs: Effects on Chromatin Structure and Transcription

- HOTAIR is a long ncRNA found in humans that regulates transcription by guiding PRC2 and LSD1 to particular genes. These complexes produce histone modifications that silence the genes (see Figure 18.2).

18.3 Non-Coding RNAs: Effects on Translation and mRNA Degradation

- Fire and Mello showed that double-stranded RNA is more potent at silencing mRNA than is antisense RNA (see Figure 18.3).
- RNA interference is a mechanism of RNA silencing in which miRNA or siRNA becomes part of an RNA-induced silencing complex (RISC) that inhibits the translation of a specific mRNA or causes its degradation (see Figure 18.4).

18.4 Non-Coding RNAs and Protein Targeting

- Signal recognition particle (SRP), which is composed of one or more proteins and an ncRNA, plays a role in the targeting of proteins to the plasma membrane of prokaryotic cells or to the ER membrane of eukaryotic cells (see Figure 18.5).

18.5 Non-Coding RNAs and Genome Defense

- The CRISPR-Cas system found in bacteria and archaea provides defense against bacteriophages, plasmids, and transposons. The defense mechanism has three phases: adaptation, expression, and interference (see Figure 18.6).
- PIWI-interacting RNAs (piRNAs) silence transposable elements in animals (see Figure 18.7).

18.6 Role of Non-Coding RNAs in Human Diseases

- Abnormalities in the expression of ncRNAs have been associated with many diseases, including cancer, neurological disorders, and cardiovascular diseases (Table 18.2).
- In the future, therapies to treat certain diseases may involve the inhibition or restoration of ncRNA function.

PROBLEM SETS & INSIGHTS

More Genetic TIPS

1. An ncRNA may have the following functions: scaffold, guide, alterer of protein function or stability, ribozyme, blocker, and/or decoy. Which of these functions are mediated by each type of ncRNA listed next? Note: A single ncRNA may have more than one function.

   A. tRNA

   B. rRNA

   C. SRP RNA

   D. piRNA

   Topic: *What topic in genetics does this question address?*

   The topic is the functions of specific types of ncRNAs.

   Information: *What information do you know based on the question and your understanding of the topic?*

   In the question, you are given a list of six general functions of ncRNAs and four specific examples of ncRNAs. From your understanding of the topic, you may recall the functions of each of these four examples.

   Problem-Solving Strategy: *Compare and contrast.*

   To solve this problem, you need to compare the six general functions of ncRNAs with your understanding of the functions of each of the four specific ncRNAs.

   Answer:

   A. Guide: A tRNA binds to a complementary codon in an mRNA and carries the specific amino acid to be added to a polypeptide.

   B. Scaffold, guide, and ribozyme: rRNA functions as a scaffold for the binding of ribosomal proteins; in bacteria, a segment of rRNA acts as a guide for the binding of the Shine-Dalgarno sequence in mRNA; and the rRNA in peptidyl transferase functions as a ribozyme that catalyzes peptide bond formation.

   C. Scaffold and alterer of protein function: SRP RNA acts as a scaffold for the binding of SRP proteins, and it affects the GTPase activity of certain proteins in SRP and in the SRP receptor.

   D. Guide: piRNA guides PIWI proteins to genes or to mRNAs.

2. With regard to the CRISPR-Cas system that defends bacteria against bacteriophage attack, what happens during the adaptation, expression, and interference phases? When a bacterium is exposed to a particular bacteriophage, is the adaptation phase always necessary?

   Topic: *What topic in genetics does this question address?*

   The topic is the CRISPR-Cas system that provides bacteria with defense against bacteriophages. More specifically, the question asks you to sort out what happens during each phase and decide whether or not the first phase is always needed.

   Information: *What information do you know based on the question and your understanding of the topic?*

   In the question, you are reminded that the CRISPR-Cas system defends bacteria against bacteriophages and that the defense mechanism occurs in three phases. From your understanding of the topic, you may recall what happens during each phase.

   Problem-Solving Strategy: *Describe the steps.*

   To solve this problem, one strategy is to sort out the steps of this genome defense mechanism.

   Answer: During adaptation, a portion of bacteriophage DNA is inserted into the *Crispr* gene. This phase requires the proteins Cas1 and Cas2. During the expression phase, tracrRNA, crRNA, and Cas9 are produced. Finally, during the interference phase, tracrRNA, crRNA, and Cas9 come together and cleave the bacteriophage DNA, thereby inactivating it.

   If the bacterium or one of its ancestors was already exposed to the type of bacteriophage that is currently infecting it, the adaptation phase is not necessary. Prior exposure to the bacteriophage may have resulted in the insertion of a portion of the bacteriophage DNA into the *Crispr* gene. This alteration would be passed on to daughter cells. Therefore, if a bacterium already had a portion of the bacteriophage DNA in its *Crispr* gene, it would not have to go through the adaptation phase; it would already be adapted to defend itself against that bacteriophage.

Conceptual Questions

C1. List and briefly describe four types of molecules that can bind to an ncRNA.

C2. An ncRNA may have the following functions: scaffold, guide, alterer of protein function or stability, ribozyme, blocker, and/or decoy. Which of those functions is/are mediated by each of the ncRNAs listed next? (Note: A single ncRNA may have more than one function.)

   A. HOTAIR

   B. RNA of RNaseP

   C. microRNA

   D. crRNA

C3. Explain how HOTAIR plays a role in the transcriptional regulation of particular genes.

C4. What is the phenomenon of RNA interference (RNAi)? Explain how double-stranded RNA is processed during RNAi and how the mechanism leads to the silencing of a complementary mRNA.

C5. With regard to RNAi, what are three possible sources for double-stranded RNA?

C6. What is the difference between an miRNA and an siRNA? How do these ncRNAs affect mRNAs?

C7. Describe the structure of SRP in eukaryotes, and outline its role in targeting proteins to the ER membrane.

C8. Referring to Figure 18.5, predict what would happen if the SRP RNA was unable to stimulate the GTPase activities of the GTP-binding proteins within SRP and the SRP receptor.

C9. Compare and contrast the roles of crRNA and tracrRNA in the defense against bacteriophages provided by the CRISPR-Cas system.

C10. In the CRISPR-Cas system, does the tracrRNA act as a scaffold, guide, ribozyme, blocker, decoy, and/or alterer of protein function or stability?

C11. What are the roles of Cas1, Cas2, and Cas9 proteins in bacterial genome defense?

C12. Outline the steps that occur when piRISCs silence transposable elements by repressing transcription and by directly inhibiting TE RNAs. What is the role of piRNAs in this process?

C13. List five types of cancer in which ncRNAs can be involved.

C14. Explain how the miR-200 family of miRNAs behave as tumor-suppressor genes. What happens when their expression is blocked or decreased?

Experimental Questions

E1. A protein called trypsin, which plays a role in digestion, is made by pancreatic cells and secreted from those cells. Starting with a sample of pancreatic cells, a researcher modified the gene that encodes trypsin by mutating the ER signal sequence so it was no longer recognized by SRP. How would this mutation affect the targeting of trypsin?

E2. In the experiment shown in Figure 18.3, were Fire and Mello injecting pre-miRNA or pre-siRNA? Explain.

E3. Explain how the data of Fire and Mello suggested that double-stranded RNA is responsible for the silencing of *mex-3* mRNA.

E4. As described in Chapter 20, the CRISPR-Cas system has been modified so it can be used as a gene mutagenesis tool (look ahead to Figure 20.18). Describe how the gene mutagenesis tool works, and explain how the natural CRISPR-Cas system is altered to produce this tool.

E5. Compare and contrast anti-miRNA oligonucleotides, locked nucleic acids (LNAs), and antagomirs, which may eventually be used to treat certain forms of cancer.

E6. What is miRNA replacement therapy? Describe three examples of this treatment approach.

Questions for Student Discussion/Collaboration

1. According to the concept of an RNA world, RNA was the first macromolecule found in primordial cells. As evolution progressed, DNA and proteins eventually came into existence. Discuss which ncRNAs described in Table 18.1 may have arisen during the RNA world, and which probably arose after the modern DNA/RNA/protein world came into existence.

2. Go to the PubMed website and do a search using the words *non-coding RNA* and *disease*. Scan through the journal articles you retrieve and make a list of the roles that ncRNAs may play in human diseases.

Answers to Comprehension Questions

18.1: e, b

18.2: d

18.3: d

18.4: d

18.5: d, c, d

18.6: d, c

Note: All answers appear in Connect; the answers to the even-numbered questions and all of the Concept Check questions are in Appendix B.

19

CHAPTER OUTLINE

- 19.1 Effects of Mutations on Gene Structure and Function
- 19.2 Random Nature of Mutations
- 19.3 Spontaneous Mutations
- 19.4 Induced Mutations
- 19.5 DNA Repair

The effects of a mutation. A mutation during embryonic development caused this sheep to have a black spot on its side.
©Robert Brooker

GENE MUTATION AND DNA REPAIR

The primary function of DNA is to store information for the synthesis of cellular proteins. A key aspect of the gene expression process is that the DNA itself does not normally change. This stability allows DNA to function as a permanent storage unit. However, on relatively rare occasions, a mutation can occur. The term **mutation** refers to a heritable change in the genetic material. When a mutation occurs, the structure of DNA is changed permanently, and this alteration can be passed from mother to daughter cells during cell division. If a mutation occurs in reproductive cells, it may also be passed from parent to offspring.

The phenomenon of mutation is centrally important in all fields of genetics, including molecular genetics, Mendelian inheritance, and population genetics. Mutations provide the allelic variation that we have discussed throughout this textbook. For example, phenotypic differences, such as tall versus dwarf pea plants, are due to mutations that alter the expression of particular genes. With regard to their phenotypic effects, mutations can be beneficial, neutral, or detrimental. On the positive side, mutations are essential to the continuity of life. They provide the variation that enables species to change and adapt to their environment. Mutations are the foundation for evolutionary change. On the negative side, however, new mutations are much more likely to be harmful than beneficial to the individual. The genes within each species have evolved to work properly. They have functional promoters, coding sequences, terminators, and so on, that allow them to be expressed. Random mutations are more likely to disrupt these sequences than to improve their function. For example, many inherited human diseases result from mutated genes. In addition, diseases such as skin and lung cancer can be caused by environmental agents that are known to cause DNA mutations. For these and many other reasons, investigating the molecular nature of mutations is a deeply compelling area of research. In this chapter, we will consider the nature of mutations and their consequences for gene expression at the molecular level.

Species have developed several ways to repair damaged DNA, and thereby prevent mutations. DNA repair systems reverse DNA damage before it results in a mutation that could potentially have negative consequences. DNA repair systems have been studied extensively in many organisms, particularly *Escherichia coli*, yeast, mammals, and plants. A variety of systems repair different types of DNA lesions. We will conclude the chapter by examining how several of these DNA repair systems operate.

19.1 EFFECTS OF MUTATIONS ON GENE STRUCTURE AND FUNCTION

Learning Outcomes:
1. Define *point mutation*.
2. Describe how a mutation within the coding sequence of a gene may alter a polypeptide's structure and function.
3. Explain how a mutation within a non-coding sequence may alter gene function.
4. Compare and contrast intragenic and intergenic suppressors.
5. Explain how changes in chromosome structure may affect gene expression.
6. Distinguish between germ-line and somatic mutations.

How do mutations affect phenotype? To answer this question, we must appreciate how changes in DNA structure can ultimately affect DNA function. Much of our understanding of mutations has come from the study of experimental organisms, such as bacteria, yeast, and *Drosophila*. Researchers can expose these organisms to environmental agents that cause mutations and then study the consequences of the induced mutations. In addition, because these organisms have a short generation time, researchers can investigate the effects of mutations when they are passed from cell to cell and from parent to offspring.

As discussed in Chapter 8, changes in chromosome structure and number are important occurrences within natural populations of eukaryotic organisms. These types of changes are considered to be mutations because the genetic material has been altered in a way that can be inherited. Changes in chromosome structure and number usually affect the expression of many genes. In comparison, a gene mutation is a relatively small change in DNA structure that affects a single gene. In this section, we will be primarily concerned with the ways that mutations can affect the molecular and phenotypic expression of single genes. We will also consider why the timing of mutations during an organism's development has important consequences.

Gene Mutations Are Molecular Changes in the DNA Sequence of a Gene

A gene mutation occurs when the sequence of the DNA within a gene is altered in a permanent way. It can involve a base substitution in the sequence within a gene or a removal or addition of one or more base pairs.

A **point mutation** is a change in a single base pair within the DNA. For example, the DNA sequence shown here has been altered by a **base substitution,** in which one base is substituted for another.

↓
5′–AACGC**T**AGATC–3′   5′–AACGC**G**AGATC–3′
3′–TTGCG**A**TCTAG–5′ → 3′–TTGCG**C**TCTAG–5′

A change of a pyrimidine to another pyrimidine, such as C to T, or a purine to another purine, such as A to G, is called a **transition.** This type of mutation is more common than a **transversion,** in which purines and pyrimidines are interchanged. The example just shown is a transversion (a change from T to G, in the top strand, and from A to C, in the bottom strand).

Another way a gene mutation may occur is if a short sequence of base pairs is deleted from or added to the chromosomal DNA:

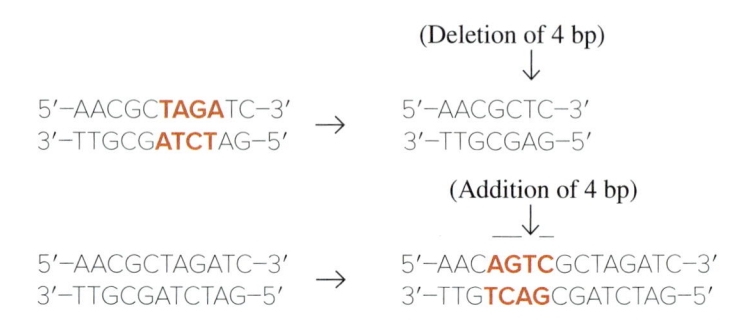

(Deletion of 4 bp)
↓
5′–AACGC**TAGA**TC–3′   5′–AACGCTC–3′
3′–TTGCG**ATCT**AG–5′ → 3′–TTGCGAG–5′

(Addition of 4 bp)
↓ __ __
5′–AACGCTAGATC–3′   5′–AAC**AGTC**GCTAGATC–3′
3′–TTGCGATCTAG–5′ → 3′–TTG**TCAG**CGATCTAG–5′

As we will see next, small deletions or additions to the sequence of a gene can significantly affect the function of the encoded protein.

Gene Mutations Can Alter the Coding Sequence of a Gene

How might a mutation within the coding sequence of a protein-encoding gene affect the amino acid sequence of the polypeptide that is encoded by the gene? **Table 19.1** describes the possible effects of point mutations.

- **Silent mutations** are those that do not alter the amino acid sequence of the polypeptide even though the base sequence has changed. Because the genetic code is degenerate, silent mutations can occur in certain bases within a codon, such as the third base, and the specific amino acid is not changed.
- **Missense mutations** are base substitutions for which an amino acid change does result. An example of a missense mutation is the one that causes the human disease known as sickle cell disease. This disease involves a mutation in the β-globin gene, which alters the polypeptide sequence such that the sixth amino acid is changed from glutamic acid to valine. This single amino acid substitution alters the structure and function of the hemoglobin protein. One consequence of this alteration is that under conditions of low oxygen, the red blood cells assume a sickle shape (**Figure 19.1**). In this case, a single amino acid substitution has a profound effect on the phenotype of cells and even causes a serious disease.
- **Nonsense mutations** involve a change from a normal codon to a stop codon. This change terminates the translation of the polypeptide earlier than expected, producing a truncated polypeptide (see Table 19.1).
- **Frameshift mutations** involve the addition or deletion of a number of nucleotides that is not divisible by three. Because the codons are read in multiples of three, this type of mutation shifts the reading frame. The translation of the mRNA results in a different amino acid sequence downstream from the mutation.

19.1 EFFECTS OF MUTATIONS ON GENE STRUCTURE AND FUNCTION   433

TABLE 19.1

Consequences of Point Mutations Within a Coding Sequence

Type of Change	Mutation in the DNA	Example*	Amino Acids Altered	Likely Effect on Protein Function
None	None	5'–A-T-G–A-C-C–G-A-C–C-C-G–A-A-A–G-G-G–A-C-C–3' Met – Thr – Asp – Pro – Lys – Gly – Thr –	None	None
Silent	Base substitution	5'–A-T-G–A-C-C–G-A-C–C-C-C–A-A-A–G-G-G–A-C-C–3' Met – Thr – Asp – Pro – Lys – Gly – Thr –	None	None
Missense	Base substitution	5'–A-T-G–C-C-C–G-A-C–C-C-G–A-A-A–G-G-G–A-C-C–3' Met – Pro – Asp – Pro – Lys – Gly – Thr –	One	Neutral or inhibitory
Nonsense	Base substitution	5'–A-T-G–A-C-C–G-A-C–C-C-G–T-A-A–G-G-G–A-C-C–3' Met – Thr – Asp – Pro – STOP!	Many	Inhibitory
Frameshift	Addition/deletion	5'–A-T-G–A-C-C G A C–G-C-C–G-A-A–A-G-G–G-A-C-C–3' Met – Thr – Asp – Ala – Glu – Arg – Asp –	Many	Inhibitory

*DNA sequence in the coding strand. Note that this sequence is the same as the mRNA sequence except that the RNA contains uracil (U) instead of thymine (T). The three-base codons are shown in alternating black and red. Mutations are shown in green. Changes in the amino acid sequence are shown in blue.

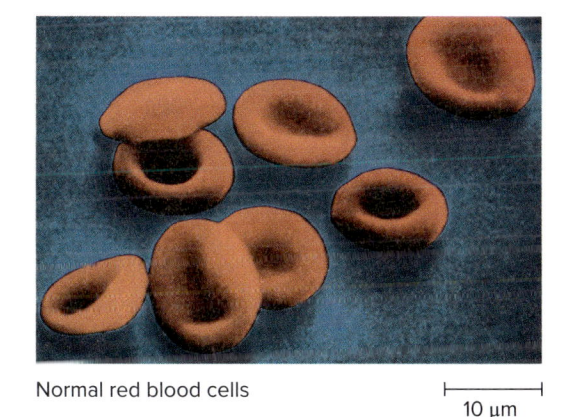

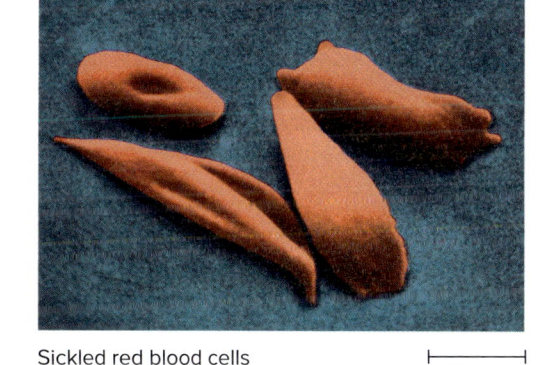

Normal red blood cells |—| 10 µm

Sickled red blood cells |—| 10 µm

(a) Micrographs of red blood cells

NORMAL: NH₂ – VALINE – HISTIDINE – LEUCINE – THREONINE – PROLINE – GLUTAMIC ACID – GLUTAMIC ACID...

SICKLE CELL: NH₂ – VALINE – HISTIDINE – LEUCINE – THREONINE – PROLINE – VALINE – GLUTAMIC ACID...

(b) A comparison of the amino acid sequence between normal β globin and sickle cell β globin

FIGURE 19.1 **Missense mutation in sickle cell disease.** (a) Normal red blood cells (left) and sickled red blood cells (right). (b) A comparison of the amino acid sequence of the normal β-globin polypeptide and the polypeptide encoded by the sickle cell allele. This figure shows only a portion of the polypeptide sequence, which is 146 amino acids long. In β globin, the first methionine at the amino terminus is removed after the polypeptide is made. As seen here, a missense mutation changes the sixth amino acid from a glutamic acid to a valine.

Genes ⟶ Traits A missense mutation alters the structure of β globin, which is a subunit of hemoglobin, the oxygen-carrying protein in the red blood cells. When an individual is homozygous for the sickle cell allele, this missense mutation causes the red blood cells to sickle under conditions of low oxygen concentration. The sickling phenomenon is a description of the trait at the cellular level. At the organism level, the sickled cells can clog the capillaries, causing painful episodes that can result in organ damage. The shortened life span of the red blood cells leads to symptoms of anemia.

©Science History Images/Alamy Stock Photo

Except for silent mutations, new mutations are more likely to produce polypeptides that have reduced rather than enhanced function. For example, when nonsense mutations produce polypeptides that are substantially shorter, the shorter polypeptides are unlikely to function properly. Likewise, frameshift mutations dramatically alter the amino acid sequence of polypeptides and are thereby likely to disrupt function. Missense mutations are less likely to alter function because they involve a change of a single amino acid within polypeptides that typically contain hundreds of amino acids. When a missense mutation has no detectable effect on protein function, it is referred to as a **neutral mutation.** A missense mutation that substitutes an amino acid with a chemistry similar to that of the original amino acid is likely to be neutral or nearly neutral. For example, a missense mutation that substitutes a glutamic acid for an aspartic acid is likely to be neutral because both amino acids are negatively charged and have similar side chain structures. Silent mutations are also considered a type of neutral mutation.

Mutations can occasionally produce a polypeptide with an enhanced ability to function. Although these favorable mutations are relatively rare, they may result in an organism with a greater likelihood of surviving and reproducing. If this is the case, natural selection may cause such a favorable mutation to increase in frequency within a population. This topic will be discussed in Chapter 23.

Gene Mutations Can Occur Outside of a Coding Sequence and Influence Gene Expression

Thus far, we have focused our attention on mutations in the coding regions of genes and their effects on polypeptide structure and protein function. In previous chapters, we learned how various sequences outside of coding sequences play important roles during the process of gene expression. A mutation can occur within a noncoding sequence, thereby affecting gene expression (**Table 19.2**). For example, a mutation may alter the sequence within the core promoter of a gene.

- Promoter mutations that increase transcription are termed **up promoter mutations.** Mutations that make a sequence more like the consensus sequence are likely to be up promoter mutations.

TABLE 19.2
Possible Consequences of Gene Mutations Outside of a Coding Sequence

Sequence	Effect of Mutation
Promoter	May increase or decrease the rate of transcription
Regulatory element/operator site	May disrupt the ability of the gene to be properly regulated
5'-UTR/3'-UTR	May alter the ability of mRNA to be translated; may alter mRNA stability
Splice recognition sequence	May alter the ability of pre-mRNA to be properly spliced

- A **down promoter mutation** causes the promoter to become less like the consensus sequence, decreasing its affinity for transcription factors and decreasing the transcription rate.

In Chapter 16, we considered how mutations could affect regulatory elements. For example, mutations in the *lac* operator site, called *lacO^C* mutations, prevent the binding of the lac repressor protein. This causes the *lac* operon to be constitutively expressed even in the absence of lactose. Bacteria strains with *lacO^C* mutations are at a selective disadvantage compared with wild-type *E. coli* strains because they waste their energy expressing the *lac* operon even when the proteins related to lactose metabolism are not needed. As noted in Table 19.2, mutations can also occur in other non-coding regions of a gene and alter gene expression in a way that may affect phenotype. For example, mutations that affect the untranslated regions of mRNA—the 5'-UTR and 3'-UTR—may affect gene expression if they alter the ability of the mRNA to be translated or its stability. In addition, mutations in eukaryotic genes can alter splice junctions and affect the order or number of exons contained within an mRNA.

Gene Mutations Are Also Given Names That Describe How They Affect the Wild-Type Genotype and Phenotype

Thus far, several genetic terms have been introduced that describe the molecular effects of mutations. Genetic terms are also used to describe the effects of mutations relative to a wild-type genotype or phenotype.

- In a natural population, a **wild type** is a relatively prevalent genotype or phenotype. Many or most genes exist as multiple alleles, so a population may have two or more wild-type alleles.
- A mutation may change a wild-type genotype by altering the DNA sequence of a gene. When such a mutation is rare in a population, the result is generally referred to as a **mutant allele.**
- A reverse mutation, more commonly called a **reversion,** changes a mutant allele back to a wild-type allele.

Another way to describe a mutation is based on its influence on the wild-type phenotype. Mutants are often characterized by their differential ability to survive.

- A neutral mutation does not alter protein function, so it does not affect survival or reproductive success.
- A **deleterious mutation,** however, decreases the chances of survival and reproduction.
- The extreme example of a deleterious mutation is a **lethal mutation,** which results in the death of a cell or organism.
- On the other hand, a **beneficial mutation** enhances the survival or reproductive success of an organism.

In some cases, an allele may be either deleterious or beneficial, depending on the genotype and/or the environmental conditions. An example is the sickle cell allele. In the homozygous state, the sickle cell lessens the chances of survival. However, an individual who is heterozygous for the sickle cell and wild-type alleles has an increased chance of survival due to malarial resistance.

Finally, some mutations result in **conditional mutants** in which the phenotype is affected only under a defined set of conditions. Geneticists often study conditional mutants in microorganisms; a common example is a temperature-sensitive (*ts*) mutant. A bacterium that has a *ts* mutation grows normally in one temperature range—the permissive temperature range—but exhibits defective growth at a different temperature range—the nonpermissive range. For example, an *E. coli* strain carrying a *ts* mutation may be able to grow in the range 33°C—38°C but not in the range 40°C—42°C, whereas a wild-type strain can grow in either temperature range.

Suppressor Mutations Reverse the Phenotypic Effects of Another Mutation

A second mutation sometimes affects the phenotypic expression of a first mutation. As an example, let's consider a mutation that causes an organism to grow very slowly. A second mutation at another site in the organism's DNA may restore the normal growth rate, converting the mutant back to the wild-type phenotype. Geneticists call these second-site mutations **suppressors**, or **suppressor mutations**. This name is meant to indicate that this type of mutation acts to suppress the phenotypic effects of another mutation. A suppressor mutation differs from a reversion, because it occurs at a DNA site that is distinct from the location of the first mutation. Suppressor mutations are classified according to their relative locations with regard to the mutation they suppress (**Table 19.3**).

Intragenic Suppressors When the second mutation site is within the same gene as the first, the mutation is termed an **intragenic suppressor.** This type of suppressor often produces a change in protein structure that compensates for an abnormality in protein structure caused by the first mutation. Researchers may isolate suppressor mutations to obtain information about protein structure and function. For example, Robert Brooker and colleagues have isolated many intragenic suppressors in the *lacY* gene of *E. coli*, which encodes lactose permease, as described in Chapter 16. This protein must undergo conformational changes to transport lactose across the cell membrane. These researchers began with single mutations that altered amino acids on transmembrane regions, which inhibited this conformational change, thereby preventing growth on media containing lactose. Suppressor mutations were then isolated that restored transport function and allowed growth on lactose. By analyzing the locations of these suppressor mutations, the researchers were able to determine that certain transmembrane regions in the protein are critical for conformational changes required for lactose transport.

Intergenic Suppressors Alternatively, a suppressor mutation can occur in a different gene from the first mutation—an **intergenic suppressor.** How do intergenic suppressors work? These suppressor mutations usually involve a change in the expression of one gene that compensates for a loss-of-function mutation affecting another gene (see Table 19.3). For example, a first mutation may cause one protein to be partially or completely defective. An intergenic suppressor in a different protein-encoding gene might overcome this defect by altering the structure of a second protein so that it can take over the functional role the first protein cannot perform. Alternatively, intergenic suppressors may affect proteins that participate in a common cellular pathway. When a first mutation affects the activity of a protein, a suppressor mutation could alter the function of a second protein involved in this pathway, thereby overcoming the defect in the first.

In some cases, intergenic suppressors have effects on multimeric proteins, in which each subunit is encoded by a different gene. A mutation in one subunit that inhibits function may be compensated by a mutation in another subunit. Another type of intergenic suppressor involves mutations in genetic regulatory proteins such as transcription factors. When a first mutation causes a protein to be defective, a suppressor mutation may occur in a gene that encodes a transcription factor. The mutant transcription factor transcriptionally activates another gene that can compensate for the loss-of-function mutation in the first gene.

Changes in Chromosome Structure Can Affect the Expression of a Gene

Thus far, we have considered small changes in the DNA sequence of particular genes. A change in chromosome structure can also be associated with an alteration in the expression of single genes. Quite commonly, an inversion or translocation has no obvious phenotypic consequence. However, in 1925, Alfred Sturtevant was the first to recognize that chromosomal rearrangements in *Drosophila* can influence phenotypic expression (namely, eye morphology). In some cases, a chromosomal rearrangement may affect a gene because a chromosomal **breakpoint**—a region where two chromosome pieces break and rejoin with other chromosome pieces—occurs within a gene. A breakpoint in the middle of a gene is very likely to inhibit gene function because it separates the gene into two pieces.

In other cases, a gene may be left intact, but its expression may be altered when it is moved to a new location. When this occurs, the change in gene location is said to have a **position effect.** How do position effects alter gene expression? Researchers have discovered two common explanations. **Figure 19.2** depicts schematic examples in which a piece of one chromosome has been inverted or translocated to a different chromosome. One possibility is that a gene may be moved next to regulatory sequences for a different gene, such as silencers or enhancers, that influence the expression of the relocated gene (Figure 19.2a). Alternatively, a chromosomal rearrangement may reposition a gene from a less condensed, or euchromatic chromosome, where it is active, to a very highly condensed, or heterochromatic chromosome. When the gene is moved to a heterochromatic region, its expression may be turned off (Figure 19.2b). This second type of position effect may produce a variegated phenotype in which the expression of the gene is variable. For genes that affect pigmentation, this produces a mottled appearance rather than an even color. **Figure 19.3** shows a position effect that alters eye color in *Drosophila*. Figure 19.3a depicts a normal red-eyed fruit fly, and Figure 19.3b shows a mutant fly that has inherited a chromosomal rearrangement in which a gene affecting eye color has been relocated to a heterochromatic chromosome. The variegated

TABLE 19.3 Examples of Suppressor Mutations

Type	No Mutation	First Mutation	Second Mutation	Description
Intragenic	Transport can occur	Transport inhibited	Transport can occur	A first mutation disrupts normal protein function, and a suppressor mutation affecting the same protein restores function. In this example, the first mutation inhibits lactose transport function, and the second mutation restores it.
Intergenic Redundant function	Functional enzyme / (second enzyme)	Nonfunctional enzyme / (second enzyme)	(nonfunctional) / Gain of a new functional enzyme	A first mutation inhibits the function of a protein, and a second mutation alters a different protein to carry out that function. In this example, the proteins function as enzymes.
Common pathway	Precursor → Fast → Intermediate → Slow → Product	Precursor → Slow → Intermediate → Slow → Little product	Precursor → Slow → Intermediate → Fast → Product	Two or more different proteins may be involved in a common pathway. A mutation that causes a defect in one protein may be compensated for by a mutation that alters the function of a different protein in the same pathway.
Multimeric protein	Active	Inactive	Active	A mutation in a gene encoding one protein subunit that inhibits function may be suppressed by a mutation in a gene that encodes a different subunit. The double mutant has restored function.
Transcription factor	Normal function	Loss of function	(see below)	A first mutation causes loss of function of a particular protein. A second mutation may alter a transcription factor and cause it to activate the expression of another gene. This other gene encodes a protein that can compensate for the loss of function caused by the first mutation.

Transcription factor

Mutant transcription factor turns on a gene that compensates for the loss of function.

Causes expression of this protein

Compensates for inactive protein so function is restored

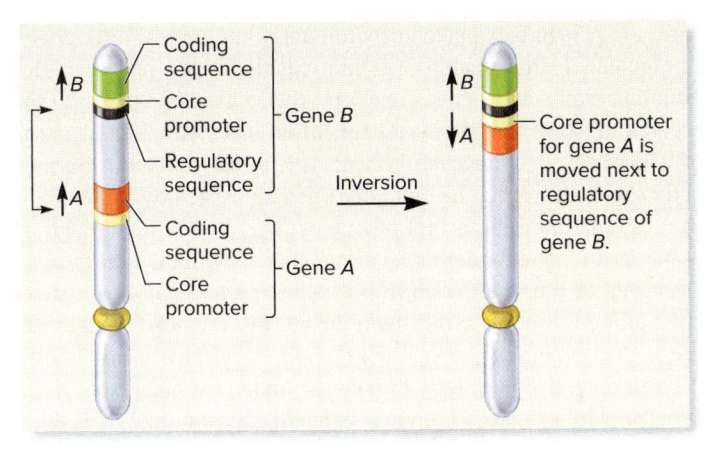

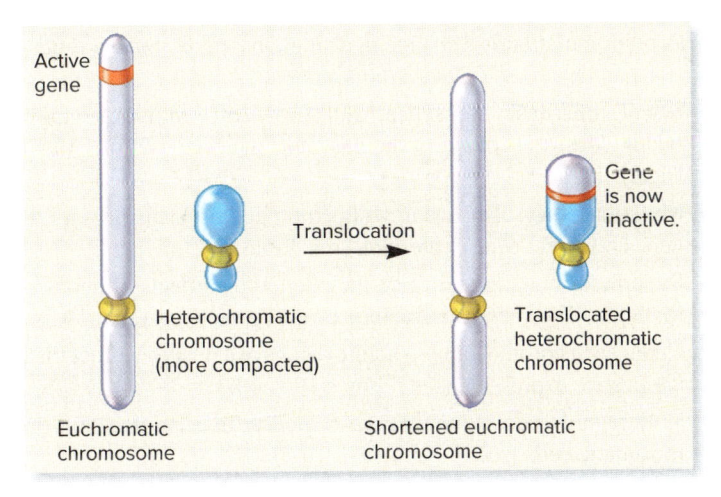

FIGURE 19.2 Causes of position effects. (a) A chromosomal inversion has repositioned the core promoter of gene A next to the regulatory sequences for gene B. Because regulatory sequences are often bidirectional, the regulatory sequences for gene B may regulate the transcription of gene A. (b) A translocation has moved a gene from a euchromatic to a heterochromatic chromosome. This type of position effect inhibits the expression of the relocated gene.

Concept Check: Explain what the term position effect means.

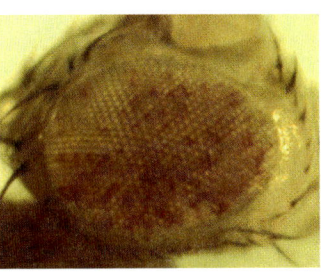

appearance of the eye occurs because the degree of heterochromatin formation varies across different regions of the eye. In cells where heterochromatin formation has turned off the eye color gene, a white phenotype occurs, but other cells allow this same region to remain euchromatic and produce a red phenotype.

Mutations Can Occur in Germ-Line or Somatic Cells

In this section, we have considered many different ways that mutations affect gene expression. For multicellular organisms, the timing of mutations also plays an important role. A mutation can occur very early in life, such as in a gamete or a fertilized egg, or it may occur later in life, such as in the embryonic or adult stages.

FIGURE 19.3 A position effect that alters eye color in *Drosophila*. (a) A normal red eye. (b) An eye in which an eye color gene has been relocated to a heterochromatic chromosome. This inactivates the gene in some cells and produces a variegated phenotype.

Genes → Traits Variegated eye color occurs because the degree of heterochromatin formation varies throughout different regions of the eye. In some cells, heterochromatin formation occurs and turns off the eye color gene, thereby leading to the white phenotype. In other cells, the region containing the eye color allele remains euchromatic, yielding a red phenotype.
©Dr. Jack R. Girton

Concept Check: Has the DNA sequence of the eye color gene been changed in part (b) compared with part (a)? How do we explain the phenotypic difference?

The exact time when mutations occur can be important with regard to the severity of the genetic effect and whether the mutations are passed from parent to offspring. Geneticists classify the cells of animals into two types: the germ line and the somatic cells. The term **germ line** refers to cells that give rise to the gametes such as eggs and sperm.

Germ-Line Mutations A **germ-line mutation** can occur directly in a sperm or egg cell, or it can occur in a precursor cell that produces the gametes. If a mutant gamete participates in fertilization, all cells of the resulting offspring will contain the mutation (**Figure 19.4a**). Likewise, when an individual with a germ-line mutation produces gametes, the mutation may be passed along to future generations of offspring.

Somatic Mutations The **somatic cells** comprise all cells of the body excluding the germ-line cells. Examples include muscle cells, nerve cells, and skin cells. Mutations can also happen within somatic cells at early or late stages of development. **Figure 19.4b** illustrates the consequences of a mutation that took place during the embryonic stage. In this example, a **somatic mutation** has occurred within a single embryonic cell. As the embryo grows, this single cell is the precursor for many cells of the adult organism. Therefore, in the adult, a portion of the body contains the mutation. The size of the affected region depends on the timing of the mutation. In general, the earlier the mutation occurs during development, the larger the affected region. An individual that has somatic regions that differ genotypically from each other is called a **genetic mosaic**.

Figure 19.5 is a photo of an individual with a somatic mutation that occurred during an early stage of development. In this case, the person has a patch of white hair, but the rest of the hair is pigmented. Presumably, this individual initially had a single mutation occur in an embryonic cell that ultimately gave rise to a patch of scalp that produced the white hair. Although a patch of white hair is not a harmful phenotypic effect, mutations during early stages of life can be quite detrimental, especially if they disrupt essential developmental processes. Therefore, even though it is smart to avoid environmental agents that cause mutations during all stages of life, the possibility of somatic mutations is a rather compelling reason to avoid them during the very early stages of life such as embryonic development, infancy, and early childhood. For example, the possibility of somatic mutations in an embryo is a reason why women are advised to avoid exposure to X-rays during pregnancy.

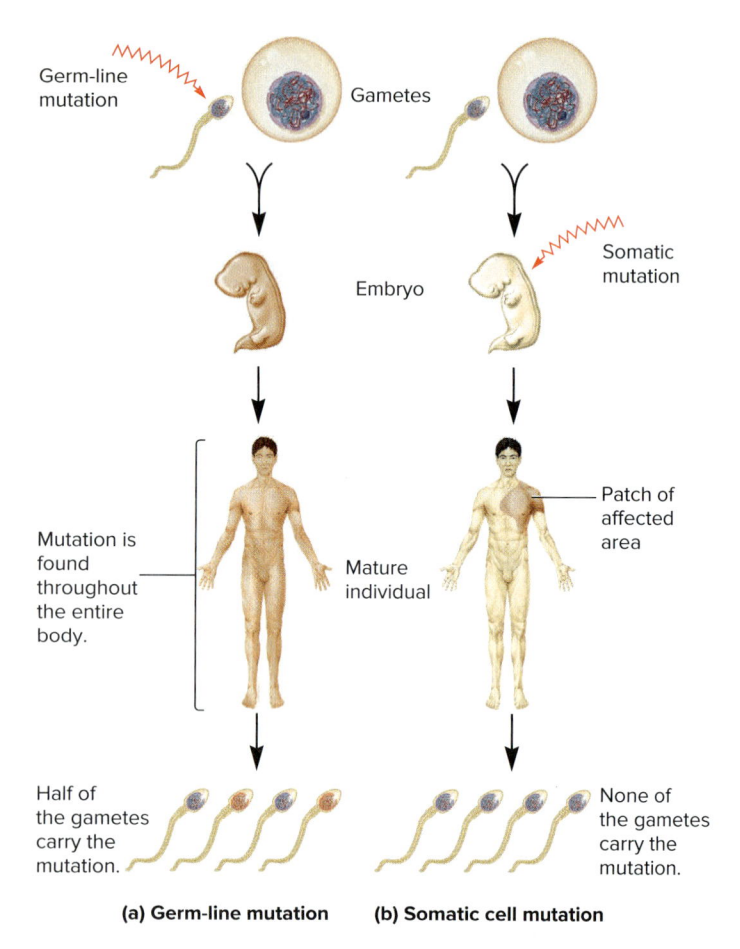

FIGURE 19.4 The effects of germ-line versus somatic mutations.

FIGURE 19.5 Example of a somatic mutation.
Genes → Traits This person has a patch of white hair because a somatic mutation that occurred in a single cell during embryonic development prevented pigmentation of the hair. This cell continued to divide, producing a patch of white hair.
©Otero/gtphoto

Concept Check: Can this trait be passed to offspring?

19.1 REVIEWING THE KEY CONCEPTS

- A point mutation is a change in a single base pair. Such a mutation can be a transition or a transversion.
- Silent, missense, nonsense, and frameshift mutations may occur within a coding sequence of a gene (see Table 19.1, Figure 19.1).
- Mutations may also occur within non-coding sequences of a gene and affect gene expression (see Table 19.2).
- Suppressor mutations reverse the phenotypic effects of another mutation. They can be intragenic or intergenic (see Table 19.3).
- Changes in chromosome structure can have a position effect that alters gene expression (see Figures 19.2, 19.3).
- With regard to timing, mutations can occur in germ-line cells or in somatic cells (see Figures 19.4, 19.5).

19.1 COMPREHENSION QUESTIONS

1. A mutation changes a codon that specifies tyrosine into a stop codon. This type of mutation is a
   a. missense mutation.
   b. nonsense mutation.
   c. frameshift mutation.
   d. neutral mutation.

2. A down promoter mutation causes the promoter of a gene to be _____ like the consensus sequence and _____ transcription.
   a. less, stimulates
   b. more, stimulates
   c. less, inhibits
   d. more, inhibits

3. A mutation in one gene that reverses the effects of a mutation in a different gene is
   a. an intergenic suppressor.
   b. an intragenic suppressor.
   c. an up promoter mutation.
   d. a position effect.

4. Which of the following is an example of a somatic mutation?
   a. A mutation in an embryonic muscle cell
   b. A mutation in a sperm cell
   c. A mutation in an adult nerve cell
   d. Both a and c are examples of somatic mutations.

19.2 RANDOM NATURE OF MUTATIONS

Learning Outcome:

1. Analyze the results obtained by the Lederbergs, and explain how they are consistent with the random mutation theory.

For a couple of centuries, biologists had questioned whether heritable changes occur as a result of behavior or exposure to particular environmental conditions or are events that happen randomly. In the nineteenth century, the naturalist Jean-Baptiste Lamarck proposed that physiological events—such as the use or disuse of muscles—determine whether traits are passed along to offspring. For example, his hypothesis suggested that an individual who practiced and became adept at a physical activity and developed muscular legs would pass that characteristic on to the next generation. The alternative point of view is that genetic variation exists in a population as a matter of random chance, and natural selection results in the differential reproductive success of organisms that are better adapted to their environments. Those individuals who, by chance, happen to have beneficial mutations will be more likely to survive and pass these genes to their offspring. These opposing ideas of the nineteenth century—one termed *physiological adaptation* and the other termed *random mutation*—were tested in bacterial studies in the 1940s and 1950s.

One such study was carried out by Joshua and Esther Lederberg, who were interested in the relationship between mutation and the environmental conditions under which they occur. To distinguish between the physiological adaptation and random mutation hypotheses, they developed a technique known as **replica plating** in the 1950s. As shown in **Figure 19.6**, they plated a large number of bacteria onto a master plate that did not contain any selective agent (namely, no T1 phage). A sterile piece of velvet cloth was lightly touched to this plate in order to pick up a few bacterial cells from each colony. This replica was then transferred to two secondary plates that contained an agent that selected for the growth of bacterial cells with a particular genotype.

In the example shown in Figure 19.6, the secondary plates contained T1 bacteriophages. On these plates, only those mutant cells that are resistant to T1 (ton^r) could grow. On the secondary plates, a few colonies were observed. Strikingly, they occupied the same location on each plate. These results indicated that the mutations conferring ton^r occurred randomly while the cells were growing on the nonselective master plate. The presence of the T1 phage in the secondary plates simply selected for the growth of previously occurring ton^r mutants. These results supported the random mutation hypothesis. In contrast, the physiological adaptation hypothesis would have predicted that ton^r bacterial mutants would occur after exposure to the selective agent on the secondary plates. If that had been the case, the colonies would be expected to arise in different locations on the two secondary plates.

The results of the Lederbergs supported the random mutation hypothesis, now known as the **random mutation theory**. According to this theory, mutations are a random process—they can occur in any gene and do not involve exposure of an organism to a particular condition that causes specific types of mutations to happen.

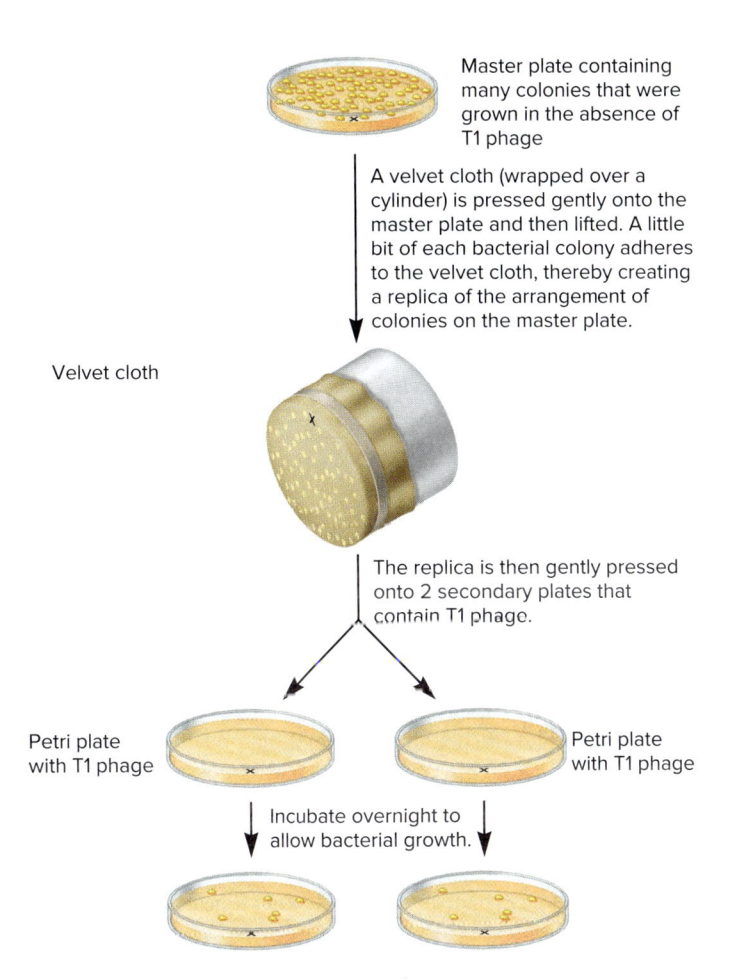

FIGURE 19.6 Replica plating. Bacteria were first plated on a master plate under nonselective conditions. A sterile velvet cloth was used to make a replica of the master plate. This replica was gently pressed onto two secondary plates that contained a selective agent, which was T1 bacteriophage. Only those mutant cells that are ton^r (resistant to T1) could grow to form visible colonies. Note: The black × indicates the alignment of the velvet and the plates.

Source: Adapted from Lederberg, J., & Lederberg, E. M. (1952) Replica Plating and Indirect Selection of Bacterial Mutants, *Journal of Bacteriology*, vol. 63, 399-406.

Concept Check: Why are these results inconsistent with the physiological adaptation hypothesis?

In some cases, a random mutation may provide a mutant organism with an advantage, such as resistance to T1 phage. Although such mutations occur as a matter of random chance, environmental conditions may select for organisms that happen to carry them.

As researchers have learned more about mutation at the molecular level, the view that mutations are a totally random process has required some modification. Within the same individual, some genes mutate at a much higher rate than other genes. Why does this happen? Some genes are larger than others, which provides a greater chance for mutation. Also, the relative locations of genes within a chromosome may cause some genes to be more susceptible to mutation than others. Even within a single gene, **hot spots** are usually found—certain regions of a gene that are more likely to mutate than other regions.

19.2 REVIEWING THE KEY CONCEPTS

- The replica plating experiments performed by Lederberg and Lederberg were consistent with the random mutation theory (see Figure 19.6).

19.2 COMPREHENSION QUESTION

1. In the replica plating experiments of the Lederbergs, bacterial colonies appeared at the same locations on each of two secondary plates because
   a. T1 phage caused the mutations to happen.
   b. the mutations occurred on the master plate prior to T1 exposure and prior to replica plating.
   c. Both a and b are true.
   d. Neither a nor b is true.

19.3 SPONTANEOUS MUTATIONS

Learning Outcomes:
1. Distinguish between spontaneous and induced mutations.
2. List examples of spontaneous mutations.
3. Outline how mutations arise by depurination, deamination, and tautomeric shifts.
4. Describe how reactive oxygen species alter DNA structure and cause mutations.
5. Explain the mechanism of trinucleotide repeat expansion.

Mutations can have a wide variety of effects on the phenotypic expression of genes. For this reason, geneticists have expended a great deal of effort identifying the causes of mutations. This task has been truly challenging because a staggering number of agents can alter the structure of DNA and thereby cause mutations. Geneticists categorize the causes of mutations in one of two ways: **Spontaneous mutations** are changes in DNA structure that result from natural biological or chemical processes, whereas **induced mutations** are caused by environmental agents (**Table 19.4**). Many causes of spontaneous mutations are examined in other chapters throughout this text.

Overall, a distinguishing feature of spontaneous mutations is that their underlying cause originates within the cell. By comparison, the cause of induced mutations originates outside the cell. In this section, we will explore mechanisms, such as depurination, deamination, and tautomeric shifts, by which some spontaneous mutations can occur, and consider how oxidative stress and trinucleotide repeat expansions cause mutations. In the following section, we will consider induced mutations.

Spontaneous Mutations Can Arise by Depurination, Deamination, and Tautomeric Shifts

How can molecular changes in DNA structure cause mutation? Our first examples concern changes that can occur spontaneously, albeit at a low rate.

TABLE 19.4
Causes of Mutations

Common Causes of Mutations	Description
Spontaneous	
Aberrant recombination	Abnormal crossing over may cause deletions, duplications, translocations, and inversions (see Chapter 8).
Aberrant segregation	Abnormal chromosomal segregation may cause aneuploidy or polyploidy (see Chapter 8).
Errors in DNA replication	A mistake by DNA polymerase may cause a point mutation (see Chapter 13).
Transposable elements	Transposable elements can insert themselves into the sequence of a gene (see Chapter 12).
Depurination	On rare occasions, the linkage between a purine (i.e., adenine or guanine) and deoxyribose can spontaneously break. If not repaired, this can lead to mutation.
Deamination	Cytosine or 5-methylcytosine can spontaneously deaminate to create uracil or thymine, respectively.
Tautomeric shifts	Spontaneous changes in base structure can cause mutations if they occur immediately prior to DNA replication.
Toxic metabolic products	The products of normal metabolic processes, such as reactive oxygen species, may be chemically reactive agents that can alter the structure of DNA.
Induced	
Chemical agents	Chemical substances may cause changes in the structure of DNA.
Physical agents	Physical phenomena such as UV light and X-rays can damage DNA.

Depurination The most common type of chemical change that occurs naturally is **depurination**, which involves the removal of a purine (adenine or guanine) from the DNA. The covalent bond between deoxyribose and a purine base is somewhat unstable and occasionally undergoes a spontaneous reaction with water that releases the base from the sugar, thereby creating an **apurinic site** (**Figure 19.7a**). In a typical mammalian cell, approximately 10,000 purines are lost from the DNA in a 24-hour period at 37°C. The rate of loss is higher if the DNA is exposed to agents that cause certain types of base modifications such as the attachment of alkyl groups (methyl or ethyl groups). Fortunately, apurinic sites are recognized by DNA repair systems that fix the problem. If the repair systems fail, however, a mutation may result during subsequent rounds of DNA replication. What happens at an apurinic site during DNA replication? Because a complementary base is not present to specify the incoming base for the new strand, any of the four bases are added to the new strand in the region that is opposite the apurinic site (**Figure 19.7b**). This may produce a new mutation.

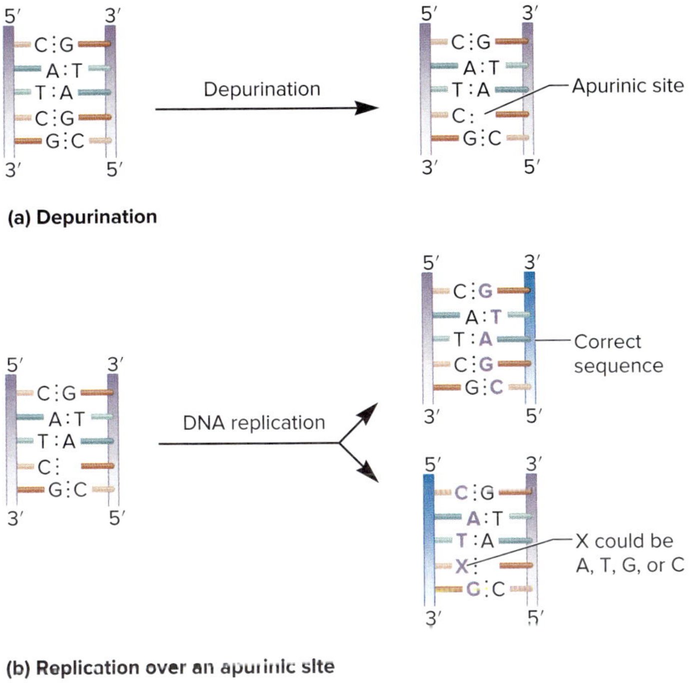

(a) Depurination

(b) Replication over an apurinic site

FIGURE 19.7 Spontaneous depurination. (a) The bond between guanine and deoxyribose is broken, thereby releasing the base. This leaves an apurinic site in the DNA. (b) If an apurinic site remains in the DNA as it is being replicated, any of the four nucleotides can be added to the newly made strand.

Concept Check: When DNA replication occurs over an apurinic site, what is the probability that a mutation will occur?

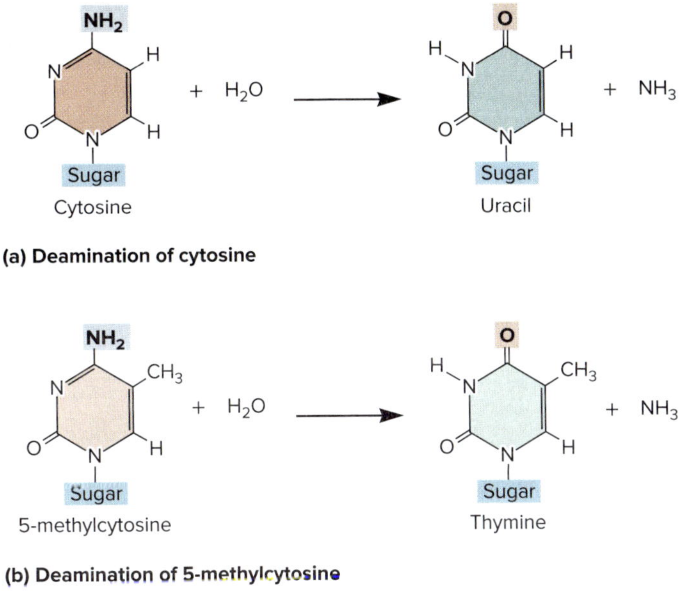

Deamination A second spontaneous chemical change that may occur in DNA is the **deamination** of a cytosine base. The other bases are not readily deaminated. As shown in **Figure 19.8a**, deamination involves the removal of an amino group from cytosine. This produces uracil. DNA repair systems can recognize uracil as an inappropriate base within DNA and subsequently remove it. However, if such a repair does not take place, a mutation may result because uracil hydrogen bonds with adenine during DNA replication. Therefore, if a DNA template strand has uracil instead of cytosine, a newly made strand will incorporate adenine instead of guanine.

Figure 19.8b shows the deamination of 5-methylcytosine. As discussed in Chapter 17, the methylation of cytosine occurs in many eukaryotic species, forming 5-methylcytosine. This methylation also occurs in prokaryotes. If 5-methylcytosine is deaminated, the resulting base is thymine, which is a normal constituent of DNA. Therefore, this poses a problem for DNA repair because DNA repair systems cannot distinguish which is the incorrect base—the thymine that was produced by deamination or the guanine in the opposite strand that originally base-paired with the methylated cytosine. For this reason, methylated cytosine bases tend to produce hot spots for mutation. As an example, researchers analyzed 55 spontaneous mutations that occurred within the *lacI* gene of *E. coli* and determined that 44 of them involved changes at sites that were originally occupied by a methylated cytosine base.

(a) Deamination of cytosine

(b) Deamination of 5-methylcytosine

FIGURE 19.8 Spontaneous deamination of cytosine and 5-methylcytosine. (a) The deamination of cytosine produces uracil. (b) The deamination of 5-methylcytosine produces thymine.

Concept Check: Which of these two changes is more difficult for DNA repair systems to fix correctly? Explain why.

Tautomeric Shift A third way that mutations may arise spontaneously involves a temporary change in base structure called a **tautomeric shift**. In this case, the **tautomers** are bases, which exist in keto and enol forms or amino and imino forms. The two forms in each of these pairs can interconvert by a chemical reaction that involves the migration of a hydrogen atom and a switch of a single bond and an adjacent double bond. The common, stable form of guanine and thymine is the keto form; the common form of adenine and cytosine is the amino form (**Figure 19.9a**). At a low rate, G and T can interconvert to an enol form, and A and C can change to an imino form. Though the amounts of the enol and imino forms of these bases are relatively small, their presence can cause a mutation because these rare forms do not conform to the AT/GC rule of base pairing. Instead, if one of the bases is in the enol or imino form, hydrogen bonding will promote TG and CA base pairs, as shown in **Figure 19.9b**.

How does a tautomeric shift cause a mutation? The answer is that the shift must occur immediately prior to DNA replication. When DNA is in a double-stranded condition, the base pairing usually holds the bases in their more stable forms. After the strands unwind, however, a tautomeric shift may occur. In the example shown in **Figure 19.9c**, a thymine base in the template strand has undergone a tautomeric shift just prior to the replication of the complementary daughter strand. During replication, the daughter strand incorporates a guanine opposite this thymine, creating a base mismatch. This mismatch could be repaired via the proofreading function of DNA polymerase or via a mismatch repair system (discussed later in this chapter). However, if these repair mechanisms do not occur, the next round of DNA replication will

442   CHAPTER 19 :: GENE MUTATION AND DNA REPAIR

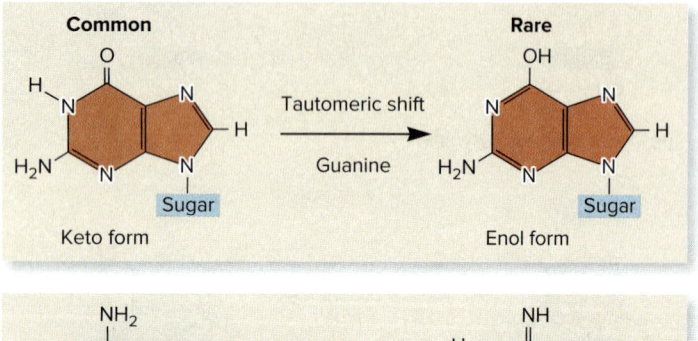

(a) Tautomeric shifts that occur in the 4 bases found in DNA

(b) Base pair mismatches due to tautomeric shifts

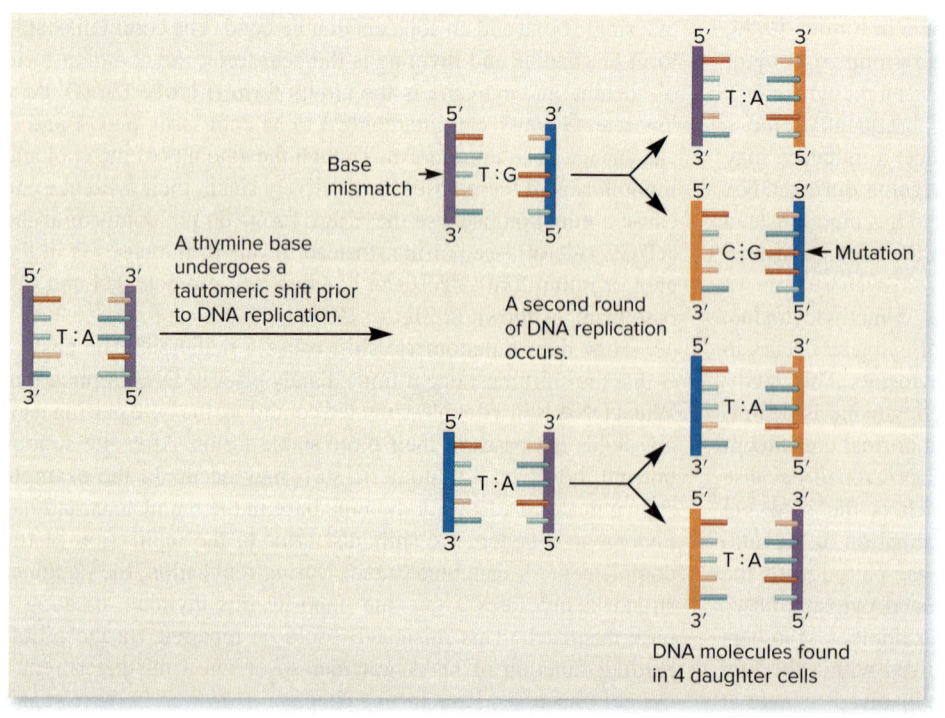

(c) A tautomeric shift just prior to DNA replication can produce a mutation.

FIGURE 19.9 Tautomeric shifts and their ability to cause mutation. (a) The common forms of the bases are shown on the left, and the rare forms produced by tautomeric shifts are shown on the right. (b) On the left, the rare enol form of thymine pairs with the common keto form of guanine (instead of adenine); on the right, the rare imino form of cytosine pairs with the common amino form of adenine (instead of guanine). (c) A tautomeric shift occurred in a thymine base just prior to replication, causing the formation of a TG base pair. If not repaired, a second round of replication will lead to the formation of a permanent CG mutation. Note: A tautomeric shift is a very temporary situation. During the second round of replication, the thymine base that shifted prior to the first round of DNA replication is likely to have shifted back to its normal form. Therefore, during the second round of replication, an adenine base is found opposite this thymine.

produce a double helix with a CG base pair, whereas the correct base pair should be TA. As shown in the right side of Figure 19.9c, one of four daughter cells inherits this CG mutation.

Oxidative Stress May Lead to DNA Damage and Mutation

Aerobic organisms use oxygen as a terminal acceptor of their electron transport chains. **Reactive oxygen species (ROS),** such as hydrogen peroxide, superoxide, and hydroxyl radical, are products of oxygen metabolism in all aerobic organisms. In eukaryotes, ROS are naturally produced as unwanted by-products of energy production in mitochondria. They may also be produced during certain types of immune responses and by a variety of detoxification reactions in the cell. If ROS accumulate, they can damage cellular molecules, including DNA, proteins, and lipids. To prevent this from happening, cells use a variety of enzymes, such as superoxide dismutase and catalase, to prevent the buildup of ROS. In addition, small molecules, such as vitamin C, may act as antioxidants. Certain foods contain other chemicals that act as antioxidants. Colorful fruits and vegetables, including grapes, blueberries, cranberries, citrus fruits, spinach, broccoli, beets, beans, red peppers, carrots, and strawberries, are usually high in antioxidants. In humans, the overaccumulation of ROS has been implicated in a wide variety of medical conditions, including cardiovascular disease, Alzheimer disease, chronic fatigue syndrome, and aging. However, the production of ROS is not always harmful. ROS are produced by the immune system as a means of killing pathogens. In addition, some ROS are used in cell signaling.

Oxidative stress refers to an imbalance between the production of ROS and an organism's ability to break them down. If ROS overaccumulate, one particularly harmful consequence is **oxidative DNA damage,** which refers to changes in DNA structure that are caused by ROS. DNA bases are very susceptible to oxidation. Guanine bases are particularly vulnerable to oxidation, which can lead to several different oxidized products. The most thoroughly studied guanine oxidation product is 7,8-dihydro-8-oxoguanine, which is commonly known as 8-oxoguanine (8-oxoG) (**Figure 19.10**). Researchers often measure the amount of 8-oxoG in a sample of DNA to determine the extent of oxidative stress. Why are oxidized bases harmful? In the case of 8-oxoG, it base pairs with adenine during DNA replication, causing mutations in which a GC base pair becomes a TA base pair. This is a transversion mutation.

Although oxidative DNA damage can occur spontaneously, it also results from environmental agents, such as ultraviolet light, X-rays, and many chemicals, including those found in cigarette smoke. Later in this chapter, we will discuss such environment agents and their abilities to cause mutation.

DNA Sequences Known as Trinucleotide Repeats Are Hotspots for Mutation

Researchers have discovered that several human genetic diseases are caused by an unusual form of mutation known as **trinucleotide repeat expansion (TNRE).** The term refers to the phenomenon in which a repeated sequence of three nucleotides can readily increase in number from one generation to the next. In humans and other species, certain genes and chromosomal locations contain regions where trinucleotide sequences are repeated in tandem. These sequences are usually transmitted normally from parent to offspring without mutation. However, in persons with TNRE disorders, the length of a trinucleotide repeat has increased above a certain critical size and thereby causes disease symptoms.

Table 19.5 describes several human diseases that involve these types of expansions: spinal and bulbar muscular atrophy (SBMA), Huntington disease (HD), spinocerebellar ataxia (SCA1), fragile X syndromes (FRAXA and FRAXE), and myotonic muscular dystrophy (also called dystrophia myotonica, DM). In some cases, the expansion is within the coding sequence of a gene. Typically, such an expansion involves CAG repeats. Because CAG encodes glutamine, these repeats cause the encoded proteins to contain long tracts of glutamine. The presence of glutamine tracts causes the proteins to aggregate. This aggregation of proteins or protein fragments carrying glutamine repeats is correlated with the progression of the disease. However, recent evidence suggests that the aggregated proteins may not directly cause the disease symptoms. In other TNRE disorders, the expansions are located in non-coding regions of genes. In the case of the two fragile X syndromes, the expansions produce CpG islands that become methylated. As discussed in Chapter 17, DNA methylation can silence gene transcription. For DM, it has been hypothesized that these expansions cause abnormal changes in RNA structure, which produce disease symptoms.

Some TNRE disorders have the unusual feature of progressively worsening severity in future generations—a phenomenon called **anticipation.** An example is depicted here, where the repeat of triplet CAG has expanded from 11 tandem copies to 18.

CAGCAGCAGCAGCAGCAGCAGCAGCAGCAGCAG $n = 11$

to

CAGCAGCAGCAGCAGCAGCAGCAGCAGC
AGCAGCAGCAGCAGCAGCAGCAGCAG $n = 18$

However, anticipation does not occur with all TNRE disorders and usually depends on whether the disease is inherited from the mother or father. In the case of HD, anticipation is likely to occur if the mutant gene is inherited from the father. In contrast,

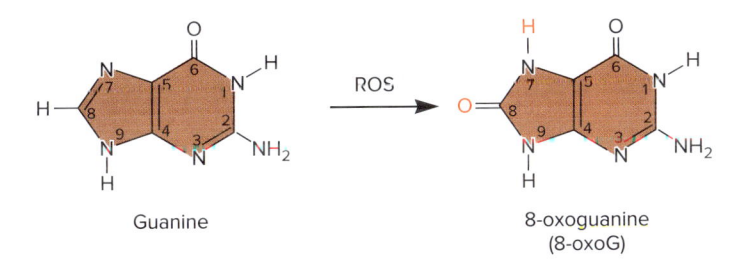

FIGURE 19.10 **Oxidation of guanine to 8-oxoguanine by a reactive oxygen species (ROS).**

Concept Check: What is a reactive oxygen species?

TABLE 19.5
TNRE Disorders

Disease	SBMA	HD	SCA1	FRAXA	FRAXE	DM
Repeated Triplet	CAG	CAG	CAG	CGG	GCC	CTG
Location of Repeat	Coding sequence	Coding sequence	Coding sequence	5′-UTR	5′-UTR	3′-UTR
Number of Repeats in Unaffected Individuals	11–33	6–37	6–44	6–53	6–35	5–37
Number of Repeats in Affected Individuals	36–62	27–121	43–81	>200	>200	>200
Pattern of Inheritance	X-linked	Autosomal dominant	Autosomal dominant	X-linked	X-linked	Autosomal dominant
Disease Symptoms	Neuro-degenerative	Neuro-degenerative	Neuro-degenerative	Mental impairment	Mental impairment	Muscle disease
Anticipation*	None	Male	Male	Female	None	Female

*Indicates the parent in which anticipation occurs.
SBMA, spinal and bulbar muscular atrophy; HD, Huntington disease; SCA1, spinocerebellar ataxia; FRAXA and FRAXE, fragile X syndromes; DM, dystrophia myotonica (myotonic muscular dystrophy).

DM is more likely to get worse if the gene is inherited from the mother. These results suggest that TNRE happens more frequently during oogenesis or spermatogenesis, depending on the particular gene involved. The phenomenon of anticipation makes it particularly difficult for genetic counselors to advise couples about the likely severity of these diseases if they are passed to their children.

How does TNRE occur? Though it may occur in more than one way, researchers have determined that a key aspect of TNRE is that the triplet repeat can form a hairpin, also called a stem-loop. A consistent feature of the triplet sequences associated with TNRE is they contain at least one C and one G (see Table 19.5). As shown in **Figure 19.11a**, such a sequence can form a hairpin due to the formation of CG base pairs.

The formation of a hairpin during DNA replication can lead to an increase in the length of a DNA region if it occurs in the newly made daughter strand (**Figure 19.11b**).

1. DNA replication proceeds just past the TNRE.
2. When the hairpin forms, DNA polymerase temporarily slips off the template strand.
3. Next, DNA polymerase "backs up" and hops back onto the template strand and resumes DNA replication from the end of the hairpin. When this occurs, DNA polymerase is synthesizing most of the hairpin region twice.
4. Hairpins within DNA are short-lived. The hairpin can spread out, which leaves a gap in the opposite strand.
5. This gap is later filled in by DNA polymerase and ligase as shown by the black DNA in Figure 19.11b. The end result is that the TNRE has become longer.

When the trinucleotide repeat sequence is abnormally long, such expansions may frequently occur during gamete formation, and therefore offspring in successive generations may have trinucleotide repeat sequences that are even longer than those in their parents.

19.3 REVIEWING THE KEY CONCEPTS

- Spontaneous mutations result from natural biological and chemical processes, whereas induced mutations are caused by environmental agents (see Table 19.4).
- Three common ways that mutations can arise spontaneously is by depurination, deamination, and tautomeric shifts (see Figures 19.7–19.9).
- Reactive oxygen species (ROS) can cause spontaneous mutations by oxidizing bases in DNA (see Figure 19.10).
- In individuals with a trinucleotide repeat expansion (TNRE), the number of repeats of a trinucleotide sequence increases above a critical level and becomes prone to frequent expansion. This type of mutation is responsible for certain types of human diseases. The repeats can expand due to hairpin formation during DNA replication (see Table 19.5, Figure 19.11).

19.3 COMPREHENSION QUESTIONS

1. Which of the following is *not* an example of a spontaneous mutation?
   a. A mutation caused by an error in DNA replication
   b. A mutation caused by a tautomeric shift
   c. A mutation caused by UV light
   d. All of the above are spontaneous mutations.
2. A point mutation could be caused by
   a. depurination.
   b. deamination.
   c. tautomeric shift.
   d. any of the above.
3. One way that TNRE may occur involves the formation of _____ that disrupts _____.
   a. a double-strand break, chromosome segregation
   b. an apurinic site, DNA replication
   c. a hairpin, DNA replication
   d. a free radical, DNA structure

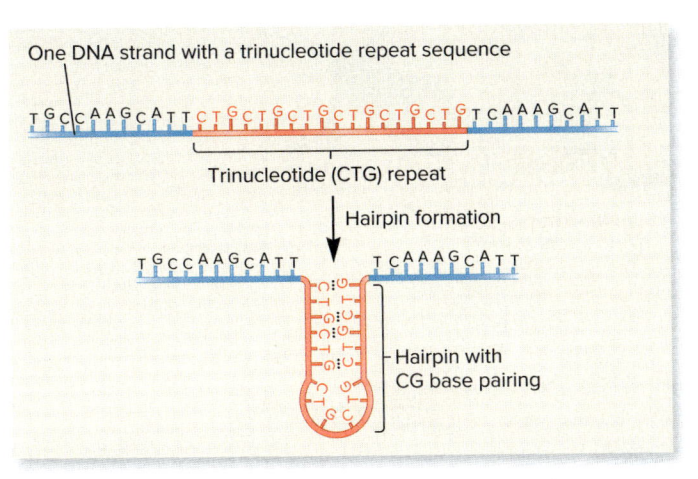

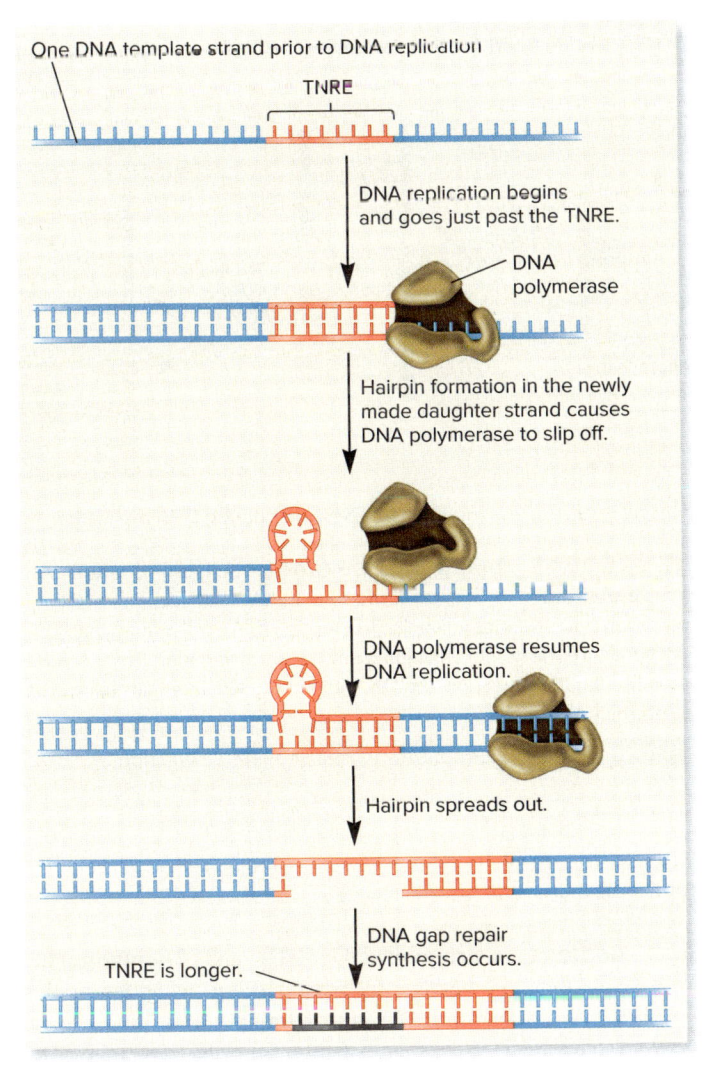

19.4 INDUCED MUTATIONS

Learning Outcomes:
1. Define *mutagen*.
2. Distinguish between chemical and physical mutagens, and provide examples.
3. Define *mutation rate*.
4. Analyze the results of an Ames test.

As we have seen, spontaneous mutations can occur in a wide variety of ways. They result from natural biological and chemical processes. In this section, we turn our attention to induced mutations that are caused by environmental agents. These agents can be either chemical or physical. They enter cells and lead to changes in DNA structure. Agents known to alter the structure of DNA in ways that lead to mutations are called **mutagens.** In this section, we will explore several mechanisms by which mutagens alter the structure of DNA. We will also consider laboratory tests that can identify potential mutagens.

Mutagens Alter DNA Structure in Different Ways

Over the past few decades, researchers have found that an enormous array of agents act as mutagens that permanently alter the structure of DNA. We often hear in the news media that we should avoid these agents in our foods and living environment. For example, we use products such as sunscreens to help us avoid the mutagenic effects of ultraviolet (UV) light. The public is concerned about mutagens for two important reasons. First, mutagenic agents are often involved in the development of human cancers. In addition, because new mutations may be deleterious, people want to avoid mutagens to prevent gene mutations that may have harmful effects on their future offspring.

Mutagenic agents are usually classified as chemical or physical mutagens. Examples of both types of agents are listed in **Table 19.6**. Let's first consider some ways that chemical mutagens alter DNA structure.

Base Modification Some chemical mutagens act by covalently modifying the structure of bases. For example, **nitrous acid** (HNO_2) replaces amino groups with keto groups ($-NH_2$ to $=O$), a process called deamination. Deamination changes cytosine to uracil and adenine to hypoxanthine. When the altered DNA replicates, the modified bases do not pair with the appropriate bases in the newly made strand. Instead, uracil pairs with adenine, and hypoxanthine pairs with cytosine (**Figure 19.12**).

Other chemical mutagens also disrupt the appropriate pairing between nucleotides by alkylating bases within the DNA. During alkylation, methyl or ethyl groups are covalently attached to the bases. Examples of alkylating agents include **nitrogen mustard** (a type of mustard gas) and **ethyl methanesulfonate (EMS).** Mustard gas was used as a chemical weapon during World War I. Such agents severely damage the skin, eyes, mucous membranes, lungs, and blood-forming organs.

FIGURE 19.11 Proposed mechanism of trinucleotide repeat expansion (TNRE). (a) Trinucleotide repeats can form hairpin structures due to CG base pairing. (b) Formation of a trinucleotide repeat expansion.

TABLE 19.6
Examples of Mutagens

Mutagen	Effect(s) on DNA Structure
Chemical	
Nitrous acid	Deaminates bases
Nitrogen mustard	Alkylating agent
Ethyl methanesulfonate	Alkylating agent
Proflavin	Intercalates within DNA helix
5-Bromouracil	Base analog
2-Aminopurine	Base analog
Physical	
X-rays	Cause base deletions, single-strand breaks in the DNA backbone, crosslinking, and chromosomal breaks
UV light	Promotes formation of pyrimidine dimers, such as thymine dimers

(a) Base pairing of 5BU (a thymine analog) with adenine or guanine

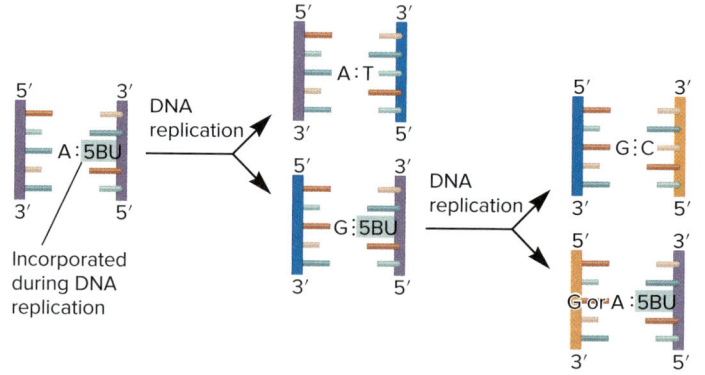

(b) How 5BU causes a mutation in a base pair during DNA replication

FIGURE 19.13 **Base pairing of 5-bromouracil and its ability to cause mutation.** (a) In its keto form, 5BU bonds with adenine; in its enol form, it bonds with guanine. (b) During DNA replication, guanine may be incorporated into a daughter strand by pairing with 5BU. After a second round of replication, the DNA contains a GC base pair instead of the original AT base pair.

Concept Check: Does 5-bromouracil cause a transition or a transversion?

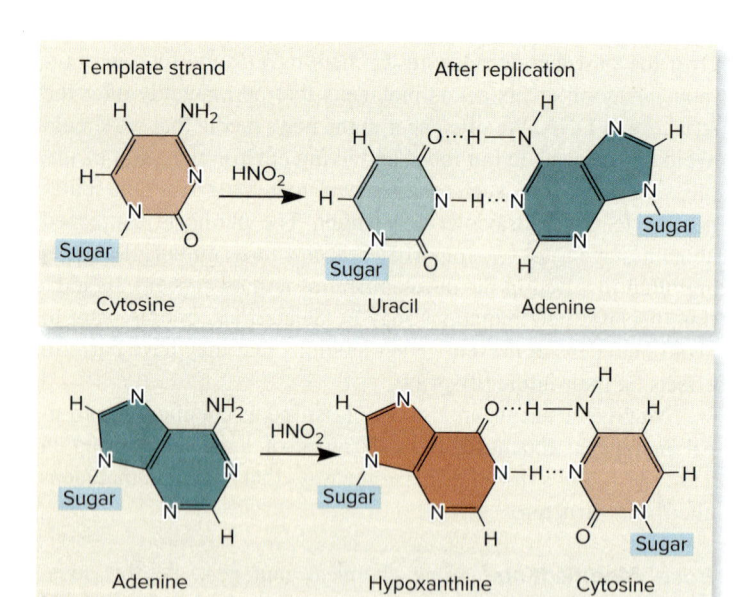

FIGURE 19.12 **Mispairing of modified bases that have been deaminated by nitrous acid.** Nitrous acid converts cytosine to uracil, and adenine to hypoxanthine. During DNA replication, uracil pairs with adenine, and hypoxanthine pairs with cytosine. This incorrect pairing creates mutations in the newly replicated strand during DNA replication.

Intercalating Agents Some mutagens exert their effects by directly interfering with the DNA replication process. For example, **acridine dyes** such as **proflavin** contain flat structures that intercalate, or insert themselves, between adjacent base pairs, thereby distorting the helical structure. When DNA containing these mutagens is replicated, single-nucleotide additions and/or deletions can occur in the newly made daughter strands, creating frameshift mutations.

Base Analogs **5-bromouracil (5BU)** and **2-aminopurine** are base analogs that become incorporated into daughter strands during DNA replication. 5BU is a thymine analog that can be incorporated into DNA instead of thymine. Like thymine, 5BU base-pairs with adenine. However, at a relatively high rate, 5BU undergoes a tautomeric shift and base-pairs with guanine (**Figure 19.13a**). When this occurs during DNA replication, 5BU causes a mutation in which a TA base pair is changed to a 5BU-G base pair (**Figure 19.13b**). This mutation is a transition, because the adenine has been changed to a guanine, both of which are purines. During the next round of DNA replication, the template strand containing the guanine base produces a GC base pair. In this way, 5BU promotes a change of an AT base pair into a GC base pair.

Compounds like 5BU are sometimes used in chemotherapy for cancer. The rationale is that these compounds are incorporated only into the DNA of actively dividing cells such as cancer cells. When incorporated, these compounds tend to cause many mutations in the cells, leading to the death of cancer cells. Unfortunately, other actively dividing cells, such as those in the skin and the lining of the digestive tract, also incorporate 5BU, which leads to unwanted side effects of chemotherapy, such as hair loss and a diminished appetite.

DNA molecules are also sensitive to physical agents such as radiation, including both ionizing and nonionizing radiation.

Ionizing Radation Radiation of short wavelength and high energy, known as ionizing radiation, can alter DNA structure. This type of radiation includes X-rays and gamma rays. Ionizing radiation can penetrate deeply into biological materials, where it produces chemically reactive molecules known as free radicals. These molecules alter the structure of DNA in a variety of ways. Exposure to high doses of ionizing radiation results in base deletions, oxidized bases, single-strand breaks in the DNA backbone, crosslinking, and even chromosomal breaks.

Nonionizing Radiation Nonionizing radiation, such as UV light, has less energy, and so it penetrates only the surface of an organism, such as the skin. Nevertheless, UV light is known to cause DNA mutations. As shown in **Figure 19.14**, UV light causes the formation of **thymine dimers,** which are adjacent thymine bases that have become covalently linked. Thymine dimers interfere with DNA replication and therefore can produce a mutation when the DNA strand is replicated. Plants, in particular, must have effective ways of preventing UV damage because they are exposed to sunlight throughout the day. Tanning greatly increases a person's exposure to UV light, raising the potential for thymine dimers and mutation. This explains the higher incidence of skin cancer among people who have been exposed to large amounts of sunlight during their lifetime. Because of the known link between skin cancer and sun exposure, people now apply sunscreen to their skin to prevent the harmful effects of UV light. Most sunscreens contain organic compounds, such as oxybenzone, which absorb UV light, and/or opaque ingredients, such as zinc oxide, that reflect UV light.

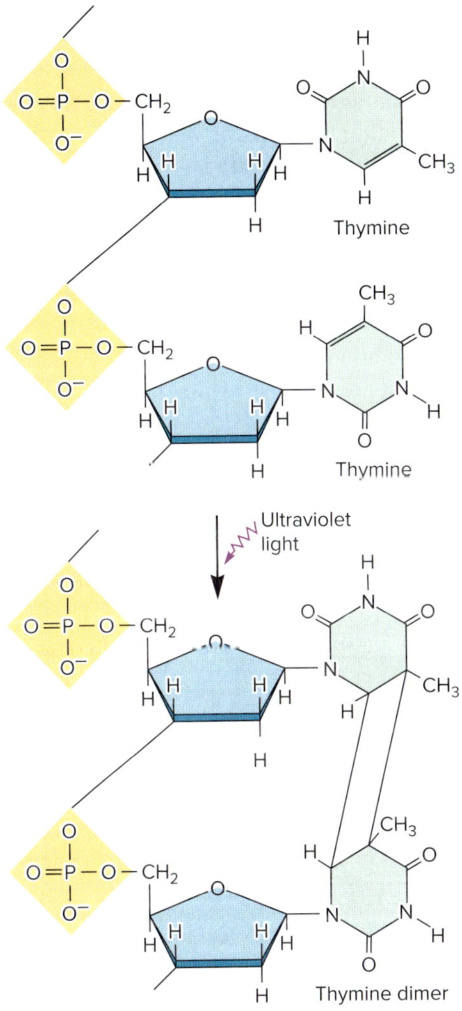

The Mutation Rate Is the Likelihood of a New Mutation

Because mutations occur spontaneously and may be induced by environmental agents, geneticists are greatly interested in learning how prevalent they are. The **mutation rate** is the likelihood that a gene will be altered by a new mutation. This rate is commonly expressed as the number of new mutations in a given gene per cell

FIGURE 19.14 Formation and structure of a thymine dimer.

Concept Check: What is a common cause of thymine dimer formation in people, and in what cell type(s) is it most likely to occur?

generation. The spontaneous mutation rate for a particular gene is typically in the range from 1 in 100,000 to 1 in 1 billion, or 10^{-5} to 10^{-9} per cell generation. In addition, mutations can occur at other sites in a genome, not only in genes. For example, people usually carry about 100 to 200 new mutations in their entire genome that were not present in their parents. Most of these are single nucleotide changes that do not occur within the coding sequences of genes. Given the human genome size of approximately 3,200,000,000 bp, these numbers tell us that a mutation is a relatively infrequent event. However, the mutation rate is not a constant number. The presence of certain environmental agents, such as X-rays, can increase the rate of induced mutations to a much higher value than the spontaneous mutation rate. In addition, mutation rates vary substantially from species to species and even within different strains of the same species. One explanation for this variation is that there are many different causes of mutations (refer back to Table 19.4).

Testing Methods Can Determine If an Agent Is a Mutagen

To determine if an agent is mutagenic, researchers use testing methods that monitor whether or not the agent increases the mutation rate. Many different kinds of tests have been used to evaluate mutagenicity. One common method is the **Ames test**, which was developed by Bruce Ames. This test uses strains of a bacterium, *Salmonella typhimurium*, that cannot synthesize the amino acid histidine. These strains contain a point mutation within a gene that encodes an enzyme required for histidine biosynthesis. The mutation renders the enzyme inactive. Therefore, the bacteria cannot grow on petri plates unless histidine has been added to the growth medium. However, a second mutation—a reversion—may occur that restores the ability to synthesize histidine. In other words, a second mutation can cause a reversion back to the wild-type condition. The Ames test monitors the rate at which this second mutation occurs, thereby indicating whether an agent increases the mutation rate above the spontaneous rate.

Figure 19.15 outlines the steps in the Ames test.

1. The suspected mutagen is mixed with a rat liver extract and a strain of *Salmonella* that cannot synthesize histidine. A mutagen may require activation by cellular enzymes, which are provided by the rat liver extract. This step improves the ability of the test to identify agents that may cause mutation in mammals. In the control tube, no mutagen is added.
2. A large number of bacteria are then plated on a growth medium that does not contain histidine.
3. The *Salmonella* strain is not expected to grow on these plates. However, if a mutation has occurred that allows the strain to synthesize histidine, it can grow on these plates to form a visible bacterial colony.

To estimate the mutation rate, the colonies that grow on the media are counted and compared with the total number of bacterial cells that were originally streaked on the plate. For example, if 10,000,000 bacteria were plated and 10 growing colonies were observed, the rate of mutation is 10 out of 10,000,000; this equals 1 in 10^6, or simply 10^{-6}. As a control, bacteria that have not been exposed to the mutagen are also tested, because a low level of spontaneous mutation is expected to occur.

How do we judge if an agent is a mutagen? Researchers compare the mutation rate in the presence and absence of the suspected mutagen. The test shown in Figure 19.15 is conducted several times. If statistical analysis reveals that the mutation rate in the experimental and control samples are significantly different, researchers may tentatively conclude that the agent may be a mutagen. Many studies have been conducted in which researchers used the Ames test to compare the urine from cigarette smokers to that from nonsmokers. This research has shown that the urine from smokers contains much higher levels of mutagens.

Mix together the suspected mutagen, a rat liver extract, and a *Salmonella* strain that cannot synthesize histidine. The suspected mutagen is omitted from the control sample.

Plate the mixtures onto petri plates that lack histidine.

Incubate overnight to allow bacterial growth.

A large number of colonies suggests that the suspected mutagen causes mutation.

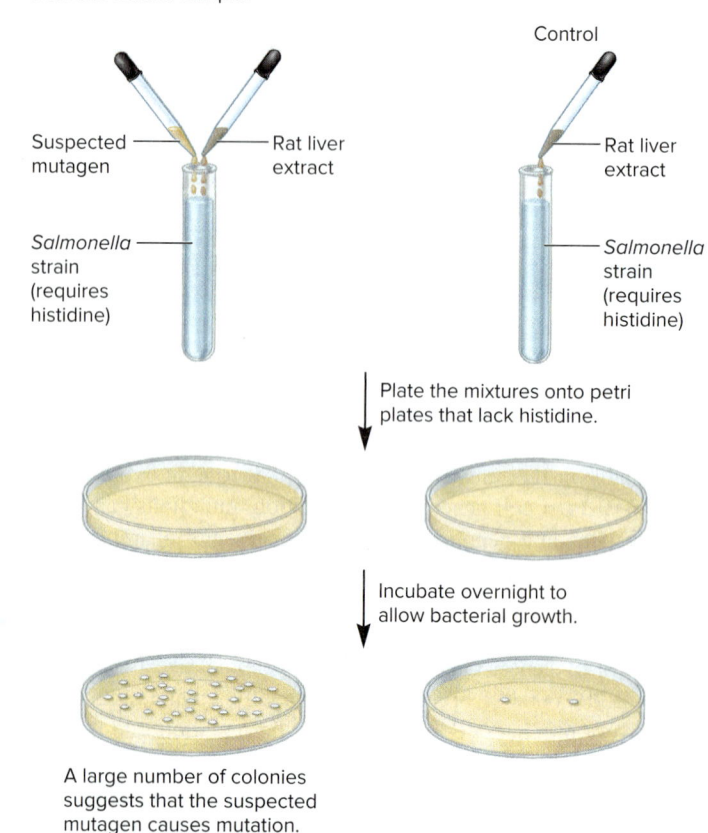

FIGURE 19.15 The Ames test for mutagenicity.

Concept Check: What is the purpose of the rat liver extract in this procedure?

Genetic TIPS

The Question: A researcher studied the effects of a suspected mutagen, called mutagen X, using the procedure for an Ames test, shown in Figure 19.15. The following data were obtained after placing 2 million cells on each plate:

Trial	Control (number of colonies)	Plus mutagen X (number of colonies)
1	3	62
2	2	77
3	5	46
4	2	55

Calculate the average mutation rate in the presence and absence of mutagen X. Conduct a *t*-test to determine if suspected mutagen X is significantly affecting the mutation rate.

Topic: What topic in genetics does this question address?

The topic is testing for mutagens. More specifically, the question is about the Ames test.

Information: What information do you know based on the question and your understanding of the topic?

In the question, you are given data regarding the outcome of four trials using the Ames test. From your understanding of the topic, you may remember that a higher number of colonies on the experimental plates may indicate that a substance is a mutagen.

Problem-Solving Strategy: Make a calculation. Analyze data.

To begin to solve this problem, you first need to calculate the mutation rate. You take the average of the four trials and then divide the average number of mutant colonies by the total number of cells applied to each plate (in this case, 2 million). You also need to conduct a *t*-test to determine if the control and experimental data are significantly different. A description of a *t*-test can be found in various statistics textbooks and the Statistics Primer available in Connect.

Answer: For the control data, the mutation rate is 1.5 in 1 million, or 1.5×10^{-6}. In the presence of the suspected mutagen, the rate is 30 in a million, or 30×10^{-6}. If you conduct a *t*-test on these data, $P < 0.01$, so you can reject the null hypothesis that the control and experimental data are not different from each other. Therefore, you can accept the hypothesis that the suspected mutagen is causing a higher mutation rate. Keep in mind that this hypothesis is not proven; you simply are able to accept it based on this statistical outcome.

19.4 REVIEWING THE KEY CONCEPTS

- A mutagen is an agent that can cause a mutation. Researchers have identified many different mutagens that change DNA structure in a variety of ways (see Table 19.6, Figures 19.12–19.14).
- Mutation rate is the likelihood that a new mutation will occur.
- The Ames test is used to determine if an agent is a mutagen (see Figure 19.15).

19.4 COMPREHENSION QUESTIONS

1. Nitrous acid replaces amino groups with keto groups, a process called
   a. alkylation.
   b. deamination.
   c. depurination.
   d. crosslinking.

2. A mutagen that is a base analog is
   a. ethyl methanesulfonate (EMS).
   b. 5-bromouracil.
   c. UV light.
   d. proflavin.

3. In an Ames test, a _____ number of colonies is observed if a substance _____ a mutagen, compared with the number of colonies for a control sample that is not exposed to the suspected mutagen.
   a. significantly higher, is
   b. significantly higher, is not
   c. significantly lower, is
   d. significantly lower, is not

19.5 DNA REPAIR

Learning Outcomes:

1. Compare and contrast the different types of DNA repair systems.
2. Describe how specialized DNA polymerases are able to synthesize DNA over a damaged region.

Because mutations are often deleterious, DNA repair systems are vital to the survival of all organisms. If DNA repair systems did not exist, spontaneous and environmentally induced mutations would be so prevalent that few species, if any, would survive. The necessity of DNA repair systems becomes evident when they are missing. Bacteria contain several different DNA repair systems. Yet, when even a single system is absent, the bacteria have a much higher rate of mutation. In fact, the rate of mutation is so high that these bacterial strains are sometimes called mutator strains. Likewise, in humans, an individual who is defective in only a single DNA repair system may manifest various disease symptoms, including a higher risk of skin cancer. This increased risk is due to the inability to repair UV-induced mutations.

Living cells have several DNA repair systems that can fix different types of DNA alterations (**Table 19.7**). Each repair system is composed of one or more proteins that play specific roles in the repair mechanism. In most cases, DNA repair is a multistep process. First, one or more proteins in the DNA repair system detect an irregularity in DNA structure. Next, the abnormality is removed by the action of DNA repair enzymes. Finally, normal DNA is synthesized via DNA replication enzymes. In this section, we will examine some of these different repair systems, which have been characterized in bacteria, yeast, mammals, and plants. Their diverse ways of repairing DNA underscore the necessity for the proper maintenance of DNA structure.

Damaged Bases Can Be Directly Repaired

In a few cases, the covalent modification of nucleotides by mutagens can be reversed by specific cellular enzymes. As discussed earlier in this chapter, UV light causes the formation of thymine dimers. Bacteria, fungi, most plants, and some animals produce an enzyme called **photolyase** that recognizes thymine dimers and splits them, which returns the DNA to its original condition (**Figure 19.16**). Photolyase contains two light-sensitive cofactors. The repair mechanism itself requires light and is known

CHAPTER 19 :: GENE MUTATION AND DNA REPAIR

TABLE 19.7
Common Types of DNA Repair Systems

System	Description
Direct repair	An enzyme recognizes an incorrect alteration in DNA structure and directly converts the structure back to the correct form.
Base excision repair and nucleotide excision repair	An abnormal base or nucleotide is first recognized and removed from the DNA, and a segment of DNA in this region is excised, and then the complementary DNA strand is used as a template to synthesize a normal DNA strand.
Mismatch repair	Similar to excision repair except that the DNA defect is a base pair mismatch in the DNA, not an abnormal nucleotide. The mismatch is recognized, and a segment of DNA in this region is removed. The parental strand is used as a template to synthesize a normal daughter strand of DNA.
Homologous recombination repair	Occurs at double-strand breaks or when DNA damage causes a gap in synthesis during DNA replication. The strands of a normal sister chromatid are used to repair a damaged sister chromatid.
Nonhomologous end joining	Occurs at double-strand breaks. The broken ends are recognized by proteins that keep the ends together; the broken ends are eventually rejoined.

as **photoreactivation.** This process directly restores the structure of DNA. Because plants are exposed to sunlight throughout the day, photolyase is a critical DNA repair enzyme for many plant species.

Nucleotide Excision Repair Systems Remove Segments of Damaged DNA

An important general process for DNA repair is **nucleotide excision repair (NER)**, which is often used to repair bulky, helix-distorting lesions. This type of system can repair many different types of DNA damage, including thymine dimers, chemically modified bases, missing bases, and certain types of crosslinks. In NER, several nucleotides in the damaged strand are removed from the DNA, and the intact strand is used as a template for resynthesis of a normal complementary strand. NER is found in all eukaryotes and prokaryotes, although its molecular mechanism is better understood in prokaryotic species.

In *E. coli*, the NER system requires four key proteins, designated UvrA, UvrB, UvrC, and UvrD, plus DNA polymerase and DNA ligase. The four Uvr proteins recognize and remove a short segment of a damaged DNA strand. These proteins are named Uvr because they are involved in ultraviolet light repair of thymine dimers, although they are also important in repairing chemically damaged DNA.

Figure 19.17 outlines the steps involved in the *E. coli* NER system.

1. A protein complex consisting of two UvrA molecules and one UvrB molecule tracks along the DNA in search of damaged DNA. Such DNA has a distorted double helix, which is detected by the UvrA/UvrB complex.
2. When a damaged segment is identified, the two UvrA proteins are released, and UvrC binds to the site.
3. The UvrC protein makes cuts in the damaged strand on both sides of the damaged site.
4. After this process, UvrD, which is a helicase, recognizes the region and separates the two strands of DNA. This releases a short DNA segment that contains the damaged region, and UvrB and UvrC are also released.
5. Following the excision of the damaged DNA, DNA polymerase fills in the gap, using the undamaged strand as a template. Finally, DNA ligase makes the final covalent connection between the newly made DNA and the original DNA strand.

In eukaryotes, NER systems are thought to operate similarly to those in bacteria, though more proteins are involved. Several human diseases are due to inherited defects in genes involved in NER. These include xeroderma pigmentosum (XP) and Cockayne syndrome (CS). A common symptom of both syndromes is an increased sensitivity to sunlight because of an inability to repair UV-induced lesions. **Figure 19.18** is a photograph of an individual with XP. Such individuals have pigmentation abnormalities and many premalignant lesions and are highly predisposed to developing skin cancer. They may also develop early degeneration of the nervous system.

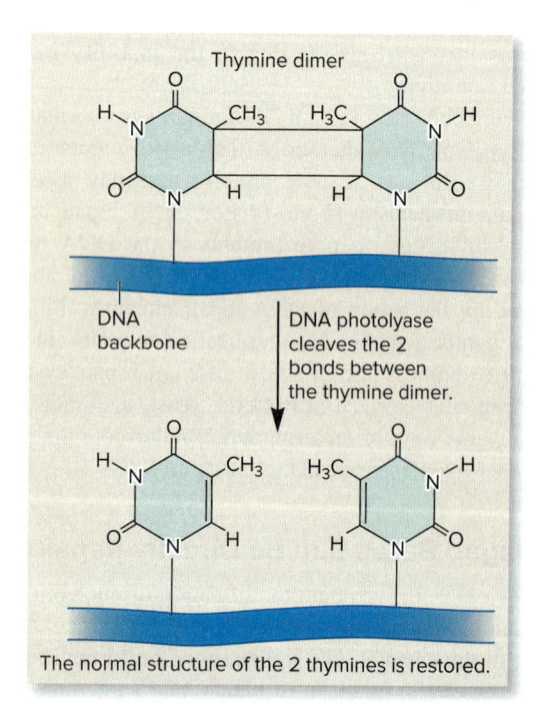

FIGURE 19.16 **Direct repair of damaged bases in DNA.** The repair of a thymine dimer by photolyase.

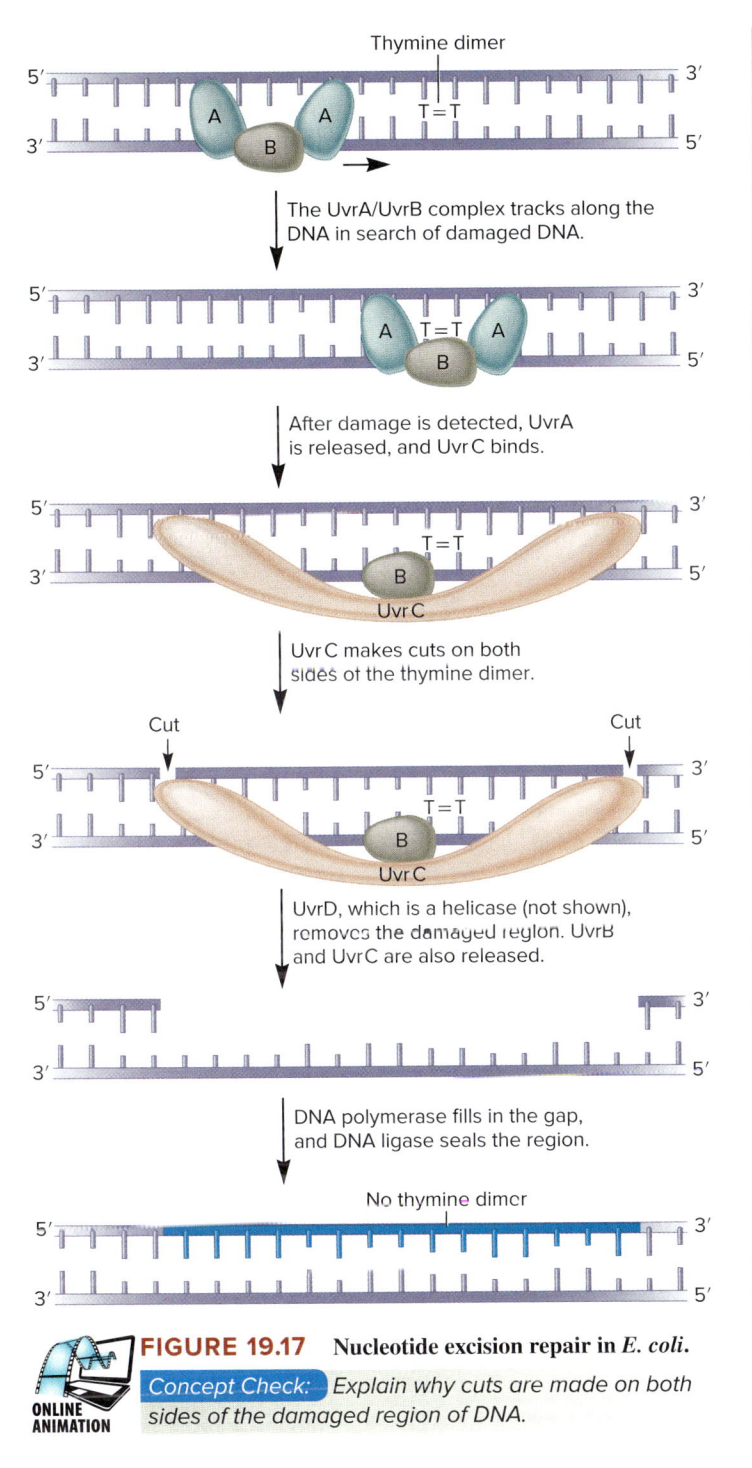

FIGURE 19.17 Nucleotide excision repair in *E. coli*.

Concept Check: *Explain why cuts are made on both sides of the damaged region of DNA.*

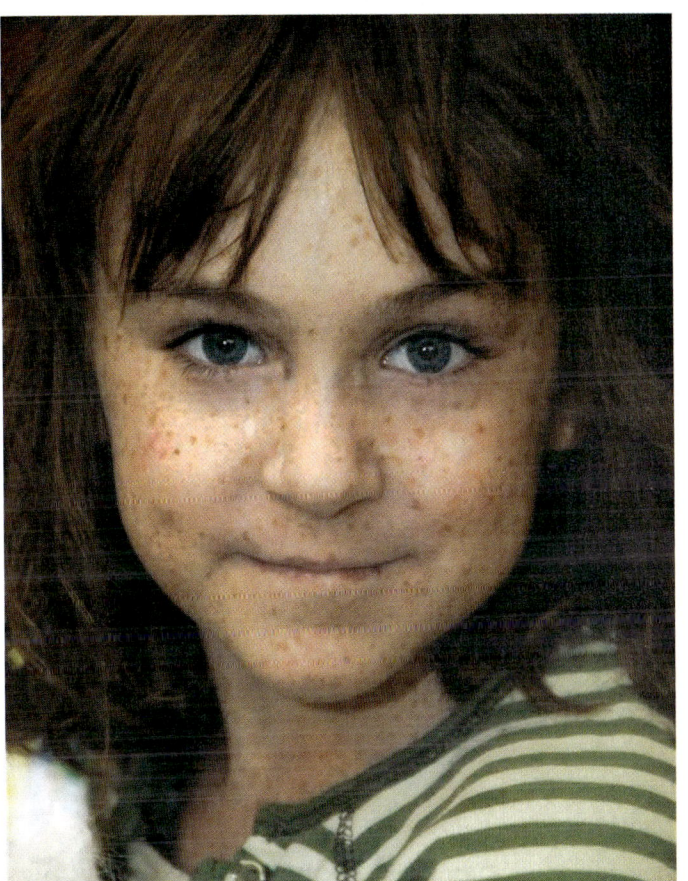

FIGURE 19.18 An individual affected with xeroderma pigmentosum.

Genes → Traits Xeroderma pigmentosum is caused by a defect in one of seven different NER genes. Affected individuals have an increased sensitivity to sunlight because of an inability to repair UV-induced DNA lesions. In addition, they may also have pigmentation abnormalities, many premalignant lesions, and a high predisposition to skin cancer.

©Barcroft Media/Getty Images

Mismatch Repair Systems Recognize and Correct a Base Pair Mismatch

Thus far, we have considered DNA repair systems that recognize abnormal nucleotide structures within DNA, including thymine dimers. Another type of abnormality that should not occur in DNA is a **base pair mismatch.** The structure of the DNA double helix obeys the AT/GC rule of base pairing. During the normal course of DNA replication, however, an incorrect nucleotide may be added to the growing daughter strand by mistake. This produces a mismatch between a nucleotide in the parental strand and one in the daughter strand. Various DNA repair mechanisms can recognize and remove this mismatch. For example, as described in Chapter 13, DNA polymerase has a 3′ to 5′ proofreading ability that detects mismatches and removes them. However, if this proofreading ability fails, cells have additional DNA repair systems that detect base pair mismatches and fix them. An interesting DNA repair system that exists in all species is the **mismatch repair system.**

In the case of a base pair mismatch, how does a DNA repair system determine which base to remove? If the mismatch is due to an error in DNA replication, the newly made daughter strand contains the incorrect base, whereas the parental strand is normal. Therefore, an important aspect of mismatch repair is that it specifically repairs the newly made strand rather than the parental template strand. Prior to DNA replication, the parental DNA has already been methylated. Immediately after DNA replication, some time must pass before a newly made strand is methylated. Therefore, newly replicated DNA is hemimethylated—only the parental DNA strand is methylated. Hemimethylation provides a way for a DNA repair system to distinguish between the parental DNA strand and the daughter strand.

The molecular mechanism of mismatch repair has been studied extensively in *E. coli*. As shown in **Figure 19.19**, three proteins, designated MutS, MutL, and MutH, detect the mismatch and direct the removal of the mismatched base from the newly made strand. These proteins are named Mut because their absence leads to a much higher mutation rate than occurs in normal strains of *E. coli*.

1. The role of MutS is to locate mismatches. Once a mismatch is detected, MutS forms a complex with MutL. MutL acts as a linker that binds to MutH, forming a loop in the DNA.
2. This loop formation stimulates MutH, which is bound to a hemimethylated site, to make a cut in the newly made, nonmethylated DNA strand. After the strand is cut, MutU, which functions as helicase, separates the strands, and an exonuclease then digests the nonmethylated DNA strand in the direction of the mismatch and proceeds beyond the mismatch site.
3. The gap in the daughter strand is repaired by DNA polymerase and DNA ligase.

The net result is that the mismatch has been corrected by removing the incorrect region in the daughter strand and then resynthesizing the correct sequence using the parental DNA as a template.

Eukaryotic species have homologs to MutS and MutL, along with many other proteins that are needed for mismatch repair. Eukaryotes do not have a MutH homolog. Instead, the eukaryotic MutL protein has the ability to make a cut in the nonmethylated DNA strand.

As with defects in nucleotide excision repair systems, mutations in the human mismatch repair system are associated with particular types of cancer. For example, mutations in two human mismatch repair genes, *hMSH2* and *hMLH1*, play a role in the development of a type of colon cancer known as hereditary nonpolyposis colorectal cancer.

Double-Strand Breaks Can Be Repaired by Homologous Recombination Repair and by Nonhomologous End Joining

Of the many types of DNA damage that can occur within living cells, the breakage of chromosomes—called a DNA double-strand break (DSB)—is perhaps the most dangerous. DSBs can be caused by ionizing radiation (X-rays or gamma rays), chemical mutagens, and certain drugs used for chemotherapy. In addition, reactive oxygen species that are the by-products of aerobic metabolism cause double-strand breaks. Surprisingly, researchers estimate that naturally occurring double-strand breaks in a typical human cell occur at a rate of 10 to 100 breaks per cell per day! Such breaks are harmful in a variety of ways. First, DSBs can result in chromosomal rearrangements such as inversions and translocations (refer back to Figure 8.2). In addition, DSBs can lead to terminal or interstitial deletions (refer back to Figure 8.3). Such genetic changes have the potential to result in detrimental phenotypic effects.

How are DSBs repaired? The two main mechanisms are **homologous recombination repair (HRR)** (see Table 19.7) and **nonhomologous end joining (NHEJ)**. During nonhomologous

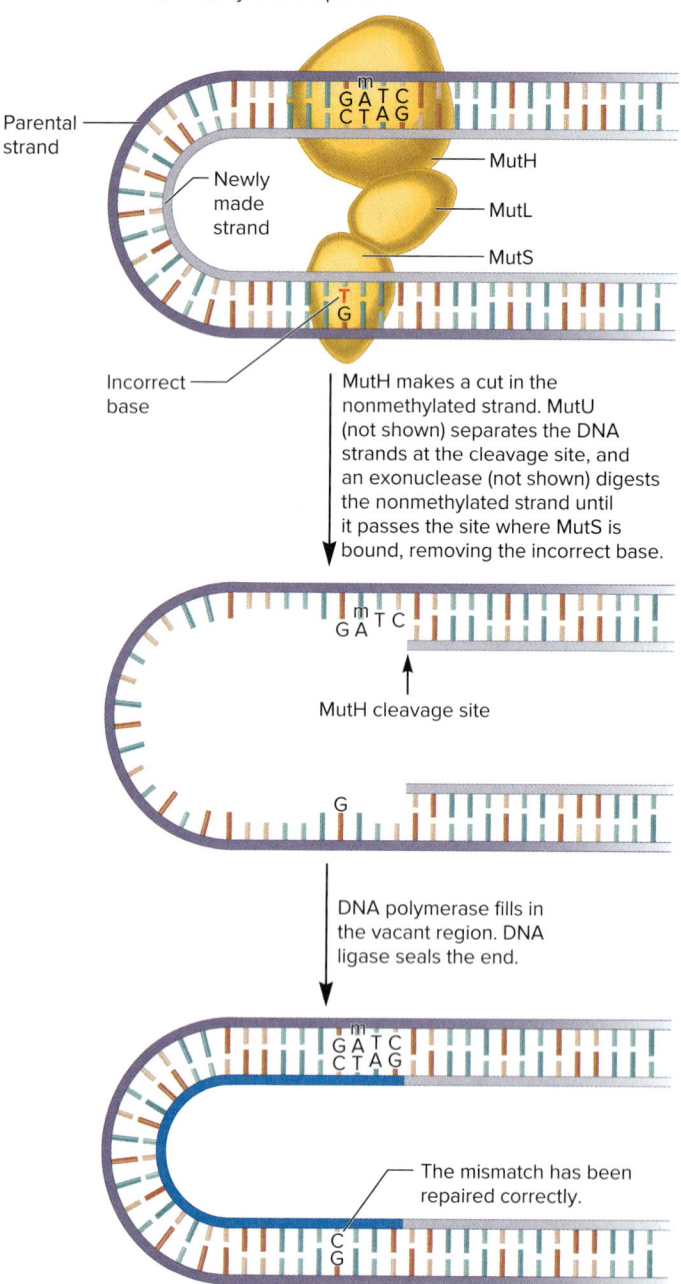

FIGURE 19.19 Mismatch repair in *E. coli*.

Concept Check: Which of the three Mut proteins is responsible for ensuring that the mismatched base in the newly made daughter strand is the one that is removed?

end joining, the two broken ends of DNA are simply pieced back together (**Figure 19.20**). This mechanism requires the participation of several proteins that play key roles in the process.

1. The DSB is recognized by end-binding proteins.
2. These proteins then recognize additional proteins that form a cross bridge that prevents the two ends from drifting apart.
3. Additional proteins are recruited to the region, and they may process the ends of the broken chromosome by digesting particular DNA strands. This processing may result in the deletion of a small amount of genetic material from the region.
4. Finally, any gaps are filled in via DNA polymerase, and the DNA ends are ligated together.

One advantage of NHEJ is that it doesn't involve a sister chromatid, so it can occur at any stage of the cell cycle. However, a disadvantage is that NHEJ may result in small deletions in the region that has been repaired.

Damaged DNA May Be Replicated by Translesion DNA Polymerases

Despite the efficient action of numerous repair systems that remove lesions in DNA in an error-free manner, it is inevitable that some lesions may escape these repair mechanisms. Such lesions may be present when DNA is being replicated. If so, replicative DNA polymerases, such as polIII in *E. coli*, which are highly sensitive to geometric distortions in DNA, are unable to replicate through DNA lesions. During the past decade, researchers have discovered that cells are equipped with specialized DNA polymerases that assist the replicative DNA polymerases during the process of **translesion synthesis (TLS)**—the synthesis of DNA over a template strand that harbors some type of DNA damage. These translesion-replicating polymerases, which are also described in Chapter 13 (see Table 13.2), contain an active site with a loose, flexible pocket that can accommodate aberrant structures in the template strand. When a replicative DNA polymerase encounters a damaged region, it is swapped with a lesion-replicating polymerase.

A negative consequence of translesion synthesis is low fidelity. Due to their flexible active site, translesion-replicating polymerases are much more likely to incorporate the wrong nucleotide into a newly made daughter strand. The mutation rate is typically in the range from 10^{-2} to 10^{-3}. Therefore, translesion synthesis is referred to as **error-prone replication.** By comparison, replicative DNA polymerases are highly intolerant of the geometric distortions imposed on DNA by the incorporation of incorrect nucleotides, and consequently, they incorporate wrong nucleotides with a very low frequency of approximately 10^{-8}. In other words, they copy DNA with a high degree of fidelity.

19.5 REVIEWING THE KEY CONCEPTS

- All species have a variety of DNA repair systems to avoid the harmful effects of mutations (see Table 19.7).
- Photolyase can directly repair thymine dimers (see Figure 19.16).
- Nucleotide excision repair removes damaged bases and damaged segments of DNA. Human inherited diseases involve defects in nucleotide excision repair (see Figures 19.17, 19.18).
- The mismatch repair system recognizes a base pair mismatch and removes a segment of the DNA strand containing the incorrect base (see Figure 19.19).
- Double-strand breaks can be repaired by homologous recombination repair (HRR) or by nonhomologous end joining (NHEJ) (see Figure 19.20).
- Damaged DNA may be replicated by translesion-replicating polymerases that are error-prone.

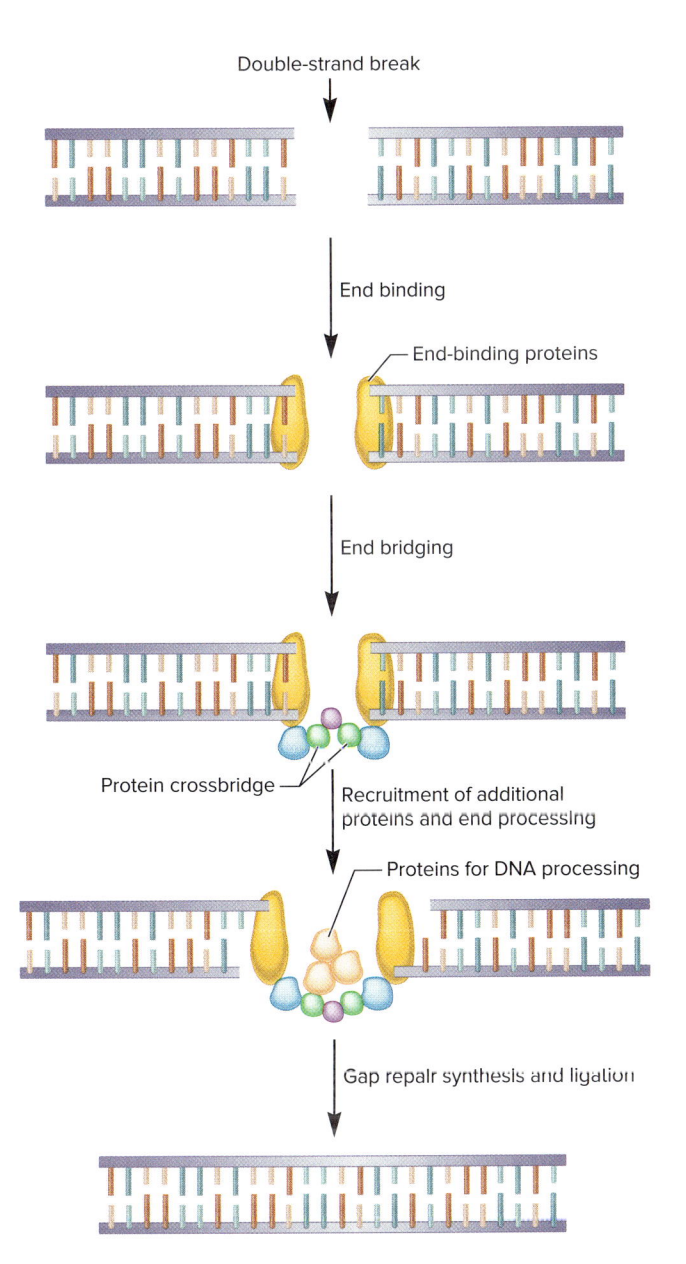

FIGURE 19.20 Repair of a double-strand break in DNA via nonhomologous end joining.

Concept Check: What is an advantage and a disadvantage of this repair system?

19.5 COMPREHENSION QUESTIONS

1. The function of photolyase is to repair
   a. double-strand breaks.
   b. apurinic sites.
   c. thymine dimers.
   d. all of the above.
2. Which of the following DNA repair systems may involve the removal of a segment of a DNA strand?
   a. Direct repair
   b. Nucleotide excision repair
   c. Mismatch repair
   d. Both b and c
3. In nucleotide excision repair in *E. coli*, the function of the UvrA/UvrB complex is to
   a. detect DNA damage.
   b. make cuts on both sides of the damage.
   c. remove the damaged piece of DNA.
   d. replace the damaged DNA with undamaged DNA.
4. Double-strand breaks can be repaired by
   a. homologous recombination repair (HRR).
   b. nonhomologous end joining (NHEJ).
   c. nucleotide excision repair (NER).
   d. either a or b.
5. An advantage of translesion-replicating polymerases is that they can replicate _____, but a disadvantage is that they _____ .
   a. very quickly, have low fidelity
   b. over damaged DNA, have low fidelity
   c. when resources are limited, are very slow
   d. over damaged DNA, are very slow

KEY TERMS

Page 431. mutation
Page 432. point mutation, base substitution, transition, transversion, silent mutations, missense mutations, nonsense mutations, frameshift mutations
Page 434. neutral mutation, up promoter mutations, down promoter mutation, wild type, mutant allele, reversion, deleterious mutation, lethal mutation, beneficial mutation
Page 435. conditional mutants, suppressor (suppressor mutation), intragenic suppressor, intergenic suppressor, breakpoint, position effect
Page 437. germ line, germ-line mutation, somatic cells, somatic mutation, genetic mosaic
Page 439. replica plating, random mutation theory, hot spots
Page 440. spontaneous mutations, induced mutations, depurination, apurinic site
Page 441. deamination, tautomeric shift, tautomers
Page 443. reactive oxygen species (ROS), oxidative stress, oxidative DNA damage, trinucleotide repeat expansion (TNRE), anticipation
Page 445. mutagens, nitrous acid, nitrogen mustard, ethyl methanesulfonate (EMS)
Page 446. acridine dyes, proflavin, 5-bromouracil (5BU), 2-aminopurine
Page 447. thymine dimers, mutation rate
Page 448. Ames test
Page 449. photolyase
Page 440. photoreactivation, nucleotide excision repair (NER)
Page 451. base pair mismatch, mismatch repair system
Page 452. homologous recombination repair (HRR), nonhomologous end joining (NHEJ)
Page 453. translesion synthesis (TLS), error-prone replication

CHAPTER SUMMARY

- A mutation is a heritable change in the genetic material.

19.1 Effects of Mutations on Gene Structure and Function

- A point mutation is a change in a single base pair. Such a mutation can be a transition or a transversion.
- Silent, missense, nonsense, and frameshift mutations may occur within a coding sequence of a gene (see Table 19.1, Figure 19.1).
- Mutations may also occur within non-coding sequences of a gene and affect gene expression (see Table 19.2).
- Suppressor mutations reverse the phenotypic effects of another mutation. They can be intragenic or intergenic (see Table 19.3).
- Changes in chromosome structure can have a position effect that alters gene expression (see Figures 19.2, 19.3).
- With regard to timing, mutations can occur in germ-line cells or in somatic cells (see Figures 19.4, 19.5).

19.2 Random Nature of Mutations

- The replica plating experiments performed by Lederberg and Lederberg were consistent with the random mutation theory (see Figure 19.6).

19.3 Spontaneous Mutations

- Spontaneous mutations result from natural biological or chemical processes, whereas induced mutations are caused by environmental agents (see Table 19.4).
- Three common ways that mutations can arise spontaneously is by depurination, deamination, and tautomeric shifts (see Figures 19.7–19.9).

- Reactive oxygen species (ROS) can cause spontaneous mutations by oxidizing bases in DNA (see Figure 19.10).
- In individuals with a trinucleotide repeat expansion (TNRE), the number of repeats of a trinucleotide sequence increases above a critical level and becomes prone to frequent expansion. This type of mutation is responsible for certain types of human diseases. The repeats can expand due to hairpin formation during DNA replication (see Table 19.5, Figure 19.11).

19.4 Induced Mutations

- A mutagen is an agent that can cause a mutation. Researchers have identified many different mutagens that change DNA structure in a variety of ways (see Table 19.6, Figures 19.12–19.14).
- Mutation rate is the likelihood that a new mutation will occur.
- The Ames test is used to determine if an agent is a mutagen (see Figure 19.15).

19.5 DNA Repair

- All species have a variety of DNA repair systems to avoid the harmful effects of mutations (see Table 19.7).
- Photolyase can directly repair thymine dimers (see Figure 19.16).
- Nucleotide excision repair removes damaged bases and damaged segments of DNA. Human inherited diseases involve defects in nucleotide excision repair (see Figures 19.17, 19.18).
- The mismatch repair system recognizes a base pair mismatch and removes a segment of the DNA strand containing the incorrect base (see Figure 19.19).
- Double-strand breaks can be repaired by homologous recombination repair (HRR) or by nonhomologous end joining (NHEJ) (see Figure 19.20).
- Damaged DNA may be replicated by translesion-replicating polymerases that are error-prone.

PROBLEM SETS & INSIGHTS

More Genetic TIPS

1. If the rate of mutation is 10^{-5} per gene per cell generation, how many new mutations per gene would you expect in a population of 1 million bacteria?

Topic: What topic in genetics does this question address?

The topic is the mutation rate in a population of bacteria.

Information: What information do you know based on the question and your understanding of the topic?

From the question, you know the mutation rate and are asked to predict the number of new mutations per gene in a populaton composed of 1 million bacteria. From your understanding of the topic, you may remember that the mutation rate is the number of new mutations per cell generation.

Problem-Solving Strategy: Make a calculation.

To solve this problem, you multiply the mutation rate times the number of bacteria.

Answer: The mutation rate times the number of bacteria is $10^{-5} \times 10^{6}$, which equals 10. Therefore, you would expect about 10 new mutations in this gene per million bacteria.

2. A reversion is a mutation that returns a mutant codon back to a codon that gives a wild-type phenotype. At the DNA level, this type of mutation can be an exact reversion or an equivalent reversion.

GAG (glutamic acid)	First mutation	→ GTG (valine)	Exact reversion	→ GAG (glutamic acid)
GAG (glutamic acid)	First mutation	→ GTG (valine)	Equivalent reversion	→ GAA (glutamic acid)
GAG (glutamic acid)	First mutation	→ GTG (valine)	Equivalent reversion	→ GAT (aspartic acid)

An equivalent reversion produces a protein that is equivalent to the wild-type protein in structure and function. This outcome can occur in two ways. In some cases, the reversion produces the wild-type amino acid (in this case, glutamic acid), but it uses a different codon than the wild-type gene. Alternatively, an equivalent reversion may substitute an amino acid structurally similar to the wild-type amino acid. In our example, an equivalent reversion has changed valine to an aspartic acid. Because aspartic and glutamic acids are structurally similar—they are acidic amino acids—this type of reversion can restore wild-type structure and function.

Here is the question: The template strand within the coding sequence of a gene has the following sequence:

3′–TACCCCTTCGACCCCGGA–5′

This template produces the following mRNA:

5′–AUGGGGAAGCUGGGGCCA–3′

The mRNA encodes a polypeptide with the following sequence:

methionine–glycine–lysine–leucine–glycine–proline

A mutation changes the template strand to this sequence:

3′–TACCCCTACGACCCCGGA–5′

After the first mutation, another mutation occurs to change this sequence again. Is each of the following second mutations an exact reversion, an equivalent reversion, or neither?

A. 3′–TACCCCTCCGACCCCGGA–5′

B. 3′–TACCCCTTGACCCCGGA–5′

C. 3′–TACCCGACGACCCCGGA–5′

Topic: What topic in genetics does this question address?

The topic is about mutations called reversions. More specifically, the question is about the effects of such mutations on the coding sequence of a gene.

Information: What information do you know based on the question and your understanding of the topic?

From the question, you know the difference between an exact reversion and an equivalent reversion. From your understanding of the topic, you may remember that the template strand is used to make a complementary strand of mRNA. You should be able to look codons up in Table 15.1 and determine whether a particular mutation will alter the coding sequence. You may also recall that AUG is the start codon.

Problem-Solving Strategy: Relate structure and function. Predict the outcome.

To begin solving this problem, you first need to write out the sequence of the mRNA that will be produced from the wild-type, mutant, and reversion sequences. Note: Every other codon is shown in red, beginning with the AUG start codon:

Wild-type:	5′–AUGGGGAAGCUGGGGCCA–3′
Mutant:	5′–AUGGGGAUGCUGGGGCCA–3′
Reversion A:	5′–AUGGGGAGGCUGGGGCCA–3′
Reversion B:	5′–AUGGGGAAGCUGGGGCCA–3′
Reversion C:	5′–AUGGGGCUGCUGGGGCCA–3′

The mutations occur in the third codon. You need to look up the codons in the codon table. Finally, you need to decide if a change in amino acid sequence is likely to affect protein function. In general, when an amino acid is substituted with a closely related amino acid, the change is less likely to inhibit protein function.

Answer:

A. This is probably an equivalent reversion. The third codon, which encodes a lysine in the wild-type gene, is now an arginine codon. Arginine and lysine are both basic amino acids, so the polypeptide would probably function normally.

B. This is an exact reversion.

C. The third codon, which is a lysine in the wild-type gene, has been changed to a leucine codon. It is difficult to say if this would be an equivalent reversion or not. Lysine is a basic amino acid, and leucine is a nonpolar, aliphatic amino acid. The protein may still function normally with a leucine at the third codon, or it may function abnormally. You would need to test the function of the protein to determine if this was an equivalent reversion or not.

Conceptual Questions

C1. Is each of the following mutations a transition, transversion, addition, or deletion? The sequence in the original DNA strand is

5′–GGACTAGATAC–3′

(Note: Only the coding DNA strand is shown.)

A. 5′–GAACTAGATAC–3′
B. 5′–GGACTAGAGAC–3′
C. 5′–GGACTAGTAC–3′
D. 5′–GGAGTAGATAC–3′

C2. A gene mutation changes a base pair from AT to GC. This change causes a gene to encode a truncated protein that is nonfunctional. An organism that carries this mutation cannot survive at high temperatures. Make a list of all the genetic terms that could be used to describe this type of mutation.

C3. What does a suppressor mutation suppress? What is the difference between an intragenic and an intergenic suppressor?

C4. How would each of the following types of mutations affect protein function or the amount of functional protein that is expressed from a gene?

A. Nonsense mutation
B. Missense mutation
C. Up promoter mutation
D. Mutation that affects splicing

C5. X-rays strike a chromosome in a living cell and ultimately cause the cell to die. Did the X-rays produce a mutation? Explain why or why not.

C6. Lactose permease is encoded by the *lacY* gene of the *lac* operon. Suppose a mutation occurred at codon 64 that changed the normal glycine codon to a valine codon. The mutant lactose permease is unable to function. However, a second mutation, which changes codon 50 from an alanine codon to a threonine codon, is able to restore function. Is each of the following terms appropriate or inappropriate to describe this second mutation?

A. Reversion
B. Intragenic suppressor
C. Intergenic suppressor
D. Missense mutation

C7. Is each of the following mutations a silent, missense, nonsense, or frameshift mutation? The sequence in the original DNA strand is 5′–ATGGGACTAGATACC–3′. (Note: Only the coding strand is shown; the first codon is methionine.)

A. 5′–ATGGGTCTAGATACC–3′
B. 5′–ATGCGACTAGATACC–3′
C. 5′–ATGGGACTAGTTACC–3′
D. 5′–ATGGGACTAAGATACC–3′

C8. In Chapters 14 through 17, we discussed many sequences that are outside of a coding sequence but are important for gene expression. Look up two of these sequences and write them out. Explain how a mutation could change these sequences, thereby altering gene expression.

C9. Explain two ways that a chromosomal rearrangement can cause a position effect.

C10. Is a random mutation more likely to be beneficial or harmful? Explain your answer.

C11. Which of the following mutations could be appropriately described as a position effect?

A. A point mutation at the –10 position in the promoter region prevents transcription.

B. A translocation places the coding sequence for a muscle-specific gene next to an enhancer that is turned on in nerve cells.

C. An inversion flips a gene from the long arm of chromosome 17 (which is euchromatic) to the short arm (which is heterochromatic).

C12. As discussed in Chapter 22, most forms of cancer are caused by environmental agents that produce mutations in somatic cells. Is an individual with cancer considered a genetic mosaic? Explain why or why not.

C13. Discuss the consequences of a germ-line versus a somatic mutation.

C14. Make a drawing that shows how alkylating agents alter the structure of DNA, and explain the process.

C15. Explain how a mutagen can interfere with DNA replication to cause a mutation. Give two examples.

C16. What type of mutation (transition, transversion, or frameshift) would you expect each of the following mutagens to cause?

A. Nitrous acid

B. 5-Bromouracil

C. Proflavin

C17. Explain what happens to the sequence of DNA during trinucleotide repeat expansion (TNRE). If someone was mildly affected with a TNRE disorder, what issues would be important when considering whether to have offspring?

C18. Distinguish between spontaneous and induced mutations. Which are more harmful? Which are avoidable?

C19. Are mutations random events? Explain your answer.

C20. Give an example of a mutagen that can change cytosine to uracil. Which DNA repair system(s) would be able to repair this defect?

C21. If a mutagen causes bases to be removed from nucleotides within DNA, what repair system could fix this damage?

C22. Trinucleotide repeat expansions (TNREs) are associated with several different human inherited diseases. Certain types of TNREs produce a long stretch of the amino acid glutamine within the encoded protein. When a TNRE exerts its detrimental effect by producing a glutamine stretch, are the following statements true or false?

A. The TNRE is within the coding sequence of the gene.

B. The TNRE prevents RNA polymerase from transcribing the gene properly.

C. The trinucleotide sequence is CAG.

D. The trinucleotide sequence is CCG.

C23. With regard to TNRE, what is meant by the term *anticipation*?

C24. Achondroplasia is a rare form of dwarfism. It is caused by an autosomal dominant mutation within a single gene. Among 1,422,000 live births, the number of babies born with achondroplasia was 31. Among those 31 babies, 18 of them had one parent with achondroplasia. The remaining babies had two unaffected parents. What is the mutation rate for achondroplasia?

C25. A segment of DNA has the following sequence:

TTGGATGCTGG
AACCTACGACC

A. What would the sequence be immediately after reaction with nitrous acid? Let H represent hypoxanthine and U represent uracil.

B. Let's suppose this DNA was treated with nitrous acid. The nitrous acid was then removed, and the DNA was replicated for two generations. What would be the sequences of the DNA products after the DNA had replicated two times? (Note: Hypoxanthine pairs with cytosine.) Your answer should contain the sequences of four double strands.

C26. In the treatment of cancer, the basis for many types of chemotherapy and radiation therapy is that mutagens are more effective at killing dividing cells than nondividing cells. Explain why. What are possible harmful side effects of chemotherapy and radiation therapy?

C27. An individual has a somatic mutation that changes a lysine codon to a glutamic acid codon. Prior to acquiring this mutation, the individual had been exposed to UV light, proflavin, and 5-bromouracil. Which of these three agents would be the most likely to have caused this somatic mutation? Explain your answer.

C28. Which of the following examples is likely to be caused by a somatic mutation?

A. A purple flower has a small patch of white tissue.

B. One child, in a family of seven, is an albino.

C. One apple tree, in a very large orchard, produces its apples 2 weeks earlier than any of the other trees.

D. A 60-year-old smoker develops lung cancer.

C29. How would nucleotide excision repair be affected if one of the following proteins was missing? Describe the condition of the DNA on which repair was attempted in the absence of the protein.

A. UvrA

B. UvrC

C. UvrD

D. DNA polymerase

C30. During mismatch repair, why is it necessary to distinguish between the template strand and the newly made daughter strand? How is this accomplished?

C31. What are the two main mechanisms by which cells repair double-strand breaks?

C32. What is the underlying genetic defect that causes xeroderma pigmentosum? How can the symptoms of this disease be explained by the genetic defect?

C33. In *E. coli*, a methyltransferase enzyme encoded by the *dam* gene recognizes the sequence 5′–GATC–3′ and attaches a methyl group to the nitrogen at position 6 of adenine. *E. coli* strains that have the *dam* gene deleted are known to have a higher spontaneous mutation rate than normal strains. Explain why.

C34. Discuss the similarities and differences between nucleotide excision repair and the mismatch repair system.

Application and Experimental Questions

E1. Explain how the technique of replica plating supports the random mutation theory but conflicts with the physiological adaptation hypothesis.

E2. Outline how you would use the technique of replica plating to show that antibiotic resistance is due to random mutations.

E3. From an experimental point of view, is it better to use haploid or diploid organisms for mutagen testing? Consider the Ames test when preparing your answer.

E4. How would you modify the Ames test to evaluate physical mutagens? Would it be necessary to add the rat liver extract? Explain why or why not.

E5. During an Ames test, bacteria were exposed to a potential mutagen. Also, as a control, another sample of bacteria was not exposed to the mutagen. In both cases, 10 million bacteria were plated and the following results were obtained:

No mutagen: 17 colonies

With mutagen: 2017 colonies

Calculate the mutation rate in the presence and absence of the mutagen. How much does the mutagen increase the rate of mutation?

E6. Richard Boyce and Paul Howard-Flanders conducted an experiment that provided biochemical evidence that thymine dimers are removed from DNA by a DNA repair system. In their studies, bacterial DNA was radiolabeled so the amount of radioactivity reflected the amount of thymine dimers. The DNA was then subjected to UV light, causing the formation of thymine dimers. When radioactivity was found in the soluble fraction, thymine dimers had been excised from the DNA by a DNA repair system. But when the radioactivity was in the insoluble fraction, the thymine dimers had been retained within the DNA. The following table illustrates some of the experimental results involving a normal strain of *E. coli* and a second strain that was very sensitive to killing by UV light:

Strain	Treatment	Radioactivity in the Insoluble Fraction (cpm*)	Radioactivity in the Soluble Fraction (cpm)
Normal	No UV	<100	<40
Normal	UV-treated, incubated 2 hours at 37°C	357	940
Mutant	No UV	<100	<40
Mutant	UV-treated, incubated 2 hours at 37°C	890	<40

Source: Adapted from Boyce, R. P., & Howard-Flanders, P. (1964) Release of Ultraviolet Light-Induced Thymine Dimers from DNA in E. coli K-12," *Proceedings of the National Academy of Sciences of the United States of America*, vol. 51, 293-300.
*The abbreviation cpm stands for "counts per minute," which is a measure of the number of radioactive emissions from the sample.

Explain the results found in this table. Why is the mutant strain sensitive to UV light?

Questions for Student Discussion/Collaboration

1. In *E. coli*, a variety of mutator strains have been identified in which the spontaneous rate of mutation is much higher than in normal strains. Make a list of the types of abnormalities that could cause a strain of bacteria to become a mutator strain. Which abnormalities do you think would give the highest rate of spontaneous mutation?

2. Discuss the times in a person's life when it is most important to avoid mutagens. Which parts of a person's body should be the most highly protected from mutagens?

3. A large amount of research is aimed at studying mutation. However, there is not an infinite amount of research funding. Where would you put your money for mutation research?

   A. Testing of potential mutagens

   B. Investigating molecular effects of mutagens

   C. Investigating DNA repair mechanisms

   D. Some other topic

Answers to Comprehension Questions

19.1: b, c, a, d

19.2: b

19.3: c, d, c

19.4: b, b, a

19.5: c, d, a, d, b

Note: All answers appear in Connect; the answers to even-numbered questions and all Concept Check questions are in Appendix B.

PART V GENETIC TECHNOLOGIES

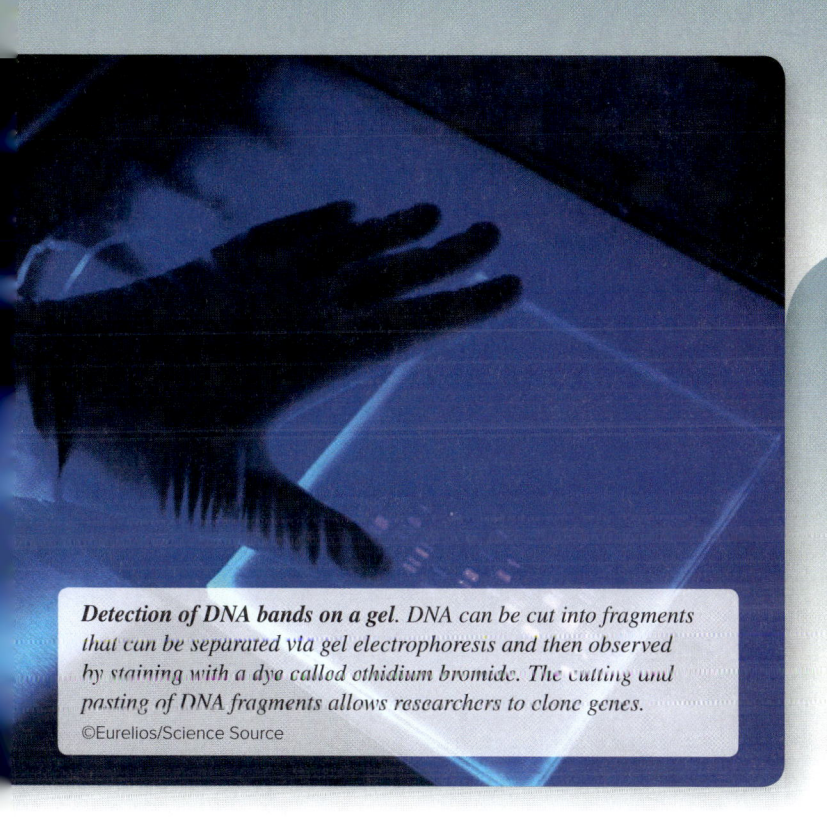

20

CHAPTER OUTLINE

- 20.1 Gene Cloning Using Vectors
- 20.2 Polymerase Chain Reaction
- 20.3 Reproductive Cloning and Stem Cells
- 20.4 DNA Sequencing
- 20.5 Gene Editing
- 20.6 Blotting Methods to Detect Gene Products
- 20.7 Analyzing DNA- and RNA-Binding Proteins

Detection of DNA bands on a gel. DNA can be cut into fragments that can be separated via gel electrophoresis and then observed by staining with a dye called ethidium bromide. The cutting and pasting of DNA fragments allows researchers to clone genes.
©Eurelios/Science Source

MOLECULAR TECHNOLOGIES

In this chapter, we will focus on methods that are used for manipulating and analyzing DNA and gene products at the molecular level. In some cases, these technologies involve the cutting and pasting of DNA segments to produce new arrangements. **Recombinant DNA technology** is the use of in vitro molecular techniques that manipulate fragments of DNA to produce new arrangements. In the early 1970s, the first successes in making recombinant DNA molecules were accomplished independently by two groups at Stanford University: David Jackson, Robert Symons, and Paul Berg; and Peter Lobban and A. Dale Kaiser. Both groups were able to isolate and purify pieces of DNA in a test tube and then covalently link DNA fragments from two different sources. In other words, they constructed molecules called **recombinant DNA molecules.** Shortly thereafter, researchers were able to introduce such recombinant DNA molecules into living cells. Once inside a host cell, the recombinant molecules are replicated to produce many identical copies of a gene—a process called gene cloning.

DNA technologies, such as gene cloning, have enabled geneticists to investigate relationships between gene sequences and phenotypic consequences, and thereby have been widely used to further increase our understanding of gene structure and function. Researchers in molecular genetics employ different methods to study gene structure and gene expression. Also, many practical applications of DNA technologies have been developed, including exciting advances such as gene therapy, screening for human diseases, recombinant vaccines, and the production of transgenic plants and animals in agriculture, in which a cloned gene from one species is transferred to some other species. Transgenic organisms have also been important in basic research.

In this chapter, we will focus primarily on the use of DNA technologies as a way to further our understanding of gene structure and function. We will look at the materials and molecular techniques used in gene cloning and explore polymerase chain reaction (PCR), which can make many copies of DNA within a defined region. We will also consider cloning at the level of whole organisms, an approach called reproductive cloning, and explore the properties and uses of stem cells. We then examine how scientists analyze and alter DNA sequences through the techniques of DNA sequencing (a method that enables researchers to determine the base sequence of a DNA strand)

and gene editing (a procedure that allows researchers to make mutations within a segment of DNA). Finally, we will discuss techniques for identifying specific gene products, as well as methods for detecting the binding of proteins to DNA or RNA sequences.

20.1 GENE CLONING USING VECTORS

Learning Outcomes:
1. Outline the procedure for cloning a gene into a vector.
2. Describe how cDNA is made.
3. Compare and contrast a genomic library with a cDNA library.

Molecular biologists want to understand how the molecules within living cells contribute to cell structure and function. Because proteins are the workhorses of cells and because they are the products of genes, many molecular biologists focus their attention on the structure and function of proteins or the genes that encode them. Researchers may focus their efforts on the study of just one or perhaps a few different genes or proteins. At the molecular level, this poses a daunting task. In eukaryotic species, any given cell can express thousands of different proteins, making the study of any single gene or protein akin to a "needle in a haystack" exploration. To overcome this truly formidable obstacle, researchers frequently take the approach of cloning the genes that encode their proteins of interest. The term **gene cloning** refers to the phenomenon of making many copies of a gene. The laboratory methods necessary to clone a gene were devised during the early 1970s. Since then, many technical advances have enabled gene cloning to become a widely used procedure among scientists, including geneticists, cell biologists, biochemists, plant biologists, microbiologists, evolutionary biologists, clinicians, and biotechnologists.

Table 20.1 summarizes some of the common uses of gene cloning. In modern molecular biology, the diversity of uses for gene cloning is remarkable. For this reason, gene cloning has provided the foundation for critical technical advances in a variety of disciplines, including molecular biology, genetics, cell biology, biochemistry, and medicine. In this section and the following section, we will examine the two general strategies used to make copies of a gene—the insertion of a gene into a vector that is then propagated in living cells, and cloning via polymerase chain reaction.

Cloning Experiments May Involve Two Kinds of DNA Molecules: Chromosomal DNA and Vector DNA

If a scientist wants to clone a particular gene, a common source of the gene is the chromosomal DNA of the species that carries the gene. For example, if the goal is to clone the rat β-globin gene, this gene is found within the chromosomal DNA of rat cells. In this case, the rat's chromosomal DNA is one type of DNA needed in a cloning experiment. To prepare chromosomal DNA, an experimenter first obtains cellular tissue from the organism of interest. The preparation of chromosomal DNA then involves the breaking open of cells and the extraction and purification of the DNA using biochemical techniques such as chromatography and centrifugation (see Appendix A for a description of these techniques).

Let's begin our discussion of gene cloning by considering a DNA technology in which a gene is removed from its native site within a chromosome and inserted into a smaller segment of DNA known as a **vector**—a small DNA molecule that replicates independently of the chromosomal DNA and produces

TABLE 20.1

Some Uses of Gene Cloning

Technique	Description
DNA sequencing	Cloned genes provide enough DNA to subject the gene to DNA sequencing (described in Section 20.4). The sequence of the gene can reveal the gene's promoter, regulatory sequences, and coding sequence. Gene sequencing is also important in the identification of alleles that cause cancer and inherited human diseases.
Gene editing	A gene can be manipulated to change its DNA sequence. Mutations within genes help to identify gene sequences such as promoters and regulatory elements. The study of a mutant gene can also help to elucidate its normal function and to discover how its expression may affect the roles of other genes. Mutations in the coding sequence may reveal which amino acids are important for a protein's structure and function.
DNA probes	Labeled DNA strands from a cloned gene can be used as probes for identifying RNA. This method, known as Northern blotting, is described in Section 20.6.
Expression of cloned genes	Cloned genes can be introduced into a different cell type or different species. The expression of cloned genes has many uses:

Research
1. The expression of a cloned gene can help to elucidate its cellular function.
2. The coding sequence of a gene can be placed next to an active promoter and then introduced into a culture of cells that will express a large amount of the protein. This greatly aids in the purification of large amounts of protein that may be needed for biochemical or biophysical studies.

Biotechnology
1. Cloned genes can be introduced into bacteria to make pharmaceutical products such as insulin.
2. Cloned genes can be introduced into plants and animals to make transgenic species with desirable traits.

Clinical trials
1. Cloned genes have been used in clinical trials involving gene therapy.

many identical copies of an inserted gene. The purpose of vector DNA is to act as a carrier of the DNA segment to be cloned. In cloning experiments, a vector may carry a small segment of chromosomal DNA, perhaps only a single gene. By comparison, a chromosome carries many more genes, perhaps a few hundred or thousand. Like a chromosome, a vector is replicated within a living cell; a cell that harbors a vector is called a **host cell.** When a vector is replicated within a host cell, the DNA that it carries is also replicated. The vectors commonly used in gene-cloning experiments were derived originally from two natural sources: plasmids or viruses.

Plasmids Most vectors are **plasmids,** which are small, circular pieces of DNA. As discussed in Chapter 9, plasmids are found naturally in many strains of bacteria and occasionally in eukaryotic cells.

- Plasmids contain a DNA sequence, known as an **origin of replication,** that is recognized by the replication enzymes of the host cell, which allows the plasmid to be replicated.
- Commercially available plasmids have been genetically engineered for effective use in cloning experiments. They contain unique sites where geneticists insert pieces of DNA.
- Another useful feature of plasmids used as vectors is they often contain resistance genes that provide host cells with the ability to grow in the presence of a toxic substance. Such a gene is called a **selectable marker,** because the expression of the gene selects for the growth of the host cells. For example, the gene amp^R encodes an enzyme known as β-lactamase. This enzyme degrades ampicillin, an antibiotic that normally kills bacteria. However, bacteria carrying the amp^R gene can grow and form visible colonies on media containing ampicillin, because they can degrade it.

Viral Vectors An alternative type of vector used in cloning experiments is a viral vector. As discussed in Chapter 10, viruses infect living cells and propagate themselves by taking control of the host cell's metabolic machinery. When a chromosomal gene is inserted into a viral genome, the gene is replicated whenever the viral DNA is replicated. Therefore, viruses can be used as vectors to carry other pieces of DNA.

Molecular biologists may choose from hundreds of different vectors to use in their cloning experiments. **Table 20.2** provides a general description of several types of vectors that are used to clone small segments of DNA. In addition, other types of vectors, such as cosmids, bacterial artificial chromosomes (BACs), and yeast artificial chromosomes (YACs), are used to clone large pieces of DNA.

Restriction Enzymes Cut DNA into Pieces and DNA Ligase Joins the Pieces Together

A key step in a cloning experiment is the insertion of chromosomal DNA into a plasmid or viral vector. This requires the cutting and pasting of DNA fragments. To cut DNA, researchers

TABLE 20.2

Some Vectors Used in Cloning Experiments

Example	Type	Description
pBluescript	Plasmid	A type of vector like the one shown in Figure 20.2. It is used to clone small segments of DNA and propagate them in *E. coli*.
YEp24	Plasmid	This plasmid is an example of a shuttle vector, which can replicate in two different host species, *E. coli* and *Saccharomyces cerevisiae*. It carries origins of replication for both species.
λgt11	Viral	This vector is derived from bacteriophage λ, which is described in Chapter 10. λgt11 also contains a promoter from the *lac* operon. When fragments of DNA are cloned next to this promoter, the DNA is expressed in *E. coli*. This is an example of an expression vector. An expression vector is designed to clone the coding sequences of genes so that they are transcribed and translated correctly.
SV40	Viral	This virus naturally infects mammalian cells. Genetically altered derivatives of the SV40 viral DNA are used as vectors for the cloning and expression of genes in mammalian cells that are grown in the laboratory.
Baculovirus	Viral	This virus naturally infects insect cells. In a laboratory, insect cells can be grown in liquid media. Unlike many other types of eukaryotic cells, insect cells often express large amounts of proteins that are encoded by cloned genes. When researchers want to make a large amount of a protein, they can clone the gene that encodes the protein into baculovirus and then purify the protein from insect cells.

use enzymes known as **restriction endonucleases,** or **restriction enzymes.** The restriction enzymes used in cloning experiments bind to a specific base sequence and then cleave the DNA backbone at two defined locations, one in each strand. Proposed by Werner Arber in the 1960s and discovered by Hamilton Smith and Daniel Nathans in the 1970s, restriction enzymes are made naturally by many different species of bacteria. They protect bacterial cells from invasion by foreign DNA, particularly that of bacteriophages.

Figure 20.1 shows the role of a restriction enzyme, called *Eco*RI, in producing a recombinant DNA molecule.

1. Certain types of restriction enzymes are useful in cloning because they cut the DNA into fragments with "sticky ends." As shown in Figure 20.1, the sticky ends are single-stranded regions of DNA that can hydrogen bond to a complementary sequence of DNA from a different source.
2. The ends of two different DNA pieces hydrogen bond to each other because of their complementary sticky ends.

462   CHAPTER 20 :: MOLECULAR TECHNOLOGIES

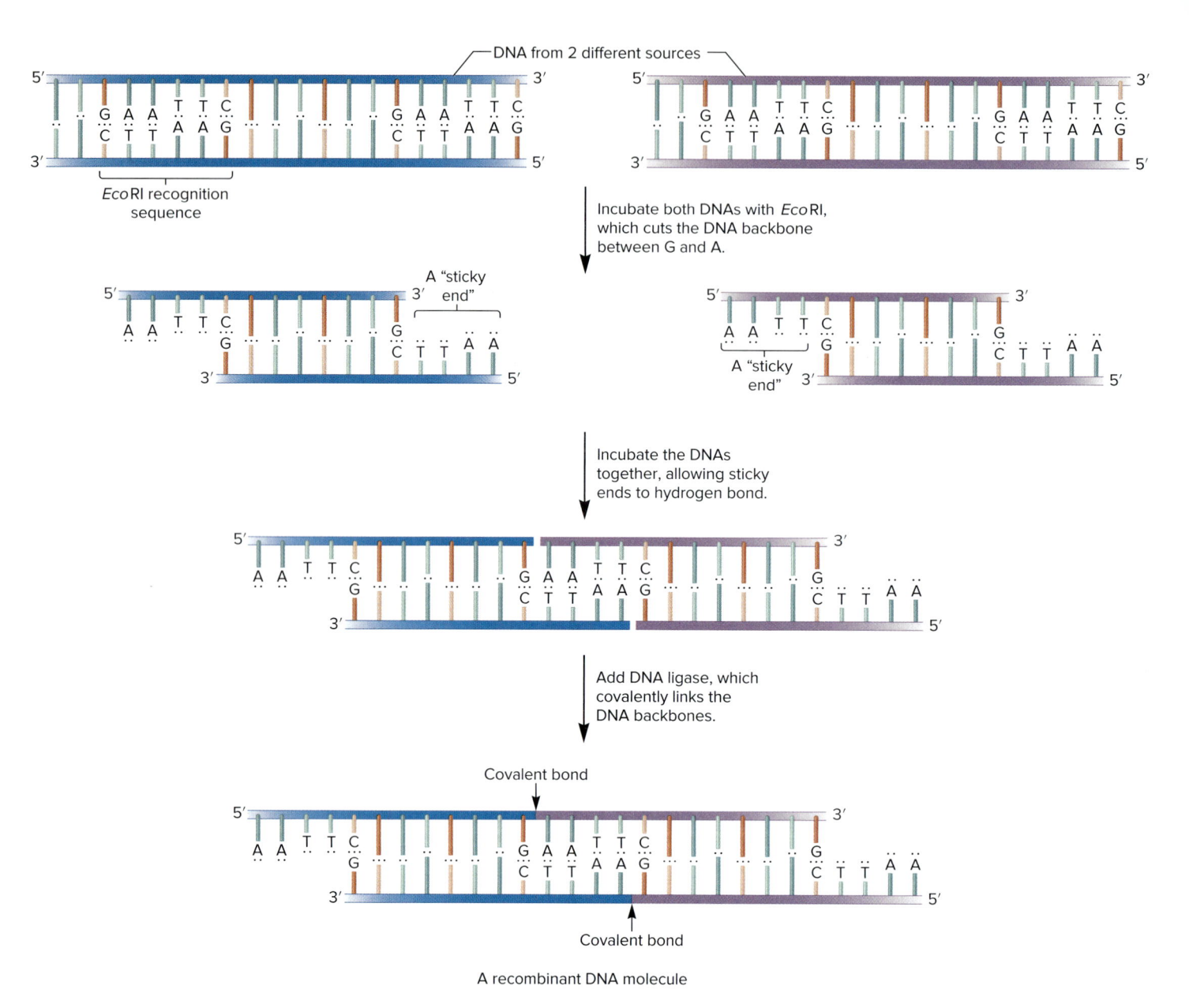

FIGURE 20.1 **The action of a restriction enzyme and the production of recombinant DNA.** The restriction enzyme *Eco*RI binds to a specific sequence, in this case 5′–GAATTC–3′. It then cleaves the DNA backbone between G and A, producing DNA fragments. The single-stranded ends of different DNA fragments can hydrogen bond with each other, because they have complementary sequences. DNA ligase then catalyzes the formation of covalent bonds in the DNA backbones of the fragments.

Concept Check: Prior to the action of DNA ligase, how many hydrogen bonds are holding these two DNA fragments together?

3. The hydrogen bonding between the sticky ends of DNA fragments promotes a temporary interaction between the two fragments. However, this interaction is not stable because it involves only a few hydrogen bonds between complementary bases. How can this interaction be made more permanent? The answer is that the sugar-phosphate backbones within the DNA strands must be covalently linked together. Experimentally, this linkage is catalyzed by the addition of **DNA ligase.**

Currently, several hundred different restriction enzymes from various bacterial species have been identified and are available commercially to molecular biologists. **Table 20.3** gives a few examples. Restriction enzymes usually recognize sequences that are **palindromic;** that is, the sequence in one strand is the same as that in the complementary strand when read in the opposite direction. For example, the sequence recognized by *Eco*RI is 5′–GAATTC–3′ in the top strand. Read in the opposite direction in the bottom strand, this sequence is also 5′–GAATTC–3′.

TABLE 20.3
Some Restriction Enzymes Used in Gene Cloning

Restriction Enzyme*	Bacterial Source	Sequence Recognized†
BamHI	Bacillus amyloliquefaciens H	5′–GGATCC–3′ 3′–CCTAGG–5′
Sau3A I	Staphylococcus aureus 3A	5′–GATC–3′ 3′–CTAG–5′
EcoRI	E. coli RY13	5′–GAATTC–3′ 3′–CTTAAG–5′
NaeI	Nocardia aerocolonigenes	5′–GCCGGC–3′ 3′–CGGCCG–5′
PstI	Providencia stuartii	5′–CTGCAG–3′ 3′–GACGTC–5′

*Restriction enzymes are named according to the species in which they are found. The first three letters are italicized because they indicate the genus and species names. Because a species may produce more than one restriction enzyme, the enzymes are designated I, II, III, and so on, to indicate the order in which they were discovered in a given species. Some restriction enzymes, like EcoRI, produce a sticky end with a 5′ overhang (see Figure 20.1), whereas others, such as PstI, produce a sticky end with a 3′ overhang. However, not all restriction enzymes cut DNA to produce sticky ends. For example, the enzyme NaeI cuts DNA to produce blunt ends.

†The arrows show the locations in the upper and lower DNA strands where the restriction enzymes cleave the DNA backbone.

Gene Cloning Involves the Insertion of DNA Fragments into Vectors, Which Are Then Propagated Within Host Cells

Now that you are familiar with the materials, let's outline the general strategy that is followed in a typical cloning experiment. In the procedure shown in **Figure 20.2**, the goal is to clone a gene of interest into a plasmid vector that already carries the amp^R gene.

1. To begin this experiment, the chromosomal DNA is isolated and digested with a restriction enzyme. This enzyme cuts the chromosomes into many small fragments. The plasmid DNA is also cut with the same restriction enzyme. However, the plasmid has only one unique site for the restriction enzyme.

2. After cutting, the plasmid has two ends that are complementary to the sticky ends of the chromosomal DNA fragments. The digested chromosomal DNA and plasmid DNA are mixed together and incubated under conditions that promote the binding of these complementary sticky ends. DNA ligase is then added to catalyze the covalent linkage between adjacent DNA backbones. In some cases, the two ends of the vector simply ligate back together, restoring the vector to its original structure. This is called a recircularized vector. In other cases, a fragment of chromosomal DNA may become ligated to both ends of the vector. In this way, a segment of chromosomal DNA has been inserted into the vector. The vector containing a piece of chromosomal DNA is a **recombinant vector.**

3. Following ligation, the DNA is introduced into living cells treated with agents that render them permeable to DNA molecules. Cells that can take up DNA from the extracellular medium are called **competent cells.** This step in the procedure is commonly called **transformation.** Only a very small percentage of bacterial cells actually take up a plasmid.

4. How can an experimenter distinguish between bacterial cells that have taken up a plasmid and those that have not? In the experiment shown in Figure 20.2, a plasmid is introduced into bacterial cells that were originally sensitive to ampicillin. The bacteria are then streaked onto plates containing bacterial growth media and ampicillin. A bacterium that has taken up a plasmid carrying the amp^R gene continues to divide and forms a bacterial colony containing tens of millions of cells. Because each cell within a single colony is derived from the same original cell, all cells within a colony contain the same type of plasmid DNA.

In the experiment shown in Figure 20.2, how can the experimenter distinguish between colonies of bacteria that have a recircularized vector and those that have a recombinant vector carrying a piece of chromosomal DNA? As shown here, the chromosomal DNA has been inserted into a region of the vector that contains the *lacZ* gene, which encodes the enzyme β-galactosidase (see Chapter 16). The insertion of chromosomal DNA into the vector disrupts the *lacZ* gene so it no longer produces a functional enzyme. By comparison, a recircularized vector has a functional *lacZ* gene. The functionality of *lacZ* is determined by providing the growth medium with a colorless compound, X-Gal (5-bromo-4-chloro-3-indolyl-β-D-galactoside), which is converted by β-galactosidase into a blue dye. Bacteria grown in the presence of X-Gal and IPTG (an inducer of the *lacZ* gene) form blue colonies if they have a functional *lacZ* gene and white colonies if they do not. In this experiment, therefore, bacterial colonies containing recircularized vectors form blue colonies, whereas colonies containing recombinant vectors are white.

In the example of Figure 20.2, one of the white colonies contains cells with a recombinant vector that carries a human gene of interest; the segment containing the human gene is shown in red. The goal of gene cloning is to produce an enormous number of copies of a recombinant vector that carry the gene of interest. During transformation, a single bacterial cell usually takes up a single copy of a recombinant vector. However, two subsequent events lead to the amplification of the cloned gene. First, because the vector has an origin of replication, the bacterial host cell replicates the recombinant vector to produce many identical copies per cell. Second, the bacterial cells divide approximately every 20 minutes.

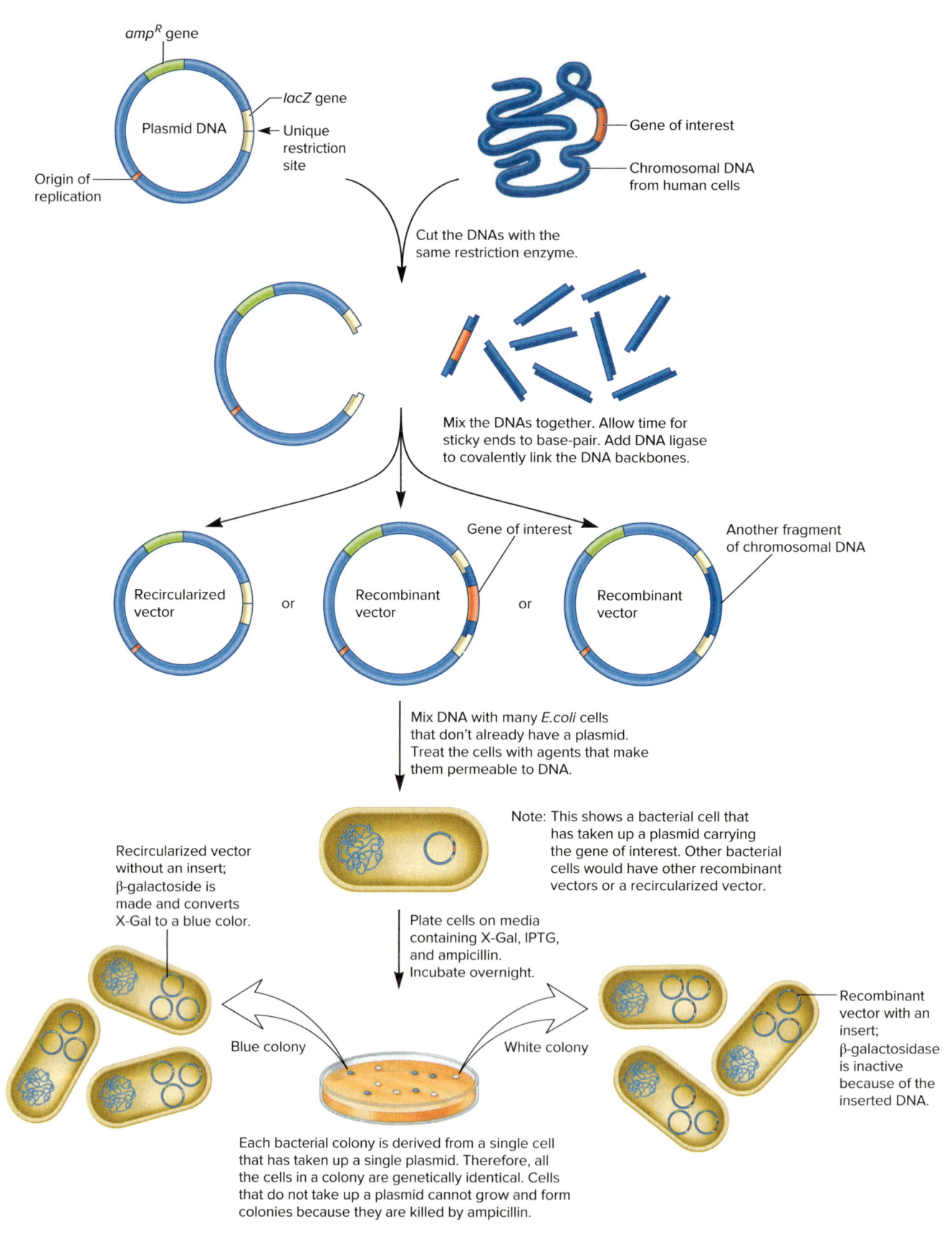

FIGURE 20.2 **The steps in gene cloning.** Note: X-Gal refers to the colorless compound 5-bromo-4-chloro-3-indolyl-β-D-galactoside. It is converted by β-galactosidase into a blue dye. IPTG is an abbreviation for isopropyl-β-D-thiogalactopyranoside, which is a nonmetabolizable lactose analog that induces the *lac* promoter.

Concept Check: Explain the role of the selectable marker gene in this experiment.

After overnight growth, a bacterial colony may be composed of 10 million cells, with each cell containing 50 copies of the recombinant vector. Therefore, this bacterial colony would contain 500 million copies of the cloned gene!

cDNA Can Be Made from mRNA via Reverse Transcriptase

In the example of gene cloning in Figure 20.2, chromosomal DNA and plasmid DNA were used as the material to clone genes. Alternatively, a sample of RNA can provide a starting point for cloning DNA. As described in Chapter 10, the enzyme **reverse transcriptase** uses RNA as a template to make a complementary strand of DNA. This enzyme is encoded in the genome of retroviruses and provides a way for retroviruses to copy their RNA genome into DNA molecules that then integrate into the host cell's chromosomes. Likewise, reverse transcriptase is encoded in some retrotransposons and is needed in the retrotransposition of such elements.

Researchers can use purified reverse transcriptase in a strategy for cloning genes, using mRNA as the starting material (**Figure 20.3**).

1. To begin this experiment, mRNAs, which naturally contain a polyA tail at their 3′ ends, are purified from a sample of cells. The mRNAs are mixed with primers composed of a string of thymine-containing nucleotides. This short strand of DNA, or **oligonucleotide**, is called a poly-dT primer. The poly-dT primers are complementary to the 3′ end of mRNAs.
2. Reverse transcriptase and deoxyribonucleotides (dNTPs) are then added to make a DNA strand that is complementary to the mRNA.
3. One way to make the other DNA strand is to use RNaseH, which partially digests the RNA, generating short RNAs that are used as primers.
4. DNA polymerase uses these primers to make segments of DNA strands that are complementary to the strand made by reverse transcriptase. During this process, it removes the RNA primers, except the one at the 5′ end. Finally, DNA ligase connects the DNA segments made by DNA polymerase I, producing a continuous DNA strand, as shown at the bottom of Figure 20.3.

When DNA is made using an RNA template, the DNA is called **complementary DNA (cDNA)**. The term originally referred to the single strand of DNA that is complementary to the RNA template. However, cDNA now refers to any DNA, whether it is single- or double-stranded, that is made using RNA as the starting material.

Why is cDNA cloning useful? From a research perspective, an important advantage of cDNA is that it lacks introns, which are often found in eukaryotic genes. Because introns can be quite large, it is simpler to insert cDNAs into vectors if researchers want to focus their attention on the coding sequence of a gene. For example, if the primary goal is to determine the coding sequence of a protein-encoding gene, a researcher inserts cDNA into a vector and then determines the DNA sequence of the insert, as described later in this chapter. Similarly, if a scientist wants to express an encoded protein of interest in a cell that does not splice out the introns properly (e.g., in a bacterial cell), it is necessary to make cDNA clones of the gene that codes for that protein.

A DNA Library May Be Constructed Using Genomic DNA or cDNA

In a typical cloning experiment that involves the use of vectors (see Figure 20.2), the treatment of the chromosomal DNA with restriction enzymes yields tens of thousands of different DNA fragments. Therefore, after the DNA fragments are

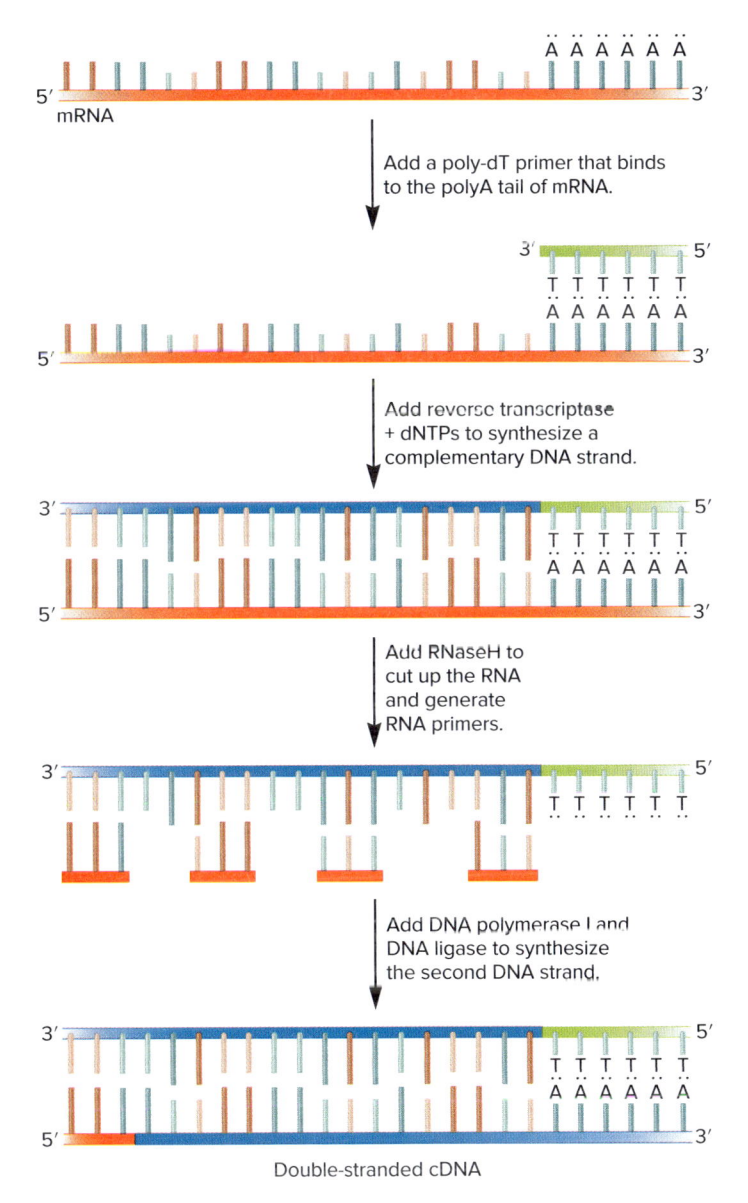

FIGURE 20.3 **Synthesis of cDNA.** A poly-dT primer anneals to the 3′ end of eukaryotic mRNAs. Reverse transcriptase then catalyzes the synthesis of a complementary DNA strand (cDNA). RNaseH digests the mRNA into short pieces that are used as primers by DNA polymerase I to synthesize the second DNA strand. The 5′ to 3′ exonuclease function of DNA polymerase I removes all of the RNA primers except the one at the 5′ end (because there is no primer upstream from this site). This RNA primer can be removed by the subsequent addition of an RNase. After the double-stranded cDNA is made, it can then be inserted into a vector, as shown in Figure 20.4b.

Concept Check: *Explain the meaning of the name reverse transcriptase.*

ligated individually to vectors, the researcher has a collection of recombinant vectors, with each vector containing a particular fragment of chromosomal DNA. A collection of recombinant vectors is known as a **DNA library.** When the starting material is chromosomal DNA, the library is called a **genomic library** (Figure 20.4a).

Researchers may also make a **cDNA library** that contains recombinant vectors with cDNA inserts. Figure 20.4b illustrates how cDNAs are made and inserted into vectors. As shown in Figure 20.3, cDNA is first made via reverse transcriptase. To insert the cDNAs into vectors, short oligonucleotides called linkers are attached to the cDNAs via DNA ligase. The linkers contain DNA sequences with a unique site for a restriction enzyme. After the linkers are attached to the cDNAs, the cDNAs and the vectors are cut with restriction enzymes and then ligated to each other and transformed into bacteria to produce a cDNA library.

20.1 REVIEWING THE KEY CONCEPTS

- Gene cloning has many uses, including gene sequencing, gene editing, the use of genes as probes, and the expression of cloned genes (see Table 20.1).
- Cloning vectors are derived from plasmids or viruses (see Table 20.2).
- Restriction enzymes, also called restriction endonucleases, cut chromosomal DNA and vector DNA to produce sticky ends that hydrogen bond with each other. DNA ligase then makes a covalent link between the DNA backbones (see Figure 20.1, Table 20.3).
- Gene cloning using vectors involves the insertion of the gene into a vector and then its propagation in a living cell such as *E. coli* (see Figure 20.2).
- Complementary DNA (cDNA) is made via reverse transcriptase starting with mRNA (see Figure 20.3).
- A DNA library is a collection of recombinant vectors. The inserts can be chromosomal DNA or cDNA (see Figure 20.4).

20.1 COMPREHENSION QUESTIONS

1. Which of the following may be used as a vector in a gene cloning experiment?
   a. mRNA
   b. Plasmid
   c. Virus
   d. Either b or c

2. The restriction enzymes used in gene-cloning experiments _____, which generates sticky ends that can _____.
   a. cut the DNA, enter bacterial cells
   b. cut the DNA, hydrogen bond with complementary sticky ends
   c. methylates DNA, enter bacterial cells
   d. methylates DNA, hydrogen bond with complementary sticky ends

3. What is the proper order of the following steps in a gene-cloning experiment involving vectors?
   1. Add DNA ligase.
   2. Incubate the chromosomal DNA and the vector DNA with a restriction enzyme.
   3. Introduce the DNA into living cells.
   4. Mix the chromosomal DNA and vector DNA together.
   a. 1, 2, 3, 4
   b. 2, 3, 1, 4
   c. 2, 4, 1, 3
   d. 1, 2, 4, 3

4. The function of reverse transcriptase is to
   a. copy RNA into DNA.
   b. copy DNA into RNA.
   c. translate RNA into protein.
   d. translate DNA into protein.

5. A collection of recombinant vectors that carry fragments of chromosomal DNA is called
   a. a genomic library.
   b. a cDNA library.
   c. a Northern blot.
   d. either a or b.

20.2 POLYMERASE CHAIN REACTION

Learning Outcomes:
1. Describe the three steps of a PCR cycle.
2. Explain how reverse-transcriptase PCR is carried out.
3. Outline the method of real-time PCR, and discuss why it is used.

In the method of gene cloning discussed in Section 20.1, the DNA of interest is inserted into a vector, which then is introduced into a host cell. The replication of the vector within the host cell, followed by the proliferation of the host cells, leads to the production of many copies of the DNA. Another way to copy DNA, without the aid of vectors and host cells, is a technique called **polymerase chain reaction (PCR),** which was developed by Kary Mullis in 1985. In this section, we begin with a general description of PCR and then examine how it can be used to quantitate the amount of DNA or RNA in a biological sample.

Each Cycle of PCR Involves Three Steps: Denaturation, Primer Annealing, and Primer Extension

The PCR method is used to make large amounts of DNA that is located in a defined region flanked by two **primers.** The primers are oligonucleotides, which are short segments of DNA, usually about 15 to 20 nucleotides in length. As shown in Figure 20.5a, the starting material of a PCR experiment can be a complex mixture of DNA. The two primers bind to specific sites in the DNA because their bases are complementary at these sites. The end result of PCR is that the region that is flanked by the primers, which contains the gene of interest, is amplified. The term *amplification* means that many copies of the region have been made. In other words, the region between the two primers has been cloned.

The primers are key reagents in a PCR experiment. They are made chemically, typically not in research or clinical laboratories, but instead at university or industrial facilities. Researchers or clinicians

20.2 POLYMERASE CHAIN REACTION   467

FIGURE 20.4 **The construction of a DNA library. (a)** To make a genomic library, chromosomal DNA is first digested to produce many fragments. The fragment containing the gene of interest is highlighted in red. Following ligation, each vector carries a different piece of chromosomal DNA. The library shown here is composed of four colonies. An actual genomic library would contain thousands of different bacterial colonies, each one carrying a different piece of chromosomal DNA. **(b)** To make a cDNA library, oligonucleotide linkers that contain a restriction site are attached to the cDNAs, so they can be inserted into vectors.

Concept Check: What is an advantage of making a cDNA library rather than a genomic library?

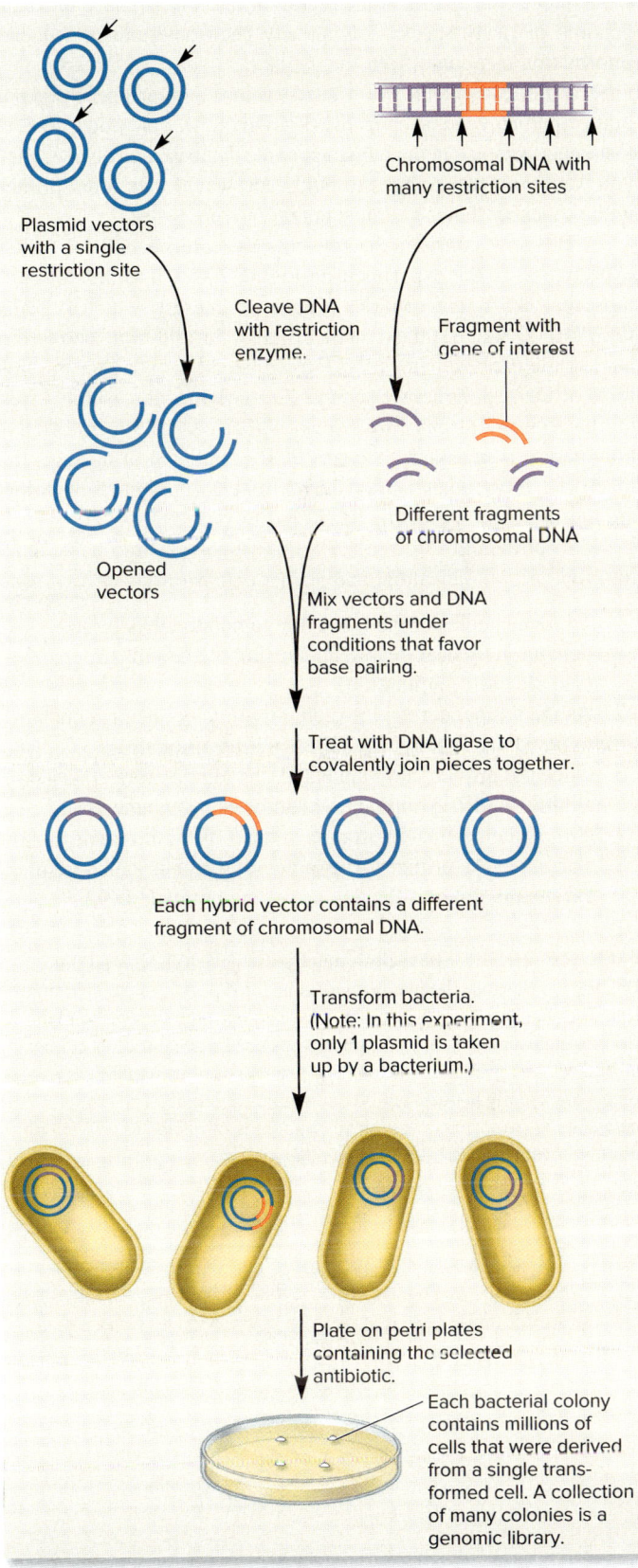

(a) Making a genomic library

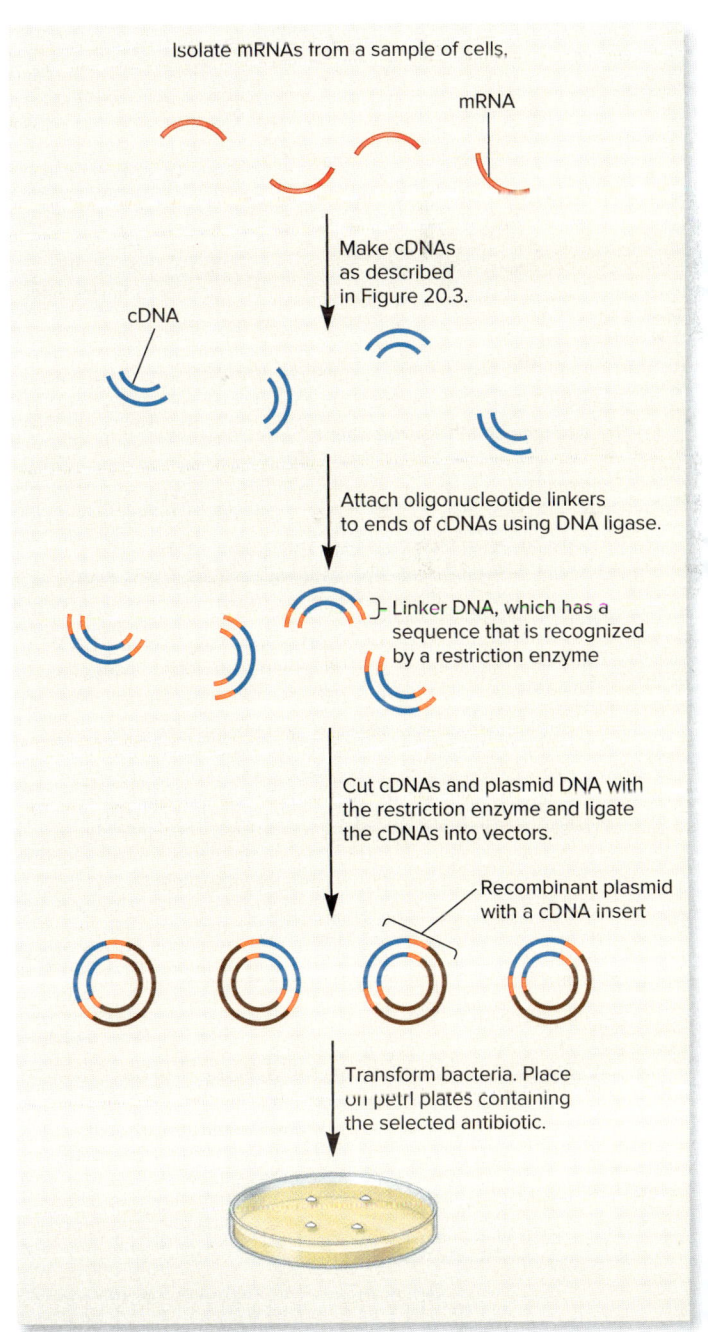

(b) Making a cDNA library

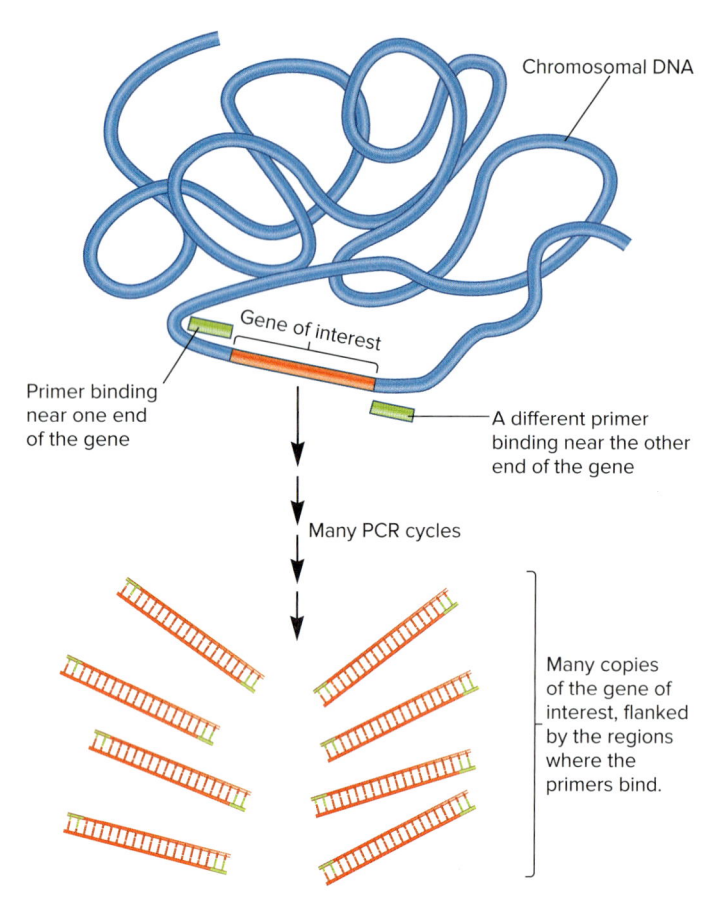

468   CHAPTER 20 :: MOLECULAR TECHNOLOGIES

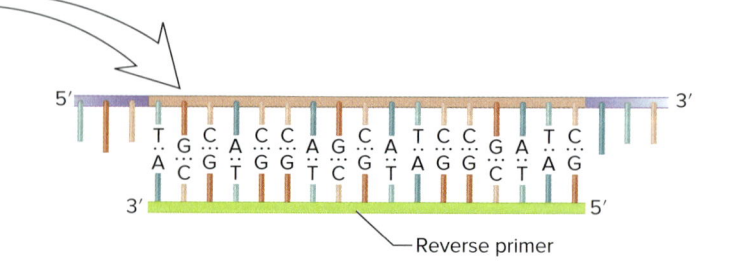

simply order the primers they need with a specified sequence. How does someone choose the primer sequences? In most cases, PCR is conducted on DNA samples in which a scientist already knows the DNA sequences that flank the region of interest. Because the entire genome of many species has already been determined, sequence information can come from the genome database of the species of interest. For example, if you wanted to clone a human gene or a portion of human gene, such as the β-globin gene, you would look up that gene sequence in the human genome database and decide where you want your two primers to bind. You would then order primers with sequences that are complementary to those sites.

Let's now consider the reagents that are needed for a PCR experiment. In the experiment in Figure 20.5a, a sample of chromosomal DNA that contains a gene of interest, which is called **template DNA,** is mixed with primers. In addition, deoxyribonucleoside triphosphates (dNTPs) are added, as is a thermostable form of DNA polymerase such as ***Taq* polymerase,** isolated from the bacterium *Thermus aquaticus*. A thermostable form of DNA polymerase is necessary because PCR includes heating steps that inactivate most other natural forms of DNA polymerase (which are thermolabile, or readily denatured by heat).

As outlined in **Figure 20.5b**, PCR involves three steps: denaturation, primer annealing, and primer extension.

1. To make copies of the DNA, the template DNA is first denatured by heat treatment, causing the strands to separate (see the online animation).
2. As the temperature is lowered, oligonucleotide primers bind to the DNA in a process called **primer annealing.**
3. Once the primers have annealed, the temperature is raised slightly, and *Taq* polymerase catalyzes the synthesis of complementary DNA strands in the 5′ to 3′ direction, starting at the primers. This process, which is called **primer extension,** doubles the amount of the template DNA.

FIGURE 20.5 The technique of polymerase chain reaction (PCR). (a) Because the two primers anneal to specific sites, PCR amplifies a defined region of DNA that is located between the two primers. (b) During each cycle of PCR, three steps occur: denaturation, primer annealing, and primer extension. The primers used in actual PCR experiments are usually about 15 to 20 nucleotides in length. The region between the two primers is typically hundreds of nucleotides in length, not just several nucleotides as shown here.

(a) The outcome of a PCR experiment

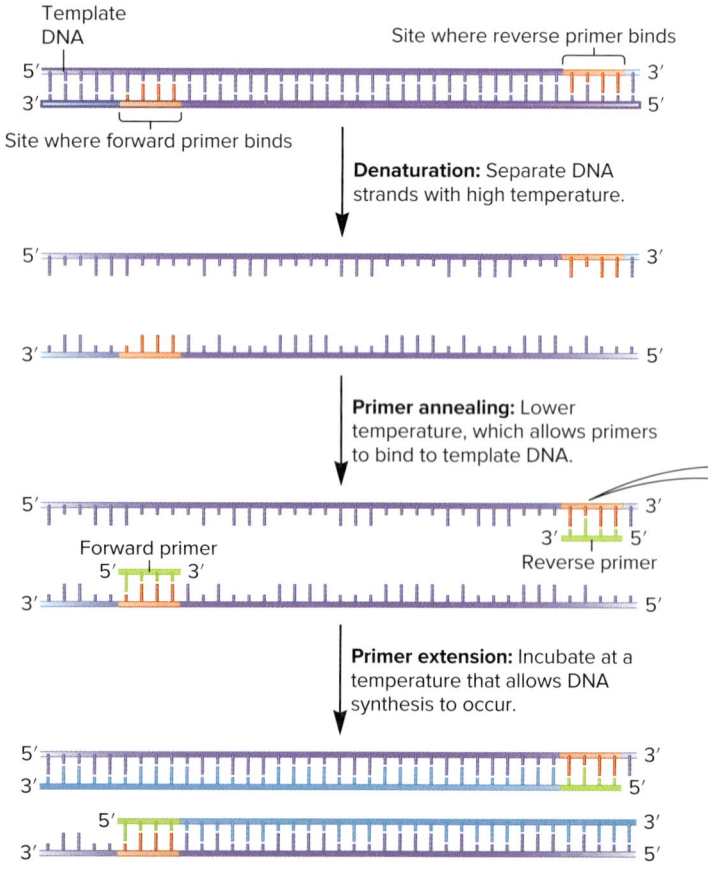

(b) The 3 steps of a PCR cycle

20.2 POLYMERASE CHAIN REACTION 469

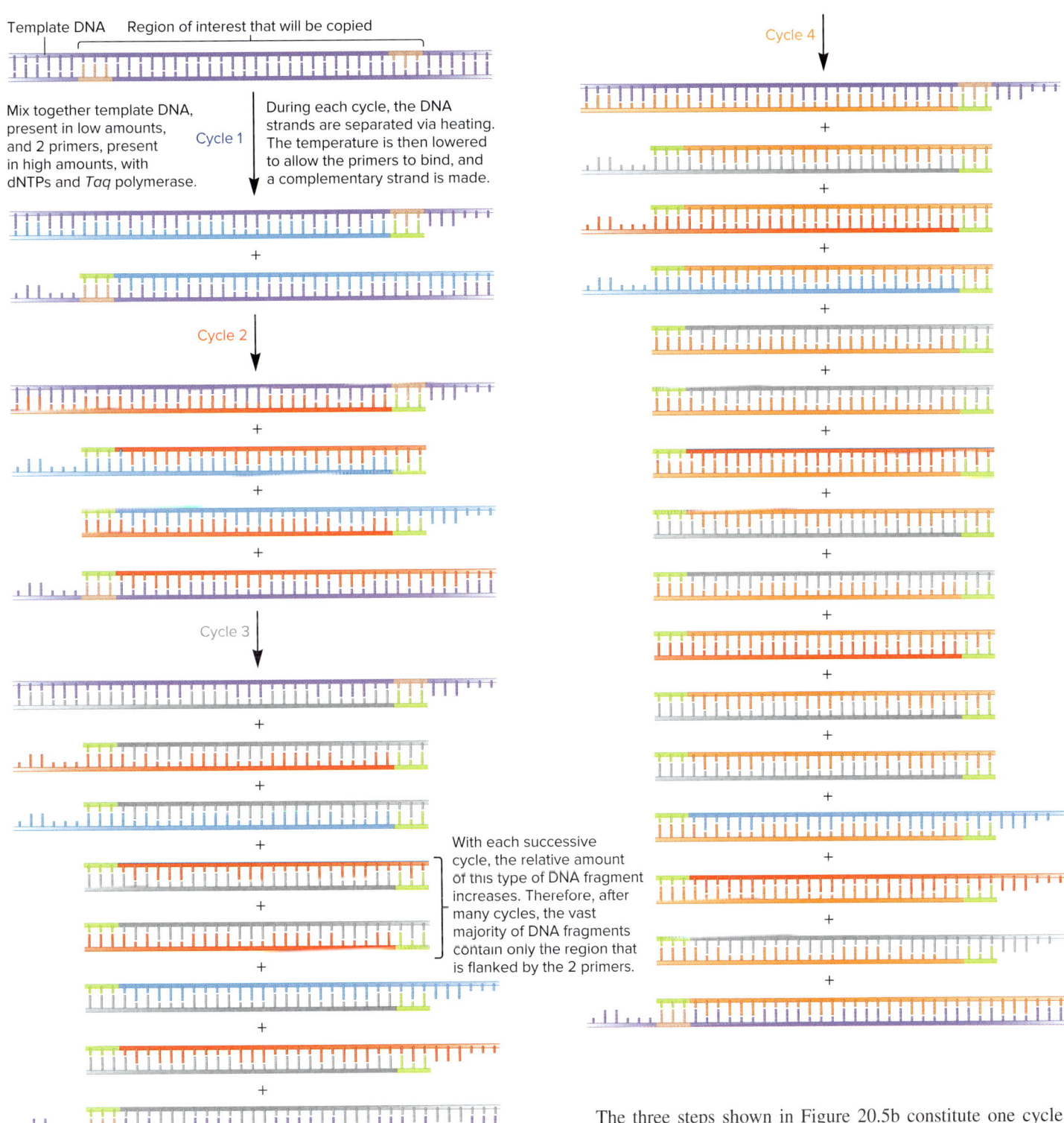

FIGURE 20.6 **The technique of PCR carried out for four cycles.** During each cycle, oligonucleotides (green) that are complementary to the ends of the targeted DNA sequence bind to the DNA and act as primers for the synthesis of this DNA region. Note: The DNA template strands are purple, and those strands made during the first, second, third, and fourth cycles are blue, red, gray, and orange, respectively.

Concept Check: After four cycles of PCR, which products predominate? Explain why.

The three steps shown in Figure 20.5b constitute one cycle of a PCR reaction. The three-step cycle of denaturation–primer annealing–primer extension is repeated for many cycles to double the amount of template DNA many times in a row. This process is called a chain reaction because the products of each previous reaction (the newly made DNA strands) are used as reactants (as the template strands) in subsequent reactions.

Figure 20.6 follows a PCR experiment through four cycles. Because the region of interest in the template DNA is doubled with each cycle, the end result of four cycles is 2^4, or 16, copies of that region. Since the starting material contains long strands of chromosomal DNA, some of the products of a PCR experiment contain the region of interest plus some additional DNA at either end. However, after many cycles, the products that contain only the region of interest greatly predominate in the mixture.

PCR is carried out in a machine, known as a **thermocycler,** that automates the timing of temperature changes in each cycle. The experimenter mixes the DNA sample, dNTPs, *Taq* polymerase, and an excess amount of primers together in a single tube. The tube is placed in a thermocycler, and the experimenter sets the machine to operate within a defined temperature range and number of cycles. During each cycle, the thermocycler increases the temperature to denature the DNA strands and then lowers the temperature to allow annealing and extension to take place. Typically, each cycle lasts 2 to 3 minutes and is then repeated. A typical PCR run is likely to involve 20 to 30 cycles of replication and takes a couple of hours to complete. The PCR technique can amplify the amount of DNA by a staggering amount. Assuming 100% efficiency, the intervening region between the two primers increases 2^{20}-fold after 20 cycles. This is approximately a million-fold increase in the original amount of DNA!

Reverse Transcriptase PCR Is Used to Amplify RNA

PCR is also used to detect and quantitate the amount of specific RNAs in living cells. To accomplish this goal, RNA is isolated from a sample and mixed with reverse transcriptase, a primer that binds near the 3′ end of the RNA of interest, and deoxynucleotides (**Figure 20.7**). The product is a single-stranded cDNA, which then can be used as template DNA in a conventional PCR reaction. The end result is that the RNA has been amplified to produce many copies of DNA.

This method, called **reverse transcriptase PCR,** is extraordinarily sensitive. Reverse transcriptase PCR can detect the expression of small amounts of RNA from a single cell! As discussed next, certain modifications to PCR allow researchers to observe the accumulation of PCR products and to quantitate the amount of DNA or RNA in a biological sample.

Real-Time PCR Is Used to Quantitate the Amount of a Specific Gene or mRNA in a Sample

In some applications of PCR, the goal is to obtain large amounts of a DNA region. To determine if PCR is successful, a researcher typically runs a sample of DNA on a gel, stains the DNA with ethidium bromide (EtBr), and then observes the gel under UV light, which causes EtBr to fluoresce. If a band of the correct size is seen, the experiment is likely to have been successful. This PCR approach is sometimes called endpoint analysis because the success of the experiment is judged after PCR is completed.

By comparison, the method of **real-time PCR** allows a researcher to follow the amount of a specific PCR product in real time, as PCR is taking place in a thermocycler. Because the PCR products are ultimately derived from the template DNA that was initially added to the reaction, this approach allows researchers to determine how much DNA, such as the DNA that encodes a specific gene, was originally in the sample before PCR was conducted.

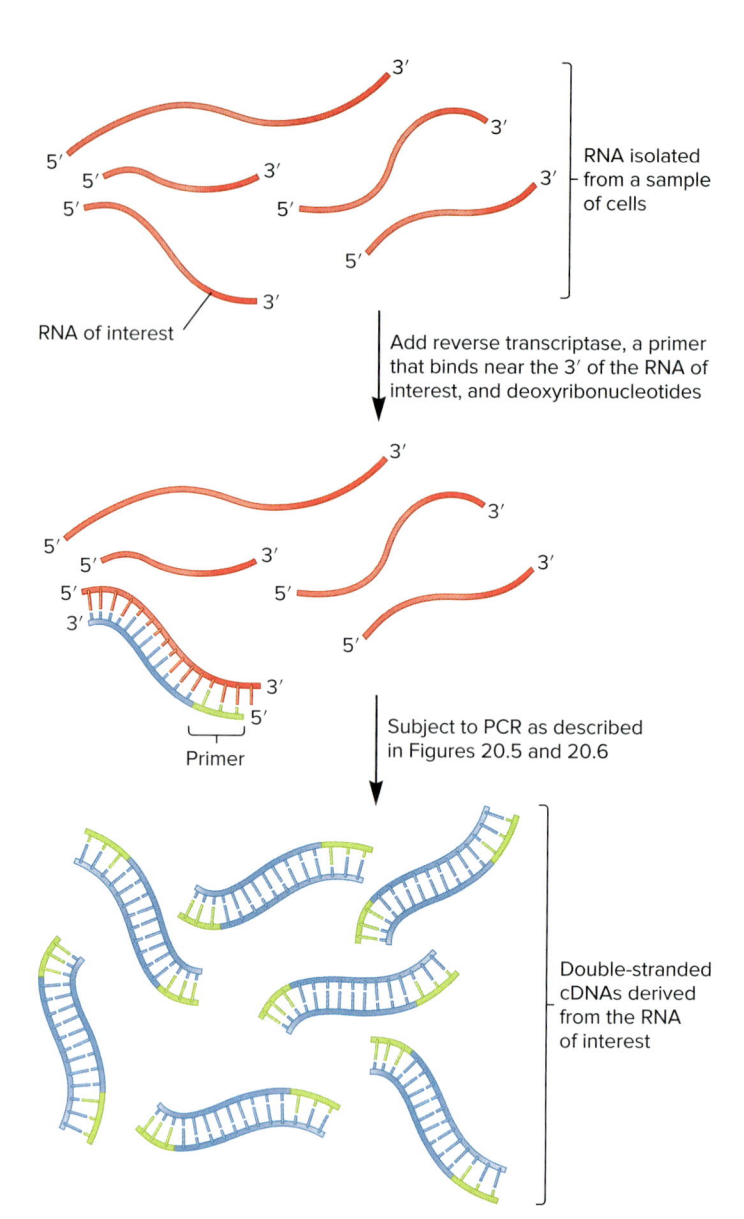

FIGURE 20.7 The technique of reverse transcriptase PCR.

Similarly, if the starting material is mRNA that is reverse transcribed into DNA, real-time PCR can be used to determine how much mRNA from a specific gene was in the sample. This provides a way to quantitatively measure gene expression.

How do researchers determine the amount of a PCR product during real-time PCR? The procedure is carried out in a thermocycler that has the capacity to measure changes in the level of fluorescence that is emitted from probes that are added to the PCR mixture. The fluorescence given off by the probes depends on the amount of the PCR product. Several probes have been developed. We will consider one type called TaqMan.

The TaqMan probe is an oligonucleotide that has a reporter molecule at one end and a quencher molecule at the other (**Figure 20.8a**). The oligonucleotide is complementary to a site within the PCR product of interest. The reporter molecule emits fluorescence at a certain wavelength, but that fluorescence is largely absorbed by the nearby quencher. Therefore, the close

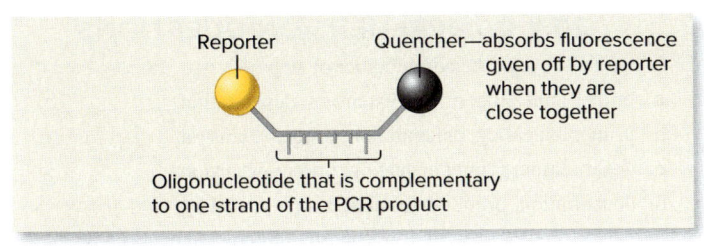

(a) TaqMan probe

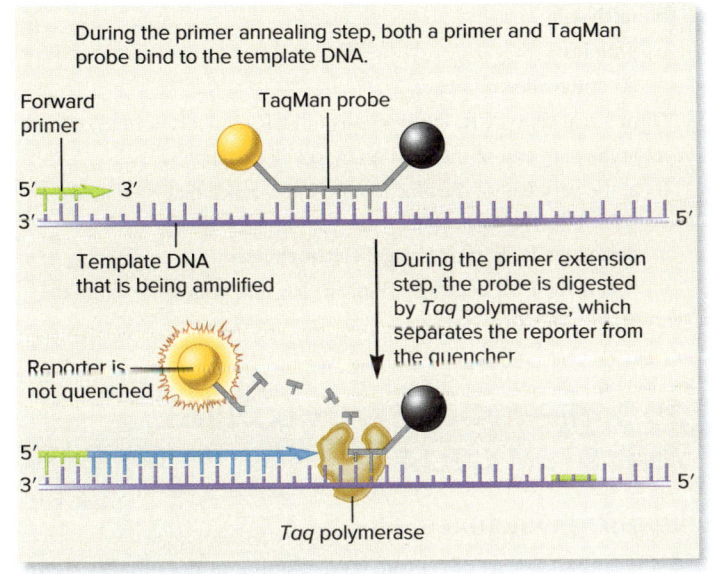

(b) Use of a TaqMan probe in real-time PCR

FIGURE 20.8 An example of a probe, called TaqMan, which is used in a real-time PCR experiment.

Concept Check: What needs to happen so that fluorescence emitted by the reporter is not quenched?

proximity of the reporter molecule to the quencher molecule prevents the detection of fluorescence from the reporter molecule.

As shown in **Figure 20.8b**, a primer and the TaqMan probe both anneal to the template DNA during the primer annealing step. During the primer extension step, the 5′ to 3′ exonuclease activity of *Taq* polymerase cleaves the oligonucleotide in the TaqMan probe into individual nucleotides, thereby separating the reporter from the quencher. This allows the reporter to emit (unquenched) fluorescence that can be measured within the thermocycler. As PCR products accumulate, more and more of the TaqMan probes are digested, and therefore the level of fluorescence increases.

Stages of Real-Time PCR **Figure 20.9a** considers a real-time PCR experiment as it occurs over the course of many cycles, such as 20 to 40. Real-time PCR goes through three main phases. Initially, when the amount of PCR product is small and reagents are not limiting, the synthesis of the PCR product occurs with close to 100% efficiency—the amount of product nearly doubles with each cycle. This exponential accumulation is difficult to detect in the earliest cycles because the amount of PCR products is small. Therefore, the amount of TaqMan that is cleaved is also small. The second phase is approximately linear as PCR products continue to

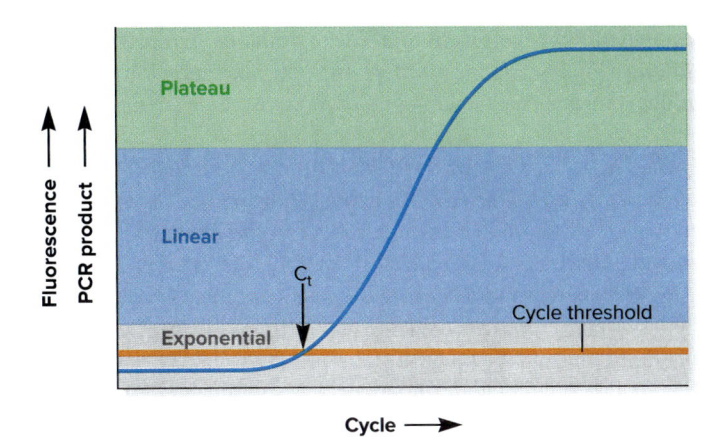

(a) Phases of PCR

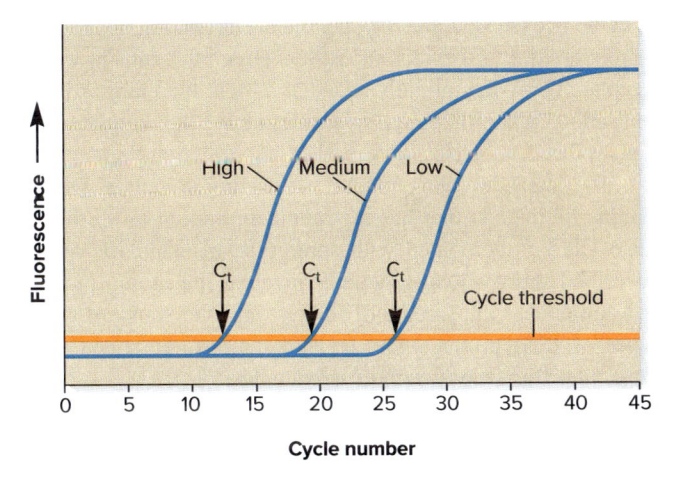

(b) Real-time PCR at high, medium, and low concentrations of the starting template DNA

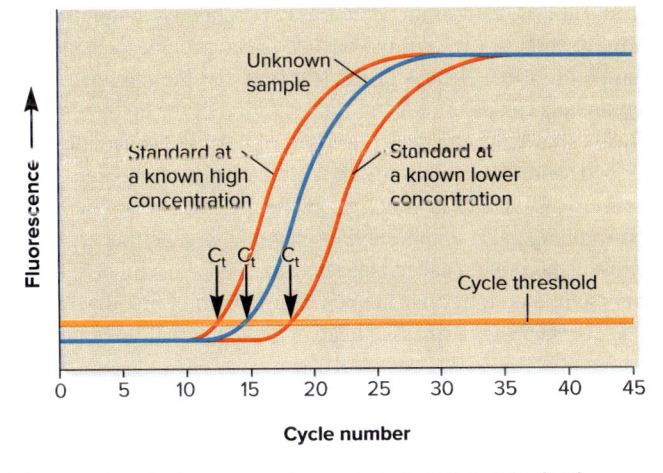

(c) A comparison between an unknown sample and standards of known concentrations

FIGURE 20.9 Examples of data that are obtained from a real-time PCR experiment. (a) The three phases that occur in a typical PCR experiment. (b) PCR carried out at three different starting concentrations of the template DNA. (c) A comparison between a sample with an unknown DNA concentration and a standard at a high and low concentration.

accumulate, but the reaction efficiency falls as reagents become limiting. Finally, in the third phase, the accumulation of PCR products reaches a plateau as one or more reagents are used up.

Cycle Threshold Method for Measuring the Initial Template Concentration During the exponential phase of PCR, the amount of PCR product is proportional to the amount of starting template that was initially added. Therefore, the exponential phase of PCR is analyzed to quantitate the initial template concentration. The commonly used method for this analysis is called the **cycle threshold method (C_t method)**. The cycle threshold (C_t) is reached when the accumulation of fluorescence is significantly greater than the background level. During the early cycles, the fluorescence signal due to the background level of fluorescence is greater than that derived from the amplification of the PCR product. Once the C_t value is exceeded, the exponential accumulation of product can be measured. **Figure 20.9b** considers three PCR runs in which the template DNA was initially added at a high, medium, or low concentration. When the initial concentration of the template DNA is higher, the C_t is reached at an earlier amplification cycle.

To determine the amount of starting template DNA, the sample of interest that has an unknown amount of starting template DNA is compared with some type of standard, which are described shortly. Real-time PCR involves the coamplification of two templates: the sample of interest and the standard. The two types of PCR products can be monitored simultaneously using molecules that fluoresce in different colors.

Using Standards to Determine the Initial Concentration of Template DNA What types of standards are commonly used? One possibility is a standard of known concentration that is added to the PCR mixture. For example, plasmid DNA carrying a specific gene can be added to the PCR mixture in known amounts, and the amplification of this plasmid-encoded gene provides a standard. By comparing the C_t values of the standard and the unknown sample, researchers can determine the concentration of the unknown sample. This is shown schematically in **Figure 20.9c**, but is actually accomplished with computer software. Alternatively, researchers may use an internal standard in which another gene that is already present in the sample is also amplified. For example, a gene that encodes the cytoskeletal protein called actin may be used. This relative quantitation method is somewhat simpler. The amount of unknown template DNA of interest is expressed relative to the internal standard.

20.2 REVIEWING THE KEY CONCEPTS

- Polymerase chain reaction (PCR) uses oligonucleotide primers to copy a specific region of DNA. Each cycle of PCR involves three steps: denaturation, primer annealing, and primer extension (see Figures 20.5, 20.6).
- Reverse transcriptase PCR begins with mRNA and is used to study gene expression (see Figure 20.7).
- Real-time PCR monitors PCR as it occurs in a thermocycler. It is used to quantitate the amount of starting DNA or mRNA in a sample. Real-time PCR uses fluorescent molecules to follow the PCR reaction. A sample that has an unknown amount of DNA is compared with a standard (see Figures 20.8, 20.9).

20.2 COMPREHENSION QUESTIONS

1. In one PCR cycle, the correct order of steps is
   a. primer annealing, primer extension, denaturation.
   b. primer annealing, denaturation, primer extension.
   c. denaturation, primer annealing, primer extension.
   d. denaturation, primer extension, primer annealing.
2. In reverse transcriptase PCR, the starting biological material is
   a. chromosomal DNA.
   b. mRNA.
   c. proteins.
   d. all of the above.
3. During real-time PCR, the synthesis of PCR products is analyzed
   a. at the very end of the reaction by gel electrophoresis.
   b. at the very end of the reaction by fluorescence that is emitted within the thermocycler.
   c. during the PCR cycles by gel electrophoresis.
   d. during the PCR cycles by fluorescence that is emitted within the thermocycler.

20.3 REPRODUCTIVE CLONING AND STEM CELLS

Learning Outcomes:

1. Outline the steps of reproductive cloning in mammals.
2. Define *stem cells,* and describe their two key properties.
3. Distinguish between totipotent, pluripotent, multipotent, and unipotent stem cells.
4. List examples of potential uses of stem cells to treat human diseases.

The previous sections focused on the cloning of segments of DNA, using either vectors or PCR. In this section, we will consider another form of cloning: the cloning of whole organisms, a methodology called reproductive cloning. We will also examine the properties of stem cells and their potential uses in treating human diseases. Both reproductive cloning and stem cells have received enormous public attention due to the complex ethical issues they raise.

Researchers Have Succeeded in Cloning Mammals from Somatic Cells

The term *cloning* has more than one meaning. Earlier in this chapter, we discussed gene cloning, which involves methods that produce many copies of a gene. The cloning of an entire organism is a different matter. **Reproductive cloning** refers to methods that produce two or more genetically identical individuals. This happens occasionally in nature; identical twins are genetic clones that began from the same fertilized egg. Similarly, researchers can take mammalian embryos at an early stage of development (e.g., the two-cell to eight-cell stage), separate the cells, implant them into the uterus, and obtain multiple births of genetically identical individuals.

In the case of plants, cloning is an easier undertaking. Plants can be cloned from somatic cells. In most cases, it is relatively easy to take a cutting from a plant, expose it to growth hormones, and obtain a separate plant that is genetically identical to the original. However, this approach has not been possible with mammals. For several decades, scientists believed that chromosomes within the somatic cells of mammals had incurred irreversible genetic changes that render them unsuitable for cloning. However, this hypothesis has proven to be incorrect. In 1997, Ian Wilmut and his colleagues at the Roslin Institute in Scotland announced that a sheep, which they named Dolly, had been cloned using the genetic material from somatic cells.

Figure 20.10 shows how Dolly was produced.

1. The researchers removed mammary cells from an adult female sheep and grew them in the laboratory. The researchers also extracted the nucleus from an egg cell of a different sheep.
2. They then used electrical pulses to fuse a diploid mammary cell with the enucleated egg cell.
3. After fusion, the zygote began embryonic development.
4. The resulting embryo was implanted into the uterus of a surrogate mother sheep.
5. One hundred and forty-eight days later, Dolly was born.

Although Dolly was clearly a clone of the donor female sheep, tests conducted when she was 3 years old suggested that she was genetically older than her actual age. As mammals age, chromosomes in somatic cells tend to shorten from the telomeres—the ends of eukaryotic chromosomes. Therefore, older individuals have shorter chromosomes in their somatic cells than younger ones do. This shortening does not seem to occur in the cells of the germ line, however. When researchers analyzed the chromosomes in Dolly's somatic cells when she was about 3 years old, the lengths of her chromosomes were consistent with a sheep that was significantly older, say, 9 or 10 years old. The sheep that donated the somatic cell that produced Dolly was 6 years old, and her mammary cells had been grown in culture for several cell doublings before a mammary cell was fused with an oocyte. This led researchers to postulate that Dolly's shorter telomeres were a result of chromosome shortening in the somatic cells of the sheep that donated the nucleus.

With regard to telomere length, research in mice and cattle has shown different results; the telomeres of these cloned animals appear to be the correct length. For example, cloning was conducted on mice using the method described in Figure 20.10 for six consecutive generations. The cloned mice of the sixth generation had normal telomeres. Further research is necessary to determine if cloning via somatic cells has an effect on the length of telomeres in subsequent generations. However, other studies in mice point to various types of genetic flaws in cloned animals. For example, Rudolf Jaenisch and his colleagues used DNA microarray technology (described in Chapter 21) to analyze the transcription patterns of over 10,000 genes in cloned mice. As many as 4% of those genes were not expressed normally. Furthermore, research has shown that cloned mice die at a younger age than their naturally bred counterparts.

Mammalian cloning is still at an early stage of development. Nevertheless, the breakthrough of creating Dolly has shown that it is technically possible. In recent years, cloning from somatic cells

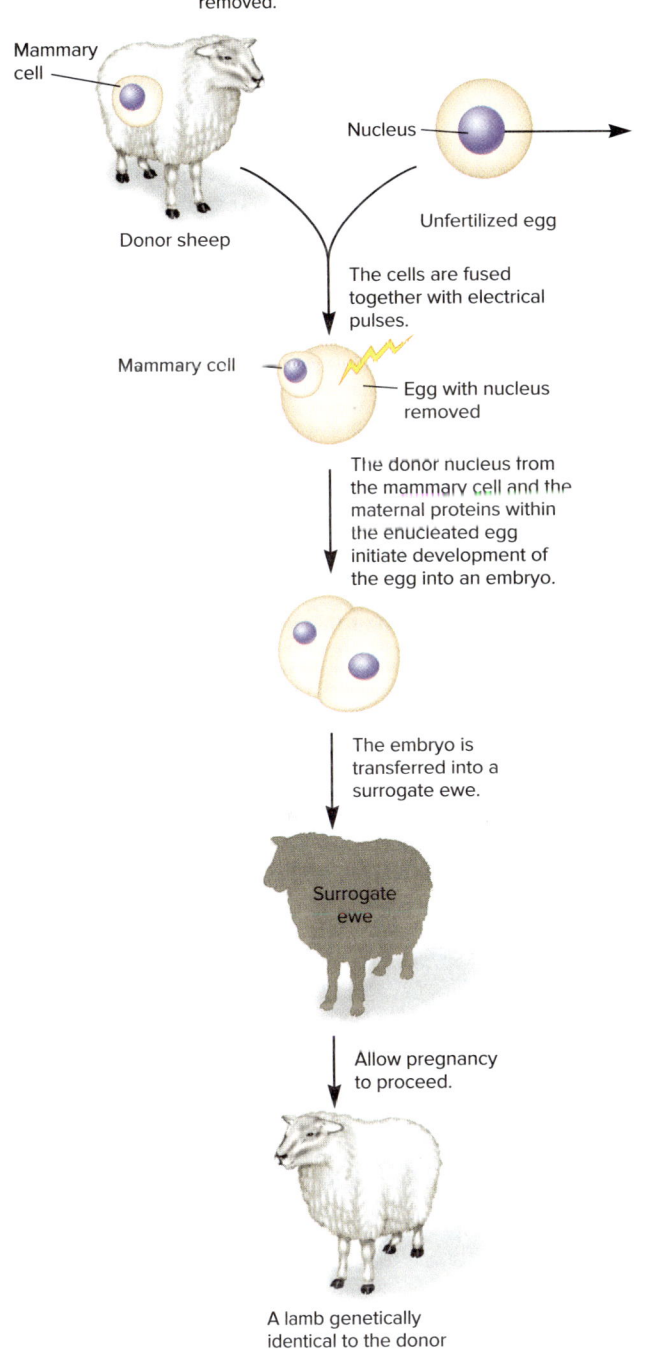

FIGURE 20.10 Protocol for the successful cloning of a sheep.

Genes → Traits Dolly was (almost) genetically identical to the sheep that donated a mammary cell to create her. Dolly and the donor sheep were (almost) genetically identical in the same way that identical twins are; they carried the same set of genes and looked remarkably similar. However, they may have had minor genetic differences due to possible variation in their mitochondrial DNA and may have exhibited some phenotypic differences due to maternal effect or imprinted genes.

Concept Check: In the protocol, why is the nucleus of the oocyte removed?

has been achieved in several other mammalian species, including cattle, mice, goats, and pigs. In 2002, the first pet was cloned. She was named Carbon Copy, but also called Copy Cat (**Figure 20.11**). Mammalian cloning may potentially have many practical applications. Cloning livestock would enable farmers to use the somatic cells from their best individuals to create genetically homogeneous herds, which could increase agricultural yield. However, animals in a genetically homogeneous herd may be more susceptible to rare diseases.

Though some people are concerned about the use of cloning with agricultural species, a majority have become very concerned about the possibility of human cloning. This prospect has raised a host of serious ethical questions. For example, some people feel that cloning humans is morally wrong and threatens the basic fabric of parenthood and family. Others feel that it is a technology that could offer a new avenue for reproduction, one that could be offered to infertile couples, for example. In the public sector, the sentiment toward human cloning has been generally negative. Many countries have issued an all-out ban on human cloning, but others permit limited research in this area. Because the technology for cloning exists, our society will continue to wrestle with the legal and ethical aspects of cloning, not only of animals but also of people.

Stem Cells Have the Ability to Divide and Differentiate into Different Cell Types

Stem cells supply the cells that construct our bodies from a fertilized egg. In adults, stem cells also replenish worn-out or damaged cells. To accomplish this task, stem cells have two common characteristics.

- They have the capacity to divide.
- They can differentiate into one or more specialized cell types.

As shown in **Figure 20.12**, the two daughter cells produced from the division of a stem cell can have different fates. One of the cells may remain an undifferentiated stem cell, while the other daughter cell can differentiate into a specialized cell type. With this type of asymmetrical division/differentiation pattern, the population of stem cells remains relatively constant, while providing a population of specialized cells. In the adult, this type of mechanism is needed to replenish cells that have a finite life span, such as skin epithelial cells and red blood cells.

In mammals, stem cells are commonly categorized according to their developmental stage and their ability to differentiate (**Figure 20.13**).

- A fertilized egg is considered **totipotent,** because it can give rise to all of the cell types in the adult organism.
- The early mammalian embryo contains **embryonic stem cells (ES cells),** which are found in the inner cell mass of the blastocyst. The blastocyst is the stage of embryonic development prior to uterine implantation—the preimplantation embryo. ES cells are **pluripotent,** which means they can differentiate into every or almost every cell type of the body. However, a single ES cell has lost the ability to produce an entire, intact individual. During the early fetal stage of development, the germ-line cells found in the gonads also are pluripotent. These cells are called **embryonic germ cells (EG cells).**
- A **multipotent** stem cell can differentiate into several cell types but far fewer than an ES cell. For example,

FIGURE 20.11 **Carbon Copy, the first cloned pet.** The animal shown here was produced using a procedure similar to the one shown in Figure 20.10.
©Texas A&M University/Getty Images

Concept Check: Is Carbon Copy a transgenic animal?

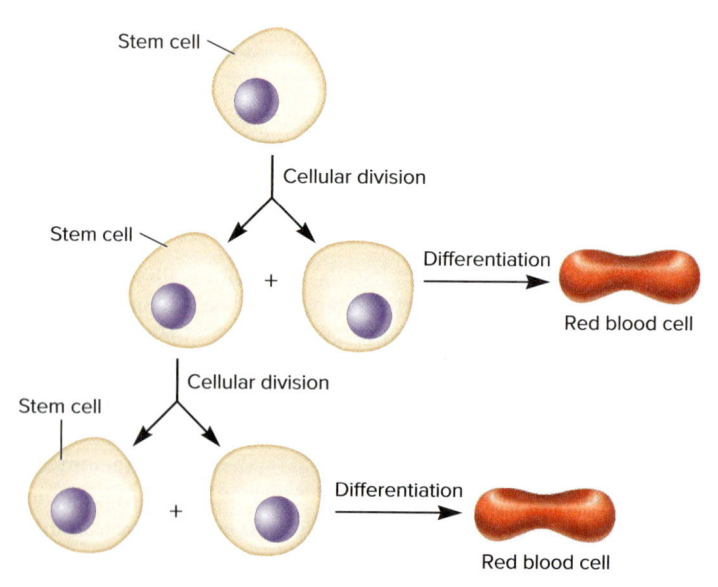

FIGURE 20.12 Growth pattern of stem cells. The two main traits that stem cells exhibit are an ability to divide and an ability to differentiate. When a stem cell divides, one of the two cells remains a stem cell, and the other daughter cell differentiates into a specialized cell type.

Concept Check: Explain why stem cells are not depleted during the life of an organism.

cells may help us to understand basic genetic mechanisms that underlie the process of multicellular development. A second compelling reason why people have become interested in stem cells is their potential to treat human diseases or injuries that cause cell and tissue damage. This application has already become a reality in certain cases. For example, bone marrow transplantation is used to treat patients with certain forms of cancers. Such patients may be given radiation treatments that destroy their immune systems. When these patients are injected with bone marrow from a healthy person, the stem cells within the transplanted marrow have the ability to proliferate and differentiate within their bodies and provide a functioning immune system.

Renewed interest in the use of stem cells in the potential treatment of many other diseases was fostered in 1998 by two separate studies, directed by James Thomson and John Gearhart, showing that embryonic cells, either ES or EG cells, can be successfully propagated in the laboratory. As shown in **Table 20.4**, embryonic cells could potentially be used to treat a wide variety of diseases associated with cell and tissue damage. By comparison, it would be difficult, based on our modern knowledge, to treat these diseases with adult stem cells because of the inability to locate most types of adult stem cells within the body and successfully grow them in the laboratory. In addition, with the exception of stem cells in the blood, other types of stem cells in the adult body are difficult to remove in sufficient numbers for transplantation. By comparison, ES and EG cells are easy to identify and have the great advantage of rapid growth in the laboratory. For these reasons, ES and EG cells offer a greater potential for transplantation, based on our current knowledge of stem cell biology.

For ES or EG cells to be used in transplantation, researchers need to derive methods that cause the cells to differentiate into the appropriate type of tissue. For example, if the goal is to repair a spinal cord injury, ES or EG cells need the appropriate cues to cause them to differentiate into neural tissue. At present, much research is needed to understand and potentially control the fate of ES or EG cells. Currently, researchers speculate that a complex array of factors determine the developmental fates of stem cells. These include internal factors within the stem cells themselves, as well as external factors such as the properties of neighboring cells and the presence of hormones and growth factors in the environment.

From an ethical perspective, the primary issue that prompts debate is the source of the stem cells for research and potential treatments. Most ES cells have been derived from human embryos that were produced from in vitro fertilization and were subsequently not used. Most EG cells are obtained from aborted fetuses. Some feel that it is morally wrong to use such tissue in research and/or the treatment of disease, or they fear such use could lead to intentional abortions for the sole purpose of obtaining fetal tissues for transplantation. Alternatively, others feel that the embryos and fetuses that provide the ES and EG cells are not going to become living individuals, and therefore, it is beneficial to study these cells and use them in a positive way to treat human diseases and injury. It is not clear whether

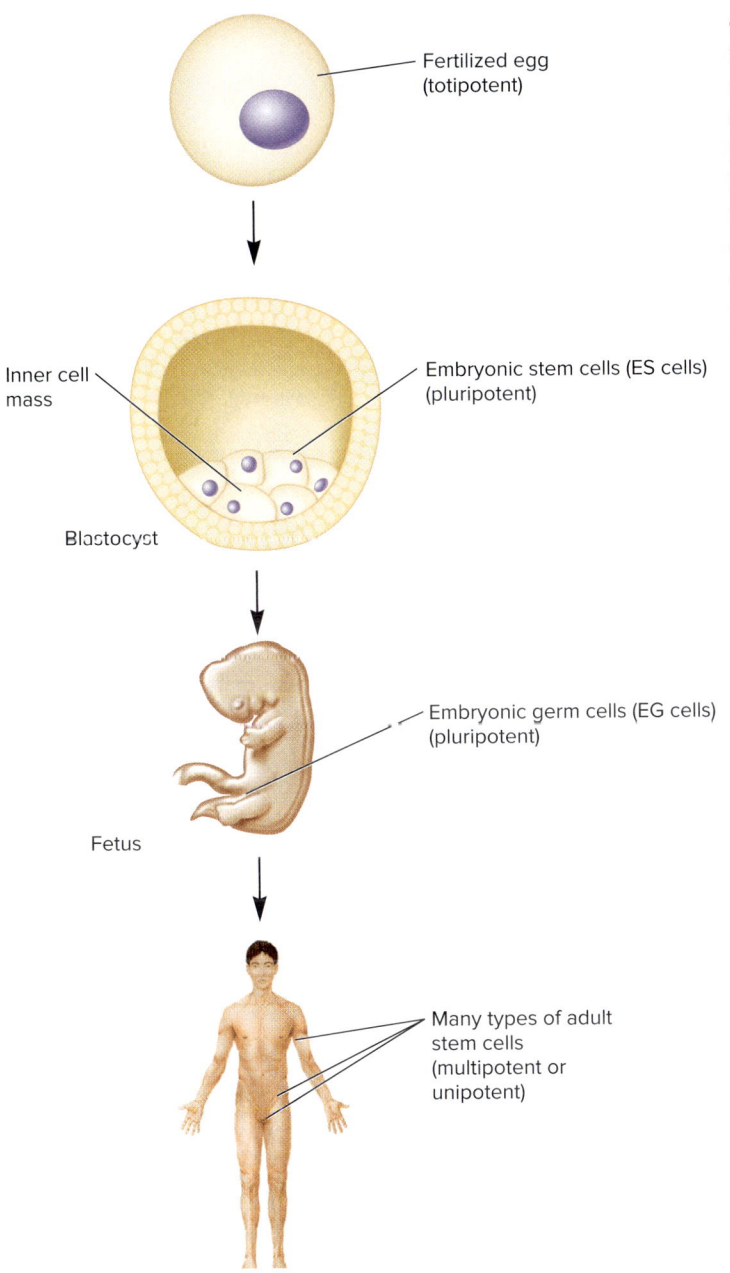

FIGURE 20.13 Occurrence of stem cells at different stages of human development.

hematopoietic stem cells (HSCs) found in the bone marrow can supply cells that populate two different tissues, namely, blood and lymphoid tissue (**Figure 20.14**).
- Other stem cells found in the adult seem to be **unipotent**. For example, primordial germ cells in the testis differentiate only into a single cell type, the sperm.

Stem Cells Have the Potential to Treat a Variety of Diseases

What are the potential uses of stem cells? Interest in stem cells centers on two main areas. Because stem cells have the capacity to differentiate into multiple cell types, the study of stem

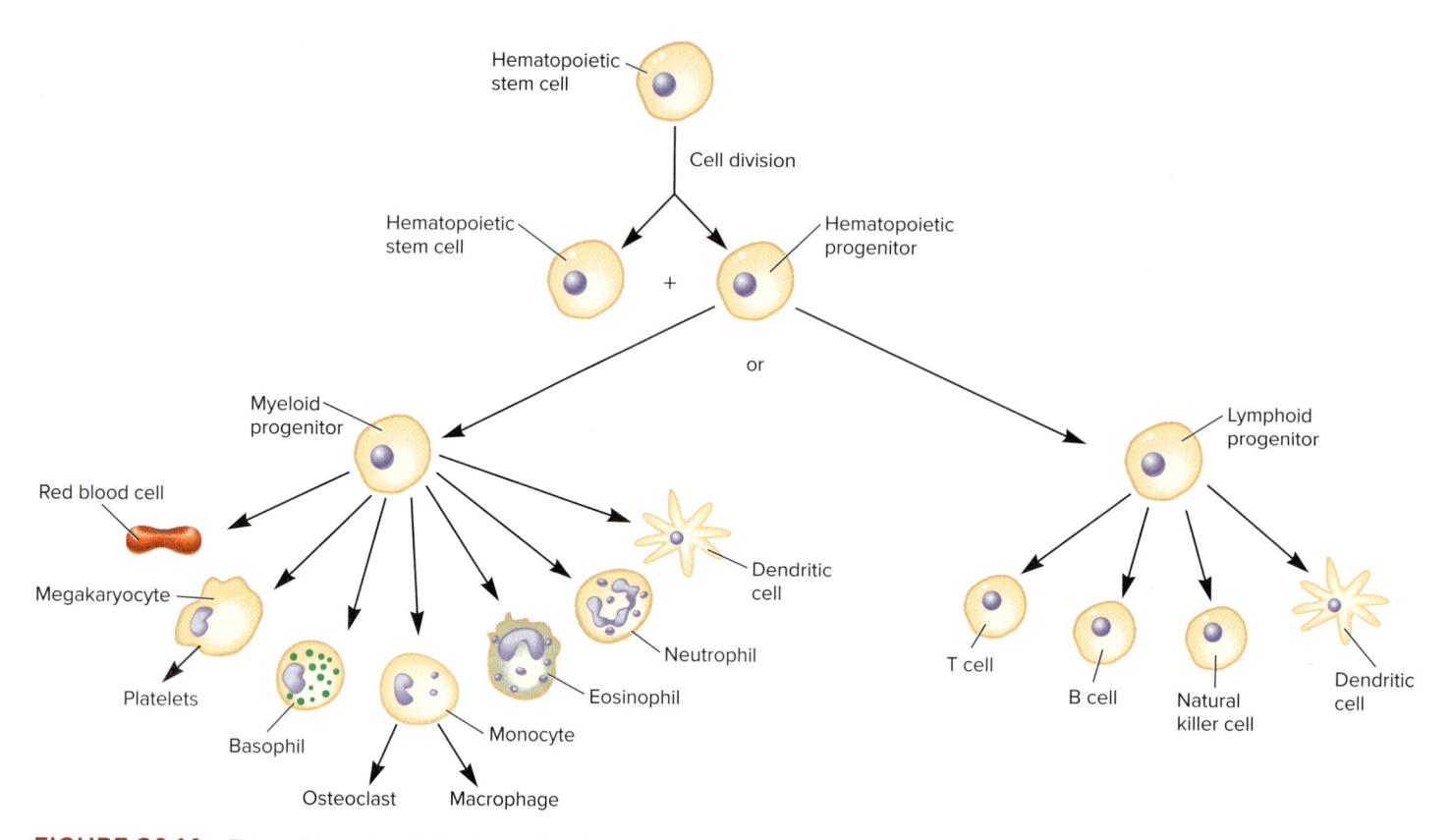

FIGURE 20.14 **Fates of hematopoietic stem cells.** Multipotent HSCs can follow a pathway in which cell division produces a myeloid progenitor cell, which can then differentiate into a red blood cell, megakaryocyte, basophil, monocyte, eosinophil, neutrophil, or dendritic cell. Alternatively, an HSC can follow a path in which it becomes a lymphoid progenitor cell, which then differentiates into a T cell, B cell, natural killer cell, or dendritic cell.

Concept Check: Are hematopoietic stem cells unipotent, multipotent, or pluripotent?

TABLE 20.4
Potential Uses of Stem Cells to Treat Diseases

Cell/Tissue Type	Disease Treatment
Neural	Implantation of cells into the brain to treat Parkinson disease
	Treatment of neurological injuries such as those to the spinal cord
Skin	Treatment of burns and other types of skin disorders
Cardiac	Repair of heart damage associated with heart attacks
Cartilage	Repair of joints damaged by injury or arthritis
Bone	Repair of damaged bone or replacement with new bone
Liver	Repair or replacement of liver tissue that has been damaged by injury or disease
Skeletal muscle	Repair or replacement of damaged muscle

these two opposing viewpoints can reach a common ground. As a compromise, some governments have enacted laws that limit or prohibit the use of embryos or fetuses to obtain stem cells, yet permit the use of stem cell lines that are already available in research laboratories.

If stem cells could be obtained from adult cells and propagated in the laboratory, an ethical dilemma may be avoided because most people do not have serious moral objections to current procedures such as bone marrow transplantation. In 2006, work by Shinya Yamanaka and colleagues showed that adult mouse fibroblasts (a type of connective tissue cell) could become pluripotent via the injection of four different genes that encode transcription factors. In 2007, Yamanaka's laboratory and two other research groups showed that such induced pluripotent stem cells (iPS cells) can differentiate into all cell types when injected into mouse blastocysts and grown into baby mice. Though further research is still needed, these recent results indicate that adult cells can be reprogrammed to become embryonic stem cells.

20.3 REVIEWING THE KEY CONCEPTS

- Mammalian reproductive cloning can be achieved by fusing somatic cells to oocytes with their nuclei removed (see Figures 20.10, 20.11).
- Stem cells have the ability to divide and differentiate. Stem cells may be totipotent, pluripotent, multipotent, or unipotent (see Figures 20.12–20.14).
- Stem cells have the potential to treat a variety of human diseases (see Table 20.4).

20.3 COMPREHENSION QUESTIONS

1. During mammalian reproductive cloning, _____ is fused with _____.
   a. a somatic cell, a stem cell
   b. a somatic cell, an egg cell
   c. a somatic cell, an enucleated egg cell
   d. an enucleated somatic cell, an egg cell

2. Which of the following is a key feature of stem cells?
   a. They have the ability to divide.
   b. They have the ability to differentiate.
   c. They are always pluripotent.
   d. Both a and b are true of stem cells.

20.4 DNA SEQUENCING

Learning Outcome:
1. Outline the steps in automated DNA sequencing via the dideoxy method.

As we have seen throughout this text, our knowledge of genetics can be largely attributed to an understanding of DNA structure and function. The feature that underlies all aspects of inherited traits is the DNA sequence. For this reason, analyzing DNA sequences is a powerful approach for understanding genetics. A technique called **DNA sequencing** enables researchers to determine the base sequence of DNA found in genes and other chromosomal regions. It is one of the most important tools for exploring genetics at the molecular level. Molecular geneticists often want to determine DNA base sequences as a first step toward understanding the function and expression of genes. For example, the investigation of genetic sequences has been vital to our understanding of promoters, regulatory elements, and the genetic code itself. Likewise, an examination of sequences has facilitated our understanding of origins of replication, centromeres, telomeres, and transposable elements.

During the 1970s, two methods for DNA sequencing were devised. One method, developed by Allan Maxam and Walter Gilbert, involved the base-specific chemical cleavage of DNA. Another method, developed by Frederick Sanger and colleagues, is known as **dideoxy sequencing**. Because it became the more popular method of DNA sequencing, we will consider the dideoxy method here. In addition, Chapter 21 considers some newer methods of sequencing DNA that are not based on the Sanger dideoxy method. These newer methods are commonly used in projects aimed at determining the DNA sequence of an entire genome.

The dideoxy method of DNA sequencing is based on our knowledge of DNA replication but includes a clever twist. As described in Chapter 13, DNA polymerase connects adjacent deoxyribonucleotides by catalyzing a covalent bond between the 5′ phosphate on one nucleotide and the 3′ —OH group on the previous nucleotide (refer back to Figure 13.13). However, chemists can synthesize deoxyribonucleotides that are missing the —OH group at the 3′ position (**Figure 20.15**). These synthetic nucleotides

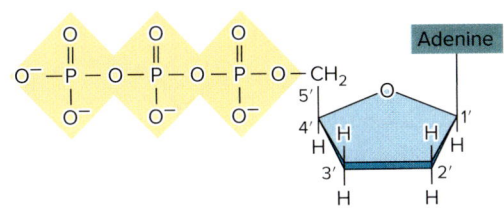

2′, 3′-Dideoxyadenosine triphosphate (ddA)

FIGURE 20.15 **The structure of a dideoxyribonucleotide.** Note that the 3′ group is a hydrogen rather than an —OH group. For this reason, another nucleotide cannot be attached at the 3′ position.

are called **dideoxyribonucleotides (ddNTPs)**. (Note: The prefix *dideoxy-* indicates that two [di] oxygens [oxy] have been removed [de] from this sugar; in comparison, ribose has —OH groups at both the 2′ and 3′ positions.) Sanger reasoned that if a dideoxyribonucleotide is added to a growing DNA strand, the strand can no longer grow because the dideoxyribonucleotide is missing the 3′ —OH group. The incorporation of a dideoxyribonucleotide into a growing strand is therefore referred to as **chain termination**.

To detect the incorporation of dideoxynucleotides during DNA replication, the newly made DNA strands must be labeled in some way. When dideoxy sequencing was first invented, researchers labeled the nucleotides with radioisotopes, which allowed newly made DNA strands to be detected via autoradiography. This traditional DNA sequencing method was replaced by **automated DNA sequencing** in which each type of dideoxynucleotide is labeled with a different colored fluorescent dye. For example, a common way to fluorescently label ddNTPs is the following: ddA is green, ddT is red, ddG is yellow, and ddC is blue.

For automated DNA sequencing, the segment of DNA to be sequenced is obtained in large amounts. This can be accomplished using gene cloning, which was described earlier in this chapter. In the example in **Figure 20.16**, the segment of DNA to be sequenced, which we will call the target DNA, was cloned into a vector at a defined location. The target DNA was inserted next to a site in the vector where a primer will bind, which is called the annealing site. The aim of the experiment is to determine the base sequence of the target DNA next to the annealing site. In the experiment shown in Figure 20.16, the recombinant vector DNA has been previously denatured into single strands, usually via heat treatment. Only the strand needed for DNA sequencing is shown here.

Let's now consider the steps that are involved in DNA sequencing (Figure 20.16).

1. A sample containing many copies of the single-stranded recombinant vector is mixed with many primers that will bind to the annealing site. The primer binds to the DNA because it is complementary to the annealing site.
2. All four types of deoxyribonucleotides are added at a high concentration and all four dideoxyribonucleotides (ddA, ddT, ddG, and ddC), which are fluorescently labeled, are added at a low concentration.
3. DNA polymerase is then added, which causes the synthesis of strands that are complementary to the target DNA sequence.

478   CHAPTER 20 :: MOLECULAR TECHNOLOGIES

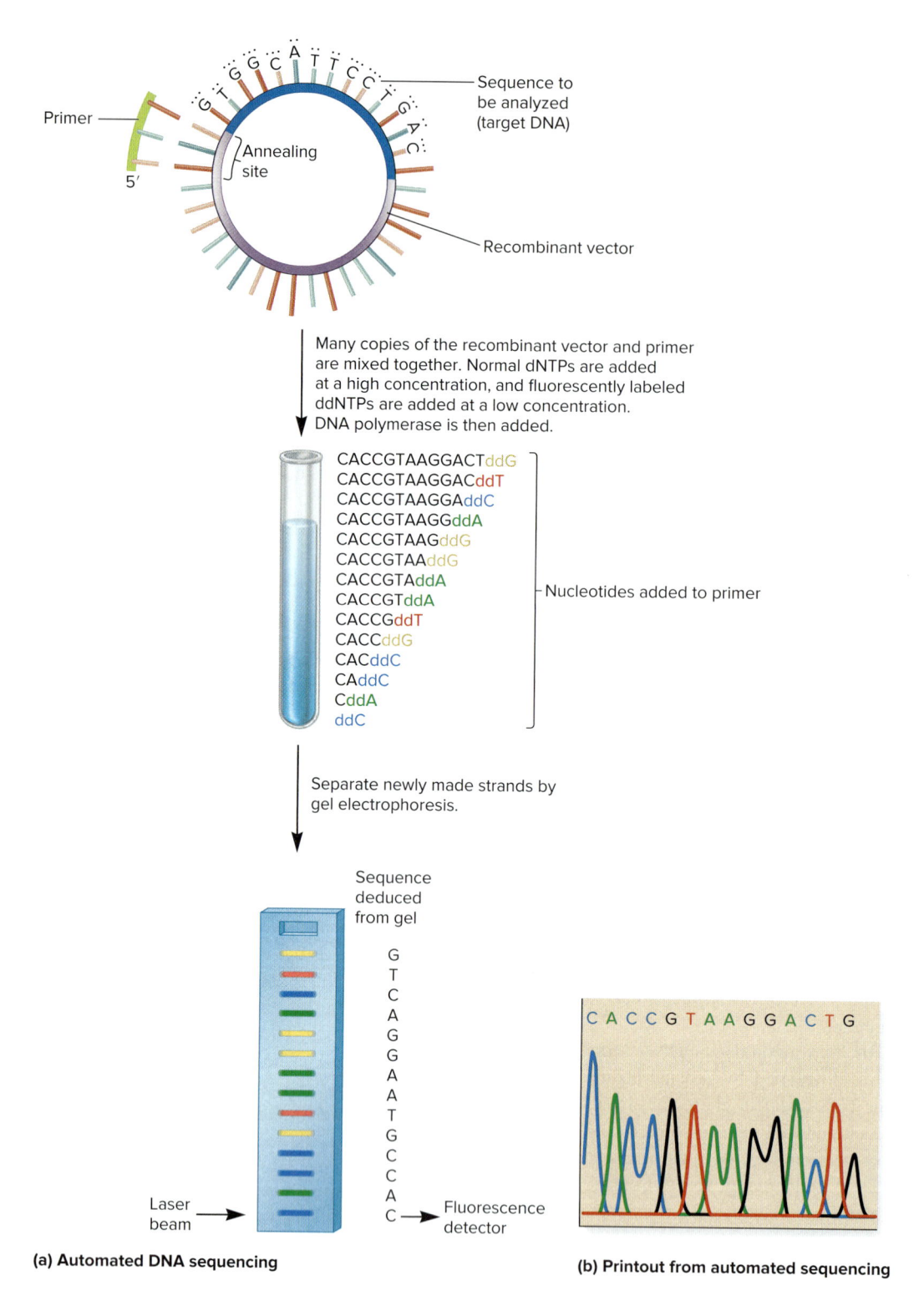

FIGURE 20.16 The protocol for DNA sequencing by the dideoxy method. (a) The method begins with a recombinant vector into which the target DNA has been inserted. For the primer to bind, the recombinant vector must be denatured into single-stranded DNA at the beginning of the experiment. Only the strand needed for DNA sequencing is shown here. This diagram schematically depicts a series of bands on a gel; the four colors of the bands occur because each type of dideoxynucleotide is labeled with a different colored fluorescent dye. As each band passes a laser, the fluorescent dye is excited by the laser beam, and the fluorescence emission is recorded by a fluorescence detector. The detector reads the level of fluorescence at four wavelengths corresponding to the four dyes. (b) As shown in the printout, the peaks of fluorescence correspond to the DNA sequence that is complementary to the target DNA.

DNA synthesis continues until a dideoxynucleotide is incorporated into a growing strand. For example, chain termination can occasionally occur at the sixth or thirteenth position of the newly synthesized DNA strand if a ddT becomes incorporated at either of these sites. Note that the complementary A base is found at the sixth and thirteenth positions in the target DNA. Therefore, we expect to obtain DNA strands that terminate at the sixth or thirteenth positions and have a ddT at their ends. Because these DNA strands contain a ddT, they are fluorescently labeled in red.

4. After the samples have been incubated for several minutes, mixtures of DNA strands of different lengths are made, depending on the number of nucleotides attached to the primer. These DNA strands are separated according to their lengths by running them on a slab gel or more commonly by running them through a gel-filled capillary tube. The shorter strands move to the bottom of the gel more quickly than the longer strands.

Because we know the color with which each dideoxyribonucleotide is labeled, we also know which base is at the very end of each DNA strand separated on the gel. Therefore, we can deduce the DNA sequence that is complementary to the target DNA by reading which base is at the end of every DNA strand and matching this sequence with the length of the strand. Reading the base sequence, from bottom to top, is much like climbing a ladder of bands. For this reason, the sequence obtained by this method is referred to as a **sequencing ladder.**

Theoretically, it is possible to read this base sequence directly from the gel. From a practical perspective, however, it is faster and more efficient to automate the procedure using a laser and fluorescence detector. As the gel is running, each band passes the laser and the laser beam excites the fluorescent dye. The fluorescence detector records the amount of fluorescence emission from the excited dye. The detector reads the level of fluorescence at four wavelengths, corresponding to the four different colored dyes. An example of the printout from the fluorescence detector is shown in **Figure 20.16b**. As seen here, the peaks of fluorescence correspond to the DNA sequence that is complementary to the target DNA. Note that ddG is usually labeled with a yellow dye, but it is converted to black ink on the printout shown in Figure 20.16b for ease of reading. Though improvements in automated sequencing continue to be made, a typical sequencing run can provide a DNA sequence that is approximately 700 to 900 bases long, or perhaps even longer.

20.4 REVIEWING THE KEY CONCEPTS

- A method of DNA sequencing, called the dideoxy method, uses fluorescently labeled dideoxynucleotides that cause chain termination (see Figures 20.15, 20.16).

20.4 COMPREHENSION QUESTION

1. When a dideoxynucleotide is incorporated into a growing DNA strand,
   a. the strand elongates faster.
   b. the strand cannot elongate.
   c. the strand becomes more susceptible to DNase I cleavage.
   d. none of the above events occurs.

20.5 GENE EDITING

Learning Outcomes:
1. Describe the methods of site-directed mutagenesis and CRISPR-Cas technology.
2. List a few reasons why these methods are useful.

To understand how the genetic material functions, researchers often analyze mutations that alter the normal DNA sequence, thereby affecting the expression of genes and the outcome of traits. For example, geneticists have discovered that many inherited human diseases, such as sickle cell disease and hemophilia, involve mutations within specific genes. These mutations provide insight into the function of the genes in unaffected individuals. Hemophilia, for example, is caused by mutations in genes that encode blood clotting factors.

Because the analysis of mutations can provide important information about normal genetic processes, researchers often wish to produce mutant organisms. As we discussed in Chapter 19, mutations can arise spontaneously or can be induced by environmental agents. Mendel's pea plants are a classic example of strains with different phenotypes that arose from spontaneous mutations. In addition, experimental organisms can be treated with mutagens that increase the rate of mutations. In this section, we will consider newer methods that allow researchers to produce specific mutations within cloned genes or within genes in living cells. This approach is called **gene editing** or **gene mutagenesis**.

Site-Directed Mutagenesis Alters the Sequence of Cloned Genes

A technique known as **site-directed mutagenesis** allows a researcher to produce a mutation at a specific site within a cloned DNA segment. For example, if a DNA sequence is 5′–AAATTTCTTTAAA–3′, a researcher can use site-directed mutagenesis to change it to 5′–AAATTTGTTTAAA–3′. In this case, the researcher deliberately changed the seventh base from a C to a G. Why is this method useful? The site-directed mutation can then be introduced into a living organism to see how the mutation affects the expression of a gene, the function of a protein, and the phenotype of the organism.

Figure 20.17 describes the general steps in the procedure. Prior to this experiment, the DNA was denatured into single strands; only the single strand needed for site-directed mutagenesis is shown. As in PCR, this single-stranded DNA is referred to as the template DNA, because it is used as a template to synthesize a complementary strand.

1. An oligonucleotide primer is allowed to hybridize or anneal to the template DNA. The primer, typically 20 or so nucleotides in length, is synthesized chemically. (A shorter version of the primer is shown in Figure 20.17 for simplicity.) The researcher designs the base sequence of the primer. The primer has two important characteristics. First, most of the sequence of the primer is complementary to the site

in the DNA where the mutation is to be made. However, a second feature is that the primer contains a region of mismatch where the primer and template DNA are not complementary. The mutation occurs in this mismatched region. For this reason, site-directed mutagenesis is sometimes referred to as oligonucleotide-directed mutagenesis.

2. After the primer and template have annealed, the complementary strand is synthesized by adding deoxyribonucleoside triphosphates (dNTPs), DNA polymerase, and DNA ligase. The result is a double-stranded molecule that contains a mismatch only at the desired location.

3. This double-stranded DNA is then introduced into a bacterial cell. Within the cell, the DNA mismatch is likely to be repaired (see Chapter 19). Depending on which base is replaced, this may produce the mutant sequence or the original sequence. Clones containing the desired mutation are identified by DNA sequencing and used for further studies.

After a site-directed mutation has been made within a cloned gene, its consequences are analyzed by introducing the mutant gene into a living cell or organism. As described earlier, recombinant vectors containing cloned genes can be introduced into bacterial cells. After transformation of such a vector into a bacterium, a researcher can study the differences in function between the mutant and wild-type genes and the proteins they encode. Similarly, mutant genes made via site-directed mutagenesis can be introduced into plants and animals.

Genes in Living Cells Can Be Altered Using CRISPR-Cas Technology

In Chapter 18, we considered how the Crispr-Cas system provides bacteria with a defense against invasion by bacteriophages (refer back to Figure 18.6). Researchers realized that the components of this system can be used to mutate genes in living cells. This application is called **CRISPR-Cas technology.** In the natural (type II) system found in bacteria, two different non-coding RNAs (tracrRNA and crRNA) play key roles. The tracrRNA binds to the Cas9 protein and also to crRNA. The crRNA binds to a target DNA, such as a DNA segment within a bacteriophage. These binding interactions guide the Cas9 protein to the bacteriophage DNA, and then Cas9 makes a double-strand break.

Researchers have made a modification to this system to make it efficient for gene editing. They create a single RNA in which the tracrRNA and crRNA are linked to each other (**Figure 20.18a**). This molecule is called the single guide RNA (sgRNA). The spacer region of the sgRNA is designed to be complementary to one of the strands of the gene that is to be mutated. The sgRNA binds to Cas9 and guides it to the gene of interest. Cas9 then makes a double-strand break in this gene.

Following this break, two different repair events are possible. If the break is repaired by nonhomologous end joining (NHEJ), the region may incur a small deletion (see left side of **Figure 20.18b**). This deletion may inactivate the gene, particularly if it causes a

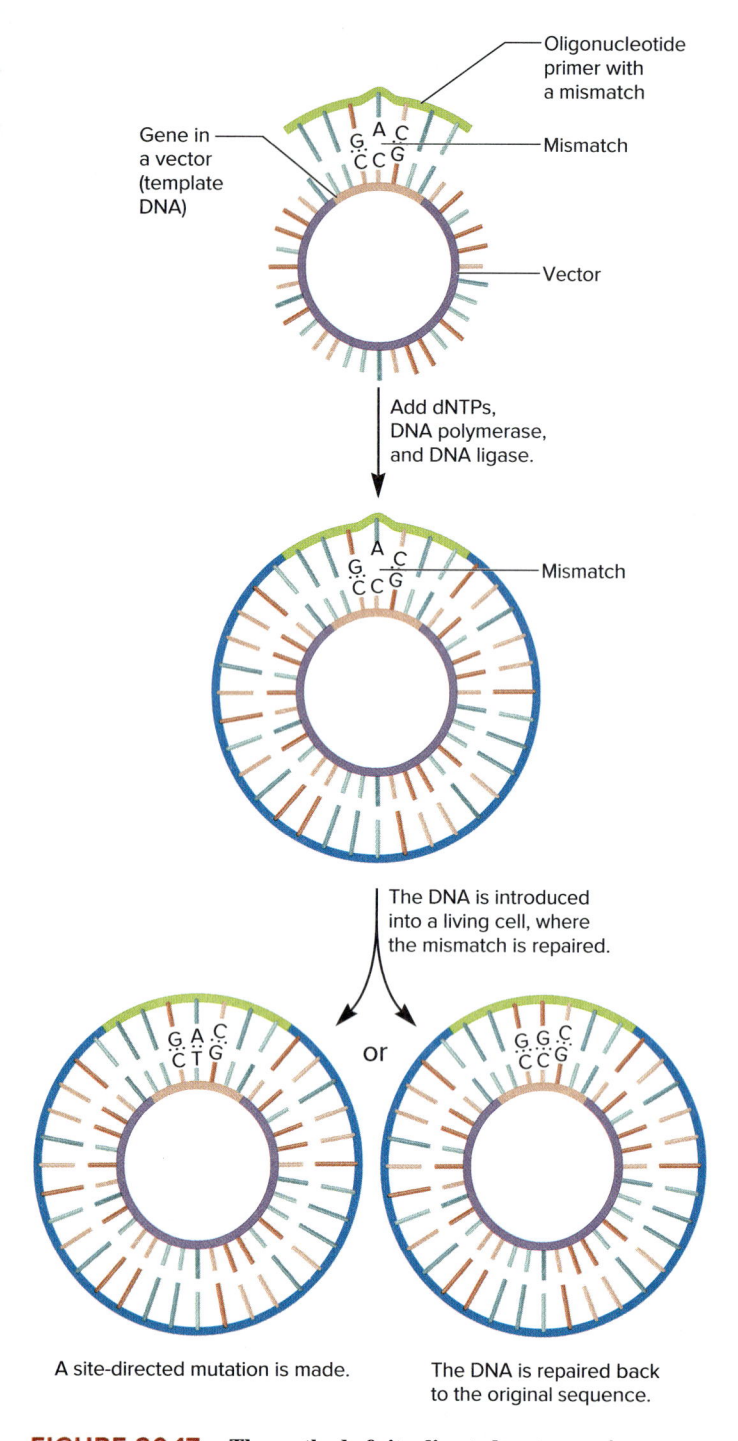

FIGURE 20.17 The method of site-directed mutagenesis.

Genes → Traits To examine the relationship between genes and traits, researchers can alter gene sequences via site-directed mutagenesis. The altered gene is introduced into a living cell or organism to examine how the mutation affects the organism's traits. For example, a researcher could produce a nonsense mutation in the middle of the *lacY* gene in the *lac* operon. If this site-directed mutant gene was then introduced into an *E. coli* bacterium that did not a have a normal copy of the *lacY* gene, the bacterium would be unable to use lactose. This result would indicate that a functional *lacY* gene is necessary for bacteria to have the trait of lactose utilization.

Concept Check: List three possible uses of site-directed mutagenesis.

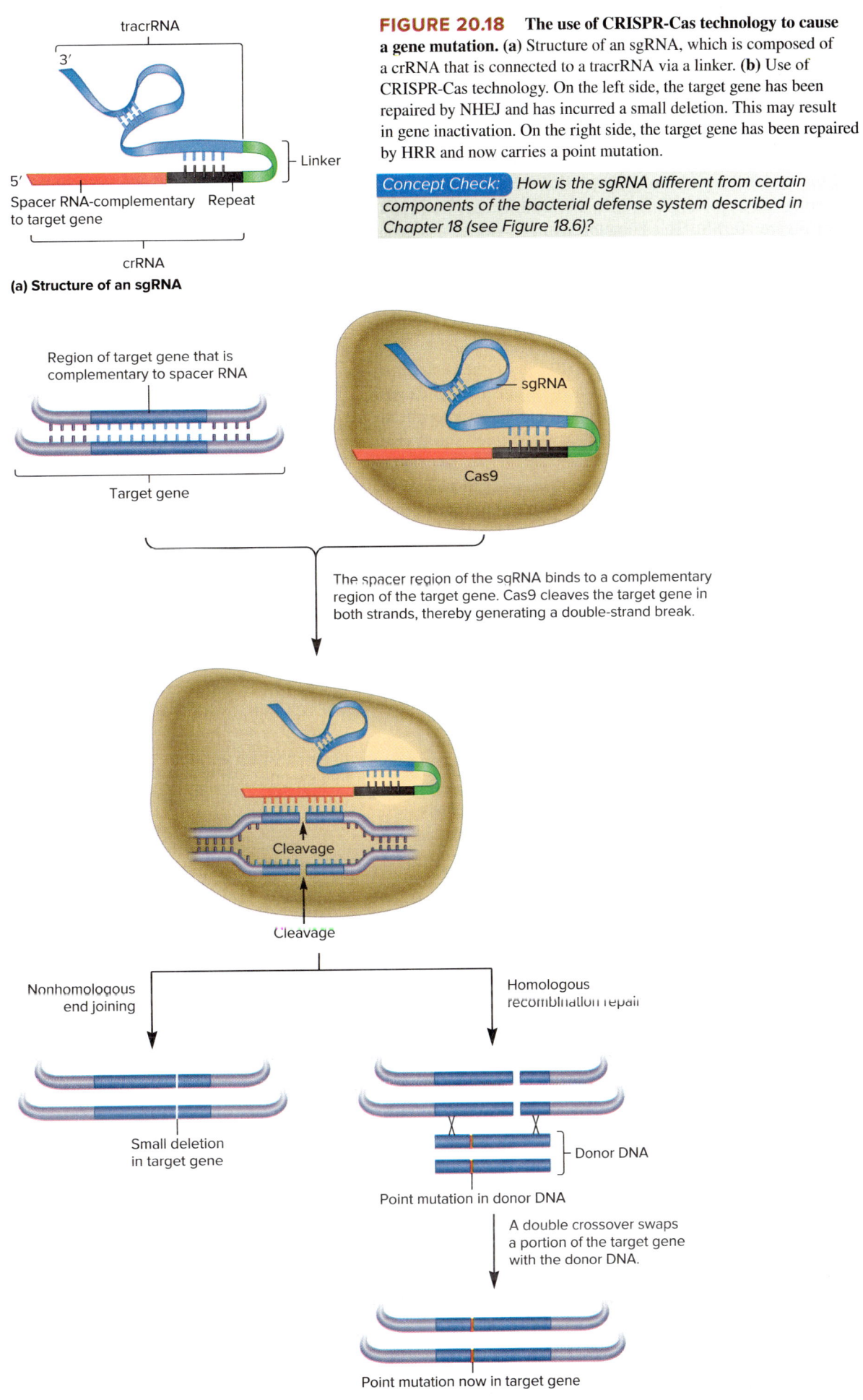

FIGURE 20.18 The use of CRISPR-Cas technology to cause a gene mutation. (a) Structure of an sgRNA, which is composed of a crRNA that is connected to a tracrRNA via a linker. (b) Use of CRISPR-Cas technology. On the left side, the target gene has been repaired by NHEJ and has incurred a small deletion. This may result in gene inactivation. On the right side, the target gene has been repaired by HRR and now carries a point mutation.

Concept Check: *How is the sgRNA different from certain components of the bacterial defense system described in Chapter 18 (see Figure 18.6)?*

(a) Structure of an sgRNA

The spacer region of the sgRNA binds to a complementary region of the target gene. Cas9 cleaves the target gene in both strands, thereby generating a double-strand break.

A double crossover swaps a portion of the target gene with the donor DNA.

(b) Use of CRISPR-Cas technology to inactivate a gene or create a point mutation frameshift mutation in the coding sequence. Alternatively, the double-strand break can be repaired by homologous recombination repair (HRR). Both of these repair mechanisms are described in Chapter 19. If HRR is desired, a researcher can also include a double-stranded segment of DNA, called the donor DNA, that is homologous to the region where the break occurs (see right side of Figure 20.18b). This homologous DNA can be designed to carry a particular mutation, such as a point mutation, that the researcher wants to make. The HRR system swaps in the donor DNA by a double crossover. In this way, a researcher can mutate a gene, similar to the result of site-directed mutagenesis shown in Figure 20.17.

The procedure presented in Figure 20.18b can be performed on different cell types and even on whole organisms. For example, this type of experiment could be carried out on a fertilized mouse oocyte. In this case, a researcher would inject a segment of DNA that encodes an sgRNA and Cas9 protein into the oocyte. After these two genes are expressed, the sgRNA-Cas9 complex would either carry out gene inactivation or produce a point mutation if the researcher also injected donor DNA. Compared to site-directed mutagenesis, a major advantage of CRISPR-Cas technology is that it can be directly conducted on living cells, such as mouse oocytes. It has also been applied to mouse embryos, adult mice, human cell lines, roundworms, and a variety of cells from different species of plants.

20.5 REVIEWING THE KEY CONCEPTS

- In the method of site-directed mutagenesis, an oligonucleotide with a region of mismatch directs a mutation into a specific location in the DNA (see Figure 20.17).
- In CRISPR-Cas technology, components of a bacterial genome defense system are used to mutate genes in living cells (see Figure 20.18).

20.5 COMPREHENSION QUESTION

1. The purpose of gene editing is to
   a. determine if a protein binds to a DNA segment.
   b. determine the sequence of a segment of DNA.
   c. alter the sequence of a segment of DNA.
   d. determine if a gene is expressed.

20.6 BLOTTING METHODS TO DETECT GENE PRODUCTS

Learning Outcome:

1. Explain how Northern blotting and Western blotting are used to detect specific RNAs and proteins, respectively.

Thus far, this chapter has largely focused on the analysis and manipulation of DNA. Scientists also need techniques for the identification of gene products, such as the RNA that is transcribed from a particular gene or the protein that is encoded by an mRNA. In this section, we will consider the methodologies for the detection of RNAs and proteins.

Northern Blotting Is Used to Detect a Specific RNA

As discussed earlier in this chapter, reverse transcriptase PCR and real-time PCR are used to detect and quantitate the amount of RNA in a biological sample. Another approach, known as **Northern blotting,** is also used to identify a specific RNA within a mixture of many RNA molecules.* Researchers employ Northern blotting to investigate the transcription of genes at the molecular level. This method can determine if a specific gene is transcribed in a particular cell type, such as nerve or muscle cells, or at a particular stage of development, such as in fetal or adult cells. Also, Northern blotting can reveal if a pre-mRNA transcript is alternatively spliced into two or more mRNAs of different sizes.

To conduct a Northern blotting experiment, RNA is extracted and purified from a sample of living cells. This RNA can be isolated from a particular cell type under a given set of conditions or during a particular stage of development. Any particular cell produces thousands of different types of RNA molecules, because cells express many genes at any given time. After the RNA is extracted from cells and purified, it is loaded onto an agarose gel that separates the RNA transcripts according to their size. (The technique of gel electrophoresis is described in Appendix A.) The RNAs within the gel are then blotted onto a nylon membrane and probed with a labeled fragment of DNA from a cloned gene. RNAs that are complementary to the DNA probe are detected as labeled bands.

Figure 20.19 shows the results of a Northern blotting experiment in which the DNA probe was complementary to an mRNA that encodes a protein called tropomyosin. In lane 1, the mRNA was isolated from smooth muscle cells, in lane 2 from striated muscle cells, and in lane 3 from brain cells. As seen here, smooth and striated muscle cells contain a large amount of this mRNA.

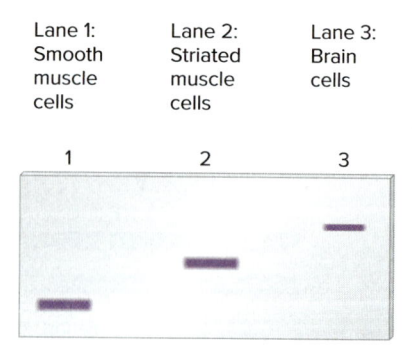

FIGURE 20.19 **The results of Northern blotting.** RNA is isolated from a sample of cells and then separated by gel electrophoresis. The separated RNA bands are blotted into a nylon membrane and then placed in a solution containing a labeled DNA probe that is complementary to tropomyosin mRNA. RNA molecules that are complementary to the probe appear as labeled bands (shown in purple).

*A technique called Southern blotting was named after Edwin Southern. The name *Northern blotting* arose because this method is similar to Southern blotting, which can detect DNA fragments but is no longer widely used.

This result is expected because tropomyosin plays a role in the regulation of cell contraction. By comparison, brain cells have much less of this mRNA. In addition, we see from the locations of the bands that the molecular masses of the three mRNAs differ among the three cell types. This observation indicates that the pre-mRNA is alternatively spliced to contain different combinations of exons.

Western Blotting Is Used to Detect a Specific Protein

For protein-encoding genes, the end result of gene expression is the synthesis of proteins. A particular protein within a mixture of many different protein molecules can be identified by **Western blotting.** This method can determine if a specific protein is made in a particular cell type or at a particular stage of development.

In a Western blotting experiment, proteins are extracted from living cells. Just as with RNAs, any given cell produces many different proteins at any time, because it is expressing many different genes. After the proteins have been extracted from the cells, they are loaded onto a gel that separates them by molecular mass. To perform the separation step, the proteins are first dissolved in sodium dodecyl sulfate (SDS), a detergent that denatures proteins and coats them with negative charges. The negatively charged proteins are then separated in a gel made of polyacrylamide. This method of separating proteins is called SDS-PAGE (polyacrylamide gel electrophoresis).

Following SDS-PAGE, the proteins within the gel are blotted onto a nylon membrane. The next step is to use a probe that recognizes a specific protein of interest. An important difference between Western blotting and Northern blotting is that Western blotting uses an **antibody** as a probe, rather than a labeled DNA strand. Antibodies bind to sites known as **epitopes.** An epitope has a three-dimensional structure that is recognized by an antibody. The term **antigen** refers to any molecule that is recognized by an antibody. An antigen contains one or more epitopes. In the case of proteins, an epitope is a short sequence of amino acids. Because the amino acid sequence is a unique feature of each protein, any given antibody specifically recognizes a particular protein. In Western blotting, this antibody is called the primary antibody for that protein.

After the primary antibody has been given sufficient time to recognize the protein of interest, any unbound primary antibody is washed away, and a secondary antibody is added. A secondary antibody is an antibody that specifically recognizes and binds to a region in the primary antibody. Secondary antibodies, which may be labeled or conjugated to an enzyme, are used for convenience, because secondary antibodies are available commercially. In general, it is easier for researchers to obtain these antibodies from commercial sources rather than labeling their own primary antibodies. In a Western blotting experiment, the secondary antibody provides a way to detect the protein of interest in a gel blot. For example, it is common for the secondary antibody to be linked to the enzyme alkaline phosphatase. When the colorless compound XP (5-bromo-4-chloro-3-indolyl-phosphate) is added to the blotting solution, alkaline phosphatase converts the compound to a dark purple dye (**Figure 20.20a**). Because the secondary antibody binds to the primary antibody, a protein band that is recognized by the primary antibody becomes dark purple (nearly black).

In the Western blot shown in **Figure 20.20b**, the primary antibody recognized β globin. In lane 1, proteins were isolated from mouse red blood cells. The labeled band indicates that the β-globin polypeptide is made in these cells. By comparison, the proteins loaded into lanes 2 and 3 were from brain cells and intestinal cells, respectively. The absence of any bands indicates that these cell types do not synthesize β globin.

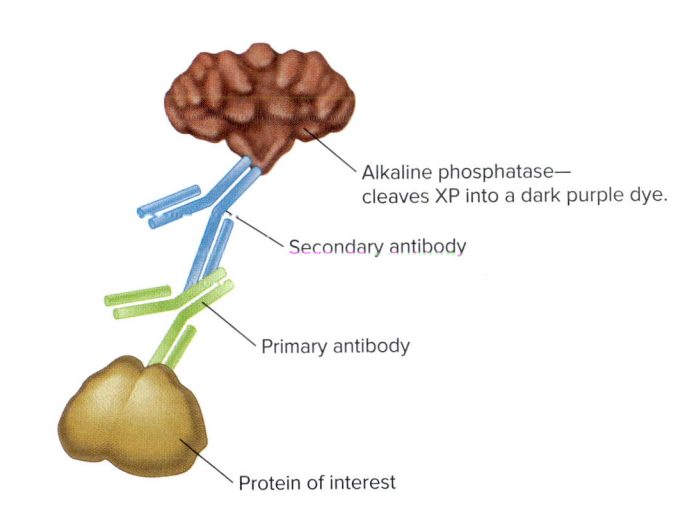

(a) Interactions between the protein of interest and antibodies

(b) Results from a Western blotting experiment

FIGURE 20.20 Western blotting. (a) After blotting, a primary antibody is added that binds to the protein of interest. Then a secondary antibody is added that binds to the primary antibody. In this example, the secondary antibody is also attached to an enzyme called alkaline phosphatase. When the colorless compound XP (5-bromo-4-chloro-3-indolyl phosphate) is added, alkaline phosphatase converts it to a dark purple dye. (b) The dark purple band indicates where the primary antibody has recognized the protein of interest. In lane 1, proteins were isolated from mouse red blood cells. As seen here, the β-globin polypeptide is made in these cells. By comparison, lanes 2 and 3 were samples of proteins from brain cells and intestinal cells, which do not synthesize β globin.

Concept Check: What is the purpose of using a secondary antibody?

Genetic TIPS

The Question: In the Western blot shown here, proteins were extracted from red blood cells obtained from tissue samples at different stages of mouse development. An equal amount of cellular proteins was added to each lane. The primary antibody recognizes the β-globin polypeptide that is found in the hemoglobin protein. Explain these results.

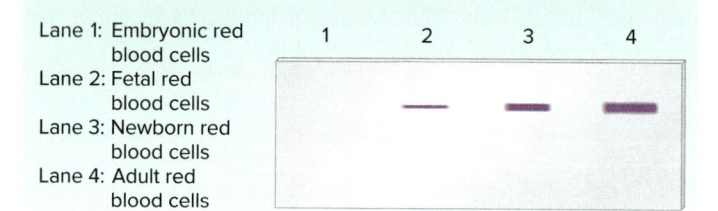

Lane 1: Embryonic red blood cells
Lane 2: Fetal red blood cells
Lane 3: Newborn red blood cells
Lane 4: Adult red blood cells

Topic: What topic in genetics does this question address?

The topic is the use of Western blotting to study gene expression in different cell types.

Information: What information do you know based on the question and your understanding of the topic?

From the question, you know the results of a Western blotting experiment. From your understanding of the topic, you may remember that the thickness of the band reflects the amount of a specific polypeptide, in this case, the β-globin polypeptide.

Problem-Solving Strategy: Compare and contrast.

One strategy to solve this problem is to look at the results, and compare and contrast the bands that are found in different lanes on the Western blot.

Answer: As the blot shows, the amount of β globin increases during development. No detectable β globin is produced during embryonic development. The amount increases significantly during fetal development and becomes maximal in the adult. These results indicate that the β-globin gene is "turned on" in later stages of development, leading to the synthesis of the β-globin polypeptide. This experiment illustrates how a Western blot can provide information concerning the relative amount of a specific protein within living cells.

20.6 REVIEWING THE KEY CONCEPTS

- Northern blotting uses a labeled DNA probe to detect a specific RNA within a mixture of many different RNAs (see Figure 20.19).
- Western blotting uses antibodies to detect a specific protein within a mixture of many different proteins (see Figure 20.20).

20.6 COMPREHENSION QUESTIONS

1. Which of the following methods uses a labeled DNA probe?
   a. Site-directed mutagenesis
   b. Northern blotting
   c. Western blotting
   d. Both a and b

2. Which of the following methods is used to detect a specific RNA within a mixture of many different RNAs?
   a. Site-directed mutagenesis
   b. Northern blotting
   c. Western blotting
   d. None of the above

3. During Western blotting, the primary antibody recognizes
   a. the secondary antibody.
   b. the protein of interest.
   c. an mRNA of interest.
   d. a specific fragment of chromosomal DNA.

20.7 ANALYZING DNA- AND RNA-BINDING PROTEINS

Learning Outcome:

1. Describe how an electrophoretic mobility shift assay is used to determine if a protein binds to DNA or RNA.

Researchers often want to study the binding of proteins to specific sites on a DNA or RNA molecule. For example, the molecular investigation of transcription factors requires methods for identifying interactions between transcription factor proteins and specific DNA sequences. A technically simple, widely used method for identifying DNA- or RNA-binding proteins is the **electrophoretic mobility shift assay (EMSA)**. This technique was used originally to study interactions between specific proteins and rRNA molecules and quickly became popular after its success in studying protein-DNA interactions in the *lac* operon. Now it is commonly used as a technique for detecting interactions between RNA-binding proteins and mRNAs and between transcription factors and DNA regulatory elements.

In the case of DNA-binding proteins, the technical basis for an EMSA is that the binding of a protein to a DNA fragment slows down the fragment's ability to move within a polyacrylamide or agarose gel. During electrophoresis, DNA fragments are pulled through the gel matrix toward the bottom of the gel by a voltage gradient. Smaller molecules migrate more quickly through a gel matrix than do larger ones. A key aspect of this procedure is that the binding of a protein to a DNA fragment slows down the DNA's rate of movement through the gel matrix, because a protein-DNA complex has a higher mass. When comparing a DNA fragment alone to that same DNA fragment when it is within a protein-DNA complex, the complex is shifted to a higher band compared to the DNA alone, because the complex migrates more slowly to the bottom of the gel

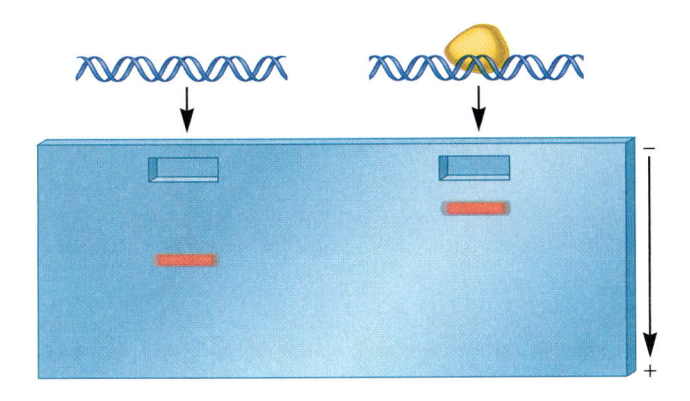

FIGURE 20.21 **The results of an electrophoretic mobility shift assay.** The binding of a protein to a labeled fragment of DNA slows down the fragment's rate of movement through a gel. For the results shown in the lane on the right, if the concentration of the DNA fragment was higher than the concentration of the protein, there would be two bands: one band with protein bound (at a higher molecular mass) and one band without protein bound (corresponding to the band found in the left lane).

(Figure 20.21). The bands are visualized by staining the DNA with a dye such as EtBr. To increase the sensitivity of an EMSA, the DNA can also be labeled with a fluorescent molecule.

An EMSA must be carried out under non-denaturing conditions. This means that the buffers and gel cannot cause the unfolding of proteins or the separation of the DNA double helix. Avoiding denaturation is necessary so that the proteins and DNA retain their proper structure and are thereby able to bind to each other. The nondenaturing conditions of an EMSA differ from those of the more common SDS gel electrophoresis, in which the proteins are denatured by the detergent SDS.

20.7 REVIEWING THE KEY CONCEPTS

- An electrophoretic mobility shift assay (EMSA) can determine if a protein binds to a specific DNA fragment or RNA molecule because the binding of the protein slows down the movement of the DNA or RNA through a gel (see Figure 20.21).

20.7 COMPREHENSION QUESTION

1. In an EMSA, the binding of a protein to DNA
   a. prevents the DNA from being digested with a restriction enzyme.
   b. causes the DNA to migrate more slowly through a gel.
   c. causes the DNA to migrate more quickly through a gel.
   d. inhibits the expression of any genes within the DNA.

KEY TERMS

Page 459. recombinant DNA technology, recombinant DNA molecules
Page 460. gene cloning, vector
Page 461. host cell, plasmids, origin of replication, selectable marker, restriction endonucleases, restriction enzymes
Page 462. DNA ligase, palindromic
Page 463. recombinant vector, competent cells, transformation
Page 465. reverse transcriptase, oligonucleotide, complementary DNA (cDNA)
Page 466. DNA library, genomic library, cDNA library, polymerase chain reaction (PCR), primer
Page 468. template DNA, *Taq* polymerase, primer annealing, primer extension
Page 470. thermocycler, reverse transcriptase PCR, real-time PCR
Page 472. cycle threshold method (C_t method), reproductive cloning
Page 474. stem cells, totipotent, embryonic stem cells (ES cells), pluripotent, embryonic germ cells (EG cells), multipotent
Page 475. unipotent
Page 477. DNA sequencing, dideoxy sequencing, dideoxyribonucleotides (ddNTPs), chain termination, automated DNA sequencing
Page 479. sequencing ladder, gene editing, gene mutagenesis, site-directed mutagenesis
Page 480. CRISPR-Cas technology
Page 482. Northern blotting
Page 483. Western blotting, antibody, epitopes, antigen
Page 484. electrophoretic mobility shift assay (EMSA)

CHAPTER SUMMARY

- Recombinant DNA technology is the use of in vitro molecular techniques to manipulate fragments of DNA and produce new arrangements.

20.1 Gene Cloning Using Vectors

- Gene cloning has many uses, including gene sequencing, site-directed mutagenesis, the use of genes as probes, and the expression of cloned genes (see Table 20.1).
- Cloning vectors are derived from plasmids or viruses (see Table 20.2).
- Restriction enzymes, also called restriction endonucleases, cut chromosomal DNA and vector DNA to produce sticky ends that will hydrogen bond with each other. DNA ligase then makes a covalent link between the DNA backbones (see Figure 20.1, Table 20.3).
- Gene cloning using vectors involves the insertion of the gene into a vector and then its propagation in a living cell such as *E. coli* (see Figure 20.2).
- Complementary DNA (cDNA) is made via reverse transcriptase starting with mRNA (see Figure 20.3).
- A DNA library is a collection of recombinant vectors. The inserts can be chromosomal DNA or cDNA (see Figure 20.4).

486 CHAPTER 20 :: MOLECULAR TECHNOLOGIES

20.2 Polymerase Chain Reaction

- Polymerase chain reaction (PCR) uses oligonucleotide primers to copy a specific region of DNA. Each cycle of PCR involves three steps: denaturation, primer annealing, and primer extension (see Figures 20.5, 20.6).
- Reverse transcriptase PCR begins with mRNA and is used to study gene expression (see Figure 20.7).
- Real-time PCR monitors PCR as it occurs in a thermocycler. It is used to quantitate the amount of starting DNA or mRNA in a sample. Real-time PCR uses fluorescent molecules to follow the PCR reaction. A sample that has an unknown amount of DNA is compared with a standard (see Figures 20.8, 20.9).

20.3 Reproductive Cloning and Stem Cells

- Mammalian reproductive cloning can be achieved by fusing somatic cells to oocytes with their nuclei removed (see Figures 20.10, 20.11).
- Stem cells have the ability to divide and differentiate. Stem cells may be totipotent, pluripotent, multipotent, or unipotent (see Figures 20.12–20.14).
- Stem cells have the potential to treat a variety of human diseases (see Table 20.4).

20.4 DNA Sequencing

- A method of DNA sequencing, called the dideoxy method, uses fluorescently labeled dideoxynucleotides that cause chain termination (see Figures 20.15, 20.16).

20.5 Gene Editing

- In the method of site-directed mutagenesis, an oligonucleotide with a region of mismatch directs a mutation into a specific location in DNA (see Figure 20.17).
- In CRISPR-Cas technology, components of a bacterial genome defense system are used to mutate genes in living cells (see Figure 20.18).

20.6 Blotting Methods to Detect Gene Products

- Northern blotting uses a labeled DNA probe to detect a specific RNA within a mixture of many different RNAs (see Figure 20.19).
- Western blotting uses antibodies to detect a specific protein within a mixture of many different proteins (see Figure 20.20).

20.7 Analyzing DNA- and RNA-Binding Proteins

- An electrophoretic mobility shift assay (EMSA) can determine if a protein binds to a specific DNA fragment or RNA molecule because the binding of the protein slows down the movement of the DNA or RNA through a gel (see Figure 20.21).

PROBLEM SETS & INSIGHTS

More Genetic TIPS

1. RNA was isolated from four different cell types and probed with labeled DNA strands from a cloned gene that is called gene X. The results are shown here.

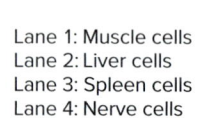

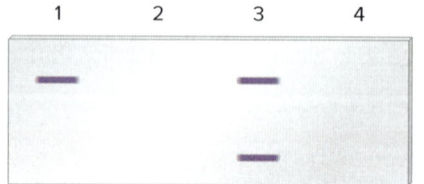

Lane 1: Muscle cells
Lane 2: Liver cells
Lane 3: Spleen cells
Lane 4: Nerve cells

Explain the results of this experiment.

Topic: **What topic in genetics does this question address?**

The topic is the use of Northern blotting to study gene expression in different cell types.

Information: **What information do you know based on the question and your understanding of the topic?**

From the question, you know the results of a Northern blotting experiment. From your understanding of the topic, you may remember that the thickness of the band reflects the amount of RNA that is made from a specific gene. Also, if the locations of the bands indicate that the RNAs have different molecular masses, alternative splicing has occurred.

Problem-Solving **S**trategy: **Analyze data. Compare and contrast.**

One strategy to solve this problem is to look at the results of this experiment, and compare and contrast the relative thicknesses and locations of the bands that are found in different lanes of the gel.

Answer: In this Northern blot, a labeled band appears in those lanes where RNA was isolated from muscle and spleen cells but not from liver and nerve cells. These results indicate that the muscle and spleen cells express a significant amount of RNA from gene X, but the liver and nerve cells do not. The muscle cells show a single band, whereas the spleen cells show this band plus a second band of lower molecular mass. An interpretation of these results is that the spleen cells can alternatively splice the RNA to produce a second RNA containing fewer exons.

2. The sequence of a region of a DNA strand is 3′–ATACGAC-TAGTCGGGACCATATC–5′. If the primer in a dideoxy sequencing experiment anneals just to the left of this sequence, draw the sequencing ladder that will be obtained.

Topic: What topic in genetics does this question address?

The topic is DNA sequencing. More specifically, the question is about predicting the banding pattern on a DNA sequencing gel.

Information: What information do you know based on the question and your understanding of the topic?

From the question, you know the sequence of a region of a template strand that is adjacent to an annealing site. From your understanding of the topic, you may remember that each dideoxynucleotide is labeled with a different color and that the strand that is made has a sequence that is complementary to the template strand. The bases that are closer to the annealing site will be contained in smaller DNA fragments and will move more quickly to the bottom of the gel.

Problem-Solving Strategy: Make a drawing.

A strategy to solve this problem is to write out the complementary sequence, 5′–TATGCTGATCAGCCCTGGTATAG–3′. The first T (at the 5′ end) will be near the bottom of the gel, the next A will be slightly higher in the gel, the third base, a T, will be slightly higher, and so on.

Answer:

G = Yellow
A = Green
T = Red
C = Blue

Conceptual Questions

C1. Discuss three important advances that have resulted from gene cloning.

C2. What is a restriction enzyme? What structure does it recognize? What type of chemical bond does it cleave? Be as specific as possible.

C3. Write a double-stranded DNA sequence that is 20 base pairs in length and is palindromic.

C4. What is cDNA? In eukaryotes, how does cDNA differ from genomic DNA?

C5. As described in Chapter 6, not all inherited traits are determined by nuclear genes (i.e., genes located in the cell nucleus) that are expressed during the life of an individual. In particular, maternal effect genes and mitochondrial genes are notable exceptions. With these ideas in mind, let's consider the cloning of a sheep (e.g., Dolly).

A. With regard to maternal effect genes, is the phenotype of such a cloned animal determined by the animal that donated the enucleated egg or by the animal that donated the somatic cell nucleus? Explain.

B. Does the cloned animal inherit extranuclear traits from the animal that donated the egg or from the animal that donated the somatic cell? Explain.

C. In what ways would you expect the cloned animal to be similar to or different from the animal that donated the somatic cell? Is it accurate to call such an animal a "clone" of the animal that donated the nucleus?

C6. Draw the structural feature of a dideoxyribonucleotide that causes chain termination. Explain how it does this.

Application and Experimental Questions

E1. What is the functional significance of sticky ends in a cloning experiment? What type of bonding makes the ends sticky?

E2. Table 20.3 describes the cleavage sites of five different restriction enzymes. After these restriction enzymes have cleaved the DNA, four of them produce sticky ends that can hydrogen bond with complementary sticky ends, as shown in Figure 20.1. The efficiency with which sticky ends bind together depends on the number of hydrogen bonds; more hydrogen bonds makes the ends "stickier" and more likely to stay attached. Rank the four restriction enzymes from Table 20.3 (from best to worst) with regard to the efficiency of their sticky ends binding to each other.

E3. Describe the important features of cloning vectors. Explain the purpose of selectable markers in cloning experiments.

E4. How does gene cloning produce many copies of a gene?

E5. In your own words, describe the series of steps necessary to clone a gene.

E6. What is a recombinant vector? How is a recombinant vector constructed? Explain how X-Gal is used in a method of identifying recombinant vectors that contain segments of chromosomal DNA.

E7. What is a DNA library? Do you think this name is appropriate?

E8. Some vectors used in cloning experiments contain bacterial promoters that are adjacent to unique cloning sites. This makes it possible to insert a gene sequence next to the bacterial promoter and express the gene in bacterial cells. These vectors are called expression vectors. If you wanted to express a eukaryotic protein in bacterial cells, would you insert genomic DNA or cDNA into the expression vector? Explain your choice.

E9. Why is a thermostable form of DNA polymerase (e.g., *Taq* polymerase) used in PCR? Is it necessary to use a thermostable form of DNA polymerase in the dideoxy method or site-directed mutagenesis?

E10. Starting with a sample of RNA that contains the mRNA for the β-globin gene, explain how you could create many copies of the β-globin cDNA using reverse transcriptase PCR.

E11. What type of probe is used for real-time PCR? Explain how the level of fluorescence correlates with the amount of PCR product.

E12. What phase of PCR (exponential, linear, or stationary) is analyzed to quantitate the amount of DNA or RNA in a sample? Explain why this phase is chosen.

E13. Evidence (see P. G. Shiels, A. J. Kind, K. H. Campbell, et al. [1999], Analysis of telomere lengths in cloned sheep, *Nature* 399, 316–17) shows that Dolly may have been genetically older than her actual age. As mammals age, the chromosomes in somatic cells tend to shorten from the telomeres. Therefore, older individuals have shorter chromosomes in their somatic cells than do younger ones. When researchers analyzed the chromosomes in the somatic cells of Dolly when she was about 3 years old, the lengths of her chromosomes were consistent with a sheep that was significantly older, say, 9 or 10 years old. (Note: As described in this chapter, the sheep that donated the somatic cell that produced Dolly was 6 years old, and her mammary cells had been grown in culture for several cell doublings before a mammary cell was fused with an oocyte.)

　A. Suggest an explanation why Dolly's chromosomes seemed so old.

　B. Let's suppose that Dolly at age 11 gave birth to a lamb named Molly; Molly was produced naturally (by mating Dolly with a normal male). When Molly was 8 years old, a sample of somatic cells was analyzed. How old would you expect Molly's chromosomes to appear, based on the phenomenon of telomere shortening? Explain your answer.

　C. Discuss how the observation of chromosome shortening, which was observed in Dolly, might affect the popularity of reproductive cloning.

E14. What is reproductive cloning? Are identical twins in humans considered to be clones? With regard to agricultural species, what are some potential advantages of reproductive cloning?

E15. DNA sequencing can help us to identify mutations within genes. The following data are derived from an experiment in which a normal gene and a mutant gene have been sequenced:

G = Yellow
A = Green
T = Red
C = Blue

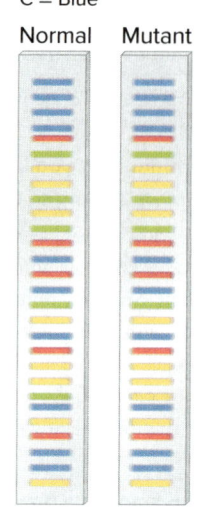

Locate and describe the mutation.

E16. A sample of DNA was subjected to automated DNA sequencing, and the printout is shown here.

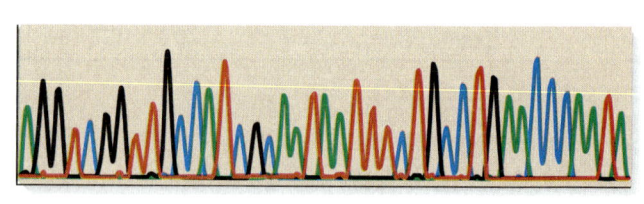

G = Black   T = Red
A = Green   C = Blue

What is the sequence of this DNA segment?

E17. A portion of the coding sequence of a cloned gene is shown here:

5′–GCCCCCGATCTACATCATTACGGCGAT–3′
3′–CGGGGGCTAGATGTAGTAATGCCGCTA–5′

This portion of the gene encodes a polypeptide with the amino acid sequence alanine–proline–aspartic acid–leucine–histidine–histidine–tyrosine–glycine–aspartic acid. Using the method of site-directed mutagenesis, a researcher wants to change the leucine codon into an arginine codon, using an oligonucleotide that is 19 nucleotides long. What is the sequence of the oligonucleotide that should be used? Designate the 5′ and 3′ ends of the oligonucleotide in your answer. Note: The mismatch should be in the middle of the oligonucleotide, and a 1-base mismatch is preferable over a 2- or 3-base mismatch. Use the bottom strand as the template strand for this site-directed mutagenesis experiment.

E18. Let's suppose you want to use site-directed mutagenesis to investigate a DNA sequence that functions as a response element for hormone binding. From previous work, you have narrowed down the response element to a sequence of DNA that is 20 bp in length with the following sequence:

5′–GGACTGACTTATCCATCGGT–3′
3′–CCTGACTGAATAGGTAGCCA–5′

As a strategy to pinpoint the actual response element sequence, you decide to make 10 different site-directed mutants and then analyze their effects by an electrophoretic mobility shift assay. What mutations would you make? What results would you expect to obtain?

E19. Gene editing is also used to explore the structure and function of proteins. For example, changes can be made to the coding sequence of a gene to determine how alterations in the amino acid sequence affect the function of a protein. Let's suppose that you are interested in the functional importance of a particular glutamic acid (an amino acid) within a protein you are studying. By gene editing, you make mutant proteins in which the glutamic acid codon has been changed to other codons. You then test the encoded mutant proteins for functionality. The results are as follows:

	Functionality
Normal protein	100%
Mutant proteins containing	
Tyrosine	5%
Phenylalanine	3%
Aspartic acid	94%
Glycine	4%

From these results, what would you conclude about the functional significance of this glutamic acid within the protein?

E20. In Northern and Western blotting, what is the purpose of gel electrophoresis?

E21. What is the purpose of a Northern blotting experiment? What types of information can it tell you about the transcription of a gene?

E22. Suppose an X-linked gene in mice exists as two alleles, B and b. X-chromosome inactivation, a process in which one X chromosome is turned off, occurs in the somatic cells of female mammals (see Chapter 4). Allele B encodes an mRNA that is 900 nucleotides long, whereas allele b contains a small deletion that shortens the mRNA to a length of 825 nucleotides. Draw the expected Northern blot that will be obtained using mRNA isolated from somatic tissue of the following mice:

Lane 1. mRNA from an $X^b Y$ male mouse

Lane 2. mRNA from an $X^b X^b$ female mouse

Lane 3. mRNA from an $X^B X^b$ female mouse. Note: The sample taken from the female mouse is not from a clone of cells.

E23. The method of Northern blotting is used to determine the amount and size of a particular RNA transcribed in a given cell type. Alternative splicing (discussed in Chapter 14) produces mRNAs of different lengths from the same gene. The Northern blot shown here was obtained using a DNA probe that is complementary to the mRNA encoded by a particular gene. The mRNA in lanes 1 through 4 was isolated from different cell types, and equal amounts of total cellular mRNA were added to each lane.

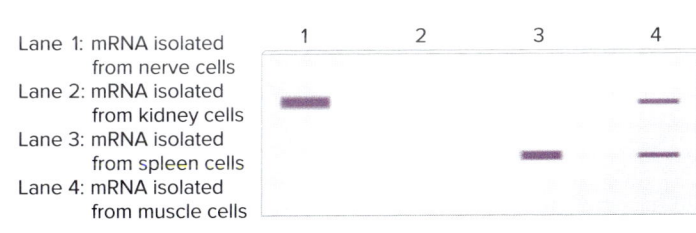

Lane 1: mRNA isolated from nerve cells
Lane 2: mRNA isolated from kidney cells
Lane 3: mRNA isolated from spleen cells
Lane 4: mRNA isolated from muscle cells

Explain these results.

E24. The technique of Northern blotting depends on the phenomenon in which a probe binds to mRNA. Explain why this binding occurs.

E25. In the Western blot shown here, proteins were isolated from red blood cells and muscle cells from two different individuals. One individual was unaffected, and the other suffered from a disease known as thalassemia, which involves a defect in hemoglobin. The blot was exposed to an antibody that recognizes β globin, which is one of the polypeptides that constitute hemoglobin. Equal total amounts of cellular proteins were added to each lane.

Lane 1: Proteins isolated from normal red blood cells
Lane 2: Proteins isolated from the red blood cells of a thalassemia patient
Lane 3: Proteins isolated from normal muscle cells
Lane 4: Proteins isolated from the muscle cells of a thalassemia patient

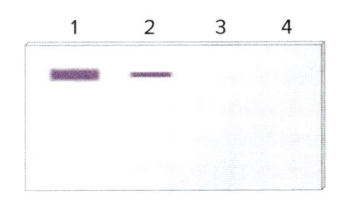

Explain these results.

E26. Let's suppose a researcher was interested in the effects of mutations on the expression of a protein-encoding gene for a protein that is 472 amino acids in length. This protein is expressed in leaf cells of *Arabidopsis thaliana*. It has a molecular mass of approximately 56,640 Da. Make a drawing that shows the expected results of a Western blot using proteins isolated from the leaf cells that were obtained from the following plants:

Lane 1. A plant homozygous for a nonmutant gene

Lane 2. A plant homozygous for a deletion that removes the promoter for this gene

Lane 3. A heterozygous plant in which one gene is nonmutant and the other gene has a mutation that introduces an early stop codon at codon 112

Lane 4. A plant homozygous for a mutation that introduces an early stop codon at codon 112

Lane 5. A plant homozygous for a mutation that changes codon 108 from a phenylalanine codon into a leucine codon

E27. Explain the basis for using an antibody as a probe in a Western blotting experiment.

E28. A cloned gene fragment contains a regulatory element that is recognized by a regulatory transcription factor. Previous experiments have shown that the presence of a hormone results in transcriptional activation by this transcription factor. To study this effect, you conduct an electrophoretic mobility shift assay and obtain the following results:

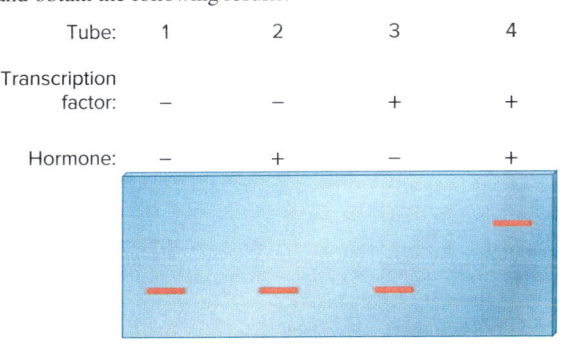

Tube:	1	2	3	4
Transcription factor:	–	–	+	+
Hormone:	–	+	–	+

Explain the action of the hormone.

E29. Describe the rationale behind an electrophoretic mobility shift assay.

E30. An electrophoretic mobility shift assay can be used to study the binding of proteins to a segment of DNA. In the results shown here, an electrophoretic mobility shift assay was used to examine the requirements for the binding of RNA polymerase II (from eukaryotic cells) to the promoter of a protein-encoding gene. The assembly of general transcription factors and RNA polymerase II at the core promoter is described in Chapter 14 (Figure 14.14). In this experiment, the segment of DNA containing a promoter sequence was 1100 bp in length. The fragment was mixed with various combinations of proteins and then subjected to an electrophoretic mobility shift assay.

Explain which proteins (TFIID, TFIIB, or RNA polymerase II) are able to bind to this DNA fragment by themselves. Which transcription factors (i.e., TFIID or TFIIB) are needed for the binding of RNA polymerase II?

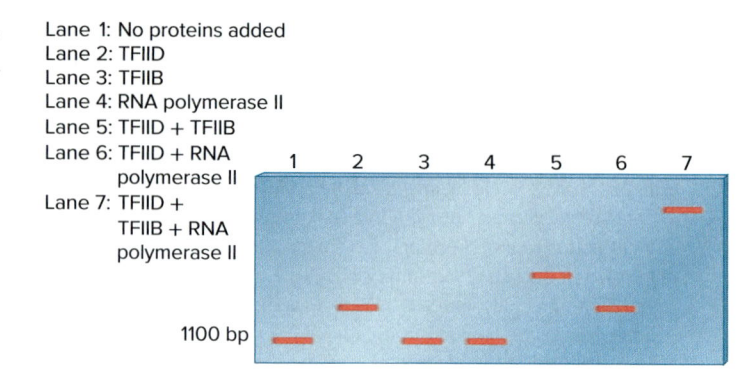

Lane 1: No proteins added
Lane 2: TFIID
Lane 3: TFIIB
Lane 4: RNA polymerase II
Lane 5: TFIID + TFIIB
Lane 6: TFIID + RNA polymerase II
Lane 7: TFIID + TFIIB + RNA polymerase II

Questions for Student Discussion/Collaboration

1. Discuss and make a list of some of the reasons why determining the amount of a particular gene product would be useful to a geneticist. Use specific examples of known genes (e.g., the gene for β globin and other genes) when making your list.

2. Make a list of possible research questions that could be answered using site-directed mutagenesis or CRISPR-Cas technology.

Answers to Comprehension Questions

20.1: d, b, c, a, a

20.2: c, b, d

20.3: c, d

20.4: b

20.5: c

20.6: b, b, b

20.7: b

Note: All answers appear in Connect; the answers to even-numbered questions and all Concept Check questions are in Appendix B.

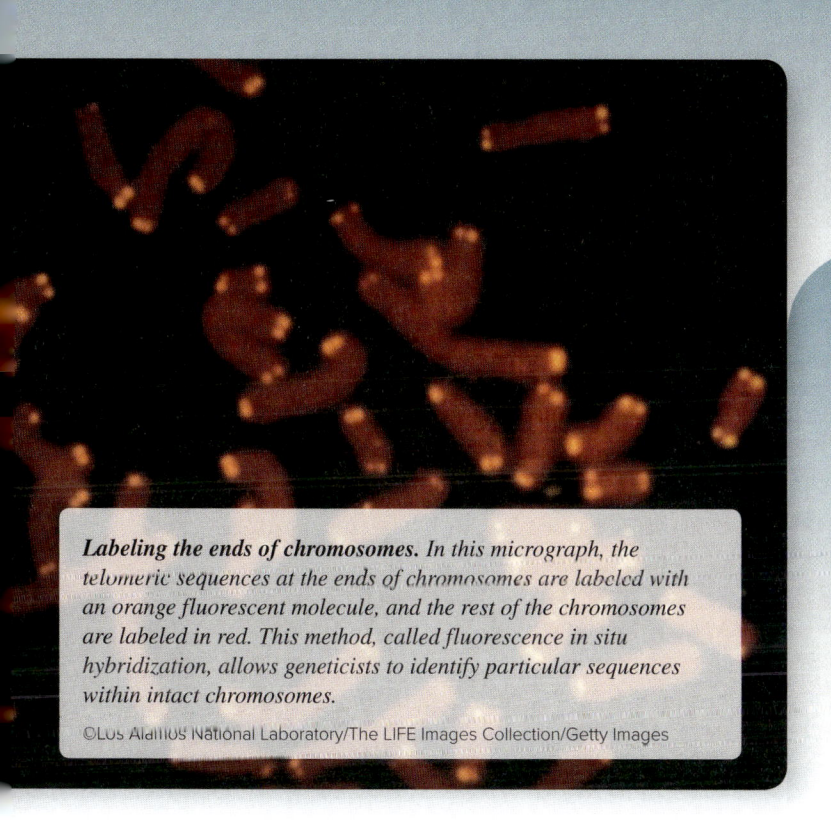

21

CHAPTER OUTLINE

- 21.1 Overview of Chromosome Mapping
- 21.2 In Situ Hybridization
- 21.3 Molecular Markers
- 21.4 Genome-Sequencing Projects
- 21.5 Metagenomics
- 21.6 Functional Genomics

Labeling the ends of chromosomes. In this micrograph, the telomeric sequences at the ends of chromosomes are labeled with an orange fluorescent molecule, and the rest of the chromosomes are labeled in red. This method, called fluorescence in situ hybridization, allows geneticists to identify particular sequences within intact chromosomes.

©Los Alamos National Laboratory/The LIFE Images Collection/Getty Images

GENOMICS

The term **genome** refers to the total genetic composition of an organism or species. For example, the nuclear genome of humans is composed of 22 different autosomes, and an X and (in males) a Y chromosome. In addition, human cells have a mitochondrial genome composed of a single circular chromosome.

As genetic technology has progressed over the past few decades, researchers have gained an increasing ability to analyze the composition of genomes as whole units. The term **genomics** refers to the molecular analysis of the entire genome of a species. Genome analysis is a molecular dissection process applied to a complete set of chromosomes. Segments of chromosomes are analyzed in progressively smaller pieces, the locations of which are known on the intact chromosomes. This is the mapping phase of genome analysis. The mapping of the genome ultimately progresses to the determination of the complete DNA sequence for all of a species' chromosomes.

In 1995, a team of researchers headed by Craig Venter and Hamilton Smith obtained the first complete DNA sequence of the genome from the bacterium *Haemophilus influenzae*. Its genome consists of a single circular chromosome that is composed of 1.83 million base pairs (bp) of DNA and contains approximately 1743 genes. In 1996, the first entire DNA sequence of a eukaryote, *Saccharomyces cerevisiae* (baker's yeast), was completed.

In this chapter, we will focus on methods aimed at elucidating the organization of the sequences within a species' genome. This process may begin with the mapping of regions along the species' chromosomes. The process is finished when the complete DNA sequence has been determined. We will first consider three mapping strategies—cytogenetic mapping, linkage mapping, and physical mapping. Then we explore genome-sequencing projects—research endeavors that have the ultimate goal of determining the sequence of DNA bases of the entire genome of a given species. We will examine the methods, goals, and results of these large undertakings, which include the Human Genome Project.

Once a genome sequence is known, researchers can examine, at the level of many genes, how the numerous genes that make up a genome interact to produce the traits of an organism. This research area is called **functional genomics.** In this chapter, we will also consider some recent advances in functional genomics.

21.1 OVERVIEW OF CHROMOSOME MAPPING

Learning Outcomes:
1. Define *mapping*.
2. Distinguish between cytogenetic, linkage, and physical mapping.

In genetics, the term **mapping** refers to the experimental process of determining the relative locations of genes or other segments of DNA on individual chromosomes. Researchers may follow any of three general approaches to mapping a chromosome: cytogenetic, linkage, or physical mapping.

- **Cytogenetic mapping** (also called cytological mapping) is aimed at determining the locations of specific sequences, such as gene sequences, within chromosomes that are viewed microscopically. When stained, each chromosome of a given species has a characteristic banding pattern, and genes are mapped cytogenetically relative to a band location.
- **Linkage mapping**, which is discussed in Chapter 7, uses the frequency of recombination between different genes to determine their relative spacing and order along a chromosome. In eukaryotes, linkage mapping involves crosses among organisms that are heterozygous for two or more genes. The number of recombinant offspring provides a relative measure of the distance between genes, which is expressed in map units (mu).
- **Physical mapping** involves DNA-cloning and/or DNA-sequencing techniques to determine the location of and distance between genes and other DNA regions. In a physical map, the distances are given as the number of base pairs between genes.

A **genetic map**, or **chromosome map**, is a diagram that shows the relative locations of genes or other DNA segments on a chromosome. The term **locus** (plural, **loci**) refers to the site within a genetic map where a specific gene or other DNA segment is found. **Figure 21.1** compares genetic maps that show the loci for two X-linked genes, *sc* (for "scute," a gene affecting bristle morphology) and *w* (a gene affecting eye color), in *Drosophila melanogaster*.

- In the cytogenetic map at the top, the *sc* gene is located at band 1A8, and the *w* gene is located at band 3B6.
- In the linkage map in the middle, genetic crosses indicate that the two genes are approximately 1.5 map units (mu) apart.
- The physical map at the bottom shows that the two genes are approximately 2.4×10^6 bp apart on the X chromosome.

Correlations between cytogenetic, linkage, and physical maps often vary from species to species and from one region of the chromosome to another. For example, a distance of 1 mu may correspond to 1 to 2 million bp in one region of the chromosome, but other regions may recombine at a much lower rate, so a distance of 1 mu may be a much longer physical segment of DNA.

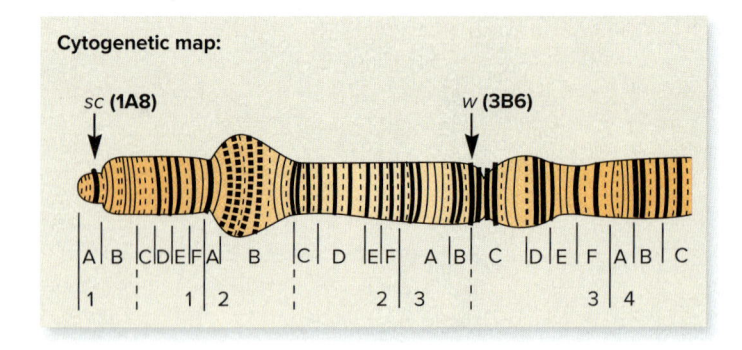

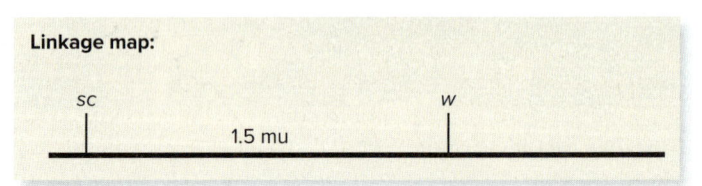

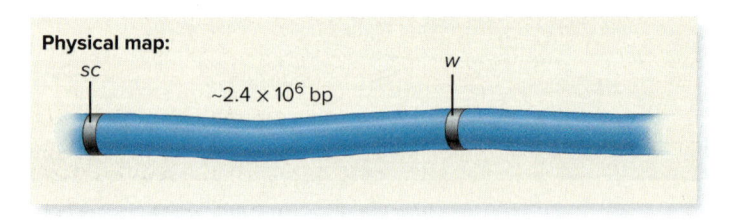

FIGURE 21.1 A comparison of cytogenetic, linkage, and physical maps. Each of these maps shows the distance between the *sc* and *w* genes on the X chromosome in *Drosophila melanogaster*. The cytogenetic map is that of the polytene chromosome.

Concept Check: What is a genetic map?

21.1 REVIEWING THE KEY CONCEPTS

- The term *mapping* refers to the experimental process of determining the relative locations of genes or other DNA segments on a chromosome, thereby producing a diagram called a genetic map. The process may involve cytogenetic, linkage, or physical mapping techniques (see Figure 21.1).

21.1 COMPREHENSION QUESTION

1. What type of chromosome mapping relies on crosses?
   a. Cytogenetic mapping
   b. Linkage mapping
   c. Physical mapping
   d. All of the above rely on crosses.

21.2 IN SITU HYBRIDIZATION

Learning Outcomes:
1. Outline the method of fluorescence in situ hybridization.
2. Define *chromosome painting*.

The technique of **in situ hybridization** is used to cytogenetically map the locations of genes or other DNA sequences within large

FIGURE 21.2 The technique of fluorescence in situ hybridization (FISH). The probe hybridizes to the denatured chromosomal DNA only at a specific, complementary site in the genome. Note that the chromosomes are highly condensed metaphase chromosomes that have already replicated. These are sister chromatids. Therefore, each X-shaped chromosome actually contains two copies of a particular gene. Because the sister chromatids are identical, a probe that recognizes a site on one sister chromatid will also bind to the same site on the other.

Concept Check: Why does the probe bind to a specific site on a chromosome?

eukaryotic chromosomes. The term *in situ* (from the Latin for "in place") indicates that the procedure is conducted on chromosomes that are being held in place—adhered to a surface. The term **hybridization** refers to the fact that a labeled strand of DNA forms a hybrid with an intact chromosome via base pairing.

To map a gene via in situ hybridization, researchers use a labeled probe to detect the location of the gene within a set of chromosomes. If the gene of interest has been cloned previously, as described in Chapter 20, the DNA of the cloned gene can be used as a probe. Because a DNA strand from a cloned gene hybridizes only to its complementary sequence on a particular chromosome, this technique provides the ability to localize the gene of interest. For example, let's consider the gene that causes the white-eye phenotype in *Drosophila* when it carries a loss-of-function mutation. This gene has already been cloned. If a single-stranded piece of this cloned DNA is mixed with *Drosophila* chromosomes in which the DNA has been denatured, the cloned piece will bind only to the X chromosome at the location corresponding to the site of the eye color gene.

The most common method of in situ hybridization uses fluorescently labeled DNA probes and is referred to as **fluorescence in situ hybridization (FISH)**. **Figure 21.2** shows the steps of the FISH procedure.

1. The cells, which are prepared using a technique that keeps the chromosomes intact, are treated with agents that cause them to swell, and their contents are fixed to the slide.
2. The chromosomal DNA is then denatured.
3. A DNA probe is added. For example, the added DNA probe might be a cloned piece of single-stranded DNA that is complementary to a specific gene. In this case, the goal of the FISH experiment is to determine the location of the gene within a set of chromosomes. The probe binds to a site in the chromosomes where the gene is located because the probe and chromosomal gene line up and hydrogen bond with each other.
4. To detect where the probe has bound to a chromosome, the probe is subsequently tagged with a fluorescent molecule. Tagging may be accomplished by first incorporating biotin-labeled nucleotides into the probe. Biotin, a small, nonfluorescent molecule, has a very high affinity for a protein called avidin. Fluorescently labeled avidin is then added, and it binds tightly to the biotin and thereby labels the probe as well.

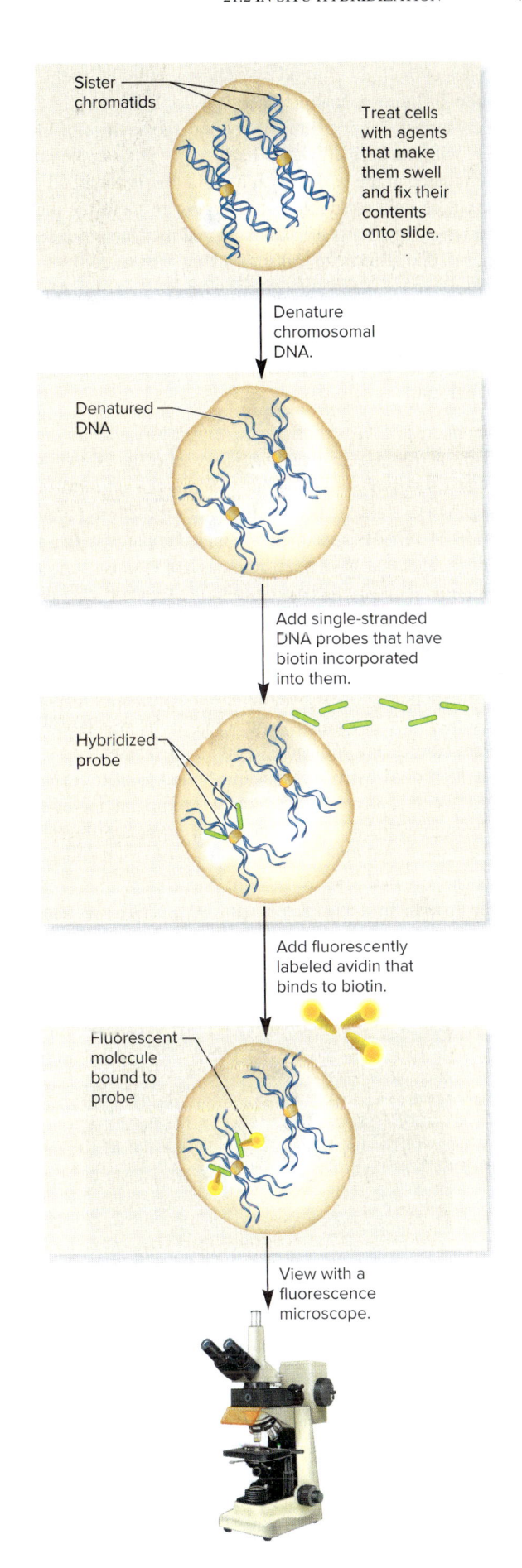

5. To detect the light emitted by a fluorescently labeled probe, a fluorescence microscope is used. Such a microscope contains filters that allow the passage of light only within a defined wavelength range. The sample is illuminated at the wavelength of light that is absorbed by the fluorescent molecule. The fluorescent molecule then emits light at a longer wavelength and the fluorescence microscope allows the transmission of the emitted light. Because filters prevent the transmission of light of other wavelengths, only the emitted light is viewed, and the background of the sample is dark. Therefore, the fluorescence is seen as a brightly glowing color on a dark background.

For many FISH experiments, chromosomes are counterstained by a fluorescent dye that is specific for DNA. A commonly used dye is DAPI (4′,6-diamidino-2-phenyl-indol), which is excited by UV light. This dye gives all of the DNA a blue color. The results of a FISH experiment are then compared with a sample of chromosomes that have been stained with Giemsa stain to produce banding. This comparison allows the location of a probe to be mapped relative to the banding pattern.

Figure 21.3 illustrates the results of an experiment involving six different DNA probes. The six probes were strands of DNA corresponding to six different DNA segments within human chromosome 5. In this experiment, each probe was labeled with a different fluorescent molecule. This enabled researchers to distinguish the probes when they became bound to their corresponding locations on chromosome 5. Computer-imaging methods were used to assign each fluorescently labeled probe a different color. This technique is called **chromosome painting.** In this way, FISH discerns the sites along chromosome 5 corresponding to the six different probes. In a visual, colorful way, FISH was used here to determine the order and relative distances between six specific sites along a single chromosome. FISH is commonly used in genetics and cell biology research, and its use has become more widespread in clinical applications. For example, clinicians may use FISH to detect changes in chromosome structure such as deletions, duplications, and translocations, which may occur in patients with genetic disorders.

Genetic TIPS

The Question: The disease called phenylketonuria (PKU) is a recessive disorder in humans that is due to a loss-of-function mutation involving the gene that encodes the enzyme phenylalanine hydroxylase. Some people with the disorder carry a point mutation that causes the loss of function, whereas other individuals have been shown to have a deletion of the entire gene. Explain how you could use fluorescence in situ hybridization to distinguish a point mutation from a deletion.

Topic: What topic in genetics does this question address? The topic is the use of FISH to distinguish a point mutation from a deletion.

Information: What information do you know based on the question and your understanding of the topic?

From the question, you know that some people with PKU have a point mutation in the gene that encodes phenylalanine hydroxylase, whereas others have experienced a deletion of that gene. From your understanding of the topic, you may remember that FISH is used to detect a gene's location within an intact set of chromosomes.

Problem-Solving Strategy: Design an experiment.

One strategy to solve this problem is to design an experiment using FISH.

Answer: You basically want to follow the procedure in Figure 21.2.

1. Obtain a blood sample from the affected individual. As a control, obtain a blood sample from an unaffected and unrelated individual. Treat the cells with agents that cause them to swell, and fix their contents to a slide.
2. Denature the chromosomal DNA.
3. Add a fluorescently labeled probe. In this case, the probe would be a strand of DNA that is complementary to one of the strands of the phenylalanine hydroxylase gene.
4. View under a fluoresence microscope.

Expected results: In the control, you would see bright spots where the phenylalanine hydroxylase gene is located. If the affected individual had a point mutation, you would still see the spots. If there had been a deletion of both copies of the gene, you would not see any fluorescently labeled spots.

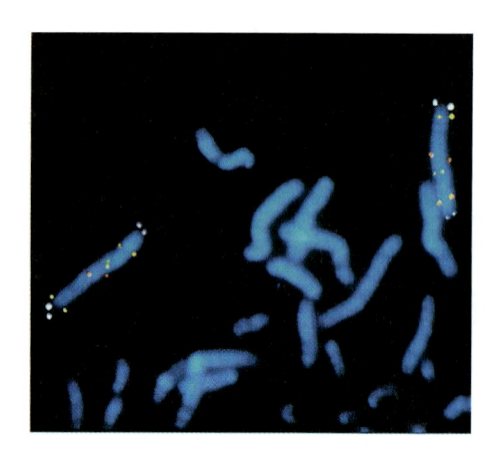

FIGURE 21.3 **Chromosome painting via fluorescence in situ hybridization.** In this experiment, six different probes were used to locate six different sites along human chromosome 5. The colors are due to computer imaging of the fluorescence emission; they are not the actual colors of the fluorescent labels. Two spots are usually seen at each site because the probe binds to both sister chromatids.

From: Ried, T., Baldini, A., Rand, T.C., and Ward, D.C. (1992) Simultaneous visualization of seven different DNA probes by in situ hybridization using combinatorial fluorescence and digital imaging microscopy. *PNAS* 89: 4.1388-1392. Courtesy Thomas Ried

21.2 REVIEWING THE KEY CONCEPTS

- Fluorescence in situ hybridization (FISH) is a method commonly used to map genes and other segments on an intact chromosome (see Figures 21.2, 21.3).

21.2 COMPREHENSION QUESTION

1. The technique of fluorescence in situ hybridization involves the use of a _____ that hybridizes to a _____.
   a. radiolabeled probe, band on a gel
   b. radiolabeled probe, specific site on an intact chromosome
   c. fluorescent probe, band on a gel
   d. fluorescent probe, specific site on an intact chromosome

21.3 MOLECULAR MARKERS

Learning Outcomes:
1. Define *molecular marker*.
2. Explain the use of microsatellites in mapping studies.

In Chapter 7, we explored how allelic differences between genes can be used to map the relative locations of those genes along a chromosome by conducting testcrosses. In this section, we will focus on the use of molecular markers to map genes. These markers can be used in both linkage mapping and physical mapping.

Molecular Markers Are Short DNA Segments That Are Identified Using Molecular Techniques

As an alternative to relying on allelic differences between genes, geneticists have realized that regions of DNA that do not encode genes can be used as markers along a chromosome. A **molecular marker** is a segment of DNA that is found at a specific site along a chromosome and has properties that enable it to be uniquely recognized using molecular tools, such as polymerase chain reaction (PCR) and gel electrophoresis.

Like alleles, molecular markers may be **polymorphic;** that is, within a population, they may vary from individual to individual. Therefore, the distances between linked molecular markers can be determined from the outcomes of crosses. Using molecular techniques, researchers have found it easier to identify many molecular markers within a given species' genome rather than identifying allelic differences that affect traits. For this reason, geneticists have increasingly turned to molecular markers as points of reference along genetic maps. As **Table 21.1** indicates, many different kinds of molecular markers are used by geneticists.

Researchers have constructed detailed genetic maps in which a series of many molecular markers have been identified along each chromosome of certain species. These species include humans, model organisms, agricultural species, and many others. Why are molecular markers useful? One key reason is that molecular markers can be used to determine the approximate location of an unknown gene that causes a human disease. Clinical geneticists sometimes follow the transmission patterns of polymorphic molecular markers in family pedigrees to locate genes that cause human disease when they are mutant. The discovery of a particular marker in those who have the disease can indicate that the marker is close to the disease-causing allele (see Chapter 22, Figures 22.6 and 22.7).

In addition, molecular markers may help researchers identify the locations of genes involved in quantitative traits, such as fruit yield and meat weight, that are valuable in agriculture. The use of molecular markers to identify such genes is described in Chapter 24 (see Figure 24.5). Genetic maps with a large number of markers are used by evolutionary biologists to determine patterns of genetic variation within a species and the evolutionary relatedness of different species.

As an example, **Figure 21.4** shows a simplified RFLP map of two chromosomes found in the plant *Arabidopsis thaliana* (a small plant in the mustard family), which is one of the favorite model organisms of plant molecular geneticists. Many RFLPs, which are described in Table 21.1, have been mapped to different locations on the *Arabidopsis* chromosomes.

TABLE 21.1

Common Types of Molecular Markers

Marker	Description
Restriction fragment length polymorphism (RFLP)	A site in a genome where the distance between two restriction sites varies among different individuals. These sites are identified by restriction enzyme digestion of chromosomal DNA and the identification of DNA fragments that vary in length.
Amplified restriction fragment length polymorphism (AFLP)	This type of marker is the same as an RFLP except that the fragment is amplified via PCR instead of isolating the chromosomal DNA.
Microsatellite	A site in the genome that contains many short sequences that are repeated many times in a row. The total length is usually in the range 50–200 bp, and the length of a given microsatellite may be polymorphic within a population. Microsatellites are isolated via PCR. They are also called short tandem repeats (STRs) and simple sequence repeats (SSRs).
Single-nucleotide polymorphism (SNP)	A site in a genome where a single nucleotide is polymorphic among different individuals. These sites occur commonly in all genomes, and they are becoming more widely used in the mapping of disease-causing alleles and of genes that contribute to quantitative traits that are valuable in agriculture (see Chapter 24).
Sequence-tagged site (STS)	This is a general term to describe any molecular marker that is found at a unique site in a genome and is amplified by PCR. AFLPs, microsatellites, and SNPs can provide sequence-tagged sites within a genome.

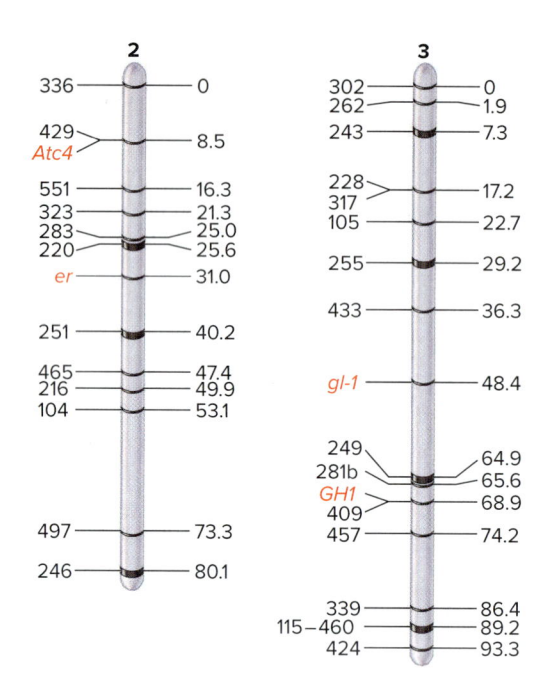

FIGURE 21.4 An RFLP linkage map of *Arabidopsis thaliana.* This plant has five different chromosomes, but only the maps of chromosomes 2 and 3 are shown here. These maps show the locations of many RFLP markers. The numbers along the left side of each chromosome designate the locations of those markers. The numbers along the right side of each chromosome are the map distances in map units. For example, the RFLPs at map positions 16.3 and 40.2 on chromosome 2 are designated 551 and 251, respectively, and are 23.9 mu apart. The top marker at the end of each chromosome was arbitrarily assigned as the starting point (zero) for that chromosome. In addition, the map shows the locations of a few known genes (shown in red): *Atc4* = actin, *er* = erecta, *gl-1* = glabra-1, and *GH1* = acetolactate synthase.

Genes → Traits As a first step in mapping the locations of an organism's genes, researchers may initially determine the sites of RFLPs along the chromosomes. In this case, researchers determined the locations of many of these sites along the *Arabidopsis* chromosomes. Mapping the RFLP sites makes it easier to locate genes within the *Arabidopsis* genome. The identification of genes helps researchers to understand the relationship between genes and traits.

Linkage Mapping Commonly Uses Molecular Markers

To make a highly refined linkage map of a genome, researchers must identify many different polymorphic sites and follow their transmission from parent to offspring over many generations. RFLPs were among the first molecular markers studied by geneticists. More recently, other molecular markers have been used because they are easier to generate via PCR. As an example, let's consider **microsatellites,** which are short, repetitive sequences that are abundantly interspersed throughout a species' genome and tend to vary in length among different individuals. Microsatellites usually contain di-, tri-, tetra-, or pentanucleotide sequences that are repeated many times in a row. For example, the most common microsatellite encountered in humans is a dinucleotide sequence $(CA)_n$, where n ranges from 5 to more than 50. In other words, this dinucleotide sequence can be tandemly repeated 5 to 50 or more times. The $(CA)_n$ microsatellite is found,

on average, about every 10,000 bases in the human genome. Researchers have identified thousands of different DNA segments that contain $(CA)_n$ microsatellites, located at many distinct sites within the human genome.

How do researchers identify a specific microsatellite within a chromosome? The procedure is shown in **Figure 21.5**.

1. The starting material is a sample that contains all of the chromosomes of an organism.
2. Primers complementary to the unique DNA sequences that flank a specific region are used to amplify a particular microsatellite by PCR. In other words, the PCR primers copy only a particular microsatellite, but not the thousands of others that are interspersed throughout the genome. If a pair of PCR primers copies a single site within a set of chromosomes, the amplified region is called a **sequence-tagged site (STS).**
3. The PCR products are then run on a gel to identify their size(s). In a diploid species, an individual has two copies of a given STS. When an STS contains a microsatellite, the two PCR products may be identical and result in a single band on a gel if the region is the same length in both copies (i.e., if the individual is homozygous for the microsatellite). However, if an individual has two copies that differ in the number of repeats in the microsatellite sequence (i.e., if the individual is heterozygous for the microsatellite), the two PCR products obtained will be different in length (as in the results shown in Figure 21.5). The DNA fragments in the two bands in this figure were made via PCR, using primers that flank a particular microsatellite on chromosome 2. The DNA fragment in the higher band is longer, because it has more repeat sequences than the lower band does.

When microsatellites have length polymorphisms, researchers can follow their transmission from parent to offspring. PCR amplification of particular microsatellites provides a strategy for the genetic analysis of human pedigrees, as shown in **Figure 21.6**. Prior to this analysis, a unique segment of DNA containing a microsatellite had been identified. Using PCR primers complementary to this microsatellite's unique flanking segments, two parents and their three offspring were tested for the inheritance of this microsatellite. A small sample of cells was obtained from each individual and subjected to PCR amplification, as described in Figure 21.5. The amplified PCR products were then analyzed by high-resolution gel electrophoresis, which detects small differences in the lengths of DNA fragments. The mother's PCR products were 154 and 150 bp in length; the father's were 146 and 140 bp. Their first offspring inherited the 154-bp product from the mother and the 146-bp one from the father, the second inherited the 150 bp from the mother and the 146 bp from the father, and the third inherited the 150 bp from the mother and the 140 bp from the father. As shown in the figure, the transmission of polymorphic microsatellites is relatively easy to follow from generation to generation.

The simple pedigree analysis shown in Figure 21.6 illustrates the general method used to follow the transmission of a single microsatellite that is polymorphic for length. In linkage studies,

21.3 MOLECULAR MARKERS   497

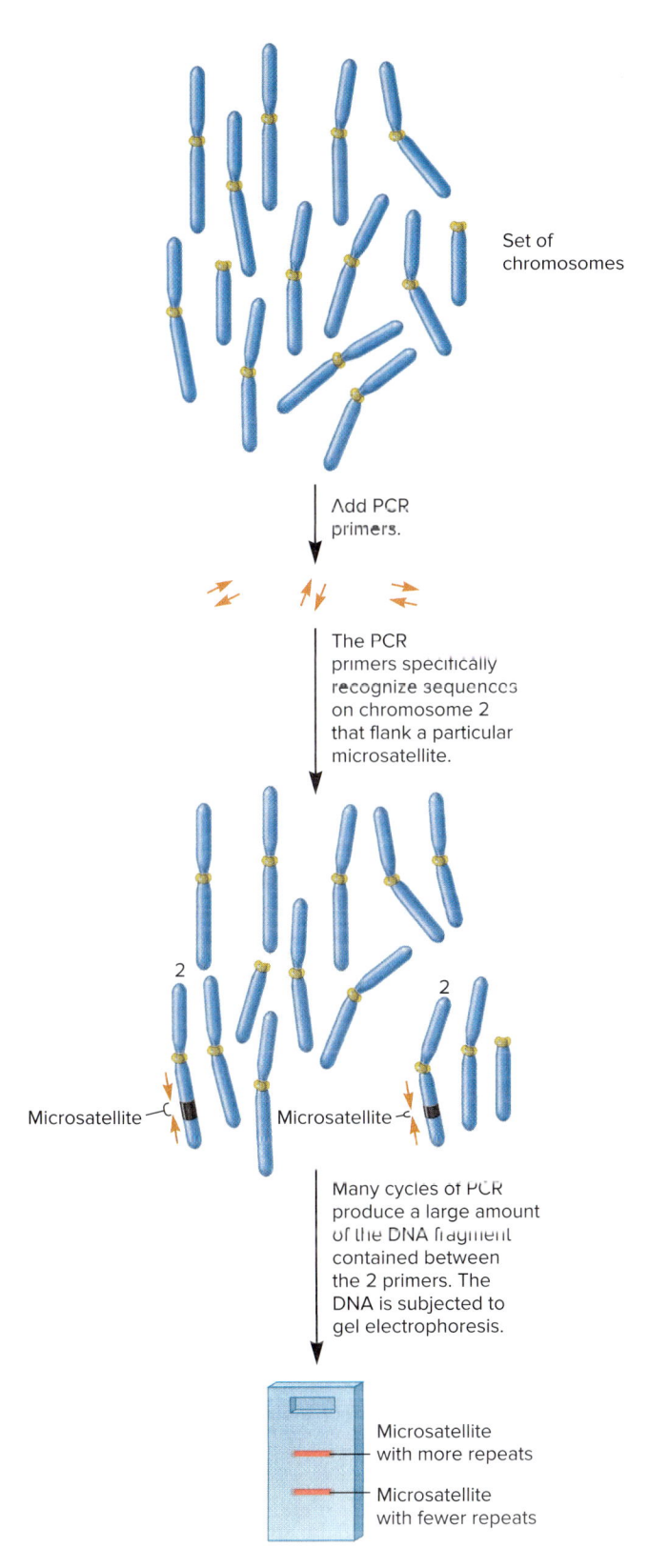

FIGURE 21.5 **Identifying a microsatellite using PCR primers.** The individual shown here is heterozygous for a microsatellite on chromosome 2. The two copies of this microsatellite differ in length.

Concept Check: *What causes microsatellites to be polymorphic?*

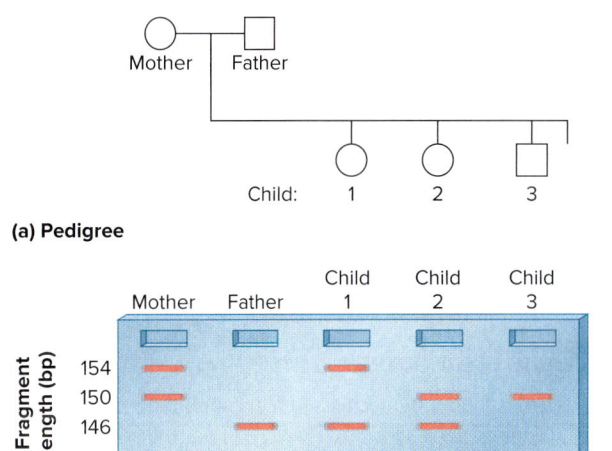

(a) Pedigree

(b) Electrophoretic gel of PCR products for a polymorphic microsatellite found in the family in (a)

FIGURE 21.6 **Inheritance pattern of a polymorphic microsatellite in a human pedigree.** This microsatellite is autosomal.

the goal is to follow the transmission of many different microsatellites to determine which ones are linked along the same chromosome and which ones are not. Those that are not linked will independently assort from generation to generation. Those that are linked tend to be transmitted together to the same offspring. In a large pedigree, it is possible to identify cases in which linked microsatellites have segregated due to crossing over. The frequency of crossing over provides a measure of the map distance, in this case between the different microsatellites. This approach can help researchers obtain a finely detailed linkage map of the human chromosomes without having to depend on alleles of closely linked genes that affect phenotype.

21.3 REVIEWING THE KEY CONCEPTS

- Linkage mapping and physical mapping often rely on molecular markers to map locations on chromosomes (see Table 21.1).
- Restriction fragment length polymorphisms (RFLPs), microsatellites, and single-nucleotide polymorphisms (SNPs) are used as molecular markers (see Figures 21.4–21.6).

21.3 COMPREHENSION QUESTIONS

1. A molecular marker is a _____ that is found at a specific site on a chromosome and has properties that allow it to be _____.
   a. colored dye, visualized via microscopy
   b. colored dye, visualized on a gel
   c. segment of DNA, uniquely identified using molecular tools
   d. segment of DNA, visualized via microscopy

2. Which of the following is an example of a molecular marker?
   a. RFLP
   b. Microsatellite
   c. Single-nucleotide polymorphism
   d. All of the above are types of molecular markers.

3. To map the distance between molecular markers via testcrossses, the markers must be
   a. polymorphic.
   b. monomorphic.
   c. fluorescently labeled.
   d. on different chromosomes.

21.4 GENOME-SEQUENCING PROJECTS

Learning Outcomes:
1. Describe the shotgun method for sequencing an entire genome.
2. List the goals of the Human Genome Project.
3. Outline different methods of DNA sequencing.
4. Compare and contrast the sizes of different species' genomes.

Genome-sequencing projects are research endeavors that have the ultimate goal of determining the sequence of DNA bases of the entire genome of a given species. Such projects involve many participants, including scientists who isolate DNA and perform DNA-sequencing reactions, as well as theoreticians who gather the DNA sequence information and assemble it into a complete DNA sequence for each chromosome. For bacteria and archaea, which usually have just one chromosome, the genome sequence is that of a single chromosome. For eukaryotes, each chromosome must be sequenced. Thus, for humans, the genome sequence includes sequences of 22 autosomes, 2 sex chromosomes, and the mitochondrial genome.

In just a couple of decades, our ability to map and sequence genomes has improved dramatically. As of 2017, the complete genome sequences have been obtained from many different species, including over 4000 prokaryotes and 300 eukaryotes. Considering that the first genome sequence was generated in 1995, the progress of genome-sequencing projects since then has been truly remarkable! In this section, we will examine the approaches that researchers follow when tackling such large projects. We will also survey some of the general goals of the Human Genome Project and compare the results from the genome sequencing of various species.

Venter, Smith, and Colleagues Sequenced the First Genome in 1995

The first genome to be entirely sequenced was that of the bacterium *Haemophilus influenzae*. This bacterium causes a variety of diseases in humans, including respiratory illnesses and bacterial meningitis. *H. influenzae* has a relatively small genome: approximately 1.8 Mb (i.e., 1.8 million bp) of DNA in a single circular chromosome.

When sequencing an entire genome, researchers must consider factors such as genome size, the efficiency of the methods used to sequence DNA, and the costs of the project. Since genome-sequencing projects began in the 1990s, researchers have learned that the most efficient and inexpensive way to sequence genomes is via an approach called **shotgun sequencing,** in which DNA fragments to be sequenced are randomly generated from larger DNA fragments. In this method, genomic DNA is isolated and broken into smaller DNA fragments, typically 1500 bp or more in length. The researchers then randomly sequence such fragments from the genome. As a matter of chance, some of the fragments overlap, as shown schematically in **Figure 21.7**. The DNA sequences in two different fragments are identical in the overlapping region. This allows researchers to order them as they are found in the intact chromosome. An advantage of shotgun DNA sequencing is that it does not require extensive mapping, which can be very time-consuming and expensive. A disadvantage is that researchers waste some time sequencing the same region of DNA more times than needed.

To obtain a complete sequence of a genome with the shotgun approach, how do researchers decide how many fragments to sequence? We can calculate the probability that a base will not be sequenced using this approach with the following equation:

$$P = e^{-m}$$

where

P is the probability that a base will be left unsequenced
e is the base of the natural logarithm: $e \approx 2.72$
m is the number of bases sequenced divided by the total genome size

For example, in the case of *H. influenzae* with a genome size of 1.8 Mb, if researchers sequence 9 Mb, $m = 5$ (i.e., 9.0 Mb divided by 1.8 Mb), and the calculated probability is:

$$P = e^{-m} = e^{-5} = 0.0067, \text{ or } 0.67\%$$

This result means that if we randomly sequence 9.0 Mb, which is five times the length of a single genome, we are likely to miss only 0.67% of the genome. With a genome size of 1.8 Mb, we will miss about 12,000 nucleotides out of approximately 1,800,000. Such missed sequences are typically on small DNA fragments that—as a matter of random chance—did not happen to be sequenced. The missing links in the genome can be sequenced later using mapping methods.

The general protocol followed by J. Craig Venter, Hamilton Smith, and colleagues in this discovery-based investigation is presented in **Figure 21.8**. This is a shotgun DNA-sequencing approach. The researchers isolated chromosomal DNA from *H. influenzae* and used sound waves to break the DNA into small fragments approximately 2000 bp long. These fragments were

```
                         Overlapping region
                         ⌠‾‾‾‾‾‾‾‾‾‾‾‾‾‾‾⌡
    TTACGGTACCAGTTACAAATTCCAGACCTAGTACC
    AATGCCATGGTCAATGTTTAAGGTCTGGATCATGG
                         GACCTAGTACCGGACTTATTCGATCCCCAATTTTGCAT
                         CTGGATCATGGCCTGAATAAGCTAGGGGTTAAAACGTA
```

FIGURE 21.7 A comparison of two DNA fragments that contain an overlapping region.

randomly cloned into vectors, allowing the DNA to be propagated in *E. coli*. Each *E. coli* clone carried a vector with a different piece of DNA from *H. influenzae*. The researchers then subjected many of these clones to DNA sequencing. They sequenced a total of approximately 10.8 Mb of DNA.

▶ **THE GOAL**

The goal was to obtain the entire genome sequence of *H. influenzae*. This information reveals the genome's size as well as the genes the organism has.

▶ **ACHIEVING THE GOAL** — **FIGURE 21.8** The use of shotgun DNA sequencing to determine the first genome sequence of a bacterial species.

Starting materials: A strain of *H. influenzae*.

Experimental level | **Conceptual level**

1. Purify DNA from a strain of *H. influenzae*. This involves breaking the cells open by adding phenol and chloroform. Most protein and lipid components go into the phenol-chloroform phase. DNA remains in the aqueous (water) phase, which is removed and used in step 2.

2. Sonicate the DNA to break it into small fragments about 2000 bp long.

3. Clone the DNA fragments into vectors. The procedures for cloning are described in Chapter 20. This produces a DNA library.

   Refer back to Figure 20.2.

4. Subject many clones to the procedure of dideoxy DNA sequencing, also described in Chapter 20. A total of 10.8 Mb was sequenced.

   Refer back to Figure 20.16.

   Produces a large number of overlapping sequences.

5. Use computer methods to identify various types of genes in the genome.

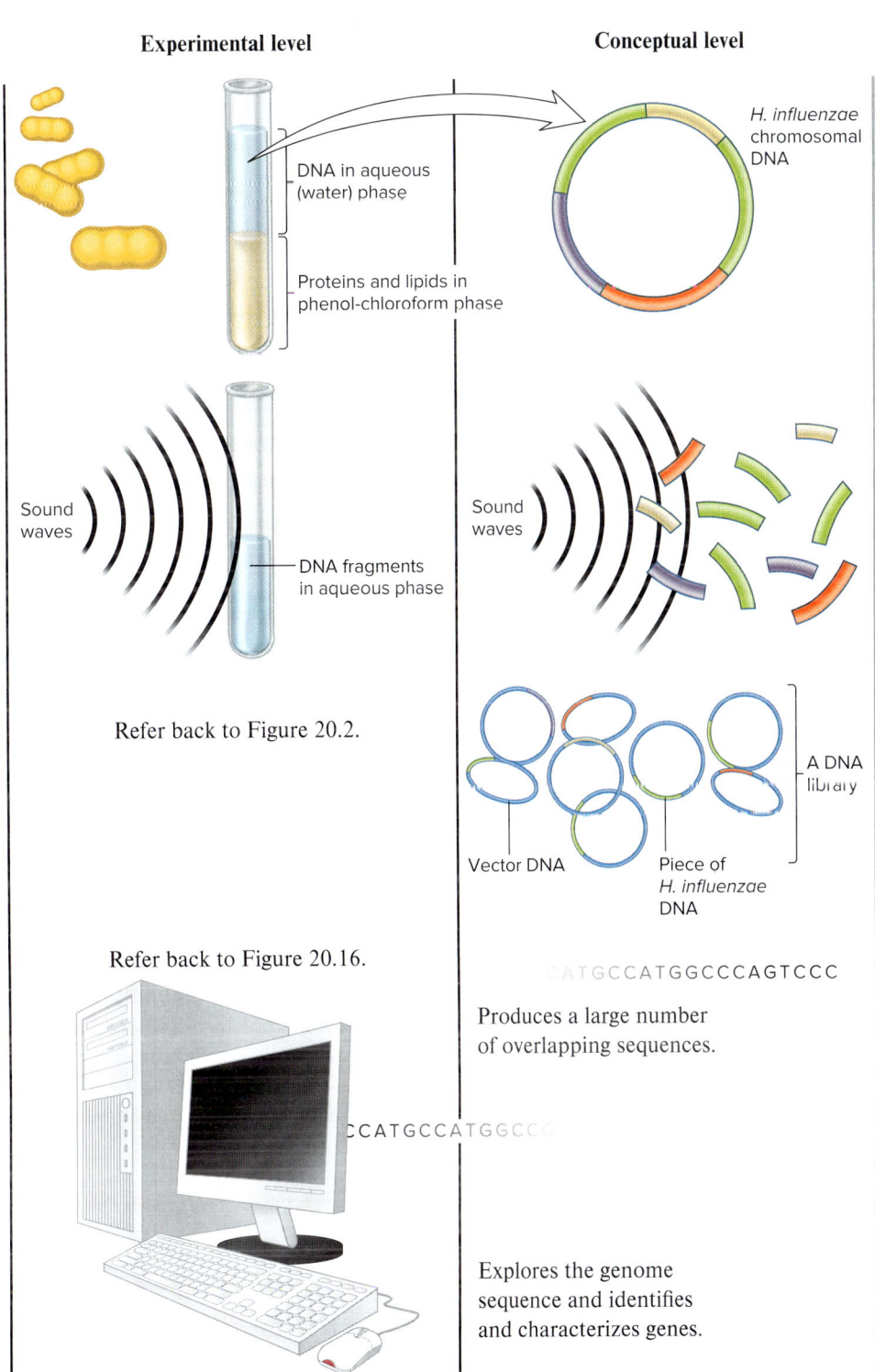

   Explores the genome sequence and identifies and characterizes genes.

THE DATA

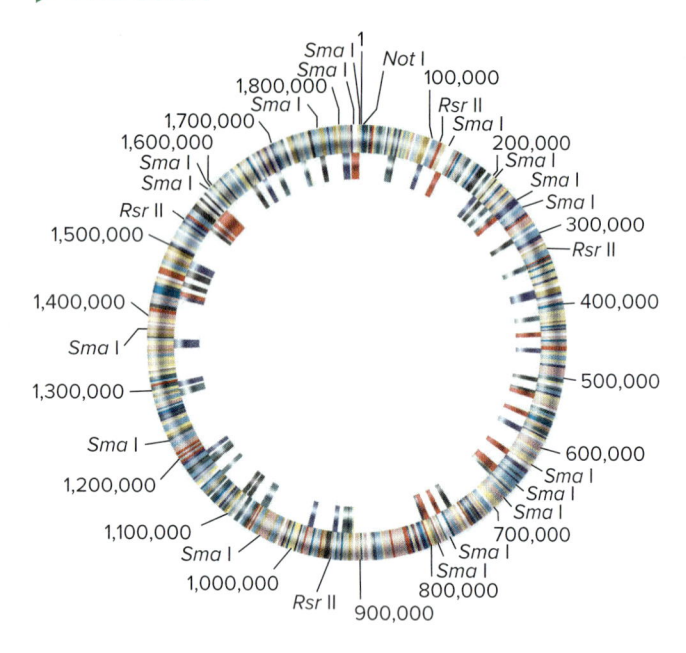

Functions of Proteins Encoded by Genes

% of genome	
6.8	Amino acid biosynthesis
5.4	Biosynthesis of cofactors, prosthetic groups, carriers
8.3	Cell envelope
5.3	Cellular processes
3.0	Central intermediary metabolism
10.4	Energy metabolism
2.5	Fatty acid/phospholipid metabolism
5.3	Purines, pyrimidines, nucleosides, and nucleotides
6.3	Regulatory functions
8.6	Replication
12.2	Transport-binding proteins
14.0	Translation
2.7	Transcription
9.2	Other categories
	Hypothetical
	Unknown

INTERPRETING THE DATA

The outcome of this genome-sequencing project was a very long DNA sequence. In 1995, Venter, Smith, and colleagues published the entire DNA sequence of *H. influenzae*. The researchers then analyzed the genome sequence using a computer to obtain information about the properties of the genome. They asked: How many genes does the genome contain and what are the likely functions of those genes? The data of Figure 21.8 summarize the results that the researchers obtained. The *H. influenzae* genome is composed of 1,830,137 bp of DNA. The computer analysis predicted 1743 genes. Based on their similarities to known genes in other species, the researchers also predicted the functions of nearly two-thirds of these genes. The diagram displaying the data obtained by Venter, Smith, and colleagues places genes into various categories based on the predicted functions of their encoded proteins. These results gave the first comprehensive genome picture of a living organism!

The Human Genome Project Was the Largest Genome-Sequencing Project in History

Due to the large size of the human genome, sequencing it was an enormous undertaking. Scientists had been discussing how to tackle this project since the mid-1980s. In 1988, the National Institutes of Health established an Office of Human Genome Research, with James Watson as its first director. The **Human Genome Project,** which officially began on October 1, 1990, was a 13-year effort coordinated and funded by the U.S. Department of Energy and the National Institutes of Health. The human DNA that was used in the Human Genome Project was obtained from several volunteers; their identity was not revealed to protect their privacy. From its outset, the Human Genome Project had the following goals:

1. To obtain a linkage map of the human genome. This involved the identification of millions of genetic markers and their localization along the autosomes and sex chromosomes.
2. To obtain a physical map of the human genome.
3. To obtain the DNA sequence of the entire human genome. The first (nearly complete) sequence was published in February 2001. This was considered a first draft. A second draft was published in 2003, and the completed maps and sequences for all of the human chromosomes were published by 2006. The entire genome is approximately 3 billion bp in length. If the entire human genome were typed in a textbook like this, it would be nearly 1 million pages long!
4. To develop technology for the management of human genome information. The amount of information obtained from this project is staggering, to say the least. The Human Genome Project developed user-friendly tools to provide scientists easy access to up-to-date information obtained from the project. The Human Genome Project also developed analytical tools for interpreting genome information.
5. To analyze the genomes of model organisms. These include bacterial species (e.g., *Escherichia coli* and *Bacillus subtilis*), *Drosophila melanogaster* (fruit fly), *Caenorhabditis elegans* (a nematode), *Arabidopsis thaliana* (a flowering plant), and *Mus musculus* (mouse).
6. To develop programs focused on understanding and addressing the ethical, legal, and social implications of the results obtained from the Human Genome Project. The Human Genome Project sought to identify the major genetic issues that will affect members of our society and to develop policies to address these issues. For example, what is an individual's right to privacy regarding genetic information? Some people are worried that their medical insurance company may discriminate against them if it is found that they carry a disease-causing or otherwise deleterious gene.

7. To develop technological advances in genetic methodologies. Some of the efforts of the Human Genome Project involved improvements in molecular genetic technology such as gene cloning and DNA sequencing. The project also developed computer technology for data processing, storage, and analysis of sequence information.

A great benefit expected from the characterization of the human genome is the ability to identify and study the sequences of our genes. Mutations in many different genes are known to be correlated with human diseases, which include cancer, heart disease, and many other abnormalities. The identification of mutant genes that cause inherited diseases was a strong motivation for the Human Genome Project. A detailed genetic and physical map has made it profoundly easier for researchers to locate such genes. Furthermore, a complete DNA sequence of the human genome provides researchers with insight into the types of proteins encoded by these genes. The cloning and sequencing of disease-causing alleles is expected to play an increasingly important role in the diagnosis and treatment of disease.

In 2008, a more massive undertaking, called the 1000 Genomes Project, was launched with the goal of achieving a detailed understanding of human genetic variation. In this international project, researchers set out to determine the DNA sequence of at least 1000 anonymous participants from around the globe. In 2012, the sequencing of 1092 genomes was announced in the journal *Nature*. Since then, thousands more human genomes have been sequenced.

Innovations in DNA Sequencing Have Made It Faster and Less Expensive

Since DNA sequencing was invented in the 1980s, technological advances have been aimed at making it faster and less expensive. The Human Genome Project, which began in the early 1990s, was originally estimated to cost about $3 billion to sequence a single genome. However, cost reductions due to innovations in DNA-sequencing technology drove the actual cost down to about $300 million. By the end of the project, researchers estimated that if they were starting again, they could have sequenced the genome for less than $50 million. The project took about 13 years to complete. In 2007, researchers undertook the sequencing of James Watson's genome, which cost less than $1 million. By 2011, the cost of sequencing a human genome had been reduced to about $5000. By 2017, the sequencing of a single human genome can cost as little as $1000. Such innovation will make it feasible to sequence an individual's genome as a routine diagnostic procedure.

The ability to rapidly sequence large amounts of DNA is referred to as **high-throughput sequencing.** Different types of technological advances have made this possible. First, different aspects of DNA sequencing have become automated, so samples can now be processed rapidly in a machine. For example, in Chapter 20, we considered how fluorescent labeling of nucleotides can automate the reading of a DNA sequence by a fluorescence detector. A second advance in sequencing technologies involves parallel sequencing, which allows multiple samples to be processed at once. The first parallel sequencing machines, also called platforms, could simultaneously perform many sequencing runs via multiple gel-filled capillary tubes. For example, DNA-sequencing machines produced by Applied Biosystems, which rely on Sanger's dideoxy sequencing method (described in Chapter 20), run 96 capillary tubes in parallel. Each capillary tube is capable of producing between 700 and 900 bases of a DNA sequence.

Although the Sanger dideoxy sequencing method is still in use for smaller sequencing projects, newer high-throughput platforms based on different methods of sequencing DNA are becoming more popular. **Table 21.2** describes a few of these methods, which are often referred to as **next-generation sequencing technologies** because they have superseded the Sanger dideoxy method for large sequencing projects. What sets next-generation sequencers apart from the conventional dideoxy method? One key technological advance is the ability to process thousands or even millions of sequence reads in parallel rather than only 96 at a time. This massive parallel throughput may require only one or two instrument runs to complete the sequencing of an entire prokaryotic genome. Also, next-generation sequencers are able to use samples that contain mixtures of DNA fragments that have not been subjected to the conventional vector-based cloning. By comparison, shotgun methods of DNA sequencing using capillary sequencing involve DNA-cloning steps (see Figure 21.8). The elimination of such cloning steps saves a great deal of time and money.

The newer sequencing platforms employ a complex interplay of enzymology, chemistry, high-resolution optics, and new approaches to processing the data. These instruments allow for easy sample preparation steps prior to DNA sequencing. Most of them involve strategies in which fragmented DNA is immobilized in a fixed position and repeatedly exposed to reagents. Some sequencing platforms use PCR to amplify the DNA, whereas others actually read single DNA molecules.

As an example of a next-generation sequencing technology, **Figure 21.9** presents the steps of **pyrosequencing,** which relies on the release of pyrophosphate. This method was developed by Pal Nyren and Mostafa Ronaghi in 1996 and is the basis for the Roche 454/FLX Pyrosequencer.

1. Samples, such as genomic DNA, are broken into small 300- to 800-bp fragments.
2. Short oligonucleotides called adaptors are linked to the 5′ and 3′ ends of the DNA fragments. (Note: The red adaptor contains a sequence that is complementary to an oligonucleotide attached to the beads, and the green adaptor is complementary to primers used in the DNA-sequencing reaction.) The DNA is then denatured into single strands.
3. The single-stranded DNA fragments are attached to beads via the adaptors. Initially, just one DNA strand is attached to each bead.
4. The beads are emulsified in an oil-water mixture so that each bead becomes localized into a single droplet. The mixture also contains PCR reagents, including primers that are complementary to the adaptors. During this step, the

TABLE 21.2

Examples of Next-Generation Sequencing Technologies

Technology*	DNA Preparation	Enzyme(s) Used	Detection
Roche 454/FLX Pyrosequencer	DNA fragments are bound to small beads, which are dropped into tiny wells in a picotiter plate.	DNA polymerase, ATP sulfurylase, luciferase, apyrase	Pyrophosphate release activates luciferase, which breaks down luciferin and gives off light.
Illumina/Solexa Genome Analyzer	DNA fragments are bound to a flow cell surface.	DNA polymerase	Four different fluorescently labeled nucleotides are detected.
DNA Nanoball Sequencing	Linkers are attached to small fragments of DNA that are amplified by rolling circle replication into DNA nanoballs. The nanoballs are then bound to a flow cell surface.	DNA polymerase, restriction enzymes, DNA ligase	Four different fluorescently labeled nucleotides are detected.
Single-molecule DNA sequencers			
Pacific Biosciences SMRT	DNA fragments and DNA polymerase are trapped within tiny holes on a thin metal film.	DNA polymerase	The growth of individual DNA molecules is monitored by fluorescence imaging.
Helicos Biosciences tSMS	DNA fragments are bound to a flow cell surface.	DNA polymerase	The growth of individual DNA molecules is monitored by fluorescence imaging.
ZS Genetics TEM	DNA is labeled with heavier elements, such as iodine or bromine.	None	The nucleotide sequence of single DNA molecules is read via transmission electron microscopy (TEM).
Ion Semiconductor Sequencing	DNA fragments are bound to a semiconductor chip.	DNA polymerase	When a nucleotide is incorporated into a growing strand, this releases pyrophosphate and a hydrogen ion. The release of the hydrogen ion is detected.

*Most include the company name associated with the technology.

---

single-stranded DNA on the bead becomes amplified into many identical copies, which also become attached to the beads. At this stage, each bead carries about a million copies of a particular DNA segment.

5. Each bead is placed into a well on a picotiter plate; the diameter of each well is large enough to accommodate only one bead.
6. Sequencing reagents, which include primers that are complementary to the adaptors, are added to the wells. The picotiter plate acts like a flow cell in which the beads are stuck in the wells and a pure solution of each nucleotide flows over them in a stepwise fashion. The synthesis of DNA is monitored in real time. Therefore, this type of method is referred to as **sequencing by synthesis (SBS)** because it involves the identification of each nucleotide immediately after its incorporation into a DNA strand by DNA polymerase.
7. The pyrosequencing method relies on the functions of other enzymes (ATP sulfurylase, luciferase, and apyrase), along with molecules called adenosine 5′-phosphosulfate and luciferin, which are also included in the sequencing reaction. If a nucleotide has been incorporated into a DNA strand, pyrophosphate is released. ATP sulfurylase uses that pyrophosphate along with adenosine 5′-phosphosulfate to make ATP. The ATP is then used by luciferase to break down luciferin. This reaction gives off light, which is detected by a camera in the sequencing machine. Therefore, light is given off only when a nucleotide is incorporated into a DNA strand. Unincorporated nucleotides and ATP are degraded by apyrase, and the reaction can start again by flowing a new solution containing a particular nucleotide over the picotiter plate. By sequentially adding solutions with only one of the four possible nucleotides (A, T, G, or C), the sequence of the DNA strand can be monitored in real time by determining when light is given off.

Many Genome Sequences Have Been Determined

The amazing advances in DNA-sequencing technology have enabled researchers to determine the complete genome sequences of hundreds of species. By 2017, over 4000 prokaryotic and 300 eukaryotic genomes had been sequenced, including many different mammalian genomes. Some are described in **Table 21.3**.

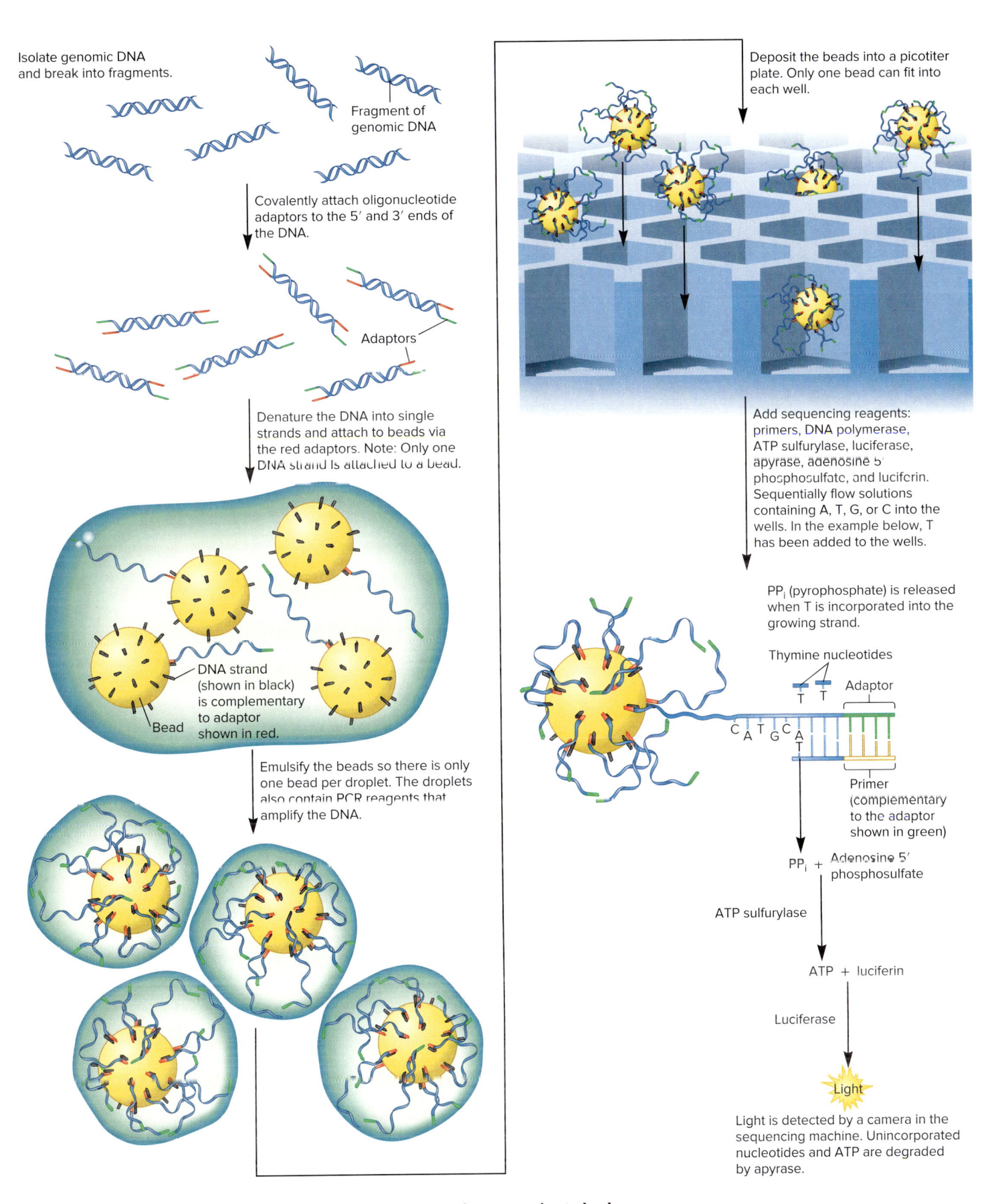

FIGURE 21.9 Pyrosequencing, an example of a next-generation sequencing technology.

Concept Check: Is this a sequencing by synthesis method? Explain.

TABLE 21.3

Examples of Genomes That Have Been Sequenced

Species	Genome Size (bp)*	Approximate Number of Protein-Encoding Genes[†]	Description
Prokaryotic genomes			
Bacteria			
Mycoplasma genitalium	580,000	521	Bacterial inhabitant of the human genital tract with a very small genome
Helicobacter pylori	1,668,000	1590	Bacterial inhabitant of the stomach that causes gastritis, peptic ulcer, and gastric cancer
Mycobacterium tuberculosis	4,412,000	4294	Bacterial species that causes tuberculosis
Escherichia coli	4,639,000	4377	Widespread bacterial inhabitant of the gut of animals; also a model research organism
Archaea			
Thermoplasma volcanium	1,580,000	1494	Archeon with an optimal growth temperature of 60°C and pH optimum of <2.0
Pyrococcus abyssi	1,760,000	1765	Archeon that was originally isolated from samples taken close to a hot spring situated 3500 m deep in the southeast Pacific; has optimal growth conditions of 103°C and 200 atm
Sulfolobus solfataricus	2,990,000	2977	Archeon found in terrestrial volcanic hot springs with optimum growth occurring at a temperature of 75–80°C and pH of 2–3
Eukaryotic genomes			
Protists			
Plasmodium falciparum	22,900,000	5268	Parasitic protist that causes malaria in humans
Entamoeba histolytica	23,800,000	9938	Amoeba that causes dysentery in humans
Fungi			
Saccharomyces cerevisiae	12,100,000	6294	Baker's yeast, a structurally simple eukaryotic species that has been extensively studied by researchers to understand eukaryotic genetics and cell biology
Neurospora crassa	40,000,000	10,082	Common bread mold, also a structurally simple eukaryotic species that has been extensively studied by researchers
Plants			
Arabidopsis thaliana	142,000,000	26,000	A small flowering plant of the mustard family used as a model organism by plant biologists
Oryza sativa	440,000,000	40,000	Rice, a cereal grain with a relatively small genome that is very important worldwide as a food crop
Populus trichocarpa	550,000,000	45,555	Balsam poplar, a tree with a relatively small genome
Animals			
Caenorhabditis elegans	97,000,000	19,000	Nematode worm that has been used as a model organism to study animal development
Drosophila melanogaster	175,000,000	14,000	Fruit fly, a model organism used to study many genetic phenomena, including development
Anopheles gambiae	278,000,000	13,683	Mosquito that carries the malaria parasite, *Plasmodium falciparum*
Canis lupus familiaris	2,400,000,000	22,000	Dog, a common house pet
Mus musculus	2,500,000,000	22,000	House mouse, a rodent and a model organism studied by researchers
Pan troglodytes	3,100,000,000	22,000	Chimpanzee, a primate and the closest living relative to humans
Homo sapiens	3,200,000,000	22,000	Human

*In some cases, the values indicate the estimated amount of DNA. For eukaryotic genomes, DNA sequencing is considered completed in the euchromatic regions. The DNA in certain heterochromatic regions cannot be sequenced, and the total amount is difficult to estimate.

[†]The numbers of genes were predicted using computer methods. These numbers should be considered estimates.

How do researchers decide which genomes to sequence? Motivation behind genome-sequencing projects comes from a variety of sources.

- Basic research scientists can greatly benefit from a genome sequence. It allows them to know which genes a given species has, and it aids in the cloning and characterization of such genes. This has been the impetus for genome projects involving model organisms such as *Escherichia coli, Saccharomyces cerevisiae, Arabidopsis thaliana, Caenorhabditis elegans, Drosophila melanogaster,* and the mouse.
- A second impetus for genome sequencing is the treatment or prevention of human diseases. As noted previously, researchers expect that the sequencing of the human genome will aid in the identification of genes that, when mutant, play a role in disease. Likewise, the decision to sequence many bacterial, protist, and fungal genomes has been related to the role of these species in infectious diseases. Thousands of microbial genomes have been sequenced. Many of them are from species that are pathogenic in humans. The sequencing of such genomes may help us understand which genes play a role in the infection process.
- In addition, the genomes of agriculturally important species have been the subject of genome sequencing. An understanding of a species' genome may aid in the development of a new strain of a livestock or plant species that has improved traits from an agricultural perspective.
- Finally, genome-sequencing projects help us to better understand the evolutionary relationships among living species. This approach, called **comparative genomics,** uses information from genome projects to understand the genetic variation among different populations.

Table 21.3 describes the results of several genome-sequencing projects that have been completed. Newly completed genome sequences are emerging rapidly, particularly those of microbial species. As we obtain more genome sequences, it becomes progressively more interesting to compare them to each other as a way to understand the process of evolution. As discussed in Section 21.6, the field of functional genomics enables researchers to study the roles of many genes as they interact to generate the phenotypic traits of the species that contain them.

21.4 REVIEWING THE KEY CONCEPTS

- Shotgun sequencing has been commonly used to sequence the DNA of many species. Overlapping regions allow researchers to determine the complete DNA sequence (see Figures 21.7, 21.8).
- The Human Genome Project resulted in the sequencing of the entire human genome.
- Newer methods of DNA sequencing can process many samples of DNA simultaneously and are superseding the Sanger dideoxy method. An example of these next-generation sequencing technologies is pyrosequencing (see Table 21.2, Figure 21.9).
- Since 1995, the genomes of thousands of species have been sequenced (see Table 21.3).

21.4 COMPREHENSION QUESTIONS

1. Shotgun sequencing is a method of DNA sequencing in which
   a. the DNA fragments to be sequenced are randomly generated from larger DNA fragments.
   b. the sequencing reactions are carried out in rapid succession.
   c. the samples to be sequenced are rapidly generated by PCR.
   d. all of the above occur.
2. Which of the following was *not* a goal of the Human Genome Project?
   a. To obtain the DNA sequence of the entire human genome
   b. To successfully clone a mammal
   c. To develop technology for the management of human genome information
   d. To analyze the genomes of model organisms
3. A prokaryotic genome is about 4 million bp in length. About how many genes would you expect it to contain?
   a. 400
   b. 4000
   c. 40,000
   d. 400,000

21.5 METAGENOMICS

Learning Outcomes:
1. Define *metagenomics*.
2. Describe the general strategy of metagenomics, and outline its uses.

Most microorganisms that exist in soil, water, and the human intestinal tract have not been successfully grown in the laboratory because researchers do not fully understand their growth requirements and because some microbes require the presence of a complex microbial community to survive. In the past, such unculturable microbes had been very difficult to study. In the 1980s and early 1990s, Norman Pace and colleagues showed that 16S rRNA genes, which are described in Chapter 15, could be analyzed from samples of different microbes using PCR. Many of the 16S rRNA genes that were sequenced from environmental samples were found to be different from 16S rRNA genes of bacteria that had been grown in the laboratory. Therefore, this work revealed that environmental samples contain an abundance of unculturable microbes. Though such studies were exciting, a limitation of PCR is that it amplifies specific genes, leaving the rest of the genome of unculturable microbes unknown.

This limitation was overcome by **metagenomics,** which is the study of a complex mixture of genetic material obtained from an environmental sample. The term **metagenome** refers to a collection of genes from a particular environmental sample. Such a sample can be analyzed in a way analogous to that used for analysis of a single genome. **Figure 21.10** outlines a general strategy for metagenomics. First, researchers obtain a sample from the environment. This may be a soil sample, a water sample from the ocean, or a fecal sample from a person. After the sample is filtered, the cells within the sample are lysed and the DNA is extracted and purified. During this procedure, the purified DNA is sheared, which breaks the DNA into fragments of different sizes. Each DNA fragment is then randomly inserted into a cloning vector and transformed into a host cell. The result is a DNA library of thousands or tens of thousands of cells, each carrying a DNA fragment from the metagenome. All of the cells together constitute the metagenomic library. The members of the library are then subjected to shotgun DNA sequencing to identify the genes they may contain. (Note: The cloning step may be skipped if next-generation DNA sequencing technologies are used.)

What are the uses of metagenomics? The following are some of the common applications of metagenomics:

- *Human medicine:* Various places in the body, such as the mouth and intestines, support a complex array of microorganisms. Metagenomics is being used to characterize these populations of organisms and to study changes in the relative compositions of microorganisms that may be associated with different diseases.
- *Agriculture:* The metagenomic analysis of soil samples has revealed an astonishing complexity of microorganisms in the soil. As researchers learn more about which microbes facilitate plant growth, such knowledge may be used to improve agricultural yields.
- *Bioremediation:* The types of microorganisms found in soil and water have a great effect on the decomposition of pollutants in the environment. Metagenomics may play a key role in the identification of microorganisms that can break down specific types of pollutants.
- *Biotechnology:* Microorganisms are capable of synthesizing a vast array of chemicals, some of which are useful to humans. An example is antibiotics that are used to treat bacterial infections. Such chemicals are made by enzymes encoded by microbial genes. Metagenomics is being used to identify such genes in soil and aquatic microorganisms as a way to discover possible antibiotics that the microorganisms can make.
- *Global change:* Microorganisms carry out about half of the photosynthesis that takes place on Earth and are key participants in the cycling of various elements such as carbon, phosphorus, and nitrogen. Metagenomics is helping us understand the complexity of microbial communities in these processes.
- *Identification of viruses:* Environmental samples are analyzed to identify viruses that infect humans and other organisms.
- *Aquatic biology:* The metagenomic analysis of water samples from rivers, freshwater lakes, and oceans has revealed a much greater complexity of microbial communities than had been expected. An example of one such study is described next.

The first extensive large-scale environmental sequencing project, called the Global Ocean Sampling Expedition, was carried out by J. Craig Venter and colleagues. It began in 2003 as a pilot project to sample the microbial population of the nutrient-limited Sargasso Sea, a region of the Atlantic Ocean close to Bermuda.

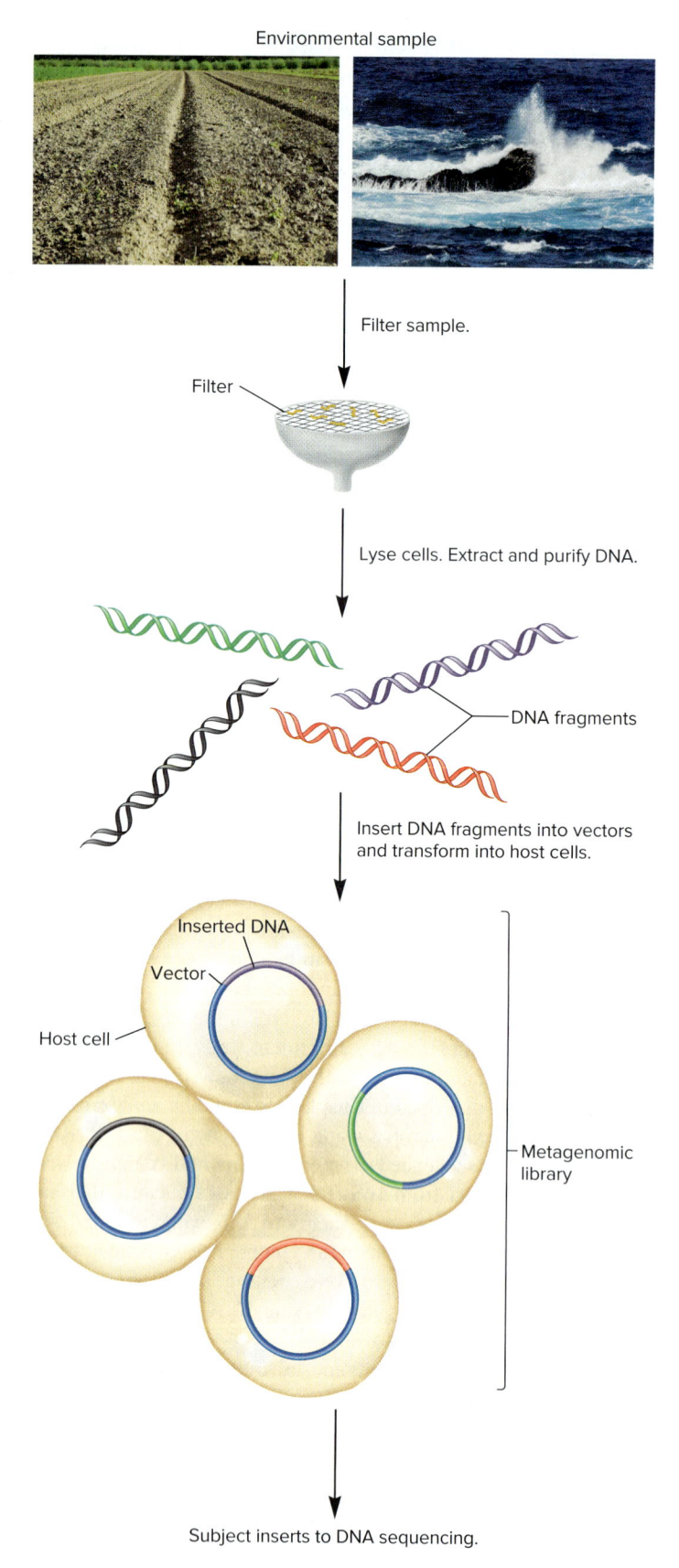

FIGURE 21.10 The general strategy of metagenomics.
©Joseph Gareri/Getty Images

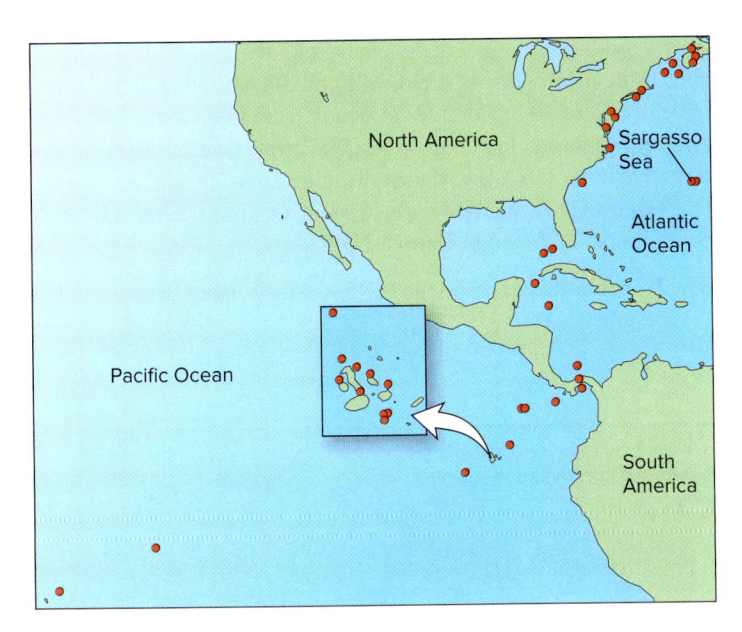

FIGURE 21.11 Sampling sites of the Global Ocean Sampling Expedition. The sampling sites are shown as red dots. The first two sites were in the Sargasso Sea. The next year the expedition started in Nova Scotia and proceeded southwest, ending in the Pacific Ocean.

This site was chosen because it was thought to contain a relatively simple population of microbes compared to other aquatic locations. The samples from the Sargasso Sea revealed several hundred previously undiscovered genes for the light-harvesting protein proteorhodopsin. This discovery may help us understand the role of this protein in energy metabolism under low nutrient conditions. In addition, over 1800 new species of microorganisms were identified.

Venter announced the full expedition in 2004. In the spirit of Darwin's travels on the *Beagle*, Venter's 95-foot sailboat (the *Sorcerer II*) was outfitted as a research vessel. Using a sample size of 200 liters, the team traveled over 32,000 miles, sampling every 200 miles; they collected about 40 different samples (**Figure 21.11**). After being frozen and shipped to a land-based research center, the samples were sequenced using shotgun techniques. A total of 7.7 million sequencing runs were performed, yielding sequence information on over 6 billion bp of DNA. Most of the DNA sequences were unique, reflecting the incredible diversity in naturally occurring microbial populations. Though the expedition yielded many exciting results, one of the key findings that emerged involved the identification of species variants, called subtypes, that exist alongside each other. As communities adapt and change, previous research had suggested that certain subtypes tend to dominate and outcompete others. In contrast, this expedition found many closely related species and subtypes that were living in the same environment. Further research will be needed to explain this paradox.

21.5 REVIEWING THE KEY CONCEPTS

- Metagenomics is the study of a complex mixture of genetic material obtained from an environmental sample (see Figures 21.10, 21.11).

21.5 COMPREHENSION QUESTION

1. Metagenomics is aimed at
   a. determining the complete genome sequence of newly identified microorganisms.
   b. mapping the genes on chromosomes of newly identified microorganisms.
   c. determining the sequence of DNA fragments in environmental samples.
   d. determining the functions of all of the genes in a given species' genome.

21.6 FUNCTIONAL GENOMICS

Learning Outcomes:
1. Describe the composition of a DNA microarray, and explain how it is used.
2. Explain how microarrays are used in conjunction with chromatin immunoprecipitation.
3. Outline the method of RNA sequencing (RNA-Seq).
4. Define *gene knockout*, and explain how a gene knockout can be experimentally useful.

Though the rapid sequencing of genomes, particularly the human genome, has generated great excitement among geneticists, many would argue that an understanding of genome function is fundamentally more interesting. In the past, our ability to study genes involved many of the techniques described in Chapter 20 such as gene cloning, Northern blotting, and gene editing. These approaches continue to provide a solid foundation for our understanding of gene function. More recently, genome-sequencing projects have enabled researchers to consider gene function at a more complex level. We now can analyze groups of many genes simultaneously to determine how they work as integrated units that produce the characteristics of cells and the traits of complete organisms. In this section, we examine the use of DNA microarrays and RNA sequencing, which enable researchers to monitor the expression of thousands of genes simultaneously. We will also discuss why researchers are producing gene knockout collections in which each gene within a given species is inactivated in order to understand gene function at the genomic level.

A Microarray Can Identify Genes That Are Transcribed

In the 1990s, researchers developed a new technology called a **DNA microarray** (also called a **gene chip**) that makes it possible to monitor the expression of thousands of genes simultaneously. A DNA microarray is a small silica, glass, or plastic slide that is dotted with many different sequences of DNA, each corresponding to a short sequence within a known gene. For example, one spot in a microarray may correspond to a sequence within the β-globin gene, whereas another could correspond to a gene that encodes actin, which is a cytoskeletal protein. A single slide may contain tens of thousands of different spots in an area the size of a postage stamp. The relative location of each gene represented in the array is known.

How are microarrays made? Some are produced by spotting different samples of DNA onto a slide, much like the way an inkjet printer works. Different DNA fragments, which are made synthetically (e.g., by PCR), are individually spotted onto the slide. The DNA fragments are typically 500 to 5000 nucleotides in length, and a few thousand to tens of thousands are spotted to make a single array. Alternatively, other microarrays contain shorter DNA segments—oligonucleotides—that are directly synthesized on the surface of the slide. In this case, the DNA sequence at a given spot is produced by selectively controlling the growth of the oligonucleotide using narrow beams of light. Such oligonucleotides are typically 25 to 30 nucleotides in length. Hundreds of thousands of different spots can be found on a single array. Overall, the technology of making DNA microarrays is quite amazing.

Once a DNA microarray has been made, it is used as a hybridization tool, as shown in **Figure 21.12**.

1. To begin this experiment, mRNA was isolated from a sample of cells.

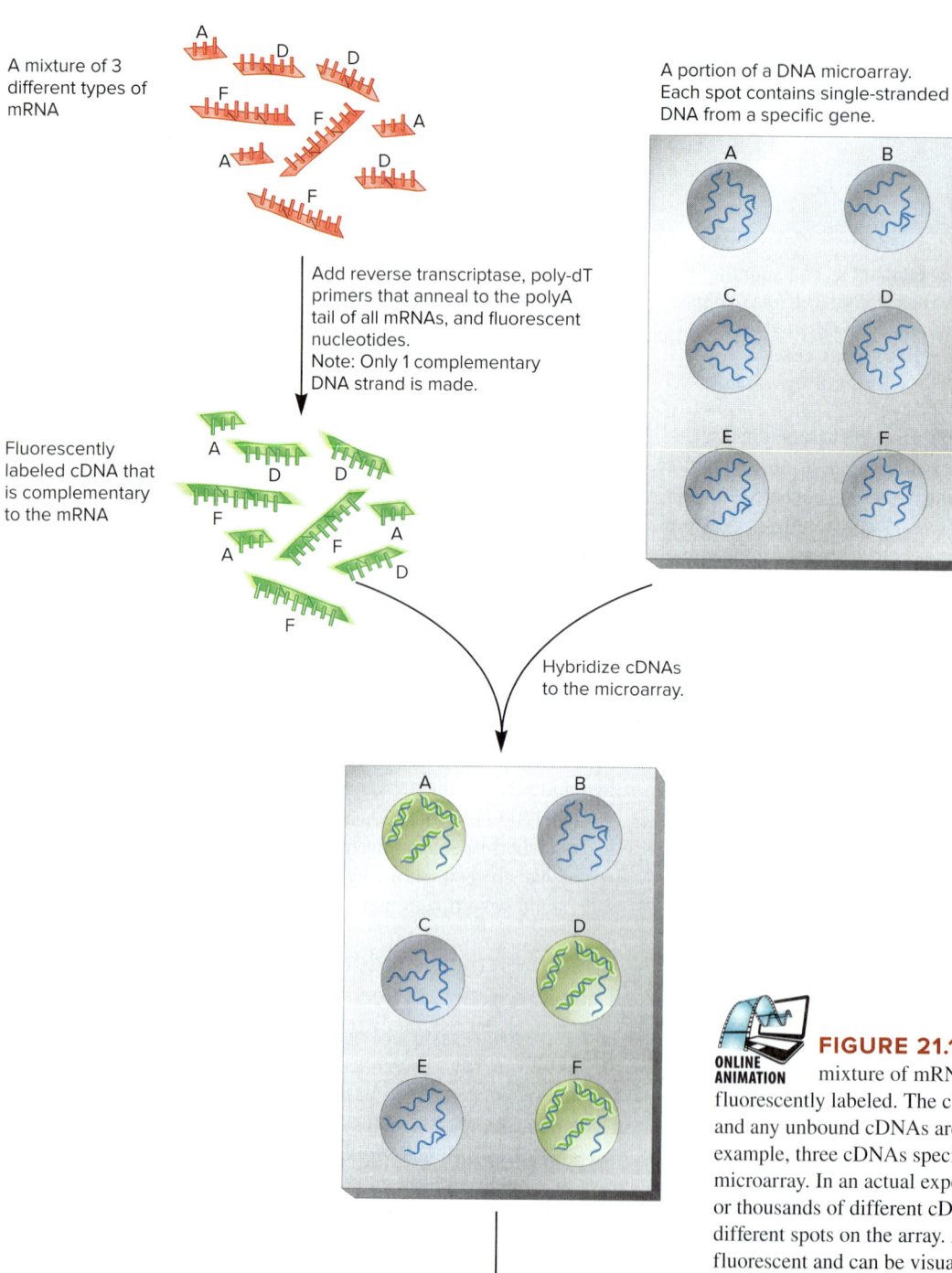

FIGURE 21.12 **A DNA microarray.** A mixture of mRNAs is used to create cDNAs that are fluorescently labeled. The cDNAs are applied to the microarray, and any unbound cDNAs are washed away. In this simplified example, three cDNAs specifically hybridize to spots on the microarray. In an actual experiment, there are typically hundreds or thousands of different cDNAs and tens of thousands of different spots on the array. After hybridization, the spots become fluorescent and can be visualized using a laser scanner.

Concept Check: Explain how this experiment provides information regarding the expression of genes.

2. The mRNA was mixed with fluorescently labeled nucleotides and reverse transcriptase to make fluorescently labeled cDNA.
3. The labeled cDNAs were then layered onto a DNA microarray. Those cDNAs that are complementary to the DNAs in the microarray hybridize, thereby becoming bound to the microarray.
4. The array is then washed with a buffer to remove any unbound cDNAs and placed in a device called a laser scanner, which produces higher resolution images than a conventional optical microscope. The device scans each pixel—the smallest element in a visual image—and after correction for local background, the final fluorescence intensity for each spot is obtained by averaging across the pixels in each spot. This results in a group of fluorescent spots at defined locations in the microarray.

High fluorescence intensity in a particular spot means that a large amount of cDNA in the sample hybridized to the DNA at that location. Because the DNA sequence of each spot is already known, a fluorescent spot identifies cDNAs that are complementary to those DNA sequences. Furthermore, because the cDNA was generated from mRNA, this technique identifies RNAs that have been made in a particular cell type under a given set of conditions.

The technology of DNA microarrays has found many important uses (**Table 21.4**). Thus far, its most common use is for studying gene expression patterns. Such studies help us to understand how genes are regulated in a cell-specific manner and how environmental conditions can induce or repress the transcription of genes. In some cases, microarrays can even help to elucidate the genes encoding proteins that participate in a complicated metabolic pathway. Microarrays can also be used as identification tools. For example, gene expression patterns can aid in the categorization of tumor types. Such identification can be important in the treatment of a disease. Instead of using labeled cDNA, researchers can also hybridize labeled genomic DNA to a microarray. This technique can be used to identify mutant alleles in a population and to detect deletions and duplications. In addition, it is useful in correctly identifying closely related bacterial species and subspecies. Finally, microarrays can be used to study DNA-protein interactions, as described next.

DNA Microarrays Can Be Used to Analyze DNA-Protein Binding at the Genome Level

As discussed throughout this text, the binding of proteins to specific DNA sites is critical for a variety of molecular processes, including gene transcription and DNA replication. To study these processes, researchers have devised a variety of techniques for identifying whether or not specific proteins bind to particular sites in DNA. For example, the electrophoretic mobility shift assay (EMSA), which is described in Chapter 20, is used for this purpose (see Figure 20.21).

More recently, a newer approach called **chromatin immunoprecipitation (ChIP)** has gained widespread use in the analysis of DNA-protein interactions. This method can determine whether proteins bind to particular sites in DNA. An advantage of this method is that it analyzes DNA-protein interactions as they occur in the chromatin of living cells. In contrast, an EMSA is an in vitro technique, which typically uses cloned DNA and purified proteins. **Figure 21.13** shows the steps of chromatin immunoprecipitation when it is used in conjunction with a DNA microarray or without one.

1. Proteins in living cells, which are noncovalently bound to DNA, are more tightly attached to the chromatin by the addition of formaldehyde or some other agent that covalently crosslinks the protein to the DNA.
2. Following crosslinking, the cells are lysed, and the DNA is broken by sonication into pieces approximately 200 to 1000 bp long.
3. Next, an antibody is added that is specific for the protein of interest. To conduct a ChIP assay, a researcher must suspect that a particular protein binds to the DNA and previously made or obtained an antibody that recognizes that protein. Antibodies specifically bind to antigens. In this case, the antigen is the protein of interest that is thought to be a DNA-binding protein.

TABLE 21.4

Applications of DNA Microarrays

Application	Description
Cell-specific gene expression	A comparison of microarray data using cDNAs derived from RNA of different cell types can identify genes that are expressed in a cell-specific manner.
Gene regulation	Environmental conditions play an important role in gene regulation. A comparison of microarray data may reveal genes that are induced under one set of conditions and repressed under another.
Elucidation of metabolic pathways	Genes that encode proteins that participate in a common metabolic pathway are often expressed in a parallel manner. This can be revealed from a microarray analysis. This application overlaps with the study of gene regulation via microarrays.
Tumor profiling	Different types of cancer cells exhibit striking differences in their profiles of gene expression. Such a profile can be revealed by a DNA microarray analysis. This approach is gaining widespread use as a method of subclassifying tumors that are sometimes morphologically indistinguishable. Tumor profiling may provide information that can improve a patient's clinical treatment.
Genetic variation	A mutant allele may not hybridize to a spot on a microarray as well as a wild-type allele. Therefore, microarrays are gaining widespread use as a tool for detecting genetic variation. They have been used to identify disease-causing alleles in humans and mutations that contribute to quantitative traits in plants and other species. In addition, microarrays are used to detect chromosomal deletions and duplications.
Microbial strain identification	Microarrays can distinguish between closely related bacterial species and subspecies.
DNA-protein binding	Chromatin immunoprecipitation, which is illustrated in Figure 21.13, can be used with DNA microarrays to determine where in the genome a particular protein binds to the DNA.

510   CHAPTER 21 :: GENOMICS

4. In Figure 21.13, the antibodies are attached to heavy beads. The antibodies bind to DNA-protein complexes and cause the complexes to form a pellet following centrifugation. (See Appendix A for a description of centrifugation.) Because an antibody is made by the immune system of an animal, this step is called immunoprecipitation.
5. The next step is to identify the DNA to which the protein is covalently crosslinked. To do so, the protein is removed by treatment with chemicals that break the covalent crosslinks.
6. Because the DNA is usually present in very low amounts, researchers must amplify the DNA to analyze it. This is done using PCR (described in Chapter 20).

- If researchers already suspect that a protein binds to a known DNA region, they can use PCR primers that specifically flank the DNA region (see bottom left of Figure 21.13). If they obtain a PCR product, this means that the protein of interest must have been bound (either directly or through other proteins) to this DNA site in living cells.
- Alternatively, a researcher may want to determine where the protein of interest binds across the whole genome. In this case, a DNA microarray can be used (see bottom right of Figure 21.13). Because a DNA microarray is created on a chip, this form of chromatin immunoprecipitation is called a **ChIP-chip assay.** The ends of the precipitated DNA are first ligated to short DNA pieces called linkers. PCR primers are then added that are complementary to the linkers and, therefore, amplify the DNA regions between the linkers. During PCR, the DNA is fluorescently labeled. The labeled DNA is then denatured and hybridized to a DNA microarray. Because the DNA was isolated using an antibody specific to the protein of interest, the fluorescent spots on the microarray identify sites in the genome where the protein binds. In this way, researchers may be able to determine where a protein binds to locations in the genome, even if those site(s) had not been previously determined by other methods.

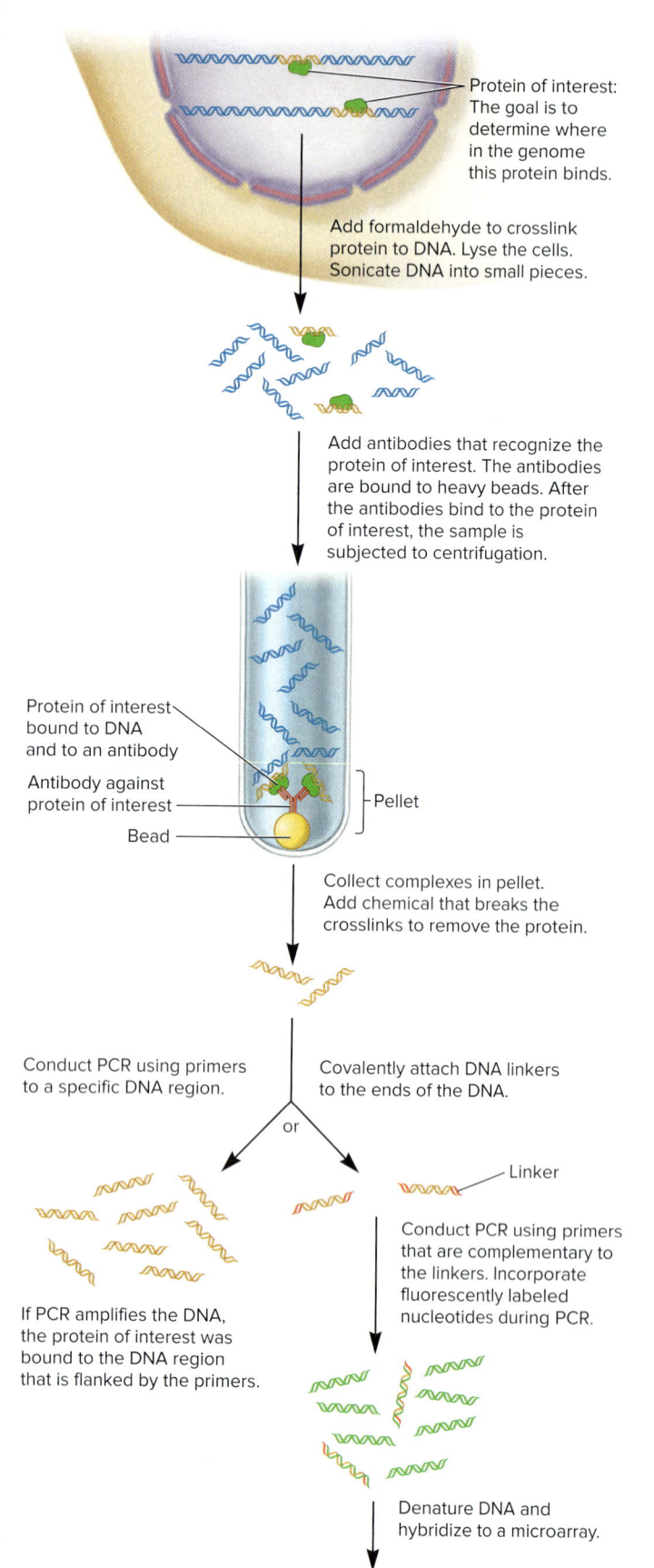

RNA Sequencing (RNA-Seq) Is a Newer Method for Identifying Expressed Genes

The **transcriptome** is the set of all RNA molecules, including mRNAs and non-coding RNAs, that are transcribed in one cell or a population of cells. Researchers may focus on the identification of each RNA molecule and also on their relative concentrations. The invention of next-generation sequencing technologies, described in Section 21.4, has changed the way in which transcriptomes are studied. A particularly exciting advance is **RNA sequencing (RNA-Seq)**, which was developed by Michael Snyder

FIGURE 21.13 Chromatin immunoprecipitation (ChIP). This method can determine whether proteins bind to a particular region of DNA found within the chromatin of living cells. If a microarray is used, as indicated in the bottom right, this is called a ChIP-chip assay.

Concept Check: *Why is an antibody used in this experiment?*

21.6 FUNCTIONAL GENOMICS

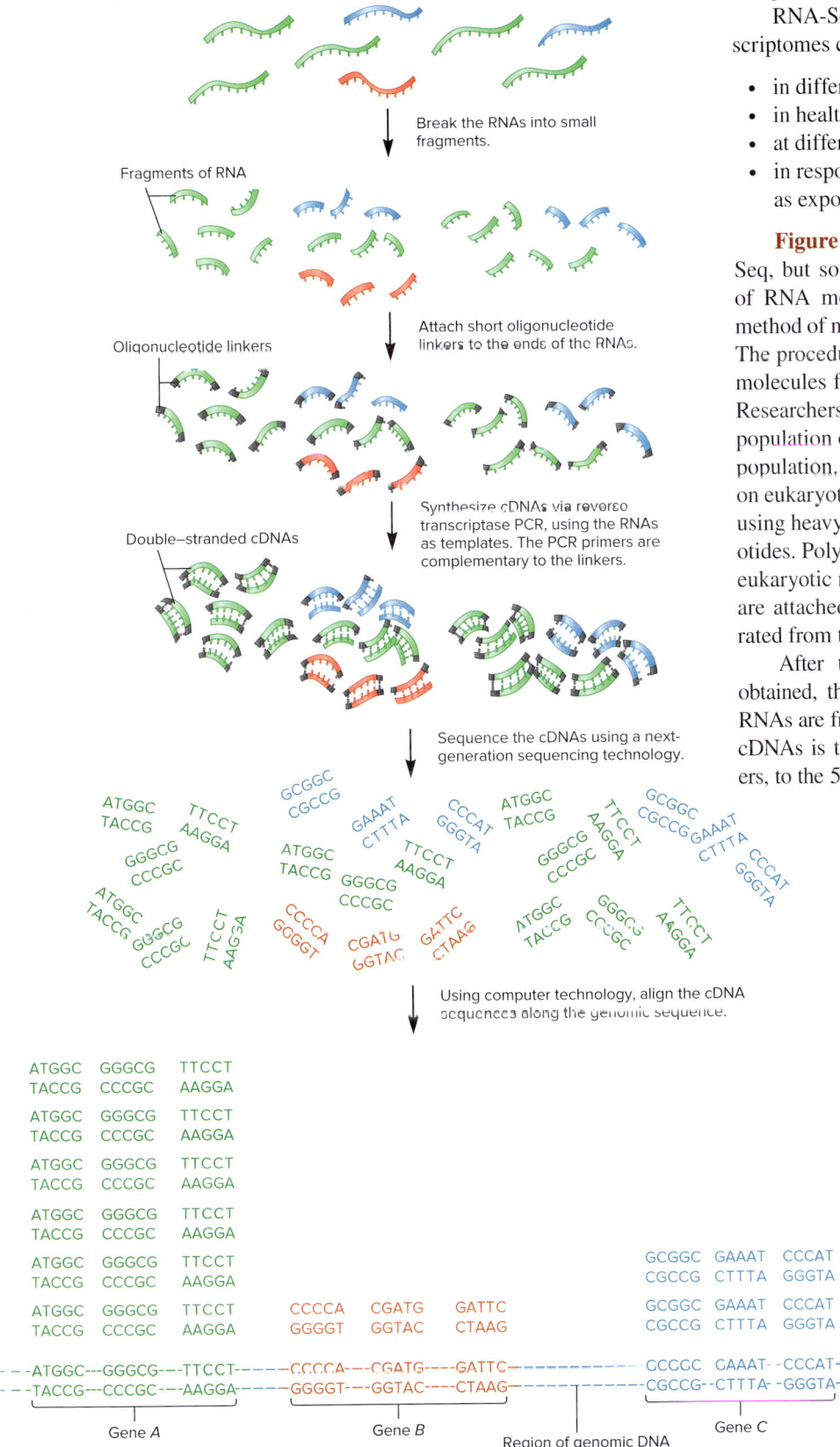

and colleagues in 2008. This method involves the sequencing of complementary DNAs (derived from RNAs) using next-generation DNA sequencing methods.

RNA-Seq has several important applications. Transcriptomes can be compared

- in different cell types;
- in healthy versus diseased cells;
- at different stages of development; and
- in response to different environment agents, such as exposure to a hormone or to toxic chemicals.

Figure 21.14 outlines a general strategy for RNA-Seq, but some steps may vary depending on the types of RNA molecules that are being sequenced and the method of next-generation DNA sequencing that is used. The procedure always begins with the isolation of RNA molecules from a sample of one or more types of cells. Researchers or clinicians may want to analyze the entire population of RNAs, or they may want to analyze a subpopulation. For example, if researchers wanted to focus on eukaryotic mRNAs, they could "pull out" mRNAs by using heavy beads that are attached to polyT oligonucleotides. PolyT oligonucleotides bind to the polyA tails of eukaryotic mRNAs. Because the polyT oligonucleotides are attached to heavy beads, the mRNAs can be separated from the rest of the RNAs by centrifugation.

After the desired population of RNAs has been obtained, the next step is to produce cDNAs. First, the RNAs are fragmented into small pieces. One way to make cDNAs is to attach short segments of DNA, called linkers, to the 5′ and 3′ ends of the RNA fragments. After the

FIGURE 21.14 The technique of RNA sequencing (RNA-Seq). This simplified example involves a population of three different RNA molecules, each shown in a different color. Protein-encoding genes in complex eukaryotes typically contain introns, which are spliced out of the pre-mRNAs. The alignment at the bottom of the figure corresponds to the sequences of the mature mRNAs. The gaps between the cDNA sequences in the alignment indicate the locations of introns. (Note: In actual RNA-Seq experiments, the RNA fragments are much longer than the ones shown here. Though the optimal length varies depending on the method of DNA sequencing used, a common length for the RNA fragments and corresponding cDNA sequences is 100–300 nucleotides.)

attachment of linkers, primers are added that are complementary to the linkers, and cDNAs are made via reverse transcriptase PCR, which is described in Chapter 20. The population of cDNAs is then subjected to next-generation DNA sequencing (see Section 21.4). This DNA sequencing produces a diverse collection of cDNA sequences.

The next phase of RNA-Seq is to compare the collection of cDNA sequences with the already known genome sequence of the organism from which the RNA was isolated. In other words, the cDNA sequences are aligned with the genomic DNA sequence (see bottom of Figure 21.14). This phase is accomplished with computer technology. When a cDNA sequence aligns with a gene sequence within the genome, this result means that the gene was expressed, because each cDNA sequence is derived from an RNA molecule. Also, as discussed in Chapter 14, eukaryotic pre-mRNAs often undergo splicing and alternative splicing (refer back to Figures 14.20 and 14.21). The alignment of cDNAs with the genomic DNA allows researchers to determine the pattern of RNA splicing that is found in a particular cell type under a given set of conditions.

Compared to the use of microarrays, RNA-Seq has several advantages:

- RNA-Seq is more accurate at quantifying the amount of each RNA transcript; the number of times that a particular cDNA sequence aligns with a gene is a measurement of the gene's expression level.
- It is superior in detecting RNA transcripts that are in low abundance.
- It identifies the precise boundaries between exons and introns and allows researchers to discover new splice variants.
- It identifies the 5′ and 3′ ends of RNA transcripts and aids in the identification of transcriptional start sites.

Gene Knockout Collections Allow Researchers to Study Gene Function at the Genomic Level

One broad goal of functional genomics is to determine the functions of all the genes in a species' genome. Because each species has thousands of different genes, this is a very complicated task. One approach to achieving this goal is to produce collections of organisms of the same species in which each strain has one of its genes knocked out. For example, in *E. coli*, which has 4377 different genes, a complete knockout collection would be composed of 4377 different strains, with a different gene knocked out in each strain. A **gene knockout** is the alteration of a gene in a way that inactivates its function.

Why are knockout collections useful? Consider, for example, a phenotype produced by a particular gene knockout in mice, which cause deafness. Such a result suggests that the function of the normal gene is to promote hearing. Geneticists may also produce knockouts involving two or more genes to understand how the protein products of genes participate in a particular cellular pathway or contribute to a complex trait.

Knockout collections can be made in different ways. One way is via transposable elements, which are described in Chapter 12. When a transposable element hops into a gene, it often inactivates the gene's function. Another way to produce knockouts is via CRISPR-Cas technology (see Chapter 20). In 2006, the National Institutes of Health (NIH) launched the Knockout Mouse Project. The goal of this program is to build a comprehensive and publicly available resource comprising a collection of mouse embryonic stem cells (ES cells) containing a loss-of-function mutation in every gene in the mouse genome. The NIH Knockout Mouse Project collaborates with other large-scale efforts including one in Canada, called the North American Conditional Mouse Mutagenesis Project (NorCOMM), and one in Europe, called the European Conditional Mouse Mutagenesis Program (EUCOMM). The collective goal of these programs is to create at least one loss-of-function mutation in each of the approximately 22,000 genes in the mouse genome.

Knockout collections are currently available for several other species, including *E. coli*, *S. cerevisiae*, and *C. elegans*. The use of these knockouts has already shed considerable light on the functions of many different genes in these species and will continue to facilitate our understanding of gene function at the genomic level in the future.

Genetic TIPS

The Question: A sample of liver cells was collected from a healthy donor and from an individual with liver cancer. mRNA was isolated from both samples of cells and subjected to DNA microarray analysis. In the results from the two samples, 77 spots on the microarry from the cancer cells were much brighter compared to those for the cells from the healthy donor. How would you interpret these results? Explain their meaning with regard to the growth of the cancer cells. (Note: Assume that each spot corresponds to a different gene.)

Topic: What topic in genetics does this question address?

The topic is DNA microarray analysis. More specifically, the question is about comparing healthy cells with cancer cells.

Information: What information do you know based on the question and your understanding of the topic?

From the question, you know the results from a DNA microarray analysis that compares healthy cells and cancer cells. From your understanding of the topic, you may remember that the brightness of a spot on a microarray indicates how much of the cDNA, which was reverse transcribed from the cells' mRNA, hybridized to the known DNA sequence at that location.

Problem-Solving **S**trategy: Compare and contrast.

One strategy to solve this problem is compare the results from the cancer and healthy cells and relate them to the transcription levels in the cells.

Answer: The cancer cells are overexpressing 77 genes that the normal cells are not overexpressing. The overexpression of some of these genes is likely to contribute to the cancerous growth.

21.6 REVIEWING THE KEY CONCEPTS

- The goal of functional genomics is to understand the role of genetic sequences in a given species.
- A microarray is a slide dotted with many DNA sequences. It can be used to study the expression of many genes simultaneously, and it also has other uses (see Figure 21.12, Table 21.4).

- In a ChIP-chip assay, chromatin immunoprecipitation is used in conjunction with a microarray to study DNA-protein interactions at the genomic level (see Figure 21.13).
- In the method known as RNA sequencing (RNA-Seq), RNA is isolated from cells, converted to cDNA, and then sequenced using a next-generation sequencing technology. The cDNA sequences are aligned with the genomic sequence (Figure 21.14).
- Researchers are producing gene knockout collections for certain species such as mice to determine the functions of genes at the genomic level.

21.6 COMPREHENSION QUESTIONS

1. A DNA microarray is a slide that is dotted with
   a. mRNAs from a sample of cells.
   b. fluorescently labeled cDNAs.
   c. known sequences of DNA.
   d. known cellular proteins.

2. To study gene expression in a particular sample of cells, what material is usually hybridized to a DNA microarray?
   a. Fluorescently labeled mRNAs
   b. Fluorescently labeled cDNAs
   c. Radiolabeled mRNAs
   d. Radiolabeled cDNAs

3. The purpose of a ChIP-chip assay is to determine
   a. the expression levels of particular genes in a genome.
   b. the amount of a specific protein that is made in a given cell type.
   c. the sites in a genome where a particular protein binds.
   d. all of the above.

4. A gene knockout is a gene
   a. whose function has been inactivated.
   b. that has been transferred to a different species.
   c. that has been moved to a new location in the genome.
   d. that has been eliminated from a species during evolution.

KEY TERMS

Page 491. genome, genomics, functional genomics
Page 492. mapping, cytogenetic mapping, linkage mapping, physical mapping, genetic map (chromosome map), locus (loci), in situ hybridization
Page 493. hybridization, fluorescence in situ hybridization (FISH)
Page 494. chromosome painting
Page 495. molecular marker, polymorphic
Page 496. microsatellites, sequence-tagged site (STS)
Page 498. genome-sequencing projects, shotgun sequencing
Page 500. Human Genome Project
Page 501. high-throughput sequencing, next-generation sequencing technologies, pyrosequencing
Page 502. sequencing by synthesis (SBS)
Page 505. comparative genomics, metagenomics, metagenome
Page 507. DNA microarray, gene chip
Page 509. chromatin immunoprecipitation (ChIP)
Page 510. ChIP-chip assay, transcriptome, RNA sequencing (RNA seq)
Page 512. gene knockout

CHAPTER SUMMARY

- The genome is the total genetic composition of an organism or species. Genomics refers to the molecular analysis of the entire genome of a species. Functional genomics is aimed at studying the expression of genes that make up a genome.

21.1 Overview of Chromosome Mapping

- The term *mapping* refers to the experimental process of determining the relative locations of genes or other DNA segments on a chromosome, thereby producing a diagram called a genetic map. The process may involve cytogenetic, linkage, or physical mapping techniques (see Figure 21.1).

21.2 In Situ Hybridization

- Fluorescence in situ hybridization (FISH) is a method commonly used to map genes and other segments on an intact chromosome chromosome (see Figures 21.2, 21.3).

21.3 Molecular Markers

- Linkage mapping and physical mapping often rely on molecular markers to map locations on chromosomes (see Table 21.1).

- Restriction fragment length polymorphisms (RFLPs), microsatellites, and single-nucleotide polymorphisms (SNPs) are used as molecular markers (see Figures 21.4–21.6).

21.4 Genome-Sequencing Projects

- Shotgun sequencing has been commonly used to sequence the DNA of many species. Overlapping regions allow researchers to determine the complete DNA sequence (see Figures 21.7, 21.8).
- The Human Genome Project resulted in the sequencing of the entire human genome.
- Newer methods of DNA sequencing can process many samples of DNA simultaneously and are superseding the Sanger dideoxy method. An example of these next-generation sequencing technologies is pyrosequencing (see Table 21.2, Figure 21.9).
- Since 1995, the genomes of thousands of species have been sequenced (see Table 21.3).

21.5 Metagenomics

- Metagenomics is the study of a complex mixture of genetic material obtained from an environmental sample (see Figures 21.10, 21.11).

21.6 Functional Genomics

- The goal of functional genomics is to understand the role of genetic sequences in a given species.
- A microarray is a slide dotted with many DNA sequences. It can be used to study the expression of many genes simultaneously, and it also has other uses (see Figure 21.12, Table 21.4).
- In a ChIP-chip assay, chromatin immunoprecipitation is used in conjunction with a microarray to study DNA-protein interactions at the genome level (see Figure 21.13).
- In the method known as RNA sequencing (RNA-Seq), RNA is isolated from cells, converted to cDNA, and then sequenced using a next-generation sequencing technology. The cDNA sequences are aligned with the genomic sequence (see Figure 21.14).
- Researchers are producing gene knockout collections for certain species, such as mice, to determine the functions of genes at the genome level.

PROBLEM SETS & INSIGHTS

More Genetic TIPS

1. Does a molecular marker have to be polymorphic to be useful in physical mapping studies? Does a molecular marker have to be polymorphic to be useful in linkage mapping (i.e., involving family pedigree studies or genetic crosses)? Explain why or why not.

Topic: What topic in genetics does this question address?

The topic is molecular markers. More specifically, the question is about whether molecular markers need to be polymorphic to be used in physical or linkage mapping.

Information: What information do you know based on the question and your understanding of the topic?

The question asks you to consider polymorphisms in molecular markers. From your understanding of the topic, you may remember that physical mapping involves the cloning of DNA segments, whereas linkage mapping involves crosses.

Problem-Solving Strategy: Compare and contrast.

One strategy to solve this problem is to compare and contrast the two approaches. In physical mapping, molecular markers are used as points of reference. In linkage mapping, researchers analyze individuals who are heterozygous for molecular markers to determine the number of recombinant offspring and thereby compute map distances.

Answer: A molecular marker does not have to be polymorphic to be useful in physical mapping studies. Many sequence-tagged sites (STSs) that are used in physical mapping studies are monomorphic. Monomorphic markers can provide landmarks in such mapping studies.

In linkage mapping studies, however, a marker must be polymorphic to be useful. Polymorphic molecular markers can be RFLPs, microsatellites, or SNPs. To compute map distances in linkage analysis, researchers must study individuals that are heterozygous for two or more markers (or genes). For experimental organisms, heterozygotes are testcrossed to homozygotes, and then the numbers of recombinant offspring and nonrecombinant offspring are determined. For markers that do not assort independently (i.e., linked markers), the map distance is computed as the number of recombinant offspring divided by the total number of offspring times 100.

2. The distance between two molecular markers that are linked along the same chromosome can be determined by analyzing the outcomes of crosses. This can be done in humans by analyzing a family's pedigree. However, the accuracy of linkage mapping with human pedigrees is fairly limited because the number of people in most families is relatively small. As an alternative, researchers can analyze a population of sperm, produced from a single male, and compute linkage distance in this manner. As an example, let's suppose a male is heterozygous for two polymorphic STSs. STS-1 exists in two sizes: 234 bp and 198 bp. STS-2 also exists in two sizes: 423 bp and 322 bp. A sample of sperm was collected from this man, and individual sperm were placed into 40 separate tubes. In other words, there was one sperm in each tube. Believe it or not, PCR is sensitive enough to allow analysis of DNA in a single sperm! Into each of the 40 tubes were added the primers that amplify STS-1 and STS-2, and then the samples were subjected to PCR. The following results were obtained:

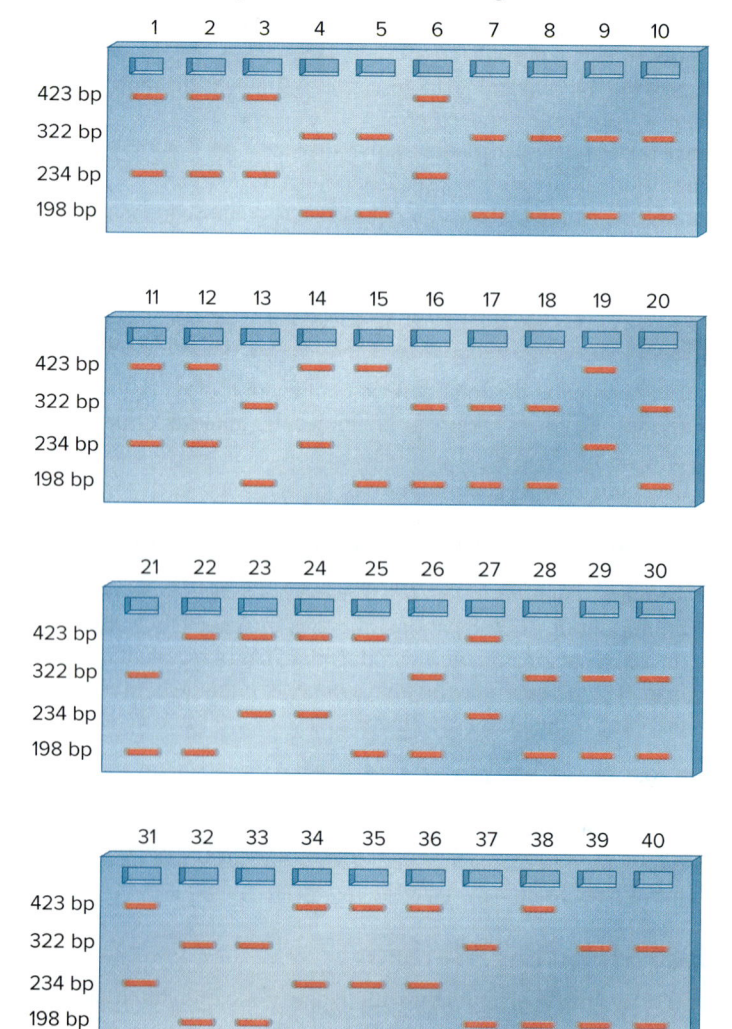

A. What is the arrangement of these two STSs in this individual?

B. What is the map distance between STS-1 and STS-2?

Topic: What topic in genetics does this question address?

The topic is linkage mapping. More specifically, the question is about using DNA in sperm to compute the map distance between two molecular markers.

Information: What information do you know based on the question and your understanding of the topic?

From the question, you know the patterns of STS-1 and STS-2 in a population of 40 sperm. From your understanding of the topic, you may remember that linkage mapping involves the identification of recombinants, which are produced by crossovers. The map distance is the percentage of recombinants.

Problem-Solving Strategy: Compare and contrast. Make a calculation.

One strategy to begin to solve this problem is to compare and contrast the lanes on the gel. When markers are linked, the nonrecombinant pattern of bands will be more common than the recombinant pattern. Map distance is computed as the number of recombinants divided by the total number of sperm analyzed, times 100.

Answer: Keep in mind that mature sperm are haploid, so they have only one copy of STS-1 and one copy of STS-2.

A. If you look at the 40 lanes, most of them (i.e., 36) have either the 234-bp and 423-bp STSs or the 198-bp and 322-bp ones. This is the arrangement of STSs in this male. One chromosome has STS-1 that is 234 bp and STS-2 that is 423 bp, and the homologous chromosome has STS-1 that is 198 bp and STS-2 that is 322 bp.

B. There are four recombinant sperm, shown in lanes 15, 22, 25, and 38.

$$\text{Map distance} = \frac{4}{40} \times 100$$

$$\text{Map distance} = 10.0 \text{ mu}$$

Note: This is a relatively easy experiment compared with a pedigree analysis, which would involve contacting lots of relatives and collecting samples from each of them.

Conceptual Questions

C1. A person with a rare genetic disease has a sample of her chromosomes subjected to fluorescence in situ hybridization using a probe that is known to recognize band p11 on chromosome 7. Even though her chromosomes look cytologically normal, the probe does not bind to this person's chromosomes. How would you explain these results?

C2. For each of the following, decide if it could be appropriately described as a genome:

A. The *E. coli* chromosome

B. Human chromosome 11

C. A complete set of 10 chromosomes in corn

D. A copy of the single-stranded RNA packaged into human immunodeficiency virus (HIV)

C3. Which of the following statements about molecular markers are true?

A. All molecular markers are segments of DNA that carry specific genes.

B. A molecular marker is a segment of DNA that is found at a specific location in a genome.

C. We can follow the transmission of a molecular marker by analyzing the phenotype (i.e., the physical characteristics) of offspring

D. We can follow the transmission of molecular markers using molecular techniques such as gel electrophoresis.

E. An STS is a molecular marker.

Application and Experimental Questions

E1. In an in situ hybridization experiment, what is the relationship between the base sequence of the probe DNA and the site on the chromosomal DNA where the probe binds?

E2. Describe the technique of fluorescence in situ hybridization. Explain how it can be used to map genes.

E3. The cells from a person's malignant tumor were subjected to fluorescence in situ hybridization using a probe that recognizes a unique sequence on chromosome 14. The probe was detected only once in each of the cells. Explain this result, and speculate on its significance with regard to the malignant characteristics of these cells.

E4. Figure 21.2 illutstrates the technique of FISH. Why is it necessary to fix the cells (and the chromosomes inside of them) to the slides? What does it mean to fix them? Why is it necessary to denature the chromosomal DNA?

E5. Explain how DNA probes with different fluorescence emission wavelengths can be used in a single FISH experiment to map the locations of two or more genes. This method is called chromosome painting. Explain why this is an appropriate term.

E6. A researcher is interested in a gene found on human chromosome 21. Describe the expected results of a FISH experiment using a probe that is complementary to this gene. How many spots would you see if the probe was used on a sample from an individual with 46 chromosomes versus an individual with Down syndrome?

E7. A woman had five children with two different men. This group of seven individuals is analyzed with regard to three different STSs: STS-1 is 146 bp and 122 bp; STS-2 is 102 bp and 88 bp; and STS-3 is 188 bp and 204 bp. The mother is homozygous for all three STSs: STS-1 = 122, STS-2 = 88, and STS-3 = 188. Father 1 is homozygous for STS-1 = 122 and STS-2 = 102, and heterozygous for STS-3 = 188 and 204. Father 2 is heterozygous for STS-1 = 122 and 146, STS-2 = 88 and 102, and homozygous for STS-3 = 204. The five children show the following results:

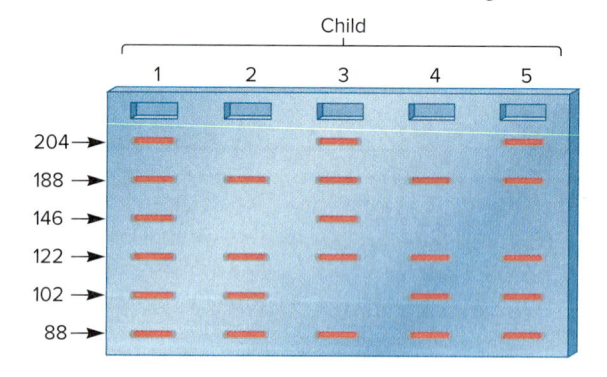

Which children can you definitely assign to father 1 and father 2?

E8. In the Human Genome Project, researchers collected linkage data from many crosses in which the male was heterozygous for molecular markers and many crosses where the female was heterozygous for the same markers. The distance between these two markers, computed in map units, is different between males and females. In other words, the linkage maps for human males and females are not the same. Propose an explanation for this discrepancy. Do you think the sizes of chromosomes (excluding the Y chromosome) in human males and females are different? How could physical mapping resolve this discrepancy?

E9. Take a look at question 2 in More Genetic TIPS. Let's suppose a male is heterozygous for two polymorphic sequence-tagged sites. STS-1 exists in two sizes: 211 bp and 289 bp. STS-2 also exists in two sizes: 115 bp and 422 bp. A sample of sperm was collected from this man, and individual sperm were placed into 30 separate tubes. Into each of the 30 tubes were added the primers that amplify STS-1 and STS-2, and then the samples were subjected to PCR. The following results were obtained:

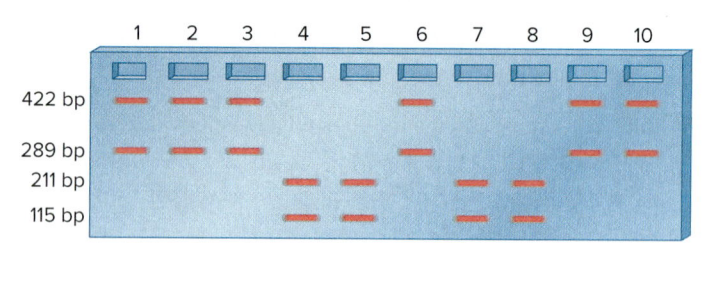

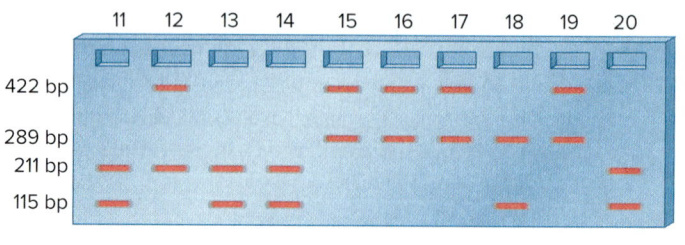

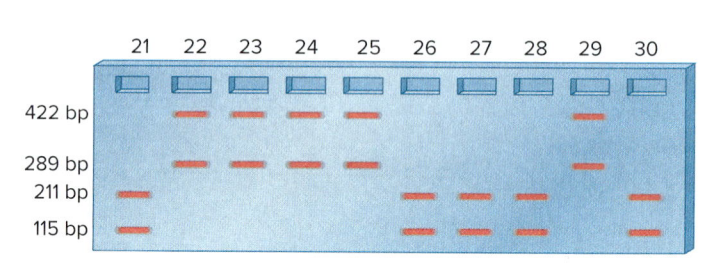

A. What is the arrangement of these STSs in this individual?

B. What is the map distance between STS-1 and STS-2?

C. Could this approach to analyzing a population of sperm be applied to RFLPs?

E10. What is an STS? How are STSs generated experimentally? What are the uses of STSs? Explain how a microsatellite can be a polymorphic STS.

E11. A bacterium has a genome size of 4.4 Mb. If a researcher carries out shotgun DNA sequencing and sequences a total of 19 Mb, what is the probability that a base will be left unsequenced? What percentage of the total genome will be left unsequenced?

E12. Discuss the advantages of next-generation sequencing technologies.

E13. What is meant by *sequence by synthesis*?

E14. Outline the general strategy used in metagenomics.

E15. With regard to DNA microarrays, answer the following questions:

A. What is attached to the slide? Be specific about the number of spots, the lengths of DNA fragments, and the origin of the DNA fragments.

B. What is hybridized to the spots on the microarray?

C. How is hybridization detected?

E16. In the procedure called RNA sequencing (RNA-Seq), what type of molecule is actually sequenced?

Questions for Student Discussion/Collaboration

1. What is a molecular marker? Give two examples. Discuss why it is generally easier to locate and map molecular markers rather than functional genes.

2. Which goals of the Human Genome Project do you think are the most important? Why? Discuss the types of ethical problems that might arise as a result of identifying all of our genes.

Answers to Comprehension Questions

21.1: b
21.2: d
21.3: c, d, a
21.4: a, b, b
21.5: c
21.6: c, b, c, a

Note: All answers appear in Connect; the answers to even-numbered questions and all Concept Check questions are in Appendix B.

22

CHAPTER OUTLINE

22.1 Inheritance Patterns of Genetic Diseases
22.2 Detection of Disease-Causing Alleles via Haplotypes
22.3 Genetic Testing and Screening
22.4 Overview of Cancer
22.5 Oncogenes
22.6 Tumor-Suppressor Genes
22.7 Role of Epigenetics in Cancer
22.8 Personalized Medicine

PART VI GENETIC ANALYSIS OF INDIVIDUALS AND POPULATIONS

Cigarette smoking and lung cancer. Cigarette smoke contains chemicals that are known to mutate genes in the cells of a person's lungs, thereby leading to lung cancer. Lung cancer remains the top cause of deaths from cancer in the United States, and 87% of deaths due to lung cancer are linked to smoking.
©UK Stock Images Ltd./Alamy Stock Photo

MEDICAL GENETICS AND CANCER

Insight into genetics is expected to bring about revolutionary changes in medical practices. In fact, changes are already underway. Currently, several hundred genetic tests are in clinical use, with many more under development. Most of these tests detect mutations associated with rare genetic disorders that follow Mendelian inheritance patterns. These include Duchenne muscular dystrophy, cystic fibrosis, sickle cell disease, and Huntington disease. In addition, there are now genetic tests to detect the predisposition to develop certain forms of cancer. As discussed in Chapter 21, DNA-sequencing technologies are progressing to the point where the sequencing of a person's entire genome will be inexpensive enough to be done as a routine diagnostic procedure. Many people in the medical field expect that this advance will usher in the era of personalized medicine—the use of information about a patient's genotype and other clinical data in order to select a medication, therapy, or preventative measure that is specifically suited to that patient. We will explore this topic at the end of this chapter.

Thousands of genetic diseases are known to afflict people. Most of the genetic disorders discussed in the first part of this chapter are the direct result of a mutation in one gene. However, many diseases have a complex pattern of inheritance involving several genes. These include common medical disorders such as diabetes, asthma, and certain forms of mental illness. In these cases, a single mutant gene does not determine whether a person has a disease. Instead, a number of genes may each make a subtle contribution to a person's susceptibility to a disease. Unraveling these complexities will be a challenge for some time to come. The availability of the human genome sequence, discussed in Chapter 21, will be of great help.

In this chapter, we will focus our attention on ways that mutant genes contribute to human disease. In the first part of the chapter, we will explore the molecular basis of several genetic disorders and their patterns of inheritance. We will also examine how genetic testing can determine if an individual carries a defective allele. We will then consider cancer, a disease that involves the uncontrolled growth of somatic cells. We will examine the underlying genetic basis for cancer and discuss the roles that many different genes may play in the development of this disease.

22.1 INHERITANCE PATTERNS OF GENETIC DISEASES

Learning Outcomes:
1. List seven observations that suggest a disease may have a genetic component.
2. Analyze human pedigrees, and be able to distinguish autosomal recessive, autosomal dominant, X-linked recessive, and X-linked dominant patterns.
3. Define *locus heterogeneity*, and explain how it can confound pedigree analysis.

Human genetics is a topic that is hard to resist. Almost everyone who looks at a newborn is tempted to speculate whether the baby resembles the mother, the father, or perhaps a more distant relative. In this section, we will focus primarily on the inheritance of human genetic diseases rather than common traits found in the general population. Even so, the study of human genetic diseases often provides insights regarding such common traits. The disease hemophilia illustrates this point. Hemophilia is a condition in which the blood does not clot properly. By analyzing people with this disorder, researchers have identified genes that participate in the process of blood clotting. The study of hemophilia has helped to identify a clotting pathway involving several different proteins. Therefore, as with the study of mutants in model organisms such as *Drosophila*, mice, and yeast, when we study the inheritance of genetic diseases, we often learn a great deal about the genetic basis for normal physiological processes as well.

Because thousands of human diseases have an underlying genetic basis, human genetic analysis is of great medical importance. In this section, we will examine the causes and inheritance patterns of human genetic diseases that result from defects in single genes. As you will learn, mutant genes that cause diseases often follow simple Mendelian inheritance patterns.

A Genetic Basis for a Human Disease May Be Suggested by a Variety of Observations

When we view the characteristics of people, we usually think that some traits are inherited, whereas others are caused by environmental factors. For example, when the facial features of two related individuals look strikingly similar, we think that this similarity has a genetic basis. The profound resemblance between identical twins is an obvious example. By comparison, other traits have environmental causes. If we see a person with purple hair, we likely suspect that he or she has used hair dye as opposed to showing an unusual genetic trait.

For human diseases, geneticists would like to know the relative contributions from genetics and the environment. Is a disease caused by a pathogenic microorganism, a toxic agent in the environment, or a faulty gene? Unlike the case with experimental organisms, we cannot conduct human crosses to determine the genetic basis for diseases. Instead, we must rely on analyzing the occurrence of a disease in families that already exist. As described in the following list, several observations are consistent with the idea that a disease is caused, at least in part, by the inheritance of mutant genes. When the occurrence of a disease correlates with several of these observations, a geneticist becomes increasingly confident that the disease has a genetic basis.

1. *When an individual exhibits the disease, it is more likely to occur in that person's genetic relatives than in someone in the general population.* For example, someone with cystic fibrosis is more likely to have relatives with this disease than would a randomly chosen member of the general population.

2. *Identical twins share the disease more often than nonidentical twins.* Identical twins, also called **monozygotic (MZ) twins,** are genetically identical to each other, because they were formed from the same sperm and egg. By comparison, nonidentical twins, also called fraternal, or **dizygotic (DZ) twins,** are formed from separate pairs of sperm and egg cells. Fraternal twins share, on average, 50% of their genetic material. When a disorder has a genetic component, a pair of identical twins is more likely to exhibit the disorder than is a pair of fraternal twins.

    Geneticists evaluate a disorder's **concordance,** the degree to which it is inherited, by calculating the percentage of twin pairs in which both twins exhibit the disorder relative to pairs where only one twin shows the disorder. Theoretically, for diseases caused by a single gene, concordance among identical twins should be 100%. For fraternal twins, concordance for dominant disorders is expected to be 50%, assuming only one parent is heterozygous for the disease. For recessive diseases, concordance among fraternal twins is expected to be 25% if we assume both parents are heterozygous carriers. However, the actual concordance values observed for most single-gene disorders are usually less than such theoretical values for a variety of reasons. For example, some disorders are not completely penetrant, meaning that the symptoms associated with the disorder are not always produced. Also, one twin may have a disorder due to a new mutation that occurred after fertilization; it would be very unlikely for the other twin to have the same mutation.

3. *The disease does not spread to individuals sharing similar environmental situations.* Inherited disorders cannot spread from person to person. The only way genetic diseases can be transmitted is from parent to offspring.

4. *Different populations tend to have different frequencies of the disease.* Because mutations are rare events, they may arise in one population but not another. Also, each population is exposed to its own unique set of environmental conditions that may influence the prevalence of a given allele. Therefore, the frequencies of genetic diseases due to mutant alleles usually vary among different populations of humans. For example, the frequency of sickle cell disease is highest among certain African and Asian populations and relatively low in other parts of the world (see Chapter 23, Figure 23.9).

5. *The disease tends to develop at a characteristic age.* Many genetic disorders exhibit a characteristic **age of onset** at

which the symptoms of the disease appear. Some mutant genes exert their effects during embryonic and fetal development, so their effects are apparent at birth. Other genetic disorders tend to develop much later in life.

6. *The human disorder may resemble a disorder that is already known to have a genetic basis in an animal.* In animals, on which we can conduct experiments, various traits are known to be governed by genes. For example, the albino phenotype is found in humans as well as in many animals (**Figure 22.1**).

7. *A correlation is observed between a disease and a mutant human gene or a chromosomal alteration.* A particularly convincing piece of evidence that a disease has a genetic basis is the identification of altered genes or chromosomes that occur only in people exhibiting the disorder. When comparing two individuals, one with a disease and one without, we expect to see differences in their genetic material if the disorder has a genetic component. Alterations in gene sequences are determined by DNA-sequencing techniques (see Chapter 20). Also, changes in chromosome structure and number can be detected by the microscopic examination of chromosomes (see Chapter 8).

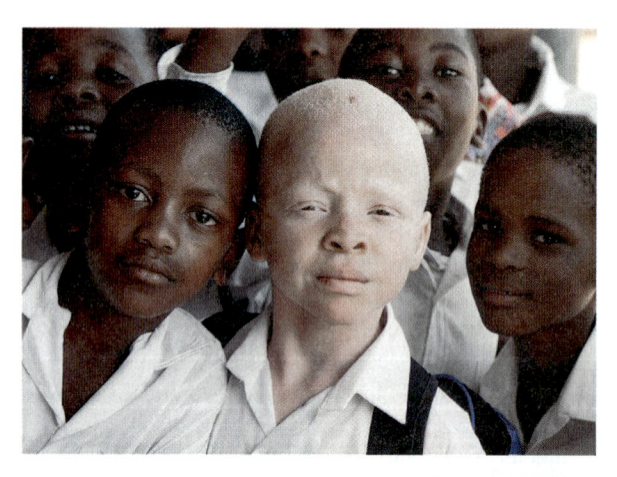

FIGURE 22.1 The albino phenotype in a human and a wildebeast.

Genes → Traits Certain enzymes (encoded by genes) are necessary for the production of pigment. A homozygote with two defective alleles in one of these pigmentation genes exhibits an albino phenotype. This phenotype can occur in humans and other animals.

(t) ©Friedrich Stark/Alamy Stock Photo; (b) ©Mitch Reardon/Science Source

Inheritance Patterns of Human Diseases May Be Determined via Pedigree Analysis

When a human disorder is caused by a mutation in a single gene, the pattern of inheritance can be deduced by analyzing human pedigrees. How is this accomplished? A geneticist must obtain data from many large pedigrees containing several individuals who exhibit the disorder and then follow its pattern of inheritance from generation to generation. To appreciate the basic features of pedigree analysis, we will examine a few pedigrees that involve diseases inherited in different ways. You may wish to review Chapter 3 (see Figure 3.14) on the organization and symbols of pedigrees.

Autosomal Recessive Inheritance The pedigree shown in **Figure 22.2** concerns a genetic disorder called Tay-Sachs disease (TSD), first described by Warren Tay, a British ophthalmologist, and Bernard Sachs, an American neurologist, in the 1880s. Affected individuals appear healthy at birth but then develop neurodegenerative symptoms at 4 to 6 months of age. The primary characteristics are cerebral degeneration, blindness, and loss of motor function. Individuals with TSD typically die in the third or fourth year of life. This disease is particularly prevalent in Ashkenazi (eastern European) Jewish populations, in which it has a frequency of about 1 in 3600 births, which is over 100 times more frequent than in most other human populations.

At the molecular level, the mutation that causes TSD is in a gene that encodes the enzyme hexosaminidase A (HexA). HexA is responsible for the breakdown of a category of lipids called G_{M2}-gangliosides, which are prevalent in the cells of the central nervous system. A defect in the ability to break down these lipids leads to their excessive accumulation in neurons and eventually causes the neurodegenerative symptoms characteristic of TSD.

As illustrated in Figure 22.2, Tay-Sachs disease is inherited in an autosomal recessive manner. Four common features of autosomal recessive inheritance are as follows:

1. *Frequently, an affected offspring has two unaffected parents.* For rare recessive traits, the parents are usually unaffected, meaning they do not exhibit the disease. For deleterious alleles that cause early death or infertility, the two parents must be unaffected. This is always the case in TSD.
2. *When two unaffected heterozygotes have children, the percentage of affected children is (on average) 25%.*
3. *Two affected individuals have 100% affected children.* This observation can be made only when a recessive trait produces fertile, viable individuals. In the case of TSD, the affected individual dies in early childhood, and so it is not possible to observe crosses between two affected people.
4. *The trait occurs with the same frequency in both sexes.*

Autosomal recessive inheritance is a common mode of transmission for genetic disorders, particularly those that involve defective enzymes. Human recessive alleles are often caused by mutations that result in a loss of function in the encoded enzyme. In the case of Tay-Sachs disease, a heterozygous carrier has approximately 50% of the functional enzyme, which is sufficient for a normal phenotype. However, for other genetic diseases, the

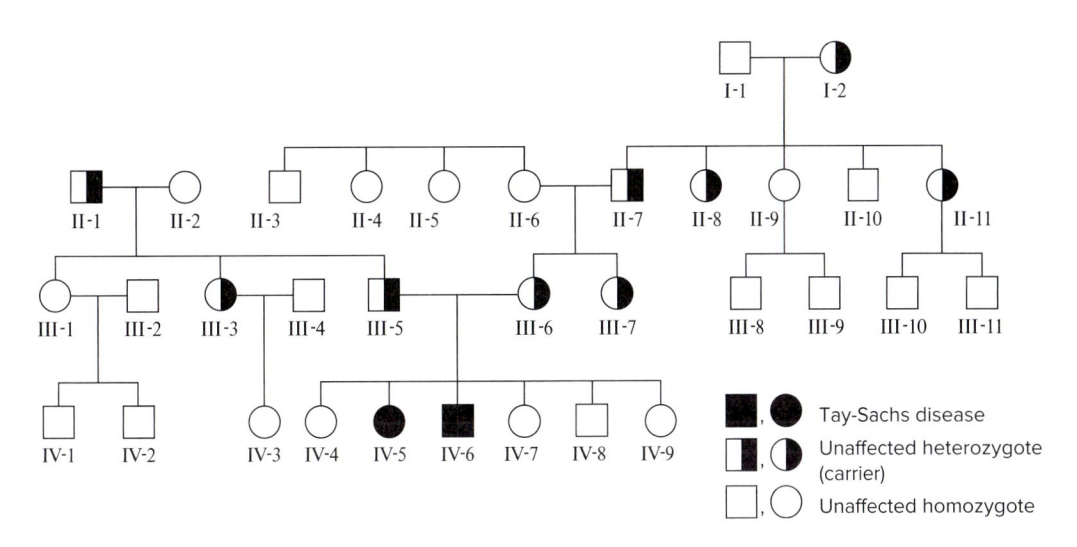

FIGURE 22.2 A family pedigree for Tay-Sachs disease, indicating autosomal recessive inheritance. Heterozygous carriers were determined by genetic testing.

Concept Check: What feature(s) of this pedigree indicate(s) autosomal recessive inheritance?

---

level of functional protein may vary due to effects of gene regulation. Thousands of human genetic diseases are inherited in a recessive manner, and in many cases, the mutant genes have been identified. A few of these diseases are described in **Table 22.1**.

Autosomal Dominant Inheritance Now let's examine a human pedigree involving an autosomal dominant disease (**Figure 22.3**). In this example, the affected individuals have a disorder called Huntington disease. The major symptoms of this disease, which usually develops during middle age, are due to the degeneration of certain types of neurons in the brain, leading to personality changes, dementia, and early death. In 1993, the gene involved in Huntington disease was identified and sequenced. It encodes a protein called huntingtin that is expressed in neurons but is also found in some cells not affected in Huntington disease.

In persons with this disorder, a mutation called a trinucleotide repeat expansion produces a polyglutamine tract—many glutamines in a row—within the huntingtin protein. This causes an aggregation of the protein in neurons. However, additional research is needed to understand the molecular relationship between the abnormality in the huntingtin protein and the disease symptoms. Five common features of autosomal dominant inheritance are as follows:

1. *An affected offspring usually has one or two affected parents.* However, this is not always the case. Some dominant traits show incomplete penetrance (see Chapter 5), so a heterozygote may not exhibit the trait even though it may be passed to offspring who do exhibit the trait. Also, a dominant mutation may occur during gametogenesis, so two unaffected parents may produce an affected offspring.

TABLE 22.1
Examples of Human Disorders Inherited in an Autosomal Recessive Manner

Disorder	Chromosomal Location of Gene	Gene Product	Effects of Disease-Causing Allele
Adenosine deaminase deficiency	20q	Adenosine deaminase	Defective immune system and skeletal and neurological abnormalities
Albinism (type I)	11q	Tyrosinase	Inability to synthesize melanin, resulting in very light skin, hair, etc.
Cystic fibrosis (CF)	7q	Cystic fibrosis transmembrane conductance regulator	Water imbalance in tissues of the pancreas, intestine, sweat glands, and lungs due to impaired ion transport; leads to lung damage
Phenylketonuria (PKU)	12q	Phenylalanine hydroxylase	Foul-smelling urine, neurological abnormalities, mental impairment; may be remedied by diet modification starting at birth

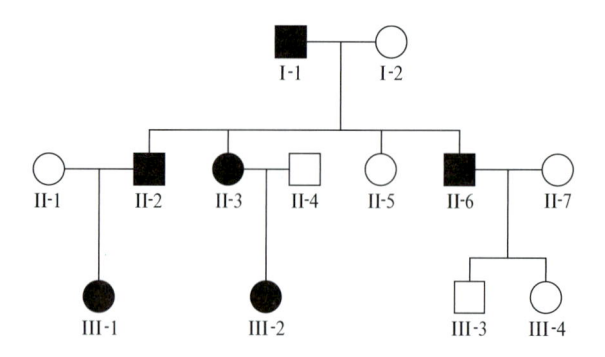

FIGURE 22.3 A family pedigree for Huntington disease, indicating autosomal dominant inheritance. Because the dominant allele is rare, most affected individuals are heterozygotes. Rare cases of people who are homozygous for the disease-causing allele have been reported. Such individuals tend to have more severe symptoms.

Concept Check: What feature(s) of this pedigree indicate(s) autosomal dominant inheritance?

2. An affected individual with only one affected parent is expected to produce 50% affected offspring (on average).
3. Two affected, heterozygous individuals have (on average) 25% unaffected offspring.
4. The trait occurs with the same frequency in both sexes.
5. For most dominant, disease-causing alleles, the homozygote is more severely affected with the disorder. In some cases, a dominant allele may be lethal in the homozygous condition.

Numerous autosomal dominant diseases have been identified in humans (**Table 22.2**). The three common explanations for dominant disorders are haploinsufficiency, a gain-of-function mutation, or a dominant-negative mutation.

- **Haploinsufficiency** refers to the phenomenon in which a person has only a single functional copy of a gene, and that single functional copy does not produce a normal phenotype. In these disorders, 50% of the functional protein is not sufficient to produce a normal phenotype. Haploinsufficiency shows a dominant pattern of inheritance because a heterozygote (with one functional allele and one inactive allele) has the disease. An example is aniridia, which is a rare disorder that results in an absence of the iris of the eye. Aniridia leads to visual impairment and blindness in severe cases.

- **Gain-of-function mutations** change the gene product so it gains a new or abnormal function. An example is achondroplasia, which is characterized by abnormal bone growth that results in short stature with relatively short arms and legs. This disorder is caused by a mutation that occurs in the gene that encodes fibroblast growth factor receptor-3. In achondroplasia, the mutant form of the receptor is overactive. This overactivity disrupts a signaling pathway and leads to severely shortened bones. Also, the allele that causes Huntington disease is thought to be an example of a gain-of-function mutation. The aggregation of the mutant protein is thought to interfere with neural function.

- **Dominant-negative mutations** alter the gene product in a way that acts antagonistically to the normal gene product. In humans, Marfan syndrome, which is due to a mutation in the *fibrillin-1* gene, is an example. The *fibrillin-1* gene encodes a glycoprotein that is a structural component of the extracellular matrix that provides structure and elasticity to tissues. The mutant gene encodes a glycoprotein that opposes the effects of the normal protein, thereby weakening the elasticity of certain body parts. For example, the walls of the major arteries such as the aorta, the large artery that leaves the heart, are often affected.

X-Linked Recessive Inheritance Let's now turn to another inheritance pattern common in humans that is called X-linked

TABLE 22.2

Examples of Human Disorders Inherited in an Autosomal Dominant Manner

Disorder	Chromosomal Location of Gene	Gene Product	Effects of Disease-Causing Allele
Aniridia	11p	Pax6 transcription factor	An absence of the iris of the eye, leading to visual impairment and sometimes blindness
Marfan syndrome	15q	Fibrillin-1	Tall and thin individuals with abnormalities in the skeletal, ocular, and cardiovascular systems due to a weakening in the elasticity of certain body parts
Osteoporosis	7q	Collagen (type 1α_2)	Brittle, weakened bones
Familial hypercholesterolemia	19p	LDL receptor	Very high serum levels of low-density lipoprotein (LDL), a predisposing factor in heart disease
Huntington disease	4p	Huntingtin	Neurodegeneration that occurs relatively late in life, usually in middle age

TABLE 22.3
Examples of Human Disorders Inherited in an X-Linked Recessive Manner

Disorder	Gene Product	Effects of Disease-Causing Allele
Duchenne muscular dystrophy	Dystrophin	Progressive degeneration of muscles that begins in early childhood
Hemophilia A	Clotting factor VIII	Defect in blood clotting
Hemophilia B	Clotting factor IX	Defect in blood clotting
Androgen insensitivity syndrome	Androgen receptor	Missing male steroid hormone receptor; XY individuals have external features that are feminine but internally have undescended testes and no uterus

recessive inheritance (Table 22.3). X-linked recessive inheritance of diseases poses a special problem for males. Why are males more likely to be affected? Most X-linked genes lack a counterpart on the Y chromosome. Males are hemizygous—have a single copy—for these genes. Therefore, a female heterozygous for an X-linked recessive allele passes this trait on to 50% of her sons, as shown in the following Punnett square for hemophilia. In this example, X^H carries the wild-type allele, whereas X^{h-A} is the X chromosome that carries a mutant allele causing hemophilia.

	♂ X^H	Y
♀ X^H	$X^H X^H$ Unaffected female	$X^H Y$ Unaffected male
X^{h-A}	$X^H X^{h-A}$ Carrier female	$X^{h-A} Y$ Male with hemophilia

As mentioned previously, hemophilia is a disorder in which the blood cannot clot properly when a wound occurs. For individuals with this trait, a minor cut may bleed for a very long time, and small injuries can lead to large bruises, because internal broken capillaries may leak blood profusely before they are repaired. For hemophiliacs, common injuries pose a threat of severe internal or external bleeding. Hemophilia A, also called classical hemophilia, is caused by a loss-of-function mutation in an X-linked gene that encodes the protein clotting factor VIII. This disease has also been called the "royal disease," because it has affected many members of European royal families. The pedigree shown in **Figure 22.4** illustrates the prevalence of hemophilia A among the descendants of Queen Victoria of England. The pattern of X-linked recessive inheritance is revealed by the following observations:

1. *Males are much more likely to exhibit the trait.*
2. *The mothers of affected males often have brothers or fathers who are affected with the same trait.*
3. *The daughters of affected males produce, on average, 50% affected sons.*

X-Linked Dominant Inheritance Relatively few genetic disorders in humans follow an X-linked dominant inheritance pattern. In most cases, males are more severely affected than females, probably because females carry an X chromosome with a normal copy of the gene in question. With most of the X-linked dominant disorders listed in **Table 22.4**, male embryos die at an early stage of development, so most individuals exhibiting the disorder are females. Also, due to their dominant nature and severity, persons with some of the disorders listed in Table 22.4 do not reproduce. Therefore, these dominant disorders, which include Rett syndrome and Aicardi syndrome, are not passed from generation to generation. Instead, they are caused by new mutations that occur during gamete formation or early embryogenesis. For those X-linked dominant disorders in which the offspring can reproduce, the following pattern is often observed:

1. *Only females exhibit the trait when it is lethal to males.*
2. *Affected mothers have a 50% chance of passing the trait to daughters.* Note: Affected mothers also have a 50% chance of passing the trait to sons, but for many of these disorders, affected sons are not observed because of lethality.

Many Genetic Disorders Exhibit Locus Heterogeneity

Hemophilia, which we considered earlier in this chapter, illustrates another concept in genetics called **locus heterogeneity.** This term refers to the phenomenon in which a particular type of disease may be caused by mutations in two or more different genes. For example, blood clotting involves the participation of several different proteins that take part in a cellular cascade that leads to the formation of a clot. Hemophilia is usually caused by a defect in one of three different clotting factors. In hemophilia A, a protein called factor VIII is missing. Hemophilia B is a deficiency in a different clotting factor, called factor IX. Factors VIII and IX are encoded by different genes on the X chromosome. These two types of hemophilia show an X-linked recessive pattern of inheritance. By comparison, hemophilia C is due to a factor XI deficiency. The gene encoding factor XI is found on chromosome 4, and this form of hemophilia follows an autosomal recessive pattern of inheritance.

Unfortunately, locus heterogeneity may greatly confound pedigree analysis. For example, a human pedigree might contain individuals with X-linked hemophilia and other individuals with hemophilia C. A geneticist who assumed that all affected individuals had defects in the same gene would be unable to explain the resulting pattern of inheritance. For disorders such as hemophilia, pedigree analysis is not a major problem because the biochemical basis for this disease is well understood. However, for rare diseases that are poorly understood at the molecular level, locus heterogeneity may obscure the pattern of inheritance.

524   CHAPTER 22 :: MEDICAL GENETICS AND CANCER

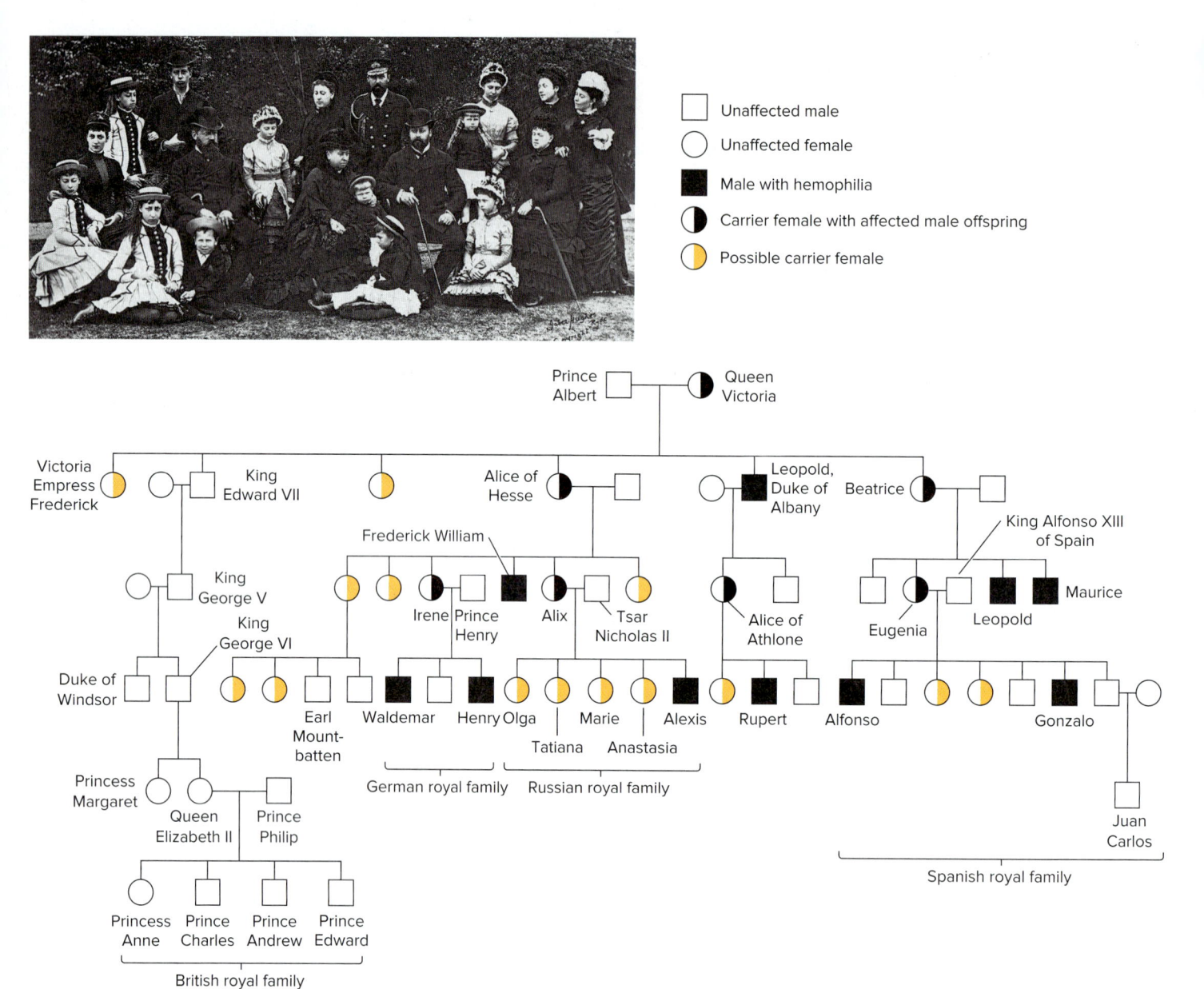

FIGURE 22.4 A family tree for hemophilia A in the royal families of Europe, indicating an X-linked recessive inheritance. Pictured are Queen Victoria and Prince Albert of Great Britain with some of their descendants.
©Hulton-Deutsch Collection/Corbis/Getty Images

Concept Check: What feature(s) of this pedigree indicate(s) X-linked recessive inheritance?

TABLE 22.4

Examples of Human Disorders Inherited in an X-Linked Dominant Manner

Disorder	Gene Product	Effects of Disease-Causing Allele
Vitamin D-resistant rickets	Metallopeptidase	Defects in bone mineralization at the sites of bone growth or remodeling, leading to bone deformity and stunted growth in children
Rett syndrome	Methyl-CpG-binding protein-2	A neurodevelopmental disorder that includes a deceleration of head growth and small hands and feet; fatal in males
Aicardi syndrome	Unknown	Characterized by the partial or complete absence of a key structure in the brain called the corpus callosum, and the presence of retinal abnormalities; fatal in males
Incontinentia pigmenti	NFκβ essential modulator	Characterized by morphological and pigmentation abnormalities in the skin, hair, teeth and nails; fatal in males

22.1 REVIEWING THE KEY CONCEPTS

- A genetic basis for a human disease may be suggested by a variety of different observations (e.g., see Figure 22.1).
- Thousands of human genetic diseases follow simple Mendelian patterns of inheritance. These patterns include autosomal recessive, autosomal dominant, X-linked recessive, and X-linked dominant inheritance (see Figures 22.2–22.4, Tables 22.1–22.4).
- Recessive diseases are usually caused by loss-of-function mutations, whereas dominant diseases may be caused by haploinsufficiency, gain-of-function mutations, or dominant-negative mutations.
- Many genetic diseases exhibit locus heterogeneity, which means that they may be caused by mutations in more than one gene.

22.1 COMPREHENSION QUESTIONS

1. Which of the following would not be consistent with the idea that a disorder has a genetic component?
   a. The disorder is more likely to occur among an affected person's relatives than in the general population.
   b. The disorder can spread to individuals sharing similar environments.
   c. The disorder tends to develop at a characteristic age.
   d. A correlation is observed between the disorder and a mutant gene.
2. Assuming complete penetrance, which type of inheritance pattern is consistent with the pedigree shown here?

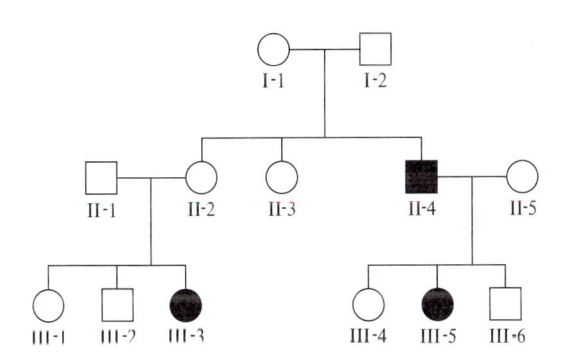

   a. Autosomal recessive
   b. Autosomal dominant
   c. X-linked recessive
   d. X-linked dominant
3. Which of the following is *not* a common explanation for a dominant disorder?
   a. Haploinsufficiency
   b. A change in chromosome number
   c. A gain-of-function mutation
   d. A dominant-negative mutation
4. Locus heterogeneity means that a genetic disorder
   a. has a heterogeneous phenotype.
   b. is caused by mutations in two or more different genes.
   c. involves a structural change in multiple chromosomes.
   d. is inherited from both parents.

22.2 DETECTION OF DISEASE-CAUSING ALLELES VIA HAPLOTYPES

Learning Outcomes:
1. Define *haplotype*.
2. Explain how haplotypes are analyzed to identify disease-causing alleles in humans.

Because mutant genes are known to play a role in thousands of diseases, researchers have devoted great effort to identifying alleles associated with genetic diseases. In this section, we will explore how geneticists analyze pedigrees to identify mutant alleles that cause disease.

Haplotypes Exhibit Genetic Variation

To identify disease-causing alleles, researchers often rely on the known chromosomal locations of genes and molecular markers that have been characterized in human populations. A disease-causing allele may be identified due to its proximity to another known gene or its proximity to molecular markers.

As discussed in Chapter 21, researchers can characterize chromosomes at the molecular level and determine the precise locations of genes and molecular markers on each chromosome. During the course of evolution, new mutations arise that alter the DNA sequences of genes and molecular markers. For this reason, homologous chromosomes exhibit gene differences (i.e., allelic variation) and show variation in their molecular markers.

As an example, **Figure 22.5** considers a pair of homologous chromosomes from two different individuals and focuses on four sites (called 1, 2, 3, and 4) that occur at particular locations on those chromosomes. These sites could be within particular genes or they could be molecular markers used in mapping studies. In this drawing, each site is also given a letter designation (A, B, or C) depending on the variation in the DNA sequence at the site. In individual 1, sites 1, 2, and 4 differ at one base pair between the homologs. In individual 2, all four sites differ at one or two base pairs. Also note that individuals 1 and 2 differ with regard to some of these sites.

The term **haplotype**, which is a contraction of haploid genotype, refers to the linkage of alleles or molecular markers on a single chromosome. In Figure 22.5, the haplotypes for these four sites are shown at the bottom of each chromosome. For example, the haplotype of the left homolog in individual 2 is 1A 2B 3B 4C.

Because mutations are rare events, haplotypes do not dramatically change from one generation to the next due to new mutations. By comparison, haplotypes are more likely to change over the course of a few generations due to crossing over. However, the likelihood of changing a haplotype depends on the distance between the alleles or molecular markers. If two sites are far apart, a crossover is more likely to alter their pattern than if they are close together. If sites 1, 2, 3, and 4 were very close together along this chromosome, the haplotypes shown in this figure would be likely to stay the same after a few generations. For example, a great-great-great grandchild of individual 2 might inherit the haplotype 1A 2B 3B 4C or 1C 2C 3C 4A. In contrast, the inheritance of either haplotype would be much less likely if the sites were far apart and could frequently recombine by crossing over.

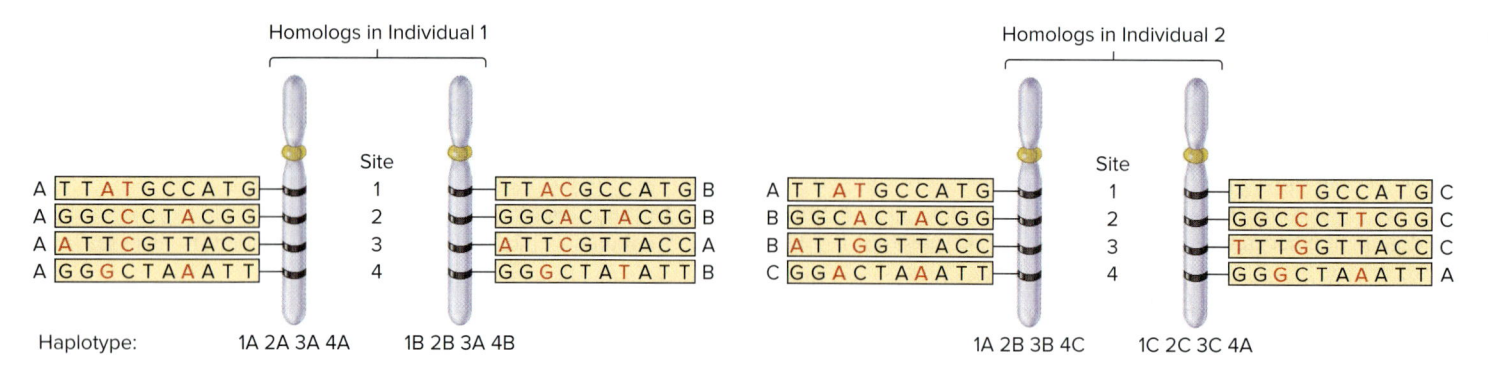

FIGURE 22.5 A schematic representation of haplotypes on a human chromosome. This example considers four sites on a chromosome that may exist in different versions, designated A, B, or C. Bases that differ are shown in red. (Note: At each site, only one of the two DNA strands is shown.)

Concept Check: What is a haplotype?

Haplotype Association Studies Are Conducted to Identify Disease-Causing Alleles

How do geneticists identify genes that cause disease when they are mutant? Although a variety of approaches may be followed, the hunt often begins with family pedigrees. The goal is to localize a disease-causing allele to a small region on a chromosome that is distinguished by its haplotype. This approach is based on two assumptions:

- The disease-causing allele had its origin in a single individual known as a **founder,** who lived many generations ago. Since that time, the allele has spread throughout portions of the human population.
- When the disease-causing allele originated in the founder, it occurred in a region of a chromosome with a particular haplotype. The haplotype is not likely to have changed over the course of several generations if the disease-causing allele and markers in this region are very close together.

By comparing the transmission patterns of many molecular markers with the occurrence of an inherited disease, researchers can pinpoint particular markers that are closely linked to the disease-causing mutant allele.

To help you understand this concept, **Figure 22.6** shows a situation in which an individual—a founder—has incurred a new mutation that results in a disease-causing allele. The molecular markers designated 1A, 2C, 3B, and 4B are very close to the location of this mutant allele. In succeeding generations, the disease-causing allele would be more likely to be present in individuals with haplotype 1A 2C 3B 4B than in those without this haplotype. Furthermore, people who have inherited the disease-causing allele would be particularly likely to inherit the 2C marker because it is the closest to the disease-causing allele. Due to their close proximity, it would be unlikely that a crossover would separate the 2C marker and the disease-causing allele from each other and create a different haplotype. When alleles and molecular markers are associated with each other at a frequency that is significantly higher than expected by random chance, they are said to exhibit **linkage disequilibrium.** The phenomenon of linkage disequilibrium is common when a disease-causing allele arises in a founder and the allele is closely linked to other markers on the same chromosome.

Haplotype Association Studies Identified the Mutant Gene That Causes Huntington Disease

As an example of haplotype association studies, let's consider Huntington disease. Nancy Wexler, whose own mother died of this disorder, has studied Huntington disease among a population of related individuals in Venezuela. Over 15,000 individuals have been analyzed. **Figure 22.7** shows a simplified version of a pedigree for a large family from this Venezuelan population in which many individuals were affected with Huntington disease. A specific molecular marker, called G8, is found in four different versions, named A, B, C, or D. In this Venezuelan population, pedigree analysis revealed that the G8-C marker, which is located near the tip of the short arm of chromosome 4, is almost always associated with the mutant gene causing Huntington disease. In other words, the G8-C marker is closely linked to the *huntingtin* allele.

Once a disease-causing mutant gene has been localized to a short chromosomal region, the next step is to determine which gene in the region is responsible for the disease. Modern haplotype association studies typically localize a gene to a chromosome region that is about 1 Mb (1 million bp) in length. One way to identify a disease-causing allele is chromosome walking, a technique in which a researcher starts at a particular molecular marker, such as G8-C, and constructs a series of clones until the gene of interest is reached. This is how the *huntingtin* gene was identified. However, the technique

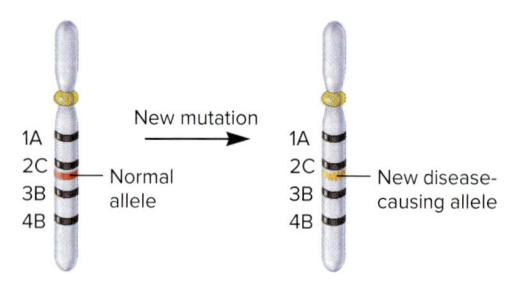

FIGURE 22.6 The occurrence of a new mutation in a founder. In this example, the new mutation occurred in a region with a specific haplotype: 1A 2C 3B 4B. Due to the close linkage between the 2C marker and the mutant allele, the same haplotype is likely to be found in individuals of succeeding generations who have inherited the mutant allele.

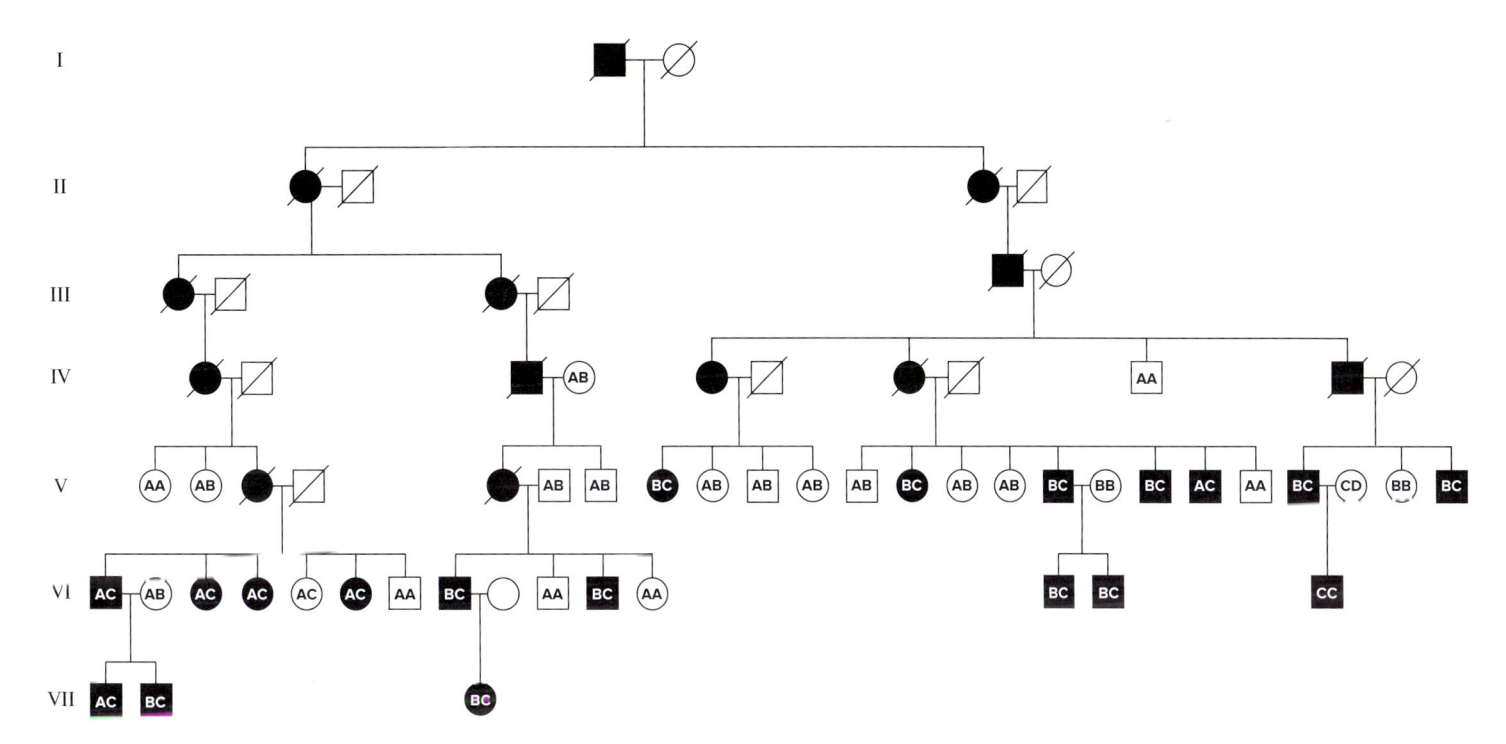

FIGURE 22.7 **The transmission pattern of a molecular marker for Huntington disease.** Affected individuals are shown with black symbols. The letters within each symbol indicate the forms of the G8 marker (A, B, C, or D) the individual carries. Affected individuals always carry the C version of this marker. In rare cases, an unaffected individual may also carry the G8-C marker. Symbols with slashes indicate deceased individuals. [Data from Gusella J. F., Wexler N. S., Conneally P. M., et al. (1983), A polymorphic DNA marker genetically linked to Huntington's disease. *Nature* 306, 234–238.]

Concept Check: Explain the connection between the founder and the G8-C marker.

of chromosome walking is no longer widely used because the entire human genome has been sequenced, and most genes have been identified. Therefore, researchers can analyze the 1-Mb region to which a gene has been mapped to determine if a mutant gene in this region is responsible for a disease. In the human genome, a 1-Mb region usually contains about 5 to 10 different genes, though the number can vary greatly. For this reason, mapping does not definitively tell researchers which gene may play a role in human disease, but it usually narrows down the list to a few candidate genes.

How does a researcher determine which of the candidate genes is the correct one? To further reduce the list, researchers may also consider biological function. As the scientific community explores the functions of genes experimentally, the data are published in the research literature and placed into databases. In some cases, this information may help to narrow down the list of candidate genes. For example, if the disease of interest is neurological, researchers may discover that only certain genes in the mapped region are expressed in neurons. Also, researchers may compare data from other organisms. If a mutant gene in a mouse causes neurological problems and a human homolog of the mouse gene is found in the mapped region, this human gene would be a good candidate for being responsible for the human disease symptoms.

After researchers have reduced the number of candidate genes to as short a list as possible, the next phase is to sequence the candidate gene(s) from many affected and unaffected individuals, using the DNA-sequencing methods described in Chapters 20 and 21. The goal is to identify a gene in which affected individuals always carry a mutation. Identifying a gene that has a genetic change found in affected individuals is strong evidence that the candidate gene causes the disease symptoms.

Why is it useful to identify disease-causing alleles? In many cases, the identification of these alleles helps us to understand how genes contribute to pathogenesis and may even aid in developing strategies aimed at the treatment of the disease. As described in Section 22.3, the identification of such genes may also result in the development of genetic tests that can enable people to determine if they are carrying disease-causing alleles.

Genome-Wide Association Studies Can Identify Disease-Causing Alleles in Human Populations

As we have seen, haplotypes can be examined to identify disease-causing alleles in humans. With the advent of DNA microarrays and next-generation sequencing technologies, described in Chapter 21, researchers can now identify disease-causing alleles on a broader scale, because genetic variation can be determined at the whole genome level. A **genome-wide association study** (**GWA study,** or **GWAS**) is an examination of a genome-wide set of genetic variants among many different individuals to see if any variant is associated with a disease or other type of trait. A GWA study usually tries to find an association between one or more single-nucleotide polymorphisms (SNPs) and a disease or other trait. (SNPs are described in Chapter 21.) Many GWA studies have focused on finding associations between SNPs and human diseases, but they can also be directed at other types of genetic variation (e.g., deletions or duplications) and at other species.

How is a GWA study carried out? One common approach is to compare the SNPs of two large groups of individuals—a control group that does not have a disease and another group in which all of the individuals are affected with the same disease, such as type 1 diabetes. Although the number of SNPs that are analyzed varies from study to study, researchers commonly compare a million or more. The goal is to identify SNPs for which the frequency is significantly different between the control group and the group of affected individuals. To identify such SNPs, researchers determine the odds ratio, which is the odds of disease for individuals having a specific SNP versus the odds of disease for individuals who do not have that same SNP. A P value for the significance of the odds ratio is typically calculated using a test that is similar to a chi square test, which is described in Chapter 3. In a standard chi square test, when $P < 0.05$, the results are considered significant. However, in a GWA study, researchers often analyze millions of SNPs, so using such a criterion for the P value would result in many false positives. It is common in GWA studies to set the level of significance at 0.05/(1,000,000 SNPs). In other words, a value of $P < 5 \times 10^{-8}$ is viewed as statistically significant.

After odds ratios and P values have been calculated for all SNPs, the data are used to produce a Manhattan plot. This name was chosen because the plot looks somewhat like the skyline of Manhattan in New York City. The data are plotted as the negative logarithm of the P value for a given SNP as a function of its location in the genome. **Figure 22.8** shows a Manhattan plot for a GWA study that looked for an association between SNPs and type 1 diabetes. In this plot, the SNPs with significant associations are indicated by stacks of green points. These results indicate that SNPs on certain chromosomes are associated with type I diabetes.

How do we interpret the results of a GWA study? As discussed later in this chapter, an association does not necessary imply a cause-and-effect relationship. Therefore, further work is usually needed to understand why a SNP is associated with a human disease. In some cases, a SNP may not be involved in a cause-and-effect relationship but is closely linked to a particular disease-causing allele due to the founder effect, described earlier in this section (see Figure 22.6). In other cases where a cause-and-effect relationship does exist, a SNP may have a variety of effects, including:

- a change in the coding sequence of a gene;
- a change in the promoter;
- a change in a regulatory sequence, such as an enhancer.

Such changes can alter gene expression and thereby contribute to disease symptoms.

> ### Genetic TIPS
>
> **The Question:** Figure 22.6 shows the location of a disease-causing allele that arose in a founder. The haplotypes of five grandchildren of this founder are as follows:
>
> Grandchild 1: 1A 2B 3C 4B/1A 2B 3A 4C
>
> Grandchild 2: 1A 2C 3B 4B/1B 2A 3B 4A
>
> Grandchild 3: 1B 2A 3A 4B/1C 2B 3A 4A
>
> Grandchild 4: 1B 2C 3B 4B/1A 2B 3A 4C
>
> Grandchild 5: 1A 2B 3A 4C/1B 2A 3B 4A
>
> Which of these grandchildren is/are likely to have inherited the disease-causing allele and which are less likely based on their haplotypes?
>
> **T**opic: What topic in genetics does this question address?
>
> The topic is the use of haplotypes to predict the likelihood that an individual may be carrying a disease-causing allele.
>
> **I**nformation: What information do you know based on the question and your understanding of the topic?
>
> From the question, you know the location of a disease-causing allele in a founder, and the haplotypes of five grandchildren. From your understanding of the topic, you may remember that markers that are closely linked to the disease-causing allele are likely to be found in individuals who have inherited the disease-causing allele.
>
> **P**roblem-Solving **S**trategy: Compare and contrast.
>
> One strategy to solve this problem is to compare the haplotypes and look for those that resemble the founder. In particular, the 2C marker is closely linked to the disease-causing allele.
>
> **Answer:** One of the chromosomes of grandchild 2 has the same haplotype as the founder, and so that grandchild is the most likely to carry the disease-causing allele. Also, grandchild 4 has the 2C marker and is relatively likely to carry the allele. The other three grandchildren are relatively unlikely to have the disease-causing allele because they do not carry the 2C marker.

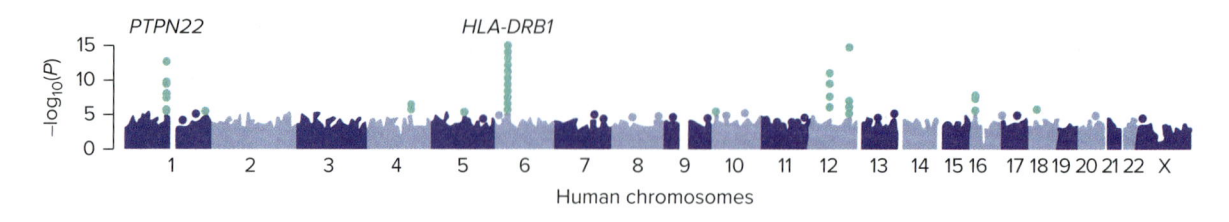

FIGURE 22.8 A GWA study of type 1 diabetes in humans. The y-axis indicates the P values of SNPs, and the x-axis shows their locations on the human chromosomes. Those SNPs that are not statistically significant are shown in dark and light blue, whereas those that are statistically significant are green. Many statistically significant SNPs are found in the genes *PTPN22* and *HLA-DRB1*. *PTPN22* encodes a protein called tyrosine phosphatase that affects the responsiveness of T-cell and B-cell receptors; mutations in this gene are associated with increases or decreases in risks of autoimmune diseases. The *HLA-DRB1* gene is part of a family of genes that encode proteins that are present on the surface of certain immune system cells. Source: The Wellcome Trust Case Control Consortium (2007) *Nature* 447, 661-678.

22.2 REVIEWING THE KEY CONCEPTS

- The term *haplotype* refers to the linkage of alleles or molecular markers on a single chromosome (see Figure 22.5).
- Disease-causing mutations may originate in a founder with a specific haplotype (see Figure 22.6).
- Researchers may identify a disease-causing allele by determining its location within a chromosome via haplotype association studies (see Figure 22.7).
- Genome-wide association studies (GWA studies) are conducted to find associations between genetic variation and human diseases (see Figure 22.8).

22.2 COMPREHENSION QUESTIONS

1. What is a haplotype?
   a. A species with one set of chromosomes
   b. A cell with one set of chromosomes
   c. The linkage of alleles or molecular markers on a chromosome
   d. All of the above describe a haplotype.
2. Haplotype association studies are aimed at the identification of a particular _____ based on _____.
   a. chromosome, an abnormality in its structure
   b. chromosome, the arrangement of molecular markers
   c. gene, its linkage to other genes or molecular markers
   d. gene, chromosomal rearrangements

22.3 GENETIC TESTING AND SCREENING

Learning Outcomes:
1. Compare and contrast genetic testing and genetic screening.
2. List different methods used to test for genetic abnormalities.
3. Describe how genetic testing can be conducted before birth.

Because genetic abnormalities occur in the human population at a significant level, people have sought ways to determine whether individuals carry disease-causing alleles or other types of genetic abnormalities. The term **genetic testing** refers to the use of testing methods to determine if an individual carries a genetic abnormality. By comparison, the term **genetic screening** refers to population-wide genetic testing. In this section, we will examine both approaches.

Genetic Testing Is Used to Identify Many Inherited Human Diseases

Table 22.5 describes several different genetic testing methods. In many cases, single-gene mutations that affect the function of proteins can be examined at the protein level. If a gene encodes an enzyme, biochemical assays to measure that enzyme's activity may be available. As mentioned earlier, Tay-Sachs disease is due to a defect in the enzyme hexosaminidase A (HexA). Enzymatic assays for this enzyme involve the use of an artificial substrate in which 4-methylumbelliferone (MU) is covalently linked to

TABLE 22.5

Testing Methods for Genetic Abnormalities

Method	Description
Protein Level	
Biochemical	As described in this chapter for Tay-Sachs disease, the enzymatic activity of a protein can be assayed in vitro.
Immunological	The presence of a protein can be detected using antibodies that specifically recognize that protein. Western blotting is an example of this type of technique (see Chapter 20).
DNA or Chromosomal Level	
DNA sequencing	If the gene associated with a disease has already been identified and sequenced, it is possible to design PCR primers that can amplify the gene from a sample of cells. The amplified DNA segment can then be subjected to DNA sequencing, as described in Chapter 20.
In situ hybridization	A DNA probe that hybridizes to a particular gene or gene segment can be used to determine if the gene is present, absent, or altered in an individual. The technique of fluorescence in situ hybridization (FISH) is described in Chapter 21 (see Figure 21.2).
Karyotyping	The chromosomes from a sample of cells can be stained and then analyzed microscopically for abnormalities in chromosome structure and number (see Figure 2.2).
DNA microarrays	Microarrays are used to detect polymorphisms found in the human population that are associated with diseases.

N-acetylglucosamine (GlcNAc). HexA cleaves this covalent bond and releases MU, which is fluorescent.

$$\text{MU–GlcNAc} \xrightarrow{\text{HexA}} \text{MU} + \text{GlcNAc}$$
$$(\text{nonfluorescent}) \qquad\qquad (\text{fluorescent})$$

To perform this assay, a small sample of cells is collected and incubated with MU–GlcNAc, and the fluorescence is measured. Samples from individuals affected with Tay-Sachs, who do not produce the HexA enzyme, produce little or no fluorescence, whereas samples from individuals who are homozygous for the normal *hexA* allele produce a high level of fluorescence. Samples from heterozygotes, who have 50% HexA activity, produce intermediate levels of fluorescence.

An alternative and more common approach is to detect single-gene mutations at the DNA level. To apply this testing strategy, researchers must have previously identified the mutant gene using molecular techniques. The identification of many human genes, such as those involved in Duchenne muscular dystrophy, cystic fibrosis, and Huntington disease, has made it possible to test for affected individuals or those who may be carriers of these diseases. Table 22.5 describes several ways to test for gene mutations.

Many human genetic abnormalities involve changes in chromosome number and/or structure. In fact, changes in chromosome number are a common class of human genetic abnormality. Most of these result in spontaneous abortions. However, approximately 1 in 200 live births are aneuploid—have an abnormal number of chromosomes (see Chapter 8, Table 8.1). About 5% of infant and childhood deaths are related to such genetic abnormalities. Changes in chromosome number and many changes in chromosome structure can be detected by karyotyping the chromosomes with a light microscope.

Genetic Screening Identifies Genetic Abnormalities at the Population Level

In the United States, genetic screening for certain disorders has become common medical practice. For example, pregnant women older than 35 often have tests conducted to see if their fetuses are carrying chromosomal abnormalities. As discussed in Chapter 8, these tests are indicated because the rate of such defects increases with the age of the mother. Another example is the widespread screening for phenylketonuria (PKU). An inexpensive test can determine if newborns have this disease. Those who test positive can then be given a low-phenylalanine diet to avoid PKU's devastating effects.

Genetic screening has also been conducted on specific populations in which a genetic disease is prevalent. For example, in 1971, community-based screening for heterozygous carriers of Tay-Sachs disease was begun among specific Ashkenazi Jewish populations. With the use of this screening, over the course of one generation, the incidence of births of children with Tay-Sachs disease in these populations was reduced by 90%. For most rare genetic abnormalities, however, genetic screening is not routine practice. Rather, genetic testing is performed only when a family history reveals a strong likelihood that a couple may produce an affected child. Typically, such a couple already has an affected child or has other relatives with a genetic disease.

Genetic testing and screening are medical practices with many social and ethical dimensions. For example, people must decide whether or not they want to make use of available tests, particularly when the disease in question has no cure. For example, Huntington disease typically does not affect people until their 50s and can last 20 years. People who learn they are carriers of genetic diseases such as Huntington disease can be devastated by the news. Some argue that people have a right to know about their genetic makeup; others assert that it does more harm than good. Another issue is privacy. Who should have access to personal genetic information, and how could it be used? Could routine genetic testing lead to discrimination by employers or medical insurance companies? In the coming years, we will gain ever-increasing awareness of our genetic makeup and the underlying causes of genetic diseases. As a society, establishing guidelines for the uses of genetic testing will be a necessary, yet very difficult, task.

Genetic Testing Can Be Performed Prior to Birth

Genetic testing can be performed during pregnancy, which may affect a woman's decision to terminate that pregnancy. The two common ways of obtaining cellular material from a fetus for the purpose of genetic testing are amniocentesis and chorionic villus sampling.

- In **amniocentesis,** a doctor removes amniotic fluid containing fetal cells, using a needle that is passed through the abdominal wall (**Figure 22.9**). Because relatively few

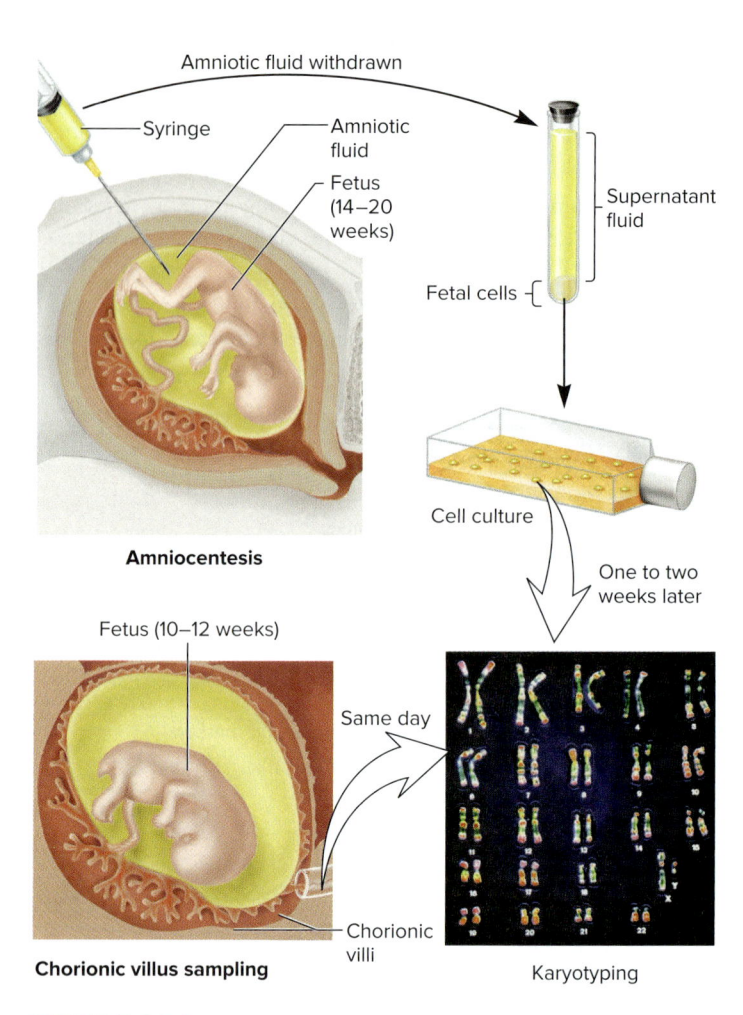

FIGURE 22.9 **Techniques to determine genetic abnormalities during pregnancy.** In amniocentesis, amniotic fluid is withdrawn, and fetal cells are collected by centrifugation. The cells are then allowed to grow in a laboratory culture medium for 1–2 weeks prior to karyotyping. In chorionic villus sampling, a small piece of the chorion is removed. These cells can be prepared directly for karyotyping.
©CNRI/SPL/Science Source

fetal cells are found in the amniotic fluid, the cells are grown in the laboratory for 1–2 weeks and then karyotyped to determine the number of chromosomes per cell and whether changes in chromosome structure have occurred.

- In **chorionic villus sampling,** a small piece of the chorion (the fetal part of the placenta) is removed, and a karyotype is prepared directly from the collected cells. Chorionic villus sampling can be performed earlier during pregnancy than amniocentesis, usually around the tenth to twelfth week, compared to the fifteenth to twentieth week for amniocentesis, and results from chorionic villus sampling are available sooner. Weighed against these advantages, however, is that this procedure may pose a slightly greater risk of causing a miscarriage.

Another method of genetic testing performed prior to birth is called **preimplantation genetic diagnosis (PGD).** This approach, which is conducted before pregnancy even occurs, involves the genetic testing of embryos that have been produced by **in vitro fertilization (IVF)**—a process in which sperm and egg are combined outside of the mother's body. The testing is typically done to check for a specific genetic abnormality, such as the allele that causes Huntington disease. PGD can also determine if an embryo has the correct number of chromosomes (called aneuploidy screening).

PGD is done by removing one or two cells usually at about the eight-cell stage, which occurs 3 days after fertilization. This process is called embryo biopsy, or blastomere biopsy. Molecular techniques described in Table 22.5 are then conducted on the removed cell(s) to either check for a particular genetic disease or determine the chromosome composition. The testing can usually be completed in a day or so. Depending on the outcome of the results, a decision can be made whether or not to transfer the embryo into the uterus of the prospective mother in hopes of implantation and the eventual birth of a baby. In most cases, only embryos that do not show genetic abnormalities are used. As with the genetic screening of adults, the testing of embryos and fetuses raises many ethical questions.

22.3 REVIEWING THE KEY CONCEPTS

- Genetic testing for abnormalities can be performed in a variety of ways (see Table 22.5).
- Genetic screening is population-wide genetic testing.
- Genetic testing by means of amniocentesis, chorionic villus sampling, or preimplantation genetic diagnosis (PGD) can be done prior to birth (see Figure 22.9).

22.3 COMPREHENSION QUESTIONS

1. Which of the following is *not* a method used in genetic testing?
   a. Chromosome walking
   b. DNA sequencing
   c. In situ hybridization
   d. DNA microarrays
2. Which of the following prenatal genetic testing methods is done in conjunction with in vitro fertilization?
   a. Amniocentesis
   b. Chorionic villus sampling
   c. Preimplantation genetic diagnosis
   d. All of the above are usually performed in conjunction with in vitro fertilization.

22.4 OVERVIEW OF CANCER

Learning Outcomes:
1. Describe the key characteristics of cancer.
2. Compare and contrast oncogenes and tumor-suppressor genes.

Cancer is a disease characterized by uncontrolled cell division. It is a genetic disease at the cellular level. More than 100 kinds of human cancers have been identified, and they are classified according to the type of cell that has become cancerous. Though cancer is a diverse collection of many diseases, some characteristics are common to all cancers.

- Most cancers originate in a single cell. This single cell and its line of daughter cells undergo a series of genetic changes that accumulate during cell division. In this regard, a cancerous growth can be considered to be **clonal** in origin. A hallmark of a cancer cell is that it divides to produce two daughter cancer cells.
- At the cellular and genetic levels, cancer is usually a multistep process that begins with a precancerous genetic change—a **benign** growth—that is followed by additional genetic changes that lead to cancerous cell growth (**Figure 22.10**).
- When cells have become cancerous, the growth is described as **malignant**. Cancer cells are **invasive**—they can invade healthy tissues—and **metastatic**—they can migrate to other parts of the body and cause secondary tumors.

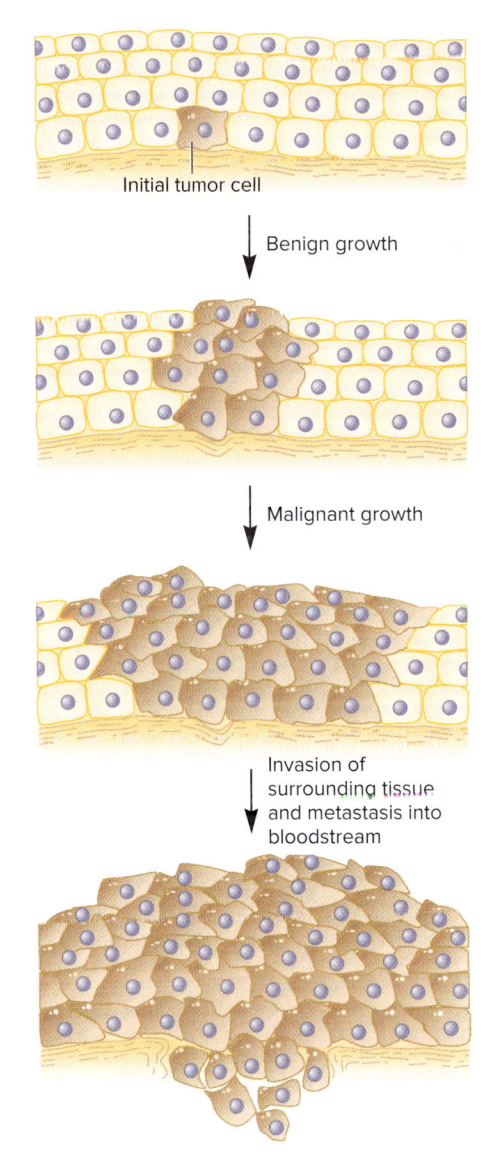

FIGURE 22.10 Progression of cellular growth leading to cancer.

Genes → Traits In a healthy individual, one or more gene mutations convert a normal cell into a tumor cell. This tumor cell divides to produce a benign tumor. Additional genetic changes in the tumor cells may occur, leading to malignant growth. At a later stage in malignancy, the tumor cells invade surrounding tissues, and some malignant cells may metastasize by traveling through the bloodstream to other parts of the body, where they can grow and cause secondary tumors. As a trait, cancer can be viewed as a series of genetic changes that eventually lead to uncontrolled cell growth.

In the United States, approximately 1 million people are diagnosed with cancer each year, and about half that number will die from the disease. In 5–10% of all cases, a predisposition to develop the cancer is an inherited trait. We will examine some inherited forms of cancer later in this chapter. Most cancers, though, perhaps 90–95%, are not passed from parent to offspring. Rather, cancer is usually an acquired condition that typically occurs later in life. Although some cancers are caused by spontaneous mutations and viruses, at least 80% of all human cancers are related to exposure to agents that promote genetic changes in somatic cells. These environmental agents, such as UV light and certain chemicals, are mutagens that alter DNA in a way that affects the function of normal genes. If the DNA is permanently modified in somatic cells, such changes will be transmitted during cell division. These DNA alterations can lead to effects on gene expression that ultimately affect cell division and thereby lead to cancer. An environmental agent that causes cancer in this manner is called a **carcinogen.**

Both genetic and epigenetic changes can promote cancerous growths. Most genetic changes that cause cancer involve mutations in particular genes. The effects of these mutations are placed into two broad categories:

- An **oncogene** is a mutant gene that is overexpressed and contributes to cancerous growth.
- A **tumor-suppressor gene** is a gene that prevents cancers. A loss-of-function mutation in a tumor-suppressor gene can allow cancerous growth to occur.

Therefore, oncogenes and tumor suppressor genes are distinguished by whether a cancer-causing mutation increases or decreases gene expression.

As discussed in Chapter 17, genes are also regulated by epigenetic mechanisms, such as DNA methylation and histone modifications. These epigenetic changes can promote cancer by increasing the expression of oncogenes or inhibiting the expression of tumor-suppressor genes. In Sections 22.5 and 22.6, we will explore how mutations promote cancer by affecting oncogenes and tumor-suppressor genes, respectively, and in Section 22.7, we will consider epigenetic effects.

22.4 REVIEWING THE KEY CONCEPTS

- Cancer is clonal in origin. It usually develops in a multistep process that involves several mutations, begins with benign growth, and progresses to growth that is invasive and metastatic (see Figure 22.10).
- Cancer can be caused by mutations that overexpress oncogenes or that inhibit the expression of tumor-suppressor genes. Epigenetic changes can also promote cancer.

22.4 COMPREHENSION QUESTION

1. A mutant gene that promotes cancer when it is overexpressed is called
   a. a tumor-suppressor gene.
   b. an oncogene.
   c. both a and b.
   d. neither a nor b.

22.5 ONCOGENES

Learning Outcomes:

1. Describe the genetic changes that convert proto-oncogenes into oncogenes.
2. Summarize the different ways in which oncogenes can lead to cancer.

As noted in the preceding section, an oncogene is a mutant gene that is overexpressed and contributes to the development of cancer. A normal, nonmutated gene that has the potential to become an oncogene is termed a **proto-oncogene.** To become an oncogene, a proto-oncogene must incur a mutation that causes its expression to be abnormally high. Such a mutation is another example of a gain-of-function mutation, which we discussed earlier in this chapter in the context of dominant mutations. A gain-of-function mutation that produces an oncogene typically has one of three possible effects:

- The amount of the encoded protein is greatly increased.
- A change occurs in the structure of the encoded protein that causes it to be overly active.
- The encoded protein is expressed in a cell type where it is not normally expressed.

In this section, we will explore the possible ways that oncogenes affect cell growth and examine the types of genetic changes that convert proto-oncogenes into oncogenes. We will also consider how viruses can introduce oncogenes into cells.

Oncogenes Have Gain-of-Function Mutations That May Affect Proteins Involved in Cell Division Pathways

As described in Chapter 2, eukaryotic cells destined to divide advance through a series of stages known as the cell cycle (refer back to Figure 2.5). The phases consist of G_1 (first gap), S (synthesis of DNA, the genetic material), G_2 (second gap), and M phase (mitosis and cytokinesis). The G_1 phase is a period in a cell's life when it may become committed to divide. Depending on the conditions, a cell in the G_1 phase may accumulate molecular changes that cause it to advance through the rest of the cell cycle. When this occurs, cell biologists say that a cell has reached a special control point called the restriction point. The commitment to divide is based on a variety of factors. For example, environmental conditions, such as the presence of sufficient nutrients, are important for cell division. In addition, multicellular organisms rely on signaling molecules to coordinate cell division throughout the body. These signaling molecules are often called **growth factors** because they promote cell division.

As researchers began to investigate cancer at the molecular level, they wanted to understand how mutant genes promote abnormal cell division. In parallel with cancer research, cell biologists have studied the roles that normal cellular proteins play in cell division. As mentioned, the cell cycle is regulated in part by growth factors that bind to cell surface receptors and initiate a cascade of cellular events that lead eventually to cell division.

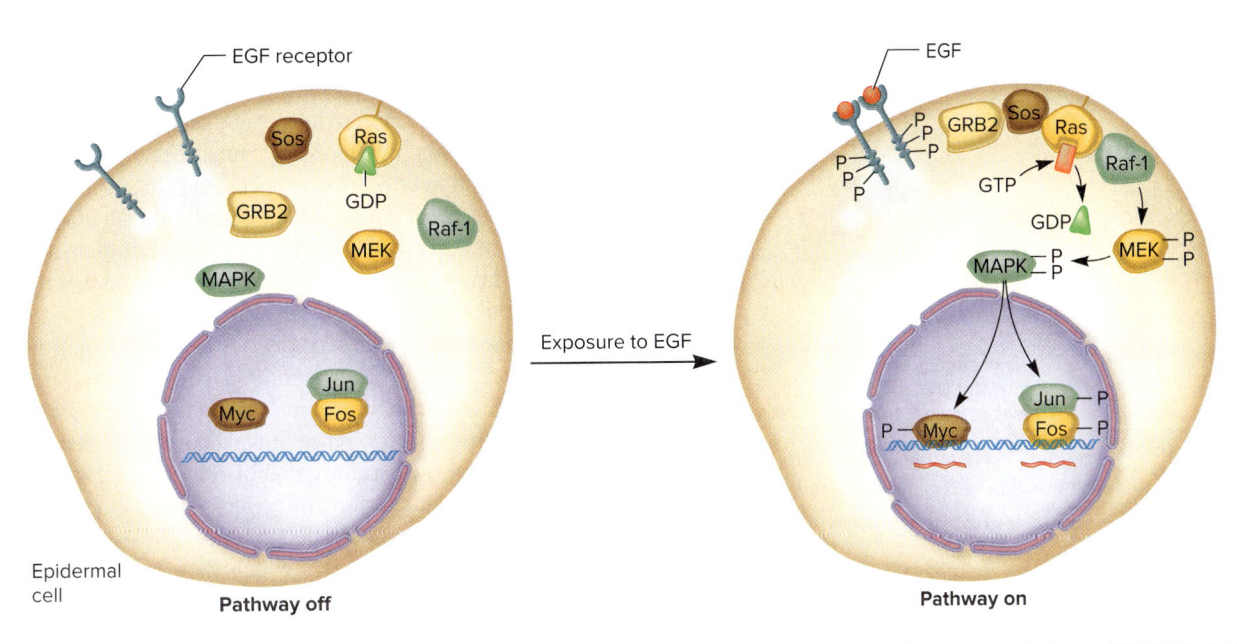

FIGURE 22.11 **The activation of a cell-signaling pathway by a growth factor.** In this example, epidermal growth factor (EGF) binds to two EGF receptors, causing them to dimerize and phosphorylate each other. An intracellular protein called GRB2 is attracted to the phosphorylated EGF receptor, and it is subsequently bound by another protein called Sos. The binding of Sos to GRB2 enables Sos to activate a protein called Ras. This activation involves the release of GDP and the binding of GTP. The activated Ras/GTP complex then activates Raf-1, which is a protein kinase. Raf-1 phosphorylates MEK, and then MEK phosphorylates MAPK. More than one MAPK may be involved. Finally, the phosphorylated form of MAPK activates transcription factors, such as Myc, Jun, and Fos. This activation leads to the transcription of genes, which encode proteins that promote cell division.

Figure 22.11 considers a hormone called epidermal growth factor (EGF) that is secreted from endocrine cells and stimulates epidermal cells, such as skin cells, to divide.

1. EGF binds to its receptor.
2. This binding results in the activation of an intracellular signaling pathway.
3. This pathway, also known as a signal cascade, leads to a change in gene transcription. In other words, the transcription of specific genes is activated in response to the growth factor.
4. Once these genes are transcribed and translated, the gene products function to promote the progression through the cell cycle.

Figure 22.11 is just one example of a pathway between a growth factor and gene activation. Eukaryotic species produce many different growth factors, and the signaling pathways are often more complex than the one shown here.

The mutations that convert proto-oncogenes into oncogenes have been analyzed in many types of cancers. Oncogenes commonly encode proteins that function in cell signaling pathways related to cell division (**Table 22.6**). These proteins include growth factors, growth factor receptors, intracellular signaling proteins, and transcription factors. As an example, mutations that alter the amino acid sequence of the Ras protein have been shown to cause functional abnormalities. The Ras protein is a GTPase, which hydrolyzes GTP to GDP + P_i (**Figure 22.12**). Therefore, after it has been activated, the Ras protein returns to its inactive state by hydrolyzing GTP. Certain mutations that convert the normal *ras* gene into an oncogene decrease the ability of the Ras protein to

TABLE 22.6

Examples of Proto-Oncogenes That Can Mutate into Oncogenes*

Gene	Cellular Function of Encoded Protein
Growth factors	
sis	Platelet-derived growth factor
int-2	Fibroblast growth factor
Growth factor receptors	
erbB	Growth factor receptor for EGF (epidermal growth factor)
fms	Growth factor receptor for NGF (nerve growth factor)
Intracellular signaling proteins	
ras	GTP/GDP-binding protein
raf	Serine/threonine kinase
src	Tyrosine kinase
abl	Tyrosine kinase
Transcription factors	
myc	Transcription factor
jun	Transcription factor
fos	Transcription factor

*The genes included in this table are found in humans as well as other vertebrate species. Many of these genes were initially identified in retroviruses. Most of the genes have been given three-letter names that are abbreviations for the type of cancer the oncogene causes or the type of virus in which the gene was first identified.

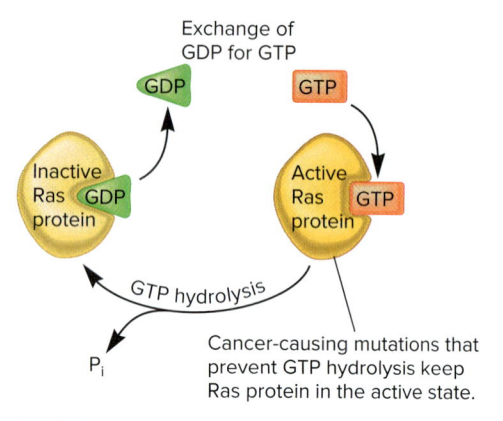

FIGURE 22.12 **Functional cycle of the Ras protein.** The binding of GTP to Ras activates the function of Ras and promotes cell division. The hydrolysis of GTP to GDP and P_i converts the active form of Ras to an inactive form.

Concept Check: How would a mutation that prevents the Ras protein from hydrolyzing GTP affect the cell-signaling pathway in Figure 22.11?

hydrolyze GTP. This results in a greater amount of the active GTP-bound form of the Ras protein. In this way, such mutations keep the signaling pathway turned on, thereby stimulating the cell to divide.

Genetic Changes in Proto-Oncogenes Convert Them to Oncogenes

How do specific genetic alterations convert proto-oncogenes into oncogenes? By isolating and studying oncogenes at the molecular level, researchers have discovered four main ways this occurs. These changes can be categorized as missense mutation, gene amplification (i.e., an increase in copy number), chromosomal translocation, or viral integration.

Missense Mutation As mentioned previously, changes in the structure of the Ras protein can cause it to become permanently activated. These changes are caused by a missense mutation in the *ras* gene. The human genome contains four different but evolutionarily related *ras* genes: *ras*H, *ras*N, *ras*K-4a, and *ras*K-4b. All four homologous genes encode proteins with very similar amino acid sequences consisting of 188 or 189 amino acids. Missense mutants in these normal *ras* genes are associated with particular forms of cancer. For example, a missense mutation in *ras*H that changes the twelfth amino acid in the protein from a glycine to a valine is responsible for the conversion of *ras*H into an oncogene. Experimentally, chemical carcinogens have been shown to cause this missense mutation and thereby lead to cancer.

Gene Amplification Another genetic event that may convert proto-oncogenes to oncogenes is gene amplification, or an abnormal increase in the copy number of a proto-oncogene. An increase in gene copy number is expected to increase the amount of the encoded protein, thereby contributing to malignancy.

Gene amplification does not normally happen in mammalian cells, but it is a common occurrence in cancer cells. Robert Gallo and Mark Groudine discovered that a gene called c-*myc* was amplified in a human leukemia cell line. Many human cancers are associated with the amplification of particular oncogenes. In such cases, the extent of oncogene amplification may be correlated with the progression of tumors to increasing malignancy. These include the amplification of N-*myc* in neuroblastomas and *erbB-2* in breast carcinomas.

Chromosomal Translocation A third type of genetic alteration that can lead to cancer is a chromosomal translocation. Abnormalities in chromosome structure are common in cancer cells, and very specific types of chromosomal translocations have been identified in certain types of tumors. In 1960, Peter Nowell and David Hungerford discovered that chronic myelogenous leukemia (CML) is correlated with the presence of a shortened version of chromosome 22, which they called the Philadelphia chromosome after the city where it was discovered. Rather than being due to a deletion, this shortened chromosome is the result of a reciprocal translocation between chromosomes 9 and 22. Later studies revealed that this translocation activates a proto-oncogene, *abl*, in an unusual way (**Figure 22.13**). The reciprocal translocation involves breakpoints within the *abl* and *bcr* genes. Following the reciprocal translocation, the coding sequence of the *abl* gene fuses with the promoter and coding sequence of the *bcr* gene.

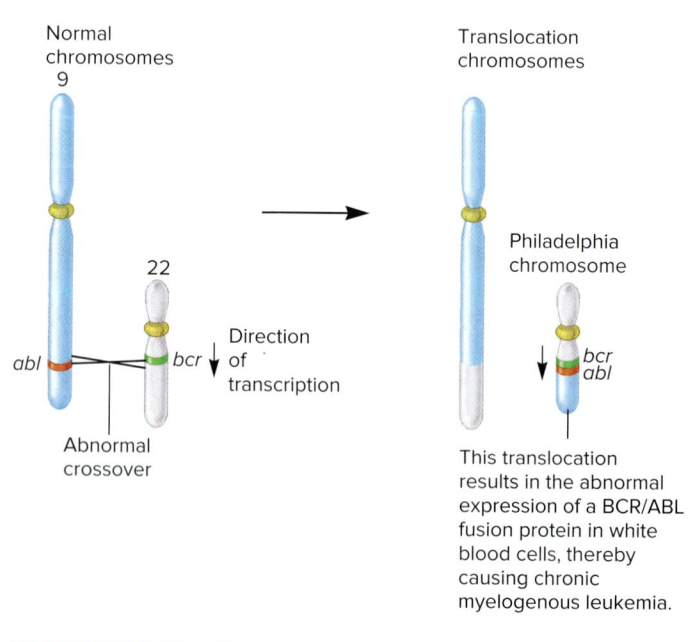

FIGURE 22.13 **The reciprocal translocation commonly found in people with chronic myelogenous leukemia.**

Genes → Traits In healthy individuals, the *abl* gene is located on chromosome 9, and the *bcr* gene is on chromosome 22. In certain forms of myelogenous leukemia, a reciprocal translocation causes the *abl* gene to fuse with the *bcr* gene. This combined gene, under the control of the *bcr* promoter, encodes an abnormal fusion protein that overexpresses the tyrosine kinase function of *the* ABL protein in white blood cells and leads to leukemia.

Concept Check: Why does this translocation cause leukemia rather than cancer in a different tissue type, such as the lung?

This yields an oncogene that encodes an abnormal fusion protein, which contains the polypeptide sequences encoded from both genes. The *abl* gene encodes a tyrosine kinase enzyme, which uses ATP to attach phosphate groups onto target proteins. This phosphorylation activates certain proteins involved with cell division. Normally, the *abl* gene is highly regulated. However, in the Philadelphia chromosome, the fusion gene is controlled by the *bcr* promoter, which is active in white blood cells. This explains why this fusion causes a type of cancer called a leukemia, which is characterized by a proliferation of white blood cells.

Interestingly, the study of the *abl* gene has led to an effective treatment for CML. Until recently, the only successful treatment was to destroy the patient's bone marrow and then restore blood-cell production by infusing stem cells from the bone marrow of a healthy donor. Applying knowledge about the function of the ABL protein, researchers developed the drug imatinib mesylate (Gleevec) that appears to dramatically improve survival. This molecule fits into the active site of the ABL protein, preventing ATP from binding there. Without ATP, the ABL protein cannot phosphorylate its target proteins. This prevents the ABL protein from stimulating cell division. In a clinical trial, almost 90% of the CML patients treated with the drug showed no further progression of their disease! In 2001, Gleevec was given FDA approval and is now used to treat CML and certain other types of cancers.

Viral Integration A fourth way that oncogenes can arise is via viral integration. As part of their reproductive cycle, certain viruses integrate their genomes into the chromosomal DNA of their host cell. If the integration occurs next to a proto-oncogene, a viral promoter or enhancer sequence may cause the proto-oncogene to be overexpressed. For example, in certain lymphomas that occur in birds, the genome of the avian leukosis virus has been found to be integrated next to the *c-myc* gene, enhancing its level of transcription.

Certain Viruses Cause Cancer by Carrying Oncogenes into Cells

As mentioned, most types of cancers are caused by mutagens that alter the structure and expression of genes. A few viruses, however, are known to cause cancer in plants, animals, and humans. Many of these viruses also infect normal laboratory-grown cells and convert them into malignant cells. In cancer biology, the process of converting a normal cell into a malignant cell is called **transformation**.

Most cancer-causing viruses are not very potent at inducing cancer. An organism must be infected for a long time before a tumor actually develops. Furthermore, most viruses are inefficient at transforming or are unable to transform normal cells grown in the laboratory. By comparison, a few types of viruses rapidly induce tumors in animals and efficiently transform cells in culture. These are called **acutely transforming viruses (ACTs)**. About 40 such viruses have been isolated from chickens, turkeys, mice, rats, cats, and monkeys. The first virus of this type, the Rous sarcoma virus (RSV), was isolated from chicken sarcomas by Peyton Rous in 1911.

TABLE 22.7

Examples of Viruses That May Cause Cancer

Virus	Description
Retroviruses	
Rous sarcoma virus (RSV)	Causes sarcomas in chickens
Hardy-Zuckerman-4 feline sarcoma virus	Causes sarcomas in cats
DNA Viruses	
Hepatitis B, SV40, polyomavirus	Causes liver cancer in several species, including humans
Papillomavirus	Causes benign tumors and malignant carcinomas in several species, including humans; causes cervical cancer in humans
Herpesvirus	Causes carcinoma in frogs and T-cell lymphoma in chickens. A human herpesvirus, Epstein-Barr virus, is a causative agent in Burkitt lymphoma, which occurs primarily in immunosuppressed individuals such as AIDS patients.

During the 1970s, RSV research led to the identification of the first gene known to cause cancer. Researchers investigated RSV by infecting chicken fibroblast cells, which are a type of connective tissue cell, in the laboratory. This causes the chicken fibroblasts to grow like cancer cells. During the course of their studies, researchers identified mutant RSV strains that infected and proliferated within chicken cells without transforming them into malignant cells. These RSV strains were determined to contain a defective viral gene. In contrast, in other strains where this gene is functional, cancer occurs. This viral gene was designated *src* for sarcoma, the type of cancer it causes. The *src* gene was the first example of an oncogene, a mutant gene that promotes cancer.

Since these early studies of RSV, many other viruses carrying oncogenes have been investigated. The characterization of such oncogenes has led to the identification of several genes with cancer-causing potential. In addition to retroviruses, several viruses with DNA genomes cause tumors, and some of these are known to cause cancer in humans (**Table 22.7**). Researchers estimate that up to 15% of all human cancers are associated with viruses.

22.5 REVIEWING THE KEY CONCEPTS

- Oncogenes often exert their effects via intracellular signaling pathways that control the cell cycle (see Figure 22.11).
- Oncogenes arise due to gain-of-function mutations in proto-oncogenes. An example is a mutation in the *ras* gene that prevents the Ras protein from hydrolyzing GTP (see Table 22.6, Figure 22.12).
- Different types of genetic changes that can convert proto-oncogenes to oncogenes include missense mutation, gene amplification, chromosomal translocation, and viral integration (see Figure 22.13).
- Certain viruses cause cancer by carrying oncogenes into cells (see Table 22.7).

22.5 COMPREHENSION QUESTION

1. Which of the following is a type of genetic change that may produce an oncogene?
   a. Missense mutation
   b. Gene amplification
   c. Chromosomal translocation
   d. All of the above may produce an oncogene.

22.6 TUMOR-SUPPRESSOR GENES

Learning Outcomes:

1. Describe the genetic changes that inactivate tumor-suppressor genes.
2. Summarize the different ways in which the loss of tumor-suppressor genes can lead to cancer.

In the previous section, we considered how oncogenes promote cancer as a result of gain-of-function mutations. An oncogene is an abnormally activated proto-oncogene whose overexpression leads to uncontrolled cell growth. We now turn our attention to a second category of genes called tumor-suppressor genes. As the name suggests, the role of a tumor-suppressor gene is to prevent cancerous growth. Therefore, when a tumor-suppressor gene becomes inactivated by mutation, cancer is more likely to occur. It is a loss-of-function mutation in a tumor-suppressor gene that promotes cancer. In this section, we will explore the possible ways that tumor-suppressor genes affect cell growth and examine the types of genetic changes that inactivate them.

Tumor-Suppressor Genes Play a Role in Preventing the Proliferation of Cancer Cells

The first identification of a human tumor-suppressor gene involved studies of retinoblastoma, a tumor that occurs in the retina of the eye. Some people have inherited a predisposition to develop this disease within the first few years of life. By comparison, the noninherited form of retinoblastoma tends to occur later in life but only rarely.

Based on these differences, in 1971, Alfred Knudson proposed a "two-hit" model for retinoblastoma. According to this model, retinoblastoma requires two mutations to occur. People with the hereditary form already have received one mutant gene from one of their parents. They need only one additional mutation in the other copy of this tumor-suppressor gene to develop the disease. Because the retina has millions of cells, it is relatively likely that a mutation may occur in one or more of these cells at an early age, leading to the disease. However, people with the noninherited form of the disease must have two mutations in the same retinal cell to cause the disease. Because two rare events are much less likely to occur than a single such event, the noninherited form of this disease is expected to occur much later in life and only rarely. Therefore, this hypothesis explains the different populations typically affected by the inherited and noninherited forms of retinoblastoma.

Since Knudson's original hypothesis, molecular studies have confirmed the two-hit hypothesis for the occurrence of retinoblastoma. The gene in which mutations occur is designated *rb* (for retinoblastoma). This tumor-suppressor gene is found on the long arm of chromosome 13. Most people have two functional copies of the *rb* gene. Persons with hereditary retinoblastoma have inherited one functional and one defective copy. In nontumorous cells throughout the body, they have one functional copy and one defective copy of *rb*. However, in retinal tumor cells, the functional *rb* gene has also suffered the second hit (i.e., a mutation), which renders it defective. Without the tumor-suppressor ability, cells are allowed to grow and divide in an unregulated manner, which ultimately leads to retinoblastoma.

More recent studies have revealed how the Rb protein suppresses the proliferation of cancer cells (**Figure 22.14**). The Rb protein regulates a transcription factor called E2F, which activates genes required for cell cycle progression. (The eukaryotic cell cycle is described in Chapter 2; see Figure 2.5.) The binding of the Rb protein to E2F inhibits its activity and prevents the cell from progressing through the cell cycle. As discussed later in this chapter, when a normal cell is supposed to divide, cellular proteins called cyclins bind to cyclin-dependent protein kinases (CDKs). This activates the kinases, which then phosphorylate the Rb protein. The phosphorylated form of the Rb protein is released from E2F, thereby allowing E2F to activate genes needed to advance through the cell cycle. What happens when both copies of the *rb* gene are rendered inactive by mutation? The answer is that the E2F protein is always active, which explains why uncontrolled cell division occurs.

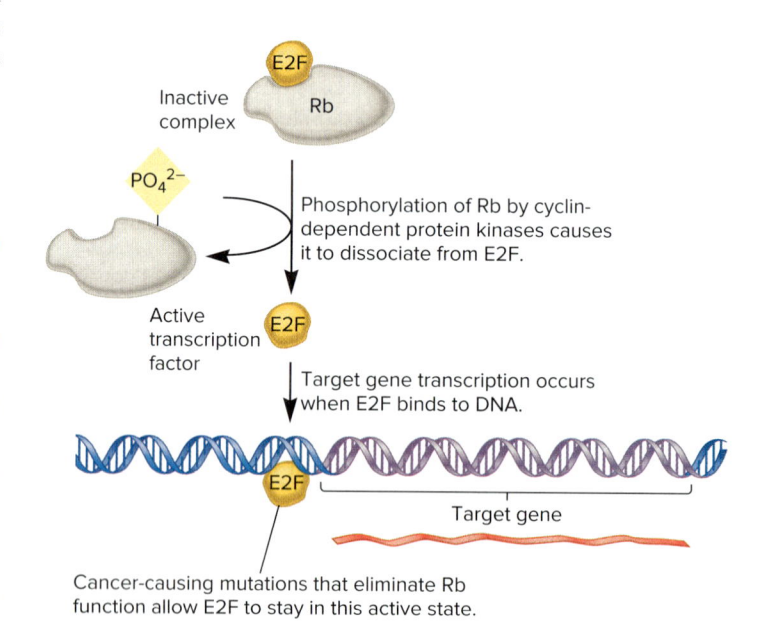

FIGURE 22.14 Interactions between the Rb and E2F proteins. The binding of the Rb protein to the transcription factor E2F inhibits the ability of E2F to function. This prevents cell division. For cell division to occur, cyclins bind to cyclin-dependent protein kinases, which then phosphorylate the Rb protein. The phosphorylated Rb protein is released from E2F. The free form of E2F can activate target genes needed to advance through the cell cycle.

Concept Check: If a cell cannot make any Rb protein, how will this affect the function of E2F?

The Vertebrate *p53* Gene Is a Master Tumor-Suppressor Gene That Senses DNA Damage

After the *rb* gene, the second tumor-suppressor gene discovered was the *p53* gene. The *p53* gene is the most commonly altered gene in human cancers. About 50% of all human cancers are associated with defects in *p53*. These include malignant tumors of the lung, breast, esophagus, liver, bladder, and brain as well as sarcomas, lymphomas, and leukemias. For this reason, an enormous amount of research has been aimed at elucidating the function of the p53 protein.

A primary role of the p53 protein is to determine if a cell has incurred DNA damage. The expression of the *p53* gene is caused by the formation of damaged DNA. The inducing signal appears to be double-strand DNA breaks. The p53 protein functions as a transcription factor. It contains a DNA-binding domain and a transcriptional activation domain. If damage is detected, p53 can promote three types of cellular pathways aimed at preventing the proliferation of cells with damaged DNA:

1. When confronted with DNA damage, the expression of *p53* activates genes involved with DNA repair. This may prevent the accumulation of mutations that activate oncogenes or inactivate tumor-suppressor genes.
2. If a cell is in the process of dividing, it can arrest itself in the cell cycle. By stopping the cell cycle, a cell gains time to repair its DNA and avoid producing two mutant daughter cells. For this to happen, the p53 protein stimulates the expression of another gene termed *p21*. The p21 protein inhibits the formation of cyclin/CDK complexes that are needed to advance from the G_1 phase of the cell cycle to the S phase.
3. The most drastic event is that the expression of *p53* can initiate a series of events called **apoptosis**, or programmed cell death. In response to DNA-damaging agents, a cell may self-destruct. Apoptosis is an active process that involves cell shrinkage, chromatin condensation, and DNA degradation. This process is facilitated by proteases known as **caspases.** These types of proteases are sometimes called the "executioners" of the cell. Caspases digest selected cellular proteins such as microfilaments, which are components of the intracellular cytoskeleton. This causes the cell to break apart into small vesicles that are eventually phagocytized by cells of the immune system. In this way, an organism can eliminate cells with cancer-causing potential.

Tumor-Suppressor Genes Can No Longer Inhibit Cancer When Their Function Is Lost

During the past three decades, researchers have identified many tumor-suppressor genes that, when defective, contribute to the development and progression of cancer (**Table 22.8**). What are the general functions of these genes? They tend to fall into two broad categories: genes that negatively regulate cell division and genes that maintain genome integrity.

Negative Regulators of Cell Division Some tumor-suppressor genes encode proteins that have direct effects on the regulation of

TABLE 22.8
Functions of Selected Tumor-Suppressor Genes

Gene	Function
Genes that negatively regulate cell division	
rb	The Rb protein is a negative regulator of E2F (see Figure 22.14). The inhibition of E2F prevents the transcription of certain genes required for DNA replication and cell division.
p16	The protein kinase p16 negatively regulates cyclin-dependent kinases and thereby controls the transition from the G_1 phase of the cell cycle to the S phase.
NF1	The NF1 protein stimulates Ras to hydrolyze its GTP to GDP. Loss of NF1 function causes the Ras protein to be overactive, which promotes cell division.
APC	APC is a negative regulator of a cell-signaling pathway that leads to the activation of genes that promote cell division.
Genes that maintain genome integrity	
p53	p53 is a transcription factor that acts as a checkpoint protein and positively regulates a few specific target genes and negatively regulates others in a general manner. It acts as a sensor of DNA damage. It can prevent advancement through the cell cycle and also can promote apoptosis.
BRCA-1, BRCA-2	BRCA1 and BRCA2 proteins are both involved in the cellular defense against DNA damage. These proteins facilitate DNA repair and can promote apoptosis if repair is not achieved.

cell division. The *rb* gene is an example. As mentioned earlier in this section, the Rb protein negatively regulates E2F. If both copies of the *rb* gene are inactivated, the growth of cells is accelerated. Therefore, loss of function of negative regulators has a direct effect on the abnormal cell division rates seen in cancer cells. In other words, when the Rb protein is lost, a cell becomes more likely to divide.

Maintenance of Genome Integrity Alternatively, other tumor-suppressor genes play a role in the proper maintenance of the integrity of the genome. The term **genome maintenance** refers to cellular mechanisms that either prevent mutations from occurring and/or prevent mutant cells from surviving or dividing. The proteins encoded by the genes that participate in genome maintenance help to ensure that gene mutations or changes in chromosome structure and number do not occur and are not transmitted to daughter cells. Such proteins can be subdivided into two classes: checkpoint proteins and those involved directly with DNA repair.

Figure 22.15 shows a simplified diagram of cell cycle control. Proteins called cyclins and cyclin-dependent protein kinases (CDKs) are responsible for advancing a cell through the four phases of the cell cycle. For example, an activated G_1-cyclin/CDK complex is necessary to advance from the G_1 to the S phase. Human cells produce several types of cyclins and CDKs. The formation of activated cyclin/CDK complexes is regulated by a variety of factors. One way is via **checkpoint proteins** that monitor the state of the cell and stop the progression through the cell cycle if abnormalities, such as DNA breaks and improperly

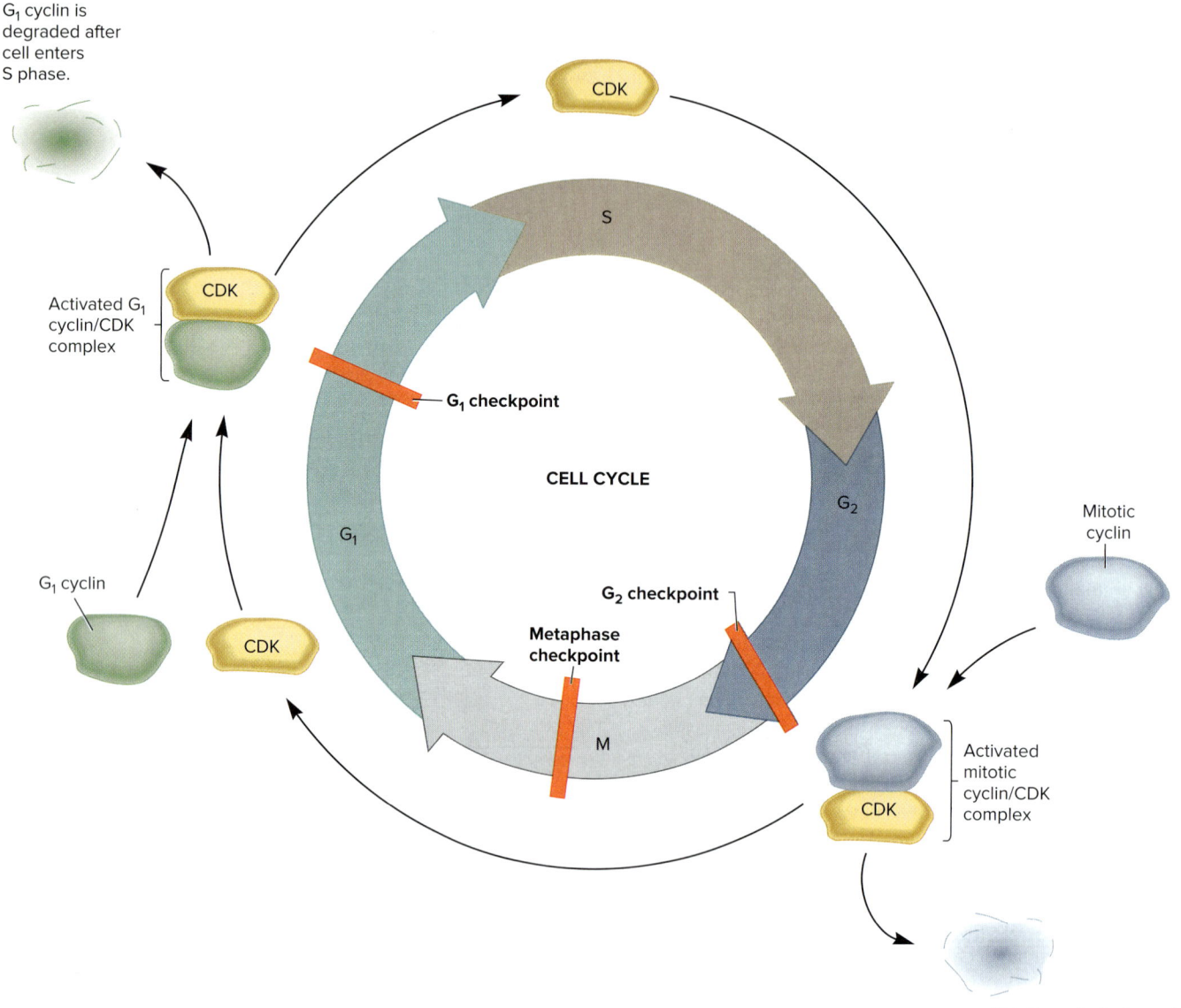

FIGURE 22.15 Cell cycle control. Progression through the cell cycle requires the formation of activated cyclin/CDK complexes. A cell may produce several different types of cyclin proteins, which are typically degraded after the cell has advanced to the next phase. The formation of activated cyclin/CDK complexes is regulated by checkpoint proteins. If these proteins detect DNA damage, they will prevent the formation of activated cyclin/CDK complexes. In addition, other checkpoint proteins can detect chromosomes that are incorrectly attached to the spindle apparatus and prevent further progression through metaphase. Note that this is a simplified diagram of the cell cycle of humans.

Concept Check: What is a checkpoint?

segregated chromosomes, are detected. These proteins are called checkpoint proteins because their role is to check the integrity of the genome and prevent cells from advancing past a certain point in the cell cycle if genetic abnormalities are detected. Therefore, these proteins provide a mechanism to stop the accumulation of genetic abnormalities that could produce cancer cells within the body. When checkpoint genes are lost, cell division may not be directly accelerated. However, the loss of checkpoint protein function makes it more likely that undesirable genetic changes occur that could cause cancerous growth.

Several checkpoint proteins regulate the cell cycle of human cells. Figure 22.15 shows three of the major checkpoints where these proteins exert their effects. Both the G_1 and G_2 checkpoints involve the functions of proteins that can sense if the DNA has incurred damage. If so, these checkpoint proteins, such as p53, prevent the formation of active cyclin/CDK complexes. This stops the progression of the cell cycle. A checkpoint also exists in metaphase. This checkpoint is monitored by proteins that can sense if a chromosome is not correctly attached to the spindle apparatus, making it likely that it will be improperly segregated.

In addition to checkpoint proteins, a second class of proteins involved with genome maintenance consists of DNA repair proteins, which were discussed in Chapter 19 (refer back to Table 19.7). The genes encoding such proteins are inactivated in certain forms of cancer. The loss of a DNA repair protein makes it more likely for a cell to accumulate mutations that could create an oncogene or eliminate the function of a tumor-suppressor gene. In Chapter 19, we considered how defects in the nucleotide excision repair process are responsible for the disease called xeroderma pigmentosum, which results in a predisposition to developing skin cancer (refer back to Figure 19.18). As discussed later in this section, defects in DNA mismatch repair proteins can contribute to colorectal cancer. In these cases, the loss of a DNA repair protein contributes to a higher mutation rate, which makes it more likely for other genes to incur cancer-causing mutations.

Tumor-Suppressor Genes Can Be Silenced by Loss-of-Function Mutations and Chomosome Loss

Thus far, we have considered the functions of proteins that are encoded by tumor-suppressor genes. Cancer biologists also want to understand how tumor-suppressor genes are inactivated, because this knowledge may aid in the prevention of cancer. Researchers have identified two different types of genetic changes that diminish the function of tumor-suppressor genes:

- A mutation can occur specifically within a tumor-suppressor gene to inactivate its function. For example, a mutation could inactivate the promoter of a tumor-suppressor gene or introduce an early stop codon in the coding sequence. Either of these would prevent the expression of a functional protein.
- Many types of cancer are associated with aneuploidy. As discussed in Chapter 8, aneuploidy involves the loss or addition of one or more chromosomes, so the total number of chromosomes is not an even multiple of a set. In some cases, chromosome loss may contribute to the progression of cancer because the lost chromosome carries one or more tumor-suppressor genes.

Most Forms of Cancer Involve Multiple Genetic Changes Leading to Malignancy

The discovery of oncogenes and tumor-suppressor genes, along with molecular techniques that can detect genetic alterations, has enabled researchers to study the progression of certain forms of cancer at the molecular level. Many cancers begin with a benign genetic alteration that, over time and with additional mutations, progresses to malignancy. Furthermore, a malignancy can continue to accumulate genetic changes that make it even more difficult to treat. For example, some tumors may acquire mutations that cause them to be resistant to chemotherapeutic agents.

In 1990, Eric Fearon and Bert Vogelstein proposed a series of genetic changes that lead to colorectal cancer, the second most common cancer in the United States. As shown in **Figure 22.16**, colorectal cancer is derived from cells in the mucosa of the colon.

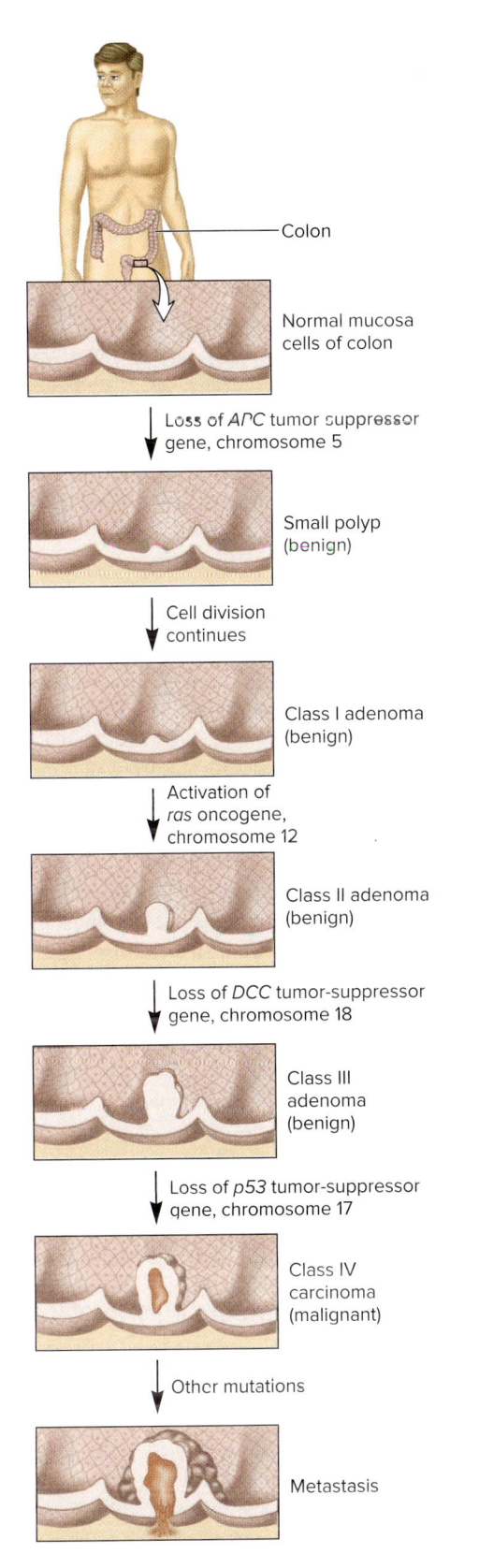

FIGURE 22.16 Multiple genetic changes leading to colorectal cancer.

The loss of function of *APC*, a tumor-suppressor gene on chromosome 5, leads to an increased proliferation of mucosal cells and the development of a benign polyp, a noncancerous growth. Additional genetic changes involving the loss of other tumor-suppressor genes and the activation of an oncogene (namely, *ras*) lead eventually to the development of a carcinoma. In Figure 22.16, the genetic changes that lead to colon cancer are portrayed as occurring in an orderly sequence. However, it is the total number of genetic changes, not their exact order, that is important.

Among different types of tumors, researchers have identified a large number of genes that are mutated in cancer cells. Though not all of these mutant genes have been directly shown to affect the growth rate of cells, such mutations are likely to be found in tumors because they provide some type of growth advantage for the cell population from which the cancer developed. For example, certain mutations may enable cells to metastasize to neighboring locations. These mutations may not affect growth rate, but they provide an advantage in that the cancer cells are not limited to growing in a particular location, but can migrate to new locations.

Researchers have estimated that about 300 different protein-encoding genes may play a role in the development of human cancer. Since the approximate genome size of humans is about 22,000 protein-encoding genes, this observation indicates that over 1% of those genes have the potential to promote cancer if their function is altered by a mutation. More recently, researchers have discovered that mutations in non-coding RNAs are also associated with certain forms of cancer. This topic is discussed in Chapter 18.

In addition to mutations within specific genes, other common genetic changes associated with cancer are alterations in chromosome structure and number. **Figure 22.17** compares the chromosome composition of a normal female cell and a tumor cell taken from the same person. The chromosome composition of the tumor cell is quite bizarre. For example, some chromosomes are missing. If tumor-suppressor genes are on these missing chromosomes, their function is lost as well. Figure 22.17 also shows that chromosome 7 is present in three copies in the tumor cell. If this chromosome carries proto-oncogenes, the expression of those genes may be overactive. Finally, tumor cells often have chromosomes with translocations (these are designated as marker chromosomes in the figure). Such translocations may create fused genes (as in the case of the Philadelphia chromosome discussed earlier in this section), or they may place two genes close together so that the regulatory sequences of one gene affect the expression of the other.

Inherited Forms of Cancers May Be Caused by Defects in Tumor-Suppressor Genes

Before we end our discussion of the genetic basis of cancer, let's consider which genes are most likely to be affected in inherited forms of the disease. As mentioned previously, about 5–10% of all cases of cancer involve inherited (germ-line) mutations. These familial forms of cancer occur because people have inherited mutations from one or both parents that give them an increased susceptibility to developing cancer. Inheriting the mutations does not mean they will definitely get cancer, but they are more likely to develop the disease than are individuals in the general population.

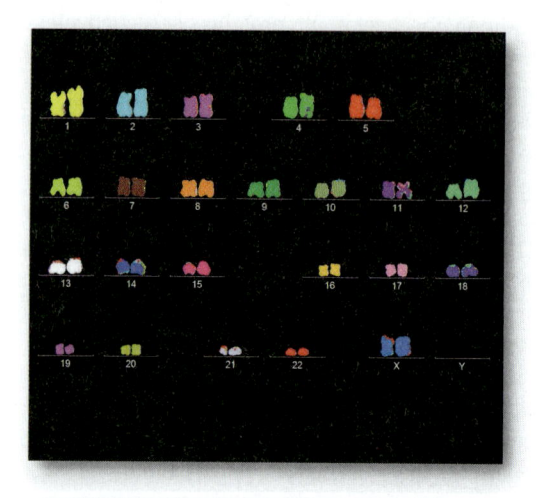

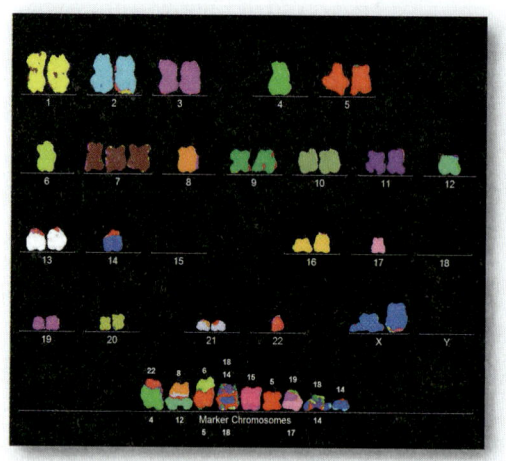

FIGURE 22.17 **A comparison between chromosomes found in a normal human cell and in a tumor cell from the same person.** The bottom set, from a tumor cell, is highly abnormal, with missing copies of some chromosomes and an extra copy of chromosome 7. Translocated chromosomes (designated marker chromosomes in this figure) are made of fused pieces of different chromosomes.
Courtesy of the Duesberg Lab, University of California, Berkeley

When individuals have family members who have developed certain forms of cancer, they may be tested to determine if they also carry a mutant gene. For example, von Hippel-Lindau disease and familial adenomatous polyposis are examples of syndromes for which genetic testing to identify at-risk family members is considered the standard of care.

What types of genes are mutant in familial cancers? Most inherited forms of cancer involve a defect in a tumor-suppressor gene (**Table 22.9**). In these cases, the individual is heterozygous, with one normal and one inactive allele.

At the level of a human pedigree, a predisposition for developing cancer is inherited in a dominant fashion because a heterozygote exhibits this predisposition. **Figure 22.18a** shows a pedigree for familial breast cancer. In this case, individuals with the disorder have inherited a loss-of-function mutation in one *BRCA-1* gene. As seen in the pedigree, the development of breast cancer

22.6 TUMOR-SUPPRESSOR GENES

TABLE 22.9
Inherited Mutant Genes That Confer a Predisposition to Developing Cancer

Gene*	Type of Cancer
Tumor-suppressor genes†	
VHL	von Hippel-Lindau disease, which is typically characterized by a clear-cell renal carcinoma
APC	Familial adenomatous polyposis and familial colon cancer
rb	Retinoblastoma
p53	Li-Fraumeni syndrome, which is characterized by a wide spectrum of tumors, including soft-tissue and bone sarcomas, brain tumors, adenocortical tumors, and premenopausal breast cancers
BRCA-1	Familial breast cancer
BRCA-2	Familial breast cancer
NF1	Neurofibromatosis
MLH1	Nonpolyposis colorectal cancer
XP-A to XP-G	UV sensitive forms of cancer such as basal cell carcinoma
Oncogenes	
RET	Multiple endocrine neoplasia type 2

*Many of the genes included in this table are mutated in more than one type of cancer. The cancers listed are those in which it has been firmly established that a predisposition to develop the disease is commonly due to germ-line mutations in the designated gene.
†MLH1 and XP-A to XP-G encode proteins that are involved in DNA repair.

shows a dominant pattern of inheritance with incomplete penetrance. Most affected individuals have an affected parent.

At the cellular level, however, the actual development of cancer is recessive. A cell must be homozygous for a loss-of-function allele to become cancerous. How does this occur? An individual initially is heterozygous, but then a somatic mutation in the normal *BRCA-1* gene eliminates its function. This somatic mutation makes the cell homozygous for two loss-of-function alleles (**Figure 22.18b**). This phenomenon is called **loss of heterozygosity** (**LOH**)—the loss of function of a normal allele when the other allele was already inactivated.

22.6 REVIEWING THE KEY CONCEPTS

- Knudson proposed a two-hit model to explain the occurrence of retinoblastoma (see Figure 22.14).
- The vertebrate p53 gene senses DNA damage and prevents damaged cells from dividing. Its expression may result in apoptosis.
- Most tumor-suppressor genes encode proteins that are negative regulators of cell division or play a role in genome maintenance (see Table 22.8).
- Checkpoint proteins prevent cells that may harbor genetic abnormalities from advancing through the cell cycle (see Figure 22.15).
- Tumor-suppressor genes can be silenced by loss-of-function mutations or chromosome loss.
- Most forms of cancer involve multiple genetic changes (see Figure 22.16).

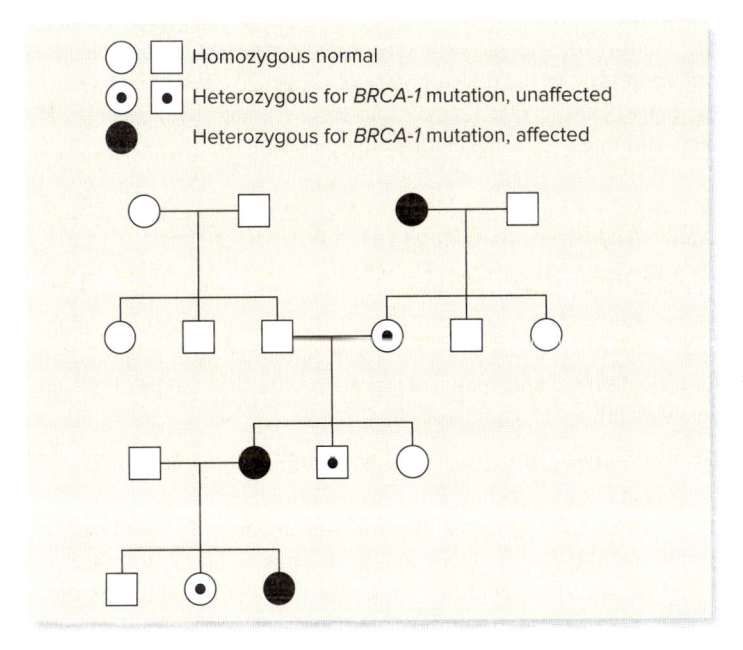

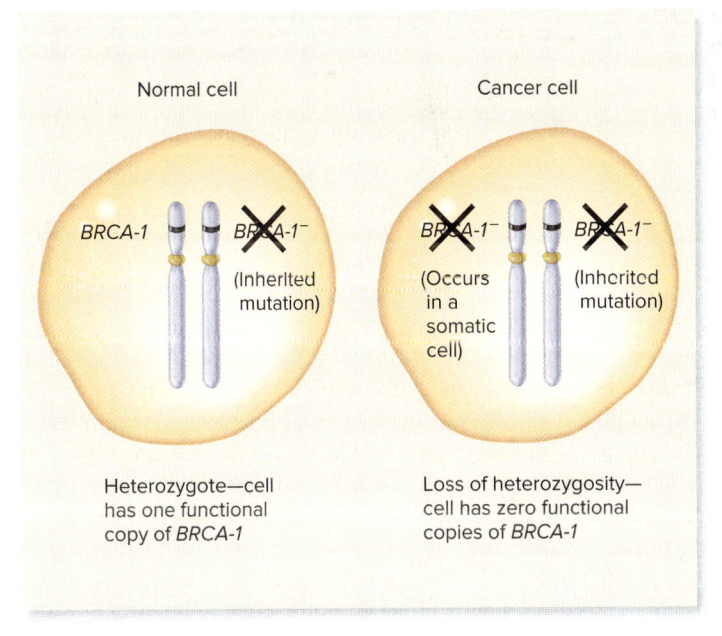

(a) Pedigree for familial breast cancer

(b) Development of cancer at the cellular level

FIGURE 22.18 **Inheritance pattern of familial breast cancer.** (a) A family pedigree that involves a loss-of-function mutation in the *BRCA-1* gene. The individuals in this pedigree were tested to determine if they carry a mutation in the *BRCA-1* gene. Affected individuals are shown with filled-in symbols. The disorder follows a dominant pattern of inheritance. (Note: Males can occasionally develop breast cancer, though it is much more common in females.) (b) Breast cancer at the cellular level. Normal cells in an affected individual are heterozygous, whereas cancer cells in the same individual have lost their heterozygosity. Therefore, at the cellular level, cancer is recessive because both alleles must be inactivated for it to occur.

Concept Check: Explain why familial breast cancer shows a dominant pattern of inheritance in a pedigree even though it is recessive at the cellular level.

- Alterations in chromosome structure and number are common in cancer cells (see Figure 22.17).
- An inherited predisposition to develop cancer is usually caused by a mutation in a tumor-suppressor gene (see Table 22.9).
- Familial breast cancer exhibits a dominant pattern of inheritance. At the cellular level, loss of heterozygosity (LOH) promotes cancer (see Figure 22.18).

22.6 COMPREHENSION QUESTIONS

1. Tumor-suppressor genes promote cancer when
   a. they are overexpressed.
   b. they are expressed in the wrong cell type.
   c. their function is inactivated.
   d. they are expressed at the wrong stage of development.
2. Normal (nonmutant) tumor-suppressor genes often function
   a. as negative regulators of cell division.
   b. in the maintenance of genome integrity.
   c. in the stimulation of cell division.
   d. Both a and b are functions of normal tumor-suppressor genes.
3. Most forms of cancer involve
   a. the activation of a single oncogene.
   b. the inactivation of a single tumor-suppressor gene.
   c. the activation of multiple oncogenes.
   d. the activation of multiple oncogenes and the inactivation of multiple tumor-suppressor genes.

22.7 ROLE OF EPIGENETICS IN CANCER

Learning Outcomes:
1. Explain how genetics and epigenetics play a role in cancer.
2. Describe the underlying causes of epigenetic changes associated with cancer.

One of the most active fields in genetics involves the study of how epigenetic changes contribute to human diseases. We are probably seeing only the "tip of the iceberg" in our current understanding of this topic. Many medical studies have identified correlations between epigenetic changes and particular diseases. In Chapter 24, we will examine how researchers can calculate a **correlation coefficient** by comparing two variables to see if they are related to each other. For example, a research study may compare one variable, such as the level of DNA methylation of a specific gene, to a second variable, such as the severity of a disease. If a high level of DNA methylation (hypermethylation) is associated with an increase in disease severity, this is a positive correlation. Researchers analyze the data to decide if such a correlation is statistically significant.

When a statistically significant correlation coefficient is obtained, how do we interpret its meaning? Such a result suggests a true **association**—changes in the two variables follow a pattern. For example, in a positive correlation, when one variable increases, the other variable also increases. However, such an association does not necessarily imply a cause-and-effect relationship.

When considering the role of epigenetic changes and human disease, an association can arise in three common ways:

- The epigenetic changes directly contribute to the disease symptoms. There is a cause-and-effect relationship.
- Conversely, the disease symptoms may arise first, and then they cause subsequent epigenetic changes to happen. This is also a cause-and-effect relationship, but in the opposite direction.
- The association is indirect because a third factor is involved. For example, a toxic agent in the environment may cause a disease, and it may also cause particular types of epigenetic changes even though those epigenetic changes do not contribute to the disease.

In general, correlation coefficients are quite useful in identifying associations between two variables. Caution is necessary, however, because a statistically significant correlation coefficient, by itself, cannot prove that the association is due to cause and effect. Even so, research studies that identify associations are very useful because they provide the motivation to carry out further research to determine if a cause-and-effect relationship exists.

Researchers have identified many examples in which epigenetic changes are associated with a particular disease. These include Alzheimer disease, cardiovascular diseases, diabetes, multiple sclerosis, and asthma. For these diseases, further research is needed to determine if these epigenetic changes are directly contributing to the disease symptoms. The role of epigenetics and disease has been most extensively studied with regard to cancer. For some forms of cancer, the evidence suggests a cause-and-effect relationship. In this section, we will consider how epigenetics may play an important role in the development of cancer. We will first explore the types of epigenetic changes that are associated with cancer, and then examine how abnormalities in epigenetic changes may arise.

Abnormalities in Chromatin Modification Are Common in Cancer Cells

Three general types of chromatin modifications are often found to be abnormal in cancer cells. These abnormalities in chromatin modification are epigenetic changes, because they are passed from a cancer cell to its daughter cells:

- *DNA methylation.* A particularly common alteration in cancer cells is a change in the level of DNA methylation. For example, hypermethylation—an abnormally high level of methylation, typically at CpG islands, is often observed. This hypermethylation may promote cancer by inhibiting the expression of tumor-suppressor genes.
- *Covalent modification of histones.* As described in Chapter 17, the covalent modification of histones can affect the expression of genes, either activating or inhibiting transcription depending on the pattern of modification. The histone tails are subject to a variety of modifications, including the attachment of acetyl, methyl, and phosphate groups (refer back to Figure 17.8). Variation in the covalent modification of histones has been shown to occur at specific genes in cancer

cells. Depending on the specific type of modification, such changes could increase the expression of oncogenes or inhibit the expression of tumor-suppressor genes.

- **Chromatin remodeling.** Another important chromatin modification is chromatin remodeling, in which the locations of nucleosomes are changed. Abnormalities in the locations of nucleosomes have been frequently found in cancer cells. Depending on how the nucleosomes are rearranged, such changes could increase the expression of oncogenes or inhibit the expression of tumor-suppressor genes.

Epigenetic Changes Associated with Cancer May Arise Because Mutations or Environmental Agents Alter the Functions of Chromatin-Modifying Proteins

Thus far, we have considered three general types of epigenetic changes that contribute to cancer. Researchers want to understand how these changes arise. In other words, what causes chromatin modifications to become abnormal and promote cancer? The underlying cause of the abnormality can be placed into two general categories:

1. Mutations may occur in genes that encode chromatin-modifying proteins.
2. Environmental agents may alter the functions of chromatin-modifying proteins.

Mutations in Genes That Encode Chromatin-Modifying Proteins Chromatin modifications are carried out by an array of different proteins. Mutations in the genes that encode these proteins are a common occurrence in many or perhaps most types of cancer (**Table 22.10**). The mutation is a genetic event, but it has an epigenetic effect because it alters the function of a chromatin-modifying protein. For example, a mutation may occur in a gene that encodes DNA methyltransferase, thereby inhibiting its function. This inhibition will decrease the level of DNA methylation, and this epigenetic change can be passed from one cancer cell to its daughter cells. As described in Table 22.10, this type of mutation is often found in cells that give rise to acute myeloid leukemia.

In some cases, the mutations described in Table 22.10 increase the function of the chromatin-modifying proteins, but it is more common for a mutation to inhibit their function. In either case, the mutation may have a widespread effect on gene expression. As discussed in Chapter 17, chromatin modifications can convert chromatin from a closed (transcriptionally inactive) to an open (transcriptionally active) state, and vice versa. When the function of chromatin-modifying proteins is altered by mutation, this may cause other genes to be overexpressed by converting chromatin to an open conformation or it may inhibit gene expression by converting chromatin to a closed conformation. Such changes may increase the expression of oncogenes or inhibit the expression of tumor-suppressor genes, respectively. Both types of change can contribute to cancer.

Environmental Agents That Alter the Functions of Chromatin-Modifying Proteins A second way that chromatin modification can be abnormally altered is by exposure to environmental agents. Such agents may directly alter the functions of chromatin-modifying proteins or they may initiate a series of cellular changes that ultimately affect the functions of chromatin-modifying proteins. **Table 22.11** lists several examples of environmental agents that are known to cause epigenetic changes involving abnormalities in DNA methylation, histone modification, and/or chromatin remodeling. These agents are also associated with particular forms of cancer.

For some of the examples listed Table 22.11, scientific evidence indicates that the association is causative. For example, certain agents in tobacco smoke have been shown to cause cellular changes that underlie lung cancer. Furthermore, some of the chemicals in tobacco smoke, such as polycyclic aromatic hydrocarbons, are both mutagenic and epimutagenic, which makes them particularly potent at promoting changes that may lead to cancer. Alternatively, some of the agents listed in Table 22.11 show an association with particular cancers, but researchers are still trying to determine if the epigenetic changes caused by these agents actually promote changes that result in cancer.

TABLE 22.10
Mutations in Genes That Encode Chromatin-Modifying Proteins and Their Occurrence in Different Types of Cancers

Type of Modification	Type of Protein Encoded by Mutant Gene	Protein Function	Particular Cancer(s) in Which Mutant Gene Is Observed
DNA methylation	DNA methyltransferase	Methylates DNA	Acute myeloid leukemia
Histone modification	Histone acetyltransferase	Attaches acetyl groups to histones	Colorectal, breast, and pancreatic cancer
Histone modification	Histone methyltransferase	Attaches methyl groups to histones	Renal and breast cancer
Histone modification	Histone demethylase	Removes methyl groups from histones	Multiple myeloma and esophageal cancer
Histone modification	Histone kinase	Attaches phosphate groups to histones	Medulloblastoma, glioma
Chromatin remodeling	SWI/SNF complex	Alters the positions of histones	Lung, breast, prostate, and pancreatic cancer

TABLE 22.11
Environmental Agents That Are Associated with Cancer and Are Known to Cause Epigenetic Changes

Environmental Agent	Occurrence	Particular Cancers Associated with Agent
Polycyclic aromatic hydrocarbons	Tobacco smoke, automobile exhaust, charbroiled food	Lung, breast, stomach, and skin cancer
Benzene	Tobacco smoke, automobile exhaust	Leukemia, lymphoma, multiple myeloma
Endocrine disruptors (e.g., diethylstilbestrol)	Insecticides, fungicides, herbicides, and some types of plastic	Breast, prostate, and thryoid cancer
Cadmium	Tobacco products, production of batteries	Lung and breast cancer
Nickel	Occupational exposure in mining, welding, and electroplating, and in the manufacturing of jewelry, stainless steel, and battery production	Lung and nasal cancer
Arsenic	Lead alloy, feed additive in agriculture, insecticides	Skin, bladder, kidney, and liver cancer

Cancer Treatments May Be Aimed at Epigenetic Changes

A potentially exciting application of the research on epigenetic changes associated with cancer is the development of new drugs that may reverse these changes, thereby providing another treatment option for cancer patients. Researchers are actively investigating drugs that may inhibit cancer cells by affecting either DNA methylation or covalent histone modifications. For example, inhibitors of DNA methyltransferase, the enzyme that attaches methyl groups to DNA, are being developed to treat certain forms of cancer including leukemia—a cancer of white blood cells. Drugs such as 5-azacytidine and decitabine, which inhibit DNA methyltransferase, have shown some promising results when used in conjunction with other anticancer drugs. In some cases, clinical improvement in patients with leukemia has been associated with a decrease in DNA methylation. Although the specific mechanisms for patient improvement are not understood at the molecular level, one possibility is that the lower level of DNA methylation has reversed the inhibition of tumor suppressor genes.

22.7 REVIEWING THE KEY CONCEPTS

- Epigenetic changes associated with cancer may arise because mutations occur in genes that encode chromatin-modifying proteins or because environmental agents alter the functions of chromatin-modifying proteins (see Tables 22.10, 22.11).

22.7 COMPREHENSION QUESTIONS

1. Which of the following types of epigenetic changes may promote cancer?
    a. DNA methylation
    b. Covalent modification of histones
    c. Chromatin remodeling
    d. All of the above
2. The underlying cause(s) of epigenetic changes associated with cancer may be
    a. mutations in genes that encode chromatin-modifying proteins.
    b. environmental agents that alter the function of chromatin-modifying proteins.
    c. mutations in genes that encode proteins that directly accelerate cell growth.
    d. all of the above.
    e. both a and b.

22.8 PERSONALIZED MEDICINE

Learning Outcomes:
1. Define *personalized medicine* and *molecular profiling*.
2. Describe specific ways in which personalized medicine affects patient care.

Personalized medicine is the use of information about a patient's genotype and other clinical data in order to select a medication, therapy, or preventative measure that is specifically suited to that patient. As we gain a better understanding of human genes and disease states, researchers expect that personalized medicine will become an increasingly important aspect of health care. In this section, we will begin by examining how personalized medicine can affect treatment options for cancer patients. We will then consider how personalized medicine may play a role in determining the correct dosage for certain types of drugs.

Molecular Profiling Is Increasingly Used to Classify Tumors and Improve Treatment Options

Traditionally, different types of tumors have been classified according to their appearance under a microscope. Although this approach is useful, a major drawback is that two tumors may have a very similar microscopic appearance but yet have very different underlying genetic changes and clinical outcomes. For this reason, researchers and clinicians are turning to methods that enable them to understand the molecular changes that occur in diseases such as cancer. This general approach is called **molecular profiling.**

In cancer biology, molecular profiling involves the identification of the genes that play a role in the development of cancer. Why is this useful? First, molecular profiling can distinguish between tumors that look very similar under the microscope. Second, researchers are optimistic that molecular profiling may lead to improved treatment options. As we gain a better understanding of the genetic changes associated with particular types of cancers, researchers may be able to develop drugs that specifically target the proteins that are encoded by cancer-causing gene mutations. As discussed earlier in this chapter, the drug imatinib mesylate, which is used to treat chronic myelogenous leukemia, was developed in this way. As another example, about 70% of all breast cancers exhibit an overexpression of the estrogen receptor. These types of breast cancer are better treated with drugs that either block the estrogen receptor or block the synthesis of estrogen. Therefore, the drug tamoxifen, which is an antagonist of the estrogen receptor, is used to treat tumors in which the estrogen receptor is overexpressed.

DNA Microarrays Are Used in the Molecular Profiling of Tumors

DNA microarrays, described in Chapter 21, are often used as a tool in the molecular profiling of tumors. In the study of cancer, researchers can compare cancer cells to normal cells and identify groups of genes that are turned on in the cancer cells and off in the normal cells, as well as other groups of genes that are turned off in the cancer cells and on in the normal cells. Likewise, researchers can compare two different types of tumors and identify groups of genes that show different patterns of expression.

As an example, **Figure 22.19a** shows a computer-generated image that presents the results of a microarray analysis of 47 samples, most of which came from the tumors of patients with a type of cancer called diffuse large B-cell lymphoma (DLBCL). Each column represents the expression pattern of a set of genes from a particular sample. Genes that are expressed are shown in red; those that are not expressed are shown in green. During the course of these studies, the researchers identified two different patterns of gene expression. The tumor samples on the left side showed a set of genes (next to the orange bar) that tended to be turned on in the tumor and another set of genes (next to the blue bar) that tended to be turned off in the tumor. This pattern of

FIGURE 22.19 The use of DNA microarrays to classify types of tumors. (a) Forty-seven samples, mostly from patients with diffuse large B-cell lymphoma (DLBCL), were subjected to a DNA microarray analysis. The figure shown here is a graphical compilation of these microarray analyses. Each column represents one sample; each row represents the expression of a particular gene. The names of some of the genes are shown along the right side. (Note: The rows and columns are not easily resolved in this illustration.) Genes highly expressed are shown in red; those not expressed are shown in green. One group of samples had an expression pattern similar to that found in germinal center B cells; the other group had an expression pattern typical of activated B cells. (b) Survival rates of the patients with DLBCL. (Reprinted by permission from Macmillan Publishers Ltd. A.A. Alizadeh, M.B. Eisen, R.E. Davis, et al. (2000), Distinct types of diffuse large B-cell lymphoma identified by gene expression profiling. *Nature 403*, 6769, 503–511. Image courtesy of Ash Alizadeh.)

©Pat Brown

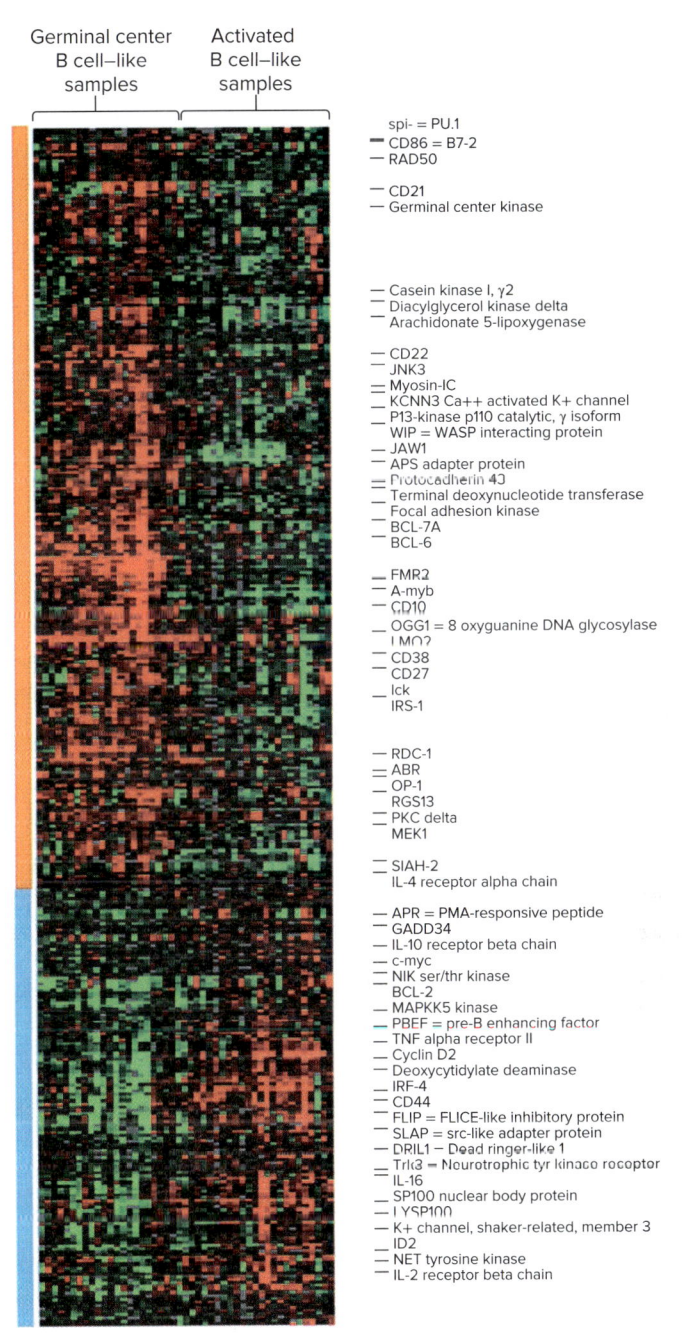

(a) Cluster analysis

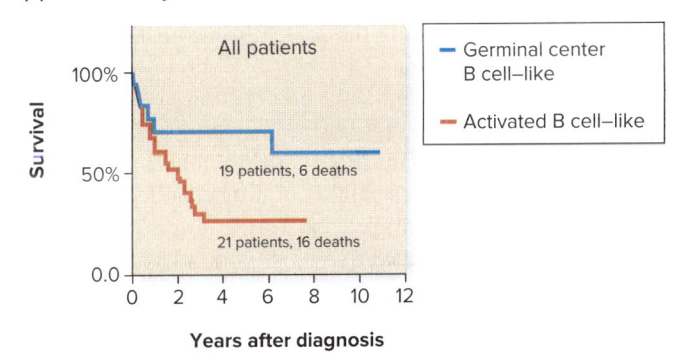

(b) Patient outcomes

gene expression was similar to the pattern found in a type of B cell called germinal center B cells. In contrast, the tumor samples on the right side showed the opposite pattern. The upper genes tended to be turned off in these patients, and the lower genes were turned on. These samples showed a gene expression pattern found in activated B cells. These results suggest that the two groups of tumors may have originated in B cells at different stages of development—those on the left originated in germinal center B cells, whereas those on the right originated in activated B cells.

Furthermore, the patients from whom these tumors were derived also appeared to have very different clinical outcomes (**Figure 22.19b**). The patients whose tumors had a pattern of gene expression similar to that in activated B cells had a significantly lower overall survival rate than did the other patients.

A Patient's Genotype Is Important in Determining the Proper Dosage of Certain Drugs

Pharmacogenetics is the study or clinical testing of genetic variation that causes differing responses to drugs. The proper dosage for any given drug depends on a variety of factors. Some key factors include the following:

- For drugs that are taken orally, the rate of transport of the drug from the digestive tract into the bloodstream
- The rate of transport of the drug into the appropriate cells where the drug exerts its effect
- The ability of the drug to affect the function of a specific target protein
- The ability of the drug to be taken up and metabolized by the liver
- The rate of excretion of the drug from the body

These five factors are affected by genetics because proteins, which are encoded by genes, are directly involved. For example, transport proteins are often required for the uptake of drugs into the bloodstream, into specific cell types, and into liver cells. Enzymes, which are proteins, are involved in metabolizing drugs into inactive products. Finally, nearly all drugs exert their effects by binding to specific target proteins and altering their function. For example, acetylsalicylic acid, more commonly known as aspirin, binds to a protein called cyclo-oxygenase, thereby inhibiting the protein's enzymatic function and reducing inflammation.

Why is genetics important in the proper dosage of drugs? The answer is that genetic variation in the human population often affects the function of proteins that are involved with drug transport, drug metabolism, and the ability of drugs to affect their target proteins.

To understand how genetic variation can affect the proper dosage of a specific drug, let's consider drug metabolism by the liver. Many drugs are broken down in the liver by a family of related enzymes called cytochrome P450. An example is warfarin (coumarin) that is used clinically as an anticoagulant, but requires periodic monitoring and is associated with adverse side affects. This drug is metabolized by a cytochrome P450 enzyme designated CYP2C9. Researchers have identified over 80 variants (SNPs) in the gene that encodes CYP2C9 in human populations. Some of these variants affect CYP2C9 function. Clinically, the variable activity of CYP2C9 can result in four different levels of warfarin metabolism, which are described as ultrarapid, extensive, intermediate, and poor.

Currently, the dose of warfarin given to patients is either a "one size fits all" approach or it may take into consideration characteristics such as gender, age, and liver function. Adjustments to the dosage are made based on periodic blood tests that measure the level of blood coagulation. Even so, over-coagulation and under-coagulation remain a problem in a significant number of patients. Recently, genetic tests have become available that help doctors determine the proper warfarin dosage for their patients. For example, a person with an ultrarapid metabolizer genotype requires a higher dosage because the drug tends to be rapidly broken down in that person's body. By comparison, someone with a poor metabolizer genotype requires a lower dosage. In the future, such genetic tests may be routinely used by doctors to determine the proper drug dosage.

As discussed in Chapter 21, DNA-sequencing technologies are progressing to the point where the sequencing of a person's entire genome will be inexpensive enough to be used as a routine diagnostic procedure. Therefore, many clinicians are predicting that it will become common practice for a patient's genome sequence to be determined and analyzed to improve care. As we gain a better understanding of how genetic variation affects drug action, transport, and metabolism, pharmacogenetics is expected to play an increasing role in personalizing health care.

22.8 REVIEWING THE KEY CONCEPTS

- Personalized medicine is the use of information about a patient's genotype and other clinical data in order to select a medication, therapy, or preventative measure that is specifically suited to that patient.
- Molecular profiling is used to classify tumors, which may affect treatment options. Methods used in such profiling include DNA microarrays (see Figure 22.19).
- Pharmacogenetics, which is the study or clinical testing of genetic variation that causes differing responses to drugs, is likely to play an increasing role in determining the proper dosages of drugs given to patients.

22.8 COMPREHENSION QUESTION

1. Personalized medicine may be used
    a. to characterize types of tumors.
    b. to predict the outcome of certain types of cancers.
    c. to determine the proper dosage of drugs.
    d. in all of the above.

KEY TERMS

Page 519. monozygotic (MZ) twins, dizygotic (DZ) twins, concordance, age of onset
Page 522. haploinsufficiency, gain-of-function mutations, dominant-negative mutations
Page 523. locus heterogeneity
Page 525. haplotype
Page 526. founder, linkage disequilibrium
Page 527. genome-wide association study (GWA study, or GWAS)
Page 529. genetic testing, genetic screening
Page 530. amniocentesis, chorionic villus sampling, preimplantation genetic diagnosis (PGD)
Page 531. in vitro fertilization (IVF), cancer, clonal, benign, malignant, invasive, metastatic
Page 532. carcinogen, oncogene, tumor-suppressor gene, proto-oncogene, growth factors
Page 535. transformation, acutely transforming viruses (ACTs)
Page 537. apoptosis, caspases, genome maintenance, checkpoint proteins
Page 541. loss of heterozygosity (LOH)
Page 542. correlation coefficient, association
Page 544. personalized medicine, molecular profiling
Page 546. pharmacogenetics

CHAPTER SUMMARY

- Thousands of genetic diseases are known to afflict people.

22.1 Inheritance Patterns of Genetic Diseases

- A genetic basis for a human disease may be suggested by a variety of different observations (e.g., see Figure 22.1).
- Thousands of human genetic diseases follow simple Mendelian patterns of inheritance. These patterns include autosomal recessive, autosomal dominant, X-linked recessive, and X-linked dominant inheritance (see Figures 22.2–22.4, Tables 22.1–22.4).
- Recessive diseases are usually caused by loss-of-function mutations, whereas dominant diseases may be caused by haploinsufficiency, gain-of-function mutations, or dominant-negative mutations.
- Many genetic diseases exhibit locus heterogeneity, which means that they may be caused by mutations in more than one gene.

22.2 Detection of Disease-Causing Alleles via Haplotypes

- The term *haplotype* refers to the linkage of alleles or molecular markers on a single chromosome (see Figure 22.5).
- Disease-causing mutations may originate in a founder with a specific haplotype (see Figure 22.6).
- Researchers may identify a disease-causing allele by determining its location within a chromosome via haplotype association studies (see Figure 22.7).
- Genome-wide association studies (GWA studies) are conducted to find associations between genetic variation and human diseases (see Figure 22.8).

22.3 Genetic Testing and Screening

- Genetic testing for abnormalities can be performed in a variety of ways (see Table 22.5).
- Genetic screening is population-wide genetic testing.
- Genetic testing by means of amniocentesis, chorionic villus sampling, or preimplantation genetic diagnosis (PGD) can be done prior to birth (see Figure 22.8).

22.4 Overview of Cancer

- Cancer is clonal in origin. It usually develops via a multistep process that involves several mutations, begins with benign growth, and progresses to growth that is invasive and metastatic (see Figure 22.10).
- Cancer can be caused by mutations that overexpress oncogenes or inhibit the expression of tumor-suppressor genes. Epigenetic changes can also promote cancer.

22.5 Oncogenes

- Oncogenes often exert their effects via intracelllular signaling pathways that control the cell cycle (see Figure 22.11).
- Oncogenes arise due to gain-of-function mutations in proto-oncogenes. An example is a mutation in the *ras* gene that prevents the Ras protein from hydrolyzing GTP (see Table 22.6, Figure 22.12).
- Different types of genetic changes that can convert proto-oncogenes to oncogenes include missense mutation, gene amplification, chromosomal translocation, and viral integration (see Figure 22.13).
- Certain viruses cause cancer by carrying oncogenes into cells (see Table 22.7).

22.6 Tumor-Suppressor Genes

- Knudson proposed a two-hit model to explain the occurrence of retinoblastoma (see Figure 22.14).
- The vertebrate *p53* gene senses DNA damage and prevents damaged cells from dividing. Its expression may result in apoptosis.
- Most tumor-suppressor genes encode proteins that are negative regulators of cell division or play a role in genome maintenance (see Table 22.8).
- Checkpoint proteins prevent cells that may harbor genetic abnormalities from advancing through the cell cycle (see Figure 22.15).
- Tumor-suppressor genes can be silenced by loss-of-function mutations or chromosome loss.

548   CHAPTER 22 :: MEDICAL GENETICS AND CANCER

- Most forms of cancer involve multiple genetic changes (see Figure 22.16).
- Alterations in chromosome structure and number are common in cancer cells (see Figure 22.17).
- An inherited predisposition to develop cancer is usually caused by a mutation in a tumor-suppressor gene (see Table 22.9).
- Familial breast cancer exhibits a dominant pattern of inheritance. At the cellular level, loss of heterozygosity (LOH) promotes cancer (see Figure 22.18).

22.7 Role of Epigenetics in Cancer

- Epigenetic changes associated with cancer may arise because mutations occur in genes that encode chromatin-modifying proteins or because environmental agents alter the functions of chromatin-modifying proteins (see Tables 22.10, 22.11).

22.8 Personalized Medicine

- Personalized medicine is the use of information about a patient's genotype and other clinical data in order to select a medication, therapy, or preventative measure that is specifically suited to that patient.
- Molecular profiling is used to classify tumors, which may affect treatment options. Methods used in such profiling include DNA microarrays (see Figure 22.19).
- Pharmacogenetics, which is the study or clinical testing of genetic variation that causes differing responses to drugs, is likely to play an increasing role in determining the proper dosages of drugs given to patients.

PROBLEM SETS & INSIGHTS

More Genetic TIPS

1. The pedigree presented here shows the inheritance of a human disease known as familial hypercholesterolemia.

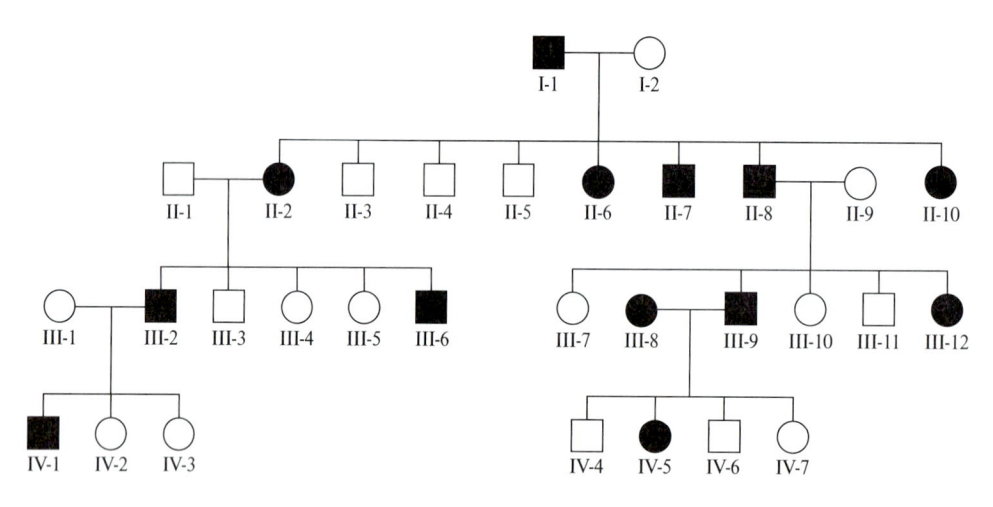

This disorder is characterized by an elevated level of serum cholesterol in the blood. Though relatively rare, this genetic abnormality can be a contributing factor to heart attacks. At the molecular level, this disease is caused by a mutant gene that encodes a protein called low-density lipoprotein receptor (LDLR). In the bloodstream, serum cholesterol is bound to a carrier protein known as low-density lipoprotein (LDL). LDL binds to LDLR, which enables cells to absorb cholesterol. When LDLR is defective, it becomes more difficult for the cells to absorb cholesterol. This explains why the blood level of cholesterol remains high. Based on the pedigree, what is the most likely pattern of inheritance of this disorder?

Topic: What topic in genetics does this question address?

The topic is inheritance patterns in humans. More specifically, the question is about identifying the inheritance pattern of familial hypercholesterolemia.

Information: What information do you know based on the question and your understanding of the topic?

In the question, you are given a pedigree to analyze. From your understanding of the topic, you may remember that single-gene disorders can follow an autosomal recessive, autosomal dominant, X-linked recessive, or X-linked dominant pattern of inheritance.

Problem-Solving **S**trategy: Predict the outcome. Compare and contrast.

One strategy to solve this problem is to consider what the different patterns of inheritance predict with regard to the traits found in parents and offspring. You can compare and contrast the four patterns described in Section 22.1, and see if you can rule any of them out.

Answer: The pedigree is consistent with an autosomal dominant pattern of inheritance. An affected individual always has an affected parent. We can rule out the other possible patterns as follows:

The pattern can't be autosomal recessive, because two affected parents (III-8 and III-9) produced some unaffected offspring.

It can't be X-linked recessive, because an affected mother, III-8 (who would have to be homozygous), produced unaffected sons.

It can't be X-linked dominant, because an affected father, III-2, produced unaffected daughters.

2. Oncogenes sometimes result from genetic rearrangements (e.g., translocations) that produce gene fusions. An example occurs with the Philadelphia chromosome, in which a reciprocal translocation between chromosomes 9 and 22 leads to fusion of the first part of the *bcr* gene with the *abl* gene. Suggest two different reasons why a gene fusion can create an oncogene.

Topic: *What topic in genetics does this question address?*

The topic is chromosomal rearrangements and how they may affect gene expression. More specifically, the question is about the translocation that produces a Philadelphia chromosome.

Information: *What information do you know based on the question and your understanding of the topic?*

From the question, you know that the Philadelphia chromosome has a fusion of the *bcr* and *abl* genes due to a chromosomal translocation. From your understanding of the topic, you may remember that the promoter controls the transcription of a gene and that the coding sequence determines the structure and function of the protein.

Problem-Solving Strategy: *Relate structure and function. Predict the outcome.*

One strategy to solve this problem is to relate the structural change in the gene to possible changes in gene expression and protein function.

Answer: An oncogene arises from a genetic change that abnormally activates the expression of a gene that plays a role in cell division. When a genetic change creates a gene fusion, this can abnormally activate the expression of a gene in two ways.

The first way is at the level of transcription. The promoter and part of the coding sequence of one gene may become fused with the coding sequence of the other gene. In this example, the promoter and part of the coding sequence of the *bcr* gene may fuse with the coding sequence of the *abl* gene. After this has occurred, the *abl* gene is under the control of the *bcr* promoter, rather than its normal promoter. Because the *bcr* promoter is turned on in different cells than the *abl* promoter is, overexpression of *abl* occurs in certain cell types where its gene product is not normally found.

A second way that a gene fusion can cause abnormal activation is at the level of protein structure. A fusion protein has parts of two different polypeptides. One portion of a fusion protein may affect the structure of the second portion in such a way that the second portion becomes abnormally active, or vice versa.

Conceptual Questions

C1. With regard to pedigree analysis, make a list of the observations that distinguish recessive, dominant, and X-linked patterns of inheritance.

C2. Explain, at the molecular level, why human genetic diseases often follow simple Mendelian patterns of inheritance, whereas most normal traits, such as the shape of your nose or the size of your head, are governed by multiple gene interactions.

C3. Many genetic disorders exhibit locus heterogeneity. Define and give two examples of locus heterogeneity. How does locus heterogeneity confound a pedigree analysis?

C4. In general, why do changes in chromosome structure or number tend to affect an individual's phenotype? Explain why some changes in chromosome structure, such as reciprocal translocations, do not.

C5. We often speak of diseases such as phenylketonuria (PKU) and achondroplasia as having a genetic basis. Explain whether the following statements are accurate with regard to the genetic basis of any human disease (not just PKU and achondroplasia).

    A. An individual must inherit two copies of a mutant allele to have disease symptoms.

    B. A genetic predisposition means that an individual has inherited one or more alleles that make it more likely that he or she will develop disease symptoms than other individuals in a population.

    C. A genetic predisposition to develop a disease may be passed from parents to offspring.

    D. The genetic basis for a disease is always more important than the effect of the environment.

C6. Figure 22.1 illustrates albinism in two different species. Describe two other genetic disorders found in both humans and animals.

C7. Discuss why a genetic disease might have a particular age of onset. Would an infectious disease have an age of onset? Explain why or why not.

C8. Gaucher disease (type I) is due to a defect in a gene that encodes a protein called acid β glucosidase. This enzyme plays a role in carbohydrate metabolism within lysosomes. The gene is located on the long arm of chromosome 1. People who inherit two defective copies of this gene exhibit Gaucher disease, the major symptoms of which include an enlarged spleen, bone lesions, and changes in skin pigmentation. Let's suppose a phenotypically unaffected woman, whose father had Gaucher disease, has a child with a phenotypically unaffected man, whose mother had Gaucher disease.

    A. What is the probability that this child will have the disease?

    B. What is the probability that this child will have two normal copies of this gene?

    C. If this couple has five children, what is the probability that one of them will have Gaucher disease and four will be phenotypically unaffected?

C9. Ehler-Danlos syndrome is a rare disorder caused by a mutation in a gene that encodes a protein called collagen (type 3 A1). Collagen is found in the extracellular matrix that plays an important role in the formation of skin, joints, and other connective tissues. People with Ehler-Danlos syndrome have extraordinarily flexible skin and very loose joints. The following pedigree contains several individuals affected with this syndrome, shown with black symbols. Based on this pedigree, does the syndrome follow autosomal recessive, autosomal dominant, X-linked recessive, or X-linked dominant inheritance? Explain your reasoning.

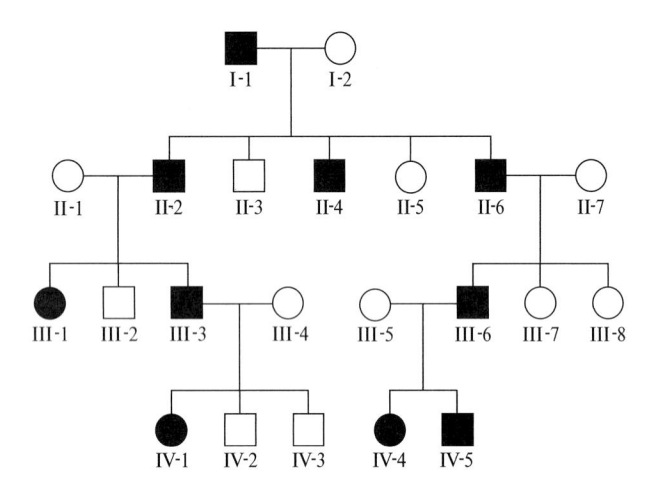

C10. Hurler syndrome is due to a mutation in a gene that encodes a protein called α-L-iduronidase. This protein functions within lysosomes as an enzyme that breaks down mucopolysaccharides (a type of polysaccharide that has many acidic groups attached). When this enzyme is defective, excessive amounts of the mucopolysaccharides dermatan sulfate and heparin sulfate accumulate within the lysosomes, especially in liver cells and connective tissue cells. This accumulation leads to symptoms such as an enlarged liver and spleen, bone abnormalities, corneal clouding, heart problems, and severe neurological problems. The following pedigree contains three individuals affected with Hurler syndrome, indicated with black symbols. Based on this pedigree, does this syndrome appear to follow autosomal recessive, autosomal dominant, X-linked recessive, or X-linked dominant inheritance? Explain your reasoning.

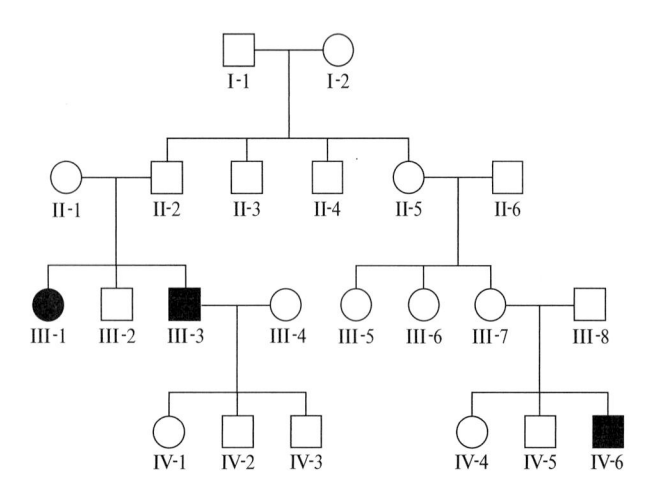

C11. Like Hurler syndrome, Fabry disease involves an abnormal accumulation of substances within lysosomes. However, the lysosomes of individuals with Fabry disease show an abnormal accumulation of lipids. The defective enzyme is α-galactosidase A, which is a lysosomal enzyme that functions in lipid metabolism. The enzymatic defect causes cell damage, especially to the kidneys, heart, and eyes. The gene that encodes α-galactosidase A is found on the X chromosome. Let's suppose a phenotypically unaffected couple produces two sons with Fabry disease and one phenotypically unaffected daughter. What is the probability that the daughter will have an affected son?

C12. Achondroplasia is a rare form of dwarfism caused by an autosomal dominant mutation that affects the gene that encodes a fibroblast growth factor receptor. Among 1,422,000 live births, the number of babies born with achondroplasia was 31. Among those 31 babies, 18 of them had one parent with achondroplasia. The remaining babies had two unaffected parents. How do you explain those 13 babies, assuming that the mutant allele has 100% penetrance? What are the odds that these 13 individuals will pass this mutant gene to their offspring?

C13. Lesch-Nyhan syndrome is due to a mutation in a gene that encodes a protein called hypoxanthine-guanine phosphoribosyltransferase (HPRT). HPRT is an enzyme that functions in purine metabolism. People afflicted with this syndrome have severe neurodegeneration and loss of motor control. The following pedigree contains several individuals with Lesch-Nyhan syndrome, shown with black symbols. Based on this pedigree, does this syndrome appear to be inherited via an autosomal recessive, autosomal dominant, X-linked recessive, or X-linked dominant pattern? Explain your reasoning.

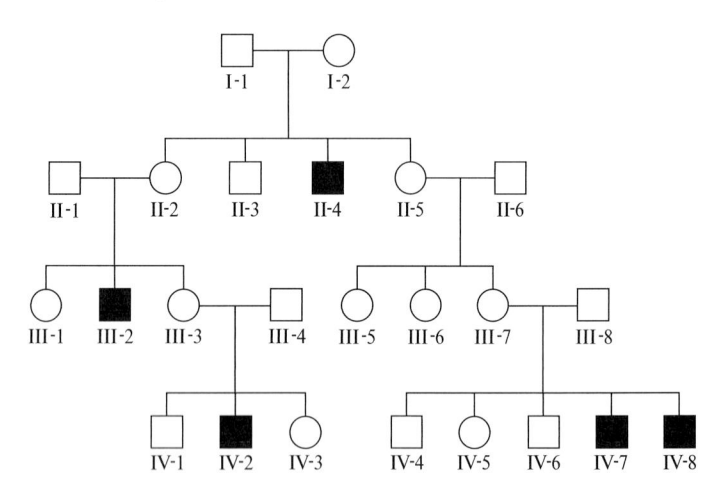

C14. Marfan syndrome is due to a mutation in a gene that encodes a protein called fibrillin-1. The syndrome is inherited as a dominant trait. The fibrillin-1 protein is the main constituent of extracellular microfibrils. These microfibrils can exist as individual fibers or associate with a protein called elastin to form elastic fibers. People with the disorder tend to be unusually tall with long limbs, and they may have defects in their heart valves and aorta. Let's suppose a phenotypically unaffected woman has a child with a man who has Marfan syndrome.

A. What is the probability this child will have the disease?

B. If this couple has three children, what is the probability that none of them will have Marfan syndrome?

C15. Sandhoff disease is due to a mutation in a gene that encodes a protein called hexosaminidase B. This disease has symptoms that are similar to those of Tay-Sachs disease. Weakness begins in the first 6 months of life. Individuals exhibit early blindness and progressive mental and motor deterioration. The following pedigree contains three individuals with Sandhoff disease, indicated with black symbols.

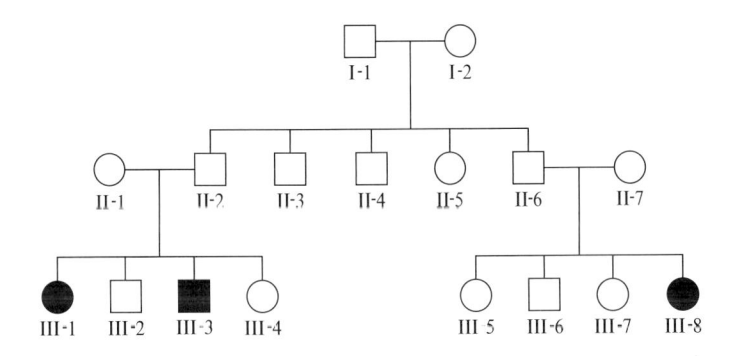

A. Based on this pedigree, does this syndrome appear to follow autosomal recessive, autosomal dominant, X-linked recessive, or X-linked dominant inheritance? Explain your reasoning.

B. What is the likelihood that II-1, II-2, II-3, II-4, II-5, II-6, and II-7 carry a mutant allele of the gene encoding hexosaminidase B?

C16. Describe the two assumptions that underlie the identification of disease-causing alleles via haplotypes.

C17. What is the difference between an oncogene and a tumor-suppressor gene? Give two examples of each type of gene.

C18. What is a proto-oncogene? What are the typical functions of proteins encoded by proto-oncogenes? At the level of protein function, what are the general ways that proto-oncogenes can be converted to oncogenes?

C19. A genetic predisposition to developing cancer is usually inherited as a dominant trait. At the level of cellular function, are the alleles involved actually dominant? Explain why some individuals who have inherited these dominant alleles do not develop cancer during their lifetimes.

C20. Describe the types of genetic changes that commonly convert a proto-oncogene to an oncogene. Give three examples. Explain how the genetic changes are expected to alter the activity of the gene product.

C21. Relatively few inherited forms of cancer involve the inheritance of mutant oncogenes. Instead, most inherited forms of cancer are defects in tumor-suppressor genes. Give two or more reasons why inherited forms of cancer seldom involve activated oncogenes.

C22. The *rb* gene encodes a protein that inhibits E2F, a transcription factor that activates several genes involved in cell division. Mutations in *rb* are associated with certain forms of cancer, such as retinoblastoma. Under each of the following conditions, would you expect cancer to occur?

A. One copy of the *rb* gene is defective; both copies of the *E2F* gene are functional.

B. Both copies of the *rb* gene are defective; both copies of the *E2F* gene are functional.

C. Both copies of the *rb* gene are defective; one copy of the *E2F* gene is defective.

D. Both copies of the *rb* gene and the *E2F* gene are defective.

C23. A knockout mouse in which both copies of the *p53* gene are defective has been produced by researchers. This type of mouse appears normal at birth. However, it is highly sensitive to UV light. Based on your knowledge of *p53*, explain the normal appearance at birth and the high sensitivity to UV light.

C24. With regard to cancer cells, which of the following statements are *true*?

A. Cancer cells are clonal, which means they are derived from a single mutant cell.

B. To become cancerous, cells usually accumulate multiple genetic changes that eventually result in uncontrolled growth.

C. Most cancers are caused by oncogene-carrying viruses.

D. Cancer cells have lost the ability to properly regulate cell division.

C25. When the DNA of a human cell becomes damaged, the *p53* gene is activated. What is the general function of the p53 protein? Is it an enzyme, transcription factor, cell cycle protein, or something else? Describe three ways in which the synthesis of the p53 protein affects cellular function. Why is it beneficial for these three things to happen when a cell's DNA has been damaged?

C26. How can environmental agents that do not cause gene mutations contribute to cancer? Are epigenetic changes that are associated with cancer passed to offspring?

Application and Experimental Questions

E1. Which of the following experimental observations suggest that a disease has a genetic basis?

A. The frequency of the disease is lower in relatives who live apart than in relatives who live together.

B. The frequency of the disease is unusually high in a small group of genetically related individuals who live in southern Spain.

C. The disease symptoms usually begin around the age of 40.

D. It is more likely that both monozygotic twins will be affected by the disease than both dizygotic twins will be.

E2. Section 22.1 described seven types of experimental observations that suggest that a disease is inherited. Which of these observations do you find the least convincing? Which do you find the most convincing? Explain your answer.

E3. What is meant by the term *genetic testing*? How do testing at the protein level and testing at the DNA level differ? Describe five different techniques used in genetic testing.

E4. A particular disease occurs within a group of South American Indians. During the 1920s, many of these people migrated to Central America. In the Central American group, the disease is never observed. Discuss whether or not you think the disease has a genetic component. What types of further observations would you make?

E5. Chapter 20 describes a method known as Western blotting that can be used to detect a polypeptide that is translated from a particular mRNA. In this method, a particular polypeptide is detected by an antibody that specifically recognizes a segment of the polypeptide's amino acid sequence. After the antibody binds to the polypeptide within a gel, a secondary antibody (which is labeled) is used to visualize the polypeptide as a dark band. For example, an antibody that recognizes α-galactosidase A could be used to specifically detect the amount of α-galactosidase A on a gel. The enzyme α-galactosidase A is defective in individuals with Fabry disease, which follows an X-linked recessive pattern of inheritance. Amy, Nan, and Pete are siblings, and Pete has Fabry disease. Aileen, Jason, and Jerry are brothers and sister, and Jerry has Fabry disease. Amy, Nan, and Pete are not related to Aileen, Jason, and Jerry. Amy, Nan, and Aileen are concerned that they could be carriers of a defective α-galactosidase A gene. A sample of cells was obtained from each of these six individuals and subjected to Western blotting, using an antibody against α-galactosidase A. Samples were also obtained from two unrelated and unaffected individuals (lanes 7 and 8). The results are shown here.

Samples from:
Lane 1. Amy
Lane 2. Nan
Lane 3. Pete
Lane 4. Aileen
Lane 5. Jason
Lane 6. Jerry
Lane 7. Unaffected male
Lane 8. Unaffected female

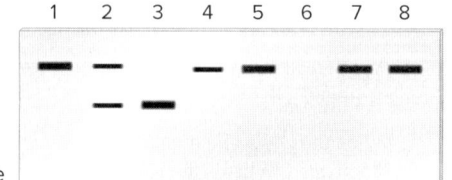

Note: Due to X-chromosome inactivation in females, the amount of expression of genes on the single X chromosome in males is equal to the amount of expression from genes on both X chromosomes in females.

A. Explain the type of mutation (e.g., missense, nonsense, promoter, etc.) that caused Fabry disease in Pete and Jerry.

B. What would you tell Amy, Nan, and Aileen regarding the likelihood that they are carriers of the mutant allele and the probability of having affected offspring?

E6. An experimental assay for the blood clotting protein called factor IX is available. A blood sample was obtained from each individual of the following pedigree. The amount of factor IX protein, shown within each symbol on the pedigree, is expressed as a percentage of the average amount observed in individuals who do not carry a mutant copy of the gene.

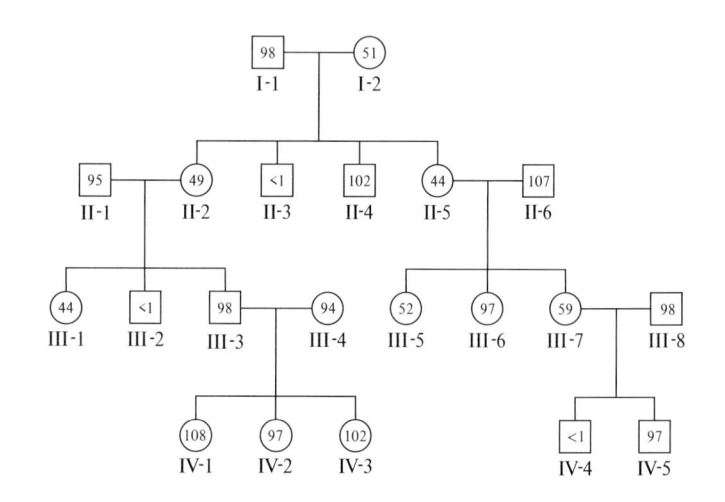

What is the likely genotype of each person in this pedigree?

E7. Discuss ways to distinguish whether a particular form of cancer involves an inherited predisposition or is due strictly to (postzygotic) somatic mutations. In your answer, consider that only one mutation may be inherited, but the cancer might develop only after several somatic mutations.

E8. The codon change (Gly-12 to Val-12) in human *ras*H that converts it to oncogenic *ras*H has been associated with many types of cancers. For this reason, researchers would like to develop drugs to inhibit oncogenic *ras*H. Based on your understanding of the Ras protein, what types of drugs might you develop? In other words, what would be the structure of the drugs, and how would they inhibit Ras protein? How would you test the efficacy of the drugs? What might be some side effects?

E9. Explain how DNA microarrays are used in molecular profiling of cancerous tumors.

Questions for Student Discussion/Collaboration

1. Make a list of the benefits that may arise from genetic testing as well as possible negative consequences. Discuss the items on your list.

2. Our government has finite funds to devote to cancer research. Discuss which of the following areas of cancer research you think should receive the most funding.

A. Identifying and characterizing oncogenes and tumor-suppressor genes

B. Identifying agents in our environment that cause cancer

C. Identifying viruses that cause cancer

D. Devising methods aimed at killing cancer cells in the body

E. Informing the public of the risks involved in exposure to carcinogens

In the long run, in which of these areas would you expect successful research to be the most effective in decreasing human mortality due to cancer?

3. Go to the PubMed website and search using the words *epigenetic* and *cancer*. Scan through the journal articles you retrieve, and make a list of environmental agents that may cause epigenetic changes that contribute to cancer.

Answers to Comprehension Questions

22.1: b, a, b, b

22.2: c, c

22.3: a, c

22.4: b

22.5: d

22.6: c, d, d

22.7: d, e

22.8: d

Note: All answers appear in Connect; the answers to even-numbered questions and all Concept Check questions are in Appendix B.

23

CHAPTER OUTLINE

- 23.1 Genes in Populations and the Hardy-Weinberg Equation
- 23.2 Overview of Microevolution
- 23.3 Natural Selection
- 23.4 Genetic Drift
- 23.5 Migration
- 23.6 Nonrandom Mating
- 23.7 Sources of New Genetic Variation

The African cheetah. This species has a relatively low level of genetic variation because the population was reduced to a small size approximately 10,000 to 12,000 years ago.
©Riccardo Vallini Pics/Getty Images

POPULATION GENETICS

Until now, we have primarily focused our attention on genes within individuals and their related family members. In this chapter and Chapter 24, we turn to the study of genes in a population or species. The field of **population genetics** is concerned with changes in genetic variation within a group of individuals over time. Population geneticists want to know the extent of genetic variation within populations, why it exists, and how it changes over the course of many generations. The field of population genetics emerged as a branch of genetics in the 1920s and 1930s. Its mathematical foundations were developed by theoreticians who extended the principles of Gregor Mendel and Charles Darwin by deriving formulas to explain the occurrence of genotypes within populations. These foundations can be largely attributed to three scientists: Ronald Fisher, Sewall Wright, and J. B. S. Haldane. As we will see, support for their mathematical theories was provided by several researchers who analyzed the genetic composition of natural and experimental populations. More recently, population geneticists have used techniques to probe genetic variation at the molecular level. In addition, staggering advances in computer technology have aided population geneticists in the analysis of their genetic theories and data. In this chapter, we will explore the genetic variation that occurs in populations and consider the reasons why the genetic composition of a population may change over the course of several generations.

23.1 GENES IN POPULATIONS AND THE HARDY-WEINBERG EQUATION

Learning Outcomes:
1. Define *gene pool* and *population*.
2. Describe the extent of polymorphism in natural populations.
3. Use the Hardy-Weinberg equation to calculate allele and genotype frequencies.

In the field of population genetics, the focus shifts away from the individual and onto the population of which the individual is a member. Population genetics may seem like a significant departure from other topics in this text, but it is a direct extension of our understanding of Mendel's laws of inheritance, molecular genetics, and the ideas of Darwin. Conceptually, all of the alleles of every gene in a population make up the **gene pool.** In this regard, each member of the population is viewed as receiving its genes from its parents, which, in turn, are members of the gene pool. Furthermore, individuals that reproduce contribute to the gene pool of the next generation. Population geneticists study the genetic variation within the gene pool and how such variation changes from one generation to the next. The emphasis is often on allelic variation. In this section, we will examine some of the general features of populations and gene pools.

A Population Is a Group of Interbreeding Individuals That Share a Gene Pool

In genetics, the term *population* has a very specific meaning. With regard to sexually reproducing species, a **population** is a group of individuals of the same species that occupy the same region and can interbreed with one another. Many species occupy a wide geographic range and are divided into discrete populations. For example, distinct populations of a given species may be located on different continents, or populations on the same continent may be divided by a geographical feature such as a large mountain range.

A large population usually is composed of smaller groups called **local populations.** The members of a local population are far more likely to breed among themselves than with members of a more distant population. Local populations are often separated from each other by moderate geographic barriers. As an example, the large ground finch (*Geospiza magnirostris*) is found on a small volcanic island called Daphne Major, which is one of the Galápagos Islands (**Figure 23.1**). Daphne Major is located northwest of the much larger Santa Cruz Island. The population of large ground finches on Daphne Major constitutes a local population of this species. Breeding is much more apt to occur among members of a local population than between members of neighboring populations. On relatively rare occasions, however, a bird may fly from Daphne Major to Santa Cruz Island, which would make breeding between the two different local populations possible.

(a) Large ground finch

(b) A view of Daphne Major (the small island in the distance) from Santa Cruz Island

FIGURE 23.1 A local population of the large ground finch. **(a)** The large ground finch (*Geospiza magnirostris*) on Daphne Major. **(b)** A view of Daphne Major, one of the Galápagos Islands, from Santa Cruz Island.

(a) ©LABETAA Andre/Shutterstock; (b) ©Deborah Freund

Concept Check: What does the term local population mean?

At the Population Level, Some Genes May Be Monomorphic, but Most Are Polymorphic

In population genetics, the term **genetic polymorphism,** or simply **polymorphism** (meaning "many forms"), refers to the observation that many inherited traits display variation within a population. Historically, genetic polymorphism first referred to variation in inherited traits that are observable with the naked eye. Polymorphisms in color and pattern have long attracted the attention of population geneticists. Some of the well-studied variations include yellow and red varieties of the elder-flowered orchid and brown, pink, and yellow shells in land snails, which are discussed later in this chapter. **Figure 23.2** illustrates a striking example of polymorphism in the Hawaiian happy-face spider (*Theridion grallator*). The three individuals shown in this figure are from the same species, but they differ in alleles that affect color and pattern.

FIGURE 23.2 Polymorphism in the Hawaiian happy-face spider.

Genes → Traits These three spiders are members of the same species and carry the same genes. However, several genes that affect pigmentation patterns are polymorphic, meaning that more than one allele occurs in each gene within the population. This polymorphism within the Hawaiian happy-face spider population produces members that look quite different from each other.
©Geoff Oxford

Concept Check: Are polymorphisms common or rare in natural populations?

What is the underlying cause of polymorphism? At the DNA level, polymorphism may be due to two or more alleles that influence the phenotype of the individual that inherits them. In other words, it is due to genetic variation. Geneticists also use the term **polymorphic** to describe a gene that commonly exists as two or more alleles in a population. By comparison, a **monomorphic** gene exists predominantly as a single allele in a population. By convention, when a single allele is found in at least 99% of all cases, the gene is considered monomorphic.

At the level of a particular gene, polymorphism may involve various types of changes such as a deletion of a significant region of the gene, a duplication of a region, or a change in a single nucleotide. This last phenomenon is called a **single-nucleotide polymorphism (SNP)**. SNPs are the smallest type of genetic change that can occur within a given gene and are also the most common. In humans, for example, SNPs represent 90% of all the variation in DNA sequences that occurs among different people. SNPs are found very frequently in genes. In the human population, a gene that is 2000 to 3000 bp in length contains, on average, 10 different sites that are polymorphic. The high frequency of SNPs indicates that polymorphism is the norm for most human genes. Likewise, relatively large, healthy populations of nearly all species exhibit a high level of genetic variation as evidenced by the occurrence of SNPs within most genes.

Within a population, the alleles of a given gene may arise through different types of genetic changes. **Figure 23.3** considers a gene that exists in multiple forms in humans. This example is a short segment of DNA found within the human β-globin gene. The top sequence is an allele designated Hb^A, whereas the middle sequence is called Hb^S. These alleles differ from each other by a single nucleotide, so they provide an example of a SNP. As discussed in Chapter 5, the Hb^S allele causes sickle cell disease in a homozygote. The bottom sequence contains a short, 5-bp deletion compared with the other two alleles. This deletion results in a nonfunctional β-globin polypeptide. Therefore, the bottom sequence is an example of a loss-of-function allele.

Population Genetics Is Concerned with Allele and Genotype Frequencies

As we have seen, population geneticists want to understand the prevalence of polymorphic genes within populations. Their goal is to identify the causative factors that govern changes in genetic variation. Much of their work evaluates the frequency of alleles in a quantitative way. Calculations of two fundamental values are central to population genetics: **allele frequencies** and **genotype frequencies.** The allele and genotype frequencies are defined as

$$\text{Allele frequency} = \frac{\text{Number of copies of an allele in a population}}{\text{Total number of all alleles for that gene in a population}}$$

$$\text{Genotype frequency} = \frac{\text{Number of individuals with a particular genotype in a population}}{\text{Total number of individuals in a population}}$$

Though these two frequencies are related, a clear distinction between them must be kept in mind. As an example, let's consider a hypothetical population of 100 frogs with the following genotypes:

64 dark green frogs with genotype GG

32 medium green frogs with genotype Gg

4 light green frogs with genotype gg

When calculating an allele frequency, homozygous individuals have two copies of an allele, whereas heterozygotes have only one. For example, in tallying the g allele, each of the 32 heterozygotes has one copy of the g allele, and each light green frog has two copies. The allele frequency for g equals

$$g = \frac{32 + 2(4)}{2(64) + 2(32) + 2(4)}$$

$$= \frac{40}{200} = 0.2, \text{ or } 20\%$$

This result tells us that the allele frequency of g is 20%. In other words, 20% of the alleles for this gene in the population are the g allele.

Let's now calculate the genotype frequency of gg (light green) frogs:

$$gg = \frac{4}{64 + 32 + 4}$$

$$= \frac{4}{100} = 0.04, \text{ or } 4\%$$

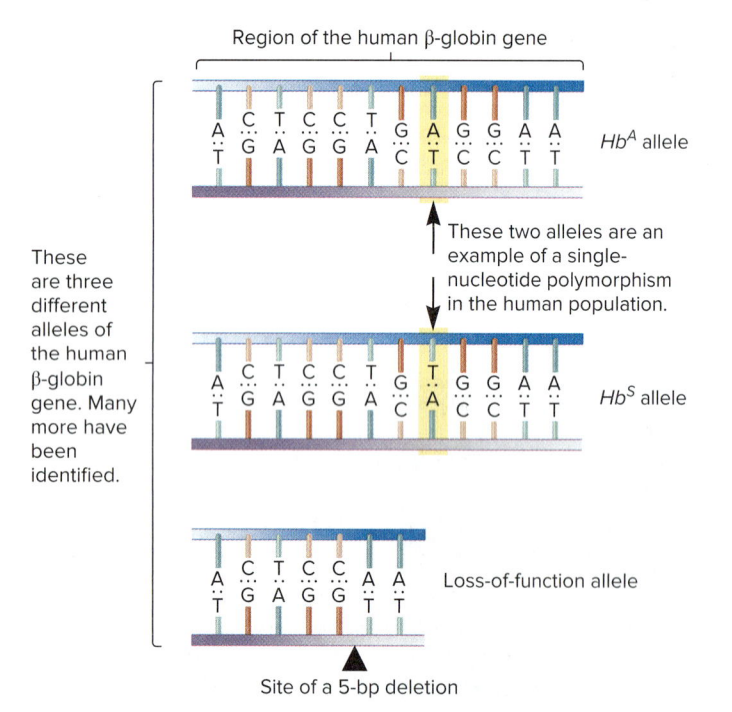

FIGURE 23.3 **The relationship between alleles and various types of mutations.** The DNA sequence shown here is a small portion of the β-globin gene in humans. Mutations have altered the gene to create the three different alleles in this figure. The top two alleles differ by a single base pair and thus exhibit what is known as a single-nucleotide polymorphism (SNP). The bottom allele has a 5-bp deletion that begins right after the arrowhead. The deletion results in a nonfunctional polypeptide, so the allele is a loss-of-function allele.

We see that 4% of the individuals in this population are light green frogs.

Allele and genotype frequencies are always less than or equal to 1 (i.e., ≤100%). If a gene is monomorphic, the allele frequency for the single allele will equal or be close to 1.0. For polymorphic genes, if we add up the frequencies for all of the alleles in the population, we should obtain a value of 1.0. In our frog example, the allele frequency of g equals 0.2. The frequency of the other allele, G, equals 0.8. If we add the two together, we obtain a value of $0.2 + 0.8 = 1.0$.

The Hardy-Weinberg Equation Can Be Used to Calculate Genotype Frequencies Based on Allele Frequencies

In 1908, a British mathematician, Godfrey Harold Hardy, and a German physician, Wilhelm Weinberg, independently derived a simple mathematical expression that predicts stability of allele and genotype frequencies from one generation to the next. The maintenance of stability of these frequencies is called **Hardy-Weinberg equilibrium,** because (under a given set of conditions, described later in this section) the allele and genotype frequencies do not change over the course of many generations.

Why is Hardy-Weinberg equilibrium a useful concept? An equilibrium is a null hypothesis, which suggests that evolutionary change is not occurring. In reality, however, populations rarely achieve an equilibrium. Therefore, the main usefulness of Hardy-Weinberg equilibrium is that it provides a framework that can be used to understand changes in allele and genotype frequencies within a population when such an equilibrium is violated.

To appreciate Hardy-Weinberg equilibrium, let's return to our hypothetical frog example in which a gene is polymorphic and exists as two different alleles: G and g. If the allele frequency of G is denoted by the variable p and the allele frequency of g is denoted by q, then

$$p + q = 1$$

For example, if $p = 0.8$, then q must be 0.2. In other words, if the allele frequency of G equals 80%, the remaining 20% of alleles must be g, because together the allele frequencies must equal 100%.

The **Hardy-Weinberg equation** is used to relate allele frequencies and genotype frequencies. For a diploid species, each individual inherits two copies of most genes. The Hardy-Weinberg equation assumes that the alleles for the next generation for any given individual are chosen randomly and independently of each other. Therefore, we can use the product rule (see Chapter 3) and multiply the sum, $p + q$, by itself. Because $p + q = 1$, we know that the product of these sums also equals 1:

$$(p + q)(p + q) = 1$$
$$p^2 + 2pq + q^2 = 1 \text{ (Hardy-Weinberg equation)}$$

This equation applies to a gene in a diploid species that is found in only two alleles, which exist at frequencies designated p and q.

The Hardy-Weinberg equation predicts an equilibrium—unchanging allele and genotype frequencies from generation to generation—if certain conditions are met in a population. With regard to a particular gene of interest, these conditions are as follows:

1. *No new mutations:* The gene of interest incurs no new mutations.
2. *No genetic drift:* The population is so large that allele frequencies do not change due to random fluctuations.
3. *No migration:* Individuals do not travel between different populations.
4. *No natural selection:* All of the genotypes have the same reproductive success.
5. *Random mating or breeding:* With respect to the gene of interest, the members of the population reproduce with each other randomly, without regard to their phenotypes and genotypes. (Note: The term *mating* is used to describe sexual reproduction among animals. In the case of plants, the term *breeding* is commonly used.)

If the Hardy-Weinberg equation is applied to our hypothetical frog population, in which a gene exists in alleles designated G and g, then

p^2 equals the genotype frequency of GG
$2pq$ equals the genotype frequency of Gg
q^2 equals the genotype frequency of gg

If $p = 0.8$ and $q = 0.2$ and if the population is in Hardy-Weinberg equilibrium, then

$$GG = p^2 = (0.8)^2 = 0.64$$
$$Gg = 2pq = 2(0.8)(0.2) = 0.32$$
$$gg = q^2 = (0.2)^2 = 0.04$$

In other words, if the allele frequency of G is 80% and the allele frequency of g is 20%, the genotype frequency of GG is 64%, Gg is 32%, and gg is 4%.

To illustrate the relationship between allele frequencies and genotypes, **Figure 23.4** compares the values obtained with the Hardy-Weinberg equation with the way that gametes combine randomly with each other to produce offspring. In a population, the

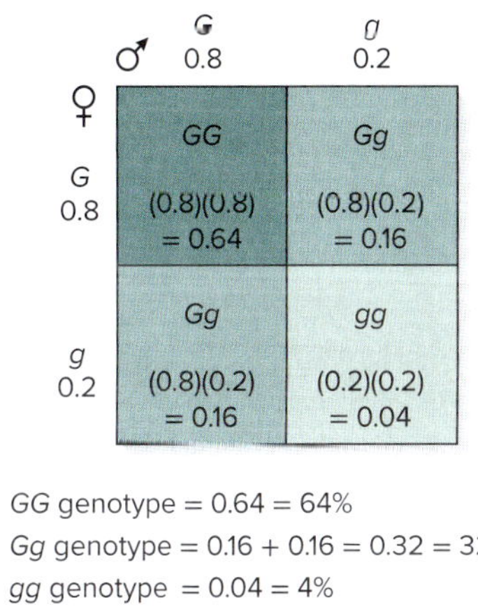

	♂ G 0.8	♂ g 0.2
♀ G 0.8	GG $(0.8)(0.8) = 0.64$	Gg $(0.8)(0.2) = 0.16$
♀ g 0.2	Gg $(0.8)(0.2) = 0.16$	gg $(0.2)(0.2) = 0.04$

GG genotype $= 0.64 = 64\%$
Gg genotype $= 0.16 + 0.16 = 0.32 = 32\%$
gg genotype $= 0.04 = 4\%$

FIGURE 23.4 A comparison between allele frequencies and the union of alleles in a Punnett square. In a population in Hardy-Weinberg equilibrium, the frequency of gametes carrying a particular allele is equal to the allele frequency in the population.

frequency of a gamete carrying a particular allele is equal to the allele frequency in that population. In this example, the frequency of a gamete carrying the *G* allele equals 0.8.

We can use the product rule to determine the frequency of genotypes. For example, the frequency of producing a *GG* homozygote is $0.8 \times 0.8 = 0.64$, or 64%. The probability of inheriting both *g* alleles is $0.2 \times 0.2 = 0.04$, or 4%. As seen in Figure 23.4, heterozygotes can be produced in two different ways. An offspring could inherit the *G* allele from its father and *g* from its mother, or *G* from its mother and *g* from its father. Therefore, the frequency of heterozygotes is $pq + pq$, which equals $2pq$; in our example, this is $2(0.8)(0.2) = 0.32$, or 32%.

The Hardy-Weinberg equation provides a quantitative relationship between allele and genotype frequencies in a population. **Figure 23.5** describes this relationship for different allele frequencies of *g* and *G*. As expected, when the allele frequency of *g* is very low, the *GG* genotype predominates; when the *g* allele frequency is high, the *gg* homozygote is most prevalent in the population. When the allele frequencies of *g* and *G* are intermediate in value, the heterozygote predominates.

In reality, no population satisfies Hardy-Weinberg equilibrium completely. Nevertheless, in large natural populations with little migration and negligible natural selection, Hardy-Weinberg equilibrium may be nearly approximated for certain genes.

The Hardy-Weinberg equation can be expanded to include three or more alleles. For example, let's consider a situation in which a gene exists as three alleles: *A1*, *A2*, and *A3*. The allele frequency of *A1* is designated by the letter *p*, *A2* by the letter *q*, and *A3* by the letter *r*. Under these circumstances, the Hardy-Weinberg equation becomes

$$(p + q + r)^2 = 1$$
$$p^2 + q^2 + r^2 + 2pq + 2pr + 2qr = 1$$

where

p^2 is the genotype frequency of *A1A1*

q^2 is the genotype frequency of *A2A2*

r^2 is the genotype frequency of *A3A3*

$2pq$ is the genotype frequency of *A1A2*

$2pr$ is the genotype frequency of *A1A3*

$2qr$ is the genotype frequency of *A2A3*

The Hardy-Weinberg Equation Can Be Used to Detect Evolutionary Change

As discussed in Chapter 3, the chi square test is used to evaluate the agreement between observed and expected data. We can use a chi square test to determine whether a population exhibits Hardy-Weinberg equilibrium for a particular gene. If a population is not in equilibrium, this result suggests evolutionary change. To carry out this calculation, it is necessary to distinguish between homozygotes and heterozygotes, either phenotypically or at the molecular level. This distinction is necessary so that we can determine both the allele and genotype frequencies. As an example, let's consider a human blood type called the MN type. In this case, the blood type is determined by two codominant alleles, *M* and *N*. In an Inuit population in East Greenland, it was found that among 200 people, 168 were *MM*, 30 were *MN*, and 2 were *NN*. We can use these observed data to calculate the expected number of each genotype based on the Hardy-Weinberg equation:

$$\text{Allele frequency of } M = \frac{2(168) + 30}{400} = 0.915$$

$$\text{Allele frequency of } N = \frac{2(2) + 30}{400} = 0.085$$

Expected frequency of $MM = p^2 = (0.915)^2 = 0.837$

Expected number of *MM* individuals = $0.837 \times 200 = 167.4$
(or 167 if rounded to the nearest individual)

Expected frequency of $NN = q^2 = (0.085)^2 = 0.007$

Expected number of *NN* individuals = $0.007 \times 200 = 1.4$
(or 1 if rounded to the nearest individual)

Expected frequency of $MN = 2pq = 2(0.915)(0.085) = 0.156$

Expected number of *MN* individuals = $0.156 \times 200 = 31.2$
(or 31 if rounded to the nearest individual)

$$\chi^2 = \frac{(O_1 - E_1)^2}{E_1} + \frac{(O_2 - E_2)^2}{E_2} + \frac{(O_3 - E_3)^2}{E_3}$$

$$\chi^2 = \frac{(168 - 167)^2}{167} + \frac{(30 - 31)^2}{31} + \frac{(2 - 1)^2}{1}$$

$$\chi^2 = 1.04$$

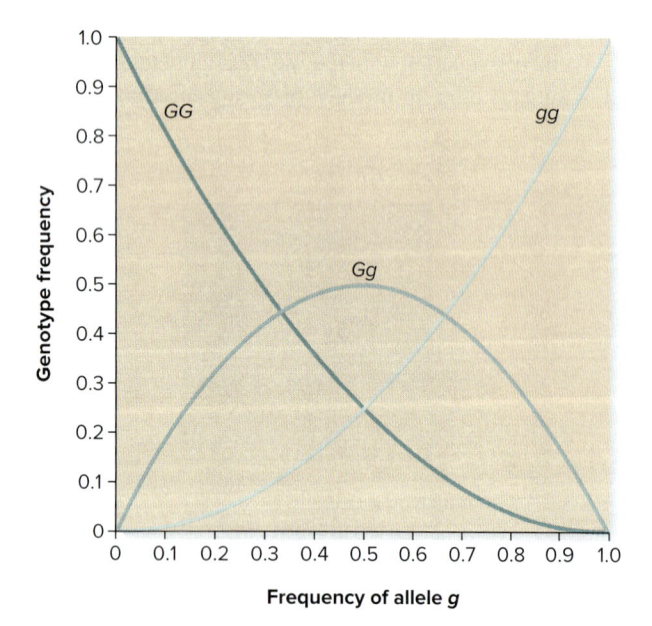

FIGURE 23.5 The relationship between allele frequencies and genotype frequencies according to the Hardy-Weinberg equation. This graph assumes that *g* and *G* are the only two alleles for this gene.

We need to refer back to Table 3.2 in Chapter 3 to evaluate the calculated chi square value. Because the gene exists in two alleles, the degrees of freedom equals 1. For a chi square value of 1.04, the P value is between 0.5 and 0.2, which is well within the acceptable range. Therefore, we fail to reject the null hypothesis. In this case, the alleles for this gene appear to be in Hardy-Weinberg equilibrium.

When researchers have investigated other genes in various populations, a high chi square value is sometimes obtained, and the hypothesis that the allele and genotype frequencies are in Hardy-Weinberg equilibrium is rejected. In these cases, we would say that the population is in **disequilibrium.** Deviation from Hardy-Weinberg equilibrium indicates evolutionary change. Factors such as natural selection, genetic drift, migration, and nonrandom mating may disrupt Hardy-Weinberg equilibrium. Therefore, when population geneticists discover that a population is not in equilibrium, they try to determine which evolutionary factors are at work.

Genetic TIPS

The Question: One particularly useful feature of the Hardy-Weinberg equation is that it allows us to estimate the frequency of heterozygotes for recessive genetic diseases, assuming that Hardy-Weinberg equilibrium exists. As an example, let's consider cystic fibrosis, which is a human genetic disease involving a gene that encodes a transmembrane transporter of chloride ions (Cl^-). Persons afflicted with this disorder have an irregularity in salt and water balance. One of the symptoms is thick mucus in the lungs that can contribute to repeated lung infections. In populations of Northern European descent, the frequency of affected individuals is approximately 1 in 2500. Because cystic fibrosis is a recessive disorder, affected individuals are homozygotes. Assuming that the population is in Hardy-Weinberg equilibrium, what is the frequency of individuals who are heterozygous carriers?

Topic: What topic in genetics does this question address? The topic is predicting the frequency of heterozygotes in a population. More specifically, the question is about predicting the frequency of heterozygotes carrying the recessive allele that causes cystic fibrosis.

Information: What information do you know based on the question and your understanding of the topic? From the question, you know the frequency of homozygotes that have the disease. From your understanding of the topic, you may realize that you can use the Hardy-Weinberg equation to determine allele and genotype frequencies.

Problem-Solving Strategy: Make a calculation. One strategy to solve this problem is to use the Hardy-Weinberg equation to first determine the allele frequencies for the disease-causing allele and the non-disease-causing allele, and then use these allele frequencies to calculate the genotype frequency of heterozygotes.

If q represents the allele frequency of the disease-causing allele, then

$q^2 = 1/2500$

$q^2 = 0.0004$

We take the square root to determine q:

$q = \sqrt{0.0004}$

$q = 0.02$

If p represents the non-disease-causing allele,

$p = 1 - q$

$p = 1 - 0.02 = 0.98$

Answer: The frequency of heterozygous carriers is

$2pq = 2(0.98)(0.02) = 0.0392$, or 3.92%

23.1 REVIEWING THE KEY CONCEPTS

- All of the alleles of every gene in a population constitute the population's gene pool.
- For sexually reproducing organisms, a population is a group of individuals of the same species that occupy the same region and can interbreed with one another (see Figure 23.1).
- In population genetics, polymorphism refers to inherited traits or genes that exhibit variation in a population (see Figure 23.2).
- Single-nucleotide polymorphisms (SNPs) are the most common type of variation among genes (see Figure 23.3).
- Geneticists analyze genetic variation by determining allele and genotype frequencies.
- The Hardy-Weinberg equation can be used to calculate genotype frequencies based on allele frequencies (see Figures 23.4, 23.5).
- Deviation from Hardy-Weinberg equilibrium indicates that evolutionary change is occurring.
- The Hardy-Weinberg equation can be used to estimate the frequency of heterozygous carriers.

23.1 COMPREHENSION QUESTIONS

1. A gene pool is
   a. all of the genes in a single individual.
   b. all of the genes in the gametes from a single individual.
   c. all of the genes in a population of individuals.
   d. the random mixing of genes during sexual reproduction.
2. In natural populations, most genes are
   a. polymorphic.
   b. monomorphic.
   c. recessive.
   d. both a and c.

3. A gene exists in two alleles, designated *D* and *d*. If 48 copies of this gene are the *D* allele and 152 are the *d* allele, what is the allele frequency of *D*?
   a. 0.24
   b. 0.32
   c. 0.38
   d. 0.76

4. The allele frequency of *C* is 0.4 and that of *c* is 0.6. If the population is in Hardy-Weinberg equilibrium, what is the frequency of heterozygotes?
   a. 0.16
   a. 0.24
   a. 0.26
   a. 0.48

23.2 OVERVIEW OF MICROEVOLUTION

Learning Outcomes:
1. Define *microevolution*.
2. Explain the role of mutation in microevolution.
3. List the mechanisms that may cause allele and genotype frequencies to significantly change from one generation to the next.

The genetic variation in natural populations typically changes over the course of many generations. The term **microevolution** describes changes in a population's gene pool from generation to generation. Such change is rooted in two related phenomena (**Table 23.1**). First, the introduction of new genetic variation into a population is one essential aspect of microevolution. As discussed in Section 23.7, gene variation can originate by a variety of mechanisms. For example, new alleles of preexisting genes can arise by random mutations. Such events provide a continuous source of new variation to populations. However, new mutations are relatively rare. For example, a common rate of mutation for a given gene may be on the order of 1 new mutation per 1 million copies of the gene per generation. Therefore, even though new mutations are a vital source of genetic variation, they do not, by themselves, act as a major factor in promoting widespread changes in a population.

Microevolution also involves the action of evolutionary mechanisms that alter the prevalence of a given allele or genotype in a population. These mechanisms are natural selection, genetic drift, migration, and nonrandom mating (see Table 23.1). The collective contributions of these evolutionary mechanisms over the course of many generations have the potential to promote widespread genetic changes in a population. In the following sections, we will examine how these mechanisms can affect the type of genetic variation that occurs when a gene exists in two or more alleles in a population. As you will learn, these mechanisms may cause a particular allele to be favored, or they may create a balance where two or more alleles are maintained in a population.

TABLE 23.1
Factors That Govern Microevolution

Source of New Allelic Variation*

Mutation	Throughout most of this chapter, we consider allelic variation. Random mutations within preexisting genes introduce new alleles into populations, but at a very low rate. New mutations may be beneficial, neutral, or deleterious. For new alleles to rise to a significant percentage in a population, evolutionary mechanisms (i.e., natural selection, genetic drift, and/or migration) must operate on them.

Mechanisms That Alter Existing Genetic Variation

Natural selection	The phenomenon in which certain phenotypes have greater reproductive success compared to other phenotypes. For example, natural selection may be related to the survival of members to reproductive age.
Genetic drift	A change in genetic variation from generation to generation due to random fluctuations. Allele frequencies may change as a matter of chance from one generation to the next. Genetic drift tends to have a greater effect in a small population.
Migration	Migration can occur between two different populations. The introduction of migrants into a recipient population may change the allele frequencies of that population.
Nonrandom mating	The phenomenon in which individuals select mates based on their phenotypes or genetic lineage. Nonrandom mating can alter the relative proportions of homozygotes and heterozygotes predicted by the Hardy-Weinberg equation but does not change allele frequencies.

*Allelic variation is just one source of new genetic variation. Section 23.7 considers a variety of mechanisms through which new genetic variation can occur.

23.2 REVIEWING THE KEY CONCEPTS
- Microevolution refers to changes in a population's gene pool from generation to generation.
- Mutations are the source of new genetic variation. However, the occurrence of new mutations does not greatly change allele frequencies because it happens at a very low rate. Other factors, such as natural selection, genetic drift, migration, and nonrandom mating, may alter allele and/or genotype frequencies (see Table 23.1).

23.2 COMPREHENSION QUESTION
1. Which of the following is a factor that, by itself, does *not* promote widespread changes in allele or genotype frequencies?
   a. New mutation
   b. Natural selection
   c. Genetic drift
   d. Migration
   e. Nonrandom mating

23.3 NATURAL SELECTION

Learning Outcomes:
1. Explain the process of natural selection.
2. Compare and contrast directional, stabilizing, disruptive, and balancing selection.

In the 1850s, Charles Darwin and Alfred Russel Wallace independently proposed the theory of evolution by **natural selection.** According to this theory, phenotypes may vary with regard to reproductive success. Because the phenotypes of individuals are largely determined by the alleles they carry, those individuals with higher reproductive success are more likely to produce offspring, and therefore pass certain alleles to the next generation. Natural selection acts on phenotypes, which are governed by individuals' genotypes.

What factors contribute to reproductive success? Some individuals may have characteristics that make them better adapted to their environment. These individuals are more likely to survive to reproductive age and contribute offspring to the next generation. In sexually reproducing species, the ability to find a mate and fertility are also key factors that contribute to an individual's reproductive success.

A modern restatement of the principles of natural selection can relate our knowledge of molecular genetics to the phenotypes of individuals.

1. Within a population, allelic variation arises in various ways, such as through random mutations that cause differences in DNA sequences. A mutation that creates a new allele may alter the amino acid sequence of the encoded protein, which, in turn, may alter the function of the protein.
2. Some alleles may encode proteins that enhance an individual's survival or reproductive capability compared with other members of the population. For example, an allele may produce a protein that is more efficient at a higher temperature, conferring on the individual a greater probability of survival in a hot climate.
3. Individuals with beneficial alleles have phenotypes that make them more likely to reproduce and contribute to the gene pool of the next generation.
4. Over the course of many generations, allele frequencies of many different genes may change due to natural selection, thereby significantly altering the characteristics of a population or species. The net result of natural selection is a population that is better adapted to its environment and more successful at reproduction. Even so, it should be emphasized that species are not perfectly adapted to their environments, because mutations are random events and because the environment tends to change from generation to generation.

Fisher, Wright, and Haldane developed mathematical formulas to explain the theory of natural selection. As our knowledge of natural selection has increased, it has become apparent that it operates in many different ways. In this section, we will consider a few examples of how natural selection may occur.

Darwinian Fitness Is a Measure of Reproductive Success

To begin our discussion of natural selection, we must examine the concept of **Darwinian fitness**—the relative likelihood that one genotype will contribute to the gene pool of the next generation rather than other genotypes. As mentioned, natural selection acts on phenotypes that are derived from individuals' genotypes. Although Darwinian fitness often correlates with physical fitness, the two concepts are not identical. Darwinian fitness is a measure of reproductive success. An extremely fertile genotype may have a higher Darwinian fitness than a less fertile genotype that appears more physically fit.

To consider Darwinian fitness, let's use the example of a gene existing in A and a alleles. If the three genotypes have the same level of mating success and fertility, we can assign a fitness value to each genotype class based on the likelihood of individuals with that genotype surviving to reproductive age. For example, let's suppose that the relative survival to adulthood of each of the three genotype classes is as follows: For every five AA individuals that survive, four Aa individuals survive, and one aa individual survives. By convention, the genotype with the highest reproductive ability is given a fitness value of 1.0. Values indicating **relative fitness** are denoted by the variable w. The genotypes other than the one with the highest reproductive ability are assigned fitness values relative to this 1.0 value:

$$\text{Fitness of } AA: w_{AA} = 1.0$$
$$\text{Fitness of } Aa: w_{Aa} = 4/5 = 0.8$$
$$\text{Fitness of } aa: w_{aa} = 1/5 = 0.2$$

Differences in reproductive success among genotypes may be due to various factors. In this example, the fittest genotype is more likely to survive to reproductive age. In other situations, the fittest genotype is more likely to mate. For example, a bird with brightly colored feathers may have an easier time attracting a mate than a bird with duller plumage. Finally, a third possibility is that the fittest genotype may be more fertile. The individual with that genotype may produce a higher number of gametes or gametes that are more successful at fertilization.

Also keep in mind that the preceding discussion of relative fitness presents the simplified case in which a single gene affects reproductive success. However, most traits are affected by allelic variation involving multiple genes. In Chapter 24, we will examine quantitative traits, such as weight and height, which are determined by alleles of many different genes. When natural selection acts on quantitative traits, relative fitness values are determined by allelic variation of multiple genes, not just one.

By studying species in their native environments, population geneticists have discovered that natural selection can occur in several ways. The patterns of natural selection depend on the relative fitness values of different genotypes and on the variation of environmental effects. The four patterns of natural selection that we will consider are called directional, stabilizing, disruptive, and balancing selection. In most of the examples described in this section, natural selection leads to adaptation that makes members of a population better able to survive to reproductive age.

Directional Selection Favors the Extreme Phenotype

Directional selection favors individuals at one extreme of a phenotypic distribution that are more likely to survive and reproduce in a particular environment. In some cases, directional selection may act on phenotypes that are largely determined by the alleles of a single gene. For example, the level of resistance to an insecticide in a mosquito population may be determined by alleles of a single gene. In other cases, directional selection may act on phenotypes that are determined by multiple genes. For example, body weight in mammals is influenced by alleles of many different genes. If directional selection favored higher body weight, it would affect the allele frequencies of many different genes.

Different phenomena may initiate the process of directional selection.

- One way that directional selection may arise is that a new allele may be introduced into a population by mutation, and the new allele may promote a higher fitness in individuals that carry it (**Figure 23.6**). If the homozygote carrying the favored allele has the highest fitness value, directional selection may cause this favored allele to eventually become the predominant allele in the population, perhaps even giving rise to a monomorphic gene.

- Another possibility is that a population may be exposed to a prolonged change in the environment in which it lives. Under the new environmental conditions, the relative fitness values may change to favor one or more genotypes, which will promote the elimination of other genotypes. As an example, let's suppose a population of finches on the mainland already has genetic variation in beak size. A small number of birds migrate to an island where the seeds are generally larger than they are on the mainland. In this new environment, birds with larger beaks have a higher relative fitness because they are better able to crack open the larger seeds and thereby survive to reproductive age. Over the course of many generations, directional selection produces a population of birds carrying alleles that promote larger beak size.

For traits for which fitness is largely determined by alleles of a single gene, we can calculate how directional selection changes allele frequencies in a step-by-step, generation-per-generation way. To appreciate how this occurs, let's take a look at how fitness affects Hardy-Weinberg equilibrium and allele frequencies. Again,

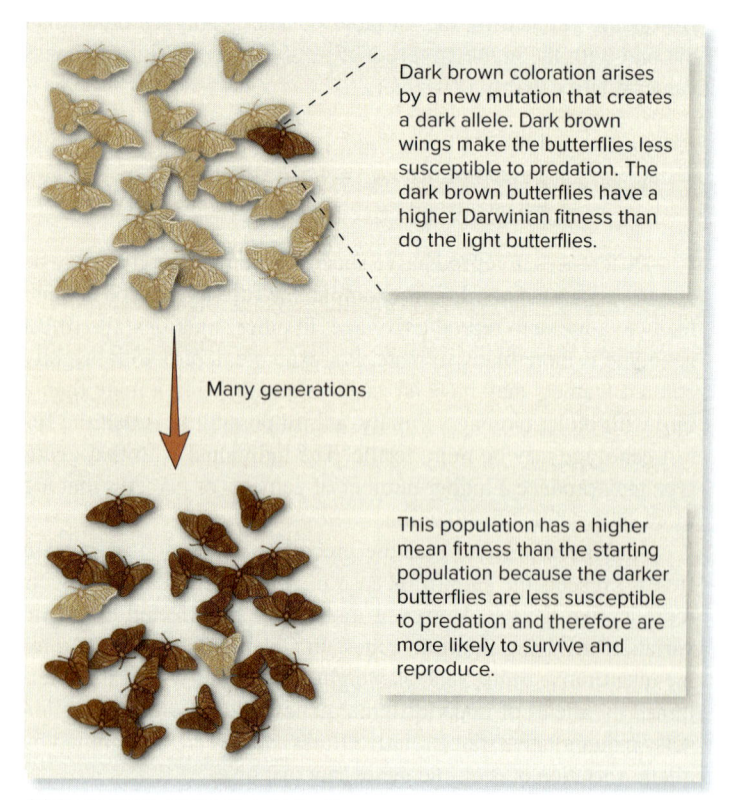

(a) An example of directional selection

Dark brown coloration arises by a new mutation that creates a dark allele. Dark brown wings make the butterflies less susceptible to predation. The dark brown butterflies have a higher Darwinian fitness than do the light butterflies.

Many generations

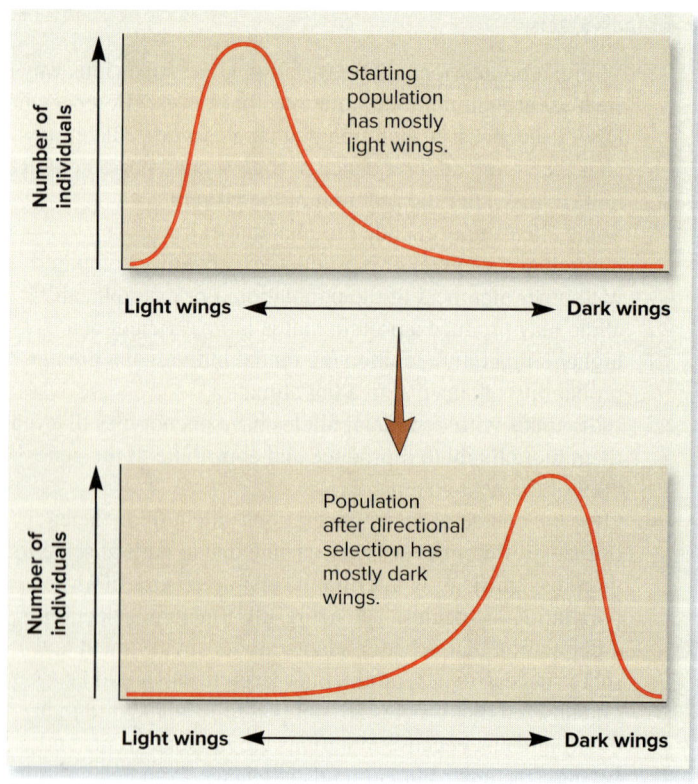

This population has a higher mean fitness than the starting population because the darker butterflies are less susceptible to predation and therefore are more likely to survive and reproduce.

(b) Graphical representation of directional selection

Starting population has mostly light wings.

Population after directional selection has mostly dark wings.

FIGURE 23.6 **Directional selection.** (a) A new mutation arises in a population that confers higher Darwinian fitness. In this example, butterflies with dark wings are more likely to survive and reproduce. Over many generations, directional selection favors the prevalence of darker individuals. (b) A graphical representation of directional selection.

Concept Check: In this pattern of natural selection, explain the meaning of the word *directional*.

let's suppose a gene exists in two alleles: *A* and *a*. The three relative fitness values, which are based on relative survival levels, are

$$w_{AA} = 1.0$$
$$w_{Aa} = 0.8$$
$$w_{aa} = 0.2$$

In the next generation, we expect that Hardy-Weinberg equilibrium is modified in the following way due to directional selection:

Frequency of *AA*: $p^2 w_{AA}$
Frequency of *Aa*: $2pq w_{Aa}$
Frequency of *aa*: $q^2 w_{aa}$

In a population that is changing due to natural selection, these three terms may not add up to 1.0, as they do in Hardy-Weinberg equilibrium. Instead, the three terms sum to a value known as the **mean fitness of the population** ($\overline{w}$):

$$p^2 w_{AA} + 2pq w_{Aa} + q^2 w_{aa} = \overline{w}$$

Dividing both sides of this equation by the mean fitness of the population gives,

$$\frac{p^2 w_{AA}}{\overline{w}} + \frac{2pq w_{Aa}}{\overline{w}} + \frac{q^2 w_{aa}}{\overline{w}} = 1$$

Using terms from this equation, we can calculate the expected genotype and allele frequencies after one generation of directional selection:

Frequency of *AA* genotype: $\dfrac{p^2 w_{AA}}{\overline{w}}$

Frequency of *Aa* genotype: $\dfrac{2pq w_{Aa}}{\overline{w}}$

Frequency of *aa* genotype: $\dfrac{q^2 w_{aa}}{\overline{w}}$

Allele frequency of *A*: $p_A = \dfrac{p^2 w_{AA}}{\overline{w}} + \dfrac{pq w_{Aa}}{\overline{w}}$

Allele frequency of *a*: $q_a = \dfrac{q^2 w_{aa}}{\overline{w}} + \dfrac{pq w_{Aa}}{\overline{w}}$

As an example, let's suppose that the starting allele frequencies are $A = 0.5$ and $a = 0.5$, and use fitness values of 1.0, 0.8, and 0.2 for the three genotypes, *AA*, *Aa*, and *aa*, respectively. We begin by calculating the mean fitness of the population:

$$p^2 w_{AA} + 2pq w_{Aa} + q^2 w_{aa} = \overline{w}$$
$$\overline{w} = (0.5)^2 (1) + 2(0.5)(0.5)(0.8) + (0.5)^2 (0.2)$$
$$\overline{w} = 0.25 + 0.4 + 0.05 = 0.7$$

After one generation of directional selection,

Frequency of *AA* genotype: $\dfrac{p^2 w_{AA}}{\overline{w}} = \dfrac{(0.5)^2(1)}{0.7} = 0.36$

Frequency of *Aa* genotype: $\dfrac{2pq w_{Aa}}{\overline{w}} = \dfrac{2(0.5)(0.5)(0.8)}{0.7} = 0.57$

Frequency of *aa* genotype: $\dfrac{q^2 w_{aa}}{\overline{w}} = \dfrac{(0.5)^2(0.2)}{0.7} = 0.07$

Allele frequency of *A*: $p_A = \dfrac{p^2 w_{AA}}{\overline{w}} + \dfrac{pq w_{Aa}}{\overline{w}}$

$$= \dfrac{(0.5)^2(1)}{0.7} + \dfrac{(0.5)(0.5)(0.8)}{0.7} = 0.64$$

Allele frequency of *a*: $q_a = \dfrac{q^2 w_{aa}}{\overline{w}} + \dfrac{pq w_{Aa}}{\overline{w}}$

$$= \dfrac{(0.5)^2(0.2)}{0.7} + \dfrac{(0.5)(0.5)(0.8)}{0.7} = 0.36$$

After one generation, the allele frequency of *A* has increased from 0.5 to 0.64, and that of *a* has decreased from 0.5 to 0.36. These changes have occurred because the *AA* genotype has the highest fitness, whereas the *Aa* and *aa* genotypes have lower fitness values. Another interesting feature of natural selection is that it raises the mean fitness of the population. If we assume that the individual fitness values are constant, the mean fitness of this next generation is

$$\overline{w} = p^2 w_{AA} + 2pq w_{Aa} + q^2 w_{aa}$$
$$= (0.64)^2(1) + 2(0.64)(0.36)(0.8) + (0.36)^2(0.2)$$
$$= 0.80$$

The mean fitness of the population has increased from 0.7 to 0.8.

What are the consequences of natural selection at the population level? After one generation, the population is better adapted to its environment than it was in the preceding generation. Another way of viewing this calculation is that the subsequent population has a greater reproductive potential than the preceding one. We could perform the same types of calculations to find the allele frequencies and mean fitness value in the next generation. If we assume the individual fitness values remain constant, the frequencies of *A* and *a* in the next generation are 0.85 and 0.15, respectively, and the mean fitness increases to 0.931. As we can see, the general trend is to increase *A*, decrease *a*, and increase the mean fitness of the population.

In the previous example, we considered the effect of natural selection by beginning with allele frequencies at intermediate levels (namely, $A = 0.5$ and $a = 0.5$). **Figure 23.7** illustrates what would happen if a new mutation introduced the *A* allele into a population that was originally monomorphic for the *a* allele. As before, the *AA* homozygote has a relative fitness value of 1.0, the *Aa* heterozygote

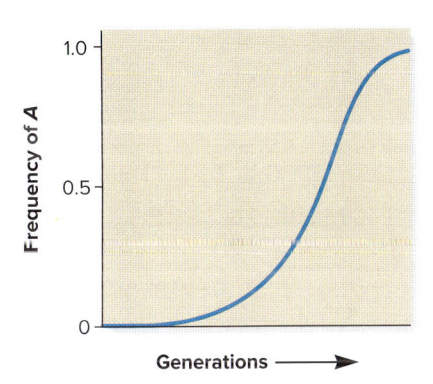

FIGURE 23.7 **The fate of a beneficial allele that is introduced into a population as a new mutation.** A new allele (*A*) is beneficial in the homozygous condition: $w_{AA} = 1.0$. The heterozygote, *Aa* ($w_{Aa} = 0.8$), and the homozygote, *aa* ($w_{aa} = 0.2$), have lower fitness values.

has a value of 0.8, and the recessive *aa* homozygote has a value of 0.2. Initially, the *A* allele is at a very low frequency in the population. If it is not lost initially due to genetic drift, its frequency slowly begins to rise and then, at intermediate values, rises much more rapidly.

Eventually, directional selection may lead to fixation of a beneficial allele. However, a new beneficial allele is in a precarious situation when its frequency is very low. As discussed later in this chapter, genetic drift is likely to eliminate new mutations, even beneficial ones, as a result of chance fluctuations.

Researchers have identified many examples in nature of directional selection affecting traits that are determined by single genes, such as resistance to pesticides. For example, the pesticide DDT (dichlorodiphenyltrichloroethane) began to be used in the 1940s as a way to decrease the populations of mosquitoes and other insects. Subsequently, certain insect species have become resistant to DDT by a dominant mutation in a single enzyme-encoding gene. The resulting mutant enzyme detoxifies DDT, making it harmless to the insect. **Figure 23.8** shows the results of an experiment in which mosquito larvae (*Aedes aegypti*) were exposed to DDT over the course of seven generations. The starting population showed a low level of DDT resistance, as evidenced by the low percentage of survivors after exposure to DDT. By comparison, in seven generations, nearly 100% of the population was DDT-resistant. These results illustrate the power of directional selection in promoting change in a population. Since the 1950s, resistance to nearly every known insecticide has evolved within 10 years of its commercial introduction!

Balanced Polymorphisms May Arise as a Result of Heterozygote Advantage or Negative Frequency-Dependent Selection

A common misperception is that natural selection always eliminates "weaker alleles" from a population. Researchers have discovered certain patterns of natural selection that actually favor the maintenance of two or more alleles in a population. One example is called **balancing selection.**

Heterozygote Advantage For genetic variation involving a single gene, balancing selection may arise when the heterozygote has a higher fitness than either corresponding homozygote, a situation called **heterozygote advantage.** In this case, an equilibrium is reached in which both alleles are maintained in the population. If the relative fitness values are known for each of the genotypes, the allele frequencies at equilibrium can be calculated. To do so, we must consider the **selection coefficient (*s*),** which measures the degree to which a genotype is selected against:

$$s = 1 - w$$

By convention, the genotype with the highest fitness has an *s* value of zero. Genotypes at a selective disadvantage have *s* values that are greater than 0 but less than or equal to 1.0. An extreme case is a recessive lethal allele, which would have an *s* value of 1.0 in the homozygote, while the *s* value in the heterozygote could be 0.

Let's consider genotypes with the following relative fitness values:

$$w_{AA} = 0.7$$
$$w_{Aa} = 1.0$$
$$w_{aa} = 0.4$$

The selection coefficients are

$$s_{AA} = 1 - 0.7 = 0.3$$
$$s_{Aa} = 1 - 1.0 = 0$$
$$s_{aa} = 1 - 0.4 = 0.6$$

The population reaches an equilibrium when

$$s_{AA}p = s_{aa}q$$

If we take this equation, let $q = 1 - p$, and then solve for *p*, we obtain

$$p = \text{Allele frequency of } A = \frac{s_{aa}}{s_{AA} + s_{aa}}$$
$$= \frac{0.6}{0.3 + 0.6} = 0.67$$

If we let $p = 1 - q$ and then solve for *q*, we get

$$q = \text{Allele frequency of } a = \frac{s_{AA}}{s_{AA} + s_{aa}}$$
$$= \frac{0.3}{0.3 + 0.6} = 0.33$$

In this example, balancing selection maintains the two alleles in the population at frequencies of 0.67 for *A* and 0.33 for *a*.

Heterozygote advantage can sometimes explain the high frequency of alleles that are harmful in a homozygous condition. A classic example is the Hb^S allele of the human β-globin gene. A homozygous $Hb^S Hb^S$ individual exhibits sickle cell disease, a disorder characterized by the sickling of the red blood cells. The $Hb^S Hb^S$ homozygote has a lower fitness than a homozygote with

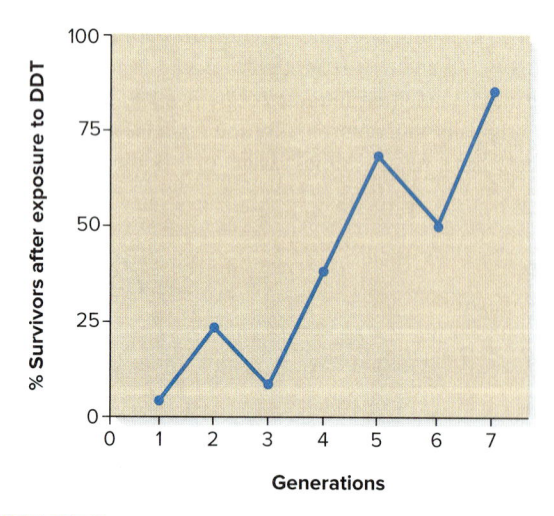

FIGURE 23.8 **Directional selection for DDT resistance in a mosquito population.** In this experiment, mosquito larvae (*Aedes aegypti*) were exposed to 10 mg/L of DDT. The percentage of survivors was recorded, and then the survivors of each generation were used as parents for the next generation.

Concept Check: In this example, is directional selection promoting genetic diversity? Explain.

two copies of the more common β-globin allele, Hb^AHb^A. However, the heterozygote, Hb^AHb^S, has a higher level of fitness than either homozygote in areas where malaria is endemic (**Figure 23.9**). Compared with Hb^AHb^A homozygotes, heterozygotes have a 10–15% higher chance of survival if infected by the malarial parasite, *Plasmodium falciparum*. Therefore, the Hb^S allele is maintained in populations in areas where malaria is prevalent, even though the allele is detrimental in the homozygous state.

In addition to sickle cell disease, other gene mutations that cause human disease in the homozygous state are thought to be prevalent because of heterozygote advantage. For example, the high prevalence of the allele causing cystic fibrosis may be related to this phenomenon, but the advantage that a heterozygote may possess has not been determined.

Negative Frequency-Dependent Selection A second mechanism of balancing selection is **negative frequency-dependent selection**. In this pattern of natural selection, the fitness of a genotype decreases when its frequency becomes higher. In other words, rare individuals have a higher fitness than more common individuals. Therefore, rare individuals are more likely to reproduce, whereas common individuals are less likely, thereby producing a balanced polymorphism in which no genotype becomes too rare or too common.

An interesting example of negative frequency-dependent selection involves the elder-flowered orchid, *Dactylorhiza sambucina* (**Figure 23.10**). Throughout the range of this plant in central and southern Europe, both yellow- and red-flowered individuals are prevalent. A proposed explanation for this polymorphism is related to the orchid's pollinators, which are mainly bumblebees such as *Bombus lapidarius* and *B. terrestris*. The pollinators increase their visits to the flowers of one color of *D. sambucina* as that color becomes less common in a given area. One reason why this may occur is that *D. sambucina* is a rewardless flower; that is, it does not provide its pollinators with any reward for visiting, such as sweet nectar. Pollinators learn that the *D. sambucina* flowers of the more common color in a given area do not offer a reward, and they increase their visits to the less-common flower. Thus, the relative fitness of the less-common flower increases.

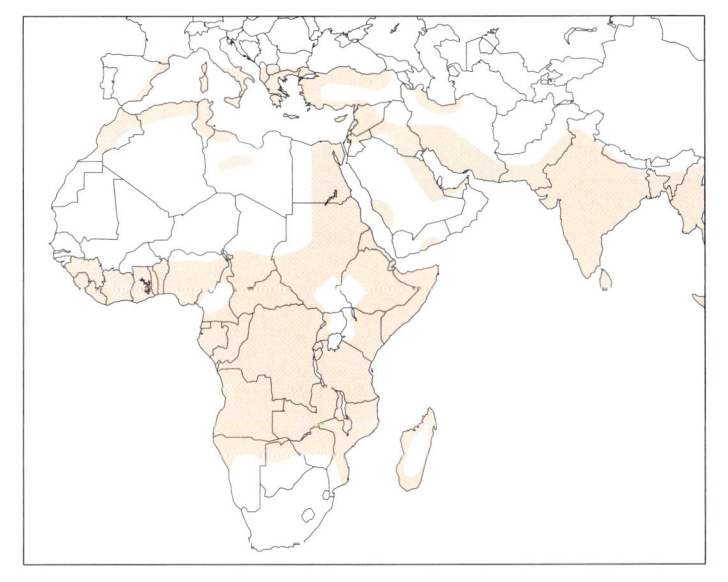

(a) Malaria prevalence

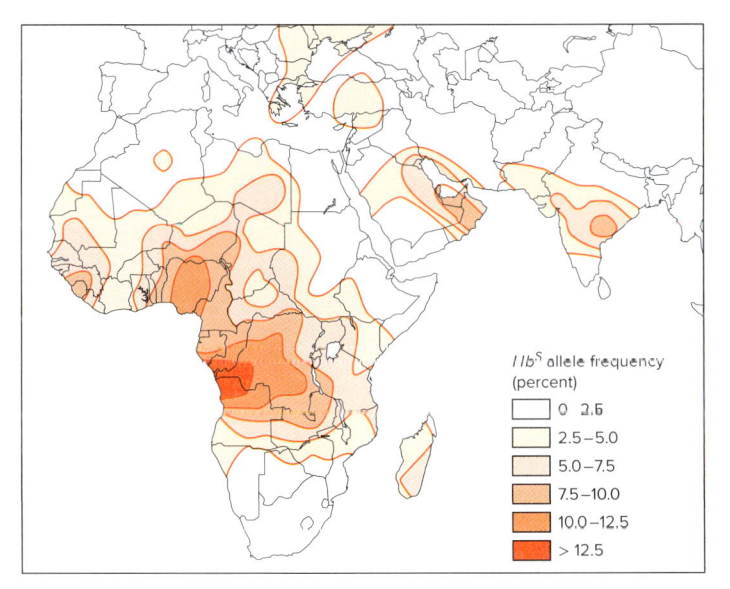

(b) Hb^S allele frequency

FIGURE 23.9 **The geographic relationship between malaria and the frequency of the sickle cell allele in human populations.** (a) The geographic prevalence of malaria in Africa and surrounding areas. (b) The frequency of the Hb^S allele in the same areas.

Genes → Traits The sickle cell allele of the β-globin gene is maintained in human populations by balancing selection. In areas where malaria is endemic, the heterozygote carrying one copy of the Hb^S allele has a greater fitness than either of the corresponding homozygotes (Hb^AHb^A and Hb^SHb^S). Therefore, even though the Hb^SHb^S homozygotes suffer the detrimental consequences of sickle cell disease, this negative outcome is balanced by the beneficial effects of malarial resistance in the heterozygotes.

Concept Check: Explain why the Hb^S allele is prevalent in certain regions even though it is detrimental in the homozygous condition.

FIGURE 23.10 **The two color variations found in the elder-flowered orchid, *Dactylorhiza sambucina*.** The two colors are maintained in the population due to negative frequency-dependent selection.
©Paul Harcourt Davies/SPL/Science Source

Concept Check: Explain how negative frequency-dependent selection works.

Disruptive Selection Favors Multiple Phenotypes in Heterogeneous Environments

As we have just seen, polymorphisms may arise due to balancing selection. Researchers have discovered another pattern of natural selection that favors the maintenance of two or more alleles in heterogeneous environments. This pattern, called **disruptive selection,** also known as diversifying selection, favors the survival of two or more different phenotypes (**Figure 23.11**). This pattern of selection typically acts on traits that are determined by multiple genes. In disruptive selection, the fitness values of particular genotypes are higher in one environment and lower in a different one. Disruptive selection is likely to occur in populations that occupy diverse environments; some members of the species are more likely to survive and reproduce in each type of environmental condition.

As an example, **Figure 23.12a** shows a photograph of land snails, *Cepaea nemoralis*, which live in woods and open fields. This snail is polymorphic with respect to color and banding patterns of the shell. In 1954, Arthur Cain and Philip Sheppard found that shell color was correlated with the environment. As shown in **Figure 23.12b**, the highest frequency of brown shell color was found in snails in the beech woods, which have wide expanses of dark soil. The frequency of brown shells was substantially less in other environments. By comparison, pink-shelled snails are most common in the leaf litter of both beech woods and deciduous woods, and the yellow-shelled snails are most abundant in the sunny, grassy areas of hedgerows and rough herbage. Researchers have suggested that this disruptive selection can be explained by different levels of predation by thrushes. Depending on the environment, certain snail phenotypes may be more easily seen by their predators than others. Migration can occasionally occur between the snail populations, which keeps the polymorphism in balance among these different environments.

Stabilizing Selection Favors Individuals with Intermediate Phenotypes

In **stabilizing selection,** the extreme phenotypes for a trait are selected against, and those individuals with intermediate phenotypes have the highest fitness values. Stabilizing selection is typically directed at quantitative traits, such as body weight and offspring number, which are determined by multiple genes. Stabilizing selection tends to decrease genetic diversity for genes affecting such traits because it eliminates alleles that cause a greater variation

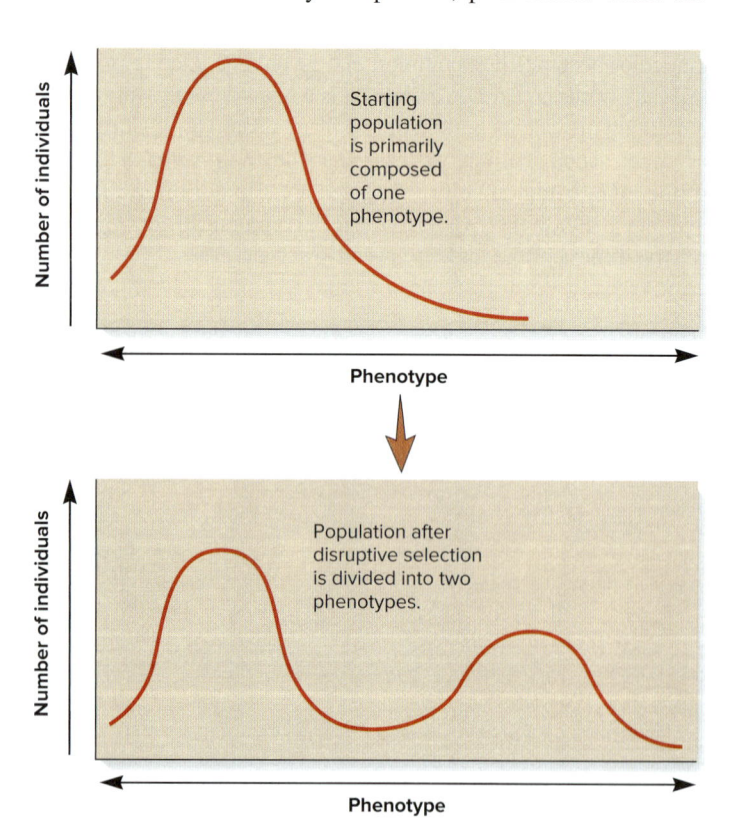

FIGURE 23.11 **Disruptive selection.** Over time, this pattern of selection favors two or more phenotypes due to heterogeneous environments.

Concept Check: Does disruptive selection favor polymorphism? Explain why or why not.

(a) Land snails

(b) Frequency of shell color

Habitat	Brown	Pink	Yellow
Beech woods	0.23	0.61	0.16
Deciduous woods	0.05	0.68	0.27
Hedgerows	0.05	0.31	0.64
Rough herbage	0.004	0.22	0.78

FIGURE 23.12 **Polymorphism in the land snail, *Cepaea nemoralis*.** (a) This species of snail can can have a shell with several different colors and banding patterns. (b) Coloration of the shells is correlated with the specific environments where the snails are located.

Genes → Traits Shell coloration is an example of genetic polymorphism due to heterogeneous environments; the genes governing shell coloration are polymorphic. The predation of snails is correlated with their ability to be camouflaged in their natural environment. Snails with brown shells are most prevalent in beech woods, where the soil is dark. Pink-shelled snails are most abundant in the leaf litter of beech woods and deciduous woods. Yellow-shelled snails are most prevalent in sunnier locations, such as hedgerows and rough herbage.

©R. Koenig/age fotostock

in phenotypes. In 1947, David Lack proposed that stabilizing selection may apply to clutch size in birds. Under stabilizing selection, birds that lay too many or too few eggs have lower fitness values than those that lay an intermediate number (**Figure 23.13**). Laying too many eggs may cause many offspring to die due to inadequate parental care and food. In addition, the strain on the parents themselves may decrease their likelihood of survival and therefore their ability to produce more offspring. Having too few offspring, on the other hand, does not contribute many individuals to the next generation. Therefore, the most successful parents are those that produce an intermediate clutch size. In the 1980s, Swedish evolutionary biologist Lars Gustafsson and colleagues examined the phenomenon of stabilizing selection in the collared flycatcher, *Ficedula albicollis*, on the island of Gotland, which is southeast of the mainland of Sweden. They discovered that Lack's hypothesis that clutch size is subject to the action of stabilizing selection appears to be true for this species.

23.3 REVIEWING THE KEY CONCEPTS

- Natural selection is a process that influences allele frequencies from one generation to the next based on fitness, which is a measure of the relative reproductive success of different genotypes.
- Directional selection favors the extreme phenotype (see Figures 23.6–23.8).
- Balancing selection results in stable polymorphism. It may arise as a result of heterozygote advantage or negative frequency-dependent selection (see Figures 23.9, 23.10).
- Disruptive selection favors multiple phenotypes in heterogeneous environments (see Figures 23.11, 23.12).
- Stabilizing selection favors individuals with intermediate phenotypes (see Figure 23.13).

23.3 COMPREHENSION QUESTIONS

1. Darwinian fitness is a measure of
   a. survival.
   b. reproductive success.
   c. heterozygosity of the gene pool.
   d. polymorphisms in a population.
2. Within a particular population, darkly colored rats are more likely to survive than more lightly colored individuals. This situation is likely to result in
   a. directional selection.
   b. stabilizing selection.
   c. disruptive selection.
   d. balancing selection.
3. A population occupies heterogeneous environments in which the fitness of some individuals is higher in one environment and the fitness of other individuals is higher in another environment. This situation is likely to result in
   a. directional selection.
   b. stabilizing selection.
   c. disruptive selection.
   d. balancing selection.
4. A gene exists in two alleles and the heterozygote has the highest fitness. This situation is likely to result in
   a. directional selection.
   b. stabilizing selection.
   c. disruptive selection.
   d. balancing selection.

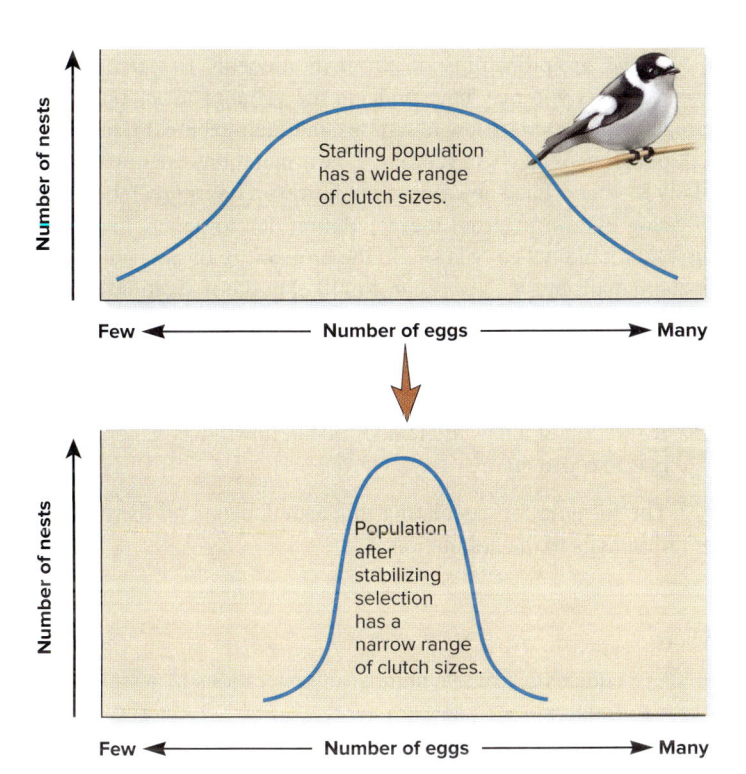

FIGURE 23.13 Stabilizing selection. In this pattern of natural selection, the extremes of a phenotypic distribution are selected against. Those individuals with intermediate traits have the highest fitness. This results in a population with less diversity and more uniform traits.

Concept Check: In general, why does stabilizing selection decrease genetic diversity?

23.4 GENETIC DRIFT

Learning Outcomes:
1. Define *genetic drift*.
2. Explain how population size affects genetic drift, and calculate the probabilities of the outcomes of this process.
3. Compare and contrast the bottleneck effect and the founder effect.

In the 1930s, geneticist Sewall Wright played a key role in developing the concept of **random genetic drift,** or simply, **genetic drift,** which refers to changes in allele frequencies in a population due to random fluctuations. As a matter of chance, the frequencies of alleles found in gametes that unite to form zygotes vary from generation to generation. Over the long run, genetic drift usually results in either the loss of an allele or its fixation at 100%

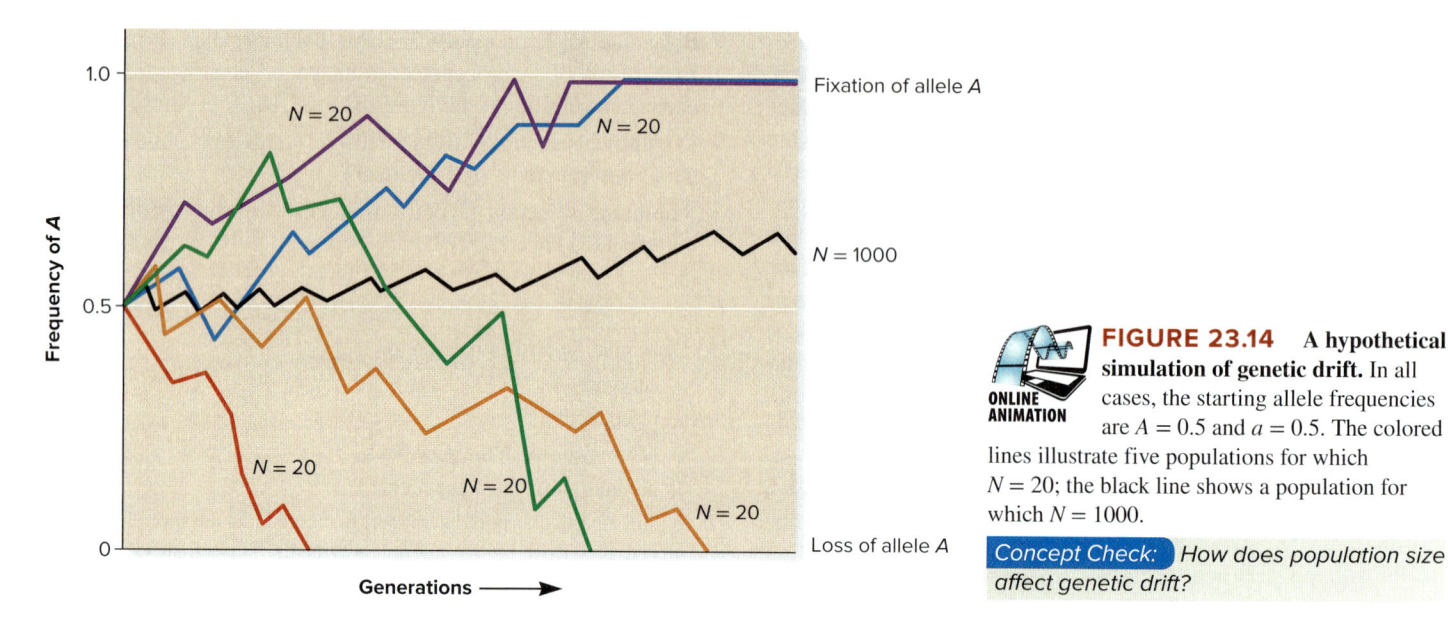

FIGURE 23.14 A hypothetical simulation of genetic drift. In all cases, the starting allele frequencies are $A = 0.5$ and $a = 0.5$. The colored lines illustrate five populations for which $N = 20$; the black line shows a population for which $N = 1000$.

Concept Check: How does population size affect genetic drift?

in the population. The process is random with regard to particular alleles. Genetic drift can lead to the loss or fixation of deleterious, neutral, or beneficial alleles. The rate at which this occurs depends on the population size and on the initial allele frequencies.

Figure 23.14 illustrates the potential consequences of genetic drift in one large ($N = 1000$) and five small ($N = 20$) populations. At the beginning of this hypothetical simulation, all of these populations have identical allele frequencies: $A = 0.5$ and $a = 0.5$. In the five small populations, this allele frequency fluctuates substantially from generation to generation. Eventually, one of the alleles is eliminated and the other is fixed at 100%. At this point, the allele has become monomorphic and does not fluctuate any further. By comparison, the allele frequencies in the large population fluctuate much less, because random sampling error is expected to have a smaller effect. Nevertheless, genetic drift leads to allele fixation even in large populations, but it takes many more generations for the effect to occur.

Now let's ask two questions:

1. How many new mutations do we expect in a natural population?
2. How likely is it that any new mutation will be either fixed in or eliminated from a population due to genetic drift?

With regard to the first question, the average number of new mutations depends on the mutation rate (μ), which is described later in this chapter, and on the number of individuals in a population (N). If each individual has two copies of the gene of interest, the expected number of new mutations in this gene is

$$\text{Expected number of new mutations} = 2N\mu$$

From this equation, we see that a new mutation is more likely to occur in a large population than in a small one. This makes sense, because the larger population has more copies of the gene to be mutated.

With regard to the second question, the probability of fixation of a newly arising allele due to genetic drift is

$$\text{Probability of fixation} = \frac{1}{2N} \text{ (assuming equal numbers of males and females contribute to the next generation)}$$

In other words, the probability of fixation is the same as the initial allele frequency in the population. For example, if $N = 20$, the probability of fixation of a new allele equals $1/(2 \times 20)$, or 2.5%.

Conversely, a new allele may be lost from the population.

$$\text{Probability of elimination} = 1 - \text{probability of fixation}$$
$$= 1 - \frac{1}{2N}$$

If $N = 20$, the probability of elimination equals $1 - 1/(2 \times 20)$, or 97.5%. As you may have noticed, the value of N has opposing effects with regard to new mutations and their eventual fixation in a population. When N is very large, new mutations are much more likely to occur. Each new mutation, however, has a greater chance of being eliminated from the population due to genetic drift. On the other hand, when N is small, the probability of new mutations is also small, but if they occur, the likelihood of fixation is relatively large.

Now that we appreciate the phenomenon of genetic drift, we can ask a third question:

3. If fixation of a new allele does occur, how many generations is it likely to take?

The formula for calculating this also depends on the number of individuals in the population:

$$\bar{t} = 4N$$

where

$\bar{t}$ equals the average number of generations to achieve fixation

N equals the number of individuals in the population, assuming that males and females contribute equally to each succeeding generation

As you may have expected, allele fixation takes much longer in large populations. If a population has 1 million breeding members, it takes, on average, 4 million generations, perhaps an insurmountable period of time, to reach fixation. In a small group of 100 individuals, however, fixation takes only 400 generations, on average.

In nature, allele frequencies in small populations are more susceptible to genetic drift. This susceptibility commonly arises as a result of the bottleneck effect or the founder effect.

Bottleneck Effect Changes in population size may influence genetic drift via the **bottleneck effect** (**Figure 23.15**). In nature, a population can be reduced dramatically in size by events such as earthquakes, floods, droughts, or human destruction of habitat. Such events may randomly eliminate most of the members of the population without regard to their genetic composition. The bottleneck may initiate genetic drift because the population of survivors may have allele frequencies that differ from those of the original population. In addition, allele frequencies are expected to drift substantially during the generations when the population size is small. In extreme cases, alleles may even be eliminated. Eventually, the population that experienced the bottleneck may regain its original size. However, the new population will have less genetic variation than the original large population. As an example, the African cheetah population lost a substantial amount of its genetic variation due to a bottleneck effect. DNA analysis by population geneticists has suggested that a severe bottleneck occurred approximately 10,000 to 12,000 years ago, when the population size was dramatically reduced. The population eventually rebounded, but the bottleneck effect significantly decreased the genetic variation.

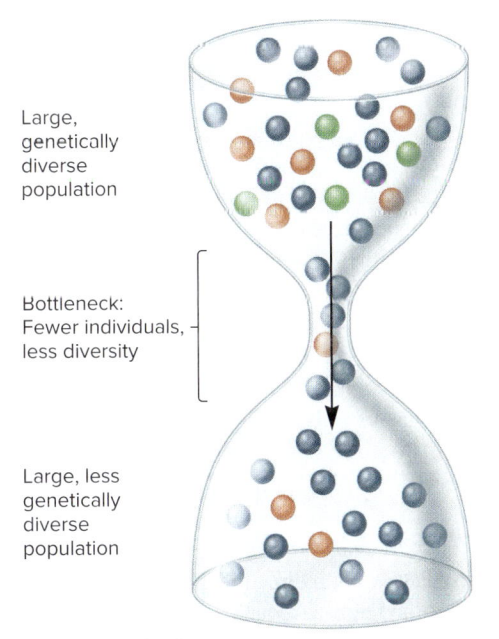

Founder Effect Geography and population size may also influence genetic drift via the **founder effect.** The key difference between the bottleneck effect and the founder effect is that the founder effect involves a geographical change; a small group of individuals separates from a larger population and establishes a colony in a new location. For example, a few individuals may move from a large continental population and become the founders of an island population. The founder effect has two important consequences.

- The founding population is expected to have less genetic variation than the original population from which it was derived.
- As a matter of chance, the allele frequencies in the founding population may differ markedly from those of the original population.

Population geneticists have studied many examples of isolated populations that were started from a few members of another population. In the 1960s, Victor McKusick studied allele frequencies in the Old Order Amish of Lancaster County, Pennsylvania. At that time, this was a group of about 8000 people, descended from just three couples who immigrated to the United States in the 1700s. Among this population of 8000, a genetic disease known as Ellis-van Creveld syndrome (a recessive form of dwarfism) was found at a frequency of 0.07, or 7%. By comparison, this disorder is extremely rare in other human populations, even the population from which the founding members had originated. The high frequency of dwarfism in the Lancaster County population is a chance occurrence due to the founder effect. The recessive allele can be traced back to one couple who came to the area in 1744.

FIGURE 23.15 **The bottleneck effect, an example of genetic drift.** (a) A representation of the bottleneck effect. Note that the genetic variation denoted by the green balls has been lost. (b) The African cheetah. The modern species has low genetic variation due to a bottleneck that is thought to have occurred about 10,000 to 12,000 years ago.
©Ingram Publishing

Concept Check: *What is happening at the bottleneck? Describe the effect of genetic drift during the bottleneck.*

23.4 REVIEWING THE KEY CONCEPTS

- Genetic drift involves changes in allele frequencies due to random fluctuations. Over the long run, it often results in allele fixation or loss. The effect of genetic drift is greater in small populations (see Figure 23.14).
- Two mechanisms that can influence genetic drift are the bottleneck effect and the founder effect (see Figure 23.15).

23.4 COMPREHENSION QUESTIONS

1. Genetic drift is
   a. a change in allele frequencies due to random fluctuations.
   b. likely to result in allele loss or fixation over the long run.
   c. more pronounced in smaller populations.
   d. all of the above.
2. Which of the following influences on genetic drift involve the movement of a population from one location to another?
   a. The bottleneck effect
   b. The founder effect
   c. Both a and b
   d. None of the above

23.5 MIGRATION

Learning Outcome:

1. Explain how migration affects allele frequencies between neighboring populations, and calculate the magnitude of such a change.

We have just seen how the movement of a relatively small group to a new location can lead to genetic drift, resulting in a population with altered allele frequencies. In addition, migration between two different established populations can alter allele frequencies. For example, a species of birds may occupy two geographic regions that are separated by a large body of water. On rare occasions, the prevailing winds may allow birds from the western population to fly over this body of water and become members of the eastern population. If the two populations have different allele frequencies and if a sufficient number of western birds migrate, the migration may alter the allele frequencies in the eastern population.

After migration has occurred, the new (eastern) population is called a **conglomerate.** To calculate the allele frequencies in the conglomerate, we need two kinds of information. First, we must know the original allele frequencies in the donor and recipient populations. Second, we must know what proportion of the conglomerate population the migrants represent. With these data, we begin by calculating the change in allele frequency in the conglomerate population using the following equation:

$$\Delta p_C = m(p_D - p_R)$$

where

Δp_C is the change in allele frequency in the conglomerate population

p_D is the allele frequency in the donor population

p_R is the allele frequency in the original recipient population

m is the proportion of migrants in the conglomerate population, that is,

$$m = \frac{\text{number of migrants in the conglomerate population}}{\text{total number of individuals in the conglomerate population}}$$

As an example, let's suppose the allele frequency of A is 0.7 in the donor population and 0.3 in the recipient population. A group of 20 individuals migrates and joins the recipient population, which originally had 80 members. Thus,

$$m = \frac{20}{20 + 80}$$
$$= 0.2$$

$$\Delta p_C = m(p_D - p_R)$$
$$= 0.2(0.7 - 0.3)$$
$$= 0.08$$

We can now calculate the allele frequency in the conglomerate population:

$$p_C = p_R + \Delta p_C$$
$$= 0.3 + 0.08 = 0.38$$

Therefore, in the conglomerate population, the allele frequency of A has changed from 0.3 (its value in the recipient population before migration occurred) to 0.38. This increase in allele frequency arises from the higher allele frequency of A in the donor population.

Gene flow is the phenomenon in which individuals migrate from one population to another population and the migrants are able to breed successfully with the members of the recipient population. Gene flow depends not only on migration, but also on the ability of the migrants' alleles to be passed to subsequent generations.

In our example, we considered the consequences of a unidirectional migration from a donor to a recipient population. In nature, it is common for individuals to migrate in both directions. What are the main consequences of bidirectional migration? Depending on its rate, such migration tends to reduce differences in allele frequencies between neighboring populations. Populations that frequently mix their gene pools via migration tend to have similar allele frequencies, whereas isolated populations are expected to be more disparate. By comparison, genetic drift, which we considered earlier, tends to make local populations more disparate from each other.

Migration can also enhance genetic diversity within a population. As mentioned, new mutations are relatively rare events. Therefore, a particular mutation may arise in only one population. Migration may then introduce this new allele into neighboring populations.

23.5 REVIEWING THE KEY CONCEPTS

- Migration can alter allele frequencies. It tends to reduce differences in allele frequencies between neighboring populations and increase genetic diversity within a population.

23.5 COMPREHENSION QUESTION

1. Gene flow depends on
   a. migration.
   b. the ability of migrant alleles to be passed to subsequent generations.
   c. genetic drift.
   d. both a and b.

23.6 NONRANDOM MATING

Learning Outcomes:
1. Define *assortative mating, inbreeding,* and *outbreeding.*
2. Calculate the inbreeding coefficient for individuals in a pedigree.
3. Explain how inbreeding affects the Hardy-Weinberg equilibrium.

As mentioned earlier in this chapter, one of the conditions required to establish Hardy-Weinberg equilibrium is random mating (or random breeding), which means that individuals reproduce with each other irrespective of their genotypes and phenotypes. In many cases, particularly in human populations, this condition is violated frequently.

When sexual reproduction is nonrandom in a population, the process is called **assortative mating.**

- Positive assortative mating occurs when individuals with similar phenotypes reproduce with each other.
- The opposite situation, where dissimilar phenotypes preferentially reproduce, is called negative assortative mating.
- Individuals may reproduce with members that are part of the same genetic lineage. Reproduction between two genetically related individuals, such as cousins, is called **inbreeding.** Inbreeding sometimes occurs in human societies and is more likely to take place in nature when population size becomes very limited.
- **Outbreeding** involves reproduction between unrelated individuals. It can create hybrids that are heterozygous for many genes.

In the absence of other evolutionary processes, inbreeding and outbreeding do not affect allele frequencies in a population. However, these patterns of mating do disrupt the balance of genotypes that is predicted by the Hardy-Weinberg equation. Let's consider inbreeding in a family pedigree. **Figure 23.16** presents a human pedigree involving a mating between cousins. Individuals III-2 and III-3 are cousins and have produced the daughter labeled IV-1. She is said to be inbred, because her parents are genetically related to each other.

Inbreeding involves a smaller gene pool because the reproducing individuals are related genetically. In the 1940s, Gustave Malécot developed methods to quantify the degree of inbreeding. The **inbreeding coefficient (F)** is the probability that two alleles for a given gene in a particular individual will be identical because both copies are due to descent from a common ancestor. An inbreeding coefficient (F) can be computed by analyzing the degree of relatedness within a pedigree.

As an example, let's determine the inbreeding coefficient for individual IV-1. To begin this calculation, we must first identify all of this individual's common ancestors. A common ancestor is anyone who is an ancestor to both of an individual's parents. In Figure 23.16, IV-1 has one common ancestor, I-2, her great-grandfather. I-2 is the grandfather of III-2 and III-3.

Our next step is to determine the inbreeding paths. An inbreeding path for an individual is the shortest path through the pedigree that includes both parents and the common ancestor. In a pedigree, there is an inbreeding path for each common ancestor. The length of each inbreeding path is calculated by adding together all of the individuals in the path except the individual of interest. In this case, there is only one path because IV-1 has only one common ancestor. To add the members of the path, we begin with individual IV-1, but we do not count her. We then move to her father (III-2); to her grandfather on her father's side (II-2); to I-2, her great-grandfather (the common ancestor); back down to her grandmother on her mother's side (II-3); and finally to her mother (III-3). This path has five members. Finally, to calculate the inbreeding coefficient, we use the following formula:

$$F = \Sigma(1/2)^n(1 + F_A)$$

where

  F is the inbreeding coefficient of the individual of interest
  n is the number of individuals in the inbreeding path, excluding the inbred offspring
  F_A is the inbreeding coefficient of the common ancestor
  Σ indicates that we add together $(1/2)^n(1 + F_A)$ for each inbreeding path

In this case, there is only one common ancestor and, therefore, only one inbreeding path. Also, we do not know anything about the heritage of the common ancestor, so we assume that F_A is zero. Thus, for the example in Figure 23.16,

$$F = \Sigma(1/2)^n(1 + 0)$$
$$= (1/2)^5 = 1/32 = 3.125\%$$

What does this value mean? The inbreeding coefficient, 3.125%, tells us the probability that a gene in the inbred individual (IV-1) is homozygous due to its inheritance from a common ancestor (I-2). In this case, each gene in individual IV-1 has a 3.125% chance of being homozygous because the individual has inherited the same allele twice from her great-grandfather (I-2), once through each parent.

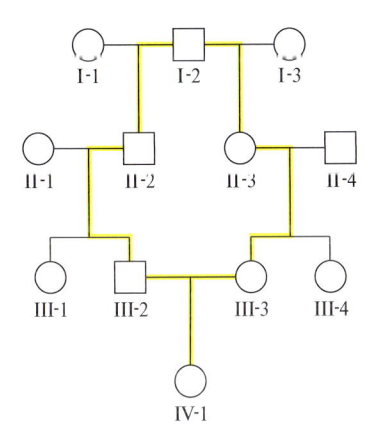

FIGURE 23.16 **A human pedigree containing inbreeding.** Individual IV-1 is the result of inbreeding because her parents are related. The inbreeding path is highlighted in yellow.

Concept Check: How does inbreeding affect the likelihood that recessive traits will be expressed? Explain.

As an example, let's suppose that the common ancestor (I-2) is heterozygous for the gene involved in cystic fibrosis (CF). His genotype is *Cc*, where *c* is the recessive allele that causes CF. Because *F* is 3.125%, the probability is 3.125% that the inbred individual (IV-1) is homozygous (*CC* or *cc*) for this gene because she has inherited both copies from her great-grandfather. She has a 1.56% probability of inheriting both normal alleles (*CC*) and a 1.56% probability of inheriting both mutant alleles (*cc*). The inbreeding coefficient is denoted by the letter *F* (for fixation) because it is the probability that an allele will be fixed in the homozygous condition. The term *fixation* signifies that the homozygous individual can pass only one type of allele to their offspring.

In other pedigrees, an individual may have two or more common ancestors. In this case, the inbreeding coefficient *F* is calculated using the sum of the number of individuals in the inbreeding paths. (Such a situation is described in question 3 of More Genetic TIPS at the end of this chapter.)

What are the consequences of inbreeding in a population? Inbreeding disrupts Hardy-Weinberg equilibrium, resulting in a higher proportion of homozygotes. From an agricultural viewpoint, this outcome may be desirable. For example, an animal breeder may use inbreeding to produce animals that are larger because they have become homozygous for alleles promoting larger size. On the negative side, many genetic diseases are inherited in a recessive manner (see Chapter 22). For these disorders, inbreeding increases the likelihood that an individual will be homozygous for the recessive allele and therefore afflicted with the disease.

Also, in natural populations, inbreeding lowers the mean fitness of the population if homozygous offspring have lower relative fitness values. The lowering of mean fitness can be a serious problem as natural populations become smaller due to human habitat destruction. As a population shrinks, inbreeding becomes more likely because individuals have fewer potential mates from which to choose. The inbreeding, in turn, produces homozygotes that are less fit, thereby decreasing the reproductive success of the population. This phenomenon is called **inbreeding depression.** Conservation biologists sometimes try to circumvent this problem by introducing individuals from one population into another. For example, the endangered Florida panther (*Felis concolor coryi*) suffers from inbreeding-related defects, which include poor sperm quality and quantity and morphological abnormalities. To help alleviate these effects, panthers of the same species from Texas have been introduced into the Florida population.

23.6 REVIEWING THE KEY CONCEPTS

- Nonrandom mating may alter the genotype frequencies that are predicted by the Hardy-Weinberg equation. Inbreeding results in a higher proportion of homozygotes in a population (see Figure 23.16).

23.6 COMPREHENSION QUESTION

1. Inbreeding is sexual reproduction between individuals that are
   a. homozygous.
   b. heterozygous.
   c. part of the same genetic lineage.
   d. both a and c.

23.7 SOURCES OF NEW GENETIC VARIATION

Learning Outcomes:
1. List different sources of genetic variation.
2. Define *mutation rate*, and calculate how it affects allele frequencies in populations.
3. Define *horizontal gene transfer*, and explain how it produces genetic variation.
4. Describe the occurrence of repetitive sequences, and explain how they are used in DNA fingerprinting.

In the previous sections, we primarily focused on genetic variation in cases when a single gene exists in two or more alleles. These simplified scenarios allow us to appreciate the general principles behind evolutionary mechanisms. As researchers have analyzed genetic variation at the molecular, cellular, and population levels, however, they have come to understand that new genetic variation occurs in many ways (**Table 23.2**). Among eukaryotic species, sexual reproduction is an important way that new genetic variation occurs among offspring. In Chapters 3 and 7, we considered how independent assortment and crossing over during sexual reproduction may produce new combinations of alleles in various genes, thereby producing new genetic variation in the resulting offspring. Though prokaryotic species reproduce asexually, they also possess mechanisms for gene transfer, such as conjugation, transduction, and transformation (see Chapter 9). These mechanisms are important for fostering genetic variation among bacterial and archaeal populations.

Rare mutations in DNA may also give rise to new types of variation (see Table 23.2). As discussed earlier in this chapter (see Figure 23.3) and in Chapter 19, mutations may occur within a particular gene to create new alleles of that gene. Such allelic variation is common in natural populations. Also, as described in Chapter 8, gene duplications may create a gene family; each family member acquires independent mutations and often evolves more specialized functions. An example is the globin gene family (refer back to Figure 8.7). In this section, we will examine some additional mechanisms through which an organism can acquire new genetic variation. These include horizontal gene transfer and changes in repetitive sequences. The diversity of mechanisms for fostering genetic variation underscores its profound importance in the evolution of species that are both well adapted to their native environments and successful at reproduction.

Mutations Provide the Source of Genetic Variation

As discussed in Chapters 8 and 19, mutations involve changes in gene sequences, chromosome structure, and/or chromosome number. Mutations are random events that occur spontaneously at a low rate or are caused by mutagens at a higher rate. In 1926, the Russian geneticist Sergei Chetverikov was the first to suggest that mutational variability provides the raw material for evolution but does not constitute a significant evolutionary change. In

TABLE 23.2 Sources of New Genetic Variation That Occur in Populations

Type	Description
Independent assortment	The independent segregation of different chromosomes may give rise to new combinations of alleles in offspring (see Chapter 3).
Crossing over	Recombination (crossing over) between homologous chromosomes can also produce new combinations of alleles that are located on the same chromosome (see Chapter 7).
Interspecies crosses	On occasion, members of different species may breed with each other to produce hybrid offspring.
Prokaryotic gene transfer	Prokaryotic species possess mechanisms of genetic transfer such as conjugation, transduction, and transformation (see Chapter 9).
New alleles	Point mutations can occur within a gene to create single-nucleotide polymorphisms (SNPs). In addition, genes can be altered by small deletions and additions. Gene mutations are discussed in Chapter 19.
Gene duplications	Events, such as misaligned crossovers, can add additional copies of a gene into a genome and lead to the formation of gene families (see Chapter 8).
Chromosome structure and number	Chromosome structure may be changed by deletions, duplications, inversions, and translocations. Changes in chromosome number result in aneuploid, polyploid, and alloploid offspring. These mechanisms are discussed in Chapter 8.
Exon shuffling	New genes can be created when exons of preexisting genes are rearranged to make a gene that encodes a protein with a new combination of domains.
Horizontal gene transfer	Genes from one species can be introduced into another species and become incorporated into that species' genome.
Changes in repetitive sequences	Short repetitive sequences are common in genomes due to the occurrence of transposable elements and tandem arrays. The numbers and lengths of repetitive sequences tend to show considerable variation in natural populations.

mutations such as frameshift mutations, missense mutations, and nonsense mutations all may cause a gene to express a protein that is nonfunctional or less functional than the wild-type protein. Also, mutations in non-coding sequences can alter gene expression (refer back to Table 19.2). Neutral mutations, which are not acted upon by natural selection, occur in several different ways. For example, a neutral mutation can change the base in the wobble position without affecting the amino acid sequence of the encoded protein, or a neutral mutation can be a missense mutation that has no effect on protein function. Such point mutations happen at specific sites within the coding sequence. Neutral mutations can also occur within introns, the non-coding sequences of genes. By comparison, beneficial mutations are relatively uncommon. To be advantageous, a new mutation might alter the amino acid sequence of a protein to yield a better-functioning product. Although such mutations do occur, they are expected to be very rare for a population in a stable environment.

The **mutation rate** is defined as the probability that a gene will be altered by a new mutation. The rate is typically expressed as the number of new mutations in a given gene per generation. A common value for the mutation rate is in the range of 1 in 100,000 to 1 in 1,000,000, or 10^{-5} to 10^{-6} per generation. However, mutation rates vary depending on species, cell type, chromosomal location, and gene size. Furthermore, in experimental studies, the mutation rate is usually measured by following the change of a normal (functional) gene to a deleterious (nonfunctional) allele. The mutation rate producing beneficial alleles is expected to be substantially less.

It is clear that new mutations provide genetic variability, but population geneticists also want to know how much the mutation rate affects the allele frequencies in a population. Can random mutations have a large effect on allele frequencies over time? To answer this question, let's consider this simple case: A gene exists as an allele, A; the allele frequency of A is denoted by the variable p. A mutation can convert the A allele into a different allele called a. The allele frequency of a is designated by q. The conversion of the A allele into the a allele by mutation occurs at a rate that is designated μ. If we assume that the rate of the reverse mutation (a to A) is negligible, the increase in the frequency of the a allele after one generation is

$$\Delta q = \mu p$$

For example, let's consider the following conditions:

$p = 0.8$ (i.e., frequency of A is 80%)

$q = 0.2$ (i.e., frequency of a is 20%)

$\mu = 10^{-5}$ (i.e., the mutation rate for the conversion of A to a)

$\Delta q = (10^{-5})(0.8) = (0.00001)(0.8) = 0.000008$

Therefore, in the next generation (designated $n + 1$),

$q_{n+1} = 0.2 + 0.000008 = 0.200008$

$p_{n+1} = 0.8 - 0.000008 = 0.799992$

As we can see from this calculation, new mutations do not significantly alter the allele frequencies in a single generation.

other words, mutation can produce new alleles in a population but does not substantially alter allele frequencies. Chetverikov proposed that populations in nature absorb mutations like a sponge and retain them in a heterozygous condition, thereby providing a source of variability that may lead to future change.

Population geneticists often consider how new mutations affect the survival and reproductive potential of the individual that inherits them. A new mutation may be deleterious, neutral, or beneficial, depending on its effect. For genes that encode proteins, the effects of new mutations depend on their influence on protein function. Deleterious and neutral mutations are far more likely to occur than beneficial ones. For example, alleles can be altered in many different ways that render an encoded protein defective. As discussed in Chapter 19, deletions and point

We can use the following equation to calculate the change in allele frequency after any number of generations:

$$(1 - \mu)^t = \frac{p_t}{p_0}$$

where

μ is the mutation rate for the conversion of A to a

t is the number of generations

p_0 is the allele frequency of A in the starting generation

p_t is the allele frequency of A after t generations

As an example, let's suppose that the allele frequency of A is 0.8, $\mu = 10^{-5}$, and we want to know what the allele frequency will be after 1000 generations ($t = 1000$). Plugging these values into the preceding equation and solving for p_t gives

$$(1 - 0.00001)^{1000} = \frac{p_t}{0.8}$$

$$p_t = 0.792$$

Therefore, after 1000 generations, the frequency of A has dropped only from 0.8 to 0.792. Again, these results point to how slowly the occurrence of new mutations changes allele frequencies. In natural populations, the rate of new mutation is rarely a significant catalyst in shaping allele frequencies. Instead, other processes such as natural selection, genetic drift, and migration have far greater effects on allele frequencies.

New Genes Are Acquired via Horizontal Gene Transfer

Species also accumulate genetic changes by a process called **horizontal gene transfer,** in which an organism incorporates genetic material from another organism without being the offspring of that organism. This process often involves the exchange of genetic material between different species. **Figure 23.17** illustrates one possible mechanism for horizontal gene transfer. In this example, a eukaryotic cell has engulfed a bacterium by endocytosis. During the degradation of the bacterium, a bacterial gene escapes to the nucleus of the cell, where it is inserted into one of the chromosomes. In this way, a gene has been transferred from a bacterial species to a eukaryotic species. By analyzing gene sequences among many different species, researchers have discovered that horizontal gene transfer is a common phenomenon. This process can occur from prokaryotes to eukaryotes, from eukaryotes to prokaryotes, between different species of prokaryotes, and between different species of eukaryotes.

Gene transfer among bacterial species is relatively widespread. As discussed in Chapter 9, bacterial species may carry out three natural mechanisms of gene transfer known as conjugation, transduction, and transformation. By analyzing the genomes of bacterial species, scientists have determined that many genes within a given bacterial genome are derived from genes acquired from other species via horizontal gene transfer. Genome studies have suggested that as much as 20% to 30% of the variation in the genetic composition of modern prokaryotic species can be attributed to this process. For example, roughly 17% of the genes of *E. coli* and *Salmonella typhimurium* have been acquired from other species via horizontal gene transfer during the past 100 million years. The roles of these acquired genes are quite varied, though they commonly involve functions that are readily acted on by natural selection. These functions include antibiotic resistance, the ability to degrade toxic compounds, and pathogenicity (the ability to cause disease).

Genetic Variation Is Produced via Changes in Repetitive Sequences

Another source of genetic variation involves changes in **repetitive sequences**—short sequences, typically a few base pairs to a few thousand base pairs long, that are repeated many times within a species' genome. Repetitive sequences usually come from two types of sources. First, transposable elements (TEs) are DNA sequences that can move from place to place in a species' genome (see Chapter 12). The prevalence and movement of TEs provide a great deal of genetic variation between species and within a single species. In certain eukaryotic species, TEs have become fairly abundant (see Table 12.2).

A second type of repetitive sequence is nonmobile and consists of short sequences that are tandemly repeated. Such tandem repeat sequences are called satellites because they usually sediment away from the rest of the chromosomal DNA during equilibrium density centrifugation. In a **microsatellite** (also called a short tandem repeat, STR), the repeat unit is usually 1 to 6 bp long, and the whole tandem repeat is less than a few hundred base pairs in length. For example, the most common microsatellite encountered in humans is the sequence $(CA)_N$, where N may range from 5 to more than 50. In other words, this dinucleotide sequence can be tandemly repeated 5 to 50 or more times. The $(CA)_N$ microsatellite is found, on average, about every 10,000 bases in the human genome. In a **minisatellite,** the repeat unit is typically 6 to 80 bp in length, and the size of the minisatellite ranges from 1 kbp to 20 kbp. An example of a minisatellite in humans is telomeric DNA. In a human sperm cell, for example, the repeat unit is 6 bp and the size of a telomere is about 15 kbp.

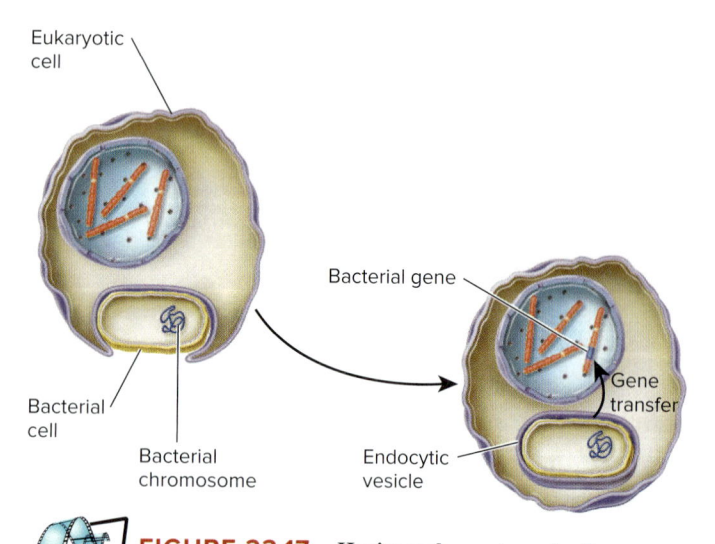

FIGURE 23.17 Horizontal gene transfer from a bacterium to a eukaryote. In this example, a bacterium is engulfed by a eukaryotic cell, and a bacterial gene is transferred to one of the eukaryotic chromosomes.

Tandem repetitive sequences such as microsatellites and minisatellites tend to undergo mutations in which the number of tandem repeats changes. For example, a microsatellite with a 4-bp repeat unit and a length of 64 bp may undergo a mutation that adds three more repeat units and becomes 76 bp long. Micro- and minisatellites may change in length by different mechanisms. One common mechanism to explain this phenomenon is slippage of DNA polymerase off the template strand during DNA replication. As discussed in Chapter 19, this mechanism can explain trinucleotide repeat expansion (TNRE) (refer back to Figure 19.11).

Because repetitive sequences tend to vary within a population, they have become a common tool that geneticists use in a variety of ways. For example, as described in Chapters 21 and 22, microsatellites are used as molecular markers to map the locations of genes (refer back to Figure 22.7). Also, population geneticists analyze microsatellites or minisatellites to study variation at the population level and to determine the relationships among individuals and between neighboring populations. The sizes of microsatellites and minisatellites found in closely related individuals tend to be more similar than the sizes of those in unrelated individuals. As described next, this phenomenon is the basis for DNA fingerprinting.

DNA Fingerprinting Is Used for Identification and Relationship Testing

The technique of **DNA fingerprinting,** also known as **DNA profiling,** analyzes individuals based on the occurrence of repetitive sequences in their genome. When subjected to traditional DNA fingerprinting, the chromosomal DNA gives rise to a series of bands on a gel (**Figure 23.18**). The sizes and order of bands is an individual's DNA fingerprint. Like the human fingerprint, the DNA of each individual has a distinctive pattern. It is the unique patterns of these bands that make it possible to distinguish individuals.

A comparison of the DNA fingerprints among different individuals has found two applications:

- DNA fingerprinting is used as a method of identification. In forensics, DNA fingerprinting can identify a crime suspect. In medicine, the technique can identify the type of bacterium that is causing an infection in a particular patient.
- DNA fingerprinting is also used for relationship testing. Closely related individuals have more similar fingerprints than do distantly related ones. In humans, such similarity is useful for paternity testing. In population genetics, DNA fingerprinting can provide evidence regarding the degree of relatedness among members of a population. Such information may help geneticists determine if a population is likely to be suffering from inbreeding depression.

The development of DNA fingerprinting has relied on the identification of DNA sequences that vary in length among members of a population. This naturally occurring variation causes each individual to have a unique DNA fingerprint. In the 1980s, Alec Jeffreys and his colleagues found that certain minisatellites within human chromosomes are particularly variable in their lengths. As discussed earlier, minisatellites tend to vary within populations due to changes in the number of tandem repeats at each site.

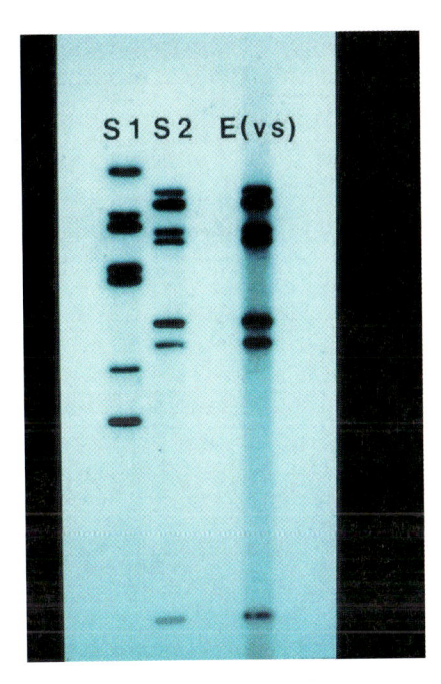

FIGURE 23.18 A comparison of two DNA fingerprints. The chromosomal DNA from two different individuals (suspect 1 is S1, and suspect 2 is S2) was subjected to DNA fingerprinting. The DNA evidence from a crime scene, E(vs), was also subjected to DNA fingerprinting. Following hybridization of a labeled molecule, the DNA appears as a series of bands on a gel. The dissimilarity in the pattern of these bands distinguishes different individuals, much as the differences in physical fingerprint patterns can be used for identification. As seen here, the DNA from S2 matches the DNA found at the crime scene.
©Leonard Lessin

Concept Check: *What are two common applications of DNA fingerprinting?*

In the past decade, the technique of DNA fingerprinting has become automated, much like the automation that changed the procedure of DNA sequencing, described in Chapter 20. DNA fingerprinting is now done using the technique of polymerase chain reaction (PCR), which amplifies microsatellites. Like minisatellites, microsatellites are found in multiple sites in the genome of humans and other species and vary in length among different individuals. The results of one example of DNA fingerprinting are shown in **Figure 23.19**.

1. In this procedure, the microsatellites from a sample of DNA are amplified by PCR using primers that flank several different repetitive regions.
2. The amplified microsatellite fragments are fluorescently labeled.
3. They are then separated by gel electrophoresis according to their molecular masses.
4. As each microsatellite approaches the end of the gel, a laser excites the fluorescent molecules within a microsatellite, and a detector records the amount of fluorescence emission for each microsatellite.

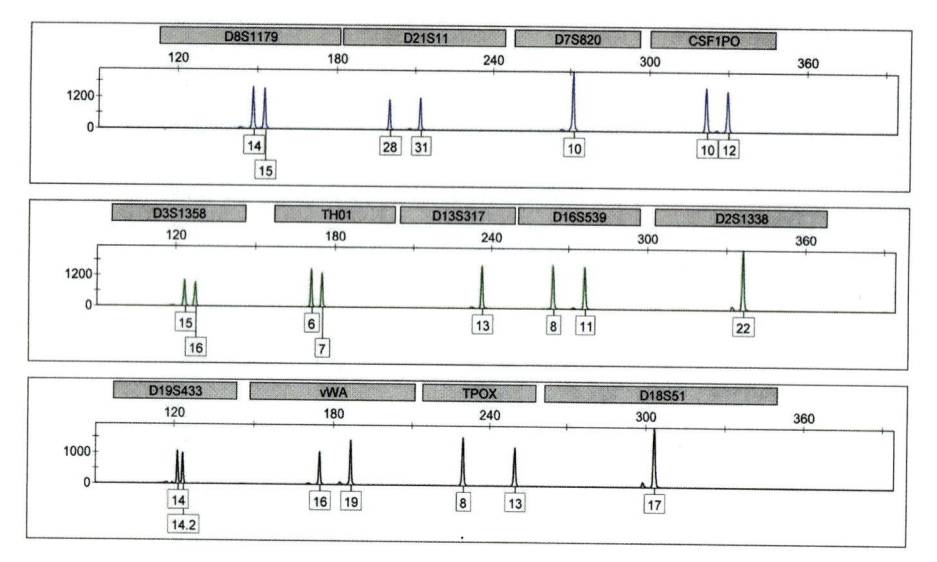

FIGURE 23.19 Automated DNA fingerprinting. In automated DNA fingerprinting, a sample of DNA is amplified, using primers that recognize the ends of microsatellites. The microsatellite fragments are fluorescently labeled and then separated by gel electrophoresis. The fluorescent molecules within each microsatellite are excited with a laser, and the amount of fluorescence is measured via a fluorescence detector. A printout from the detector is shown here. The gray boxes indicate the names of specific microsatellites. The peaks show the relative amounts of each microsatellite. The boxes beneath each peak indicate the number of tandem repeats in a given microsatellite. In this example, the individual is heterozygous for certain microsatellites (e.g., D8S1179) and homozygous for others (e.g., D7S820).

5. As shown in Figure 23.19, this type of DNA fingerprint yields a series of peaks, each peak corresponding to a characteristic molecular mass. In this automated approach, the pattern of peaks rather than bands constitutes an individual's DNA fingerprint.

23.7 REVIEWING THE KEY CONCEPTS

- A variety of different mechanisms can bring about genetic variation (see Table 23.2).
- The mutation rate is the probability that a gene will be altered by a new mutation.
- A species may acquire a new gene from another species via horizontal gene transfer (see Figure 23.17).
- A common source of genetic variation in populations involves changes in repetitive sequences, such as microsatellites.
- DNA fingerprinting is a technique that relies on variation in repetitive sequences within a population. It is used as a means of identification and in relationship testing (see Figures 23.18, 23.19).

23.7 COMPREHENSION QUESTIONS

1. The mutation rate is
   a. the likelihood that a new mutation will occur in a given gene.
   b. too low to substantially change allele frequencies in a population.
   c. lower for mutations that create beneficial alleles.
   d. All of the above are true of the mutation rate.

2. The transfer of an antibiotic resistance gene from one bacterial species to a different species is an example of
   a. exon shuffling.
   b. horizontal gene transfer.
   c. genetic drift.
   d. migration.

3. DNA fingerprinting analyzes the DNA from individuals on the basis of the occurrence of _____ in their genomes.
   a. repetitive sequences
   b. abnormalities in chromosome structure
   c. specific genes
   d. viral insertions

KEY TERMS

Page 554. population genetics, gene pool
Page 555. population, local populations, genetic polymorphism, polymorphism
Page 556. polymorphic, monomorphic, single-nucleotide polymorphism (SNP), allele frequencies, genotype frequencies
Page 557. Hardy-Weinberg equilibrium, Hardy-Weinberg equation
Page 559. disequilibrium,
Page 560. microevolution
Page 561. natural selection, Darwinian fitness, relative fitness
Page 562. directional selection
Page 563. mean fitness of the population ($\bar{w}$)
Page 564. balancing selection, heterozygote advantage, selection coefficient (s)
Page 565. negative frequency-dependent selection
Page 566. disruptive selection, stabilizing selection
Page 567. random genetic drift, genetic drift
Page 569. bottleneck effect, founder effect
Page 570. conglomerate, gene flow
Page 571. assortative mating, inbreeding, outbreeding, inbreeding coefficient (F)
Page 572. inbreeding depression
Page 573. mutation rate
Page 574. horizontal gene transfer, repetitive sequences, microsatellite, minisatellite
Page 575. DNA fingerprinting, DNA profiling

CHAPTER SUMMARY

- Population genetics is concerned with changes in genetic variation within a group of individuals over time.

23.1 Genes in Populations and the Hardy-Weinberg Equation

- All of the alleles of every gene in a population constitute the population's gene pool.
- For sexually reproducing organisms, a population is a group of individuals of the same species that occupy the same region and can interbreed with one another (see Figure 23.1).
- In population genetics, polymorphism refers to inherited traits or genes that exhibit variation in a population (see Figure 23.2).
- Single-nucleotide polymorphisms (SNPs) are the most common type of variation among genes (see Figure 23.3).
- Geneticists analyze genetic variation by determining allele and genotype frequencies.
- The Hardy-Weinberg equation can be used to calculate genotype frequencies based on allele frequencies (see Figures 23.4, 23.5).
- Deviation from Hardy-Weinberg equilibrium indicates that evolutionary change is occurring.
- The Hardy-Weinberg equation can be used to estimate the frequency of heterozygous carriers.

23.2 Overview of Microevolution

- Microevolution refers to changes in a population's gene pool from generation to generation.
- Mutations are the source of new genetic variation. However, the occurrence of new mutations does not greatly change allele frequencies because it happens at a very low rate. Other factors, such as natural selection, genetic drift, migration, and nonrandom mating, may alter allele and/or genotype frequencies (see Table 23.1).

23.3 Natural Selection

- Natural selection is a process that influences allele frequencies from one generation to the next based on fitness, which is a measure of the relative reproductive success of different genotypes.
- Directional selection favors the extreme phenotype (see Figures 23.6–23.8).

- Balancing selection results in stable polymorphism. It may arise as a result of heterozygote advantage or negative frequency-dependent selection (see Figures 23.9, 23.10).
- Disruptive selection favors multiple phenotypes in heterogeneous environments (see Figures 23.11, 23.12).
- Stabilizing selection favors individuals with intermediate phenotypes (see Figure 23.13).

23.4 Genetic Drift

- Genetic drift involves changes in allele frequencies in a population due to random fluctuations. Over the long run, it often results in allele fixation or loss. The effect of genetic drift is greater in small populations (see Figure 23.14).
- Two mechanisms that can influence genetic drift are the bottleneck effect and the founder effect (see Figure 23.15).

23.5 Migration

- Migration can alter allele frequencies. It tends to reduce differences in allele frequencies between neighboring populations and increase genetic diversity within a population.

23.6 Nonrandom Mating

- Nonrandom mating may alter the genotype frequencies that are predicted by the Hardy-Weinberg equation. Inbreeding results in a higher proportion of homozygotes in a population (see Figure 23.16).

23.7 Sources of New Genetic Variation

- A variety of different mechanisms can bring about genetic variation (see Table 23.2).
- The mutation rate is the probability that a gene will be altered by a new mutation.
- A species may acquire a new gene from another species via horizontal gene transfer (see Figure 23.17).
- A common source of genetic variation in populations involves changes in repetitive sequences, such as microsatellites.
- DNA fingerprinting is a technique that relies on variation in repetitive sequences within a population. It is used as a means of identification and in relationship testing (see Figures 23.18, 23.19).

PROBLEM SETS & INSIGHTS

More Genetic TIPS

1. In a particular population, the phenotypic frequency of people who cannot taste phenylthiocarbamide (PTC) is approximately 0.3. The inability to taste this bitter substance is due to homozygosity of a recessive allele. If we assume there are only two alleles in the population (namely, tasting, *T*, and nontasting, *t*) and that the population is in Hardy-Weinberg equilibrium, calculate the frequencies of these two alleles.

Topic: What topic in genetics does this question address?

The topic is predicting allele frequencies in a population. More specifically, the question asks you to predict the frequencies of alleles that affect people's ability to taste phenylthiocarbamide (PTC).

Information: What information do you know based on the question and your understanding of the topic?

From the question, you know the frequency of homozygotes who are nontasters. From your understanding of the topic, you may realize that you can use the Hardy-Weinberg equation to determine allele frequencies if you know the genotype frequencies.

Problem-Solving Strategy: Make a calculation.

One strategy to solve this problem is to use the Hardy-Weinberg equation to determine the allele frequencies.

If q represents the allele frequency of the recessive allele (t) that confers nontasting, then

$$q^2 = 0.3$$

We take the square root of both sides of the equation to determine q:

$$q = \sqrt{0.3}$$
$$q = 0.55$$

If p represents the tasting allele (T),

$$p = 1 - q$$
$$p = 1 - 0.55 = 0.45$$

Answer: The frequency of the nontasting allele is 0.55, or 55%, and that of the tasting allele is 0.45, or 45%.

2. Let's suppose that pigmentation in a species of insect is controlled by a single gene that exists as two alleles, D for dark and d for light. The heterozygote, Dd, is intermediate in color. In a heterogeneous environment, the allele frequencies are $D = 0.7$ and $d = 0.3$. This polymorphism is maintained because the environment has some dimly lit forested areas and some sunny fields. During a hurricane, a group of 1000 insects is blown to a completely sunny area. In this environment, the relative fitness values are $DD = 0.3$, $Dd = 0.7$, and $dd = 1.0$. What are the predicted allele frequencies in the next generation?

Topic: What topic in genetics does this question address?

The topic is about directional selection. More specifically, the question is about directional selection in an insect population that moved to a new location.

Information: What information do you know based on the question and your understanding of the topic?

In the question, you are given the relative fitness values for three genotypes and the starting allele frequencies. From your understanding of the topic, you may remember that geneticists have derived equations to calculate directional selection from one generation to the next.

Problem-Solving Strategy: Make a calculation.

To solve this problem you first need to calculate the mean fitness of the population and then calculate the allele frequencies in the next generation.

Calculating the mean fitness of the population:

$$p^2 w_{DD} + 2pq w_{Dd} + q^2 w_{dd} = \overline{w}$$
$$\overline{w} = (0.7)^2(0.3) + 2(0.7)(0.3)(0.7) + (0.3)^2(1.0)$$
$$= 0.15 + 0.29 + 0.09 = 0.53$$

After one generation of selection, the allele frequencies will be:

Allele frequency of D: $p_D = \dfrac{p^2 w_{DD}}{\overline{w}} + \dfrac{pq w_{Dd}}{\overline{w}}$

$$= \dfrac{(0.7)^2(0.3)}{0.53} + \dfrac{(0.7)(0.3)(0.7)}{0.53}$$
$$= 0.55$$

Allele frequency of d: $q_d = \dfrac{q^2 w_{dd}}{\overline{w}} + \dfrac{pq w_{Dd}}{\overline{w}}$

$$= \dfrac{(0.3)^2(1.0)}{0.53} + \dfrac{(0.7)(0.3)(0.7)}{0.53}$$
$$= 0.45$$

Answer: After one generation, the allele frequency of D has decreased from 0.7 to 0.55, and that of d has increased from 0.3 to 0.45.

3. Using the pedigree shown next, answer the following questions with regard to individual VII-1:

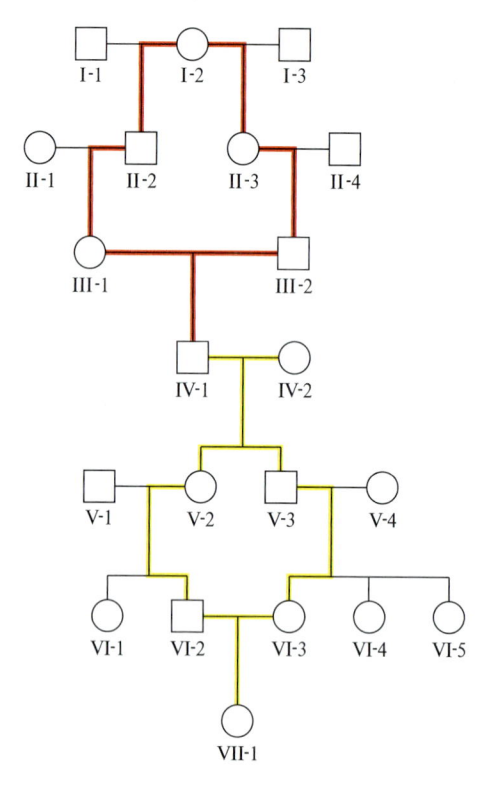

A. Who are the common ancestors of her parents?

B. What is the inbreeding coefficient for this individual?

Topic: What topic in genetics does this question address?

The topic is inbreeding. More specifically, the questions ask you to identify the common ancestor(s) in a pedigree and calculate the inbreeding coefficient for a particular individual.

Information: What information do you know based on the question and your understanding of the topic?

In the question, you are given a family pedigree. From your understanding of the topic, you may remember that a common ancestor is someone who is an ancestor of both of a person's parents. You may also remember that geneticists have derived an equation to calculate the inbreeding coefficient.

Problem-Solving Strategy: Compare and contrast. Make a calculation.

To solve this problem, you first need to compare the members of the pedigree and identify which of them are common ancestors of both of the individual's parents. You then can use the inbreeding equation to calculate her inbreeding coefficient.

Answer:

A. The common ancestors are IV-1 and IV-2. They are the grandparents of VI-2 and VI-3, who are the parents of VII-1. (Follow the yellow-highlighted lines.)

B. The inbreeding coefficient is calculated using the formula

$$F = \Sigma(1/2)^n(1 + F_A)$$

In this case, the two common ancestors are IV-1 and IV-2. Also, IV-1 is inbred, because I-2 is a common ancestor to both of IV-1's parents. The first step is to calculate F_A, the inbreeding coefficient for IV-1. The inbreeding path for IV-1, which is highlighted in red, contains five people: III-1, II-2, I-2, II-3, and III-2. Therefore,

$$n = 5$$
$$F_A = (1/2)^5 = 0.03$$

Now we can calculate the inbreeding coefficient for VII-1. Each inbreeding path for VII-1, highlighted in yellow, contains five people: VI-2, V-2, IV-1, V-3, and VI-3; and VI-2, V-2, IV-2, V-3, and VI-3. Thus,

$$F = (1/2)^5 (1 + 0.03) + (1/2)^5 (1 + 0)$$
$$= 0.032 + 0.031 = 0.063$$

4. An important application of DNA fingerprinting is relationship testing. Persons who are related genetically have some bands or peaks in common. The number they share depends on the closeness of their genetic relationship. For example, an offspring is expected to receive half of his or her minisatellites from one parent and the rest from the other.

The following diagram schematically shows traditional DNA fingerprints of an offspring, the mother, and two potential fathers. In paternity testing, the offspring's DNA fingerprint is first compared with that of the mother. The bands that the offspring has in common with the mother are shown in purple. The bands that are not similar between the offspring and the mother must have been inherited from the father. These bands are shown in red.

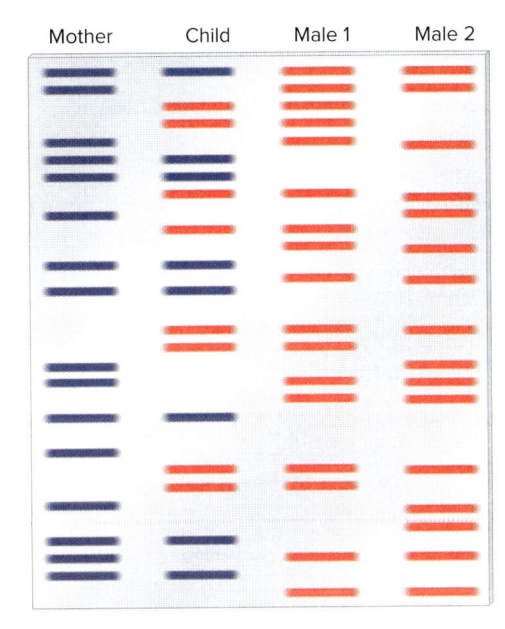

Which male could be the father?

Topic: What topic in genetics does this question address?

The topic is DNA fingerprinting. More specifically, the question is about using DNA fingerprinting to determine paternity.

Information: What information do you know based on the question and your understanding of the topic?

From the question, you know the pattern of bands in the DNA fingerprints of a mother, an offspring, and two potential fathers. From your understanding of the topic, you may remember that an offspring shares 50% of its bands with its mother and 50% with its father.

Problem-Solving Strategy: Compare and contrast.

One strategy to solve this problem is to compare the bands of the offspring with each of the fathers.

Answer: Male 2 does not have many of the red (paternal) bands seen in the offspring's fingerprint. Therefore, he can be excluded as being the father of this child. However, male 1 has all of the paternal bands. He is very likely to be the father.

Note: Geneticists can calculate the likelihood that the matching bands between the offspring and a prospective father could occur as a matter of random chance. To do so, they analyze the frequency of each band in a reference population (e.g., people of Northern European descent living in the United States). For example, let's suppose that DNA fingerprinting analyzed 40 bands. Of these, 20 bands matched with the mother and 20 bands matched with a prospective father. If the probability of each of these bands in a reference population was 1/4, the likelihood of such a match occurring by random chance would be $(1/4)^{20}$, or roughly 1 in 1 trillion. Therefore, a match between two samples is rarely a matter of random chance.

Conceptual Questions

C1. What is the gene pool? How is a gene pool described in a quantitative way?

C2. In genetics, what does the term *population* mean? Pick any species you like and describe how its population might change over the course of many generations.

C3. What is genetic polymorphism? What is the source of genetic variation?

C4. Identify each of the following as an example of allele, genotype, and/or phenotype frequency:

  A. Approximately 1 in 2500 individuals of Nothern European descent is born with cystic fibrosis.

  B. The percentage of carriers of the sickle cell allele in West Africa is approximately 13%.

  C. The number of new mutations per generation resulting in achondroplasia, a genetic disorder, is approximately 5×10^{-5}.

C5. The term *polymorphism* can refer to both genes and traits. Explain what is meant by a polymorphic gene and a polymorphic trait. If a gene is polymorphic, does the trait that the gene affects also have to be polymorphic? Explain why or why not.

C6. Cystic fibrosis (CF) is a recessive autosomal disorder. In certain populations of Northern European descent, the number of people born with this disorder is about 1 in 2500. Assuming Hardy-Weinberg equilibrium for this trait:

  A. What are the frequencies for the common (non-disease-causing) allele and the mutant (disease-causing) allele.

  B. What are the genotype frequencies of homozygous unaffected, heterozygous, and homozygous affected individuals?

  C. Assuming random mating, what is the probability that two phenotypically unaffected heterozygous carriers will choose each other as mates?

C7. For a gene existing in two alleles, what are the allele frequencies when the heterozygote frequency is at its maximum value, assuming Hardy-Weinberg equilibrium? What if there are three alleles?

C8. In a population, the frequencies of two alleles are $B = 0.67$ and $b = 0.33$. The genotype frequencies are $BB = 0.50$, $Bb = 0.37$, and $bb = 0.13$. Do these numbers suggest inbreeding? Explain why or why not.

C9. The ability to roll your tongue is inherited as a recessive trait. The frequency of the rolling allele is approximately 0.6, and the dominant (nonrolling) allele is 0.4. What is the frequency of individuals who can roll their tongues?

C10. What evolutionary factors can cause allele frequencies to change and possibly lead to genetic polymorphism? Discuss the relative importance of each type of process.

C11. What is the difference between a neutral and an adaptive evolutionary process? Describe two or more examples of each. At the molecular level, explain how mutations can be neutral or adaptive.

C12. What is Darwinian fitness? What types of characteristics can promote high fitness values? Give several examples.

C13. What is the intuitive meaning of the mean fitness of a population? How does its value change in response to natural selection?

C14. Describe the similarities and differences among directional, balancing, disruptive, and stabilizing selection.

C15. Is each of the following examples due to directional, disruptive, balancing, or stabilizing selection?

  A. Polymorphisms in color and banding pattern of the shells of land snails as described in Figure 23.12

  B. Thick fur among mammals living in cold climates

  C. Birth weight in humans

  D. Sturdy stems and leaves among plants exposed to windy climates

C16. With regard to the term *genetic drift*, what is drifting? Why is this an appropriate term to describe this phenomenon?

C17. Why is genetic drift more significant in small populations? Why does it take longer for genetic drift to cause allele fixation in large populations than in small ones?

C18. A group of four birds flies to a new location and starts a new colony. Three of the birds are homozygous AA, and one bird is heterozygous Aa.

  A. What is the probability that the a allele will become fixed in the population via genetic drift?

  B. If fixation of the a allele occurs, how long will it take?

  C. How will the growth of the population, from generation to generation, affect the answers to parts A and B? Explain.

C19. Describe what happens to allele frequencies as a result of the bottleneck effect. Discuss the relevance of this effect with regard to species that are approaching extinction.

C20. With regard to genetic drift, is each of the following statements *true* or *false*? If a statement is false, explain why.

  A. Over the long run, genetic drift leads to allele fixation or loss.

  B. When a new mutation occurs within a population, genetic drift is more likely to cause the loss of the new allele rather than the fixation of the new allele.

  C. Genetic drift promotes genetic diversity in large populations.

  D. Genetic drift is more significant in small populations.

C21. When two populations frequently intermix due to migration, what are the long-term consequences with regard to allele frequencies and genetic variation?

C22. Two populations of antelope are separated by a mountain range. The antelope occasionally migrate from one population to the other. Migration can occur in either direction. Explain how migration affects the following phenomena:

  A. Genetic diversity in the two populations

  B. Allele frequencies in the two populations

  C. Genetic drift in the two populations

C23. Does inbreeding affect allele frequencies? Why or why not? How does it affect genotype frequencies? With regard to rare recessive diseases, what are the consequences of inbreeding in human populations?

C24. Using the pedigree shown here, answer the following questions for individual VI-1.

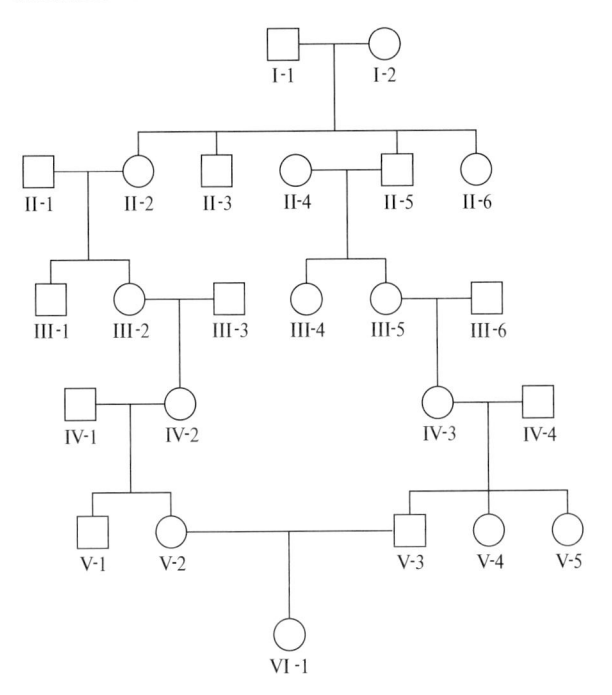

A. Is this individual inbred?

B. If so, who is/are her parents' common ancestor(s)?

C. Calculate the inbreeding coefficient for VI-1.

D. Are the parents of VI-1 inbred?

C25. A family pedigree is shown here.

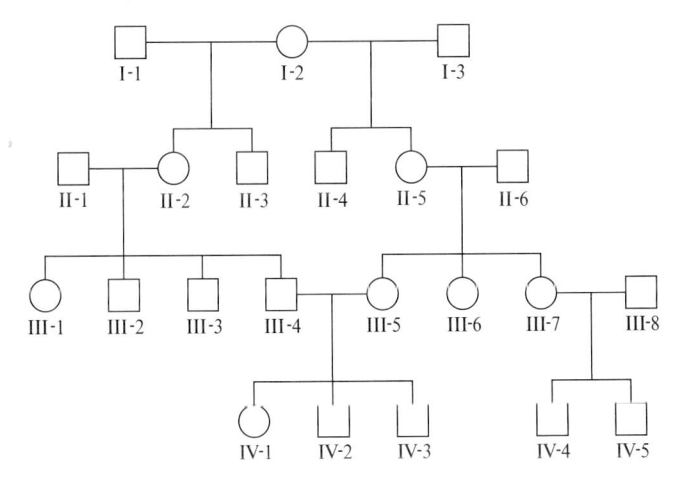

A. What is the inbreeding coefficient for individual IV-3?

B. Based on the data shown in this pedigree, is individual IV-4 inbred?

C26. A family pedigree is shown here.

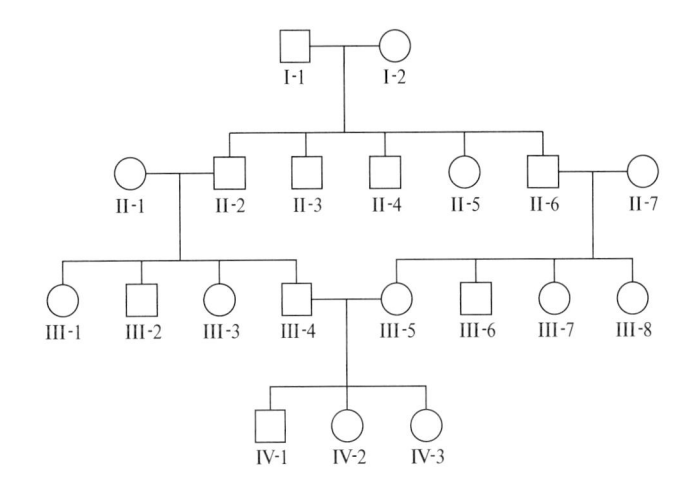

A. What is the inbreeding coefficient for individual IV-2? Who is/are her parents' common ancestor(s)?

B. Based on the data shown in this pedigree, is individual III-4 inbred?

C27. Antibiotics are commonly used to combat bacterial and fungal infections. During the past several decades, however, antibiotic-resistant strains of microorganisms have become alarmingly prevalent. This resistance has undermined the effectiveness of antibiotics in treating many types of infectious disease. Discuss how the following processes that alter allele frequencies may have contributed to the emergence of antibiotic-resistant strains:

A. Random mutation

B. Genetic drift

C. Natural selection

C28. Let's suppose the mutation rate for converting a B allele into a b allele is 10^{-4} per generation. The current allele frequencies are $B = 0.6$ and $b = 0.4$. How long will it take for the allele frequencies to equal each other, assuming that no genetic drift takes place?

Application and Experimental Questions

E1. You need to be familiar with the techniques described in Chapter 20 to answer this question. Gene polymorphisms can be detected using a variety of cellular and molecular techniques. Which techniques would you use to detect gene polymorphisms at the following levels?

A. DNA level

B. RNA level

C. Polypeptide level

E2. A gene for coat color in rabbits exists in four alleles designated C (full coat color), c^{ch} (chinchilla), c^h (Himalayan), and c (albino). In a population of rabbits in Hardy-Weinberg equilibrium, the allele frequencies are

$C = 0.34$

$c^{ch} = 0.17$

$c^h = 0.44$

$c = 0.05$

Assume that C is dominant to the other three alleles, c^{ch} is dominant to c^h and c, and c^h is dominant to c.

  A. What is the frequency of albino rabbits?

  B. Among 1000 rabbits, how many would you expect to have a Himalayan coat color?

  C. Among 1000 rabbits, how many would be heterozygotes with a chinchilla coat color?

E3. In a large herd of 5468 sheep, 76 animals have yellow fat, and the rest of the members of the herd have white fat. Yellow fat is inherited as a recessive trait. This herd is assumed to be in Hardy-Weinberg equilibrium.

  A. What are the frequencies of the white and yellow fat alleles in this population?

  B. Approximately how many sheep with white fat are heterozygous carriers of the yellow allele?

E4. The human MN blood group is determined by two codominant alleles, M and N. The following data were obtained from five human populations:

Genotype Frequencies (%)*

Population	Place	MM	MN	NN
Inuit	East Greenland	83.5	15.6	0.9
Navajo Indians	New Mexico	84.5	14.4	1.1
Finns	Karajala	45.7	43.1	11.2
Russians	Moscow	39.9	44.0	16.1
Aborigines	Queensland	2.4	30.4	67.2

*Data from E. B. Speiss (1990), *Genes in populations*, 2nd ed. Wiley-Liss, New York.

  A. Calculate the allele frequencies in these five populations.

  B. Which populations appear to be in Hardy-Weinberg equilibrium?

  C. Which populations do you think have experienced significant intermixing due to migration?

E5. In an island population, the following data were obtained for the numbers of people with each of the four blood types:

  Type O 721
  Type A 932
  Type B 235
  Type AB 112

  Is this population in Hardy-Weinberg equilibrium? Explain your answer.

E6. In a donor population, the allele frequencies for the common (Hb^A) and sickle cell alleles (Hb^S) are 0.9 and 0.1, respectively. A group of 550 individuals from this donor population migrates to a recipient population containing 10,000 individuals; in the recipient population, the allele frequencies are $Hb^A = 0.99$ and $Hb^S = 0.01$.

  A. Calculate the allele frequencies in the conglomerate population.

  B. Assuming the donor and recipient populations are each in Hardy-Weinberg equilibrium, calculate the genotype frequencies in the conglomerate population prior to any mating between the donor and recipient populations.

  C. What will be the genotype frequencies of the conglomerate population in the next generation, assuming it achieves Hardy-Weinberg equilibrium in one generation?

E7. A recessive lethal allele has achieved a frequency of 0.22 due to genetic drift in a very small population. Based on natural selection, how would you expect the allele frequencies to change in the next three generations? (Note: Your calculation can assume that genetic drift is not altering allele frequencies in either direction.)

E8. Among a large population of 2 million gray mosquitoes, one mosquito is heterozygous for a body color gene; this mosquito has one gray allele and one blue allele. There is no selective advantage or disadvantage for either gray or blue body color. All of the other mosquitoes carry the gray allele.

  A. What is the probability of fixation of the blue allele?

  B. If fixation happens to occur, how many generations is it likely to take?

  C. Qualitatively, how would the answers to parts A and B be affected if the blue allele conferred a slight survival advantage?

E9. Resistance to the poison warfarin is a genetically determined trait in rats. Homozygotes carrying the resistance allele ($W^R W^R$) have a lower fitness because they suffer from vitamin K deficiency, but heterozygotes ($W^R W^S$) do not have this deficiency. However, the heterozygotes are still resistant to warfarin. In an area where warfarin is applied, a heterozygote has a survival advantage. Due to warfarin resistance, a heterozygote is also more fit than a homozygote ($W^S W^S$) that is sensitive to warfarin. If the relative fitness values for $W^R W^S$, $W^R W^R$, and $W^S W^S$ individuals are 1.0, 0.37, and 0.19, respectively, in areas where warfarin is applied, calculate the allele frequencies at equilibrium. How would this equilibrium be affected if the rats were no longer exposed to warfarin?

E10. Describe, in as much experimental detail as possible, how you would test the hypothesis that the distribution of shell color among land snails is due to predation.

E11. Look at question 4 in More Genetic TIPS before answering this question. Here are traditional DNA fingerprints of five people: a child, the mother, and three potential fathers:

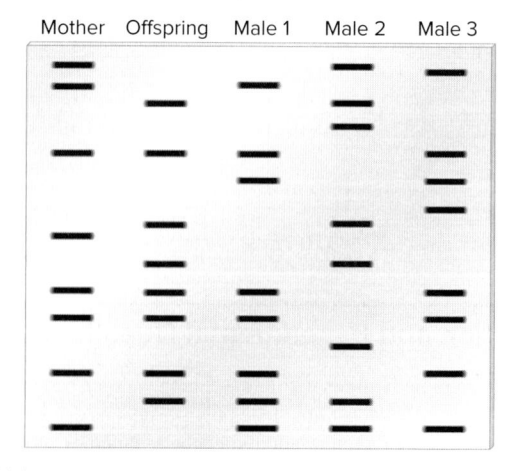

Which males can be ruled out as being the father? Explain your answer. If one of the males could be the father, explain the general strategy for calculating the likelihood that his DNA fingerprint matches the offspring's by chance alone.

E12. What is DNA fingerprinting? How can it be used in human identification?

E13. When analyzing the automated DNA fingerprints of a father and his biological daughter, a technician examined 50 peaks and found that 30 of them were a perfect match. In other words, 30 out of 50 peaks, or 60%, were a perfect match. Is this percentage too high, or would you expect a value of only 50%? Explain why or why not.

E14. What would you expect to be the minimum percentage of matching peaks in an automated DNA fingerprint for the following pairs of individuals?

A. Mother and son

B. Sister and brother

C. Uncle and niece

D. Grandfather and grandson

Questions for Student Discussion/Collaboration

1. Discuss examples of positive and negative assortative mating in natural populations, human populations, and agriculturally important species.

2. Discuss the role of mutation in the origin of genetic polymorphisms. Suppose that a genetic polymorphism involves two alleles that have frequencies of 0.45 and 0.55. Describe three different scenarios to explain these observed allele frequencies. You can propose that the mutations that produced the polymorphism are neutral, beneficial, or deleterious.

3. Most new mutations are detrimental, yet rare beneficial mutations can be adaptive. With regard to the fate of new mutations, discuss whether you think it is more important for natural selection to select against detrimental alleles or to select in favor of beneficial ones. Which do you think is more significant in human populations?

Answers to Comprehension Questions

23.1: c, a, a, d

23.2: a

23.3: b, a, c, d

23.4: d, b

23.5: d

23.6: c

23.7: d, b, a

Note: All answers appear in Connect; the answers to even-numbered questions and all Concept Check questions are in Appendix B.

24

CHAPTER OUTLINE

- 24.1 Overview of Complex and Quantitative Traits
- 24.2 Statistical Methods for Evaluating Quantitative Traits
- 24.3 Polygenic Inheritance
- 24.4 Identification of Genes That Control Quantitative Traits
- 24.5 Heritability
- 24.6 Selective Breeding

Domesticated wheat. The color of wheat ranges from dark red to white, which is an example of a complex or quantitative trait.
©Robert Glusic/Getty Images

QUANTITATIVE GENETICS

In this chapter, we will examine **complex traits**—characteristics that are determined by several genes and are significantly influenced by environmental factors. Most of the complex traits that we will consider are also called **quantitative traits** because they can be described numerically. In humans, quantitative traits include height, the shape of the nose, and the rate of food metabolism, to name a few examples.

The field of genetics that studies the mode of inheritance of complex and quantitative traits is called **quantitative genetics.** In agriculture, most of the key characteristics of interest to plant and animal breeders are quantitative traits. These include traits such as weight, fruit size, resistance to disease, and the ability to withstand harsh environmental conditions. As we will see later in this chapter, genetic techniques have improved our ability to develop strains of agriculturally important species with desirable quantitative traits. In addition, many human diseases, such as asthma and diabetes, are viewed as complex traits because they are influenced by several genes. Quantitative genetics is also important in the study of evolution. Many of the traits that allow a species to adapt to its environment are quantitative. Examples include the swift speed of the cheetah and the sturdiness of tree branches in windy climates.

In this chapter, we will examine how genes and the environment contribute to the phenotypic expression of quantitative traits.

We will begin with an overview of quantitative traits and the use of statistical methods to analyze them. We will then look at the inheritance of polygenic traits and at quantitative trait loci (QTLs)—locations on chromosomes containing genes that affect the outcome of quantitative traits. Advances in genetic mapping strategies have enabled researchers to identify these genes. Last, we will discuss heritability, which is the genetic variation that affects phenotypic variation, and how to calculate it.

24.1 OVERVIEW OF QUANTITATIVE TRAITS

Learning Outcomes:
1. List examples of complex and quantitative traits.
2. Explain how quantitative traits may be described with a frequency distribution.

When we compare characteristics among members of the same species, the differences may be complex or quantitative rather than qualitative. Humans, for example, have the same basic anatomical features (two eyes, two ears, and so on), but they differ in quantitative ways. People vary with regard to height, weight, the shape of facial features, pigmentation, and many other characteristics.

TABLE 24.1
Types of Complex and Quantitative Traits

Type of Trait	Examples
Anatomical traits	Height, weight, number of bristles in *Drosophila*, ear length in corn, and degree of pigmentation in flowers and skin
Physiological traits	Metabolic traits, speed of running and flight, ability to withstand harsh temperatures, and milk production in mammals
Behavioral traits	Mating calls, courtship rituals, ability to learn a maze, and ability to grow or move toward light
Diseases	Heart disease, hypertension, cancer, diabetes, asthma, and arthritis

As shown in **Table 24.1**, quantitative traits can be categorized as anatomical, physiological, and behavioral. In addition, many human diseases are viewed as complex traits because they are influenced by many genes and environmental factors. Three of the leading causes of death worldwide—heart disease, cancer, and diabetes—are considered complex traits.

Quantitative traits are described numerically. Height and weight can be measured in centimeters (or inches) and kilograms (or pounds), respectively. The number of bristles on a fruit fly's body can be counted, and metabolic rate can be assessed as the amount of glucose burned per minute. Behavioral traits can also be quantified. A mating call can be evaluated with regard to its duration, sound level, and pattern. The ability to learn a maze can be described as the time and/or repetitions it takes to master the skill.

- Quantitative traits, such as height and weight, are viewed as **continuous traits**—traits that do not fall into discrete categories.
- Some, such as bristle number in *Drosophila*, are **meristic traits**—traits that can be counted and expressed in whole numbers.
- Certain diseases, such as diabetes, are viewed as complex traits even though the disease itself can be considered qualitative—either you have it or you don't. The reason why such diseases are considered complex traits is that the alleles of several different genes contribute to the likelihood that an individual will develop the disease. A certain threshold must be reached in which the number of disease-causing alleles results in the development of the disease. Such diseases are referred to as **threshold traits**—traits that are inherited due to the contribution of many genes, but are expressed qualitatively.

Many Quantitative Traits Exhibit a Continuum of Phenotypic Variation That Follows a Normal Distribution

In Part II of this text, we discussed many traits that fall into discrete categories. For example, fruit flies might have white or red eyes, and pea plants may produce wrinkled or smooth seeds. The alleles that govern these traits affect the phenotype in a qualitative way. In analyzing crosses involving these types of traits, each offspring can be put into a particular phenotypic category. Such attributes are called **discontinuous traits.**

In contrast, quantitative traits show a continuum of phenotypic variation within a group of individuals. For such traits, it may be impossible to place organisms into a discrete phenotypic class. For example, **Figure 24.1a** is a photograph showing the range of heights of 82 college students. Though height is found at minimum and maximum values, the range of heights between these values is fairly continuous.

How do geneticists describe traits that show a continuum of phenotypes? Because most quantitative traits do not naturally fall into a small number of discrete categories, an alternative way to describe them is to use a **frequency distribution.** To construct a frequency distribution, the trait is divided arbitrarily into a number of convenient, discrete phenotypic categories. For example, in Figure 24.1, the range of heights is partitioned into 1-inch intervals. Then a graph is made that shows the number of individuals found in each of the categories.

Figure 24.1b shows a frequency distribution for the heights of the students pictured in Figure 24.1a. The measurement of height is plotted along the x-axis, and the number of individuals who exhibit that phenotype is plotted on the y-axis. The values along the x-axis are divided into the discrete 1-inch intervals that define the phenotypic categories, even though height is essentially continuous within a group of individuals. For example, in Figure 24.1a, 9 students were between 65.5 and 66.5 inches in height, which is plotted as the point (66 inches, 9 students) on the graph in Figure 24.1b. This type of analysis can be conducted on any group of individuals who vary with regard to a quantitative trait.

The line in the frequency distribution depicts a **normal distribution,** a distribution for a large sample in which the trait of interest varies in a symmetrical way around an average value. The distribution of measurements of many biological characteristics is approximated by a symmetrical bell curve like that in Figure 24.1b. Normal distributions are common when the phenotype is determined by the cumulative effect of many small, independent factors.

24.1 REVIEWING THE KEY CONCEPTS

- Quantitative traits can be categorized as anatomical, physiological, or behavioral (see Table 24.1).
- Quantitative traits often exhibit a continuum of phenotypic variation that follows a normal distribution (see Figure 24.1).

24.1 COMPREHENSION QUESTIONS

1. Which of the following is an example of a quantitative trait?
   a. Height
   b. Rate of glucose metabolism
   c. Ability to learn a maze
   d. All of the above are quantitative traits.

586   CHAPTER 24 :: QUANTITATIVE GENETICS

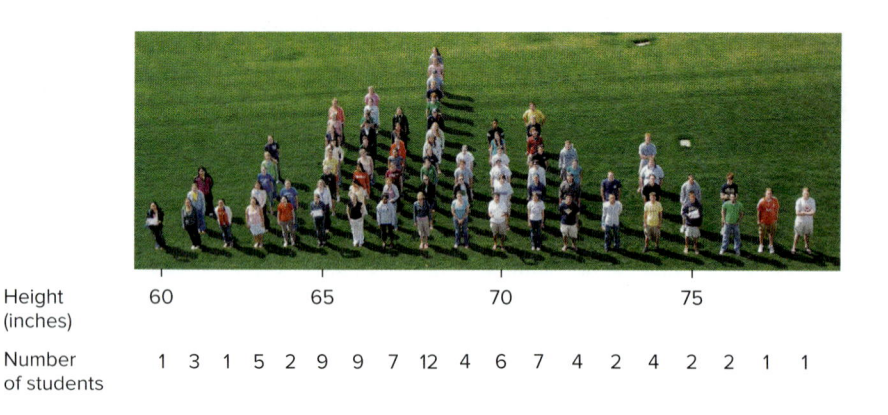

Height (inches)	60					65					70					75			
Number of students	1	3	1	5	2	9	9	7	12	4	6	7	4	2	4	2	2	1	1

(a)

(b)

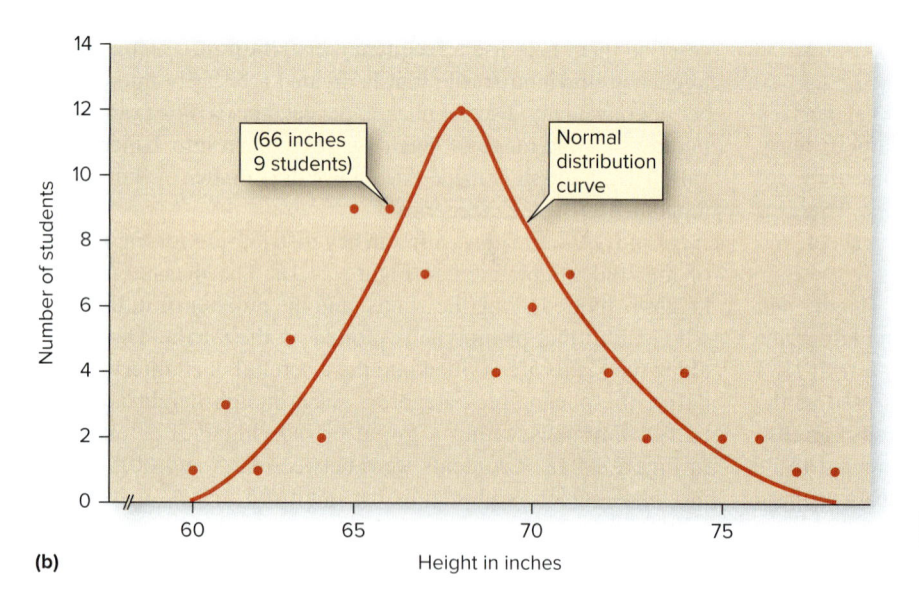

FIGURE 24.1 Normal distribution of a quantitative trait. (a) The distribution of heights of 82 college students. (b) A frequency distribution for the heights of students shown in part (a).
©McGraw-Hill Education/David Hyde/Wayne Falda, photographer

Concept Check: *Is height a discontinuous (discrete) trait, or does it exhibit a continuum?*

2. Saying that a quantitative trait exhibits a continuum means that
   a. the numerical value for the trait increases with the age of the individual.
   b. environmental effects are additive.
   c. the phenotypes for the trait are continuous and do not fall into discrete categories.
   d. the trait continuously changes during the life of an individual.

24.2 STATISTICAL METHODS FOR EVALUATING QUANTITATIVE TRAITS

Learning Outcome:
1. Calculate the mean, variance, standard deviation, and correlation coefficient for quantitative traits, and explain the meanings of these statistics.

In the early 1900s, Francis Galton and his student Karl Pearson showed that many traits in humans and domesticated animals are quantitative in nature. To understand the underlying genetic basis of these traits, they founded what became known as the **biometric field** of genetics, which involves the statistical study of biological traits. During this period, Galton and Pearson developed various statistical tools for studying the variation of quantitative traits within groups of individuals. Many of these tools are still in use today. In this section, we will examine how statistical tools are used to analyze the variation of quantitative traits within groups.

Statistical Methods Are Used to Evaluate a Frequency Distribution Quantitatively

Statistical tools can be used to analyze a normal distribution in a number of ways. One measure you are probably familiar with is a parameter called the **mean,** which is the sum of all the values in a group divided by the number of individuals in that group. The mean is computed using the following formula:

$$\overline{X} = \frac{\Sigma X}{N}$$

where
$\overline{X}$ is the mean
ΣX is the sum of all the values in the group
N is the number of individuals in the group

A more generalized form of this equation can be used:

$$\overline{X} = \frac{\Sigma f_i X_i}{N}$$

where

- $\overline{X}$ is the mean
- $\Sigma f_i X_i$ is the sum of all the values in the group; each value in the group is multiplied by its frequency (f_i) in the group
- N is the number of individuals in the group

For example, suppose a group of corn ears have the following lengths (rounded to the nearest centimeter): 15, 14, 13, 14, 15, 16, 16, 17, 15, and 15. Then

$$\overline{X} = \frac{4(15) + 2(14) + 13 + 2(16) + 17}{10}$$

$$\overline{X} = 15 \text{ cm}$$

In genetics, we are often interested in the amount of phenotypic variation that exists in a group. As we will see later in this chapter, genetic variation lies at the heart of breeding experiments. Without variation, selective breeding is not possible, and natural selection cannot favor one phenotype over another. A common way to evaluate variation within a population is with a statistic called the **variance**, which is a measure of the variation around the mean. The variance is the sum of the squared deviations from the mean divided by the degrees of freedom (df equals $N - 1$; refer back to Chapter 3 for a review of degrees of freedom).

$$V_X = \frac{\Sigma f_i (X_i - \overline{X})^2}{N - 1}$$

where

- V_X is the variance
- $X_i - \overline{X}$ is the difference between each value and the mean
- N is the number of observations

For example, if we use the values given previously for the lengths of ears of corn, the variance in length is calculated as follows:

$$\Sigma f_i (X_i - \overline{X})^2 = 4(15 - 15)^2 + 2(14 - 15)^2 +$$
$$(13 - 15)^2 + 2(16 - 15)^2 + (17 - 15)^2$$
$$\Sigma f_i (X_i - \overline{X})^2 = 0 + 2 + 4 + 2 + 4$$
$$\Sigma f_i (X_i - \overline{X})^2 = 12 \text{ cm}^2$$
$$V_X = \frac{\Sigma f_i (X_i - \overline{X})^2}{N - 1}$$
$$V_X = \frac{12 \text{ cm}^2}{9}$$
$$V_X = 1.33 \text{ cm}^2$$

Although variance is a measure of the variation around the mean, it is a statistic that may be difficult to understand intuitively because the variance is computed from squared deviations. For example, weight can be measured in grams; the corresponding variance is measured in grams squared. Even so, variances are centrally important in the analysis of quantitative traits because they are additive under certain conditions. This means that the variances for different factors that contribute to a quantitative trait, such as genetic and environmental factors, can be added together to predict the total variance for that trait. Later, we will examine how this property is useful in predicting the outcome of genetic crosses.

To gain a more intuitive grasp of variation, we can take the square root of the variance. This statistic is called the **standard deviation (SD)**. Again, using the example of the lengths of corn ears, the standard deviation is

$$SD = \sqrt{V_X} = \sqrt{1.33}$$

$$SD = 1.15 \text{ cm}$$

If the values in a population follow a normal distribution, it is easier to appreciate the amount of variation by considering the standard deviation. **Figure 24.2** illustrates the relationship between the standard deviation and the percentages of individuals that deviate from the mean. Approximately 68% of all individuals in a population have values within one standard deviation from the mean, in either the positive or the negative direction. About 95% are within two standard deviations, and 99.7% are within three standard deviations. When a quantitative characteristic follows a normal distribution, less than 0.3% of the individuals have values that are more or less than three standard deviations from the mean of the population. In our corn example, three standard deviations equals 3.45 cm. Therefore, we expect approximately 0.3% of the ears of corn have lengths less than 11.55 cm or greater than 18.45 cm, assuming that ear length follows a normal distribution.

Some Statistical Methods Compare Two Variables with Each Other

In many biological problems, it is useful to compare two different variables. For example, we may wish to compare the occurrence

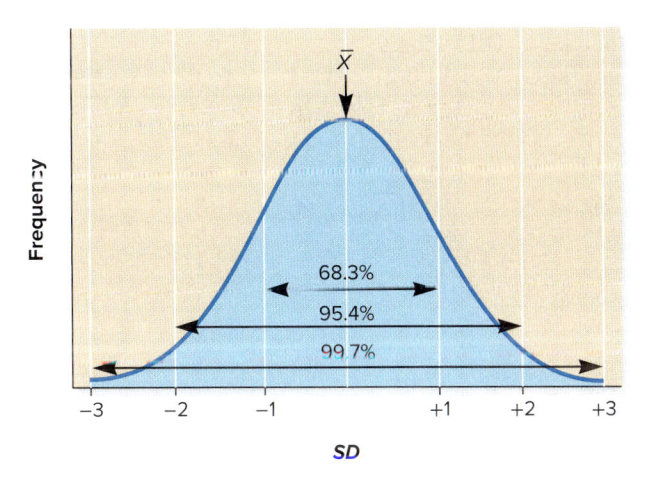

FIGURE 24.2 The relationship between the standard deviation and the proportions of individuals in a normal distribution. For example, approximately 68% of the individuals in a population are between the mean and one standard deviation (1 SD) above or below the mean.

Concept Check: What percentage of individuals fall more than two standard deviations above the mean?

of two different phenotypic traits. Do obese animals have larger hearts? Are brown eyes more likely to occur in people with dark skin pigmentation? A second type of comparison is between traits and environmental factors. Does insecticide resistance occur more frequently in areas that have been exposed to insecticides? Is heavy body weight more prevalent in colder climates? Finally, a third type of comparison is between traits and genetic relationships. Do tall parents tend to produce tall offspring? Do women with diabetes tend to have brothers with diabetes?

To gain insight into such questions, a statistic known as the correlation coefficient is often applied. To calculate this statistic, we first need to determine the **covariance,** which describes the degree of variation between two variables within a group of individuals. The covariance is similar to the variance, except that we multiply together the deviations of two different variables rather than squaring the deviations from a single factor.

$$CoV_{(X,Y)} = \frac{\Sigma f_i[(X_i - \overline{X})(Y_i - \overline{Y})]}{N - 1}$$

where

$CoV_{(X,Y)}$ is the covariance between X and Y values

X_i represents the values for one variable, and $\overline{X}$ is the mean value in the group

Y_i represents the values for another variable, and $\overline{Y}$ is the mean value in the group

N is the total number of pairs of observations

As an example, let's compare the weights of cows and those of their adult female offspring. A farmer might be interested in this relationship to determine if genetic variation plays a role in the weight of cattle. The following data give the weights at 5 years of age for 10 different cows and their female offspring.

Mother's Weight (kg)	Offspring's Weight (kg)	$X_i - \overline{X}$	$Y_i - \overline{Y}$	$(X_i - \overline{X})(Y_i - \overline{Y})$
570	568	−26	−30	780
572	560	−24	−38	912
599	642	3	44	132
602	580	6	−18	−108
631	586	35	−12	−420
603	642	7	44	308
599	632	3	34	102
625	580	29	−18	−522
584	605	−12	7	−84
575	585	−21	−13	273
$\overline{X} = 596$	$\overline{Y} = 598$			$\Sigma = 1373$

$SD_X = 21.1$ $SD_Y = 30.5$

$$CoV_{(X,Y)} = \frac{\Sigma f_i[(X_i - \overline{X})(Y_i - \overline{Y})]}{N - 1}$$

$$CoV_{(X,Y)} = \frac{1373}{10 - 1}$$

$$CoV_{(X,Y)} = 152.6$$

After we calculate the covariance, we can evaluate the strength of the association between the two variables by calculating a **correlation coefficient (r).** This value, which ranges between −1 and +1, indicates how two factors vary in relation to each other. The correlation coefficient is calculated using the following formula:

$$r_{(X,Y)} = \frac{CoV_{(X,Y)}}{SD_X SD_Y}$$

A positive r value means that two factors tend to vary in the same way relative to each other; as one factor increases, the other increases with it. A value of zero indicates that the two factors do not vary in a consistent way relative to each other; the values of the two factors are not related. Finally, a negative correlation, in which the correlation coefficient is negative, indicates that the two factors tend to vary in opposite ways to each other; as one factor increases, the other decreases.

Let's use the data on 5-year weights for cows and their offspring to calculate a correlation coefficient:

$$r_{(X,Y)} = \frac{152.6}{(21.1)(30.5)}$$

$$r_{(X,Y)} = 0.237$$

The result is a positive correlation between the 5-year weights of mothers and offspring. In other words, the positive correlation value suggests that heavy mothers tend to have heavy offspring and lighter mothers have lighter offspring.

How do we evaluate the value of r? After a correlation coefficient has been calculated, we must consider whether the r value represents a true association between the two variables or if it could simply be due to chance. To accomplish this, we can test the hypothesis that there is no real correlation (i.e., the null hypothesis, $r = 0$). The null hypothesis is that the observed r value differs from zero due only to random sampling error. We followed a similar approach in the chi square analysis described in Chapter 3. Like the chi square value, the significance of the correlation coefficient is directly related to sample size and the degrees of freedom (df). In testing the significance of correlation coefficients, df equals $N - 2$ because two variables are involved; N equals the number of paired observations. We will reject the null hypothesis if the chi square value results in a probability that is less than 0.05 (less than 5%) or if the probability is less than 0.01 (less than 1%). These are called the 5% and 1% significance levels, respectively. **Table 24.2** shows the relationship between the r values and degrees of freedom at the 5% and 1% significance levels.

The use of Table 24.2 is valid only if certain assumptions are met. First, the values of X and Y in the study must have been obtained by an unbiased sampling of the entire population. In addition, this approach assumes that the values of X and Y follow a normal distribution, like that in Figure 24.1, and that the relationship between X and Y is linear.

To illustrate the use of Table 24.2, let's consider the correlation we have just calculated for 5-year weights of cows and their female offspring. In this case, we obtained a value of 0.237 for r,

TABLE 24.2
Values of r at the 5% and 1% Significance Levels

Degrees of Freedom (df)	5%	1%	Degrees of Freedom (df)	5%	1%
1	0.997	1.000	24	0.388	0.496
2	0.950	0.990	25	0.381	0.487
3	0.878	0.959	26	0.374	0.478
4	0.811	0.917	27	0.367	0.470
5	0.754	0.874	28	0.361	0.463
6	0.707	0.834	29	0.355	0.456
7	0.666	0.798	30	0.349	0.449
8	0.632	0.765	35	0.325	0.418
9	0.602	0.735	40	0.304	0.393
10	0.576	0.708	45	0.288	0.372
11	0.553	0.684	50	0.273	0.354
12	0.532	0.661	60	0.250	0.325
13	0.514	0.641	70	0.232	0.302
14	0.497	0.623	80	0.217	0.283
15	0.482	0.606	90	0.205	0.267
16	0.468	0.590	100	0.195	0.254
17	0.456	0.575	125	0.174	0.228
18	0.444	0.561	150	0.159	0.208
19	0.433	0.549	200	0.138	0.181
20	0.423	0.537	300	0.113	0.148
21	0.413	0.526	400	0.098	0.128
22	0.404	0.515	500	0.088	0.115
23	0.396	0.505	1000	0.062	0.081

Note: df equals N − 2.
Source: Spence, J. T. and Underwood, B. J., *Elementary Statistics*. Englewood Cliffs, NJ: Prentice-Hall, 1970.

and the value of N was 10. Under these conditions, df equals 10 (the number of paired observations) minus 2, which equals 8. To be valid at a 5% significance level, the value of r would have to be 0.632 or higher. Because the value we obtained is much less than this, it is fairly likely that this value could have occurred because of random sampling error. In this case, we cannot reject the null hypothesis, and, therefore, we cannot conclude that the positive correlation is due to a true association between the weights of cows and their female offspring.

In an actual experiment, however, a researcher would examine many more pairs of cows and offspring, perhaps 500 to 1000. If a correlation of 0.237 was observed for N = 1000, the value would be significant at the 1% level. In this case, we would reject the null hypothesis that weights are not associated with each other. Instead, we would conclude that a real association occurs between the weights of cows and their offspring. In fact, these kinds of experiments have been done for cattle weights, and the correlations between mothers and offspring have often been found to be significant.

If a statistically significant correlation is obtained, how do we interpret its meaning? An r value that is statistically significant suggests a true association, but it does not necessarily imply a cause-and-effect relationship. When parents and offspring display a significant correlation for a trait, we should not jump to the conclusion that genetics is the underlying cause of the positive association. In many cases, parents and offspring share similar environments, so the positive association might be rooted in environmental factors. In general, correlations are quite useful in identifying positive or negative associations between two variables. We should use caution, however, because this statistic, by itself, cannot prove that the association is due to cause and effect.

Genetic TIPS

The Question: In a population of 100 male fruit flies, the mean abdomen length is 2.0 mm and the standard deviation is 0.3 mm. If you assume that abdomen length follows a normal distribution, what percentage of male flies will have an abdomen length equal to or greater than 2.6 mm?

Topic: What topic in genetics does this question address?

The topic is the use of values of the mean and standard deviation to determine the percentage of individuals that deviate a certain amount from the mean for a population.

Information: What information do you know based on the question and your understanding of the topic?

From the question, you know the mean and standard deviation for abdomen length in a population of fruit flies. From your understanding of the topic, you may remember that a relationship exists between the standard deviation and the proportions of individuals in a normal distribution (see Figure 24.2).

Problem-Solving Strategy: Make a calculation.

To begin to solve this problem, you first need to consider how far a male with an abdomen length of 2.6 mm deviates from the mean. The mean of the population is 2.0 mm, so this male deviates 0.6 mm from the mean. If you divide 0.6 mm by the standard deviation, which is 0.3 mm, this individual deviates 2 SDs from the mean.

According to Figure 24.2, about 95.4% of all flies are within 2 SDs from the mean, which means that 4.6% fall outside of 2 SDs. Half of these will be 2 SDs or more below the mean, and half of them, or 2.3% of them, will be 2 SDs or more above it.

Answer: Only 2.3% of the flies in this population will have an abdomen length equal to or greater than 2.6 mm.

24.2 REVIEWING THE KEY CONCEPTS

- Statistical methods are used to analyze quantitative traits. These methods include calculations of the mean, variance, standard deviation, covariance, and correlation (see Figure 24.2, Table 24.2).

24.2 COMPREHENSION QUESTIONS

1. The variance is
   a. a measure of the variation around the mean.
   b. computed as a squared deviation.
   c. higher when there is less phenotypic variation.
   d. Both a and b are correct.
2. Which of the following statistics is used to compare two variables?
   a. Mean
   b. Correlation
   c. Variance
   d. Standard deviation

24.3 POLYGENIC INHERITANCE

Learning Outcome:
1. Describe how polygenic inheritance may result in a continuum of phenotypes.

Thus far, we have seen that quantitative traits tend to show a continuum of variation and can be analyzed with various statistical tools. In this section, we begin to focus on the genetic basis of such traits. Quantitative traits are usually **polygenic**, which means they are controlled by multiple genes. The term **polygenic inheritance** refers to the transmission of any trait that is governed by two or more different genes. In this section, we will consider how the number of genes involved and environmental factors influence the continuum of variation that is exhibited by quantitative traits.

For Quantitative Traits That Are Polygenic, Each Gene May Contribute to the Trait in an Additive Way

The first experiment demonstrating polygenic inheritance was conducted by Herman Nilsson-Ehle in 1909. He studied the inheritance of red pigment in the hull of bread wheat, *Triticum aestivum* (**Figure 24.3a**). When true-breeding plants with white hulls were crossed to a variety with red hulls, the F_1 generation had an intermediate color. When the F_1 generation was allowed to self-fertilize, great variation in redness was observed in the F_2 generation, ranging over various shades: dark red, medium red, intermediate red, light red, and white. An undiscriminating observer might conclude that this F_2 generation displayed a continuous variation in hull color. However, as shown in **Figure 24.3b**, Nilsson-Ehle carefully categorized the colors of the hulls and discovered that they exhibited a 1:4:6:4:1 ratio. He concluded that this species has two different genes that control hull color, each gene existing in a red or white allele. He hypothesized that these two loci must contribute additively to the color of the hull. In this case, each red allele carried by a given plant contributed to the red color of the hull. Plants carrying more red alleles had a deeper red color.

Polygenic Inheritance and Environmental Factors May Produce a Continuum of Phenotypes

As we have just seen, Nilsson-Ehle categorized genotypes for wheat hull colors into five discrete phenotypic classes, ranging from dark red to white. However, for many polygenic traits, it is difficult or impossible to place individuals into discrete phenotypic classes. In general, as the number of genes controlling a trait increases and the variation due to the environment becomes greater, the categorization of genotypes into discrete phenotypic classes becomes increasingly problematic. Therefore, a Punnett square cannot be used to analyze most quantitative traits. Instead, statistical methods, which are described later in this chapter, must be employed.

Figure 24.4 illustrates how gene number and the environment affect the ability of genotypes to produce discrete phenotypic classes for a quantitative trait, in this case, seed weight. Before we consider these graphs, let's discuss the possible effects of the environment:

- First, the environment may or may not have great variation—some plants may be exposed to much more sunlight than others or they may be planted in better soil or receive more rain.
- Second, such environmental variation may or may not have a great effect on seed weight. For example, in one species of plant, receiving low or high amounts of rain may have a great effect on seed weight, whereas in another species, such variation in rainfall may have a minimal effect.

Figure 24.4a shows a situation in which seed weight is controlled by one gene with light (*w*) and heavy (*W*) alleles. A heterozygous plant (*Ww*) is allowed to self-fertilize. When seed weight is only slightly influenced by environmental variation, as seen on the left, the *ww*, *Ww*, and *WW* genotypes result in well-defined phenotypic classes. When environmental variation has a greater effect on seed weight, as shown on the right, more phenotypic variation is found in seed weight for each genotype. The variation in the frequency distribution on the right is much higher. Even so, most genotypes can still be categorized into the three main classes.

By comparison, **Figure 24.4b** illustrates a situation in which seed weight is governed by three genes instead of one, each existing in light and heavy alleles. A cross between two heterozygotes is expected to produce seven genotypes in a 1:6:15:20:15:6:1 ratio. When the variation in environmental factors is low and/or plays a minor role in the outcome of this trait, as shown in the upper graph in Figure 24.4b, nearly all individuals fall within a phenotypic

24.3 POLYGENIC INHERITANCE   591

(a) Wheat with red and white hulls

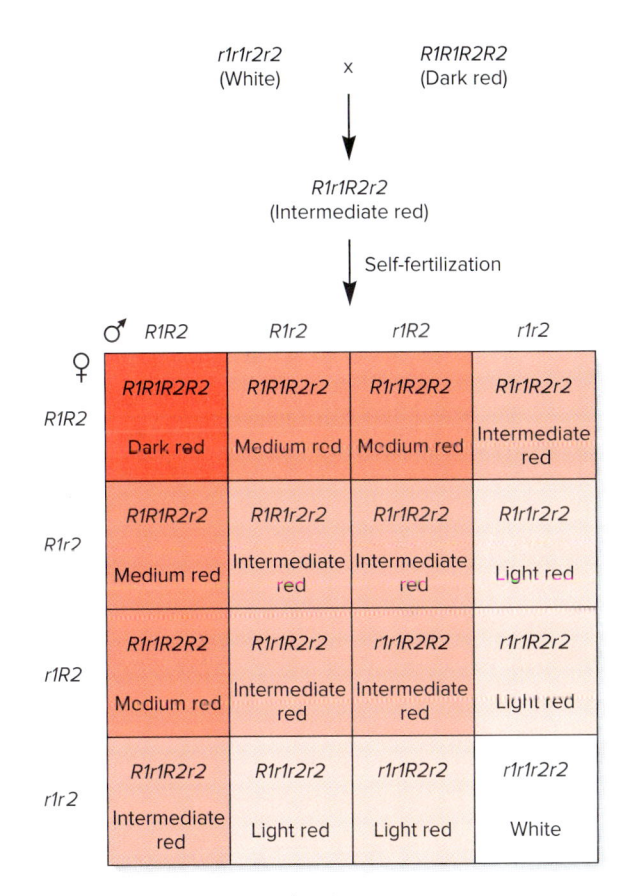

(b) Inheritance of hull color in wheat

r1r1r2r2 (White) × R1R1R2R2 (Dark red)

↓

R1r1R2r2 (Intermediate red)

↓ Self-fertilization

♂ ♀	R1R2	R1r2	r1R2	r1r2
R1R2	R1R1R2R2 Dark red	R1R1R2r2 Medium red	R1r1R2R2 Medium red	R1r1R2r2 Intermediate red
R1r2	R1R1R2r2 Medium red	R1R1r2r2 Intermediate red	R1r1R2r2 Intermediate red	R1r1r2r2 Light red
r1R2	R1r1R2R2 Medium red	R1r1R2r2 Intermediate red	r1r1R2R2 Intermediate red	r1r1R2r2 Light red
r1r2	R1r1R2r2 Intermediate red	R1r1r2r2 Light red	r1r1R2r2 Light red	r1r1r2r2 White

FIGURE 24.3 The Nilsson-Ehle experiment on the relationship of continuous variation to polygenic inheritance in wheat. (a) Red (top) and white (bottom) varieties of wheat, *Triticum aestivum*. (b) Nilsson-Ehle carefully categorized the colors of the hulls in the F_2 generation and discovered that they exhibited a 1:4:6:4:1 ratio. This pattern occurs because the contributions of the red alleles are additive.

Genes → Traits In this example, two genes, each with two alleles (red and white), govern hull color. Offspring can display a range of colors, depending on how many copies of the red allele they inherit. If an offspring is homozygous for the red allele of both genes, it will have very dark red hulls. By comparison, if it carries three red alleles and one white allele, it will be medium red (which is not quite as deep in color). Thus, this polygenic trait can exhibit a range of phenotypes from dark red to white.

(a) (top) ©Nigel Cattlin/Science Source; (a) (bottom) ©irin-k/age fotostock

Concept Check: When we say that alleles are additive, what does this mean?

class that corresponds to their genotype. When the environment has a greater effect on phenotype, as shown in the lower graph, the situation becomes more ambiguous. For example, individuals with one *w* allele and five *W* alleles have a phenotype that overlaps with that of individuals having six *W* alleles or two *w* alleles and four *W* alleles. Therefore, it becomes difficult to categorize each genotype into a unique phenotypic class. Instead, the trait displays a continuum ranging from light to heavy seed weight.

24.3 REVIEWING THE KEY CONCEPTS

- Polygenic inheritance refers to the transmission of any trait that is governed by two or more genes.

- Polygenic inheritance and environmental factors may produce a continuum of phenotypes for a quantitative trait (see Figures 24.3, 24.4).

24.3 COMPREHENSION QUESTION

1. For many quantitative traits, genotypes and phenotypes tend to overlap because
   a. the trait changes over time.
   b. the trait is polygenic.
   c. environmental variation affects the trait.
   d. Both b and c are true.

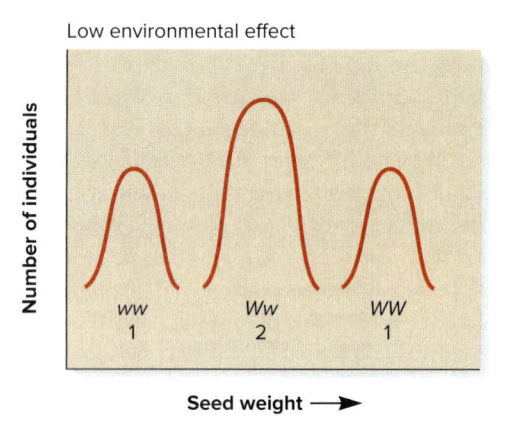

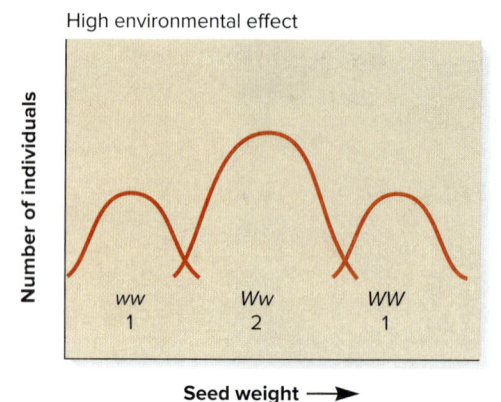

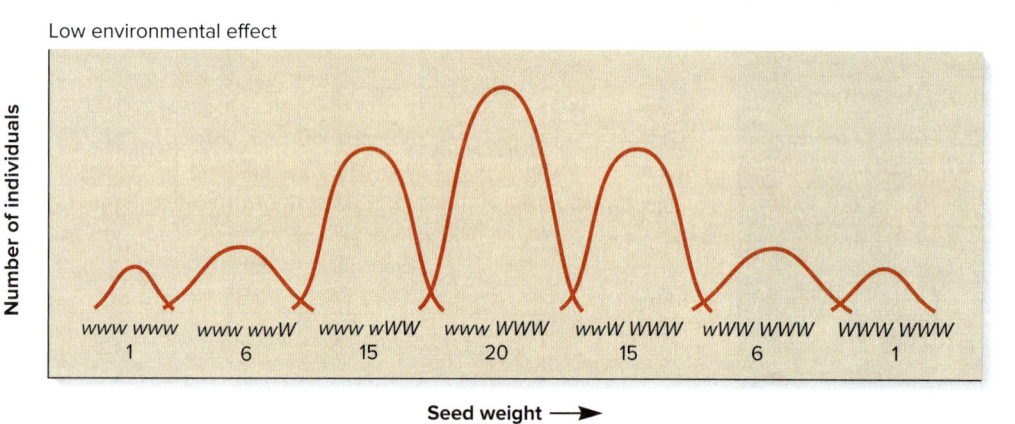

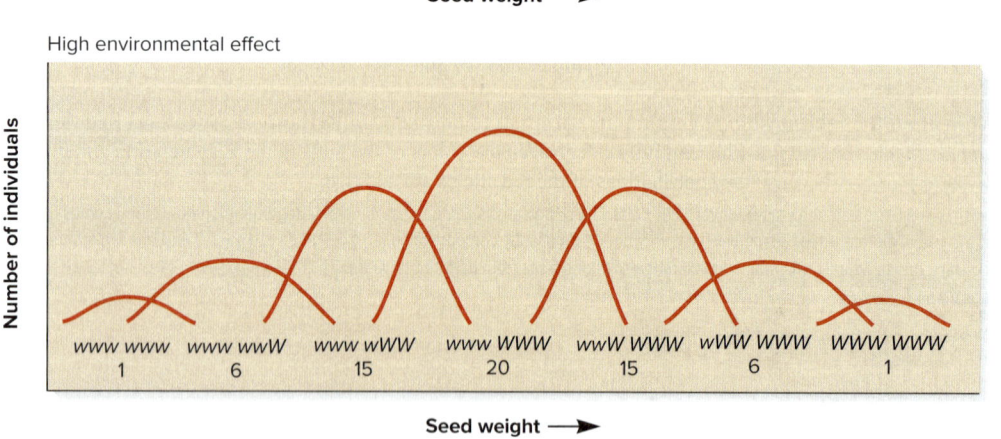

FIGURE 24.4 **How genotypes and phenotypes may overlap for polygenic traits.** (a) Situations in which seed weight is controlled by one gene, existing in light (*w*) and heavy (*W*) alleles. (b) Situations in which seed weight is governed by three genes instead of one, each existing in light and heavy alleles. (Note: The 1:2:1 and 1:6:15:20:15:6:1 ratios were derived by using a Punnett square and assuming a cross between individuals that are both heterozygous for three different genes.)

Genes → Traits The ability of geneticists to correlate genotype and phenotype depends on how many genes are involved and how much the environment causes the phenotype to vary. In part (a), a single gene influences seed weight. In the graph on the left side, environmental variation does not cause much variation in seed weight. No overlap in seed weight is observed for the *ww*, *Ww*, and *WW* genotypes. In the graph on the right side, environmental variation has a greater effect on seed weight. In this case, a few individuals with the *ww* genotype produce seeds having the same weight as seeds of a few individuals with the *Ww* genotype; and a few individuals with the *Ww* genotype produce seeds having the same weight as seeds from individuals with the *WW* genotype. As shown in part (b), it becomes even more difficult to distinguish genotype based on phenotype when three genes are involved. The overlaps are slight when environmental variation does not cause much seed weight variation. However, when environmental variation has a greater effect on phenotype, the overlaps between genotypes and phenotypes are very pronounced, and the trait appears to follow a continuum of variation.

Concept Check: Explain how gene number and environmental variation affect the overlaps between phenotypes and different genotypes.

24.4 IDENTIFICATION OF GENES THAT CONTROL QUANTITATIVE TRAITS

Learning Outcomes:
1. Define *quantitative trait locus*.
2. Explain how quantitative trait loci are mapped on chromosomes using molecular markers.

A goal of researchers working in the field of quantitative genetics is to identify the genes that are associated with complex and quantitative traits. This can be a challenging endeavor because such traits are usually polygenic. In the past few decades, the development of many molecular approaches has made it easier for researchers to identify these genes. In this section, we will consider an approach to mapping the locations of genes that control quantitative traits that is based on the locations of molecular markers.

The location on a chromosome that harbors one or more genes that affect the outcome of a quantitative trait is called a **quantitative trait locus (QTL)**. QTLs are chromosomal regions identified by genetic mapping. Because such mapping usually locates a QTL within a relatively large chromosomal region, a QTL may contain a single gene or it may contain two or more closely linked genes that affect a quantitative trait.

To identify genes associated with certain traits, researchers may determine their locations by identifying their linkage to molecular markers or by genome-wide association studies. Both of these approaches are described in Chapter 22 (see Figures 22.5–22.8). Similarly, the basis of **QTL mapping** is the association between phenotypes for quantitative traits and molecular markers such as restriction fragment length polymorphisms (RFLPs), microsatellites, and single-nucleotide polymorphisms (SNPs).

One strategy for QTL mapping is presented in **Figure 24.5**, which shows two different strains of a diploid plant species with four chromosomes per set. The strains are highly inbred, which means that they are homozygous for most molecular markers and genes. They differ in two important ways. First, the two strains differ with regard to many molecular markers; the sites of many markers on each chromosome are already known. These markers are designated 1A and 1B, 2A and 2B, and so forth. The markers 1A and 1B mark the same chromosomal location in this species, namely, the upper tip of chromosome 1. However, the two markers are distinguishable in the two strains at the molecular level. For example, 1A might be a microsatellite that is 148 bp, whereas 1B might be a microsatellite that is 212 bp. Second, the two strains differ in a quantitative trait of interest. In this example, the strain on the left produces large fruit, whereas the strain on the right produces small fruit. The unknown genes affecting this trait are designated with a black or blue QTL label. A black QTL indicates a site that harbors one or more alleles that promote large fruit. A blue QTL is at the same site but carries alleles that promote small fruit. Prior to conducting their crosses, researchers would not know the chromosomal locations of the QTLs shown in this figure. The purpose of the experiment is to determine their locations.

As shown in Figure 24.5, the QTL mapping strategy begins by crossing the two inbred strains to each other and then backcrossing the F_1 offspring to both parental strains. This produces a second generation with a great degree of variation. The offspring from these backcrosses are then characterized in two ways: First, they are examined for their fruit size; second, a cell sample from each individual is analyzed to determine which molecular markers are found in their chromosomes. The goal is to find an association between particular molecular markers and fruit size. For example, 2A is strongly associated with large size, whereas 2B is strongly associated with small size. By comparison, 9A and 9B are not associated with large or small size, because a QTL affecting this trait is not found on this chromosome. Also, markers 14A and 14B, which are fairly far away from a QTL, are not strongly associated with any particular QTL. Markers that are on the same chromosome but far away from a QTL are often separated from the QTL by crossing over that occurs during meiosis in the F_1 heterozygote. Only closely linked markers are strongly associated with a particular QTL.

Overall, QTL mapping involves the analysis of a large number of markers and offspring. The data are analyzed by computer programs that can statistically associate the phenotype (e.g., fruit size) with particular markers. Markers found throughout the genome of a species provide a way to identify the locations of several different genes that possess allelic differences that may affect the outcome of a quantitative trait.

In an early investigation using QTL mapping, in 1988, Andrew Paterson and his colleagues examined inheritance of quantitative traits in tomato plants. They studied an agricultural strain of tomato and a South American green-fruited variety. These two strains differed in their RFLPs, and they also exhibited dramatic differences in three agriculturally important characteristics: fruit mass, soluble solids content, and fruit pH. The researchers crossed the two strains and then backcrossed the offspring to the domestic tomato. The researchers then examined 237 plants with regard to 70 known RFLP markers. In addition, 5–20 tomatoes from each plant were analyzed for fruit mass, soluble solids content, and fruit pH. Using this approach, the researchers were able to map genes contributing much of the variation in these traits to particular sites on the tomato chromosomes. They identified six loci causing variation in fruit mass, four affecting soluble solids content, and five with effects on fruit pH.

More recently, the DNA sequence of the entire genome of many species has been determined. In such cases, the mapping of QTLs to a defined chromosomal region may allow researchers to analyze the DNA sequence in that region and to identify one or more genes that influence a quantitative trait of interest.

24.4 REVIEWING THE KEY CONCEPTS

- The locations on a chromosome that contain one or more genes affecting a quantitative trait are called quantitative trait loci (QTLs).
- QTLs can be identified by their proximity to known molecular markers (see Figure 24.5).

24.4 COMPREHENSION QUESTIONS

1. A QTL is a _____ where one or more genes affecting a quantitative trait are _____.
   a. site in a cell, located
   b. site on a chromosome, located
   c. site in a cell, expressed
   d. site on a chromosome, expressed

2. To map QTLs, researchers cross strains that differ with regard to
   a. a quantitative trait.
   b. molecular markers.
   c. a quantitative trait and molecular markers.
   d. a quantitative trait and a discontinuous trait.

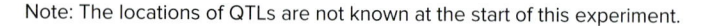

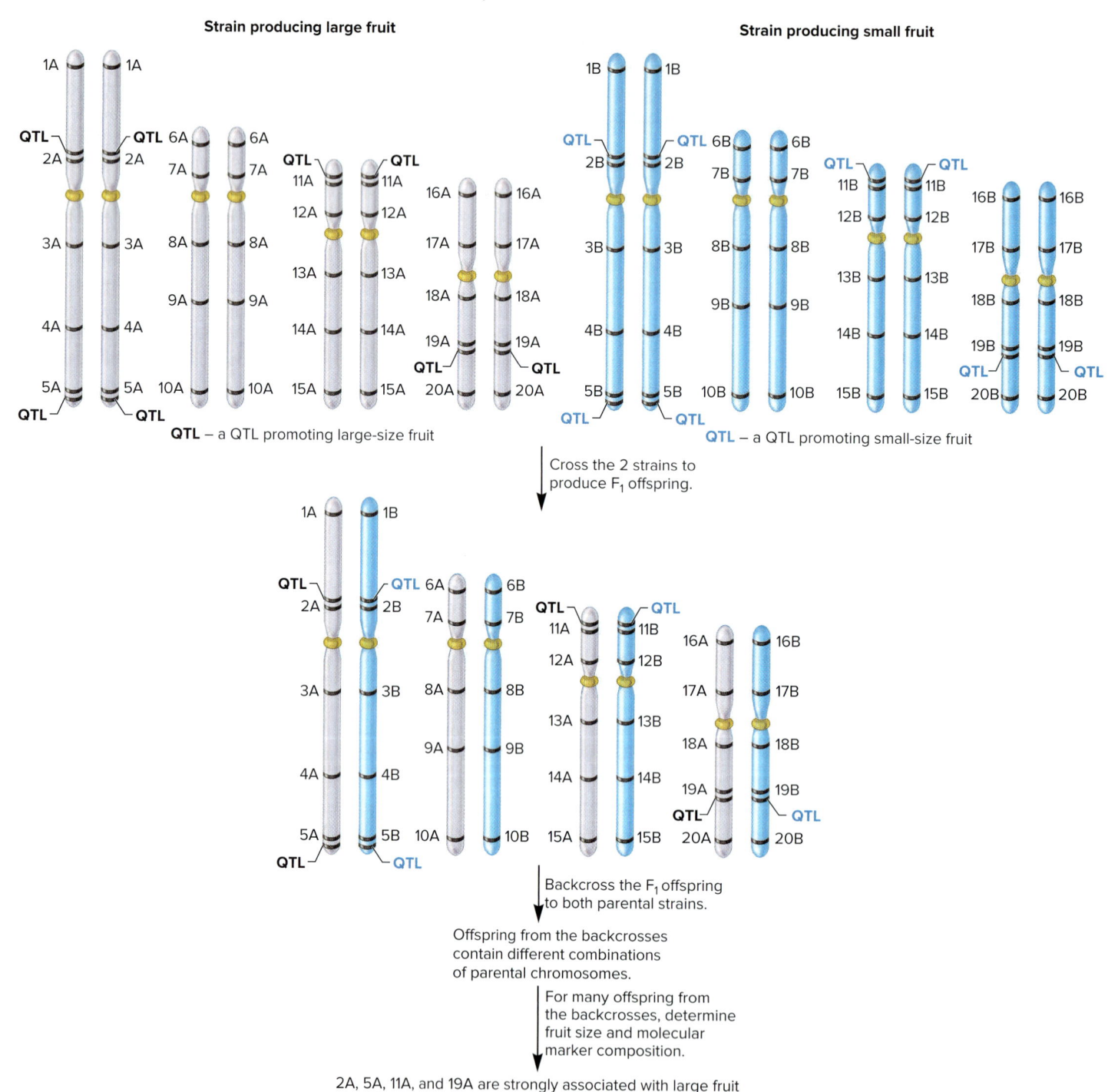

FIGURE 24.5 **The general strategy for QTL mapping using molecular markers.** Two different inbred strains have four chromosomes per set. The strain on the left produces large fruit, and the strain on the right produces small fruit. The goal of this mapping strategy is to locate the unknown genes affecting this trait, which are designated with a QTL label. A black QTL indicates a site carrying one or more alleles that promote large fruit, and a blue QTL is a site that carries alleles that promote small fruit. The two strains differ with regard to many molecular markers designated 1A and 1B, 2A and 2B, and so forth. The two strains are crossed, and then the F₁ offspring are backcrossed to both parental strains. Many offspring from the backcrosses are then examined to assess their fruit size and to determine which molecular markers are found in their chromosomes. The data are analyzed by computer programs that can statistically associate the phenotype (e.g., fruit size) with particular markers. Markers found throughout the genome of this species provide a way to locate several different genes that may affect the outcome of a single quantitative trait. In this case, the analysis predicts that four QTLs promoting heavier fruit weight are linked to regions of the chromosomes with the following markers: 2A, 5A, 11A, and 19A.

Concept Check: What are the two ways that strains A and B differ?

24.5 HERITABILITY

Learning Outcomes:
1. Explain the relationship between phenotypic variance, genetic variance, and environmental variance using an equation.
2. Describe how interactions and associations between genotype and environmental factors may affect phenotypic variance.
3. Define *heritability*, and distinguish between broad-sense heritability and narrow-sense heritability.
4. Calculate narrow-sense heritability using correlation coefficients.

As we have just seen, recently developed approaches in molecular mapping have enabled researchers to identify the genes that contribute to quantitative traits. The other key factor that affects the phenotypic outcomes of quantitative traits is the environment. All traits of organisms are influenced by both genetics and the environment, and this kind of interaction is particularly pertinent in the study of quantitative traits. Researchers want to understand how variation, both genetic and environmental, affects the phenotypic results. In this section, we will examine how geneticists analyze the genetic and environmental components that affect quantitative traits.

Both Genetic Variance and Environmental Variance May Contribute to Phenotypic Variance

Earlier, we examined the amount of phenotypic variation within a group by calculating the variance. Geneticists partition quantitative trait variation into components that are attributable to the following different causes:

Genetic variation (V_G)

Environmental variation (V_E)

Variation due to interactions between genetic and environmental factors ($V_{G \times E}$)

Variation due to associations between genetic and environmental factors ($V_{G \leftrightarrow E}$)

Let's begin by considering a simple situation in which V_G and V_E are the only factors that determine phenotypic variance, and the genetic and environmental factors are independent of each other. If this is the case, then the total variance for a trait in a group of individuals is

$$V_P = V_G + V_E$$

where

V_P is the total phenotypic variance. It is the amount of variation that is measured at the phenotypic level

V_G is the relative amount of variance due to genetic variation

V_E is the relative amount of variance due to environmental variation

Why is this equation useful? The partitioning of variance into genetic and environmental components allows us to estimate their relative importance in influencing the variation within a group. If V_G is very high and V_E is very low, genetics plays the greater role in promoting variation within a group. Alternatively, if V_G is low and V_E is high, environmental factors underlie much of the phenotypic variation. As described later in this chapter, a livestock breeder might want to apply selective breeding if V_G for an important (quantitative) trait is high. In this way, the characteristics of the herd may be improved. Alternatively, if V_G is negligible, it would make more sense to investigate and manipulate the environmental causes of phenotypic variation.

With experimental and domesticated species, one possible way to determine V_G and V_E is by comparing the variation in traits between genetically identical and genetically disparate groups. For example, researchers have used the practice of **inbreeding** to develop genetically homogeneous strains of mice. Inbreeding in mice involves many generations of brother-sister matings, which eventually produces strains that are **monomorphic** (carry the same allele) for all of their genes. Within such an inbred strain of mice, V_G equals zero. Therefore, all phenotypic variation is due to V_E. When studying quantitative traits such as weight, an experimenter might want to know the genetic and environmental variance for a different, genetically heterogeneous group of mice. To do so, genetically homogeneous and genetically heterogeneous mice could be raised under the same environmental conditions and their weights measured (in grams). The phenotypic variance for weight could then be calculated as described earlier. Let's suppose we obtained the following results:

$V_P = 15$ g^2 for the group of genetically homogeneous mice

$V_P = 22$ g^2 for the group of genetically heterogeneous mice

In the case of the homogeneous mice, $V_P = V_E$, because V_G equals 0. Therefore, V_E equals 15 g^2. To estimate V_G for the heterogeneous group of mice, we could assume that V_E (i.e., the environmentally produced variance) is the same for them as it is for the homogeneous mice, because the two groups were raised in identical environments. This assumption allows us to calculate the genetic variance in weight for the heterogeneous mice.

$$V_P = V_G + V_E$$
$$22 \text{ g}^2 = V_G + 15 \text{ g}^2$$
$$V_G = 7 \text{ g}^2$$

This result tells us that some of the phenotypic variance in the genetically heterogeneous group is due to the environment (namely, 15 g^2) and some (7 g^2) is due to genetic variation in alleles that affect weight.

Phenotypic Variation May Also Be Influenced by Interactions and Associations Between Genotype and the Environment

Thus far, we have considered the simple situation in which genetic variation and environmental variation are independent of each other and affect the phenotypic variation in an additive way.

As another example, let's suppose that three genotypes, *HH*, *Hh*, and *hh*, affect height, producing tall, intermediate, and short plants, respectively. Greater sunlight makes the plants grow taller regardless of their genotypes. In this case, our assumption that $V_P = V_G + V_E$ is reasonably valid.

Genotype-Environment Interaction For the *HH*, *Hh*, and *hh* plants, let's now consider a different environmental factor: minerals in the soil. As a hypothetical example, let's suppose the *h* allele decreases the function of a protein involved with mineral uptake from the soil. In this case, the *Hh* and *hh* plants are shorter because they cannot take up a sufficient supply of certain minerals to support maximal growth, whereas the *HH* plants are not limited by mineral uptake. According to this hypothetical scenario, adding minerals to the soil enhances the growth rate of *hh* plants by a large amount and the *Hh* plants by a smaller amount, because minerals are more available to be taken up by the plants (**Figure 24.6**). The height of *HH* plants is not affected by mineral supplementation. When the environmental effects on phenotype differ according to genotype, this phenomenon is called a **genotype-environment interaction**. As noted earlier, variation due to interactions between genetic and environmental factors is termed $V_{G \times E}$.

Interactions between genetic and environmental factors are common. As an example, **Table 24.3** shows results from a study conducted in 2000 by Cristina Vieira, Trudy Mackay, and colleagues in which they investigated the genotype-environment interaction for QTLs affecting life spans in *Drosophila melanogaster*. The data in the table compare the life span in days of male and female flies from two different strains of *D. melanogaster* raised at different temperatures. Because males and females differ in their sex chromosomes and gene expression patterns, they can be viewed as having different genotypes. The effects of environmental changes depended greatly on the strain and the sex of the flies.

- Under standard culture conditions, the females of strain A had the longest life span, whereas females of strain B had the shortest.
- In strain A, high temperature increased the longevity of males and decreased the longevity of females. In contrast, under hotter conditions, the longevity of males of strain B was dramatically reduced, whereas females of this same strain were not significantly affected.
- Lower growth temperature also had different effects in these two strains. Although low temperature increased the longevity of both strains, the effects were most dramatic in the males of strain A and the females of strain B.

Taken together, these results illustrate the potential complexity of the effects of genotype-environmental interaction on a quantitative trait such as life span.

Genotype-Environment Association Another complication confronting geneticists is that genotypes may not be randomly distributed in all possible environments. When certain genotypes are preferentially found in particular environments, this phenomenon is called a **genotype-environment association** ($V_{G \leftrightarrow E}$). When such an association occurs, the effects of genotype and environment are not independent of each other, and the association needs to be considered when determining the effects of genetic and environmental variation on the total phenotypic variation. Genotype-environment associations are very common in human genetics,

TABLE 24.3

Longevity of Two Strains of *Drosophila melanogaster* *

	Strain A		Strain B	
Temperature	Male	Female	Male	Female
Standard	33.6	39.5	37.5	28.9
High	36.3	33.9	23.2	28.6
Low	77.5	48.3	45.8	77.0

*Longevity was measured in the mean number of days of survival. Strains A and B were inbred strains of *D. melanogaster* called Oregon and 2b, respectively. The standard-, high-, and low-temperature conditions were 25°C, 29°C, and 14°C, respectively.

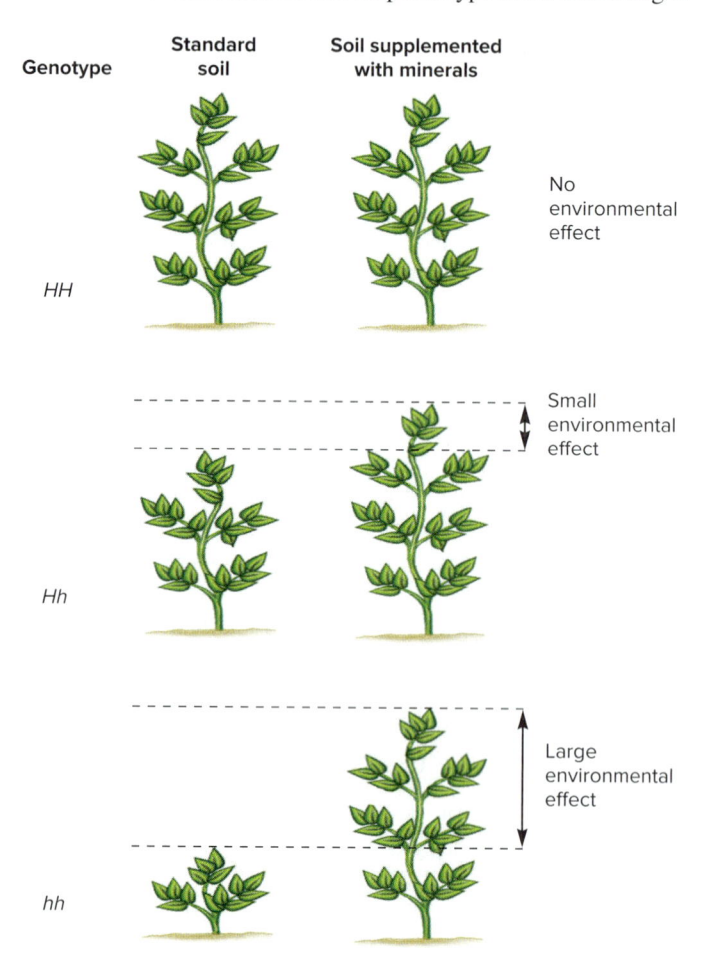

FIGURE 24.6 **A schematic illustration of a genotype-environment interaction.** When grown in standard soil, plants with the three genotypes *HH*, *Hh*, and *hh* show tall, intermediate, and short heights, respectively. When the soil is supplemented with minerals, a large effect is seen in the plants with the *hh* genotype and a smaller effect in those with the *Hh* genotype. The plants with the *HH* genotype are unaffected by the environmental change.

because members of large families tend to have more similar environments than do members of the population as a whole. One way to evaluate this type of effect is to compare individuals who have different genetic relationships, such as identical versus fraternal twins. We will examine this approach later in this section. Another strategy that geneticists might follow is to analyze siblings that have been adopted by different parents at birth. Their environmental conditions tend to be more disparate, and this may help to minimize the effects of genotype-environment association.

Heritability Is the Relative Amount of Phenotypic Variation That Is Due to Genetic Variation

Another way to view variance is to focus on the genetic contribution to phenotypic variation. The term **heritability** refers to the amount of phenotypic variation due to genetic variation within a specific group of individuals raised in a particular environment. Both genes and the environment are essential to produce the traits of an organism. However, variation of a trait in a population may be due entirely to environmental variation, entirely to genetic variation, or more commonly to a combination of the two.

If all of the phenotypic variance in a group is due to genetic variance, the heritability has a value of 1. If all of the phenotypic variation is due to environmental effects, the heritability equals 0. For most groups of organisms, the heritability for a given trait lies between these two extremes. For example, both genes and diet affect the size an individual will attain. In a given population, some individuals inherit alleles that tend to make them larger, and a proper diet also promotes larger size. Other individuals inherit alleles that make them small, and an inadequate diet may contribute to small size. Taken together, both genetics and the environment affect the amount of phenotypic variation for a trait such as size.

If we assume that environment and genetics are independent and the only two factors affecting phenotype, then

$$h_B^2 = V_G/V_P$$

where

h_B^2 is the heritability in the broad sense
V_G is the variance due to genetics
V_P is the total phenotypic variance, which equals $V_G + V_E$

The heritability defined here, h_B^2, called **broad-sense heritability**, takes into account different types of genetic variation that may affect the phenotype. As we have seen throughout this text, genes can affect phenotypes in various ways.

- The Nilsson-Ehle experiment described earlier in this chapter showed that the alleles determining hull color in wheat affect the phenotype in an additive way. A heterozygote exhibits a phenotype that is intermediate between those of the respective homozygotes.
- Alternatively, alleles affecting other traits may have a dominant/recessive relationship. In this case, the alleles are not strictly additive, because the heterozygote has a phenotype closer to, or perhaps the same as, that of the homozygote containing two copies of the dominant allele.

For example, Mendel discovered that both PP and Pp pea plants have purple flowers.
- In addition, another complicating factor is epistasis (discussed in Chapter 5), in which the alleles for one gene can mask the phenotypic expression of the alleles of another gene.

To account for these differences, geneticists usually subdivide V_G into these three different genetic categories:

$$V_G = V_A + V_D + V_I$$

where

V_A is the variance due to the additive effects of alleles
V_D is the variance due to the effects of alleles that follow a dominant/recessive pattern of inheritance
V_I is the variance due to the effects of alleles that interact in an epistatic manner

When analyzing quantitative traits, geneticists may focus on V_A and ignore the contributions of V_D and V_I. They do this for scientific as well as practical reasons. For some quantitative traits, the additive effects of alleles may play a primary role in the phenotypic outcome. In addition, when the alleles behave additively, we can predict the outcomes of crosses based on the quantitative characteristics of the parents. The heritability of a trait due to the additive effects of alleles is called **narrow-sense heritability**:

$$h_N^2 = V_A/V_P$$

For many quantitative traits, the value of V_A may be relatively large compared with V_D and V_I. In such cases, the determination of the narrow-sense heritability provides an estimate of the broad-sense heritability.

How can the narrow-sense heritability be determined? In this chapter, we will consider two common ways. As discussed in Section 24.6, one way to calculate the narrow-sense heritability involves selective breeding practices, which are done with agricultural species. A second common strategy for determining narrow-sense heritability involves the measurement of a quantitative trait among groups of genetically related individuals. For example, agriculturally important traits, such as egg weight in poultry, can be analyzed in this way. To calculate the heritability, a researcher determines the observed egg weights between individuals whose genetic relationships are known, such as a mother and her female offspring. These data can then be used to compute a correlation between the phenotypes of parent and offspring, using the methods described earlier. The narrow-sense heritability is then calculated as

$$h_N^2 = r_{obs}/r_{exp}$$

where

r_{obs} is the observed phenotypic correlation between related individuals
r_{exp} is the expected correlation based on the known genetic relationship

In our example, r_{obs} is the observed phenotypic correlation between parent and offspring. In actual research studies, the observed phenotypic correlation for egg weights between mothers and female offspring has been found to be about 0.3 (although this varies among strains). The expected correlation, r_{exp}, is based on the known genetic relationship. A parent and offspring share 50% of their genetic material, so r_{exp} equals 0.50, which gives

$$h_N^2 = r_{obs}/r_{exp}$$
$$= 0.3/0.50$$
$$= 0.60$$

(Note: For siblings, $r_{exp} = 0.50$; for identical twins, $r_{exp} = 1.0$; and for an aunt-niece relationship, $r_{exp} = 0.25$.)

According to this calculation, about 60% of the phenotypic variation in egg weight is due to additive genetic variation; the other 40% is due to the environment.

When calculating heritabilities from correlation coefficients, keep in mind that such a computation assumes that genetics and the environment are independent variables. However, this is not always the case. The environments of parents and offspring are often more similar to each other than are the environments of unrelated individuals. As mentioned earlier, there are different ways to minimize this confounding factor. First, in human studies, researchers may analyze the heritabilities from correlations between adopted children and their biological parents. Alternatively, they can examine a variety of relationships (aunt and niece, identical twins versus fraternal twins, and so on) and see if the heritability values are roughly the same in all cases. This approach was applied in the study that is described next.

Heritability of Dermal Ridge Count in Human Fingerprints Is Very High

Fingerprints are inherited as a quantitative trait. It has long been known that identical twins have fingerprints that are very similar, whereas fraternal twins show considerably less agreement. Galton was the first researcher to study fingerprint patterns, but this trait became more amenable to genetic studies in the 1920s, when Kristine Bonnevie, a Norwegian geneticist, developed a method for counting the number of ridges within a human fingerprint.

As shown in **Figure 24.7**, human fingerprints can be categorized as having an arch, loop, or whorl, or a combination of these patterns. The primary difference among these patterns is the number of triple junctions, each known as a triradius (Figure 24.7b and c). At a triradius, a ridge emanates in three different directions. An arch has zero triradii, a loop has one, and a whorl has two. In Bonnevie's method of counting, a line is drawn from a triradius to the center of the fingerprint. The ridges that touch this line are then counted. (Note: The triradius ridge itself is not counted, and the last ridge is not counted if it forms the center of the fingerprint.) With this method, one can obtain a ridge count for all 10 fingers. Bonnevie conducted a study on a small population and found that ridge count correlations were relatively high in genetically related individuals.

Sarah Holt, who was also interested in the inheritance of this quantitative trait, carried out a more exhaustive study of ridge counts by examining the fingerprint patterns of a large group of people and their close relatives. In the experiment shown in **Figure 24.8**, the ridge counts for pairs of related individuals were determined by the method described in Figure 24.7. The correlation coefficients for ridge counts were then calculated among the pairs of related or unrelated individuals. To estimate the narrow-sense heritability, the observed correlations were then divided by the expected correlations based on the known genetic relationships.

▶ THE HYPOTHESIS

Dermal ridge count has a genetic component. The goal of this experiment is to determine the contribution of genetics to the variation in dermal ridge counts.

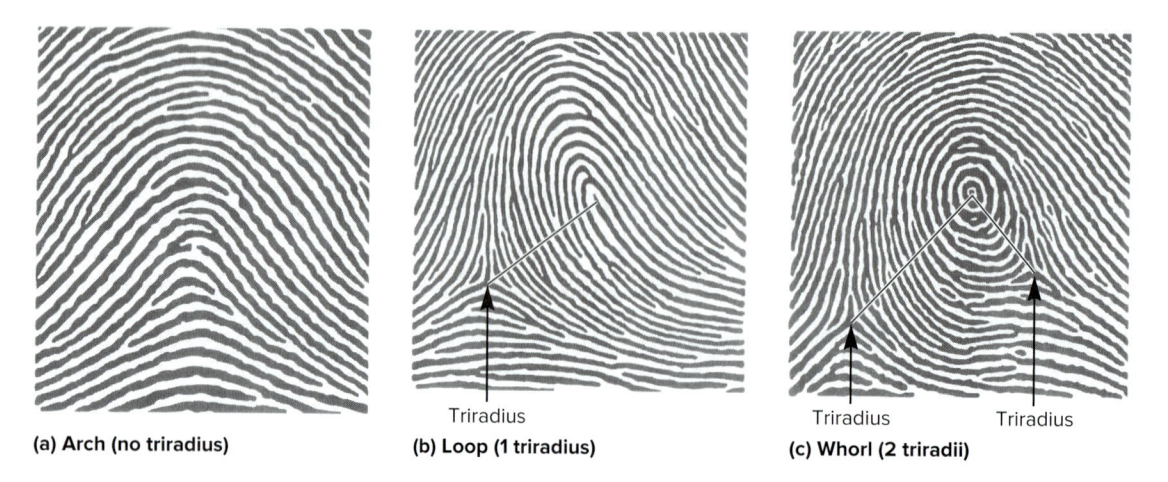

(a) Arch (no triradius)   **(b) Loop (1 triradius)**   **(c) Whorl (2 triradii)**

FIGURE 24.7 Human fingerprints and the ridge count method of Bonnevie. (a) This print has an arch rather than a triradius. The ridge count is zero. (b) This print has one triradius. A straight line is drawn from the triradius to the center of the print. The number of ridges dissecting this straight line is 13. (c) This print has two triradii. Straight lines are drawn from both triradii to the center. There are 16 ridges touching the left line and 7 touching the right line, giving a total ridge count of 23.

24.5 HERITABILITY

▶ **TESTING THE HYPOTHESIS** — **FIGURE 24.8** Heritability of human fingerprint patterns.

Starting material: A group of human subjects from Great Britain.

	Experimental level	Conceptual level
1. Take a person's finger and blot it onto an ink pad.		
2. Roll the person's finger onto a recording surface to obtain a print.		This is a method to measure a quantitative trait.
3. With a low-power binocular microscope, count the number of ridges, using the method described in Figure 24.7.		
4. Calculate the correlation coefficients between different pairs of individuals, as described earlier in this chapter.	See the data.	The correlation coefficient provides a way to determine the heritability for the quantitative trait.

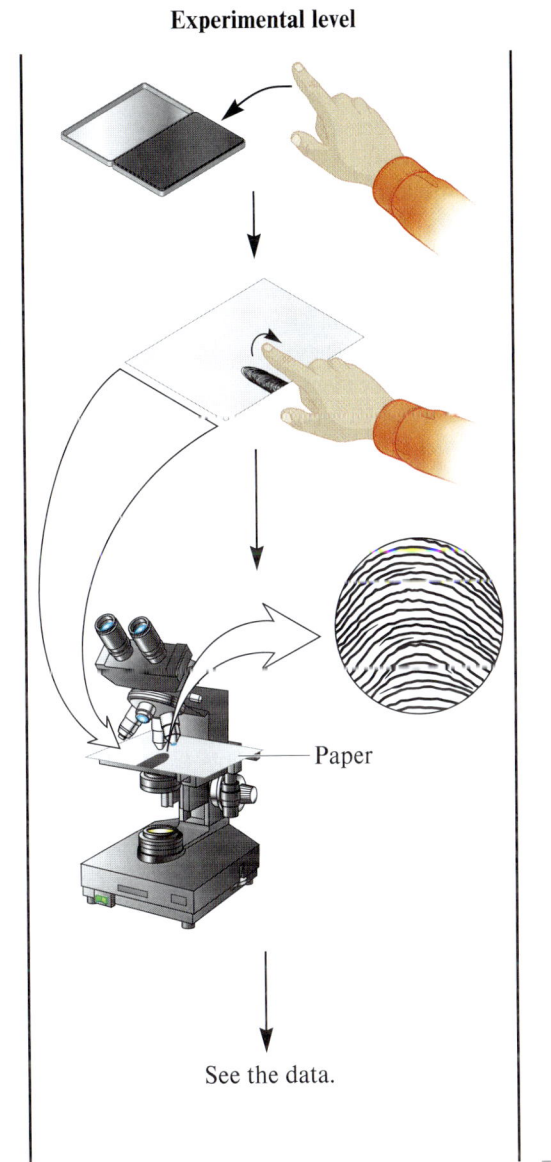

▶ **THE DATA**

Type of Relationship	Number of Pairs Examined	Correlation Coefficient (r_{obs})	Heritability r_{obs}/r_{exp}
Parent-child	810	0.48 ± 0.04*	0.96
Parent-parent	200	0.05 ± 0.07	—†
Sibling-sibling	642	0.50 ± 0.04	1.00
Identical twins	80	0.95 ± 0.01	0.95
Fraternal twins	92	0.49 ± 0.08	0.98
			0.97 average heritability

*The value following is the standard error of the mean.
†Note: We cannot calculate a heritability value in this case because the value for r_{exp} is not known. Nevertheless, the value for r_{obs} is very low, suggesting that there is a negligible correlation between unrelated individuals.
Adapted from S. B. Holt (1961) Quantitative genetics of fingerprint patterns. *Br Med Bull 17*, 247–250.

▶ INTERPRETING THE DATA

As seen in the data, the results indicate that genetics plays the major role in explaining the variation in this trait. Genetically unrelated individuals (namely, the parent-parent relationships) have a negligible correlation for this trait. By comparison, individuals who are genetically related have a substantially higher correlation. When the observed correlation coefficient is divided by the expected correlation coefficient based on the known genetic relationships, the average heritability value is 0.97, which is very close to 1.0.

What do these high heritability values mean? They indicate that nearly all of the variation in fingerprint pattern is due to genetic variation. Significantly, fraternal and identical twins have substantially different observed correlation coefficients, even though we expect they have been raised in very similar environments. These results support the idea that genetics is playing the major role in promoting variation and the results are not biased heavily by environmental similarities that may be associated with genetically related individuals. From an experimental viewpoint, the results show us how the determination of correlation coefficients between related individuals can provide insight regarding the relative contributions of genetics and environment to the variation of a quantitative trait.

24.5 REVIEWING THE KEY CONCEPTS

- Heritability is the amount of phenotypic variation due to genetic variation within a group of individuals raised in a particular environment.
- Genetic variance and environmental variance may contribute additively to phenotypic variance.
- Genetic variance and environmental variance may exhibit interactions and associations (see Figure 24.6, Table 24.3).
- Broad-sense heritability takes into account different types of genetic variation that may affect phenotype, including the additive effects of alleles, effects due to dominant/recessive relationships, and effects due to epistatic interactions.
- Narrow-sense heritability is heritability that is due to the additive effects of alleles.
- Holt determined that dermal ridge count has a very high heritability value in humans (see Figures 24.7, 24.8).

24.5 COMPREHENSION QUESTIONS

1. In a population of squirrels in North Carolina, the heritability for body weight is high. This means that
   a. body weight is primarily controlled by genes.
   b. the environment has little influence on body weight.
   c. the variation in body weight is mostly due to genetic variation.
   d. Both a and b are correct.
2. If two or more different genotypes are not affected by environmental variation in the same way, this outcome is due to
   a. a genotype-environment association.
   b. a genotype-environment interaction.
   c. the additive effects of alleles.
   d. both a and b.
3. One way to estimate narrow-sense heritability for a given trait is to compare _____ for _____.
   a. variances, related pairs of individuals
   b. correlation coefficients, related pairs of individuals
   c. variances, unrelated pairs of individuals
   d. correlation coefficients, unrelated pairs of individuals

24.6 SELECTIVE BREEDING

Learning Outcomes:
1. Describe the effects of selective breeding.
2. Calculate heritability from the results of selective breeding experiments.

The term **selective breeding** refers to programs and procedures designed to modify phenotypes in species of economically important plants and animals. This approach, also called **artificial selection,** is related to natural selection, discussed in Chapter 23. In forming his theory of natural selection, Charles Darwin was influenced by his observations of selective breeding by pigeon fanciers and other breeders. The primary difference between artificial and natural selection is how the parents are chosen. Natural selection is due to natural variation in reproductive success. In artificial selection, the breeder chooses individuals with traits that are desirable from a human perspective. In this section, we will examine the effects of selective breeding and consider its relationship to heritability.

Selective Breeding of Species Can Alter Quantitative Traits Dramatically

For centuries, humans have been practicing selective breeding to obtain domestic species with interesting or agriculturally useful characteristics. A good example is the dog, which is a common house pet. All domestic dogs are derived from the gray wolf (*Canis lupus*). The various breeds of dogs have been obtained by selective breeding strategies that typically focus on morphological traits (size, fur color, etc.) and behavioral traits (ability to hunt, friendly to humans, etc.). As shown in **Figure 24.9**, modification of quantitative traits in a species by selective breeding can have striking results. When comparing a greyhound with a bulldog, the magnitude of the differences is amazing. They hardly look like members of the same species.

A QTL study in 2007 by Nathan Sutter and colleagues indicated that the size of dogs is determined, in part, by alleles of the gene that encodes Igf1, a growth hormone called insulin-like growth factor 1. A particular allele of this gene was found to be common to all small breeds of dogs and nearly absent from very large breeds, suggesting that this allele is one of several genes that influences body size in small breeds of dogs.

Likewise, most of the food we eat is obtained from species that have been modified profoundly by selective breeding strategies. These food products include grains, fruits, vegetables, meat, milk, and juices. **Figure 24.10** illustrates how certain characteristics in

24.6 SELECTIVE BREEDING

Greyhound

German shepherd

Bulldog

Cocker spaniel

FIGURE 24.9 Some common breeds of dogs that have been obtained by selective breeding.

Genes → Traits By selecting parents carrying the alleles that influence certain quantitative traits in a desired way, dog breeders have produced breeds with distinctive sets of traits. For example, the bulldog has alleles that produce short legs and a flat face. By comparison, the corresponding genes in a German shepherd have alleles that produce longer legs and a more pointy snout. All of the dogs shown in this figure carry the same kinds of genes (e.g., many genes that affect their sizes, shapes, and fur color). However, the alleles for many of these genes are different among these dogs, thereby producing breeds with strikingly different phenotypes.

(a) ©Xseon/Shutterstock; (b) ©Pavel Shlykov/Shutterstock; (c) ©Willee Cole/Getty Images; (d) ©Labrador Photo Video/Shutterstock

Concept Check: What are the similarities and differences between natural selection and selective breeding?

the wild mustard plant (*Brassica oleracea*) have been modified by selective breeding to create several varieties of important domesticated crops. The wild plant is native to Europe and Asia, and plant breeders began to modify its traits approximately 4000 years ago. As the figure shows, certain quantitative traits in the domestic

FIGURE 24.10 Crop plants developed by selective breeding of the wild mustard plant (*Brassica oleracea*).

Genes → Traits The wild mustard plant carries a large amount of genetic (i.e., allelic) variation, which was used by plant breeders to produce modern strains that are agriculturally desirable and economically important. For example, by selecting for alleles that promote the formation of large lateral leaf buds, the strain of Brussels sprouts was created. By selecting for alleles that alter the leaf morphology, kale was developed. Although these six agricultural plants look quite different from each other, they carry many of the same alleles as the wild mustard. However, they differ in alleles affecting the formation of stems, leaves, flower buds, and leaf buds.

(a) ©Steven P. Lynch; (b) ©Sarawut Chainawarat/Shutterstock; (c) ©David Acosta Allely/Shutterstock; (d) ©Cheryl Moulton/123RF; (e) ©David Acosta Allely/Shutterstock; (f) ©L. Mouton/PhotoAlto; (g) ©Pixtal/age fotostock

Concept Check: Identify the types of traits that have been subjected to selective breeding to develop these crop plants.

Strain	Modified trait
	Wild mustard plant
Kohlrabi	Stem
Kale	Leaves
Broccoli	Flower buds and stem
Brussels sprouts	Lateral leaf buds
Cabbage	Terminal leaf bud
Cauliflower	Flower buds

strains, such as stems and lateral buds, differ considerably from those of the original wild species.

The phenomenon that underlies selective breeding is genetic variation. Within a group of individuals, allelic variation may affect the outcome of quantitative traits. The fundamental strategy of the selective breeder is to choose parents that will pass on to their offspring alleles that produce desirable phenotypic characteristics. For example, if a breeder wants large cattle, the largest members of the herd are chosen as parents for the next generation. These large cattle will transmit an array of alleles to their offspring that confer large size. The breeder often chooses genetically related individuals (e.g., brothers and sisters) as the parental stock. As mentioned previously, mating between genetically related individuals is known as inbreeding. Some of the consequences of inbreeding are also described in Chapter 23.

What is the outcome when selective breeding is conducted to modify a quantitative trait? **Figure 24.11a** shows the results of a program begun at the Illinois Agricultural Experiment Station in 1896, before the rediscovery of Mendel's laws. This experiment began with 163 ears of corn with an oil content ranging from 4% to 6%. In each of 80 succeeding generations, corn plants were divided into two separate groups. In one group, members with the highest oil content were chosen as parents of the next generation. In the other group, members with the lowest oil content were chosen. After 80 generations, the oil content in the first group rose to over 18%; in the other group, it dropped to less than 1%. These results show that selective breeding can modify quantitative traits in a very directed manner.

When comparing the curves in Figure 24.11, keep in mind that quantitative traits are often at an intermediate value in unselected populations. Therefore, artificial selection can increase or decrease the magnitude of the trait. In this case, oil content can go up or down. Artificial selection tends to be the most rapid and effective in changing the frequency of alleles that are within an intermediate range in a starting population, such as 0.2 to 0.8.

Figure 24.11 also shows the phenomenon known as a **selection limit**—after several generations a plateau is reached where artificial selection is no longer effective. A selection limit may occur for two reasons:

- Presumably, the starting population possesses a large amount of genetic variation, which contributes to the diversity in phenotypes. By carefully choosing the parents, each succeeding generation has a higher proportion of the desirable alleles. However, after many generations, the population may be monomorphic for all or most of the desirable alleles that affect the trait of interest. At this point, additional selective breeding will have no effect.
- A second reason for a selection limit is related to fitness. Some alleles that accumulate in a population due to artificial selection may have a negative influence on the population's overall fitness. A selection limit is reached in which the desired effects of artificial selection are balanced by the negative effects on fitness.

Selective Breeding Provides a Way of Estimating Heritability

With artificial selection experiments, the response to selection is a common way to estimate the narrow-sense heritability in a starting population. The narrow-sense heritability measured in this way is also called the **realized heritability.** It is calculated as

$$h_N^2 = \frac{R}{S}$$

where

R is the response in the offspring to selection, or the difference between the mean of the offspring and the mean of the starting population

S is the selection differential in the parents, or the difference between the mean of the parents and the mean of the starting population

Here,

$$R = \overline{X}_O - \overline{X}$$
$$S = \overline{X}_P - \overline{X}$$

where

$\overline{X}$ is the mean of the starting population
$\overline{X}_O$ is the mean of the offspring
$\overline{X}_P$ is the mean of the parents

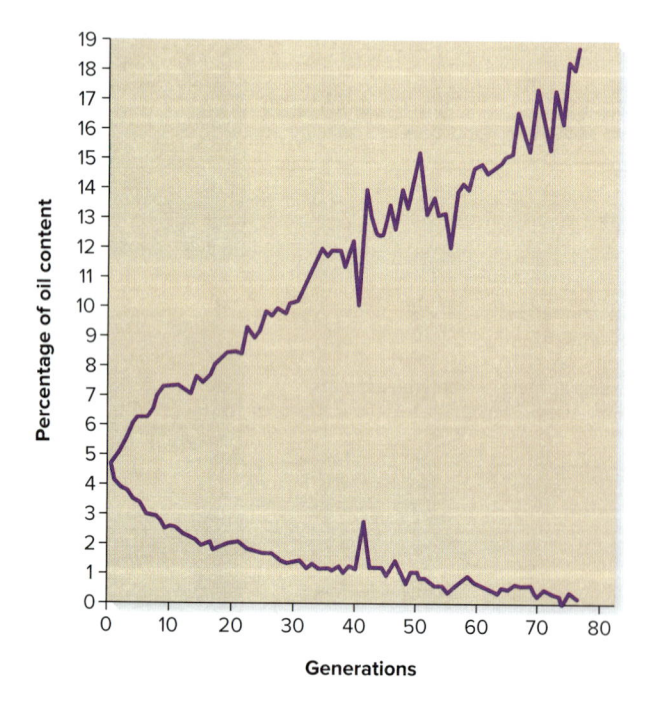

FIGURE 24.11 Common results of selective breeding for a quantitative trait. The graph shows the results of selective breeding for high and low oil content in corn.

Concept Check: What are two reasons why a selection limit is reached at which artificial selection no longer has an effect?

So,

$$h_N^2 = \frac{\overline{X}_O - \overline{X}}{\overline{X}_P - \overline{X}}$$

The narrow-sense heritability is the proportion of the variance in phenotype that can be used to predict changes in the population mean when selection is practiced.

As an example, let's consider the trait of bristles in fruit flies, which are hairlike structures that protrude from the body. Suppose that we began with a population of fruit flies in which the average bristle number was 37.5. The parents chosen from this population had an average bristle number of 40. The offspring of the next generation had an average bristle number of 38.7. With these values, the realized heritability is

$$h_N^2 = \frac{38.7 - 37.5}{40 - 37.5}$$

$$h_N^2 = \frac{1.2}{2.5}$$

$$h_N^2 = 0.48$$

This result tells us that about 48% of the phenotypic variation is due to the additive effects of alleles that affect bristle number.

As we have just seen, selective breeding can be used to predict narrow-sense heritability. Alternatively, if we already know the narrow-sense heritability, we can predict the mean phenotypes of offspring. If we rearrange the realized heritability equation as follows:

$$R = h_N^2 S$$

$$\overline{X}_O - \overline{X} = h_N^2 (\overline{X}_P - \overline{X})$$

The last equation is called the breeder's equation, because it is used to calculate the mean phenotypes of offspring based on the means of the parents, the means of the starting population, and the heritability. An example of the use of this equation is presented next.

Genetic TIPS

The Question: The narrow-sense heritability (h_N^2) for potato weight in a starting population of potato plants is 0.42, and the mean weight is 1.4 pounds. If a breeder crosses plants with average potato weights of 1.9 and 2.1 pounds, respectively, what is the predicted average weight of potatoes from the offspring?

Topic: What topic in genetics does this question address?

The topic is heritability. More specifically, the question is about using heritability to predict the phenotypes of offspring.

Information: What information do you know based on the question and your understanding of the topic?

From the question, you know the narrow-sense heritability for potato weight in a population, and you know the mean potato weight for the population and the weights of potatoes from selected parents. From your understanding of the topic, you may remember that $R = h_N^2 S$.

Problem-Solving Strategy: Make a calculation. Predict the outcome.

To solve this problem, you first need to calculate the mean weight of potatoes produced from the parents and then use the equation described above. The mean potato weight for the parental potatoes is 2.0 pounds. To solve for the mean weight for the offspring:

$$R = h_N^2 S$$

$$\overline{X}_O - \overline{X} = h_N^2 (\overline{X}_P - \overline{X})$$

$$\overline{X}_O - 1.4 = 0.42 (2.0 - 1.4)$$

Answer: $\overline{X}_O = 1.65$ pounds

24.6 REVIEWING THE KEY CONCEPTS

- Selective breeding refers to programs and procedures designed to modify phenotypes in economically important species of plants and animals (see Figures 24.9, 24.10).
- Starting with a genetically diverse population, selective breeding can usually modify a trait in a desired direction until a selection limit is reached (see Figure 24.11).

24.6 COMPREHENSION QUESTIONS

1. For selective breeding to be successful, the starting population must
    a. have genetic variation that affects the trait of interest.
    b. be very large.
    c. be amenable to phenotypic variation caused by environmental effects.
    d. have very little phenotypic variation.
2. The mean weight of cows in a population is 520 kg. Animals with a mean weight of 540 kg are used as parents and produce offspring that have a mean weight of 535 kg. What is the narrow-sense heritability (h_N^2) for body weight in this population of cows?
    a. 0.25
    b. 0.5
    c. 0.75
    d. 1.0

KEY TERMS

Page 584. complex traits, quantitative traits, quantitative genetics
Page 585. continuous traits, meristic traits, threshold traits, discontinuous traits, frequency distribution, normal distribution
Page 586. biometric field, mean
Page 587. variance, standard deviation (*SD*)
Page 588. covariance, correlation coefficient (*r*)
Page 590. polygenic, polygenic inheritance
Page 593. quantitative trait locus (QTL), QTL mapping
Page 595. inbreeding, monomorphic

Page 596. genotype-environment interaction, genotype-environment association

Page 597. heritability, broad-sense heritability, narrow-sense heritability

Page 600. selective breeding, artificial selection

Page 602. selection limit, realized heritability

CHAPTER SUMMARY

- Quantitative genetics is the field of genetics concerned with complex and quantitative traits.

24.1 Overview of Quantitative Traits

- Quantitative traits can be categorized as anatomical, physiological, or behavioral (see Table 24.1).
- Quantitative traits often exhibit a continuum of phenotypic variation that follows a normal distribution (see Figure 24.1).

24.2 Statistical Methods for Evaluating Quantitative Traits

- Statistical methods are used to analyze quantitative traits. These methods include calculations of the mean, variance, standard deviation, covariance, and correlation (see Figure 24.2, Table 24.2).

24.3 Polygenic Inheritance

- Polygenic inheritance refers to the transmission of any trait that is governed by two or more genes.
- Polygenic inheritance and environmental factors may produce a continuum of phenotypes for a quantitative trait (see Figures 24.3, 24.4).

24.4 Identification of Genes That Control Quantitative Traits

- The locations on a chromosome that contain one or more genes affecting a quantitative trait are called quantitative trait loci (QTLs).
- QTLs can be identified by their proximity to known molecular markers (see Figure 24.5).

24.5 Heritability

- Heritability is the amount of phenotypic variation due to genetic variation within a group of individuals raised in a particular environment.
- Genetic variance and environmental variance may contribute additively to phenotypic variance.
- Genetic variance and environmental variance may exhibit interactions and associations (see Figure 24.6, Table 24.3).
- Broad-sense heritability takes into account different types of genetic variation that may affect phenotype, including the additive effects of alleles, effects due to dominant/recessive relationships, and effects due to epistatic interactions.
- Narrow-sense heritability is heritability that is due to the additive effects of alleles.
- Holt determined that dermal ridge count has a very high heritability value in humans (see Figures 24.7, 24.8).

24.6 Selective Breeding

- Selective breeding refers to programs and procedures designed to modify phenotypes in economically important species of plants and animals (see Figures 24.9, 24.10).
- Starting with a genetically diverse population, selective breeding can usually modify a trait in a desired direction until a selection limit is reached (see Figure 24.11).

PROBLEM SETS & INSIGHTS

More Genetic TIPS

1. The following data describe the 6-week weights (in grams) of mice and their offspring of the same sex:

Parent's Weights (g)	Offspring's Weights (g)
24	26
21	24
24	22
27	25
23	21
25	26
22	24
25	24
22	24
27	24

Calculate the correlation coefficient.

Topic: What topic in genetics does this question address?

The topic is calculating a correlation coefficient.

Information: What information do you know based on the question and your understanding of the topic?

From the question, you know the 6-week weights of mice and their same-sex offspring. From your understanding of the topic, you may remember that you first need to calculate the mean weights and standard deviations of the parents and offspring, and then the covariance. The correlation coefficient is computed as the covariance divided by the product of the standard deviations.

Problem-Solving **S**trategy: **Make a calculation.**

To begin to solve this problem, you first need to calculate the means and standard deviations for each group:

$$\bar{X}_{parents} = \frac{24 + 21 + 24 + 27 + 23 + 25 + 22 + 25 + 22 + 27}{10} = 24$$

$$\bar{X}_{offspring} = \frac{26 + 24 + 22 + 25 + 21 + 26 + 24 + 24 + 24 + 24}{10} = 24$$

$$SD_{parents} = \sqrt{\frac{0 + 9 + 0 + 9 + 1 + 1 + 4 + 1 + 4 + 9}{9}} = 2.1$$

$$SD_{offspring} = \sqrt{\frac{4 + 0 + 4 + 1 + 9 + 4 + 0 + 0 + 0 + 0}{9}} = 1.6$$

Next, you calculate the covariance.

$$CoV_{(parents, offspring)} = \frac{\Sigma[(X_P - \bar{X}_P)(X_O - \bar{X}_O)]}{N - 1}$$
$$= \frac{0 + 0 + 0 + 3 + 3 + 2 + 0 + 0 + 0 + 0}{9}$$
$$= 0.9$$

Finally, you calculate the correlation coefficient:

$$r_{(parents, offspring)} = \frac{CoV_{(P,O)}}{SD_P SD_O}$$

$$r_{(parents, offspring)} = \frac{0.9}{(2.1)(1.6)}$$

Answer: $r_{(parent, offspring)} = 0.27$

2. A farmer wants to increase the average body weight in a herd of cattle. She begins with a herd having a mean weight of 595 kg and chooses individuals to breed that have a mean weight of 625 kg. Twenty offspring are obtained, having the following weights in kilograms: 612, 587, 604, 589, 615, 641, 575, 611, 610, 598, 589, 620, 617, 577, 609, 633, 588, 599, 601, and 611. Calculate the realized heritability for body weight in this herd.

Topic: **What topic in genetics does this question address?**

The topic is heritability. More specifically, the question is about calculating the realized heritability based on the outcomes of crosses.

Information: **What information do you know based on the question and your understanding of the topic?**

From the question, you know the mean weight of a herd of cattle and the mean weight of selected parents. You also know the weights of 20 offspring. From your understanding of the topic, you may remember that $h_N^2 = \frac{R}{S}$.

Problem-Solving **S**trategy: **Make a calculation.**

To solve this problem, you first need to calculate the mean weight of the offspring and then use the equation above.

$$\bar{X}_O = \frac{\text{Sum of the offspring's weights}}{\text{Number of offspring}}$$

$$\bar{X}_O = 604 \text{ kg}$$

$$h_N^2 = \frac{R}{S}$$
$$= \frac{\bar{X}_O - \bar{X}}{\bar{X}_P - \bar{X}}$$

$$h_N^2 = \frac{604 - 595}{625 - 595}$$

Answer: $h_N^2 = 0.3$

3. Are the following statements regarding heritability true or false?

A. Heritability applies to a specific population raised in a particular environment.

B. Heritability in the narrow sense takes into account all types of genetic variance.

C. Heritability is a measure of the amount that genetics contributes to the outcome of a trait.

Topic: **What topic in genetics does this question address?**

The topic is heritability.

Information: **What information do you know based on the question and your understanding of the topic?**

In the question, you are given statements regarding heritability. From your understanding of the topic, you may remember the definitions of heritability and narrow-sense heritability.

Problem-Solving **S**trategy: **Define key terms.**

One strategy to solve this problem is to recall the definitions of heritability and narrow-sense heritability.

Answer:

A. True

B. False. Narrow-sense heritability considers only the effects of additive alleles.

C. False. Heritability is a measure of the amount of phenotypic variation that is due to genetic variation; it applies to the variation of a specific population raised in a particular environment.

Conceptual Questions

C1. Give several examples of quantitative traits. How are these quantitative traits described within groups of individuals?

C2. At the molecular level, explain why quantitative traits often exhibit a continuum of phenotypes within a population. How does the environment help produce this continuum?

C3. What is a normal distribution? Discuss this curve with regard to quantitative traits within a population. What is the relationship between the standard deviation and the normal distribution?

C4. Explain the difference between a continuous trait and a discontinuous trait. Give two examples of each. Are quantitative traits likely to be continuous or discontinuous? Explain why.

C5. What is a frequency distribution? Explain how such a graph is plotted for a quantitative trait that is continuous.

C6. The variance for weight in a particular herd of cattle is 484 pounds^2. The mean weight is 562 pounds. How heavy would an animal have to be if it was in the top 2.5% of the herd? The bottom 0.13%?

C7. Two different varieties of potato plants produce potatoes with the same mean weight of 1.5 pounds. One variety has a very low variance for potato weight, and the other has a much higher variance.

    A. Discuss the possible reasons for the differences in variance.

    B. If you were a potato farmer, would you rather raise a variety with a low or high variance? Explain your answer from a practical point of view.

    C. If you were a potato breeder and you wanted to develop potatoes with a heavier weight, would you choose the variety with a low or high variance? Explain your answer.

C8. If $r = 0.5$ and $N = 4$, would you conclude that a positive correlation exists between the two variables? Explain your answer. What if $N = 500$?

C9. What does it mean when a correlation coefficient is negative? Can you think of examples?

C10. When a correlation coefficient is statistically significant, what do you conclude about the two variables? What do the results mean with regard to cause and effect?

C11. What is polygenic inheritance? Discuss the issues that make polygenic inheritance difficult to study.

C12. What is a quantitative trait locus (QTL)? Does a QTL contain one gene or multiple genes? What method is commonly used to identify QTLs?

C13. Let's suppose that weight in a species of mammal is a polygenic trait, and each gene exists as a heavy and light allele. If the allele frequencies in the population are equal for both types of alleles (i.e., 50% heavy alleles and 50% light alleles), what percentage of individuals will be homozygous for the light alleles in all of the genes affecting this trait, if the trait was determined by the following number of genes?

    A. Two

    B. Three

    C. Four

C14. The broad-sense heritability for a trait equals 1.0. In your own words, explain what this value means. Would you conclude that the environment is unimportant in the outcome of this trait? Explain your answer.

C15. From an agricultural point of view, discuss the advantages and disadvantages of selective breeding. It is common for plant breeders to take two different, highly inbred strains, which are the product of many generations of selective breeding, and cross them to make hybrids. How does this approach overcome some of the disadvantages of selective breeding?

C16. Many beautiful varieties of roses have been produced, particularly in the last few decades. These newer varieties often have very striking and showy flowers, making them desirable as horticultural specimens. However, breeders and novices alike have noticed that some of these newer varieties do not have very fragrant flowers compared with the older, more traditional varieties. From a genetic point of view, suggest an explanation why some of these newer varieties with superb flowers are not as fragrant.

C17. In your own words, explain the meaning of the term *heritability*. Why is a heritability value valid only for a particular population of individuals raised in a particular environment?

C18. What is the difference between broad-sense heritability and narrow-sense heritability? Why is narrow-sense heritability such a useful concept in the field of agricultural genetics?

C19. The heritability for egg weight in a group of chickens on a farm in Maine is 0.95. Are the following statements regarding this heritability true or false? If a statement is false, explain why.

    A. The environment in Maine has very little effect on the outcome of this trait.

    B. Nearly all of the phenotypic variation for this trait in this group of chickens is due to genetic variation.

    C. The trait is polygenic and likely to involve a large number of genes.

    D. Based on the observation of the heritability in the Maine chickens, it is reasonable to conclude that the heritability for egg weight in a group of chickens on a farm in Montana is also very high.

C20. In a fairly large population of people living in a commune in the southern United States, everyone cares about good nutrition. All of the members of this population eat very nutritious foods, and their diets are very similar to each other. How do you think the heights of individuals in this commune population would compare with those of the general population in the following categories?

    A. Mean height

    B. Heritability for height

    C. Genetic variation for alleles that affect height

C21. When artificial selection is practiced over many generations, eventually a plateau is reached in which further selection has little effect on the outcome of the trait. This phenomenon is illustrated in Figure 24.11. Explain why it occurs.

C22. Discuss whether a natural population of wolves or a domesticated population of German shepherds is more likely to have a higher heritability for the trait of size.

Application and Experimental Questions

E1. Here are data for height and weight among 10 male college students.

Height (cm)	Weight (kg)
159	48
162	50
161	52
175	60
174	64
198	81
172	58
180	74
161	50
173	54

   A. Calculate the correlation coefficient for height and weight for this group.

   B. Is the correlation coefficient statistically significant? Explain.

E2. The abdomen length (in millimeters) was measured in 15 male *Drosophila,* and the following data were obtained: 1.9, 2.4, 2.1, 2.0, 2.2, 2.4, 1.7, 1.8, 2.0, 2.0, 2.3, 2.1, 1.6, 2.3, and 2.2. Calculate the mean, standard deviation, and variance for this population of male fruit flies.

E3. You conduct an RFLP analysis of head weight in one strain of cabbage and determine that seven QTLs affect this trait. In another strain of cabbage, you find that only four QTLs affect this trait. Both strains of cabbage are from the same species, although they may have been subjected to different degrees of inbreeding. Explain how one strain can have seven QTLs and another strain four QTLs for exactly the same trait. Is the second strain missing three genes?

E4. Let's suppose that two strains of pigs differ in 500 RFLPs. One strain is much larger, on average, than the other. The pigs are crossed to each other, and the members of the F_1 generation are also crossed among themselves to produce an F_2 generation. Three distinct RFLPs are associated with F_2 pigs that are larger. How would you interpret these results?

E5. Outline the steps you would follow to determine the number of genes that influence the yield of rice. Describe the results you might get if rice yield is governed by variation in six different genes.

E6. In a wild strain of tomato plants, the phenotypic variance for tomato weight is 3.2 g^2. In another strain of highly inbred tomatoes raised under the same environmental conditions, the phenotypic variance is 2.2 g^2. With regard to the wild strain,

   A. Estimate V_G.

   B. What is h_B^2?

   C. Assuming that all of the genetic variance is additive, what is h_N^2?

E7. The average thorax length in a *Drosophila* population is 1.01 mm. You want to practice selective breeding to make larger *Drosophila.* To do so, you choose 10 parents (5 males and 5 females) of the following sizes: 0.97, 0.99, 1.05, 1.06, 1.03, 1.21, 1.22, 1.17, 1.19, and 1.20. You mate them and then determine the thorax lengths of 30 offspring (half male and half female):

0.99, 1.15, 1.20, 1.33, 1.07, 1.11, 1.21, 0.94, 1.07, 1.11, 1.20, 1.01, 1.02, 1.05, 1.21, 1.22, 1.03, 0.99, 1.20, 1.10, 0.91, 0.94, 1.13, 1.14, 1.20, 0.89, 1.10, 1.04, 1.01, 1.26

Calculate the realized heritability of thorax length in this group of flies.

E8. In a strain of mice, the average 6-week body weight is 25 g and the narrow-sense heritability for this trait is 0.21.

   A. What would be the average weight of the offspring if parents with a mean weight of 27 g were chosen?

   B. What parental mean weight would you have to choose to obtain offspring with an average weight of 26.5 g?

E9. A danger in computing heritability values from studies involving genetically related individuals is the possibility that these individuals share more similar environments than do unrelated individuals. In the experiment shown in Figure 24.8, which data are the most compelling evidence that ridge count is not caused by genetically related individuals sharing common environments? Explain.

E10. A large, genetically heterogeneous group of tomato plants was used as the original breeding stock by two different breeders, named Mary and Hector. Each breeder was given 50 seeds and began a selective breeding strategy, much like that described in Figure 24.11. The seeds were planted, and the breeders selected the 10 plants with the highest mean tomato weights as the breeding stock for the next generation. This process was repeated over the course of 12 growing seasons, and the following data were obtained:

Mean Weight of Tomatoes (pounds)

Year	Mary's Tomatoes	Hector's Tomatoes
1	0.7	0.8
2	0.9	0.9
3	1.1	1.2
4	1.2	1.3
5	1.3	1.3
6	1.4	1.4
7	1.4	1.5
8	1.5	1.5
9	1.5	1.5
10	1.5	1.5
11	1.5	1.5
12	1.5	1.5

   A. Explain these results.

   B. Another tomato breeder, named Martin, got some seeds from Mary's and Hector's tomato strains (after 12 generations), grew the plants, and then crossed them to each other. The mean weight of the tomatoes in these hybrids was about 1.7 pounds. For a period of 5 years, Martin subjected these hybrids to the same selective breeding strategy that Mary and Hector had followed, and he obtained the following results:

Mean Weight of Tomatoes (pounds)

Year	Martin's Tomatoes
1	1.7
2	1.8
3	1.9
4	2.0
5	2.0

Explain Martin's data. Why was Martin able to obtain tomatoes heavier than 1.5 pounds, whereas Mary's and Hector's strains appeared to plateau at this weight?

E11. For each of the following relationships, the correlations for height were determined for 15 pairs of individuals:

Mother/daughter: 0.36

Mother/granddaughter: 0.17

Sister/sister: 0.39

Sister/sister (fraternal twins): 0.40

Sister/sister (identical twins): 0.77

What is the average heritability for height in this group of females?

E12. An animal breeder had a herd of sheep with a mean weight of 254 pounds at 3 years of age. He chose animals with a mean weight of 281 pounds as parents for the next generation. When these offspring reached 3 years of age, their mean weight was 269 pounds.

A. Calculate the narrow-sense heritability for weight in this herd.

B. Using the heritability value that you calculated in part A, what mean weight would you have to choose for the parents to get offspring that weigh 275 pounds on average (at 3 years of age)?

E13. The trait of blood pressure in humans has a frequency distribution that is similar to a normal distribution. The following graph shows the ranges of blood pressures for a selected population of people. The red line depicts the frequency distribution of the systolic pressures for the entire population. Several individuals with high blood pressure were identified, and the blood pressures of their relatives were determined. This frequency distribution is depicted with a blue line. (Note: The blue line does not include the people who were identified with high blood pressure; it includes only their relatives.)

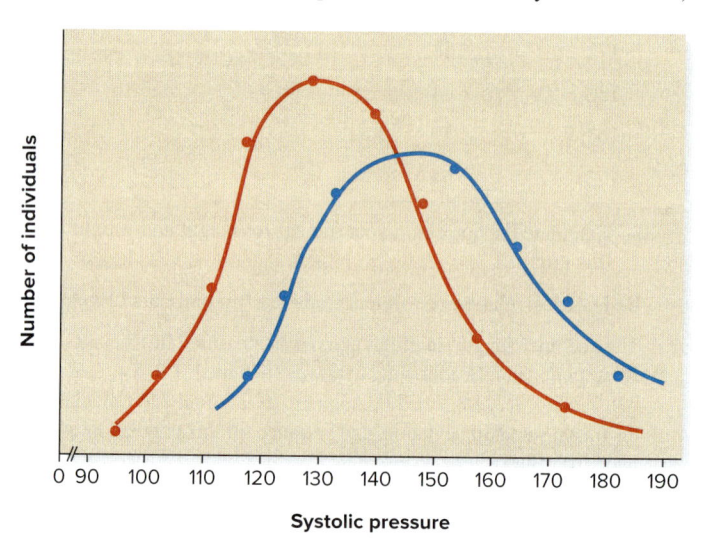

What do these data suggest with regard to a genetic basis for high blood pressure? What statistical approach could you use to determine the heritability for this trait?

Questions for Student Discussion/Collaboration

1. Explain why heritability is an important phenomenon in agriculture. Discuss how it is misunderstood.

2. From a biological viewpoint, speculate as to why many traits seem to fit a normal distribution. Students with a strong background in math and statistics may want to explain how a normal distribution is generated, and what it means. Can you think of biological examples that do not fit a normal distribution?

Answers to Comprehension Questions

24.1: d, c

24.2: d, b

24.3: d

24.4: b, c

24.5: c, b, b

24.6: a, c

Note: All answers appear in Connect; the answers to even-numbered questions and all Concept Check questions are in Appendix B.

APPENDIX
EXPERIMENTAL TECHNIQUES

OUTLINE

A.1 Methods for Growing Cells

A.2 Microscopy

A.3 Separation Methods

A.4 Methods for Measuring Concentrations of Molecules and Detecting Radioisotopes

A.1 METHODS FOR GROWING CELLS

Researchers often grow cells in a laboratory as a way to study their properties. A population of cells grown in the laboratory is known as a **cell culture.** Cell culturing offers several technical advantages. The primary advantage is that the growth medium is defined and can be controlled. Minimal growth medium contains the bare essentials for cell growth: salts, a carbon source, an energy source, essential vitamins, amino acids, and trace elements. In their experiments, geneticists often compare strains that can grow in minimal media and mutant strains that cannot grow unless the medium is supplemented with additional components. A rich growth medium contains many more components than are required for growth.

Researchers also add other substances to the culture medium for experimental reasons. For example, radioactive isotopes can be added to the culture medium to radiolabel cellular macromolecules. In addition, an experimenter might add a hormone to the growth medium and then monitor the cells' response to the hormone. In all of these cases, cell culturing is advantageous because the experimenter can control and vary the composition of the growth medium.

The first step in creating a cell culture is the isolation of a cell population that the researchers wish to study. For bacteria, such as *Escherichia coli,* and eukaryotic microorganisms, such as yeast and *Neurospora,* researchers simply obtain a sample of cells from a colleague or a stock center. For animal or plant tissues, the procedure is a bit more complicated. When cells are contained within a complex tissue, they must first be dispersed by treating the tissue with agents that separate it into individual cells to create a cell suspension.

Once a desired population of cells has been obtained, researchers can grow them in a laboratory (i.e., in vitro) either suspended in a liquid growth medium or attached to a solid surface such as agar. Both methods have been commonly used in the experiments considered throughout this text. Liquid culture is often used when researchers want to obtain a large quantity of cells and isolate individual cellular components, such as nuclei or DNA. **Figure A.1** shows animal cells and bacteria cells that are grown on solid growth media. In gene-cloning experiments with bacteria and yeast, solid media are also used. Each colony of cells is

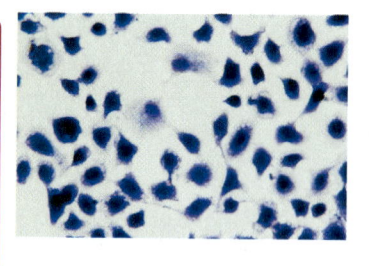

FIGURE A.1 Growth of cells on solid growth media. (a) This micrograph shows fibroblasts growing as a monolayer on a solid growth medium. (b) Bacterial cells form colonies that are a clonal population of cells derived from a single cell.
©De Agostini Picture Library/Age fotostock, ©Science Photo Library-CNRI/SPL/Getty Images

a clone of cells that is derived from a single cell that divided to produce many cells (**Figure A.1b**). As discussed in Chapter 20, a solid medium is used in the isolation of individual clones that contain a desired gene.

A.2 MICROSCOPY

Microscopy is a technique for observing things that cannot be seen (or can hardly be seen) with the unaided eye. A key aspect in microscopy is **resolution,** which is the minimum distance between two objects that enables them to be seen as separate from each other. The ability to resolve two points as being separate depends on several factors, including the wavelength of the illumination source (light or electron beam), the medium in which the sample is immersed, and the structural features of the microscope (which are beyond the scope of this text).

As shown in **Figure A.2**, two widely used kinds of microscopes are the optical (light) microscope and the transmission electron microscope (TEM). The light microscope is used to resolve cellular structures to a resolution limit of approximately 0.3 μm. (For comparison, a typical bacterium is about 1 μm long.) At this resolution, the individual cell organelles in eukaryotic cells can be discerned easily, and chromosomes are also visible. Karyotyping is accomplished via light microscopy after the chromosomes have been treated with stains. A variation of light microscopy known as fluorescence microscopy is often used to highlight a particular feature of a chromosome or cellular structure. The technique of fluorescence in situ hybridization (FISH; see Chapter 21) makes use of this type of microscope. Also, optical modifications in certain light microscopes (e.g., phase contrast and differential interference) can be used to exaggerate the differences in densities between neighboring cells or cell structures. These kinds of light microscopes are useful in monitoring cell division in living (unstained) cells or in transparent worms.

The structural details of large macromolecules such as DNA and ribosomes are not observable by light microscopy. The coarse topology of these macromolecules can be determined by electron microscopy. Electron microscopes have a limit of resolution of about 2 nm, which is about 100 times finer than the best light microscopes. The primary advantage of electron microscopy over light microscopy is its better resolution. Disadvantages include a much higher expense and more extensive sample preparation. In transmission electron microscopy, the sample is bombarded with an electron beam. This requires that the sample be dried, fixed, and usually coated with a heavy metal that absorbs electrons.

A.3 SEPARATION METHODS

Biologists often wish to take complex systems and separate them into less complex components. For example, the cells within a complex tissue can be separated into individual cells, or the macromolecules within cells can be separated from the other cellular components. In this section, we will focus primarily on methods aimed at separating and purifying macromolecules.

Disruption of Cellular Components

In many experiments described in this text, researchers have obtained a sample of cells and then wish to isolate particular components from the cells. For example, a researcher may want to purify a protein that functions as a transcription factor. To do so, he or she would begin with a sample of cells that synthesize this protein and then break open the cells using one of the methods described in **Table A.1**. In eukaryotes, the breakage of cells releases the soluble proteins from the cells; it also dissociates the cell organelles that are bounded by membranes. This mixture of proteins and cell organelles can then be isolated and purified by centrifugation and chromatographic methods, which are described next.

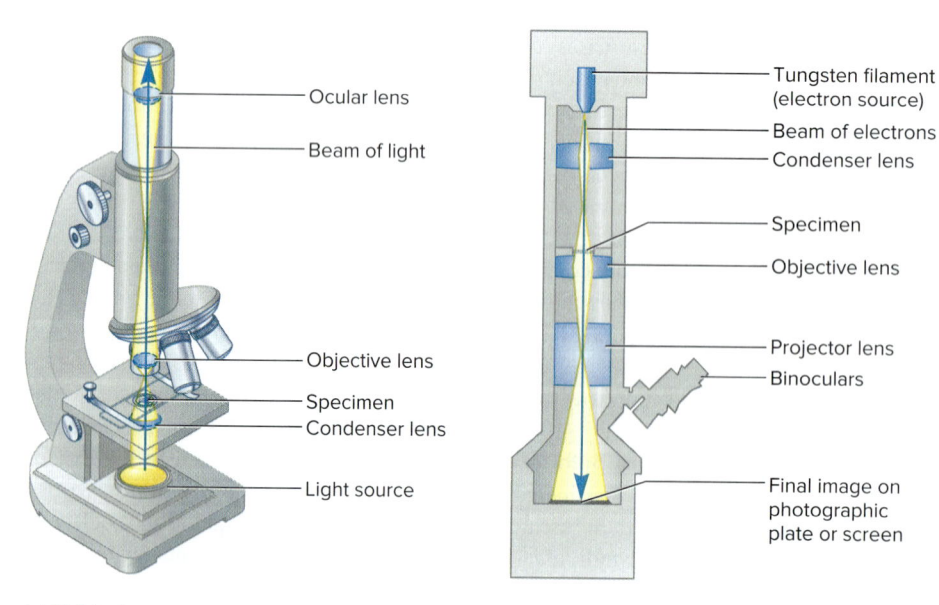

(a) Light microscope
(b) Transmission electron microscope

FIGURE A.2 Design of (a) optical (light) and (b) transmission electron microscopes.

TABLE A.1 Common Methods of Cell Disruption

Method	Description
Sonication	The exposure of cells to intense sound waves, which breaks the cell membranes.
French press	The passage of cells through a small aperture under high pressure, which breaks the cell membranes and cell wall.
Homogenization	Cells are placed in a tube that contains a pestle. When the pestle is spun, the cells are squeezed through the small space between the pestle and the glass wall of the tube, thereby breaking them.
Osmotic shock	The transfer of cells into a hypo-osmotic medium. The cells take up water and eventually burst.

Centrifugation

Centrifugation is a method commonly used to separate cell organelles and macromolecules. A **centrifuge** contains a motor that causes a rotor holding centrifuge tubes to spin very rapidly. As the rotor spins, particles move toward the bottom of each centrifuge tube; the rate at which they move depends on several factors, including their densities, sizes, and shapes, and the viscosity of the medium. The rate at which a macromolecule or cell organelle sediments to the bottom of a centrifuge tube is called its **sedimentation coefficient,** which is normally expressed in Svedberg units (S): $1\ S = 1 \times 10^{-13}$ second.

When a sample contains a mixture of macromolecules or cell organelles, it is likely that different components will sediment at different rates. This phenomenon, which is the basis for the method of **differential centrifugation,** is shown in **Figure A.3**. As seen here, particles with large sedimentation coefficients reach the bottom of the tube more quickly than those with smaller coefficients. Researchers can follow two different strategies that use differential centrifugation as a separation technique. One way is to separate the **supernatant** from the **pellet** following centrifugation. The pellet is a collection of particles found at the bottom of the tube, and the supernatant is the liquid found above the pellet. In Figure A.3, after the experimenter subjects the sample to a low-speed spin, most of the particles with large sedimentation coefficients are found in the pellet, whereas most of the particles with small and intermediate coefficients are found in the supernatant. A high-speed spin of the supernatant then separates the small and intermediate particles. Therefore, differential centrifugation provides a way of segregating these three types of particles.

A second way to separate particles using centrifugation is to collect fractions. A **fraction** is a portion of the liquid contained within a centrifuge tube. The collection of fractions is done when the solution within the centrifuge tube contains a gradient. For example, as shown in **Figure A.4**, the solution at the top of the

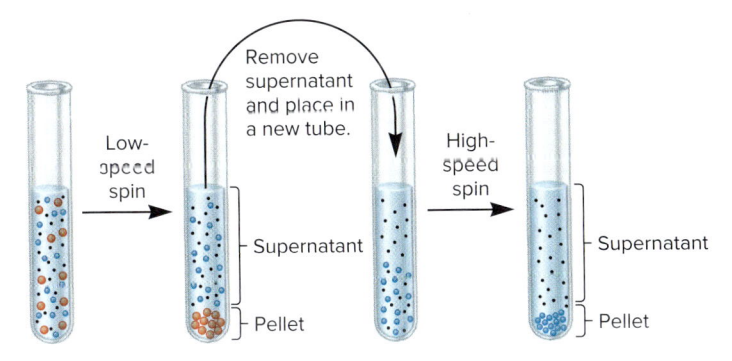

- Particles with large sedimentation coefficients
- Particles with intermediate sedimentation coefficients
- Particles with small sedimentation coefficients

FIGURE A.3 **The method of differential centrifugation.** A sample containing a mixture of particles with different sedimentation coefficients is placed in a centrifuge tube. The tube is subjected to a low-speed spin that pellets the particles with large sedimentation coefficients. After a high-speed spin of the supernatant, the particles with an intermediate sedimentation coefficient are found in the pellet, and those with a small sedimentation coefficient are in the liquid supernatant.

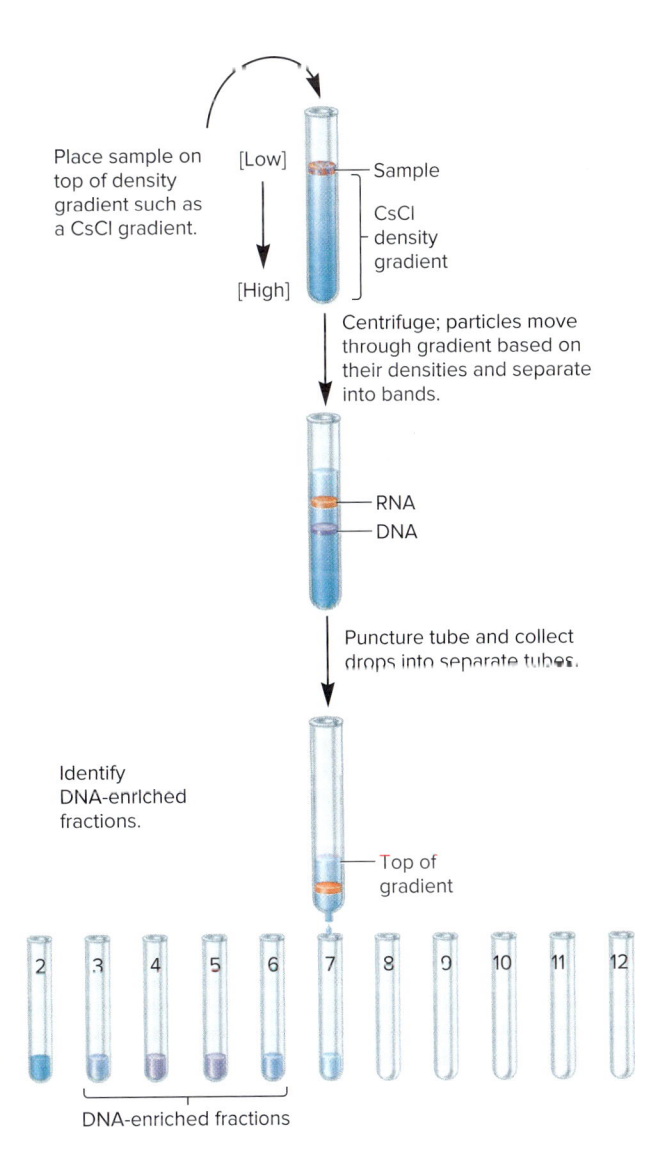

FIGURE A.4 **Gradient centrifugation and the collection of fractions.**

tube has a lower concentration of cesium chloride (CsCl) than that at the bottom. In this experiment, a sample is layered on the top of the gradient and then centrifuged. In this example, the DNA and RNA separate from each other, because they have different sedimentation coefficients. The experimenter then punctures the bottom of the tube and collects fractions. The DNA fragments, which are heavier, come out of the tube in the earlier fractions; the RNA molecules are collected in later fractions.

A type of gradient centrifugation that may also be used to separate macromolecules and organelles is **equilibrium density centrifugation.** In this method, the particles sediment through the gradient, reaching a position where the density of the particle matches the density of the solution. At this point, the particle is at equilibrium and does not move any farther toward the bottom of the tube.

Chromatography and Gel Electrophoresis

Chromatography is a method of separating different macromolecules or smaller molecules based on their chemical and physical properties. In this method, a sample is dissolved in a liquid solvent and exposed to some type of matrix, such as a column containing beads or a thin strip of paper. The degree to which the molecules interact with the matrix depends on their chemical and physical characteristics. For example, a positively charged molecule binds tightly to a negatively charged matrix, but a neutral molecule does not.

Figure A.5 illustrates how column chromatography can be used to separate molecules that differ with respect to charge. Prior to this experiment, a column is packed with beads that are positively charged. There is plenty of space between the beads for molecules to flow from the top of the column to the bottom. However, if the molecules are negatively charged, they will spend some of their time binding to the positive charges on the surface of the beads. In the example shown in Figure A.5, the red proteins are positively charged and, therefore, flow rapidly from the top of the column to the bottom. They emerge in the fractions that are collected early in this experiment. The blue proteins, however, are negatively charged and tend to bind to the beads. The binding of the blue proteins to the beads can be disrupted by changing the ionic strength or pH of the solution that is added to the column. Eventually, the blue proteins will be eluted (i.e., leave the column) in later fractions.

Researchers use many variations of chromatography to separate molecules and macromolecules. The type shown in Figure A.5 is called ion-exchange chromatography, because its basis for separation depends on the charge of the molecules. In another type of column chromatography, known as gel filtration chromatography, the beads are porous. Small molecules are temporarily trapped within the beads, whereas large molecules flow between the beads. In this way, gel filtration separates molecules on the basis of size. To separate different types of macromolecules, such as proteins, researchers may use another type of bead; this bead has a preattached molecule that binds specifically to the protein they want to purify. For example, if a transcription factor binds a particular DNA sequence as part of its function, the beads within a column may have this DNA sequence preattached to them. Therefore, the transcription factor binds tightly to the DNA attached to these beads, whereas all other proteins are eluted rapidly from the column. This form of chromatography is called affinity chromatography, because the beads have a special affinity for the macromolecule of interest.

Besides column chromatography, in which beads are packed into a column, a matrix can be made in other ways. In paper chromatography, molecules pass through a matrix composed of paper. The rate of movement of molecules through the paper depends on their degree of interaction with the solvent and paper. In thin-layer chromatography, a matrix is spread out as a very thin layer on a rigid support such as a glass plate. In general, paper and thin-layer chromatography are effective at separating small molecules, whereas column chromatography is used to separate macromolecules such as DNA fragments or proteins.

Gel electrophoresis combines chromatography and electrophoresis to separate molecules and macromolecules. As its name suggests, the matrix used in gel electrophoresis is composed of a gel. As shown in **Figure A.6**, samples are loaded into wells at one end of the gel, and an electric field is applied across the gel. This electric field causes charged molecules to migrate from one side of the gel to the other. The migration of molecules in response to an electric field is called **electrophoresis.** In the examples of gel electrophoresis found in this text, the macromolecules within the sample migrate toward the positive end of the gel. In most forms of gel electrophoresis, a mixture of macromolecules is separated according to their molecular masses. Small proteins or DNA fragments move to the bottom of the gel more quickly than larger ones. Because the samples are loaded in rectangular wells at the top of the gel, the molecules within the sample are separated into bands within the gel. These bands of separated macromolecules can be visualized with stains. For example, ethidium bromide is a stain that binds to DNA and RNA and can be seen under ultraviolet light.

The two most commonly used gels are polymers made from acrylamide or agarose. Proteins typically are separated on

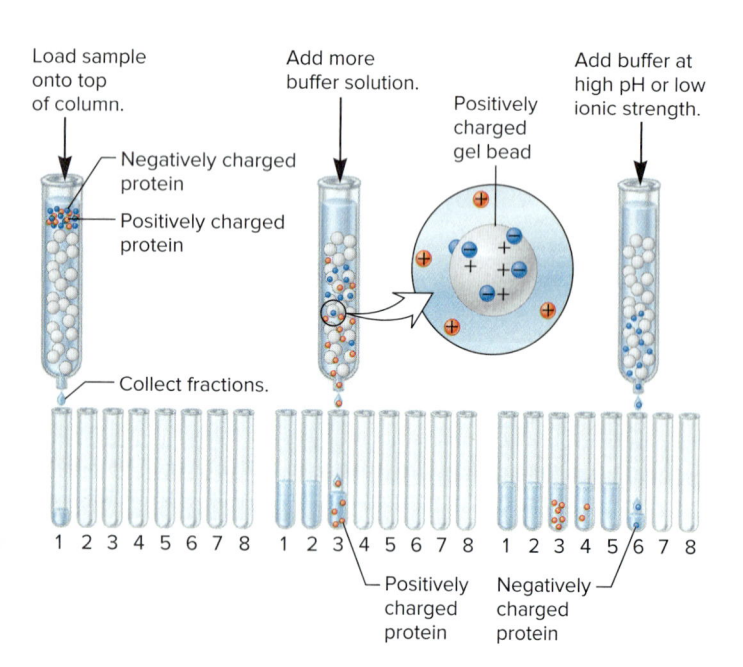

FIGURE A.5 Ion-exchange chromatography.

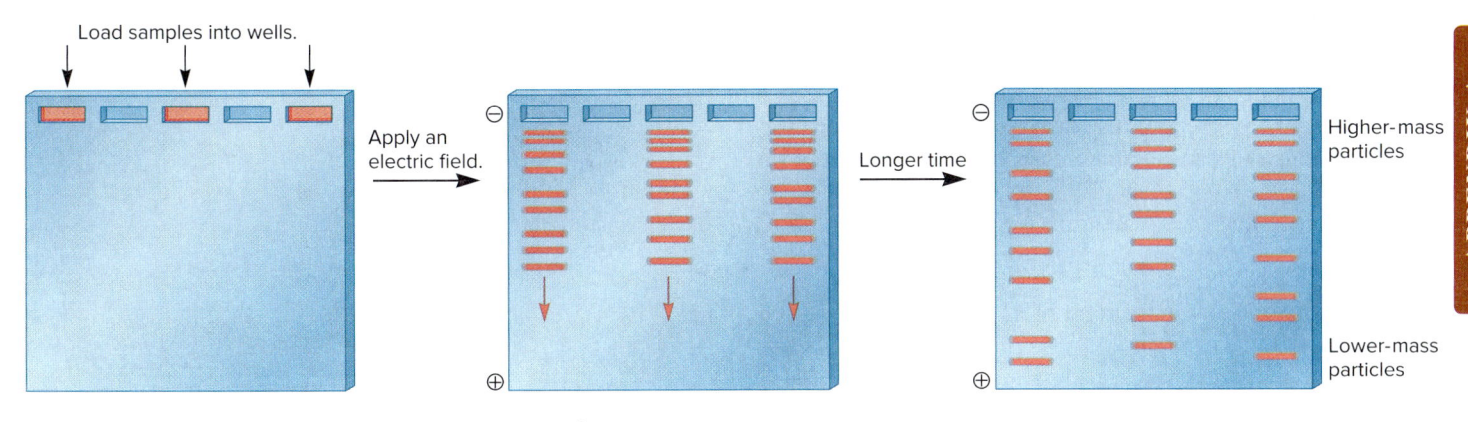

(a) Separation of a mixture of particles by gel electrophoresis

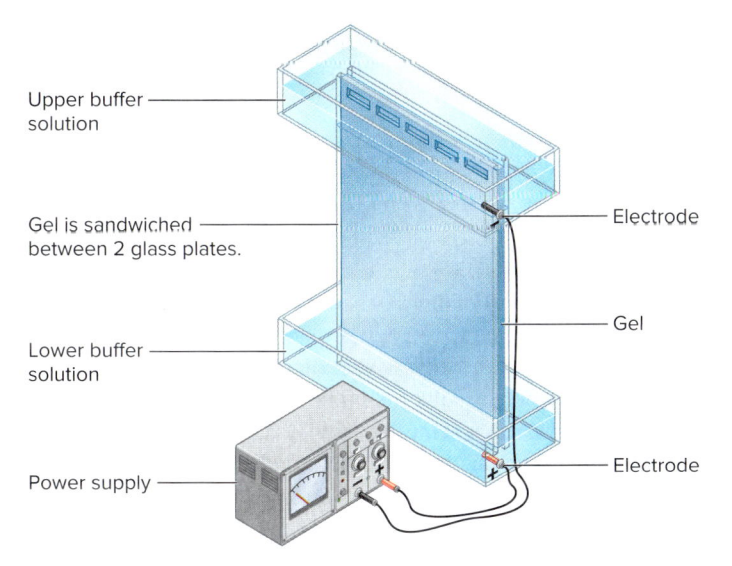

(b) Apparatus used in gel electrophoresis

FIGURE A.6 Gel electrophoresis of DNA fragments.

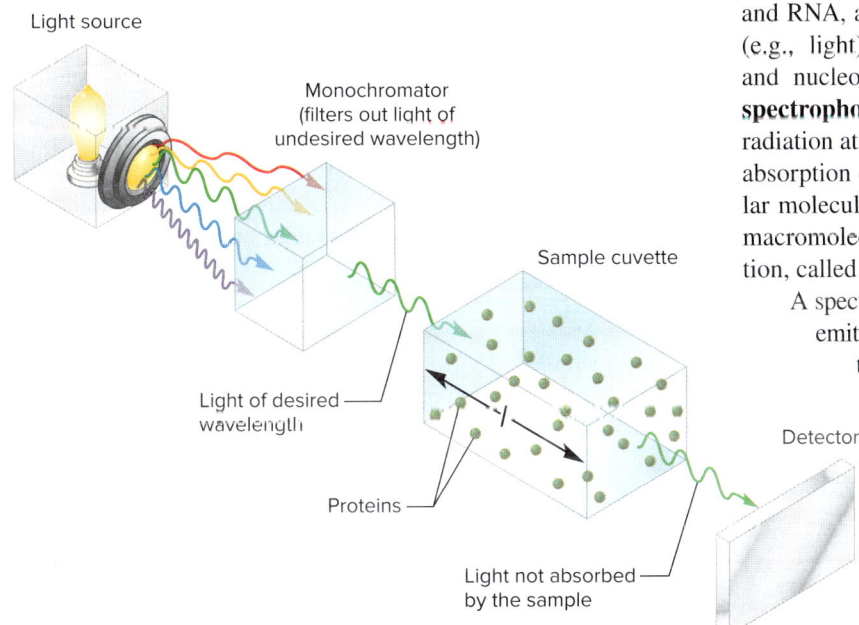

FIGURE A.7 Design of a spectrophotometer.

polyacrylamide gels, whereas DNA fragments are separated on agarose gels. Occasionally, researchers use polyacrylamide gels to separate DNA fragments that are relatively small (i.e., shorter than 1000 bp).

A.4 METHODS FOR MEASURING CONCENTRATIONS OF MOLECULES AND DETECTING RADIOISOTOPES

To understand the structure and function of cells, researchers often need to detect the presence of molecules and macromolecules and to measure their concentrations. In this section, we consider a variety of methods for detecting and measuring the concentrations of biological molecules and macromolecules.

Spectroscopy

Macromolecules found in living cells, such as proteins, DNA, and RNA, are fairly complex molecules that can absorb radiation (e.g., light). Likewise, small molecules such as amino acids and nucleotides can also absorb light. A device known as a **spectrophotometer** is used by researchers to determine how much radiation at various wavelengths a sample absorbs. The amount of absorption can be used to determine the concentration of particular molecules within a sample, because each type of molecule or macromolecule has its own characteristic wavelength(s) of absorption, called its absorption spectrum.

A spectrophotometer typically has two light sources, which can emit ultraviolet or visible light. As shown in **Figure A.7**, the light is passed through a monochromator, which allows the passage of light at a desired range of wavelengths and filters out undesired wavelengths. The light of desired wavelengths then strikes a sample contained within a cuvette. Some of the incident light is absorbed, and some is not. The amount and wavelengths of light that are absorbed depend on the concentration and structures of the molecules and macromolecules in the cuvette. The unabsorbed light passes through the sample and is detected by the spectrophotometer. The amount of light that strikes the detector is subtracted from the amount of incident light, yielding the measure of absorption. In this way, the spectrophotometer provides an absorption reading for the sample. This reading can be used to calculate the concentration of particular molecules or macromolecules in a sample.

Detection of Radioisotopes

A radioisotope is an unstable form of an atom that decays to a more stable form by emitting α-, β-, or γ-rays, which are types of ionizing radiation. In research, radioisotopes that are β and/or γ emitters are commonly used. A β-ray is an emitted electron, and a γ-ray is an emitted photon. Some radioisotopes commonly used in genetics experiments are shown in **Table A.2**.

Experimentally, radioisotopes are used because they are easy to detect. Therefore, if a particular compound is radiolabeled, its presence can be detected specifically throughout the course of an experiment. For example, if a nucleotide is radiolabeled with ^{32}P, a researcher can determine whether the isotope becomes incorporated into newly made DNA or whether it remains as the free nucleotide. Researchers commonly use two different methods of detecting radioisotopes: scintillation counting and autoradiography.

The technique of **scintillation counting** permits a researcher to count the number of radioactive emissions from a sample containing a population of radioisotopes. In this approach, the sample is dissolved in a solution (called the scintillant) that contains organic solvents and one or more compounds known as fluors. When radioisotopes emit ionizing radiation, energy is absorbed by the fluors in the solvent. This excites the fluor molecules, causing their electrons to be boosted to higher energy levels. The excited electrons return to lower, more stable energy levels by releasing photons of light. When a fluor is struck by ionizing radiation, it also absorbs the energy and then releases a photon of light within a particular wavelength range. A device known as a scintillation counter counts the photons of light emitted by the fluors. **Figure A.8** shows a scintillation counter. To use this device, a researcher dissolves a sample in a scintillant and then places the solution in a scintillation vial. The vial is then placed in the scintillation counter, which detects the amount of radioactivity. The scintillation counter has a digital meter that displays the amount of radioactivity in the sample and provides a printout of the amount of radioactivity in counts per minute. A scintillation counter contains several rows for the loading and analysis of many scintillation vials. After they have been loaded, the scintillation counter counts the amount of radioactivity in each vial and provides the researcher with a printout showing those amounts.

FIGURE A.8 A scintillation counter.
©Nigel Cattlin/Science Source

A second way of detecting radioisotopes is via **autoradiography.** This technique is not as quantitative as scintillation counting, because it does not provide the experimenter with a precise measure of the amount of radioactivity in counts per minute. However, autoradiography has the great advantage that it can detect radioisotopes in their actual locations in macromolecules or cells. For example, autoradiography is used to detect a particular band on a gel or to map the location of a gene within an intact chromosome.

To conduct autoradiography, a sample containing a radioisotope is fixed and usually dried. If it is a cellular sample, it also may be thin-sectioned. The sample is then pressed next to X-ray film (in the dark) and placed in a lightproof cassette. When a radioisotope decays, it emits a β- or γ-ray, which may strike a thin layer of photoemulsion next to the film. The photoemulsion contains silver salts such as AgBr. When a radioactive particle is emitted and strikes the photoemulsion, a silver grain is deposited on the film. This produces a dark spot on the film, which correlates with the original location of the radioisotope in the sample. In this way, the dark image on the film reveals the location(s) of the radioisotopes in the sample.

TABLE A.2
Some Isotopes Used in Genetics Experiments

Isotope	Stable or Radioactive	Emission	Half-Life
^2H	Stable		
^3H	Radioactive	β	12.3 years
^{13}C	Stable		
^{14}C	Radioactive	β	5730 years
^{15}N	Stable		
^{18}O	Stable		
^{24}Na	Radioactive	β (and γ)	15 hours
^{32}P	Radioactive	β	14.3 days
^{35}S	Radioactive	β	87.4 days
^{45}Ca	Radioactive	β	164 days
^{59}Fe	Radioactive	β (and γ)	45 days
^{131}I	Radioactive	β (and γ)	8.1

ANSWERS TO CONCEPT CHECKS AND EVEN-NUMBERED PROBLEMS

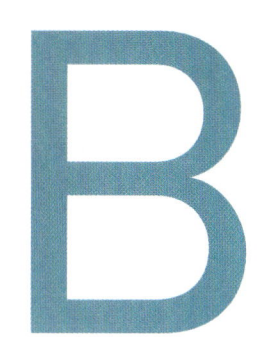

::

CHAPTER 1

Note: The answers to the Comprehension Questions are at the end of the chapter.

Concept Check Questions (in figure legends)

FIGURE 1.1 Understanding our genes may help with diagnoses of inherited diseases. It may also lead to the development of drugs to combat diseases. Other answers are possible.

FIGURE 1.2 Many ethical issues are associated with human cloning. Is it the wrong thing to do? Does it conflict with an individual's religious views? And so on.

FIGURE 1.3 Sorting mosquitoes allows sterile males to be released into the environment, which limits mosquito reproduction because females mate only once.

FIGURE 1.4 DNA is a macromolecule.

FIGURE 1.5 DNA and proteins are found in chromosomes. RNA may also be associated with chromosomes.

FIGURE 1.6 The information to make a polypeptide is stored in DNA.

FIGURE 1.7 The dark-colored butterfly has a more active pigment-synthesizing enzyme.

FIGURE 1.8 Genetic variation is the reason the frogs look different.

FIGURE 1.9 These are examples of variation in chromosome number.

FIGURE 1.10 If this girl had been given a standard diet, she would have developed the harmful symptoms of PKU, which include mental impairment and foul-smelling urine.

FIGURE 1.11 A corn gamete contains 10 chromosomes. (The leaf cells are diploid.)

FIGURE 1.12 The horse populations have become adapted to their environment, which has changed over the course of many years.

FIGURE 1.13 There are several possible examples of other model organisms, including rats and frogs.

End-of-Chapter Questions

Conceptual Questions

C2. At the molecular level, a gene (a sequence of DNA) is first transcribed into RNA. The genetic code within the RNA is used to synthesize a protein with a particular amino acid sequence. This second process is called translation.

C4. Genetic variation is the occurrence of genetic differences within members of the same species or different species. Within any population, variation may occur in the genetic material. Variation may occur in particular genes, so some individuals carry one allele and other individuals carry a different allele. Examples include differences in coat color among mammals or flower color in plants. At the molecular level, this type of genetic variation is caused by changes in the DNA sequences of genes. There may also be variation in chromosome structure and number.

C6. You can pick almost any trait. For example, flower color in petunias would be an interesting choice. Some petunias are red and others are purple. There must be different alleles in a flower color gene that affect this trait in petunias. In addition, the amount of sunlight, fertilizer, and water also affects the intensity of flower color.

C8. A DNA sequence is a sequence of nucleotides. Each nucleotide may have one of four different bases (i.e., A, T, G, or C). When we speak of a DNA sequence, we focus on the sequence of those bases.

C10. A. A gene is a segment of DNA. For most genes, the expression of the gene results in the production of a functional protein. The functioning of proteins within living cells largely determines the traits of an organism.

B. A gene is a segment of DNA that usually encodes the information for the production of a specific polypeptide. Genes are found within chromosomes. Many genes are found within a single chromosome.

C. An allele is an alternative version of a particular gene. For example, suppose a plant has a flower color gene. One allele could produce a white flower, while a different allele could produce an orange flower. The white allele and orange allele are alleles of the flower color gene.

D. A DNA sequence is a sequence of nucleotides. The information within a DNA sequence (which is transcribed into an RNA sequence) specifies the amino acid sequence within a polypeptide.

C12. A. How genes and traits are transmitted from parents to offspring

B. How the genetic material functions at the molecular and cellular levels

C. Why genetic variation exists in populations, and how it changes over the course of many generations

Application and Experimental Questions

E2. A genetic cross involves breeding two different individuals.

E4. You would see 47 chromosomes instead of 46. There would be three copies of chromosome 21 instead of two copies.

E6. You need to follow the scientific method. You can take a look at an experiment in another chapter to see how the scientific method is followed.

CHAPTER 2

Note: The answers to the Comprehension Questions are at the end of the chapter.

Concept Check Questions (in figure legends)

FIGURE 2.1 Compartmentalization means that cells have membrane-bound compartments.

FIGURE 2.2 The chromosomes would not be spread out very well and would probably be overlapping. It would be difficult to see individual chromosomes.

FIGURE 2.3 Homologs are similar in size and banding pattern, and they carry the same types of genes. However, the alleles of a given gene may be different.

FIGURE 2.4 FtsZ assembles into a ring at the future site of the septum and recruits to that site other proteins that produce a cell wall between the two daughter cells.

FIGURE 2.5 The G_1 phase is a phase of the cell cycle when a cell may become committed to cell division. By comparison, the G_0 phase is a phase in which a cell is either not advancing through the cell cycle or has committed to never divide again.

FIGURE 2.6 Homologs are genetically similar; one is inherited from the mother and the other from the father. By comparison, chromatids are the product of DNA replication. The chromatids within a pair of sister chromatids are genetically identical.

FIGURE 2.7 One end of a kinetochore microtubule is attached to a kinetochore on a chromosome. The other end is within the centrosome.

FIGURE 2.8 Anaphase

FIGURE 2.9 Ingression occurs because myosin motor proteins shorten the contractile ring, which is formed from actin proteins.

FIGURE 2.10 The end result of crossing over is that homologous chromosomes have exchanged pieces.

FIGURE 2.11 The cells at the end of meiosis are haploid, whereas the mother cell is diploid.

FIGURE 2.12 In mitosis, each pair of sister chromatids is attached to both poles, whereas in metaphase of meiosis I, each pair of sister chromatids is attached to just one pole.

FIGURE 2.13 Polar bodies are small cells that are produced during oogenesis and then degenerate.

FIGURE 2.14 All of the nuclei in the embryo sac are haploid. The central cell has two haploid nuclei, and all of the other cells, including the egg, have just one.

End-of-Chapter Questions

Conceptual Questions

C2. The term *homolog* refers to the members of a chromosome pair. Homologs are usually the same size and carry the same types and order of genes. They may differ in that the genes they carry may be different alleles.

C4. Metaphase is the organization phase, and anaphase is the separation phase.

C6. In metaphase I of meiosis, each pair of chromatids is attached to only one pole via the kinetochore microtubules. In metaphase of mitosis, there are two attachments (i.e., to both poles). If the attachment is lost, a chromosome will not migrate to a pole and may not become enclosed in a nuclear membrane after telophase. If left out in the cytosol, it would eventually be degraded.

C8. The reduction occurs because there is a single DNA replication event but two cell divisions. Because of the nature of separation during anaphase I, each cell receives one copy of each type of chromosome.

C10. It means that the maternally derived and paternally derived chromosomes are randomly aligned along the metaphase plate during metaphase I.

C12. There are three pairs of chromosomes. The number of different random alignments equals 2^n, where n equals the number of chromosomes per set. So the possible number of arrangements equals 2^3, which is 8.

C14. The probability would be much lower because pieces of maternal chromosomes would be incorporated into the paternal chromosomes. Therefore, a gamete would be unlikely to carry a chromosome that was completely paternally derived.

C16. During interphase, the chromosomes are greatly extended. In this conformation, they might get tangled up with each other and not sort properly during meiosis and mitosis. The condensation process probably occurs so that the chromosomes easily align along the equatorial plate during metaphase without getting tangled up.

C18. During prophase II, your drawing should show four replicated chromosomes (i.e., four structures that look like Xs). Each chromosome is one homolog. During prophase of mitosis, there should be eight replicated chromosomes (i.e., eight Xs). During prophase of mitosis, there are pairs of homologs. The main difference is that prophase II has a single copy of each of the four chromosomes, whereas prophase of mitosis has four pairs of homologs. At the end of meiosis I, each daughter cell has received only one copy of a homologous pair, not both. This is due to the alignment of homologs during metaphase I and their separation during anaphase I.

C20. DNA replication does not take place during interphase II. The chromosomes at the end of telophase I have already replicated (i.e., they are found in pairs of sister chromatids). During meiosis II, the sister chromatids separate from each other, yielding individual chromosomes.

C22. A. 20 B. 10 C. 30 D. 20

C24. Male gametes are usually small and mobile. Animal and some plant male gametes often have flagella, which make them motile. The mobility of the male gamete makes it likely that it will come in contact with the female gamete. Female gametes are usually much larger and contain nutrients to help the growth of the embryo after fertilization occurs.

C26. During oogenesis in humans, the cells are arrested in prophase I of meiosis for many years until selected primary oocytes advance through the rest of meiosis I and begin meiosis II. If fertilization occurs, meiosis II is completed.

Application and Experimental Questions

E2. During interphase, the chromosomes are longer, thinner, and much harder to see. In metaphase, they are highly condensed, which makes them thicker and shorter.

Questions for Student Discussion/Collaboration

2. A major advantage of sexual reproduction is that it fosters genetic diversity in future populations. A major disadvantage is that to reproduce, each individual must find a mate of the opposite sex.

CHAPTER 3

Note: The answers to the Comprehension Questions are at the end of the chapter.

Concept Check Questions (in figure legends)

FIGURE 3.2 The male gamete is found within pollen.

FIGURE 3.3 The white flower is providing the sperm and the purple flower is providing the eggs.

FIGURE 3.4 A true-breeding strain maintains the same trait over the course of many generations.

FIGURE 3.6 Segregation means that the T and t alleles separate from each other so that a haploid cell receives one of them, but not both.

FIGURE 3.7 In this hypothesis, two different genes are linked. The alleles of the same gene are not linked.

FIGURE 3.9 Independent assortment allows for new combinations of alleles among different genes to be found in future generations of offspring.

FIGURE 3.10 Such a parent could make two types of gametes, Ty and ty, in equal proportions.

FIGURE 3.12 Homologous chromosomes separate at anaphase of meiosis I.

FIGURE 3.13 These chromosomes could line up in four different ways.

FIGURE 3.14 Horizontal lines connect two individuals that have offspring together, and they connect all of the offspring that are produced by the same two parents.

End-of-Chapter Questions

Conceptual Questions

C2. For plants, cross-fertilization occurs when the pollen and eggs come from different plants; in self-fertilization they come from the same plant.

C4. A true-breeding organism is a homozygote that has two copies of the same allele.

C6. Diploid organisms contain two copies of each type of gene. When they make gametes, only one copy of each gene is found in a gamete. Two alleles cannot stay together within the same gamete.

C8. Genotypes: 1 TT : 1 tt
Phenotypes: 1 tall : 1 dwarf

C10. In this cross, c is the recessive allele for constricted pods; Y is the dominant allele for yellow color. The cross is $ccYy \times CcYy$. Follow the directions for setting up a Punnett square, as described in Section 3.3. The genotypic ratio is 2 $CcYY$: 4 $CcYy$: 2 $Ccyy$: 2 $ccYY$: 4 $ccYy$: 2 $ccyy$. This 2:4:2:2:4:2 ratio can be reduced to a 1:2:1:1:2:1 ratio.

The phenotypic ratio is 6 smooth pods, yellow seeds : 2 smooth pods, green seeds: 6 constricted pods, yellow seeds : 2 constricted pods, green seeds. This 6:2:6:2 ratio could be reduced to a 3:1:3:1 ratio.

C12. Offspring with a nonparental phenotype are consistent with the idea of independent assortment. If two different traits were always transmitted together as a unit, it would not be possible to get nonparental phenotypic combinations. For example, if a true-breeding parent had two dominant traits and was crossed to a true-breeding parent having the two recessive traits, the F_2 generation could not have offspring with one recessive and one dominant trait. However, because independent assortment can occur, it is possible for F_2 offspring to have one dominant and one recessive trait.

C14. A. Barring a new mutation during gamete formation, the probability is 100%. They must be heterozygotes in order to produce a child with a recessive disorder.

B. Construct a Punnett square. There is a 50% chance of heterozygous offspring.

C. Use the product rule. The chance of being phenotypically unaffected is 0.75 (i.e., 75%), so the answer is $0.75 \times 0.75 \times 0.75 = 0.422$, which is 42.2%.

D. Use the binomial expansion equation, where $n = 3$, $x = 2$, $p = 0.75$, $q = 0.25$. The answer is 0.422, or 42.2%.

C16. First construct a Punnett square. The chances are 75% of producing a solid pup and 25% of producing a spotted pup.

A. Use the binomial expansion equation, where $n = 5$, $x = 4$, $p = 0.75$, $q = 0.25$. The answer is 0.396, or 39.6%.

B. You can use the binomial expansion equation for each litter. For the first litter, $n = 6$, $x = 4$, $p = 0.75$, $q = 0.25$; for the second litter, $n = 5$, $x = 5$, $p = 0.75$, $q = 0.25$. Because the litters are in a specified order, we use the product rule and multiply the probability of the first litter times the probability of the second litter. The answer is 0.070, or 7.0%.

C. To calculate the probability of the first litter, we use the product rule and multiply the probability of the first pup (0.75) times the probability of the remaining four. We use the binomial expansion equation to calculate the probability of the remaining four, where $n = 4$, $x = 3$, $p = 0.75$, $q = 0.25$. The probability of the first litter is 0.316. To calculate the probability of the second litter, we use the product rule and multiply the probability of the first pup (0.25) times the probability of the second pup (0.25) times the probability of the remaining five. To calculate the probability of the remaining five, we use the binomial expansion equation, where $n = 5$, $x = 4$, $p = 0.75$, $q = 0.25$. The probability of the second litter is 0.025. To get the probability of these two litters occurring in this order, we use the product rule and multiply the probability of the first litter (0.316) times the probability of the second litter (0.025). The answer is 0.008, or 0.8%.

D. Because this is a specified order, we use the product rule and multiply the probability of the firstborn (0.75) times the probability of the second born (0.25) times the probability of the remaining four. We use the binomial expansion equation to calculate the probability of the remaining four pups, where $n = 4$, $x = 2$, $p = 0.75$, $q = 0.25$. The answer is 0.040, or 4.0%.

C18. A. Use the product rule: $(1/4)(1/4) = 1/16$

B. Use the binomial expansion equation:
$n = 4, p = 1/4, q = 3/4, x = 2$
$P = 0.21$, or 21%

C. Use the product rule:
$(1/4)(3/4)(3/4) = 0.14$, or 14%

C20. A. 1/4   B. 1, or 100%

C. $(3/4)(3/4)(3/4) = 27/64 = 0.42$, or 42%

D. Use the binomial expansion equation, where
$n = 7, p = 3/4, q = 1/4, x = 3$
$P = 0.058$, or 5.8%

E. The probability that the first plant is tall is 3/4. To calculate the probability that among the next four, any two will be tall, we use the binomial expansion equation, where $n = 4$, $p = 3/4$, $q = 1/4$, $x = 2$.

The probability P equals 0.21.

To calculate the overall probability of these two outcomes:

$(3/4)(0.21) = 0.16$, or 16%

C22. It violates the law of segregation because two copies of one gene are in the gamete. The two alleles for the A gene did not segregate from each other.

C24. Based on this pedigree, it is likely to be dominant inheritance because an affected child always has an affected parent. In fact, Marfan syndrome is a dominant disorder.

C26. It is impossible for the F_1 individuals to be true-breeding because they are all heterozygotes.

C28. 2 *TY*, *tY*, 2 *Ty*, *ty*, *TTY*, *TTy*, 2 *TtY*, 2 *Tty*

It may be tricky to think about, but you get 2 *TY* and 2 *Ty* because either of the two *T* alleles could combine with *Y* or *y*. Also, you get 2 *TtY* and 2 *Tty* because either of the two *T* alleles could combine with *t* and then combine with *Y* or *y*.

C30. The genotype of the F_1 plants is *Tt Yy Rr*. According to the laws of segregation and independent assortment, the alleles of each gene will segregate from each other, and the alleles of different genes will randomly assort into gametes. A *Tt Yy Rr* individual could make eight types of gametes: *TYR, TyR, Tyr, TYr, tYR, tyR, tYr,* and *tyr,* in equal proportions (i.e., 1/8 of the gametes will be of each type). To determine genotypes and phenotypes, you could make a large Punnett square that contains 64 boxes. You would need to line up the eight possible gametes across the top and along the side, and then fill in the 64 boxes. Alternatively, you could use the multiplication method or the forked-line method. The genotypes and phenotypes are as follows:

1 *TT YY RR*
2 *TT Yy RR*
2 *TT YY Rr*
2 *Tt YY RR*
4 *TT Yy Rr*
4 *Tt Yy RR*
4 *Tt YY Rr*
8 *Tt Yy Rr*   = 27 tall, yellow, round
1 *TT yy RR*
2 *Tt yy RR*
2 *TT yy Rr*
4 *Tt yy Rr*   = 9 tall, green, round
1 *TT YY rr*
2 *TT Yy rr*
2 *Tt YY rr*
4 *Tt Yy rr*   = 9 tall, yellow, wrinkled
1 *tt YY RR*
2 *tt Yy RR*
2 *tt YY Rr*
4 *tt Yy Rr*   = 9 dwarf, yellow, round
1 *TT yy rr*
2 *Tt yy rr*   = 3 tall, green, wrinkled
1 *tt yy RR*
2 *tt yy Rr*   = 3 dwarf, green, round
1 *tt YY rr*
2 *tt Yy rr*   = 3 dwarf, yellow, wrinkled
1 *tt yy rr*   = 1 dwarf, green, wrinkled

C32. The woolly haired male is a heterozygote, because he has the trait and his mother did not. (He must have inherited the normal allele from his mother.) Therefore, he has a 50% chance of passing the woolly allele to his offspring; his offspring have a 50% of passing the allele to their offspring; and these grandchildren have a 50% chance of passing the allele to their offspring (the woolly haired man's great-grandchildren). Because this is an ordered sequence of independent outcomes, we use the product rule: $0.5 \times 0.5 \times 0.5 = 0.125$, or 12.5%. Because no other Scandinavians are on the island, the chance is 87.5% for the offspring being normal (because they could not inherit the woolly hair allele from anyone else). We use the binomial expansion equation to determine the likelihood that one out of eight great-grandchildren will have woolly hair, where $n = 8$, $x = 1$, $p = 0.125$, $q = 0.875$. The answer is 0.393, or 39.3%, of the time.

C34. Use the product rule. If the woman is heterozygous, there is a 50% chance of having an affected offspring: $(0.5)^7 = 0.0078$, or 0.78%, of the time. This is a pretty small probability. If the woman has an eighth child who is unaffected, however, she has to be a heterozygote, because it is a dominant trait. She would have to pass a normal allele to an unaffected offspring. The answer is 100%.

Application and Experimental Questions

E2. The experimental difference depends on where the pollen comes from. In self-fertilization, the pollen and eggs come from the same plant. In cross-fertilization, they come from different plants.

E4. According to Mendel's law of segregation, the genotypic ratio should be 1 homozygote dominant : 2 heterozygotes : 1 homozygote recessive. The data table considers only the plants with a dominant phenotype. The genotypic ratio should be 1 homozygote dominant : 2 heterozygotes. The homozygote dominants would be true-breeding while the heterozygotes would not be true-breeding. This 1:2 ratio is very close to what Mendel observed.

E6. All three offspring had black coats. The ovaries from the albino female could only produce eggs with the dominant black allele (because they were obtained from a true-breeding black female). The actual phenotype of the albino mother does not matter. Therefore, all offspring were heterozygotes (Bb) with black coats.

E8. If we construct a Punnett square according to Mendel's laws, we expect a 9:3:3:1 ratio. Because a total of 556 offspring were observed, the expected numbers of offspring are

$556 \times 9/16 = 313$ round, yellow

$556 \times 3/16 = 104$ wrinkled, yellow

$556 \times 3/16 = 104$ round, green

$556 \times 1/16 = 35$ wrinkled, green

If we put the observed and expected values into the chi square equation, we get a value of 0.51. With four categories, our degrees of freedom are $n - 1$, or 3. If we look up the value of 0.51 in the chi square table (see Table 3.2), we see that it falls between the P values of 0.80 and 0.95. This means that the probability is 80% to 95% that a value equal to or greater than 0.51 is expected to occur due to random sampling error. Therefore, we accept the hypothesis. In other words, the results are consistent with the law of independent assortment.

E10. A. If we let c^+ represent straight wings and c represent curved wings, and e^+ represent gray body and e represent ebony body:

Parental cross: $cce^+e^+ \times c^+c^+ee$

F_1 generation is heterozygous: c^+ce^+e

An F_1 offspring crossed to a fly with curved wings and ebony body is

$c^+ce^+e \times ccee$

The F_2 offspring will have this genotypic ratio:

$c^+ce^+e : c^+cee : cce^+e : ccee$

B. The phenotypic ratio of the F_2 flies is 1:1:1:1, as follows:

straight wings, gray body: straight wings, ebony bodies : curved wings, gray bodies : curved wings, ebony bodies

C. From part B, we expect 1/4 of each category. There are a total of 444 offspring. The expected number of each category is $1/4 \times 444$, which equals 111.

$\chi^2 = \frac{(114 - 111)^2}{111} + \frac{(105 - 111)^2}{111} + \frac{(111 - 111)^2}{111} + \frac{(114 - 111)^2}{111}$

$\chi^2 = 0.49$

With 3 degrees of freedom, a value of 0.49 or greater is likely to occur between 80% and 95% of the time. Therefore, we accept our hypothesis.

E12. Follow the basic chi square strategy. We expect a 3:1 ratio, or 3/4 of the dominant phenotype and 1/4 of the recessive phenotype.

The observed and expected values are as follows (rounded to the nearest whole number):

Observed*	Expected	$\frac{(O - E)^2}{E}$
5,474	5,493	0.066
1,850	1,831	0.197
6,022	6,017	0.004
2,001	2,006	0.012
705	697	0.092
224	232	0.276
882	886	0.018
299	295	0.054
428	435	0.113
152	145	0.338
651	644	0.076
207	215	0.298
787	798	0.152
277	266	0.455

$\chi^2 = 2.15$

*Due to rounding, the observed and expected values may not add up to precisely the same number

Because $n = 14$, there are 13 degrees of freedom. If we look up this value in the chi square table, we have to look between 10 and 15 degrees of freedom. In either case, we expect a value of 2.15 or greater to occur more than 99% of the time. Therefore, we accept the hypothesis.

E14. The dwarf parent with terminal flowers must be homozygous for both genes, because it is expressing these two recessive traits: $ttaa$, where t is the recessive dwarf allele and a is the recessive allele for terminal flowers. The phenotype of the other parent is dominant for both traits. However, because this parent was able to produce dwarf offspring with axial flowers, it must have been heterozygous for both genes: $TtAa$.

E16. You need to make crosses to understand the pattern of inheritance of traits (determined by genes) from parents to offspring. And you need to microscopically examine cells to understand the pattern of transmission of chromosomes. The correlation between the pattern of transmission of chromosomes during meiosis and Mendel's laws of segregation and independent assortment is what led to the chromosome theory of inheritance.

Questions for Student Discussion/Collaboration

2. If you construct a Punnett square, the following probabilities will be obtained:

tall with axial flowers: 3/8

dwarf with terminal flowers: 1/8

The probability of being tall with axial flowers or dwarf with terminal flowers is then calculated as follows:

$3/8 + 1/8 = 4/8 = 1/2$

You use the product rule to calculate the probability of the ordered outcome of the first three offspring being tall/axial or dwarf/terminal and the fourth offspring being tall/axial:

$(1/2)(1/2)(1/2)(3/8) = 3/64 = 0.047 = 4.7\%$

CHAPTER 4

Note: The answers to the Comprehension Questions are at the end of the chapter.

Concept Check Questions (in figure legends)

FIGURE 4.1 In the X-Y system, the presence of the Y chromosome causes maleness, whereas in the X-0 system, the ratio between the number of X chromosomes and number of sets of autosomes determines sex. A ratio of 0.5 is male and 1.0 is female.

FIGURE 4.2 A male bee is not produced by sexual reproduction because it is produced from an unfertilized egg. Sexual reproduction involves the union of gametes.

FIGURE 4.3 A higher average temperature would favor a high percentage of female alligators. This might limit population size if there are insufficient numbers of males for reproduction. On the other hand, a high number of females might favor a larger population size if males are not limiting.

FIGURE 4.4 The sporophytes are the opposite sexes in dioecious plants.

FIGURE 4.5 The Barr body is more brightly stained because it is very compact.

FIGURE 4.6 X-chromosome inactivation initially occurs during embryonic development.

FIGURE 4.7 Only the maintenance phase occurs in an adult female.

FIGURE 4.9 In the F_2 generation, only the males had white eyes.

FIGURE 4.10 The key pedigree feature that points to X-linked inheritance is that only males are affected with the disorder. Also, carrier females often have affected brothers.

FIGURE 4.11 The reason why the reciprocal cross yields a different result is because females carry two copies of an X-linked gene whereas males have only one.

End-of-Chapter Questions

Conceptual Questions

C2. A. Dark males and light females; reciprocal: all dark offspring
     B. All dark offspring; reciprocal: dark females and light males
     C. All dark offspring; reciprocal: dark females and light males
     D. All dark offspring; reciprocal: dark females and light males

C4. A. Female; there are no Y chromosomes.
     B. Female; there are no Y chromosomes.
     C. Male; the Y chromosome determines maleness.
     D. Male; the Y chromosome determines maleness.

C6. A Barr body is a mammalian X chromosome that is highly condensed. It is found in somatic cells with two or more X chromosomes. Most genes on the Barr body are inactive.

C8. X-chromosome inactivation in heterozygous females produces a mosaic pattern of gene expression. During early embryonic development, some cells have the maternal X chromosome inactivated and other cells have the paternal X chromosome inactivated; these embryonic cells will divide and produce billions of cells. In the case of a female that is heterozygous for a gene that affects pigmentation of the fur, this produces a variegated pattern of coat color. Because it is a random process in any given animal, two female cats will vary as to where the orange and black patches occur. A variegated coat pattern could not occur in female marsupials due to X-chromosome inactivation because the paternal X chromosome is always inactivated in the somatic cells of females.

C10. The male is XXY. The person is male due to the presence of the Y chromosome. Because two X chromosomes are counted, one of the X chromosomes is inactivated to produce a Barr body.

C12. A. In females, one of the X chromosomes is inactivated. When the X chromosome that is inactivated carries the normal allele, only the defective color-blindness allele will be expressed. Therefore, on average, about half of a female's eye cells are expected to express the common (not color blind) allele. Depending on the relative amounts of cells expressing the common versus the color-blindness allele, the end result may be partial color blindness.

     B. In this female, as a matter of chance, X-chromosome inactivation in the right eye always, or nearly always, inactivated the X chromosome carrying the normal allele. In the left eye, the chromosome carrying the color-blindness allele was often inactivated.

C14. The spreading phase is when the X chromosome is inactivated (i.e., condensed) in a wave that spreads outward from the X-inactivation center (Xic). The condensation spreads from Xic to the rest of the X chromosome. The *Xist* gene is transcribed from the inactivated X chromosome. It encodes an RNA that coats the X chromosome, which subsequently attracts proteins that are responsible for the compaction.

C16. First set up the following Punnett square:

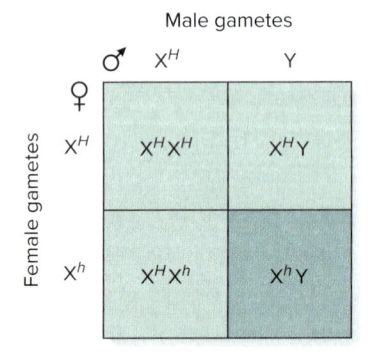

	X^H	Y
X^H	$X^H X^H$	$X^H Y$
X^h	$X^H X^h$	$X^h Y$

Male gametes across top, Female gametes down side. There is a 1/4 probability of each type of offspring.

A. 1/4   B. (3/4)(3/4)(3/4)(3/4) = 81/256   C. 3/4

D. The probability of an affected offspring is 1/4, and the probability of an unaffected offspring is 3/4. For this problem, you use the binomial expansion equation, where $x = 2$, $n = 5$, $p = 1/4$, and $q = 3/4$. The answer is 0.26, or 26%, of the time.

Application and Experimental Questions

E2. In general, you cannot distinguish between autosomal and pseudoautosomal inheritance from a pedigree analysis. Mothers and fathers have an equal probability of passing the alleles to sons and daughters. However, if an offspring had a chromosomal abnormality, you might be able to tell. For example, in a family tree involving the *Mic2* allele, an offspring that was X0 would have less of the gene product and an offspring that was XXX or XYY or XXY would have an extra amount of the gene products.

This may lead you to suspect that the gene is located on the sex chromosomes.

E4. Actually, his data are consistent with this hypothesis. To rule out a Y-linked allele, he could have crossed an F_1 female with a red-eyed male rather than an F_1 male. The same results would be obtained. Because the red-eyed male would not have a white allele, this would rule out Y-linkage.

E6. To be a white-eyed female, a fly must inherit two X chromosomes and both must carry the white-eye allele. This could occur only if both X chromosomes in the female stayed together and the male gamete contained a Y chromosome. The white-eyed females would be XXY. To produce a red-eyed male, a female gamete lacking any sex chromosomes could unite with a normal male gamete carrying the X^{w+}. This would produce an X0, red-eyed male.

E8. The rare female flies would be XXY. Both X chromosomes would carry the white-eye allele and the miniature allele. These female flies would have miniature wings because they would have inherited both X chromosomes from their mother.

Questions for Student Discussion/Collaboration

2. One possibility is fertilization of an abnormal female gamete. The white-eyed male parent could make sperm carrying the X chromosome that could fertilize a female gamete without any sex chromosomes. This would produce an X0, white-eyed male. This is the most likely explanation.

   Another possibility is that the flies in Morgan's lab were not completely true-breeding. The allele creating the white phenotype may have occurred several generations earlier in a female fly. Perhaps, among the many red-eyed sisters, one of them could have already been a heterozygote. The rare sister could produce white-eyed male offspring. New mutations are a third, but unlikely possibility, because we already know that the mutation rate is very low. It took Morgan 2 years to get one white-eyed male.

CHAPTER 5

Note: The answers to the Comprehension Questions are at the end of the chapter.

Concept Check Questions (in figure legends)

FIGURE 5.1 Both colors are considered wild type because both are prevalent in natural populations.

FIGURE 5.2 Yes. The *PP* homozygote probably makes twice the amount of protein that is needed for purple pigment formation.

FIGURE 5.3 Individual III-2 shows the effect of incomplete penetrance.

FIGURE 5.4 Genes and the environment determine an organism's traits.

FIGURE 5.5 50% of the functional protein is not enough to give a red color.

FIGURE 5.6 It is often easier to observe incomplete dominance at the molecular/cellular level.

FIGURE 5.7 In this case, the heterozygote is resistant to malaria.

FIGURE 5.8 The scenario shown in part (a) explains overdominance in the case of the sickle cell allele.

FIGURE 5.9 The *i* allele is a loss-of-function allele.

FIGURE 5.10 A heterozygous female does not have scurs.

FIGURE 5.11 Certain traits are expressed only in males or females, possibly due to differences in the levels of sex hormones or other factors that differ between the sexes.

FIGURE 5.12 The heterozygote has one normal copy of the gene, which allows for development to proceed in a way that is not too far from normal. Having two mutant copies of the gene probably adversely affects development to a degree that is incompatible with survival.

FIGURE 5.14 Epistasis means that the alleles of one gene mask the phenotypic effects of the alleles of a different gene. Complementation occurs when two strains exhibiting the same recessive trait produce offspring that show the dominant (wild-type) trait. This outcome usually means that the alleles for the recessive trait are in two different genes.

FIGURE 5.15 The two genes are redundant. Having one functional copy of either gene produces a triangular capsule. If both genes are nonfunctional, an ovate capsule is produced.

End-of-Chapter Questions

Conceptual Questions

C2. Sex-influenced traits are affected by the sex of the individual even though the gene that governs the trait is autosomally inherited. Scurs in cattle is an example. The expression of a sex-limited trait is limited to one sex. For example, colorful plumage in certain species of birds is limited to the male sex. Sex-linked inheritance involves traits whose genes are found on the sex chromosomes. Examples in humans include hemophilia and color blindness.

C4. If the functional allele is dominant, then one copy of the gene produces a sufficient amount of the protein. Having twice as much of this protein, as in the normal homozygote, does not alter the phenotype. If the allele is incompletely dominant, this means that one copy of the normal allele does not produce the same trait as the homozygote.

C6. The ratio would be 1 normal : 2 star-eyed individuals.

C8. Types O and AB provide an unambiguous genotype. Type O can only be *ii*, and type AB can only be $I^A I^B$. It is possible for a couple to produce children with all four blood types. The couple would have to be $I^A i$ and $I^B i$. If you construct a Punnett square, you will see that they can produce children with AB, A, B, and O blood types.

C10. A. 1/4   B. 0   C. (1/4)(1/4)(1/4) = 1/64

   D. Use the binomial expansion equation:

   $$P = \frac{n!}{x!(n-x)!} p^x q^{(n-x)}$$

   $n = 3, p = 1/4, q = 1/4, x = 2$

   $P = 3/64 = 0.047$, or 4.7%

C12. All of the F_1 generation will be white because they have inherited the dominant white allele from their Leghorn parent. Construct a Punnett square. Let *W* and *w* represent one gene, where *W* is dominant and causes a white phenotype. Let *A* and *a* represent the second gene, where the recessive allele causes a white phenotype in the homozygous condition. The genotype of the F_1 birds is *WwAa*. The phenotypic ratio of the

F_2 generation will be 13 white : 3 brown. The only brown birds will be 2 *wwAa* and 1 *wwAA*.

C14. We know that the parents must be heterozygotes for both genes.

The genotypic ratio of their offspring is 1 *ScSc* : 2 *Scsc* : 1 *scsc*

The phenotypic ratio depends on sex: 1 *ScSc* male with scurs : 1 *ScSc* female with scurs : 2 *Scsc* males with scurs : 2 *Scsc* females, no scurs : 1 *scsc* male, no scurs : 1 *scsc* female, no scurs.

A. 50%   B. 1/8, or 12.5%
C. (3/8)(3/8)(3/8) = 27/512 = 0.05, or 5%

C16. You would look at the pattern within families over the course of many generations. For a recessive trait, 25% of the offspring within a family are expected to be affected if both parents are unaffected carriers, and 50% of the offspring are expected to be affected if one parent is affected. You could look at many families and see if these 25% and 50% values are approximately true. Incomplete penetrance would not necessarily yield such numbers. Also, for very rare alleles, incomplete penetrance would probably have a much higher frequency of affected parents producing affected offspring. For rare recessive disorders, it is most likely that both parents are heterozygous carriers. Finally, the most informative pedigrees would be those in which two affected parents produce children. If they can produce an unaffected offspring, this indicates incomplete penetrance. If all of their offspring were affected, this would be consistent with recessive inheritance.

C18. The probability of a heterozygote passing the allele to his/her offspring is 50%. The probability of an affected offspring expressing the trait is 80%. We use the product rule to determine the likelihood of these two independent outcomes.

(0.5)(0.8) = 0.4, or 40% of the time

C20. This pattern is an example of incomplete dominance. The heterozygous horses are palominos. For example, if *C* represents chestnut and *c* represents cremello, the chestnut horses are *CC*, the cremello horses are *cc*, and the palominos are *Cc*.

Application and Experimental Questions

E2. Mexican hairless dogs are heterozygous for a dominant allele that is lethal when homozygous. In a cross between two Mexican hairless dogs, we expect 1/4 to be normal, 1/2 to be hairless, and 1/4 to die.

E4. The first offspring must be homozygous for the horned allele. The father's genotype is still ambiguous; he could be heterozygous or homozygous for the horned allele. The mother's genotype must be heterozygous because her phenotype is polled (she cannot be homozygous for the horned allele) but she produced a horned daughter (who must have inherited a horned allele from its mother).

E6. It is a sex-limited trait, where *W* (white) is dominant but expressed only in females. In the cross of two yellow butterflies, the male is *Ww* but is still yellow because the white phenotype is limited to females. The female is *ww* and yellow. The offspring would be 50% *Ww* and 50% *ww*. All males would be yellow. Half of the females would be white (*Ww*) and half would be yellow (*ww*). Overall, this would yield 50% yellow males, 25% yellow females, and 25% white females.

E8. The yellow squash has to be *wwgg*. It has to be *ww* because it is colored, and it has to be *gg* because it is yellow. Because the cross produced 50% white and 50% green offspring, the other parent (i.e., the white squash) must be *WwGG*. This cross would produce 50% offspring that are *WwGg* (white) and 50% that are *wwGg* (green).

E10. For this cross, you expect a 9:7 ratio of red to white flowers. In other words, 9/16 will be red and 7/16 will be white. Because there are a total of 345 plants, the expected values are

$9/16 \times 345 = 194$ red

$7/16 \times 345 = 151$ white

$$\chi^2 = \sum \frac{(O - E)^2}{E}$$

$$\chi^2 = \frac{(201 - 194)^2}{194} + \frac{(144 - 151)^2}{151}$$

$$\chi^2 = 0.58$$

With 1 degree of freedom, this chi square value is too small to reject your hypothesis. Therefore, you accept that it may be correct.

Questions for Student Discussion/Collaboration

2. Let's refer to the alleles as *B* dominant, *b* recessive and *G* dominant, *g* recessive.

The parental cross is *BBGG* × *bbgg*.

All of the F_1 offspring are *BbGg*.

Make a Punnett square for *BbGg* × *BbGg*.

According to your Punnett square, the genotypes that are homozygous for the *b* allele and have at least one copy of the dominant *G* allele are gray. To explain this phenotype, you could hypothesize that the *B* allele encodes an enzyme that can make lots of pigment, whether or not the *G* allele is present. Therefore, you get a black phenotype when one *B* allele is inherited. The *G* allele encodes a somewhat redundant enzyme, but maybe it does not function quite as well or its pigment product may not be as dark. Therefore, in the absence of a *B* allele, the *G* allele will give a gray pigment.

CHAPTER 6

Note: The answers to the Comprehension Questions are at the end of the chapter.

Concept Check Questions (in figure legends)

FIGURE 6.1 A nucleoid is not surrounded by a membrane, whereas the cell nucleus is.

FIGURE 6.2 A reciprocal cross is a cross in which the sexes and phenotypes of the parents are reversed compared to a first cross.

FIGURE 6.3 No. Once a patch of tissue is white, it has lost all of the normal chloroplasts, so it could not produce a patch of green tissue unless there was a new mutation.

FIGURE 6.4 One mitochondrial nucleoid usually contains multiple copies of the mitochondrial chromosome.

FIGURE 6.5 Mitochondria need rRNAs and tRNAs to synthesize polypeptides within the mitochondrial matrix.

FIGURE 6.6 Chloroplast and mitochondrial genomes have lost most of their genes during evolution. Many of these have been transferred to the cell nucleus.

FIGURE 6.7 All of the offspring would be normal size because they would inherit a functional allele from their father.

FIGURE 6.8 Erasure allows eggs to transmit unmethylated copies of the gene to the offspring.

Figure 6.9. Maintenance methylation is automatic methylation that occurs when a methylated gene replicates and is transferred to daughter cells. It occurs in somatic cells. De novo methylation is the methylation of a gene that is not already methylated. It usually occurs in germ-line or embryonic cells.

FIGURE 6.10 The offspring are all dextral because all of the F_1 mothers are *Dd*, and the genotype of the mother determines the phenotype of the offspring.

FIGURE 6.11 The egg cell will receive both *D* and *d* gene products.

End-of-Chapter Questions

Conceptual Questions

C2. Extranuclear inheritance does not always occur via the female gamete. Sometimes it occurs via the male gamete. Even in species in which maternal inheritance is prevalent, paternal leakage may occur; that is, the paternal parent occasionally provides mitochondria via the sperm. Maternal inheritance is the most common form of extranuclear inheritance because the female gamete is relatively large and more likely to contain cell organelles.

C4. The mitochondrial and chloroplast genomes are composed of a circular chromosome found in one or more copies in a region of the organelle known as the nucleoid. The number of genes per chromosome varies from species to species. Chloroplast genomes tend to be larger than mitochondrial genomes. See Tables 6.1 and 6.3 for examples of the variation among mitochondrial and chloroplast genomes.

C6. A. Yes. B. Yes.
 C. No, it is determined by a gene in the chloroplast chromosome.
 D. No, it is determined by a gene in the mitochondrial chromosome.

C8. Biparental extranuclear inheritance might resemble Mendelian inheritance in that offspring could inherit alleles of a given gene from both parents. The patterns differ, however, when considered from the perspective of heterozygotes. For a Mendelian trait, the law of segregation tells us that a heterozygote passes one allele for a given gene to an offspring, but not both. In contrast, if a parent has a mixed population of mitochondria (e.g., some carrying a mutant gene and some carrying a functional gene), that parent could pass both types of genes to a single offspring, because more than one mitochondrion could be contained within a sperm or egg cell.

C10. Genomic imprinting is a modification that occurs to a nuclear gene and alters gene expression, but is not permanent. Examples of imprinting include the *Igf2* gene and the pattern of X-chromosome inactivation in certain species.

C12. The process of de novo methylation will not occur in adult cells but may occur in germ-line cells.

C14. A. Genotypes: All *Nn*
   Phenotypes: All nonfunctional
 B. Genotypes: All *Nn*
   Phenotypes: All functional
 C. Genotypes: 1 *NN* : 2 *Nn* : 1 *nn*
   Phenotypes: All functional (because the mother is heterozygous and *N* is dominant)

C16. A. For an imprinted gene, you need to know whether the mutant *Igf2* allele is inherited from the mother or the father.
 B. For a maternal effect gene, you need to know the genotype of the mother.
 C. For maternal inheritance, you need to know whether the mother has the trait, because the offspring will inherit her mitochondria.

C18. The mother is *hh*. We know this because it is a maternal effect gene and all of its offspring have small heads. The offspring are all *hh* because their mother is *hh* and their father is *hh*.

Application and Experimental Questions

E2. Let's first consider the genotypes of male A and male B. Male A must have two functional copies of the *Igf2* gene. We know this because male A's mother was *Igf2 Igf2*; the father of male A must have been a heterozygote *Igf2 Igf2⁻* because half of the litter that contained male A also contained dwarf offspring. But because male A was not dwarf, it must have inherited the functional allele from its father. Therefore, male A must be *Igf2 Igf2*. We cannot be completely sure of the genotype of male B. It must have inherited the functional *Igf2* allele from its father because male B is normal size. We do not know the genotype of male B's mother, but she could be either *Igf2⁻ Igf2⁻* or *Igf2 Igf2⁻*. In either case, the mother of male B could pass the *Igf2⁻* allele to an offspring, but we do not know for sure if she did. So, male B could be either *Igf2 Igf2⁻* or *Igf2 Igf2*.

For the *Igf2* gene, we know that the maternal allele is inactivated. Therefore, the genotypes and phenotypes of females A and B are irrelevant. The phenotype of the offspring is determined only by the allele that is inherited from the father. Because we know that male A has to be *Igf2 Igf2*, we know that it can produce only normal size offspring. Because females A and B both produced dwarf offspring, male A cannot be the father. In contrast, male B could be either *Igf2 Igf2* or *Igf2 Igf2⁻*. Because both females gave birth to dwarf babies (and because male A and male B were the only two male mice in the cage), we conclude that male B must be *Igf2 Igf2⁻* and is the father of both litters.

E4. A haploid egg should express either the shorter or the longer mRNA, but not both (because it has only one copy of the gene). The nurse cells, however, can express both types of mRNAs if the female is heterozygous. Therefore, if we begin with heterozygous females, we could separate the nurse cells from the eggs. We could then isolate mRNA from the nurse cells and (in a separate tube) isolate mRNA from eggs. As described in Chapter 20, the mRNA would then be run on a gel and subjected to Northern blotting, using a probe that is complementary to both types of mRNAs. According to our knowledge of maternal effect genes, we would expect the egg to contain both types of mRNAs, because it receives them from the nurse cells. Both types of mRNAs would also be found in the nurse cells.

E6. Mate the female to a *dd* male. If all of the offspring coil to the left, you know the female must be *dd*. If they all coil to the right, she could be either *DD* or *Dd*. If the F_1 offspring coil to the right, you can let them mate with each other to produce an F_2 generation. If the original mother was *Dd*, then half of the F_1 female offspring will be *Dd* and half will be *dd*. Therefore, half of the F_2 snails will coil to the right and half to the left. In contrast, if the original mother was *DD*, all of the F₁ female offspring will be *Dd*. In this case, all of the F₂ snails will coil to the right.

Questions for Student Discussion/Collaboration

2. Obviously, you cannot maintain a population of flies that are homozygous for a recessive lethal allele. However, heterozygous females can produce viable offspring and these can be crossed to heterozygous or homozygous males to produce homozygous females. An experimenter would routinely have to make crosses and determine that the recessive allele was present in a population of flies by identifying homozygous females that were unable to produce any viable offspring. Maintaining a population of flies that carry a lethal recessive allele can be much easier if the recessive lethal allele is closely linked to a dominant allele that is not lethal, such as one affecting eye color or some other trait. If a fly exhibits the dominant trait, it is likely that it is also carrying the recessive lethal allele.

CHAPTER 7

Note: The answers to the Comprehension Questions are at the end of the chapter.

Concept Check Questions (in figure legends)

FIGURE 7.1 The offspring found in excess are those with purple flowers, long pollen and those with red flowers, round pollen.

FIGURE 7.2 No, such a crossover would not change the arrangement of these alleles.

FIGURE 7.3 A single crossover can produce offspring with these four phenotypic combinations: gray body, red eyes, miniature wings; gray body, white eyes, miniature wings; yellow body, red eyes, long wings; and yellow body, white eyes, long wings.

FIGURE 7.4 When genes are relatively close together, a crossover is relatively unlikely to occur between them. Therefore, the nonrecombinant offspring are more common.

FIGURE 7.5 The reason is that the *w* and *m* genes are farther apart than the *y* and *w* genes.

FIGURE 7.6 Crossing over can change the combination of kernel phenotypes, and it can also change the morphologies of the chromosomes compared to the original chromosomes.

FIGURE 7.7 Genetic maps are useful because they help us to: (1) understand the complexity and genetic organization of a species; (2) understand the underlying basis of inherited traits; (3) clone genes; (4) understand evolution; (5) diagnose and treat diseases; (6) predict the likelihood of a couple having offspring with genetic diseases; (7) breed livestock and crops.

FIGURE 7.8 Crossing over occurred during oogenesis in the female parent of the recombinant offspring.

FIGURE 7.9 Multiple crossovers prevent the maximum percentage of recombinant offspring from exceeding 50%.

FIGURE 7.10 Mitotic recombination occurs in somatic cells.

End-of-Chapter Questions

Conceptual Questions

C2. An independent assortment hypothesis is proposed because it enables us to calculate the expected values based on Mendel's ratios. Using the observed and expected values, we can calculate whether or not the deviations between the observed and expected values are too large to occur as a matter of chance. If the deviations are very large, we reject the hypothesis of independent assortment.

C4. The single crossover produces *ABC, Abc, aBC,* and *abc*.

A. Between 2 and 3, between genes B and C

B. Between 1 and 4, between genes *A* and *B*

C. Between 1 and 4, between genes *B* and *C*

D. Between 2 and 3, between genes *A* and *B*

C6. The likelihood of scoring a basket will be greater if the basket is larger. Similarly, the chances of a crossover initiating in a region between two genes is proportional to the size of the region between the two genes. There are a finite number (usually a few) of crossovers that occur between homologous chromosomes during meiosis, and the likelihood that a crossover will occur in a region between two genes depends on how big that region is.

C8. The pedigree suggests linkage between the dominant allele causing nail-patella syndrome and the I^B allele of the ABO blood type gene. In every case, the individual who inherits the I^B allele also inherits the disorder.

C10. <u>*Ass-1* 43 *Sdh-1* 5 *Hdc* 9 *Hao-1* 6 *Odc-2* 8 *Ada-1*</u>

C12. The inability to detect double crossovers causes the map distance to be underestimated. In other words, more crossovers occur in the region than we realize. When there is a double crossover, there are no recombinant offspring (in a two-factor cross). Therefore, the second crossover cancels out the effects of the first crossover.

C14. Mitotic recombination is crossing over between homologous chromosomes during mitosis in somatic cells. Mitotic recombination is one explanation for the blue patch. Following mitotic recombination, the two chromosomes carrying the *b* allele could segregate into the same cell and produce the blue color. Another explanation could be chromosome loss; the chromosome carrying the *B* allele could be lost during mitosis.

Application and Experimental Questions

E2. They could have used a strain with two abnormal chromosomes. In this case, the recombinant chromosomes would either look normal or have abnormalities at both ends.

E4. The rationale behind a testcross is to determine if recombination has occurred during meiosis in the heterozygous parent. The other parent is usually homozygous recessive, so we cannot tell if crossing over has occurred in the recessive parent. It is easier to interpret the data if a testcross uses a completely homozygous recessive parent. However, in the other parent, it is not necessary for all of the dominant alleles to be on one chromosome and all of the recessive alleles on the other. The parental generation provides us with information concerning the original linkage pattern between the dominant and recessive alleles.

E6. The reason why the percentage of recombinant offspring is more accurate when the genes are close together is because fewer double crossovers occur. The inability to detect double crossovers causes the map distance to be underestimated. If two genes are very close together, very few double crossovers occur, so underestimation due to double crossovers is minimized.

E8. Morgan determined this by analyzing the data in gene pairs. This analysis revealed that there were fewer recombinants between certain gene pairs (e.g., body color and eye color) than between other gene pairs (e.g., eye color and wing length). From this comparison, he hypothesized that genes that are close together on the same chromosome will produce fewer recombinants than genes that are farther apart.

E10. We consider the genes in pairs: There should be 10% offspring due to crossing over between genes *A* and *B*, and 5% due to crossing over between *A* and *C*.

  A. This genotype is due to a crossover between *B* and *A*. The nonrecombinants are *Aa bb Cc* and *aa Bb cc*. The 10% recombinants are *Aa Bb Cc* and *aa bb cc*. If we assume an equal number of both types of recombinants, 5% are *Aa Bb Cc*.

  B. This genotype is due to a crossover between *A* and *C*. The nonrecombinants are *Aa bb Cc* and *aa Bb cc*. The 5% recombinants are *aa Bb Cc* and *Aa bb cc*. If we assume an equal number of both types of recombinants, 2.5% are *aa Bb Cc*.

  C. This genotype is also due to a crossover between *A* and *C*. The nonrecombinants are *Aa bb Cc* and *aa Bb cc*. The 5% recombinants are *aa Bb Cc* and *Aa bb cc*. If we assume an equal number of both types of recombinants, 2.5% are *Aa bb cc*.

E12. A. One basic strategy to solve this problem is to divide the data into gene pairs and determine the map distance between two genes.

  184 tall, smooth
  13 tall, peach
  184 dwarf, peach
  12 dwarf, smooth

  $$\text{Map distance} = \frac{13 + 12}{184 + 13 + 184 + 12} = 6.4 \text{ mu}$$

  153 tall, normal
  44 tall, oblate
  155 dwarf, oblate
  41 dwarf, normal

  $$\text{Map distance} = \frac{44 + 41}{153 + 44 + 155 + 41} = 21.6 \text{ mu}$$

  163 smooth, normal
  33 smooth, oblate
  31 peach, normal
  166 peach, oblate

  $$\text{Map distance} = \frac{33 + 31}{163 + 33 + 31 + 166} = 16.3 \text{ mu}$$

  Use the two shortest distances to compute the map:
  tall, dwarf 6.4 smooth, peach 16.3 normal, oblate

E14. The two nonrecombinant types are homozygotes that cannot make either enzyme and heterozygotes that can make both enzymes. The recombinants can make one enzyme but not both. Because the two genes are 12 mu apart, 12% of the offspring will be recombinants and 88% will be nonrecombinant types. Because two nonrecombinant types are produced in equal numbers, we expect 44% of the mice to be unable to make either enzyme.

E16. A. If we use *B* (bushy tail) and *b* (normal tail) for one gene and *Y* (yellow) and *y* (white) for the second gene:

Parental generation: *BBYY* × *bbyy*

F$_1$ generation: All *BbYy* (Note: if the two genes are linked, *B* would be linked to *Y* and *b* would be linked to *y*.)

Testcross: F$_1$ *BbYy* × *bbyy*

Nonrecombinant offspring from testcross: *BbYy* and *bbyy*

BbYy males—bushy tails, yellow
bbyy males—normal tails, white
BbYy females—normal tails, yellow
bbyy females—normal tails, white

Recombinant offspring from testcross: *Bbyy* and *bbYy*

Bbyy males—bushy tails, white
bbYy males—normal tails, yellow
Bbyy females—normal tails, white
bbYy females—normal tails, yellow

We cannot use the data on female offspring because we cannot tell if females are recombinant or nonrecombinant: All females have normal tails. However, we can tell if male offspring are recombinant.

If we use the data on male offspring to conduct a chi square analysis, we expect a 1:1:1:1 phenotypic ratio. Because there are 197 male offspring in total, we expect 1/4, or 49 (rounded to the nearest whole number), of each of the four possible phenotypes. Computing the chi square value:

$$\chi^2 = \frac{(28-49)^2}{49} + \frac{(72-49)^2}{49} + \frac{(68-49)^2}{49} + \frac{(29-49)^2}{49}$$

$$\chi^2 = 9.0 + 10.8 + 7.4 + 8.2$$

$$\chi^2 = 35.4$$

If we look up the value of 35.4 in the chi square table, with 1 degree of freedom, the value lies far beyond the 0.01 probability level. Therefore, it is very unlikely to get such a large deviation if the hypothesis of independent assortment is correct. Therefore, we reject our hypothesis and conclude that the genes are linked.

  B. Computing the map distance:

  $$\frac{28 + 29}{28 + 72 + 68 + 29} \times 100 = 28.9 \text{ mu}$$

E18. A.

  Parent   Parent
  b 7 A 4 C   ×   *B 7 a 4 c*
  b 7 A 4 C   *B 7 a 4 c*
              ↓
          Offspring
          b 7 A 4 C
          B 7 a 4 c

  B. A heterozygous F$_2$ offspring would have to inherit a chromosome carrying all of the dominant alleles. In the F$_1$ parent (of the F$_2$ offspring), a crossover in the region between genes *b* and *A* (and between *B* and *a*) would yield a chromosome that was *BAC* and *bac*. If an F$_2$ offspring inherited the *BAC* chromosome from its F$_1$ parent and the *bac* chromosome from the homozygous parent, it would be heterozygous for all three genes.

  C. From part B, a crossover between genes *b* and *A* (and between *B* and *a*) would yield *BAC* and *bac* chromosomes.

If an offspring inherited the *bac* chromosome from its F₁ parent and the *bac* chromosome from its homozygous parent, it would be homozygous for all three genes. The chances of a crossover in this region are 7%. However, half of the crossovers yield chromosomes that are *BAC* and the other half yield chromosomes that are *bac*. Therefore, the chances are 3.5% of getting homozygous F₂ offspring.

Questions for Student Discussion/Collaboration

2. The X and Y chromosomes are not completely distinct linkage groups. One might describe them as overlapping linkage groups having some but not most of their genes in common.

CHAPTER 8

Note: The answers to the Comprehension Questions are at the end of the chapter.

Concept Check Questions (in figure legends)

FIGURE 8.1 The staining of chromosomes results in banding patterns that make it easier to distinguish chromosomes that are similar in size and have similar centromeric locations.

FIGURE 8.2 Deletions and duplications alter the total amount of genetic material.

FIGURE 8.3 A chromosomal fragment that does not contain a centromere will not segregate properly. If it remains outside the nucleus, it will be degraded.

FIGURE 8.5 Nonallelic homologous recombination occurs in this example because of the pairing of homologous sites that duplicated within the chromosomes. This pairing causes the chromosomes to be misaligned.

FIGURE 8.10 The chromosomes form an inversion loop so that the homologous regions can pair with each other. For the inverted and noninverted regions to pair, a loop must form.

FIGURE 8.11 The mechanism shown in part (b) may occur if transposable elements are found in different chromosomes. These elements may promote the pairing between nonhomologous chromosomes, and a subsequent crossover could occur.

FIGURE 8.13 These chromosomes form a translocation cross because homologous regions are pairing with each other.

FIGURE 8.14 aneuploid, monosomic, monosomy 3.

FIGURE 8.15 The genes on chromosome 2 are present in single copies, whereas the genes on the other chromosomes are present in two copies. The expression of genes on chromosome 2 would be less (perhaps 50%) relative to a normal individual. This creates an imbalance between genes on chromosome 2 and those on the other chromosomes.

FIGURE 8.18 About 512.

FIGURE 8.19 Polyploid plants are often larger and more robust. They may have larger flowers and produce more fruit.

FIGURE 8.20 During meiosis in a triploid individual, the homologs cannot pair properly. This results in highly aneuploid gametes, which are usually nonviable. Also, if aneuploid gametes participate in fertilization, the offspring are usually nonviable.

FIGURE 8.21 Nondisjunction means that pairs of chromosomes are not separating from each other properly during meiosis or mitosis.

FIGURE 8.23 In autopolyploidy, multiple sets of chromosomes come from the same species. In allopolyploidy, multiple sets of chromosomes come from at least two different species.

End-of-Chapter Questions

Conceptual Questions

C2. Small deletions and duplications are less likely to affect phenotype simply because they usually involve fewer genes. If a small deletion did have a phenotypic effect, you could conclude that a gene or genes in the region of the deletion are required to have a normal phenotype.

C4. A gene family is a group of genes that are derived from the process of gene duplications. They have similar sequences, but the sequences have some differences due to the accumulation of mutations over many generations. The members of a gene family usually encode proteins with similar but specialized functions. The specialization may occur in different cells or at different stages of development.

C6. It has a pericentric inversion.

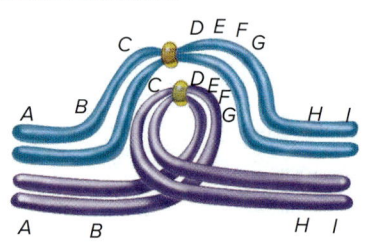

C8. There are four possible products: One is a normal chromosome and one contains the inversion shown in the drawing in conceptual question C8. The other two chromosomes will be dicentric or acentric with the following order of genes:

centromere   centromere

$\underline{A \downarrow B C D E F G H I J D C B \downarrow A}$ Dicentric
$\underline{M L K J I H G F E K L M}$ Acentric

C10. In the absence of crossing over, alternate segregation will yield half of the cells with two normal chromosomes and half with a balanced translocation. With adjacent-1 segregation, all cells will be unbalanced. Two cells will be

$\underline{A B C D E} + \underline{A I J K L M}$

And the other two cells will be

$\underline{H B C D E} + \underline{H I J K L M}$

C12. One of the parents may carry a balanced translocation between chromosomes 5 and 7. The phenotypically abnormal offspring has inherited an unbalanced translocation due to the segregation of translocated chromosomes during meiosis.

C14. A deletion and an unbalanced translocation are more likely to have phenotypic effects because they create genetic imbalances. With a deletion, there are too few copies of several genes, and for an unbalanced translocation, there are too many.

C16. The chromosome is due to a crossover within the inverted region. Your drawing should show the inversion loop. The crossover occurred between *P* and *U*.

C18. This person has a total of 46 chromosomes. However, the person would be considered aneuploid rather than euploid, because one of the sets is missing a sex chromosome and one set has an extra copy of chromosome 21.

C20. Imbalances due to aneuploidy, deletions, and duplications are related to the copy number of genes. For many genes, the level of gene expression is directly related to the number of genes per cell. If there are too many copies, as in trisomy, or too few, as in monosomy, the level of gene expression will be too high or too low, respectively. It is difficult to say why deletions and monosomies are more detrimental, although one could speculate that having too little of a gene product causes more cellular problems than having too much of a gene product. In addition, monosomies may unmask rare recessive alleles that are detrimental.

C22. One explanation is that one lizard is diploid and the other is a closely related tetraploid species. Their offspring would be triploid, which would explain the sterility. Another possibility is that one of the lizards may carry a large inversion.

C24. Mosaicism is a condition in which an organism contains a subset of cells that are genetically different from those of the rest of the organism. It may result from mitotic nondisjunction in which sister chromatids separate improperly, so one daughter cell has three copies of a chromosome while the other has only one, or in which the sister chromatids separate, but one chromosome does not attach to a spindle and is therefore lost.

C26. Polyploid plants are often more robust than their diploid counterparts. Such plants may produce a greater yield of fruits and vegetables. In the field, they tend to be more resistant to harsh environmental conditions. When polyploid plants have an odd number of sets, they are typically seedless. This can be a desirable trait for certain fruit-producing crops such as bananas and watermelons.

C28. Polyploid, triploid, and euploid should not be used.

C30. The odds of producing a euploid gamete are $(1/2)^{n-1}$, which equals $(1/2)^5$, or 1 in 32. The chance of producing an aneuploid gamete is therefore 31 out of 32 gametes. We use the product rule to determine the chances of getting a euploid offspring, because a euploid individual is produced from two euploid gametes: $1/32 \times 1/32 = 1/1024$. In other words, if this plant self-fertilized, only 1 in 1024 offspring would be euploid. The euploid offspring could be diploid, triploid, or tetraploid.

C32. The nondisjunction usually occurs during meiosis I when the homologs synapse to form bivalents. During meiosis II and mitosis, the homologs do not synapse. Instead, the chromosomes align randomly along the metaphase plate and then the centromeres separate.

C34. In meiotic nondisjunction, the bivalents are not separating correctly during meiosis I. During mitotic nondisjunction, the sister chromatids are not separating properly.

C36. Complete nondisjunction occurs during meiosis I, and one nucleus receives all the chromosomes and the other nucleus does not get any. The nucleus with all of the chromosomes then proceeds through a normal meiosis II to produce two haploid sperm cells.

Application and Experimental Questions

E2. Colchicine interferes with the mitotic spindle apparatus and thereby causes nondisjunction. At high concentrations, it can cause complete nondisjunction and produce polyploid cells.

E4. A. The F_1 offspring of this cross would probably be phenotypically normal, because they would carry the correct number of genes.

B. The F_1 offspring would have lowered fertility because they would be inversion heterozygotes. Because this is a large inversion, crossing over is fairly likely in the inverted region. When crossing over occurs, it will produce deletions and duplications that will probably be lethal in the resulting F_2 offspring.

E6. First, you would cross the two strains together. It is difficult to predict the phenotype of the offspring. Nevertheless, you would keep crossing offspring to each other and backcrossing them to the parental strains until you obtained a great-tasting tomato strain that was resistant to heat and the viral pathogen. You could then make this strain tetraploid by treatment with colchicine. If you crossed the tetraploid strain with your great-tasting diploid strain that was resistant to heat and the viral pathogen, you may get a triploid that had these characteristics. This triploid would probably be seedless.

E8. A polytene chromosome is formed when a chromosome replicates many times, and the chromatids lie side by side, as shown in Figure 8.18. The homologous chromosomes also lie side by side. Therefore, if one chromosome carries a deletion, there will be a loop in its homolog. The loop is the segment that is not deleted from one of the two homologs.

Questions for Student Discussion/Collaboration

2. There are lots of possibilities. You could look in agriculture and botany books to find many examples. In the insect world, there are interesting examples of euploidy affecting gender determination. Among amphibians and reptiles, there are also several examples of closely related species that have euploid variation.

4. 1. Polyploid plants are often more robust and disease resistant.
   2. Allopolyploids may have useful combinations of traits.
   3. Hybrids are often more vigorous; they can be generated from monoploids.
   4. Strains with an odd number of chromosome sets (e.g., triploids) are usually seedless.

CHAPTER 9

Note: The answers to the Comprehension Questions are at the end of the chapter.

Concept Check Questions (in figure legends)

FIGURE 9.1 To grow, those colonies must have functional copies of all five genes. This could occur by the transfer of the met^+ and bio^+ genes to the met^- bio^- thr^+ leu^+ thi^+ strain or the transfer of the thr^+, leu^+, and thi^+ genes to the met^+ bio^+ thr^- leu^- thi^- strain.

FIGURE 9.2 Because bacteria are too large to pass through the filter, the U-tube apparatus can be used to determine if direct cell-to-cell contact is necessary for gene transfer to occur.

FIGURE 9.3 It would be found in an F$^+$ cell.

FIGURE 9.4 Relaxase is a part of the relaxosome, which is needed for the cutting of the F factor and its transfer to the recipient cell. The coupling factor guides the DNA strand to the exporter, which transports it to the recipient cell.

FIGURE 9.5 An F′ factor carries a portion of the bacterial chromosome, whereas an F factor does not.

FIGURE 9.6 Because conjugation occurred for a longer period of time, *pro*$^+$ was transferred in the conjugation experiment to the recipient cell at the bottom right.

FIGURE 9.8 This type of map is based on the time of gene transfer in conjugation experiments, which is determined in minutes.

FIGURE 9.9 The *lacZ* gene is closer to the origin of transfer; its transfer began at 16 minutes.

FIGURE 9.10 The normal process is for bacteriophage DNA to be incorporated into a phage coat. In transduction, a segment of bacterial chromosomal DNA is incorporated into a phage coat.

FIGURE 9.11 If the two genes were very far apart, the *arg*$^+$ gene would never be cotransduced with the *met*$^+$ gene.

FIGURE 9.12 If the recipient cell did not have a *lys*$^-$ gene, the *lys*$^+$ gene could be incorporated by nonhomologous recombination.

End-of-Chapter Questions

Conceptual Questions

C2. Conjugation is not a form of sexual reproduction, in which two distinct parents produce gametes that unite to form a new individual. However, conjugation is similar to sexual reproduction in that the genetic material from two cells is somewhat mixed. In conjugation, there is not the mixing of two genomes, one from each gamete. Instead, there is a transfer of genetic material from one cell to another. This transfer can alter the combination of genetic traits in the recipient cell.

C4. An F$^+$ strain contains a separate, circular piece of DNA that has its own origin of transfer. An Hfr strain has its origin of transfer integrated into the bacterial chromosome. An F$^+$ strain can transfer only the DNA contained on the F factor. If given enough time, an Hfr strain can actually transfer the entire bacterial chromosome to the recipient cell.

C6. Sex pili promote the binding of donor and recipient cells.

C8. Cotransduction is the transduction of two or more genes. The distance between the genes determines the frequency of cotransduction. When two genes are close together, the cotransduction frequency will be higher than for two genes that are relatively farther apart.

C10. If a site that frequently incurred a breakpoint is between two genes, the cotransduction frequency of these two genes will be much less than expected. This is because the breakage will separate the two genes from each other.

C12. The transfer of conjugative plasmids such as F factor DNA does not require recombination.

C14. A. Transformation is the most likely mechanism because conjugation does not usually occur between different species, particularly distantly related species, and different species are not usually infected by the same bacteriophages.

B. The genetic transfer could occur in a single step, but it may be more likely to have involved multiple steps.

C. The use of antibiotics selects for the survival of bacteria that have resistance genes. If a population of bacteria is exposed to an antibiotic, those carrying resistance genes will survive and their relative numbers will increase in subsequent generations.

Application and Experimental Questions

E2. Mix the two strains together and then put some of them on plates containing streptomycin and some of them on plates without streptomycin. If mated colonies are present on both types of plates, then the *thr*$^+$, *leu*$^+$, and *thi*$^+$ genes were transferred to the *met*$^+$ *bio*$^+$ *thr*$^-$ *leu*$^-$ *thi*$^-$ strain. If colonies are found only on the plates that lack streptomycin, then the *met*$^+$ and *bio*$^+$ genes were transferred to the *met*$^-$ *bio*$^-$ *thr*$^+$ *leu*$^+$ *thi*$^+$ strain. This answer assumes a one-way transfer of genes from a donor to a recipient strain.

E4. An interrupted mating experiment is a procedure in which two bacterial strains are allowed to conjugate, and then the conjugation is interrupted at various time points. The interruption occurs by agitation of the solution in which the bacteria are found. This type of study is used to map the locations of genes. It is necessary to interrupt conjugation so that you can vary the time and obtain information about the order of transfer: which gene was transferred first, second, and so on.

E6. Conjugate unknown strains A and B to the F$^-$ strain in your lab that is resistant to streptomycin and cannot use lactose. Do this in two separate tubes (i.e., strain A plus your F$^-$ strain in one tube, and strain B plus your F$^-$ strain in the other tube). Plate the conjugated cells on growth media containing lactose plus streptomycin. If colonies grow, the unknown strain had to be strain A, the F$^+$ strain that had lactose utilization genes on its F factor.

E8. A. If we extrapolate the lines back to the x-axis, the line for the *hisE* gene intersects the axis at about 3 minutes and the line for the *pheA* gene intersects it at about 24 minutes. These are the values for the times of entry. Therefore, the distance between these two genes is 21 minutes (i.e., 24 minus 3).

B. *hisE*   *pabB*   *pheA*
      4   17

E10. One possibility is that you could treat the P1 lysate with DNase I, an enzyme that digests DNA. (Note: If DNA were digested with DNase I, the function of any genes within the DNA would be destroyed.) If the DNA were within a P1 phage, it would be protected from DNase I digestion. This would allow you to distinguish between transformation (which would be inhibited by DNase I) versus transduction (which would not be inhibited by DNase I). Another possibility is that you could try to fractionate the P1 lysate. Naked DNA will be smaller than a P1 phage carrying DNA. You could try to filter the lysate to remove naked DNA, or you could subject the lysate to centrifugation and remove the lighter fractions that contain naked DNA.

E12. Cotransduction frequency = $(1 - d/L)^3$

For the normal strain:

Cotransduction frequency = $(1 - 0.7/2)^3 = 0.275$, or 27.5%

For the new strain:

Cotransduction frequency = $(1 - 0.7/5)^3 = 0.64$, or 64%

The experimental advantage is that you could map genes that are farther than 2 minutes apart. You could map genes that are up to 5 minutes apart.

E14. Cotransduction frequency = $(1 - d/L)^3$

$$0.53 = (1 - d/2 \text{ minutes})^3$$
$$(1 - d/2 \text{ minutes}) = \sqrt[3]{0.53}$$
$$(1 - d/2 \text{ minutes}) = 0.81$$
$$d = 0.38 \text{ minute}$$

E16. A. We first need to calculate the cotransformation frequency, which equals 2/70, or 0.029.

Cotransformation frequency = $(1 - d/L)^3$
$$0.029 = (1 - d/2 \text{ minutes})^3$$
$$d = 1.4 \text{ minutes}$$

B.
Cotransformation frequency = $(1 - d/L)^3$
$$= (1 - 1.4/4)^3$$
$$= 0.27$$

As you may have expected, the cotransformation frequency is much higher when the transformation involves larger pieces of DNA.

Questions for Student Discussion/Collaboration

2. Conjugation requires direct cell contact, which is mediated by proteins that are found in the same species. So, it is less likely to occur between different species unless they are very closely related evolutionarily. Similarly, gene transfer via transduction involves bacteriophages that are usually species-specific. Gene transfer via transformation is most likely to occur between different species. Gene transfer has several possible consequences, including antibiotic resistance and the ability to survive under new growth conditions.

CHAPTER 10

Note: The answers to the Comprehension Questions are at the end of the chapter.

Concept Check Questions (in figure legends)

FIGURE 10.1 One feature that varies among viruses is the genome. Viral genomes vary with regard to size, RNA versus DNA, and single-stranded versus double-stranded. The structures of viruses also vary with regard to the complexity of their capsids and whether or not they have an envelope.

FIGURE 10.3 A virus may remain latent following its integration into the genome of the host cell.

FIGURE 10.4 The lytic cycle produces new phage particles.

FIGURE 10.6 An emerging virus is one that has arisen recently and is more likely than previous strains to cause disease.

End-of-Chapter Questions

Conceptual Questions

C2. All viruses have a nucleic acid genome and a capsid composed of protein. Some eukaryotic viruses are surrounded by an envelope that consists of a membrane with embedded proteins.

C4. A viral envelope is a membrane with embedded proteins. It is made when the virus buds from the host cell, taking with it a portion of the host cell's plasma membrane.

C6. The attachment step usually involves the binding of the virus to a specific protein on the surface of the host cell. Only certain cell types will make that protein.

C8. Reverse transcriptase is used to copy the viral RNA into DNA so it can be integrated into a chromosome of the host cell.

C10. A temperate phage can follow either the lytic or lysogenic cycle, whereas a virulent phage can follow only the lytic cycle.

Application and Experimental Questions

E2. Electron microscopy must be used to visualize a virus.

E4. The two traits that Fraenkel-Conrat and Singer analyzed were the lesions on leaves that the viruses caused, and the amino acid composition of the viral coat proteins. When the nucleic acid of the reconstituted virus came from the wild-type strain, the lesions and the protein composition of newly made viruses were wild-type. Alternatively, if the RNA of the reconstituted virus came from the Holmes ribgrass strain, the lesions and protein composition were like those of the Holmes ribgrass strain.

E6. The drugs prevent the assembly of new viruses. Therefore, if new viruses cannot be made, the virus cannot spread.

Questions for Student Discussion/Collaboration

2. There are many possibilities. One difference is that bacterial diseases can be treated with antibiotics, whereas viral diseases cannot. You could try to isolate a sample (e.g., of blood) from an infected mouse and see if there are bacteria present. (Bacteria are visible via light microscopy, whereas viruses are not.) You could try to filter the infected blood through filters that allow the passage of viruses (which are smaller) but do not allow the passage of bacteria, and then determine if the filtrate is able to infect other mice. Other answers are possible.

CHAPTER 11

Note: The answers to the Comprehension Questions are at the end of the chapter.

Concept Check Questions (in figure legends)

FIGURE 11.1 In this experiment, the type R bacteria had taken up genetic material from the heat-killed type S bacteria, which converted the type R bacteria into type S. This enabled them to proliferate within the mouse and kill it.

FIGURE 11.2 RNase or protease were added to the DNA extract to rule out the possibility that small amounts of contaminating RNA or protein were responsible for converting the type R bacteria into type S.

FIGURE 11.4 Ribose and uracil are not found in DNA.

FIGURE 11.7 Deoxyribose and phosphate form the backbone of a DNA strand.

FIGURE 11.8 Modeling is useful because it shows how atoms can fit together in a three-dimensional structure of a complex molecule.

FIGURE 11.11 Hydrogen bonding between base pairs and base stacking hold the DNA strands together.

FIGURE 11.12 The major and minor grooves are the indentations where atoms in the bases are in contact with water in the cellular fluid. The major groove is wider than the minor groove.

FIGURE 11.13 B DNA is a right-handed helix and the backbone is helical, whereas Z DNA is a left-handed helix and the backbone appears to zigzag slightly. Z DNA has the bases tilted relative to the central axis, whereas they are perpendicular to that axis in B DNA. The two forms also differ in the number of base pairs per turn.

FIGURE 11.14 Covalent bonds hold nucleotides together in an RNA strand.

FIGURE 11.15 A hydrogen bonds with U and G hydrogen bonds with C.

End-of-Chapter Questions

Conceptual Questions

C2. The transformation process is described in Chapter 9.
1. A fragment of DNA binds to the cell surface.
2. It penetrates the cell membrane.
3. It enters the cytoplasm.
4. It recombines with the chromosome.
5. The genes within the DNA are expressed (i.e., transcription and translation).
6. The gene products create a capsule. That is, they are enzymes that synthesize a capsule using cellular molecules as building blocks.

C4. The building blocks of a nucleotide are a sugar (ribose or deoxyribose), a nitrogenous base, and a phosphate group. In a nucleotide, the phosphate is already linked to the 5′ position on the sugar. When two nucleotides are linked together, a phosphate on one nucleotide forms a covalent bond with the 3′ hyrdroxyl group on another nucleotide.

C6. The structure is a phosphate group connecting two sugars at the 3′ and 5′ positions, as shown in Figure 11.7.

C8. 3′–CCGTAATGTGATCCGGA–5′

C10. A drawing of a DNA helix with 10 bp per turn looks like Figure 11.11. To make 15 bp per turn, you would have to add five more base pairs, but the helix should still make only one complete turn.

C12. The bases occupy the major and minor grooves. Phosphates and sugars are found in the backbone. If a DNA-binding protein does not recognize a nucleotide sequence, it probably is not binding in the grooves, but instead is binding to the DNA backbone (i.e., sugar-phosphate sequence). DNA-binding proteins that recognize a base sequence must bind into the major or minor groove of the DNA, which is where the bases are accessible to a DNA-binding protein. Most DNA-binding proteins that recognize a base sequence fit into the major groove. By comparison, other DNA-binding proteins such as histones, which do not recognize a base sequence, bind to the DNA backbone.

C14. The structure is shown in Figure 11.5. You begin numbering at the carbon that is to the right of the ring oxygen and continue to number the carbon atoms in a clockwise direction. Antiparallel means that the backbones are running in the opposite direction. In one strand, the sugar carbons are oriented in a 3′ to 5′ direction, while in the other strand they are oriented in a 5′ to 3′ direction.

C16. The structures are similar in that RNA and DNA double helices are helical and antiparallel and base pairing is due to complementarity. The shapes of the helices are slightly different with regard to the number of base pairs per turn. Another difference is that RNA base pairing involves A with U, whereas DNA base pairing involves A with T.

C18. Its base sequence

C20. G = 32%, C = 32%, A = 18%, T = 18%

C22. Lysines and arginines and polar amino acids

C24. They always run parallel.

C26. The viral genetic material is probably double-stranded RNA because the amount of A equals that of U and the amount of G equals that of C. Therefore, this molecule could be double stranded and obey the AU/GC rule. However, it is possible that it is merely a coincidence that A happens to equal U and G happens to equal C, and the genetic material is really single stranded.

C28. There are 10^8 base pairs in the chromosome. In a double helix, a single base pair has a length of about 0.34 nm, which equals 0.34×10^{-9} m. If we multiply the two values together:

$10^8(0.34 \times 10^{-9}) = 0.34 \times 10^{-1}$ m, or 0.034 m, or 3.4 cm

The answer is 3.4 cm, which equals 1.3 inches! That is enormously long considering that a typical human cell is only 10 to 100 μm in diameter. As described in Chapter 12, the DNA has to be greatly compacted to fit into a living cell.

C30. Yes, a stem-loop could form, as long as there are sequences that are complementary and antiparallel to each other. These would be similar to the complementary double-stranded regions observed in RNA molecules (e.g., see Figures 11.15 and 11.16).

C32. Region 1 cannot form a stem-loop with region 2 and region 3 at the same time. Complementary regions of RNA form base pairs, not base triplets. The region 1/region 2 interaction would be slightly more stable than the region 1/region 3 interaction because it is one nucleotide longer, and it has a higher amount of GC base pairs. Remember that GC base pairs form three hydrogen bonds, whereas AU base pairs form two hydrogen bonds. Therefore, helices with a higher GC content are more stable.

Application and Experimental Questions

E2. A. These are possible reasons why most of the cells were not transformed.
1. Most of the cells did not take up any of the type S DNA.
2. The type S DNA was usually degraded after it entered the type R bacteria.
3. The type S DNA was usually not expressed in the type R bacteria.

B. The antibody/centrifugation steps were used to remove the bacteria that had not been transformed. It enabled the researchers to determine the phenotype of the bacteria that had been transformed. If this step was omitted, there would have been so many colonies on the plate it would have been difficult to identify any transformed bacterial colonies, because they would have represented a very small proportion of the total number of bacterial colonies.

C. The researchers were trying to determine that it was really the DNA in their DNA extract that was the genetic material. It was possible that the extract was not entirely pure and could contain contaminating RNA or protein. However, treatment with RNase and protease did not prevent transformation, indicating that RNA and protein were not the genetic material. In contrast, treatment with DNase blocked transformation, confirming that DNA is the genetic material.

E4. Here are five advantages of modeling on a computer:
1. You can create lots of different shapes.
2. You can move things around very quickly with a mouse.
3. You can use mathematical formulas to fit things together in a systematic way.
4. Computers process information very rapidly.
5. You can store the information you have obtained from model building in a computer file.

Questions for Student Discussion/Collaboration

2. There are many possibilities. You could use a DNA-specific chemical and show that it causes heritable mutations. Perhaps you could inject an oocyte with a piece of DNA and produce a mouse with a new trait.

CHAPTER 12

Note: The answers to the Comprehension Questions are at the end of the chapter.

Concept Check Questions (in figure legends)

FIGURE 12.1 The sequences that comprise genes constitute most of a bacterial genome.

FIGURE 12.2 Two

FIGURE 12.5 Strand separation is needed for certain processes such as DNA replication and RNA transcription.

FIGURE 12.6 Eukaryotic chromosomes have centromeres and telomeres, but bacterial chromosomes do not. Also, eukaryotic chromosomes typically have many more repetitive sequences.

FIGURE 12.7 One reason for variation in genome size is that the number of genes among different eukaryotes varies. A second reason is that the number of repetitive sequences varies.

FIGURE 12.9 Retrotransposition always causes the TE to increase in number. Transposition by DNA transpososons, by itself, does not increase the TE number. However, such transposition can increase the TE number if it occurs around the time of DNA replication.

FIGURE 12.13 Reverse transcriptase uses RNA as a template to make a strand of DNA.

FIGURE 12.14 11 nm at its widest point

FIGURE 12.16 The solenoid model shows the nucleosomes in a repeating, spiral arrangement, whereas the zigzag model has a more irregular and dynamic arrangement of nucleosomes.

FIGURE 12.17 The nuclear matrix helps to organize and compact the chromosomes within the cell nucleus.

FIGURE 12.18 A chromosome territory is a discrete region in the cell nucleus that is occupied by a single chromosome.

FIGURE 12.19 Active genes are found in the less compacted regions of euchromatin.

FIGURE 12.20 Converstion of the diameter from 300 nm to 700 nm results when the radial loop domains become more tightly packed.

End-of-Chapter Questions

Conceptual Questions

C2. A bacterium with two nucleoids is similar to a diploid eukaryotic cell in that it has two copies of each gene. The bacterium is different, however, with regard to alleles. A eukaryotic cell can have two different alleles for the same gene. For example, a cell from a pea plant could be heterozygous, Tt, for the gene that affects height. By comparison, a bacterium with two nucleoids has two identical chromosomes. Therefore, such a bacterium is homozygous for its chromosomal genes. Note: As discussed in Chapter 9, a bacterium can contain another piece of DNA, called an F' factor, that can carry a few genes. The alleles on an F' factor can be different from the alleles on the bacterial chromosome.

C4. DNA is a double helix. The helix is a coiled structure. Supercoiling involves additional coiling of a structure that is already a coil. Positive supercoiling is called overwinding because it adds additional twists in the same direction as the DNA double helix, that is, in a right-handed direction. Negative supercoiling is in the opposite direction. Z DNA is a left handed helix. Positive supercoiling in Z DNA is in a left-handed direction, whereas negative supercoiling is in the right-handed direction (the opposite of the directions of positive and negative supercoiling in B DNA).

C6. A. The three twists would create either three fewer or three more turns for a total of seven or 13, respectively.

B. If the helix now has seven turns, it was left-handed. The three right-handed twists you made would cause three fewer turns in a left-handed helix. If the helix now has 13 turns, it was right-handed. The three right-handed twists you made would add three more turns to a right-handed helix (compare Figures 12.4a and d).

C. The twisting would probably not make supercoils because the two strings are not tightly interacting with each other. It's easy for the two strings to change the number of coils.

D. If the strings were glued together with rubber cement, the three additional twists would probably make supercoils. A glued pair of strings is more like the DNA double helix. In a double helix, the two strands are hydrogen bonding to each other. The hydrogen bonding is like the glue. Additional twists tend to create supercoils rather than alter the number of coils.

C8. Topoisomers are different with regard to the number of supercoils they contain. They are identical with regard to the number of base pairs in the double helix.

C10. Centromeres are structures found in eukaryotic chromosomes that provide an attachment site for kinetochore proteins so the chromosomes are sorted (i.e., segregated) during mitosis and meiosis. They are most important during M phase.

C12. Retrotransposons have the greatest potential for proliferation because the element is transcribed into RNA as an intermediate. Many copies of this RNA could be transcribed and then copied into DNA by reverse transcriptase. Theoretically, many copies of the element could be inserted into the genome in a single generation.

C14. A. LTR and non-LTR retrotransposons
B. Insertion elements and simple transposons
C. All of them are flanked by direct repeats.
D. Insertion elements and simple transposons

C16. An autonomous transposable element has the genes that are necessary for transposition. For example, a cut-and-paste TE that is autonomous has the transposase gene. A nonautonomous transposable element does not have all of the genes that are necessary for transposition. However, if a cell contains an autonomous element and a nonautonomous element of the same type, the nonautonomous element can move. For example, if a *Drosophila* cell contained two P elements, one autonomous and one nonautonomous, the transposase expressed from the autonomous P element could recognize the nonautonomous P element and catalyze its transposition.

C18. A nucleosome is composed of double-stranded DNA wrapped 1.65 times around an octamer of histones. In a 30-nm fiber, histone H1 helps to compact the nucleosomes. The solenoid and three-dimensional zigzag models describe how this compaction may occur to form a 30-nm fiber. The solenoid model has a more regular, spiral arrangement of nucleosomes. By comparison, the arrangement of nucleosomes in the zigzag model appears somewhat random (zigzagging) and is thought to be more dynamic.

C20.

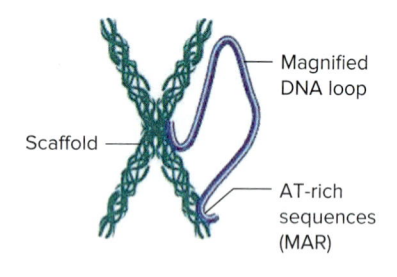

[Diagram showing a chromosome with labels: Magnified DNA loop, Scaffold, AT-rich sequences (MAR)]

C22. During interphase, the chromosomes are found within the cell nucleus. They are less tightly packed and are transcriptionally active. Segments of chromosomes are anchored to the nuclear matrix. During M phase, the chromosomes become highly condensed, and the nuclear membrane is fragmented into vesicles. The chromosomal DNA remains anchored to a scaffold formed from the nuclear matrix. The chromosomes eventually become attached to the spindle apparatus via microtubules that are attached to the kinetochore, which is attached to the centromere.

C24. There are 146 bp around the core histones. If the linker region is 54 bp long, we expect 200 bp of DNA (i.e., 146 + 54) for each nucleosome and linker region. If we divide 46,000 bp by 200 bp, we get 230. Because there are two molecules of H2A for each nucleosome, we expect there will be 460 molecules of H2A in a 46,000-bp sample of DNA.

C26. A. There are 10^8 bp in this human chromosome. In a double helix, a single nucleotide is about 0.34 nm long, which equals 0.34×10^{-3} μm. If we multiply the two values together: $10^8 (0.34 \times 10^{-3}) = 0.34 \times 10^5$ μm, or 34,000 μm

B. The 30-nm fiber is about 49 times shorter than a linear double helix. (Note: The 11-nm fiber compacts the DNA about seven times. The 30-nm fiber compacts the DNA an additional seven times. Therefore, compared to linear DNA, the 30-nm fiber is $7 \times 7 = 49$ times more compact.) If we divide 34,000 μm by 49 we get 694 μm.

C. The 30-nm fiber would not fit inside the nucleus if it were stretched out in a linear manner because the diameter of the nucleus is much smaller than 694 μm. However, the 30-nm fiber is very thin and is compacted by many radial loop domains.

Application and Experimental Questions

E2. Supercoiled DNA appears curled up into a relatively compact structure. You could add different purified topoisomerases and use microscopy to see how they affect the structure. For example, DNA gyrase relaxes positive supercoils, while topoisomerase I relaxes negative supercoils. If you added topoisomerase I to a DNA preparation and it became less compacted, then the DNA was negatively supercoiled.

E4. With a salt solution of moderate concentration, the nucleosome structure is still preserved, so the same pattern of results would be observed. DNase I would cut the linker region and produce fragments of DNA that would be in multiples of 200 bp. However, with a highly concentrated salt solution, the core histones would be lost, and DNase I could cut anywhere. On the gel, you would see fragments of almost any size. Because there would be a continuum of fragments of many different sizes, the lane on the gel would probably look like a smear rather than having a few prominent bands of DNA.

E6. There are many possibilities. You could digest the DNA with DNase I and see if that yields multiples of 200 bp or so. You could try to purify proteins from the sample and see if eukaryotic proteins or bacterial proteins are present.

E8. A. Because the *Alu*I sequence is interspersed throughout all of the chromosomes, many brightly colored spots would be seen along all chromosomes.
B. Only the centromeric region of the X chromosome would be brightly colored.

Questions for Student Discussion/Collaboration

2. This is a matter of opinion. Having so much DNA that has no obvious function seems like a waste of energy. Perhaps it has a function that we don't know about yet. On the other hand, evolution does allow bad things to accumulate within genomes, such as genes that cause diseases, etc. Perhaps this is just another example of the negative consequences of evolution.

CHAPTER 13

Note: The answers to the Comprehension Questions are at the end of the chapter.

Concept Check Questions (in figure legends)

FIGURE 13.1 The two features that allow DNA to be replicated are its double-stranded structure and the base pairing between A and T and between G and C.

FIGURE 13.2 The semiconservative model is correct.

FIGURE 13.5 The DnaA boxes are recognized by DnaA proteins, which bind to them and cause the DNA strands to separate at the AT-rich region.

FIGURE 13.6 Two replication forks are formed at the origin of replication.

FIGURE 13.7 Primase is needed for DNA replication because DNA polymerase cannot initiate DNA replication on a bare template strand.

FIGURE 13.8 The template strand is read in the 3′ to 5′ direction.

FIGURE 13.10 The leading strand is made as one long, continuous strand in the same direction that the replication fork is moving. The lagging strand is made as Okazaki fragments in the direction away from the replication fork.

FIGURE 13.13 The oxygen in the newly made ester bond comes from the sugar.

FIGURE 13.16 Because eukaryotic chromosomes are so large, they need multiple origins so that the DNA can be replicated in a reasonable length of time.

FIGURE 13.21 Six times (36 divided by 6)

FIGURE 13.22 An advantage of genetic recombination is that it may foster genetic diversity, which may produce organisms that have reproductive advantages over other organisms.

FIGURE 13.23 A heteroduplex region may be produced after branch migration because the two strands are not perfectly complementary.

FIGURE 13.24 During recombination, a D-loop is formed when a single DNA strand from one homolog invades the homologous region of the other homolog.

FIGURE 13.26 During gene conversion, the region of the gene that contained the genetic variation that created the *b* allele was digested away. The same region from the homolog, which carried the *B* allele, was then used as a template to make another copy of the *B* allele.

End-of-Chapter Questions

Conceptual Questions

C2. Bidirectional replication refers to DNA replication in both directions starting from one origin.

C4. A. TTGGHTGUTGG
        HHUUTHUGHUU

   B.   TTGGHTGUTGG
                     HHUUTHUGHUU
                          ↓
          TTGGHTGUTGG   CCAAACACCAA
          AACCCACAACC   HHUUTHUGHUU
                          ↓
   TTGGHTGUTGG TTGGGTGTTGG CCAAACACCAA CCAAACACCAA
   AACCCACAACC AACCCACAACC GGTTTGTGGTT HHUUTHUGHUU

C6. Let's assume that there are 4,600,000 bp of DNA and that DNA replication is bidirectional and occurs at a rate of 750 nucleotides per second.

   If there were just a single replication fork:

   4,600,000/750 = 6133 seconds, or 102.2 minutes

   Because replication is bidirectional: 102.2/2 = 51.1 minutes

   Actually, this is an average value based on a variety of growth conditions. Under optimal growth conditions, replication can occur substantially faster.

With regard to errors, if we assume an error rate of one mistake per 100,000,000 nucleotides:

4,600,000 × 1000 bacteria = 4,600,000,000 nucleotides of replicated DNA

4,600,000,000/100,000,000 = 46 mistakes

This is pretty amazing. In this population, DNA polymerase would cause only 46 single mistakes in a total of 1000 bacteria, each containing 4.6 million bp of DNA.

C8. DNA polymerase would slide from right to left. The sequence of the new strand would be

3′–CTAGGGCTAGGCGTATGTAAATGGTCTAGTGGTGG–5′

C10. A. In Figure 13.5, the first, second, and fourth DnaA boxes are running in the same direction, and the third and fifth are running in the opposite direction. Once you realize that, you can see that the sequences are very similar to each other.

   B. According to the direction of the first DnaA box, the consensus sequence is

   TGTGGATAA
   ACACCTATT

   C. The consensus sequence is nine nucleotides long. Because there are four kinds of nucleotides (i.e., A, T, G, and C), the chance of this sequence occurring by random chance is 4^{-9}, or once every 262,144 nucleotides. Because the *E. coli* chromosome is more than 10 times longer than this, it is fairly likely that this consensus sequence occurs elsewhere. The reason why there are not multiple origins, however, is because the origin has five copies of the consensus sequence very close together. The chance of having five copies of this consensus sequence occurring close together (as a matter of random chance) is very small.

C12. 1. It recognizes the origin of replication.
   2. It initiates the formation of a replication bubble.
   3. It recruits helicase to the region.

C14. An Okazaki fragment is a short segment of newly made DNA to be incorporated in the lagging strand. It is necessary to make these short fragments because in the lagging strand, the replication fork is exposing nucleotides in a 5′ to 3′ direction, but DNA polymerase is sliding along the template strand in a 3′ to 5′ direction away from the replication fork. Therefore, the newly made lagging strand is synthesized in short pieces that are eventually attached to each other.

C16. A processive enzyme is one that remains clamped to one of its substrates. DNA polymerase remains clamped to the template strand as it makes a new daughter strand. This processivity ensures a fast rate of DNA synthesis.

C18. The opposite strand is made in the conventional way by DNA polymerase, using the strand made via telomerase as a template.

C20. The ends labeled B and C could not be replicated by DNA polymerase. DNA polymerase makes a strand in the 5′ to 3′ direction using a template strand that is running in the 3′ to 5′ direction. Also, DNA polymerase requires a primer. At the ends labeled B and C, there is no place (upstream) for a primer to be made.

C22. As shown in Figure 13.21, the first step of replication involves the binding of telomerase to the telomere. The 3′ overhang binds to the complementary RNA in telomerase. For this reason, a 3′ overhang is necessary for telomerase to replicate the telomere.

C24. Branch migration will not create a heteroduplex during SCE because the sister chromatids are genetically identical. There should not be any mismatches between the complementary strands. Gene conversion cannot take place because the sister chromatids carry alleles that are already identical to each other.

C26. A recombinant chromosome is one that has been derived from a crossover and contains a combination of alleles that is different from those in the original chromosomes. A recombinant chromosome is a hybrid of the original chromosomes.

C28. Gene conversion occurs when a pair of different alleles is converted to a pair of identical alleles. For example, a pair of *Bb* alleles could be converted to *BB* or *bb*.

C30. Gene conversion is likely to take place near the breakpoint in a recombinant chromosome. According to the double-strand break model, a gap may be created by the digestion of one DNA strand in the double helix. Gap repair synthesis may result in gene conversion. A second way that gene conversion can occur is by mismatch repair. A heteroduplex may be created after DNA strand migration and may be repaired in such a way as to result in gene conversion.

Application and Experimental Questions

E2. A. The researcher would probably still see a band of DNA, but only see a heavy band.

B. The researcher would probably not see a band because the DNA would not be released from the bacteria. The bacteria would sediment to the bottom of the tube.

C. The researcher would not see a band. UV light is needed to see the DNA, which absorbs light in the UV region.

E4. For a DNA strand to grow, a phosphoester bond is formed between the 3′ —OH group on one nucleotide and the innermost 5′ phosphate group on the incoming nucleotide (see Figure 13.13). If the —OH group is missing, such a bond cannot form.

E6. The more likely mechanism in this case is DNA gap repair synthesis. At any given base, mismatch repair has a 50% chance of causing gene conversion. The odds of converting all six bases to those found in the *B* allele would therefore be $(1/2)^6$, or 1/64. Because this is a relatively small number, gap repair synthesis is perhaps more likely because it could remove a large region that contains all six bases and use a DNA strand from the *B* allele as a template to make a new strand.

Questions for Student Discussion/Collaboration

2. This is a matter of opinion. DNA replication needs to be fast so that life can be sustained on a reasonable time scale. That it is error-free is very important. DNA stores information, and that information needs to be maintained with high fidelity. Finally, the regulation of DNA replication during cell division is necessary to ensure the proper amount of DNA per cell.

CHAPTER 14

Note: The answers to the Comprehension Questions are at the end of the chapter.

Concept Check Questions (in figure legends)

FIGURE 14.2 The mutation would not affect the length of the mRNA, because it would not terminate transcription. However, the encoded polypeptide would be shorter.

FIGURE 14.5 For a group of related sequences within DNA, the consensus sequence consists of the most common base found at each location within those sequences.

FIGURE 14.6 Parts of sigma factor must fit into the major groove so that the protein can recognize a base sequence of a promoter.

FIGURE 14.7 The −10 sequence is an AT-rich region and thus has fewer hydrogen bonds than a region with a lot of G-C base pairs.

FIGURE 14.10 Such a mutation would prevent ρ-dependent termination of transcription.

FIGURE 14.11 NusA helps RNA polymerase to pause, which facilitates transcriptional termination.

FIGURE 14.13 The TATA box provides a precise starting point for the transcription of eukaryotic protein-encoding genes.

FIGURE 14.14 The phosphorylation of CTD allows RNA polymerase to proceed to the elongation phase of transcription.

FIGURE 14.17 An endonuclease can cleave within a strand, whereas an exonuclease digests a strand, one nucleotide at a time, starting at one end.

FIGURE 14.18 Splicing via a spliceosome is very common in eukaryotes.

FIGURE 14.20 snRNPs are involved in recognizing the intron boundaries, cutting out the intron, and connecting the two adjacent exons together.

FIGURE 14.22 Exon 4 would be spliced out. Exon 3 would be retained in the mRNA.

FIGURE 14.23 The 7-methylguanosine cap is important for the proper splicing of pre-mRNA, the exit of mRNA from the nucleus, and the binding of mRNA to a ribosome.

FIGURE 14.25 A functional consequence of RNA editing is that the amino acid sequence of an encoded polypeptide may be changed, which can affect its function.

End-of-Chapter Questions

Conceptual Questions

C2. The release of sigma factor marks the end of the initiation stage of transcription.

C4. GGCATTGTCA

C6. The most highly conserved positions are the first, second, and sixth. In general, when promoter sequences are conserved, they are more likely to be important for binding. That explains why changes do not occur at these positions; if a mutation altered a conserved position, the promoter would probably not work very well. By comparison, changes are tolerated occasionally at the fourth position and frequently at the third and fifth positions. The positions that tolerate changes are less important for binding by sigma factor.

C8. This mutation will not affect transcription. However, it will affect translation by preventing the initiation of polypeptide synthesis.

C10. Sigma factor can slide along the major groove of the DNA and recognize base sequences that are exposed in the groove. When it encounters a promoter sequence, hydrogen bonding between the bases and the sigma factor protein promotes a tight and specific interaction.

C12.

DNA	RNA
G	C
C	G
A	U
T	A

The template strand is 3′–CCGTACGTAATGCCGTAG TGTGATCCCTAG–5′ and the coding strand is 5′–GGCA TGCATTACGGCATCACACTAGGGATC–3′. The promoter is to the left of where the template and coding strands are located (e.g., to the left of the 5′ end of the coding strand).

C14. When transcriptional termination occurs, the hydrogen bonds are broken between the DNA and the part of the newly made RNA transcript that is located in the open complex.

C16. DNA helicase and ρ protein bind to a nucleic acid strand and travel in the 5′ to 3′ direction. When they encounter a double-stranded region, they break the hydrogen bonds between complementary strands. The ρ protein is different from DNA helicase in that it moves along an RNA strand, whereas DNA helicase moves along a DNA strand. The function of DNA helicase is to promote DNA replication, whereas the function of ρ protein is to promote transcriptional termination.

C18. A. Mutations that alter the uracil-rich region by introducing Gs and Cs, and mutations that prevent the formation of the stem-loop structure

B. Mutations that alter the termination sequence, and mutations that alter the ρ recognition site

C. Eventually, somewhere downstream from the gene, another transcriptional termination sequence would be found, and transcription would terminate there. This second termination sequence might be positioned randomly or it might be at the end of an adjacent gene.

C20. Eukaryotic promoters are somewhat variable with regard to the pattern of sequence elements. In the case of protein-encoding genes that are transcribed by RNA polymerase II, it is common to have a TATA box, which is about 25 bp upstream from a transcriptional start site. The TATA box is important in the identification of the transcriptional start site and the assembly of RNA polymerase and various transcription factors. The transcriptional start site defines where transcription actually begins.

C22. The two models are described in Figure 14.15. The allosteric model is more like ρ-independent termination, and the torpedo model is more like ρ-dependent termination. In the torpedo model, a protein knocks RNA polymerase off the DNA, much like the effect of ρ protein.

C24. Hydrogen bonding is usually the predominant type of interaction in an assembly and disassembly process. In addition, ionic bonding and hydrophobic interactions could occur. Covalent interactions would not occur. High temperature and high salt concentrations tend to break hydrogen bonds and thus would inhibit assembly and stimulate disassembly.

C26. In bacteria, the 5′ end of the tRNA is cleaved by RNase P. The 3′ end is cleaved by a different endonuclease, and then a few nucleotides are digested away by an exonuclease that removes nucleotides until it reaches a CCA sequence.

C28. A ribozyme is a catalyst that is composed of RNA. Examples are RNase P and self-splicing group I and II introns. It is thought that the spliceosome may contain ribozymes as well.

C30. Self-splicing means that an RNA molecule can splice itself without the aid of a protein. Group I and II introns are self-splicing, although proteins can enhance the rate at which it proceeds.

C32. In alternative splicing, variation in the pattern of splicing results in mRNAs that contain alternative combinations of exons. The biological significance is that two or more different proteins can be produced from a single gene. This is a more efficient use of the genetic material. In multicellular organisms, alternative splicing is often used in a cell-specific manner.

C34. As shown at the left side of Figure 14.18, the guanosine, which binds to the guanosine-binding site, does not have a phosphate group attached to it. This guanosine is the nucleoside that winds up at the 5′ end of the intron. Therefore, the intron does not have a phosphate group at its 5′ end.

C36. U5

Application and Experimental Questions

E2. The 1100-nucleotide band would be observed for a normal individual (lane 1). A deletion that removed the –50 to –100 region would greatly diminish transcription, so the homozygote would produce hardly any of the transcript (just a faint amount, as shown in lane 2), and the heterozygote would produce roughly half as much of the 1100-nucleotide transcript (lane 3) compared to a normal individual. A nonsense codon would not have an effect on transcription; it affects only translation. So the individual with this mutation would produce a normal amount of the 1100-nucleotide transcript (lane 4). A mutation that removed the splice acceptor site would prevent splicing. Therefore, this individual would produce a 1550-nucleotide transcript (actually, 1547 nucleotides to be precise, 1550 minus 3). The Northern blot is shown here:

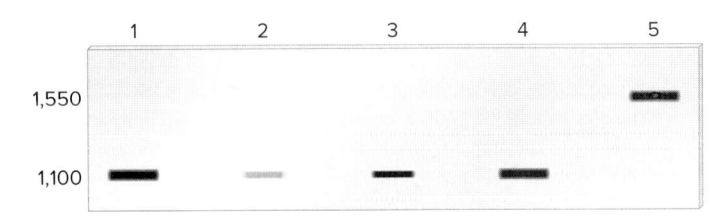

E4. A. Migration would not be shifted upwards because ρ protein would not bind to the mRNA that is encoded by a gene that is terminated in a ρ-independent manner. The mRNA from such genes does not contain the sequence near the 3′ end that acts as a recognition site for the binding of ρ protein.

B. Migration would be shifted upwards because ρ protein would bind to the mRNA.

C. Migration would be shifted upwards because U1 would bind to the pre-mRNA.

D. Migration would not be shifted upwards because U1 would not bind to mRNA that has already had its introns removed. U1 binds only to pre-mRNA.

Questions for Student Discussion/Collaboration

2. RNA transcripts come in two basic types: those that function as RNA (e.g., tRNA, rRNA, etc.) and those that are translated

CHAPTER 15

Note: The answers to the Comprehension Questions are at the end of the chapter.

Concept Check Questions (in figure legends)

FIGURE 15.1 Alkaptonuria occurs when homogentisic acid oxidase is defective.

FIGURE 15.2 The strain 2 mutants are unable to convert O-acetylhomoserine into cystathionine.

FIGURE 15.3 The role of DNA is to store the information that specifies the amino acid sequence of a polypeptide.

FIGURE 15.5 Tryptophan and phenylalanine are the least soluble in water.

FIGURE 15.6 Hydrogen bonding promotes the formation of secondary structures in proteins.

FIGURE 15.8 Only one radiolabeled amino acid was found in each sample. If radioactivity was trapped on the filter, this meant that the RNA triplet specified that particular amino acid.

FIGURE 15.9 A tRNA has an anticodon that recognizes a codon in the mRNA. It also has a 3′ acceptor site where the correct amino acid is attached.

FIGURE 15.11 A charged tRNA has an amino acid attached to it.

FIGURE 15.12 The wobble rules allow for a smaller population of tRNAs to recognize all of the possible mRNA codons.

FIGURE 15.14 A site in mRNA promotes the binding of the mRNA to the ribosome during formation of the initiation complex. During elongation, the codons in mRNA specify the polypeptide sequence. The stop codon of mRNA is needed to terminate transcription.

FIGURE 15.16 The Shine-Dalgarno sequence in bacterial mRNA is complementary to a region in the 16S rRNA within the small ribosomal subunit. These complementary sequences hydrogen bond with each other.

FIGURE 15.17 Peptidyl transferase catalyzes peptide bond formation between amino acids in the growing polypeptide.

FIGURE 15.18 The structures of release factors, which are proteins, resemble the structures of tRNAs.

End-of-Chapter Questions

Conceptual Questions

C2. When we say that the genetic code contains degeneracy, it means that more than one codon can specify the same amino acid. For example, GGG, GGC, GGA, and GGU all specify glycine. In general, the genetic code is nearly universal, because it is used in the same way by viruses, prokaryotes, fungi, plants, and animals. As shown in Table 15.2, there are a few exceptions, which occur primarily in protists and in yeast and mammalian mitochondria.

C4. A. The mutant tRNA would recognize glycine codons in the mRNA but would put in tryptophan where glycine was supposed to be in the polypeptide.

B. This mutation indicates that the aminoacyl-tRNA synthetase is primarily recognizing other regions of the tRNA molecule besides the anticodon region. In other words, tryptophanyl-tRNA synthetase (the aminoacyl-tRNA synthetase that attaches tryptophan) primarily recognizes other regions of the tRNA^{Trp} sequence (that is, other than the anticodon region). If aminoacyl-tRNA synthetases recognized only the anticodon region, we would expect glycyl-tRNA synthetase to recognize this mutant tRNA and attach glycine. That is not what happens.

C6. A. Three tRNAs. There are six leucine codons: UUA, UUG, CUU, CUC, CUA, and CUG. The anticodon AAU recognizes UUA and UUG. Two more tRNAs are needed to efficiently recognize the other four leucine codons. These could be GAG and GAU or GAA and GAU.

B. One tRNA. There is only one codon, AUG, so only one tRNA with the anticodon UAC is needed.

C. Three tRNAs. There are six serine codons: AGU, AGC, UCU, UCC, UCA, and UCG. One tRNA can recognize AGU and AGC. This tRNA could have the anticodon UCG or UCA. Two more tRNAs are needed to efficiently recognize the other four tRNAs. These could be AGG and AGU or AGA and AGU.

C8. 3′–CUU–5′ or 3′–CUC–5′

C10. It can recognize 5′–GGU–3′, 5′–GGC–3′, and 5′–GGA–3′. All of these specify glycine.

C12. All tRNA molecules have a cloverleaf structure with three stem-loop structures. The second stem-loop contains the anticodon sequence that recognizes the codon sequence in mRNA. At the 3′ end, there is an acceptor stem, with the sequence CCA, that serves as an attachment site for an amino acid. Most tRNAs also have base modifications within their nucleotide sequences.

C14. The role of aminoacyl-tRNA synthetase is to specifically recognize tRNA molecules and attach the correct amino acid to them. This ability is sometimes described as the second genetic code because the specificity of the attachment is a critical step in deciphering the genetic code. For example, if a tRNA has a 3′–GGG–5′ anticodon, it will recognize a 5′–CCC–3′ codon, which should specify proline. The aminoacyl-tRNA synthetase known as prolyl-tRNA-synthetase will recognize this tRNA and attach proline to the 3′ end. Other aminoacyl-tRNA synthetases will not recognize this tRNA.

C16. Bases that have been chemically modified can occur at various locations throughout tRNA molecules. The significance of all of these modifications is not entirely known. However, within the anticodon region, base modification alters base pairing to allow the anticodon to recognize two or more different bases within the codon.

C18. No, it is not. Due to the wobble rules, the 5′ base in the anticodon of a tRNA can recognize two or more bases in the third (3′) position of the mRNA. Therefore, any given cell type can synthesize fewer than 61 types of tRNAs.

C20. The assembly process is very complex at the molecular level. In eukaryotes, 33 proteins and one rRNA assemble to form a 40S subunit, and 49 proteins and three rRNAs assemble to form a 60S subunit. This assembly occurs within the nucleolus.

C22. A. Within the 30S subunit and at the interface between the two subunits

B. Within the 50S subunit

C. Within the 50S subunit

D. Within the 30S subunit

C24. Bacterial mRNAs contain a Shine-Dalgarno sequence, which is necessary for the binding of the mRNA to the small ribosomal subunit. This sequence, UUAGGAGGU, is complementary to a sequence in the 16S rRNA. Due to this complementarity, these sequences will hydrogen bond to each other during the initiation stage of translation.

C26. The ribosome binds at the 5' end of the mRNA and then scans in the 3' direction in search of an AUG start codon. If it finds one that obeys Kozak's rules, it will begin translation at that site. Aside from an AUG start codon, two other important features are a guanosine at the +4 position and a purine at the −3 position.

C28. The A (aminoacyl) site is the location where a tRNA carrying a single amino acid initially binds. The only exception is the initiator tRNA, which binds to the P (peptidyl) site. The growing polypeptide is removed from the tRNA in the P site and transferred to the amino acid attached to the tRNA in the A site. The ribosome translocates in the 3' direction, with the result that the two tRNAs in the P and A sites are moved to the E (exit) and P sites, and the uncharged tRNA in the E site is released.

C30. The initiation phase involves the binding of the Shine-Dalgarno sequence to the rRNA in the 30S subunit. The elongation phase involves the binding of anticodons in tRNA to codons in mRNA.

C32. A. E site and P site (Note: A tRNA without an amino acid attached is only briefly found in the P site, just before translocation occurs.)

B. P site and A site (Note: A tRNA with a polypeptide chain attached is only briefly found in the A site, just before translocation occurs.)

C. Usually the A site, except the initiator tRNA, which can be found in the P site.

C34. The tRNAs bind to the mRNA because their anticodon and codon sequences are complementary. When the ribosome translocates in the 5' to 3' direction, the tRNAs remain bound to their complementary codons, and the two tRNAs shift from the A site and P site to the P site and E site. If the ribosome moved in the 3' direction, it would have to dislodge the tRNAs and drag them to a new position where they would not (necessarily) be complementary to the mRNA.

C36. 52

Application and Experimental Questions

E2. The initiation phase of translation is very different in bacteria and in eukaryotes. A bacterial mRNA would not be translated very efficiently in a eukaryotic translation system, because it lacks a cap structure attached to its 5' end. A eukaryotic mRNA would not have a Shine-Dalgarno sequence near its 5' end, so it would not be translated very efficiently in a bacterial translation system.

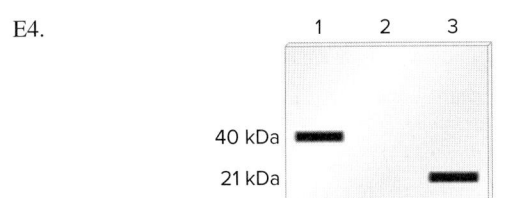

E4.

E6. A. If codon usage were significantly different between kangaroo and yeast cells, this would inhibit the translation process. For example, if the preferred leucine codon in kangaroos was CUU, translation would probably be slow in a yeast translation system. We would expect the cell-free translation system from yeast cells to primarily contain leucine tRNAs with the anticodon sequence AAC, because this tRNA^{Leu} would match the preferred yeast leucine codon, which is UUG. In a yeast translation system, there probably would not be a large amount of tRNA with the anticodon sequence GAA, which would match the preferred leucine codon, CUU, of kangaroos. For this reason, kangaroo mRNA would not be translated very well in a yeast translation system, but it probably would be translated to some degree.

B. The advantage of codon bias is that a cell can rely on a smaller population of tRNA molecules to efficiently translate its proteins. A disadvantage is that mutations, which do not change the amino acid sequence but do change a codon (e.g., UUG to UUA), may inhibit the production of a polypeptide if a preferred codon is changed to a nonpreferred codon.

Questions for Student Discussion/Collaboration

2. There are similarities along several lines:

    1. There is a lot of molecular recognition going on, either between two nucleic acid molecules or between proteins and nucleic acid molecules.

    2. There is biosynthesis going on in both processes. Small building blocks are being connected, which requires an input of energy.

    3. There are genetic signals that determine the beginning and ending of these processes.

    There are also many differences:

    1. Transcription produces an RNA molecule with a similar structure to the DNA, whereas translation produces a polypeptide with a structure that is very different from RNA.

    2. Depending on your point of view, it seems that translation is more biochemically complex, requiring more proteins and RNA molecules to accomplish the task.

CHAPTER 16

Note: The answers to the Comprehension Questions are at the end of the chapter.

Concept Check Questions (in figure legends)

FIGURE 16.1 Gene regulation is energy-efficient. A cell does not waste energy making RNAs and proteins it does not need.

FIGURE 16.2 Activators and repressors are regulatory proteins. Inducers, corepressors, and inhibitors are small effector molecules.

FIGURE 16.3 The *lacZ*, *lacY*, and *lacA* genes are under the control of the *lac* promoter.

FIGURE 16.5 The binding of lac repressor to the *lac* operon occurs when lactose is not present in the environment—when allolactose is not bound to the repressor.

FIGURE 16.8 The repressor protein allows the bacterial cell to avoid turning on the operon in the absence of lactose. The activator protein allows the cell to choose between glucose and lactose.

FIGURE 16.9 The data for the cases in which O_2 and O_3 are deleted indicate that O_1, by itself, is not very strong in repressing the *lac* operon.

FIGURE 16.11 Tryptophan acts as a corepressor that causes trp repressor to bind to the *trp* operon and repress transcription.

FIGURE 16.12 Hydrogen bonding between complementary sequences causes stem-loops to form.

FIGURE 16.13 The presence of tryptophan causes the *trpL* gene to be translated to its stop codon. This blocks both regions 1 and 2, which allows a 3–4 stem-loop to form. The 3–4 stem-loop causes termination of transcription.

FIGURE 16.14 The *micF* antisense RNA binds to the *ompF* mRNA and inhibits its translation.

FIGURE 16.15 Feedback inhibition prevents a bacterium from overproducing the product of a metabolic pathway.

FIGURE 16.16 The RNA conformation with an antiterminator stem-loop favors transcription.

FIGURE 16.17 The RNA conformation with the Shine-Dalgarno antisequestor favors translation.

End-of-Chapter Questions

Conceptual Questions

C2. Gene regulation greatly enhances the efficiency of cell growth. It takes a lot of energy to transcribe and translate genes. Therefore, a cell is much more efficient and better at competing in its environment if it expresses a gene only when the gene product is needed. For example, a bacterium will express only the genes that are necessary for lactose metabolism when the bacterium is exposed to lactose. When the environment is missing lactose, these genes are turned off. Similarly, when tryptophan levels are high within the cytoplasm, the genes that are required for tryptophan biosynthesis are repressed.

C4. A. Regulatory protein B. Effector molecule
     C. DNA segment   D. Effector molecule
     E. Regulatory protein F. DNA segment
     G. Effector molecule

C6. A mutation that has a *cis*-effect is within a genetic regulatory sequence, such as an operator site, that affects the binding of a genetic regulatory protein. A *cis*-effect mutation affects only the adjacent genes that the genetic regulatory sequence controls. A mutation having a *trans*-effect is usually in a gene that encodes a genetic regulatory protein. A *trans*-effect mutation can be complemented in a merozygote experiment by the introduction of a normal gene that encodes the regulatory protein.

C8. A. No transcription would take place. The *lac* operon could not be expressed.
     B. No regulation would take place. The operon would be continuously turned on.
     C. The rest of the operon would function normally but no transacetylase would be made.

C10. Diauxic growth refers to a phenomenon due to gene regulation in which a cell first uses up one type of sugar (such as glucose) before it begins to metabolize a second sugar (such as lactose). When a bacterial cell is exposed to both sugars, the uptake of glucose causes the cAMP levels in the cell to fall. When this occurs, the catabolite activator protein (CAP) is removed from the *lac* operon so it is not able to be activated by CAP.

C12. A mutation that prevented lac repressor from binding to the operator would make the *lac* operon constitutive only in the absence of glucose. However, this mutation would not be entirely constitutive because transcription would be inhibited in the presence of glucose. The disadvantage of constitutive expression of the *lac* operon is that the bacterial cell would waste a lot of energy transcribing the genes and translating the mRNA when lactose was not present.

C14. A. Attenuation will not occur because a 2–3 stem-loop will form.
     B. Attenuation will occur because a 2–3 stem-loop cannot form, so a 3–4 one will form.
     C. Attenuation will not occur because a 3–4 stem-loop cannot form.
     D. Attenuation will not occur because a 3–4 stem-loop cannot form.

C16. The addition of Gs and Cs into the U-rich sequence would prevent attenuation. The U-rich sequence promotes the dissociation of the mRNA from the DNA, when the terminator stem-loop forms. This causes RNA polymerase to dissociate from the DNA and thereby causes transcriptional termination. The UGGUUGUC sequence would probably not dissociate because of the Gs and Cs. Remember that G-C base pairs have three hydrogen bonds and are more stable than A-U base pairs, which have only two hydrogen bonds.

C18. It takes a lot of cellular energy to translate mRNA into a protein. A cell wastes less energy if it prevents the initiation of translation rather than preventing a later stage such as elongation or termination.

C20. One possible mechanism is that histidine acts as a corepressor that shuts down the transcription of the histidine synthetase gene. A second mechanism is that histidine acts as an inhibitor via feedback inhibition. A third possibility is that histidine inhibits the ability of the mRNA encoding histidine synthetase to be translated. Perhaps it induces a gene that encodes an antisense RNA. If the amount of histidine synthetase protein is identical in the presence and absence of extracellular histidine, a feedback inhibition mechanism is favored, because this affects only the activity of the histidine synthetase enzyme, not the amount of the enzyme. The other two mechanisms would diminish the amount of this protein.

C22. The two proteins are similar in that both bind to a segment of DNA and repress transcription. They are different in three ways: (1) They recognize different effector molecules (i.e., lac repressor recognizes allolactose, and trp repressor recognizes tryptophan. (2) Allolactose causes lac repressor to release from the operator, whereas tryptophan causes trp repressor to bind to its operator. (3) The sequences of the operator sites that these two proteins recognize are different from each other. Otherwise, lac repressor could bind to the *trp* operator, and trp repressor could bind to the *lac* operator.

Application and Experimental Questions

E2. In samples loaded in lanes 1 and 4, we expect the repressor to bind to the operator because no lactose is present. In the sample loaded into lane 4, the CAP protein could still bind cAMP because there is no glucose. However, there is no difference between lanes 1 and 4, so CAP does not appear to activate transcription when lac repressor is bound. If we compare samples loaded into lanes 2 and 3, lac repressor is not bound in either case, and CAP is not bound in the sample loaded into lane 3. There is less transcription in lane 3 compared to lane 2, but because there is some transcription observed in lane 3, we can conclude that the removal of CAP (because cAMP levels are low) is not entirely effective at preventing transcription. Overall, the results indicate that the binding of lac repressor is more effective at preventing transcription of the *lac* operon than is the removal of CAP.

E4. A. Yes, with no sonication, β-galactosidase will not be released from the cell, and not much yellow color will be observed. (Note: A little yellow color may be seen because some of the β-ONPG may be taken into the cell.)

B. No, yellow color should be observed in the first two tubes even if the researcher forgot to add lactose because the unconjugated strain does not have a functional lac repressor.

C. Yes, if the researcher forgot to add β-ONPG, there will be no yellow color because the cleavage of β-ONPG by β-galactosidase is what produces the color.

E6. You could mate a strain that has an F′ factor carrying a normal *lac* operon and a normal *lacI* gene to the mutant strain. Because the mutation is in the operator site, expression of β-galactosidase would continue to occur, even in the absence of lactose.

Questions for Student Discussion/Collaboration

2. A DNA loop may inhibit transcription by preventing RNA polymerase from recognizing the promoter. Or, it may inhibit transcription by preventing the formation of the open complex. Alternatively, a bend may enhance transcription by exposing the base sequence that the sigma factor of RNA polymerase recognizes. The bend may expose the major groove in such a way that this base sequence is more accessible to binding by sigma factor and RNA polymerase.

CHAPTER 17

Note: The answers to the Comprehension Questions are at the end of the chapter.

Concept Check Questions (in figure legends)

FIGURE 17.1 Gene regulation is most energy-efficient at the level of transcription, because a cell avoids wasting energy making RNA or proteins.

FIGURE 17.3 If TFIID cannot bind to the TATA box, RNA polymerase will not be recruited to the core promoter, and therefore transcription will not begin.

FIGURE 17.4 When an activator interacts with mediator, it causes mediator to phosphorylate CTD, which causes RNA polymerase to proceed to the elongation phase of transcription.

FIGURE 17.6 The glucocorticoid receptor binds specifically next to genes that have an adjacent GRE.

FIGURE 17.7 Nucleosome eviction may allow certain proteins access to bind to particular sites in the DNA.

FIGURE 17.8 Histone modifications may directly affect the interaction between histones and the DNA, or they may affect the binding of other proteins to the chromatin.

FIGURE 17.9 An NFR is needed at the core promoter so that transcription factors can recognize enhancers and the preinitiation complex can form.

FIGURE 17.10 Histone eviction or displacement is needed for elongation to occur because RNA polymerase cannot transcribe through nucleosomes. It needs to unwind the DNA for transcription to take place.

FIGURE 17.12 In part (a), DNA methylation blocks an activator protein from binding to the DNA. This prevents transcriptional activation.

FIGURE 17.13 De novo methylation occurs on unmethylated DNA, whereas maintenance methylation occurs on hemimethylated DNA.

FIGURE 17.16 The choice occurs only during embryonic development.

FIGURE 17.17 A knot-like structure would prevent proteins such as TFIID and RNA polymerase from binding to the gene and would thereby inhibit transcription of the gene.

FIGURE 17.19 At the level of DNA sequences, queen and worker bees are not different from each other. However, differences in gene expression due to epigenetics can explain the morphological differences between the two types of female bees.

FIGURE 17.21 The mutation would cause the overproduction of ferritin, because ferritin synthesis would occur even if iron levels were low.

End-of-Chapter Questions

Conceptual Questions

C2. Regulatory elements are relatively short genetic sequences that are recognized by regulatory transcription factors. After the regulatory transcription factor has bound to the regulatory element, it will affect the rate of transcription, either increasing or decreasing it, depending on the action of the regulatory protein. Regulatory elements are typically located in the upstream region near the promoter, but they can be located almost anywhere (i.e., upstream and downstream) and even quite far from the promoter.

C4. Transcriptional activation occurs when a regulatory transcription factor binds to a regulatory element and activates transcription. These transcriptional activator proteins may interact with TFIID and/or mediator to promote the assembly of RNA polymerase and general transcription factors at the promoter region. They may also alter the structure of chromatin so that RNA polymerase and transcription factors are able to gain access to the promoter. Transcriptional inhibition occurs when a regulatory transcription factor inhibits transcription. Such repressors may interact with TFIID and/or mediator to inhibit RNA polymerase.

C6. For the glucocorticoid receptor to bind to a GRE, a steroid hormone must first enter the cell. The hormone then binds to the glucocorticoid receptor, which releases HSP90. The release of HSP90 exposes a nuclear localization signal (NLS) within the receptor, which enables it to dimerize and then enter the nucleus. Once inside the nucleus, the dimer binds to a GRE, which activates transcription of the adjacent genes.

C8. Eventually, the glucocorticoid hormone will be degraded by the cell. The glucocorticoid receptor binds the hormone with a certain affinity. The binding is a reversible process. Once the concentration of the hormone falls below the affinity of the hormone for the receptor, the receptor will no longer have the glucocorticoid hormone bound to it. When the hormone is released, the glucocorticoid receptor will change its conformation, and it will no longer bind to the DNA.

C10. ATP-dependent chromatin remodeling complexes may change the positions of nucleosomes, evict histones, and/or replace histones with histone variants.

C12. The attraction between DNA and histones occurs because the histones are positively charged and the DNA is negatively charged. The covalent attachment of acetyl groups decreases the amount of positive charge on the histone proteins and thereby may decrease the binding of the DNA. In addition, histone acetylation may attract proteins to the region that loosen chromatin compaction.

C14. A nucleosome-free region (NFR) is a location on a chromosome where nucleosomes are missing. These regions are typically found at the beginnings and ends of genes. An NFR at the beginning of a gene is thought to be important for activation of the gene. An NFR at the end of a gene may be important for its proper termination.

C16. The attachment of the acetyl group (—COCH$_3$) eliminates the positive charge on the lysine side chain, thereby disrupting the electrostatic attraction between the histone protein and the negatively charged DNA backbone. This makes it easier for the histones to be displaced from the DNA.

C18. Perhaps the DNA methyltransferase is responsible for methylating and inhibiting a gene that causes a cell to become a muscle cell. The methyltransferase is inactivated by the mutation.

C20. Epigenetics is the study of mechanisms that lead to changes in gene expression that can be passed from cell to cell and are reversible, but do not involve a change in the sequence of DNA. Not all epigenetic changes are passed from parent to offspring. For example, those that occur in somatic cells, such as lung cells, would not be passed to offspring.

C22. During oogenesis, the ICR and DMR are not methylated. This allows CTCF to bind to these sites and to each other to form a loop. This loop prevents a nearby enhancer from activating the *Igf2* gene. During sperm formation, methylation of these sites prevents CTCF binding and thereby prevents loop formation. The nearby enhancer activates the *Igf2* gene. Therefore, the copy of the *Igf2* gene inherited from the father via sperm is active.

C24. See Figure 17.16. The choice of which X chromosome remains active occurs when the pluripotency factors and CTCFs, which were previously bound to both X chromosomes, shift entirely to one of the X chromosomes. Because the pluripotency factors stimulate the expression of the *Tsix* gene, the X chromosome to which they shift is chosen as the active X chromosome. By comparison, on the other X chromosome, the *Tsix* gene is not expressed, which permits the expression of the *Xist* gene.

C26. Epigenetic modifications lead to changes in gene expression that persist from early development to adulthood. These changes allow some genes to be expressed in certain cell types but not in others. The trithorax and polycomb group complexes are involved in causing such epigenetic changes.

C28. See Figure 17.17.

C30. Dietary factors can affect the occurrence of epigenetic changes, particularly those that happen during early stages of development. For example, dietary factors that affect the level of DNA methylation appear to be responsible or partly responsible for altering coat color in mice and affecting female bee development. These dietary factors may affect the level of S-adenosylmethionine and/or the function of DNA methyltransferase and thereby influence whether particular genes are methylated or not. If genes are methylated, this typically inactivates them.

C32. This person would be unable to make ferritin, because the IRP would always be bound to the IRE. The amount of transferrin receptor mRNA would be high, even in the presence of high amounts of iron, because the IRP would remain bound to the IRE, stabilizing the transferrin receptor mRNA. Such a person would not have any problem taking up iron into his or her cells and, in fact, would take up a lot of iron via the transferrin receptor, even when the iron concentrations were high. Therefore, this person would not need more iron in the diet. However, excess iron in the diet would be very toxic for two reasons. First, the person cannot make ferritin, which prevents the toxic buildup of iron in the cytosol. Second, when iron levels are high, the person would continue to synthesize the transferrin receptor, which functions in the uptake of iron.

Application and Experimental Questions

E2. If the DNA fragment is 3800 bp, it means that the site is unmethylated, because it has been cut by *Not*I (and *Not*I cannot cut methylated DNA). If the DNA fragment is 5300 bp, the DNA is not cut by *Not*I, so we assume that it is methylated. For the sample in lane 4, the gene is isolated from root tissue, and the DNA is not methylated because the sample runs at 3800 bp. This suggests that gene *T* is expressed in root tissue. In the other samples, the DNA runs at 5300 bp, indicating that the DNA is methylated. The pattern of methylation seen here is consistent with the known function of gene *T*. We would expect it to be expressed in root cells because it functions in the uptake of phosphate from the soil. It would be silenced in the other parts of the plant via methylation.

E4. The coat color relies on the expression of the *Agouti* gene. In this case, the *Agouti* gene is under the control of a promoter that is found in a transposable element, causing it to be overexpressed. When the promoter is not methylated and the gene is overexpressed, the coat color is yellow because the *Agouti* gene product promotes the synthesis of yellow pigment. When the promoter is methylated, this inhibits the *Agouti* gene and coat color is darker because less yellow pigment is made.

E6. The agent in cigarette smoke may have caused the methylation of CpG islands near the promoter of the *p53* gene, thereby inhibiting transcription. Another possibility is that it may affect covalent histone modifications or chromatin remodeling in a way that causes the gene to be in a closed conformation.

E8. The blot shown in (a) is correct for ferritin mRNA regulation. The presence and absence of iron does not affect the amount of mRNA; it affects the ability of the mRNA to be translated. The blot shown in (b) is correct for transferrin receptor mRNA. In the absence of iron, the IRP binds to the IRE in the mRNA and stabilizes it, so the mRNA level is high. In the presence of iron, the iron binds to IRP and IRP is released from the IRE, and the transferrin receptor mRNA is degraded.

Questions for Student Discussion/Collaboration

2. The list could be a long one; it could include agents in cigarette smoke, pesticides, herbicides, heavy metals, etc.

CHAPTER 18

Note: The answers to the Comprehension questions are at the end of the chapter.

Concept Check Questions (in figure legends)

FIGURE 18.1 DNA, RNAs, proteins, and small molecules can bind to an ncRNA.

FIGURE 18.4 RISC binds to a specific mRNA because the small RNA within RISC is complementary to that mRNA. The bonding is hydrogen bonding between complementary bases.

FIGURE 18.5 The hydrolysis of GTP promotes the release of SRP, which allows translation to resume.

FIGURE 18.6 The crRNA binds to one of the strands of the phage DNA due to complementary base pairing.

FIGURE 18.7 piRNAs and PIWI proteins cause TE RNAs to be degraded and prevent TEs from being transcribed.

End-of-Chapter Questions
Conceptual Questions

C2. A. Scaffold and guide
B. Ribozyme
C. Guide
D. Guide

C4. RNA interference refers to the phenomenon in which the presence of a double-stranded RNA molecule results in the silencing of a complementary mRNA. First, the double-stranded RNA is processed by dicer into small RNA fragments (miRNAs or siRNAs). The miRNA or siRNA associates with a complex called the RNA-induced silencing complex (RISC). RISC then binds to a complementary mRNA. This either leads to mRNA degradation or inhibits translation.

C6. MicroRNAs (miRNAs) are ncRNAs that are transcribed from endogenous eukaryotic genes—genes that are normally found in the genome. They play key roles in regulating gene expression, particularly during embryonic development in animals and plants. Most commonly, a single type of miRNA inhibits the translation of several different mRNAs. By comparison, small-interfering RNAs (siRNAs) are ncRNAs that usually originate from exogenous sources, which means that they are not normally made by cells. They usually promote mRNA degradation. The sources of siRNAs can be viruses that infect a cell, or researchers can make siRNAs in order to study gene function experimentally.

C8. If the SRP RNA did not stimulate the GTPase activities, SRP would remain bound to the polypeptide and translation would not resume.

C10. tracrRNA is a guide that guides crRNA to Cas9.

C12. See Figure 18.7. The role of piRNAs is to guide the PIWI proteins to TE RNA, where they either cause transcription to be inhibited or cause degradation of the TE RNA.

C14. The miR-200 family plays an essential role in tumor suppression by inhibiting an event called the epithelial-mesenchymal transition (EMT), which is the initiating step of metastasis. During the EMT, cells lose their adhesion to neighboring cells. This loss of adhesion is associated with a decrease in expression of E-cadherin, which is a membrane protein that adheres adjacent cells to each other. When E-cadherin levels are low, cells can more easily move to new sites in the body. In an adult, when the miR-200 family of miRNAs is expressed at normal levels, these miRNAs inhibit the EMT. This inhibition maintains a normal level of E-caderin and thereby prevents metastasis. However, if miR-200 miRNAs are expressed at low levels, EMT is not inhibited, and metastasis can occur.

Application and Experimental Questions

E2. The researchers were injecting pre-siRNA. In their experiment, it was a perfect match to the *mex-3* RNA and caused its degradation.

E4. The CRISPR-Cas system is used to alter genes by causing a small deletion within a gene or producing a change in the sequence of a gene. In this system, tracrRNA and crRNA form one RNA called a single guide RNA (sgRNA). The sgRNA recognizes both the Cas9 protein and a target gene that a researcher wants to mutate. After the sgRNA-Cas9 complex binds to the target gene, Cas9 cleaves the gene. Following nonhomologous end joining, the result is often a deletion of a small segment of the target gene, or if donor DNA is added, a double crossover can introduce a point mutation into the target gene.

E6. miRNA replacement therapy is aimed at restoring the function of miRNAs that have been down-regulated in cancer cells. One example of this treatment approach is the use of a viral vector to restore a particular miRNA. Another example is to increase the levels of miRNAs by targeting the miRNA-processing machinery via drugs such as enoxacin, which enhances the miRNA-processing machinery in certain cancer cells and thereby results in an increase of many miRNAs. A third example of miRNA replacement therapy is the use of DNA demethylating agents and histone deacetylase inhibitors. These compounds reverse the epigenetic silencing of miRNAs that behave as tumor suppressors.

Questions for Student Discussion/Collaboration

2. The search will lead to many examples of roles played by ncRNAs that are not discussed in this text. It should be easy and interesting to make such a list.

CHAPTER 19

Note: The answers to the Comprehension Questions are at the end of the chapter.

Concept Check Questions (in figure legends)

FIGURE 19.2 A position effect has occurred when the expression of a gene has been altered due to a change in its position along a chromosome.

FIGURE 19.3 In part (b), the DNA sequence of the eye color gene has not been changed. The change in its expression is due to a position effect.

FIGURE 19.5 A somatic mutation is not passed from parent to offspring.

FIGURE 19.6 Because the colonies on the secondary plates appear in the same locations, the mutations occurred on the master plate before exposure to T1. Therefore, the mutations did not occur as a physiological adaptation to T1 exposure.

FIGURE 19.7 The probability is 75%.

FIGURE 19.8 The deamination of methylcytosine is more difficult to repair because thymine is a normal base found in DNA. This makes it difficult for DNA repair enzymes to distinguish between the correct and altered base.

FIGURE 19.10 A reactive oxygen species is a molecule that contains oxygen and may react with cellular molecules in a way that may cause harm.

FIGURE 19.13 The base analog 5-bromouracil causes a transition.

FIGURE 19.14 Thymine dimer formation is often the result of exposure to UV light. It most commonly occurs in skin cells.

FIGURE 19.15 Enzymes within the rat liver extract may convert non-mutagenic molecules into mutagenic agents. Adding the extract allows researchers to identify molecules that may be mutagenic in mammals.

FIGURE 19.17 Cuts are made on both sides of the damaged region so that it can be removed by UvrD.

FIGURE 19.19 MutH distinguishes between the parental strand and the newly made daughter strand, which ensures that the daughter strand is repaired.

FIGURE 19.20 An advantage of NHEJ as a DNA repair system is that it can occur at any stage of the cell cycle. A disadvantage is that it may be imprecise and result in a short deletion in the DNA.

End-of-Chapter Questions

Conceptual Questions

C2. It is a gene mutation, a point mutation, a base substitution, a transition mutation, a deleterious mutation, a mutant allele, a nonsense mutation, a conditional mutation, and a temperature-sensitive lethal mutation.

C4. A. A nonsense mutation would probably inhibit protein function, particularly if it was not near the end of the coding sequence.

B. A missense mutation may or may not affect protein function, depending on the nature of the amino acid substitution and whether the substitution is in a critical region of the protein.

C. An up promoter mutation would increase the amount of functional protein.

D. Such a mutation may affect protein function if the alteration in splicing changes an exon in the mRNA that results in a protein with a different structure.

C6. A. Not appropriate, because the second mutation occurred in a different codon.

B. Appropriate.

C. Not appropriate, because the second mutation affected the same gene as the first mutation.

D. Appropriate.

C8. Here are two possible examples:

The consensus sequences for many bacterial promoters are –35: 5′–TTGACA–3′ and –10: 5′–TATAAT–3′. Most mutations that alter the consensus sequence would be expected to decrease the rate of transcription. For example, a mutation that changed the –35 region to 5′–GAGACA–3′ would decrease transcription.

The sequence 5′–TATAAT–3′ is recognized by the transcription factor TFIID. If this sequence was changed to 5′–TGTAAT–3′, TFIID would not recognize it very well and the adjacent gene would probably show a lower rate of transcription.

C10. Random mutations are more likely to be harmful than beneficial. The genes within each species have evolved to work properly. They have functional promoters, coding sequences, terminators, and so on, that allow them to be expressed. Mutations are more likely to disrupt these sequences. For example, mutations within the coding sequence may produce early stop codons, frameshift mutations, and missense mutations that result in a nonfunctional polypeptide. On rare occasions, however, mutations are beneficial; they may produce a gene that is expressed better than the original gene or produce a polypeptide that functions better.

C12. Yes, a person with cancer is a genetic mosaic. The cancerous tissue contains gene mutations that are not found in noncancerous cells of the body.

C14. The drawing should show the attachment of a methyl or ethyl group to a base within the DNA. The presence of the alkyl group disrupts the proper base pairing between the alkylated base and the normal base in the opposite DNA strand.

C16. A. Nitrous acid causes mutations that change A to G or C to T, which are transition mutations.

B. 5-bromouracil causes mutations that change G to A, which are transitions.

C. Proflavin causes small additions or deletions, which may result in frameshift mutations.

C18. A spontaneous mutation originates within a living cell. It may be due to spontaneous changes in nucleotide structure, errors in DNA replication, or products of normal metabolism that may alter the structure of DNA. The causes of induced mutations originate from outside the cell. They may be physical agents, such as UV light or X rays, or chemicals that act as mutagens. Both spontaneous and induced mutations may cause a harmful phenotype such as cancer. In many cases, induced mutations are avoidable if the individual can prevent exposure to the environmental agent that acts as a mutagen.

C20. Nitrous acid can change a cytosine to uracil. Nucleotide excision repair could remove the defect and replace it with the correct base.

C22. A. True

B. False; the TNRE is not within the promoter, it is within the coding sequence.

C. True; CAG is a codon for glutamine.

D. False; CCG is a codon for proline.

CHAPTER 20 A-35

C24. The mutation rate is the number of new mutations per generation. There are 13 babies who did not have a parent with achondroplasia; thus, 13 is the number of new mutations. To calculate the mutation rate as the number of new mutations per generation, we divide 13 by 2,844,000. The mutation rate is 4.6×10^{-6}.

C26. The effects of mutations are cumulative. If one mutation occurs in a cell, this mutation will be passed to the daughter cells. If a mutation occurs in the daughter cell, it will then have two mutations. These two mutations will be passed to the next generation of daughter cells, and so forth. The accumulation of many mutations eventually kills the cells. That is why mutagens are more effective at killing dividing cells than nondividing cells. It is because the number of mutations accumulates to a lethal level.

There are two main side effects to these treatments. First, some normal (noncancerous) cells of the body, particularly skin cells and intestinal cells, are actively dividing. These cells are also killed by chemotherapy and radiation therapy. Secondly, it is possible that the therapy may produce mutations that will cause noncancerous cells to become cancerous. For these reasons, there is a maximal recommended dose for chemotherapy or radiation therapy.

C28. A. Yes
B. No, the albino trait affects the entire individual.
C. No, the early apple-producing trait affects the entire tree.
D. Yes

C30. Mismatch repair is aimed at eliminating mismatches that may have occurred during DNA replication. In this case, the wrong base is in the newly made strand. The binding of MutH, which occurs on a hemimethylated sequence, provides a means of distinguishing between the unmethylated and methylated strands. In other words, MutH binds to the hemimethylated DNA in a way that allows the mismatch repair system to distinguish which strand is methylated and which is not.

C32. The underlying genetic defect that causes xeroderma pigmentosum is a defect in one of the genes that encode a polypeptide involved with nucleotide excision repair. Individuals with this disease are defective at repairing DNA abnormalities such as thymine dimers and abnormal bases. Therefore, they are very sensitive to environmental agents such as UV light, which is more likely to cause mutations in these people compared to unaffected individuals. For this reason, people with XP develop pigmentation abnormalities and premalignant lesions, and they have a high predisposition to skin cancer.

C34. Both types of repair systems recognize an abnormality in the DNA and excise the abnormal strand. The normal strand is then used as a template to synthesize a complementary strand of DNA. The systems differ in the types of abnormalities they detect. The mismatch repair system detects base pair mismatches, while the nucleotide excision repair system recognizes thymine dimers, chemically modified bases, missing bases, and certain types of crosslinks. The mismatch repair system operates immediately after DNA replication, allowing it to distinguish between the daughter strand (which contains the wrong base) and the parental strand. The nucleotide excision repair system can operate at any time in the cell cycle.

Application and Experimental Questions

E2. To show that antibiotic resistance is due to random mutations, you could follow the same basic strategy as Lederberg and Lederberg did, except the secondary plates would contain the antibiotic instead of T1 phage. If the antibiotic resistance arose as a result of random mutation on the master plate, you would expect the antibiotic-resistant colonies to appear at the same locations on two different secondary plates.

E4. You would expose the bacteria to the physical agent. You could also expose the bacteria to the rat liver extract, but it is probably not necessary for two reasons. First, a physical mutagen is not something that a person would eat. Therefore, the actions of digestion via the liver are probably irrelevant if you are concerned that the agent might be a mutagen. Second, the rat liver extract would not be expected to alter the properties of a physical mutagen.

E6. The results suggest that the strain is defective in nucleotide excision repair. If we compare the normal and mutant strains that have been incubated for 2 hours at 37°C, much of the radioactivity in the normal strain has been transferred to the soluble fraction because it has been excised. In the mutant strain, however, less of the radioactivity has been transferred to the soluble fraction, suggesting that this strain is not as efficient at removing thymine dimers.

Questions for Student Discussion/Collaboration

2. The worst time to be exposed to mutagens is at very early stages of embryonic development. An early embryo is most sensitive to mutagens because any mutation will affect a large region of the body. Adults must also worry about mutagens for several reasons. Mutations in somatic cells can cause cancer, a topic that is discussed in Chapter 22. Also, adults should be careful to avoid mutagens that may affect the ovaries or testes because these mutations could be passed to offspring.

CHAPTER 20

Note: The answers to the Comprehension Questions are at the end of the chapter.

Concept Check Questions (in figure legends)

FIGURE 20.1 Eight

FIGURE 20.2 In this experiment, the selectable marker gene selects for the growth of bacteria that have taken up a plasmid.

FIGURE 20.3 The name *reverse transcriptase* refers to the idea that this enzyme catalyzes the opposite of transcription. It uses an RNA template to make DNA, whereas during transcription a DNA template is used to make RNA.

FIGURE 20.4 The advantage of a cDNA library is that the recombinant vectors lack introns. This is useful if researchers want to focus their attention on the coding sequence of a gene or if they want to express the gene in cells that do not splice out introns.

FIGURE 20.6 The products that predominate are the DNA fragments that are flanked by primers. This occurs because these products can be used as templates to make products only like themselves.

FIGURE 20.8 The probe must be cleaved to separate the quencher from the reporter.

FIGURE 20.10 The nucleus of the oocyte is removed so that the resulting organism contains (nuclear) genetic material only from the somatic cell.

FIGURE 20.11 No. Carbon Copy did not receive genetic material from a different species.

FIGURE 20.12 When stem cells divide, they produce one cell that remains a stem cell and another cell that differentiates. This pattern maintains a population of stem cells.

FIGURE 20.14 Multipotent

FIGURE 20.17 Some examples are to study the core promoter of a gene; to study regulatory elements; to identify splice sites; and to study the importance of particular amino acids with regard to protein function.

FIGURE 20.18 In the bacterial defense system, tracrRNA and crRNA are separate molecules. In gene editing using CRIPSR-Cas technology, tracrRNA and crRNA are covalently linked to form a single guide molecule (sgRNA).

FIGURE 20.20 The secondary antibody is labeled, which provides a way to detect the protein of interest.

End-of-Chapter Questions

Conceptual Questions

C2. A restriction enzyme recognizes and binds to a specific DNA sequence and then cleaves a (covalent) phosphoester bond in each of two DNA strands.

C4. cDNA is DNA that is made using RNA as the starting material. Unlike genomic DNA, it lacks introns.

C6. A dideoxynucleotide is missing the —OH group at the 3′ position. When the 5′ end of a dideoxynucleotide binds to a growing strand of DNA, another phosphoester bond cannot be formed at the 3′ position. Therefore, the dideoxynucleotide prevents any further addition of nucleotides to the growing strand of DNA.

Application and Experimental Questions

E2. Remember that AT base pairs form two hydrogen bonds, while GC base pairs form three hydrogen bonds. The order (from stickiest to least sticky) is as follows:

*Bam*HI = *Pst*I = *Sau*3AI > *Eco*RI > *Nae*I

E4. In conventional gene cloning, many copies are made because the vector replicates to a high copy number within the cell, and the cells divide to produce many more cells. In PCR, the replication of the DNA to produce many copies is facilitated by primers, deoxyribonucleoside triphosphates (dNTPs), and *Taq* polymerase.

E6. A recombinant vector is a vector that has a piece of foreign DNA inserted into it. The foreign DNA may be from the chromosomal DNA of some organism. To construct a recombinant vector, the vector and the source of foreign DNA are digested with the same restriction enzyme. The complementary ends of the fragments are allowed to hydrogen bond to each other (i.e., sticky ends are allowed to bind), and then DNA ligase is added to create covalent phosphoester bonds. In some cases, a piece of the foreign DNA will become ligated to the vector, thereby creating a recombinant vector. In other cases, the two ends of the vector ligate back together, restoring the vector to its original structure.

As shown in Figure 20.2, the insertion of foreign DNA can be detected using X-Gal. The insertion of the foreign DNA causes the inactivation of the *lacZ* gene. The *lacZ* gene encodes the enzyme β-galactosidase, which converts the colorless compound X-Gal to a blue compound. If the *lacZ* gene is inactivated by the insertion of foreign DNA, the enzyme will not be produced, and the bacterial colonies will be white. If the vector has simply recircularized, and the *lacZ* gene remains intact, the enzyme will be produced and the colonies will be blue.

E8. It would be necessary to use cDNA so that the gene would not carry any introns. Bacterial cells do not contain spliceosomes (which are described in Chapter 14). To express a eukaryotic protein in bacteria, a researcher would clone cDNA into the bacteria, because the cDNA does not contain introns.

E10. Initially, the mRNA would be mixed with reverse transcriptase and nucleotides to create a complementary strand of DNA. Reverse transcriptase also needs a primer. This could be a primer that is known to be complementary to the β-globin mRNA. Alternatively, mature mRNAs have a polyA tail, so you could add a primer that consists of many Ts, called a poly-dT primer. After the complementary DNA strand has been made, the sample would then be mixed with primers, *Taq* polymerase, and nucleotides and subjected to the standard PCR protocol. (Note: the PCR reaction would have two kinds of primers. One primer would be complementary to the 5′ end of the mRNA and would be unique to the β-globin sequence. The other primer would be complementary to the 3′ end. This second primer could be a poly-dT primer or it could be a unique primer that would bind slightly upstream from the polyA-tail region.)

E12. The exponential phase of PCR is analyzed because it is during this phase that the amount of PCR product is proportional to the amount of the original DNA in the sample.

E14. Reproductive cloning is the cloning of entire multicellular organisms. With plants, this is easy. Most species of plants can be cloned by asexual cuttings. In animals, cloning occurs naturally to produce identical twins. Identical twins are genetic replicas of each other because they begin from the same fertilized egg. (Note: Some somatic mutations could occur in identical twins that would make them slightly different.) Recently, as in the case of Dolly, reproductive cloning has been accomplished by fusing somatic cells with enucleated eggs. The advantage, from an agricultural point of view, is that reproductive cloning would allow farmers to choose the best animal in a herd and make many clones from it. Breeding would no longer be necessary. Also, breeding may be less reliable because the offspring inherit traits from both the mother and father.

E16. AGGTCGGTTGCCATCGCAATAATTTCTGCC
TGAACCCAATA

E18. There are many different strategies you could follow. For example, you could mutate every other base and see what happens. It would be best to make very nonconservative mutations such as a purine for a pyrimidine or a pyrimidine for a purine. If the mutation prevents protein binding in the EMSA, then the mutation is probably within the response element. If the mutation has no effect on protein binding, it probably is outside the response element.

E20. The purpose of gel electrophoresis is to separate the many RNA molecules or proteins that were obtained from the sample being analyzed. This separation is based on molecular mass and allows a researcher to identify the molecular mass of the RNA molecule or protein that is being recognized by the labeled probe.

E22. The Northern blot is shown here. The female mouse expresses the same total amount of this mRNA as the male does. For the heterozygous female, there is 50% of the 900 bp band and 50% of the 825 bp band.

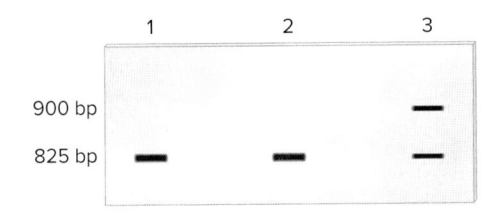

E24. Binding occurs due to the hydrogen bonding between complementary sequences. Due to the chemical properties of DNA and RNA strands, they form double-stranded regions when the base sequences are complementary.

E26. The Western blot is shown here. The sample in lane 2 came from a plant that was homozygous for a mutation that prevented the expression of this polypeptide. Therefore, no protein was observed in this lane. The sample in lane 4 came from a plant that is homozygous for a mutation that introduces an early stop codon into the coding sequence; thus, the polypeptide is shorter than normal (13.3 kDa). The sample in lane 3 was from a heterozygote that expresses about 50% of each type of polypeptide. Finally, the sample in lane 5 came from a plant that is homozygous for a mutation that changed one amino acid to another amino acid. This type of mutation, termed a missense mutation, may not be detectable on gel. However, a single amino acid substitution could affect polypeptide function.

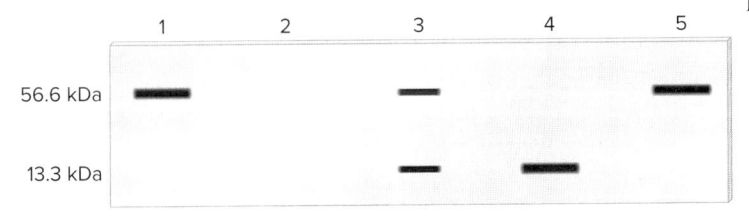

E28. In this case, the transcription factor binds to the response element when the hormone is present. Therefore, the hormone promotes the binding of the transcription factor to the DNA and thereby promotes transcriptional activation.

E30. TFIID can bind to this DNA fragment by itself, as seen in lane 2. However, TFIIB and RNA polymerase II cannot bind to the DNA by themselves (lanes 3 and 4). As seen in lane 5, TFIIB can bind, if TFIID is also present, because the mobility shift is higher than with TFIID alone (compare lanes 2 and 5). In contrast, RNA polymerase II cannot bind to the DNA when only TFIID is present. The mobility shift in lane 6 is the same as in lane 2, indicating that only TFIID is bound. Finally, in lane 7, when all three components are present, the mobility shift is higher than when both TFIIB and TFIID are present (compare lanes 5 and 7). These results mean that all three proteins are bound to the DNA. Taken together, the results indicate that TFIID can bind by itself, TFIIB needs TFIID to bind, and RNA polymerase II needs both proteins to bind to the DNA.

Questions for Student Discussion/Collaboration

2. Here are just a few possible research questions:

   Does a particular amino acid within a protein sequence play a critical role in the protein's structure or function?

   Does a DNA sequence function as a promoter?

   Does a DNA sequence function as a regulatory site?

   Does a DNA sequence function as a splicing junction?

   Is a sequence important for correct translation?

   Is a sequence important for RNA stability?

CHAPTER 21

Note: The answers to the Comprehension Questions are at the end of the chapter.

Concept Check Questions (in figure legends)

FIGURE 21.1 A genetic map is a diagram that shows the relative locations of genes or other DNA sequences on a chromosome.

FIGURE 21.2 The probe binds to a specific site because the site has a complementary sequence.

FIGURE 21.5 Microsatellites are polymorphic when the number of repeat sequences varies.

FIGURE 21.9 Yes. The sequence is determined as the DNA is synthesized.

FIGURE 21.12 A key point is that mRNA is made only when a gene is expressed. In this experiment, mRNA is first isolated and then used to make cDNA, which is fluorescently labeled. The fluorescent spots indicate which genes have been transcribed into mRNA.

FIGURE 21.13 The antibody recognizes the protein of interest that is covalently crosslinked to DNA. The antibody is also attached to a heavy bead, which makes it easy to separate it from the rest of the cellular components by centrifugation. This procedure provides an easy way to purify the protein of interest along with its attached DNA.

End-of-Chapter Questions

Conceptual Questions

C2. A. Yes. B. No, this is only one chromosome in the human genome.
     C. Yes. D. Yes.

Application and Experimental Questions

E2. Fluorescence in situ hybridization is a cytological method of mapping genes. A probe that is complementary to a chromosomal sequence is used to locate the gene microscopically within a mixture of many different chromosomes. Thus, this method can be used to cytologically map the location of a gene sequence. When more than one probe is used, the order of genes along a particular chromosome can be determined.

E4. The term *fixing* refers to procedures that chemically freeze cells and prevent degradation. After fixation has occurred, the contents of the cells do not change their morphology. In a sense, they are frozen in place. In a FISH experiment, fixing keeps all of the chromosomes within a cell in the vicinity of each other; they cannot float around on the slide and get mixed up with chromosomes from other cells. A group of chromosomes observed in a FISH experiment comes from a single cell.

It is necessary to denature the chromosomal DNA so that the probe can bind to it. The probe is a segment of DNA that is complementary to the DNA of interest. The strands of chromosomal DNA must be separated (i.e., denatured) so that the probe can bind to complementary sequences.

E6. If the sample was from an unaffected individual, two spots (one on each copy of chromosome 21) would be observed. Three spots would be observed if the sample was from a person with Down syndrome, because that person would have three copies of chromosome 21.

E8. The discrepancy arises because the rate of recombination between homologous chromosomes is different during oogenesis compared to spermatogenesis. Physical mapping measures the actual distance (in bp) between markers. The physical mapping of chromosomes from males and females reveals that they are the same lengths (excluding the Y chromosome). Therefore, the sizes of chromosomes in males and females are the same. The differences observed in linkage maps are due to differences in the rates of recombination during oogenesis versus spermatogenesis.

E10. A sequence-tagged site (STS) is a segment of DNA, usually quite short (e.g., 100 to 400 bp in length), that is a unique site in a genome. STSs are identified using primers in a PCR reaction. STSs serve as molecular markers in genetic mapping studies. Sometimes the region within an STS may contain a microsatellite. A microsatellite is a short DNA segment that is variable in length, usually due to a short repetitive sequence. When a microsatellite is within an STS, the length of the STS varies among different individuals or even in the same individual if the individual is heterozygous for the STS. Such an STS is polymorphic. Polymorphic STSs can be used in linkage analysis, because their transmission can be followed in family pedigrees and through crosses of experimental organisms.

E12. The overall advantage of next-generation sequencing technologies is that a large amount of DNA can be sequenced in a short period of time and at a lower cost. Also, subcloning fragments of DNA into vectors is no longer necessary. Further, some of the methods allow parallel analysis of an enormous number of samples simultaneously.

E14. First, researchers obtain a sample from the environment, which could be a soil sample, a water sample from the ocean, or a fecal sample from a person. After the sample is filtered, the cells within the sample are lysed and the DNA is extracted and purified. During this procedure, the purified DNA is sheared into fragments of different sizes. The DNA fragments are then randomly inserted into cloning vectors and transformed into host cells. The result is a DNA library of thousands or tens of thousands of cells, each carrying a DNA fragment from the metagenome. All of these cells together constitute the metagenomic library. The members of the library are then subjected to shotgun DNA sequencing to identify the genes they may contain. (Note: The cloning step may be skipped if certain types of DNA sequencing methods, such as pyrosequencing, are used.)

E16. Complementary DNA (cDNA) is subjected to a next-generation sequencing method.

Questions for Student Discussion/Collaboration

2. This is a matter of opinion. Many people would say that the ability to identify many human genetic diseases is the most important goal. In addition, we will better understand how humans are constructed at the molecular level. As we gain a greater understanding of our genetic makeup, some people are worried that this may lead to greater discrimination. Insurance or health companies could refuse people who are known to carry genetic abnormalities. Similarly, employers could make their employment decisions based on the genetic makeup of potential employees rather than their past accomplishments. At the family level, genetic information may affect how people choose mates, and whether or not they decide to have children.

CHAPTER 22

Note: The answers to the Comprehension Questions are at the end of the chapter.

Concept Check Questions (in figure legends)

FIGURE 22.2 The key feature is that affected offspring have both parents who are unaffected by the disease. The parents are heterozygous carriers.

FIGURE 22.3 The key feature is that affected offspring have an affected parent.

FIGURE 22.4 The key feature is that all of the affected individuals are males. Furthermore, these males all have mothers who were descendants of Queen Victoria.

FIGURE 22.5 A haplotype is a linkage of alleles or molecular markers on a single chromosome.

FIGURE 22.7 In this example, the original mutation that caused the *huntingtin* allele occurred in a germ-line cell or gamete in which the allele was linked to the G8-C marker.

FIGURE 22.12 Such a mutation would keep the Ras protein in an active state and thereby promote cancerous growth. The cell-signaling pathway would be turned on.

FIGURE 22.13 The *bcr* gene is expressed in white blood cells. This translocation causes the *abl* gene to be under the control of the *bcr* promoter. The abnormal expression of *abl* in white blood cells causes leukemia.

FIGURE 22.14 E2F will be active all of the time, and this will lead to cancer.

FIGURE 22.15 A checkpoint is a point in the cell cycle at which proteins detect if the cell is in the proper condition to divide. If an abnormality such as DNA damage is detected, the checkpoint proteins will halt the cell cycle.

FIGURE 22.18 An individual with a predisposition to develop familial breast cancer inherits only one copy of the mutant allele. Therefore, this disease shows a dominant pattern of inheritance. However, for breast cancer to actually develop, the other allele must become mutant in somatic cells.

End-of-Chapter Questions

Conceptual Questions

C2. When a disease-causing allele affects a trait, it is causing a deviation from normality, but the gene involved is not usually the only gene that governs the trait. For example, an allele causing hemophilia prevents the normal blood-clotting pathway from operating correctly. It follows a simple Mendelian pattern because a single gene affects the phenotype. Even so, it is known that normal blood clotting is due to the actions of many genes.

C4. Changes in chromosome number and unbalanced changes in chromosome structure tend to affect phenotype because they create an imbalance of gene expression. For example, in Down syndrome, there are three copies of chromosome 21 and therefore three copies of all the genes on that chromosome. The result is a relative overexpression of genes that are located on chromosome 21 compared to the other chromosomes. Balanced translocations and inversions often lack phenotypic consequences because the total amount of genetic material is not altered, and the level of gene expression is not significantly changed.

C6. Here are a few possible answers: Dwarfism occurs in people and dogs. Breeds like the dachshund and basset hound are examples of dwarfism in dogs. There are diabetic people and mice. There are forms of inherited obesity in people and mice. Hip dysplasia affects people and dogs.

C8. A. Because a person must inherit two defective copies of this gene and it is known to be on chromosome 1, the mode of transmission is autosomal recessive. Both parents must be heterozygous because they had to have one affected parent (who transmitted the mutant allele to them) and their phenotypes are unaffected (so they must have received the normal allele from their other parent). Because both parents are heterozygotes, there is a 1/4 chance of producing an affected child (a homozygote) with Gaucher disease. If we let G represent the nonmutant allele and g the mutant allele, we obtain the following Punnett square:

	G	g
G	GG Normal	Gg Normal
g	Gg Normal	gg Gaucher's

B. From this Punnett square, we can also see that there is a 1/4 chance of producing a homozygote with both normal copies of the gene.

C. We need to apply the binomial expansion equation (see Chapter 3 for a description of this equation) with $n = 5$, $x = 1$, $p = 0.25$, $q = 0.75$. The answer is 0.396, or 39.6%.

C10. The pattern of inheritance is autosomal recessive. All of the affected individuals do not have affected parents. Also, the disorder is found in both males and females. If the inheritance pattern were X-linked recessive, individual III-1 would have to have an affected father, which she does not.

C12. The 13 babies have acquired a new mutation. In other words, during spermatogenesis or oogenesis, or after the egg was fertilized, a new mutation occurred in the fibroblast growth factor gene. These 13 individuals have the same chances of passing the mutant allele to their offspring as the 18 individuals who inherited the mutant allele from a parent. The chance is 50%.

C14. Because this is a dominant trait, the mother must have two normal copies of the gene, and the father (who is affected) is most likely to be a heterozygote. (Note: the father could be a homozygote, but this is extremely unlikely because the dominant allele is very rare.) If we let M represent the mutant Marfan allele and m the normal allele, we can construct the following Punnett square:

	M	m
m	Mm Marfan	mm Normal
m	Mm Marfan	mm Normal

A. There is a 50% chance that this couple will have an affected child

B. We use the product rule. The odds of having an unaffected child are 50%. So we multiply $0.5 \times 0.5 \times 0.5$, which equals 0.125, or a 12.5% chance of having three unaffected offspring.

C16. 1. The disease-causing allele had its origin in a single individual known as a founder, who lived many generations ago. Since that time, the allele has spread throughout portions of the human population.

2. When the disease-causing allele originated in the founder, it occurred in a region with a particular haplotype. The haplotype is not likely to have changed over the course of several generations if the disease-causing allele and markers in this region are very close together.

C18. A proto-oncogene is a normal cellular gene that typically plays a role in cell division. It can be altered by mutation to become an oncogene and thereby cause cancer. At the level of protein function, a proto-oncogene can become an oncogene by synthesizing too much of a protein or synthesizing the same amount of a protein that is abnormally active.

C20. Conversion of a proto-oncogene to an oncogene can occur by missense mutation, gene amplification, chromosomal translocation, and viral integration. Examples are given in Table 22.6. The genetic changes are expected to increase the amount of the encoded protein or alter its function in a way that makes it more active.

C22. A. No, because E2F is inhibited.
B. Yes, because E2F is not inhibited.
C. Yes, because E2F is not inhibited.
D. No, because there is no E2F.

C24. A. True   B. True
C. False, most cancer cells are caused by mutations that result from environmental mutagens.
D. True

C26. Though they don't change the DNA sequence, epigenetic modifications due to environmental agents can affect gene expression. Such changes could increase gene expression and thereby result in oncogenes, or they could inhibit the expression of tumor suppressor genes. Either type of change could contribute to cancer. For example, DNA methylation of a tumor suppressor gene could promote cancer. Epigenetic changes associated with cancer occur in somatic cells and cannot be passed from parents to offspring.

Application and Experimental Questions

E2. Perhaps the least convincing observation is the higher incidence of a disease in particular populations. Because populations living in specific geographic locations are exposed to their own unique environments, it is difficult to distinguish genetic from environmental causes for a particular disease. The most convincing evidence might be the higher incidence of a disease in related individuals or the ability to correlate a disease with the presence of a mutant gene. Overall, however, the conclusion that a disease has a genetic component should be based on as many observations as possible.

E4. You would probably conclude that the disease does not have a genetic component. If it were rooted primarily in genetics, it would be likely to be found in the Central American population. Of course, there is a chance that very few or none of the people who migrated to Central America were carriers of the mutant gene, but that is somewhat unlikely for a large migrating population. By comparison, you might suspect that an environmental agent that is present in South America but not present in Central America may underlie the disease. Researchers could try to search for this environmental agent (e.g., a pathogenic organism).

E6. Males I-1, II-1, II-4, II-6, III-3, III-8, and IV-5 have a normal copy of the gene. Males II-3, III-2, and IV-4 are hemizygous for an inactive mutant allele. Females III-4, III-6, IV-1, IV-2, and IV-3 have two normal copies of the gene, whereas females I-2, II-2, II-5, III-1, III-5, and III-7 are heterozygous carriers of a mutant allele.

E8. One possible category of drugs would be GDP analogues (i.e., compounds whose structures resemble that of GDP). Perhaps you could find a GDP analogue that binds to the Ras protein and locks it in the inactive conformation.

One way to test the efficacy of such a drug would be to incubate the drug with a type of cancer cell that is known to have an overactive Ras protein, and then plate the cells on solid media. If the drug locked the Ras protein in the inactive conformation, it should inhibit the formation of malignant growth or malignant foci.

There are possible side effects of such drugs. First, they might block the growth of normal cells, because Ras protein plays a role in normal cell proliferation. Second (as you might know from a cell biology course), there are many GTP/GDP-binding proteins in cells, and the drugs could somehow inhibit cell growth and function by interacting with these proteins.

Questions for Student Discussion/Collaboration

2. There isn't a clearly correct answer to this question, but it should stimulate interesting discussion.

CHAPTER 23

Note: The answers to the Comprehension Questions are at the end of the chapter.

Concept Check Questions (in figure legends)

FIGURE 23.1 A local population is a group of individuals that are more likely to interbreed with each other than with individuals in more distant populations.

FIGURE 23.2 Polymorphisms are very common in nearly all natural populations.

FIGURE 23.6 The word *directional* means that selection is favoring a particular phenotype. Selection is moving the population in the direction in which that phenotype will predominate.

FIGURE 23.8 No, because directional selection is eliminating individuals that are sensitive to DDT.

FIGURE 23.9 The Hb^S allele is an advantage in the heterozygous condition because it confers resistance to malaria. The heterozygote advantage outweighs the homozygote disadvantage.

FIGURE 23.10 With negative frequency-dependent selection, the rarer phenotype has a higher fitness, which improves its reproductive success.

FIGURE 23.11 Yes, disruptive selection fosters polymorphism. The fitness values of phenotypes depend on the environment. Some phenotypes are the fittest in one environment, whereas other phenotypes are the fittest in another.

FIGURE 23.13 Stabilizing selection decreases genetic diversity because it eliminates individuals that carry alleles that promote more extreme phenotypes.

FIGURE 23.14 Genetic drift tends to have a greater effect in small populations, where it can lead to the rapid loss or fixation of an allele.

FIGURE 23.15 At the bottleneck, genetic diversity may be lower because the population consists of fewer individuals. Also, during the bottleneck, genetic drift may promote the loss of certain alleles and the fixation of other alleles, thereby diminishing genetic diversity.

FIGURE 23.16 Inbreeding increases the likelihood of homozygosity, and therefore tends to increase the likelihood that an individual will exhibit a recessive trait. Homozygosity becomes more likely because an individual can inherit both copies of the same allele from a common ancestor.

FIGURE 23.18 DNA fingerprinting is used for identification and for relationship testing.

End-of-Chapter Questions

Conceptual Questions

C2. A population is a group of interbreeding individuals. An example is a squirrel population in a forested area. Over the course of many generations, several things could happen to this population. A forest fire, for example, could dramatically decrease the number of individuals and thereby cause a bottleneck. This would decrease the genetic diversity of the population. A new predator may enter the region, and natural selection may favor the survival of squirrels that are best able to evade the predator. Another possibility is that a group of squirrels within the population may migrate to a new region and found a new squirrel population.

C4. A. Phenotype frequency and genotype frequency

B. Genotype frequency

C. Allele frequency

C6. A. The genotype frequency for the *CF* homozygote is 1/2500, or 0.004. This value equals q^2. The allele frequency is the square root of this value, which equals 0.02. The frequency of the corresponding non-disease-causing allele equals $1 - 0.02 = 0.98$.

B. The frequency for the *CF* homozygote is 0.004; for the unaffected homozygote, $(098)^2 = 0.96$; and for the heterozygote, $2(0.98)(0.02)$, which equals 0.039.

C. If a person is known to be a heterozygous carrier, the chances that this particular person will happen to choose another as a mate is equal to the frequency of heterozygous carriers in the population, which is 0.039, or 3.9%. The chances that two randomly chosen individuals will choose each other as mates equals $0.039 \times 0.039 = 0.0015$, or 0.15%.

C8. If we apply the Hardy-Weinberg equation:

$BB = (0.67)^2 = 0.45$, or 45%

$Bb = 2(0.67)(0.33) = 0.44$, or 44%

$bb = (0.33)^2 = 0.11$, or 11%

The actual data show a higher percentage of homozygotes (compare 45% with 50% and 11% with 13%) and a lower percentage of heterozygotes (compare 44% with 37%) than expected. Therefore, these data are consistent with inbreeding, which increases the percentage of homozygotes and decreases the percentage of heterozygotes.

C10. Migration, genetic drift, and natural selection are the main factors that alter allele frequencies within a population. Natural selection acts to eliminate harmful alleles and promote beneficial alleles. Genetic drift involves random changes in allele frequencies that may eventually lead to elimination or fixation of alleles. It is thought to be important in the establishment of neutral alleles in a population. Migration is important because it introduces new alleles into neighboring populations.

C12. Darwinian fitness is the relative likelihood that a genotype will survive and contribute to the gene pool of the next generation as compared to other genotypes. The genotype with the highest reproductive ability is given a value of 1.0. Characteristics that promote survival, ability to attract a mate, or enhanced fertility are expected to increase Darwinian fitness. Examples are the thick fur of a polar bear, which helps it to survive in a cold climate; the bright plumage of male birds, which helps them to attract a mate; and the high number of gametes released by certain species of fish, which enhances their fertility.

C14. In all cases, these patterns of natural selection favor one or more phenotypes because such phenotypes have a reproductive advantage. However, the patterns differ with regard to whether a single phenotype or multiple phenotypes are favored, and whether the phenotype that is favored is in the middle of the phenotypic range or at one or both extremes. Directional selection favors one phenotype at a phenotypic extreme. Over time, natural selection is expected to favor the fixation of alleles that cause these phenotypic characteristics. Disruptive selection favors two or more phenotypic categories. It will lead to a population with a balanced polymorphism for the trait. Balancing selection may occur because of heterozygote advantage or negative frequency-dependent selection. These promote a stable polymorphism in a population. Stabilizing selection favors individuals with intermediate phenotypes. It tends to decrease genetic diversity because alleles that favor extreme phenotypes are eliminated.

C16. In genetic drift, allele frequencies are drifting. Genetic drift is an appropriate term because the word *drift* implies a random process. Nevertheless, drift can be directional. A boat may drift from one side of a lake to another. It would not drift in a straight path, but the drifting process will alter its location. Similarly, allele frequencies can drift up and down and eventually lead to the elimination or fixation of particular alleles within a population.

C18. A. Probability of fixation = $1/2N = 1/2(4) = 1/8$, or 0.125

B. $\bar{t} = 4N = 4(4) = 16$

C. The preceding calculations assume a constant population size. If the population grows after it has been founded by the four individuals, the probability of fixation will be lower and the time to reach fixation will be longer.

C20. A. True B. True

C. False; it causes allele loss or fixation, which results in less diversity.

D. True

C22. A. Migration will increase the genetic diversity in both populations. A random mutation could occur in one population to create a new allele. This new allele could be introduced into the other population via migration.

B. The allele frequencies between the two populations will tend to be similar to each other, due to the intermixing of their alleles.

C. Genetic drift depends on population size. When two populations intermix, this has the effect of increasing the overall population size. In a sense, the two smaller populations behave like one big population. Therefore, the effects of genetic drift are lessened when the individuals in two populations can migrate. The net effect is that allele loss and allele fixation are less likely to occur due to genetic drift.

C24. A. Yes

B. The common ancestors are I-1 and I-2.

C.

$F = \Sigma(1/2)^n (1 + F_A)$

$F = (1/2)^9 + (1/2)^9$

$F = 1/512 + 1/512 = 2/512 = 0.0039$

D. Based on the data shown in the pedigree, the parents are not inbred.

C26. A. The inbreeding coefficient is calculated using the formula

$F = \Sigma(1/2)^n (1 + F_A)$

In this case, there are two common ancestors, I-1 and I-2. Because we have no prior history on I-1 or I-2, we assume they are not inbred, which makes $F_A = 0$. The two inbreeding loops for IV-2 contain five people: III-4, II-2, I-1, II-6, III-5; and III-4, II-2, I-2, II-6, III-5. Therefore, $n = 5$ for both loops.

$F = (1/2)^5 (1 + 0) + (1/2)^5 (1 + 0)$

$F = 0.031 + 0.031 = 0.062$

B. Based on the data shown in the pedigree, individual III-4 is not inbred.

C28. We can use the following equation to calculate the change in allele frequency after any number of generations:

$(1 - u)^t = \dfrac{p_t}{p_o}$

$(1 - 10^{-4})^t = 0.5/0.6 = 0.833$

$(0.9999)^t = 0.833$

$t = 1,827$ generations

Application and Experimental Questions

E2. In this case:

$(p + q + r + s)^2 = 1$

$p^2 + q^2 + r^2 + s^2 + 2pq + 2qr + 2qs + 2rp + 2rs + 2sp = 1$

Let $p = C$, $q = c^{ch}$, $r = c^h$, and $s = c$.

A. The frequency of albino rabbits is s^2:

$s^2 = (0.05)^2 = 0.0025$, or 0.25%

B. Himalayan is dominant to albino but recessive to full and chinchilla. Therefore, Himalayan rabbits would be represented by r^2 and by $2rs$:

$r^2 + 2rs = (0.44)^2 + 2(0.44)(0.05) = 0.24$, or 24%

Among 1000 rabbits, about 240 would have a Himalayan coat color.

C. Chinchilla is dominant to Himalayan and albino but recessive to full coat color. Therefore, heterozygotes with chinchilla coat color would be represented by $2qr$ and by $2qs$.

$2qr + 2qs = 2(0.17)(0.44) + 2(0.17)(0.05) = 0.17$, or 17%

Among 1000 rabbits, about 170 would be heterozygotes with chinchilla fur.

E4. A.

Inuit	$M = 0.913$	$N = 0.087$
Navajo	$M = 0.917$	$N = 0.083$
Finns	$M = 0.673$	$N = 0.327$
Russians	$M = 0.619$	$N = 0.381$
Aborigines	$M = 0.176$	$N = 0.824$

B. To determine if these populations are in equilibrium, we can use the Hardy-Weinberg equation and calculate the expected number of individuals with each genotype. For example:

Inuit $MM = (0.913)^2 = 83.3$

$MN = 2(0.913)(0.087) = 15.9$

$NN = (0.087)^2 = 0.76$

In general, the values agree pretty well with an equilibrium. The same is true for the other four populations.

C. Based on similar allele frequencies, the Inuit and Navajo Indians seem to have interbred as well as the Finns and Russians.

E6. A. $\Delta p_C = m(p_D - p_R)$

With regard to the sickle cell allele:

$\Delta p_C = (550/10{,}550)(0.1 - 0.01) = 0.0047$

$p_C = p_R + \Delta p_C = 0.01 + 0.0047 = 0.0147$

B. We need to calculate the genotypes separately:

For the 550 migrating individuals,

$Hb^A Hb^A = (0.9)^2 = 0.81$ We expect $(0.81)550 = 445.5$.

$Hb^A Hb^S = 2(0.9)(0.1) = 0.18$ We expect $(0.18)550 = 99$.

$Hb^S Hb^S = (0.1)^2 = 0.01$ We expect $(0.01)550 = 5.5$.

For the original recipient population,

$Hb^A Hb^A = (0.99)^2 = 0.98$ We expect 9801 individuals to have this genotype.

$Hb^A Hb^S = 2(0.99)(0.01) = 0.0198$ We expect 198 with this genotype.

$Hb^S Hb^S = (0.01)^2 = 0.0001$ We expect 1 with this genotype.

For the conglomerate population,

$(445.5 + 9801)/10{,}550 = 0.971$ $Hb^A Hb^A$ homozygotes

$(99 + 198)/10{,}550 = 0.028$ heterozygotes

$(5.5 + 1)/10{,}550 = 0.00062$ $Hb^S Hb^S$ homozygotes

C. In the next generation, the allele frequencies in the conglomerate population (calculated in part A), should yield the expected genotype frequencies according to Hardy-Weinberg equilibrium.

Allele frequency of $Hb^S = 0.0147$, so $Hb^A = 0.985$

$Hb^A Hb^A = (0.985)^2 = 0.97$

$Hb^A Hb^S = 2(0.985)(0.0147) = 0.029$

$Hb^S Hb^S = (0.0147)^2 = 0.0002$

E8. A. Probability of fixation = $1/2N$ (assuming equal numbers of males and females contributing to the next generation)

Probability of fixation = $1/2(2{,}000{,}000)$

= 1 in 4,000,000 chance

B. $\bar{t} = 4N$

where $\bar{t}$ = the average number of generations to achieve fixation

N = the number of individuals in a population, assuming that males and females contribute equally to each succeeding generation

$\bar{t} = 4(2 \text{ million}) = 8$ million generations

C. If the blue allele had a slight survival advantage, the value calculated in part A would be slightly larger; there would be a higher chance of allele fixation. The value calculated in part B would be smaller; it would take a shorter period of time to reach fixation.

E10. You could mark snail shells with a dye and release equal numbers of dark-shelled and light-shelled snails into dimly lit forested regions and sunny fields. At a later time, recapture the snails and count them. It would be important to have a method of unbiased recapture because it would be easier to find the light-shelled snails in a forest and the dark-shelled snails in a field. Perhaps you could bait the region with something that the snails like to eat and only collect snails that are at the bait. In addition to this type of experiment, you could also observe predation as it occurs.

E12. DNA fingerprinting is a method of identification based on the properties of DNA. Minisatellites and microsatellite sequences vary in size in natural populations. This variation can be seen when DNA fragments are subjected to gel electrophoresis. Within a population, any two individuals (except for identical twins) will display a different pattern of DNA fragments; that pattern is called their DNA fingerprint.

E14. The minimum percentage of matching bands is based on the genetic relationships.

A. 50%

B. 50% (on average, but it could be less or more)

C. 25% (on average, but it could be less or more)

D. 25% (on average, but it could be less or more)

Questions for Student Discussion/Collaboration

2. Mutation is responsible for creating new alleles, but the rate of new mutations is so low that it cannot explain allele frequencies in this range. Let's call the two alleles B and b and assume that B was the original allele and b is a more recent allele that arose as a result of mutation. Three scenarios to explain the allele frequencies are as follows:

   1. The b allele is neutral and reached its present frequency by genetic drift. It hasn't reached elimination or fixation yet.
   2. The b allele is beneficial and its frequency is increasing due to natural selection. However, there hasn't been enough time to reach fixation.
   3. The Bb heterozygote is at a selective advantage leading to a balanced polymorphism.

CHAPTER 24

Note: The answers to the Comprehension Questions are at the end of the chapter.

Concept Check Questions (in figure legends)

FIGURE 24.1 In most populations (like this one), height exhibits a continuum.

FIGURE 24.2 About 95.4% are within two standard deviations, which means that 4.6% are outside this range. Half of them, or 2.3%, fall more than two standard deviations above the mean.

FIGURE 24.3 When alleles are additive, this means they contribute in an incremental way to the outcome of a trait. Having three heavy alleles will make an individual heavier than having two heavy alleles.

FIGURE 24.4 Increases in gene number and in environmental variation tend to cause greater overlaps between different genotypes and the same phenotype.

FIGURE 24.5 These two strains differ with regard to a quantitative trait and in their molecular markers.

FIGURE 24.9 Both natural selection and selective breeding affect allele frequencies due to differences in reproductive success. In the case of natural selection, reproductive success is determined by environmental conditions. For selective breeding, reproductive success is determined by the people who choose the parents for breeding.

FIGURE 24.10 The traits involve variation in the stem, leaves, flower buds, lateral leaf buds, and terminal bud.

FIGURE 24.11 A selection limit may be reached because a population has become monomorphic for all of the desirable alleles. A second reason is because the desired effects of artificial selection are balanced by the negative effects on fitness.

End-of-Chapter Questions

Conceptual Questions

C2. At the molecular level, quantitative traits often exhibit a continuum of phenotypic variation because they are usually influenced by multiple genes that exist as multiple alleles. A large amount of environmental variation will also increase the overlap between genotypes and phenotypes for polygenic traits.

C4. A discontinuous trait is one that falls into discrete categories. Examples include brown eyes versus blue eyes in humans and purple versus white flowers in pea plants. A continuous trait is one that does not fall into discrete categories. Examples include height in humans and fruit weight in tomatoes. Most quantitative traits are continuous; the trait exhibits a range of values. The reason why quantitative traits are continuous is because they are usually polygenic and greatly influenced by the environment. As shown in Figure 24.4b, this tends to create ambiguities between genotypes and a continuum of phenotypes.

C6. The top 2.5% is about two standard deviation units above the mean. If we take the square root of the variance, the standard deviation is 22 lb. To be in the top 2.5%, an animal has to be at least 44 lb heavier than the mean, that is, have a weight of 606 lb. To be in the bottom 0.13%, an animal has to be three standard deviations lighter, or at least 66 lb lighter than the mean and thus weigh 496 lb.

C8. There is a positive correlation, but it could have occurred as a matter of chance alone. According to Table 24.2, this value could be due to random sampling error. You would need more data to determine if there is a significant correlation. If $N = 500$, the correlation is statistically significant, and you would conclude that the correlation did not occur as a matter of random chance. However, you could not claim cause and effect.

C10. When a correlation coefficient is statistically significant, it means that the association is likely to have occurred for reasons other than random sampling error. Statistical significance may indicate cause and effect but not necessarily. For example, large parents may have large offspring due to genetics (cause and effect). However, the correlation may be related to the sharing of similar environments rather than cause and effect.

C12. A quantitative trait locus is a site within a chromosome that contains genes that affect a quantitative trait. It is possible for a QTL to contain one gene, or it may contain two or more closely linked genes. QTL mapping, which involves linkage to known molecular markers, is commonly used to determine the locations of QTLs.

C14. If the broad-sense heritability equals 1.0, it means that all of the variation in the population is due to genetic variation rather than environmental variation. It does not mean that the environment is unimportant in the outcome of the trait. Under another set of environmental conditions, the trait may have turned out quite differently.

C16. When a species is subjected to selective breeding, the breeder is focusing on improving one particular trait. In this case, rose breeders focused on the size and quality of the flowers. Because a breeder usually selects a small number of individuals (e.g., the ones with best flowers) as the breeding stock for the next generation, this may lead to a decrease in the allelic diversity of other genes. For example, several genes affect flower fragrance. In an unselected population, these genes may exist as "fragrant" alleles and "nonfragrant" alleles. After many generations of breeding for large flowers, the fragrant alleles may be lost from the population, just as a matter of random chance. This is a common problem of selective breeding: Selecting for an improvement in one trait may inadvertently diminish the quality of an unselected trait.

Others have suggested that the lack of fragrance may be related to flower structure and function. Perhaps the amount of energy that a flower uses to make beautiful petals somehow diminishes its capacity to make fragrance.

C18. Broad-sense heritability takes into account all genetic factors that affect the phenotypic variation in a trait. Narrow-sense heritability considers only alleles that behave in an additive fashion. In many cases, the alleles affecting quantitative traits appear to behave additively. More importantly, if a breeder assumes that the heritability of a trait is due to the additive effects of alleles, it is possible to predict the outcome of selective breeding, which is also termed the realized heritability.

C20. A. Because of their good nutrition, individuals in the commune would likely be taller.

B. If the environment is rather homogeneous, then heritability values tend to be higher because the environment contributes less to the amount of variation in the trait. Therefore, in the commune, the heritability might be higher, because the members uniformly practice good nutrition. On the other hand, because the commune is a smaller size than the general population, the amount of genetic variation might be less, so this would make the heritability lower. However, because the problem states that the commune population is large, we would probably assume that the amount of genetic variation is similar to that in the general population. Overall, the best guess would be that the heritability in the commune population is higher because of the uniform nutrition standards.

C. As stated in part B, the amount of variation would probably be similar, because the commune population is large. As a general answer, larger populations tend to have more genetic variation. Therefore, the general population probably has a bit more variation.

C22. A natural population of animals is more likely to have a higher genetic diversity than a domesticated population. This is because domesticated populations have been subjected to many generations of selective breeding, which decreases genetic diversity. Therefore, V_G is likely to be higher for the natural population. The other issue is the environment. It is difficult to say which group would have a more homogeneous environment. In general, natural populations tend to have a more heterogeneous environment, but not always. If the environment is more heterogeneous, this tends to cause more phenotypic variation, which makes V_E higher.

$$\text{Heritability} = V_G / V_T$$
$$= V_G / (V_G + V_E)$$

When V_G is high, this increases heritability. When V_E is high, this decreases heritability. In the natural wolf population, we would expect that V_G would be high. In addition, we would guess that V_E might be high as well (but that is less certain). Nevertheless, if this were the case, the heritability for the wolf population might be similar to that for the domestic population, because the high V_G in the wolf population is balanced by its high V_E. On the other hand, if V_E is not that high in the wolf population, or if it is fairly high in the domestic population, then the wolf population would have a higher heritability for this trait.

Application and Experimental Questions

E2. To calculate the mean, we add the values together and divide by the total number:

$$\text{Mean} = \frac{1.9 + 2(2.4) + 2(2.1) + 3(2.0) + 2(2.2) + 1.7 + 1.8 + 2(2.3) + 1.6}{15}$$
$$= 2.1$$

The variance is the sum of the squared deviations from the mean divided by $N - 1$. The mean value of 2.1 must be subtracted from each value, and then the square is taken. These 15 values are added together and then divided by 14 (which is $N - 1$):

$$\text{Variance} = \frac{0.85}{14}$$
$$= 0.061$$

The standard deviation is the square root of the variance:

$$\text{Standard deviation} = 0.25$$

E4. These results suggest there are (at least) three different genes that influence the size of pigs. This is a minimum estimate because a QTL may have two or more closely linked genes. Also, it is possible that the large and small strains have the same RFLP band that is associated with one or more of the genes that affect size.

E6. A. If we assume that the highly inbred strain has zero genetic variance:

V_G (for the wild strain) $= 3.2\text{ g}^2 - 2.2\text{ g}^2 = 1.0\text{ g}^2$

B. $h_B^2 = 1.0\text{ g}^2/3.2\text{ g}^2 = 0.31$

C. It is the same as h_B^2, so it also equals 0.31.

E8. A.
$$h_N^2 = \frac{\overline{X}_O - \overline{X}}{\overline{X}_P - \overline{X}}$$
$0.21 = (\overline{X}_O - 25\text{ g})/(27\text{ g} - 25\text{ g})$
$\overline{X}_O - 25\text{ g} = 2\text{ g }(0.21)$
$\overline{X}_O = 25.42\text{ g}$

B.
$0.21 = (26.5\text{ g} - 25\text{ g})/(\overline{X}_P - 25\text{ g})$
$(\overline{X}_P - 25\text{ g})(0.21) = 1.5\text{ g}$
$\overline{X}_P = 32.14\text{ g}$ (parental mean weight)

However, because this value is so far from the mean, there may not be parents of that weight in the population of mice that you have available.

E10. A. After six or seven generations, the selective breeding seems to have reached a plateau. This suggests that the tomato plants have become monomorphic for the alleles that affect tomato weight.

B. Because Martin's first generation has a mean tomato weight of 1.7 lb, which is heavier than either Mary's or Hector's tomatoes, these results suggest that heterozygosity may increase tomato weight. This partially explains why Martin has obtained tomatoes that are heavier than 1.5 lb. However, this is not the whole story; it does not explain why Martin obtained tomatoes that weigh 2 lb. Even though Mary's and Hector's tomatoes were selected for heavier weight, they may not have all of the heavy alleles for each gene that controls weight. For example, let's suppose there are 20 genes that affect weight, with each gene existing in a light and heavy allele. During the early stages of selective breeding, when Mary and Hector picked their 10 plants as seed producers for the next generation, as a matter of random chance, some of these plants may have been homozygous for the light alleles at a few of the 20 genes that control weight. Therefore, just as a matter of chance, they probably lost a few of the heavy alleles that affect weight. So, after 12 generations of breeding, they have predominantly heavy alleles but also have light alleles for some of the genes. If we represent each heavy allele with a capital letter and each light allele with a lowercase letter, Mary's and Hector's strains could have the following genotypes:

Mary's strain: *AA BB cc DD EE FF gg hh II JJ KK LL mm NN OO PP QQ RR ss TT*

Hector's strain: *AA bb CC DD EE ff GG HH II jj kk LL MM NN oo PP QQ RR SS TT*

As we see here, Mary's strain is homozygous for the heavy allele at 15 of the genes but carries the light allele at the other five. Similarly, Hector's strain is homozygous for the heavy allele at 15 genes and carries the light allele at the other five. It is important to note, however, that the light alleles in Mary's and Hector's strains are not in the same genes. Therefore, when Martin crosses them together, he will initially get

Martin's F_1 offspring: *AA Bb Cc DD EE Ff Gg Hh II Jj Kk LL Mm NN Oo PP QQ RR Ss TT*

If the alleles are additive and contribute equally to the trait, we would expect about the same mean weight (1.5 lb), because this hybrid has a total of 10 light alleles. However, genes (which were homozygous recessive in Mary's and Hector's strains) are heterozygous in the F_1 offspring, and this may make the plants healthier and contribute to a higher weight. If Martin's F_1 strain is subjected to selective breeding, the 10 genes that are heterozygous in the F_1 offspring may eventually become homozygous for the heavy allele. This would explain why Martin's tomatoes achieved a weight of 2.0 pounds after five generations of selective breeding.

E12. A. $h_N^2 = \dfrac{\overline{X}_O - \overline{X}}{\overline{X}_P - \overline{X}}$

$h_N^2 = \dfrac{269 - 254}{281 - 254} = 0.56$

B. $0.56 = \dfrac{275 - 254}{\overline{X}_P - 254}$

$\overline{X}_P = 291.5$ lb

Questions for Student Discussion/Collaboration

2. Most traits depend on the influence of many genes. Also, genetic variation is a common phenomenon in most populations. Therefore, most individuals have a variety of alleles that contribute to a given trait. For quantitative traits, some alleles may alter the sizes of body parts, such as the size of fruit. If a population contains many different genes and alleles that govern a quantitative trait, most individuals will have an intermediate phenotype because they will have inherited some alleles that confer larger size and some that confer smaller size. Fewer individuals will inherit a predominance of alleles for larger size or a predominance of alleles for smaller size. An example of a quantitative trait that does not fit a normal distribution is snail shell pigmentation. The dark snails and light snails are favored rather than the intermediate colors because they are less susceptible to predation.

GLOSSARY

A

A an abbreviation for adenine.

acentric fragment a fragment of a chromosome that lacks a centromere.

acquired antibiotic resistance the acquisition of antibiotic resistance by a bacterium that has taken up a gene or plasmid from another bacterium that has such resistance.

acridine dye a type of chemical mutagen that causes frameshift mutations.

acrocentric describes a chromosome with the centromere significantly off center, but not at the very end.

activator a transcriptional regulatory protein that increases the rate of transcription.

acutely transforming virus (ACT) a virus that readily transforms normal cells into malignant cells, when grown in a laboratory.

adaptor hypothesis a hypothesis that proposes that a tRNA has two functions: recognizing a three-base codon sequence in mRNA and carrying an amino acid that is specific for that codon.

adenine a purine base found in DNA and RNA. It base-pairs with thymine in DNA.

age of onset the time of life at which symptoms of a genetic disease appear.

alkaptonuria a human genetic disorder involving the accumulation of homogentisic acid due to a defect in homogentisic acid oxidase.

allele an alternative form of a specific gene.

allele frequency the number of copies of a particular allele in a population divided by the total number of all alleles for that gene in the population.

allelic variation genetic variation in a population that involves the occurrence of two or more different alleles for a particular gene.

allodiploid describes an organism that contains one set of chromosomes from two different species.

alloploid describes an organism that contains sets of chromosomes from two or more different species.

allopoidy the phenomenon in which a cell or organism contains sets of chromosomes from two or more different species.

allopolyploid describes an organism that contains two (or more) sets of chromosomes from two (or more) species.

allosteric enzyme an enzyme that contains two binding sites: a catalytic site and a regulatory site.

allosteric regulation the phenomenon in which an effector molecule binds to a noncatalytic site on a protein and causes a conformational change that regulates the protein's function.

allosteric site the site on a protein where a small effector molecule binds to regulate the function of the protein.

allotetraploid describes an organism that contains two sets of chromosomes from two different species.

α helix a type of secondary structure found in proteins.

alternative exon an exon that is not always found in mRNA. It is only found in certain types of alternatively spliced mRNAs.

alternative splicing the phenomenon in which a pre-mRNA can be spliced in more than one way.

Ames test a test using strains of a bacterium, *Salmonella typhimurium*, to determine if a substance is a mutagen.

amino acid a building block of polypeptides and proteins. It contains an amino group, a carboxyl group, and a side chain.

aminoacyl site (A site) a site on the ribosome where a charged tRNA initially binds.

aminoacyl-tRNA a tRNA molecule that has an amino acid covalently attached to its 3′ end.

aminoacyl-tRNA synthetase an enzyme that catalyzes the attachment of a specific amino acid to the correct tRNA.

2-aminopurine a base analog that acts as a chemical mutagen.

amino-terminal end the location of the first amino acid in a polypeptide.

amniocentesis a method of obtaining cellular material from a fetus for the purpose of genetic testing.

AMO see *anti-miRNA oligonucleotide*.

anaphase the fourth stage of M phase. As anaphase proceeds, half of the chromosomes move to one pole, and the other half move to the other pole.

aneuploid not euploid; a variation in chromosome number such that the total number of chromosomes is not an exact multiple of a set or *n* number. Aneuploidy is the condition of being aneuploid.

antagomer a type of anti-miRNA oligonucleotide having one or more base modifications that promote stronger binding to an miRNA.

anther the structure in flowering plants that gives rise to pollen grains.

antibiotic a chemical that inhibits the growth of microorganisms.

antibodies proteins produced by the B cells of the immune system that recognize foreign substances (namely, viruses, bacteria, and so forth) and target them for destruction.

anticipation the phenomenon in which the severity of an inherited disease tends to get worse in successive generations.

anticodon a three-nucleotide sequence in tRNA that is complementary to a codon in mRNA.

antigens foreign substances that are recognized by antibodies.

anti-miRNA oligonucleotide (AMO) an RNA that is complementary to an miRNA and inhibits its function.

antiparallel the opposite orientation of the two strands of a DNA molecule with regard to their 3′ and 5′ ends.

antisense RNA an RNA strand that is complementary to a strand of mRNA.

apoptosis programmed cell death.

apurinic site a site in DNA that is missing a purine base.

ARS elements DNA sequences found in yeast that function as origins of replication.

artificial selection see *selective breeding*.

artificial transformation transformation of bacteria that occurs via experimental treatments.

asexual reproduction a form of reproduction that does not involve the union of gametes; at the cellular level, a preexisting cell divides to produce two new cells.

A site see *aminoacyl site*.

association a relationship between variables in which changes in them follow a pattern.

assortative mating breeding in which individuals preferentially mate with each other based on their phenotypes.

AT/GC rule in DNA, the observation that an adenine base in one strand always hydrogen bonds with a thymine base in the opposite strand, and a guanine base always hydrogen bonds with a cytosine.

ATP-dependent chromatin remodeling see *chromatin remodeling*.

AT-rich region a region at a bacterial origin of replication that has a high percentage of A and T base pairs and easily separates so that replication forks can form.

attenuation a mechanism of genetic regulation, as with the *trp* operon, in which a short RNA is made but its synthesis is terminated before RNA polymerase can transcribe the rest of the operon.

attenuator sequence a sequence found in certain operons (e.g., *trp* operon) in bacteria that stops transcription soon after it has begun.

automated DNA sequencing the use of fluorescently labeled dideoxyribonucleotides and a fluorescence detector to sequence DNA.

autonomous element a transposable element that contains all of the information necessary for transposition or retroposition to take place.

autopolyploid a polyploid produced within a single species as a result of nondisjunction.

autoradiography a technique that involves the use of X-ray film to detect the location of radioisotopes in macromolecules or cells. It is used to detect a particular band on a gel or to map the location of a gene within an intact chromosome.

autosomes chromosomes that are not sex chromosomes.

auxotroph a bacterial strain that cannot synthesize a particular nutrient and needs that nutrient supplemented in its growth medium.

B

backbone the portion of a DNA or RNA strand that is composed of covalently linked phosphates and sugar molecules.

bacteriophages (or **phages**) viruses that infect bacteria.

balanced translocation a translocation, such as a reciprocal translocation, for which the total amount of genetic material is normal or nearly normal.

balancing selection a pattern of natural selection that favors the maintenance of two or more alleles and may arise as a result of heterozygote advantage or negative frequency-dependent selection.

Barr body a structure in the interphase nuclei of somatic cells of female mammals that is a highly condensed X chromosome.

basal transcription in eukaryotes, a low level of transcription via the core promoter. The binding of transcription factors to enhancer elements may increase transcription above the basal level.

base a nitrogen-containing molecule that is a portion of a nucleotide in DNA or RNA. Examples of bases are adenine, thymine, guanine, cytosine, and uracil.

base excision repair (BER) a type of DNA repair in which a modified base is removed from a DNA strand. Following base removal, a short region of the DNA strand is removed, which is then resynthesized using the complementary strand as a template.

base pair the structure consisting of two nucleotides in opposite strands of DNA linked by hydrogen bonds. For example, an AT base pair is a structure in which an adenine-containing nucleotide in one DNA strand hydrogen bonds with a thymine-containing nucleotide in the complementary strand.

base pair mismatch the situation in which two bases opposite each other in a double helix do not conform to the AT/GC rule. For example, if A is opposite C, that is a base mismatch.

base stacking the arrangement in DNA in which the bases are oriented so their flattened regions face each other; this stabilizes the double helix by excluding water.

base substitution a point mutation in which one base is substituted for another.

B DNA the predominant form of DNA in living cells. It is a right-handed DNA helix with 10 bp per turn.

behavioral trait a trait that involves behavior. An example is the ability to learn a maze.

beneficial mutation a mutation that has a beneficial effect on phenotype.

benign describes a noncancerous tumor that is not invasive and cannot metastasize.

β sheet a type of secondary structure found in proteins.

bidirectional in gene regulation, refers to the ability of regulatory elements to function in either direction.

bidirectionally in DNA replication, describes the movement of the two replication forks in opposite directions outward from the origin of replication.

bidirectional replication the phenomenon in which two DNA replication forks move in opposite directions from an origin of replication.

binary fission the physical process whereby a bacterial cell divides into two daughter cells. During this event, the two daughter cells become divided by the formation of a septum.

binomial expansion equation an equation used to solve genetic problems involving two types of unordered events.

biological evolution the accumulation of genetic changes in a species or population from one generation to the next.

biometric field a field of genetics that involves the statistical study of biological traits.

bivalent a structure in which two pairs of homologous sister chromatids have synapsed (i.e., aligned) with each other.

blending hypothesis of inheritance an early, incorrect hypothesis of heredity. According to this view, the seeds that dictate hereditary traits are able to blend together from generation to generation. The blended traits would then be passed to the next generation.

bottleneck effect a potential cause of genetic drift that occurs when most members of a population are eliminated without any regard to their genetic composition.

bp see *base pair*.

branch migration the lateral movement of a Holliday junction.

breakpoint the region where two chromosome pieces break and rejoin with other chromosome pieces.

broad-sense heritability heritability that takes into account all genetic factors.

5-bromouracil (5BU) a base analog that acts as a chemical mutagen.

C

C an abbreviation for cytosine.

cAMP see *cyclic-AMP*.

cancer a disease characterized by uncontrolled cell division.

CAP an abbreviation for catabolite activator protein, a genetic regulatory protein found in bacteria.

capping the covalent attachment of a 7-methylguanosine to the 5′ end of mRNA in eukaryotes.

capsid the protein coat of a virus.

CAP site the sequence of DNA that is recognized by CAP.

capsomer a protein subunit of a viral capsid.

carbohydrates organic molecules with the general formula $C_n(H_2O)_n$. An example of a simple carbohydrate is the sugar glucose. Large carbohydrates are composed of multiple sugar units.

carboxyl-terminal end the location of the last amino acid in a polypeptide.

carcinogen an agent that can cause cancer.

caspases proteolytic enzymes that play a role in apoptosis.

catabolite activator protein see *CAP*.

catabolite repression the phenomenon in which a catabolite (e.g., glucose) represses the expression of certain genes (e.g., the *lac* operon).

cDNA see *complementary DNA*.

cDNA library a DNA library consisting of a collection of cDNAs.

cell culture the growth of cells in a laboratory.

cell cycle in eukaryotic cells, a series of stages through which a cell advances in order to divide. The phases are G for gap, S for synthesis (of the genetic material), and M for mitosis. There are two G phases, G_1 and G_2.

cell-free translation system an experimental mixture that can synthesize polypeptides.

cell plate the structure that forms between two daughter plant cells and leads to the separation of the cells by formation of an intervening cell wall.

cellular level with regard to gene expression, the level of observation at which genes affect the traits of cells.

centiMorgan (cM) (same as a map unit) a unit of map distance obtained from genetic crosses. Named in honor of Thomas Hunt Morgan.

central dogma of genetics the idea that the usual flow of genetic information is from DNA to RNA to polypeptide (protein). In addition, DNA replication serves to copy the information so that it can be transmitted from cell to cell and from parent to offspring.

centrifugation a method used to separate cell organelles and macromolecules in which samples are placed in tubes and spun very rapidly. The rate at which particles move toward the bottom of the tube depends on their densities, sizes, shapes, and the viscosity of the medium.

centrifuge a machine that contains a motor, which causes a rotor holding centrifuge tubes to spin very rapidly.

centrioles in animal cells, a pair of cylindrically shaped structures found at the centrosome.

centromere a region of a eukaryotic chromosome that provides an attachment site for the kinetochore.

centrosome a cellular structure from which microtubules emanate.

chain termination the stoppage of growth of a DNA strand, RNA strand, or polypeptide.

chaperone a protein that aids in the folding of polypeptides.

character in genetics, a general characteristic of an organism.

Chargaff's rule the observation that in DNA, the amount of A equals that of T, and the amount of G equals that of C.

charged tRNA a tRNA that has an amino acid covalently attached to its 3′ end.

checkpoint protein a protein that monitors the conditions of DNA and chromosomes and may prevent a cell from advancing through the cell cycle if an abnormality is detected.

chiasma (pl. **chiasmata**) the site where crossing over occurs between two chromosomes. It resembles the Greek letter chi, χ.

ChIP-chip assay a form of chromatin immunoprecipitation that utilizes a microarray to determine where in the genome a particular protein binds.

chi square (χ^2) test a commonly used statistical method for determining the goodness of fit. This method can be used to analyze population data in which the members of the population fall into different categories.

chloroplast DNA (cpDNA) the genetic material found within a chloroplast.

chorionic villus sampling a method for obtaining cellular material from a fetus for the purpose of genetic testing.

chromatid following chromosomal replication in eukaryotes, the two copies that remain attached to each other as sister chromatids.

chromatin the complex between DNA and proteins found within chromosomes.

chromatin immunoprecipitation (ChIP) a method for determining whether proteins bind to a particular region of DNA. This method analyzes DNA-protein interactions as they occur in the chromatin of living cells.

chromatin remodeling a change in chromatin structure that alters the composition of histones or the spacing of nucleosomes (or both).

chromatography a method of separating different macromolecules and small molecules based on their chemical and physical properties. A sample is dissolved in a liquid solvent and exposed to some type of matrix, such as a gel, a column containing beads, or a thin strip of paper.

chromocenter the central point where chromosomes making up a polytene chromosome aggregate.

chromosome the structures within living cells that contain the genetic material. Genes are physically located within the structure of chromosomes. Biochemically, chromosomes contain a very long segment of DNA, which is the genetic material, and proteins, which are bound to the DNA and give it structure.

chromosome map see *genetic map*.

chromosome painting the use of probes to identify particular regions of chromosomes. The probes are usually assigned a computer-generated color.

chromosome territory a distinct region in the cell nucleus occupied by a particular chromosome.

chromosome theory of inheritance the theory of Sutton and Boveri that the inheritance patterns of traits can be explained by the transmission patterns of chromosomes during gametogenesis and fertilization.

***cis*-acting element** a sequence of DNA, such as a regulatory element, that exerts a *cis*-effect.

***cis*-effect** an effect on gene expression due to genetic sequences that are within the same chromosome and often are immediately adjacent to the gene of interest.

cleavage furrow a constriction that precedes the division of two animal cells during cytokinesis.

clonal refers to something related to cloning or a clone. For example, a clonal population of cells is a group of cells that are derived from the same cell.

clone or **cloning** see *gene cloning*.

closed complex the complex between transcription factors, RNA polymerase, and a promoter before the DNA has denatured to form an open complex.

closed conformation a tightly packed conformation of chromatin that cannot be transcribed.

cM an abbreviation for centiMorgan; also see *map unit*.

coactivator a protein that does not bind directly to the DNA, but plays a role in the activation of transcription.

coding strand the strand of DNA within a protein-encoding gene that has the same sequence as the mRNA except that T replaces U in the DNA.

codominance a pattern of inheritance in which two alleles are both expressed in the heterozygous condition. For example, a person with the genotype $I^A I^B$ has the blood type AB and expresses both surface antigens A and B.

codon a sequence of three nucleotides in mRNA that functions in translation. A start codon, which usually specifies methionine, initiates translation, and a stop codon terminates translation. The other codons specify the amino acids within a polypeptide sequence according to the genetic code.

coefficient of inbreeding (F) see *inbreeding coefficient*.

colinearity the correspondence between the sequence of codons in the DNA coding strand and the amino acid sequence of a polypeptide.

combinatorial control the phenomenon common in eukaryotes in which the combination of many factors determines the expression of any given gene.

comparative genomics the field of study that uses information from genome projects to understand the genetic variation between different populations and evolutionary relationships among different species.

competence factors proteins that are needed for bacterial cells to become naturally transformed by extracellular DNA.

competent cells cells that can be transformed by extracellular DNA.

complementary describes base sequences in two DNA strands that match each other according to the AT/GC rule. For example, if one strand has the sequence ATGGCG-GATT, then the complementary strand must be TACCGCCTAAA.

complementary DNA (cDNA) DNA that is made from an RNA template by the action of reverse transcriptase.

complementation the phenomenon in which the presence of two different mutant alleles in the same organism produces a wild-type phenotype. It usually happens because the two mutations are in different genes, so the organism carries one copy of each mutant allele and one copy of each wild-type allele.

complete nondisjunction event that may occur during meiosis or mitosis when all of the chromosomes fail to disjoin and remain in one of the two daughter cells.

complex traits characteristics that are determined by several genes and are significantly influenced by environmental factors.

concordance in genetics, the degree to which pairs of individuals (e.g., identical twins or fraternal twins) exhibit the same trait.

condense to form a more compact structure, as chromatids do during prophase.

conditional lethal allele an allele that is lethal, but only under certain environmental conditions.

conditional mutant a mutant whose phenotype depends on the environmental conditions, such as a temperature-sensitive mutant.

conglomerate a population composed of members of an original population plus new members that have migrated from another population.

conjugation a form of genetic transfer between bacteria that involves direct physical interaction between two bacterial cells. One bacterium acts as donor and transfers genetic material to a recipient cell.

conjugation bridge a connection between two bacterial cells that provides a passageway for DNA during conjugation.

consensus sequence the most commonly occurring bases within a sequence element.

conservative model an incorrect model that proposed that both strands of parental DNA remain together following DNA replication.

constitutive exon an exon that is always found in mRNA following splicing.

constitutive gene a gene that is not regulated and has essentially constant levels of expression over time.

continuous traits traits for which the phenotype exhibits a continuum.

control element see *regulatory sequence*.

copy number variation (CNV) variation in the copy number of a gene within a species.

core enzyme the subunits of an enzyme that are needed for catalytic activity, as in the core enzyme of RNA polymerase.

corepressor a small effector molecule that binds to a repressor protein, thereby causing the repressor protein to bind to DNA and inhibit transcription.

core promoter in eukaryotes, a DNA sequence that is absolutely necessary for transcription to take place. It provides the binding site for general transcription factors and RNA polymerase.

correlation coefficient (*r*) a statistic with a value that ranges between −1 and 1. It describes how two factors vary relative to each other.

cotransduction the transfer during bacterial transduction of a piece of DNA carrying two closely linked genes.

covariance a statistic that describes the degree of variation between two variables within a group.

cpDNA an abbreviation for chloroplast DNA.

CpG island a group of CG sequences that may be clustered near a promoter of a gene. The methylation of the cytosine bases usually inhibits transcription.

CRISPR-Cas system a system found in prokaryotes that is composed of non-coding RNAs and proteins, and defends the prokaryotes against bacteriophages, plasmids, and transposons.

CRISPR-Cas technology a method of gene editing based on the CRISPR-Cas system.

cross a breeding between two distinct individuals. An analysis of their offspring may be conducted to understand how traits are passed from parent to offspring.

cross-fertilization the process of breeding (crossing) two particular plants in which the pollen from one is placed on the stigma of the other.

crossing over a physical exchange of chromosome pieces that most commonly occurs during prophase of meiosis I.

C-terminus see *carboxyl-terminal end*.

cycle threshold method (C_t method) in real-time PCR, a method of determining the starting amount of DNA based on a threshold level at which the accumulation of fluorescence is significantly greater than the background level.

cyclic-AMP (cAMP) in bacteria, a small effector molecule that binds to CAP (catabolite activator protein). In eukaryotes, cAMP functions as a second messenger in a variety of intracellular signaling pathways.

cytogeneticist a scientist who studies chromosomes using microscopy.

cytogenetic mapping the mapping of genes or genetic sequences using microscopy.

cytogenetics the field of genetics that involves the microscopic examination of chromosomes.

cytokinesis the division of a single cell into two cells. The two nuclei produced in M phase are segregated into separate daughter cells during cytokinesis.

cytoplasmic inheritance see *extranuclear inheritance*.

cytosine a pyrimidine base found in DNA and RNA. It base-pairs with guanine in DNA.

D

Darwinian fitness the relative likelihood that a genotype will contribute to the gene pool of the next generation as compared with other genotypes.

daughter strand in DNA replication, the newly made strand of DNA.

deamination the removal of an amino group from a molecule. For example, the removal of an amino group from cytosine produces uracil.

decondensed the state of chromosomes in which they are less tightly compacted.

deficiency the condition in which a segment of chromosomal material is missing.

degeneracy the characteristic of the genetic code that more that one codon specifies the same amino acid. For example, the codons GGU, GGC, GGA, and GGG all specify the amino acid glycine.

degrees of freedom in a statistical analysis, the number of categories that are independent of each other.

deleterious mutation a mutation that has a detrimental effect on phenotype.

deletion the condition in which a segment of DNA is missing.

de novo methylation the methylation of DNA that has not been previously methylated. This is usually a highly regulated event.

deoxyribonucleic acid (DNA) the genetic material. It is a double-stranded structure, with each strand composed of repeating units of deoxyribonucleotides.

deoxyribose the sugar found in DNA.

depurination the removal of a purine base from DNA.

diakinesis the fifth stage of prophase of meiosis I.

diauxic growth the sequential use of two sugars by a bacterium.

dicentric describes a chromosome with two centromeres.

dicentric bridge the region between the two centromeres in a dicentric chromosome.

dideoxyribonucleotide (ddNTP) a nucleotide used in DNA sequencing that is missing the 3′ —OH group. If a dideoxyribonucleotide is incorporated into a DNA strand, it stops further growth of the strand.

dideoxy sequencing a method of DNA sequencing that uses dideoxyribonucleotides to terminate the growth of DNA strands.

differential centrifugation a form of centrifugation involving a series of centrifugation steps in which the supernatant or pellet is used in each subsequent step.

differentially methylated region (DMR) in imprinting, a site that is methylated during spermatogenesis or oogenesis, but not both.

dimeric DNA polymerase a complex of two DNA polymerase holoenzymes that moves as a unit during DNA replication.

dioecious refers to a species of plants in which some individuals produce only male gametophytes and others produce only female gametophytes.

diploid refers to an organism or cell that contains two sets of chromosomes.

diplotene the fourth stage of prophase of meiosis I.

directionality in DNA and RNA, the 5′ to 3′ arrangement of nucleotides in a strand; in proteins, the linear arrangement of amino acids from the N-terminus to the C-terminus.

directional selection natural selection that favors an extreme phenotype. This usually leads to the fixation of the favored allele.

direct repeats short DNA sequences that flank transposable elements and in which the DNA sequence is repeated in the same direction.

discontinuous trait a trait for which each offspring can be put into a particular phenotypic category.

discovery-based science the collection and analysis of data without a preconceived hypothesis. In some cases, the goal is to collect data to be able to formulate a hypothesis.

disequilibrium in population genetics, refers to a population that is not in Hardy-Weinberg equilibrium.

dispersive model an incorrect model for DNA replication that proposed that segments of parental DNA and newly made DNA are interspersed in both strands following the replication process.

disruptive selection a pattern of natural selection that favors different phenotypes in different local environments. This results in a balanced polymorphism.

dizygotic (DZ) twins also known as fraternal twins; twins formed from separate pairs of sperm and egg cells.

DMR see *differentially methylated region*.

DNA the abbreviation for deoxyribonucleic acid.

DnaA box a recognition site for the binding of the DnaA protein, which is involved in the initiation of bacterial DNA replication.

DnaA protein a protein that binds to the DnaA box at the origin of replication in bacteria and initiates DNA replication.

DNA fingerprinting a technology for identifying a particular individual based on the properties of his or her DNA.

DNA gap repair synthesis the synthesis of DNA in a region where a DNA strand has been previously removed, usually by a DNA repair enzyme or by an enzyme involved in homologous recombination.

DNA gyrase also known as topoisomerase II; an enzyme that introduces negative supercoils into DNA using energy from ATP. Gyrase can also relax positive supercoils when they occur.

DNA helicase an enzyme that separates the two strands of DNA.

DNA library a collection of many recombinant vectors, each carrying a particular fragment of DNA from a larger source. For example, each recombinant vector in a DNA library might carry a small segment of chromosomal DNA from a particular species.

DNA ligase an enzyme that catalyzes the formation of a covalent bond between two DNA fragments.

DNA methylation the phenomenon in which an enzyme covalently attaches a methyl group (—CH_3) to a base (adenine or cytosine) in DNA.

DNA methyltransferase the enzyme that attaches methyl groups to adenine or cytosine bases.

DNA microarray a small silica, glass, or plastic slide that is dotted with many different sequences of DNA, corresponding to short sequences within known genes.

DNA polymerase an enzyme that catalyzes the covalent attachment of nucleotides to form a strand of DNA.

DNA profiling see *DNA fingerprinting*.

DNA replication the process in which original DNA strands are used as templates for the synthesis of new DNA strands.

DNA replication licensing in eukaryotes, the process in which MCM helicase is bound at an origin, enabling the formation of two replication forks.

DNase an enzyme that cuts the sugar-phosphate backbone in DNA.

DNA sequencing a method for determining the base sequence in a segment of DNA.

DNA supercoiling the formation of additional coils in DNA due to twisting forces.

DNA translocase a type of motor protein that can move along DNA or cause DNA to move.

dominant describes an allele that determines the phenotype in the heterozygous condition. For example, if a plant is *Tt* and has a tall phenotype, the *T* (tall) allele is dominant over the *t* (dwarf) allele.

dominant-negative mutation a mutation that produces an altered gene product that acts antagonistically to the normal gene product. Shows a dominant pattern of inheritance.

dosage compensation the phenomenon that in species with sex chromosomes, one of the sex chromosomes is altered so that males and females have similar levels of gene expression, even though they do not contain the same complement of sex chromosomes.

double helix a helical structure formed when two strands of DNA (and sometimes RNA) bind to each other.

double-strand break model a model for homologous recombination in which the event that initiates recombination is a double-strand break in one of the double helices.

down promoter mutation a mutation in a promoter that inhibits the rate of transcription.

down regulation genetic regulation that leads to a decrease in gene expression.

duplication repetition of a segment of DNA more than once within a genome.

dyad a pair of sister chromatids.

E

egg cell also known as an egg or ovum; a female gamete that is usually very large and nonmotile.

electrophoresis the migration of ions or molecules in response to an electric field.

elongation (1) in transcription, the synthesis of RNA using DNA as a template; (2) in translation, the synthesis of a polypeptide using the information within mRNA.

electrophoretic mobility shift assay (EMSA) a technique used to determine if a protein binds to a fragment of DNA; the protein shifts the mobility of the DNA fragment during gel electrophoresis.

embryonic germ cell (EG cell) a type of pluripotent stem cell found in the gonads of the fetus.

embryonic stem cell (ES cell) a type of pluripotent stem cell found in the early blastocyst.

embryo sac in flowering plants, the female gametophyte that contains an egg cell.

emerging virus a virus that has arisen recently and has the potential to cause widespread disease.

empirical approach a strategy in which experiments are designed to determine quantitative relationships as a way to derive laws that govern biological, chemical, or physical phenomena.

empirical laws laws that are discovered using an empirical (observational) approach.

EMSA see *electrophoretic mobility shift assay*.

endonuclease an enzyme that can cut in the middle of a DNA strand.

endopolyploidy in a diploid individual, the phenomenon in which certain cells of the body are polyploid.

endosymbiosis a symbiotic relationship in which the symbiont actually lives inside (endo) the larger of the two species.

endosymbiosis theory the theory that the ancient origin of plastids and mitochondria was the result of certain species of bacteria taking up residence within primordial eukaryotic cells.

enhancer a DNA sequence that functions as a regulatory element. The binding of a regulatory transcription factor to the enhancer increases the level of transcription.

environment the surroundings an organism experiences.

enzyme a protein that functions to accelerate chemical reactions within the cell.

enzyme adaptation the phenomenon in which a particular enzyme appears within a living cell after the cell has been exposed to the substrate for that enzyme.

epigenetic inheritance the passing of an epigenetic change from parent to offspring.

epigenetics the study of mechanisms that lead to changes in gene expression that can be passed from cell to cell and are reversible, but do not involve a change in the sequence of DNA.

epimutation a heritable change in gene expression that does not alter the DNA sequence.

episome a segment of bacterial DNA that can exist as a plasmid and also integrate into the chromosome.

epistasis an inheritance pattern in which one gene can mask the phenotypic effects of a different gene.

epitope the structure on the surface of an antigen that is recognized by an antibody.

equilibrium density centrifugation a form of centrifugation in which the particles sediment through the gradient, reaching a position where the density of each particle matches the density of the solution.

error-prone replication a form of DNA replication that is carried out by translesion-replicating DNA polymerases and results in a high rate of mutation.

ER signal sequence a short amino acid sequence near the amino terminus of some proteins in eukaryotes that directs the protein to the ER membrane.

E site see *exit site*.

essential gene a gene that is essential for survival of the organism.

ethyl methanesulfonate (EMS) a type of chemical mutagen that alkylates bases (i.e., attaches methyl or ethyl groups).

euchromatin DNA that is not highly compacted and may be transcriptionally active.

eukaryotes (Greek, "true nucleus") one of the three domains of life. A defining feature of these organisms is that their cells contain a nucleus bounded by a membrane. Some simple eukaryotic species are single-celled protists and yeast; more complex multicellular species include fungi, plants, and animals.

euploid describes an organism in which the chromosome number is an exact multiple of a chromosome set.

evolution see *biological evolution*.

exit site (E site) a site on a ribosome from which an uncharged tRNA exits.

exon a segment contained within the RNA after splicing has occurred. In mRNA, the coding sequence of a polypeptide is contained within the exons.

exon shuffling the phenomenon in which exons are transferred between different genes during evolution.

exon skipping the splicing of an exon out of a pre-mRNA.

exonuclease an enzyme that digests an RNA or DNA strand from the end.

expressivity the degree to which a trait is expressed. For example, flowers with deep red color have a high expressivity of the allele for red flower color.

extranuclear inheritance (also known as cytoplasmic inheritance) the inheritance of genetic material that is not found within the nucleus.

F

F see *inbreeding coefficient*.

feedback inhibition the phenomenon in which the final product of a metabolic pathway inhibits an enzyme that acts early in the pathway.

feedback loop the phenomenon in which activation of a gene encodes a transcription factor that continues to stimulate its own expression.

fertilization the union of gametes (e.g., sperm and egg) to begin the life of a new organism.

F factor a fertility factor found in certain strains of bacteria in addition to their circular chromosome. Strains of bacteria that contain an F factor are designated F^+; strains without F factors are F^-.

F′ factor an F factor that also carries genes derived from the bacterial chromosome.

F_1 generation the offspring produced from a cross of individuals of the parental generation.

F_2 generation the offspring produced from a cross of individuals in the F_1 generation.

fidelity the accuracy of a process. If there are few mistakes, a process has a high fidelity.

fitness see *Darwinian fitness*.

flap endonuclease an endonuclease found in eukaryotes that removes RNA or DNA flaps that are generated during DNA replication.

fluorescence in situ hybridization (FISH) a form of in situ hybridization in which the probe is fluorescently labeled.

fork see *replication fork*.

forked-line method a method to solve independent assortment problems in which lines are drawn to connect particular genotypes.

founder with regard to genetic diseases, an individual who lived many generations ago and in whom a genetic disease originated.

founder effect changes in allele frequencies that occur when a small group of individuals separates from a larger population and establishes a colony in a new location.

fraction following centrifugation, a portion of the liquid contained within a centrifuge tube.

frameshift mutation a mutation that involves the addition or deletion of nucleotides not in a multiple of three and thereby shifts the reading frame of the codon sequence downstream from the mutation.

frequency distribution a graph that shows the numbers of individuals that are found in each of several phenotypic categories.

functional genomics the study of gene function at the genome level. It involves the study of many genes simultaneously.

G

G an abbreviation for guanine.

gain-of-function mutation a mutation that causes a gene to be expressed in an additional place where it is not normally expressed or during a stage of development when it is not normally expressed or at a level that is higher than normal.

gamete a reproductive cell (usually haploid) that can unite with another reproductive cell to create a zygote. Sperm and egg cells are types of gametes.

gametogenesis the production of gametes (e.g., sperm or egg cells).

gametophyte the haploid generation of a plant.

GATC methylation site a DNA sequence in bacteria that is methylated; it plays a role in preventing DNA replication from happening too early.

G bands the chromosomal banding pattern that is observed when the chromosomes have been treated with the dye called Giemsa.

gel electrophoresis a method that combines chromatography and electrophoresis to separate molecules and macromolecules. Samples are loaded into wells at one end of the gel, and an electric field is applied across the gel that causes charged molecules to migrate from one side of the gel to the other.

gene a unit of heredity that may influence the outcome of an organism's traits. At the molecular level, a gene contains the information to make a functional product, either RNA or protein.

gene chip see *DNA microarray*.

gene cloning the production of many copies of a gene using molecular methods such as PCR or the introduction of a gene into a vector that replicates in a host cell.

gene conversion the phenomenon in which one allele is converted to another due to genetic recombination and DNA repair.

gene duplication an increase in the copy number of a gene. Can lead to the evolution of gene families.

gene expression the process in which the information within a gene is accessed, first to synthesize RNA and usually proteins, and eventually to affect the phenotype of the organism.

gene family two or more different genes within a single species that are homologous because they were derived from the same ancestral gene.

gene flow changes in allele frequencies due to migration.

gene interaction the phenomenon in which two or more different genes influence the outcome of a single trait.

gene knockout the inactivation of both copies of a normal gene.

gene mutation a relatively small mutation that affects only a single gene.

gene pool all of the genes within a particular population.

general transcription factor (GTF) one of several proteins that are necessary for basal transcription at the core promoter in eukaryotes.

gene redundancy the phenomenon in which an inactive gene is compensated for by another gene with a similar function.

gene regulation the variation in the level of gene expression under different conditions.

genetic approach in research, the study of mutant genes that have abnormal function. By studying mutant genes, researchers may better understand normal genes and normal biological processes.

genetic code the correspondence between a codon (i.e., a sequence of three bases in an mRNA molecule) and the functional role that the codon plays during translation. Each codon specifies a particular amino acid or the end of translation.

genetic cross the breeding of two individuals and the analysis of their offspring in an attempt to understand how traits are passed from parent to offspring.

genetic drift random changes in allele frequencies due to sampling error.

genetic linkage see *linkage*.

genetic map (also referred to as a genetic linkage map) a diagram that describes the relative locations of genes or other DNA segments on a chromosome.

genetic mapping any method used to determine the linear order and distance of separation of genes that are linked to each other on the same chromosome.

genetic mosaic see *mosaicism*.

genetic polymorphism the occurrence of two or more wild-type alleles in a population; each allele is found at a frequency of 1% or higher.

genetic recombination (1) the process in which chromosomes are broken and then rejoined to form a new genetic combination; (2) the process in which alleles are assorted and passed to offspring in combinations that are different from the parents.

genetics the study of heredity.

genetic screening the use of testing methods to determine if an individual is a heterozygous carrier for or has a genetic disease.

genetic testing the analysis of individuals with regard to their genes or gene products. In many cases, the goal is to determine if an individual carries a mutant gene.

genetic transfer the physical transfer of genetic material from one bacterial cell to another.

genetic variation genetic differences among members of the same species or among different species.

genome all of the chromosomes and DNA sequences that an organism or species can possess.

genome maintenance cellular mechanisms that either prevent mutations from occurring and/or prevent mutant cells from surviving or dividing.

genome-sequencing projects research endeavors that have the ultimate goal of determining the sequence of DNA bases of the entire genome of a given species.

genome-wide association study (GWA study, or GWAS) an examination of a genome-wide set of genetic variants among many different individuals to see if any variant is associated with a disease or other type of trait.

genomic imprinting a pattern of inheritance that involves a change in a single gene or chromosome during gamete formation. Depending on whether the modification occurs during spermatogenesis or oogenesis, imprinting governs whether an offspring will express a gene that has been inherited from its mother or father.

genomic library a DNA library made from chromosomal DNA fragments.

genomics the molecular analysis of the entire genome of a species.

genotype the genetic composition of an individual, especially in terms of the alleles for particular genes.

genotype-environment association phenomenon in which certain genotypes are preferentially found in particular environments.

genotype-environment interaction phenomenon in which the environmental effects on phenotype differ according to genotype.

genotype frequency the number of individuals with a particular genotype in a population divided by the total number of individuals in the population.

germ cells the gametes (i.e., sperm and egg cells).

germ line a lineage of cells that gives rise to gametes.

germ-line mutation a mutation in a cell of the germ line.

glucocorticoid receptor a type of steroid receptor that functions as a regulatory transcription factor.

goodness of fit the degree to which observed data and expected data are similar to each other. If the observed and expected data are very similar, the goodness of fit is high.

G_1 phase a gap phase of the eukaryotic cell cycle during which a cell is making a decision to divide; it precedes the S phase.

G_2 phase a gap phase of the eukaryotic cell cycle during which a cell is preparing to divide; it precedes the M phase.

group I intron a type of intron found in self-splicing RNA that uses free guanosine in its splicing mechanism.

group II intron a type of intron found in self-splicing RNA that uses an adenine nucleotide within itself in its splicing mechanism.

growth factors protein factors that influence cell division.

guanine a purine base found in DNA and RNA. It base-pairs with cytosine in DNA.

GWAS see *genome-wide association study*.

gyrase see *DNA gyrase*.

H

haplodiploid system a mechanism of sex determination found in some species, such as bees, in which one sex is haploid (e.g., male) and the other sex is diploid (e.g., female).

haploid containing half the genetic material found in somatic cells. For a species that is diploid, a haploid gamete contains a single set of chromosomes.

haploinsufficiency the phenomenon in which a person has only a single functional copy of a gene, which does not produce a normal phenotype. Shows a dominant pattern of inheritance.

haplotype the linkage of particular alleles or molecular markers on a single chromosome.

haplotype association study a study in which disease-causing alleles are identified due to their linkage to particular markers along a chromosome.

Hardy-Weinberg equation $p^2 + 2pq + q^2 = 1$.

Hardy-Weinberg equilibrium the phenomenon in which, under certain conditions, allele frequencies are maintained in a stable condition and genotypes can be predicted according to the Hardy-Weinberg equation.

helicase see *DNA helicase*.

helix-turn-helix motif a structure found in transcription factor proteins that promotes binding to the major groove of DNA.

hemizygous describes the single copy of an X-linked gene in a male. A male mammal is said to be hemizygous for X-linked genes.

heritability the amount of phenotypic variation within a particular group of individuals that is due to genetic factors.

heterochromatin highly compacted DNA. It is usually transcriptionally inactive.

heterodimer when two polypeptides encoded by different genes bind to each other to form a dimer.

heteroduplex a double-stranded region of DNA that contains one or more base mismatches.

heterogametic sex in species with two types of sex chromosomes, the heterogametic sex produces two types of gametes. For example, in mammals, the male is the heterogametic sex, because a sperm can contain either an X or a Y chromosome.

heterogamous describes a species that produces two morphologically different types of gametes (i.e., sperm and eggs).

heterogeneity see *locus heterogeneity*.

heteroplasmy the condition of a cell that contains variation in a particular type of organelle. For example, a plant cell could contain some chloroplasts that make chlorophyll and other chloroplasts that do not.

heterozygote an individual who is heterozygous.

heterozygote advantage a pattern of inheritance in which a heterozygote has a reproductive advantage compared to either of the corresponding homozygotes.

heterozygous describes a diploid individual that has two different alleles of the same gene.

Hfr strain (for high frequency of recombination) a bacterial strain in which an F factor has been integrated into the bacterial chromosome. During conjugation, an Hfr strain can transfer segments of the bacterial chromosome.

highly repetitive sequences sequences that are found tens of thousands or even millions of times throughout a genome.

high-throughput sequencing DNA sequencing technologies that can process a large number of samples and thereby produce a large amount of DNA sequence information in a relatively short period of time.

histone acetyltransferase an enzyme that attaches acetyl groups to the amino-terminal ends of histone proteins.

histone code hypothesis the hypothesis that the pattern of histone modification acts much like a language or code in specifying alterations in chromatin structure.

histone variants histones that have amino acid sequences slightly different from those of the core histones and play specialized roles in chromatin structure and function.

histone proteins a group of proteins involved in forming the nucleosome structure of eukaryotic chromatin.

holandric gene a gene on the Y chromosome.

Holliday junction a site where an unresolved crossover has occurred between two homologous chromosomes.

Holliday model a model to explain the molecular mechanism of homologous recombination.

holoenzyme an enzyme containing all of its subunits, such as the holoenzyme of RNA polymerase, which contains σ factor along with the core enzyme.

homodimer a protein formed when two polypeptides encoded by the same gene bind to each other to form a dimer.

homogametic sex in species with two types of sex chromosomes, the homogametic sex produces only one type of gamete. For example, in mammals, the female is the homogametic sex, because an egg can only contain an X chromosome.

homolog one of the chromosomes in a pair of homologous chromosomes.

homologous (1) describes two genes that are derived from the same ancestral gene and have similar DNA sequences; (2) the two homologs of a chromosome pair are said to be homologous to each other.

homologous recombination the exchange of DNA segments between homologous chromosomes.

homologous recombination repair (HRR) also called homology-directed repair, occurs when the DNA strands from a sister chromatid are used to repair a lesion in the other sister chromatid.

homozygous describes a diploid individual that has two identical alleles of a particular gene.

horizontal gene transfer the transfer of genes from one organism to another organism that is not its offspring.

host cell a cell that is infected with a virus or bacterium.

host range the spectrum of host species that a virus or other pathogen can infect.

hot spots sites within a gene that are more likely to be mutated than other locations.

housekeeping gene a gene that encodes a protein required in most cells of a multicellular organism.

Human Genome Project a worldwide collaborative project that provided a detailed map of the human genome and obtained its complete DNA sequence.

human immunodeficiency virus (HIV) a relatively recent emerging virus that infects humans and other primates and causes acquired immunodeficiency syndrome (AIDS).

hybrid (1) an offspring obtained from a hybridization experiment; (2) a cell produced from a cell fusion experiment in which the two separate nuclei have fused to make a single nucleus.

hybrid dysgenesis the production of defective *Drosophila* offspring due to the phenomenon that P elements can transpose freely.

hybridization (1) the mating of two organisms of the same species with different characteristics; (2) the phenomenon in which two single-stranded molecules renature together to form a hybrid molecule.

hypothesis testing using statistical tests to determine if the data from experimentation are consistent with a hypothesis.

I

ICR see *imprinting control region*.
imprinting see *genomic imprinting*.
imprinting control region (ICR) a DNA region that is differentially methylated and plays a role in genomic imprinting.
inborn error of metabolism a genetic disease that involves a defect in a metabolic enzyme.
inbreeding reproduction by genetically related individuals.
inbreeding coefficient (*F*) the probability that two alleles of a given gene in a particular individual will be identical because both copies are due to descent from a common ancestor.
inbreeding depression the phenomenon in which inbreeding produces homozygotes that are less fit, thereby decreasing the reproductive success of a population.
incomplete dominance a pattern of inheritance in which a heterozygote that carries two different alleles exhibits a phenotype that is intermediate to those of the corresponding homozygous individuals. For example, an *Rr* heterozygote may be pink, whereas the *RR* and *rr* homozygotes are red and white, respectively.
incomplete penetrance a pattern of inheritance in which an allele does not always control the phenotype of the individual. For example, an individual carrying a dominant allele may not exhibit the dominant trait.
induced describes a gene that has been transcriptionally activated by an inducer.
induced mutation a mutation caused by environmental agents.
inducer a small effector molecule that binds to a genetic regulatory protein and thereby increases the rate of transcription.
inducible gene a gene that is regulated by an inducer, which is a small effector molecule that causes transcription to increase.
inhibitor a small effector molecule that binds to an activator protein, causing the protein to be released from the DNA, thereby inhibiting transcription.
initiation (1) in transcription, the stage that involves the initial binding of RNA polymerase to the promoter in order to begin RNA synthesis; (2) in translation, the formation of a complex between mRNA, the initiator tRNA, and the ribosomal subunits.
initiator tRNA during translation, the tRNA that binds to the start codon.
insertion element the simplest transposable element, which is commonly found in bacteria.
in situ hybridization a technique used to cytogenetically map the locations of genes or other DNA sequences within large eukaryotic chromosomes. In this method, a complementary probe is used to detect the location of a gene within a set of chromosomes.
integrase an enzyme that functions in the integration of viral DNA or retroelements into a chromosome.
intergenic region in a chromosome, a region of DNA that lies between two different genes.
intergenic suppressor a suppressor mutation that is in a different gene from the gene that contains the first mutation.
internal nuclear matrix a network of irregular protein fibers and other proteins that is connected to the nuclear lamina and fills the interior of the nucleus.
interphase the series of phases G_1, S, and G_2, during which a cell spends most of its life.
interrupted mating a method used in conjugation experiments in which the length of time that the bacteria spend conjugating is ended by a blender treatment or other type of harsh agitation.
interstitial deletion loss of an internal segment from a linear chromosome.
intervening sequence also known as an intron; a segment of RNA that is removed during RNA splicing.
intragenic suppressor a suppressor mutation that is within the same gene as the mutation that it suppresses.
intrinsic termination transcriptional termination that does not require the rho protein.
introns intervening sequences that are found between exons. Introns are spliced out of the RNA prior to translation.
invasive refers to a tumor that can invade surrounding tissue.
inversion a change in the direction of genetic material along a chromosome in which a segment is flipped so it runs in the reverse order.
inversion heterozygote a diploid individual that carries one normal chromosome and a homologous chromosome with an inversion.
inversion loop the loop structure that is formed when the homologous chromosomes of an inversion heterozygote attempt to align themselves (i.e., synapse) during meiosis.
inverted repeats DNA sequences found in transposable elements that are identical (or very similar) but run in opposite directions.
in vitro fertilization (IVF) in the case of humans, the fertilization of an egg outside of a female's body.
iron regulatory protein (IRP) a translational regulatory protein that recognizes iron response elements that are found in specific mRNAs. It may inhibit translation or stabilize the mRNA.
iron response element (IRE) an RNA sequence that is recognized by the iron regulatory protein.
IRs see *inverted repeats*.
IS element see *insertion sequence element*.
isoacceptor tRNAs two different tRNAs that can recognize the same codon.
isogamous describes a species that makes morphologically similar gametes.

K

karyotype an organized representation of a micrograph of all the chromosomes within a cell. It reveals how many chromosomes are found within an actively dividing somatic cell.
kinetochore a group of cellular proteins that attach to the centromere during meiosis and mitosis.
knockout see *gene knockout*.
Kozak's rules a set of rules that describes the most favorable types of bases that flank a start codon in complex eukaryotes.

L

lac repressor a protein that binds to the operator site of the *lac* operon and inhibits transcription.
lagging strand a strand that is synthesized during DNA replication as short Okazaki fragments in the direction away from the replication fork.
latent refers to a virus that exists in a dormant state for a long period of time before producing new virus particles.
law of independent assortment see *Mendel's law of independent assortment*.

law of segregation see *Mendel's law of segregation*.
lncRNA see *long non-coding RNA*.
leading strand a strand that is synthesized during DNA replication continuously toward the replication fork.
leptotene the first stage of prophase of meiosis I.
lethal allele an allele that may cause the death of an organism.
lethal mutation a mutation that produces an allele that causes the death of an organism.
library see *DNA library*.
ligase see *DNA ligase*.
linkage the occurrence of two or more genes on the same chromosome.
linkage disequilibrium the association of alleles and molecular markers with each other at a frequency that is significantly higher than expected by random chance.
linkage group a group of genes that are linked together because they are found on the same chromosome.
linkage mapping the mapping of genes or other genetic sequences on a chromosome by analyzing the outcome of crosses.
lipid a general name given to an organic molecule that is insoluble in water. Cell membranes contain a large amount of lipids.
LNA see *locked nucleic acid*.
local population a segment of a population that is somewhat isolated. Members of a local population are more likely to breed with each other than with members that are outside of the local population.
locked nucleic acid (LNA) a type of anti-miRNA oligonucleotide that has a locked conformation and binds more tightly to an miRNA.
locus (pl. **loci**) the physical location of a gene within a chromosome.
locus heterogeneity the phenomenon in which a particular type of disease or trait may be caused by mutations in two or more different genes.
LOH see *loss of heterozygosity*.
long non-coding RNA (lncRNA) a non-coding RNA that is longer than 200 nucleotides.
long terminal repeats (LTRs) base sequences containing many short segments that are tandemly repeated. They are found in retroviruses and viral-like retroelements.
loss-of-function allele an allele of a gene that encodes an RNA or protein that is nonfunctional or compromised in function.
loss-of-function mutation a change in a genetic sequence that creates a loss-of-function allele.
loss of heterozygosity (LOH) a genetic change that inactivates the single functional allele in a heterozygous somatic cell.

LTR retrotransposon a type of retrotransposon that is derived from a virus and has long terminal repeats.
LTRs see *long terminal repeats*.
Lyon hypothesis a hypothesis to explain the pattern of X-chromosome inactivation seen in mammals. Initially, both X chromosomes are active. However, at an early stage of embryonic development, one of the two X chromosomes is randomly inactivated in each somatic cell of a female.
lysis cell breakage.
lysogenic cycle a type of growth cycle for a phage in which the phage integrates its genetic material into the chromosome of a bacterium. This integrated phage DNA can exist in a dormant state for a long time, during which no new bacteriophages are made.
lysogeny a dormant state of a virus that has integrated its DNA into a host cell's chromosome.
lytic cycle a type of growth cycle for a phage in which the phage directs the synthesis of many copies of its genetic material and coat proteins. These components then assemble to make new phages. When synthesis and assembly are completed, the bacterial host cell is lysed, and the newly made phages are released into the environment.

M

macromolecule a large organic molecule composed of smaller building blocks. Examples include DNA, RNA, proteins, and large carbohydrates.
maintenance methylation the methylation of hemimethylated DNA following DNA replication.
major groove a wide indentation in the DNA double helix in which the bases have access to water.
malignant describes a tumor composed of cancerous cells.
map distance the relative distance between sites (e.g., genes) on a single chromosome.
mapping the experimental process of determining the relative locations of genes or other segments of DNA on individual chromosomes.
mapping functions approaches to determine the quantitative relationship between recombination frequencies and physical distances along chromosomes.
map unit (mu) a unit of map distance obtained from genetic crosses. One map unit is equivalent to 1% recombinant offspring in a testcross.
marker see *molecular marker*.
MAR see *matrix-attachment region*.

maternal effect an inheritance pattern for certain nuclear genes in which the genotype of the mother directly determines the phenotypic traits of her offspring.
maternal inheritance inheritance of DNA that occurs through the cytoplasm of the egg.
matrix-attachment region (MAR) a site in the chromosomal DNA that is anchored to proteins in the nuclear matrix.
maturase a protein that enhances the rate of splicing of group I and group II introns.
MCM helicase a group of eukaryotic proteins needed to complete a process called DNA replication licensing, which is necessary for the formation of two replication forks at an origin of replication.
mean ($\bar{X}$) the sum of all the values in a group divided by the number of individuals in the group.
mean fitness of the population ($\bar{w}$) the average fitness of a population that is calculated by considering the frequencies and fitness values for all genotypes.
mediator a protein complex that interacts with RNA polymerase II and various regulatory transcription factors. Depending on its interactions with regulatory transcription factors, mediator may stimulate or inhibit RNA polymerase II.
meiosis a form of nuclear division in which the sorting process results in the production of haploid cells from a diploid cell.
meiotic nondisjunction the event in which chromosomes do not segregate equally during meiosis.
Mendelian inheritance the common pattern of inheritance observed by Mendel, which involves the transmission of eukaryotic genes that are located on the chromosomes found within the cell nucleus.
Mendel's law of independent assortment two different genes randomly assort their alleles during the formation of haploid cells (if they are not linked).
Mendel's law of segregation the two copies of a gene segregate from each other during transmission from parent to offspring.
meristic traits traits that can be counted and expressed in whole numbers.
merozygote a partial diploid strain of bacteria containing F′ factor genes.
messenger RNA (mRNA) a type of RNA that contains the information for the synthesis of a polypeptide.
metacentric describes a chromosome with the centromere in the middle.
metagenome a collection of genes from a particular environmental sample.
metagenomics the study of a complex mixture of genetic material obtained from an environmental sample.

metaphase the third stage of M phase. The chromosomes align along the center of the spindle apparatus.

metaphase plate the plane at which chromosomes align during metaphase.

metastatic describes cancer cells that migrate to other parts of the body.

methylation see *DNA methylation*.

methyl-CpG-binding protein a protein that binds to a CpG island when it is methylated.

methyl-directed mismatch repair see *mismatch repair system*.

methyltransferase see *DNA methyltransferase*.

microarray see *DNA microarray*.

microdomains loops of bacterial chromosomal DNA, typically 10 kbp in length, that emanate from a central core.

microevolution changes in the gene pool with regard to particular alleles that occur over the course of many generations.

microRNA (miRNA) an ncRNA that usually originates from an endogenous eukaryotic gene and silences an mRNA via RNA interference.

microsatellite short tandem repeats (typically a couple hundred base pairs in length) that are interspersed throughout a genome and are quite variable in length among different individuals. They can be amplified by PCR.

microscopy the use of a microscope to view cells or subcellular structures.

microtubule-organizing center (MTOC) a site in a cell where microtubules begin to grow.

minimal medium a type of growth medium for microorganisms that contains a mixture of reagents that are required for growth, with nothing additional added.

minisatellite a repetitive sequence that was formerly used in DNA fingerprinting. Its use has been largely superseded by smaller repetitive sequences called microsatellites.

minor groove a narrow indentation in the DNA double helix in which the bases have access to water.

minute a unit of measure in bacterial conjugation experiments. This unit refers to the relative time it takes for genes to first enter a recipient strain during conjugation.

mismatch repair system a DNA repair system that detects a mismatch and specifically removes the segment from the newly made daughter strand.

missense mutation a base substitution that leads to a change in the amino acid sequence of the encoded polypeptide.

mitochondrial DNA (mtDNA) the DNA found within mitochondria.

mitosis a type of nuclear division into two nuclei, such that each daughter cell receives the same complement of chromosomes.

mitotic nondisjunction an event in which chromosomes do not segregate equally during mitosis.

mitotic recombination crossing over that occurs during mitosis.

mitotic spindle apparatus (also known as the mitotic spindle) the structure that organizes and separates the chromosomes during M phase of the eukaryotic cell cycle.

model organism an organism studied by many researchers so that they can more easily compare their results and begin to understand the properties of a given species.

moderately repetitive sequences sequences that are found a few hundred to several thousand times in the genome.

molecular genetics an examination of DNA structure and function at the molecular level.

molecular level with regard to gene expression, the level of observation at which genes affect the molecular properties of an organism.

molecular marker a segment of DNA that is found at a specific site in the genome and has properties that enable it to be uniquely recognized using molecular tools such as gel electrophoresis.

molecular profiling methods that enable researchers to understand the molecular changes that occur in diseases such as cancer.

monad a single chromatid within a pair of sister chromatids; can also refer to a chromosome that has not replicated.

monoallelic expression the phenomenon in which only one of the two alleles of a given gene is transcriptionally expressed because the other allele has been silenced due to imprinting.

monohybrid a heterozygous individual produced from a single-factor cross.

monomorphic describes a gene that is found as only one allele in a population.

monosomic refers to a diploid cell or organism that is missing a chromosome (i.e., $2n-1$).

monozygotic (MZ) twins twins that are genetically identical because they were formed from the same sperm and egg.

morph a form or phenotype in a population. For example, red eyes and white eyes are different eye color morphs.

morphological trait a trait that affects the morphology (physical form) of an organism. An example is eye color.

mosaicism the condition in which the cells of part of an organism differ genetically from the rest of the organism.

M phase a general name for nuclear and cellular division during mitosis or meiosis. Nuclear division is divided into prophase, prometaphase, metaphase, anaphase, and telophase. Nuclear division is usually followed by cellular division or cytokinesis.

mRNA see *messenger RNA*.

mtDNA an abbreviation for mitochondrial DNA.

MTOC see *microtubule-organizing center*.

multinomial expansion equation an equation used to solve genetic problems involving three or more types of unordered events.

multiple alleles two or more alleles of the same gene existing within a population.

multiplication method a method for solving independent assortment problems in which the probabilities of the outcome for each gene are multiplied together.

multipotent describes a type of stem cell that can differentiate into several different types of cells.

mutagen an agent that causes alterations in the structure of DNA.

mutant alleles alleles that have been created by random mutation of wild-type alleles.

mutation a permanent change in the genetic material that can be passed from cell to cell or from parent to offspring.

mutation rate the likelihood that a gene will be altered by a new mutation.

N

n an abbreviation that stands for the number of chromosomes in a set. In humans, $n = 23$, and a diploid cell has $2n = 46$ chromosomes.

NAPs see *nucleoid-associated proteins*.

narrow-sense heritability heritability that takes into account only those genetic factors that are additive.

natural selection the process whereby differential fitness acts on the gene pool. When a mutation creates a new allele that is beneficial, the allele may become prevalent within future generations because the individuals possessing the allele are more likely to reproduce and pass it to their offspring.

natural transformation a natural process of transformation that occurs in certain strains of bacteria.

ncRNA see *non-coding RNA*.

negative control transcriptional regulation by a repressor protein.

negative frequency-dependent selection a mechanism giving rise to balancing selection in which the fitness of a genotype decreases when its frequency becomes higher.

neutral mutation a mutation that has no detectable effect on protein function or no detectable effect on the phenotype of the organism.

next-generation sequencing technologies newer DNA sequencing technologies that are more rapid and inexpensive.

NFR see *nucleosome-free region*.

nitrogen mustard an alkylating agent that can cause mutations in DNA.

nitrous acid a type of chemical mutagen that deaminates bases, thereby changing amino groups to keto groups.

nonallelic homologous recombination recombination that occurs at nonhomologous sites within chromosomes due to the occurrence of repetitive sequences.

nonautonomous element a transposable element that lacks a gene that encodes a protein (such as transposase or reverse transcriptase) that is necessary for transposition.

non-coding RNA (ncRNA) an RNA that does not encode a polypeptide.

nondisjunction event in which chromosomes do not segregate properly during mitosis or meiosis.

nonessential genes genes that are not absolutely required for survival, although they are likely to be beneficial to the organism.

nonhomologous end-joining (NHEJ) protein a protein that joins the ends of DNA fragments that are not homologous.

nonhomologous recombination the exchange of DNA between nonhomologous segments of chromosomes or plasmids.

non-LTR retrotransposon a type of retrotransposon that does not have long terminal repeats.

nonparental see *recombinant*.

nonrecombinant in a testcross, a phenotype or arrangement of alleles on a chromosome that occurs in the offspring but is not found in the parental generation.

nonsense codon a stop codon.

nonsense mutation a mutation that involves a change from a sense codon to a stop codon.

normal distribution a distribution for a large sample in which the trait of interest varies in a symmetrical way around an average value.

norm of reaction the effects of environmental variation on an individual's traits.

Northern blotting a technique used to detect a specific RNA within a mixture of many RNA molecules.

N-terminus see *amino-terminal end*.

nuclear genes genes that are located on chromosomes found in the nucleus of eukaryotic cells.

nuclear lamina a collection of fibers that line the inner nuclear membrane.

nuclear matrix a group of proteins that anchor the loops found in eukaryotic chromosomes.

nucleic acid RNA or DNA. A macromolecule that is composed of repeating nucleotide units.

nucleoid a darkly staining region that contains the genetic material of mitochondria, chloroplasts, or bacteria.

nucleoid-associated proteins (NAPs) a set of DNA-binding proteins that are found in bacteria and facilitate compaction and organization of the chromosome.

nucleolus a region within the nucleus of eukaryotic cells where the assembly of ribosomal subunits occurs.

nucleoside structure in which a base is attached only to a sugar, with no phosphate attached to the sugar.

nucleosome the repeating structural unit within eukaryotic chromatin. It is composed of double-stranded DNA wrapped around an octamer of histone proteins.

nucleosome-free region (NFR) a region within a eukaryotic chromosome where nucleosomes are not found.

nucleotide the repeating structural unit of nucleic acids, composed of a sugar, one or more phosphates, and base.

nucleotide excision repair (NER) a DNA repair system in which several nucleotides in the damaged strand are removed and the undamaged strand is used as a template to resynthesize a normal strand.

nucleus a membrane-bound organelle in eukaryotic cells where sets of chromosomes are found.

null hypothesis a hypothesis that assumes there is no real difference between the observed and expected values.

O

Okazaki fragments short segments of DNA that are synthesized to form the lagging strand during DNA replication.

oligonucleotide a short strand of DNA, typically a few or a few dozen nucleotides in length.

oncogene a gene that promotes cancer due to a gain-of-function mutation.

one-gene/one-enzyme hypothesis the idea, which later needed to be expanded, that one gene encodes one enzyme.

oogenesis the production of egg cells.

open complex the region of separation of two DNA strands produced by RNA polymerase during transcription.

open conformation a loosely packed chromatin structure that is capable of transcription.

operator (or **operator site**) a sequence of nucleotides in bacterial DNA that provides a binding site for a genetic regulatory protein.

operon a group of two or more genes that are transcribed from a single promoter.

ORC see *origin recognition complex*.

organelle a large, specialized structure within a cell, which is surrounded by a single or double membrane.

organism level the level of observation or experimentation that involves a whole organism.

orientation-independent refers to certain types of genetic regulatory elements that can function in the forward or reverse direction. Certain enhancers are orientation-independent.

origin of replication a nucleotide sequence that functions as an initiation site for the assembly of several proteins required for DNA replication.

origin of transfer the location on an F factor or within the chromosome of an Hfr strain that is the initiation site for the transfer of DNA from one bacterium to another during conjugation.

origin recognition complex (ORC) a complex of six proteins found in eukaryotes that is necessary to initiate DNA replication.

outbreeding sexual reproduction by genetically unrelated individuals.

ovary (1) in plants, the structure in which the ovules develop; (2) in animals, the structure that produces egg cells and female hormones.

overdominance an inheritance pattern in which a heterozygote has greater reproductive success than either of the corresponding homozygotes.

ovule the structure in higher plants where the female gametophyte (i.e., embryo sac) is produced.

ovum a female gamete, also known as an egg cell.

oxidative DNA damage changes in DNA structure that are caused by reactive oxygen species (ROS).

oxidative stress an imbalance between the production of reactive oxygen species (ROS) and an organism's ability to break them down.

P

p an abbreviation for the short arm of a chromosome.

pachytene the third stage of prophase of meiosis I.

palindromic having the same sequence when read in the forward and reverse directions.

pangenesis an incorrect hypothesis that suggested that hereditary traits could be modified depending on the lifestyle of the individual. For example, it was believed that a person who practiced a particular skill would produce offspring that would be better at that skill.

paracentric inversion an inversion in which the centromere is found outside of the inverted region.

paralogs homologous genes within a single species that constitute a gene family.

parental generation (P generation) in a genetic cross, the first generation in the experiment. In Mendel's studies, the parental generation was true-breeding with regard to particular traits.

parental strand in DNA replication, the DNA strand that is used as a template.

particulate theory of inheritance a theory proposed by Mendel that states that traits are inherited as discrete units that remain unchanged as they are passed from parent to offspring.

PcG see *polycomb group*.

paternal leakage the phenomenon in which species where maternal inheritance is generally observed, the male parent may, on rare occasions, provide mitochondria or chloroplasts to the zygote.

PCR see *polymerase chain reaction*.

pedigree analysis genetic analysis using information contained within family trees. In this approach, the aim is to determine the type of inheritance pattern that a gene follows.

pellet a collection of particles found at the bottom of a centrifuge tube following centrifugation.

peptide bond a covalent bond formed between the carboxyl group in one amino acid in a polypeptide and the amino group in the next amino acid.

peptidyl site (P site) a site on the ribosome that carries a tRNA along with a polypeptide.

peptidyl transfer the step during the elongation stage of translation in which the polypeptide is removed from the tRNA in the P site and transferred to the amino acid at the A site.

peptidyl transferase a complex that functions during translation to catalyze the formation of a peptide bond between the amino acid in the A site of the ribosome and the growing polypeptide.

pericentric inversion an inversion in which the centromere is located within the inverted region of the chromosome.

personalized medicine the application of genetic or molecular data in the treatment of disease.

PGD see *preimplantation genetic diagnosis*.

P generation the parental generation in a genetic cross.

phage see *bacteriophage*.

phage λ a bacteriophage that infects *E. coli*.

pharmacogenetics the study or clinical testing of genetic variation that causes differing responses to drugs.

phenotype the observable traits of an organism.

phenylketonuria (PKU) a human genetic disorder involving a defect in the enzyme phenylalanine hydroxylase.

phosphodiester linkage in a DNA or RNA strand, a linkage in which a phosphate group connects two sugar molecules.

photolyase an enzyme found in yeast and plants that can repair thymine dimers by splitting them, which returns the DNA to its original condition.

photoreactivation a mechanism of DNA repair of thymine dimers via photolyase that requires light.

physical mapping the mapping of genes or other genetic sequences using DNA cloning methods.

physiological trait a trait that affects a cellular or body function. An example is the rate of glucose metabolism.

piRISC see *piRNA-induced silencing complex*.

piRNA see *PIWI-interacting RNA*.

piRNA-induced silencing complex (piRISC) a complex composed of a piRNA and PIWI proteins that prevents the movement of transposable elements in animal cells.

PIWI-interacting RNA (piRNA) a non-coding RNA that associates with PIWI proteins and prevents the movement of transposable elements in animal cells.

PKU see *phenylketonuria*.

plasmid a circular piece of DNA that exists independently of the chromosomal DNA. Some plasmids are used as vectors in cloning experiments.

pleiotropy the multiple effects of a single gene on the phenotype of an organism.

pluripotent describes a type of stem cell that can differentiate into all or nearly all the types of cells of the adult organism.

pluripotency factors transcription factors that influence which of the two X chromosomes in female mammals stays active in somatic cells.

point mutation a change in a single base pair within DNA.

pollen grain the male gametophyte of flowering plants.

polyadenylation the process of attaching a string of adenine nucleotides to the 3' end of eukaryotic mRNAs.

polyA tail the string of adenine nucleotides at the 3' end of eukaryotic mRNAs.

polycistronic mRNA an mRNA transcribed from an operon that encodes two or more proteins.

polycomb group (PcG) a group of protein complexes that regulate development in animals by promoting epigenetic changes that inhibit gene expression.

polycomb response element (PRE) a site on DNA that is recognized by a polycomb group such as PRC2.

polygenic refers to traits that are governed by two or more different genes.

polymerase chain reaction (PCR) the method for amplifying a DNA region involving the sequential use of oligonucleotide primers and *Taq* polymerase.

polymorphic describes a trait or gene that is found in two or more forms in a population.

polymorphism (1) the prevalence of two or more phenotypic forms in a population; (2) the phenomenon in which a gene exists in two or more alleles within a population.

polypeptide a sequence of amino acids that is the product of mRNA translation. One or more polypeptides fold and associate with each other to form a functional protein.

polyploid describes an organism or cell with three or more sets of chromosomes.

polyribosome an mRNA transcript that has many bound ribosomes in the act of translation.

polysome see *polyribosome*.

polytene chromosome a type of chromosome found in certain cells, such as *Drosophila* salivary cells, that consists of many copies of a chromosome that lie side by side.

population a group of individuals of the same species that are capable of interbreeding with one another.

population genetics the field of genetics that is primarily concerned with the prevalence of genetic variation within populations.

population level the level of observation or experimentation that involves a population of organisms.

position effect a change in phenotype that occurs when the position of a gene is changed from one chromosomal site to a different one.

positive control genetic regulation by activator proteins.

posttranslational describes events that occur after translation is completed.

posttranslational covalent modification the covalent attachment of a molecule to a protein after it has been synthesized via ribosomes.

posttranslational regulation functional control of proteins already present in a cell.

preimplantation genetic diagnosis (PGD) a form of genetic testing in which an embryo obtained via in vitro fertilization is tested for genetic abnormalities.

preinitiation complex the stage of transcription in which the assembly of RNA polymerase and general transcription factors occurs at the core promoter, but the DNA has not yet started to unwind.

PRE see *polycomb response element*.

pre-mRNA in eukaryotes, a long transcript produced within the cell nucleus by the transcription of protein-encoding genes.

The pre-mRNA is usually altered by splicing and other modifications before it exits the nucleus.

prereplication complex (preRC) in eukaryotes, an assembly of at least 14 different proteins, including a group of 6 proteins called the origin recognition complex (ORC), that acts as the initiator of eukaryotic DNA replication.

Pribnow box the TATAAT sequence that is often found at the −10 site of bacterial promoters.

primary structure with regard to proteins, the linear sequence of amino acids.

primase an enzyme that synthesizes a short RNA primer during DNA replication.

primer a short segment of DNA or RNA that initiates DNA replication.

primer annealing in PCR, the process in which an oligonucleotide primer binds to a complementary segment of DNA.

primer extension in PCR, the step during which DNA is made by extending the length of a DNA primer.

primosome a multiprotein complex composed of DNA helicase, primase, and several accessory proteins.

probability the chance that an event will occur in the future.

processing body (P-body) a cellular structure that stores mRNA molecules, such as those that have been silenced by RNA interference.

processive enzyme an enzyme, such as RNA or DNA polymerase, that glides along the DNA and does not dissociate from the template strand as it catalyzes the covalent attachment of nucleotides.

product rule the probability that two or more independent events will occur is equal to the products of their individual probabilities.

proflavin a type of chemical mutagen that causes frameshift mutations.

prokaryotes (Greek, "prenucleus") another name for bacteria and archaea. The term refers to the fact that their chromosomes are not contained within a separate nucleus in the cell.

prometaphase the second stage of M phase. During this stage, the nuclear membrane vesiculates, and the mitotic spindle is completely formed.

promoter a sequence within a gene that initiates (i.e., promotes) transcription.

proofreading the ability of DNA polymerase to remove mismatched bases from a newly made strand.

prophage phage DNA that has been integrated into a bacterial chromosome.

prophase the first stage of M phase. The chromosomes have already replicated and begin to condense. The mitotic spindle starts to form.

protandrous hemaphrodite an organism that first becomes a male and can later transform into a female.

protease an enzyme that digests the polypeptide backbone in proteins.

protein a functional unit composed of one or more polypeptides.

protein-encoding gene a gene that encodes a polypeptide; also called a structural gene.

proteome the collection of all proteins that a given cell or species can make.

proto-oncogene a normal cellular gene that does not cause cancer but may incur a gain-of-function mutation or become incorporated into a viral genome and thereby lead to cancer.

prototroph a bacterial strain that does not need a particular nutrient supplemented in its growth medium.

provirus viral DNA that has been incorporated into the chromosome of a host cell.

pseudoautosomal genes genes that are located in the regions found on both the X and Y chromosome.

pseudoautosomal inheritance the inheritance pattern of genes that are found on both the X and Y chromosomes. Even though such genes are located physically on the sex chromosomes, their pattern of inheritance is identical to that of autosomal genes.

P site see *peptidyl site*.

Punnett square a diagram in which the gametes that two parents can produce are arranged on a square grid as a way to predict the types of offspring the parents will produce and in what proportions.

purine a type of nitrogenous base that has a double-ring structure. Examples are adenine and guanine.

P value in a chi square table, the probability that the deviations between observed and expected values are due to random chance.

pyrimidine a type of nitrogenous base that has a single-ring structure. Examples are cytosine, thymine, and uracil.

pyrosequencing a type of next-generation DNA sequencing technology.

pyrrolysine a nonstandard amino acid that may be incorporated into polypeptides during translation.

Q

q an abbreviation for the long arm of a chromosome.

QTL mapping the determination of the location of QTLs using mapping methods such as genetic crosses coupled with the analysis of molecular markers.

QTLs see *quantitative trait loci*.

quantitative genetics the area of genetics concerned with traits that can be described in a quantitative way.

quantitative trait a trait, usually polygenic in nature, that can be described with numbers.

quantitative trait loci (QTLs) the locations on chromosomes where the genes that influence quantitative traits reside.

quaternary structure two or more polypeptides bound to each other to form a protein.

R

radial loop domains organization of chromatin during interphase into loops, often 25,000 to 200,000 bp in size, which are anchored to the nuclear matrix at matrix-attachment regions.

random genetic drift see *genetic drift*.

random mutation theory the idea that mutations are a random process—they can occur in any gene and do not involve exposure of an organism to a particular condition that selects for specific types of mutations.

random sampling error the deviation between the observed and expected outcomes that is due to chance.

reactive oxygen species (ROS) products of oxygen metabolism in all aerobic organisms that can damage cellular molecules, including DNA, proteins, and lipids.

reading frame a series of codons beginning at a start codon as a frame of reference.

realized heritability a form of narrow-sense heritability that is observed when selective breeding is practiced.

real-time PCR a method of PCR in which the synthesis of DNA is monitored in real time. It can quantitate the starting amount of DNA in a sample.

recessive refers to a trait or gene that is masked by the presence of a dominant trait or gene.

recessive epistasis a form of epistasis in which an individual must be homozygous for a recessive allele for a particular phenotype to be masked.

reciprocal crosses a pair of crosses in which the traits of the two parents differ with regard to sex. For example, one cross could be a red-eyed female fly and a white-eyed male fly, and the reciprocal cross would be a red-eyed male fly and a white-eyed female fly.

reciprocal translocation a translocation in which two different chromosomes exchange pieces.

recombinant (1) refers to combinations of alleles or traits that are not found in the true-breeding parental generation; (2) describes DNA molecules that are produced by molecular techniques in which segments of DNA are joined to each other in ways that differ from their original arrangement. The cloning of DNA into vectors is an example.

recombinant DNA molecules molecules that are produced in a test tube by covalently linking DNA fragments from two different sources.

recombinant DNA technology the use of in vitro molecular techniques to isolate and manipulate pieces of DNA.

recombinant vector a vector that contains an inserted fragment of DNA, such as a gene from a chromosome.

recombination see *genetic recombination*.

redundancy see *gene redundancy*.

regulatory sequence (regulatory element or control element) a sequence of DNA (or possibly RNA) that binds a regulatory protein and thereby influences gene expression. Bacterial operator sites and eukaryotic enhancers and silencers are examples.

regulatory transcription factor a protein or protein complex that binds to a regulatory element and influences the rate of transcription via RNA polymerase.

relative fitness the relative likelihood that one genotype will contribute to the gene pool of the next generation compared to other genotypes.

relaxosome a protein complex that recognizes the origin of transfer in F factors and other conjugative plasmids, cuts one DNA strand, and aids in the transfer of the T DNA.

release factor a protein that recognizes a stop codon and promotes translational termination and the release of the completed polypeptide.

repetitive sequences DNA sequences that are present in many copies in the genome.

replica plating a technique in which a replica of bacterial colonies is transferred to a new petri plate.

replication see *DNA replication*.

replication fork the region where two DNA strands have separated and new strands are being synthesized.

replisome a complex that contains a primosome and dimeric DNA polymerase.

repressible gene a gene that is regulated by a corepressor or inhibitor, which are small effector molecules that cause transcription to decrease.

repressor a regulatory protein that binds to DNA and inhibits transcription.

reproduction the production of daughter cells from a mother cell or the production of offspring from parents.

reproductive cloning the cloning of a eukaryotic organism, which may involve the use of genetic material from somatic cells.

resolution (1) the last stage of homologous recombination, in which the entangled DNA strands become resolved into two separate structures; (2) in microscopy, the ability to distinguish two adjacent objects as separate from each other.

restriction endonuclease (or restriction enzyme) an endonuclease that cleaves DNA. The restriction enzymes used in cloning experiments bind to specific base sequences and then cleave the DNA backbone at two defined locations, one in each strand.

restriction point a point in the G_1 phase of the cell cycle at which a cell becomes committed to cell division.

retroelement see *retrotransposon*.

retrotransposition a form of transposition in which the element is transcribed into RNA. The RNA is then used as a template via reverse transcriptase to synthesize a DNA molecule that is integrated into a new region of the genome via integrase.

retrotransposon a type of transposable element that moves via an RNA intermediate.

reverse mutation see *reversion*.

reverse transcriptase an enzyme that uses an RNA template to make a complementary strand of DNA.

reverse transcriptase PCR a modification of PCR in which the first round of replication involves the use of RNA and reverse transcriptase to make a complementary strand of DNA.

reversion a mutation that returns a mutant allele back to the wild-type allele.

R group the side chain of an amino acid.

rho (ρ) protein a protein that is involved in transcriptional termination for certain bacterial genes.

rho-dependent termination transcriptional termination that requires the function of the rho protein.

rho-independent termination transcription termination that does not require the rho protein. It is also known as intrinsic termination.

ribonucleic acid (RNA) a nucleic acid that is composed of ribonucleotides. In living cells, RNA is synthesized via the transcription of DNA.

ribose the sugar found in RNA.

ribosomal RNA (rRNA) RNA that is a component of a ribosome.

ribosome a large macromolecular structure that acts as the catalytic site for polypeptide synthesis. The ribosome allows the mRNA and tRNAs to be positioned correctly as a polypeptide is made.

ribosome-binding site a sequence in bacterial mRNA that binds to a ribosome to initiate translation.

riboswitch a form of genetic regulation in which an RNA can exist in two different secondary conformations.

ribozyme an RNA molecule with catalytic activity.

RISC see *RNA-induced silencing complex*.

RNA see *ribonucleic acid*.

RNA editing a change in the nucleotide sequence of an RNA molecule that involves additions or deletions of particular bases or a conversion of one type of base to a different type.

RNA-induced silencing complex (RISC) the complex that mediates RNA interference.

RNA interference the phenomenon that double-stranded RNA targets complementary RNAs within the cell for silencing or degradation.

RNA polymerase an enzyme that synthesizes a strand of RNA using a DNA strand as a template.

RNA primer a short strand of RNA, made by primase, that is used to elongate a strand of DNA during DNA replication.

RNase an enzyme that cuts the sugar-phosphate backbone in RNA.

RNA-Seq see *RNA sequencing*.

RNA sequencing (RNA-Seq) a strategy for determining which genes are transcribed at the genome level.

RNA splicing the process in which pieces of RNA are removed and the remaining pieces are covalently attached to each other.

Robertsonian translocation the structure produced when two telocentric chromosomes fuse at their short arms.

rRNA see *ribosomal RNA*.

S

SBS see *sequencing by synthesis*.

scaffold a collection of proteins that holds the DNA in place and gives chromosomes their characteristic shapes.

scaffold-attachment region (SAR) a site in the chromosomal DNA that is anchored to the nuclear matrix.

SCE see *sister chromatid exchange*.

science a way of knowing about our natural world. The science of genetics allows us to understand how the expression of genes produces the traits of an organism.

scientific method a process that scientists typically follow so that they may reach verifiable conclusions about the natural world.

scintillation counting a technique that permits a researcher to count the number of radioactive emissions from a sample containing a population of radioisotopes.

secondary structure a regular repeating pattern of molecular structure, such as the DNA double helix or the α helix and β sheet found in proteins.

sedimentation coefficient a measure of centrifugation that is normally expressed in Svedberg units (S): $1\ S = 1 \times 10^{-13}$ second.

segmental duplication a tandem duplication of a small segment of a chromosome.

segregate to put two things in separate locations. For example, homologous chromosomes segregate into different gametes.

selectable marker a gene that provides a selectable phenotype in a cloning experiment. Many selectable markers are genes that confer antibiotic resistance.

selection coefficient (*S*) the degree to which a genotype is selected against; equal to one minus the fitness value.

selection limit the phenomenon in which several generations of artificial selection result in a plateau where artificial selection is no longer effective.

selective breeding programs and procedures designed to modify the phenotypes in economically important species of plants and animals.

selenocysteine a nonstandard amino acid that may be incorporated into polypeptides during translation.

self-fertilization fertilization that involves the union of male and female gametes derived from the same parent.

selfish DNA hypothesis the idea that transposable elements exist because they possess characteristics that allow them to multiply within the host cell DNA and inhabit the host without offering any selective advantage.

self-splicing the ability of some RNA molecules to remove their own introns without the aid of other proteins or other RNAs.

semiconservative model the correct model for DNA replication that proposes that the newly made double-stranded DNA contains one parental strand and one daughter strand.

semilethal alleles lethal alleles that kill some individuals but not all.

semisterility the condition in which an individual has a lowered fertility.

sense codon a codon that specifies an amino acid.

sequence complexity the number of times a particular base sequence appears throughout the genome of a given species.

sequence element a DNA sequence with a specialized function.

sequence-tagged site (STS) a short segment of DNA, usually between 100 and 400 bp long, whose base sequence is found to be unique within an entire genome. Sequence-tagged sites are identified by PCR.

sequencing see *DNA sequencing*.

sequencing by synthesis (SBS) a next-generation DNA sequencing technology in which the synthesis of DNA is directly monitored to deduce the base sequence.

sequencing ladder a series of bands on a gel that can be followed in order (e.g., from the bottom of the gel to the top of the gel) to determine the base sequence of DNA.

sex chromosomes a pair of chromosomes (e.g., X and Y in mammals) that determines sex in a species.

sex determination the process in which certain factors govern whether an individual develops into a male or female organism.

sex-influenced inheritance an inheritance pattern in which an allele is dominant in one sex but recessive in the opposite sex. Scurs in certain breeds of cattle is an example of a sex-influenced trait.

sex-limited inheritance an inheritance pattern in which a trait is found in only one of the two sexes. An example is beard development in men.

sex-linked gene a gene that is located on one of the sex chromosomes.

sex pilus (pl. **pili**) a structure on the surface of bacterial cells that acts as an attachment site to promote binding between bacteria.

sexual dimorphism the phenomenon in which the males and females of a species are morphologically distinct.

sexual reproduction the process whereby parents make gametes (e.g., sperm and egg) that fuse with each other in the process of fertilization to begin the life of a new organism.

Shine-Dalgarno sequence a sequence in bacterial mRNAs that functions as a ribosomal binding site.

short-interfering RNAs (siRNAs) see *small-interfering RNA*.

shotgun sequencing a genome sequencing strategy in which DNA fragments to be sequenced are randomly generated from larger DNA fragments.

side chain in an amino acid, the chemical structure that is attached to the carbon atom (i.e., the α carbon) that is located between the amino group and carboxyl group.

sigma (σ) factor a transcription factor that recognizes bacterial promoter sequences and facilitates the binding of RNA polymerase to the promoter.

signal recognition particle (SRP) a complex between a non-coding RNA and one or more proteins that directs newly made polypeptides to the ER membrane in eukaryotes or to the plasma membrane in prokaryotes.

silencer a DNA sequence that functions as a regulatory element. The binding of a regulatory transcription factor to the silencer decreases the level of transcription.

silent mutation a mutation that does not alter the amino acid sequence of the encoded polypeptide even though the nucleotide sequence has changed.

simple Mendelian inheritance an inheritance pattern involving a simple, dominant/recessive relationship that produces observed ratios in the offspring that readily obey Mendel's laws.

simple translocation a translocation in which a piece of a chromosome becomes attached to a different chromosome.

simple transposition a cut-and-paste mechanism for transposition in which a transposable element is removed from one site and inserted into another.

simple transposon a transposon that moves via a DNA intermediate and carries additional genes that are not required for transposition, such as an antibiotic resistance gene.

single-factor cross a cross in which an experimenter is following the outcome of only a single trait.

single-nucleotide polymorphism (SNP) a genetic polymorphism within a population in which two alleles of a gene differ by a single base pair.

single-strand binding protein a protein that binds to both of the single strands of DNA during DNA replication and prevents them from re-forming a double helix.

siRNA see *small-interfering RNAs*.

sister chromatid exchange (SCE) the phenomenon in which crossing over occurs between sister chromatids, which thereby exchange identical genetic material.

sister chromatids pairs of replicated chromosomes that are attached to each other at the centromere. Sister chromatids are genetically identical.

site-directed mutagenesis a technique that enables scientists to change the sequence of cloned DNA segments.

small-interfering RNA (siRNA) an ncRNA that usually originates from an exogenous source and silences an mRNA via RNA interference.

small nuclear riboprotein see *snRNP*.

small regulatory RNA a non-coding RNA that is shorter than 200 nucleotides.

snRNP a complex containing small nuclear RNAs and a set of proteins, which are components of the spliceosome.

somatic cell any cell of the body except for germ-line cells that give rise to gametes.

somatic mutation a mutation in a somatic cell.

species a group of organisms that maintains a distinctive set of attributes in nature.

spectrophotometer a device used by researchers to determine how much radiation at various wavelengths a sample absorbs.

spermatogenesis the production of sperm cells.

sperm cell also known as a sperm; a male gamete. Sperm are small and usually travel relatively far distances to reach the female gamete.

S phase a phase of the eukaryotic cell cycle during which the DNA is replicated.

spindle see *mitotic spindle apparatus*.

spindle pole during cell division in eukaryotes, one of two sites in the cell where microtubules originate.

spliceosome a complex that functions in the splicing of eukaryotic pre-mRNA.

splicing see *RNA splicing*.

splicing factor a protein that regulates the process of RNA splicing.

spontaneous mutation a change in DNA structure that results from random abnormalities in biological processes.

sporophyte the diploid generation of a plant.

SRP see *signal recognition particle*.

SR protein a type of splicing factor.

stabilizing selection a pattern of natural selection that favors individuals with an intermediate phenotype.

standard deviation (SD) a statistic that is computed as the square root of the variance.

start codon a three-base sequence in mRNA that initiates translation. It is usually 5′-AUG-3′ and encodes methionine.

stem cell a cell that has the capacity to divide and to differentiate into one or more specific cell types.

stigma the structure in flowering plants on which the pollen grains land and the pollen tubes start to grow so that sperm cells can reach the egg cells.

stop codon a three-base sequence in mRNA that signals the end of translation of a polypeptide. The three stop codons are 5′-UAA-3′, 5′-UAG-3′, and 5′-UGA-3′.

strain a variety that continues to exhibit the same characteristic after several generations.

strand in DNA or RNA, the long linear polymer formed of nucleotides covalently linked together.

structural gene see *protein-encoding gene*.

STS see *sequence-tagged site*.

submetacentric describes a chromosome in which the centromere is slightly off center.

subunit a part of a larger structure or complex. In a protein, each subunit is a single polypeptide.

supercoiling see *DNA supercoiling*.

supernatant following centrifugation, the fluid that is found above the pellet.

suppressor (or **suppressor mutation**) a mutation at a second site that suppresses the phenotypic effects of another mutation.

synapsis the event in which homologous chromosomes recognize each other and then align themselves along their entire lengths.

synaptonemal complex a complex of proteins that promotes the interconnection between homologous chromosomes during meiosis.

synonymous codons two different codons that specify the same amino acid.

synteny a group of genes that are found in the same order on the chromosomes of different species.

T

T an abbreviation for thymine.

tandem array (or **tandem repeat**) a short nucleotide sequence that is repeated many times in a row.

***Taq* polymerase** a thermostable form of DNA polymerase used in PCR experiments.

target-site duplications see *direct repeats*.

TATA box a sequence found within eukaryotic core promoters that determines the starting site for transcription. The TATA box is recognized by a TATA-binding protein, which is a component of TFIID.

tautomeric shift a change in chemical structure such as an alternation between the keto and enol forms of the bases that are found in DNA.

tautomers the chemically similar forms of certain small molecules, such as bases, which can spontaneously interconvert.

T DNA a segment of DNA found within a Ti plasmid that is transferred from a bacterium to infected plant cells. The T DNA from the Ti plasmid becomes integrated into the chromosomal DNA of the plant cell by recombination.

TE see *transposable element*.

telocentric describes a chromosome with its centromere at one end.

telomerase the protein/RNA complex that recognizes telomeric sequences at the ends of eukaryotic chromosomes and synthesizes additional numbers of telomeric repeat sequences.

telomerase reverse transcriptase (TERT) the enzyme within telomerase that uses RNA as a template to make DNA.

telomerase RNA component (TERC) a segment of RNA within telomerase that contains a sequence complementary to the telomeric repeat sequence.

telomeres specialized DNA sequences found at the ends of linear eukaryotic chromosomes.

telophase the fifth stage of M phase. The chromosomes have reached their respective poles and decondense.

temperate phage a bacteriophage that may follow the lysogenic cycle.

temperature-sensitive allele an allele for which the resulting phenotype depends on the environmental temperature.

temperature-sensitive (ts) lethal allele an allele that is lethal at a certain environmental temperature.

template DNA a strand of DNA that is used to synthesize a complementary strand of DNA or RNA.

template strand a strand of DNA that is used to synthesize a complementary strand of DNA or RNA.

TERC see *telomerase RNA component*.

terminal deficiency see *terminal deletion*.

terminal deletion loss of a segment from the end of a linear chromosome.

termination (1) in transcription, the release of the newly made RNA transcript and RNA polymerase from the DNA; (2) in translation, the release of the polypeptide and the last tRNA and the disassembly of the ribosomal subunits and mRNA.

termination codon see *stop codon*.

terminator a sequence within a gene that signals the end of transcription.

TERT see *telomerase reverse transcriptase*.

tertiary structure the three-dimensional structure of a macromolecule, such as a polypeptide.

testcross an experimental cross between a recessive individual and an individual whose genotype the experimenter wishes to determine.

tetrad the association among four sister chromatids during meiosis.

tetraploid having four sets of chromosomes (i.e., 4*n*).

TFIID a type of general transcription factor in eukaryotes that is required for the function of RNA polymerase II. It binds to the TATA box and recruits RNA polymerase II to the core promoter.

thermocycler a device that automates the timing of temperature changes in each cycle of a PCR experiment.

30-nm fiber the association of nucleosomes to form a more compact structure that is 30 nm in diameter.

three-factor cross a cross in which an experimenter follows the outcome of three different traits.

threshold traits traits that are inherited quantitatively due to the contribution of many genes, but are expressed qualitatively.

thymine a pyrimidine base found in DNA. It base-pairs with adenine in DNA.

thymine dimer two adjacent thymine bases in a DNA strand that have become covalently linked; may result in a mutation.

tissue-specific gene a gene that is highly regulated and is expressed in a particular cell type.

TNRE see *trinucleotide repeat expansion*.

topoisomerase an enzyme that alters the degree of supercoiling in DNA.

topoisomers DNA conformations that differ only with regard to supercoiling.

totipotent having the genetic potential to produce an entire individual. A somatic plant cell or a fertilized egg is totipotent.

trait the specific properties of a character. Morphological traits affect the appearance of an organism. Physiological traits affect the ability of an organism to function. Behavioral traits are those that affect an organism's behavior.

***trans*-acting factor** a regulatory protein that binds to a regulatory element in the DNA and exerts a *trans*-effect.

transcription the process of synthesizing RNA from a DNA template.

transcriptional start site the site in a gene where transcription begins.

transcription factors a broad category of proteins that influence the ability of RNA polymerase to transcribe DNA into RNA.

transcriptome all of the RNA molecules that are transcribed in a given cell or group of cells.

transduction a form of genetic transfer between bacterial cells in which a bacteriophage transfers bacterial DNA from one bacterium to another.

***trans*-effect** an effect on gene expression that occurs even though two DNA segments are not physically adjacent to each other. *Trans*-effects are mediated through diffusible genetic regulatory proteins.

transfer RNA (tRNA) a type of RNA used in translation that carries an amino acid. The anticodon in tRNA is complementary to a codon in the mRNA.

transformation (1) introduction of a plasmid vector or segment of chromosomal DNA into a bacterial cell; (2) conversion of a normal cell into a malignant cell.

transgenerational epigenetic inheritance see *epigenetic inheritance*.

transition a point mutation involving a change of a pyrimidine to another pyrimidine (e.g., C to T) or a purine to another purine (e.g., A to G).

translation the synthesis of a polypeptide using the information contained in the codons of an mRNA.

translational regulatory protein a protein that regulates translation.

translational repressor a protein that binds to mRNA and inhibits its ability to be translated.

translesion-replicating polymerase a type of DNA polymerase that can replicate over a DNA region that contains an abnormal structure (i.e., a lesion).

translesion synthesis (TLS) the synthesis of DNA over a template strand that harbors some type of DNA damage. This occurs via translesion-replicating polymerases.

translocation (1) the phenomenon in which one segment of a chromosome breaks off and becomes attached to a different chromosome; (2) movement of a ribosome from one codon in an mRNA to the next codon.

translocation cross the structure that is formed when the chromosomes of a reciprocal translocation attempt to synapse during meiosis. This structure contains two normal (nontranslocated chromosomes) and two translocated chromosomes. A total of eight chromatids are found within the cross.

transposable element (TE) a small genetic element that can move to multiple locations within the chromosomal DNA.

transposase the enzyme that catalyzes the movement of transposons.

transposition the phenomenon of transposon movement.

transposon a transposable element that moves via a DNA intermediate.

transversion a point mutation in which a purine is interchanged with a pyrimidine, or vice versa.

trithorax group (TrxG) a group of protein complexes that regulate development in animals by promoting epigenetic changes that activate gene expression.

trimethylation attachment of three methyl groups to an amino acid such as lysine.

trinucleotide repeat expansion (TNRE) a type of mutation that involves an increase in the number of tandemly repeated trinucleotide sequences.

triploid describes an organism or cell that contains three sets of chromosomes.

trisomic refers to a diploid cell or organism with one extra chromosome (i.e., $2n + 1$).

tRNA see *transfer RNA*.

trp repressor a protein that binds to the operator site of the *trp* operon and inhibits transcription.

true-breeding line a strain of a particular species that continues to exhibit the same trait after several generations of self-fertilization (in plants) or inbreeding.

TrxG see *trithorax group*.

tumor-suppressor gene a gene that functions to inhibit cancerous growth.

two-factor cross a cross in which an experimenter follows the outcome of two different traits.

U

U an abbreviation for uracil.

unbalanced translocation a translocation that results in a cell having too much genetic material compared with a normal cell.

unipotent refers to a type of stem cell that can differentiate into only a single type of cell.

3′-untranslated region the region of an mRNA molecule that follows the stop codon.

5′-untranslated region the region of an mRNA molecule that precedes the start codon.

up promoter mutation a mutation in a promoter that increases the rate of transcription.

up regulation genetic regulation that leads to an increase in gene expression.

uracil a pyrimidine base found in RNA.

UTR an abbreviation for the untranslated region of mRNA.

V

variance the sum of the squared deviations from the mean divided by the degrees of freedom.

variants (1) versions of a trait; (2) individuals of the same species that exhibit different traits, such as tall and dwarf pea plants.

vector a small segment of DNA that is used as a carrier of another segment of DNA. Vectors are used in DNA cloning experiments.

viral envelope a structure composed of a membrane and viral spike proteins that surrounds the capsid of certain eukaryotic viruses.

viral genome the genetic material of a virus.

viral reproductive cycle the series of steps that lead to the production of new viruses; the cycle may also include a latent phase, such as the lysogenic cycle in bacteriophages, in which no viruses are made.

virulent phage a phage that follows only the lytic cycle.

virus a small infectious particle that contains nucleic acid as its genetic material, surrounded by a capsid of proteins. Some viruses also have an envelope consisting of a membrane embedded with spike proteins.

W

Western blotting a technique used to detect a specific protein among a mixture of proteins.

wild type a genotype or phenotype that is common in a natural population.

wild-type allele an allele that is prevalent in a natural population, generally found in greater than 1% of the population. For polymorphic genes, there is more than one wild-type allele.

wobble base the base in an anticodon that pairs with the third base in a codon; this base can vary without affecting recognition between codon and anticodon during translation.

wobble rules rules that govern the specificity of binding between the third base in a codon and the corresponding base in an anticodon.

X

X-chromosome inactivation (XCI) a process in which mammals equalize the expression of X-linked genes by randomly turning off one X chromosome in the somatic cells of females.

X-inactivation center (Xic) a site on the X chromosome that appears to play a critical role in X-chromosome inactivation.

X-linked genes (alleles) genes (or alleles of genes) that are physically located on the X chromosome.

X-linked inheritance an inheritance pattern in certain species that involves genes that are located only on the X chromosome.

X-linked recessive inheritance pattern an inheritance pattern for an allele or trait in which the gene is found on the X chromosome and the allele is recessive relative to a corresponding dominant allele.

Y

Y-linked genes (alleles) genes (or alleles of genes) that are located only on the Y chromosome.

Z

Z DNA a left-handed DNA double helix that is found occasionally in living cells.

zygote a cell formed from the union of a sperm and egg.

zygotene the second stage of prophase of meiosis I.

INDEX

A

Abl gene, 534–535
ABO blood type, multiple alleles, 98, 99
Accuracy, of DNA replication, 279
Ac element, 249
Acentric fragment, 165
Achondroplasia, 522
Acquired antibiotic resistance, 201
Acridine dyes, 446
Acrocentric centromere, 156, 158
Activator protein, 357
Activator, regulatory protein, 356, 357, 364–365
Acutely transforming viruses (ACTs), 535
Adaptation (CRISPR-Cas system), 423–424
Adaptor hypothesis, 338, 341
Adenine (A), 5, 225, 226, 229, 231, 234, 266
Adenosin deaminase (ADA) deficiency, 521
Adenovirus, 208, 209
Adjacent-1 segregation, 167
Adjacent-2 segregation, 167
Aedes aegypti (mosquito larvae), 564
Age of onset
    genetic disorders, 519–520
    lethal alleles, 101
Agouti gene, 402–403
Agriculture
    and inbreeding, 572
    and metagenomics, 506
    quantitative traits in, 584
    selective breeding, 600–601
Aicardi syndrome, 524
AIDS (acquired immune deficiency syndrome), 217–218
Alanine (Ala), 329, 332, 336, 339
Albinism, 92, 521
Albino phenotype, 520
Alkaptonuria, 326
Allele frequencies, 556–558, 561, 562–564, 567–568
    and migration, 570
    mutation rate affecting, 573–574
Alleles, 7, 21
    codominance, 98–99
    definition, 47
    detection of disease-causing, 525–528
    dominant, 21–22, 91–93
    dominant mutant, 92
    gene pool, 554–555
    genes existing in multiple, 98
    and haplotypes, 525
    human recessive, 520
    law of segregation and, 47–48
    lethal, 90, 101–102
    linkage and, 133
    multiple, 98–99
    mutant, 91
    and polymorphism, 555–556
    recessive, 91
    relationship between mutations and, 556
    semilethal, 101
    wild-type, 91, 95
    X-linked, 82
Allelic variation, 156
Allis, C. David, 388
Allodiploid species, 177–178
Allolactose, 359, 360, 361, 370
Alloploid organism, 177
Alloploidy, 177
Allopolyploidy, 177
Allosteric enzyme, 373
Allosteric model (transcriptional termination), 308, 309
Allosteric regulation, 359
Allosteric site, 359, 373
Allotetraploid plants, 177–178
Alopex lagopus (Arctic fox), 94
Alternate segregation, 167
Alternative exons, 315
Alternative splicing, 314–315, 316
Alu family of sequences, 247, 249, 251
Alzheimer's disease, 427, 542
American holly, 75
Ames, Bruce, 448
Ames test, 448–449
Amino acids, 3, 5
    attaching to a tRNA, 339
    function of, 4
    incorporated into polypeptides during translation, 331–332
Amino acid sequence, 5
    gene mutations effecting, 432, 433
Aminoacyl site (A site), 343
Aminoacyl-tRNA, 340
Aminoacyl-tRNA synthetases, 339–340
Amino-terminal end, 331
Amino-terminal tail, 253, 388, 389
Amniocentesis, 530
AMP (adenosine monophosphate), 226
Amplification, 466
Amplification (PCR), 496
Amplified restriction fragment length polymorphism (AFLP), 495
Amyotrophic lateral sclerosis (ALS), 427
Anaphase, 28
    meiosis I, 32, 33, 167, 174
    meiosis II, 32
Androgen insensitivity syndrome, 523
Androgen receptor gene, 99
Aneuploid births, 530
Aneuploid organisms, 169
Aneuploidy, 169, 170–172, 174, 178
    and cancer, 539
Angelman syndrome, 122–123
Angiosperms (flowering plants)
    gametophyte development, 36
    pollination and fertilization in, 43

I-1

Animal cell
  cytokinesis in, 29
  mitosis in, 28
  mitotic spindle, 27, 29
  stages of meiosis in, 32
Animals
  cloning, 1
  gamete production, 37
  gametogenesis in, 35
  genetic technologies modifying traits of, 2–3
  genome sequencing, 504
  origin of DNA replication in, 280
  sperm and egg cells made in, 35–36
  variations in euploidy, 172–173
Aniridia, 522
Anopheles gambiae, 504
Anopheles mosquito, 96
Antagomirs, 427
Anthers, 43
Antibiotic resistance, 186, 200–201
Antibiotics, and bacterial translation, 348, 349
Antibody, Western blotting using, 483
Anticipation, 443–444
Anticodon pairing, 340–341
Anticodons, 328, 339
  adaptor hypothesis, 339
  definition, 328
  tRNA, 235, 330, 338–339
Antigens, 483
Anti-miRNA oligonucleotides (AMOs), 427
Antiparallel arrangement
  DNA strands, 231
  RNA double helices, 234
Antisense RNA, 372, 415
Antisense strand, 301
APC gene, 537, 540
Apoptosis, 537
Apurinic site, 440
Aquatic biology, and metagenomics, 506
Arabidopsis thaliana (flowering plant), 495, 496, 504, 505
  epigenetic effects, 395, 397
  Human Genome Project, 500
  model organism, 12
  RFLP map, 495, 496
Arber, Werner, 461
Arctic fox *(Alopex lagopus),* color change in, 94
Arginine (Arg), 329, 332, 336, 337
Arginines (R), 315
Aromatic amino acids, 332
ARS (autonomously replicating sequence) elements, 280
Artificial chromosomes, 461
Artificial transformation, 199
Asexual reproduction, 23, 24–25
Ashkenazi Jewish population, 520, 530
Asparagine (Asn), 332, 336, 337
Aspartic acid (Asp), 329, 332, 336
Aspergiullus nidulans, 30
Association, correlation coefficient and, 542
Association studies
  genome-wide, 527–528
  identifying mutant gene causing Huntington disease, 526–527
Assortative mating, 571
Aster microtubules, 27
Asthma, and epigenetic changes, 542
AT/GC rule, 231, 266–267, 278
ATP, and myosin, 29
ATP-dependent chromatin remodeling, 386–387
ATP synthesis, 21

AT-rich region, 271
Attachment, in viral reproductive cycle, 212
Attenuation, 356, 368–369, 370, 371
Attenuator sequence, 368, 369
AUG codon, 328, 329
Automated DNA sequencing, 477
Autonomous elements, 249
Autopolyploid, 177
Autoradiography, A-6
Autosomal dominant inheritance, 521–522
Autosomal recessive inheritance, 520–521
Autosomes, 73
Auxotroph, 187–188
Avery, Amos, 178
Avery, Oswald, 223, 224

B

Bacillus subtilis, 242, 374, 500
Backbone, DNA, 227, 229, 232
Bacteria
  asexual reproduction, 24
  gene regulation in, 356
  genome sequencing, 504
  Hfr strain, 190–191
  homologous recombination, 286
  plasmids in, 189–190
  and replica plating, 439
  riboswitches in, 373–375
  transduction in, 196–199
  transposable elements, 251
Bacterial artificial chromosomes (BACs), 461
Bacterial cells, ribosome composition in, 341–342
Bacterial chromosomes
  chromosomal loops, 241–242
  CRISPR-Cas system in, 422–423
  general features, 241
  Hfr strains, 191–192
  homologous recombination, 199–200
  microdomains, 242
  organization of functional sites along, 240–241
  origin of replication, 271
  structure of, 241–244
  supercoiling, 242–244
Bacterial conjugation, 186, 187–192, 362
Bacterial DNA replication, 266
  accuracy of, 279
  chemistry of, 278–279
  key features of a bacterial origin of, 270–272
  synthesis of new DNA strands, 273–277
Bacterial gene regulation. *See* Gene regulation (bacteria)
Bacterial species, gene transfer among, 574
Bacterial transcription, 299–303
Bacterial transduction, 186, 196–199
Bacterial transformation, 186, 199–200, 223
Bacterial translation
  antibiotics inhibiting, 348, 349
  beginning before transcription is completed, 347
  comparison with eukaryotic translation, 348, 349
  coupling between transcription and, 347
  elongation stage, 347
  initiation stage, 344–345
  protein factors, 346
  ribosome composition, 341–342, 343
  termination stage, 347, 348
Bacteriophage reproductive cycle, 196, 197

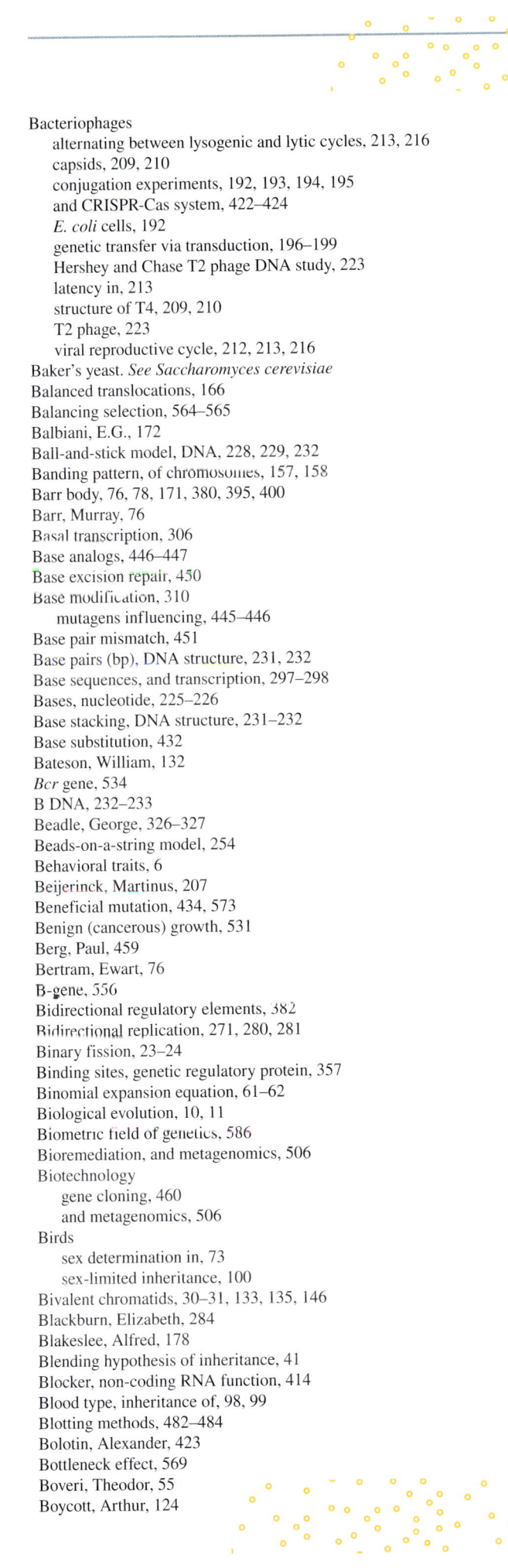

INDEX

Bacteriophages
    alternating between lysogenic and lytic cycles, 213, 216
    capsids, 209, 210
    conjugation experiments, 192, 193, 194, 195
    and CRISPR-Cas system, 422–424
    E. coli cells, 192
    genetic transfer via transduction, 196–199
    Hershey and Chase T2 phage DNA study, 223
    latency in, 213
    structure of T4, 209, 210
    T2 phage, 223
    viral reproductive cycle, 212, 213, 216
Baker's yeast. *See Saccharomyces cerevisiae*
Balanced translocations, 166
Balancing selection, 564–565
Balbiani, E.G., 172
Ball-and-stick model, DNA, 228, 229, 232
Banding pattern, of chromosomes, 157, 158
Barr body, 76, 78, 171, 380, 395, 400
Barr, Murray, 76
Basal transcription, 306
Base analogs, 446–447
Base excision repair, 450
Base modification, 310
    mutagens influencing, 445–446
Base pair mismatch, 451
Base pairs (bp), DNA structure, 231, 232
Base sequences, and transcription, 297–298
Bases, nucleotide, 225–226
Base stacking, DNA structure, 231–232
Base substitution, 432
Bateson, William, 132
Bcr gene, 534
B DNA, 232–233
Beadle, George, 326–327
Beads-on-a-string model, 254
Behavioral traits, 6
Beijerinck, Martinus, 207
Beneficial mutation, 434, 573
Benign (cancerous) growth, 531
Berg, Paul, 459
Bertram, Ewart, 76
B-gene, 556
Bidirectional regulatory elements, 382
Bidirectional replication, 271, 280, 281
Binary fission, 23–24
Binding sites, genetic regulatory protein, 357
Binomial expansion equation, 61–62
Biological evolution, 10, 11
Biometric field of genetics, 586
Bioremediation, and metagenomics, 506
Biotechnology
    gene cloning, 460
    and metagenomics, 506
Birds
    sex determination in, 73
    sex-limited inheritance, 100
Bivalent chromatids, 30–31, 133, 135, 146
Blackburn, Elizabeth, 284
Blakeslee, Alfred, 178
Blending hypothesis of inheritance, 41
Blocker, non-coding RNA function, 414
Blood type, inheritance of, 98, 99
Blotting methods, 482–484
Bolotin, Alexander, 423
Bottleneck effect, 569
Boveri, Theodor, 55
Boycott, Arthur, 124

Branch migration, 286
BRCA-1 gene, 537, 540, 541
BRCA-2 gene, 537
Breeding true, 44
Brenner, Sydney, 299
Bridges, Calvin, 55, 84
Broad-sense heritability, 597
Brooker, Robert, 435
Bulge loop, RNA structure, 234

C

Caenorhabditis elegans, 12, 76, 417, 500, 504, 505, 512
Cain, Arthur, 566
Calico cats, 76, 77, 78–79
Cancer
    associated with viruses, 535
    definition, 531
    key characteristics of, 531–532
    and multiple genetic changes, 539–540
    and non-coding RNAs, 426–427
    and oncogenes, 532–535
    progression of cellular growth leading to, 531
    role of epigenetics in, 542–544
    and tumor-suppressor genes, 536
Cancer treatment, aimed at epigenetic changes, 544
Canis lupus familiaris, 504
Capping, 310, 316–317, 319
Capsella bursa-pastoris, 105–106
Capsids, 208, 209, 210
CAP site, 358, 359, 365, 366
Capsomers, 208
Carbohydrates, 4
Carbon Copy (cloned pet), 474
Carboxyl terminal domain (CTD), 308
Carboxyl-terminal end, 331
Cardiovascular diseases, 427
    and epigenetic changes, 542
    and non-coding RNAs, 427
Carothers, E. Eleanor, 55
Cartilage-hair hypoplasia (CHH), 426
Cas1 gene, 423
Cas2 gene, 423
Cas9 genes, 424
Cas genes, 423
Caspases, 537
Catabolite activator protein (CAP), 358, 359
Catabolite repression, 364, 365
Cattle, sex-influenced inheritance and, 99, 100
Cavalli-Sforza, Luca, 191
CCL3 gene, 163
cDNA. *See* Complementary DNA (cDNA)
cDNA library, 466, 467
Cell(s)
    basic organization of, 20
    composed of biochemicals, 3–4
    molecular organization of, 4
    prokaryotes *vs.* eukaryotes, 20–21
    protein function and, 4–5
Cell culture, A-1
Cell cycle
    checkpoint regulating, 537–538
    eukaryotic cells, 24–25
    growth factors, 532–533
Cell cycle control, 537, 538
Cell differentiation, and epigenetic changes, 396, 400–402

Cell disruption, A-2–A-3
Cell division, 19, 23–26
    cytogenetics and, 21, 22
    cytokinesis, 29
    epigenetic gene regulation, 400
    eukaryotic chromosomes during, 259
    and gene regulation, 355
    growth factors promoting, 532
    meiosis, 30–33
    mitosis, 27–29
    and tumor-suppressor genes, 537
Cell division pathways, and oncogenes, 532–534
Cell-free translation system, 334, 335–336
Cell plate, 29
Cell-to-cell recognition and signaling, 5
Cellular level
    proteins functioning at the, 7
    relationship between genes and traits at the, 7
Cellular molecules, 4
CENP-A, 246
CentiMorgans (cM), 141
Central dogma of genetics, 296, 297
Centrifugation, A-3–A-4
Centrifuge, A-3
Centrioles, 27
Centromeres, 25, 26, 38
    eukaryotic, 245–246
    and inversions, 164
    paracentric inversion, 164, 165
    reciprocal translocations, 167, 168
    Robertsonian translocation, 166
Centromeric locations, 157, 158
Centrosomes, 27
Cepaea nemoralis (land snail), 566
Cerevisiae
    knockout programs, 512
Cesium chloride (CsCl) gradient, 268
CFTR (cystic fibrosis transmembrane conductance regulator), 59
Chain termination, 477
Chalfie, Martin, 3
Chaperones, 333
Characters
    definition, 44
    Mendel's study of seven, 44, 45, 46
    Mendel's study of two-factor crosses, 50–52
    not assorting independently, 132
Charcot-Marie-Tooth disease, 161
Chargaff, Erwin, 228–229
Chargaff's rule, 229
Charged tRNA, 340
Chase, Martha, 223
Checkpoint proteins, 537–538
Chemical bonds, 4
Chemistry, of DNA replication, 278–279
Chemotherapy, 447
Chetverikov, Sergei, 572–573
Chiasma(ta), 31
Chickenpox, 213
ChIP-chip assay, 510
Chi square test
    to distinguish between linkage and independent assortment, 136–138
    and Hardy-Weinberg equation, 558, 559
    to test validity of a genetic hypothesis, 62–64
Chi square value, 63–64, 588
Chloroplast DNA (cpDNA), 114
Chloroplasts
    circular chromosomes in a nucleoid, 114
    endosymbiotic origin of, 119, 120
    extranuclear inheritance, 114–116
    genetic composition of, 114
    inheritance patterns varying among different species, 116
    plant cell, 21
    transmission among different organisms, 116
Chorionic villus sampling, 530
Chromatids, 25, 29
    banding patterns, 157, 158
Chromatin, 19
    definition, 253
    of eukaryotic cells, 253
    structure during interphase, 258
Chromatin immunoprecipitation (ChIP), 509, 510
Chromatin modifications, abnormal in cancer cells, 542–543
Chromatin remodeling, 386–387, 543
Chromatins
    histone variants role in structure and function of, 387–388
    HOTAIR altering structure of, 416–417
Chromatography, A-4
Chromocenter, 172
Chromosomal DNA
    using vectors, 460–461
Chromosomal rearrangement, 435–436, 452
Chromosomal translocation, 534–535
Chromosome compaction
    chromosomal loops, 241–242
    supercoiling, 242–243
Chromosome map, 492. *See also* Genetic maps/mapping
Chromosome painting, 494
Chromosomes, 5. *See also* Eukaryotic chromosomes
    bacterial (*See* Bacterial chromosomes)
    cancer development and alterations in, 540
    changes in number of, 168–169
    changes in structure of, 159–168
    classified by banding pattern, 157
    classified by centromeric location, 157
    classified by size, 158
    comparison of homologous, 23
    crossing over, 31
    definition, 19, 240
    deletion, 159, 160
    diploid cells, 21
    DNA and, 2
    duplication, 159, 161–163
    eukaryotic (*See* Eukaryotic chromosomes)
    following DNA replication, 26
    general features of, 19–23
    genetic linkage on, 131–132
    in human gamete, 34
    inversions, 164–166
    karyotpes, 21
    mechanisms producing variation in number of, 175–178
    in meiosis, 30
    meiosis I, 31, 33
    meiosis II, 33
    in Mendel's law of independent assortment, 52
    in Mendel's law of segregation, 47
    micrograph of, 5
    microscopy preparation, 21, 22
    microscropic study of, 156–158
    observed under microscope, 21, 22
    three-dimensional structure of DNA in, 225
    translocations, 166–167
    transmission during mitosis, 27–29
    usual number of, 156
    variation in number of sets of, 172–175
    variation in number within a set, 170–172
Chromosome sorting, 24, 25

INDEX I-5

Chromosome structure
    altering gene expression, 435, 437
    consequences of transposition on, 252
Chromosome territory, 257–258
Chromosome theory of inheritance, 55–58
Chromosome variation, 8
Chromosome walking, 526–527
Chronic myelogenous leukemia (CML), 534–535
Ciprofloxacin (Cipro), 244
Cis-acting elements, 306, 364
Cis-effect, 364, 367
Classical hemphilia, 523
Cleavage furrow, 29
Clonal (cancerous growth), 531
Cloning. *See also* Gene cloning
    human, 1
    mammalian, 1, 2, 473
    of mammals, 1, 2
    reproductive, 472–474
Cloning experiments, vectors used in, 461
Closed complex, 301
Closed conformation, 386
Cloverleaf pattern, tRNA structure, 339
Clownfish, 74
C-*myc* gene, 534, 535
Coactivators, 383
Cockayne syndrome (CS), 450
Coding strand, 302
Codominance inheritance pattern, 90, 98–99
Codons, 5, 339
    anticodon pairing, 340–341
    bacterial transcription, 297
    codon-anticodon recognition process, 340
    copolymer, 337
    and exceptions to the genetic code, 330
    initiation stage of translation, 345–346
    mRNA, 328–330, 331
Colchicine, polyploidy in plants, 178
Colinearity of gene expression, 309–310
Collared flycatcher (*Ficedula albicollis*), 567
Colorectal cancer, 539–540
Col-plasmids, 190
Column chromatography, A-4
Combinatorial control, 382
Compaction, chromosome, 241–243
Comparative genomics, 505
Competence factors, 199
Competent cells, 199, 463
Complementarity of DNA strands, 266–267
Complementary DNA (cDNA), 465, 508
    DNA microarray experiment, 508, 509
    RNA sequencing, 511–512
Complementary sequences, 231
Complementation, two gene inheritance pattern, 104, 105
Complete nondisjunction, 175
Complex traits, 584
    examples of, 584–585
Concordance, disease heritability, 519
Condense (chromatids), 27
Conditional lethal alleles, 101
Conditional mutants, 435
Congenital analgesia, 60
Conglomerate, 570
Conjugation (bacterial mating)
    and F factor, 188–189, 190
    genetic mapping experiments, 192–196
    genetic transfer of bacteria through, 186, 187–190
    and Hfr strains, 191–192
    requiring direct physical contact, 188
Conjugation bridge, 189
Consensus sequence, 300, 303–304, 313
Conservative model (DNA replication), 268
Constitutive exons, 315
Constitutive (unregulated) genes, 355
Continuous traits, 585
Control elements, gene regulation, 381
Copolymers, 336–337
Copy Cat (cloned pet), 474
Copy number variation (CNV), 162–163
Core enzyme, 300, 301
Core histone proteins, 253, 388, 389
Corepressor, gene regulation, 308, 356, 357, 370
Core promoter, 306, 307
Corn plant
    crossover studies, 138, 139
    micrographs of chromosomes, 157, 158
Correlation coefficient *(r)*, 542, 588, 598
Correns Carl, 42, 55
Correns, Carl, 95, 114–115
Cotransduction, 197–199
Coumarins, 244
Covalent histone modification, 395
Covalent modification, 386, 389
    of histones, 542–543
    transcription factor modulation, 384, 385
Covariance, 588
CpG islands, 392
Creighton, Harriet, 138
Cremer, Christoph, 257
Cremer, Thomas, 257
Crick, Francis, 228, 229, 230, 296, 338, 340
Cri-du-chat syndrome, 160
Crisanti, Andrea, 2
CRISPR-Cas system, 422–424, 512
CRISPR-Cas technology, 480–482
Crispr gene, 424
CRISPR locus. *See* CRISPR-Cas system
CRISPR RNA (crRNA), 415
Crop plants, selective breeding, 601
Crossed, definition, 42
Cross-fertilization, 44, 45, 47
Crossing over
    dependent on distance, 135–136
    explained, 133
    and inversion heterozygotes, 164–165
    meiosis, 31, 133
    relationship between linkage and, 133–139
crRNA, 480
CTC-binding factor (CTCF), 397–398, 399, 400
C (carboxyl group)-terminus, 331
Cut-and-paste mechanism, 248
Cyanobacterium, 119, 120
Cycle threshold method (C_t method), 472
Cyclic-AMP (cAMP), 364–365
Cyclin-dependent protein kinases (CDKs), 537
Cyclins, 537
Cysteine (Cys), 329, 332, 336, 337
Cystic fibrosis (CF), 59, 92, 102–103, 518, 521, 529
Cytogeneticists, 21, 156
Cytogenetic mapping, 492
Cytogenetics, 21
Cytokinesis, 28, 29, 33, 36–37
Cytological mapping, 492
Cytoplasmic inheritance, 114

Cytosine (C), 5, 225, 226, 229, 231, 234, 266
   deamination of, 441
   DNA methylation on, 392

D

Dactylorhiza sambucina (elder-flowered orchid), 91
Daphne Major, ground finches on, 555
DAPI dye, 494
Darwin, Charles, 10, 13, 554, 561
Darwinian fitness, 561
Daughter cells, 23, 24, 25, 27, 28, 29, 30, 33
Daughter strands, DNA, 267
Davis, Bernard, 188
Deamination, spontaneous mutations by, 441
Decondensed chromosomes, 27
Decoy, non-coding RNA function, 414
Deficiency, chromosome change through, 159
Degeneracy, 329, 340
Degradative plasmids, 190
Degrees of freedom, chi square test, 63
Deleterious mutation, 434, 573
Deletion, chromosome change through, 159, 160
Denaturation, 468, 469
De novo (new) methylation, 122, 393
Deoxyribonucleic acid (DNA)
   bacterial replication, 270–279
   base content in, 229
   B DNA and Z DNA alternatives, 232–233
   chromosomes and, 19
   double helix model (*See* DNA double helix)
   in a eukaryotic chromosome, 253
   eukaryotic replication, 280–284
   extrachromosomal, 21
   extranuclear, 21
   gene expression and, 5–6
   Human Genome Project and, 1
   levels of structure, 225
   macromolecule, 4
   mutagens altering structure of, 445–447
   origin of *nucleic acid* term, 224
   protein synthesis and, 5
   strand structure, 227
   structure, 225–226
Deoxyribonucleoside triphosphates (dNTPs), 278, 465, 468, 480
Deoxyribonucleotides (dNTPs), 465, 470
Deoxyribose, 226, 440, 441
Depurination, spontaneous mutations by, 440, 441
Dermal ridge count, 598–599
Development, 397
DeVries, Hugo, 42, 55
Diabetes
   and epigenetic changes, 542
Diakinesis stage, 31
Diauxic growth, 364
Dicentric bridge, 165–166
Dicentric chromosome, 165
Dideoxyribonucleotides (ddNTPs), 477
Dideoxy sequencing, 477, 478, 479
Diet, and epigenetic changes, 396, 397
Differential centrifugation, A-3
Differentially methylated region (DMR), 397, 398
Diffuse large B-cell lymphoma (DLBCL), 545
Dimeric DNA polymerase, 276

Dimer, protein subunit, 97
Dioecious plant species, 75
Diplococcus pneumoniae, 199
Diploid/diploid cells, 133
   definition, 9
   eukaryotic species, 21
   and haploid cells, 34
   meiosis, 30, 33
   meiosis versus mitosis, 33
   mitotic cell division, 29
   oogenesis, 35, 36
   process of mitois in, 27, 28
Diploid species, and Hardy-Weinberg equation, 557
Diploid sporophyte, 36
Diplotene stage of meiosis, 31
Directionality, 227
   polypeptide synthesis, 331
Direct (DNA) repair, 449–450
Direct repeats (DRs), 248, 249, 250
Discovery-based science
   definition, 15
   Morgan's fruit fly eye color study, 81–82
   Wolman and Jacob's *E. coli* mapping study, 194–195
Disease resistance, and heterozygote advantage, 97
Disequilibrium, 559
Dispersive model (DNA replication), 268
Displacement loop (D-loop), 288
Disruptive selection, 566
Distance
   and crossing over, 135–136
   and recombination, 141, 143
Dizygotic (DZ) twins, 519
DNA. *See* Deoxyribonucleic acid (DNA)
DnaA proteins, 271, 272
DNA-binding proteins, 484–485, 537
DNA double helix, 225
   ball-and-stick model of, 232
   core histone proteins, 253
   discovery of, 228–230
   forming different types of structures, 232–233
   space-filling model of, 232
   structures features, 230–232
   unwinding, and DNA replication, 273
DNA double-strand break (DSB), 452
DNA fingerprinting, 575–576
DNA gap repair synthesis, 288, 289–290
DNA gyrase, 243–244, 273
DNA helicase (DnaB protein), 271, 273
DNA library, 465–466
DNA ligase, 276, 462
DNA methylation, 122, 391–393
   Agouti gene, 403
   dietary effects on development, 403, 404
   gene mutations, 543
   heritability, 392–393
   inhibiting gene transcription, 392
   molecular model for, 394
   spermatogenesis, 398
DNA methyltransferase, 392
   cancer treatment, 544
DNA microarray analysis
   analyze transcription patterns in cloned mice, 473
   applications of, 509
   compared with RNA sequencing, 512
   identifying disease-causing alleles, 527
   identifying transcribed genes, 507–509

testing method for genetic abnormalities, 529
used to analyze DNA-protein binding at genome level, 509–510
used to classify types of tumors, 545–546
DNA mismatch repair, 288–289
DNA polymerase, 273, 274, 276
    chemistry of DNA replication, 278, 279
    eukaryotic, 281, 282
    proofreading function of, 279
DNA polymerase III, 278–279
DNA profiling, 575
DNA repair proteins, 539
DNA repair systems
    direct repair, 449–450
    homologous recombination repair (HRR), 450, 452
    mismatch repair system, 451–452
    nonhomologous end joining (NHEJ), 452–453
    nucleotide excision repair (NER), 450, 451
    translesion synthesis (TLS), 453
DNA replication
    accuracy of, 279
    bacterial, 270–279
    central dogma of genetics, 297
    chemistry of, 278–279
    DNA polymerase, 274–275, 281–282
    eukaryotic, 280–284
    explained, 266
    homologous recombination, 286
    key features of a bacterial origin of, 270–272
    key proteins involved with bacterial, 273
    leading and lagging strands, 274, 275–276
    origin of replication, 271, 272, 280–281
    replication fork, 273–275
    RNA primers, 274, 282–283
    structural basis for, 266–267
    synthesis of new strands, 273–277
    tautomeric shifts occurring prior to, 441–442
    telomerase, 283–284
    three-dimensional view of, 276, 277
    three models describing end result of, 268–270
DNA replication licensing, 281
DNase, 223
DNase I, 254–256
DNA sequence(s), 5
    1000 Genomes Project, 1
    Human Genome Project and, 1
    of *Saccharomyces cerevisiae*, 491
    transposable elements, 248–249
DNA sequencing, 477–479, 529. *See also* Genome-sequencing projects
    dideoxy method, 477–479
    function of, 477
    protocol for dideoxy method, 478
    steps involved in, 477–479
    technological advances in, 501–502
DNA strand(s)
    acting as templates for synthesis of new strands, 267
    structural features of, 227
DNA supercoiling, 242–243
DNA translocase, 387
DNA viruses, 535
Dogs
    selective breeding of, 600, 601
    white-spotting phenotype in, 103
    X-linked muscular dystrophy in, 83
Dolly (cloned mammal), 2, 473

Dominant alleles, 21–22, 91, 92, 96
Dominant mutant alleles, 92
Dominant-negative mutations, 92, 522, 922
Dominant/recessive relationship, 89, 90, 91, 92
Dominant trait, and genetic disease, 59
Dominant variant, 47
Dosage compensation, 75–76
Double crossover, 136, 137, 143, 144–145
Double helices, RNA, 234
Double helix. *See* DNA double helix
Double-strand break model, 286–288
Double-stranded RNA, 417–419
Down, John Langdon, 171
Down promoter mutation, 434
Down regulation, eukaryotic gene regulation, 382
Down syndrome (trisomy 21), 8, 171
Drosophila melanogaster (fruit fly)
    chromosomal rearrangements in, 435
    chromosomes, 169, 172, 173
    DNA methylation, 391
    dosage compensation, 76
    environmental factors, 94–95
    eye color mutations, 76
    gene affecting eye color, 80–81
    genetic linkage map of, 140
    genetic map, 492, 493
    genome sequencing, 504, 505
    lethal alleles in, 101
    longevity of two strains, 596
    meristic trait, 585
    mitotic recombination, 146, 147
    model organism, 12
    Morgan's crossover study, 134
    PIWI-interacting RNAs (piRNAs), 424
    polycombe response element (PRE), 402
    position effect, 435, 437
    repetitive sequences, 247
    RNA editing, 318
    sex chromosomes, 73
    in situ hybridization, 493
    testcross, 142
    transposable elements, 251, 252
    TrxG and PcG complexes in, 400
    wild-type alleles in, 91
Drugs, genotype determining proper dosage of, 546
Ds element, 249
Duchenne, Guillaume, 82
Duchenne muscular dystrophy (DMD), 82, 518, 523, 529
Duplication, chromosome change through, 159, 161–163
Dwarfism, 569
Dyad, 25, 33
Dystrophia myotonica (DM), 443
Dystrophin, 82, 83

E

E2F protein, 536, 537
E. coli chromosome, 197
*Eco*RI, 461
Edward syndrome, 171
Effector molecule, transcription factor modulation, 384, 385
Egg cell, 35
Eggs (female gamete), 43
Electrophoresis, A-4–A-5

Electrophoretic mobility shift assay (EMSA), 484–485, 509
Ellis-van Creveld syndrome, 569
Elongation (transcription), 298, 301, 319
Elongation factors, translation, 346
Elongation stage of translation, 344, 346–347
Embryogenesis, 120, 121, 123, 125
Embryonic development, 103
Embryonic germ cells (EG cells), 474, 475
Embryonic stem cells (ES cells), 474, 475
Embryo sac, 36–37
Emerging viruses, 217–218
Empirical approach, 45
Empirical laws, 45
EMSA (electrophoretic mobility shift assay), 484–485, 509
Endonuclease, 311, 317–318
Endopolyploidy, 173
Endosymbiosis theory, 119–120
Enhancers, 306, 308
    and regulatory transcription factors, 381, 382
    splicing, 316
Entamoeba histolytica, 504
Enterococcus faecalis, 201
Entry, in viral reproductive cycle, 212
Environmental stress, and gene regulation, 355, 380
Environment/environmental factors
    altering function of chromatin-modifying proteins, 543
    and epigenetic changes, 402–404
    epigenetics and, 395, 396, 397, 402–404
    and gene expression, 93–95, 96
    human disease and, 519
    induced mutations, 440, 447
    lac operon regulation, 360
    and lethal alleles, 101
    oxidative DNA damage, 443
    and phenotypic variance, 595–597
    phenylketonuria (PKU), 9
    polygenic inheritance, 590–591
    and random mutation, 439
    traits and, 8–9
Enzyme adaptation, 358
Enzymes, 5
    genes encoding, 326–328, 327–328
Epidermal growth factor (EGF), 533
Epigenetic gene regulation
    development of body parts, 400
    environmental agents, 395, 396, 397, 402–404
    genomic imprinting, 397–398
    molecular changes underlying, 395
    non-coding RNAs, 395, 396
    programmed changes during development, 395, 396
    transcription factors, 395, 396
    trithorax and polycomb groups promoting, 400–402
    X-chromosome inactivation (XCI), 398–400
Epigenetic inheritance, 113, 120–123, 394
Epigenetics
    and change in gene expression, 394
    definition, 394
    role in cancer, 542–544
Epimutation, 394
Episomes, 190, 213
Epistasis, 104–105
Epithelial-mesenchymal transition (EMT), 429
Epitopes, 483
Epstein-Barr virus, 208
Equilibrium density centrifugation, A-4
Equus, modern horse evolution, 11
Error-prone replication, 453
ER signal sequence, 421

Escherichia coli, 213
    anticodon sequence, 341
    attachment in viral reproductive cycle, 212
    bacterial chromosomes, 241
    bacterial transcription, 299
    bacteria reproduction, 24
    basic organization of cells, 20
    conjugation experiments, 192–196
    core enzyme, 300
    cotransduction experiment, 197–199
    deamination, 441
    DNA replication, 268, 269, 274, 276
    DNA replication in, 269, 271
    DNA supercoiling, 243
    gene knockout, 512
    gene mutations, 434, 435
    gene regulation, 358, 359
    gene sequences, 300, 504, 505
    genome sequencing, 504, 505
    Hershey and Chase T2 phage DNA study, 223
    Hfr strain/F factor, 191
    homologous recombination, 288
    horizontal gene transfer, 574
    Human Genome Project, 500
    intragenic suppressor, 435
    lac operon, 364
    lactose metabolism, 358, 359
    microdomains, 242
    mismatch repair in, 452
    model organism, 12
    NER system, 450, 451
    ompF gene, 372
    origin of replication in, 271, 272
    phages, 213
    processing of precursor tRNA molecule in, 311
    replicative DNA polymerases, 453
    riboswitches, 374, 375
    site-directed mutagenesis, 480
    supercoiling and, 243
    T2 phage, 223
    transcription, 299–300, 303
    transcription termination, 302, 303
    transfer of F factor, 190
    transfer of genetic material, 187–188
    trp operon, 368
Essential gene, 101
Ester bond, 278
Estradiol, 74
Ethyl methanesulfonate (EMS), 445
Euchromatin, 258
Eukaryotes
    about, 20–21
    binary fission, 23–24
    having two copies of most genes, 47
    horizontal gene transfer from a bacterium to, 574
    signal recognition particle (SRP) in, 421
Eukaryotic cell ribosomes, 341–342
Eukaryotic cells, 4, 21
    cell cycle, 24–25
    cell division, 532–533
    chromatin, 19
    dividing by cytokinesis, 29
    dividing by mitosis, 27–29
    dividing meiosis, 30–33
    endosymbiosis theory, 119–120
    meiosis and, 30
    Mendelian inheritance pattern, 113, 114
    microtubule-organizing centers (MTOCs), 27

mitosis and, 27–28
ribosome composition in, 341–342
Eukaryotic chromosomes
during cell division, 259, 260
crossing over between, 285–286
cytogenetics and, 21
general features of, 245
inherited in sets, 21–23
microscopic examination of, 156–158
mitotic spindle apparatus and, 26–27
nondividing cells, 253–258
nuclear matrix, 256–257
nucleosome structure, 253–256
organization of functional sites along, 245–246
radial loop domains, 256–257, 258, 259, 260
relicated by telomerese, 283–284
repetitive sequence, 248
30-nm fiber, 256
variation in genome size, 246–247
Eukaryotic DNA replication
DNA polymerases, 281–282
multiple origins of replication, 280–281
RNA primers, 282–283
similarities and differences with bacterial DNA replication, 280
telomerase, 283–284
Eukaryotic gene regulation. See Gene regulation (eukaryotic)
Eukaryotic genomes, sequencing, 504
Eukaryotic initiation factor (eIF2), 345
Eukaryotic ribosome, 342
Eukaryotic species
extranuclear inheritance, 114
multicellularity, 304–305
retrotransposition, 248
Eukaryotic transcription, 304–309
Eukaryotic translation
comparison of bacterial translation and, 348, 349
ribosomes, 341–342
stages of translation, 345–347
Euploid organisms/individuals, 169, 170
Euploidy, 172, 173–174, 177
European Conditional Mouse Mutagenesis Program (EUCOMM), 512
Evolution, 10
changes in horse genus *Equus*, 10, 11
and gene duplication, 161–162
of the globin gene family in humans, 162
and Hardy-Weinberg, deviation from, 559
and natural selection, 561–562
and quantitative genetics, 584
retrotransposons, 249
and transposable elements, 251–252
Exit site (E site), 343
Exons, 313, 314, 315
coding sequences within, 310
Exon shuffling, 252
Exon skipping, 316
Exonuclease, 309, 311
Experimental techniques, A-1–A-6
Experiments on Plant Hybrids (Mendel), 42
Expression (CRISPR-Cas system), 424
Expression phage, 424
Expressivity, 93
Extrachromosomal DNA, 21
Extranuclear DNA, 21
Extranuclear inheritance
chloroplasts, 114–117
definition, 114
mitochondria, 117–119

F

F_1 generation, 45, 46, 47, 51–52, 63, 89, 124
F_2 generation, 45, 46, 47, 51, 63, 89, 124–125, 132, 134, 135, 137, 141–142, 143, 144
F_3 generation, 125
Familial breast cancer, 540–541
Familial forms of cancer, 540–541
Familial hypercholesterolemia, 522
F^- cell, 191, 192–193
Feedback inhibition, 372–373
Feedback loops, 395
Felis concolor coryi (Florida panther), 572
Female gametes (eggs), 43
Ferritin mRNA, 406
Fertility plasmids, 190
Fertilization, Mendel's study of pea plants, 43
F (fertility) factor, 188–189, 190, 191, 192, 362
F′ factors (F prime factors), 191, 192, 362
Fibrillin-1 gene, 522
Fidelity, bacterial DNA replication, 279
Fingerprints, dermal ridge count in, 598–599
Fire, Andrew, 417, 418, 419
Fisher, Ronald, 554, 561
5-bromouracil (5BU), 446–447
5′ cap/capping, 310, 316–317
5′ direction, 302, 310, 312, 313, 314
5′ end of exon, 312, 313
5-methylcytosine, 392, 441
5′-untranslated region, 328
Fixation, 572
Flap endonuclease, eukaryotic DNA replication, 282–283
Flemming, Walther, 55
Florida panther *(Felis concolor coryi)*, 572
Flower color, 95
Flowering plants
gamete production, 36, 37
reproduction, 35
Fluorescence in situ hybridization (FISH), 493–494
Fol, Hermann, 55
Forked-line method, for independent-assortment problems, 53, 54
Founder, disease-causing allele, 526, 528
Founder effect, 569
Four-o'clock plant. See *Mirabilis jalapa* (four-o'clock plant)
Fraction, centrifugation, A-3
Fraenkel-Conrat, Heinz, 210
Fragile X syndromes (FRAXA/FRAXE), 443
Frameshift mutations, 432, 433, 573
Franklin, Rosalind, 228, 229
Fraternal twins, 519, 598, 600
Free radicals, and mtDNA, 118
French press, cell disruption, A-3
Frequency distribution, quantitative traits, 585, 586–587
Frog, euploidy in, 172
Fruit flies, 80, 169
micrographs of chromosomes, 157, 158
FtsZ (protein), 24
Full methylation, 392
Functional genomics, 506–512
definition, 491
DNA microarrays, 507–510
RNA sequencing, 510–512
Functional sites
bacterial chromosomes, 240–241
eukaryotic chromosomes, 245–246

G

G_0 phase of cell cycle, 24
G_1 phase of cell cycle, 24–25
G_2 phase of cell cycle, 24, 25
G8-C marker, 526, 527
Gabrb gene, 123
Gain-of-function mutations, 92, 522
Gallo, Robert, 534
Galton, Francis, 586, 598
Gamete formation, genomic imprinting during, 397–398
Gametes, 9, 10, 21, 34
   definition, 42
   Mendel's law of segregation and, 47
   Punnett square, 48–49
Gametogenesis, 34, 121–122, 521
Gametophytes, 36–37
GA-rich region, 416
Garrod, Archibald, 326
GATC methylation sites, 271
G bands, 157
Gearhart, John, 475
Gel electrophoresis, A-4–A-5
Gel filtration chromatography, A-4
Gellert, Martin, 243
Gene(s)
   definition, 47, 296
   encoding enzymes, 326–328
   and the environment, 8–9
   following Mendelian inheritance patterns, 113
   Mendel's law of segregation and, 47
   molecular expression of, 3–6
   passed from parent to offspring, 9–10
   relationship between traits and, 6–10
   tumor-suppressor, 536–541
   variation (*See* Genetic variation)
Gene amplification, 534
Gene chip, 507. *See also* DNA microarray analysis
Gene cloning
   complementary DNA (cDNA), 465
   definition, 460
   DNA library, 465–466, 467
   restriction enzymes used in, 461–462, 463
   reverse transcriptase, 465
   steps in, 463–465
   uses of, 460
   viral vectors, 461
Gene conversion, 286, 288–290
Gene duplication, 161
Gene editing, 479–482
Gene expression, 5, 6, 296
   aneuploidy causing imbalance in, 170–172
   base sequence roles during, 297–298
   changes in chromosome structure altering, 435, 437
   colinearity of, 309
   consequences of transposition on, 252
   and dosage compensation, 75–76
   environmental effects on, 93–95
   and epigenetic effect, 394
   gene mutations influencing, 434, 435, 437
   incomplete penetrance, 92–93
   levels of, subject to regulation, 381
   multiple phenoytic effects, 102–103
   regulation, 372
Gene families, 161–162

Gene flow, 570
Gene interaction, 104–106
Gene knockout, 121, 512
Gene knockout collections, 512
Gene mutagenesis, 479–482
Gene mutations, 8
   altering the coding sequence of a gene, 432–433
   as molecular changes in DNA sequence of a gene, 432
   molecular genetics and, 13
   within a noncoding sequence, 434
   pleiotropy, 102–103
   wild-type genotype/phenotype, 434–435
Gene order, and three-factor crosses, 143
Gene pool, 554–555
General transcription factors (GTFs), 306, 307–308, 381
Gene redundancy, 104, 105–106
Gene regulation (bacteria)
   in bacteria, 356
   dominant alleles and, 91
   explained, 355
   inducible *vs.* repressible, 370
   lac operon, 358–366
   posttranslational, 372–373
   riboswitches, 373–374
   translational, 372
   trp operon, 368–371
Gene regulation (eukaryotic). *See also* Epigenetic gene regulation
   chromatin remodeling, 386–387
   DNA methylation, 391–393
   epigenetic, 394–397
   explained, 380
   histone code, 388–389
   histone modifications, 390–391
   histone variants, 387–388
   levels of gene expression subject to, 381
   nucleosome-free region (NFR), 389–390
   regulatory transcription factors, 381–385
Gene repression, and HOTAIR, 416–417
Gene silencing, 122, 123, 400–402
Genetic abnormalities
   genetic screening identifying, 530
   testing methods for, 529
Genetic approach, 13
Genetic code, 5
   codons, 328–329
   exceptions to, 330
   experimental determination of, 334–338
Genetic cross(es), 12–13
   chi square test, 62
   first systematic studies of, 41
   probability of outcomes, 59–60
   Punnett square used to determine outcome of, 48–49
Genetic disease(s). *See also* Human disease
   association studies, 526–528
   autosomal dominant inheritance, 521–522
   autosomal recessive inheritance, 520–521
   inheritance patterns of, 519–524
   mutant alleles, 91
   ncRNA gene associated with, 426
   pedigree analysis and, 59
   X-linked dominant inheritance, 523
   X-linked recessive inheritance, 522–523
Genetic drift, 564, 567–569
   bottleneck effect, 569
   definition, 567
   founder effect, 569
   and Hardy-Weinberg equation, 557

hypothetical simulation of, 568
and microevolution, 560
migrations vs., 571
and population size, 568–569
Genetic linkage
overview, 131–132
relationship between crossing over and, 133–139
Genetic linkage map, 140
Genetic maps/mapping, 131, 491, 492
calculating map distance, 141–143
comparison of types of, 492
conjugation studies, 192–196
cytogenetic, 492
definition, 492
linkage, 492, 494, 496–497
molecular markers for, 495–497
physical, 492
purpose of, 139–140
in situ hybridization, 492–494
three-factor crosses used for, 143–145
transduction studies, 197–199
usefulness of, 140
Genetic material
DNA identification as, 221–224
primary function of, 240
T2 phage, 223–224
Genetic mosaic, 437
Genetic polymorphism, 91
Genetic recombination, 56–57, 133, 286
Genetics
as an experimental science, 14
definition, 3
fields of, 12–14
levels of organization, 7
overview, 1–10
science of, 14–15
Genetic screening, 529
Genetic testing/tests, 518
amniocentesis, 530
chorionic villus sampling, 530
definition, 529
performed prior to birth, 530–531
preimplantation genetic diagnosis (PGD), 530–531
Genetic transfer
explained, 186
three mechanisms for, 186–187
through bacterial transduction, 196–199
through conjugation, 187–192
Genetic variation
explained, 8, 156
haplotypes exhibiting, 525
inherited differences in traits due to, 8
and microevolution, 560
and mutations, 572–573, 572–574
1000 Genomes Project and, 1
phenotypic variance, 595
and polymorphism, 556
and population genetics, 554, 555
population genetics and, 13
sources of new, 573
through repetitive sequences, 574–576
Gene transcription. *See* Transcription
Genital herpes, 213
Genome, 1
definition, 240, 491
viral, 210

Genome maintenance, 537–538
Genome-sequencing projects, 498–505
first genome sequenced (*H. influenza*, 1995), 498–500
Human Genome Project, 500–501
motivation behind, 505
next-generation sequencing technologies, 501–502
pyrosequencing, 501–502, 503
results of, 504
shotgun sequencing, 498–499
Genome-wide association study (GWA study), 527–528
Genomic imprinting, 120–123
epigenetic changes, 396, 397–398
epigenetic gene regulation, 395
example, in a mouse, 120, 121
examples of human genes imprinted, 123
during gametogenesis, 121–122
and human diseases, 122–123
marking process, 122
Genomic library, 466, 467
Genomics. *See also* Genetic maps/mapping
chromosome mapping, 492
definition, 491
functional genomics, 507–512
molecular markers, 495–497
in situ hybridization, 492–493
Genotype
definition, 48
importance for proper dosage of drugs, 546
interaction with the environment, and phenotypic class, 590–591
overlapping with phenotypes for polygenic traits, 592
sex-influenced inheritance, 99
Genotype-environment association, 596–597
Genotype-environment interaction, 596
Genotype frequencies, 556–558, 563
Geospiza agnirostris (ground finch), 555
Germ-line mutations, 437
Giemsa, Gustav, 157
Giemsa stain, 157, 158
Gierer, Alfred, 210
Gilbert, Walter, 477
Global change, and metagenomics, 506
Global Ocean Sampling Expedition, 506–507
Globin genes, 162
Glucocorticoid hormones, 384, 386
Glucocorticoid receptor, 385, 386
Glucose, 4, 364–365
Glutamic acid (Glu), 329, 332, 336, 337
Glutamine (Gln), 332, 336
Glycine (Gly), 329, 332, 336
Goodness of fit, 62
Gotland, collared flycatcher on, 567
Gradient centrifugation, 268, A-3–A-4
Gram-negative nonsulfur purple bacteria, 120
Green fluorescent protein (GFP), 2, 3
Greider, Carol, 284
Griffith, Frederick, 222–223
Groudine, Mark, 534
Ground finch *(Geospiza agnirostris)*, 555
Group II introns, 312
Group I introns, 312
Growing cells methods, A-1–A-2
Growth factors, 532–533
Guanine (G), 5, 225, 226, 229, 231, 234, 266
Guillain-Barré syndrome, 217
Gustafsson, Lars, 567

H

H1 (histone protein), 253
H1N1 (swine flu), 218
H19 gene, 397
Haemophilus influenzae, 241, 491, 498–500
Hairpins (stem-loop), RNA structure, 234
Haldane, J.B.S., 554, 561
Haplodiploid species, 172
Haplodiploid system, 73
Haploid bacteria, 286
Haploid cells, 30, 33, 133
Haploid, definition, 9
Haploid egg, 36
Haploid gametes, 34, 56
Haploid gametophyte, 36
Haploid genome sizes, 246
Haploinsufficiency, 92, 522
Haplotype (haploid genotype), 528
    association studies, 526–527
    definition, 525
    exhibiting genetic variation, 525–526
    schematic representation, on human chromosome, 526
Hardy-Weinberg equation, 557–559, 563, 571
Hardy-Zuckerman-4 feline sarcoma virus, 535
Hawaiian happy-face spider, 555
Hayes, William, 191
Heat shock proteins (HSP), 385
Helicobacter pylori, 504
α helix, 333
Helix-turn-helix motif, 300
Hellcos Biosciences tSMS, 502
Helper T cells, 212, 217, 218
Hematopoietic stem cells (HSCs), 474–475, 476
Hemimethylation, 392, 451–452
Hemizygous, X-linked gene in male, 80
Hemophilia, 519, 523
Hemophilia A, 523
Hemophilia B, 523
Hepatitis B, 525
Hereditary traits, 41
Heritability, 595–600
    broad-sense, 597
    definition, 597
    dermal ridge count in human fingertips, 598–599
    DNA methylation, 392–393, 394
    and environmental interactions and associations, 595–597
    narrow-sense, 597–598
    realized, 602–603
    and selective breeding, 602–603
Herpes simplex type I, 213
Herpes simplex type II, 208, 213
Herpes viruses, 213, 535
Hershey, Alfred, 223
Hertwig, Oscar, 55
Heterochromatin, 258
Heteroduplex, 200, 286
Heterogametic sex, 73
Heterogamous organisms, 34
Heterogamous species, 116
Heterogeneous environments, and disruptive selection, 566
Heteroplasmy, 115, 119
Heterozygote advantage, 96–98, 564–565
Heterozygotes, 52, 58, 59, 91, 558, 559
Heterozygous individuals, 23, 47, 52, 56, 59
HexA (hexosaminidase A), 520, 529

Hfr (high frequency of recombination) strains, 191–193, 192, 194, 202
Highly repetitive sequences, 247
High-throughput sequencing, 501
Hippocrates, 41
Histidine (His), 332, 336, 337
Histone acetyltransferase, 388
Histone code hypothesis, 388–389
Histone eviction, chromatin-remodeling, 387
Histone H3, 246
Histone modifications. *See also* Covalent histone modification
    effect on nucleosome structure, 389
    effect on transcription, 416–417
    epigenetic changes, 541
    and gene transcription, 388–389
    and transcriptional activation, 390
Histone proteins, 253, 254, 261
Histones, covalent modification of, 542–543
Histone variants, 387–388, 391, 395
HIV (human immunodefiency virus), 208, 212–213, 217
Hoagland, Mahlon, 338
Holandric genes, 80
Holley, Robert, 339
Holliday junction, 286, 287
Holliday model, 286, 287, 288
Holliday, Robin, 286
Holoenzyme, 300, 301
Holt, Sarah, 598
Homodimer, 97
Homogametic sex, 73
Homogenization, cell disruption, A-3
Homologous chromosomes, 21, 23, 26, 47, 56–57
    crossing over and, 133, 134, 136, 138
    double-strand break model, 286, 288
Homologous genes, 162
Homologous recombination, 199–200, 266
    explained, 285–286
    gene conversion, 286, 288–290
    Holliday Model, 286, 287
    proteins playing a role in, 288
Homologous recombination repair (HRR), 450, 452
Homologs, 9, 21–22
Homozygotes, 558, 564–565, 572
Homozygous individual, 23, 47, 91
Honeybees *(Apis mellifora)*
    dietary effects on development, 403, 405
    sex differences in, 73
Horizontal gene transfer, 200–201, 574
Hormone levels, sex determination and, 74–75
Horowitz, Rachel, 256
Horse, evolution of modern-day, 10, 11
Horvath, Philippe, 423
Host cell, 461
    definition, 208
    viral envelope, 208, 209
    viral reproductive cycle, 212, 213
Host range, 208
HOTAIR, 415, 416–417, 429
Hot spots, gene mutation, 439
Housekeeping genes, 392
*Hox*C genes, 416
Hox genes, 416
Hox transcript antisense intergenic RNA. *See* HOTAIR
Huberman, Joel, 280
Hull color, bread wheat, 491, 590
Human chromosomes, 21, 22
    classified according to centromeric location, 157
    complement of, in somatic cells and gametes, 10

deletions in human chromosome 1, 160
G bands, 157, 158
Giemsa stain, 158
haplotypes on, 525, 526
human genome, 2
Human cloning, 1
Human disease
  caused by mitochondrial mutations, 118–119
  complex traits, 585
  examples of recessive, 92
  genetic tests identifying, 529–530
  and genomic imprinting, 122–123
  pedigree analysis and, 59
  potential uses of stem cells for, 475–476
  role of non-coding RNAs in, 426–428
  threshold traits, 585
  and trinucleotide repeat expansion (TNRE), 443, 444
  X-linked recessive inheritance pattern, 82, 83
Human disorders. *See also* Genetic disease(s)
  inherited in an autosomal dominant manner, 522
  inherited in an autosomal recessive manner, 521
  inherited in an X-linked recessive manner, 523
Human fingerprint patterns, heritability of, 597–598
Human genome
  unique and repetitive DNA sequence in, 247
Human Genome Project, 1, 2, 500–501
Human medicine
  and metagenomics, 506
Human pedigree
  inbreeding in, 571
  symbols used in, 58–59
Humans
  studying inheritance patterns in, 58–59
  transposable elements, 251
Human viruses, latency in, 213, 215
Human virus, first, 207
Hungerford, David, 534
Huntington disease, 101, 443, 518, 521–522, 529
  association studies identifying mutant gene causing, 526–527
Hybridization, 42, 493
Hybrids, 42
Hyla chrysoscelis, 172
Hyla versicolor, 172
Hypothesis testing, 14
  chi square test, 62–63, 136, 138
  lacI gene encoding a diffusible repressor protein, 362–364
  probability calculations used in, 60

I

Identical twins, 519, 598, 600
Igf2 gene, 121, 122, 123, 397–398
Igf2R gene, 123
Ilex opaca (American holly), 75
Imprinting control region (ICR), 122, 123, 397, 398
Imprinting, genomic, 120–123
Inborn error of metabolism, 326
Inbreeding, 595
Inbreeding coefficient *(F),* 571–572
Inbreeding depression, 572, 575
Incoming deoxyribonucleoside triphosphates, 274
Incomplete dominance, 90, 95–96
Incomplete penetrance, 90, 92–93, 521
Incontinentia pigmenti, 524
Independent assortment

chi square test used to distinguish between linkage and, 136–138
and genetic recombination, 56–57
Mendel's law of, 50–53, 56, 57
not occurring, 132
Induced (operon), 359
Induced mutations, 440, 445–448
Inducer, in gene regulation, 308, 356, 357
Inducible genes, 308, 356, 357
Infant deaths, and genetic abnormalities, 530
Influenza virus, 208, 209, 217
Information, genetic material criteria, 221, 223
Inheritance, 9
  of blood type, 98, 99
  pseudoautosomal, 80
Inheritance, particulate theory of, 47
Inheritance patterns. *See also* Mendelian inheritance
  in chromosome theory of inheritance, 55
  codominance, 90, 98–99
  dominant/recessive inheritance, 47, 91–93
  of familial breast cancer, 541
  human disease, 519–524
  incomplete dominance, 90, 95–96
  incomplete penetrance, 90, 92–93
  lethal alleles, 90, 101–102
  maternal effect, 124–126
  Mendel's experiments, 41–42
  overdominance, 90, 96–98
  overview of simple, 89–90
  sex-influenced inheritance, 90, 99, 100
  sex-limited inheritance, 90, 100
  Simple Mendelian, 89, 90
  studying in humans (pedigree analysis), 58–59
  X-linked, 80–83, 90
  X-linked recessive, 82, 522–523
  Y-linked, 80
Inhibitor, in gene regulation, 308, 356, 357, 370
Initiation (transcription), 298, 308, 319
Initiation phase, X-chromosome inactivation, 78, 79
Initiation stage of translation, 344–345, 346
Initiator tRNA, 344
Insects, sex determination in, 73
Insertion element (IS element), 248
In situ hybridization, 492–494, 529
Integrase, 212, 250
Integration, in viral reproductive cycle, 212
Intercalating agents, 446
Interference (CRISPR-Cas system), 424
Intergenic regions, 241
Intergenic suppressors, 435, 436
Internal loop, RNA structure, 234
Internal nuclear matrix, 257
Interphase, 24, 28
  chromosomes in cell nucleus during, 257–258
Interphase chromatin compaction, 257–258
Interrupted mating experiment, 192
Interstitial deletion, 160
Intervening sequences, 310. *See also* Introns
Intragenic suppressors, 435, 436
Intrinsic termination, 303
Introns, 245, 310, 312–313, 314, 315, 318
Invasive cancer cells, 531
Inversion, chromosome change through, 159, 164–166
Inversion heterozygote, 164–165
Inversion loop, 164, 165
Inversions, 159, 164–166
Inverted repeats (IRs), 248, 249
In vitro fertilization (IVF), 529–530

In vitro translation system, 334
Ion-exchange chromatography, A-4
Ionizing radiation, 447
Ion semiconductor sequencing, 502
Iron assimilation, 405–406
Iron regulatory proteins (IRP), 406
Iron response element (IRE), 405, 406
IS1 transposable element, 251
Isoacceptor tRNAs, 340
Isogamous organisms, 34
Isoleucine (Ile), 329, 332, 336, 337
Isotopes, useful in genetics, A-6
Ivanovski, Dmitri, 207

J

Jackson, David, 459
Jacob, François, 192–195, 358, 360, 364
Jacobs syndrome, 171
Jaenisch, Rudolf, 473
Janssens, Frans Alfons, 134
Jeffreys, Alec, 575
Jellyfish, 2, 3
Jirtle, Randy, 403, 404
Johannsen, Wilhelm, 47

K

Kaiser, A. Dale, 459
Karyotype, 21, 22, 157, 158
Karyotyping, 529
Khorana, H. Gobind, 334, 336–337
Kinetochore, 25, 26, 27, 33, 246
Kinetochore microtubules, 27, 28, 29, 30, 31, 33
Klinefelter syndrome, 171
Knockout Mouse Project, 512
Knudson, Alfred, 536
Kölreuter, Joseph, 41
Kornberg, Roger, 253, 308
Kozak, Marilyn, 345
Kozak's rules, 346

L

L1 transposable element, 251
LacA gene, 358
LacI mutation, 359–360, 360–362, 361, 362
LacIS mutations, 359–360
Lack, David, 568
LacI gene, 359, 360, 362
LacI mutants, 376
Lac operon, 358–366
  catabolite repression, 364–365
  cycle of induction and repression, 361
  cyclic-AMP (cAMP), 364–365
  inducible and under negative control, 358–359
  inducible form of regulation, 370
  mechanism of induction of the, 360
  operator sites, 365–366, 367
  regulated by a repressor protein, 358–360
Lac operon promoter, 303, 304
Lac repressor, 358–360, 361, 364, 365–366, 367
Lactose
  catabolite activator protein (CAP), 365
  and catabolite expression, 364
  E. coli metabolism, 358
  lacI gene experiment, 362–364
  lac operon, 360, 361
  and *lac* operon, 359–360
Lactose metabolism, 358–359
LacY gene, 358
LacZ gene, 358, 463
Lagging strand, bacterial DNA replication, 273, 274, 275–276
Lamarck, Jean-Baptiste, 439
Lancaster County, Pennsylvania, 569
Land snail *(Cepaea nemoralis)*, 566
Latency
  in bacteriophages, 213
  in human viruses, 213
Latent, definition, 213
Law of independent assortment, 50–54, 56, 131
Law of inheritance, 59, 60
Law of segregation, 45–49, 56
Laws of inheritance, Mendel's, 9
Leading strand, bacterial DNA replication, 273, 274, 275–276
Leber hereditary optic naturopathy (LHON), 118, 119
Lederberg, Esther, 188, 439
Lederberg, Joshua, 187, 191, 439
Leder, Philip, 337
Left-handed helix, 232
Lejeune, Jérôme, 171
Leptotene stage of meiosis, 30, 31
Lesch-Nyhan syndrome, 92
Lethal alleles, 90, 101–102
Lethal mutation, 434
Leucine (Leu), 329, 332, 336, 337
Light microscope, A-2
Linkage
  overview, 131–132
  relationship between crossing over and, 133–139
Linkage disequilibrium, 526
Linkage groups, 132
Linkage mapping, 492, 496–497, 514. *See also* Genetic maps/mapping
  genetic linkage map, 140
Linker histone, 253, 254
Linker region, nucleosome structure, 252–253, 256
Lipids, 4
Livestock, cloning of, 1
Lobban, Peter, 459
Local populations, 555
Locked nucleic acids (LNAs), 427
Locus heterogeneity, 523
Locus/loci, 23, 492
Long non-coding RNAs (lncRNAs), 415, 427
Long terminal repeats (LTRs), 249, 250
Loop domains, radial, 246–257, 258, 259, 260
Loss-of-function allele, 13, 92, 105–106
Loss-of-function mutation, 13, 101, 185
Loss of heterozygosity (LOH), 541
Lou Gehrig disease, 427
LSD1, 416
LTR retrotransposons, 249
Lymnaea peregra, 124
Lyon hypothesis, 76–77
Lyon, Mary, 76
Lyosozyme, 215
Lysine (Lys), 329, 332, 336
Lysis, 196, 198
Lysogenic cycle, 212, 213, 214
Lysogeny, 213
Lytic cycle, 213, 215, 216

M

Mackay, Trudy, 596
MacLeod, Colin, 223
Macrodomains, 242
Macromolecules, 4, 5
Maintenance methylation, 393
Maintenance phase, X-chromosome inactivation, 78, 79
Major groove, DNA double helix model, 232
Malaria, 96, 565
Male gametes, 34–35, 43
Maleszka, Ryszard, 403
Malignant growth, 531
Mammalian cloning, 1, 2, 472–474
Mammalian genome size, 246
Mammals
    dosage compensation in, 76
Manhattan plot, 528
Manx cat, semilethal alleles, 101, 102
Map distance, 141, 143, 145
Mapping, definition, 492. *See also* Genetic maps/mapping
Mapping functions, 141
Mapping studies, 131
Map units (mu), 141
Marfan syndrome, 522
Marking process, genomic imprinting, 120, 122
Maternal age, Down syndrome and, 171–172
Maternal effect, 113, 124–126
Maternal inheritance pattern
    chloroplasts, 115
    mitochondria, 118
    and transmission of mitochondria, 118
Matrix-attachment regions (MARs), 257
Matthaei, J. Heinrich, 335
Maturases, 313
Maxam, Allan, 477
Mayer, Adolf, 207
McCarthy, Maclyn, 223
McClintock, Barbara, 138, 248
McClung, Clarence, 55, 73
McKusick, 569
MCM (minichromosome maintenance), 281
MCM helicase, eukaryotic DNA replication, 281
Mean fitness of the population, 563, 572
Mean, statistical method of, 586–587
Mediator, 307, 308
    and regulatory transcription factors, 383, 384
Medicine
    and metagenomics, 506
    personalized, 544–545
Megaspores, 36
Meiosis, 30–33
    crossing over during, 133
    gamete production and, 36, 37
    Mendel's law of segregation and, 47–48, 56
    Mendel's laws of inheritance and, 50–53, 55, 56–57
    metaphase of, 31–32
    nondisjunction, 175, 176
    reciprocal translocations, 167
    similarity to mitosis, 30
    stages of prophase of, 31
Meiosis I, 167, 175
    anaphase, 32, 33, 167, 174
    compared with mitosis and meiosis II, 33
    metaphase, 31, 32, 33
    nondisjunction, 171, 176
    phases of, 32
    prometaphase of, 31
    prophase of, 30–31
    spermatocyte, 35
    telophase, 32
Meiosis II, 167, 175, 176
    compared with mitosis and meiosis I, 33
    nondisjunction, 176
    phases of, 32
    secondary oocyte and, 36
    sorting events occurring during, 33
    spermatocyte, 35
Meiotic nondisjunction, 175
Meiotic segregation, 167, 168
Mello, Craig, 417, 418, 419
Mendel, Gregor, 9
    early life, 42
    population genetics, 554
    study of pea plants, 42–45, 96, 104
    transmission genetics and, 12
Mendelian inheritance. *See also* Inheritance patterns
    chromosome theory of inheritance, 55–58
    definition, 89
    and hypothesis testing, 62
    inheritance patterns involving two genes, 104–106
    Mendel's law of independent assortments, 52–54
    Mendel's law of segregation, 47–49
    pea plants, study of, 42–45
    popular views before, 41
    rules related to Mendelian inheritance patterns, 113
    simple Mendelian inheritance, 89, 90
    single-factor cross experiments, 45–47
    two-factor cross experiments, 50–52
Mendelian inheritance patterns, 518, 519
Mendel's 3:1 phenotypic ratio, 47–48
Mendel's law of independent assortment, 52–54, 55, 56–57, 131
Mendel's seven characters, 44, 45, 47, 91
Meristic traits, 585
Merozygote, 362, 364, 367
Meselson, Matthew, 268, 270
Messenger RNA (mRNA), 5. *See also* Translation
    alternative splicing, 315, 316
    bacterial gene sequences, 297
    bacterial transcription, 299
    base sequences within, 298, 328
    binding to a ribosome, 374–375
    central dogma of genetics, 296, 297
    codons in, 328–330
    and colinearity, 309–310
    DNA microarray experiment, 508–509
    eukaryotic modification, 308
    and experimental determination of the genetic code, 334
    functions, 325–326
    organization of base sequences, 297, 298
    polyA tail, 317–318
    polycistronic, 358
    recognition between tRNAs and, 338
    ribosomes, 343
    RNA editing, 318
    RNA polymerase II, 305
    RNA sequencing, 511
    RNAs inhibiting, 417–419
    RNA splicing, 310, 312
    during translation, 296, 309, 325
    translational repressors binding to, 372
    trp operon, 370
Metabolism, and gene regulation, 355
Metacentric centromere, 156, 158
Metagenome, 505

Metagenomics
   definition, 505
   general strategy of, 506
   Global Ocean Sampling Expedition, 506–507
   strategy of, 505–506
   uses of, 506
Metaphase, 28, 29
   meiosis I, 31, 32, 33
   meiosis II, 32
Metaphase chromosomes, 258, 259, 260, 261
Metaphase plate, 29
Metastasis, and miR-200 family, 429
Metastatic cancer cells, 531
Methionine (Met), 329, 332, 336
Methylation process, 123, 124. *See also* DNA methylation
Methyl-CpG-binding proteins, 392, 393
Mex-3 gene, 417, 418–419
Mic2 gene, 80
MicF antisense RNA, 372
Microarrays. *See* DNA microarray analysis
Microdomains, 242
Microevolution, 560
Microphthalmia-associated transcription factor (MITF), 103
MicroRNAs (miRNAs), 414, 415, 419, 427–428
Microsatellites, 574–575, 593
   molecular markers, 495, 496–497
Microscopic study of chromosomes, 156–158
Microscopy, A-2
Microsporocytes, 36
Microtubule-organizing centers (MTOCs), 27
Microtubules, mitotic spindle, 27, 29, 30
Miescher, Friedrich, 224
Migration, 570
   and Hardy-Weinberg equation, 557
Miniature allele, 85
Minimal medium, definition, 187
Minisatellite, 574–575
Minor groove, DNA double helix model, 232
Minutes, bacterial genetic map, 196
MiR-200 family, 426–427
Mirabilis jalapa (four-o'clock plant), 114–115
Mismatch repair, 450
Mismatch repair system, 451–452
Missense mutations, 432, 433, 434, 534, 573
Mitochondria, 21
   circular chromosomes in a nucleotide, 117
   endosymbiotic origin of, 119, 120
   extranuclear inheritance, 114, 117–119
   genetic composition of, 117
   and human disease, 118–119
   maternal inheritance pattern, 118
Mitochondrial DNA (mtDNA), 117–118
Mitochondrial genome, 117, 118
Mitosis, 25, 27–29
   compared with meiosis I and meiosis II, 33
   similarity to meiosis, 30
Mitotic nondisjunction, 175, 176
Mitotic recombination, 146, 147
Mitotic spindle, 26–28, 29–30, 32
MN blood type, 558
Model organisms, definition, 12
Moderately repetitive DNA sequence, 247
Modifying genes, 104
Molecular genetics, 13, 90, 221
Molecular level, genes expressed at the, 3, 5–6, 7
Molecular markers, 495–497
   genetic variation, 525
   for Huntington disease, transmission pattern, 527
   quantitative trait locus, 593, 594
Molecular mechanisms, gene regulation, 380, 387, 395
Molecular model, inheritance of DNA, 394
Molecular profiling, 544–545
Monad, 25, 29, 31, 32
Monoallelic expression, 120
Monod, Jacques, 358, 360, 364
Monohybrids, 45
Monomorphic gene, definition, 556
Monomorphic strains, 595
Monosomic, copy of chromosomes, 169
Monozygotic (MZ) twins, 519
Montgomery, Thomas, 55
Morgan, Thomas Hunt, 55, 80, 134, 135–136
Morphological traits, 6
Morphs, definition, 8
Mosaicism, 175, 177
Mosquitoes, GFP expression, 2–3
Mosquito larvae *(Aedes aegypti)*, 564
Mouse. *See Mus musculus* (mouse)
M (mitosis) phase of cell cycle, 25, 259
MRNA. *See* Messenger RNA (mRNA)
MRNA-interfering complementary RNA *(mic)*, 372
MRSA strains, 201
Muller, Hermann, 76
Mullis, Kary, 466
Multibranched junction, RNA structure, 234
Multicellularity, 23
   and gene regulation, 380
Multinomial expansion equation, 62
Multiple alleles, 98
Multiple genetic changes, leading to malignancy, 539–540
Multiple sclerosis, 427, 542
Multiplication method, for independent-assortment problems, 53, 54
Multipotent stem cells, 474–475
Muscular dystrophy, in dogs, 83
Mus musculus (mouse), 12
   genome sequencing, 504, 505
   imprinting and dwarfism, 120–121
Mutagens, 445–448
Mutant alleles, 91, 434
   definition, 91
   detection of disease-causing, 525–526
   frequencies of disease, 519
   X-linked muscular dystrophy, 83
Mutant genes, 13
   and human disease, 519–520
   predisposition to cancer, 541
Mutation rate, 447, 573
Mutations. *See also* Gene mutations
   causes of, 440
   changes in chromosome structure, 159
   explained, 431
   in a founder, 526
   frameshift, 432, 433
   gain-of-function, 92, 522
   in genes encoding chromatin-modifying proteins, 543
   and gene variation, 560
   genome maintenance preventing, 537–539
   germ-line, 437
   haplotypes, 525, 526
   and Hardy-Weinberg equation, 557
   human disease caused by mitochondrial, 118–119
   inactivating function of tumor-suppressor gene, 539
   induced, 440, 445–448
   loss-of-function, 13, 101, 185

missense, 432, 433, 434, 534
nonsense, 432, 433, 434
random nature of, 438–439
relationship between alleles and, 556
retinoblastoma, 536
somatic, 437–438
as source for genetic variation, 572–573
spontaneous, 440–444
study of induced, 432
suppressor, 435, 436
Mycobacterium tuberculosis, 504
Mycoplasma genitalium, 504
Myosin, 5
Myotonic muscular dystrophy, 443

N

Nägeli, Carl, 55, 222
Narrow-sense heritability, 597–598, 602–603
Nathans, Daniel, 461
National Institutes of Health, Office of Human Genome Research, 500
Natural selection, 10, 561–567
  balancing selection, 564–565
  directional selection, 562–564
  disruptive selection, 566
  and Hardy-Weinberg equation, 557
  negative frequency-dependent selection, 564–565
  principles of, 561
  reproductive success, 561–562
  stabilizing selection, 566–567
Natural transformation, 199
Nature (journal), 1, 229
NcRNAs. *See* Non-coding RNAs (ncRNAs)
Negative control, transcription regulation, 356
Negative frequency-dependent selection, 565
Negative supercoiling, 243–244
Neurological disorders, and non-coding RNAs, 426
Neurospora crassa (bread mold), 286, 327, 504
Neutral mutation, 434, 573
Next-generation sequencing technologies, 501–502
NF1 gene, 537
Nierhaus, Knud, 343
NIH Knockout Mouse Project, 512
Nilsson-Ehle, Herman, 590, 591
Nirenberg, Marshall, 334, 335, 337
Nitrous acid (HNO_2), chemical mutagens, 445, 446
Noll, Markus, 254
Nonallelic homologous recombination, 161, 163
Nonautonomous element, 249
Non-coding RNAs (ncRNAs)
  binding to other molecules, 413–414
  CRISP-Cas technology, 480
  effects on chromatin structure and transcription, 416–417
  effects on translation and mRNA degradation, 417–419
  epigenetic gene regulation, 395, 396
  functions of, 414–415
  and genome defense, 422–425
  importance of, 412
  movement of transposable elements, 252
  and mRNA degradation, 417–419
  and protein targeting, 421
  role in human diseases, 426–428
Non-coding sequences, mutations in, 434, 573
Nondisjunction, 171

meiotic, 175, 176
mitotic, 176
and polyploidy, 178
Nondividing cells
  structure of eukaryotic chromosomes in, 253–258
Nonessential gene, 101
Nonhomologous chromosomes, 56
Nonhomologous crossover, translocations and, 166
Nonhomologous end joining (NHEJ), 450, 452–453, 480
Nonhomologous recombination, bacterial transformation, 200
Nonionizing radiation, 447
Non-LTR retrotransposons, 249
Non-Mendelian pattern of inheritance
  extranuclear inheritance, 114–119
  imprinting, 120
  maternal effect, 124–126
Nonparentals, 51–52
Nonpolar amino acids, 332
Nonrandom mating, 560, 571–572
Nonrecombinant cells, 133
Nonrecombinant offspring, 134, 135, 141, 142
Nonsense codons, 328
Nonsense mutations, 432, 433, 434, 573
Nonstandard amino acids, 332
Normal distribution, 585
Norm of reaction, 8–9, 94–95
North American Conditional Mouse Mutagenesis Project (NorCOMM), 512
Northern blotting, 303–304, 482–483
Nowell, Peter, 534
N (amino)-terminus, 331
Nuclear genes, 114
Nuclear genome, 240
Nuclear lamina, eukaryotic chromosomes, 256
Nuclear localization signal (NLS), 385, 386
Nuclear matrix, eukaryotic chromosomes, 256–257, 259
Nucleic acids, 4, 224. *See also* Deoxyribonucleic acid (DNA); Ribonucleic acid (RNA)
Nucleoid, 20
  bacterial chromosome, 241, 242
  within a chloroplast, 114
  definition, 114
  mitochondria, 117
Nucleoid-associated proteins (NAPs), 242
Nucleoid- structuring (H-NS), 242
Nucleolus, rRNA assembly, 311
Nucleoside, 226
Nucleoside diphosphates (NDPs), 334
Nucleoside triphosphates (NTPs), 305
Nucleosome(s)
  30-nm fiber, 256
  beads-on-a-string model, 254–256
  chromatin-remodeling complexes altering, 387
  and histone variants, 388, 389
  model of, 253–254
  structures of, 253, 254
  zigzag model, 256
Nucleosome-free region (NFR), 389–390
Nucleotide excision repair (NER), 450
Nucleotides, 4, 5
  components of, 225
  short strand of DNA containing, 227
  structure of, 225–226
Nucleus, 21, 253
Null hypothesis, 62, 63, 136, 138, 588
Numbering system, of promoters, 299
Nurse cells, 125, 126
Nyren, Pal, 501

O

OCA2 gene, 21–22
Ochoa, Severo, 334
Ohno, Susumu, 76
Okazaki fragments, 273, 275, 276
Okazaki, Reiji, 276
Okazaki, Tsuneko, 276
Old Order Amish study, 569
Oligonucleotide, 465
OmpF mRNA, 372
Oncogene(s)
    definition, 532
    proto-oncogenes converted to, 532–535
One-gene/one-enzyme hypothesis, 327–328
1000 Genomes Project, 1000
Oogenesis, 35–36, 397–398
Open complex, 298, 299, 301, 302, 307, 308
Open conformation, 386
Operator sites, 358, 359, 360, 362, 364, 365–366, 367
Operon, 358. *See also Lac* operon
Organelles, 4, 20–21
Organism level, 7, 8
oriC region, 271, 272
Orientation-independent, eukaryotic gene regulation, 382
Origin of recognition complex (ORC), 281
Origin of replication, 271, 461
    bacterial chromosomes, 241
Origin of transfer, bacterial conjugation, 189
Oryza sativa, 504
Osmoregulation, 372
Osmotic shock, cell disruption, A-3
Osteoporosis, 522
Outbreeding, 571
Ovaries, 43
Overdominance inheritance pattern, 90, 96–98
Overwinding, supercoiling and, 243
Ovules, 43
Ovum, 35
Oxidative DNA damage, 443
Oxidative phosphorylation, 118
Oxidative stress, 443

P

P1 phage, 197
P16 gene, 537
P22 phage, 197
P53 gene, 537
Pace, Norman, 505
Pachytene stage, meiosis, 30–31
Pacific Biosclenes SMRT, 502
Painter, Theophilus, 172
Palindromic sequences, 462
Pangenesis, definition, 41
Pan troglodytes, 504
Papillomavirus, and cancer, 535
Paracentric inversion, 164–165
Paralogs, 105, 161
    definition, 162
Pardee, Arthur, 360, 361, 362
Parental generation (P generation), 45
Parental strands, DNA, 267
Particulate theory of inheritance, 47

Parvovirus, 208
Patau, Klaus, 78
Patau syndrome, 171
Paternal leakage, 118
Paterson, Andrew, 593
Pauling, Linus, 228
P cross, 45
Pea plants *(Pisum sativum)*
    cross-fertilization, 44
    flower structure and pollination in, 43
    and genetic cross, 12–13
    law of segregation, 47
    Mendel's study of, 9, 41, 42–45, 104
    three-factor crosses, 53
    two-factor crosses, 50, 52, 53
Pearson, Karl, 586
Pedigree analysis, 58–59
    Duchenne muscular dystrophy, 83
    familial breast cancer, 540–541
    hemophilia, 524
    Huntington disease, 522
    inheritance patterns of human disease determined via, 520–523
    and locus heterogeneity, 523
    polydactyly, 92, 93
    symbols, 58–59
    Tay-Sachs disease, 521
P elements, 51, 252
Pellet, centrifugation, A-3
Penrose, L.S., 171
Pentaploids, 174
Peptide bond, 331, 346
Peptidyl site (P site), 343
Peptidyl transfer, 346
Peptidyl transferase, 346
Pericentric inversion, 164–165
Personalized medicine, 518, 544–546
Pets, cloning of, 474
Phage fd, 208
Phage λ (lambda), 208, 212, 213, 214
Phage Qβ, 208
Phages. *See* Bacteriophages
Phage T4, 208
Pharmacogenetics, 546
Phenotype, 48
    albino, 520
    and aneuploidy, 170–171
    complementation, 105
    deletions affecting, 160
    and directional selection, 562
    disease resistance, 97
    epistatic interaction, 104–105
    F_2 offspring, 141
    gene redundancy, 105
    incomplete dominance and, 95, 96
    incomplete penetrance and, 92, 93
    and inversions, 164
    lethal, 101
    mutations in single genes, 102–103
    nonrandom mating, 571
    relationship between environment and, 94
    sex-influenced inheritance, 99, 100
    and stabilizing selection, 566–567
    variegated, 115–116
    white-spotting, in dogs, 103
Phenotype(s)
    multiple, and disruptive selection, 566
    polygenic inheritance resulting in a continuum of, 590–591

Phenotypic variation
    genetic variance and environmental variance contributing to, 595
    heritability, 597
    interactions/association between genotype and environment, 595–597
    quantitative traits, 585
Phenylalanine (Phe), 326, 329, 332, 336, 337, 338, 339
    metabolic pathway of, 326
    tRNA, 234
Phenylalanine hydroxylase, 9
Phenylketonuria (PKU), 9, 92, 94, 97, 521
    fluorsenscence in situ hybridization (FISH) and, 494
    screening for, 530
    tuberculosis resistance, 97
Phosphodiester linkage, 227, 278
Photolyase, 449–450
Photoreactivation, 449–450
Physical mapping, 492
Physiological adaptation, 439
Physiological traits, 6
piRNA-induced silencing complexes (piRISCs), 424, 425
PIWI-interacting RNA (piRNA), 415
PIWI-interacting RNAs (piRNAs), 424
Plant cells
    cytokinesis in, 29
    DNA, 21
    molecular organization, 4
Plant fertilization, 37
Plants
    alternation of generations between haploid sporophyte and diploid gametophyte, 36–37
    chromosome variation in, 8
    gamete production, 36, 37
    genome sequencing, 504
    polyploidy, 173–174
    sex chromosomes in, 76
    sperm and egg cells made in, 35–36
    transposable elements, 251
Plasmids, 189–190
    used as vectors, 461
Plasmodium, 96, 97, 504
Pleiotropy, 102–103
Plethodon larselli, genome size, 246–247
Plethodon richmondi, genome size, 246
Pluripotency factors, 398, 400
Pluripotent stem cells (iPS cells), 474, 476
Pneumococcus, 199
    Griffith's experiments on genetic transformation in, 222–223
Pneumocystis jiroveci, 217
Point centromere, 245
Point mutations, 432, 480, 481, 482
Polar amino acids, 332
Polar microtubules, 27
Pollen grains, 36, 43
Pollination, 43
Polyadenylation, 308, 309, 310, 317
PolyA tail, 317, 318
Polycistronic mRNA, 358
Polycomb group (PcG), 400–402
Polycomb response element (PRE), 402
Polydactyly, 92, 93
Polygenic (quantitative traits), 590
Polygenic inheritance, 590–592
Polymerase chain reaction (PCR)
    cycle threshold method, 472
    and environmental samples, 505
    four-step cycle, 469–470
    microsatellites, 496, 497
    real-time, 470–472
    reverse transcriptase, 470
    steps of, 466, 468
    thermocycler, 470
    three-step cycle, 468
Polymorphic, definition, 556
Polymorphic molecular markers, 495, 497
Polymorphism, 555–556
Polynucleotide phosphorylase, 334
Polyomavirus, 535
Polypeptides, 3, 5
    alternative splicing, 314, 315
    amino acid sequence, 328–330, 331–334
    central dogma of genetics, 297
    directionality, 331
    gene mutations effecting amino acid sequence of, 432, 433, 434
    primary structure, 332–333
    and proteins, 328
    and protein structure and function, 331–334
    quaternary structure, 333–334
    referring to structure, 328
    secondary structure, 333
    sequence of amino acids, 297, 309
    tertiary structure, 333
    in transcription, 298
    during translation, 296
Polypeptide synthesis, 296. *See also* Translation
Polyploid organisms, 169, 171
Polyploid plant, 174
Polyploidy, 169, 172, 173–174, 175, 178
Polyribosome, 342
Polysome, 342
Polytene chromosome, 172, 173
Population
    allele and genotype frequencies, 556–557
Population, definition, 555
Population genetics
    explained, 13
    gene pool, 554–555
    genetic drift, 567–569
    Hardy-Weinberg equation, 557–559
    microevolution, 560
    migration, 570
    natural selection, 561–567
    new genetic variation sources, 572–576
    nonrandom mating, 571–572
    overview, 554
    polymorphism, 555–556
Population level, relationship between genes and traits at the, 7, 8
Populus trichocarpa, 504
Position effect, 164, 435, 437
Positive control, 356
Positive supercoiling, 243
Posttranslational regulation, 372–373
Posttranslational covalent modification, 373
Posttranslational regulation, 372
PRC2 (polycomb repressive complex 2), 400, 401, 402, 416
PRE-binding protein, 402
Pre-crRNAs, 423, 424
Preimplantation genetic diagnosis (PGD), 530–531
Preinitiation complex, eukaryotic transcription, 307
Pre-mRNA
    capping, 316–317
    polyA tail, 317
    splicing, 312, 313–315

INDEX

Prereplication complex (preRC), 281
Pribnow box, 300
Pribnow, David, 300
Primary spermatocyte, 35, 36
Primary structure, polypeptide, 332–333
Primase, 274
Primer annealing, 468, 469
Primer extension, 468, 469
Primers
    identifying a microsatellite using PCR, 496, 497
    in PCR method, 466
    self-directed mutagenesis, 479–480
Primosome, 276
Probability calculations, 60–62
Probability, definition, 60
Processing body (P-body), 419
Processive enzyme, 279
Processivity, 279
Product rule, 60–61, 557, 558
Proflavin, 446
Programmed epigenetic changes, 395, 396
Prokaryotes
    cellular differences between eukaryotic species and, 20
    genomes sequenced, 504
Prokaryotic cells, 20
Prokaryotic species, new genetic variation, 572
Proline (Pro), 332, 336, 338
Prometaphase, 27, 28
    meiosis I, 31, 32
    meiosis II, 32
Promoters, 319
    bacterial transcription, 297, 299–300
    eukaryotic, 306
    operon, 358
Proofreading, bacterial DNA replication, 279
Prophage, 212
Prophase, 27, 28
    meiosis I, 30–31, 32
    meiosis II, 32
Protandrous hermaphrodites, 74
Protease, 223
Protein(s)
    blotting methods for detection of, 482–484
    denoting function, 328
    DNA storing information needed for synthesis of, 5
    and enzyme adaptation, 358
    functions of, 4–5
    levels of structures formed in, 33
    and polypeptides, 328
    regulatory, 356, 357, 364, 372
Protein-encoding genes, 241, 296, 305, 306–307, 312, 313, 315, 325, 328
    bacterial chromosomes, 241
    role in cancer development, 540
Protein-protein interaction, transcription factor modulation, 384, 385
Protein scaffold, eukaryotic chromosomes during cell division, 259, 261
Protein synthesis
    genetic basis for, 325–328
    relationship between genetic code and, 328–334
Proteome, 4
Proto-oncogenes, 532–535
Prototroph, 187, 188
Provirus, 212, 213
Pseudoautosomal genes, 80, 171
Pseudoautosomal inheritance, 80
Punnett, Reginald, 48, 132
Punnett square
    allele frequencies and union of alleles in a, 557
    inheritance pattern of X-linked traits, 82, 83
    and quantitative traits, 590
    single-factor cross, 48–49
    for a two-factor cross, 52–53
Purines, 226, 440, 441
P values, 63
Pyrimidines, 226
Pyrococcus abyssi, 504
Pyrophosphate (PP_i), 278
Pyrosequencing, 501–502, 503
Pyrrolysine (Pyl), 330, 332

Q

QTL mapping, 593, 594
Quantitative analysis, 45, 47
Quantitative genetics
    definition, 584
Quantitative trait locus (QTL), 593, 600
Quantitative traits
    continuum of phenotypic variation, 585
    examples of, 584–585
    explained, 584
    heritability, 595–600
    identification of genes that control, 592–594
    normal distribution of, 585, 586
    polygenic inheritance, 590–591
    selective breeding, 600–602
    selective breeding altering, 600–602
    statistical methods for evaluating, 586–589
Quaternary structure, polypeptides, 333–334

R

Radial loop domains, 256–257, 258, 259, 260
Radiation, ionizing, 447
Radiation, nonionizing, 447
Radioisotope detection, A-6
Random breeding, and Hardy-Weinberg equation, 557
Random genetic drift, 567. *See also* Genetic drift
Random mutations, 560
Random mutation theory, 439
Random sampling error, 60
Rare mutations, 572
Ras protein, 533–534
Rb gene, 536
Rb protein, 536
Reactive ends, broken chromosomes, 166
Reactive oxygen species (ROS), 443
Reading frame, mRNA, 329–330
Real-time PCR technology, 470–472
Recessive alleles, 56, 57, 59, 60, 62, 91, 564
Recessive epistasis, 104–105
Recessive trait, and genetic disease, 59
Recessive variants, 47
Reciprocal cross, 83, 115
Reciprocal translocation, 159, 166, 167, 168
Recombinant bacterium, 197
Recombinant cells, 133
Recombinant chromosomes, 146
Recombinant DNA molecules, 459, 461–462
Recombinant DNA technology, 459
Recombinant offspring

and crossing over, 133, 134–136, 135, 137, 138
and genetic mapping, 140, 141, 142–143
Recombinant vector, 463
Recombination. *See* Genetic recombination
Recombination, bacterial chromosomes (F factors), 191
Regional centromeres, 245
Regulatory elements, 381
Regulatory proteins, 356, 357, 364, 372
Regulatory sequence, 297, 298
Regulatory transcription factors, 381, 382
Relationship testing, DNA fingerprinting used for, 575–576
Relative fitness, 562
Relaxosome, bacterial conjugation, 189, 190
Release factors, translation, 346, 347, 348
Release, in viral reproduction cycle, 213
Repetitive sequences, 161, 574–575
    bacterial chromosomes, 241
    eukaryotic chromosomes, 246
    transposable elements, 251
    transposition, 247–248
Replica plating, 439
Replication fork, 271, 273
Replication, genetic material criteria, 222, 223
Replisome, 276
Repressible genes, 308, 356, 357
Repressor, regulatory protein, 356
    lacI gene, 360–363, 364
    lac operon, 358–359
    regulatory protein, 356, 357
    trp operon regulated by, 368
Repressors, regulatory transcription factors acting as, 381–382
Reproduction. *See also* Sexual reproduction
    asexual, 23, 24–25
    and binary fission, 23–24
    definition, 19
    factors contributing to success of, 561
    genetic variation and, 9–10
    meiosis, 30–34
    mitosis and cytokinesis, 26–29
    traits passed from parents to offspring through, 9
Reproductive cloning, 472–474
Resistance plasmids, 190
Resistant strains, 200–201
Resolution, 286
    microscopy, A-2
Restriction endonucleases, 461
Restriction enzymes, 461, 462, 463
Restriction fragment length polymorphism (RFLP), 495, 496, 593
Restriction point, 24–25
Retinoblastoma, 536
Retroelements, 248
Retrotransposition, 248, 249, 250–251
Retrotransposons, 248
Retroviruses, 535
Rett syndrome, 523, 524
Reverse mutation, 434
Reverse transcriptase, 212, 217–218, 250–251, 465
    gene cloning, 465
Reverse transcriptase PCR, 470
Reversion (reverse mutation), 434
RFLP linkage map, 495, 496
R group, amino acids, 332
Rheinberger, Hans-Jorg, 343
Rho (r)-dependent termination, bacterial transcription, 302, 303
Rho (r)-independent termination, bacterial transcription, 302

Rho (r) protein, bacterial transcription, 302
Ribbon model, RNA, 235
Ribonucleic acid (RNA), 5, 6, 222. *See also* Messenger RNA (mRNA); Non-coding RNAs (ncRNAs)
    analyzing binding of proteins to, 484
    blotting methods for detection of, 482–484
    nucleotide structure, 224–225
    origin of *nucleic acid* term, 224
    retrotransposition, 248
    ribbon model, 235
    space-filling model, 235
    structure, 225–226, 234–235
    synthesis of, 337
    synthesizing, 336–337
    tobacco mosaic virus, 210–211
    viruses, 208, 209, 210–211, 212–213, 217–218, 223
Ribose, 226
Ribosomal RNA (rRNA), 342
    description and role of, 415
    moderately repetitive sequence, 247
    processed in eukaryotes, 311
Ribosome
    binding tRNA, 337
    composition in bacteria and eukaryotes, 342, 349
    elongation stage of translation, 346
    structural features of, 341–343
    structure, 343
    subunits, 342–343
    termination stage of translation, 347
Ribosome-binding site, bacterial transcription, 297, 298
Riboswitches
    regulation of transcription and translation, 373–375
    types of, 374
Ribozymes, 311, 313
    non-coding RNAs functioning as, 414
Rich, Alexander, 232
Richmond, Timothy, 253, 256
Riggs, Arthur, 280
Right-handed, double helix model, 231, 232
Right-handed, RNA double helices, 234
RNA-binding proteins, 405–406
RNA component of signal recognition particle (SRP RNA), 415
RNA editing, 310, 318, 319
RNA induced silencing complex (RISC), 419
RNA interference (RNAi), 419, 420
RNA modification, 309–318, 319
RNA polymerase
    bacterial transcription, 300–302
    differences between transcription and RNA modification in, 319
    eukaryotic transcription, 305, 306, 308, 309
    and regulatory transcription factors, 381, 383
    schematic structure of, 305
    transcription, 297–298
    transcription factors, 381
RNA polymerase holoenzyme, 300
RNA polymerase I, 305
RNA polymerase II, 305, 306–307, 308, 309
RNA polymerase III, 305
RNA primers, 273, 274, 276
    removal by flap endonuclease, 282
RNase, 223
RNaseH, 465
RNaseMRP, 426
RNaseP, 311
RNaseP RNA, 311, 415
RNA sequencing (RNA-Seq), 510–512
RNA splicing, 305, 310, 312–316

RNA transcript
    cleaved into smaller functional transcripts, 310, 311
    dissociate from DNA, 298, 299, 302–303
    splicing/alternate splicing, 310, 314–315
    synthesized during elongation stage, 298, 301–302
Robertsonian translocation, 166–167
Robertson, William, 166
Roche 454/FLX Pyrosequencer, 501, 502
Ronaghi, Mostafa, 501
Rous, Peyton, 535
Rous sarcoma virus (RSV), 535
Roux, Wilhelm, 55
Royal families, hemophilia in, 523
Royal jelly, and dietary effects on honeybee development, 403, 405
rRNA. *See* Ribosomal RNA (rRNA)

S

Saccharomyces cerevisiae, 245
    genome, 491
    genome sequencing, 504, 505
    model organism, 12
    and splicing, 314–315
    transcription, 304
Saccharomyces cerevisiae, 504
Sachs, Bernard, 520
Salamander, 246
Salmonella typhimurium, 448
    horizontal gene transfer, 574
Sandhoff disease, 92
Sanger, Frederick, 477
Santa Cruz Island, 555
Sargasso Sea, 506–507
Scaffold-attachment regions (SARs), 257
Schimper, Andreas, 119
Schizosaccharamoyces pombe
    meiosis and, 30
Schramm, Gerhard, 210
Scientific method, 14
Scintillation counting, A-6
Scurs, in cattle, 99, 100
Secondary oocyte, 36
Secondary structure of tRNA, 339
Secondary structure, polypeptides, 333
"Second genetic code," 340
Sedimentation coefficient, A-3
Segmental duplication, chromosomes, 162
Segregate, of genes, 47
Selectable marker, 461
Selection coefficient *(s)*, 564
Selective breeding, 600–603
Selenocysteine (Sec), 330, 332
Selenocysteine insertion sequence (SECIS), 330
Self-assembling virus, 213
Self-fertilization, 43, 45
    Mendel's law of independent assortment and, 52
    Punnett square used to predict outcome of, 48–49
Selfish DNA hypothesis, 251–252
Self-splicing, 312–313
Semiconservative model (DNA replication), 268, 269, 270
Semilethal alleles, 101, 102
Semisterility, 167
Sense codons, 328, 339
Sense RNA, 417
Sense strand, 302

Separation methods, A-2–A-5
Sequence complexity, 247, 248
Sequence elements, bacterial transcription, 299–300
Sequence-tagged sites (STS), 495, 496
Sequencing by synthesis (SBS), 502
Sequencing ladder, 479
Serine (Ser), 329, 332, 336, 337
Serines (S), 315
7SL RNA gene, 249
Sex chromosomes, 9, 72–73, 75
    and aneuploid conditions, 171
    cytogenetics and, 21
Sex determination
    definition, 72
    of deoecious plant species, 75
    dependent on sex chromosomes, 72–73
    dosage compensation, 75–77
    environmental factors, 73–75
    haplodiploid system, 73
    X and Y chromosomal properties, 79–80
    X-chromosome inactivation center (Xic), 76–79
    X-linked inheritance, 80–83
    X-Y system of, 73
    Y-linked genes, 80
Sex-influenced inheritance pattern, 90, 99, 100
Sex-limited inheritance pattern, 90, 100
Sex-linked genes, 79
Sex pili (pilus), 189
Sexual dimorphism, 100
Sexual reproduction, 9–10, 19
    assortative mating, 571
    gametes, 21, 34–35
    oogenesis, 35–36
    outbreeding, 571
    spermatogenesis, 35
sgRNA, 480, 481, 482
sgRNA-Cas9, 482
Sheet, 333
Shepherd's purse, 105–106
Sheppard, Philip, 566
Shimomura, Osamu, 3
Shine-Dalgarno sequence, 346, 372
Shingles, 213
Short ncRNAs, 415
Shotgun sequencing, 498–500
Shull, George, 105–106
Sickle cell disease, 96, 518, 564–565
    missense mutation in, 433
Side chain, 332
Sigma (σ) factor, bacterial transcription, 300
Signal recognition particle (SRP), 421
Signal recognition particle, RNA component of (SRP RNA), 415
Silencers, 306, 308, 382
Silencing, of transposable elements, 424, 425
Silene latifolia (white campion), 75
Silent mutations, 432, 433
Simple Mendelian inheritance, 89, 90
Simple translocation, 159
Simple transposition, 248, 249
Simple transposons, 248–249
Singer, Beatrice, 210
Single-factor crosses, 45–46
Single genes
    disease cause by, 518, 519
Single genes, multiple effects of (pleiotropy), 102–103
Single-nucleotide polymorphism (SNP), 495, 556, 593
    genome-wide association study, 527–528

Single-strand binding proteins, 275
Sister chromatid exchange (SCE), 286
Sister chromatids, 25, 26, 27, 31, 33, 285–286
Site-directed mutagenesis, 479–480
Size, chromosomes numbered by their, 157
Small-interfering RNAs (siRNAs), 419
Small nuclear riboproteins (snRNPs), 313
Small nuclear RNA, 313, 415
Small regulatory RNAs, 415
Smith, Hamilton, 491, 498–500
Snail coiling, inheritance pattern of, 124–125
Snail coiling, maternal effect in, 126
snRNPs. *See* Small nuclear riboproteins (snRNPs)
SNRPN gene, 123
Snyder, Michael, 510
Solenoid model, eukaryotic nucleosome structure, 256
Solid growth media, A-1–A-2
Somatic cells, 9, 10, 21, 437
　cloning from, 474
　mammals maintaining one active X chromosome in, 78
Somatic mutations, 437–438
Sonication, cell disruption, A-3
Sorcerer II (sailboat), 507
Space-filling model of the double helix, 232
Species
　and biological evolution, 10
　definition, 7
Spectrophotometer, A-5
Spectroscopy, A-5–A-6
Sperm, 35, 43
Spermatogenesis, 31, 35, 398
Sperm cells, 34–35
S (synthesis) phase of cell cycle, 25, 280, 281
Spinal and bulbar muscular atrophy (SBMA), 443
Spindle pole, 27
Spinocerebellar ataxia (SCA1), 443
Spliceosome, 312, 313–314
Splicing factors, 315–316
Splicing, RNA, 305, 310, 312–316
Spontaneous abortions, 170, 171, 529–530
Spontaneous mutations, 440–444
Sporophyte, 36
Spreading phase, X-chromosome inactivation, 78, 79
SR proteins, 315
SRY gene, 73, 79
Stabilizing selection, 566–567
Stahl, Franklin, 268, 270
Staining chromosomes, 157, 158
Standard deviation, 587
Staphylococcus aureus, 201
　antibiotic resistance, 201
Start codons, 297, 298, 299, 318, 328, 329, 330, 344, 345–346
Statistical methods, 62–63
Statistical methods, quantitative traits evaluation, 586–589
Stem cells
　ability to divide and differentiate, 474–475
　definition, 474
　growth pattern of, 474
　occurrence at human development stages, 475
　potential to treat diseases, 475–476
Stem-loops
　attenuation, 368–369
　RNA structure, 234
　Shine-Dalgarno antisequestor, 374–375
　TPP riboswitch, transcription regulation, 374
　trpL mRNA, 371
Sternbach, Hans, 343

Stern, Curt, 146
Steroid hormones, and gene transcription, 384–385
Stigma, 43
Stop codons, 297, 298, 299, 318, 328, 330, 344, 347
Strahl, Brian, 388
Strain, definition, 44
Strand (DNA or RNA), 225
Strasburger, Eduard, 55
Streptococcus pneumoniae, 199
　Avery, MacLeod and McCarty transformation study, 224
　Griffith's genetic material study, 222
Structural genes, 241, 325. *See also* Protein-encoding genes
Structural maintenance of chromosomes (SMCs), 242
Sturtevant, Alfred, 435
S. typhimurium chromosome, 197
Submetacentric centromere, 156, 158
Subunits (polypeptides), 334
Subunits (ribosomal), 342–343
Sugars, nucleotide, 226
Sulfolobus solfataricus, 504
Supercoiling, bacterial chromosome, 242–244
Supernatant, centrifugation, A-3
Super-repressor mutations, 359–360
Suppressor mutations, 435
Sutton, Walter, 55
SV40, 535
Svedberg, Theodor, 342
Svedberg units (S), 342
Swine flu, 217
Symons, Robert, 459
Synapsis, 30
Synaptonemal complex, 30
Synonymous codons, 329
Synteny, 132
Synthesis of viral components, in viral reproduction cycle, 212–213
Synthetic RNA, used to determine the genetic code, 334–338

T

T2 phage, 223–224
T4 bacteriophage, 209, 213
Tailing, 319
Tandem array/tandem repeat, 247
TaqMan probe, 470–471
Taq polymerase, 468
Target-site duplications, 248
TATA-binding protein (TBP), 307
TATA box, 306, 307
Tatum, Edward, 187, 188, 326–327
Tautomeric shift, spontaneous mutation by, 441–443
Tautomers, 441
Tay-Sachs disease, 92, 97, 520, 529
Tay, Warren, 520
TBP-associated factors (TAFs), 307
T DNA (transferred DNA), 189
Telocentric centromere, 156, 158
Telomerase, 283–284
Telomerase reverse transcriptase (TERT), 284
Telomerase RNA component (TERC), 284, 415
Telomeres, 166, 246, 283, 284
Telophase, 28, 29
　meiosis I, 32
　of meiosis I and cytokinesis, 33
　meiosis II, 32
Temperate phage, 213, 216

INDEX

Temperature, and epigenetic changes, 395, 396
Temperature-sensitive (ts) lethal alleles, 101
Template DNA, 468
Template strands, 267
    bacterial transcription, 301, 302
    RNA polymerase, 305
TE promoter, 403, 404
Terminal deletion, 160
Termination
    stage of translation, 344, 347, 348
    transcription, 298, 320
Termination codons, 328
Terminator
    bacterial transcription, 297
    operon, 358
Tertiary structure, polypeptides, 333
Testcross, 81, 82, 141–143
Testosterone, 74
Tetrads, 30–31
Tetraploid organisms, 169
Therman, Eeva, 78
Thermocycler, 470
Thermoplasma volcanium, 504
Thiamine, 374
Thiamin pyrophosphate (TPP), 374
30-nm fiber, 256
Thomson, James, 475
3-1 phenotypic ratio, 82
3:1 phenotypic ratio, 47–48
3′ polyadenylation, 310
3′ tail, 315
3′-untranslated region, 328
Three-factor crosses, 132, 134, 143–145
Threonine (Thr), 329, 332, 336, 337
Threshold traits, 585
Thymine (T), 5, 225, 226, 229, 231, 266
Thymine dimers, 447
Tissue-specific genes, 392
TMV. *See* Tobacco mosaic virus (TMV)
Tn10 transposable element, 251
Tobacco mosaic virus (TMV), 207, 208, 209, 210–211, 212
Tonegawa, Susumu, 383
Topoisemerase, 274
Topoisomerase I, 243
Topoisomerase II, 243, 273, 277
Topoisomers, 243
Torpedo model (transcriptional termination), 308, 309
Totipotent (fertilized egg), 474
Toxins, and epigenetic changes, 396, 397
TPP (thiamin pyrophosphate), 374–375
Tracr gene, 423, 424
TracrRNA, 423, 424, 480
Trait(s)
    categories of, 6
    definition, 44
    environmental factors, 93–95
    the environment and, 8–9
    explained, 3
    expressivity, 93
    passed from parents to offspring, 9
    relationship between genes and, 6–10
    skipping generations, 92
Trans-acting factors, 306, 364
Transcription, 5, 6
    bacterial, 299–303
    bacterial promoter, 299–300
    base sequences, 297–298
    central dogma of genetics, 296, 297
    comparison between bacterial and eukaryotic, 319
    definition, 296
    and DNA methylation, 392
    eukaryotic, 304–309
    RNA modification, 309–318
    RNA polymerase hooenzyme, 300–301
    role of histones in eukaryotic, 391
    termination of, 302–303, 308, 309
    three stages of, 298–299
Transcription (gene), 7
Transcriptional activation, of eukaryotic gene, 390–391
Transcriptional regulation, 355–356, 358–359, 364, 372
    activators and repressors, 356–357
    catabolite repression, 364–365
    chromatin remodeling, 386–387
    histone modification, 388–389, 390
    lac operon, 358–359
    lactose metabolism, 358
    nucleosome-free region (NFR), 389–390
    regulatory transcription factors, 381–385
    riboswitches, 373–374
    trp operon, 370
Transcriptional start site, 299, 306
Transcription factor IIB (TFIB), 307, 308
Transcription factor IID (TFID), 307, 383, 402
Transcription factor IID (TFIID), 307
Transcription factor IIE (TFIIE), 307
Transcription factor IIF (TFIF), 307
Transcription factor IIH (TFIIH), 307, 308
Transcription factors, 298, 299, 306–307, 381
Transcriptomes, 510, 511
Transduction, 186, 197, 198–199
Trans-effect, 364, 366, 367
Transferrin receptor mRNA, 405–406, 406–407
Transfer RNA (tRNA), 235, 311, 312, 328
    and aminoacyl-tRNA synthetases, 339–340
    function, 338–339
    functions during translation, 338–339
    initiation stage of translation in bacteria, 344
    processing of, 311
    ribosome binding, 337–338
    role and description of, 415
    space-filling model of, 235
    structural features, 339
    wobble rules, 340–341
Transformation
    bacterial, 199–200, 223, 224
    gene cloning, 463
    and genetic transfer, 186
    normal to malignant cell, 535
Transgenerational epigenetic inheritance, 394
Transition, gene mutation, 432
Translation, 5, 6, 7, 296
    adaptor hypothesis, 338, 341
    antibiotics inhibiting bacterial, 348, 349
    central dogma of genetics, 297
    codons in mRNA, 328–330
    comparison of bacterial and eukaryotic, 346
    directionality, 331
    effect of non-coding RNAs on, 417–419, 420
    explained, 325
    ribosome structure and assembly, 341–343
    riboswitches, 374–375
    RNA-binding protein, 405–407
    stages of, 343–348, 349
    tRNA structure and function, 338–341
Translational regulatory protein, 372
Translational repressors, 372

Translesion-replicating polymerases, 282
Translesion synthesis (TLS), 453
Translocation, 138, 139
    and cancer formation, 534–535
    chromosome change through, 159
    involving exchange between chromosomes, 166–167
    producing abnormal gametes, 167
    Robertsonian, 166–167
Translocation cross, 167
Transmission electron microscope, A-2
Transmission, genetic material criteria, 221, 223
Transmission genetics, 12–13
Transposable elements (TE), 247
    different transposition pathways, 248–249
    examples of, 251
    in genomes of selected species, 252
    increase in number of, 251
    influences on mutation and evolution, 251–252
    organizations of DNA sequences in, 248–249
    PIWI-interacting RNAs silencing, 424–425
    and repetitive sequences, 241, 247–248
    transposition pathways, 248
Transposase, 248, 250
Transposition, 247–248
    different mechanisms of, 248–249
    possible consequences of, 252
    process, 249–250
Transposons, 248, 250
Transversion, gene mutation, 432
Trimethylation, 400
Trinucleotide repeat expansion (TNRE), 443–444, 575
Triplet-binding assay, 337–338
Triple X syndrome, 171
Triploid, chromosome set, 174–175
Triploid fruit flies, 169
Trisomic/trisomy, 169, 171
Trisomy 1, 171
Trisomy 13, 171
Trisomy 18, 171
Trisomy 21, 171
Trisomy X individuals, 171
Trithorax group (TrxG), 400
Triticum aestivum, 174, 590, 591
tRNAs. *See* Transfer RNA (tRNA)
TrpL gene, 368, 369, 370
Trp operon, 368–371
    attenuation, 368–369, 370
    regulated by repressor protein, 368, 369
    stem-loops, 368–369
Trp repressor, 368, 369, 370
True-breeding line (strain), 44, 46, 51, 80–81, 104
Tryptophan (Trp), 329, 332, 336
Tsien, Roger, 3
Tsix gene, 399, 400
Tumors
    DNA microarrays used in molecular profiling of, 545–546
    molecular profiling used to classify, 544–545
Tumor-suppressor genes
    definition, 532
    functions of, 537–539
    inherited forms of cancers caused by defects in, 540–541
    p53, 537
    predisposition to developing cancer, 541
    rb gene, 536
    role in preventing proliferation of cancer cells, 536
    sensing DNA damage, 537
    silenced, 539
Turner, Bryan, 388
Turner syndrome, 171
Two-factor crosses, 50–51, 52, 53, 62–63, 132, 136, 143
Two-hit model for retinoblastoma, 536
Tyrosine (Tyr), 326, 329, 332, 336

U

UAA codon, 328
UBE3A gene, 123
UGA codon, 328
Ultraviolet (UV) radiation, 380
Unbalanced translocation, 166
Underwinding, supercoiling and, 243
Unipotent stem cells, 475
Unregulated (constitutive) genes, 355
Up regulation, eukaryotic gene regulation, 382
Uracil (U), 225, 226, 234
U-tube apparatus, bacterial contact studies, 188

V

Valine, 337, 340, 345
Valine (Val), 329, 332, 336, 337
Van Beneden, Edouard, 55
Variables, statistical methods for comparing, 587–589
Variance, 587
Variant(s)
    crossing, 45
    definition, 44
    dominant, 47
    recessive, 47
Variation, genetic material criteria, 222, 223
Varicella-zoster virus, 213
Variegated phenotype, 115–116
Vectors, gene cloning using, 460–461
Venter, J. Craig, 491, 498–500, 506, 507
Vieira, Cristina, 596
Viral assembly, in viral reproduction cycle, 213
Viral envelope, 208, 209
Viral genome, 210
Viral integration, 535
Viral reproductive cycle, 212–216
Viral vectors, 461
Virulence plasmids, 190
Virulent phages, 213
Viruses
    causing cancer, 535
    definition, 208
    first to be discovered, 207
    genome differences in, 210
    host range differences in, 208
    metagenomics and identification of, 506
    reproductive cycles, 212–215
    spread of emerging viruses, 217–218
    structural differences in, 208–210
Vitamin D-resistant rickets, 524
Von Tschermak, Erich, 42, 55

W

Waddington, Conrad, 394
Wallace, Alfred Russel, 561
Waterland, Robert, 403, 404

Water snail *(Lymnaea peregra)*, 113, 124
Watson, James, 228, 229, 230, 343
Weinberg, Wilhelm, 557
Weismann, August, 55, 222
Western blotting, 483, 484
Wexler, Nancy, 526
White campion, 75
Wild mustard plant, and selective breeding, 601
Wild-type alleles, 91, 95
Wild-type genotype, 434
Wild-type homozygote, 96
Wild-type mice *(AA)*, 402–403
Wild-type phenotype, 91, 104, 105, 140, 434
Wild-type strain, 327
Wilkins, Maurice, 228, 229
Wilmut, Ian, 1, 473
Wobble base, tRNA, 329
Wobble position, 341
Wobble rules, 340–341
Wollman, Elie, 192–195
Woodcock, Christopher, 256
Worker honeybees, 403–404
Wright, Sewall, 554, 561, 568
Wu, Tai Te, 198

X

X-0 system of sex determination, 73
X-chromosome inactivation (XCI), 76–79
    epigenetic changes, 396, 398–400
    phases, 400
X chromosomes, 9, 10, 72, 73
    dosage compensation, 76
    eye color alleles, 82
    properties in mammals, 79–80
    reciprocal cross, 83
    variation in number of, 171

Xeroderma pigmentosum (XP), 450, 451
X-inactivation center (Xic), 78
X inactive-specific transcript (Xist RNA), 400, 415
Xist gene, 398, 400
Xist RNA, 400, 415
X-linked alleles, 82
X-linked dominant inheritance, 523
X-linked genes, 79, 83
    transmission patterns for, 80–83
X-linked inheritance pattern, 80–84, 90, 134
X-linked recessive, 82, 83
X-linked recessive inheritance, 522–523
X-ray crystallography, 253, 254
X-ray diffraction, 228, 229
X-Y system of sex determination, 73

Y

Yamanaka, Shinya, 476
Y chromosome, 9, 10, 72, 73, 75, 78, 79–80
Yeast artificial chromosomes (YACs), 461
Yeast, transposable elements, 251, 252
Y-linked genes, 80

Z

Zamecnik, Paul, 338
Z DNA, 232–233
Zickler, H., 286
Zigzag model, eukaryotic nucleosomes structure, 256
Zika virus, 217
ZS Genetics TM, 502
Z-W system of sex determination, 73
Zygotene stage of meiosis, 30, 31